Electronics Engineer's Reference Book

Sixth Edition

Electronics Engineer's Reference Book

Sixth Edition

Edited by
F F Mazda
DFH, MPhil, CEng, FIEE, DMS, MBIM

With specialist contributors

Butterworths
London · Boston · Singapore · Sydney
Toronto · Wellington

 PART OF REED INTERNATIONAL P.L.C.

First published 1989

Butterworth International Edition, 1989
ISBN 0-408-05430-1

British Library Cataloguing in Publication Data

Electronics engineer's reference book—6th ed
1. Electronic equipment
I. Mazda, F.F. (Fraidoon Framroz), 1939–
621.381

ISBN 0-408-00590-4

Library of Congress Cataloging in Publication Data

Electronics engineer's reference book/edited by F.F. Mazda; with specialist contributors.
p. cm.
Bibliography: p.
Includes index.
ISBN 0-408-00590-4:
1. Electronics—Handbooks, manuals, etc. I. Mazda, F.F.
TK7825.E36 1989
621.381—dc 19

Typeset by Mid-County Press Ltd, 2a Merivale Road, London SW15 2NW
Printed and bound by Hartnolls Ltd., Bodmin, Cornwall

Preface

On taking over as editor of the fifth edition of *The Electronics Engineer's Reference Book*, I did what all new editors do; out came my broom and I started sweeping! The changes from the fourth edition to the fifth were dramatic, and it has been grafitying to note the success which the book has subsequently had on both sides of the Atlantic. The whole text was laid out in a larger, more convenient to use format; the information was grouped together into five parts; and much of the old material was removed in order to make way for thirty-two new chapters.

Having completed, what I considered to be, such an Herculean task, I naturally expected to rest on my laurels! So when the time came around to produce the sixth edition, I brought out my pen in readiness to dot the odd i and cross the occasional t. Alas, how mistaken I was! As someone famous (whose name escapes me for the moment) once said, 'Time and Technology wait for no man'. So the son of Hercules was recruited to work on the sixth edition!

For this new edition I have retained the same five groupings, which have proved so successful in the fifth edition. As before the first part contains a synopsis of mathematical and electrical techniques used in the analysis of electronic systems. Part two covers physical phenomena, such as electricity, light and radiation, often met with in electronic systems. The third part of the book contains chapters on basic electronic components and materials, the building blocks of any electronic design. The information presented covers a wide spectrum of devices, from the humble resistor to the glamorous microprocessor.

Part four has chapters on electronic circuit design and on instrumentation. A range of design techniques are covered from linear to digital circuits, and from signal power levels to those in the megawatt region. The fifth part of the book contains topics, such as radar and computers, which form well recognised application areas of electronics.

All the chapters have been revised in going from the fifth to the sixth edition, many of them being extensively rewritten to ensure that they are up to date. Inevitably some of the older material has had to be deleted, to make room for new text. Six new chapters have been added, on topics which have gained considerably in importance since the last edition. These are on: application specific integrated circuits, such as gate arrays and standard cells, which have rapidly become an essential element in any economical electronic design; computer aided design techniques, involving the full range of design tools and methodology from designs on silicon chips to those on printed circuit boards; digital system analysis, covering the instrumentation and measurement techniques used to analyse digital circuits, including those based on microprocessors; software engineering which, with the importance of software in all major projects, plays a vital part in electronic designs; local area networks, describing the latest developments in networks based on the ISO open system interconnect (OSI) standards; and integrated services digital network (ISDN), a topic which no self-respecting modern book on electronics could be without!

A large number of sub-editors from Butterworths worked on the sixth edition of *The Electronics Engineer's Reference Book*. To them go my thanks for the dedication which they have shown in labouring tirelessly in the background, and even though I do not mention them by name they will know the genuine gratitude I have for their hard work and support.

And finally, as always, the last word must be reserved for the many, many authors who made this book possible. Thank you all for the excellent manuscripts you have produced; for accepting the tight schedules you have had to work to; and for the skilful excuses you have employed when late manuscripts were being chased by the sub-editors!

FFM
Bishop's Stortford
March 1989

Contents

Part 4 · Electronic Design

Part 5 · Applications

Acknowledgements

The production of this reference book would have been impossible without the good will, help and cooperation of the electronics industry, the users of electronic equipment and members of the educational profession. Bare acknowledgements are very inadequate but the editor wishes to thank the following firms and organisations which so readily made available information and illustrations and permitted members of their specialist staffs to write contributions:

AMP of Great Britain Ltd
AMPEX Great Britain Ltd
Atomic Energy Establishment
British Aerospace
British Broadcasting Corporation
British Railways Board, Research and Development Division
British Telecom International
British Telecom plc
Cambridge Scientific Instruments Ltd
Cavendish Laboratory, Cambridge
Clinical Research Centre
Cranfield Information Technology Institute
Department of Trade & Industry
Digital Equipment Co. Ltd
Duracell Batteries Ltd
E-Sil Components Ltd
English Electric Valve Company Ltd
Ever Ready Ltd
Ever Ready Special Batteries Ltd
Exacta Circuits Ltd
Fujitsu Microelectronics Ltd
GEC-Marconi Electronics Ltd
Gould Advance Ltd
Greater London Council, Department of Planning and Transportation
Heriot-Watt University, Edinburgh
Hewlett-Packard Ltd
Hitachi Europe Ltd
International Rectifier Ltd
LSI Logic Ltd
Lucas Electrical Company Ltd
Marconi Company Ltd
Meteorological Office
Morgan Materials Technology Ltd
Mostek Corporation
Motorola Ltd
Mullard Ltd
Murata Manufacturing Co.
North East Thames Regional Health Authority
Nortronic Associates Ltd
N.V. Philips Gloeilampenfabrieken
Ovum Ltd
Plessey Company Ltd
Polytechnic of North London
Post Office
Racal-Redac Ltd
Rank Xerox Ltd
RCA
Rediffusion Ltd
Rutherford and Appleton Laboratories, Science Research Council
SAFT (UK) Ltd
SEEQ Corporation
Siemens Ltd
Standard Telephones and Cables Ltd
STC plc
STC Technology Ltd
STC Telecommunications Ltd
Texas Instruments Incorporated
University of Aston
University of Cambridge
University of Leeds
University of Manchester Institute of Science and Technology
University of Oxford
University of Sussex
University of Technology, Loughborough
Xerox Corporation

Acknowledgement is made to the Director of the International Radio Consultative Committee (CCIR) for permission to use information and reproduce diagrams and curves from the CCIR Documents.

Extracts from British Standards publications are reproduced by permission of the British Standards Institution.

List of Contributors

P Aloisi
Motorola Semiconductors Ltd

S W Amos, BSc(Hons), CEng, MIEE
Freelance Technical Editor and Author

J Barron, BA, MA(Cantab)
Cambridge University

J Berry, BSc(Hons)
LSI Logic Ltd

C J Bowry
Ever Ready Ltd

P A Bradley, BSc, MSc, CEng, MIEE
Rutherford Appleton Laboratory
Science and Engineering Research Council

G H Browton
STC Technology Ltd

D K Bulgin, PhD
Cambridge Instruments Ltd

M Burchall, CEng, FIEE
REL Ltd

J M Camarata
Electronics Division, Xerox Corporation

J R Cass
STC Telecommunications Ltd

P M Chalmers, BEng, AMIEE
English Electric Valve Co. Ltd

A Clark, DipEE, CEng, MIEE
Electrical Engineering Consultant

G T Clayworth, BSc
High Power Klystron Department
English Electric Valve Co. Ltd

H W Cole
Marconi Radar Systems Ltd

A P O Collis, BA CEng, MIEE
Broadcast Tubes Division, English Electric Valve Co. Ltd

P J Cottam, BSc
Eltek Semiconductors Ltd

J Curtis
Electronic Components Group, Siemens Ltd

J A Dawson, CChem, MRSC
Rank Xerox Ltd

V A Downes
Ovum Ltd

J P Duggan
Mullard Ltd

G W A Dummer, MBE, CEng, FIEE, FIEEE, FIERE
Electronics Consultant

J L Eaton, BSc, MIEE, CEng
Consultant (Broadcasting)

M D Edwards, BSc, MSc, PhD
The University of Manchester Institute of Science and Technology

M Ewing
SAFT (UK) Ltd

T C Fleming, BSc
Liddicon Camera Tubes
English Electric Valve Co. Ltd

P A Francis
Racal-Redac UK Ltd

R M Gibb, MA, DPhil
STC Optical Devices Division

C L S Gilford, MSc, PhD, FInstP, MIEE, FIOA
Independent Acoustic Consultant

A Goodings, PhD, DIC, BSc, ARCS
Atomic Energy Establishment

C J Goodings
The Cavendish Laboratory

M W Gray
Hewlett-Packard GmbH

V J Green, BSc
STC plc

P G Hamer
STC Technology Ltd

J E Harry, BSc(Eng), PhD, DSc, CEng, FIEE
Loughborough University of Technology

P Hawker
Formerly, Independent Broadcasting Authority

D R Heath, BSc, PhD
British Aerospace Ltd

E W Herold, BSc, MSc, DSc, FIEE

D W Hill, MSc, PhD, DSc, FInstP, FIEE
North East Thames Regional Health Authority

P H Hitchcock, PhD
Ever Ready Special Batteries Ltd

S R Hodge
Racal-Redac Systems Ltd

J Houldsworth
International Computers Ltd, UK

P J Howard, BSc, CEng, MIEE
STC Telecommunications Ltd

R S Hurst
STC Technology Ltd

M Jones, BSc
Hitachi Europe Ltd

P Jones, BSc
Fujitsu Microelectronics Ltd

J Kempster, BSc(Hons), MBCS
Digital Equipment Co. Ltd

C Kindell
AMP of Great Britain Ltd

T Kingham
AMP of Great Britain Ltd

P Kirkby, PhD
STC Technology Ltd

P R Knott, MA, CPhys, MInstP, MICeram
Morgan Materials Technology Ltd

J A Lane, DSc, CEng, FIEE
Radio Communications Division
Department of Trade and Industry

I G Lang
Exacta Circuits Ltd

J R G Lloyd, CEng, MIMechE, MIQA
Nortronic Associates Ltd

R W Lomax, MSc, CEng, FIEE, CPhys, FInstP
Power Components Division, STC Components Ltd

C E Longhurst
Duracell Batteries Ltd

P Longland, PhD
STC Hybrids Division

S Lowe
Ampex Great Britain Ltd

P G Lund
Oxford University

R C Marshall, MA, FIEE, MInstMC
Protech Instruments & Systems Ltd

F F Mazda, DFH, MPhil, CEng, FIEE, DMS, MBIM
STC Telecommunications Ltd

A D Monk, MA, CEng, MIEE
Consultant Engineer

D Ord
E-Sil Components Ltd

T Oswald, BSc, MIEE
STC Telecommunications Ltd

S C Pascall, BSc, PhD, CEng, MIEE
Telecommunications Policies Division
Commission of the European Communities

A M Pope, BSc
RR Electronics Ltd

S C Redman, AMIEE
STC Telecommunications Ltd

M Remfry, CEng, MIEE, MBIM, MIIM
STC Components Group

J Riley
AMP of Great Britain Ltd

I Robertson
Digital Equipment Corporation

M J Rose, BSc(Eng)
Mullard Ltd

M G Say, PhD, MSc, CEng, ACGI, DIC, FIEE, FRSE
Heriot-Watt University

M J B Scanlan, BSc, ARCS
Marconi Research Laboratories

A Shewan, BSc
Motorola Ltd

S F Smith, BSc(Eng), CEng, FIEE
STC Telecommunications Ltd

C R Spicer, DipEE, AMIEE
BBC Design and Equipment Department

K R Sturley, BSc, PhD, FIEE, FIEEE
Telecommunications Consultant

C J Tully
Cranfield Information Technology Institute

L W Turner, CEng, FIEE, FRTS
Consultant Engineer

K Wakino
Murata Manufacturing Co.

D B Waters, BSc(Eng)
STC Telecommunications Ltd

F J Weaver, BSc, CEng, MIEE
English Electric Valve Co. Ltd

F Welsby, BSc, CEng, MIEE
British Telecom plc

J D Weston, BSc(Eng)
STC Technology Ltd

R C Whitehead, CEng, MIEE
Polytechnic of North London

Part 1

Techniques

1 Trigonometric Functions and General Formulae

J Barron BA, MA (Cantab)
University of Cambridge

Contents

1.1 Mathematical signs and symbols

Sign, symbol	*Quantity*
$=$	equal to
$\mp$ $\neq$	not equal to
$\equiv$	identically equal to
Δ	corresponds to
$\approx$	approximately equal to
$\rightarrow$	approaches
$\simeq$	asymptotically equal to
$\sim$	proportional to
∞	infinity
$<$	smaller than
$>$	larger than
$\leq$ $\leqslant$ $\leqq$	smaller than or equal to
$\geq$ $\geqslant$ $\geqq$	larger than or equal to
$\ll$	much smaller than
$\gg$	much larger than
$+$	plus
$-$	minus
$.$ $\times$	multiplied by
$\frac{a}{b}$ a/b	a divided by b
$\lvert a\rvert$	magnitude of a
a^n	a raised to the power n
$a^{1/2}$ $\sqrt{a}$	square root of a
$a^{1/n}$ $\sqrt[n]{a}$	nth root of a
$\bar{a}$ $\langle a\rangle$	mean value of a
$p!$	factorial p, $1\times2\times3\times\ldots\times p$
$\binom{n}{p}$	binomial coefficient, $\frac{n(n-1)\ldots(n-p+1)}{1\times2\times3\times\ldots\times p}$
Σ	sum
Π	product
$f(x)$	function f of the variable x
$[f(x)]_a^b$	$f(b)-f(a)$
$\lim\limits_{x\to a} f(x)$; $\lim_{x\to a} f(x)$	the limit to which $f(x)$ tends as x approaches a
Δx	delta x = finite increment of x
δx	delta x = variation of x
$\frac{df}{dx}$; df/dx; $f'(x)$	differential coefficient of $f(x)$ with respect to x
$\frac{d^nf}{dx^n}$; $f^{(n)}(x)$	differential coefficient of order n of $f(x)$
$\frac{\partial f(x,y,\ldots)}{\partial x}$; $\left(\frac{\partial f}{\partial x}\right)_{y\ldots}$	partial differential coefficient of $f(x,y,\ldots)$ with respect to x, when $y,\ldots$ are held constant
df	the total differential of f
$\int f(x)\,dx$	indefinite integral of $f(x)$ with respect to x
$\int_a^b f(x)\,dx$	definite integral of $f(x)$ from $x=a$ to $x=b$
e	base of natural logarithms
e^x; $\exp x$	e raised to the power x
$\log_a x$	logarithm to the base a of x
$\lg x$; $\log x$; $\log_{10} x$	common (Briggsian) logarithm of x
$\text{lb}\,x$; $\log_2 x$	binary logarithm of x
$\sin x$	sine of x
$\cos x$	cosine of x
$\tan x$; $\text{tg}\,x$	tangent of x
$\cot x$; $\text{ctg}\,x$	cotangent of x
$\sec x$	secant of x
$\text{cosec}\,x$	cosecant of x
$\arcsin x$	arc sine of x
$\arccos x$	arc cosine of x
$\arctan x$, $\text{arctg}\,x$	arc tangent of x
$\text{arccot}\,x$, $\text{arcctg}\,x$	arc cotangent of x
$\text{arcsec}\,x$	arc secant of x
$\text{arccosec}\,x$	arc cosecant of x
$\sinh x$	hyperbolic sine of x
$\cosh x$	hyperbolic cosine of x
$\tanh x$	hyperbolic tangent of x
$\coth x$	hyperbolic cotangent of x
$\text{sech}\,x$	hyperbolic secant of x
$\text{cosech}\,x$	hyperbolic cosecant of x
$\text{arsinh}\ x$	inverse hyperbolic sine of x
$\text{arcosh}\ x$	inverse hyperbolic cosine of x
$\text{artanh}\ x$	inverse hyperbolic tangent of x
$\text{arcoth}\ x$	inverse hyperbolic cotangent of x
$\text{arsech}\ x$	inverse hyperbolic secant of x
$\text{arcosech}\,x$	inverse hyperbolic cosecant of x
i, j	imaginary unity, $i^2=-1$
$\text{Re}\,z$	real part of z
$\text{Im}\,z$	imaginary part of z
$\lvert z\rvert$	modulus of z
$\arg z$	argument of z
z^*	conjugate of z, complex conjugate of z
$\bar{A}$, A', A^t	transpose of matrix A
A^*	complex conjugate matrix of matrix A
A^+	Hermitian conjugate matrix of matrix A
A, **a**	vector
$\lvert\mathbf{A}\rvert$, A	magnitude of vector
$\mathbf{A}\cdot\mathbf{B}$	scalar product
$\mathbf{A}\times\mathbf{B}$, $\mathbf{A}\wedge\mathbf{B}$	vector product
∇	differential vector operator
$\nabla\varphi$, grad φ	gradient of φ
$\nabla\cdot A$, div **A**	divergence of **A**
$\nabla\times\mathbf{A}$, $\nabla\wedge\mathbf{A}$, curl **A**, rot **A**	curl of **A**
$\nabla^2\varphi$, $\Delta\varphi$	Laplacian of φ

1.2 Trigonometric formulae

$$\sin^2 A+\cos^2 A=\sin A\,\text{cosec}\,A=1$$

$$\sin A=\frac{\cos A}{\cot A}=\frac{1}{\text{cosec}\,A}=(1-\cos^2 A)^{1/2}$$

$$\cos A=\frac{\sin A}{\tan A}=\frac{1}{\sec A}=(1-\sin^2 A)^{1/2}$$

$$\tan A=\frac{\sin A}{\cos A}=\frac{1}{\cot A}$$

$$1+\tan^2 A=\sec^2 A$$

$$1+\cot^2 A=\text{cosec}^2 A$$

$$1-\sin A=\text{coversin}\,A$$

$$1-\cos A=\text{versin}\,A$$

$$\tan\tfrac{1}{2}\theta=t;\quad \sin\theta=2t/(1+t^2);\quad \cos\theta=(1-t^2)/(1+t^2)$$

$\cot A = 1/\tan A$

$\sec A = 1/\cos A$

$\operatorname{cosec} A = 1/\sin A$

$\cos(A \pm B) = \cos A \cos B \mp \sin A \sin B$

$\sin(A \pm B) = \sin A \cos B \pm \cos A \sin B$

$$\tan(A \pm B) = \frac{\tan A \pm \tan B}{1 \mp \tan A \tan B}$$

$$\cot(A \pm B) = \frac{\cot A \cot B \mp 1}{\cot B \pm \cot A}$$

$\sin A \pm \sin B = 2 \sin \frac{1}{2}(A \pm B) \cos \frac{1}{2}(A \mp B)$

$\cos A + \cos B = 2 \cos \frac{1}{2}(A + B) \cos \frac{1}{2}(A - B)$

$\cos A - \cos B = 2 \sin \frac{1}{2}(A + B) \sin \frac{1}{2}(B - A)$

$$\tan A \pm \tan B = \frac{\sin(A \pm B)}{\cos A \cos B}$$

$$\cot A \pm \cot B = \frac{\sin(B \pm A)}{\sin A \sin B}$$

$\sin 2A = 2 \sin A \cos A$

$\cos 2A = \cos^2 A - \sin^2 A = 2\cos^2 A - 1 = 1 - 2\sin^2 A$

$\cos^2 A - \sin^2 B = \cos(A + B) \cos(A - B)$

$\tan 2A = 2 \tan A/(1 - \tan^2 A)$

$$\sin \tfrac{1}{2}A = \left(\frac{1 - \cos A}{2}\right)^{1/2}$$

$$\cos \tfrac{1}{2}A = \pm\left(\frac{1 + \cos A}{2}\right)^{1/2}$$

$$\tan \tfrac{1}{2}A = \frac{\sin A}{1 + \cos A}$$

$\sin^2 A = \frac{1}{2}(1 - \cos 2A)$

$\cos^2 A = \frac{1}{2}(1 + \cos 2A)$

$$\tan^2 A = \frac{1 - \cos 2A}{1 + \cos 2A}$$

$$\tan \tfrac{1}{2}(A \pm B) = \frac{\sin A \pm \sin B}{\cos A + \cos B}$$

$$\cot \tfrac{1}{2}(A \pm B) = \frac{\sin A \pm \sin B}{\cos B - \cos A}$$

1.3 Trigonometric values

Angle	*0°*	*30°*	*45°*	*60°*	*90°*	*180°*	*270°*	*360°*
Radians	0	$\pi/6$	$\pi/4$	$\pi/3$	$\pi/2$	π	$3\pi/2$	2π
Sine	0	$\frac{1}{2}$	$\frac{1}{2}\sqrt{2}$	$\frac{1}{2}\sqrt{3}$	1	0	−1	0
Cosine	1	$\frac{1}{2}\sqrt{3}$	$\frac{1}{2}\sqrt{2}$	$\frac{1}{2}$	0	−1	0	1
Tangent	0	$\frac{1}{3}\sqrt{3}$	1	$\sqrt{3}$	∞	0	∞	0

1.4 Approximations for small angles

$\sin\theta = \theta - \theta^3/6$; $\cos\theta = 1 - \theta^2/2$; $\tan\theta = \theta + \theta^3/3$;
(θ in radians)

1.5 Solution of triangles

$$\frac{\sin A}{a} = \frac{\sin B}{b} = \frac{\sin C}{c} \qquad \cos A = \frac{b^2 + c^2 - a^2}{2bc}$$

$$\cos B = \frac{c^2 + a^2 - b^2}{2ca} \qquad \cos C = \frac{a^2 + b^2 - c^2}{2ab}$$

where A, B, C and a, b, c are shown in *Figure 1.1*. If $s = \frac{1}{2}(a + b + c)$,

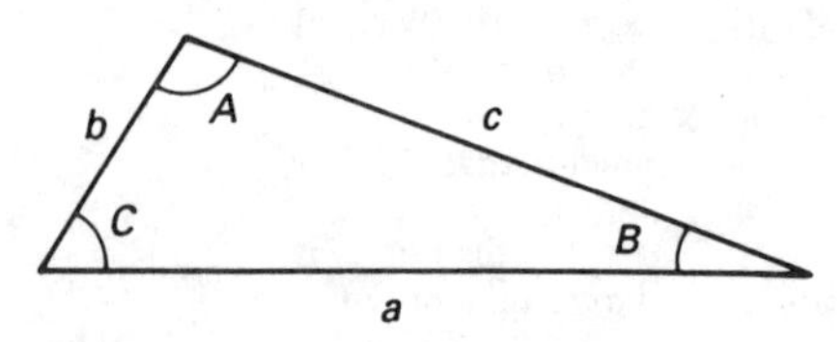

Figure 1.1 Triangle

$$\sin\frac{A}{2} = \sqrt{\frac{(s-b)(s-c)}{bc}} \qquad \sin\frac{B}{2} = \sqrt{\frac{(s-c)(s-a)}{ca}}$$

$$\sin\frac{C}{2} = \sqrt{\frac{(s-a)(s-b)}{ab}}$$

$$\cos\frac{A}{2} = \sqrt{\frac{s(s-a)}{bc}} \qquad \cos\frac{B}{2} = \sqrt{\frac{s(s-b)}{ca}}$$

$$\cos\frac{C}{2} = \sqrt{\frac{s(s-c)}{ab}}$$

$$\tan\frac{A}{2} = \sqrt{\frac{(s-b)(s-c)}{s(s-a)}} \qquad \tan\frac{B}{2} = \sqrt{\frac{(s-c)(s-a)}{s(s-b)}}$$

$$\tan\frac{C}{2} = \sqrt{\frac{(s-a)(s-b)}{s(s-c)}}$$

1.6 Spherical triangle

$$\frac{\sin A}{\sin a} = \frac{\sin B}{\sin b} = \frac{\sin C}{\sin c}$$

$\cos a = \cos b \cos c + \sin b \sin c \cos A$

$\cos b = \cos c \cos a + \sin c \sin a \cos B$

$\cos c = \cos a \cos b + \sin a \sin b \cos C$

where A, B, C and a, b, c are now as in *Figure 1.2*.

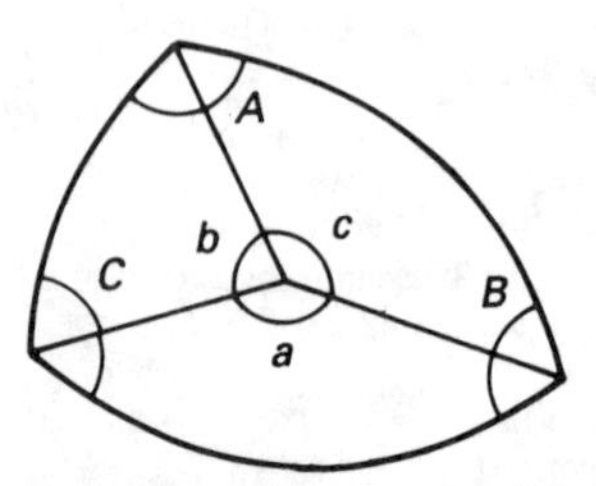

Figure 1.2 Spherical triangle

1.7 Exponential form

$$\sin\theta=\frac{e^{i\theta}-e^{-i\theta}}{2i}\qquad \cos\theta=\frac{e^{i\theta}+e^{-i\theta}}{2}$$

$$e^{i\theta}=\cos\theta+i\sin\theta\qquad e^{-i\theta}=\cos\theta-i\sin\theta$$

1.8 De Moivre's theorem

$$(\cos A+i\sin A)(\cos B+i\sin B)$$
$$=\cos(A+B)+i\sin(A+B)$$

1.9 Euler's relation

$$(\cos\theta+i\sin\theta)^n=\cos n\theta+i\sin n\theta=e^{in\theta}$$

1.10 Hyperbolic functions

$$\sinh x=(e^x-e^{-x})/2\qquad \cosh x=(e^x+e^{-x})/2$$

$$\tanh x=\sinh x/\cosh x$$

Relations between hyperbolic functions can be obtained from the corresponding relations between trigonometric functions by reversing the sign of any term containing the product or implied product of two sines, e.g.:

$$\cosh^2 A-\sinh^2 A=1$$

$$\cosh 2A=2\cosh^2 A-1=1+2\sinh^2 A$$
$$=\cosh^2 A+\sinh^2 A$$

$$\cosh(A\pm B)=\cosh A\cosh B\pm\sinh A\sinh B$$

$$\sinh(A\pm B)=\sinh A\cosh B\pm\cosh A\sinh B$$

$$e^x=\cosh x+\sinh x\qquad e^{-x}=\cosh x-\sinh x$$

1.11 Complex variable

If $z=x+iy$, where x and y are real variables, z is a complex variable and is a function of x and y. z may be represented graphically in an Argand diagram (*Figure 1.3*).

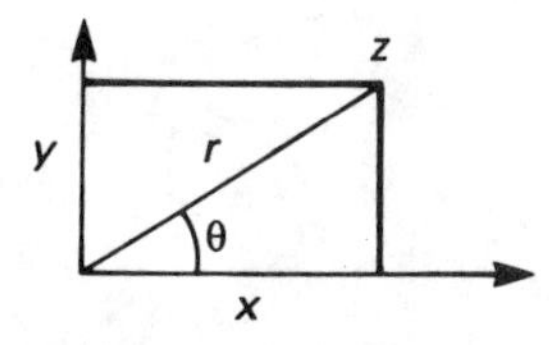

Figure 1.3 Argand diagram

Polar form:

$$z=x+iy=|z|\,e^{i\theta}=|z|(\cos\theta+i\sin\theta)$$

$$x=r\cos\theta\qquad y=r\sin\theta$$

where $r=|z|$.

Complex arithmetic:

$$z_1=x_1+iy_1\qquad z_2=x_2+iy_2$$

$$z_1\pm z_2=(x_1\pm x_2)+i(y_1\pm y_2)$$

$$z_1\,.\,z_2=(x_1x_2-y_1y_2)+i(x_1y_2+x_2y_1)$$

Conjugate:

$$z^*=x-iy\qquad z\,.\,z^*=x^2+y^2=|z|^2$$

Function: another complex variable $w=u+iv$ may be related functionally to z by

$$w=u+iv=f(x+iy)=f(z)$$

which implies

$$u=u(x,y)\qquad v=v(x,y)$$

e.g.,

$$\cosh z=\cosh(x+iy)=\cosh x\cosh iy+\sinh x\sinh iy$$
$$=\cosh x\cos y+i\sinh x\sin y$$

$$u=\cosh x\cos y\qquad v=\sinh x\sin y$$

1.12 Cauchy–Riemann equations

If $u(x,y)$ and $v(x,y)$ are continuously differentiable with respect to x and y,

$$\frac{\partial u}{\partial x}=\frac{\partial v}{\partial y}\qquad \frac{\partial u}{\partial y}=-\frac{\partial v}{\partial x}$$

$w=f(z)$ is continuously differentiable with respect to z and its derivative is

$$f'(z)=\frac{\partial u}{\partial x}+i\frac{\partial v}{\partial x}=\frac{\partial v}{\partial y}-i\frac{\partial u}{\partial y}=\frac{1}{i}\left(\frac{\partial u}{\partial y}+i\frac{\partial v}{\partial y}\right)$$

It is also easy to show that $\nabla^2u=\nabla^2v=0$. Since the transformation from z to w is conformal, the curves $u=$constant and $v=$constant intersect each other at right angles, so that one set may be used as equipotentials and the other as field lines in a vector field.

1.13 Cauchy's theorem

If $f(z)$ is analytic everywhere inside a region bounded by C and a is a point within C

$$f(a)=\frac{1}{2\pi i}\int_C\frac{f(z)}{z-a}\,dz$$

This formula gives the value of a function at a point in the interior of a closed curve in terms of the values on that curve.

1.14 Zeros, poles and residues

If $f(z)$ vanishes at the point z_0 the Taylor series for z in the region of z_0 has its first two terms zero, and perhaps others also: $f(z)$ may then be written

$$f(z)=(z-z_0)^n g(z)$$

where $g(z_0)\neq 0$. Then $f(z)$ has a *zero* of order n at z_0. The reciprocal

$$q(z)=1/f(z)=h(z)/(z-z_0)^n$$

where $h(z)=1/g(z)\neq 0$ at z_0. $q(z)$ becomes infinite at $z=z_0$ and is said to have a *pole* of order n at z_0. $q(z)$ may be expanded in the form

$$q(z)=c_{-n}(z-z_0)^n+\ldots+c_{-1}(z-z_0)^{-1}+c_0+\ldots$$

where c_{-1} is the *residue* of $q(z)$ at $z=z_0$. From Cauchy's theorem, it may be shown that if a function $f(z)$ is analytic throughout a region enclosed by a curve C except at a finite number of poles, the integral of the function around C has a value of $2\pi i$ times the sum of the residues of the function at its poles within C. This fact can be used to evaluate many definite integrals whose indefinite form cannot be found.

1.15 Some standard forms

$$\int_0^{2\pi} e^{\cos\theta}\cos(n\theta-\sin\theta)\,d\theta=2\pi/n!$$

$$\int_0^{\infty}\frac{x^{a-1}}{1+x}dx=\pi\operatorname{cosec}a\pi$$

$$\int_0^{\infty}\frac{\sin\theta}{\theta}d\theta=\frac{\pi}{2}$$

$$\int_0^{\infty}x\exp(-h^2x^2)\,dx=\frac{1}{2h^2}$$

$$\int_0^{\infty}\frac{x^{a-1}}{1-x}dx=\pi\cot a\pi$$

$$\int_0^{\infty}\exp(-h^2x^2)dx=\frac{\sqrt{\pi}}{2h}$$

$$\int_0^{\infty}x^2\exp(-h^2x^2)dx=\frac{\sqrt{\pi}}{4h^3}$$

1.16 Coordinate systems

The basic system is the rectangular Cartesian system (x, y, z) to which all other systems are referred. Two other commonly used systems are as follows.

1.16.1 Cylindrical coordinates

Coordinates of point P are (x,y,z) or (r,θ,z) (see *Figure 1.4*), where

$x=r\cos\theta \qquad y=r\sin\theta \qquad z=z$

In these coordinates the volume element is $r\,dr\,d\theta\,dz$.

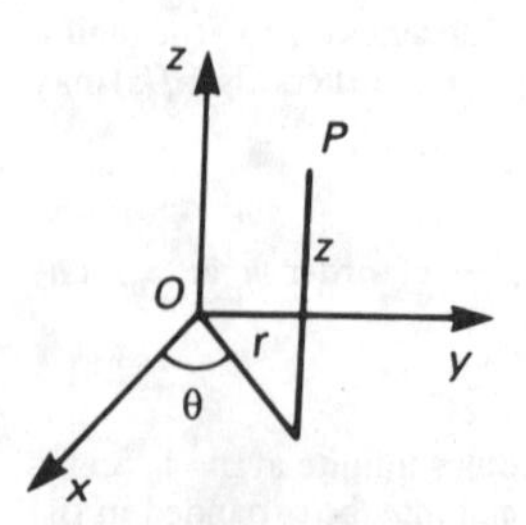

Figure 1.4 Cylindrical coordinates

1.16.2 Spherical polar coordinates

Coordinates of point P are (x,y,z) or (r,θ,φ) (see *Figure 1.5*), where

$x=r\sin\theta\cos\phi \qquad y=r\sin\theta\sin\phi \qquad z=r\cos\theta$

In these coordinates the volume element is $r^2\sin\theta\,dr\,d\theta\,d\phi$.

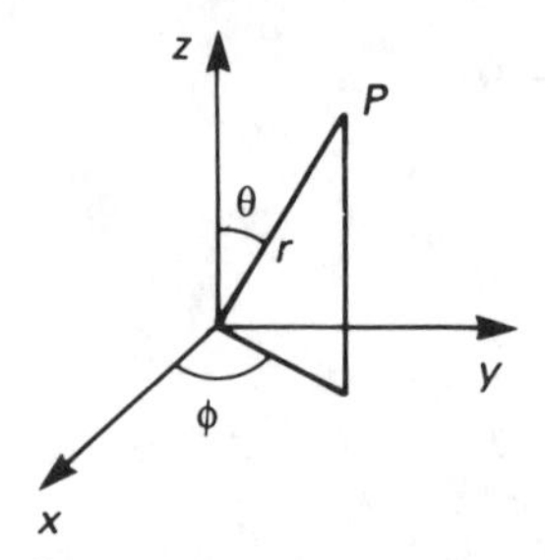

Figure 1.5 Spherical polar coordinates

1.17 Transformation of integrals

$$\iiint f(x,y,z)\,dx\,dy\,dz=\iiint\varphi(u,v,w)|J|\,du\,dv\,dw$$

where

$$J=\begin{vmatrix}\frac{\partial x}{\partial u} & \frac{\partial y}{\partial u} & \frac{\partial z}{\partial u}\\ \frac{\partial x}{\partial v} & \frac{\partial y}{\partial v} & \frac{\partial z}{\partial v}\\ \frac{\partial x}{\partial w} & \frac{\partial y}{\partial w} & \frac{\partial z}{\partial w}\end{vmatrix}=\frac{\partial(x,y,z)}{\partial(u,v,w)}$$

is the Jacobian of the transformation of coordinates. For Cartesian to cylindrical coordinates, $J=r$, and for Cartesian to spherical polars, it is $r^2\sin\theta$.

1.18 Laplace's equation

The equation satisfied by the scalar potential from which a vector field may be derived by taking the gradient is Laplace's equation, written as:

$$\nabla^2\phi=\frac{\partial^2\phi}{\partial x^2}+\frac{\partial^2\phi}{\partial y^2}+\frac{\partial^2\phi}{\partial z^2}=0$$

In cylindrical coordinates:

$$\nabla^2\phi=\frac{1}{r}\frac{\partial}{\partial r}\left(r\frac{\partial\phi}{\partial r}\right)+\frac{1}{r^2}\frac{\partial^2\phi}{\partial\theta^2}+\frac{\partial^2\phi}{\partial z^2}$$

In spherical polars:

$$\nabla^2\phi=\frac{1}{r^2}\frac{\partial}{\partial r}\left(r^2\frac{\partial\phi}{\partial r}\right)+\frac{1}{r^2\sin\theta}\frac{\partial\phi}{\partial\theta}\Bigg)+\frac{1}{r^2\sin^2\theta}\frac{\partial^2\phi}{\partial\phi^2}$$

The equation is solved by setting

$$\phi=U(u)V(u)W(w)$$

in the appropriate form of the equation, separating the variables and solving separately for the three functions, where (u, v, w) is the coordinate system in use.

In Cartesian coordinates, typically the functions are trigonometric, hyperbolic and exponential; in cylindrical coordinates the function of z is exponential, that of θ trigonometric and that of r is a Bessel function. In spherical polars, typically the function of r is a power of r, that of φ is trigonometric, and that of θ is a Legendre function of $\cos\theta$.

1.19 Solution of equations

1.19.1 Quadratic equation

$$ax^2+bx+c=0$$

$$x=-\frac{b}{2a}\pm\frac{\sqrt{b^2-4ac}}{2a}$$

In practical calculations if $b^2>4ac$, so that the roots are real and unequal, calculate the root of larger modulus first, using the same sign for both terms in the formula, then use the fact that $x_1x_2=c/a$ where x_1 and x_2 are the roots. This avoids the severe cancellation of significant digits which may otherwise occur in calculating the smaller root.

For polynomials other than quadratics, and for other functions, several methods of successive approximation are available.

1.19.2 Bisection method

By trial find x_0 and x_1 such that $f(x_0)$ and $f(x_1)$ have opposite signs (see *Figure 1.6*). Set $x_2=(x_0+x_1)/2$ and calculate $f(x_2)$. If $f(x_0)f(x_2)$ is positive, the root lies in the interval (x_1, x_2); if negative in the interval (x_0, x_2); and if zero, x_2 is the root. Continue if necessary using the new interval.

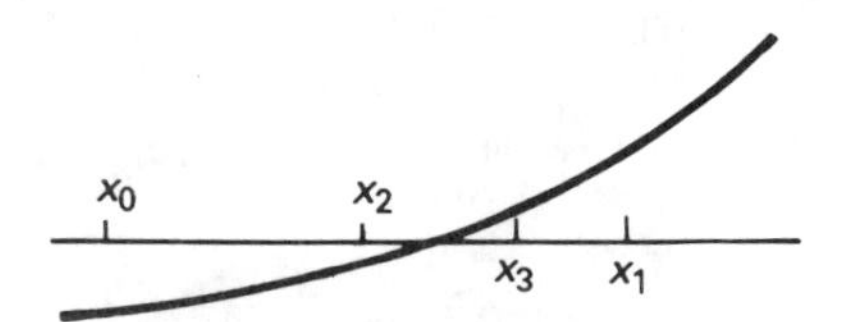

Figure 1.6 Bisection method

1.19.3 Regula falsi

By trial, find x_0 and x_1 as for the bisection method; these two values define two points $(x_0, f(x_0))$ and $(x_1, f(x_1))$. The straight line joining these two points cuts the x-axis at the point (see *Figure 1.7*)

$$x_2=\frac{x_0f(x_1)-x_1f(x_0)}{f(x_1)-f(x_0)}$$

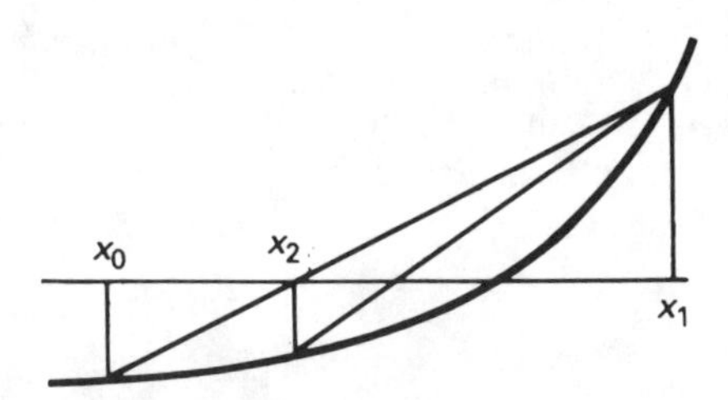

Figure 1.7 Regula falsi

Evaluate $f(x_2)$ and repeat the process for whichever of the intervals (x_0, x_2) or (x_1, x_2) contains the root. This method can be accelerated by halving at each step the function value at the retained end of the interval, as shown in *Figure 1.8*.

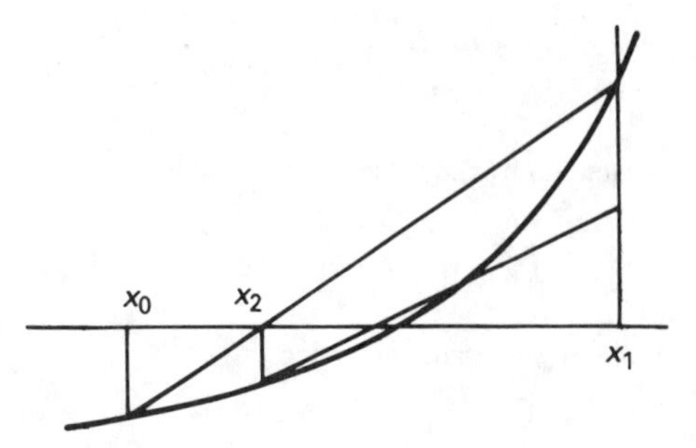

Figure 1.8 Accelerated method

1.19.4 Fixed-point iteration

Arrange the equation in the form

$$x=f(x)$$

Choose an initial value of x by trial, and calculate repetitively

$$x_{k+1}=f(x_k)$$

This process will not always converge.

1.19.5 Newton's method

Calculate repetitively (*Figure 1.9*)

$$x_{k+1}=x_k-f(x_k)/f'(x_k)$$

This method will converge unless: (a) x_k is near a point of inflexion of the function; or (b) x_k is near a local minimum; or (c) the root is multiple. If one of these cases arises, most of the trouble can be overcome by checking at each stage that

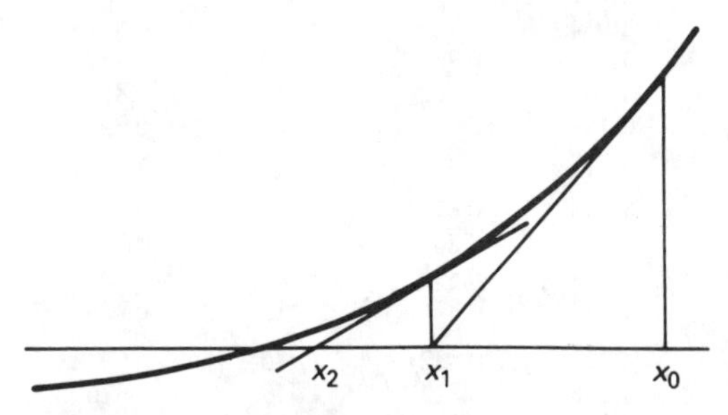

Figure 1.9 Newton's method

$$f(x_{k+1})<f(x_k)$$

and, if not, halving the preceding value of $|x_{k+1}-x_k|$.

1.20 Method of least squares

To obtain the best fit between a straight line $ax+by=1$ and several points $(x_1, y_1), (x_2, y_2), \ldots, (x_n, y_n)$ found by observation, the coefficients a and b are to be chosen so that the sum of the squares of the errors

$$e_i = ax_i + by_i - 1$$

is a minimum. To do this, first write the set of inconsistent equations

$$ax_1 + by_1 - 1 = 0$$

$$ax_2 + by_2 - 1 = 0$$

$$\vdots$$

$$ax_n + by_n - 1 = 0$$

Multiply each equation by the value of x it contains, and add, obtaining

$$a \sum_{i=1}^{n} x_i^2 + b \sum_{i=1}^{n} x_i y_i - \sum_{i=1}^{n} x_i = 0$$

Similarly multiply by y and add, obtaining

$$a \sum_{i=1}^{n} x_i y_i + b \sum_{i=1}^{n} y_i^2 - \sum_{i=1}^{n} y_i = 0$$

Lastly, solve these two equations for a and b, which will be the required values giving the least squares fit.

1.21 Relation between decibels, current and voltage ratio, and power ratio

$$\text{dB} = 10 \log \frac{P_1}{P_2} = 20 \log \frac{V_1}{V_2} = 20 \log \frac{I_1}{I_2}$$

dB	I_1/I_2 or V_1/V_2	I_2/I_1 or V_2/V_1	P_1/P_2	P_2/P_1
0.1	1.012	0.989	1.023	0.977
0.2	1.023	0.977	1.047	0.955
0.3	1.035	0.966	1.072	0.933
0.4	1.047	0.955	1.096	0.912
0.5	1.059	0.944	1.122	0.891
0.6	1.072	0.933	1.148	0.871
0.7	1.084	0.923	1.175	0.851
0.8	1.096	0.912	1.202	0.832
0.9	1.109	0.902	1.230	0.813
1.0	1.122	0.891	1.259	0.794
1.1	1.135	0.881	1.288	0.776
1.2	1.148	0.871	1.318	0.759
1.3	1.162	0.861	1.349	0.741
1.4	1.175	0.851	1.380	0.724
1.5	1.188	0.841	1.413	0.708
1.6	1.202	0.832	1.445	0.692
1.7	1.216	0.822	1.479	0.676
1.8	1.230	0.813	1.514	0.661
1.9	1.245	0.804	1.549	0.645
2.0	1.259	0.794	1.585	0.631
2.5	1.334	0.750	1.778	0.562
3.0	1.413	0.708	1.995	0.501
3.5	1.496	0.668	2.24	0.447
4.0	1.585	0.631	2.51	0.398
4.5	1.679	0.596	2.82	0.355
5.0	1.778	0.562	3.16	0.316
5.5	1.884	0.531	3.55	0.282
6.0	1.995	0.501	3.98	0.251
6.5	2.11	0.473	4.47	0.224
7.0	2.24	0.447	5.01	0.200
7.5	2.37	0.422	5.62	0.178
8.0	2.51	0.398	6.31	0.158
8.5	2.66	0.376	7.08	0.141
9.0	2.82	0.355	7.94	0.126
9.5	2.98	0.335	8.91	0.112
10.0	3.16	0.316	10.00	0.100
10.5	3.35	0.298	11.2	0.0891
11.0	3.55	0.282	12.6	0.0794
15.0	5.62	0.178	31.6	0.0316
15.5	5.96	0.168	35.5	0.0282
16.0	6.31	0.158	39.8	0.0251
16.5	6.68	0.150	44.7	0.0224
17.0	7.08	0.141	50.1	0.0200
17.5	7.50	0.133	56.2	0.0178
18.0	7.94	0.126	63.1	0.0158
18.5	8.41	0.119	70.8	0.0141
19.0	8.91	0.112	79.4	0.0126
19.5	9.44	0.106	89.1	0.0112
20.0	10.00	0.1000	100	0.0100
20.5	10.59	0.0944	112	0.00891
21.0	11.22	0.0891	126	0.00794
21.5	11.88	0.0841	141	0.00708
22.0	12.59	0.0794	158	0.00631
22.5	13.34	0.0750	178	0.00562
23.0	14.13	0.0708	200	0.00501
23.5	14.96	0.0668	224	0.00447
24.0	15.85	0.0631	251	0.00398
24.5	16.79	0.0596	282	0.00355
25.0	17.78	0.0562	316	0.00316
25.5	18.84	0.0531	355	0.00282
26.0	19.95	0.0501	398	0.00251
26.5	21.1	0.0473	447	0.00224
27.0	22.4	0.0447	501	0.00200
27.5	23.7	0.0422	562	0.00178
28.0	25.1	0.0398	631	0.00158
28.5	26.6	0.0376	708	0.00141
29.0	28.2	0.0355	794	0.00126
29.5	29.8	0.0335	891	0.00112
30.0	31.6	0.0316	1000	0.00100
31.0	35.5	0.0282	1260	7.94×10^{-4}
32.0	39.8	0.0251	1580	6.31×10^{-4}
33.0	44.7	0.0224	2000	5.01×10^{-4}
34.0	50.1	0.0200	2510	3.98×10^{-4}
35.0	56.2	0.0178		3.16×10^{-4}
36.0	63.1	0.0158	3980	2.51×10^{-4}
37.0	70.8	0.0141	5010	2.00×10^{-4}

2 Calculus

J Barron BA, MA (Cantab)
University of Cambridge

Contents

2.1 Derivative

$$f'(x)=\lim_{\delta x\to 0}\frac{f(x+\delta x)-f(x)}{\delta x}$$

If u and v are functions of x,

$$(uv)'=u'v+uv'$$

$$\left(\frac{u}{v}\right)'=\frac{u'v-uv'}{v^2}$$

$$(uv)^{(n)}=u^{(n)}v+nu^{(n-1)}v^{(1)}+\ldots+{}^nC_pu^{(n-p)}v^{(p)}+\ldots+uv^{(n)}$$

where

$${}^nC_p=\frac{n!}{p!\,(n-p)!}$$

If $z=f(x)$ and $y=g(z)$, then

$$\frac{\mathrm{d}y}{\mathrm{d}x}=\frac{\mathrm{d}y}{\mathrm{d}z}\frac{\mathrm{d}z}{\mathrm{d}x}$$

2.2 Maxima and minima

$f(x)$ has a stationary point wherever $f'(x)=0$: the point is a maximum, minimum or point of inflexion according as $f''(x)<$, $>$ or $=0$.

$f(x, y)$ has a stationary point wherever

$$\frac{\partial f}{\partial x}=\frac{\partial f}{\partial y}=0$$

Let (a, b) be such a point, and let

$$\frac{\partial^2 f}{\partial x^2}=A,\qquad \frac{\partial^2 f}{\partial x\,\partial y}=H\qquad \frac{\partial^2 f}{\partial y^2}=B$$

all at that point, then:

If $H^2-AB>0$, $f(x, y)$ has a saddle point at (a, b).
If $H^2-AB<0$ and if $A<0$, $f(x, y)$ has a maximum at (a, b), but if $A>0$, $f(x, y)$ has a minimum at (a, b).
If $H^2=AB$, higher derivatives need to be considered.

2.3 Integral

$$\int_a^b f(x)\,\mathrm{d}x=\lim_{N\to\infty}\sum_{n=0}^{N-1} f\left(a+\frac{n(b-a)}{N}\right)\left(\frac{b-a}{N}\right)$$
$$=\lim_{N\to\infty}\sum_{n=1}^{N} f(a+(n-1)\,\delta x)\,\delta x$$

where $\delta x=(b-a)/N$.

If u and v are functions of x, then

$$\int uv'\,\mathrm{d}x=uv-\int u'v\,\mathrm{d}x \quad \text{(integration by parts)}$$

2.4 Derivatives and integrals

y	$\frac{\mathrm{d}y}{\mathrm{d}x}$	$\int y\,\mathrm{d}x$
x^n	nx^{n-1}	$x^{n+1}/(n+1)$
$1/x$	$-1/x^2$	$\ln(x)$
e^{ax}	$a\,e^{ax}$	e^{ax}/a
$\ln(x)$	$1/x$	$x[\ln(x)-1]$
$\log_a x$	$\frac{1}{x}\log_a e$	$x\log_a\left(\frac{x}{e}\right)$
$\sin ax$	$a\cos ax$	$-\frac{1}{a}\cos ax$
$\cos ax$	$-a\sin ax$	$\frac{1}{a}\sin ax$
$\tan ax$	$a\sec^2 ax$	$-\frac{1}{a}\ln(\cos ax)$
$\cot ax$	$-a\,\mathrm{cosec}^2\,ax$	$\frac{1}{a}\ln(\sin ax)$
$\sec ax$	$a\tan ax\sec ax$	$\frac{1}{a}\ln(\sec ax+\tan ax)$
$\mathrm{cosec}\,ax$	$-a\cot ax\,\mathrm{cosec}\,ax$	$\frac{1}{a}\ln(\mathrm{cosec}\,ax-\cot ax)$

y	$\dfrac{dy}{dx}$	$\int y\,dx$
$\arcsin(x/a)$	$1/(a^2-x^2)^{1/2}$	$x\arcsin(x/a)+(a^2-x^2)^{1/2}$
$\arccos(x/a)$	$-1/(a^2-x^2)^{1/2}$	$x\arccos(x/a)-(a^2-x^2)^{1/2}$
$\arctan(x/a)$	$a/(a^2+x^2)$	$x\arctan(x/a)-\frac{1}{2}a\ln(a^2+x^2)$
$\text{arccot}(x/a)$	$-a/(a^2+x^2)$	$x\,\text{arccot}(x/a)+\frac{1}{2}a\ln(a^2+x^2)$
$\text{arcsec}(x/a)$	$a(x^2-a^2)^{-1/2}/x$	$x\,\text{arcsec}(x/a)-a\ln[x+(x^2-a^2)^{1/2}]$
$\text{arccosec}(x/a)$	$-a(x^2-a^2)^{-1/2}/x$	$x\,\text{arccosec}(x/a)+a\ln[x+(x^2-a^2)^{1/2}]$
$\sinh ax$	$a\cosh ax$	$\frac{1}{a}\cosh ax$
$\coth ax$	$a\sinh ax$	$\frac{1}{a}\sinh ax$
$\tanh ax$	$a\,\text{sech}^2\,ax$	$\frac{1}{a}\ln(\cosh ax)$
$\coth ax$	$-a\,\text{cosech}^2\,ax$	$\frac{1}{a}\ln(\sinh ax)$
$\text{sech}\,ax$	$-a\tanh ax\,\text{sech}\,ax$	$\frac{2}{a}\arctan(e^{ax})$
$\text{cosech}\,ax$	$-a\coth ax\,\text{cosech}\,ax$	$\frac{1}{a}\ln\left(\tanh\frac{ax}{2}\right)$
$\text{arsinh}(x/a)$	$(x^2+a^2)^{-1/2}$	$x\,\text{arsinh}(x/a)-(x^2+a^2)^{1/2}$
$\text{arcosh}(x/a)$	$(x^2-a^2)^{-1/2}$	$x\,\text{arcosh}(x/a)-(x^2-a^2)^{1/2}$
$\text{artanh}(x/a)$	$a(a^2-x^2)^{-1}$	$x\,\text{artanh}(x/a)+\frac{1}{2}a\ln(a^2-x^2)$
$\text{arcoth}(x/a)$	$-a(x^2-a^2)^{-1}$	$x\,\text{arcoth}(x/a)+\frac{1}{2}a\ln(x^2-a^2)$
$\text{arsech}(x/a)$	$-a(a^2-x^2)^{-1/2}/x$	$x\,\text{arsech}(x/a)+a\arcsin(x/a)$
$\text{arcosech}(x/a)$	$-a(x^2+a^2)^{-1/2}/x$	$x\,\text{arcosech}(x/a)+a\,\text{arsinh}(x/a)$
$(x^2\pm a^2)^{1/2}$		$\frac{1}{2}x(x^2\pm a^2)^{1/2}\pm\frac{1}{2}a^2\,\text{arsinh}(x/a)$
$(a^2-x^2)^{1/2}$		$\frac{1}{2}x(a^2-x^2)^{1/2}+\frac{1}{2}a^2\arcsin(x/a)$
$(x^2\pm a^2)^p x$		$\begin{cases}\frac{1}{2}(x^2\pm a^2)^{p+1}/(p+1) & (p\neq-1)\\ \frac{1}{2}\ln(x^2\pm a^2) & (p=-1)\end{cases}$
$(a^2-x^2)^p x$		$\begin{cases}-\frac{1}{2}(a^2-x^2)^{p+1}/(p+1) & (p\neq-1)\\ -\frac{1}{2}\ln(a^2-x^2) & (p=-1)\end{cases}$
$x(ax^2+b)^p$		$\begin{matrix}(ax^2+b)^{p+1}/2a(p+1) & (p\neq-1)\\ [\ln(ax^2+b)]/2a & (p=-1)\end{matrix}$
$(2ax-x^2)^{-1/2}$		$\arccos\left(\frac{a-x}{a}\right)$
$(a^2\sin^2 x+b^2\cos^2 x)^{-1}$		$\frac{1}{ab}\arctan\left(\frac{a}{b}\tan x\right)$
$(a^2\sin^2 x-b^2\cos^2 x)^{-1}$		$-\frac{1}{ab}\,\text{artanh}\left(\frac{a}{b}\tan x\right)$
$e^{ax}\sin bx$		$e^{ax}\frac{a\sin bx-b\cos bx}{a^2+b^2}$
$e^{ax}\cos bx$		$e^{ax}\frac{(a\cos bx+b\sin bx)}{a^2+b^2}$

y	$\int y\,dx$	
$\sin mx \sin nx$	$\frac{1}{2}\frac{\sin(m-n)x}{m-n}-\frac{1}{2}\frac{\sin(m+n)x}{m+n}$	$(m\neq n)$
	$\frac{1}{2}\left(x-\frac{\sin 2mx}{2m}\right)$	$(m=n)$
$\sin mx \cos nx$	$-\frac{1}{2}\frac{\cos(m+n)x}{m+n}-\frac{1}{2}\frac{\cos(m-n)x}{m-n}$	$(m\neq n)$
	$-\frac{1}{2}\frac{\cos 2mx}{2m}$	$(m=n)$
$\cos mx \cos nx$	$\frac{1}{2}\frac{\sin(m+n)x}{m+n}+\frac{1}{2}\frac{\sin(m-n)x}{m-n}$	$(m\neq n)$
	$\frac{1}{2}\left(x+\frac{\sin 2mx}{2m}\right)$	$(m=n)$

2.5 Standard substitutions

Integral a function of	*Substitute*
a^2-x^2	$x=a\sin\theta$ or $x=a\cos\theta$
a^2+x^2	$x=a\tan\theta$ or $x=a\sinh\theta$
x^2-a^2	$x=a\sec\theta$ or $x=a\cosh\theta$

2.6 Reduction formulae

$$\int \sin^m x\,dx=-\frac{1}{m}\sin^{m-1}x\cos x+\frac{m-1}{m}\int\sin^{m-2}x\,dx$$

$$\int \cos^m x\,dx=\frac{1}{m}\cos^{m-1}x\sin x+\frac{m-1}{m}\int\cos^{m-2}x\,dx$$

$$\int \sin^m x\cos^n x\,dx=\frac{\sin^{m+1}x\cos^{n-1}x}{m+n}+\frac{n-1}{m+n}\int\sin^m x\cos^{n-2}x\,dx$$

If the integrand is a rational function of $\sin x$ and/or $\cos x$, substitute $t=\tan\frac{1}{2}x$, then

$$\sin x=\frac{1}{1+t^2},\qquad \cos x=\frac{1-t^2}{1+t^2},\qquad dx=\frac{2dt}{1+t^2}$$

2.7 Numerical integration

2.7.1 Trapezoidal rule (*Figure 2.1*)

$$\int_{x1}^{x2} y\,dx=\tfrac{1}{2}h(y_1+y_2)+O(h^3)$$

2.7.2 Simpson's rule (*Figure 2.1*)

$$\int_{x1}^{x2} y\,dx=2h(y_1+4y_2+y_3)/6+O(h^5)$$

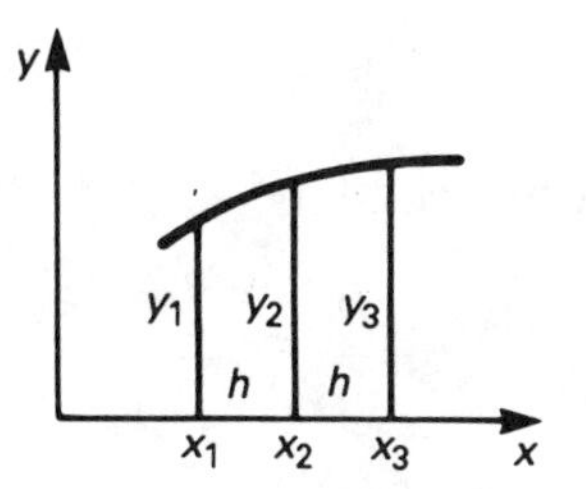

Figure 2.1 Numerical integration

2.7.3 Change of variable in double integral

$$\iint f(x,y)\,dx\,dy=\iint F(u,v)|J|\,du\,dv$$

where

$$J=\frac{\partial(x,y)}{\partial(u,v)}=\begin{vmatrix}\frac{\partial x}{\partial u} & \frac{\partial x}{\partial v}\\ \frac{\partial y}{\partial u} & \frac{\partial y}{\partial v}\end{vmatrix}=\begin{vmatrix}\frac{\partial x}{\partial u} & \frac{\partial y}{\partial u}\\ \frac{\partial x}{\partial v} & \frac{\partial y}{\partial v}\end{vmatrix}$$

is the Jacobian of the transformation.

2.7.4 Differential mean value theorem

$$\frac{f(x+h)-f(x)}{h}=f'(x+\theta h)\qquad 0<\theta<1$$

2.7.5 Integral mean value theorem

$$\int_a^b f(x)g(x)\,dx=g(a+\theta h)\int_a^b f(x)\,dx$$

$h=b-a,\ 0<\theta<1$

2.8 Vector calculus

Let $s(x,y,z)$ be a scalar function of position and let

$\mathbf{v}(x,y,z)=\mathbf{i}v_x(x,y,z)+\mathbf{j}v_y(x,y,z)+\mathbf{k}v_z(x,y,z)$

be a vector function of position. Define

$$\nabla=\mathbf{i}\frac{\partial}{\partial x}+\mathbf{j}\frac{\partial}{\partial y}+\mathbf{k}\frac{\partial}{\partial z}$$

so that

$$\nabla.\nabla=\nabla^2=\frac{\partial^2}{\partial x^2}+\frac{\partial^2}{\partial y^2}+\frac{\partial^2}{\partial z^2}$$

then

$$\text{grad}\,s=\nabla s=\mathbf{i}\frac{\partial s}{\partial x}+\mathbf{j}\frac{\partial s}{\partial y}+\mathbf{k}\frac{\partial s}{\partial z}$$

$$\text{div}\,\mathbf{v}=\nabla.\mathbf{v}=\frac{\partial v_x}{\partial x}+\frac{\partial v_y}{\partial y}+\frac{\partial v_z}{\partial z}$$

$$\text{curl}\,\mathbf{v}=\nabla\times\mathbf{v}=\mathbf{i}\left(\frac{\partial v_z}{\partial y}-\frac{\partial v_y}{\partial z}\right)+\mathbf{j}\left(\frac{\partial v_x}{\partial z}-\frac{\partial v_z}{\partial x}\right)+\mathbf{k}\left(\frac{\partial v_y}{\partial x}-\frac{\partial v_x}{\partial y}\right)$$

The following identities are then true:

$\text{div}(s\mathbf{v})=s\,\textit{div}\,\mathbf{v}+(\text{grad}\,s).\mathbf{v}$

$\text{curl}(s\mathbf{v}) = s\,\text{curl}\,\mathbf{v} + (\text{grad}\, s) \times \mathbf{v}$

$\text{div}(\mathbf{u} \times \mathbf{v}) = \mathbf{v}\,.\,\text{curl}\,\mathbf{u} - \mathbf{u}\,.\,\text{curl}\,\mathbf{v}$

$\text{curl}(\mathbf{u} \times \mathbf{v}) = \mathbf{u}\,\text{div}\,\mathbf{v} - \mathbf{v}\,\text{div}\,\mathbf{u} + (\mathbf{v}\,.\,\nabla)\mathbf{u} - (\mathbf{u}\,.\,\nabla)\mathbf{v}$

$\text{div grad}\, s = \nabla^2 s$

$\text{div curl}\,\mathbf{v} = 0$

$\text{curl grad}\, s = 0$

$\text{curl curl}\,\mathbf{v} = \text{grad}(\text{div}\,\mathbf{v}) - \nabla^2\mathbf{v}$

where ∇^2 operates on each component of **v**.

$\mathbf{v} x\,\text{curl}\,\mathbf{v} + (\mathbf{v}\,.\,\nabla)\mathbf{v} = \text{grad}\,\frac{1}{2}\mathbf{v}^2$

Potentials:

If $\text{curl}\,\mathbf{v} = 0$, $\mathbf{v} = \text{grad}\,\varphi$ where φ is a scalar potential.
If $\text{div}\,\mathbf{v} = 0$, $\mathbf{v} = \text{curl}\,\mathbf{A}$ where **A** is a vector potential.

3 Series and Transforms

J Barron BA, MA (Cantab)
University of Cambridge

Contents

3.1 Arithmetic series

Sum of n terms,

$$S_n = a + (a+d) + (a+2d) + \ldots + [a+(n-1)d]$$
$$= n[2a+(n-1)d]/2$$
$$= n(a+l)/2$$

3.2 Geometric series

Sum of n terms,

$$S_n = a + ar + ar^2 + \ldots + ar^{n-1} = a(1-r^n)/(1-r)$$

$(|r|<1)$

$$S_\infty = a/(1-r)$$

3.3 Binomial series

$$(1+x)^p = 1 + px + \frac{p(p-1)}{2!}x^2 + \frac{p(p-1)(p-2)}{3!}x^3 + \ldots$$

If p is a positive integer the series terminates with the term in x^p and is valid for all x; otherwise the series does not terminate, and is valid only for $-1<x<1$.

3.4 Taylor's series

Infinite form

$$f(x+h) = f(x) + hf'(x) + \frac{h^2}{2!}f''(x) + \ldots$$
$$+ \frac{h^n}{n!}f^{(n)}(x) + \ldots$$

Finite form

$$f(x+h) = f(x) + hf'(x) + \frac{h^2}{2!}f''(x) + \ldots$$
$$+ \frac{h^n}{n!}f^{(n)}(x) + \frac{h^{n+1}}{(n+1)!}f^{(n+1)}(x+\lambda h)$$

where $0 \leqslant \lambda \leqslant 1$.

3.5 Maclaurin's series

$$f(x) = f(0) + xf'(0) + \frac{x^2}{2!}f''(0) + \ldots + \frac{x^n}{n!}f^{(n)}(0) + \ldots$$

Neither of these series is necessarily convergent, but both usually are for appropriate ranges of values of h and of x respectively.

3.6 Laurent's series

If a function $f(z)$ of a complex variable is analytic on and everywhere between two concentric circles centre a, then at any point in this region

$$f(z) = a_0 + a_1(z-a) + \ldots + b_1/(z-a) + b_2/(z-a)^2 + \ldots$$

This series is often applicable when Taylor's series is not.

3.7 Power series for real variables

	Math	*Comp*
$e^x = 1 + x + \frac{x^2}{2!} + \ldots$	all x	$\lvert x\rvert \leqslant 1$
$\ln(1+x) = x - \frac{x^2}{2} + \frac{x^3}{3} - \frac{x^4}{4} + \ldots$	$-1 < x \leqslant 1$	
$\sin x = x - \frac{x^3}{3!} + \frac{x^5}{5!} - \frac{x^7}{7!} + \ldots$	all x	$\lvert x\rvert \leqslant 1$
$\cos x = 1 - \frac{x^2}{2!} + \frac{x^4}{4!} - \frac{x^6}{6!} + \ldots$	all x	$\lvert x\rvert \leqslant 1$
$\tan x = x + \frac{x^3}{3} + \frac{2x^5}{15} + \frac{17x^7}{315} + \ldots$		$\lvert x\rvert < \frac{\pi}{2}$
$\arctan x = x - \frac{x^3}{3} + \frac{x^5}{5} - \frac{x^7}{7} + \ldots$		$\lvert x\rvert \leqslant 1$
$\sinh x = x + \frac{x^3}{3!} + \frac{x^5}{5!} + \frac{x^7}{7!} + \ldots$	all x	$\lvert x\rvert \leqslant 1$
$\cosh x = 1 + \frac{x^2}{2!} + \frac{x^4}{4!} + \frac{x^6}{6!} + \ldots$	all x	$\lvert x\rvert \leqslant 1$

The column headed 'Math' contains the range of values of the variable x for which the series is convergent in the pure mathematical sense. In some cases a different range of values is given in the column headed 'Comp', to reduce the rounding errors which arise when computers are used.

3.8 Integer series

$$\sum_{n=1}^{N} n = 1+2+3+4+\ldots+N = N(N+1)/2$$

$$\sum_{n=1}^{N} n^2 = 1^2+2^2+3^2+4^2+\ldots+N^2 = N(N+1)(2N+1)/6$$

$$\sum_{n=1}^{N} n^3 = 1^3+2^3+3^3+4^3+\ldots+N^3 = N^2(N+1)^2/4$$

$$\sum_{n=1}^{\infty} \frac{(-1)^{n+1}}{n} = 1 - \frac{1}{2} + \frac{1}{3} - \frac{1}{4} + \ldots = \ln(2) \quad \text{(see } \ln(1+x)\text{)}$$

$$\sum_{n=1}^{\infty} \frac{(-1)^{n+1}}{2n-1} = 1 - \frac{1}{3} + \frac{1}{5} - \frac{1}{7} + \ldots = \frac{\pi}{4} \quad \text{(see } \arctan x\text{)}$$

$$\sum_{n=1}^{\infty} \frac{1}{n^2} = 1 + \frac{1}{4} + \frac{1}{9} + \frac{1}{16} + \ldots = \frac{\pi^2}{6}$$

$$\sum_{n=1}^{N} n(n+1)(n+2)\ldots(n+r)$$
$$= 1.2.3\ldots + 2.3.4\ldots + 3.4.5\ldots + \ldots$$
$$+ N(N+1)(N+2)\ldots(N+r)$$
$$= \frac{N(N+1)(N+2)\ldots(N+r+1)}{r+2}$$

3.9 Fourier series

$$f(\theta) = \tfrac{1}{2}a_0 + \sum_{n=1}^{\infty}(a_n \cos n\theta + b_n \sin n\theta)$$

with

$$a_n = \frac{1}{\pi}\int_0^{2\pi} f(\Theta)\cos n\Theta\,\mathrm{d}\Theta$$

$$b_n = \frac{1}{\pi}\int_0^{2\pi} f(\Theta)\sin n\Theta\,\mathrm{d}\Theta$$

or

$$f(\theta) = \sum_{n=-\infty}^{\infty} c_n \exp(jn\theta)$$

with

$$c_n = \frac{1}{2\pi}\int_0^{2\pi} f(\Theta)\exp(-jn\Theta)\,\mathrm{d}\Theta = \begin{cases}\frac{1}{2}(a_n + jb_n) & n<0\\ \frac{1}{2}(a_n - jb_n) & n>0\end{cases}$$

The above expressions for Fourier series are valid for functions having at most a finite number of discontinuities within the period 0 to 2 of the variable of integration.

3.10 Rectified sine wave

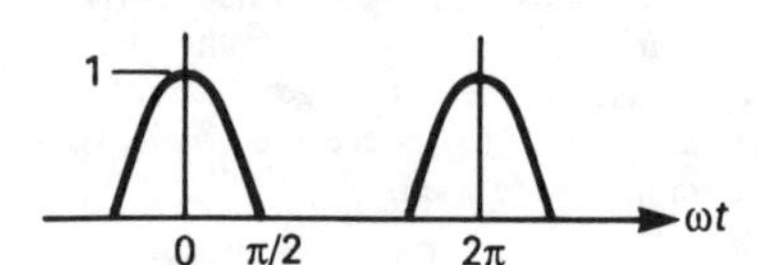

Figure 3.1 Half wave

$$f(\omega t) = \frac{1}{\pi} + \frac{1}{2}\cos\omega t + \frac{2}{\pi}\sum_{n=1}^{\infty}(-1)^{n+1}\frac{\cos 2n\omega t}{4n^2-1}$$

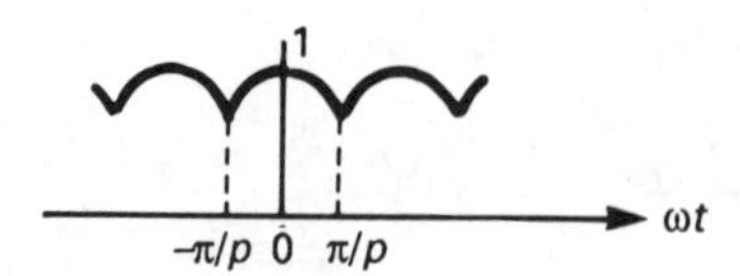

Figure 3.2 p-phase

$$f(\omega t) = \frac{\sin(\pi/p)}{\pi/p} + \frac{2p}{\pi}\sin\left(\frac{\pi}{p}\right)\sum_{n=1}^{\infty}(-1)^{n+1}\frac{\cos np\omega t}{p^2n^2-1}$$

3.11 Square wave

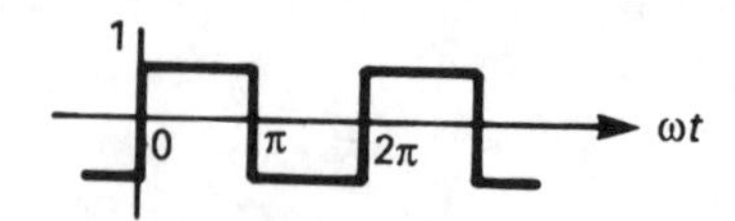

Figure 3.3 Square wave

$$f(\omega t) = \frac{4}{\pi}\sum_{n=1}^{\infty}\frac{\sin(2n-1)\omega t}{(2n-1)}$$

3.12 Triangular wave

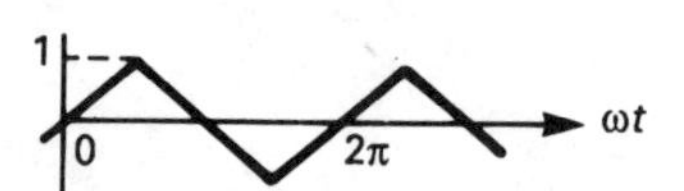

Figure 3.4 Triangular wave

$$f(\omega t) = \frac{8}{\pi^2}\sum_{n=1}^{\infty}(-1)^{n+1}\frac{\sin(2n-1)\omega t}{(2n-1)^2}$$

3.13 Sawtooth wave

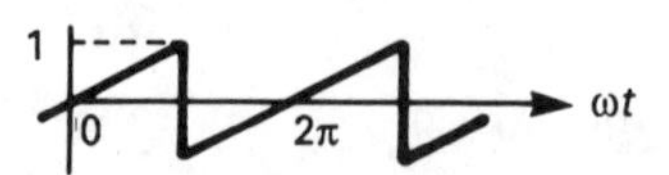

Figure 3.5 Sawtooth wave

$$f(\omega t) = \frac{2}{\pi}\sum_{n=1}^{\infty}(-1)^{n+1}\frac{\sin n\omega t}{n}$$

3.14 Pulse wave

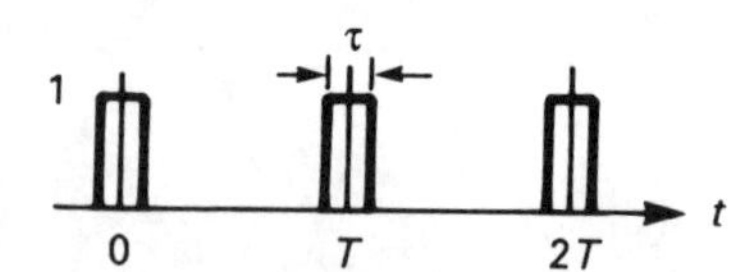

Figure 3.6 Pulse wave

$$f(t) = \frac{\tau}{T} + \frac{2\tau}{T}\sum_{n=1}^{\infty}\frac{\sin(n\omega\tau/T)}{n\pi\tau/T}\cos\left(\frac{2n\pi t}{T}\right)$$

3.15 Fourier transforms

Among other applications, these are used for converting from the time domain to the frequency domain.

Basic formulae:

$$\int_{-\infty}^{\infty} U(f)\exp(j2\pi ft)\,\mathrm{d}f = u(t) \leftrightarrows U(f) = \int_{-\infty}^{\infty} u(t)\exp(-j2\pi ft)\,\mathrm{d}t$$

Change of sign and complex conjugates:

$u(-t) \leftrightarrows U(-f)$, $u^*(t) \leftrightarrows U^*(-f)$

Time and frequency shifts (τ and φ constant):

$u(t-\tau) \leftrightarrows U(f)\exp(-j2\pi f\tau)\exp(j2\pi\varphi t)u(t) \leftrightarrows U(f-\varphi)$

Scaling (T constant):

$u(t/T) \leftrightarrows TU(fT)$

Products and convolutions:

$u(t)\dagger v(t) \leftrightarrows U(f)V(f), \qquad u(t)v(t) \leftrightarrows U(f)\dagger V(f)$

Differentiation:

$u'(t) \leftrightarrows j2\pi f U(f), \qquad -j2\pi t u(t) \leftrightarrows U'(f)$

$\partial u(t,\alpha)/\partial\alpha \leftrightarrows \partial(U-f,\alpha)/\partial\alpha$

Integration ($U(0)=0$, a and b real constants):

$$\int_{-\infty}^{t} u(\tau)\,d\tau \leftrightarrows U(f)/j2\pi f$$

$$\int_{a}^{b} v(t,\alpha)\,d\alpha \leftrightarrows \int_{a}^{b} V(f,\alpha)\,d\alpha$$

Interchange of functions:

$U(t) \leftrightarrows u(-f)$

Dirac delta functions:

$\delta(t) \leftrightarrows 1 \qquad \exp(j2\pi f_0 t) \leftrightarrows \delta(f-f_0)$

Rect(t) (unit length, unit amplitude pulse, centred on $t=0$):

$\mathrm{rect}(t) \leftrightarrows \sin \pi f/\pi f$

Gaussian distribution:

$\exp(-\pi t^2) \leftrightarrows \exp(-\pi f^2)$

Repeated and impulse (delta function) sampled waveforms:

$$\sum_{-\infty}^{\infty} u(t-nT) \leftrightarrows (1/T)U(f) \sum_{-\infty}^{\infty} \delta(f-n/T)$$

$$u(t) \sum_{-\infty}^{\infty} \delta(t-nT) \leftrightarrows (1/T) \sum_{-\infty}^{\infty} U(f-n/T)$$

Parseval's lemma:

$$\int_{-\infty}^{\infty} u(t)v^*(t)\,dt = \int_{-\infty}^{\infty} U(f)V^*(f)\,df$$

$$\int_{-\infty}^{\infty} |u(t)|^2\,dt = \int_{-\infty}^{\infty} |U(f)|^2\,df$$

3.16 Laplace transforms

$$\bar{x}_s = \int_0^{\infty} x(t)\exp(-st)$$

Function	*Transform*	*Remarks*
$e^{-\alpha t}$	$\dfrac{1}{s+\alpha}$	
$\sin \omega t$	$\dfrac{\omega}{s^2+\omega^2}$	
$\cos \omega t$	$\dfrac{s}{s^2+\omega^2}$	
$\sinh \omega t$	$\dfrac{\omega}{s^2-\omega^2}$	
$\cosh \omega t$	$\dfrac{s}{s^2-\omega^2}$	
t^n	$n!/s^{n+1}$	
1	$1/s$	
$H(t-\tau)$	$\dfrac{1}{s}\exp(-s\tau)$	Heaviside step function
$x(t-\tau)H(t-\tau)$	$\exp(-s\tau)\bar{x}(s)$	Shift in t
$\delta(t-\tau)$	$\exp(-s\tau)$	Dirac delta function
$\exp(-\alpha t)\,x(t)$	$\bar{x}(s+\alpha)$	Shift in s
$\exp(-\alpha t)\sin \omega t$	$\dfrac{\omega}{(s+\alpha)^2+\omega^2}$	
$\exp(-\alpha t)\cos \omega t$	$\dfrac{(s+\alpha)}{(s+\alpha)^2+\omega^2}$	
$tx(t)$	$-\dfrac{d\bar{x}(s)}{ds}$	
$\dfrac{dx(t)}{dt}=x'(t)$	$s\bar{x}(s)-x(0)$	
$\dfrac{d^2x(t)}{dt^2}=x''(t)$	$s^2\bar{x}(s)-sx(0)-x'(0)$	
$\dfrac{d^nx(t)}{dx^n}=x^{(n)}(t)$	$s^n\bar{x}(s)-s^{n-1}x(0)-s^{n-2}x'(0)\ldots -sx^{(n-2)}(0)-x^{(n-1)}(0)$	

Convolution integral

$$\int_0^t x_1(\sigma)x_2(t-\sigma)\,d\sigma \to \bar{x}_1(s)\bar{x}_2(s)$$

4 Matrices and Determinants

J Barron BA, MA (Cantab)
University of Cambridge

Contents

4.1 Linear simultaneous equations

The set of equations

$a_{11}x_1 + a_{12}x_2 + \ldots + a_{1n}x_n = b_1$

$a_{21}x_1 + a_{22}x_2 + \ldots + a_{2n}x_n = b_2$

$\ldots$

$a_{n1}x_1 + a_{n2}x_2 + \ldots + a_{nn}x_n = b_n$

may be written symbolically

$\mathbf{Ax} = \mathbf{b}$

in which $\mathbf{A}$ is the *matrix* of the coefficients a_{ij}, and $\mathbf{x}$ and $\mathbf{b}$ are the *column matrices* (or vectors) $(x_1 \ldots x_n)$, and $(b_1 \ldots b_n)$. In this case the matrix $\mathbf{A}$ is square $(n \times n)$. The equations can be solved unless two or more of them are not independent, in which case

$\det \mathbf{A} = |\mathbf{A}| = 0$

and there then exist non-zero solutions x_i only if $\mathbf{b} = 0$. If $\det \mathbf{A} \neq 0$, there exist non-zero solutions only if $\mathbf{b} \neq 0$. When $\det \mathbf{A} = 0$, $\mathbf{A}$ is *singular*.

4.2 Matrix arithmetic

If $\mathbf{A}$ and $\mathbf{B}$ are both matrices of m rows and n columns they are *conformable*, and

$\mathbf{A} \pm \mathbf{B} = \mathbf{C}$ where $C_{ij} = A_{ij} \pm B_{ij}$

4.2.1 Product

If $\mathbf{A}$ is an $m \times n$ matrix and $\mathbf{B}$ an $n \times l$, the product $\mathbf{AB}$ is defined by

$$(\mathbf{AB})_{ij} = \sum_{k=1}^{n} (\mathbf{A})_{ik}(\mathbf{B})_{kj}$$

In this case, if $l \neq m$, the product $\mathbf{BA}$ will not exist.

4.2.2 Transpose

The transpose of $\mathbf{A}$ is written $\mathbf{A}'$ or $\mathbf{A}^t$ and is the matrix whose rows are the columns of $\mathbf{A}$, i.e.

$(\mathbf{A}^t)_{ij} = (\mathbf{A})_{ji}$

A square matrix may be equal to its transpose, and it is then said to be *symmetrical*. If the product $\mathbf{AB}$ exists, then

$(\mathbf{AB})^t = \mathbf{B}^t\mathbf{A}^t$

4.2.3 Adjoint

The *adjoint* of a square matrix $\mathbf{A}$ is defined as $\mathbf{B}$, where

$(\mathbf{B})_{ij} = (A)_{ji}$

and A_{ji} is the *cofactor* of a_{ji} in $\det \mathbf{A}$.

4.2.4 Inverse

If $\mathbf{A}$ is non-singular, the *inverse* $\mathbf{A}^{-1}$ is given by

$\mathbf{A}^{-1} = \operatorname{adj} \mathbf{A}/\det \mathbf{A}$ and $\mathbf{A}^{-1}\mathbf{A} = \mathbf{AA}^{-1} = \mathbf{I}$

the *unit* matrix.

$(\mathbf{AB})^{-1} = \mathbf{B}^{-1}\mathbf{A}^{-1}$

if both inverses exist. The original equations $\mathbf{Ax} = \mathbf{b}$ have the solutions $\mathbf{x} = \mathbf{A}^{-1}\mathbf{b}$ if the inverse exists.

4.2.5 Orthogonality

A matrix $\mathbf{A}$ is orthogonal if $\mathbf{AA}^t = \mathbf{I}$. If $\mathbf{A}$ is the matrix of a coordinate transformation $\mathbf{X} = \mathbf{AY}$ from variables y_i to variables x_i, then if $\mathbf{A}$ is orthogonal $\mathbf{X}^t\mathbf{X} = \mathbf{Y}^t\mathbf{Y}$, or

$$\sum_{i=1}^{n} x_i^2 = \sum_{i=1}^{n} y_i^2$$

4.3 Eigenvalues and eigenvectors

The equation

$\mathbf{Ax} = \lambda\mathbf{x}$

where $\mathbf{A}$ is a square matrix, $\mathbf{x}$ a column vector and λ a number (in general complex) has at most n solutions $(\mathbf{x}, \lambda)$. The values of λ are *eigenvalues* and those of $\mathbf{x}$ *eigenvectors* of the matrix $\mathbf{A}$. The relation may be written

$(\mathbf{A} - \lambda\mathbf{I})\mathbf{x} = 0$

so that if $\mathbf{x} \neq 0$, the equation $\mathbf{A} - \lambda\mathbf{I} = 0$ gives the eigenvalues. If $\mathbf{A}$ is symmetric and real, the eigenvalues are real. If $\mathbf{A}$ is symmetric, the eigenvectors are orthogonal. If $\mathbf{A}$ is not symmetric, the eigenvalues are complex and the eigenvectors are not orthogonal.

4.4 Coordinate transformation

Suppose $\mathbf{x}$ and $\mathbf{y}$ are two vectors related by the equation

$\mathbf{y} = \mathbf{Ax}$

when their components are expressed in one orthogonal system, and that a second orthogonal system has unit vectors $\mathbf{u}_1, \mathbf{u}_2, \ldots, \mathbf{u}_n$ expressed in the first system. The components of $\mathbf{x}$ and $\mathbf{y}$ expressed in the new system will be $\mathbf{x}'$ and $\mathbf{y}'$, where

$\mathbf{x}' = \mathbf{U}^t\mathbf{x}, \qquad \mathbf{y}' = \mathbf{U}^t\mathbf{y}$

and $\mathbf{U}^t$ is the orthogonal matrix whose rows are the unit vectors $\mathbf{u}_1^t$, $\mathbf{u}_2^t$, etc. Then

$\mathbf{y}' = \mathbf{U}^t\mathbf{y} = \mathbf{U}^t\mathbf{Ax} = \mathbf{U}^t\mathbf{A}\mathbf{x}' = \mathbf{U}^t\mathbf{AUx}'$

or

$\mathbf{y}' = \mathbf{A}'\mathbf{x}'$

where

$\mathbf{A}' = \mathbf{U}^t\mathbf{AU}$

Matrices $\mathbf{A}$ and $\mathbf{A}'$ are *congruent*.

4.5 Determinants

The determinant

$$D = \begin{vmatrix} a_{11} & a_{12} & \cdots & a_{1n} \\ a_{21} & a_{22} & \cdots & a_{2n} \\ \vdots & \vdots & \vdots & \vdots \\ a_{n1} & a_{n2} & \cdots & a_{nn} \end{vmatrix}$$

is defined as follows. The first suffix in a_{rs} refers to the row, the second to the column which contains a_{rs}. Denote by M_{rs} the determinant left by deleting the rth row and sth column from D, then

$$D=\sum_{k=1}^{n}(-1)^{k+1}a_{1k}M_{1k}$$

gives the value of D in terms of determinants of order $n-1$, hence by repeated application, of the determinant in terms of the elements a_{rs}.

4.6 Properties of determinants

If the rows of $|a_{rs}|$ are identical with the columns of $|b_{sr}|$, $a_{rs}=b_{sr}$ and

$$|a_{rs}|=|b_{sr}|$$

that is, the *transposed* determinant is equal to the original.

If two rows or two columns are interchanged, the numerical value of the determinant is unaltered, but the sign will be changed if the permutation of rows or columns is odd.

If two rows or two columns are identical, the determinant is zero.

If each element of one row or one column is multiplied by k, so is the value of the determinant.

If any row or column is zero, so is the determinant.

If each element of the pth row or column of the determinant c_{rs} is equal to the sum of the elements of the same row or column in determinants a_{rs} and b_{rs}, then

$$|c_{rs}|=|a_{rs}|+|b_{rs}|$$

The addition of any multiple of one row (or column) to another row (or column) does not alter the value of the determinant.

4.6.1 Minor

If row p and column q are deleted from $|a_{rs}|$, the remaining determinant M_{pq} is called the *minor* of a_{pq}.

4.6.2 Cofactor

The *cofactor* of a_{pq} is the minor of a_{pq} prefixed by the sign which the product $M_{pq}a_{pq}$ would have in the expansion of the determinant, and is denoted by A_{pq}:

$$A_{pq}=(-1)^{p+q}M_{pq}$$

A determinant a_{ij} in which $a_{ij}=a_{ji}$ for all i and j is called *symmetric*, whilst if $a_{ij}=-a_{ji}$ for all i and j, the determinant is *skew-symmetric*. It follows that $a_{ii}=0$ for all i in a skew-symmetric determinant.

4.7 Numerical solution of linear equations

Evaluation of a determinant by direct expansion in terms of elements and cofactors is disastrously slow, and other methods are available, usually programmed on any existing computer system.

4.7.1 Reduction of determinant or matrix to upper triangular or to diagonal form

The system of equations may be written

$$\begin{bmatrix} a_{11} & a_{12} & \cdots & a_{1n} \\ a_{21} & a_{22} & \cdots & a_{2n} \\ \vdots & \vdots & \vdots & \vdots \\ a_{n1} & a_{n2} & \cdots & a_{nn} \end{bmatrix} \begin{matrix} \\ x_1 \\ \vdots \\ \end{matrix} \begin{bmatrix} x_1 \\ x_2 \\ \vdots \\ x_n \end{bmatrix} = \begin{bmatrix} b_1 \\ b_2 \\ \vdots \\ b_n \end{bmatrix}$$

The variable x_1 is eliminated from the last $n-1$ equations by adding a multiple $-a_{i1}/a_{11}$ of the first row to the ith, obtaining

$$\begin{bmatrix} a_{11} & a_{12} & \cdots & a_{1n} \\ 0 & a'_{22} & \cdots & a'_{2n} \\ \vdots & \vdots & \cdots & \vdots \\ 0 & 0 & \cdots & a''_{nn} \end{bmatrix} \begin{matrix} \\ x_1 \\ \vdots \\ \end{matrix} \begin{bmatrix} x_1 \\ x_2 \\ \vdots \\ x_n \end{bmatrix} = \begin{bmatrix} b_1 \\ b'_1 \\ \vdots \\ b''_n \end{bmatrix}$$

where primes indicate altered coefficients. This process may be continued by eliminating x_2 from rows 3 to n, and so on. Eventually the form will become

$$\begin{bmatrix} a_{11} & a_{12} & \cdots & a_{1n} \\ 0 & a'_{22} & \cdots & a'_{2n} \\ \vdots & \vdots & \cdots & \vdots \\ 0 & 0 & \cdots & a''_{nn} \end{bmatrix} \begin{matrix} \\ x_1 \\ \vdots \\ \end{matrix} \begin{bmatrix} x_1 \\ x_2 \\ \vdots \\ x_n \end{bmatrix} = \begin{bmatrix} b_1 \\ b'_2 \\ \vdots \\ b''_n \end{bmatrix}$$

x_n can now be found from the nth equation, substituted in the $(n-1)$th to obtain x_{n-1} and so on.

Alternatively the process may be applied to the system of equations in the form

$$\mathbf{Ax}=\mathbf{Ib}$$

where $\mathbf{I}$ is the unit matrix, and the same operations carried out upon $\mathbf{I}$ as upon $\mathbf{A}$. If the process is continued after reaching the upper triangular form, the matrix $\mathbf{A}$ can eventually be reduced to diagonal form. Finally, each equation is divided by the corresponding diagonal element of $\mathbf{A}$, thus reducing $\mathbf{A}$ to the unit matrix. The system is now in the form

$$\mathbf{Ix}=\mathbf{Bb}$$

and evidently $\mathbf{B}=\mathbf{A}^{-1}$. The total number of operations required is $O(n^3)$.

5

Electric Circuit Theory

P G Lund
Department of Engineering Science,
Oxford University

Contents

5.1 Types of source

If we were to measure the terminal voltage of a source as an increasing current was drawn we should find the relationships shown in *Figure 5.1(a)* where A is a line of constant voltage V_0 obtained when the generator is perfect and without internal impedance, and where B shows the practical case in which there is internal impedance Z and $V=V_0-IZ$. The corresponding graphs for a constant current generator are shown in *Figure 5.1(b)* where A shows the perfect case and B the imperfect case. Apart

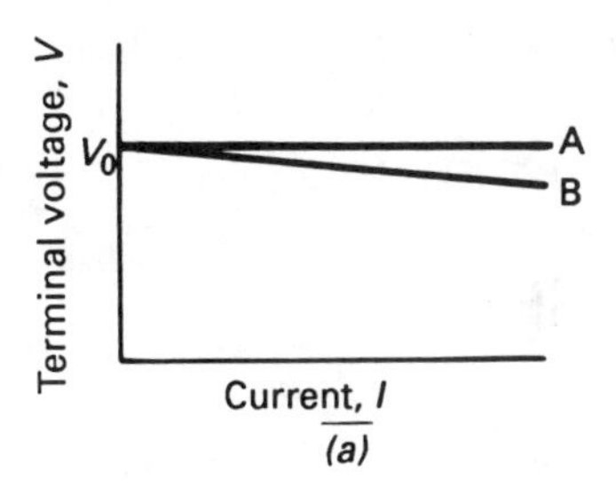

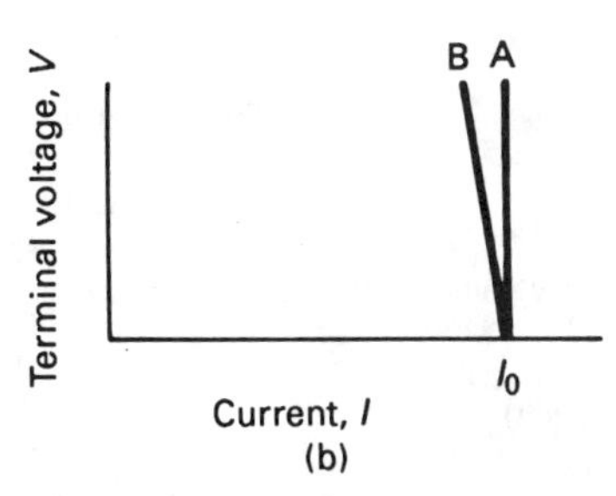

Figure 5.1 Types of sources: (a) constant voltage; (b) constant current

from the lack of familiarity the difficulty with a constant current generator is that the perfect case calls for an infinite impedance in parallel in any equivalent circuit and a practical case requires an impedance which is large but not infinite. This can best be seen by considering the equations which represent line B in *Figure 5.1(b)*, i.e. $I=I_0-VY$ or $I=I_0-V/Z$, where Y is the admittance of an element and equals $1/Z$ where Z is the impedance.

The equivalent circuits for the two types of generator are usually shown as in *Figures 5.2(a)* and *(b)*.

In practice the constant current generator is a useful aid towards the understanding of many transistors in which, crudely but often sufficiently accurately, the output is a current constant over a range of loads.

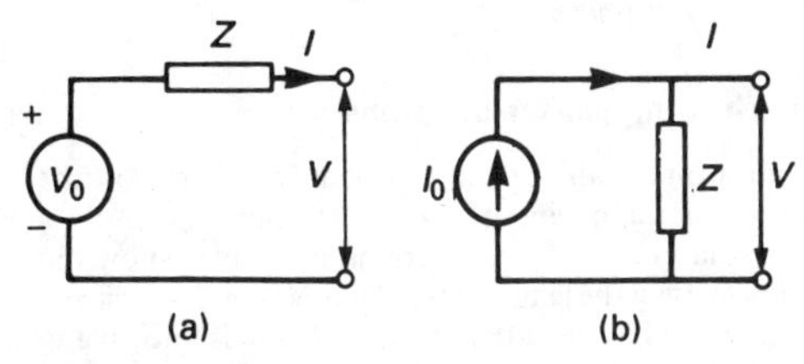

Figure 5.2 Equivalent circuit: (a) voltage source; (b) current source

5.2 Alternating current theory

Because it is possible to analyse all periodic functions of time into series of sinusoids and because of the mathematical properties of sinusoids we always consider currents and voltages varying in such a way that

$$\left.\begin{array}{l} i=I\sin\omega t=I\sin 2\pi ft \\ v=V\sin\omega t=V\sin 2\pi ft \end{array}\right\} \tag{5.1}$$

or

$$\left.\begin{array}{l} i=I\cos\omega t=I\sin 2\pi ft \\ v=V\cos\omega t=V\cos 2\pi ft \end{array}\right\} \tag{5.2}$$

where

i, v = instantaneous values of current, voltage
I, V = maximum or peak values of current, voltage
ω = angular frequency (in radians/s)
f = frequency (in cycles/s or hertz).

An alternative approach is to make use of complex number ideas based on de Moivre's theorem which states that

$$e^{j\theta}=\cos\theta+j\sin\theta \tag{5.3}$$

where $j=\sqrt{-1}$, so that we can write

$$v=V\,\mathrm{Re}\,(e^{j\omega t}) \tag{5.4}$$

for $V\cos\omega t$, or

$$v=V\,\mathrm{Im}\,(e^{j\omega t}) \tag{5.5}$$

for $V\sin\omega t$, where Re and Im stand for the real and imaginary parts of $e^{j\omega t}$. The convenient mathematical properties mentioned above are the simple forms of the differentials:

$$dv/dt=V\omega\cos\omega t \tag{5.6}$$

if $v=V\sin\omega t$, and

$$dv/dt=\mathrm{Re}\text{ or }\mathrm{Im}\,(j\omega V e^{j\omega t}) \tag{5.7}$$

dependent on whether v was assumed to be the real or the imaginary part of $V e^{j\omega t}$.

Integration produces similar expressions:

$$\int v\,dt=-(V/\omega)\cos\omega t \tag{5.8}$$

for $v=V\sin\omega t$, and

$$\int v\,dt=\mathrm{Re}\text{ or }\mathrm{Im}\left(\frac{V}{j\omega}e^{j\omega t}\right)=\mathrm{Re}\text{ or }\mathrm{Im}\left(-\frac{jV}{\omega}e^{j\omega t}\right) \tag{5.9}$$

The nature of these variations is shown in *Figure 5.3*. Other properties of sinusoids and various trigonometrical relations will be developed in later sections.

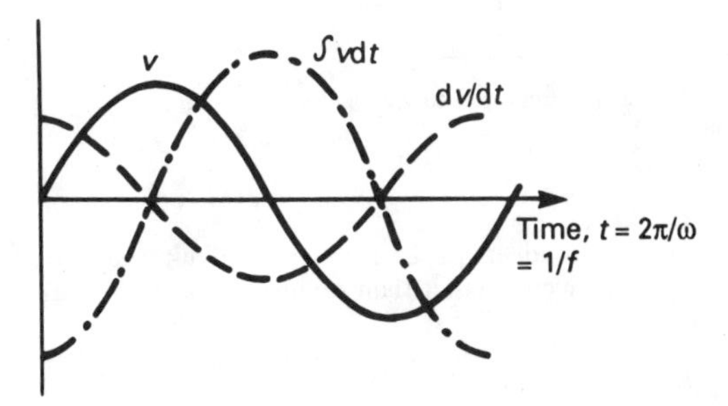

Figure 5.3 Sinusoidal variation

5.3 Resistance, inductance, capacitance and related quantities

Resistance, inductance and capacitance are the three main elements that either absorb or store electrical energy. They constitute the passive as opposed to the active parts of a circuit such as voltage and current sources. The relationship between voltage and current for each is of fundamental importance.

5.3.1 Resistance

Ohm's law states that

$$v = Ri \tag{5.10}$$

The unit of resistance is the ohm (symbol Ω).

5.3.2 Inductance

$$v = L\frac{di}{dt} \tag{5.11}$$

The unit of inductance is the henry (symbol H).

This equation is sometimes found with a minus sign, but then the 'v' is voltage induced in the inductor which has to be overcome by an applied voltage. Because our purpose is to study circuits we think in terms of the applied voltage.

5.3.3 Capacitance

The basic relation for a capacitor is

$$q = vC \tag{5.12}$$

where q is charge in coulombs (symbol C) and C is capacitance in farads (symbol F). Differentiation leads to the more usual expressions

$$\frac{dq}{dt} = i = C\frac{dv}{dt} \tag{5.13}$$

or

$$v = \frac{1}{C}\int i\,dt \tag{5.14}$$

In some circumstances it is convenient to use the reciprocal quantities, the main occasion being when many elements are in parallel, as in *Figure 5.4*. If we introduce the conductance G (equal

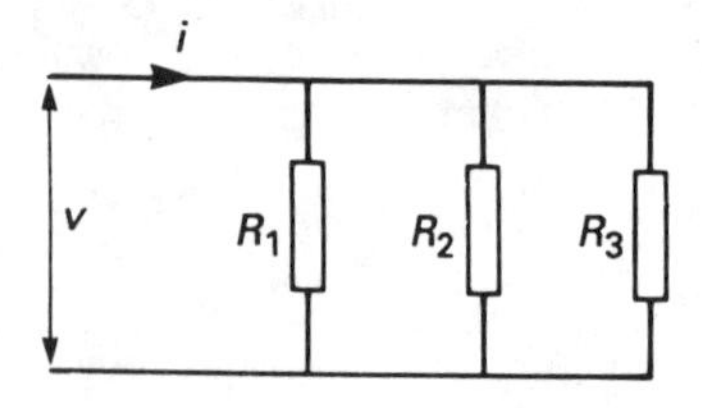

Figure 5.4 Parallel circuit to show advantage of using conductances

to $1/R$) then the conductance of the whole circuit becomes $G = G_1 + G_2 + G_3$ which is easier than solving

$$\frac{1}{R} = \frac{1}{R_1} + \frac{1}{R_2} + \frac{1}{R_3}$$

for R. Of course Ohm's law now takes the form $i = Gv$.

The unit of conductance is the siemen (symbol S). The phrase 'reciprocal ohms' (symbol ℧) is sometimes used.

Reciprocals for inductance and capacitance are not used in the same way but related quantities will be introduced when discussing a.c. circuits. Similarly the idea of impedance and its reciprocal, admittance, will be introduced later.

5.4 AC analysis of electric circuits

5.4.1 Mathematical approach

Consider a circuit such as that shown in *Figure 5.5* in which the

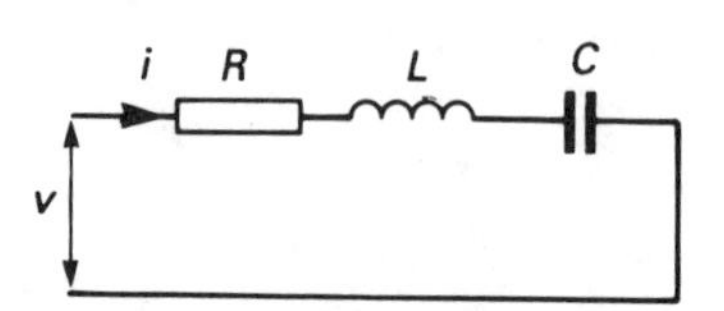

Figure 5.5 Alternating current series circuit

elements are connected in series. The applied voltage must be

$$v = Ri + L\frac{di}{dt} + \frac{1}{C}\int i\,dt \tag{5.15}$$

If it is assumed that v and i vary sinusoidally i can be expressed as $I\exp(j\omega t)$ and Equation (5.15) becomes

$$\begin{aligned} v &= \left(RI + j\omega LI + \frac{1}{j\omega C}I\right)\exp(j\omega t) \\ &= \left(R + j\omega L - \frac{j}{\omega C}\right)I\exp(j\omega t) \\ &= \left[R + j\left(\omega L - \frac{1}{\omega C}\right)\right]I\exp(j\omega t) \end{aligned} \tag{5.16}$$

The expression $R + j(\omega L - 1/\omega C)$ can be written

$$\sqrt{R^2 + \left(\omega L - \frac{1}{\omega C}\right)^2}\exp(j\varphi)$$

where

$$\varphi = \tan^{-1}\left(\frac{\omega L - 1/\omega C}{R}\right)$$

The quantity $\sqrt{R^2 + (\omega L - 1/\omega C)^2}$ is known as the impedance, Z. Equation (5.16) can now be written

$$v = Z\exp(j\varphi)I\exp(j\omega t) = ZI\exp[j(\omega t + \varphi)] \tag{5.17}$$

The product ZI is the magnitude V of this applied voltage and angle φ is termed the phase angle.

5.4.2 Approach using phasor diagrams

Sinusoidal variations can be understood in terms of the projection onto a straight line of a point that moves round a circle, or of a radial line that sweeps round a circle as shown in *Figure 5.6*. If $\theta = \omega t$ then the lengths OA, OB represent $V\sin\omega t$ and $V\cos\omega t$ respectively. The line OP represents V and is known as a phasor. Such lines used to be called vectors but this is now regarded as misleading because vectors, as understood in

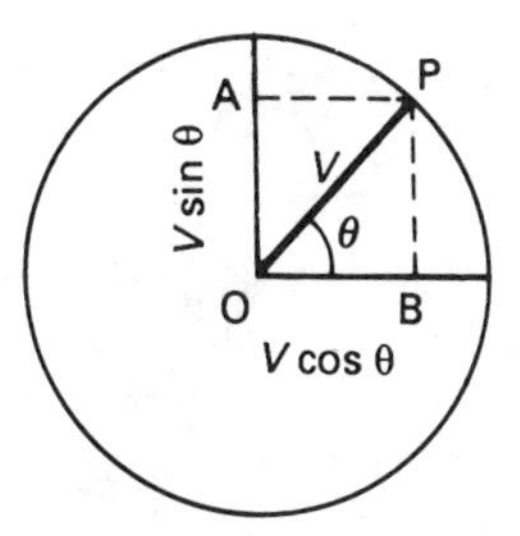

Figure 5.6 Phasor representation of sinusoidal variation

mechanics, represent *space*-dependent quantities such as force and momentum but the electrical quantities which we represent in a similar graphical manner are *time* dependent.

Figure 5.7 shows the phasor diagram for the circuit of *Figure 5.5*.

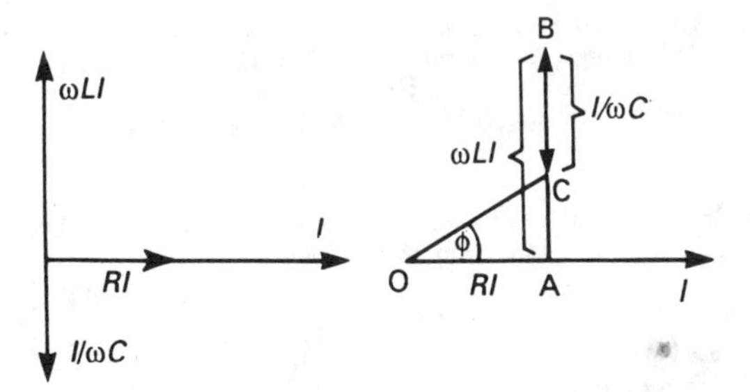

Figure 5.7 Phasor diagram for the series circuit of *Figure 5.5*

It will be realised that the triangle OAC is a graphical method of displaying and calculating the quantities Z and φ introduced in the mathematical approach. If $i = I\sin\omega t$ we can write $v = ZI\sin(\omega t + \varphi)$.

5.4.3 Summary

For a.c. circuits in which the frequency is ω the impedance to current flow of a simple series combination of R, L and C is denoted by $Z = \sqrt{R^2 + (\omega L - 1/\omega C)^2}$ so that, confining our attention to magnitudes, $V = IZ$ or $I = V/Z$. There also exists a phase shift because the voltage and current will not, in general, be in phase, i.e. their maxima, minima, and zero values will not occur at the same instant of time.

Figure 5.8 shows how voltage and current vary with time. In this case the current lags behind the voltage or the voltage leads

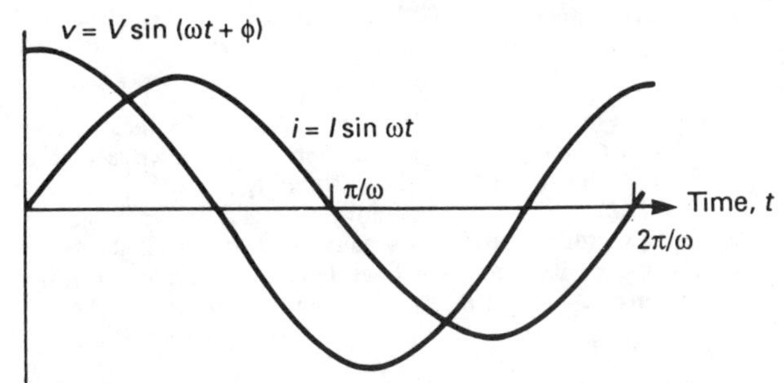

Figure 5.8 Voltage and current in a circuit in which current lags behind voltage

the current. If the effect of the capacitor were to exceed that of the inductor (i.e. if $I/\omega C > \omega LI$) the current would lead the voltage.

It should be appreciated that although diagrams could be drawn to scale and the answer for, say, the impedance could be measured, it is much more common to sketch the diagrams and use trigonometry to find the answer.

5.5 Impedance, reactance, admittance and susceptance

The impedance Z is best regarded as the quantity used in a.c. circuits analogous to resistance in d.c. circuits. In general it is a complex quantity consisting of two parts: the real part is resistance and represents energy dissipated in the circuit; while the imaginary part represents the energy stored and ultimately returned to the circuit. In this case, for an LCR series circuit,

$$Z = R + \mathrm{j}\left(\omega L - \frac{1}{\omega C}\right)$$

The terms representing the energy storage elements are $(\omega L - 1/\omega C)$ and we introduce the term reactance, X, to represent the energy storage element. Consequently $Z = R + \mathrm{j}X$. The units of Z and X must be the same as those of R, that is ohms. The inverse of impedance Z is the admittance Y and that of reactance X is susceptance B. As the reciprocal of resistance R is conductance G we can write $Y = G + \mathrm{j}B$.

5.6 Technique for a.c. circuits

Drawing on the example of the series LCR circuits we can form general rules. Considering a frequency ω we write $\mathrm{j}\omega L$ for every inductance, $1/\mathrm{j}\omega C$ for every capacitance and proceed as for a d.c. circuit containing only resistances. For example, consider the circuit in *Figure 5.9*, having written in $\mathrm{j}\omega L$ and $1/\mathrm{j}\omega C$ as necessary:

$$Z = \frac{R_2(R_1 + \mathrm{j}\omega L)}{R_2 + R_1 + \mathrm{j}\omega L} + \frac{1}{\mathrm{j}\omega C_1} + \frac{R_3/\mathrm{j}\omega C_2}{R_3 + 1/\mathrm{j}\omega C_2}$$

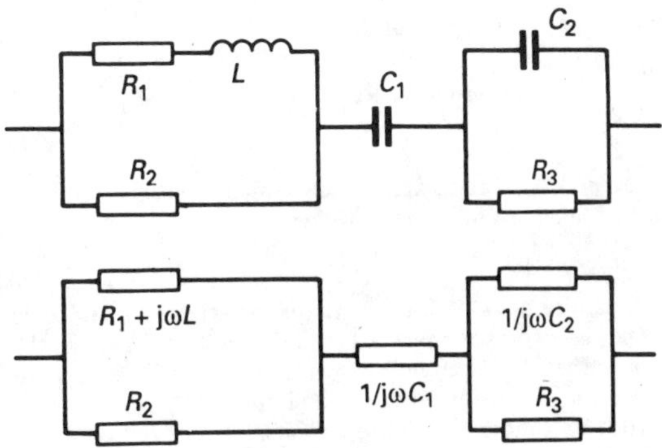

Figure 5.9 Circuit for analysis and first step in solution

This will reduce to the form $R + \mathrm{j}X$ enabling the relationship between overall voltage and current to be established. It is possible to achieve the same result using phasor diagrams, but when combinations of series and parallel elements are needed the work becomes tricky. The principles are that parts in series carry the same current whereas those in parallel experience the same voltage.

5.7 Average and r.m.s. values

So far we have dealt with the instantaneous values of voltage v and current i and the peak or maximum V and I. The true average of any quantity which is varying sinusoidally is zero. However, circumstances exist, such as full wave rectification, where we have waveforms like that shown in *Figure 5.10*. If the original voltage waveform was $v = V \sin \omega t$ the average of the fully rectified wave is $V_{av} = (2/\pi)V$.

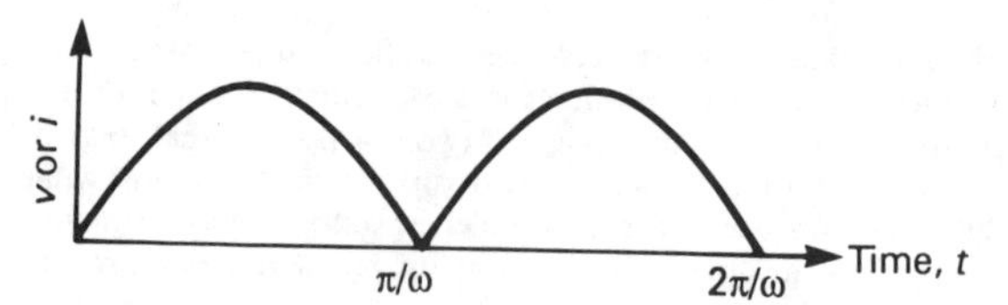

Figure 5.10 Full wave rectification

Another form of average is that associated with the current's ability to deliver power to a resistor R. If the current waveform is $i = I \sin \omega t$ then the power delivered is $i^2R = (I \sin \omega t)^2 R$ averaged over a cycle. We introduce I_{rms} to represent this average such that $I_{rms} = (1/\sqrt{2})I$. The subscript stands for root mean square (r.m.s.) and those are the operations which, in the order square, mean and root, have been performed on the current.

Similarly we have $V_{rms} = (1/\sqrt{2})V$. It is the r.m.s. value of an a.c. quantity that is usually quoted and, unless stated to the contrary, it is to be assumed that a given voltage or current is the r.m.s. value.

5.8 Power, power factor

Instantaneously the power delivered to a circuit is given by $w = vi$ where w is the power and v and i the instantaneous voltage and current. The average power delivered is the mean of vi; therefore

$$W = \frac{\omega}{2\pi} \int_0^{2\pi/\omega} vi \, dt$$

If $v = V \sin \omega t$ and $i = I \sin(\omega t + \varphi)$

$$W = \frac{\omega}{2\pi} \int_0^{2\pi/\omega} VI \sin \omega t \sin(\omega t + \varphi) \, dt = \tfrac{1}{2} VI \cos \varphi \qquad (5.18)$$

In this expression V and I are the peak values and φ is the phase angle. Introducing r.m.s. values we write

$$W = V_{rms} I_{rms} \cos \varphi \qquad (5.19)$$

The quantity $\cos \varphi$ is known as the power factor and φ is sometimes known as the power factor angle. This form has the advantage that, for a circuit in which $\varphi = 0$, such as a pure resistance, the expression for power is simply $V_{rms} I_{rms}$ as would be expected from simple d.c. considerations.

Figure 5.11 shows how power varies through a cycle for different power factors. When current and voltage have the same polarity, either positive or negative, power is supplied to the circuit, and is shown shaded. When current and voltage have opposite polarities power is recovered from the circuit and is shown cross-hatched. The shaded areas represent power and it will be seen that power is always positive, i.e. being absorbed by the resistor in case I. In case II the shaded and cross-hatched areas are equal, indicating that energy is alternately stored and recovered with no net consumption. Case III is intermediate and

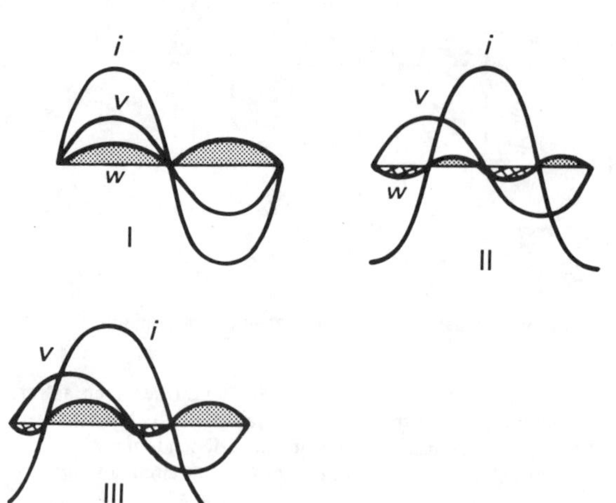

Figure 5.11 Power transfer in a.c. circuit. Case I: load entirely resistive, v and i in phase, maximum power w. Case II: load entirely inductive, i lags v by 90°, no net power. Case III load partially inductive, i lags v by less than 90°, some net power

it will be noticed that the shaded areas exceed the cross-hatched ones indicating a net consumption of power.

5.9 Network laws and theorems: Kirchhoff's laws

The basis of systematic analysis of any circuit other than the most elementary is the two laws attributed to Kirchhoff. The first is the current law which states that the sum of all currents flowing into a node must be zero so in *Figure 5.12(a)*

$$i_1 + i_2 + i_3 + i_4 = 0$$

The second is the voltage law which states that there is no net change of voltage round a closed loop so in *Figure 5.12(b)* $V = iR_1 + iR_2$ when V is the rise of voltage from A to B and $iR_1 + iR_2$ is the fall of voltage from B through C and D to A.

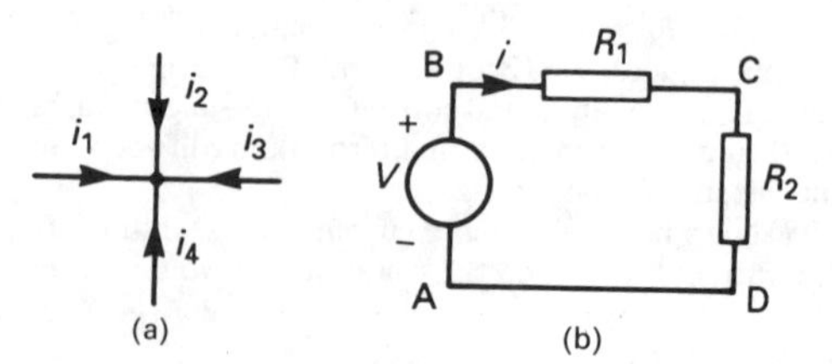

Figure 5.12 Kirchhoff's laws: (a) current law; (b) voltage law

The voltage law is usually used explicitly, for example a possible method of solving a circuit such as that shown in *Figure 5.13* would be to write in the currents i_1, i_2, $i_1 - i_2$, i_3 and $i_1 - i_2 - i_3$ in that order, implicitly making use of Kirchhoff's current law, and then to form as many equations as are necessary by the use of Kirchhoff's voltage law. Considering the loop comprising V, R_1, and R_3 we can write

$$V = i_1 R_1 + i_2 R_3$$

Considering R_2, R_3 and R_4

$$0 = (i_1 - i_2)R_2 + i_3 R_4 - i_2 R_3$$

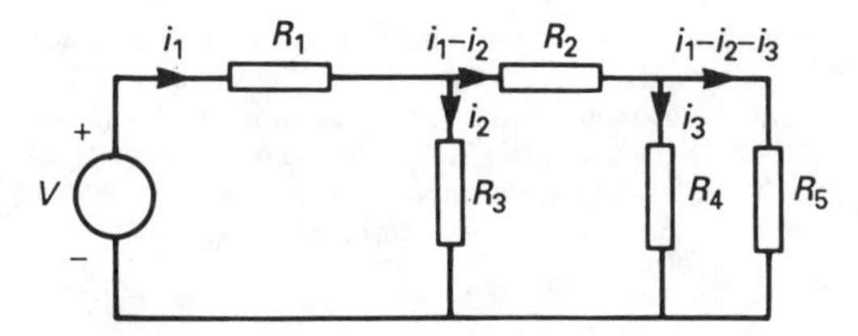

Figure 5.13 Illustration of Kirchhoff's laws

Considering R_4 and R_5

$$0=(i_1-i_2-i_3)R_5-i_3R_4$$

There are alternatives such as that obtained by considering V, R_1, R_2 and R_5 which would give

$$V=i_1R_1+(i_1-i_2)R_2+(i_1-i_2-i_3)R_5$$

However, this is not independent of the previous set of three equations as can be seen by adding these equations.

5.9.1 Loop or mesh currents

A device that is often helpful is to consider the current flowing round a loop rather than that actually in a wire. *Figure 5.14* shows the method. The current in any part is the sum of the loop currents which flow in the adjacent loops. If we apply the idea to the circuit of *Figure 5.13* we will consider loop currents i_4, i_5, i_6 as shown in *Figure 5.15*. In this case $i_1=i_4$, $i_2=i_4-i_5$, $i_3=i_5-i_6$.

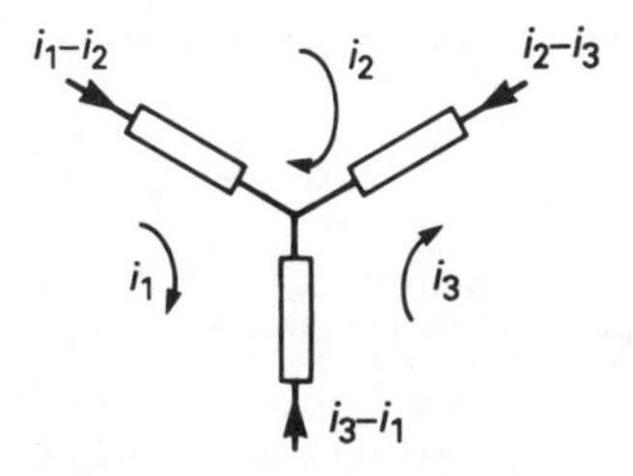

Figure 5.14 Loop currents

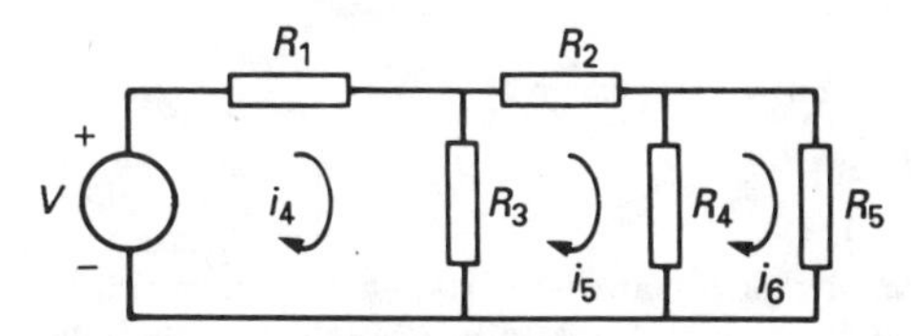

Figure 5.15 Illustration of loop currents

Writing the voltage loop equation we have the following equations. Considering loop i_4

$$V=i_4R_1+(i_4-i_5)R_3$$

Considering loop i_5

$$0=i_5R_2+(i_5-i_6)R_4+(i_5-i_4)R_3$$

Considering loop i_6

$$0=i_6R_5+(i_6-i_5)R_4$$

These can be arranged in a systematic manner which makes checking easy and leads to a matrix solution:

$$V=(R_1+R_3)i_4-R_3i_5$$

$$0=-R_3i_4 \quad +(R_2+R_3+R_4)i_5-R_4i_6$$

$$0=-R_4i_5 \quad +(R_4+R_5)i_6$$

The expression is symmetrical about the leading diagonal and the terms of the leading diagonal are the resistances as one traverses the appropriate loop.

5.9.2 Superposition

If a circuit is linear and contains more than one source the principle of superposition may be useful. Linearity means that currents are proportional to voltages. This means that diodes etc., which do not conduct equally for both directions of applied voltage, and devices (such as incandescent light bulbs) where the resistance changes, and so the current depends on the voltage to some power other than one, are excluded. Consider a circuit such as that shown in *Figure 5.16(a)*. Supposing we require the current I in R_2, the principle of superposition tells us that it is the sum of

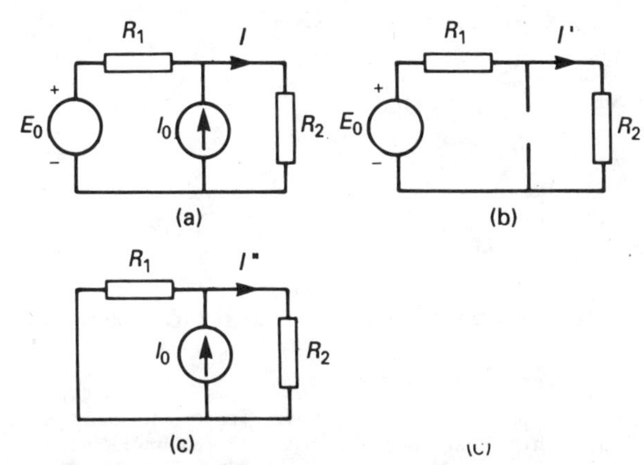

Figure 5.16 Illustration of superposition: (a) complete circuit; (b) current source deactivated or suppressed; (c) voltage source deactivated or suppressed

the currents I' and I'' where I' is the result of deactivating the current source and calculating the current due to the voltage source and I'' is the result of deactivating the voltage source and calculating the current due to the current source. A deactivated voltage source offers no impedance to the flow of current and so becomes a short circuit, whereas a deactivated current source offers an infinite impedance to the flow of current and so becomes an open circuit. Therefore *Figures 5.16(b)* and *(c)* show the two constituent parts and we can see that

$$I'=\frac{E_0}{R_1+R_2} \qquad I''=I_0\left(\frac{R_1}{R_1+R_2}\right)$$

and so

$$I=I'+I''=\frac{E_0+I_0R_1}{R_1+R_2}$$

5.9.3 Star–delta and delta–star transformations

These transformations may be useful in simplifying a circuit and are illustrated in *Figures 5.17* and *5.18*:

$$Y_{AB}=\frac{Y_AY_B}{Y_A+Y_B+Y_C} \tag{5.20}$$

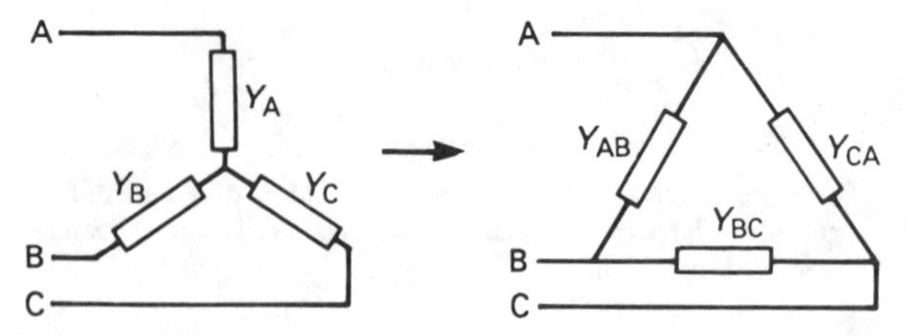

Figure 5.17 Star–delta transformation

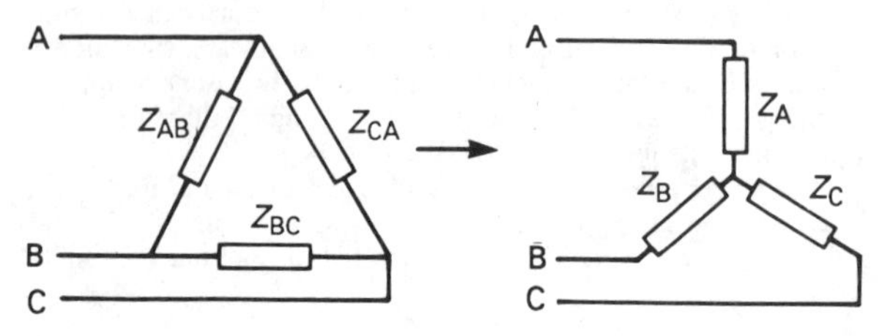

Figure 5.18 Delta–star transformation

Similarly for Y_{BC} and Y_{CA}.

$$Z_A = \frac{Z_{AB}Z_{CA}}{Z_{AB}+Z_{BC}+Z_{CA}} \tag{5.21}$$

Similarly for Z_B and Z_C.

5.10 Thevenin's theorem and Norton's theorem

Acceptance of these theorems enables one to analyse circuits consisting of one or more 'black boxes'. The so-called 'black box' approach encourages one to regard a piece of equipment in terms of what is observable at its terminals without regard to what is going on inside. Thevenin's and Norton's theorems give the quantities which must be used in any analysis. Previous sections have introduced current generators alongside voltage generators and the basic difference between Thevenin and Norton is that Thevenin expresses the circuit in terms of an equivalent voltage generator whereas Norton gives an equivalent current generator. Thevenin's theorem states that any two-terminal linear network can be represented by an ideal voltage source V_T in series with an impedance Z_T. The value of V_T is the voltage observed between the terminals when on open circuit and the value of Z_T is the impedance measured between the terminals with independent sources of voltage and current deactivated. The implications of deactivation were explained in Section 5.9.2. There is an alternative approach to the calculation of Z_T; it is that Z_T can be expressed as $Z_T = V_{oc}/I_{sc}$ when $V_{oc} = V_T$ and I_{sc} is the current that would flow in a short circuit placed across the terminals. Norton's theorem states that any two-terminal linear network can be represented by an ideal current generator I_N in parallel with an impedance. The value of the impedance is the same as that introduced with the Thevenin equivalent circuit but now it is in parallel with the source. The value of I_N equals I_{sc}, namely the short circuit current.

5.11 Resonance, 'Q' factor

If a circuit containing inductance and capacitance is subjected to a voltage of constant amplitude but varying frequency the current will have a maximum or minimum value for a particular frequency. Alternatively if the frequency is held fixed and the value of one of the components varied the same maximum or minimum will be observed. This phenomenon is known as resonance. To develop the important relations we will consider a simple circuit as shown in *Figure 5.19(a)* containing capacitance

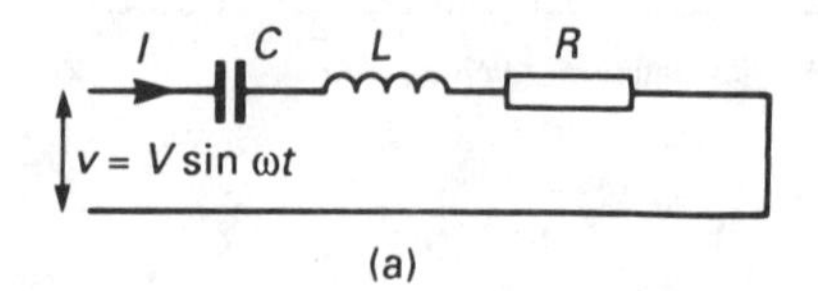

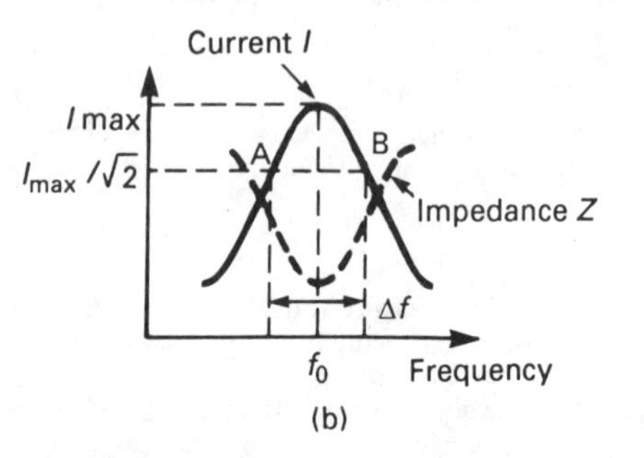

Figure 5.19 Resonance: (a) series circuit; (b) resonance curves

C, inductance L and resistance R. Using the techniques already developed for a.c. circuit analysis we write

$$v = \left(R + j\omega L + \frac{1}{j\omega C}\right) I \exp(j\omega t) \tag{5.22}$$

where the instantaneous current i is given by $I \exp(j\omega t)$. The instantaneous voltage v can be similarly represented by

$$v = V \exp[j(\omega t + \varphi)] \tag{5.23}$$

where φ is the angle by which V leads I. It follows that V, the peak value or modulus of the voltage, is related to I, the peak value or modulus of the current, by the relationship

$$V = \sqrt{R^2 + (\omega L - 1/\omega C)^2}\, I \tag{5.24}$$

and the phase angle φ is given by

$$\tan \varphi = \frac{\omega L - 1/\omega C}{R} \tag{5.25}$$

If we consider the usual elementary situation in which the circuit is supplied with a voltage of constant amplitude but varying frequency then the current I is given by

$$I = \frac{V}{\sqrt{R^2 + (\omega L - 1/\omega C)^2}} \tag{5.26}$$

and a possible shape is sketched in *Figure 5.19(b)*.

The current has a maximum value at a frequency f_0 $(=\omega_0/2\pi)$ known as the resonant frequency. Its value is given by putting $\omega_0^2 = 1/LC$ in Equation (5.24). At this frequency it should be noticed that the phase angle φ is zero. If $\omega > \omega_0$ the phase angle is positive, that is, the voltage leads the current as is typical of an inductance; and if $\omega < \omega_0$ the phase angle is negative as is found with a capacitance. *Figure 5.19(b)* also shows the manner in which the impedance Z varies.

Resonance circuits, because of their frequency-dependent characteristics, are used for tuning purposes; that is, to pick out signals of one particular frequency from a range of signals, e.g. the tuning stages of a radio receiver. In these cases, of course, one of the circuit elements—normally the capacitor C—is varied until the resonant frequency equals that of the desired signal. For good selectivity the peak of the resonance curve needs to be sharp. The measure of the sharpness is the resonant frequency f_0 divided by the width Δf at some particular level. The level that is usually chosen is that at which the current has a value $1/\sqrt{2}$ of its maximum value. Bearing in mind that the power developed by a current in a resistor is I^2R it will be realised that if the current is $1/\sqrt{2}$ of its maximum then the power will be 1/2 of the maximum. These points are therefore known as the half power points (A and B in *Figure 5.19(b)*).

The symbol Q, defined as $f_0/\Delta f$ or $\omega_0/\Delta\omega$, is used for the sharpness as it measures the quality of the circuit when used as a tuning device. If we return to the simple series circuit shown in *Figure 5.19(a)* and if we draw the full phasor diagram as in *Figure 5.20* it will be seen that at resonance the voltages across both the

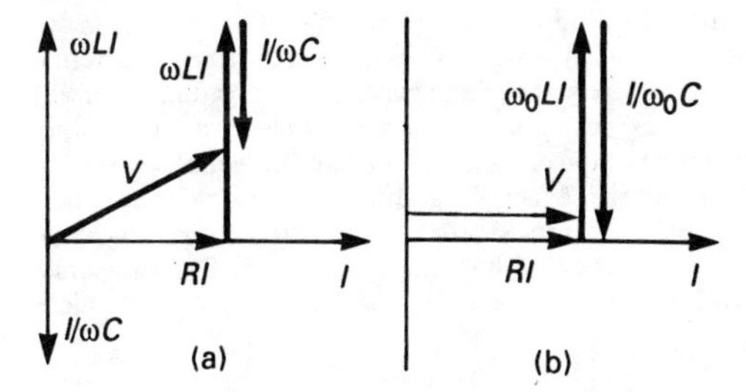

Figure 5.20 Phasor diagram for the circuit of *Figure 5.19(a)*: (a) circuit not in resonance; (b) circuit in resonance

inductor and the capacitor are equal. This leads to an alternative definition of Q, based on voltage magnification:

$$\frac{\text{voltage across inductor}}{\text{total voltage}}=\frac{\omega_0 LI}{RI}=\frac{\omega_0 L}{R} \tag{5.27}$$

$$\frac{\text{voltage across capacitor}}{\text{total voltage}}=\frac{I}{\omega_0 CRI}=\frac{1}{\omega_0 CR} \tag{5.28}$$

So we can define Q as $\omega_0 L/R$ or $1/\omega_0 CR$.

One should point out that, although voltage has been magnified, there has been no magnification of power. The voltages across both the inductor and the capacitor are in quadrature with the current so no power is available. Looked at from the energy point of view, relatively large amounts of energy are stored in the inductor and the capacitor but they are always returned to the source.

This section has, so far, only considered series circuits; the results for a parallel circuit are similar except that whereas the series circuit has a minimum impedance at resonance the parallel one has a maximum. The analysis is complicated by the fact that a true parallel circuit (R, L and C in parallel) is not a correct representation of a real inductor connected in parallel with a real capacitor. It will be realised that to obtain a high Q in a series circuit it is necessary to have a small R. A real good quality capacitor has negligible resistance but a real inductor, made of a coil of wire, must possess some resistance. However, it can be shown that the practical situation of a real capacitor in parallel with a real inductor complete with parasitic resistance, r (*Figure 5.21(a)*) is equivalent to a true parallel circuit (*Figure 5.21(b)*) provided that Q is large and we put $C=C'$, $L=L'$ and $R=Q^2r$. In other words, the small parasitic r in series is equivalent to a large R in parallel.

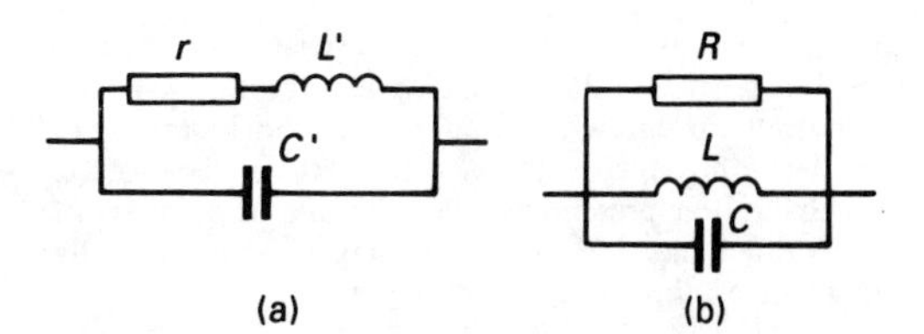

Figure 5.21 To show equivalence of real parallel circuit (a) with circuit (b) which is easier to analyse

5.12 Mutual inductance

This is the phenomenon whereby a changing current in one circuit produces a voltage in another. It is usually explained by appealing to the ideas of lines of magnetic flux which, when they change, produce an electromotive force in a circuit. Consider two coils as shown in *Figure 5.22*. If current in coil 1 is changing at a

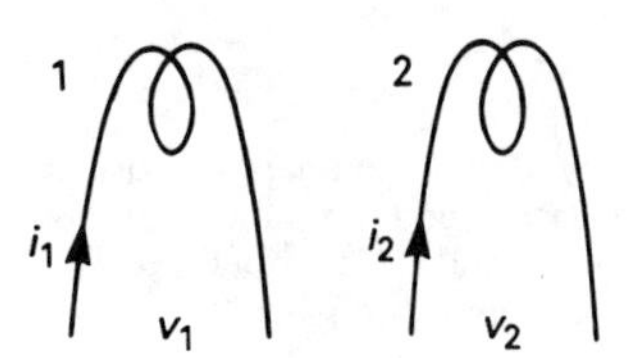

Figure 5.22 Illustration of mutual inductance (air cored)

rate di_1/dt then a voltage proportional to di_1/dt will be induced in coil 2. The coefficient of proportionality is known as the coefficient of mutual inductance M so that

$$v_2=M\,di_1/dt \tag{5.29}$$

and

$$v_1=M\,di_2/dt \tag{5.30}$$

The fact that the same value of M is used in both equations will probably be intuitively obvious; it can be proved by considering the energy stored in the coupled circuits when first one and then the other current is switched on. For two coils such as those shown in *Figure 5.22* the value of M will be much smaller than it would be if the coils were closer together to reduce leakage and if they were wound on an iron former. Much of the flux emanating from, say, coil 1 will not go through coil 2 and there is said to be a lot of leakage; this could be reduced by bringing the coils together. If, however, we were to wind the coils on an iron former the value of M would be much greater because more flux is produced due to ferromagnetism.

When analysing a circuit containing mutual inductance it is often helpful to redraw the mutual inductance adding generators as shown in *Figure 5.23*. The directions shown for positive current flow have the merit of symmetry but if one is considering

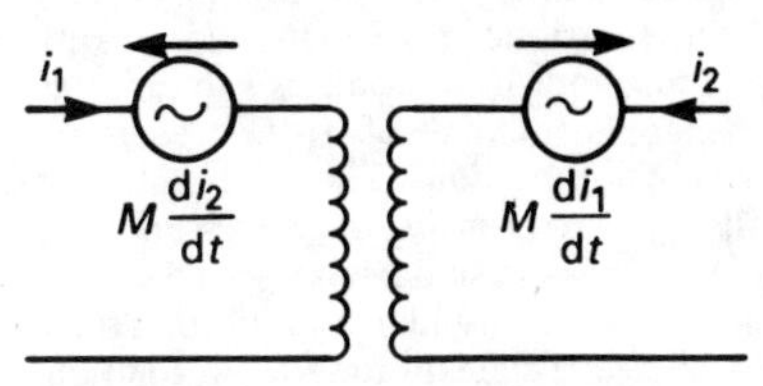

Figure 5.23 Circuit representation of mutual inductance

a transformer, the most commonly occurring example of mutual inductance, it is normal to think of an input current i, as shown, and of an output current, which would be $-i_2$ in *Figure 5.23.*

In line with the introduction to L and C and the analysis of a.c. circuits which has been presented earlier it should be pointed out that, if, as is often the case, we are dealing with sinusoidally varying quantitives, then di/dt becomes $j\omega i$.

5.13 Differential equations and Laplace transforms

Equations containing integral and differential expressions are known as differential equations and many books have been devoted to their solution. To illustrate the process we will consider in more detail the equation developed in Section 5.4. The basic equation is (see Equation (5.15))

$$v=Ri+L\frac{di}{dt}+\frac{1}{C}\int i\,dt$$

An alternative form which may appear simpler, because it contains only differential coefficients and not a mixture of differential coefficients and integrals, is obtained by differentiating throughout and rearranging to give

$$L\frac{d^2i}{dt^2}+R\frac{di}{dt}+\frac{i}{C}=\frac{dv}{dt} \tag{5.31}$$

There are two parts to the solution of such an equation and the complete solution is the sum of both. The term on the right-hand side (dv/dt) is known as the forcing or driving function. In this case it is the voltage which is applied to the circuit even though in this analysis it is its rate of change that is used. Without a driving function nothing would happen unless, say, there was an initial charge on the capacitor in which case a solution is required to describe what happens as a result of this initial charge. The two parts of the solution referred to above are the complementary function and the particular integral.

(*1*) *The complementary function* is the solution to the equation with the independent variable, which is usually written on the right-hand side, put equal to zero. The complementary function describes the result of any initial charges or currents. In a stable system it decays to zero and represents the transient behaviour.

(*2*) *The particular integral* requires the inclusion of the independent variable, the driving voltage in the case we are considering. The particular integral is any solution to the full equation resulting from the inclusion of the forcing function. In our case it cannot be found until the amplitude and frequency of the driving voltage are known.

Traditionally, for this type of equation, the method of finding these two solutions goes as follows. For the complementary function we let $i=A\,e^{mt}$ and substitute into Equation (5.31) with the right-hand side equal to zero, which produces values for m. For each m there will be a different value of A; these arbitrary constants depend on the initial conditions. The solution for the particular integral calls for an element of guesswork and intuition. If the driving function is sinusoidal it is a sensible guess that the response will also be sinusoidal and of the same frequency. It represents the steady-state, or long-term, solution.

All the guesswork can be taken out of the solution of these differential equations by the use of the Laplace transform. The effect of this is to transform the differential (and integral) expressions into straightforward algebraic equations. Once this has been done to the complete equation the transform of the dependent variable, in our case the current i, is expressed in terms of the other quantities and the use of the inverse transform produces the solution we require.

The Laplace transform of a function of time $f(t)$ is denoted by $F(s)$ and is defined as

$$F(s)=\int_0^{\infty} f(t)\,e^{-st}\,dt$$

The inverse transform is

$$f(t)=\frac{1}{2\pi j}\int_{\sigma-j\infty}^{\sigma+j\infty} F(s)\,e^{st}\,ds$$

However, tables are frequently used rather than these formulae.

Use of the Laplace transform method quickly leads to the realisation that the equations can easily be set up by considering the impedance of an inductor to be sL and that of a capacitor to be $1/sC$. This is satisfactory as long as there are no initial currents or voltages respectively. The term generalised impedance is used for impedances expressed in this manner. The advantage of the method is that any type of input can be handled equally easily as long as the transform exists. In addition to sinusoids the other inputs that are encountered are step functions, repeated step functions or square waves and ramp functions. For sinusoids replace s with $j\omega$ and we arrive back at the equation with which we started.

5.14 Transients and time constants

As the word transient implies, any phenomenon to which it is applied is short-lived and the time constant is a measure of its duration. A simple circuit that is often used to illustrate these ideas is that of a capacitor connected to a source of voltage through a switch and a resistor as shown in *Figure 5.24(a).*

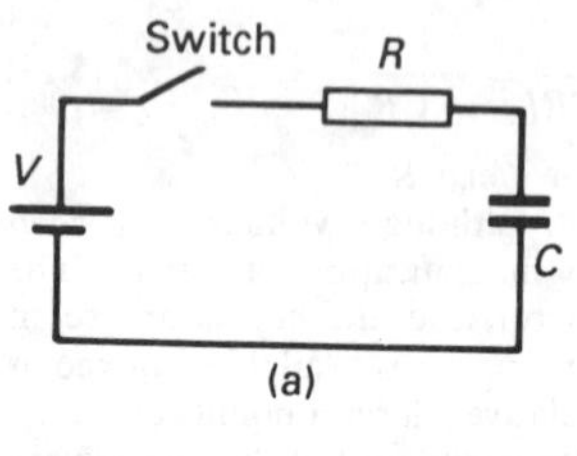

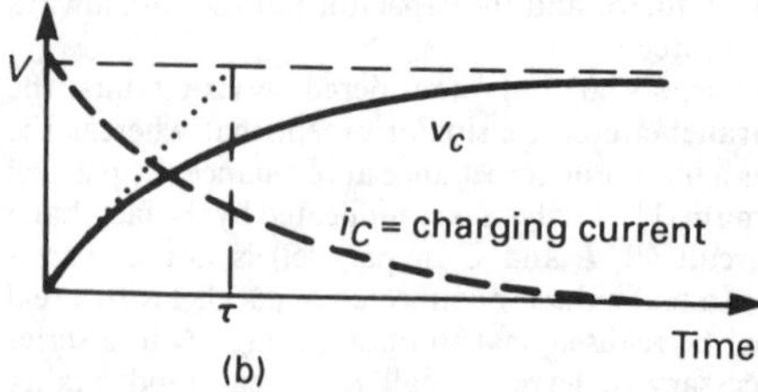

Figure 5.24 Illustration of transient phenomena: (a) circuit; (b) voltage and current graphs

If the capacitor is initially uncharged it can be shown that the voltage v_C across the capacitor at time t after closing the switch is given by

$$v_C=V[1-\exp(-t/RC)] \tag{5.32}$$

The current in the circuit decays according to the expression

$$i=(V/R)\exp(-t/RC) \tag{5.33}$$

The time constant τ can be regarded in two ways. Firstly it is defined, for this circuit, as being RC which means that it is the time it takes for the voltage to rise to $1-1/e=0.632$ of its final value. The second approach is to calculate how long it would take for v_C to reach V if the initial rate of change of v_C were to be maintained. The dotted line in *Figure 5.24(b)* is drawn tangential to the curve of v_C at the origin and illustrates this approach to the time constant.

If a circuit were to contain an inductor L instead of the capacitor the time constant would be L/R, and the current in a circuit containing an inductor and a resistor in series would be

$$i=\frac{V}{R}[1-\exp(-Rt/L)] \tag{5.34}$$

The final current must be V/R because there will be no voltage across the inductor in the ultimate steady state. If one is faced with a circuit containing both inductance and capacitance the results are more complicated and, in the absence of resistance, there will be no steady long-term solution. The solution can best be approached by setting up the differential equation and using Laplace transforms to solve it. The closing of a switch to apply a constant voltage is to apply a 'step function' and this can be handled easily using Laplace transforms because the transform of a step is $1/s$ times the size of the step.

5.14.1 Pulses and square waves

The response of a circuit to a pulse is often easily understood by considering the response to two step functions of opposite signs and one delayed relative to the other, as shown in *Figure 5.25(a)*. The shape depends on the relative sizes of the time constant τ and the delay interval T. *Figure 5.25(c)* shows the situation when $\tau \simeq T$.

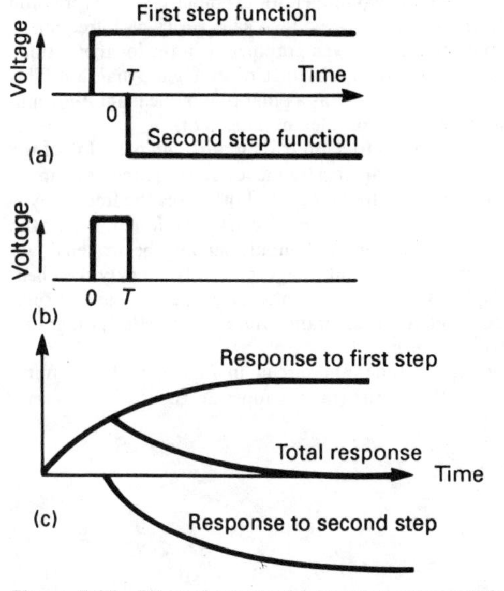

Figure 5.25 Effect of a pulse: (a) two-step representation; (b) pulse; (c) circuit response

If $\tau \ll T$ then the response would be much closer to the square wave input. If $\tau \gg T$ the output voltage would hardly change because the second and negative pulse would come so soon after the first and positive one.

Square waves are a series of such pulses and the response can be built up accordingly. The time gap between pulses and its relation to the time constant is the important quantity.

5.15 Three-phase circuits

Alternating current has many advantages over direct current when considering generation and transmission; no commutator is needed in the generator and the voltage can easily be changed to the level appropriate for economical transmission. However, single-phase a.c. systems are not as efficient as d.c. when considering the quantity of material required to transmit a given amount of energy. A further drawback to single-phase a.c. motors is that they are not inherently self-starting. A three-phase system overcomes both these disadvantages. Imagine a generator with three coils—R (red), Y (yellow) and B (blue)—on the stator and a magnet turning inside. The magnet could be a permanent one but is more likely to be an electromagnet fed from a d.c. source. If the pole faces are shaped so that the distribution of flux is sinusoidal then sinusoidal waves of voltage will be induced in the coils (see *Figure 5.26*).

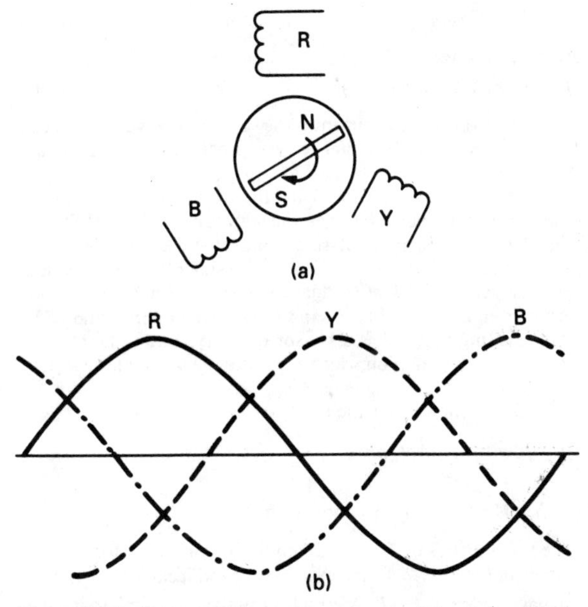

Figure 5.26 Elementary three-phase generator (a) and waveforms (b)

5.15.1 Star or Y connection

It is not necessary to carry all six wires from the three coils to the load. The easiest way to connect them together is to join one end of each coil to a common point and call that point the star or neutral point. This is known as a star connection as shown in *Figure 5.27*.

There are two ways in which the voltage can be expressed, namely the phase value or the line value. The three-phase values V_R, V_Y and V_B are equal in magnitude (V_ϕ) but differ in phase by 120°; similarly the line values V_{RB}, V_{BY} and V_{YR} are also equal in magnitude (V_L), but differ in phase by 120°. The relationship between V_L and V_ϕ is

$$V_L=\sqrt{3}\,V_\phi \tag{5.35}$$

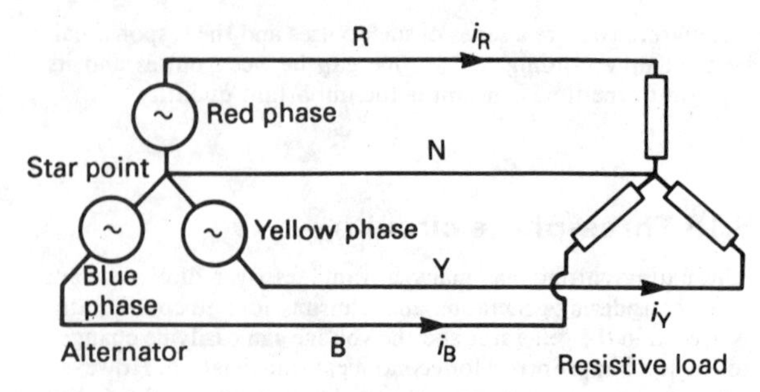

Figure 5.27 Star or Y connected alternator and load

If we consider the red phase to be the reference one we can express the driving voltages as follows:

$$v_R = V \sin \omega t \quad (5.36)$$

$$v_Y = V \sin(\omega t + 120°) \quad (5.37)$$

$$v_B = V \sin(\omega t + 240°) \quad (5.38)$$

If the impedances are of magnitude Z and produce a phase shift θ then the currents are

$$i_R = (V/Z)\sin(\omega t - \theta) \quad (5.39)$$

$$i_Y = (V/Z)\sin(\omega t + 120° - \theta) \quad (5.40)$$

$$i_B = (V/Z)\sin(\omega t + 240° - \theta) \quad (5.41)$$

In the neutral wire the current is the sum of these values and can easily be shown to be equal to zero, since the loads are equal (balanced).

If the loads are not balanced and there is no neutral conductor there will be a voltage between the star point of the supply and that of the load. In any analysis one introduces an unknown such as V_N to represent the drop between the star point of the source and the neutral point of the load. The voltages across the loads are written in terms of v_R, v_Y and v_B as defined above and of V_N remembering that this is a phasor quantity. The value of V_N is found from the equation which expresses the fact that the sum of the three line currents equals zero.

It can be shown that the total power consumed is

$$3V_\phi I_\phi \cos\theta = \sqrt{3}\, V_L I_L \cos\theta \quad (5.42)$$

5.15.2 Delta or mesh connection

The three phases of *Figure 5.26* could be connected together as shown in *Figure 5.28*. In this case $V_L = V_\phi$ and it can be shown that in magnitude $I_L = \sqrt{3}\, I_\phi$. Methods of analysis are similar and it is often helpful to use the delta–star transformation described in Section 5.9.3 to find the equivalent star-connected load. In general, as far as the supply is concerned, one does not know (or care) if it is star or delta connected and one is free to choose whichever is more convenient, which is usually star.

The result for the power consumed is the same for delta as for star, namely $\sqrt{3}\, V_L I_L \cos\theta$.

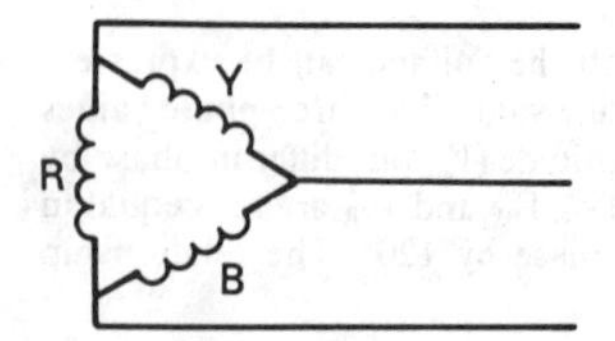

Figure 5.28 Delta or mesh connection

5.16 The decibel

When using the bel (symbol B), or more frequently the decibel (symbol dB), as a unit in measuring such electrical quantities as voltage and current it must be remembered that it is the power that the voltage or current could produce that is being considered. This power is proportional to the square of the voltage or current, consequently the expression in decibels is

$$20\log_{10}(V/V_{ref}) \quad (5.43)$$

and similarly for current ratios.

These expressions assume the existence of a reference value of voltage V_{ref} or current I_{ref} and also that both the voltage or current being measured and the reference values are developing power in the same resistor. In electrical work, these have been chosen as 1 mW and 600 Ω which implies that $V_{ref} = \sqrt{0.6} = 0.775$. Information to this effect is often found in small figures somewhere on the dial of a decibel meter. If the same meter or one with the same value of V_{ref} is used for measuring both the input voltage (V_1) and the output voltage (V_2), the value of (V_2/V_1) dB is obtained by subtracting V_1 expressed in dB from V_2 expressed in dB. The following explains why:

$$(V_2 \text{ in dB}) - (V_1 \text{ in dB}) = 20\log_{10}(V_2/V_{ref}) - 20\log_{10}(V_1/V_{ref})$$

$$= 20\log_{10}\left(\frac{V_2}{V_{ref}} \cdot \frac{V_{ref}}{V_1}\right)$$

$$= V_2/V_1 \text{ in dB}$$

It will be seen that V_{ref} cancels.

5.17 Frequency response and Bode diagrams

This is a most important description of the behaviour of a circuit; in the case of a voltage amplifier, for example, it is an expression of the manner in which the voltage gain varies with frequency. The information is often given graphically in the form of a Bode diagram, or occasionally in the form of an Argand diagram. This latter can be regarded either as a polar plot of the magnitude and phase relation or as a complex number plot.

There are two parts to a Bode diagram; one is a plot of the amplitude (in decibels) against frequency and the other is a plot of the phase shift against frequency. In both cases the frequency is plotted on a logarithmic scale. The transfer function (i.e. the expression for output in terms of input) can often be broken down into simple terms whose Bode diagrams can be quickly sketched. As the complete transfer function is the *product* of several such terms the complete Bode diagram is the *sum* of the Bode diagrams of the individual terms.

For example, consider the circuit in *Figure 5.29*; assuming sinusoidal conditions, the transfer function is:

$$\frac{e_2}{e_1}(j\omega) = \frac{Z_2}{Z_1 + Z_2}$$

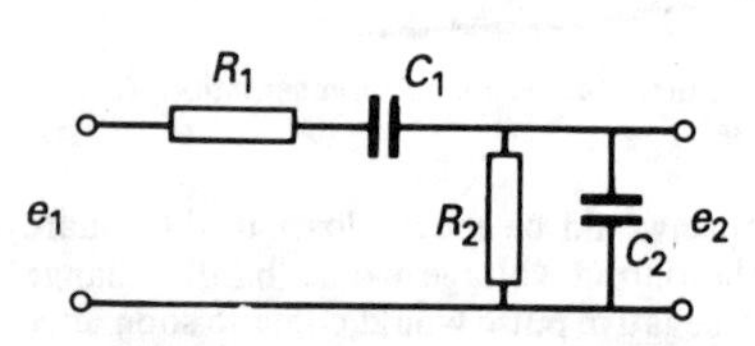

Figure 5.29 Circuit used to illustrate Bode diagrams

where

$$Z_2 = \frac{R_2}{1 + j\omega C_2 R_2} \quad \text{and} \quad Z_1 = R_1 + \frac{1}{j\omega C_1}$$

Analysis shows that, provided $C_1 R_2 \ll C_1 R_1 + C_2 R_2$,

$$\frac{e_2}{e_1}(j\omega) = A \frac{j\omega C_1 R_1}{(1 + j\omega C_1 R_1)(1 + j\omega C_2 R_2)}$$

where A is a numerical factor independent of frequency. Alternatively, this can be expressed as

$$\frac{e_2}{e_1}(j\omega) = A \frac{j\omega/\omega_1}{(1 + j\omega/\omega_1)(1 + j\omega/\omega_2)}$$

where $\omega_1 = 1/R_1 C_1$ and $\omega_2 = 1/R_2 C_2$. If $R_1 C_1 \gg R_2 C_2$ then $\omega_1 \ll \omega_2$.

To draw the Bode diagram for the amplitude of this transfer function we express e_2/e_1 in decibels as follows:

$$\left(\frac{e_2}{e_1}\right) \text{dB} = 20 \log_{10} A + 20 \log_{10} \left(\frac{\omega/\omega_1}{\sqrt{1 + (\omega/\omega_1)^2}}\right)$$

$$- 20 \log_{10} \sqrt{1 + (\omega/\omega_2)^2}$$

The Bode diagrams of the three terms are shown in *Figure 5.30*.

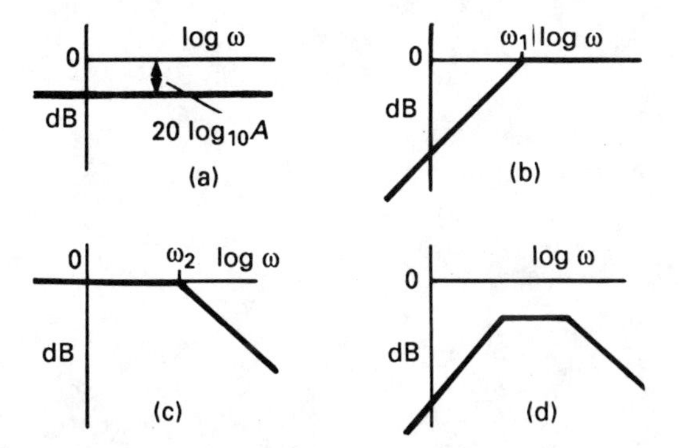

Figure 5.30 Bode diagrams for amplitude of transfer function of circuit shown in *Figure 5.29*

Part (a) shows $20 \log_{10} A$ as a negative number because A will be less than 1. Part (b) shows

$$20 \log_{10} \left(\frac{\omega/\omega_1}{\sqrt{1 + (\omega/\omega_1)^2}}\right)$$

which is best regarded as

$$20 \log_{10}(\omega/\omega_1) - 20 \log_{10} \sqrt{1 + (\omega/\omega_1)^2}$$

Part (c) shows

$$- 20 \log_{10} \sqrt{1 + (\omega/\omega_2)^2}$$

Part (d) shows the whole transfer function made by adding the constituent parts. The sloping lines rise or fall at a rate of 20 dB for every decade, a decade being a ten-fold change of frequency.

The other part of the Bode diagram showing phase is built up in a similar manner as shown in *Figure 5.31*. The first term

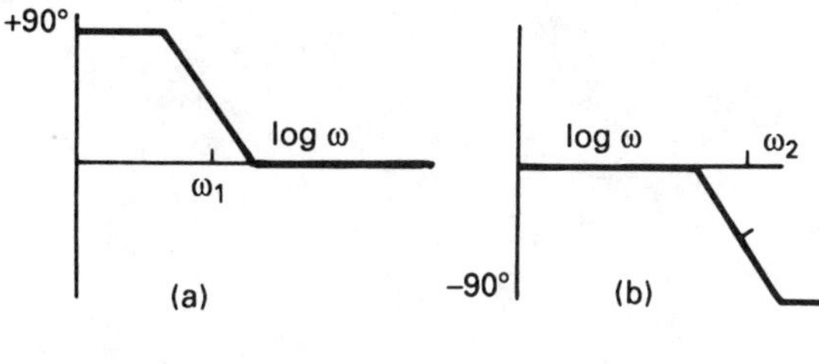

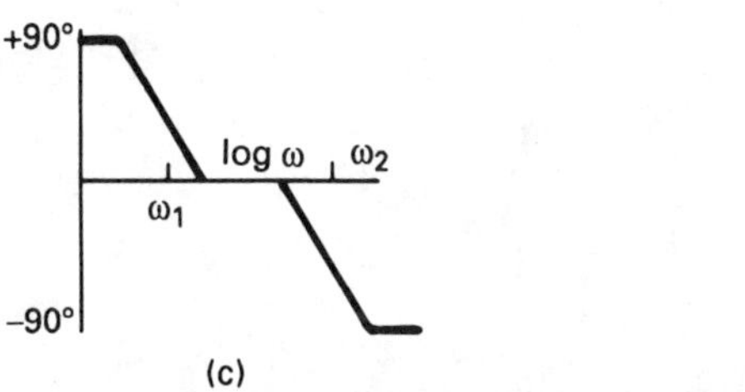

Figure 5.31 Bode diagram for phase shift of transfer function of circuit shown in *Figure 5.29*

introduces no phase shift, the second one a shift of $90° - \arctan(\omega/\omega_1)$ (*Figure 5.31(a)*) and the third a shift of $-\arctan(\omega/\omega_2)$.

In *Figures 5.30* and *5.31* the lines shown are the asymptotes to which the true curves approach.

Statistics

F F Mazda
DFH, MPhil, CEng, FIEE, DMS, MBIM
STC Telecommunications Ltd

Contents

6.1 Introduction

Data are available in vast quantities in all branches of electronic engineering. This chapter presents the more commonly used techniques for presenting and manipulating data to obtain meaningful results.

6.2 Data presentation

Probably the most common method used to present engineering data is by tables and graphs. For impact, or to convey information quickly, pictograms and bar charts may be used. Pie charts are useful in showing the different proportions of a unit.

A strata graph shows how the total is split amongst its constituents. For example, if a voltage is applied across four parallel circuits, then the total current curve may be as in *Figure 6.1*. This shows that the total current is made up of currents in the four parallel circuits, which vary in different ways with the applied voltage.

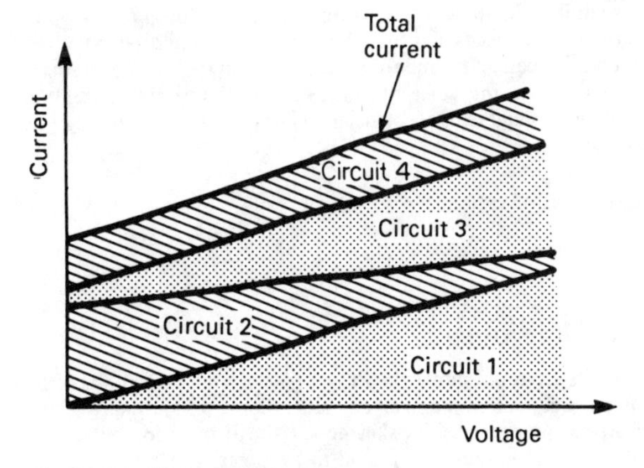

Figure 6.1 Illustration of a strata graph

Logarithmic or ratio graphs are used when one is more interested in the change in the ratios of numbers rather than their absolute value. In the logarithmic graph, equal ratios represent equal distances.

Frequency distributions are conveniently represented by a histogram as in *Figure 6.2*. This shows the voltage across a batch

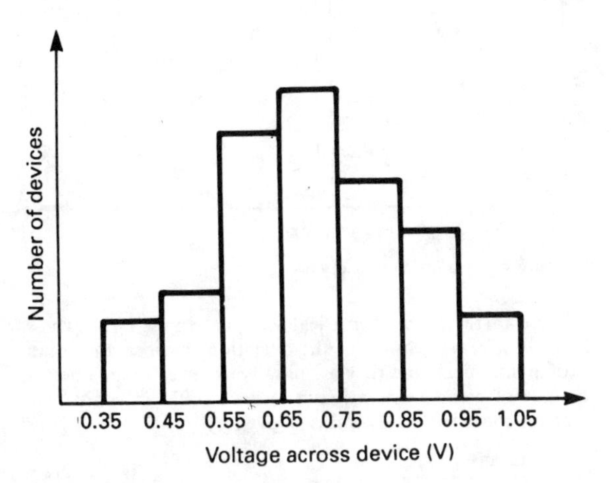

Figure 6.2 A histogram

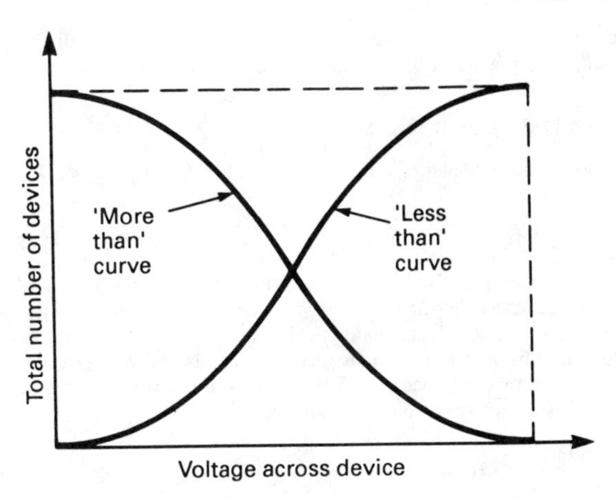

Figure 6.3 Illustration of ogives

of diodes. Most of the batch had voltage drops in the region 0.65 to 0.75 V, the next largest group being 0.55 to 0.65 V. In a histogram, the areas of the rectangles represent the frequencies in the different groups. Ogives, illustrated in *Figure 6.3*, show the cumulative frequency occurrences above or below a given value. From this curve it is possible to read off the total number of devices having a voltage greater than or less than a specific value.

6.3 Averages

6.3.1 Arithmetic mean

The arithmetic mean of n numbers $x_1, x_2, x_3, \ldots, x_n$ is given by

$$\bar{x} = \frac{x_1 + x_2 + x_3 + \cdots + x_n}{n}$$

or

$$\bar{x} = \frac{\sum_{r=1}^{n} x_r}{n} \tag{6.1}$$

The arithmetic mean is easy to calculate and it takes into account all the figures. Its disadvantages are that it is influenced unduly by extreme values and the final result may not be a whole number, which can be absurd at times, e.g. a mean of $2\frac{1}{2}$ men.

6.3.2 Median and mode

Median or 'middle one' is found by placing all the figures in order and choosing the one in the middle, or if there are an even number of items, the mean of the two central numbers. It is a useful technique for finding the average of items which cannot be expressed in figures, e.g. shades of a colour. It is also not influenced by extreme values. However, the median is not representative of all the figures.

The mode is the most 'fashionable' item, that is, the one which appears the most frequently.

6.3.3 Geometric mean

The geometric mean of n numbers $x_1, x_2, x_3, \ldots, x_n$ is given by

$$x_g = \sqrt[n]{(x_1 \times x_2 \times x_3 \times \ldots \times x_n)} \tag{6.2}$$

This technique is used to find the average of quantities which follow a geometric progression or exponential law, such as rates of changes. Its advantage is that it takes into account all the numbers, but is not unduly influenced by extreme values.

6.3.4 Harmonic mean

The harmonic mean of n numbers $x_1, x_2, x_3, \ldots, x_n$ is given by

$$x_h = \frac{n}{\sum_{r=1}^{n} (1/x_r)} \quad (6.3)$$

This averaging method is used when dealing with rates or speeds or prices. As a rule when dealing with items such as A per B, if the figures are for equal As then use the harmonic mean but if they are for equal Bs use the arithmetic mean. So if a plane flies over three equal distances at speeds of 5 m/s, 10 m/s and 15 m/s the mean speed is given by the harmonic mean as

$$\frac{3}{\frac{1}{5}+\frac{1}{10}+\frac{1}{15}} = 8.18 \text{ m/s}$$

If, however, the plane were to fly for three equal times, of say, 20 seconds at speeds of 5 m/s, 10 m/s and 15 m/s, then the mean speed would be given by the arithmetic mean as $(5+10+15)/3 = 10$ m/s.

6.4 Dispersion from the average

6.4.1 Range and quartiles

The average represents the central figure of a series of numbers or items. It does not give any indication of the spread of the figures in the series from the average. Therefore, in *Figure 6.4*, both curves, A and B, have the same average but B has a wider deviation from the average than curve A.

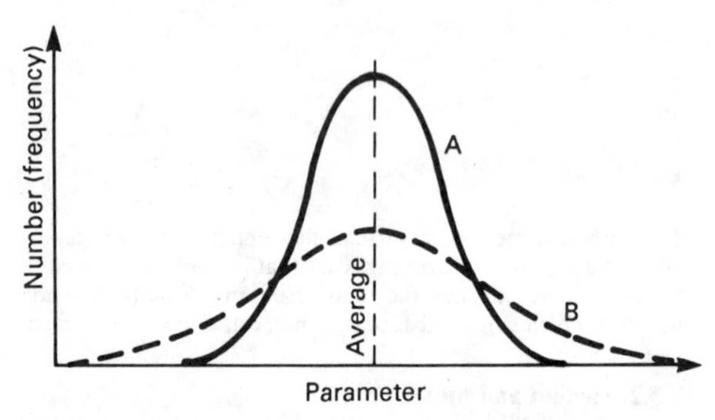

Figure 6.4 Illustration of deviation from the average

There are several ways of stating by how much the individual numbers in the series differ from the average. The range is the difference between the smallest and largest values. The series can also be divided into four quartiles and the dispersion stated as the interquartile range, which is the difference between the first and third quartile numbers, or the quartile deviation which is half this value.

The quartile deviation is easy to use and is not influenced by extreme values. However, it gives no indication of distribution between quartiles and covers only half the values in a series.

6.4.2 Mean deviation

This is found by taking the mean of the differences between each individual number in the series and the arithmetic mean, or median, of the series. Negative signs are ignored.

For a series of n numbers $x_1, x_2, x_3, \ldots, x_n$ having an arithmetic mean of $\bar{x}$ the mean deviation of the series is given by

$$\frac{\sum_{r=1}^{n} |x_r - \bar{x}|}{n} \quad (6.4)$$

The mean deviation takes into account all the items in the series, but it is not very suitable since it ignores signs.

6.4.3 Standard deviation

This is the most common measure of dispersion. For this the arithmetic mean must be used and not the median. It is calculated by squaring deviations from the mean, so eliminating their sign, adding the numbers together and then taking their mean and then the square root of the mean. Therefore, for the series in Section 6.4.2, the standard deviation is given by

$$\sigma = \left(\frac{\sum_{r=1}^{n} (x_r - \bar{x})^2}{n} \right)^{1/2} \quad (6.5)$$

The unit of the standard deviation is that of the original series. So if the series consists of the heights of a group of children in metres, then the mean and standard deviation are in metres. To compare two series having different units, such as the height of children and their weights, the coefficient of variation is used, which is unitless:

$$\text{coefficient of variation} = \frac{\sigma}{\bar{x}} \times 100 \quad (6.6)$$

6.5 Skewness

The distribution shown in *Figure 6.4* is symmetrical since the mean, median and mode all coincide. *Figure 6.5* shows a skewed distribution with positive skewness. If the distribution bulges the other way, the skewness is said to be negative.

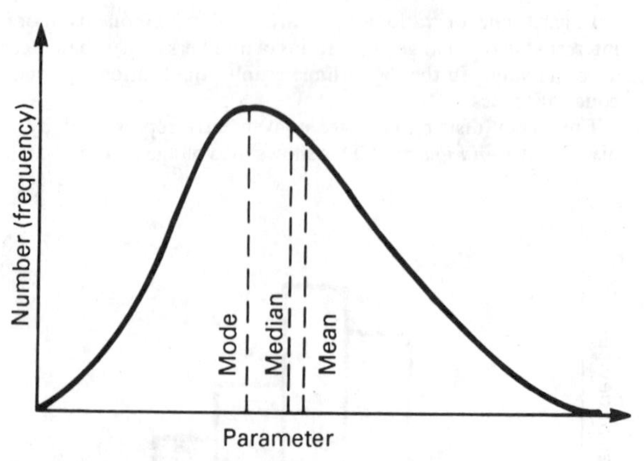

Figure 6.5 Illustration of skewness

There are several mathematical ways for expressing skewness. They all give a measure of the deviation between the mean, median and mode and they are usually stated in relative terms, for ease of comparison between series of different units. The Pearson coefficient of skewness is given by

$$P_k = \frac{\text{mean} - \text{mode}}{\text{standard deviation}} \quad (6.7)$$

Since the mode is sometimes difficult to measure this can also be stated as

$$P_k = \frac{3\ (\text{mean} - \text{median})}{\text{standard deviation}} \quad (6.8)$$

6.6 Combinations and permutations

6.6.1 Combinations

Combinations are the number of ways in which a proportion can be chosen from a group. Therefore the number of ways in which two letters can be chosen from a group of four letters A, B, C, D is equal to 6, i.e. AB, AC, AD, BC, BD, CD. This is written as

$$^4C_2 = 6$$

The factorial expansion is frequently used in combination calculations where

$$n! = n \times (n-1) \times (n-2) \times \cdots \times 3 \times 2 \times 1$$

Using this the number of combinations of r times from a group of n is given by

$$^nC_r = \frac{n!}{r!\,(n-r)!} \quad (6.9)$$

6.6.2 Permutations

Combinations do not indicate any sequencing. When sequencing within each combination is involved the result is known as a permutation. Therefore the number of permutations of two letters out of four letters A, B, C, D is 12, i.e. AB, BA, AC, CA, AD, DA, BC, CB, BD, DB, CD, DC. The number of permutations of r items from a group of n is given by

$$^nP_r = \frac{n!}{(n-r)!} \quad (6.10)$$

6.7 Regression and correlation

6.7.1 Regression

Regression is a method for establishing a mathematical relationship between two variables. Several equations may be used to establish this relationship, the most common being that of a straight line. *Figure 6.6* shows the plot of seven readings. This is called a scatter diagram. The points can be seen to lie approximately on the straight line AB.

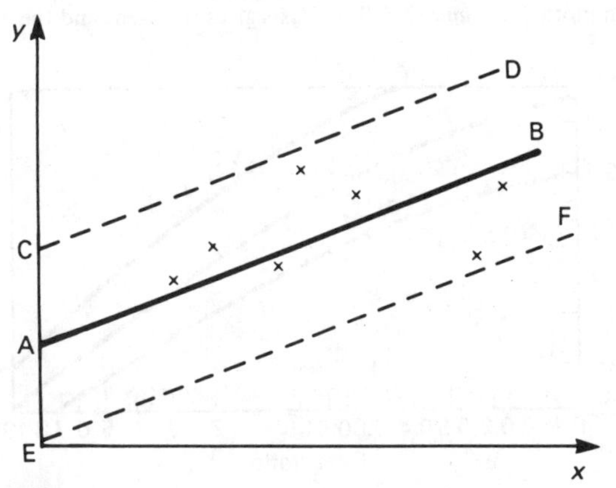

Figure 6.6 A scatter diagram

The equation of a straight line is given by

$$y = mx + c \quad (6.11)$$

where x is the independent variable, y the dependent variable, m the slope of the line and c its intercept on the y-axis. c is negative if the line intercepts the y-axis on its negative part and m is negative if the line slopes the other way to that shown in *Figure 6.6*.

The best straight line to fit a set of points is found by the method of least squares as

$$m = \frac{\sum xy - (\sum x \sum y)/n}{\sum x^2 - (\sum x)^2/n} \quad (6.12)$$

and

$$c = \frac{\sum x \sum xy - \sum y \sum x^2}{(\sum x)^2 - n \sum x^2} \quad (6.13)$$

where n is the number of points. The line passes through the mean values of x and y, i.e. $\bar{x}$ and $\bar{y}$.

6.7.2 Correlation

Correlation is a technique for establishing the strength of the relationship between variables. In *Figure 6.6* the individual figures are scattered on either side of a straight line and although one can approximate them by a straight line it may be required to establish if there is correlation between the x- and y-readings.

Several correlation coefficients exist. The product moment correlation coefficient (r) is given by

$$r = \frac{\sum (x-\bar{x})(y-\bar{y})}{n\sigma_x\sigma_y} \quad (6.14)$$

or

$$r = \frac{\sum (x-\bar{x})(y-\bar{y})}{[\sum (x-\bar{x})^2 \sum (y-\bar{y})^2]^{1/2}} \quad (6.15)$$

The value of r varies from $+1$, when all the points lie on a straight line and y increases with x, to -1, when all the points lie on a straight line but y decreases with x. When $r=0$ the points are widely scattered and there is said to be no correlation between x and y.

The standard error of estimation in r is given by

$$S_y = \sigma_y(1-r^2)^{1/2} \quad (6.16)$$

In about 95% of cases, the actual values will lie within plus or minus twice the standard error of estimated values given by the regression equation. This is shown by lines CD and EF in *Figure 6.6*. Almost all the values will be within plus or minus three times the standard error of estimated values.

It should be noted that σ_y is the variability of the y-values, whereas S_y is a measure of the variability of the y-values as they differ from the regression which exists between x and y. If there is no regression then $r=0$ and $\sigma_y = S_y$.

It is often necessary to draw conclusions from the order in which items are ranked. For example, two judges may rank contestants in a beauty contest and we need to know if there is any correlation between their rankings. This may be done by using the Rank correlation coefficient (R) given by

$$R = 1 - \frac{6\sum d^2}{n^3 - n} \quad (6.17)$$

where d is the difference between the two ranks for each item and n is the number of items. The value of R will vary from $+1$ when the two ranks are identical to -1 when they are exactly reversed.

6.8 Probability

If an event A occurs n times out of a total of m cases then the probability of occurrence is stated to be

$$P(\mathrm{A})=n/m \tag{6.18}$$

Probability varies between 0 and 1. If $P(\mathrm{A})$ is the probability of occurrence then $1-P(\mathrm{A})$ is the probability that event A will not occur and it can be written as $P(\bar{\mathrm{A}})$.

If A and B are two events then the probability that either may occur is given by

$$P(\mathrm{A\ or\ B})=P(\mathrm{A})+P(\mathrm{B})-P(\mathrm{A\ and\ B}) \tag{6.19}$$

A special case of this probability law is when events are mutually exclusive, i.e. the occurrence of one event prevents the other from happening. Then

$$P(\mathrm{A\ or\ B})=P(\mathrm{A})+P(\mathrm{B}) \tag{6.20}$$

If A and B are two events then the probability that they may occur together is given by

$$P(\mathrm{A\ and\ B})=P(\mathrm{A})\times P(\mathrm{B|A}) \tag{6.21}$$

or

$$P(\mathrm{A\ and\ B})=P(\mathrm{B})\times P(\mathrm{A|B}) \tag{6.22}$$

$P(\mathrm{B|A})$ is the probability that event B will occur assuming that event A has already occurred and $P(\mathrm{A|B})$ is the probability that event A will occur assuming that event B has already occurred. A special case of this probability law is when A and B are independent events, i.e. the occurrence of one event has no influence on the probability of the other event occurring. Then

$$P(\mathrm{A\ and\ B})=P(\mathrm{A})\times P(\mathrm{B}) \tag{6.23}$$

Bayes' theorem on probability may be stated as

$$P(\mathrm{A|B})=\frac{P(\mathrm{A})P(\mathrm{B|A})}{P(\mathrm{A})P(\mathrm{B|A})+P(\bar{\mathrm{A}})P(\mathrm{B}|\bar{\mathrm{A}})} \tag{6.24}$$

As an example of the use of Bayes' theorem suppose that a company discovers that 80% of those who bought its product in a year had been on the company's training course. 30% of those who bought a competitor's product had also been on the same training course. During that year the company had 20% of the market. The company wishes to know what percentage of buyers actually went on its training course, in order to discover the effectiveness of this course.

If B denotes that a person bought the company's product and T that he went on the training course then the problem is to find $P(\mathrm{B|T})$. From the data $P(\mathrm{B})=0.2$, $P(\bar{\mathrm{B}})=0.8$, $P(\mathrm{T|B})=0.8$, $P(\mathrm{T}|\bar{\mathrm{B}})=0.3$. Then from Equation (6.24)

$$P(\mathrm{B|T})=\frac{0.2\times0.8}{0.2\times0.8+0.8\times0.3}=0.4$$

6.9 Probability distributions

There are several mathematical formulae with well defined characteristics and these are known as probability distributions. If a problem can be made to fit one of these distributions then its solution is simplified. Distributions can be discrete when the characteristic can only take certain specific values, such as 0, 1, 2, etc., or they can be continuous where the characteristic can take any value.

6.9.1 Binomial distribution

The binomial probability distribution is given by

$$(p+q)^n=q^n+{}^nC_1pq^{n-1}+{}^nC_2p^2q^{n-2}+\cdots+{}^nC_xp^xq^{n-x}+\cdots+p^n \tag{6.25}$$

where p is the probability of an event occurring, $q\ (=1-p)$ is the probability of an event not occurring and n is the number of selections.

The probability of an event occurring m successive times is given by the binomial distribution as

$$p(m)={}^nC_mp^mq^{n-m} \tag{6.26}$$

The binomial distribution is used for discrete events and is applicable if the probability of occurrence p of an event is constant on each trial. The mean of the distribution $B(M)$ and the standard deviation $B(S)$ are given by

$$B(M)=np \tag{6.27}$$

$$B(S)=(npq)^{1/2} \tag{6.28}$$

6.9.2 Poisson distribution

The Poisson distribution is used for discrete events and, like the binomial distribution, it applies to mutually independent events. It is used in cases where p and q cannot both be defined. For example, one can state the number of goals which were scored in a football match, but not the goals which were not scored.

The Poisson distribution may be considered to be the limiting case of the binomial when n is large and p is small. The probability of an event occurring m successive times is given by the Poisson distribution as

$$p(m)=(np)^m\frac{e^{-np}}{m!} \tag{6.29}$$

The mean $P(M)$ and standard deviation $P(S)$ of the Poisson distribution are given by

$$P(M)=np \tag{6.30}$$

$$P(S)=(np)^{1/2} \tag{6.31}$$

Poisson probability calculations can be done by the use of probability charts as shown in *Figure 6.7*. This shows the probability that an event will occur at least m times when the mean (or expected) value np is known.

6.9.3 Normal distribution

The normal distribution represents continuous events and is shown plotted in *Figure 6.8*. The x-axis gives the event and the

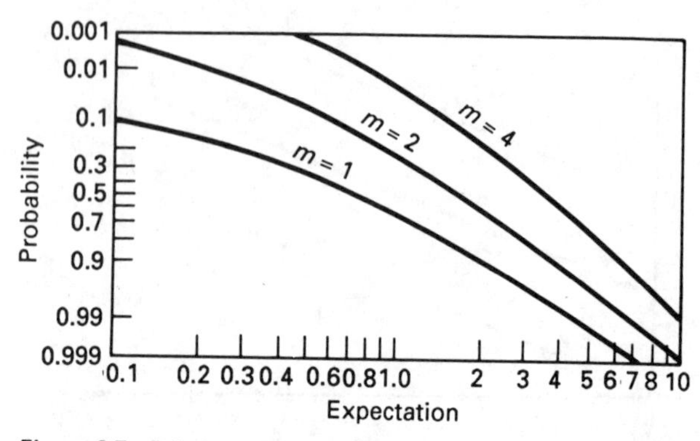

Figure 6.7 Poisson probability paper

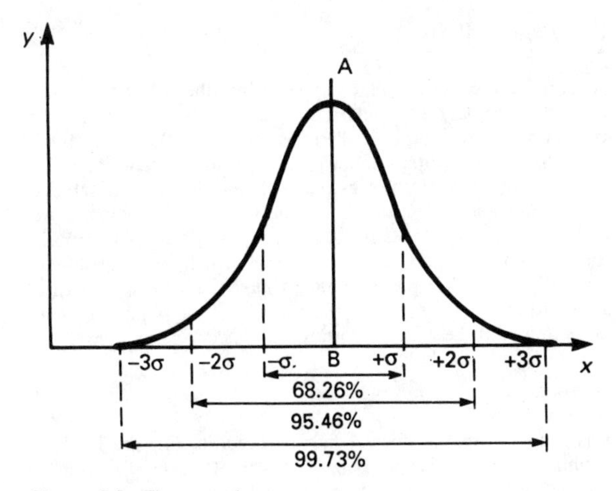

Figure 6.8 The normal curve

y-axis the probability of the event occurring. The curve shows that most of the events occur close to the mean value and this is usually the case in nature. The equation of the normal curve is given by

$$y=\frac{1}{\sigma(2\pi)^{1/2}}\exp[-(x-\bar{x})^2/2\sigma^2] \tag{6.32}$$

where $\bar{x}$ is the mean of the values making up the curve and σ is their standard deviation.

Different distributions will have varying mean and standard deviations but if they are distributed normally their curves will all follow Equation (6.32). These distributions can all be normalised to a standard form by moving the origin of their normal curve to their mean value, shown as B in *Figure 6.8*. The deviation from the mean is now represented on a new scale of units given by

$$\omega=\frac{x-\bar{x}}{\sigma} \tag{6.33}$$

The equation for the standardised normal curve now becomes

$$y=\frac{1}{(2\pi)^{1/2}}\exp(-\omega^2/2) \tag{6.34}$$

The total area under the standardised normal curve is unity and the area between any two values of ω is the probability of an item from the distribution falling between these values. The normal curve extends infinitely in either direction but 68.26% of its values (area) fall between $\pm\sigma$, 95.46% between $\pm 2\sigma$, 99.73% between $\pm 3\sigma$ and 99.994% between $\pm 4\sigma$.

Table 6.1 gives the area under the normal curve for different values of ω. Since the normal curve is symmetrical the area from $+\omega$ to $+\infty$ is the same as from $-\omega$ to $-\infty$. As an example of the use of this table, suppose that 5000 street lamps have been installed in a city and that the lamps have a mean life of 1000 hours with a standard deviation of 100 hours. How many lamps will fail in the first 800 hours? From Equation (6.33)

$$\omega=(800-1000)/100=-2$$

Ignoring the negative sign, Table 6.1 gives the probability of lamps not failing as 0.977 so that the probability of failure is $1-0.977$ or 0.023. Therefore 5000×0.023 or 115 lamps are expected to fail after 800 hours.

Table 6.1 Area under the normal curve from $-\infty$ to ω

ω	0.00	0.02	0.04	0.06	0.08
0.0	0.500	0.508	0.516	0.524	0.532
0.1	0.540	0.548	0.556	0.564	0.571
0.2	0.579	0.587	0.595	0.603	0.610
0.3	0.618	0.626	0.633	0.640	0.648
0.4	0.655	0.663	0.670	0.677	0.684
0.5	0.692	0.700	0.705	0.712	0.719
0.6	0.726	0.732	0.739	0.745	0.752
0.7	0.758	0.764	0.770	0.776	0.782
0.8	0.788	0.794	0.800	0.805	0.811
0.9	0.816	0.821	0.826	0.832	0.837
1.0	0.841	0.846	0.851	0.855	0.860
1.1	0.864	0.869	0.873	0.877	0.881
1.2	0.885	0.889	0.893	0.896	0.900
1.3	0.903	0.907	0.910	0.913	0.916
1.4	0.919	0.922	0.925	0.928	0.931
1.5	0.933	0.936	0.938	0.941	0.943
1.6	0.945	0.947	0.950	0.952	0.954
1.7	0.955	0.957	0.959	0.961	0.963
1.8	0.964	0.966	0.967	0.969	0.970
1.9	0.971	0.973	0.974	0.975	0.976
2.0	0.977	0.978	0.979	0.980	0.981
2.1	0.982	0.983	0.984	0.985	0.985
2.2	0.986	0.987	0.988	0.988	0.989
2.3	0.989	0.990	0.990	0.991	0.991
2.4	0.992	0.992	0.993	0.993	0.993
2.5	0.994	0.994	0.995	0.995	0.995
2.6	0.995	0.996	0.996	0.996	0.996
2.7	0.997	0.997	0.997	0.997	0.997
2.8	0.997	0.998	0.998	0.998	0.998
2.9	0.998	0.998	0.998	0.998	0.999
3.0	0.999	0.999	0.999	0.999	0.999

6.9.4 Exponential distribution

The exponential probability distribution is a continuous distribution and is shown in *Figure 6.9*. It has the equation

$$y=\frac{1}{x}\exp(-x/\bar{x}) \tag{6.35}$$

where $\bar{x}$ is the mean of the distribution. Whereas in the normal distribution the mean value divides the population in half, for the exponential distribution 36.8% of the population is above the average and 63.2% below the average. *Table 6.2* shows the area under the exponential curve for different values of the ratio $K=x/\bar{x}$, this area being shown shaded in *Figure 6.9*.

As an example suppose that the time between failures of a piece

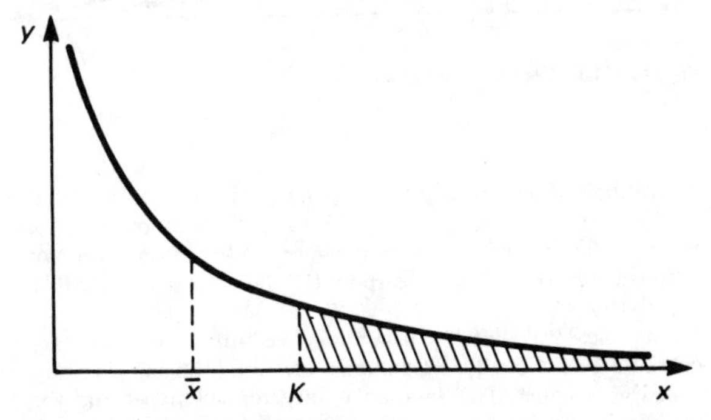

Figure 6.9 The exponential curve

Table 6.2 Area under the exponential curve from K to $+\infty$

K	0.00	0.02	0.04	0.06	0.08
0.0	1.000	0.980	0.961	0.942	0.923
0.1	0.905	0.886	0.869	0.852	0.835
0.2	0.819	0.803	0.787	0.771	0.776
0.3	0.741	0.726	0.712	0.698	0.684
0.4	0.670	0.657	0.644	0.631	0.619
0.5	0.607	0.595	0.583	0.571	0.560
0.6	0.549	0.538	0.527	0.517	0.507
0.7	0.497	0.487	0.477	0.468	0.458
0.8	0.449	0.440	0.432	0.423	0.415
0.9	0.407	0.399	0.391	0.383	0.375

of equipment is found to vary exponentially. If results indicate that the mean time between failures is 1000 hours, then what is the probability that the equipment will work for 700 hours or more without a failure? Calculating K as $700/1000=0.7$ then from *Table 6.2* the area beyond 0.7 is 0.497 which is the probability that the equipment will still be working after 700 hours.

6.9.5 Weibull distribution

This is a continuous probability distribution and its equation is given by

$$y=\alpha\beta(x-\gamma)^{\beta-1}\exp[-\alpha(x-\gamma)^{\beta}] \tag{6.36}$$

where α is called the scale factor, β the shape factor and γ the location factor.

The shape of the Weibull curve varies depending on the value of its factors. β is the most important, as shown in *Figure 6.10*, and

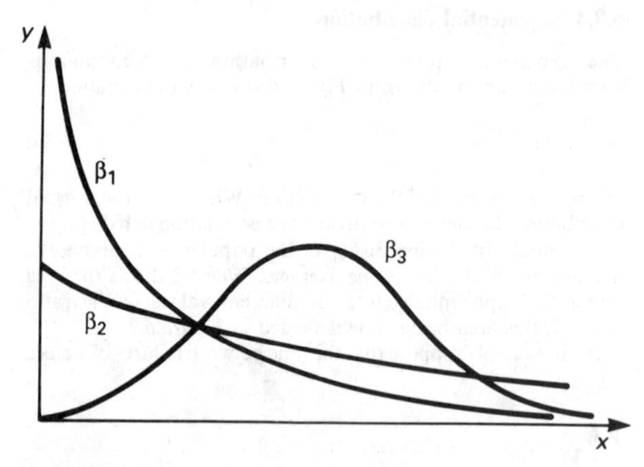

Figure 6.10 Weibull curves ($\alpha=1$)

the Weibull curve varies from an exponential ($\beta=1.0$) to a normal distribution ($\beta=3.5$). In practice β varies from about $\frac{1}{3}$ to 5. Because the Weibull distribution can be made to fit a variety of different sets of data, it is popularly used for probability distributions.

Analytical calculations using the Weibull distribution are cumbersome. Usually predictions are made using Weibull probability paper. The data are plotted on this paper and the probability predictions read from the graph.

6.10 Sampling

A sample consists of a relatively small number of items drawn from a much larger population. This sample is analysed for certain attributes and it is then assumed that these attributes apply to the total population, within a certain tolerance of error.

Sampling is usually associated with the normal probability distribution and, based on this distribution, the errors which arise due to sampling can be estimated. Suppose a sample of n_s items is taken from a population of n_p items which are distributed normally. If the sample is found to have a mean of μ_s with a standard deviation of σ_s then the mean μ_p of the population can be estimated to be within a certain tolerance of μ_s. It is given by

$$\mu_p=\mu_s\pm\frac{\gamma\sigma_s}{n_s^{1/2}} \tag{6.37}$$

γ is found from the normal curve depending on the level of confidence we need in specifying μ_p. For $\gamma=1$ this level is 68.26%; for $\gamma=2$ it is 95.46% and for $\gamma=3$ it is 99.73%.

The standard error of mean σ_e is often defined as

$$\sigma_e=\frac{\sigma_s}{n_s^{1/2}} \tag{6.38}$$

so Equation (6.37) can be rewritten as

$$\mu_p=\mu_s\pm\gamma\sigma_e \tag{6.39}$$

As an example suppose that a sample of 100 items, selected at random from a much larger population, gives their mean weight as 20 kg with a standard deviation of 100 g. The standard error of the mean is therefore $100/(100)^{1/2}=10$ g and one can say with 99.73% confidence that the mean value of the population lies between $20\pm3\times0.01$ or 20.03 kg and 19.97 kg.

If in a sample of n_s items the probability of occurrence of a particular attribute is p_s, then the standard error of probability p_e is defined as

$$p_e=\left(\frac{p_s q_s}{n_s}\right)^{1/2} \tag{6.40}$$

where $q_s=1-p_s$.

The probability of occurrence of the attribute in the population is then given by

$$p_p=p_s\pm\gamma p_e \tag{6.41}$$

where γ is again chosen to cover a certain confidence level.

As an example suppose a sample of 500 items shows that 50 are defective. Then the probability of occurrence of the defect in the sample is $50/500=0.1$. The standard error of probability is $(0.1\times0.9/500)^{1/2}$ or 0.0134. Therefore we can state with 95.46% confidence that the population from which the sample was drawn has a defect probability of $0.1\pm2\times0.0134$, i.e. 0.0732 to 0.1268; or we can state with 99.73% confidence that this value will lie between $0.1\pm3\times0.0134$, i.e. 0.0598 to 0.1402.

If two samples have been taken from the same population and these give standard deviations of σ_{s1} and σ_{s2} for sample sizes of n_{s1} and n_{s2} then Equation (6.38) can be modified to give the standard error of the difference between means as

$$\sigma_{de}=\left(\frac{\sigma_{s1}^2}{n_{s1}}+\frac{\sigma_{s2}^2}{n_{s2}}\right)^{1/2} \tag{6.42}$$

Similarly Equation (6.40) can be modified to give the standard error of the difference between probabilities of two samples from the same population as

$$p_{de}=\left(\frac{p_{s1}q_{s1}}{n_{s1}}+\frac{p_{s2}q_{s2}}{n_{s2}}\right)^{1/2} \tag{6.43}$$

6.11 Tests of significance

In taking samples we often obtain results which deviate from the expected. Tests of significance are then used to determine if this deviation is real or if it could have arisen due to sampling error.

6.11.1 Hypothesis testing

In this system a hypothesis is set up and is then tested at a given confidence level. For example, suppose a coin is tossed 100 times and it comes up heads 60 times. Is the coin biased or is it likely that this falls within a reasonable sampling error? The hypothesis is set up that the coin is not biased. Therefore one would expect that the probability of heads is 0.5, i.e. $p_s = 0.5$. The probability of tails, q_s, is also 0.5. Using Equation (6.40) the standard error of probability is given by $p_e = (0.5 \times 0.5/100)^{1/2}$ or 0.05. Therefore from Equation (6.41) the population probability at the 95.45% confidence level of getting heads is $0.5 + 2 \times 0.05 = 0.6$. Therefore it is highly likely that the coin is not biased and the results are due to sampling error.

The results of any significance test are not conclusive. For example, is 95.45% too high a confidence level to require? The higher the confidence level the greater the risk of rejecting a true hypothesis, and the lower the level the greater the risk of accepting a false hypothesis.

Suppose now that a sample of 100 items of production shows that five are defective. A second sample of 100 items is taken from the same production a few months later and gives two defectives. Does this show that the production quality is improving? Using Equation (6.43) the standard error of the difference between probabilities is given by $(0.05 \times 0.95/100 + 0.02 \times 0.98/100)^{1/2} = 0.0259$. This is less than twice the difference between the two probabilities, i.e. $0.05 - 0.02 = 0.03$, therefore the difference is very likely to have arisen due to sampling error and it does not necessarily indicate an improvement in quality.

6.11.2 Chi-square test

This is written as χ^2. If O is an observed result and E is the expected result then

$$\chi^2 = \sum \frac{(O-E)^2}{E} \qquad (6.44)$$

The χ^2 distribution is given by tables such as *Table 6.3*, from which the probability can be determined. The number of degrees of freedom is the number of classes whose frequency can be assigned independently. If the data are presented in the form of a table having V vertical columns and H horizontal rows then the degrees of freedom are usually found as $(V-1)(H-1)$.

Returning to the earlier example, suppose a coin is tossed 100 times and it comes up heads 60 times and tails 40 times. Is the coin biased? The expected values for heads and tails are 50 each so that

$$\chi^2 = \frac{(60-50)^2}{50} + \frac{(40-50)^2}{50} = 4$$

The number of degrees of freedom is one since once we have fixed the frequency for heads that for tails is defined. Therefore entering *Table 6.3* with one degree of freedom the probability level for $\chi^2 = 4$ is seen to be above 2.5%, i.e. there is a strong probability that the difference in the two results arose by chance and the coin is not biased.

As a further example suppose that over a 24-hour period the average number of accidents which occur in a factory is seen to be as in *Table 6.4*. Does this indicate that most of the accidents occur during the late night and early morning periods? Applying the χ^2 tests the expected value, if there was no difference between the time periods, would be the mean of the number of accidents, i.e. 5.

Table 6.3 The chi-square distribution

Degrees of freedom	*Probability level* 0.100	0.050	0.025	0.010	0.005
1	2.71	3.84	5.02	6.63	7.88
2	4.61	5.99	7.38	9.21	10.60
3	6.25	7.81	9.35	11.34	12.84
4	7.78	9.49	11.14	13.28	14.86
5	9.24	11.07	12.83	15.09	16.75
6	10.64	12.59	14.45	16.81	18.55
7	12.02	14.07	16.01	18.48	20.28
8	13.36	15.51	17.53	20.09	21.96
9	14.68	16.92	19.02	21.67	23.59
10	15.99	18.31	20.48	23.21	25.19
12	18.55	21.03	23.34	26.22	28.30
14	21.06	23.68	26.12	29.14	31.32
16	23.54	26.30	28.85	32.00	34.27
18	25.99	28.87	31.53	34.81	37.16
20	28.41	31.41	34.17	37.57	40.00
30	40.26	43.77	46.98	50.89	53.67
40	51.81	55.76	59.34	63.69	66.77

Table 6.4 Frequency distribution of accidents in a factory during 24 hours

Time (24 hour clock)	*Number of accidents*
0–6	9
6–12	3
12–18	2
18–24	6

Therefore from Equation (6.44)

$$\chi^2 = \frac{(9-5)^2}{5} + \frac{(3-5)^2}{5} + \frac{(2-5)^2}{5} + \frac{(6-5)^2}{5} = 6$$

There are three degrees of freedom, therefore from *Table 6.3* the probability of occurrence of the result shown in *Table 6.4* is seen to be greater than 10%. The conclusion would be that although there is a trend, as yet there are not enough data to show if this trend is significant or not. For example, if the number of accidents were each three times as large, i.e. 27, 9, 6, 18 respectively, then χ^2 would be calculated as 20.67 and from *Table 6.3* it is seen that the results are highly significant since there is a very low probability, less than ½%, that it can arise by chance.

6.11.3 Significance of correlation

The significance of the product moment correlation coefficient of Equations (6.14) or (6.15) can be tested at any confidence level by means of the standard error of estimation given by Equation (6.16). An alternative method is to use the Student t test of significance. This is given by

$$t = \frac{r(n-2)^{1/2}}{(1-r^2)^{1/2}} \qquad (6.45)$$

where r is the correlation coefficient and n the number of items. Tables are then used, similar to *Table 6.3*, which give the probability level for $(n-2)$ degrees of freedom.

The Student t for the rank correlation coefficient is given by

$$t = R[(n-2)/(1-R^2)]^{1/2} \quad (6.46)$$

and the same Student t tables are used to check the significance of R.

Further reading

BESTERFIELD, D. H., *Quality Control*, Prentice Hall (1979)
CAPLEN, R. H., *A Practical Approach to Quality Control*, Business Books (1982)
CHALK, G. O. and STICK, A. W., *Statistics for the Engineer*, Butterworths (1975)
COHEN, S. S., *Practical Statistics*, Edward Arnold (1988)
DAVID, H. A., *Order Statistics*, Wiley (1981)
DUDEWICZ, E. J. and MISHRA, S. N., *Modern Mathematical Statistics*, Wiley (1988)
DUNN, R. A. and RAMSING, K. D., *Management Science, a Practical Approach to Decision Making*, Macmillan (1981)
FITZSIMMONS, J. A., *Service Operations Management*, McGraw-Hill (1982)
GRANT, E. I. and LEAVENWORTH, R. S., *Statistical Quality Control*, McGraw-Hill (1980)
HAHN, W. C., *Modern Statistical Methods*, Butterworths (1979)
JONES, M. E. M., *Statistics*, Schofield & Sims (1988)
MAZDA, F. F., *Quantitative Techniques in Business*, Gee & Co. (1979)
SIEGEL, A. F., *Statistics and Data Analysis*, Wiley (1988)

Part 2

Physical Phenomena

7 Quantities and Units

L W Turner CEng, FIEE, FRTS
Consultant Engineer

Contents

7.1 International unit system

The International System of Units (SI) is the modern form of the metric system agreed at an international conference in 1960. It has been adopted by the International Standards Organisation (ISO) and the International Electrotechnical Commission (IEC) and its use is recommended wherever the metric system is applied. It is now being adopted throughout most of the world and is likely to remain the primary world system of units of measurement for a very long time. The indications are that SI units will supersede the units of existing metric systems and all systems based on Imperial units.

SI units and the rules for their application are contained in *ISO Resolution* R1000 (1969, updated 1973) and an informatory document *SI-Le Système International d'Unités*, published by the Bureau International des Poids et Mesures (BIPM). An abridged version of the former is given in British Standards Institution (BSI) publication PD 5686 *The use of SI Units* (1969, updated 1973) and BS 3763 *International System (SI) Units*; BSI (1964) incorporates information from the BIPM document.

The adoption of SI presents less of a problem to the electronics engineer and the electrical engineer than to those concerned with other engineering disciplines as all the practical electrical units were long ago incorporated in the metre-kilogram-second (MKS) unit system and these remain unaffected in SI.

The SI was developed from the metric system as a fully coherent set of units for science, technology and engineering. A coherent system has the property that corresponding equations between quantities and between numerical values have exactly the same form, because the relations between units do not involve numerical conversion factors. In constructing a coherent unit system, the starting point is the selection and definition of a minimum set of independent 'base' units. From these, 'derived' units are obtained by forming products or quotients in various combinations, again without numerical factors. Thus the base units of length (metre), time (second) and mass (kilogram) yield the SI units of velocity (metre/second), force (kilogram-metre/second-squared) and so on. As a result there is, for any given physical quantity, only one SI unit with no alternatives and with no numerical conversion factors. A single SI unit (joule = kilogram metre-squared/second-squared) serves for energy of any kind, whether it be kinetic, potential, thermal, electrical, chemical..., thus unifying the usage in all branches of science and technology.

The SI has seven base units, and two supplementary units of angle. Certain important derived units have special names and can themselves be employed in combination to form alternative names for further derivations.

Each physical quantity has a quantity-symbol (e.g., *m* for mass) that represents it in equations, and a unit-symbol (e.g., kg for kilogram) to indicate its SI unit of measure.

7.1.1 Base units

Definitions of the seven base units have been laid down in the following terms. The quantity-symbol is given in italics, the unit-symbol (and its abbreviation) in roman type.

Length: *l*; metre (m). The length equal to 1 650 763.73 wavelengths in vacuum of the radiation corresponding to the transition between the levels $2p_{10}$ and $5d_5$ of the krypton-86 atom.

Mass: *m*; kilogram (kg). The mass of the international prototype kilogram (a block of platinum preserved at the International Bureau of Weights and Measures at Sèvres).

Time: *t*; second (s). The duration of 9 192 631 770 periods of the radiation corresponding to the transition between the two hyperfine levels of the ground state of the caesium-133 atom.

Electric current: *i*; ampere (A). The current which, maintained in two straight parallel conductors of infinite length, of negligible circular cross-section and 1 m apart in vacuum, produces a force equal to 2×10^{-7} newton per metre of length.

Thermodynamic temperature: *T*; kelvin (K). The fraction 1/273.16 of the thermodynamic (absolute) temperature of the triple point of water.

Luminous intensity: *I*; candela (cd). The luminous intensity in the perpendicular direction of a surface of 1/600 000 m^2 of a black body at the temperature of freezing platinum under a pressure of 101 325 newtons per square metre.

Amount of substance: *Q*; mole (mol). The amount of substance of a system which contains as many elementary entities as there are atoms in 0.012 kg of carbon-12. The elementary entity must be specified and may be an atom, a molecule, an ion, an electron, etc., or a specified group of such entities.

7.1.2 Supplementary angular units

Plane angle: α, β...; radian (rad). The plane angle between two radii of a circle which cut off on the circumference an arc of length equal to the radius.

Solid angle: Ω; steradian (sr). The solid angle which, having its vertex at the centre of a sphere, cuts off an area of the surface of the sphere equal to a square having sides equal to the radius.

Force: The base SI unit of electric current is in terms of force in newtons (N). A force of 1 N is that which endows unit mass (1 kg) with unit acceleration (1 m/s^2). The newton is thus not only a coherent unit; it is also devoid of any association with gravitational effects.

7.1.3 Temperature

The base SI unit of thermodynamic temperature is referred to a point of 'absolute zero' at which bodies possess zero thermal energy. For practical convenience two points on the Kelvin temperature scale, namely 273.15 K and 373.15 K, are used to define the Celsius (or Centigrade) scale (0°C and 100°C). Thus in terms of temperature *intervals*, 1 K = 1°C; but in terms of temperature *levels*, a Celsius temperature θ corresponds to a Kelvin temperature $(\theta + 273.15)$ K.

7.1.4 Derived units

Nine of the more important SI derived units with their definitions are given

Quantity	*Unit name*	*Unit symbol*
Force	newton	N
Energy	joule	J
Power	watt	W
Electric charge	coulomb	C
Electrical potential difference and EMF	volt	V
Electric resistance	ohm	Ω
Electric capacitance	farad	F
Electric inductance	henry	H
Magnetic flux	weber	Wb

Newton That force which gives to a mass of 1 kilogram an acceleration of 1 metre per second squared.

Joule The work done when the point of application of 1 newton is displaced a distance of 1 metre in the direction of the force.

Watt The power which gives rise to the production of energy at the rate of 1 joule per second.

Coulomb The quantity of electricity transported in 1 second by a current of 1 ampere.

Volt The difference of electric potential between two points of a conducting wire carrying a constant current of 1 ampere, when the power dissipated between these points is equal to 1 watt.

Ohm The electric resistance between two points of a conductor when a constant difference of potential of 1 volt, applied between these two points, produces in this conductor a current of 1 ampere, this conductor not being the source of any electromotive force.

Farad The capacitance of a capacitor between the plates of which there appears a difference of potential of 1 volt when it is charged by a quantity of electricity equal to 1 coulomb.

Henry The inductance of a closed circuit in which an electromotive force of 1 volt is produced when the electric current in the circuit varies uniformly at a rate of 1 ampere per second.

Weber The magnet flux which, linking a circuit of one turn, produces in it an electromotive force of 1 volt as it is reduced to zero at a uniform rate in 1 second.

Some of the simpler derived units are expressed in terms of the seven basic and two supplementary units directly. Examples are listed in *Table 7.1*.

Table 7.1 Directly derived units

Quantity	*Unit name*	*Unit symbol*
Area	square metre	m^2
Volume	cubic metre	m^3
Mass density	kilogram per cubic metre	kg/m^3
Linear velocity	metre per second	m/s
Linear acceleration	metre per second squared	m/s^2
Angular velocity	radian per second	rad/s
Angular acceleration	radian per second squared	rad/s^2
Force	kilogram metre per second squared	kg m/s^2
Magnetic field strength	ampere per metre	A/m
Concentration	mole per cubic metre	mol/m^3
Luminance	candela per square metre	cd/m^2

Units in common use, particularly those for which a statement in base units would be lengthy or complicated, have been given special shortened names (see *Table 7.2*). Those that are named from scientists and engineers are abbreviated to an initial capital letter: all others are in small letters.

Table 7.2 Named derived units

Quantity	*Unit name*	*Unit symbol*	*Derivation*
Force	newton	N	kg m/s^2
Pressure	pascal	Pa	N/m^2
Power	watt	W	J/s
Energy	joule	J	N m, W s
Electric charge	coulomb	C	A s
Electric flux	coulomb	C	A s
Magnetic flux	weber	Wb	V s
Magnetic flux density	tesla	T	Wb/m^2
Electric potential	volt	V	J/C, W/A
Resistance	ohm	Ω	V/A
Conductance	siemens	S	A/V

Table 7.2 *continued*

Quantity	*Unit name*	*Unit symbol*	*Derivation*
Capacitance	farad	F	A s/V, C/V
Inductance	henry	H	V s/A, Wb/A
Luminous flux	lumen	lm	cd sr
Illuminance	lux	lx	lm/m^2
Frequency	hertz	Hz	l/s

The named derived units are used to form further derivations. Examples are given in *Table 7.3*.

Table 7.3 Further derived units

Quantity	*Unit name*	*Unit symbol*
Torque	newton metre	N m
Dynamic viscosity	pascal second	Pa s
Surface tension	newton per metre	N/m
Power density	watt per square metre	W/m^2
Energy density	joule per cubic metre	J/m^3
Heat capacity	joule per kelvin	J/K
Specific heat capacity	joule per kilogram kelvin	J/(kg K)
Thermal conductivity	watt per metre kelvin	W/(m K)
Electric field strength	volt per metre	V/m
Magnetic field strength	ampere per metre	A/m
Electric flux density	coulomb per square metre	C/m^2
Current density	ampere per square metre	A/m^2
Resistivity	ohm metre	Ω m
Permittivity	farad per metre	F/m
Permeability	henry per metre	H/m

Names of SI units and the corresponding EMU and ESU CGS units are given in *Table 7.4*.

Table 7.4 Unit names

Quantity	*Symbol*	*SI*	*EMU & ESU*
Length	l	metre (m)	centimetre (cm)
Time	t	second (s)	second
Mass	m	kilogram (kg)	gram (g)
Force	F	newton (N)	dyne (dyn)
Frequency	f, v	hertz (Hz)	hertz
Energy	E, W	joule (J)	erg (erg)
Power	P	watt (W)	erg/second (erg/s)
Pressure	p	newton/metre^2 (N/m^2)	dyne/centimetre^2 (dyn/cm^2)
Electric charge	Q	coulomb (C)	coulomb
Electric potential	V	volt (V)	volt
Electric current	I	ampere (A)	ampere
Magnetic flux	Φ	weber (Wb)	maxwell (Mx)
Magnetic induction	B	tesla (T)	gauss (G)
Magnetic field strength	H	ampere turn/ metre (At/m)	oersted (Oe)
Magnetomotive force	F_m	ampere turn (At)	gilbert (Gb)
Resistance	R	ohm (Ω)	ohm
Inductance	L	henry (H)	henry
Conductance	G	mho (Ω^{-1}) (siemens)	mho
Capacitance	C	farad (F)	farad

Table 7.5 Commonly used units of measurement

	SI (absolute)	*FPS (gravitational)*	*FPS (absolute)*	*cgs (absolute)*	*Metric technical units (gravitational)*
Length	metre (m)	ft	ft	cm	metre
Force	newton (N)	lbf	poundal (pdl)	dyne	kgf
Mass	kg	lb or slug	lb	gram	kg
Time	s	s	s	s	s
Temperature	°C K	°F	°F °R	°C K	°C K
Energy {mech.	joule*	ft lbf	ft pdl	dyn cm = erg	kgf m
Energy {heat		Btu	Btu	calorie	kcal
Power {mech.	watt	hp	hp }	erg/s	metric hp
Power {elec.		watt	watt }		watt
Electric current	amp	amp	amp	amp	amp
Pressure	N/m^2	lbf/ft^2	pdl/ft^2	dyn/cm^2	kgf/cm^2

* 1 joule = 1 newton metre or 1 watt second.

7.1.5 Gravitational and absolute systems

There may be some difficulty in understanding the difference between SI and the Metric Technical System of units which has been used principally in Europe. The main difference is that while mass is expressed in kg in both systems, weight (representing a force) is expressed as kgf, a gravitational unit, in the MKSA system and as N in SI. An absolute unit of force differs from a gravitational unit of force because it induces unit acceleration in a unit mass whereas a gravitational unit imparts gravitational acceleration to a unit mass.

A comparison of the more commonly known systems and SI is shown in *Table 7.5.*

7.1.6 Expressing magnitudes of SI units

To express magnitudes of a unit, decimal multiples and submultiples are formed using the prefixes shown in *Table 7.6.* This method of expressing magnitudes ensures complete adherence to a decimal system.

Table 7.6 The internationally agreed multiples and submultiples

Factor by which the unit is multiplied		*Prefix*	*Symbol*	*Common everyday examples*
One million million (billion)	10^{12}	tera	T	
One thousand million	10^{9}	giga	G	gigahertz (GHz)
One million	10^{6}	mega	M	megawatt (MW)
One thousand	10^{3}	kilo	k	kilometre (km)
One hundred	10^{2}	hecto*	h	
Ten	10^{1}	deca*	da	decagram (dag)
UNITY	1			
One tenth	10^{-1}	deci*	d	decimetre (dm)
One hundredth	10^{-2}	centi*	c	centimetre (cm)
One thousandth	10^{-3}	milli	m	milligram (mg)
One millionth	10^{-6}	micro	μ	microsecond (μs)
One thousand millionth	10^{-9}	nano	n	nanosecond (ns)
One million millionth	10^{-12}	pico	p	picofarad (pF)
One thousand million millionth	10^{-15}	femto	f	
One million million millionth	10^{-18}	atto	a	

* To be avoided wherever possible.

7.1.7 Auxiliary units

Certain auxiliary units may be adopted where they have application in special fields. Some are acceptable on a temporary basis, pending a more widespread adoption of the SI system. *Table 7.7* lists some of these.

Table 7.7 Auxiliary units

Quantity	*Unit symbol*	*SI equivalent*
Day	d	86 400 s
Hour	h	3600 s
Minute (time)	min	60 s
Degree (angle)	°	$\pi/180$ rad
Minute (angle)	′	$\pi/10\,800$ rad
Second (angle)	″	$\pi/648\,000$ rad
Are	a	$1\ dam^2 = 10^2\ m^2$
Hectare	ha	$1\ hm^2 = 10^4\ m^2$
Barn	b	$100\ fm^2 = 10^{-28}\ m^2$
Standard atmosphere	atm	101 325 Pa
Bar	bar	$0.1\ MPa = 10^5\ Pa$
Litre	l	$1\ dm^3 = 10^{-3}\ m^3$
Tonne	t	$10^3\ kg = 1\ Mg$
Atomic mass unit	u	$1.660\,53 \times 10^{-27}$ kg
Angström	Å	$0.1\ nm = 10^{-10}$ m
Electron-volt	eV	$1.602\,19 \times 10^{-19}$ J
Curie	Ci	$3.7 \times 10^{10}\ s^{-1}$
Röntgen	R	2.58×10^{-4} C/kg

7.1.8 Nuclear engineering

It has been the practice to use special units with their individual names for evaluating and comparing results. These units are usually formed by multiplying a unit from the cgs or SI system by a number which matches a value derived from the result of some natural phenomenon. The adoption of SI both nationally and internationally has created the opportunity to examine the practice of using special units in the nuclear industry, with the object of eliminating as many as possible and using the pure system instead.

As an aid to this, ISO draft Recommendations 838 and 839 have been published, giving a list of quantities with special names, the SI unit and the alternative cgs unit. It is expected that as SI is increasingly adopted and absorbed, those units based on cgs will go out of use. The values of these special units illustrate the fact that a change from them to SI would not be as revolutionary as might be supposed. Examples of these values together with the SI units which replace them are shown in *Table 7.8*.

Table 7.8 Nuclear engineering

Special unit			*SI replacement*
Name		*Value*	
Angström	(Å)	10^{-10} m	m
Barn	(b)	10^{-28} m^2	m^2
Curie	(Ci)	3.7×10^{10} s^{-1}	s^{-1}
Electron-volt	(eV)	$(1.602\,189\,2\pm.000\,004\,6)\times10^{-19}$ J	J
Röntgen	(R)	2.58×10^{-4} C/kg	C/kg

7.2 Universal constants in SI units

Table 7.9 Universal constants

The digits in parentheses following each quoted value represent the standard deviation error in the final digits of the quoted value as computed on the criterion of internal consistency. The unified scale of atomic weights is used throughout (^{12}C = 12). C = coulomb; G = gauss; Hz = hertz; J = joule; N = newton; T = tesla; u = unified nuclidic mass unit; W = watt; Wb = weber. For result multiply the numerical value by the SI unit.

Constant	*Symbol*	*Numerical value*	*SI unit*
Speed of light in vacuum	c	2.997 925(1)	10^8 m/s
Gravitational constant	G	6.670(5)*	10^{-11} N m^2 kg^2
Elementary charge	e	1.602 10(2)	10^{-19} C
Avogadro constant	N_A	6.022 52(9)	10^{26} kmol^{-1}
Mass unit	u	1.660 43(2)	10^{-27} kg
Electron rest mass	m_e	9.109 08(13)	10^{-31} kg
		5.485 97(3)	10^{-4} u
Proton rest mass	m_p	1.672 52(3)	10^{-27} kg
		1.007 276 63(8)	u
Neutron rest mass	m_n	1.674 82(3)	10^{-27} kg
		1.008 665 4(4)	u
Faraday constant	F	9.684 70(5)	10^4 C/mol
Planck constant	h	6.625 59(16)	10^{-34} J s
	$h/2\pi$	1.054 494(25)	10^{-34} J s
Fine-structure constant	α	7.297 20(3)	10^{-3}
	$1/\alpha$	137.038 8(6)	
Charge-to-mass ratio for electron	e/m_e	1.758 796(6)	10^{11} C/kg
Quantum of magnetic flux	hc/e	4.135 56(4)	10^{-11} Wb
Rydberg constant	R_∞	1.097 373 1(1)	10^7 m^{-1}
Bohr radius	a_0	5.291 67(2)	10^{-11} m
Compton wavelength of electron	h/m_ec	2.426 21(2)	10^{-12} m
	$\lambda C/2\pi$	3.861 44(3)	10^{-13} m
Electron radius	$e^2/m_ec^2=r_e$	2.817 77(4)	10^{-15} m
Thomson cross-section	$8\pi r_e^2/3$	6.651 6(2)	10^{-29} m^2
Compton wavelength of proton	$\lambda_{C,p}$	1.321 398(13)	10^{-15} m
	$\lambda_{C,p}/2\pi$	2.103 07(2)	10^{-16} m
Gyromagnetic ratio of proton	γ	2.675 192(7)	10^8 rad/(s T)
	$\gamma/2\pi$	4.257 70(1)	10^7 Hz/T
(uncorrected for diamagnetism of H_2O)	γ'	2.675 123(7)	10^8 rad/(s T)
	$\gamma'/2\pi$	4.257 59(1)	10^7 Hz/T
Bohr magneton	μ_B	9.273 2(2)	10^{-24} J/T
Nuclear magneton	μ_N	5.050 50(13)	10^{-27} J/T
Proton magnetic moment	μ_p	1.410 49(4)	10^{-26} J/T
	μ_p/μ_N	2.792 76(2)	
(uncorrected for diamagnetism in H_2O sample)	μ'_p/μ_N	2.792 68(2)	
Gas constant	R_0	8.314 34(35)	J/K mol
Boltzmann constant	k	1.380 54(6)	10^{-23} J/K
First radiation constant ($2\pi hc^2$)	c_1	3.741 50(9)	10^{-16} W/m^2
Second radiation constant (hc/k)	c_2	1.438 79(6)	10^{-2} m K
Stefan–Boltzmann constant	σ	5.669 7(10)	10^{-8} W/m^2 K^4

* The universal gravitational constant is not, and cannot in our present state of knowledge, be expressed in terms of other fundamental constants. The value given here is a direct determination by P. R. Heyl and P. Chrzanowski, *J. Res. Natl. Bur. Std.* (*U.S.*) 29, 1 (1942).

The above values are extracts from *Review of Modern Physics* Vol. 37 No. 4 October 1965 published by the American Institute of Physics.

7.3 Metric to Imperial conversion factors

Table 7.10 Conversion factors

SI units	*British units*
SPACE AND TIME	
Length:	
1 μm (micron)	$=39.37 \times 10^{-6}$ in
1 mm	=0.039 370 1 in
1 cm	=0.393 701 in
1 m	=3.280 84 ft
1 m	=1.093 61 yd
1 km	=0.621 371 mile
Area:	
1 mm^2	$=1.550 \times 10^{-3}$ in^2
1 cm^2	=0.155 0 in^2
1 m^2	=10.763 9 ft^2
1 m^2	=1.195 99 yd^2
1 ha	=2.471 05 acre
Volume:	
1 mm^3	$=61.023 7 \times 10^{-6}$ in^3
1 cm^3	$=61.023 7 \times 10^{-3}$ in^3
1 m^3	=35.314 7 ft^3
1 m^3	=1.307 95 yd^3
Capacity:	
10^6 m^3	$=219.969 \times 10^6$ gal
1 m^3	=219.969 gal
1 litre (l)	=0.219 969 gal
	=1.759 80 pint
Capacity flow:	
10^3 m^3/s	$=791.9 \times 10^6$ gal/h
1 m^3/s	$=13.20 \times 10^3$ gal/min
1 litre/s	=13.20 gal/min
1 m^3/kW h	=219.969 gal/kW h
1 m^3/s	=35.314 7 ft^3/s (cusecs)
1 litre/s	$=0.588 58 \times 10^{-3}$ ft^3/min (cfm)
Velocity:	
1 m/s	=3.280 84 ft/s = 2.236 94 mile/h
1 km/h	=0.621 371 mile/h
Acceleration:	
1 m/s^2	=3.280 84 ft/s^2
MECHANICS	
Mass:	
1 g	=0.035 274 oz
1 kg	=2.204 62 lb
1 t	=0.984 207 ton = 19.684 1 cwt
Mass flow:	
1 kg/s	=2.204 62 lb/s = 7.936 64 klb/h
Mass density:	
1 kg/m^3	=0.062 428 lb/ft^3
1 kg/litre	=10.022 119 lb/gal
Mass per unit length:	
1 kg/m	=0.671 969 lb/ft = 2.015 91 lb/yd
Mass per unit area:	
1 kg/m^2	=0.204 816 lb/ft^2
Specific volume:	
1 m^3/kg	=16.018 5 ft^3/lb
1 litre/tonne	=0.223 495 gal/ton
Momentum:	
1 kg m/s	=7.233 01 lb ft/s
Angular momentum:	
1 kg m^2/s	=23.730 4 lb ft^2/s
Moment of inertia:	
1 kg m^2	=23.730 4 lb ft^2

Table 7.10 *continued*

SI units	*British units*
MECHANICS	
Force:	
1 N	=0.224 809 lbf
Weight (force) per unit length:	
1 N/m	=0.068 521 lbf/ft
	=0.205 566 lbf/yd
Moment of force (or torque):	
1 N m	=0.737 562 lbf ft
Weight (force) per unit area:	
1 N/m^2	=0.020 885 lbf/ft^2
Pressure:	
1 N/m^2	$=1.450 38 \times 10^{-4}$ lbf/in^2
1 bar	=14.503 8 lbf/in^2
1 bar	=0.986 923 atmosphere
1 mbar	=0.401 463 in H_2O
	=0.029 53 in Hg
Stress:	
1 N/mm^2	$=6.474 90 \times 10^{-2}$ $tonf/in^2$
1 MN/m^2	$=6.474 90 \times 10^{-2}$ $tonf/in^2$
1 hbar	=0.647 490 $tonf/in^2$
Second moment of area:	
1 cm^4	=0.024 025 in^4
Section modulus:	
1 m^3	=61 023.7 in^3
1 cm^3	=0.061 023 7 in^3
Kinematic viscosity:	
1 m^2/s	=10.762 75 ft^2/s = 10^6 cSt
1 cSt	=0.038 75 ft^2/h
Energy, work:	
1 J	=0.737 562 ft lbf
1 MJ	=0.372 5 hph
1 MJ	=0.277 78 kW h
Power:	
1 W	=0.737 562 ft lbf/s
1 kW	=1.341 hp = 737.562 ft lbf/s
Fluid mass:	
(Ordinary) 1 kg/s	=2.204 62 lb/s = 793 6.64 lb/h
(Velocity) 1 kg/m^2 s	=0.204 815 lb/ft^2s
HEAT	
Temperature:	
(Interval) 1 K	=9/5 deg R (Rankine)
1°C	=9/5 deg F
(Coefficient) $1°R^{-1}$	=1 deg F^{-1} = 5/9 deg C
$1°C^{-1}$	=5/9 deg F^{-1}
Quantity of heat:	
1 J	$=9.478 17 \times 10^{-4}$ Btu
1 J	=0.238 846 cal
1 kJ	=947.817 Btu
1 GJ	$=947.817 \times 10^3$ Btu
1 kJ	=526.565 CHU
1 GJ	$=526.565 \times 10^3$ CHU
1 GJ	=9.478 17 therm
Heat flow rate:	
1 W(J/s)	=3.412 14 Btu/h
1 W/m^2	=0.316 998 Btu/ft^2 h
Thermal conductivity:	
1 W/m °C	=6.933 47 Btu in/ft^2 h °F
Coefficient and heat transfer:	
1 W/m^2 °C	=0.176 110 Btu/ft^2 h °F
Heat capacity:	
1 J/°C	$=0.526 57 \times 10^{-3}$ Btu/°R

Table 7.10 *continued*

SI units	*British units*
HEAT	
Specific heat capacity:	
1 J/g °C	=0.238 846 Btu/lb °F
1 kJ/kg °C	=0.238 846 Btu/lb °F
Entropy:	
1 J/K	$=0.52657\times10^{-3}$ Btu/°R
Specific entropy:	
1 J/kg °C	$=0.238846\times10^{-3}$ Btu/lb °F
1 J/kg K	$=0.238846\times10^{-3}$ Btu/lb °R
Specific energy/specific latent heat:	
1 J/g	=0.429 923 Btu/lb
1 J/kg	$=0.429923\times10^{-3}$ Btu/lb
Calorific value:	
1 kJ/kg	=0.429 923 Btu/lb
1 kJ/kg	=0.773 861 4 CHU/lb
1 J/m^3	$=0.0268392\times10^{-3}$ Btu/ft^3
1 kJ/m^3	=0.026 839 2 Btu/ft^3
1 kJ/litre	=4.308 86 Btu/gal
1 kJ/kg	=0.009 630 2 therm/ton
ELECTRICITY	
Permeability:	
1 H/m	$=10^7/4\pi\ \mu_0$
Magnetic flux density:	
1 tesla	$=10^4$ gauss = 1 Wb/m^2
Conductivity:	
1 mho	=1 reciprocal ohm
1 siemens	=1 reciprocal ohm
Electric stress:	
1 kV/mm	=25.4 kV/in
1 kV/m	=0.025 4 kV/in

7.4 Symbols and abbreviations

Table 7.11 Quantities and units of periodic and related phenomena (based on ISO Recommendation R31)

Symbol	*Quantity*
T	periodic time
τ, (T)	time constant of an exponentially varying quantity
f, ν	frequency
η	rotational frequency
ω	angular frequency
λ	wavelength
σ ($\tilde{\nu}$)	wavenumber
k	circular wavenumber
$\log_e (A_1/A_2)$	natural logarithm of the ratio of two amplitudes
$10 \log_{10} (P_1/P_2)$	ten times the common logarithm of the ratio of two powers
δ	damping coefficient
Λ	logarithmic decrement
α	attenuation coefficient
β	phase coefficient
γ	propagation coefficient

Table 7.12 Symbols for quantities and units of electricity and magnetism (based on ISO Recommendation R31)

Symbol	*Quantity*
I	electric current
Q	electric charge, quantity of electricity
ρ	volume density of charge, charge density (Q/V)
σ	surface density of charge (Q/A)
E, (K)	electric field strength
V, (φ)	electric potential
U, (V)	potential difference, tension
E	electromotive force
D	displacement (rationalised displacement)
D'	non-rationalised displacement
ψ	electric flux, flux of displacement (flux of rationalised displacement)
ψ'	flux of non-rationalised displacement
C	capacitance
ε	permittivity
ε_0	permittivity of vacuum
ε'	non-rationalised permittivity
ε'_0	non-rationalised permittivity of vacuum
ε_r	relative permittivity
χ_e	electric susceptibility
χ'_e	non-rationalised electric susceptibility
P	electric polarisation
p, (p_e)	electric dipole moment
J, (S)	current density
A, (α)	linear current density
H	magnetic field strength
H'	non-rationalised magnetic field strength
U_m	magnetic potential difference
F, (F_m)	magnetomotive force
B	magnetic flux density, magnetic induction
Φ	magnetic flux
A	magnetic vector potential
L	self-inductance
M, (L)	mutual inductance
k, (x)	coupling coefficient
σ	leakage coefficient
μ	permeability
μ_0	permeability of vacuum
μ'	non-rationalised permeability
μ'_0	non-rationalised permeability of vacuum
μ_r	relative permeability
k, (χ_m)	magnetic susceptibility
k', (χ'_m)	non-rationalised magnetic susceptibility
m	electromagnetic moment (magnetic moment)
H_i, (M)	magnetisation
J, (B_i)	magnetic polarisation
J'	non-rationalised magnetic polarisation
w	electromagnetic energy density
S	Poynting vector
c	velocity of propagation of electromagnetic waves *in vacuo*
R	resistance (to direct current)
G	conductance (to direct current)
ρ	resistivity
γ, σ	conductivity
R, R_m	reluctance
Λ, (P)	permeance
N	number of turns in winding
m	number of phases
p	number of pairs of poles
φ	phase displacement
Z	impedance (complex impedance)

Table 7.12 *continued*

Symbol	*Quantity*
$[Z]$	modulus of impedance (impedance)
X	reactance
R	resistance
Q	quality factor
Y	admittance (complex admittance)
$[Y]$	modulus of admittance (admittance)
B	susceptance
G	conductance
P	active power
S, (P_s)	apparent power
Q, (P_q)	reactive power

Table 7.13 Symbols for quantities and units of acoustics (based on ISO Recommendation R31)

Symbol	*Quantity*
T	period, periodic time
f, ν	frequency, frequency interval
ω	angular frequency, circular frequency
λ	wavelength
k	circular wavenumber
ρ	density (mass density)
P_s	static pressure
p	(instantaneous) sound pressure
ε, (x)	(instantaneous) sound particle displacement
u, v	(instantaneous) sound particle velocity
a	(instantaneous) sound particle acceleration
q, U	(instantaneous) volume velocity
c	velocity of sound
E	sound energy density
P, (N, W)	sound energy flux, sound power
I, J	sound intensity
Z_s, (W)	specific acoustic impedance
Z_a, (Z)	acoustic impedance
Z_m, (w)	mechanical impedance
L_p, (L_N, L_w)	sound power level
L_p, (L)	sound pressure level
δ	damping coefficient
Λ	logarithmic decrement
α	attenuation coefficient
β	phase coefficient
γ	propagation coefficient
δ	dissipation coefficient
r, τ	reflection coefficient
γ	transmission coefficient
α, (α_a)	acoustic absorption coefficient
R	{ sound reduction index { sound transmission loss
A	equivalent absorption area of a surface or object
T	reverberation time
L_N, (Λ)	loudness level
N	loudness

Table 7.14 Some technical abbreviations and symbols

Quantity	*Abbreviation*	*Symbol*
Alternating current	a.c.	
Ampere	A or amp	
Amplification factor		μ
Amplitude modulation	a.m.	
Angular velocity		ω
Audio frequency	a.f.	
Automatic frequency control	a.f.c.	
Automatic gain control	a.g.c.	
Bandwidth		Δf
Beat frequency oscillator	b.f.o.	
British thermal unit	Btu	
Cathode-ray oscilloscope	c.r.o.	
Cathode-ray tube	c.r.t.	
Celsius	C	
Centi-	c	
Centimetre	cm	
Square centimetre	cm^2 or sq cm	
Cubic centimetre	cm^3 or cu cm or c.c.	
Centimetre-gram-second	c.g.s.	
Continuous wave	c.w.	
Coulomb	C	
Deci-	d	
Decibel	dB	
Direct current	d.c.	
Direction finding	d.f.	
Double sideband	d.s.b.	
Efficiency		η
Equivalent isotropic radiated power	e.i.r.p.	
Electromagnetic unit	e.m.u.	
Electromotive force instantaneous value	e.m.f.	E or V, e or v
Electron-volt	eV	
Electrostatic unit	e.s.u.	
Fahrenheit	F	
Farad	F	
Frequency	freq.	f
Frequency modulation	f.m.	
Gauss	G	
Giga-	G	
Gram	g	
Henry	H	
Hertz	Hz	
High frequency	h.f.	
Independent sideband	i.s.b.	
Inductance-capacitance		L-C
Intermediate frequency	i.f.	
Kelvin	K	
Kilo-	k	
Knot	kn	
Length		l
Local oscillator	l.o.	
Logarithm, common		log or $\log_{10}$
Logarithm, natural		ln or $\log_e$
Low frequency	l.f.	
Low tension	l.t.	

Table 7.14 *continued*

Quantity	*Abbreviation*	*Symbol*
Magnetomotive force	m.m.f.	F or M
Mass		m
Medium frequency	m.f.	
Mega-	M	
Metre	m	
Metre-kilogram-second	m.k.s.	
Micro-	μ	
Micromicro-	p	
Micron		μ
Milli-	m	
Modulated continuous wave	m.c.w.	
Nano-	n	
Neper	N	
Noise factor		N
Ohm		Ω
Peak to peak	p–p	
Phase modulation	p.m.	
Pico-	p	
Plan-position indication	PPI	
Potential difference	p.d.	V
Power factor	p.f.	
Pulse repetition frequency	p.r.f.	
Radian	rad	
Radio frequency	r.f.	
Radio telephony	R/T	
Root mean square	r.m.s.	
Short-wave	s.w.	
Single sideband	s.s.b.	
Signal frequency	s.f.	
Standing wave ratio	s.w.r.	
Super-high frequency	s.h.f.	
Susceptance		B
Travelling-wave tube	t.w.t.	
Ultra-high frequency	u.h.f.	
Very high frequency	v.h.f.	
Very low frequency	v.l.f.	
Volt	V	
Voltage standing wave ratio	v.s.w.r.	
Watt	W	
Weber	Wb	
Wireless telegraphy	W/T	

Table 7.15 Greek alphabet and symbols

Name	*Symbol*		*Quantities used for*
alpha	A	α	angles, coefficients, area
beta	B	β	angles, coefficients
gamma	Γ	γ	specific gravity
delta	Δ	δ	density, increment, finite difference operator
epsilon	E	ε	Napierian logarithm, linear strain, permittivity, error, small quantity
zeta	Z	ζ	coordinates, coefficients, impedance (capital)
eta	H	η	magnetic field strength, efficiency
theta	Θ	θ	angular displacement, time
iota	I	ι	inertia
kappa	K	κ	bulk modulus, magnetic susceptibility
lambda	Λ	λ	permeance, conductivity, wavelength
mu	M	μ	bending moment, coefficient of friction, permeability
nu	N	ν	kinematic viscosity, frequency, reluctivity
xi	Ξ	ξ	output coefficient
omicron	O	o	
pi	Π	π	circumference ÷ diameter
rho	P	ρ	specific resistance
sigma	Σ	σ	summation (capital), radar cross-section, standard deviation
tau	T	τ	time constant, pulse length
upsilon	Y	u	
phi	Φ	φ	flux, phase
chi	X	χ	reactance (capital)
psi	Ψ	ψ	angles
omega	Ω	ω	angular velocity, ohms

References

1 COHEN, E. R. and TAYLOR, B. N., *Journal of Physical and Chemical Reference Data*, vol. 2, 663, (1973).
2 'Recommended values of physical constants', CODATA (1973).
3 McGLASHAN, M. L., *Physicochemical quantities and units*, London: The Royal Institute of Chemistry, (1971).

Electricity

M G Say
PhD, MSc, CEng, ACGI, DIC, FIEE, FRSE
Professor Emeritus of Electrical Engineering,
Heriot-Watt University, Edinburgh

Contents

8.1 Introduction

Most of the observed electrical phenomena are explicable in terms of electric *charge* at rest, in motion and in acceleration. Static charges give rise to an *electric field* of force; charges in motion carry an electric field accompanied by a *magnetic field* of force; charges in acceleration develop a further field of *radiation*.

Modern physics has established the existence of elemental charges and their responsibility for observed phenomena. Modern physics is complex: it is customary to explain phenomena of engineering interest at a level adequate for a clear and reliable concept, based on the electrical nature of matter.

8.2 Molecules, atoms and electrons

Material substances, whether solid, liquid or gaseous, are conceived as composed of very large numbers of *molecules*. A molecule is the smallest portion of any substance which cannot be further subdivided without losing its characteristic material properties. In all states of matter molecules are in a state of rapid continuous motion. In a *solid* the molecules are relatively closely 'packed' and the molecules, although rapidly moving, maintain a fixed mean position. Attractive forces between molecules account for the tendency of the solid to retain its shape. In a *liquid* the molecules are less closely packed and there is a weaker cohesion between them, so that they can wander about with some freedom within the liquid, which consequently takes up the shape of the vessel in which it is contained. The molecules in a *gas* are still more mobile, and are relatively far apart. The cohesive force is very small, and the gas is enabled freely to contract and expand. The usual effect of heat is to increase the intensity and speed of molecular activity so that 'collisions' between molecules occur more often; the average spaces between the molecules increase, so that the substance attempts to expand, producing internal pressure if the expansion is resisted.

Molecules are capable of further subdivision, but the resulting particles, called *atoms*, no longer have the same properties as the molecules from which they came. An atom is the smallest portion of matter that can enter into chemical combination or be chemically separated, but it cannot generally maintain a separate existence except in the few special cases where a single atom forms a molecule. A molecule may consist of one, two or more (sometimes many more) atoms of various kinds. A substance whose molecules are composed entirely of atoms of the same kind is called an *element*. Where atoms of two or more kinds are present, the molecule is that of a chemical *compound*. At present 102 atoms are recognised, from combinations of which every conceivable substance is made. As the simplest example, the atom of hydrogen has a mass of 1.63×10^{-27} kg and a molecule (H_2), containing two atoms, has twice this mass. In one gram of hydrogen there are about 3×10^{23} molecules with an order of size between 1 and 0.1 nm.

Electrons, as small particles of negative electricity having apparently almost negligible mass, were discovered by J. J. Thomson, on a basis of much previous work by many investigators, notably Crookes. The discovery brought to light two important facts: (1) that atoms, the units of which all matter is made, are themselves complex structures, and (2) that electricity is atomic in nature. The atoms of all substances are constructed from particles. Those of engineering interest are: *electrons*, *protons* and *neutrinos*. Modern physics concerns itself also with *positrons*, *mesons*, *neutrons* and many more. An *electron* is a minute particle of negative electricity which, when dissociated from the atom (as it can be) indicates a purely electrical, nearly mass-less nature. From whatever atom they are derived, all electrons are similar. The electron charge is $e = 1.6 \times 10^{-19}$ C, so that $1\ C = 6.3 \times 10^{18}$ electron charges. The apparent rest mass of an electron is 1/1850 of that of a hydrogen atom, amounting to $m = 9 \times 10^{-28}$ g. The meaning to be attached to the 'size' of an electron (a figure of the order of 10^{-13} cm) is vague. A *proton* is electrically the opposite of an electron, having an equal charge, but positive. Further, protons are associated with a mass the same as that of the hydrogen nucleus. A *neutron* is a chargeless mass, the same as that of the proton.

8.3 Atomic structure

The mass of an atom is almost entirely concentrated in a nucleus of protons and neutrons. The simplest atom, of hydrogen, comprises a nucleus with a single proton, together with one associated electron occupying a region formerly called the K-shell. Helium has a nucleus of two protons and two neutrons, with two electrons in the K-shell. In these cases, as in all normal atoms, the sum of the electron charges is numerically equal to the sum of the proton charges, and the atom is electrically balanced. The neon atom has a nucleus with 10 protons and 10 neutrons, with its 10 electrons in the K- and L-shells.

The *atomic weight* A is the total number of protons and neutrons in the nucleus. If there are Z protons there will be $A-Z$ neutrons: Z is the *atomic number*. The nuclear structure is not known, and the forces that keep the protons together against their mutual repulsion are conjectural.

A nucleus of atomic weight A and atomic number Z has a charge of $+Ze$ and is normally surrounded by Z electrons each of charge $-e$. Thus copper has 29 protons and 35 neutrons ($A = 64$, $Z = 29$) in its nucleus, electrically neutralised by 29 electrons in an enveloping cloud. The atomic numbers of the known elements range from 1 for hydrogen to 102 for nobelium, and from time to time the list is extended. This multiplicity can be simplified: within the natural sequence of elements there can be distinguished groups with similar chemical and physical properties (see *Table 8.1*). These are the *halogens* (F 9, Cl 17, Br 35, I 53); the *alkali metals* (Li 3, Na 11, K 19, Rb 37, Cs 55); the *copper* group (Cu 29, Ag 47, Au 79); the *alkaline earths* (Be 4, Mg 12, Ca 20, Sr 38, Ba 56, Ra 88); the *chromium* group (Cr 24, Mo 42, W 74, U 92); and the *rare gases* (He 2, Ne 10, A 18, Kr 36, Xe 54, Rn 86). In the foregoing the brackets contain the chemical symbols of the elements concerned followed by their atomic numbers. The difference between the atomic numbers of two adjacent elements within a group is always 8, 18 or 32. Now these three bear to one another a simple arithmetical relation: $8 = 2 \times 2 \times 2$, $18 = 2 \times 3 \times 3$ and $32 = 2 \times 4 \times 4$. Arrangement of the elements in order in a periodic table beginning with an alkali metal and ending with a rare gas shows a remarkable repetition of basic similarities. The periods are I, 1–2; II, 3–10; III, 11–18; IV, 19–36; V, 37–54; VI, 55–86; VII, 87–?.

An element is often found to be a mixture of atoms with the same chemical property but different atomic weights (*isotopes*). Again, because of the convertibility of mass and energy, the mass of an atom depends on the energy locked up in its compacted nucleus. Thus small divergences are found in the atomic weights which, on simple grounds, would be expected to form integral multiples of the atomic weight of hydrogen. The atomic weight of oxygen is arbitrarily taken as 16.0, so that the mass of the proton is 1.0076 and that of the hydrogen atom is 1.0081.

Atoms may be in various energy states. Thus the atoms in the filament of an incandescent lamp may emit light when excited, e.g. by the passage of an electric heating current, but will not do so when the heater current is switched off. Now heat energy is the kinetic energy of the atoms of the heated body. The more vigorous impact of atoms may not always shift the atom as a whole, but may shift an electron from one orbit to another of higher energy level within the atom. This position is not normally stable, and the electron gives up its momentarily acquired

Table 8.1 Elements

Period	*Atomic number*	*Name*	*Symbol*	*Atomic weight*
I	1	Hydrogen	H	1.008
	2	Helium	He	4.002
II	3	Lithium	Li	6.94
	4	Beryllium	Be	9.02
	5	Boron	B	10.82
	6	Carbon	C	12.00
	7	Nitrogen	N	14.008
	8	Oxygen	O	16.00
	9	Fluorine	F	19.00
	10	Neon	Ne	20.18
III	11	Sodium	Na	22.99
	12	Magnesium	Mg	24.32
	13	Aluminium	Al	26.97
	14	Silicon	Si	28.06
	15	Phosphorus	P	31.02
	16	Sulphur	S	32.06
	17	Chlorine	Cl	35.46
	18	Argon	A	39.94
IV	19	Potassium	K	39.09
	20	Calcium	Ca	40.08
	21	Scandium	Sc	45.10
	22	Titanium	Ti	47.90
	23	Vanadium	V	50.95
	24	Chromium	Cr	52.01
	25	Manganese	Mn	54.93
	26	Iron	Fe	55.84
	27	Cobalt	Co	58.94
	28	Nickel	Ni	58.69
	29	Copper	Cu	63.57
	30	Zinc	Zn	65.38
	31	Gallium	Ga	69.72
	32	Germanium	Ge	72.60
	33	Arsenic	As	74.91
	34	Selenium	Se	78.96
	35	Bromine	Br	79.91
	36	Krypton	Kr	83.70
V	37	Rubidium	Rb	85.44
	38	Strontium	Sr	87.63
	39	Yttrium	Y	88.92
	40	Zirconium	Zr	91.22
	41	Niobium	Nb	92.91
	42	Molybdenum	Mo	96.00
	43	Technetium	Tc	99.00
	44	Ruthenium	Ru	101.7
	45	Rhodium	Rh	102.9
	46	Palladium	Pd	106.7
	47	Silver	Ag	107.9
	48	Cadmium	Cd	112.4
	49	Indium	In	114.8
	50	Tin	Sn	118.0
	51	Antimony	Sb	121.8
	52	Tellurium	Te	127.6
	53	Iodine	I	126.9
	54	Xenon	Xe	131.3
VI	55	Caesium	Cs	132.9
	56	Barium	Ba	137.4
	57	Lanthanum	La	138.9
	58	Cerium	Ce	140.1
	59	Praseodymium	Pr	140.9
	60	Neodymium	Nd	144.3

Table 8.1 *continued*

Period	*Atomic number*	*Name*	*Symbol*	*Atomic weight*
VI	61	Promethium	Pm	147
	62	Samarium	Sm	150.4
	63	Europium	Eu	152.0
	64	Gadolinium	Gd	157.3
	65	Terbium	Tb	159.2
	66	Dysprosium	Dy	162.5
	67	Holmium	Ho	163.5
	68	Erbium	Er	167.6
	69	Thulium	Tm	169.4
	70	Ytterbium	Yb	173.0
	71	Lutecium	Lu	175.0
	72	Hafnium	Hf	178.6
	73	Tantalum	Ta	181.4
	74	Tungsten	W	184.0
	75	Rhenium	Re	186.3
	76	Osmium	Os	191.5
	77	Iridium	Ir	193.1
	78	Platinum	Pt	195.2
	79	Gold	Au	197.2
	80	Mercury	Hg	200.6
	81	Thallium	Tl	204.4
	82	Lead	Pb	207.2
	83	Bismuth	Bi	209.0
	84	Polonium	Po	210
	85	Astatine	At	211
	86	Radon	Rn	222
VII	87	Francium	Fr	223
	88	Radium	Ra	226.0
	89	Actinium	Ac	227
	90	Thorium	Th	232.1
	91	Protoactinium	Pa	234
	92	Uranium	U	238.1
	93	Neptunium	Np	239
	94	Plutonium	Pu	242
	95	Americium	Am	243
	96	Curium	Cm	243
	97	Berkelium	Bk	245
	98	Californium	Cf	246
	99	Einsteinium	Es	247
	100	Fermium	Fm	256
	101	Mendelevium	Md	256
	102	Nobelium	No	—

potential energy by falling back into its original level, releasing the energy as a definite amount of light, the *light-quantum* or *photon*.

Among the electrons of an atom those of the outside peripheral shell are unique in that, on account of all the electron charges on the shells between them and the nucleus, they are the most loosely bound and most easily *removable*. In a variety of ways it is possible so to excite an atom that one of the outer electrons is torn away, leaving the atom *ionised* or converted for the time into an *ion* with an effective positive charge due to the unbalanced electrical state it has acquired. Ionisation may occur due to impact by other fast-moving particles, by irradiation with rays of suitable wavelength and by the application of intense electric fields.

The three 'structures' of *Figure 8.1* are based on the former 'planetary' concept, now modified in favour of a more complex

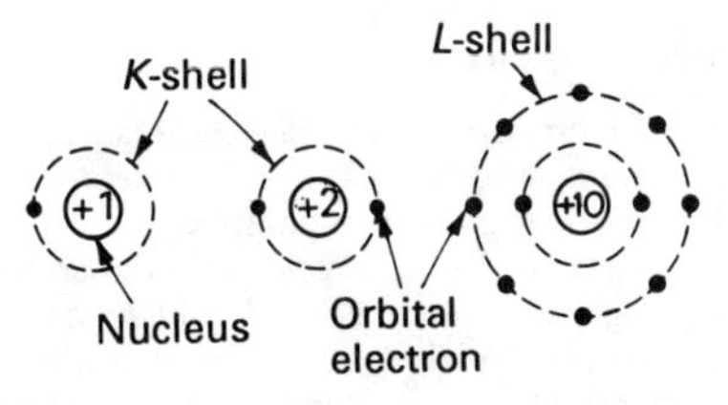

Figure 8.1 Atomic structure. The nuclei are marked with their positive charges in terms of total electron charge. The term 'orbital' is becoming obsolete. Electron: mass $m=9\times10^{28}$ g, charge $e=-1.6\times10^{-19}$ C. Proton: mass $=1.63\times10^{-24}$ g, charge $=+1.6\times10^{-19}$ C. Neutron = mass as for proton; no charge

idea derived from consideration of wave mechanics. It is still true that, apart from its mass, the chemical and physical properties of an atom are given to it by the arrangement of the electron 'cloud' surrounding the nucleus.

8.4 Wave mechanics

The fundamental laws of optics can be explained without regard to the nature of light as an electromagnetic wave phenomenon, and photoelectricity emphasises its nature as a stream or ray of corpuscles. The phenomena of diffraction or interference can only be explained on the wave concept. *Wave mechanics* correlates the two apparently conflicting ideas into a wider concept of 'waves of matter'. Electrons, atoms and even molecules participate in this duality, in that their effects appear sometimes as corpuscular, sometimes as of a wave nature. Streams of electrons behave in a corpuscular fashion in photoemission, but in certain circumstances show the diffraction effects familiar in wave action. Considerations of particle mechanics led de Broglie to write several theoretic papers (1922–6) on the parallelism between the dynamics of a particle and geometrical optics, and suggested that it was necessary to admit that classical dynamics could not interpret phenomena involving energy quanta. Wave mechanics was established by Schrödinger in 1926 on de Broglie's conceptions.

When electrons interact with matter they exhibit wave properties: in the free state they act like particles. Light has a similar duality, as already noted. The hypothesis of de Broglie is that a particle of mass m and velocity u has wave properties with a wavelength $\lambda=h/mu$, where h is the Planck constant, $h=6.626\times10^{-34}$ J s. The mass m is relativistically affected by the velocity.

When electron waves are associated with an atom, only certain fixed-energy states are possible. The electron can be raised from one state to another if it is provided, by some external stimulus such as a photon, with the necessary energy-difference Δw in the form of an electromagnetic wave of wavelength $\lambda=hc/\Delta w$, where c is the velocity of free-space radiation (3×10^8 m/s). Similarly, if an electron falls from a state of higher to one of lower energy, it emits energy Δw as radiation. When electrons are raised in energy level, the atom is *excited*, but not ionised.

8.5 Electrons in atoms

Consider the hydrogen atom. Its single electron is not located at a fixed point, but can be anywhere in a region near the nucleus with some probability. The particular region is a kind of shell or cloud, of radius depending on the electron's energy state.

With a nucleus of atomic number Z, the Z electrons can have several possible configurations. There is a certain radial pattern of electron probability cloud distribution (or shell pattern). Each electron state gives rise to a cloud pattern, characterised by a definite energy level, and described by the series of quantum numbers n, l, m_l and m_s. The number n ($=1,2,3\ldots$) is a measure of the energy level; l ($=0, 1, 2\ldots$) is concerned with angular momentum; m_l is a measure of the component of angular momentum in the direction of an applied magnetic field; and m_s arises from the electron spin. It is customary to condense the nomenclature so that electron states corresponding to $l=0$, 1, 2 and 3 are described by the letters s, p, d and f and a numerical prefix gives the value of n. Thus boron has 2 electrons at level 1 with $l=0$, two at level 2 with $l=0$, and one at level 3 with $l=1$: this information is conveyed by the description $(1s)^2(2s)^2(2p)^1$.

The energy of an atom as a whole can vary according to the electron arrangement. The most stable state is that of minimum energy, and states of higher energy content are *excited*. By Pauli's *exclusion principle* the maximum possible number of electrons in states 1, 2, 3, 4 ... n are 2, 8, 18, 32, ..., $2n^2$ respectively. Thus only 2 electrons can occupy the 1s state (or K-shell) and the remainder must, even for the normal minimum-energy condition, occupy other states. Hydrogen and helium, the first two elements, have respectively 1 and 2 electrons in the 1-quantum (K) shell; the next, lithium, has its third electron in the 2-quantum (L) shell. The passage from lithium to neon (*Figure 8.1*) results in the filling up of this shell to its full complement of 8 electrons. During the process, the electrons first enter the 2s subgroup, then fill the 2p subgroup until it has 6 electrons, the maximum allowable by the exclusion principle (see *Table 8.2*).

Table 8.2 Typical atomic structures

Element and atomic number		*Principal and secondary quantum numbers*									
		1s	*2s*	*2p*	*3s*	*3p*	*3d*	*4s*	*4p*	*4d*	*4f*
H	1	1									
He	2	2									
Li	3	2	1								
C	6	2	2	2							
N	7	2	2	3							
Ne	10	2	2	6							
Na	11	2	2	6	1						
Al	13	2	2	6	2	1					
Si	14	2	2	6	2	2					
Cl	17	2	2	6	2	5					
A	18	2	2	6	2	6					
K	19	2	2	6	2	6		1			
Mn	25	2	2	6	2	6	5	2			
Fe	26	2	2	6	2	6	6	2			
Co	27	2	2	6	2	6	7	2			
Ni	28	2	2	6	2	6	8	2			
Cu	29	2	2	6	2	6	10	1			
Ge	32	2	2	6	2	6	10	2	2		
Se	34	2	2	6		6	10	2	4		
Kr	36	2	2	6		6	10	2	6		
		1	2	3	*4s*	*4p*	*4d*	*4f*	*5s*	*5p*	
Rb	37	2	8	18	2	6			1		
Xe	54	2	8	18	2	6	10		2	6	

Very briefly, the effect of the electron-shell filling is as follows. Elements in the same chemical family have the same number of electrons in the subshell that is incompletely filled. The rare gases (He, Ne, A, Kr, Xe) have no uncompleted shells. Alkali metals (e.g. Na) have shells containing a single electron. The alkaline earths have two electrons in uncompleted shells. The good conductors (Ag, Cu, Au) have a single electron in the uppermost quantum state. An irregularity in the ordered sequence of filling (which holds consistently from H to A) begins at potassium (K) and continues to Ni, becoming again regular with Cu, and beginning a new irregularity with Rb.

8.5.1 Energy levels

The electron of a hydrogen atom, normally at level 1, can be raised to level 2 by endowing it with a particular quantity of energy most readily expressed as 10.2 eV. (1 eV = 1 electron volt = 1.6×10^{-19} J is the energy acquired by a free electron falling through a potential difference of 1 V, which accelerates it and gives it kinetic energy.) 10.2 V is the *first excitation potential* for the hydrogen atom. If the electron is given an energy of 13.6 eV it is freed from the atom, and 13.6 V is the *ionisation potential*. Other atoms have different potentials in accordance with their atomic arrangement.

8.6 Conduction

Conduction is the name given to the movement of electrons, or ions, or both, giving rise to the phenomena described by the term *electric current*. The effects of a current include a redistribution of charges, heating of conductors, chemical changes in liquid solutions, magnetic effects, and many subsidiary phenomena.

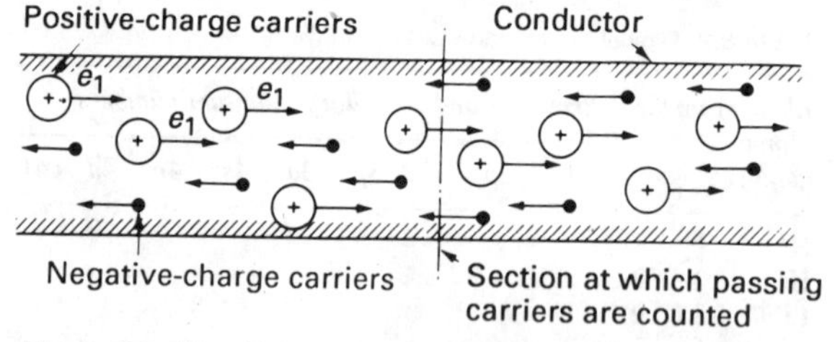

Figure 8.2 Electric current as the result of moving charges

If at some point on a conductor (*Figure 8.2*) n_1 carriers of electric charge (they can be water-drops, ions, dust particles, etc.) each with a positive charge e_1 arrive per second, and n_2 carriers (such as electrons) each with a negative charge e_2 arrive in the opposite direction per second, the total rate of passing of charge is $n_1e_1 + n_2e_2$, which is the charge per second or *current*. A study of conduction concerns the kind of carriers and their behaviour under given conditions. Since an electric field exerts mechanical forces on charges, the application of an electric field (i.e. a potential difference) between two points on a conductor will cause the movement of charges to occur, i.e. a current to flow, so long as the electric field is maintained.

The discontinuous particle nature of current flow is an observable factor. The current carried by a number of electricity carriers will vary slightly from instant to instant with the number of carriers passing a given point in a conductor. Since the electron charge is 1.6×10^{-19} C, and the passage of one coloumb per second (a rate of flow of one *ampere*) corresponds to $10^{19}/1.6 = 6.3 \times 10^{18}$ electron charges per second, it follows that the discontinuity will be observed only when the flow comprises the very rapid movement of a few electrons. This may happen in gaseous conductors, but in metallic conductors the flow is the very slow drift (measureable in mm/s) of an immense number of electrons.

A current may be the result of a two-way movement of positive and negative particles. Conventionally the direction of current flow is taken as the same as that of the positive charges and against that of the negative ones.

8.6.1 Conductors and insulators

In substances called *conductors*, the outer-shell electrons can be more or less freely interchanged between atoms. In copper, for example, the molecules are held together comparatively rigidly in the form of a 'lattice'—which gives the piece of copper its permanent shape—through the interstices of which outer electrons from the atoms can be interchanged within the confines of the surface of the piece, producing a random movement of free electrons called an 'electron atmosphere'. Such electrons are responsible for the phenomenon of electrical conductivity.

In other substances called *insulators* all the electrons are more or less firmly bound to their parent atoms so that little or no relative interchange of electron charges is possible. However, a potential difference applied to a perfect insulator affects the atoms by a 'stretching' or 'rotation' which displaces the electrical centres of negative and positive in opposite directions. This polarisation of the dielectric insulating material may be considered as taking place in the manner indicated in *Figure 8.3*.

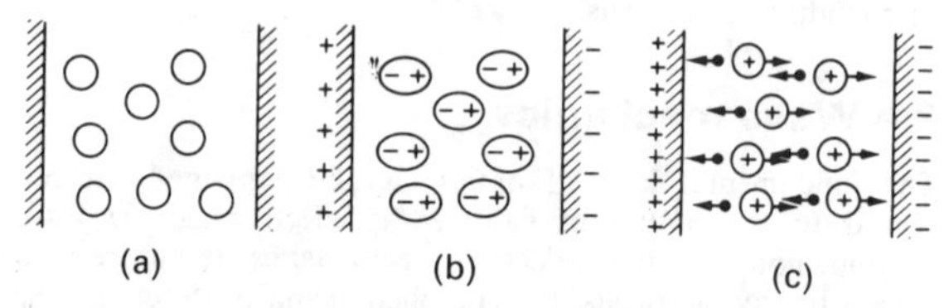

Figure 8.3 Polarisation, displacement and breakdown in a dielectric material: (a) no electric field; atoms unstrained; (b) electric field applied; polarisation; (c) intensified electric field; atoms ionised

Before the electric field is applied, in (*a*), the atoms of the insulator are neutral and unstrained; (*b*) as the potential difference is raised the electric field exerts opposite mechanical forces on the negative and positive charges and the atoms become more and more highly strained. On the left face the atoms will all present their negative charges at the surface: on the right face, their positive charges. These surface polarisations are such as to account for the effect known as *permittivity*. The small displacement of the electric charges is an electron shift, i.e. a *displacement current* flows while the polarisation is being established. *Figure 8.3(c)* shows that under conditions of excessive electric field atomic disruption or ionisation may occur, converting the insulator material into a conductor, resulting in *breakdown*.

8.6.2 Semiconductors

Intrinsic semiconductors (i.e. materials between the good conductors and the good insulators) have a small spacing of about 1 eV between their permitted bands, which affords a low conductivity, strongly dependent on temperature and of the order of one-millionth that of a conductor.

Impurity semiconductors have their low conductivity provided by the presence of minute quantities of foreign atoms (e.g. 1 in 10^8) or by deformations in the crystal structure. The impurities 'donate' electrons of energy-level that can be raised into a conduction band (n-type); or they can attract an electron from a filled band to leave a 'hole', or electron deficiency, the movement of which corresponds to the movement of a positive charge (p-type).

8.7 Conduction in various media

The phenomena are here discussed in terms adequate for simple explanation.

8.7.1 Metals

Reference has been made above to the 'electron atmosphere' of electrons in random motion within a lattice of comparatively rigid molecular structure in the case of copper, which is typical of the class of good metallic conductors. The random electronic motion, which intensifies with rise in temperature, merges into an average shift of charge of almost (but not quite) zero continuously (*Figure 8.4*). When an electric field is applied along the length of a conductor (as by maintaining a potential difference across its ends), the electrons have a *drift* towards the positive end superimposed upon their random digressions. The drift is slow, but such great numbers of electrons may be involved that very large currents, entirely due to electron drift, can be produced by this means. In their passage the electrons are

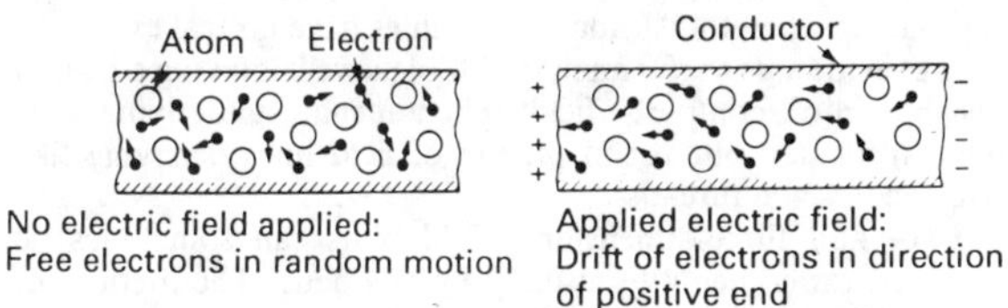
Atom
Electron
Conductor
No electric field applied:
Free electrons in random motion
Applied electric field:
Drift of electrons in direction of positive end

Figure 8.4 Electronic conduction in metals

impeded by the molecular lattice, the collisions producing heat and the opposition called *resistance*. The conventional direction of current flow is actually opposite to that of the drift of charge, which is exclusively electronic.

8.7.2 Crystals

When atoms are brought together to form a crystal, their individual sharp and well defined energy levels merge into energy *bands*. These bands may overlap, or there may be gaps in the energy levels available, depending on the lattice spacing and interatomic bonding. Conduction can take place only by electron migration into an empty or partly filled band: filled bands are not available. If an electron acquires a small amount of energy from the externally applied electric field, and can move into an available empty level, it can then contribute to the conduction process.

8.7.3 Liquids

Liquids are classified according to whether they are *non-electrolytes* (non-conducting) or *electrolytes* (conducting). In the former the substances in solution break up into electrically balanced groups, whereas in the latter the substances form ions, each a part of a single molecule, with either a positive or a negative charge. Thus common salt, NaCl, in a weak aqueous solution breaks up into sodium and chlorine ions. The sodium ion Na^+ is a sodium atom less one electron, the chlorine ion Cl^- is a chlorine atom with one electron more than normal. The ions attach themselves to groups of water molecules. When an electric field is applied the sets of ions move in opposite directions, and since they are much more massive than electrons the conductivity produced is markedly inferior to that in metals. Chemical actions take place in the liquid and at the electrodes when current passes. Faraday's electrolysis law states that the mass of an ion deposited at an electrode by electrolyte action is proportional to the quantity of electricity which passes and to the *chemical equivalent* of the ion.

8.7.4 Gases

Gaseous conduction is strongly affected by the pressure of the gas. At pressures corresponding to a few centimetres of mercury gauge, conduction takes place by the movement of positive and negative ions. Some degree of ionisation is always present due to stray radiations (light, etc.). The electrons produced attach themselves to gas atoms and the sets of positive and negative ions drift in opposite directions. At very low gas pressures the electrons produced by ionisation have a much longer free path before they collide with a molecule, and so have scope to attain high velocities. Their motional energy may be enough to *shock-ionise* neutral atoms, resulting in a great enrichment of the electron stream and an increased current flow. The current may build up to high values if the effect becomes cumulative, and eventually conduction may be effected through a *spark* or *arc*.

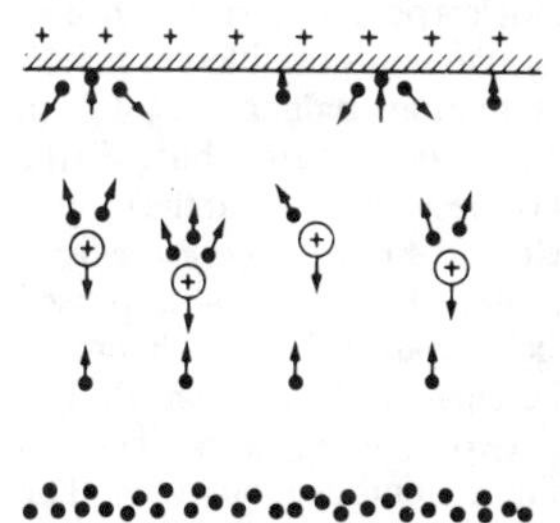
Anode to which electrons flow by influence of electric field
Electrons striking anode surface may produce secondary emission
Gas atoms ionised by collision: increased electron flow to anode and drift of positive ions towards cathode
Electrons moving to anode
Electron space charge
Cathode from which primary electrons are emitted

Figure 8.5 Electrical conduction in gases at low pressure

Some of the effects described above are illustrated in *Figure 8.5*. At the bottom is an electrode, the *cathode*, from the surface of which electrons are emitted, generally by heating the cathode material. At the top is a second electrode, the *anode*, and an electric field is established between anode and cathode, which are enclosed in a vessel which contains a low-pressure inert gas. The electric field causes electrons emitted from the cathode to move upwards. In their passage to the anode these electrons will encounter gas molecules. If conditions are suitable, the gas atoms are ionised, becoming in effect positive charges associated with the nuclear mass. Thereafter the current is increased by the detached electrons moving upwards and by the positive ions moving more slowly downwards. In certain devices (such as the mercury-arc rectifier) the impact of ions on the cathode surface maintains its emission. The impact of electrons on the anode may be energetic enough to cause the *secondary emission* of electrons from the anode surface. If the gas molecules are excluded and a vacuum established, the conduction becomes purely electronic.

8.7.5 Vacua

This may be considered as purely electronic, in that any electrons present (there can be no molecular *matter* present if the vacuum is perfect) are moved in accordance with the forces exerted on them by an applied electric field. The number of electrons is always small, and although high speeds may be reached the currents conducted in vacuum tubes are generally measurable only in milli- or micro-amperes.

8.8 Energy conversion

Charge separation involves displacement against a force and the development of potential difference (p.d.), indicating a system of potential energy. In some electrophysical phenomena a conversion of potential into a different form can occur. Some common cases of energy conversion are listed below.

8.8.1 Contact and friction effects

Pairs of dissimilar materials brought into contact exhibit a contact p.d. if their abilities to diffuse electrons differ. In a Zn/Cu pair, the zinc more readily transfers electrons to the copper, producing a small *contact* p.d. at the junction. *Friction* can build up potential differences of several kilovolts, e.g. the metal body of a vehicle can be raised to a high voltage to earth by reason of friction between the rubber tyres and the surface of the road.

8.8.2 Thermoelectric effects

Irreversible production of heat occurs at a rate I^2R when a conductor of resistance R carries a conduction current I: this is the *Joule effect*. Reversible thermoelectric conversion can also occur. In a closed loop comprising two wires of different metals joined at their ends, the contact potentials balance out if both junctions are held at the same temperature; but if the temperatures differ, there is a drift of electrons around the loop, a direct conversion of thermal to electric energy (*Seebeck effect*). The converse is also true: if a conduction current is passed through a dissimilar junction, heat is absorbed or liberated depending on the direction of the current (*Peltier effect*). The *Thomson effect* embraces the two above, and includes electrothermal conversion in a single homogeneous wire. The Seebeck effect is employed in devices for the measurement of temperature.

8.8.3 Photoelectric effects

These arise from the physical changes in photosensitive cells when irradiated by light within the spectrum ultraviolet/infrared. The result may be (1) electron emission from the cathode of a *photoemissive* cell, (2) the generation of a p.d. in a *photovoltaic* cell, or (3) a change in the conductivity of a *photoconductive* cell.

8.8.4 Mechanical-electric effects

The *piezoelectric* effect occurs in crystalline materials such as quartz and tourmaline. When mechanically stressed, a charge separation occurs, making some regions positively and others negatively charged. The converse is also true: application of an electric field causes the crystal to expand or contract. The crystal acts as a reversible electromechanical energy converter.

8.8.5 Chemical-electric effects

The ionic transfer of charge in electrolytes (*Section 8.7.3*) depends on the nature of the electrolyte. Chemically basic solutions of sodium and similar hydroxides produce hydrogen and oxygen gases at the cathode and anode respectively; acid solutions yield products determined by the nature of the electrodes. With solutions of metal salts and appropriate electrodes, electrodeposition is achieved. Ores of copper, zinc and cadmium can be treated to deposit the metal at the cathode, with oxygen gas released at the anode. Electrorefining may be employed with copper, tin, nickel, etc.—the impure metal at the anode is dissolved and re-deposited at the cathode, leaving impurities at the bottom of the cell. Electroplating resembles electrorefining, except that pure metal or alloy is used as the anode. Commercially important applications of electrochemistry are the irreversible primary ('dry') cell and the reversible secondary ('accumulator') battery.

8.9 Fields

Many phenomena of electrical and electronic interest must be explained in subatomic terms, but others are adequately taken as mass effects on a macroscopic basis. Thus, for example, conduction in a metal conductor can often be expressed in terms of Ohm's law, and 'action at a distance' in terms of a *field of force*. The term 'field' implies an effect distributed throughout a region of space, in contrast with one that is concentrated at a point. The field concept was postulated by Faraday for the effects in the neighbourhood of an electric charge, and later those describing the magnetic effect of a current. Maxwell formulated analytic expressions for field phenomena, and with the 'displacement current' hypothesis predicted energy radiation by electromagnetic waves, subsequently confirmed experimentally by Hertz. A brief outline of field philosophy is given in *Section 8.1*.

The geometric pattern of a field of force in a plane can be depicted by lines along which the force acts. Maxwell postulated that a force line represented the line of action of two mechanical stresses, tensile along the line and compressive in all directions at right-angles thereto. Although such lines have no real existence, they provide a powerful analytical tool as well as giving a helpful insight into complex field phenomena, underlying the unsophisticated but useful picture of field lines behaving like stretched elastic threads.

Two- and three-dimensional field plots can sometimes be obtained experimentally, usually by analogue. The method of 'finite differences' enables complex fields to be computed.

8.9.1 Electric field

When two conducting electrodes are separated in a dielectric medium and raised by equal charges $+q$ and $-q$ respectively to a potential difference V, a static electric field of force exists between them; *Figure 8.6(a)*. The broken lines depict the field structure: they emerge from the positive and terminate on the negative electrode. The Maxwell tensile-stress concept shows that there is, between electrodes, a force of mutual attraction. If both charges have the same polarity, say positive as in *Figure 8.6(b)*, the force lines emerge from both electrodes, to terminate on corresponding negative charges elsewhere in the system. From the Maxwell compressive-stress concept, there is now a force of mutual repulsion between the electrodes.

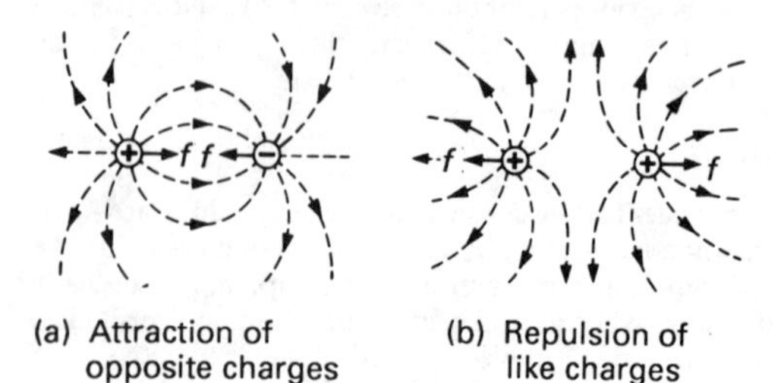

Figure 8.6 Electric field between charged electrodes

The potential difference V between electrodes is distributed along the field lines. Joining points on neighbouring field lines having the same potential forms *equipotential* lines. The field and equipotential lines cross orthogonally.

The field can be taken as an electric flux q between the electrodes. At a given point the electric flux density is D (coulomb per square metre) and the electric field strength, or voltage gradient, is E (volt per metre). These are related by $D=\varepsilon E$, where $\varepsilon=\varepsilon_0\varepsilon_r$. Here ε_0 is a basic property of free space (the *electric constant*), and ε_r is the relative permittivity of the dielectric medium, a number usually between 1 and 10, but occasionally larger (e.g. 80 for pure water). The system in *Figure 8.6(a)* is, in technological terms, a capacitor of capacitance $C=q/V$ (coulomb per volt, or farad).

8.9.2 Magnetic field

The region surrounding a steady current is occupied by a magnetic field of force proportional to the current, and a corresponding magnetic flux. Both are circuital, i.e. they encircle the current in closed loops. *Figure 8.7* gives an indication of the flux distribution, in free space or in air, for three simple cases, showing the directional conventions adopted. In (*a*) for a long straight isolated current, the flux has circular paths concentric with the conductor. In (*b*) there is a concentration of flux within the current loop. In (*c*) the flux is greatly increased by use of a coil of N turns, because much of the flux links the current N times. The flux linkage is an important parameter: its value per unit current is the inductance L of the coil.

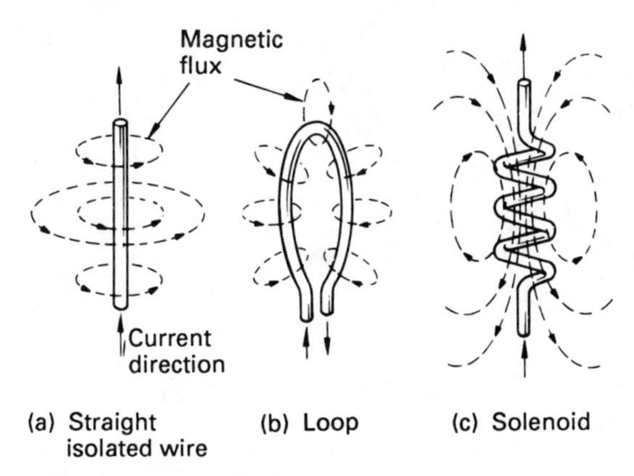

Figure 8.7 Simple magnetic fields

The closed path of a flux element is its magnetic circuit. If the total effective current (taking account of multiple linkages) is regarded as a magnetomotive force, $F=i$ (ampere or ampere-turn). The m.m.f. is distributed around the magnetic circuit to produce, at a given point, a magnetic field intensity H (ampere per metre). Let the ambient medium be free space or air; then H establishes a magnetic flux density $B=\mu_0 H$ (weber per square metre, or tesla). Here μ_0 is the absolute permeability of free space, termed the *magnetic space constant*. If, however, the medium is ferromagnetic, then $B=\mu_0\mu_r H$, where μ_r is the relative permeability of the medium. It varies over a wide range, typically 10 to 10^5, depending on the magnetic properties of the medium, and it is a nonlinear function of the ratio B/H as a result of 'magnetic saturation'.

Electromagnetic induction A magnetic flux is a store of energy. When the flux is increased or decreased, so is the stored energy. Where the energy transfer is derived from, or returned to, a linked electric circuit, it takes the form of a current i that flows by reason of an induced electromotive force e for the duration t of the change: e, i and t are the quantities that determine the electrical energy converted. The relative directions of e and i depend on the direction of energy transfer. Faraday's law of electromagnetic induction states that the induced e.m.f. is proportional to the rate of change of linkage in the form $e=-(\mathrm{d}\phi/\mathrm{d}t)$. Typical cases are the *transformer effect* (a varying flux in a stationary circuit) and the *generator effect* (the movement of a circuit through a constant flux).

8.9.3 Electromagnetic field

On the basis of experimental work by Ampere, Coulomb and Faraday, a comprehensive theory of the electromagnetic field, both static and dynamic, was established by Maxwell. He saw that a useful symmetry is attained if the rate of change of an electric flux ψ is taken as equivalent to a displacement current. (The same idea underlies the treatment of a capacitor in standard circuit analysis.) In simple terms, the substance of the Maxwell equations is the following.

(1) The magnetomotive force around a closed path linking a current I is equal to $I=I_c+I_d$, the sum of the conduction and displacement currents.
(2) The electric flux from a charge Q is equal to Q.
(3) The electric force around a closed path linking a magnetic flux ϕ is equal to the rate of change of the flux linkage.
(4) The magnetic flux is circuital.

These are expressed symbolically as below. The letter O by the integral sign means summation around a closed path of elements dl, and S means summation over a surface of elements ds.

(1) $\int_O H\,\mathrm{d}l=I_c+I_d$

(2) $\int_S D\,\mathrm{d}s=Q$

(3) $\int_O E\,\mathrm{d}l=-(\partial\phi/\partial t)$

(4) $\int_S B\,\mathrm{d}s=0$

To these are added the constitutive equations

$D=\varepsilon E$

$B=\mu H$

$J=J_c+J_d=\sigma E+(\partial\psi/\partial t)$

relating densities and field intensities. J is the volume current density, and σ is the volume conductivity.

Light

D R Heath BSc, PhD, CPhys, MInstP
British Aerospace plc

Contents

9.1 Introduction

In recent years the growth of the field of opto-electronics has required the engineer to furnish himself with a knowledge of the nature of optical radiation and its interaction with matter. The increase in the importance of measurements of optical energy has also necessitated an introduction into the somewhat bewildering array of terminologies used in the hitherto specialist fields of radiometry and photometry.

9.2 The optical spectrum

Light is electromagnetic radiant energy and makes up part of the electromagnetic spectrum. The term *optical spectrum* is used to described the *light* portion of the electromagnetic spectrum and embraces not only the visible spectrum (that detectable by the eye) but also the important regions in optoelectronics of the ultraviolet and infrared.

The electromagnetic spectrum, classified into broad categories according to wavelength and frequency, is given in *Figure 10.1*, Chapter 10. It is observed that on this scale the optical spectrum forms only a very narrow region of the complete electromagnetic spectrum. *Figure 9.1* is an expanded diagram

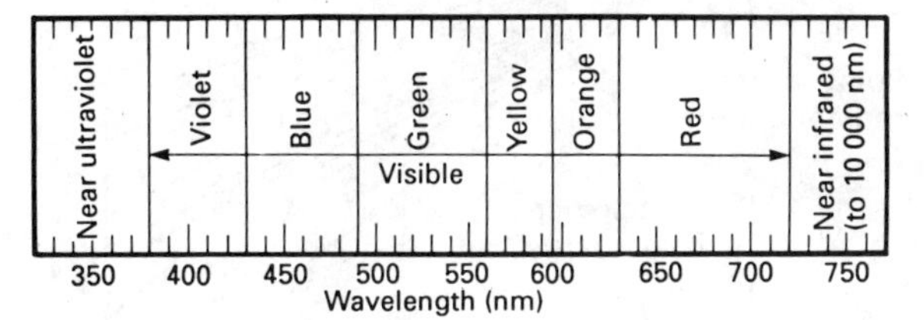

Figure 9.1 The visible spectrum

showing more detail of the ultraviolet, visible and infrared regions. By convention, optical radiation is generally specified according to its wavelength. The wavelength can be determined from a specific electromagnetic frequency from the equation:

$$\lambda = c/f \tag{9.1}$$

where λ is the wavelength (m), f is the frequency (Hz) and c is the speed of light in a vacuum ($\sim 2.99 \times 10^8$ m/s). The preferred unit of length for specifying a particular wavelength in the visible spectrum is the nanometre (nm). Other units are also in common use, namely the angström (Å) and the micrometre or micron. The relation of these units is as follows:

1 nanometre (nm) $= 10^{-9}$ metre
1 angström (Å) $= 10^{-10}$ metre
1 micron (μm) $= 10^{-6}$ metre

The micron tends to be used for describing wavelengths in the infrared region and the nanometre for the ultraviolet and visible regions.

The wavenumber (cm^{-1}) is the reciprocal of the wavelength measured in centimetres, i.e. $1/\lambda$ (cm) = wavenumber (cm^{-1}).

9.3 Basic concepts of optical radiation

In describing the measurement of light and its interaction with matter, three complementary properties of electromagnetic radiation need to be invoked: ray, wave and quantum. At microwave and longer wavelengths it is generally true that radiant energy exhibits primarily wave properties while at the shorter wavelengths, X-ray and shorter, radiant energy primarily exhibits ray and quantum properties. In the region of the optical spectrum, ray, wave, and quantum properties will have their importance to varying degrees.

9.4 Radiometry and photometry

Radiometry is the science and technology of the measurement of radiation from all wavelengths within the optical spectrum. The basic unit of power in radiometry is the watt (W).

Photometry is concerned only with the measurement of light detected by the eye, i.e. that radiation which falls between the wavelengths 380 nm and 750 nm. The basic unit of power in photometry is the lumen (lm).

In radiometric measurements the ideal detector is one which has a flat response with wavelength whereas in photometry the ideal detector has a spectral response which approximates to that of the average human eye. To obtain consistent measurement techniques the response of the average human eye was established by the Commission Internationale de l'Eclairage (CIE) in 1924. The response known as the photopic eye response is shown in *Figure 9.2* and is observed to peak in the green/yellow part of the visible spectrum at 555 nm. The curve indicates that it takes approximately ten times as many units of blue light as green light to produce the same visibility effect on the average human eye.

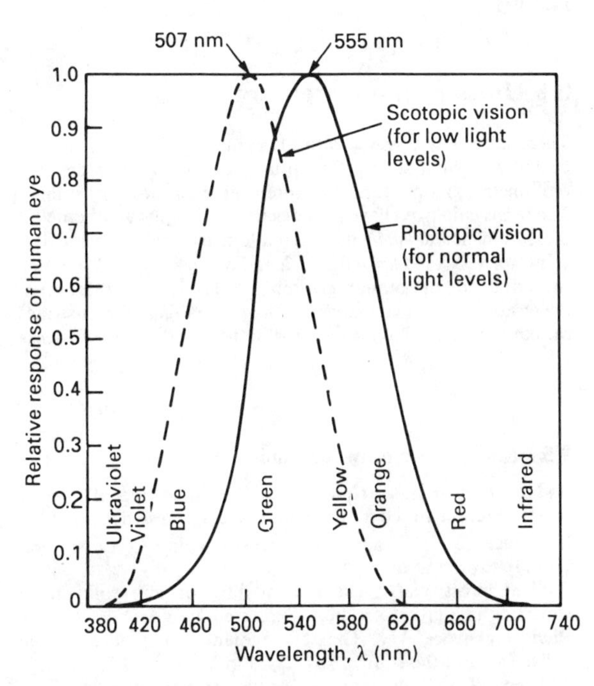

Figure 9.2 The photopic and scotopic eye responses

The broken curve in *Figure 9.2* with a peak at 507 nm is termed the scotopic eye response. The existence of the two responses arises out of the fact that the eye's spectral response shifts at very low light levels. The retina of the human eye has two types of optical receptors, cones and rods. Cones are mainly responsible for colour vision and are highly concentrated in a 0.3 mm diameter spot, called the fovea, at the centre of the field of vision. Rods are not present in the fovea but have a very high density in the peripheral regions of the retina. They do not give rise to colour response but at low light levels are significantly more sensitive than cones. At normal levels of illumination (photopic response) the eye's response is determined by the cones in the retina whilst at very low light levels the retina's rod receptors take over and cause a shift in the response curve to the scotopic response.

In normal circumstances photometric measurements are based on the CIE photopic response and all photometric instruments must have sensors which match this response. At the peak wavelength of 555 nm of the photopic response one watt of radiant power is defined as the equivalent of 680 lumens of luminous power.

In order to convert a radiometric power measurement into photometric units both the spectral response of the eye and the spectral output of the light source must be taken into account. The conversion is then achieved by multiplying the energy radiated at each wavelength by the relative lumen/watt factor at that wavelength and summing the results. Note that in the ultraviolet and infrared portions of the optical spectrum although one may have high output in terms of watts the photometric value in lumens is zero due to lack of eye response in those ranges. However, it should be said that many observers can see the 900 nm radiation from a GaAs laser or the 1.06 μm radiation from an Nd:YAG laser since in this instance the intensity can be sufficiently high to elicit a visual response. Viewing of these sources in practice is not to be recommended for safety reasons and the moderately high energy densities at the eye which are involved.

9.5 Units of measurement

There are many possible measurements for characterising the output of a light source. The principles employed in defining radiometric and photopic measurement terms are very similar. The terms employed have the adjective *radiant* for a radiometric measurement and *luminous* for a photometric measurement. The subscript e is used to indicate a radiometric symbol and the subscript v for a photometric symbol. A physical visualisation of the terms to be defined is given in *Figure 9.3*. *Figure 9.4* illustrates the concept of solid angle required in the visualisation of *Figure 9.3*.

9.5.1 Radiometric terms and units

Radiant flux or **radiant power,** Φ_e The time rate of flow of radiant energy emitted from a light source. Expressed in J/s or W.

Irradiance, E_e The radiant flux density incident on a surface. Usually expressed in W/cm^2.

Radiant intensity, I_e The radiant flux per unit solid angle travelling in a given direction. Expressed in W/sr.

Radiant exitance, M_e The total radiant flux divided by the surface area of the source. Expressed in W/cm^2.

Radiance, L_e The radiant intensity per unit area, leaving, passing through, or arriving at a surface in a given direction. The surface area is the projected area as seen from the specified direction. Expressed in $W/(cm^2\ sr)$.

9.5.2 Photometric terms and units

The equivalent photometric terminologies to the radiometric ones defined above are as follows:

Luminous flux or **power,** Φ_v The time rate of flow of luminous energy emitted from a light source. Expressed in lm.

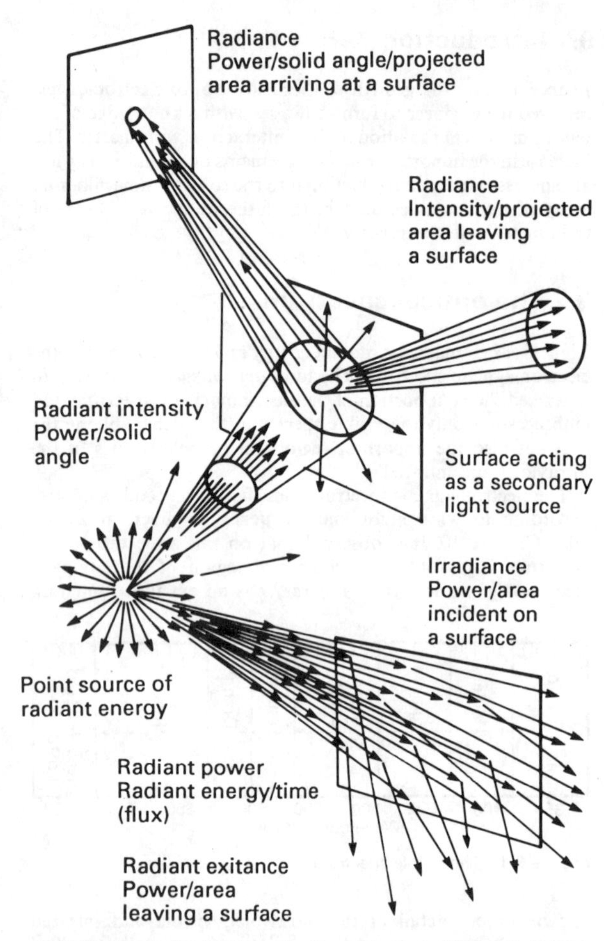

Figure 9.3 A visualisation of radiometric terms (from reference 1)

Illuminance or **illumination,** E_v The density of luminous power incident on a surface. Expressed in lm/cm^2. Note the following:

$1\ lm/cm^2 = 1$ phot
$1\ lm/m^2 = 1$ lux
$1\ lm/ft^2 = 1$ footcandle

Luminous intensity, I_v The luminous flux per unit solid angle, travelling in a given directon. Expressed in lm/sr. Note that 1 lm/sr = 1 cd.

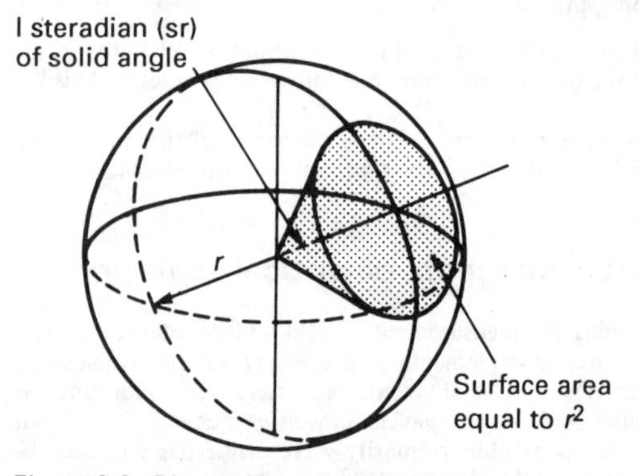

Figure 9.4 Diagram illustrating the steradian (from reference 1)

Luminous exitance, M_v The total luminous flux divided by the surface area of the source. Expressed in lm/cm^2.

Luminance, L_v The luminous intensity per unit area, leaving, passing through or arriving at a surface in a given direction. The surface area is the projected area as seen from the specified direction. Expressed in lm/(cm^2 sr) or cd/cm^2.

Mathematically if the area of an emitter has a diameter or diagonal dimension greater than 0.1 of the distance of the detector it can be considered as an area source. Luminance is also called the photometric **brightness**, and is a widely used quantity. In *Figure 9.5* the projected area of the source, A_p, varies directly as the cosine of θ, i.e. is a maximum at 0° or normal to the surface and minimum at 90°. Thus

$$A_p = A_s \cos\theta \tag{9.2}$$

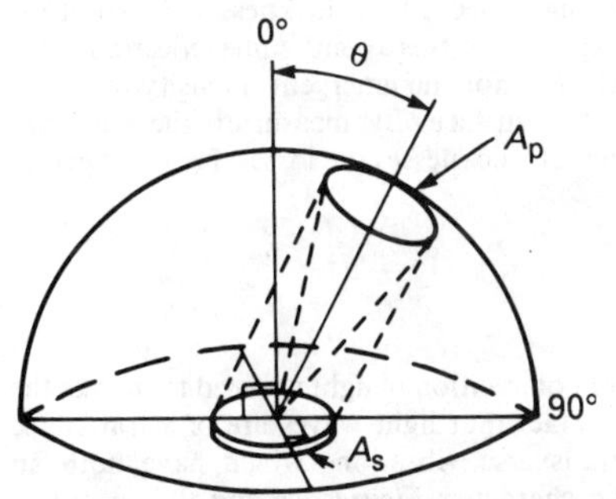

Figure 9.5 Diagram illustrating the projected area

Luminance is then the ratio of the luminous intensity (I_v) to the projected area of the source (A_p):

$$\text{luminance} = \frac{\text{luminous intensity}}{\text{projected area}} = \frac{I_v}{A_p}$$

$$= \frac{I_v}{A_s \cos\theta} \text{ lm/sr per unit area}$$

since 1 lm/sr = 1 cd, depending on the units used for the area we have

1 cd/cm^2 = 1 stilb

$1/\pi$ cd/cm^2 = 1 lambert

$1/\pi$ cd/ft^2 = 1 footlambert

Table 9.1 provides a summary of the radiometric and photometric terms with their symbols and units.

Table 9.1 Radiometric and photometric terms

Quantity	*Symbol*	*Unit(s)*
Radiant flux	Φ_e	W
Luminous flux	Φ_v	lm
Irradiance	E_e	W/cm^2
Illuminance	E_v	lm/cm^2 = phot lm/m^2 = lux lm/ft^2 = footcandle
Radiant intensity	I_e	W/sr
Luminous intensity	I_v	lm/sr = cd
Radiant exitance	M_e	W/cm^2
Luminous exitance	M_v	lm/cm^2
Radiance	L_e	W/(cm^2 sr)
Luminance (Photometric brightness)	L_v	lm (cm^2 sr) cd/cm^2 = stilb $1/\pi$ cd/cm^2 = lambert $1/\pi$ cd/ft^2 = footlambert

Some typical values of natural scene illumination expressed in units of lm/m^2 and footcandles are given in *Table 9.2*. *Table 9.3* gives some approximate values of luminance for various sources.

Table 9.2 Approximate levels of natural scene illumination (from reference 1)

	Footcandles	lm/m^2
Direct sunlight	1.0–1.3 × 10^4	1.0–1.3 × 10^5
Full daylight	1–2 × 10^3	1–2 × 10^4
Overcast day	10^2	10^3
Very dark day	10	10^2
Twilight	1	10
Deep twilight	10^{-1}	1
Full moon	10^{-2}	10^{-1}
Quarter moon	10^{-3}	10^{-2}
Starlight	10^{-4}	10^{-3}
Overcast starlight	10^{-5}	10^{-4}

Table 9.3 Approximate levels of luminance for various sources (from reference 1)

	Footlamberts	cd/m^2
Atomic fission bomb (0.1 ms after firing, 90 ft diameter ball)	6 × 10^{11}	2 × 10^{12}
Lightning flash	2 × 10^{10}	6.8 × 10^{10}
Carbon arc (positive crater)	4.7 × 10^6	1.6 × 10^7
Tungsten filament lamp (gas-filled, 16 lm/W)	2.6 × 10^5	8.9 × 10^5
Sun (as observed from the earth's surface at meridian)	4.7 × 10^3	1.6 × 10^4
Clear blue sky	2300	7900
Fluorescent lamp (T-12 bulb, cool white, 430 mA medium loading)	2000	6850
Moon (as observed from earth's surface)	730	2500

9.6 Practical measurements

A wide variety of commercial instruments is available for carrying out optical radiation measurements.

The radiometer is an instrument which will normally employ a photodiode, phototube, photomultiplier or photoconductive cell as its detector. Each of these detectors has a sensitivity which varies with wavelength. It is therefore necessary for the instrument to be calibrated over the full range of wavelengths for which it is to be used. For measurement of monochromatic radiation the instrument reading is simply taken and multiplied by the appropriate factor in the detector sensitivity at the given wavelength. A result in units of power or energy is thereby obtained.

For the characterisation of broadband light sources, where the output is varying with wavelength, it is necessary to measure the source in narrow band increments of wavelength. This can be achieved by using a set of calibrated interference filters.

The spectroradiometer is specifically designed for broadband measurements and has a monochromator in front of the detector which performs the function of isolating all the wavelengths of interest. These can be scanned over the detector on a continuous basis as opposed to the discrete intervals afforded by filters.

The photometer is designed to make photometric measurements of sources. It usually consists of a photoconductive cell, silicon photodiode or photomultiplier with a filter incorporated to correct the total system response to that of the standard photopic eye response curve.

Thermopiles, bolometers and pyrometers generate signals which can be related to the incident power as a result of a change in temperature which is caused by absorption of the radiant energy. They have an advantage that their response as a function of wavelength is almost flat (constant with wavelength), but are limited to measurement of relatively high intensity sources and normally at wavelengths greater than 1 μm.

Calibration of most optical measuring instruments is carried out using tungsten lamp standards and calibrated thermopiles. The calibration accuracy of these lamp standards varies from approximately ±8% of absolute in the ultraviolet to ±5% of absolute in the visible and near infrared. Measurement systems calibrated with these standards will generally have accuracies of 8–10% of absolute. It is important to realise that the accuracy of optical measurements is rather poor compared to other spheres of physics. To obtain an accuracy of 5% in a measurement is very difficult; a good practitioner will be doing well to keep errors to between 10 and 20%.

9.7 Interaction of light with matter

Light may interact with matter by being reflected, refracted, absorbed or transmitted. Two or more of these are usually involved.

9.7.1 Reflection

Some of the light impinging on any surface is reflected away from the surface. The reflectance varies according to the properties of the surface and the wavelength of the impinging radiation.

Regular or *specular reflection* is reflection in accordance with the laws of reflection with no diffusion (surface is smooth compared to the wavelength of the impinging radiation).

Diffuse reflection is diffusion by reflection in which on the microscopic scale there is no regular reflection (surface is rough when compared to the wavelength of the impinging radiation).

Reflectance (ρ) is the ratio of the reflected radiant or luminous flux to the incident flux.

Reflection (*optical*) *density* (D) is the logarithm to the base ten of the reciprocal of the reflectance.

$$D(\lambda)=\log_{10}\left(\frac{1}{\rho(\lambda)}\right) \tag{9.3}$$

where $\rho(\lambda)$ is the spectral reflectance.

9.7.2 Absorption

When a beam of light is propagated in a material medium its speed is less than its speed in a vacuum and its intensity gradually decreases as it progresses through the medium. The speed of light in a material medium varies with the wavelength and this variation is known as *dispersion*. When a beam traverses a medium some of the light is scattered and some is absorbed. If the absorption is true absorption the light energy is converted into heat. All media show some absorption—some absorb all wavelengths more or less equally, others show selective absorption in that they absorb some wavelengths very much more strongly than others. The phenomena of scattering, dispersion and absorption are intimately connected.

Absorption coefficient

Lambert's law of absorption states that equal paths in the same absorbing medium absorb equal fractions of the light that enters them. If in traversing a path of length $\mathrm{d}x$ the intensity is reduced from I to $I-\mathrm{d}I$ then Lambert's law states that $\mathrm{d}I/I$ is the same for all elementary paths of length $\mathrm{d}x$. Thus

$$\frac{\mathrm{d}I}{I}=-K\,\mathrm{d}x$$

where K is a constant known as the absorption coefficient. Therefore $\log I=-Kx+C$ where C is a constant. If $I=I_0$ at $x=0$, $C=\log I_0$ and so

$$I=I_0\,\mathrm{e}^{-Kx} \tag{9.4}$$

Note that in considering a medium of thickness x, I_0 is not the intensity of incident light due to there being some reflection at the first surface. Similarly I is not the emergent intensity owing to reflection at the second surface. By measuring the emergent intensity for two different thicknesses the losses due to reflection may be eliminated.

9.7.3 Polarisation

For an explanation of polarisation of light we need to invoke the wave concept and the fact that light waves are of a transverse nature possessing transverse vibrations which have both an electric and magnetic character. *Figures 9.6* and *9.7* set out to illustrate the meaning of unpolarised and linearly polarised light.

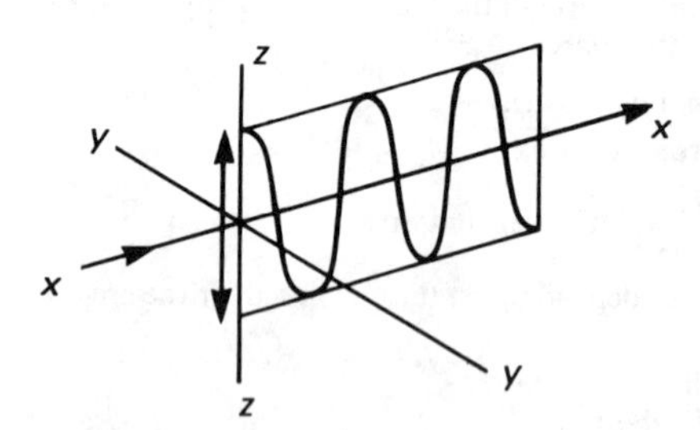

Figure 9.6 Linearly polarised light

In *Figure 9.6* a wave is propagating in the x direction with the vibrations in a single plane. Any light which by some cause possesses this property is said to be linearly polarised. Ordinary light, such as that received from the sun or incandescent lamps, is unpolarised and in this case the arrangement of vibrations is in all possible directions perpendicular to the direction of travel, as in *Figure 9.7*.

There are numerous methods for producing linearly polarised light, those most widely known being birefringence or double refraction, reflection, scattering and dichroism. Double reflection

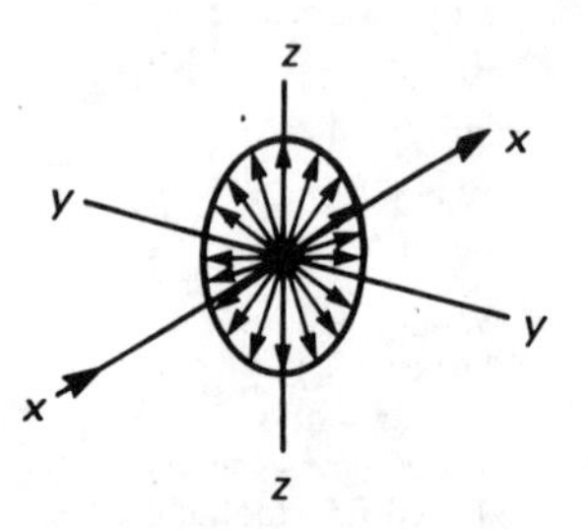

Figure 9.7 Unpolarised light

occurs in certain types of natural crystal such as calcite and quartz and will divide a beam of unpolarised light into two separate polarised beams of equal intensity. By eliminating one of the polarised beams a very efficient linear polariser can be made.

Dichroic polarisers make up the great majority of commercially produced synthetic polarisers. They exhibit dichroism, the property of absorbing light to different extents depending on the polarisation form of the incident beam.

The light emerging from a linear polariser can be given a 'twist' so that the vibrations are no longer confined to a single plane but instead form a helix. This is achieved by inserting a sheet of double-refracting material into the polarised beam which divides the beam into two beams of equal intensity but with slightly different speeds, one beam being slightly retarded. The light is said to be circularly polarised.

The application and uses of polarised light are very considerable—liquid crystal displays, control of light intensity, blocking and prevention of specular glare light, measuring optical rotation, measuring propagation of stress and strain are some notable ones.

References

1 ZAHA, M. A., 'Shedding some needed light on optical measurements', *Electronics*, 6 Nov., 91–6 (1972)

Further reading

BRUENING, R. J., 'Spectral irradiance scales based on filtered absolute silicon photodetectors'. *Appl. Optics*, **26**, No. 6 (1987)

CLAYTON, R. K., *Light and Living Matter*, Vol. 2, *The Biological Part*, McGraw-Hill, New York (1971)

GRUM, F. and BECHENER, R. J., *Optical Radiation Measurements*, Vol. 1, *Radiometry*, Academic Press, London (1979)

JENKINS, F. A. and WHITE, F. E., *Fundamentals of Optics*, 3rd edn, McGraw-Hill, New York (1957)

KEYS, R. J., *Optical and infrared detectors*, Springer Verlag (1980)

LAND, E. H., 'Some aspects on the development of sheet polarisers', *J. Opt. Soc. Am.*, **41**, 957 (1951)

LERMAN, S., *Radiant Energy and the Eye*, Macmillan (1980)

LONGHURST, R. S., *Geometrical and Physical Optics*, Longman, Green & Co., London (1957)

McEWEN, R. S., 'Liquid crystals, displays and devices for optical processing'. *J. Phys. E, Sci. Instrum.*, **20**, 364 (1987)

MAYER-ARENDT, J. R., *Introduction to Classical and Modern Optics*, Prentice Hall, Englewood Cliffs, NJ (1972)

RCA Electro-Optics Handbook, RCA Commercial Engineering, Harrison, NJ (1974)

SMITH, W. J., *Modern Optical Engineering*, McGraw-Hill, New York (1966)

TAYLOR, J. H., 'Radiation exchange', *Appl. Optics*, **26**, No. 4, 619 (1987)

WALSH, J. W. T., *Photometry*, Dover, New York (1965)

10 Electromagnetic and Nuclear Radiation

A Goodings PhD, DIC, BSc, ARCS
Head of Applied Physics,
Instrumentation & Electronics Branch,
Atomic Energy Establishment, Winfrith, Dorset

C J Goodings BA (Oxon)
Microelectronics Research Group,
The Cavendish Laboratory, Cambridge

Contents

10.1 Principal symbols and constants[1]

A	Atomic weight
c_0	Velocity of electromagnetic waves in free space ($=(\mu_0\varepsilon_0)^{-1/2}=2.977\,924\,580\times 10^8$ m/s)
$\lvert e\rvert$	Electronic charge ($1.602\,189\times 10^{-19}$ C)
E_0	Particle energy at rest
h	Planck's constant ($=6.626\,176\times 10^{-34}$ J/Hz$=4.135\,701\times 10^{-15}$ eV/Hz)
j	$(-1)^{1/2}$
m_0	Particle mass at rest
Z	Atomic number
ε	$=\varepsilon_r\varepsilon_0$
ε_r	Relative permittivity of a medium
ε_0	Permittivity of free space ($8.854\,818\times 10^{-12}$ F/m)
λ	Wavelength
μ	$=\mu_0\mu_r$
μ_r	Relative permeability
μ_0	Permeability of free space ($4\pi\times 10^{-7}$ H/m)
ν	Frequency
σ	Reaction cross-section per atom or per nucleus
ω	Angular frequency ($2\pi\nu$)

10.2 Types of radiation

This chapter is concerned with the energetic elementary particles that emerge from nuclear reactions and with electromagnetic radiation, both of which involve the transmission of energy through space, in a manner which does not require the presence of a transmitting medium. They can, therefore, propagate in vacuum. Both types occur in ionising and non-ionising forms. In passing through matter, ionising radiation displaces electrons (or larger charged particles), leaving a trail of free electrons and ions (charged atoms). Attention is concentrated in this chapter on the similarities and differences between the ionising and non-ionising interactions and on ionisation phenomena in general.

All radiation can be described in terms of waves or particles. Electromagnetic radiation is often thought of as photons (quanta), whilst the transmission of particles is often discussed in terms of their de Broglie waves. Circumstances dictate the more convenient concept and the decision usually involves the ratio of the relevant wavelength to the size of the object with which the radiation is interacting.

10.2.1 Electromagnetic radiation

10.2.1.1 Basic concepts

Electromagnetic radiation is generated by the acceleration of electric charge and is propagated according to Maxwell's equations:[2–5]

$$\nabla\times\bar{D}=\rho \quad (10.1)$$

$$\nabla\times\bar{B}=0 \quad (10.2)$$

$$\nabla\times\bar{E}=-\delta\bar{B}/\delta t \quad (10.3)$$

$$\nabla\times\bar{H}=\bar{J}+\delta\bar{D}/\delta t \quad (10.4)$$

where $\bar{B}$ is the magnetic induction field, $\bar{D}$ the electric displacement field, $\bar{E}$ the electric field, $\bar{H}$ the magnetic field, ρ the density of free charges, and J the conduction current density.

These are the fundamental laws of electromagnetism. They are justified by continued agreement with experiment and arise respectively from a number of empirical results: Coulomb's law via Gauss' law of electrostatics; the absence of magnetic monopoles; Faraday's and Lenz's laws of electromagnetic induction; and Ampere's law plus the concept of displacement current.

For linear isotropic media:

$$\bar{D}=\varepsilon\bar{E} \quad (10.5)$$

and

$$\bar{B}=\mu\bar{H} \quad (10.6)$$

The simplest solution of Maxwell's equations applies to linear, isotropic, homogeneous media with no free charge. It leads to:

$$\nabla^2\bar{E}-(1/c^2)\delta^2\bar{E}/\delta t^2=0 \quad (10.7)$$

and

$$\nabla^2\bar{H}-(1/c^2)\delta^2\bar{H}/\delta t^2=0 \quad (10.8)$$

These equations represent electric and magnetic waves propagating together with velocity c. The $\bar{E}$ and $\bar{H}$ vectors are at right angles to each other and to the direction of propagation. Conventionally, the plane of the $\bar{E}$ vector is taken as the plane of polarisation. The velocity of propagation in media other than vacuum, c, is less than c_0. At optical frequencies the ratio c_0/c is known as the refractive index.

Energy is shared equally by the $\bar{E}$ and $\bar{H}$ fields and energy transmission is described by Poynting's vector, $\bar{E}\times\bar{H}$. The flow of energy per unit area due to a plane wave is εcE^2 W/m^2. In free space this becomes $2.65\times 10^{-3}E^2$ W/m^2. The total outward energy flow from a given volume is the integral of the vector over the surface of the volume. In a.c. complex notation, Poynting's theorem can be written:[6]

$$-\nabla\cdot(\bar{E}\times\bar{H}^*)=\bar{E}\cdot\bar{J}^*+\mathrm{j}\omega(\mu_r\mu_0\bar{H}\cdot\bar{H}^*-\varepsilon_r\varepsilon_0\bar{E}\cdot\bar{E}^*) \quad (10.9)$$

where * represents a complex conjugate and $\bar{J}$ is conduction current density.

In Equation (10.9), the left-hand term represents the energy flux, the first on the right is the energy dissipation into the medium and the last represents rate of change of energy in the fields.

Since the propagation velocity is fixed by the permittivity and permeability of the medium, wavelength depends on the frequency of the originating disturbance. The range of possible wavelengths is therefore very wide. It is known as the electromagnetic spectrum and extends from hundreds of kilometres to less than 10^{-14} m. *Figure 10.1* illustrates this range. The visible region extends from about 750 nm (red) to about 400 nm (violet). The rest of the spectrum is invisible to the human eye.

10.2.1.2 Interference

Maxwell's equations are linear. The principle of superposition applies and interference between different wave trains is possible. It requires coherent sources (with the same frequency, the same polarisation and constant phase difference) and will only be observed if the waves have similar amplitudes. The amplitude of the resultant signal is obtained by summation of all possible contributions from the source(s), phase difference being taken into account.

In many cases the interfering waves are derived from a common source either by division of amplitude (e.g. by partial reflection) or by division of wavefront (e.g. by slits). At radio frequencies interference is used to produce directional aerials and similar devices. At optical wavelengths, interference is employed, for example, to reduce surface reflections in lenses and to generate precise measuring scales. It is fundamental to the action of lasers.

10.2.1.3 Diffraction

Diffraction describes the propagation of waves into a geometric shadow and the interference-like effects which take place when

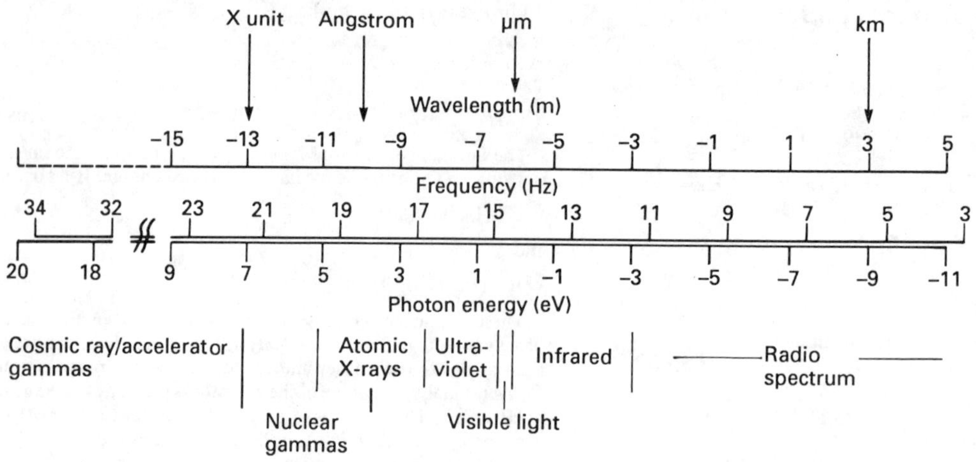

Figure 10.1 The electromagnetic spectrum in vacuum

waves impinge upon structures whose edges are smooth to dimensions comparable with the wavelength. Distinction is made between two types:[7,8]

Fraunhofer diffraction This is the more common type and assumes that the incident and diffracted radiation are approximated by plane waves. It sets the ultimate limit to the performance of optical instruments and electron microscopes and describes the spreading of radio waves round obstructions.

The amplitude of the diffraction pattern can be obtained by means of a Fourier transform. For example, for any plane aperture with dimensions x, y:

$$A(l, m) = K \int_y \int_x F(x, y) \exp[2\pi j(lx \times my)/\lambda] \, dx \, dy \qquad (10.10)$$

where l and m are the sines of output angles measured with respect to a line at right angles to the aperture, $F(x, y)$ is the complex amplitude of the wave across the aperture, and K is a constant. For a single long slit, the diffracted wave amplitude becomes:

$$A = A_0(\sin \beta)/\beta \qquad (10.11)$$

$$\beta = \pi W(\sin \psi + \sin \theta)/\lambda \qquad (10.12)$$

where A_0 is the incident amplitude, W is the width of the slit, and ψ and θ are the incident and output angles measured to the normal to the plane of the slit. The output intensity is A^2.

Essentially, diffraction gratings comprise many such slits. They depend on interference between the multiple diffraction patterns:

$$A = A_0 \frac{\sin (N\delta)}{\sin \delta} \cdot \frac{\sin \beta}{\beta} \qquad (10.13)$$

where $\delta = D\beta/W$, D is the slit spacing, and N is the number of slits. The resolving power of a grating is given by:

$$\lambda/\Delta\lambda = ND(\sin \psi + \sin \theta)/\lambda \qquad (10.14)$$

Fresnel diffraction This occurs when curvature of the wave front across the transmitting aperture is significant compared with the wavelength.

10.2.1.4 Photons

Electromagnetic radiation can also be considered in particulate terms as a flow of photons. The energy of each photon is related to its frequency:

$$E_P = h\nu \qquad (10.15)$$

where h is Planck's constant, and has dimensions of energy per unit frequency. It is often expressed in terms of the energy given to an electron by a change in potential of 1 V, where 1 eV equals $1.602\,189 \times 10^{-19}$ J.

The wavelength (or frequency) and the photon energy both uniquely describe a particular radiation. Photon energy is usually used for this purpose when it is greater than a few eV.

Visible electromagnetic radiation, with an energy of the order of 1 eV, is usually generated by the transition of electrons from one outer orbit to another within an atom. X-rays, with energies of a few keV, are generated by electron beams and by changes amongst inner atomic orbits. Electromagnetic photons with energies greater than a few keV are known as gamma radiation and are produced in large accelerators or, more usually, by transitions within atomic nuclei. In the latter case energies up to a few MeV may be expected. Photons in the cosmic ray flux may reach perhaps 10^{20} eV (i.e. a few joules!).

10.2.2 Particulate radiation

10.2.2.1 Basic concepts

Particles are produced when atomic nuclei or other high-velocity particles interact with each other or when atomic nuclei decay from an excited to a more stable state. Each has an energy, E_0, corresponding to its mass at rest, m_0:

$$E_0 = m_0 c_0^2 \qquad (10.16)$$

It will also have energy due to its motion and, in an engineering context, this is usually quoted as the particle's energy. The rest energy, although large, is not normally of direct interest. Energy of motion cannot always be treated conventionally as kinetic energy. Low-mass particles, such as electrons, often have velocities approaching c_0 and are influenced by relativity. The observed mass, m, of a particle with velocity v (with respect to the observer) then becomes:[9]

$$m = m_0[1 - (v/c_0)^2]^{-1/2} \qquad (10.17)$$

Its total energy is then mc_0^2.

The motion of particles can be described approximately with conventional mechanics but a rigorous treatment requires quantum mechanics. They are then said to be described by so-called de Broglie waves whose frequency and length are given respectively by:[10]

$$\nu_d = mc^2/h \tag{10.18}$$

$$\lambda_d = h/mv \tag{10.19}$$

These are the waves that explain instruments such as electron microscopes. The wavelength of a non-relativistic electron with an energy V eV is

$$\lambda_d = 1.226\,431 \times 10^{-9}/\sqrt{V} \text{ m} \tag{10.20}$$

Particles may interact with material through which they pass and the probability of this happening is usually expressed as a cross-section, σ.[11,12] Cross-sections are conceptual areas with sizes proportional to the interaction probability and there will be one for every possible process. They are ascribed to each target atom or particle so that, if an incident particle flux ϕ m^{-2} s^{-1} impinges on unit volume of material containing N targets, the rate of the event associated with σ is given by:

$$R = \phi N \sigma \text{ s}^{-1} \tag{10.21}$$

An incident parallel beam will be attenuated exponentially:

$$\phi = \phi_0 \exp -(N\sigma x) \tag{10.22}$$

where ϕ_0 is the initial value and x is distance travelled (m).

Cross-sections are small and are usually expressed in barns where 1 barn equals 10^{-28} m^2. Particles have magnetic moments and these can be relevant in interactions, particularly those involving neutrons.

10.2.2.2 Types of particle

A very large number of different particles have been identified[13] but, in addition to the photon, only five are normally of interest.

Beta particles (β) are energetic electrons. They have rest mass, m_e, and a negative charge, $-|e|$, where $m_e = 9.109\,534 \times 10^{-13}$ kg and $E_0 = 0.511$ MeV.

The energies of the β particles from a given nuclear decay are distributed from zero to a unique maximum, E_{max}, the difference between the actual β energy and E_{max} being carried away by an unobserved particle, the neutrino. The β energy distribution can often be approximated by a half cosine function. In many cases, β decay from a given parent leads to a daughter in one of several excited states which then reach the ground state by photon emission. Decay of this type generates a number of independent β energy distributions, each being accompanied by one or more γ-rays and having a certain probability of occurrence. These different routes are known as decay channels. β-particles from nuclei have maximum energies less than a few MeV.

Positrons are positive particles with the same mass as the electron but with charge $+|e|$. They are created, as a pair with an electron, by photons which have energy greater than the combined rest energy. After a short life, positrons turn back into electromagnetic radiation by mutual annihilation with another electron. During their lives they behave in much the same way as electrons and are usually discussed in the same context.

Protons (p) are the nuclei of hydrogen atoms. They have a rest mass, m_p, and carry a charge $+|e|$, where $m_p = 1.672\,648 \times 10^{-27}$ kg and $E_0 = 938.279$ MeV.

Neutrons (n) have approximately the same mass, m_n, as protons but are uncharged, where $m_n = 1.674\,954 \times 10^{-27}$ kg and $E_0 = 939.573$ MeV.

Protons and neutrons occur in approximately equal numbers in most nuclei. Proton radiation is relatively rare (except in an accelerator context) but many free neutrons exist in nuclear reactors. They are produced when heavy nuclei, such as uranium or plutonium, fission and can also be generated by the action of α-particles or γ-rays on certain nuclear isotopes. Some unstable nuclei decay by the emission of a neutron.

Alpha (α) particles are a stable combination of two protons and two neutrons. They are the nuclei of helium atoms, carry charge $+2|e|$ and are produced in the decay of many unstable heavy nuclei.

Fission fragments are generated two at a time in nuclear fission. They are the atoms into which the fissile material splits, can take many forms and are relatively heavy (of order $100m_p$). When first generated, they are highly ionised, i.e. their outer, atomic electron shells are far from complete. However they quickly acquire these electrons and become normal atoms. Fission fragment nuclei are usually unstable and decay with the emission of further radiation, particularly β-particles and γ-rays.

10.3 The interaction of radiation with matter

10.3.1 Non-ionising electromagnetic radiation

When the wavelength is long compared with the mean spacing between electrons in the material, the wave will interact with the electrons in bulk. It will not ionise but will follow Maxwell's equations. These must be solved for the particular medium and the appropriate boundary conditions.

For transverse waves in linear, isotropic, homogeneous media:[2]

$$(2\pi/\lambda)^2 = (K_r + jK_i)^2 = \varepsilon_r \mu_r (\omega/c)^2 (1 - js/\omega\varepsilon) \tag{10.23}$$

where s is the conductivity of the medium. K_r is a travelling wave solution, whilst K_i yields an exponential attenuation in the medium. Attenuation increases with the ratio of conduction to displacement current, $s/\omega\varepsilon$. In the general case, s and ε may be functions of ω.

Skin depth, δ, is the distance in which the wave is attenuated by the factor e as it passes into the medium:

$$\delta = 1/K_i \tag{10.24}$$

For materials with high conductivity, $s/\omega\varepsilon \gg 1$:

$$\delta = \sqrt{2/\omega s \mu} \tag{10.25}$$

When an electromagnetic wave is incident on the boundary between two media, the general result is a transmitted wave plus a reflected one. Their amplitudes can be derived from the wave equations constrained such that the tangential components of E and H and the normal components of D and B are continuous across the boundary. It can be shown that reflection is efficient from good conductors ($s/\omega\varepsilon \gg 1$) and that it can be reduced at a particular frequency by the use of a $\lambda/4$ thick surface layer of an appropriate second medium. Arguments of this type apply to ionised gases as well as to solid materials. They also govern the detailed behaviour of transmission lines and waveguides.[14]

Maxwell's equations imply that the properties of a medium can be expressed in terms of a characteristic impedance, Z_m, and this is often a convenient concept. In general:

$$Z_m = E/H = \sqrt{\mu/\varepsilon}$$

$$Z_{\text{free space}} = 376.7\ \Omega$$

Thus, if a wave travels through medium 1 towards a boundary with medium 2, the fractional reflection in amplitude is given by:

$$R = \frac{Z_2 - Z_1}{Z_2 + Z_1}$$

Negative reflection means a reversal of phase, and the transmission, T, through the boundary follows from:

$$R+T=1$$

These results are particularly useful when applied to junctions between transmission lines or to the termination of a single line. In these cases, Z_1 will be the characteristic impedance of the first line whilst Z_2 becomes that of the second or the value of the terminating impedance.

Many of the optical formulae can be derived from Maxwell's equations. However, at the shorter optical wavelengths, photons have enough energy to eject a conduction electron from the medium (photoelectric emission). The minimum energy required is known as the work function of the material[15] and the probability of the process is very sensitive to the material, its surface condition, photon energy and temperature. It is considered non-ionising because it involves only a conduction electron.

10.3.2 Ionising electromagnetic radiation

At higher photon energies (shorter wavelengths) the photon can interact only with single electrons and ionisation occurs by the photoelectric or the Compton process.[11] At the highest energies electron/positron pair production dominates.

In the photoelectric process the photon knocks an orbital electron out of the material and this carries away all of the difference between the original photon energy and that required to free the electron. Ejection is usually from an inner (K or L) shell so that the latter is much more than the work function. The cross-section for this interaction is a strong function of atomic number and shows a series of discontinuities with photon energy because of the different ionisation energies for different shells. For incident photon energies, E_g greater than the K absorption edge of the material (of order 10 keV), σ varies as $Z^5 \cdot E_g^{7/2}$ whilst at high E_g it varies inversely to E_g. Since the atom is left in an excited state, photoelectric ionisation is usually followed by the emission of additional photons but these will not be related to the originating one and may be in the near-visible region.

The Compton interaction also produces a free energetic electron but this time the photon does not vanish. It is scattered through an angle θ and its energy is reduced to E'_g:

$$E'_g=\frac{0.51E_g}{0.51+E_g(1-\cos\theta)} \quad \text{(energies in MeV)}$$

$E_g-E'_g$ is carried by the electron and is distributed from zero to a well-defined maximum (the Compton edge) corresponding to $\theta=\pi$ (electron projected forwards). For energies measured in MeV:

$$E_{max}=\frac{4E_g^2}{1+4E_g} \tag{10.26}$$

The cross-section varies approximately inversely to $Z\sqrt{E_g}$.

Pair production has an energy threshold of 1.02 MeV. Its cross-section varies as Z^2 and increases rapidly with E_g above the threshold.

All three effects are illustrated in *Figure 10.2*.

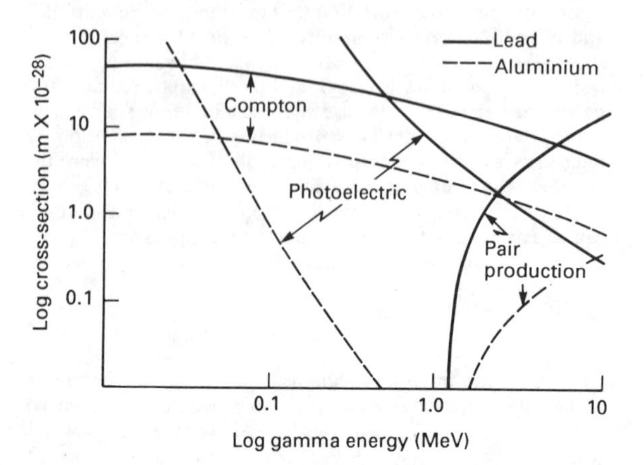

Figure 10.2 Cross-sections for the absorption of γ-rays

10.3.3 Particles interacting with atoms

Low-energy particles will interact with crystals and other regular structures to produce diffraction and other wave-related phenomena but this process will cease when the de Broglie wavelength becomes short compared with atomic spacings (of order 10^{-10} m). Particles with energies greater than a few eV can, however, interact directly with single atoms.

In addition to nuclear reactions, there are three main processes by which an energetic particle can lose energy:

(1) *Elastic collisions* Slow, heavy particles near the end of their range can transfer appreciable amounts of energy to whole atoms by simple collisions.
(2) *Inelastic collisions* The particle interacts with the electronic shell of the target atom. This is the main process for a particle with energy not great compared with its rest energy.
(3) *Radiative loss or bremsstrahlung* The particle is deflected by coulomb forces when it passes near nuclei, and photons are produced by the resultant acceleration. This process applies when the energy is much greater than the rest energy.

Thus, for charged particles and in most practical cases, inelastic collisions dominate. A single electron in the absorber feels an impulse as the particle passes and gains energy which may knock it out of its parent atom or may only be enough to raise it into a new orbit with subsequent emission of a photon. An incoming particle will meet many electrons in this way and in some cases the electron will gain enough kinetic energy to become a secondary, ionising particle in its own right. Such secondary electrons are called δ-rays.

The linear stopping power of an absorber is its power to extract energy from the incident particle:

$$S=-\mathrm{d}E/\mathrm{d}x \quad \text{in appropriate units} \tag{10.27}$$

To a good approximation, for a heavy particle which cannot lose a large fraction of its energy in a single collision with the light electrons:[16]

$$S=-\frac{0.031Zz^2}{A\beta^2}\{\ln[1.02\times10^6\beta^2/ZI(1-\beta^2)]-\beta^2\} \quad \text{MeV kg}^{-1}\,\text{m}^2 \tag{10.28}$$

where z_e is the charge on the incoming particle; β is the velocity of the incoming particle as a fraction of c; Z and A apply to the target; I is an average excitation potential obtained by summation of the various electron transitions in the atom (of the order of 11.5 eV depending on the target and the projectile); and kg/m^2 has dimensions of density times length to allow for the fact that stopping power depends on density.

For a given particle velocity in a given material, S depends on z^2. Thus, α-particles ($z=2$) will deposit energy more densely than protons ($z=1$) but less densely than heavily ionised fission fragments. Residual range is obtained from the integral of S and is the distance that a particle with a given energy will travel. *Figure 10.3* shows values of this function.

An incident electron can lose up to 50% of its energy in a single collision and the above formula does not apply. S for electrons is relatively small, i.e. they deposit little energy per unit length of track. Their attenuation is characterised by prolific δ-formation,

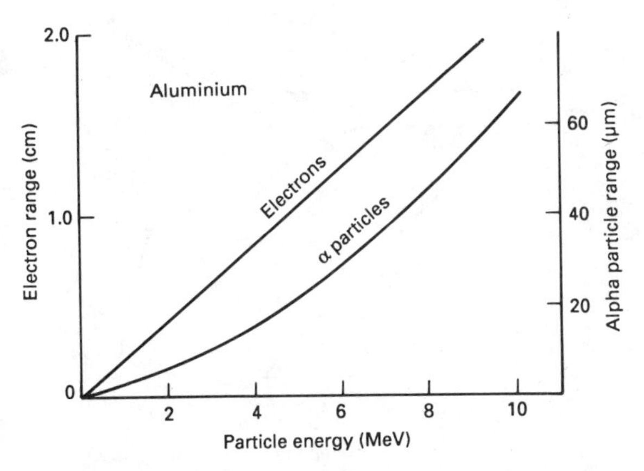

Figure 10.3 Approximate range/energy relationships

the tracks having multiple branches and no well-defined range. Backscatter can occur. Because of this, a beam of electrons suffers roughly exponential absorption with an end-point depending on the initial energy. *Figure 10.3* gives ranges, but they are of a different type from those of α-particles. Electrons with energies greater than about $\frac{1}{2}$ MeV show significant relativistic effects and bremsstrahlung may be important.

All of these processes are subject to statistical variation.

10.3.4 Nuclear interactions

Particles (including photons) with high energy or no charge may avoid the orbital electrons and interact directly with a nucleus. Energy will be released or absorbed and the interaction may eject another particle. The atomic number of the target may be changed, altering the chemical properties of the target, which may be left in a radioactive state. Each of these processes will have an energy-dependent cross-section.

In practice, the main agents for this type of process are neutrons, particularly neutrons which are in thermal equilibrium with their surroundings, having an energy of about $\frac{1}{40}$ eV at room temperature. Typical reactions are:

$^{10}B(n,\alpha)^{7}Li$	$\sigma = 3820$ b	Used in detectors
$^{13}C(n,\gamma)^{14}C$	$\sigma = 0.5$ mb	Used in dating
$^{55}Mn(n,\gamma)^{56}Mn$	$\sigma = 13.2$ b	Activation of steel
$^{113}Cd(n,\gamma)^{114}Cd$	$\sigma = 20\,000$ b	Used in control rods

Neutrons that impinge upon certain heavy nuclei can cause fission and the release of significant energy as well as more neutrons to continue the process.

$^{235}U(n,n)$ fission fragments + 195 MeV

10.3.5 The effect of radiation on materials

This follows from the above processes. Non-ionising electromagnetic radiation will deposit energy as heat and will do so efficiently if it can couple effectively. Ionising but non-nuclear events generate free ions and promote chemical reactions which would either not take place at all or would only occur under very different circumstances, possibly at much higher temperatures. These reactions proceed at a rate set by the ionisation density and described by their *G* values.

G equals the number of events of a given type per 100 eV deposited. Several *G* values may be necessary to describe competing chemical processes and they may depend on the ionisation density and type. They may also depend, for example, on whether oxygen is available to support a given chemical route. Thus irradiation under one gas (or vacuum) may produce different results from irradiation under another and there may be synergism with, for example, temperature. Materials processed in this way will not display residual radioactivity because the nuclei are unchanged.

High-energy heavy particles such as neutrons at a few MeV can damage materials by colliding with constituent atoms and creating crystal defects. This can alter dimensions and strengths, change electrical properties, etc. Properties that are sensitive to structure can change relatively quickly.

10.4 The detection of ionising radiation

10.4.1 Techniques[16–19]

Radiation detectors depend on the deposition of ionisation energy and its subsequent availability in a useful form. Neutrons are converted into ionising particles and their presence inferred. γ-rays are detected because they produce photoelectric or Compton electrons.

Detectors may be classified by the ways in which these processes are carried out.

Films and similar devices Ionising radiation will produce a latent image in photographic film and the fog density after development is a measure of the radiation intensity. The thermoluminescent dosimeter (TLD) is analogous because it stores energy for release as light when the dosimeter is subsequently warmed. In this case the quantity of light is a measure of dose.

Detectors of this type can be made very sensitive. They are fabricated into special badges containing materials which extend the sensitive range of the film to, for example, neutrons.

Scintillation detectors Radiation produces excited atomic states and, in a suitable phosphor, these decay by emission of a photon in the near-visible part of the spectrum which is detected with a photomultiplier. A phosphor with high atomic number will yield a high proportion of photoelectric interactions and, if big enough to absorb the photoelectrons, will permit energy spectrometry. In this context, problems arise from Compton interactions and from pair production.

Solid-state detectors Various techniques are available, which, in materials such as silicon and germanium, will produce regions containing relatively few free carriers. The carriers generated by ionisation can then be detected. Detectors based on germanium are cooled to reduce thermal noise and have virtually replaced scintillators for γ-energy spectrometry. Radiation damage can be a problem and sometimes limits the application of solid-state detectors.

Pulse ionisation chambers An ionising event can be made to deposit energy in a gas between electrodes and this ionisation may be collected as a pulse by an applied field. The charge depends on the deposited energy and on the gas. Typically:

$$\text{charge collected} = 1.6 \times 10^{-19} \times \text{energy deposited (eV)}/30\ \text{C} \quad (10.29)$$

Because the electrons produced by γ-rays are lightly ionising, γ-events deposit only a few keV and are difficult to resolve against noise. In contrast, neutrons are often detected via fissile coatings and ionisation from fission fragments. These will deposit perhaps 30 MeV to give a charge greater than 10^{-13} C. This charge will appear in about 100 ns as a 1 μA current pulse.

Counters can be used with pulse height discriminators which respond only to large pulses and γ-rays can be rejected. However, depending on the system resolving time, counts will be lost at high rates and this limits the dynamic range. Corrections for this can be quite complex but, in a simple case, dominated by a resetting discriminator with fixed dead-time, τ, the effective counting rate is given by:

$$N' = N \exp(-N\tau) \text{ s}^{-1}$$

where N is the true counting rate.

Proportional and Geiger counters Ionisation electrons in a pulse chamber make multiple collisions with the filling gas as they move towards the anode and, in high collecting fields, they can collect enough energy to ionise the atoms with which they collide. This secondary ionisation enhances the original signal.

In the proportional counter the magnitude of the final output is still proportional to that of the primary ionisation—it is merely multiplied by the gas gain. In the Geiger counter, excitation photons from the multiplication process produce yet more carriers from the chamber walls and all proportionality is lost. Proportional and Geiger counters generate relatively large outputs from small primary ionisations and overcome amplifier signal-to-noise problems in an elegant way. They are well suited to the detection of γ-photons. Gas gain in proportional counters is of the order of 100 and pulse output from Geiger counters is of the order of 10^{-9} C.

DC ionisation chambers In high radiation fields, chamber pulses can be integrated into an effective direct current and this may be used as a signal. Such chambers can work at the highest flux levels but have the disadvantage that pulse height discrimination is no longer available.

The quartz fibre electroscope is a form of d.c. chamber which directly indicates charge lost as a result of ionisation leakage.

Current fluctuation or campbelling chambers The noise component generated by the random arrival of pulses contains information and the rms current output from an ionisation chamber is given by:

$$i^2 = 2NB\overline{(Q^2)} \qquad (10.30)$$

where N is the mean event rate (s^{-1}), B is the bandwidth (Hz), and $\overline{(Q^2)}$ is the mean square charge per event (C^2). The latter two parameters are instrument constants and i^2 can be calibrated against N. The advantage of this process is that the terms in $\overline{Q^2}$ generates bias against events with low Q and, for example, the detection of neutrons in the presence of very large γ-fields becomes possible. These systems operate in bandwidths of about 100 kHz centred at about 200 kHz.

10.4.2 Special electronic requirements

Radiation detection involves the measurement of very small charges, sometimes in wide bandwidths. Chamber pulses can be integrated on the first stage of high-input impedance amplifiers but the output then varies inversely as the parasitic capacitance and signal-to-noise limits tend to be reached with cables as short as 10 m. An alternative is to use capacitive feedback (charge-sensitive amplification) but this restricts the upper limit counting rate. A third possibility is to use low-noise amplifiers that have an input impedance equal to that of the counter cable. The pulse is not reflected and is handled as current rather than charge.

The amplifier frequency response must optimise signal to thermal noise ratio and yet be consistent with the required system resolving time. When the latter is important, a first approximation to optimum is obtained with equal differentiate and integrate time constants, both being equal to the counter electron collecting time.[20] When resolving time is not important, better S/N ratio can usually be obtained by integrating for a few microseconds.

Because of the small signals, electrical interference is important, particularly in installations containing long cables

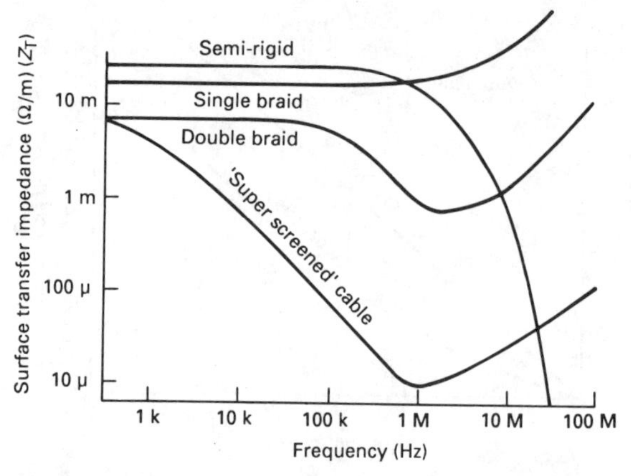

Figure 10.4 The surface transfer impedances of four types of detector cable

and in current fluctuation systems. Normal electromagnetic compatibility techniques are employed and it is additionally necessary to protect against so-called earth coupling. In this mode, disturbances on the earth are coupled directly or capacitively to detector cable screens. Screen currents as large as 100 mA appear at frequencies governed by screen resonances and these currents couple to the chamber signal via the cable surface transfer impedence (Z_t). *Figure 10.4* shows values of Z_t for typical chamber cables and illustrates the advantage of so-called superscreened designs containing magnetic material.

10.5 Health physics

Radiation has biological effects.[21–22] The following notes provide background but cannot, within the space available, be other than superficial. As in all safety matters, the reader should seek qualified local advice when necessary.

Non-ionising radiation causes heating in tissue. Limits on power density are specified and must be imposed. There are said to be other mechanisms by which non-ionising radiation produces biological effects but there is currently no general agreement on the existence or magnitude of such effects and no limits are available.

Ionising radiation affects tissue by the same mechanisms as it damages other materials. It creates chemical radicals and unwanted chemical activity. Different G values apply to different damage routes but on average the total damage will be dependent on the total dose, the 'effectiveness' of the radiation type in causing damage and the sensitivity of the particular tissue under irradiation. *Table 10.1* emphasises the difference between

Table 10.1 Relationships between radiation units

Quantity of radioactivity	Curie *Becquerel	The source term
Radiation exposure	Rontgen	Gives effect of X- and γ-rays in air (historic)
Absorbed dose	Rad *Gray	Energy absorption from exposure (any material)
Dose equivalent	Rem *Sievert	Absorbed dose times quality
Absorbed dose rate Dose equivalent rate	Units/h	

* SI units

quantity of radioactivity (the source term), radiation exposure (the effect of that source in the absence of the absorber), absorbed dose (the energy deposited in the absorber by a given exposure) and the effect of the absorbed dose on tissue.

Table 10.2 Definitions of radiation units

Curie (Ci)	3.7×10^{10} disintegrations/s
Becquerel (Bq)	1 disintegration/s
Rontgen (R)	1 esu in 1.293 mg of air (or 258 μC in 1 kg)
Rad (rad)	100 erg/g (applies to any material)
Gray (Gy)	1 J/kg (= 100 rad)
Rem (rem)	rad $\times Q \times N$ (*Table 10.3*)
Sievert (Sv)	Gy $\times Q \times N$ (= 100 rem)

Table 10.2 gives definitions of the various units and clarifies the relationship between the old and new versions. The quantity Q allows for the fact that different types of radiation produce different degrees of damage. For example, fast neutrons generate knock-on protons from tissue hydrogen. These are heavily ionising and 1 Gy of fast neutrons will do as much damage as 10 Gy of γ-radiation. Hence the quality factor (or relative biological effectiveness, RBE) of fast neutrons will be 10. A second modifier, N, has, to date, been set to unity. *Table 10.3* gives values of Q.

Table 10.3 Quality factors for ionising radiation

Radiation type	*Q value*
X-rays, γ-rays and electrons	1
Thermal (slow) neutrons	2.3
Fast neutrons and protons	10
α-particles	20

Radiation injury can be placed in two categories:

(1) Somatic—damage to the irradiated individual (immediate or delayed).
(2) Hereditary—damage to offspring.

The radiation levels at which such damage may become serious are assessed by the International Commission on Radiological Protection. The ICRP has no legislative function and is independent of governments but its recommendations form the basis of regulations in many countries. They have identified three primary objectives for radiological protection:

(1) All particles involving radiation exposure should be justified, with the benefits outweighing the detriment.
(2) Radiation doses should be as low as reasonably achievable, economic and social factors being taken into account.
(3) All radiation exposures should be within the recommended dose limits.

Objective (2) established the principle of ALARA which has been expressed as a regulatory requirement to make all doses as low as reasonably practical (ALARP). The limits noted in (3) distinguish external dose from that delivered by ingested radioactive material (orally and through wounds), whole-body dose from that to certain parts of the body, and dose to the general public from that to radiation workers. They have also been formulated as national legal requirements and take into account the fact that some 98% of the dose to the public arises from natural background and medical applications.

References

1 NATIONAL PHYSICAL LABORATORY, *Fundamental Physical Constants and Energy Conversion Factors*, HMSO, London (1974)
2 LORRAIN, P. and CORSON, D. R., *Electromagnetic Fields and Waves*, Freeman & Co (1970)
3 RAMO, S., WHINNERY, J. R. and VAN DUZER, T., *Fields and Waves in Communication Electronics*, Wiley, Chichester (1967)
4 SOLYMAR, L., *Lectures on Electromagnetic Theory—A Short Course for Engineers*, Oxford University Press, Oxford (1976)
5 JACKSON, J. D., *Classical Electrodynamics*, Wiley, Chichester (1962)
6 HUGHES, W. F. and GAYLORD, E. W., *Basic Equations of Engineering Science*, Schaum, New York (1964)
7 FOWLES, G. R., *Introduction to Modern Optics*, Holt, Rinehart & Winston (1968)
8 SMITH, F. J. and THOMPSON, J. H., *Optics*, Wiley, Chichester (1971)
9 FRENCH, A. P., *Special Relativity*, Nelson (1968)
10 RAE, A. I. M., *Quantum Mechanics*, Adam Hilger (1986)
11 MEYERHOF, W. E., *Elements of Nuclear Physics*, McGraw-Hill, Maidenhead (1967)
12 BOWLER, M. G., *Nuclear Physics*, Pergamon Press, Oxford (1973)
13 HUGHES, I. S., *Elementary Particles*, Penguin, Harmondsworth (1972)
14 CONNOR, F. R., *Wave Transmission*, Edward Arnold (1972)
15 ROSENBERG, H. M., *The Solid State*, Oxford University Press, Oxford (1978)
16 SHARPE, J., *Nuclear Radiation Detectors*, Methuen (1955)
17 COOPER, P. N., *Introduction to Nuclear Radiation Detectors*, Cambridge University Press, Cambridge (1986)
18 KNOLL, G. F., *Radiation Detection and Measurement*, Wiley, Chichester (1979)
19 PRICE, W. J., *Nuclear Radiation Detection*, McGraw-Hill, Maidenhead (1958)
20 GILLESPIE, A. B., *Signal, Noise and Resolution in Nuclear Counter Amplifiers*, Pergamon Press, Oxford (1953)
21 MARTIN, A. and HARBISON, S. A., *An Introduction to Radiation Protection*, Chapman & Hall (1986)
22 MARSHALL, W. (Ed.), *Nuclear Power Technology*, Vol. 3, *Nuclear Radiation*, Clarendon Press, Oxford (1983)

11 The Ionosphere and the Troposphere

P A Bradley BSc, MSc, CEng, MIEE
Rutherford Appleton Laboratory,
Science and Engineering Research Council
(Sections 11.1–11.6)

J A Lane DSc, CEng, FIEE
Radio Communication Division,
Department of Trade and Industry
(Sections 11.7–11.9)

Contents

11.1 The ionosphere

The ionosphere is an electrified region of the Earth's atmosphere situated at heights of from about fifty kilometres to several thousand kilometres. It consists of ions and free electrons produced by the ionising influences of solar radiation and of incident energetic solar and cosmic particles. The ionosphere is subject to marked geographic and temporal variations. It has a profound effect on the characteristics of radio waves propagated within or through it. By means of wave refraction, reflection or scattering it permits transmission over paths that would not otherwise be possible, but at the same time it screens some regions that could be illuminated in its absence (see *Figure 11.1*).

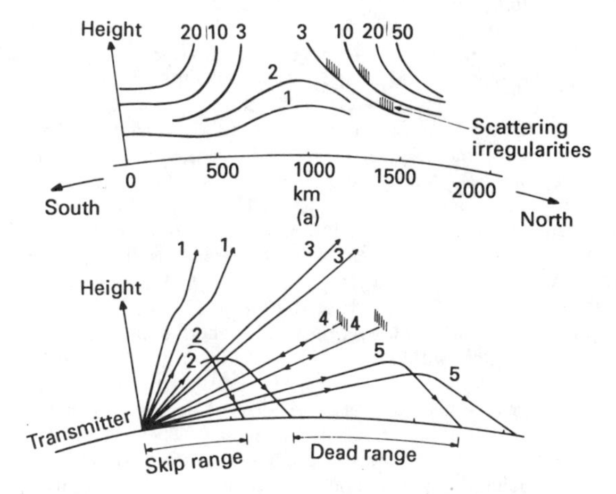

Figure 11.1 High-frequency propagation paths via the ionosphere at high latitudes: (a) sample distribution of electron density (arbitrary units) in northern hemisphere high-latitude ionosphere (adapted from Buchau[1]); (b) raypaths for signals of constant frequency launched with different elevation angles

The ability of the ionosphere to refract, reflect or scatter rays depends on their frequency and elevation angle. Ionospheric refraction is reduced at the higher frequencies and for the higher elevation angles, so that provided the frequency is sufficiently great rays 1 in *Figure 11.1* escape whereas rays 2 are reflected back to ground. Rays 3 escape because they traverse the ionosphere at latitudes where the electron density is low (the Muldrew trough[2]). Irregularities in the F-region are responsible for the direct backscattering of rays 4. The low-elevation rays 5 are reflected to ground because of the increased ionisation at the higher latitudes. Note that for this ionosphere and frequency there are two ground zones which cannot be illuminated.

The ionosphere is of considerable importance in the engineering of radio communication systems because:

(1) It provides the means of establishing various communication paths, calling for system-design criteria based on a knowledge of ionospheric morphology.
(2) It requires specific engineering technologies to derive experimental probing facilities to assess its characteristics, both for communication-systems planning and management, and for scientific investigations.
(3) It permits the remote monitoring by sophisticated techniques of certain distant natural and man-made phenomena occurring on the ground, in the air and in space.

11.2 Formation of the ionosphere and its morphology

There is widespread interest in the characteristics of the ionosphere by scientists all over the world. Several excellent general survey books have been published describing the principal known features[3,4] and other more specialised books concerned with aeronomy, and including the ionosphere and magnetosphere, are of great value to the research worker.[5–7] Several journals in the English language are devoted entirely, or to a major extent, to papers describing investigations into the state of the ionosphere and of radio propagation in the ionosphere.

The formation of the ionosphere is a complicated process involving the ionising influences of solar radiation and solar and cosmic particles on an atmosphere of complex structure. The rates of ion and free-electron production depend on the flux density of the incident radiation or particles, as well as on the ionisation efficiency, which is a function of the ionising wavelength (or particle energies) and the chemical composition of the atmosphere. There are two heights where electron production by the ionisation of molecular nitrogen and atomic and molecular oxygen is a maximum. One occurs at about 100 km and is due to incident X-rays with wavelengths less than about 10 nm and to ultraviolet radiation with wavelengths near 100 nm; the other is at about 170 km and is produced by radiation of wavelengths 20–80 nm.

Countering this production, the free electrons tend to recombine with the positive ions and to attach themselves to neutral molecules to form negative ions. Electrons can also leave a given volume by diffusion or by drifting away under the influences of temperature and pressure gradients, gravitational forces or electric fields set up by the movement of other ionisation. The electron density at a given height is given from the so-called *continuity equation* in terms of the balance between the effects of production and loss.

Night-time electron densities are generally lower than in the daytime because the rates of production are reduced. *Figure 11.2* gives examples of a night-time and a daytime height distribution

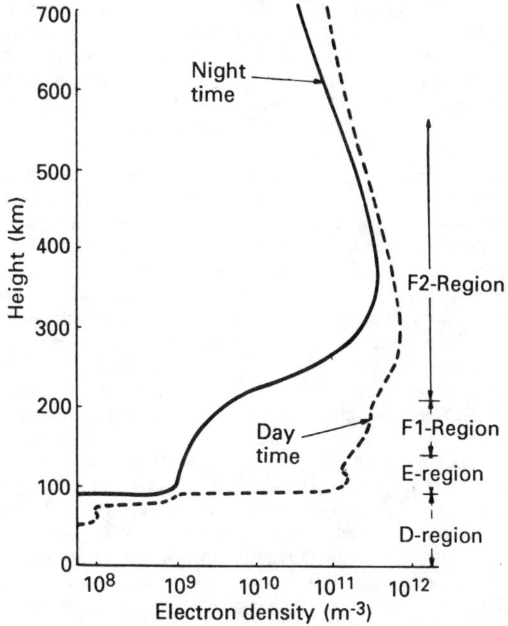

Figure 11.2 Sample night-time and daytime height distributions of electron density at mid-latitudes in summer

of electron density. The ionisation is continuous over a wide height range, but there are certain height regions with particular characteristics, and these are known, following E. V. Appleton, by the letters D, E and F. The E-region is the most regular ionospheric region, exhibiting a systematic dependence of maximum electron density on solar-zenith angle, leading to predictable diurnal, seasonal and geographical variations. There is also a predictable dependence of its electron density on the changes in solar radiation which accompany the long-term fluctuations in the state of the sun. Maximum E-region electron density is approximately proportional to sunspot number, which varies over a cycle of roughly 11 years.

In the daytime the F-region splits into two, with the lower part known as the F1-region and the upper part as the F2-region. This splitting arises because the principal loss mechanism is an ion–atom interchange process followed by dissociative recombination, the former process controlling the loss rates in the F2-region and the latter in the F1-region. Although maximum production is in the F1-region, maximum electron density results in the F2-region, where the loss rates are lower. The maximum electron density of the F1-region closely follows that of the E-region, but there are significant and less predictable changes in its height. The maximum electron density and height of the F2-region are subject to large changes which have important consequences to radio wave propagation. Some of these changes are systematic but there are also major day-to-day variations. It seems likely that the F2-region is controlled mainly by ionisation transport to different heights along the lines of force of the Earth's magnetic field under the influence of thermospheric winds at high and middle latitudes[8] and by electric fields at low latitudes.[9] These effects, taken in conjunction with the known variations in atmospheric composition, can largely explain characteristics of the F2-region which have in the past been regarded as anomalous by comparison with the E-region—namely, diurnal changes in the maximum of electron density in polar regions in the seasons of complete darkness, maximum electron densities at some middle-latitude locations at times displaced a few hours from local noon with greater electron density in the winter than the summer, and at low latitudes longitude variations linked more to the magnetic equator than to the geographic equator, with a minimum of electron density at the magnetic equator and maxima to the north and south where the magnetic dip is about 30°. At all latitudes electron densities in the F2-region, like those in the E- and F1-regions, increase with increase of sunspot number. The electron densities at heights above the maximum of the F2-region are controlled mainly by diffusion processes.

The D-region shows great variability and fine structure, and is the least well understood part of the ionosphere. The only ionising radiations that can penetrate the upper regions and contribute to its production are hard X-rays with wavelengths less than about 2 nm and Lyman-α radiation at 121.6 nm. Chemical reactions responsible for its formation principally involve nitric oxide and other minor atmospheric constituents.

The D-region is mainly responsible for the absorption of radio waves because of the high electron-collision frequencies at such altitudes (see below). While the electron densities in the upper part of the D-region appear linked to those in the E-region, leading to systematic latitudinal, temporal and solar-cycle variations in absorption, there are also appreciable irregular day-to-day absorption changes. At middle latitudes anomalously high absorption is experienced on some days in the winter. This is related to warmings of the stratosphere and is probably associated with changes in D-region composition. In the lower D-region at heights below about 70 km the ionisation is produced principally by energetic cosmic rays, uniformly incident at all times of day. Since the free electrons thereby generated tend to collide and become attached to molecules to form negative ions by night, but are detached by solar radiation in the daytime, the lower D-region ionisation, like that in the upper D-region, is much greater by day than night. In contrast, however, electron densities in the lower D-region, being related to the incidence of cosmic rays, are reduced with increase in the number of sunspots. Additional D-region ionisation is produced at high latitudes by incoming particles, directed along the lines of force of the Earth's magnetic field. Energetic electrons, probably originating from the Sun, produce characteristic auroral absorption events over a narrow band of latitudes about 10° wide, associated with the visual auroral regions.[10]

From time to time disturbances occur on the Sun known as solar flares. These are regions of intense light, accompanied by increases in the solar far ultraviolet and soft X-ray radiation. Solar flares are most common at times of high sunspot number. The excess radiation leads to sudden ionospheric disturbances (SIDs), which are rapid and large increases in ionospheric absorption occurring simultaneously over the whole sunlit hemisphere. These persist for from a few minutes to several hours giving the phenomena of short-wave fadeouts (SWFs), first explained by Dellinger. Accompanied by solar flares are eruptions from the Sun of energetic protons and electrons. These travel as a column of plasma, and depending on the position of the flare on the Sun's disc and on the trajectory of the Earth, they sometimes impinge on the ionosphere. Then, the protons, which are delayed in transit from fifteen minutes to several hours, produce a major enhancement of the D-region ionisation in polar regions that can persist for several days. This gives the phenomenon of polar cap absorption (PCA) with complete suppression of h.f. signals over the whole of both polar regions.[11] Slower particles, with transit times of 20–40 hours, produce ionospheric storms. These storms, which result principally from movements in ionisation, take the form of depressions in the maximum electron density of the F2-region.[12] They can last for several days at a time with effects which are progressively different in detail at different latitudes. Since the sun rotates with a period of about 28 days, sweeping out a column of particles into space when it is disturbed, there is a tendency for ionospheric storms to recur after this time interval.

Additional ionisation is sometimes found in thin layers, 2 km or less thick, embedded in the E-region at heights between 90 and 120 km. This has an irregular and patchy structure, a maximum electron density which is much greater than that of the normal E-region, and is known as sporadic-E (or Es) ionisation because of its intermittent occurrence. It consists of patches up to 2000 km in extent, composed of large numbers of individual irregularities each less than 1 km in size. Sporadic-E tends to be opaque to the lower h.f. waves and partially reflecting at the higher frequencies. It results from a number of separate causes and may be classified into different types,[13] each with characteristic occurrence and other statistics. In temperate latitudes sporadic-E arises principally from wind shear, close to the magnetic equator it is produced by plasma instabilities and at high latitudes it is mainly due to incident energetic particles. It is most common at low latitudes where it is essentially a daytime phenomenon.

Irregularities also develop in the D-region due to turbulence and wind shears and other irregularities are produced in the F-region. The F-region irregularities can exist simultaneously over a wide range of heights, either below or above the height of maximum electron density, and are referred to as spread-F irregularities. They are found at all latitudes but are particularly common at low latitudes in the evenings where their occurrence is related to rapid changes in the height of the F-region.[14]

11.3 Ionospheric effects on radio signals

A radio wave is specified in terms of five parameters: its amplitude, phase, direction of propagation, polarisation and frequency. The principal effects of the ionosphere in modifying these parameters are considered as follows.

11.3.1 Refraction

The change in direction of propagation resulting from the traverse of a thin slab of constant ionisation is given approximately by Bouger's law in terms of the refractive index and the angle of incidence. A more exact specification including the effects of the Earth's magnetic field is given by the Haselgrove equation solution.[15] The refractive index is determined from the Appleton–Hartree equations of the magnetoionic theory[16,17] as a function of the electron density and electron-collision frequency, together with the strength and direction of the Earth's magnetic field, the wave direction and the wave frequency. The dependence on frequency leads to wave dispersion of modulated signals. Since the ionosphere is a doubly refracting medium it can transmit two waves with different polarisations (see below). The refractive indices appropriate to the two waves differ. Refraction is reduced at the greater wave frequencies, and at v.h.f. and higher frequencies it is given approximately as a function of the ratio of the wave and plasma frequencies, where the plasma frequency is defined in terms of a universal constant and the square root of the electron density.[16] *Table 11.1* lists the magnitude of the refraction and of other propagation parameters for signals at a frequency of 100 MHz which traverse the whole ionosphere.

11.3.2 Change in phase-path length

The phase-path length is given approximately as the integral of the refractive index with respect to the ray-path length. Ignoring spatial gradients, the change in phase-path length introduced by passage through the ionosphere to the ground of signals at v.h.f. and higher frequencies from a spacecraft is proportional to the total electron content. This is the number of electrons in a vertical column of unit cross-section.

11.3.3 Group delay

The group and phase velocities of a wave differ because the ionosphere is a dispersive medium. The ionosphere reduces the group velocity and introduces a group delay which for transionospheric signals at v.h.f. and higher frequencies, like the phase-path change, is proportional to the total electron content.

11.3.4 Polarisation

Radio waves that propagate in the ionosphere are called characteristic waves. There are always two characteristic waves known as the ordinary wave and the extraordinary wave; under certain restricted conditions a third wave known as the Z-wave can also exist.[16] In general the ordinary and extraordinary waves are elliptically polarised. The polarisation ellipses have the same axial ratio, orientations in space that are related such that under many conditions they are approximately orthogonal, and electric vectors which rotate in opposite directions.[16] The polarisation ellipses are less elongated the greater the wave frequency. Any wave launched into the ionosphere is split into characteristic ordinary and extraordinary wave components of appropriate power. At m.f. and above these components may be regarded as travelling independently through the ionosphere with polarisations which remain related, but continuously change to match the changing ionospheric conditions. The phase paths of the ordinary and extraordinary wave components differ, so that in the case of transionospheric signals when the components have comparable amplitudes, the plane of polarisation of their resultant slowly rotates. This effect is known as Faraday rotation.

11.3.5 Absorption

Absorption arises from inelastic collisions between the free electrons, oscillating under the influence of the incident radio wave, and the neutral and ionised constituents of the atmosphere. The absorption experienced in a thin slab of ionosphere is given by the Appleton–Hartree equations[16] and under many conditions is proportional to the product of electron density and collision frequency, inversely proportional to the refractive index and inversely proportional to the square of the wave frequency. The absorption is referred to as non-deviative or deviative depending on whether it occurs where the refractive index is close to unity. Normal absorption is principally a daytime phenomenon. At frequencies below 5 MHz it is sometimes so great as to completely suppress effective propagation. The absorptions of the ordinary and extraordinary waves differ, and in the range 1.5–10 MHz the extraordinary wave absorption is significantly greater.

11.3.6 Amplitude fading

If the ionosphere were unchanging the signal amplitude over a fixed path would be constant. In practice, however, fading arises as a consequence of variations in propagation path, brought about by movements or fluctuations in ionisation. The principal causes of fading are:

(1) Variations in absorption.
(2) Movements of irregularities producing focusing and defocusing.
(3) Changes of path length among component signals propagated via multiple paths.
(4) Changes of polarisation, such as for example due to Faraday rotation.

These various causes lead to different depths of fading and a range of fading rates. The slowest fades are usually those due to

Table 11.1 Effect of one-way traverse of typical mid-latitude ionosphere at 100 MHz on signals with elevation angle above 60 degrees[22]

Effect	*Day*	*Night*	*Frequency dependence, f*
Total electron content	$5\times10^{13}/\text{cm}^2$	$5\times10^{12}/\text{cm}^2$	
Faraday rotation	15 rotations	1.5 rotations	f^{-2}
Group delay	12.5 μs	1.2 μs	f^{-2}
Change in phase-path length	5.2 km	0.5 km	f^{-2}
Phase change	7500 radians	750 radians	f^{-2}
Phase stability (peak-to-peak)	±150 radians	±15 radians	f^{-1}
Frequency stability (r.m.s.)	±0.04 Hz	±0.004 Hz	f^{-1}
Absorption	0.1 dB	0.01 dB	f^{-2}
Refraction	$\leqslant1°$	—	f^{-2}

absorption changes which have a period of about 10 minutes. The deepest and most rapid fading occurs from the beating between two signal components of comparable amplitude propagated along different paths. A regularly reflected signal together with a signal scattered from spread-F irregularities can give rise to so-called *flutter* fading, with fading rates of about 10 Hz. A good general survey of fading effects, including a discussion of fading statistics, has been produced.[18] On operational communication circuits fading may be combated by space diversity or polarisation-diversity receiving systems and by the simultaneous use of multiple-frequency transmissions (frequency diversity).

11.3.7 Frequency deviations

Amplitude fading is accompanied by associated fluctuations in group path and phase path, giving rise to time and frequency-dispersed signals. When either the transmitter or receiver is moving, or there are systematic ionospheric movements, the received signal is also Doppler-frequency shifted. Signals propagated simultaneously via different ionospheric paths are usually received with differing frequency shifts. Frequency shifts for reflections from the regular layers are usually less than 1 Hz, but shifts of up to 20–30 Hz have been reported for scatter-mode signals at low latitudes.[19]

11.3.8 Reflection, scattering and ducting

The combined effect of refraction through a number of successive slabs of ionisation can lead to ray reflection. This may take place over a narrow height range as at l.f. or rays may be refracted over an appreciable distance in the ionosphere as at h.f. Weak incoherent scattering of energy occurs from random thermal fluctuations in electron density, and more efficient aspect-sensitive scattering from ionospheric irregularities gives rise to direct backscattered and forward-scatter signals. Ducting of signals to great distances can take place at heights of reduced ionisation between the E- and F-regions, leading in some cases to round-the-world echoes.[20] Ducting can also occur within regions of field-aligned irregularities above the maximum of the F-region.

11.3.9 Scintillation

Ionospheric irregularities act as a phase-changing screen on transionospheric signals from sources such as Earth satellites or radiostars. This screen gives rise to diffraction effects with amplitude, phase and angle-of-arrival scintillations.[21]

11.4 Communication and monitoring systems relying on ionospheric propagation

Ionospheric propagation is exploited for a wide range of purposes, the choice of system and the operating frequency being largely determined by the type and quantities of data to be transmitted, the path length and its geographical position.

11.4.1 Communication systems

Radio communication at very low frequencies (v.l.f.) is limited by the available bandwidth, but since ionospheric attenuation is very low, near world-wide coverage can be achieved. Unfortunately the radiation of energy is difficult at such frequencies and complex transmitting antenna systems, coupled with large transmitter powers, are needed to overcome the high received background noise from atmospherics—the electromagnetic radiation produced by lightning discharges. Because of the stability of propagation, v.l.f. systems are used for the transmission of standard time signals and for c.w. navigation systems which rely on direction-finding techniques, or on phase comparisons between spaced transmissions as in the Omega system (10–14 kHz).[23] At low frequencies (l.f.) increased propagation losses limit area coverage, but simpler antenna systems are adequate and lower transmitter powers can be employed because of the reduced atmospheric noise. Low frequency systems are used for communication by on–off keying and frequency-shift keying. Propagation conditions are more stable than at higher frequencies because the ionosphere is less deeply penetrated. Low frequency signals involving ionospheric propagation are also used for communication with submarines below the surface of the sea, with receivers below the ground and with space vehicles not within line-of-sight of the transmitter. Other l.f. systems[24] relying principally on the ground wave, which are sometimes detrimentally influenced by the sky wave at night, include the Decca c.w. navigation system (70–130 kHz), the Loran C pulse navigation system (100 kHz) and long-wave broadcasting.

At medium frequencies (m.f.) daytime absorption is so high as to completely suppress the sky wave. Some use is made of the sky wave at night-time for broadcasting, but generally medium frequencies are employed for ground-wave services. Despite the advent of reliable multichannel satellite and cable systems, high frequencies continue to be used predominantly for broadcasting, fixed, and mobile point-to-point communications, via the ionosphere—there are still tens of thousands of such circuits.

Very high frequency (v.h.f.) communication relying on ionospheric scatter propagation between ground-based terminals is possible. Two-way error-correcting systems with scattering from intermittent meteor trains can be used at frequencies of 30–40 MHz over ranges of 500–1500 km.[25] Bursts of high-speed data of about 1 s duration with duty cycles of the order of 5% can be achieved, using transmitter powers of about 1 kW. Meteor-burst systems find favour in certain military applications because they are difficult to intercept, since the scattering is usually confined to 5–10 degrees from the great-circle path. Forward-scatter communication systems at frequencies of 30–60 MHz, also operating over ranges of about 1000 km, rely on coherent scattering from field-aligned irregularities in the D-region at a height of about 85 km.[26] They are used principally at low and high latitudes. Signal intensities are somewhat variable, depending on the incidence of irregularities. During magnetically disturbed conditions signals are enhanced at high latitudes, but are little affected at low latitudes. Directional transmitting and receiving antennas with intersecting beams are required. Particular attention has to be paid to avoiding interference from signal components scattered from sporadic-E ionisation or irregularities in the F-region. Special frequency-modulation techniques involving time division multiplex are used to combat Doppler effects. Systems with 16 channels, automatic error correction and operating at 100 words per minute, now exist.

11.4.2 Monitoring systems

High frequency (h.f.) signals propagated obliquely via the ionosphere and scattered at the ground back along the reverse path may be exploited to give information on the characteristics of the scattering region.[27] Increased scatter results from mountains, from cities and from certain sea waves. Signals backscattered from the sea are enhanced when the signal wavelength is twice the component of the sea wavelength along the direction of incidence of the signal, since round-trip signals reflected from successive sea-wave crests then arrive in phase to give coherent addition. The Doppler shift of backscatter returns from sea waves due to the sea motion usually exceeds that imposed on the received signals by the ionosphere, so that Doppler filtering enables the land- and sea-scattered signals to be examined separately. This permits studies of distant land–sea boundaries,[28] and since the wavelength of a sea wave is directly related to its velocity, provides a means of synoptic monitoring of distant ocean waves. Doppler filtering can also resolve signals reflected or scattered back along ionosphere paths from aircraft, rockets, rocket trails or ships.

Studies of the Doppler shift of stable frequency, h.f., c.w. signals propagated via the ionosphere between ground-based terminals provide important information about infrasonic waves in the F-region originating from nuclear explosions,[29] earthquakes,[30] severe thunderstorms[31] and air currents in mountainous regions.

High-altitude nuclear explosions lead to other effects which may be detected by radio means.[32] An immediate wideband electromagnetic pulse is produced which can be monitored throughout the world at v.l.f. and h.f. Also, enhanced D-region ionisation, lasting for several days over a wide geographical area, produces an identifiable change in the received phase of long-distance v.l.f. ionospheric signals. Other more localised effects which can be detected include the generation of irregularities in the F-region.

Atmospherics may be monitored at v.l.f. out to distances of several thousand kilometres and by recording simultaneously the arrival azimuths at spaced receivers the locations of thunderstorms may be determined and their movements tracked as an aid to meteorological warning services.

11.5 Ionospheric probing techniques

There are a wide variety of methods of sounding the state of the ionosphere involving single- and multiple-station ground-based equipments, rocket-borne and satellite probes (see *Table 11.2*). A comprehensive survey of the different techniques has been produced by a Working Group of the International Union of Radio Science (URSI).[33] Some of the techniques involve complex analysis procedures and require elaborate and expensive equipment and antenna systems (*Figure 11.3*), others need only a single radio receiver.

The swept-frequency ground-based *ionosonde* consisting of a co-located transmitter and receiver was developed for the earliest of ionospheric measurements and is still the most widely used probing instrument. The transmitter and receiver frequencies are swept synchronously over the range from about 0.5–1 MHz to perhaps 20 MHz depending on ionospheric conditions, and short pulses typically of duration 100 μs with a repetition rate of 50/s are transmitted. Calibrated film or computer digitised records of the received echoes give the group path and its frequency dependence.

In practice these records require expert interpretation because: (1) multiple echoes occur, corresponding to more than one traverse between ionosphere and ground (so-called *multiple-hop* modes) or when partially reflecting sporadic-E or F-region irregularities are present; (2) the ordinary and extraordinary waves are sometimes reflected from appreciably different heights; and (3) oblique reflections occur when the ionosphere is tilted. Internationally agreed procedures for scaling ionosonde records (ionograms) have been produced.[34]

Table 11.2 Principal ionospheric probing techniques

Height range	*Technique*	*Parameters monitored*	*Site*
Above 100 km	vertical-incidence sounding	up to height of maximum ionisation—electron density	ground
	topside sounding	from height of satellite to height of maximum ionisation—electron density; electron and ion temperatures; ionic composition	satellite
	incoherent scatter	up to few thousand km—electron density; electron temperature; ion temperature; ionic composition; collision frequencies; drifts of ions and electrons	ground
	Faraday rotation and differential Doppler	total electron content	satellite-ground
	in-situ probes	wide range of parameters	satellite
	c.w. oblique incidence	solar flare effects; irregularities; travelling disturbances; radio aurorae	ground
	pulse oblique incidence	oblique modes by ground backscatter and oblique sounding; meteors; radio aurorae; irregularities and their drifts	ground
	whistlers	out to few Earth radii—electron density; ion temperature; ionic composition	ground or satellite
Below 100 km	vertical-incidence sounding	absorption	ground
	riometer	absorption	ground
	c.w. and pulse oblique incidence	electron density and collision frequency	ground
	wave fields	electron density and collision frequency	rocket–ground
	in situ probes	electron density and collision frequency; ion density; composition of neutral atmosphere	rocket
	cross-modulation	electron density and collision frequency	ground
	partial reflection	electron density and collision frequency	ground
	lidar	neutral air density; atmospheric aerosols; minor constituents	ground

Figure 11.3 The EISCAT radar antenna (Photograph courtesy of Rutherford Appleton Laboratory)

Since reflection takes place from a height where the sounder frequency is equal to the ionospheric plasma frequency, and since the group path can be related to that height provided the electron densities at all lower heights are known, the data from a full frequency sweep can be used to give the true-height distribution of electron density in the E- and F-regions up to the height of maximum electron density of the F2-region. The conversion of group path to true height requires assumptions regarding missing data below the lowest height from which echoes are received and over regions where the electron density does not increase monotonically with height. The subject of true-height analysis is complex and controversial.[35] Commercially manufactured pulse sounders use transmitter powers of about 1 kW. Sounders with powers of around 100 kW and a lower frequency limit of a few kHz have been operated successfully in areas free from m.f. broadcast interference, to study the night-time E-region—this has a maximum plasma frequency of around 0.5 MHz.

Pulse-compression systems[36] and c.w. chirp sounders[37] offer the possibility of improved signal/noise ratios and echo resolution. In the chirp sounder system, originally developed for use at oblique incidence, the transmitter and receiver frequencies are swept synchronously so that the finite echo transit time leads to a frequency modulation of the receiver i.f. signals. These signals are then spectrum-analysed to produce conventional ionograms. Receiver bandwidths of only a few tens of hertz are needed so that transmitter powers of a few watts are adequate. Other ionosondes have been produced and used operationally which record data digitally on magnetic tape.[34] An ionosonde has been successfully flown in an aircraft to investigate geographical changes in electron density at high latitudes.[38] Over 100 ground-based ionosondes throughout the world make regular soundings each hour of each day; data are published at monthly intervals.[39]

Since 1962 swept-frequency ionosondes have been operated in satellites orbiting the earth at altitudes of around 1000 km. These are known as *topside sounders* and they give the distributions of electron density from the satellite height down to the peak of the F2-region. They also yield other plasma-resonance information, together with data on electron and ion temperatures and ionic composition. Fixed-frequency topside sounders are used to study the spatial characteristics of spread-F irregularities and other features with fine structure.

Many different monitoring probes are mounted in satellites orbiting the earth at altitudes above 100 km, to give direct measurements of a range of ionospheric characteristics. These include r.f. impedance, capacitance and upper-hybrid resonance probes for local electron density, modified Langmuir probes for electron temperature, retarding potential analysers and sampling mass spectrometers for ion density, quadrupole and monopole mass spectrometers for ion and neutral-gas analysis and retarding potential analysers for ion temperature measurements.

Ground measurements of satellite beacon signals permit studies of total-electron content, either from the differential Doppler frequency between two harmonically related h.f./v.h.f. signals,[40] or from the Faraday rotation of a single v.h.f. transmission.[41] Beacons on geostationary satellites are valuable for investigations of temporal variations. The scintillation of satellite signals at v.h.f. and u.h.f. gives information on the incidence of ionospheric irregularities, their heights and sizes.[42]

A powerful tool for ionospheric investigations up to heights of several thousand kilometres is the vertical-incidence incoherent-scatter radar. The technique makes use of the very weak scattering from random thermal fluctuations in electron density which exist in a plasma in quasi-equilibrium. Several important parameters of the plasma affect the scattering such that each of these can be determined separately. The power, frequency spectrum and polarisation of the scattered signals are measured and used to give the height distributions of electron density, electron temperature, ionic composition, ion-neutral atmosphere and ion–ion collision frequencies, and the mean plasma-drift velocity. Tristatic receiving systems enable the vertical and horizontal components of the mean plasma drift to be determined. Radars operate at frequencies of 40–1300 MHz using either pulse or c.w. transmissions. Transmitter peak powers of the order of 1 MW, complex antenna arrays (*Figure 11.3*), and sophisticated data processing procedures are needed. Ground clutter limits the lowest heights that can be investigated to around 100 km.

Electron densities, ion temperatures and ionic composition out to several Earth radii may be studied using naturally occurring whistlers originating in lightning discharges. These are dispersed audio-frequency trains of energy, ducted through the ionosphere and then propagated backwards and forwards along the Earth's magnetic-field lines to conjugate points in the opposite hemisphere. Whistler dispersions may be observed either at the ground or in satellites.[43]

Continuous wave and pulsed signals, transmitted and received at ground-based terminals, may be used in a variety of ways to study irregularities or fluctuations in ionisation. Cross-correlation analyses of the amplitudes on three spaced receivers, of pulsed signals of fixed frequency reflected from the ionosphere at near vertical incidence, give the direction and velocity of the horizontal component of drift.[44] The heights, patch sizes and incidence of F-region irregularities responsible for oblique-path forward-scatter propagation at frequencies around 50 MHz may be investigated by means of highly-directional antennas and from signal transit times.[14] Measurements of the Doppler frequency variations of signals from stable c.w. transmitters may be used to study: (1) ionisation enhancements in the E- and F-regions associated with solar flares; (2) travelling ionospheric disturbances;[45] and (3) the frequency-dispersion component of the ionospheric channel-scattering function.[46]

Sporadic-E and F-region irregularities associated with visual aurorae may be examined by pulsed-radar techniques over a wide range of frequencies from about 6 MHz to 3000 MHz. They may also be investigated using c.w. bistatic systems in which the transmitter and receiver are separated by several hundred kilometres. Since the irregularities are known to be elongated and aligned along the direction of the Earth's magnetic field and since at the higher frequencies efficient scattering can only occur under restricted conditions, the scattering centres may readily be located. Using low-power v.h.f. beacon transmitters, this technique has proved very popular with radio amateurs. Pulsed meteor radars incorporating Doppler measurements indicate the properties and movements of meteor trains.[25]

Two other oblique-path techniques, giving information on the regular ionospheric regions, are high-frequency ground backscatter sounding and variable-frequency oblique sounding. The former uses a nearby transmitter and receiver, and record interpretation generally involves identifying the skip distance (see *Figure 11.1*) where the signal returns are enhanced because of ray convergence. It is important to use antennas with azimuthal beamwidths of only a few degrees to minimise the ground area illuminated. Long linear antenna arrays with beam slewing, and circularly-disposed banks of log-periodic antennas with monopulsing are used. Oblique-incidence sounders are adaptions of vertical-incidence ionosondes with the transmitter and receiver controlled from stable synchronised sources. Atlases of characteristic records obtained from the two types of sounder under different ionospheric conditions have been produced. Mean models of the ionosphere over the sounding paths may be deduced by matching measured data with ray-tracing results.[47]

So far, no mention has been made of the height region below about 100 km. As already noted, the D-region is characterised by a complex structure and high collision frequencies which lead to large daytime absorption of h.f. and m.f. waves. This absorption may be measured using fixed-frequency vertical-incidence pulses,[48] or by monitoring c.w. transmissions at ranges of 200–500 km, where there is no ground-wave component and the dominant signals are reflected from the E-region by day and the sporadic-E layer by night. There is then little change in the raypaths from day to night so that, assuming night absorption can be neglected, daytime reductions in amplitude are a measure of the prevailing absorption. Multifrequency absorption data give information on the height distributions of electron density.[49] Auroral absorption is often too great to be measured in such ways, but special instruments known as riometers can be used.[50] These operate at a frequency around 30 MHz and record changes in the incident cosmic noise at the ground caused by ionospheric absorption.

D-region electron densities and collision frequencies may be inferred from oblique or vertical-path measurements of signal amplitude, phase, group-path delay and polarisation at frequencies of 10 Hz to 100 kHz, with atmospherics as the signal sources at the lower frequencies. Vertically radiated signals in the frequency range 1.5–6 MHz suffer weak partial reflections from heights of 75–90 km. Measurements of the reflection coefficients of both the ordinary and extraordinary waves, which can be of the order of 10^{-5}, enable electron density and collision-frequency data to be deduced.[51] Pulsed signals with high transmitter antenna systems and very sensitive receivers are needed. As well as *in-situ* probes in rockets, there are a wide range of other schemes for determining electron density and collision frequency, involving the study of wave-fields radiated between the ground and a rocket. These use combinations of frequencies in the v.l.f.–v.h.f. range and include the measurement of differential-Doppler frequency, absorption, differential phase, propagation time and Faraday rotation.

Theory shows that signals propagated via the ionosphere can become cross-modulated by high-power interfering signals which heat the plasma electrons through which the wanted signals pass. This heating causes the electron-collision frequency, and therefore the amplitude of the wanted signal to fluctuate at the modulation frequency of the interfering transmitter. Investigations of this phenomenon (known as the Luxembourg effect after the first identified interfering transmitter) usually employ vertically transmitted and received wanted pulses, modulated by a distant disturbing transmitter radiating synchronised pulses at half the repetition rate. Changes in signal amplitude and phase between successive pulses are measured, and by altering the relative phase of the two transmitters, the height at which the cross-modulation occurs can be varied. Such data enable the height distributions of electron density in the D-region to be determined.[52]

Using a laser radar (lidar), the intensity of the light back-scattered by the atmospheric constituents at heights above 50 km gives the height distributions of neutral-air density and the temporal and spatial statistics of high-altitude atmospheric aerosols. Minor atmospheric constituents may be detected with tunable dye lasers from their atomic and molecular-resonance scattering.

11.6 Propagation prediction procedures

Long-term predictions based on monthly median ionospheric data are required for the circuit planning of v.l.f.–h.f. ground-based systems. Estimates of raypath launch and arrival angles are needed for antenna design, and of the relationship between transmitter power and received field strength at a range of frequencies, so that the necessary size of transmitter and its frequency coverage can be determined. Since there are appreciable day to day changes in the electron densities in the F2-region, in principle short-term predictions based on ionospheric probing measurements or on correlations with geophysical indices should be of great value for real-time frequency management. In practice, however, aside from the technical problems of devising schemes of adequate accuracy: (*a*) not all systems are frequency agile (e.g. broadcasting); (*b*) effective schemes may require two-way transmissions; and (*c*) only assigned frequencies may be used. An alternative to short-term predictions is real-time channel sounding; certain procedures involve a combination of the two techniques.

11.6.1 Long-term predictions

The first requirement of any long-term prediction is a model of the ionosphere. At v.l.f. waves propagate between the Earth and the lower boundary of the ionosphere at heights of 70 km by day and 90 km by night as if in a two-surface waveguide. Very low frequency field-strength predictions are based on a full-wave theory that includes diffraction and surface-wave propagation. For paths beyond 1000 km range only three or fewer waveguide modes need be considered. A general equation gives field strength as a function of range, frequency, ground-electrical properties and the ionospheric reflection height and reflection coefficients.[53] Unfortunately the reflection coefficients vary in a complex way with electron density and collision frequency, the direction and strength of the Earth's magnetic field, wave frequency and angle of incidence, so that in the absence of accurate D-region electron-density data, estimates are liable to appreciable error. At l.f. propagation is more conveniently described by wave-hop theory in terms of component waves with different numbers of hops. As at v.l.f. reflection occurs at the base of the ionosphere and the accuracy of the field-strength prediction is largely determined by uncertainties in ionospheric models and reflection coefficients. Medium-frequency signals penetrate the lower ionosphere and are usually reflected from heights of 85–100 km, except over distances of less than 500 km by night when reflection may be from the F-region. Large absorptions occur near the height of reflection and daytime signals are very weak. It is now realised that because of the uncertainties in ionospheric models, signal-strength predictions are best based on empirical equations fitted to measured signal-strength data for other oblique paths.

Prediction schemes for h.f. tend to be complicated because they must assess the active modes and elevation angles; these vary markedly with ionospheric conditions and transmitter frequency. Equations are available for the raypaths at oblique incidence through ionospheric models composed of separate layers, each with a parabolic distribution of electron density with height.[54] They are employed in one internationally used prediction scheme,[55] with the parameters of the parabolas determined from numerical prediction maps of the vertical-incidence ionospheric characteristics, as given by data from the world network of ionosondes[56] (see *Figure 11.4*). Calculations over a fixed path for a range of frequencies indicate the largest frequency (the basic m.u.f. or maximum usable frequency) that propagates via a given mode. Assuming some statistical law for the day to day variability of the parameters of the model they also give the *availability*, which is the fraction of days that the mode can exist. Received signal strengths are then determined in terms of the transmitter power and a number of transmission loss and gain factors. These include transmitting and receiving antenna gain, spatial attenuation, ray convergence gain, absorption, intermediate-path ground-reflection losses and polarisation-coupling losses. Predictions may be further extended by including estimates of the day to day variability in signal intensity. Calculations are prohibitively lengthy without computing aids and a number of computerised prediction schemes have been produced. By means of estimates of background noise intensity, and from the known required signal/noise ratio, the mode reliability may also be determined. This is the fraction of the days that the signals are received with adequate strength. For some systems involving fast data transmission, predictions of the probability of multipath, with two or more modes of specified comparable amplitude with propagation delays differing by less than some defined limit, are also useful and can be made.

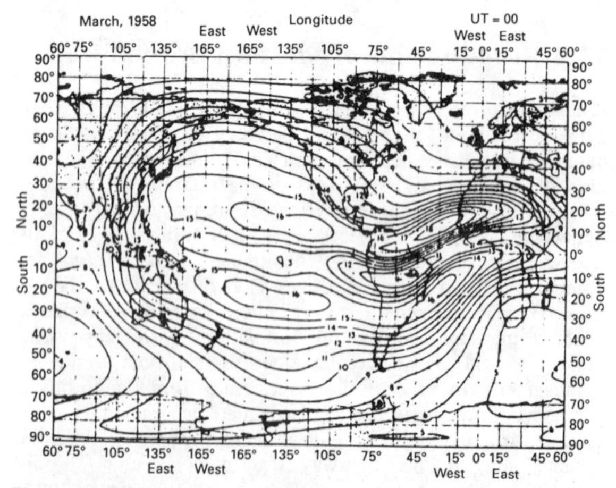

Figure 11.4 Predicted median, foF2, MHz for 00h, UT in March 1958 (Reproduced by permission of the Institute for Telecommunication Sciences, Boulder, USA)

11.6.2 Short-term predictions and real-time channel sounding

Some limited success has been achieved in the short-term prediction of the ionospheric characteristics used to give the parameters of the ionospheric models needed for h.f. performance assessment. Schemes are based either on spatial or temporal extrapolation of near real-time data or on correlations with magnetic activity indices. Regression statistics have been produced for the change in the maximum plasma frequency of the F2-region (foF2) with local magnetic activity index K, and other work is concerned with producing joint correlations with K and with solar flux.

In principle at h.f. the most reliable, although costly, way of ensuring satisfactory propagation over a given path and of optimising the choice of transmission frequency involves using an oblique-incidence sounder over the actual path; in practice, however, sounder systems are difficult to deploy operationally, require expert interpretation of their data, lead to appreciable spectrum pollution, and give much redundant information. Some schemes involve low-power channel monitoring of the phase-path stability on each authorised frequency, to ensure that at all times the best available is used. Real-time sounding on one path can aid performance predictions for another. Examples include ray tracing through mean ionospheric models simulated from measured backscatter or oblique-incidence soundings. Many engineers operating established radio circuits prefer, for frequency management, to rely on past experience, rather than to use predictions. This is not so readily possible for mobile applications. Real-time sounding schemes involving ground transmissions on a range of frequencies to an aircraft, but only single-frequency transmission in the reverse direction, have proved successful.[57]

11.7 The troposphere

The influence of the lower atmosphere, or troposphere, on the propagation of radio waves is important in several respects. At all frequencies above about 30 MHz refraction and scattering, caused by local changes in atmospheric structure, become significant—especially in propagation beyond the normal horizon. In addition, at frequencies above about 5 GHz, absorption in oxygen and water vapour in the atmosphere is important at certain frequencies corresponding to molecular absorption lines. An understanding of the basic characteristics of these effects is thus essential in the planning of very high frequency communication systems. The main features of tropospheric propagation are summarised from a practical point of view as follows.

There are two general problems: firstly, the influence of the troposphere on the *reliability* of a communication link. Here attention is concentrated on the weak signals which can be received for a large percentage of the time, say 99.99%. Secondly, it is necessary to consider the problem of *interference* caused by abnormal propagation and unusually strong, unwanted signals of the same frequency as the wanted transmission.

In both these aspects of propagation, the radio refractive index of the troposphere plays a dominant role. This parameter depends on the pressure, temperature and humidity of the atmosphere. Its vertical gradient and local fluctuations about the mean value determine the mode of propagation in many important practical situations. Hence the interest in the subject of radio meteorology, which seeks to relate tropospheric structure and radio-wave propagation. In most ground-to-ground systems the height range 0–2 km above the Earth's surface is the important region, but in some aspects of Earth–space transmission, the meteorological structure at greater heights is also significant.

11.7.1 Historical background

Although some experiments on ultra-short-wave techniques were carried out by Hertz and others more than 90 years ago, it was only after about 1930 that any systematic investigation of tropospheric propagation commenced. For a long time it was widely believed that at frequencies above about 30 MHz transmission beyond the geometric horizon would be impossible. However, this view was disputed by Marconi as early as 1932. He demonstrated that, even with relatively low transmitter powers, reception over distances several times the optical range was possible. Nevertheless, theoreticians continued for several years to concentrate on studies of diffraction of ultra-short waves around the Earth's surface. However, their results were found to over-estimate the rate of attenuation beyond the horizon. To correct for this disparity, the effect of refraction was allowed for by assuming a process of diffraction around an Earth with an effective radius of 4/3 times the actual value. In addition, some experimental work began on the effect of irregular terrain and the diffraction caused by buildings and other obstacles.

However, it was only with the development of centimetric radar in the early years of World War II that the limitations of earlier concepts of tropospheric propagation were widely recognised. For several years attention was concentrated on the role of unusually strong refraction in the surface layers, especially over water, and the phenomenon of trapped propagation in a *duct*. It was shown experimentally and theoretically that in this mode the rate of attenuation beyond the horizon was relatively small. Furthermore, for a given height of duct or surface layer having a very large, negative, vertical gradient of refractive index, there was a critical wavelength above which trapping did not occur; a situation analogous to that in waveguide transmission.

Again however it became apparent that further work was required to explain experimental observations. The increasing use of v.h.f., and later u.h.f., for television and radio communication emphasised the need for a more comprehensive approach on beyond-the-horizon transmission. The importance of refractive-index variations at heights of the order of a kilometre began to be recognised and studies of the correlation between the height-variation of refractive index and field strength began in several laboratories.

With the development of more powerful transmitters and antennae of very high gain it proved possible to establish communication well beyond the horizon even in a 'well-mixed' atmosphere with no surface ducts or large irregularities in the height-variation of refractive index. To explain this result, the concept of *tropospheric scatter* was proposed. The trans-horizon field was assumed to be due to incoherent scattering from the random, irregular fluctuations in refractive index produced and maintained by turbulent motion. This procedure has dominated much of the experimental and theoretical work since 1950 and it certainly explains some characteristics of troposphere propagation. However, it is inadequate in several respects. It is now known that some degree of stratification in the troposphere is more frequent than was hitherto assumed. The possibility of reflection from a relatively small number of layers or *sheets* of large vertical gradient must be considered, especially at v.h.f. At u.h.f. and s.h.f. strong scattering from a 'patchy' atmosphere, with local regions of large variance in refractive index filling only a fraction of the common volume of the antenna beams, is probably the mechanism which exists for much of the time.

The increasing emphasis on microwaves for terrestrial and space systems has recently focused attention on the effects of precipitation on tropospheric propagation. While absorption in atmospheric gases is not a serious practical problem below 40 GHz, the attenuation in rain and wet snow can impair the performance of links at frequencies of about 10 GHz and above. Moreover, scattering from precipitation may prove to be a significant factor in causing interference between space and terrestrial systems sharing the same frequency. The importance of interference-free sites for Earth stations in satellite links has also stimulated work on the shielding effect of hills and mountains. In addition, the use of large antennae of high gain in space systems requires a knowledge of refraction effects (especially at low angles of elevation), phase variations over the wavefront, and the associated effects of scintillation fading and gain degradation. Particularly at the higher microwave frequencies, thermal noise radiated by absorbing regions of the troposphere (rain, clouds, etc.) may be significant in space communication. Much of the current research is therefore being directed towards a better understanding of the spatial structure of precipitation.

In addition, there has been a recent revival of interest in the effects of terrain (hills, buildings, etc.) at v.h.f., u.h.f. and s.h.f., especially in relation to the increasing requirements of the mobile services.

11.8 Survey of propagation modes in the troposphere

Figure 11.5 illustrates qualitatively the variation of received power with distance in a homogeneous atmosphere at frequencies above about 30 MHz. For antenna heights of a

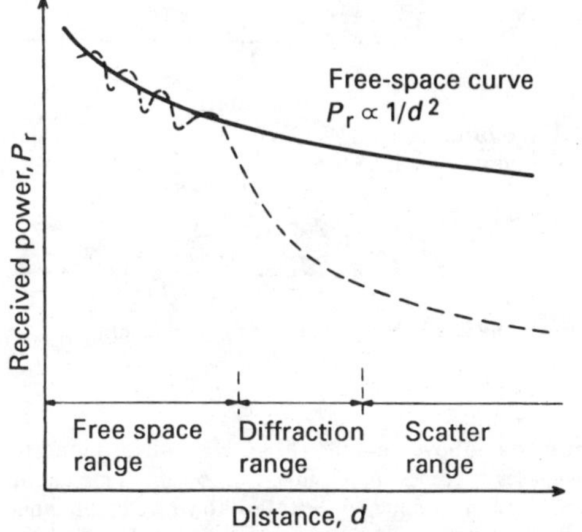

Figure 11.5 Tropospheric attenuation as a function of distance in a homogeneous atmosphere. Direct and ground-reflected rays interfere in the free-space range; obstacle-diffraction effects predominate in the diffraction range; and refractive-index variations are important in the scatter range

wavelength or more, the propagation mode in the free-space range is a *space wave* consisting of a direct and a ground-reflected ray. For small grazing angles the reflected wave has a phase change of nearly 180° at the Earth's surface, but imperfect reflection reduces the amplitude below that of the direct ray. As the path length increases, the signal strength exhibits successive maxima and minima. The most distant maximum will occur where the path difference is $\lambda/2$, where λ is the wavelength.

The range over which the space-wave mode is dominant can be determined geometrically allowing for refraction effects. For this purpose we can assume that the refractive index, n, decreases linearly by about 40 parts in 10^6 (i.e. 40 N units) in the first kilometre. This is the equivalent to increasing the actual radius of the Earth by a factor of 4/3 and drawing the ray paths as straight lines. The horizon distance d, from an antenna at height h above an Earth of effective radius a is

$$d=(2ah)^{1/2} \qquad (11.1)$$

For two antennae 100 m above ground the total range is about 82 km, 15% above the geometric value.

Beyond the free-space range, diffraction around the Earth's surface and its major irregularities in terrain is the dominant mode, with field strengths decreasing with increasing frequency and being typically of the order of 40 dB below the free space value at 100 km at v.h.f. for practical antenna heights. As the distance increases, the effect of reflection or scattering from the troposphere increases and the rate of attenuation with distance decreases. In an actual inhomogeneous atmosphere the height-variation of n is the dominant factor in the *scatter zone* as illustrated in *Figure 11.6*. However, in practice the situation is rarely as simple as that indicated by these simple models.

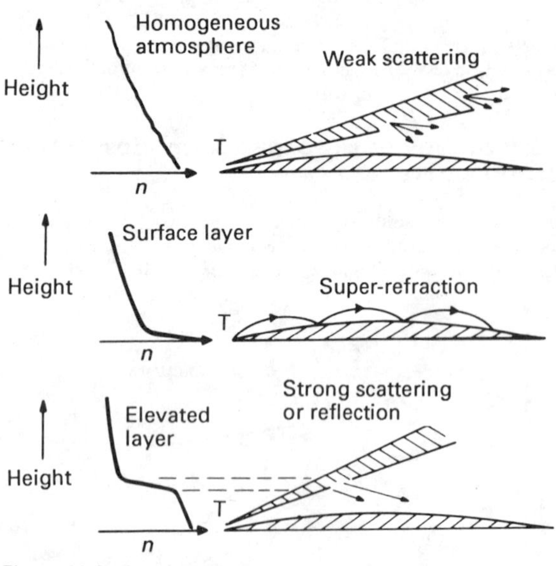

Figure 11.6 Tropospheric propagation modes and height-variation index, n

At frequencies above about 40 GHz, absorption in atmospheric gases becomes increasingly important. This factor may determine system design and the extent to which co-channel sharing is possible; for example, between terrestrial and space communication services. There are strong absorption lines due to oxygen at 60 and 119 GHz, with values of attenuation, at sea level, of the order of 15 and 2 dB km^{-1} respectively. At 183 GHz, a water vapour line has an attenuation of about 35 dB km^{-1}. Between these lines there are 'windows' of relatively low attenuation; e.g. around 35, 90 and 140 GHz. These are the preferred bands for future exploitation of the millimetric spectrum; for example, for short-range communication systems or radar. Further details are given in Report 719 of the CCIR.

11.8.1 Ground-wave terrestrial propagation

When the most distant maximum in *Figure 11.5* occurs at a distance small compared with the optical range, it is often permissible to assume the Earth flat and perfectly reflecting, particularly at the low-frequency end of the v.h.f. range. The space-wave field E at a distance d is then given by

$$E=(90W^{1/2}h_t h_r)/\lambda d^2 \qquad (11.2)$$

where W is the power radiated from a $\lambda/2$ dipole, and h_t and h_r are the heights of the transmitting and receiving antennae respectively.

The effects of irregular terrain are complex. There is some evidence that, for short, line-of-sight links, a small degree of surface roughness increases the field strength by eliminating the destructive interference between the direct and ground-reflected rays. Increasing the terrain irregularity then reduces the field strength, particularly at the higher frequencies, as a result of shadowing, scattering and absorption by hills, buildings and trees. However, in the particular case of a single, obstructing ridge visible from both terminals it is sometimes possible to receive field strengths greater than those over level terrain at the same distance. This is the so-called *obstacle gain*.

In designing microwave radio-relay links for line-of-sight operation it is customary to so locate the terminals that, even with unfavourable conditions in the vertical gradient of refractive index (with *sub-refraction* effects decreasing the effective radius of the Earth), the direct ray is well clear of any obstacle. However, in addition to multipath fading caused by a ground-reflected ray, it is possible for line-of-sight microwave links to suffer fading caused by multi-path propagation via strong scattering or abnormal refraction in an elevated layer in the lower troposphere. This situation may lead to a significant reduction in usable bandwidth and to distortion, but the use of spaced antennae (space diversity) or different frequencies (frequency diversity) can reduce these effects. Even in the absence of well defined layers, scintillation-type fading may occasionally occur at frequencies of the order of 30–40 GHz on links more than say 10 km long. The development of digital systems has further emphasised the importance of studies of the effect of refractive index variation on distortion, bandwidth and error rate.

As a guide to the order of magnitude of multipath fading, Report 338 of the CCIR gives the following values for a frequency of 4 GHz, for the worst month of the year, for average rolling terrain in northwest Europe:

0.01% of time; path length 20 km; 11 dB or more below free space
path length 40 km; 23 dB or more below free space

Over very flat, moist ground the values will be greater and, for a given path length, will tend to increase with frequency. But at about 10 GHz and above, the effect of precipitation will generally dominate system reliability.

The magnitude of attenuation in rain can be estimated theoretically and the reliability of microwave links can then be forecast from a knowledge of rainfall statistics. But the divergence between theory and experiment is often considerable. This is partly due to the variation which can occur in the drop-size distribution for a given rainfall rate. In addition, many difficulties remain in estimating the intensity and spatial characteristics of rainfall for a link. This is an important practical problem in relation to the possible use of *route diversity* to minimise the effects of absorption fading. Experimental results show that for very high reliability (i.e. for all rainfall rates less than say 50–100 mm/h in temperate climates) any terrestrial link operating at frequencies much above 30 GHz must not exceed say 10 km in length. It is possible, however, to design a system with an alternative route so that by switching between the two links the worst effects of localised, very heavy rain can be avoided. The magnitude of attenuation in rain is shown in *Figure 11.7(a)* and the principle of route diversity is illustrated in *Figure 11.7(b)*. For temperate climates, the diversity gain (i.e. the difference between the attenuation, in dB, exceeded for a specified, small percentage of time on a single link and that exceeded simultaneously on two parallel links) varies as follows:

(1) It tends to decrease as the path length increases from 12 km, for a given percentage of time and for a given lateral path separation.
(2) It is generally greater for a spacing of 8 km than for 4 km, though an increase to 12 km does not provide further improvement.
(3) It is not strongly dependent on frequency in the range 20–40 GHz for a given geometry.

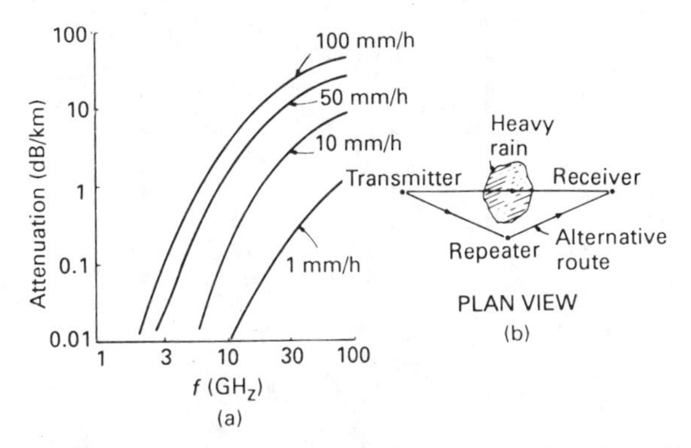

Figure 11.7 (a) Attenuation in rain. (b) The application of route diversity to minimise effects of fading

The main problem in the v.h.f. and u.h.f. broadcasting and mobile services (apart from prediction of interference) is to estimate the effect of irregularities in terrain and of varying antenna height on the received signal. The site location is of fundamental importance. Prediction of received signal strengths, on a statistical basis, has been made using a parameter, Δh, which characterises terrain roughness (see CCIR Recommendation 370 and Reports 239 and 567). However, there is considerable path-to-path variability, even for similar Δh values. Especially in urban areas, screening and multipath propagation due to buildings are important. Moreover, in such conditions—and especially for low heights of receiving antenna—depolarization effects can impair performance of orthogonally polarized systems sharing a common frequency. At the higher u.h.f. frequencies, attenuation due to vegetation (e.g. thick belts of trees) is beginning to be significant.

11.8.2 Beyond-the-horizon propagation

Although propagation by surface or elevated layers (see *Figure 11.6*) cannot generally be utilised for practical communication circuits, these features remain important as factors in co-channel interference. Considerable theoretical work, using waveguide mode theory, has been carried out on duct propagation and the results are in qualitative agreement with experiment. Detailed comparisons are difficult because of the lack of knowledge of refractive index structure over the whole path, a factor common to all beyond-the-horizon experiments. Nevertheless, the theoretical predictions of the maximum wavelength trapped in a duct are in general agreement with practical experience. These values are as follows:

λ *(max) in cm*	*Duct height in m*
1	5
10	25
100	110

Normal surface ducts are such that complete trapping occurs only at centimetric wavelengths. Partial trapping may occur for the shorter metric wavelengths. Over land the effects of irregular terrain and of thermal convection (at least during the day) tend to inhibit duct formation. For a ray leaving the transmitter horizontally, the vertical gradient of refractive index must be steeper than -157 parts in 10^6 per kilometre.

Even when super-refractive conditions are absent, there remains considerable variability in the characteristics of the received signal usable in the 'scatter' mode of communication. This variability is conveniently expressed in terms of the transmission loss, which is defined as $10 \log (P_t/P_r)$, where P_t and P_r are the transmitted and received powers respectively. In scatter propagation, both slow and rapid variations of field strength are observed. Slow fading is the result of large-scale changes in refractive conditions in the atmosphere and the hourly median values below the long-term median are distributed approximately log-normally with a standard deviation which generally lies between 4 and 8 decibels, depending on the climate. The largest variations of transmission loss are often seen on paths for which the receiver is located just beyond the diffraction region, while at extreme ranges the variations are less. The slow fading is not strongly dependent on the radio frequency. The rapid fading has a frequency of a few fades per minute at lower frequencies and a few hertz at u.h.f. The superposition of a number of variable incoherent components would give a signal whose amplitude was Rayleigh-distributed. This is found to be the case when the distribution is analysed over periods of up to five minutes. If other types of signal form a significant part of that received, there is a modification of this distribution. Sudden, deep and rapid fading has been noted when a frontal disturbance passes over a link. In addition, reflections from aircraft can give pronounced rapid fading.

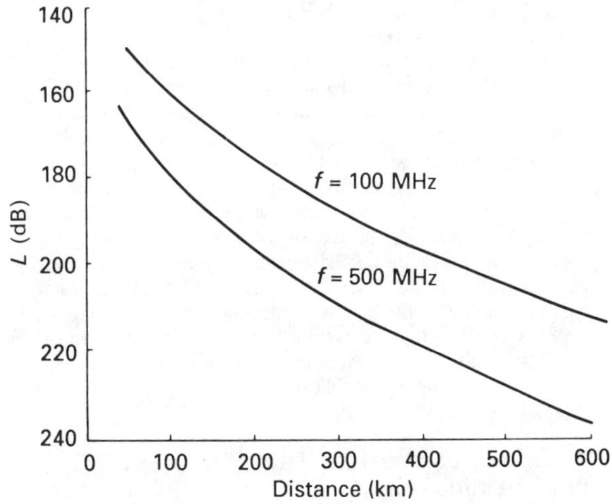

Figure 11.8 Median transmission loss, L, between isotropic antennas in a temperate climate and over an average rolling terrain. The height of the transmitting antenna is 40 m, and the height of the receiving antenna is 10 m

The long-term median transmission loss relative to the free-space value increases approximately as the first power of the frequency up to about 3 GHz. Also, for most temperate climates, monthly median transmission losses tend to be higher in winter than in summer, but the difference diminishes as the distance increases. In equatorial climates, the annual and diurnal variations are generally smaller. The prediction of transmission loss, for various frequencies, path lengths, antenna heights, etc., is an important practical problem. An example of the kind of data required is given in *Figure 11.8*.

At frequencies above 10 GHz, the heavy rain occurring for small percentages of the time causes an additional loss due to absorption, but the accompanying scatter from the rain may partly offset the effect of absorption.

11.8.3 Physical basis of tropospheric scatter propagation

Much effort has been devoted to explaining the fluctuating trans-horizon field in terms of scattering theory based on statistical models of turbulent motion. The essential physical feature of this approach is an atmosphere consisting of irregular *blobs* in random motion which in turn produce fluctuations of refractivity about a stationary mean value. Using this concept, some success has been achieved in explaining the approximate magnitude of the scattered field but several points of difficulty remain. There is now increasing evidence, from refractometer and radar probing of the troposphere, that some stratification of the troposphere is relatively frequent.

By postulating layers of varying thickness, horizontal area and surface roughness, and of varying lifetime it is possible in principle to interpret many of the features of tropospheric propagation. Indeed, some experimental results (e.g. the small distance-dependence of v.h.f. fields at times of anomalous propagation) can be explained by calculating the reflection coefficient of model layers of constant height and with an idealised height-variation of refractive index such as half-period sinusoidal, exponential, etc. The correlation between field strength and layer height has also been examined and some results can be explained qualitatively in terms of *double-hop* reflection from extended layers. Progress in ray-tracing techniques has also been made. Nevertheless, the problems of calculating the field strength variations on particular links remain formidable, and for many practical purposes statistical and empirical techniques for predicting link performance remain the only solution. (See CCIR Report 238.)

Other problems related to fine structure are space and frequency diversity. On a v.h.f. scatter link with antennae spaced normal to the direction of propagation, the correlation coefficient may well fall to say 0.5 for spacings of 5–30λ in conditions giving fairly rapid fading. Again, however, varying meteorological factors play a dominant role. In frequency diversity, a separation of say 3 or 4 MHz may ensure useful diversity operation in many cases, but occasionally much larger separations are required. The irregular structure of the troposphere is also a cause of gain degradation. This is the decrease in actual antenna gain below the ideal free-space value. Several aspects of the irregular refractive-index structure contribute to this effect and its magnitude depends somewhat on the time interval over which the gain measurement is made. Generally, the decrease is only significant for gains exceeding about 50 dB.

11.9 Tropospheric effects in space communications

In space communication, with an Earth station as one terminal, several problems arise due to refraction, absorption and scattering effects, especially at microwave frequencies. For low angles of elevation of the Earth station beam, it is often necessary to evaluate the refraction produced by the troposphere, i.e. to determine the error in observed location of a satellite. The major part of the bending occurs in the first two kilometres above ground and some statistical correlation exists between the magnitude of the effect and the refractive index at the surface. For high-precision navigation systems and very narrow beams it is often necessary to evaluate the variability of refraction effects from measured values of the refractive index as a function of height. A related phenomenon important in tracking systems is the phase distortion in the wave-front due to refractive index fluctuations, a feature closely linked with gain degradation. This phase distortion also affects the stability of frequencies transmitted through the troposphere.

Absorption in clear air may affect the choice of frequencies, above about 40 GHz, to minimise co-channel interference. *Figure 11.9* shows the zenith attenuation from sea level for an average clear atmosphere as a function of frequency. It illustrates the 'window' regions mentioned in the survey of propagation modes.

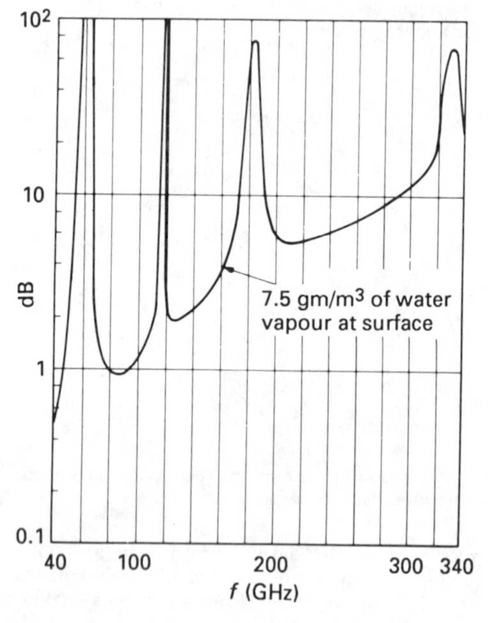

Figure 11.9 Zenith attenuation (dB) in clear air

From an altitude of 4 km, the values would be about one-third of those shown. This indicates the potential application of frequencies above 40 GHz for communication on paths located above the lower layers of the troposphere.

Clouds produce an additional loss which depends on their liquid-water content. Layer-type cloud (stratocumulus) will not cause additional attenuation of more than about 2 dB, even at 140 GHz. On the other hand, cumulonimbus will generally add several decibels to the total attenuation, the exact value depending on frequency and cloud thickness.

Absorption in precipitation (see *Figure 11.7(a)*) has already been mentioned in relation to terrestrial systems. Water drops attenuate microwaves both by scattering and by absorption. If the wavelength is appreciably greater than the drop-size then the attenuation is caused almost entirely by absorption. For rigorous calculations of absorption it is necessary to specify a drop-size distribution; but this, in practice, is highly variable and consequently an appreciable scatter about the theoretical value is found in experimental measurements. Moreover, statistical information on the vertical distribution of rain is very limited. This makes prediction of the reliability of space links difficult and emphasises the value of measured data. Some results obtained using the Sun as an extraterrestrial source are shown in *Figure 11.10*.

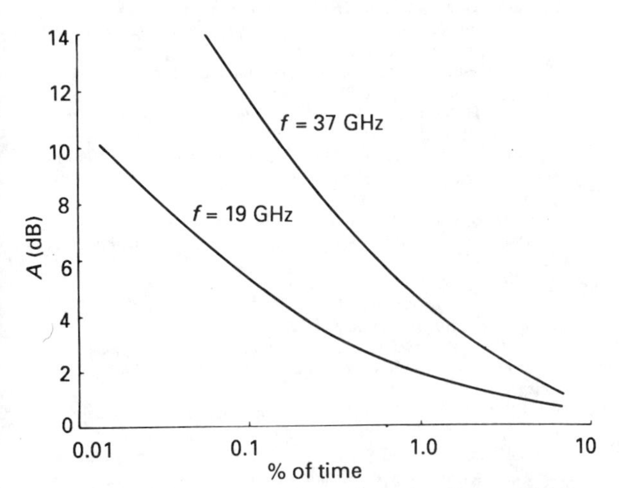

Figure 11.10 Measured probability distribution of attenuation A on earth–space path at 19 GHz and 37 GHz (southeast England: Elevation angles 5° to 40°: data from solar-tracking radiometers)

Scatter from rain (and ice crystals at and above the freezing level in the atmosphere) can cause significant interference on co-channel terrestrial and space systems even when the beams from the two systems are not directed towards each other on a great-circle path; such scattering being – to a first approximation – isotropic. It may also be significant in the case of two Earth Stations with beams elevated well above the horizon: for example, with one a feeder link transmitter to a broadcasting satellite and the other a receiver in the fixed-satellite service. This mode of interference may be dominant when hills or other obstacles provide some 'site shielding' against signals arriving via a ducting mode.

Because precipitation (and to a smaller extent the atmospheric gases) absorb microwaves, they also radiate thermal-type noise. It is often convenient to specify this in terms of an *equivalent black-body temperature* or simply *noise temperature* for an antenna pointing in a given direction. With radiometers and low-noise receivers it is now possible to measure this tropospheric noise and assess its importance as a factor in limiting the performance of a microwave Earth–space link. For a complete solution, it is necessary to consider not only direct radiation into the main beam but also ground-reflected radiation, and emission from the ground itself, arriving at the receiver via side and back lobes. From the meteorological point of view, radiometer soundings (from the ground, aircraft, balloons or satellites) can provide useful information on tropospheric and stratospheric structure. Absorption in precipitation becomes severe at frequencies above about 30 GHz and scintillation effects also increase in importance in the millimetre range. However, for space links in or near the vertical direction the system reliability may be sufficient for practical application even at wavelengths as low as 3–4 mm. Moreover, spaced receivers in a site-diversity system can be used to minimise the effects of heavy rain.

In recent years, extensive studies of propagation effects (attenuation, scintillation, etc.) have been carried out by direct measurements using satellite transmissions. Special emphasis has been given to frequencies between 10 and 30 GHz, in view of the effect of precipitation on attenuation and system noise. Details are given in Report 564 of the CCIR. For typical elevation angles of 30°–45°, the total attenuation exceeded for 0.01% of the time has values of the following order:

f=12 GHz; 5 dB in temperate climates (e.g. northwest Europe)
20 dB in tropical climates (e.g. Malaysia)
10 dB in East Coast USA (Maryland)
f=20 GHz; 10 dB (northwest Europe)

Site-diversity experiments using satellite transmissions show that site spacings of the order of 5–10 km can give a useful improvement in reliability. However, the improvement may depend on the site geometry and on topographical effects. At frequencies above say 15 GHz, the advantage of site diversity may be quite small if the sites are so chosen that heavy rain in, for example, frontal systems tends to affect both sites simultaneously.

Frequency re-use is envisaged, in space telecommunication systems, by means of orthogonal polarisation. But this technique is restricted by depolarisation due to rain and ice clouds and—to a lesser extent—by the system antennas. Experimental data on polarisation distortion, obtained in satellite experiments, are given in CCIR Report 564.

11.10 Techniques for studying tropospheric structure

The importance of a knowledge of the structure of the troposphere in studies of propagation is clearly evident in the above sections. The small-scale variations in refractive index and in the intensity of precipitation are two important examples. They form part of the general topic of *tropospheric probing*.

Much useful information on the height-variation of refractive index can be obtained from the radio-sondes carried on free balloons and used in world-wide studies of meteorological structure. However, for many radio applications these devices do not provide sufficient detail. To obtain this detail instruments called refractometers have been developed, mainly for use in aircraft, on captive balloons or on tall masts. They generally make use of a microwave cavity for measuring changes in a resonance frequency, which in turn is related to the refractive index of the enclosed air. Such refractometers are robust, rapid-response instruments which have been widely used as research tools, though they have yet to be developed in a form suitable for widespread, routine use.

High-power, centimetric radar is also a valuable technique. By its use it is possible to detect layers or other regions of strong scatter in the troposphere, and to study their location and structure. Joint radar-refractometer soundings have proved of special interest in confirming that the radar does indeed detect irregularities in clear-air structure. The application of radar in precipitation studies is, of course, a well-known and widely used technique in meteorology; although to obtain the detail and precision necessary for radio applications requires careful refinements in technique.

Optical radar (lidar) and acoustic radar have also been used to probe the troposphere, although the information they provide is only indirectly related to radio refractive index.

The millimetre and sub-millimetre spectrum, as yet not exploited to any significant degree for communications, is nevertheless a fruitful region for tropospheric probing. In particular, the presence of several absorption lines (in water vapour, oxygen and minor constituents such as ozone) makes it possible to study the concentration and spatial distribution of these media. Near the ground, direct transmission experiments are feasible; for example, to study the average water-vapour concentration along a particular path. In addition, it is possible to design radiometers for use on the ground, in an aircraft or in a satellite, which will provide data on the spatial distribution of absorbing atmospheric constituents by measurement of the emission noise they radiate. This topic of *remote probing* is one exciting considerable current interest in both radio and meteorology.

References

1 BUCHAU, J., 'Instantaneous versus averaged ionosphere', *Air Force Surveys in Geophysics No. 241* (*Air Force Systems Command, United States Air Force*), **1** (1972)
2 MULDREW, D. B., 'F-layer ionisation troughs deduced from Alouette data', *J. Geophys. Res.*, **70**, 2635 (1965)
3 DAVIES, K., *Ionospheric Radio Propagation*, Monograph 80, National Bureau of Standards, Washington (1965)
4 RATCLIFFE, J. A., *Sun, Earth and Radio—an Introduction to the Ionosphere and Magnetosphere*, Weidenfeld and Nicolson, London (1970)
5 RISHBETH, H. and GARRIOTT, O. K., *Introduction to Ionospheric Physics*, Academic Press, London (1969)
6 DAVIES, K., *Ionospheric Radio Waves*, Blaisdell, Waltham, Mass. (1969)
7 RATCLIFFE, J. A., *An Introduction to the Ionosphere and Magnetosphere*, Cambridge University Press, Cambridge (1972)
8 RISHBETH, H., 'Thermospheric winds and the F-region: a review', *J. Atmosph. Terr. Phys.*, **34**, 1 (1972)
9 DUNCAN, R. A., 'The equatorial F-region of the ionosphere', *J. Atmosph. Terr. Phys.*, **18**, 89 (1960)
10 HARTZ, T. R., 'The general pattern of auroral particle precipitation and its implications for high latitude communication systems', *Ionospheric Radio Communications*, ed. K. Folkestad, Plenum, New York, 9 (1968)
11 BAILEY, D. K., 'Polar cap absorption', *Planet. Space Sci.*, **12**, 495 (1964)
12 MATSUSHITA, S., 'Geomagnetic disturbances and storms', *Physics of Geomagnetic Phenomena*, ed. Matsushita, S. and Campbell, W. H., Academic Press, London, 793 (1967)
13 SMITH, E. K. and MATSUSHITA, S., *Ionospheric Sporadic-E*, Macmillan, New York (1962)
14 COHEN, R. and BOWLES, K. L., 'On the nature of equatorial spread-F', *J. Geophys. Res.*, **66**, 1081 (1961)
15 HASELGROVE, J., 'Ray theory and a new method for ray tracing', *Report on Conference on Physics of Ionosphere*, Phys. Soc. London, 355 (1954)
16 RATCLIFFE, J. A., *The Magnetoionic Theory*, Cambridge University Press, Cambridge (1959)
17 BUDDEN, K., *Radio Waves in the Ionosphere*, Cambridge University Press, Cambridge (1961)
18 CCIR REPORT 266-6. 'Ionospheric propagation and noise characteristics pertinent to terrestrial radiocommunication systems design and service planning', *Documents of XVIth Plenary Assembly*, Geneva, ITU (1986)
19 NIELSON, D. L., 'The importance of horizontal F-region drifts to transequatorial VHF propagation', *Scatter Propagation of Radio Waves*, ed. Thrane, E., *AGARD Conference Proceedings No. 37*, NATO, Neuilly-sur-Seine, France (1968)
20 FENWICK, F. B. and VILLARD, O. G., 'A test of the importance of ionosphere reflections in long distance and around-the-world high frequency propagation', *J. Geophys. Res.*, **68**, 5659 (1963)
21 RATCLIFFE, J. A., 'Some aspects of diffraction theory and their application to the ionosphere', *Reports on Progress in Physics*, Phys. Soc., London, **19**, 188 (1956)
22 CCIR REPORT 263-6. 'Ionospheric effects upon Earth-space propagation', *Documents of XVIth Plenary Assembly, Geneva*, ITU (1986)
23 PIERCE, J. A., 'OMEGA', *IEEE Trans. Aer. and Elect. Syst.*, **1**, 206 (1965)
24 STRINGER, F. S., 'Hyperbolic radionavigation systems', *Wireless World*, **75**, 353 (1969)
25 SUGAR, G. R., 'Radio propagation by reflection from meteor trails', *Proc. IEEE*, **52**, 116 (1964)
26 BAILEY, D. K., BATEMAN, R. and KIRBY, R. C., 'Radio transmission at VHF by scattering and other processes in the lower ionosphere', *Proc. IRE*, **43**, 1181 (1955)
27 CROFT, T. A., 'Skywave backscatter: a means for observing our environment at great distances', *Rev. Geophys. and Space Physics*, **10**, 73 (1972)
28 BLAIR, J. C., MELANSON, L. L. and TVETEN, L. H., 'HF ionospheric radar ground scatter map showing land-sea boundaries by a spectral separation technique', *Electronics Letters*, **5**, 75 (1969)
29 BAKER, D. M. and DAVIES, K., 'Waves in the ionosphere produced by nuclear explosions', *J. Geophys. Res.*, **73**, 448 (1968)
30 DAVIES, K. and BAKER, D. M., 'Ionospheric effects observed around the time of the Alaskan earthquake of March 28 1964', *J. Geophys. Res.*, **70**, 2251 (1965)
31 BAKER, D. M. and DAVIES, K., 'F2-region acoustic waves from severe weather', *J. Atmosph. Terr. Phys.*, **31**, 1345 (1969)
32 PIERCE, E. T., 'Nuclear explosion phenomena and their bearing on radio detection of the explosions', *Proc. IEEE*, **53**, 1944 (1965)
33 SMITH, E. K., 'Electromagnetic probing of the upper atmosphere', ed. U.R.S.I. Working Group, *J. Atmosph. Terr. Phys.*, **32**, 457 (1970)
34 PIGGOTT, W. R. and RAWER, K., *U.R.S.I. Handbook of Ionogram Interpretation and Reduction*, 2nd edn, Rep. UAG-23, Dept. of Commerce, Boulder, USA (1972)
35 BEYNON, W. J. G., 'Special issue on analysis of ionograms for electron density profiles', ed. URSI Working Group, *Radio Science*, **2**, 1119 (1967)
36 COLL, D. C. and STOREY, J. R., 'Ionospheric sounding using coded pulse signals', *Radio Science*, **68D**, 1155 (1964)
37 FENWICK, R. B. and BARRY, G. H., 'Sweep frequency oblique ionospheric sounding at medium frequencies', *IEEE Trans. Broadcasting*, **12**, 25 (1966)
38 WHALEN, J. A., BUCHAU, J. and WAGNER, R. A., 'Airborne ionospheric and optical measurements of noontime aurora', *J. Atmosph. Terr. Phys.*, **33**, 661 (1971)
39 Ionospheric Data—Series FA—published monthly for National Geophysical and Solar-Terrestrial Data Centre, Boulder, USA
40 GARRIOTT, G. K. and NICHOL, A. W., 'Ionospheric information deduced from the Doppler shifts of harmonic frequencies from earth satellites', *J. Atmosph. Terr. Phys.*, **22**, 50 (1965)
41 ROSS, W. J., 'Second-order effects in high frequency transionospheric propagation', *J. Geophys. Res.*, **70**, 597 (1965)
42 AARONS, J., 'Total-electron content and scintillation studies of the ionosphere', ed. *AGARDograph 166*, NATO, Neuilly-sur-Seine, France (1973)
43 HELLIWELL, R. A., *Whistlers and Related Ionospheric Phenomena*, Stanford University Press, Stanford, California (1965)
44 MITRA, S. N., 'A radio method of measuring winds in the ionosphere', *Proc. IEE*, **46**, Pt. III, 441 (1949)
45 MUNRO, G. H., 'Travelling disturbances in the ionosphere', *Proc. Roy. Soc.*, **202A**, 208 (1950)
46 BELLO, P. A., 'Some techniques for instantaneous real-time measurements of multipath and Doppler spread', *IEEE Trans. Comm. Tech.*, **13**, 285 (1965)
47 CROFT, T. A., 'Special issue on ray tracing', ed. *Radio Science*, **3**, 1 (1968)
48 APPLETON, E. V. and PIGGOTT, W. R., 'Ionospheric absorption measurements during a sunspot cycle', *J. Atmosph. Terr. Phys.*, **5**, 141 (1954)
49 BEYNON, W. J. G. and RANGASWAMY, S., 'Model electron density profiles for the lower ionosphere', *J. Atmosph. Terr. Phys.*, **31**, 891 (1969)
50 HARGREAVES, J. K., 'Auroral absorption of H. F. radio waves in the ionosphere—a review of results from the first decade of riometry', *Proc. IEEE*, **57**, 1348 (1969)
51 BELROSE, J. S. and BURKE, M. J., 'Study of the lower ionosphere using partial reflections', *J. Geophys. Res.*, **69**, 2799 (1964)
52 FEJER, J. A., 'The interaction of pulsed radio waves in the ionosphere', *J. Atmosph. Terr. Phys.*, **7**, 322 (1955)
53 CCIR REPORT 265-6, 'Sky-wave propagation and circuit performance at frequencies between about 30 kHz and 500 kHz', *Documents of XVIth Plenary Assembly, Geneva*, ITU (1986)
54 APPLETON, E. V. and BEYNON, W. J. G., 'The application of ionospheric data to radio communications', *Proc. Phys. Soc.*, **52**, 518 (1940); and **59**, 58 (1947)

55 CCIR REPORT 252–2, 'CCIR interim method for estimating sky-wave field strength and transmission loss at frequencies between the approximate limits of 2 and 30 MHz', *Documents of XIIth Plenary Assembly, New Dehli*, ITU, Geneva (1970)
56 CCIR REPORT 340-5, 'CCIR atlas of ionospheric characteristics', *Documents of XVIth Plenary Assembly, Geneva*, ITU (1986)
57 STEVENS, E. E., 'The CHEC sounding system', *Ionospheric Radio Communications*, ed. K. Kolkestad, Plenum, New York, 359 (1968)
58 BEAN, B. R. and DUTTON, E. J., *Radio Meteorology, Monograph 92*, US Government Printing Office, Washington (1966)
59 CASTEL, F. Du., *Tropospheric Radiowave Propagation Beyond the Horizon*, Pergamon Press, Oxford (1966)
60 CCIR 16th PLENARY ASSEMBLY, Vol. V, *Propagation in Non-Ionized Media*, International Telecommunication Union, Geneva (1986)
61 HALL, M. P. M., *Effects of the Troposphere on Radio Communication*, Peter Peregrinus (for IEE), London (1979)
62 *IEE Conference Publications in Propagation*; No. 48, London (1968); No. 98, London (1973); No. 169, London (1978); No. 195, York (1981); Norwich (1983); No. 248, London (1985)
63 SAXTON, J. A. (Ed.) *Advances in Radio Research*, Academic Press, London (1964)
64 URSI Commission F, *Colloquium Proceedings*; La Baule, France (CNET, Paris, 1977). Also, at Lennoxville, Canada (Proceedings edited by University of Bradford, England, 1980)
65 BOITHIAS, L., *Radio-Wave Propagation*, North Oxford Academic Publishers, London (1988)

Part 3

Materials and Components

12 Resistive Materials and Components

G W A Dummer
MBE, CEng, FIEE, FIEEE, FIERE
Electronics Consultant
(Sections 12.1–12.2 and 12.7)

D Ord
E-Sil Components Ltd
(Sections 12.3–12.4)

M J Rose BSc (Eng)
Mullard Ltd
(Sections 12.5–12.6)

Contents

12.1 Basic laws of resistivity

Ohm's law states that: the ratio of the potential difference E between the ends of a conductor to the current I flowing in it is a constant, provided the physical conditions of the conductor are unaltered. Ohm defined the resistance, R, of the given conductor as the ratio E/I. The practical units in which the law is expressed are resistance in ohms, potential difference in volts and current in amperes.

The resistance of a given conductor is proportional to its length l and inversely proportional to its area of cross-section a. Thus R varies as l/a and $R=\rho(l/a)$ where ρ is a constant of the material known as its *specific resistance* or *resistivity*. When l and a are unity, $R=\rho$; thus the resistivity of a given material may be defined as the resistance of *unit cube* of the material. In the SI system ρ may be expressed in *ohms per metre cube* or simply *ohm-metres*.

It will be seen that resistivity may be expressed in many terms, of which the following are those in general use:

(1) Microhms per centimetre cube ($\mu\Omega$ cm)
(2) Ohms square mil per foot (Ω sq mil/foot)
(3) Ohms circular mil per foot (Ω cir mil/foot).

In connection with the above terms, there are three points which should be noted:

(1) A mil is an expression used to denote one thousandth of an inch (0.001 in) and should not be confused with millimetres.
(2) Microhm cm is sometimes expressed as microhms/cm^3.
(3) Resistivity must always be qualified by reference to the temperature at which it is measured, usually 20°C.

12.2 Resistance materials

12.2.1 Mass and volume resistivities

The units of mass resistivity and volume resistivity are interrelated through the density. For copper wires this is stated in the IACS (International Annealed Copper Standard) as 8.89 g/cm^3 at 20°C. The volume resistivity, ρ, the mass resistivity, δ, and density, d, are related in the formula

$$\delta=\rho d \tag{12.1}$$

The IACS in various units of mass and volume resistivity, all at 20°C, is:

0.153 28	ohm gram/metre^2
875.20	ohm pound/mile^2
0.017 241	ohm millimetre^2/metre
1.724 1	microhm centimetre
0.678 79	microhm inch
10.371	ohm circular mil/foot

12.2.2 Temperature coefficient of resistance

At a temperature of 20°C, the coefficient of variation of resistance with temperature of standard annealed copper—measured between two potential points rigidly fixed to the wire—the metal being allowed to expand freely, is given as 0.003 93 = 1/254.45 per degree Celsius. The temperature coefficient of resistance of a copper wire of constant mass and conductivity of 100% at 20°C is therefore 0.003 93 per degree Celsius.

For other conductivities of copper, the 20°C temperature of resistance is given by multiplying the decimal number expressing the per cent conductivity by 0.003 93.

The conductivity of copper may be calculated from the coefficient of resistance. The temperature coefficient of resistance, α, for different initial Celsius temperatures and different conductivities may be calculated from the formula,

$$\alpha_{t_1}=\frac{1}{\dfrac{1}{\eta(0.003\,93)}+(t_1-20)} \tag{12.2}$$

in which η is the conductivity expressed decimally, and α_{t_1} is the temperature coefficient at t_1.

The temperature coefficient of resistivity is generally taken to be 0.0068 microhm cm per °C. Expressed in other values, it is

0.000 597	ohm gram/metre^2
3.31	ohm pound/mile^2
0.000 681	microhm centimetre
0.002 68	microhm inch
0.040 9	ohm circular mil/foot

The Fahrenheit equivalent for these constants may be found by dividing them by 1.8. The change of resistivity per degree Fahrenheit is

0.001 49 microhm-inch

12.2.3 Temperature coefficients and resistivities for metals and alloys

These are given for many elements, metals and alloys in *Table 12.1.*

12.3 Fixed resistors

12.3.1 Marking

Resistors are marked either by colour bands or by print marking. With the colour code system it is popular to use the four-band or five-band system. The four-band system is used for resistance values having two significant figures and the five-band coding is usually used for values having three significant figures. An explanation of the method is given in *Figure 12.1* and *Table 12.2*.

Resistance values are made according to a standard series, i.e. E24 (24 values per decade), E48, E96 and E192. *Table 12.3* lists these preferred values. Using values other than these may incur a cost premium from the manufacturer and also a minimum purchase quantity.

12.3.2 General characteristics of discrete fixed resistors

Fixed resistors are generally placed in three categories—Precision, Semi-Precision and General Purpose. Precision types include metal film and wirewound. Semi-Precision are metal film and metal oxide. General purpose are carbon film, carbon composition and wirwound.

The following are some of the factors that influence the choice of a resistor for a particular application. These are summarised in *Table 12.4*.

12.3.2.1 Size

In general metal film and carbon resistors dissipate less power than wirewound resistors for the same resistance value and size. Metal film and metal oxide resistors have, in most professional applications, replaced carbon film and carbon composition resistors. This is because of their better stability characteristics and environmental performance.

Table 12.1 Resistivities and temperature coefficients of metals and alloys

Material	*Composition*	*Temperature (°C)*	*Resistivity (Ω m × 10^{-8})*	*Temperature (°C)*	*Temperature coefficient of resistance (per °C)*	*Authority*
Advance* (See Constantan)	Ni 30, Cr 5, Fe 65			20	+0.000 72	
Alumel*	Ni 94, Mn 2.5, Fe 4.5, Al 2, Si 1			0	0.001 2	
Aluminium				0	0.001 2	
				18	+0.003 9	
				25	0.003 4	
				100	0.004	
				500	0.005	
annealed, highest purity				0–100	0.004 45	
Aluminium-bronze	Cu 97, Al 3				0.001 02	
	Cu 90, Al 10				0.003 20	
	Cu 6, Al 94				0.003 80	
Antimony				20	0.003 6	
Argentan*	Cu 61.6, Ni 15.8, Zn 22.6			0–160	0.000 387	
Arsenic				20	0.004 2	
Bismuth				20	0.004	
				0–100	0.004 46	
Brass				20	0.002	
	Cu 66, Zn 34			15	0.002	
	Cu 60, Zn 40			15	0.001	
Brightray B				20–500	+0.000 14	
Brightray C				20–500	+0.000 079	
Brightray F				20–500	+0.000 25	
Brightray H				20–500	+0.000 084	
Brightray S				20–500	+0.000 061	
Bronze	Cu 88, Sn 12			20	0.000 5	
Cadmium				20	0.003 8	
drawn				0–100	0.004 24	
annealed, pure				0	0.004 2	
Carbon		0	3500		−0.005	
		500	2700			
		100	2100			
		2000	1100			
		2500	900			
Chromax*	Ni 30, Cr 20, Fe 50			20–500	0.000 31	
Chromel A*	Ni 80, Cr 20			20–500	0.000 13	
Chromel C*	Ni 60, Cr 16, Fe 24			20–500	0.000 17	
Chromel D*	Ni 30, Cr 20, Fe 50			20–500	0.000 32	
Climax				20	+0.000 7	
Cobalt				0	0.003 3	
				0–100	0.006 58	
Constantan	Cn 60, Ni 40	20	49	12	0.000 008	Bureau of Standards
		−200	42.4	25	0.000 002	Nicolai
		−150	43.0	100	0.000 033	Nicolai
		−100	43.5	200	0.000 02	Nicolai
		−50	43.9			Nicolai
		0	44.1			Nicolai
		+100	44.6			Nicolai
		400	44.8			Nicolai
Copper, commercial:		20	1.7241*	20	0.003 93	Bureau of Standards
annealed		20	1.77	20	0.003 82	Bureau of Standards
hard drawn		20	1.692	100	0.003 8	Wolff, Delinger, 1910
pure, annealed		−258.6	0.014	400	0.004 2	Nicolai
		−206.6	0.163	1000	0.006 2	Nicolai
		−150	0.567			Nicolai
		−100	0.904			Nicolai
		+100	2.28			Northrup, 1914
		200	2.96			Northrup, 1914
		500	5.08			Northrup, 1914

Table 12.1 *continued*

Material	*Composition*	*Temperature (°C)*	*Resistivity (Ω m × 10⁻⁸)*	*Temperature (°C)*	*Temperature coefficient of resistance (per °C)*	*Authority*
		1000	9.42			Northrup, 1914
electrolytic				0	0.0041	
pure, annealed				0–100	0.00433	
Copper–manganese	Mn 0.98	0	4.83			Munker, 1912
	Mn 1.49	0	6.66			Munker, 1912
	Mn 4.2	20	17.9			Sebast & Gray, 1916
	Mn 7.4	20	19.7			Sebast & Gray, 1916
	Mn 15	20	50			Klein, 1924
Copper–manganese	Cu 96.5, Mn 3.5				0.00022	
	Cu 95, Mn 5				0.000026	
	Cu 70, Mn 30				0.00004	
Copper–manganese–iron	Cu 91, Mn 7.1, Fe 1.9	0	20	0	0.00012	Blood
	Cu 70.6, Mn 23.2, Fe 6.2	0	77	0	0.000022	Blood
Copper–manganese–nickel	Cu 73, Mn 24, Ni 3	0	48	0	−0.00003	Feussner, Lindeck
Copper–nickel	Cu 60, Ni 40			0	±0.00002	
Eureka*		0	47	0	+0.00005	Drysdale, 1907
Evanohm*	Cr 20, Al 2.5, Cu 2.5, Ni bal.			−50 to +100	±0.00002	
Excello				20	0.00016	
Ferry*				20–100	±0.00002	
German silver	Cu 60.16, Zn 25.37	−200	27.9			Dewar, Fleming
	Ni 14.03, Fe 0.3	−100	29.3			
	Co and Mn trace	+100	33.1			
German-silver	Ni 18			20	0.00004	
	Cu 60, Zn 25, Ni 15			0	0.00036	
Gold				20	0.0034	
				100	0.0035	
				500	0.0035	
				1000	0.0049	
Gold–copper–silver	Au 58.3, Cu 26.5, Ag 15.2			0	0.000574	
	Au 66.5, Cu 15.4, Ag 18.1			0	0.000529	
	Au 7.4, Cu 78.3, Ag 14.3			0	0.001830	
Gold–silver	Au 90, Ag 10			0	0.0012	
	Au 67, Ag 33			0	0.00065	
Graphite‡		0	800			
		500	830			
		1000	870			
		2000	1000			
		2500	1100			
Indium				0	0.0047	
Iridium				0–100	0.00411	
Iron				20	0.0050	
				0	0.0062	
				25	0.0052	
				100	0.0068	
				500	0.0147	
				1000	0.005	
Karma*	Ni 80, Cr 20			20	0.00016	
Lead				18	0.0043	
pure				0–100	0.00422	
Lithium				0	0.0047	
				230	0.0027	
Lohm*	Ni 6, Cu 94			20–100	0.00071	

Material	Composition	Temperature (°C)	Resistivity (Ω m × 10^{-8})	Temperature (°C)	Temperature coefficient of resistance (per °C)	Authority
Magnesium				20	0.004	
				0	0.003 8	
				25	0.005 0	
				100	0.004 5	
				500	0.003 6	
				600	0.010 0	
Mancoloy*				200	0.000 5	
Manganese–copper	Mn 30, Cu 70	0	100	0	0.000 040	Feussner, Lindeck
Manganese–nickel	Mn 2, Ni 98			20–100	0.004 5	
Manganin	Cu 84, Mn 12, Ni 4	20	44	12	0.000 006	Bureau of Standards
		22.5	45	25	0.000 000	Kimura, Sakamaki
		−200	37.8	100	−0.000 042	Nicolai
		−100	38.5	250	−0.000 052	Nicolai
		−50	38.7	475	0.000 000	Nicolai
		0	38.8	500	+0.000 11	Nicolai
		100	38.9			Nicolai
		400	38.3			Nicolai
Mercury				20	0.000 89	
				0	0.000 88	
Midohm*	Ni 23, Cu 77			20–100	0.000 18	
				40	0.000 000	
Minalpha*				25	+0.003 3	
Molybdenum				100	0.003 4	
				1000	0.004 8	
Monel-metal*				20	0.002	
Nichrome*	Ni 61, Cr 14, Fe 24			20–500	0.000 17	
Nichrome V*	Ni 80, Cr 20			20–500	0.000 13	
Nickel				20	0.006	
				0	0.006	
				25	0.004 3	
				100	0.004 3	
				500	0.003	
				1000	0.003 7	
pure, annealed				0–100	0.006 75	
Nickel–chromium	Ni 80, Cr 20	20	100			Bureau of Standards
Nickel		20	7.8			Bureau of Standards
pure		−182.5	1.44			Fleming, 1900
		−78.2	4.31			Fleming, 1900
		0	6.93			Fleming, 1900
		94.9	11.1			Nicolai, 1907
		400	60.2			
Nickel–copper–zinc	Ni 12.84, Cu 30.59 Zn 6.57 by volume	0	20.3			Matthiessen
Ohmax*				20–500	0.000 066	
Palladium				20	0.003 3	
pure				0–100	0.003 77	
pure				0	0.003 5	
Phosphor–bronze				0	0.004–0.003	
Palladium–gold	Pd 50, Au 50			0–100	0.000 36	
Palladium–silver	Pd 60, Ag 40			0–100	0.000 04	
Palladium		20	11			Bureau of Standards
		−183	2.78			Dewar, Fleming
		−78	7.17			Dewar, Fleming
		0	10.21			Dewar, Fleming
		98.5	13.79			Dewar, Fleming
Palladium–copper	Pd 72, Cu 28	20	47			Johansson, Linde
Palladium–gold	Pd 50, Au 50	20	27.5			Sedstrom, Wise
Palladium–silver	Pd 60, Ag 40	20	42			Sedstrom & Svensson
Platinum				20	0.003	

Table 12.1 *continued*

Material	*Composition*	*Temperature (°C)*	*Resistivity (Ω m × 10^{-8})*	*Temperature (°C)*	*Temperature coefficient of resistance (per °C)*	*Authority*
Platinum				0	0.0037	
				0–100	0.00392	
Platinum–iridium	Pt 90, Ir 10			0	0.0012	
	Pt 80, Ir 20			0	0.0008	
Platinum–rhodium	Pt 90, Rh 10			0	0.0013	
Platinum–silver	Pt 33, Ag 67			0	0.00024	
Platinum–gold	Pt 40, Au 60			20	0.0006	
	Pt 20, Au 80			20	0.0025	
Platinum–copper				20	0.0003	
	Pt 75, Cu 25			0	0.0055	
Potassium liquid				100	0.0042	
Platinum		20	10			Bureau of Standards
		−203.1	2.44			Dewar, Fleming
		−97.5	6.87			Dewar, Fleming
		0	10.96			Dewar, Fleming
		+100	14.85			Dewar, Fleming
		400	26			Nicolai
		−265	0.10			Nernst
		−253	0.15			Nernst
		−233	0.54			Nernst
		−153	4.18			Nernst
		−73	7.82			Nernst
		0	11.05			Nernst
		+100	14.1			Pirrani
		200	17.9			Pirrani
		400	25.4			Pirrani
		800	40.3			Pirrani
		1000	47.0			Pirrani
		1200	52.7			Pirrani
		1400	58.0			Pirrani
		1600	63.0			Pirrani
Platinum–gold	Au 60, Pt 40	20	42.0			Johansson, Linde
	Au 80, Pt 20	20	25.0			Johansson, Linde
Platinum–iridium	Pt 90, Ir 10	0	24			Barnes
	Pt 80, Ir 20	0	31			Barnes
	Pt 65, Ir 35	20	36			Geibel, Carter and Nemilow
Platinum–rhodium	Pt 90, Rh 10	−200	14.40			Dewar, Fleming
		−100	18.05			Dewar, Fleming
		0	21.14			Dewar, Fleming
		+100	24.2			Dewar, Fleming
	Pt 80, Rh 20	20	20			Acken, Nemilow, Voronow and Carter
Platinum–silver	Pt 67, Ag 33	0	24.2			Kurnakow and Nemilow
	Pt 55, Ag 45	20	61			
Platinum–copper	Pt 75, Cu 25	20	92			Sedstron
Rheotan	Cu 53.28, Ni 25.31 Zn 16.80, Fe 4.46 Mn 0.37	0	53	0	0.0004	Feussner, Lindeck
Rose metal	Bi 49, Pb 28	0	64	0	0.002	
	Sn 23			0	0.006	
Rhodium				0–100	0.00443	
Rubidium				0	0.006	
Silchrome*	Si, Cr, Fe			20	0.000025	
Silicon bronze				0	0.0038–0.0023	
				20	0.0038	
				25	0.003	
				100	0.0036	

Material	*Composition*	*Tempera-ture* (°C)	*Resistivity* (Ω m ×10⁻⁸)	*Tempera-ture* (°C)	*Temperature coefficient of resistance* (per °C)	*Authority*
				500	0.004 4	
				0–100	0.004 1	
				9	+0.004 4	
				120	0.003 3	
Silver		−200	0.357			Nicolai
		−100	0.916			Nicolai
		0	1.506			Nicolai
		+100	2.15			Northrup
		+750	6.65			Northrup
pure, annealed						
Sodium liquid						
Steel						
aluminium	Al 5, C 0.2	20	65			Portevin, 1909
	Al 15, C 0.9	20	88			Portevin, 1909
chromium	Cr 13, C 0.7	20	60			Portevin, 1909
	Cr 40, C 0.8	20	71			Portevin, 1909
Invar	35 Ni	20	81			Bureau of Standards
manganese		20	70			Bureau of Standards
nickel	Ni 10, C 0.1	20	29			
	Ni 25, C 0.1	20	39			
	Ni 80, C 0.1	20	82			Portevin, 1909
		20	18			
Siemens–Martin	Si 2.5	20	45			Bureau of Standards
silicon	Si 4	20	62			
tempered glass-hard			45.7			Stronhal, Barnes
tempered yellow			27			Stronhal, Barnes
tempered blue			20.5			Stronhal, Barnes
tempered soft			15.9			Stronhal, Barnes
titanium	Ti 2.5, C 0.15	20	16			Portevin, 1909
tungsten	W 5, C 0.2	20	20			Portevin, 1909
	W 20, C 0.2	20	24			Portevin, 1909
vanadium	V 5, C 1.1	20	121			Portevin, 1909
Steel						
Invar	Ni 36, C 0.2			0	0.002	
piano wire				0	0.003 2	
Siemens-Martin				20	0.003	
silicon	Si 4			20	0.000 8	
tempered glass-hard				0	0.001 6	
tempered blue				0	0.003 3	
Tantalum				20	0.003 1	
				0–100	0.003 47	
Thalium				0	0.004	
Therlo*				20	0.000 01	
Thorium				20–1800	0.002 1	
Tin				20	0.004 2	
Tungsten				18	0.004 5	
				500	0.005 7	
				1000	0.008 9	
pure, annealed				0–100	0.004 65	
Tellurium‡		19.6	200 000			Matthiessen
Tin–bismuth	Sn 90, Bi 9.5	12	16			
	Sn 2, Bi 98	0	244			
Vacrom*	Ni 80, Cr 20			0–500	0.000 06	
Wood's metal*				0	0.002	
Zinc				20	0.003 7	
				0	0.004	
				0–100	0.004 15	

* Trade name.
† N. B. Polycrystalline; the resistivity of a single crystal in the plane of the hexagonal network is about 60×10^{-8} Ω m at 20°C.
‡ Resistivity is greatly dependent on purity of the specimen.

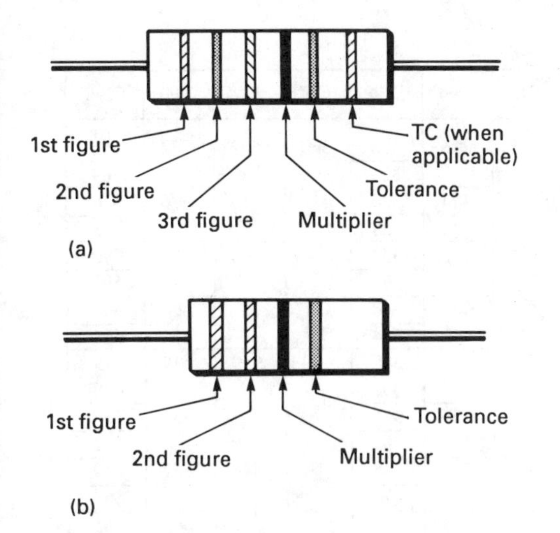

Figure 12.1 Resistor colour coding system: (a) E48, E96 and E192 series; (b) E12 and E24 series

Table 12.2 Colour code marking

Colour	*Figure*	*Multiplier*	*Tolerance* (%)	*Temperature coefficient*
Silver		× 0.01 Ω	±10	
Gold		× 0.1 Ω	±5	
Black	0	× 1 Ω		
Brown	1	× 10	±1	
Red	2	× 100 Ω	±2	$\pm 50 \times 10^{-6}$/K
Orange	3	× 1 kΩ		$\pm 15 \times 10^{-6}$/K
Yellow	4	× 10 kΩ		$\pm 25 \times 10^{-6}$/K
Green	5	× 100 kΩ	±0.5	
Blue	6	× 1 MΩ	±0.25	
Violet	7	× 10 MΩ	±0.1	
Grey	8	× 100 MΩ		
White	9			

12.3.2.2 Power handling capacity

In general, metal film and metal oxide resistors are available up to 10 W. Special types are available up to 3000 W and are usually used as dummy antennas. Cracked carbon resistors are available up to 2 W. All of these ratings have to be reduced when resistors are used at high ambient temperatures. Wirewound resistors are invariably used when high powers are to be dissipated. Some vitreous enamelled wirewound types will handle power as great as 1000 W. It is important to remember that the surface temperatures reached by the resistors when dissipating these wattages can be very high – of the order of several hundred degrees Celsius. In pulse operation (particularly when the duty cycle is low) only the mean power is effective in raising the internal temperature of a resistor. As the power is supplied in short pulses, very high peak ratings are possible but the mean power should not be exceeded. Where the resistor may be subjected to long pulses over a long period of time, the manufacturer should be consulted concerning the long-term performance. A useful test can be made as follows to assess how the resistor may be affected:

(1) Measure and record the resistor value to two decimal places.
(2) Subject the resistor to the pulse conditions for as long as possible. Allow the resistor to cool for 1 hour.
(3) Repeat step 1.
(4) Calculate the change in resistance.

For metal oxide and metal film resistors a change of 0.5% would indicate that the resistor was coping adequately with the pulse conditions.

12.3.2.3 Load life or endurance

Resistors are tested for their change in resistance when dissipating full rated power at 70°C. The standard test times are 1000 hours and 8000 hours after which the product performance is assessed.

In general, metal film and metal oxide film resistors have replaced carbon composition and to some extent carbon film in professional applications. This is because of their better stability (<1%) and lower temperature coefficients of resistance. Carbon film and carbon composition types still prove effective for most domestic applications having typical respective stabilities of 1 and 5%. The highest stability with the lowest temperature coefficient can be obtained from wirewound and metal film resistors.

12.3.2.4 Tolerance (or accuracy)

General-purpose metal film/metal oxide and carbon film resistors are available with tolerances of 1, 2 and 5%. Carbon composition types are made to approximate target values and then selected to 5, 10 or 20%. Special wirewound and metal film resistors are available with tolerances as low as 0.005%.

12.3.2.5 Maximum operating temperature

Carbon-composition resistors are seriously affected by ambient temperatures over 100°C, mainly by changes in the structure of the binder used in the resistor mixture. The maximum recommended surface temperature is about 110–115°C. This is the total working temperature produced by the power dissipated inside the resistance, the heat from associated components, and the ambient temperature in which the resistor is operating. Cracked-carbon resistors can be operated up to a maximum surface temperature of 150°C under the same conditions, metal films up to 175°C, and oxide films up to 200–250°C, approximately. Some special metal and metal-oxide-film power resistors can operate at 500–600°C when no limiting protective coating is applied.

Wirewound resistors are generally lacquered or vitreous enamelled for protection of the windings. For both types the safe upper limit is set by the protective coating. For lacquered types the maximum recommended temperature is 175°C (some will work up to 450°C).

Free circulation of air should be allowed, and the ends of tubular resistors should not be placed flat against the chassis. If the resistors are badly mounted, or if several resistors are placed together, derating is necessary.

12.3.2.6 Maximum operating voltage

The maximum operating voltage is determined mainly by the physical shape of the resistor and by the resistance value (which determines the maximum current through the resistor and therefore the voltage for a given wattage), that is, the 'critical value' referred to previously.

Commercial ratings at room temperature are some 25–50% higher than military ratings, and reference should always be made to the resistor manufacturer for the maximum voltage rating.

Table 12.3 Preferred resistor values

E24	E48	E96	E192	E24	E48	E96	E192	E24	E48	E96	E192	E24	E48	E96	E192
100	100	100	100		178	178	178		316	316	316	560	562	562	562
			101	180			180				320				569
		102	102			182	182			324	324			576	576
			104				184	330			328				583
	105	105	105		187	187	187		332	332	332		590	590	590
			106				189				336				597
		107	107			191	191			340	340			604	604
			109				193				344				612
110	110	110	110		196	196	196		348	348	348	620	619	619	619
			111				198				352				626
		113	113	200		200	200			357	357			634	634
			114				203	360			361				642
	115	115	115		205	205	205		365	365	365		649	649	649
			117				208				370				657
		118	118			210	210			374	374			665	665
120			120				213				379				673
	121	121	121		215	215	215		383	383	383	680	681	681	681
			123	220			218				388				690
		124	124			221	221	390		392	392			698	698
			126				223				397				706
	127	127	127		226	226	226		402	402	402		715	715	715
			129				229				407				723
130		130	130			232	232			412	412			732	732
			132				234				417				741
	133	133	133		237	237	237		422	422	422	750	750	750	750
			135	240			240	430			427				759
		137	137			243	243			432	432			768	768
			138				246				437				777
	140	140	140		249	249	249		442	442	442		787	787	787
			142				252				448				796
		143	143			255	255			453	453			806	806
			145				258				459	820			816
	147	147	147		261	261	261		464	464	464		825	825	825
			149				264	470			470				835
150		150	150	270		267	267			475	475			845	845
			152				271				481				856
	154	154	154		274	274	274		487	487	487		866	866	866
			156				277				493				876
		158	158			280	280			499	499			887	887
160			160				284				505				898
	162	162	162		287	287	287	510	511	511	511	910	909	909	909
			164				291				517				920
		165	165			294	294			523	523			931	931
			167				298				530				942
	169	169	169	300	301	301	301		536	536	536		953	953	953
			172				305				542				965
		174	174			309	309			549	549			976	976
			176				312				556				988

12.3.2.7 Frequency range

On a.c. carbon-composition resistors (up to about 10 kΩ in value) behave as pure resistors up to frequencies of several MHz. At higher frequencies the self-capacitance of the resistor becomes predominant, and the impedance falls. The inductance of carbon-composition resistors does not usually cause trouble below 100 MHz (except in special cases such as in attenuator resistors).

Cracked-carbon resistors specially manufactured with little or no spiral grinding can be operated at frequencies of many hundreds of MHz, but methods of mounting and connection become important at these frequencies. Other film-type resistors such as metal film and metal-oxide film are also suitable for use at high frequencies, and the effect of spiralling of the film is relatively unimportant below 50 MHz.

Table 12.4 Summary of the electrical characteristics of fixed resistors

Resistor type	*Load life* (1000 h) (±%)	*Max. noise* (μV/V)	*Temperature coeff.* (ppm/°C)	*Voltage coeff.* (%/V)	*Max. resistor temp.* (°C)
Metal film	1	0.2	±50	None	175
Ultra-precision	0.03	None	±1.0	None	150
Metal oxide film	2	0.2	±300	0.0005	300
Cracked carbon	2	2.0	−200 to −1500	0.0005	150
Moulded carbon composition	+4 −6	6.0	±1200	0.05	115
Wirewound precision type	0.01 (if hermetically sealed)	None	(Ni/Cr)+70 (Cu/Ni)+70	None	70
general-purpose type	1	None	±200	None	320

For wirewound resistors, the inductance of single-layer windings becomes appreciable, and *Ayrton–Perry* or *back-to-back* windings are often used for so-called non-inductive resistors. At high frequencies the capacitive rather than the inductive effect limits the frequency of operation. For example, the reactance of a typical resistor of 6 kΩ with an Ayrton–Perry winding becomes capacitance at 3 MHz.

In all measurements on resistors at high frequencies, the method of mounting the resistor is important. The direct end-to-end capacitance of the resistor and the capacitance of the two leads to the resistor body are included in the total capacitance being measured, and the resistor should therefore be mounted as near as possible as it is to be mounted in use. Ideally the mounting fixtures should be standardised for comparison measurements.

To summarise, for a resistor to be suitable for operation at high frequencies it should meet the following general requirements:

(1) Its dimensions should be as small as possible.
(2) It should be low in value.
(3) It should be of the film type.
(4) A long thin resistor has a better frequency characteristic than a short fat one.
(5) All connections to the resistor should be made as short as possible.
(6) There should be no sudden geometrical discontinuity along its length.

12.3.2.8 Noise

Carbon-composition resistors generate noise of two types: thermal agitation, or *Johnson* noise, which is common to all resistive impedances, and *current* noise, which is caused by internal changes in the resistor when current is flowing through it. The latter is peculiar to the carbon-composition resistor and other non-metallic films and does not occur in good quality wirewound resistors. Cracked-carbon resistors generate noise in a similar fashion to the carbon-composition types but at a very much lower level. For low values of resistance (where the film is thick) the noise is difficult to measure. Metal-film and metal-oxide-film resistors generate noise at a very low level indeed.

Measurements have shown that for carbon-composition resistors current noise increases linearly with current up to about 15 μA. With greater currents the noise curve approximates to a parabola.

12.3.2.9 Temperature coefficient of resistance

A resistor measured at 70°C will have a different resistance value from that at 20°C; the change in value at differing temperatures can be calculated from the temperature coefficient for each class of resistor. Approximate maximum values are given in *Table 12.4*.

12.3.2.10 Voltage coefficient

When a voltage is applied across a carbon resistor, there is an immediate change in resistance, usually a decrease. The change, which is not strictly proportional to the voltage, is usually measured at values not less than 100 kΩ. In carbon-composition resistors, the change in resistance value due to the applied voltage is usually within 0.02%/V d.c. With cracked-carbon resistors, particularly the larger sizes, the effect is negligible for low values of resistance, certainly being less than 0.001%. On the higher values it can rise to 0.002%, and on very small resistors, where the stress is clearly much greater, the maximum values may approach 0.005%/V. The voltage coefficient is frequently quoted at too high a figure because of the difficulty in separating it from effects due to temperature coefficient. Wirewound resistors do not show this effect, provided they are free from leakage between turns. Metal-film resistors have voltage coefficients from 0.0001%/V to 0.003%/V depending on wattage, whereas metal-oxide-film resistors approximate from 0.0001%/V to 0.0005%/V.

12.3.2.11 Solderability

Manufacturers routinely check the solderability of their components by using either a solder bath or the solder globule test.

When resistors are soldered into a equipment the resistance value will experience a small permanent change due to the heating effects of the solder. For film resistors this change is typically ±0.2%.

12.3.2.12 Shelf life

There is a change in the resistance of most types of resistor during storage. Carbon composition resistors exhibit the biggest change and may be up to 5% after one year. Wirewound and film resistors are very stable and may change by up to 0.2%. This change usually occurs in the first year and thereafter the change in resistance value is minimal.

12.4 Surface-mounted resistors

The desire for more automated production lines and a growing demand for smaller sized electronic equipment are twin elements spurring on the rise in popularity of surface-mounted resistors. Although the phrase 'surface mounted' is a relatively new term in the electronics vocabulary, it represents a concept which has been developing for more than a decade.

12.4.1 Types

The range of resistors in chip form is continually expanding and includes metal glaze, metal film and wirewound resistors. Metal glaze resistors are usually regarded as general-purpose types but, where tolerances of 1% are required with low temperature coefficients, metal film types are usually preferred.

12.4.2 Format

Two types are commonly available—flat chip and MELF (metal electrode face bonding). The former are manufactured in standard sizes and their case sizes reflect the length and width in inches, i.e. 1206 = 0.12 in × 0.060 in. The most common sizes available are 0603, 0805 and 1206 (see *Figure 12.2*).

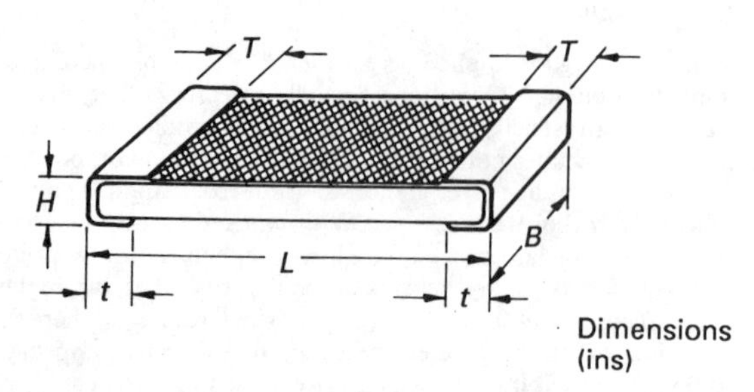

Dimensions	*CR 0805*	*CR 1206*
L	0.080	0.12
B	0.050	0.060
H	0.018	0.024
T	0.016	0.018
t	0.012	0.012

Figure 12.2 Flat chip surface-mounted resistor format

The MELF types have been derived from conventional film resistor technology and are available as 0.125 W or 0.25 W (see *Figure 12.3*).

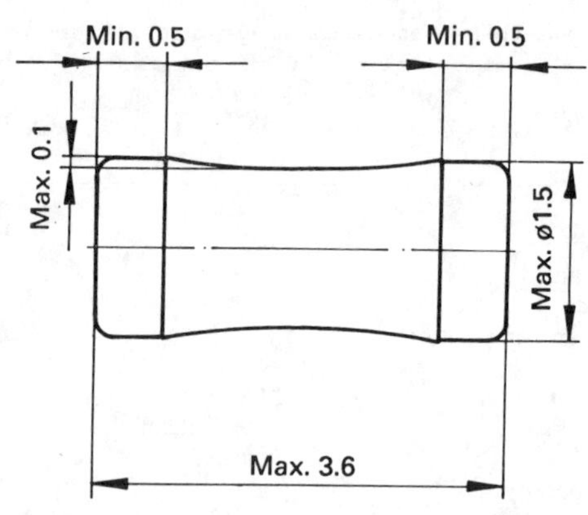

Figure 12.3 MELF surface-mounted resistor (SMM 0204), dimensions in millimetres

12.4.3 Electrical characteristics

A summary of the typical electrical characteristics of surface-mounted resistors is given in *Table 12.5*.

In surface mounting technology the most critical operation is soldering for which the definition of the soldering pad dimensions is vital to prevent major soldering problems.

The pad dimensions need to be adapted to component size, soldering method and packaging density. Reference to the product manufacturer is essential to obtain their recommendations. *Table 12.6* gives a list of typical pad dimensions based on *Figure 12.4*.

12.5 Thermistors

The name thermistor is an acronym for *therm*ally sensitive res*istor*. This name is applied to resistors made from semiconductor materials whose resistance value depends on the temperature of the material. This temperature is determined by both the ambient temperature in which the thermistor is operating and the temperature rise caused by the power dissipation within the thermistor itself. The thermistors first developed had a negative temperature coefficient of resistance (n.t.c.), that is the resistance value decreased as the temperature increased. This is in contrast to the behaviour of most metals and therefore of wirewound and metal-film resistors whose resistance increases with an increase of temperature. Subsequently thermistors with a positive temperature coefficient of resistance (p.t.c.) were developed, with the advantage over normal resistors

Table 12.5 Characteristics of commonly used surface-mounted resistors

	Flat chip metal glaze	*Flat chip metal film*	*MELF metal film*	*Wirewound*
Power (W)	1/16–1/4	1/16–1/4	1/8–1/4	1–5
Resistance range	2R2–10M	50R–100K	1R–5M1	0.001R–2K7
Tolerance (%)	1, 2, 5	0.1–0.5	0.5–2	0.1–10
Temperature coefficient (ppm/°C)	100–600	25–50	50	20–90
Thermal shock (%)	±0.5	±0.05	±0.25	±0.2
Short time overload (%)	±0.5	±0.05	±0.25	±0.2
Resistance to solder heat (%)	±0.25	±0.05	±0.25	±0.2
Moisture resistance (%)	±0.5	±0.05	±0.5	±0.2
Load life (1000 h) (%)	±0.2	±0.1	±0.5	±0.5

Table 12.6 Surface-mounted resistor pad dimensions

Style	*Dimensions* (mm)	*Wave soldering*	*Reflow soldering*
Flat chip			
0805	A	1.6	1.5
	B	1.0	0.7
	C	1.0	1.0
	D	3.0	2.4
1206	A	2.0	1.9
	B	1.2	1.0
	C	1.8	1.8
	D	4.2	3.8
0204(MELF)	A	1.6	1.6
	B	1.2 min	1.0
	C	2.2	2.2
	D	4.6	4.2
0207(MELF)	A	1.7–2.3	2.8
	B	1.9 min	2.0
	C	3.8 nom	3.8
	D	7.6 min	7.8

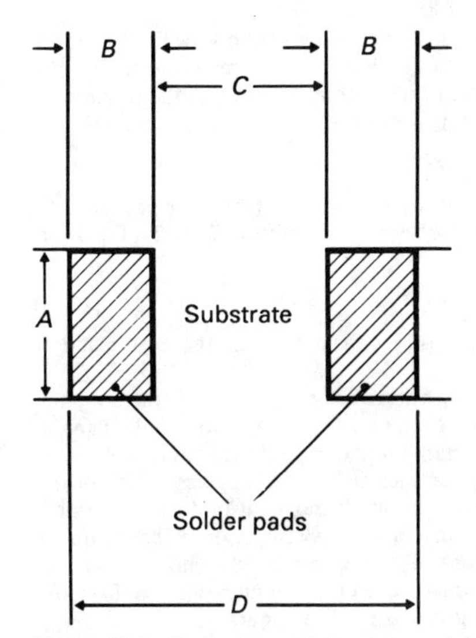

Figure 12.4 Surface-mounted resistor pad dimensions (see *Table 12.6*)

of having a larger temperature coefficient. Thermistors are used in applications as temperature sensors or as stabilising elements to compensate the effects of temperature changes in a circuit.

12.5.1 Electrical characteristics of n.t.c. thermistors

The variation of resistance with temperature for various types of n.t.c. thermistor has the form shown in *Figure 12.5*. The relationship between the resistance and temperature can be expressed as

$$R_T = A\,e^{B/T} \tag{12.3}$$

where R_T is the resistance in ohms at an absolute temperature T in kelvin; e is the base of natural logarithms (2.718); and A and B are constants.

The value of B for a particular thermistor material can be found by measuring the resistance at two values of absolute

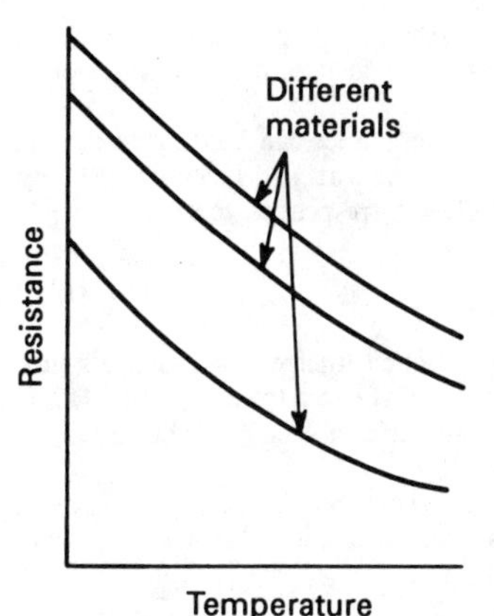

Figure 12.5 Resistance/temperature characteristic for n.t.c. thermistors

temperature T_1 and T_2 and using Equation (12.3):

$$B = 2.303\left(\frac{T_1 T_2}{T_2 - T_1}\right)(\log_{10} T_{T_1} - \log_{10} R_{T_2}) \tag{12.4}$$

When B is calculated from Equation (12.4), it is found that in practice it is not a true constant but slight deviations occur at high temperatures. More exact expressions for the variation of resistance with temperature have been suggested to replace Equation (12.3). These include $R_T = AT^C e^{B/T}$ where C is a small positive or negative constant which may sometimes be zero, or $R_T = A\,e^{B/(T+K)}$, where K is a constant.

The value of B for practical thermistor materials lies between 2000 and 5500; the unit is the kelvin.

Equation (12.4) can be rearranged in terms of R_{T_1} to give

$$\log_{10} R_{T_1} = \log_{10} R_{T_2} + B\left(\frac{T_2 - T_1}{T_1 T_2}\right)\log_{10} e \tag{12.5}$$

If the resistance of the thermistor at temperature T_2 is known, and the value of B for the thermistor material is known, the resistance value at any temperature in the working range can be calculated from Equation (12.5).

Curves relating the ratio of R_T, the resistance at temperature T, and R_{25}, the resistance at 25°C which is taken as a *standard* value for comparing different types of thermistor, to the value of B with temperature as parameter are sometimes given in published data.

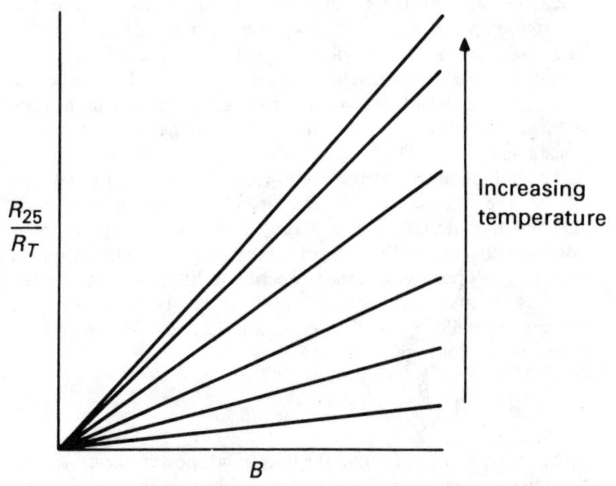

Figure 12.6 Ratio of R_{25}/R_T plotted against B value with temperature as parameter

From these curves, shown in *Figure 12.6*, and the known values of R_{25} and B for the thermistor type given in the data, the value of R_T can be determined.

A temperature coefficient of resistance can be derived for the thermistor material. This coefficient α is obtained by differentiating Equation (12.3) with respect to temperature:

$$\alpha = \frac{1}{R}\frac{dR}{dT} = -\frac{B}{T^2} \tag{12.6}$$

At 25°C, the value of α for practical thermistor materials lies, typically, between 2.5 and 7%. It can be seen from Equation (12.6) that the temperature coefficient varies inversely as the square of the absolute temperature.

The voltage/current characteristic for an n.t.c. thermistor is shown in *Figure 12.7*. The characteristic relates the current

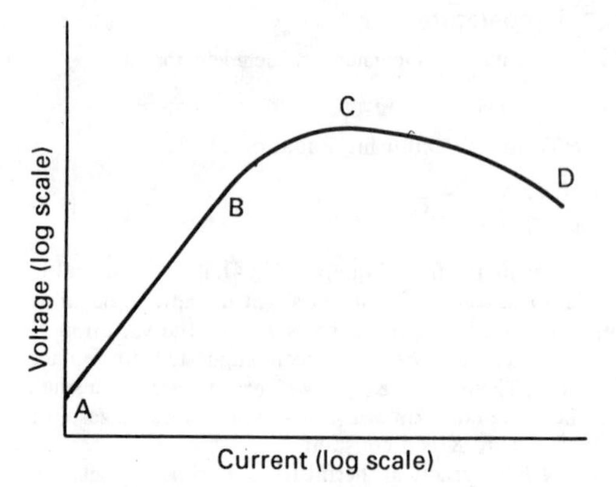

Figure 12.7 Voltage/current characteristic for n.t.c. thermistor

through the thermistor to the voltage drop across it after thermal equilibrium has been established in a constant ambient temperature. This static characteristic is plotted on logarithmic scales.

Over the low-current part of the characteristic, part A–B, the power input to the thermistor is too low to cause any rise in temperature by internal heating. The thermistor therefore acts as a linear resistor. Above point B, however, the current causes sufficient power dissipation within the thermistor to produce a rise in temperature and therefore a fall in resistance. The resistance value is therefore lower than would be expected for a linear resistor. As the current is increased further, the power dissipation within the thermistor causes a progressively larger fall in resistance, so that at some current value the voltage drop across the thermistor reaches a maximum, point C. Above this current value, the fall in resistance caused by the internal heating is large enough to give the thermistor a negative incremental resistance, part C–D.

The temperature corresponding to the maximum voltage across the thermistor can be calculated for a particular thermistor material. If it is assumed that the temperature is constant throughout the body of the thermistor, and that the heat transfer is proportional to the temperature difference between the thermistor and its surroundings (which is true for low temperature differences), then Equation (12.3) can be rewritten in terms of natural logarithms as

$$\log_e R_T = \log_e A + \frac{B}{T} \tag{12.7}$$

At thermal equilibrium, the electrical power input to the thermistor is equal to the heat dissipated; that is

$$VI = D(T - T_{amb}) \tag{12.8}$$

where V is the voltage drop across the thermistor, I is the current through it, T is the temperature of the thermistor body, T_{amb} is the ambient temperature, and D is the dissipation constant (the power required for unit temperature rise). The dissipation constant may also be represented by the symbol δ.

Substituting $R_T = V/I$ in Equation (12.7), taking logarithms of Equation (12.8) and adding gives

$$\log_e V = \frac{1}{2}\log_e AD + \frac{1}{2}\log_e (T - T_{amb}) + \frac{B}{2T} \tag{12.9}$$

At the maximum voltage point, Equation (12.9) will be a maximum. This gives

$$T_{V(max)} = \frac{B}{2} \pm \left(\frac{B^2}{4} - BT_{amb}\right)^{1/2} \tag{12.10}$$

The value of $T_{V(max)}$ corresponding to the maximum voltage across the thermistor is

$$T_{V(max)} = \frac{B}{2} - \left(\frac{B^2}{4} - BT_{amb}\right)^{1/2} \tag{12.11}$$

It can be seen from Equation (12.10) that a solution is possible only if $B > 4T_{amb}$. Also, the temperature corresponding to the maximum voltage is determined by the B value only and not by the resistance value. For practical thermistor materials, $T_{V(max)}$ lies between 45°C and 85°C.

In many applications, it is necessary to know the time taken for the thermistor to reach equilibrium. Assuming again that the temperature is constant throughout the body of the thermistor, the cooling in a time dt is given by

$$-H\,dT = D(T - T_{amb})\,dt \tag{12.12}$$

where H is the thermal capacity of the thermistor in joules/°C. When the thermistor cools from temperature T_1 to T_2, Equation (12.12) gives

$$(T_1 - T_{amb}) = (T_2 - T_{amb})\,e^{-t/\tau} \tag{12.13}$$

where τ is the thermal time constant of the thermistor and is equal to H/D.

In practice, the temperature is not constant throughout the body of the thermistor, the surface cooling more rapidly than the interior. The time constant quoted in published data is defined as the time required by the thermistor to change by 63.2% of the total change between the initial and final body temperatures when subjected to an instantaneous temperature change under zero power conditions. Cooling curves showing the rise in resistance with time when the electrical input is removed from the thermistor are often given in published data.

12.5.2 Electrical characteristics of p.t.c. thermistors

The variation of resistance with temperature for a p.t.c. thermistor is more complex than that for an n.t.c. thermistor shown in *Figure 12.5*. Because the temperature coefficient of resistance is positive only over part of the temperature range, a typical resistance/temperature characteristic for a p.t.c. thermistor may have the form shown in *Figure 12.8*. Over parts A–B (temperature T_1 to T_2) and C–D (temperature T_3 to T_4) the coefficient is negative; it is only over part B–C, corresponding to the temperature range T_2 to T_3, that the required positive coefficient is obtained. Over this range, the relationship between resistance and temperature can be expressed (approximately) as

$$R_T = A + C\,e^{BT} \tag{12.14}$$

where R_T is the resistance in ohms at an absolute temperature T in kelvin, e is the base of natural logarithms, A, B and C are constants, and T is restricted in value to $T_2 < T < T_3$.

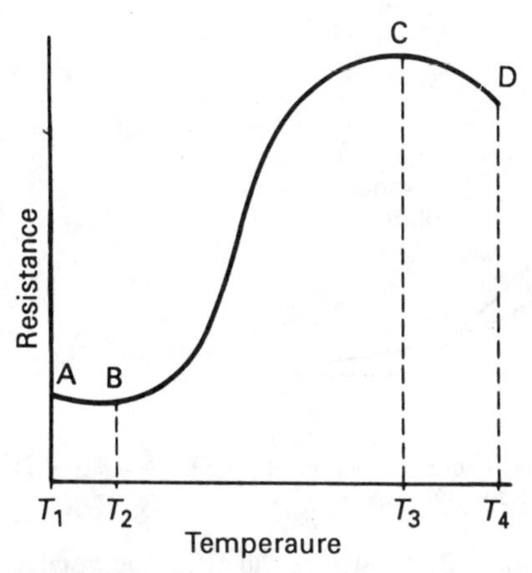

Figure 12.8 Resistance/temperature characteristic for p.t.c. thermistor

Equation (12.14) can be differentiated to give the temperature coefficient of resistance α:

$$\alpha=\frac{1}{R}\frac{dR}{dT}=\frac{BC\,e^{BT}}{A+C\,e^{BT}} \qquad (12.15)$$

In practice, unfortunately, the resistance/temperature characteristic can seldom be described by such a simple relationship as that of Equation (12.14). Unlike n.t.c. thermistors where Equation (12.3) provides a reasonable approximation to the practical behaviour, any attempt to modify Equation (12.14) to a more accurate form results in complicated expressions. For this reason, graphical methods are often used for design calculations. A quantity called the switch temperature, T_s, is often quoted in published data. This temperature is the one at which a p.t.c. thermistor begins to have a usable positive temperature coefficient of resistance. T_s is defined as the higher of two temperatures at which the resistance of the thermistor is twice its minimum value.

The voltage/current characteristic for a p.t.c. thermistor is shown in *Figures 12.9* and *12.10*. In both figures the voltage and current axes are interchanged with respect to those of the characteristic for an n.t.c. thermistor shown in *Figure 12.7*. The characteristic shown in *Figure 12.9* is plotted on linear scales, and characteristics for different ambient temperatures are included to show the effect on the thermistor. When the voltage across the thermistor is low, the power dissipation within the thermistor is insufficient to heat it above the ambient temperature. The thermistor behaves as a linear resistor, part A–B. When the voltage is increased, the power dissipation causes the thermistor temperature to rise above the switch temperature T_s, point C. The

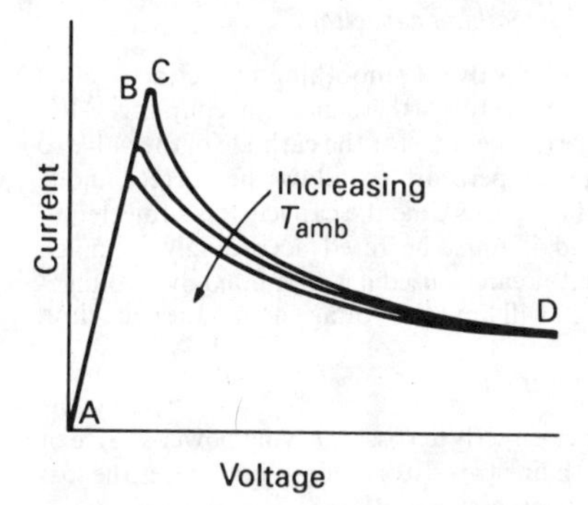

Figure 12.9 Voltage/current characteristic for p.t.c. thermistor with ambient temperature as parameter (linear scales)

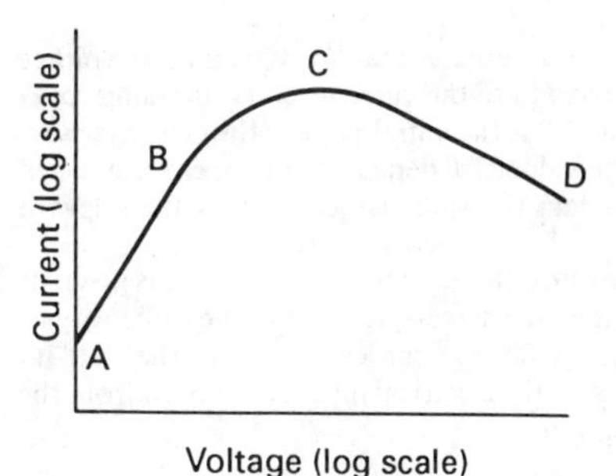

Figure 12.10 Voltage/current characteristic for p.t.c. thermistor (log scales)

resistance of the thermistor rises, and the current falls. Any further increase in voltage results in a progressive fall in current, part C–D. The characteristic for a constant ambient temperature plotted on logarithmic scales is shown in *Figure 12.10*.

At higher voltages, p.t.c. thermistors show a voltage dependency, the resistance value being determined by the voltage across the thermistor as well as by the temperature. The behaviour of the thermistor under these conditions can be represented by an equivalent circuit consisting of an ideal p.t.c. thermistor (no voltage dependency) in parallel with an ideal voltage dependent resistor (the voltage and current being related by the expression: $V \propto I^{\beta}$). The voltage/current characteristics of the components of this equivalent circuit are shown in *Figure 12.11* compared with the characteristic of a normal p.t.c.

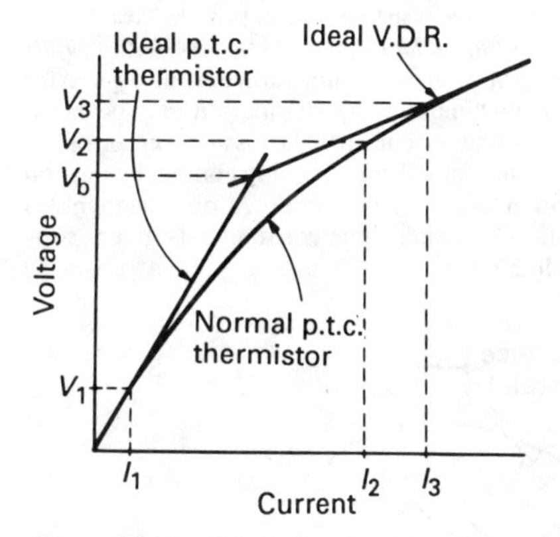

Figure 12.11 Voltage/current characteristic of p.t.c. thermistor compared with *ideal* components of equivalent circuit

thermistor. The *normal* characteristic is measured at constant ambient temperature under pulse conditions to avoid self-heating of the thermistor. At low voltages, the characteristic of the normal p.t.c. thermistor coincides with the characteristic of the ideal thermistor. At the higher voltages where the voltage dependency becomes effective, the characteristic of the normal p.t.c. thermistor coincides with the characteristic of the ideal thermistor. At the higher voltages where the voltage dependency becomes effective, the characteristic of the normal thermistor coincides with that of the voltage dependent resistor. The point of intersection of the two 'ideal' characteristics, where the currents through the two components of the equivalent circuit are equal, defines the balance voltage V_b. The value of balance voltage for a specified ambient temperature is given in published data. The voltage dependency of the thermistor β can be calculated from the expression

$$\beta=\frac{\log V_3-\log V_2}{\log(I_3R-V_3)-\log(I_2R-V_2)} \qquad (12.16)$$

where V_2 is a pulse voltage greater than V_b, V_3 is a pulse voltage greater than V_2; I_2 and I_3 are the currents corresponding to V_2 and V_3 respectively; and R is the initial slope of the characteristic, given by V_1/I_1. As the value of β depends on temperature, when quoted in published data the value is qualified by the relevant temperature.

As with n.t.c. thermistors, a thermal time constant τ is given in published data for p.t.c. thermistors, defined in the same way as previously described. Cooling curves showing the fall in resistance with time after the electrical input is removed from the thermistor are also given.

12.5.3 Choice of thermistor for application

When n.t.c. and p.t.c. thermistors are used in circuits, the designer must consider certain requirements before he can select the correct thermistor for his purpose. These requirements include the resistance value, temperature coefficient of resistance and temperature range; the power dissipation required of the thermistor; and the thermal time constant.

Most of these factors have been discussed in the previous sections describing the characteristics of n.t.c. and p.t.c. thermistors. In the published data a resistance value, usually at 25°C, is given, supplemented by resistance values at other temperatures, a temperature coefficient of resistance (α or B value), and resistance/temperature curves for various ambient temperatures. Sometimes an operating temperature range or a maximum operating temperature is given. For p.t.c. thermistors, the switch temperature is given. The voltage/current characteristic of the thermistor sometimes has resistance and power axes superimposed, as shown in *Figure 12.12*. The balance voltage V_b and voltage dependency β at a specified temperature are also given for p.t.c. thermistors. A maximum power dissipation may be specified, or a maximum voltage or current. The dissipation factor δ is often given in two forms: for still air assuming cooling by natural convection and radiation, and when mounted on a heatsink to increase the cooling. A thermal time constant τ is given, supplemented by cooling curves.

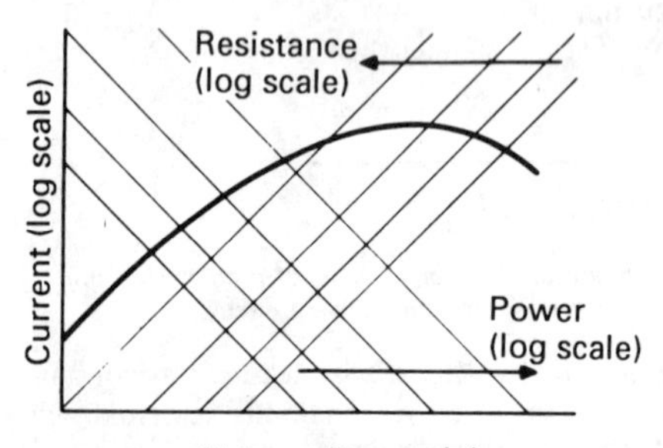

Figure 12.12 Voltage/current characteristic of thermistor with resistance and power axes superimposed

From these data, the designer is able to choose a suitable thermistor. For some applications, however, a single thermistor may not meet the requirements, and it is possible to combine a thermistor with a series/parallel combination of linear resistors so that the overall characteristic corresponds to that required. For p.t.c. thermistors, it should be noted that with a series resistance three working points are possible. In *Figure 12.13*, the voltage/current characteristic of a p.t.c. thermistor plotted on linear scales is shown with a resistive loadline superimposed. The resistive loadline intersects the voltage axis at the supply voltage V_s, and the current axis at I_s where I_s is given by V_s/R. Of the three working points given by the intersection of the characteristic and

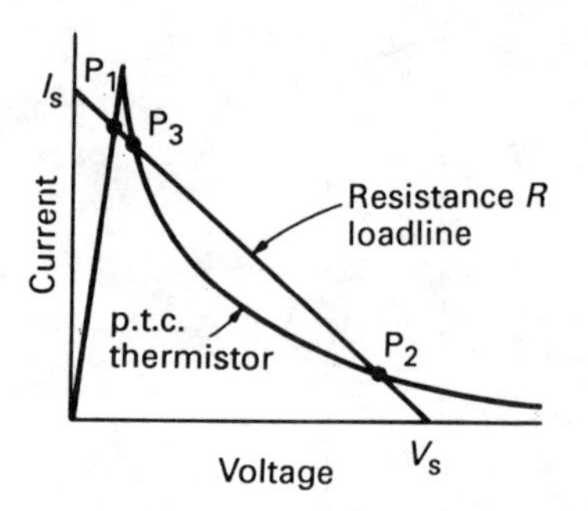

Figure 12.13 Voltage/current characteristic of p.t.c. thermistor with resistive loadlines superimposed

loadline, points P_1 and P_2 are stable, while P_3 is an unstable working point. When the supply voltage is first applied, equilibrium will be established at working point P_1 at which the load current is relatively high. Working point P_2 can only be reached if the supply voltage increases, the ambient temperature increases, or the resistance in series with the thermistor decreases. In each case, the displacement of the loadline or the change in the thermistor characteristic with temperature must be large enough to allow the peak of the thermistor characteristic to lie under the loadline.

12.6 Voltage-sensitive resistors

These are formed by dry-pressing silicon carbide with a ceramic binder into discs or rods and firing at about 1200°C. The ends of the rods, or the sides of the discs, are sprayed with metal (usually brass) to which connections are soldered. They are often known as voltage-dependent resistors, the current through the resistor being given by

$$I = KE^n \qquad (12.17)$$

where K is a constant equal to the current in amperes at $E = 1$ V; and n is a constant dependent on voltage, varying between 3 and 7 for common mixes. It is usually between 4.0 and 5.0.

12.6.1 Applications of voltage-sensitive resistors

12.6.1.1 Suppressing voltage surges and quenching contact sparks

The resistor is connected across inductive loads and prevents voltage surges. The space required by the resistor is small and the current normally passing through it is quite small. A similar application is in avoiding sparking of relay contacts although the current permissible is limited to about 0.2 A per unit.

12.6.1.2 Protection of smoothing capacitors

In the anode circuit of a valve a smoothing capacitor is often connected from the anode to earth to prevent coupling. After switching on it requires some time for the cathode of the valve to reach a high enough temperature to allow the correct anode current to flow and during this time the capacitor is subjected to full input voltage and it must be rated accordingly. A much 'lighter' type of capacitor can be used if it is shunted by a voltage-sensitive resistor which will limit the voltage across the capacitor.

12.6.1.3 Voltage stabilisation

A resistor is used either directly across a varying power source or in a bridge circuit with linear resistors, but in both cases the loss of energy is somewhat excessive.

Other uses suggested include lightning arrestors, shunting of rectifiers, arc furnaces, thyratrons, armatures, etc.

12.7 Variable resistors (potentiometers)

12.7.1 Types of variable resistor

There are two general classes of variable resistor: general purpose and precision. The general-purpose resistors may be sub-divided into wirewound and carbon-composition types. The precision resistors, which are always wirewound, usually follow linear, sine–cosine or other mathematical laws. Linearities as high as 0.01% (for linear) and 0.1% (for sine–cosine and other laws) are obtainable. The general-purpose types usually follow a linear law, but some follow a logarithmic law. They have overall resistance tolerances of 10% for the wirewound types (although much closer tolerances can be obtained) and 20% for the carbon-composition types.

The metal-film and the high-quality moulded-track types can also be considered to be in the precision category. Linearities of 0.5% in the moulded type and 0.1% in metal-film types (with the aid of trimming) are obtainable.

12.7.2 Electrical characteristics of variable resistors

A summary of the electrical characteristics of most types is given in *Table 12.7*.

Table 12.7 Summary of electrical characteristics of variable resistors

	Mfg. tolerance (%)	*Selection tolerance* (%)	*Overall stability (after climatic tests)* (%)	*Linearity* (%)	*Resolution (in degrees)*	*Life (number of sweeps)*
Carbon composition Coated-track types	±20	±20	±20	±15 (at 50% rotation)	(Stepless)	20 000 minimum Max. depends on construction
Carbon construction Moulded solid-track types	±20	±20	±5	±15 (at 50% rotation)	(Stepless)	20 000 minimum Max. depends on construction
Cermet (ceramic/metal)	±20	±20	±20	±15 (at 50% rotation)	(Stepless)	20 000 minimum
Conductive plastic	±20	±20	±10	±15 (at 50% rotation)	(Stepless)	Up to 25 million
Rotary wirewound general-purpose types	±10	±10	±2	1.0	1.0	20 000 minimum Max. depends on construction
Wirewound precision linear types	±5	Not applicable	High if sealed	Average 0.5 (can be 0.01)	Average 0.1 (depends on size, wire, etc.)	50 000 minimum Max. several millions
Wirewound precision toroidal types	±5	Not applicable	High if sealed	Average 0.1 (can be higher)	Average (depends on size, wire, etc.)	50 000 minimum Max. several millions
Wirewound precision helical types	±5	Not applicable	High if sealed	Average 0.25 (can be higher)	0.01 (for 10-turn pot.) (depends on wire, etc.)	50 000 minimum Max. several millions
Sine–cosine potentiometer Card-wound types	±5	Not applicable	High if sealed	Average 0.5 (can be higher)	Varies with slider position	50 000 minimum Max. several millions

12.7.2.1 Resistance value

For precision-variable resistors the upper limit of resistance value is about 100 kΩ; above this the element size may exceed 150 mm in diameter. General-purpose types are made in values up to 500 kΩ (wirewound) and 5 MΩ (carbon). The lower limit is about 1 Ω for wirewound resistors and about 10 Ω for carbon-composition types.

12.7.2.2 Resistance law

The resistance law is the law relating the change of resistance to the movement of the wiper, and it may be linear, logarithmic, log–log, sine–cosine, secant and the like, depending on the requirement for which the variable resistor is designed.

12.7.2.3 Linearity

There is often confusion between the terms *linearity*, *resolution*, *discrimination* and *accuracy* in discussing variable resistors. An ideal linear variable resistor has a constant resistance change for each equal increment in angular rotation (or linear movement) of the slider. In practice, this relationship is never achieved, and the linearity, or linear accuracy, is the amount by which the actual resistance at any point on the winding varies from the expected straight line of a 'resistance *vs* rotation' graph in a rotary variable resistor or 'resistance *vs* movement' graph in a linear variable resistor. For example, a 1 kΩ variable resistor held to a linearity of ±0.1% would not vary more than 1 Ω on either side of the line of zero error.

The terms *resolution* and *discrimination* are synonymous. Resolution, or discrimination, is the resistance per turn of resistance wire and is thus a function of the number of turns on the variable resistor. For example, a resistor of 100 Ω containing 100 turns of wire has a resolution of 1 Ω. Resolution may be defined more accurately as *resistance resolution*, *voltage resolution* or *angular resolution*. Resistance resolution is the resistance per turn; voltage resolution is the voltage per turn; and angular resolution is the minimum change in slider angle necessary to produce a change in resistance. In general, the resistance resolution is one-half of the linearity. For example, if the linearity (linear accuracy) of a 1 kΩ resistor is to be held to within 0.1%, the resistance resolution should be 0.05% or less, and the winding should have at least 2000 turns. The word *accuracy*, unqualified, has no meaning in defining a variable resistor.

12.7.2.4 Stability

Stability concerns the change of resistance with time, or under severe climatic conditions, as well as the behaviour under normal load conditions. For general-purpose carbon-composition variable resistors the stability tolerance is 25% and for general-purpose wirewound resistors it is 2%. The stability of the precision types of wirewound variable resistor is much higher, as these are usually sealed to exclude moisture and dust.

12.7.2.5 Minimum effective resistance

All variable resistors have some method of *ending off* the resistance element so that the slider goes into a *dead* position at each end, although it may rotate a few degrees more. There is a small jump in resistance, known as the *hop-off* resistance, as the slider touches the element. For general-purpose wirewound types, this should be less than 3% of the nominal resistance, and for carbon composition types it should be less than 5%.

12.7.2.6 Effective angle of rotation

The *dead* positions mentioned in the previous paragraph are known as the *hop-off* angles. For military use the hop-off angle must not exceed 10% of the total angular rotation at either end for general-purpose wirewound resistors and 30% for carbon-composition types. The effective angle of rotation is 360° less the sum of the hop-off angles at the ends and the space allowed for terminations.

12.7.2.7 Life under given conditions

Service specifications require that both wirewound and carbon-composition types should withstand 10 000 sweeps at 30 cycles/min with full-load current through the resistance element, totalling 20 000 cycles. After the test the change in resistance should not be more than 2% for the wirewound types and 5% for the composition types. Precision-variable resistors designed for long life—for example, with low brush pressure and carrying little current—have much longer lives—up to two million sweeps and sometimes up to ten million.

12.7.2.8 Performing under various climatic conditions

The most frequent causes of failure in variable resistors are corrosion of the metal parts and swelling and distortion of plastic parts such as track mouldings, cases and the like, due to moisture penetration. To combat these problems, the variable resistor should embody metal parts made from non-corroding metals, which may be difficult to fabricate, or be sealed in a container with a rotating seal for the spindle. The wattage rating is sometimes lowered slightly because of the sealing, but the life of the component is increased by many times. Some present types are made with solid-moulded carbon-composition tracks and bases that resist the effects of humidity.

12.7.2.9 Performance under vibration

In variable resistors difficulties may be experienced due to open circuit or intermittent contact if the slider vibrates off the track or due to change of resistance if the slider moves along the track. In general, the second is much more serious, particularly if the vibration occurs sideways to the potentiometer. Resonant frequencies vary between 100 and 300 Hz for the small 45 mm diameter wirewound potentiometer and at amplitudes of 5–10 g. The shaft-length and knob-weight also affect the resonant frequency. Reduction in shaft length to 6 mm may raise the resonant frequency to 1000 Hz or more, and the knob should be as light and as small as possible.

12.7.2.10 Noise

Electrical noise in carbon-composition variable resistors is usually due to poor or intermittent contact between the slider and the track. Variations of pressure, or the presence of dust or metal particles, cause changes in contact resistance, resulting in noise. Sealing or at least dust-proofing is necessary to avoid trouble due to dust contact variation. In wirewound variable resistors there are several types of noise contact resistance or constriction resistance noise, loading noise, resolution noise and vibrational noise due to slip-rings (if these are used).

Further reading

ABE, O., *et al.*, 'Effect of substrate thermal expansion coefficient on the physical and electrical properties of thick film resistors', *Thin Solid Films*, **162** (August 1988)

BROWN, J. A., 'Metal oxide film resistors', *Electronic Equipment News*, 10 (August 1970)

DUMMER, G. W. A., *Fixed Resistors*, 2nd edn, Pitman, London (1967)

DUMMER, G. W. A., *Variable Resistors*, 2nd edn, Pitman, London (1963)

DUMMER, G. W. A., *Materials for Conductive and Resistive Functions*, Hayden Book Co., New York (1970)

EDWARDS, L., 'Selecting NTC thermistors for control applications', *Sensor Rev.*, **8**(4) (October 1988)
KARP, H. R., 'Trimmers take a turn for the better', *Electronics*, **79** (17 Jan. 1972)
KHATER, F., *et al.*, 'Temperature coefficient of resistivity of double-layer thin metallic films for high-performance thin-film resistors', *J. Mater. Sci. Lett.*, **7**(10) (October 1988)
MODINE, F. A. *et al.*, 'New varistor materials', *J. Appl. Phys.*, **64**(8) (15 October 1988)
NEALE, L., 'Manufacturing and testing precision potentiometers', *Electronics & Power*, **497** (27 June 1974)
RAGAN, R., 'Power rating calculations for variable resistors', *Electronics*, **129** (19 July 1973)

13 Dielectric Materials and Components

J R G Lloyd CEng, MIMechE, MIQA
Technical Director
Nortronic Associates Ltd
(Sections 13.1–13.2.3, 13.2.5–13.3.6)

K Wakino
Senior Executive Director
Murata Manufacturing Co.
(Section 13.2.4)

Contents

13.1 Characteristics of dielectric materials

13.1.1 General characteristics

Dielectric materials used for capacitors can be grouped into the following five main classes:

(1) Mica, low-loss ceramic, glass, etc.: used for capacitors from a few pF to a few thousand pF.
(2) High-permittivity ceramic: used for capacitors from a few hundred pF to some μF.
(3) Paper and metallised paper: used for capacitors from a few thousand pF up to some μF.
(4) Electrolytic (oxide film): used for capacitors from just under 1 μF to many thousands of μF. Aluminium electrolytic and tantalum electrolytic.
(5) Dielectrics such as polystyrene, polyethylene terephthalate (polyester), polycarbonate, polypropylene, etc.: range of use from a few hundred pF to a few hundred μF.

Many factors affect the dielectric properties of a material when it is used in a capacitor. Among these are the permittivity, dissipation or power factor, insulation resistance, dielectric absorption, dielectric strength, operating temperature, and temperature coefficient of capacitance.

13.1.2 Summary of properties of capacitor dielectrics

A table of the main characteristics of some dielectric materials used in capacitors is given in *Table 13.1*.

13.1.3 Permittivity (dielectric constant)

The permittivity, dielectric constant or specific inductive capacity of any material used as a dielectric is equal to the ratio of the capacitance of a capacitor using the material as a dielectric, to the capacitance of the same capacitor using vacuum as a dielectric. The permittivity of dry air is approximately equal to one. A capacitor with solid or liquid dielectric of higher permittivity (ε) than air or vacuum can therefore store ε times as much energy for equal voltage applied across the capacitor plates. A few typical figures for capacitor dielectrics are shown in *Table 13.2*.

Table 13.2

	Permittivity (ε)
Vacuum	1.0
Dry air	1.000 59
Plastic film (polyester, polystyrene, etc.)	2.0–3.5
Impregnated paper	2.5–6.0
Mica and glass	4.0–7.0
Ceramic	
stable	up to 100
semi-stable	up to 1200
high K or high ε*	200–12 000
Aluminium oxide	7+
Tantalum oxide	11+

* The high permittivity in high-ε ceramic capacitors comes from the fact that the electric charges in the molecular structure of the material are very loosely bound and can move almost freely under the polarising voltage, resulting in high total capacitance.

Dielectrics can be classified in two main groups—polar and non-polar materials. Polar materials have a permanent unbalance in the electric charges within the molecular structure. The dipoles within the structure consist of molecules whose ends are oppositely charged. These dipoles therefore tend to align themselves in the presence of an alternating electric field (if the frequency is not too high). The resultant oscillation causes a large loss at certain frequencies and at certain temperatures. In non-polar materials the electric charges within the molecular structure are in balance and the dipoles do not rotate under applied fields, although they may distort. No sharp loss peaks with frequency and temperature therefore exist. Consequently dielectrics which are non-polar in structure make the best a.c. working capacitors, particularly at higher frequencies.

13.1.4 Losses in dielectrics

Losses occur due to current leakage, dielectric absorption, etc., depending on the frequency of operation. For a good non-polar dielectric the curve relating loss with frequency takes the approximate shape given in *Figure 13.1(a)*. For a polar material the loss–frequency curve may be shown approximately as in *Figure 13.1(b)*.

The variation of permittivity with frequency is negligible so long as the loss is low. Increased losses occur when the process of alignment cannot be completed, owing to molecular collisions, and in these regions there is a fall in permittivity. Viscous drag in the molecular structure limits the frequency at which full alignment can be carried out. If the applied frequency is comparable with the limiting frequency losses still become high.

Equivalent circuits showing series and parallel loss resistance can be given, but are greatly dependent on the system of measurement at any particular frequency. The important criterion is the ratio:

$$\frac{\text{power wasted per cycle}}{\text{power stored per cycle}}$$

This is the power factor of the material.

13.1.5 Absorption

If a capacitor were completely free from dielectric absorption the initial charging or polarisation current when connected to a d.c. supply would be

$$I=(V/R)\exp(-t/CR)$$

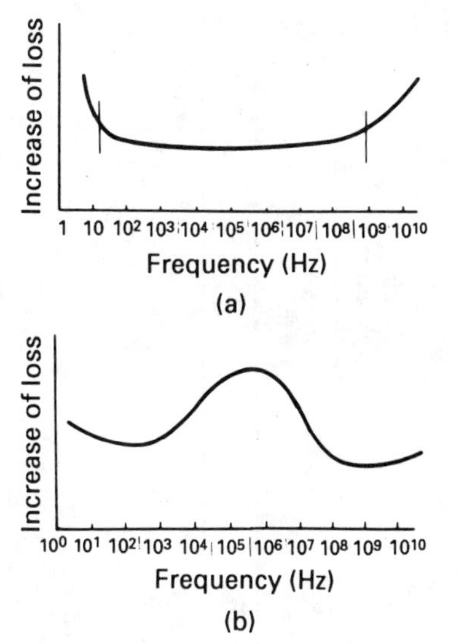

Figure 13.1 Loss–frequency curve for (a) non-polar dielectric, (b) polar dielectric

Table 13.1 Properties of some capacitor dielectric materials

Material	*Loss (at room temperature)*			*Permittivity (over operating frequency range)*	*Dielectric strength (V/mil) (Breakdown)*	*Temperature limits (°C)*		*Remarks (and some (registered) Trade Names)*
	Limiting frequency of operation		*Power factor (Loss angle tan δ) at 1 kHz*			*Approx. min.*	*Approx. max.*	
	Approx. min.	*Approx. max.*						
1	*2*	*3*	*4*	*5*	*6*	*7*	*8*	*9*
Kraft paper (capacitor tissue)	Poor below 100 Hz (can be d.c.)	1 MHz	0.01 to 0.03	4.5	500 to 1000 (depends on impregnant)	No limit but impregnant freezes	85 to 100	Capacitor properties depend greatly on impregnant
with mineral oil	—	—	0.003 5	2.23	—	−55	105	Effect of various impregnants
with castor oil	—	—	0.007	4.7	—	−25	65	
with silicone oil	—	—	0.003 5	2.6	—	−60	125	
with polyisobutylene	—	—	0.003	2.2	—	−55	125	
with epoxy resin	—	—	0.010	4.7	—	−40	100	
Mica (Ruby)	Precision work 100 Hz (can be d.c.)	10 000 MHz	0.000 5	7.0	1000	No limit	+200	Properties vary according to source of origin
Ceramic								
(Low-permittivity types)	10 Hz (can be d.c.)	10 000 MHz	0.001	Up to 100	200 to 300	No limit	+150 to 200	Class I ceramic—very stable, no ageing, capacitance and d.f. not affected by voltage and frequency
(Medium-permittivity types)	10 Hz (can be d.c.)	5000 MHz	0.02	Up to 1200	100 to 150	No limit	+120	Class II ceramic—capacitance and dissipation factor affected by time (ageing), a.c. and d.c. volts and frequency. Non-linear TCC (applies to both medium- and high-permittivity types)
(High-permittivity types)	10 Hz (can be d.c.)	1000 MHz	0.03	Approx. 1000 to over 12 000	100	−100	+120	

Table 13.1—*contd.*

Material	*Loss (at room temperature)*			*Permittivity (over operating frequency range)*	*Dielectric strength (V/mil) (Breakdown)*	*Temperature limits (°C)*		*Remarks (and some (registered) Trade Names)*
	Limiting frequency of operation		*Power factor (Loss angle tan δ) at 1 kHz*			*Approx. min.*	*Approx. max.*	
	Approx. min.	*Approx. max.*						
1	*2*	*3*	*4*	*5*	*6*	*7*	*8*	*9*
Glass (Soft lead–soda)	200 Hz	10 000 MHz	0.001	6.5 to 6.8	<500	No limit	+200	
Glass (Hard, borosilicate)	100 Hz	10 000 MHz	0.001	4.0	<500	No limit	+200	
Quartz (Fused)	100 Hz	>10 000 MHz	0.000 2	3.8	1000	No limit	+300	
Plastics								
Polyethylene	d.c.	10 000 MHz	0.000 2	2.3	1000	Becomes brittle −40	+70	Polythene, Alkathene, Telcothene
Polystyrene film	d.c.	10 000 MHz	0.000 2	2.5	2000	−40	+85	Distrene, Lustron, Styron, Styroflex
Polycarbonate	d.c.	10 000 MHz	0.001	2.9	1800	−40	+125	Makrofol
Polymonochlorotrifluoro-ethylene (PCTFE) film	d.c.	10 000 MHz	0.01 to 0.05	2.3 to 2.8	3000 to 5000	−195	+180	Kel-F, Hostaflon
Polyethylene terephthalate film	d.c.	1000 MHz	0.005	3.3	4000 to 5000	−40	+130	Mylar, Melinex, Terphane, Hostophan
Polypropylene	d.c.	1000 MHz	0.000 5	2.2	3000 to 5000	−40	+100	Bexphane, Moplefan
Paraffin wax	d.c.	10 000 MHz	0.000 1	2.2	1000	No limit but hardens	60	Capacitor impregnant, coil potting, etc.
Liquid insulants								
Paraffin	d.c.	10 000 MHz	0.001	2.2	1000	No limit but solidifies	50	Ozokerite (slightly plastic). Used for coil potting
Transformer oil	d.c.	1 MHz	0.001	2.4	1000	−40	120	
Silicone oil	d.c.	10 000 MHz	0.000 1	2.8	1000	−50	250	
Nitro-benzene	(see remarks)	10 000 MHz	0.000 5	40	50 to 1000	Freezes at +3 then low permittivity	150	Note high permittivity. Cannot be used on d.c. because of leakage. Has large negative temperature coefficient. Has specialised use as capacitor dielectric

* Can be d.c. but slow capacitance change possible.

† ppm/°C = parts per million per degree Celsius.

where

I = current flowing after a time t
V = applied voltage
R = capacitor series resistance
C = capacitance
e = base of Napierian logs (2.718)

and the polarisation current would die off asymptotically to zero. If R is small, this takes place in a very short time and the capacitor is completely charged. In all solid-dielectric capacitors it is found that, after a fully charged capacitor is momentarily discharged and left open-circuited for some time, a new charge accumulates within the capacitor, because some of the original charge has been 'absorbed' by the dielectric. This produces the effect known as dielectric absorption. A time lag is thus introduced in the rate of charging and of discharging the capacitor which reduces the capacitance as the frequency is increased and also causes unwanted time delays in pulse circuits.

13.1.6 Leakage currents and time constants of capacitors

Losses due to leakage currents when a capacitor is being used on d.c. prevent indefinite storage capacity being realised, and the charge acquired will leak away once the source is removed. The time in which the charge leaks away to 1/e, or 36.8%, of its initial value is given by RC, where R is the leakage resistance and C is the capacitance. If R is measured in megohms and C in microfarads the time constant is in seconds. This can also be expressed as megohm microfarads or as ohm farads. Some typical time constants for various dielectrics used in capacitors are:

Film (polystyrene–polypropylene)	several days
Impregnated paper	several hours
Tantalum–electrolytic capacitors	one or two hours
High-permittivity capacitors (ceramic)	several minutes
Etched foil aluminium electrolytic capacitors	several minutes

It should be borne in mind that below capacitance values of about 0.1 μF the time constant is generally determined by the structure, leakage paths, etc., of the capacitor assembly itself rather than the dielectric material. Leakage current increases with increase of temperature (roughly exponentially). In good dielectrics at room temperature it is too small to measure, but at higher temperatures the current may become appreciable, even in good dielectrics.

13.1.7 Insulation resistance (or insulance)

An ideal dielectric would allow no electrons to flow or make their way from one electrode plate to the other through the dielectric. There is no ideal, practical dielectric available, so a current of electrons does flow always from one plate to the other, resulting in 'leakage current' for the capacitor. Hence the insulation resistance for a capacitor is related to leakage current by Ohm's law ($V=IR$). It is usual to use leakage current for electrolytic capacitors, all other types being expressed in insulation resistance as the leakage current is low. The insulation resistance of a dielectric material may be measured in terms of surface resistivity in ohms or megohms, or as volume resistivity in ohm centimetres.

The insulation resistance of the assembled capacitor is important in circuit use. The insulation resistance of any capacitor will be lowered in the presence of high humidity (unless it is sealed), and will be reduced when operated in high ambient temperatures (whether sealed or not).

For perfectly sealed capacitors used under conditions of high humidity there should be no deterioration, but for imperfectly sealed capacitors the drop in insulation resistance will be roughly inversely proportional to the effectiveness of the sealing. Unsealed capacitors will show a large and rapid drop in insulation resistance under these conditions.

The effect of the insulation resistance value, both its magnitude and how it varies with temperature and humidity, is quite critical in circuitry where leakage of current through the capacitor can cause malfunction or undesirable effects to occur.

A method of measurement is given in BS 9070:1970.

13.1.8 Dielectric strength

The ultimate dielectric strength of a material is determined by the voltage at which it breaks down. The stress in kilovolts per metre (or volts per mm) at which this occurs depends on the thickness of the material, the temperature, the frequency and the waveform of the testing voltage, and the method of application, etc., and therefore comparisons between different materials should ideally be made on specimens equal in thickness and under identical conditions of measurement. The ultimate dielectric strength is measured by applying increasing voltage through electrodes to a specimen with recessed surfaces (to ensure that the region of maximum stress is as uniform as possible). Preparation of the specimen is important and their previous histories should be known.

The dielectric strength of a material is always reduced when it is operated at high temperatures or if moisture is present. Few materials are completely homogeneous and breakdown may take the form of current leak along certain small paths through the material; these become heated and cause rapid deterioration, or flashover along the surface and permanent carbonisation of the surface of organic materials. Inorganic materials such as glass and mica are usually resistant to this form of breakdown. The time for which the voltage is applied is important; most dielectrics will withstand a much higher voltage for brief periods. With increasing frequency the dielectric strength is reduced, particularly at radio frequencies, depending on the power factor, etc., of the material.

13.1.9 Effect of frequency

At very low frequencies, also at very high frequencies, there is an increase of loss which sets a limit to the practical use of a capacitor with any given dielectric. At very low frequencies various forms of leakage in the dielectric material have time to become apparent, such as d.c. leakage currents and long time constant effects, which have no effect at high frequencies. At very high frequencies some of the processes contributing to dielectric polarisation do not have time to become effective and therefore cause loss. These losses might be simply and approximately represented as in *Figure 13.2*.

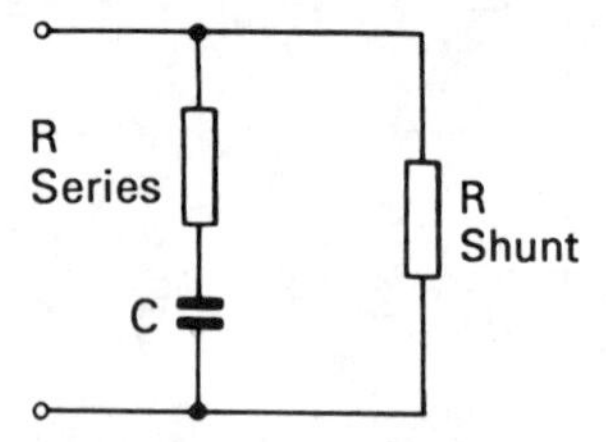

Figure 13.2 Losses in a capacitor

At very low frequencies the circuit is entirely resistive, all the current passing through the shunt resistance (d.c. leakage resistance, etc.). At very high frequencies the current passes through the capacitance C but all the volts are dropped across the series resistance and again the circuit is lossy. These limit the upper frequency independently of the dielectric material used. Similarly, leakage across the case containing the dielectric may limit the lower frequency so that not all the useful range of the dielectric itself may be realised.

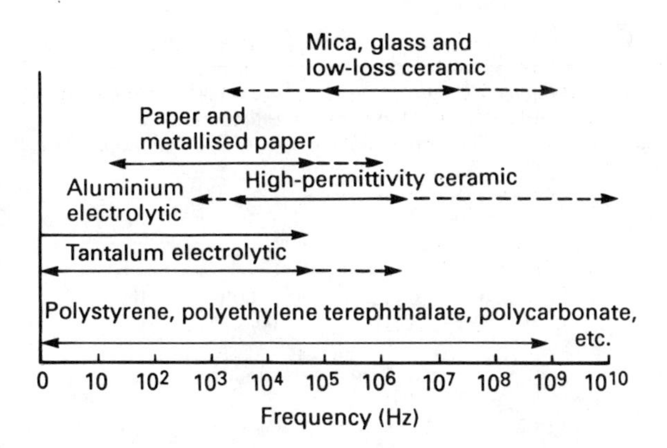

Figure 13.3 Frequency coverage of different classes of capacitor

The chart in *Figure 13.3* shows the approximate usable frequency ranges for capacitors with various dielectric materials. The construction of the capacitor assembly will affect the frequency coverage to some extent, so that the chart should be regarded as a guide only.

13.1.10 The impedance of a capacitor

The current (amperes) in a capacitor when an alternating voltage is applied is given by

$I = 2\pi fCV$

where

C = capacitance in farads
V = voltage in volts
f = frequency in hertz

The capacitive reactance (ohms) is given by

$X_C = 1/2\pi fC$

and the inductive reactance (ohms) is given by

$X_L = 2\pi fL$

An ideal capacitor would have entirely negative reactance but the losses, described previously, due to dielectric, case and leads, preclude this. In addition, inductance is also present in varying amounts and therefore as the frequency is increased the inductive or positive reactance increases, and above a critical frequency the capacitor will behave as an inductor. At the resonant frequency the impedance of the capacitor is controlled by its effective resistance or equivalent series resistance (ESR), which in turn is made up of the losses described. Every capacitor will resonate at some given frequency (depending on its construction) and, having inductance and resistance, will exhibit a complex impedance, capacitive in one range of frequencies, resistive in another and inductive in still another.

The impedance (Z) of a capacitor is expressed as follows:

$$Z = \sqrt{\left[\mathrm{ESR}^2 + \left(2\pi fL - \frac{1}{2\pi fC}\right)^2\right]}$$

It therefore follows that, when a capacitor is operated at a frequency higher than its resonant frequency, the inductance of the device becomes the major consideration.

13.1.11 Temperature effect on capacitors

Temperature variations have an effect on capacitance and dissipation factor of all types of capacitor. Some react only slightly, others will react moderately, while some react very considerably. Polystyrene is very stable on capacitance with a change in value of less than 2% from −55°C to +85°C. Polypropylene and polycarbonate are also classed as stable but polyethylene terephthalate although stable from 0°C to +50°C varies by as much as 10% in capacitance between −55°C and +100°C. The ceramic capacitor ranges have a very large variety of temperature coefficient of capacitance characteristics from very stable to very unstable. Almost any characteristic desired can be achieved depending on the dielectric mixture used, processing, method of assembly and stabilisation techniques following manufacture. Aluminium and tantalum electrolytic capacitors are classed as very poor, aluminium varying 30% between −10°C and +85°C and tantalum approximately 15%.

The capacitance change with temperature (TCC) characteristic is of critical importance in a good many circuits where the correct functioning of the circuit depends on the capacitance value remaining within very tight tolerance limits over a prescribed temperature range.

13.2 General characteristics of discrete fixed capacitors

13.2.1 Summary of electrical characteristics

Capacitors are generally divided into classes according to their dielectric, e.g. film, paper, ceramic, mica, aluminium oxide. It is useful to a designer to know the chief characteristics of these classes of capacitor and the main characteristics are briefly outlined in the following paragraphs. It is important to remember that capacitance is never constant, except under certain fixed conditions. It changes with temperature, frequency and age, and the capacitance value marked on the capacitor strictly applies only at room temperature and at low frequencies. A brief summary of their electrical characteristics is given in *Table 13.2*.

13.2.2 Impregnated-paper capacitors

This type of capacitor was the most widely used of capacitor types, made by rolling paper as insulation between two metal foils and filling with an impregnant, the whole being encapsulated in a suitable housing. Gradually the design was replaced by impregnated metallised paper and subsequently by the plastic film family of capacitors, film and foil, and metallised film. Owing to the very excellent self-healing characteristic of metallised paper (very low percentage of free carbon and high oxygen content), this type of capacitor has remained dominant worldwide in a.c. mains type applications. In the metallised paper capacitor, one side of the paper is metallised before rolling (winding). The main characteristics are the excellent self-healing under voltage stress, very good pulse handling capability (high dv/dt), and small size. If the paper is punctured due to stress of any kind, the thin metallising quickly evaporates in the area of the puncture, preventing a short circuit and allowing the capacitor to continue normal operation. This is a very desirable feature, when as many as ten transients per day can exist on mains supplies with amplitudes up to 1200 V and pulse steepness between 200 and 2000 V/ms. The paper of the capacitor is impregnated with a suitable impregnant, one of the best ones being epoxy resin. Such a.c. capacitors are very strictly controlled throughout the world by national and international bodies (BSI, VDE, SEV, UL, CSA, to name a few) and have to meet very stringent tests before being granted approval for use. The capacitors have approval categories—X1, X2, Y—which they have to meet depending on application, as in some instances a capacitor failing in service can result in danger to human life.

Table 13.2 Summary of the electrical characteristics of fixed capacitors

Capacitor type	*Capacitor stability (after climatic tests)* (%)	*Normal manu-facturing tolerance on capacitance* (%)	*Best manu-facturing accuracy* (%)	*Permit-tivity* (ε)	*Power factor* (*at* 1 kHz)	*Temperature coefficient* (ppm/°C)	*Maximum capacitor temperature for long life* (°C)
Impregnated-paper (rect. metal-cased and tubular capacitors)	5	±20	5	approx. 5	0.005 to 0.01	+100 to 200	100
Metallised-paper tubular capacitors	5	±25	5	5	0.005 to 0.01	+150 to 200	85
Moulded stacked-mica capacitors	2	±20 or ±10	5	4 to 7	0.001 to 0.005	±200	Up to 120 Depends on casing
Moulded metallised-mica capacitors	1	±10	±2	4 to 7	0.001 to 0.005	±60	Up to 120 Depends on casing
Glass-dielectric capacitors	1	±10	(±1 down to 10 pF)	approx. 8	0.001	+150	200
Glaze-dielectric capacitors	1	±10	1	5 to 10	0.001	+120	150
Ceramic-tubular, normal-ε capacitors	1	±10	±1	6 to 15 80 to 90	0.001	+100 −30 −750 (according to mix)	150
Ceramic-tubular, high-ε capacitors	20	±20	5 (low values) 1 (high values)	1500 and 3000 (may be higher)	0.01 to 0.02 (Varies with tem-perature, etc.)	−1500, varies. Non-linear	100
Ceramic class I (low permittivity type)	1	±10	±1	up to 100	0.001	0+30	125
Ceramic class II (medium permittivity type)	5	±20	±10	up to 1200	0.02	$\Delta C \pm 15\%$ max. (−55°C to +125°C)	125
Ceramic class II (high permittivity type)	20	−20% +80%	±20%	1000 to 12 000	0.03	ΔC+22–56% max. (from +10°C to +85°C)	85
Polystyrene-film, tubular and rectangular capacitors	>1	±10	±1	2.3	0.000 5	−150	85
Polyethylene terephthalate (Melinex) capacitors	5	±10	5	2 to 5, depending on frequency	0.01 at 1 kHz varies with tempera-ture and frequency	Varies with temperature	125
Polycarbonate	2	±10	±1	2.9	0.001	150	125
Polypropylene	2	±10	±1	2.2	0.0005	−170	100
Electrolytic (normal) capacitors	10	−20 to +50	10	—	0.02 to 0.05	+1000 to 2000 approx.	105
Electrolytic, tantalum-pellet	5	±20	10	—	0.05	+100 to 200	125
Electrolytic, tantalum-foil capacitors	10	±10	5	—	0.05	+500	85
Precision-type air-dielectric capacitors	±0.01	—	0.01	1	0.000 01	+10	20

13.2.3 Mica-dielectric capacitors

The main characteristics of this type of capacitor are low power factor, high voltage operation and excellent long-term stability at room temperature. Mica, being a natural mineral and adapted to use without physical or chemical alteration, is completely inert both dimensionally and electrically. Mica capacitors have been manufactured in two common constructions, the earliest being accompanied by the classic technique of interleaving alternate layers of insulating mica and conducting material (foil). This construction has been largely supplanted by silvered mica capacitors, where a thin layer of silver replaces the foil and is screened and fired onto the surface of the mica. The extract screening process provides an area of the surface of the mica which is predictable in size and position to a degree not possible in a foil construction. This coupled to the plate being physically bonded to the dielectric, allowing no relative motion to occur in the presence of physical, electrical or environmental stresses, makes the capacitor extremely stable.

The stability of the silvered-mica plate type is about 1% under normal conditions of use and that of the stacked-mica-plate type about 2%. They are sealed in suitable housing to prevent moisture, etc., from affecting the stability. The temperature coefficient is low, between ± 60 ppm/°C. The silvered-plate capacitor has a better temperature coefficient than the stacked-plate tupe. Both types, especially the stacked-plate capacitor, show slight non-cyclic capacitance shifts during temperature cycling and, in most of the types available, the temperature/capacitance curve is not entirely linear. There is also a wide spread of mean temperature coefficients between different specimens, even of the same batch.

The power factor of mica is approximately 0.0003 at 1 kHz, but can be as low as 0.0001 when specially selected and very dry. At radio frequencies the permittivity is about 7. The current-carrying capacity of the silvered plate imposes a limit to radio frequency and pulse loading. The silvered-plate capacitor is therefore less suitable for heavy current work than the stacked-plate type, although the latter is less stable and cannot be made to such a close selection tolerance as the silvered-plate capacitor.

13.2.4 Ceramic-dielectric capacitors

Ceramic capacitors are categorised into three classes by their dielectrics and two types by their construction.

13.2.4.1 Dielectrics

Class 1 These dielectrics are called temperature compensating ceramics because their capacitance changes almost linearly with temperature and is cyclic in behaviour. Capacitors with these dielectrics are used for temperature stabilisation of resonant circuits or for applications where high *Q*-factor and stability are design criteria. The permittivity usually ranges from 6 to 400, and the temperature coefficient is between $+120$ and -4700 ppm/°C. The most commonly specified temperature coefficient is NPO (0 ± 30 ppm/°C). The dielectrics are composed of paraelectric titanates.

Class 2 Capacitors of this class are used for bypass and decoupling applications or frequency discrimination where *Q*-factor and stability are not of major importance. This class is divided into two sub-classes composed of barium titanate.

In the medium-*K* class, the permittivity is in the range from 500 to about 4000 and is relatively stable over a wide temperature range of -55 to $+125$°C with a maximum capacitance change of ± 15%.

In the high-*K* class, the high permittivity ranging from 5000 to 30 000 is obtained by shifting the Curie point of the ferroelectrics such as barium titanate and lead-based relaxor to near room temperature. The capacitance is, however, fairly sensitive to temperature and voltage stress.

Class 3 Thin effective dielectric layers are formed at the ceramic surface or the grain boundary of the semiconductive barium titanate or strontium titanate ceramics, resulting in very high capacitance values. Apparent permittivity reaches 200 000. The capacitors are suited for decoupling or bypassing where low leakage resistance and voltage sensitivity can be tolerated.

13.2.4.2 Construction

Single-layer ceramic capacitors Disc, rectangular plate and tube types are commonly manufactured by highly automated processes. Disc types are common in these capacitors because they are most inexpensive to construct. To have a high volumetric efficiency, the dielectric thickness is designed as thin as possible with high-*K* materials. High voltage disc capacitors, which have rating voltage from 500 V to 30 kV, are composed of strontium titanate based materials that have stable dielectric properties under high voltage stress.

Multilayer ceramic capacitors The capacitors are made from a pile of many thin ceramic layers with inner electrodes separating each of the dielectric layers. The electrodes are exposed at alternate ends of the capacitor so that a paralleling of the individual capacitor elements is achieved. This type of capacitor, particularly the chip type, has become extremely popular because of its wide capacitance range, high volumetric efficiency, suitability for surface mounting technology (SMT) and high performance at high-frequency applications.

13.2.5 Glass-dielectric capacitors

These capacitors are formed of very thin glass sheets (approximately 12 μm thick) which are extruded as foil. The sheets are interleaved with aluminium foil and fused together to form a solid block. Their most important characteristics are the high working voltages obtainable and their small size compared with encased mica capacitors.

Glass-dielectric capacitors (glass capacitors) have a positive temperature coefficient of about 150 ppm/°C, and their capacitance stability and *Q*-factor are remarkably constant. The processes involved in the manufacture of glass can be accurately controlled, ensuring a product of constant quality, whereas mica, which is a natural product, may vary in quality. As the case of a glass capacitor is made of the same material as the dielectric, the *Q*-factor maintains its value at low capacitances, while the low-inductance direct connections to the plates maintain the *Q*-factor at high capacitances.

These capacitors are capable of continuous operation at high temperatures and can be operated up to 200°C. They are also used as high-voltage capacitors in transmitters.

13.2.6 Glaze- or vitreous-enamel-dielectric capacitors

Glaze- or vitreous-enamel-dielectric capacitors (glaze- or vitreous-enamel capacitors) are formed by spraying a vitreous lacquer on metal plates which are stacked and fired at a temperature high enough to *vitrify* the glaze. Capacitors made in this way have excellent r.f. characteristics, exceedingly low loss and can be operated at high temperatures, 150°C to 200°C. As they are *vitrified* into a monolithic block they are capable of withstanding high humidity conditions and can also operate over a wide temperature range. The total change of capacitance over a temperature range of -55°C to $+200$°C is of the order of 5,9

The temperature coefficient is about +120 ppm/°C and the cyclic or retrace characteristics are excellent. As in the glass capacitor the encasing material is the same as the dielectric material and therefore all corona at high voltages is within the dielectric. They are extremely robust and the electrical characteristics cannot normally change unless the capacitor is physically broken.

13.2.7 Plastic-dielectric capacitors

In plastic-dielectric capacitors the dielectric consists of thin films of synthetic polymer material. The chief characteristic of plastic-film capacitors is their very high insulation resistance, small physical size and high reliability, particularly with the metallised film types. There are many different materials and each one is used because of some particular characteristic or combination of characteristics required for specific circuit applications. The main synthetic polymer films used as capacitor dielectrics are as given below.

13.2.7.1 Polyethylene terephthalate (polyester)

This is a tough polymer with high tensile strength, free from pinholes and with good insulating properties over a reasonably wide temperature range. The film, which has a polar molecular structure, is available for capacitor products in a thickness range from 100 μm down to 1.5 μm. Its combination of good insulation properties over a wide temperature range and high mechanical strength are the outstanding characteristics. It also has a low moisture absorption. The high mechanical strength and high softening point (approximately 210°C) make polyester ideally suited for metallising by vacuum evaporation. The resulting capacitors exhibit the very beneficial self-healing properties associated with this type of construction, and polyester consists of sufficient oxygen to produce good self-healing. The insulation resistance of capacitors manufactured with polyester is a factor of 50 or more higher than that of equivalent impregnated capacitors, but is still lower than that of polycarbonate, polystyrene and polypropylene film capacitors. Owing to the good all-round mechanical and electrical properties and the high dielectric constant of polyester film, capacitors made using this film are the most widely used of any film or paper capacitor. They offer the widest selection of capacitance values, voltage ranges, sizes and styles. Polyester capacitors are normally used for coupling/decoupling applications at low and medium frequencies but find many other applications in general and power electronics as well as in certain a.c. applications. Polyester is typically ideal for d.c. applications because of its high dielectric strength and resistivity. Polyester is known under a variety of trade names such as Melinex (ICI), Mylar (DuPont), Hostophan (Germany) and Terphane (France).

13.2.7.2 Polycarbonate

This is a polyester of carbonic acid and bisphenols. It combines good physical properties with a lower loss (or dissipation factor) than polyethylene terephthalate. It has a temperature characteristic nearer to zero and is available in film form down to 1.5 μm in thickness. The dielectric constant is 2.9 at +20°C and 1000 Hz. The film softening point is approximately 155°C—this coupled to good mechanical strength allows polycarbonate to be metallised by vacuum evaporation. The self-healing properties of polycarbonate are not quite as good as for polyester and polypropylene film, as it does not have as much available oxygen. The film, which has a polar molecular structure, has a higher insulation resistance than polyester but finds greatest use when small physical size coupled with a flat and tight temperature coefficient of capacitance, between 0°C and 70°C, is required. Polycarbonate allows the manufacture of close-tolerance (typically ±1%), stable, long-life capacitors and is predominantly used for d.c. applications. Because of its low dissipation factor it can be used for a.c. applications.

13.2.7.3 Polystyrene

This is a hydrocarbon material and has a lower permittivity than the previous two dielectrics. However, it has a better dissipation factor. It has excellent dielectric properties due to its non-polar molecular structure. Its tensile strength is much lower than that of the other three films listed, and this, coupled to its relatively low softening temperature (90°C) and lack of available oxygen for the self-healing action, makes polystyrene unsuitable for metallising. Consequently film thicknesses below 6 μm are not commercially required or available. After winding with 5 μm foils, the capacitor elements are carefully heat-treated. This makes the capacitor extremely stable in capacitance owing to the slight contraction of the film that takes place. Polystyrene film and foil capacitors have a very high insulation resistance and a very low and stable temperature coefficient of capacitance. This makes them ideally suited for use in equipment that must meet exact requirements, although they do not have the self-healing properties of the other three metallised films discussed.

13.2.7.4 Polypropylene

This is a low-price material and has a very low dissipation factor. It is obtained from a non-polar polymer and its most advantageous electrical properties are due to the fact that the dielectric properties are very little affected by temperature and frequency changes. The film, which is available down to a thickness of 4 μm, has a softening point of approximately 145°C. This together with its good mechanical strength allows the material to be metallised by vacuum evaporation. Hence, in a similar way to polyester and polycarbonate, polypropylene also provides the very desirable self-healing property associated with metallised dielectrics. Polypropylene has a higher negative temperature coefficient and a lower dielectric constant than polystyrene; however, as it is available in metallised form, much smaller physical sizes are obtained. The non-polar nature of the film, coupled with good frequency response and all the other good dielectric properties it exhibits, makes polypropylene ideally suited for a.c., high-pulse and high-frequency applications. Capacitors are available in a very wide variety of values and configurations. Frequently, metallised polypropylene is combined with impregnated paper to provide capacitors with high a.c. and d.c. voltage capabilities, ideally suited for power electronic applications.

A comparison chart of some of the main characteristics of the four film materials is given in *Table 13.3*.

13.2.8 Electrolytic capacitors

The most notable characteristic of these capacitors is the large capacitance obtainable in a given volume, especially if the working voltage is low. Generally, electrolytic capacitors can be divided into two basic types by the nature of the base-oxidisable metal used. This is usually either aluminium or tantalum. There are other usable metals but these two yield relatively large capacitance per unit volume at an economical price. Electrolytic capacitors are very widely used in electronic/electrical circuits

Table 13.3 Film material properties

	Polyethylene terephthalate	*Polycarbonate*	*Polystyrene*	*Polypropylene*
Permittivity (at 1 kHz, 20°C)	3.3	2.9	2.5	2.2
Dissipation factor (at 1 kHz, 20°C)	0.005	0.001	0.0002	0.0005
Dielectric strength (V/μm) (at 20°C)	304	184	200	204
Insulance resistance (Ω – F) (at 20°C)	3×10^4	6×10^4	1×10^6	1×10^5
Dielectric absorption (%) (at 20°C)	0.2	0.05	0.01	0.01
Max. capacitor operating temperature (°C)	125	125	85	100
Film molecular structure	Polar	Polar	Non-polar	Non-polar
Metallisation	Available	Available	Not available	Available

and also in power electronic applications. They are found in filters, time constant circuits, bypass, coupling/decoupling and smoothing applications to name a few. Electrolytic capacitors are manufactured in many shapes and configurations from as low as 0.1 μF up to 1.0 F, in voltages up to 500 V d.c. Generally they are polarised and are used on d.c. only. They are available in a non-polarised form but usage is limited (motor-start being a typical application).

Aluminium electrolytic capacitors are generally made by a wet foil process (aluminium foil anode and liquid electrolyte), a more recent development being a solid aluminium version (aluminium anode and a solid semiconductor electrolyte).

Tantalum electrolytics are available in several forms: wet foil type (tantalum foil anode and liquid electrolyte), wet slug type (tantalum 'sintered slug' anode and liquid electrolyte), dry slug type (tantalum 'sintered slug' anode and solid semiconductor electrolyte). Wet foil aluminium and dry slug tantalum electrolytic capacitors are the most widely used types.

Construction of all types of aluminium and tantalum electrolytics, although widely different in detail, does tend to follow the same basic principle. A super-pure anode (99.99%) is produced with a very enhanced surface area (e.g. etched foil). This anode is treated electrochemically to form a very thin oxide of the anode material on the anode. The oxide is the dielectric of the capacitor and the thickness of the oxide determines the capacitor voltage rating. Since the oxide is so thin and the surface area so great, the capacitance of electrolytic capacitors is very large. Next to the oxide is a spacer of some form (paper and electrolyte in the wet aluminium) which becomes the other conducting plate of the capacitor. Some form of conductor is placed next to the spacer (cathode foil of the wet aluminium) to enable contact to be made with the external negative terminal.

It is most important in modern electrical applications that high temperature and good long-life performance are obtained from electrolytic capacitors. (Typically available are 125°C capacitors, giving 20 years and more of life at 40–50°C.) Capacitors are available which can perform well at frequencies up to 100 kHz. Permanent changes in capacitance, power factor and leakage current are used to determine end of life.

13.2.8.1 Capacitance

Electrolytic capacitors used for their very high capacitance in a given volume are not noted for their capacitance stability. Indeed, in the applications in which these are used, it is not particularly required. There is generally a slight increase in capacitance (about 15%) when the temperature is raised from 20°C to 85°C, and a gradual decrease as the temperature is reduced to −30°C. A more rapid capacitance decrease takes place below −30°C. There is a marked difference between aluminium and tantalum electrolytics in their temperature coefficient of capacitance. There is also a difference between capacitors of similar types manufactured by different manufacturers. If temperature coefficient of capacitance is of great importance in a circuit, manufacturers' literature should always be consulted. Electrolytic capacitors are manufactured to be within a selection tolerance. At 100 Hz, there is a slight decrease in capacitance as the applied frequency is increased (typically a 10% reduction at 10 kHz).

13.2.8.2 Power factor

The power factor (PF) is the ratio of the ESR to the total impedance of the capacitor. The dissipation factor (DF) of a capacitor is the ratio of the ESR to the capacitive reactance. For low losses the power factor is equal to dissipation factor. In effect, power factor indicates the total losses in the capacitor but the energy loss due to the d.c. leakage current is so small that it can be neglected, all of the energy loss being attributed to the a.c. or ripple current component present. As power factor can be considered to be a resistive component in the capacitor, any factor that adds to this resistive effect will considerably affect power factor. In electrolytic capacitors one such factor is the resistance of the electrolyte, and this resistance is the major contribution of power factor to an electrolytic capacitor. Since the resistance of the electrolyte rises rapidly with a fall in temperature, it will be seen that power factor is very susceptible to temperature changes and at temperatures of the order of −30°C can be approaching 100%. This effect can be modified to a certain extent by modification of the electrolytes. Modern non-aqueous electrolytes give a much smaller change of power factor with temperature and can be operated over a much wider range of temperature.

Power factor is affected by operating frequency, the change being greater with high-voltage electrolytes than with low. For example, a 25 V capacitor with a power factor of 7% at 50 Hz could have a power factor of 40% at 1000 Hz.

13.2.8.3 Leakage current

This is normally considered instead of insulation resistance in electrolytic capacitors. All electrolytic capacitors allow a small amount of d.c. leakage current to pass through when rated polarised d.c. voltage is applied. This current will vary with temperature and will rise with increased applied voltage. In general the amount of leakage current is indicative of the immediate quality of the electrolytic capacitor; the lower the better. At low temperatures (typically −30°C) the leakage current is quite a low value, but at +85°C it can be more than ten times the +20°C value. Initially leakage current increases with applied voltage, being very high when the load voltage is first applied, but then it falls rapidly and after about five minutes tends to reach a stable value.

13.2.8.4 Impedance/ESR

There is a gradual increase in impedance as the temperature is reduced and at −30°C and 100 Hz this can be over five times the +20°C impedance. At still lower temperatures a more rapid increase occurs; at temperatures above +20°C much smaller variations occur. The impedance falls rapidly with increase in frequency until a minimum value is reached at a point called the *resonance point* of the capacitor. The absolute value of impedance and the frequency at which it occurs vary considerably from type to type and also style to style of capacitor. (For example, a foil type aluminium electrolytic varies considerably from a sintered slug tantalum.) The resonance point is typically between 10 kHz and 100 kHz, and this can always be obtained from manufacturer's literature. At resonance the impedance of an electrolytic capacitor is influenced by the equivalent series resistance only, capacitive and inductive reactances cancelling each other out. At frequencies below resonant frequency the capacitive reactance plays the most important role along with ESR, while above the resonant frequency the inductive reactance plays an ever-increasing role. Consequently it is most important that electrolytic capacitors, used more and more for high-frequency applications, are designed, manufactured and selected with their ESR and equivalent series inductance (ESL) as low as possible. Once again we see the importance of capacitors with low impedance working at high frequencies.

13.2.8.5 Life expectancy

Very long lives can now be obtained from a great number of electrolytic capacitor types (aluminium and tantalum). Modern technology allows capacitors to be designed for certain specific applications (low leakage, high temperature, high frequency, long life, etc.) and combinations thereof. A circuit designer is therefore able to select the electrolytic capacitor that specifically meets the design conditions. End of life usually comes about with a deterioration in one or more of the major parameters—drop in capacitance, increase in power factor, leakage current or ESR. It is well to note that high temperatures reduce life expectancy, and great care should be taken regarding this point when designing in electrolytic capacitors by either selecting the highest working temperature types available or taking action to reduce working temperatures. It is now possible with computer-aided design techniques to design in electrolytic capacitors which exactly suit the design and working parameters, giving the optimum technical and commercial solution to a given application.

13.2.9 Air-dielectric capacitors

Air-dielectric capacitors are used mainly as laboratory standards of capacitance for measurement purposes. With precision construction and use of suitable materials, they can have a permanence of value of 0.01% over a number of years for large capacitance values.

13.2.10 Vacuum and gas-filled capacitors

Vacuum capacitors are used mainly as high-voltage capacitors in airborne radio transmitting equipment and as blocking and decoupling capacitors in large industrial and transmitter equipments. They are made in values up to 500 pF for voltages up to 12 000 V peak. Gas-filled types are used for very high voltages—of the order of 250 000 V. Clean dry nitrogen may be used at pressures up to 10^5 kg/m^2. They are specially designed for each requirement.

13.3 Variable capacitors

Variable capacitors may be grouped into five general classes: precision types, general-purpose types, transmitter types, trimmers and special types.

13.3.1 Precision types

Precision types have been developed for many years mainly as laboratory sub-standards of capacitance in bridge and resonant circuits and numerous measuring instruments have been designed around them. Various laws are available and capacitances up to 5000 pF can be obtained in one swing. Capacitance tolerances are of the order of one part in ten thousand and long term stabilities under controlled conditions of 0.02% over as many years as possible.

13.3.2 General-purpose radio types

General-purpose radio types are used as tuning capacitors in broadcast receivers. They have developed from large single capacitors to compact four- or five-gang units which can have a standard capacitance tolerance of within 1% or 1 pF to a stated law. The power factor of a modern air-dielectric variable capacitor, at 1 MHz, varies between 0.03% (at the minimum capacitance setting) and 0.6% (at the maximum capacitance setting). They are available in many laws, e.g. straight-line frequency, straight-line wavelength, straight-line percentage frequency, so that they can be used in test equipment and receivers of many types. The normal capacitance swing of this type of capacitor is about 400–500 pF, but components can be obtained in capacitance swings (in ranges) from 10 to about 600 pF.

13.3.3 Transmitter types

Transmitter types are basically similar in design but have wider spacing between vanes to allow safe operation at much higher voltages. The capacitance swing usually ranges up to about 1000 pF. The edges of the vanes are rounded and polished to avoid flashover, and special attention is paid to shape and mounting for high-voltage operation. The most common laws are square-law capacitance (or linear frequency) and straight-line capacitance. Special split-stator constructions are also used for push–pull circuits. Oil filling increases the capacitance and working voltage from two to five times, depending on the dielectric constant of the oil used. Compressed gas variable capacitors use nitrogen under pressures up to 1.5×10^6 kg/m^2 for broadcast transmitters. Bellows-type variable capacitors are made in the USA, variable from 10 to 60 pF, which are only 3 in (7.62 cm) in diameter and 5 in (12.7 cm) long. These can operate at 20 000 r.f. peak volts and 10 A maximum r.m.s. current at 20 MHz.

13.3.4 Trimmer capacitors

With the wide variety of trimmer capacitors currently available, there is uncertainty and at times confusion when selecting the correct trimmer for a particular application. Whilst a wide variety of trimmers can often be used, there are usually only a small number that will ideally meet all the requirements of a particular application (technically, commercially and environmentally). Under the pressure of the trend towards miniaturisation, trimmers are now available in very small sizes for printed circuit board mounting.

Trimmer capacitors come in many classes, some of the main ones being: air-spaced rotary types; compression types (usually mica); ceramic, glass, or quartz dielectric, rotary and tubular types; and PTFE dielectric rotary types. The capacitance range for rotary air-spaced types covers from approximately 1.6 pF to 100 pF. This is effected in a very wide variety of swings, namely 1.7–3.8 pF; 2.4–21 pF; 3.0–32 pF; etc. Rotary air-spaced trimmers come in a wide variety of sizes, shapes, mounting, etc. (sub-miniature, standard, low-profile) and can consist of both rotor and stator assemblies, being precision machined from solid metal. This results in the elimination of practically all soldered joints and particularly improves mechanical rigidity. This consequently improves electrical stability and consistency of capacitance. Working voltages go up to 400 V d.c., insulation resistance can be greater than 5×10^3 MΩ, temperature coefficient of capacitance is 50 ppm/°C, and temperature range is −55°C to 125°C, with 56 days humidity classification.

Compression types cover a much wider total range from 1.5 pF to 2000 pF, again in a wide variety of stages.

Ceramic-dielectric capacitance ranges are usually smaller, from 0.9 pF to 2.0 pF, ranging up to 5–100 pF, depending on the temperature coefficient required (typically N1500, N700, N500, NPO, P100, etc.). Voltages are typically 50 V d.c., 100 V d.c., etc., with the maximum being 500 V d.c. Insulation resistance is typically greater than 10^3 MΩ and self-resonant frequency is from 100 MHz to 1000 MHz depending on style (disc, tubular rotating piston, tubular non-rotating piston, etc.). The ceramic-dielectric types have a temperature range of −40°C to +85°C. Ceramic trimmers are used successfully by the million and have two distinct advantages; low price and small size. Should a higher grade device be required, a glass or quartz dielectric tubular trimmer has traditionally been used. This has enabled greater excursions of temperature (+125°C), better temperature coefficient of capacitance (typically ±50 ppm/°C), higher working voltage (1250 V d.c.) and higher, higher insulation resistance (10^6 MΩ), and in general a greater stability, better linearity, etc.

However, now available is the PTFE dielectric rotary trimmer, which can also be classed as a high-reliability high-grade trimmer. This type uses the PTFE fluorocarbon as the dielectric, which ensures high stability under severe operating conditions. The PTFE is specially processed from high-grade dispersion polymer and machined from solid to enhance the extremely low loss and hydrophobic nature of the dielectric. Working voltages range up to 600 V d.c. and climatic category can be −55°C to +125°C, with 56 days damp heat. Insulation resistance is greater than 10^4 MΩ and temperature coefficient of capacitance is 50 ppm/°C. Self-resonant frequency is of the order of 1–4 GHz and capacitance values range from 0.2 pF to 29 pF, in the usual stages.

13.3.5 Special types

Special types of variable capacitor are sometimes required, such as differential, split-stator, or phase-shifting capacitors. Phase-shifting capacitors are used mainly in radar systems for the accurate measurement of time intervals, and in high-speed sweep-scanning circuits. In addition, specialised multi-gang capacitors designed for transmitter/receivers are also produced.

13.3.6 Summary of electrical charcteristics

A summary of main electrical characteristics of variable capacitors is given in *Table 13.4*.

Further reading

APPS, L. T. and PLASKETT, J. A., 'Capacitors and fixed resistors—a continuing evolution', *Electronics & Power (GB)*, **22**, No. 7, p. 429 (July 1976)

ATKINSON, A. D., 'Fixed and variable capacitors—past, present and future', *New Electron. (GB)*, **12**, No. 6, p. 102 (20 March 1979)

BELL, R. W., 'Tuning devices. Variable capacitors', *Electronic Equipment (GB)*, **17**, No. 9, p. 69 (September 1978)

BOGATIN, E., 'Design rules for microstrip capacitance', *IEEE Trans. (CHMT)*, **11**(3), September (1988)

Table 13.4 Summary of electrical characteristics of variable capacitors

Type	*Capacitance law*	*Approximate capacitance swing* (pF)	*Power factor*	*Approximate operating voltage* (V d.c.)	*Temperature coefficient* (ppm/°C)
Single unit, precision	SLC	from 100 to 1500	0.000 01	1000	+10 (best)
Multi-gang, precision	SLF	320	0.000 05	500	+20 (−15 with compensating vane)
Single unit, general-purpose	SLC and exp.	15–100 and 350–550	0.001	750	+120
Multi-gang, general-purpose	SLC and exp.	15–100 and 350–550	0.001	750	+120
Multi-gang miniature, general-purpose	SLC	300–350	0.001	500	+150
Trimmers, air-dielectric, vane-type	SLC	Range from 2 to 100	0.001	500 to 1250	+50 to +120
Concentric trimmers, air-dielectric	SLC	6.0 and 27	0.007	300	+300
Mica compression trimmers	Non-linear	Range from 8 to 650 (rotary type to 3000)	0.001	350–500 —rotary type 3000 (up to 1000 pF)	Poor
Ceramic trimmers	Approx. SLC	from 5 to 100	0.002 to 0.005	500	Depends on ceramic
Plastic concentric trimmers	SLC	29	0.001	600	Depends on plastic (PTFE 50)

CAMPBELL, D. S., 'Electrolytic capacitors', *The Radio & Electronic Engineer*, **41**, No. 1 (Jan. 1971)
'CAPACITORS—A COMPREHENSIVE EDN REPORT', *Electronic Design News, USA*, p. 139 (May 1966)
CAPACITORS (Supplement), *Electronic Weekly*, p. 14 (21 March 1973)
CAPACITOR AND RESISTOR TECHNOLOGY SYMPOSIUM (CARTS), Europe and USA (1987)
CAPACITOR AND RESISTOR TECHNOLOGY SYMPOSIUM (CARTS), Europe and USA (1988)
DEKKER, E., 'SAL capacitors for automotive applications', *Electronic Components and Applications*, **8**(4) (1988)
DUMMER, G. W. A., *Variable Capacitors and Trimmers*, Pitman, London (1957)
DUMMER, G. W. A., *Fixed Capacitors*, 2nd edn, Pitman, London (1964) (Contains comprehensive bibliography up to 1962)
DUMMER, G. W. A. and NORDENBERG, H. M., *Fixed and Variable Capacitors*, McGraw-Hill, New York (1960)
EVANS, D., 'Fixed ceramic capacitors', *New Electron. (GB)*, **8**, No. 9, p. 86 (29 April 1975)
GIRLING, D. S., 'Quality control in capacitor-production and testing', *The Radio & Electronic Engineer*, **40**, No. 4, p. 173 (Oct. 1970)
LLOYD, J. R. G., 'Fail safe ceramic capacitors', *Electronic Engineering* (June 1984)
'MINIATURE CAPACITOR PROGRESS', *Electron*, p. 15 (29 June 1972)
PAMPLIN, B. F., 'Capacitor selection—facts and figures', *Electronic Equipment News*, p. 44 (Dec. 1969)
'RESISTORS AND CAPACITORS', *Electronics Weekly*, p. 17 (14 June 1978)
RIFA CAPACITORS, *RFI Suppression, Basic Information* (1988)
VON HIPPEL, A., *Dielectrics & Waves*, Chapman & Hall, London (1954)
VON HIPPEL, A., *Dielectric Materials & Application*, Chapman & Hall, London (1954)
YOSHIDA, A. *et al.*, 'Aluminium collector electrodes formed by the plasma spraying method for electric double-layer capacitors', *IEEE Trans. (CHMT)*, **11**(3), September (1988)

14 Magnetic Materials

P R Knott MA, CPhys, MInstP, MICeram
Morgan Materials Technology Ltd

Contents

14.1 Basic magnetic properties

14.1.1 The origins of magnetism

Permanent magnets, in the shape of lodestones, were known to have been in use around 2000 BC for direction finding. The dipolar nature of a magnet was apparent to the ancient Greeks, with its north seeking or N pole and S pole. They also knew that pieces of iron are attracted by a magnet. Gilbert, in the late 16th century, summarised the forces of interaction between two magnets in the famous law 'like poles repel, unlike poles attract'. It is convenient to think of a *magnetic field* existing wherever magnetic effects are experienced, whose direction at any point is given by the N–S axis of a small permanent magnet freely suspended at that point. This is the basis of the familiar compass-needle method of field plotting. Lines of force for a bar magnet are shown in *Figure 14.1*.

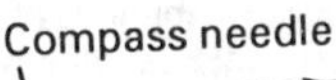

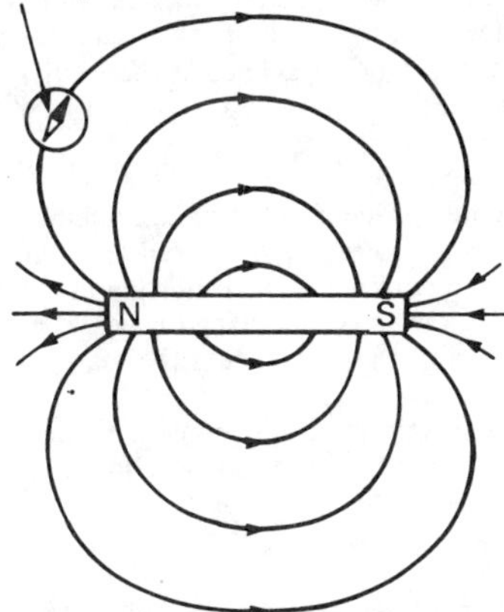

Figure 14.1 Lines of force for a bar magnet (After Duffin,[2] reproduced by permission of the publisher)

In the early 1820s Ampère showed that two coils carrying electric currents, or one such coil and a magnet, exerted forces and couples on each other. At separations which are large compared with the dimensions of coils or magnet, these effects are similar, and analogous to the force and couple exerted by one electric dipole on another. The corollary is that the fields of a small permanent dipole and a small current loop should be identical, which is confirmed experimentally. Lines of force can be plotted for a solenoid (*Figure 14.2*); the similarity in patterns outside the bar magnet and solenoid is obvious. Thus an *electric current* is a source of magnetic field.

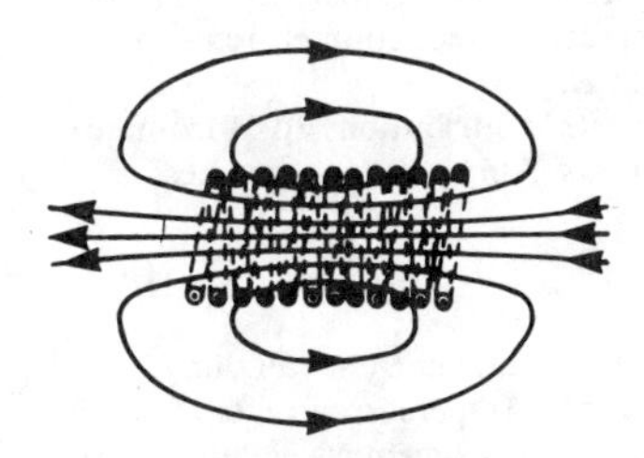

⊗ Current in

⊙ Current out

Figure 14.2 Lines of force for a solenoid (After Duffin,[2] reproduced by permission of the publisher)

By analogy with electric dipoles, magnetic dipole moments m_1, m_2 may be defined for pairs of small coplanar dipoles 1, 2 of either type, such that the force and couple each exerts on the other are *always* proportional to the product $m_1 m_2$. The effective moment of either dipole is found to vary with orientation; thus m is a vector quantity, denoted by $\mathbf{m}$. Experiments with various current loops show that $\mathbf{m}$ is proportional to the current I and loop area A. For a small single turn loop, by definition

$$\mathbf{m} = I\mathbf{A} \tag{14.1}$$

$\mathbf{m}$ and $\mathbf{A}$ are now *axial* vectors (*Figure 14.3*), the positive direction of each being defined by the sense of current flow. $\mathbf{m}$ is in units of ampere metre² (A m²).

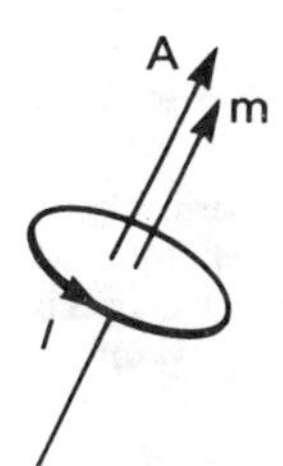

Figure 14.3 Current loop dipole (After Duffin,[2] reproduced by permission of the publisher)

Ampère subsequently proposed that the origin of all magnetism in materials lay in *small circulating currents* associated with each atom. They would each give rise to a magnetic dipole moment, the vector sum of which over all atoms would be the bulk magnetic moment. This explains why isolated magnetic poles are never observed. Even on the atomic scale only dipoles exist, which are due to electric currents.

In modern atomic theory the elementary currents of Ampère's theory are replaced by the closed orbits of negatively charged electrons about the atomic nuclei. However, the electron is now known to have an intrinsic 'spin' moment. Thus an atom may have a resultant moment due to both *orbital* and *spin* contributions.

14.1.2 Magnetic fields *in vacuo*

Using a test current loop dipole of known $\mathbf{m}$, the field quantity $\mathbf{B}$, called the magnetic flux density, may be defined at a point as follows. The *direction* of $\mathbf{B}$ is the equilibrium direction of $\mathbf{m}$; the *magnitude* of $\mathbf{B}$ is the couple per unit moment required to keep the test dipole at 90° to $\mathbf{B}$. In general if $\mathbf{m}$ makes an angle θ with $\mathbf{B}$ (*Figure 14.4*) then

$$C = mB \sin\theta \quad \text{or} \quad \mathbf{C} = \mathbf{m} \times \mathbf{B} \tag{14.2}$$

The unit of $\mathbf{B}$ is the tesla (T) defined as kg/(s² A) or equivalently as N/(A m). While Equation (14.2) holds for a permanent magnetic dipole, it is then a definition of $\mathbf{m}$, not $\mathbf{B}$, since $\mathbf{m}$ is independently defined only for a current loop.

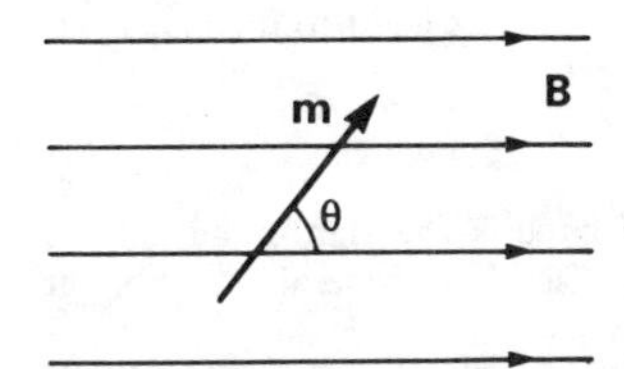

Figure 14.4 Dipole moment **m** in flux density **B** (After Duffin,[2] reproduced by permission of the publisher)

The *total* flux of **B** passing through a surface S is defined as the sum of all the products of the normal components of **B** and elements of area d**S**:

$$\varphi = \iint_S \mathbf{B}.\mathrm{d}\mathbf{S} \qquad (14.3)$$

φ is in units of tesla metre² (T m²), also called the weber (Wb), and **B** is sometimes given the unit Wb/m². By convention, B lines of force are said to cross a unit area in a region of flux density **B** so that φ represents the total number of lines crossing S.

By analogy with electromotive force (e.m.f.) which is the line integral of the electric field **E** round a closed circuit C, the magnetomotive force (m.m.f.) is defined as follows:

$$\mathcal{M} = \oint_C \frac{\mathbf{B}.\mathrm{d}\mathbf{s}}{\mu_0} \qquad (14.4)$$

where d**s** is an element of length and μ_0 is called the permeability of free space. In SI units $\mu_0 = 4\pi \times 10^{-7}$ N/A² by definition, so the unit for $\mathcal{M}$ is the ampere. The choice of $\mathbf{B}/\mu_0$ rather than **B** is for convenience as explained in Section 14.1.3. Now in an electrostatic field

$$\oint \mathbf{E}.\mathrm{d}\mathbf{s} = 0$$

At large distances the flux density due to a magnetic dipole takes the *same* form as the electric field caused by an electric dipole, so that for a circuit which is far from any dipole

$$\oint_C \mathbf{B}.\mathrm{d}\mathbf{s} = 0 \qquad (14.5)$$

Figure 14.5 shows the closed lines of force outside a current-carrying wire. They are called a *vortex* of **B**. Evidently any line

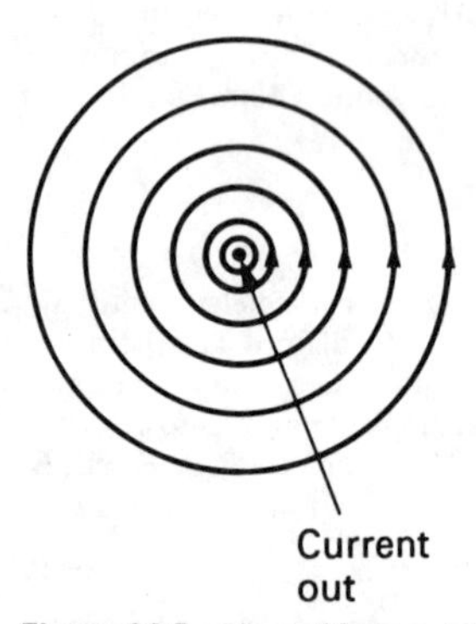

Figure 14.5 Lines of force outside a wire (After Duffin,[2] reproduced by permission of the publisher)

integral of **B** around the wire cannot vanish. In fact a closed path C encircling a current I once contributes just I to the m.m.f., i.e.

$$\oint_C \mathbf{B}.\mathrm{d}\mathbf{s} = \mu_0 I \qquad (14.6)$$

This is Ampère's law. For N circuits the right-hand side of Equation (14.6) becomes $\mu_0 NI$. Clearly in general $\mathcal{M}$ is neither path-independent nor single-valued.

The total flux of **B** through any closed surface S containing or intersecting magnetic dipoles is given by the *magnetic* analogue of Gauss' law. Just as in electrostatics

$$\oint\!\!\oint_S \mathbf{E}.\mathrm{d}\mathbf{S} = 0$$

because any number of dipoles contribute zero net charge, so here

$$\oint\!\!\oint_S \mathbf{B}.\mathrm{d}\mathbf{S} = 0 \qquad (14.7)$$

This is Gauss' law. Equations (14.6) and (14.7) are the general laws summarising the properties of steady magnetic fields in a vacuum.

14.1.3 Magnetisable media

A circuit in a vacuum in which a current I is maintained has a certain flux density $\mathbf{B}_0$ crossing it. The introduction of a magnetisable medium *changes* $\mathbf{B}_0$ to some new value $\mathbf{B}'$, where in general both $\mathbf{B}_0$ and $\mathbf{B}'$ vary from point to point in magnitude and direction. Ampère's theory provides a model in which the effect of the medium is found by replacing it with suitable distributions of *surface* and *volume* currents. Specifically, when the vacuum round the circuit is *completely* filled with the medium, the *mean* flux density $\bar{B}_0$ changes by a factor μ_r which is defined as the *relative permeability* of the medium, i.e.

$$\mu_r = \bar{B}'/\bar{B}_0 \qquad (14.8)$$

Evidently the same result would ensue if the vacuum remained and instead I changed to $\mu_r I$. According to the model the increased flux density is due to currents $(\mu_r - 1)I$ flowing in the surface next to I. Hereafter the currents producing the fields are called *conduction currents* I_C and those replacing the material *Amperean currents* I_A.

The magnetisation at a point is the magnetic moment per unit volume **M** at that point, given by the dipole moment d**m** of an elementary volume $\mathrm{d}\tau$ surrounding the point, i.e.

$$\mathbf{M} = \mathrm{d}\mathbf{m}/\mathrm{d}\tau \qquad (14.9)$$

where **M** is in units of ampere per metre (A/m) and $\mathrm{d}\tau$ is assumed small enough for **M** to be constant. Let the element have length $\mathrm{d}l$ parallel to **M**, and cross-sectional area $\mathrm{d}S$, i.e. $\mathrm{d}\tau = \mathrm{d}l\mathrm{d}S$. If it is replaced by loops of total *surface* current $\mathrm{d}I_A$, then by Equation (14.1)

$$\mathrm{d}\mathbf{m} = \mathbf{M}\,\mathrm{d}l\,\mathrm{d}S = \mathrm{d}I_A\,\mathrm{d}S = J_{SA}\,\mathrm{d}l\,\mathrm{d}S$$

where J_{SA} is the surface current density, i.e.

$$J_{SA} = |\mathbf{M}| \qquad (14.10)$$

This result holds for a macroscopic piece of material provided **M** is uniform throughout. Adjacent 'interior' Amperean currents cancel out as regards magnetic effect and only those on the outer surface contribute. $J_{SA} = M\cos\theta$ at a general point where the tangent to the surface is at angle θ to the direction of **M**. A *volume* current distribution J_{VA} need only be invoked when **M** is not uniform, i.e. the interior currents do not balance. This may happen if the medium itself carries conduction currents, or it is *non-linear* (μ_r is a function of **B**).

Since there are now separate contributions to **B** from conduction and Amperean currents, Ampère's law becomes

$$\oint_C \mathbf{B}.\mathrm{d}\mathbf{s} = \mu_0(I_C + I_A) \qquad (14.11)$$

where C is a circuit linking conduction currents totalling I_C and magnetised material equivalent to Amperean currents I_A. Let d**s** be a line element of C cutting an elementary volume of the material at some arbitrary angle to **M**. As before, $\mathrm{d}I_A = J_{SA}\,\mathrm{d}l = \mathbf{M}.\mathrm{d}\mathbf{s}$. So Equation (14.11) may be rearranged to give

$$\oint_C (\mathbf{B}/\mu_0 - \mathbf{M}).\mathrm{d}\mathbf{s} = I_C$$

and the magnetic field strength **H**, in ampere per metre (A/m), is defined by

$$\mathbf{H}=\mathbf{B}/\mu_0-\mathbf{M} \tag{14.12}$$

Ampère's law, previously defined *in vacuo* by Equation (14.6) can now be generalised:

$$\oint_C \mathbf{H}.d\mathbf{s}=I_C \tag{14.13}$$

Likewise the m.m.f. ($\mathcal{M}$) round a closed path simply equals the total conduction current linked. The utility of **H** lies in the fact that it is generated *solely* by conduction currents and is therefore unaffected by the presence of a medium.

The magnetic susceptibility χ_m is defined at a point by the ratio of magnetic moment per unit volume to magnetic field strength:

$$\mathbf{M}=\chi_m\mathbf{H} \tag{14.14}$$

χ_m is dimensionless and is a *scalar* for an isotropic homogeneous medium, otherwise a tensor, and is independent of **H** *only* for linear media. From Equations (14.12) and (14.14),

$$\mathbf{B}=\mu_0(1+\chi_m)\mathbf{H}$$

so

$$\mu_r=1+\chi_m \tag{14.15}$$

$$\mathbf{B}=\mu_0\mu_r\mathbf{H} \tag{14.16}$$

which now define the relative permeability μ_r *at a point*.

The notion of *magnetic poles* is familiar. It is simple in essence because of the similarity with electric charges. Much of electrostatic theory can be used by analogy to account for magnetic phenomena. While it is often useful in calculations involving permanent magnet materials, the magnetic pole model has several drawbacks:

(1) *Isolated* poles do not exist, whereas atomic magnetic moments can be satisfactorily related to electron angular momenta via current loops.
(2) While magnetic moment is a measurable quantity, pole location and strength are *ill-defined*.
(3) There is no analogue of charge conduction.
(4) With other than *permanent* magnet materials, problems are generally more difficult to solve using the pole concept.
(5) *Practical* permanent dipoles do not exist. The properties of any magnet are affected by stray fields, including the proximity of other magnetic material. By contrast the moment of a current loop can be maintained constant.

This section has dealt with media filling the whole space in which magnetic fields exist, i.e. effectively *infinite* media. The presence of boundaries may affect **B**, **H** and **M** considerably. In particular demagnetising effects have important practical consequences for ferromagnetic materials. This is considered in Sections 14.2.2 and 14.4.1.

Two excellent though contrasting treatments of the fundamentals of magnetism are given by Bleaney and Bleaney[1] and Duffin.[2]

14.2 Ferromagnetic media

Ferromagnetic substances are always solids and usually metals. χ_m is positive, large (often $\gg 1$), nonlinear and dependent on previous history (showing hysteresis). Most materials in this class can exhibit a *finite and perhaps a large moment when* $\mathbf{H}=0$. There is an abrupt change of properties at a characteristic temperature known as the Curie temperature (T_C). Above this, behaviour is simpler, the material becomes *paramagnetic*, i.e. χ_m is still positive but small ($\ll 1$), and it obeys the *Curie–Weiss law*. At absolute temperature T

$$\chi_m=\frac{C}{T-T_p} \tag{14.17}$$

where C is the Curie constant and T_p is the paramagnetic Curie temperature. T_p and T_C usually differ by some 10 or 20°C.

14.2.1 Hysteresis properties

Many of the characteristic properties of a ferromagnetic material are displayed in curves of B (or M) against H. For ease of interpretation the sample used is preferably a toroid wound with primary and secondary coils. **B**, **H** and **M** are uniform and parallel, and there are no demagnetising effects (see Section 14.2.2). A current I in the primary, of n turns per unit length, generates a field $H=nI$ (by Ampère's law) and B is derived from flux linking the secondary. If the material is initially unmagnetised, then increasing H from zero gives an S-shape *magnetisation curve* OABC in *Figure 14.6*, of which the portion OA is reversible. Beyond A, reduction of H from some point B causes the path BB′ to be followed. If H is continuously reversed

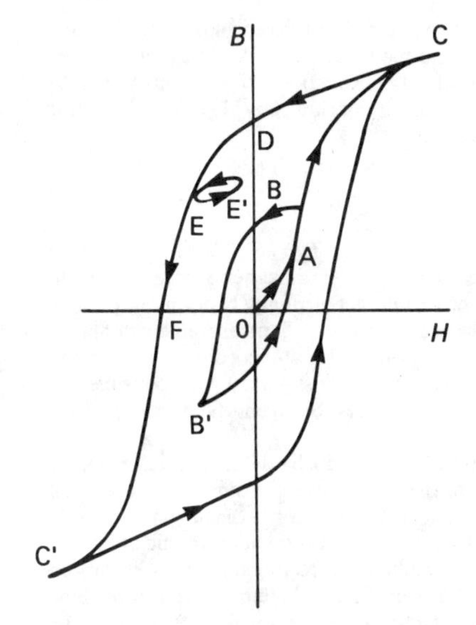

Figure 14.6 Typical hysteresis loops (After Duffin,[2] reproduced by permission of the publisher)

through amplitude BB′, a *minor hysteresis loop* is described. The biggest loop CC′, called simply *the hysteresis loop*, is obtained when the amplitude of H is large enough to cause saturation of the magnetisation. Beyond the points CC′, B increases as H simply, as Equation (14.12) predicts. If at any point E small reversals of H are made, *subsidiary loops* like EE′ are followed. The *demagnetisation curve* DEF gives the basic information required for permanent magnet design. The intercept OD on the B axis (i.e. $H=0$) is called the *remanence* B_r, and the intercept on the $-H$ axis where $B=0$ is the *coercivity* H_{cB}. The *relative permeability* $\mu_r=B/\mu_0 H$ evidently goes through a maximum near point B on the magnetisation curve as H increases from zero, then gradually falls towards unity. The *incremental permeability* μ_{inc}, often of greater practical importance, is defined for small ranges of B and H by $\mu_{inc}=\delta B/\mu_0\delta H$. The *initial permeability* μ_i clearly equals μ_{inc} for small excursions of H about the origin. The *recoil permeability* μ_{rec} is the slope of the chord joining the tips of a subsidiary loop such as EE′; this is also of relevance in permanent magnet design. A graph of M against H shows similar features but as M saturates the extremities of the major loop are of zero slope. The intercepts here are the *remanent magnetisation* M_r ($\mu_0 M_r = B_r$) and *intrinsic coercivity* (or coercive field) H_{cM}. Since numerically $H_{cB} = M(B=0)$ and $H_{cM} = B/\mu_0$ $(M=0)$, $H_{cM} \neq H_{cB}$ in general but is arbitrarily related. Finally, application of a reverse field equal to H_{cB} does not permanently demagnetise the sample as when H is returned to zero, a *recoil line* parallel to EE′ is followed and the sample is left in some remanent state $M_r' < M_r$.

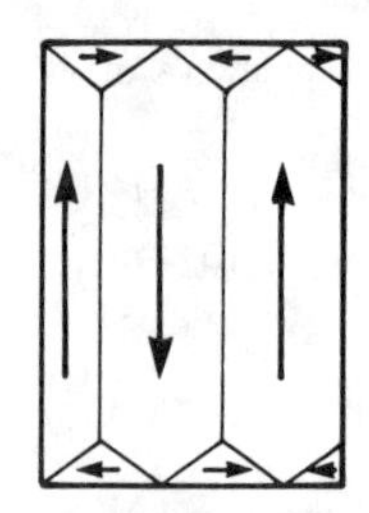

Figure 14.8 Typical domain closure pattern (After Duffin,[2] reproduced by permission of the publisher)

14.2.2 Microscopic theory

Ferromagnetism can be accounted for via the theory of paramagnetism, since the two are clearly intimately related. *Paramagnetism* is exhibited when the atoms or ions in a material possess a resultant moment $\mathbf{m}_p$ which is the vector sum of all the electron moments. While free atoms often have net moments, atoms in combination and ions usually do not. These substances possess *weak* moments in an applied field which disappear when the field is removed. Typically $\chi_m \sim +10^{-3}$ at room temperature. According to Curie's law

$$\chi_m = C/T \tag{14.18}$$

which holds quite well for gases, solutions and some solids. Many liquids and solids obey the modified Curie–Weiss law of Equation (14.17). It is assumed that a molecular field $\lambda\mathbf{M}$ is contributed by interactions between atoms. Replacing $\mathbf{H}$ by $\mathbf{H} + \lambda\mathbf{M}$ in Equation (14.14) gives to first order

$$\chi_m = \frac{C}{T}(1 + \lambda\chi_m) \quad \text{or} \quad \chi_m = \frac{C}{T - \lambda C} \tag{14.19}$$

which is the Curie–Weiss law with $T_p = \lambda C$. λ is called the *Weiss constant.* Weiss suggested that in ferromagnetic materials neighbouring atomic moments interact strongly, resulting in parallel alignment over considerable regions. T_C measurements confirm that λ is very large, of order 10^4. According to quantum mechanics *electron spins* are almost entirely responsible for ferromagnetism, the large forces between them being called *exchange interactions.*

That ferromagnetic materials can be demagnetised at all by external fields is due to the existence of *domains.* These are small regions, generally 10^{-6} to 10^{-2} cm^3 in size, each spontaneously magnetised to saturation in a definite crystallographic direction, adjacent domains having different directions. They are separated by *Bloch walls* of order 5 nm thick wherein the spin directions change progressively (*Figure 14.7*). They are formed to lower the energy stored in internal and external (flux leakage) fields. Subdivision into a number of domains takes place with broadly opposing magnetisations, with small closure domains at the surfaces (*Figure 14.8*). The process does not go on indefinitely because Bloch walls have a certain energy arising from *magnetocrystalline anisotropy*. In a crystalline material there are preferred directions of magnetisation dictated by the crystal symmetry. The *anisotropy energy* W_A is that required to rotate the magnetisation from an 'easy' to a 'hard' direction. W_A is thought to be due both to the electric field of the ionic charges and spin–orbit coupling whereby the atomic spin direction is coupled to the lattice via the orbit. The wall contribution to W_A arises since inevitably most of the spins in the wall will not be parallel to the easy direction. Domain size is thus largely determined by the balance of field and wall anisotropy energies. The shape of the magnetisation curve (*Figure 14.6*) is now explained by reference to *Figure 14.9.* In unmagnetised material, domains are quasi-randomly orientated and there is no net moment (a). At small **H**, more favourably magnetised domains grow at the expense of others (b), so in the initial part OA of the curve magnetisation proceeds via small reversible displacements. The steeper part **AB** corresponds to larger irreversible displacements (c) where wall movements are hindered by crystal structure defects. At still higher **H**, between B and C, magnetisation increases by rotation of domains from easy directions towards the field (d). Finally, at C, all domain walls are swept away, and alignment with **H** is virtually complete (e).

The irreversible wall movements at moderate **H** can be detected audibly by voltages induced in loudspeaker coils wound on a specimen, as a wall jumps across a defect (Barkhausen effect). Domain patterns may be observed on the surface of a material by applying colloidal ferromagnetic powder (Bitter patterns).

14.2.3 Energy losses in ferromagnetic materials

Two sources of loss of major practical importance are possible when the material is subjected to *alternating* magnetic fields.

14.2.3.1 Hysteresis

During a cycle, domain rotation is opposed by anisotropy effects, the energy to overcome this being derived from the magnetising field and appearing ultimately as heat. It can be shown that the energy lost per cycle or the net work done by the field (in J/m^3 per cycle) is

$$W_h = \oint \mathbf{H} . d\mathbf{B} \tag{14.20}$$

which is just *the area of the BH loop.* Note that this is essentially a d.c. loop. With alternating magnetisation, non-hysteresis losses result in a phase angle between B and H which in turn modifies

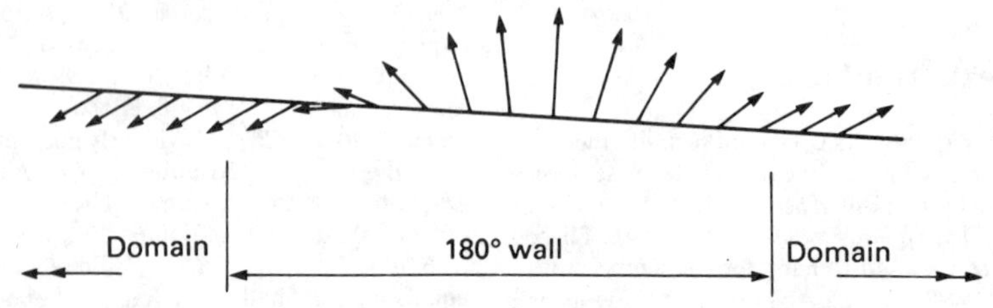

Figure 14.7 Change of spin orientation in a Bloch wall (after Snelling,[3] reproduced by permission)

(a) 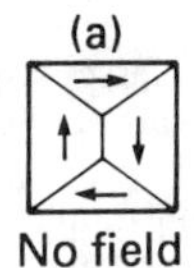 (b)  (c) 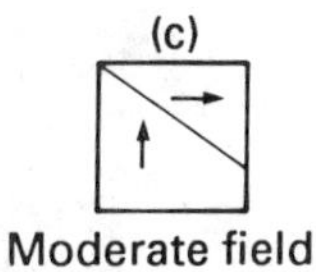 (d) 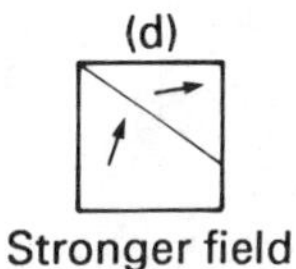 (e) 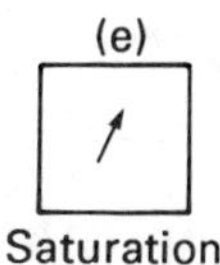

No field — Moderate field — Stronger field — Saturation

Figure 14.9 Effect of an external field on domains (After Duffin,[2] reproduced by permission of the publisher)

loop shape. By making various assumptions about loop properties, or appealing to experiment, power series for $B(H)$ and $W_h(H)$ may be derived which converge for low amplitudes of H. The first term for W_h in one such approach, due to Peterson, is given by

$$W_h \simeq \tfrac{8}{3}\mu_0 a_1 \hat{H}^3 \qquad (14.21)$$

where a_1 is a coefficient with unit m/A and $\hat{H}$ is the maximum field strength. These aspects of hysteresis loss are discussed by Snelling.[4]

14.2.3.2 Eddy current losses

An alternating flux in a conductive medium induces *eddy currents* which also result in energy loss. The magnitude of the loss depends partly on medium shape and size, and is reduced by subdivision into electrically insulated regions, e.g. laminations or grains. At low frequencies f where eddy current inductive effects are negligible, and for a peak flux density $\hat{B}$ and bulk resistivity ρ, the power loss per unit volume (in W/m^3) in a lamination of thickness d is given by

$$P_E = \frac{(\pi \hat{B} f d)^2}{6\rho} \qquad (14.22)$$

where $\hat{B}$ is assumed to lie in the plane of the lamination. Equation (14.22) is quoted by Snelling.[5]

14.2.4 Boundaries and demagnetising factors

At the boundary of a medium of relative permeability μ_r, **B** and **H** cannot both be continuous. Specifically:

(1) for a boundary parallel to **H**, **B** is discontinuous;
(2) for a boundary normal to **B**, **H** is discontinuous.

It is useful to interpret discontinuities in **H** as due to an opposing specimen field $\mathbf{H}_d$. μ_r is so large for ferromagnetic materials that the net internal **H** may oppose **B** and **M**. $\mathbf{H}_d$ is called an *internal demagnetising field*. In case 2, as **B** is continuous, $\mathbf{H}_d = \mathbf{M}$. For a specimen of arbitrary shape, **M** and $\mathbf{H}_d$ vary from point to point in a complex way. A ferromagnetic ellipsoid is the only shape which when placed in a *uniform* field $\mathbf{H}_a$ has the same magnetisation and demagnetising field at every point. In this case $\mathbf{H}_d = [N_d]\mathbf{M}$, where $[N_d]$ is a *demagnetisation tensor*. If the ellipsoid is orientated with its axes along rectangular coordinate axes x, y, z then $[N_d]$ is diagonalised and the *demagnetising factors* N_x, N_y, N_z along the respective axes are related very simply by

$$N_x + N_y + N_z = 1 \qquad (14.23)$$

It is still difficult to calculate individual demagnetisation factors except in limiting cases by appealing to symmetry. For instance:

(1) Sphere: $N_x = N_y = N_z = 1/3$
(2) Thin plate in yz plane: $N_x \simeq 1$, $N_y = N_z \simeq 0$
(3) Long rod, axis in x direction, circular section: $N_x \simeq 0$, $N_y = N_z \simeq 1/2$

Demagnetising effects are of fundamental importance in *any circuit containing air gaps*. Demagnetising factors need to be known in calculating the working point of a permanent magnet, for instance, or the condition for gyromagnetic resonance of a microwave ferrite.

14.2.5 Soft magnetic materials

Soft magnetic materials have small or moderate remanence, low coercivity and hysteresis loss, and high permeability. Ease of magnetisation and demagnetisation implies *freely moving* domain walls. To achieve this a low anisotropy is required, and materials are usually annealed to remove strain and local atomic disorder, which inhibit wall motion. Materials of technological importance are all *iron alloys* (apart from ferrites). Iron itself is often used in electromagnets, including relays and valves, but is sensitive to light element impurities such as carbon, sulphur, nitrogen and oxygen. *Silicon–iron* is extensively used in transformers and large electromagnetic machinery generally. It shows lower eddy current and hysteresis losses, and higher permeabilities than iron. Commercial 3.2% SiFe sheet used for making transformer laminations has enhanced magnetic properties in the rolling direction. Hot-rolled strip is cold reduced in several passes to its final thickness. During subsequent annealing at about 1100°C, uniformly sized crystallites are formed which are highly orientated in the rolling direction. SiFe alloys, like many other soft magnetic materials, are magnetostrictive, and magnetic losses are increased (reduced) by external compressive (tensile) stresses. Sheets of 3.2% SiFe have a glassy coating which provides interlaminar insulation and also puts the material under tensile stress, partly overcoming the compressive stresses that occur in transformer construction. Higher permeability and lower losses are achieved in *nickel–iron* alloys (~40–80% Ni). They are extensively used for magnetic screening, but they saturate at lower fields than SiFe and are relatively expensive, limiting their use otherwise to small transformers and chokes. High saturation magnetisations are achieved in *iron–cobalt* alloys which may contain low levels of vanadium and nickel to improve malleability and mechanical strength. They are used in aircraft generators where a large power/weight ratio is critical. There is much interest in the *amorphous metals*, also known as *metallic glasses* and *rapidly solidified metals*, of general formula $T_{80}M_{20}$, where T is one or more transition metals (Fe, Co, Ni, Cr, Mn) and M is one or more metalloids (B, C, Si, P). Ribbons up to ~60 μm thick and up to ~75 mm wide are simply and economically prepared using the *drum quenching technique* (*Figure 14.10*). The melt is forced through an orifice by gas pressure and impinges on two rotating drums. The extremely rapid cooling (~10^6 °C/s) ensures that the disordered state of the melt is retained. Metallic glasses are unique in that they are hard and strong and at the same time tough and ductile; they are highly corrosion resistant and possess excellent soft magnetic properties. Applications include low-loss transformer cores, magnetic shielding, recording heads and magnetic springs. The origin of ferromagnetism in amorphous substances is not well understood at present. *Soft ferrites*, though not actually ferromagnetic, behave like soft magnetic alloys in many respects. These are described in *Section 14.3.2*.

Properties of representative soft ferromagnetic materials are given in *Table 14.1*. A recent review of amorphous metals has been made by Pye.[6] Soft magnetic alloys in general are discussed by Chin and Wernick.[7]

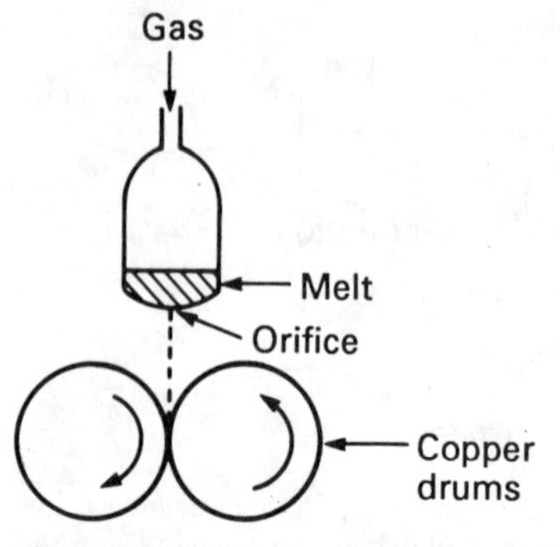

Figure 14.10 Drum quenching technique for manufacture of amorphous ribbon alloys (after Overshott,[8] reproduced by permission)

14.2.6 Hard magnetic materials

Hard magnetic materials usually have high coercivity and remanence, with relatively low permeability. Magnetisation and demagnetisation are now deliberately made difficult and there are various ways of introducing structural defects in order to *inhibit* domain wall motion. Some materials may be prepared containing effectively single particle domains in which there are few or no Bloch walls. Magnetisation can only occur by domain rotation requiring high fields. Suitable alloy and ferrite materials, their manufacture and properties, are described in Section 14.4. *Hard ferrites*, like the soft materials, are not strictly ferromagnetic.

14.3 Ferrites

Ferrites are mixed metal oxide ceramics, black or dark grey in colour, and very hard and brittle. They have high electrical resistivities, typically in the range 10–10^{12} Ω cm (see values for soft magnetic alloys in *Table 14.1*). Their strong magnetic properties are due to *ferrimagnetism* in which magnetic ions (principally Fe^{3+}) on different crystallographic sites have spin moments coupled antiparallel via oxygen ions by so-called *superexchange interactions.* Opposing sets of moments or magnetic sublattices are formed, which are not equal in magnitude, resulting in a large magnetisation. Like ferromagnetic materials, ferrites have a well-defined magnetic transition temperature called the Curie temperature (or sometimes the Néel temperature).

14.3.1 Crystal structures

The three technically important classes are: (a) spinels; (b) garnets; and (c) hexaferrites. Spinels and garnets are magnetically soft, while the hexaferrites are hard.

The general formula of an unsubstituted spinel is MFe_2O_4 where M is one or more divalent cations such as Mg, Mn, Co, Ni, Cu, or Zn. The metal ions occupy two types of interstitial site (called A and B) in a cubic close-packed array of oxygen ions. B sublattice moments usually dominate as there are twice as many B as A sites. The site preference of a cation depends on its size and electronic structure, and the nature of the other cation(s) present. The generalised distribution for a cation M with no exclusive preference is:

$(M_\alpha Fe_{1-\alpha})$	$[M_{1-\alpha}Fe_{1+\alpha}]O_4$	(partially inverted spinel)
A sites	B sites	
spin up	spin down	

If M is non-magnetic, the net moment is due to an imbalance of $(1+\alpha)-(1-\alpha)$ or $2\alpha Fe^{3+}$. A practical example is magnesium ferrite, $(Mg_{0.1}Fe_{0.9})[Mg_{0.9}Fe_{1.1}]O_4$ $(\alpha=0.1)$. Zinc ferrite, $(Zn)[Fe_2]O_4$ $(\alpha=1)$, is a *normal* spinel and is not ferrimagnetic as all the Fe^{3+} are on B sites. By contrast nickel ferrite, $(Fe)[NiFe]O_4$ $(\alpha=0)$, is *inverted* and ferrimagnetism arises only because the Ni ions are coupled magnetically. Clearly no net moment arises from equal numbers of Fe^{3+} ions in the two sublattices.

The general formula of an unsubstituted garnet is $R_3Fe_5O_{12}$, where R is trivalent yttrium (Y) or a rare earth (or again a mixture of ions). The structure is also cubic but here the cations occupy three types of interstitial site, called c, a, and d. With one or two important exceptions, each cation in a garnet is exclusively confined to one type of site, mainly because now the preferences of ionic species on three types of site have to be mutually compatible. The generalised distribution is

$\{R_3\}$	$[Fe_2]$	$(Fe_3)O_{12}$	(rare earth iron garnet)
c sites	a sites	d sites	
spin up	spin up	spin down	

The moment of the d sublattice dominates that of the a sublattice in the ratio 3:2. Y^{3+} is not magnetic but rare earth ions in the c sublattice couple antiparallel with the d sublattice, which interactions predominate at low temperatures. The most important unsubstituted materials are yttrium iron garnet (YIG), $Y_3Fe_5O_{12}$, and mixed yttrium gadolinium iron garnets (YGdIG), $Y_{3-a}Gd_aFe_5O_{12}$. Barium hexaferrite ($BaFe_{12}O_{19}$) is by far the most important of a number of complex ferrites with *hexagonal* as opposed to cubic symmetry. The structure is a close-packed array of the larger O^{2-} and Ba^{2+} ions with the smaller Fe^{3+} in

Table 14.1 Properties of typical soft magnetic alloys (After Duffin[2] and Brailsford[16])

Material and approximate composition	μ_i	μ_{max}	H_c (A/m)	B_{sat} (T)	ρ ($\mu\Omega$ cm)
Iron, motor grade (99.6% Fe)	250	5×10^3	80	2.1	14
Silicon–iron, cold reduced, grain orientated (3.2% Si, 96.8% Fe)	2×10^3	7×10^4	8	2.0	48
Rhometal (36% Ni, 64% Fe)	1.8×10^3	7×10^3	12	0.9	85
Supermalloy (79% Ni, 15% Fe, 5% Mo, 1% Mn)	10^5	10^6	0.2	0.8	65
Permendur (49% Fe, 49% Co, 2% V)	800	5×10^3	180	2.4	28

Table 14.2 Main properties with typical values of NiZn and MnZn ferrites (After Snelling[3])

Property	*Symbol or expression*	*Measurement conditions*	*Range of values*		*Units*
			NiZn	*MnZn*	
Initial permeability	μ_i	$\hat{B}\to 0$ $f<10$ kHz	10–2000	300–10^4	—
Saturation flux density	B_{sat}	$\hat{H}<10$ A mm^{-1}	0.1–0.45	0.3–0.55	T
Temperature factor	$\frac{\mu_2-\mu_1}{\mu_2\mu_1(T_2-T_1)}$	$T_1=25°C$, $T_2=55°C$ $\hat{B}\to 0$ $f<10$ kHz	-10 to $+40$	-1.0 to $+3.0$	$°C^{-1}\times 10^{-6}$
Disaccommodation factor	$\frac{\mu_1-\mu_2}{\mu_1^2\log(t_2/t_1)}$	$\hat{B}\to 0$ $f<10$ kHz	—	1–10	10^{-6}
Residual loss factor	$\frac{\tan\delta_r}{\mu}$	$\hat{B}\to 0$ $f<100$ kHz $f>100$ kHz	— 20–2000	1–100 —	10^{-6}
Hysteresis loss coefficient	$\frac{\tan\delta_h}{\mu\hat{B}}$	$f=10$ kHz $\hat{B}=1$–3 mT	2–4000	0.1–2.0	$mT^{-1}\times 10^{-6}$
Resistivity	ρ	d.c.	10^5–10^9	10–2000	Ω cm

no less than five different types of site. Per formula unit, eight Fe^{3+} are 'spin up' and four are 'spin down'. The symmetry bestows a ***high uniaxial anisotropy*** the material being magnetically hard with the easy direction of magnetisation along the hexagonal axis. $BaFe_{12}O_{19}$ and the very similar strontium compound are considered with other permanent magnet materials in Section 14.4.

14.3.2 Soft ferrites

The term 'soft ferrites' covers those spinel materials used broadly in inductor and transformer applications at frequencies up to about 500 MHz. Among all ferrites they represent the largest application category in terms of tonnes produced per annum.

Soft ferrites are chiefly *manganese–zinc* (MnZn) and *nickel–zinc* (NiZn). Substitution of Zn^{2+} (non-magnetic) for Mn^{2+} or Ni^{2+} (magnetic) is on A sites only which initially increases the net moment, and so the magnetisation, and reduces the anisotropy energy. The major practical effect is to increase the initial permeability μ_i and reduce the Curie temperature T_C.

Representative values of the main properties of NiZn and MnZn ferrites are given in *Table 14.2*. The complementary *Table 14.3* lists the main applications. Comprehensive reviews of soft ferrites and their applications have been made by Snelling[3] and Slick.[9]

14.3.3 Microwave ferrites

These are used basically in the frequency range 100 MHz to 100 GHz and above. An electromagnetic wave of frequency ω, propagating through a ferrite in the direction of an external biasing d.c. magnetic field H_a, has *two circularly polarised* components whose behaviour is described by the complex permeability $\mu_\pm' - j\mu_\pm''$ (*Figure 14.11*). The electron spin moments precess about H_a in sympathy with the positive sense component. Precession is sustained by energy coupled from the microwave field. *Gyromagnetic resonance* (peak in μ_+'') occurs when

$$\omega=\gamma H_0 \text{ (Larmor precession)} \tag{14.24}$$

where H_0 is the *internal* field for resonance, i.e. H_a corrected for shape demagnetising effects and the effective anisotropy field. γ is the gyromagnetic ratio (~0.035 MHz m/A). Little energy is absorbed for precession in the anti-Larmor sense (μ_-',μ_-''), i.e. for signal propagation in the reverse direction, or a reversed H_a.

These non-reciprocal effects give rise to the isolation and phase shift necessary for the operation of most microwave devices, wherein the material is housed in a suitable waveguide, coaxial or stripline transmission line. The dispersive and dissipative character of μ is exploited in different ways, depending on where the ferrite is biased, and details of device geometry and construction.

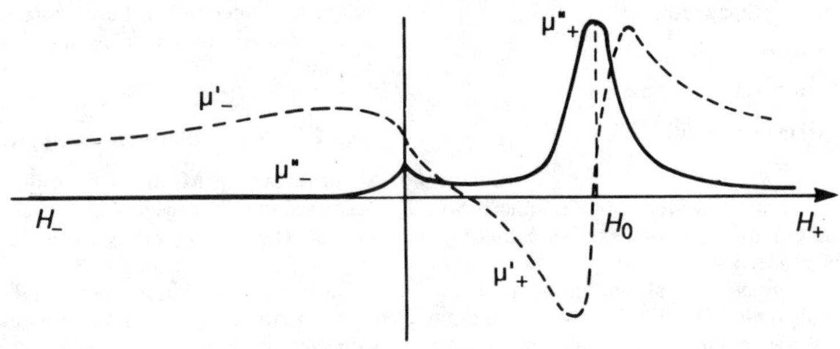

Figure 14.11 Real and imaginary parts of complex permeability versus d.c. magnetic field at fixed frequency

Table 14.3 Summary of non-microwave ferrite applications (after Slick[9])

Ferrite chemistry	Device	Device function	Frequencies	Desired ferrite properties
Linear B/H, low flux density				
MnZn, NiZn	Inductor	Frequency selection network Filtering and resonant circuits	$\leqslant$1 MHz (MnZn) ~1–100 MHz (NiZn)	High μ, high μQ, high stability of μ with temperature and time
MnZn, NiZn	Transformer (pulse and wide band)	V and I transformation Impedance matching	Up to 500 MHz	High μ, low hysteresis losses
NiZn	Antenna rod	Electromagnetic wave receival	Up to 15 MHz	High μQ, high resistivity
MnZn	Loading coil	Impedance loading	Audio	High μ, high B_s, high stability of μ with temperature, time and d.c. bias
Nonlinear B/H, medium to high flux density				
MnZn, NiZn	Flyback transformer	Power converter	<100 kHz	High μ, high B_s, low hysteresis losses
MnZn	Deflection yoke	Electron beam deflection	<100 kHz	High μ, high B_s
MnZn, NiZn	Suppression bead	Block unwanted a.c. signals	Up to 250 MHz	Mod. high μ, high B_s, high hysteresis losses
MnZn, NiZn	Choke coil	Separate a.c. from d.c. signals	Up to 250 MHz	Mod. high μ, high B_s, high hysteresis losses
MnZn, NiZn	Recording head	Information recording	Up to 10 MHz	High μ, high density, high μQ, high wear resistant
MnZn	Power transformer	Power converter	<60 kHz	High B_s, low hysteresis losses
Nonlinear B/H, rectangular loop				
MnMg, MnMgZn, MnCu, MnLi, etc.	Memory cores	Information storage	Pulse	High squareness, low switching coefficient and controlled coercive force
MnMgZn, MnMgCd	Switch cores	Memory access transformer	Pulse	High squareness, controlled coercive force
MnZn	Magnetic amplifiers			

Table 14.4 Typical ranges of properties for microwave ferrite systems

Type	General formula	M_s (kA/m)	T_C (°C)	ΔH (kA/m)	ε'	$\tan\delta$ ($\times 10^{-4}$)
Garnets						
YAlFe	$Y_3Fe_{5-y}Al_yO_{12}$	140–16	280–95	4.5–2.5	15.5–14	<2
YGdFe	$Y_{3-a}Gd_aFe_5O_{12}$	140–55	280	4.5–19	15.5	<2
Spinels						
NiAlFe	$NiFe_{2-y}Al_yO_4$	250–80	580–400	24–40	13–11	<10
NiZnFe	$Ni_{1-x}Zn_xFe_2O_4$	250–400	580–375	24–14	13–13.5	<10
MgAlFe	$MgFe_{2-y}Al_yO_4$	190–70	275–120	30–12	12.5–11	<5
LiTiFe	$Li_{0.5(1+y)}Fe_{2.5-1.5y}Ti_yO_4$	295–80	620–300	40–24	15.5–18	<5
LiZnFe	$Li_{0.5(1-x)}Fe_{2.5-0.5x}Zn_xO_4$	295–385	620–420	40–16	15.5–15	<5

The important properties of microwave ferrites are:

(1) Magnetisation: saturation magnetisation, M_s; temperature coefficient, $\dfrac{1}{M_s}\dfrac{dM_s}{dT}$

(2) Magnetic losses: resonance linewidth, ΔH; spinwave linewidth, ΔH_K

(3) Dielectric properties: dielectric constant, ε'; loss tangent, $\tan\delta_\varepsilon = \varepsilon''/\varepsilon'$

(4) *BH* loop properties: remanence ratio, B_r/B_{sat}; coercivity, H_{cB}

Both garnet and spinel types are widely used as microwave ferrite. Substitution of several different ions can be made to get the best combination of properties.

Major substituents in *garnets* are aluminium (Al^{3+}) and gadolinium (Gd^{3+}). Small quantities of ions such as indium (In^{3+}), holmium (Ho^{3+}), dysprosium (Dy^{3+}) and manganese (Mn^{3+}) are included to control specific properties.

In *spinels*, major substituents in *nickel* (Ni^{2+}) and *magnesium* (Mg^{2+}) ferrites are aluminium and zinc (Zn^{2+}); in *lithium* (Li^+) ferrites they are titanium (Ti^{4+}) and zinc. Again, small amounts of cobalt (Co^{2+}) and manganese control specific properties.

Typical ranges of properties for microwave ferrite systems are given in *Table 14.4*. The effects of individual ions on ferrite properties are summarised in *Table 14.5*. Von Aulock[10] has edited a sourcebook of these materials, while a concise review with emphasis on more recent advances has been made by Nicolas.[11]

14.3.4 Ferrites for memory application

Small toroids in MgMn, CuMn or LiNi spinel materials have been used in very large numbers as elements for rapid access data storage. High B_r/B_{sat} ratios are needed, with controlled values of H_{cB} to increase switching speeds.

Garnets are used in *magnetic bubble memories*. If a thin single crystal plate (~5 μm thick) with orthogonal easy direction is subjected to an increasing magnetic field H_a also normal to the plate, the antiparallel domains shrink until, over a narrow range

Table 14.5 Effect of substituent ions on the properties of microwave ferrites

		Garnets					*Spinels*			
Property	*Ion:* *Site:*	Al^{3+} *a, d*	Gd^{3+} *c*	In^{3+} *a*	Ho^{3+}, Dy^{3+} *c*	Mn^{3+} *a*	Al^{3+} *A, B*	Zn^{2+} *A*	Ti^{4+} *B*	Co^{2+} *B*
M_s		↓	↓	↑	→	↗	↓	↑	↓	→
T_C		↓	→	↓	→	↘	↓	↓	↓	→
$\frac{1}{M_s}\frac{dM_s}{dT}$ (20°C)		↗	↓	↑	→	↗	↗	↑	↗	→
ΔH		↘	↑	↓	↑	↘	→	↘	→	↘
ΔH_K		→	↗	↘	↑	→	→	↘	→	↑
$\tan \delta_\varepsilon$		→	→	↘	→	↘	→	→	↘	↘
B_r/B_{sat}		↘	→	↘	→	↑	↘	↘	→	↘

Key: ↑ = up, ↓ = down, ↗, ↘ = marginal, → = unchanged

of $\boldsymbol{H}_a$, small stable cylindrical domains ('magnetic bubbles') are left (*Figure 14.12*), each of which carries one bit of information. Typical bubble diameters are 2–5 μm. To ensure their stability and mobility, the plate must be free from imperfections, with faces as flat and parallel as possible. A promising technique is to grow a magnetic film by *liquid phase epitaxy* on a non-magnetic single crystal substrate. A suitable film composition is the garnet $Y_{2.9}La_{0.1}Fe_{3.8}Ga_{1.2}O_{12}$ using a substrate of $Gd_3Ga_5O_{12}$ (gadolium gallium garnet or GGG).The use of garnet and other ferrite films for bubble memories has been reviewed by Eschenfelder.[12]

14.3.5 Manufacture of ferrites

Most ferrites are prepared as polycrystalline materials by ceramic processing. Metal oxides are reacted chemically in a *powder preparation stage* to form a suitably reactive ferrite powder of correct composition. For instance, YIG is formed by reacting yttrium and iron oxides as follows:

$$3Y_2O_3 + 5Fe_2O_3 \rightarrow 2Y_3Fe_5O_{12}$$

Densification is achieved by green-forming the powder and sintering to make a dense solid with the right mechanical, electrical and magnetic properties.

To prepare the powder suitable grades of oxides (or carbonates, citrates, etc.) are weighed out in the correct amounts and wet ground in a ball mill. Other types of mill, e.g. attritors or colloid mills, are used or sometimes a dry grinding process such as air cyclone milling. The main object is to achieve *very intimate mixing*. After drying the material is usually *presintered* (calcined) at between 800 and 1300°C, below the final firing temperature, converting the fine oxide mixture to a coarse ferrite powder. This reduces shrinkage of the final shape and improves homogeneity and batch repeatability. A *more intensive milling* is usually needed to reduce the reacted oxide to a particle size that is ceramically workable. *Chemical solution methods*, e.g. hydroxide coprecipitation and solution freeze-drying, are used to make extremely reactive materials, where mixing occurs on a molecular scale. These are often difficult to control, and are less flexible than oxide milling processes.

The powder is green-formed as required, the material then having some 40–60% of the final density. The usual methods are die pressing, isostatic pressing or extrusion. Powder binders and lubricants are often necessary, and sometimes plasticisers as well. These are usually organic, and are added before the powder is granulated to improve flow properties. Mouldings are sintered to the final density by firing at between 1000 and 1500°C, depending on composition. Heating up is gradual, to eliminate smoothly any moisture and organic additives. *Microwave ferrites* are usually fired in pure oxygen to achieve the high densities required, whereas *MnZn ferrites* are fired, and more particularly cooled, in atmospheres low in oxygen, otherwise free Fe_2O_3 appears in the material, causing cracking and poor magnetic properties. Peak firing temperature and duration determine the final density, grain size and residual porosity. One special technique is *hot pressing* which is used to make very dense fine

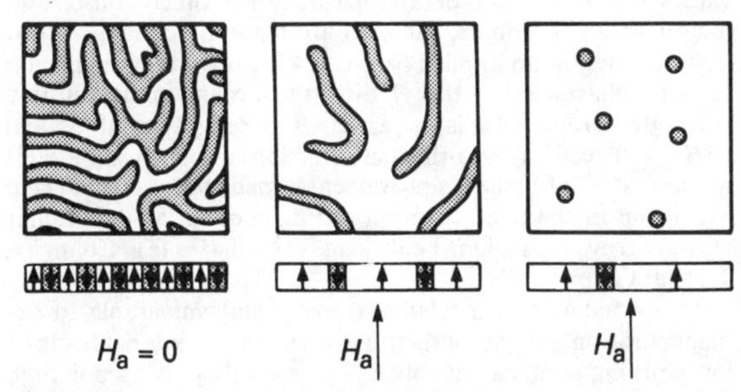

Figure 14.12 Domains in a thin plate of magnetic material with orthogonal easy direction (after Pistorius *et al.*,[13] reproduced by permission)

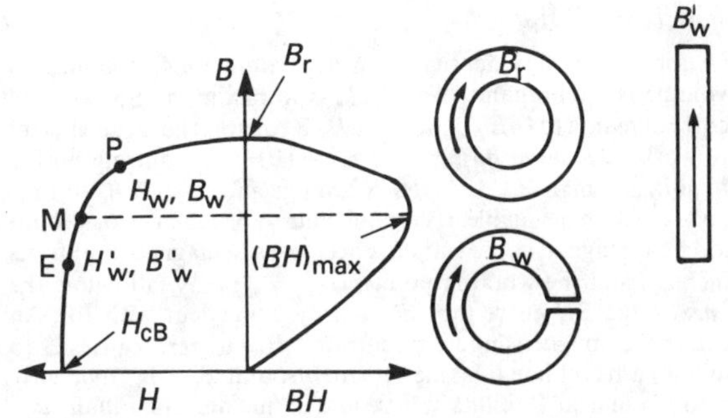

Figure 14.13 Permanent magnet design (after Brailsford,[16] reproduced by permission)

Table 14.6 Properties of typical hard magnetic materials (After Duffin,[2] Brailsford[16] and Buschow[19])

Material and approximate composition	H_{cB} (kA/m)	B_r (T)	$(BH)_{max}$ (kJ/m^3)	
Carbon steel (1% C, 99% Fe)	4	0.95	1.4	I
Chromium steel (5% Cr, 0.7% C, 94.3% Fe)	5	0.94	2.4	I
Alnico (10% Al, 17% Ni, 12% Co, 6% Cu, 55% Fe)	40	0.80	13.6	I
Alcomax III (8% Al, 13.5% Ni, 24% Co, 3% Cu, 1% Nb, 50.5% Fe)	56	1.3	46.5	A
Columax (as Alcomax III)	60	1.35	62	AC
Samarium cobalt 1–5 ($SmCo_5$)	600	0.78	130	AC
Samarium cobalt 2–17 (Sm_2Co_{17})	400	1.2	300	AC
Feroba I ($BaFe_{12}O_{19}$)	140	0.21	8	I
Ferroxdure 380 ($SrFe_{12}O_{19}$)	265	0.39	28.5	AC
Neomax 35 ($Nd_{15}Fe_{77}B_8$)	891	1.22	279	AC

Key: I, isotropic; A, anisotropic; C, crystal orientated.

grain materials used in some high peak power microwave applications.

Ferrite manufacturing processes are considered in some detail by Snelling,[14] Slick[9] and Wang.[15]

14.4 Permanent magnets

14.4.1 Magnetic circuits

If a closed ring of permanent magnet material is magnetised to saturation and the magnetising coil is removed, the internal flux density is the remanence B_r. A small air gap of length g in the ring sets up a demagnetising field H_w causing the internal B to fall to some value B_w at P on the demagnetisation curve (*Figure 14.13*). P is called the *working point*. If the gap field and flux density are H_g and B_g, the pole face area is A_g, and the magnet has length l and cross-section A then:

$$\varphi = B_w A = B_g A_g \quad \text{(assuming no flux leakage)}$$
$$H_w l = -H_g g \quad \text{(by Ampères law)}$$

confirming that H_w and H_g are opposed. Hence

$$B_w H_w l A = -B_g H_g g A_g \tag{14.25}$$

i.e. for a given flux density B_g in gap volume gA_g, the magnet volume is a minimum when $B_w H_w$ is a maximum. So point M corresponding to $(BH)_{max}$ on the BH–B curve is the most efficient to work at. A large B_g implies a large $(BH)_{max}$ requiring both a high B_r and high H_{cB}. In fact flux leakage affects both B_g and B_w, and is seldom negligible. It certainly has to be taken into account for a bar magnet: as the ratio l/A decreases demagnetising effects increase and the working point E(H_w', B_w') may fall below the *knee* of the *BH* curve (not necessarily coincident with M). An *external* demagnetising field returning H_w' to zero causes B to follow a recoil line to some remanent point $B_r' < B_r$. Evidently recoil from near P results in little loss in flux density. Finally *any* element of a magnetic circuit of length l, cross-section A and permeability μ_r has a *reluctance* (in A/Wb) defined by

$$R_m = \frac{l}{\mu_0 \mu_r A} \tag{14.26}$$

Magnetic reluctance is analogous to electrical resistance and reluctances in series and in parallel may be combined using the same laws.

14.4.2 Ferrous alloys

Martensitic steels contain 0.5–1.0% carbon and were the first man-made permanent magnets. Properties are improved by adding Cr, Co and W. After alloying they are quench-hardened in oil or water; magnetic properties are stabilised by annealing at 100°C. The relatively high H_{cB} is due to internal strains which impede domain wall motion.

Dispersion hardened alloys are the major group of alloys. They contain basically Al, Ni, Fe and Co and are cast from a melt at 1100–1200°C and annealed at 550–650°C. *Dispersion hardening* occurs during cooling when small ferromagnetic particles of FeCo precipitate in a non-ferromagnetic NiAl matrix. Controlled cooling is vital for correct dispersion of the FeCo particles, which are probably essentially single domains to account for the high values of H_{cB}. Materials of this type are called Alnico. *Anisotropic* materials (e.g. Alcomax, Ticonal) are made by cooling Alnico from the melt in an applied field ($\gtrsim$24 kA/m). Recrystallisation of both phases along the H direction occurs for favourably orientated grains. H_{cB} is increased, B_r is roughly doubled and $(BH)_{max}$ tripled in the principal direction and correspondingly reduced at 90°. Further improvement is made by *differential heat extraction* in the field direction during cooling, and columnar crystals grow with aligned cube edge crystal axes (e.g. Columax, Ticonal GX).

These materials are relatively costly and vulnerable to demagnetisation, obviating their use where large external fields or low working points are involved. Typically they are used in high field applications, e.g. loudspeakers, chucks, clamps, magnetrons, and generally in professional applications requiring high stability of B_r and H_{cB} and/or high environmental temperatures.

Ferrous alloys in general are discussed by Brailsford[16] and Alnico/Alcomax in particular by McCurrie.[17] Interesting new alloys include the *ductile* Fe–Co–Cr materials (see Zijlstra[18]) which are magnetically similar to Columax-type materials, and the remarkable *neodymium iron boron* compounds based on $Nd_2Fe_{14}B$, which have extremely high H_{cB} and $(BH)_{max}$ and very high B_r values, and are *the most powerful permanent magnet materials* yet made. These are the subject of intense research and development at present.[19]

14.4.3 Non-ferrous alloys

Samarium cobalt 1–5 is the most important of various rare earth cobalt alloys like NdFeB having *extremely high* H_{cB} and $(BH)_{max}$, and high B_r values. It has a linear *BH* (demagnetisation) characteristic and magnetisation and demagnetisation are both difficult. Manufacture involves fairly complex powder metallurgical techniques. Raw materials and processing are expensive, but as $(BH)_{max}$ is so high $SmCo_5$ is often more cost effective than Alcomax, and the very low working points achievable make for remarkable savings in weight and space.

14.4.4 Hard ferrites

The $BaFe_{12}O_{19}$ structure was described in Section 14.3.1. Powder preparation of the important isotropic (barium) and anisotropic (barium and strontium) compounds is conventional, but must ensure grain sizes under 1 μm in the final ceramic to achieve the *very high* H_{cB} values of which they are capable. B_r by contrast is moderately high. *Isotropic* $BaFe_{12}O_{19}$, also formed and fired in conventional ways, has a linear demagnetisation curve and is used in general applications. The *anisotropic* materials are made by dry or wet pressing the powder in a field of order 500 kA/m. The crystallites are thus partially aligned, and remarkably alignment is improved on sintering. B_r is roughly doubled and $(BH)_{max}$ tripled in the principal direction (cf. Alnico and Alcomax). The *BH* curve shows a pronounced knee for anistrotropic $BaFe_{12}O_{19}$ precluding its use in severe demagnetising environments and/or at low temperatures where the working point may move below the knee. It is best suited to high working point *static* applications, e.g. loudspeakers, transducers. By contrast the anisotropic $SrFe_{12}O_{19}$ is particularly resistant to demagnetisation. It has a linear *BH* characteristic, and its H_{cB} is about 50% higher, though B_r and $(BH)_{max}$ are slightly lower, compared with anisotropic $BaFe_{12}O_{19}$. The Sr material finds use in *dynamic* applications with large external demagnetising fields, e.g. d.c. motor stators, torque drives.

Ba and Sr hexaferrites represent some 80% of all world permanent magnet production. Raw materials and processing are fairly cheap and the good B_r/H_{cB} characteristics make flat and compact assemblies possible. The extremely low eddy current losses are of value in r.f. work and some types of motor. One limitation is that the temperature variations of B_r and H_{cB} are an order higher than in Alnico/Alcomax.

A comprehensive review of hard ferrites has been made by Van den Broek and Stuijts.[20] Properties of representative permanent magnet materials are given in *Table 14.6*.

References

1 BLEANEY, B. I. and BLEANEY, B., *Electricity and Magnetism*, 3rd edn, Oxford University Press, Oxford (1976)
2 DUFFIN, W. J., *Electricity and Magnetism*, 3rd edn, McGraw-Hill (1980)
3 SNELLING, E. C., *Soft Ferrites*, Iliffe Books, London (1969)
4 SNELLING, E. C., *Soft Ferrites*, p. 23, Iliffe Books, London (1969)
5 SNELLING, E. C., *Soft Ferrites*, p. 27, Iliffe Books, London (1969)
6 PYE, A., 'Amorphous metals by rapid solidification', *Design Engineering*, **37** (March 1983)
7 CHIN, G. Y. and WERNICK, J. H., *Ferromagnetic Materials*, Vol. 2, *Soft Magnetic Metallic Materials*, North-Holland, Amsterdam, 55 (1980)
8 OVERSHOTT, K. J., *Electronics and Power*, November/December, 768 (1976)
9 SLICK, P. I., *Ferromagnetic Materials*, Vol. 2, *Ferrites for Non-Microwave Applications*, North-Holland, Amsterdam, 189 (1980)
10 VON AULOCK, W. H., Ed., *Handbook of Microwave Ferrite Materials*, Academic Press, London (1965)
11 NICOLAS, J., *Ferromagnetic Materials*, Vol. 2, *Microwave Ferrites*, North-Holland, Amsterdam, 243 (1980)
12 ESCHENFELDER, A. H., *Ferromagnetic Materials*, Vol. 2, *Crystalline Films for Bubbles*, North-Holland, Amsterdam, 297 (1980)
13 PISTORIUS, J. A., ROBERTSON, J. M. and STACY, W. T., *Philips Tech. Rev.*, **35**, 1 (1975)
14 SNELLING, E. C., *Soft Ferrites*, Iliffe Books, London, 5 (1969)
15 WANG, F. F. Y., Ed., *Treatise on Materials Science and Technology*, Vol. 9, *Ceramic Fabrication Processes*, Academic Press, London (1976)
16 BRAILSFORD, F., *Physical Principles of Magnetism*, Van Nostrand Reinhold, Wokingham, 243 (1966)
17 McCURRIE, R. A., *Ferromagnetic Materials*, Vol. 3, *The Structure and Properties of Alnico Permanent Magnet Alloys*, North-Holland, Amsterdam, 107 (1982)
18 ZIJLSTRA, H., *IEEE Trans. Magnetics*, **MAG-14**, 661 (1978)
19 BUSCHOW, K. H. J., *Materials Science Reports*, **1**, 1 (1986)
20 VAN DEN BROEK, C. A. M. and STUIJTS, A. L., *Philips Tech. Rev.*, **37**, 157 (1977)

15 Inductors and Transformers

H C Remfry CEng, MIEE, MBIM, MIIM
formerly with
STC Components Group

Contents

15.1 General characteristics of transformers

The performance of transformer components may be considered in terms of their impedance characteristics and their equivalent circuit, which allows the analysis of transient and transfer functions. The ideal transformer has infinite turns and zero losses. It can therefore represent perfect coupling between input and output circuit or primary to secondary windings. The equivalent circuit includes the parameters shown in *Figure 15.1*.

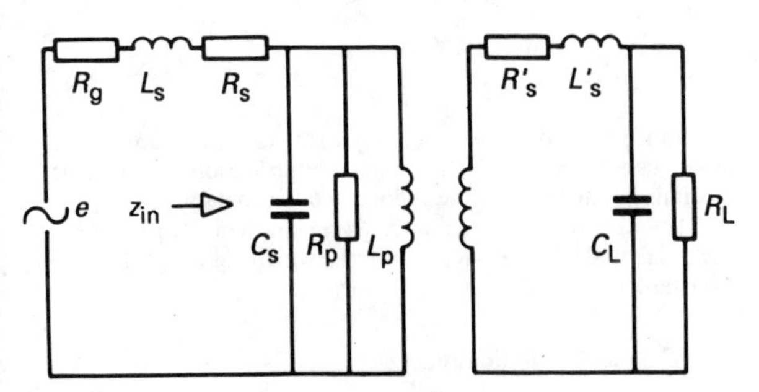

Figure 15.1 Transformer equivalent circuit

The frequency response dependent parameters at low frequencies are the resistive and shunt inductive elements which will attenuate the output voltage as the reactance of the shunt inductance decreases with reducing frequency. At midband frequency the resistance and the reactance of the leakage inductance define the output characteristic and at high frequency the series inductance and shunt capacitance define a stop band. The input impedance Z_{in} is affected by resonance points in the frequency range and these occur due to shunt inductance and self-capacitance acting in current resonance, i.e. apparent infinite impedance, and the self-capacitance with the leakage reactance acting in voltage resonance, i.e. apparent zero impedance. The shunt resonance defines a change in input impedance from inductive to capacitive reactance. Other resonance points occur due to the distributed nature of the reactances, but these will normally lie outside the operating band.

For a step function input the delay parameters are leakage inductance and self-capacitance. Depending on their magnitude either one or the other will be dominant, that is *L/R* or *CR* will delay the rise-time or the fall-time and overswing may be present depending on the degree of damping. The equivalent circuit shows discrete component parameters which allow analysis by the use of linear differential equations, the physical model however resembles a transmission line with distributed parameters where non-linear differential equations prevent satisfactory solution[1].

15.2 Transformer design practice

Design practice involves using a series of trials to achieve convergence to a desired performance. A summary of the steps involved is:

(1) Analysis of performance to define parameter specification.
(2) Selection of available material to satisfy parameter specification.
(3) From the defined parameter specification determine the approximate core excitation, that is minimum inductance or *Q*, or maximum magnetising current or core loss. The number of turns *N* and conductor size may then be provisionally allocated.
(4) With starting value of *N* turns and the determined ratio of additional windings, the allocation of conductor size, turns and insulation for subsequent windings may be made with reference to the rating or current capacity.
(5) The geometry of the coil will allow calculation of mean turn, conductor length, winding depth, weight and resistance.
(6) From the estimated dimensions and parameters found in (5), the resistive drop and regulation can be calculated, together with estimates of the leakage inductance, the winding and interwinding capacitance, the voltage gradients, dissipation and thermal gradient.
(7) The mathematical model can then be checked against the performance requirement with corrective action taken on deviations.

15.3 Transformer materials

The principal winding material is round insulated copper conductor which may be solid, stranded or bunched. Variations are round insulated aluminium over a limited size range; insulated rectangular copper or aluminium which may be solid or stranded; sheet or foil in anoidised aluminium or film coated copper.

Published round wire tables are available from manufacturers, and the corresponding British Standards give quality assurance and qualifying limit performance data.

Insulation coverings are normally synthetic enamels, but variations include single, double or multiple coverings in paper, glass, rayon or silk, where increased spacings to reduce capacitance or arduous duty requirements have to be met.

15.4 Transformer design parameters

15.4.1 Windings

Using the perimeter length of the tube or spool faces on which the windings is to be wound as a base turn dimension (bt), calculation of the mean turn (MT) of each winding can be made for *Figure 15.2* as,

$$MT_1(W_1) = \pi d_1 + bt \quad (15.1)$$

$$MT_2(W_2) = 2\pi(d_1 + t_1 + \tfrac{1}{2}d_2) + bt \quad (15.2)$$

$$MT_N(W_N) = 2\pi(d_1 + t_1 + d_2 + t_2 + \tfrac{1}{2}d_N) + bt \quad (15.3)$$

where d_1, d_2 and d_N are winding depths (heights) and t_1, t_2 are thicknesses of insulation. The MT value may then be used to calculate the active material weight *W* and the resistance *R* as follows:

$$W = MT \times N \times a \times S_d \quad (15.4)$$

$$R = \rho l/a = \rho N MT/a \quad (15.5)$$

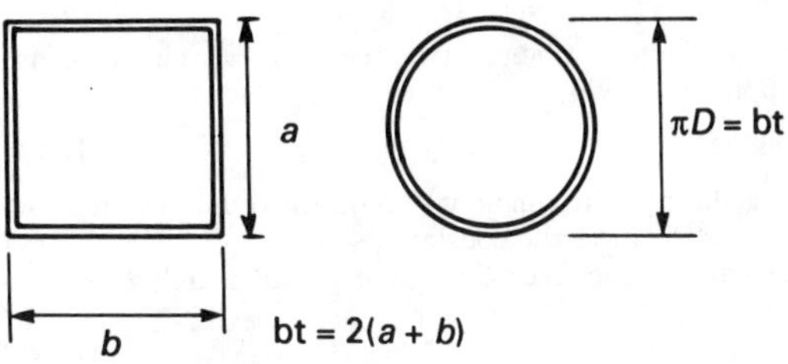

Figure 15.2 Base turn dimension

where N is the number of winding turns, a is the conductor area, S_d is the wire density (9.6 g/cm^3 for copper) and ρ is the resistance per unit length (1.73 μΩ/mm at 20°C for copper).

For small coils an average mean turn of all the windings will give small error, since the error cancels on the sum of all the windings, the inner windings being overstated, the outer windings understated.

With larger coils it is wise to restrict the error and to compute the individual winding dimensions as above.

The sum of the winding depths, $d_1+d_2+d_N$, must be less than the total winding space d, by a winding space factor which is dependent on the winding practices and skills employed. Variations in factors are great particularly for toroidal, transposed, random or orthocyclic windings.

15.4.2 Capacitance

Capacitive elements are present at low value and consequently become dominant only at high frequency or fast wavefronts. They have significance in terms of voltage stress, r.f. interference, crosstalk and common-mode coupling between circuits.

The capacitive elements affect and may override the inductive voltage distribution interval to the winding. *Figure 15.3* shows distributed capacitance elements present in windings.

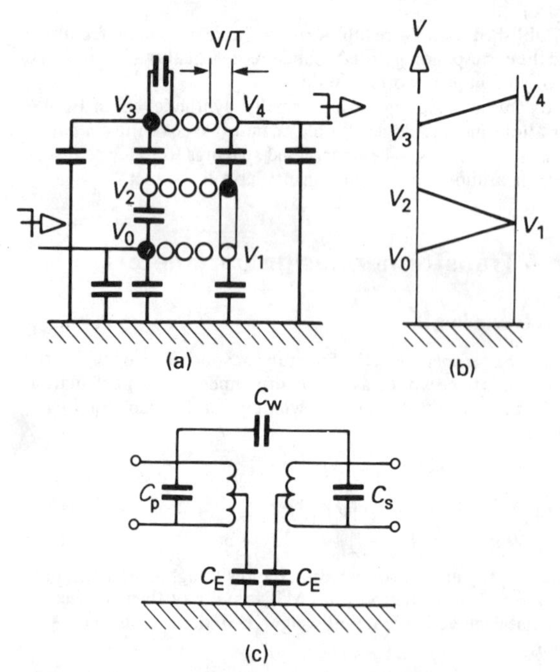

Figure 15.3 Winding capacitance: (a) distributed capacitance in a two-winding coil; (b) voltage gradient; (c) capacitive impedance balance

Considering the first winding layer, the layer capacitance C_l is given approximately by

$$C_l=0.008\,854A\varepsilon/t \text{ pF} \tag{15.6}$$

where A is the layer area (in mm^2), t is the dielectric thickness (in mm) and ε is the dielectric constant (~3–5).

The winding capacitance C_s is given approximately by

$$C_s=\frac{4C_l}{3N}\left(1-\frac{1}{N}\right)\text{pF} \tag{15.7}$$

where N is the number of layers in the winding. Computation of capacitance value is approximate and measured values may need to be established. For reference see MacFadyen,[2] Snelling,[3] *Mullard Technical Handbook*,[4] and Siemens.[8]

The interwinding capacitance C_W may usually be measured directly on *LCR* bridges whilst the self-capacitance requires a *Q* bridge to determine two values of resonating *C* for two resonant frequency points:

$$\frac{(f_1)^2}{(f_2)^2}=\frac{C_1}{C_2} \tag{15.8}$$

and

$$C_s=\frac{(f_1)^2}{(f_2)^2}C_1-C_2 \tag{15.9}$$

Alternatively a dip meter measurement for minimum current point can be made to determine the self-resonant frequency assuming that the winding inductance is constant.

The interwinding capacitance C_W is interactive with the core materials and particularly so for ferrite which has a high dielectric constant.

15.4.3 Leakage inductance

The series leakage inductance is calculated in the following general form with reference to *Figure 15.4*. It is dependent on the

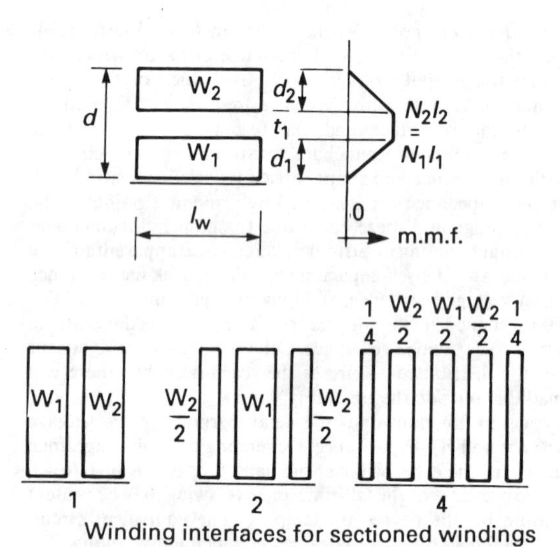

Figure 15.4 Winding leakage inductance

winding arrangement and it should be noted that the leakage inductance is not frequency dependent, the system acting as an air-cored inductor.

$$L_s=4\pi\times10^{-4}N_1^2\frac{\text{MT}}{l_w}\left(\frac{d_1+d_2}{3}+t_1\right)\mu\text{H} \tag{15.10}$$

where MT is the average of two windings, l_w is the length of winding traverse, and MT, l_w, d_1, d_2 and t_1 are in mm.

The core excitation for transformers may be considered as a shunt inductance and resistance in parallel with the load, and corresponds to the core magnetisation and loss currents. Values depend on the working flux density *B*, the frequency and the material used in the core. The definition of self-inductance *L* is the induced e.m.f. *E* per unit rate of change of current *I* where $E=-L\mathrm{d}I/\mathrm{d}t$.

The energy required to establish a current I in an inductance L can be regarded as stored in the magnetic field, so the stored energy

$$W=\int_0^I LI\,\mathrm{d}t \tag{15.11}$$

and if L is constant the total stored energy is $\frac{1}{2}LI^2$.

The flux density B is related to E by $\mathrm{d}\varphi/\mathrm{d}t$ and $Et=BNA\times10^{-8}$ where A is a uniform core section area in cm^2, N is the number of turns and t the time in seconds.

The m.m.f. H is equal to the work done in carrying a unit pole N times around the current in the winding where

$$H=\frac{4\pi NI}{10l} \tag{15.12}$$

The permeability μ is the ratio B/H, giving

$$\frac{E}{I}=Z=R+\mathrm{j}X=\mathrm{j}\omega\frac{4\pi N^2A}{10^9 l_e}\mu \tag{15.13}$$

and

$$L=\frac{4\pi N^2A\mu}{10^9 l_e} \tag{15.14}$$

where l_e is the magnetic path length in cm and $\omega=2\pi f$.

15.4.4 Losses

The core loss resistance R may be expressed as

$$R=\frac{4\pi N^2Ak}{10^9 l_e} \tag{15.15}$$

where k is an experimentally determined factor for the material and flux condition. The permeability μ identifies with values of flux density B. Hence at low flux density μ remains constant[5] and is known as initial permeability μ_i or μ_0. When B has the greatest rate of change with H the permeability is μ_{max}. At operating B_{max} it is called the amplitude or apparent permeability μ_a and for a composite permeability, such as a core interrupted by an air gap space, it is the effective permeability μ_e. Incremental permeability $\Delta\mu$ refers to the shape of a cyclic magnetisation superimposed onto a static magnetisation. For a specific size of core the permeability may be expressed as an inductance factor A_L, usually in nanohenry per turn. Permeability coefficients are evaluated for temperature change, frequency and time.

The loss resistance k is frequency and flux dependent and consists of hysteresis and eddy current loss together with stray or residual loss. It has the form $W=W_h+W_e+C$ where at low induction W_h varies as B^3f, or at high induction as $NB^{1.6}f$, where N is the Steinmetz factor. W_e varies as B^2f^2 or at high induction as

$$\frac{\pi tB^2}{(\mu\rho)^{1/2}}f^{3/2}$$

C represents an energy loss of unknown origin, t is the thickness and ρ the resistivity.[5]

The effect of eddy current circulation is such as to oppose the field producing them and to prevent the creation of a uniform field within the lamination or core thickness. The effective permeability is therefore dependent on this skin effect, and at very high frequency 95% of the true permeability may be lost. Welsby[6] defines a critical depth of penetration equal to half the lamination thickness occurring at critical frequency f_c where

$$f_c=\frac{Q}{\pi^2f^2\mu_0}\times10^9 \quad \text{for } Q>6 \tag{15.16}$$

The eddy current skin effect is present also in the winding conductor where the eddy currents oppose the flow of current in the centre of the conductor and effectively increase the resistance of the conductor. A similar action is produced by the proximity effect of the alternating magnetic fields of nearby conductors. To limit this increase in losses the use of Litz, bunched or transposed, conductors is necessary, usually at high frequency or in larger windings with high non-active winding space.

15.4.5 Heating

Losses produced within the core and windings create heat and raise the temperature of the materials. Because of the uneven source of heating the structure has thermal gradients recognised as hot spot, and average and surface temperature. With cooling by natural convection the heat is transferred to the surrounding ambient and a steady-state temperature will be achieved under constant load conditions, and after a time period dependent on the size and shape of the component. Assessment of thermal gradient for air natural cooling can be made only with extreme simplification. Representation of the heat flow may be made by considering an equivalent electrical circuit as in *Figure 15.5*, but experimental data obtained from physical models are usually necessary to validate a final design.

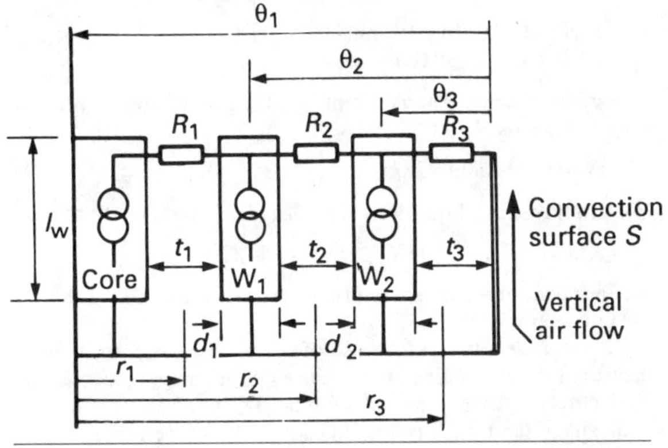

	Thermal conductivity (10^3 μW/(mm °C))
Copper	385
Iron	13
Aluminium	220
Ferrite	3.5
Filled epoxy	0.4
Plastic nylon	0.25
Paper	0.1

Figure 15.5 Radial heat flow gradient

The schematic[3] represents radial heat flow from heat generators (shown as current sources) passing through low conductivity insulating material (shown as resistors) towards the convection surface where the heat is transferred to the ambient air. The surface areas of the low conduction barriers are calculated from the winding length l_w and a mean turn circumference for the barrier of $2\pi r$. Using known values of thermal conductivity a calculation of temperature gradients may be made. For example, using the following values from the outline design for a full-wave centre-tap rectifier and capacitive-input (FWCT-CIF) mains transformer:

$l_w=38.8$ mm

$r_1=14.29$ mm	$r_2=18.29$ mm	$r_3=25.29$ mm
$t_1=1.0$ mm	$t_2=0.25$ mm	$t_3=0.25$ mm
$P_1=1.87$ W	$P_2=3.74$ W	$P_3=3.61$ W

(assuming round limb), the heat flow through R_3 is given by

$$\frac{P_1+P_2+P_3}{2\pi r_3 l_w}=\frac{1.87+3.74+3.61}{6.28+25.29+\times 38.8}=0.0015\ \text{W/mm}^2$$

The thermal resistance of R_3 is given by t_3/λ, where $\lambda=200\ \mu\text{W/(mm °C)}$ for tape. Hence

$t_3/\lambda=0.25/200=0.001\,25°\text{C mm}^2/\mu\text{W}$

Thus the temperature drop θ_3 across R_3 is given by

$$\theta_3=0.001\,25\times 1500=1.87°\text{C} \qquad (15.17)$$

The heat flow through R_2 is

$$\frac{P_1+P_2}{2\pi r_2 l_w}=\frac{5.61}{4457}=0.001\,26\ \text{W/mm}^2$$

The thermal resistance of R_2 is t_2/λ and so

$t_2/\lambda=0.25/200=0.001\,25°\text{C mm}^2/\mu\text{W}$

Thus the temperature drop $(\theta_3-\theta_2)$ across R_2 is

$$\theta_3-\theta_2=0.001\,25\times 1260=1.57°\text{C} \qquad (15.18)$$

Performing the same calculation for R_1 we have

$P_1/2\pi r_1 l_w=0.000\,501\,9\ \text{W/mm}^2$
$t_1/\lambda=1.0/250=0.004°\text{C mm}^2/\mu\text{W}$

where λ for plastic is $250\ \mu\text{W/(mm °C)}$. Thus the temperature drop $(\theta_1-\theta_2)$ across R_1 is

$$\theta_1-\theta_2=0.004\times 502=2.0°\text{C} \qquad (15.19)$$

We can now solve Equations (15.17), (15.18) and (15.19) to give

$\theta_3=1.87°\text{C} \qquad \theta_2=3.44°\text{C} \qquad \theta_1=5.44°\text{C}$

The temperatures calculated are the temperatures above that of the convection surface *S*.

The average temperature rise of the winding is conveniently measured by the change in resistance whilst carrying a constant load current. After a period of time 3*T*, where *T* is the time constant of the transformer, a steady-state temperature will be reached. The percentage change in resistance R_0 (initial) to R_T (steady state) will give the temperature rise of the winding using the temperature coefficient of the resistivity of copper of 0.393°C/% change.

The resistance measurement is the average resistance of the winding and includes increments of resistance higher than average (hotter) and also of lower resistance (cooler). For coils with height greater than a few centimetres the surface temperatures at the base and at the top of the coil will be different. For coils of 50 cm height the temperature difference becomes extreme. The radial temperature gradient is also distorted at various levels of coil height and for large units the use of cooling ducts becomes necessary to prevent excessive temperature values. The presence and type of varnish, also the degree of penetration and retention have a marked effect on reducing thermal gradient and significant improvements result in the use of liquid cooling.

Safety requirements involve limiting the temperature rise under fault conditions, which may be a continuously applied short circuit of the load. A limit may also apply to rated load so that a dissipation curve may need to be followed closely.

15.4.6 Vector presentation

The input impedance of a transformer connected load may be represented by the magnitude and vector of the constituent parameter losses. The method is used to determine the accuracy of ratio and phase errors in precision measuring current and voltage transformers. It is also useful to consider the performance

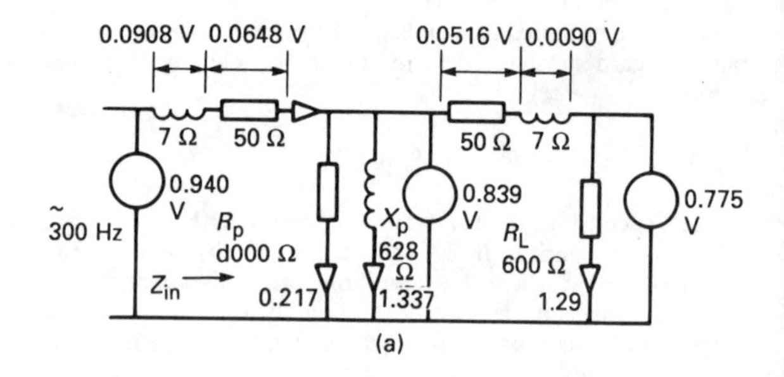

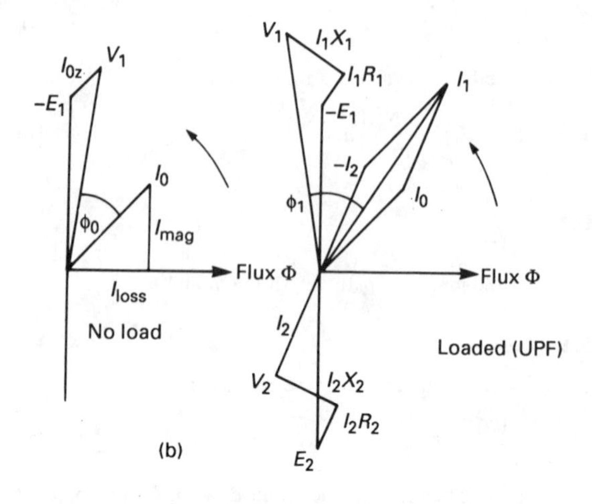

Figure 15.6 (a) Equivalent circuit and (b) vector diagram of an audio transformer

of audio transformers at a given frequency (*Figure 15.6*). Using the equivalence of a unity ratio transformer the effects on the load of the series and shunt parameters are illustrated in the vector diagrams. The parameters have been treated as partially distributed. If the secondary winding leakage and resistance are transferred to the primary winding values and lumped to them, there is a variation in performance:

	Lumped parameter	*Partial distributed parameter*
$Z_{in}=E_{in}/I_{in}$	469 Ω	467 Ω
Insertion loss $20\log(E_o/E_{in})$	−1.35 dB	−1.68 dB
Insertion loss $20\log\left(\frac{Z_{in}+R_L}{Z_{in}-R_L}\right)$	−12.23 dB	−18.09 dB

Care is therefore necessary in the allocation of parameters to an equivalent circuit and recognition that the distribution may not be entirely defined. The equivalence of equivalent circuits is given by MacFadyen.[2] Vector representations of loads with lead and lag current in the secondary winding are illustrated in *Figure 15.7*.

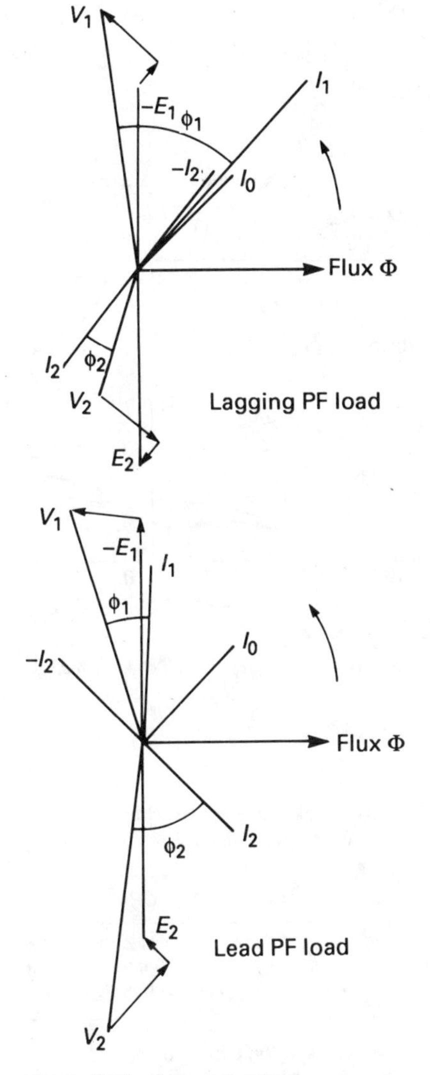

Figure 15.7 Parameter vector

15.5 Inductors

The ability to store energy in the form of electromagnetic charge is a characteristic of inductors and reactors and complements the electrostatic charge energy storage present in capacitance. The inductor has low impedance at zero frequency and high impedance at high frequency. The capacitor has high impedance at zero frequency, low impedance at high frequency. The characteristic equations for the voltage across and the current through these passive components are:

Inductance $\quad v=-L\dfrac{\mathrm{d}i}{\mathrm{d}t} \qquad i=\dfrac{1}{L}\displaystyle\int_0^t v\,\mathrm{d}t$

Capacitance $\quad v=\dfrac{1}{C}\displaystyle\int_0^t i\,\mathrm{d}t \qquad i=C\dfrac{\mathrm{d}v}{\mathrm{d}t}$

The inductor is normally used in series mode and is current driven, whilst the capacitor is used in parallel or shunt mode and is voltage fed. They have linear characteristics but have limit value at overcurrent (saturation) or overvoltage (breakdown).

The inductor dissipation losses are due to the winding conductor resistance and to the loss generated when under magnetic excitation, and a relationship $Q=2\pi fL/R$ may be used to define the inductor performance. It also expresses the loss angle or power factor presented to the circuit.

The use of air gaps in the magnetic path extends the parametric range of inductive components and enables highly stable linear components to be constructed. Provision of moveable core material within the air gap space allows adjustment of inductance. Inductor sizes vary from very small air cored to very large protection reactors where they may be embedded in concrete to withstand electromagnetic forces produced by fault currents.

Their application in electronic circuits however is mainly in suppression, timing and smoothing duties. Here the core requires a large flux range and the silicon–iron materials are usually most suitable to accommodate the dual excitation of high d.c. ampere turns and the superimposed smaller a.c. excitation. In multi-phase rectifier circuits a centre-tapped reactor is often used to solve commutation problems and limit the large circulating current that would otherwise flow between the conduction phases. For power applications the design may be limited by thermal rating due to either winding loss or to the core loss present at high a.c. excitation or high frequency.

Inductors required in frequency selective circuits and particularly in carrier telephony networks are characterised by the need for high stability performance at very low signal amplitude. The construction normally requires ferrite material in pot core form with high shielding from crosstalk coupling. The inductors will be provided with an adjustment to permit setting the value accurately. To assist in the selection of core size it is possible to calculate and plot a range of inductance and loss values against frequency in the form of iso-Q curves. A line joining equal Q-value points provides a Q-contour from which inductance, frequency and turns may be selected. (See *Figure 15.8*.)

At high frequency the filter inductor is designed to have maximum Q-factor at a given frequency. At lower frequencies requiring high inductance value, which in turn requires smaller gap dimension and high μ_i, the variability encountered may require the use of a lower Q-factor with less sharp cut-off frequency. The need to work at low induction requires correspondingly high turns giving increased winding resistance and, at the higher frequencies, increased a.c. resistance. The presence of the distributed capacitance within the winding produces dielectric losses which, because of the small signal amplitude, are no longer negligible and contribute to the terminal impedance.

The choice of conductor size becomes complex, as for instance the use of stranded 0.071 mm diameter conductors with higher eddy current loss may give lower overall loss than stranded 0.040 mm conductor which may have higher capacitance and dielectric loss. The performance is therefore dependent on careful manufacture and model validation, before design release, although the use of computer programs is offered by some of the manufacturers of the core material.

Powdered nickel and iron dust cores are currently used for inductors and smoothing chokes at high frequency. The powdered nature of the core material, held together by a binding agent, provides a distributed air gap with low electromagnetic interference field. In normal form the cores are toroidal and are available with toleranced A_L value, stabilised and with stated temperature coefficients. A wide range of values may be produced usually with single layer windings using wire size suited to the current rating. The saturation flux density for this material is twice that of ferrite. The use of nickel core material in lamination form giving high permeability and low hysteresis loss is common in telecommunication components at speech frequency and particularly where superimposed direct current is present—the high inductance and low loss requirement being necessary to

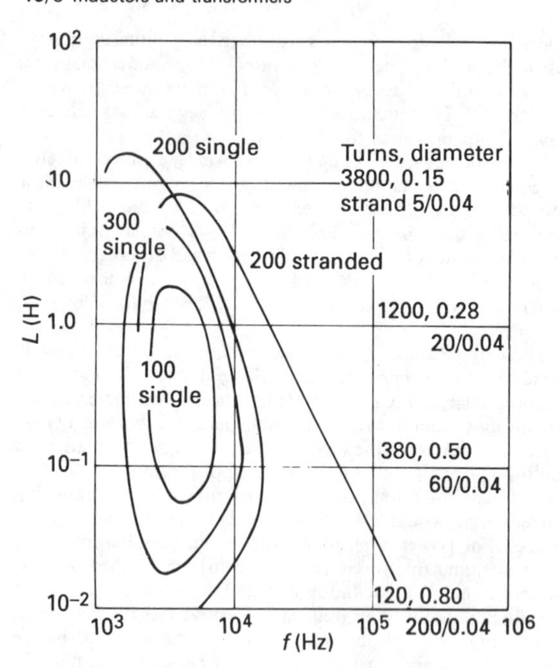

Figure 15.8 Typical full winding Q-factor: 45 mm pot core, $\mu = 100$ (from *Mullard Technical Handbook*[8])

Figure 15.9 Full-wave rectification (from O. H. Schade, *Proc. IRE, July 1943*)

reduce attenuation of low frequency signals.

A wide range of nickel alloy materials are available with permeability figures of 260 000 being achieved. The effect of orientation and cold reduction in the preparation of nickel strip allows square loop materials with the ratio of remanence flux density to saturation flux density greater than 95%, and which will operate in free running oscillators and have ideal characteristics for saturating reactors or transducers.

15.6 Design examples

Design examples are presented in the form of working notes and establish the scenario, but they need further work to complete the design. The parameters involved in the examples are described briefly in the text and covered by data presented in the illustrations. Magnetisation data have not been included and it is necessary to refer to manufacturers' handbooks for further information.

15.6.1 50 Hz rectifier mains transformer

It is desired to provide a smoothed direct current output of 40 V at 1 A full load and a current of 0.1 A light load, using a full-wave centre-tap rectifier and capacitive input filter with values (see *Figure 15.9*): $R_{FL} = 40\,\Omega$, $R_{LL} = 400\,\Omega$, $N = 2$ (number of rectifiers), R_s = source resistance. Using $2\pi fCR_L$ of 0.3 and $R_s/R_L = 10\%$ gives $C = 24\,\mu F$, $E_0 = 0.57E_{pk}$, E_{ripple} is almost 100%, $T = CR =$ 0.96 ms. With $2\pi fCR_L$ of 100 and $R_s/R_L = 10\%$ then $C = 8000\,\mu F$, $E_0 = 0.76E_{pk}$, $E_{ripple} = 1.3\%$, $CR = 320$ ms. The first has too low a conversion factor and little smoothing, the second gives high conversion but excessive peak currents. Checking alternative $2\pi fCR_L$ values we have (in obvious units):

$2\pi fCR_L$	C	E_0	E_{pk}	E_{rms}	E_{ripple}	CR	I_{rms}	I_{pk}
13	1035	0.75	53.3	37.7	10	41.4	2.2	6.4
28	2230	0.76	53.3	37.7	5	89.2	2.3	6.4
150	12000	0.76	53.3	37.7	1	48.0	2.3	6.4

Satisfactory utilisation occurs at $2\pi fCR_L = 30$.

Figure 15.9 for the circuit shows the approximate current conduction for the sine wave period, the relative discharge time constant to $T = 3$ periods or 30×10^{-3} seconds from a 50 Hz supply frequency, e.g. 1.5 cycles. The time constant of 39.2 ms corresponds to a ripple percentage of E_0 of approximately 5%.

The secondary rating for this unit is then 76 V centre-tap at 1.6 A = 60.8 volt ampere/phase.

Referring to the table of lamination sizes (*Table 15.1*) the pattern 196 in square stock configuration has a turns per volt value of 4.05 at a working flux density of 1.5 T at 50 Hz. Allowing for a winding regulation of 10%, $V_{secondary} = 83.6$ V at no load.

The design would proceed as follows:

Primary turns, N_1	240 V × turns/V = 972
Secondary turns, N_2	83.6 V × turns/V = 338 tap at 169
Winding area, A_N	4.65 cm^2 say space factor (SF) 0.7 Allow $\frac{1}{3}$ winding space for primary and $\frac{2}{3}$ space for secondary
Conductor for N_1	$972 \times (\frac{1}{3} \times 4.65 \times 0.7)$ turns/cm^2 = 895 turns/cm^2 Bare diameter 0.280 mm from wire table has 896 turns/cm^3, grade 2 covering

Table 15.1 Core parameters

No. (8)	Plan L (cm)	Plan W (cm)	A_e (cm³)	l_e (cm)	Wt. of Fe (kg)	l_w (mm)	d (mm)	bt (mm)	A_N (mm²)	Wt. of Cu (kg)	A_L (3)	A_R (4)	$LI/R^{1/2}$ (5)	Rating (VA) (2)	Turns/V (1)	Comment
35	57.2	47.6	3.26	11.5	0.296	25.4	7.5	86.5	190	0.075	3572	23.1	0.031	14	9.2	DIN 60 near
29	76.2	63.5	5.8	15.3	0.695	33.8	10.2	117	345	0.209	4776	15.6	0.069	45	5.1	DIN 75 near
196	85.7	71.4	7.34	17.2	0.985	38.8	12	135	465	0.382	5377	11.8	0.101	75	4.05	DIN 84 near
248	133	111	17.8	26.7	3.72	63.5	19.1	203	1210	1.46	8400	7.1	0.344	250	1.65	DIN 135 near
0405	23.8	33.3	0.58	7.45	0.032	12.7	5.5	44.5	70.5	0.018	981	32.9		1	49.5	Single HWR C core
1008	57.2	54	1.98	12.4	0.177	33.3	9.9	79.5	330	0.154	2012	11.4	0.05	5	14.9	Double HWR C core
5014	95.3	88.9	5.25	20.6	0.827	57.3	19	118	1089	0.867	3211	5.1	0.25	75	5.35	Double HWR C core
9024	146	127	13.05	30.2	3.12	79.5	31	184	2465	2.89	5440	3.6	190	300	2.19	Double HWR C core
6/3	28.6	19	0.206	7.48	0.012	56.5	4.5	23	191	0.027	347	8.18	—	0.4	129	HWT ring DEF5193
10/6	47.6	31.8	0.715	12.47	0.068	96.7	7.7	38.8	559	0.142	722	4.41	—	8	37	HWT L=OD, W=ID
12/6	57.2	38.1	0.824	14.96	0.099	116	9.3	42.1	810	0.257	744	3.1	—	16	32.1	HWT L=OD, W=ID
18/9	85.7	57.1	1.93	22.45	0.332	182	14.5	59.2	1990	0.935	1083	1.84	—	150	13.7	HWT L=OD, W=ID
C118	18	11	0.43	2.6	0.006	6	3.05	27.3	16	0.002	2084	93.9			4.3[7]	IEC 2 slot BS 4061
RM10	24.7	18.7	0.842	4.13	0.020	11	4.2	39.8	46.2	0.010	2569	44.4	0.002		8.4[7]	DIN 41980
K45	45	29.2	3.62	6.7	0.109	16.84	6.7	72.5	113	0.041	6808	31.8	0.014		36.2[7]	BS 4061 Range 1
EC35	34.5	34.6	0.865	10.3	0.040	28.8	7.5	51.8	216	0.018	1432	21.7	0.003		8.6[7]	
EC70	70	69	2.83	14.1	0.180	41.4	11.3	62.8	468	0.214	2529	6.6	0.026		28.3[7]	

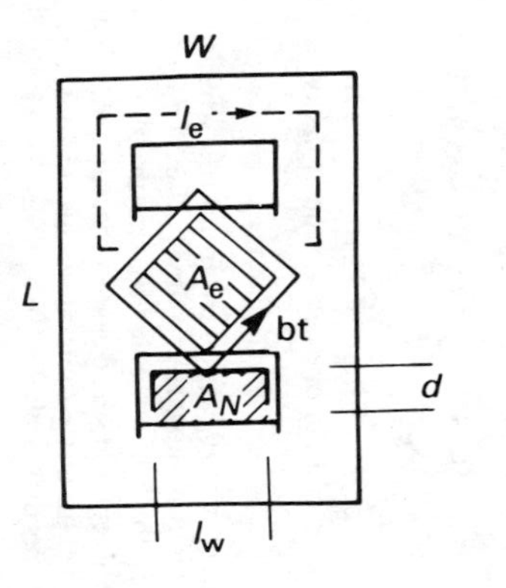

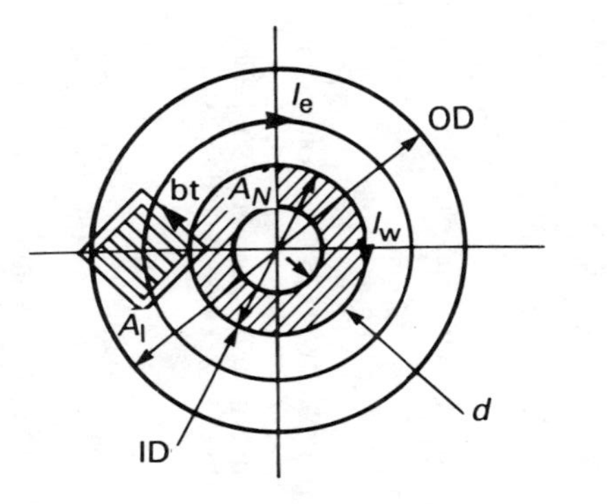

1 Turns/V for $B_{max}=1.5$ T, 50 Hz
2 Load rating, normal ambient for turns/V
3 $L=N^2A_L$ in nH at $\mu_R=1000$
4 $R=N^2A_R$ in $\mu\Omega$/20°C
5 $LI/R^{1/2}$ constant, incremental excitation at 10 mT 50 Hz with IDC magnetisation
6 Stacking factor 0.9—square/near-square section
7 $E\times t$ in μV s for $N=1$, $B=100$ mT
8 Pattern number or designation.

Conductor for N_2 — $338/(\frac{2}{3}\times 4.65\times 0.7)$ turns/cm^2 = 156 turns/cm^2
Bare diameter 0.750 mm from wire table has 144.4 turns/cm^2, grade 2 covering

Winding 1 (primary) — Turns per layer = 116, layer length (traverse) = 38.8 mm
Layers = 8.4; depth, $d_1 = 3.0$ mm
$MT_1 = \pi d_1 + bt = 3\pi + 135 = 144.4$ mm
$a_1 = \pi(0.280/2)^2 = 0.061\,58$ mm^2
$Wt_1 = MT_1 \times N_1 \times a_1 \times S_d$
$= 144.4 \times 972 \times 0.061\,58 \times 9.6/1000$
$= 86.2$ g

$$R_1 = \frac{\rho l}{a_1} = \frac{17.3\times 10^{-6}\times 144.4\times 972}{0.061\,58}$$
$= 39.4\ \Omega$

Winding 2 (secondary) — Turns per layer = 46, layer length (traverse) = 38.8 mm
Layers = 7.4; depth, $d_2 = 6.7$ mm
$MT_2 = 2\pi(d_1 + t_1 + \frac{1}{2}d_2) + bt$
$= 2\pi(3.0 + 0.25 + 3.35) + 135$
$= 176.4$ mm
$Wt_2 = MT_2 \times N_2 \times a_2 \times S_d$

$= 176.4 \times 338 \times 0.441\,8 \times 9.6/1000$
$= 252.8$ g

$$R_2 = \frac{\rho l}{a_2} = \frac{17.3\times 10^{-6}\times 176.4\times 338}{0.441\,8}$$
$= 2.33\ \Omega$

Winding depth — $d_1 + t_1 + d_2 + t_2 = 3 + 0.25 + 6.7 + 0.25 = 10.2$ mm

Calculating losses — $d = 12$ mm, space factor (SF) = 85%

$R_1 = 39.4\ \Omega/20°\text{C} = 47.67\ \Omega/75°\text{C}$ reference temperature at full load
$I_1 = 60.8 \times 1.10/240 = 0.28$ A, assuming 90% efficiency
$R_2 = 2.33\ \Omega/20°\text{C} = 2.82\ \Omega/75°\text{C} = 1.41\ \Omega$/half

Then:

Winding 1 (primary) — 240 V, $I_1 = 0.28$ A, $R_1 = 47.67\ \Omega$, $(I^2R) = 3.74$ W, $(IR)_1 = 5.56\%$

Winding 2 (secondary) — 41.8 V at no load, $I_2 = 1.6$ A, $R_2 = 1.41\ \Omega$, $(I^2R)_2 = 3.61$ W, $(IR)_2 = 5.39\%$

Hence $(I^2R)_{total} = 7.35$ W, $(IR)_{total} = 10.95\%$

Regulation — $41.8 \times 0.89 = 37.2$ V at full load

Core loss at 1.5 T, 50 Hz and 0.985 kg core weight

Material 800/0.50, 6 W/kg, 19 V A/kg = 5.91 W 78 mA
M6/0.35, 1.9 W/kg, 17 V A/kg = 1.87 W 70 mA

Total loss = copper + iron = 7.35 + 5.91 = 13.26 W, $\eta = 88.1\%$, $\theta_r = 74°$C or for grain-oriented M6 core: total loss = 7.35 + 1.87 = 9.22 W, $\eta = 86.8\%$, $\theta_r = 57°$C

Temperature rise θ_r is taken from the average conductor temperature rise shown for pattern 196 in *Figure 15.10*. For ambient up to 45°C a permissible temperature rise of $\theta_r = 55°$C would be acceptable; the design is therefore marginal. It should be noted that the dissipation rate at this steady-state temperature is flat, at approximately 6°C/W.

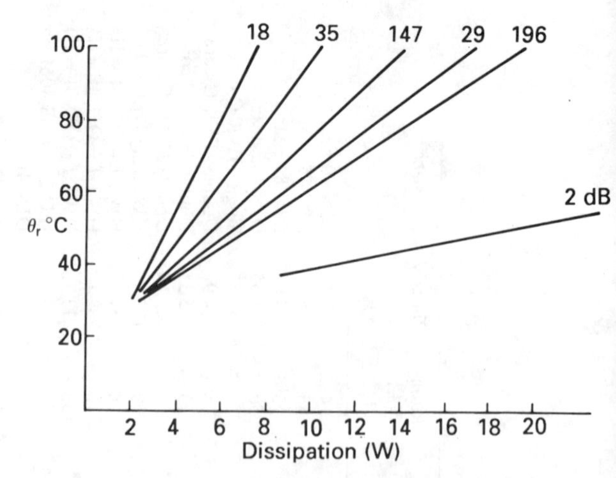

Figure 15.10 Temperature rise curves for a rectifier mains transformer

The light load condition of 0.1 A and $R_{LL} = 400\ \Omega$ modifies the conversion to $2\pi fCR_{LL}$ at 300 with reduction in ripple and a rise in E_0. The choice of capacitor value is of course related to available standard values.

Insulations of polyurethane wire, nylon bobbin and polyester tape have continuous temperature category class A to BS 2757 which permits continuous operation at 110°C and therefore allows a hot spot temperature of +10°C above average. For hostile ambients above 45°C an uprating in insulation class would be necessary or an increase in transformer frame size.

15.6.2 50 Hz safety isolating transformer

Rating — 240 V to 17.8 V, 4.9 V A

Heating — (a) Temperature rise to be limited to 85°C at FL × 1.1 V in
(b) Temperature rise to be limited to 135°C at SC × 1.1 V in

Regulation — Less than 100%, $V_2 > 0.5E_2$

No Load — Current less than 10% of rated output, <2 mA
Core loss less than 10% of rated output, <0.49 W

Isolation — 4 kV and 6 mm creepage

A selected core would be pattern 35 in square stack configuration from *Table 15.1*. The core area A_e is 3.26 cm^2. After trial calculations a flux density of 660 mT is chosen giving a turns per volt of 20.83. To achieve the specified isolation a two-section spool is chosen giving $l_w = 12$ mm, $d = 7.5$ mm, $A_N = 90$ mm^2 per section, with a separating centre cheek of thickness 1 mm.

The design proceeds as follows:

Winding 1 (primary) — Volts 240 V: turns, $N_1 = 240 \times 20.83 = 5000$

Winding 2 (secondary) — Volts 35.6 V: turns, $N_2 = 35.6 \times 20.83 = 740$

The available winding space per section is 90 mm^2 or with 80% SF = 0.72 cm^2

Conductor for N_1 — 5000/0.72 = 6944 turns/cm^2, diameter = 0.100, grade 2 (from wire table)

Conductor for N_2 — 740/0.72 = 1028 turns/cm^2, diameter = 0.250, grade 2 (from wire table)

Winding 1 (primary) — Turns per layer = 1.2 cm × 77.5 turns/cm = 93

Layers = 5000/93 = 54; depth $d_1 = 54 \times 0.129 = 7$ mm

$MT_1 = \pi d_1 + bt = 7\pi + 86.5 = 108.5$ mm

$W_1 = MT_1 \times N_1 \times a_1 \times S_d$
$= 108.5 \times 5000 \times 0.05^2\pi \times 9.6 \times 10^{-6}$
$= 40.9$ g

$$R_1 = \frac{\rho l}{a_1} = \frac{17.3 \times 108.5 \times 5000}{0.05^2\pi}$$

$= 1195\ \Omega$

Winding 2 (secondary) Turns per layer = 1.2 cm × 33.2 turns/cm = 39

Layers = 740/39 = 19; depth, $d_2 = 19 \times 0.301 = 5.72$ mm

$MT_2 = \pi d_2 + bt = 5.72\pi + 86.5 = 104.5$ mm

$W_2 = MT_2 \times N_2 \times a_2 \times S_d$
$= 104.5 \times 740 \times 0.125^2\pi \times 9.6 \times 10^{-6}$
$= 36.4$ g

$$R_2 = \frac{\rho l}{a_2} = \frac{17.3 \times 10^{-6} \times 104.5 \times 740}{0.125^2\pi}$$

$= 27.25\ \Omega$

Under full load condition the losses are calculated as follows:

$I_1 = 5.8$ V A/240 V = 0.024 A
$I_2 = 4.9$ V A/17.8 V = 0.275 A allowing for losses
$R_1 = 1195\ \Omega/20°C = 1446\ \Omega/75°C$
$R_2 = 27.25\ \Omega/20°C = 33\ \Omega/75°C$

Core loss for 800/50 material at 660 mT, 50 Hz, 0.296 kg. From manufacturers' data 1.5 W/kg = 0.44 W, 1.4 A/kg = 0.44 VA = 1.85 mA

$V_1 = 240$ V, $I_1 = 0.024$ A, $R_1 = 1446\ \Omega$
$(I^2R)_1 = 0.833$ W, $(IR)_1 = 14.46\%$
$V_2 = 35.6$ V, $I_2 = 0.275$ A, $R_2 = 33\ \Omega$
$(I^2R)_2 = 2.49$ W, $(IR)_2 = 25.49\%$
Loss = $W_1 + W_2 + \text{Fe} = 0.833 + 2.49 + 0.44 = 3.76$ W, $\eta = 56.6\%$, $\theta_r = 47°C$ (*Figure 15.5*)
Regulation $14.46 + 25.4 = 39.95\%$ $V_2 = 21.36$ V (17.8 V)

Correction is required to $N_2 = \frac{17.8}{0.6} \times 20.83$

$= 618$ turns, 30.4 g, 22.75 Ω

Revised V_2 values:

$V_2 = 29.7$ V, $I_2 = 0.275$ A, $R_2 = 27.5\ \Omega$,
$(I^2R)_2 = 2.08$ W, $(IR)_2 = 25.46\%$
Loss = $0.833 + 2.08 + 0.44 = 3.35$ W, $\eta = 54.4\%$, $\theta_r = 43°C$
Regulation = $14.46 + 25.46 = 39.92\%$ $V_2 = 17.82$ V

The FL temperature rise is within the specified 85°C and will remain so at $1.1V_1$.

Short-circuit load condition entails a possible winding temperature of 135°C, the winding resistance should therefore be corrected to this temperature:

$R_1 = 1195\ \Omega/20°C = 1735\ \Omega/135°C$
$R_2 = 22.75\ \Omega/20°C = 33\ \Omega/135°C$

With the load shorted the output voltage $V_2 = 0$, the transfer voltage E_2 is therefore developed across the internal winding impedance, mainly R. The condition approximates to:

$V_1 = 264$ V, $I_1 = 0.091$ A, $R_1 = 1735\ \Omega$
$(I^2R)_1 = 14.46$ W, $(IR)_1 = 60\%$
$V_2 = 29.7$ V, $I_2 = 0.36$ A, $R_2 = 33\ \Omega$
$(I^2R)_2 = 4.28$ W, $(IR)_2 = 40\%$
Loss = $14.46 + 4.28 + 0.44 = 19.18$ W, $\eta = 0\%$
Regulation = $60 + 40 = 100\%$ $V_2 = 0$

θ_r at 19.18 W has an indicated dissipation rate of 10°C/W giving a temperature of 192°C (see *Figure 15.11*). The cooling is modified by the heat distribution and with heat flow into the core and to W_2, a value of near 135°C would be found on a physical model.

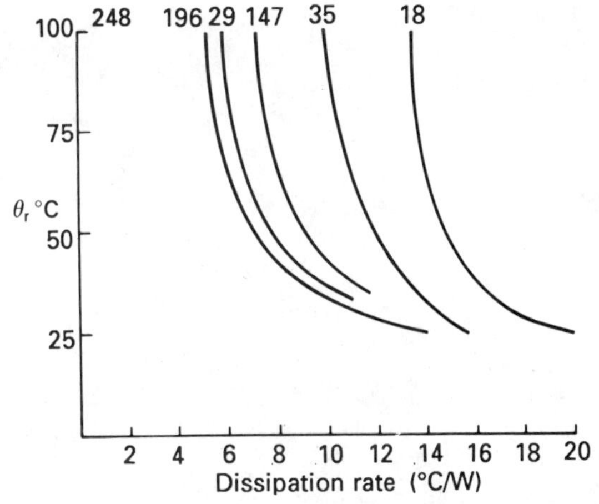

Figure 15.11 Temperature rise curve for a safety isolating transformer

References

1 CHENG, D. K., *Analysis of Linear Systems*, Addison-Wesley, New York (1977)
2 MACFADYEN, R. A., *Small Transformers and Inductors*, Chapman and Hall, London (1954)
3 SNELLING, E. C., *Soft Ferrites*, Iliffe Books, London (1969)
4 *Mullard Technical Handbook*, Book Three, Part 3, 'Vinkor inductor cores'
5 *Encyclopaedia Brittanica*, 'Magnetism', p. 594 (1968)
6 WELSBY, W. G., *The Theory and Design of Inductance Coils*, 2nd edn, Macdonald & Co., London (1960)
7 HECK, C., *Magnetic Materials and their Application*, Butterworths, London (1974)
8 SIEMENS, *Data Book*, 'Soft-magnetic siferrit'
9 MATTIN, R., 'The evolution of the toroidal transformer', *Electronic Engineering*, (June 1982)

16

Relays

R W Lomax
MSc, CEng, FIEE, CPhys, FInstP
Power Components Division
STC Components Ltd

Contents

16.1 Relay characteristics

Most electromechanical relays are operated electromagnetically. Some can operate thermally or electrostatically, but these are unusual. Electromagnetic relays have a low reluctance, high flux density magnetic circuit. In a gap in this circuit is an armature providing mechanical movement. By passing current through an electric coil on the magnetic circuit, magnetic flux is generated around the circuit. The armature moves to close the gap and operates electric contacts via an insulated actuator.

Consider the changeover contact spring set in *Figure 16.1*. With the relay unenergised the lever contact presses on the break contact. When the relay is energised, the armature moves against a restraining backstop force and the actuator moves through a pre-travel to touch the lever spring. With increased energisation upon overcoming the break contact force, the break contact is opened. The actuator then moves the lever spring through the contact gap. The lever contact touches the make contact and the lever spring is pushed through an overtravel by the actuator to provide a make contact force.

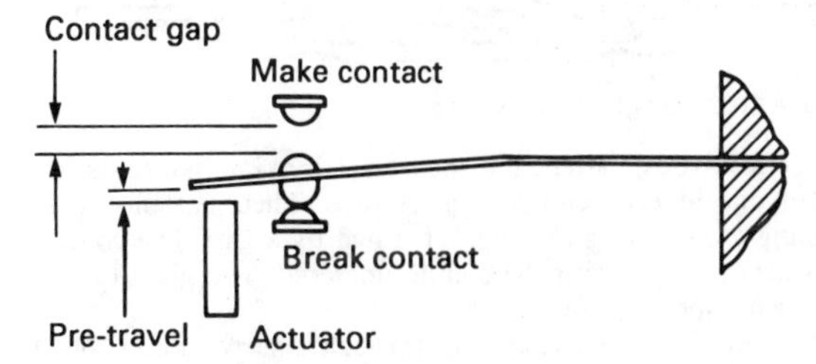

Figure 16.1 The changeover contact

Once operated, coil current change has no effect until it is made so low that it cannot hold the relay operated. Then the armature falls right back to its backstop.

As the armature has mass, external forces may move it, causing transient contact opening. The likelihood of this can be reduced by reducing armature mass, by increasing armature restraining forces and by increasing forces holding contacts together. The relay operate time is largely determined by the coil rise time. Increasing the applied coil voltage reduces the operate time, but increases the coil dissipation. The timing of a changeover contact is shown in *Figures 16.2*(*a*) and (*b*). *Figure 16.3* shows that the coil voltage range, which varies with ambient temperature, is limited. However, the operate time is reduced by applying a higher coil voltage along with a series resistance.

Upon open circuiting the coil the relay release time is short, but can cause a surge voltage across the coil. Suppression elements can be added, as described in Section 16.7. With a diode across the coil, the release time is increased by perhaps five times.

Upon operation the contact may bounce. Bounce on contact opening is small, but not so on contact closing. The armature has kinetic energy. Its dissipation can cause contacts to bounce, reducing their lifetime. Contact bounce is reduced by buffer springs and air damping.

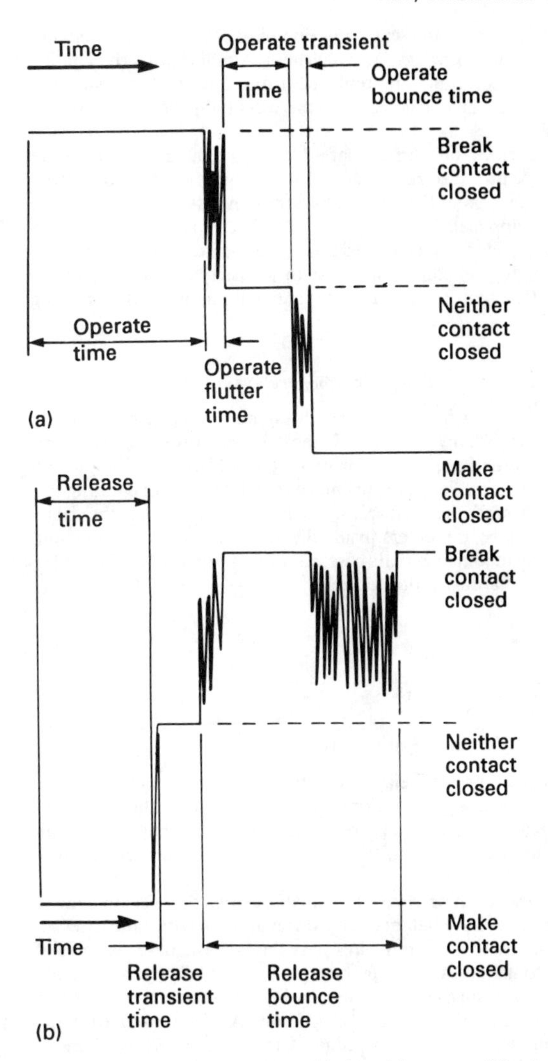

Figure 16.2 Changeover contact timing: (a) operate; (b) release

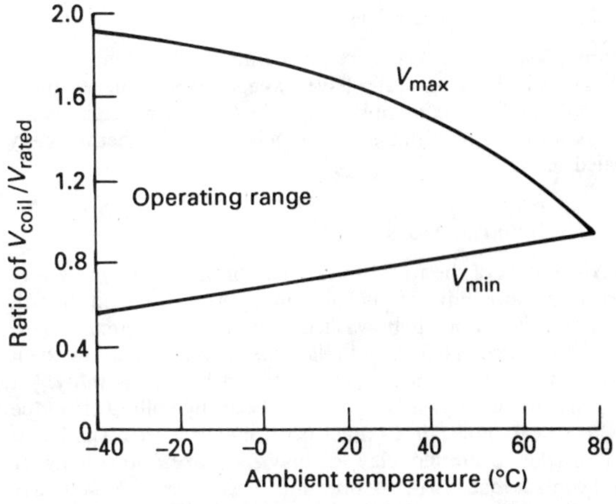

Figure 16.3 Coil voltage operating range

16.2 Relay constructions

16.2.1 Unsealed relays

16.2.1.1 Dust-protected relays

Many relays in use are dust protected. They are inside a case, which is open to the air at gaps around tags and join lines. This is the most important feature of these relays. Any fumes generated inside the case can escape and do not contaminate the contacts. Oxygen and other gases, which may combine with the contact during operation, are replaced from outside the case. The atmosphere inside the case remains essentially the same as the air outside.

Impurities on surfaces are likely to remain oxidised and produce little electric leakage. However, gases and vapours from the air may enter the case to affect the contacts. This can limit the relay's application.

Versions of dust protected relays are available for low, medium and high level contact loads, with a variety of contact actions, mixed in many combinations, and with numerous mounting methods.

16.2.1.2 Pin sealed or 'flux-proof' relays

Many relays, which are for mounting on printed circuit boards, are made with the pins sealed in the base with an epoxy resin. This prevents the flux, used to promote the soldering of component leads to the tracks on the printed circuit board, from entering the relay around the base pins. Such flux can change the relay performance, if it enters inside the relay case. Air can still enter the relay at join lines in its case, so contact performance is similar to that of a dust-protected relay.

16.2.2 Sealed relays

16.2.2.1 Hermetically sealed relays

Hermetically sealed relays are primarily for military use. They are usually for low and medium level contact loads. They have low contact resistance and are quite sensitive. They have high isolation between circuits and operate over a wide temperature range.

Typical of these relays are crystal can sized electromagnetic relays. They have balanced armatures and can withstand external forces. The termination wires pass through the base, as cores of glass to metal seals. Inside the relay mounted on the wires, close to the base, are the contact assemblies. The magnetic circuit and coil are held above in a strong frame. A metal cover over the assembly is sealed to the edge of the base. There is often an exhaust hole in the base. The relay is vacuum baked and back filled with dry nitrogen after which the exhaust hole is sealed. The relay has a low leak rate, which is what is usually meant by the term hermetically sealed.

16.2.2.2 Plastic sealed relays

Many plastic-cased relays are made and are fully sealed by an epoxy resin. They naturally have leakage rates about ten times greater than the maximum rates specified for hermetically sealed relays. Contact performance approaches that of a hermetically sealed relay.

16.2.2.3 Washtight relays

In some parts of the world the climate promotes the growth of mould on any flux left behind on a printed circuit board. Therefore the flux has to be washed from the printed circuit board assembly. Some plastic-cased relays are so sealed to allow them to be washed while on the printed circuit board assembly, yet they may be subsequently opened to the air by pulling off a tape seal or breaking a pip. Contact performance is then similar to that of a dust-protected relay and may allow operation on higher level contact loads than possible in the sealed relay. Such relays are called washtight relays.

16.2.3 Sealed contact unit relays

In these types of relays the actual contact is sealed, whereas the electromagnetic actuating mechanism is not. The moving parts are within the sealed contact units. The contact action within one contact unit is simple, either a single make action or a single changeover action. More complex action relays are made with several contact units operated by one electromagnetic mechanism.

Three types of relays are described in this class, reed relays, mercury-wetted relays and diaphragm relays.

In the reed relay the sealed contact unit is the reed contact. A single make action version is shown in *Figure 16.4*. Two long thin

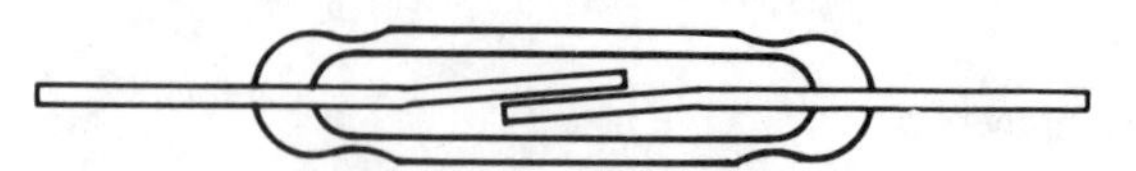

Figure 16.4 Single make reed contact unit

nickel–iron rods, flattened at their ends, looking like reeds, are sealed one at each end of a glass tube. Their flattened ends forming the contacts overlap, separated by a gap. The contact areas are covered with the contact material. The glass tube is filled with special gases.

By passing an electric current through a surrounding coil, a magnetic flux is generated through the reed contact. The flux passes through the contact gap, so attracting one reed blade onto the other. The deflection of the blades produces the required release force.

The reed contact has a relatively high magnetic reluctance. Its blades are long and thin since short and thick blades would require large release forces. In practice, for a reasonable coil power, this limits the contact force. This is compensated by optimising the contact material and the filling gases.

The mercury-wetted contact unit is long lived, of stable low resistance and exhibits no electrical bounce. Its contacts are wetted by a film of mercury which is at the upper end of a blade, which moves under the action of magnetic flux excited through the contact unit. The contacts are sealed in a glass tube filled with high-pressure hydrogen. A single changeover mercury-wetted contact unit is shown in *Figure 16.5*. A relay is completed by a surrounding electric coil and an outer magnetic shield.

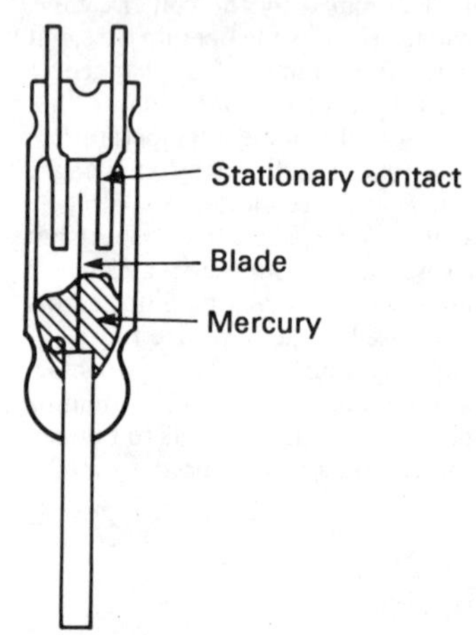

Figure 16.5 Single changeover mercury-wetted contact unit

A changeover contact unit may be magnetically biased to remain on one contact, the break contact, when the relay is unenergised. Upon opening a contact the mercury film in between is stretched until it breaks rapidly. A small globule of mercury falls away during this opening, encouraging fresh mercury to rise by capillary action up the blade from a pool at the lower end of the tube. The contact unit is position-sensitive; it cannot be up-ended. The rapid breaking of the connection may produce large surge voltages. Suitable suppression circuit elements are necessary across the contacts, as described in Section 16.7.

In diaphragm relays the sealed contact unit is the diaphragm switch. A single make diaphragm switch is shown in *Figure 16.6*. It is designed around a glass to metal seal, of which the core forms the stationary contact. In front of this is suspended a thin nickel–iron diaphragm, which is the moving contact. When magnetic flux is excited through the switch from the core, the diaphragm is attracted onto it, closing the switch. The switch is of low magnetic reluctance giving a large magnetic flux. This produces a high contact force for relatively low coil power.

The coil is assembled on the switch core, around which the magnetic circuit is completed by a cover, end cap and tube. There are two types of switch, the single make and the single change-over. Multiples of these switches are built into relays. They are small and are used for printed circuit board mounting.

The contacts are plated with hard gold giving low contact resistances over long lifetimes, on low and medium level loads. *Figure 16.7* shows a section through a single make diaphragm relay.

16.2.4 'New generation' DIL relays

Small printed circuit board mounting plastic cased relays with a dual-in-line (DIL) pin configuration are now readily available. They are primarily for telecommunication applications. They usually have only up to a two-changeover contact action. They are often fully sealed and have good low-level load contact performance. Many have the restraining backstop force on the armature produced magnetically by incorporating a permanent magnet. This can reduce the size of the magnetic circuit and increase the relay sensitivity.

The contacts may have a multilayer design, which, as explained in *Section 16.5*, can allow good switching performance upon various contact loads.

16.2.5 Solid-state relays

At present the relay market is dominated by electromechanical relays (e.m.r.'s). These have some disadvantages such as contact erosion, arcing, limited operating speed and susceptibility to external forces.

Solid-state relays (s.s.r.'s) can overcome these disadvantages. They are of three types: firstly the hybrid e.m.r., which has a solid-state input and a mechanically switched output; secondly the more common hybrid s.s.r., which has an electromechanical input and a solid-state switched output; and thirdly the true s.s.r., which contains no moving parts.

The hybrid relays are usually cheaper than true s.s.r.'s. The hybrid e.m.r. can provide cheap multipole outputs. The hybrid s.s.r. provides good input isolation with the solid-state switching removing the adverse contact effects and is usually used for a.c. loads. It can incorporate trigger circuits, which switch the output when the load voltage is near zero, so that inrush currents and radio frequency interference are reduced.

S.s.r.'s are reliable with a long lifetime. They give arc-free switching, are less susceptible to external forces, and they give noiseless operation. However, they have certain disadvantages. Except for single pole relays of output greater than 10 A, they may be more expensive than e.m.r.'s, multipole outputs being proportionally more expensive. They have no physical break in the output circuit, and a relay with full four terminal isolation is complex. They also run hot, requiring a heatsink for outputs greater than about 5 A, and the heatsink may be large.

16.3 Comparison of relay types

Table 16.1 shows a comparison of different relay types with regard to parameters of interest. The larger the number of asterisks the better is the relay for that particular parameter.

The reliability of a relay could be said to be the probability of making a connection through a contact. A dust-protected relay is probably the poorest in this respect, as it can be affected by the surrounding atmosphere, although within its contact load range it is good. A reed relay has better reliability as it has a sealed contact. A hermetically sealed relay is even better, as it generally has a greater contact force than a reed relay. However, a mercury wetted relay is best because of its wetted contact. A diaphragm relay is as good, because it has a very high contact force.

The resistance of a relay to external forces, which cause it to be accelerated, can be considered. A mercury wetted relay is probably the poorest, as it is somewhat fragile. A dust-protected relay is generally not good, as it often does not have a balanced armature. A reed relay is a little better. A hermetically sealed relay

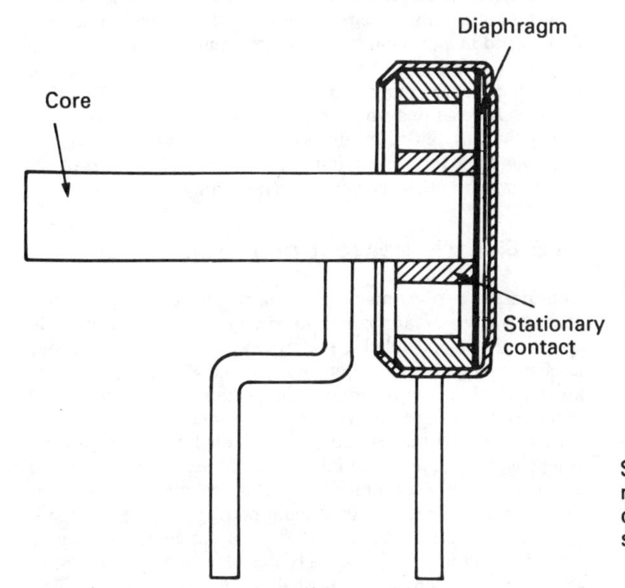

Figure 16.6 Single make diaphragm switch

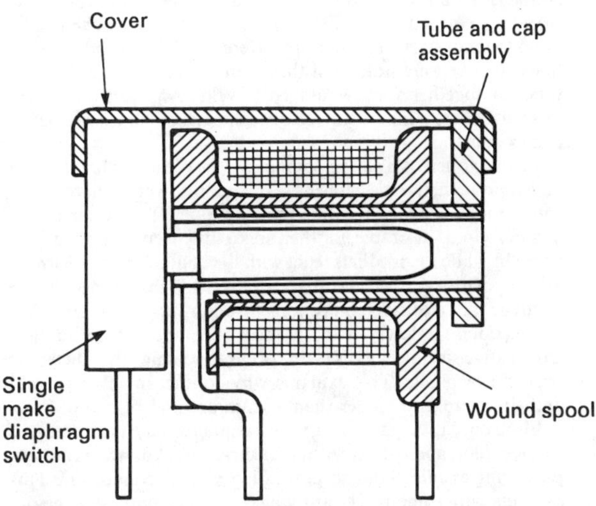

Figure 16.7 Single make diaphragm relay

Table 16.1 Comparison of relay types

	Unsealed relay	*Hermetically sealed relay*	*Plastic sealed relay*	*Reed relay*	*Mercury wetted relay*	*Diaphragm relay*
Reliability	*	***	***	**	****	****
Life	**	*	**	**	****	***
Resistance to external forces	**	***	**	**	*	****
Maximum contact load	****	***	***	***	*	**
Sealing	unsealed	****	**	***	***	**
Position sensitive	no	no	no	no	yes	no
Price	****	**	****	****	*	**
Multiple contact actions	***	**	**	*	*	*

is good, as it usually has a balanced armature. The diaphragm relay is the best from this point of view, as it has a very high contact force coupled with an armature of very low mass.

A dust-protected relay can probably deal more easily with the large contact loads, because it is open to the atmosphere, and this can be used to advantage to reduce the probability of contact welding. Hermetically sealed relays and reed relays are good in this respect. A diaphragm relay may be a little more likely to weld on large contact loads because of its high contact force in a protected atmosphere.

Considering relay sealing, a hermetically sealed relay offers an assured level of sealing with a leak rate below a given limit. Often a reed relay or a mercury-wetted relay is practically as well sealed.

The mercury-wetted relay suffers because it is position sensitive. To a first approximation other relays are not position sensitive.

Dust-protected relays are usually cheap, when the number of contact actions available is considered. Reed relays are also cheap. The complexity of a mercury-wetted relay is usually reflected in its price.

Dust-protected relays usually offer the largest range of contact actions on a relay. Reed relays, mercury-wetted relays and diaphragm relays only offer multiple contact actions by incorporating a multiple of contact units within a relay.

16.4 Electric contact phenomena

All relays except solid-state relays rely on the switching of electric contacts. Contact phenomena are complex. All surfaces in the world are covered with a film of deposits, which may chemically bond to them. Metal surfaces are covered with films about 10 nm thick. If they were not, i.e. if they were atomically clean, when touched together they would cold weld. An electric contact formed between metal surfaces needs surface films, so that it comes apart.

A voltage across a closed contact causes dielectric breakdown of the surface film. This happens at thin points, where microscopic contact metal spikes protrude into the film. This explains why contact wipe is desirable, for then spikes dig deeper into the film. Some breakdown products react with the contact metal. With an electropositive contact metal, its atoms dope the surface film to produce an electric connection through it: the more such connections the lower is the contact resistance. With an electronegative metal anodisation occurs, sealing the dielectric breakdown points. This explains why contact metals like gold and silver are used rather than aluminium and titanium.

Electronegative gases in the atmosphere may permeate the surface film and react with dielectric breakdown products, producing anodisation and preventing electric connection. This explains why some relays are sealed and are outgassed before sealing.

The greater the force between two contact surfaces, the larger is the number of metal spikes protruding into the surface film and the more are the electric connections through it, producing a low contact resistance.

The doping of the surface films enhances their adhesion to the positively charged metal contact. On opening the films may be torn from the negatively charged surface, causing erosion. So the making of an electric contact cannot be divorced from its erosion.

The conditions to make an electric connection are the same as to maintain an electric current after establishing a connection. This explains why a 'wetting' current is often desired. Each parallel connection through the surface film appears to conduct well or not at all, and abruptly changes between these states. Therefore conduction is noisy, with a large high-frequency component. The smaller the total contact current the less reliable is the connection and the larger is its percentage noise. The wider the frequency response of any measuring device, the greater appears the peak contact resistance. Hence the measured contact resistance depends entirely on the method and equipment used.

The making of an electric connection of two metal surfaces and their electric resistance spot welding are just two degrees of the same process. Good electrical contact operation lies between the extremes of too high a contact resistance developing and the contacts welding up.

When the electric contact is open, but closing, so the surfaces are near, with a high open circuit voltage between them, a large electric field exists. This field exists in the surface films, because they are not metals. Where the film on the negative contact has been doped in a particular way electron emission can be excited into the contact gap. This can initiate an arc. An arc can be similarly produced during opening of a contact, particularly if an inductively generated surge voltage appears across it. To reduce arcing, the initial electron emission needs reducing. Some contact materials dope the surface films in a way less likely to produce arcing. Impurities in the contact can enhance arcing.

16.5 Electric contact materials

Consideration of contact phenomena suggests that certain contact materials are better for particular applications. No contact gives good reliable performance for all loads. On low-level contact loads pure electropositive metal contacts like gold are required, perhaps working in protected atmospheres, as obtained in sealed relays and with high contact forces. However, with high contact loads there is a high probability of welding. To compensate for this, material may be added in the contact to enhance any surface film. This could be nickel in gold or cadmium oxide in silver. An amount of oxygen in the surrounding atmosphere may be beneficial, so unsealed relays may be better for high contact loads. A silver or silver alloy contact with a gold flash has often been suggested as a universal contact. This has given problems and was forbidden by the German Post Office in 1971. It appears that, if silver sulphide grows on silver, it is heavily doped with silver and so conducts. With an additional gold flash, silver sulphide grows through pinholes in the gold to cover the contact surface. The gold then acts as a bottleneck between the underlying silver contact and the silver sulphide. The excess silver in the silver sulphide is reduced, which reduces the electric conduction through the surface.

Table 16.2 lists different types of contact materials, the properties each possess, their resistance to atmospheric corrosion and their application. The gold alloys contain several per cent of another transition metal, such as nickel or cobalt. The silver alloys have about 10% or more of another metal or a metal oxide such as nickel or copper or cadmium oxide. The platinum alloys have about 10% of another metal such as ruthenium.

Palladium or platinum in an electrical contact can give rise to the phenomenon of 'brown powder' occurring on the contact. This is because palladium or platinum naturally polymerises gases or vapours in the atmosphere, which are adhering to the contact surface. However, the growth of the resulting polymer film on the contact is self-limiting. The polymer film is usually so thin that it is unseen, until any contact wipe breaks it up and shows it as a 'brown powder'. A further polymer film develops, where the original film was wiped off, and in this way 'brown powder' continues to be produced.

Recently it has been recognised that, once a relay has been put into service on a certain contact load, it is unlikely to be subsequently used on a different contact load. Many contacts are now made with a thin top layer suitable for use on a low-level contact load. Below this top layer is a contact material suitable for a high-level contact load. On a low-level contact load there is little wear and switching occurs on the thin top layer. On a high-level contact load the thin top layer is quickly worn away and switching occurs on the lower contact material.

16.6 Relay reliability

From the results of endurance tests on relays, it is found that coil failures due to wire breakages and poor solder joints are far fewer than contact failures. The failure rate of an electric contact determines the failure rate of the relay.

From the contact phenomena given in *Section 16.4* it can be said that the probability of making an electric connection at a contact, through a film on it, depends upon:

(1) The open circuit voltage.
(2) The inverse of the film thickness of the contact.
(3) The electropositivity of the contact metal.
(4) The inverse of the partial pressure of electronegative gas in the atmosphere.
(5) The power dissipated at the contact, i.e. a function of the current.
(6) The inverse of the heat capacity of the contact.
(7) The thermal resistance of the contact to its surroundings.
(8) The ambient temperature of its surroundings.

A large number of such parallel connections are made, so the contact resistance is inversely proportional to the connection probability. However, the probability of welding a contact is directly proportional to this connection probability. Additionally, contact erosion depends on this probability. The connection probability for a given design in a constant environment at a low-level load is small; contact resistance may be high and the consequent failure rate, say λ_r, may be high. This is region 1 shown in the diagram of contact failure rate against contact load (*Figure 16.8*).

If just the contact load is increased to a large value, the probability of contact welding and erosion may be large. The consequent failure rate, say λ_w, may be high. This is region 2 of *Figure 16.8*. At a contact load between the low level and large value the sum of the two above failure rates, say λ_b, is least and then contact life is greatest. This is region 3, sometimes called the self-cleaning region.

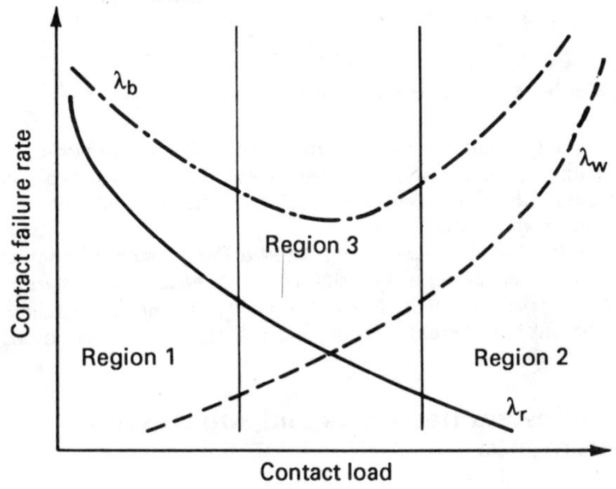

Figure 16.8 Variation of contact failure rate with load

Table 16.2 Types of contact material

Material type	*Properties*	*Atmospheric corrosion*	*Load level*
Fine gold	tends to cold weld	excellent corrosion resistance	low
Gold alloys	greater hardness; greater wear resistance than fine gold	good corrision resistance; may oxidise	low to medium
Silver	most popular contact material; tends to weld; relatively high erosion	tends to form sulphide in sulphurous atmosphere	medium to high
Silver alloys	higher hardness; resistant to arcing; reduced tendency to weld	tends to form sulphide in sulphurous atmosphere; may oxidise	high
Palladium	resistant to arcing	corrosion resistance; tends to form 'brown powder'	high
Palladium–silver alloys	hard; resistant to arcing; popular in telecommunications	essentially corrosion resistant	medium
Platinum alloys	very low erosion; extreme demands in high reliability applications	good corrosion resistance; tends to form 'brown powder'	high

The contact design in terms of shape, material and its operating environment determines the value of the minimum failure rate and at what load it occurs. Consequently it is difficult to generalise on relay reliability due to the many different relay designs available. However, *Figure 16.9* and *Table 16.3* between them give in orders of magntitude the areas of operation and the corresponding failure rates of contacts of various types of materials, in different general-purpose relays. The values given can only be taken as rule of thumb figures. One of the reasons for giving this information is to emphasise that different types of contact have different levels of performance. There is no 'universal' contact.

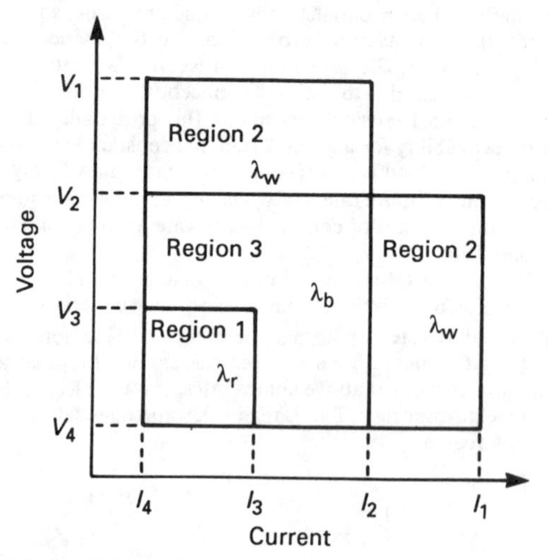

Figure 16.9 Approximate areas of contact operation

For all contacts there are suggested lower limits for contact voltages and currents. The values given in the table are generally for d.c. resistive loads. The a.c. capability of contacts is greater. The effect of different types of loads on the contact is considered in the next section.

16.7 Switched loads and suppression elements

Electrical loads switched by the relay contacts can be placed in the following groups:

(1) *Resistive load* This is the more easily understood. The load current and voltage should be within that recommended for the contact. This may be approximately as given in *Table 16.3*.

(2) *Capacitive d.c. load* On closing contacts to a capacitive load there is a large initial current. This may cause contact welding.

(3) *Tungsten lamp load* The cold resistance of a tungsten filament lamp is about one tenth of its hot resistance. This produces a large initial current.

(4) *Inductive d.c. load* This is a difficult load to switch. A large transient voltage is generated on breaking the load current. This voltage is of opposite polarity to the supply voltage. It may damage some devices, which are sensitive to the polarity of the voltage applied. It may break down insulation, and it may initiate an arc between the opening contacts which can radiate radio frequency interference (r.f.i.) and increase contact erosion.

A number of circuit elements may be incorporated to suppress these effects. Some such circuit elements are given in *Table 16.4*.

(5) *Inductive a.c. load* This may be a motor or transformer or solenoid. On closing this can cause an initial current of about five times the steady current value. A large transient voltage may also be generated on breaking the load current. This produces adverse effects similar to those described in (4). Two suppression circuits are given in *Table 16.5*.

16.8 Glossary of useful relay terms

All-or-nothing relay A relay intended to be energised either higher than that at which it operates or lower than that at which it releases.

Bistable relay A relay able to stay in either its operated or unoperated condition after the coil energisation has been removed.

Bounce time Time between when a contact first closes and finally closes during an operation (see *Figure 16.2*).

Break-before-make contact A changeover contact which on relay operation opens a common connection from a break contact before closing to make contact. This is the usual form of changeover contact.

Break contact A contact which is open when the relay is operated and closed when the relay is released.

Bridging time The time in which both contact circuits of a make-before-break changeover contact are closed.

Changeover contact Contact which on relay operation opens a common connection from a break contact and closes it to

Table 16.3 Approximate current and voltage limits and failure rates for various contacts in general-purpose relays

Relay type	*Contact material*	*Approximate voltages* (V)				*Approximate currents* (A)				*Contact failure rates* ($\times 10^{-9}$)		
		V_1	V_2	V_3	V_4	I_1	I_2	I_3	I_4	λ_w	λ_b	λ_r
Unsealed	gold alloy	100	10	1	0.1	0.5	0.1	0.01	0.001	30	8	10
Unsealed	silver	250	50	10	3	5	1	0.2	0.05	250	10	10
Unsealed	palladium–silver	100	50	15	5	2	1	0.1	0.01	100	30	10
Unsealed	silver alloy	250	100	30	10	10	1	0.5	0.1	1000	200	300
Sealed	gold alloy	100	10	1	0.01	0.5	0.1	0.01	0.001	100	2	5
Sealed	silver	100	30	3	1	3	1	0.1	0.005	200	5	5
Sealed	silver alloy	100	50	10	3	5	1	0.3	0.05	1000	200	200
Sealed	mercury	200	20	1	0.01	0.5	0.3	0.1	0.001	300	0.5	1

Table 16.4 Suppression circuit elements for use in switching inductive d.c. loads

Suppression elements	*Advantages*	*Disadvantages*	*Comments*
Load; V; R_s; L; R	suppresses surge voltage; low cost reduces contact erosion	extra current drain	if load is a relay coil it lengthens release time
Load; V; D_s; L; R	limits surge voltage to 1 V suppresses contact erosion and r.f.i.	greatly lengthens relay release time if load is relay coil	
Load; V; Z_s; D_s; L; R	reduces surge voltage suppresses contact erosion	moderately lengthens relay release time if load is relay coil	costly
Load; V; R_s; C_s; L; R	good contact protection	large surge voltage makes this poor for use with solid-state devices	C_s (in μF)$=I_L$ (in A) R_s must prevent capacitance current welding contacts
Load; V; R_s; C_s; D_s; I_L; L; R	good comprehensive suppression	costly	peak forward current of $D_s=I_L$ $R_s \geqslant R$

a make contact. This can be either a break-before-make contact or a make-before-break contact.

Contact circuit resistance The resistance of a closed contact, when measured by a four pole method at the appropriate relay tags with a given applied voltage and current.

Contact force The force between two contact points when closed.

Contact gap The gap between two contact points when open.

Dielectric test *See* Voltage proof test.

Table 16.5 Suppression circuit elements for use in switching inductive a.c. loads

Suppression elements	*Advantages*	*Disadvantages*	*Comments*
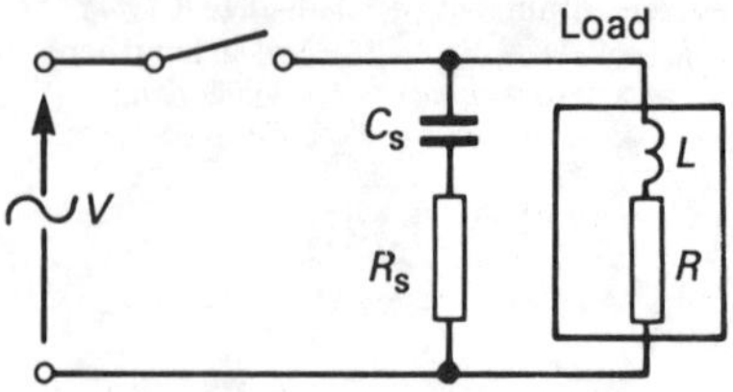	suppresses surge voltage reduces contact erosion	extra current drain	$R_s \simeq R$ $C_s \simeq L/R^2$
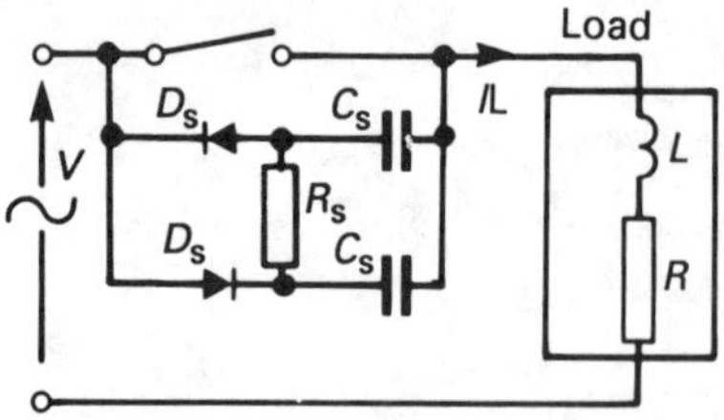	good comprehensive suppression	costly	$R_s \simeq 100\ k\Omega$ peak forward current of $D_s = I_L$ $C_s \simeq L/R^2$

Drop-out *See* Release.

Drop-out energisation *See* Release energisation.

Drop-out time *See* Release time.

Duty cycle The ratio of the coil energisation time to the total period of one cycle of operation, often given as a percentage. This is of interest if a number of similar cycles of operation are performed.

Electromechanical relay A relay in which the intended operation arises through movement produced by energisation.

Energisation Current or voltage or power applied to the relay input, usually its coil.

Hold energisation The energisation at a given temperature, which will hold the relay operated, when energisation is reduced from a larger value.

Hybrid electromechanical relay A relay in which electro-mechanical and solid-state devices combine to give a mechanically switched output.

Hybrid solid-state relay A relay in which electromechanical and solid-state devices combine to give a solid-state switched output.

Make-before-break contact A changeover contact which on relay operation closes a common connection to a make contact before opening from a break contact.

Make contact A contact which is closed when the relay is operated and open when the relay is released.

Monostable relay A relay which operates upon energisation and releases after energisation is removed.

Non-operate energisation The energisation at a given temperature which will not operate the relay.

Operate energisation The energisation at a given temperature which will operate the relay.

Operate time Time between the application of a given coil energisation and the first closing or opening of a given contact. The operate time is greatly dependent on the value of coil energisation (see *Figure 16.2*).

Pick-up *See* Operate.

Pick-up energisation *See* Operate energisation.

Pick-up time *See* Operate time.

Polarised relay A relay in which the change of condition depends on the sign of the applied energisation.

Pull-in *See* Operate.

Pull-in energisation *See* Operate energisation.

Pull-in time *See* Operate time.

Release energisation The energisation at a given temperature at which the relay will release, when energisation is reduced from a larger value.

Release time The time between the removal of a given coil energisation and the first opening or closing of a given contact. As usually defined this does not include bounce time. The release time is dependent on the value of coil energisation (see *Figure 16.2*).

Solid-state relay A relay whose output switching is achieved using solid-state devices and without the use of any moving parts.

Thermoelectric electromotive force The electromotive force (e.m.f.) generated by a closed contact of an energised relay at an elevated temperature.

Transfer time *See* Transit time.

Transit time The time during which both contact circuits of a changeover contact are open (see *Figure 16.2*).

Further reading

ANDREIEV, N., 'Power relays: solid state vs. electromechanical', *Control Engineering*, **20** (1), 46–9 (1973)

BEDDOE, S., 'High-performance miniature relays: a design review', *Design Electronics*, **8**, December, 38 (1970)

CAPP, A. O., 'Electrical contact considerations', *Electronics World*, April, 49 (1967)

DUMMER, G. W. A., 'Failure rates, long term changes and failure mechanisms of electronic components: part 1: failure rates and the influence of environments', *Electronic Components*, **5** (10), 835–84 (1964)

GAYFORD, M. L., *Modern Relay Techniques: STC Monograph No. 1*, Newnes–Butterworths, Sevenoaks (1969)

GROSSMAN, M., 'Focus on reed relays', *Electronic Design*, **20** (14), 50–9 (1972)

GROSSMAN, M., 'Focus on relays', *Electronic Design*, No. 26, December, 119 (1978)

HYDE, N., 'Electromechanical relays: part 3: the contact problem', *Electronic Components*, **5** (8), 665–9 (1964)

HYDE, N., 'Electromechanical relays: part 4: reliability testing', *Electronic Components*, **5** (10), 865–72 (1964)

JORDAN, J. S., 'Supressing relay coil transients', *Electronic Industries*, **24** (4), 73–4 (1965)

KODA, A. J., 'Mercury-wetted relays', *Electronics World*, April, 56 (1967)

KOSCO, J. C., 'Fundamentals of electrical contact materials and selection factors', *Insulation/Circuits*, **17** (8), 27–34 (1971)

LOMAX, R. W., 'The design and development of sealed contact change over diaphragm relay', *Proc. 20th NARM Conf., Stillwater, Oklahoma, USA, 18–19 April* (1972)

LYONS, R. E. 'Relays: electromechanical vs. solid state', *Mechanical Engineering*, **94** (10), 43–6 (1972)

NAPLES, J. G., 'Relay reliability in surface mounting', *New Electronics*, February (1988)

ROSENBERG, R. L., 'Reed relays', *Electronics World*, April, 41 (1967)

ROSINE, L. L., 'Miniature relays', *Electrotechnology*, **84** (1), 47–55 (1969)

ROVNYAK, R. M., 'Arc, surge and noise suppression', *Electronics World*, May, 46 (1967)

SAUER, H., 'Modern relay technology', 2nd edn, Huethig, Heidelberg (1986)

WOODHEAD, H. S., 'The diaphrgm relay', *Electronic Components*, **9** (3), 290–4 (1968)

17 Piezoelectric Materials and Components

M J B Scanlan BSc, ARCS
Marconi Research Laboratories

Contents

17.1 Introduction

Piezoelectricity, discovered by the brothers Curie in 1880, takes its name from the Greek piezein, to press, but piezoelectric materials will in general react to any mechanical stress by producing electric charge: in the converse effect, an electric field results in mechanical strain. As a simple introduction, compare small cubes of crystal quartz (a piezoelectric material) and fused quartz (non-piezoelectric). If the former is mechanically stressed, in an appropriate way, electric charge will appear on its faces: there will be some elastic change in dimensions, but the amount of this change will depend on the electric boundary conditions, i.e. whether electrodes on the faces of the cube are open—or short-circuited. Conversely, if a voltage be applied to these electrodes of the unstressed cube, the cube will not only be charged electrically but strained mechanically: as in the first case, the amount of charge stored electrically will depend on whether the cube is clamped or free mechanically. In contrast, the cube of fused quartz behaves purely elastically under mechanical stress, and as a simple dielectric in an electric field: there is no coupling between the two effects.

As in this example, piezoelectric materials are crystalline in structure and anisotropic in many properties (e.g. thermal conductivity, thermal expansion, dielectric constant) while the non-piezoelectrics are often amorphous and isotropic. It is also worth noting that the same crystalline structure which results in piezoelectricity results also in optical activity[1] (the linear electro-optic, or Pockels effect is possible with all piezoelectric materials, and only with them) and in pyroelectricity, the generation of electricity by heat (every pyroelectric crystal is piezoelectric, and every ferroelectric crystal, but not every piezoelectric crystal, is pyroelectric). Ferroelectricity, an important sub-class of piezoelectricity, will be further discussed below.

Piezoelectricity is a much more common phenomenon than is generally realised. Cady[2] reports that over 1000 piezoelectric crystals have been identified; this is perhaps not so surprising when it is remembered that only 11 out of the 32 crystal classes cannot, because of their symmetrical crystal structure, have any piezoelectric members.

A caution should be given here about notation, both as regards piezoelectricity itself and on crystal structure. Most modern texts follow the IRE *Standards on Piezoelectric Crystals*, dated 1949 and later,[3] but the first edition of Cady[4] (1946) does not, of course, and the second edition[2] (1964) only outlines the standards rather briefly in an appendix. Hueter and Bolt's[5] *Sonics* use Voight's notation, dating from 1890, and explains that conversion into other notations can be made as explained in the standards.

On crystal structure, Cady[2] gives a cross-reference between the various crystal notations, which will be useful to those not familiar with the subject. A recent paper[6] gives a good short review of the whole topic of piezoelectricity including theory and materials as well as devices, and has a good bibliography.

17.2 The crystalline basis of piezoelectricity

Piezoelectricity is essentially a phenomenon of the crystalline state and the two subjects are dealt with together in Kittel,[7] Cady,[2] Berlincourt *et al.*,[8] Mason[9] and Nye.[10]

Crystals are classified into seven systems and 32 classes. Of these, eleven have a centre of symmetry and cannot be piezoelectric: the 21 classes lacking a centre of symmetry all have piezoelectric members. It can therefore be stated as a necessary condition for piezoelectricity that the crystal lacks a centre of symmetry. Only if this is the case will there be a shift of the electrical centre of gravity when the crystal is stressed, and a resultant generation of charge on the crystal faces.

An important sub-class of piezoelectrics is that of the ferroelectrics, which have two or more stable asymmetric states, between which they can be switched by a sufficiently strong electric field. Ferroelectrics have a domain structure and a Curie temperature, and show hysteresis; hence their name, by analogy with the ferromagnetics. Cady[2] deals at some length with Rochelle salt, one of the earliest ferroelectrics, and Kittel[7] has a chapter on barium titanate, which has the perovskite structure typical of one family of ferroelectrics.

Another important family (lithium niobate, lithium tantalate, etc.) is of considerable importance for electro-optic purposes as well as (especially lithium niobate) piezoelectricity.

Ferroelectricity can be exploited not only in single crystal material, but also in polycrystalline form, and this gives the commercially important class of piezoelectric ceramics. If the polycrystalline material is *poled*, i.e. subjected to a strong electric field as it is cooled through its Curie point, the domains in the crystals are aligned in the direction of the field. One can in this way have a piezoelectric material whose size, shape and piezoelectric qualities can, within limits, be made to order. These commercially available materials are often solid solutions of lead zirconate and lead titanate with various additives to modify the behaviour. Such a range is marketed under the general name of PZT, with distinguishing numbers, by the Clevite Corporation USA, and under the general name PXE, again with a number, by Mullard in the UK. A good account of these materials (and of piezoelectric materials in general) is given in Berlincourt *et al.*[8] The Mullard materials are listed in a booklet,[11] together with some application notes.

Finally, it is worth noting that some piezoelectric materials, notably cadmium sulphide, CdS, and zinc oxide, ZnO, can be evaporated or sputtered on to a suitable substrate, on which the deposit is polycrystalline, but oriented.

These materials, 'grown' to a thickness of a half wavelength, can be used as electroacoustic transducers to launch elastic waves into the substrate at much higher frequencies than can be achieved by the alternative (low frequency) technique of bonding a halfwave thickness of a crystal transducer (quartz, lithium niobate, ceramic) to the substrate.

17.3 Piezoelectric constants

In a dielectric medium which is elastic but not piezoelectric (e.g. fused quartz, glass), independent relations exist giving its behaviour in an electric field E, and under a mechanical stress T. These are

$$D=\varepsilon E \tag{17.1}$$

$$S=sT \tag{17.2}$$

where D is the electric displacement, ε the dielectric constant, S the strain in the material and s the elastic compliance (the reciprocal of Young's modulus, if the stress is tension or compression).

In a piezoelectric medium, the strain or the displacement depend linearly on both the stress and the field, and the equations for D and S become

$$D=dT+\varepsilon^T E \tag{17.3}$$

$$S=s^E T+dE \tag{17.4}$$

where d is a piezoelectric constant characteristic of the material, and the superscripts T and E denote that ε and s are to be measured at constant stress and constant electric field respectively.

In words, d may be defined as the displacement (charge per unit area) per unit applied stress (the electric field being constant) or as the strain per unit applied field (the stress being constant). From the first definition, d is the piezoelectric charge constant.

The piezoelectric voltage constant g can be introduced by an equivalent pair of equations

$$E = -g^T + D/\varepsilon^T \tag{17.5}$$

$$S = s^D T + gD \tag{17.6}$$

By substituting for D from Equation (17.3) into Equation (17.5),

$$d = g\varepsilon^T \tag{17.7}$$

Another very important piezoelectric quantity is the electromechanical coupling coefficient, k, the significance of which is seen as follows. The mechanical energy stored in the piezoelectric medium is, from Equation (17.4),

$$U_M = \tfrac{1}{2}ST = \tfrac{1}{2}s^E T^2 + \tfrac{1}{2}dET \tag{17.8}$$

and the electrical energy stored is, from Equation (17.3),

$$U_E = \tfrac{1}{2}DE = \tfrac{1}{2}dTE + \tfrac{1}{2}\varepsilon^T E^2$$

U_M contains a mechanical term ($\tfrac{1}{2}s^E T^2$) and a mixed term ($\tfrac{1}{2}dET$): similarly U_E contains an electrical term ($\tfrac{1}{2}\varepsilon^T E^2$) and the same mixed term. By analogy with a transformer, in which the coupling factor is the ratio of the mutual energy to the square root of the product of the primary and secondary energies, k is the ratio of the mixed term to the square root of the products of electrical and mechanical terms, i.e.

$$k = \frac{\tfrac{1}{2}dET}{\sqrt{\tfrac{1}{4}\varepsilon^T s^E E^2 T^2}}$$

or

$$k^2 = \frac{d^2}{\varepsilon^T s^E} \tag{17.9}$$

At low frequency, i.e. below the frequency of any mechanical resonance of the material, k^2 is a measure of how much of the energy supplied in one form (electrical or mechanical) is stored in the other form: it is not an efficiency since the energy applied is all stored in one form or the other.

17.4 Piezoelectric notations

For the sake of simplicity, the discussion so far of the piezoelectric constants has implied that there is only one piezoelectric axis, and that the stress or electric field is applied along this axis and the resulting strain or charge is also measured in this direction. This situation is appropriate, for instance, to a cylinder of a ceramic piezoelectric material, poled along its axis, and with the stress or field also along this axis. It is not appropriate to a single crystal piezoelectric, with its anisotropic properties, to which stress is applied in an arbitrary direction. To deal with this situation, a more complicated notation must be set up (IRE Standards, 1949).[3]

Figure 17.1 shows a right-handed orthogonal set of axes X, Y, Z, together with a small cube with faces *1, 2, 3*, perpendicular to X, Y and Z respectively. Any stress applied to the cube of material can be resolved into forces *1, 2, 3*, along the axes, or shears *4, 5, 6*, acting about the axes. Any electric field can be resolved into its components along the axes, and any charge resulting from the field or from piezoelectric stress will appear on the faces *1, 2, 3*, as shown. As for stress, there are six ways in which a strain can appear, i.e. tension or compression along the three axes, or shear about the axes.

Hence for the general case, there are six permitivities ε_{ij} (i and j, 1–3), 21 elastic compliances s_{ij} (i and j, 1–6) and 18 piezoelectric constants d_{ij} (i, 1–3; j, 1–6). Fortunately, most crystals are symmetrical enough that the number of independent piezoelectric coefficients is greatly reduced: quartz for instance has only two, d_{11} connecting a field along the X axis with strain in the same direction, and d_{14}, connecting a field along an axis normal to Z and a shear strain in the plane normal to the field.

Berlincourt *et al.*[8] and Nye[10] give elasto-optic matrices for all 32 of the crystal classes. These are 9×9 matrices, symmetrical about the diagonal, so giving the possible 45 elasto-optic constants listed in the previous paragraph. Such matrices are also given in the IRE Standards (1958).[3]

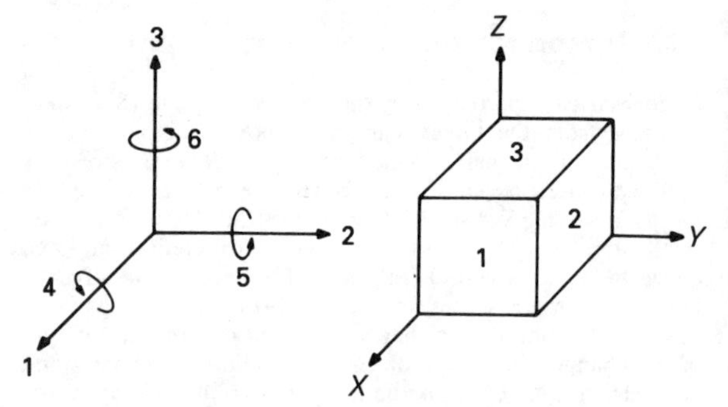

Figure 17.1 Diagram of axes (1, 2, 3), shears (4, 5, 6) and surfaces (1, 2, 3) for piezoelectric notation

17.5 Applications of piezoelectric materials

17.5.1 Frequency stabilisation

The use of quartz as a frequency standard is still the most important application of piezoelectricity technically and commercially. This use of quartz relies on its high Q as a mechanical resonator: Mason[9] shows curves of Q against frequency which imply that the internal Q (i.e. that due to internal frictional losses, and excluding mounting losses and air losses) of an *AT*-cut shear mode crystal can be over 10^7 at 1 MHz. Contributory factors are its high quality as a dielectric, its low dielectric constant, and the relative ease of cutting and polishing. The processes involved in quartz technology are described in some detail by Cady.[2] It is because of the increasing difficulty of mining sufficient high-quality natural quartz that the hydrothermal growth of quartz has increased in the last 10 or 15 years. The synthetic quartz can be made with at least as high a Q as natural quartz.

The most important cuts of the crystal for frequency standards at frequencies of 1 MHz upwards are the *AT* and *BT* cuts, shown in *Figure 17.2*. A more comprehensive diagram, showing the

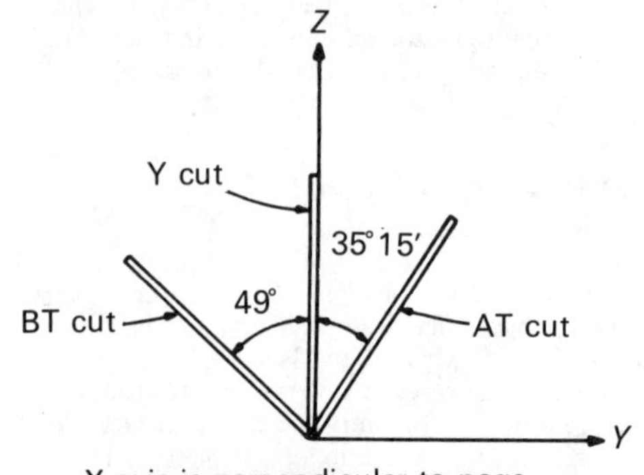

Figure 17.2 Diagram of crystal axes in quartz, illustrating some of the more important crystal cuts

position of a large number of the more important cuts, is in Mason:[9] *AT* and *BT* cuts are *rotated Y-cuts*, i.e. they are obtained by rotating the *Y* plane (perpendicular to *Y*) about the *X* axis, as shown in the diagram. These cuts, which vibrate in a shear mode, are important because they give a zero temperature coefficient of frequency, as shown in *Figure 17.3* (*BT* cut) and *17.4* (*AT* cut).

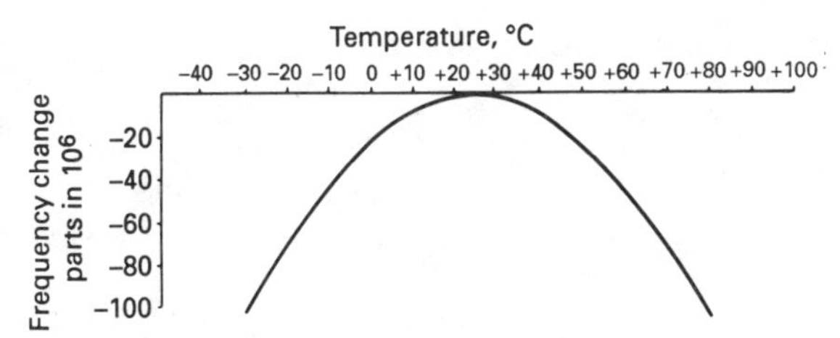

Figure 17.3 The variation of frequency with temperature for a BT cut quartz crystal

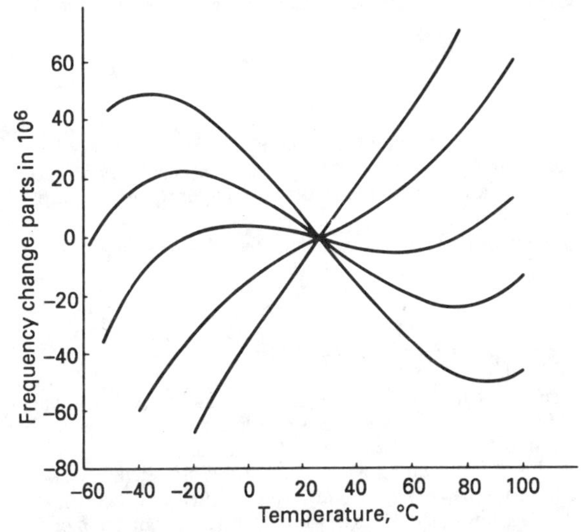

Figure 17.4 Typical family of frequency variation with temperature of AT cut quartz crystals. The curves shown cover an angular range of about 20 minutes of arc

The *BT* temperature coefficient curve is parabolic and is of zero slope at only one temperature: this temperature is a function of the exact angle, and changes by about 1.5°C per minute of arc. *BT* cuts are preferred to *AT* at frequencies above 10 MHz, since they are more stable against changes in drive level and load capacitance.

The *AT* cut, on the other hand, can be chosen to give two temperatures of zero coefficient, and a wide temperature range (the middle curve of *Figure 17.4*) over which the coefficient is very small. The five curves in *Figure 17.4* are spread over about 20 minutes of arc. For best performance as a frequency standard, contoured (i.e. convex on one or both surfaces) *AT* cut plates are used, often operating at a harmonic frequency (e.g. a 5th overtone crystal, abbreviated to 5th *OT*). The crystal, despite its low temperature coefficient, will be in a temperature-controlled oven, or, for the highest quality, a double oven. A double oven will reduce temperature fluctuations by a factor of 10^4 or more, i.e. the temperature of the crystal will not vary by more than a few millidegrees.

The frequency of a newly made quartz crystal oscillator will drift quite rapidly for a few months, a process known as ageing.[12] After a few months at constant temperature the ageing rate will be about 1 part in 10^{10} per month or less and the frequency will continue to drift at this, or a very slowly decreasing rate, indefinitely. If conditions are disturbed, e.g. the oscillator or its oven is switched off for a few hours, ageing will recommence at near the initial rate. Hence a quartz frequency standard should never be switched off, nor run at a power output above a few microwatts. The specification of a commercial quartz frequency standard,[13] using a 5th *OT AT* crystal, would include a frequency stability better than 15 parts in 10^{12} at constant voltage and temperature measured over 1 second and an ageing rate of less than 1 part in 10^9 per month after three months. These are of course maximum values, and do not contradict the lower ageing rate quoted above.

Because they are not absolute standards (i.e. they require calibration) and because of ageing, quartz crystals have been superseded as frequency/time standards by atomic clocks (rubidium, caesium, hyrogen). However, the short-term (1 second) stability of the best quartz crystals is better than that of the atomic standards (except hydrogen) which therefore consist of an atomic clock, used to control and correct the ageing of a good quartz oscillator. It is from the quartz oscillator that the output frequencies are delivered.

Quartz crystal oscillators of high stability are well reviewed in Gerber and Sykes,[12] in an issue of *Proceedings IEEE* devoted to frequency stability: the issue therefore also reviews atomic frequency standards.

Quartz crystals can be made to any frequency from 1 kHz to 750 kHz, and from 1.5 MHz to 200 MHz, the latter being high harmonic overtone crystals. The frequencies between 750 kHz and 1.5 MHz are difficult to cover as strong secondary resonances occur. There is a tendency to use a higher frequency crystal counted down rather than a very low frequency crystal, which tends to be larger and less stable than is desirable. The units may be packaged in glass envelopes, or in a solder-sealed or a cold-welded metal can: this container may also contain a heater

Figure 17.5 A micro-miniature quartz crystal oscillator mounted on a TO5 header (courtesy The Marconi Company Ltd)

and temperature sensor. *Figure 17.5* shows the interior of a microminiature crystal oscillator (Marconi Type F 3187), providing frequencies in the range 6–25 MHz over the temperature range −55°C to +90°C to an accuracy of ±50 ppm.

A new family of quartz crystal cuts, known successively as TS (thermal shock),[14] SC (stress compensated)[15] and TTC (thermal transient compensated)[16] have been introduced recently: the name SC cut seems to have prevailed. This is a doubly rotated cut (unlike the simple single rotations shown in *Figure 17.2*) and is therefore intrinsically more difficult to manufacture: moreover, since SC cut crystals are stiffer (less easily pulled) than AT cuts, they must be made to a tighter tolerance (on the positive side, they are also more tolerant of circuit changes). Commercial units at 5 and 10 HMz as frequency standards, and at 10.23 MHz for GPS (Global Positioning System) are now available, at a significantly greater cost (~ × 10) than for AT or BT cuts. However, these units offer improved stability over a wide temperature range, improving ageing (<1 in 10^7 per year), very good phase noise, very small change of frequency with thermal transients, and good stability against vibration and acceleration. Some early results are given in Burgoon and Wilson.[17]

17.5.2 Crystal filters

Quartz crystals can also be used, and for the same reasons, in very precise filters, used for defining a passband very accurately, and for rejecting adjacent channel interference. *Figure 17.6* shows a very narrow bandpass filter, with a passband of about 50 Hz at a centre frequency of 100 kHz: the 60 dB bandwidth is about 300 Hz. This filter might be used for telegraphy. *Figure 17.7*, on the other hand, shows the passband of a filter for a reasonable quality speech circuit. The passband is 5 kHz wide at 3 dB, and less than 11 kHz at 60 dB. Such filters can be very useful in conditions of severe congestion, or for s.s.b. working.

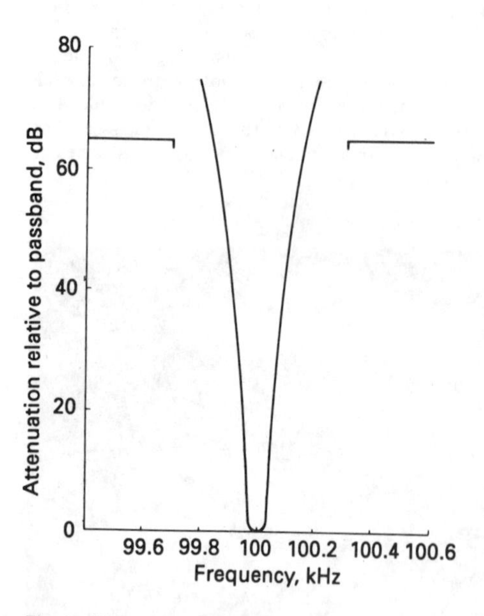

Figure 17.6 Response curve of a very narrow passband crystal filter

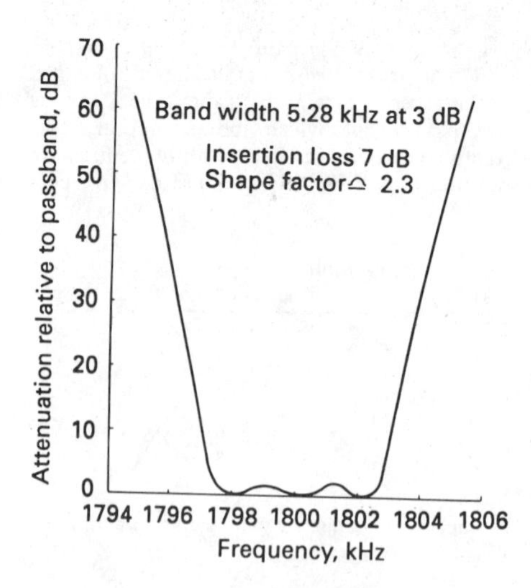

Figure 17.7 Response curve of a multi-element narrow passband crystal filter

A more recent development is the monolithic crystal filter, i.e. a single quartz plate on which a number of mechanically coupled resonators are printed, the whole assembly forming a multiple element filter.

Filters using surface wave techniques are discussed in *Section 17.5.5*.

17.5.3 Bulk delay lines[18–20]

Piezoelectric transducers (quartz, lithium niobate, ceramics) are used to launch ultrasonic waves into liquid (water, mercury) or solid (fused quartz, glass, sapphire, spinel) media, in which the velocity of propagation is about 10^5 times slower than that of radio waves in free space. Shear waves are often used, because the velocity is lower and because, with the correct polarisation of the shear axis, the wave is reflected from a boundary as a pure shear wave. The transducers are either ground to thickness (half a wavelength at the required frequency) and then bonded to the medium, or evaporated or sputtered (cadmium sulphide, zinc oxide) to the required thickness. Whereas ground transducers are limited to, say, 100 MHz by their fragility at this frequency, evaporated or sputtered transducers can be deposited to resonate at frequencies up to 10 GHz. Long delays at such frequencies are, however, precluded by high losses in the available media,[21] and by the fact that the transducers can launch only longitudinal waves. Mode conversion from longitudinal to shear mode can be used for intermediate frequencies and delay times.

Bulk delay lines are used in colour television receivers (~62 μs at the colour subcarrier frequency),[11] in MTI (moving target indication) for the removal of radar clutter from the display, in vertical aperture correction for television cameras, in APEGs (artificial permanent echo generators) for radar performance monitoring, and for a variety of other purposes. Mercury and quartz delay lines were used as data memories in early computers, but are now obsolete in this application. Most bulk wave delay lines are non-dispersive, but dispersive lines can also be made for pulse compression and other purposes (see *Figure 17.8* and *Section 17.5.5*).

Acoustic waves in bulk media can also be used for the deflection and modulation of light.[22]

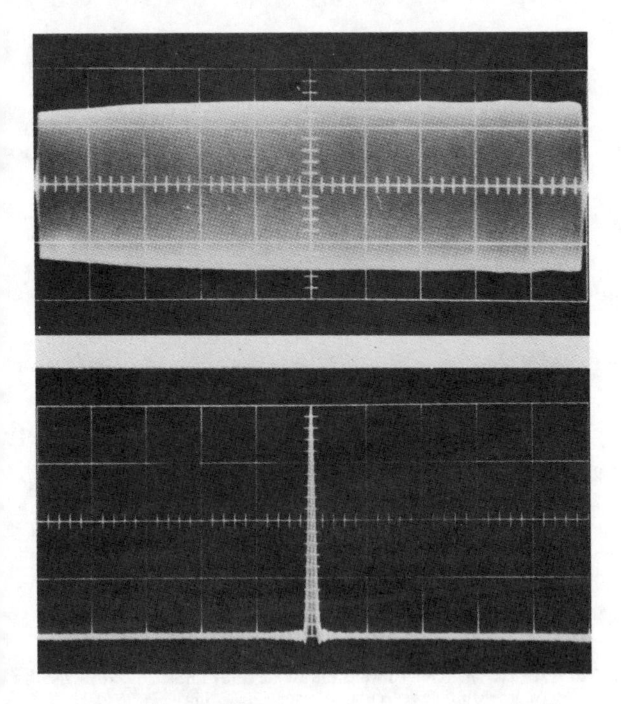

Figure 17.8 Pulse compression from a dispersive ultrasonic delay line. Input (top) is a 5 μs pulse, swept over 25 MHz: the compressed pulse (below) is 52 ns wide, i.e. the compression ratio is about 100 to 1 (courtesy WS Mortley, The Marconi Company Ltd)

17.5.4 Acoustic amplifiers

Some well known semiconductor materials (gallium arsenide, GaAs, cadmium sulphide, CdS, etc.) are also piezoelectric. It was discovered by Hutson *et al.*[23] that if electrons were made to travel by means of a d.c. electric field at the same velocity as an acoustic wave in the material, there could be an interaction between the wave and the electrons, resulting in amplification. Since the acoustic wavelength is small, the gain per unit length of material could be high. These devices do not seem to have fulfilled their early promise, doubtless because of their very low efficiency, which makes it difficult to operate them in a c.w. mode.

More recently, surface wave amplifiers have been made, using a single semiconducting piezoelectric material, or two materials, one piezoelectric (e.g. lithium niobate) the other semiconducting (e.g. silicon), in close proximity.[24] It is too early to say whether these amplifiers will be more successful than the bulk wave versions.

17.5.5 Surface acoustic waves

By means of a suitable transducer geometry, a shear or longitudinal wave in a bulk medium may be transferred to the surface of another medium: alternatively, if interdigital transducers (i.e. arrays of fingers one half wavelength apart) are laid down on a piezoelectric substrate and driven electrically, a surface wave is again excited.[25] The wave excited in these cases is the Rayleigh wave, travelling as a surface deformation which dies away very rapidly below the surface of the medium. It can be shown fairly simply that the velocity is a little lower than that of a shear wave in the medium, and that the wave is non-dispersive. Moreover, it travels on the surface and so is accessible as required. The velocity, and therefore the wavelength, is about 10^5 times lower than that of the radio wave in free space, giving the possibility, at least, of truly microelectronic circuits at radio frequencies.

This technology of surface acoustic waves (SAW) is now maturing rapidly; early work,[26–28] and a more recent review of applications[29] covers such topics as signal processors, code generators, filters, stabilised oscillators, TV filters, convolvers, correlators, etc. As examples, *Figure 17.8* shows a chirp or pulse compression in which an f.m. pulse 5 μs long, swept over 25 MHz, is compressed to a pulse 50 ns long. (This result was actually achieved with a bulk wave device, but SAW techniques are now almost universal in this application.) Another relatively simple example is that of a Bragg cell, in which an acoustic wave in SAW[30] (or in a bulk material[31]) is used as a light deflector. Since the deflection angle is proportional to the acoustic frequency, the system can be used as a spectrum analyser which will simultaneously measure the frequencies of all signals present over a bandwidth of 2 GHz or more.[32] The principles of Bragg cells in SAW and bulk wave versions are shown in *Figures 17.9* and *17.10.*

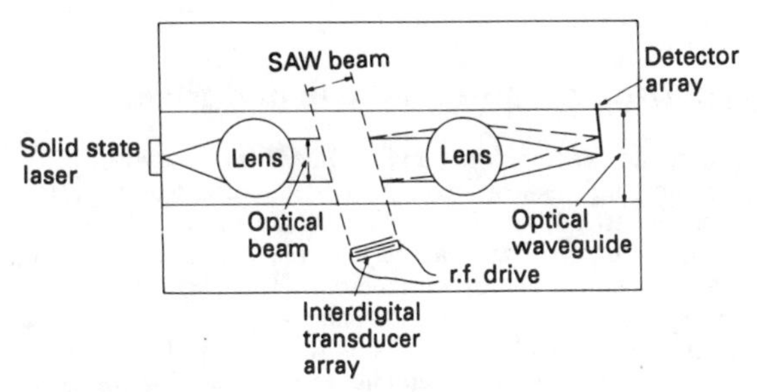

Figure 17.9 A SAW Bragg cell (courtesy of The Marconi Company Ltd)

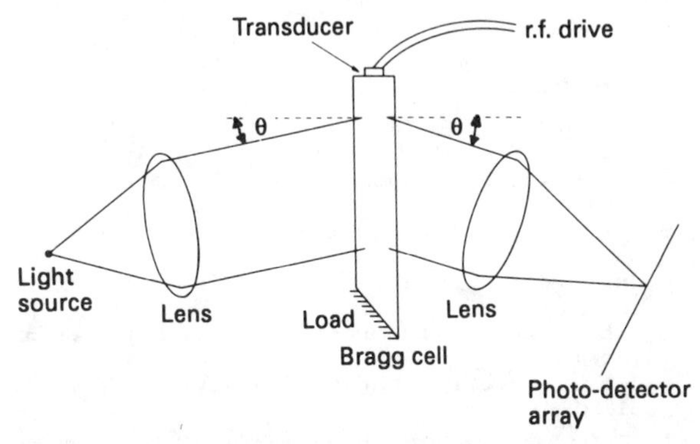

Figure 17.10 A bulk wave Bragg cell (courtesy of The Marconi Company Ltd)

Bragg cells are an example of a new class of acousto-optic devices in which a SAW (or bulk wave) is used to interact with a light beam, often from a laser. Further and more complex examples are given in the reference already quoted.[26]

17.5.6 Piezoelectric plastic films

Considerable interest has been shown in the applications of piezoelectric plastic films, especially polyvinylidene fluoride, PVF_2: films of this material, after polarisation, are strongly piezoelectric (5 to 10 times stronger than quartz) and pyroelectric.

Its structure and applications have been reviewed,[33,34] the applications include microphones, hydrophones (the material couples well to water), pressure switches and pyroelectric radiation detectors: there may also be useful medical applications.

17.5.7 Miscellaneous piezoelectric applications

Besides the major areas dealt with above, piezoelectric materials (usually ceramic materials because of their relatively low cost, ease of fabrication and high coupling coefficient) are used in a wide variety of other applications. These include piezoelectric ('ceramic') pickups for record players[11] (bi-morphs or multi-morphs), gas ignition systems, echo sounders, ultrasonic cleaners and soldering irons and ultrasonic drills. (As an alternative to piezoelectric operation, such systems, which rely on transmitting high ultrasonic powers, i.e. drills, cleaners and soldering irons, may be driven magnetostrictively.)

Useful accounts of these miscellaneous and high-power applications may be found in books by Crawford[35] and Hueter and Bolt.[5]

17.6 Data on piezoelectric materials

Because of their complexity and volume, it is impossible to give here numerical data on even the most important piezoelectric materials: instead, sources of such data are quoted.

Berlincourt *et al.*[8] give a review of the sources (p. 171) and tables of the more important materials (Tables II–VI inclusive (pp. 180–4) covering single crystal materials, and Tables VII and VIII (pp. 195 and 202) on ceramic materials). The Mullard Booklet[11] covers the PXE ceramics (pp. 14–15). *The American Institute of Physics Handbook*[37] lists the piezoelectric strain constants of some 50 or 60 materials (Section 9–97), and the temperature coefficients of a few of these. Pointon[6] gives useful data on about 20 of the more commonly used crystals.

References

1 KAMINOW, I. P. and TURNER, E. H., 'Electro-optic light modulator', *Proc. IEEE*, **54**, 1374 (1966)
2 CADY, W. G., *Piezoelectricity*, Dover edition, Dover Publications (1964)
3 IRE Standards on Piezoelectric Crystals (1949), *Proc. IRE*, **37**, 1378 (1949)
IRE Standards on Piezoelectric Crystals (1957), *Proc. IRE*, **45**, 354 (1957)
IRE Standards on Piezoelectric Crystals (1958), *Proc. IRE*, **46**, 765 (1958)
IRE Standards on Piezoelectric Crystals (1961), *Proc. IRE*, **49**, 1162 (1961)
4 CADY, W. G., *Piezoelectricity*, 1st edn, McGraw-Hill, Maidenhead (1946)
5 HUETER, T. F. and BOLT, R. H., *Sonics*, John Wiley, Chichester (1955)
6 POINTON, A. J., 'Piezoelectric devices', *Proc. IEE*, **129**, Pt. A., No. 5 (July 1982)
7 KITTEL, C., *Introduction to Solid State Physics*, Chapman and Hall, London (1953)
8 BERLINCOURT, D. A., CURRAN, D. R. and JAFFE, H., 'Piezoelectric and piezomagnetic materials', in *Physical Acoustics*, Vol. 1, Part A (ed. Mason, W. P.), Academic Press, London (1964)
9 MASON, W. P., *Piezoelectric Crystals and Their Applications to Ultrasonics*, Van Nostrand Reinhold, London (1950)
10 NYE, J. F., *The Physical Properties of Crystals*, Oxford, Oxford University Press (1957)
11 VAN RANDERAAT, J., ed., *Piezoelectric Ceramics*, Mullard Ltd. (1968)
12 GERBER, E. A. and SYKES, R. A., 'State of the art—quartz crystal units and oscillators', *Proc. IEEE*, **54**, 103 (1966)
13 MARCONI TYPE F.3160, taken from *Catalogue of Quartz Crystal Oscillators, Ovens and Filters*, Marconi Specialised Components Division
14 HOLLAND, R., 'Nonuniformly heated anisotropic plates. 1. Mechanical distortion and relaxation', *IEEE Trans.*, **SU-21**, 171–8 (1974)
15 EERNISSE, E. P., 'Quartz resonator frequency shifts arising from electrode stress', *Proc. 29th AFCS* (1975)
16 KUSTERS, J., 'Transient thermal compensation for quartz resonators', *IEEE Trans.* **SU-23**, 273–6 (1975)
17 BURGOON, R. and WILSON, R. L., 'Performance results of an oscillator using the SC-cut crystal', *Proc. 33rd Annual Symp. on Frequency Control*, p. 406 (1979)
18 BROCKELSBY, C. F., PALFREEMAN, J. S. and GIBSON, R. W., *Ultrasonic Delay Lines*, Iliffe Books, London (1963)
19 EVELETH, J. E., 'A survey of ultrasonic delay lines below 100 MHz', *Proc. IEEE*, **53**, 1406
20 MAY, J. E., 'Guided wave ultrasonic delay lines', and MASON, W. P., 'Multiple reflection ultrasonic delay lines', in *Physical Acoustics*, Vol. 1, Part A (ed. Mason, W. P.), Academic Press (1964)
21 KING, D. G., 'Ultrasonic delay lines for frequencies above 100 MHz', *Marconi Review*, **34**, 314 (1971)
22 GORDON, E. L., 'A review of acoustic optical deflection and modulation devices', *Proc. IEEE*, **54**, 1391
23 HUTSON, A. R., MCFEE, J. H. and WHITE, D. L., *Phys. Rev. Lett.*, **1**, 237 (1961)
24 LAKIN, K. M. and SHAW, H. J., 'Surface wave delay line amplifiers', *IEEE Trans.*, **MTT 17**, 912 (Nov. 1969)
25 STERN, E., 'Microsound components, circuits and applications', *IEEE Trans.*, **MTT 17**, 912, p. 835 (Nov. 1969)
26 *IEEE Trans.*, **MTT 17** (Nov. 1969)
27 *IEEE Trans.*, **MTT 21** (April 1973)
28 MAINES, J. D. and PAIGE, E. G. S., 'Surface-acoustic-wave components, devices and applications', *IEE Reviews, Proc. IEE*, **120**, No. 10R, 1078 (1973)
29 *IEEE Trans.*, **MTT-29** (5 May 1981); **SU-28** (3 May 1981)
30 ANDERSON, D. B., 'Integrated optical spectrum analyzer: an imminent chip', *IEEE Spectrum*, 22 (Dec. 1978)
31 CARTER, R. W. and WILLATS, T. F., 'The design and peformance of a bulk wave Bragg cell spectrum analyzer', *Marconi Rev.*, **44**, 57 (1981)
32 BAGSHAW, J. M. and WILLATS, T. F., 'Aspects of the performance of broadband anisotropic Bragg cells', *GEC J. Res.*, **3**, 4 (1985)
33 GALLANTRAE, H. R. and QUILLIAM, R. M., 'Polarized polyvinylidene fluoride—its application to pyroelectric and piezoelectric devices', *Marconi Rev.*, **39**, 189 (1976)
34 SUSSNER, H., 'The piezoelectric polymer PVF_2 and its applications', *Proc. IEEE Ultrasonics Symp.*, 491 (1979)
35 CRAWFORD, A. R., *Ultrasonics for Industry*, Iliffe Books, London (1969)
36 CRAWFORD, A. H., ed., *High Power Ultrasonics–International Conference Proceedings*, IPC Press (1972)
37 *American Institute of Physics Handbook*, 2nd edn, McGraw-Hill, Maidenhead (1963)

18

Connectors

Written by the Product Managers
of AMP of Great Britain Ltd,
including

C Kindell
T Kingham
J Riley

Contents

18.1 Connector housings

Connector housings are of different shapes, sizes and form, being able to satisfy requirements for a range of applications and industries—commercial, professional, domestic and military. In a commercial low-cost connector the insulator material can be nylon which can be used in a temperature range of −40°C to +105°C and is also available with a flame retardant additive.

The connector can be wire to wire using crimp snap-in contacts, wire to printed circuit board and also printed circuit board to printed circuit board. In the automotive industry it is necessary to have waterproof connectors to prevent any ingress of water thrown up by moving wheels and the velocity of the vehicle through rain. Wire seals are placed at the wire entry point in the connector. These seals, made from neoprene, grip the insulation of the wire very tightly thus preventing any water ingress. When the two halves of the connector are mated it is necessary to have facial seal thus precluding any water ingress between these two parts and preventing any capillary action of the water.

In the professional and military fields the housing material needs to be very stable and to counteract any attack by fluids. One type of material in this form is diallyl phthalate. The primary advantages of diallyl phthalate are exceptional dimensional stability, excellent resistance to heat, acids, alkalies and solvents, low water absorption and good dielectric strength. This combination of outstanding properties makes diallyl phthalate the best choice of plastics for high-quality connectors.

Most connectors which are available in diallyl phthalate are also available in phenolic. While phenolic does not have outstanding resistance to acids, alkalies and solvents it nevertheless has many characteristics which make it a good choice for connector housings. Among these characteristics are excellent dimensional stability, good dielectric strength and heat resistance. In addition there are a number of fillers which can be added to phenolic to obtain certain desired properties.

18.2 Connector contacts

The contacts that are used with the majority of connectors are made from brass or phosphor bronze, with a variety of platings from tin through to gold. The most common type of brass used in the manufacture of contacts is cartridge brass which has a composition of 70% copper and 30% zinc. This brass possesses good spring properties and strength, has excellent forming qualities and is a reasonably good conductor. Phosphor bronze alloys are deoxidised with phosphorus and contain from about 1 to 10% tin. These alloys are primarily used when a metal is needed with mechanical properties superior to those of brass and where the slightly reduced conductivity is of little consequence.

One extremely important use of phosphor bronze is in locations where the terminal may be exposed to ammonia. Ammonia environment cause stress corrosion cracking in cartridge brass terminals. On the other hand, phosphor bronze terminals are approximately 250 times more resistant to this type of failure. Associated with the materials used in the manufacture of the contacts is a variety of platings. Plating is a thin layer of metal applied to the contact by electrodeposition.

Corrosion is perhaps the most serious problem encountered in contacts and the plating used is designed to eliminate or reduce corrosion. Corrosion can spread uniformly over the surface of the contact covering it with a low-conductivity layer, with the thickness of this layer being dependent upon environmental conditions, length of exposure and the type of metal being used. Brass contacts that are unplated and have been in service for a period of time have a reddish brown appearance rather than the bright yellow colour of cartridge brass. This reddish colour is a tarnish film caused by oxidation of the metal. Although this film may not impair conductivity at higher voltages, it does at the very least destroy the appearance of the contact. To eliminate the problem of this tarnish film the contact is usually tin plated. Although oxides form on tin they are the same colour as the tin and the appearance remains the same. In addition tin is relatively soft, and if it is to be used as a contact plating, most of the oxides will be removed during mating and unmating of the contact.

Tin is the least expensive of the platings and is used primarily for corrosion protection and appearance on contacts which operate at a fairly high voltage. Another important feature of tin is that it facilitates soldering. Gold plating is always used on contacts that operate in low voltage level circuitry and corrosive environments. The presence of films caused by the combination of sulphur or oxygen with most metals can cause open circuit conditions in low-voltage equipment. Since gold will not combine with sulphur or oxygen there is no possibility of these tarnish films forming.

18.3 Connector shapes and sizes

A family of connectors would need to include a range of rectangular and circular connectors associated with a variety of size-16 contacts including signal and power contacts, fibre optic contacts and subminiature coaxial contacts. The connectors shown in *Figure 18.1* are rectangular for in-line, panel mounting or rack and panel applications, and circular either for environmentally sealed application or for the commercial portion of MIL-C-5015 connector areas. The housings are available in nylon, diallyl phthalate or phenolic with, in certain areas, a metal shell. The contacts are available in brass or phosphor bronze with tin or gold plating finishes. Associated are a variety of accessories such as strain reliefs, pin hoods, jackscrews, guide pins and pin headers to give greater versatility to the connector ranges.

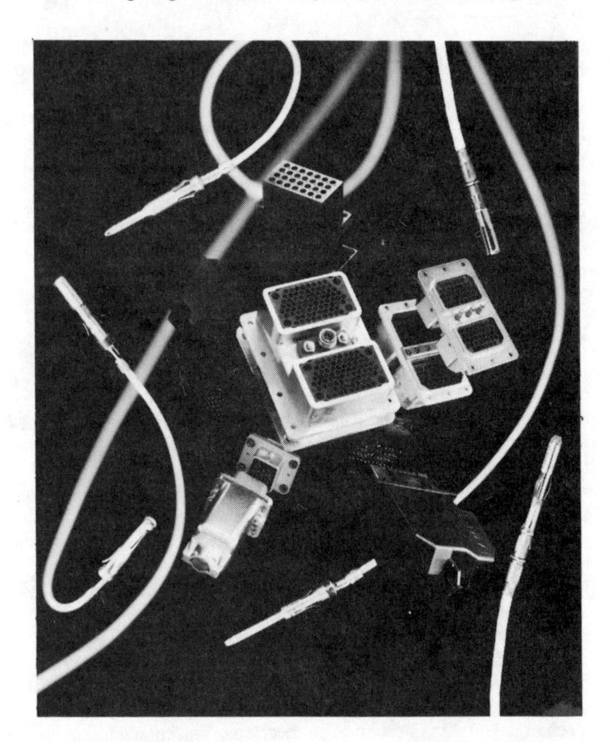

Figure 18.1 Multi-way pin and socket connectors (courtesy AMP of Great Britain Ltd)

18.4 Connector terminations

Crimping has long been recognised as an electrically and mechanically sound technique for terminating wires. Since crimping is a strictly mechanical process, it is relatively easy to automate. Because of this automation capability, crimping has become the accepted terminating technique in many industries.

Terminals or contacts designed for speed crimping in automatic or semi-automatic machines are often significantly different from those designed for handtool assembly, although most machine-crimpable terminals and contacts can also be applied with handtools.

Those designed for hand application cannot normally be used for automatic machinery. The selection of a crimping method is determined by a combination of five factors: (a) access of wire; (b) wire size; (c) production quantity; (d) power availability; and (e) terminal or contact design.

There are, of course, other factors which must be considered; for example, if the finished leads are liable to be roughly handled, as in the appliance or automotive industries, the conductor insulation and the terminal or contact will have to be larger than electrically necessary in order to withstand misuse.

The user must remember the importance of maintaining the proper combination of wire terminal and tool; only then can the optimum crimp geometry and depth be obtained. In this respect it is best to follow the manufacturer's recommendations since most terminals have been designed for a specific crimp form. The effects of crimp depth are shown in *Figure 18.2*.

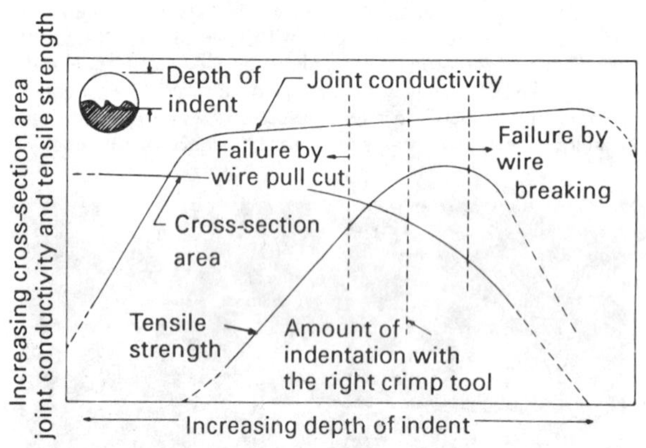

Figure 18.2 Effects of crimp depth

Tensile strength and electrical conductivity increase in proportion to the crimp depth. When the deformation is too great, tensile strength and conductivity suffer because of the reduced cross-sectional area. There is an optimum crimped depth for tensile strength and another for conductivity and, in general, these peaks do not coincide. Thus a design compromise is required to achieve the best combination of properties. However, the use of improper tooling can void the entire design.

Merely selecting the proper wire terminal and tool combination is not enough. The wire must be stripped to the recommended dimension, without nicking, and inserted into the terminal to the correct depth before crimping. A properly crimped terminal is shown in *Figure 18.3*.

It is possible to determine the relative quality of a crimp joint by measuring crimp depth in accordance with manufacturers' suggestions. Tensile strength also provides a relative indication of mechanical quality of crimped connections. This factor has been utilised as a 'user control test' in British and other international specifications.

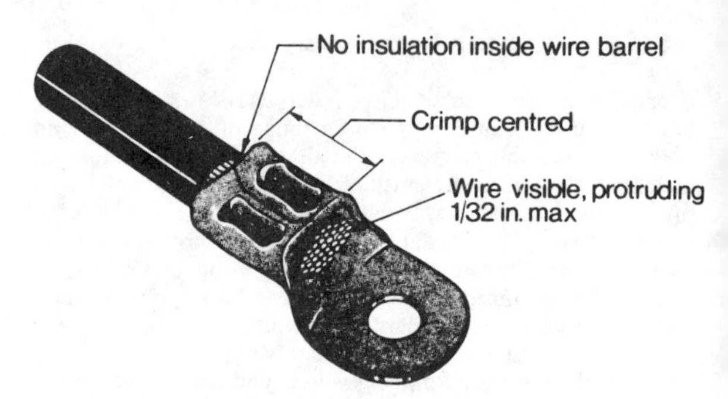

Figure 18.3 A correctly crimped un-insulated terminal

Crimped terminations do not require pre-soldering, and are designed to be used with untinned wires. Soldering may damage the crimp, burn the wire and produce a bad joint. Also the wire could stiffen from wicked solder, and break off later because of vibration. Soldering can affect the characteristics designed into the crimp and seriously affect the performance.

Many terminal designs have some means of supporting the wire insulation. These features are divided into two main categories: (a) insulation gripping; and (b) insulation supporting.

Insulation gripping terminals prevent wire flexing at the termination point and deter movement of the insulation. This feature improves the tensile strength of the crimp for applications where severe vibration is present.

Insulation-supporting types of terminals have insulation that extends beyond the crimping barrel and over the wires own insulation. This only provides support and does not grip the insulation in a permanent manner. The latest feature for this type of termination is the funnel entry type. This, as its name implies, has a funnel form on the inside of the insulation sleeve, which aids in the correct placing of the stripped wire in the barrel of the terminal. It has long been possible to snag strands of wire on the edge of the wire barrel while putting the stripped wire into the terminal. This, depending on the number of snagged strands, could impair the electrical characteristics and cause 'hot spots' or under-crimping of the terminal.

All these problems are resolved with the introduction of funnel entry, as the wire is able to go straight into the wire barrel without damage. Minimal operator skills, increased production rates and added benefits are possible.

Finally there is the introduction of insulation displacement or slotted beam termination, which has enjoyed wide use in the telecommunication and data systems industries. With the advent of connectors or terminal blocks designed for mass termination, this concept has spread and it is rapidly finding use in many other applications (see Section 18.6).

Although the appearance and materials in insulated displacement connectors vary, the design of the slotted areas is basically the same in each.

Insulated wire fits loosely into the wider portion of the V-shaped slot and, as the wire is pushed deeper into the terminal, the narrowing slot displaces the insulation and restricts the conductor as in *Figure 18.4*. Additional downward movement of the insertion tool forces the conductor into the slot where electrical contact is made.

Insulation displacement has recently been applied to terminal blocks. One side has the insulation displacement contact whilst the other side would have the conventional screw terminal accepting ring tongues or bare wire. It would accept a wide range of wire sizes or stranded wire 0.3 mm^2 to 2 mm^2.

Wire termination is accomplished by two simple screwdriver type handtools with different insertion depths. The first is used to

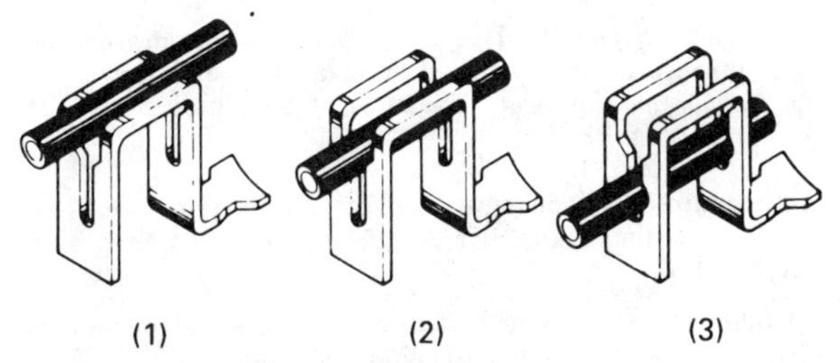

Figure 18.4 Insulation displacement contacts

insert a single wire into the terminal, the other for inserting the second wire.

18.5 Tooling

Crimping tools or machines should be selected after a thorough analysis, as with any other production system.

Generally, the following rates can be achieved with various types of tooling:

manual tools	100–175 per hour
power tools	150–300 per hour
semi-automatic machines	100–4000 per hour
lead-making machine	up to 11 000 per hour

As these figures show, manual tools are intended for repair and maintenance, while powered tools and machines are designed for production applications. To the manufacturer whose output in wire terminations is relatively small, automated tooling is not necessary. While the economics of automated wire terminations may vary with different applications, an output of more than one million terminations a year can be taken as a guide for considering semi-automatic tooling.

The basics of a good crimp connection are the same whether the tool to be used is a simple plier type or a fully automatic lead-making machine.

The basic type of crimping tool is the simple plier type. It is used for repair, or where very few crimps are to be made. These are similar in construction to ordinary pliers except that the jaws are specifically machined to form a crimp. Most of these tools are dependent upon the operator to complete the crimp properly by closing the pliers until the jaws bottom together. Many of the tools may be used for several functions such as wire stripping, cutting, and crimping a wide range of terminal sizes and types. Tools of this type are in wide use.

Other more sophisticated tooling is available, such as cycle controlled tools. This type normally contains a ratchet mechanism which prevents the tool opening before the crimp has been properly completed. This ratchet action produces a controlled uniform crimp every time regardless of operator skill. However, operator fatigue is normally a limiting factor in production with any manual tool.

Powered handtools, either pneumatic or electronic controlled, can be semi-portable or bench mounted. When larger production quantities of terminations are required, the need for this form of tooling is essential. They not only yield high rates of output at low installation cost but also give high standards of quality and are repeatable throughout the longest production run.

These tools offer the opportunity for the introduction of tape-mounted products. A variety of tape mounted terminations are available in either reel or boxed form. Advanced tooling, with interchangeable die sets, gives a fast changeover with minimum downtime. During the crimping cycle the machine will automatically break the tape bonds and free the crimp product for easy extraction, at the same time indexing the termination into position for the next crimp operation.

18.6 Mass termination connectors

Mass termination is a method of manufacturing harnesses by taking wires directly to a connector and eliminating the steps of wire stripping, crimping and contact insertion into housings. It employs a connection technique known as insulation displacement, an idea developed many years ago for the telecommunications industry.

An unstripped insulated wire is forced into a slot which is narrower than the conductor diameter as in *Figure 18.3*. Insulation is displaced from the conductor. The sides of the slot deflect like a spring member and bear against the wire with a residual force that maintains high contact pressure during the life of the termination.

This is the basic principle, but it must be realised that each slot is carefully designed to accept dimensional changes without reducing contact forces. This is accomplished by designing enough deflection into the slot to compensate for creep, stress relaxation and differential thermal expansion.

The force required to push the wire into the slot is approximately 25 times less than for a conventional crimp and it is this factor, in conjunction with the facilitity of not having to strip the wire, that makes insulation displacement readily acceptable for mass termination, by taking several wires to the connector and terminating them simultaneously.

A typical system would employ a pre-loaded connector, with the receptacle having dual slots offering four regions of contact to the wire. The exit of the wire from the connector is at 90° to the mating pin, and can have a maximum current rating of up to 7.5 A.

The average tensile strength of the displacement connection when pulled along the axis of the wire is 70% of the tensile strength of the wire and 20% when pulled on axis parallel to the mating pin. Therefore, plastic strain ears are moulded into the connector to increase the wire removal force in this direction.

The different systems have been developed to accept a wide range of wire including 28 to 22 AWG (0.08 to 0.33 mm^2) wire and 26 to 18 AWG (0.13 to 0.82 mm^2) wire. The connectors are colour coded for each wire gauge since the dimensional difference of the slot width cannot be readily identified.

The pin headers for most systems are available for vertical and right angle applications in flat style for economical wire to post applications, polarised for correct mating and alighnment of housings, and polarised with friction lock for applications in a vibration environment.

18.6.1 Types of tooling

To obtain all the benefits for harness manufacture, a full range of tooling is available, from simple 'T' handle tools to cable makers.

The 'T' handle tool would be used only for maintanance and repair. For discrete wires a self-indexing handtool, either manually or air operated, would be used for intermediate volumes.

For terminating ribbon cable, there are small bench presses for relatively low volumes of harnesses, and electric bench presses for higher production needs.

However, it is the innovation of the harness board tool and the cable maker that offers highest production savings. The harness board tool allows connectors to be mass terminated directly onto a harness board. The equipment consists of three parts: power tool, applicator and board mounted comb fixtures. The wires are routed on the harness board and placed through the appropriate comb fingers. The power tool and applicator assembly is placed on the combs to cut and insert the wires into the connectors. After binding with cable ties the harness can be removed from the board.

The cable maker, either double end or single end, will accept up to 20 wires which can be pulled from drums or reels on an appropriate rack. The individual wires can all be the same length, or variable, with a single connector on one end of the cable and multiple connectors on the other end.

In general, a complete cycle would take approximately 15–20 s according to how many connectors are being loaded. However, three double-ended cables, six-way at each end, able to be produced on the machine in one cycle, would be using the machine to its maximum capacity and the overall time would be expected to be longer.

A comparison can be made between an automatic cut, strip and terminating machine and the cable maker mentioned earlier. This comparison is on 100 000, six-way connectors:

Standard method

Cut, strip and terminate	3400/h		176 h
Manually insert contacts	900/h		666 h
		Total	842 h

New method (mass termination)

Cable maker: assume conservative figure of two single-ended cables every 20 s	360 cables/h	Total	278 h

It can clearly be seen that labour savings of 67% are not unrealistic, which must be the major benefit from using mass termination techniques. Other such benefits include no strip control, no crimp control, reduced wiring errors, no contact damage and reduced tooling wear.

18.6.2 Ribbon cable connectors

Connectors for 0.050″ pitch ribbon cable can also be considered as mass termination types. The basic four types of ribbon cable are extruded, bonded, laminated and woven, with extruded offering the best pitch tolerance and 'tearability'. The connectors, normally loaded with gold-plated contacts, are available in a standard number of ways up to 64, these being 10, 14, 16, 20, 26, 34, 40, 44, 50, 60 and 64.

There are various types of connectors used:

(1) Receptacle connectors, 0.100 grid for plugging to a header.
(2) Card edge connectors, 0.100 pitch to connect to the edge of a PCB.
(3) Pin connectors, to mate with receptacle connectors and offer ribbon-to-ribbon facility.
(4) Transition connectors, for soldering direct to PCB.
(5) DIL plug, 0.100 × 0.300 grid for either soldering to PCB or connecting to a DIP header.

The normal rating for these types of connector is 1 A with an operating temperature range of −55°C to 105°C and a dielectric withstanding voltage of 500 V r.m.s.

18.7 Fibre optics connectors

When joining fibres light losses will occur in four ways:

(1) *Surface finish* The ends of the fibres must be square and smooth and this is usually accomplished by polishing the cut ends.
(2) *End separation* Ideally the fibre ends should touch but this could cause damage and so they are normally held between 0.001″ and 0.005″ apart.
(3) *Axial misalignment* This causes the highest loss and must be controlled to within 50% of the smaller fibre diameter.
(4) *Angular misalignment* The ends of the fibres should be parallel to within 2%.

Any connector system must therefore hold the fibre ends to within these limits, and several different variations have been developed:

(1) *Tube method* This method uses a metal jack and plug which are usually held together by a threaded coupling. The fit of the plug into the jack provides the primary alignment and guides the fibre in the jack into a tapered alignment hole in the plug. The depth of engagement must be accurately controlled to ensure correct end separation. These connectors are normally made from turned metal parts and have to be produced to close tolerances (*Figure 18.5*).

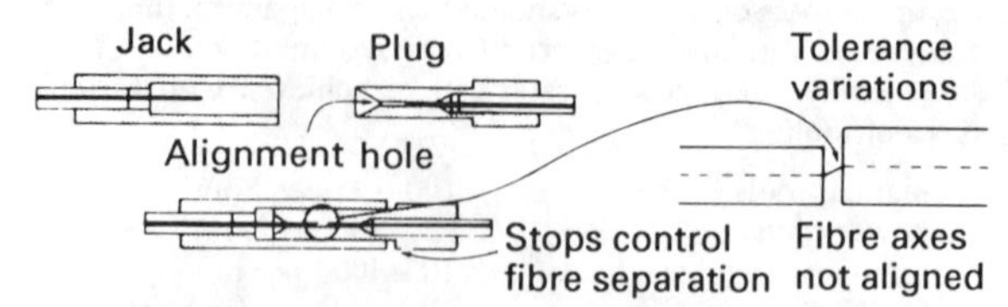

Figure 18.5 Fibre optic tube alignment connector

(2) *Straight sleeve method* A precision sleeve is used to mate two plugs which are often designed similar to the SMA coaxial connectors and made from very tightly toleranced metal turned parts and, due to the design, concentricity needs to be very good (*Figure 18.6*).

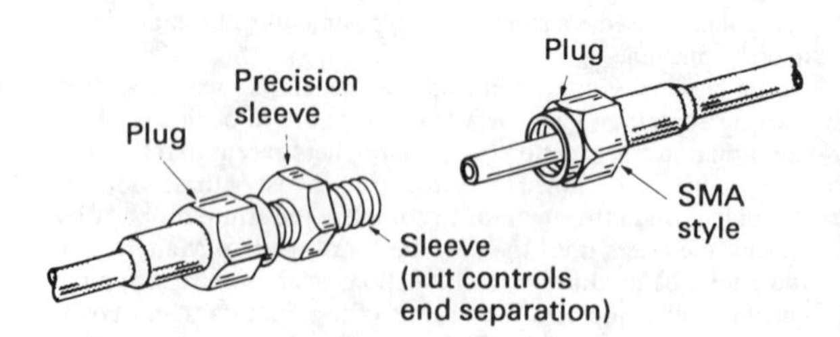

Figure 18.6 Fibre optic straight sleeve connector

(3) *Double eccentric method* Here the fibres are mounted within two eccentrics which are then mated. The eccentrics are then rotated to bring the fibre axes into very close alignment and locked. This produces a very good coupling with much looser manufacturing tolerances but the adjustment can be cumbersome and must usually be done with some test equipment to measure maximum adjustment (*Figure 18.7*).

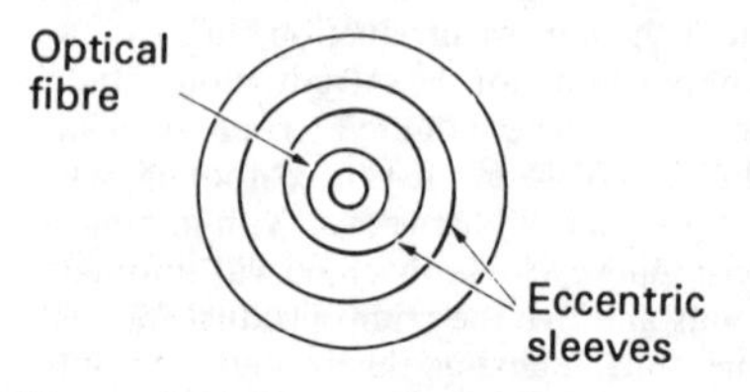

Figure 18.7 Fibre optic double eccentric connector

(4) *Three-rod method* Three rods can be placed together such that their centre space is the size of the fibre to be joined. The rods, all of equal diameter, compress and centre the fibres radially and usually have some compliancy to absorb fibre variations. With this design it is important that the two mating parts overlap to allow both members to compress each fibre. The individual parts in this design can be moulded plastic but need to be well toleranced (*Figure 18.8*).

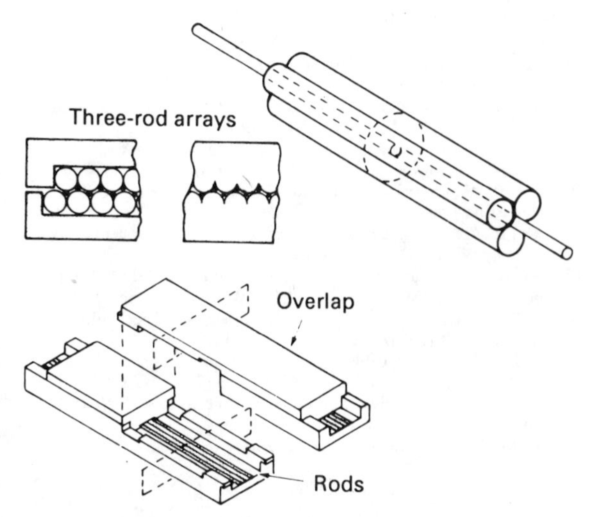

Figure 18.8 Fibre optic three-rod connector

(5) *Four-pin method* Four pins can be used to centre a fibre and the pins are held in a ferrule. This method is sometimes used with the straight sleeve design when the pins are used to centre the fibre and the sleeve is used to align the mating halves. These parts are normally turned metal and held to tight tolerances (*Figure 18.9*).

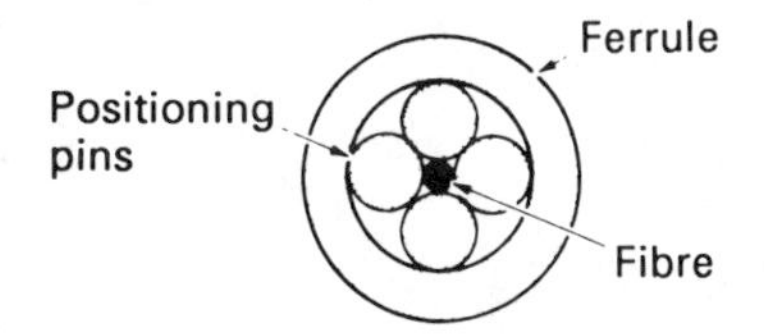

Figure 18.9 Fibre optic four-pin method of connection

(6) *Resilient ferrule* This method utilises a ferrule and a splice bushing. The front of the ferrule is tapered to match a similar taper in the bush and the two parts are compressed together with a screw-on cap which forces the two tapers together. This moves the fibre in the ferrule on centre and provides a sealed interface between the two parts preventing foreign matter from entering the optical interface. The compression feature accommodates differences in fibre size and enables manufacturing tolerances to be considerably relaxed. The parts are plastic mouldings and typically produce a connector loss of less than 2 dB per through-way at very low cost (*Figure 18.10*).

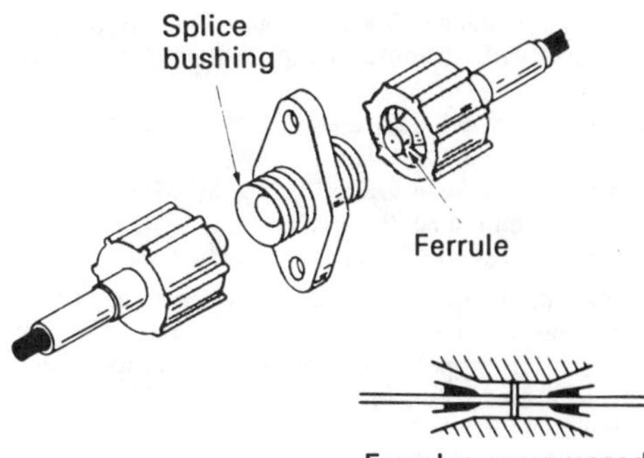

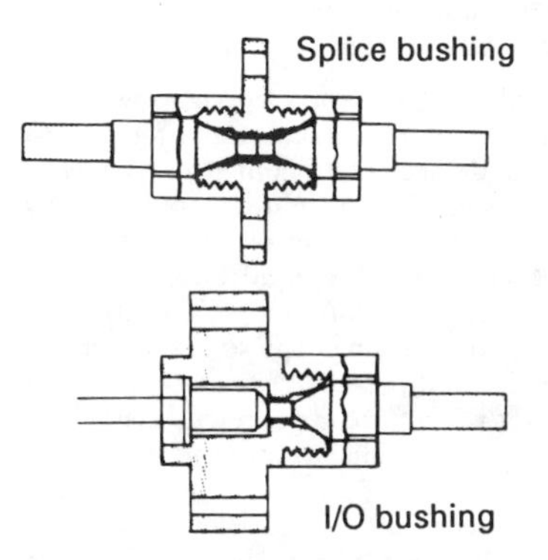

Figure 18.10 Fibre optic resilient ferrule alignment mechanism

18.8 Radio frequency connectors

Radio frequency connectors are used for terminating radio frequency transmission lines which have to be run in coaxial cables. These cables are available in sizes ranging from less than 3 mm diameter for low power applications of around 50 W to over 76.2 mm diameter for powers of 100 000 W. In addition to power-handling capabilities cables are also available for high-frequency applications, high and low temperature applications, severe environmental applications and many other specialised uses which all require mating connectors.

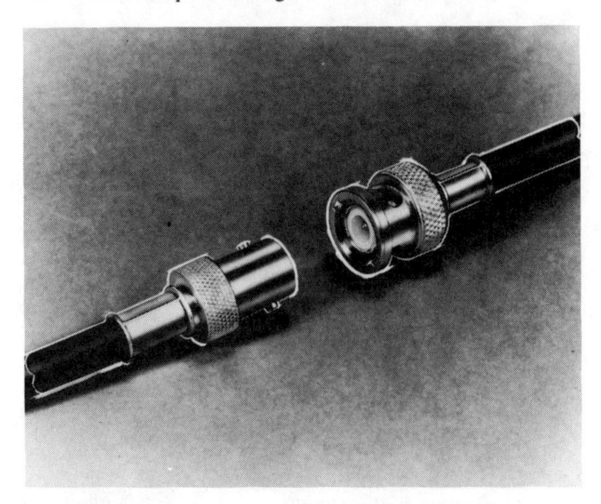

Figure 18.11 A typical r.f. connector (courtesy AMP of Great Britain Ltd)

Some of the more popular ranges are grouped in approximate cable diameter size with their operating frequency ranges as follows:

3 mm diameter	SMA 0–12.4 GHz, SMB 0–1 GHz, SMC 0–1 GHz
5 mm diameter	BNC 0–4 GHz, TNC 0–11 GHz, min. u.h.f. 0–2 GHz
7 mm diameter	N 0–11 GHz, C 0–11 GHz, u.h.f.

The design and construction of the range of connectors is very similar throughout as they all have to terminate a centre conductor and a woven copper braid screen. The variations are in size and materials. For example, to meet the MIL-C-39012 specification it is necessary to use high-quality materials such as brass, silver plated for the shells, and teflon for the dielectric with gold-plated copper centre contacts. This is mainly due to the requirement for a temperature range of −65°C to +165°C. There are, however, three distinct types of termination:

(1) With the *soldering and clamping* type of design the centre contact is soldered to the centre conductor and the flexible braid is then clamped to the shell of the connector by a series of tapered washers and nuts. The biggest advantage of this type of connector is that it is field repairable and replaceable without the use of special tools. The disadvantages are the possibility of a cold solder joint through underheating, or melting the dielectric by overheating. Any solder which gets onto the outside of the centre contact must be removed otherwise the connector will not mate properly. It is easy to assemble the connector wrongly due to the large number of parts involved.
These connectors are used in large numbers by the military.

(2) In the *crimping* design the centre contact is crimped to the centre conductor and the flexible braid is then crimped between the connector shell and a ferrule. There are versions which require two separate crimps, normally to meet the MIL specification, and versions which can have both crimps made together. The advantages of this method of termination are speed and reliability together with improved electrical performance. Testing has shown that the SWR of a crimped connector is lower than the soldered and clamped version. The crimp is always repeatable and does not rely upon operator skill. The disadvantages of this design are that a special crimp tool is required and the connectors are not field repairable.

(3) With the *soldering and crimping* type of design the centre contact is soldered to the centre connector and the flexible braid is crimped. Obviously all the advantages and disadvantages of the previous two methods are involved with this design.

Further reading

BEARD, T., 'Technologies combine in connectors', *Electronic Product Design* (Dec. 1987)

BISHOP, P. J., 'The connector design costs of three tracks between pads', *Electronic Product Design* (June 1988)

BROWNE, J., 'Protecting and connecting RF and microwave hardware', *Microwave & RF*, **27**(8) (Aug. 1988)

BURCH, C., 'Stacking connectors for fast circuits', *Electronic Product Design* (June 1988)

CLARK, R., 'The critical role of connectors in modern system design', *Electronics & Power* (Sept. 1981)

EVANS, C. J., 'Connector finishes: tin in place of gold', *IEE Trans. Comps, Hybrid & Manf. Technol.*, **CHMT-3**, No. 2 (June 1980)

JOWETT, S., 'The way ahead in connector design and technology', *New Electronics* (1 Sept. 1987)

KINDELL, C., 'Ribbon cable review', *Electronic Production* (Nov./Dec. 1980)

McDERMOTT, J., 'Hardware and interconnect devices', *EDN* (July 1980)

McDERMOTT, J., 'Flat cable and its connector systems', *EDN* (Jan. 1981)

McDERMOTT, J., 'Flat cable and connectors', *EDN* (Aug. 1982)

MILNER, J., 'LDCs for IDCs', *New Electronics* (Jan. 1982)

PEEL, M., 'Material for contact integrity', *New Electronics* (Jan. 1980)

ROELOFS, J. A. M. and SVED, A., 'Insulation displacement connections', *Electronic Components and Applications*, **4**, No. 2 (Feb. 1982)

SAVAGE, J. and WALTON, A., 'The UK connector scene—a review', *Electronic Production* (Sept. 1982)

TANAKA, T., 'Connectors for low-level electronic circuitry', *Electronic Engineering* (Feb. 1981)

19 Printed Circuits

I G Laing
Technical Director
Exacta Circuits Ltd

Contents

19.1 Introduction

19.1.1 The role of the printed circuit

The semiconductor technology of integrated circuits is the driving force behind electronic systems development. However, the printed circuit board (PCB) is an essential part of an overall system since equipment is built up by connecting active devices together into manageable, cost-effective blocks.

The printed circuit is an interconnection system which is multilevel, highly conductive, consistently reproduceable, and has a low to medium dielectric constant. These attributes have made the PCB the standard, almost universal, method of construction for practically all electronic systems. It is also a convenient field replacement unit.

The printed circuit provides mechanical support as well as functional electrical interconnection for the components, and some thermal management.

19.1.2 History of the printed circuit

Dr Paul Eisler patented the first 'modern' printed circuit boards in 1943 but development of the technology only really took shape with the invention of the transistor in the late 1950s. The printed circuit board industry grew in the 1960s and 1970s and is now considered mature. Multilayer products date essentially from 1961 with the Hazeltyne patent.

19.1.3 Basic manufacturing

There are many possible ways of manufacturing a PCB but only two types of processing have real commercial use. These are:

(1) The *subtractive process*, where metal, usually copper, is etched from the surface of an insulating substrate to produce the required pattern.
(2) The *additive process*, where conductive material is added to the surface of an insulating substrate. This is generally achieved by either electroless plating or conductive ink printing.

The majority of circuits today are produced by the subtractive process.

19.1.4 Types of printed circuit

Circuit boards may be rigid, flexible or a combination of both. They may have interconnection patterns which can typically achieve 50 cm of interconnection per square centimetre of signal layer. They can be made in large panels—typically 61 cm × 46 cm —with larger panels requiring more specialised, yet available, manufacturing equipment.

Interconnection between layers can be achieved by a wide variety of methods. *Figures 19.1–19.9* show examples of various

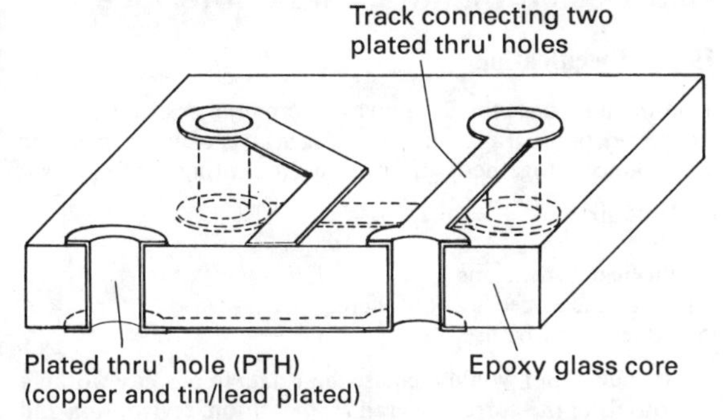

Figure 19.1 Double-sided plated thru' hole printed circuit board

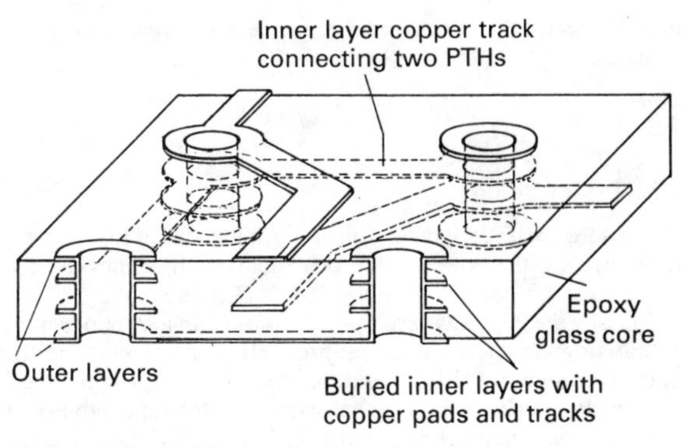

Figure 19.2 Multilayer printed circuit board

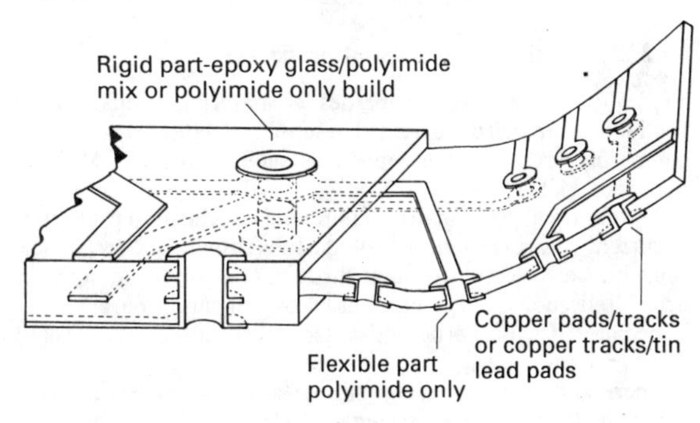

Figure 19.3 Flexi-rigid multilayer printed circuit board

Anodised aluminium or nickel-plated copper surface-mounted heat sink

Heat sink bonded to PCB using no flow prepreg

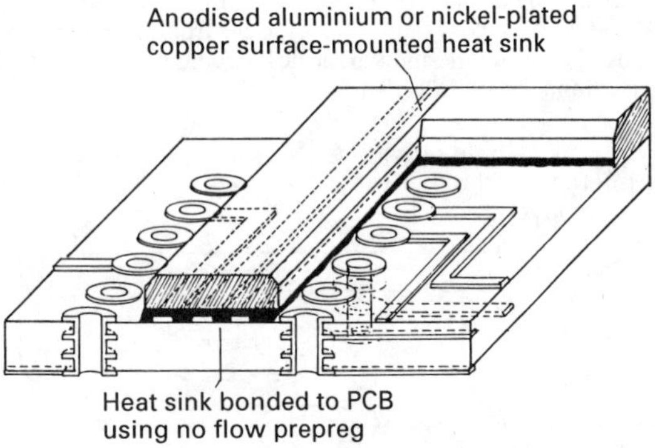

Figure 19.4 Multilayer printed circuit board with surface-mounted heat sink

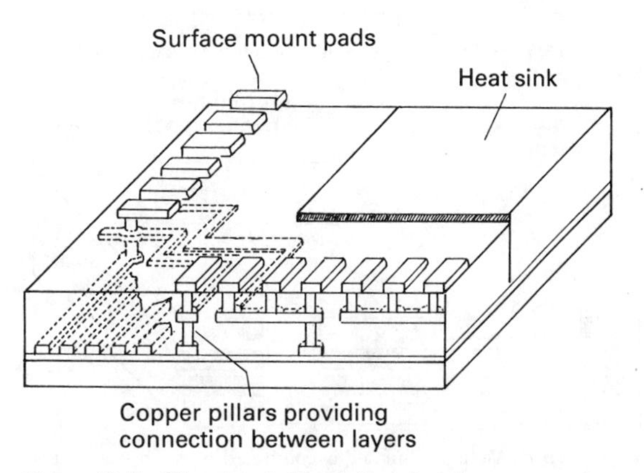

Figure 19.5 Pillar plated printed circuit board

printed circuits, and illustrate the variety of types which are available.

19.2 Design

19.2.1 Design aim

Good electronic design sets out to achieve the required performance at the lowest cost consistent with achieving the required standards of reliability and maintainability.

Initially the PCB was developed to produce an interconnection technology which facilitated mass production and mass assembly, and gave economy of weight and space. Close control over circuit electrical parameters is easily achievable by the repeatability of the dielectric contant and dielectric thickness and the consistency of line width. The current-carrying capacity is that of copper.

19.2.2 Electrical and thermal properties

Typically the electrical properties of materials used in PCB production are well documented in suppliers' data sheets, and the commonly used laminate standards such as the US Military Standard MIL P 13949.

The electrical performance of the circuit is related to properties of current-carrying capacity, voltage breakdown, line resistance, capacitance, dielectric constant through control of propagation delay, reflections, surge currents, crosstalk and characteristic impedance. The properties of printed circuit boards are covered in several references.[1–5]

Thermal design is frequently important, particularly for very dense, high-performance systems. Once it is decided how the system is to be partitioned, it is possible to move through the schematic or logic description of the circuitry that requires to be packaged, to the layout stage, where the electronics components parts list, board outline and type critical circuit information and specification all come together.

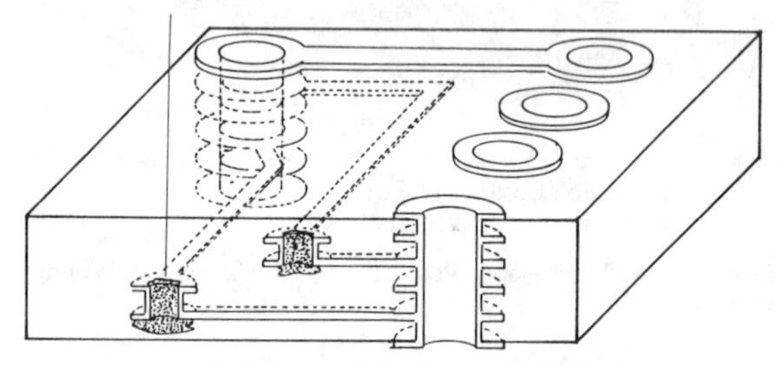

Figure 19.6 Buried via hole multilayer printed circuit board

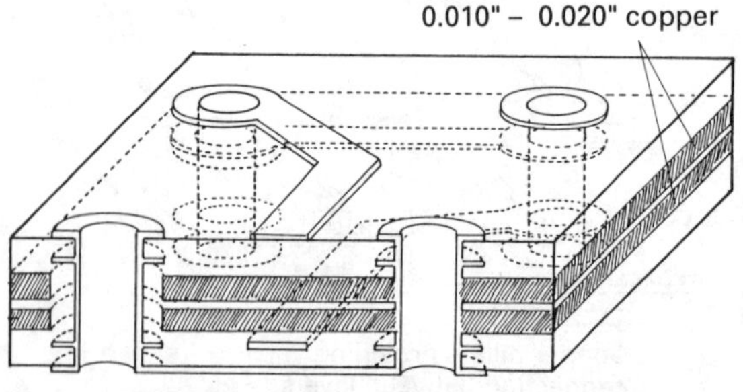

Figure 19.7 Multilayer printed circuit board with internal heat sink

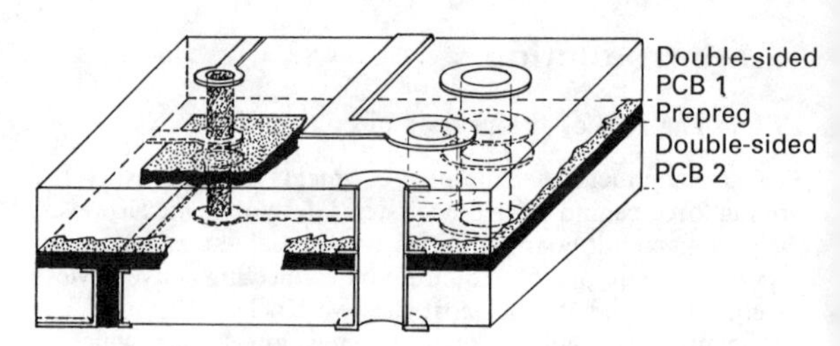

Figure 19.8 Double double-sided plated thru' hole printed circuit board

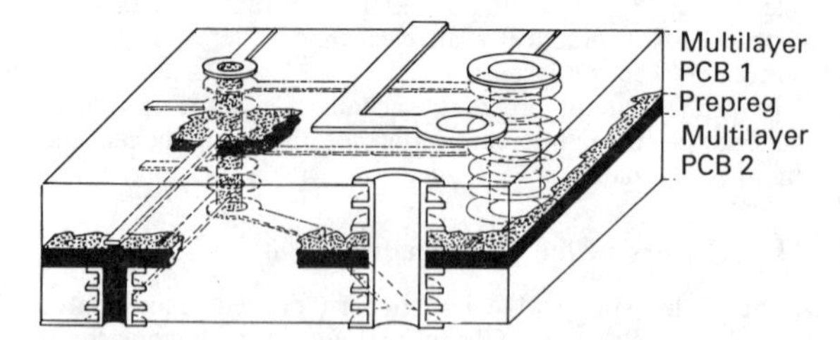

Figure 19.9 Double multilayer printed circuit board

19.2.3 Layout

The usual pattern is to set out the components in a suitable electrical array, minimising interconnection lengths between key components or key functional areas. Occasionally component layout has to take prime account of thermal distribution.

Once the component layout and therefore footprint is set, the required interconnections may be made. Manual taping to a scaled-up footprint pattern is still in fairly common usage. Routing is made easier by using at least two signal layers, orthogonal to each other. Interconnection between layers is freely available. For dense packaging, more than one or two signal layers may be necessary.

19.2.4 Use of CAD

More commonly today the majority of tracks are generated by computer aided design (CAD) systems. Here autorouting of tracks to a high percentage success rate of 80–95% is possible within the given design guidelines of number of layers, size of board, track size, gap size, pad size and hole size. The final track routings are usually operator assisted in an interactive mode. These layouts are then plotted onto stable photographic film. In these stages, humidity and temperature control are essential for medium to complex work.

19.3 Design/manufacturing interface

19.3.1 Specification

Traditionally, once the PCB layout (commonly called design) is completed, photographic masters of each layer of interconnection are produced, together with drawings indicating the following:

(1) Hole sizes
(2) Hole positions
(3) Profiled dimensions
(4) Acceptance specification reference
(5) Materials to be used, etc.

The manufacturer would then use these data to produce working phototools of the correct overall size, position, registration and feature size to allow for economic manufacturing processing and tolerancing. Similarly, a numerical control (NC) tape would be produced to drive automatic drilling or profiling machines. The phototools would also be used to determine the ends of electrical nets such that a bed of nails electrical test head can be designed and built for end-product acceptance. Frequently individual circuits can be made on a multiple image panel.

19.3.2 CADEM

Today many of these activities can be aided by computer aided design, engineering and manufacture (CADEM) systems, where the design data are accessed while still in digital format and the information is manipulated by computer with reference to the manufacturer's process rules and plant size. Direct machine drivers would then be output for manufacturing use, with the minimum of delay and the maximum of accuracy. *Figure 19.10* shows the route that would normally be followed.

Each layer of interconnection has to match all other layers and each has to be easily and accurately registered one to another in various stages of the manufacture.

The working phototools would conveniently be directly plotted on to prepunched photographic film pinned in register under conditions of controlled humidity and temperature.

19.4 Materials and processing

19.4.1 Laminates

The most common base laminate is epoxy resin impregnated woven glass fibre. There are several glass cloth styles, from the thickest individual ply, designated 7628 and 0.17 mm thick, to the thinnest of 104 style at 0.04 mm thick.

These plies are built up in various combinations, usually with outer copper foils to produce working core thicknesses of typically 0.1 mm up to 3 mm. These double-sided copper-clad cores are used to produce the required circuit patterns. The copper foils range from 5 μm through 9 μm, 17 μm, 35 μm and further steps to 175 μm. Typically the thinner foils are used for the finest track widths, with 17 μm and 35 μm being used for conventional logic circuitry and the wider copper tracks for power application. Other foils would include nickel alloys for the production of printed and etched resistor elements.

Other common materials include phenolic resin-impregnated paper which finds application particularly in consumer products. Chopped strand glass is cheaper than woven glass and this can be impregnated with epoxy, phenolic, polyester or other resins, and used on its own or with facing sheets of resin-impregnated woven glass. These materials are commonly called composite woven glass.

Filled polyimide resin and cyanoacrylate, although more expensive than epoxy, are used in cases where there are continuous operating temperatures of 150°C. Polyimide or polyester can also be used on its own and these are used for flexible circuit applications.

Polytetrafluoroethylene (PTFE) impregnated glass is used for microwave and other high-frequency applications benefitting from its relatively low dielectric constant.

A recent development is an expanded PTFE fibre, which is woven into sheets and then filled and bonded together with epoxy or other resins, providing a more easily processable microwave application material.

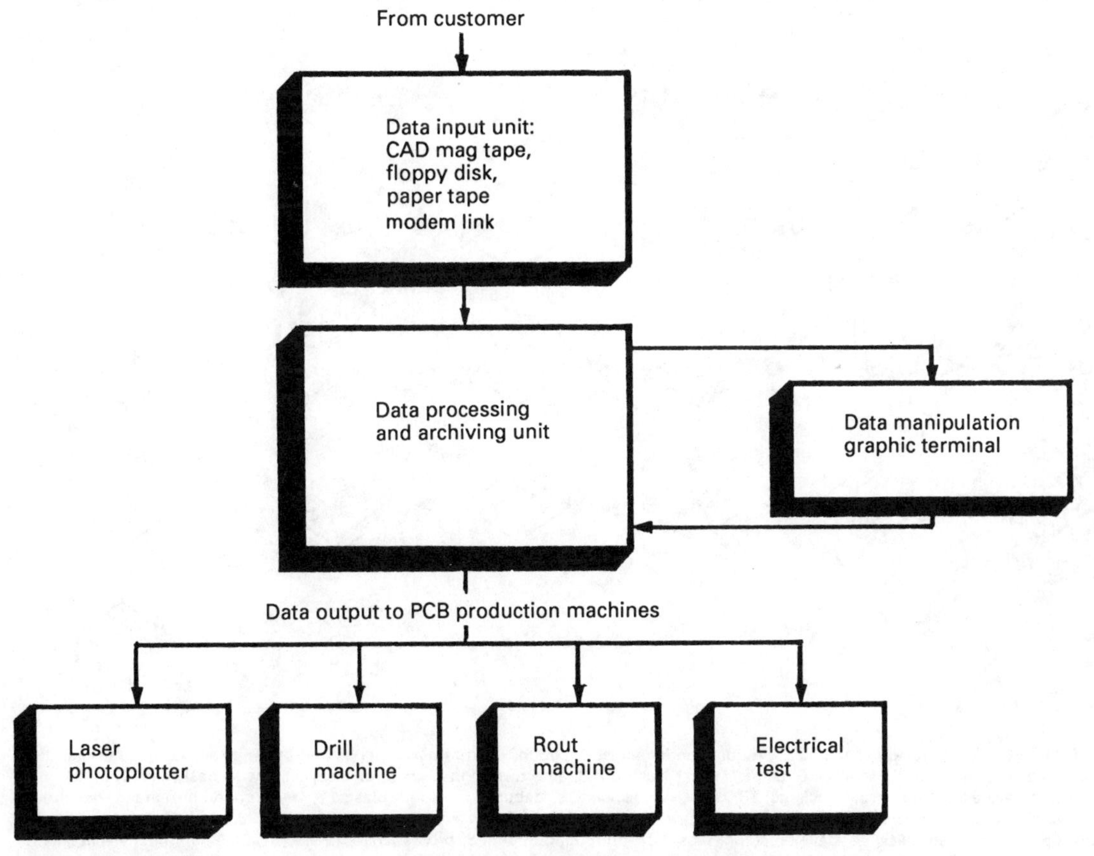

Figure 19.10 Block diagram of a typical computerised artwork and data generation system

Polyethersulphone and other related resins can be used to produce moulded circuits where the mechanical features are moulded in three dimensions with the current pattern commonly being subsequently generated by additive means. This type of circuit would suit high volume applications.

Generally, dielectric constants range from about 2 to about 5.

19.4.2 Finishes

19.4.2.1 Metallic

Copper is used to produce the interconnection pattern with tin/lead for solderability protection. Nickel, gold and palladium/nickel are used for contact areas. Nickel alloys are also used to produce resistive elements.

19.4.2.2 Non-metallic

Benzotriazole is used to preserve the solderability of clean copper, as are solderable flux lacquers. Carbon-loaded pastes provide lower cost contact areas, and various pastes called polymer thick film are used, which are printed and dried to provide conducting or insulating patterns depending on the composition of the paste.

Epoxy or acrylic resins are used as permanent solder masks on finished boards. Peelable solder masks are also practical for temporary protection of contact pads during soldering or as part of a two-stage assembly and soldering operation.

19.4.3 Processing

19.4.3.1 Influence of sequence

The actual format of any particular design of circuit board determines the selection and sequencing of particular processes. For instance, hole generation operations are the common starting point for plated through-hole double-sided boards, whereas it is commonly the last operation for non-plated through boards, and is in the middle of the sequence for multilayer boards. *Figure 19.11* shows a typical sequence of manufacturing for a multilayer printed circuit.

19.4.3.2 Registration

For all types of printed circuit, registration has to be considered and set at the beginning. Most often used are a series of holes and/or slots, but edge location is frequently seen. The principle is to be able to have all the required features on and through each

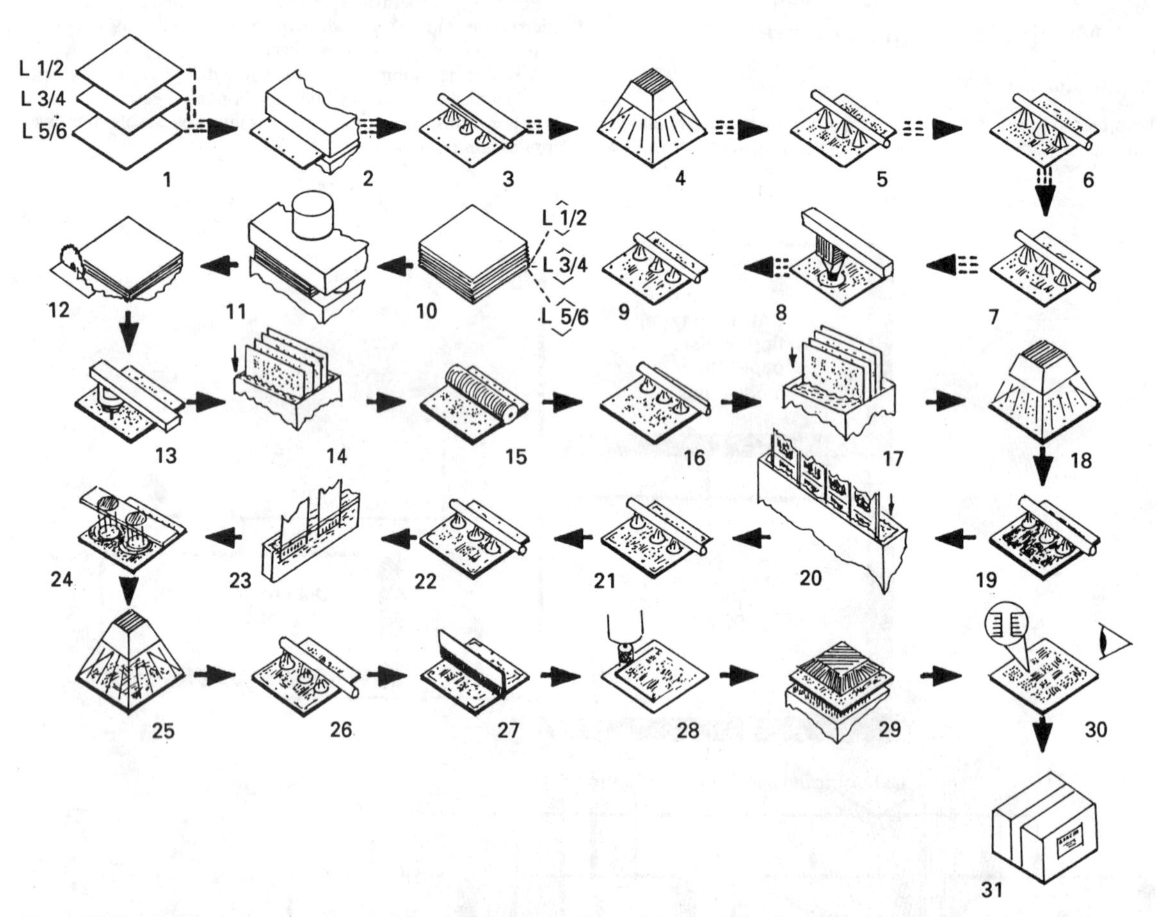

Figure 19.11 Multilayer process sequence: 1, requisition laminate; 2, punch tooling holes; 3, chemical/brush clean; 4, image layers GND/volt/logic (inner layers); 5, develop; 6, etch; 7, resist wash off; 8, auto optical inspect; 9, oxidise; 10, lay up package; 11, bond package; 12, remove excess resin/split package; 13, drill; 14, desmear; 15, deburr; 16, vapour blast; 17, clean/condition/catalyse/electroless copper plate; 18, image outer layers; 19, develop; 20, copper and tin/lead plate; 21, resist wash off; 22, etch; 23, gold edge connector plate; 24, reflow; 25, image solder mask; 26, develop solder mask; 27, print annotation; 28, profile; 29, auto electrical test; 30, final inspect/environmental test; 31, dispatch

layer, and each board in a fixed and consistent position with respect to each other in order to:

(1) Provide a basis for design rules.
(2) Produce an electrically and mechanically consistent product.
(3) Ease manufacturability.
(4) Facilitate assembly and test.

19.4.3.3 Hole generation

For round holes, mechanical drilling using tungsten carbide drill bits is preferred on fibre-filled materials, whereas paper phenolic is usually punched. Routing is used for unusual shaped holes. Recently lasers have been developed for holes smaller than about 200 μm in diameter. Today virtually all these machines are NC, driven by tapes or cassettes, or computer numerical control (CNC), driven directly from a central computer database. These machines have automatic tool change facilities and commands for cutter speed and feed control as well as positional control.

19.4.3.4 Pattern generation

Patterns are usually created using a selective resist followed by etching or by plating and etching as in *Figures 19.12* and *19.13*. Screen printing, through stainless steel or polyester mesh, is used for high volume work, where features are greater than about 250 μm.

Photoimageable resists are in extensive use for higher technology work, producing better resolution and more accurate registration, and needing less skill. These resists can be liquid where they would be applied by screen, curtain, roller, electrostatic spray, or electrophoretic or other coating. Their use is not as common as that of dry film resists where the precoated and dried photoimageable resist is rolled with heat and pressure on to the surface to be imaged.

Once applied, the resist is imaged using ultraviolet light in the 300–500 nm range causing a chemical change to take place, selectively hardening the resist. The unpolymerised resist is then washed away using chlorothene, in the case of solvent-processable

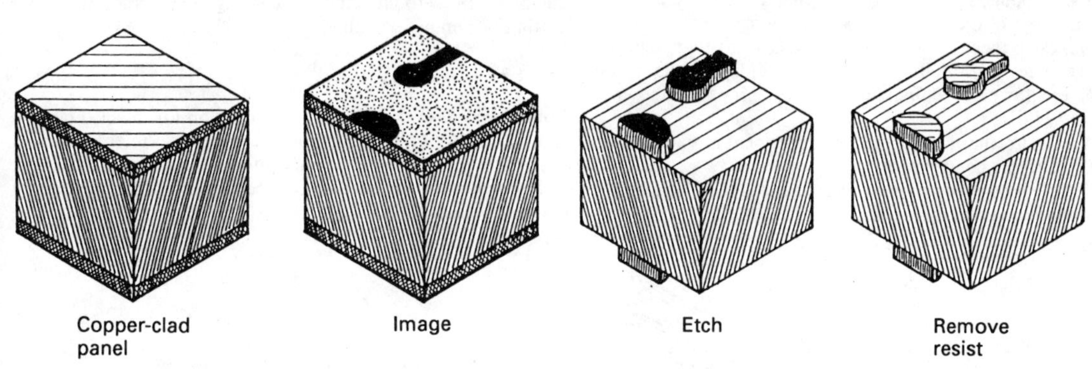

Figure 19.12 Print and etch process from imaging to resist removal

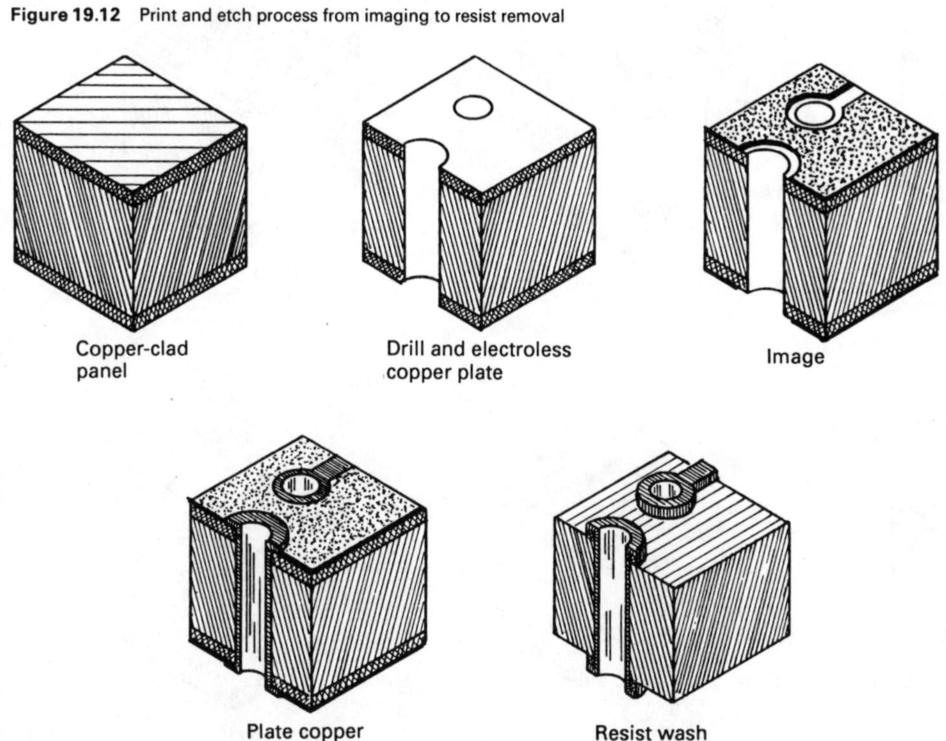

Figure 19.13 Print, plate and etch process

resists, or a weak solution of sodium carbonate for aqueous processable resists. This type of resist is positive working as shown in *Figure 19.14*. Negative-working resists are also used but these are more commonly found in liquid forms.

Etchants used are acidic cupric chloride, sulphuric/peroxide or sulphuric/persulphate systems, most of which are fully regenerable. In cases where tin/lead is the etch resist (i.e. the print, plate, ink wash off, etch process), ammoniacal cupric chloride needs to be used in place of the acidic version. Chrome/phosphoric etch is used for nickel foils.

19.4.3.5 Hole conditioning and reinforcement

Interconnection between layers is achieved by the metallisation of holes, commonly termed 'plated through hole'. The initial coverage of the hole is normally performed by electroless copper plating, although vacuum sputtering is possible. The initial deposition of the electroless copper is reinforced by electroplated copper to a thickness of about 25 μm. This reinforcement takes place after the reverse image of the desired track pattern is created on the surface. Tin/lead is then electroplated and this layer is subsequently used as an etch resist. At a later stage the tin/lead layer is fused to form a solder alloy prior to final finishing. The process is shown in *Figure 19.15*.

In many circuits, the entire copper reinforcement is made with the same electroless copper plating.

The electroless copper process covers the stages of hole and surface conditioning, catalysing and activating the hole and the chemical deposition of the copper itself on to the catalysed surfaces. In some cases the laminate itself can be precatalysed and often a pattern print operation is placed between catalysing and deposition to create selective copper deposition on to unclad dielectric surfaces and holes, commonly called additive processing. This is illustrated in *Figure 19.16*.

Where the circuit is multilayered, reliable connections need to be made to buried inner layers. In this case the hole is frequently preconditioned by etching some resin from the hole to ensure that any resin which may have been smeared over the interconnection point by the drilling process is removed prior to the electroless copper step.

In some cases both the resin and reinforcement are etched and this is common with US military specifications.

19.4.3.6 Profiling

Normally the circuit is made on a larger manufacturing panel and this needs to be cut to the size of the finished circuit for the customer. Press tooling and NC routing are common with laser cutting becoming available.

19.4.3.7 Verification

Point-to-point electrical testing is common for finished circuits using a 'bed of nails' technique. Access to the interconnection pattern was once on 2.54 mm pitch, but newer component technologies now mean that connection points may be within 0.5 mm of each other.

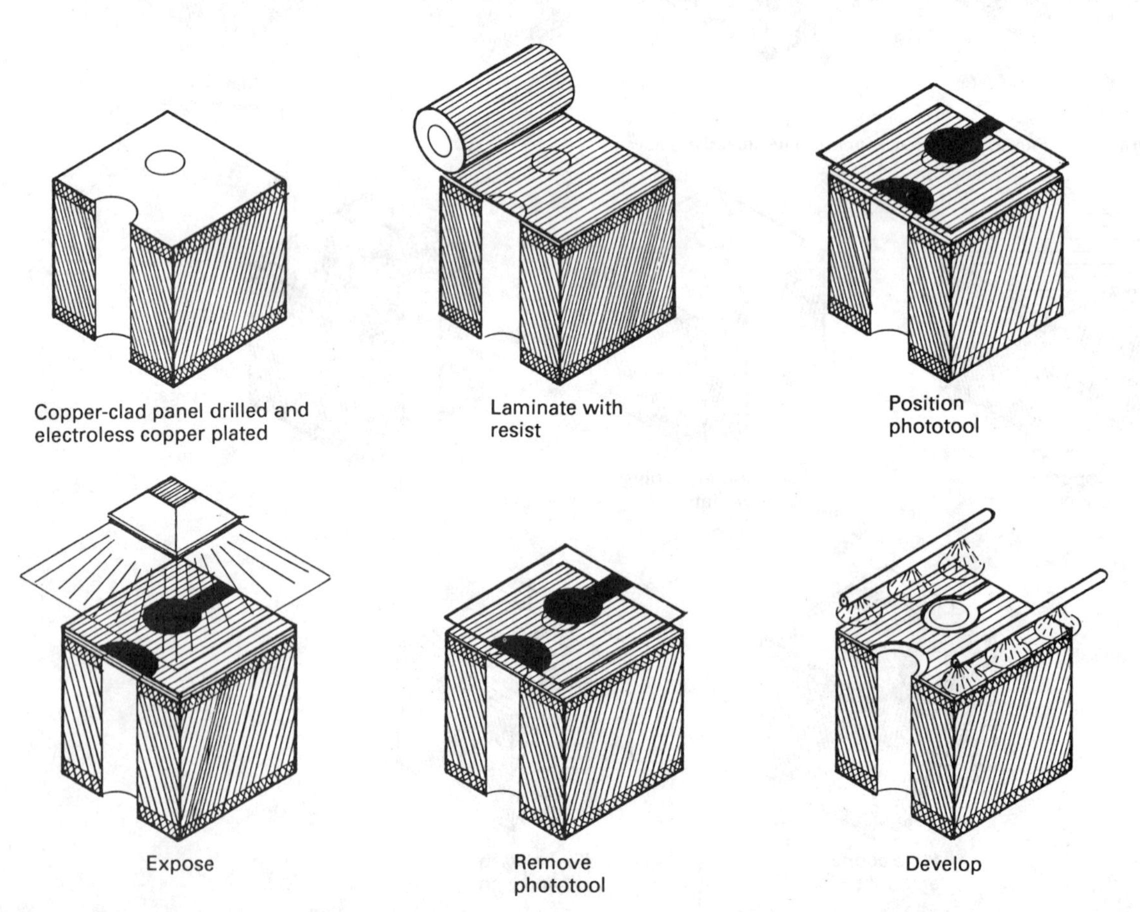

Figure 19.14 Positive working imaging process

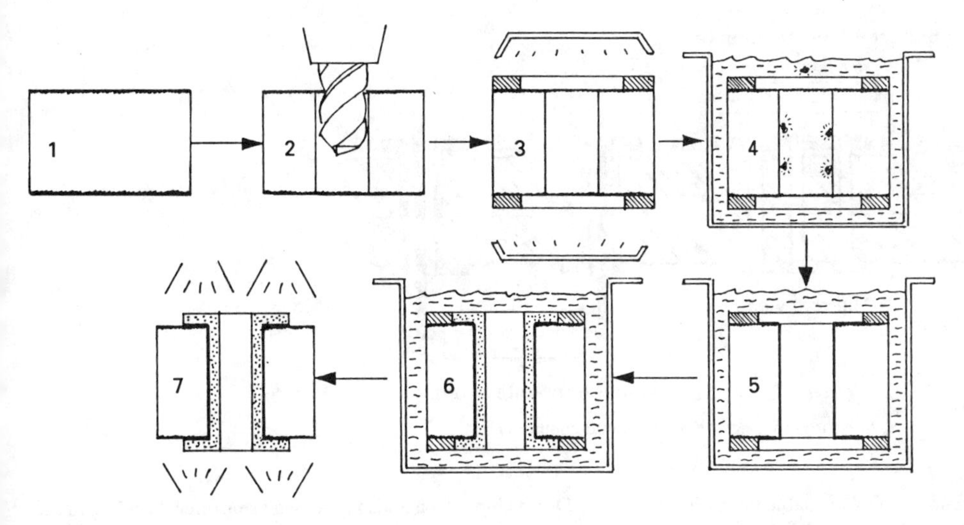

Figure 19.15 Schematic diagram showing the creation of a plated through hole using a subtractive process: 1, copper-clad laminate; 2, drill; 3, clean/condition; 4, catalysis/accelerate; 5, plate electroless copper; 6, image; 7, plate electrolytic copper; 8, plate electrolytic tin/lead; 9, resist removel; 10, etch

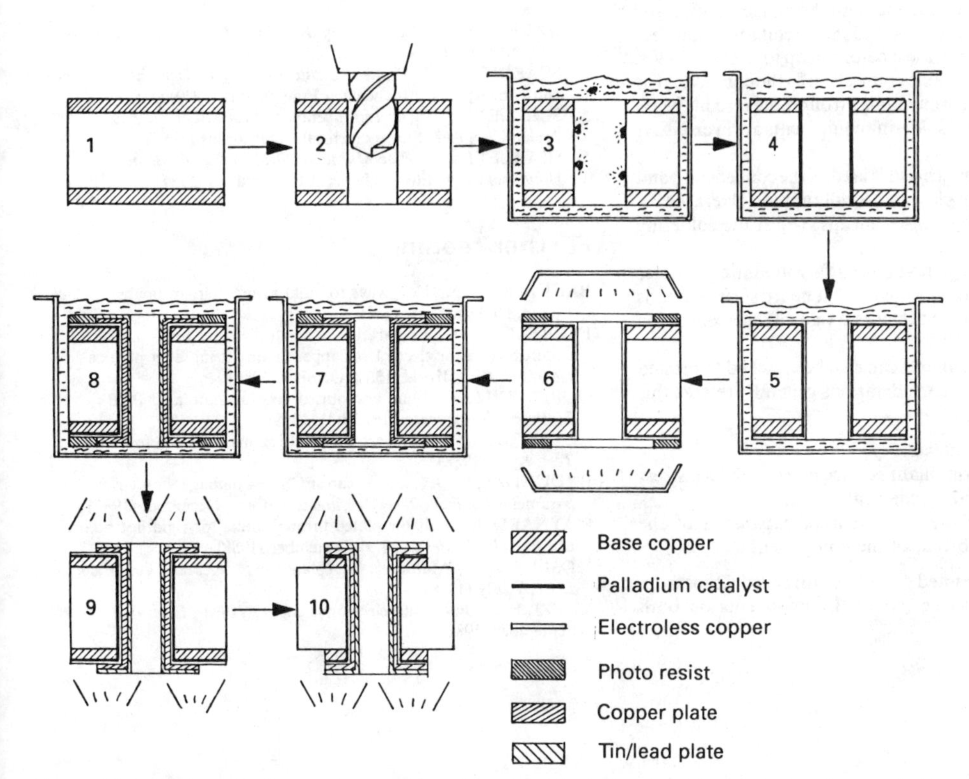

Figure 19.16 Schematic diagram showing the additive method of printed circuit manufacture; 1, unclad adhesive-coated laminate; 2, drill; 3, image; 4, clean/condition; 5, catalysis/accelerate; 6, plate electroless copper; 7, resist removal

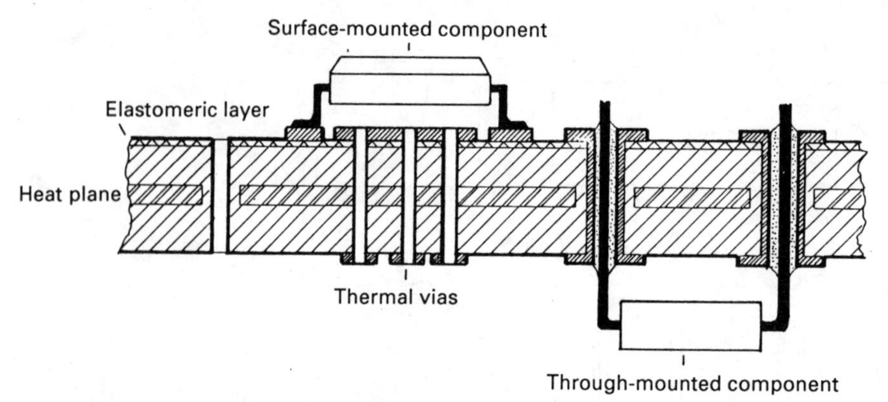

Figure 19.17 Diagram of Chipstrate with through- and surface-mounted components

Automatic optical inspection is now available for both inner layers and outer layers using computer integrated CCD cameras.

Solderability, thermal shock, insulation resistance, properties after environmental conditioning, dimensions, flatness and interconnection continuity are the most frequently applied test procedures.

19.5 Assembly

The printed circuit provides the interconnection pattern for discrete electrical components to interact as a subsystem. Packaged components are generally leaded and both through mounting and surface mounting of leads into or on to the circuit are employed.

Through mounting was the most common up to the mid-1980s. The hole provided a good reliable connection to the component as well as providing the interconnection through the circuit. Mass NC-controlled assembly and soldering equipments are in extensive use.

In surface mounting, a surface pad is used as the connection point and the component is held in place by a locally applied glue, a locally applied tacky solder paste or other means, whilst the soldering operation is taking place.

It is possible to mix through hole and surface mounting. Surface mounting technology reduced printed circuit board surface area as holes for component fixing no longer take up valuable interconnection space.

Leadless packaged components can also be attached to printed circuit boards but reliability considerations generally restrict this application to:

(1) Relatively small components.
(2) Use of boards with a compliant surface layer to reduce stress levels from thermal and mechanical strain.
(3) Use of boards where the coefficient of expansion of the substrate is matched to that of the component.

Figure 19.17 illustrates a printed circuit board, called Chipstrate, with through hole and surface mounted components on both sides.

Direct chip attach is also practised using direct bonding to the substrate from the chip or by the intermediate vehicle of tape automated bonding (TAB).

Acknowledgement

I would like to thank Tony Baillie of Exacta Circuits for the diagrams.

References

1 COOMBS, D. F., *Printed Circuits Handbook*, McGraw-Hill, New York (1967)
2 HARPER, C. A., *Handbook of Electronic Packaging*, McGraw-Hill, New York (1969)
3 SCARLETT, J. S., *Printed Circuit Boards for Micro Electronics*, 2nd edn, Ayr, Electrochemical Publications (1980)
4 SCARLETT, J. S., *An Introduction to Printed Circuit Board Technology*, Ayr, Electrochemical Publications (1985)
5 SCARLETT, J. S., *The Multilayer Printed Circuit Board Handbook*, Ayr, Electrochemical Publications (1985)

Further reading

BAHNIUK, D. E., 'New ways to build printed circuit boards', *Mach. Design*, **60**(3), 6 Oct. (1988)
FARRELL, D. M., 'Preparation of small diameter holes in printed circuit boards for electroless copper plating using a jet pumice machine', *Circ. World*, **15**(1), October (1988)
GOBERECHT, A., 'Blind and buried vias for multilayer PCBs', *Electron. Components Appl.*, **8**(4) (1988)
GOTHARD, A., 'Multilayers—a maturing market', *Electron. Manuf. Test*, April (1988)
HROUNDAS, G., 'Eliminate bare-PCB test damage with strict parameter control', *Test Measurement World*, December (1987)
MAYNARD, B., 'Reliable plated through holes for rigid flex boards', *Electron. Pack. Prod.*, **28**(9), September (1988)
MOSLEY, J. D., 'Multilayer backplanes require careful design specs', *EDN*, 7 July (1988)
NUZZI, F. J., 'Full build electroless copper', *Print. Circ. Fab.*, **11**(9), September (1988)

20

Power Sources

F F Mazda
DFH, MPhil, CEng, FIEE, DMS, MBIM
STC Telecommunications Ltd
(Sections 20.1, 20.2.6, 20.3.1, 20.3.3, 20.5, 20.6)

C J Bowry
Ever Ready Ltd
(Sections 20.2.1, 20.2.2)

C E Longhurst
Duracell Batteries Ltd
(Sections 20.2.3, 20.2.4)

M Ewing
SAFT (UK) Ltd
(Section 20.2.5)

P H Hitchcock PhD
Ever Ready Special Batteries Ltd
(Sections 20.3.2, 20.4)

Contents

20.1 Cell characteristics

Power sources or batteries are made from several cells which are connected together to give a higher voltage. There are two types of cells:

(1) A *primary cell* is used once, until it is discharged, and then thrown away. Examples of primary cells described in this chapter are zinc–carbon, zinc chloride, alkaline manganese, mercuric oxide, zinc–lithium and zinc–air.
(2) A *secondary cell* needs to be charged, after it is made, before use. Once discharged the cell can be recharged and used again. Examples of secondary cells are lead–acid, nickel–cadmium and zinc–air.

Several parameters are important in the choice of a cell for a given application. These parameters are usually determined by the electrochemistry of the material used within a cell.

Open circuit voltage This is the voltage at the terminals of the cell. It depends on several factors such as the history of the cell (*Figure 20.1*) and the amount of energy which it has supplied (*Figure 20.2*). Cells are usually designed to have a relatively flat discharge curve until the cell is almost totally discharged, and then the terminal voltage decays rapidly.

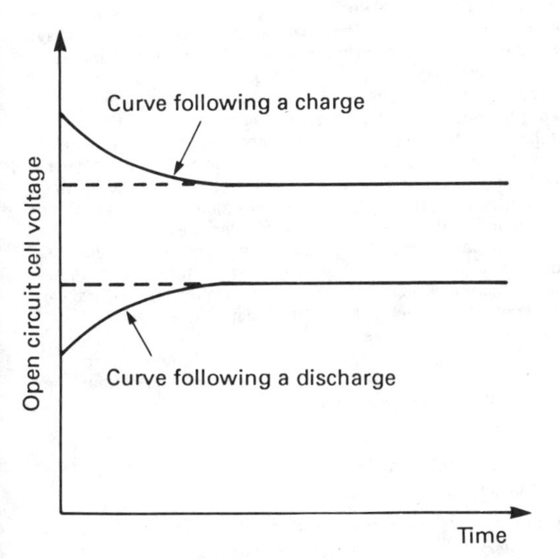

Figure 20.1 Effect of cell history on open circuit cell voltage

Cell capacity This is the amount of energy, usually stated in ampere hours, which the cell can provide without its terminal voltage falling below a given value. This is also specified as its '*C*' rate, which is the rate at which a fully charged cell would be discharged in one hour. So a cell with a capacity of 5 A h has a '*C*' rate of 5 A.

Depth of discharge This is the percentage of the cell's capacity by which it has been discharged. So a 50 A h cell which has been discharged by 10 A h has a depth of discharge of 20%.

Charge retention or shelf life This is a measure of the ability of the cell to maintain its charge when stored. It is affected by the state of charge of the cell and the storage temperature.

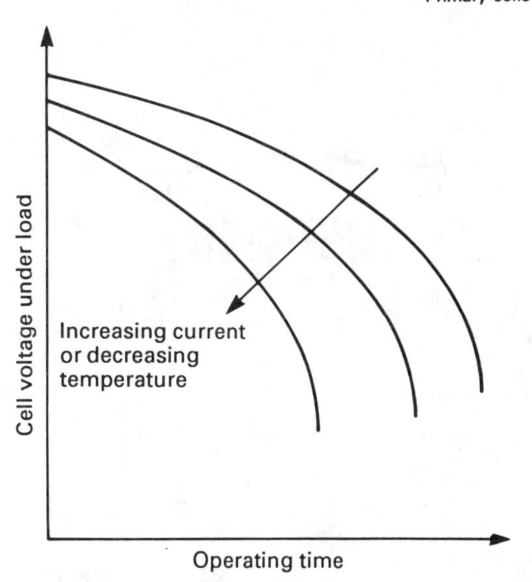

Figure 20.2 Effect of load or temperature on the cell voltage

Charge rate For a secondary cell this is the charging current expressed as a function of the cell's capacity.

Cycle life This is the number of charge–discharge cycles which a secondary cell can go through before failure occurs.

Charge acceptance This is the ability of a secondary cell to accept energy. It is measured as the proportion of the charge input which the cell can give out again, without its voltage falling below a specified value.

Charge voltage This is the voltage developed across a secondary cell when it is under charge. This voltage can be up to 50% higher than the rated discharge voltage of the cell. It increases with the charge rate and at low temperatures.

20.2 Primary cells

20.2.1 The Leclanché (zinc–carbon) cell

The Leclanché (zinc–carbon) cell is available in two distinct forms: the round cell and the layer cell. The former (*Figure 20.3*) is marketed as a single unit or as a multicell battery, whilst the latter (*Figure 20.4*) is sold only as a multicell (layer stack) battery.

20.2.1.1 Theory of operation

The initial open circuit voltage or e.m.f. of a Leclanché cell is the difference between the electrode potentials of a zinc electrode and a manganese dioxide electrode immersed in a solution of battery electrolyte. For a fresh cell it is usually near 1.6 V falling to 1.4–1.2 V for a fully discharged cell.

The internal resistance of a cell is a very complicated property and depends principally on the construction. It is commonly calculated from a measurement of the e.m.f. and the short circuit current read instantaneously on an ammeter whose resistance including leads does not exceed 0.01 Ω:

$$r = \frac{E}{C} - 0.01 \qquad (20.1)$$

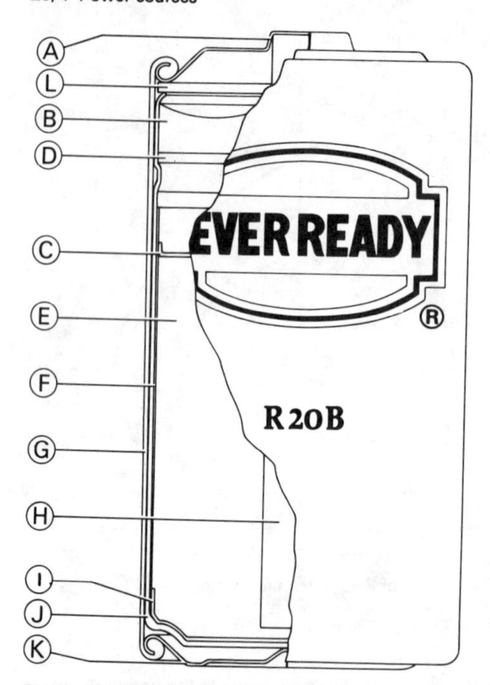

Figure 20.3 The cylindrical type, typified by the R20 battery, of which the main construction elements are: A, metal top cap to secure the best possible electrical contact between cells; B, sub-seal of a soft bitumen compound, applied to seal the cell; C, top washer spacer situated between the depolariser mix and the top collar; D, top collar centralising the carbon rod and supporting the bitumen sub-seal; E, cathode made from thoroughly mixed, high-quality materials, containing manganese dioxide to act as the electrode material and carbon black or graphite for conductivity, with ammonium chloride and zinc chloride which are also necessary ingredients; F, absorbent paper lining impregnated with electrolyte, to act as a separator; G, jacket, outside the cell and carrying the printed design, to resist bulging, breakage and leakage, and hold all the components firmly together; H, the positive pole, a rod made of highly conductive carbon, functioning as a current collector and remaining unaltered by the reactions occurring within the cell; I, bottom washer separating the depolariser from the zinc cup; J, anode of zinc metal, extruded to form a seamless cup which holds all the other constituents, making the article clean, compact and easily portable—when the cell is discharged, part of the cup is consumed to produce electrical energy; K, metal bottom cover made of tin plate, in contact with the bottom of the zinc cup, giving an improved negative contact in the torch or other equipment, and sealing off the cell to increase its leakage resistance; L, insulating washer providing electrical insulation between the zinc cup and top cap, thus preventing the cell short circuiting

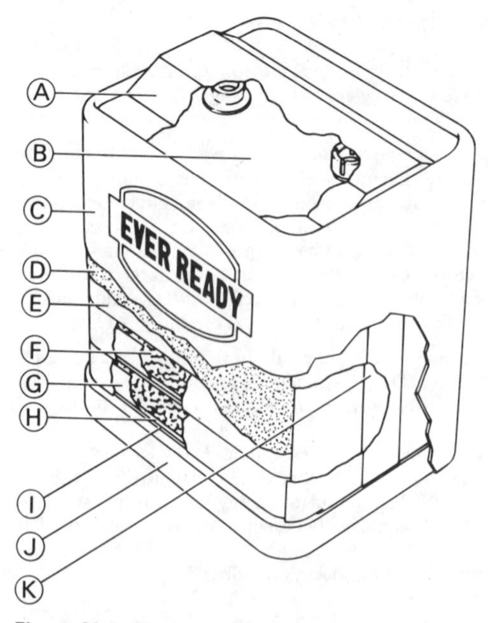

Figure 20.4 The layer cell battery: A, protector card to protect the terminals—torn away before use; B, plastic top plate carrying the snap fastener connectors and closing the top of the battery; C, metal jacket, crimped on to the outside of the battery and carrying the printed design, helping to resist bulging, breakage and leakage, and holding all the components firmly together; D, wax coating sealing any capillary passages between cells and the atmosphere, preventing loss of moisture; E, plastic cell container—a plastic band holding together all the components of a single cell; F, positive electrode—a flat cake containing a mixture of manganese dioxide as the electrode material and carbon black or graphite for conductivity, plus ammonium chloride and zinc chloride, which are also necessary ingredients; G, paper tray acting as a separator between the mix cake and the zinc electrode; H, duplex electrode—a zinc plate coated with a thin layer of highly conductive carbon which is impervious to electrolyte; I, electrolyte-impregnated paper which contains the electrolyte and acts as an additional separator between the mix cake and the zinc; J, plastic bottom plate closing the bottom of the battery; K, conducting strip in contact with the negative zinc plate at the base of the stack and the negative socket at the other end

where
r = internal resistance (in Ω)
E = e.m.f. (in V)
C = short circuit current (in A)

The value obtained is only approximate because it is impossible to obtain a truly instantaneous reading of the current. More accurate measurements can be made with an oscilloscope. The voltage can be measured within a few microseconds of shunting the cell with a known resistance. The internal resistance can then be calculated using the formula:

$$r = \frac{E - V}{V} R \qquad (20.2)$$

where
r = internal resistance (in Ω)
E = e.m.f. (in V)
R = shunt resistance (in Ω)
V = instantaneous voltage with shunt (in V)

Note that the internal resistance of a cell will vary according to the load placed on it; the higher the shunt resistance the higher the internal resistance.

In the Leclanché cell the actual reactions which occur when the cell is discharged through an external circuit are very complicated. *Figure 20.5* represents chemically what happens on discharge.

At the surface of the negative electrode, zinc atoms ionise. The zinc ion goes into solution while electrons travel to the positive electrode through the external load and the carbon rod. In the positive electrode the manganese is reduced from a four valent state in MnO_2 to a trivalent state in MnOOH.

The hydroxyl ions, OH, produced in the cathode reaction and the zinc ions produced in the anode reaction, participate in one of the alternative reactions (a) or (b) (*Figure 20.5*). If reaction (a) takes place ammonium chloride from the electrolyte also takes part. Zinc diammine chloride, $Zn(NH_3)_2Cl_2$, and water are the products. If reactions (b) occurs the trivalent compound of manganese, MnOOH, reacts with zinc and hydroxyl ions to form hetaerolite, $ZnO.Mn_2O_3$, and water. Analyses of discharged cells show that both reactions can occur side by side in one cell

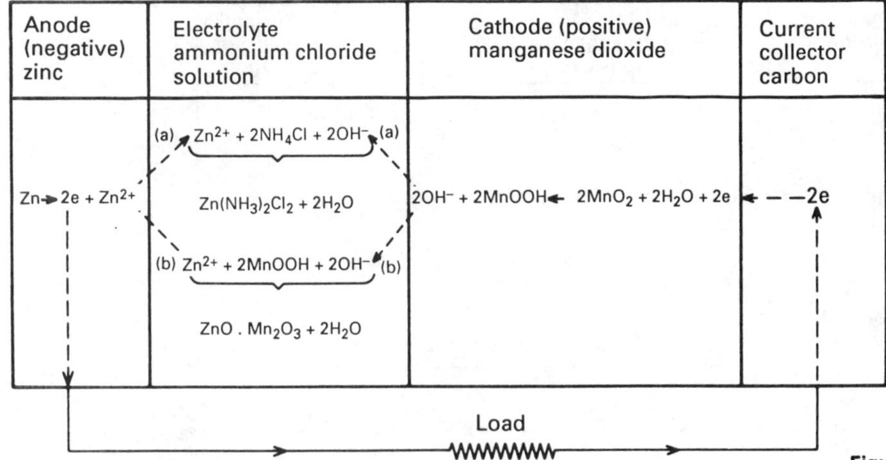

Figure 20.5 Chemical reactions in the Leclanché cell

because both zinc diammine chloride and hetaerolite are found, but reaction (a) usually predominates.There are, however, alternative reactions and the total zinc involved in the reactions is the same as the total zinc produced in the anode reaction. The total water produced by the two reactions is equal to the water used in the cathode process. Consequently, the cell neither gains nor loses water as a result of normal discharge.

If the cell is discharged beyond its useful life, zinc oxychloride, $ZnCl_2.4Zn(OH)_2$, may be formed as a white precipitate. Water is consumed in its formation.

Electroneutrality of the solution is maintained by the process of electrolytic conductivity. Some of the positive ions from the vicinity of the anode migrate towards the cathode and some of the negative ions from the vicinity of the cathode migrate in the opposite direction in such a manner that the sum of negative charges always equals the sum of positive charges in any volume of solution. Since the hydroxyl ions and zinc ions are removed in chemical reactions, most of the conductivity is due to ammonium and chloride ions. Some ammonium chloride is also removed in reaction (a), so excess of the component has to be provided. For convenience it is incorporated in the cathode mix.

20.2.1.2 Performance and use

The quantity of electricity which can be obtained from a Leclanché dry cell depends on a number of factors, of which the more important are:

(1) The physical size of the cell
(2) The rate at which the cell is discharged
(3) The daily duty period
(4) The end-point voltage
(5) The method of construction and skill of the manufacturer
(6) The temperature
(7) The age of the cell

For cells of the same grade, structure and electrochemical system, the quantity of electricity realised is dependent on cell size. *Figure 20.6* shows the effect of the end-point voltage on the service life of an LR14 when discharged through 6.8 Ω for 1 hour per day. If the end point voltage is reduced from 1.2 to 1.1 V an increase of 125% in service life is obtained.

The effect of temperature on the discharge life of a Leclanché cell is complicated. If the discharge is completed over a fairly short period (about 4 weeks), raising the temperature from that of IEC test (20°C) may give significant increases in service life. If full discharge takes much longer, the deleterious effects of high-temperature storage may be greater than the benefits due to increased efficiency of the discharge reactions. At −20°C the Leclanché cell is virtually inoperative and at −10°C little useful service will be obtained except at low current drains. It is advantageous to keep battery operated equipment in a warm environment before use in subzero temperatures.

The Leclanché cell is designed to store satisfactorily at 20°C but storage at lower temperatures (−10°C to +10°C) is beneficial provided the cells are enclosed in sealed containers which should be retained to protect them from condensation when warmed to ambient temperature.

20.2.2 Zinc chloride cell

A variant of the Leclanché cell is the so-called zinc chloride cell which contains little or no ammonium chloride. Such a cell has

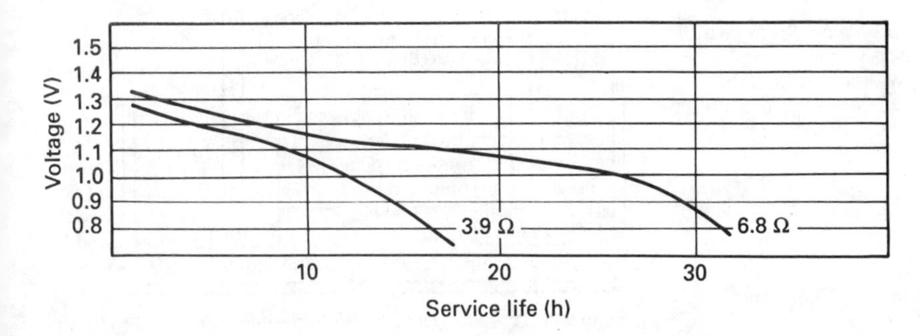

Figure 20.6 Service life curve: discharge period 1 h/day

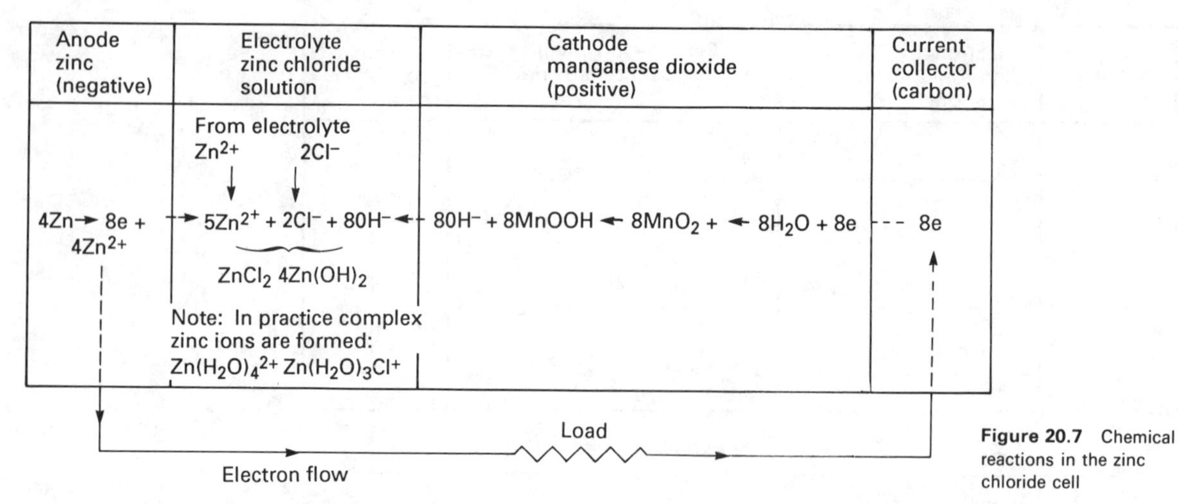

Figure 20.7 Chemical reactions in the zinc chloride cell

a better high-rate discharge capability and has better leakage resistance on long-term, abusive over discharge.

The discharge reactions in a zinc chloride cell are also complex but may be characterised by the schematic diagram of *Figure 20.7*.

It will be noted that in this type of cell, unlike the Leclanché cell, water is consumed during discharge.

Figures 20.8–20.10 show the effect of discharge rate, intermittency of use and temperature on the performance of typical Leclanché and zinc chloride cells.

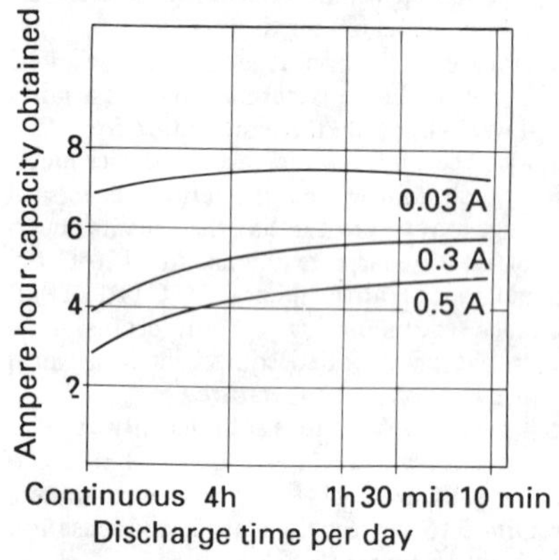

Figure 20.8 Discharge rate and intermittency effect. R20S battery discharged at 0.03 A (radios), 0.3 A (cassette players) and 0.5 A (torches and toys)

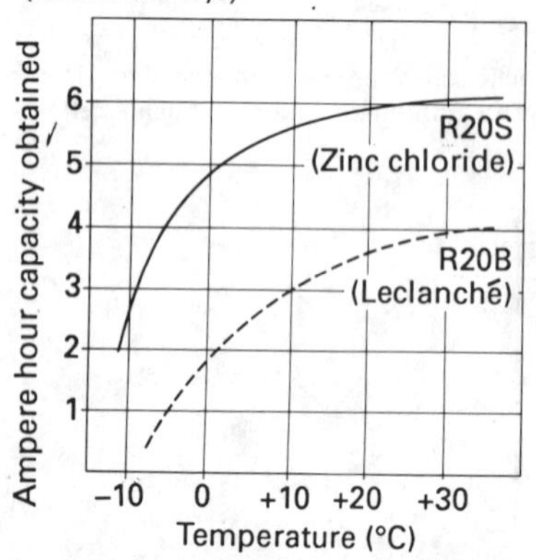

Figure 20.9 Temperature effect on discharge. R20 battery discharged at 0.3 A for 30 min/day to 0.9 V

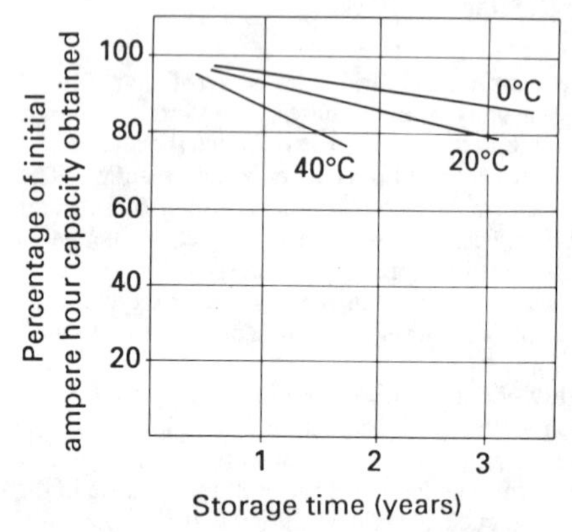

Figure 20.10 Capacity retention on storage. R20 zinc chloride battery discharged at 0.3 A, 30 min/day to 0.9 V

20.2.3 Mercuric oxide cell

20.2.3.1 Theory of operation

In the mercury cell (*Figure 20.11*), the anode is the element zinc (Zn) and the cathode is the compound mercuric oxide (HgO). As current is drawn from the cell the anode is oxidised (that is, the zinc anode attracts the oxygen from the cathode), becoming the compound zinc oxide (ZnO), and the cathode is reduced (or de-

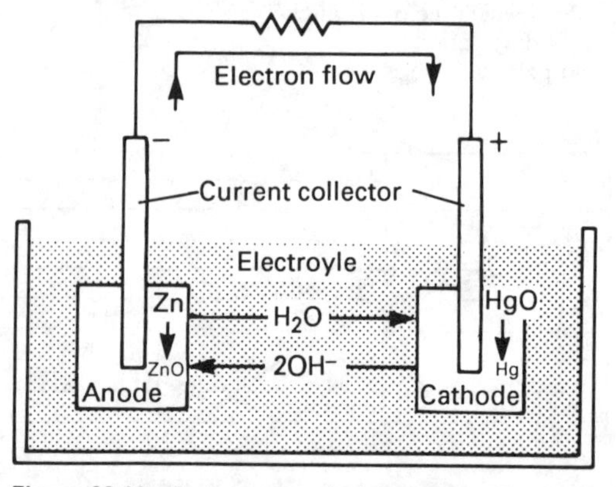

Figure 20.11 Basic reactions in the mercury cell

oxidised), giving up its oxygen to become pure mercury (Hg). Mercury is quite low on the electromotive scale, lying immediately above silver, platinum and gold, and for this reason mercuric oxide will readily yield its oxygen to the zinc.

Each electrode is attached to a current collector, and both are immersed in an electrolyte. When the current collectors of the cell are connected, the strong affinity of the zinc in the anode for the oxygen that is weakly bound to mercury in the cathode is 'felt' over the circuit, and the mercuric oxide begins to break down and give up oxygen. In the reaction that results, the water component of the electrolyte also plays a critical role.

The two electrons are provided by the current collector of the cathode, and attach themselves to the mercury ion, replacing the electrons it donated to its oxygen partner. Thus neutralised, the mercury separates from the oxygen, leaving:

$O^{2-} + (H^+)_2O^{2-}$

Or, two oxygen and hydrogen ions which can be written as:

$2(H^+O^{2-})$

Or, cancelling a *positive* and *negative*:

$2HO^-$

The hydroxyl ion is usually written as:

OH^-

The end-products of the reaction at the cathode are mercury, plus a pair of hydroxyl ions, which are released into the electrolyte. The electrolyte *already* has a plentiful supply of these. Consequently there is a 'pressure' exerted across the electrolyte (the 'domino effect') which makes a corresponding pair of ions available at the anode, to react with the zinc. If the two ions are written separately, the 'confrontation' that now occurs at the anode can be expressed as follows:

$$Zn + \begin{matrix} OH^- \\ OH^- \end{matrix}$$

Given its very strong affinity for oxygen, the zinc will now snatch away one of the two oxygen atoms, and, like mercury, it has two outer electrons for doing this. But the oxygen aleady has two electrons, so to combine with the zinc it must release them. Left behind will be two hydrogen atoms and one oxygen atom, which can then form a simple water molecule. We can now describe the reaction at the anode in simple terms:

$Zn + 2OH^- \rightarrow ZnO + H_2O + 2e^-$ (the two released electrons)

We can do the same for the events at the cathode:

$HgO + H_2O + 2e^-$ ('borrowed' from the anode) $\rightarrow Hg + 2OH^-$

Everything balances. The water molecule used at the cathode is replaced by one produced at the anode, and the two electrons 'borrowed' at the cathode are restored by the anode. The end result is a steady flow of electrons (electricity) from anode to cathode as a consequence of the sophisticated and dynamic electrochemical processes within the power cell. In the simplest of all terms, then:

$Zn + HgO \rightarrow ZnO + Hg$

20.2.3.2 Construction

The effective voltage range is typically 1.3 V down to 1.0 V per cell depending on load and temperature. The mercury system appears in two variants—one with a well defined no-load voltage between 1.35 V and 1.36 V, the second with a voltage which can vary between 1.36 V and 1.55 V. The voltage difference is only important during the first 5–10% of discharge, and only becomes a design consideration when maximum voltage stability is required.

Mercury cells are produced in both cylindrical and button types. Electrochemically both are identical and differ only in can design and internal arrangement. The anode is formed from high purity amalgamated zinc. The cathode is a compressed mercuric oxide/manganese dioxide graphite mixture separated from the anode by an ion permeable barrier. The electrolyte is a solution of an alkali metal hydroxide whose ions act as carriers for the chemical reactions in the cell, but are not part of the reaction. In operation this combination produces metallic mercury which does not inhibit the current flow within the cell. The inside of the cell top is electrochemically compatible with the zinc anode. The cell can is made from nickel plated steel and does not take part in the chemical reaction. *Figure 20.12* illustrates a mercury button cell.

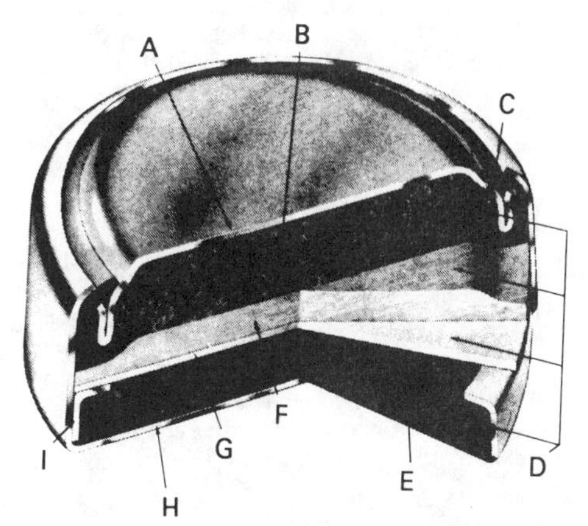

Figure 20.12 Mercury button cell: A, cell top (negative terminal), single type, of steel coated with copper inside and with nickel and gold externally; B, anode of powdered zinc (amalgamated), together with gelled electrolyte; C, nylon grommet coated with sealant to ensure freedom from leakage; D, electrolyte—alkaline solution in anode, cathode and separators; E, cathode of mercuric oxide with graphite, highly compacted; F, absorbent separator of felted fabric (cotton or synthetic), preventing direct contact between anode and cathode and holding the electrolyte; G, barrier separator membrane, permeable to electrolyte but not to dissolved cathode components; H, cell can (positive terminal), nickel or steel coated on both sides with nickel; I, sleeve, nickel-coated steel, supporting grommet pressure and aiding in consolidating the cathode

20.2.3.3 Characteristics

Voltage stability The uniform voltage of the mercury cell is due to the efficient nature of the cathode. Over a long period of discharge a regulation within 1% can be sustained, and for short-term operation higher stability can be achieved. This flat voltage characteristic may be utilised as a reference source or in other applications where voltage stability is essential.

Capacity The ability of the mercury system to withstand both continuous and intermittent discharge with relatively constant ampere hour output allows the capacity rating to be specified. Rest periods as for ordinary cells are not required.

Shelf life The capacity retention of the mercury system is excellent in storage and the voltage characteristic is not affected. Mercury cells can be stored for periods of up to 3 years, according to type. It is not, however, good practice to store cells for unnecessarily long periods.

Mechanical strength The mercury cell can withstand severe vibration, shock and acceleration forces.

Vacuum and pressure High vacuum has no detectable effect and maximum permitted pressure depends on the size of the cell.

Corrosion Mercury cells normally have an excellent resistance to corrosive atmospheres and high relative humidity conditions.

Leakage resistance The use of nickel-plated steel cans and precision moulded seals are part of meticulous design aimed at eliminating leakage under the most adverse conditions.

Temperature range The mercury system provides a stable voltage over the temperature range −30 to +70°C. Special cells based on a wound anode construction are capable of efficient operations at temperatures 15°C below what is possible in standard cells. Although successful performance of some cells above 120°C for short periods has been reported it is recommended that +70°C should not be exceeded.

20.2.4 Alkaline manganese cells

20.2.4.1 Theory of operation

The basic electrode reactions in the alkaline manganese cell are similar to those occurring in the zinc–mercuric oxide cell. The electrochemically active components of the cell are a zinc metal anode, a strong alkaline electrolyte (KOH) and a manganese dioxide (MnO_2) cathode.

The reaction at the zinc anode is an oxidation process as previously described for the zinc–mercuric oxide cell and this can be simply described by the following chemical equation:

$$Zn + 2OH^- \rightarrow ZnO + H_2O + 2e^-$$

The electrode potential for this reaction is approximately −1.35 V (with reference to a standard hydrogen electrode) and remains relatively constant during discharge of the cell.

At the manganese dioxide cathode, the reaction is more complicated and the reduction process occurs in two steps corresponding to $MnO_2 \rightarrow MnO_{1.5}$ (or Mn_2O_3) $\rightarrow MnO$. During this reduction process, the manganese oxide chemical stoichiometry changes in a continuous fashion, i.e. a homogeneous reaction in contrast to the heterogeneous reaction for mercuric oxide being reduced to mercury metal which does not involve intermediate oxide compositions other than HgO. The consequence of this homogeneous reaction is a change of cathode electrode potential during the discharge life.

Initially, the cathode electrode potential is approximately +0.23 V (with reference to a standard hydrogen electrode) and this falls steadily over the first step of discharge to about −0.5 V. The second step occurs at a more constant potential, around −0.5 V, and then falls sharply in the composition region $MnO_{1.2}$ to $MnO_{1.1}$ to below −0.9 V.

The actual chemical species present during the two-step reduction process have been described as:

$$MnO_2 \rightarrow \underset{\text{(amorphous)}}{MnOOH} \rightarrow Mn(OH)_2$$

This implies incorporation of part of the electrolyte into the cathode material during discharge but in simple terms the cathode reaction (first step) can be described by the following chemical equation:

$$2MnO_2 + H_2O + 2e^- \rightarrow 2OH^- + 2MnO_{1.5} \text{ (or } Mn_2O_3\text{)}$$

and the resultant overall cell reaction as:

$$Zn + 2MnO_2 \rightarrow ZnO + 2MnO_{1.5} \text{ (or } Mn_2O_3\text{)}$$

Combination of the two electrode potentials gives an approximate open circuit voltage of 1.58 V.

Most of the useful life of the cell is given by the first step of the manganese dioxide discharge and, by the time this is completed, the cell voltage will have fallen in a fairly continuous manner to 0.85 V. This is in contrast to the relatively constant voltage discharge of the zinc–mercuric oxide cell, which occurs until all of the active components have been used, when the voltage falls sharply to a low value. The rate of fall of voltage during discharge is also dependent on the discharge load and will increase with increasing load (i.e. higher current) due to other factors such as electrode polarisation, as is the case for all cell types.

20.2.4.2 Construction

To achieve performance which is far superior to ordinary zinc–carbon cells in most applications, alkaline manganese cells require higher quality and more expensive materials, and a more sophisticated construction. Here the anode, which does not have to double as part of the cell structure, is formed of zinc powder. The zinc particles are of carefully controlled size, shape, and purity, and are amalgamated (i.e. combined with mercury) to supress gassing and to maximise performance at all discharge rates. The present tendency is to reduce the mercury content of the amalgamation by adding alternative heavy metals in order to maintain gassing suppression.

The alkaline electrolyte which gives this cell its popular designation is a solution of potassium hydroxide (KOH), which is highly conductive. The electrolyte is diffused throughout the powdered zinc and in intimate contact with its granules, ensuring that the anode material is almost completely oxidised by the time the cell's stored energy is exhausted.

Similar design and engineering refinements apply to the cathode of the alkaline manganese cell. The basic cathode material is what is known as *electrolytic manganese dioxide*. This material is produced synthetically by electrolysis. Derived this way it is a much purer oxide than that found in the natural ore, and also has a greater oxygen content per unit volume. This additional oxygen in the cathode material provides increased reactivity and so significantly extends the capacity of the cell.

The supply of reactive oxygen is still further increased because the cathode material is highly compressed, forcing more of it into the available space than it would hold ordinarily. As in the anode, some electrolyte is absorbed into the cathode material during manufacture, assuring good contact with electrolyte throughout the complete cell system. The use of a more conductive electrolyte, and the higher quality of both anode and cathode materials, results in a system that has considerably more usable electrochemical energy stored within it than could possibly be contained in a zinc–carbon cell of a comparable size.

In structural design (*Figure 20.13*) the typical alkaline manganese cell is in many ways the exact opposite of the zinc–carbon configuration. Here the anode is on the inside and the cathode is on the outside, instead of the reverse. The alkaline cell has a central, nail-like anode collector inserted from the bottom, instead of a carbon-rod cathode collector extending down from the top.

In the alkaline manganese system, the entire cell is enclosed in a steel case, which provides a considerably stronger and more secure container than the zinc can of zinc–carbon cells. Just inside the steel case, and in intimate contact with it, is the cathode material. With this arrangement the can becomes the current collector for the cathode, with its positive terminal formed by a protrusion at the top of the case. Lining the thick, cylindrical cathode is a sleeve of absorbent material, which acts as a separator between cathode and anode, as in the zinc–carbon cell. Within the sleeve is the zinc anode. Anode, cathode and separator are all infused with electrolyte for maximum capacity and conductivity.

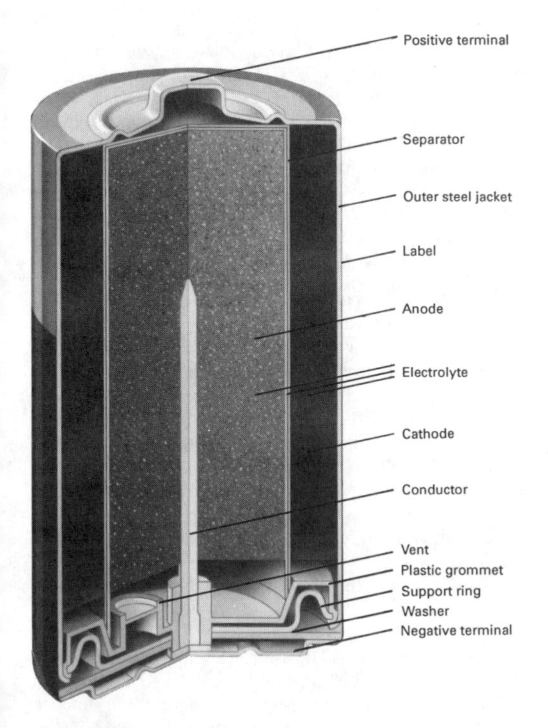

Figure 20.13 Typical alkaline manganese cell

At the core of the cell, in direct contact with the anode, is the 'nail'. This forms the current collector for the anode and is welded to a cap at the bottom of the can to form the negative terminal of the cell. Because of this somewhat 'reverse' arrangement of design in the alkaline cell, the top terminal is positive just as it is in the zinc–carbon cell (because of the carbon rod), and the bottom terminal is negative (provided by the bottom of the can in the zinc–carbon system).

20.2.4.3 Characteristics

Voltage range The operating voltage range is 1.3 to 0.8 V per cell under most conditions of load and temperature. Maximum open circuit voltage is typically 1.56 V per cell. The recommended end voltage for single cell operation at room temperature is 0.8 V increasing to 1 V per cell when six or more series cells are used.

Load currents The alkaline battery excels on continuous heavy loads. There is no distinct upper load limit, and the system is typically capable of supplying intermittent loads up to 2 A at room temperature.

Leakproofness Alkaline cells are leakproof under all normal conditions. The following situations should be avoided whenever possible:

(1) Cell insertion with the wrong polarity
(2) External short circuit
(3) Reverse drive of series cells
(4) Charging

Environmental The alkaline system operates efficiently between −30°C and +70°C subject to load and duty cycle regimes. High relative humidity creates no particular problems, and the battery is tolerant to both high pressure and vacuum.

Storage life Alkaline cells may be stored for prolonged periods at room temperature without significant losses in capacity. Typical capacity retention exceeds 85% after 2½ years storage at 20°C. Short-term exposure to temperatures above 45°C is permissible. Long-term storage at elevated temperatures causes a progressive deterioration in both capacity and high rate properties and should therefore be avoided whenever possible.

20.2.5 Lithium cells

The commercial benefits associated with the utilisation of lithium cells are now being realised in full by many major industrial sectors. From the proliferation of lithium systems which existed at the beginning of the decade a considerable degree of product rationalisation has taken place with many of the early products being either discarded or limited by application and cost effectiveness. The major lithium cell technologies now being produced on a highly industrialised basis are typically 3 V and 1.5 V systems classified by cathode construction, the most important of which are listed in *Table 20.1*.

Table 20.1 Major lithium cell technologies

Electrochemical system	*Cathode*	*Nominal operating voltage* (V)
Lithium–copper oxyphosphate ($Li/Cu_4O(PO_4)_2$)	Solid	2.8
Lithium–carbon monofluoride (Li/CF_x)	Solid	2.8
Lithium–iodine (Li/I_2)	Solid	2.8
Lithium–manganese dioxide (Li/MnO_2)	Solid	3.0
Lithium–sulphur dioxide (Li/SO_2)	Liquid	2.8
Lithium–thionyl chloride ($Li/SOCl_2$)	Liquid	3.4
Lithium–copper oxide (Li/CuO)	Solid	1.5
Lithium–iron disulphide (Li/FeS_2)	Solid	1.5

20.2.5.1 Cell technology and construction

Apart from being the lightest metal known, lithium has the highest electrode potential of any metallic element (3.045 V) with a theoretical electrochemical equivalent capacity of 3860 (A h)/kg.

In general, the chemical energy available in a cell is directly converted into electrical energy through a controlled process represented as the sum of the anode and cathode reaction.

In a lithium cell, the anode (lithium) is the site where oxidation and release of electrons and positive ions occurs simultaneously with the reduction of the cathode (oxidiser) which accepts the electrons and reacts with the positive ions from the anode. This process occurs when the cell, also containing a lithium electrolyte salt, is connected through an external circuit where the electrons flow from anode to cathode allowing ionic transfer through the completed current path inside the cell.

Careful matching with compatible electropositive products has made it possible to develop a number of different lithium couples, although those listed in *Table 20.1* are emerging as favourites for the future.

Lithium is highly reactive with water and must be processed under strictly controlled moisture-free conditions. Whilst it is malleable and easy to manipulate special techniques have had to be developed for it to be precision worked.

Lithium cells feature a high-purity lithium anode, compatible cathode and either non-aqueous organic or inorganic electrolytes. The cells are usually hermetically sealed or crimped to prevent leakage of electrolyte, cathode materials and byproducts precipitated during discharge by the mainstream electrochemical reaction. Hermetic sealing also inhibits moisture ingress and ensures that cells will function safely even if they are subjected to mechanical and electrical abuse.

The cross-sectional arrangements of the three most popular cell constructions are illustrated in *Figure 20.14*. The internal configurations of different lithium cells, apart from possible differences in electrochemistry, influence capacity discharge performance. In cylindrical cells the bobbin construction type consists of an annular core of lithium as the anode inside an outer concentric cathode ring. Interfacial electrode surface area is limited. The spiral wound configuration consists of thinner flat electrodes rolled up in a jelly roll configuration inserted into a can. For a cell of a given size the bobbin construction type contains more active material and so has a greater capacity, whereas the spiral wound construction with its larger interfacial electrode area exposed to the reduction–oxidation process enables the available energy to be delivered at a higher rate.

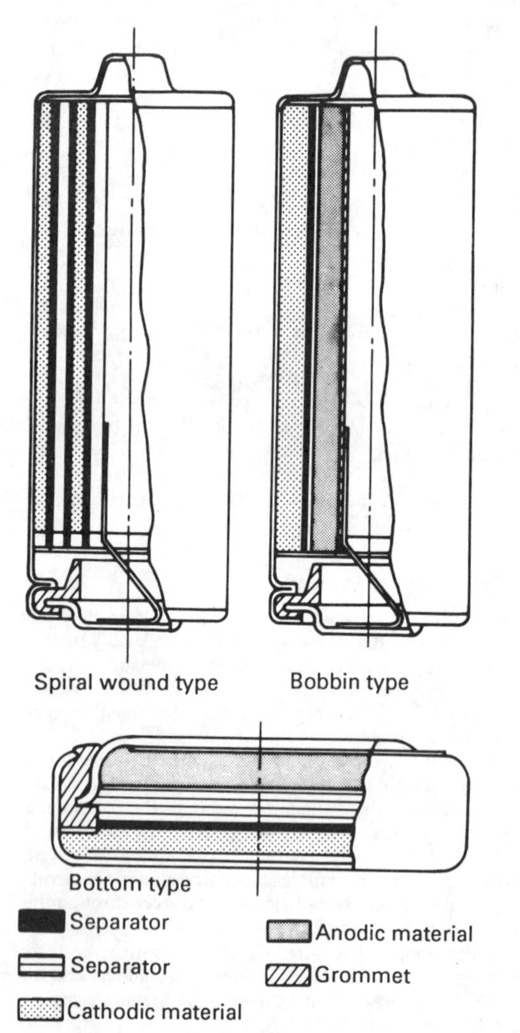

Figure 20.14 Various cell configurations used in lithium primary batteries (Courtesy of SAFT (UK) Ltd)

For exceedingly low-rate, long-life, low-capacity applications (C-MOS RAM memory backup including stand-by, watches and calculators) button cell technology admirably fulfils the requirements, particularly if serious space constraints also have to be considered.

20.2.5.2 Characteristics of lithium cells

Voltage Unlike zinc–carbon and alkaline manganese cells, whose voltage falls progressively during discharge, most lithium cells exhibit stable on-load voltage characteristics for up to 90% of their useful life. Discharge profiles of ten major primary and secondary battery systems illustrate key differences to designers (see *Figure 20.15*). For lithium cells, typified by the lithium–thionyl chloride, lithium–sulphur dioxide and lithium–manganese dioxide systems, the voltage–time profiles are effectively flat and stable, indicating constancy of voltage for virtually their entire service lives.

In those cells with open circuit voltage significantly higher than nominal on-load voltage the initial high voltage can be artificially suppressed by 'burn-in', the use of zener diodes or voltage regulators.

Operating temperature range The use of non-aqueous inorganic and organic electrolytes gives lithium cells their ability to function over extremely wide temperature ranges.

Lithium–sulphur dioxide and lithium–thionyl chloride cells will function efficiently down to −40°C, whilst lithium–copper oxide cells will, in their encapsulated form, operate quite satisfactorily up to 175°C.

The $Li/SOCl_2$ system is capable of discharging well over 60% of its rated capacity at −40°C under specific load conditions, and the $Li/Cu_4O(PO_4)_2$ over 90% of its rated capacity at 175°C under ideal load conditions.

Current drain The ability of a lithium cell to sustain continuous and intermittent discharge is influenced by cathode resistivity, electrolyte, cell construction and subsequent internal impedance growth characteristics. Liquid cathode spiral wound cells with their large electrode surface areas are able to deliver significantly greater currents than their solid cathode lithium cell counterparts.

Shelf life Lithium batteries are being adopted for the same reasons as those which persuaded industry to use alkaline manganese dioxide in preference to traditional zinc–carbon batteries. Systems such as lithium–manganese dioxide, lithium–thionyl chloride, lithium–sulphur dioxide and lithium–carbon monofluoride have proven 10 year plus shelf lives with over 85% retained capacity at room temperature. Accumulated real time and accelerated storage testing completed at ambient and elevated temperatures now conclusively supports the claims of most reputable manufacturers.

Load response time This characteristic, although of little or no significance in low-rate continuous discharge applications, is of considerable importance to users and specifiers of lithium batteries where high-rate performance is required, particularly when the batteries may be required to be stored for up to 10 years at elevated temperatures before use. Applications falling into this category are emergency standby power supplies, military and paramilitary man pack radios, and computer backup.

Solid anode/solid cathode systems tend not to exhibit the voltage delay phenomenon which manifests itself as a significant reduction in voltage upon the application of a high initial load and is particularly pronounced if the cell or battery has been stored for a prolonged period at elevated temperature. However,

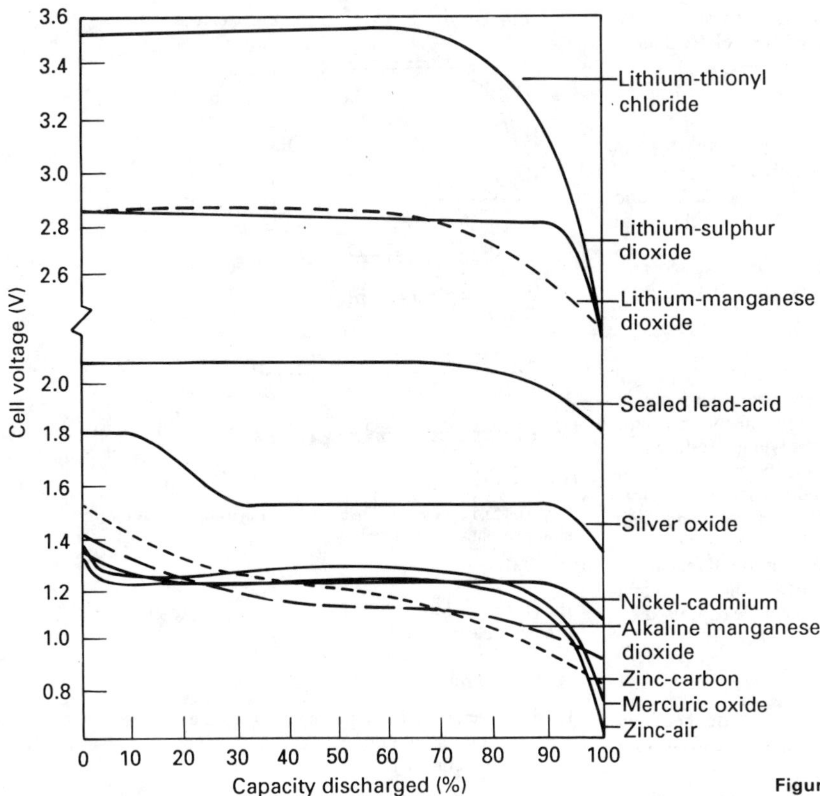

Figure 20.15 Discharge curves for various cells

this is not the case with solid anode/liquid cathode technologies such as lithium–sulphur dioxide (Li/SO_2) and lithium–thionyl chloride ($Li/SOCl_2$). The reformulation of electrolytes and the use of additives progressively continues to represent the two main avenues for voltage delay alleviation by limiting the build-up of the protective film which develops on the anode and which conversely imparts to liquid cathode lithium cells their excellent shelf life properties.

Reliability The nature of the manufacturing processes which have had to be specially developed, the care with which all cell components are selected, and the hospitalised production facilities alll contribute to ensure that cells are less susceptible to failure than any of their predecessors.

High reliability is epitomised by the Li/I_2 and Li/Ag_2CrO_4 batteries used by the medical industry. In over 10 years no reports of Li/Ag_2CrO_4 cells failing prematurely have been received. Reliability levels are greater than 0.7×10^{-8} and increasing.

Safety The lithium battery industry has directed much of its research effort to addressing the question of safety typified by the problems associated with the early designs of lithium–sulphur dioxide cells, and most lithium technologies which are now in use adequately satisfy acknowledged international safety standards. Underwriters Laboratories with mandate to assess electronic components including batteries has not only recognised but granted user replacement status to many manufacturers' lithium cell and battery products, and most equipment manufacturers of renown are now using lithium batteries in many of their consumer products.

20.2.5.3 Theory of operation

Lithium–sulphur dioxide cells (Li/SO_2) These cells are available in spiral form only. Lithium foil, polypropylene separator and teflon bonded carbon cathode pressed on to a support grid are rolled together to give the required active surface area. Sulphur dioxide under pressure acts as the cathode/depolarising agent.

The overall discharge mechanism is as follows:

$2Li + 2SO_2 \rightarrow Li_2S_2O_4$ (lithium dithionite)

Lithium–thionyl chloride cells ($Li/SOCl_2$) Considered by many to be the natural successor to the Li/SO_2 couple the spiral wound version has the highest theoretical energy density and drain capability of any commercially produceable lithium battery.

In place of pressurised SO_2, the cell relies upon $SOCl_2$ which doubles as depolarising agent and cathode.

Several discharge mechanisms have been proposed although it is now considered by most experts that the discharge reaction is adequately described by the equation:

$4Li + 2SOCl_2 \rightarrow S + SO_2 + 4LiCl$

Under normal discharge negligible amounts of gas are generated. Maximum pressure developed under use consistent with Raoult's law predicted to be 3.8 bar (55 psi).

Lithium–manganese dioxide cells (Li/MnO_2) This was the first solid anode/solid cathode 3 V lithium system to be manufactured commercially. The use of inexpensive materials makes it especially suitable for powering low-drain, long-life, consumer applications.

Unlike the Li/SO_2 and $Li/SOCl_2$ cells this couple uses an organic lithium perchlorate dioxolane based electrolyte. Overall discharge reaction can be summarised as follows:

$$Li + MnO_2 \rightarrow LiMnO_2$$

Lithium–copper oxide cells (Li/CuO) One of the first lithium couples to be released onto the market, this was originally developed as a replacement for the traditional zinc–carbon and alkaline manganese systems. It is available in button, bobbin and spiral configurations.

Cells consist of lithium anode, cupric oxide cathode and lithium perchlorate dioxolane based electrolyte. The overall discharge reaction is as follows:

$$2Li + CuO \rightarrow Li_2O + Cu$$

Lithium–carbon monofluoride cells ($LiCF_x$) The lithium anode is composed of flat lithium sheet pressed on to a metal connector such as nickel or stainless steel. The electrolyte is produced by dissolving an alkali metal salt in a non-aqueous organic solvent to produce an electrolyte with ionic conductivity. An organic solvent with high dielectric constant, low viscosity and high boiling point is used for the electrolyte.

The cathode $(CF_x)_n$, namely fluorocarbon, is an intercalation compound produced through the reaction of carbon powder and fluorine gas. The general discharge reaction can be expressed by the following equation:

$$mLi(CF_x)_m \rightarrow mLiF_x + mC$$

Lithium–iodine cells (Li/I_2) As the cell uses solid-state chemistry it is suitable only for low drain devices because of current rate limitations.

The cathode consists of iodine and a complex organic conductive charge transfer complex (CTC) which in contact with lithium produces lithium iodine. Continuous discharge promotes precipitation of lithium iodine increasing internal impedance and diminishing terminal voltage.

The active reaction materials are hermetically sealed in laser welded cases complete with tinned nickel ceramic corrosion resistant terminals to prevent diffusion of moisture and gas into the battery, and leakage.

The overall discharge mechanism is as follows:

$$2Li + CTC\ nI_2 \rightarrow CTC\ (n-1)I_2 + 2LiI$$

The volume of cell products within the cell remain unchanged during discharge. Solid electrolyte with restricted ionic conductivity prevents leakage.

20.2.6 Zinc–air cell

20.2.6.1 Construction

The cell uses zinc as the anode and oxygen from the air as the cathode material. Since the space usually taken by the cathode oxidising agent can now be allocated to the anode this cell has a high energy to weight ratio.

The key to the zinc–air cell is the construction, which lets air into the cell without allowing any of the electrolyte to leak out. *Figure 20.16* shows the construction of a button cell. The anode is the zinc top and the potassium hydroxide electrolyte is contained within this. An insulating gasket separates the anode and cathode. The air cathode arrangement, shown in *Figure 20.16(b)* is made up of a system of separators and is only about 0.5 mm thick. The oxygen from the air combines with the hydroxide from the electrolyte, under the catalytic action of the carbon, to form water. The metallic mesh gives mechanical support and carries

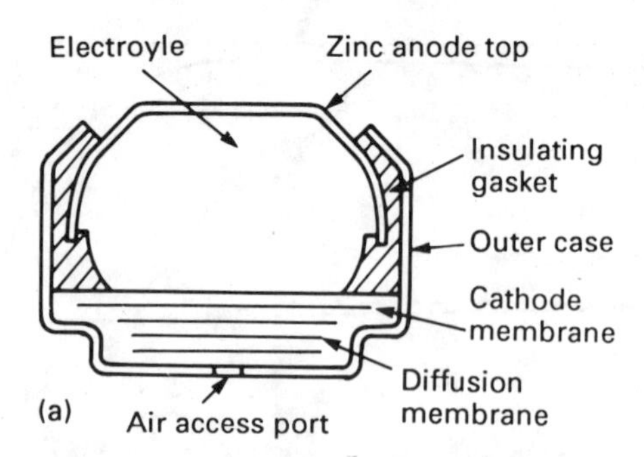

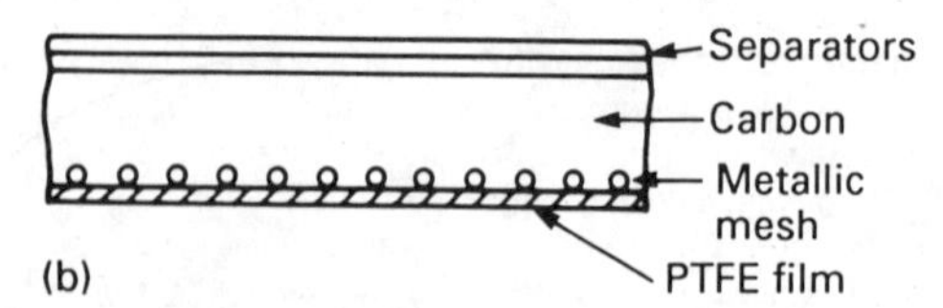

Figure 20.16 Zinc–air cells: (a) button cell; (b) cathode arrangement of button cell

the current, whilst the PTFE film allows air to enter the cell but prevents electrolyte from escaping.

20.2.6.2 Theory of operation

During operation the zinc reacts with the electrolyte leaving electrons on the anode, which give it a negative charge. The chemical reaction is:

$$Zn + 2OH \rightarrow ZnO + H_2O + 2e^-$$

Oxygen from the air combines with the water in the electrolyte, taking electrons from the cathode and replacing the hydroxyl ions lost at the anode. Electrons removed from the cathode leave it positively charged. The reaction is:

$$\tfrac{1}{2}(O_2) + H_2O + 2e^- \rightarrow 2OH^-$$

The current rating of the cell is determined by the rate of air flow and by the cathode surface area. The ampere hour rating depends on the weight of the zinc anode. The cell will be exhausted when all the zinc material is used up.

20.2.6.3 Characteristics

The zinc–air cell has a no load voltage of 1.4 V, with a higher energy to weight ratio, and a higher current than the alkaline or mercuric oxide cells. The zinc–air cell is used in applications which need to work at high currents for long periods.

The internal resistance of the zinc–air cell is low and this resistance is primarily determined by the oxygen diffusion rate. The cell can operate over a wide temperature range of −40°C to +60°C.

The shelf life of a zinc–air cell is very good, the average loss of capacity during storage being about 2% per year. This is because one of the cell's reactants is oxygen, and this can be excluded during storage by airtight wrapping around the cell.

The zinc–air cell is inherently safe in operation since any gasses developed can escape via the air diffusion path. The short circuit current of the cell is limited by the rate at which the cell can absorb oxygen.

The main problem with the zinc–air cell is that it is affected by atmospheric conditions. The water vapour pressure in the cell is equivalent to 55% relative humidity at 20°C, so on wet days the cell will gain moisture and on dry days it will lose moisture. This affects the aqueous electrolyte, which is usually 30% potassium hydroxide, so the cell can fail if operated for long periods at extremes of atmospheric conditions. The carbon dioxide from the air also reacts with the electrolyte to form potassium carbonate, and this increases the internal resistance of the cell.

20.3 Secondary cells

20.3.1 Lead–acid cells

20.3.1.1 Construction

Although the lead–acid cell was developed by Gaston Plante in 1860 it was not until much later, when it was adopted by the automobile industry, that it gained in popularity. The open type of construction used in automobiles is not suitable for use in electronic equipment since the cell must be mounted upright, there is risk of spillage, and the cell needs frequent topping up.

For electronic applications a gelled electrolyte cell is used. This is sealed, does not need topping up with electrolyte and can be mounted in any position. In this type of construction both plates of the cell are made from lead in the form of a grid. The positive plate is filled with lead dioxide and the negative plate with spongy lead. The plates are formed into thin metal sheets and are interleaved with layers of porous fibreglass separators and wound into a cylinder. This is sealed in a chemically stable polypropylene case and then put into a metal case for strength.

To achieve a low internal resistance, low polarisation and long life the material of the metal plates is kept close to the surface by using thin plates and a spiral wound construction. The separator electrically separates the positive and negative plates and holds the electrolyte, distributing it over the working surface.

The electrolyte is a dilute solution of sulphuric acid. The quantity of electrolyte is such that it is retained by the plates and separators and none of it is free to leak. This gives the cell good gas diffusion and oxygen recombination, and avoids large gas pressures during overcharging. A safety vent is provided in the cell case to allow for the escape of gas.

20.3.1.2 Theory of operation

During the discharge period the spongy lead in the negative plate, and the lead dioxide in the positive plate, react with the sulphuric acid to give lead sulphate crystals and water. The lead sulphate crystals grow on the lead dioxide and if they are excessive the operation of the cell will slow down. Charging reverses the process. The equations are as follows:

$$PbO_2 + Pb + 2H_2SO_4 \rightleftarrows 2PbSO_4 + 2H_2O$$

During overcharge the charging current electrolyses the water in the electrolyte and forms oxygen and hydrogen. This can lead to a build-up of gas pressure.

20.3.1.3 Characteristics

Voltage The nominal cell voltage is 2.1 V. The voltage droops during discharge due to loss in the internal resistance of the cell.

Current capacity The lead–acid cell can withstand high charge and discharge rates, and is specially good on pulsed operation where it can deliver large currents for a short time. During rest periods acid diffuses from the separator back to the working areas of the plates, and allows a greater working capacity. There is a drop in capacity at high discharge currents because of insuffficient ion diffusion caused by the depletion of electrolyte near the active material of the plates. This effect is called concentration polarisation.

Temperature range The cell operates over a temperature range of −60°C to +60°C although the optimum operating temperature is +20°C. The capacity and terminal voltage of the cell decrease at low temperatures due to a reduction of the ionic diffusion rate.

Shelf life The shelf life of the cell is reduced by internal electrochemical discharge, which is worse at high temperatures. The loss of charge per day varies from 0.01% for a nearly discharged cell at 0°C to 2% for a fully charged cell at 45°C.

Chemically the effects are similar during self-discharge as that which occurs during normal operation. However, the lead sulphate crystals are now large and completely surround the active plate material. This is known as sulphation and if it is allowed to continue it can prevent the cell from accepting charge. To avoid sulphation the lead–acid cell should be stored at low temperature and recharged at periodic intervals.

20.3.2 Nickel–cadmium cells

20.3.2.1 Construction

There are two types of construction which are of interest to the electronics engineer: the cylindrical cell and the button cell. Before we consider the two types in turn we should summarise a few important differences. Cylindrical cells can be made larger, they have pressure-release safety vents and can give very high discharge currents; button cells are by comparison more compact, give lower currents, but have the ability to hold charge for longer periods. *Figures 20.17* and *20.18* illustrate typical constructions.

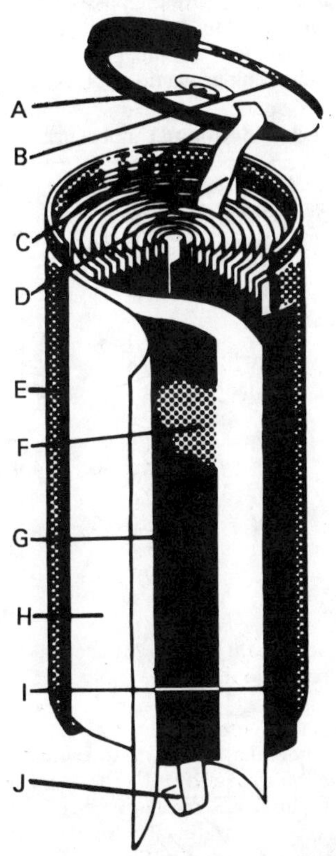

Figure 20.17 Typical cylindrical nickel–cadmium cell construction: A, resealing safety vent; B, nickel-plated steel top plate (positive); C, nylon sealing grommet; D, positive connectors; E, nickel-plated steel can (negative); F, support; G, sintered negative electrode; H, separator; I, sintered positive electrode; J, negative connector

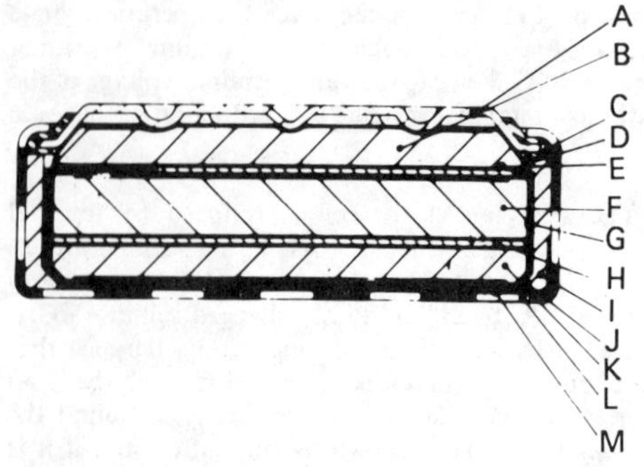

Figure 20.18 Typical button nickel–cadmium cell construction: A, cell lid; B, negative electrode (A); C, grommet; D, negative connecting strip; E, negative electrode wrappers (2 off); F, positive electrode; G, positive electrode wrapper; H, separator; I, insulating sleeve; J, insulating cup; K, negative electrode (B); L, cell case; M, positive connecting strip

Cylindrical cells are generally made to comply with the standards BS 5932:1980 and IEC 285:1983. *Table 20.2* shows some sizes available. The dimensions given include plastic insulating sleeves. Three configurations are specified, in relation to solder tags, as shown in *Figure 20.19* and HB is the most commonly used. It is strongly recommended to solder the cells into the circuits especially as high currents are often required, or perhaps the reliability of electrical contact after years on standby charging. Solder tags are generally fitted at no extra cost. Soldering directly to cell cases or top caps is likely to damage the plastic insulating materials inside the cell with the possibility of internal short circuits at the time or during subsequent service.

Cells or batteries can be encapsulated in resin. This is often required, for example, for mechanical protection for rugged applications, or electrical and thermal insulation for 'intrinsically safe' or similar uses. It is important that steps are taken to avoid sealing up the safety vent on the cells. Various techniques are employed, such as drilling small holes down to voids round the vents, or using a foam with interconnecting pores as the encapsulant. Any external resistors introduced for intrinsic safety reasons can also be housed conveniently within the encapsulation. Care must be taken to avoid overheating of encapsulated cells during discharge or overcharge.

Figure 20.19 Solder tag styles

20.3.2.2 Theory of operation

We must consider the functions of the four basic components of cells: the positive electrode, the negative electrode, the separator and the electrolyte. During charging the chemical compositions of the two electrodes are progressively changed and during discharge the process is reversed, the alkaline electrolyte allowing the transport of charge through the porous separator.

The 'active materials' in the uncharged electrodes consist of nickel hydroxide in the positive and cadmium hydroxide in the negative and these react in the following way during the passage of the charging and then discharging currents:

$$\underset{\text{discharged}}{2Ni(OH)_2 + Cd(OH)_2} \rightleftarrows \underset{\text{charged}}{Cd + 2NiOOH + 2H_2O}$$

Table 20.2 Nickel–cadmium cylindrical sealed cells

IEC designation	*Size (ANSI)*	*Nominal capacity* (A h)	*Max. dimensions* (mm) A	B	C	*Approx. wt* (g)	*Charge rate for* 16 h (mA)	*Typical internal resistance* (mΩ)
KR 11/45	AAA	0.18	43.0	10.5	44.5	10.0	18	80
KR 15/18	⅓AA	0.11	16.1	14.1	17.0	8.0	12	43
KR 15/29	½AA	0.24	27.1	14.3	28.1	14.0	24	39
KR 16/29	½A	0.45	27.1	16.7	28.1	19.0	45	24
KR 15/51	AA	0.50	49.3	14.3	50.3	25.0	50	26
KR 17/51	super AA	0.60	49.0	15.6	50.0	30.0	60	25
KR 23/43	RR	1.40	41.0	22.6	42.6	50.0	140	16
KR 27/50	C	2.20	46.4	26.0	49.0	70.0	220	14
KR 35/44	½D	2.60	42.6	32.5	43.7	100.0	260	11
KR 35/62	D	4.00	58.0	32.5	61.3	140.0	400	6
KR 35/62	D	4.50	58.4	33.8	61.0	150.0	450	6
KR 35/92	F	7.00	89.7	33.8	91.0	225.0	770	5
KR 44/91	super F	10.00	89.9	41.5	91.0	345.0	1000	3

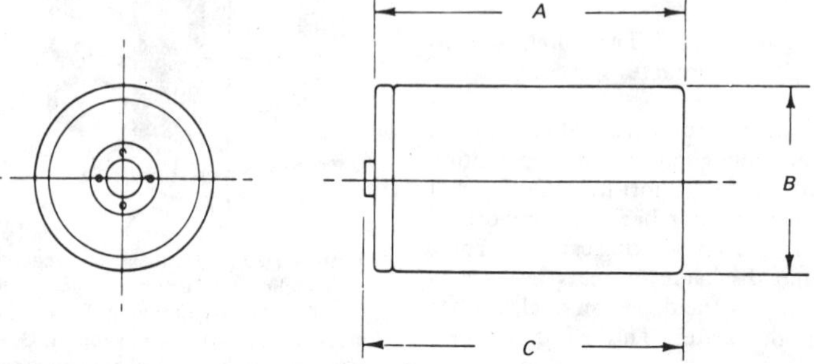

In order to understand how the maintenance-free, completely sealed system is achieved we must consider what happens on overcharge and in particular the oxygen recombination reaction, which enables the necessarily evolved oxygen gas to be continuously absorbed and reused inside the cell in accordance with the chemical equations:

$$O_2 + 2H_2O + 2Cd \rightarrow 2Cd(OH)_2$$

Thus, during overcharge, parts of the negative electrode are being continuously charged to metallic cadmium and discharged back again to cadmium hydroxide.

The oxygen is given off at the overcharged positive nickel electrode and passes through the fine porous separator and is very quickly absorbed at the cadmium negative electrode. To speed this reaction within the cell the two electrodes are mounted in close proximity to each other, insulated only by the thin layer of separator. The cell is designed so that the negative electrode is electrically bigger than the positive electrode to avoid its becoming fully charged and therefore it cannot be overcharged and cannot evolve hydrogen. To absorb the energy of overcharge at least one of these two gases must be produced and as hydrogen cannot be recombined and oxygen easily can, a convenient and reliable cell system presents itself.

20.3.2.3 Characteristics of cylindrical cells

Capacity The ampere hour capacity of a cell is defined as the product of the discharge current and the time in hours for which it can be drawn. It is somewhat dependent on the actual rate of discharge and it is common commercial practice to give the figure at the 5 hour rate, the end of discharge being taken as 1 V per cell. It is a convenient general practice to use common terminology for currents for all sizes of cell or battery. We take the current expected to discharge the cell to 1.0 V in 1 hour as the 'C' rate. It is obviously equal arithmetically to the ampere hour capacity of the cell. Other currents of charge and discharge are given as multiples and sub-multiples of this. Thus the $C/5$ rate will discharge in 5 hours and the $5C$ rate in 12 minutes. *Figure 20.20* illustrates this by showing the voltage–time discharge curves for $C/5$, $C/1$ and $5C$ rates and it will be seen that the capacity obtained decreases somewhat with increasing current. It will also be seen from this figure that the voltage curve is very flat for a large part of the time. This is a consequence of the low and stable internal resistance and these cylindrical cells are particularly useful when large discharge currents are required. In many applications differing discharge currents are drawn in sequence and to a very good approximation the net capacity taken can be obtained by adding up all the individual current × time pulses. All these cylindrical cells can be charged indefinitely at the $C/10$ rate regardless of their initial state of charge. It should be noted, however, that to allow for a charge efficiency factor of about 1.4, a 'flat' or discharged cell will need at least 14 hours on charge at this '10 hour' rate to attain full capacity. By way of an example, consider a cylindrical nickel–cadmium cell of the 'AA' or 'penlight' size. It has a capacity of about 500 mA h, can be left permanently on charge at 50 mA and its internal resistance is such that it can deliver 10 A for 30 s, 5 A for 3 min, 500 mA for 1 h or 50 mA for 10 h. For many applications it can replace a penlight R6 zinc–carbon or LR6 alkaline manganese battery.

Nickel–cadmium 'secondary' cells can deliver their full capacity in one continuous discharge without disadvantage as no time is needed for 'depolarising'. Because of this it is very difficult to give a meaningful comparison of the ampere hour capacity per cycle of nickel–cadmium cells and their primary cell equivalents. As the latter give their optimum performance only with intermittent use, and the former can be recharged many times, most manufacturers quote expected cycle lives of 300–1000 recharges, at normal temperatures and conditions.

Safety These cylindrical cells are fitted with a resealing vent that relieves any excess internal pressure caused by abuse. It opens, typically, at 14 atm and closes at 12 atm. Abuse conditions could be overcharging at too high a current, or reverse charging. Batteries should not be disposed of by incineration as this may cause a rapid pressure rise. To facilitate the internal chemical reactions only a small amount of electrolyte is present and the cells are said to be 'starved' of this component. Thus on venting only a small amount of gas should be released. A full guide to the safe operation of all types of nickel–cadmium cell is given in British Standard BS 6132:1983.

Temperature Nickel–cadmium batteries give their optimum preformance at about 25°C and *Figure 20.21* shows the deviations to be expected at other temperatures. Note that there

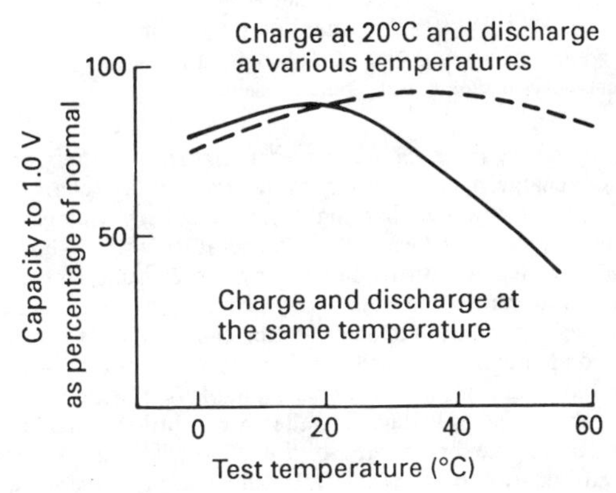

Figure 20.21 Capacity variation with temperature of nickel–cadmium cells

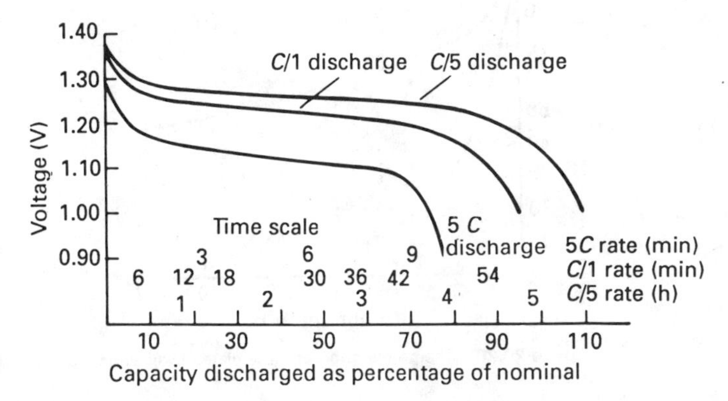

Figure 20.20 Discharge voltage curves of nickel–cadmium cells at 20°C

are two curves. The broken curve gives the changes in capacity, in this case after charging at the $C/10$ rate for 14 hours at a 'room temperature' of 20°C and discharging, after conditioning at the test temperature. The full curve is obtained when one charges and discharges at the test temperatures. It can be inferred from this graph that the efficiency of the charging process improves at lower temperatures but can fall off markedly at elevated temperatures.

Shelf life Batteries are by nature chemical entities and are thus affected by temperature, and an important temperature effect is the influence on the ability of the cell to hold charge during storage. All secondary batteries gradually lose their charge and the nickel–cadmium cell is no exception although button cells are better in this respect than cylindrical cells. *Figure 20.22* shows the big effect that temperature has on this property.

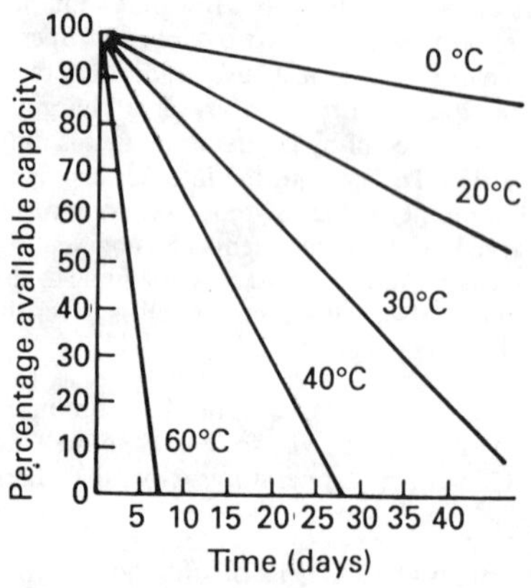

Figure 20.22 Charge retention as a function of temperature for cylindrical nickel–cadmium cells

Effects of temperature on charge characteristics When on continuous charge, once the battery has achieved its fully charged state all the energy (i.e. charging current × battery voltage = power) is converted into heat and the temperature will rise above ambient to a degree largely dictated by the amount of heat insulation and ventilation round the battery. An undesirable effect of this is the amount of deterioration caused in the various plastics components of the cells, and in recent years cells with more stable separators have become available (see below).

Whilst the discharge voltages are altered very little by changes in temperature, the on-charge voltage of a cell is markedly influenced as shown in *Figure 20.23*. It will be seen that there is a decrease with an increase in temperature and the slope of the line is -4 mV/°C. It is largely due to the influence of this phenomenon that sealed nickel–cadmium batteries are almost invariably charged from a constant current source. Control of charge current by cell voltage, or parallel charging, can lead to unstable conditions when the battery temperature rises with a falling voltage and more current passes with consequent further temperature rise and so on. It is possible for batteries, or parts of batteries, in parallel systems to be completely destroyed by this sequence of events. The process is called 'thermal runaway'. A clear distinction must be made in this respect between 'open' or 'vented' cells and the 'sealed' cells being considered here. This unstable condition is much more likely to arise with the sealed cells; hence the recommendation for constant current charging, and never connecting sealed cells in parallel.

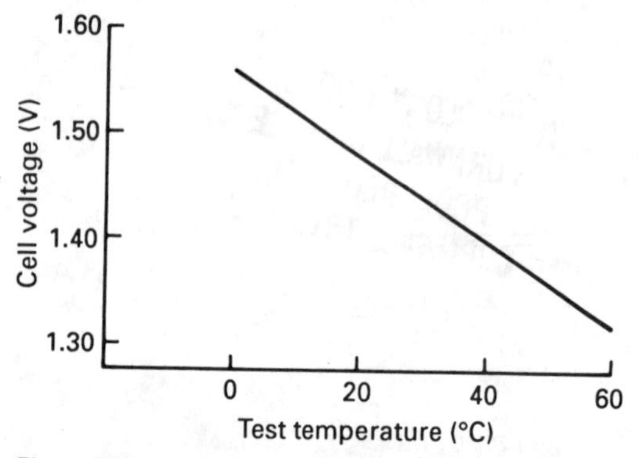

Figure 20.23 Charge–voltage variation with temperature

Internal resistance Typical values are shown in *Table 20.2*. The very low internal resistances of these cells is illustrated by the chart (*Figure 20.24*) of short circuit currents of various cylindrical cells. These currents do not harm the cells for short periods but are not normally used as it would be a very inefficient use of the cells as little external work is being done.

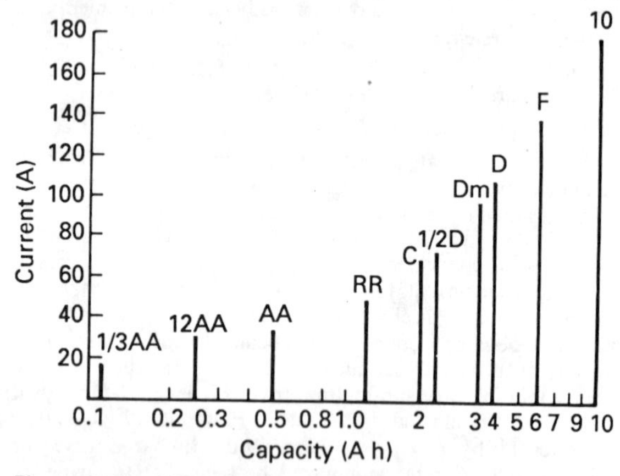

Figure 20.24 Short circuit currents for various cylindrical cells

20.3.2.4 Characteristics of button cells

Closely related as they are to cylindrical cells, sealed nickel–cadmium button cells have features which make them especially suitable for certain applications. Their capacities range from 30 mA h to 600 mA h and details of some sizes available are given

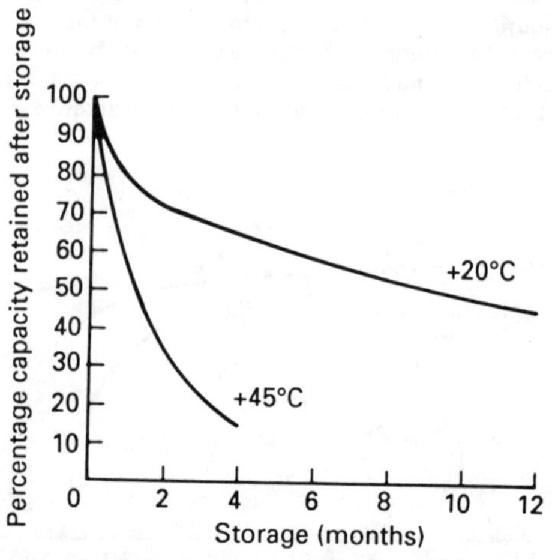

Figure 20.25 Charge retention of mass plate nickel–cadmium button cells

Table 20.3 Nickel–cadmium button cells

IEC designation	*Reference*	*Capacity* (mA h)	*Voltage* (V)	*Maximum diameter* (mm)	*Maximum thickness* (mm)	*Approx. weight* (g)	*C/10 charge rate* (mA)	*Typical internal resistance* (mΩ)
KB 12/6	NCB 3 Z	30	1.2	12	5.6	2	3	700
KB 16/7	NCB 6 Z	60	1.2	16	6.3	4	6	300
KB 23/6	NCB 11 Z	110	1.2	23	4.9	6	11	170
KB 26/7	NCB 18 Z	180	1.2	26	6.0	9	18	120
KB 26/10	NCB 28 Z	280	1.2	26	9.7	13.5	28	120
KB 26/10	NCB 28 D	280	1.2	26	9.7	13.5	28	80
KB 35/11	NCB 60 Z	600	1.2	35.1	10.5	30	60	70
KB 35/11	NCB 60 V	600	1.2	35.1	10.5	30.5	60	30

in *Table 20.3*. Some cells have electrodes constructed on the so-called mass plate systems and these are characterised by their low rate of self-discharge. *Figure 20.25* illustrates this for a commercially available range (see also *Figure 20.22*).

Various combinations of electrodes are used. Some cells have one positive and one negative electrode (Z in *Table 20.3*); some one positive and two negative (D types); and some two positive and two negative (V types). As will be seen, extra electrodes lower the internal resistance and that has a marked influence on the maximum pulse and maximum discharge currents which can be drawn (see *Tables 20.3* and *20.4* and *Figure 20.26*). It will also be seen from *Figure 20.26* that extra electrodes mean that the capacity given is less dependent on the discharge current.

Table 20.4 Maximum discharge currents of some button nickel–cadmium cells

Reference	*Maximum continuous current* (mA)	*Maximum 2 s pulse* (A)
NCB 3 Z	40	0.3
NCB 6 Z	75	0.7
NCB 11Z	165	1.5
NCB 18Z	250	2.0
NCB 28Z	250	2.0
NCB 28D	500	3.0
NCB 60Z	600	4.0
NCB 60V	1500	8.0

The trickle-charge current is less than that possible for cylindricals and should be limited to $C/100$. Thus for the 60 mA h size the maximum continuous current is 0.6 mA. The charge efficiency is very good and even this small current will keep the battery fully charged and the capacity can be expected to be still over 60% of nominal after four years overcharge at normal temperatures.

As with cylindricals, cell temperature affects the capacity available and this is shown in *Figure 20.27*. The optimum performance is obtained at normal temperatures and again the comparison is made between the two and three electrode types.

20.3.3 Silver–zinc cells

The silver–zinc cell is made in a sealed button form as in *Figure 20.28*. The positive electrode consists of a silver mesh which is coated with silver oxide. The negative electrode uses a perforated metal plate, made from silver or silver-coated copper, which is then covered with zinc oxide. The electrodes are saturated with potassium hydroxide which acts as the electrolyte.

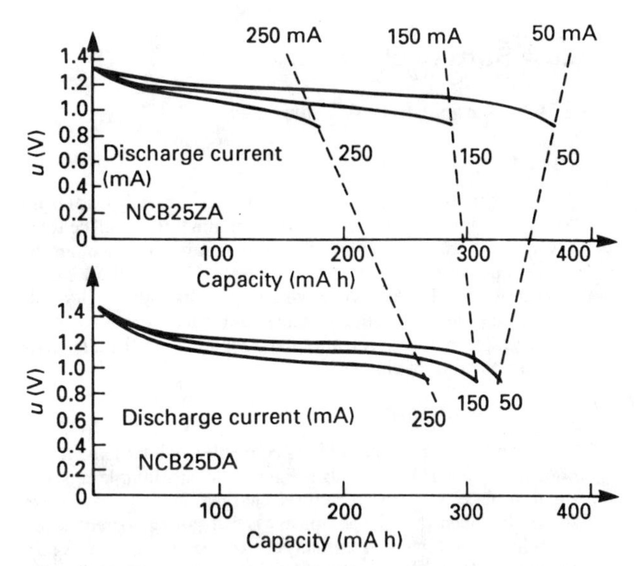

Figure 20.26 Comparison of some button nickel–cadmium cells

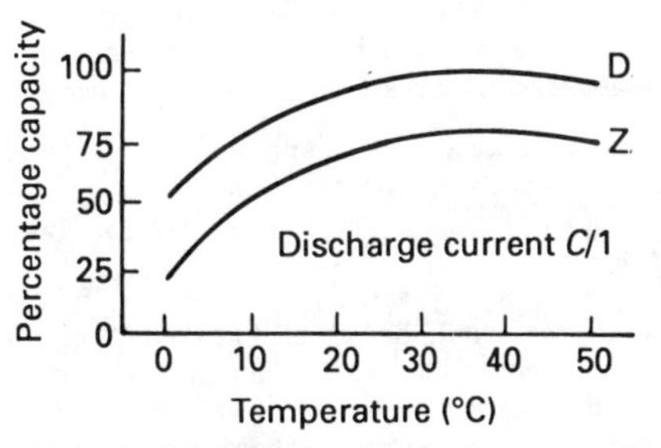

Figure 20.27 Capacity variation with temperature for two button nickel–cadmium cells

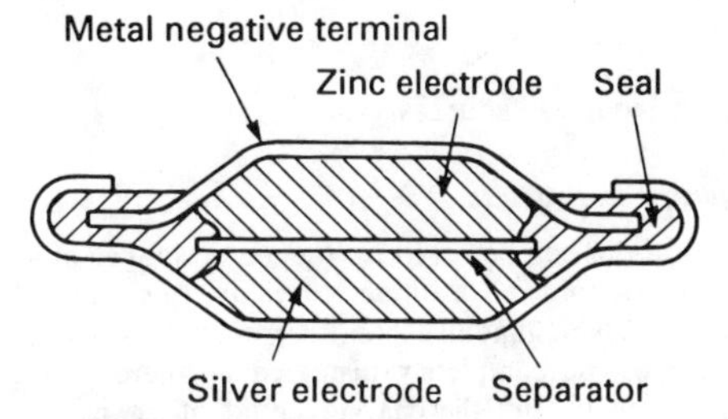

Figure 20.28 The silver–zinc cell

The separator consists of cellulose material. It should not become hydrated by the electrolyte as this will cause it to expand and press against the zinc anode. The cellulose material can also oxidise and degenerate. Several new types of material are being developed for the separator.

Silver–zinc cells have more than double the capacity of nickel–cadmium cells of the same size. Their terminal voltage is 1.5 V and the discharge curve is almost flat. During charging the terminal voltage rises to 1.85 V when the cell reaches 95% of its charge and this serves as a control mechanism for the charging circuit.

The silver–zinc cell has a charge efficiency of 95%, and this is useful when it is being charged from low-energy sources such as solar cells. The leakage current of silver–zinc cells is also low. The disadvantage of the cell is that it is relatively expensive, primarily due to the cost of the silver, and currently available cells have a low charge–discharge cycle life of about 100 cycles.

20.4 Battery chargers

20.4.1 Charging from d.c. sources

20.4.1.1 Dry cells

This is the simplest form of charging and can be merely leads from a bank of dry cells, a technique used sometimes for small cells for toy applications, when connection to the mains supply might be undesirable for both safety and cost reasons. The method is only suitable for small batteries of low-capacity cells and uses dry cells in series/parallel connection having about twice the voltage of the rechargeable battery. The internal resistance of the dry cells controls the rate of charge.

20.4.1.2 Vehicle batteries

Charging from vehicle electrics is a convenient method and this is often accomplished via the cigar-lighter socket. Hand lamps and small transceivers are sometimes charged in this way. Care should be taken that the equipment is not taking current when the vehicle is not in use as this may drain the main battery. Again the battery being charged should have about half the voltage of the supply, but here a resistance is essential in series with the battery.

20.4.1.3 Solar power chargers

A single solar cell can produce a voltage of up to about 0.5 V and will deliver a current dependent on the Sun's intensity (the 'insolation'). Thus, for a given insolation, they can be regarded as constant current devices.

However, the current delivered varies during the day and the maximum in clear weather would be expected at noon. Some of the time these currents could be greater than that needed or recommended for the battery being charged. Under these circumstances it is advisable to control the charge current with a simple constant current circuit, eliminating the high current peaks and ensuring more accurate and evenly spread charging.

20.4.2 Charging from a.c. sources

20.4.2.1 Transformer-less circuits

In its simplest form this can be based on the a.c. reactance of a non-electrolytic capacitor, and a diode to rectify. A circuit is given in *Figure 20.29*. Note that all current-carrying parts of this type of charger must be completely insulated as there is no isolation from the mains and the charger could otherwise be dangerous.

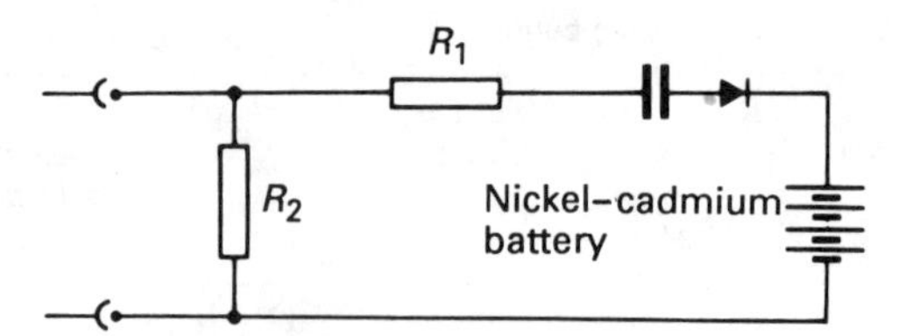

Figure 20.29 A transformer-less circuit for charging small batteries

Capacitor charging circuits are usually found in appliances using either button cells or small cylindricals as their motive force; e.g. electric razors, toothbrushes, etc., which can have the completely isolated circuitry.

20.4.2.2 Transformer circuits

A step-down transformer, diode and resistance are the bare essentials of a simple safe charger (see *Figure 20.30*). It is common practice to choose a transformer with a secondary r.m.s. voltage approximately twice that of the battery to be charged. This is to produce a reasonably constant current supply. If all measurements are made using r.m.s. values then calculation of the resistance R for a charge current I is given by Ohm's law:

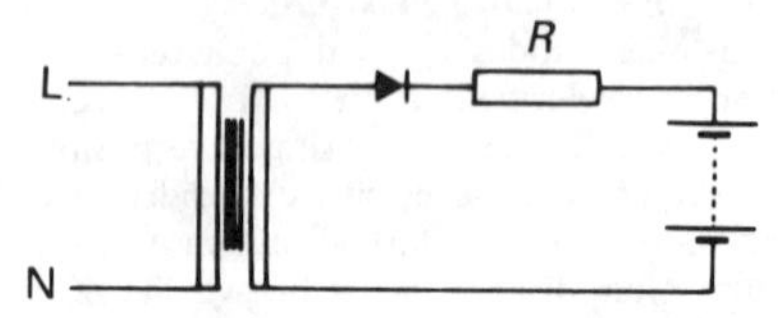

Figure 20.30 A simple charging circuit

$$R = \frac{V_s - V_b}{I}$$

where V_s is the transformer secondary voltage and V_b is the battery on-charge voltage (1.4 × No. of cells).

Improvements may be made by adding a smoothing capacitor and rectifying to full wave with a bridge rectifier.

A transistorised circuit gives a more stable constant current and its simplest form is given in *Figure 20.31*.

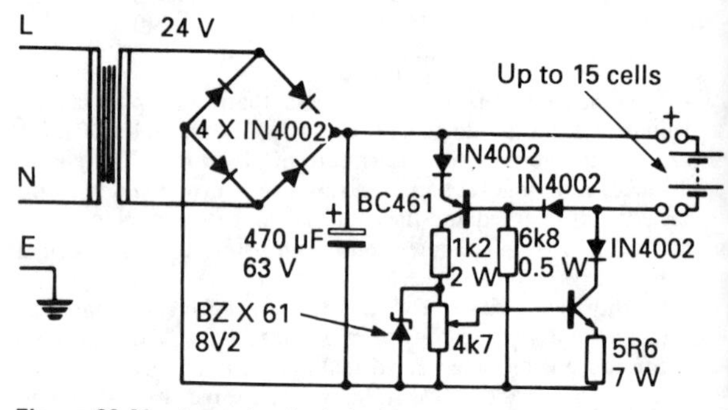

Figure 20.31 A transistorised charger

20.4.3 Fast charging

An easy way to control fast charging is first to ensure the battery is completely discharged either as a consequence of its service routine or by incorporating a discharge circuit into the charger. Charge can then be put back by timing an accurately known

Table 20.5 Fast charge currents and times for fully discharged nickel–cadmium batteries

Charge current	*Normal charge*	*Maximum charge*
$C/8$	12 h	indefinite
$C/4$	5 h	6 h
$C/2$	$2\frac{1}{4}$ h	$2\frac{1}{2}$ h
$C/1$	1 h	$1\frac{1}{4}$ h
$2C$	27 min	30 min
$4C$	12 min	12 min
$8C$	5 min	5 min

current (±5% or less) and the times given in *Table 20.5* are a guide.

The fast charging can also be controlled by sensing of the voltage rise associated with the attainment of full charge. This technique can replace about 80% or more of the nominal capacity in a few minutes. It is not necessary to discharge the battery first nor know the state of charge. A circuit is shown in *Figure 20.32* for charging a 12 V, 0.5 A h battery to about 80% of nominal in about 12 min.

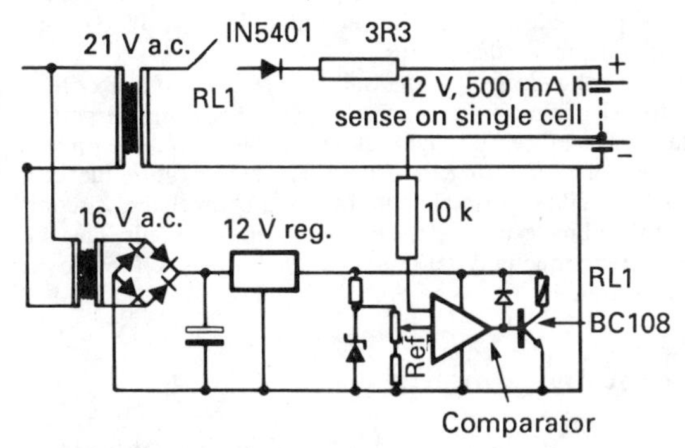

Figure 20.32 A fast-charge circuit

The voltage is sensed by the comparator IC and a usual voltage cut-off is 1.54 V per cell. Inclusion of a negative temperature thermistor in the potential divider network would enable charging to be carried out between 0°C and 45°C. The variation in cut-off must be adjusted to −4 mV/°C per cell, reaching 1.62 V per cell at 0°C. Often a timer is added to the circuit as an additional precaution.

20.4.4 Battery management systems

Recently highly sophisticated battery management systems have become available for nickel–cadmium batteries. A typical system consists of a fast charger, usually a pulse charger, which can charge a battery in 30 min or less, and detector circuits which indicate the battery's state of charge and general condition. However, the high cost of such systems currently limits their use to military equipment and some camera and broadcast equipment.

20.5 Battery selection

The first choice which faces an equipment designer is whether to use primary or secondary batteries. If the application requires a low initial cost, or the life of the equipment is short, then primary batteries are the obvious choice. Secondary batteries can prove cheaper in the long run as they can be recharged and reused, but they require a charger, so the initial cost of the battery system is higher.

Table 20.6 compares the parameters of a few types of primary cells. Zinc–carbon is cheap and readily available. It is primarily used in application having a light, intermittent duty cycle. The battery has a low shelf life and a drooping discharge curve.

For heavy-duty applications needing continuous operation at high currents, alkaline manganese dioxide batteries are preferred to zinc–carbon. They have 50–100% more energy for the same weight, and a longer shelf life, but they are more expensive at currents below about 200 mA.

The mercuric oxide battery has a flat discharge curve and a high energy to weight ratio. It is often used in voltage reference applications. Lithium batteries are used in applications which require a long shelf life, or a wide operating temperature range, or a very high energy density.

Table 20.7 summarises the parameters of a few secondary batteries. Lead–acid cells are in wide general use as they cost about one half as much as nickel–cadmium cells, this price difference being more marked at higher current ratings. Nickel–cadmium batteries are used in applications requiring a flat discharge characteristic, or where many charge–discharge cycles are involved. They have a lower weight and volume than lead–acid batteries, and below about 0.5 A h ratings they are competitive on price.

20.6 Fuel cells

The fuel cell converts chemical energy from the oxidation of a fuel into electrical energy. *Figure 20.33* shows the operation of

Table 20.6 Comparison of primary cells

Parameter	*Carbon–zinc*	*Zinc chloride*	*Alkaline manganese dioxide*	*Mercuric oxide*	*Lithium*	*Zinc–air*
Nominal cell voltage (V)	1.5	1.5	1.5	1.35 or 1.40	3.0	1.4
Energy output ((W h)/kg)	40	90	60	100	300	200
Shelf life at 20°C (years)	1.5	2	3	3	5	5
Operating temperature range (°C)	+5 to +60	−10 to +60	−10 to +60	−20 to +100	−40 to +80	−40 to +60
Flatness of discharge curve (comparative)*	3	3	3	2	1	1
Cost (comparative)*	4	3	3	2	1	1

* 1 = highest or best, 4 = lowest or worst

Table 20.7 Comparison of secondary cells

Parameter	*Lead–acid*	*Nickel–cadmium*	*Silver–zinc*
Nominal cell voltage (V)	2.1	1.2	1.5
Energy output ((W h))/kg)	20	30	110
Shelf life at 20°C (years)	1.0	0.2	0.3
Cycle life (number of operations)	500	2000	100
Operating temperature range (°C)	−60 to +60	−40 to +60	−20 to +80
Flatness of discharge curve (comparative)*	2	1	1
Cost (comparative)*	3	2	1

* 1=highest or best, 4=lowest or worst

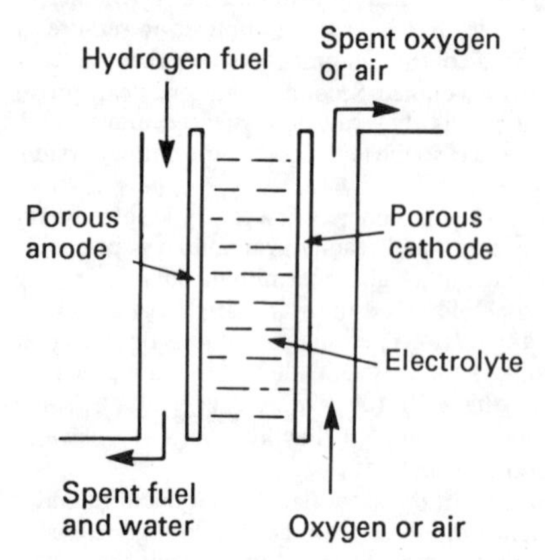

Figure 20.33 Operation of a hydrogen–oxygen fuel cell

a hydrogen–oxygen fuel cell. The cell reaction is as follows:

$H_2 \rightarrow 2H^+ + 2e^-$ at the anode

$O_2 + 4H^+ + 4e^- \rightarrow 2H_2O$ at the cathode

$2H_2 + O_2 \rightarrow 2H_2O$ overall

This reaction gives a difference of potential between the electrodes of the cell, causing a flow of current in a load connected between them.

The fuel cell works isothermally. Its thermal efficiency is high, of the order of 60–80%. In low temperature cells the electrodes are made of finely divided platinum, in the form of wire screens, and the electrolyte is potassium hydroxide.

Figure 20.34 shows the characteristic of a hydrogen–oxygen cell. The power density, measured in milliwatts per square centimetre, is greater when pure oxygen is used since it is a better oxidant than air.

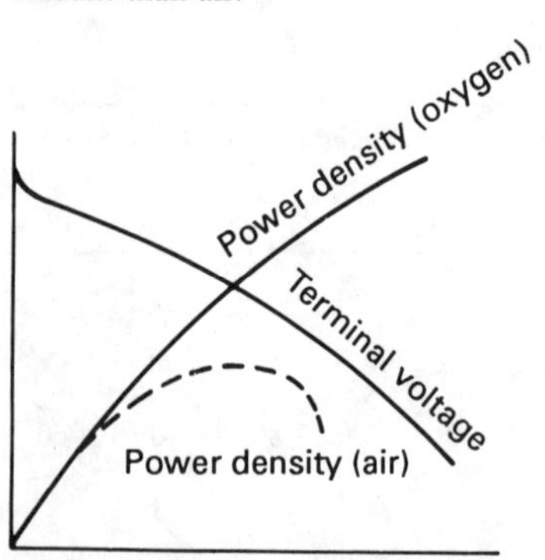

Figure 20.34 Characteristics of a hydrogen–oxygen fuel cell

Many other types of fuel cells have been developed. Methanol is a good fuel since it has a low carbon content, it is cheap, and it can easily be handled and stored. The chemical reactions of the cell are as follows:

$CH_3OH + H_2O \rightarrow CO_2 + 6H^+ + 6e^-$ at the anode

$O_2 + 4H^+ + 4e^- \rightarrow H_2O$ at the cathode

Liquid electrolytes present several problems in fuel cells. The walls must be porous to hold the solution, and the chemical reaction occurs in the capillaries of the walls where there is contact between the gas, solution and solid. It is difficult and expensive to make these walls.

With a solid electrolyte the wall surface is no longer critical, and the cell can be run at higher temperatures, giving larger cell voltages. The power to weight ratio is also improved since a very thin layer of electrolyte can be used. The key to future fuel cells lies in the ability to obtain suitable solid electrolytes. Zirconia doped with lime is used as the electrolyte in some cells. This has high ion conductivity at 1000°C but it falls off at lower temperatures.

Acknowledgements

Sections 20.1, 20.2.2, 20.2.6, 20.3.1, 20.3.3, 20.5 and 20.6 are based on material published in *Discrete Electronic Components* by F F Mazda, published by Cambridge University Press, 1981. British Ever Ready Ltd are thanked for granting permission to reproduce extracts from *Modern Portable Electricity* and *Battery Data Book*.

Further reading

BATTLES, J. E., SMAGA, J. A. and MYLES, K. M., 'Materials requirements for high performance secondary batteries', *Metallurgical Trans. A*, **9A**, February (1978)

BEAUSSART, L., 'Thermal batteries and their applications', *Communications International*, February (1978)

BERKOVITCH, I., 'What prospects for photovoltaic generators?', *Electronics & Power*, August (1980)

BRODD, R. J., KOZAWA. A. and KORDESCH, K. V., 'Primary batteries 1951–1976', *J. Electrochem. Soc.*, July (1978)

BUTTERFIELD, P. N., 'The development, theory and use of nickel–cadmium batteries', *Radio Communication*, May (1978)

DAY, M., 'Batteries—power packed', *Communicate*, May (1988)

DONNELLY, W., 'Battery technology—uses and abuses', *Electronics Industry*, May (1982)

GABANO, J. P., 'Lithium batteries' (1983)

HARRISON, A. I., 'Lead–acid standby-power batteries in telecommunications', *Electronics & Power*, July/August (1981)

HOLMES, L., 'Lithium primary batteries—an expanding technology, *Electronics & Power*, August (1980)

JAY, M.A., 'Lithium–thionyl chloride cells', *Electronics & Power*, July/August (1981)

KNUTSEN, J. E., 'A new generation of battery systems', *New Electronics*, July 14 (1981)
KORDESCH, K. V., '25 years of fuel cell development (1951–1976)', *J. Electrochem. Soc.*, March (1978)
KORDESCH, K. V. and TOMANTSCHGER, K., 'The physics teacher', January (1981)
KUWANO, Y., 'Amorphous silicon solar cell seen as power source of future', *JEE*, June (1982)
McDERMOTT, J., 'Manganese-dioxide, lithium power sources sub for high-priced silver-oxide cells', *EDN*, February 3 (1982)
MEEDS, P., 'Lithium–sulphur dioxide batteries for long-term memory back up', *New Electronics*, July 13 (1982)
MORRIS, R. J., 'Lithium cells', *New Electronics*, July 14 (1981)
MULLER, N. and VEBELHART, H., 'Lithium power modules for better memory back-up', *New Electronics*, April (1988)
NABESHIMA, T., 'Sealed nickel–cadmium cells at their limits', *New Electronics*, June 10 (1980)
ORMOND, T., 'Rechargeable batteries', *EDN*, December 8 (1988)
ROGERSON, S., 'Alkaline batteries to take 50% of the market by 1990', *New Electronics*, January (1988)
RUETSCHI, P., 'Review on the lead–acid battery science and technology', *J. Power Sources*, **2**, 3–24 (1977/78)
SALKIND, A. J., FERRELL, D. T. and HEDGES, A. J., 'Secondary batteries: 1952–1977', *J. Electrochem. Soc.*, August (1978)
SHIRLAND, F. A. and RAI-CHOUDHURY, P., 'Materials for low cost solar cells', *REP. Prog. Phys.*, **41** (1978)
SPENCER, E. W., 'Lithium batteries: new technology and new problems', *Prof. Safety*, January (1981)
STIRRUP, B. N., 'Charge regimes for nickel–cadmium and lead–acid stationary batteries', *Electronics & Power*, July/August (1982)
SUBBARAO, E. C., 'Fuel cell as a direct energy conversion device', *Trans. of the SAEST*, **11**, No. 4 (1976)
UMEO, Y., 'Batteries play major role in miniaturization of electronics equipment', *JEE*, November (1981)
WALKER, D. and GILHAM, D. 'Rechargeable nickel–cadmium cells, their construction, reliability and use', *Electron. Eng., September* (1979)

21 Discrete Semiconductors

P Aloisi
Motorola Semiconductors Ltd

Contents

21.1 p–n junctions

If part of a single crystal of silicon or germanium is formed into p-type material, and part formed into n-type material, the abrupt interface between the two types of material is called a p–n junction. As soon as the junction is formed, majority carriers will diffuse across it. Initially both types of material are electrically neutral. Holes will diffuse from the p-type material into the n-type material, and electrons from the n-type material into the p-type. Thus the p-type material is losing holes and gaining electrons, and so acquires a negative charge. The n-type material is losing electrons and gaining holes, and so acquires a positive charge. The negative charge on the p-type material prevents further electrons crossing the junction, and the positive charge on the n-type material prevents further holes crossing the junction. This space charge creates an internal potential barrier across the junction. Because of this internal potential barrier the region close to the junction is free of majority carriers and this region is called a depletion layer. This induced voltage is inversely linked to the intrinsic carrier concentration (N_i). For example, for germanium, $N_i(\text{Ge}) = 2.4 \times 10^{13}$ cm^{-3} and for silicon, $N_i(\text{Si}) = 1.4 \times 10^{10}$ cm^{-3}, so junction voltage V_j is $V_j(\text{Ge}) = 200$ mV and $V_j(\text{Si}) = 560$ mV.

An external battery can be connected to the p–n junction in one of two ways:

(1) The positive terminal can be connected to the p-type material. The junction is forward biased and results in a decrease of the potential barrier and depletion layer, giving

$$I_F = K \exp((V_a - V_g)/U) \qquad (21.1)$$

where

V_a = supply voltage
$U = kT/q$ (26 mV for 300 K)
V_g = the forbidden energy layer potential (0.72 V) for germanium and 1.1V for silicon)

So a small increase in forward voltage will cause a large increase in forward current.

The forward current will increase with junction temperature with the following law:

$$I_F = KT^2 \exp[(V_a - V_g)/U] \qquad (21.2)$$

for the germanium and

$$I_F = KT^{1.4} \exp[(V_a - V_g)/U] \qquad (21.3)$$

for silicon. So a large increase in temperature increases the forward current, mainly for the germanium semiconductor.

(2) The external battery is connected to the p–n junction with the positive terminal connected to the n-type material. The junction is reverse biased and results in an increase of the potential barrier and consequently an increase in the depletion region, mainly in the weakest doped region, which is normally the n-region. The leakage current is very small, caused by the minority carriers, and if the p-region is the highest doped region the leakage current is only related to the minority carriers, P_n, in the n-type region, so

$$I_r = KP_n/L_p \qquad (21.4)$$

where L_p is the hole diffusion length, such that

$$L_p = \sqrt{(D_p \tau_p)} \qquad (21.5)$$

where

D_p = the hole diffusion constant
τ_p = the hole lifetime

To control I_r we can increase the hole lifetime or decrease P_n by increasing the n-material resistivity.

This leakage current is called a 'volume' leakage current. In reality the leakage current is given by:

$$I_R = I_{r(\text{volume})} + I_{(\text{surface})} + I_{(\text{generation})} \qquad (21.6)$$

$I_{(\text{surface})}$ is quite small for a good dice process. $I_{(\text{generation})}$ is due to carrier generation in the depletion region, and is large in reverse bias. $I_{(\text{generation})}$ is important for the silicon material up to a junction temperature of 175°C:

$$I_{(\text{generation})} = qN_i x_m A/\tau \quad \text{for } V_r > 1\text{ V} \qquad (21.7)$$

where

x_m = depletion region width $= K\sqrt{V_r}$
τ = effective carrier lifetime

In silicon, the leakage current doubles for every increase of the junction temperature of about 10°C.

We can consider the diffused carriers at both sides of the depletion layer as a capacitor with an area of plate equal to the p–n junction area and the distance between two plates equal to x_m. Since x_m is proportional to the square root of V_r, the junction capacitance is inversely proportional to the square root of the reverse voltage. Constructive use of this effect is made in variable-capacitance (varactor) diodes. On the other hand, the capacitance can limit the performance of switching diodes at high frequencies.

If the applied reverse voltage is increased, the critical electrical field in the p–n junction (E_{cr}) will eventually be reached and breakdown will occur. The mechanism of avalanche breakdown is essentially ionisation; the electric field is large enough to give the free electrons enough velocity to dislodge other electrons from the atoms of the crystal lattice and so on. A sudden build-up of electron–hole pairs will provide a large current flow, only limited by the external impedance.

To control the breakdown voltage V_{br} of the diode, the doping profile of the n-region must be controlled. Increasing the resistivity will increase V_{br}.

The junction temperature also has a detrimental effect on the diode lifetime, thermal runaway being given by the increase in internal losses:

$$V_f I_F + V_r I_R \qquad (21.8)$$

$V_f I_F$ decreases slightly with the temperature but I_R increases considerably. Between 50°C and 150°C, for a silicon diode, I_R will be multiplied by 1000, so losses increase, junction temperature increases and so on until the device is destroyed. Every increase of 10°C in junction temperature will decrease the MTBF (mean time before failure) by two. This is known as *Arrhenius' law*.

So to give an acceptable industrial lifetime, semiconductor manufacturers give a limit for the junction temperature of around 75°C for germanium diodes and around 200°C for p–n junction diodes.

The properties of the p–n junction can be exploited in many ways. Besides the normal rectifier diode, other types such as small signal diodes, high-speed switching diodes, zener or avalanche diodes, voltage regulator, transient suppressors and varactors are available.

21.2 Small-signal diodes

Diodes which are used for demodulation and switching applications, where only small amounts of power are involved, can be conveniently grouped together as small-signal diodes. As such diodes dissipate a low energy they can be wired directly on the board, without any heatsink. The usual construction is a glass or plastic envelope with axial connecting wires as in *Figure 21.1*.

The first small-signal junction diodes were introduced in the mid-1950s, and were manufactured from germanium by an alloy

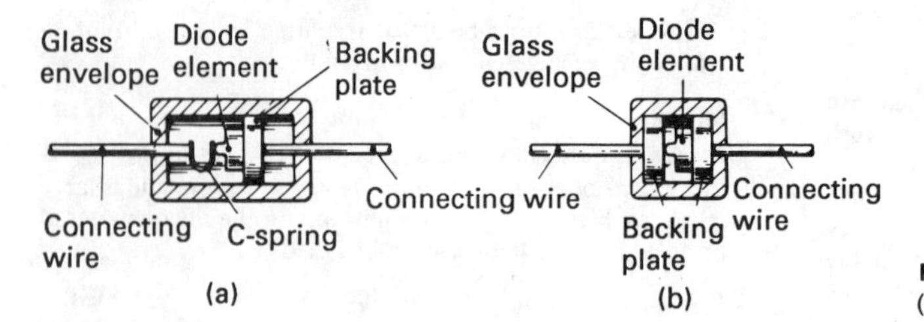

Figure 21.1 Construction of small-signal diodes: (a) spring-contact, (b) whiskerless

junction process with a small pellet of indium. During the late 1950s silicon started to replace germanium as the generally used material for semiconductor devices, thanks to the better reverse current and better allowable junction temperature given by this material. The introduction of planar technology in 1960 enabled planar and planar epitaxial diodes to be manufactured.

A slice of n-type monocrystalline silicon is coated, a window is cut and boron diffused into the silicon through the window to form a region of p-type silicon. The coating is etched away and connecting leads attached to the p-type and n-type regions, and the assembly is then encapsulated.

When a diode is used for switching at high frequency, two important parameters are needed; capacitance associated with the junction, which for the circuit designer means a trade-off with the allowable forward current, and recovery time. Heavy materials such as gold or platinum can be diffused into the p–n junction to reduce the lifetime of the minority carriers.

Fast small-signal diodes can give switching times of about a few nanoseconds, and better switching times can be obtained by using gallium arsenide instead of silicon.

21.3 Rectifier diodes

High-power rectifier diodes were developed during the 1960s for use in rectifier systems operating from the mains supply. The aim was to replace existing systems using thermionic valves or mercury arc rectifiers. Such diodes are required to conduct high currents and withstand a large reverse voltage. Cooling, therefore, is an important requirement. Because high-power diodes need to be mounted on heatsinks, a stud mounting construction or large plastic packages such as case 340 (TO-218 or TO-3P) or 221 (TO-220) is used. Thermal resistance between junction and case is normally given in the manufacturer's data sheet, and is expressed as temperature change with power (°C/W). Like electrical resistance, thermal resistance can be considered connected in series. Thus a chain of thermal resistance exists from the diode junction to the cooling element (air or liquid) considered as heat well:

$$R_{th(j-c)} + R_{th(c-h)} + R_{th(h-a)} \qquad (21.9)$$

So the circuit designer has to choose the appropriate thermal resistance for the heatsink and the interface between the case and the heatsink (quality of surfaces: grease or insulator), in order to evacuate the losses in the rectifier and to keep the junction temperature below the maximum allowable junction temperature.

In addition to the normal data for a diode, such as average forward current, reverse current and forward voltage drop for high-power rectifier diodes, the circuit designer requires information on the transient performances. These include repetitive and non-repetitive peak reverse voltage, and RMS reverse voltage. Similar ratings are given for forward current, i.e. a repetitive peak forward current I_{FRM} and a non-repetitive peak forward current I_{FSM}. Thermal characteristics, such as thermal resistance (junction to case), operating junction temperature range and transient thermal resistance, are also given.

21.4 Commutating diodes

This type of diode is used as a switch in the system, with natural commutation. However, each transient needs a time to be performed. The turn-on and turn-off processes are as follows:

(1) *Turn-on* To obtain the forward characteristics, the diode has to build up its internal minority carrier level. Before this build-up takes place, the diode internal resistance remains high, mainly for high-voltage diodes, so when a high turn-on gradient of current (dI/dt) is imposed an overvoltage will appear at the diode terminals. This overvoltage is also given by the internal parasitic inductances ($L\,dI/dt$), *Figure 21.2*. From this figure we can define a peak voltage, called also V_{dyn}, and a turn-on time t_{fr}.

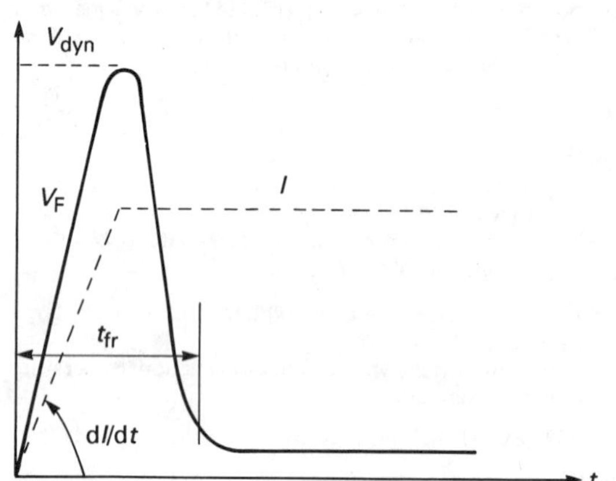

Figure 21.2 Turn-on process of a diode

(2) *Turn-off* To recover its reverse characteristics, consisting of high impedance, low leakage current and high reverse voltage, the minority carriers have to be removed, by the current itself and by internal recombination. When the current reaches zero, not all the carriers have yet been removed from the structure and the current will reverse until a depletion region is built up in order to sustain the field of the reverse voltage.

From *Figure 21.3*, we can define the reverse recovery current I_{RM}, the recovery time t_{rr} and the recovery charge Q_{rr}.

We can define another parameter called the softness, which represents the shape of the current when it returns to zero (the recovery part of the current). If the diode recovery is hard it will generate a lot of electrical noise.

Fast commutating diodes, with t_{rr} less than 500 ns, are built in the same way as rectifier diodes; this technology is called double diffused. The starting material is a high-resistivity material (n^-), such as silicon, to sustain the reverse voltage of the diode. Boron is diffused into one side of this pellet to create the p^+ anode, and phosphorus (n^+) is simultaneously diffused into the other side to provide a very low resistivity cathode connection. The speed of

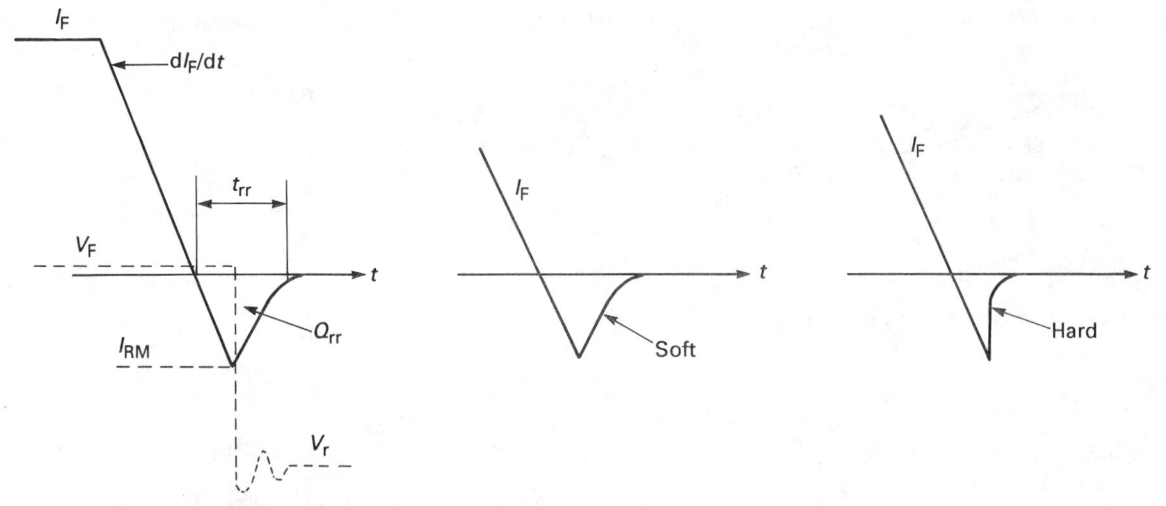

Figure 21.3 Turn-off process of a diode

this diode can be improved by diffusing heavy materials, such as gold or platinum, into the structure to create recombination centres, or by irradiation of the pellet for the same purpose.

Ultrafast diodes use a new technology called epitaxial diode or fast recovery epitaxial diode (FRED) technology. The manufacturer starts with a low resistivity (n^+) silicon die, and an n^- epitaxial layer with a chosen resistivity and thickness is grown to obtain the desired reverse voltage; a p^+ anode is diffused into this layer (*Figure 21.4*).

This technology can obtain very precise control of the electrical characteristics such as reverse voltage, leakage current, forward voltage and switching time, and a big improvement in switching times; for example, for epitaxial diodes under 200 V, $t_{rr}=35$ ns and for diodes around 1000 V, $t_{rr}=100$ ns.

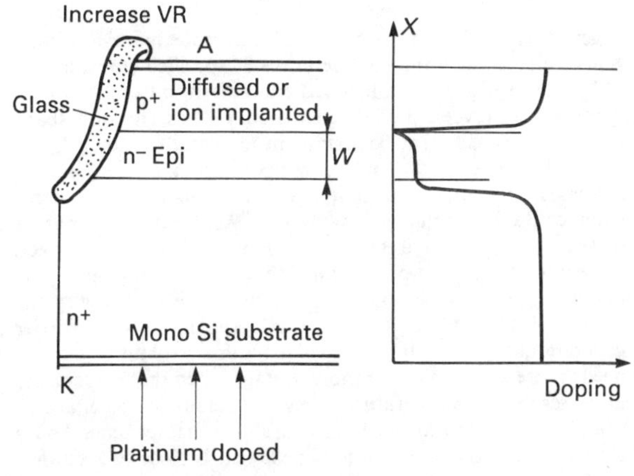

Figure 21.4 Epitaxial technology

21.5 Schottky diodes or metal–semiconductor barrier diodes

Figure 21.5 shows an energy diagram for a metal and semiconductor junction. ϕ_m is the metal work function, E_F is the Fermi level of the metal and the semiconductor, and E_c and E_v are the energy levels of the conduction and valence bands of the semiconductor.

The electrons of the conduction band of the n material, which are more free than the metal ones, can diffuse through the barrier into the metal and create an electric field which stops this diffusion. This produces a majority carrier device (n^{++} n^-). The thermionic current is given by:

$$J_F=RT^2\exp(-q\phi_b/kT)\exp(qV_a/kT) \qquad (21.10)$$

where

R = Richardson's constant ≈ 110 A/(cm^2 K^2) for silicon
T = absolute temperature in K
V_a = applied voltage
ϕ_b = the barrier level in electron-volts (eV), which depends on the metal and is between 0.5 eV for chromium and 0.87 eV for platinum
k = Boltzmann's constant
q = electron charge

The forward static characteristics versus barrier metal types and temperature are given in *Figure 21.6*.

The theoretical Schottky leakage current is given by the formula:

$$J_r=J_s=-RT^2\exp(-\phi_b q/kT) \qquad (21.11)$$

Schottky diodes are mainly used in low-voltage applications because of their low forward characteristic, and high-frequency ($\geqslant 100$ kHz) power systems because this majority carrier device has inherently no reverse recovery time.

Designers would like a Schottky with the highest possible ratio V_F/V_R, so many different systems are in development to meet this requirement. Examples are junction barrier Schottky (JBS), MOS barrier controlled Schottky (MBS) and diodes using semiconductor materials like gallium arsenide (GaAs), whose carrier mobility is about 12 times higher than that of silicon.

21.6 Zener diodes

The zener family represents two distinctive phenomena—zener breakdown and avalanche breakdown, but their static characteristics are the same. Their uses are also identical; voltage regulator or voltage reference, and transient or overvoltage surge suppressor.

In the p–n junction if there is a high doping profile, i.e. a low resistivity zone, the space charge width, d, is very narrow and in spite of the low applied reverse voltage ($V_R \approx 5$ V) the electric field

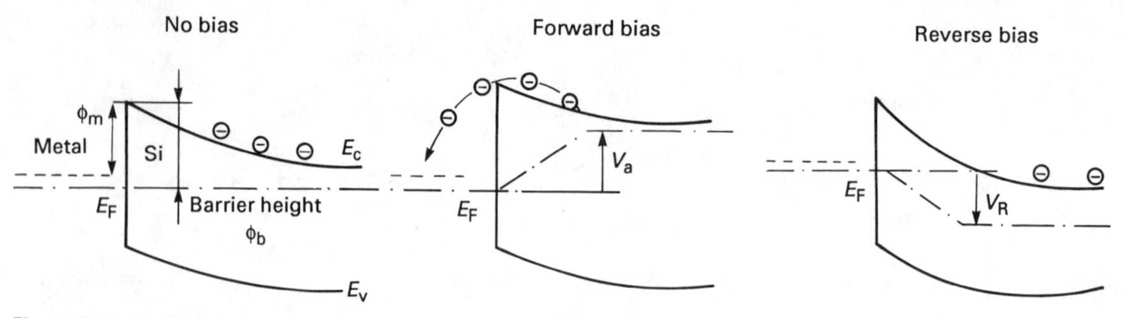

Figure 21.5 Metal–semiconductor barrier energy diagram

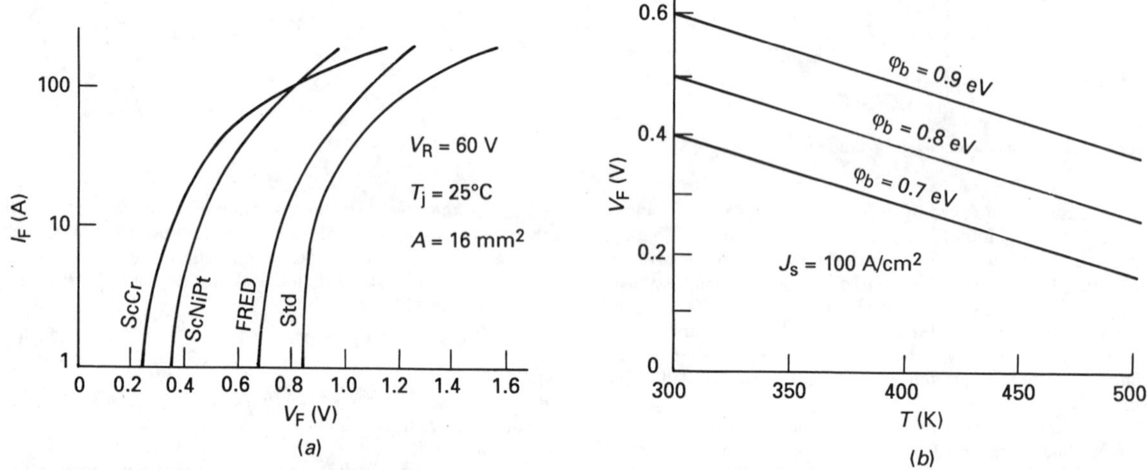

Figure 21.6 Schottky forward static characteristic: (a) current–voltage, (b) voltage–temperature

$E=V_R/d$ could be very high. If $E \geqslant 3\times10^5$ V/cm, the force on the covalent bonds is such that a lot of electrons will be freed and a high current can result. The pure zener breakdown effect can exist only for zener diodes under 5–6 V.

Now, if the p–n junction has a low doping level, a larger depletion layer will occur in the junction under reverse bias and the electric field will not reach the critical zener breakdown level. However, the space charge region will be large enough to allow the free carriers to take sufficient velocity and, if the travelling distance is higher than the mean free distance for the carrier at the given temperature, there will be a high probability of collisions with the atoms of the crystal and the creation of an electron–hole pair, and so on. The space charge region becomes ionised, and this ionisation will stop when the electric field has decreased. The number of created free carriers is practically infinite and the avalanche current is limited only by the external impedance. Avalanche breakdown occurs for the diode over 8 V, and for the devices between 6 and 8 V there is a mix of the zener and avalanche phenomena.

It is possible, therefore, to manufacture a range of zener diodes with different breakdown voltages. The value of the breakdown voltage is determined principally by the doping level of the semiconductor material from which the diode is made. The range of voltages obtained is typically from 4V to 200V.

Diodes operating at breakdown voltages below 10V usually have the junctions manufactured by alloying, and those operating at higher voltages have diffused junction in the epitaxial layer. Because of tolerances during the manufacturing process, there is also a tolerance on the value on the zener voltage. This tolerance is generally $\pm5\%$, but special selection can be made to obtain zener voltages with tighter tolerances at specified currents.

The zener diode has a forward characteristic like a 'normal' diode and in reverse mode (zener mode) it has almost a sharp knee. It has a different behaviour in temperature; for a 'pure' zener, V_z decreases when the temperature increases but V_z increases with temperature for an avalanche diode. This phenomenon could be explained as follows. When the temperature increases, the valence electrons will gain some energy and need less energy from the supply to break free. On the other hand, the possible number of carrier recombinations increases with the square of the leakage current, while the number of carrier generations increase linearly, so the same avalanche current needs more electric field or more voltage when the temperature increases. So the temperature coefficient versus zener voltage V_z can be drawn as in *Figure 21.7*, which shows that for zener diodes close to 5–6 V this temperature coefficient is close to zero.

The characteristic $I_z=f(V_z)$ is not strictly parallel to the y-axis, therefore we can define a dynamic characteristic $Z_z=\Delta V_z/\Delta I_z$. Zener published data give two values of the zener impedance, Z_{zk} for the knee and Z_{zt} for $I_{zt}=I_{zm}/4$, I_{zm} being the maximum zener current. These parameters are measured with ΔI_z(peak to peak)$=0.282I_c$ ($\pm10\%$ I_c(RMS)).

Among their possible uses, zener diodes can be used as a voltage reference. Typically this device can offer good voltage stability in time; less than 500×10^{-6} (500 parts per million or ppm) for 1000 hours working time. Precision-reference zener diodes can have a time stability of a few ppm for 1000 h. These types of zener diodes are made with a string of several diodes in

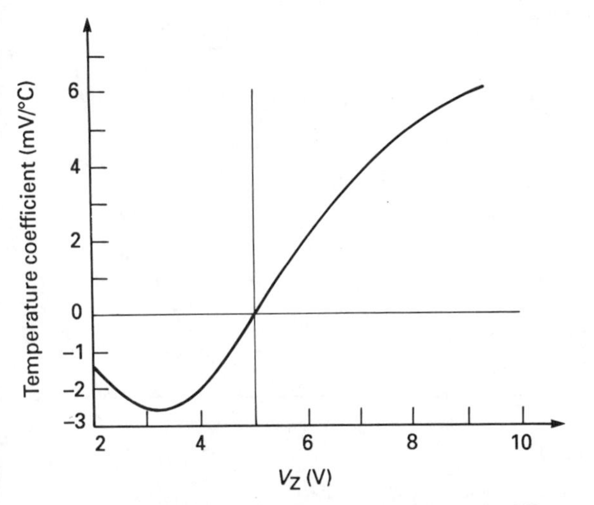

Figure 21.7 Temperature coefficient of zener voltage for different zener voltages

series, because a p–n junction has a negative temperature coefficient in the forward direction and a positive temperature coefficient in the reverse direction ($V_r \geqslant 8$ V). These diodes must be mounted in the same package.

The emergence of ultra-fast switches such as power MOSFETs requires the use of very fast zeners which can sustain high transient powers, for voltage suppressors. The normal zener exhibits a parasitic inductance so its turn-on time is quite slow for some applications. Transient suppressors should be much faster; the device structure should be fired very quickly and optimised to sustain a large power surge. A cellular technology, like the power MOSFET, answers these requirements. Products with a turn-on time in the region of the nanosecond and which can sustain a transient current around 200 A and a power of 1500 W for 1 W for 1 ms are available. These ratings are non-repetitive ratings, and are given for the device without a heatsink. In applications where the transients are repetitive a heatsink should be used.

21.7 Symbols for main electrical parameters of semiconductor diodes

C_d	diode capacitance (reverse bias)
I_F	continuous (d.c.) forward current
$I_{F(AV)}$	average forward current
I_{FRM}	repetitive peak forward current
I_{FSM}	non-repetitive (surge) peak forward current
I_O	average output current
I_{OSM}	non-repetitive (surge) output current
I_R	continuous reverse leakage current
I_{RRM}	repetitive peak reverse current
I_{RSM}	non-repetitive peak reverse current
I_Z	voltage regulator (zener) diode continuous (d.c.) operating current
I_{ZM}	voltage regulator (zener) diode peak current
P_{tot}	total power dissipated within device
Q_s	recovered (stored) charge
$R_{th(h)}$	thermal resistance of heatsink
$R_{th(i)}$	contact thermal resistance
$R_{th(j-amb)}$	thermal resistance, junction to ambient
$R_{th(j-mb)}$	thermal resistance, junction to mounting base
r_Z	voltage regulator (zener) diode differential (dynamic) resistance
T_{amb}	ambient temperature
T_{jmax}	maximum permissible junction temperature
t_p	pulse duration
t_{rr}	reverse recovery time
$V_{(CL)R}$	surge suppressor diode clamping voltage
V_F	continuous (d.c.) forward voltage
V_{IRM}	repetitive peak input voltage
$V_{I(RMS)}$	RMS input voltage
V_{ISM}	non-repetitive (surge) input voltage
V_{IWM}	crest working input voltage
V_O	average output voltage
V_R	d.c. reverse voltage
V_{RM}	peak reverse voltage
V_{RRM}	repetitive peak reverse voltage
V_{RSM}	non-repetitive (surge) peak reverse voltage
V_{RWM}	crest working reverse voltage
V_Z	voltage regulator (zener) diode operating voltage

21.8 Bipolar junction transistors

The transistor is a three-terminal device consisting of collector, emitter and base, built with a sandwich of three layers of semiconductor, npn or pnp.

The characteristics of this device are normally given in the published data as the maximum ratings such as:

(1) Breakdown voltages, BV_{ceo}, BV_{cbo}, BV_{ces}, BV_{ebo}. The symbols are self-explanatory for the breakdown conditions. BV_{ceo} is, for example, the breakdown voltage between collector and emitter, base open.
(2) Sustained, continuous, peak or overload collector current, I_C.
(3) Leakage currents, I_{cbo}, I_{ceo} $(=I_{cbo} \times H_{FE})$, I_{ebo}.
(4) Thermal resistance between junction and case, $R_{th(j-c)}$ (°C/W).
(5) Maximum junction temperature, T_j.

Static characteristics are also given, such as d.c. current gain H_{FE} *vesus* collector current I_c, collector emitter voltage saturation V_{CEsat} *versus* I_C, mutual or transfer characteristic as I_c versus emitter voltage V_{be} and the capacitances seen at the terminals C_{ib}, C_{ob}, C_{re} *versus* bias voltage.

It is possible to find some a.c. or dynamic characteristics as the small-signal current gain at given frequency h_{fe}, current-gain bandwidth product F_t versus I_C, the noise figures or the switching times. Normally, these parameters are given at the ambient temperature: 25°C. Temperature variations of some parameters are available. See *Figure 21.8* for the typical shape of these parameters.

In addition, mainly for the power transistors, published data include safe operating areas, such as FBSOA (forward bias safe operating area), RBSOA (reverse bias SOA) or E_{SB} (energy of second breakdown).

It is possible to connect this product in three configurations: common emitter, common base and common collector, as in *Figure 21.9*.The first configuration is used as a current amplifier. The output current (I_c) equals the input current (I_b) times the current gain (β) of transistor. β is also called the static common emitter forward current transfer ratio:

$$H_{FE}=I_c/I_b \qquad (21.12)$$

The second configuration is used as an impedance adaptor. The input has a low impedance and the output a higher one, so that even if the current gain, α, or more precisely the static comon base forward transfer ratio, given by Equation (21.13), is close to unity, a power gain between input and output is obtained.

$$H_{FB}=I_c/I_E \qquad (21.13)$$

The common collector configuration is also called the emitter follower. The static common collector forward current transfer ratio H_{FC} is high, the voltage gain is little less than unity and the

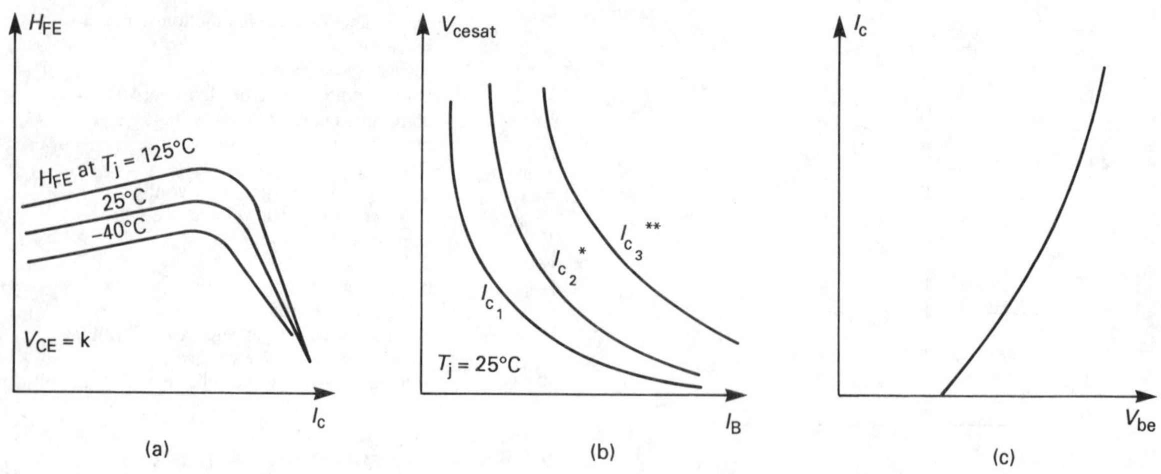

Figure 21.8 Bipolar junction transistor characteristics: (a) gain, (b) saturation voltage, (c) collector current

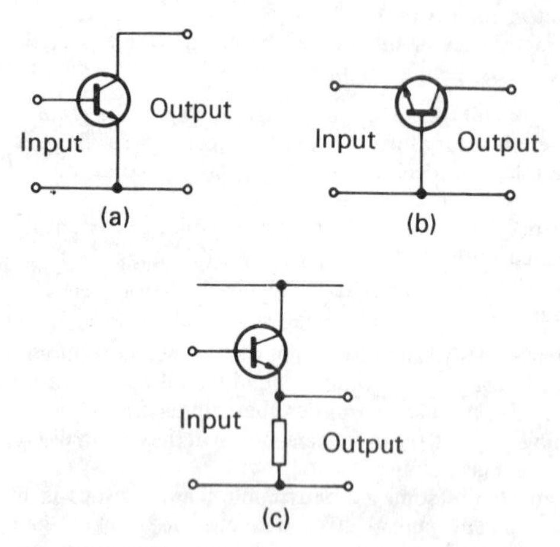

Figure 21.9 Transistor configurations: (a) common emitter, (b) common base, (c) common collector

power gain low, but the main interest of this circuit is the impedance adaptation from a high input impedance to a low load impedance.

To summarise the main characteristics of these configurations, the common emitter configuration is a current amplifier and the others are impedance adaptors.

21.8.1 Small-signal transistors

To enable the behaviour of a transistor under a.c. conditions to be analysed, an equivalent circuit is required. Among the complex suggested circuits three are useful; they are applicable only to small signals.

Figure 21.10 shows these equivalent circuits. These models are given for the common emitter configuration. R_b, R_c and R_e are the base, collector and emitter resistances. The current generator provides a current $h_{fe}I_b$ to represent the current gain of the transistor.

The T circuit applies to low and medium frequencies only. The π equivalent circuit can be used for any frequency because the internal junction capacitances are added, the gain of the transistor being represented by the forward conductance g_m.

The third equivalent circuit does not represent the physical structure of the transistor, but it is close to the published h parameters. This circuit contains both voltage and current generators. The current generator produces a current h_f times the input current, representing the forward gain of the transistor. The voltage generator produces a voltage h_r times the output voltage, representing the feedback effect of the transistor. The basic equations for the network are:

$$v_i = h_i i_i + h_r v_o \quad (21.14)$$

$$i_o = h_f i_i + h_o v_o \quad (21.15)$$

If the output is short circuited, $v_o = 0$, giving:

$$v_i = h_i i_i \quad (21.16)$$

$$h_i = v_i / i_i \quad (21.17)$$

h_i is the input impedance of the transistor. Similarly:

$$i_o = h_f i_i \quad (21.18)$$

$$h_f = i_o / i_i \quad (21.19)$$

that is, h_f is the current gain of the transistor.

If the input terminals are open circuit, $i_i = 0$, giving:

$$h_r = v_i / v_o \quad (21.20)$$

which is the voltage feedback ratio.

Similarly:

$$h_o = i_o / v_o \quad (21.21)$$

which is the output admittance of the transistor.

Subscripts can be added to the h parameters to indicate the transistor configuration: b for the common base, c for the common collector and e for the common emitter. Thus h_{fe} is the forward current transfer ratio in the common emitter configuration.

To summarise the h parameter definitions, we have:

h_{ie}	input impedance	$\lvert\partial V_{be}/\partial I_b\rvert V_{ce}$
h_{oe}	output admittance	$\lvert\partial I_c/\partial V_{ce}\rvert I_b$
h_{fe}	forward current transfer ratio	$\lvert\partial I_c/\partial I_b\rvert V_{ce}$
h_{re}	voltage feedback ratio	$\lvert\partial V_{be}/\partial V_{ce}\rvert I_b$

The values will change with changes in I_b and V_{ce}. Curves of the h parameters are given in the published data. The h parameters are easily measured for actual transistors, and this gives them considerable advantage over other systems of parameters that have been suggested.

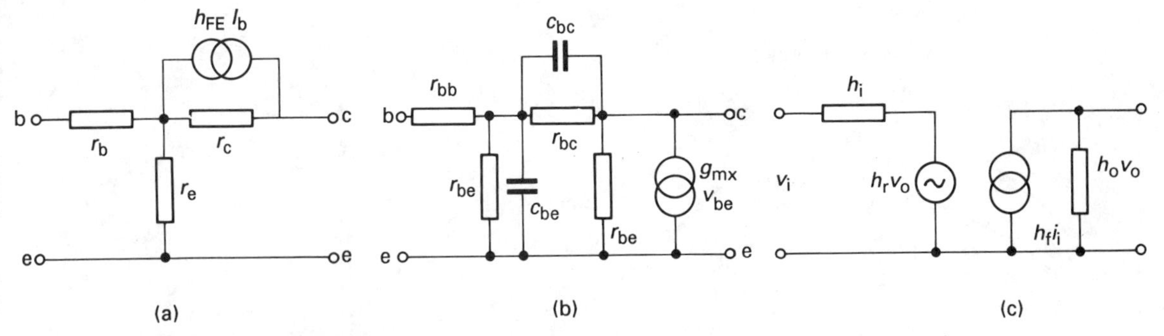

Figure 21.10 Small signal transistor equivalent circuits: (a) T circuit, (b) π circuit, (c) hybrid parameter network

The performance of the transistor is limited at high frequencies by the capacitances of the junctions. Various frequency values have been suggested as 'figures of merit' for high-frequency performance, for example $f_h f_b$ or f_{hfe} at which the current gain has fallen to 0.7 times its low frequency value, and f_1, the frequency at which the current gain has fallen to unity in the common emitter configuration. This parameter is better known as the common emitter gain bandwidth product F_t, the product of the frequency at which the value of h_{fe} falls off at the rate of 6 dB per octave and the value of h_{fe} at this frequency. F_t varies with the collector current, so the curves of F_t *versus* I_c at the relevant collector voltage is normally published.

21.8.2 Power transistors

Power transistors are mainly used in power amplifiers or as a power switch. For both of the applications the main parameters are the controlled power, the thermal dissipation and the switching speed. These are determined by the SOA, RBSOA, E_{SB}, V_{CEsat} and switching speeds.

21.8.2.1 Forward bias safe operating area

Figure 21.11 gives the first parameter for the choice of the power amplifier. It can be broken into four parts:

(1) Limit of the maximum current in the output connections.
(2) Maximum power dissipation, $P=VI=$constant, a hyperbola or a straight line with a slope of -1 in log × log coordinates.
(3) Limitations by 'hot spots', instabilities, second breakdown.
(4) Limitation by the avalanche, BV_{ceo}.

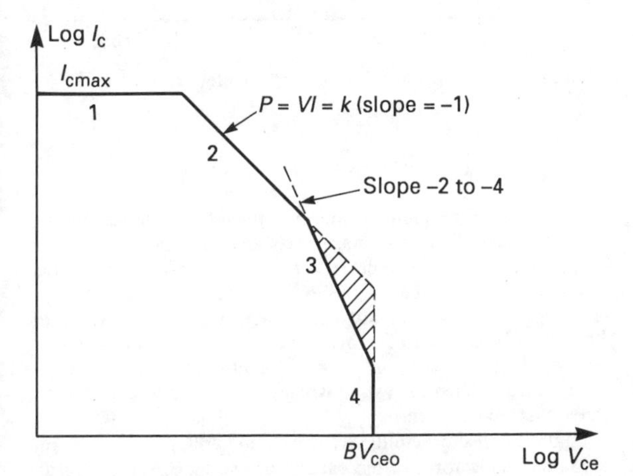

Figure 21.11 Forward biased safe operating area (FBSOA)

The effect of base drive on the breakdown is shown in *Figure 21.12*. The area is limited for the worst case drive which is open base BV_{ceo}.

The physical phenomenon which limits the part 3 of the area is related to second breakdown and hot spots: if the transistor is driven into the avalanche with a collector current more than a critical current, I_{sb}, a sudden collapse of voltage is observed and the current through the transistor is only limited by the external impedance.

A high current density exists under the edges of the emitter fingers during the turn-on and conduction time of the transistor. This effect is due to the transverse electric field in the base layer, and the electric field between base and collector, this junction being reverse biased during the conduction. The higher the electric field, the higher the 'defocalisation' and the higher the probability of getting small plasmas in the high current density region, which are known as 'hot spots'. These plasmas have a very low resistance giving a sudden decrease in voltage known as second breakdown.

The safe operating area can be extended for pulse operation. Published data give the maximum junction temperature T_j and the continuous junction-to-case thermal resistance $R_{th(j-c)}$, in addition to other related thermal resistances; case to heatsink $R_{th(c-h)}$ and heatsink to ambient $R_{th(h-a)}$. We can calculate the maximum power dissipation (P_d) as:

$$T_j - T_a = \Delta T_j = P_d(R_{th(j-c)} + R_{th(c-h)} + R_{th(h-a)}) \qquad (21.22)$$

Data sheets also give the graph of effective normalised transient thermal resistance $r(t)$ *versus* pulse width T, and duty cycle $D=t_{on}/T$. With this thermal factor we can calculate the new allowed power dissipation in the device:

$$\Delta T_j = P_{d(T,D)}(\sum R_{th})r(t) \qquad (21.23)$$

21.8.2.2 Reverse bias safe operating area

When a power transistor is used as the switch, during turn-off the physical behaviour in the structure is different from that during turn-on and in conduction. Now the electric field due to the base layer and the base current is reversed, so the collector current is 'focalised' under the axis of the emitter finger, and the safe operating area at turn-off is as shown in *Figure 21.13*.

21.8.2.3 Turn-on safe operating

Normally at turn-on the FBSOA should be used, but with the fast devices hot spots cannot take place during a short time such as hundreds of nanoseconds because they are related to the thermal aspects. So a turn-on safe operating area such as that given in *Figure 21.14* could be used.

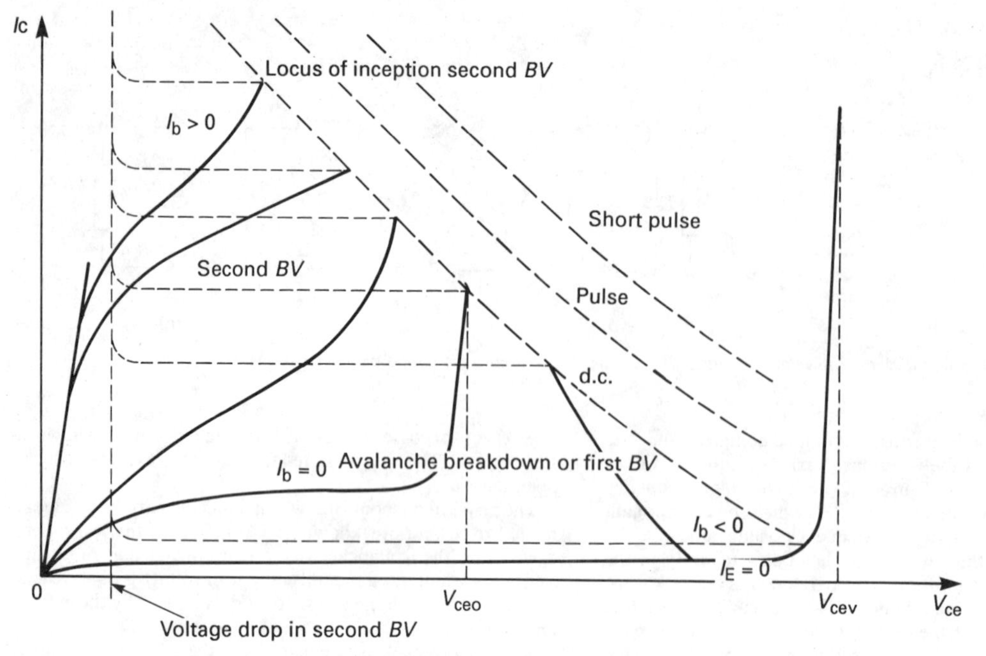

Figure 21.12 Avalanche breakdown characteristic of power transistors

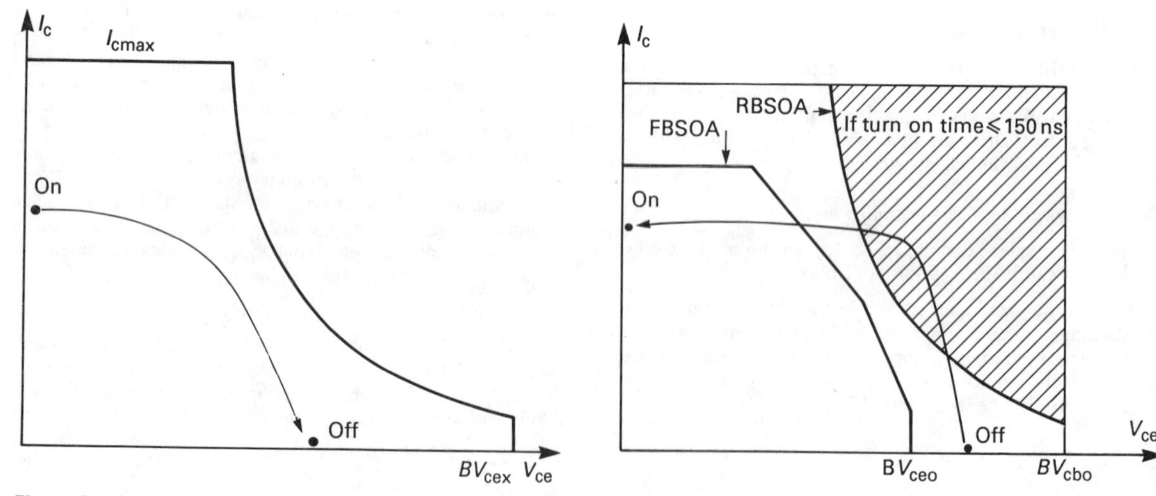

Figure 21.13 Reverse-biased safe operating area (RBSOA) for a power transistor

Figure 21.14 Turn-on safe operating area for fast devices

21.8.2.4 Energy of second breakdown

For some devices which could dissipate energy stored in the load inductance, $0.5LI^2$, during avalanche, semiconductor manufacturers' data sheets may give a maximum energy E_{SB} that the power transistor switch can sustain before going into second breakdown and being destroyed.

21.8.2.5 Saturation voltage

For power transistors used in a linear mode, the d.c. current gain is important in order to have a high transfer ratio in power or impedance. In a conduction mode, collector-to-emitter voltage saturation is of first importance because the continuous losses are a product of the load current and the transistor saturation voltage. During switch-on, the voltage at the transistor collector–emitter terminals does not decrease immediately, due to the transistor internal capacitances, and a certain amount of energy is stored in the structure. Semiconductor manufacturers give a parameter called $V_{CEsat(dyn)}$ which can be used to calculate the energy stored at turn-on.

Static collector–emitter saturation voltage V_{CEsat} and dynamic collector–emitter saturation voltage $V_{CEsat(dyn)}$ are improved by a high interdigitated emitter–base structure and an optimised breakdown voltage; the higher the breakdown voltage the higher the saturation voltage.

21.8.2.6 Speed

Speed is an important parameter for a power transistor used as the switch because switching losses are the main part of the total losses. If a fast turn-on time is required, a high pulse of base current is needed to charge the structure quickly. However, for fast turn-off times it is better to have the minimum of carriers stored in the base, and in the collector of the power transistor, which needs a trade-off with a fast turn-on time. The more difficult carriers to extract at turn-off are the ones stored in the collector far from the base connection, so it is better to avoid any hard saturation when the base–collector junction is forward biased.

21.8.3 Transistor technologies

21.8.3.1 Alloyed transistors

The first transistors commercially available, in the early 1950s, were made by the germanium alloy-junction process, but because of the high leakage current of this semiconductor material, which limits the sustaining voltage, semiconductor manufacturers changed to silicon.

The alloying techniques developed for germanium were applied for silicon: pnp transistors were manufactured by using a slice of n-type silicon (which formed the base) and two pellets of aluminium, which would form the emitter and the collector, were placed on opposite sides of the monocrystalline silicon slice. The whole assembly was heated and some aluminium dissolved into the silicon to form a p-type region, giving two p–n junctions. This technique was extensively used for high-frequency transistors until it was superseded by the silicon planar transistor.

21.8.3.2 Mesa technique

This principle consists of etching the edges of the transistor structure to decrease the junction area and so reduce the capacitances. The resulting shape is a 'mesa' or plateau as shown in *Figure 21.15*.

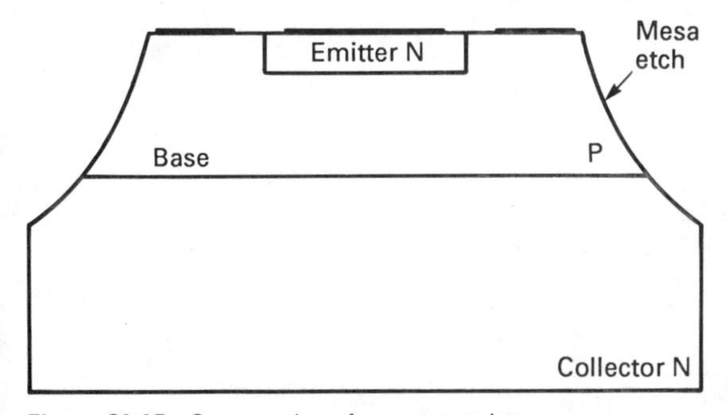

Figure 21.15 Cross-section of a mesa transistor

The principle of mesa etching is extensively used in many types of transistor. It allows the edges of the transistor to be controlled, and in particular allows the collector–base junction to be clearly defined and passivated. 'Passivation' or 'glassivation' consists of depositing silicon oxide or glass into the mesa etch to prevent contamination of the open junction, which sustains a high electric field. This guarantees the reliability of the device. A big improvement to this technique is the 'planar' technique.

21.8.3.3 Planar technique

The discovery that thermally grown silicon oxide on the surface of the slice could form a barrier to diffusion, and so could be used to define the impurity regions, led to a breakthrough in transistor manufacture: for the first time a mass production technique could be applied. Operation at frequencies up to gigahertz (microwave region) became possible, and the planar process made possible the most important step in semiconductor devices: the integrated circuit.

The principle of the planar process is the diffusion of impurities into areas of silicon slice defined by windows in a covering oxide layer, as described in Chapter 27.

The combination of diffusion and epitaxy growth has given rise to a family of transistors such as single diffused, double diffused, triple diffused, epitaxy, epibase double diffused, and so on. All these constructions give various advantages in different applications as summarised in *Table 21.1*.

21.8.4 The Darlington

The Darlington, from the name of its inventor, is a cascade combination of two transistors. *Figure 21.16* shows three main arrangements: the npn Darlington made with two npn transistors in cascade; the MosBip Darlington, consisting of a MOSFET transistor as the driver and a bipolar transistor as the power source; and a composite Darlington, or Sziklaï, a pnp-like Darlington built with a pnp driver and an npn power transistor.

The main advantage of a Darlington is the overall current gain:

$$H_{FE(D)} \approx H_{FE(driver)} \times H_{FE(power)} \qquad (21.24)$$

for the bipolar structure, and

$$G \text{ (Darlington } G_{fs} = T_{out}/V_{in}) = G_{fs(MOS)} \times H_{FE(power)} \qquad (21.25)$$

The disadvantage is a high saturation voltage of the driver.

$$V_{CEsat} = V_{BEsat} + V_{CEsat} \qquad (21.26)$$

Most Darlingtons are usually monolithic, being built in the same silicon substrate by the manufacturer. A resistor between base and emitter is sometimes needed in order to avoid any leakage current amplification. When the Darlington is monolithic, these resistors can also be introduced in the structure, using the base layer resistance. When the turn-off time of the Darlington is of importance, a small 'speed-up diode' (SUD) is introduced between the two bases of the Darlington in order to extract the carriers stored in the base of the power transistor. This diode can be monolithically built or added as a hybrid.

21.8.5 Transistor packaging

Bipolar transistors may be sealed in metal packages (TO-18, TO-39, TO-3); in plastics packages (TO-92, TO-126, 'D pack', TO-220, TO-218, depending on the die size); in 'surface-mounted' packages (SOT-23, SOT-89); or very big package assemblies with multiple connections.

Normally, the back of the chip (collector) contains a multilayer structure in titanium, nickel or silver. This chip is bonded on the package header, made with a copper–nickel alloy, with various types of soldering material such as molybdenum, eutectics (Au–Ge, Au–Sn, Au–Si), or 'soft' solder like lead, tin, silver, indium, antimony or a mixture of these materials. This die-bonding is done in a reducing atmosphere or forming-gas (5% hydrogen, 95% nitrogen). For the future this attachment could be made through an insulated material like beryllium oxide (BeO) or alumina (Al_2O_3).

Emitter and base connections are made with aluminium or aluminium–magnesium wires ultrasonic bonded on the die-pads and the external connections.

Plastic packages are increasing in use because the plastic material epoxy has a very good hermeticity and the mounting system is very easy to use. This plastic material can sometimes be the insulator for the package heatsink.

Table 21.1 Various bipolar transistor technologies

Technology	*Doping profile*	*Advantages*	*Drawbacks*
Homobase	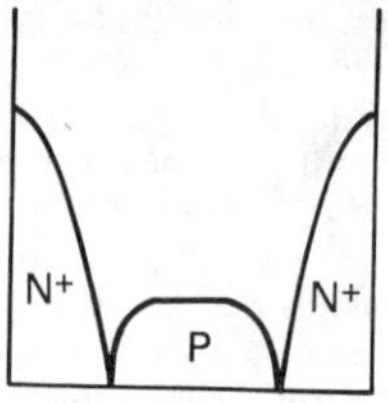 N+ P N+	Good ruggedness	Only pnp Slow Low BV_{ce} Large electrical characteristics distribution
Triple diffused	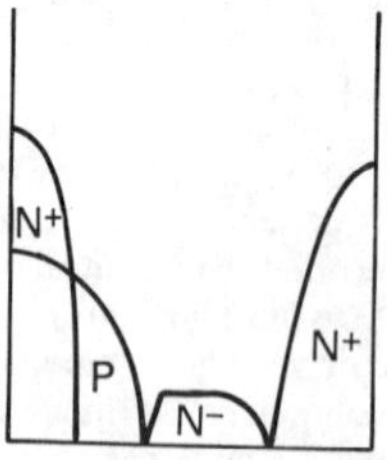 N+ P N− N+	High BV_{ce} Good switching speeds Good RMSOA	Moderate V_{CEsat} Only npn Large electrical characteristics distribution Fragile
Epicollector	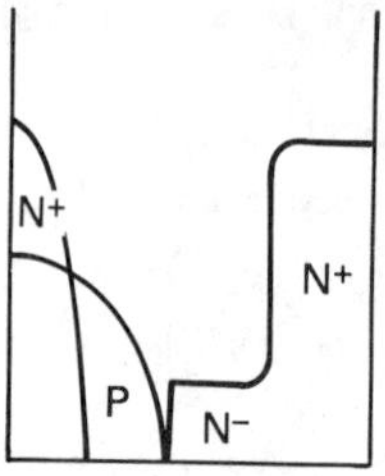 N+ P N− N+	npn/pnp Good V_{CEsat}, BV_{ces}, speed, ruggedness Narrow distribution	Limited BV_{ceo} (< 1200 V)
Epibase	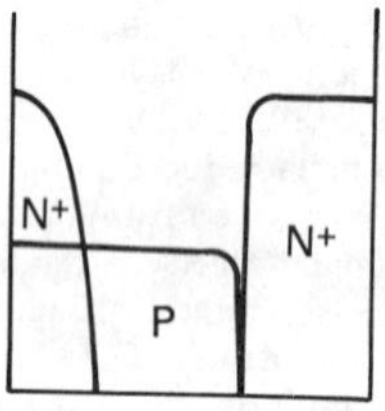 N+ P N+	npn/pnp Good SOA Quite fast	BV <250 V
Multiepitaxial	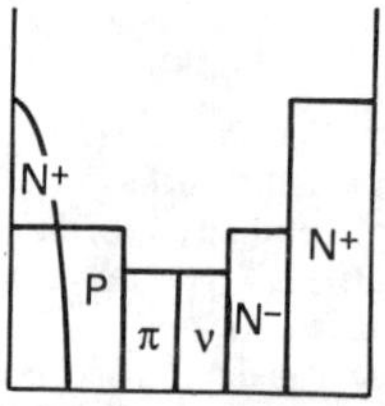 N+ P π ν N− N+	npn/pnp High ruggedness Good V_{CEsat}, H_{FE}, speed	Quite expensive

Simple packages like TO-3 or TO-218 are limited to 1500 V and around 40 A, so with the increase of power electronics the need for higher current is stronger and new big packages emerge. They are normally insulated and in plastic material. They may contain a Darlington or triple Darlington, but most of the time they are an inverter leg made with two Darlingtons in series, two diodes as freewheeling diodes, and all the external connections needed. Today there are products on the market with a working voltage of 1200 V and a working current of 600 A!

21.8.6 The diac

The word diac, defined by GE, signifies a diode alternating current switch, also called a bidirectional diode thyristor. A diac is in fact a three-layer device (npn) as the transistor structure, without the base terminal and with a symmetric doping profile. The device symbol and the static characteristic are given in *Figure 21.17*.

When an increasing voltage is applied between its two terminals, the breakdown voltage BV_{ceo} of the device is reached and the device switches on. When the external voltage is removed the device switches off.

A diac is used as a trigger device, mostly to drive a triac in a power relaxation oscillator.

21.9 Unipolar transistors

Instead of using both available carriers in the material, these types of device use only one carrier type, the majority carrier.

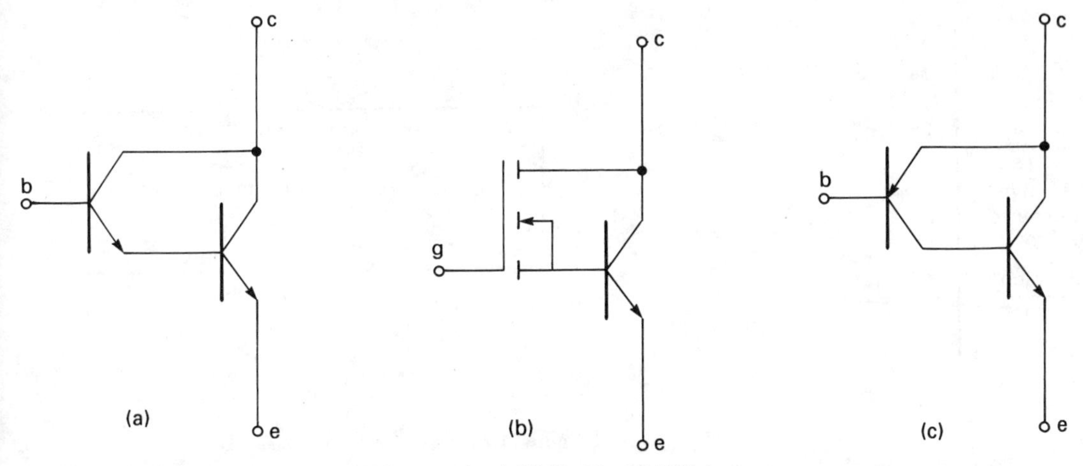

Figure 21.16 Darlington arrangements: (a) conventional, (b) MosBip, (c) Sziklai

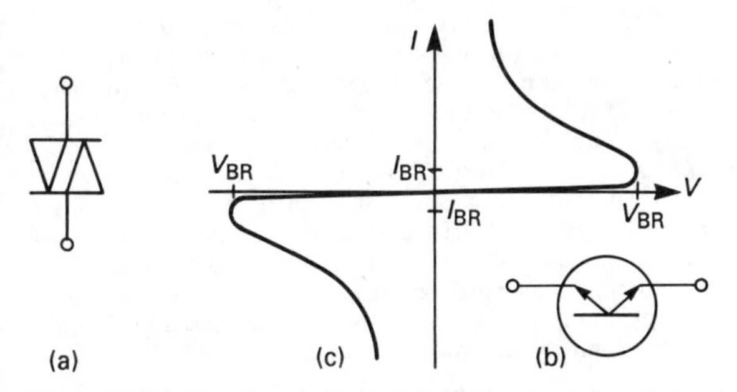

Figure 21.17 The diac: (a, b) symbols, (c) static characteristic

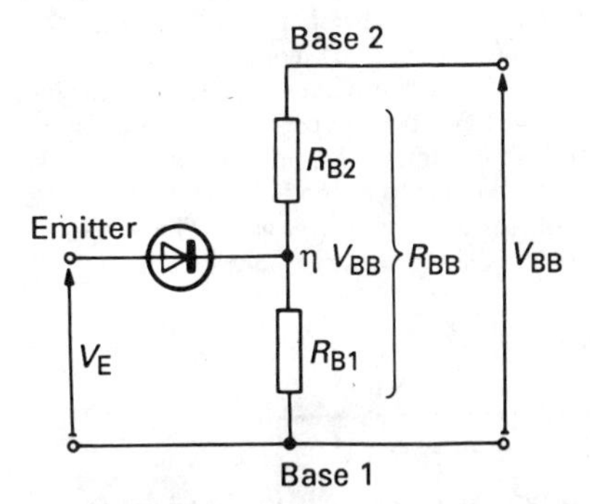

Figure 21.19 Equivalent circuit of unijunction transistor

21.9.1 Unijunction transistors

As the name implies, a unijunction transistor contains only one junction although it is a three-terminal device. The junction is formed by alloying p-type impurity at a point along the length of a short bar-shaped n-type silicon slice. This p-type region is called the emitter. Non-rectifying contacts are made at the ends of the bar to form the base 1 and base 2 connections. The structure of a unijunction transistor is shown in *Figure 21.18(a)*, and the circuit symbol in *Figure 21.18(b)*.

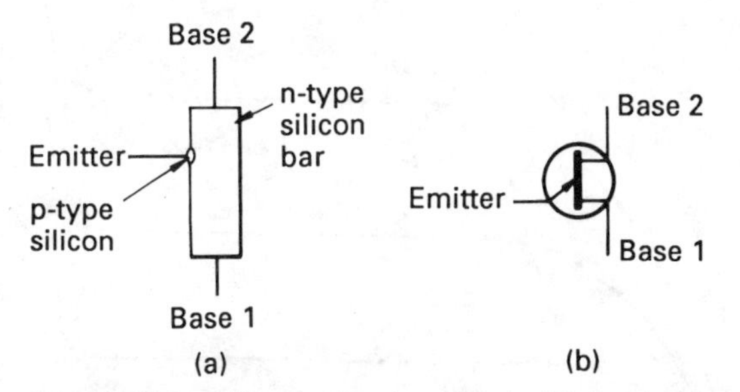

Figure 21.18 Unijunction transistor: (a) simplified structure, (b) circuit symbol

The resistance between the base 1 and base 2 connections will be that of the silicon bar. This is shown on the equivalent circuit in *Figure 21.19* as R_{BB}, and has a typical value between 4 kΩ and 12 kΩ. A positive voltage is applied across the base, the base 2 contact being connected to the positive terminal. The base acts as a voltage divider, and a proportion of the positive voltage is applied to the emitter junction. The value of this voltage depends on the position of the emitter along the base, and is related to the voltage across the base, V_{BB}, by the intrinsic stand-off ratio η. The value of η is determined by the relative values of R_{B1} and R_{B2}, and is generally between 0.4 and 0.8.

The emitter p–n junction is represented in the equivalent circuit by the diode. When the emitter voltage V_E is zero, the diode is reverse biased by the voltage ηV_{BB}. Only the small reverse current flows. If the emitter voltage is gradually increased, a value is reached where the diode becomes forward-biased and starts to conduct. Holes are injected from the emitter into the base, and are attracted to the base 1 contact. The injection of these holes reduces the value of R_{B1} so that more current flows from the emitter to base 1, reducing the value of R_{B1} further. The unijunction transistor therefore acts as a voltage-triggered switch, changing from a high 'off' resistance to a low 'on' resistance at a voltage determined by the base voltage and the value of η.

The voltage/current characteristic for a unijunction transistor is shown in *Figure 21.20*. It can be seen that, after the device has been triggered, there is a negative resistance region on the characteristic. This enables the unijunction transistor to be used in oscillator circuits as well as in simple trigger circuits.

21.9.2 Junction field-effect transistors

There are two types of field effect transistor, the junction field-effect transistor (JFET) and metal oxide semiconductor field-effect tran-

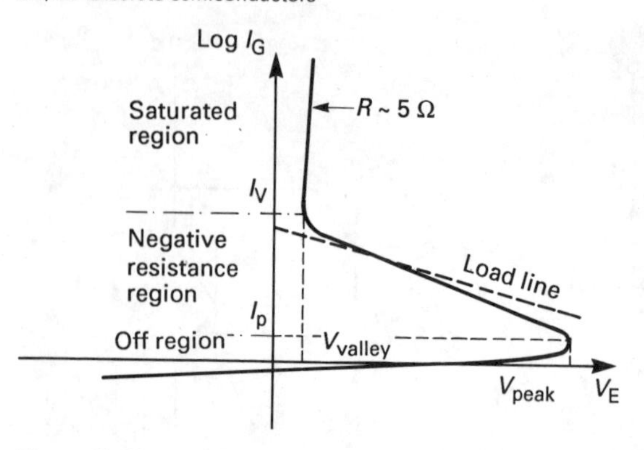

Figure 21.20 Unijunction transistor characteristic

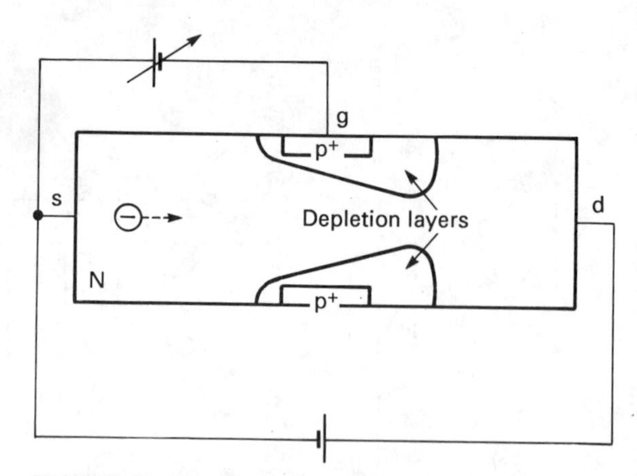

Figure 21.22 Operation of an n-channel JFET

sistor (MOSFET). The principles on which these devices operate are that current is carried by majority carriers (electrons in n-type semiconductors, holes in p-type semiconductors) and controlled by an electric field. The primary difference between them is in the method by which the control element is made.

The n-channel JFET is formed from a bar-shaped slice of n-type monocrystalline silicon into which two p-type junctions are diffused. Because of this the JFET is sometimes called a double-base unipolar transistor. Connections are made to the ends of the channel (the source and the drain), and to the p-regions (the gate). The p-channel JFET is formed by an opposite process (*Figure 21.21*).

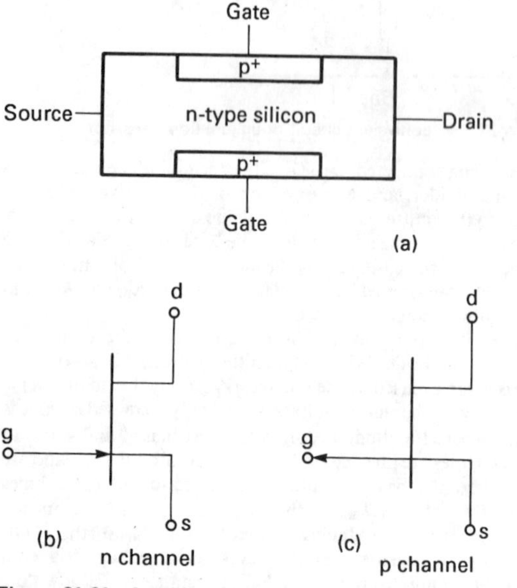

Figure 21.21 Schematic of a JFET: (a) structure, (b, c) symbols

21.9.2.1 Operation of JFETs

When the gate voltage is zero, if a poitive voltage supply is connected to the drain with the source grounded, a flow of current will take place in the silicon bar. As in any resistor, this current will cause a linear drop in voltage along the length of the channel. A gate-to-source voltage will start in the reverse direction, and a larger reverse bias at the gate-to-drain end. So the depletion layers given by the reverse biased p–n junctions will have the wedge shape shown in *Figure 21.22*.

As the drain-to-source voltage is increased, the drain current will increase linearly. However, the gate-to-channel voltage will also increase, so that the depletion layer will penetrate further into the channel. A point will be reached where the decrease in channel thickness has a greater effect than the increase in drain-to-source voltage. The characteristic of the drain current I_D plotted against drain-to-source voltage V_{DS} will therefore form a knee.

As V_{DS} is increased further, a value will be reached above which virtually no further increase in drain current will occur. This is because the depletion layer extends across the whole of the channel at the drain end of the gate and the channel is pinched off.

The drain current is maintained by electrons being swept through the depletion layer. Eventually as V_{DS} is increased further, the gate-to-channel breakdown value will be reached.

21.9.2.2 JFET characteristics

The operation just described is for zero gate voltage. If the gate-to-source voltage V_{GS} has a small constant negative value, the depletion layer will extend into the channel even with no drain current flowing. Thus the knee of the characteristic curve will occur at lower drain current, and the voltage at which the channel is pinched off occurs at lower value, V_{DS}. A higher negative value of V_{GS} will decrease the knee and pinch-off value further, as in *Figure 21.23*.

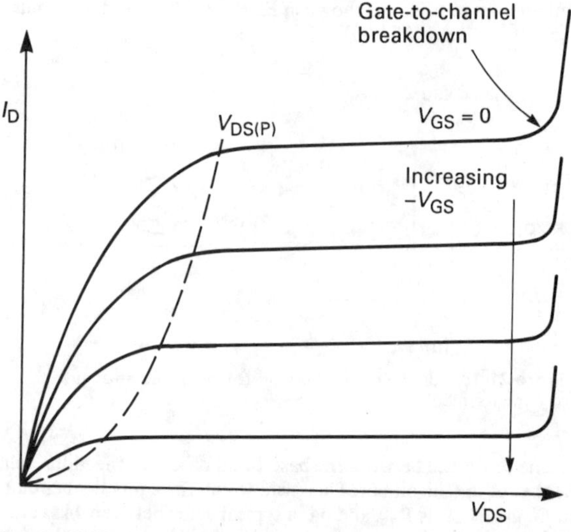

Figure 21.23 JFET drain current characteristics

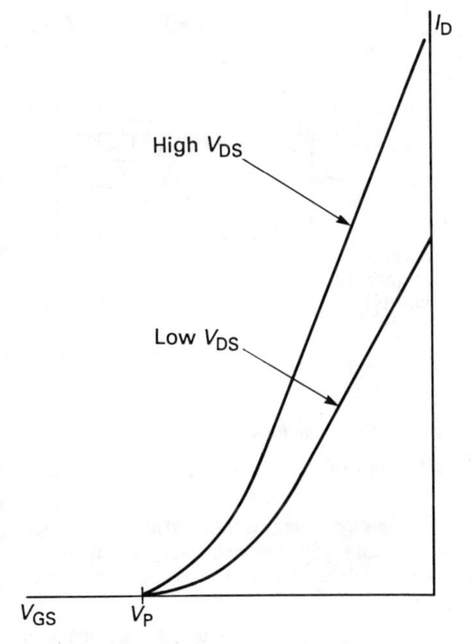

Figure 21.24 JFET transfer characteristics

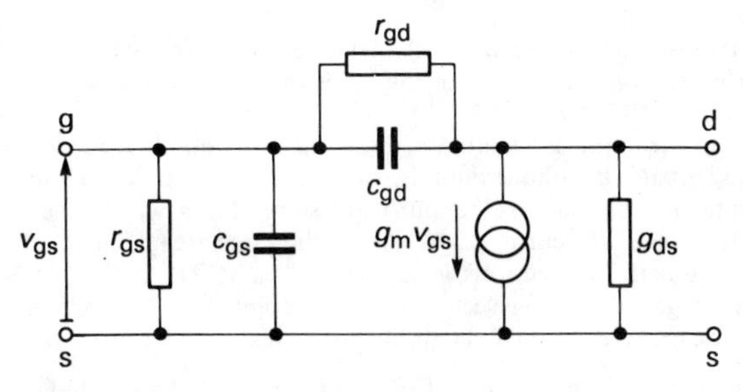

Figure 21.25 Small-signal equivalent circuit for a JFET

The part of the characteristic where V_{DS} is greater than the pinch-off value is called the pinch-off or saturation region.

The characteristic I_{out} versus V_{in} or transfer characteristic ($I_D = f(V_{GS})$ is given in *Figure 21.24*. The slope of this characteristic is the transconductance of the JFET, g_m. The value of g_m is given by

$$g_m = |\partial I_d / \partial V_{GS}| V_{DS} \tag{21.27}$$

Maximum ratings are given in the published data as maximum drain-to-gate voltage ($V_{DGO(max)}$) the maximum drain current I_d which will define the safe operating area of the JFET, in conjunction with the maximum power dissipated by the device, and the maximum gate current $I_{G(max)}$. The characteristics of the JFET are also temperature dependent.

An equivalent circuit for the JFET in the common source configuration under small signal conditions is shown in *Figure 21.25*. The gain is represented by the current generator $g_m V_{GS}$. The input resistance ($r_{GS} \approx 10^{11}$ Ω) is high because of the reverse bias of the gate. The input capacitance, c_{GS}, is the gate-to-channel capacitance formed by the depletion layer between the source and the pinch-off point. Feedback capacitance, or Miller capacitance, c_{GD}, is the capacitance of the depletion layer beyond the pinch-off point to the drain. Resistance, r_{GD}, is the gate-to-drain resistance which has a similar value to r_{GS}. The output conductance g_{DS} is the drain-to-source conductance. Both C_{GS} and C_{GD} are voltage dependent.

21.9.3 Metal oxide semiconductor field-effect transistors

The MOSFET planar technology presented by Khang and Atalla in June 1960 resulted in a reliable structure, as in *Figure 21.26*. The structure of the insulated gate FET differs from that of the JFET in two respects:

(1) The gate is not a diffused region but a thin layer of metal insulated from the rest of the FET by a layer, usually of oxide.
(2) A current carrying channel is formed by the accumulation of charges beneath the gate electrode acting as an electrostatic pilot.

21.9.3.1 Operation of MOSFET

With no voltage applied to the gate, only the depletion layers about the p–n junctions are present. Even if a large voltage is applied between source and drain, no current will flow; the output impedance is very high.

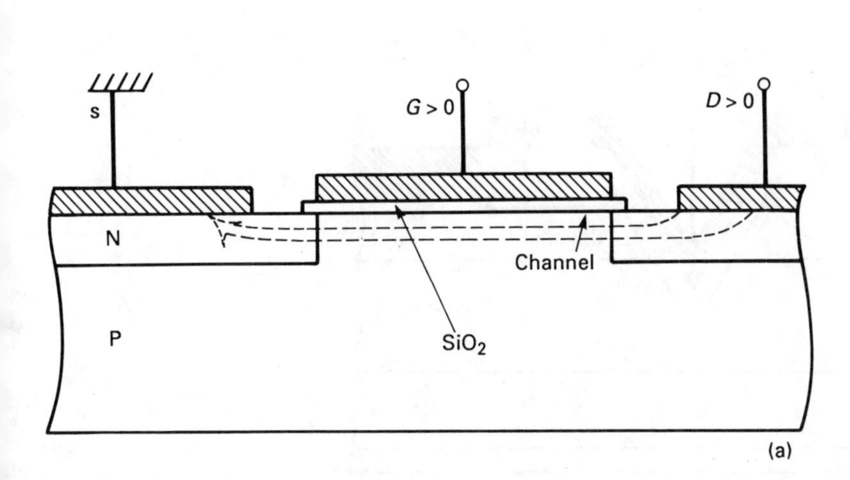

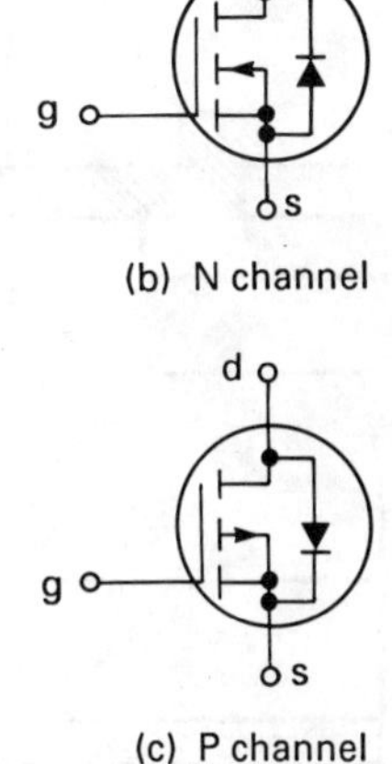

Figure 21.26 Simplified n channel MOSFET: (a) structure, (b, c) symbols

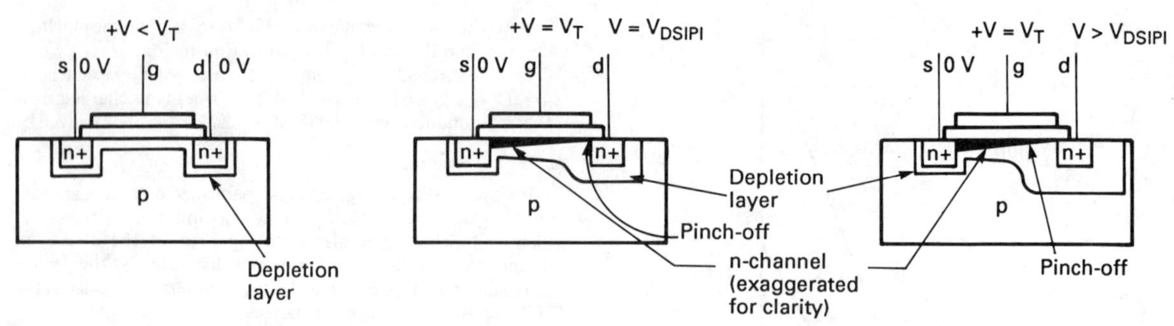

Figure 21.27 Operations of an n channel MOSFET

If a small positive voltage, with respect to the source, is applied to the gate, free holes in the substrate are repelled from the surface and a depletion layer is formed beneath the gate. If the gate voltage is increased further, free electrons from the souce will be attracted to the region under the gate oxide. In this way, a layer of electrons will be formed in the surface of p-type substrate. This layer is called an inversion layer. The gate to source voltage at which the inversion layer is just formed is called the threshold voltage V_{th}. The electrons in the inversion layer can be used as charge carriers by applying a drain-to-source voltage. A drain current will flow whose magnitude is determined by the drain-to-source voltage and the conductivity of the channel which in its turn depends on the charge by unit area in the channel. This charge density is proportional to the voltage difference between the gate and the channel.The flow of drain current will produce a voltage drop along the channel, leading to a reduction voltage between gate and channel towards the drain. The voltage distribution in the channel is then indicated in *Figure 21.27*, in which the thickness of the depletion layer between channel and substrate is seen to increase towards the drain.

This type of MOSFET, normally off with no voltage on the gate and where a positive voltage is needed to put it in a conduction mode, is called enhancement mode MOSFET. A normally on n channel MOSFET (like JFET) can exist, it is called depletion mode MOSFET. Very few MOSFETs in this mode are available on the market, so only the enhancement mode MOSFET will be described.

21.9.3.2 Power MOSFET technologies

Currently, three families of power MOSFET are in use, as described below.

Coplanar structure This technology is still used for its very good transconductance linearity and complementary capability in HiFi amplifiers. A field plate is built to allow working voltages to around 200 V.

Non-planar structure, vertical channel The main drawback of the previous structure is a quite small current density that could be carried by this technology. So, in order to increase the silicon efficiency (higher current density), the use of the vertical part of the silicon became mandatory and the drain connection was made beneath the chip.

To make a quite vertical channel an anisotropic etching was performed on the surface of the chip: the V groove was born (Sillconix 1975), *Figure 21.28*.

This structure is capable of high current, mobility carriers being greater in volume than the surface. Each chemical etching creates a V shape across the source and so produces two channels which double the current capability of the same area. Control of the geometric dimensions is simple and easy. The major disadvantage is the strong electric field at the point of the V; this high electric field creates rollability problems on high-voltage products.

Horizontal channel, vertical flow of current This cellular structure, the most widely used one today on the market, was

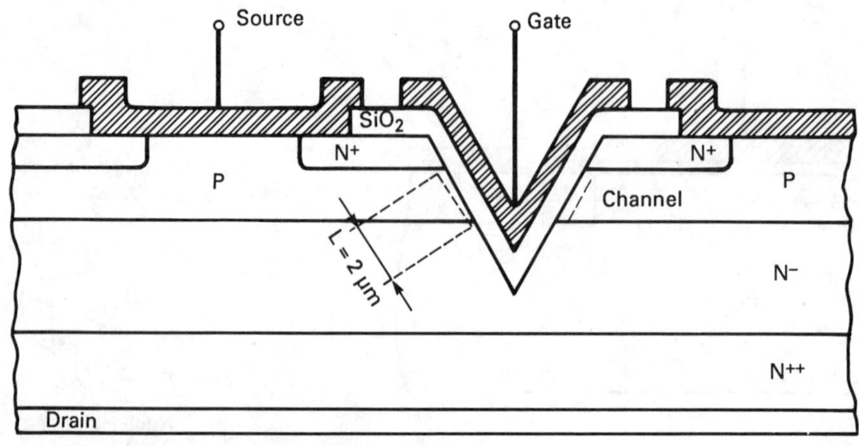

Figure 21.28 Vertical structure of power MOSFET (Sillconix)

created by International Rectiflor in 1979, followed by a lot of manufacturers like Motorola and Siemens in 1980.

The advantages of this structure are:

(1) The gate is above the low-doped drain region.
(2) The source metallisation is above the chip which acts as a field plate.
(3) The cellular structure of the internal parasitic diode will act as the shield field for the gate-to-drain region and allows a very high voltage device to be built with a low maximum gate-to-drain voltage $V_{GDmax} \approx 50$ V.
(4) Using a fairly simple diffusion process, without too much masking, the polysilicon gate enabling self-alignment of source windows, and results in the best R_{DSon}/BV_{DSS} compromise.

21.9.3.3 *MOSFET transistor characteristics*

The practical model of low frequency MOSFET is given in *Figure 21.29* and its characteristics in *Figure 21.30*. This model will give a quick overview of the main characteristics of the MOSFET transistor.

Seen from the input, the MOSFET is like a capacitance: C_{GS} the geometrical capacitance gate-to-source and C_{GD} the feedback dynamic capacitance drain-to-gate, also called the Miller capacitance.

Seen from the output the MOSFET is like a resistance R_{DSon}, a current source $G_{fs} \times V_{DS}$ and a capacitance C_{DS} given by the internal parasitic diode.

In *Figure 21.29*, another equivalent model is given, which explains some of the VDMOS characteristics by the JFET in series with the MOSFET itself. This JFET is given by the depletion layer created by the parasitic internal diode. When the drain-to-source voltage V_{DS} increases this depletion layer extends and pinches the part of the drain as in the JFET, which explains why the current drain remain constant when the drain voltage increases.

R_{DSon} characteristic The internal R_{DSon} resistance can be broken in three parts: the channel resistance R_{ch}, the resistance R_{ac} between two cells and the drain resistance R_d, the n-epitaxial resistive part of the drain. R_{ch} is linearly related to the die geometrical factors, channel length L and distance between two cells d, and not to the breakdown voltage of the device. R_{ac} is linearly related to the breakdown voltage of the MOSFET.

The low-voltage MOSFET (< 100 V) has a low R_{DSon} by unit area and directly related to the cell density with an optimum given by the JFET limit phenomenon, and the high-voltage MOSFET is limited only by the p–n junction (parasitic diode) breakdown voltage.

R_{DSon} increases when the temperature increases, so any loss calculations have to take into account the R_{DSon} at the working junction temperature.

Internal capacitances These internal capacitances are related to the drain-to-source (V_{DS}) and gate-to-source (V_{GS}) voltages.

Leakage currents below the threshold voltage V_{th} These currents, gate-to-body and drain-to-source, are very low (less than 10^{-9} A).

Forward characteristic The curves $I_D = f(V_{DS})|V_{GS}$ and $V_{DS} = f(V_{GS})|I_D$ are given in *Figure 21.30*. When drain source increases, at the constant gate-to-source voltage, the drain current remains constant (*Figure 21.30(a)*). *Figure 21.30(b)* shows the three main MOSFET working regions—the off part, the active or linear amplifier part, and the saturated or switching part.

Transconductance $G_{fs} = \partial I_D/\partial V_{GS}$ The gain or the transconductance of the MOSFET is the change of I_{DS} with V_{GS}. When $V_{th} < V_{GS} < (V_{GS} + 2V)$, the transconductance G_{fs} is in quadratic form:

$$I_D = k(V_{GS} - V_{th})^2 \quad (21.28)$$

After that, G_{fs} is constant.

The 'gain' of the power MOSFET decreases when the temperature increases, so a MOSFET does not exhibit thermal runaway.

Safe operating areas of a MOSFET The forward safe operating area of a MOSFET is only limited by $R_{DSon} \times I_{DS}$, the maximum current in the connecting wires, the maximum dissipated power in the device, and the breakdown voltage BV_{DSS}. A MOSFET is very rugged for sustaining a short circuit current, because the short circuit current is self-limited at $I_{SC} = G_{fs} \times V_{GS}$.

MOSFETs were susceptible to destruction by BV_{DSS} breakdown phenomenon. In 1986, a lot of improvements were made in the structure in order to sustain the avalanche energy. Among these improvements the internal parasitic diode was built in such a fashion that its breakdown voltage protects the MOSFET itself, like a zener diode.

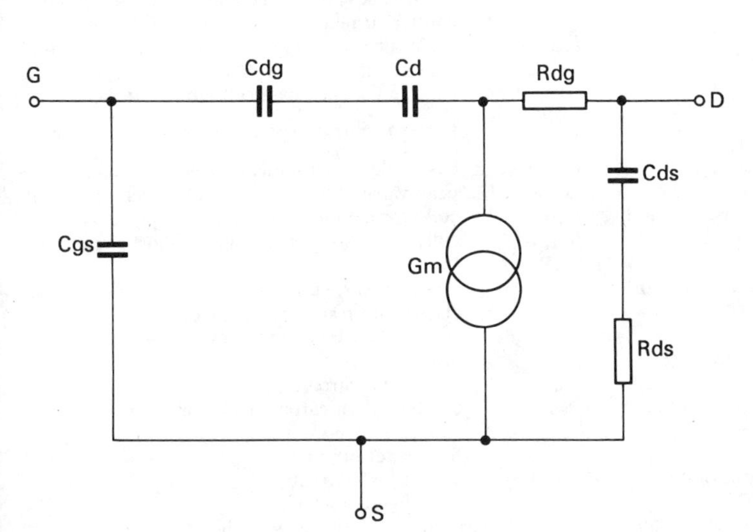

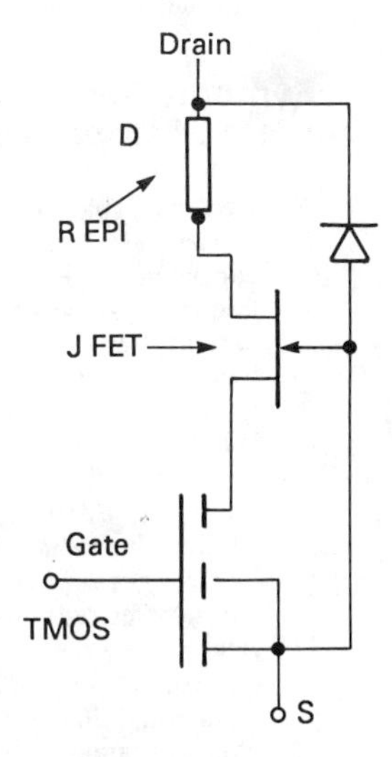

Figure 21.29 Low voltage equivalent circuit of a VD MOSFET

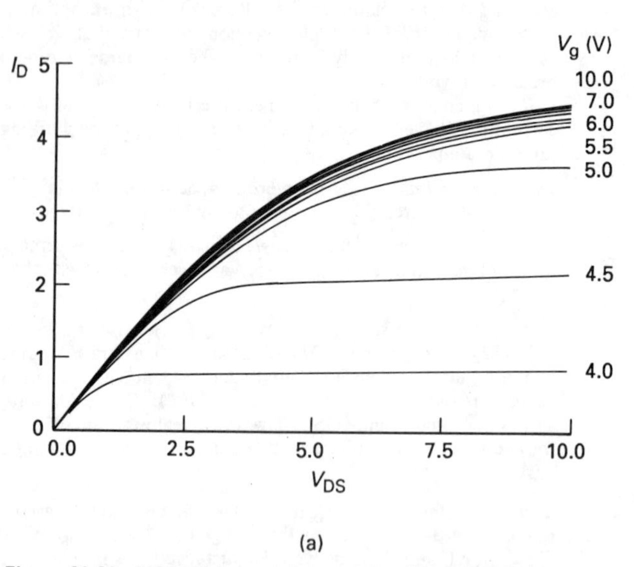

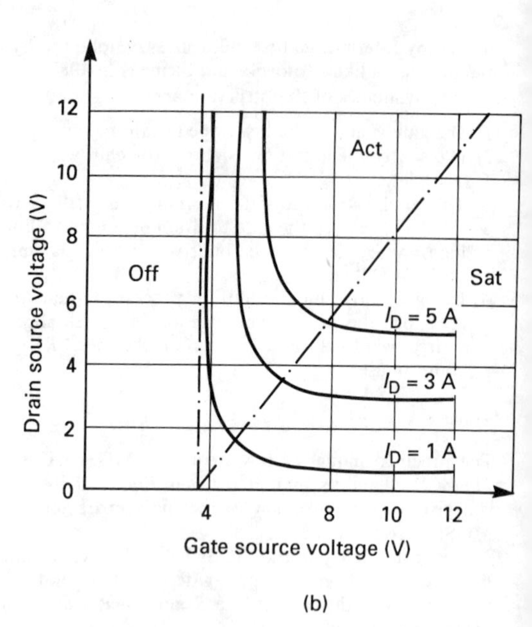

Figure 21.30 MOSFET forward characteristics (Motorola MTM 5N40): (a) I_D *versus* V_{DS}, (b) V_{DS} *versus* V_{GS}

The gate oxide thickness is very low, around 100 nm. This oxide can sustain around 50 V and published data give a maximum V_{GS} at 20 V, so a fast zener must be mounted in parallel with gate–source terminals to protect the gate oxide against any overvoltage.

Speed-gate drive The model of MOSFETs shows that this product is like a capacitance seen from the input, so the commutation speed of the MOSFET is only related to the speed of the capacitance charge:

$$t = CV/I_g \qquad (21.29)$$

C being the total input capacitances of the MOSFET, V the V_{GS} driving voltage and I_g the gate charge current.

Tens of nanoseconds transient switching can be obtained, so a MOSFET is a good candidate for high working frequency applications, in systems between 100 Hz and 1 MHz. Junction temperature has a very small influence on the MOSFET internal capacitances and switching speed.

21.10 Symbols for main electrical parameters of transistors

C_{rs}	feedback capacitance in field-effect transistor
C_{Tc}	capacitance of collector depletion layer
C_{Te}	capacitance of emitter depletion layer
f_{hfb}	frequency at which the common-base forward current transfer ratio has fallen to 0.7 × low-frequency value
f_{hfe}	frequency at which the common-emitter forward current transfer ratio has fallen to 0.7 × low-frequency value
f_T	transition frequency (common-emitter gain–bandwidth product)
f_1	frequency at which common-emitter forward current transfer ratio has fallen to 1
G	gain
g_m	transconductance of field-effect transistor
h_{fb}, h_{fc}, h_{fe}	small-signal forward current transfer ratio for transistor configuration indicated by second subscript, output voltage held constant
h_{FB}, h_{FC}, h_{FE}	static forward current transfer ratio for transistor configuration indicated by second subscript, output voltage held constant
h_{ib}, h_{ic}, h_{ie}	small-signal input impedance for transistor configuration indicated by second subscript, output short-circuited to alternating current
h_{ob}, h_{oc}, h_{oe}	small-signal output impedance for transistor configuration indicated by second subscript, input open-circuit to alternating current
h_{rb}, h_{rc}, h_{re}	small-signal reverse voltage transfer ratio (voltage feedback ratio) for transistor configuration indicated by second subscript, output voltage held constant
i_b, i_c, i_e	instantaneous value of varying component of base/collector/emitter current
i_B, i_C, i_E	instantaneous value of total base/collector/emitter current
I_b, I_c, I_e	r.m.s. value of varying component of base/collector/emitter current
I_{bm}, I_{cm}, I_{em}	peak value of varying component of base/collector/emitter current
I_B, I_C, I_E	continuous (d.c.) base/collector/emitter current
$I_{B(AV)}$, $I_{C(AV)}$, $I_{E(AV)}$	average value of base/collector/emitter current
I_{BEX}, I_{CEX}	base/collector cut-off current in specified circuit
I_{BM}, I_{CM}, I_{EM}	peak value of total base/collector/emitter current
I_{CBO}	collector cut-off current, emitter open-circuit
I_{CBS}, I_{CES}	collector cut-off current, emitter short-circuited to base
I_{CBX}	collector current with both junctions reverse biased with respect to base
I_{CEO}	collector cut-off current, base open-circuit
I_D	drain current
I_{DM}	peak drain current
I_{EBO}	emitter cut-off current, collector open-circuit
I_G	gate current
I_{GM}	peak gate current
I_{GSS}	gate cut-off current
N	noise figure
P_{tot}	total power dissipated within device
r_{DSoff}	drain-to-source off resistance

r_{DSon}	drain-to-source on resistance
$R_{th(j\text{-}amb)}$	thermal resistance, junction to ambient
$R_{th(j\text{-}case)}$	thermal resistance, junction to case
$R_{th(j\text{-}mb)}$	thermal resistance, junction to mounting base
T_{amb}	ambient temperature
T_{jmax}	maximum permissible junction temperature
T_{mb}	mounting base temperature
T_{stg}	storage temperature
t_{off}	turn-off time
t_{on}	turn-on time
V_{BE}	base-to-emitter d.c. voltage
V_{BEsat}	base-to-emitter saturation voltage
V_{CB}	collector-to-base d.c. voltage
V_{CBO}	collector-to-base voltage, emitter open-circuit
V_{CC}	collector d.c. supply voltage
V_{CE}	collector-to-emitter d.c. voltage
V_{CEK}	collector knee voltage
V_{CEO}	collector-to-emitter voltage, base open-circuit
V_{CEsat}	collector-to-emitter saturation voltage
V_{DB}	drain-to-substrate voltage
V_{DG}	drain-to-gate voltage
V_{DGM}	peak drain-to-gate voltage
V_{DS}	drain-to-source voltage
V_{DSM}	peak drain-to-source voltage
V_{EB}	emitter-to-base d.c. voltage
V_{EBO}	emitter-to-base voltage, collector open-circuit
V_{GB}	gate-to-substrate voltage
V_{GBM}	peak gate-to-substrate voltage
V_{GS}	gate-to-source voltage
V_{GSM}	peak gate-to-source voltage
$V_{GS(P)}$	gate-to-source cut-off voltage
V_{GSO}	gate-to-source voltage, drain open-circuit
V_P	pinch-off voltage in field-effect transistor
V_{SB}	source-to-substrate voltage
V_T	threshold voltage in field-effect transistor
η	intrinsic stand-off ratio in unijunction transistor

21.11 The thyristor family

The name thyristor is an analogy to the thyratron valve, and is the generic name for a whole family of semiconductors having three junctions (four layers).

These devices, which can have two, three or four external terminals, are bistable (fully off or fully on) in operation, and may be uni- or bi-directional.

Table 21.2 shows the whole thyristor family, with their symbols and main characteristics.

21.11.1 The thyristor

The original name for this structure was the reverse blocking triode thyristor. As it is the most commonly used device in the thyristor, it has generally become known simply as the thyristor.

It is convenient to regard the thyristor as a silicon controlled rectifier (SCR). Like a rectifier diode, the thyristor will conduct a load current in one direction only, but the current can flow only when the thyristor has been triggered. It is this property of being rapidly switched from the non-conducting to the conducting state that enables the thyristor to be used for the control of electrical power.

21.11.2 Operation of thyristors

The schematic structure and the analogy of the thyristor are given in *Figure 21.31*. It is a four-layer structure with three terminals called the anode, cathode and gate.

Table 21.2 The thyristor family

Initials	*Names and typical ratings*	*Symbol*	*Static characteristics*
SCR	Silicon controlled rectifier 5 kV, 5 kA		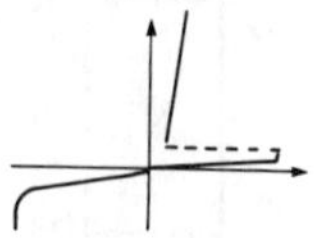
ASCR	Asymmetrical silicon controlled rectifier 2 kV, 2 kA		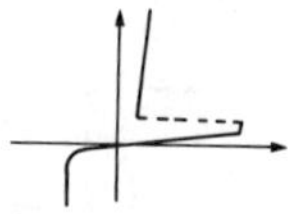
Triac	Triode a.c. switch 800 V, 50 A		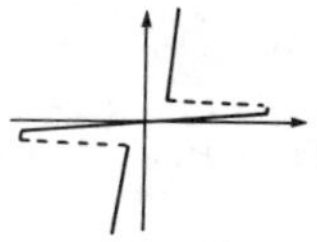
GTO	Gate turn-off thyristor 4 kV, 4 kA		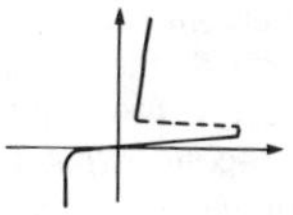
SIDAC	Switching diode SIDAC Breakover diode 400 V, 1 A		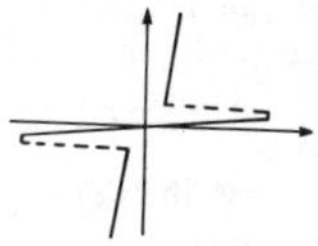
PUT	Programmable unijunction transistor 40 V, 15 A		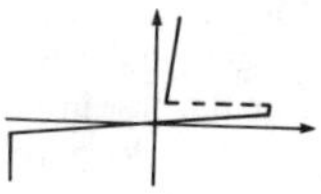
SUS	Silicon unilateral switch 30 V, 1 A	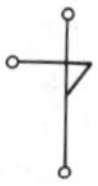	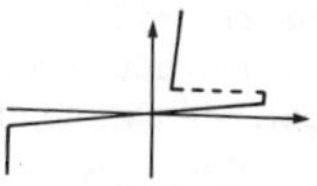
SBS	Silicon bilateral switch 30 V, 1 A	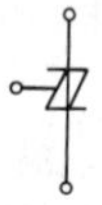	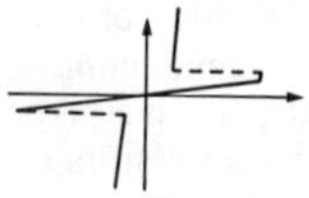
SCS	Silicon controlled switch 100 V, 1 A	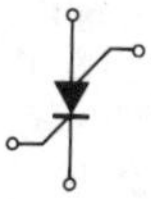	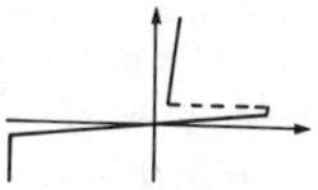

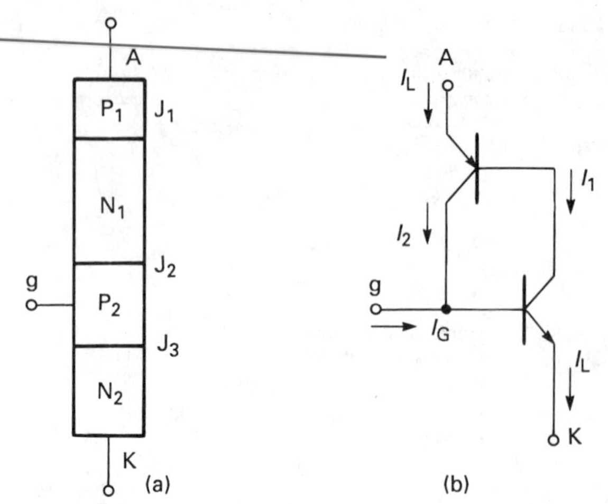

Figure 21.31 Schematic structure and analogy of a thyristor: (a) structure, (b) two-transistor analogy

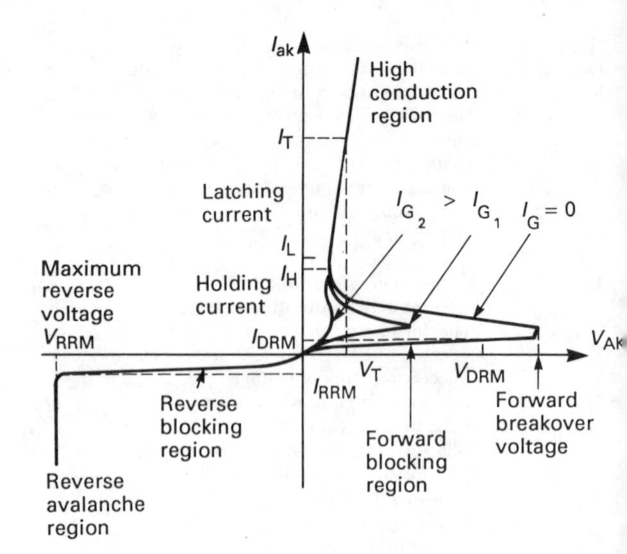

Figure 21.32 Static characteristic of a thyristor

The four-layer structure may be considered as forming two interconnected transistors: with only the anode to cathode voltage applied, the current through the thyristor is the small reverse current flowing across junction J_2. The thyristor is in the non-conducting forward blocking state. When the small positive voltage is applied to the gate, electrons flow from the cathode to the gate. This cathode–gate current forms the emitter base current of the npn transistor, and by normal transistor action some of the electrons cross junction J_2 to enter the n collector region. Because the gate region is thicker than the base region of the transistor, the current gain will be less than the unity.

The electron flow across junction J_2 causes the depletion layer to become narrower. The proportion of the anode to cathode voltage across the depletion layer is reduced, and the forward voltage across junctions J_1 and J_3 is increased. The flow of holes from the anode to the n-region (emitter–base current of the pnp transistor) is increased by this increase in forward bias across junction J_1. By normal transistor action, some of these holes cross junction J_2 to the gate (collector) region. Again, because the n region is thicker than the base in a transistor, the current gain is less than unity. The flow of holes across junction J_2 further reduces the width of the depletion layer. Consequently the proportion of the anode-to-cathode voltage across the depletion layer is reduced further, the forward voltage across junctions J_1 and J_3 is increased further, and the electron flow in the npn transistor is inreased further.

This cumulative action, once initiated by the application of the gate voltage, continues until the depletion layer at junction J_2 collapses. The anode-to-cathode impedance becomes very small, and a large current flows through the thyristor.

This current flow is self-sustaining, the thyristor is now in the forward conducting state, and the gate voltage can be removed. The value of current at which the thyristor changes from the non-conducting to the conducting state is called the latching current, and this occurs when the product of the gain current of the two transistors reaches the unity.

The input current for the npn transistor is the gate current I_G. The collector current for this transistor is I_1 as shown in *Figure 21.31(b)*, and this is the base current for the pnp transistor. The collector current of the pnp transistor I_2 is fed back to the base of the npn transistor. If the current gain of the npn transistor is H_{FE1}, then:

$$I_1 = H_{FE1}(I_G + I_2) \tag{21.30}$$

If the gain of the pnp transistor is H_{FE2}, then:

$$I_2 = H_{FE2} I_1 \tag{21.31}$$

The load current through the thyristor, I_L, is the sum of I_1 and I_2 and can be found from the previous equations as:

$$I_L = \frac{H_{FE1} I_G (1 + H_{FE2})}{1 - H_{FE1} H_{FE2}} \tag{21.32}$$

The load current becomes infinite, that is the value will depend on the load resistance outside the thyristor rather than on the thyristor itself, when the denominator in the last equation is zero. That is:

$$H_{FE1} H_{FE2} = 1 \tag{21.33}$$

This is the condition for the thyristor to turn on and remain in the forward conduction state.

The thyristor is turned off by reducing the current through the device below a value called the holding current. This is often done by making the anode of the thyristor negative with respect to the cathode.

21.11.3 Thyristor characteristics and ratings

21.11.3.1 Static characteristic of the thyristor

The static characteristic of the thyristor is shown in *Figure 21.32*. With a reverse voltage applied, the voltage–current characteristic is similar to that of any reverse-biased semiconductor diode. A small reverse current flows as the reverse voltage is increased until the reverse avalanche region is reached. In the forward direction, if the thyristor is not triggered, a small forward leakage current flows until the forward avalanche region is reached. This is the non-conducting forward blocking state. In the avalanche region, the leakage current increases until, at the forward breakover voltage V_{BO}, the thyristor switches rapidly to the conduction state. If the thyristor is triggered by a positive voltage applied to the gate when the anode-to-cathode voltage is less than V_{BO}, the thyristor again switches rapidly to the conducting state. In the conducting state, the thyristor behaves as a forward-biased semiconductor diode.

The trigger pulse applied to the gate must remain until the current through the thyristor exceeds the latching value I_L. This is very important with a highly inductive load. If the trigger voltage is removed before the latching value has been reached, the thyristor will stop conducting when the pulse ends. In practice, thyristors are triggered by trains of pulses rather than a single pulse.

Also shown on the static characteristic is the holding current. If the load current through the thyristor falls below this value, the thyristor rapidly switches off to the non-conducting state.

21.11.3.2 Gate control

A gate characteristic is often given in published data. The form of this characteristic is shown in *Figure 21.33*. The two solid curves are the limits (min–max) of the total distribution of one family of devices. The curve in broken lines defines the maximum permissible gate dissipation. The operating point defined by the resistance of the drive loop which intersects V_{GS}/I_g, and its variation with the temperature, should be well below this limit.

The minimum voltage and minimum current to initiate turn-on are also shown. These minima are temperature dependent, and the area of uncertain triggering will decrease as the junction temperature increases.

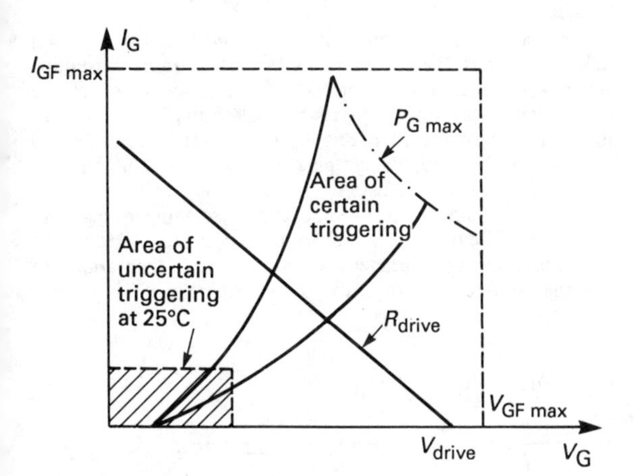

Figure 21.33 Thyristor gate characteristic

21.11.3.3 Dynamic characteristics of the thyristor

When a gate drive voltage is applied, the positive anode voltage with respect to the cathode of the thyristor starts to decrease at the end of a time called the 'delay time', t_d (see *Figure 21.34*). Then the thyristor switches rapidly to the conduction state. The rise time, t_r, of the anode current is also defined by the time needed by the anode-to-cathode voltage to fall from 90% to the 10% level.

The rate of rise and the value of the current trigger pulse appreciably affect the turn-on time of the thyristor. At the end of the turn-on time t_d, only a part of the chip close to the gate is conducting. From here conduction spreads throughout the silicon at a rate of approximately 100 μm/μs.

So for each type of thyristor it is essential to define a critical limit of increase in load current dI/dt. If the gradient of the load current is greater than the permitted maximum, the value can be limited by connecting a small inductance in series with the thyristor.

The thyristor is turned off by reducing the current through the device to below the holding value, I_H. In practical circuits, the thyristor is often reverse biased to ensure that it is reliably turned

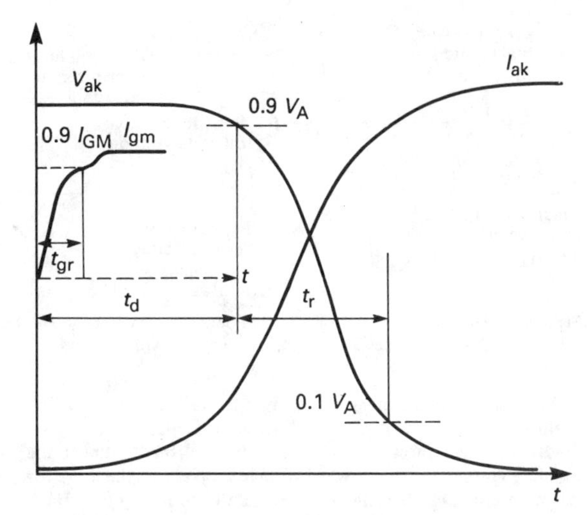

Figure 21.34 Switching characteristic of a thyristor

off. The reverse bias must be applied for a minimum time called the turn-off time, t_q.

The turn off time is the time needed by the minority carriers to recombine, mainly in the central layers. The forward voltage can be reapplied only when these excess carriers are recombined. The value of turn-off time increases with the junction temperature, so for fast switching it is necessary to reduce the lifetime of minority carriers. Heavy materials or irradiation can be used to create a recombination centre.

When a fast rate of change of anode-to-cathode voltage dV/dt is applied to the anode of the thyristor a capacitive current, due to the reverse-biased centre junction capacitance, may be injected to the gate, as follows:

$$I_G = C\frac{dV}{dt} \tag{21.34}$$

If this current is sufficiently large the device may be inadvertently turned on.

This dV/dt limitation depends upon junction temperature and state of the product before its application of a stable blocking state, or reapplied voltage.

The spurious triggering given by a high dV/dt can be prevented in practice by using 'snubber' components such as capacitance and resistance networks connected between anode and cathode of the thyristor to slow down the rate of rise of the applied voltage.

21.11.3.4 Maximum ratings

It is convenient to group the ratings of thyristors into anode-to-cathode voltage ratings, current ratings, temperature ratings and gate ratings.

Because thyristors can sustain reverse and forward voltages in non-conducting states, reverse and forward repetitive (or continuous) and non-repetitive voltage ratings can be specified. Sometimes reverse and forward characteristics are symmetric, so the published data give V_{DRM} or V_{RRM} for the repetitive peak voltage and V_{RSM} for the non-repetitive reverse blocking voltage. The repetitive and the non-repetitive ratings are determined by the breakover voltage V_{BO} for the forward direction and by the avalanche breakdown for the reverse one. These ratings are also determined by the energy carried by these transients and therefore by the maximum related junction temperature.

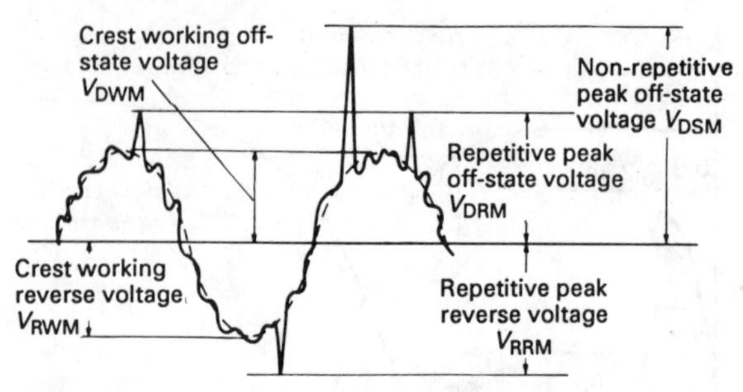

Figure 21.35 Diagrammatic relationship between thyristor voltage ratings and typical mains supply waveform with transients

The relationship between the voltage ratings and a typical main supply waveform is shown in *Figure 21.35*.

Other voltage ratings usually given in published data are the on-state voltage V_T, the forward voltage drop across the thyristor when conducting a specified current, and the maximum dV/dt rating.

As with the voltage ratings, the current that can be conducted by a thyristor is specified by the on-state current $I_{T(RMS)}$, the average current $I_{T(AV)}$ and the peak non-repetitive surge current I_{TSM}. This surge current is given for a half sinusoid of 60 or 50 Hz (8.3 or 10 ms). It may be applied once when the junction temperature is at its maximum.

Two other current ratings are given in published data, the I^2t and the dI/dt rating. The surge current capability I^2t is required for selecting fuses to protect the thyristor against excessive current being drawn from the device if a short circuit occurs in a load circuit. Other data tied to the surge current capability sometimes given as ratings or charts are I_{max} and I_{peak} *versus* time or mains supply cycles.

The power that can be dissipated with a thyristor (as with all semiconductor devices) is limited by the maximum permissible junction temperature.

It is usual to mount the device on a heatsink to increase the cooling performance. The total thermal resistance of the thyristor mounted on its heatsink is

$$\sum R_{th} = (T_{jmax} - T_{amb})/P_{tot} \quad (21.35)$$

Curves giving the various thermal resistances, $R_{th(j-c)}$ and $R_{th(h-a)}$, are published. Maximum operating junction temperature is available, so the maximum power dissipated by the thyristor can be calculated by Equation (21.35). For a particular application, the average current and the conduction angles are known, and so is the total dissipated power.

The principal gate ratings are the peak gate power, P_{GM}, the average gate power, $P_{AG(AV)}$, the peak gate current, I_{GM}, the minimum instantaneous gate current to initiate the turn-on, I_{GT}, and the minimum instantaneous gate voltage to initiate the turn-on, V_{GT}.

21.11.4 Manufacture of thyristors

Three manufacturing processes are used for thyristors; the diffused-alloyed, the all-diffused and the planar processes. The diffused-alloyed process is the older, and was used for the first devices.

The manufacture of both diffused-alloyed and all-diffused pellets starts with the pnp structure. An n-type monocrystalline silicon slice is chosen to provide the characteristics of the intermediate n-layer of the thyristor; sustain the forward breakdown voltage of the thyristor, V_{BO}, but with a thickness to prevent 'punch-through'; and give the lowest possible forward drop voltage, V_T. A diffusion process is used to form p-type regions on both sides of the n-type slice, thus forming the gate and the anode of the thyristor as shown in *Figure 21.36*. Depending on the final device characteristics, an n-layer is then added to form the cathode of the thyristor, or the original wafer is cut into

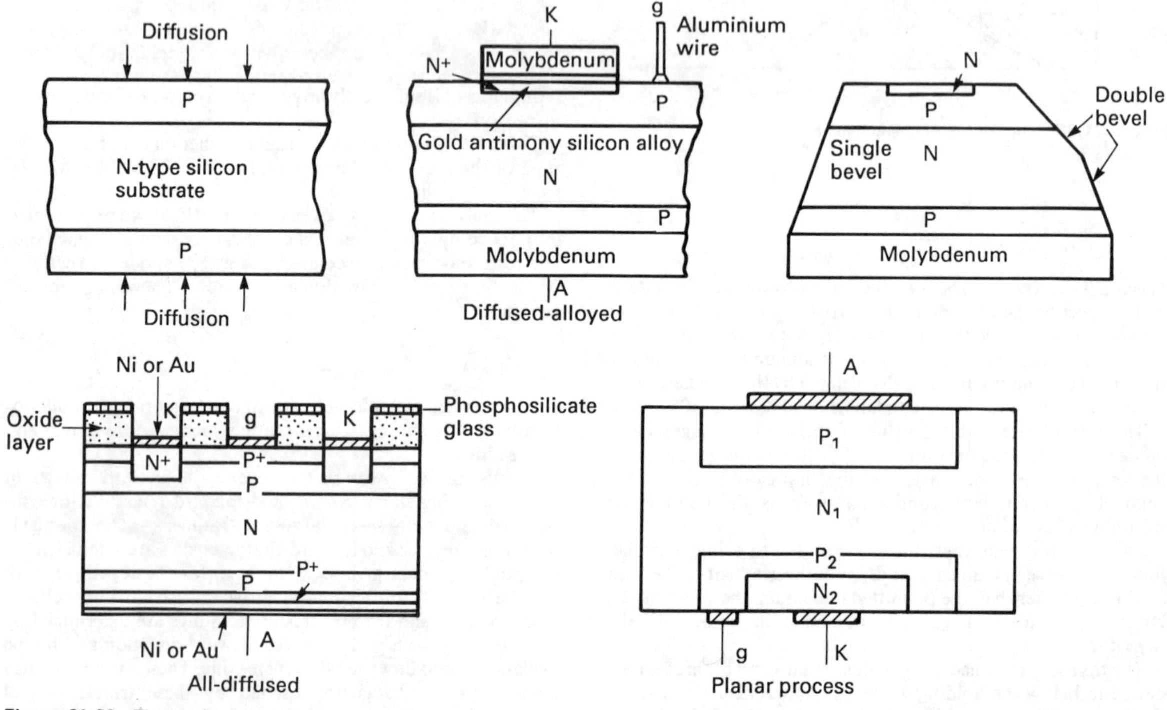

Figure 21.36 Thyristor constructions

a round slice whose diameter is determined by the required current capability. For the highest thyristor characteristics, several kilovolts and several kiloamperes, only one thyristor is processed in one wafer with, presently, a diameter as large as 75 mm.

For the diffused-alloyed process the n-layer is given by a preform of gold–antimony which is fused onto the p-cathode layer. For the diffused process, one side of the pnp pellet is selectively masked and subsequently diffused with n-type impurities through the windows of the mask.

A 'planar' structure describes a type of pellet where all the junctions rise to the upper surface of the wafer. The principal advantage of this technology is the thin layer of silicon dioxide grown over the pellet which prevents contamination of the silicon surface, giving low leakage currents and high reliability. The disadvantage of this technology is lower silicon efficiency, since more silicon is required per ampere of current-carrying capability. Planar structures are best suited to low current devices where many products can be obtained from a single wafer.

With other techniques than the planar, the edges of the dice can be bevelled or double bevelled to stretch the electric field on the surface, which is very importnt in sustaining the breakdown voltage and from the reliability point of view. A glass passivation will protect the open junctions.

Thyristor structure improvements To allow a highest possible dI/dt, the gate geometry is of importance, and the interdigitated gate, centre gate or spiral gate is used for this purpose.

For large thyristors, these improvements are not sufficient to fire the silicon in a short time, so an 'amplifying gate' can be implemented. This consists of a small pilot thyristor, which will quickly control the large main structure.

For the highest possible dV/dt it is useful to place a resistance between gate and cathode to give the parasitic current another path than the gate cathode injecting junction. This is called 'shorted emitter' technology.

The asymmetrical thyristor Continuing technological improvements on thyristors have resulted in reducing t_q, thus increasing operating frequency. Increasing the forward current density through efficient use of silicon, reduces the on voltage V_T, and thus reduces losses, but to the detriment of reverse blocking voltage; hence the name asymmetric thyristor (ASCR). This is not of great importance, since for most applications a reverse current is passed through an anti-parallel diode, or reverse voltage can be blocked with a series diode.

In the ASCR, a highly doped layer is added at the junction J_1 with the aim of stopping the extension of electric field. This is shown as the cross-hatched portion of *Figure 21.37*. For the same V_{DM} and N_1 resistivity, t_q and switching losses may be reduced without increasing on-losses.

Thyristor packaging Several types of package are used for the thyristor family. Small plastic packages such as TO-92, TO-220 or TO-218 are limited to 1500 V and 55 A.

The assembly process for this type of package is the same as that used for the bipolar transistor; soft solder (Pb–Sn–Sb) for the die attach (anode) and aluminium wires ultrasonic bonded for the gate and the cathode.

Thermal fatigue occurs in thyristors caused by the heavy currents that have to be conducted and the consequent large amount of heat developed, coupled with the power cycling that they have to sustain. Different temperature coefficients of expansion occur for the silicon and the copper used for the package slug. To avoid this, molybdenum is the most used material for hard soldering the dice to the heatsink. Hard solder is also used for the joint between cathode and output. The gate connection to the thyristor element is made with an aluminium wire to ensure a non-rectifying contact. The wire is welded ultrasonically to the exposed p-layer. The device is encapsulated in a ceramic and metal top-cap which is welded to the header as shown in *Figure 21.38(a)*. Unlike the rectifier, it is difficult to connect the cathode of the heatsink because of the presence of the gate, so thyristors usually have the anode connected to the heatsink.

Another method of encapsulation involves replacement by pressure contacts of the solders between the silicon pellet and the molybdenum or tungsten back-up plates (press-pack packages). This is a practical method to encapsulate the large pellets, more than 25 mm in diameter. The force required to develop the necessary pressure to ensure adequate electrical and thermal contact at the joints may be either retained internal to the encapsulating housing or retained externally to the encapsulation, see *Figure 21.38(b)*.

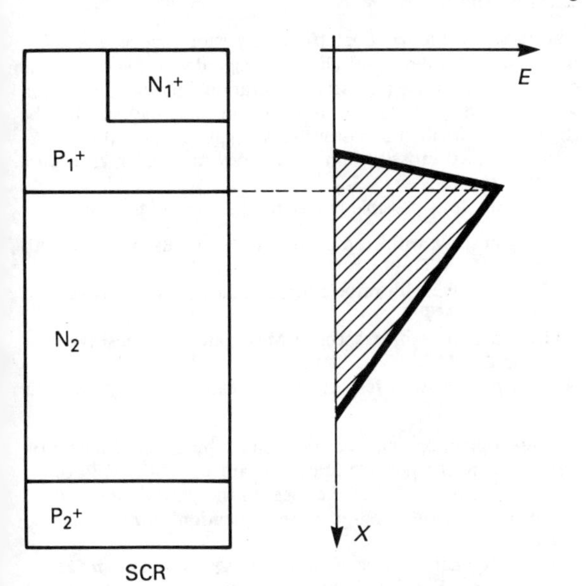

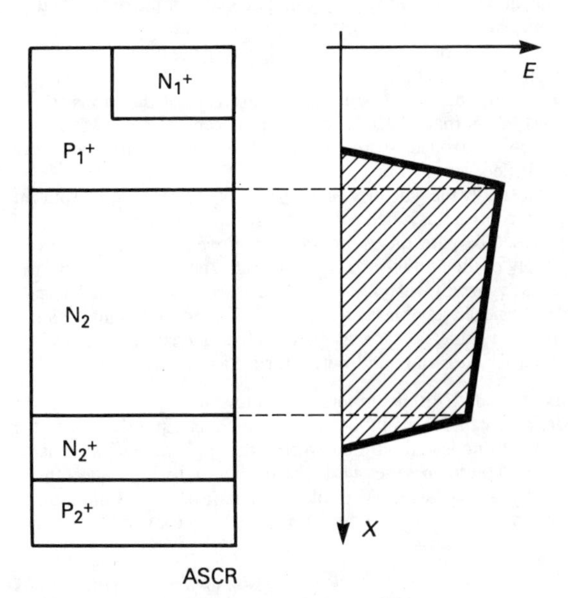

Figure 21.37 Comparison between conventional SCR and asymmetrical SCR. Blocking capability = area of electric field

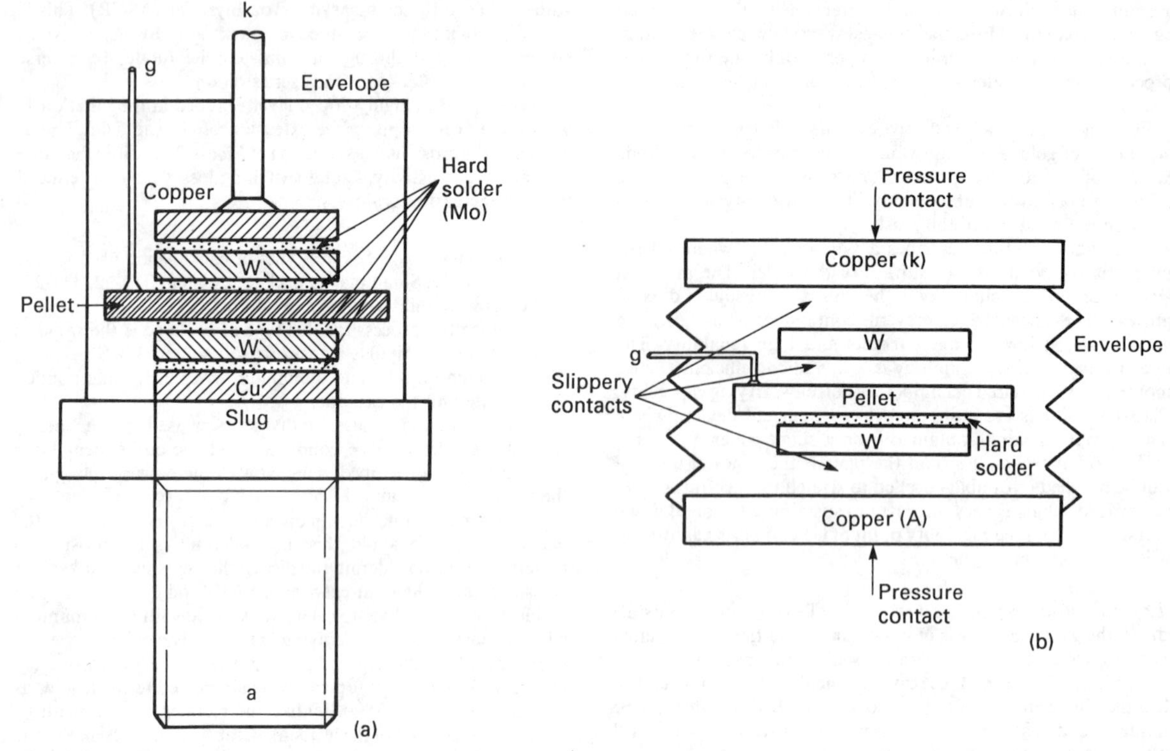

Figure 21.38 Thermal fatigue resistant thyristor constructions: (a) hand solder technique, (b) press-pack technique

21.11.5 Thyristor utilisation

Drive techniques The thyristor family structure can be turned on by several means. Light can have sufficient energy to start the lower npn transistor in the equivalent structure, as in a whole family of LASCRs, LAPUTs, LASCSs, etc. A forward V_{AK} voltage greater than the forward breakdown V_{BO} can also be used to turn it on, as in the family of 'breakover' diodes or 'SIDACs'. The rate of change of applied voltage dV/dt , or the over-junction temperature, is only used for protective systems.

The most common method of triggering the thyristor is to pass current into the base of the npn transistor. The base of the pnp transistor could also be used, but the gain of this transistor is always less than the gain of the npn transistor. However, the silicon controlled switch (SCS) allows the control of both equivalent transistors.

Normally the gate drive can be removed after turn-on, but for good control it is much better to drive the thyristor by a train of pulses. The gate drive should be removed during the reverse supply of the anode with the respect to the cathode. If not, the leakage current will increase with the associated losses. The gate can be negatively biased during the off condition. This bias has some advantages, such as better blocking voltage capability, better dV/dt rating, and better turn-off time, t_q.

dV/dt and dI/dt limitations To limit the system dV/dt the designer can use a snubber circuit, such as a parallel capacitor between anode and cathode, to limit the applied dV/dt. A resistor is also placed in series with the capacitor to limit the rate of capacitor discharge. To limit the system dI/dt, an inductor is placed in series with the thyristor. This inductor can be a normal or saturable device.

Parallel or series connection of thyristors Very large thyristors are available today and since paralleling is not very easy to implement due to current sharing problem, paralleling thyristors is not often used.

Because thyristors are used to control large energy systems connected to high-voltage mains, series connection of thyristors is extensively used. Due to the wide distribution of leakage current, a balancing network is now essential.

21.11.6 The triac

The triac, or bidirectional triode thyristor (*tri*ode *a.c.* semiconductor), is a device which can be regarded as two thyristors connected in inverse parallel configuration, but with a common gate electrode. Unlike the thyristor, however, the triac can be triggered with either a positive or a negative gate pulse.

A simplified cross-section of this device is given in *Figure 21.39*.

There are four possible modes of operation of a triac:

(1) MT_2 positive with respect to MT_1, gate pulse positive with respect to MT_1.
(2) MT_2 positive with respect to MT_1, gate pulse negative with respect to MT_1.
(3) MT_1 positive with respect to MT_2, gate pulse positive with respect to MT_1.
(4) MT_1 positive with respect to MT_2, gate pulse negative with respect to MT_1.

These operations can be represented by a coordinate axis system where the positive directions are $V_G - MT_1$ voltage on the x-axis and $MT_2 - MT_1$ voltage on the y-axis. Operation 1 above is called the 'first-quadrant operation' and so on (see *Figure 21.40*).

The static characteristic of this device is shown in *Figure 21.41*. As would be expected, it consists of two thyristors' forward voltage–current characteristics combined. The dynamic character-

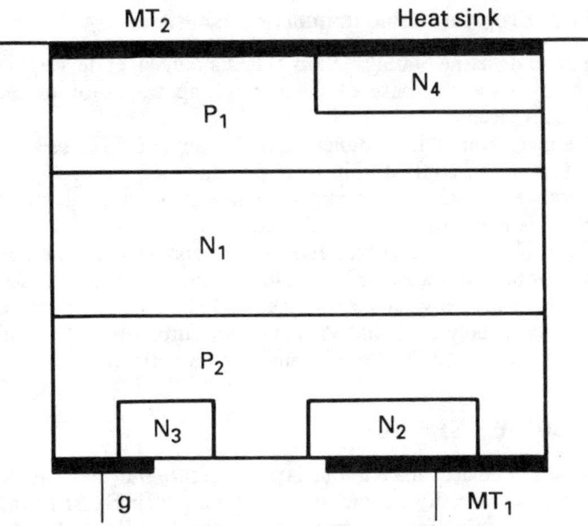

Figure 21.39 Simplified cross-section of a triac.
First SCR: $P_1N_1P_2N_2$. Second SCR (opposite): $P_2N_1P_1N_4$

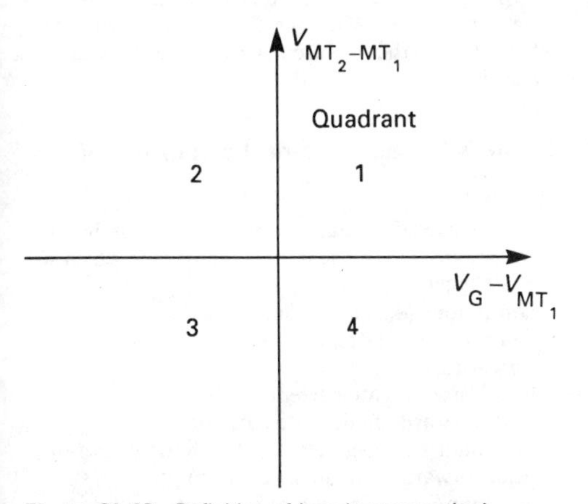

Figure 21.40 Definition of 'quadrant operation'

istics of the triac are the same as the thyristor's characteristics, but the triac has some severe limitations in dV/dt and dI/dt; 5 V/μs instead of 500 V/μs in dV/dt, for example.

The best switching performance occurs in quadrant 1 and the worst in quadrant 4. Quadrants 2 and 3 are far more uniform, so a negative gate drive with respect to the MT_1 terminal is commonly used in order to obtain a similar switch-on characteristic when a triac is used to control an a.c. system.

Because of the triac dV/dt sensitivity it is mandatory to put a snubber network across it: 0.1 μF in series with 100 Ω is commonly used.

21.11.7 Gate turn-off thyristors

GTOs can also be called GCOs (gate cut-off) or GSCs (gate controlled switch). The gate turn-off thyristor is a thyristor specially built to be turned off by its gate terminal as it is for a transistor. This device should therefore combine the thyristor advantages, such as high silicon efficiency, with the ease of gate drive of the bipolar transistor.

The schematic structure of the thyristor given in *Figure 21.31* can be used to explain the GTO operation. If enough gate current (I_{GRmin}) is extracted, the thyristor can be turned off, so a gate

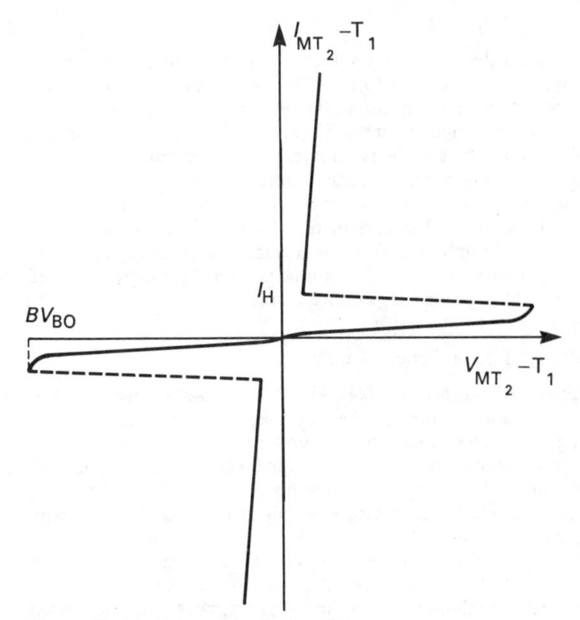

Figure 21.41 Triac static characteristic

current gain limit can be defined:

$$G_{limit} = \frac{\text{anode current}}{I_{GRmin}} \tag{21.36}$$

This ratio is mainly directed by the common base gain current of the pnp transistor in the structure (α):

$$G_{limit} = \frac{1}{\alpha} \tag{21.37}$$

The pnp transistor is therefore very important in a GTO structure. α is tied to the carrier lifetime and the thickness of the base layer (n layer).

21.11.7.1 Static characteristic of GTOs

The static characteristic is given in *Figure 21.42*. It is similar to the thyristor static characteristic, but if the gate current I_G is less than a critical value called I_{GFcrit} the I_{AK} *versus* V_{AK} characteristic looks like the transistor I_C *versus* V_{CE} characteristic. Normally a GTO cannot sustain a reverse voltage, only 10–20 V, as is common for a bipolar transistor.

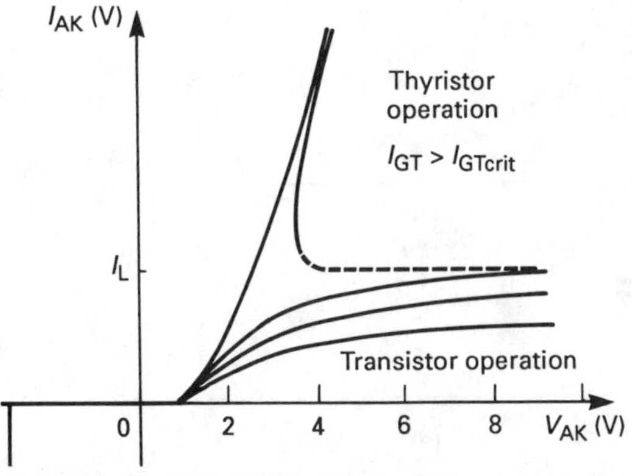

Figure 21.42 Static characteristic of a GTO

21.11.7.2 *GTO drive*

A high gate current is needed for a good turn-on, as it is for thyristor or transistor turn-on. It is important to maintain the forward gate current during the on-time and to check that this current is enough to put the GTO in thyristor mode, because the GTO latching current is quite high. The reverse gate current should be higher than the minimum reverse gate current I_{GRmin}; the limit gain previously defined is between 5 and 10 in the available GTO. The current turn-off time, mainly the storage time, is strongly related to the reverse gate current, I_{GR}. These ratings, turn-on time, turn-off time and limit gain, are degraded when the junction temperature increases.

21.11.7.3 *Manufacture of GTOs*

The manufacture of a GTO is like the manufacture of a thyristor, but the doping profile of the four layers are optimised to obtain a good common base current gain for the npn transistor ($\alpha \approx 1$), and a low common base current gain for the pnp transistor. This requires a large thickness of the upper n layer (base of the pnp transistor) and heavy material doping, in order to kill to minority carrier lifetime.

The turn-off behaviour of the GTO is like that of the bipolar transistor. It is a high current density under the cathode layer (current pinch-off), so hot spots can appear with associated second breakdown. The first manufacturing solution to improve this behaviour was implementation of a thin low resistivity p^+ buffer in the gate p layer structure. The extracted gate current creates a lower lateral field and less pinch-off phenomenon.

A better solution is obtained by using a ramified or spiralled gate structure, because this structure will decrease the load current density by increasing the cathode area and will increase the dI/dt rating of the device by quickly firing the silicon.

21.11.8 SIDAC or breakover diodes

The breakover diode is a bilateral switch operating at medium voltage (under 500 V at present). The static characteristic is the same as the triac static characteristic, but this device is only driven by an overvoltage seen at its terminals. It is a fast device with a very low turn-on time and a low t_q (under 1 μs).

This switch can be used as overvoltage protection (OVP) on d.c. supply in conjunction with a fast fuse (crowbar). It can also limit the lightning induced transients on communication lines in conjunction with a SLIC (subscriber loop interface circuit).

21.11.9 Programmable unijunction transistors

In spite of its name of unijunction, this is a four-layer device, but the gate drives the base of the upper pnp transistor in the equivalent circuit.

The programmable unijunction transistor (PUT) is used as a UJT, but the control voltage is programmed externally by a resistance–capacitor network, as in *Figure 21.43*, with the static characteristic obtained in this circuit.

The PUT is more sensitive, faster, more versatile and generally more economical than a UJT and often replaces it. It can be used in clock circuits between 0.0003 Hz and 2 kHz. This clock can be accurately set between 2 and 3% in temperature, between 25 and 60°C, by using a series diode compensation network.

21.11.10 SUS-SBS-SCS

These small devices have a four-layer structure but they can be controlled as for a thyristor for the unilateral switch (SUS), by the base of the equivalent npn transitor; as for the SIDAC for the bilateral switch (SBS), by the overvoltage seen at its terminals; and by the two bases of the equivalent transistors for the silicon controlled switch (SCS).

They have the same statics characteristics as shown in *Table 21.2*. They have much lower ratings than their counter-part, thyristor or SIDAC, and they are used in low-power applications or as trigger devices.

21.11.11 Symbols for main electrical parameters of thyristors

di/dt	rate of rise of on-state current after triggering
dV/dt	maximum rate of rise of off-state voltage which will not trigger any device
I_D	continuous (d.c.) off-state current
I_{DM}	peak off-state current
I_{FG}	forward gate current
I_{FGM}	peak forward gate current
I_{GaM}	peak forward anode–gate current
I_{GaT}	minimum anode–gate current to initiate turn-on
I_{GkM}	peak forward cathode–gate current
I_{GkT}	minimum cathode–gate to initiate turn-on
I_{GT}	minimum instantaneous gate current to initiate turn-on
I_{GQ}	gate turn-off current

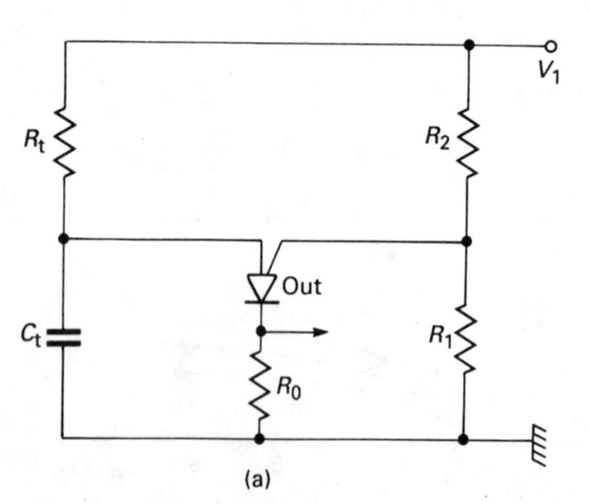

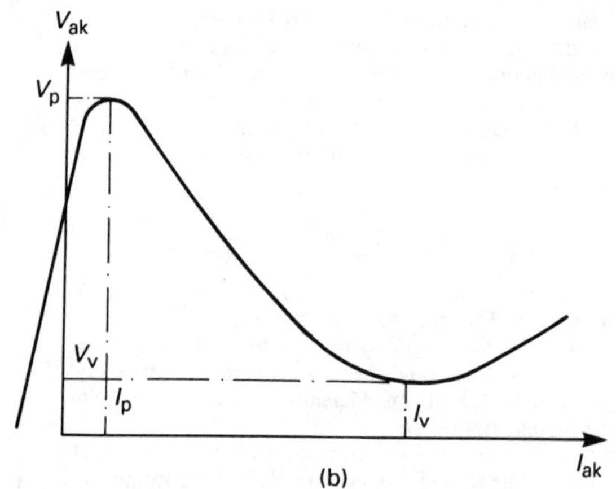

Figure 21.43 PUT application: (a) oscillator circuit, (b) characteristic

I_H holding current
I_L latching current
I_{RG} reverse gate current
I_{RGM} peak reverse gate current
I_T continuous (d.c.) on-state current
$I_{T(AV)}$ average on-state current
$I_{T(ov)}$ overload mean on-state current
$I_{T(RMS)}$ r.m.s. on-state current
I_{TRM} repetitive peak on-state current
I_{TSM} non-repetitive peak on-state current
$P_{G(AV)}$ average gate power
P_{GM} peak gate power
$R_{th(h)}$ thermal resistance of heatsink
$R_{th(i)}$ contact thermal resistance at specified torque
$R_{th(j-amb)}$ thermal resistance, junction to ambient
$R_{th(j-h)}$ thermal resistance, junction to heatsink
$R_{th(j-mb)}$ thermal resistance, junction to mounting base
$R_{th(mb-h)}$ thermal resistance, mounting base to heatsink
T_{amb} ambient temperature
$T_{j,max}$ maximum permissible junction temperature
T_{mb} mounting-base temperature
T_{stg} storage temperature
t_d delay time
t_{gt} gate-controlled turn-on time
t_{off} circuit-commutated turn-off time
t_{on} turn-on time for SCS
t_q circuit-commutated turn-off time
t_r rise time
V_{AK} forward on-state voltage of SCS
V_{BO} breakover voltage
V_D continuous off-state voltage
V_{DRM} repetitive peak off-state voltage
V_{DSM} non-repetitive peak off-state voltage
V_{DWM} crest working off-state voltage
V_{FG} forward gate voltage
V_{FGM} peak forward gate voltage
V_{GaM} peak reverse anode-gate-to-anode voltage
V_{GaT} minimum anode-gate voltage that will initiate turn-on
V_{GD} maximum continuous gate voltage which will not initiate turn-on
V_{GkM} peak reverse cathode-gate-to-cathode voltage
V_{GkT} minimum cathode-gate voltage to initiate turn-on
V_{GT} minimum instantaneous trigger voltage to initiate turn-on
V_{RG} reverse gate voltage
V_{RGM} peak reverse gate voltage
V_{RRM} repetitive peak reverse voltage
V_{RSM} non-repetitive peak reverse voltage
V_{RWM} crest working reverse voltage
V_T continuous (d.c.) on-state voltage

Bibliography

ALOISI, P., *Power Switch, Application Manual*, Motorola DLE 401/D (1986)
ALOISI, P., 'Utilizzazione di MOS di potenza', *Elettronica*, February (1987)
ALOISI, P. and CORDONNIER, C. E., 'A medium power switch using a GEMFET/bipolar Darlington combination', *Electronic Engineering*, May (1987)
ALOISI, P. and DUBBERKE, D., 'Leistungstransistoren in automobilen', *Elektroniker*, No. 3 (1984)
BALIGA, B. J., *Modern Power Devices*, Wiley, New York (1987)
CHANTE, J. P., 'Les composants de puissance et le genie electrique', *Club EEA, IDN*, 26–27 March (1987)
CORDINGLEY, B. V., 'Pin fast power diode', *PEVD IEE*, London, p. 19 (1984)
COULTHARD, N. C. and PEZZANI, R., 'Understanding gate assisted turn off of an interdigitated ultra fast, asymmetrical power thyristor', *Proc. PCI* (1981)
FAY, G., 'Power ICs features and future', *Powertechnics*, June (1986)
FAY, G., 'Current mirror FETs cut costs and sensing losses', *EDN*, September 4 (1986)
FILANOVSKY, I. M., 'Switching characteristic of MOS differential pair with mismatched transistors', *Int. J. Electron.*, **65**, No. 5 (1988)
FRANK, R. and JANIKOWSKI, R., 'Trends in power ICs development', *Power Conversion and Intelligent Motion*, April (1986)
GAMBOA-ZURRIGA, M., *MOS de puissance, relaxation thermique*, Thèse de Docteur-Ingenieur, Toulouse (1980)
GAUEN, K., *Designing with TMOS Power MOSFETs*, Motorola AN 913 (1983)
GAUEN, K., *Paralleling Power MOSFETs in Switching Applications*, Motorola AN 918 (1984)
GAUEN, K., *Insuring Reliable Performance from MOSFETs*, Motorola AN 929 (1984)
GENERAL ELECTRIC, *SCR manual*, 6th edn (1979)
GUEGAN, G., *Etudes des propriétés dynamiques du transistor MOS a canal vertical*, Doctorat de sciences, Toulouse (1979)
HARTMAN, C. and HILLEN, M., 'Redesign of small-signal switching diodes in SOT packages', *Electro. Components Appl.*, **8**, No. 4 (1988)
LEE, S. K. *et al.*, 'Reliability study on rapid thermal processed metal–oxide–semiconductor field effect transistors', *Solid State Electron.*, **31**, No. 10 (1988)
LETURQ, P., 'Composants de puissance a semiconducteurs, évolutions et perspectives', *Colloque sur l'électronique de puissance, Centre de perspective et d'evaluation*, Paris (1982)
LINDEMAYER and WRIGLEY, *Fundamentals of Semiconductor Devices*, D. Van Nostrand, New York (1966)
LOCCI, N., 'Measurement of instantaneous losses in switching power devices', *IEEE Trans. (IM)*, **37**, No. 4 (1988)
MASDURAUD, J. M. and ROGER, B., 'A new technology for high voltage, high speed Darlington', *PCI*, Geneva (1981)
MIYAMOTO, M. *et al.*, 'Trade off between cut off frequency and breakdown voltage for power MOSFETs', *Solid State Electron.*, **31**, No. 11 (1988)
MOTOROLA, *UJT Theory and Characteristics*, AN 293 (1972)
MOTOROLA, *UJT Applications*, AN 294 (1972)
MOTOROLA, *UJT Circuits for Thyristors*, AN 413 (1975)
MOTOROLA, *PUT Theory and Applications*, AN 527 (1976)
NAKATANI, Y. and KURUYU, I., 'An ultra high speed, large safe operating area, switching power transistor with new fine emitter structure', *4 INTELEC 83*, **18**, No. 6, 500 (1983)
PHAM-PHAM, T., *Le compromis entre la resistance a l'état passant et la tenue en tension dans les transistors MOS de puissance*, Thèse de 3e cycle, Toulouse (1982)
PHILLIPS, A.-B., *Transistor Engineering*, McGraw-Hill (1962)
PSHAENICH, A., *New Power Bipolars Compare Favorably with FETs for Switching Efficiency*, Motorola AN 845 (1981)
RCA, *Solid State Power Circuits* (1971)
REYNES, J. M., *Relations entre performances et parametres structuraux des transistors bipolaires de puissance*, Thèse de Docteur Ingenieur, INSA, Toulouse (1986)
ROBERTS, D. and SPREADBURY, P., 'Assessing zener diodes for calibration standards', *Test Measurement World*, October (1988)
SCHULTZ, W. and ALBERKRACK, J., *A New High Performance Current Mode Controller Teams up with Current Sensing Power MOSFETs*, Motorola AN 976 (1986)
SILBER, D., 'Improved dynamic properties of GTO and diodes by proton implantation', *IEDM*, p. 162-6.6 (1985)
SINGER, P. H., 'The transistor: 40 years later', *Semiconductor International*, January (1988)
TRAN DUC, H., 'The on state resistance versus breakdown voltage', *ESSDERC*, Toulouse (1981)
VAN DE WOW, T., 'Reliability of encapsulated glass passivated high voltage transistors', *PCI*, Munich (1980)
WOODWORTH, A., 'High voltage, glass passivated, gate turn-off thyristor', *Proc. PCI* (1980)
WOODWORTH, F. A., 'High voltage epitaxial rectifier', *IEDM IEE*, London, p. 38 (1984)
ZAREMBA, D., *The Operation and Application of the SMARTMOS Overvoltage and Overtemperature Protectors*, Motorola AN 956 (1985)

22 Microwave Semiconductor Devices

M W Gray
Hewlett-Packard GmbH

Contents

22.1 Introduction

The requirements of high frequency performance impose the greatest possible demands on semiconductor technology. The substrate material must be of the greatest purity so that there will be no imperfections in the active structure. Also, the material itself may need to be a more difficult substance to work with, such as gallium arsenide, because the high mobility of its charge carriers means improved performance at maximum frequencies. The required complexity of the device makes it difficult to define and special techniques such as ion implantation are required. Repeatability of characteristics becomes difficult and production yields are relatively low.

In this section on microwave semiconductors each device is described in terms of its construction and characteristics and the various types are mentioned. The most important parameters are defined and applications examples are given.

22.2 Schottky diodes

Standard pn diodes are limited at higher frequencies by the capacitance provided by the lifetime of the minority carrier. This means that such diodes will not switch on and off quickly compared to the frequency of the signal they are trying to process, resulting in reduced performance in, for example, mixing and detecting in a radio receiver. Schottky diodes overcome this problem by employing a metal–semiconductor barrier at the active junction so that majority carriers only participate in the diode action. It should be noted that Schottky diodes may also be termed 'hot carrier' diodes, 'hot electron' diodes and 'Schottky barrier' diodes.

22.2.1 Schottky diode types

Typical Schottky diodes belong to one of the three types: passivated, hybrid or mesh.

22.2.1.1 Passivated diodes

Figure 22.1 shows a typical passivated diode. The n-type silicon and a metal, such as NiCr (nickel chromium) form the Schottky junction. The surface of the semiconductor is passivated

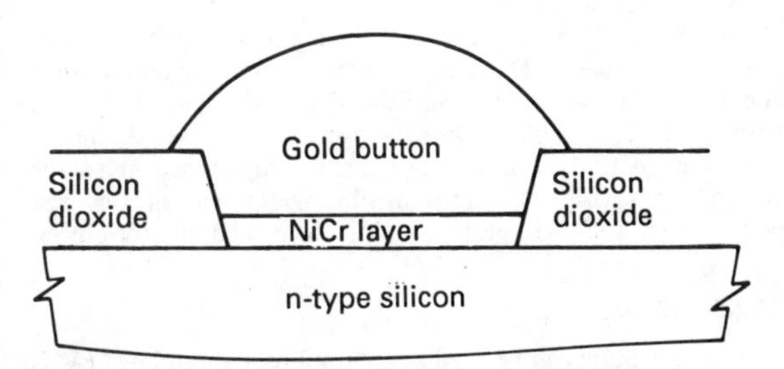

Figure 22.1 Passivated Schottky diode cross-section

(protected) by a layer of silicon dioxide against outside contaminations. Sometimes a layer of silicon nitride is used in addition to the oxide. The thick gold layer (called gold button due to its shape) connects to the outside world as one terminal of the diode, the other terminal being the semiconductor itself.

Usually passivated diodes have small Schottky junction areas, therefore, they have low junction capacitance and are suitable for operation up to 18 GHz. However, under small reverse bias voltages (about 5 V), large electric fields occur where the Schottky junction meets the silicon dioxide, causing voltage breakdown.

22.2.1.2 Hybrid diodes

The hybrid Schottky diode has a higher reverse breakdown voltage than the passivated diode. In addition to the Schottky junction, a p–n junction is located at the oxide–Schottky interface as in *Figure 22.2*. In this case 'hybrid' indicates the presence of both Schottky and p–n junctions in one diode. This

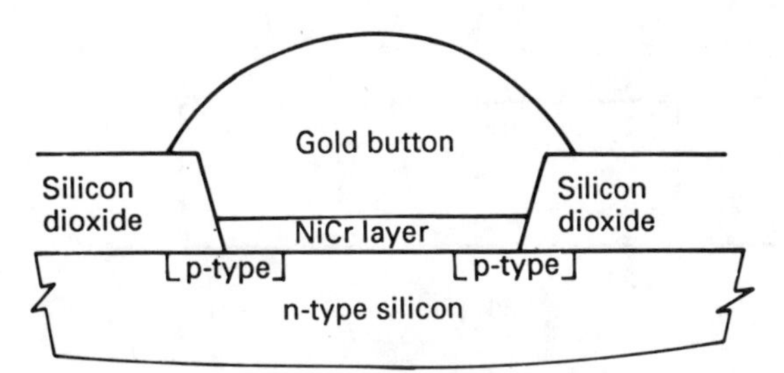

Figure 22.2 Hybrid Schottky diode cross-section

arrangement does not allow high electric fields to build up at the interface, and a reverse breakdown voltage as high as 70 V is achieved. The penalty is high junction capacitance, limiting operating frequency to 4 GHz.

22.2.1.3 Mesh diodes

A mesh diode has an unpassivated Schottky junction. *Figure 22.3* is the top view of a mesh diode chip. Each chip has about 80 pads, each of which is a diode itself. The pads are defined on the

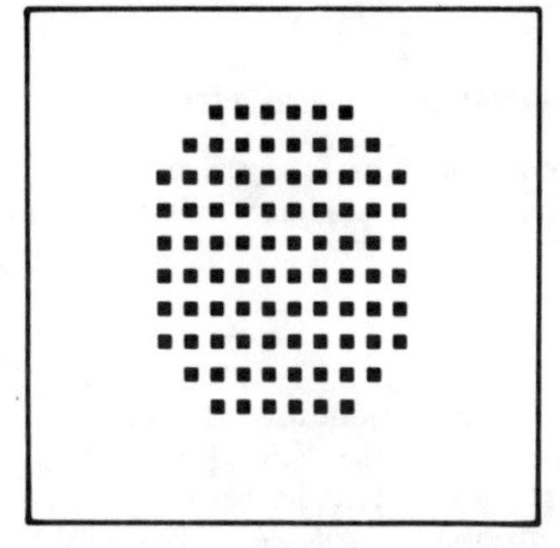

Figure 22.3 Top view of a mesh diode chip

semiconductor chip by a mesh mask. The distance from the centre of one pad to another is about 50 μm (two thousandths of an inch).

Figure 22.4 shows the cross-section of a single pad on a mesh diode. During assembly the diode chip is put into a package and a metal whisker is placed on the chip. A randomly achieved contact of the whisker with one of the pads is indicated on an oscilloscope. This assembly technique does not lend itself to high volume production.

The mesh diodes usually have medium size Schottky junction areas, and the resulting average junction capacitances limit operation frequencies to about 4 GHz. The breakdown voltage is about 30 V, and the diodes have low flicker ($1/f$) noise.

The Schottky junction in the mesh diode is between the nickel and n-type silicon (*Figure 22.4*). The metal whisker is for diode lead connection only and does not contribute to the formation of

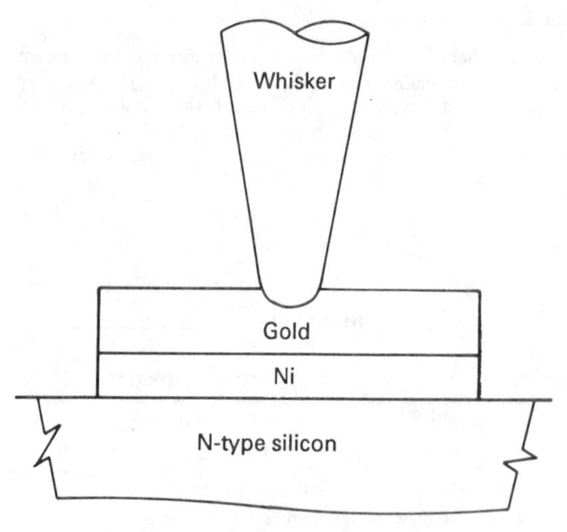

Figure 22.4 Mesh diode cross-section

the junction, so the diode is completely reliable and uniform in its performance. In contrast the junction of a point contact diode is formed by the contact of the metal whisker and the semiconductor.

N-type silicon and nickel chromium (or nickel) are not the only combination of semiconductor and metal which can be used to make a Schottky junction. An alternative semiconductor is gallium arsenide (GaAs) and suitable metals include molybdenum, tungsten, aluminium, titanium and platinum.

Silicon Schottky diodes are capable of operations up to 40 GHz. Gallium arsenide Schottky diodes, due to the higher electron mobility in this semiconductor, are made to cover applications up to 100 GHz and above.

22.2.2 Applications and performance parameters

The three main areas of applications for Schottky diodes are switching (including clipping and sampling), mixing and detecting.

22.2.2.1 Switching

The most important characteristic for a diode used in any form of switching application is speed, and the Schottky diode is unsurpassed in this respect. However, several other parameters are also important and these are given in *Table 22.1*. The forward voltage at 1 mA is also known as turn-on voltage. Schottky diodes made with different metal–semiconductor combinations offer different forward voltages.

Table 22.1 Switching diode performance comparison

Parameters	*Diode*		
	Silicon Schottky	*Silicon p–n*	*Germanium p–n*
Forward voltage drop (mV at 1 mA)	150–450	600	300
Reverse leakage (nA)	<100	<100	>1000
Breakdown voltage (V)	5–70	>1000	>1000
Carrier lifetime (ps)	<100	500	2000

The minority carrier lifetime is applicable to the p–n junction diode only because its operation involves both majority and minority carriers. The Schottky diode has only majority carriers although their parasitic reactances give rise to an effect similar to minority carrier lifetime. The 'lifetime' is an indication of the switching speed of the diodes; the lower the lifetime, the faster the switching speed.

22.2.2.2 Mixing

Any non-linear element will mix but the Schottky diode is particularly effective for this because of its nearly square law characteristics and low-noise performance. The parameters concerned with mixing are noise figure, input admittance, and intermediate frequency noise and impedance.

Noise figure is defined as the ratio of the input signal to noise ratio to the output signal to noise ratio: the lower the noise figure of the mixer diode, the better the performance of the mixer under conditions of weak r.f. signal input. A typical noise figure for a mixer diode is 6 to 7 dB.

The noise figure of the mixer diode is dependent on the local oscillator power (P_{LO}). There is a range of local oscillator power over which the diode exhibits the best noise figure. The noise figure is also dependent on the frequency of the input.

The input admittance of a mixer diode indicates how the diode can be matched to a certain r.f. circuit for best performance. It is plotted on a Smith chart. Since the input admittance varies with local oscillator power and d.c. bias these conditions are always indicated. Often a single standing wave ratio (SWR) is quoted for the test frequency, showing how well the diode matches a fixed tuned test system.

Intermediate frequency (i.f.) noise is also known as flicker noise. Its amplitude increases with decreasing frequency. The mixer diode noise has two components. One is broadband noise which is constant in amplitude at all frequencies and the other is i.f. noise. At high frequencies the former dominates and at low frequencies the latter. Intermediate frequency noise is measured and expressed as noise temperature ratio (T_N) in dB. The higher the number, the noisier the diode. Schottky diodes have low i.f. noise. Of the three types of Schottky diodes the mesh diodes have the lowest i.f. noise, and the passivated diodes the highest. In contrast germanium and silicon point contact diodes have i.f. noise which is 20–30 dB worse.

The intermediate frequencies of some mixer applications are low. For example, in Doppler radar the i.f. is in the audio range (d.c. to 10 kHz) and the mesh Schottky diode, with its low i.f. noise, is the logical choice as a Doppler radar mixer diode.

Similar to the r.f. admittance the i.f. impedance helps the designer put the diode into a suitable i.f. circuit for best performance. It is often plotted against the local oscillator power.

22.2.2.3 Detecting

A Schottky diode can be used as an amplitude modulated (AM) detector or for detecting the presence of any signal. Again, the Schottky diode is particularly useful in this application because of its excellent characteristics of tangential sensitivity, voltage sensitivity, video resistance and i.f. noise.

The tangential sensitivity (TS) of a detector diode describes the performance of the diode under low signal level conditions. It is a subjective measurement using an oscilloscope. The TS is the input signal level which produces a detected signal which is just above the noise floor. At this point the signal to noise ratio is 8 dB. The tangential sensitivity is measured as an input power level in dB m: the smaller the value, the more sensitive the detector diode. For example, the TS for the 5082-2755 detector diode is

−55 dB m for a 2 MHz video bandwidth, at 10 GHz and 20 μA bias. A small amount of d.c. bias usually improves the TS by reducing the output resistance of the diode at low signal levels; thus causing the majority of the detected voltage to appear across the load resistance.

The voltage sensitivity (γ) is a measure of the output voltage of the detector diode for a given input power. It is measured in mV/μW. The value of γ is useful to the circuit designer in determining how much voltage can be obtained at the video output of the diode, allowing him or her to design a video amplifier of the right gain.

The video resistance is simply the output resistance of the diode at video frequencies, and effectively forms a potential divider with the load resistance. Applying a small d.c. bias, or increasing the signal power (thus inducing a rectified current within the diode), will decrease the video resistance and proportionally increase the voltage across the load.

Zero-bias Schottky detectors are also available. These are specifically designed to have a low video resistance with no applied bias (typically 100 kΩ compared with 40 MΩ for a conventional detector). A number of different contact metals can be used, resulting in diodes with a range of barrier heights. A low barrier height results in a diode with a low video resistance for low signal powers (because of the rectified current induced in the diode), but a high video resistance for high signal powers (because of the higher residual series resistance). Thus the performance characteristics of a low barrier diode are good TS and noise performance, but poor voltage sensitivity.

If the frequency of the video signal is low (below 1 MHz) i.f. noise, as in the case of the mixer diode, becomes the main source of detector diode noise. The low i.f. noise of the Schottky diode is therefore a definite advantage over that of the point contact diode.

22.3 PIN diodes

The most important feature of the PIN diode is its basic property of being an almost pure resistor at r.f. frequencies, whose resistance value can be varied from about 10^4 Ω to less than 1 Ω by the control current flowing through it. Most diodes exhibit this characteristic to some degree, but the PIN diode is specially optimised in design to achieve a relatively wide resistance range, good linearity, low distortion, and low current drive. The characteristics of the PIN diode make it suitable for use in switches, attenuators, modulators, limiters, phase shifters and other signal control circuits.

The PIN diode is a silicon semiconductor consisting of a layer of intrinsic material sandwiched between regions of highly doped p and n material, hence the abbreviation. Under reverse bias, the I-layer is depleted of mobile charges and the PIN diode appears essentially as a capacitor. When forward bias is applied across the PIN diode positive charge (holes) from the p region and negative charge (electrons) from the n region are injected into the I-layer, therefore increasing its conductivity and lowering its resistance.

The conductance (G) of the PIN diode is proportional to the stored charge (Q) in the I-layer, which in turn is proportional to the d.c. bias current. Resistance (R) is therefore inversely proportional to the d.c. bias current.

An important parameter of the PIN diode is the lifetime (τ), which essentially defines the length of time required for the stored charge to deplete by recombination once the forward bias is removed. There is a frequency, f_0, which is related to τ by the expression,

$$f_0 = 1/2\pi\tau \qquad (22.1)$$

used for determining the low frequency limit of the PIN diode.

Since charge depletes for times greater than τ, a signal with cycle period greater than τ will experience varying conductance. This defines a lower frequency limit, given by Equation (22.1), for linear performance of a diode. For r.f. signal frequencies below f_0 the PIN diode rectifies the signal much like an ordinary p–n diode, and considerable output distortion occurs. At f_0, there is some rectification with resulting distortion. At frequencies above f_0, less and less rectification occurs as charge storage due to r.f. current (I_{rf}) diminishes according to the relationship:

$$Q = \frac{I_{rf}}{2\pi f} \qquad (22.2)$$

Since I_{rf} is dependent on the power absorbed by the diode, distortion is also dependent on the r.f. signal level, r.f. frequency, and d.c. bias current. For most applications the minimum frequency of operation should be about $10f_0$ or

$$f_{min} = \frac{10}{2\pi\tau} = \frac{1.59}{\tau} \qquad (22.3)$$

This restriction is not important in switching applications, where the diode is normally biased either completely off or completely on. In those states, since most of the power is either reflected or transmitted, the effect of r.f. current on the total charge is small and distortion is not a problem.

At frequencies much higher than f_0, the PIN diode with forward bias behaves essentially as a pure resistor. The resistance of the PIN diode is related to the bias current and characteristics of the diode as follows:

$$R = W^2/\mu I\tau \qquad (22.4)$$

where W is the I-layer thickness, μ the combined electron and hole mobilities, τ the lifetime, and I the bias current.

The resistance is inversely proportional to both lifetime and bias current. However, since long lifetime is usually associated with thick I-layer and resistance is proportional to the square of the I-layer thickness, a diode with long lifetime has higher resistance. Exceptions to this case are increases in lifetime by other means without widening the I-layer.

The voltage dependency (or independency) of capacitance is related to the dielectric relaxation frequency, f_D. The undepleted portion of the I-layer has the resistance (R) shunting its capacitance (C). f_D is the frequency where the capacitive reactance is equal to the resistance, i.e.:

$$1/2\pi f_D C = R \qquad (22.5)$$

For an undepleted layer of thickness (W) and area (A)

$$R = \rho W/A \qquad (22.6)$$

$$C = \varepsilon A/W \qquad (22.7)$$

Thus, solving for the dielectric relaxation frequency gives

$$f_D = 1/2\pi\rho\varepsilon \qquad (22.8)$$

where ρ and ε are, respectively, the resistivity and permittivity of the I-layer silicon, with ε being approximately 10^{-12} F/cm. At frequencies much higher than f_D, since the reactance of the undepleted layer is much less than its resistance, total capacitance is independent of bias. At frequencies much lower than f_D, the resistance of the undepleted layer shorts out the capacitance, and sufficient reverse bias is required to deplete the I-layer of charge.

22.3.1 PIN diode construction

Most PIN diodes are derived from the four basic chip structures illustrated in *Figure 22.5*.

22.3.1.1 Epitaxial mesa PIN

Because of its small I-layer (both in thickness and area) the epitaxial mesa structure has little charge storage. Therefore it is

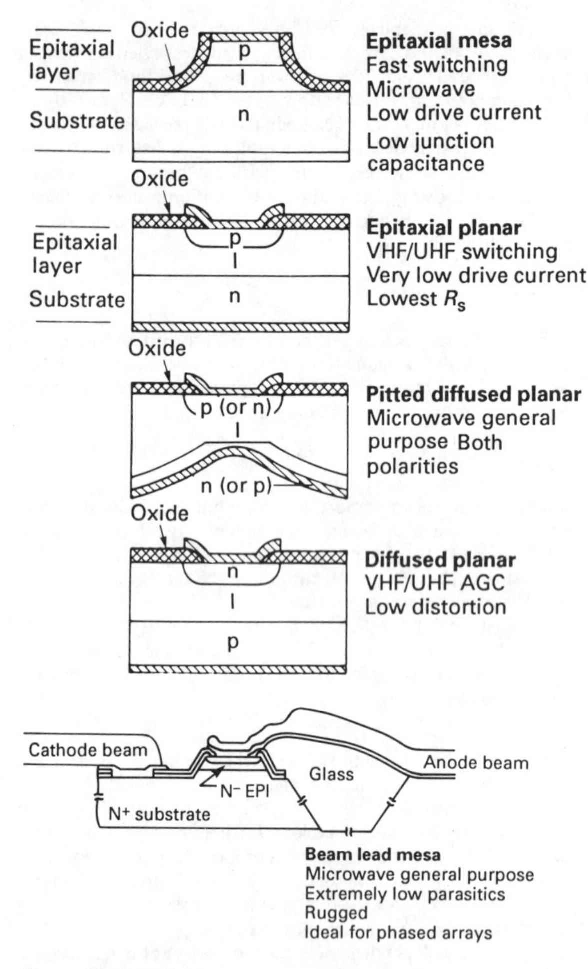

Figure 22.5 PIN chip structures

capable of fast switching. Also because of its thin I-layer, relatively low current drive is required for a given resistance. The mesa structure results in low junction capacitance, and thus enables good frequency response through the microwave bands.

22.3.1.2 Planar PIN

The planar structure (compared to the mesa) has a somewhat larger junction area, and a correspondingly higher junction capacitance. Two types of planar PIN chip are available: epitaxial or diffused. The large junction area and thin I-layer of the epitaxial PIN results in very low resistance, and very low bias currents are required. This type of diode is well suited to switching applications in the VHF/UHF range. The thick I-layer structure of the diffused planar PIN is used for long lifetime diodes. These are most suitable for low-frequency, low-distortion applications, e.g. automatic gain control (AGC) in the VHF/UHF range. Both epitaxial and diffused types of diode can be made in either PIN or NIP polarity.

22.3.1.3 Pitted diffused planar PIN

The diffused planar structure with a pit etched on the reverse side has low junction capacitance, and is therefore suited for switching and general purpose use in the microwave frequencies. Both polarities are available, i.e. NIP and PIN.

22.3.1.4 Beam lead mesa PIN

This is of similar electrical design to the epitaxial mesa chip, but with metalised beams for anode and cathode connections deposited as an integral part of the device. The resulting structure offers extremely low parasitic capcitances and inductances. This gives exceptional high frequency performance (low capacitance Schottky beam lead diodes are used at frequencies above 100 GHz), as well as improved performance at lower frequencies compared with chips.

Beam lead diodes are small (typically 0.7 mm × 0.25 mm) and, once bonded in a circuit, are supported completely by the leads. This eliminates the die attach process (attaching the base of the chip to the conductor substrate), and the relatively fragile wire bond from the anode of a chip. Thus, ruggedness, reliability and ease of bonding are all enhanced.

22.3.2 Applications and diode parameters

PIN diodes are used principally for the control of r.f. and microwave signals from frequencies below 1 MHz to 20 GHz. Applications include switching, attenuating, modulating, limiting and phase shifting. Certain diode requirements are common to all these control functions, while others are more important in a particular type of usage.

22.3.2.1 Switching

Because of its high r.f. resistance when unbiased and its low r.f. resistance when biased with a d.c. current of only 10 mA or less a PIN diode is an ideal circuit element for use in r.f. switches. They may be used as single diode switches in either series or parallel configurations, or for more demanding requirements in combinations using two or more diodes.

The series switch is on when the diode is forward biased, and it is off when the diode is zero or reverse biased. The opposite is true for the shunt switch where zero or reverse bias turns the switch on, and forward bias turns the switch off. For a good aproximation, the PIN diode in a switch is essentially a resistor in the forward-biased state and a capacitor in the reverse-biased state.

The loss of signal atrributed to the diode when the switch is on (transmission state) is insertion loss. The insertion loss is primarily determined in a series switch by the forward-biased resistance of the diode and in a shunt switch both by the diode capacitance and signal frequency. In either case it is the diode impedance in relation to the source and load impedance (generally 50 Ω). For low insertion loss, low resistance is needed in a series switch. Low capacitance (particularly at high frequencies) is needed in a shunt switch.

Isolation is the measure of r.f. leakage between the input and output when the switch is off. For high isolation, low capacitance (especially at high frequencies) is required in a series switch. Low resistance is required in a shunt switch.

Reverse recovery time is a measure of switching time, and is dependent on the forward and reverse bias applied. With forward bias current, I_F, charge is stored in the I-layer. When reverse biased, reverse current, I_R, will flow for a short period of time, known as delay time, t_d. When sufficient number of carriers have been removed, the current begins to decrease. The time required for the reverse current to decrease from 90% to 10% is called the transition time, t_t. The sum, $t_d + t_t$, is the reverse recovery time, which is a good indication of the time it takes to switch the diode from on to off.

Standing wave ratio (SWR) which is a measure of the r.f. impedance match, is particularly important in high frequency applications. Since the SWR of most package styles depends on the mounting arrangement, it is only specified for diodes in 50 Ω stripline packages.

The r.f. power (CW or pulse) that can be handled safely by a diode switch is limited by two factors: the breakdown voltage of the diode, and thermal considerations, which involve maximum diode junction temperature and the thermal resistance of the diode and packaging. Other factors affecting power handling capability are ambient temperature, frequency, attenuation level (diode resistance) pulse width, and pulse duty cycle.

When the maximum isolation requirements are greater than that which can be achieved with a single diode switch, multiple diode arrangements can be used. Two diodes connected closely in series or shunt will only increase the isolation by 6 dB over that of a single diode. If n diodes are spaced a quarter wavelength apart, the resulting isolation will be n times that of a single diode plus 6 dB. Where $\lambda/4$ spacing is impractical, isolation can be increased by using a series–shunt combination. This arrangement will give the isolation of a series diode and that of a shunt diode plus 6 dB. The power handling capability is not improved in a pair of diodes spaced $\lambda/4$ apart. The lead diode absorbs a much larger percentage of the incident power than in the case of a closely spaced pair of diodes.

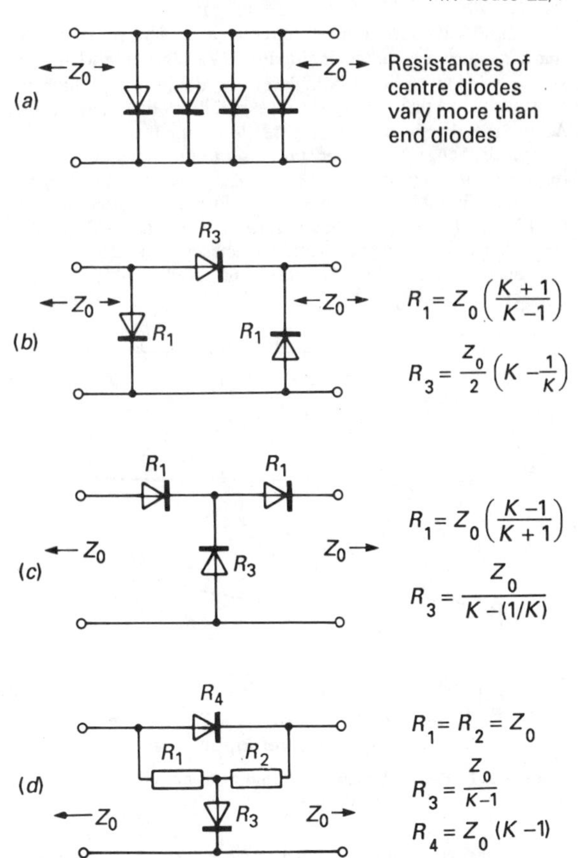

Figure 22.6 Constant impedance absorption attenuators: (a) resistive line; (b) π; (c) T; (d) bridged-T

22.3.2.2 *Phase shifters*

In a phase shifting circuit the PIN diode serves basically as a switch to transfer a transmission line from one electrical length to another, therefore resulting in a phase shift of the transmitted signal.

In each circuit, forward bias applied at one of the bias ports will produce a phase shift in the r.f. output signal. Because of its fast switching speeds and low on and high off impedance, the PIN diode is well suited for phase shifter applications.

22.3.2.3 *Attenuators*

Whereas a switch is used only in its maximum on or off state, an attenuator is operated throughout its dynamic range (or resistance range in the case of a diode attenuator). Although the basic series or shunt diode switch can be used as an attenuator, it cannot offer constant input and output impedance, which is required for good source and load matching in most attenuator applications.

The resistive line attenuator is distributed in structure with a large number of diodes. The diodes in the centre of the structure vary more in impedance than the ones near the ports, resulting in constant impedance at the ports. This structure is capable of large attenuation range, but is not useful at low frequencies because of large size (*Figure 22.6*). The π, T, and bridged-T attenuators are more compact in structure. Both the π and T configurations use three diodes each, while the bridged-T circuit contains two diodes and two resistors as the basic circuit elements. For a particular value of attenuation, K, the design equations determine the values of resistance to which the series and shunt diodes must be biased in order for each attenuator to achieve constant impedance match.

While certain switching parameters are applicable for attenuators, other diode and circuit considerations are more important for attenuator performance. The range from maximum attenuation (isolation) to minimum attenuation (insertion loss) defines the dynamic range of an attenuator. The theoretical dynamic range achievable by the π, T, and bridged-T attenuators are respectively 52, 40, and 34 dB. The linearity of attenuation versus bias is important in most attenuator applications, and is directly attributable to the basic linearity of resistance versus control current of the diode. In many applications, particularly where more than one diode is driven by a single bias supply, power drain is of prime concern. The resistance requirements of good PIN diodes are achieved with low bias currents. Diode and circuit parasitics limit high frequency performance. Proper choice of circuit components and careful circuit layout will help to minimise limitations and sustain performance at high frequencies. Distortion due to rectification limits low frequency performance. The two principal types of distortion in PIN attenuators are intermodulation and cross-modulation.

22.3.2.4 *Modulators*

A PIN diode constant impedance attenuator can be used to modulate an r.f. signal by applying the modulation signal to the bias port. For performance with minimum distortion the carrier frequency should be much greater than f_0 given by Equation 22.1) and the modulation frequency much less than f_0. For most applications a compromise in these requirements may be necessary.

22.3.2.5 *Limiters*

Many microwave systems contain sensitive amplifiers, mixers, and detectors, etc., which are subject to burnout by inadvertent high-level transient signals. Protection for these systems can be provided by a limiter. A PIN diode limiter is essentially an attenuator that uses self-bias rather than externally applied bias.

A basic passive limiter circuit and its performance characteristics are shown in *Figure 22.7*. When the r.f. input signal is below the threshold level, the diode is very high in resistance, and the output very closely follows the input signal. Above the threshold level, the r.f. is rectified and the diode resistance changes to a lower state. As a result, much of the r.f. is reflected, allowing only a small, almost constant output with increase in input. For this circuit to be efficient, it is essential that the PIN diode has fast switching time (low lifetime). A PIN with thin I-layer will help rectification efficiency and result in low resistance. Another diode requirement is good heat transfer characteristics.

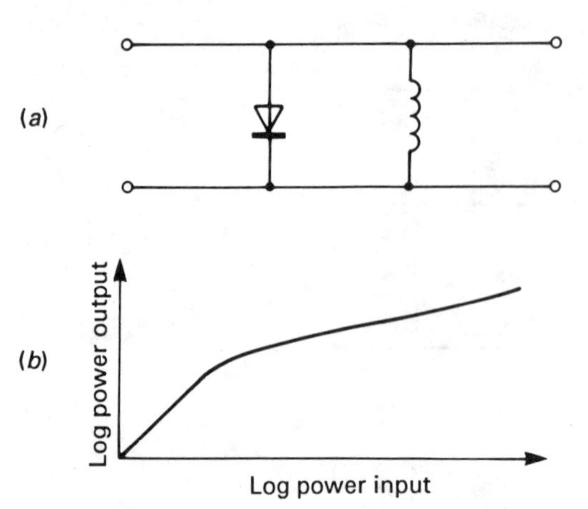

Figure 22.7 Passive limiter circuit and performance characteristics: (a) circuit; (b) characteristic

In a quasi-active limiter, self-rectified current is not needed. Part of the incoming r.f. signal is detected by a Schottky detector diode and the rectified current used to forward bias the PIN limiter diode. This type of circuit enhances the turn-on capability of the limiter. Thus the PIN I-layer thickness may be increased and power handling capability improved.

22.4 Step recovery diodes

The step recovery diode is a charge-controlled switch. A forward bias stores charge, a reverse bias depletes this stored charge, and when fully depleted the diode ceases to conduct current. The action of turning off, or ceasing current conduction, takes place so fast that the diode can be used to produce an impulse. If this is done cyclically, a train of impulses is produced. A periodic series of impulses in the time domain converts to an infinitude of frequencies (all multiples of the basic exciting frequency) in the frequency domain. If these impulses are used to excite a resonant circuit, much of the total power in the spectrum can be concentrated into a single frequency. Thus input power at one frequency can be converted to output power at a higher frequency.

Figure 22.8 is a representation of step recovery diode structure. The 'intrinsic' region is actually very lightly doped n-type silicon. It is in this layer that charge is stored.

Figure 22.9 is a simple equivalent circuit consisting of an extremely fast switch and two capacitors. When in forward conduction, and until the stored charge is totally depleted by reverse conduction, the switch is connected to the large capacitor, C_{FWD}. When the charge is depleted the switch changes instantaneously (60–400 ps) to the other capacitor, C_{DEP}.

A simple circuit, shown in *Figure 22.10*, is the 'impulse generator'. This circuit will help to explain the step recovery

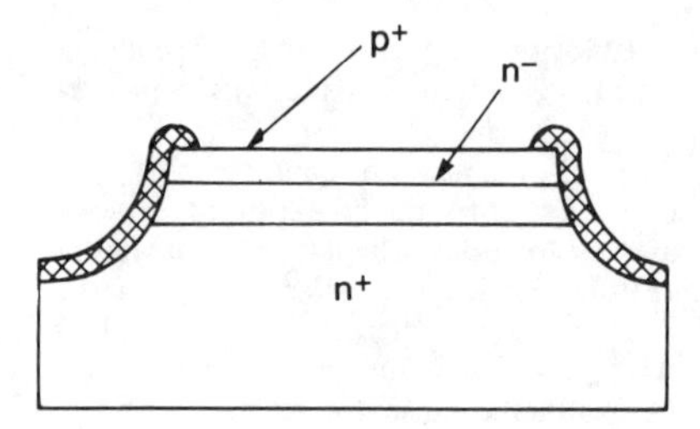

Figure 22.8 Step recovery diode cross-section

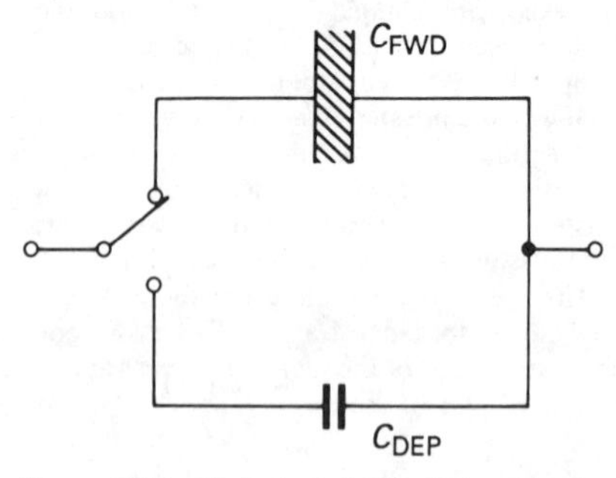

Figure 22.9 Equivalent circuit of step recovery diode

diode action. As the signal source overcomes the reverse bias set by the d.c., the diode becomes forward biased by the input source. It then conducts in the forward direction and charge is stored in the intrinsic layer much as it is in a PIN diode. Under reverse bias, conduction continues until all the charge is depleted. (Recall from basic electronics that it is impossible to change the current through an inductor instantaneously without producing a voltage impulse.) So we now have a circuit whose output is a series of impulses occurring once each cycle, as shown in *Figure 22.11*. A series of impulses contains a spectrum of frequencies. For

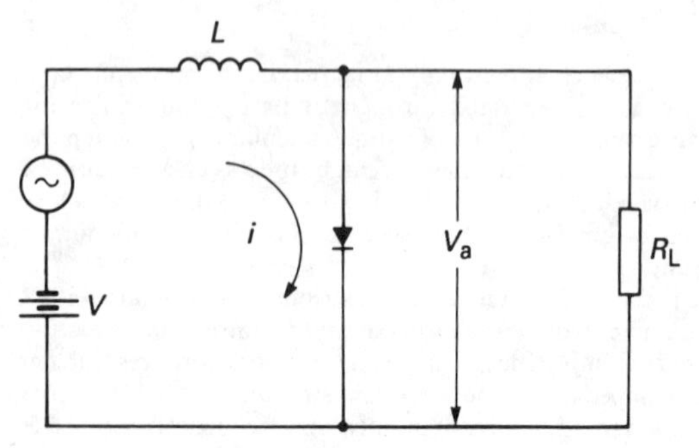

Figure 22.10 Impulse generator circuit

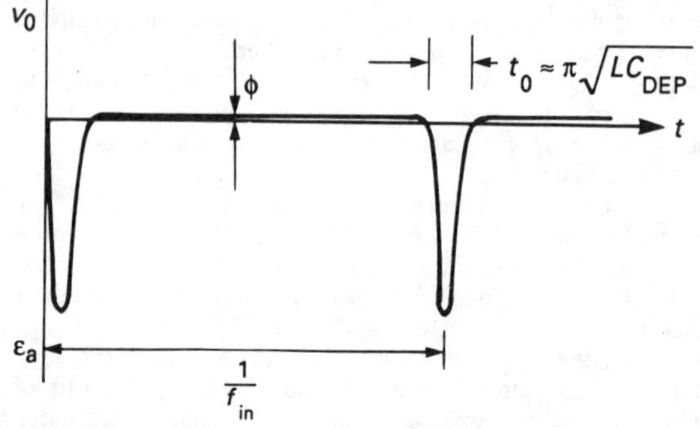

Figure 22.11 Output of impulse generator

practical reasons we cannot produce perfect impulses, so the power contained in the higher harmonics drops off.

22.4.1 Applications and parameters

Applications can be approached in two ways:

(1) The way the diode is used
(2) The equipment in which it is used

The step recovery diode can be made to produce very sharp and narrow pulses, and these in turn contain a virtual infinitude of harmonics of the exciting frequency.

A circuit which exploits the step recovery diode's production of a multitude of frequency components is called a *comb generator*. Comb generators are used in measurement equipment such as spectrum analysers to produce locking signals.

Another type of circuit picks out a single harmonic and optimises the power output around that harmonic. This circuit is called a *multiplier*. The end result of a multiplier is output power at some multiple ($2f_i$, $3f_i$, etc.) of the input frequency. The efficiency of the conversion is high enough to make this a very practical scheme for multiplying up from a readily available transistor oscillator at 600 MHz to get a 2.4 GHz signal ($\times 4$). Multipliers are used as local oscillators, low power transmitters, or transmitter drivers in radar, telemetry, telecommunications, and instrumentation.

Two specifications that determine the *total power output* in any given multiplier mode are *maximum junction temperature* and *thermal resistance*. It may also be necessary to know the *efficiency* of conversion. Efficiency depends heavily on the design of the multiplier, so we do not specify it.

Two other specifications affect the power output by determining the maximum energy in the impulse. The *reverse voltage breakdown* limit, V_{BR}, limits the pulse height and therefore the energy in the pulse. The *reverse bias capacitance*, C_V, does two things. First, it determines the energy which can be put into the pulse, and secondly it sets the impedance level of the impulse circuit.

22.5 Silicon bipolar transistors

Bipolar transistors are commonly used in amplifiers up to 6 GHz and in oscillators up to 12 GHz. The techniques used in bipolar circuit design are well formulated. A bipolar transistor is made of semiconductor materials such as germanium or silicon. Microwave bipolar transistors are almost exclusively silicon.

Figure 22.12 is a simplified model of a bipolar transistor for gain behaviour at microwave frequencies. There are four important time constants involved. They are:

τ_e = emitter charging time
τ_b = base transit time
τ_d = collector depletion layer transit time
τ_c = collector charging time

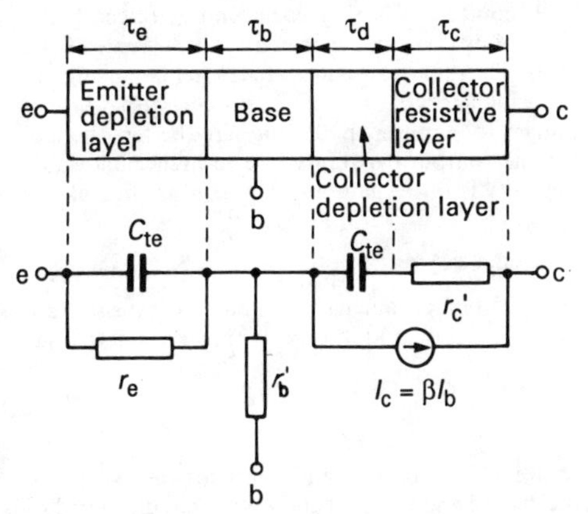

Figure 22.12 Microwave transistor model

The low frequency limit of the exciting signal is set by *minority carrier lifetime*, τ, and the ability to form an effective impulse at the higher frequencies is determined by the *transition time*, t_r. As with PIN diodes, the minority carrier lifetime is the time required for 63% of the charge carriers to recombine. Minority carrier lifetime sets the lower input frequency limit because, as the frequency gets lower and lower, more and more of the charge is dissipated by recombination during a cycle. We do not want the charge to dissipate by recombination because that reduces the energy in the impulse. The input frequency should be higher than a number related to τ by:

$$f_{in} \geqslant \frac{10}{2\pi\tau} \qquad (22.9)$$

The emitter charging time is caused by the parallel combination of the base–emitter space charge resistance r_e and base–emitter transition capacitance C_{te}. The base transit time is the time taken by a carrier to cross the neutral base region mainly by diffusion. The collector depletion layer transit time is the time taken by a carrier to cross the base–collector depletion region by electric field and by diffusion. The collector charging time is due to the RC combination of the collector resistance R_c and base–collector transition capacitance C_{tc}.

22.5.1 Microwave transistor construction

Figure 22.13 is the top view of a simplified bipolar transistor. The patterns are the metal contacts which connect the outside circuit to the working parts of the transistor. Each length of the pattern is

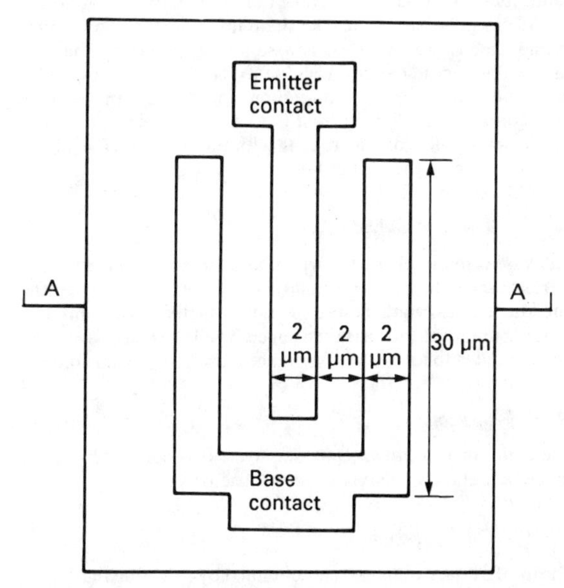

Figure 22.13 Top view of a typical microwave bipolar transistor

called a finger. An actual transistor may have more fingers. The collector is the back of the transistor, underneath the emitter and base.

Figure 22.14 is the cross-section of *Figure 22.13* along A–A. In both figures the relevant dimensions are indicated. A microwave

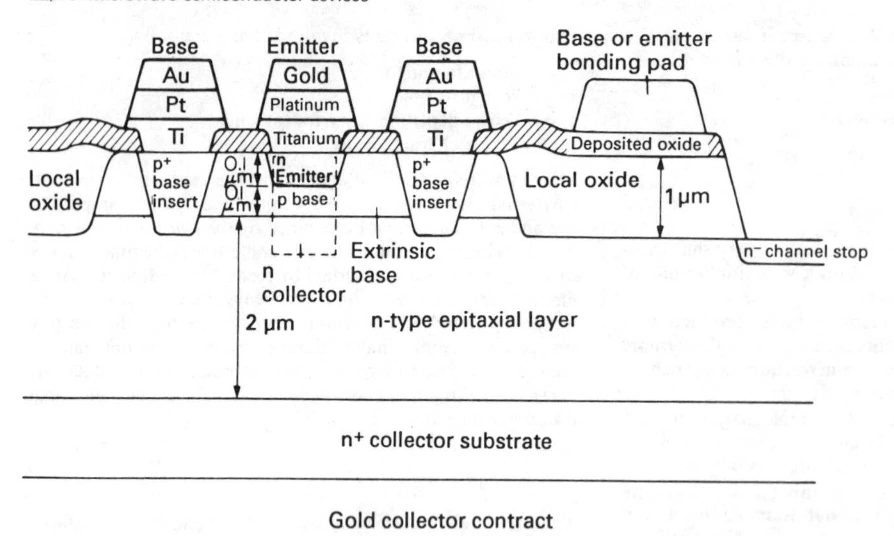

Figure 22.14 Cross-section of a bipolar transistor

transistor is physically very small. The active surface of some low noise transistors is less than 0.025 mm square.

In *Figure 22.14* the functional parts of the bipolar transistor are outlined by the dotted box. This npn transistor has an n-type emitter, p-type base and an n-type collector. The rest of the device serves the purpose of connecting the transistor to the outside circuit, with a minimum of unwanted resistance and reactance.

22.5.1.1 The emitter and metal contact

The emitter is ion implanted n-type silicon material. The connection to the outside is by the titanium–platinum–gold (Ti–Pt–Au) metallic sandwich. The titanium provides a low ohmic resistance contact and good adhesion to the silicon. The gold carries the electric current well due to its low resistivity; and it does not suffer metal migration problems under high temperature and high current density. The platinum layer prevents the gold from interacting with the titanium and silicon and is known as the barrier layer.

22.5.1.2 The base and base insert

The base is ion implanted p-type silicon. It is connected via the extrinsic base, the base insert and the same metal system to the outside. The extrinsic base does not contribute to transistor action, but rather adds base resistance. The base insert is heavily ion implanted to form a low resistance path to the metal contact.

22.5.1.3 The collector

The collector is a grown epitaxial layer. It is connected to the collector metal contact via the n^+ collector substrate.

22.5.1.4 The local oxide

The local oxide is an oxide layer obtained by oxidising the silicon locally, as opposed to deposited oxide. It is a thick oxide layer onto which the emitter and base bonding pads are located. The thickness minimises the bonding pad to collector capacitance: the smaller the capacitance, the better the microwave performance of the transistor. In the making of a transistor, photomasking is used to define the areas of the silicon to be worked on. For best results the mask should be in intimate contact with the surface. Local oxide is made so that it is flush with the rest of the surface, while deposited oxide creates steps on the surface. So the local oxide allows better masking, leading to finer geometry and precision in the end product.

22.5.1.5 Channel stop

Under conditions of elevated temperature and a reverse biased base–collector junction, a phenomenon known as inversion (or channeling) could occur. The surface of the n-type epitaxial layer may become a p-type layer, joining with the base and causing an undesirable extension of the base. A heavily n-doped surface prevents this from happening. This is the channel stop.

22.5.1.6 Emitter ballasting

Emitter ballasting is useful in providing thermal stability to microwave transistors which operate at collector currents of 30 mA or higher.

A power transistor has many emitter fingers and acts like a large transistor made up of many small transistors. At high current levels the emitters get hot. One will always be the hottest, therefore the lowest resistance. Without emitter ballasting the hottest emitter will let more current through it and this thermal runaway effect will eventually destroy this emitter, thus the whole transistor. With emitter ballasting each emitter finger has a feedback resistor of its own, and each emitter is prevented from thermal runaway. The same transistor with emitter ballasting can be operated at a higher current than the version without.

The advantages of emitter ballasting are better thermal stability and higher output capability due to higher allowable current. The penalty is lower gain and higher noise figure.

22.5.2 Radio frequency parameters

The three most useful r.f. parameters for a bipolar transistor are noise gain, power output and the scattering parameters (*S*-parameters).

22.5.2.1 Noise gain

The minimum noise figure of the transistor is specified when the transistor is d.c. biased and its input and output circuits tuned to achieve minimum noise from the device.

Three noise parameters predict the noise performance of a transistor under input matching conditions other than that for minimum noise figure. The set of three quantities are:

Γ_0 optimal input reflection coefficient
R_n noise resistance
F_{min} minimum noise figure

22.5.2.2 Gain

The maximum available gain is power gain achieved at the given frequency when the input and output circuits are tuned for maximum gain and it is unconditionally stable.

Associated gain is the gain achieved when the transistor is tuned and biased for the minimum noise figure. Associated gain is always lower than the maximum available gain and both are frequency and d.c. bias dependent.

22.5.2.3 Power output at 1 dB compression

This is the power handling parameter, and the output power achievable for 1 dB compression in input/output power ratio (*Figure 22.15*).

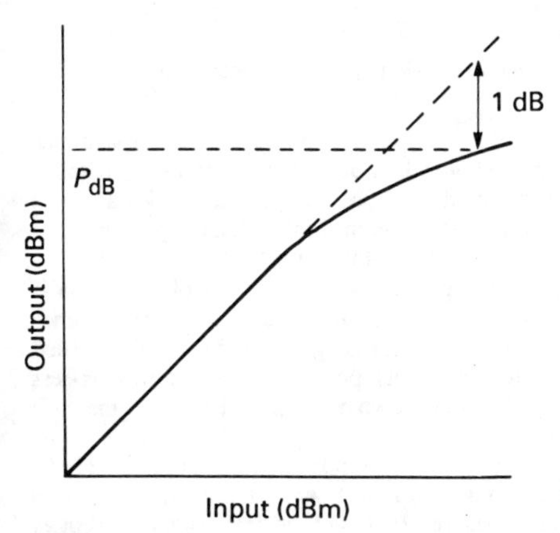

Figure 22.15 Output compression

22.5.2.4 Scattering parameters

S- parameters consist of four vectors (magnitude and angle): S_{11}, S_{21}, S_{12} and S_{22}. They are input reflection coefficient (S_{11}), forward transmission gain (S_{21}), reverse transmission isolation (S_{12}) and output reflection coefficient (S_{22}). They are almost always measured in a 50 Ω system. They are dependent on the d.c. bias conditions and the signal power of the transistor. For the bipolar transistors the *S*-parameters are usually given for small input power (e.g. −20 dB m) and at the d.c. bias condition at which they are expected to perform best.

22.5.3 Applications

As for transistors at lower frequencies, microwave silicon bipolar transistors may be used for amplification up to 4 GHz and are typically configured in a stripline or microstrip circuit. Design is complex and often accomplished with the aid of a computer and most transistor data sheets contain a wealth of data to achieve effective matching characteristics to minimise noise and maximise gain over the desired frequency band.

Computer aided design is used for microwave transistor design to such an extent that the characteristics of many popular devices are contained within powerful proprietary software packages.

Bipolar microwave transistors may be used to make fundamental oscillators up to 6 GHz. They may also be used in applications such as mixing, logic switching, gating, signal detection and sampling.

22.6 Gallium arsenide field-effect transistors

Microwave field-effect transistors are made with gallium arsenide (GaAs) material with a reverse-biased Schottky junction as the gate. Due to the metal-to-semiconductor gate it is often called a metal–semiconductor field-effect transistor (MESFET). The higher electron mobility in gallium arsenide in comparison to that in silicon and the unipolar FET principle make the GaAs MESFET (or GaAs FET) a superior transistor to either silicon bipolar transistors or silicon FETs at high microwave frequencies.

Figure 22.16 is a top view of the GaAs FET chip. The long thin structure is the gate, and the source and drain are labelled

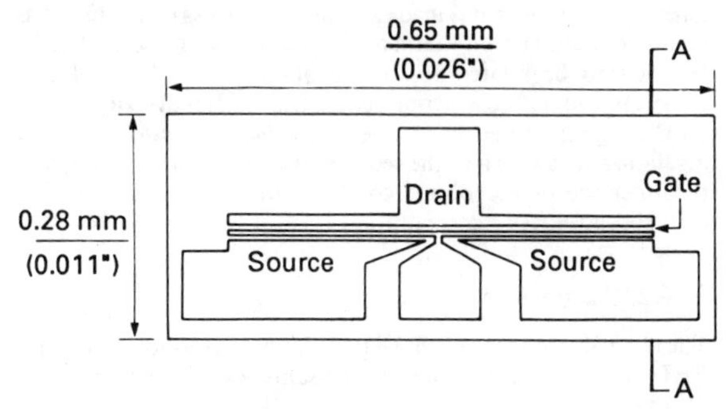

Figure 22.16 GaAs FET chip (top view)

accordingly. By convention, the long dimension of the gate is called its width and the short dimension its length. The cross-section of the FET at section AA of *Figure 22.16* is shown in *Figure 22.17*. It gives an indication of the minute device dimensions necessary for good high-frequency performance.

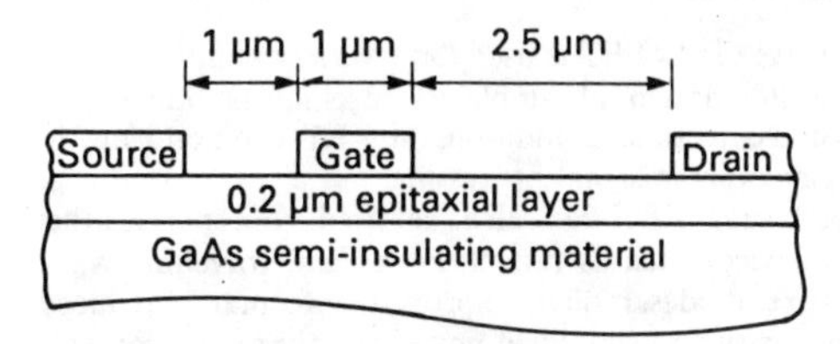

Figure 22.17 GaAs FET cross-section view

22.6.1 Device construction

Figure 22.18 shows the construction of the GaAs FET. The functional layer of the FET is the semiconducting epitaxial layer. It is grown on top of the building block, the semi-insulating substrate. The former is precisely doped for its semiconducting properties and the latter made as pure as possible to act as an insulator. They are both of GaAs material so that one can be grown on top of the other with a perfect crystal interface.

The gate of the FET is the aluminium (Al)–GaAs Schottky barrier, with a gold (Au) layer for wire bonding separated by a chrome (Cr)–platinum (Pt) interface which prevents the migration between gold and aluminium. The chrome is for layer adhesion and the platinum is the actual migration barrier.

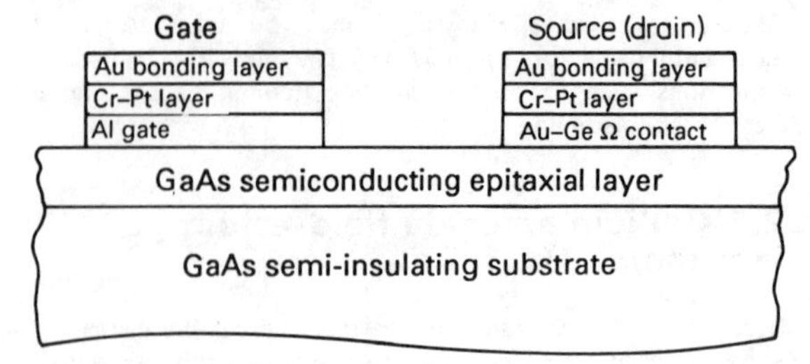

Figure 22.18 GaAs FET construction

The source and drain of the FET are of the same construction. They have a gold–germanium (Ge) ohmic contact with the same gold bonding layer and the chrome–platinum interface. The source, gate and drain surface areas of the FET are covered with a dielectric layer (not shown in *Figure 22.18*) for mechanical scratch protection.

The performance of the FET is controlled largely by the material from which it is made and by the process with which it is manufactured. The long-term performance stability of the FET is believed to be related to the quality of the semi-insulating substrate and semiconducting epitaxial layer. It is also dependent on the degree of perfection of the interface between the semi-insulating substrate and the semiconducting epitaxial layer, and the interface of the ohmic contact with the semiconducting epitaxial layer.

22.6.2 Parameters

The d.c. parameters are defined in the same way as low frequency FETs and may be measured in the same way. These are:

I_{DSs} saturation drain current
V_{GSp} pinch-off voltage
g_m transconductance

The important r.f. specifications for an FET are gain, noise figure, power output and *S*-parameters, and these are defined in the same way as for microwave bipolar transistors.

22.6.3 Applications

The microwave GaAs FET is used for a variety of applications, such as in the design of amplifiers, oscillators, mixers, i.f. amplifiers and detectors. In addition, the FET also performs the switching functions in signal processing.

The microwave FET can also be used to replace other microwave devices such as tunnel diodes and travelling wave tubes and mixer diodes. It offers improved performance, reduced circuit complexity and simplified power supply requirements.

22.7 Tuning varactor diodes

Tuning varactor diodes are p–n junction devices which, when reverse biased, exhibit a capacitance which varies inversely with applied reverse bias voltage. The phenomenon occurs because a reverse bias widens the depletion layer, so decreasing the capacitance of the device. Tuning varactors may be used in a variety of applications where voltage variable capacitance is required such as tuning tank circuits, AFC loops and filters. It should be noted that signal levels are small so that no signal contribution to bias is evident and the device is never used in the forward-bias configuration. Varactor diodes can be made from both silicon and gallium arsenide.

The capacitance–voltage characteristic is shown for a typical diode in *Figure 22.19*. The curve is of a log/linear type and we define:

$$\text{Tuning ratio} = \frac{C_1 V_1}{C_2 V_2} \qquad (22.10)$$

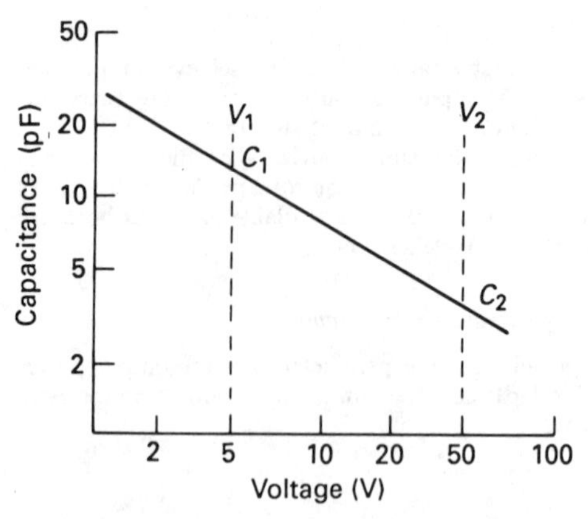

Figure 22.19 Capacitance–voltage for a varactor diode

The tuning ratio determines the magnitude of capacitance change available. It should be noted that the total capacitance includes fixed elements due to package characteristics and this will affect the tuning ratio since, in some cases, the package may contribute up to 50% of the total capacitance.

When the space charge across the n region of the diode is totally depleted the varactor action ceases and the capacitance will charge no further with increasing reverse bias. This is called punch-through. Normally this punch-through voltage is less than or equal to the breakdown voltage and the parameter is specified for the designer.

Q-factor is one of the most important characteristics of the varactor diode so that losses in its circuit are kept as low as possible. For high-*Q* the reactances of the device must be kept as low as possible. Obviously the capacitance of the device contributes to the reactance so the *Q*-factor will increase as the reverse bias is increased and the capacitance falls.

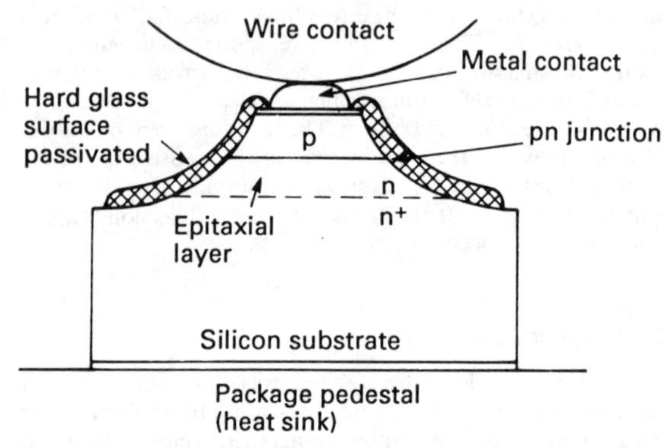

Figure 22.20 Varactor diode construction

For high-*Q* the series resistance must be kept as low as possible and resistance can come from the chip itself, bonding contact and package contacts.

Figure 22.20 shows a commonly used mesa construction of the varactor diode.

Further reading

BARRERA, J. S., 'GaAs field effect-transistors', *Microwave Journal*, February, 28–36 (1976)

BERSON, B., 'Semiconductors prove fruitful for microwave power devices', *Electronics*, Jan. 22, 83–90 (1976)

CASTERLINE, E. T. and BENJAMIN, J. A., 'Trends in microwave power transistors', *Solid State Technol.*, April, 51–56 (1975)

DAVIES, R., NEWTON, B. H. and SUMMERS, J. G., 'The TRAPATT oscillator', *Phillips Technical Review*, **40,** No. 4 (1982)

DUH, K. *et al.*, 'Ultra low noise characteristics of millimetre-wave high electron mobility transistors', *IEEE Electron. Dev. Lett.*, **9**, No. 10 (1988)

HAMMERSCHMITT, J., 'Microwave semiconductors for SMT', *Siemens Comps.*, **23**, No. 2 (1988)

MAAS, S. A., *Microwave Mixers*, Artech House (1986)

Microwave and RF Designer's Catalogue, Hewlett-Packard (1987)

WEARDEN, T., 'Varactor diodes' *Electron. Product Des.*, December, 81, 82 (1980)

23 Optical Sources and Detectors

R M Gibb, MA, DPhil
Technical Manager
STC Optical Devices Division
(Section 23.1)

J Curtis
Distributor Manager
Electronic Components Group
Siemens Ltd
(Section 23.2)

Contents

23.1 Optical sources

This chapter considers optical sources which emit photons by the transition from an excited to a ground state as distinct from incandescence. The wavelength of emitted radiation is given by the equation

$$E = 1.239/\lambda$$

where E is the energy difference between the two states in electron-volts, and λ is the wavelength in nanometres. Such devices should, in theory, be highly efficient with quantum efficiencies approaching 100%. However, it will be seen that device performance is limited by material purity and the fraction of radiation which can be extracted from the chip. Two classes of devices will be described. The first class is spontaneous radiation emitters comprising light and infrared emitting diodes (LEDs and IREDs), and the second is the wide ranging group of devices known as lasers.

23.1.1 LEDs and IREDs

23.1.1.1 Mechanism

Radiation is emitted from electroluminescent p–n junctions as a result of radiative recombination of electrons and holes whose concentrations exceed those statistically permitted at thermal equilibrium. Excess carrier concentrations are obtained in a forward-biased p–n junction through minority carrier injection: the lowering of the potential barrier of the junction under forward bias allows conduction band electrons from the n side and valence band holes from the p side to diffuse across the junction. These injected carriers significantly increase the minority carrier concentrations and recombine with the oppositely charged majority carriers. This recombination, which tends to restore the equilibrium carrier densities, can result in the emission of photons from the junction. The recombination process, hence the amount as well as the wavelength of the light generated, is a strong function of the physical and electrical properties of the material.

Both energy and momentum must be conserved when an electron and a hole recombine to emit a photon. Since the photon has considerable energy but very small momentum ($h\nu/c$), simple recombination only occurs in direct bandgap materials, that is, where the conduction band minimum and valence band maximum both lie at the zero momentum position. This condition was assumed to be a prerequisite for efficient electroluminescence until about 1964 when Grimmeiss and Scholz[1] demonstrated reasonably efficient electroluminescence in the indirect bandgap material GaP.

In an indirect bandgap material where the valence band maximum and conduction band minimum lie at different values of momentum, recombination can only occur when a third momentum-conserving particle is involved; phonons (i.e. lattice vibrations) serve this purpose. The probability of an electron–hole recombination involving both a photon and a phonon is considerably smaller than the simpler process involving only a photon in a direct bandgap material. This is clearly illustrated by the differences in recombination coefficients (R) for various materials: this coefficient (R), which relates the radiative recombination rate (r) to the excess minority (Δn) and majority (p) carrier concentrations

$$r = R(\Delta n)p \qquad (23.1)$$

is of the order of 10^{-14} to 10^{-15} cm^3/s for indirect bandgap materials (such as Si, Ge and GaP) while it is of the order of 10^{-10} to 10^{-11} cm^3/s for direct bandgap materials (such as GaAs, GaSb, InAs or InSb).

LEDs and IREDs can, in principle, be made from any semiconducting compound containing a p–n junction, and having a sufficiently wide bandgap. Despite considerable effort on other materials to achieve efficient luminescence, currently only III–V compounds are of practical interest.

23.1.1.2 Doping

Donor and acceptor impurities determine the magnitude and type of conductivity and also play a major role in the radiative recombination processes. In direct bandgap III–V compounds, donor impurities are usually chosen from group VI of the periodic table (Te, Se, S), while acceptor impurities usually belong to group II (Zn, Cd, Mg). Sometimes group IV impurities (Ge, Si, Sn) are used either as acceptors or donors depending on which of the available lattice sites (III or V) they occupy.

Optimum impurity concentrations are best determined experimentally. High doping levels, i.e. large majority carrier concentrations, are desirable to lower the bulk resistivity, and consequently to minimise heating and voltage drops at high current densities; and also because the recombination probability, which is directly proportional to the carrier concentration, should be as large as possible. The doping concentrations are practically limited, however, by the formation of precipitates and other crystallographic imperfections which introduce competing non-radiative recombination centres and therefore reduce the electroluminescence efficiency. Donor concentrations of 10^{17}–10^{18} per cm^3 and acceptor concentrations of 10^{17}–10^{19} per cm^3 are typical.

Although the luminescence efficiencies in n- and p-type materials at optimum doping concentrations are essentially equal, the radiation emitted from the p-region normally dominates for several reasons:

(1) Electron injection into the p-region is favoured over hole injection into the n-region because of the high electron to hole mobility ratio in most III–V compounds.
(2) The Fermi level is slightly higher than the intrinsic energy gap in n-type material,while it is lower in p-type material. The resulting heterojunction effect further favours electron injection into the p-region over hole injection into the n-region.
(3) Radiation generated in the n-region is usually of shorter wavelength (higher energy) than that generated in the p-region. Therefore n-generated radiation is strongly absorbed in the p-region while p-generated radiation passes through the n-region with reduced absorption losses. This latter fact is usually taken into account in the design of efficient devices.

In indirect bandgap materials, e.g. GaP, recombination across the gap requires the participation of a momentum-conserving phonon and is therefore inefficient. More efficient recombination can occur when a charged carrier is first trapped at a neutral impurity centre and then used to attract the oppositely charged carrier. Momentum conservation in this case is more easily satisfied because the carrier trapped at a neutral impurity centre is highly localised in space and consequently has a wide range of crystal momentum.

Only a limited number of impurities have been found which enhance the recombination in GaP. Near bandgap ($\sim$2.23 eV) green emission is increased by nitrogen substitution for phosphorus in GaP, but competing non-radiative recombination processes limit the internal quantum efficiency to about 1%. Red luminescence ($\sim$1.79 eV) is improved by incorporation of zinc and oxygen centres. The internal quantum efficiency in GaP:Zn,O is higher (10–20%) than for green luminescence, but the emission saturates at relatively low current densities ($\lesssim$10 A/cm^2) due to the limited concentration of Zn–O centres ($<10^{17}$ per cm^3). Green luminescence in GaP:N does not readily saturate since nitrogen concentrations of 10^{19} per cm^3 are practical.

23.1.1.3 Quantum efficiency and brightness

Since most of the applications of LEDs involve an observer, the response of the human eye at the emitted wavelength is of primary importance. The eye sensitivity for normal (photopic) vision extends from about 400 nm to 700 nm. It peaks in the green (555 nm) at 680 lm/W and falls off towards the red and blue regions of the spectrum (*Figure 9.2*). The brightness, and to a large extent the visibility, of an LED is proportional to the product of its external quantum efficiency and the sensitivity of the eye at the emitted wavelength.

The external quantum efficiency of an LED is the ratio of emitted photons to number of electrons passing through the diode, and it is typically 0.1–7% at room temperature. The external quantum efficiency is always less than the internal quantum efficiency because all the light generated cannot exit from the diode.

The internal quantum efficiency is highly dependent on the perfection of the material near the p–n junction. Various defects, contaminants or dislocations reduce the internal quantum efficiency by producing deep recombination centres, which lead to long wavelength radiation, or by enhancing non-radiative recombination. It is thought that the superiority of GaP:N diodes grown by liquid phase epitaxy (LPE) compared to those grown by vapour-phase epitaxy (VPE) may be due to the lower density of Ga vacancies in LPE material.

Poor quality substrates are a major cause of defects. Additional imperfections can be introduced during the growth of the epitaxial layer, especially when the lattice mismatch is relatively large. These latter imperfections can be reduced by grading the composition of the film during growth from that of the substrate to the composition desired at the junction. Defects are also sometimes introduced during the diffusion process used to form the p–n junction.

The high index of refraction of III–V compounds (typically ~3.5) leads to additional light losses. Much of the generated light suffers total internal reflection. This increases the optical path length inside the diode thereby increasing internal optical absorption which is particularly high for near bandgap emission. Thus the external quantum efficiency can be 50 to 100 times smaller than the internal efficiency.

Practically, these losses can be reduced by increasing the transmissivity of the surface. Antireflection coatings can be applied, the diode can be shaped in the form of a hemisphere, or a hemispherical epoxy or acrylic lens can be used to increase the efficiency by a factor of 2–3 typically. The internal absorption is reduced at longer emission wavelengths, which are obtained by incorporating deeper acceptors (e.g. silicon instead of zinc in GaAs), or in indirect bandgap materials, where the emission is much below the energy gap. In indirect gap materials, the reduced internal efficiency (10% to 20% for GaP:Zn,O as compared to ~50% for GaAs:Zn) sets an upper bound for the external efficiency of the device.

The brightness (B) of an LED in cd/m^2 is given by

$$B=\frac{3940\eta_{\text{ext}}KJ}{\lambda}\left(\frac{A_{\text{j}}}{A_{\text{s}}}\right) \tag{23.2}$$

where

η_{ext} = external quantum efficiency
K = luminous efficiency of the eye (in lm/W)
J = junction current density (in A/cm^2)
$A_{\text{j}}/A_{\text{s}}$ = ratio of junction area to observed emitting surface
λ = emission wavelength (in μm)

Commercial LEDs are available with luminous intensities as high as 500 mcd at 20 mA for red emitters.

23.1.1.4 Ternary and quaternary compounds

The use of binary compounds limits the range of bandgap energies which can be used. As a result considerable work has been spent on the development of ternary and even quaternary alloys as emitters. Ternary alloy systems, composed of narrow direct gap and wider indirect gap materials (e.g. GaAsP, GaAlAs, InGaP and InAlP), provide a monotonically varying direct energy gap as the relative concentration of the indirect material is increased up to a critical composition x_c, where the energy gap becomes indirect. This is shown in *Figure 23.1* for the GaAsP system.

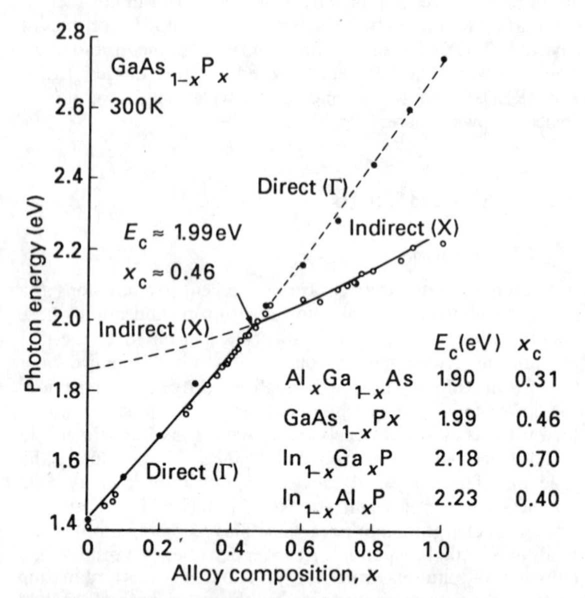

Figure 23.1 Direct (Γ) and indirect (×) conduction band minima for GaAs$_{1-x}$ as a function of alloy composition, x. Closed data points are from electroreflectance measurements; open points are from electroluminescence spectra (Reproduced with permission from Smith[2])

The majority of visible LEDs are now fabricated from the GaAsP system. Red emitting diodes utilise an alloy composition close to the direct/indirect transition at a value of x=0.4. To achieve orange emission at around 600 nm the indirect transition is enhanced by the addition of nitrogen.

The range of IREDs covers emitters utilising the alloy GaAlAs in the range 900–700 nm and the quaternary alloy GaInAsP for emission between 1200 and 1600 nm.

23.1.1.5 Materials growth

III–V semiconductor devices are fabricated from epitaxial material deposited on single crystal substrates. The choice of substrates depends on the lattice constant and the thermal expansion coefficient of the epitaxial layers being grown. Since GaAs and AlAs have roughly the same lattice constant, all compositions of GaAlAs can be grown on GaAs substrates without inducing strain or misfit dislocations. Certain GaAsP compositions require a graded layer between the substrate and the emitting layer. For InGaAsP grown on InP substrates careful control of the composition is required to prevent misfit dislocations forming.

Four techniques of epitaxy are widely used: liquid phase epitaxy (LPE), vapour phase epitaxy (VPE), metal organic chemical vapour deposition (MOCVD) and molecular beam epitaxy (MBE). Each has its own advantage for certain types of material.

In LPE, originally developed by Nelson[3], a layer of material is grown from a solvent comprising one of the constituent metals, i.e. Ga or In. At the growth temperatures, which are in excess of 500°C, these metals are liquid and are saturated with the elements to be deposited plus the dopants. This 'melt' is contained in a graphite boat. The substrate is positioned under the melt while the boat is cooled and the layer is epitaxially deposited. The layer thickness is controlled by the cooling rate and the time. Using a multi-well boat a multilayer structure can be grown in a single run.

This technique has been largely used for GaAlAs materials. However, it suffers from being of poor uniformity and requires long flushing times to remove residual oxygen. It has also been used to produce the most efficient GaP:N green LEDs and high-reliability long-wavelength lasers in InGaAsP.

VPE is widely used in the commercial preparation of red $GaAs_{1-x}P_x$ LEDs. A typical VPE system, as originally described by Tietjen and Amick,[4] is shown in *Figure 23.2* Here the gallium is

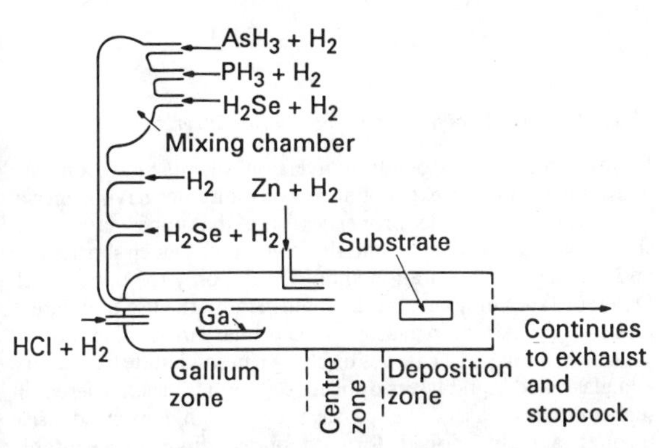

Figure 23.2 Schematic diagram of typical vapour-phase epitaxy growth apparatus used for the deposition of $GaAs_{1-x}P_x$ layers on GaAs substrates (Reproduced with permission from Smith[2])

transported by flowing hydrogen chloride gas over the molten metal. Arsenic and phosphorus are obtained from the thermal decomposition of arsine and phosphine, respectively. Hydrogen is normally used as the carrier gas.

As the gases pass over the substrate, an epitaxial layer of $GaAs_{1-x}P_x$ forms whose composition is determined by the composition of the gases; these can be controlled accurately over wide ranges with precision flowmeters and valves. Varying the gas composition during growth provides the ability to slowly grade the composition of the epitaxial layer, thereby minimising the lattice mismatch and resultant strains.

Doping is also provided by introducing suitable gases. H_2S and H_2Se are normally used to incorporate sulphur and selenium donors, and $(CH_3)_2Te$ and SiH_3 have sometimes been used to obtain tellurium or silicon doping, respectively. Acceptor impurities can also be incorporated at the appropriate moment, e.g. by adding zinc vapour and hydrogen, but more often a postgrowth diffusion of zinc metal is used to form the p-layer.

In addition to the original horizontal VPE systems, much larger vertical reactors have been developed for mass production. These systems process many wafers simultaneously ($\sim 300\,cm^2$ of substrates at a time), usually rotating the substrates during growth to improve uniformity. The availability of multi-wafer growth equipment and high-quality GaAs substrates, coupled with the fact that the epitaxial layers require no postgrowth polishing, have contributed significantly to the commercialisation and reduction in cost of LEDs.

VPE is also used[5] for preparation of $In_{1-x}Ga_xP$ but is less suited for $Al_xGa_{1-x}As$ because of the reactivity of aluminium-containing gases.

MOCVD is related to VPE in that it is a vapour phase transport system and can be automated with flow controllers in the same way. Fully computerised equipments are now commercially available for the growth of GaAlAs, while InGaAsP devices grown by this technique are in development.

This method uses liquid metal alkyls such as trimethyl gallium as the metal source. These compounds are stable at low temperature under dry hydrogen and have well defined vapour pressures. When mixed with gaseous hydrides, e.g. phosphine or arsine, and passed over a heated substrate, the metal alkyl is cracked and the semiconductor is deposited. Dopants in the form of hydrides or alkyls are also used.

This technique has made a major impact in the manufacture of certain types of devices since it allows growth to be achieved uniformly on 2-inch wafers. It can also achieve very abrupt interfaces between layers of different composition and has particularly been used for the growth of quantum well structures. Reviews of MOCVD have been written by Ludowise[6] and Dapkus.[7]

MBE has been used in the research laboratory to grow a wide range of semiconductor materials, but has never found commercial usage. It uses heated elemental or gas souces to form a beam of atoms which impinge on a substrate held in ultra-high vacuum. Alternate layers of group III and group V elements can be directed at the substrate by the use of shutters. In this way some of the most abrupt interfaces and thinnest layers have been grown. This technique is well reviewed by Chang[8] and by Cho and Arthur.[9]

23.1.1.6 *Devices*

Two principal device types exist; surface and edge emitters. The former comprise all LEDs and the majority of IREDs.

In a simple form, a p–n junction is produced in an epitaxial layer either during the growth process or by diffusion. The metalisation covers a minimum area of the surface to allow low contact resistance without obscuring the radiation. For devices emitting by transition through an impurity centre, the whole chip 'glows'. For direct bandgap devices, the radiation is strongly absorbed. This phenomenon can be used in forming addressable monolithic arrays of emitting regions, e.g. a seven-segment numeric array.

More complex devices entail the growth of heterostructures. These comprise epitaxially grown layers of different composition. In 900 nm diodes, a layer of GaAs is sandwiched between two GaAlAs layers. The wider bandgap GaAlAs confines the injected electrons and holes within the GaAs region and thereby improves the recombination efficiency. However, if the recombination region is lightly doped with aluminium to tune the wavelength, the GaAs substrate and the surface contact layer are strongly absorbing. This effect can be overcome by etching away the absorbing layer in the emitting region. In the Burrus[10] diode a well is etched in the substrate down to the GaAlAs cladding layer. The diode is mounted substrate up and the emission is from the etched well. Output powers as high as 15 mW are quoted for such devices at 100 mA drive current, but more normally powers of a few milliwatts are achieved.

The problem of substrate absorption does not occur in emitters at 1300 nm where the InP substrate is transparent. Efficient surface emitters are produced by etching a curved surface onto the diode which increases the effective critical angle and, hence, the extraction efficiency. Output powers of up to 500 μW can be achieved at 100 mA and rise times <2 ns have been recorded.

More efficient devices can be produced by confining the carriers parallel to the junction and by forming localised contact regions. The diode is cleaved perpendicular to the junction so that the radiation is emitted from the edge. These produce high-intensity sources. An example of this is shown in *Figure 23.3*. Here the contact stripe is limited to one end of the chip. Absorption in the unpumped region is sufficient to prevent stimulated emission from predominating. These devices form ideal sources for fibre optic systems since the output can be efficiently coupled to an optical fibre. At 850 nm 1 mW can be launched into a 50 μm core fibre, while at 1300 nm 10 μW is readily achieved with single mode fibre.

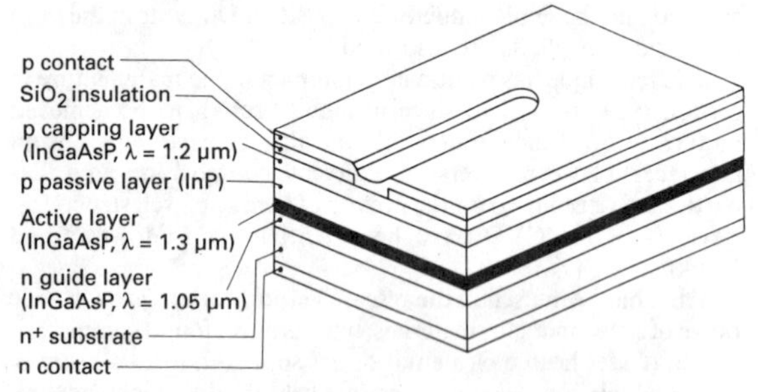

Figure 23.3 Edge-emitting IRED (Courtesy of STC plc)

Most simple devices are packaged in transistor headers or on lead frames moulded in transparent epoxy. However, complex arrays and displays involve a range of plastic and metal package types.

23.1.2 Lasers

The word laser is an acronym for *L*ight *A*mplification by the *S*timulated *E*mission of *R*adiation. The advent of the laser has enabled visible light and infrared radiation sources to be produced with a spectral purity and stability as good as, or better than, can be achieved in the radio spectrum, e.g. a HeNe laser can be made with a linewidth of a few kHz or less at a carrier frequency of about 5×10^{14} Hz. Such light sources are said to have a high temporal or longitudinal coherence. The meaning of this is illustrated by the fact that the light from the source mentioned above could still interfere with another beam split off from the same source with a path difference between the two beams of up to 300 km. The corresponding figure for the best conventional light source is only 10 cm.

Another property of laser beams is a high degree of transverse coherence, i.e. if one takes any section perpendicular to a laser beam, then there is a very well defined relationship between the phase of all points on the plane at all times, no matter how far from the source (this is only exactly true in free space). With normal light sources the phase relationship between different points on the source is completely random for distances greater than at most a few μm. This high transverse coherence is responsible for the very slow spreading of laser beams.

The unique properties of the laser are due to the process of stimulated emission, first postulated by Einstein in 1917 in order to explain Planck's radiation law. The idea of stimulated emission is that an atom in an excited state can be 'stimulated' to emit radiation by placing it in a radiation field of the same frequency, as it will normally emit in going to one of the lower excited states or to the ground state. It can be seen that such a process can result in amplification, if the process of stimulated emission from atoms in the upper state is more probable than that of absorption by atoms in the lower state. Einstein showed that, in general, these processes have equal probabilities for a single atom in the radiation field. Thus in order to obtain net amplification there must be more atoms in the upper state than in the lower state. This is an inversion of the normal energy distribution, and thus we speak of population inversion being necessary to achieve amplification.

It can also be shown that the light produced by the stimulated emission process is in phase with the stimulating light, i.e. it is coherent with it. However, at optical frequencies as distinct from microwaves, spontaneous emission, i.e. the random de-excitation of upper state atoms, is a highly probable process which makes the achievement of inversion difficult. All things being equal the probability of spontaneous emission is inversely proportional to the wavelength cubed, thus inversion is much more difficult to achieve in the ultraviolet than the infrared, and extremely difficult in the X-ray region. Of course spontaneous emission is the dominant process in ordinary light sources and is responsible for the poor coherence properties mentioned above since the excited atoms emit in a quite independent and random manner, resulting in poor transverse coherence, and the emission is in the form of very short pulses, resulting in poor longitudinal coherence.

23.1.2.1 Methods of achieving population inversion

In order to achieve population inversion, we must have a means of excitation and a system of atoms or molecules having energy levels with favourable properties, e.g. the upper laser level should have a long lifetime and the lower level a much shorter one and the means of excitation should excite only the upper level. Different types of lasers are distinguished by the use of different modes of excitation. In gas lasers the primary mode of excitation is electron impact in a gaseous discharge, in solid-state (crystalline and glass) and liquid lasers optical excitation is usual, whereas in semiconductor lasers the passage of a high current density through a highly doped, forward biased diode of a suitable material is the preferred mode of excitation. In most cases the excitation process is a multi-step one. For instance in many gas lasers a mixture of gases is used, one of the gases is chosen to have a metastable excited state with a high probability of excitation, whose energy coincides with that of the upper laser level, so that de-excitation by collision with a ground state atom or molecule of the other species is highly probable. Thus helium in the HeNe laser and nitrogen in the 'CO_2' laser (which usually uses a mixture of carbon dioxide, helium and nitrogen) both have suitable metastable levels for the excitation of neon atoms and carbon dioxide molecules respectively.

In solid-state lasers, the light absorbed excites an ion to a high level which then decays very rapidly by a non-radiative process to the upper laser level which usually has a lifetime of at least several hundred microseconds, thus making inversion relatively easy to achieve on a transient basis, i.e. pulsed excitation by a flashtube. In fact solid-state laser materials are rather special phosphors, chosen for the narrow linewidth of their emission. Similarly de-excitation of the lower laser level is not usually a simple radiative process. In gas lasers, more often than not, collisions with the walls of the discharge tube (HeNe) or with another species of atom or molecule (helium in the CO_2 laser) are important in de-excitation. In some solid-state lasers (so-called four level materials such as neodymium doped materials) de-excitation is achieved by radiationless transfer of energy to the atoms of the lattice in which the Nd^{3+} is embedded. In the case of 'three level' materials such as ruby, the lower laser level is the ground state of the Cr^{3+} ions in an Al_2O_3 crystalline matrix. Thus 'de-excitation' is automatic, but inversion is difficult to achieve since more than half of all the Cr^{3+} ions must be excited in order to achieve inversion.

23.1.2.2 Oscillators and modes

If we simply have a medium in which population inversion is maintained, then it will act as a tuned amplifier resonant at a frequency corresponding to the transition concerned, with 3 dB points determined by the linewidth of spontaneous emission. If the overall amplification is very high, then amplification of spontaneous emission or super-radiance can depopulate the inversion very rapidly, with some line narrowing.

To make an oscillator, we require positive feedback; this is provided by a pair of mirrors placed at each end of the amplifier as shown in *Figure 23.4*. Usually one of these is as near 100% reflecting as possible, and the other one has a transmission typically of a few per cent or so. At longer infrared wavelengths a hole in the output mirror may be used to couple power out. To obtain oscillation GR_1R_2 must be greater than unity (G is the gain of the medium; R_1,R_2 are the mirror reflectivities). This arrangement is similar to a microwave resonator, except that there are no reflecing walls between the mirrors, and is therefore spoken of as an 'open' resonator.

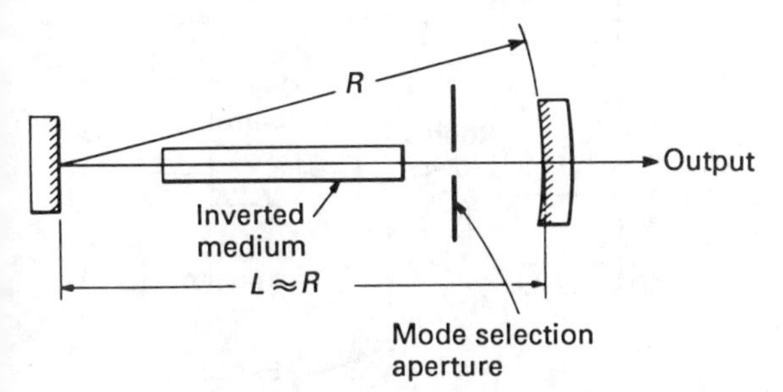

Figure 23.4 Laser with hemispherical resonator incorporating transverse mode selection

Since the distance between the mirrors is typically more than 10^5 wavelengths, and the width of the medium is usually at least 10^3 wavelengths, it can be seen that many more modes of oscillation are possible than in a microwave resonator. The behaviour of such modes has been described by Fox and Li[11] and others.

The modes in an open resonator of this type are what is known as transverse electromagnetic or t.e.m. waves. They are specified by three subscript mode numbers p, l, q (cylindrical symmetry) or m, n, q (rectangular). The last subscript in each case is the longitudinal mode number, i.e. a very large number. The other two refer to the transverse directions, and in practice here the low order modes (<5) are usually dominant.

Transverse mode patterns corresponding to some of the low order rectangular modes are shown in *Figure 23.5*. The 0,0q mode is of most practical importance since it has the highest degree of transverse coherence and the lowest beam divergence. The radial intensity distribution of energy in any cross-section is given by a Gaussian function. The propagation of such 'Gaussian' beams has been treated in some detail by Kogelnik.[12]

Schemes for selecting a particular mode, usually the zero order, depend on the fact that the diffraction losses of different modes propagating in a resonator are different, and by using the hemispherical resonator configuration shown in *Figure 23.4*, the absolute magnitudes of the losses for higher order modes make it easy to select the zero order mode by the use of a suitable aperture as shown in the figure. The longitudinal mode behaviour is determined by the width of the gain curve of the inverted medium and by the frequency response of the resonator. The simplest form of resonator, using a pair of relatively broadband reflecting multilayer dielectric mirrors has very many resonances, separated by a frequency $\Delta f = c/2L$ where c is the velocity of light and L is the optical path length of the resonator.

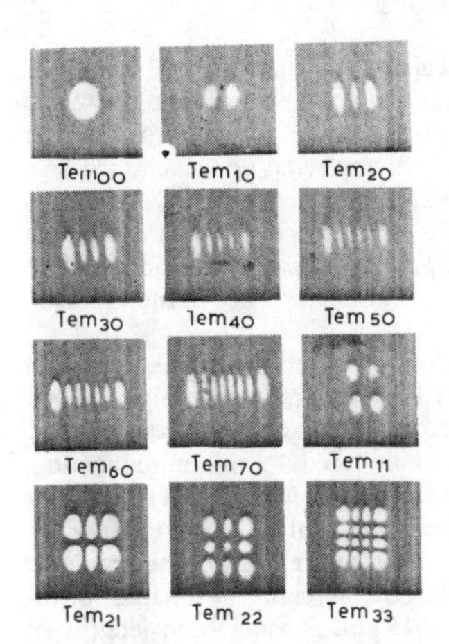

Figure 23.5 Mode patterns of a gas laser oscillator (rectangular symmetry)

In rare cases, such as the low pressure CO_2 laser, it is relatively easy to ensure single frequency operation since the linewidth of the gain curve is typically less than $c/2L$. In most types of laser there are very many resonances within the gain curve, and thus many longitudinal modes oscillate, unless a more complex resonator, having only one principal resonance within the width of the gain curves, is used. The subject of mode selection schemes has been reviewed by Smith.[2]

When many longitudinal modes oscillate simultaneously the phases are usually random. By inserting in the cavity a loss modulator or a phase modulator, tuned to the mode spacing Δf, the sidebands so produced cause the phases of all the modes to be locked together, resulting in a pulsed output with a p.r.f. Δf and a pulse length $T = 1/\Delta f_{osc}$ where Δf_{osc} is the width of the gain curve. Typically $\Delta f \geqslant 100$ MHz, $T \leqslant 1$ ns. Mode locking has also been reviewed in some detail by Smith.[13]

Another pulsed mode of laser operation of great practical importance is the so called 'Q-switched' or 'giant pulse' mode. In this case extra loss is introduced into the cavity, by means of a Pockels effect or acousto-optic modulator, while the inversion is built up to a much higher value than required to sustain oscillation in the low-loss case. The extra loss is then switched out rapidly, resulting in the stored energy being discharged in a single very short pulse (typically 10–100 ns).

A third pulsed mode operation is 'spiking', which occurs spontaneously in pulse excited solid-state lasers. The output in this case consists of a random train of submicrosecond pulses of duration comparable with the pumping pulse. This type of operation is reviewed by Roess.[14]

23.1.2.3 Gas lasers

As mentioned above, these are usually electrically excited via collisions by passing a discharge through a gas or, more often, a gas mixture. Only the most common types are described here. These are the helium–neon (HeNe) laser, the argon ion laser, the CO_2 laser and the excimer laser.

The helium–neon laser usually contains a mixture of about five parts helium with one part neon at a total pressure of a few torr. The discharge tube has a bore of 1–2 mm and a length of 20–100 cm, depending on the power output, usually in the range 0.5–50 mW. The current is usually ~10 mA with voltages up to a few kilovolts. The gain of the medium is very low and losses must be kept to a minimum. Either Brewster angle windows (resulting in plane polarised output) are used with external mirrors, or integral mirrors sealed to the discharge tube are used to ensure low loss. Although alignment is critical, and the sealed mirror method of construction is more difficult to engineer satisfactorily, this has become the preferred approach to design since it provides superior life and long-term stability. Although the usual output wavelength is ~633 nm in the red, outputs at 594 nm, 604 nm, 612 nm, 1.15 μm or 3.39 μm can be obtained by using suitable mirrors. Typical beam divergence in the visible is ~1 mrad.

In argon lasers the transitions involved are between excited levels of singly ionised argon ions. Consequently high current densities (> 100 A/cm^2) are required to achieve threshold, and plasma tubes of BeO are usually used to minimise heating, erosion and gas clean-up. Brewster angle windows of fused silica are usually used with external mirrors. Outputs of up to 20 W, on a number of lines in the blue and green regions of the spectrum simultaneously, are available commercially. Principal wavelengths are ~488 nm and ~515 nm. Plasma tube lengths of 10–100 cm with diameters of 1–2 mm are used and more gain is available than with HeNe lasers. However, the efficiency, ⩽0.1%, is very low. Typical beam divergence is ⩽1 mrad.

The CO_2 laser, which gives outputs at a number of wavelengths between 9.6 μm and 10.6 μm, is the most efficient of all the gas lasers (typically 10–20%), and is currently the one of greatest technological importance. It is the most powerful (CW powers up to 30 kW or so) and the most varied in construction since there are many different versions operating at pressures from a few torr up to above atmospheric, with pulsed or CW excitation, and even incorporating *Q*-switching. The CO_2 laser is one of the few gas lasers in which the lifetime of the upper level is great enough to make *Q*-switching worthwhile.

The most common form is still the low-pressure flowing gas (He:N_2:CO_2::8:1:1 typical) system using a plasma tube ~1–2 cm in diameter and up to several metres in length, depending on the power output required. Typical commercial systems with fast transverse or axial gas flow to improve cooling can achieve TEM_{00} CW outputs of ~250 W/m. Integral mirror or Brewster angle plus external mirror configurations are used. For high-power systems of > 100 W or so with long plasma tubes, a folded configuration is often used (*Figure 23.6*). Useful reviews of CO_2 laser systems have been given by De Maria[15] and Basov *et al.*[16]

The excimer laser (the term being a contraction of excited dimer) is the most recent powerful gas laser to be developed and its importance is due to its ultraviolet output which, depending on gas choice, lies between 157 nm (fluorine) and 350 nm (xenon fluoride). Operation, via short-pulse electrical excitation, is due to the creation of new molecular species without ground states. The short lived (⩽ 50 ns) molecules decay via disassociation from their excited states emitting ultraviolet photons. The lasing process can be efficient (typically ~1% overall) allowing ultraviolet average powers of ≳100 W to be achieved from discharge cells of ~1 M length. Pulse energies are typically < 1 J but can be up to ≳100 J in large experimental systems.

Uses of gas lasers The selection of a laser for a specific application depends primarily on its output power and wavelength. The most common gas laser is the helium–neon laser where its visible emission is commonly used for alignment and distance measurement, inspection and bar code reading.

The argon laser with a higher power blue and green emission is used by the scientific community in a broad range of applications including dye laser pumping. Commercial uses include colour separation in the printing industry and photocoagulation and retinal welding in medicine.

The CO_2 laser with its very high power CW and pulsed capability is used in cutting and welding applications. However, the penetration of its 10.6 μm line in poor weather makes low-power devices of interest to developers of thermal imaging

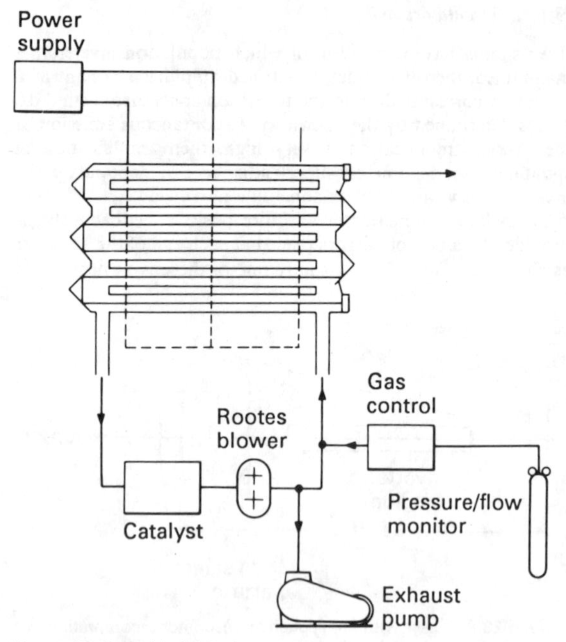

Figure 23.6 Folded CO_2 laser with gas recirculation (Courtesy Marconi-Elliott Flight Automation Ltd)

systems working at infrared wavelengths. The multiplicity of lines in its spectrum offers a wide choice for use in aerial pollution monitoring 'Lidar' systems.

Commercial applications for the excimer laser are currently being explored. Photo-assisted chemistry (e.g. in the pharmaceutical industry) and photolithography in the semiconductor industry are currently areas of keen research and development interest. The use of excimers for materials processing has recently been reviewed by Znotins *et al.*[17]

23.1.2.4 Liquid lasers

These use a solution of a fluorescent molecule, optically pumped either by another laser or by a flash tube, in a configuration similar to that used for solid-state lasers. The most common form of liquid laser is the dye laser, so called because the fluorescent molecule is a suitable dye molecule, e.g. rhodamine 6G in an organic solvent. Usually a substance is added to suppress the triplet excited state which would otherwise be formed, leading to unacceptable losses at the laser wavelength. Often a flowing solution is used to remove molecules in the triplet state from the lasing region. Dye laser outputs are usually broad band (≳10 nm) unless a grating is used in the cavity. Narrow-band tuned outputs can be obtained in this way over most of the visible and near infrared spectrum, using a number of dyes. Argon lasers are usually used to pump CW dye lasers and excimer or yttrium–aluminium–garnet (YAG) lasers are used for high-power pulsed (nanosecond) dye laser output. Outputs up to tens of watts average power have been obtained at some wavelengths as the excitation can be quite efficient (~20%).

Uses of liquid lasers The wide range of dyes available and the associated wavelengths makes the dye laser a flexible analysis tool capable of providing powerful tunable CW output or ⩾MW pulses down to only a few femtoseconds (10^{-15} s) in duration. It is used extensively in photochemical research to study many processes with specific energy absorption, matched by the laser wavelength. The two best known are probably the analyses of automobile exhaust and power station emissions.

23.1.2.5 Solid-state lasers

The first solid-state laser was the ruby laser, which is a three-level laser using chromium-doped aluminium oxide. Four-level lasers using transitions in the Nd^{3+} ions in various host materials (chiefly various glasses and yttrium–aluminium–garnet) are now more commonly used. The ruby laser emits at ~694 nm, the neodymium lasers at ~1.06 μm. All lasers in these categories are optically pumped using a xenon-filled flash tube, or a krypton-filled arc or a tungsten–iodine lamp in the case of CW Nd-YAG lasers.

A common pumping configuration is to use the laser rod in the form of a right cylinder, placed along one focal line of an elliptical cylinder, with the lamp along the other focal line. In 'normal' pulsed operation outputs of ≳100 J can be attained from a single oscillator in pulses up to a few milliseconds in duration. *Q*-switched pulses of a few joules or more can be readily obtained from oscillators in nanosecond pulses. The addition of amplifiers allows even higher energies. For example, in the case of Nd-glass, systems are now operating capable of greater than 10 kJ in nanosecond pulses for fusion studies. The wavelength spread from Nd glass lasers may be up to ~10 nm in the absence of mode selection. CW Nd-YAG systems have outputs from ~1 W using tungsten–iodine lamps for pumping. An experimental system has produced up to ~1 kW using a cascaded multiple rod configuration pumped by krypton arc lamps. Efficiencies of 2–4% are attainable. Repetitively *Q*-switched operation at p.r.f. values of ≳10 kHz is possible with continuously pumped Nd-YAG systems using acousto-optic *Q*-switching. They are often used in lasers for resistor trimming and marking. An excellent review of solid-state lasers is given by Koechner.[18]

Uses of solid-state lasers Solid-state lasers are more compact than gas or liquid designs and have found uses in precision industrial materials processing equipment. They are second only to CO_2 lasers in industrial applications. The major type (~90% of all commercial uses) is Nd-YAG, which is readily available commercially to 400 W average power and is used in industrial welding, cutting and drilling, particularly in the aerospace and electronics industries. A major use of low-power ruby and Nd-YAG lasers is in portable military range-finding equipment. The 100 kW, 20 ns pulses from a binocular-sized laser system allow accurate ranging up to many kilometres.

23.1.2.6 Semiconductor lasers

This class of components utilises the electron–hole recombination principles described in *Section 23.1.1.1*. In direct bandgap materials where high recombination velocities exist, optical gain can be achieved by creating population inversion of carriers through high-level current injection and by forming a resonant cavity. This cavity is usually produced by the high Fresnel reflectivity obtained from cleaving the material along faces perpendicular to the junction plane.

The semiconductor laser is typified by its small size (the semiconductor chip is typically 200 × 300 μm) and low drive requirement which makes it compatible with conventional semiconductor circuitry. It does, however, have particular characteristics. *Figure 23.7* shows the structure and characteristics of a typical laser device. Firstly, it has a well defined current threshold as seen from the power output *versus* drive current characteristic. Below this threshold the device emits low levels of spontaneous emission. At a limiting current density stimulated emission occurs and the emitted radiation increases linearly with drive current. Secondly, because the inverted medium and the resonant cavity are one and the same, mode locking is not readily achieved and the majority of devices show broad spectral emission with clearly defined Fabry–Perot mode peaks. Finally, the optical waveguide is of the same order of dimension as the emission wavelength, and therefore the output beam profile is diffraction limited. This results in a highly divergent beam.

Materials Any direct bandgap semiconductor can in principle support lasing. For practical room temperature operating devices,

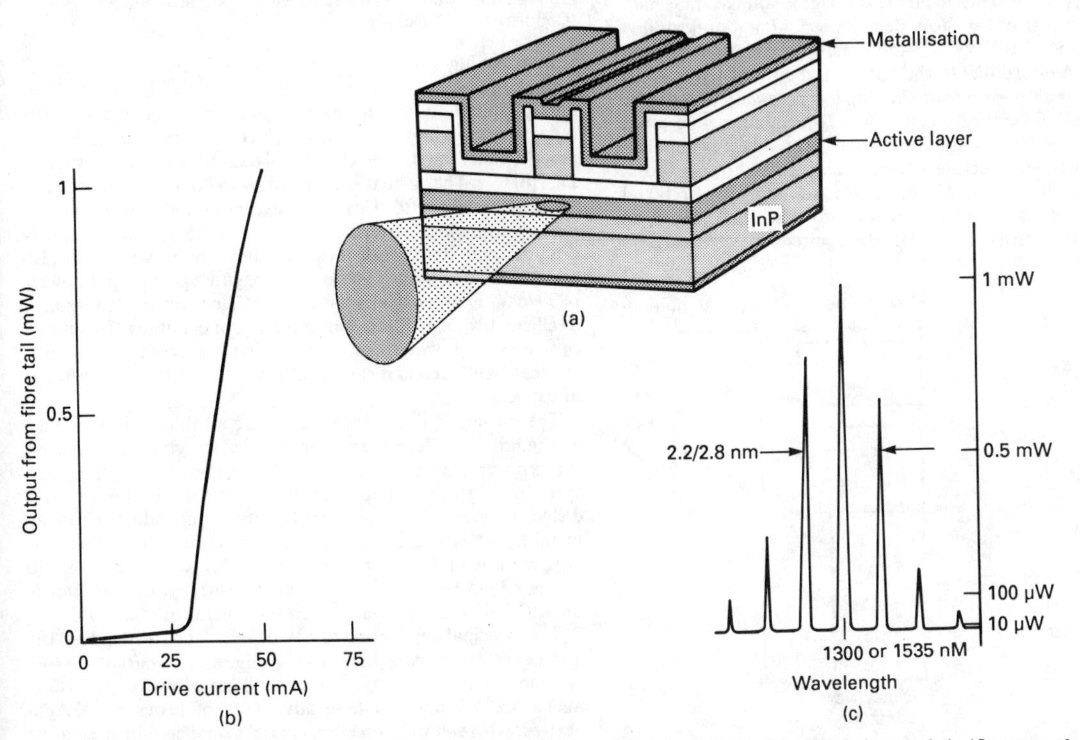

Figure 23.7 Fabry-Perot ridge laser chip: (a) construction, (b) electrical characteristic, (c) optical characteristic (Courtesy of STC plc)

however, it is necessary to provide an optical and electrical confinement region frequently referred to as the 'active region'. This typically comprises a low bandgap, high refractive index layer with cladding or confining layers of higher bandgap and lower refractive index material. These heterostructures are similar to those described in *Section 23.1.1.6*. The wafer is cleaved along the natural cleave direction of the semiconductor material to form the resonant cavity.

The principal semiconductor materials used are the GaAlAs and InGaAsP alloy systems which have been developed to cover the lasing wavelength ranges of 760–900 nm and 1200–1550 nm respectively. Both types are now well established in production. GaAlAs devices fabricated from double heterostructure material have threshold current densities of typically 1200 A/cm^2. The temperature performance of these devices is given by the characteristic temperature T_0. The lower the characteristic temperature the more sensitive the device. 850 nm emitters have T_0 values around 150 K, while 1300 nm devices are more typically 65 K. Other alloy systems are being investigated for emission at either ends of these wavelength ranges; InGaAlAs and InGaP for the low wavelength end, and antimony-containing alloys for the high wavelengths. Room temperature lasing has now been achieved at as low[19] as 646 nm and as high[20] as 2200 nm.

The layer structures necessary are grown using the liquid and vapour phase epitaxial techniques.

Devices There are a very large number of devices and designs available depending on the application and performance requirement. The power dissipation is one of the principal features that governs the ultimate performance of a device. This in turn depends on the threshold current. Thus, for pulsed applications, broad contact and wide pumped regions can offer high peak powers, while for CW operation low drive currents are achieved by reducing the pumped material to a minimum.

In designing GaAlAs lasers for high power, the device reliability has to be considered. This material, unlike InGaAsP, undergoes catastrophic facet degradation at an optical flux density of around 10^6 W/cm^2. Additionally, the life of the device is inversely proportional to the optical and current stresses. To obtain high peak powers from the chip it is necessary to separate the electrical and optical cavities while still maintaining good coupling between the photons and carriers. This is best exemplified by the multiheterostructure device. *Figure 23.8* shows a cross-section through the chip. The device is made using five epitaxial layers. The centre layer or active region is grown with a composition defined by the emission wavelength required. Injected carriers

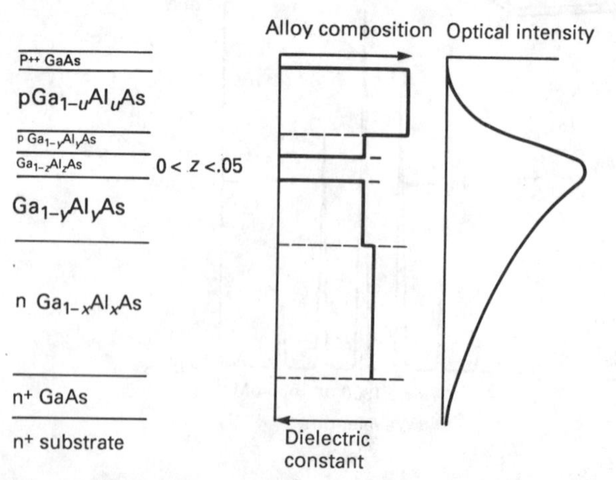

Figure 23.8 The multihet, LOC laser structure (Courtesy of STC plc)

are trapped in this region by two current-confining layers of higher aluminium content. By making the active layer less than 0.1 μm thick the threshold current can be minimised. The two current-confining layers comprise the optical cavity and are bounded by lower refractive index material which forms a waveguide. By optimising the aluminium content and layer thicknesses, output powers of up to 35 W from an emitting width of 250 μm at 40 A drive current have been achieved. Higher output powers can be achieved by stacking chips or by forming a linear array of emitting regions. The performance of these devices is limited by the heat dissipation. Thus they operate at pulse widths of around 50 ns and repetition rates of 1–10 kHz. Nevertheless, under these conditions, output powers at constant drive current can be maintained within 30% over the temperature range 0 – 125°C.

CW operation is achieved by further reduction in the heat dissipation by current and optical confinement in the plane parallel to the junction. There are two principal techniques used; the buried heterostructure and the ridge waveguide. These devices provide the lateral confinement in slightly different ways.

The buried heterostructure (*Figure 23.9*) comprises a planar structure into which two grooves are etched through the active

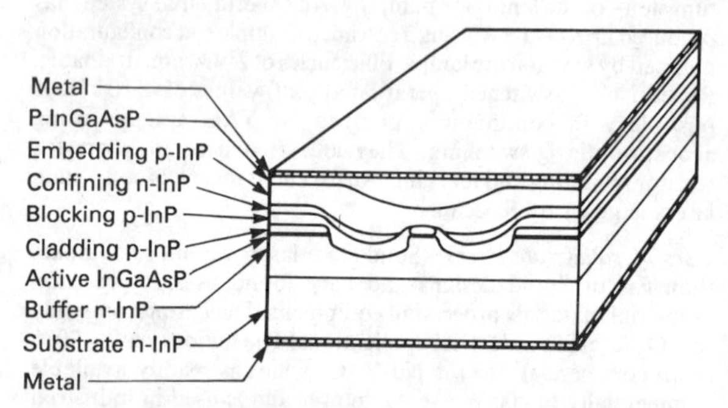

Figure 23.9 Double-channel planar buried heterostructure (Courtesy of STC plc)

region leaving a mesa of the order of 2 μm wide. A second epitaxial growth is used to fill in the grooves with low refractive index material. In some forms, this infill contains a series of layers of alternating p and n dopant which create a reverse-biased p–n junction (e.g. the double channel planar buried heterostructure DCPBH[21]). The current is confined by the thyristor structure to the mesa and the infill of low refractive index material forms a real index waveguide. Thresholds as low as 15 mA are routinely obtained with this type of device. In addition, the waveguide also allows only one optical mode to propagate up to output powers of 30 mW. Thus the device has a well defined far field and is suited to efficient launching into single mode optical fibres. Two major problems occur with this structure. First it is a complex device to fabricate, and second it can be subject to degradation by leakage of carriers.

The structure that overcomes these problems is the ridge waveguide.[22,23] Here, the planar epitaxial structure is etched, but the grooves stop short of the active region. The optical field spreading into the upper confining layer sees an effective dielectric constant change caused by the semiconductor/contact metal surface shape. This confines the optical field. Devices of this type are known for their very high reliability. The inverted rib waveguide (a related structure) has a projected degradation of only 5% after 25 years of continuous operation[24] at 50°C.

The principal application of CW devices fabricated in GaAlAs is in lasers for compact discs, data storage and xerography. Most such devices are designed to emit at 780 nm. However, recording media are available to take advantage of lower wavelength devices, where diffraction effects enable smaller spot sizes to be achieved and hence higher information packing densities. The long wavelength emitters are used in fibre optic communications. In both cases, to overcome the temperature sensitivity and diode degradation, the packaged device contains a photodetector which monitors the emission from the rear facet of the chip. If this is connected into a suitable feedback circuit, the drive current can be controlled to maintain a constant output power. A typical device for fibre optic communications will contain a laser diode, a monitoring detector, a thermoelectric element to maintain a constant diode temperature, and a thermistor. The optical fibre will be hermetically sealed into the package and aligned to the laser chip.

Most applications require analogue or digital modulation rather than CW operation. Since semiconductor diodes can be fabricated with very low capacitance, direct current modulation can be achieved in excess of 12 GHz for an 850 nm device. To achieve these modulation rates the device must be prebiassed to around threshold for frequencies up to 2 GHz and at several times threshold for the highest frequency operation.

In long-distance fibre optic communications systems, the chromatic dispersion caused by the spread of wavelengths emitted by a Fabry–Perot laser is sufficient to cause signal distortion. The most frequently used technique to obtain single frequency emission is the distributed feedback laser (DFB).[25] The device structure is similar to that described earlier, but incorporates a modulation in the interface of one of the confinement layers (*Figure 23.10*). The grating acts like a series of partial reflectors, and hence the longitudinal mode selected is that with a wavelength or multiple thereof corresponding to twice the grating spacing. For this device to operate effectively, the reflectivity of the facets must be reduced.

It has been shown that to produce a stable single frequency it is necessary to form a phase shift in the grating of a quarter wavelength. Devices of this type are less susceptible to mode hopping. The grating can be produced by holographic techniques. However, the easiest way to produce a grating with a phase shift is to write by electron beam lithography.

In double heterostructure devices, the threshold current density and emission wavelength are dependent on the active region thickness and composition, respectively. Using liquid phase epitaxy, the thickness reproducibly is limited to around 10 nm. MOCVD and MBE allow layer thicknesses to be controlled down to atomic levels When the thickness falls below 20 nm, quantum effects occur. Within the well a number of discrete states exist in both the valence and conduction band, the energy levels being dependent on the well width. Since the emission wavelength is dependent on the energy gap, in a quantum well device this is controlled by well width.

Figure 23.11 shows the effect of well width on emission wavelength for a double quantum well structure in which the well is undoped GaAs and the confinement region is GaAlAs. These devices are highly efficient, with threshold current densities of

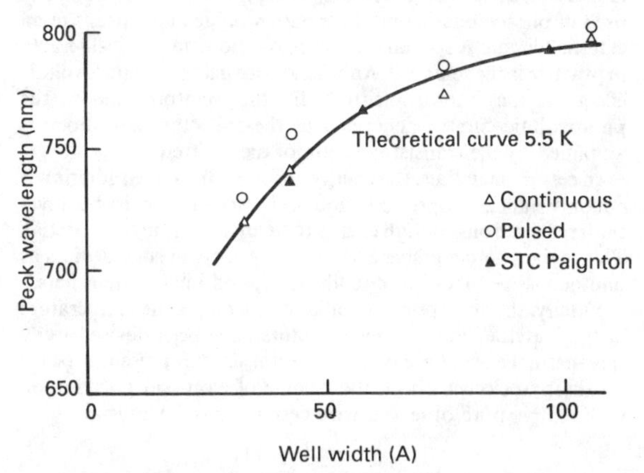

Figure 23.11 Comparison of measured PL data from multiquantum well structures with theory (Courtesy of STC plc)

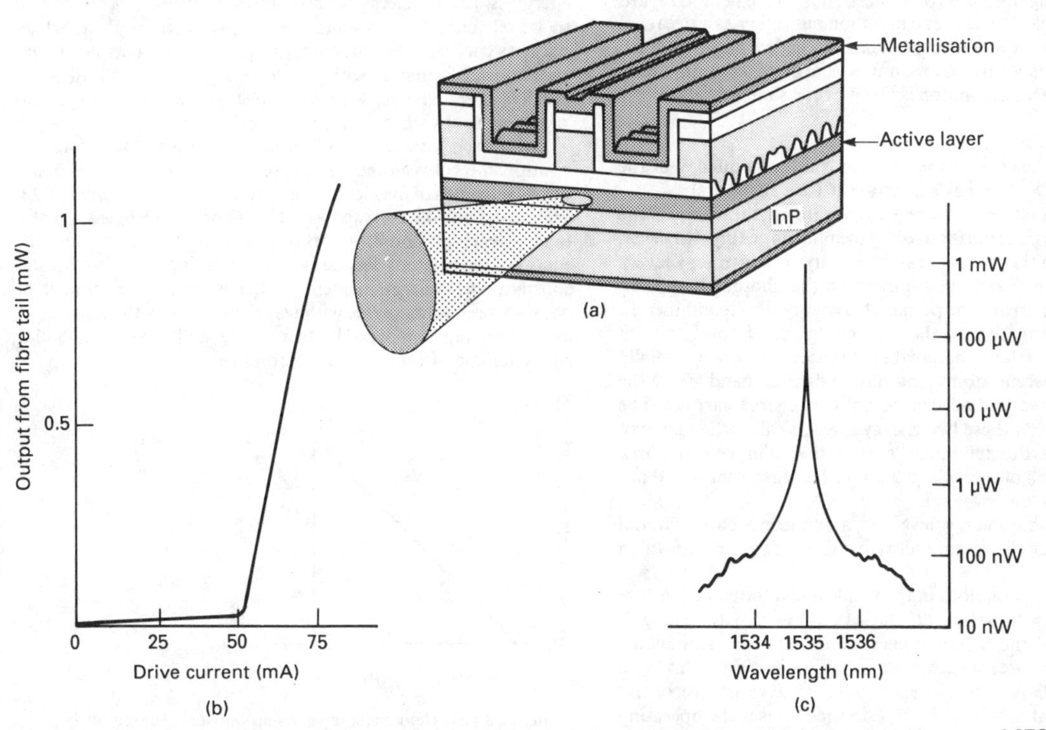

Figure 23.10 DFB (ridge) laser chip: (a) construction, (b) electrical characteristic, (c) optical characteristic (Courtesy of STC plc)

400 A/cm^2. Such devices are used principally for high-power application where power dissipation is important.

23.2 Optical detectors

Radiation falling on a semiconductor material can produce electron–hole pairs in the material which can be used as charge carriers. Thus the conductivity of the illuminated material is increased considerably. This is the photoconductive effect, and it can be used in various solid-state devices to detect visible and infrared radiation.

If the electron–hole pairs are generated in or near a p–n junction, the electrons and holes will be separated by the built-in electric field of the junction. An open-circuit voltage or a short-circuit current will be generated. This photovoltaic effect can be used in photodiodes to produce both voltage and current in an external circuit. A special form of photodiode used considerably in practice is the solar cell. Another device using the photovoltaic effect is the phototransistor. In the phototransistor, the photovoltaic current generated at the collector–base diode is amplified by the transistor action of the emitter.

In certain materials, the energy absorbed from the radiation is not only sufficient for the creation of electron–hole pairs but gives the freed electrons enough energy to be emitted from the material. This is the photoemissive effect which is used in such devices as photoemissive tubes, photomultipliers, and image intensifiers.

Finally, the absorption of radiation increases the temperature of the material, and various detectors have been devised using this thermal effect. One type of detector used at present is based on the pyroelectric effect, the change of electrical polarisation with temperature observed with certain complex crystals.

23.2.1 Photoconductive effect

When discussing optoelectronic effects, it is convenient to regard light not as an electromagnetic radiation but rather as a stream of elementary particles. These particles are called photons. Each photon contains a certain amount of energy called a quantum. The energy of a single photon is given by the simple relationship:

$$E = h\nu \qquad (23.3)$$

where E is the energy of the photon, ν is the frequency of the radiation, and h is Planck's constant (6.62×10^{-34} J s).

When the radiation is absorbed by a material, the energy of the electrons in the material is raised by the photon energy. However, the electrons in the material can have only certain energy levels, so the radiation is absorbed only when the photon energy can raise the electron from one permissible energy level to another. In semiconductor materials, the two energy levels or bands of interest are the valence band where the electrons are essentially bound to the parent atoms, and the conduction band where the electrons are free and so can be used as charge carriers. The difference between these two energy levels is called the bandgap energy, and is the minimum energy that will generate free carriers. Photons of energy less than the bandgap energy will not be absorbed by the material.

The permissible energy levels in a semiconductor material can be represented by an energy-band diagram, as shown in *Figure 23.12*.

If the energies of photons in the visible and infrared regions are calculated using Equation (23.3), it is found that the energies equal or exceed the bandgap energies of many semiconductor materials. Therefore, in such materials, illumination with visible or infrared radiation considerably lowers the resistance of the material compared to the 'dark' resistance. This is the operating principle of all solid-state photoconductive detectors. These

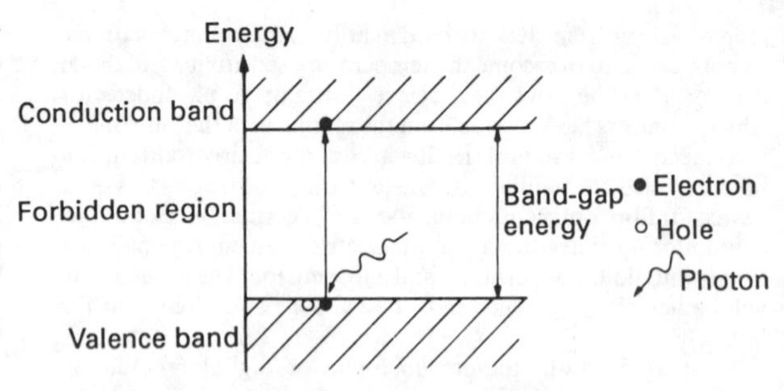

Figure 23.12 Permissible energy levels for electron in semiconductor material, generation of electron–hole pair

detectors are called quantum detectors since they depend for their operation on the quantum nature of the radiation.

It is customary to describe radiation in terms of wavelength rather than frequency. Since wavelength and frequency are related by the speed of light c:

$$\nu \times \lambda = c \qquad (23.4)$$

where λ is the wavelength, Equation (23.3) can be rewritten as

$$E = \frac{hc}{\lambda} \qquad (23.5)$$

If the minimum bandgap energy for a particular semiconductor material is represented by E_g, Equation (23.5) can be rearranged in terms of a critical wavelength λ_c corresponding to this minimum energy:

$$\lambda_c = \frac{hc}{E_g} \qquad (23.6)$$

This equation can then be used to determine the wavelength of photons whose energy corresponds to the minimum bandgap energy. Since the energy of the photon is inversely proportional to wavelength, λ_c represents the longest wavelength at which photons contain sufficient energy to produce free carriers within the material. Radiation with wavelengths greater than λ_c does not produce free carriers; with a wavelength shorter than λ_c, the radiation is absorbed and free carriers are produced.

For a fixed total energy of incident radiation, the theoretical relationship between the number of free electrons produced and the wavelengths of the incident radiation is shown in *Figure 23.13*. The sharp cut-off corresponds to λ_c. For wavelengths shorter than λ_c, each photon produces one electron–hole pair. Since the energy of the photons decreases with increasing wavelength, for a constant total energy incident on the material, the number of photons present increases with wavelength. Thus the number of free electrons produced increases, and the triangular characteristic of *Figure 23.13* is obtained.

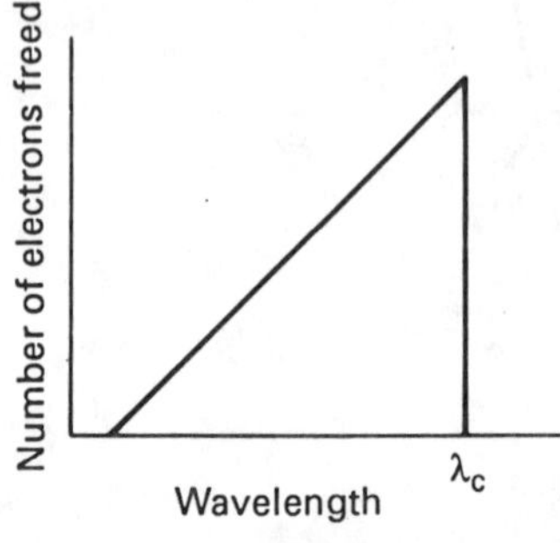

Figure 23.13 Theoretical relationship between number of free electrons produced by incident radiation and wavelength

The higher energy of the photons with shorter wavelengths may excite the electron sufficiently not only to free it from the parent atom, but also to exceed the work function of the material and so allow it to escape from the surface. Obviously such electrons cannot be used as charge carriers within the material, and so a second limiting wavelength is set for the photons, λ_m, corresponding to the energy which will cause the electron to be emitted from the surface. An equation similar to Equation (23.6) can be used to relate λ_m to this energy E_t:

$$\lambda_m = \frac{hc}{E_t} \qquad (23.7)$$

The wavelengths λ_c and λ_m represent the limiting wavelengths of the incident radiation for use with a particular material. The theoretical relationship between the number of usable charge carriers produced and the wavelength of the radiation is shown in *Figure 23.14*. This curve is called the spectral response of the material. The values of λ_c and λ_m can be calculated from Equations (23.6) and (23.7) using the known values of E_g and E_t for the material. For example, the values of λ_c for silicon and germanium are calculated as 1.13 μm and 1.77 μm respectively. The range of wavelengths for visible light is from approximately 0.40 μm (violet) to 0.70 μm (red). Therefore both silicon and germanium can be used in devices operating with visible and near-infrared radiation.

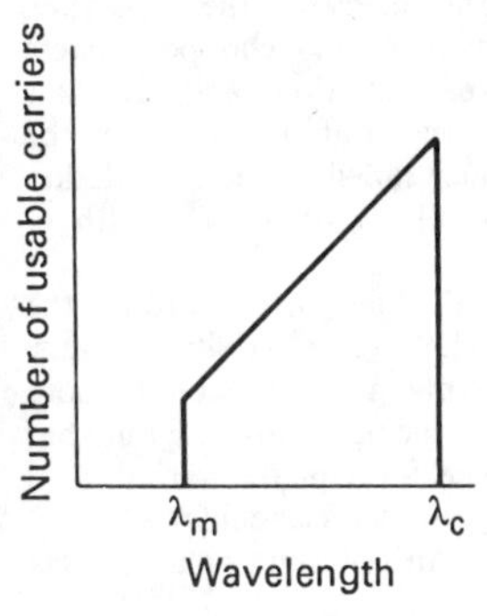

Figure 23.14 Theoretical relationship between number of usable charge carriers produced by incident radiation and wavelength (spectral response)

A practical spectral response is shown in *Figure 23.15* with the theoretical curve in broken lines for comparison. The cut-off wavelengths are the same for both curves since they are determined by the semiconductor material itself. The peak of the practical response, however, is shifted to a shorter wavelength.

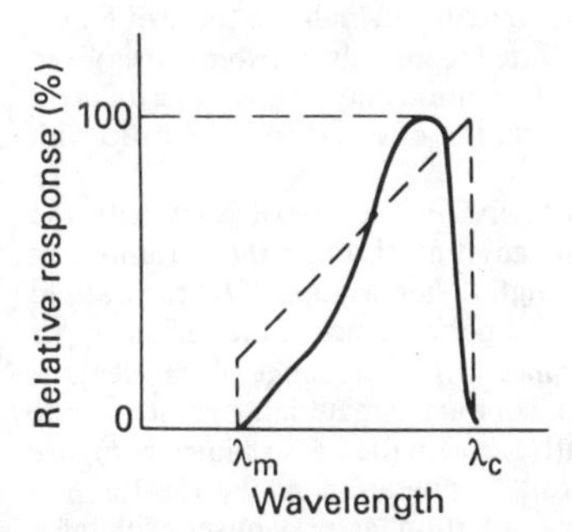

Figure 23.15 Practical spectral response (full line) compared with theoretical response (broken lines)

The shape of the practical curve differs from the theoretical one because the energy-band diagram from which the theoretical curve is derived is an oversimplification of what occurs in practice, and because of surface effects in the semiconductor material. These effects combine to produce the shape of the practical response curve of *Figure 23.15*.

The peak of the response curve can be controlled to some extent by doping the semiconductor material. Adding an impurity to the material creates an intermediate state for the electron between the valence and conduction states. This intermediate state is shown as the trapping level in the energy-band diagram of *Figure 23.16*. Thus an electron can be freed by a photon whose energy is less than the bandgap energy, but sufficient to raise the electron to the trapping level. A second photon can then impart further energy to the electron to transfer it from the trapping level to the conduction band. This two-stage freeing of an electron makes a hole available as a charge carrier before the electron is available. This energy effect combines with the effects mentioned previously to change further the theoretical response curve.

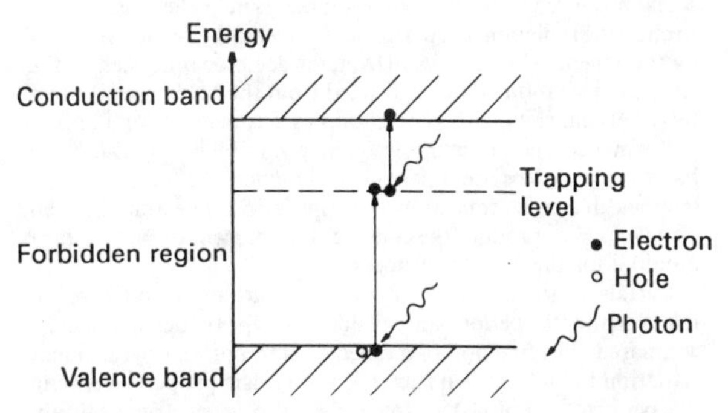

Figure 23.16 Permissible energy levels for electron in doped semiconductor material generation of electron–hole pair in two stages

If a polycrystalline semiconductor material is used, the peak of the practical response curve is flattened so that a wider peak response than that shown in *Figure 23.15* is obtained. Polycrystalline material, however, contains many traps for the generated charge carriers so that a device manufactured from such material will have a slower response time than one using monocrystalline material.

It can be seen from the preceding discussion that by choosing a particular semiconductor material, and by choosing a suitable doping level, a practical device can be constructed to respond to a particular range of wavelengths.

For radiation of a particular wavelength or fixed range of wavelengths, the greater the energy the higher is the number of photons incident on the irradiated material, and hence the higher the number of charge carriers generated within the material. For visible radiation (light), the quantity of light falling on to a given surface is called the *illumination*. The unit of illumination is the lux, defined as the illumination produced when 1 lumen of luminous flux falls on an area of 1 square metre. Thus the fall in resistance of the irradiated semiconductor material is proportional to the illumination of the incident light. For infrared radiation, it is usual to measure the radiation in terms of the power incident on an area. In this case, the fall in resistance of the irradiated material will be proportional to the radiation power per unit area.

23.2.2 Photoconductive detectors

The performance of a photoconductive detector is limited by noise. The noise in the detector itself is produced by the thermally generated carriers in the semiconductor material. At very low levels of incident radiation, the thermally generated carriers may swamp the photo-generated carriers. The effect of the thermally generated carriers can be minimised by cooling the detector, if necessary to liquid-nitrogen or liquid-helium temperatures. There is, however, a theoretical limit to the performance of the detector set by the thermal background radiation. This radiation is noisy in character, and gives rise to a limiting noise in the detector.

Another technique to improve the performance when the signal produced by the incident radiation is near the noise level is to chop the radiation at a particular frequency before it is incident on the detector. The a.c. signal from the detector produced by the chopped radiation can be extracted from the background noise by an amplifier tuned to the chopping frequency. For the best performance, the chopping frequency should be high and the bandwidth of the system narrow. In practice, the frequency and bandwidth are determined by the application. The amplifier can also be used to produce the constant bias current or bias voltage required for the detector element.

Certain characteristics of the detector are used as 'figures of merit' so that the performance of different types of detector can be compared. The function of a detector is to convert the incident radiation to an electrical signal. Thus a basic property defining the performance of a detector is the ratio of electrical output, expressed as a voltage, to the radiation input, expressed as the incident energy. This ratio is called the responsivity, and is expressed in V/W. For detectors that use a constant bias voltage, a current responsivity is used, expressed in A/W.

Various noise figures can be quoted for the detector. The noise of the detector itself is usually expressed as the RMS value of the electrical output measured at a bandwidth of 1 Hz under specified test conditions. Although a low noise enables smaller radiation levels to be detected, it can make the design of suitable amplifiers difficult because the amplifier noise must be low. It is therefore often important to know the ratio of detector noise to the value of Johnson noise in a resistor at room temperature equal in magnitude to the detector. This ratio may be quoted as a noise factor. A signal-to-noise ratio can be measured, but this will apply only for the test conditions under which it was obtained. A more useful quantity, which is often used as a figure of merit for a detector, is the noise equivalent power or NEP. This is the amount of energy that will give a signal equal to the noise in a bandwidth of 1 Hz. It is, in general, a function of the wavelength and the frequency of the measurement. The NEP is equal to the noise per unit bandwidth divided by the responsivity, and is given by the expression:

$$\text{NEP} = \frac{W}{(\Delta f)^{1/2} V_s / V_n} \tag{23.8}$$

where W is the radiation power incident on the detector (RMS value in watts), V_s is the signal voltage across the detector terminals, V_n is the noise voltage across the detector terminals, and Δf is the bandwidth of the measuring amplifier in Hz. The units of NEP are W/Hz$^{1/2}$.

Both responsivity and NEP vary with the size and the shape of the active element of the detector. It is found in practice, and supported by theory, that for similarly made detectors the NEP is often proportional to the square root of the area of the active element. A quantity called the area normalised detectivity or D^* is commonly used, which is related to NEP by the expression:

$$D^* = \frac{A^{1/2}}{\text{NEP}} \tag{23.9}$$

where A is the area of the active element of the detector in cm^2. The units of D^* are (cm Hz$^{1/2}$)/W.

When responsivity, NEP, and D^* are quoted in published data, certain figures are given in brackets after the quantity. Typical examples of this are responsivity (5.3 μm, 800, 1) and D^* (5.3 μm, 800, 1). These figures refer to the test conditions under which the value was measured. The figure 5.3 μm refers to the wavelength of the monochromatic radiation incident on the detector, 800 to the modulation (chopping) frequency in Hz of the radiation, and 1 to the electronic bandwidth in Hz. An alternative form of the test conditions is given in terms of black-body radiation. An example of this is D^* (500 K, 800, 1). The 500 K refers to the temperature of the black body from which the incident radiation was obtained. The other figures refer to the modulation frequency and bandwidth as before. Details of the other test conditions under which the quantities were measured are also given in published data. These conditions include the distance between the detector element and the source of the radiation, the aperture through which the radiation reaches the element, the operating temperature of the element, details of the chopper system producing the modulation of the radiation, and the bias conditions of the element. The electrical output signal from the detector is amplified by an amplifier tuned to the modulation frequency, which is typically 800 Hz, with a bandwidth of typically 50 Hz.

Both signal and noise vary with the bias current through the element, and so the responsivity, NEP and detectivity also vary with bias current. At high bias currents the noise increases more rapidly than the signal, and therefore the signal-to-noise ratio has a peak value at some current. The form of the variation of responsivity, noise, and detectivity with bias current for a typical detector is shown in *Figure 23.17*. An optimum value of bias current for the detector can be chosen from these curves. Variations between detectors of the same type may sometimes occur, and so for highly sensitive applications a fine adjustment of the bias current may be necessary to obtain the optimum performance from the detector. Curves similar to those of *Figure 23.17*, or of the quantities plotted separately, are sometimes given in published data. Similar information may be given in a different form; for example, as the variation of signal-to-noise ratio with bias current.

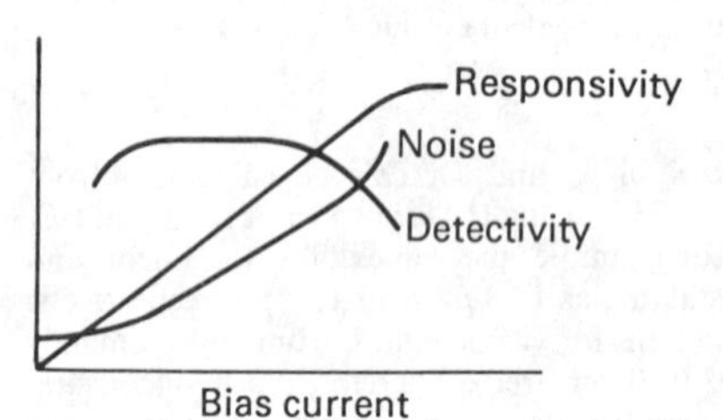

Figure 23.17 Variation of responsivity, detectivity and noise with bias current in infrared photoconductive detector

The values of responsivity, NEP and D^* will vary with the wavelength of the incident radiation through the variations of signal and noise with wavelength. The variation of D^* can be used as an indication of the spectral performance of the various types of detector, as shown in *Figure 23.18*. The range of wavelengths over which the various semiconductor materials operate can be seen, together with the relative sensitivities. Also shown in *Figure 23.18* is the theoretical limit to operation set by the thermal background radiation. This radiation, as previously explained, limits the attainable detectivity to what is called the background limited value. For wavelengths up to approximately 10 μm, the

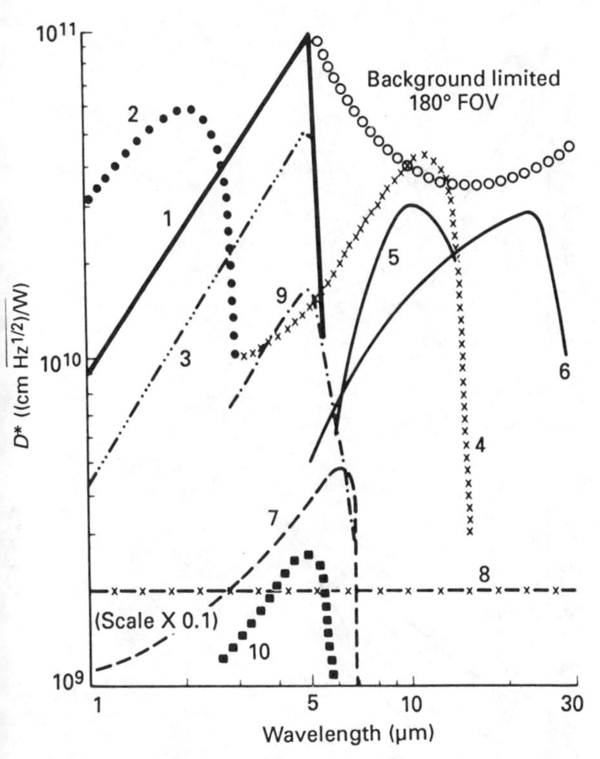

Figure 23.18 Variation of D^* with wavelength: 1, indium antimonide, 77 K, 60° FOV; 2, lead sulphide, 300 K, 180° FOV; 3, indium antimonide, 77 K, 180° FOV; 4, cadmium mercury telluride, 77 K, 60° FOV; 5, mercury-doped germanium, 35 K, 60° FOV; 6, copper-doped germanium, 4.2 K, 60° FOV; 7, indium antimonide, 300 K (scale × 0.1); 8, TGS pyroelectric detector; 9, cadmium mercury telluride, 193 K; 10, cadmium mercury telluride, 295 K

background limited value falls with increasing wavelength, and so imposes a maximum value on D^*.

One curve in *Figure 23.18* is shown operating above the theoretical limit. This is because the detector has a cooled aperture restricting the field of view and hence limiting the amount of background radiation received. Because the field of view (FOV) affects the performance, the FOV is stated as well as the temperature for the curves.

The material most suitable to detect radiation of a given wavelength is usually one with the peak spectral response equal to or slightly longer than the given wavelength. Detectors sensitive to wavelengths much longer than the given wavelength require more cooling to reduce the internal thermal noise. It has been shown previously that silicon and germanium are sensitive to radiation of wavelengths up to 1.13 μm and 1.77 μm respectively. These values lie within the infrared region, but silicon and germanium detectors are generally used for visible radiation. These detectors, in the form of photodiodes and phototransistors, are discussed separately. Other commonly used materials are shown in *Table 23.1*.

The choice of material for an application may often be influenced by the cooling required. Although detectors requiring cooling to liquid-nitrogen or liquid-helium temperatures are acceptable in scientific work, there are considerable disadvantages in their use for industrial applications. Thus, in industrial applications, a less sensitive detector that can operate at room temperature or with a thermoelectric cooler may be used in preference to a more sensitive or cheaper detector requiring greater cooling.

Table 23.1 Useful detection ranges of commonly used materials

Material		*Useful detection range* (μm)
Lead sulphide	PbS	0.6–3.0
Indium antimonide	InSb	1.0–7.0
Mercury-doped germanium	Ge:Hg	2.0–13
Cadmium telluride	CdHgTe	3.0–15
Copper-doped germanium	Ge:Cu	2.0–25
Cadmium sulphide	CdS	0.4–0.8
Cadmium selenide	CdSe	0.5–0.9

23.2.2.1 Photoconductive detectors for visible radiation

The semiconductor materials used in photoconductive detectors for visible light are cadmium sulphide (CdS) and cadmium selenide (CdSe). The spectral responses are shown in *Figure 23.19*.

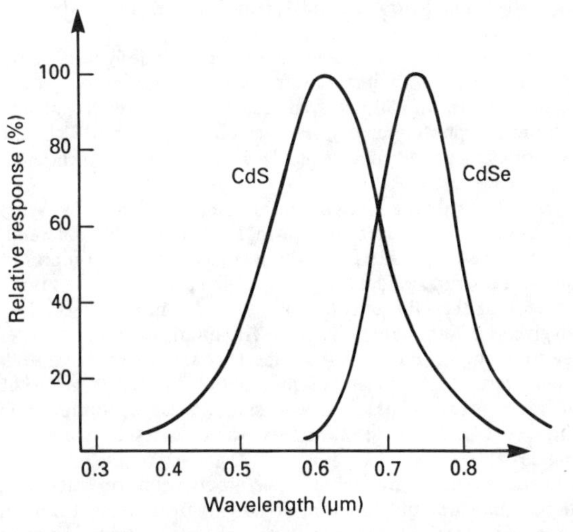

Figure 23.19 Spectral response for cadmium sulphide (CdS) and cadmium selenide (CdSe)

The detectors are two terminal devices, used as light-dependent resistors. The circuit symbol for a photoconductive detector (*Figure 23.20*) consists of a resistance symbol with incident arrows representing the radiation.

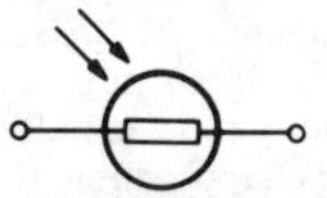

Figure 23.20 Circuit symbol for photoconductive detector (light-dependent resistor)

The active element of the detector is usually made by sintering cadmium sulphide/selenide powder onto a substrate. Interleaving finger-like contacts are deposited onto the surface of the active device. The number and geometry of contacts may be varied to give different voltage and resistance ratings. For example, a detector with a small number of widely spaced contact fingers will have higher voltage and resistance ratings than a detector with many closely spaced contact fingers.

The whole assembly is encapsulated in either a glass or a plastic envelope, and is either end or side sensitive depending on the package used.

A second and more recent manufacturing process for the sensitive element uses individual grains of cadmium sulphide, the so-called monograin process. The grains, about 40 μm in diameter, are inserted into a thin insulating sheet of a synthetic material so that each grain projects both sides of the material. Each grain is insulated from its neighbours by the sheet, but electrical connection is made by evaporating gold contact pads on to both sides of the sheet. Terminal wires are connected to these contact pads. The complete element is then encapsulated in a transparent plastic. This method of construction enables smaller detectors to be manufactured.

The electrical characteristics of cadmium sulphide detectors depend on several factors. Some of these may be considered as 'direct' factors: the illumination, the wavelength of the incident radiation, the temperature, and the device voltage and current. Other factors, however, affect the operation of the device: the time it has been kept in darkness or the time it has been operating in a circuit, and the operation of the device in the previous 24 hours. Sometimes preconditioning is specified in the component data. This consists of illumination of the detector at a specified level for a specified time before the measurements given in the data are made.

The characteristics of most interest to circuit designers are the fall of resistance with illumination and the device response time. Values of 'dark' and 'illuminated' resistance are given taking into account the 'photo memory' hysteresis of the material. The high dark resistance is given after the device has been in darkness for a specified time.

Often two values are given; one after a short time such as 20 s and one after a longer time such as 30 min. The difference between these two values of dark resistance can be considerable, the value after the longer time in the dark being at least ten times greater than the value after the shorter time. Similarly two values are given for illuminated resistance. An initial value is given after device storage in darkness for some 16 h and a second value is given after a further period of illumination such as 15 min. The difference between these values is small. Over an illumination range of 1–10 000 lx the resistance value varies by some 3–5 decades.

Response times for these detectors depend on intensity, coating thickness and doping, and can vary from 10 ms to several seconds. Curves showing response times are sometimes given in published data. The form of such curves is shown in *Figure 23.21*.

When a cadmium sulphide detector is used in a practical circuit, it can be used either as an 'on/off' device or some intermediate resistance value can be used as a trigger level. In the on/off type of operation, the device is used to detect the presence or absence of a light source. The change of resistance between the illuminated and dark values can be used to operate a relay to initiate further action. Applications of this type include alarm systems operated by the interruption of a beam of light directed on to the detector, or counting systems where objects, say on a conveyor belt, interrupt a light beam to produce a series of pulses which operates a counter. In the second type of application, the detector is used to measure the light level. The resistance of the device corresponding to a predetermined light level is used as a threshold value to trigger another circuit. An example of this type of application is a twilight switching circuit. When the daylight has faded to a given level, the corresponding resistance of the detector causes another circuit to switch on the required lights.

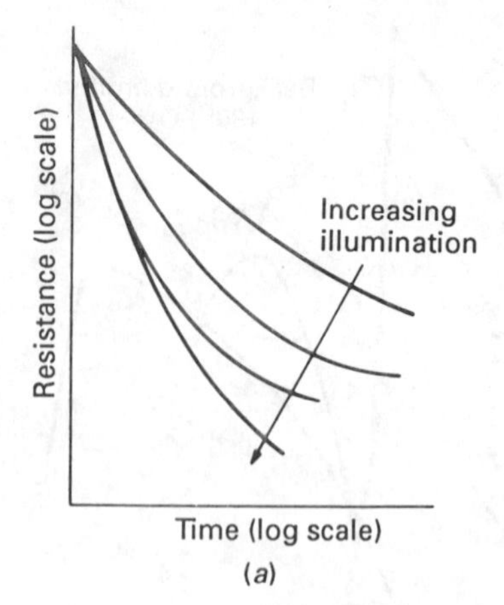

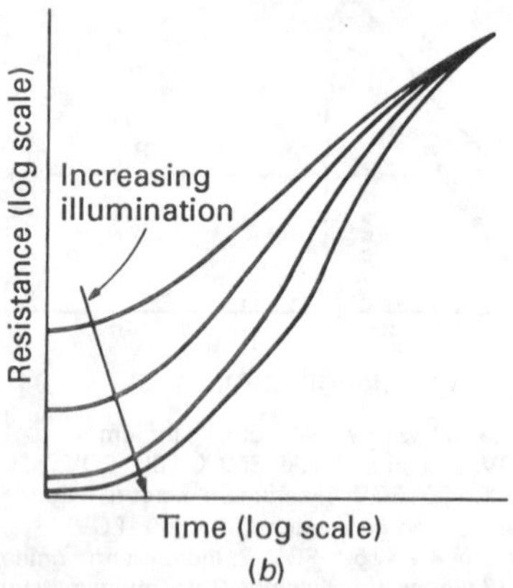

Figure 23.21 Resistance decay times (a) and rise times (b) with illumination as parameter for cadmium sulphide photoconductive detector

23.2.2.2 Photoconductive detectors for infrared radiation

The operating principle of photoconductive detectors for use with infrared radiation is the same as that of detectors for visible radiation just described. Charge carriers are generated in the illuminated material, and the consequent fall in resistance provides a measure for the incident radiation. Various semiconductor materials can be used to detect infrared radiation, enabling detectors to be chosen to suit particular wavelengths. Some detectors can be used at room temperature, while others must be cooled to low temperatures, for example with liquid nitrogen or liquid helium, to minimise the effects of thermally generated carriers (thermal noise).

Infrared photoconductive detectors are usually operated with a constant bias current through the active element so that the change in resistance when illuminated appears as a voltage change. For the more sensitive detectors, this voltage can be amplified to produce a larger output signal. Some detectors are operated with a constant bias voltage across the element. For these, the change in resistance produces a change in current.

The time constant for these types of detector is usually specified. This is defined as the time taken from application of the radiation for the detector output to fall to 63% of the peak value. Typical values range from 0.1 μs to 350 μs. Other characteristics that may be given in published data include varistors of responsivity D^* and noise with modulation frequency and the operating temperature of the detector.

The ratings given in the data are similar to those already discussed for photoconductive detectors used with visible radiation. They include the maximum detector power, maximum bias current, and the maximum operating and storage temperatures.

The construction of infrared photoconductive detectors depends on the material used for the active element, in particular whether or not it requires cooling. Detectors for operation at room temperature can be encapsulated in a metal envelope with a viewing window, or the element can be deposited on a flat plate. Detectors that require cooling can be deposited on a Dewar vessel or arranged for mounting on to a cooling vessel. For detectors that use thermoelectric cooling, the active element is mounted on the cooler in a suitable encapsulation.

Lead sulphide detectors are of two types, depending on the method used to form the active element. The element is deposited as a film either by evaporation or by a chemical reaction. The evaporated-film type can be operated at room temperature or cooled; the chemically deposited type is generally operated at room temperature only.

A typical construction used for the evaporated-film type is to form the active element on an inner surface of a Dewar vessel, as shown in *Figure 23.22*. A metal housing can be used to protect the

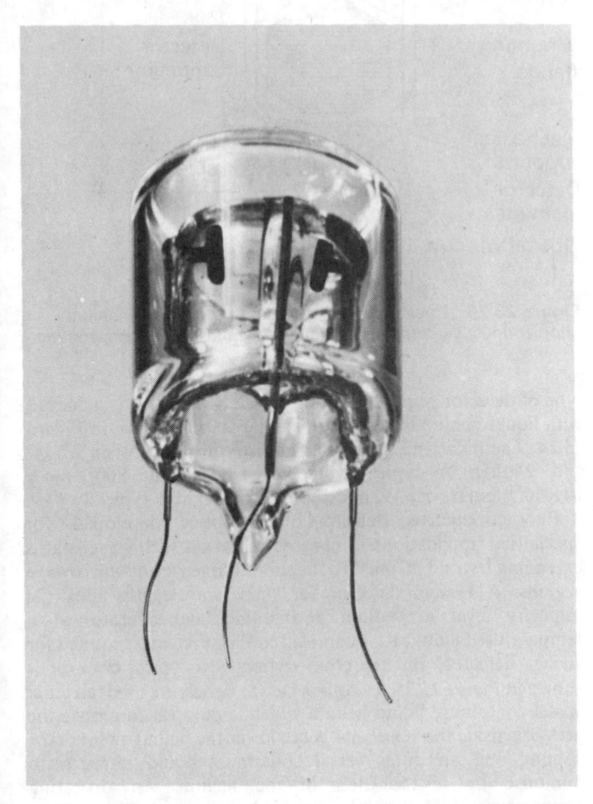

Figure 23.22 Lead sulphide photoconductive detector. The lead sulphide film is deposited on an inner surface of the Dewar vessel (Courtesy of Mullard Limited)

Dewar vessel if the detector is to be used at room temperature only, while for operation below room temperature the Dewar vessel can be filled with a suitable coolant. The operating temperature range of this type of lead sulphide detector is from 293 K to 173 K (+20°C to −100°C), while the responsivity has a peak value at 240 K and D^* has a peak value at 220 K. A typical value of responsivity (500 K, 800, 1) is 2.0×10^3 V/W and typical values of D^* (2.0 μm, 800, 1) are 4.0×10^{10} (cm $Hz^{1/2}$)/W at 293 K and 2.0×10^{11} (cm $Hz^{1/2}$)/W at 230 K. The range of wavelengths over which this detector can be used is from the visible region to 3.0 μm with a peak response at 2.3 μm. The time constant is typically 100 μs.

Because the chemically deposited lead sulphide detector is generally operated at room temperature, the construction does not have to allow for cooling. Typical constructions used for this detector are encapsulation in a small metal envelope such as the TO-5 outline, or as a 'flat pack' in which the element is deposited on an insulating substrate. Both types of construction can incorporate a filter to modify the spectral response. This type of detector is operated with a constant bias voltage, and has a typical current responsivity (2.0 μm, 800, 1) of 200 mA/W. A typical value of D^* (2.0 μm, 800,1) is 1.0×10^{10} (cm $Hz^{1/2}$)/W. The range of wavelengths is from the visible to 2.8 μm, but this can be restricted by using a germanium window to 1.5 μm to 2.8 μm. The time constant is typically 250 μs.

Both types of lead sulphide detector have a high resistance compared to detectors using other materials. A typical resistance for the evaporated-film type is 1.5 MΩ, and for the chemically deposited type 200 kΩ.

Indium antimonide detectors can be operated at room temperature, or cooled by liquid nitrogen to 77 K at which temperature a detectivity near the background limited value is obtained. The photoconductive material is a doped single crystal which is cut into thin slices from which the active element is cut. At room temperature the resistivity of indium antimonide is low, giving detectors with resistances of about 5Ω/square. The most useful forms of element are therefore long strips or 'labyrinths' made from a number of long strips laid in parallel but connected electrically in series. The effect of temperature on performance is considerable at room temperature, and so a good heatsink is required for the element. Typical constructions for room-temperature indium antimonide detectors mount the element either in a copper block or in a flat-pack encapsulation which can be provided with a viewing window (usually sapphire) if it is necessary to protect the element from dirty or corrosive atmospheres. A typical value of responsivity (6.0 μm, 800, 1) is between 1.0 V/W and 3.5 V/W, and a typical value of D^* (6.0 μm, 800, 1) is 1.5×10^8 to 3.0×10^8 (cm $Hz^{1/2}$)/W. The range of wavelengths is from the visible region to 7.0 μm, and the time constant is typically 0.1 μs.

The elements for cooled indium antimonide detectors are made in a similar way to the room-temperature types, but are mounted on the inner surface of a glass Dewar vessel. The radiation is transmitted to the element through a sapphire window. The element is cooled with liquid nitrogen either by filling the dewar vessel with the liquid or by using a miniature Joule–Thomson cooler. Cooling increases the resistance of the element to about 2 kΩ/square, and a wide range of shapes and sizes can be made for the element. Arrays as small as 0.1 mm square are possible. Because the detectors of this type are 'background limited', the use of a cooled aperture improves the detectivity. Typical values of responsivity (5.3 μm, 800, 1) and D^* (5.3 μm, 800, 1) are 1.2×10^4 V/W and 5.0×10^{10} (cm $Hz^{1/2}$)/W respectively. With a restricted field of view, the value of D^*(5.3 μm, 800, 1) can be increased to 1×10^{11} (cm $Hz^{1/2}$)/W. The range of wavelengths is from the visible to 5.6 μm, and the time constant is typically 2 μs to 5 μs.

The cross-section of a typical construction for a cooled indium antimonide detector is shown in *Figure 23.23*.

Cadmium mercury telluride is an alloy semiconductor material which forms a mixed crystal of cadmium telluride and mercury telluride. The peak spectral response of the material can be varied by the relative proportions of cadmium and mercury telluride in the crystal from 9.5 μm to 15 μm. Cadmium mercury telluride (CMT) detectors can be used to cover a range of wavelengths from approximately 3 μm to 15 μm. Detector elements are made in a similar way to those of indium antimonide just described, being cut from thin slices of a single crystal. CMT detectors can be operated at room temperature, with thermoelectric coolers at temperatures down to approximately 200 K, or with liquid nitrogen at 77 K. For operation with liquid nitrogen, a construction for the detector similar to that for the indium

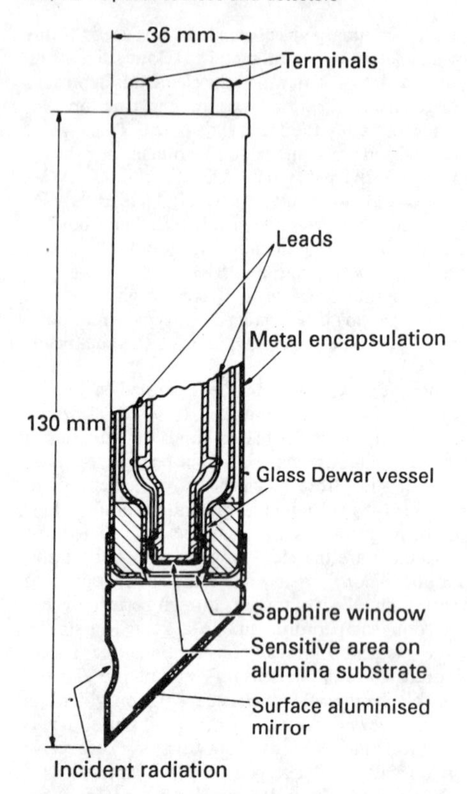

Figure 23.23 Construction of cooled indium antimonide photoconductive detector for operation at 77 K

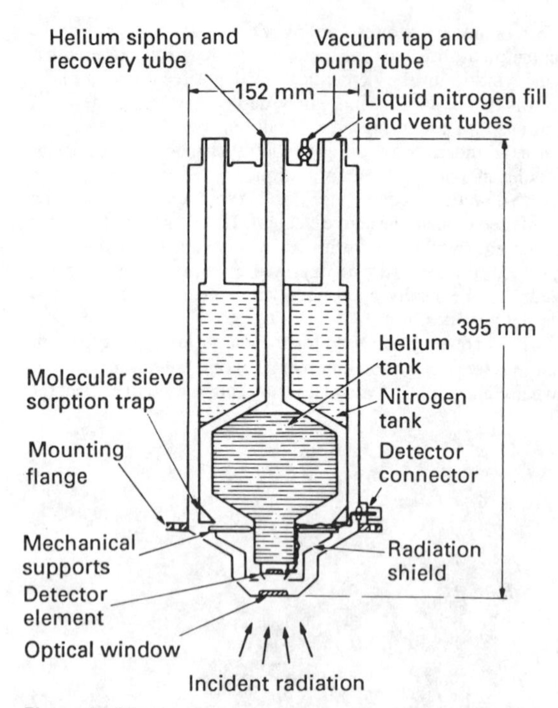

Figure 23.24 Construction of cryostat for doped-germanium photoconductive detectors operating at liquid-helium temperatures

antimonide type shown in *Figure 23.23* can be used, except that the window is made from silicon with an anti-reflective coating ('bloomed' silicon). This window material has a peak transmission between 9 μm and 11 μm. Typical values of D^* (λ_p, 800, 1), λ_p being the peak spectral response, are greater than 10^{10} (cm $Hz^{1/2}$)/W. The typical time constant is less than 1 μs.

In doped-germanium detectors, the radiation is absorbed by the electrons in the added impurity. This leads to a lower absorption coefficient, and hence the need for a thicker element. It is also essential to cool the element so that the electrons are initially in the impurity centres ready to be excited. The germanium must be extremely pure apart from the added impurity. Mercury and copper are the most widely used impurities.

Mercury-doped germanium detectors are used for wavelengths from 2.0 μm to 13 μm with a peak response at approximately 10 μm. The element requires cooling to 35 K, and this is achieved by using liquid helium either in bulk or in a Joule–Thomson cooler. The cross-section of a typical cryostat for a mercury-doped germanium detector is shown in *Figure 23.24*. The upper tank contains liquid nitrogen, and carries a radiation shield which partly surrounds the lower liquid-helium tank. This minimises evaporation of the helium, and also forms a shield for the detector element. A vacuum is maintained inside the cryostat, and the quality of the vacuum is maintained by a molecular sieve trap incorporated in the radiation shield. The cooled window through which the radiation is transmitted to the element is made of bloomed germanium. The resistance of the element is typically between 10 kΩ and 60 kΩ. A typical value of D^* (10 μm, 800, 1) is 1.3×10^{10} (cm $Hz^{1/2}$)/W, and the time constant is typically less than 1 μs.

Copper-doped germanium detectors are used for wavelengths between 2.0 μm and 25 μm with a peak response at 15 μm. This type of detector requires cooling to 4.2 K, and this is achieved with liquid helium in a cryostat similar to that shown in *Figure 23.24*. The resistance of the element is typically between 2.5 kΩ and 240 kΩ. A typical value of D^* (15 μm, 800, 1) is 10×10^{10} (cm $Hz^{1/2}$)/W, and the time constant is typically 1 μs.

Photoconductive detectors have been developed for specialised applications of infrared radiation with wavelengths extending beyond 50 μm into the sub-millimetric and microwave regions. A typical detector for these wavelengths uses the impurity level in indium antimonide, and operates at a temperature below 2 K. A complex cooling system is required for such a detector, and the cross-section of a typical cryostat is shown in *Figure 23.25*. Two glass Dewar vessels are used: an inner vessel containing liquid helium which is pumped to reduce the pressure inside the vessel and hence lower the boiling point of the helium, and an outer vessel containing liquid nitrogen to minimise the evaporation of the helium. An operating temperature of 1.6 K can be obtained in this way. The detector operates in a magnetic field produced by a superconducting solenoid immersed in the liquid helium. The incident radiation is directed on to the detector element by a light pipe fitted with a polythene window. In a typical detector of this type, the diameter of the window would be approximately 2 cm, and the length of the light pipe between the window and the element approximately 60 cm. This type of detector would operate at wavelengths, say, between 0.1 mm and 10 mm with a peak response at, typically, 1 mm. A typical value of responsivity (1 mm, 800, 1) is 2×10^3 V/W, and a typical value of D^* (1 mm, 800, 1) is 1×10^{12} (cm $Hz^{1/2}$)/W. The time constant is typically 1 μs.

The range of applications of infrared photoconductive detectors is very wide, extending from simple systems using uncooled detectors to give warning of flame failure in boilers, to complex and sophisticated systems with detectors operating at very low temperatures which are used for physical research or very precise measurements. The detection of the black-body radiation from an object can be used for such applications as

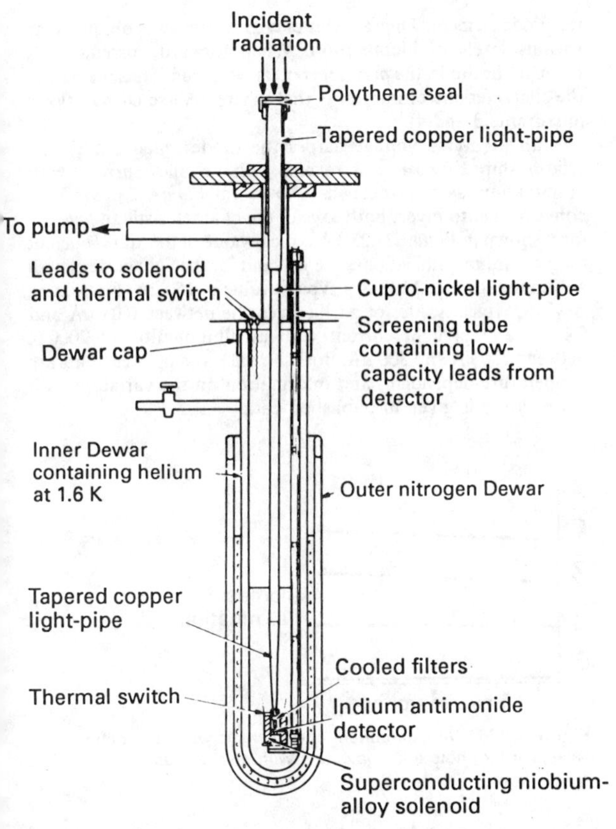

Figure 23.25 Construction of cryostat for indium antimonide photoconductive detector for sub-millimetric radiation operating at 1.6 K

detecting overheating in mechanical and electrical systems, intruder detection in security areas, temperature measurement without physical contact with the object, and thermal imaging. The varying absorption of infrared radiation by different materials can be used in such applications as chemical analysis by infrared spectroscopy or leak detection in closed systems.

23.2.2.3 Other infrared detectors

To complete this brief survey of infrared detectors, two other types are described that use photoelectronic effects other than photoconductivity.

A recently introduced material for use in infrared detectors is lead tin telluride, known as LTT. The detector uses the photovoltaic effect, a junction formed in the lead tin telluride alloy being used to separate the electron–hole pairs generated by the incident radiation. The photovoltaic effect is discussed more fully in connection with photodiodes; for the operation of the infrared detector it is sufficient to say that the separation of the charge carriers results in an open-circuit voltage or a short-circuit current. Thus an output signal can be produced by the detector without the need for an external bias voltage or current.

The peak spectral response of the detector occurs at, typically, 11 μm, enabling a range of wavelengths from approximately 8 μm to 14 μm to be covered. The detector is operated at a temperature of 77 K. Typical values of responsivity (λ_p, 800, 1) and D^* (λ_p, 800, 1) are 150 V/W and 8×10^9 (cm Hz$^{1/2}$)/W respectively. The time constant is typically less than 0.1 μs.

The second type of infrared detector uses the pyroelectric effect. This effect occurs in certain crystals with complex structures in which there is an inbuilt electrical polarisation which is a function of temperature. At temperatures above the Curie point, the pyroelectric properties disappear, but below the Curie point changes in temperature result in changes in the degree of polarisation. This change can be observed as an electrical signal if electrodes are placed on opposite faces of a thin slice of the material to form a capacitor. When the polarisation changes, the charge induced on the electrodes can flow as a current through a comparatively low impedance external circuit, or produce a voltage across the slice if the impedance of the external circuit is comparatively high. The detector produces an electrical signal only when the temperature changes.

Many crystals exhibit the pyroelectric effect, but the one most commonly used for infrared detectors at present is triglycine sulphate $(NH_2CH_2COOH)_3H_2SO_4$, known as TGS. The incident radiation on the active element of the detector produces an increase in temperature by the absorption of energy, and hence a change in the polarisation occurs. Changes in the level of the incident radiation produce an electrical signal from the detector.

The spectral response of TGS detectors, as with other thermal detectors, is wide. It extends from 1 μm, below which incident energy is not absorbed, to the millimetric region. Filters can be used to define the range of wavelengths required for particular applications. TGS detectors generally operate at low modulation frequencies (approximately 10 Hz) but can operate at higher frequencies. This is because, although the responsivity falls with increasing frequency (responsivity is inversely proportional to frequency), the noise also falls with increasing frequency. The value of NEP therefore rises only slightly up to frequencies of, say, 10 kHz. The detectivity is constant over the spectral range, D^* (500 K, 10, 1) being typically 2×10^9 (cm Hz$^{1/2}$)/W. A typical value for NEP is 2×10^{-10} W/Hz$^{1/2}$.

23.2.3 Photovoltaic devices

23.2.3.1 Photodiode

Visible or infrared radiation incident on a semiconductor material will generate electron–hole pairs by the normal photoconductive action. If these charge carriers are generated near a p–n junction, the electric field of the depletion layer at the junction will separate the electrons and holes. This is the normal action of a p–n junction which acts on charge carriers irrespective of how they are produced. However, the separation of the carriers gives rise to a short-circuit current or an open-circuit voltage, and this effect is called the photovoltaic effect. It can be used in such devices as photodiodes and phototransistors. *Figure 23.26* shows the construction of a typical photodiode.

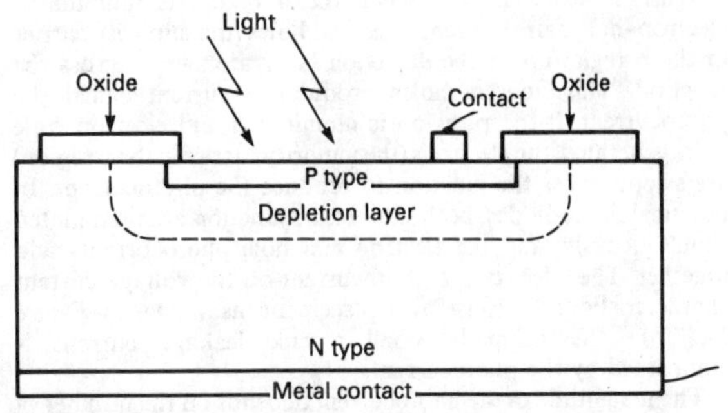

Figure 23.26 Typical construction of a p–n photodiode

In a normal p–n junction with no external bias applied, an equilibrium state is reached with an internal potential barrier across the junction. This potential barrier produces a depletion layer and prevents majority carriers crossing the junction. Minority carriers, however, can still cross the junction, and this gives rise to the small reverse leakage current of the junction diode. An external reverse voltage adds to the internal potential barrier, but the reverse current remains substantially constant because of the limited number of minority carriers available until avalanche breakdown occurs. An external forward voltage overcomes the internal potential barrier, and causes a large majority carrier current to flow. A photodiode differs from a small-signal junction diode only in that visible or infrared radiation is allowed to fall on the diode element instead of being excluded. If there is no illumination of the diode element, the photodiode acts as a normal small-signal junction diode, and has a voltage/current characteristic like that shown in *Figure 23.27(a)*.

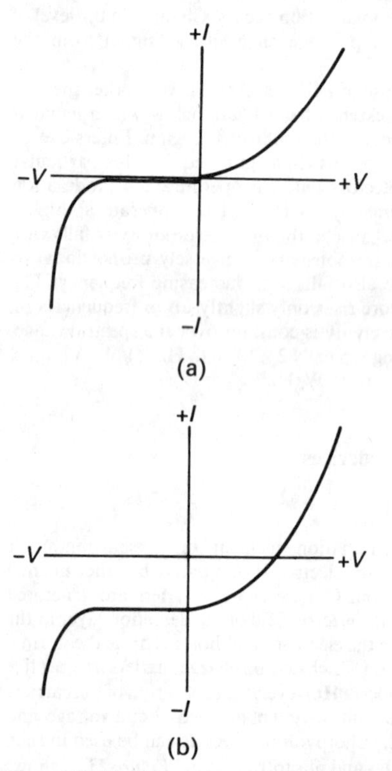

Figure 23.27 Voltage/current characteristic for photodiode: (a) not illuminated; (b) illuminated

With no external bias, when the n region is illuminated electron–hole pairs are generated and holes (the minority carrier in the n region) near the depletion layer are swept across the junction. This flow of holes produces a current called the photocurrent. If the p region is illuminated and electron–hole pairs generated, the electrons (the minority carrier in the p region) are swept across the junction to produce the photocurrent. In practical photodiodes, both sides of the junction are illuminated simultaneously, and the electron and hole photocurrents add together. The effect of the photocurrent on the voltage/current characteristic is to cause a displacement, as shown in *Figure 23.27(b)*. The normally small reverse leakage current is augmented by the photocurrent.

The magnitude of the photocurrent depends on the number of charge carriers generated, and therefore on the illumination of the diode element. Thus a series of characteristics is obtained for various levels of illumination. When forward current flows through the diode, the photocurrent is swamped. Thus the part of the characteristic of interest is that where reverse current flows (quadrants 3 and 4).

With a reverse voltage across the diode (quadrant 3), the relationship between the reverse voltage and current with illumination as parameter is shown in *Figure 23.28*. (It is conventional to invert both axes of the characteristic to give the form shown in *Figure 23.28*.) A typical value of the 'dark' current for germanium photodiodes is 6 μA, and for the light current with an illumination of 1600 lx a typical value is 150 μA. For silicon devices, typical values of dark current lie between 0.01 μA and 0.1 μA, and of light current with an illumination of 2000 lx between 250 μA and 300 μA. Both the dark and light currents are temperature dependent, and information on the variation with temperature is given in published data.

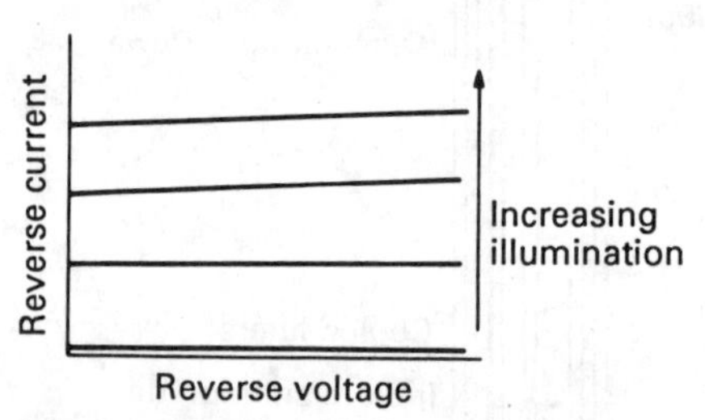

Figure 23.28 Voltage/current characteristic with illumination as parameter for photodiode operating with reverse bias

When operated with a reverse bias, the photodiode is sometimes described as being in the photoconductive mode. Although this term is deprecated since the operating principle is strictly speaking the photovoltaic effect, it is descriptive of the action of the photodiode in quadrant 3. The current in the diode is proportional to the number of carriers generated by the incident radiation. The term also provides a convenient distinction from the operation of the diode in quadrant 4.

With no bias voltage across an illuminated photodiode, a reverse current flows, the sum of the small minority-carrier thermal current and the photocurrent. As an increasing forward bias is applied, the magnitude of the reverse current decreases as the majority-carrier forward current increases. Eventually, the magnitude of the majority-carrier current equals that of the photocurrent, and no current flows through the diode. If the forward bias is increased further, a forward current flows as in a normal junction diode. Thus the limiting values in quadrant 4 are the current at zero bias, and the forward voltage at which the current is zero. These values represent the short-circuit current and open-circuit voltage which can be obtained from the photodiode. (Again, it is conventional sometimes to invert this characteristic, but this time only about the voltage axis.) With a suitable load resistance connected across the photodiode, it is possible to extract power from the diode. This is the principle of the solar cell which is described in *Section 23.2.3.4*.

Because the magnitude of the reverse current depends on the illumination, the relationship between the forward voltage and the current with illumination as parameter in quadrant 4 has the form shown in *Figure 23.29*.

A refinement of the p–n photodiode providing higher operating speeds is the PIN photodiode. Within the PIN photodiode, the space charge width is made significantly wider by sandwiching a layer of intrinsic semiconductor between the p and n layers shown in *Figure 23.26*.

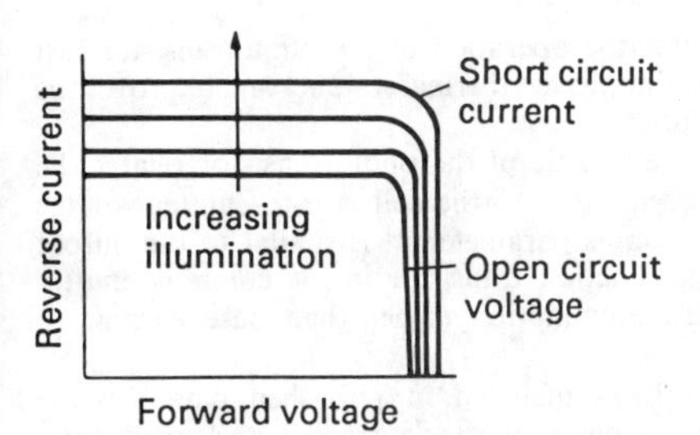

Figure 23.29 Voltage/current characteristic with illumination as parameter for photodiode with forward bias

Whereas the diffusion current is the main cause of current flow in p–n photodiodes, the drift current is the dominant component in PIN photodiodes.

A comparison of PIN and p–n photodiode is shown in *Table 23.2.*

Table 23.2 Comparison of some parameters of typical photodiodes

Parameter	*p–n photodiode*	*PIN photodiode*
Sensitivity	0.4–0.5 A/W	0.6 – 0.7 A/W
Rise time	1 μs	300 ns
Fall time	1 μs	300 ns
NEP	3×10^{-15} W/Hz$^{1/2}$	4×10^{-14} W/Hz$^{1/2}$
D	5×10^{13} (cm Hz$^{1/2}$)/W	6×10^{12} (cm Hz$^{1/2}$)/W

An alternative method of presenting the light transfer characteristics of photodiodes, sometimes used in published data, is shown in *Figure 23.30*. Sensitivities are published either relating light current to illumination (μA/lx) or to the incident monochromatic radiation (A/W).

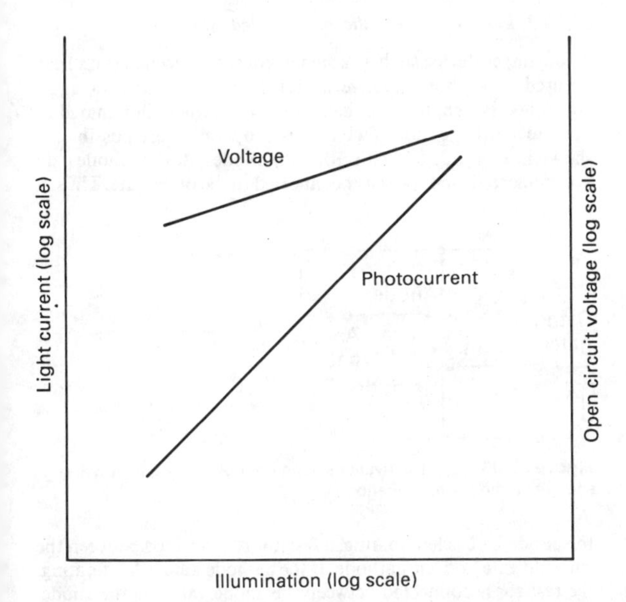

Figure 23.30 Light current and open circuit voltage plotted against illumination for photodiode

To enable a suitable external load resistance to be chosen for the photodiode, curves of current plotted against load resistance are sometimes given, as shown in *Figure 23.31*. Ratings for the photodiode given in the data include the maximum reverse voltage, and the maximum forward and reverse currents. From these ratings and the load resistance curve, a suitable external circuit to be operated by the photodiode for the particular application can be designed.

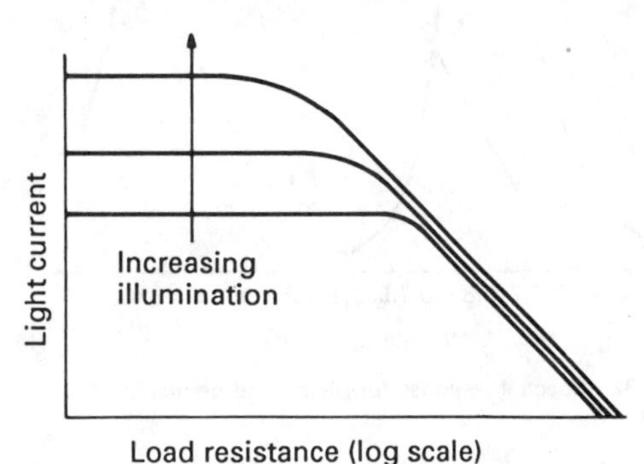

Figure 23.31 Light current plotted against load resistance with illumination as parameter for photodiode with forward bias

In many optical applications, the radiation incident on the photosensitive device is modulated or is chopped by some mechanical system. The maximum frequency to which the photodiode responds is determined by three factors: diffusion of carriers, transit time in the depletion layer, and the capacitance of the depletion layer. Carriers generated in the semiconductor material away from the depletion layer must diffuse to the edge of the layer before they can be used to form the photocurrent, and this can cause considerable delay. To minimise the time of diffusion, the junction should be formed close to the illuminated surface. The maximum amount of light is absorbed when the depletion layer is wide (with a large reverse voltage), but the layer cannot be too wide otherwise the transit time of the carriers will be too long. The depletion layer cannot be too thin, however, as the capacitance would be too high and so limit the high-frequency response of the diode. With careful design of the photodiode and choice of a suitable reverse voltage, operation up to at least 1 GHz is possible, enabling the photodiode to be used with rapidly pulsating radiation. Information on the relative response with modulation frequency, or information on the switching time expressed as the light-current rise and fall times in response to the sudden application and removal of the incident radiation, is given in published data.

Typical spectral responses for germanium and silicon photodiodes are shown in *Figure 23.32.*

Many different constructions are used for photodiodes, largely dictated by the end application. These vary from cheap moulded plastic encapsulations (used in domestic remote control applications) to metal packages with a window to allow radiation to fall onto the diode. Sometimes glass lenses are fitted to these packages. In the case of the plastic encapsulation, it is now common practice to mould filtering pigments into the epoxy resin of the package. This filters visible radiation and gives the device a black appearance.

The circuit symbol for a photodiode is shown in *Figure 23.33*.

23.2.3.2 Phototransistor

In principle, a phototransistor corresponds to a photodiode (collector–base diode) with a series connected transistor as amplifier. It is the simplest integrated photoelectric component.

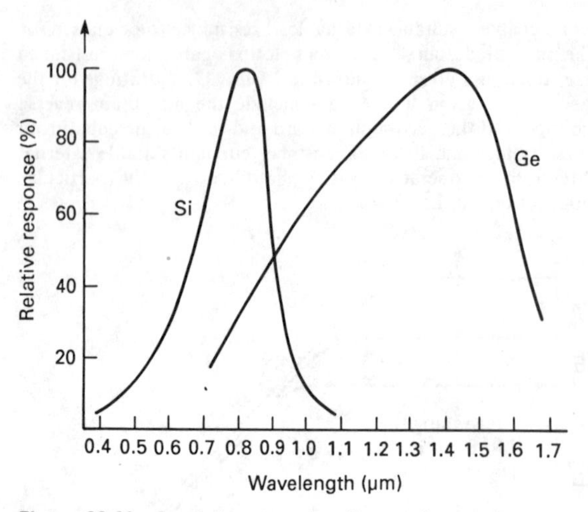

Figure 23.32 Spectral response for silicon and germanium photodiodes

Figure 23.33 Circuit symbol for photodiode

An npn phototransistor is shown in *Figure 23.34*. The base connection is open-circuit, that is the base is floating and the transistor is operating as a two-terminal device. The normal bias

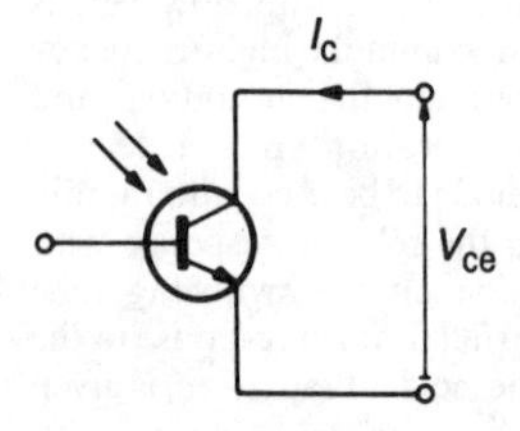

Figure 23.34 npn phototransistor

conditions are still maintained, however. With no illumination of the transistor, the current in the collector–emitter circuit is the common-emitter leakage current I_{CEO}. When the collector–base diode is illuminated, electron–hole pairs are generated and the minority-carrier photocurrent flows across the reverse-biased collector–base junction. Electrons flow out of the base, and holes flow into the base. Thus the forward bias across the base–emitter junction is increased, which in turn increases the electron flow from the emitter, across the base, into the collector. The collector current is therefore the sum of the electron photocurrent and the electron current from the emitter which is h_{FE} times the photocurrent. In other words, the photocurrent of the diode has been amplified by the current gain of the phototransistor. Thus the phototransistor is used in applications where greater sensitivity is required than can be obtained with a photodiode. In some applications, the base is not left open-circuit but a relatively high-value resistor is connected between the base and emitter. This does not affect the operation of the phototransistor just described, but gives improved thermal stability and improves the light-to-dark ratio.

The output characteristic of the phototransistor relates the light collector current $I_{CE(L)}$ to the collector-to-emitter voltage V_{CE} with illumination as parameter. It is similar to the output characteristic of a junction transistor in the common-emitter configuration with illumination rather than base current as parameter.

Other characteristics included in published data are the variation of light and dark collector currents with temperature, and the variation of collector currents with illumination.

In a phototransistor the base collector junction area available to incident radiation is large. The gain of phototransistors normally lies between 100 and 1000. However, since gain depends on current, the relationship between illumination and photocurrent is largely nonlinear. Though having more sensitivity than photodiodes, phototransistors exhibit longer rise and fall times caused by the delay due to the amplification system (Miller effect). This can cause typical rise and fall times of the order of 5–10 μs.

The spectral response of a phototransistor is very similar to that of a photodiode (see *Figure 23.32*), the peak response being around 800 μm, and is partly determined by the passivating oxide layer deposited over the light-sensitive area. This layer is necessary to prevent performance degradation. The ratings for the phototransistor given in published data are similar to those for the junction transistor. They include the maximum collector-to-base and collector-to-emitter voltages, the maximum emitter-to-collector or emitter-to-base voltages, the maximum collector current, and the maximum power dissipation. A thermal resistance is given so that a suitable heatsink can be chosen if required for the application.

There are many different encapsulations of phototransistors in common use. Dependent on application, they range from linear arrays (for punch hole reading) through epoxy resin types to small metal encapsulations (e.g. TO-18) with either plain windows or lenses. Phototransistors are used in applications where, for example, motion or solid objects need to be detected (machine safety guards, etc.) and also in flame detection systems for boilers.

23.2.3.3 Light-activated silicon controlled switch

A four-layer device such as a silicon controlled switch (s.c.s.) can be used as a photosensitive device. It can be used as a light-controlled switch, the s.c.s. being turned on when illuminated to operate another circuit. Two modes of operation are possible, as shown in *Figure 23.35*, with either the anode gate or cathode gate not connected, and a resistor connected to the other gate. Thus, if

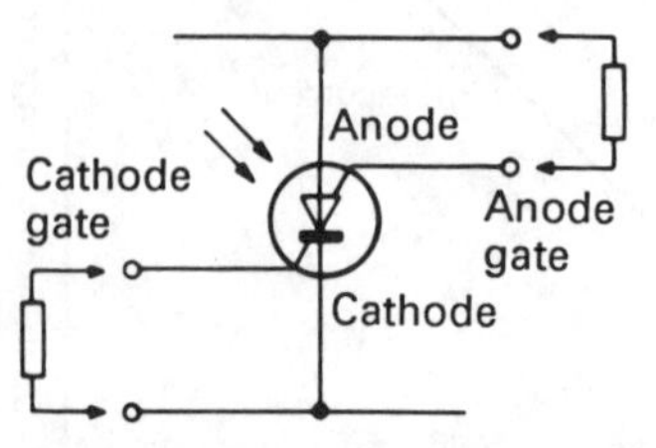

Figure 23.35 Light-activated silicon controlled switch showing alternative modes of operation

the anode gate is left floating, a resistor is connected between the cathode gate and the cathode. If the cathode gate is left floating, the resistor is connected between the anode gate and the anode.

The operation of the light-activated s.c.s. can be described in terms of the two-transistor analogue which is shown, with the

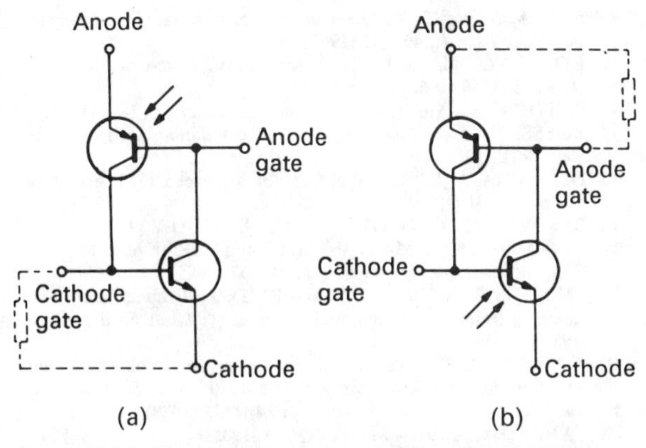

Figure 23.36 Two-transistor analogue for light-activated silicon controlled switch: (a) pnp phototransistor; (b) npn phototransistor

external resistors in broken lines, in *Figure 23.36*. When the anode gate lead is not connected, *Figure 23.36(a)*, the base of the pnp transistor is left floating, and this device acts as a phototransistor. When the s.c.s. has a positive anode-to-cathode voltage applied, and is not illuminated, only the small leakage current flows. When the device is illuminated, the photocurrent in the pnp transistor flows through the resistor between the cathode gate and cathode. A voltage is developed across this resistor so that the base of the npn transistor becomes positive with respect to its emitter, and so this transistor starts to conduct. (The development of the voltage drop across the resistor is equivalent to placing a small positive voltage on the cathode gate, which is one of the ways in which an s.c.s. can be triggered.) Cumulative action between the two transistors occurs, and the s.c.s. is turned on. If the cathode gate is not connected *Figure 23.36(b)*, the base of the npn transistor is floating and this device acts as the phototransistor. The photocurrent develops a voltage across the resistor between the anode gate and anode, so that the anode gate becomes negative with respect to the anode, and the s.c.s. is turned on. The light-activated s.c.s. is turned off, as with the normal s.c.s., by reducing the current through the device to below the holding value.

The characteristics and ratings for a light-activated s.c.s. given in published data are similar to those for a normal s.c.s. Additional information is given on the variation of cathode-gate current with illumination, and the spectral response. Two illumination levels are specified: a minimum value $E_{on,min}$ which will trigger all devices, and a maximum non-triggering value $E_{off,max}$. These values are determined by the leakage current of the device.

The usual encapsulation for the light-activated s.c.s. is a small metal envelope such as the TO-72 outline with an end-viewing window. The device is generally used in applications as a relay driver. The relay is connected in the anode circuit, and is energised when the device is turned on by the incident radiation.

23.2.3.4 Solar cell

The solar cell is a form of photodiode which is optimised for operation from the sun's radiation. The operating principle is the same as that of the photodiode, the generation of electron–hole pairs by the incident radiation, and the separation of these charge carriers by the electric field of the depletion layer at a p–n junction. The flow of minority carriers across the junction produces a short-circuit current or open-circuit voltage so that power can be extracted from the device by a suitable load resistance. The surface area of the solar cell is made as large as possible so that the maximum amount of radiation is incident on the device. The p–n junction is formed near the surface to minimise carrier recombination.

The most commonly used materials from which practical solar cells are manufactured are silicon and gallium arsenide. The cell can be constructed as a thin p-type layer on an n-type substrate, or as a thin n layer on a p substrate. Both regions of the cell are heavily doped. Non-rectifying contacts must be made to both regions. The contact of the substrate can be made on the back of the device, but the contact to the front layer must be made in such a way that the minimum surface area is obscured. Narrow contact fingers can be deposited on the front surface, as shown in the schematic structure of *Figure 23.37(a)*, or the material of the front layer can be taken round the side of the cell to the back of the substrate and the contact made there, *Figure 23.37(b)*.

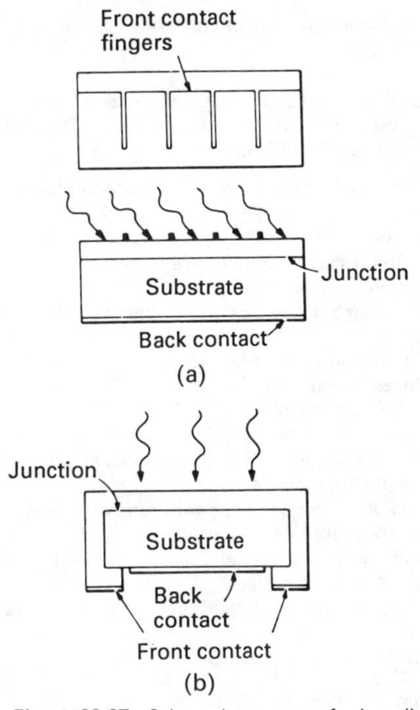

Figure 23.37 Schematic structure of solar cell: (a) contact on front surface; (b) contact on back of cell

The relationship between the voltage and photocurrent for a solar cell is shown in *Figure 23.38* with the maximum-power rectangle superimposed. This rectangle represents the maximum amount of power that can be extracted from the cell, the product $V_{mp}I_{mp}$. With a suitable load resistance, this power can be 80%

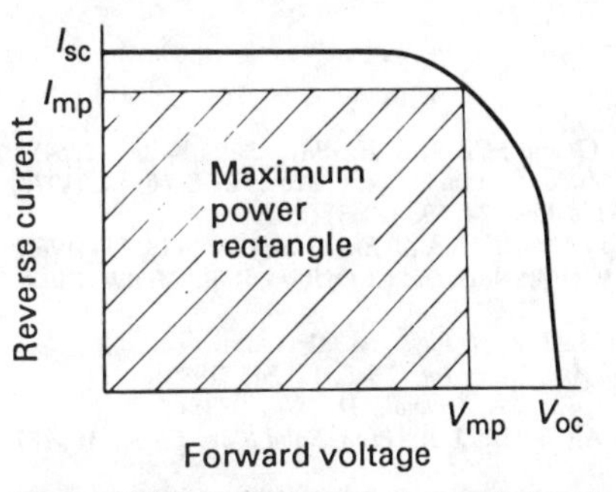

Figure 23.38 Voltage/current characteristic for solar cell with maximum power rectangle superimposed

of the product $V_{oc}I_{sc}$ where V_{oc} is the open-circuit voltage and I_{sc} is the short-circuit current. Typical values of the open-circuit voltage and short-circuit current are 0.5 V and 0.1 A respectively.

Solar cells can be connected in series to provide a higher voltage than can be obtained from a single cell, and connected in parallel to provide a higher current. The main application of solar cells is for the power supplies of space vehicles and communication satellites. Large arrays of cells are used, connected in series/parallel, to provide the required values of voltage and current to drive the electronic circuitry in these vehicles. The conversion efficiency of such batteries is between 10% and 15%. Another application of solar cells is for terrestial power supplies. In areas where extended periods of sunlight can be relied on, arrays of solar cells can be used as a power source.

23.2.4 Symbols for principal parameters of photoelectronic devices

Symbols for the main electrical and optical parameters of photoelectronic devices, other than those common to small-signal diodes and transistors, are listed below.

- D^* — area-normalised detectivity of photoconductive detector
- E — illumination
- $I_{CE(D)}$ — dark collector current of phototransistor
- $I_{CE(L)}$ — light collector current of phototransistor
- I_{cell} — current through active element of photoconductive detector
- I_{SC} — short-circuit current of solar cell
- I_{RD} — dark current of photodiode
- $I_{R(SC)}$ — short-circuit current of photodiode
- NEP — noise equivalent power
- P_{cell} — power dissipation of active element of photoconductive detector
- S, S_R — sensitivity of photodiode and phototransistor, change of current with incident radiation
- T_{tablet} — operating temperature of active element of photoconductive detector
- V_{cell} — voltage across active element of photoconductive detector
- V_{OC} — open-circuit voltage of solar cell
- λ, λ_p — peak spectral response

Acknowledgements

The assistance of Mr C. L. M. Ireland of Lumonics Ltd in the preparation of Sections 23.3.2–23.3.4 is gratefully acknowledged.

References

1 GRIMMEIS, H. G. and SCHOLZ, H., *Phys. Lett.*, **8**, 233 (1964)
2 SMITH, P. W., 'Mode selection in lasers', *Proc. IEEE*, **60**, 422 (1972)
3 NELSON, H., *RCA Rev.*, **24**, 603 (1963)
4 TIETJEN, J. J. and AMICK, J. A., *J. Electronics Soc.*, **113**, 724 (1966)
5 NUESE, C. J., RICHMAN, D. and CLOUGH, R. B., *Metall. Trans.*, **2**, 789 (1971)
6 LUDOWISE, M. J., *J. Appl. Phys.*, **58**, 31 (1985)
7 DAPKUS, P. D., *Ann. Rev. Mater. Sci.*, **12**, 243 (1982)
8 CHANG, L. L., *J. Vac. Sci. Technol.*, **31**, 120 (1983)
9 CHO, A. Y. and ARTHUR, J. R., *Prog. Solid State Chem.*, **10**, 157 (1975)
10 BURRUS, C. A. and DAWSON, R. W., *Appl. Phys. Lett.*, **17**, 97 (1970)
11 FOX, A. G. and LI, T., 'Resonant modes in a laser interferometer', *Bell Syst. Tech. J.*, **40**, 453 (1961)
12 KOGELNIK, N. and LI, T., 'Laser beams and resonators', *Appl. Optics*, **5**, 1550 (1966)
13 SMITH, P. W., 'Mode locking of lasers', *Proc. IEEE*, **58**, 1342 (1970)
14 ROESS, D., *Lasers, Light Amplifiers and Oscillators*, Academic Press, New York (1969)
15 DE MARIA, A. J., 'Review of C.W. high power CO_2 lasers', *Proc. IEEE*, **61**, 731 (1973)
16 BASOV, N. G., GLOTOV, E. P., DANILYCHEV, A. V., KERIMOV, O. M., MALYSH, M. M. and SOROKA, A. M., *IEEE J. Quant. Electron.*, **QE-21**, No. 4, 342–57 (1985)
17 ZNOTINS, T. A., PULIN, D. and REID, J., 'Excimer lasers—an emerging technology in materials processing', *Laser Focus*, May (1987)
18 KOECHNER, W., *Solid State Laser Engineering*, Vol. 1, Springer Series in Optical Sciences, Springer-Verlag, New York (1976)
19 KAWATA, S. *et al.*, *Electron. Lett.*, **24**, 1489 (1988)
20 CANEAU, C., *Electron. Lett.*, **21**, 815 (1985)
21 MITO, I. *et al.*, *Lightwave Technol.*, **LT-1**, 195 (1983)
22 KAMINOW, I. *et al.*, *Electron. Lett.*, **15**, 763 (1979)
23 KAMINOW, I. *et al.*, *Electron. Lett.*, **17**, 318 (1981)
24 JANSSEN, A. P. *et al.*, *Proc. Int. Conf. on Optical Fibre Submarine Telecoms Systems*, France, 193 (1986)
25 KOGELNIK, H. and SHANK, C. V., *Appl. Phys. Lett.*, **18**, 152 (1971)

Further reading

ALLEN, L. and JONES, D. G. C., *Principles of Gas Lasers*, Butterworth, London (1967)
ANON., 'High power semiconductor laser arrays', *ITT Application Note*
ANON., 'Laser diodes for communication', *ITT Application Note*
ANON., 'Modulation of semiconductor laser diodes', *ITT Application Note*
ANON., 'Safety aspects of laser diodes', *ITT Application Note*
ANON., 'Using semiconductor lasers', *ITT Application Note*
ARCHER, R. J., 'Light emitting diodes in III–V alloys', Paper 66, *Electrochem. Soc. Mtg, Los Angeles, California, Spring*, 1970
ARCHER, R. J., *J. Electr. Mater.*, **1**, 128 (1972)
ARECCHI, F. T. and SCHULTZ-DUBOIS, E. O. (Eds), *Laser Handbook*, Vols 1, 2, North-Holland, Amsterdam (1972)
AUZEL, F. E., 'Materials and devices using double-pumped phosphors with energy transfer', *Proc. IEEE*, **61**, 758 (1973)
BERGH, A.A. and DEAN, P. J., 'Light emitting diodes', *Proc. IEEE*, **60**, 156 (1972)
BLOOM, A. L., *Gas Lasers*, Wiley, New York (1968)
CASEY, H. C. and PANISH, M. B., *Heterostructure Lasers*, Part A: *Fundamental Principles* (1978), Part B: *Materials and Operating Characteristics* (1979) Academic Press, London
DAVIES, I. G. A. and GOODWIN, A. R., 'Reliable sources for fibre optic communication—the high power edge emitting LED', *STL/ITT Application Note*
ELECCION, M., 'The family of lasers: survey', *IEEE Spectrum*, **9**, 326 (1972)
FABIAN, M. E., *Semiconductor Lasers, A Users Handbook*, Electrochemical Publications (1981)
GOOCH, C. H. (Ed.), *Gallium Arsenide Lasers*, Wiley–Interscience, New York (1969)
GOODWIN, A. R. and PLUMB, R. G., 'Reliable sources for fibre optic communication—The 20 μm oxide insulated stripe laser', *STL/ITT Application Note*
GOODWIN, A. R. and SELWAY, P., 'Heterostructure injection lasers', *Elect. Commun.*, **47**, 149 (1972)
HARRY, J. E., *Industrial Lasers and Their Applications*, McGraw-Hill, New York (1974)
HARVEY, N. F., *Coherent Light*, Wiley, New York (1970)
JACOBS, S. D. *et al.*, 'Liquid crystal laser optics: design, fabrication and performance', *Opt. Soc. Am. J., Part B*, **5**, No. 9, September (1988)
KOBAYASHI, K. *et al.*, *Electron. Lett.*, **21**, 1162 (1985)
KRESSEL, H. and BUTLER, J. K., *Semiconductor Lasers and Heterojunction LEDs*, Academic Press, London (1978)
KIRKBY, P. A., 'Semiconductor laser sources for optical communication', *Radio Electron Engr*, **31**, No 7/8, 362–76 (1981)
LADANY, I. and KRESSEL, H., *RCA Rev.*, **33**, 517 (1972)

McCLELLAND, S., 'A clear view for fibre optic sensors', *Sensor Rev.*, **8**, No. 4, October (1988)
MARUSKA, H. P. and TIETJEN, J. J., *Appl. Phys. Lett.*, **15**, 327 (1969)
NELSON, H., *U.S. Patent* 3 565 702 (23 February 1971)
NELSON, P. *et al.*, 'Micropower laser diodes for optical interconnects', *SPIE Proc.*, **752**, Digital Optical Comp. (1988)
PAUK, Y. S., GEESNER, C. R. and SHIN, B. K., *Phys. Lett.*, **21**, 567 (1972)
RIECK, H., *Semiconductor Lasers*, MacDonald, London (1968)
SANDBANK, C. P. (Ed.), *Optical Fibre Communication Systems*, Wiley, New York (1980)
THOMPSON, G. H. B., *Physics of Semiconductor Laser Devices*, Wiley, New York (1980)
TURLEY, S. E. H., HENSHALL, G. D., GREENE, P. D., KNIGHT, V. P., MOULE, D. M. and WHEELER, S. A., 'Properties of inverted rib waveguide, lasers operating at 1.3 μm wavelength', *STL/ITT Application Note*
VECHT, A., WERRING, N. J. and SMITH, P. J. F., 'High efficiency DC electroluminescence in ZnS(Mn,Cu)', *J. Phys. D. Appl. Phys.*, **1**, 134 (1968)
VECHT, A. and WERRING, N. J., 'Direct current electroluminescence in ZnS', *J. Phys. D: Appl. Phys.*, **3**, 105 (1970)
WAGNER, W. G. and LENGYEL, B. A., 'Evolution of the giant pulse in a laser', *J. Appl. Phys.*, **34**, 2040 (1963)
WITTKE, J. P., LADANY, I. and YOCOM, P. N., *J. Appl. Phys.*, **43**, 597 (1972)
WOODALL, J. M., RUPPRECHT, H. and REUTER, W., *J. Electrochem. Soc.*, **116**, 899 (1969)

24 Displays

M Jones, BSc
Product Manager
Hitachi Europe Ltd

Contents

24.1 Light-emitting diode displays

Although the cost of LEDs has come down considerably during the last decade, largely as a result of batch fabrication techniques, the cost of a single diode is still excessive for high-resolution ($\gtrsim 10^5$ elements) displays. Nevertheless experimental television displays[1,2] using LEDs have been demonstrated. LEDs are most widely used for numeric display applications where only a few digits need to be displayed, e.g. readouts for digital equipment.

Dot matrix systems based on 5×7 dots per character and continuous graphics panels, for example 32×128 elements, are available for instrumentation. LEDs are also widely used as large-area information panels or signs, with typical panels covering several square metres and over 100 000 LEDs. Control technology allows each point to be individually addressed.

The size and geometry of the p–n junction, and the fabrication process used, depend on the intended application. Small seven-segment displays are often made in a single step on a common substrate, but the substrate cost makes this approach impractical for larger-sized displays; consequently, larger numeric LED displays are generally assembled from seven individual diodes. A seven-segment display is shown in *Figure 24.1*. From the seven segments, all the numerals and certain letters such as A, H, P and U can be formed.

In monolithic displays the contact electrodes can be evaporated or photolithographically deposited on the surface parallel to the p–n junction plane (*Figure 24.2(a)*). Light emission is predominantly from this surface. The uppermost layer is kept thin (<5 μm) to minimise self-absorption, especially in direct bandgap materials; it cannot be too thin, however, since current spreading to the whole junction area is desirable. In GaP displays the transparency of the substrate material leads to optical cross-coupling between segments; a special structure (*Figure 24.2(b)*), in which the material is etched away in between the segments and coated with absorbing or reflecting films, was developed to minimise this optical coupling.

By slicing the p–n junction into bars (*Figure 24.2(c)*) after electroding, very high line brightnesses from the edge of the junction can be obtained. Electrical connection to the small bar segments and subsequent assembly are more complicated.

Other structures have been designed which use external means (cylindrical lenses, reflecting cavities, diffusing covers—*Figure 24.2(d)*) to enlarge the area from which light is emitted; these techniques are conservative of diode area but require a hybrid process for assembly and have reduced surface brightness.

Improvements in materials and processing have led to the development of 'ultra-bright' red LEDs based on GaAlAs (gallium aluminium arsenide) using a liquid phase epitaxy process. The brightness is up to five times greater than standard red LEDs based on GaP and GaAsP materials. Blue emitting LEDs (480 nm) have also been developed, although suffering from high cost and low light output at this stage.

The reliability, long life and compatibility with low-voltage integrated driving circuitry make LEDs very attractive for small display applications. However, their power consumption of typically 5–10 mW per small digit is still considered high for many battery-operated products, e.g. portable calculators and electronic watches, hence the dominance of LCD technology.

Additional information about materials, fabrication and applications for LEDs can be found in several excellent review articles.[3–5]

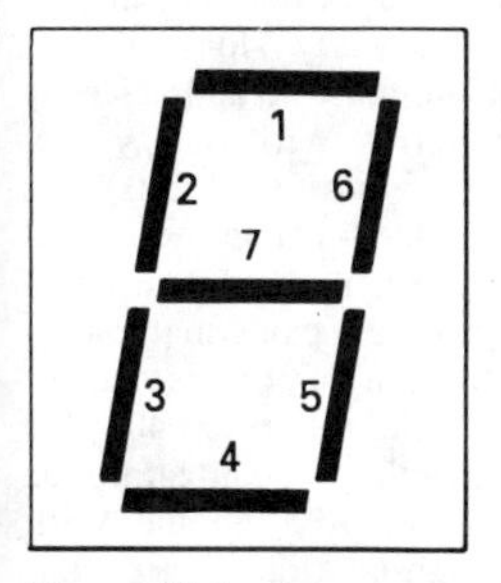

Figure 24.1 Seven-segment character display using electroluminescent diodes. 1–7, electroluminescent diodes

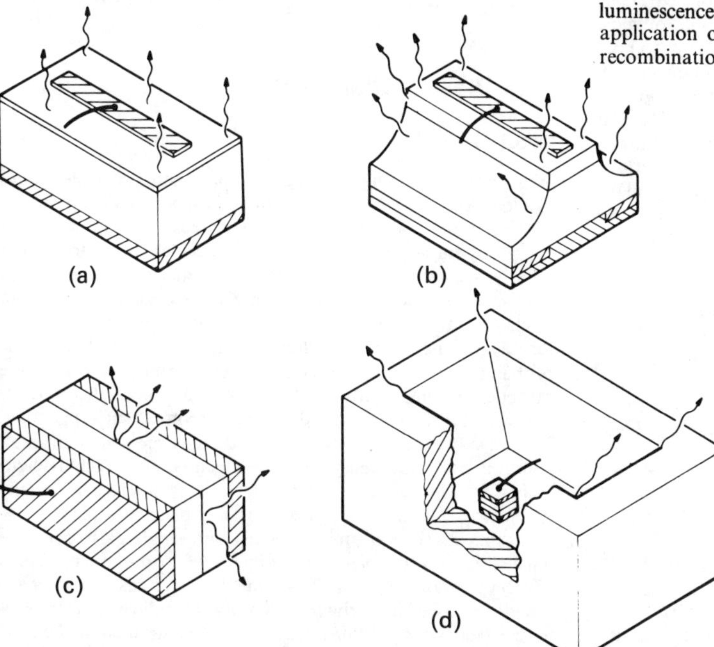

Figure 24.2 LED geometries used for the fabrication of seven-segment numeric displays: (a,b) surface emitters; (c) edge emitter; (d) cavity emitter (Courtesy of American Vacuum Society)

24.2 Electroluminescence—Destriau effect

Some phosphors emit light when a sufficiently high electric field is applied across them. This phenomenon, known as electroluminescence (EL), was first observed by Destriau[6] upon application of a changing electric field. Light is generated by recombination of electrons and holes whereby their excess energy is transferred to an emitted photon. Donors and acceptors play a major role in facilitating this radiative recombination and determining the spectral characteristics of the emitted light, i.e. the increment by which the photon energy is less than the energy gap of the phosphor material.

The mechanism by which electrons and holes are generated in the EL phosphor is not fully understood. The applied voltage is believed to be concentrated near a barrier or thin high-impedance region in the phosphor layer; the electric field strength is thereby locally increased to the point where carrier injection by field emission and possibly avalanche multiplication can occur.

24.2.1 AC electroluminescence

Most extensively studied are a.c.-EL phosphors, usually zinc sulphides or zinc sulphoselenides doped with copper or chlorine to produce yellow, green or blue emission; red EL phosphors have been made, but their brightness and efficiency are usually much lower.

In practice, a thin (~10–50 μm) layer of EL phosphor is sandwiched between two electrodes, at least one of which is transparent and through which the display is viewed. Both rigid (glass, ceramic) and flexible (plastic) substrates can be used. An a.c. excitation voltage, typically 50–5000 Hz, of several hundred volts is applied to the electrodes.

An area of early interest for a.c.-EL was the combination of large-area photoconductor layers and EL phosphors for image intensification or X-ray image storage.[7] However, saturation and trapping effects in the photoconductors and the availability of better alternatives prevented commercialisation of these devices.

Originally a.c.-EL was also believed to be of importance for large-area illumination and flat-panel television. However, in both applications luminous efficiency, average brightness and display life are of prime importance. In particular, for the television application where each display element is excited only during a small fraction of each frame time, high peak brightnesses are required in order to obtain useful average brightnesses.

Despite much effort to develop improved a.c.-EL phosphors, some fundamental materials problems remain. The brightness of EL displays increases with increasing excitation voltage and frequency. On the other hand, both life and luminous efficiency tend to decrease in the same direction. The half-life of most phosphors, i.e. the time over which the brightness drops to one-half of its initial value, varies inversely with frequency of excitation, implying a constant number of cycles. Luminous efficiencies in yellow-green phosphors of 1–5 lm/W (0.1–1%) can be achieved at 35–350 nits brightness with a typical half-life of 1000–3000 h at 400 Hz excitation.[8] The most important commercial applications are for low-power, reliable night lights, instrument panel lighting (there is usually no catastrophic failure mechanism) and backlighting of LCDs.

The technical feasibility of an a.c.-EL (ZnS,Se: Cu,Br) television panel has also been demonstrated;[9,10] however, the contrast ratio and brightness, which decrease with increasing number of scan lines due to the decreasing duty factor (~35 nits and a contrast of 10:1 for an 80-line panel *versus* ~12 nits for a 225-line panel) are not adequate for commercial television displays. Also power consumption is high due to the relatively large stray capacitance of the thin panel, and the high voltages and short pulses used to excite it.

24.2.2 DC electroluminescence

More recently, interest in d.c.-EL materials has increased with the discovery that high brightness (several hundred nits), good life (> 100 h), and modest luminous efficiency (0.1% ≈0.5 lm/W) can be achieved in ZnS doped with manganese and coated with copper sulphide. The light emitted from these phosphors originates from a thin (3–5 μm) high-impedance layer near the anode; the rest of the phosphor layer is optically inactive but acts as a resistive protection layer. The high-impedance layer has to be created by a 'forming' process (high voltage and high current) and presumably results from the diffusion of copper ions out of it.

Of particular interest has been the finding that average brightness and luminous efficiency remain high under low duty-cycle pulse excitation: for instance, an average brightness of 270 nits was achieved using 4 μs-wide 250-V pulses at 0.5% duty cycle in a panel originally formed at 60 V. This finding is of special significance for grey-scale displays, such as television where the duty cycle is low due to the large number of scanning lines. Thus an off-the-air 330-mm-diagonal flat d.c.-EL television display has been developed[11,12] with 224 × 224 elements, producing a 34-nit display with 20:1 contrast and a total power consumption of 150 W.

24.2.3 Thin film electroluminescence

The combination of long life, stable operation and high brightness under low duty cycle excitation has also been demonstrated in thin film EL panels.[13] These panels typically consist of a sputtered or electron-beam evaporated thin film EL phosphor (Mn-doped ZnS, about 0.5 μm thick) sandwiched between two insulating dielectric layers (about 0.2 μm thick), e.g. Y_2O_3 or Si_3N_4. The dielectric layers block d.c. current flow and provide the hot carriers necessary for efficient electroluminescence; the panels therefore must be addressed by high voltage pulses of alternating polarity. With this type of addressing circuitry high average brightness (200 nits) has been achieved in high-resolution, line-at-a-time addressed alphanumeric or grey-scale TV displays where the duty cycle of excitation is 0.2–1%.[14] By modifying the EL panel materials a high-resolution storage panel has also been demonstrated.[15]

High-performance, high-resolution panels of up to 640 × 200 pixels are now in production.[16,17] Generally these integrate the driver circuits and interface circuitry on board a module. With increasing yields and production automation, price and performance levels have improved. New developments are aimed at the introduction of multi-colour EL panels and 'chip-on-glass' technology for weight and size reduction.

24.3 Plasma displays

24.3.1 Operation

When a sufficiently high voltage is applied across two electrodes in a low pressure gas, a breakdown of the insulating properties of the gas is observed.[18] The sudden increase in conductivity is caused by impact ionisation of gas molecules by electrons that have been accelerated to sufficiently high energies by the applied electric field. Each electron leaving the cathode produces an electron avalanche travelling towards the anode and a positive ion avalanche travelling towards the cathode. When the probability of this ion stream regenerating an electron at the cathode becomes unity, the Townsend breakdown condition is satisfied and the discharge becomes self-supporting. The flow of electrons also excites the gas molecules through collision and this energy is subsequently given off as light emission. This emitted light, which is characteristic of the ambient gas, is the basis for a variety of plasma displays.

Plasma displays normally consist of a single glass envelope filled with a few torr of gas, typically neon for optimum luminous efficiency (orange light at 0.1–1 lm/W); other gas additives are sometimes used, in particular mercury, to reduce sputtering of the electrodes or to obtain ultraviolet emission which can be used

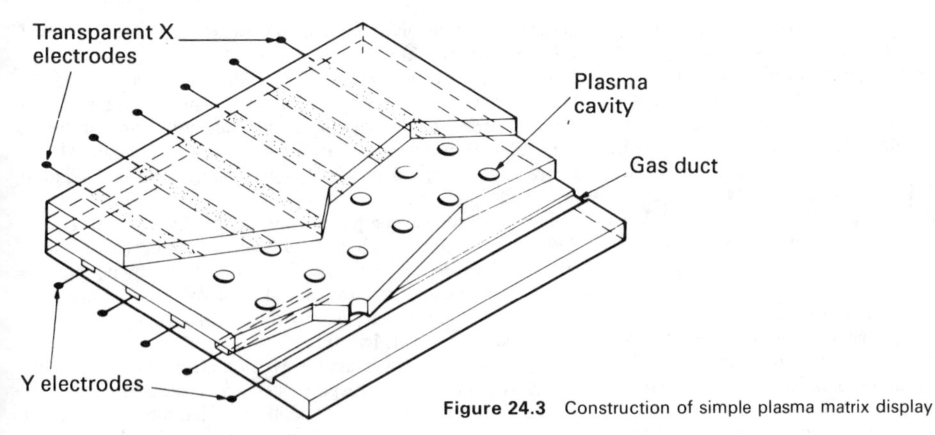

Figure 24.3 Construction of simple plasma matrix display

to excite phosphors (in order to change the colour of the display). Some internal structure is often used to isolate the display elements and to confine the discharge (*Figure 24.3*).

The extension of plasma displays to other colours can, in principle, be achieved by incorporating fluorescent phosphors which are excited by ultraviolet radiation from the discharge[19,20] or by using carriers from the discharge to bombard and excite a low-voltage cathodoluminescent phosphor directly,[21] or by using electrons extracted from the plasma, and subsequently accelerated to high voltages, to excite more or less standard cathodoluminescent phosphors.[22] Suitable cathodoluminescent colour primaries, either for low-voltage applications (i.e. phosphors with negligible 'dead-layers' to provide adequate luminous efficiency) or for high-voltage applications (i.e. phosphors resistant to inevitable ion bombardment) are not readily available however.

The high voltage needed to fire a plasma cell (100–200 V) is a disadvantage in many applications, especially the alphanumeric display field (calculators, digital clocks, digital meters) where competition from other technologies (LED, liquid crystals) is strong. Nevertheless, small and large-size plasma displays are being used because of their inherent low fabrication cost, pleasant bright appearance and reliability.

24.3.2 Matrix (multi-element) plasma displays

In larger-sized or high-resolution displays,[23,24] plasmas have many advantages. Since the current through a plasma, and therefore the light output, is negligible until breakdown occurs, there is a built-in threshold which makes plasma displays ideal for matrix (coincidence or half-select) addressing. The display elements of a matrix display are arranged at the intersection of a set of orthogonal X and Y electrodes (*Figure 24.3*), so that many elements ($n \times m$) can be addressed with relatively few ($n+m$) electrode connections.

Plasma display panels (PDPs) may be categorised into DC plasma panels, where the electrodes are in direct contact with the gas, and AC plasma panels where the electrodes are electrically isolated from the gas. DC plasma panels are mostly used for small and medium complexity commercial products, whereas AC plasma panels, due to higher costs and greater reliability, are used in professional and military markets.

24.3.3 DC plasma display panels

A conceptually simple multiplexing technique has been developed by Holz[25] which drastically reduces the number of external connections required in multi-element plasma displays. The DC PDP, shown in *Figure 24.4*, consists of two functionally separate parts:

(1) In the back of the panel, a gas discharge is shifted linearly under the influence of external clock pulses and used to address the display elements.
(2) The front of the panel contains the video-modulated display elements. These are connected to the scanning cells by means of small holes in the common electrode which are used for 'glow priming' (addressing).

The operation of the scanning part is as follows: the cathodes of a row of plasma elements are connected alternately to one of three common electrodes (*Figure 24.4*). A three-phase sinusoidal voltage whose amplitude is just below threshold is applied to these three electrodes—consequently, none of the elements will be turned 'on'. However, if a particular element, say the first, is externally turned on, then metastable ions from this discharge will diffuse to the adjacent element thereby lowering its threshold for firing. This element therefore will turn on when the voltage across it next reaches its maximum; in the meantime the voltage across the previous element is dropping and it is extinguished. In this manner the discharge is 'stepped' along by three elements for every full cycle of the sinusoidal clock voltage. In the DC PDP the first element of a row is used to start the discharge and the frequency of the clock voltage is used to determine the rate at which it propagates down the row.

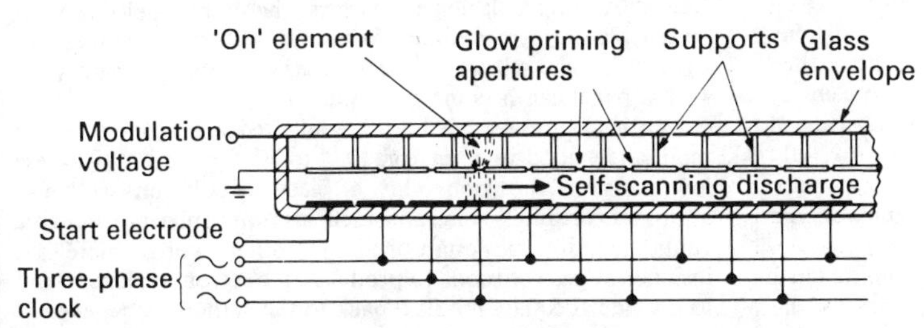

Figure 24.4 Principle of operation of the DC PDP

This linearly scanning discharge is then used to sequentially address ('glow-prime') the display elements which are interconnected to the scanning elements by means of a small hole in the central common electrode. The diffusion of carriers through these holes from the scanning discharge in the back is sufficient to remove the threshold in the brightness *vs.* voltage characteristic of the display element. Thus by modulating a common display anode, the brightness of the element in front of the scanning discharge can be determined: a single line of 256 elements can be addressed time-sequentially with less than 10 external leads. By scanning at a sufficiently high frequency (60 Hz) a steady, nonflickering display results.[26,27]

DC PDPs are supplied by a number of manufacturers, with resolutions up to 640 × 400 pixels being common.[28] Generally the units are complete modules controlled by standard TTL levels, with a typical power consumption of 20 W. Current major developments are aimed at increased production automation and driver count reduction in order to reduce costs. The longer term goal is a multi-coloured panel.

24.3.4 AC plasma display panels

A very different type of multielement plasma display was developed by Bitzer and Slottow.[29,30] The construction of this plasma panel is illustrated in *Figure 24.5*. Individual plasma elements are located at the intersection of orthogonal X-Y electrodes but these electrode strips are insulated from the gas discharge by means of a thin insulating layer—thus no direct current can flow through the panel.

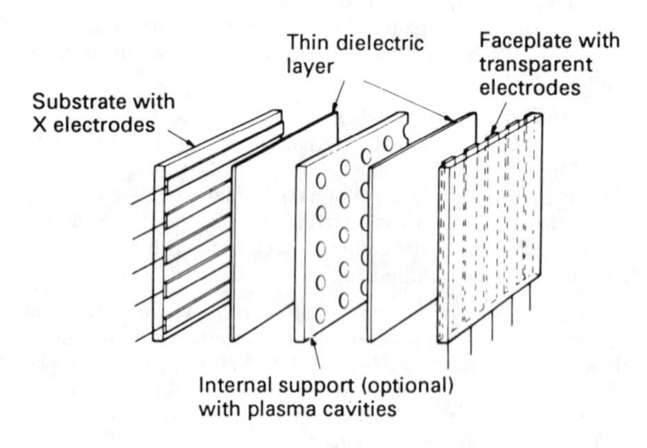

Figure 24.5 Construction of an AC PDP

When a sinusoidal voltage is applied to all electrodes, the voltage will divide capacitively across the insulating layers and the plasma element; if the voltage across the plasma is insufficient to cause breakdown, then the element remains 'off'. However, if any element is separately triggered 'on', then the current flow through the plasma will deposit an electric charge on the insulating layer covering the electrodes; this reduces the voltage across the plasma element causing it to extinguish itself. When the polarity of the applied voltage next is reversed, this charge on the insulator produces an increased voltage across the display element thus allowing it to fire again, and the charge is then transferred to the opposite electrode. In this manner, any element that has been triggered 'on' continues to fire at every half cycle of the applied voltage while all other elements remain 'off'. Therefore the AC PDP is uniquely suited for alphanumeric (on/off) display applications requiring long-term memory or slow updating.

Although the light from a single-plasma element is emitted in a very short time (10^{-7} to 10^{-6} s), the average brightness of the panel can be quite high ($\gtrsim$170 nits) with excitation frequencies of the order of 10–100 kHz.

The inherent memory of AC PDPs allowed the construction of large size, very high definition displays. The image quality, in terms of contrast ratio, viewing angle, uniformity and stability, is very high. In addition AC PDPs give a good mechanical and temperature specification and so are widely used in industrial and military markets.

The storage plasma panel is undoubtedly best suited for alphanumeric and graphics display applications where frame-storage (memory) is desirable. It has been shown possible to display grey-scale images on such a plasma using time modulation—that is, varying the fraction of a frame time during which each element is left 'on'.[31]

Video rates are now available even on large screens enabling fast data handling and image processing. On a 256 × 512 pixel AC PDP, 16 levels of grey are possible at video rates. The highest resolution currently demonstrated is 2048 × 2048 pixels on a 1.5 m diagonal with full optical quality. Development efforts are continuing on multi-colour panels. Most manufacturers except colour AC PDPs to be available by the early 1990s.[32,33]

24.4 Large-area TV displays

The concept of a large-area TV display continues to fascinate and challenge researchers around the world.[34,35] The bulk and weight of a glass envelope that can withstand the tremendous forces due to atmospheric pressure present a practical size limit for the cathode ray tube or kinescope. On the other hand the requirement in large-area consumer television displays for grey-scale, high-resolution, full-colour capability, high brightness and uniformity, and acceptable luminous efficiency has so far prevented the other previously discussed display technologies from reaching this market place. Since luminous efficiency and colour appear to be the two most difficult requirements a number of approaches have been investigated recently which would use available cathodoluminescent phosphors in a large, flat-faced, evacuated display. The tremendous atmospheric pressure (e.g. ±75 000 N on a 125 cm diameter display) requires that an internal support structure be incorporated in the panel to keep it from collapsing. The problem then is to develop a panel with an internal support structure that allows the generation, modulation and acceleration of electrons to high energies in order to excite the standard colour phosphors.

A number of different approaches have been considered recently. In one instance the discharge in a type of plasma panel is used as the source of electrons which are modulated and subsequently accelerated to several kilovolts.[22] Normally the gas pressure and electrode spacings in a plasma panel are chosen to minimise the breakdown voltage in order to simplify the addressing circuitry. At low gas pressures and small electrode spacing the breakdown strength of the plasma rises steeply.[18] Thus it is possible to matrix address a portion of the panel with moderately low voltages to establish a localised plasma, then to extract electrons from the discharge, modulate them and accelerate them towards a close-spaced phosphor anode.[22] Indeed by using a plasma discharge as the source of electrons for several picture elements or even one or more horizontal lines, and by utilising the self-scan concepts, the addressing circuitry for such a panel can become quite simple.

Another approach has been investigated which operates in a moderately good vacuum, e.g. 10^{-4} to 10^{-5} Torr.[36,37] At these low pressures the probability of electron collisions with gas atoms is exceedingly small and a self-sustaining plasma discharge cannot readily be established. In other words there are insufficient gas collisions to produce an electron avalanche and to provide adequate ion flow back to the cathode to regenerate the primary electrons. The necessary electron multiplication is obtained in this panel by incorporating electron multipliers in the form of dynodes into the supporting walls of the panel. The supporting walls, or vanes, run vertically from the cathode plate to the phosphor-coated anode plate. In addition to the dynodes other electrodes are incorporated in the vanes to extract electrons from the self-sustaining discharge, to focus, modulate and accelerate the electrons and to deflect them to the appropriate phosphor stripe of a colour display.

Although there are a large number of electrodes in this panel, many are electrically connected in parallel. Moreover the addressing requires moderate voltages only and the ion bombardment of the phosphors is minimised because of the low pressures, thus promising long life. The stabilisation of the gas pressure in this panel may be difficult, however.

Yet another large-area display panel has been studied which also uses cathodoluminescent phosphors as the source of bright, efficient primary colours.[38] Here also vertical internal support vanes are used to keep the evacuated panel from collapsing under atmospheric pressure. The modulatable area source of electrons is provided by a number of 'electron guides'. These guides use periodic electrostatic focusing to confine and steer a low-voltage electron beam over large distances. This type of periodic confinement of electrons is also used in travelling wave tubes and is the electron equivalent of an optical fibre. The electron guides in this panel have been provided with a number of discrete extraction points where the electron beam can be pushed out of the guide and accelerated towards the phosphor screen. A number of systems' options exist; normally there is one extraction point for each individual, or pair of horizontal display lines.

24.5 Vacuum fluorescent displays

The vacuum fluorescent display (VFD) tube was first commercialised in the late 1960s as the original seven-segment display for electronic calculators. Since that time, VFDs have advanced significantly and are now dominant in applications such as audio equipment, VCRs, instrumentation and automotive.[39,40]

The basic structure of a VFD is shown in *Figure 24.6*. When the cathode filament is heated to around 600°C and a voltage is applied, the filament emits thermoelectrons. When a voltage is applied to the appropriate anode and grid the thermoelectrons are accelerated towards the anode. A fluorescent material on the surface of the anode is excited and luminesces when the thermoelectrons hit the surface. By selection of suitable materials, a number of coloured emissions can be obtained ranging from blue to red. These colours can be further selected to white by appropriate filtration.[41]

Recent developments in VFD include the introduction of graphic dot matrix units employing time division of the digit grids. Units of up to 640 × 200 pixel resolutions have been demonstrated. VFDs produce a clear, high-contrast, bright display with low voltage and power consumption. Since all the critical parts of a VFD are in vacuum, it has inherently good reliability.

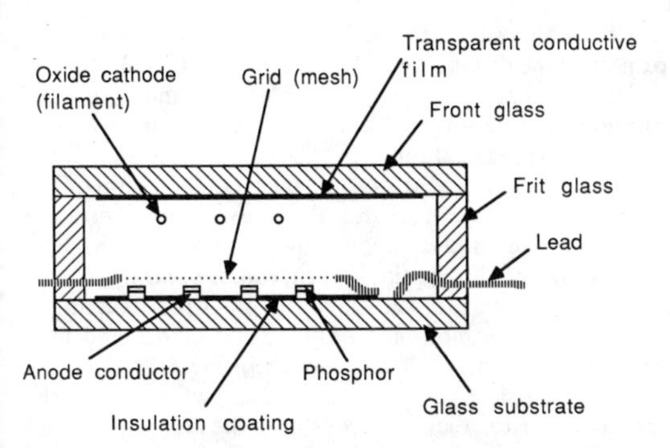

Figure 24.6 Vacuum fluorescent display structure

24.6 Passive electro-optic displays

Passive electro-optic devices modify light that is generated elsewhere. The parameter modulated may be the optical path length or the absorption, reflection or scattering parameters, or a combination of these. If the modulated quantity is wavelength dependent, then the display will be coloured. The display generally consists of a thin sheet of the material sandwiched between two planes of electrodes. The electrode configuration is such that, by applying a voltage to certain of the leads, specified areas of the display are modulated. A typical commercial display is shown in *Figure 24.7*. When only the phase of the transmitted or reflected light is modulated, a combination of polarisers is required to convert phase changes to amplitude changes which make the display visible.

Figure 24.7 Typical low-complexity twisted nematic LCD

The properties of interest of an electro-optic material, in addition to the kind and amount of light modulation, are the driving voltage and power required, whether they can be excited by a.c. or d.c., the speed of response, and the life (expressed in operating life or number of addressing cycles).

To address a simple display like that shown in *Figure 24.7*, it is practical to have a separate connection leading to each segment of one electrode plane with a common connection on the other plane. However, for more complicated displays, it is necessary to reduce the number of leads by using a grid or matrix system.[42] On one plane, the segments are interconnected in horizontal rows, on the other plane, in vertical columns, so that each segment is uniquely addressed by a combination of one horizontal and one vertical lead. For this scheme to work, it is required that the electro-optic effect have a sharp threshold, fast rise time and long decay time, so that the application of two (or three) times the threshold voltage produces a large optical change, even if that voltage is only applied for a small part of the time (low duty cycle).

Presently, the most widely used passive electro-optic materials are liquid crystals. They exhibit a considerable variety of modulation effects. Other electro-optic effects that may become important are electrophoresis and electrochromism. The properties of these materials are compared in *Table 24.1*.

24.7 Liquid crystals

Liquid crystals[43] are useful in display applications because of their very large electro-optic effects.[44] The materials themselves exist in a phase which is different from both the conventional liquid and solid phases, and exhibit properties intermediate between the two. Materials having this special phase are found in the group of organic materials with moderately large rod-like molecules. In recent years, numerous compounds and mixtures have been discovered which are liquid crystalline at room temperature and over a temperature range of as much as 100 K. This accounts for the increased practical interest in display applications.

Table 24.1

Effect	*Liquid crystals*		*Electrophoresis*	*Electrochromism*
	Hydrodynamic	*Field effect*		
Operation in transmission	T	T		
Operation in reflection	R	R	R	R
Viewing angle	narrow	medium	no restrictions	no restrictions
Operating voltage (V)	15–40	5–10	50–100	1–2
Power consumption (mW/cm²) (continuous operation)	0.1	0.002	0.4	
Energy consumption/cycle (mJ/cm²)				2–100
Switching speed (s)	0.05	0.2	1	0.01

The various material phases are distinguished by their macroscopic symmetry properties. The conventional liquid is isotropic and has no unique symmetry elements while even the lowest order crystals have complete translational symmetry. The intermediate liquid crystals retain the translational freedom of the liquid in which the molecules are free to move in varying degrees. They also exhibit long range ordering which adds certain macroscopic symmetry elements, either translational or orientational. There are presently five known liquid crystalline phases—nematic, cholesteric and three smectic phases. It is primarily the first two which are used in practical electro-optic devices.

A nematic liquid crystal is one where all the molecules align themselves approximately parallel to a unique axis while retaining the complete translational freedom. This symmetry axis is defined locally and may vary in direction in different parts of a volume of liquid. Its direction may be described by a vector (of arbitrary sense) which is called the director (*Figure 24.8*(*a*)). In the absence of external forces, the lowest energy state of a nematic is that with a uniform orientation of the director in the entire sample.

A cholesteric liquid crystal is a modification of a nematic one. In the undisturbed state, it may be described as a nematic liquid crystal which has been twisted about an axis (the helical axis) lying perpendicular to the orientation of the director at all times (*Figure 24.8*(*b*)).

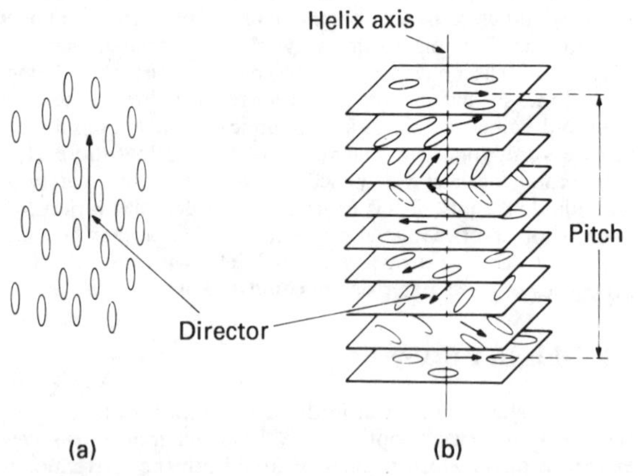

Figure 24.8 Schematic arrangement of molecules in liquid crystals: (a) nematic; (b) cholesteric. The arrow represents the director

The optical properties of these materials are derived from their symmetry. A nematic liquid crystal is optically uniaxial, so that, as in a uniaxial crystal, two different indices of refraction apply for light propagation with the electric vector parallel and perpendicular to the axis. The same is true of the dielectric constant and the conductivity which are tensors of the same order as the index of refraction. The dielectric constant anisotropy may be positive (larger for the fields parallel to the axis than for the perpendicular direction) or negative depending on the polarisation properties of the molecules. The other anisotropies have only been observed to be positive.

The optical properties of the cholesteric fluid are more complicated. It is uniaxial with a screw-type symmetry. The axis is at right angles to the axis in the nematic (or 'untwisted' cholesteric). Light propagating parallel to the helical axis experiences optical rotation, reflection and change of polarisation, depending on the ratio of the optical wavelength to the pitch of the helix.[45] In the extreme case of very small ratio, the polarisation ellipse of the light simply rotates about the axis in the same way that the director does. Light propagating at right angles to the helical axis encounters a sinusoidal variation of index along the axis with the period equal to half the pitch. This is a phase grating which causes diffraction of the light.

The electro-optic properties of a liquid crystal are very different from those of a crystalline solid. This is what causes these materials to be of great interest in display applications. In a solid, the electric field produces small changes in location of ions and in atomic polarisation and correspondingly small changes in the refractive index. In liquids, the electric field can only polarise individual molecules and the effects are even smaller. Liquid crystals respond to an electric field like solids since the molecules remain aligned by long range forces, but now it is possible to rotate the optical axis by a large amount because of the freedom of individual molecules to rearrange themselves while still maintaining the alignment. Consequently, the optical changes are much larger than those in solids and simple liquids. In addition, it is possible to destroy the long range order by changing from a uniform 'single crystal' state to a randomised polycrystalline state which exhibits strong light scattering. This change is reversible and therefore qualifies as an electro-optical effect.

The electro-optic effects can be divided into two different types with respect to the driving force. In one, the electric field acts on the dielectric properties of the molecules and produces reorientation in a uniform manner. The liquid remains static except during the reorientation. In the other, the current is the driving force and the liquid crystal must be conducting. The current is carried by ions, and the moving ions produce movement of the liquid above the electrohydrodynamic threshold.[46] The moving liquid, in turn, causes shear forces and displacements which are accompanied by optical changes. Electrohydrodynamic motion is only created by d.c. and low-frequency a.c. driving voltages. At frequencies above the dielectric relaxation frequency ($f_d = \sigma/2\pi\varepsilon$, where σ is the conductivity and ε the dielectric constant), the current becomes primarily capacitive and only field effects can be observed.

24.7.1 Practical considerations

Liquid crystal electro-optic devices are generally constructed as sandwich cells with a thin layer of the liquid placed between two conducting glass plates. Because of the large electro-optic effect, the liquid layer may be very thin (3–50 μm, typically). The cells are used in two different configurations, one where the light is transmitted through the cell, the other where the incident light is reflected back at the viewer. In the latter case, a specularly reflecting aluminium film is placed on the rear glass substrate. Both electrodes are formed by a transparent conducting layer, such as SnO_2 or In_2O_3.

For practical use, the cells must be hermetically sealed, so that the liquid crystal is confined and cannot interact with the atmosphere. The major problem is water vapour which can cause decomposition and chemical reactions between the various components of the mixture. Some kinds of liquid crystal material are more susceptible to this problem than others. Chemical reactions may also take place during operation of the devices because of the electrolytic nature of the cell. This is particularly true when direct current is applied to the cell, which causes transport of ions and electrochemical reactions at the electrodes. Consequently, the lifetime of the device under d.c. conditions is generally much shorter than under a.c. conditions, and most applications require the latter.

To produce any given electro-optic effect, it is necessary that the liquid crystals have a certain crystalline form or texture. The form is determined by the alignment of the molecules at the surfaces which may be perpendicular to the surface, parallel to it, or at some intermediate angle. In the parallel situation, the direction of the molecules in the plane may be required to be uniformly in a specified direction, or randomly oriented. The alignment is determined by appropriate treatments of the surface[47,48] in combination with chemical aligning agents.[49]

One of the important variables in the operation of the device is its speed of response. The electro-optic effects involve a reorientation of the molecules and the response times are therefore related to the viscosity of the material, decreasing, for example, with increasing temperature. Response will be faster at higher driving fields. The equilibrium condition with no external force applied is determined by the alignment at the surface, and the speed of return to that condition increases rapidly as the cell is made thinner.

24.7.2 Hydrodynamic effects

24.7.2.1 Domains and dynamic scattering

Conductivity in liquids is generally by ionic motion. At low values of conduction, the ions move through the stationary liquid but, at higher values, instabilities are produced and the liquid is set into motion.[46] It is in this region that conduction-induced electro-optic effects occur. The effects are much more important in liquid crystals than in isotropic liquids because the optical anisotropy makes the liquid motion directly visible.

The best-understood instabilities are the Williams domains.[50] They consist of rotation vortices of liquid and are observed mainly in nematic materials of negative dielectric anisotropy. A cross-section of a cell, at right angles to the vortex axes, is shown in *Figure 24.10*. The spacing between the axes is a fixed multiple of the electrode separation. The rotating fluid causes the director to tilt, as indicated. For light polarised in the plane of the figure, the optical pathlength varies depending on the location in the cell. This causes the axes and the regions separating any two vortices to become visible as an array of parallel lines. The uniform line spacing (proportional to the cell thickness) causes the cell to act as a diffraction grating.

The threshold for onset of the domain instability is determined by the interrelationship of a number of physical effects.[51] The conductivity anisotropy produces charge separation in regions where the ions are not moving parallel to the director due to the rotating liquid (*Figure 24.9*). The resulting transverse electric

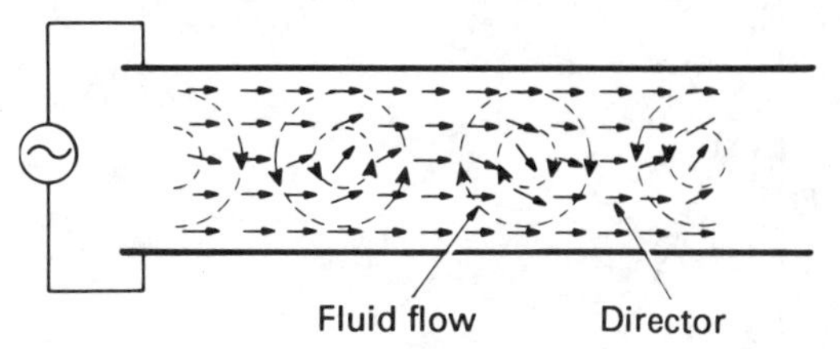

Figure 24.9 Cross-section of a nematic cell exhibiting Williams domains

fields act on the charge to produce torques in the liquid. These torques reinforce the rotation. This explains why ionic conductivity is required for domain formation.

Driving voltages can be d.c. or low frequency a.c. but, as the frequency is increased, the ionic current decreases relative to the displacement current and the instability threshold rises. Finally, near the dielectric relaxation frequency, no domains can be formed.[52] There exist other instabilities, but they are not of a hydrodynamic nature.[52]

As the driving voltage is increased beyond the threshold for domain formation, the vortices become smaller and less regular, leading finally to a turbulent state. Because the direction of the molecular axes changes rapidly from place to place, the liquid becomes highly scattering and opaque. The scattering pattern is varying continuously when observed under the microscope. This state is called *dynamic scattering*.[53]

The intensity of scattered light as a function of the applied voltage is shown in *Figure 24.10*. It can be seen that there is a

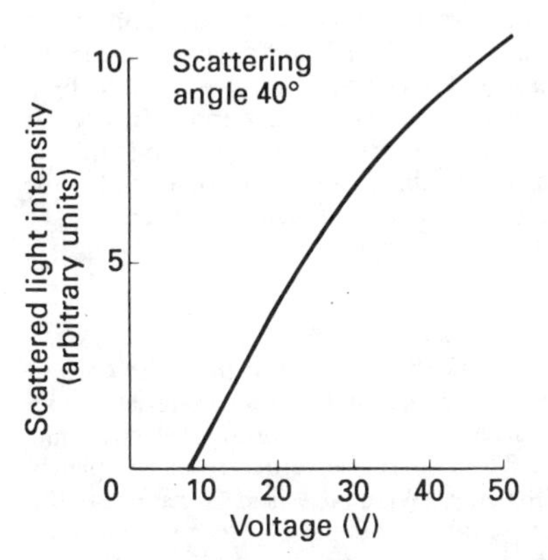

Figure 24.10 Voltage dependence of the scattered light in a dynamic scattering cell

threshold voltage, below which there is little light scattering. This voltage is slightly above the domain threshold. The amount of scattering below the threshold (off-state) depends on how well the material is aligned. Good alignment, either perpendicular or parallel to the electrodes, reduces the scattered light to a negligible quantity. The scattered light intensity in the on-state depends strongly on angle, as is shown in *Figure 24.11*. A determination of the contrast ratio of a display cell therefore depends on the conditions under which it is measured. The steady-state

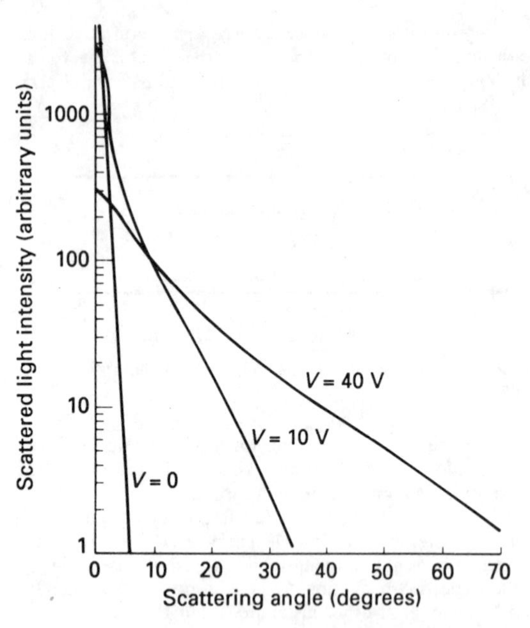

Figure 24.11 Angular dependence of the scattered light in a dynamic scattering cell

scattering properties (*Figures 24.10* and *24.11*) do not depend on the thickness of the liquid crystal film over the range of 5–10 μm. The response time to application and removal of the voltages also has a strong angular dependence. It increases proportional to the square of the film thickness if the other parameters are held constant.

Figure 24.11 shows that the scattering of light takes place predominantly in the forward direction. This produces a very good display when transmitted light is used, particularly when the geometry is arranged so that the unscattered light is not viewed by the observer. The situation is not as favourable in the reflective case because the liquid crystal must be backed by a highly reflecting mirror to produce good scattering efficiency. It becomes more difficult to prevent reflected, but unscattered, ambient light from reaching the viewer. For good results, the display must be recessed behind a set of baffles which restrict the viewing angle.

24.7.2.2 Storage effect

A related hydrodynamic effect[54] is observed in cholesterics of negative dielectric anisotropy. In loosely wound materials with a long pitch, a certain current flow will also produce the dynamic scattering instability. The cholesteric structure is completely disrupted by the turbulence. When the field is removed, the movement stops and the cholesteric forms a 'polycrystalline' state called focal-conic texture. The individual crystallites or domains are oriented randomly with respect to each other. This state is highly scattering. To make a display, a clear state is needed and this is obtained by applying a high-frequency electric field. The frequency is well above the dielectric relaxation frequency so that there is no ionic current flowing, which would re-establish the instability. Instead, the dielectric field causes the molecules to align parallel to the electrodes which, in turn, produces a uniform alignment of the helical axis at right angles to the electrodes. This cholesteric state persists after the removal of the field (storage). Thus, the sample may be switched back and forth between the scattering and clear state by applying two different kinds of driving pulses. Optically, the two states are quite similar to the on and off states of dynamic scattering but the driving voltages are higher and the transition times longer, particularly if a long-time storage is desired.

24.7.3 Field effects

24.7.3.1 Field-induced birefringence

There are a number of electric field-induced electro-optic effects in liquid crystals. The simplest one, field-induced birefringence or distortion of an aligned phase,[55,56] is considered first, and in some detail, as the basic principles apply to all field effects. The material used is a nematic liquid with negative anisotropy of dielectric constant and negligible conductivity. The properties of the cell surfaces are controlled so that the alignment of the molecules is uniform and at right angles to the surface (*Figure 24.12(a)*). When an electric field is applied in the perpendicular direction, it produces a torque on the individual molecules trying to turn them in a direction parallel to the electrodes because the lowest energy state of the unconstrained system is that in which the larger dielectric constant is oriented parallel to the field.

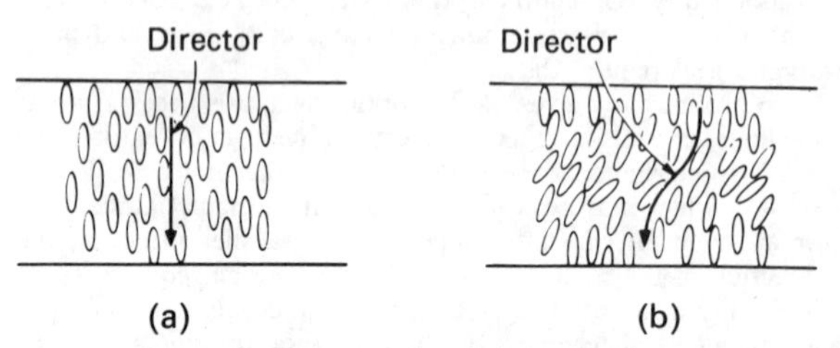

Figure 24.12 Demonstration of the field-induced birefringence effect: (a) alignment in the absence of the field; (b) when a field above the threshold field is applied between the electrodes

However, the surface alignment is permanently fixed in the perpendicular direction and, through the elastic forces of long range order, tries to keep the bulk of the material in that same orientation. This balance between electric and elastic energy is such that below a certain threshold field E_c, the alignment remains uniformly perpendicular while, above that field, it gradually distorts towards the parallel direction as shown in *Figure 24.12(b)*. In the figure, the rotation of the molecules from the perpendicular is shown in one particular direction and the long range forces tend to align the entire sample in the same direction. There is, however, no reason to prefer one direction in the plane of the cell over any other and the direction may vary gradually throughout the sample.

Both states of the nematic fluid below and above the critical field are optically clear. Light propagating through the cell is subject to different indices of refraction and, therefore, phase shifts in the two cases. These phase differences are converted to amplitude differences by positioning the cell between crossed polarisers. Then, light transmitted perpendicularly through the sample will be extinguished as long as the situation of *Figure 24.12(a)* applies (equivalent to light propagating along the axis of a uniaxial crystal). Above E_c, more and more light will be transmitted (*Figure 24.13*)[57] until a saturation is reached when all but the surface molecules are aligned to the electrodes.

The induced birefringence effect can also be created in the opposite sense by using a material of positive dielectric anisotropy and making the surface alignment parallel to the electrodes. Then, the light transmission will be maximum at low fields and reduce to zero at high fields. The largest contrast is obtained if the polariser is located at 45° to the surface aligning direction.

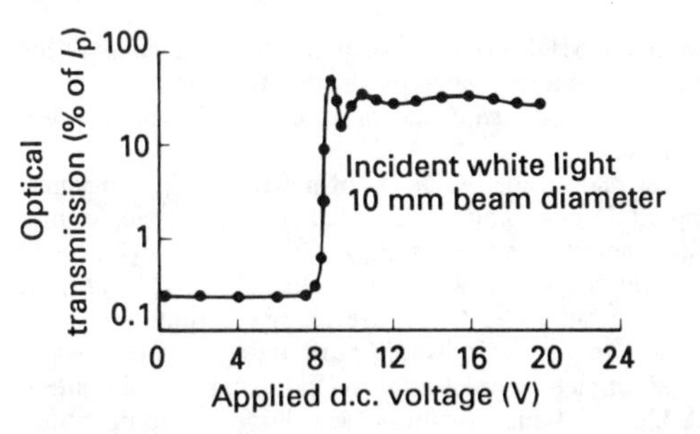

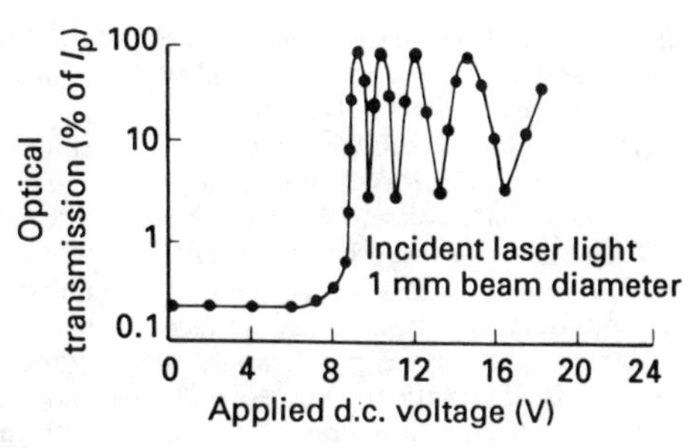

Figure 24.13 Voltage dependence of the transmitted light for a field-induced birefringence cell at normal incidence for two different kinds of light beams (Courtesy of the American Institute of Physics)

The brightness and colour of this device depend on the difference in phase between the light polarised along the crystal axis and that polarised at right angles to it. This means that the colour and intensity of the transmitted light observed by the viewer will not be uniform if the cell thickness is not uniform. A further disadvantage is that the transmitted light is strongly dependent on the angle of viewing.

24.7.3.2 Twisted nematic

A field-effect arrangement which overcomes this disadvantage is the so-called twisted nematic cell.[58] It is similar to the second case of induced birefringence with the positive anisotropy material. The alignment is parallel to the surfaces, but the directions of the alignment at the two surfaces are positioned at right angles to each other. The elastic forces cause the director in the interior of the crystal to remain in the plane of the sample but to twist gradually by 90° from one electrode to the other. This produces a structure which is identical to that of a cholesteric material with rather long pitch (four times the thickness of the sample). If plane-polarised visible light, which has a wavelength much shorter than this pitch, propagates through the cell, its plane of polarisation is rotated at the same angular rate at which the director is rotated.[45] The total angle of rotation is 90° so that, with analyser and polariser parallel to each other, no light will be transmitted. When a field above E_c is applied to the sample, the orientation will be changed to a more or less perpendicular one and the light polarisation direction no longer rotates. The light is transmitted by the analyser. If the analyser is set at right angles to the polariser, light will be transmitted by the undisturbed cell and absorbed when the field is applied.

Generally, a reflective aluminium plate is placed behind the display enabling it to be used with incident ambient lighting. The twisted nematic LCD has major advantages such as low power consumption, low voltage drive (less than 5 V) and excellent lifetime specification ($>50\,000$ h). For these reasons the twisted nematic LCD has become established as the dominant LCD effect.

24.7.3.3 Guest–host effect

Another kind of field effect in nematic liquids[59] modifies the absorption of light, rather than its phase. A pleochroic dye (in which the absorption of light varies strongly depending on whether the light is polarised parallel to the molecular axis or perpendicular to it) is added to the liquid crystal. Because such dye molecules tend to be rod-like, they orient themselves parallel to the similar-shaped liquid crystal molecules. The absorption is modulated by changing the orientation of the director and, therefore, the dye molecules. For maximum absorption by the dye, the initial alignment of a liquid crystal should be in the plane of the sample, and the incident light should be polarised parallel to the director. If the liquid crystal has a positive dielectric anisotropy, then an electric field rotates the director to a perpendicular direction where the absorption is a minimum. Note that no analyser is required. To obtain a rapid response for this effect, it is necessary that the dye molecules resemble the liquid crystal molecules as closely as possible. Also, the absorption of the dye should be as high as possible to reduce the quantity of dye required.

24.7.3.4 Cholesteric field effects

A number of field effects are possible with cholesteric materials. Some produce reorientation of the existing structure as was the case with the nematics. In addition, the field may cause two kinds of changes in the cholesteric structure. These are a change in pitch of the helix and a complete disruption of the helix with a transition to the nematic phase.

A change of pitch causes optical effects if the pitch is of the order of the wavelength of light. This is because of the wavelength-sensitive optical properties.[45] In particular, there is a region of total reflection for one sense of circularly polarised light propagating along the helical axis if the wavelength is close to the pitch of the helix. In such a case, the application of a field causes rapid and vivid changes of colour. However, the pitch also depends strongly on other variables, such as pressure and temperature, so that this is not a very practical electro-optic effect.

The field-induced phase change[60] is more important. In this case, use is made of the fact that, without any preferential surface alignment, the cholesteric material forms the focal-conic texture which is highly scattering and appears opaque. If the individual molecules have a positive dielectric anisotropy, then an applied field will tend to orient them parallel to the field. For a loosely wound cholesteric fluid, i.e. one with a large value of pitch, a moderate field will align the molecule into the induced nematic phase. This texture is optically clear and forms a large contrast to the zero-field opaque case. Upon removal of the field, the material returns to the scattering cholesteric phase. Cholesteric materials of large pitch are produced by mixing a short pitch cholesteric with a nematic material which, in this case, must have positive dielectric anisotropy.[61]

24.7.3.5 Super-twisted nematic LCD effect

The standard twisted nematic LCD, in spite of its simplicity and suitability for high-volume manufacturing, is limited in performance for high pixel density applications. As the number of pixels is increased, a multiplexing drive scheme has to be adopted which involves time scanning of the rows. However, multiplexing leads to a reduction in the contrast and viewing angle of twisted nematic LCDs due to the low electro-optical sharpness.

A modified twisted nematic LCD called super-twisted nematic (STN) has been developed which overcomes the quality problems of the standard twisted nematic technology.[62–64]

The structure of the STN display is similar to that of conventional twisted nematic but with an increased internal twist angle of the nematic liquid crystal material from front to rear substrate (*Figure 24.14*). The twist angle may be up to 270° (compared to 90° for twisted nematic). This is achieved by doping with a cholesteric liquid crystal. By modified surface treatment of the glass substrates, the surface pretilt angle is increased from 2° to 3° for the twisted nematic to in excess of 5° for the STN. In addition, the manufacturing tolerance for cell spacing is improved to better than 0.15 μm for the display area.

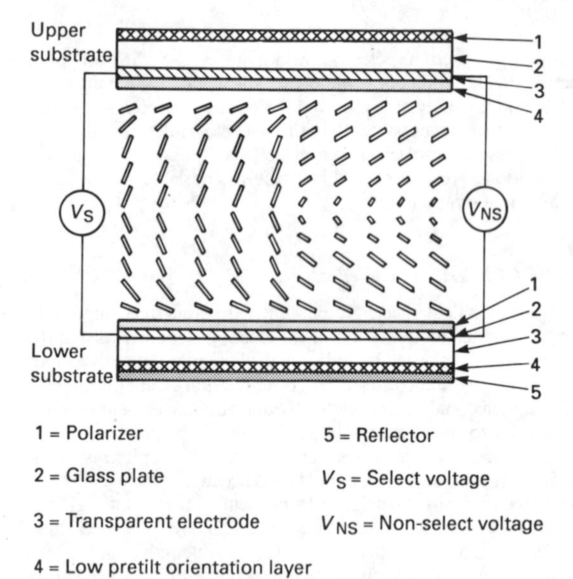

Figure 24.14 Cross-section of STN LCD

These modifications result in a display with increased electro-optical sharpness (transmission *versus* voltage characteristic). This allows multiplexed waveforms to be differentiated, resulting in higher optical performance of graphic dot matrix panels. Typically a 200% improvement in contrast and viewing angle compared to convential twisted nematic LCDs is achieved.

24.7.4 Applications

Twisted nematic LCDs are widely used in consumer, industrial and telecommunication products. The large market share is due to the low power consumption, slim profile, low weight, high reliability and CMOS compatible drive requirements. Other areas include automotive dashboards, due to wide temperature range characteristics, ease of customisation and suitability for colour filtration.

The largest market for LCDs is dot matrix module products integrating LCD glass and driver circuitry onto a single unit. Small units generally use twisted nematic technology, with larger units based on the super-twisted nematic effect. STN graphic panels of up to 640 × 400 pixels are in production. Reflective graphic panels are used in portable computers, word processors and other equipment demanding low power. By backlighting with thick film EL panels or fluorescent lamps, graphic panels are being increasingly used as computer terminals with performance levels close to those of monochrome CRT.

A new development in the LCD field is 'active drive' technology, where thin film transistors are incorporated into the cell structure. By depositing red, green and blue filters onto the display front face, a multi-colour display is possible. The thin film transistors typically made of amorphous silicon behave as on/off switches in series with every pixel. The result is an LCD with high optical performance, typically better than 20:1 contrast. Active drive colour LCDs are being used in small pocket TVs, but development of larger sized panels is limited by yield problems.

24.8 Electrophoretic displays

24.8.1 Operation

Electrophoresis—the migration of charged colloidal particles in an electric field—has been used extensively for the separation and analysis of components in biological substances.[65] It is also widely used for electrostatic coating in liquids or the development process in electrophotography.

An electrophoretic display device has been developed[66] which utilises the migration of charged pigment particles suspended in a coloured liquid to produce a reflective display. An electrophoretic suspension (25–100 μm thick), composed of pigment particles and the suspending liquid, is sandwiched between a pair of electrodes, at least one of which is transparent. This basic structure is shown in *Figure 24.15*. Under the influence of a d.c. field, the pigment particles are moved towards, and deposited on, one of the electrodes; this electrode is coloured by the pigment particles while unelectroded areas or electrodes of opposite polarity retain the colour of the suspending liquid. A high-contrast reflective display results.

The pigment particles must have the same density as the suspending liquid to avoid precipitation or separation. Organic pigment particles, e.g. hansa yellow, can be provided with the same density as that of the liquid, which consists of a mixture of solvents with different densities. Inorganic pigment particles, e.g. TiO_2, are denser than normal liquids, but their average density can be reduced by encapsulation with an inert plastic, e.g. polyethylene.

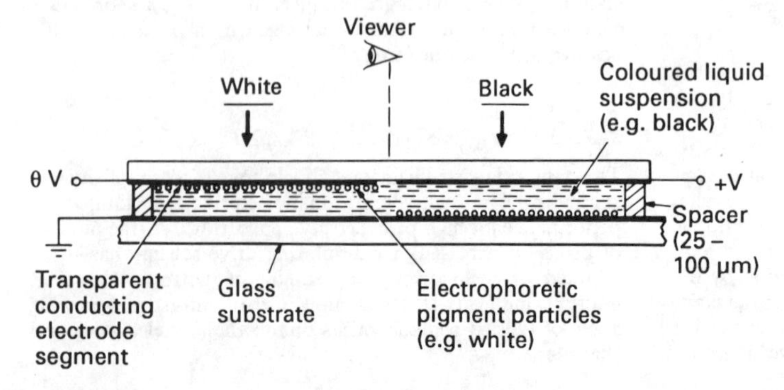

Figure 24.15 Operation of an electrophoretic display

24.8.2 Electrical characteristics and device performance

Electrophoretic displays, like liquid crystal displays, are passive and therefore attractive for viewing in high-brightness ambients. There are essentially no viewing angle restrictions, a distinct advantage over liquid crystal displays. Electrophoretic displays have memory, that is an element stays 'on' even if the power is removed, until the electric field excitation is reversed. Power dissipation (typically 0.4 mW/cm^2) is considerably higher than for liquid crystals, but low enough for moderate-sized numeric displays; the storage effect can be used to lower the average power consumption whenever the display is updated slowly.

Contrast ratios in excess of 50:1 are possible. A variety of colours is achievable by changing the pigment or suspending liquid. The addressing voltages (50–100 V), however, are a distinct disadvantage since high-voltage addressing circuitry is costly. Also, the high d.c. voltage can produce undesirable electrochemical effects which shorten the life of the display; operational life of 3000 h has been reported[66] and can be expected to improve.

24.9 Electrochromic displays

Electrochromism is the coloration or change in colour which occurs in certain materials upon the application of an electric field. The basis for this coloration is the formation of colour centres or an electrically induced oxidation or reduction whereby a new substance is formed which absorbs visible light. The ability to colour a material reversibly by means of a locally applied electric field can be used to produce a simple reflective display, suitable for viewing in high-brightness ambients.

In general, electrochromic displays require relatively high currents (high power) or long times to achieve useful display contrasts; thus they are best suited for small alphanumeric display applications, especially those requiring infrequent updating, e.g. digital clocks, calculators.

In inorganic materials, e.g. WO_3, colour centres can be formed[67] when a d.c. electric field of $\sim 10^4$ V/cm ($\sim$3 V across a 1 μm thick film) is applied at room temperature. Useful contrasts require $\sim 10^{18}$ colour centres/cm^3, corresponding to the passage of $\sim$100 mA/cm^2 for a second or more. This is comparable to the charge required to deposit an opaque layer of metal by electroplating: a 50 nm thick layer of copper requires $\sim$70 mC/cm^2 (70 mA for 1 s).

Erasibility in electrochromic displays means that irreversible electrochemical reactions must be avoided. In WO_3, electrons are trapped at colour centres reducing W^{6+} to W^{5+} while oxygen ions are given up at the anode; these ions must again be able to enter the material when the polarity is reversed. Although many electrode materials have been explored which can accept and give up oxygen ions readily, the life of these displays remains problematic.

An organic electrochromic display[68] has also been developed which produces a strongly coloured organic dye at the cathode. Since the coloration due to one dye molecule is greater than that of one metal atom or one colour centre, the electrical charge required to achieve good contrast (up to 20:1) in the organic system is considerably smaller (2 mC/cm^2). Life problems are claimed to be absent provided oxygen is excluded from the cell; 10^5 write–erase cycles have been achieved.[68] As with the inorganic systems, addressing voltages (1–2 V) are compatible with integrated circuits making them of potential use for small alphanumeric applications.

References

1 NIINA, T., KURODA, S., YONEI, H. and TAKESADA, H., 'A high brightness GaP green LED flat panel device for character and TV display', *Conf. Record of 1978 Biennial Display Research Conf.*, p. 18, October (1978)
2 NIINA, T., KURODA, S., YAMAGUCHI, T., YONEI, H., TOMIDA, Y. and YAGI, K., 'A multicolor GaP LED flat panel display device', *1981 DID Int. Symp. Digest Tech. Papers*, **12**, 140, New York, April (1981)
3 BERGH, A. A. and DEAN, P. J., 'Light-emitting diodes', *Proc. IEEE*, **60**, 156 (1972)
4 NUESE, C. J., KRESSEL, H. and LADANY, I., 'Light-emitting diodes and semiconductor materials for displays', *J. Vac. Sci. Tech.*, Sept/Oct (1973)
5 HOFFMAN, L. *et al.*, 'Silicon carbide blue light emitting diodes with improved external quantum efficiency', *J. Appl. Phys.*, **53**, 6962 (1982)
6 DESTRIAU, G. J., *J. Chem. Phys.*, **33**, 620 (1936)
7 KAZAN, B. and KNOLL, M., *Electronic Image Storage*, Academic Press, New York, pp. 418–37 (1968)
8 LARACH, S. and SHRADER, R. E., 'Electroluminescence of polycrystallites', *RCA Rev.*, **20**, 532 (1959)
9 ARAI, H., YOSHIZAWA, T., AWAZU, K., KURAHASHI, K. and IBUKI, S., 'EL panel display', *IEEE Conf. Record of 1970 IEEE Conf. on Display Devices*, p. 52, New York, 2–3 December (1970)
10 ARAI, H., 'EL panel TV', *JAEU*, **3**, 39 (1971)
11 YOSHIYAMA, M., 'Ligting the way to flat-screen TV', *Electronics*, **42**, No. 6, 114 (1969)
12 YOSHIYAMA, M., OSHIMA, N. and YAMAMOTO, R., 'A television display device utilizing DC–EL panel', *National Technical Report*, **17**, No. 6, 670 (1971) (in Japanese)
13 INOGUCHI, T., TAKEDA, M., KAKIHARA, Y., NAKATA, Y. and YOSHIDA, M., 'Stable high-brightness thin-film electroluminescent panels', *1974 SID Int. Symp. Digest Tech. Papers*, **5**, 84, San Diego, May (1974)
14 MITO, S., SUZUKI, C., KANATANI, Y. and ISE, M., 'TV imaging system using electroluminescent panels', *1974 SID Int. Symp. Digest Tech. Papers*, **5**, 86, San Diego, May (1974)
15 SUZUKI, C., KANATANI, Y., ISE, M., MIZUKAMI, E., INAZAKI, K. and MITO, S., 'Optical writing on thin-film EL panel with inherent memory', *1976 SID Int. Symp. Digest Tech. Papers*, **7**, 52, Beverley Hills, CA, May (1976)
16 FUJITA, Y. *et al.*, *Digest of Japan Display*, p. 76 (1983)
17 UEDE, H. *et al.*, *1981 SID Int. Symp.*, Los Angeles, p. 28 (1981)
18 VON HIPPEL, A. R., *Dielectrics and Waves*, p. 234, Wiley, New York (1954)
19 FORMAN, J., 'Phosphor color in gas discharge panel displays', *Proc. SID*, **13**, 14 (1972)
20 KANEKO, R., KAMEGAYA, T., YOKOZAWA, M., MATSUZAKI, H. and SUZUKI, S., 'Color TV display using 10″ planar positive-column discharge panel', *1978 SID Int. Symp. Digest Tech. Papers*, **9**, 46 (1978)
21 KRUPKA, D. C., CHEN, Y. S. and FUKUI, H., 'On the use of phosphors excited by low-energy electrons in a gas discharge flat-panel display', *IEEE Proc., Special Issue on New Materials and Devices for Displays*, 1025, July (1973)
22 CHODIL, G. J., DEJULE, M. and GLASER, D., 'Cathodo-luminescent display with hollow cathodes', *US Patent No.* 3 992 633, issued 16 November (1976)
23 DE BOER, Th. J., 'An experiment 4000-picture-element gas discharge TV display panel', *Proc. 9th National Symp. on Information Display*, p. 193, Los Angeles, May (1968)
24 AMANO, Y., 'A flat-panel color TV display system', *1974 Conf. on Display Devices and Systems, Conf. Record*, 99; *IEEE Trans. Electron. Devices*, **ED-22**, 1, January (1975)
25 HOLZ, G. E., 'The primed gas discharge cell—a cost and capability improvement for gas discharge matrix displays', *1970 SID IDEA Symp. Digest of Papers*, p. 30, May (1970)
26 CHODIL, G. J., DE JULE, M. C. and MARKIN, J., 'Good quality TV pictures using a gas discharge panel', *IEEE Conf. Record of 1972 Conf. on Display Devices*, p. 77, October (1972)
27 CHIN, Y. S. and FUKUI, H., 'A field-interlaced real-time gas-discharge flat-panel display with gray-scale', *IEEE Conf. Record of 1972 Conf. on Display Devices*, p. 70, October (1972)

28 AMANO, Y. *et al.*, 'A high resolution DC plasma display panel', *1982 SID Int. Symp. Digest Tech. Papers* (1982)
29 BITZER, D. L. and SLOTTOW, H. G., 'The plasma display panel—a digitally addressable display with inherent memory', *Proc. Fall Joint Computer Conf.*, San Francisco, November (1966)
30 BITZER, D. L. and SLOTTOW, H. G., 'Principles and applications of the plasma display panel', *Proc. 1968 Microelectronic Symp.*, *IEEE*, St. Louis (1968)
31 KURAHASHI, K., TOTTORI, H., ISOGAI, F. and TSURUTA, N., 'Plasma display with gray scale', *1973 SID Int. Symp. Digest Tech. Papers*, **4**, 72, New York, 15–17 May (1973)
32 CURRAN, P. *et al.*, 'High voltage ICs supply the driver for ac panels', *Electronics*, 130–2, April (1982)
33 SOPER, T. J. *et al.*, 'High resolution meter size display technology', *1982 SID Symp. Digest*, pp. 162–3 (1982)
34 MURAKAMI, H. *et al.*, 'An experimental TV display using a gas discharge panel with internal memory', *Proc. SID*, **12**, No. 4, 327–32 (1986)
35 YOKOZAWA, M. *et al.*, 'Colour TV display with AC PDP', *Proc. 3rd Int. Display Research Conference*, pp. 514–17, October (1983)
36 CATANESE, C. and ENDRIZ, J., 'The physical mechanisms of feedback multiplier electron sources', *1978 SID Int. Symp. Digest Tech. Papers*, **9**, 122, San Francisco, April (1978)
37 ENDRIZ, J., KENEMAN, S., CATANESE, C. and JOHNSTON, L., 'Flat TV display using feedback multipliers', *1978 SID Int. Symp. Digest Tech. Papers*, **9**, 124, San Francisco, April (1978)
38 CREDELLE, T., ANDERSON, C., MARLOWE, F., GANGE, R., FIELDS, J., FISHER, J., VAN RAALTE, J. and BLOOM, S., 'Cathodoluminescent flat panel TV using electron beam guides', *1980 SID Int. Symp. Digest Tech. Papers*, **11**, 26, San Diego, April (1980)
39 MORIMOTO, K. *et al.*, *Proceedings, Japan Display '83* (1983)
40 LE VAN, J. D., US Patent 2 226 567, 'Fluorescent Coating'
41 IWADE *et al.*, 'VF display for TV video image', *SID Int. Symp. Digest Tech. Papers*, **13**, 136 (1981)
42 HARENG, M., ASOULINE, G. and LEIBA, E., *Proc. IEEE*, **60**, 913 (1972)
43 GREY, G. W., *Molecular Structure and the Properties of Liquid Crystals*, Academic Press, New York (1962); a general reference of all early work
44 SUSSMAN, A., *IEEE Trans. Parts, Hybrids, and Packaging*, **PHP-8**, 28 (1972)
45 DEVRIES, H., *Acta Cryst.*, **4**, 219 (1951)
46 FELICI, N., *Rev. Gen. Elec.*, **78**, 717 (1969)
47 CHATELAIN, P., *Bull. Soc. Fr. Miner. Cryst.*, **66**, 105 (1943)
48 JANNING, J. L., *Appl. Phys. Lett.*, **21**, 173 (1972)
49 DREYER, F. J., in *Liquid Crystals and Ordered Fluids*, 2nd edn, JOHNSON, J. F. and PORTER, R. S. eds., Plenum Press, New York, 311 (1970)
50 WILLIAMS, R., *J. Chem. Phys.*, **39**, 384 (1963)
51 HELFRICH, W., *J. Chem. Phys.*, **51**, 4092 (1969)
52 ORSAY LIQUID CRYSTAL GROUP, *Mol. Cryst. Liq. Cryst.*, **12**, 251 (1971)
53 HEILMEIER, G. H., ZANONI, L. A. and BARTON, L. A., *Appl. Phys. Lett.*, **13**, 46 (1968)
54 HEILMEIER, G. H. and GOLDMACHER, J. E., *Proc. IEEE*, **57**, 34 (1969)
55 FREEDERICKSZ, V. and ZWETKOFF, Y., *Acta Physicochemica URSS*, **3**, 9 (1935)
56 SCHIEKEL, M. F. and FAHRENSCHON, K., *Appl. Phys. Lett.*, **19**, 390 (1971)
57 SOREF, R. A. and RAFUSE, M. J., *J. Appl. Phys.*, **43**, 2029 (1972)
58 SCHADT, M. and HELFRICH, W., *Appl. Phys. Lett.*, **18**, 127 (1971)
59 HEILMEIER, G. H. and ZANONI, L. A., *Appl. Phys. Lett.*, **13**, 91 (1968)
60 WYSOCKI, J. J., ADAMS, J. and HAAS, W., *Phys. Rev. Lett.*, **20**, 1024 (1968)
61 HEILMEIER, G. H. and GOLDMACHER, J. E., *J. Chem. Phys.*, **51**, 1258 (1969)
62 SCHEFFER, J. and NEHRING, J., *Appl. Phys. Lett.*, **45**, No. 10, 1021 (1984)
63 SCHADT, M. and LEENHOUTS, F., *Appl. Phys. Lett.*, **50**, No. 5, 236 (1987)
64 KAWASAKI, K. *et al.*, *1987 SID Int. Symp. Digest Tech. Papers*, p. 391 (1987)
65 MILAN BIER, ed., *Electrophoresis—Theory, Methods and Applications*, Academic Press, London (1959)
66 OTA, I., OHNISHI, J. and YOSHIYAMA, M., 'Electrophoretic display device', *IEEE Conf. Record of 1972 Conf. on Display Devices*, p. 46, New York, October 11–12 (1972)
67 DEB, S. K., 'A novel electrophotographic system', *Appl. Opt. Suppl.*, **3**, 192 (1969)
68 SCHOOT, C. J., PONJÉE, T. J., VAN DAM, H. T., VAN DOORN, R. A. and BOLWYN, P. T., *Appl. Phys. Lett.*, **23**, 64 (1973)

Further reading

FENGER, C. *et al.*, 'High resolution LCD panels change demands on driver electronics', *EDN*, 14 April (1988)
LIDDLE, A., 'New techniques boost resolution of vacuum fluorescent panels', *New Electron.*, May (1988)
LIEBERMAN, D., 'LCDs make the leap to display of choice for low-cost portables', *Computer Design*, July (1988)
LOWRY, J. B., 'Pulsed SCOphony laser projection system', *Opt. Laser Tech.*, **20**, No. 5, October (1988)
McMANUS, C., 'Take care when choosing controllers for flat panel displays', *EDN*, 28 April (1988)
NAGATA, M., 'Paper white LCDs use NTN technology', *Electron. Product Design*, July (1988)
WENZEL, E. H., *et al.*, 'Development of a three dimensional auditory display system', *ACM SIGCHI Bull.*, **20**, No. 2, October (1988)
WIEGAND, J., 'Innovations in monolithic display drivers improve flat panel cost/performance ratios', *EDN*, February (1988)

25 Integrated Circuit Fabrication and Packaging

F F Mazda DFH, MPhil, CEng, FIEE, DMS, MBIM
STC Telecommunications Ltd

Contents

25.1 Manufacturing processes

25.1.1 Slicon wafer preparation

The starting material for integrated circuits is a slice of single crystal silicon: the larger the slice the greater the number of identical circuits which can be made at one time. However, large slices require greater process control to ensure uniformity over the whole area. The most usual method of obtaining these slices is to pull from a silicon melt in a Czochralski puller. Impurities of p or n type can be added to the melt to give the final silicon ingots the required resistivity (typically up to about 50 Ω cm). The ingots are then cut into slices, about a millimetre thick, using a diamond impregnated saw. The saw cuts can be made along the ⟨1 1 1⟩ or ⟨1 0 0⟩ planes of the crystal. These cuts usually damage the crystal lattice near the silicon surface, resulting in poor resistivity and minority carrier lifetime. The damaged area, which is about 20 μm deep, is removed by etching in a mixture of hydrofluoric and nitric acids, and the surface is then polished with a fine diamond power to give a strain-free, highly flat region.

25.1.2 Oxide growth

Silicon oxide (SiO_2, also called silica) is grown and removed from the surface of the silicon slice many times during the manufacture of an integrated circuit. It is such a fundamental substance that one of the reasons for using silicon in integrated circuits is its ability to grow a stable oxide. The oxide layer is used for diffusion masking, for sealing and passivating the silicon surface, and for insulating the metal interconnections from the silicon. Although the oxide layer may be deposited onto the slice, as for the epitaxy layer described in *Section 25.1.4*, it is more usual to grow it using dry or wet oxygen or steam. *Figure 25.1* shows a typical arrangement of the apparatus used for oxidation (and diffusion). The silicon slices are stacked upright in a quartz boat

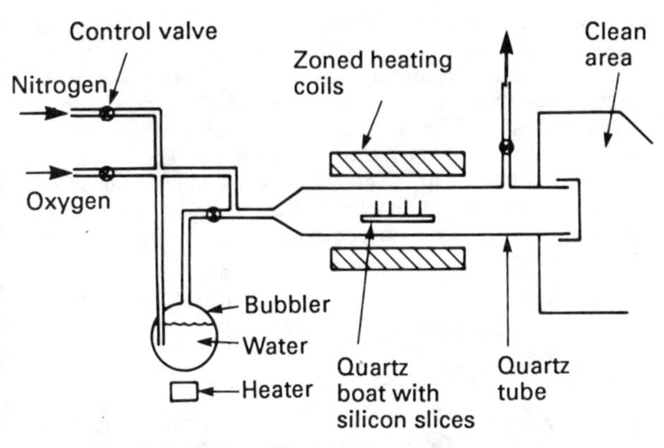

Figure 25.1 Oxidation–diffusion open tube arrangement

and inserted into a quartz tube. The tube is heated to between 1000°C and 1200°C by zoned heaters so that the boat is located in an area having a uniform temperature along a length of the tube. Nitrogen (inert atmosphere), dry oxygen, wet oxygen (oxygen bubbled into water at 95°C) or steam can be passed over the slices to grow the oxide layer. A thickness of about 1 μm takes approximately 4 h to grow and consumes between 0.4 μm and 0.5 μm of the silicon. The colour of the silicon surface changes with the thickness of the oxide layer due to the shift in the wavelength of the reflected light. This effect is used as an indication of the layer thickness.

25.1.3 Photolithography

The prime use of photolithography in integrated circuit manufacture is to selectively remove the oxide from areas of the silicon slice. To do this the surface of the oxide is first covered with a thin uniform layer of liquid photoresist. This is best obtained by holding the silicon slice in a vacuum chuck and placing a fixed amount of photoresist onto its centre. The slice is then rotated at very high speeds for about 1 min. The resist spreads over the slice, the excess flying off due to centrifugal forces. The amount left on the slice is clearly a function of the oxide properties and the viscosity of the resist. The slice is then heated for a few minutes in an oven to harden the resist.

The mask is next placed in contact with the oxide layer. This is usually a glass plate with the black emulsion areas on one face. It is essential that the mask is accurately aligned on the silicon slice using a microscope, and is close to it. Close physical contact is necessary to prevent the light spreading sideways on the photoresist when the mask is exposed. Unfortunately this requires considerable force with the result that particles on the slice can damage the mask. Also parts of the photoresist come away when the mask is removed. Consequently the mask is damaged and is only usable for an average of ten exposures. Special materials, such as chrome masks,. give about 50 exposures, but are more expensive. The modern trend is towards projection masking where an image of the mask is projected onto the silicon slice. The mask does not now make physical contact with the silicon and consequently has a much greater life.

Once the mask is in contact with the photoresist layer it is exposed to ultraviolet light for about 20 s. If negative resist material is used the exposed areas become hardened by the light. The slice is now placed in a rotating chuck and developer is dropped onto it. This dissolves the unexposed resist areas. The spinning action ensures the speedy removal of dissolved parts so that all areas have equal exposure to the developer. This part of the process usually takes about 30 s. The slice is now placed in a bath of hydrofluoric acid which dissolves the exposed oxide areas but not the silicon. The etching time must be closely controlled; if too long the etchant spreads sideways under the remaining photoresist. For applications which require most of the oxide to be removed it is more convenient to use a positive resist material, such that the areas exposed to ultraviolet light are dissolved in the developer.

25.1.4 Epitaxy

Epitaxy means growing a single crystal silicon structure on the original slice such that the new structure is essentially a molecular extension of the original silicon. Epitaxy layers can be closely controlled regarding size and resistivity (i.e. ±10%). This compares favourably with ±30% resistivity control obtained when pulling from the silicon melt. Most of the integrated circuit structure is formed in the epitaxy layer, the rest of the slice acting purely as a ground plane.

Epitaxy apparatus is very similar to the oxide growth arrangement shown in *Figure 25.1*. However, r.f. heating coils are normally used and the silicon slices are placed in a graphite boat, which may be coated with quartz to prevent the graphite contaminating the silicon. The bubbler usually contains silicon tetrachloride ($SiCl_4$) to which may be added a controlled amount of an impurity such as PCl_3. Hydrogen gas is bubbled through this mixture before entering the quartz epitaxy tube.

Initially the slices are heated to about 1200°C and pure hydrogen and hydrochloric acid vapour are passed over them to etch away any oxide or impurities which may exist on the surface of the silicon. HCl vapour is then turned off and H_2 bubbled through the $SiCl_4$, the vapour passing over the silicon slices. When this reaches the hot silicon it dissociates and silicon atoms are deposited on the slice where they rapidly establish themselves as part of the original crystal structure. It is essential to saturate the tube with $SiCl_4$ vapour to ensure a uniform layer thickness over the whole slice. The epitaxy is about 10 to 15 μm in depth and has a resistivity which varies in the region of about 10 Ω cm. The epitaxy layer may be doped by p- or n-type impurities by introducing these, in the required concentration, into the vapour stream.

25.1.5 Diffusion

In epitaxy a large area is doped by a closely controlled amount of impurity. Diffusion on the other hand enables selective areas to be doped. These areas represent those which are not covered by an oxide layer so that the photolithographic stage is normally followed by diffusion.

The diffusion furnace resembles the arrangement shown in *Figure 25.1*. The bubbler contains the impurity, and nitrogen is passed through it on the way to the boat containing the silicon slices. The furnace temperature is kept close to the melting point of silicon, i.e. 1200°C, and at this value the silicon atoms are highly mobile. Impurity atoms readily move through the silicon lattice by substitution, going from a region of high concentration to that of lower density.

Diffusion can be carried out by one of two techniques, as shown in *Figure 25.2*. In the error function or one-step process the concentration of impurities is kept fixed throughout the diffusion period, giving the curves shown. In Gaussian or two-step diffusion a fixed amount of impurity is present so that as this moves deeper into the silicon bulk the surface concentration decreases. This gives a flatter dopant distribution of higher resistivity. Generally diffusion takes place in a slightly oxidising atmosphere. This results in the formation of a glassy layer of impurity on the silicon surface as well as a slight penetration into the silicon bulk. The silicon slice can be removed to a second furnace and heated in an inert atmosphere when the dopants diffuse out of the glassy layer to give an error function distribution. The glassy layer now not only forms a diffusion source but also protects the silicon surface from evaporation, and acts as a getter for impurities from the silicon bulk. For Gaussian diffusion the glassy layer is etched off using hydrofluoric acid prior to the slice being heated in an inert atmosphere. The critical dopants just below the surface now diffuse into the silicon bulk.

Apart from the open tube arrangement shown in *Figure 25.1* it is possible to diffuse slices by putting them in a sealed quartz container with doped silicon powder and then heating the combination in a quartz furnace. The advantage of this method is that many slices can be diffused simultaneously giving a larger throughput. However, the quartz container must be broken to remove the slices after diffusion so the process can prove expensive.

The most commonly used p and n type impurities are boron and phosphorus respectively. Both reach maximum solubility at about 1200°C and have a high diffusion constant. Arsenic, on the other hand, is a n type impurity which diffuses very slowly. It is used for making the buried layer in transistors since this must not diffuse appreciably during subsequent high temperature processes. Gold is sometimes introduced as an impurity into integrated circuits. Gold is a lifetime killer and enables fast switching circuits to be built. To dope a slice with gold its rear surface is first coated with a thin film of gold using evaporation techniques and this is then heated in a diffusion furnace until the whole slice is saturated with gold. The gold atom is much smaller than a silicon atom. This means that it does not move through the silicon lattice by substitution as other impurities, but moves very rapidly in between the silicon atoms, i.e. intrinsically.

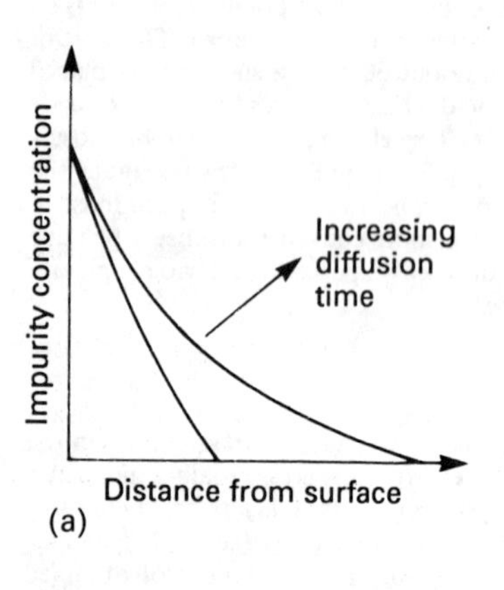

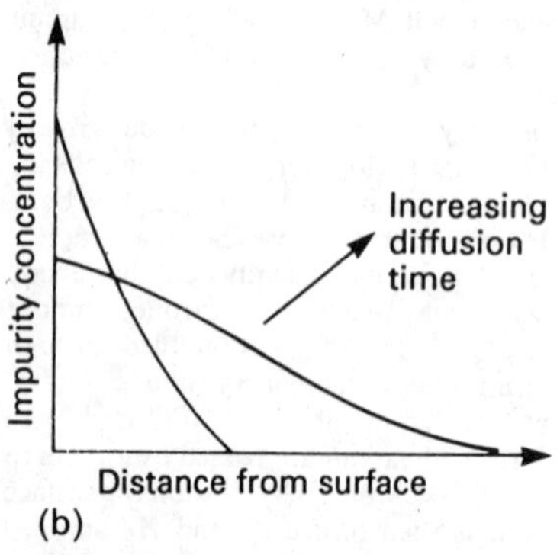

Figure 25.2 Impurity concentration during diffusion: (a) error function, (b) Gaussian

25.1.6 Vacuum deposition

During the manufacture of integrated circuits it is necessary to deposit a thin layer of metal on the silicon, for instance to form the aluminium interconnections. This is carried out using vacuum deposition in an arrangement of the type shown in *Figure 25.3*. The silicon slices are placed face down around the bell jar, with the metal source in the centre. The vacuum is lowered to below 5×10^{-6} Torr before deposition is commenced. The silicon is heated to between 100°C and 300°C which causes the deposited metal to form a chemical reaction with the silicon

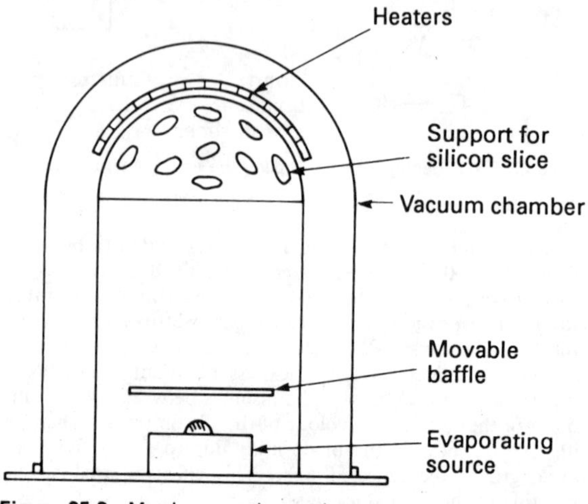

Figure 25.3 Metal evaporation equipment

oxide and adhere to the slice. Upward evaporation is also used to prevent impurities, which may be generated by the heat source, from falling onto the slices. The metal film is usually about 1 μm thick. This thickness can be monitored by including a quartz crystal oscillator in the vacuum, whose frequency changes with the amount of metal deposited on its surface.

The metal source must have a large area. Shadowing effects can arise from point sources, which would result in weaknesses in the metal film. There are two ways of heating the metal source. It can be wrapped as a wire around a resistance heated tungsten filament. This is a cheap method but there is risk of contamination due to the evaporation of the heater metal. A cleaner solution is to use electron beam evaporation. The metal source is placed in a boat which acts as an anode to an electron gun. The high energy electrons striking the source cause it to evaporate and deposit on the surrounding silicon slices.

After evaporation it is usual to heat the slices to about 500°C in an inert (nitrogen) atmosphere. This causes the metal to alloy well with the silicon surface so that the interface between the two has a low resistance. This is referred to as a low ohmic contact joint.

25.1.7 Electron beam technology

Electron beams are widely used during the processing of integrated circuits, and electron beam evaporation for metal deposition has been introduced in the last section. The second largest use of electron beams is for ion implantation, and this is covered in the present section. However, electron beams are also becoming increasingly popular for high-density circuits, formed by direct writing the integrated circuit pattern onto a silicon wafer. With this technique circuit geometries of tenths of microns are obtainable. The disadvantage of the process is that it is slow, and therefore expensive, when compared with traditional diffusion methods. It is mainly used during the prototype stage, when small quantities are needed quickly, since now no masks are required.

There are several basic requirements which must be met in any implantation system. These are:

(1) The impurity concentration must be uniform over a given slice and the process must be accurately reproducible over repeated slices.
(2) The system must have a high throughput so that the expensive implantation equipment can be amortised over a large number of devices. This means that the capacity of each batch must be high and the cycle time, which is the vacuum cycle time, since all such systems operate under a vacuum, must be short. A further requirement is that the doping current can be made large enough to reduce the time per slice.
(3) The purity of the dopant must be accurately controlled. Most ion sources produce a range of dopants in addition to the one required. The impurities must be completely removed from the ion stream before it reaches the silicon slice.
(4) The energy imparted to the ions (by the accelerating voltage) must be high enough to enable them to penetrate the maximum distance likely to be required.

These requirements are discussed with reference to the basic implantation equipment shown in *Figure 25.4*. The main parts of this arrangement are as follows:

(1) An ion source which produces an abundance of the boron or phosphorus dopant required. The source output should remain constant over many days to allow reproducible devices to be made without the necessity for constant adjustment. The source current should also be variable up to about 0.5 mA to give short implant times. There are several techniques used for producing ions, such as by a radio frequency discharge or by a heated filament discharge.
(2) A focusing system which is usually electrostatic since magnetic lens techniques, such as used for electron optics, give more bulky and expensive equipment due to the fact that the ions are now much heavier.
(3) An ion accelerator. This gives the ions the necessary penetration energy by applying a high voltage across the ions. It can vary from anything between 10 kV and 500 kV.
(4) The mass analyser is used in conjunction with the focusing system to separate out the impurities from the ion beam. First the ions are separated into different streams according to their mass and then all but the required dopant stream is blocked off before it reaches the silicon wafer chamber.
(5) The beam scanner is used to move the ion beam over the silicon surface so as to give a uniform dopant concentration. In addition it is also usual to move the slices physically past the beam for more uniformity.
(6) The doping concentration is monitored by measuring the current in the ion detector. This is quite easily done since each ion carries one positive charge unit. The dose imparted to the silicon, measured in ions/cm^2, is equal to the product of the current and the exposure time, divided by the wafer area.
(7) Finally the wafer chamber holds the silicon samples. It must be quickly accessible and large enough to hold a useful batch at each operation.

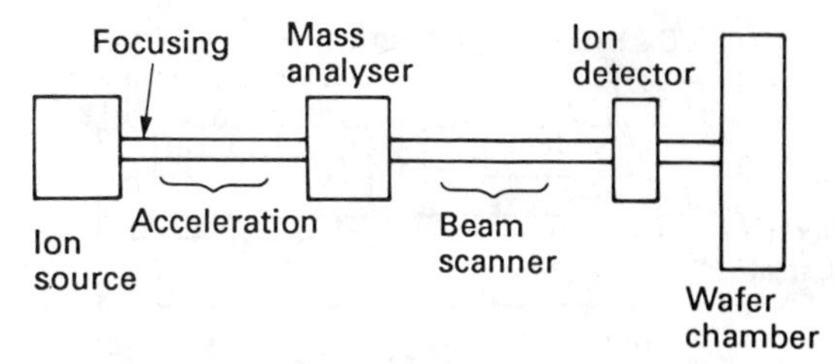

Figure 25.4 Principal parts of an ion implantation equipment

Ion implantation differs from diffusion in that the dopants are given enough energy to allow them to force their way through the silicon lattice until they reach the desired depth below the surface. The process therefore differs from diffusion where the impurity ions move gradually through the lattice and the concentration decreases away from the surface (unless epitaxy layers are grown over diffused regions, of course). Ion implantation is therefore a very precise process where the depth and concentration of impurities can be closely controlled. However, since the ions plough their way through the silicon lattice they distort the material so that implantation is usually followed by an annealing stage in which the atoms are allowed to drift into their places within the lattice.

25.2 Bipolar circuits

25.2.1 Circuit fabrication

Having looked at the different production techniques we can now see how they fit together to make an integrated circuit. The heart of any integrated circuit is its transistors. The transistor structure is very easily made in monolithic integrated circuits and it occupies relatively little space. Consequently, unlike discrete circuits, transistors are used liberally in integrated circuit designs, often to replace passive devices like resistors.

The key to integrated circuit fabrication is to build several components onto a common silicon slice without interfering with each other. There are several techniques which may be used for

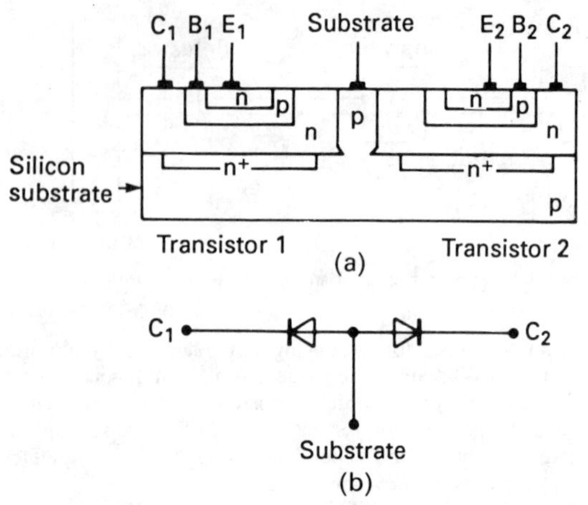

Figure 25.5 Diode of two transistors: (a) device structure, (b) isolation configuration

this in bipolar circuits. *Figure 25.5(a)* illustrates the commonest technique, known as diffusion isolation, or more correctly diode isolation. Two transistors are shown here, each completely surrounded by p-type silicon. The collector of each transistor forms a diode with this p region, as shown in *Figure 25.5(b)*. If the substrate is always connected to the most negative voltage in the system then, irrespective of the voltage on each transistor, the devices are separated from each other by two reverse biased diodes.

Apart from transistors it is possible to form other electronic components in integrated circuit structures. A p–n junction of base–collector or base–emitter forms a diode, a linear distance within the p base region or n emitter region of a transistor forms a resistor, and a reverse biased diode may be used as a capacitive element. All these elements have limitations, as will be evident later. For the present however let us consider how the circuit shown in *Figure 25.6(a)* can be built into an integrated circuit. This is illustrated in *Figure 25.6(b)*. Islands of n-type material are first formed in a p-type substrate. By taking the substrate to the most negative voltage available in the system, each island is therefore separated from the adjoining ones by the diode isolation mechanism discussed earlier. Into each of these islands it is now possible to build transistors, resistors, capacitors or diodes, as required. In *Figure 25.6(b)* the resistor is formed in what

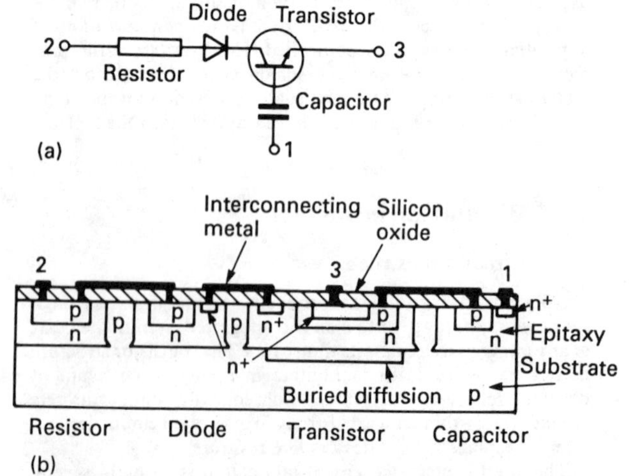

Figure 25.6 An integrated circuit with diode isolation: (a) electronic circuit, (b) integrated circuit layout

would normally be the base of a transistor and the diode utilises the base–collector of a transistor structure. The capacitor shown is a reverse diode structure although unipolar type capacitors, as described in the next section, can also be used. The various individual devices on the chip are then interconnected by metal patterns to form the required configuration shown in *Figure 25.6(a)*. The silicon oxide acts primarily as a convenient electrical insulation between the metal connectors and the rest of the silicon material.

The actual steps used in the fabrication of *Figure 25.6(b)* are shown in *Figure 25.7*. The starting material is a slice of polished silicon as in *Figure 25.7(a)*. This slice will be capable of accommodating many hundred identical circuits or dice. Only one such die is considered here. First the transistor buried layer has to be diffused. This is done by growing an oxide on the silicon surface, and by photolithographic techniques etching a window at the required location. The mask used (first mask) will clearly be transparent except for the area covering the buried layer. The slice can now be introduced into a diffusion furnace and the n^+ layer formed (*Figure 25.7(b)*). Arsenic is used for this since, as explained earlier, it diffuses slowly and so will not move much during subsequent diffusion stages.

An n epitaxy is now grown over the whole area (*Figure 25.7(c)*). The advantage of the epitaxy is evident since it allows diffusions to be formed at different planes. Then an oxide layer is grown and photolithography is used to open up windows (second mask) through which isolation regions can be diffused (*Figure 25.7(d)*). Once again an oxide is grown over the surface and this is masked (third mask) and etched for the p diffusion of the transistor base and other areas (*Figure 25.7(e)*). The doping concentration is selected to suit the transistor requirements, and capacitors, diodes and resistors must accept this limitation. A fourth masking operation is required for the transistor emitters (*Figure 25.7(f)*) and this is followed by a fifth mask which opens up the contact areas to the silicon surface. Metal (aluminium) is now deposited over the whole surface using vacuum techniques (*Figure 25.7(g)*) and finally a sixth mask is used to etch the metal to the required interconnection pattern as shown in *Figure 25.7(h)* which corresponds to *Figure 25.6(b)*.

Two important factors should be noticed from the above description. Firstly there is a lot of to and fro movement between photolithographic, oxidation and diffusion plants. Careful handling is needed at all stages and each operation should be carried out in a clean atmosphere since dirt particles can damage the slice and could act as unwanted impurity dopants. Many different diffusion furnaces are also required since in each furnace the quartz walls become contaminated with the dopant and cannot be re-used for another purpose unless it is thoroughly saturated with the new impurity for several hours. This means that for the process just described different furnaces are needed for oxidation and for deposition and drive in (two stage, Gaussian diffusion) of three different types of impurities, i.e. arsenic (buried layer), boron (p type) and phosphorus (n type). This means a bank of eight furnaces just for one process. It is therefore very common in integrated circuit production areas to see banks of diffusion furnaces.

In addition, if gold doping is used then this is usually introduced after the fourth emitter mask so that the emitter diffusion stage also drives in the gold. To prevent contamination of gold onto other devices the gold vacuum deposition plant is kept separated from that used for aluminium deposition.

The second important aspect to note is that one impurity can reverse the doping of a previous stage provided it is strong enough. Therefore the n^+ emitter diffusion is carried out into the p base diffusion. This is illustrated in *Figure 25.8* where the junctions between base, emitter and collector are defined. Note however that the transition is gradual, the impurity differences building up as one moves away from the junction. It should also

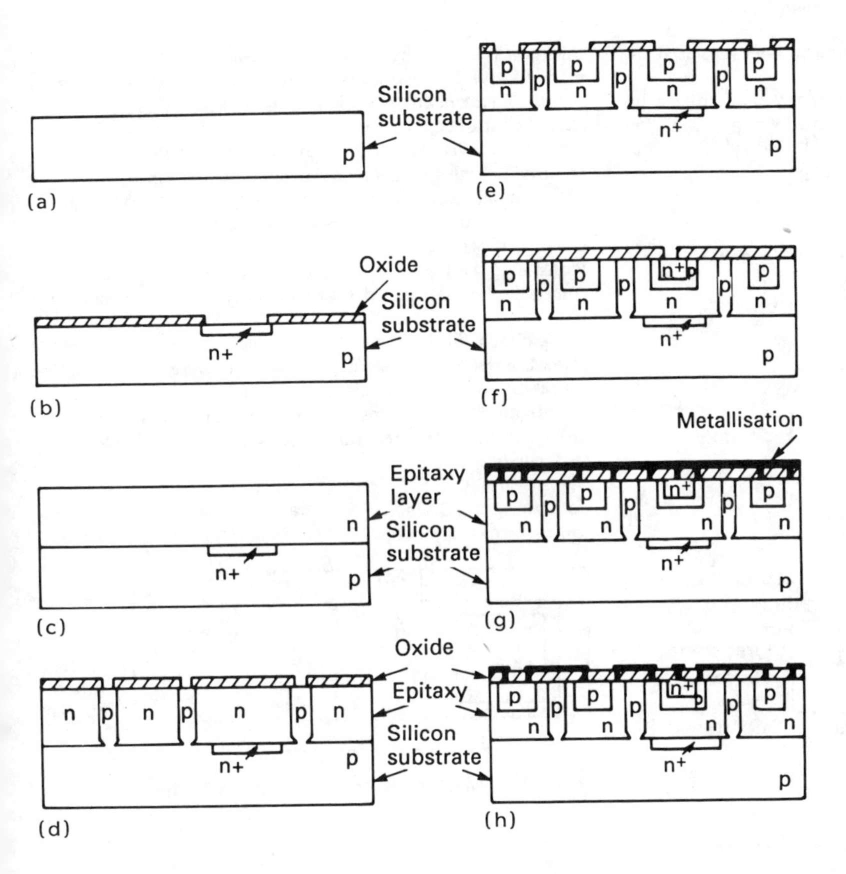

Figure 25.7 Stages in the manufacture of an integrated circuit: (a) silicon slice; (b) buried layer diffusion; (c) epitaxy; (d) isolation diffusion; (e) base diffusion; (f) emitter diffusion; (g) metal deposition; and (h) interconnection etch

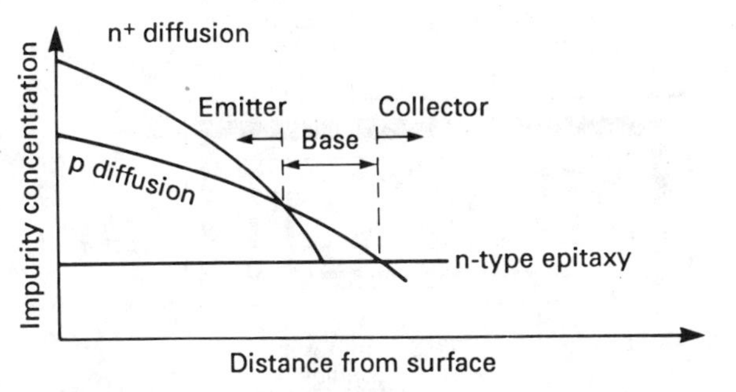

Figure 25.8 Formation of an npn transistor

be noted, as illustrated in *Figure 25.7*, that diffusions tend to spread downwards as well as sideways under the oxide masks. This must be allowed for in the mask design.

It is usual, but not always the case, to cover the completed slice with a thin layer of glass or other passive device (e.g. silicon nitride) to protect it from damage and contamination during the subsequent test and packaging stages which involve manual handling in relatively unpurified atmospheres. This step is called passivation. The glass is then etched off (seventh mask) from the bonding pad areas and in between the individual dice ready for cutting (scribing) and separation.

25.2.2 Bipolar circuit components

Most of the different types of components which are used in integrated circuits have already been briefly introduced in the previous section. We will examine here their construction and characteristics in greater detail. There are four types of components: transistors, diodes, resistors and capacitors. As mentioned before the transistor is the single most important device in integrated circuits, all other component characteristics being dependent on it.

25.2.2.1 Transistors

In a monolithic system the substrate primarily acts as a mechanical holder to the devices and is of very high resistivity. The epitaxy forms the collector and is also of high resistivity. It is usual, as shown in *Figure 25.7*, to diffuse an n^+ region, at the same time that the emitter is diffused, at the point where the metal is to connect onto the silicon surface, to make a better joint and to prevent Schottky diode action. The high series resistance present in this collector gives poor amplifier and switching performance. However, one requires a high collector resistivity at the base junction in order to give a good collector–base breakdown voltage. Both these requirements are met by a buried n^+ diffusion as described earlier. It has also been explained that diode isolation is the most common technique in use for monolithic integrated circuits. This however presents two problems. Firstly the diode in between two devices is reverse biased and, as seen in *Section 25.2.1*, this acts as a capacitor. There is therefore a relatively large valued capacitor connecting the transistor collectors to their common substrate and these capacitors are voltage dependent. For low power applications this can give rise to significant leakage currents. A second problem can arise with monolithic transistors when they are biased such that the lower three layers, including the substrate, form a parasitic pnp transistor. A large current can now flow through this device from the base of the original npn transistor to the substrate. Both these effects are overcome when other isolation techniques are used, as described in the next section.

Although npn transistors are by far the most common in monolithic circuits, pnp devices can be formed if required. There are three main types, as shown in *Figure 25.9*. The substrate pnp

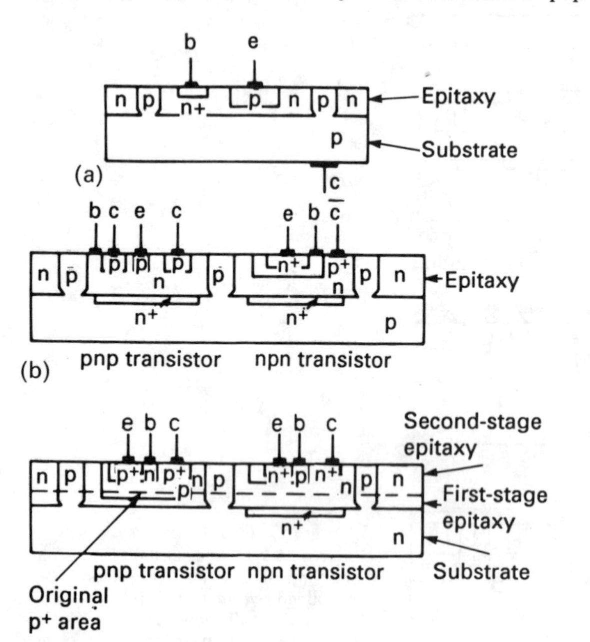

Figure 25.9 pnp transistor configurations: (a) substrate, (b) lateral, (c) two-step epitaxy

is the simplest arrangement and only requires an emitter diffusion, although an n^+ base diffusion is also usual. It has two serious faults. Firstly all transistors on the same substrate also share the same collector, so that circuit flexibility is greatly reduced. Secondly, in spite of careful process control techniques, the epitaxy layer cannot be grown to a uniform thickness over the whole slice. This means that the base width of the transistors, and consequently their gains, vary over the slice and are not predictable.

In the lateral pnp the collector is a concentric p-type diffused ring around a central emitter diffusion. The base is the epitaxy between the emitter and collector diffusions. Unfortunately this again makes the base width difficult to control since it is the difference in sideways diffusion between the emitter and collector, both of which are very dependent on processing conditions. Generally to allow for process variations, the width of the base is kept relatively wide so that the gain is low. These transistors also have a low frequency response and a parasitic pnp action down to the substrate. This effect is however reduced by the presence of the n^+ buried layer. A further limitation is that the collector–base junction occurs at the surface of the silicon, instead of in the bulk as in other arrangements. This means that it is influenced by surface effects giving a low breakdown voltage.

Truly complementary pnp and npn transistors can be obtained by the two-step epitaxy process. In this the epitaxy growth is interrupted half way through and a p^+ diffusion formed where the pnp transistors are due to occur. This is a mobile diffusant. The epitaxy growth is now resumed. When next a diffusion process occurs the p^+ buried diffusion moves up into the epitaxy to form the collector of the pnp transistor. Base and emitter diffusion occur as before. Quite clearly this type of pnp transistor requires more process steps. However, since the base thickness is formed by two lateral diffusions it can be accurately made to give pnp and npn devices of comparable performance.

Since base thickness in diffused monolithic transistors can be closely controlled it is possible to have very thin base, high gain devices. Gains of the order of 10 000 are attainable, and are used to provide high input impedance amplifiers. Thin base regions unfortunately also give lower breakdown voltages. This can be overcome by using composite transistors, which are in effect two devices in which one provides the gain whereas the other withstands the high voltage. Using monolithic techniques both transistors are fabricated simultaneously giving minimal extra cost.

Figure 25.10(a) shows the plan view of two small signal transistors separated by a p diffusion. *Figure 25.10(b)* is the schematic of a high current transistor and illustrates the interdigitated emitter construction, which is necessary to obtain a large perimeter to area ratio and so improve the frequency performance.

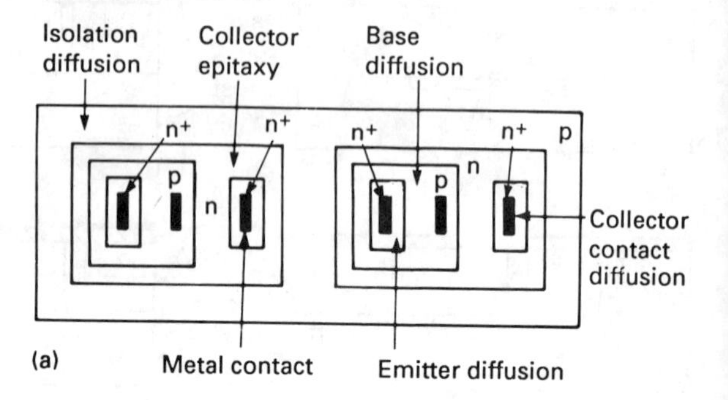

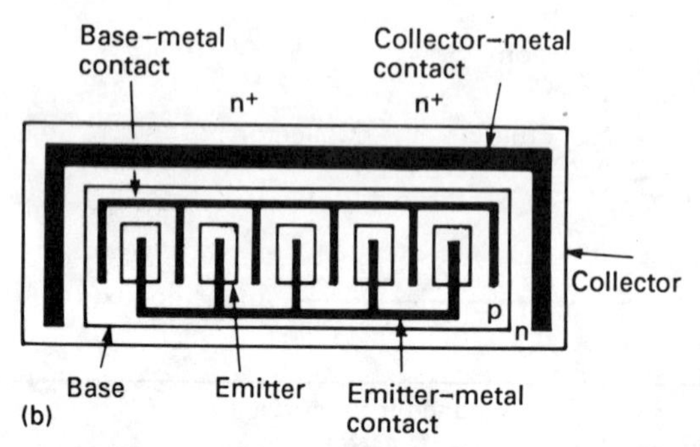

Figure 25.10 Transistor geometries: (a) two signal transistors, (b) high current transistor

25.2.2.2 Diodes

The thickness and sheet resistivity of the p and n diffusions used to make a diode are determined by those required for a transistor, since these are made at the same time as the transistor diffusions. It is also usual not to make just a p and n diffusion but to utilise two of the transistor's areas to form the diode. Clearly there are now many possibilities, and these are illustrated in *Table 25.1*. The differences between the parameters are primarily due to the variation in the minority carriers used in the arrangements. It is also seen that there are two capacitances to be considered, that of the diode itself and that of the leakage to the substrate. The fastest diode is the emitter–base arrangement since both these are heavily doped with impurities. It is also possible to obtain zener diode action, for instance by connecting the emitter and collector together and biasing it positive to the base. The zener voltage is

Table 25.1 Typical characteristics of monolithic diodes obtained from transistors

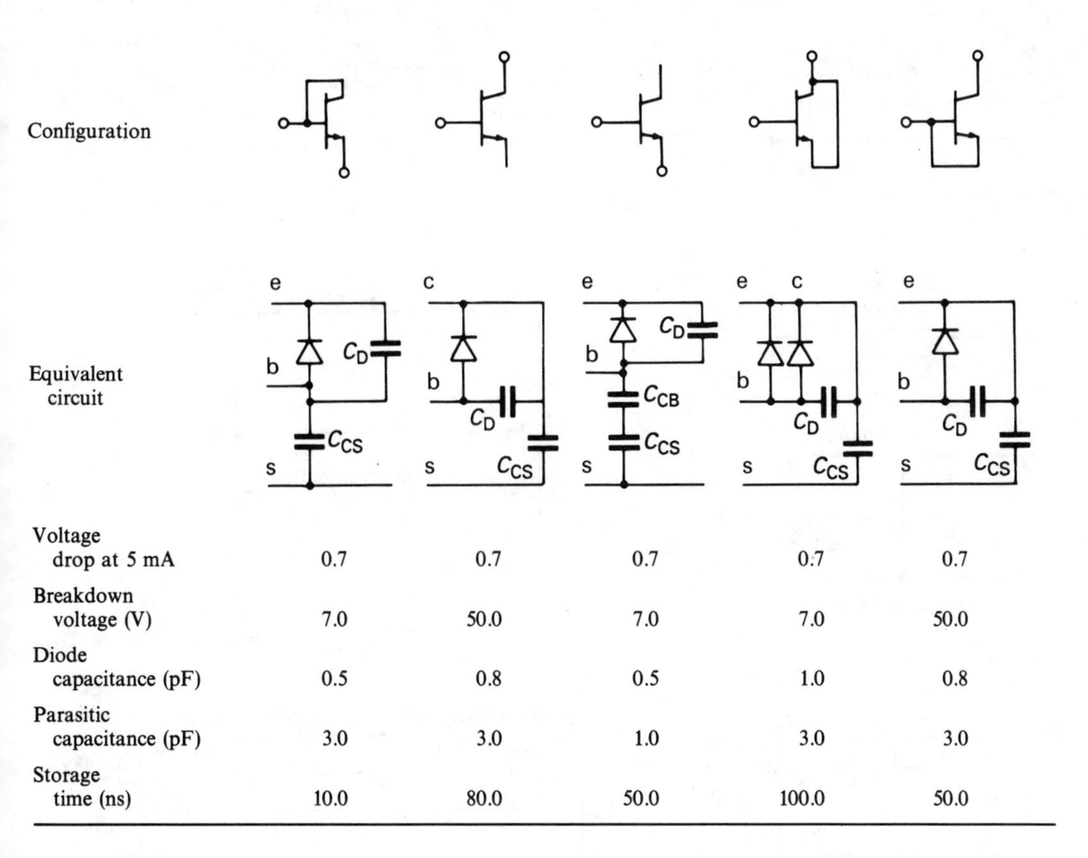

Voltage drop at 5 mA	0.7	0.7	0.7	0.7	0.7
Breakdown voltage (V)	7.0	50.0	7.0	7.0	50.0
Diode capacitance (pF)	0.5	0.8	0.5	1.0	0.8
Parasitic capacitance (pF)	3.0	3.0	1.0	3.0	3.0
Storage time (ns)	10.0	80.0	50.0	100.0	50.0

now approximately 7 V but it will vary appreciably from slice to slice due to changes in the process parameters.

A second type of diode used in monolithic circuits is the Schottky device. Essentially it consists of a metal area such as the aluminium used in the interconnection, in contact with a lightly doped n region. The excess of free electrons in the metal cause the formation of a Schottky (voltage) barrier at the junction. If the metal is now biased positive to the n region it conducts current. If it is negative it blocks, as in a traditional diode. The Schottky diode is very fast, with 1 ns typical storage time, and has a low voltage drop when conducting, typically of about 0.3 V. It is important to note, however, that the n region must be lightly doped. If it has a large impurity concentration then tunnelling takes place and the electrons in the two regions cause the formation of a good ohmic contact between the metal and the silicon. This is why an n^+ diffusion is usually made under the collector metal region.

25.2.2.3 Resistors

The resistance of a strip of diffused silicon is determined by its resistivity (sheet resistance) and dimensions. Clearly for high values of resistance the resistivity should be as large as possible. However, this implies a low impurity doping and the intrinsic carriers will now have a significant effect in reducing the temperature coefficient of the resistor. Dimensions are also limited. For instance the thinness of a line is determined by the accuracy tolerance of the masks, and the longer the resistor the greater the silicon area it occupies (long resistors are normally in the form of a zigzag pattern). Generally resistivities of 300 Ω per square are maximum for a workable temperature coefficient. The base diffusion of a transistor has a value of about 200 Ω per square and is the one most often used. The emitter has too low a resistivity and the collector too high. *Figure 25.11*(*a*) shows a diffused resistor in the base region (i.e. it is formed at the same time that the base of other transistors on the chip are diffused). The device is really a complex circuit, as shown in *Figure 25.11*(*b*), consisting of a parasitic pnp transistor, a leakage capacitor and a distributed resistor.

The absolute accuracy of a diffused resistor is limited to about 10% due primarily to the difficulty of controlling the diffusion process between slices. On a chip, however, the relative accuracy is about 3% being determined by tolerances in diffusion, masking and etching. *Figure 25.11*(*c*) shows the construction of a pinch resistor in which the emitter diffusion is used to reduce the effective width of the resistor.

25.2.2.4 Capacitors

Monolithic capacitors are not frequently used in integrated circuits since they are limited in range and performance. There are however two types available, as shown in *Figure 25.12*. The

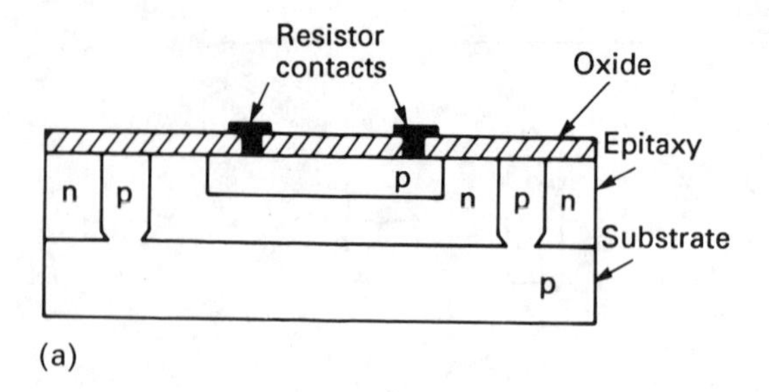

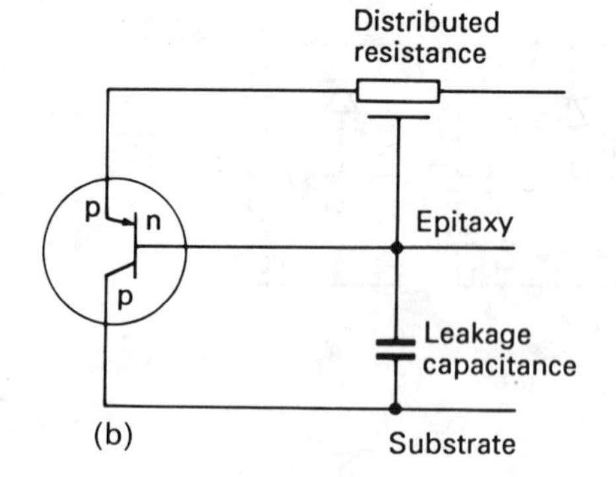

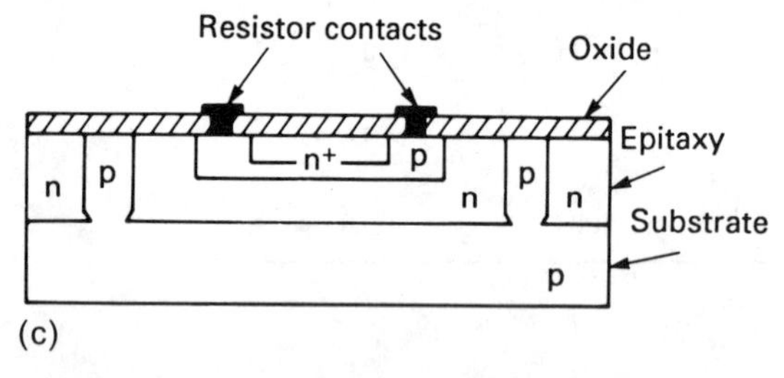

Figure 25.11 Monolithic resistor: (a) base diffused resistor, (b) equivalent circuit, (c) pinch resistor

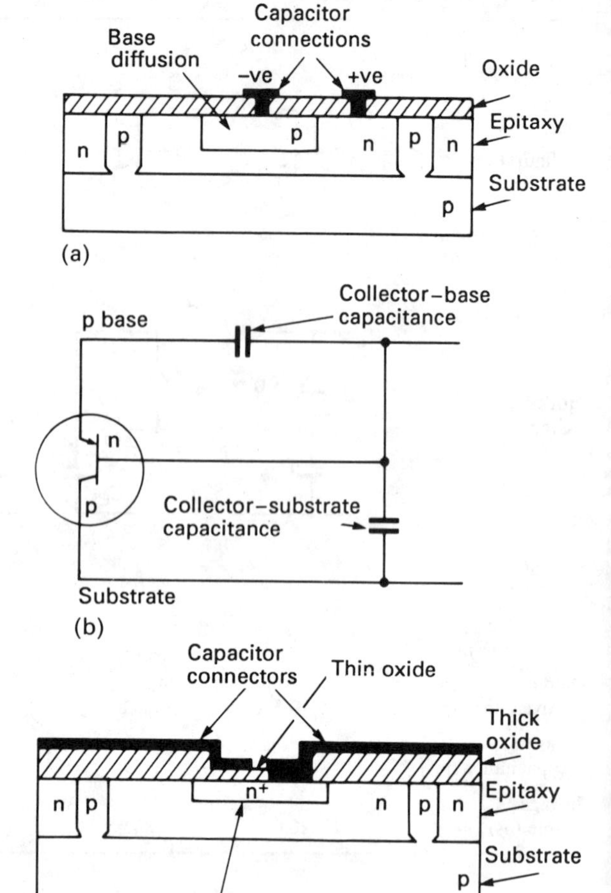

Figure 25.12 Monolithic capacitor: (a) diffused capacitor, (b) equivalent circuit, (c) metal oxide silicon capacitor

diffused capacitor is a reverse biased p–n junction, as explained in *Section 25.2.1*, formed by the collector–base or emitter–base diffusion of the transistor. The capacitance is proportional to the area of the junction and inversely proportional to the depletion thickness. This in turn varies with the resistivity of the layers so that for high impurity concentrations the capacitance is relatively large. However, the breakdown voltage of the capacitor is directly proportional to the resistivity so that one cannot simultaneously have a large capacitance value and breakdown voltage unless the junction size is made very big.

Diffused capacitors have very poor voltage coefficients since the depletion thickness is dependent on the bias voltage. Typically one can obtain a capacitance of about 1.2 nF/mm^2 using a base resistivity of 200 Ω per square and zero bias. This value is halved at a bias of 5 V. A second characteristic is low Q (equal to ratio of reactance to resistance) due to the higher resistivities involved. Generally all monolithic capacitors are inferior in this to discrete devices. *Figure 25.12(b)* shows an equivalent circuit of a diffused capacitor. In addition to the required collector–base capacitance there exists a parasitic collector–substrate capacitance and a pnp transistor. This requires the collector n region to be biased positive to cut it off.

A metal oxide silicon capacitor uses a thin layer of silicon oxide as the dielectric. One plate is the connecting metal and the other a heavily doped layer of silicon which is formed during the emitter diffusion. This capacitor has a lower leakage current and a much higher Q due to lower resistivities, and is non-directional since either plate can be biased positively. The capacitance value can be varied between about 0.3 and 0.8 nF/mm^2 and it is independent of the applied voltage provided this is below the breakdown of about 50 V. Parasitic capacitors still exist but are lower than in the case of a diffused capacitor.

25.2.3 Bipolar circuit isolation

The standard diode isolation bipolar process suffers from many faults. The most serious of these are the leakage currents, the parasitic capacitance between collector and substrate giving low speeds, and the large size of the devices due to the fact that the epitaxy layer is relatively thick and the diffusion isolations spread considerably sideways as they move down. Many new isolation techniques have been devised to overcome these disadvantages. They can all be divided into two groups: diffusion isolation and oxide isolation. In all cases these processes are backed by individual companies, therefore the company names are given in the discussion in the following sections.

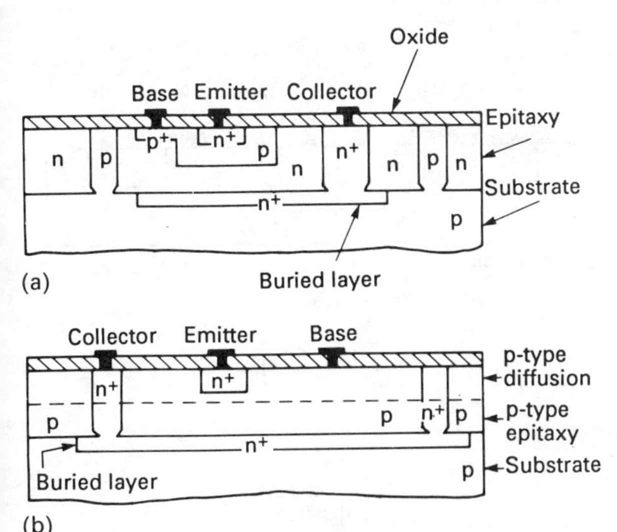

Figure 25.13 Diffusion isolated transistors: (a) Process III, (b) Collector Diffusion Isolation (CDI)

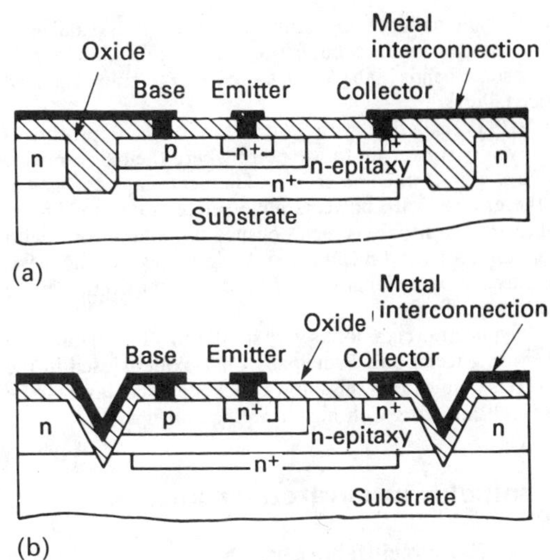

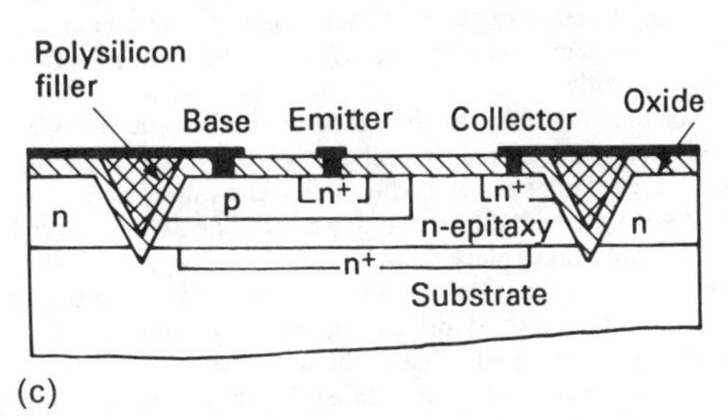

Figure 25.14 Oxide isolation techniques: (a) Isoplanar, (b) V-ATE, (c) VIP and Polyplanar

25.2.3.1 Diffusion isolation

Two different processes are illustrated in *Figure 25.13*: Process III (Plessey Semiconductors Ltd) and Collector Diffusion Isolation (CDI) (Ferranti Semiconductors Ltd). Process III is a fast system for analogue and digital devices. It uses a very thin epitaxy layer and shallow diffusions which give narrow base regions and high-frequency devices. The parasitic capacitance is also low. Diode isolation techniques are used. Process III uses an extra diffusion stage over the standard process since the collector diffusion is taken right down to the buried layer and must be made as a separate operation. This reduces the series resistance between collector junction and the buried layer, giving high frequency transitors and lower voltage drops. An extra p^+ diffusion is also used to reduce the base resistance. The thin epitaxy means that sideways diffusion is small and the devices can be closely packed.

In CDI the epitaxy is p doped and for the higher performance this may be followed by a non-mask, p^+ diffusion over the whole surface. This produces double diffused transistors different from the epitaxy base devices obtained with the usual diode isolation process. However, CDI still uses diode isolation between devices. It differs from other techniques in that the collector diffusion goes right down to the buried layer and completely surrounds the transistor. Isolation therefore occurs during the collector diffusion stage. The process therefore does not need a separate mask for the isolation diffusions. This, and the fact that the epitaxy is thin, gives very high packing densities on a chip. The thin epitaxy also gives this process, along with others which use thin layers, another advantage. Since the buried layer is close to the base region the injection efficiency of the collector is improved. This means that the transistors have high inverse gain, i.e. good gain when the roles of the collector and emitter are reversed. Not only does this result in low saturation resistance but also allows high switching speeds to be obtained. Both fast digital and linear circuits can therefore be fabricated on the same silicon chip.

25.2.3.2 Oxide isolation

Four oxide isolation processes are shown in *Figure 25.14*. These are Isoplanar (Fairchild Ltd), V-ATE (Raytheon Ltd), VIP (Motorola Ltd) and Polyplanar (Harris Ltd). All four processes depend on etching the silicon surface. This utilises the property that silicon nitride (Si_3N_4) when deposited on a silicon area prevents it from oxidising. Furthermore it is easy to remove the nitride by etching with phosphoric acid without affecting the silicon or oxide.

In the Isoplanar process a buried n^+ diffusion and a thin epitaxy layer are grown as usual. This is followed by coating the silicon surface with nitride and then oxide which, using photolithographic techniques, selectively removes the nitride from the location where grooves are to be etched. The epitaxy is then etched and a thermally grown oxide used to fill up the grooves so that they are level with the surface. Base and emitter diffusions then complete the transistor. As seen in *Figure 25.14(a)* the packing density is much higher than with diode isolation since the devices can be placed closer together. Furthermore since an oxide area prevents significant impurity diffusion there is a considerable amount of self-aligning effect in these transitors. The absence of diode isolation also reduces the junction capacitance giving higher operating speeds.

In the V-ATE process (vertical anisotropic etch) thin oxide and air are used as isolation between devices. It is important to use $\langle 1\,0\,0 \rangle$ silicon, which can be etched along the $\langle 1\,0\,0 \rangle$ plane some 30 to 40 times faster giving sloping edges of about 54° angle to the surface. Using nitride masking techniques V-grooves are cut into the epitaxy. The depth of the groove is determined by the initial mask width and can therefore be easily controlled. An oxide–nitride sandwich is then grown over the surface and a metallisation layer evaporated over this to form the interconnection patterns. Generally this process is similar to Isoplanar in performance but since the metallisation has to negotiate sharp bends in the V it is more likely to crack and lead to device failures.

The VIP (V isolation with polysilicon backfill) process is essentially very similar up to the point where the oxide–nitride sandwich is grown into the groove. The next stage consists in filling the grooves with polycrystaline silicon so as to slightly overfill them. The surface is then polished flat and finished with the usual diffusions and metallisations. Now however the metal does not need to pass over any sharp bends and the overall device is much more reliable.

The Polyplanar process looks very similar to VIP but differs in manufacturing technique. Principally only oxide is used in the grooves. It is claimed that this allows thicker epitaxy layers to be used giving transistors with higher voltage ratings.

25.3 Unipolar integrated circuits

25.3.1 Unipolar circuit fabrication

The manufacture of unipolar integrated circuits uses the same processes as bipolar systems. However, there is now no epitaxy layer and the most critical stage in the fabrication is the growth of the thin gate oxide.

For p-channel transistors the starting material is n-doped silicon which is then covered by a relatively thick oxide to a depth of about 1.5 μm. Holes are etched to define the source and drain and two p wells are next diffused in to a depth of between 2 and 4 μm. The diffusion takes place in an oxidising atmosphere so that a thin oxide layer forms over the whole surface. The oxide in the gate region is now removed and a thin, precisely controlled very pure oxide layer is regrown. This is the most important stage in the process since the gate oxide must be thin to give a low operating threshold voltage but it should also be pure, to maintain device stability, and be free from pinholes which would short the metal electrode to the silicon surface. The ohmic contact areas are then etched and metal is evaporated over the slice to a depth of 1–2 μm. The final masking stage defines the interconnection patterns as in bipolar integrated circuit systems.

Although bipolar and unipolar transistors derive their names from the number of charge types involved in their action, there are two other fundamental differences between them. Firstly the bipolar transistor operation occurs at its base region which is buried some distance under the silicon surface, i.e. in its bulk. Hence it is often called a 'bulk' transistor. In a unipolar transistor, on the other hand, conduction occurs due to channel formation on the silicon surface under the thin oxide. Hence it is a surface charge device. For this reason a unipolar device is very susceptible to contamination, and this initially presented many difficulties during manufacture, which have now been overcome. The third difference between a bipolar and unipolar transistor is evident from an examination of their cross-sectional diagrams. Whereas the base metal connects directly to the silicon of a bipolar transistor, the gate of the unipolar device is insulated by a layer of silicon oxide. This means that the unipolar transistor has a high gate input impedance.

When unipolar devices are combined together in a silicon die a fourth difference between them and bipolar circuits becomes evident. *Figure 25.15(a)* shows two unipolar transistors on a die. Comparing this with *Figure 25.5* it is evident that the unipolar system is much less complex. Part of this is accounted for by the simpler transistor structure. Another reason however is that unipolar devices are self-isolating. There already exist surrounding areas of opposite polarity around each device so that the system resembles two reverse biased diodes shown in *Figure 25.15(b)*. By taking the substrate to the most positive

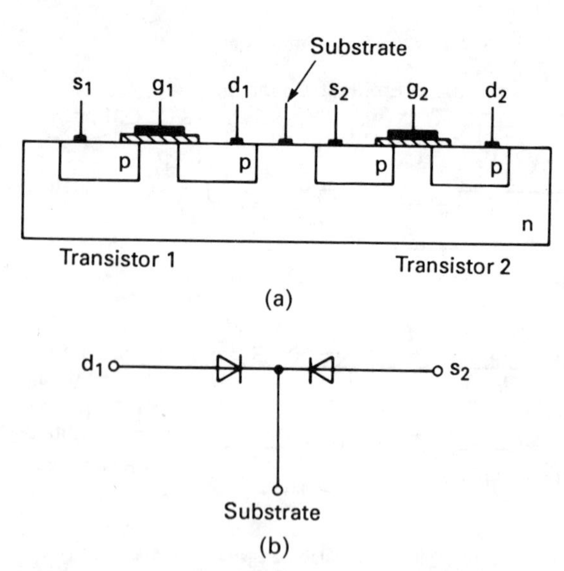

Figure 25.15 Self-isolation in unipolars: (a) two transistors on die, (b) diode isolation of (a)

potential in the system (most negative potential for n channel devices) isolation between transistors is obtained. This means that a unipolar system is usually capable of greater functional density than a bipolar system.

Figure 25.16 shows how the unipolar equivalent of *Figure 25.6* can be fabricated. The capacitor used in this instance consists of a

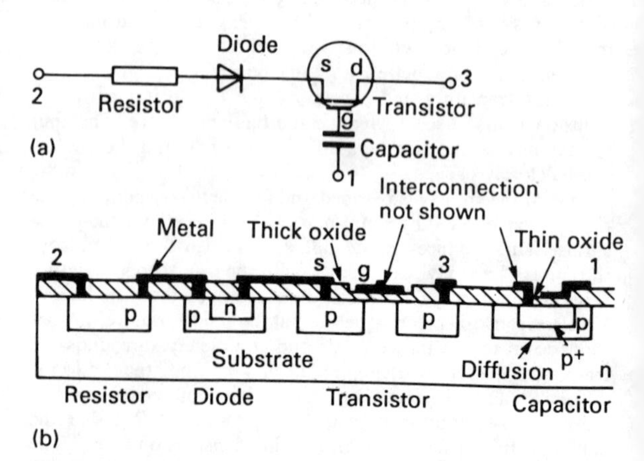

Figure 25.16 Unipolar integrated circuit: (a) functional diagram, (b) chip construction

dielectric (silicon oxide) separated by two conducting plates. The top plate is metal. The bottom plate is heavily doped silicon, which is a good conductor, and is eventually brought out to the metal interconnectors. The connection between the capacitor and transistor gate occurs round the side and is consequently not shown. Note also that thin silicon oxide is used as part of the capacitor and transistor gate whereas thick oxide forms an insulator between the metal interconnectors and the silicon surface, as in the bipolar circuit. It will be seen in the next section that the actual thickness of these two oxide layers is very important.

25.3.2 Developments in unipolars

Whereas the main battle in bipolar technology has been towards greater circuit density, unipolar systems have developed in two directions. In the early stages the prime objective was to reduce the operating voltages so that the unipolar circuit could be interfaced to traditional bipolar systems which run from 5 V supply rails. This objective having been achieved the efforts of the researchers turned towards making unipolar circuits faster. *Figure 25.17* illustrates only a few of the many systems which are currently in production.

Perhaps the easiest way of reducing the threshold voltage of a unipolar transistor, and hence reduce the operating voltage, is to use ⟨1 0 0⟩ silicon instead of the more usual ⟨1 1 1⟩ silicon. This gives a lower surface state charge and lowers the threshold from its original value of 3–5 V down to 1.5–3 V. However, the carrier mobility in ⟨1 0 0⟩ silicon is also lower so that the gain factor is seen to decrease giving lower operating speeds.

In an integrated circuit there are often metal interconnection tracks crossing two p regions, for p-channel devices, where a transistor should not exist. Transistor action is prevented by ensuring that the oxide, called field oxide, is now thick enough so that the threshold voltage needed to turn the transistor on is greater than the system voltage. For a unipolar device the gate oxide is therefore made thin while the remaining areas are kept thick. Unfortunately this thinness is limited by processing difficulties since there must be no pinholes, whereas if the oxide is too thick, say above 1.8 μm, then the aluminium conductors will have high steps which will cause them to crack. In ⟨1 1 1⟩ silicon devices the gate threshold is about 3–5 V and the field threshold 30–35 V. Going to ⟨1 0 0⟩ silicon reduces the gate threshold to 1.5–3 V but also lowers the field voltage to 15–20 V so that overall there is little benefit in this system and it is not frequently used.

An alternative unipolar structure is the metal–nitride–oxide–silicon (MNOS) device. The dielectric constant for silicon nitride is 7.5 compared to 3.9 for silicon so that when it is used as the gate material the threshold voltage is reduced to the 1.5–3 V range. Not only is the field threshold now not affected from its original 30–35 V value but the gain factor is doubled. Therefore the size of output transistors is reduced for the same current capability. A thin silicon oxide layer is generally used in these devices to prevent surface action between the silicon and the nitride. Pinholes are no longer a problem because of the relatively thick nitride layer so the oxide can be made very thin.

Another method of reducing threshold voltage is to use a polycrystalline silicon gate, as shown in *Figure 25.17(b)*. This reduces both the surface state charge (Q_{ss}) and the difference in work function (Q_{ms}) between the gate conductor, heavily doped

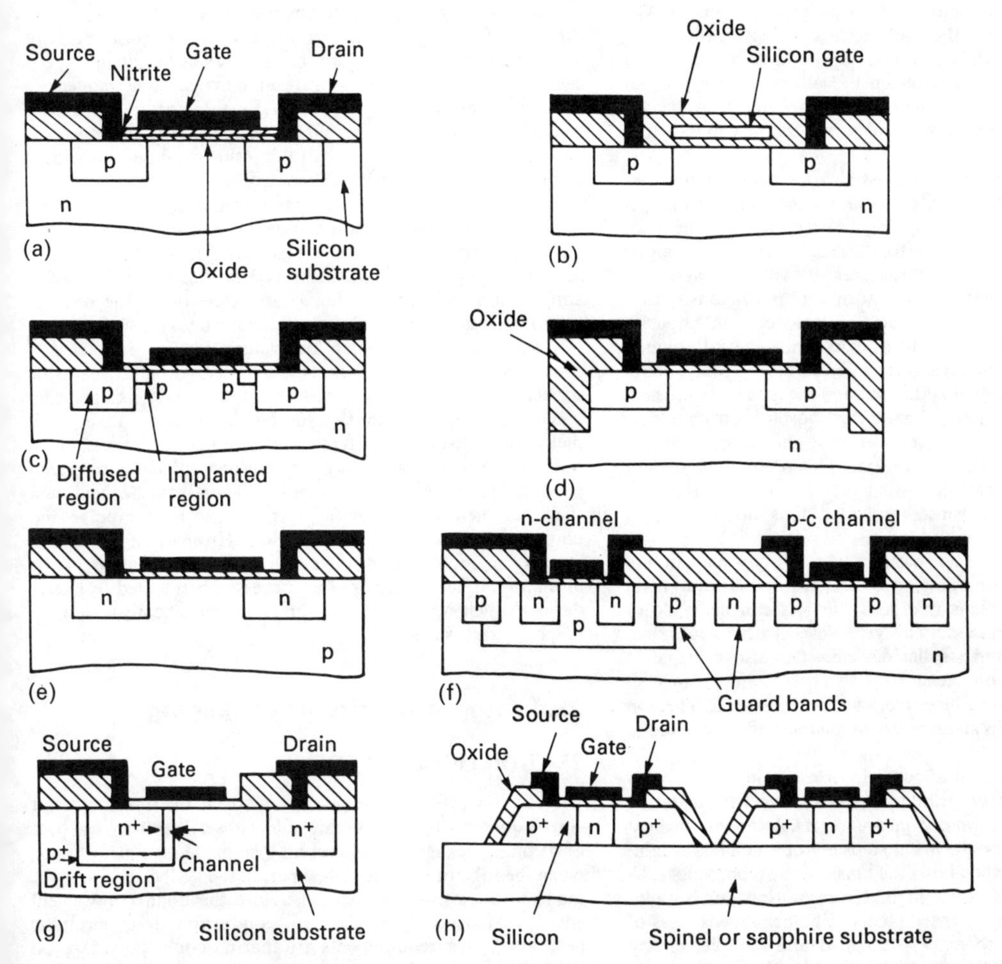

Figure 25.17 Unipolar processes: (a) MNOS, (b) silicon gate, (c) ion implant, (d) oxide isolated, (e) NMOS, (f) CMOS, (g) DMOS, (h) SOI

silicon, and the substrate. The threshold is again in the region 1.5–3 V. The doping concentration of the gate can be varied to give transistors with different thresholds if required in the circuit.

A very important advantage of the silicon gate structure is that it is a self-aligning process. To explain this let us look briefly at the production method used for the devices. Following the growth of the thick oxide a partial etch is used to form the thin oxide. A layer of silicon is then deposited over the entire surface, and oxide and silicon are then removed from the regions which are to form the source and drain. A non-mask diffusion is now carried out. Since the gate silicon and thick oxide will prevent impurity diffusion, the source and drain areas are very accurately defined. There is therefore very little overlap between the gate and these areas, giving low parasitic capacitance and fast operating speeds. For non-self-aligned processes it is necessary to build in tolerances for the fact that masks which define the diffusions and the gate metal cannot be exactly aligned. Therefore there can be considerable overlap between these areas. In the self-aligned processes the only overlap is that caused by the sideways diffusion under the gate, which always occurs.

There are other advantages to the silicon gate process. Reduced tolerances mean that devices can be packed closer together giving increased circuit density. The yield is high since the delicate gate oxide is covered up almost immediately by the silicon. The silicon gate can stand further high-temperature processes, which the aluminium gate cannot, so that unipolar and bipolar devices can be built on the same chip if required. The silicon gate also gives a third interconnection layer, the others being the metal, which is still used to interface to the outside world, and diffused conductors on the silicon surface, so that circuit density is further increased. The disadvantage of silicon gate is that it is a more complex process and can therefore be higher in cost.

Apart from diffusion it is also possible to introduce impurities into the silicon by ion implantation. In this the atoms, say boron for p type, are accelerated to a high energy so that they can penetrate the thin oxide and enter the silicon. The depth can be closely controlled by the accelerating voltage, and the concentration is adjustable by monitoring the boron ions. This is a low temperature process and since metal stops the impurity ions the system can be used to form devices with very little gate to source or drain overlap. As shown in *Figure 25.17(c)* the bulk of the source and drain are first diffused and the gate metal formed just short of these regions. High energy boron atoms are now implanted into the silicon to connect up to the original diffused areas without any overlap on the gate. This reduces the sizes of the devices and the parasitic capacitance even more than for silicon gate transistors since implanted ions do not diffuse sideways.

Ion implantation can also be used to vary the concentration of impurities in the channel region of selected transistors, metal being absent. This means that some devices can be made to operate in a depletion mode. This gives several advantages such as increased speed and smaller devices. It is also possible to implant resistor regions accurately so that a wider range of devices with a 2% absolute tolerance is attainable. The ion implantation process is also easily controllable and very suited to automation.

As in bipolar circuits it is possible to use oxide isolation in unipolar devices. Silicon nitride is used for selective oxidation giving the structure shown in *Figure 25.17(d)*. Although the field oxide is relatively thick the metal steps are of much less height. This prevents cracks and allows the masks to be placed closer to the silicon surface resulting in better resolution and smaller devices—hence reduced capacitance and higher speeds. It is of course also possible to make n channel and complementary transistors (described later in this section) by this process.

N-channel unipolar transistors (*Figure 25.17(e)*) have the advantage over p channel in that the electrons have twice the mobility of holes. This gives a higher gain factor and smaller transistors, with less capacitance and higher speeds. The thresholds are also lower giving much smaller operating voltages. However, initially the n-channel process was difficult to make. This was due to the fact that in silicon most of the mobile contaminants is positively charged. The positive gate potential used in n-channel transistors draws these to the silicon–oxide interface where they act so as to reduce threshold voltages and cause devices to behave unpredictably. In a p-channel transistor the negative gate bias moves the contaminants to the aluminium–oxide interface where they have less influence on threshold voltages. However, the n-channel processing problems have now been overcome and the device has replaced PMOS as the most frequently used unipolar transistor.

P channel and n channel transistors can be fabricated on the same slice and connected together to give complementary MOS (CMOS) devices. These have very low standby power dissipation, high operating speeds and many other circuit advantages. The disadvantage is larger size and greater processing complexity. *Figure 25.17(f)* illustrates the device. Guard band diffusions surround groups of like devices preventing leakage between them. Like NMOS and CMOS process has established itself as an industry standard for most applications.

A unipolar process which is capable of very high operating speeds is DMOS (double diffused MOS). This is shown in *Figure 25.17(g)*. A very narrow gate channel is established by double diffusion of a p^+ and n^+ region into a low-impurity drift region of the substrate. This drift region gives a very fast switching speed and together with the narrow gate enables operating frequencies well into several hundred megahertz. Furthermore it is capable of making high voltage devices if required. However, the process is relatively expensive since an extra diffusion must be used and all diffusions have to be precisely controlled.

Figure 25.17(h) shows the silicon on insulated (SOI) structure. In this a sapphire or spinel substrate, both of which have crystal structures very similar to silicon, is heated in an atmosphere of silicon-bearing gasses. The silicon is deposited on the insulating surface in a thin layer which is an extension of the original substrate. It is now possible to etch the silicon layer so as to form isolated islands into which transistors can be diffused and connected together. Clearly the leakage and parasitic capacitance of such a system is very low so that its operating speed is comparable to that of bipolar circuits. Like other dielectrically isolated systems SOI is also very resistant to radiation. A more complex system, but one which is currently in production, uses etchants to remove silicon from the back of a slice and then fills this with glass. Clearly both bipolar and unipolar devices can be built in SOI. However, it is still not possible to grow the silicon as an exact continuation of the insulator surface. Therefore the process is best suited to lateral devices which operate on the surface of the silicon, such as in unipolar transistors.

25.4 Integrated circuit packaging

25.4.1 Die bonding

Die bonding refers to the process in which the semiconductor chip is bonded down to the thick film substrate or onto the base of its final package. This bond may be made to a metal track, when the substrate of the chip is part of the system, or it may be made to an insulating area. Therefore the bond requirement could be that of good electrical conduction or of electrical isolation. Other requirements are that it should provide good mechanical strength, good heat conduction properties and should not create any unwanted p–n junctions.

There are generally three die bonding methods in use. These are eutectic alloying, soft soldering and methods using plastic adhesives.

25.4.1.1 Eutetic alloying

A eutectic alloy has a much lower melting point that that of any of its individual constituents. For die bonding a gold–silicon alloy is most commonly used. The area to which the die is to be attached is plated with gold and the silicon chip placed on it. The substrate is heated to between 400°C and 450°C and at this temperature gold and silicon form a eutectic alloy. The substrate is now removed from the heat and allowed to cool to give a strong alloy bond between the chip and the substrate. In a practical process the chip is normally held in a chuck and scrubbed ultrasonically on the bonding area in order to break down any oxides which may have formed on the silicon surface and so expose the silicon area (*Figure 25.18*). This usually takes place in an atmosphere of nitrogen since this discourages further oxide formation.

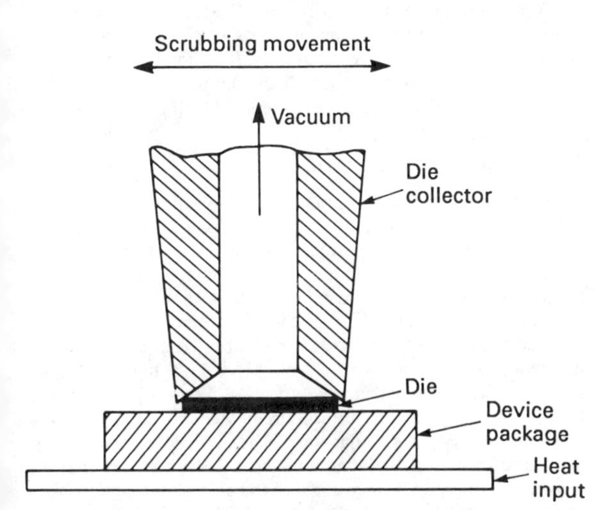

Figure 25.18 Eutectic die bonding

The gold–silicon eutectic alloy melts at 370°C (the melting point of gold is 1063°C and of silicon 1404°C). This is high enough to enable the bond to withstand subsequent processing steps, in particular the soldering of the completed device into a printed circuit board. The joint formed has excellent adhesion properties and heat conductivity. It is also capable of accommodating a slight amount of deformation. Its disadvantages are:

(1) It requires a metallisation area on the substrate.
(2) The high temperatures involved can affect any thick film resistors which have been fired on the same substrate.
(3) Due to the manual handling involved in the scrubbing process there is loss of yield caused by chips cracking or scratching.

The relatively high gold content of this alloy also makes it expensive.

In spite of its limitations the gold–silicon eutectic alloy is commonly used for die bonding. Either plain dice or devices with gold-plated backing may be used. The gold plating is expensive but results in a more tenacious bond. Apart from gold other gold alloys such as gold–silicon and gold–germanium are used in this process. Gold–gallium and gold–indium can be employed with p-type materials where it is required either to convert the die to a p type or to prevent n-type formation. Similarly gold–antimony alloys can be used with n-type dice. The melting points of all these alloys are in the region 340°C to 450°C. If a low temperature operation is required then gold–tin alloys (250°C) can be used. This also gives a very high conductivity between the chip and substrate.

25.4.1.2 Soft soldering

Lead–tin soft solders, of the type used in the assembly of printed circuit boards, can be employed to solder the silicon chip to the substrate. It is now necessary to ensure that both the pad area and the dice are gold plated for solderability. The resulting bond has good all round properties and requires a relatively low temperature, in the range 200°C to 300°C, for formation. The low temperature has the advantage of not affecting fired resistors on a thick film substrate. It does, however, mean that care must be taken during subsequent process steps not to exceed this value. The bond is also weak under thermal cycling conditions. It has the advantage that chips can be removed, with relative ease, from the substrate for replacement. Generally soldering is not used apart for large power chips.

25.4.1.3 Adhesives

Adhesives are used to stick the chip to the substrate when a low process temperature, below 200°C, is necessary. Silicon or epoxy adhesives may be used and the chip may be plain or gold backed. A measured amount of the adhesive is dispensed onto the bond area and the chip pressed on to it. Both conducting and non-conducting adhesives are available depending on whether or not an electrical contact is required to the bond area. The assembly is baked in an oven to cure the adhesive. This method of chip bonding has several advantages such as a strong bond formation, a low process temperature, absence of mechanical handling of the type required for eutectic systems and a thermal coefficient of expansion which can be readily controlled by modifying the composition of the adhesive. Its disadvantages are again the low temperature to which subsequent steps must be limited, the relatively high contact resistance to the bond, and the long curing times required. In addition there is always the risk of the adhesive contaminating the silicon dice.

25.4.2 Wire bonding

Having fixed the silicon dice to the substrate, wire bonding techniques must be used to electrically connect the metallisation pads on the dice to the hybrid conductor tracks or to the package leads. Thin gold or aluminium wires are most frequently used for this and the connections are made by thermocompression or ultrasonic bonding methods.

25.4.2.1 Thermocompression bonding

Thermocompression bonding depends on the fact that good molecular joints can be made between metals which are heated to well below their melting temperatures so long as the materials are of small dimension and sufficient pressure is applied to the junction area. Three methods are used to provide this force.

Ball or nail head bonding The basic steps involved in this process are illustrated in *Figure 25.19*. The fine gold wire, about 25 μm in diameter, is led through a heated tungsten carbide capillary tube and its tip melted by a hydrogen flame. This causes

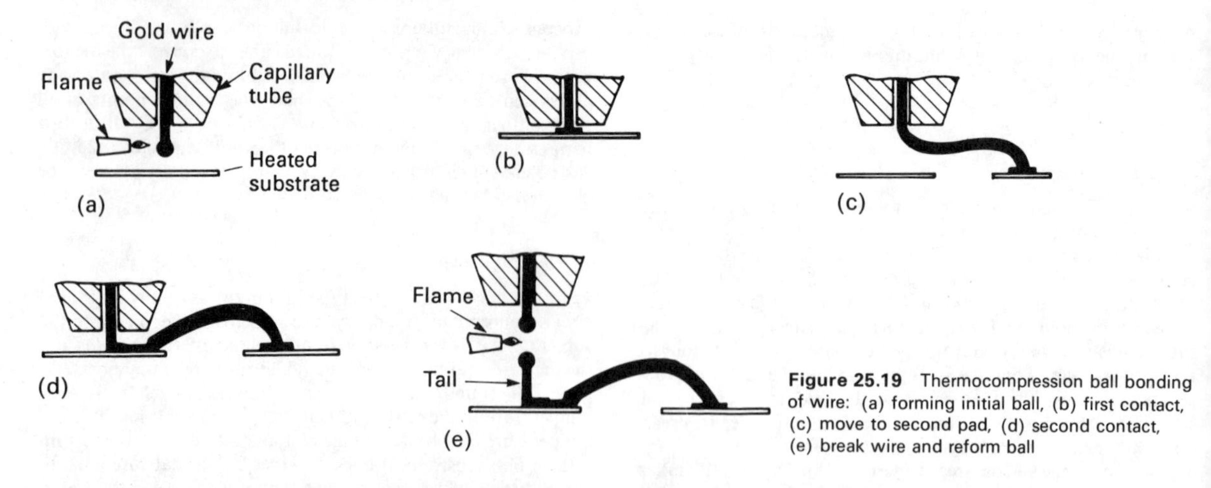

Figure 25.19 Thermocompression ball bonding of wire: (a) forming initial ball, (b) first contact, (c) move to second pad, (d) second contact, (e) break wire and reform ball

meltback to occur which results in a ball of about two to three times the original wire diameter. While the ball is still hot the wire is pulled so that the ball is held tight against the end of the capillary tip and this is then brought down hard onto the contact pad. This pad is initially maintained at a temperature of about 350°C so that the effect of heat and pressure results in a strong molecular bond with the wire. The capillary tube is now moved to the second pad, and the wire is fed through during transit. Once above the pad the tube is again brought down hard on the gold wire to deform it and form the second joint. The tube is now lifted and the hydrogen flame used to cut the wire and form a ball ready for the second wire bond. This operation leaves a pigtail of gold wire on the substrate which must be removed either manually with a tweezer or by automatic machinery.

Ball bonding results in a strong bond and enables fairly high production throughputs to be obtained. The bonds are, however, relatively large in size and the requirement of having a heated substrate can lead to deterioration of previously fired thick film devices. The operation must also be carried out in a clean environment since the capillary tube is very fine and the wires must move through it freely. Ball bonding cannot generally be used with aluminium wire since it does not form a ball when melted.

Stitch bonding In this system, the bond wire is again fed through a thin capillary tube. In the starting position this wire is hooked under the edge of the tube and contact is made by pressure on the first pad. As before the substrate is heated to about 350°C. The capillary tube then moves to the second pad and contact is made by pressure from the edge of this tube. Finally the wire is broken off by thermal shock or with a mechanical cutter. In either case there is no pigtail formation and the remaining wire is hooked under the tube and ready for the next bond. Stitch bonding is particularly suitable for making successive connections in stitch fashion without breaking the wire. It is capable of moving in any direction, as was also the case with ball bonding. Although both gold and aluminium wires may be used it is difficult by this method to break through the surface aluminium oxides and make reliable joints. Hence it is primarily used for gold wire bonding. Generally this method is not much used now, having given way to ultrasonic bonding techniques.

Wedge bonding Although similar in principle to stitch bonding this system uses a capillary tube to feed out the wire, and a separate heated wedge shaped tungsten carbide tool to provide the striking force which deforms the wire and makes the join to the pad. After the join the wire is pulled off at the bond and no pigtail is left. The substrate must once again be heated to about 350°C and gold wire is primarily used. Wedge bonding is slower than any of the other wiring methods since two independent tools need to be positioned accurately. It is, however, capable of making connection to much finer wires, down to a few micrometres, and to contact pads less than 20 μm square.

25.4.2.2 Ultrasonic bonding

Although ball bonding is commonly used for interconnections in hybrid circuits, ultrasonic bonding is more popular for packaging monolithic devices. The system has the merit of requiring no direct heat to either the capillary tube through which the wire is fed, or the substrate. It is primarily used with aluminium wires and is shown in *Figure 25.20*. The wire is pressed down on to the contact pad and the capillary tube is vibrated at an ultrasonic frequency in the region of 20–60 kHz. This high-frequency scrubbing action microscopically grinds down any unevenness between the two metal surfaces, removes surface oxides and results in a strong molecular joint. The absence of heat further deters the formation of aluminium oxides. It is important that the downward pressure of the tube and the chip clamping force are carefully controlled so as to optimise the dynamic stress pattern formed at the wire–pad interface by the sideways ultrasonic movement. The bond energy must exceed a given minimum but too large a value results in damage to the silicon dice.

Ultrasonic bonding tools are restricted in their movement so that constant realignment of the substrate may be necessary. Therefore they are less popular for use in hybrid circuits than ball bonding systems. However, they are generally capable of some 50% greater production rates than ball bonding and can form bonds which are two or three times more reliable.

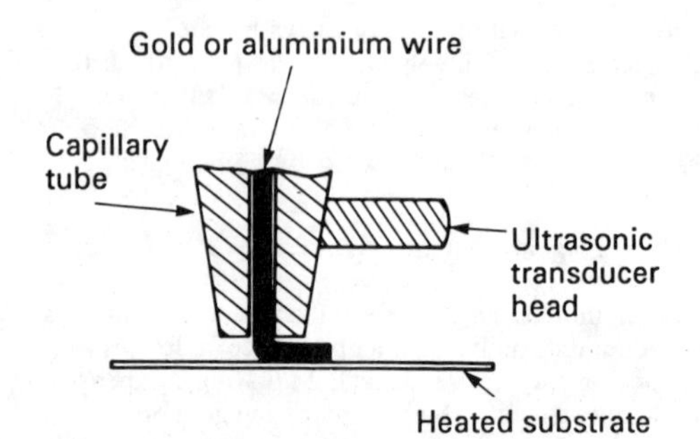

Figure 25.20 Ultrasonic bonding of wire

25.4.3 Film carrier bonding

The film carrier technique uses a continuous film of polyimide which has copper lead frames mounted on its surface. The whole system resembles a film spool. A window in the film allows the inner leads of the frame to protrude. It is to these fingers that the silicon die is bonded. After bonding the lead frame and die are wound onto a second spool for storage until required. The lead frames are then removed from the film and connected into the required circuit or package by thermocompression, ultrasonic or solder bonding.

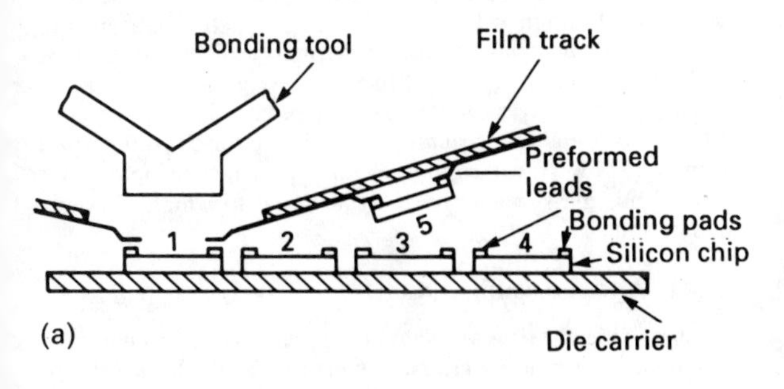

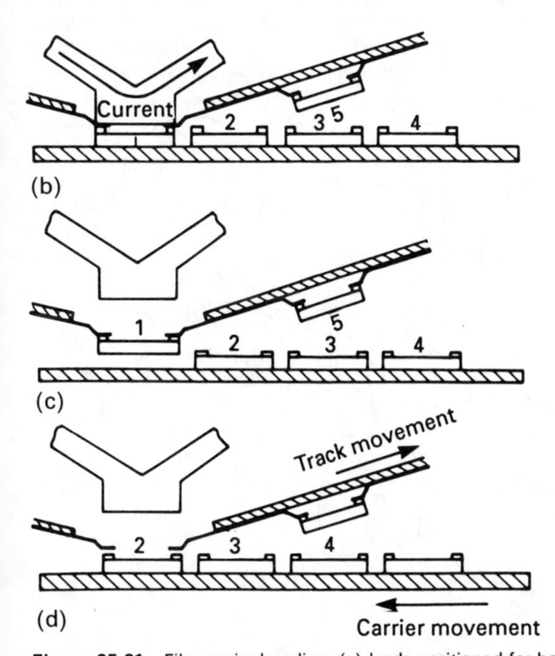

Figure 25.21 Film carrier bonding: (a) leads positioned for bond, (b) bonding operation, (c) bonded chip lifted off carrier, (d) next chip and leads in position

Figure 25.21 shows the steps used in bonding the lead frame to the semiconductor die. The inner leads are lowered onto the die in (*a*), the die having bumps as for flip chip applications and being secured to the die carrier by wax. An inert gas now flushes away the ambient air and the bonding tool descends onto the leads. A current pulse heats the tool in (*b*), forming the bond and melting the wax to release the die as in (*c*). The film carrier moves to the right as in (*d*) and the die carrier to the left so that a new lead frame and die are in position for another bond operation. Since all leads are bonded in one step this method is some ten times faster than wire bonding systems.

25.4.4 Plastic packages

The basic principle involved in a plastic package is to assemble the circuit on a substrate or metal lead frame and then to mould the entire structure, apart from the leads, in plastic to form the body of the device. The most commonly used plastic materials are epoxy, phenolic and silicone resins of which epoxy is the most popular. Several requirements are placed on the plastic material. It should adhere well to the lead frame so that it prevents moisture from creeping in along the package legs. The material must also have a thermal coefficient of expansion which is matched to the rest of the circuit in order to prevent stresses being set up in it which could damage the leads or the thin bonding wires. It is of course also important that the material does not contain or release any impurities which might contaminate the enclosed silicon.

Epoxy has relatively good characteristics for plastic packages. It is low in cost, chemically stable and capable of keeping out the effects of considerably hostile environments. In addition the material is capable of high mechanical strength, has good adhesion properties and is an excellent electrical insulator. Silicone compounds have a much higher resistance to heat than epoxy materials. However, they are structurally weaker and are attacked by some chemicals such as salts. Phenolics are structurally very strong and relatively economical to use, but they release water vapour during curing and can contain harmful chemical impurities which attack the enclosed circuit.

Plastic packages are low cost, especially when compared to hermetically sealed devices. They are also ideally suited to volume production techniques. However, their resistance to moisture and environmental contaminants is not as good and they can be damaged under thermal cycling conditions. Plastic packages also have lower heat dissipation compared to other types, especially metal packages. Heat dissipation properties can be considerably improved by wrapping the circuit in a metal surround before plastic encapsulation and in some cases bringing out part of this metal for bolting onto an external heatsink. Plastic packages are also usually characterised for operation over the industrial temperature range of 0–70°C while hermetic devices can cover the full military range from −55°C to 125°C. However, for normal industrial use, plastic devices are very suitable and with chips which have been passivated with a glass or silicon nitride layer these devices can be made to operate in fairly hostile environments.

Three methods are in use for making plastic packages.

25.4.4.1 Liquid casting

In this method a mould of metal or plastic material is used into which is placed the circuit to be packaged. A premixed solution of resin and catalyst is then poured into the mould so as to completely cover the circuit but to leave a part of its lead exposed. The mould is then baked in an oven to cure the resin. After setting the moulds may be left on the devices as permanent packages or removed and reused. It is possible in this method of moulding to introduce a degree of automation in the mixing and dispension of the encapsulating liquid. However, the method is generally relatively slow and requires long curing times. It is best suited to small batch production of a few thousand devices per week or where a long bake period is essential to fully de-gas the resin and produce higher reliability.

25.4.4.2 Transfer moulding

This is a suitable for making a wide variety of packages such as dual-in-lines and flat packs. It places no restrictions on the positions of the package leads and is capable of production rates of many thousand devices per hour. The assemblies which are to be packaged are put into the bottom half of a multicavity mould, each cavity corresponding to the position of a device. The mould is closed in a transfer moulding machine and the encapsulant material injected in fine pellet form from a nozzle. The pressure at which this injection takes place is closely controlled to prevent damage to wire bonds. Under pressure and temperature the pellets melt and flow in channels within the mould so as to fill the device cavities. The resin is now cured, while still in the mould, by the applied heat and pressure. This operation takes between 1 and 3 min, after which the mould is opened and the individual circuits separated from the lead frame assembly. If required further curing can now take place in an oven. Transfer moulding has very low operating costs although the initial price of the equipment can be high.

25.4.4.3 Fluid bed encapsulation (or conformal coating)

In this method of encapsulation only a thin layer of resin is coated over the surface and acts as the package. The coating provides a degree of protection against mechanical damage and it conforms to the outline of the encapsulated device. The system is very popularly used for low-cost hybrid circuits. The substrate is heated and dipped into a bed containing epoxy powder which is fluidised by having air blown through a porous bottom in the container. On contact with the heated substrate the powder melts and adheres to it. Several dippings are used to build up a sufficiently thick layer of epoxy over the whole surface, but this is usually done using automatic equipment with many substrates undergoing simultaneous coating. After dipping the resin is cured in an oven. This method is suitable for use with substrates having leads on one side only.

Conformal coating can also be done by another encapsulation method in which the completed substrates are dipped into a liquid bed of a thixotropic resin. Many more dippings are required before a covering of sufficient thickness is built up over the substrate. The curing period is also much longer than that required with fluid bed systems and reproducibility is poorer.

25.4.5 Hermetic packages

A semiconductor chip is sensitive to the presence of contaminants such as sodium ions, hydrogen, oxygen and water vapour. These reduce the collector–base breakdown voltage of transistors, increase the leakage current and parasitic capacitance, reduce current gain, and attack the chip metallisation and the wire bonds. Hermetic sealing aims to minimise these effects by first removing the contaminants from the package, usually by heating it to about 250°C, and then sealing it in the presence of an inert gas. During sealing the package temperature is kept to as low a value as practical.

A hermetic package usually consists of a base, a body, leads and a cover. Obtaining a seal between metal surfaces is relatively easy and is done by several methods such as welding, brazing and glass frit sealing. Glass-to-glass seals also present little problem since both surfaces can be fused at high temperatures or cemented by means of an interconnecting lower melting point glass. The problem arises when glass to metal seals need to be made, such as when metal leads pass through a glass package. The difficulties generally arise due to the unequal heat conduction and thermal coefficients of expansion of the two materials, which make the seal unreliable under temperature cycling conditions. Two methods are generally used to produce glass to metal seals. In the first method a thin oxide layer is grown on the metal surface prior to coating with glass. The surface of this oxide is dissolved into the glass and results in a smooth transition from metal to oxide to glass. This results in a good seal since the oxide in effect acts as a intermediate buffer. The second sealing technique uses a solder glass seal. This is a special composition low melting point glass which is coated onto the two surfaces to be sealed. On heating the interconnecting glass melts and wets the two surfaces, forming a good seal on cooling.

Three types of materials are used for hermetic packages: metals, ceramics and glasses. Of the many metals the most popular is a nickel–iron–cobalt alloy called kovar. It has a relatively low electrical and thermal conductivities and is compatible with standard sealing glass. Ceramics make excellent hermetic packages and are very commonly used. They have high thermal conduction and a thermal coefficient of expansion which is compatible with that of glass. Many different ceramics can be used, the most popular being alumina having a purity between 75% and 99.5%. For high-power dissipation beryllia is most common although other ceramics in production are steatite, forsterite, titanate and zircon. The thermal coefficients of all these materials are close to that of metal. Glass packages are cheap since they can give good inexpensive seals. They have poor conductivity and are generally not suitable for anything apart from very low power circuits. Hybrid packages using ceramic bases and glass walls have been used for higher power dissipations.

25.4.6 Package configurations

So many different sizes and shapes of packages have been used in the manufacture of integrated circuits that it would be folly to attempt a description of them all in this short section. Instead only a few of the more popular packages are considered here.

The TO type of package was a natural development from the transistor case. Various sizes are in use such as TO-3, TO-5, TO-

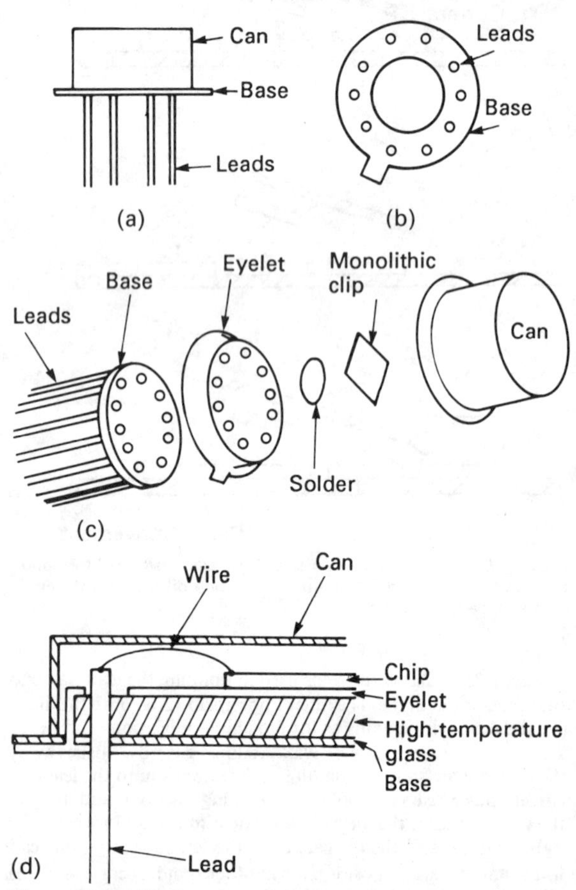

Figure 25.22 TO-5 package: (a) side view, (b) underneath view, (c) exploded view, (d) cross-sectional view

8, TO-99 and TO-100. The number of pins in these packages generally varies between 8 and 14. *Figure 25.22* shows the construction of one type of TO package. The base is a gold plated header made from kovar. The leads pass into the header via glass to metal seals. The die is attached by means of a gold–silicon eutectic solder. The can or cover is usually also made from kovar and is welded onto the header flange.

Flat pack encapsulations also come in many varieties. *Figure 25.23* shows typical structures. The die is bonded to the base by eutectic solder. The gold plated leads are embedded in a glass frame and the lid is sealed with a low-temperature glass frit. A metal flat pack uses kovar for most parts except for the walls between the ring and base which are glass. The leads pass through this. A glass package is usually made from borosilicate and consists of a one-piece base and ring assembly in which the kovar leads are sealed. This package has good thermal and mechanical properties. The lid can be made of glass, ceramic or metal and is attached by a low melting point sealing glass. Ceramic flat packs are similar to metal devices. The base, ring and lid are now usually made from alumina or, for higher power dissipation, from beryllia.

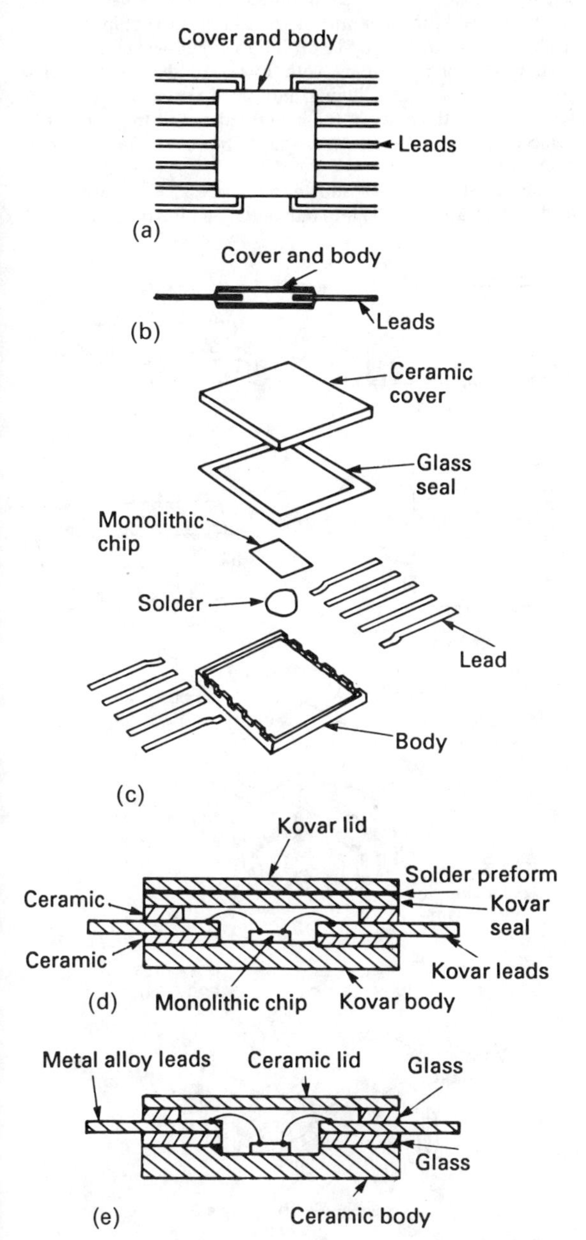

Figure 25.23 Flat package: (a) top view of 14-lead package, (b) side view, (c) exploded view of 10-lead ceramic package, (d) cross-section of metal package, (e) cross-section of ceramic package

The dual-in-line package is very popular especially for monolithic integrated circuits, since it is convenient to handle and can be readily adapted for automatic insertion into printed circuit boards. *Figure 25.24* shows a typical device, which can be made from metal or ceramic. The chip is placed in a cavity and bonded by glass frit. The leads also pass through glass frit seals in the package. The ceramic lid is initially metallised and then brazed or solder sealed to the body, or a glass frit can again be used to connect the two together. Plastic dual-in-line packages do not have a cavity for the chip. It is essential in these instances that the moulded package completely surrounds the die to protect it from the hostile environment, and all seals between the body and the leads are reliable. The overall outlines of these packages are essentially as in *Figure 25.24*.

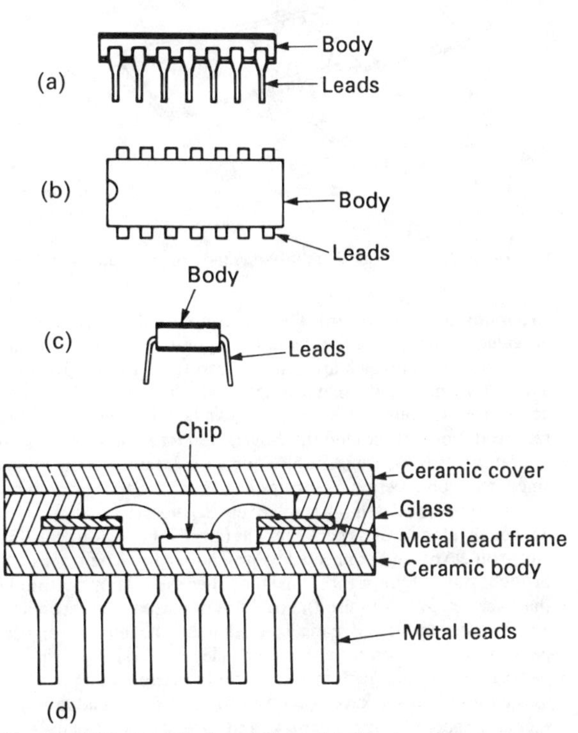

Figure 25.24 Dual-in-line package: (a) side view, (b) top view, (c) end view, (d) cross-section of ceramic package

Packages used to house LSI devices can present a problem. These chips are relatively large and generally have many outputs, typically about 40. The most popular configuration is at present still a dual-in-line, but due to the necessity of maintaining 0.25 cm spacing between pin centres the package tends to be relatively bulky. It is difficult to align the pins and to keep the device in position during insertion into a board or socket. The pins also tend to be fragile and break, and unsoldering a board to remove a device for replacement is often a very trickly operation. To overcome these limitations several different types of packages are in use. The edge mount package, shown in *Figure 25.25*, is similar

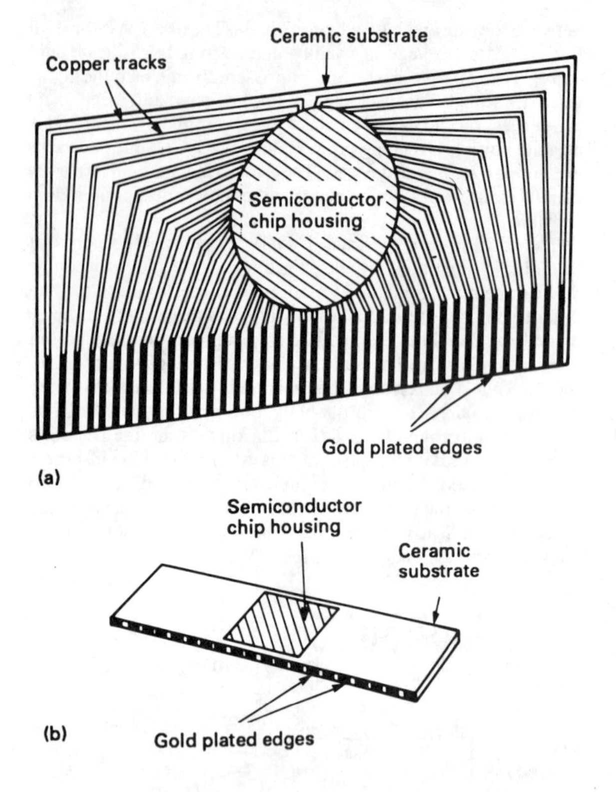

Figure 25.25 LSI packages: (a) edge mount, (b) leadless

to a miniature printed card. The chip is housed in a cavity which is sealed. Connections are made to the chip and the thick film conductor tracks which are printed onto the ceramic substrate. The tracks have gold plated fingers which plug into an edge connector. Although such a package can be quickly inserted and removed from its holder the fingers tend to wear with each operation. It is also unsuitable for many external connections since the long ceramic substrate has a tendency to bow. Furthermore the assembly is relatively expensive. An alternative approach is the leadless package. The chip is once again sealed in a central cavity and connected to printed tracks. The tracks are brought out to bumps (not pins) on either the side or bottom of the package. When in use the package is placed in a receptacle which has conducting pads located near the bumps on the package. Once the cover of the receptacle is closed it squeezes the package bumps and pads into close contact, connecting the chip into circuit. This package is claimed to have several advantages such as freedom from pin damage and ease of replacement. It is, however, still relatively expensive and not widely used in industry.

With the ever-increasing density of boards, surface-mount technology is becoming a necessity. Several packages have been introduced for this and a few of these are illustrated in *Figure 25.26*. Apart from having a footprint area which is about a third of that of comparable through-hole mounted components, surface-mounted packages are often the only economical way of handing devices with pin-outs greater than about 68. Pin grid array packages can be used, but are much more expensive.

Figure 25.26(*a*) and (*b*) illustrate the small outline integrated circuit package (SOIC). This has gull-winged leads which extend out from two sides of the package. The leads are close together, having a pitch of 1.27 mm. The package is thin and therefore takes up very little board width. The solder pads are also small and there is room for tracks to be run between the pads, when laid out on a printed circuit board. The package is self-aligning during the soldering process, the viscosity of the solder causing the leads to be pulled into place on its pads. Because the solder joints are visible they can be inspected, and manual or automatic soldering can be used, making the replacement of components easier. The disadvantage of SOICs is that the maximum number of pins is usually limited to about 28.

Placing pins on all four sides of the integrated circuit package increases the number of pin-outs. Plastic leaded chip carriers (PLCCs) can have gull-winged leads or J leads, as in *Figure 25.26*(*c*) and (*d*). *Figure 25.26*(*e*) shows the same package with the body cut away, to illustrate the bonding of the chip onto the leadframe. J leads take up less room than gull-wings, and give a high positioning accuracy with low lead deformation. The package is again self-aligning, but since the solder joints are formed under the body of the integrated circuit they cannot be inspected, and special tools must be used for replacing components.

The butt style package, shown in *Figure 25.26*(*f*), has straight leads, which are soldered into individual puddles of solder on the

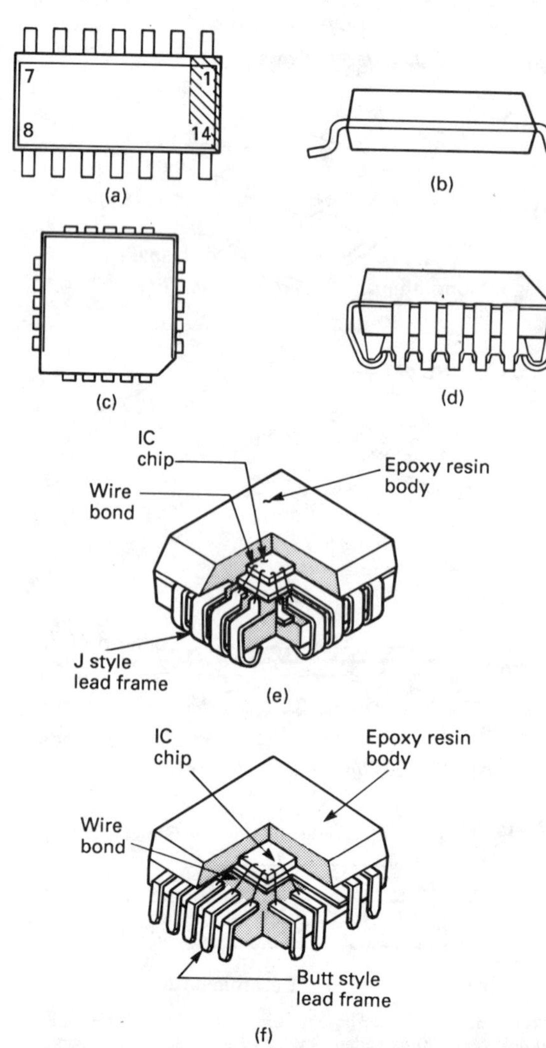

Figure 25.26 Surface-mounted IC packages: (a,b) SOIC, (c,d) PLCC, (e) PLCC sectioned, (f) butt style sectioned

pads. The package has low lead deformation and high positioning accuracy. Its solder joints can also be readily inspected. However, the package is not self-aligning, so accurate placement on a board is essential, and manual soldering is difficult to achieve.

Further reading

AHMED, A., 'Microlithography', *New Electroncis*, November 25 (1980)

AHMED, A., 'Chemical vapour deposition', *New Electronics*, December 9 (1980)

AHMED, A., 'VLSI—scaling down', *New Electronics*, March 24 (1981)

ANCEAU, F. and REIS, R. A., 'Complex integrated circuit design strategy', *IEEE J. Solid-State Circuits*, **SC-17**, No. 3, June, 459–64 (1982)

BAILEY, B., 'Ceramic chip carriers—a new standard in packaging', *Electron. Eng.*, August (1980)

BRADEN, J. S., 'Advanced surface mountable packages for VLSI devices', *Semiconductor International*, November (1987)

BRADEN, J. S., 'VLSI packaging and assembly', *Semiconductor International*, January (1988)

BURGGRAAF, P. S., 'GaAs bulk-crystal growth technology', *Semiconductor International*, June (1982)

BURGGRAAF, P. S., 'E-beam lithography: a standard tool?', *Semiconductor International*, September (1982)

BURGGRAAF, P. S., 'Magnetron sputtering systems', *Semiconductor International*, October (1982)

CHATTERJEE, P., 'Interconnections—the key to submicron VLSI', *Electronic Engineering*, February (1988)

DANCE, B., 'Below the micron threshold', *New Electronics*, June (1988)

DeSENA, A., 'Innovative packages emerge to carry faster, denser chips', *Computer Design*, 1 October (1988)

DOOLEY, A., 'Choosing a reliable IC socket', *New Electronics*, January 27 (1981)

EIDSON, J. C., HAASE, W. C. and SCUDDER, R. K., 'A precision high-speed electron beam lithography system', *Hewlett-Packard J.*, May (1981)

EL REFAIE, M., 'Chip-package substrate cushions dense, high-speed circuitries', *Electronics*, July 14 (1982)

EL REFAIE, M., 'Interconnect substrate for advanced electronic systems', *Electron. Eng.*, September (1982)

FARRELL, J. S., 'Trends towards alternative packaging', *Electron. Eng.*, Mid-September (1981)

GARRETTSON, G. A. and NEUKERMANS, A. P., 'X-ray lithography', *Hewlett-Packard J.*, August (1982)

GOULDING, M. R., 'Low *vs* high temperature epitaxial growth', *Semiconductor International*, May (1988)

GROSS, D. R., 'Packaging complex ICs', *New Electronics*, June 15 (1982)

GROSS, D. R., 'Trends in semiconductor packaging', *Electronics & Power*, October (1982)

GROSSMAN, M., 'E-beams, new processes write a powerful legacy', *Electronic Design*, June 7 (1980)

HACKE, H. J., 'Micropack packaging technology', *Siemens Res. Dev. Rep.*, **17**, No. 5 (1988)

HAYASAKA, A. and TAMAKI, Y., 'U-groove isolation technology', *JEE*, August (1982)

HAYDAMACK, W. J. and GRIFFIN, D. J., 'VLSI design strategies and tools', *Hewlett-Packard J.*, June (1981)

HESLOP, C. J., 'Reactive plasma processing in IC manufacture', *Electronic Production*, February (1980)

HOFFMAN, P. 'TAB implementation and trends', *Solid State Technology*, June (1988)

JONAS, A. W. and GARNER, L. E., 'Leadless chip carriers for LSI packaging', *Electronic Product Design*, April (1981)

JONES, G. A. C. and AHMED, H., 'Electron-beam lithography—a new approach to registration', *New Electronics*, August 12 (1980)

KOPP, R. J., 'Wafer processing and materials', *Semiconductor International*, January (1988)

LAND, I. G., 'Mounting leadless chip carriers', *New Electronics*, January 26 (1982)

LEIBSON, S. H., 'New package technology supports soaring IC and system complexity', *EDN*, 14 April (1988)

LOESCH, W., 'Custom ICS from standard cells: a design approach', *Computer Design*, May (1982)

LYMAN, J., 'Scaling the barriers to VLSIs fine lines', *Electronics*, June 19 (1980)

LYMAN, J., 'It's a three-way fight in high pin count IC packages', *Electronics*, October (1988)

McCORMICK, P., 'Designing pcbs for smt component packages', *Electronic Product Design*, June (1988)

MARCOUX, P. J., 'Dry etching: an overview', *Hewlett-Packard J.*, August (1982)

MARSTON, P., 'Future semiconductor packaging trends', *Electron. Eng.*, September (1982)

MARTIN, S. L., 'Smart power ICs get boost from new bipolar/MOS technologies', *Computer Design*, 1 May (1988)

MELLOR, P. J. T., 'Gallium arsenide integrated circuits for telecommunications systems', *Br. Telecom Tech. J.*, **5**, No. 4, October (1987)

MING LIAW, H., 'Trends in semiconductor material technologies for VLSI and VHSIC applications', *Solid State Technology*, July (1982)

MULLINS, C., 'Single wafer plasma etching', *Solid State Technology*, 88–92, August (1982)

NISHI, Y., 'Challenges in CMOS technology', *Solid State Technology*, November (1988)

NIXEN, D., 'Effects of materials and processes on package reliability', *Semiconductor Production*, September (1982)

OCHIAI, Y. *et al.*, 'Focused ion beam technology', *Solid State Technology*, November (1987)

O'LEARY, C., 'Isolation at the turning point', *New Electronics*, January (1988)

QUINNELL, R. A., 'Dielectrically isolated ICs move into high-performance applications', *EDN*, 13 October (1988)

RAPPAPORT, A., 'Automated design and simulation aids speed semiconductor IC development', *EDN*, August 4 (1982)

ROLFE, D., 'E-beam lithography for sub-micron geometries', *Electronic Product Design*, September (1981)

SAITOU, N. *et al.*, 'Electron beam direct writing technology: system and process', *Solid State Technology*, November (1987)

SHAH, G. N. *et al.*, 'Advances in wire bonding technology for high lead count, high-density devices', *IEEE Trans. (CHMT)*, September (1988)

SINGER, P., 'Today's plasma etch chemistries', *Semiconductor International*, March (1988)

SINGER, P. H., 'New directions in CMOS processing', *Semiconductor International*, April (1988)

SKIDMORE, K., 'Package trends for VLSI devices', *Semiconductor International*, June (1988)

STAFFORD, J. W., 'Reliability implications of destructive gold wire bond pull and ball bond shear testing', *Semiconductor International*, May (1982)

STENGL, G., KAITNA, R., LOSCHNER, H., RIEDER, R., WOLF, P. and SACHER, R., 'Ion projection microlithography', *Solid State Technology*, 104–9, August (1982)

TEXAS INSTRUMENTS INC., 'Technology and design challenges of MOS VLSI', *IEEE J. Solid-State Circuits*, **SC-17**, No. 3, 442–8, June (1982)

TSANTES, J., 'Leadless chip carriers revolutionize IC packaging', *EDN*, May 27 (1981)

TWADDELL, W., 'GaAs technology continues to advance, but proponents fear credibility window', *EDN*, February (1982)

VILENSKI, D. and MALTIN, L., 'Full automation of IC assembly will push productivity to new highs', *Electronics*, August 11 (1982)

WEARDEN, T. 'Applications of 111—V semiconductor materials', *Electronic Product Design*, March (1981)

WEISS, A., 'Hermetic packages and sealing techniques', *Semiconductor International*, June (1982)

WEISS, A., 'Power semiconductor packaging', *Semiconductor International*, August (1982)

WEISS, A., 'Plasma etching of aluminium: review of process and equipment technology', *Semiconductor International*, October (1982)

WEITZEL, C. E. and FRARY, J. M., 'A comparison of GaAs and Si processing technology', *Semiconductor International*, June (1982)

WINCHELL, B. and WINKLER, E. R., 'Packaging to contain the VLSI explosion', *Computer Design*, September (1982)

ZARLINGO, S. P. and SCOTT, J. R., 'Leadframe materials for packaging semiconductors', *Semiconductor International*, September (1982)

26 Hybrid Integrated Circuits

P Longland, PhD
STC Hybrids Division

Contents

26.1 Introduction

Integrated circuits can be classified into the major subdivisions of monolithic and hybrid. Hybrid circuits consist of two types: thick film and thin film. In physical appearance the two are very similar, but they differ in their manufacturing process, which determines their operating characteristics.

26.2 Manufacture of thin film circuits

There are five major steps in the manufacture of a thin film circuit. These are:

(1) Deposition of the film layer consisting of resistive, conductive or dielectric material.
(2) Patterning of the layer to form the required components and interconnections.
(3) Adjustment of the resistor and capacitor components formed in the film to give the required accuracy.
(4) Adding on the chip components (resistors, capacitors or semiconductors) and connecting them into the circuit.
(5) Packaging the completed film circuit assembly.

26.2.1 Deposition of the film

There are three major techniques currently in use to deposit thin films. These are evaporation, sputtering and ion-plating.

26.2.1.1 Evaporation

This is perhaps the most popular method and uses equipment very similar to that employed in evaporating the metal onto monolithic circuits, as described in *Chapter 25*. The substrate is placed around a source which is heated above its vaporisation temperature. A vacuum of between 10^{-5} and 10^{-7} Torr is maintained and under these conditions the mean free path of the evaporated molecules is much greater than that of the distance between the source and substrate. The molecules therefore radiate in straight lines onto the substrate. On reaching it they condense and form a thin layer. The substrate is heated by an auxiliary heater so that the adhesion between it and the film is improved. The overall temperature of the substrate is determined by these auxiliary heaters, plus the radiated heat from the source, plus the energy imparted by the arriving molecules.

There are three forms of evaporation, namely resistance heated, electron beam heated and flash heated source. In the resistance heated system the source is either in the form of a wire which is wrapped round a tungsten (or tantalum or molybdenum) heater through which current is passed, or else it is a powder which is placed in a resistance heated evaporation boat. In either case this method introduces impurities from the heater into the bell jar and it is usually limited to an evaporation temperature below about 1500°C. Both these limitations are overcome by the electron beam heated system in which a high energy beam of electrons is focused onto the source material held in a water cooled container. The beam is produced from an electron source which acts as the cathode of the system and the water cooled container is its anode. High vaporisation temperatures are attainable while keeping heater contamination to a minimum.

Neither of the two heated systems is suitable for evaporating alloys. Since the constituents of the alloy will have different vaporisation pressures they will evaporate in different amounts so that the film composition will be different from that of the source. This is overcome by using flash heating. In this a small quantity of the source material is dropped in powder form into a container which is kept at a very high temperature. The powder immediately vaporises completely and deposits onto the substrate. Another method of depositing alloys is to use a non-thermal process such as sputtering.

26.2.1.2 Sputtering

A glow discharge is formed in an atmosphere of argon at between 0.01 and 1 Torr by a high voltage between the source, which forms the cathode, and the anode, which incorporates the substrate. Argon ions are formed by this discharge and are accelerated to the cathode. On striking it they cause the release of the source molecules which acquire a negative charge. This causes them to be accelerated rapidly to the anode where they impinge onto the substrate and adhere to it. The process is known as sputtering.

Sputtering is slower than evaporation and requires more elaborate equipment. An evaporated film can be deposited at a rate of about 15 μm/min whereas the rate is limited to 1 μm/min for sputtering. Sputtering however has several advantages:

(1) It is a cold process since the source does not have to be heated to anywhere near its vaporisation temperature. Hence it can be used to deposit materials with a high vaporisation temperature such as tantalum, and also for alloys, which would give incorrect film constituents if vaporised.
(2) The molecules reaching the substrate have considerable energy so that the film density is high. The molecules on striking the substrate also ensure that all residual gas and other impurities are removed prior to and during the film formation. The adhesion is therefore very good. Both these advantages are also obtained with the ion-plated deposition method.
(3) Since negative ions are attracted by a positive substrate the three dimensional coating capability is also much greater than for vaporisation, and equal to that of ion-plating.

26.2.1.3 Ion-plating

This may be considered to be a mixture of evaporation and sputtering. A glow discharge is formed in a low pressure gas as in the sputtering method. The substrate is now the cathode. The film molecules are introduced into the discharge by evaporating the source using resistance, electron beam or flash heated techniques. These molecules are rapidly accelerated to the substrate and impact on it to form a strong bond. Since evaporation is used film formation is rapid but there is the risk of impurities being introduced. However, all the other advantages of sputtering are now applicable.

26.2.2 Patterning of films

The previous sections described how a thin film layer can be formed on the substrate. In this section the ways in which these layers can be modified to give the required pattern, whether it be resistor, conductor or dielectric, are described. Basically there are four systems in use:

26.2.2.1 Photoresist masking

This is probably the most common process. Initially the substrate is completely covered with a film layer. Photoresist is then spun over the film and it is exposed through a glass mask, in a similar process to that used for monolithic circuits. If positive photoresist is used, the exposed areas harden and are not removed in the subsequent wash. The surface is now etched to remove the film from the unprotected regions and finally the remaining photoresist is washed away. Clearly if several layers are to be formed (e.g. conductor and resistor) then these are both first evaporated on and then followed by photoresist masking and etchants which attack each layer separately.

26.2.2.2 Metal masking

This is an additive process since the starting material is the bare substrate. A metal mask with the desired film pattern cut out is put into contact with the substrate and a layer of film is evaporated on. The mask is then removed to leave the film on the substrate corresponding to the cut out regions. The metal mask must be placed in close contact with the substrate during evaporation to prevent diffused edges from occurring.

Metal masking does not involve the use of photoresists and is therefore a cheaper process. It is also useful when etchants cannot be used, for instance when SiO_2 is employed for capacitors, since etchants which dissolve this also attack the glass substrate. It is also much easier to reclaim the unused evaporated material, which may be a precious metal such as gold, from the mask than when they are part of the etchant chemicals. However, metal masking is not suitable for producing fine lines or when certain shapes are needed, for example a circle with a solid centre. They also become critical when several layers are to be formed requiring careful alignment of the successive masks.

26.2.2.3 Inverse photoresist masking

In this system the bare substrate is first coated with a layer of photoresist. This is then masked and etched to form a pattern in which photoresist is left on the areas which are to have no film. As such it forms an inverse mask. The film layer is next formed over the whole substrate. It adheres to the bare substrate regions. All the remaining photoresist is then dissolved carrying away with it the film from the unwanted areas.

Inverse masking methods are capable of very good line definitions due to the close proximity of mask and substrate. They are also used when the evaporated film is difficult to etch without attacking the substrate or the photoresist. The photoresist mask is however damaged by high temperatures and so must only be used with low-temperature deposition systems.

26.2.2.4 Inverse metal masking

This is very similar to inverse photoresist masking but uses a metal mask. The bare substrate is initially covered with a metal layer which is then etched to give the inverse metal mask. The required film layer is then deposited over this and finally the original metal mask is dissolved out with a special etchant taking with it the unwanted film.

26.3 Thin film components

This section is primarily concerned with components which are formed on a thin film substrate using one of the techniques described earlier. Devices in this class include resistors, capacitors, inductors and the interconnecting conductors between them.

26.3.1 The substrate

The starting material for any thin film circuit is obviously the substrate on which it is formed. This can be made from a variety of materials such as glass or pure glazed alumina. Thin films can also be formed on silicon dioxide which is very useful as it gives a method of combining monolithic and film circuits on a single chip.

Since thin films have relatively tight tolerances they place many demands on the requirements of the substrate. It should be clean and free from cracks and scratches. The surface of the substrate should be smooth and there should be no warping or dimensional changes with time. Such changes cause poor registration of masks and bad line definition. In addition the substrate should be a good electrical insulator since many devices are placed close to each other on its surface. For power circuits it is also important that the substrate is a good thermal conductor and in these applications it is sometimes mounted on an external heatsink.

26.3.2 Conductors

The conductor tracks on film circuits should provide a low resistance path and make low-loss contacts to both the resistor and capacitor films. Generally the minimum line widths and line spacings are limited to about 25–50 μm due to mask tolerances. The maximum thickness is determined by the available substrate area and the relatively high cost of the metal used.

By far the most popular conductor metal is gold. It has low resistivity and makes an excellent low-resistance joint to most resistor and capacitor plate materials. This joint takes the form of intermetallic bonds. However gold has very poor adhesion to the glass substrate due to the fact that no stable bonding oxide is formed at the glass–metal interface. It is therefore usual to cushion the gold from the glass by a thin (about 0.015 μm) nichrome layer, which does form a stable oxide with glass and therefore adheres strongly to it. The gold layer is usually about 0.5 μm in thickness. It will be seen in the next section that nichrome is the most popular thin film resistor material and the gold–nichrome combination is very frequently utilised. One of its further advantages is that the metals can be selectively etched to form the conductor and resistor patterns.

Aluminium is also used as a thin film conductor. Its obvious advantage is low material cost. However, apart from the disadvantage of higher resistivity, aluminium has the undesirable property of reacting with gold at relatively low temperatures. Gold wires are often used to connect add-on semiconductor devices to the conductor tracks, and if these tracks are made from aluminium then a chemical reaction occurs between them above a critical temperature. This results in the formation of a purple compound of $AuAl_2$ at the gold–aluminium interface, which eventually leads to device failure. This effect is called the 'purple plague'.

26.3.3 Resistors

Thin film resistors generally have more involved manufacturing processes than monolithic resistors but present several advantages. For instance the frequency response of a thin film resistor is several times better than that of a monolithic device. This is partly due to the lower parasitic capacitances associated with the resistors and partly due to their higher available sheet resistivity, which reduces the overall size. Thin film resistors can also be made to much tighter tolerances than monolithics since they may be adjusted to the required value after fabrication.

The most popular thin film resistor material is an alloy of about 80% nickel and 20% chromium, popularly known as nichrome. It has good adhesion properties and a low temperature coefficient, which is determined by several processing factors such as the rate of film deposition and the substrate composition. Typical resistor sizes are in the range 20 Ω to 50 kΩ. Nickel is unsuitable on its own for use as a film circuit material since it does not adhere adequately to the substrate. However, chromium has good adherence and is sometimes used for resistors. Its principal disadvantage is the tendency to form surface oxides which make it difficult to etch.

When high resistance values are required, cermets, which are compounds of a dielectric and a metal, are often used. They are usually formed by a two-step process. The first step lays down the film and the second step anneals it to the substrate. Since cermets are usually available in powder form they are ideally suited to vacuum deposition using flash techniques. It is generally difficult to control the thickness of a cermet film during deposition but it can be subsequently trimmed to give resistor values up to about 1 MΩ with a tolerance better than ±1%.

Tantalum resistors are characterised by excellent stability, a high value of dielectric field strength and high annealing temperature. Tantalum is capable of producing uniform and reproducible film layers which can also be oxidised to form a thin passivating layer. The material has a high melting point and must therefore be deposited by sputtering. Tantalum resistors can be produced with tolerances in the range ±0.1% with stability in this range over 25 years.

Table 26.1 summarises the characteristics of a few thin film resistors and compares these with monolithic devices.

Table 26.1 Characteristics of typical thin film resistors and comparison with monolithic devices

Material	*Sheet resistance** (Ω/square)	*Temperature coefficient* (ppm/°C)
Chromium	100–300	±50
Nichrome	100–400	±50
Cermets	1000	±100
Tantalum	10–500	±100
Diffused monolithic	2.5 (emitter) 100–300 (base)	500–2000

* refers to a square shaped material.

26.3.4 Capacitors

A thin film capacitor is basically a sandwich of a dielectric between two metal plates. The metal–oxide–silicon capacitor is shown in *Figure 26.1*. It can be used in monolithic or film circuits. It has the advantage over a diffused device of being insensitive to the polarity and the applied voltage and of having a relatively high Q due to its lower junction capacitances. *Figure 26.1(b)* shows its equivalent circuit. The parasitic capacitance is due primarily to the epitaxy–substrate diode, and the series resistance occurs in the diffused layer which forms the lower plate of the capacitor.

The thinner the oxide dielectric the higher the capitance value. However, the minimum thickness is usually limited to about 0.5 μm to give a uniform layer and prevent pinholes. The typical capacitance value is in the range 0.4–0.7 nF/mm² for an oxide thickness of 0.8–0.1 μm.

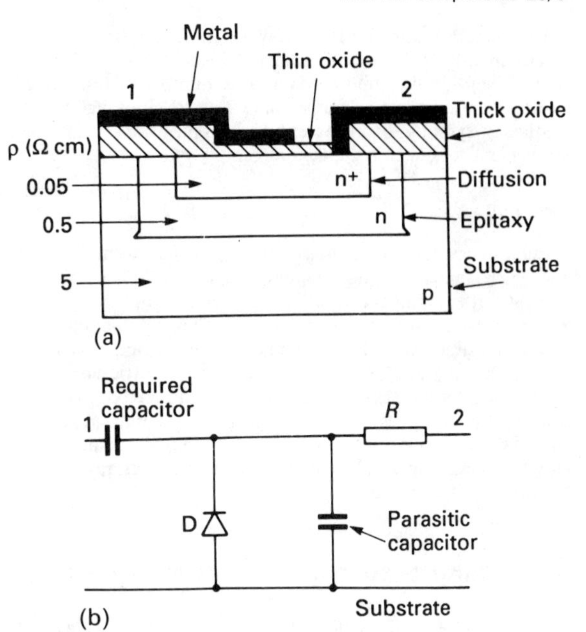

Figure 26.1 Metal–oxide–silicon capacitor: (a) cross-section, (b) equivalent circuit

Tantalum oxide (Ta_2O_5) capacitors are formed by first sputtering on a tantalum layer where the capacitor is to be located. This acts as the bottom plate. The tantalum is then oxidised to give a thin tantalum oxide dielectric layer and the capacitor is completed by sputtering on the top tantalum plate. These capacitors have very high dielectric constants and a good breakdown voltage. They are sometimes troubled by the movement of substrate faults through the metal and into the dielectric, which degrades their characteristic. This is usually corrected by an electrochemical back-etch process.

In an alumina (Al_2O_3) capacitor an aluminium layer is formed on the substrate to give the lower plate of the device. A thin layer of nickel is then evaporated onto the aluminium to prevent it migrating into the dielectric. Alumina is then deposited to form the dielectric and this is followed by a second nickel layer and then the top aluminium plate.

The properties of these thin film capacitors are summarised in *Table 26.2* and compared with a diffused monolithic device. Thin film capacitors are capable of larger values and lower parasitics than monolithic devices. They are also not dependent on the polarity and magnitude of the applied voltage provided that this

Table 26.2 Thin film capacitors and comparison with diffused monolithic devices

Dielectric material	*Dielectric constant*	*Capacitance* (nF/mm²)	*Temperature coefficient* (ppm/°C)	*Maximum operating voltage* (V)	*Q* (*at* 10 MHz)
Silicon dioxide	5.8	0.4–7	200	50	10–100
Tantalum oxide	20–25	2.0–3.5	150–300	20	50
Alumina (Al_2O_3)	9	0.6–2.0	400	20–50	10–100
Diffused monolithic	—	0.3	200	5–20	1–10

is below the breakdown value. However, thin film capacitors are not frequently used in practice, due to their requirement for more process steps, unless one needs a large number of low-valued devices on a substrate, as for example in the decoupling of high-frequency systems. In all except such applications it is more convenient to add chip capacitors.

26.3.5 Inductors

Inductors can be made quite simply in thin film technology by forming a circular or square conductor spiral on to the substrate. However, the inductance available is very low and the Q of the device is poor. Generally therefore thin film inductors are not frequently used except for some microwave hybrid applications in which devices are formed on high-quality dielectric substrates. The coils in these cases operate primarily as lumped components and have inductances of the order of a few microhenries and Q values of about 50. It is much more usual in film circuits to put tiny ferrite inductors onto the substrate as add-ons, rather than to form them as a film.

26.4 Manufacture of thick film circuits

The major stages in the manufacture of a thick-film hybrid circuit are as follows:

(1) Screen-print and fire lower conductor pattern.
(2) Screen-print and fire insulating layers.
(3) Screen-print and fire top conductor pattern.
(4) Screen-print and fire resistors.
(5) Passive laser adjustment of resistors.
(6) Add-on component attachment.
(7) Active laser adjustment of circuit.
(8) Final test.

26.4.1 Thick film screen preparation

To define the pattern geometry of printed thick film inks, three basic methods exist:

(1) Stencil screens, where a woven screen mesh, typically a plain weave, either stainless steel or polyester, has been coated with a photosensitive emulsion film and subsequently exposed to ultraviolet light to polymerise, and then developed to form the screen pattern.
(2) Metal masks, in which the pattern geometry has been directly etched in a sheet of solid metal, typically molybdenum or stainless steel, 0.05 to 0.25 mm thickness.
(3) Writing-in, where the pattern geometry is directly written onto the substrate base by the application of a computer controlled X–Y table, which moves the substrate to directly beneath a stylus. The stylus is fed from an ink reservoir. The coordinates for the X–Y table movement are generated by the direct digitising of the scaled layout.

A typical flow chart for the preparation of both stencil screens and metal masks is outlined in *Figure 26.2*.

In both the manufacture of stencil screens and metal masks, the production of 1:1 photographic masters from scaled master artwork is required. The master artwork may be prepared by a number of techniques such as applying opaque tapes to white paper or film, inking on a matt film, cut and peel off two-layer mylar backed film or the direct use of computer aided design systems. The selection of master artwork scale, typically 10:1, is controlled by the design aids such as plotters and the reduction camera in terms of lens focal length and resolution.

The process of screen printing by production equipment requires that the manufactured stencil screen or metal mask be

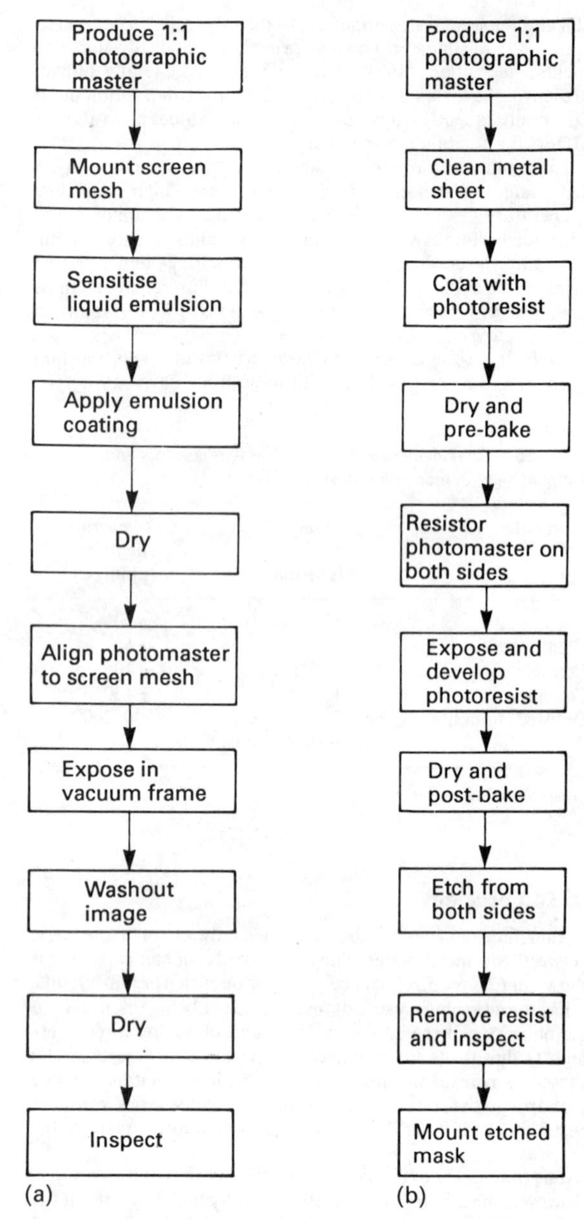

Figure 26.2 Flowchart for thick film screens: (a) stencilled screen, (b) metal mask

mounted onto a supporting frame, as indicated in *Figure 26.2*. This supporting frame may be fabricated from a number of materials, including wood, plastic, phenolics and metal castings, typically aluminium. The most commonly used frame material of those listed is cast aluminium due to the advantages of light weight, strength and stability.

Attachment of the stencil screen or metal mask to the frame is achieved by the use of one of a number of techniques including epoxy adhesive, staples, crimped tubing or metal strips. The selected dimensions of the mounting frame are totally dependent on the available printing equipment, however, sizes ranging from 7.62 cm × 7.62 cm up to 30.5 cm × 30.5 cm can be purchased for thick film printers, the most common dimensions in use being 12.7 cm × 12.7 cm and 20.3 cm × 25.4 cm.

26.4.2 Thick film deposition

Two basic methods of stencil screen or metal mask printing exist. These are: off contact printing, in which only the portion of the screen or mask directly under the squeegee is in contact with the substrate during the print cycle; and contact printing, in which the entire screen or mask remains in direct contact with the substrate during the complete pass of the squeegee and is then subsequently lifted or peeled away from the substrate. These techniques are shown in *Figure 26.3*.

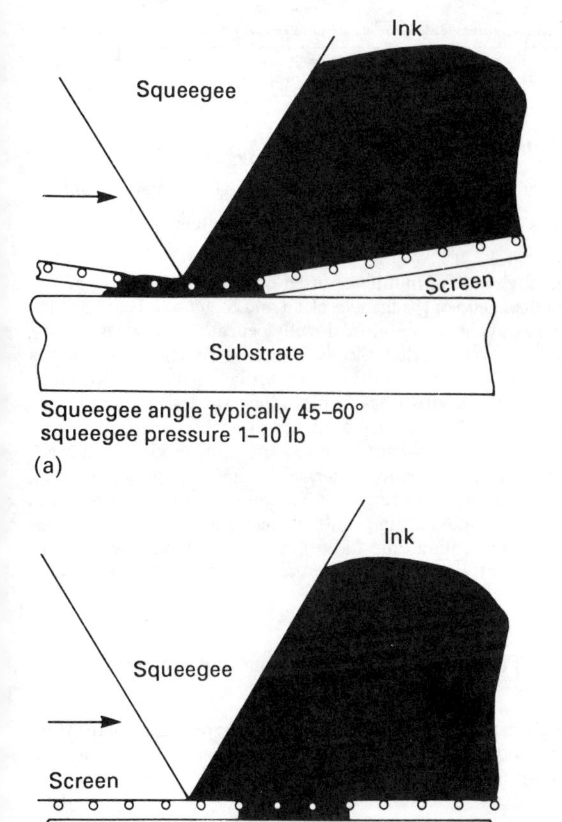

Figure 26.3 Squeegee, substrate, screen relationship during: (a) off-contact print cycle, (b) contact print cycle

The two major requirements for thick film printing process are conformity and predictability of print as shown diagrammatically in *Figure 26.4*.

The ideal print sequence would result in the complete filling of the defined pattern on the stencil screen or metal mask with ink during the squeegee pass, concluding with all the dispensed ink being transferred directly onto the substrate base in a uniform thickness and in full compliance with the defined pattern. The inherent process variables are:

(1) Ink rheology, in which the viscosity of the ink determines to a large extent the quality of the printed pattern.
(2) Screen printer settings, where print speed, squeegee pressure and angle, screen tension and screen to substrate snap-off significantly affect the print quality.

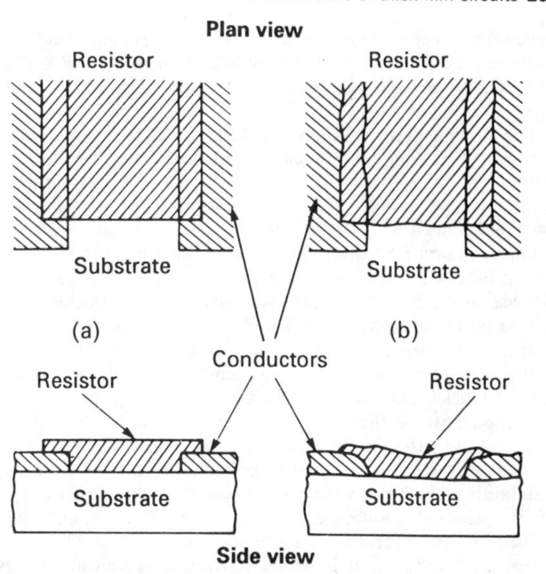

Figure 26.4 (a) Representation of an ideal thick film conductor and resistor print, (b) typical as-fired thick film conductor and resistor print

(3) Pattern geometry as defined on the stencil screen or metal mask, which affect print thickness and definition.
(4) Screen mesh size, where ink thickness and print definition become affected.
(5) Substrate physical characteristics of flatness and parallelism, which affect print thickness and definition.

These variables prohibit the ideal print characteristics being achieved. The introduction of specialised equipment such as a lunometer to calibrate the screen mesh count and viscometer to check the viscosity of the thick film inks assist in reducing the effects of process variables, but never completely remove them.

These limiting characteristics of the screen-printing process result in a minimum line width and spacing of 125 μm for conductor tracks printed directly onto the base substrate. For multilayer applications, where the number of individual conductor layers can be as many as 15, such minimum design rules must be increased to typically 250 μm to allow for the increasingly uneven surface receiving each printed layer. The use of multilayer printing techniques has increased significantly in response to the requirement for interconnection of high pin-count semiconductor devices, and five-layer multilayers involving 17 sequential print/fire operations are commonly encountered in today's hybrid manufacturing plants.

26.4.3 Thick film inks

The thick film ink has four main constituents: active material particles, binding agents, organic binders and volatile solvents. The active material consists of metal or metal oxide particles typically 2 to 5 μm in diameter, which determine the electrical characteristics of the fired film. In general, inks with metal particles are conductive in nature and inks with metal oxide particles are resistive. The selection of an ink system with a specific active particle content is based on the availability and applicability of the system to the design parameters and required processing conditions such as soldering, where the conductor leach resistance is essential, and wire bonding, where gold or aluminium wires are directly bonded to a conductor on the base substrate. Materials such as gold, gold–platinum alloy, palladium–silver, platinum–silver alloy and copper can be added to form the conductor film. The formulation of a resistor ink system can include palladium, silver–palladium oxide, and ruthenium dioxides.

The binding ingredients consists of a glass frit, typically a glass with a low melting point, such as lead borosilicate glass. The purpose of the glass frit is twofold. Firstly, the sintering or melting of the glass frit during the furnace firing provides an amorphous mass within which active particle to active particle contact can occur. Secondly, adhesion of the fired thick film ink to the base substrate is promoted by both the chemical and physical reaction of the glass frit with the substrate. Ink systems can be purchased where the glass frit has been replaced by a copper oxide, these systems being termed 'fritless' inks. They rely on the copper oxide providing a chemical bond between the active particles and the base substrate.

The purpose of the organic binder is to hold the active particles and glass frit in suspension such that the essential viscosity characteristics for screen printing are achieved, whilst maintaining a high rest viscosity to prevent the flow of the thick film ink prior to the squeegee pass and subsequently to retain the defined pattern on the substrate after the squeegee pass. The term 'thixotropic' is applied to those materials which exhibit such fluid characteristics.

The purpose of a volatile solvent is to reduce the 'yield point' of the thixotropic organic binder, the 'yield point' being the minimum force applied to the thick film ink by the squeegee at which the ink commences to flow through the stencil screen or metal mask. As either the 'yield point' is lowered or the applied force increased, the flow rate of the ink through the stencil screen or metal mask is increased. The volatility of the solvent system is one of the most important factors for the ease of screening. Excessively volatile solvents tend to dry out during use. This causes fluctuations in the obtained resistance values where resistor ink systems are being screened since the amount of ink deposited will vary as the solvent leaves the screened ink. Common resistor ink systems contain solvents such as terpineol, carbitol and cellosolve and their relatives and derivatives.

26.4.4 Thick film firing

When a film pattern has been screen printed onto the base substrate, the screened ink is dried by heating the substrate to about 125°C. This dries off the volatile solvent from the thick film ink. The screened substrate is then fired. The purpose of firing the ink is to burn off the organic binder and to allow the necessary physical and chemical changes to provide adhesion and the desired electrical properties to take place in the ink system. A temperature of about 350°C is required for the organic binder burn off and a temperature between 550°C to 1000°C is required to produce the changes in the ink system. The binding ingredient, if glass is used, must melt, wet the active particles and react with the base substrate. If a fritless ink system is used, the copper oxide must chemically react with the substrate to provide adhesion. The firing temperature profile affects the properties of the screened ink and must therefore be carefully controlled. In order to accomplish this, a means must be provided to gradually heat the screened substrate to the required temperature, maintain it at the temperature long enough for the desired physical and chemical reactions to take place and then return the substrate, gradually, back to room temperature. Within this sequence, time and temperature must be precisely controlled, the temperature typically within ±1°C. Two techniques exist for firing screen-printed thick film, the moving belt furnace, which is generally the commonest production technique, and the programmable kiln, which allows the temperature–time profile to be preset. *Figure 26.5* diagrammatically illustrates a moving belt furnace. It has a number of specific components. These include a muffle,

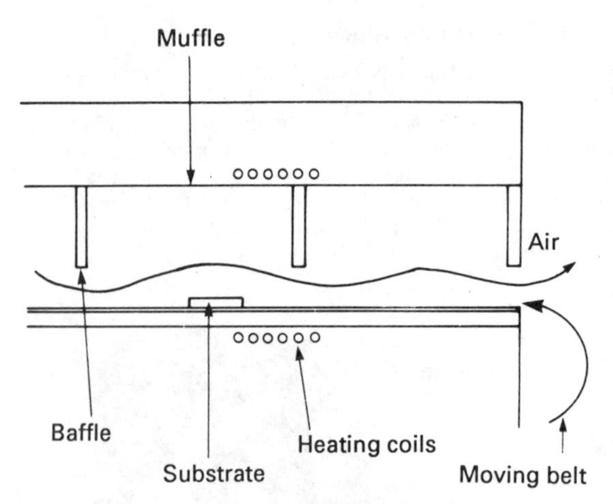

Figure 26.5 Cross-section of a moving belt furnace

generally made of mullite or fused quartz, which is designed to keep the inside of the furnace clean and to act as a heat radiator to even out the temperature directly beneath the heating coils; a continuous metal belt, typically fabricated from materials such as nichrome, which passes through the furnace at a controlled speed carrying screen substrates; baffles which are designed to provide control regarding furnace draft and atmosphere.

To achieve the essential temperature–time profile, the furnace is divided into sections, termed 'zones', each zone being surrounded by a heating coil. The minimum number of zones required for most firing profiles is four, a preheat zone, two hot zones and a cooling zone. Large furnaces may have as many as 13 independently controlled zones, in order to define the specific firing profile more accurately.

26.4.5 Thick film resistor trimming and substrate scribing

The use of thick film technology permits the designing of resistors with a wide range of ohmic values. The basic resistance, R, of a screen-printed resistor is given by:

$$R=\frac{\rho L}{WT} \tag{26.1}$$

where ρ is the bulk resistivity of the ink and L, W and T are the length, width and thickness respectively of the screen-printed resistor. Thus by choosing a resistor ink of suitable resistivity and by varying the relative values of L, W and T, resistors of the required values can be obtained. Practical limits are imposed by the minimum area which can be accurately screened using stencil screen or metal mask and by the maximum area of the substrates which can be allotted to a specific resistor. In addition, extremely low value resistors, typically 1 to 10 Ω, can be sensitive to screen-printed conductor tracks, where they are provided an interconnection pattern to the resistor. Whilst the basic resistance, R, can be determined theoretically, the inherent process variations apparent in the screening of thick film resistors result in the actual value being a variable. Typically, with the ink as purchased exhibiting a ±10% value variation and the screening process introducing an additional ±15% value variation, resistor values as fired can be within the band of ±25% of the nominal design value. This band can be reduced by additional process controls, such that resistors with a ±15% tolerance on the design value may be achieved without any additional processing subsequent to the substrates firing. In general, tolerance requiring greater accuracy than ±15% would be achieved by the trimming of the resistor. There are two basic techniques of trimming: abrasive trimming and laser trimming.

Abrasive trimming is a technique in which material is removed by using compressed air to force finely ground alumina through a small nozzle onto the fired resistor ink. The removal of resistive material increases the aspect ratio of the resistor and increases its value. *Figure 26.6* shows a sketch of an abrasively trimmed resistor. The value of the resistor is monitored during the trimming cycle by a resistance bridge. When the value of resistance is within the design technique, a signal from the bridge is sent to the valve controlling the compressed air, shutting it off and thereby stopping the trimming. Manufacturing tolerances of ±1.0% can be readily achieved by the use of air abrasive trimming. With the introduction of a two-part trimming technique which involves an initial coarse trim followed by a 24 h at 150°C stabilisation bake, prior to the second trim, manufacturing tolerances of ±0.1% can be achieved.

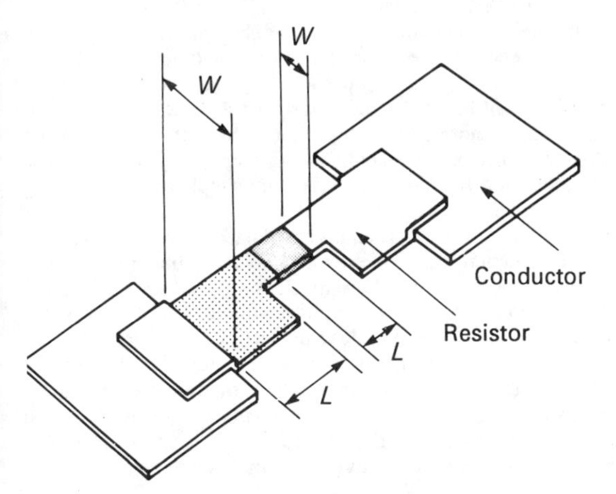

Figure 26.6 Air abrasion resistor trimming

Laser trimming is a technique in which the heat from a laser beam is utilised to evaporate resistive material, which creates a small channel or 'kerf' in the resistor, typically 20 μm wide. This re-routes the current through the resistor thereby effectively increasing the aspect ratio of the resistor and the resistance. The laser trim may consist of a single cut into the screened resistor or several cuts which could include L-shaped cuts where the rate of change in resistance introduced by each laser pulse is required to be minimal. The evaporation process melts the glass in the resistor, causing it to reflow and seal the kerf, reducing the long-term drift characteristics associated with abrasive trimming. In general, trimming lasers are YAG lasers (yttrium–aluminium–garnet), in preference to the gas lasers, such as the CO_2 lasers, which tend to produce too much power which in turn can produce excessive variation in trimmed resistor values. Manufacturing tolerances associated with laser trimming are comparable with those achieved with air abrasive trimming, however, speed of adjustment, cleanliness of working environment and resistor stability are considerably improved.

For applications requiring end-of-life resistor tolerances (typically 25 years) of less than ±1%, resistors are overprinted with a glass material prior to resistor trimming. This material holds the resistor in compression and prevents the formation and subsequent growth of micro-cracks in the resistor film during trimming.

The requirement for substrate scribing occurs when either multiples of individual circuits have been screen printed onto a single substrate base or when an individual circuit as printed is less than the base substrate in area. Typically, production hybrid houses utilise 102 mm × 102 mm, 96% aluminium oxide substrates though standard substrates ranging from 25.4 mm × 25.4 mm to 180 mm × 180 mm can be purchased. The conventional technique used for scribing the individual circuits from the base substrate is the manually operated diamond scribe which scores the substrate thereby permitting the breakout of the individual substrates. This technique is slow, however, and can be inaccurate. The use of a laser for substrate scribing, in particular a CO_2 base system, permits a rapid scribe speed and improved accuracy to be achieved, typical figures being 17.8 cm/s and ±0.01 mm respectively.

26.5 Chip components

In the previous sections of this chapter the characteristics of thick and thin film devices have been described. This section looks at the add-on components which are available for both thick and thin film circuits. These include resistors, capacitors, wound components such as inductors and miniature transformers, and semiconductors.

26.5.1 Resistors

Generally resistors are formed as films, rather than chips, in hybrid circuits. Instances when chip components are more economical are when a few devices are required in a particular resistance range which could involve a separate screen and fire operation. Generally a given resistor paste can be used to cover a resistance range of 10:1, by adjustment of its physical dimensions, before a higher or lower resistivity ink becomes necessary. Chip resistors are also used when special characteristics have to be obtained, such as high valued resistors in thin film circuits, low TCR resistors, large wattage resistors and high frequency resistors.

There are three basic types of chips which can be bought as standard products from several suppliers. These are thin film, thick film and cermet resistors.

Thin film chip resistors are formed by thin film deposition processes on an alumina (or sometimes glass) substrate. Materials used are generally chromium or nickel–chromium. Surface resistivities up to about 500 Ω per square are available giving resistors in the region of 1 MΩ with standard tolerances of 5–10% or lower if specified. These resistors have a TCR better than 75 ppm/°C and a temperature drift below 0.05% at 125°C for 1000 h.

Thick film resistors are formed on alumina substrates using screen and firing techniques with noble metal pastes. Unlike thin film resistors the thick film devices are more stable at high values of resistance since the larger quantity of binder acts as an effective sealant. A very high resistance value, however, is difficult to reproduce (unless trimmed) and the resistors are noisy. Resistivities up to 1 MΩ per square are commonly available giving a resistance range of 10 Ω to 15 MΩ with 5–10% standard tolerance without trimming. The TCR is between 30 and 300 ppm/°C with a temperature drift of 0.5% at 150°C for 1000 h. 1000 hours.

Solid cermet resistors are made from a mixture of metal, metal oxide, powdered glass and organic binder. The paste is formed into tiny blocks and fired at high temperatures to give a solid homogeneous mass. They are therefore equivalent to thick film resistors without substrates. Their electrical characteristics are very similar to those of the thick film devices described above.

Chip resistors come in a variety of different sizes and shapes, the smallest size being a block having dimensions approximately 1 mm × 1 mm × 0.2 mm. The chips can be bought with gold backing for eutectic bonding to the substrate and for interconnections into the circuit via bond wires. Alternatively they are available with metallised edges to enable them to be flow soldered. In either case the devices (or their bond wires) can often be positioned so as to provide crossovers between tracks, if these are required.

26.5.2 Capacitors

Screened capacitors are relatively difficult to produce since they require three different operations to form the two conductor plates and the intermediate dielectric. Furthermore the values of capacitance available are severely limited. For these reasons chip capacitors are very commonly used in film circuits and are consequently available in a variety of shapes.

The most popular chip capacitor dielectric material is a ceramic. The value of the dielectric constant in this type of capacitor can be varied in the range 10 to 10 000 by changing the ingredients in the ceramic while it is still in the slurry stage. Ceramic also has good physical strength and thermal tolerance to enable it to withstand the high soldering temperatures which may be attained during thick film assembly. Ceramic capacitors are basically of two types, high stability NPO (negative positive zero) and high K. NPO capacitors have dielectric constants in the region of 100, a linear temperature–capacitance curve, and a temperature coefficient of about ± 30 ppm/°C over the range -55 to $+125$°C. Capacitance values are fairly low since the physical size of the device is limited. However, by using multilayer devices, the overall plate area is increased while the dielectric thickness is kept small, giving higher capacitance values. In such a system alternate thin layers of unfired ceramic and a conducting noble metal paste are formed and the ends of the conductors connected to external solder pads. The whole stack is then pressed and fired to yield a ceramic block which looks like a homogeneous mass. The conductor thickness is now about 2 μm and the dielectric thickness about 20 μm. Although the total height of these devices is relatively large compared to printed capacitors they occupy about the same substrate area. Capacitor values between 1 pF and about 3 μF with voltage ratings of about 50 V are common.

High K ceramic capacitors are based on a dielectric of barium titanate. The dielectric constant can be as much as 8000 giving capacitance densities in the region of 10 $\mu F/cm^3$. These capacitors have a lower stability and a poorer temperature coefficient compared to NPO types.

For very large capacitance values tantalum capacitors are used. The dielectric consists of a porous tantalum slug anode which is anodised and coated with a deposit of manganese dioxide to form the cathode. Capacitor values are available from about 0.1 μF to several hundred microfarads within a voltage range of 5–50 V. Temperature coefficients are about 50–2000 ppm/°C for a range -50 to $+125$°C.

Chip capacitors can be bought with tinned metal ends for reflow soldering or with wire or ribbon leads. Alternatively they are available with metal areas for thermocompression or ultrasonic bonding. It is also possible to buy capacitors with electrodes on either side such that one end is soldered down to the conducting track whilst the second side is wire bonded.

26.5.3 Wound components

Very low value inductors, in the few nanohenry region, can be formed as metal spirals in thin film or thick film. To obtain larger value inductors or transformers it is necessary to add them in chip form. Miniature toroids are available for use as inductors or pulse and broad band transformers. These are capable of relatively high inductance values, in the region of 1 mH from a volume size 6 mm × 6 mm × 1 mm. The Q of these components is about 15 which is suitable for most applications. The main limitation of these devices is that the maximum continuous power which they can handle is of the order of a few milliwatts.

For applications requiring tunable inductors the rod type of construction is used. With adjustable metal rods of high permeability inductances in the region of 20 mH are available for rod diameters of about 1 mm. These devices have a high Q and good thermal stability. They are used in applications such as r.f. and i.f. amplifiers, discriminators, modulators and tuners.

26.5.4 Semiconductors

Perhaps the largest category of add-on components is semiconductor chips. These devices can range from small signal diodes and rectifiers to complex integrated circuits. For use in hybrid circuits all devices should be passivated to prevent contamination. This means essentially that they have to be fabricated as planar devices.

Although semiconductors can be bought in the form of miniature encapsulated components specially made for hybrid circuits this is a costlier approach than using unencapsulated chips. Its merit lies in the fact that since the device is packaged it can be tested more thoroughly prior to assembly so increasing the overall yield of the system. Packaged components can also be connected into the circuit by relatively simple reflow soldering techniques.

For minimum costs unencapsulated chips are usually connected directly into the circuit. Many techniques may be used to wire these devices. Alternatively semiconductor chips are available in a form which simplifies circuit connections. These include flip chip and beam lead devices.

In the flip chip bonding technique either the chip or the substrate to which it is connected is provided with contact bumps at the mounting regions. The chip is now inverted and bonded face down onto the substrate pads as shown in *Figure 26.7*. The chip is handled by a vacuum tube and the machine

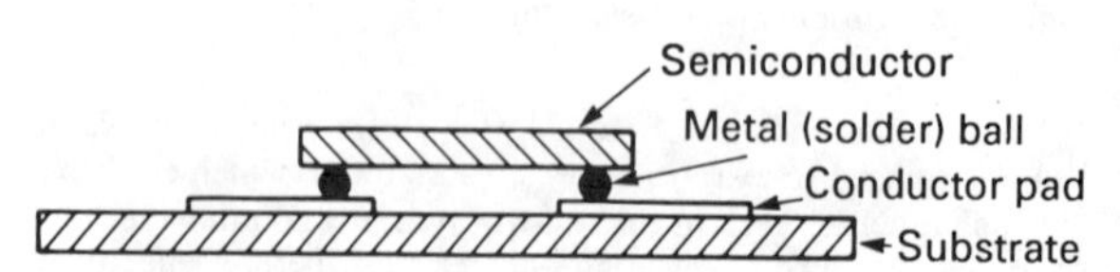

Figure 26.7 Flip chip mounted on substrate

presents to the operator a magnified split optical image of the underside of the chip, superimposed on to that of the substrate, so that alignment is simplified. The great advantage of the flip chip technique is that all bonds are made simultaneously and therefore assembly time is reduced. Furthermore electrical and mechanical connections are made in the same operation and a separate die bonding step is not required.

Flip chip bonds can be made by ultrasonic, thermocompression or solder techniques. In ultrasonic bonding the chip is pressed against the substrate and vibrated at about 60 kHz by a tungsten carbide tool. The pressure required depends on the chip size since the larger the number of bonds the greater the force. However, the chip may be damaged if this pressure is too large so that for big chips thermocompression bonding is preferred. Essentially thermocompression bonding of flip chips is similar to that used with bond wires. Pressure is applied to the chip and substrate and both are heated to about 350°C. The system needs careful matching between the temperature coefficients of expansion of chip and substrate to prevent fracture of the bond. It is generally best applied to gold-on-gold systems.

Both thermocompression and ultrasonic bonding methods apply pressure to deform the interconnecting bumps and make reliable contacts. This deformation should be due to plastic flow, and not elastic deformation, or the reliability of the contact will suffer after the pressure has been removed, and stresses will be set up in the chip which could damage it. An alternative bonding technique uses reflow soldering in which the eutectic solder is applied to the bumps or the contact pad. The chip is placed on the pad and heated to cause solder reflow. Enough flow must occur to fill all voids, but it must not be excessive or short circuits can occur between adjacent pads. Soldering occurs at low temperatures so that unequal coefficients of expansion of the chip and substrate do not present a problem. Soldered devices can be removed relatively easily for inspection or replacement. Chips connected by thermocompression or ultrasonic bonding may also be removed by rotating the devices so as to shear off the contacts. If the bumps are on the chip, however, this will result in the removal of part of the bonding pad so that if chip replacement is desired the pad must be large enough to accommodate it. Alternatively if the bumps are on the substrate then the bond between the bump and the substrate must be large enough to prevent it being removed with the chip.

In principle beam leads are very similar to flip chips except that in place of the contact bumps on the surface there is now a relatively thick wire lead which overhangs the surface of the dice. This lead is an extension of the normal aluminium metallisation of the pad, and since its thickness is of the order of 12 μm compared to 1 μm for the metallisation, it is called a beam. The beams are rigidly attached to the chip and act as electrical conductors as well as providing mechanical support. The chip is usually 'flipped' to connect it into circuit, as shown in *Figure 26.8*.

Figure 26.9 shows the cross-section through part of a beam lead chip. The processing of the devices, if it is bipolar, follows the usual steps right up to the stage where the emitter diffusions are made. The oxide layer is now etched to open up the metal contact areas. Silicon nitride is then deposited over the whole surface to provide a protective layer and the contact areas again etched through it. A thin layer of platinum is then sputtered on. This reacts with the exposed silicon to form platinum silicide which makes a low resistance contact. The unreacted platinum is then etched off. Alternative layers of first titanium and then platinum are next sputtered onto the surface. The titanium forms a strongly adherent surface to the nitride and the platinum acts as a barrier layer. The platinum is then etched to form the conducting paths and is plated with about 2 μm of gold. Selected areas of this gold are next etched to give beams about 70 μm wide and 200 μm long, which are further plated to a thickness of about 12 μm. The titanium layer is then etched off using gold as a mask and the silicon under the beams is also etched to give a wafer with each of its dice having cantilever gold contacts protruding over its edges. The wafer can now be probe tested and separated into individual dice using chemical etchants. To facilitate this process it is usual to use ⟨1 0 0⟩ oriented silicon for beam leads, instead of the more conventional ⟨1 1 1⟩ so that anisotropic etching can be employed. Beam leads can be attached to the substate in many ways, the most common being thermocompression bonding.

26.6 Hybrid packaging

The methods available for the packaging of hybrids are very similar to those described in *Chapter 25* for the packaging of integrated circuits. However, certain differences have evolved and these are described here.

For hybrid circuits containing plastic encapsulated silicon devices, i.e. SOT-23, SO-8, SO-14, etc., further encapsulation is no longer deemed necessary. Indeed the use of certain 'historically acceptable' organic encapsulants can result in a degradation of lifetime performance, and the vast majority of industrial and telecommunication hybrids are now being supplied and used 'unencapsulated'.

Where hybrid assemblies consist of a mixture of encapsulated and 'naked' semiconductor devices, encapsulation is applied locally to the 'naked' die. The encapsulation material can be either a purified epoxide resin or a silicone resin. Although providing better humidity protection to the device, silicone resins require a secondary encapsulation in situations where exposure to chlorinated solvents is involved. Such secondary encapsulation can either be an epoxide resin or a ceramic lid attached to the substrate with adhesive.

Conventional metal, hermetic, packages with glass-sealed pin-outs are still extensively used for the packaging of hybrids, containing wire-bonded semiconductor die, operating at temperatures above 70°C and below 20°C. These packages provide secure protection from the external environment for high-reliability professional space applications.

The hermetic properties of the hybrid substrate itself are now commonly utilised in a form of packaging known as 'integral substrate packaging'. The substrate is used as the base of the package and a hermetic enclosure is formed by either 'solder-sealing' or 'glass-sealing' a hermetic lid onto the substrate. Lead-outs are provided by conventional lead frame in either SIL or DIL format.

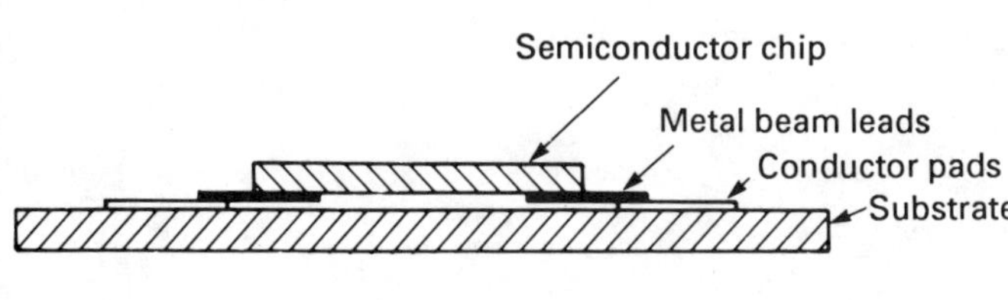

Figure 26.8 Beam lead mounted on substrate

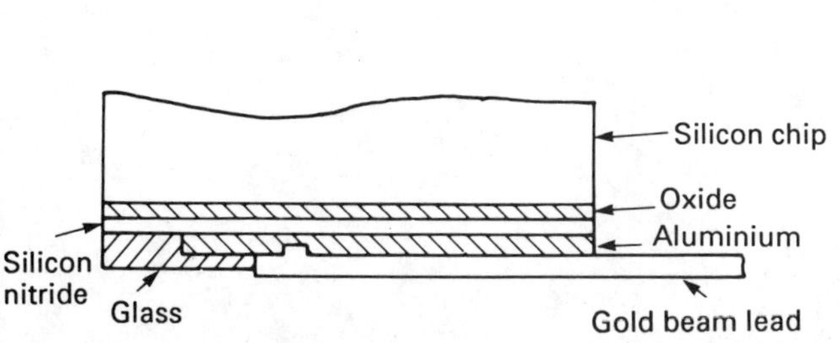

Figure 26.9 Cross-section of a beam lead

Further reading

ANON., 'Hybrid packaging technology broadens the use of new substrate structures', *Electronic Packaging and Production*, March (1983)

ANON., 'Hybrids offer alternative to higher chip densities', *Computer Design*, January 1 (1989)

BAILEY, B., 'Ceramic chip carriers—a new standard in packaging', *Electron Eng.*, November (1980)

COLEMAN, M. V., 'Ageing mechanisms and stability in thick-film resistors' *Hybrid Circuits*, **4** (1984)

DENDA, S., 'Trends in thick-film technology for electronic components', *JEE*, June (1982)

DETTMER, R., 'The silicon PCB', *IEE Rev.*, November (1988)

EMBREY, D. M., 'Thick film components', *New Electronics*, May 5 (1981)

HEID, K. K. W. and STEDDOM, C. M., 'Thick film, thin film: which to use and when', *Int. J. Hybrid Microelectronics*, **4**, No. 2, 376–8, October (1981)

HUGHES, J., 'Chips for hybrids', *New Electronics*, May 5 (1981)

KEIZER, A., 'Tape automated bonding systems', *New Electronics*, May 27 (1980)

LYMAN, J., 'Tape automated bonding meets VLSI challenge', *Electronics*, December 18 (1980)

NEIDORFF, R., 'Laser-trimmed thin-film resistors hold voltage references steady', *Electronics*, June 16 (1982)

Proceedings of the Sixth European Micro-electronics Conference, Bournemouth (1987)

ROTTIERS, L. *et al.*, 'Hot spot effects in hybrid circuits', *IEEE Trans.* (*CHMT*), **11**, No. 3, September (1988)

SERGENT, J. E., 'Understand the basics of thick-film technology', *EDN*, October 14 (1981)

SERGENT, J. E., 'Thin-film hybrids provide an alternative', *EDN*, November 25 (1981)

SINNADURAI, N., 'Trends towards miniaturisation, micro-packaging and cost effectiveness of interconnection technologies for telecoms applications', *Br. Telecom Tech. J.*, **1**, No. 1, July (1983)

THACKRAY, I., 'Using chip carriers', *Semiconductor Production*, February (1982)

TRAVIS, W., 'Thin film hybrid circuits', *New Electronics*, October 14 (1980)

WEARDEN, T., 'The technology of thick-film hybrid circuits', *Electronic Product Design*, October (1980)

27 Digital Integrated Circuits

F F Mazda DFH, MPhil, CEng, FIEE, DMS, MBIM
STC Telecommunications Ltd

Contents

27.1 Introduction

Digital circuits may be differentiated in two ways: by the function they perform and by the technology in which they are made. Therefore a CMOS shift register will perform the same function as a TTL shift register, but it will have different parameters such as speed of operation and power consumption.

In this chapter the basic parameters used to measure the performance of digital circuits are first introduced. This is followed by a description of the different digital integrated circuit technologies, and their functional systems.

27.2 Logic parameters

Although there are many parameters of interest, such as environmental operating range, availability and cost, five key parameters are generally used in comparing digital circuit families.

27.2.1 Speed

This indicates how fast a digital circuit can operate. It is usually specified in terms of gate propagation delay, or as a maximum operating speed such as the maximum clock rate of a shift register or a flip-flop.

Gate propagation delay is defined as the time between equal events on the input and output waveforms of a gate. For an AND gate this is the delay between a point on the last changing input waveform and an equal point on the output waveform. For an OR gate the first changing input waveform is used as the reference.

27.2.2 Power dissipation

This gives a measure of the power which the digital circuit draws from the supply. It is measured as the product of the supply voltage and the mean supply current for given operating conditions such as speed and output loading. Low power dissipation is an obvious advantage for portable equipment. However, since the amount of power which can be dissipated from an integrated circuit package is limited, the lower the dissipation, the greater the amount of circuit which can be built into a silicon die.

27.2.3 Speed–power product

Compromise is generally required in the speed and power dissipation since a circuit which is designed to have high speed also has a high dissipation. *Figure 27.1* shows typical speed–power curves where the speed is measured in terms of gate delay. Curve A has an overall better speed–power product than B.

27.2.4 Current source and sink

This measures the amount of current which the digital circuit can interchange with an external load. Generally digital circuits interconnect with others of the same family and this parameter is defined as a 'fan-out', which is the number of similar gates which can be driven simultaneously from one output. The 'fan-in' of a circuit is the number of its parallel inputs.

27.2.5 Noise susceptibility and generation

Digital circuits can misoperate if there is noise in the system. This may be slowly changing noise, such as drift in the power supplies, or high energy spikes of noise. Some types of logic families can tolerate more noise than others, and generally the faster the inherent speed of the logic, the more it is likely to be affected by transient noise.

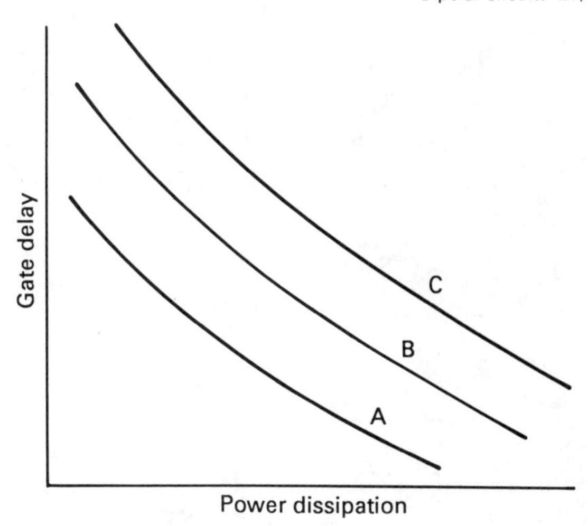

Figure 27.1 Speed–power curves

Digital circuits also generate noise when they switch and this can affect adjacent circuits. Noise generation is a direct function of the current being switched in the circuit and the speed with which it is changed.

27.3 Bipolar circuits

27.3.1 Saturating circuits

One of the earliest digital circit families was that based on the resistor–transistor logic (RTL) gate shown in *Figure 27.2(a)*. If any of the transistors TR_1 to TR_3 is turned on by a signal on its base, then the output voltage at *D* falls to a low value. The disadvantage of this circuit is that it has low noise immunity, since the noise voltage has only to overcome one base–emitter junction voltage, and it is slow since input capacitors have to be charged and discharged via the input resistors.

By putting capacitors across input resistors, R_2 to R_4, the operating speed is increased, and this is known as resistor–capacitor–transistor logic (RTCL). The disadvantage of RTL and RTCL is that they occupy a large area of silicon because of the passive components.

In diode–transistor logic (DTL), the input resistors are replaced by diodes. Some DTL circuits include a second transistor, as in *Figure 27.2(b)*, to increase the circuit gain. Diodes D_1 and D_2 increase the noise immunity. For operation in very noisy environments high noise immunity logic (HNIL) is used. This is also known as high threshold logic (HTL). It is similar to DTL except that a zener diode D_1 is introduced to increase the voltage levels. This circuit is also now largely obsolete.

The most popular digital family is transistor–transistor logic (TTL) and it is available in many configurations, a few of which are shown in *Figure 27.3*. If any of the inputs *A*, *B* or *C* is low in *Figure 27.3(a)*, then TR_3 is off. Transistors TR_1 and TR_2 form a totem-pole output stage in which the output at *D* swings between V_{cc} and 0 V. Since only one of the output transistors is on at a time, resistor R_1 can be made low valued without risk of overdissipation. This enables load capacitances to be charged and discharged rapidly, so that operating speed is increased. However, due to charge storage effects, a transistor switches off more slowly than it switches on, so that for a short time both

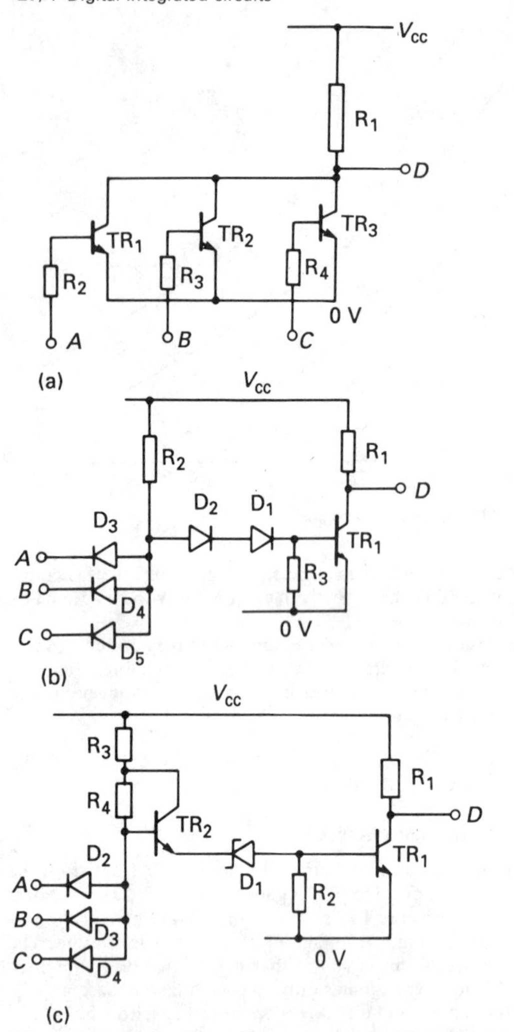

Figure 27.2 Earlier logic families: (a) RTL, (b) DTL, (c) HNIL

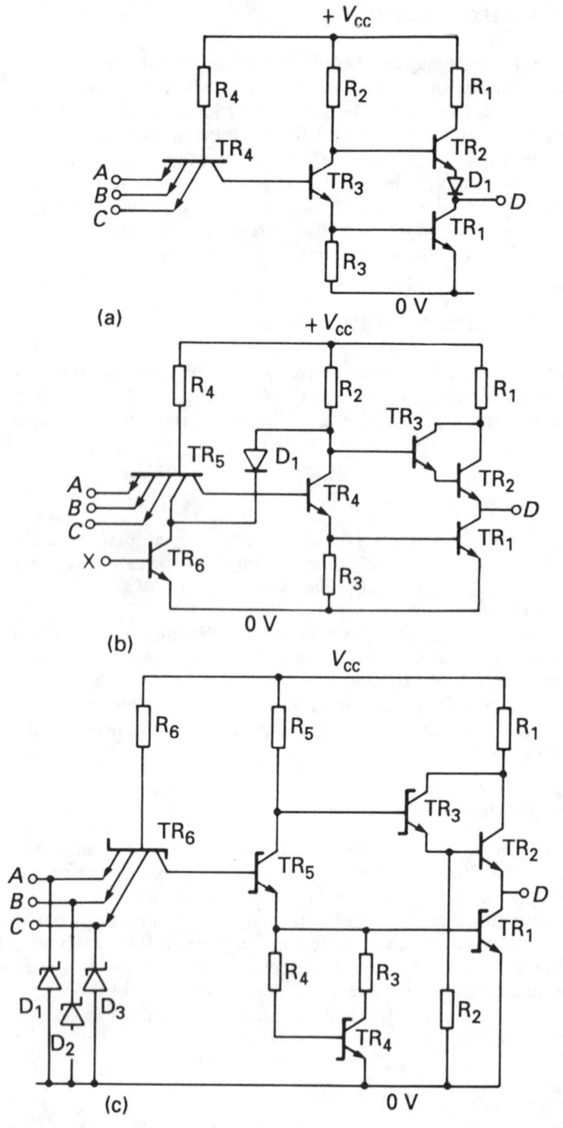

Figure 27.3 Transistor–transistor logic: (a) basic gate, (b) gate with Darlington output and tri-state control, (c) Schottky gate

transistors are on, generating current noise. This can be reduced by making R_2 and R_1 of high ohmic value, and this also reduced the power drawn from the supply. These are called low-power TTL gates (LPTTL).

Figure 27.3(b) shows a Darlington output gate with tri-state control. When input X is taken to a positive voltage, both TR_2 and TR_1 are off, so that the output is at a high impedance. This feature enables the output of several TTL circuits to be connected in parallel.

27.3.2 Non-saturating circuits

The circuits illustrated in the last section are called saturating since the transistors operate either in the off mode or fully saturated. If the transistor is prevented from saturating, then hole storage effects are reduced and the operating speed can be increased.

One method of reducing the hole storage within a device is to dope it with gold. This is expensive and can result in yield loss. A better technique is to use Schottky transistors in the TTL gate. Because of the clamping effect of the collector–base diode, a Schottky transistor can be prevented from saturating even when fully on, and it can therefore be designed to have a higher gain than a conventional transistor. However, the Schottky junction introduces capacitances so that careful selection of circuit components is required.

Schottky transistor–transistor logic (TTL-S) and LPTTL-S, which compensates for some of the speed lost in LPTTL circuits, are currently popular logic families, and are available under a variety of names from several manufacturers.

The fastest logic family commercially available is emitter-coupled logic (ECL), and it is also non-saturating. *Figure 27.4* shows the construction of a three-input ECL gate. A stable reference voltage is generated at point X by transistor TR_3 and its associated circuitry. Transistor TR_4 and the three input transistors form a differential amplifier. Depending on the input voltage relative to that at X, the current is switched through the circuit and either TR_1 or TR_2 is on.

In *Figure 27.4* none of the transistors saturate so the circuit has a high operating speed. The output transistors also have a low impedance when turned on so that they can provide high current drives and rapidly charge external circuit capacitances. The gate

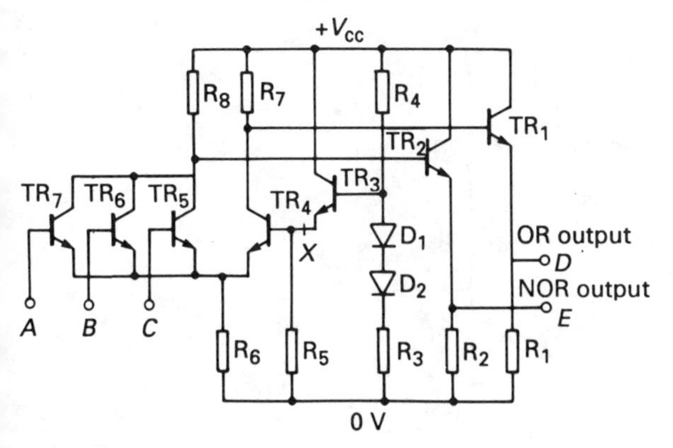

Figure 27.4 An ECL gate

is also capable of producing both OR and NOR functional outputs, and since it draws almost constant current from the supply it generates negligible switching current spikes. However, the logic voltage swings are low, usually below 1 V, so that it is difficult to interface ECL directly with other types of logic.

27.3.3 Bipolar LSI

Unipolar logic is still more popular than bipolar in large-scale integration (LSI). However, several circuits have been designed which show some advantage over unipolar in certain LSI applications. Only one such system, integrated injection logic (IIL), is described here.

In IIL all emitters go to a common point and so do the bases of transistors connected to the same resistor. Only the collectors of the transistors are separate. The driving base resistor is replaced by a current source which injects current into the output transistor, as shown in *Figure 27.5*. By using lateral transistors,

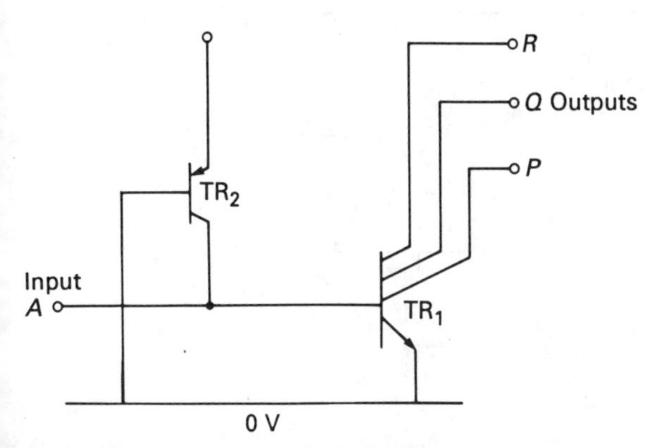

Figure 27.5 Integrated injection logic (IIL)

the pnp and npn devices can be merged together, and since no insulating regions or resistors are used the circuit density on a die is increased, which makes this logic suitable for LSI applications.

The noise immunity and speed of an IIL gate is improved as the injection current is increased. However, power dissipation also increases. The speed–power product is low, and this means that, if very high speed is not required, the power dissipation can be made low, enabling large LSI devices to be built.

Figure 27.6 shows a typical curve for the propagation delay–

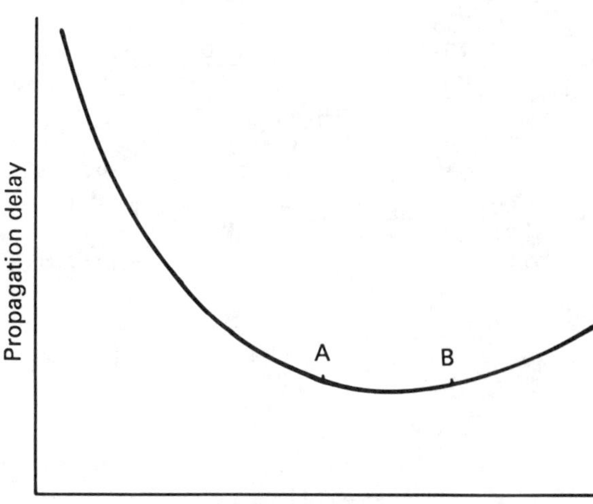

Figure 27.6 Propagation delay–power dissipation curve for IIL

power dissipation product of an IIL gate, as the injection current is increased. Up to point A this product is substantially independent of injection current, since the power dissipation increase is roughly proportional to the fall in propagation delay. Between points A and B the delay is constant since it is primarily due to active charge in the transistors. Beyond point B the base series resistance prevents rapid removal of accumulated charge so that propagation delay increases.

27.4 Unipolar logic circuits

27.4.1 CMOS logic

Complementary MOS (CMOS) is the only unipolar logic family which has found widespread use for SSI and MSI circuits. *Figure 27.7* shows the arrangement for an inverter, NAND gate and NOR gate. The circuit arrangement is similar to TTL with totem-pole outputs. Like TTL, the CMOS gate passes through a switching stage when all output transistors are on simultaneously. However, a series resistor is not now required to limit the current drawn from the power supply, since unipolar transistors have a higher inherent impedance than bipolar transistors.

CMOS logic can operate over a wide range of supply voltages. The logic 0 and logic 1 input levels are also usually guaranteed as 30% and 70% of V_{DD} so that the noise immunity band is high, as shown in *Figure 27.8*: the higher the supply voltage, the greater the absolute noise immunity.

The power dissipation of CMOS logic is low, but it increases with operating frequency since load and parasitic capacitances need to be charged and discharged.

27.4.2 Unipolar LSI logic

Unipolar circuits are generally capable of high circuit densities on a chip. They are therefore better suited to LSI than to MSI systems, where the chip size is often limited by the bonding pad areas.

Negative logic notations are normally used in discussing logic performance. *Figure 27.9* for instance shows a NAND gate in which V_{DD} is negative with respect to V_{SS} which is at ground. TR_1 is a load transistor. When inputs *A* and *B* are at logic 1 (negative), the output goes to logic 0 (positive).

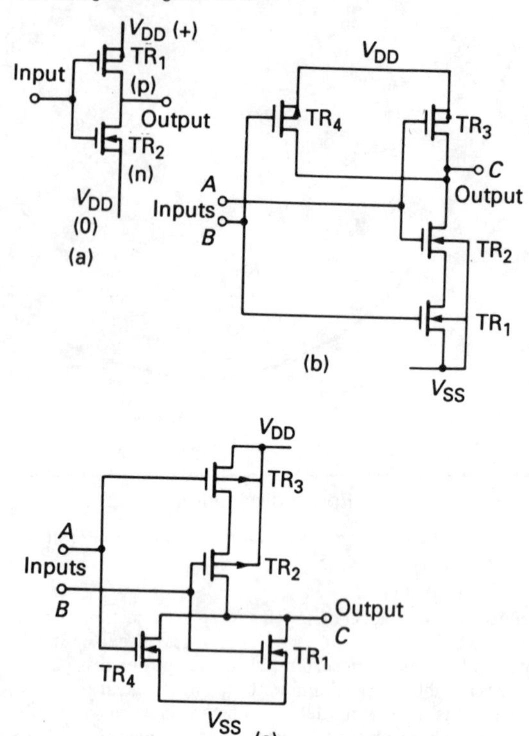

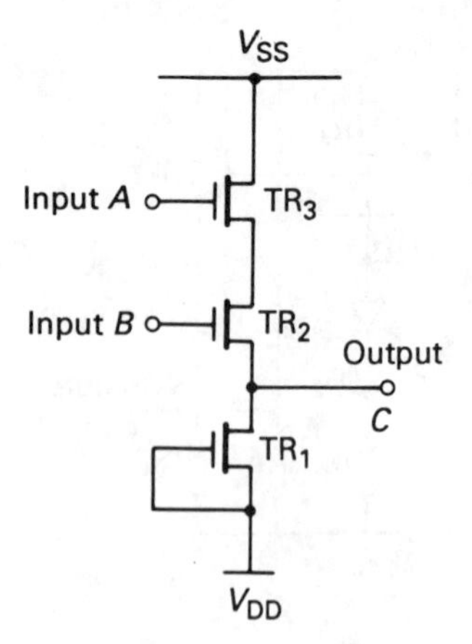

Figure 27.9 Unipolar NAND gate

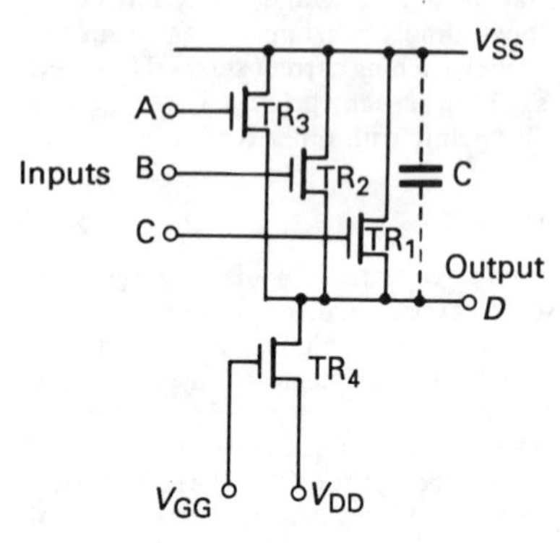

Figure 27.10 Unipolar NOR gate

Figure 27.7 CMOS logic circuit: (a) inverter, (b) NAND gate, (c) NOR gate

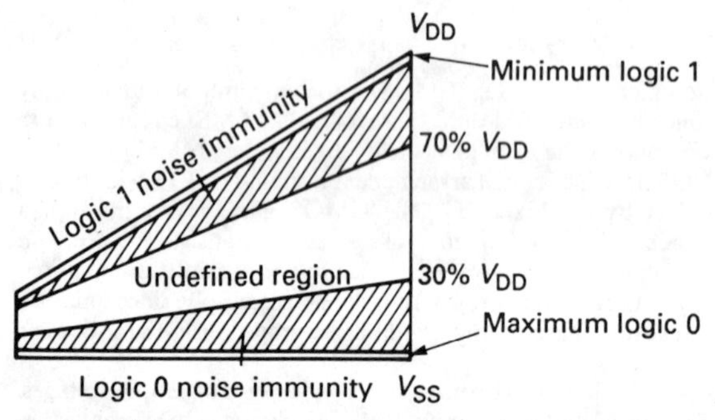

Figure 27.8 Guaranteed noise immunity bands for CMOS

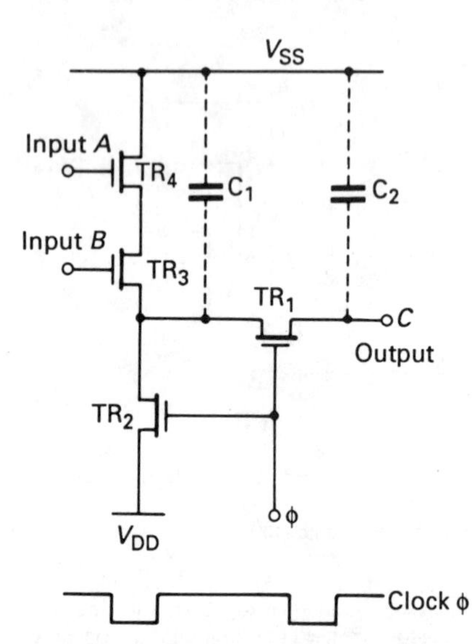

Figure 27.11 Dynamic two-phase NAND gate

A NOR gate is shown in *Figure 27.10*. A separate supply is used to bias the gate of the load transistor where $|V_{GG}| = |V_{DD} + V_T|$ to allow full logic swing on the output. The output is at logic 1 so long as none of the inputs goes to logic 1.

Both the previous figures illustrated static circuits. Dynamic circuits are used in unipolar LSI systems where the logic charge is held on parasitic capacitors. This reduces dissipation since transistors are on for a short time only, and it increases the circuit density on the die.

Figure 27.11 shows a simple dynamic NAND gate. Transistors TR_1 and TR_2 come on when the clock goes negative. Depending on the state of the inputs at *A* and *B* at this time the parasitic capacitors C_1 and C_2 will be charged or discharged. TR_1 goes off during the positive clock period and this turns TR_2 off, but the charge on C_2 is maintained, provided the off period is not long.

Unipolar transistors can be enhancement mode or depletion mode, although enhancement mode devices are used in all the circuits described so far. *Figure 27.12* shows inverters with three types of load, and *Figure 27.13* gives their characteristics. Depletion loads give larger currents and therefore shorter

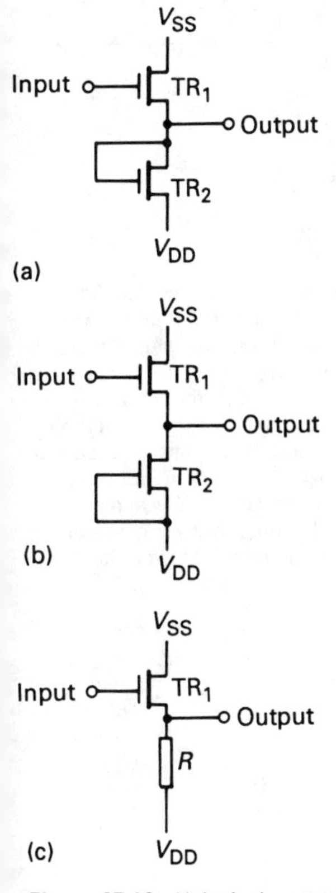

Figure 27.12 Unipolar inverters with different loads: (a) depletion, (b) enhancement, (c) resistive

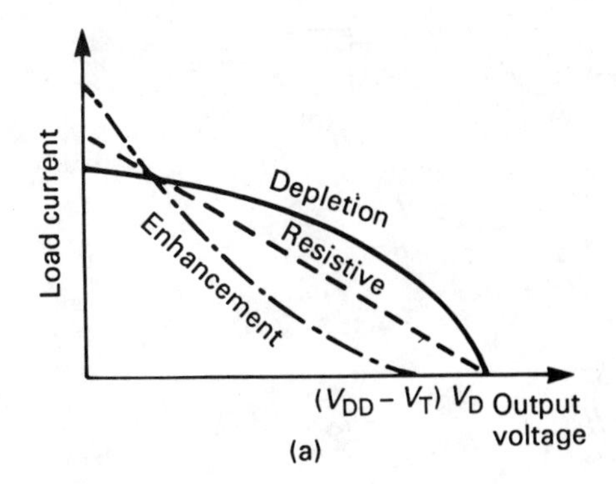

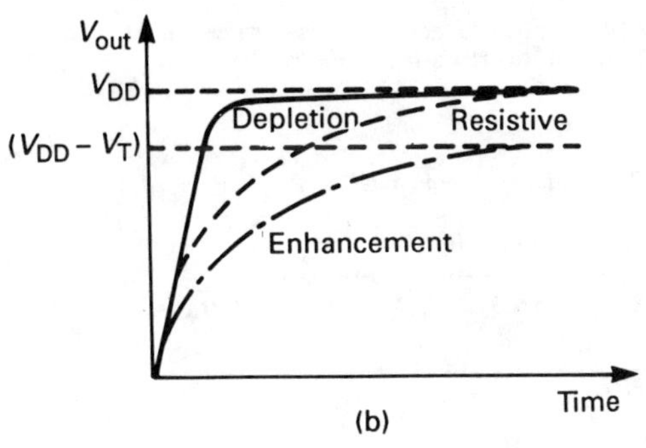

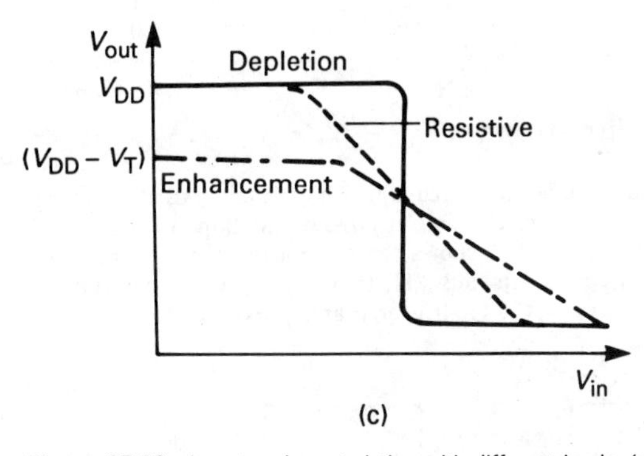

Figure 27.13 Inverter characteristics with different loads: (a) load current, (b) output risetime, (c) transfer curve

risetimes so they can generally work at higher speeds. They also have the squarest transfer characteristic, which gives the best d.c. noise immunity.

27.5 Comparisons

Table 27.1 compares the salient properties of logic families. A comparative system is used, since the absolute values can vary between vendors and in the fabrication process used. *Chapter 25* describes some of the integrated circuit processes. Currently Schottky TTL and CMOS are the most commonly used SSI and MSI families, with ECL being used for high-speed systems. NMOS is used for most large memory systems.

Table 27.1 Comparison of logic families (1 = best, 6 = worst)

Logic family	*Speed*	*Power dissipation*	*Fan-out*	*Noise immunity*	*Noise generation*
DTL	4	4	3	4	2
TTL	3	4	3	3	3
TTL-S	2	5	3	3	3
LPTTL-S	3	3	3	3	3
ECL	1	7	2	3	1
NMOS	5	2	2	2	2
CMOS	6	1	1	1	2

27.6 Gates

The circuits which are described in this and subsequent sections can be fabricated using any of the logic technologies described earlier. The electrical characteristics such as speed and power consumption will be determined by the logic family but the functional performance will be the same in all cases.

Figure 27.14 shows commonly used gates and *Table 27.2* gives the functional performance in logic 1 and 0 states. Gates are usually available as hex inverter, quad two input, treble three input, dual four input and single eight input. AND–OR–INVERT gates are also available and these are used to connect the outputs of two gates together in a wired-OR connection.

In a CMOS transmission gate a p- and an n-channel transistor are connected together. A gate signal turns on both transistors and so provides an a.c. path through them, whereas when the transistors are off the gate blocks all signals.

Gates are also made in Schmitt trigger versions. These operate as the discrete component circuits and exhibit a hysteresis between the on and off switching positions.

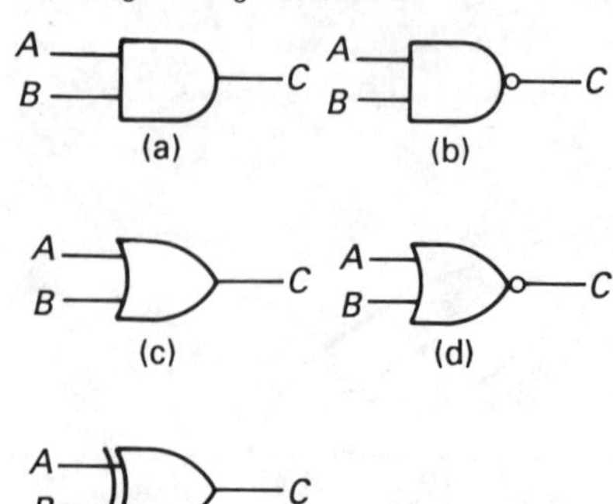

Figure 27.14 Symbols for commonly used gates: (a) AND, (b) NAND, (c) OR, (d) NOR, (e) exclusive–OR

Table 27.2 Truth table of logic gates

Inputs		*Output (C)*				
A	*B*	*AND*	*NAND*	*OR*	*NOR*	*Exclusive-OR*
0	0	0	1	0	1	0
0	1	0	1	1	0	1
1	0	0	1	1	0	1
1	1	1	0	1	0	0

27.7 Flip–flops

Flip–flops are bistable circuits which are mainly used to store a bit of information. The simpler types of flip–flops are also called latches. *Figure 27.15* shows the symbol for some of the more commonly used flip–flops. There are many variations such as master–slave *J–K*, edge-triggered and gated flip–flops.

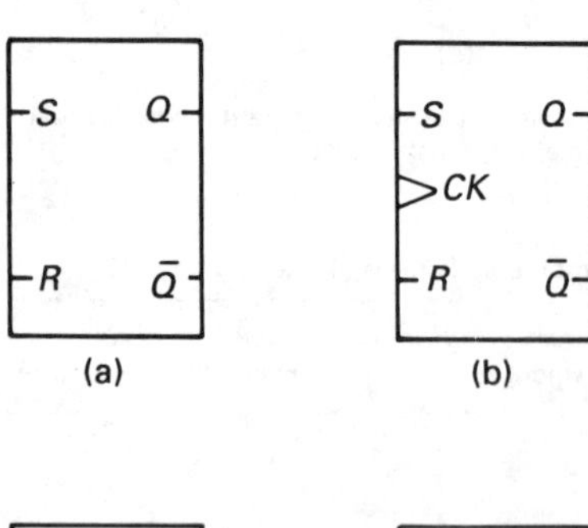

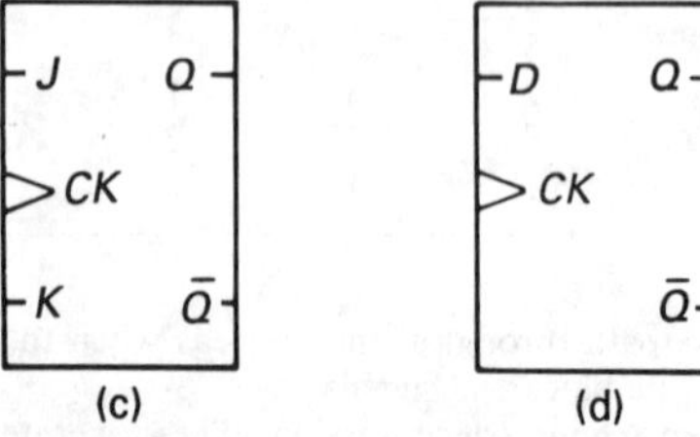

Figure 27.15 Symbols for commonly used flip–flops: (a) set–reset, (b) Master–slave set–reset, (c) *J–K*, (d) *D*-type

Table 27.3 Truth table of a set–reset flip–flop

S	*R*	Q_{n+1}	$\bar{Q}_{n+1}$
0	0	Q_n	$\bar{Q}_n$
0	1	0	1
1	0	1	0
1	1	?	?

Table 27.3 illustrates the operation of the set–rest flip–flop. Q_n represents the state of the Q output before the S and R inputs assume the logic state shown, and Q_{n+1} is the state after this event. It is seen that, with S and R both at logic zero, the outputs are unchanged. Otherwise the Q output will be set to the value of the S input and it will maintain this state even after the S and R inputs both return to zero. For S and R inputs both at logic 1, the output state will be forced to a logic 1, but when the S and R inputs return to zero the value of the Q and $\bar{Q}$ outputs is indeterminate.

In the master–slave set–reset flip–flop, whose operation is shown in *Table 27.4*, a clock pulse is required. During the rising edge of the clock information is transferred from the S and R inputs to the master part of the flip–flop. The outputs are unchanged at this stage. During the falling edge of the clock the inputs are disabled so that they can change their state without affecting the information stored in the master section. However, during this phase the information is transferred from the master to the slave of the flip–flop, so that the outputs change to the original S and R values. The flip–flop is disabled when no clock pulse is present.

Table 27.4 Truth table of a master–slave set–reset flip–flop (× = don't care state)

S	*R*	*CK*	Q_{n+1}	$\bar{Q}_{n+1}$
×	×	0	Q_n	$\bar{Q}_n$
0	0	⎍	Q_n	$\bar{Q}_n$
0	1	⎍	0	1
1	0	⎍	1	0
1	1	⎍	?	?

J–K flip–flops, illustrated in *Table 27.5*, are triggered on an edge of the clock waveform. Feedback is used internally within the logic such that the indeterminate state, when both inputs are equal to logic 1, is avoided. Now when the inputs are both at 1, the output will continually change state on each clock pulse. This is also known as toggling.

Table 27.5 Truth table of a *J–K* flip–flop (× = don't care state)

J	*K*	*CK*	Q_{n+1}	$\bar{Q}_{n+1}$
×	×	0	Q_n	$\bar{Q}_n$
0	0	↑	Q_n	$\bar{Q}_n$
0	1	↑	0	1
1	0	↑	1	0
1	1	↑	$\bar{Q}_n$	Q_n

D-type flip–flops have a single input. An internal inverter circuit provides two signals to the J and K inputs so that it operates as a *J–K* flip–flop, having only two input modes, as in *Table 27.6*.

Table 27.6 Truth table of a *D*-type flip–flop

D	*CK*	*Q*	$\bar{Q}$
0	↑	0	1
1	↑	1	0

27.8 Counters

Flip–flops can be connected together to form counters in a single package. There are primarily two types, asynchronous and synchronous. The synchronous counters may be subdivided into those with ripple enable and those with parallel or look-ahead carry.

In an asynchronous counter the clock for the next stage is obtained from the output of the preceding stage so that the command signal ripples through the chain of flip–flops. This causes a delay so asynchronous counters are relatively slow, especially for large counts. However, since each stage divides the output frequency of the previous stage by two, the counter is useful for frequency division.

In a synchronous counter, the input line simultaneously clocks all the flip–flops so there is no ripple action of the clock signal from one stage to the next. This gives a faster counter, although it is more complex since internal gating circuitry has to be used to enable only the required flip–flops to change state with a clock pulse. The enable signal may be rippled through between stages or parallel (or look-ahead) techniques may be used, which gives a faster count.

Counters are available commercially, having a binary or BCD count, and which are capable of counting up or down.

27.9 Shift registers

When stored data are moved sequentially along a chain of flip–flops, the system is called a shift register. *Figure 27.16* shows a four-bit register, although commercial devices are available in sizes from four bits to many thousands of bits.

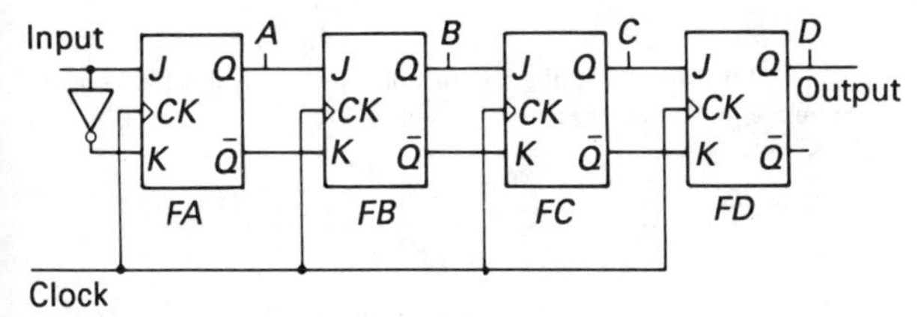

Figure 27.16 A four-bit serial-in, serial-out shift register

The shift register shown in *Figure 27.16* is serial-in, serial-out but it is possible to have systems which are parallel data input and output. The only limitation is the number of pins available on the package to accommodate the inputs and outputs. Shift registers can also be designed for left or right shift.

In the register of *Figure 27.16* input data ripple through at the clock pulse rate from the first to the last stage. Sometimes it is advantageous to be able to clock the inputs and outputs at different rates. This is achieved in a first-in, first-out (FIFO) register which is illustrated in *Figure 27.17*. Each data bit has an associated status bit and input data are automatically moved along until it reaches the last unused bit in the chain. Therefore the first data to come in will be the first to be clocked out. Data

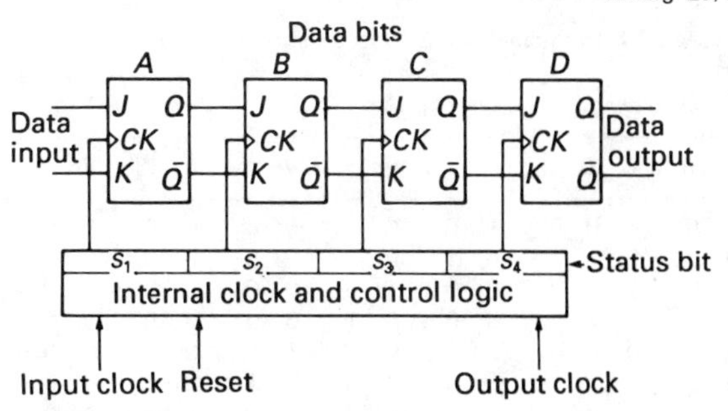

Figure 27.17 First-in, first-out shift register

can also be clocked into and out of the register at different speeds, using the two independent clocks.

27.10 Data handling

Several code converter integrated circuits are available commercially. *Figure 27.18* shows a BCD to decimal converter and *Table 27.7* gives its truth table. These converters can also be used as priority encoders. For example, *Figure 27.18* may be considered to be a ten-input priority encoder. Several of the lines

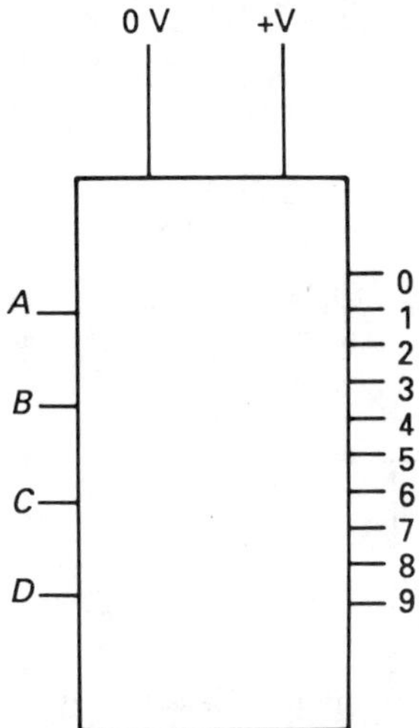

Figure 27.18 Symbol for a BCD to decimal converter

0 to 9 may be energised simultaneously but the highest number will generate a BCD output code on lines *A* to *D*.

Figure 27.19 shows an eight-channel multiplexer and *Table 27.8* gives its truth table. The channel select lines connect one of the eight data input lines to the output line using BCD code.

27.11 Timing

A variety of commercial devices are available to give monostable and astable multivibrators. Most of these incorporate control gates so that they are more versatile when used in digital systems. *Figure 27.20* illustrates a gated monostable, and its truth table, which shows the trigger mode, is given in *Table 27.9*. The monostable will trigger when the voltate at *T* goes to a logic 1.

Table 27.7 Truth table of a BCD to decimal converter

BCD code				*Outputs*									
A	*B*	*C*	*D*	*0*	*1*	*2*	*3*	*4*	*5*	*6*	*7*	*8*	*9*
0	0	0	0	1	0	0	0	0	0	0	0	0	0
0	0	0	1	0	1	0	0	0	0	0	0	0	0
0	0	1	0	0	0	1	0	0	0	0	0	0	0
0	0	1	1	0	0	0	1	0	0	0	0	0	0
0	1	0	0	0	0	0	0	1	0	0	0	0	0
0	1	0	1	0	0	0	0	0	1	0	0	0	0
0	1	1	0	0	0	0	0	0	0	1	0	0	0
0	1	1	1	0	0	0	0	0	0	0	1	0	0
1	0	0	0	0	0	0	0	0	0	0	0	1	0
1	0	0	1	0	0	0	0	0	0	0	0	0	1
1	0	1	0	0	0	0	0	0	0	0	0	0	0
1	0	1	1	0	0	0	0	0	0	0	0	0	0
1	1	0	0	0	0	0	0	0	0	0	0	0	0
1	1	0	1	0	0	0	0	0	0	0	0	0	0
1	1	1	0	0	0	0	0	0	0	0	0	0	0
1	1	1	1	0	0	0	0	0	0	0	0	0	0

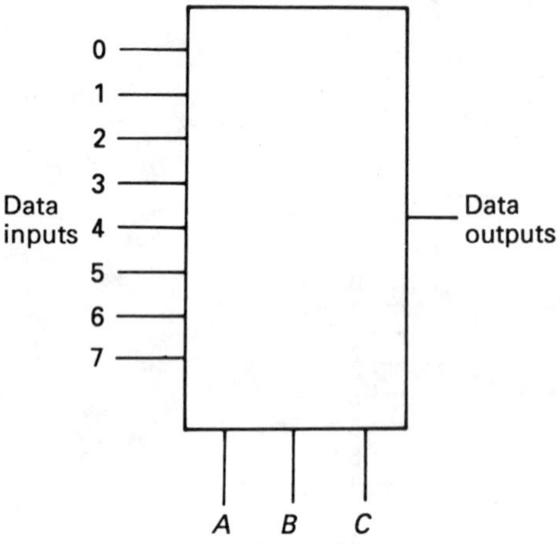

Table 27.8 Truth table of an eight-channel multiplexer (× = don't care state)

Channel select			*Data inputs*								*Data outputs*
A	*B*	*C*	*0*	*1*	*2*	*3*	*4*	*5*	*6*	*7*	
0	0	0	0	×	×	×	×	×	×	×	0
0	0	0	1	×	×	×	×	×	×	×	1
0	0	1	×	0	×	×	×	×	×	×	0
0	0	1	×	1	×	×	×	×	×	×	1
0	1	0	×	×	0	×	×	×	×	×	0
0	1	0	×	×	1	×	×	×	×	×	1
0	1	1	×	×	×	0	×	×	×	×	0
0	1	1	×	×	×	1	×	×	×	×	1
1	0	0	×	×	×	×	0	×	×	×	0
1	0	0	×	×	×	×	1	×	×	×	1
1	0	1	×	×	×	×	×	0	×	×	0
1	0	1	×	×	×	×	×	1	×	×	1
1	1	0	×	×	×	×	×	×	0	×	0
1	1	0	×	×	×	×	×	×	1	×	1
1	1	1	×	×	×	×	×	×	×	0	0
1	1	1	×	×	×	×	×	×	×	1	1

Figure 27.19 Symbol for an eight-channel multiplexer

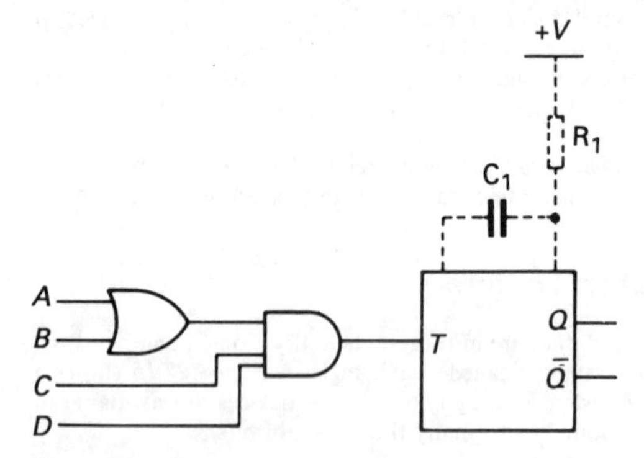

Figure 27.20 Logic diagram of a gated monostable

Table 27.9 Truth table of a monostable (↓ = 0 to 1 change, ↑ = 1 to 0 change, × = don't care)

A	*B*	*C*	*D*
↓	1	1	1
1	↓	1	1
0	×	↑	1
×	0	↑	1
0	×	1	↑
×	0	1	↑

External resistors and capacitors are used to vary the duration of the monostable pulse. By feeding the $\bar{Q}$ line back to the *D* input, this circuit can also be operated as an astable multivibrator.

Figure 27.21 shows a versatile timer circuit, which is available commercially as the 555 family. External timing components are again used, and the circuit can be interconnected to provide a variety of monostable and astable functions.

27.12 Drivers and receivers

Digital integrated circuits have limited current and voltage drive capability. To interface to power loads, driver and receiver circuits are required, which are also available in an integrated circuit package. The simplest circuit in this category is an array of transistors. Usually the emitters or collectors of the transistors are connected together inside the package to limit the package pin requirements.

Figure 27.22 shows a circuit used for driving inductive loads. The output transistors have high voltage ratings and freewheeling diodes are also provided inside the package. Four logic inputs are used for each output stage.

For digital transmission systems line drivers and receivers are available as in *Figure 27.23*. The digital input on the line driver controls an output differential amplifier stage, which can operate

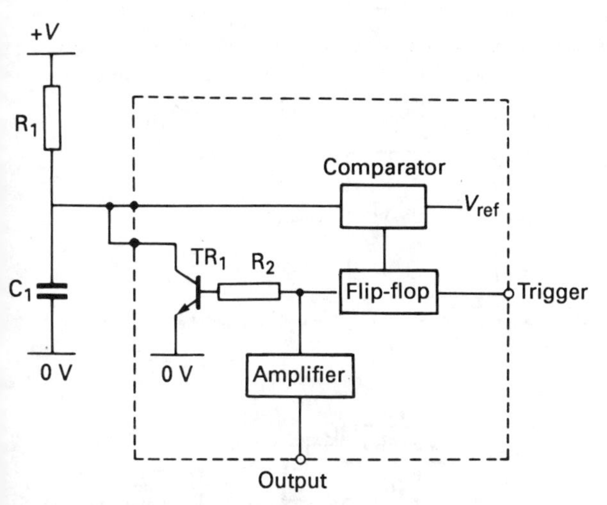

Figure 27.21 A 555 family timer

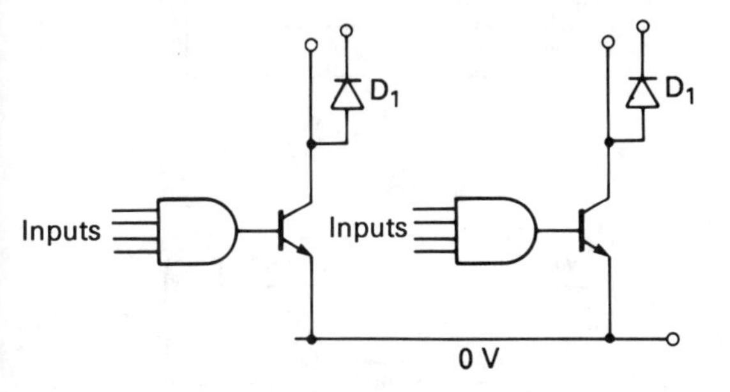

Figure 27.22 Dual driver

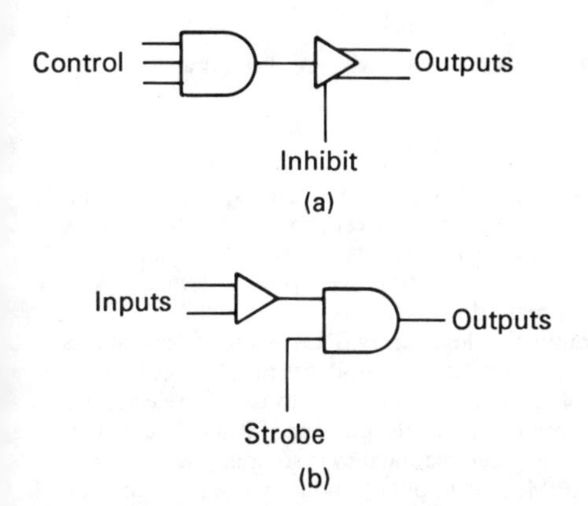

Figure 27.23 Line drivers and receivers: (a) symbol for a line driver, (b) symbol for a line receiver

into low impedance lines. The line receiver can sense low level signals via a differential input stage and provide a logic output.

27.13 Adders

Adders are the basic integrated circuit units used for arithmetic operations such as addition, subtraction, multiplication and division. A half adder adds two bits together and generates a sum

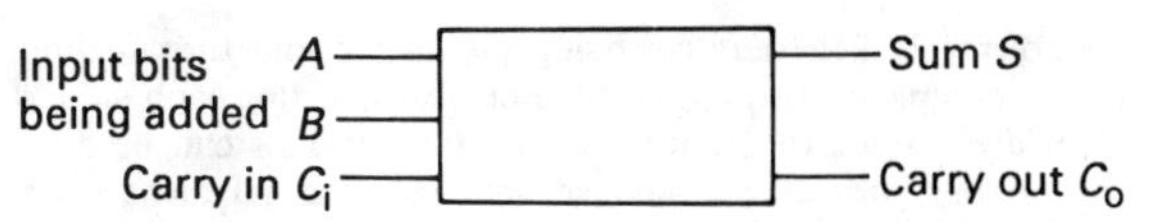

Figure 27.24 Symbol for a single bit of a full adder

Table 27.10 Truth table of a full adder

Carry in C_i	*Input bits being added* *A*	*B*	*Sum* *S*	*Carry out* C_o
0	0	0	0	0
0	0	1	1	0
0	1	0	1	0
0	1	1	0	1
1	0	0	1	0
1	0	1	0	1
1	1	0	0	1
1	1	1	1	1

and carry bit. A full adder has the facility to bring in a carry bit from a previous addition. *Figure 27.24* shows one bit of the full adder and *Table 27.10* gives its truth table.

The basic single-bit adder can be connected in several ways to add multi-bit numbers together. In a serial adder the two numbers are stored in shift registers and clocked to the *A* and *B* inputs one bit at a time. The carry out is delayed by a clock pulse and fed back to the adder as a carry in.

Serial adders are slow since the numbers are added one bit at a time. *Figure 27.25* shows a parallel adder which is faster. The

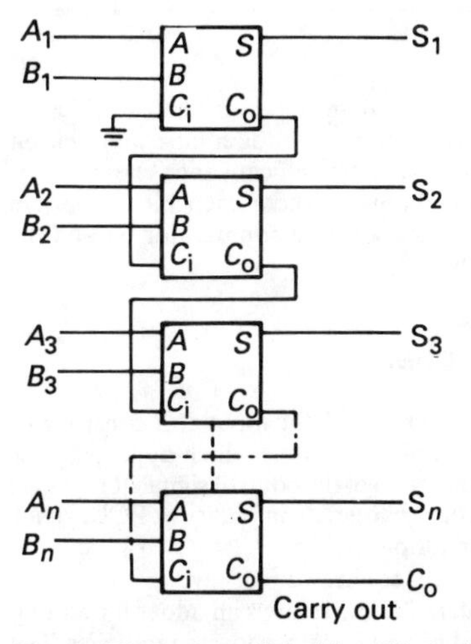

Figure 27.25 A parallel adder with ripple carry

carry output ripples through from one bit to the next so that the most significant bit cannot show its true value until the carry has rippled right through the system. To overcome this delay, look-ahead carry generators may be used. These take in the two numbers in parallel, along with the first carry-in bit, and generate the carry input for all the remaining carry bits.

Adders can be used as subtractors by taking the two's

complement of the number being subtracted and then adding. Two's complementing can be obtained by inverting each bit and then adding one to the least significant bit. This can be done within the integrated circuit so that commercial devices are available which can add or subtract, depending on the signal on the control pin.

Adders are used for multiplication by a process of shifting and addition, and division is obtained by subtraction and shifting.

27.14 Magnitude comparators

A magnitude comparator gives an output signal on one of three lines, which indicates which of the two input numbers is larger, or if they are equal. *Figure 27.26* shows one bit of a magnitude comparator and *Table 27.11* its truth table. Multi-bit numbers

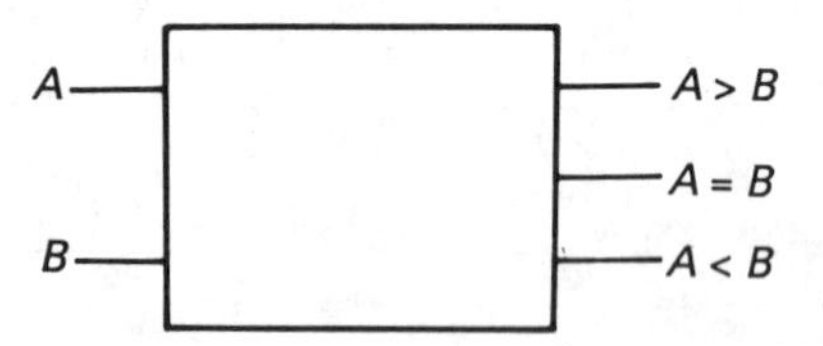

Figure 27.26 Symbol for a one-bit magnitude comparator

Table 27.11 Truth table of a one-bit magnitude comparator

A	B	$A>B$	$A<B$	$A=B$
0	0	0	0	1
0	1	0	1	0
1	0	1	0	0
1	1	0	0	1

can be compared by storing them in shift registers and clocking them to the input of the single-bit magnitude comparator, one bit at a time, starting from the most significant bit. Alternatively, parallel comparators may be used where each bit of the two numbers is fed in parallel to a separate comparator bit and the outputs are gated together.

27.15 Rate multiplier

In this device, shown in *Figure 27.27*, the input at R consists of a series of pulses and the number of pulses which appears at the output is controlled by the value of the control signal at C, where $O=RC/2^n$. Commercial devices operate in binary or BCD, giving binary or decimal rate multipliers.

Rate multipliers can be connected to give a variety of arithmetic functions. *Figure 27.28(a)* shows an adder for adding the numbers X and Y and *Figure 27.28(b)* shows a multiplier. The clock input in all cases is produced by splitting a single clock into several phases. For the adder $Z=X+Y$ and for the multiplier $Z=XY$.

27.16 Programmable logic arrays

A memory structure can be considered to be a programmable array since each cell may be programmed to a given logic state, which is subsequently read out. Read only memories (ROMs) are

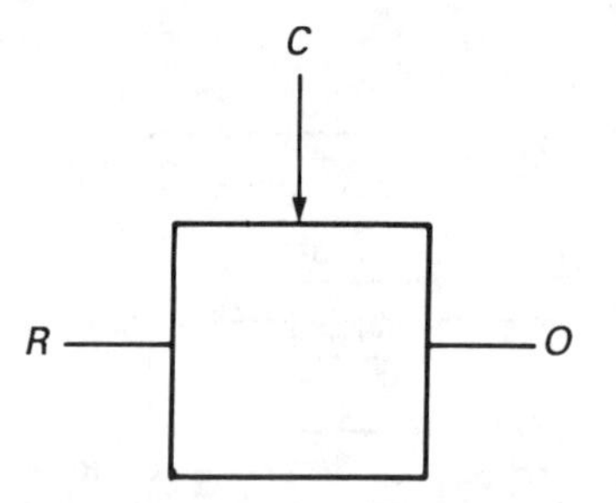

Figure 27.27 Symbol for a binary rate multiplier

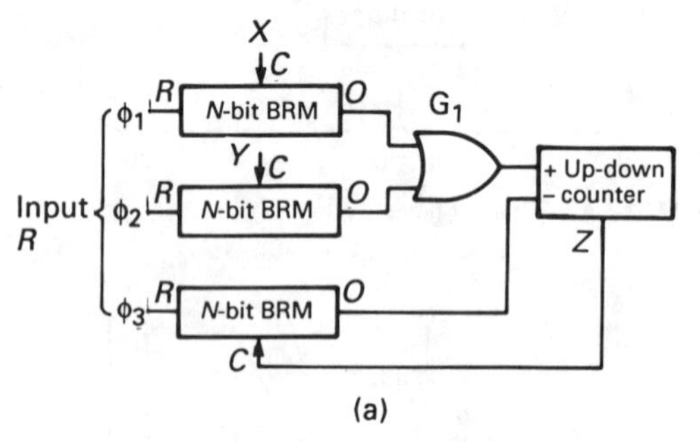

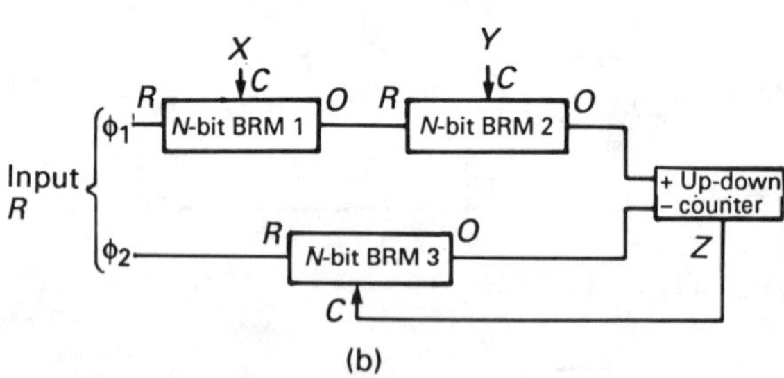

Figure 27.28 Using the binary rate multiplier: (a) addition, (b) multiplication

frequently used as such simple logic arrays, and they can be designed so that the output corresponds to an AND or an OR function of the input. All the inputs of a ROM generate a product term, and this is the case whether these product terms, which are ROM words, are used or not.

In a programmable logic array (PLA) there are two stages of encoding of the input. The first produces the product terms, and the number of terms can be designed to be any selected value below a maximum equal to the number of ROM lines. There is therefore no widespread redundancy in some applications, as can happen in a ROM. The output is produced by an OR function of the product terms in the second encoder.

There are many different forms of PLA available from several manufacturers. *Figure 27.29* shows one representation of a PLA. The logic array of *Figure 27.29(a)* is closest in form to the actual PLA layout in which the programmable junctions are represented by dots. The structure consists basically of an AND array which generates the product terms P_1 and P_2 from the inputs I_1 and I_2, and an OR array which produces outputs O_1 and O_2 from the product terms. The block diagram shows the functional operation much more clearly, whereas the gate logic diagram gives a more conventional representation.

The static form of PLA shown in *Figure 27.29* is the one most commonly used, and if required it can be combined with external storage elements for sequential applications. Many PLAs,

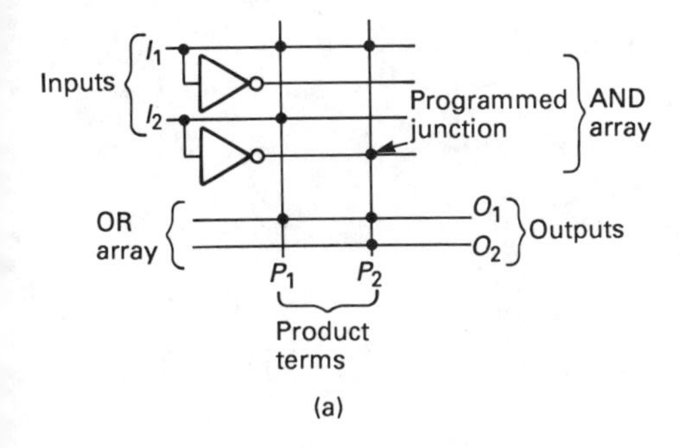

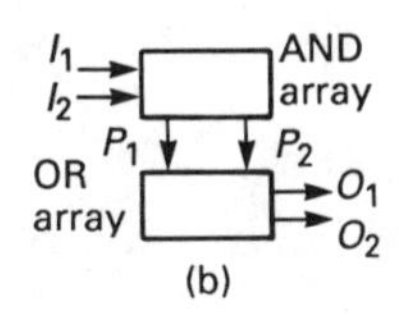

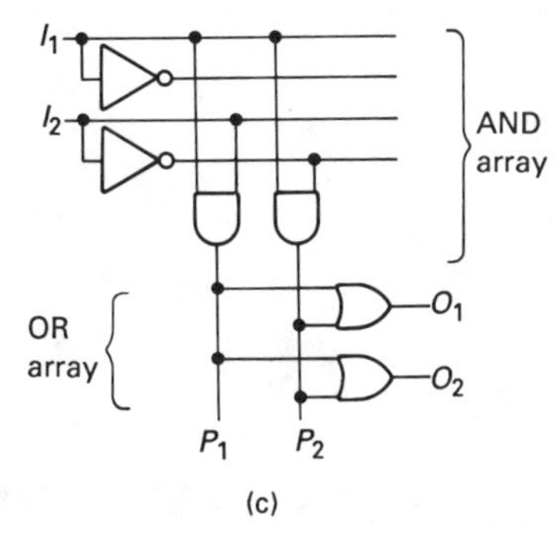

Figure 27.29 PLA representations: (a) array logic, (b) block schematic, (c) gate logic

however, include elements such as flip–flops on the silicon chip, so that they can be used for sequential applications. This is shown in *Figure 27.30*.

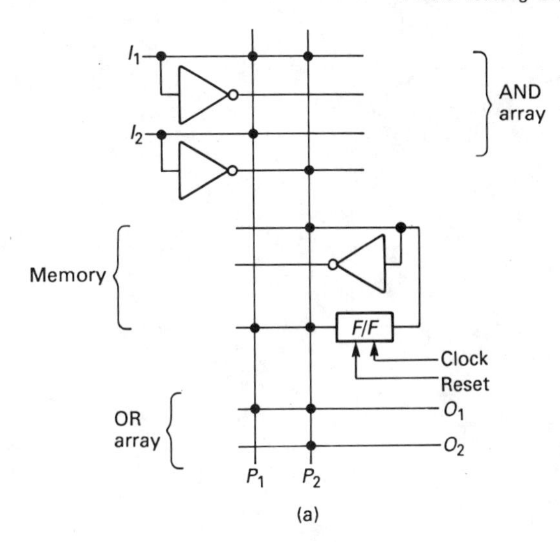

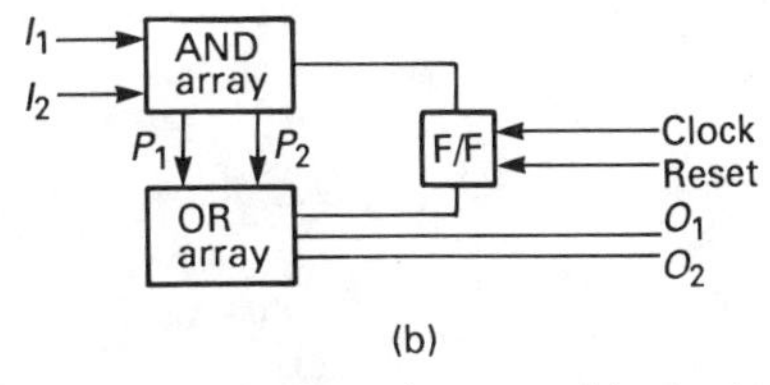

Figure 27.30 PLA with internal memory: (a) array logic, (b) block schematic

Acknowledgements

The contents of this chapter have been extracted from *Integrated Circuits* by F. F. Mazda, published by Cambridge University Press.

Further reading

CROES, R. and DE PAGTER, A., 'Advanced CMOS logic that lengthens the stride of low-power systems', *Electronic Components and Applications*, **8**, No. 2 (1987)

DIKE, C., 'Equivalent circuits model subtle traits of advanced CMOS ICs', *EDN*, 14 April (1988)

GALLANT, J., 'Gallium arsenide digital ICs complement ECL families in high speed applications', *EDN*, 3 March (1988)

MIDDLETON, R. G., *Understanding Digital Logic*, Howard W. Sams, 392 pp (1982)

NIEWIERSKI, W. J., 'Static-system design exploits low-power CMOS', *EDN*, 4 August (1988)

PATE, R. and BERG, W., 'Observe simple design rules when customizing gate arrays', *EDN*, November (1982)

PROUDFOOT, J. T., 'Programmable logic arrays', *Electronics & Power*, Nov/Dec (1980)

PRYCE, D., 'Smart-power ICs', *EDN*, 3 March (1988)

TWADDELL, W., 'Uncommitted IC logic', *EDN*, April (1980)

WALKER, R., 'CMOS logic arrays: a design direction', *Computer Design*, May (1982)

WEIGL, K. H., 'Fast MPUs direct EPLD development', *Electronic Product Design*, January (1988)

WILSON, R., 'PDL architectures vie for larger share of CPU support logic', *Computer Design*, November 1 (1987)

WILSON, R., 'CMOS VLSI sets the direction for new ICs', *Computer Design*, December (1987)

WILSON, R., 'Controversy, user doubts continue to embroil advanced CMOS logic', *Computer Design*, 1 March (1988)

28 Linear Integrated Circuits

A M Pope, BSc
Marketing Manager
RR Electronics Ltd

M Trowbridge, MA
Hitachi Europe Ltd

Contents

28.1 Introduction

The semiconductor industry today classifies almost every monolithic integrated circuit as a 'linear IC' that does not fit neatly into the categories of digital logic and memory/microprocessor circuits. Some 20 years ago, the situation was much simpler as the manufacturers were grappling with the problems of integrating basic operational amplifiers to handle analogue circuit applications in as flexible a way as possible. Then the technology in use was, of course, bipolar. Today this has expanded to embrace all the major process technologies, wherever their particular merits are advantageous, including combinations of different technologies. Hybrid integrated circuits have filled the gaps where the required technologies are impossible to combine on a single chip. The available functions have increased enormously and may be broadly categorised as operational amplifiers, interface circuits, voltage regulators, data acquisition circuits and special functions. However, the operational amplifier is still the workhorse of the analogue signal processing world and forms an excellent starting point for any discussion of linear integrated circuits (LICs).

28.2 Operational amplifiers

The operational amplifier, or op amp, was originally developed in response to the needs of the analogue computer designer. The concept is to provide a gain block whose performance is totally predictable from unit to unit and perfectly defined by the characteristics of an external feedback network. This has been achieved by LIC technology with varying degrees of accuracy, largely governed by unit cost and complexity. Nevertheless, the performance has been refined to a point where it is possible to use an op amp in almost any d.c. to 1 MHz amplifier/signal processor application. Indeed, it is probably a deliberate case of 'wheel re-invention' not to use such an approach. It is feasible to produce operational amplifier designs with various types of active device, including valves (vacuum tubes). However, IC technology is remarkably successful in offering low-cost, high-performance op amps. They have thus become the keystone of all modern signal processing and conditioning techniques.

28.2.1 Ideal operational amplifier

The 'ideal' operational amplifier of *Figure 28.1* is a differential input, single-ended output device that is operated from bipolar power supplies. As such, it can easily be used as a virtual earth

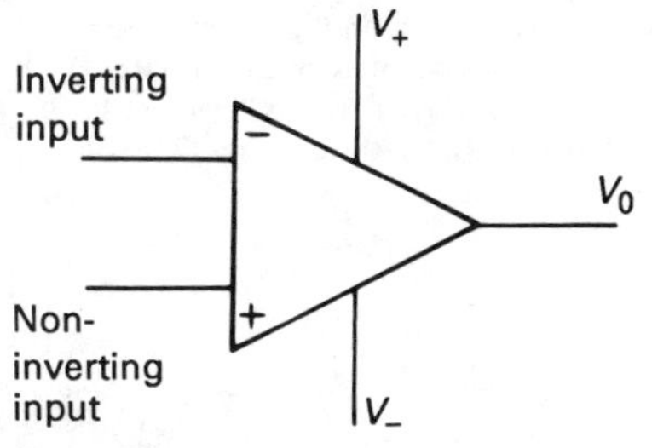

Figure 28.1 Ideal op amp

amplifier. Differential output is possible, although not common, and single supply operation will be discussed later. The ideal op amp possesses infinite gain, infinite bandwidth, zero bias current to generate a functional response at its inputs, zero offset voltage (essentially a perfect match between the input stages) and infinite input impedance. Because of these characteristics, an infinitesimally small input voltage is required at one input with respect to the other to exercise the amplifier output over its full range. It is important to realise that, when considered as open loop devices, op amp have a very small linear range. An op amp powered from ±15 V supply rails with a gain of 100 000 will achieve its maximum output voltage swing with a differential input of less than 150 μV.

Hence, if one input is held at earth, the other cannot deviate from it within the linear operational region and becomes a 'virtual earth' of the feedback theory definition. The shortfalls against the ideal of practical op amps are now considered together with their application consequences.

28.2.2 Input bias current

Input bias current affects all applications of op amps. In order for the op amp to function, it is necessary to supply a small current, from picoamperes for op amps with FET inputs to microamperes for junction transistor type inputs. To see the reason for this, *Figure 28.2* shows a typical op amp input stage, using a differential 'long-tailed pair' with a constant current sink in the emitters. The following gain stage is coupled out from the

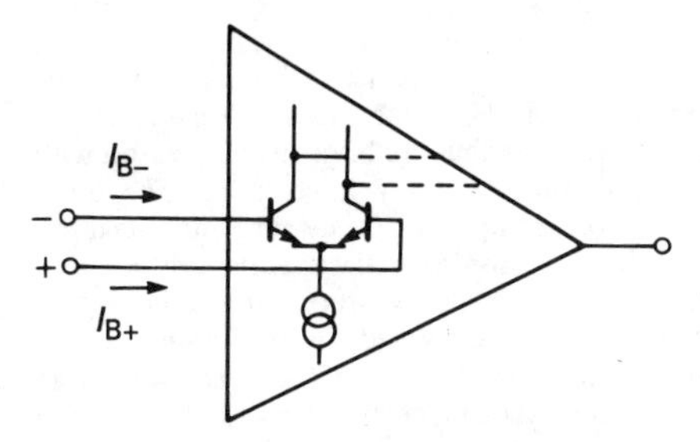

Figure 28.2 Op amp input bias current path

collector loads. The input bias current is primarily a function of the large signal current gain of the transistors and is defined as the absolute average of I_{B+} and I_{B-}, i.e.

$$I_B = \frac{|I_{B+}| + |I_{B-}|}{2} \tag{28.1}$$

One point that is immediately obvious when considering *Figure 28.2*, but is often forgotten when designing op amp circuitry on a 'black box' approach, is that a d.c. path is always required to the op amp inputs. This is mandatory even if only a.c. amplifiers are under consideration or if op amps of the FET input variety with ultra low bias current, essentially voltage operated input devices, are being used.

28.2.3 Input offset current

Op amp fabrication uses monolithic integrated circuit construction, which can produce very well matched devices for input stages, etc., by using identical geometry for a pair of devices diffused at the same time on the same chip. Nevertheless, there is always some mismatch, which gives rise to the input offset current. This is defined as the absolute difference in input bias currents, i.e.

$$I_{os} = |I_{B+} - I_{B-}| \tag{28.2}$$

28.2.4 Effect of input bias and offset current

The above indicates that the effect of input bias current will be to produce an unwanted input voltage, which can be much reduced by arranging for the bias current to each input to be delivered from an identical source resistance. *Figure 28.3* shows a simple

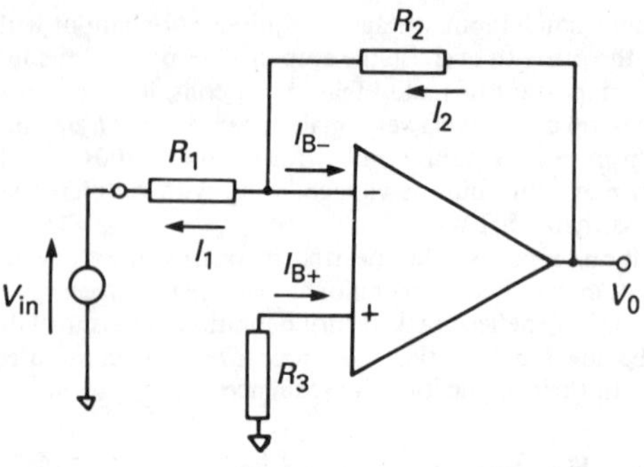

Figure 28.3 Simple inverting amplifier

inverting amplifier with a gain defined by R_2/R_1, in which R_3 is added to achieve this effect. In this example the effect of amplifier input impedance is ignored, and signal input impedance is assumed to be zero. In order to balance the bias current source resistance, R_3 is made equal to R_2 in parallel with R_1, i.e.

$$R_3 = \frac{R_1 R_2}{R_1 + R_2} \tag{28.3}$$

Obviously, the values of the feedback network resistors must be chosen so that, with the typical bias current of the op amp in use, they do not generate voltages that are large in comparison with the supplies and operating output voltage levels. The other sources of error, after balancing source resistances, are the input offset current, and the more insidious effects of the drift of input offset and bias current with temperature, time and supply voltage, which cannot easily be corrected. Op amp designers have taken great trouble to minimise the effect of these external factors on the bias and offset currents, but, of course, more sophisticated performance is only obtained from progressively more expensive devices. As might be expected, a high precision application will require an op amp with a high price tag.

The output offset voltage due to input offset current for *Figure 28.3* under zero input signal conditions can be simply derived. The voltages around the input loop must sum to zero, hence $I_{B+}R_3$ is the voltage across R_1 and the contribution to I_{B-} flowing through R_1 is $I_{B+}R_3/R_1$. The current through R_2 is then $I_{B-} - (I_{B+}R_3)/R_1$, so:

$$V_o = R_2\left[I_{B-} - \frac{I_{B+}R_3}{R_1}\right] - I_{B+}R_3 \tag{28.4}$$

where V_o is the output offset voltage. Combining Equations (28.2), (28.3) and (28.4),

$$V_o = I_{os}R_2 \tag{28.5}$$

Similarly for the condition where input source resistance balancing is ignored:

$$V_o = I_B R_2 \tag{28.6}$$

By a similar process, the effect of input bias and offset current can be calculated for other op amp circuit configurations.

28.2.5 Input offset voltage and nulling

As mentioned before, a mismatch always exists between the input stages of an op amp, and the input offset voltage (V_{os}) is the magnitude of the voltage that, when applied between the inputs, gives zero output voltage. In bipolar input op amps the major contributor to V_{os} is the V_{BE} mismatch of the differential input stage. General-purpose op amps usually have a V_{os} in the region of 1–10 mV. This also applies to the modern JFET or MOSFET input stage op amps, which achieve an excellent input stage matching with the use of ion implantation. As with the input current parameters, input offset voltage is sensitive to temperature, time and to a lesser extent input and supply voltages. Offset voltage drift with temperature is often specified as μV per mV of initial offset voltage per °C. As a general rule, the lower the offset voltage of an op amp, the lower the temperature coefficient of V_{os} will be. Precision op amps often employ isothermal design layout rules to improve temperature drift performance at the 'front end'. Usually, op amps are relatively low power dissipation devices, but it is worth remembering that some variants are capable of substantial current outputs, e.g. for low impedance line driving, and the effects of thermal feedback on the chip from output to input can be significant. This can result in pronounced input offset changes and modulation of the open loop gain.

Enhanced performance low V_{os} op amps are often produced today by correcting or 'trimming' the inherent unbalance of the input stage on chip before or after packaging the device. Techniques used are mainly laser trimming of thin film resistor networks in the input stage and a proprietary process known as 'zener zapping'.

Other op amps are available with extra pins connected for a function known as 'offset null'. Here a potentiometer is used externally by the user with its slider connected to V_+ or V_- to adjust out the unbalance of the device input stage (see *Figure 28.4*). A note of caution should be sounded here when using an op

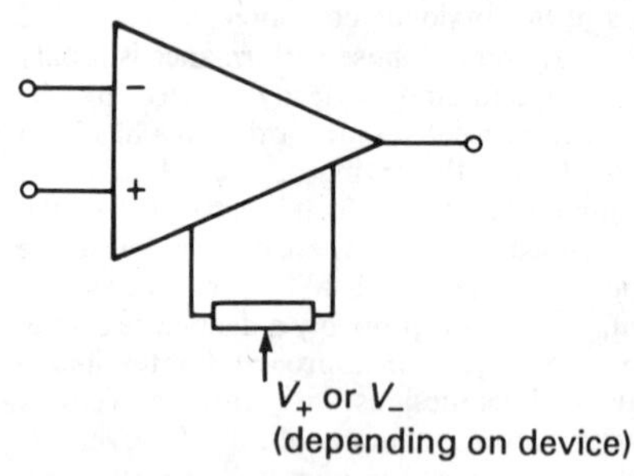

Figure 28.4 Offset nulling

amp with an offset null feature in precision circuitry; the temperature coefficient of V_{os} can be changed quite significantly in some op amp types by the nulling process.

28.2.6 Effect of input offset voltage

As with the input current parameters discussed previously, the input offset voltage will also produce an unwanted output voltage component. Using *Figure 28.3*, and ignoring the effect of input bias currents, V_{os} will appear across the inputs. With no input signal, V_{os} must appear across R_1, giving I_1 as V_{os}/R_1. Since I_1 must equal I_2 in this ideal case, the output offset voltage, V_o, is given by:

$$V_o = I_2(R_1 + R_2) \tag{28.7}$$

This reduces to:

$$V_o = V_{os}\left(1 + \frac{R_2}{R_1}\right) \tag{28.8}$$

In this example of a simple inverting amplifier, it is clear that the output offset voltage derived from the input offset voltage is $V_{os}(1 + A_{cl})$ where A_{cl} is the closed loop gain. Other configurations will respond to a similar analysis.

28.2.7 Combined effect of input offsets

If the effects of the input offset current and voltage are combined to give a 'real world' equation for the output error voltage, some actual device performance figures in a simple amplifier can be evaluated.

Combining Equations (28.6) and (28.8)

$$V_o = V_{os}\left(1+\frac{R_2}{R_1}\right)+I_{os}R_2 \quad (28.9)$$

If the values of $R_1 = 10\ k\Omega$, $R_2 = 100\ k\Omega$ and $R_3 = 9\ k\Omega$ are taken, the result is an amplifier with inverting gain of 10 (cf. *Figure 28.3*).

Using the well known μA741 op amp,

$V_{os(max)} = 6$ mV $\quad I_{os(max)} = 200$ nA
Output offset = 86 mV max

Using a precision type of amp (OP-07)

$V_{os(max)} = 250\ \mu$V $\quad I_{os(max)} = 8$ nA
Output offset = 3.55 mV max

This example shows the reduction that is possible in static errors in op amp circuits by improved grade op amps.

28.2.8 Dynamic errors

Earlier sections have covered in outline the effect of static errors on LIC op amp performance. To complete the picture in real-life circuit applications, various dynamic errors must be considered. The most important of these are mentioned below. The full treatment of stability in feedback amplifiers and the effects of noise are more properly the province of textbooks on network analysis theory and circuit design. However, the op amp concept requires the active device itself to be as tolerant as possible to external feedback effects. This results in markedly different responses to step input signals than 'normal' amplifier theory predicts when linking small signal bandwidth to rise time.

28.2.9 Open loop voltage gain

This is one parameter where the practical op amp approaches the infinite gain ideal very closely and typical gains of 250 000 and higher are quite common at zero frequency. However, it is also common for the gain to start to fall off rapidly at low frequencies, e.g. 10 Hz, with many internally compensated op amps (see *Figure 28.5*). The μA741 op amp, for instance, has its gain reduced

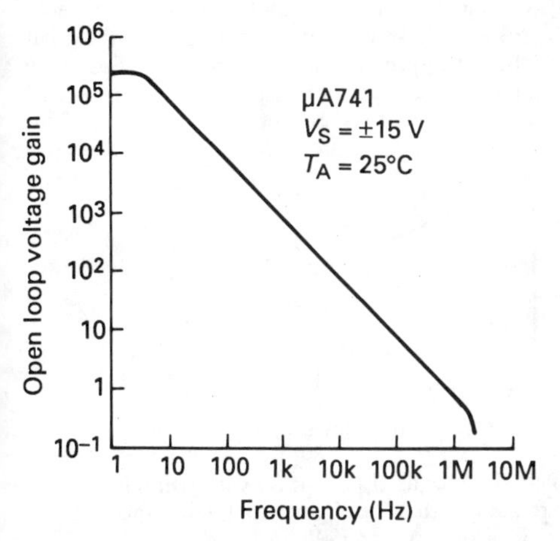

Figure 28.5 Typical op amp open loop frequency response

to unity at around 1 MHz. To assess the significance of this factor, it is necessary to develop a relationship linking the closed loop gain (A_{cl}) to the initial open loop gain (A_{ol}), in terms of accuracy. Obviously the closed loop gain is limited by the open loop gain and feedback theory indicates that the greater the difference between open and closed loop gain, the greater the gain accuracy. For the inverting amplifier of *Figure 28.3* the closed loop gain error is given by:

$$A_{cle} = \frac{100}{1+A_{ol}R_1/(R_1+R_2)}\ \% \quad (28.10)$$

Using the example of the previous section, the μA741 in an amplifier with a gain of 10 produces a remarkable 0.005% gain accuracy at d.c., with its typical A_{ol} of 2×10^5. This assumes that R_1, R_2 are exact values of course. It is a different story at 10 kHz, where, using the A_{ol} value of *Figure 28.5*, gain accuracy will be 10%. To select an op amp for a given d.c. closed loop gain of Y and a gain accuracy of X% at a given maximum signal frequency F_{max}, a criterion is found by rearranging the above expression:

$$A_{ol} \geqslant \frac{100(1+Y)}{X} - Y + 1 \quad (28.11)$$

Note that when the open loop voltage gain falls to the value of the feedback ratio $(R_2+R_1)/R_1$, the closed loop gain falls by 6 dB. For the gain of ten μA741 example of *Figure 28.3*, this occurs at around 100 kHz. There are also other factors involved in specifying op amp frequency response, the most important of these being slew rate limitations. Due attention must also be paid to the overall circuit stability of the op amp and its associated feedback network impedances. Note that open loop voltage gain is strongly dependent on supply voltage in most op amps.

28.2.10 Slew rate limiting

Slew rate limiting affects all amplifiers where capacitance on internal nodes, or as part of the external load, has to be charged and discharged as voltage levels change. Indeed, the effect is clearly seen in digital logic circuitry as well, as limitations in rise and fall time. It is significant in op amps as it determines the difference between the small signal and power bandwidth. The slew rate limit in most op amps is derived from the capability of the internal circuitry to charge and discharge the compensation capacitance network, which may either be internal or external.

A general purpose op amp such as the μA741 has a typical slew rate of 0.5 V/μs. It is also capable of at least a $\pm$10 V swing into a 2 kΩ load resistance on $\pm$15 V supplies. The slew rate controls the large signal pulse response, forcing the op amp output to change from -10 V to $+10$ V in a minimum of 40 μs. Thus the large signal square wave response cannot be greater than 12.5 kHz (50% duty cycle) and, at this frequency, the output is distorted to a triangular wave. The sine wave response is affected by slew rate such that the full peak output voltage V_{pk} is available until the frequency F_{max} at which the sine wave maximum dV/dT exceeds the slew rate. Simple differentiation of a sine function, and taking the maximum rate of change of resultant cosine, leads to the following criterion:

$$\text{slew rate (V/s)} = 2\pi F_{max} V_{pk} \quad \text{at the limit} \quad (28.12)$$

As an example, the μA741 at nominal 0.5 V/μs slew rate can produce 10V_{pk} (which is 20 V peak to peak or 7.07 V r.m.s.) up to a maximum of 8 kHz. The LM318 high-speed op amp with 50 V/μs slew rate could extend the full power bandwidth to 800 kHz. A typical general-purpose op amp output voltage capability–frequency graph is shown in *Figure 28.6*. Again note that slew rate is dependent on supply voltage to a large extent.

28.2.11 Settling time

Frequently in op amp applications, such as digital-to-analogue converters, the device output is required to acquire a new level within a certain maximum time from a step input change.

The slew rate is obviously a factor in this time, but transient effects will inevitably produce some measure of overshoot and

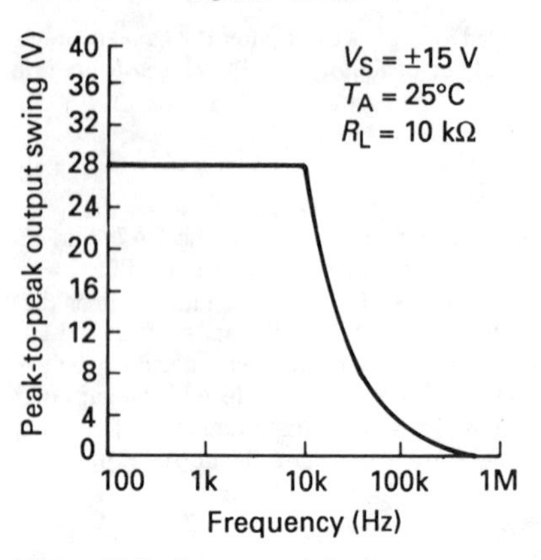

Figure 28.6 Op amp typical output swing versus frequency

possibly ringing before the final value is achieved. This time, measured to a point where the output voltage is within a specified percentage of the final value, is termed the settling time, usually measured in nanoseconds. Careful evaluation of op amp specifications is required for critical settling time applications, since high slew rate op amps may have sufficient ringing to make their settling times worst than medium slew rate devices.

28.2.12 Wide bandwidth and high slew rate

The definition of wide bandwidth and high slew rate is somewhat indistinct in LIC op amps, but, in general, greater than 10 MHz and 10 V/μs is a good standard. The problem does not lie essentially in the circuit design but rather revolves around the need for fast pnp transistors somewhere in the circuit for interstage coupling, etc. Normal junction isolation fabrication yields poor pnp transistors with gain bandwidth products of around 3 MHz (lateral pnp transistors) and 50 MHz (substrate pnp transistors), compared to standard npn figures of 400 MHz. FETs may offer a solution, but orders of magnitude increase of performance are usually achieved with dielectric isolation fabrication, which yields pnp transistors with up to 500 MHz transition frequency. Such op amps can offer 600–1000 V/μs slew rate with very little sacrifice in other performance parameters. Such devices find application in fast signal processing in areas like video and radar.

28.2.13 Output capabilities

As would be expected due to internal design limitations, op amps are unable to achieve an output voltage swing equal to the supply voltages for any significant load resistance. Some devices do achieve a very high output voltage range versus supply, notably those with CMOS FET output stages (BIMOS). All modern op amps have short circuit protection to ground and either supply built in, with the current limit threshold typically being around 25 mA. It is thus good practice to design for maximum output currents in the 5–10 mA region.

Higher current op amps are available but the development of these devices has been difficult due to the heat generated in the output stage affecting input stage drift. Monolithic devices currently approach 1 A and usually have additional circuitry built in to effect thermal overload and output device safe operating area protection, in addition to normal short circuit protection.

28.2.14 Power supply parameters

The supply consumption will usually be specified at ±15 V and possibly other supply voltages additionally. Most performance parameters usually deteriorate with reducing supply voltage. Devices are available with especially low power consumption but their performance is usually a trade-off for reduced slew rate and output capability.

Power supply rejection ratio (PSRR) is a measure of the susceptibility of the op amp to variations of the supply voltage. The definition is expressed as the ratio of the change in input offset voltage to the change in supply voltage (μV/V or dB). No self-respecting op amp ever has a PSRR of less than 70 dB. This should not be taken to indicate that supply bypassing/decoupling is unnecessary. It is good practice to decouple the supplies to general-purpose op amps at least every five devices. High-speed op amps require careful individual supply decoupling on a by-device basis for maximum stability, usually using tantalum and ceramic capacitors.

28.2.15 Common mode range and rejection

The input voltage range of an op amp is usually equal to its supply voltages without any damage occurring. However, it is required that the device should handle small differential changes at its inputs, superimposed on any voltage level in a linear fashion.

Due to design limitations, input device saturation, etc., this operational voltage range is less than the supplies and is termed the common mode input voltage range or swing. As a differential amplifier, the op amp should reject changes in its common mode input voltage completely, but in practice they have an effect on input offset and this is specified as the common mode rejection ratio (CMRR) (μV/V or dB). Note that most general-purpose op amps have a CMRR of at least 70 dB (approximately 300 μV offset change per volt of common mode change) at d.c. but this is drastically reduced as the input frequency is raised.

28.2.16 Input impedance

The differential input resistance of an op amp is usually specified together with its input capacitance. Typical values are 2 MΩ and 1.4 pF for the μA741. The input resistance of this type of device exhibits a slow decrease from about 10 kHz.

In practice, the input impedance is usually high enough to be ignored. In inverting amplifiers, the input impedance will be set by the feedback network. In non-inverting amplifiers, the input impedance is 'bootstrapped' by the op amp gain, leading to extremely high input impedance for configurations as *Figure 28.7* which shows the 100% feedback case or voltage follower. This

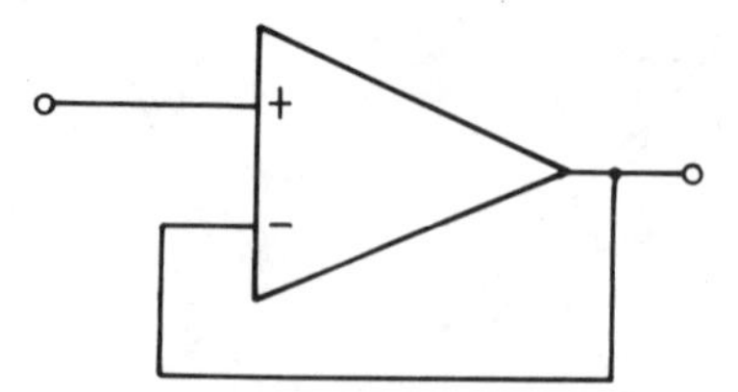

Figure 28.7 Op amp voltage follower

bootstrap effect declines as A_{ol} drops off with rising frequency, but it is safe to assume that the circuit input impedance is never worse than that of the op amp itself.

28.2.17 Circuit stability and compensation

To simplify design-in and use, many op amps are referred to as 'internally' or 'fully' compensated. This indicates that they will remain stable for closed loop gains down to unity (100% feedback). This almost certainly means that performance has been sacrificed for the sake of convenience from the point of view of users who require higher closed loop gains. To resolve this problem, many op amps are available that require all compensation components to be added externally, according to manufacturers' data and gain requirements. A compromise to this split has been the so-called 'undercompensated' op amps, which are inherently stable at closed loop gains of 5 to 10 and above.

The aim of the standard compensation is to shape the open loop response to cross unity gain before the amplifier phase shift exceeds 180°. Thus unconditional stability for all feedback connections is achieved. A method exists for increasing the bandwidth and slew rate of some uncompensated amplifiers known as feedforward compensation. This relies on the fact that the major contributors of phase shift are around the input stage of the op amp and bypassing these to provide a separate high frequency amplifying path will increase the combined frequency response (see *Figure 28.8*). Full details of the technique are available in manufacturers' literature.

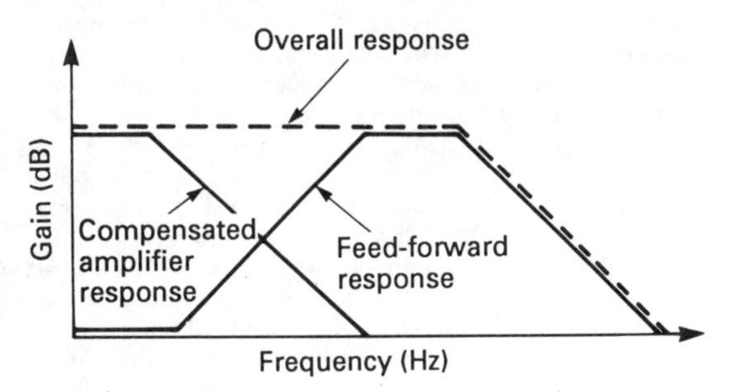

Figure 28.8 Feed-forward compensation

28.2.18 Internal circuit design

The circuit design of actual op amps is too detailed and complex a subject to enter into here. Suffice it to say that many circuit techniques not normally available to the discrete component designer are freely used in monolithic designs; in particular, the superior component matching (differential amplifiers) and Wilson current mirrors for biasing, active loads and interstage signal coupling. An approximate internal circuit diagram of a typical μA741 device is shown in *Figure 28.9*.

In brief, the device has a differential input stage, the npn transistors Q_1 and Q_2. The emitter current sinks are formed by Q_5 to Q_7. Q_3, Q_4 are lateral pnp transistors and the signal is taken from Q_4 to the main gain stage Q_{16}, Q_{17}, which has an active collector load, Q_{13}. C_1 is the compensation capacitor. The class AB output stage is formed by Q_{14} and Q_{20}, biased by Q_{18}, Q_{19}. Short circuit limiting is provided by Q_{15}, Q_{21} and Q_{23}, Q_{24} prevent latch-up under overload conditions. The rest of the transistors are used to set bias levels. The arrangement may look strange to a discrete electronics designer, but the approach is determined by the difficulty of obtaining a wide range and number of resistor and capacitor values on a monolithic IC, compared with the relative abundance of high performance npn transistors. Elements of this circuit design may be seen in all linear ICs.

28.2.19 Operational amplifier configurations

Op amps are available ranging from general purpose to ultra-precision. The low to medium specification devices are often available in duals and quads. There is a growing trend to standardise on FET input op amps for general-purpose applications, particularly those involving a.c. amplifiers, due to their much enhanced slew rate and lower noise.

Specific devices are available for high-speed and high-power output requirements. Additionally, some op amps are available as programmable devices. This means their operating characteristics (usually slew rate, bandwidth and output capability) can be traded off against power supply consumption by an external setting resistor, for instance, to tailor the device to a particular application. Often these amplifiers can be made to operate in the micropower mode, i.e. at low supply voltages and current.

28.2.20 Single supply operational amplifiers

It is entirely possible to operate any op amp on a single supply. However, the amplifier is then incompatible with bipolar d.c. signal conditioning. Single supply operation is entirely suitable for a.c. amplifiers. An example is shown in *Figure 28.10*. The circuit sets up the non-inverting input at $V_+/2$ to give the maximum output swing. Older design op amps such as the μA741 may not perform very well in such a circuit because of limited common mode input voltage range.

Most later generation op amps have been designed with single supply operation in mind and frequently use pnp differential input arrangements. They have an extended input voltage range, often including ground in single supply mode, and are referred to as 'single supply' op amps.

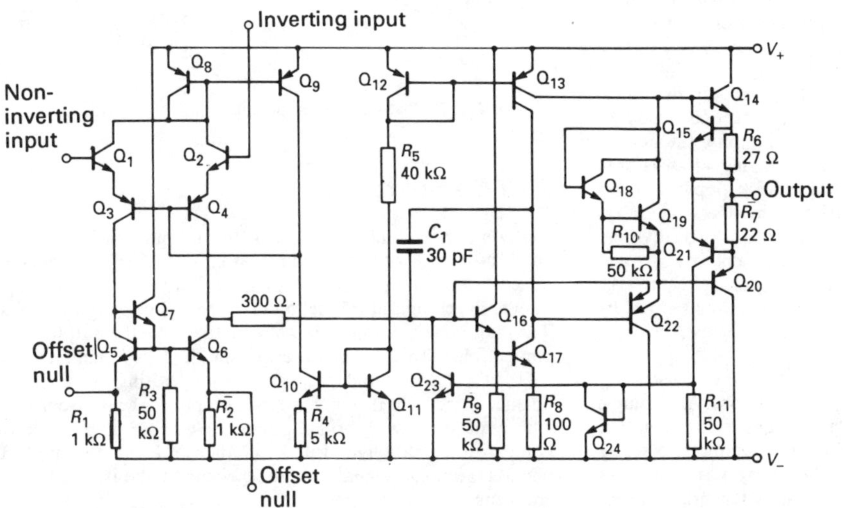

Figure 28.9 μA741 internal circuit

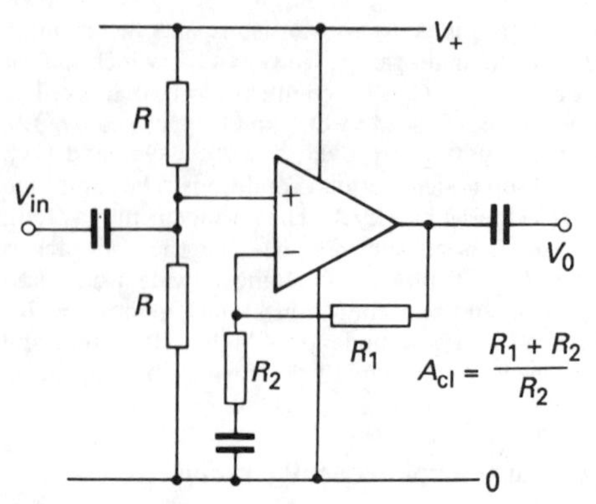

Figure 28.10 Single supply non-inverting a.c. amplifier

28.2.21 Chopper and auto zero operational amplifiers

Many attempts have been made to circumvent the offset and drift problems in op amps for precision amplifier applications. One classic technique is the chopper amplifier, in which the input d.c. signal is converted to a proportional a.c. signal by a controlled switch. It is then applied to a high gain accuracy a.c. amplifier, removing the inherent drift and offset problems. After amplification, d.c. restoration takes place using a synchronous switching action at the chopper frequency. This is available on a single monolithic device.

Other approaches have used the fact that the output is not required continually, i.e. as in analogue-to-digital converters, and have used the idle period as a self-correction cycle. This again involves the use of a switch, but in this case it is usually used to ground the input of the op amp. The subsequent output is then stored as a replica of actual device error at that moment in time and subtracted from the resultant output in the measurement part of the cycle.

28.2.22 Operational amplifier applications

Op amps are suitable for amplifiers of all types, both inverting and non-inverting, d.c. and a.c. With capacitors included in the feedback network, integrators and differentiators may be formed. Active filters form an area where the availability of relatively low cost op amps with precisely defined characteristics has stimulated the development of new circuit design techniques. Filter responses are often produced by op amps configured as gyrators to simulate large inductors in a more practical fashion. Current-to-voltage converters are a common application of op amps in such areas as photodiode amplifiers, etc.

Non-linear circuits may be formed with diodes or transistors in the feedback loop. Using diodes, precision rectifiers may be constructed, for a.c. to d.c. converters, overcoming the normal errors involved with forward voltage drops. Using a transistor in common base configuration within the feedback loop is the basic technique used for generating logarithmic amplifiers. These are extensively used for special analogue functions, such as division, multiplication, squaring, square rooting, comparing and linearisation. Linearisation may also be approached by the 'piecewise' technique, whereby an analogue function is 'fitted' by a series of different slopes (gain) taking effect progressively from adjustable breakpoints.

Signal generation is also an area where op amps are useful for producing square, triangle and sine wave functions. An example of the potential simplicity of op amp designs is shown in the triangle wave generator of *Figure 28.11*. The ramp rate is set by the integrator *RC* component values. The other op amp is used as a threshold comparator with hysteresis that switches the polarity

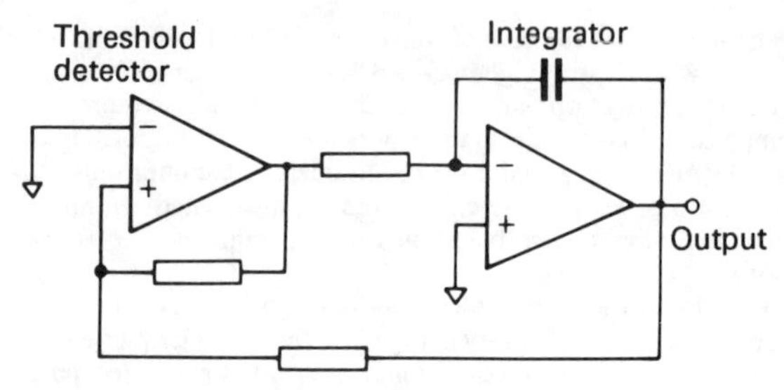

Figure 28.11 Triangle wave generator

of the integrator current input as the output reaches the positive and negative limit points.

28.2.23 Instrumentation amplifiers

Many applications in precision measurement require precise, high gain differential amplification with very high common mode rejection for transducer signal conditioning, etc. To increase the common mode rejection and gain accuracy available from a single op amp in the differential configuration, a three op amp circuit is used, which is capable of much improved performance. *Figure 28.12* shows the standard instrumentation amplifier format, which may be assembled from individual op amps or may be available as a complete (often hybrid) integrated circuit. The inputs are assigned one op amp each which may have gain or act as voltage followers. The outputs of these amplifiers are combined into a single output via the differential op amp stage. Resistor matching and absolute accuracy is highly important to maintain firstly the CMRR, and secondly the gain precision of this arrangement. Suitable resistor networks are available in thin film (hybrid) form.

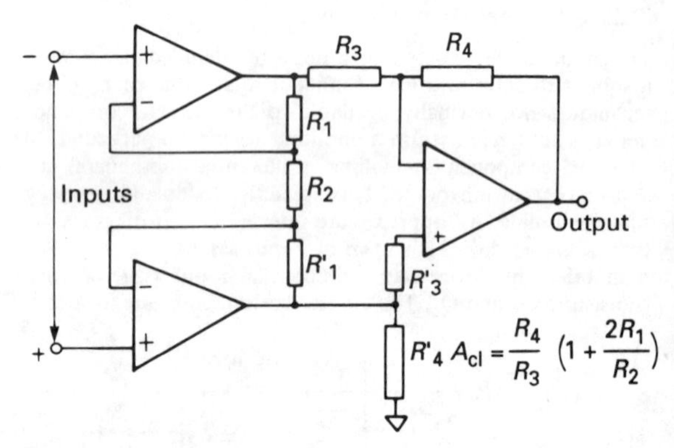

Figure 28.12 Instrumentation amplifier

28.3 Comparators

The comparator function can be performed quite easily by an op amp, but not with particularly well optimised parameters. Specific devices are available, essentially modified op amps, to handle the comparator function in a large range of applications. The major changes to the op amp are to enable the output to be compatible with the logic levels of standard logic families (e.g. TTL, ECL, CMOS) and trade off a linear operating characteristic against speed. A wide common mode input range is also useful. Thus, all the usual op amp parameters are applicable to comparators, although usually in their role as an interface between analogue signals and logic circuits there is no need for compensation.

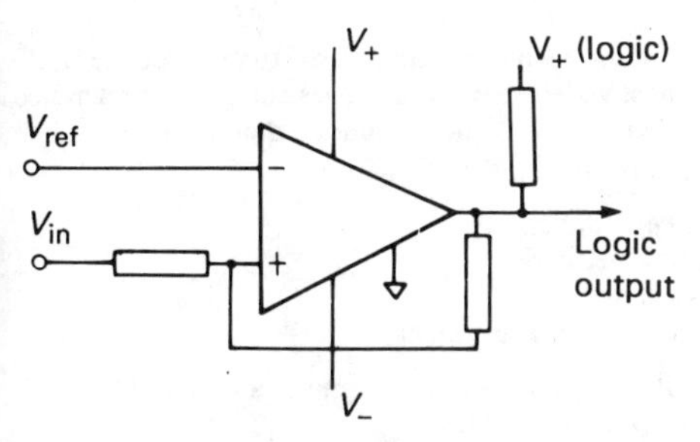

Figure 28.13 Voltage comparator using LM311

A typical comparator application is shown in *Figure 28.13*. The LM311 has a separate ground terminal to reference the output swing to ground whilst maintaining bipolar supply operation for the inputs. The output TTL compatibility is achieved by a pull-up resistor to the TTL supply rail V_{cc}, as this is an open collector type comparator. Higher speed comparators will of necessity employ 'totem-pole' output structures to maintain fast output transition times. The circuit of *Figure 28.13* produces a digital signal dependent on whether the input voltage is above or below a reference threshold (V_{ref}). To clean up the switching action and avoid oscillation at the threshold region, it is quite often necessary to apply hysteresis. This is usually relatively easy to achieve with a small amount of positive feedback from the output to input as in *Figure 28.13*.

General-purpose comparators such as the LM311 exhibit response times (time from a step input change to output crossing the logic threshold) in the order of 200 ns. Their output currents are limited, compared to op amps, being usually sufficient to drive several logic inputs. Often a strobe function is provided to disable the output from any input related response under logic signal control.

28.3.1 Comparator applications

Voltage comparators are useful in Schmitt triggers and pulse height discriminators. Analogue-to-digital converters of various types all require the comparator function and frequently the devices can be used independently for simple analogue threshold detection. Line receivers, *RC* oscillators, zero crossing detectors and level shifting circuits are all candidates for comparator use. Comparators are available in precision and high-speed versions and as quads, duals and singles of the general-purpose varieties. These latter are often optimised for single supply operation.

28.4 Analogue switches

Most FETs have suitable characteristics to operate as analogue switches. As voltage controlled, majority carrier devices, they appear as quite linear low value resistors in their 'on' state and as high resistance, low leakage path in the 'off' state.

Useful analogue switches may be produced with JFETs (junction field effect transistors) or MOSFETs (metal oxide semiconductor FETs). As a general rule, JFET switches can supply the lowest 'on' resistances and MOSFET switches the lowest 'off' leakage. Either technology can be integrated in single or multi-channel functions, complete with drivers in monolithic form.

The switched element is the channel of a FET or channels of a multiple FET array, hence the gate drive signal must be referred to the source terminal to control the switching of the drain–source resistance. It can be seen that the supply voltages of the gate driver circuit in effect must contain the allowable analogue signal input range. The FET switched element has its potential defined in both states by the analogue input voltage and additionally in the off stage by the output potential. The gate drive circuit takes an input of a ground referred, logic compatible (usually TTL) nature and converts it into a suitable gate control signal to be applied to the floating switch element. The configuration will be such that the maximum analogue input range is achieved whilst minimising any static or dynamic interaction of the switch element and its control signal. *Figure 28.14* shows a typical analogue switch in block diagram form. This is a dual SPST function. Because of their close relationship to mechanical switches in use, mechanical switch nomenclature is often used, e.g. SPST is single pole, single throw, and DPDT is double pole, double throw, both poles being activated by the same control signal. Break-before-make switching is frequently included to avoid shorting analogue signal inputs by a pair of switch elements.

The equivalent of single pole multi-way switches in monolithic form are referred to as analogue multiplexers.

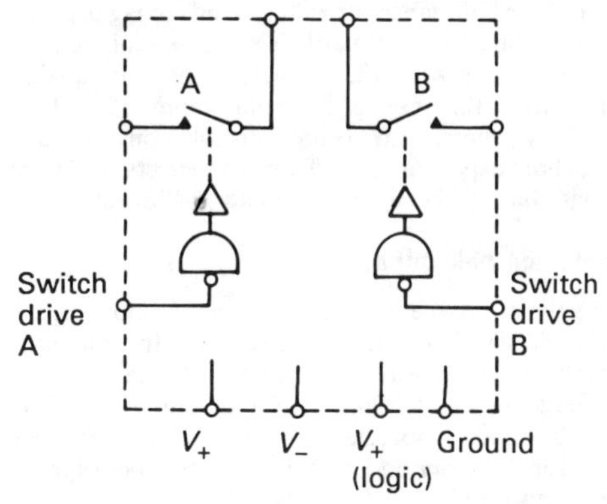

Figure 28.14 Dual SPST analogue switch (DG181)

28.4.1 Analogue switch selection

Selection is based on voltage and current handling requirements, maximum on resistance tolerable, minimum off resistance and operating speed. Precautions have to be taken in limiting input overvoltages and some CMOS switches were prone to latch up, a destructive SCR effect that occurred if the input signal remained present after the supplies had been removed. Later designs have since eliminated this hazard.

28.5 Sample and hold circuits

A sample and hold takes a 'snapshot' of an analogue signal at a point in time and holds its voltage level for a period by storing it on a capacitor. It can be made up from two op amps connected as voltage followers and an analogue switch.

This configuration block diagram is as shown in *Figure 28.15*. In operation, the input follower acts as an impedance buffer and charges the external hold capacitor when the switch is closed, so that it continually tracks the input voltage (the 'sample' mode). With the switch turned off, the output voltage is held at the level of the input voltage at the instant of switch off (the 'hold' mode). Eventually, the charge on the hold capacitor will be drained away by the input bias current of the follower. Hence the hold voltage 'droops' at a rate controlled by the hold capacitor size and the leakage current, expressed as the droop rate in mV/(s μF).

When commanded to sample, the input follower must rapidly achieve a voltage at the hold capacitor equivalent to that present at the input. This action is limited by the slew rate into the hold capacitor and settling time of the input follower and it is referred

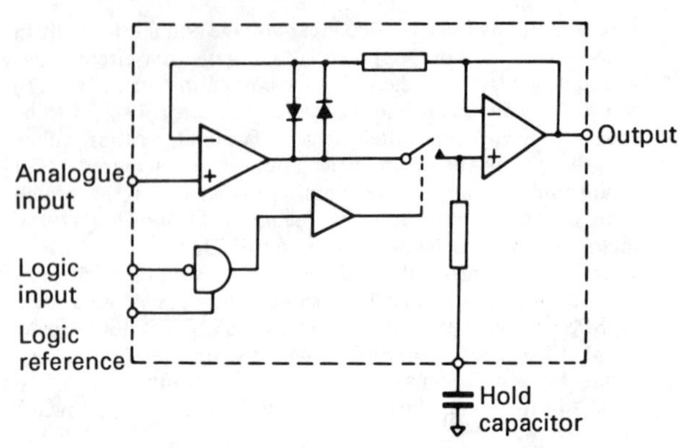

Figure 28.15 Sample and hold circuit (LF398)

to as the acquisition time. Dynamic sampling also has other sources of error. The hold capacitor voltage tends to lag behind a moving input voltage, due primarily to the on-chip charge current limiting resistor. Also, there is a logic delay (related to 'aperture' time) from the onset of the hold command and the switch actually opening, this being almost constant and independent of hold capacitor value. These two effects tend to be of opposite sign, but will rarely be completely self-cancelling.

28.5.1 Sample and hold capacitor

For precise applications, the hold capacitor needs most careful selection. To avoid significant errors, capacitors with a dielectric exhibiting very low hysteresis are required. These dielectric absorption effects are seen as changes in the hold voltage with time after sampling and are not related to leakage. They are much reduced by using capacitors constructed by polystyrene, polypropylene and PTFE.

The main applications for sample and hold circuits are in analogue-to-digital converters and the effects of dielectric absorption can often be much reduced by effecting the digitisation rapidly after sampling, i.e. in a period shorter than the dielectric hysteresis relaxation time constant.

28.6 Digital-to-analogue converters

Digital-to-analogue converters (DACs) are an essential interface from the digital world into the analogue signal processing area. They are also the key to many analogue-to-digital converter (ADC) techniques that rely on cycling a DAC in some fashion through its operating range until parity is achieved between the DAC output and the analogue input signal. All DACs conform to the block diagram of *Figure 28.16*. The output voltage is a product of the digital input word and an analogue reference voltage. The output can only change in discrete steps and the number of steps is immediately defined by the digital inputs. Should this be eight, e.g. an 8-bit DAC, the number of steps will be 256 and the full scale output will be 256 times some voltage increment related to the reference.

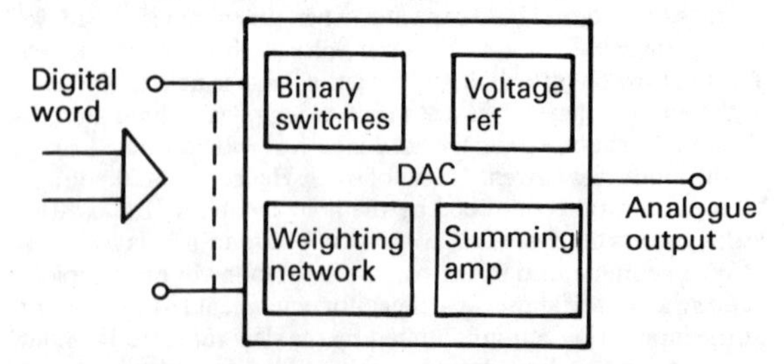

Figure 28.16 Digital-to-analog converter block diagram

Although various techniques can be used to produce the DAC function, the most widely used is variable scaling of the reference by a weighting network switched under digital control. The building blocks of this kind of DAC are:

(1) A reference voltage.
(2) A weighting network.
(3) Binary switches.
(4) An output summing amplifier.

All may be built in, but usually a minimum functional DAC will combine network and switches.

28.6.1 *R–2R* ladders

The weighting network could be binary weighted as in *Figure 28.17(a)*, with voltage switching and summation achieved with a normal inverting op amp. However, even in this simple 4-bit example, there is a wide range of resistor values to implement and speed is likely to suffer due to the charging and discharging of the network input capacitances during conversion. *Figure 28.17(b)* shows the *R–2R* ladder network employing current switching to develop an output current proportional to the digital word and a reference current. The current switching technique eliminates the transients involving the nodal parasitic capacitances.

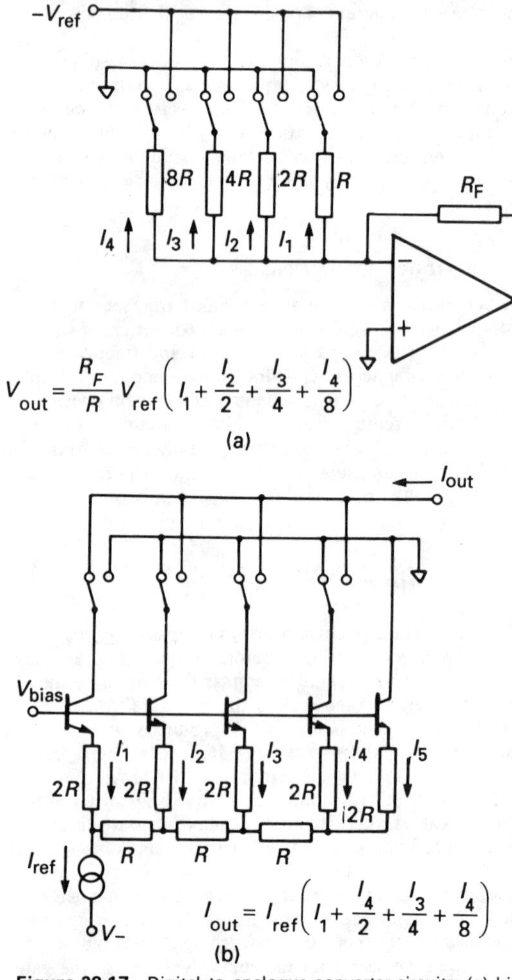

Figure 28.17 Digital-to-analogue converter circuits: (a) binary weighted ladder—voltage switching, (b) *R-2R* ladder—current switching

Significantly, only two resistance values are required and the accuracy of the converter is set by the ratio precision rather than of the absolute value. Linear IC fabrication is capable of accommodating just such resistance value constraints quite conveniently. Monolithic DACs are available in 8- and 12-bit versions and recently in 16 bit. 16-bit and above DACs are also available in hybrid form. Progress in IC DAC technology has been spurred by the demands of the telecommunication and consumer industries. PCM digital telephony systems require large volumes of D/A and A/D functions to digitise and recover speech bandwidth signals. Compact disc audio has probably presented a greater challenge in necessitating low-cost 16-bit DACs handling a 20 kHz bandwidth.

Note that the output of the DAC shown in *Figure 28.17*(*b*) is in a current form. It is converted to a voltage output quite simply by an external op amp current-to-voltage converter. *Figure 28.18* shows the standard internal layout of a typical 8-bit DAC. The op

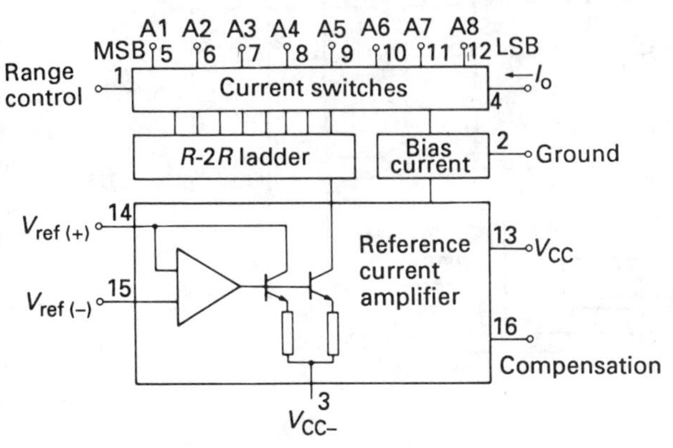

Figure 28.18 8-bit digital-to-analogue converter IC (MC1408)

amp and current mirror circuit are used to produce a current sink related to the supply from a voltage reference of either polarity referred to ground. True or complementary current outputs are available and both may be used in differential mode, in single-ended mode by grounding one output.

The output of such a DAC is directly proportional to the reference current and where it is able to operate over a significant dynamic reference current range it is referred to as a 'multiplying' DAC.

28.6.2 Resolution, linearity and monotonicity

Resolution has already been touched upon as being defined by the digital word length and reference voltage increment. However, it is important to note that it is quite possible to have, say, a 12-bit DAC in terms of resolution which is incapable of 12-bit accuracy. This is because of non-linearities in the transfer function of the DAC. It may well be accurate at full scale but deviate from an ideal straight line response due to variations in step size at other points on the characteristic. Linearity is specified as a worst case percentage of full scale output over the operating range. To be accurate to n bits, the DAC must have a linearity better than $\frac{1}{2}$LSB (LSB = least significant bit, equivalent to step size) expressed as a percentage of full scale (full scale = $2^n \times$ step size).

Differential linearity is the error in step size from ideal between adjacent steps and its worst case level determines whether the converter will be monotonic. A monotonic DAC is one in which, at any point in the characteristic from zero to full scale, an increase in the digital code results in an increase in the absolute value of the output voltage. Non-monotonic converters may actually 'reverse' in some portion of the characteristic, leading to the same output voltage for two different digital inputs. This is obviously a most undesirable characteristic in many applications.

28.6.3 Settling time

The speed of a DAC is defined in terms of its settling time, very similar to an op amp. The step output voltage change used is the full scale swing from zero and the rated accuracy band is usually $\pm\frac{1}{2}$LSB. Settling times of 100–200 ns are common with 8-bit R–$2R$ ladder monolithic DACs. The compensation of the reference op amp will have a bearing on settling time and must be handled with care to maintain a balance of stability and speed.

28.6.4 Other DAC techniques

There are many other methods that have been proposed and used for the DAC function. There is insufficient space here to make a full coverage. However, mention should be made of the pulse width technique that is frequently used in consumer ICs. The digital inputs are assigned time weighting so that in, say, a 6-bit DAC, the LSB corresponds to 1 time unit and the MSB to 32 time units. The time units are derived by variable division from a master clock and usually clocked out at a rate which is some sub-multiple of the master clock.

The duty cycle will then vary between 0 and 63/64 in its most simple form. This pulse rate is integrated by an averaging filter (usually *RC*) to produce a smooth d.c. output. The full scale is obviously dependent on the pulse height and this must be related back to a voltage reference, albeit only a regulated supply. This type of DAC is often used for generating control voltages for voltage controlled amplifiers used for volume, brightness, colour, etc. (TV application) and it is possible to combine two 6-bit pulse width ratio outputs together for 12-bit resolution (not accuracy) to drive voltage controlled oscillators, i.e. varicap diode tuners. This is one application (tuning) where non-monotonicity can be acceptable.

28.7 Analogue-to-digital converters

ADCs fall into three major categories: direct converters, DAC feedback and integrating.

28.7.1 Direct or 'flash' converters

Flash converters may have limited resolution but are essential for very high speed applications. They also have the advantage of continuous availability of the digitised value of the input signal, with no 'conversion time' waiting period. They consist of a reference voltage which is subdivided by a resistor network and applied to the inputs of a set of comparators. The other inputs of the comparator are common to the input voltage. A digital encoder would then produce, in the case of a 3-bit converter for example, a weighted 3-bit output from the eight comparator inputs. The method is highly suitable to video digitisation and is usually integrated with a high-speed digital logic technology such as ECL or advanced Schottky TTL. Flash converters in monolithic form of up to 9-bit resolution have been built.

28.7.2 Feedback ADCs

Feedback ADCs use a DAC within a self-checking loop, i.e. a DAC is cycled through its operating range until parity is achieved with the input. The digital address of the DAC at that time is the required ADC output. Feedback ADCs are accurate with a reasonably fast conversion time. Their resolution limitations are essentially those of the required DAC cost, performance and availability.

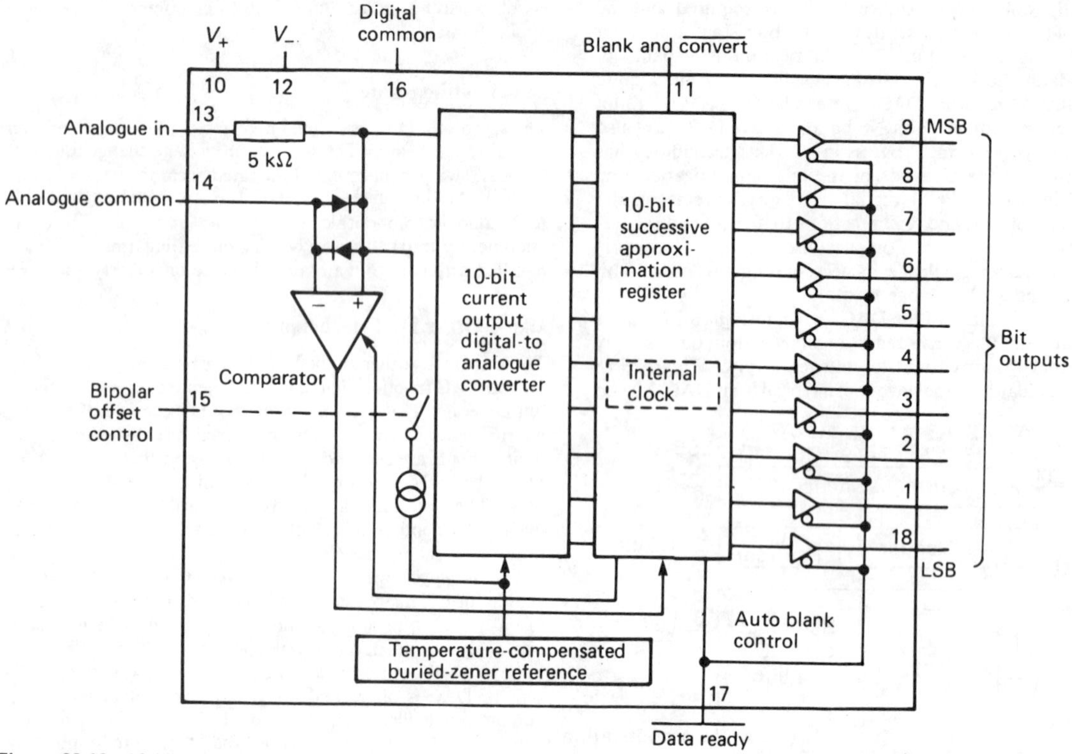

Figure 28.19 10-bit analog-to-digital converter (AD571)

Feedback ADCs require a special-purpose logic function known as a successive approximation register (SAR). The SAR uses an iterative process to arrive at parity with the ADC input voltage in the shortest possible time. It changes the DAC addresses in such a way that the amplitude of the input is checked to determine whether it is greater or smaller than the input on a bit sequential basis, commencing with the MSB. This implies continuous feedback from a comparator on the analogue input and DAC output connected to the SAR.

Figure 28.19 shows a block diagram of a 10-bit ADC available on a single chip. The 10-bit DAC is controlled by 10-bit SAR with internal clock and the DAC has an onboard reference. The digital output is fed through tri-state buffers to ease the problems of moving data onto a bus structured system. These are high impedance (blank) until the device is commanded to perform a conversion.

After the conversion time, typically 25 μs for this type of device, a data ready flag is enabled and correct data presented at the ADC output. An input offset control is provided so that the device can operate with bipolar inputs. Because of the nature of the conversion process, satisfactory operation with rapidly changing analogue inputs may not be achieved unless the ADC is preceded by a sample and hold circuit.

28.7.3 Integrating ADCs

Integrating ADCs use a variety of techniques, but all usually display high resolution and linearity. They also have a much greater ability to reject noise on the analogue input than other types of ADC. The penalty that is paid for this is a much longer conversion time which is also variable with actual input voltage. These factors make integrating ADCs very suitable for digital voltmeter, panel meter and other digital measuring applications. LICs to perform this function often have direct display (LED, LCD, etc.) drive capability built in and, because of their BCD coding and subsequent display segment/digit formatting, will be unsuitable for data acquisition systems.

28.7.4 'Dual slope' integrating ADCs

The dual slope ADC will be explained in some detail, as the understanding of this technique leads to familiarity with most other integrating converter principles. The dual slope technique has advantages in that a large number of the possible error sources are inherently self-cancelling. The basic principle consists of ramping down a capacitor from a known voltage level with a discharge current directly proportional to the input voltage for a fixed period of time. *Figure 28.20* shows the effect on the integrating capacitor. At the end of the fixed period the capacitor

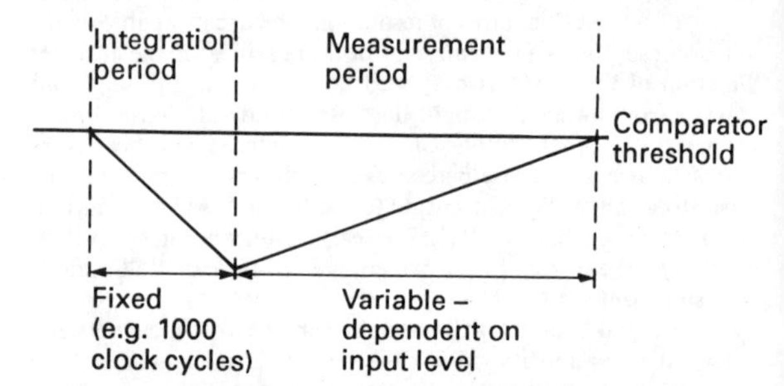

Figure 28.20 Capacitor voltage ramp—integrating ADC

is recharged at constant current defined by a voltage reference. During this second period, the clock used to generate the fixed period timing is counted until a comparator indicates that the original voltage level has been reached. This then stops the

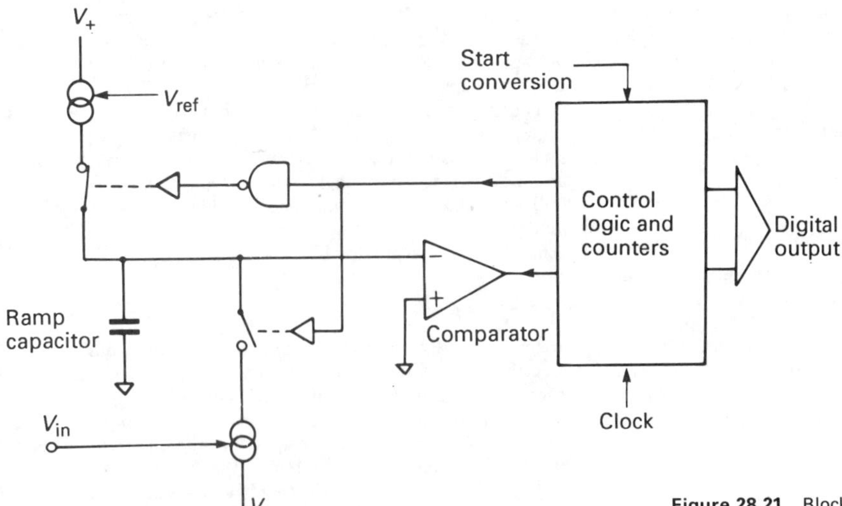

Figure 28.21 Block diagram—dual slope integrating ADC

count, which is latched and presented to the output in binary or BCD form. With suitable scaling, the BCD output can be directly in volts.

Figure 28.21 shows a block diagram of the implementation of the dual slope ADC. The switch positions will be reversed during the integration period. In a typical application, say a $3\frac{1}{2}$ digit DVM, full scale count is 1999, the control logic using BCD counters. After the previous conversion, the counters are reset to 0999. The integration period is defined from this start point to when the counter reaches its maximum count of 1999, this point being arranged to reverse the switches and the next counter state is 0000. The counter continues during the measurement period until the comparator changes state. Exceeding 1999 during measurement will latch an overload indication. Note that this example ($3\frac{1}{2}$ digit BCD) is equivalent to 11-bit binary resolution. Integrating ADCs are capable of 18-bit accuracy and resolution with careful design and application.

Dual slope ADCs can perform this accurately due mainly to the tolerance of long-term drift of various components. The clock frequency is not critical as long as its short-term stability is good, i.e. relative to the conversion period, since the integration period changes are corrected by counting the same clock during the measurement period.

The absolute value of the ramp capacitor is not important as long as it possesses short-term capacitance stability and those dielectric parameters as outlined for sample and hold circuits. The actual comparator threshold and its error parameters, e.g. offsets, are not a source of error as long as they do not change over the dynamic ramp range and in the short term. Primary error sources revolve around the reference current source (or sink) and the input amplifier, acting as a voltage-to-current converter. The op amp errors can often be corrected by auto zero techniques applied during the measurement period. This will involve switching the op amp to measure (and store for correction during integration) its own offset, during the period in which its output is redundant. The effects of noise on the input voltage are averaged out during the integration, and in mains powered equipment the conversion rate can be synchronised to the mains frequency to avoid jitter induced at mains frequency.

28.7.5 Voltage-to-frequency converters

The voltage-to-frequency converter is a form of direct ADC that produces a serial rather than parallel digital output. Converters of this type have advantages in their fast response, good linearity and continuous output that may be used for transmission of analogue values in serial digital form over cables, fibre optic and radio channels. Direct counting of the output with normal digital frequency counter techniques will produce a read-out in BCD or binary format. Additionally, most monolithic VFCs can have their functional elements configured by external connections to perform as frequency-to-voltage converters (FVCs). 12-bit resolution is possible with monolithic VFCs, but greater resolution and accuracy is obtained with hybrid devices.

Figure 28.22 shows the basic principle of a VFC operating on positive input voltages. If the input voltage is greater than the

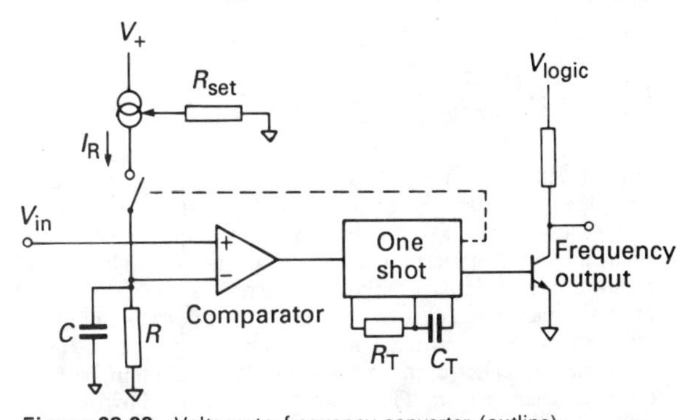

Figure 28.22 Voltage-to-frequency converter (outline)

voltage stored on the capacitor, the comparator will trigger the monostable multibrator (one-shot). The output of the one-shot switches the logic compatible output stage and also switches on the current source. The current source is on for a period determined by the one-shot *RC* time constant, and is programmed by an external resistor in conjunction with an internal voltage reference. The charge increment delivered to the comparator input capacitor will usually be sufficient to raise its voltage to a level in excess of the input voltage. At the end of the one-shot period, the current source is switched off and the voltage on the input capacitor decays at a rate determined by the parallel resistor *R*. When it falls below the input, the comparator will retrigger the one-shot again and the cycle repeats. At balance, the average current flowing into and out of the capacitor must be equal. The reference current, I_R, and the one-shot period, t, are constants, so that the average capacitor current is $I_R tF$, where F is the VFC frequency. Average capacitor current is also, to a good approximation, given by V_{in}/R. So it can be seen that frequency output is directly proportioned to input voltage, and a relatively simple VFC can be linear over a wide range.

28.8 Voltage regulators

LICs have an important part to play in power supplies for all forms of electronic circuitry, digital or analogue. They have, for a considerable period, dominated linear regulator applications and, with the increasing tendency to space and weight reduction and energy conservation, are emerging as self-contained switching regulators or control circuits for switchers. The widest usage at present in linear regulators is for series pass types with some shunt devices for special applications.

28.8.1 Series regulators

The block diagram of a series regulator is shown in *Figure 28.23*. This equates functionally to the μA723 type regulator IC with the exception of thermal and SOA protection. This was an early

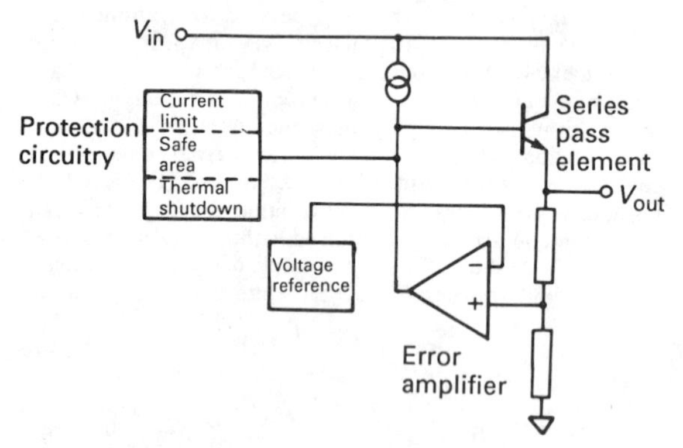

Figure 28.23 Voltage regulator block diagram

approach to a flexible 'universal' voltage regulator that remained an industry standard for many years. The essential parts are a voltage reference, an error amplifier (op amp), a series pass element, and facilities for overload protection. The μA723 contains a temperature-compensated zener diode reference and the series pass element is an npn Darlington, capable of about 150 mA collector current. An additional npn transistor is available to act as a current limiter in conjunction with an external sense resistor. The device connections allow it to be configured as a series or shunt linear regulator or various types of switching regulator using different external components to achieve the desired output voltage and boost the current handling capability. This amount of flexibility is not always the optimum solution and other fixed and adjustable concepts will be discussed.

28.8.2 Load and line regulation

The ideal voltage regulator produces its desired output voltage without any change when the input voltage and load current fluctuate. The line regulation is defined as the change in output voltage, in percentage or absolute terms, for a specified change in input voltage. Similarly, a load regulation is defined as the output voltage change for a specified change in load current. Usual applications of voltage regulator ICs will be to produce regulated d.c. lines from normal a.c. mains, i.e. regulating the raw d.c. voltage produced by a step-down transformer, rectifier and reservoir capacitor combination.

The dynamic performance of the device line regulation is important here to attenuate the mains frequency ripple present at its input. This parameter is usually specified as ripple rejection and the frequency at which it is guaranteed is usually 100–120 Hz (full wave a.c. rectification). Line and load regulation will be better than 1% and ripple rejection around 60–70 dB on a general-purpose regulator IC.

28.8.3 Drop out voltage

Drop out voltage is significant in power supply design as it identifies the minimum input/output voltage differential under which a regulator can still function. This is usually set by the saturation voltage of the series pass element and its circuit configuration. The user must ensure that the input voltage is in excess of the required output plus the worst case drop out voltage. This must allow, in the case of a mains power supply, for the troughs of the waveform of unregulated d.c. input caused by ripple effects.

28.8.4 'Three-terminal fixed' regulators

As the electronics industry has standardised on particular supply voltages, e.g. +5 V for TTL, ±15 V for analogue signal processing (op amp) circuitry, the demand has grown for fixed output regulators with minimum external components requirements and simplicity of use.

This has resulted in the families of so-called fixed voltage, three-terminal regulators. The feedback resistor network sampling the output and the current limit resistor have been integrated into these devices, along with a more substantial series pass transistor to boost current availability. This has dictated the choice of various power packages, e.g. TO-3, TO-220, TO-202, to dissipate the extra heat. It has also made mandatory the inclusion of additional protection circuitry as in *Figure 28.23* to prevent 'blowout' of the regulator in case of overdissipation (internal thermal sensing and shutdown) and simultaneous excess voltage and current on the pass element (safe operating area protection).

Fixed voltage three-terminal regulators are available, usually with 5% initial voltage tolerance, in positive and negative versions, at various voltages from 2.6 V to 24 V. Widely available current levels are 0.1 A, 0.25 A, 0.5 A and 1–1.5 A. The current rating usually decides the package options used. Monolithic fixed voltage regulators are available up to around 5 A at certain popular voltages (+5 V) and hybrid devices can achieve higher currents (10–20 A).

Figure 28.24 shows the circuit arrangement of a three terminal regulator. The quiescent (stand-by) current, I_Q, of the regulator

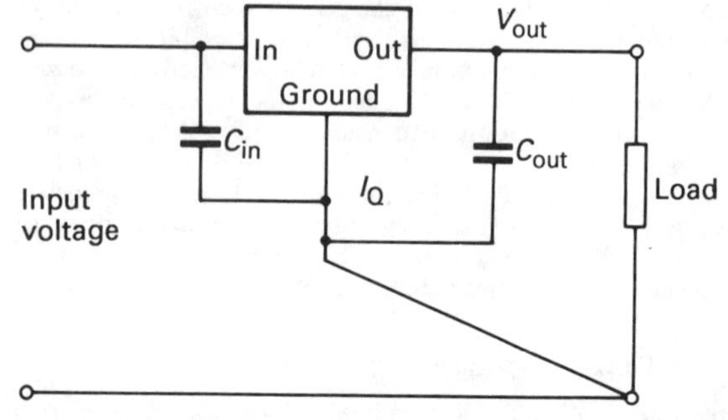

Figure 28.24 Three-terminal fixed-voltage regulator (μA7800)

flows down the ground leg and is of the order of a few milliamperes to avoid excessive power loss during no-load periods. To ensure stability, a capacitor, C_{in}, is usually required, although some regulators may tolerate its absence if the power supply reservoir capacitor is physically close enough to the device.

Negative regulators are usually much more susceptible to oscillation without complete input bypassing and the use of the recommended C_{in} (value and type) is necessary. C_{out} improves the transient response of the regulator to step load changes but is not usually essential.

A disadvantage of the three-terminal configuration is that the device can only regulate to its terminal and hence uncompensated voltage drops can occur if the load is remote. Because I_Q is low, the ground terminal can be returned to a load sense point easily. However, heavier gauge wire must be used to reduce this effect on the positive terminal, as in *Figure 28.24*.

Since the I_Q of this type of regulator is quite constant with load changes, alterations of the output voltage are possible with resistance added in the ground lead. However, three-terminal regulators specifically designed as adjustable parts are available with better performance.

28.8.5 Adjustable regulators

The proliferation of fixed voltage regulator ranges have satisfied many applications, but there is always a demand for different voltages, greater accuracy, etc. LICs such as the μA723, LM104 and LM105 can handle a variety of voltage and current combinations with a high level of precision. Specialised three-terminal adjustable regulators have been developed to offer a simple solution to both non-standard and standard voltages.

These have a 'floating' configuration as in *Figure 28.25*, and always maintain a nominal 1.2 V reference across the V_{out} and ADJ terminals. The output voltage is defined by the voltage across R_2 plus the reference voltage. Since the constant current source biasing of the reference is low, typically around 50 μA, the current through R_2 is very close to $1.25/R_1$. Errors resulting from

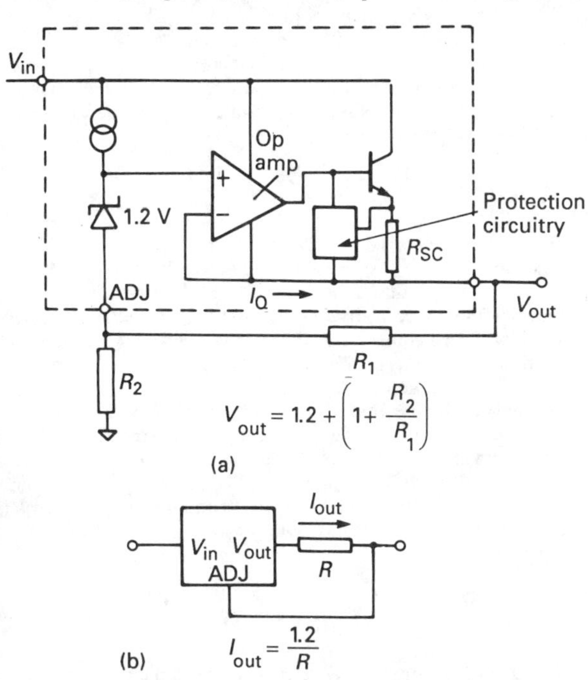

Figure 28.25 Three-terminal 'floating' adjustable regulator: (a) voltage, (b) current

the reference bias current can be swamped by selecting a much greater value for the current through R_1/R_2. This current has to be equal to or greater than the regulator quiescent current, which flows out of the V_{out} terminal, in any case. The reference used in this, as in other three-terminal regulators, is of the 'bandgap' variety. This type of three-terminal regulator also has the same comprehensive overload protection circuitry. The three-terminal adjustable regulator is also capable of implementing very simply the constant current source function as in *Figure 28.25(b)*. Versions of this type of regulator are available with MOSFET series pass elements. These can handle much higher input/output differentials than their bipolar counterparts, e.g. 125 V as compared to the normal 40 V, and hence can regulate higher supply voltages or cope with significant input transients, such as occur in automotive systems.

28.8.6 Foldback limiting

Merely limiting the current output of a regulator during overloads is rarely sufficient, since, in the case of short circuit, input/output differential increases greatly and hence so does power dissipation. To avoid exceeding the thermal design limitations, regulators using devices such as the μA723 can be arranged to progressively reduce their limit current as the output voltage is forced below its regulated level. Complete regulator subsystems, i.e. three terminals, achieve foldback by thermal sensing on the chip and progressive shutdown (output current restriction), commencing at around 125–150°C.

28.8.7 Overvoltage protection

One regulator may well be supplying many ICs in a system and it is often necessary to ensure that, in the event of a regulator failure, the output voltage cannot go significantly higher than the nominal level. TTL, in particular, has a relatively small range between normal operating (5 V $\pm$5%) and absolute maximum (7.0 V) supply levels.

Overvoltage protection uses an additional voltage reference and comparator to sample the regulator output and, should it exceed a preset (higher than nominal) limit, to rapidly fire a thyristor which short circuits or 'crowbars' the output. The regulator current limiting may also have failed, of course, and a fuse in series with the regulator input will be the terminal interruption of the fault condition. LIC devices are available containing all the elements of power supply supervisory systems, thus providing overvoltage and undervoltage protection, power on reset signals for microprocessors, remote shutdown and automatic switching to battery backup. However, the overvoltage 'crowbar' thyristor is usually an external component.

28.8.8 Voltage references

All regulator ICs contain some form of voltage reference. Originally references used in voltage regulators were zener diodes, either selected for low temperature coefficient by having a breakdown voltage close to the transition between true zener operation and avalanche operation (5.1–5.6 V), or an avalanche diode compensated with the addition of a forward biased junction (around 7 V). This latter type has been extensively used in ICs, typified by the μA723. Its disadvantages are noise and shortcomings in long-term stability. For low-voltage regulators, it dictates a minimum V_{in} of 8–9 V to activate the reference, even if the load related dropout parameter is lower.

The 'bandgap' reference produces a highly predictable, stable voltage of monimally 1.2 V with low noise and temperature coefficient. 1.2 V is the bandgap of silicon, corresponding to the V_{BE} of a silicon transistor at 0 K. The operation of the reference depends on the ability to fabricate transistors adjacent to each other within an IC whose V_{BE} difference can be controlled by area (geometry) ratio during diffusion.

Because of the number of individual components making up the reference, the bandgap can be more susceptible to thermal gradients on the chip. High-precision references have therefore tended to continue to use the zener technique with improvements being made by employing ion implantation. The subsurface breakdown thus produced has much better long-term stability, and temperature stability effects can be minimised by a constant temperature oven formed by an on-chip heater, thermal sensing and feedback control. This kind of reference is frequently employed for precision DACs and ADCs.

28.8.9 Shunt regulators

Shunt regulators are commonly seen using zener diodes to regulate supply lines. Series resistance is essential with a shunt regulator, and, if the load current falls to zero, all the power previously supplied to the load is dissipated in the regulator. This makes the shunt regulator cumbersome and inefficient for high power loads, particularly if they are variable. LIC shunt regulators are more precise, adjustable versions of the standard zener diode, with the ability to be programmed over a wide voltage range with two resistors. *Figure 28.26* shows a typical device and its basic circuit application.

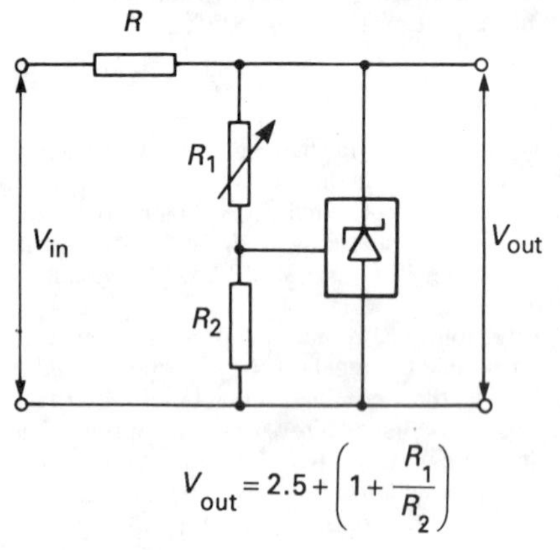

$$V_{out} = 2.5 + \left(1 + \frac{R_1}{R_2}\right)$$

Figure 28.26 Adjustable shunt regulator (μA431)

28.8.10 Switching regulators

Switching regulators differ from the series and shunt regulators discussed previously in two major areas. Firstly, they are capable of much greater efficiency since the minimum possible power is dissipated in the switching elements. Secondly, they are not limited to regulating a voltage to some lower level than its original form (step-down) but are capable of stepping up and inverting the polarity of the d.c. voltages.

LICs for switching regulators basically fall into two categories, those intended as controllers for switching regulators utilising transformer coupling from output to input, and those intended as switching inverters/converters with relatively low d.c. input voltages (around 50 V and below) and no input/output isolation. The former are often used for direct off-line (mains) switch mode power supply (SMPS) controllers.

28.8.11 SMPS control ICs

The typical off-line SMPS reduces the bulk, weight and heat dissipation of a linear supply by full wave rectifying and filtering and a.c. line input and then switching it at a frequency, say 25 kHz, which is much higher than that of the mains supply. This means a dramatic reduction in size from the normal mains frequency transformer and reservoir capacitors. The primary of the SMPS transformer is switched, either single-ended or in push–pull, by one or more high-speed bipolar or MOSFET power transistors. The output voltage is rectified at the secondary with fast recovery, e.g. Schottky rectifiers. The SMPS control circuit contains the necessary oscillator and, in addition to the normal regulator components, a comparator, pulse steering logic and drivers to interface to the power switch devices. The switching drives are pulse width modulated to regulate the output. To preserve isolation the control circuits will be coupled to the output via some form of isolator, e.g. optocouplers.

28.8.12 Single inductor switching regulators

Many useful voltage conversion/inversion and regulator functions can be achieved with LIC switching regulators of the single-ended, single-inductor variety. Such circuits allow the provision of multiple supplies (of either polarity) from a single voltage input, such as a battery, whilst ensuring low power consumption through switching efficiency. A typical LIC able to perform such d.c./d.c. translations is shown in *Figure 28.27*. The op amp is a 'bonus' which can be used to provide additional series regulated supplies as it has high current output capability. The remaining circuitry is used to compare a fraction of the output voltage with a reference (bandgap) and commit the switch to a cycle of the oscillator waveform, depending on the result.

Figure 28.28 shows the three basic switching configurations with a single inductor, and all of these, with the exception of inverting, which requires an external switch transistor, can be handled by the device of *Figure 28.27*. The oscillator free runs and the comparator enables the output switch on a cycle-by-cycle basis. During the on period, the rate of rise on inductor current will be controlled by the applied voltage and the inductance value, until either the inductor saturates (it is likely to be ferrite not air cored) or the switch element can no longer maintain a saturated state, i.e. there is insufficient base drive. This could result in switch destruction, so the risk is contained by an active current limit within the oscillator. At this point, the switch-on time is aborted, and the inductor allowed to discharge into the output load and storage capacitor. Thus the frequency of operation of this type of switching regulator is variable with load and input voltage change.

28.9 IC timers

Timers may be considered as digital circuits, since their basic functions are as astable and monostable multivibrators. However, the function is extremely useful in linear systems. Most timer circuits are derivatives of the basic NE555 type as in *Figure 28.29*, with two comparators controlling a set–reset flip–flop. In continuous (astable) operation, an external capacitor is charged and discharged between two thresholds, $\frac{2}{3}V_{cc}$ and $\frac{1}{3}V_{cc}$, via an external resistor network. In the monostable mode, the circuit times out when the external capacitor charges from zero to $\frac{2}{3}V_{cc}$. Quad and dual versions of the circuit are available and also versions in CMOS technology. LICs are also available containing a 555 type timer coupled to an *n*-bit programmable counter. These can produce long delays very accurately and repeatably.

28.10 Consumer LICs

Consumer LICs are usually devices for which the prime volume usage is within the consumer radio and TV industries. Certain functions may be usable in the professional communications,

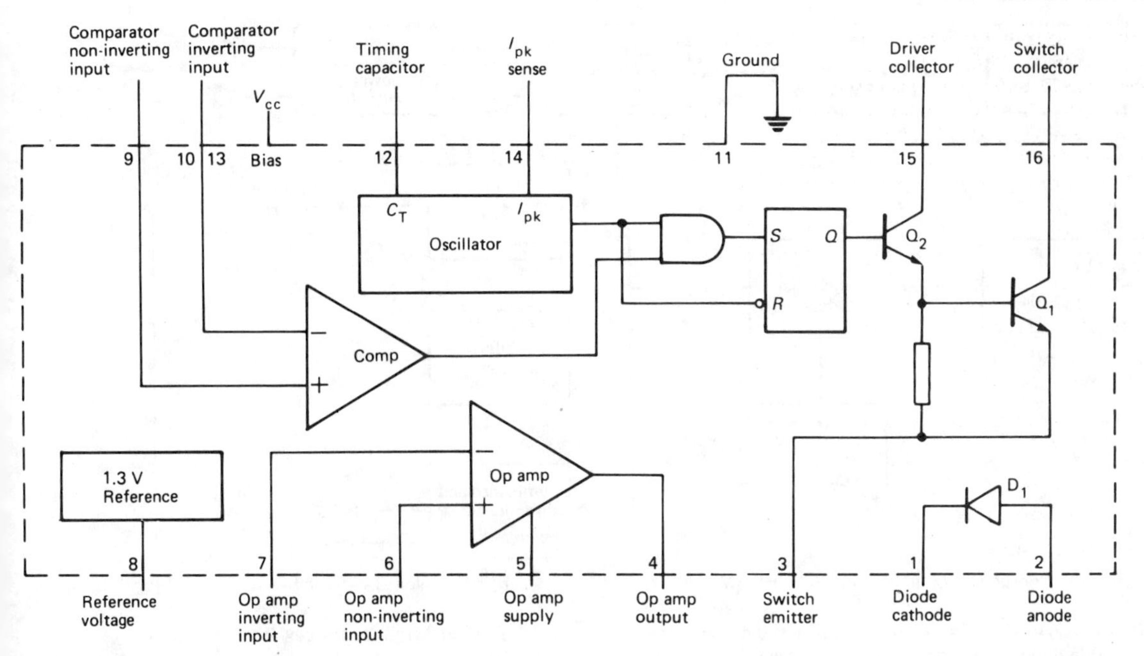

Figure 28.27 Single inductor switching regulator sub-system (μA78S40)

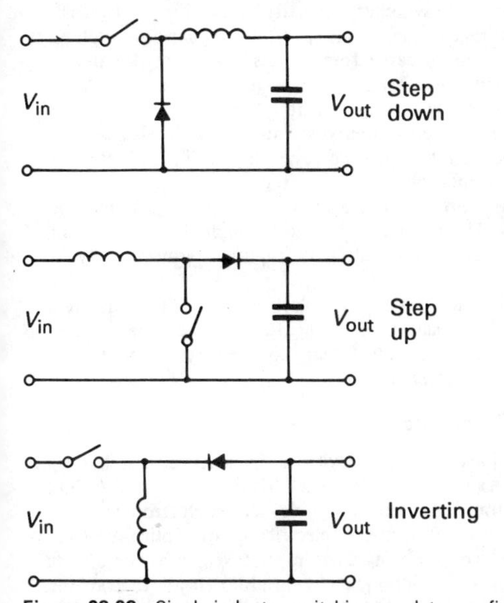

Figure 28.28 Single inductor switching regulator configurations

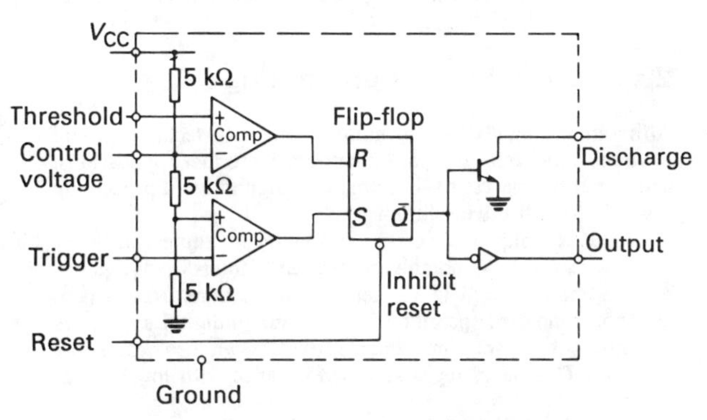

Figure 28.29 IC timer block diagram (NE555)

telecommunications and, of course, broadcasting areas. However, some functions are quite specialised and are only really useful for their designed application. A good example of this is the chrominance processing circuitry for a television receiver.

The range of available devices is very wide and cost pressures on consumer equipment have led to some of the largest scale linear ICs on the market. *Figure 28.30* gives basic outlines of a VHF/FM radio receiver and a UHF colour TV. All of the functions of the radio can be integrated and the present state of the art for minimum chip count is four, partitioning being as shown. The tuner IC would use external tuned circuits for selectivity (with varicap diode tuning) and ceramic (block) filtering prior to the 10.7 MHz limiting IF amplifier and detector IC. Multiplex stereo decoding is achieved in another IC which consists of a phase locked loop. This locks an oscillator (often 76 kHz) to the incoming 19 kHz pilot tone to regenerate the correct phase 38 kHz signal for matrix decoding. The dual audio power amplifier then drives the loudspeakers. Monolithic audio power amplifiers are currently capable of upwards of 20 W output. The complete receiver may of course require voltage regulators, tuning control circuits, etc.

The TV block diagram is an even more graphic indication of available LIC complexity. The UHF tuner is so far resistant to monolithic integration, but most other functions are heavily integrated. The video IF amplifier and detector IC handles the incoming video signal at 38–39 MHz and provides the amplitude modulation detection. Intercarrier sound at 6 MHz is handled in the sound channel IC, which additionally provides FM detection and audio power amplification. The composite video is applied to three ICs, two of which separate the synchronising signals and process the line and field scan oscillators. Field scan drive is direct, but line scan and associated EHT (25 kV) generation for the CRT are handled by an external high voltage power semiconductor. The chrominance information is decoded in another IC and the resultant red, green and blue (RGB) drives are applied to discrete video amplifiers, since ICs cannot currently achieve the breakdown voltage necessary to drive the CRT grids.

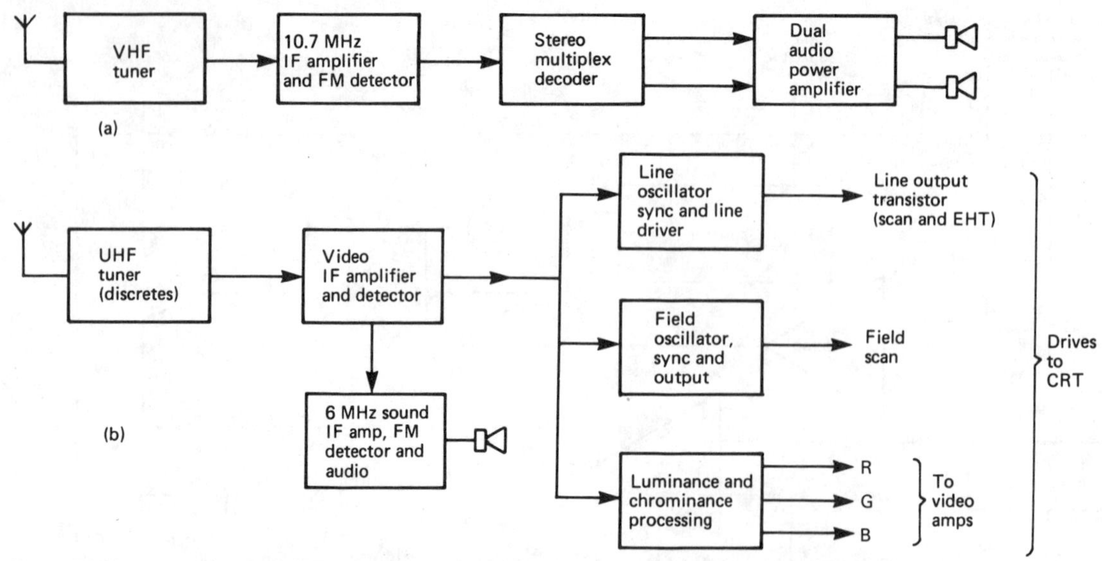

Figure 28.30 Consumer LIC applications: (a) VHF/FM stereo radio (four ICs), (b) UHF colour TV (basic five ICs)

Single chip chroma processors are probably the densest linear ICs available. Colour TVs also contain many other ICs for power supply control (SMPS controllers), tuning, remote control, etc.

28.11 Digital signal processing

Although analogue design techniques using LIC technology still have enormous relevance to contemporary product applications, considerable progress has been made in digital signal processing towards cost-effective solutions.

Digital techniques have been gradually invading the linear world for some time, notably in the area of filters. Progress has been extremely rapid in the telecommunications area, where telephone line card speech bandpass filters in digital exchanges are now exclusively implemented with switched capacitor circuitry. This offers repeatable performance with low drift at much lower cost.

The basis of digital signal processing is the conversion of analogue signals to digital and their subsequent manipulation in real time by high-speed microprocessors. These signals can then be re-converted to the analogue domain in their conditioned form via A/D converters. The development of high-speed microprocessors with enormous processing power and the improvements in speed, accuracy and cost of integrated D/A and A/D converters have made this concept a reality.

There has thus been an emergence of dedicated LSI chips containing these functions (known as DSPs), which put a vast range of signal conditioning functions under total software control, leading to significant improvements in design flexibility and product redesign time. This approach is obviously going to have great impact in analogue designs with the constant reduction in digital IC cost *versus* complexity, and should be considered for future large systems.

28.12 Charge transfer devices

A charge transfer device (CTD) is a semiconductor structure in which discrete charge packets are moved. CTDs find application in shift registers, high-speed filtering, imaging systems and dynamic memories. The two main methods of constructing CTDs result in bucket brigade devices and charge coupled devices.

28.12.1 Bucket brigade devices

Bucket brigade devices (BBDs) are formed by connecting a series of capacitors with switches, normally FET or bipolar transistors. A single storage element consists of two capacitor–switch units. Charge held on the capacitors is transferred along the device by turning odd and even switches on alternately. Modern commercial BBDs can transfer at 1 MHz.

Non-linearities encountered at small signal levels are avoided by making the zero signal correspond to a fixed offset. Charge deficit represents the signal amplitude.

FETs as switches have a slower transfer rate than bipolar transistors. The latter lose charge through the base current of each transistor. Compensating by amplifiers reduces the dynamic range of the device.

Bucket brigade devices are relatively simple to fabricate and have found application in such areas as audio delay lines to implement reverberation. Integration density and performance are better in a CCD structure.

28.12.2 Charge coupled devices

A CCD (charge coupled device) is, in essence, a shift register formed by a string of closely spaced MOS capacitors. A CCD can store and transfer analogue signals—either electrons or holes—which may be introduced electrically or optically. Charge coupled devices are being used in photosensor arrays and such signal processing components as variable delay lines, transversal filters and signal correlators.

At earlier stages in CCD development, the technology was considered to offer great potential for mass memory applications, with very low cost per bit. Structures proposed emulated the storage concept utilised in magnetic disk media, with relatively long access times by current semiconductor memory standards. As yet they have not been able to rival hard disk media, such as Winchester drives, on cost or non-volatility. The rapid increase in dynamic RAM density and decrease in cost per bit has also contributed to relegating CCD memory into a backwater. However, in analogue applications, particularly TV imaging, CCDs have made a significant impact.

28.12.3 CCD operation

To retain data, CCDs rely on dynamic charge storage on a semiconductor surface. *Figure 28.31* shows a metal oxide

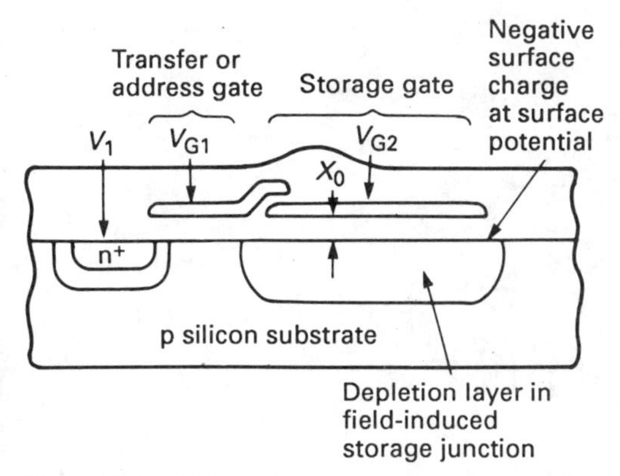

Figure 28.31 CCD construction

semiconductor structure that could be part of a CCD shift register. The storage region is the silicon surface beneath the storage gate where a field-induced junction contains a negative charge of mobile electrons.

The surface of the semiconductor behaves like a capacitive voltage divider—the upper capacitor is the fixed gate capacitance, while the lower one is the space-charge region of the field-induced junction; the connection between the capacitors is the field-induced junction itself.

The transfer or address gate forms a temporary conductive path between the two junction regions. If the voltage applied to the storage gate (V_{G2}) exceeds the threshold voltage, then the surface potential of the field-induced junction will equal the diffuse-junction potential when the transfer or address gate (V_{G1}) is turned on.

Several techniques have been utilised to move the CCD charge packet, as in *Figure 28.32*. One technique uses one electrode per

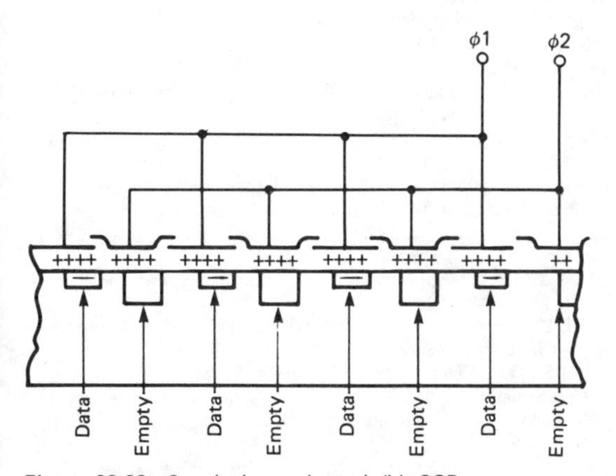

Figure 28.32 Standard two-electrode/bit CCD

bit and has one additional storage location for the entire register. This requires a data rippling effect to function. Each electrode has a unique clock that is time sequenced to all other clocks. This permits moving data one bit at a time, allowing the next location to be vacated before data is moved in. This technique almost halves the number of storage locations needed.

The primary disadvantage of this technique is that a clock decoder is required to provide the ripple through clock signals necessary. This decoder adds complexity and area to the device.

28.12.4 Charge transfer in CCDs

A voltage applied to an electrode placed on, but insulated from, the p-type silicon substrate creates a potential maximum to which electrons are attracted. It is useful to consider the potential profile as a well, and charge as water which is held in it. Clocking a series of electrodes on the surface creates a moving potential profile which 'pours' the charge along the device. This avoids the spurious leakage paths in the transistor switching of BBDs.

Losses incurred by charge being trapped while it is transferred along an interface can be reduced by moving the potential maximum away from that interface by selective diffusion. This results in the buried channel CCD (BCCD) structure. This has a faster, less noisy transfer than its surface channel equivalent. The typical efficiency of a BCCD transfer is 0.999 995 at up to 20 MHz.

Early CCDs used three-phase clocking to transfer charge while maintaining potential barriers to separate the charge packets. Doping creates an asymmetrical potential profile along the device, allowing two-phase clocking to transfer charge, while still keeping the charge packets separate (*Figure 28.32*). By applying a d.c. level to one set of electrodes charge may be transferred with only one external clock ($1\frac{1}{2}$ phase clocking). This is the basis of most modern CCD structures.

28.12.5 Charge input/output in CCDs

Charge may be injected to a CCD shift register electrically or optically. Silicon is sensitive to light in the wavelength range 400–1200 nm, which includes the visible region of the spectrum. Light falling on the depletion region of a CCD creates an electron–hole pair. The electron moves to the potential well. A charge packet is created in proportion to the intensity of the light falling on the device.

Charge sensing methods vary according to whether the charge can be read destructively, as at the end of an image sensor, or must be read without destruction, such as part-way along a tapped delay line, or in a CCD memory. In the former, charge enhances the leakage current of a reverse biased diode. In the latter, charge packets produce mirror image charges in a floating gate electrode. In both, the charges are then amplified for output.

28.12.6 CCD applications

Line scan CCD sensors In a typical image sensor (*Figure 28.33*), 2048 photosites, 13 μm square on 13 μm centres, accumulate charge which is linearly dependent on the intensity of the light entering the device and its integration time. Eight shielded elements, separated from the photosites by isolation cells, represent the random generation of charge in the device—the dark signal.

After a suitable exposure time (typically 100 μs to 100 ms) a pulse applied to the transfer clock (ϕ_x) causes charge packets to be transferred to the transport registers. By transferring odd photosites to one side, and even photosites to the other, the number of packet transfers is halved.

The d.c. voltage on V_T and the waveform on the transport clock, ϕ_t, provide the $1\frac{1}{2}$ phase clocking which moves charge in the CCD shift registers. Charge packets are delivered to the output circuit alternately to re-establish their original sequence. Here, the charge alters the potential of a precharged diode. The change is amplified, sampled, held and output. The reset clock (ϕ_R) recharges the diode.

A white reference charge is injected into the ends of the shift registers to be output with the video and used in an external automatic gain control circuit.

Two further shift registers provide an end-of-scan pulse, and reduce incursion of stray electrons into the inner shift registers.

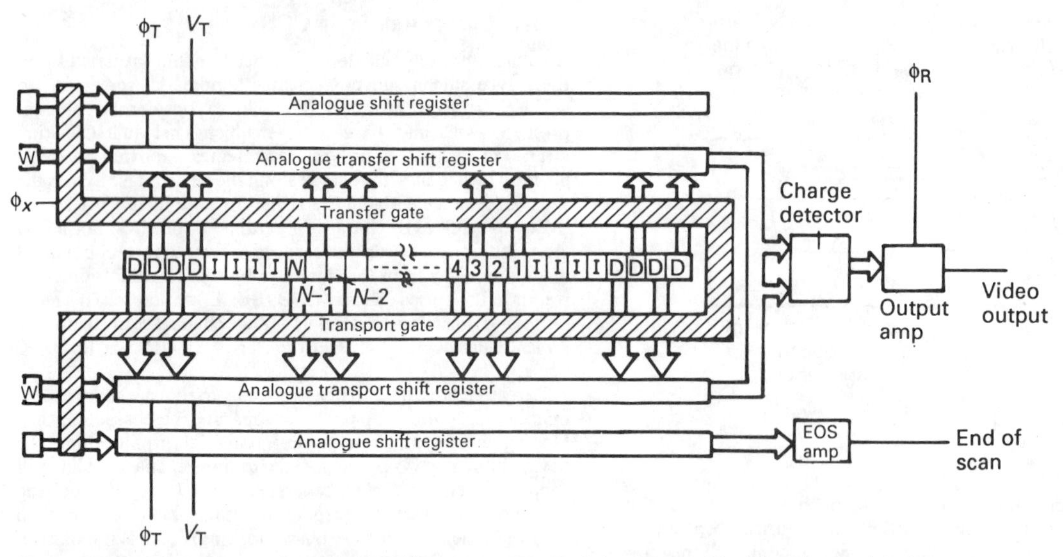

Figure 28.33 Block diagram of a typical CCD line-scan sensor: D, dark reference; W, white reference; I, isolation cell

The dynamic range of this type of device is typically 2500 to 1 with an output data rate of 2 MHz.

The precise manufacture of the photosites means that objects may be resolved accurately and repeatedly. These attributes are exploited in facsimile, telecine and industrial inspection applications. By moving objects past the sensor a two-dimensional picture is built up.

Area CCD sensors Area image sensors are fabricated with their photosites arranged in a matrix. The sensor turns a picture into precise electrical signals. Suitable clocking of the sensor results in a full interlaced video output signal. Currently available production sensors have 185 440 photosites in a 488 × 380 matrix, although development devices up to 800 × 800 elements have been produced.

The low noise, high sensitivity, low power consumption, high accuracy and inherent reliability make CCD cameras ideal for replacement of tube cameras, for applications such as broadcasting, robot vision and image analysis.

Signal processing With the charge input, shifting and output elements, a CCD delay line may be constructed. In its simplest form, an input waveform appears at the output after a delay dependent on the number of shift register elements and the clock rate. The horizontal line delay needed in PAL TV receivers can be created in this way.

In a tapped delay line, portions of a signal can be sensed after variable delays and used for adding to or subtracting from the input signal. The matched filters which may be constructed in this way find application in communications and radar to detect weak signals in high background noise. Here CCDs offer longer delays than surface acoustic wave (SAW) devices.

Further reading

BOARDMAN, C. M. and DESCURE, P., 'CCDs—injection, detection, operation and use', *Electronic Engineering*, September (1982)

DANCE, M., 'Recent developments in op amps', *Electronics Industry*, June (1981)

DEMLER, M. J., 'Understand CMOS flash ADCs to apply them effectively', *EDN*, 21 January (1988)

FAULKENBERRY, L. M., *An Introduction to Operational Amplifiers with Linear IC Applications*, Wiley, Chichester (1982)

FLEMING, T., 'Isolation amplifiers break ground loops and achieve high CMRR', *EDN*, 24 December (1987)

FULLAGAR, D. J., 'Optimising analogue to digital conversion', *Electron. Product Design*, November (1988)

HAROLD, P., 'Current feedback op amp ease high speed circuit design', *EDN*, 7 July (1988)

JUNG, W. G., 'Stable FET—input op amps achieve precision performance', *EDN*, November (1982)

KRAUS, K., 'Designing with programmable operational amplifiers', *Electronics Industry*, October (1982)

LITTLE, A. *et al.*, 'S/H amp–ADC matrimony provides accurate sampling', *EDN*, 4 February (1988)

MAYER, J. H., 'Higher speeds, CMOS designs drive flash A–D converter developments', *Comp. Design*, 15 October (1988)

MUTO, A. and NEIL, M., 'ADC dynamic performance testing', *Electronic Product Design*, June (1982)

OHR, S., 'Converters show off linear LSI processing', *Electronic Design*, June (1980)

RAHIM, Z., 'Micropower op amp offers simplicity and versatility', *EDN*, 7 January (1988)

SCHAEFER, K., 'Video op amps find use in RF and active filters', *EDN*, 1 September (1988)

SHIER, J., 'New ±5 V standard unshackles analog IC designers', *EDN*, 9 June (1988)

SHOREYS, F., 'New approaches to high-speed high-resolution analogue-to-digital conversion', *Electronics & Power*, February (1982)

TSANTES, J., 'Data converters', *EDN*, August (1982)

WIEGAND, J., 'Self-calibration and oversampling make room for more digital circuitry on monolithic ADCs', *EDN*, October 15 (1987)

ZUCH, E. L., 'Understanding and applying sample and hold circuits', Parts 1 and 2, *New Electronics*, October (1978)

29 Semiconductor Memories

P Jones, BSc
Fujitsu Microelectronics Ltd

Contents

29.1 Dynamic RAM

Dynamic RAM is the lowest cost, highest density random access memory available. Since the 4K generation, DRAM (dynamic RAM) has held a 4 to 1 density advantage over static RAM, its primary competitor. Dynamic RAM also offers low power and a package configuration which easily permits using many devices together to expand memory sizes. Today's computers use DRAM for main memory storage with memory sizes ranging from 16 kbytes to many megabytes. With these large size memories very low failure rates are required. The metal oxide semiconductor (MOS) DRAM has proven itself to be more reliable than previous memory technologies (magnetic core) and capable of meeting the failure rates required to build huge memories. Dynamic RAM is the highest volume and revenue product in the industry.

29.1.1 Cell construction

The 1K and early 4K memory devices used the three-transistor (3-T) cell shown in *Figure 29.1*. The storage of a 'one' or a 'zero'

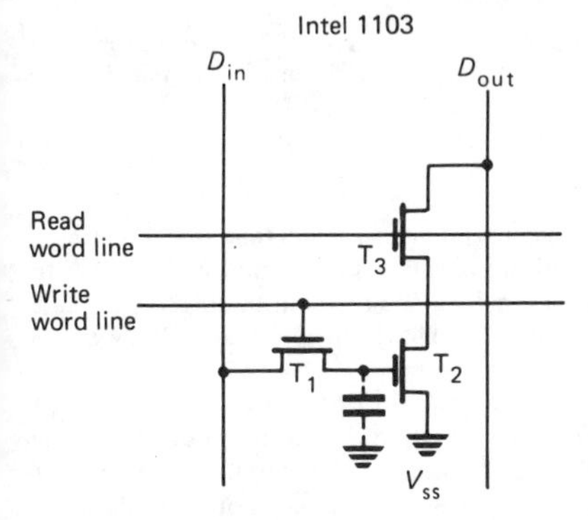

Figure 29.1 Three-transistor cell used in 1 kbit RAM device

occurred on the parasitic capacitor formed between the gate and source of T_2. Each cell had amplification thus permitting storage on the very small capacitor. Due to junction and other leakage paths the charge on the capacitor had to be replenished at fixed intervals; hence the name dynamic RAM. Called refresh, a maximum duration was specified as 2 ms.

The next evolution, the 1-T cell, was the breakthrough required to make MOS RAM a major product. The 1-T cell is shown in *Figure 29.2*. A two-level poly process is used to improve the cell density by about a factor of two at the expense of process complexity. The transistor acts as a switch between the capacitor and the bit line. The bit line carries data into and out of the cell. The transistor is enabled by the word line which is a function of the row address. The row address inputs are decoded such that one out of *N* word lines is enabled. *N* is the number of rows which is a function of density and architecture. A 16K RAM has 128 rows and 128 columns. The storage cell transistor is situated such that its source is connected to the capacitor, its drain is connected to the bit line and the gate is connected to the word line.

The technology has remained consistent through the 64K (256 rows and 256 columns), and the 256K DRAM (512 rows and 512 columns).

The 256K DRAM has seen the arrival of the first CMOS technology device in volume production, giving a reduced power consumption, and low-power standby mode. The CMOS cell will become the industry standard for all larger devices.

Although the CMOS 256K DRAM has been available, the industry standard has been an NMOS single-transistor memory cell, using a triple-layer polysilicon process allowing for even higher packing densities. Typically, the triple polysilicon

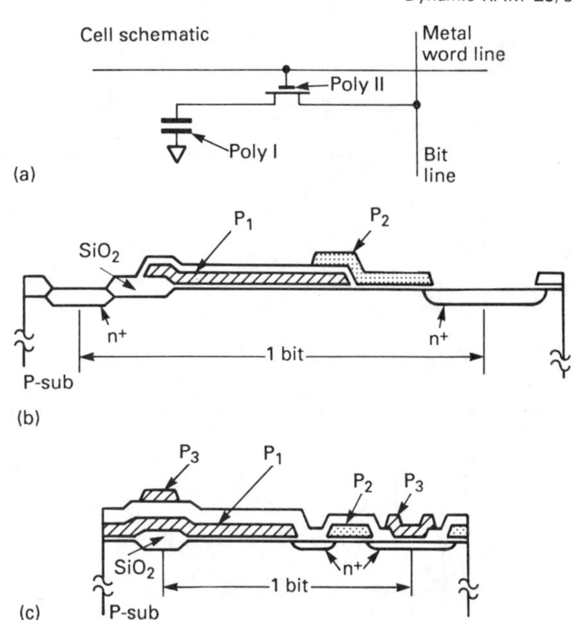

Figure 29.2 Storage cell configuration for 64K and 256K DRAM: (a) 1-T cell schematic, (b) 64K open bit line cell, (c) 256 open bit line cell

technology provides 0.75 of the cell capacity in only 0.33 of the cell area compared to the double polysilicon used on the 64K DRAM. In this process the storage capacitor and address transistor are first made using double polysilicon technology. Then another oxidised insulating layer is deposited. After making a pattern of via holes, the third layer of polysilicon is deposited and etched to form the bit lines.

29.1.2 Cell sensing

Two types of sense amplifiers have been used historically with 1-T cells. Since one approach draws significantly less power, it has become the standard of the 16K RAM generation. Sensing of the signal from a 1-T cell is explained in the following paragraphs while contrasting the two approaches.

When the row decoder enables a row, charge is transferred from each storage capacitor in that row to its respective digit/sense line, destructively reading data. Each column has its own sense amplifier, the function of which is to detect and amplify the signal caused by this charge. To maximise the signal into the sense amplifier, the bit line capacitance must be minimised. This is accomplished by cutting the bit line in half. The sense amplifier is then placed in the centre of the bit line, where it senses a differential voltage between the two halves of the line.

To each side of the sense amplifier is added one additional column of storage cells. These additional cells, commonly called 'dummy cells', have a capacitance equal to approximately one half of a data storage cell. The dummy cell is used to establish a reference voltage on the side of the sense amplifier not containing the selected cell. Since the dummy cell is half the capacitance of a storage cell, a voltage reference level halfway between the voltage difference between a '1' and a '0' is established. This reference voltage is commonly called the cell's 'trip point' and may be varied by changing the physical size of the 'dummy cell'. In actual practice this 'trip point' varies due to alignment tolerances in the manufacturing process.

Between cycles, the two halves of each bit line are equilibrated to precisely the same voltage and the 'dummy cell' is restored to 0 V (*Figure 29.3*). After the row decoder has been activated, a

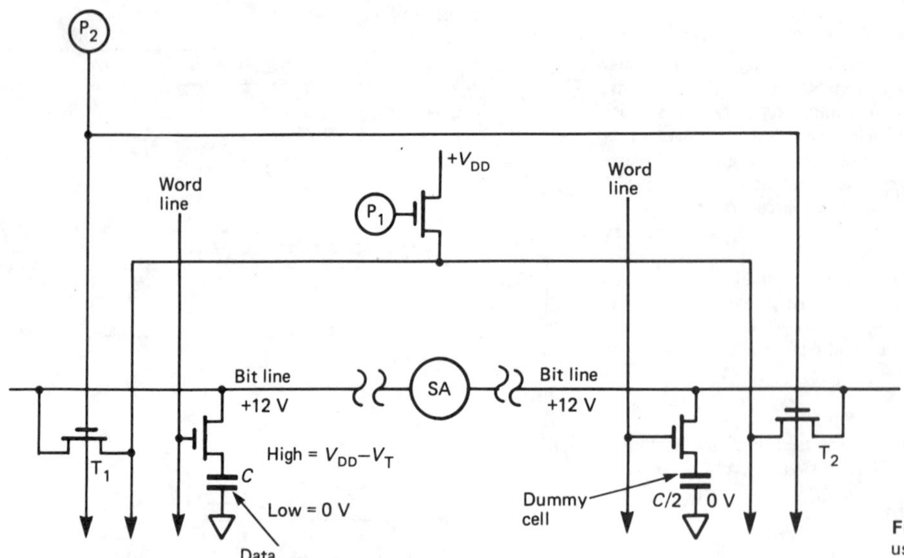

Figure 29.3 Sensing scheme uses dummy cell to establish voltage reference

single cell in each column turns on, including the dummy cell. If we assume the bit lines were initially set to 12 V then the following occurs. The 0 V stored in the dummy cell pulls the side of the sense amplifier not containing the selected storage cell to approximately 11.75 V. (This is determined by the ratio of the bit line capacitance to that of the dummy cell's capacitance.) If the storage cell contained 0 V, charge sharing between the bit line and the storage cell would bring the bit line voltage to approximately 11.5 V (remember the storage cell is twice as large as the dummy cell). A voltage difference between 11.5 and 11.75 V can now be sensed. Had 12 V been stored, the bit line would remain at 12 V and a voltage difference between 12 and 11.75 V would be sensed. In practice the storage cell stores 12 V minus the threshold voltage of the acess transistor thereby somewhat moving the bit line away from 12 V.

The sense amplifier is now turned on and via the regenerative action of the cross-coupled amplifier the lower voltage side of the sense amplifier is pulled to ground. The higher voltage side remains at its previous level. At the end of a cycle the word lines are deselected and the preceding transistors, T_1 and T_2, are turned on. This equilibrates the bit lines to the V_{DD} supply potential. The node P_1 is placed at a voltage at least one threshold above V_{DD} to permit the bit lines to reset to the V_{DD} potential.

Two types of sense amplifiers have been used in commercially available products—variations of the static design of *Figure 29.4(a)* and of the dynamic amplifier of *Figure 29.4(b)*. Both are about equal in their ability to detect and amplify small signals.

The load resistors (R_1 and R_2) employed in the static amplifiers consume a substantial amount of power, typically half or more of total chip power. Since dynamic amplifiers do not employ these resistors, their power consumption is considerably reduced. However, there are formidable design and layout problems associated with dynamic amplifiers, and many designers in the early days chose to incorporate power-consuming static sense amplifiers into their designs.

We must look at the write cycle—or more accurately the read–modify–write cycle—to understand the differing circuit requirements for static and dynamic sense amplifiers. As an example, suppose cell 64 (*Figure 29.4(a)*) had originally stored a low voltage and was read.

Upon detecting a lower voltage on node B than that on node A, the sense amplifier will drive node B to ground and node A to power supply V_{DD}. Transistor T_3 then turns on and the data from the cell becomes available to the output buffer at one end of the data bus.

Now assume we want to write the opposite data back into the cell. This requires forcing a high voltage onto node B and the storage capacitor, C_{64}. To do this, the data input buffer drives the input/output data bus to ground. T_3 then forces node A to ground, overpowering R_1. When node A goes to ground, T_2 turns off allowing R_2 to pull node B to V_{DD} as required to write the high level into the storage cell.

Without R_2, node B would remain at ground, and it would be impossible to write a high voltage into the cell. With the resistors present, data can be written into a cell in either matrix half with a single input/output data bus.

A trade-off exists in the choice of values for R_1 and R_2. Since either R_1 or R_2 dissipates power in all sense amplifiers, low resistance values result in very high power consumption. Then again, digit line capacitance of node B is quite large, and a high-value resistor would lead to excessively long write time. Unfortunately, there is no good compromise, and circuits employing static sense amplifiers suffer from high power consumption and long write times.

On paper, the dynamic amplifier solves the problem very well. Having read a low voltage from cell 64 (*Figure 29.4(b)*), assume that we again want to write a high voltage back into the cell. As before, the input buffer drives the true data bus to ground, with T_3 causing node A to follow. However, the input buffer also forces the complement data bus to V_{DD} with T_4 causing node B to follow. The complement data bus thus performs the job previously done by the resistor. With cell 64 still selected, the high voltage on node B is transferred into the cell and the write operation is complete.

Note that T_3 and T_4 function only as switches and can have very low resistances to speed up write time. This involved no speed/power trade-off. Therefore, memory designs using dynamic amplifiers consume far less power and write much faster than do designs using static sense amplifiers.

The previous was the technique used in the 4K/16K generations. This generation permitted use of a 12 V (V_{DD}) power supply and

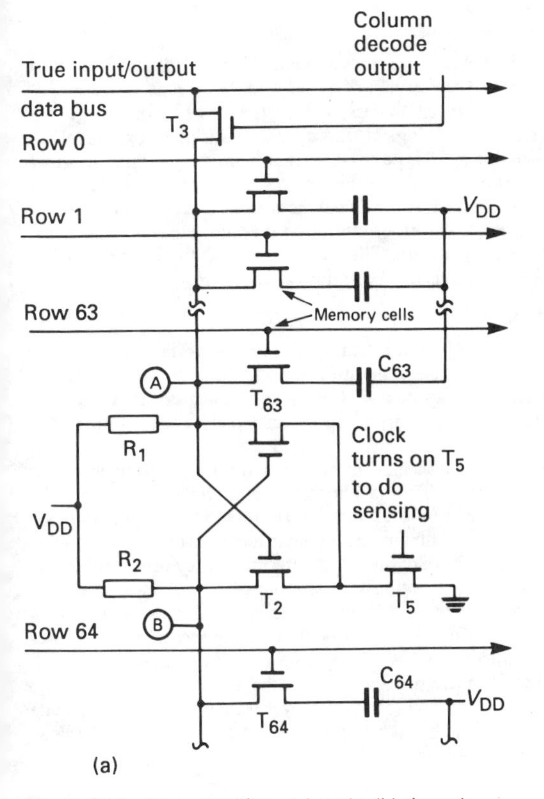

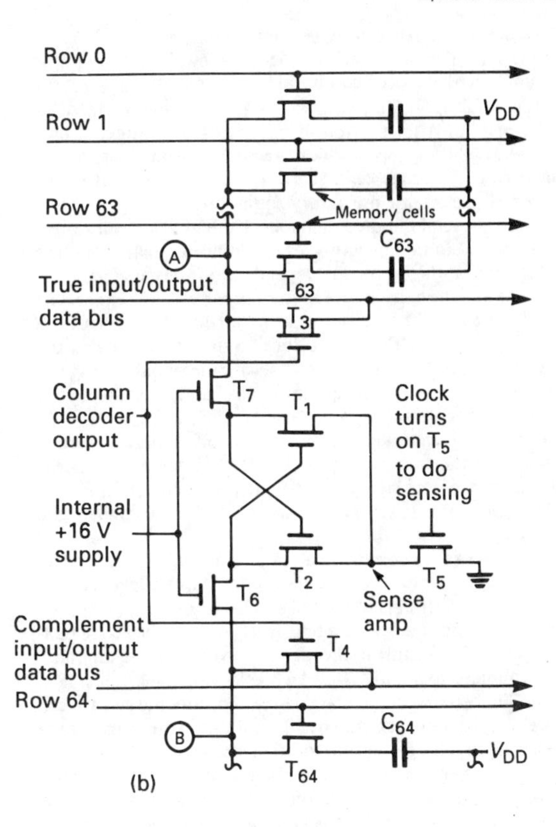

Figure 29.4 Sense amplifiers: (a) static, (b) dynamic

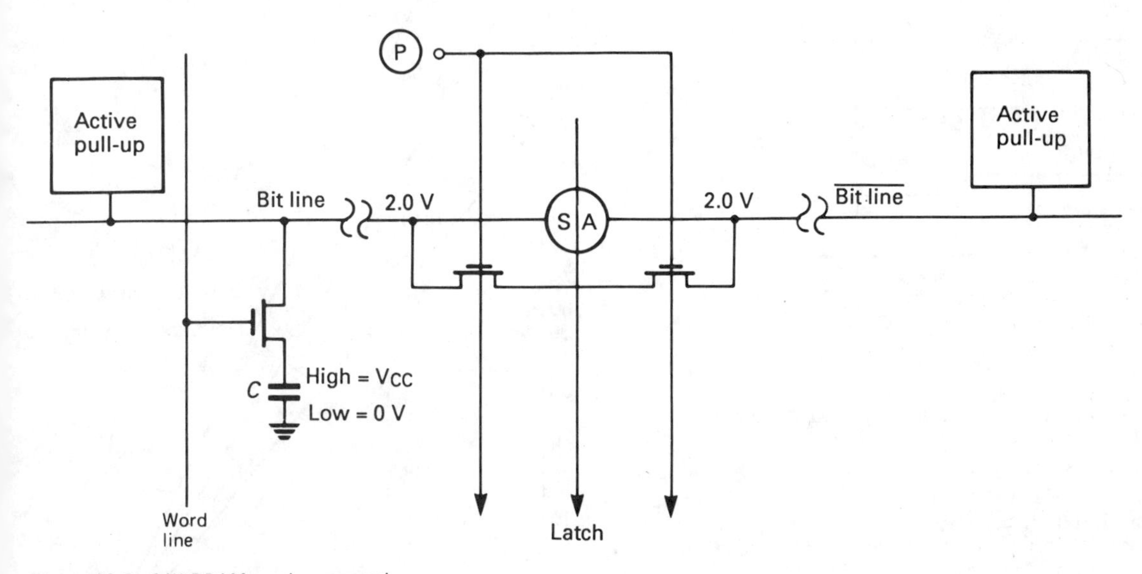

Figure 29.5 64K DRAM sensing approach

a -5 V (V_{BB}) bias supply. The 64K generation offers neither of these to the designer. The designer must generate a bias supply on chip and also has only a 5 V supply to work with. One obvious ramification is that the signal is decreased to 5/12 of the 16K if circuit techniques and geometries are held constant. To permit manufacture of a 64K RAM both dimensions and circuit techniques were changed.

Figure 29.5 illustrates one approach used at the 64K density level. The following describes the operation of this approach.

This sensing scheme equilibrates to about half the supply voltage, $V_{CC}/2$, and uses active pull-ups in order to write a full power supply level back into the cell. This method eliminates the neccessity of a dummy cell; no longer is a dummy cell required in order to establish a reference voltage to guarantee a differential voltage across the sense amplifier during sensing.

At the start of the cycle, when RAS goes low, both sides of the sense amplifier have been equilibrated to about $V_{CC}/2$. A high is stored as a full V_{CC} level (not a $V_{CC}-V_T$) and a low is stored as ground or 0 V. After the row decoder has determined which row to select, the appropriate word line is enabled turning on the selected row of storage cells. If a zero had been stored in the storage cell, the bit line previously equilibrated to $V_{CC}/2$ will be pulled down several hundred millivolts. If a high had been stored, the bit line would be pulled up several hundred millivolts. (The exact change on the bit lines is dependent on the capacitance of the bit line and the capacitance of the storage cell. The higher the ratio of storage cell capacitance to the bit line capacitance the greater the change on the bit line due to a stored high or a stored low level.) Note that in either case a differential voltage is established, without using dummy cells, and this voltage is sufficient to be detected by the sense amplifier.

When the differential voltage on the bit lines has settled following the selection of the storage cells, the sense amplifier is activated via the latch signal. The regenerative action of the sense amplifier pulls the lower side down to 0 V. The higher side remains unchanged.

At the end of the cycle, when RAS goes inactive, three events occur. First, the word line is bootstrapped up to a level greater than V_{CC} plus a V_T (threshold voltage). This assures that a full V_{CC} is stored in the storage cell for a high level. Second, the active pull-ups are enabled resulting in the higher side of the sense amplifier being reinforced to a full V_{CC} level while leaving the lower side at ground. Third, the word lines are deselected, after the storage cell has been written into, and the equilibration transistors are turned on. This allows charge sharing to occur between the bit lines on both sides of the sense amplifier to about half the supply voltage. This equilibrated level remains until the next RAS active cycle and the above process is repeated. The sequence of events described above is summarised in the timing diagram shown in *Figure 29.6*.

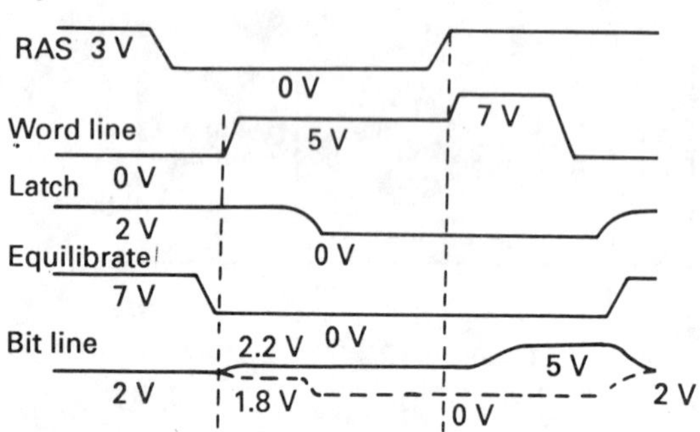

Figure 29.6 Timing diagram for the system of *Figure 29.5*

29.1.3 System design considerations using dynamic RAMs

Dynamic RAMs provide the advantages of high density, low power, and low cost. These advantages do not come for free; dynamic RAMs are considered to be more difficult to use than static RAMs. This is because dynamic RAMs require periodic refreshing in order to retain the stored data. Furthermore, although not generic to dynamics RAMs, most dynamic RAMs multiplex the address bits which requires more complex edge-activated multi-clock timing relationships. However, once the system techniques to handle refreshing, address multiplexing, and clock timing sequences have been mastered, it becomes obvious that the special care required for dynamic RAMs is only a minor inconvenience.

A synopsis of the primary topics of concern when using dynamic RAMs follows.

29.1.4 Refreshing

Dynamic RAMs use a 'sample-and-hold' principle to store binary data; the presence of a charge in the storage capacitor can represent a stored logical '1' and the absence of charge can represent a stored logical '0'. A characteristic of a stored level in a 'sample-and-hold' type circuit is that it can retain data for only a short period of time and that the retention time or storage time interval is limited by the size of the storage capacitor and leakage paths present. A further limitation on data retention time is the junction temperature. The storage time of any dynamic RAM may be expressed by the empirical equation:

$$t_s = A\exp(-BT) \qquad (29.1)$$

where T is the junction temperature, B is a variable relating the magnitude of the generation–recombination current to the junction temperature, and A is a scaling constant reflecting such variables as junction area, bulk defect density, and sense amplifier design.

Storage time typically doubles each time the junction decreases by about 10°C. This is illustrated in *Figure 29.7*. Typically, dynamic RAM vendors guarantee the retention of stored data for about 2 ms at a maximum ambient temperature of 70°C.

The architecture of the RAM determines the number of refresh cycles required to refresh all storage cells. The most common refresh requirement is 128 refresh cycles each 2 ms interval. This is accomplished by cycling through 128 row locations (addresses clocked into the RAM during RAS assertion) each and every 2 ms interval. A less popular refresh scheme which appeared at the 64K density level is 256 refresh cycles each 4 ms. This scheme requires that each cell retain the stored charge twice as long as with the previous method.

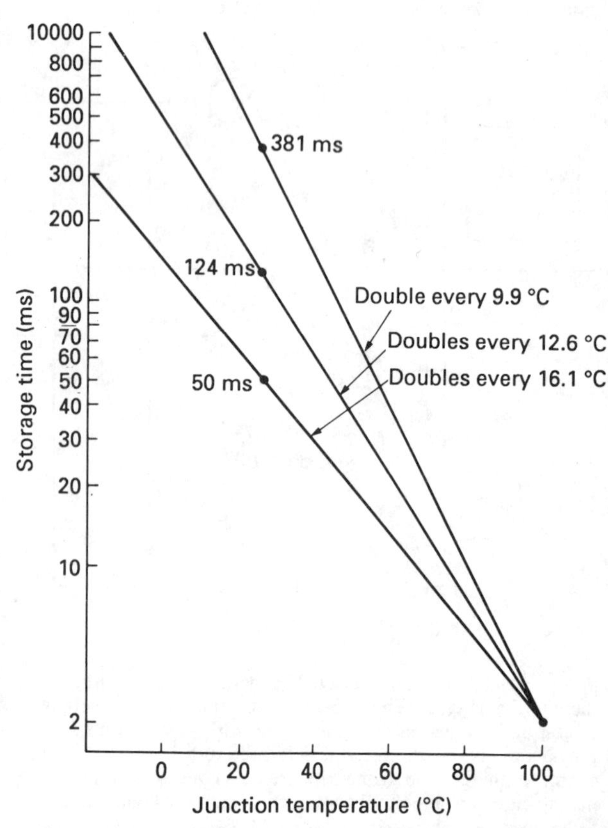

Figure 29.7 Effect of temperature on cell storage time

As indicated, dynamic RAMs require periodic refresh cycles. The refresh cycles may be performed either in a distributed fashion or in a burst mode. With distributed refreshing, single refresh cycles are scheduled evenly throughout the refresh interval. With burst refreshing, all the required refresh cycles are performed in a back-to-back fashion. Since dynamic RAMs use dynamic circuitry not only in the storage array circuitry but also in the peripheral clock circuits, these clocks must be cycled periodically or the device will not operate properly; a long interval between active RAM cycles results in degraded levels within the clock circuitry. For this reason, distributed refreshing is the preferred refreshing mode.

There are a number of methods that can be used to schedule the required periodic refresh cycles. Some microprocessors provide a signal or status bits that can easily be decoded to indicated when the refresh cycle can be safely performed.

Many systems require arbitration circuitry to select between normal memory cycles and pending refresh cycles. The arbitration task is not trivial. This circuitry must decide between either a normal access or a refresh and start a memory cycle.

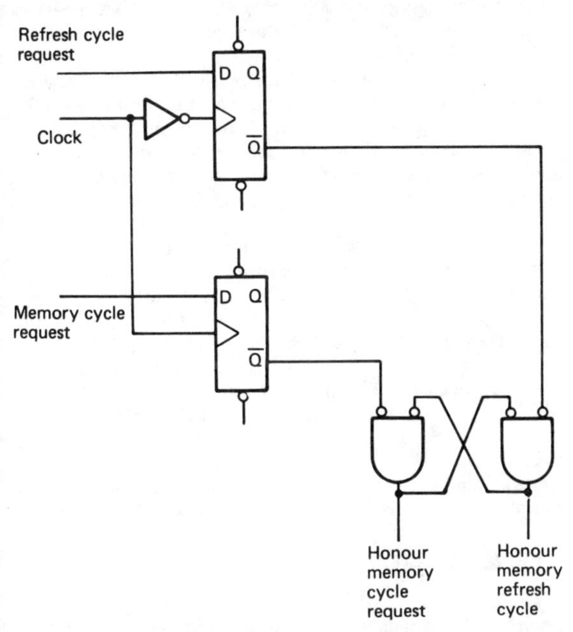

Figure 29.8 Cross-coupled flip–flop arbitration between a refresh request and a memory access request

Furthermore, the arbitration must be done such that glitching on the RAM clocks (RAS, CAS, or W) does not occur and that a memory cycle is not prematurely aborted. *Figures 29.8* and *29.9* show two different schemes that have been used successfully to arbitrate between memory access activity and refresh cycles. *Figure 29.8* indicates a technique when a memory 'start' timing signal is available with specific timing relative to a system clock which is also available to the arbitrator. *Figure 29.9* illustrates a technique which has been successful when the memory request signal is asynchronous and a synchronising clock is not available.

The purpose of the arbitration circuit is to schedule processor (or user cycle) requests and required refresh requests. Once a refresh has been given priority by the arbitrator the refresh circuitry must provide the refresh address (during row address time) to the RAM and generally a 'RAS-only' type cycle is executed. A read cycle (RAS and CAS active, W inactive) will also accomplish the refresh operation. However, when two or more devices have their data outputs connected together, 'RAS-only' type cycles should be executed in order to avoid data output contention. Note that if normal memory access cycles can guarantee that all combinations of the row addresses within the refresh field are cycled within the specified refresh interval, additional refreshing is not required.

29.1.5 Address multiplexing

The use of address multiplexing reduces the number of address lines and associated address drivers required for memory interfacing by a factor of 2. This scheme, however, requires that the RAM address space be partitioned into an X–Y matrix with half the addresses selecting the X or row address field and the other half of the addresses selecting the Y or column address field. The row addresses must be valid during the RAS clock and must remain in order to satisfy the hold requirements of the on-chip address latches. After the row address requirements have been met, the addresses are allowed to switch to the second address field, column addresses, a similar process is repeated using CAS (column address strobe) clock to latch the second (column) address field. Note that for RAS-only refreshing the column address field phase is not required. (See *Figure 29.10* for address multiplexing timing relationships for read cycles and *Figure 29.11* for write cycles.)

29.1.6 Dynamic RAM system timing generation

When using dynamic RAMs, the system performance is very much dependent on the timing relationships of the interface signals. RAM vendors have attempted to simplify the system timing problem. This is done by first reducing the time that the

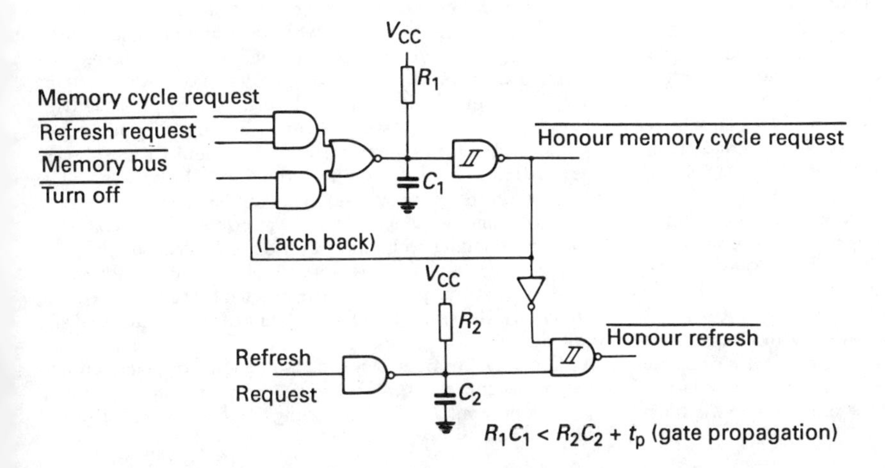

Figure 29.9 Arbitration between a synchronous memory cycle request and refresh request

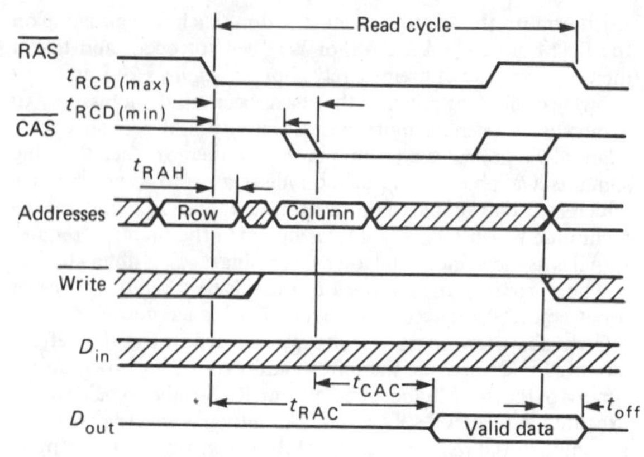

Figure 29.10 Read cycle for 18-pin multiplexed dynamic RAM

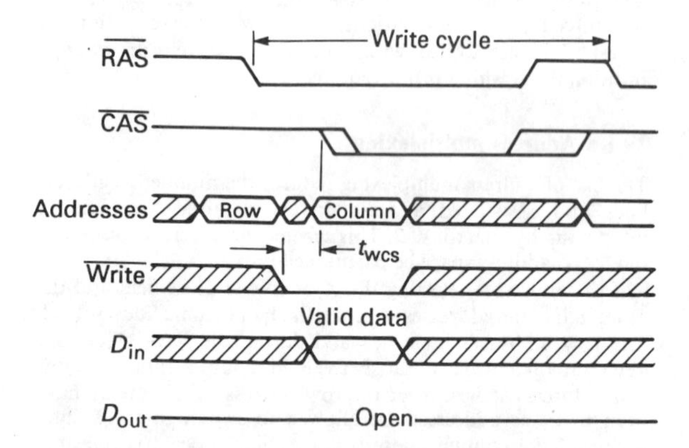

Figure 29.11 Write cycle timing for 18-pin multiplexed dynamic RAM. Note: if write command occurs before CAS, then D_{out} remains high impedance

row address must be held valid following the RAS clock going active. This allows the column addresses to be presented to the RAM as soon as practical. After the column addresses are valid, the CAS clock can be asserted. CAS can be activated at any time during the t_{RCD} minimum-to-maximum interval without affecting the worst case data access time relative to the RAS active edge. If, however, the CAS signal cannot be asserted prior to the t_{RCD} maximum point, storage errors or reading errors will not result. The only penalty that results from pulling CAS active beyond the t_{RCD} maximum limit is that the access time is pushed out such that the access time is now entirely dependent on the access time from CAS or t_{CAC} (see *Figure 29.12*).

Figures 29.13, *29.14* and *29.15* illustrate three popular schemes that are commonly used to generate the required interface timing signals. A combination of these techniques can be used. The method shown in *Figure 29.13* is generally used where low cost is the primary consideration. This method uses the inherent propagation delays of the devices themselves to generate the required timing relationships. *Figure 29.14* shows how delay lines can be used to generate the required relationships. 'Digital' delay modules have recently found widespread acceptance in this application. *Figure 29.15* shows a synchronous technique which can be used when high frequency timing edges are available. Regardless of the method used, worst case timing calculations should be made in order to ensure operation within the limits specified by the RAM vendor.

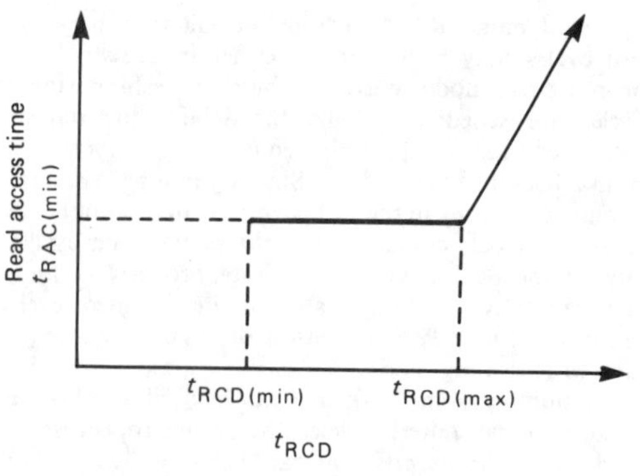

Figure 29.12 Gated CAS timing relationship

29.1.7 Dynamic RAM operational modes

At the 256K level and above, in some applications the normal read/write process is not always ideal. From 256K upwards several different operating modes are common, as follows.

Page mode This can be applied to any DRAM. The row address and first column address are applied in the normal way. To address further locations with the same row address, only the column address need be applied and strobed with CAS. In this mode the minimum time between successive accesses is the page mode cycle time, i.e. 100 ns for a 100 ns DRAM.

Nibble mode This applies to nibble mode drams only. The row address and column address are strobed, with further strobing of CAS cycles around four neighbouring address locations. In this mode the time between successive accesses is determined by the nibble mode cycle time, i.e. 45 ns for a 100 ns DRAM.

Static column mode This applies to static column mode parts only. The row address is strobed with RAS. The column address is not strobed, but applied directly as in a static RAM. A change in column address will result in a new location being accessed. The minimum time between successive accesses is determined by the static column access time, i.e. 45 ns for a 100 ns DRAM.

29.1.8 Board layout considerations

Several guidelines should be followed when laying out printed circuit boards for memory arrays using dynamic RAMs. The four areas of primary concern are power distribution, power decoupling, placement of array driver chips, and treatment of signal and clocks through the memory array.

Careful attention to power distribution and power decoupling is important since dynamic RAM current requirements are transient in nature. Dynamic RAMs require very little quiescent current during periods of inactivity. However, large current spikes occur during any active memory or refresh cycle. Due to the transient current requirements of these devices, low inductance power distribution techniques should be used within the memory array. This can be done by dedicating 'voltage' planes when using multilayer printed circuit boards or by using gridding techniques with double-sided boards. High frequency decoupling capacitors should be judiciously placed close to the memory devices and connected to the low impedance power bus. Generally, 0.1 μF decoupling capacitors are used and will adequately suppress power bus transients in the array. Additionally, low frequency capacitors should be placed around the perimeter of the memory array to provide further smoothing of the supply voltages by minimizing the effects of inductive and resistive voltage drops.

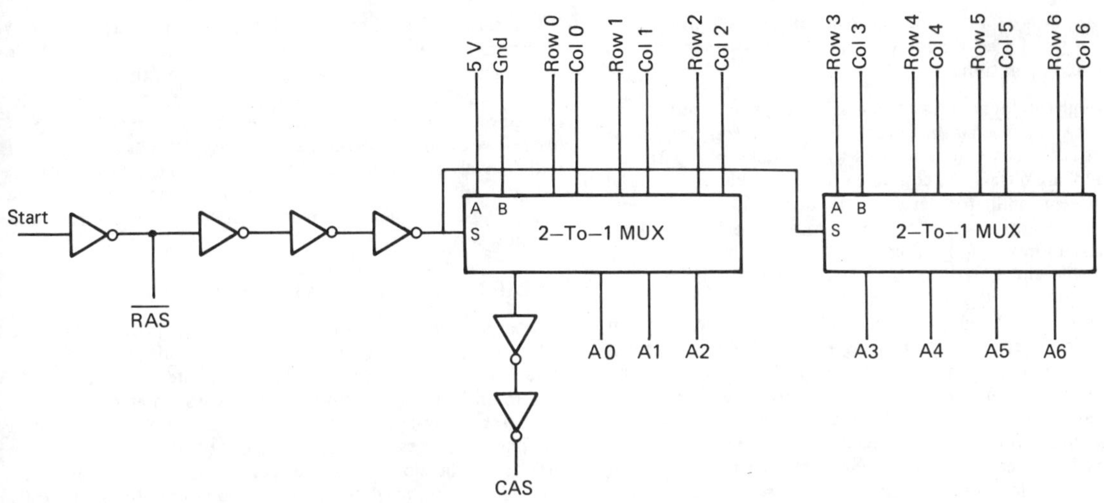

Figure 29.13 Device delays are used to generate timing

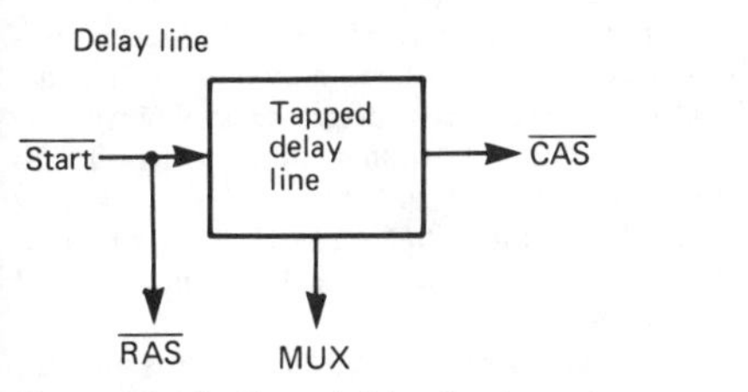

Figure 29.14 Tapped delay line is used to generate timing signals

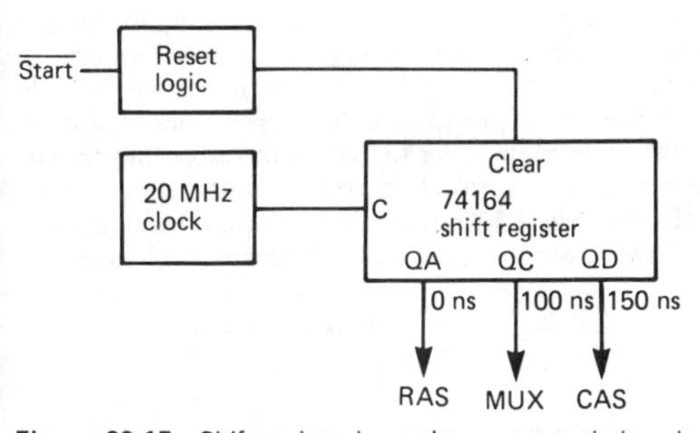

Figure 29.15 Shift register is used to generate timing signals

29.1.9 Soft errors

With dynamic RAMs, particular care must be taken to provide a friendly operating environment in order to avoid the occurrence of intermittent errors. When dynamic RAMs are operated in an 'unfriendly' environment, the occurrence of intermittent errors may exist. These errors are referred to as 'soft' errors. Board related soft errors can be prevented by making certain that all operating conditions are within the RAM manufacturer's specification limits. Some general design guidelines that should be observed when using MOS dynamic RAMs include:

(1) Always design well within the vendor's specification limits. Adequate voltage (supply and signal), timing, and temperature margins are essential for trouble free operation.
(2) Lay out arrays as densely as reasonably possible while maintaining adequate power busses and array power decoupling.
(3) Use low impedance power distribution through the array. Gridding is essential with double-sided PC boards.
(4) Place high frequency decoupling capacitors within the array matrix. At least one capacitor per critical supply for every two devices (preferably each device) is suggested. A 0.1 μF capacitor has been found to be suitable for this purpose.
(5) Where possible, place bulk decoupling capacitors around the perimeter of the memory matrix.
(6) Use transmission line terminations on all signals to the RAMS. Conventional termination techniques are effective when used with constant impedance transmission lines that have predictable impedances. Many of the single supply (+5 V) RAMs will tolerate as much as −2.0 V negative excursions below ground on input signals. This does not indicate that line terminations are not required. When excessive negative overshoot occurs, even within the specification limit, the positive excursions resulting from the damped ringing will reduce noise margins and will generally result in a violation of the $V_{IL(max)}$ specification limit. The user should be cautioned that high speed dynamic RAMs can respond to nanosecond spikes and glitches which may result in unpredictable behaviour.
(7) Avoid long parallel runs with signals that cannot tolerate crosstalk. If long runs are necessary, provide an effective ground plane by routing power or ground runners on the adjacent layer. This reduces the crosstalk by *E*-field cancellation and results in a nearly constant impedance transmission line (inclusion of a power grid yields an effective microstrip characteristic) for signals transversing the array. Also, crosstalk can be reduced by placing decoupled power ground runners between the signal traces.
(8) When possible, avoid stubbing of array signal lines. This represents an impedance discontinuity with resulting energy reflections which can reduce noise margins or possibly violate the vendor's specification limits.
(9) Locate the driver circuits as close to the array as possible with short direct routing of interface signals between the drivers and the RAMs.
(10) Place both high frequency caps and bulk decoupling caps near the array driver circuits.
(11) Provide sufficient ground bussing for drive circuits and between the drivers and the array to minimise ground bounce and ground offset conditions.
(12) Avoid using driver gates in a common package when crosstalk due to ground bounce or other internal coupling mechanisms cannot be tolerated.

(13) If any device is being used in an unconventional way, verify with the device vendor that the device will support the desired operation.

Although careful board/system design can prevent soft errors, other soft error mechanisms exist. These soft errors are due to traces of radioactive contaminants which emit alpha particles. These alpha emitters are contained in the packaging material used for semiconductor devices. The most common elements with alpha emitting isotopes are uranium, thorium, radium, polonium, bismuth, and radon.

When the alpha particle passes into the RAM it creates a positive charge in the oxide. The alpha particle continues to travel into the silicon to a depth of approximately 25 μm resulting in electron–hole pairs being generated. These electrons are attracted to the positively charged oxide. Electrons generated in depletion regions drift toward the surface and electrons generated in the substrate diffuse toward the surface. This process can result in a modification of the charge on the storage cell capacitor or can affect levels during the sensing process. In either case, a soft error can occur if the energy levels associated with the alpha particle passing through the silicon are high enough. Seldom does a single alpha particle cause more than a single location to err and the mechanism is non-catastrophic.

(a)

D_2	D_1	D_0
0	0	0
0	0	1
0	1	0
0	1	1
1	0	0
1	0	1
1	1	0
1	1	1

(b)

D_2	D_1	D_0	$P_{(even)}$
0	0	0	0
0	0	1	1
0	1	0	1
0	1	1	0
1	0	0	1
1	0	1	0
1	1	0	0
1	1	1	1

(c)

D_2	D_1	D_0	$P_{(odd)}$
0	0	0	1
0	0	1	0
0	1	0	0
0	1	1	1
1	0	0	0
1	0	1	1
1	1	0	1
1	1	1	0

Figure 29.16 Error detection. (a) Complementing any single bit results in another valid data word (Hamming distance=1). (b,c) Complementing any single bit does not result in another valid data word (Hamming distance=2)

The location affected by the alpha 'hit' will properly store new (rewritten) data and this location is no more susceptible to another soft error than any other location on the device.

Fortunately, alpha-related soft errors occur very infrequently. Device manufacturers have attempted to make these devices as insensitive to alpha radiation as possible. Methods such as using radioactively 'clean' packaging materials, protective die coatings, design improvements (for more signal margins), and new processing techniques have reduced alpha-type soft errors to a very low level.

29.1.10 Error detection and error correction

In a large memory array there is a statistical probability that a soft error(s) or device failure(s) will occur resulting in erroneous data being accessed from the memory system. The larger the system, the greater the probability of a system error.

In memory systems using dynamic RAMs organised in N by 1 configuration, the most common type of error/failure is single bit oriented. Therefore, error detection and error correction schemes with limited detection/correction capabilities are ideally suited for these applications.

Single bit detection does little more than give some level of confidence that an error has not occurred. Single bit detection is accomplished by increasing the word width by one bit. In order to understand how this provides error detection capability refer to *Figure 29.16*. In the example shown a small word length of 3 bits has been used. Note that 3 bits provides eight binary combinations as shown in *Figure 29.16(a)*. With the 3-bit word, complementing any single bit in the word results in another valid binary word combination.

By adding a parity bit and selecting the value of this bit such that the 4-bit word has an even number of 'ones' (*Figure 29.16(b)*) or an odd number of 'ones' (*Figure 29.16(c)*), complementing any single bit in the word no longer results in a valid binary combination. Note that it is necessary to complement at least two bit locations across the word in order to achieve a new valid data word with the proper number of 'ones' or 'zeros' in the word, in order to extract the error information from the stored data word. This concept was introduced by Richard Hamming in 1950. The scheme of defining the number of positions between any two valid word combinations is called the

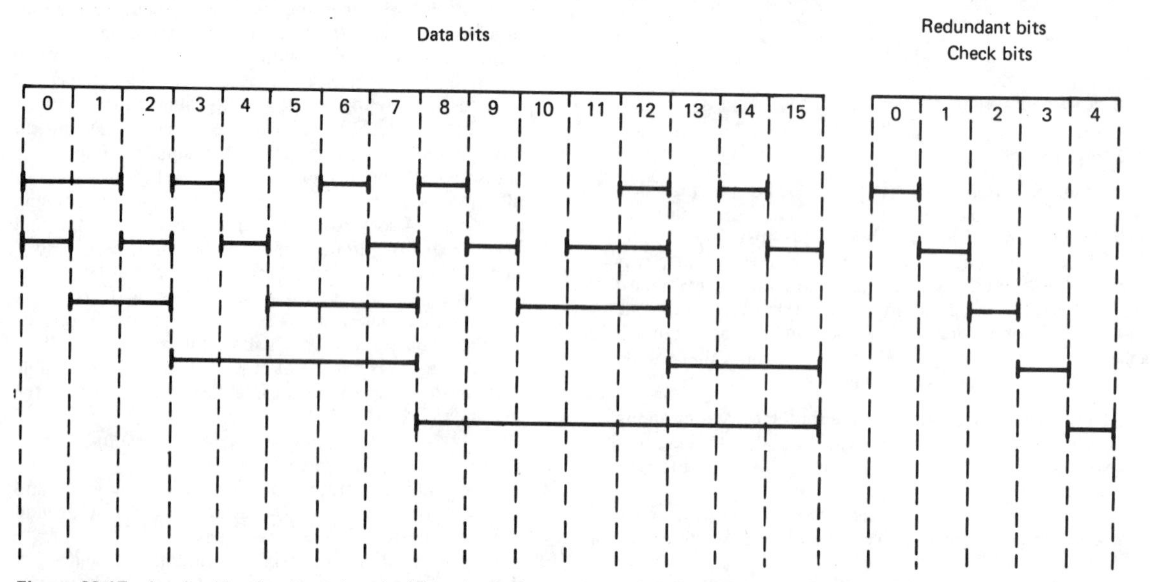

Figure 29.17 An algorithm for determing check bits (parity from subsets of data word) for providing single error bit detection and single bit correction

'Hamming distance'. The Hamming distance determines the detection/correction capability.

Correction/detection capability can be extended by additional parity bits in such a way that the Hemming distance is increased. *Figure 29.17* shows how a 16-bit word with five parity bits (commonly referred to as 'check bits') can provide single bit detection and single bit correction. In this case, a single bit error in any location across the word, including the check bits, will result in a unique combination of the parity errors when parity is checked during an access.

As indicated above, the ability to tolerate system errors is bought at the expense of using additional memory devices (redundant bits). However, the inclusion of error correction at the system level has been greatly simplified by the availability of error correction chips. These chips provide the logic functions necessary to generate the check bits during a memory write cycle and perform the error detection and correction process during a memory read cycle.

29.2 Static RAM

Static random access memory is a memory device that utilises a storage vehicle that requires only that power remain on to retain data. Typically it requires no synchronous timing edges and is easy to use. Due to its characteristics it can be an extremely fast memory device. Speed and/or ease of use are therefore the primary motivations for designing with this class of device. This contrasts with density, low cost, and lower power which are the key motives for using a dynamic RAM.

29.2.1 Organisation

Static RAMs in the 1K (K=1024) and 4K generations are organised as a square array (number of rows of cells equals the number of columns of cells). For the 1K device there are 32 rows and 32 columns of memory cells. For a 4096 bit RAM there are 64 rows and 64 columns. The RAMs of the seventies and beyond include on-chip decoders to minimise external timing requirements and number of input/output (I/O) pins.

To select a location uniquely in a 4096 bit RAM, 12 address inputs are required. The lower-order six addresses decode one of 64 to select the row while the upper-order 64 addresses decode to select one column. The intersection of the selected row and the selected column locates the desired memory cell.

Static RAMs of the late seventies and early eighties departed from the square array organisation in favour of a rectangle. This occurred for two primary reasons: performance and packaging considerations.

Static RAM has developed several application niches. These applications often require different organisations and/or performance ranges. Static RAM is currently available in 1-bit I/O, 4-bit I/O, and 8-bit I/O configurations. The 1- and 4-bit configurations are available in a 0.3 inch wide package with densities up to 64 kbits. The 8-bit configuration is available in a 0.6 inch wide package again with densities up to 64 kbits. The 8-bit I/O device is pin compatible with ROM (read only memory), and EPROM (electrically programmable ROM) devices. All of the static RAM configurations are offered in two speed ranges. Typically, 100 ns is used as the dividing line for part numbering.

29.2.2 Construction

The static random access memory (RAM) uses a six device storage cell. The cell consists of two cross-coupled transistors, two I/O transistors and two load devices. There are three techniques that have been successfully employed for implementing the load devices. These are: enhancement transistors, depletion transistors, and high impedance polysilicon load resistors.

Enhancement transistors are normally off and require a positive gate voltage (relative to the source) to turn the device on. There is a voltage drop across the device equal to the transistors' threshold. Depletion transistors are normally on and require a negative gate voltage (relative to the source) to turn them off. There is no threshold drop across this device. Polysilicon load resistors are passive loads which are always on. Formed in the poly level of the circuit they are normally undoped and have very high impedances (5000 MΩ typically). They have the advantage of very low current (nanoamperes) and occupy a relatively small area. All present large density, state of the art, static RAMs now use this technique for the cell loads. Further enhancements have been made by manufacturing a two-level poly-process thereby further decreasing cell area.

The operation of the three-cell approaches is shown in *Figure 29.18*.

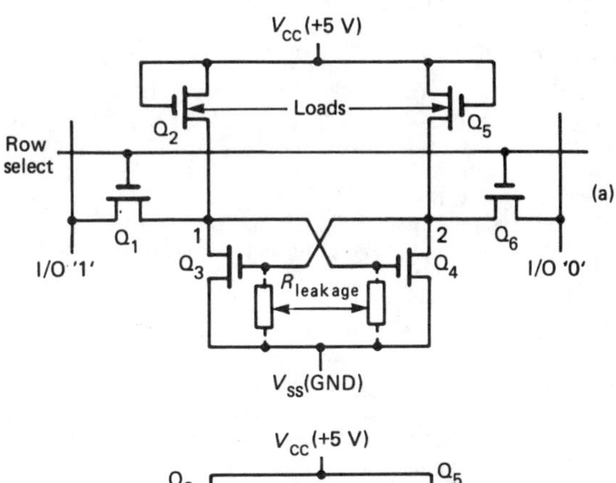

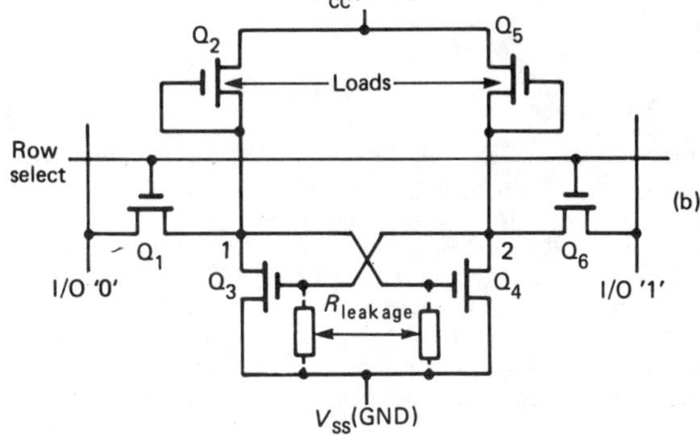

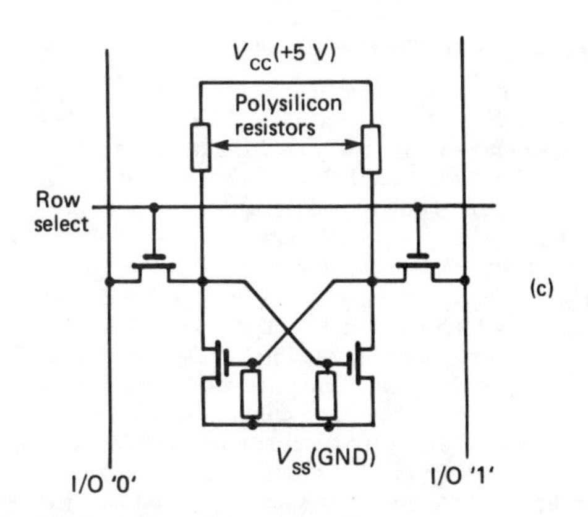

Figure 29.18 Static RAM storage cells

29.2.2.1 Enhancement cell

Consider the storage cell shown in *Figure 29.18(a)*. Data are stored as a charge on the gate of either Q_3 or Q_4 (which determines the logic state of the cell). The voltage on the charged node is approximately $V_{CC}-V_{TH}$ (where V_{TH} is the effective threshold of the load devices) and turns Q_3 or Q_4 on. By definition a logic '0' is stored in the cell if Q_3 is on and a logic '1' is stored if Q_4 is on.

If it is assumed that Q_3 is on (logic '0' stored) then current will flow from the load on Q_3 (device Q_2) through Q_3 to ground (V_{SS}). This current will cause the voltage at node 1 to assume a value near V_{SS} (the voltage is proportional to the effective on resistance of Q_2 and Q_3). The resultant low voltage on node 1 turns device Q_4 off. Device 5 maintains the charge on the gate of Q_3 by replacing charge leaked off through the high impedance parasitic leakage resistor $R_{leakage}$ (this leakage is typically in the picoampere range). The storage cell will remain in this logic state until an external forcing function is applied (write cycle).

29.2.2.2 Depletion cell

Operation of the storage cell is as follows. Assume the gate of Q_3 is high turning Q_3 on causing current to flow in Q_3 and Q_2. Since devices Q_2 and Q_3 are ratioed (that is, Q_2 has a higher impedance than Q_3) the voltage at node 1 will drop close to V_{SS}. Note that the gate of Q_2 is tied to node 1; therefore, as node 1 decreases in voltage, the voltage drive on Q_2 is reduced, making the effective impedance of Q_2 higher. This allows the voltage at node 1 to move even closer to V_{SS}.

Since node 1 is low and is tied to the gate of Q_4, device Q_4 is off. The charge on Q_3 is maintained by the load device Q_5. Note that only leakage currents flow through device Q_5 which has a minimal effect on the voltage at node 2. Since increased positive voltage at node 2 increases the voltage drive on Q_5, device Q_5 turns on hard. The voltage at node 2 is therefore equal to V_{CC} (note that there is no threshold drop across device Q_5 since it is a depletion mode device).

29.2.2.3 Resistor cell

Operation of the storage cell is as follows. Assume the gate of Q_3 is high turning Q_3 on causing current to flow in Q_3 and Q_2. Since device Q_2 has a much higher impedance than Q_3 (orders of magnitude higher) the voltage at node 1 will drop close to V_{SS}. Since node 1 is low and is tied to the gate of Q_4, device Q_4 is off. The charge on Q_3 is maintained by the load resistor Q_5. Note that only leakage currents flow through resistor Q_5 which has a minimal effect on the voltage at node 2. Since the current through Q_5 is very low the voltage at node 2 asymptotically approaches V_{CC}.

29.2.3 Accessing the storage cell

The storage cell is interrogated for a read or write operation by activating the proper row select line which turns devices Q_1 and Q_6 on (*Figure 29.18*). For a read operation, a sense amplifier connected to both the I/O '0' and I/O '1' outputs of each column detects the state of the selected storage cell in that column. If Q_3 is on (logic '0') then current will flow in the I/O '0' line. If Q_4 is on (logic '1'), current will flow in the I/O '1' line. A write buffer places a high level (approximately V_{CC}) on the I/O '0' line to write a logic '0', and a high level on the I/O '1' to write a logic '1'. For both write conditions, the opposite line is held low (V_{SS}).

Figure 29.19 is typical of circuitry used in static RAM. Data is gated to/from the appropriate columns by column select. Note that chip enable(s) gate the output data to a three state buffer and then to the output pin. Therefore, if a chip is not selected, the output pin goes to a high impedance state (allowing the output pins to be OR tied).

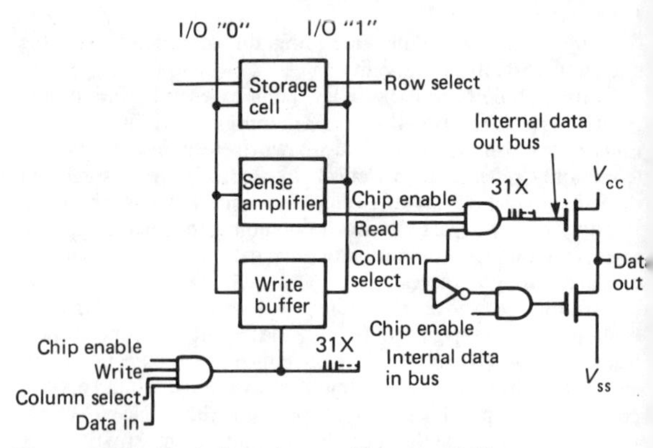

Figure 29.19 Typical circuitry used in static RAM

29.2.4 Characteristics

Static RAM is typified by ease of use. This comes at the expense of power dissipation, density, and cost. To service a broad spectrum of applications many variations of the peripheral circuitry surrounding the static cell have been implemented. This section will analyse the key device types available.

The fundamental device is often referred to as a ripple through static RAM. Its design objective is ease of use. This device offers one power dissipation mode 'on'. Therefore its selected power (active power) is equal to its deselected power (standby power). One need simply present an address to the device and select it (via chip select) to access data. Note that the write signal must be inactive. This sequence is shown in *Figure 29.20*. In this RAM type the chip select access is typically 50% of the address access.

The write cycle can be executed in a simple manner as shown in *Figure 29.21*.

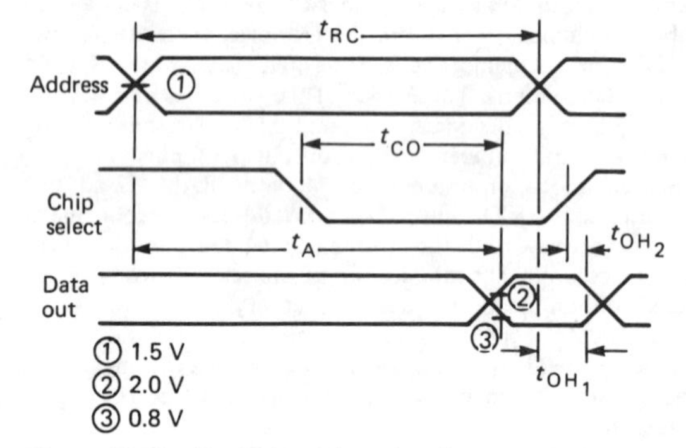

Figure 29.20 Ripple through read cycle

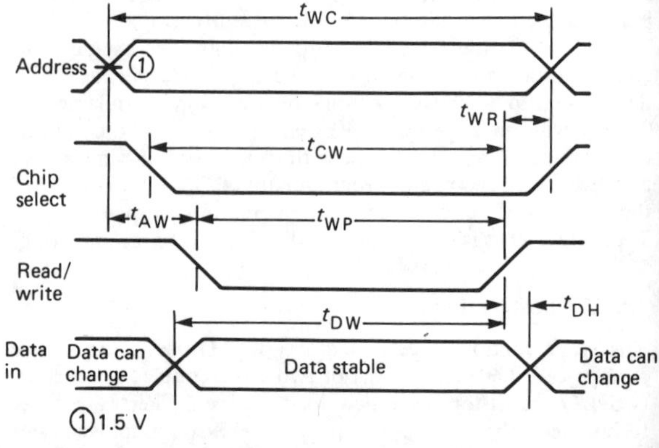

Figure 29.21 Write cycle ripple through interface

The address must be valid before and after read/write active to prevent writing into the wrong cell. Chip select gates the write pulse into the chip and must meet a minimum pulse width which is the same as the read/write minimum pulse width.

Read–modify–write cycles can be performed by this as well as other static RAMs. Read–modify–write merely means combining a read and write cycle into one operation. This is accomplished by first performing a read then activating the read/write line. The status of the data out pin during the write operation varies among vendors. The product's data sheet should therefore be read carefully.

The device achieves its ease of use by using very simplistic internal circuitry. To permit a non-synchronous interface (no timing sequence required) all circuitry within the part is active at all times. The static 'NOR' row decoder shown in *Figure 29.22* is

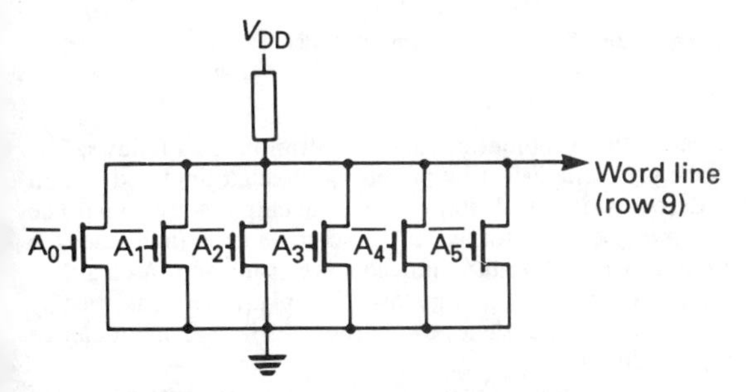

Figure 29.22 Static NOR row decoder

typical of the circuitry used. For a 4K RAM with 64 rows, this circuit is repeated 64 times with the address lines varing to identify each row uniquely. The decoder operates as follows.

In the static row decoder (*Figure 29.22*) a row is selected when all address inputs to the decoder are low making the output high. The need to keep the power consumption of this decoder at a minimum is in direct conflict with the desire to make it as fast as possible. The only way to make the decoder fast is to make the pull-up resistance small so that the capacitance of the word line can be charged quickly. However, only one row decoder's output is high, while the output of each of the other 63 decoders is low, causing the resistor current to be shunted to ground. This means that the pull-up resistance must be large in order to reduce power consumption.

Figure 29.23 is a typical input buffer. In this circuit device A and C act as loads and are always on. If the address is a logic '0'

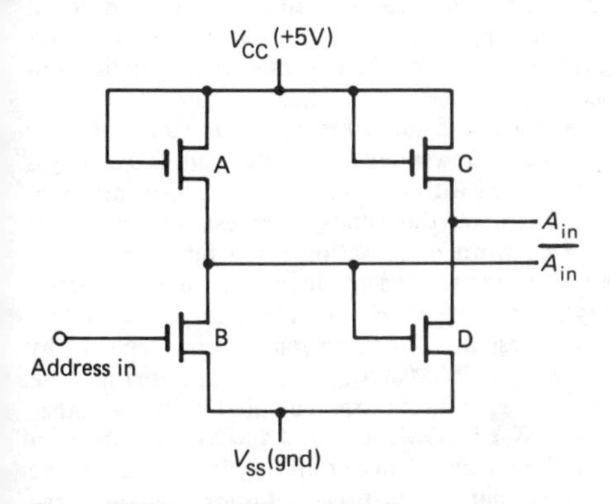

Figure 29.23 Address input buffer

then B is off and D is on; conversely, if the address is a logic '1' then B is on and D is off. There is always a current path to ground. This circuit is repeated for each address (12 times for 4K).

29.2.5 Power gated static RAM

On the previous static circuit the chip select function was used to select the device for write cycles and enable the output for read cycles. A type of static RAM, introduced at the 4K density level, uses the chip select to reduce standby power. A ratio of active to standby of 5 to 1 is achieved. The timing for this device differs from the previous RAM in that the chip select access time is now equal to address access. An anomaly in timing occurs when the chip select is inactive for a short interval of time (approximately 30 ns or less). For this situation the access time is increased by about 20 ns. This anomaly was caused by a design problem and should be eliminated as a specification requirement in future product generations.

The effect of the power down feature is clearly shown in *Figure 29.24*.

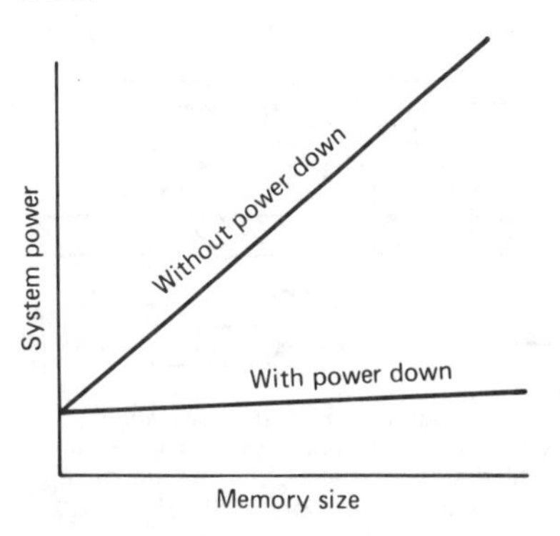

Figure 29.24 Effect of power down at the system level

This technique is also a compromise between ease of use and power. If we build a 64K × 9 system the average power is:

without power down

P_{npd} = 144 devices × 5 V × 180 mA/device = 129.6 W

where P_{npd} is power with no power down;

with power down

P_{pd} = 9 devices (active) × 5 V × 180 mA/device
+ 135 devices (standby) × 5 V × 30 mA/device
= 8.100 + 20.250 W
= 28.350 W

where P_{pd} is power with power down.

The power reduction is achieved with the minor complication of requiring chip select to be valid at the start of a cycle. If only 4K of memory is used CS can be grounded, raising power levels but simplifing the timing requirements. The internal circuitry is slightly more complicated in order to implement the power gate feature.

A radically different static RAM was designed to maintain the static cell (no periodic activation needed) and dramatically reduce the devices power consumption. The power dissipation of the static RAM can be reduced by using a 'clocked' interface. The clocked interface required synchronisation of timing and trades ease of use for power dissipation. Clocked static rams are commonly known as 'edge-activated' RAMs. The device is designed to work easily with microprocessors thereby minimising the difficulty of use. The need to synchronise timing edges to the RAM will cause a speed penalty in some applications. Therefore this RAM was designed for the microprocesser market and is not applicable to the sub-100 ns application. In microprocessor systems the system timing of the micro usually masks the speed penalty due to synchronisation.

The edge-activated device is configured as a 4K × 1. It is best used in applications where the system requires a memory more than 4K deep. Since it is a by 1 organisation, parity can be implemented easily. Parity is the addition of one extra bit per word to verify the validity of the word.

The edge-activated concept can best be explained by reference to *Figure 29.25*. The edge-activated RAM requires all address inputs to be valid prior to initiation of a negative going edge-on the chip enable input. The chip enable signal must remain valid for a specified duration, equivalent to the minimum access time of the component. A recovery time between cycles, 50% of specified access, is required for proper operation. Data out becomes available, t_{acc} after CE and remains valid until chip enable is deactivated.

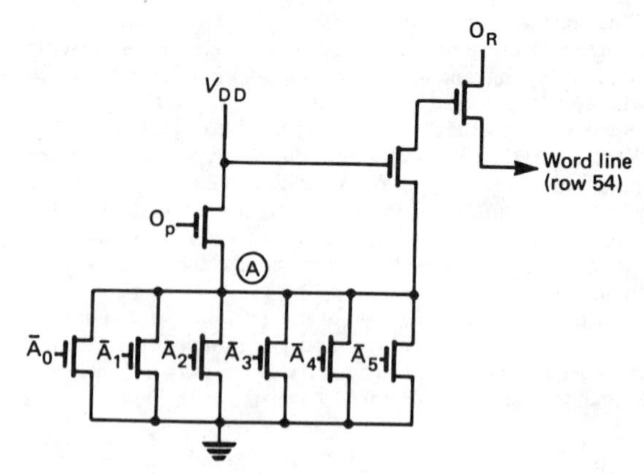

Figure 29.26 Dynamic NOR row decoder

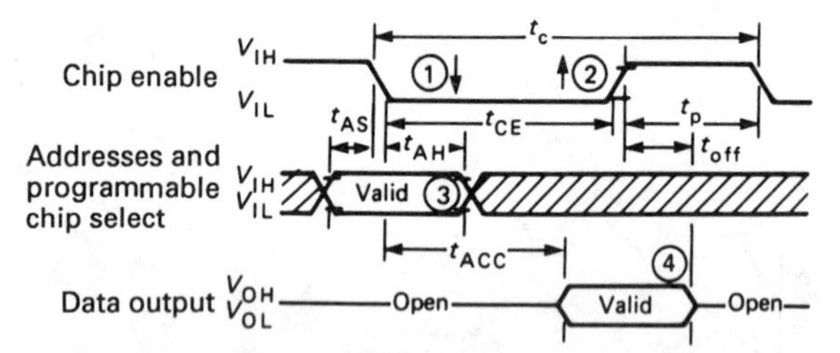

Figure 29.25 Low-voltage, edge-activated RAM/ROM family. 1, a simple high to low transition at the chip enable (CE) input activates this entire family of memory devices; 2, returning CE input to a high level is all that is required to achieve a 75% reduction in device operating power; 3, address information is strobed and latched into a set of on-chip registers; 4, you have full control of the data output, determined by the CE pulse width

The requirements of the edge-activated interface is one which is readily available from most microprocessor chips or can be readily obtained from processor/memory controller timing.

The benefit of this RAM is its very low power dissipation. Active power is typically 20% of a ripple through static's. Standby power is 1/5 to 1/30 of that achieved with other static techniques. For example, a 64K × 9 (144 devices) static RAM memory implemented with edged-activated circuits dissipates 4.7 W while a 'ripple through' static dissipates 129 W. In a similar application, the power P_{ea} required for an edge-activated static RAM is given by:

$$
\begin{aligned}
P_{ea} &= \text{devices (active)} \times \text{power (active)} \\
&\quad + \text{devices (standby)} \times \text{power (standby)} \\
&= 8 \text{ devices} \times 5 \text{ V} \times 30 \text{ mA} \\
&\quad + 135 \text{ devices} \times 5 \text{ V} \times 5 \text{ mA} \\
&= 4.7 \text{ W}
\end{aligned}
$$

$$
\begin{aligned}
P_{ripple} &= \text{devices} \times \text{power} \\
&= 144 \text{ devices} \times 5 \text{ V} \times 180 \text{ mA} \\
&= 129 \text{ W}
\end{aligned}
$$

where P_{ripple} is power ripple through static.

The low power dissipation of the edge-activated RAM is achieved by using many circuit techniques commonly associated with dynamic RAM. It is designed as is a dynamic RAM to eliminated d.c. current paths. The circuit of *Figure 29.26* is typical of the techniques used. Its operation is as follows. The decode is accomplished by presetting (precharging) node A, on the decoders, to a high state during the chip inactive time. The precharge clock is turned off at chip enable time and the addresses are strobed into the decoders causing node A on all but the one selected decoder to go low. The only power consumed by this circuit is the transient power drawn during capacitor charge time at node A.

The last technique to be described combines the best features of the static RAMs previously discussed. Called address activated, it uses many of the clocked circuitry techniques of edge activated but obtains the 'edge' from any one of the addresses or clocks (CS, WE). Since all inputs must be active between cycles, looking for a transition, the device has a power specification that is less than ripple through static but greater than edge activated. Power gate techniques can be applied to further reduce power if the application the part is designed for warrants it. This device uses clocked circuitry primarily for speed enhancement, not power.

If we analyse the static NOR row decode its performance is a function of the load device's ability to charge the word line. Also performance is a function of the transistor's (between the word line and ground) ability to discharge the word line. As we discussed earlier, in a 4K RAM, 63 out of 64 decoders are shunting current to ground to prevent selection. This means that the pull-up (load) resistance must be large in order to reduce power consumption which inpacts speed. Since the dynamic decoder has no d.c. path the trade-off between power and speed is minimised. Other areas of the circuit also benefit in speed if a clock can be used. These all add up to make address activated a useful high performance lower power circuit technique.

The timing specifications for the address activated RAM resemble that of the ripple through static RAM. In by 1 devices power gating is used while with the by 8 devices the applications do not warrant it.

Static RAM applications are segmented into two major areas based on performance. Slower devices, greater than 100 ns access time, are generally used with microprocessors. These memories are usually small (several thousands of bytes) with a general preference for wide word organisation (by 8 bit). When by 1 organised memories are used, the applications tend to be 'deeper' (larger memory), and desire lower power or parity. The by 4 bit product is an old design which is currently being replaced by 1K × 8 and 2K × 8 static RAM devices. Statics are used in these applications due to their ease of use and compatibility with other memory types (RAM, EPROM). The by 4 and by 8 devices use a common pin for data in and data out. In this configuration, one must be sure that the data out is turned off prior to enabling the data in circuitry to avoid a bus conflict (two active devices fighting each other drawing unnecessary current). Bus contention can also occur if two devices connected to the same bus are simultaneously on, fighting each other for control. The by 8 devices include an additional function called output enable (OE) to avoid this problem. The output enable function controls the output buffer only and can therefore switch it on and off very quickly. The by 4 devices do not have a spare pin to incorporate this function. Therefore, the chip select (CS) function must be used to control the output buffer. Data out turns off a specified delay after CS turns off. One problem with this approach is that it is possible for bus contention to occur if a very fast device is accessed while a slow device is being turned off. During a read–modify–write cycle the leading edge of the read write line (WE) is normally used to turn off the data out buffer to free the bus for writing.

The second segment is high performance applications. Speeds required here are typically 55–70 ns. This market is best serviced by static RAM due to their simple timing and faster circuit speeds. The requirement of synchronising two edges as required in clocked part typically costs the system designer 10–20 ns. Therefore, fast devices are always designed with a ripple through interface. The by 1 devices have been available longer, and tend to be faster than available by 8 devices. As a result the by 1 device dominates the fast static RAM market today. Applications such as cache and writeable control store tend to prefer wide word memories. The current device mix should shift over time as more suppliers manufacture faster devices.

29.2.6 Process

The pinouts and functions tend to be the same for fast and slow static devices. The speed is achieved by added process complexity and the willingness to use more power for a given process technology.

The key to speed enhancement is increasing the circuit transistor's gain, minimising unwanted capacitance and minimising circuit interconnect impedance. Advanced process technologies have been developed to achieve these goals. In general, they all reduce geometries (scaling) while pushing the state of the art in manufacturing equipment and process complexity. These process technologies have been given identification names such as HMDS and scaled poly 5. *Figure 29.27* shows several of the key parameters involved.

The figure shows the cross-section of a scaled device and lists the parameters of scaling, one of which is device gain. The slew rate of an amplifier or device is proportional to the gain. Because faster switching speeds occur with high gain, the gain is maximised for high speed. Device gain is inversely proportional to the oxide thickness (T_{ox}) and device length (l), consequently, scaling these dimensions increases the gain.

Another factor which influences performance is unwanted capacitance which appears in two forms: diffusion and Miller capacitance. Diffusion capacitance is directly proportional to the overlap length of the gate and the source (l_D). Capacitance on the input shunts the high frequency portion of the input signal so that the device can only respond to low frequencies. Secondly, capacitance from the drain to the gate forms a feedback path creating an integrator or low pass filter which degrades the high frequency performance. This effect is minimised by reducing l_D.

One of the limits on scaling is punch through voltage, which occurs when the field strength is too high, causing current to flow when the device is 'turned off'. Punch through voltage is a function of channel length (l) and doping concentration (C_B) thus channel shortening can be compensated by increasing the doping concentration. This has the additional advantage of balancing the threshold voltage which was decreased by scaling the oxide thickness for gain.

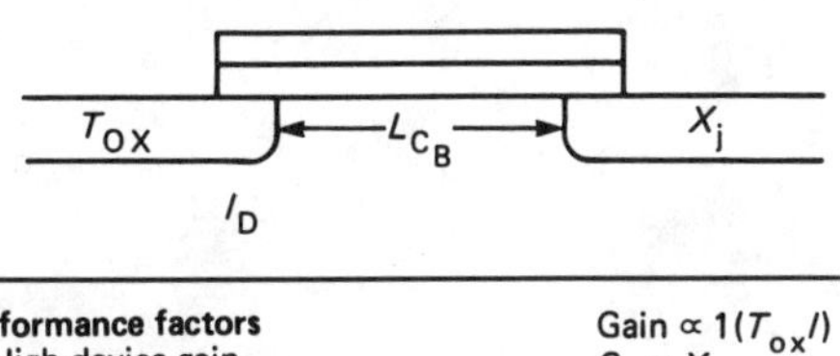

Performance factors
- High device gain
- Low diffusion capacitance
- Low Miller capacitance
- Low body effect

Gain $\propto 1(T_{ox}l)$
$C_p \propto X_j$
$C_m \propto l_D$
$\Delta V_r \propto \sqrt{C_B} T_{ox}$

Limits
- Punch through voltage
- Threshold voltage

$V_{PT} \propto C_B l^2$
$V_r \propto \sqrt{C_B} T_{ox}$

Result
- Decrease l, T_{ox}, X_j, l_D
- Increase C_B

l = Channel length
T_{ox} = Oxide thickness
X_j = Diffusion depth
l_D = Gate overlap
C_B = Concentration

Figure 29.27 HMOS scaling

29.2.7 Design techniques

29.2.7.1 By 1 statics

The by 1 static memories are offered in 18-pin or 20-pin packages for 4K and 16K densities respectively.

The 4K/16K by 1 static RAM can be easily integrated into large memory configurations in highly compact board layouts. The devices can use standard TTL logic to achieve the desired signal characteristics. A typical design for a 16K × 9 (using 4K devices) is shown in *Figure 29.28*. Chip select is decoded in order that it occurs on only one memory word simultaneously. Address lines and write enable (read/write line) go to all chips simultaneously.

Power supplies were placed in the corners of the devices to permit use of two-sided printed circuit boards. The RAM chips draw significant transient currents as well as d.c. current. To supply the transient current at the very high frequencies required it is necessary to place high frequency (ceramic) capacitors in close proximity to the memory chip. The rule of thumb is to use one 0.1 μF ceramic capacitor every other RAM.

29.2.7.2 By 8 static

The byte wide (8-bit) statics offer pin compatibility with RAM and EPROM circuits. This gives the designer the ability to utilise onboard design for RAM/ROM/EPROM needs. The pinout also permits flexibility in defining the type of memory to be implemented in a given application. The compatibility is most useful in the lower performance application geared at microprocessor memory. The pinouts chosen are expandable well beyond the 2K × 8 level. To permit this expansion a 28-pin package pinout was defined to be compatible with the 24-pin used by current generation statics, EPROMs, and ROMs.

29.3 Non-volatile memories

With today's faster, more powerful microcomputer chips emerging in abundance, and larger, more memory-intensive programs being written, semiconductor memory requirements for larger storage capacities, faster access times, and lower subsequent costs have become dominant system design factors. Basic semiconductor memory-chip technology involves variations of random access memory (RAM) and read only memory (ROM). RAM allows binary data to be written in, and to be read out. New and different programs and data can be loaded

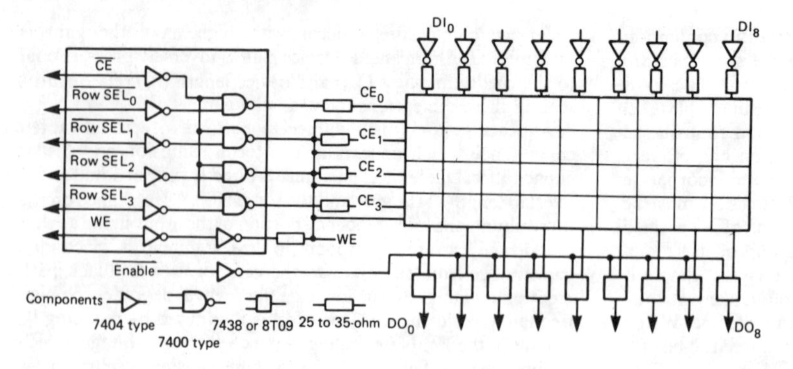

Figure 29.28 Design of 16K × 9 memory

and stored in RAM as needed by the processor. Because information is stored electrically in RAM its contents are lost whenever power goes down or off. When fixed or unchanging programs and data are needed by the processor, they are loaded into some form of ROM. In ROM, information is physically (permanently) embedded; therefore, its contents are preserved whenever power is off or interrupted momentarily.

Depending on the type and quantity of microprocessor systems to be produced, a decision has to be made as to whether ROM, PROM, or EPROM will be used for permanent program storage. If only a few systems are to be manufactured, it may be more cost-effective to use either PROM or EPROM. EPROM-based storage also allows the main program to be changed at any time, even in the field by the end-user. The PROM based system requires replacement; however, it is field programmable. If the main requirement is a minimum parts configuration and many microprocessor systems must be produced the decision should be to use ROM-based storage.

29.3.1 ROM operation

In mask-programmed ROM, the memory bit pattern is produced during fabrication of the chip by the manufacturer using a masking operation. The memory matrix is defined by row (X) and column (Y) bit-selection lines that locate individual memory cell positions.

For example, in *Figure 29.29* refer to column C_2 and row 127 as the storage cell location of interest. When the proper binary inputs on the address lines are decoded, the cell at $R_{127}C_2$ will be selected. If the drain contact of this cell is connected to bit line L_2, then L_2 will be pulled below threshold, turning off device C_2; note that devices C_0, C_1, and C_3 through C_{15} will also be off since they are not addressed. Therefore, device A pulls the OUT line to V_{CC} for a logic 1 output when cell $R_{127}C_2$ is selected.

Alternatively, when cell $R_{127}C_2$ is masked it does not have a drain contact to bit line L_2. Then when this cell is addressed, device C_2 is not connected to V_{CC} and will be turned on. Thus, the OUT line will be pulled to ground through device C_2 and will appear as a logic 0 output. To program a 1 or a 0 into a ROM storage cell, the drain contact will or will not be connected, respectively, to the particular bit line. Note that this type of programming is permanent. An alternative method of performing the same operation would be to eliminate the gate of the storage cell.

29.3.2 ROM process

For many designs, fast manufacturing turnaround time on ROM patterns is essential for fast entry into system production. This is especially true for the consumer 'games' market. Several vendors now advertise turnaround times that vary from two to six weeks for prototype quantities (typically small quantities) after data

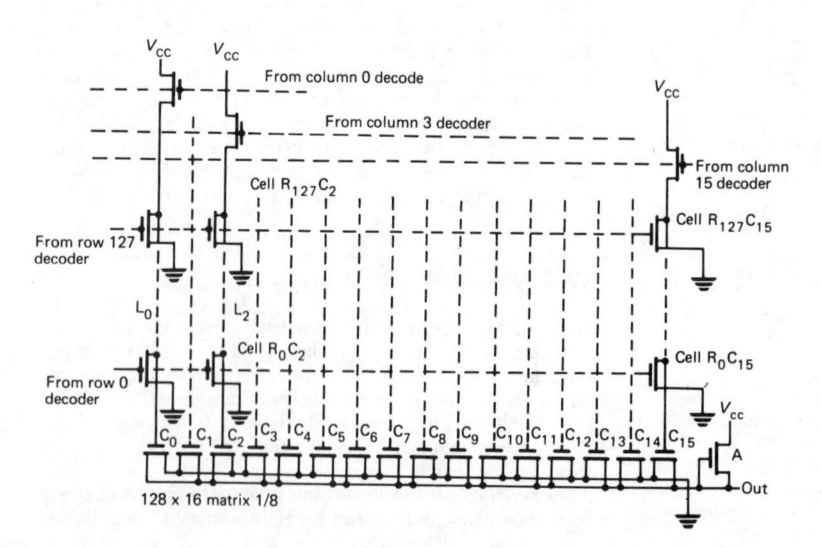

Figure 29.29 Portion of ROM matrix and output circuitry

verification. Data verification is the time when the user confirms that data have been transferred correctly into ROM in accordance with the input specifications.

Contact programming is one method that allows ROM programming to be accomplished in a shorter period of time than with gate mask programming. In mask programming, most ROMs are programmed with the required data bit pattern by vendors at the first (gate) mask level, which occurs very early in the manufacturing process. In contact programming, actual programming is not done until the fourth (contact) mask step, much later in the manufacturing process. That technique allows wafers to be processed through a significant portion of the manufacturing process, up to 'contact mask' and then stored until required for a user pattern. Some vendors go one step further and program at fifth (metal) mask per process. This results in a significantly shorter lead time over the old gate-mask programmable time of eight to ten weeks; the net effect is time and cost savings for the end-user.

29.3.3 ROM applications

Typical ROM applications include code converters, look-up tables, character generators, and non-volatile storage memories. In addition, ROMs are now playing an increasing role in microprocessor-based systems where a minimum parts configuration is the main design objective. The average amount of ROM in present microprocessor systems is in the 10–32 kbyte range. In this application, the ROM is often used to store the control program that directs CPU operation. It may also store data that will eventually be output to some peripheral circuitry through the CPU and the peripheral input/output (P I/O) device.

In a microprocessor system development cycle, several types of memory (RAM, ROM and EPROM or PROM) are normally used to aid in the system design. After system definition, the designer will begin developing the software control program. At this point, RAM is usually used to store the program, because it allows for fast and easy editing of the data. As portions of the program are debugged, the designer may choose to transfer them to PROM or EPROM while continuing to edit in RAM, thus avoiding having to reload fixed portions of the program into RAM each time power is applied to the development system.

29.3.4 Electrically erasable programmable ROM

The ideal memory is one that can perform read/write cycles at speeds meeting the needs of microprocessors (200 ns typical) and store data in the absence of power. The EEPROM (electrically erasable PROM) is a device which meets two of the three requirements. It is non-volatile (retains data in the absence of power) and can be read at speeds compatible with today's microprocessors. However, its write cycle requires typically 20 ms to execute. The EEPROM has two major advantages over the EPROM. The EEPROM can be programmed in circuit and can selectively change a byte of memory instead of all bytes. The process technology to implement the EEPROM is quite complex and the industry is currently on the verge of mastering production.

29.3.5 EEPROM theory

EEPROMs use a floating gate structure, much like the ultraviolet erasable PROM (EPROM), to achieve non-volatile operation. To achieve the ability to electrically erase the PROM a principle known as Fowler–Nordheim tunnelling was implemented. Fowler–Nordheim tunnelling predicts that under a field strength of 10 MV/cm a certain number of electrons can pass a short distance, from a negative electrode, through the forbidden gap of an insulator entering the conduction band and then flow freely toward a positive electrode. In practice the negative electrode is a polysilicon gate, the insulator is a silicon dioxide and the positive electrode is the silicon substrate.

Fowler–Nordheim tunnelling is bilateral in nature, and can be used for charging the floating gate as well as discharging it. To permit the phenomenon to work at reasonable voltages (e.g. 20 V), the oxide insulator needs to be less than 200 Å thick. However, the tunnelling area can be made very small to aid the manufacturability aspects (20 nm oxides are typically one half the previously used thickness).

Intel Corporation produces an EEPROM based on this principle. Intel named the cell structure used FLOTOX. A cross-section of the FLOTOX device is shown in *Figure 29.30*.

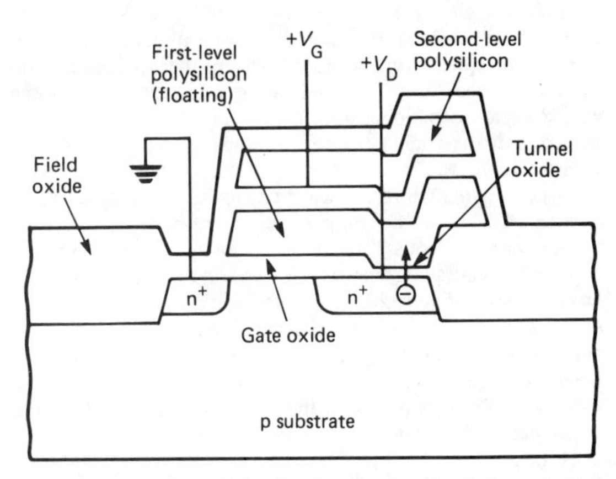

Figure 29.30 EEPROM cell using Fowler–Nordheim tunnelling mechanism

The FLOXTOX structure resembles the FAMOS structure used by Intel for EPROM devices. The primary difference is in the additional tunnel-oxide region over the drain. (See *Figure 29.31* for EPROM cell configuration.) To charge the floating

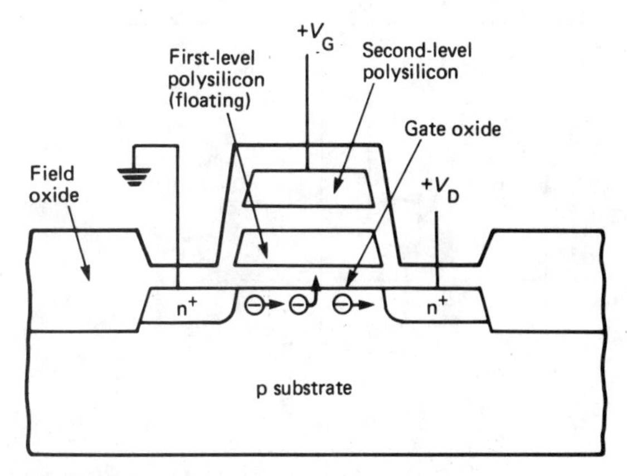

Figure 29.31 EPROM ultraviolet erasable cell. Similar to the EEPROM cell

gate of the FLOTOX structure of *Figure 29.30*, a voltage V_G is applied to the top gate and with the drain voltage V_D at 0 V the loating gate is capacitively coupled to a positive potential. Electrons will then flow to the floating gate. If a positive potential is applied to the drain and the gate is grounded the process is reversed and the floating gate is discharged.

29.3.6 EEPROM circuit operation

The electrically erasable cell is the building block for EEPROM memories. One manufacturer's approach is to assemble the Fowler–Nordheim concept cell into a memory array using two transistors per cell as shown in *Figure 29.32*. The floating gate cell (bottom device) is the actual storage device. The upper device is used as a select transistor to prevent devices on non-selected rows from discharging when a column is raised high. Before information is entered the array must be cleared. This returns all cells to a charged state as shown schematically in *Figure 29.32(a)*. To clear the memory all the select lines and program lines are raised to 20 V while all the columns are grounded. This forces electrons through the tunnel oxide to charge the floating gates on all of the selected rows. The Intel Corporation device offers the user the option of chip-clear or byte-clear. When byte-clear is initiated, only the select and program lines of an addressed byte are raised to 20 V.

To write a byte of data, the select line for the addressed byte is raised to 20 V while the program line is grounded (see *Figure 29.32(b)*). Simultaneously, the columns of the selected byte are raised or lowered according to the incoming data pattern. The byte on the left in *Figure 29.32(b)*, for example, has its column at a high voltage, causing the cell to discharge, whereas the bit on the right has its column at ground so its cell will experience no change. Reading is accomplished by applying a positive bias to the select and program lines of the addressed cell. A cell with a charged gate will remain off in this condition but a discharged cell will be turned on.

The EEPROMs being introduced today are configured externally to be compatible with ROM and EPROM standards which already exist. Devices typically use the same 24 pin pinout as the generic 2716 (2K × 8 EPROM) device. A single 5 V supply is all that is needed for read operations. For the write and clear operations an additional supply (V_{PP}) of 20 V is necessary. The device reads in the same manner as the EPROM it will eventually replace.

The timing for writing a byte is shown in *Figure 29.33*. The chip is powered up by bringing CE low. With address and data applied, the write operation is initiated with a single 10 ms, 20 V

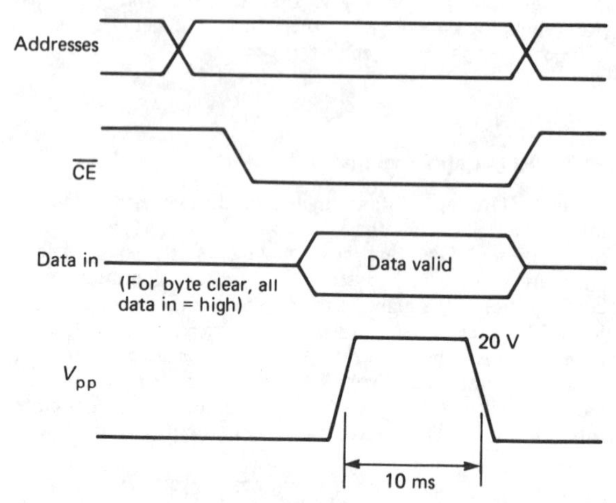

Figure 29.33 Byte write operation for an EEPROM. Two cycles are required

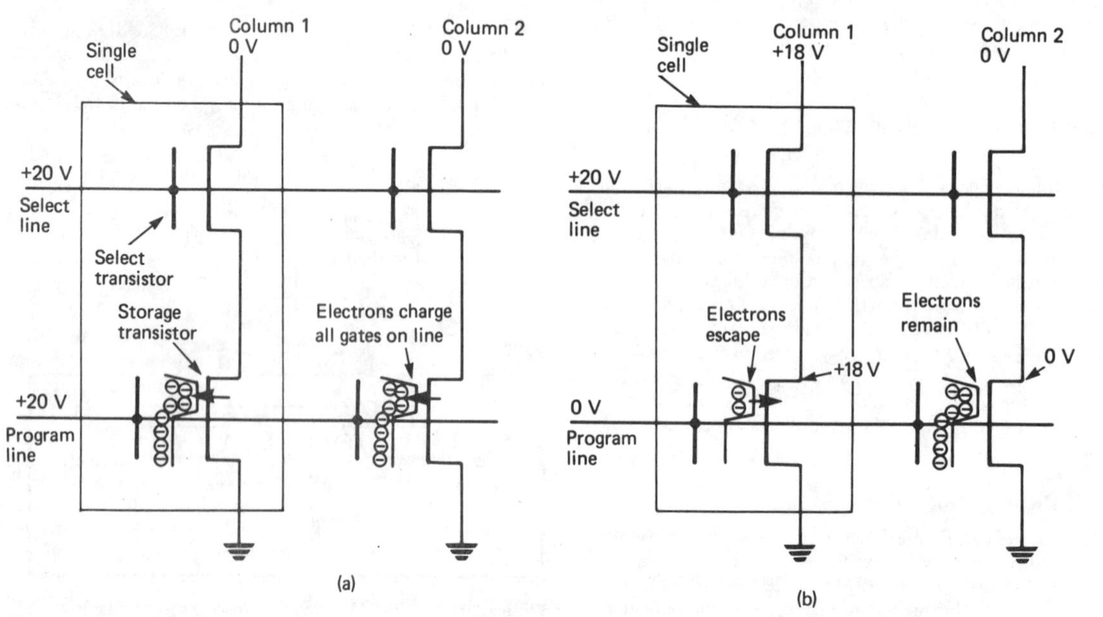

Figure 29.32 Two-transistor memory cell implementation for EEPROM using floating gate transistor

pulse applied to the V_{PP} pin. During the write operation, OE is not needed and is held high.

A byte clear is really no more than a write operation. A byte is cleared merely by being written with all 1's (high). Thus altering a byte requires nothing more than two writes to the address byte, first with the data set to all 1's and then with the desired data. This alteration of a single byte takes only 20 ms. In other non-volatile memories, changing a single byte requires that the entire contents be read out into an auxiliary memory. Then the entire memory is rewritten. This process not only requires auxilary memory; for a 2 kbyte device it takes about one thousand times as long (20 ms versus 20 s).

The only difference between byte clear and chip clear is that OE is raised to 20 V during chip clear. The entire 2 kbytes are cleared with a single 10 ms pulse. Addresses and data are not all involved in a chip-clear operation.

29.3.7 EEPROM applications

The electrically erasable PROM (EEPROM) has the non-volatile storage characteristics of core, magnetic tape, floppy and Winchester disks but is a rugged low-power solid-state device and occupies much less space. Solid-state non-volatile devices, such as ROM and EPROM, have a significant disadvantage in that they cannot be deprogrammed (ROM) or re-programmed in place (EPROM). The non-volatile bipolar PROM blows fuses inside the device to program. Once set the program cannot be changed, greatly limiting their flexibility. The EEPROM therefore has the advantages of program flexibility, small size, and semiconductor memory ruggedness (low voltages and no mechanical parts).

The advantages of the EEPROM create many applications that were not feasible before. The low power supports field programming in portable devices for communication encoding, data formatting and conversion, and program storage. The EEPROM in circuit change capability permits computer systems whose programs can be altered remotely, possibly by telephone. It can be changed in circuit to quickly provide branch points or alternate programs in interactive systems.

The EEPROMs non-volatility permits a system to be immune to power interruptions. Simple fault tolerant multiprocessor systems also become feasible. Programs assigned to a processor that fails can be reassigned to the other processors with a minimum interruption of the system. Since a program can be backed up into EEPROM in a short period of time key data can be transferred from volatile memory during power interruption and saved. The user will no longer need to either scrap parts or make service calls should a program bug be discovered in fixed memory. With EEPROM this could even be corrected remotely. The EEPROM's flexibility will create further applications as they become available in volume and people become familiar with their capabilities.

29.3.8 Erasable programmable ROM

The EPROM like the EEPROM satisfies two of our three requirements. It is non-volatile and can be read at speeds comparable with today's microprocessors. However, its write cycle is significantly slower like the EEPROM. The EPROM has the additional disadvantage of having to be removed from the circuit to be programmed, in contrast to the EEPROM's ability to be programmed in circuit.

EPROM is electrically programmable, then erasable by ultraviolet light, and programmable again. Erasability is based on the floating gate structure of an n- or p-channel MOSFET. This gate, situated within the silicon dioxide layer, effectively controls the flow of current between the source and drain of the storage device. During programming, a high positive voltage (negative if p-channel) is applied to the source and gate of a selected MOSFET, causing the injection of electrons into the floating silicon gate. After voltage removal, the silicon gate retains its negative charge because it is electrically isolated (within the silicon dioxide layer) with no ground or discharge path. This gate then creates either the presence or absence of a conductive layer in the channel between the source and the drain directly under the gate region. In the case of an n-channel circuit, programming with a high positive voltage depletes the channel region of the cell; thus a higher turn-on voltage is required than on an unprogrammed device. The presence or absence of this conductive layer determines whether the binary 1-bit or the 0-bit is stored. The stored bit is erased by illuminating the chip's surface with ultraviolet light, which sets up a photocurrent in the silicon dioxide layer which causes the charge on the floating gate to discharge into the substrate. A transparent window over the chip allows the user to perform erasing, after the chip has been packaged and programmed, in the field.

29.3.9 Programmable ROM

The PROM has a memory matrix in which each storage cell contains a transistor or diode with fusible link in series with one of the electrodes. After the programmer specifies which storage cell positions should have a 1-bit or 0-bit, the PROM is placed in a programming toll which addresses the locations designated for a 1-bit. A high current is passed through the associated transistor or diode to destroy (open) the fusible link. A closed fusible link may represent a 0-bit, while an open link may represent a 1-bit (depending on the number of data inversions done in the circuit). A disadvantage of the fusible-link PROM is that its programming is permanent; that is, once the links are opened, the bit pattern produced cannot be changed.

29.3.10 Shadow RAM

A recently introduced RAM concept is called the shadow RAM. This approach to memory yields a non-volatile RAM (data are retained even in the absence of power), by combining a static RAM cell and an electrically erasable cell into a single cell. The electrically erasable programmable read only memory (EEPROM) shadows the static RAM on a bit by bit basis. Hence the name shadow RAM. This permits the device to have read/write cycle times comparable to a static RAM, yet offer non-volatility.

The non-volatility has a limit in that the number of write cycles to the EE cell is typically one thousand to one million maximum. The device can however be read indefinitely. Currently two product types are offered. One permits selectively recalling bits stored in the non-volatile array while the other product recalls all bits simultaneously.

29.3.11 Shadow RAM operation

The static RAM portion of the shadow RAM operates the same as the depletion mode static cell discussed earlier in the static RAM section. The EEPROM portion is formed by adding a transistor with a floating gate and a capacitor as shown in *Figure 29.34*. While operating as a static RAM the capacitor and transistor have no effect. The device will automatically move the contents of the EEPROM into the static cell upon power up. This operates as follows. During switching of the internal +5 V supply, the word-select lines isolating nodes N_1 and N_2 from the bit lines are turned off. If the floating gate transistor was programmed on, N_2 will have a larger capacitance than N_1. This causes N_2 to rise more slowly than N_1, and the cell will latch with N_2 low and N_1 high. Had the floating gate transistor been programmed off, the capacitance load on N_1 would be larger than that on N_2. As the supply switches from ground to +5 V, node N_2 will rise more quickly than N_1 and N_2 will latch high with N_1 low.

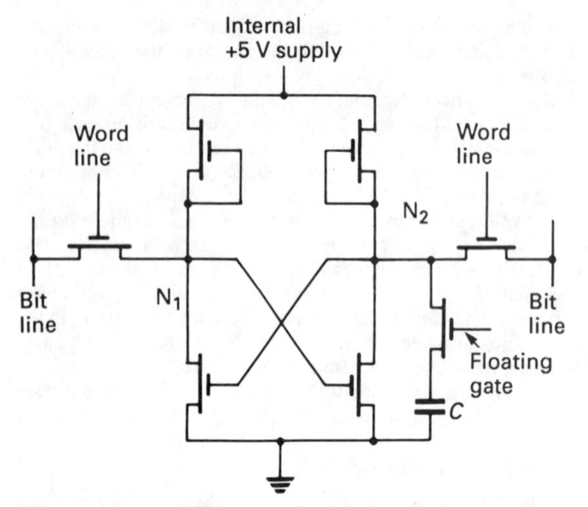

Figure 29.34 Shadow RAM storage cell comprising a classical depletion load static cell with a transistor and capacitor added

During operation data are moved between the RAM and EEPROM positions under the control of two TTL signals, recall and store. In the case of storing into the EEPROM all bits must be transferred. For recall two parts are available to give the designer the option of either recalling all bits or selectively recalling one bit at a time. The bits are recalled in less than 1 ms. The operation in non-power-up conditions is described in the next section.

29.3.12 Shadow RAM construction

The shadow RAM currently produced by Xicor Coporation uses a triple-poly-process. A schematic of the cell is shown in *Figure 29.35*. The polysilicon electrodes are separated by oxide layers

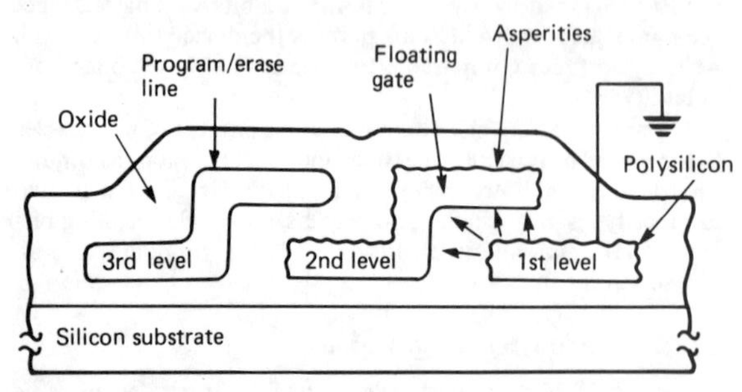

Figure 29.35 Schematic of shadow RAM cell

approximately 100 nm thick. Asperities (projections), which exist uniformly across the surface of the polysilicon layer, help to create a reproducible flow of electrons. The asperities greatly reduce the voltage potential needed. The part is programmed by bringing the program/erase line to a positive voltage thereby coupling the floating gate to a positive potential. The programming electrode (on poly 1 level) is held at ground and, with a large enough electric field, emits electrons, which are captured by the floating gate making it negative. For erasure, the floating gate is held close to ground potential while the program/erase line is brought high. To avoid the need for an extra polysilicon electrode, a switching transistor holds the gate to ground. Having established an electric field between the program/erase electrode and the floating gate, thc asperities on the gate emit electrons to the electrode so that the gate becomes positively charged.

29.3.13 Shadow RAM theory

It was observed for many years that a fairly low voltage would cause electrons to flow between polysilicon layers (electron tunnels) in unexpectedly large numbers. The enhanced electron tunnelling was found to be due to asperities. The shadow RAM uses a triple-poly-process aimed at enhancing and exploiting this tunnelling effect.

The shadow RAM uses a floating gate transistor with the ability to be left either charged or uncharged, in the absence of power, as the basis for operation of the non-volatile RAM. A floating gate is an island of conducting material surrounded by oxide and coupled capacitively to the silicon substrate to form a transistor. The presence or absence of charge on the gate determines whether the transistor is on or off. Charge is induced into the gate region by overcoming the oxide barrier by applying a relatively low voltage between the floating gate itself and the other two polysilicon layers. Since the charge is trapped in the oxide, it acts as a barrier to the flow of current, and the charge once placed on the gate should remain indefinitely as long as the charging voltage is removed. When the memory is being programmed, the floating gate is charged with electrons, turning the transistor off; during erasure, electrons are removed from the gate turning the transistor on. Hence a programmed floating gate is at a negative potential and an erased gate is positive.

Further reading

ANDREWS, W., 'Architectural and process enhancements deliver faster, more flexible PLDs', *Computer Design*, 1 January (1988)

BROWN, J. R. Jr, 'Timing peculiarities of multiplexed RAMs', *Computer Design*, July (1977)

BURSKY, D., 'Memories pace systems growth', *Electronic Design*, 27 September (1980)

BURSKY, D., 'UV EPROMs and EEPROMs crash speed and density limits', *Electronics Design*, 22 November (1980)

DONNELLY, W., 'Memories—new generations push technology forward', *Electronics Industry*, October (1982)

DROIR, J., OWEN, W. H. and SIMKO, R. T., 'Computer systems acquire both RAM and EEPROM from one chip with two memories', *Electronics Design*, 15 February (1980)

EATON, S. S. and WOOTON, D., 'Circuit advances propel 64K RAM across the 100 ns barrier', *Electronics*, 24 March (1982)

FOSS, R. C. and HARLAND, R., 'Standards for dynamic MOS RAMs', *Electronics Design*, 16 August (1977)

GOSNEY, M., 'Reappraising CCD memories: can they stand up to RAMs?', *Electronics*, 7 June (1979)

GOSNEY, M., *CCDs, Production Device or LAB Experiment*, Mostek Corporation Technology Brief

GREENE, R., 'Pinout standard amplifies variable density memory design', *Electronics Design*, 6 December (1980)

GREGORY, R., 'Caching designs eliminate wait states to relieve bottlenecks', *Computer Design*, 15 October (1988)

HNATEK, E. R., 'Semiconductor memory update: EEPROMs', *Computer Design*, December (1981)

HNATEK, E. R., 'Semiconductor memory update: DRAMs', *Computer Design*, January (1982)

INTEL CORPORATION, *Which Way For 4K . . . 16, 18, or 22 Pin?*, Ontel Corporation, Application Brief AP-11

ISLES, J., 'Flash EEPROMs bridge the gap', *New Electronics*, July/August (1988)

JOHNSON, W. S., KUHN, G. L., RENNIGER, A. L. and PERLEGOS, G., '16k EE-PROM relies on tunnelling for byte erasable program storage', *Electronics*, February 28 (1980)

JONES, F. and LAUTZENHEISER, D., 'Printed circuit board layouts for compatible dynamic RAM family', *Computer Design*, December (1980)

KLEIN, R., OWEN, W. H., SIMKO, R. T. and TCHON, W. E., '5-volt-only nonvolatile RAM owes it all to polysilicon', *Electronics*, 11 October (1979)

KOPS, P. B., 'Testability is crucial in PLD circuit design', *END*, 18 August (1988)

KURITA, K., 'Very high speed ROM using bipolar/CMOS technology', *Electron. Commun. Japan, Part 2, Electronics*, **71**, No. 7 (1988)

MERCER, C., 'The registered PROM can replace PALs in large state machines', *EDN*, 10 November (1988)

MITCHELL, P., 'Solid state mass memory now a real option', *New Electronics*, June (1988)

MOSTEK CORPORATION, *Mostek's BYTEWYDE(tm) Memory Products*, Mostek Corporation, Applications Note

MOSTEK CORPORATION, *Designing Memory Boards For RAM/ROM/EPROM Interchange*, Moste, Corporation, Applications Note

MOSTEK CORPORATION, *Resolving Microprocessor Memory Bus Contention*, Mostek Corporation, Applications Note

MOSTEK CORPORATION, *N-Channel MOS—Its Impact on Technology*, Mostek Corporation, Applications Note

OLIPHANT, J., *Designing Non-Volatile Memory Systems with Intel's 5101 RAM*, Intel Applications Note AP-12

OWEN, R., *A Testing Philosophy for 16K Dynamic Memories*, Mostek Corporation, Applications Note

PROEBSTING, R., 'Dynamic MOS RAM's. An economic solution for many system designs', *Electronic Design News*, 20 June (1977)

PROEBSTING, R. and GREEN, R., 'A TTL compatible 4096 N-channel RAM', *ISSCC*, February (1973)

PROEBSTING, R. and SCHROEDER, P., 'A 16k × 1 bit dynamic RAM', *ISCC*, February (1977)

ROCHERS, G. D., 'EEPROM eclipses other reprogrammable memories', *Electronics Design*, 22 November (1980)

'Technology focus – memory devices', *Electronic Engineering*, May (1988)

THREEWITT, B., 'A VLSI approach to cache memory', *Computer Design*, January (1982)

VOLGERS, R., 'HCMOS FIFOs and their applications', *Electron. Components Appl.*, **8**, No. 4 (1988)

WHITTIER, R. J., 'Semiconductor memories', *Mini-Micro Systems*, December (1982)

WILCOCK, J. D., 'Semiconductor memories', *New Electronics*, 17 August (1982)

WILSON, R., 'Flash EPROMs offer new alternative for code storage', *Computer Design*, 15 April (1988)

WYLAND, D. C., 'Cache tag RAM chips boost speed and simplify design', *Computer Design*, 1 November (1987)

YOUNG, S., 'Uncompromising 4-K static RAM runs fast on little power', *Electronics*, 12 May (1977)

YOUNG, S., 'Evolution of MOS technology', Presented at *Wescon* (1978)

YOUNG, S., 'Memories have not a density ceiling but new process will push through', *Electronics Design*, 25 October (1978)

30 Microprocessors

A Shewan, BSc
Motorola Ltd

Contents

30.1 Basic structure

The basic principles of operation of a microprocessor are identical to those that control the operation of any larger computer, and consist of four parts, as shown in *Figure 30.1*: the memory, the arithmetic logic unit (ALU), the input/output system, and the control unit.

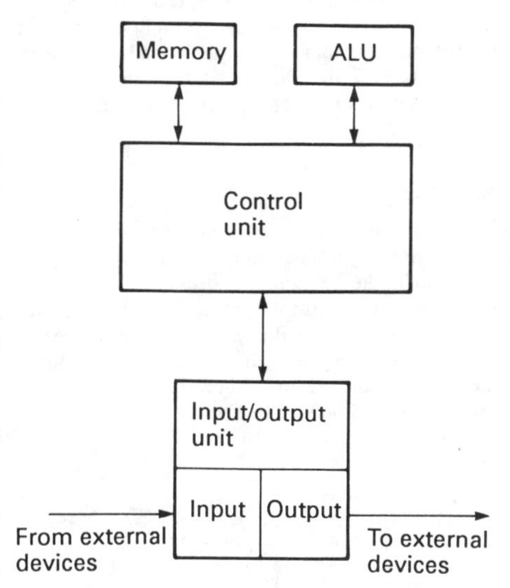

Figure 30.1 Basic structure of a microprocessor

The *memory* is an indispensable part of a microprocessor. It contains the *data* used in a program as well as the *instructions* for executing the program. A *program* is a group of instructions telling the microprocessor what to do with the data. The *arithmetic logic unit* (ALU) is part of the microprocessor that performs those arithmetic or logic operations required by the routine or program, and generates the status bits (condition codes) that are the heart of the decision-making capabilities of any computer. The *input/output system* controls communication between the microprocessor and its external devices. The *control unit* consists of a group of flip-flops and registers that regulate the operation of the microprocessor itself. The function of the control unit is to cause the proper sequence of events to occur during the execution of each computer instruction.

30.1.1 Memory concepts

The binary bits of a block of memory can be organised in various ways. It is usual for most microprocessors to be arranged so that the bits are grouped into *words* of 4, 8 or 16 bits each. In most present-day microprocessors, an 8-bit word length (or byte) is used. Modern microprocessors are *word oriented* in that they transfer one byte (or word) at a time by means of the data bus. This bus, the internal registers, and the ALU (arithmetic logic unit) are *parallel* devices that handle all the bits of a word *at the same time.*

Microprocessor instructions require 1, 2, or 3 bytes (or words) and typical routines include 5 to 50 instructions. Since IC memories are physically small and relatively inexpensive, they can include thousands of words and thereby provide very comprehensive computer programs.

Each word in memory has *two parameters*; its *address*, which locates it within memory, and the *data*, which is stored at that location. The process of accessing the contents of the memory locations requires two registers, one associated with address and one with data. The memory address register (MAR) holds the *address of the word currently being accessed*, and the memory data register (MDR) holds the *data being written into or read out of the addressed memory location*. These registers can be considered part of the memory or of the control unit.

In most microprocessors the memory consists of two parts with different memory addresses, a *ROM area* used to hold the *program, constant data*, and *tables*, and a *RAM area*, used to hold *variable data*. Generally, the data on which the program operates must be rewritten every time the system is started again (restarted), and the system must always be restarted (going through a start-up procedure) after any power failure occurs. This is not a severe drawback because data is usually invalid after a power failure. Fortunately, the program, if it is contained in ROM, can be restarted immediately because a power failure does not affect a ROM.

30.1.2 The arithmetic logic unit

The ALU performs all the arithmetic and logical operations required by the microprocessor. It accepts two operands as inputs (each operand contains as many bits as the basic word length of the computer) and performs the required arithmetic or logical operation upon them. ALUs are readily available as ICs but microprocessors contain their ALUs within their chip.

Most ALUs perform the following arithmetic or logical operations: addition, subtraction, logical OR, logical AND, EXCLUSIVE OR, complementation, and shifting.

Microprocessors are capable of performing more sophisticated arithmetic operations such as multiplication, division, extracting square roots, and taking trigonometric functions; however, in most devices these operations are performed as *software subroutines.* A multiplication command, for example, is translated by the appropriate subroutine into a series of add and shift operations that can be performed by the ALU.

30.1.3 The control section

The function of the *control section* is to regulate the operation of the microprocessor. It decodes the instructions and causes the proper events to occur in the correct order.

The control section of a microprocessor consists of a group of registers and flip-flops and the timing circuitry necessary to make them operate properly. In a rudimentary device the following registers might be part of the control section:

(1) *The memory address register (MAR)*
(2) *The program counter (PC)* This is a register that contains as many bits as the MAR. It holds the memory address of the next instruction word to be executed. It is usually *incremented* during the execution of an instruction so that it contains the address of the next instruction to be executed.
(3) *The instruction register* This register holds the instruction while it is in the process of being executed.
(4) *The instruction decoder* This decodes the instruction presently being executed. Its inputs come from the instruction register.
(5) *The accumulator* The accumulator contains the basic operand used in each instruction. In ALU operations where two operands are required, one of the operands is stored in the accumulator as a result of previous instructions. The other operand is generally read from memory. The two operands form the inputs to the ALU and the result is normally sent back to the accumulator.

The control unit usually contains several flip-flops. The flags or condition codes are flip-flops and most microprocessors also have FETCH and EXECUTE flip-flops.

These determine the state of the microprocessor. The instructions are contained in the microprocessor's memory. The microprocessor starts by fetching the instruction. This is the FETCH portion of the computer's cycle. It then executes, or performs the instruction. At this time, the computer is in EXECUTE mode. When it has finished executing the instruction, it returns to FETCH mode and reads the next instruction from memory. Thus the computer alternates between FETCH and EXECUTE modes and the FETCH and EXECUTE flip-flops determine its current mode of operation.

30.2 The Motorola 6800 microprocessor

In this chapter the Motorola 6800 microprocessor is used to describe features of operation, which are also common to many other devices currently available from other manufacturers.

Figure 30.2 shows the block diagram of a 6800 microprocessor. An 8-bit internal bus interconnects the various registers with the instruction decode and control logic. The nine control lines that communicate with the external devices can be seen on the left, with the address bus at the top and the data bus at the bottom of the figure.

30.2.1 ALU and registers

The arithmetic logic unit included in the 6800 is an 8-bit, parallel processing, two's complement device. It includes the condition code (or processor status) register. The system has a 16-bit address bus, but as seen in *Figure 30.2*, the internal bus is 8-bits wide, and the index and stack pointer registers (and the address buffers) are implemented with a high (H) and (L) byte. The two 8-bit accumulators, A and B, speed up program execution by allowing two operands to remain in the microprocessor. Instructions that can be performed using both accumulators (e.g. ABA and SBA) are very fast because they do not require additional cycles to fetch the second operand.

30.2.2 Vectored interrupts

An interrupt is a signal to the microprocessor that causes it to stop execution of the normal program and to branch (or jump) to another location that is the beginning address of an interrupt service routine. These routines are written to provide whatever action is necessary to respond to the interrupt. Four types of interrupts are provided in the 6800, and each has its unique software service routine and vector. Three of them are also implemented by pins on the 6800. The four types are:

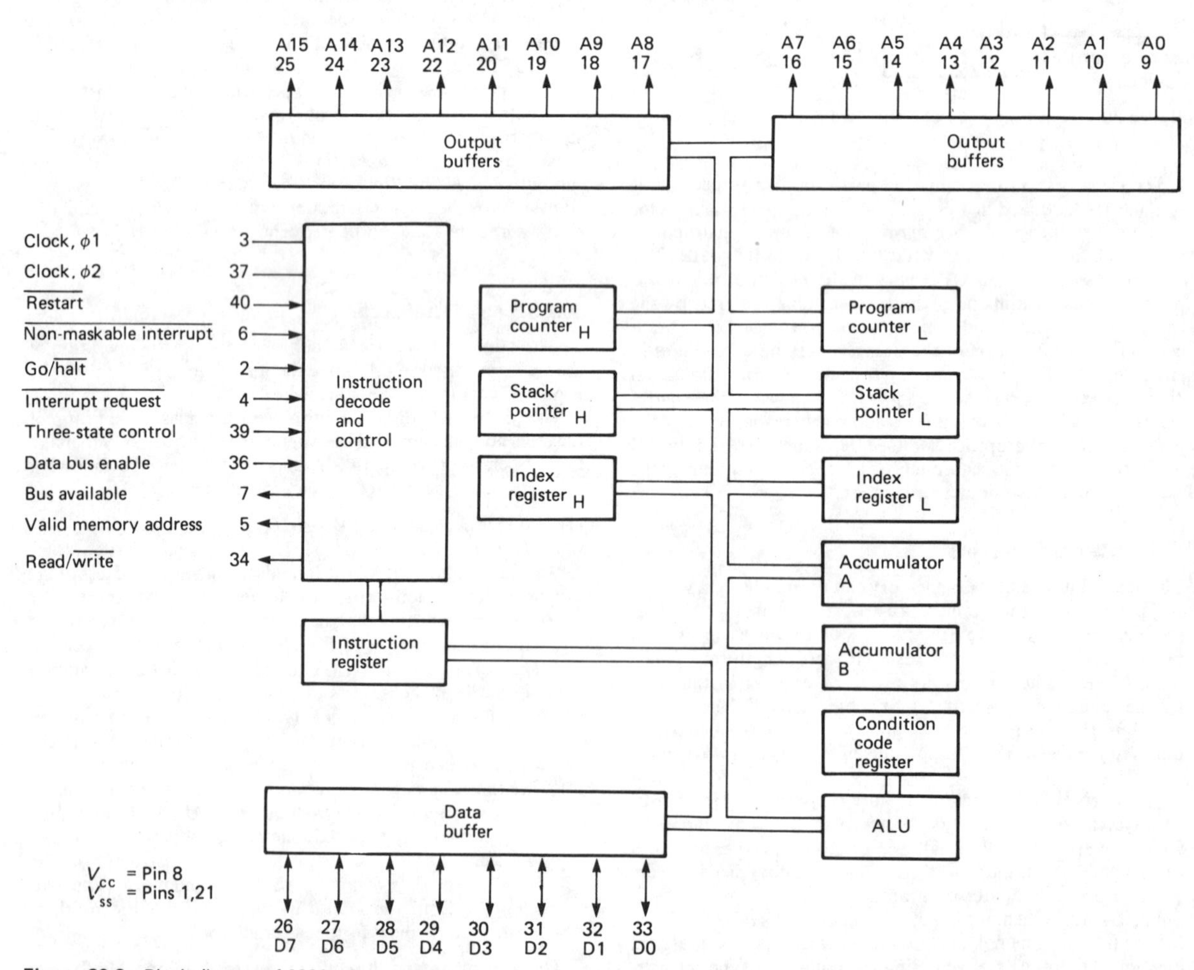

Figure 30.2 Block diagram of 6800 microprocessor

RESTART (RST)
NONMASKABLE INTERRUPT (NMI)
SOFTWARE INTERRUPT (SWI)
INTERRUPT REQUEST (IRQ)

In order for the program to be branched to the appropriate routine, the top eight locations of the ROM or PROM that is highest in memory are reserved for these interrupt vectors. The contents of these locations contain the 16-bit addresses where the service routines begin. When activated, the program is 'vectored' or pointed to the appropriate address by the microprocessor logic. This is called a 'vectored interrupt'.

30.2.3 Control lines and their functions

The hardware aspects of system design involve interconnecting the components of the system so that the data is properly transferred between them and to any external hardware that is being utilised. In addition to the obvious requirement for power and ground to each component, the address bus, and the bidirectional data bus, numerous control signal lines are required. These control lines are in effect a third bus called the control bus. The lines are identified in *Figure 30.2*, which also shows their pin numbers.

To design a system that performs properly, it is necessary to understand each signal's characteristics and function. These are described in the following paragraphs.

30.2.3.1 Read/write (R/W)

This output line is used to signal all external devices that the microprocessor is in a READ state (R/$\overline{\text{W}}$ = HIGH), or a WRITE state (R/$\overline{\text{W}}$ = LOW). The normal standby state of this line, is HIGH. This line is three-state. When three-state control (TSC) goes HIGH, the R/$\overline{\text{W}}$ line enters the high-impedance mode.

30.2.3.2 Valid memory address (VMA)

The VMA output line (when in the HIGH state) tells all devices external to the microprocessor there there is a valid address on the address bus. During the execution of certain instructions, the address bus may assume a random address because of internal calculations. VMA goes LOW to avoid enabling any device under those conditions. Note also that VMA is held LOW during HALT, TSC, or during the execution of a WAIT (WAI) instruction. VMA is not a three-state line and therefore direct memory access (DMA) cannot be performed unless VMA is externally opened (or gated).

30.2.3.3 Data bus enable (DBE)

The DBE signal enables the data bus drivers of the microprocessor when in the HIGH state. This input is normally connected to the phase 2 clock but is sometimes delayed to assure proper operation with some memory devices. When HIGH, it permits data to be placed on the bus during a WRITE cycle. During a microprocessor READ cycle, the data bus drivers within the microprocessor are disabled internally. If an external signal holds DBE LOW, the microprocessor data bus drivers are forced into their high-impedance state. This allows other devices to control the I/O bus (as in DMA).

30.2.3.4 HALT and RUN modes

When the HALT input to the 6800 is HIGH, the microprocessor is in the RUN mode and is continually executing instructions. When the HALT line goes LOW, the microprocessor halts after completing its present instruction. At that time the microprocessor is in the HALT mode. Bus available (BA) goes HIGH, VMA becomes a 0, and all three-state lines enter their high-impedance state. Note that the microprocessor does not enter the HALT mode as soon as HALT goes LOW, but does so only when the microprocessor has finished execution of its current instruction. It is possible to stop the microprocessor while it is in the process of executing an instruction by stopping or 'stretching' the clock.

30.2.3.5 Bus available (BA)

The bus available (BA) signal is a normally LOW signal generated by the microprocessor. In the HIGH state it indicates that the microprocessor has stopped and that the address bus is available. This occurs if the microprocessor is in the HALT mode or in a WAIT state as the result of the WAI instruction.

30.2.3.6 Three-state control (TSC)

TSC is an externally generated signal that effectively causes the microprocessor to disconnect itself from the address and control buses. This allows an external device to assume control of the system. When TSC is HIGH, it causes the address lines and the READ/WRITE line to go to the high impedance state. The VMA and BA signals are forced LOW. The data bus is not affected by TSC and has its own enable (DBE).

The microprocessor is a dynamic IC that must be periodically refreshed by its own clocks. Because TSC stops the clocking of the internal logic of the microprocessor, it should not be HIGH longer than 3 clock pulses, or the dynamic registers in the microprocessor may lose some of their internal data (up to 19 cycles at 2 MHz for the 68B00).

30.2.4 Clock operation

The 6800 utilises a *two-phase clock* to control its operation. The waveforms and timing of the clock are critical for the proper operation of the microprocessor and the other components of the family.

The clock synchronises the internal operations of the microprocessor, as well as all external devices on the bus. The program counter, for example, is advanced on the falling edge of phase 1 and data are latched into the microprocessor on the falling edge of phase 2. All operations necessary for the execution of each instruction are synchronised with the clock.

In a typical instruction table the number of bytes and clock cycles are listed for each instruction. The number of bytes for each instruction determines the size of the memory, and the number of cycles determines the time required to execute the program. LDA A $1234 (which is the extended addressing mode), for example, requires three bytes, one to specify the operation (OP) code and two to specify the address, but requires four cycles to execute. Often an instruction requires the processor to perform internal operations in addition to the fetch cycles. Consequently, for any instruction the number of cycles is generally larger than the number of bytes.

30.3 Hardware configuration of a microprocessor system

Several different types of devices are connected to a microprocessor to form a total control system. A few such devices are described here and illustrated by reference to the Motorola 6800 family.

The basic family of microcomputer components consists of five parts:

(1) The microprocessor.
(2) Masked programmable read only memory (ROM) (1024 bytes of 8 bits each).
(3) Static random access memory (RAM) (128 bytes of 8 bits each).

(4) Peripheral interface adapter (PIA) (for parallel data input/output).
(5) Asynchronous communications interface adapter (ACIA) (for serial data input/output).

As shown in *Figure 30.3*, a complete microcomputer can be built using the components listed above, plus a clock, which is needed to control the timing of the system. Several clock devices

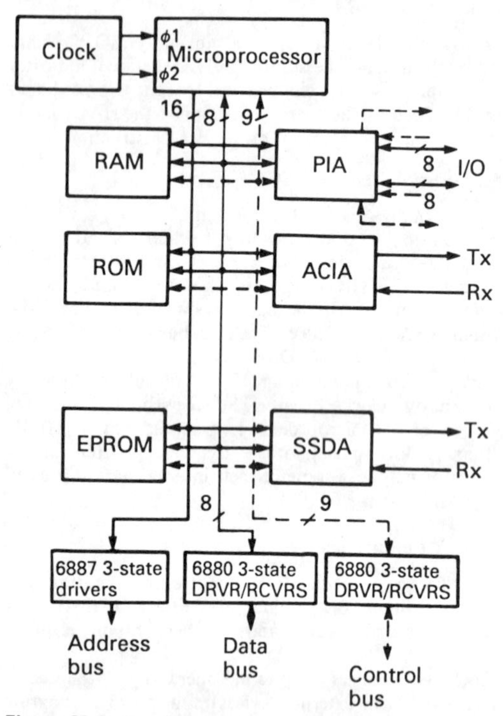

Figure 30.3 The 6800 family components

are available. These are *crystal controlled oscillators* that provide the necessary two-phase non-overlapping timing pulses and are equipped with output circuits suitable for driving MOS circuitry.

The suitability of the family of components as elements of a microcomputer system depends on several factors. First is the requirement that *all elements of a computer be present* and, second, that they be partitioned into the various packages so that they are *modular* and a variety of configurations can be easily assembled. Finally, there is a need for a simple way to interconnect them.

In the system of *Figure 30.3*, the program instructions for the system would typically be stored in ROM or EROM, and all variable data would be written into or read from RAM. The input/output (I/O) of data for the system would be done via the PIA, ACIA or SSDA.

To understand the ways these units work together, it is necessary to know the hardware features of each part. *Figure 30.3* also shows that the system components are interconnected via a 16-wire address bus, an 8-wire data bus, and a 9-wire control bus. For a component to be a member of a microcomputer family and to insure compatibility, it must meet specific system *standards*. The standards must also make it convenient for external devices to interface (or communicate) with the microprocessor. These standards, which apply to all components, are as follows:

(1) 8-bit bidirectional data bus.
(2) 16-bit address bus.
(3) Three-state bus switching techniques.
(4) TTL/DTL level compatible signals.
(5) 5 volt n-channel MOS silicon gate technology.
(6) 24 and 40 pin packages.
(7) Clock rate 100 kHz to 2 MHz (MC68B00).
(8) Temperature range of 0 to 70°C.

Since the basic word length of the 6800 is 8 bits (one byte) it communicates with other components via an 8-bit data bus. The data bus is *bidirectional*, and data are transferred into or out of the 6800 over the same bus. A read/write line (one of the control lines) is provided to allow the microprocessor to control the direction of data transfer.

An 8-bit data bus can also accommodate ASCII (American Standard Code for Information Interchange) characters and packed BCD (two BCD numbers in one byte).

A 16-bit address bus was chosen for these reasons:

(1) For programming ease, the addresses should be multiples of 8 bits.
(2) An 8-bit address bus would only provide 256 addresses but a 16-bit bus provides 65 536 distinct addresses, which is adequate for most applications.

30.3.1 Peripheral interface adapters

The peripheral interface adapter (PIA), shown in *Figure 30.4*, provides a simple means of interfacing peripheral equipment on a parallel or byte-wide basis to the microcomputer system. This device is compatible with the bus interface on the microprocessor side, and provides up to 16 I/O lines and 4 control lines on the peripheral side, for connection to external units. The PIA outputs are TTL and CMOS compatible.

30.3.2 The asynchronous communications interface adapter

The ACIA provides the circuitry to connect serial asynchronous data communications devices (such as a teletypewriter terminal TTY) to bus organised systems such as the 6800 microcomputer system.

The bus interface includes select, enable, read/write, and interrupt signal pins in addition to the 8-bit bidirectional data bus lines. The parallel data of the 6800 system is serially transmitted and received (simultaneously) with proper ASCII formatting and error checking. The control register of the ACIA is programmed via the data bus during system initialisation. It determines word length, parity, stop bits, and interrupt control of the transmit or receive functions.

The PIA and ACIA are the most used I/O components in the 6800 family. Other I/O devices are available however, and include:

(1) Synchronous serial data adapter (SSDA).
(2) Advanced data link controller (ADLC).
(3) General-purpose interface adapter (IEEE 488-1975 bus).
(4) CRT controller (CRTC).
(5) Floppy disk controller (FDC).
(6) Direct memory access controller (DMAC).

The reader should consult the manufacturers' literature for information on these peripheral controller products.

30.3.3 The synchronous serial data adapter

Asynchronous communications via the ACIA are used primarily with slow-speed terminals where information is generated on a keyboard (manually) and where the communication is not necessarily continuous. Synchronous communications are usually encountered where high-speed continuous transmission is required. This information is frequently read to MODEMs, disk or tape systems at 1200 bps or faster. Synchronous systems

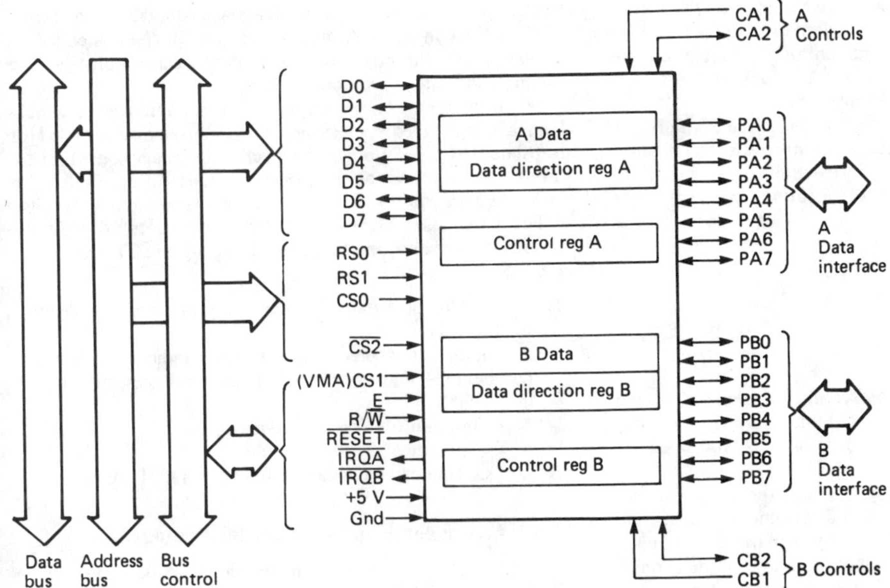

Figure 30.4 PIA bus interface and registers

transmit a steady stream of bits even when no characters are available (a *sync* character is substituted). Because there are no start and stop bits to separate the characters, care must be taken to synchronise the receiving device with the transmitted signal so that the receiver end of the circuit can determine which bit is bit 1 of the character.

Synchronous systems usually use a preamble (all 1s, for example) to establish synchronisation between the receiver and transmitter and will then maintain sync by transmitting a sync pattern until interrupted. Because start and stop bits are not needed, the efficiency of transmission is 20% better for 8-bit words (8 instead of 10 bits per character).

The SSDA provides a bidirectional serial interface for synchronous data information interchange with bus organised systems such as the 6800 microprocessor. It is a complex device containing seven registers. Although primarily designed for synchronous data communications using a 'Bi-sync' format, several of its features and, in particular, the first-in, first-out (FIFO) buffers, make it useful in other applications where data is to be transferred between devices that are not being clocked at precisely the same speed, such as tape cassettes, tape cartridges, or floppy disk systems.

30.3.4 Memory space allocation

The I/O devices such as the PIA, ACIA, and SSDA all have internal registers that contain the I/O data or control the operation of the device. Each of these registers must be allocated a unique address on the address bus and is communicated with just as if it were memory. This technique is called *memory-mapped* I/O and, in addition to allowing the use of all memory referencing instructions for the I/O functions, it eliminates the need for special I/O instructions.

The process of addressing a particular memory location includes not only selecting a cell in a chip, but also selecting that chip from among all those on the same bus. The low order address lines are generally used to address registers within the chips, and the high order lines are available to single out the desired chip. These high order lines could be connected to a decoder circuit with one output line used to enable each chip but, since most of the 6800 family devices have several chip select pins available, these are frequently connected directly to the high order address lines (A15, A14, A13, ...) and separate decoders are not needed, particularly in simple systems.

Table 30.1 Addresses required and chip select pins needed

Component	*Addresses required*	*Positive CS pins*	*Negative CS pins*
RAM	128	2	4
ROM	1024	*	*
PIA	4	2	1
ACIA	2	2	1
SSDA	2	0	1

* Four programmable enables are defined as positive or negative when the mask is made.

Table 30.1 shows the number of addresses required and the number of chip select (CS) pins available for typical 6800 family components.

A component is selected only if all its CS lines are satisfied. When unselected, a component places its outputs in a high-impedance state and is effectively disconnected from the data bus.

When setting up a microcomputer system, each component must be allocated as much memory space as it needs and must be given a unique address so that no address selects more than one component. In addition, one component (usually a ROM) must contain the vector interrupt addresses.

30.3.5 Addressing techniques

Most systems do not use all 64K of available memory space. Therefore, not all of the address lines need be used, and *redundant addresses* will occur (i.e. components will respond to two or more addresses). It is important, however, to choose the lines used so that no two components can be selected by the same address. Since the chip selects serve to turn ON the bus drivers (for a microprocessor READ), only *one* component should be on at any time or something may be damaged.

For proper system operation, most memory and peripheral devices should only transfer data when phase 2 and VMA are

HIGH. Consequently one CS line on each component is usually connected to each of these signals or a derivative of these signals. The PIA, for example, must have its E pin connected to phase 2 or it will not respond to an interrupt and can lock up the system following the execution of a WAI instruction.

When allocating memory, it is wise to place the 'scratch pad' RAMs at the bottom of the memory map, since it is then possible to use the direct address mode instructions throughout the program when referencing these RAM locations. Because 2-byte instead of 3-byte instructions are used, a saving of up to 25% in total memory requirements is possible.

30.4 Software

This section introduces the software features of microprocessors, again using the 6800 as an example, the mnemonics or assembly language concept, and then the accumulator and memory referencing instructions.

The first byte of each instruction is called the *operation code* (op code) because its bit combination determines the operation to be performed. The second and third bytes of the instruction, if used, contain address or data information. They are the *operand* part of the instruction. The op code also tells the microprocessor logic how many additional bytes to fetch for each instruction. There are 197 unique instructions in the 6800 instruction set. Since an 8-bit word has 256 bit combinations, a high percentage of the possible op codes are used.

When an instruction is fetched from the location in memory pointed to by the program counter (PC) register, and the first byte is moved into the instruction register, it causes the logic to execute the instruction. If the bit combination is 01001111 (or 4F in hex), for example, it CLEARs the A accumulator (RESETs it to all 0s), or if the combination is 01001100 (4C), the processor increments the A accumulator (adds 1 to it). These are examples of *one-byte* instructions that involve only one accumulator. An instruction byte of 10110110 (B6) is the code for the *extended addressing mode* of LOAD A. It also commands the logic to fetch the *next two bytes* and use them as the *address* of the data to be loaded into the A accumulator.

30.4.1 Assembly language

When the op codes for each instruction are expressed in their binary or hex format, as in the previous paragraph, they are known as *machine language instructions.* The bits of these op codes must reside in memory and be moved to the instruction decoder for analysis and action by the microprocessor. Programs and data can be entered into a microprocessor in machine language using data switches or keyboards.

Writing programs in machine language is very tedious, and trying to follow even a simple machine language program strains the ability of most people. As a result, various techniques have evolved in an effort to simplify the program documentation. The first step in this simplification is the use of *hexadecimal notation* to express the codes. This reduces the digits from 8 to 2 (10100011 = A3, for example) or, in the case of an address, from 16 binary digits (bits) of 4 hex digits. Even this simplification, however, is insufficient and the concept of using *mnemonic language* to describe each instruction has therefore been developed. A *mnemonic* is defined as a device to help the memory. For example, the mnemonics LDA A (load A) and STA A (store A) are used for the LOAD and STORE instructions instead of the hex OP codes of 86 and B7 because the mnemonics are descriptive and easier to remember.

Not only are mnemonics easy to remember, but the precise meaning of each one makes it possible to use a computer program to translate them to the machine language equivalent that can be used by the computer. A program for this purpose is called an *assembler* and the mnemonics that it recognises constitute an *assembly language program.*

When a program is written in mnemonic or assembly language, it is called a *source program.* After being assembled (or translated) by the computer, the machine language codes that are produced are known as the *object program.* An assembler also produces a *program listing*, which is kept as a record of the design.

The 6800 assembler uses several symbols to identify the various types of numbers that occur in a program. These symbols are:

(1) A blank or no symbol indicates the number is a *decimal* number.
(2) A $ immediately preceding a number indicates it is a *hex* number ($24, for example, is 24 in hex or the equivalent of 36 in decimal).
(3) A # sign indicates an *immediate* operand.
(4) A @ sign indicates an *octal* value.
(5) A % sign indicates a *binary* number (01011001, for example).

30.4.2 Accumulator and memory instructions

The accumulator and memory instructions can be broken down roughly into the following categories:

(1) Transfers between an accumulator and memory (LOADs and STOREs).
(2) Arithmetic operations—ADDITION, SUBTRACTION, DECIMAL ADJUST ACCUMULATOR.
(3) Logical operations—AND, OR, XOR.
(4) Shifts and rotates.
(5) Test operations—bit test and compares.
(6) Other operations—CLEAR, INCREMENT, DECRE MENT, and COMPLIMENT.

In order to reduce the number of instructions required in a typical program and, consequently, the number of memory locations needed to hold a program, microprocessors feature several ways to address them. They are called *addressing modes* and many of them only use one or two bytes.

Instructions can be addressed in one or more of the following modes: *immediate, direct, indexed, implied,* or *extended.*

These modes are explained in the following paragraphs and illustrated in *Figure 30.5*, where each instruction is assumed to start at location 10. The figure shows the op code at location 10 and the following bytes. It explains the function of each byte and gives a sample instruction for each mode.

30.4.2.1 Immediate addressing instructions

All the *immediate instructions* require two bytes. The first byte is the op code and the *second byte* contains the *operand* or *information to be used.* If, for example, the accumulator contains the number 2C and the following instruction occurs in a program:

ADD A #$23

The # indicates that the hex value $23 is to be added to the A accumulator. The immediate mode of the instruction is indicated by the # sign. After the instruction is executed, A contains 4F (23 + 2C).

Immediate instructions are used if the variable or operand is known to be programmer when coding and need not reside in memory. For example, if the programmer wants to add 5 to a variable, it is more efficient to add it immediately than to store 5 in memory and do a direct or extended ADD. When one of the operands must reside in memory, however, direct or extended instructions are required.

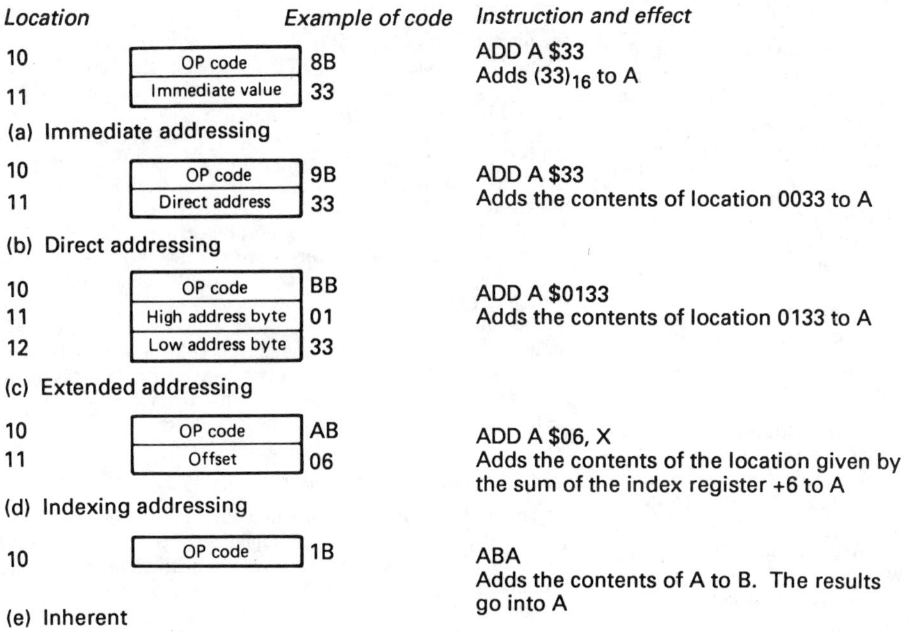

Location	Example of code		Instruction and effect
10	OP code	8B	ADD A $33
11	Immediate value	33	Adds $(33)_{16}$ to A
(a) Immediate addressing			
10	OP code	9B	ADD A $33
11	Direct address	33	Adds the contents of location 0033 to A
(b) Direct addressing			
10	OP code	BB	ADD A $0133
11	High address byte	01	Adds the contents of location 0133 to A
12	Low address byte	33	
(c) Extended addressing			
10	OP code	AB	ADD A $06, X
11	Offset	06	Adds the contents of the location given by the sum of the index register +6 to A
(d) Indexing addressing			
10	OP code	1B	ABA
			Adds the contents of A to B. The results go into A
(e) Inherent			

Figure 30.5 Examples of the various addressing modes of the 6800

30.4.2.2 Direct instructions

Like immediate instructions, direct instructions require two bytes. The second byte contains the address of the operand used in the instruction. Since the op code identifies this as a 2-byte instruction, only 8 address bits are available. The microprocessor contains 16 address lines, but for this instruction, the 8 MSBs of the address are effectively set to 0. The memory locations that can be addressed by a direct instruction are therefore restricted to 0000 to 00FF. It is often wise to place variable data in these memory locations because this data is usually referenced frequently throughout the program. The programmer can then make maximum use of direct instructions and reduce memory requirements by up to 25%.

30.4.2.3 Extended instructions

Extended instructions are 3-byte instructions. The OP code is followed by two bytes that specify the address of the operand used by the instruction. The second byte contains the 8 high order bits of the address. Because 16 address bits are available, any one of the 65 536 memory locations can be selected. Thus, extended instructions have the advantage of being able to select any memory locations, but direct instructions only require two bytes in the program instead of three. Direct instructions also require one less cycle for execution so they are somewhat faster than the corresponding extended instructions.

30.4.2.4 Indexed instructions

As its name implies, an indexed instruction makes use of the *index register* (X). For any indexed instruction, the address referred to is the sum of the number (called the OFFSET) in the second byte of the instruction, plus the contents of X. The 6800 provides instructions to LOAD, STORE, INCREMENT, and DECREMENT X as well as the stack pointer (SP).

Index registers are useful when it is necessary to relocate a program. For example, if a program that originally occupied locations 0000 to 00CF must be moved or relocated to locations 0400 to 04CF, all addresses used in the original program must be changed. In particular, direct instructions cannot be used because the program no longer occupies lower memory. If X is loaded with the base address of the program (400 in this example), then all direct instructions can be changed to indexed instructions and the program will function as before.

The index register is often used by programs that are required to perform *code conversions.* Such programs might convert one code to another (ASCII to EBCDIC, for example), or might be used for trigonometric conversions where the sine or cosine of a given angle may be required.

30.4.2.5 Implied addressing

Implied instructions are used when all the information required for the instruction is already within the CPU and no external operands from memory or from the program (in the case of immediate instructions) are needed. Since no memory references are needed, implied instructions only require one byte for their OP code. Examples of implied instructions are CLEAR, INCREMENT and DECREMENT the accumulators, and SHIFT, ROTATE, ADD, or SUBTRACT accumulators.

30.4.3 Logic instructions

AND, OR, and EXCLUSIVE OR instructions allow the programmer to perform Boolean algebra manipulations on a variable, and to SET or CLEAR specific bits in a byte. They can also be used to test specific bits in a byte, but other logic instructions such as BIT, TEST, or COMPARE may be more useful for these tests.

30.4.3.1 Setting and clearing specific bits

AND and OR instructions can be used to SET or CLEAR a specific bit or bits in an accumulator or memory location. This is very useful in those systems where each bit has a specific meaning, rather than being part of a number. In the control and status registers of the PIA or ACIA, for example, each bit has a distinct meaning.

30.4.3.2 Testing bits

In addition to being able to SET or CLEAR specific bits in a register, it is also possible to test specific bits to determine if they are 1 or 0. In the peripheral interface adapter (PIA), for example, a 1 in the MSB of the control register indicates some external event has occurred. The microprocessor can test this bit and react appropriately. Typically the results of the test sets the Z or N bit. The program then executes a conditional branch and takes one of two different paths depending on the results of the test.

Accumulator bits can be tested by the AND and OR instructions, but this modifies the contents of the accumulator. If the accumulator is to remain unchanged, the BIT TEST instruction is used. This ANDs memory (or an immediate operand) with the accumulator without changing either.

30.4.3.3 Compare instructions

A COMPARE instruction essentially subtracts a memory or immediate operand from an accumulator, leaving the contents of both memory and accumulator unchanged. The actual results of the subtraction are discarded; the function of the COMPARE is to SET the condition code bits.

There are two types of COMPARE instructions; those that involve accumulators, and those that use the index register. Effectively, the two numbers that are being compared are subtracted, but neither value is changed. The subtraction serves to SET the condition codes. In the case of the CPX instruction, only the Z bit is significant, but for the COMPARE accumulator (CMP A or B) instruction, the carry, negative, zero and overflow bits are affected and allow us to determine which of the operands is greater.

The COMPARE INDEX REGISTER instructions compare the contents of the Index Register with a 16-bit operand. They are often used to terminate loops.

30.4.3.4 The TEST instruction

The TEST (TST) instruction subtracts 0 from an operand and therefore does not alter the operand. Its effect, like that of compares or bit tests, is to set the N and Z bits. It differs in that it always CLEARs the overflow and carry bits. It is used to set the condition codes in accordance with the contents of an accumulator or memory location.

30.4.4 Branch and jump instructions

BRANCH and JUMP instructions allow programs requiring *decisions*, *branches*, and *subroutines* to be written.

30.4.4.1 Jump instructions

One of the simplest instructions is the JUMP instruction. It loads the PC with a new value and thereby transfers or jumps the program to a new location.

The JUMP instruction can be specified in one of two modes, indexed or extended. The JUMP INDEXED (JMP 0,X) is a two-byte instruction.

The second byte or *offset* is added to the index register and the sum is loaded into the PC.

Extended jumps are 3-byte instructions, where the last two bytes are a 16-bit address. Since a 16-bit address is available, they allow the program to jump to any location in memory. They are very easily understood because they require no calculations.

30.4.4.2 Unconditional branch instructions

Branch instructions are 2-byte relative address instructions. The second byte contains a displacement. Normally, when a branch instruction is executed, the contents of the PC are incremented twice to point to the address of the next instruction. When the branch is taken, the PC is altered by the displacement, and the next op code is found at the address that equals the address of the branch instruction plus two, plus the displacement. The displacement is treated as an 8-bit signed number that is added to the PC. Displacements with MSBs of 1 are negative numbers, which cause the program to branch backward. Since the maximum positive number that can be represented by an 8-bit signed byte is 127_{10} and the most negative number is -128, the program can branch to any location between PC $+129$ and PC -126, where PC is the address of the first byte of the BRANCH instruction.

The unconditional branch, BRA, causes the program to branch whenever it is encountered. It is equivalent to a JUMP instruction, but since it is only a two-byte instruction, it is used when the location being jumped to is within the range of $+129$ or -126 bytes relative to the current PC address.

30.4.4.3 Conditional branch instructions

A conditional branch instruction branches only when a particular condition code or combination of condition codes is SET or CLEAR. Therefore the results of instructions preceding the branch determine whether or not the branch is taken. Conditional branches allow the user to write programs that make decisions and give the microprocessor its ability to compute.

30.4.5 Subroutines

A *subroutine* is a small program that is generally used more than once by the main program. Multiplications, 16-bit adds, and square roots are typical subroutines.

Figure 30.6 illustrates the use of the same subroutine by two different parts of the main program. The subroutine located at 200 can be entered from either location 0011 or 00CC by placing a jump to subroutine (AD, BD, or 8D, depending on the mode) in these addresses. The PC actions, as a result of the subroutine jump at 0011, are identified by the *a* in *Figure 30.6* and *b* identifies jumps from 00CC.

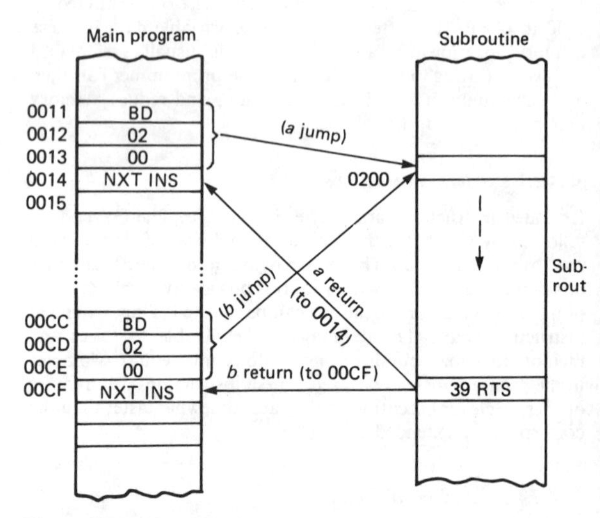

Figure 30.6 Use of a subroutine

After the subroutine is complete, the program resumes from the instruction following the location where it called the subroutine. Because the subroutine must return to one of several locations, depending on which one caused it to be entered, the original contents of the PC must be preserved so the subroutine knows where to return.

30.4.5.1 Jumps to subroutines

The JUMP TO SUBROUTINE (JSR) instruction remembers the address of the next main instruction by writing it to the stack before it takes the jump. The JSR can be executed in the indexed or extended mode. There is a branch to subroutine (BSR) that can be used if the starting address of the subroutine is within +129 to −126 locations of the program counter. The advantage of the BSR is that it requires one less byte in the main program.

30.4.5.2 Return from subroutine

The JSR instructions preserve the contents of the PC on the stack, but a return from subroutine (RTS) instruction is required to properly return. The RTS is the last instruction executed in a subroutine. It places the contents of the stack in the PC and causes the SP to be incremented twice. Because these bytes contain the address of the next main instruction in the program (put there by the JSR or BSR that initiated the subroutine), the program resumes at the place where it left off before it entered the subroutine.

30.4.5.3 Nested subroutines

In some sophisticated programs the main program may call a subroutine, which then calls on a second subroutine. The second subroutine is called a nested subroutine because it is used by and returns to the first subroutine.

30.4.5.4 Use of registers during subroutines

During the execution of a subroutine, the subroutine will use the accumulators; it may use X and it changes the contents of the CCR. When the main program is re-entered, however, the contents of these registers must often be as they were before the jump to the subroutine.

The most commonly used method of preserving register contents during a subroutine is to write the subroutine so that it PUSHes those registers it must preserve onto the stack at the beginning of the subroutine and then PULLs them at the end of the subroutine, thus restoring their contents before returning to the main program.

30.4.6 Other instructions

30.4.6.1 CLEAR, INCREMENT and DECREMENT instructions

These allow the user to alter the contents of an accumulator or memory location as specified. Those instructions referring to memory locations can be executed in extended or indexed modes only. They are simple to write, but require six or seven cycles for execution because the contents of a memory location must be brought to the CPU, modified, and rewritten to memory.

30.4.6.2 SHIFT and ROTATE instructions

The drawings in *Figure 30.7* show diagrammatically how these work. Note that they all use the carry bit, either for input, outpit, or both. Rotates are 9-bit rotates that combine the 8-bit accumulator and the carry bit.

30.4.6.3 COMPLEMENT and NEGATE instructions

The COMPLEMENT instructions invert all bits of a memory location. They are useful as logic instructions and in programs requiring complementation, such as the BCD subtraction program.

The NEGATE instructions take the two's complement of a number and therefore negate it. The negate instruction works by subtracting the operand from 00. Since the absolute value of the operand is always greater than the minuend, except when the operand itself is 00, the negate instruction SETs the carry flag for all cases, except when the operand is 00.

30.4.7 Assembler directives

When writing a program, options are available to enable the programmer to reserve memory bytes for data, specify the starting address of the program, and select the format of the assembler output. These options are called assembler directives. Assembler directives are written into the source program and interpreted during the assembly process.

30.4.8 The two-pass assembler

To convert a source program to an object program, this assembler reads the source program twice. It is known as a two-pass assembler. When it reads through the program the first time (called the first pass or pass 1), the number of bytes required for each instruction is determined and the PC is incremented accordingly. As each label or symbol is encountered, its address is stored away in the symbol table but nothing is printed (except errors if they occur). During the second reading (pass 2), the values of these labels are inserted in the object code and the offsets are calculated by the assembler for each branch instruction. It then prints the assembly listing. It also produces the object code in a form that allows it to be entered into the microcomputer.

30.4.9 Error indication

If the source program is not written in accordance with the rules specified for each type of statement, error lines will be printed. Note that each error has a number. The user's manual provides an explanation of each type of error as an aid to debugging. The most common causes of errors are improper spacing or the use of illegal characters.

Most of the errors are printed during pass 1 and allow the operator to abort the assembly to make corrections. If there are only a few errors and they are not serious, the operator may elect to let the assembler complete the listing anyway, so an object machine code file can be obtained and tested. Usually, other revisions are also required. In addition to the printing of an error number, which identifies the type of error, the original line is reprinted (unformatted) so the programmer can find any mistake.

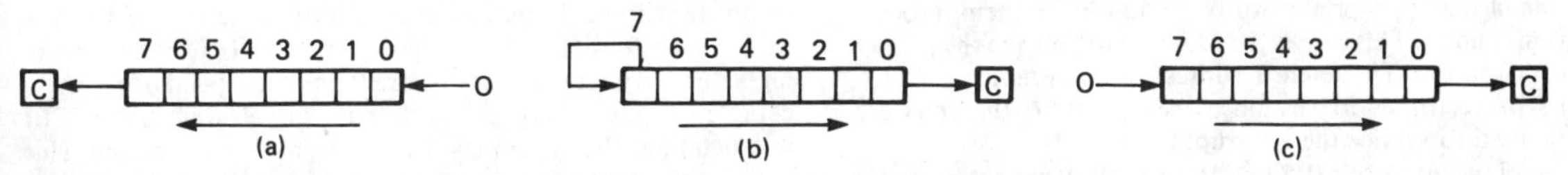

Figure 30.7 The 6800 shift instructions: (a) arithmetic shift left (ASL), (b) arithmetic shift right (ASR), (c) logical shift right

Many times the error is not obvious, and the rules may have to be reviewed before the reason is found.

A summary schematic of what an assembler does is shown in *Figure 30.8*.

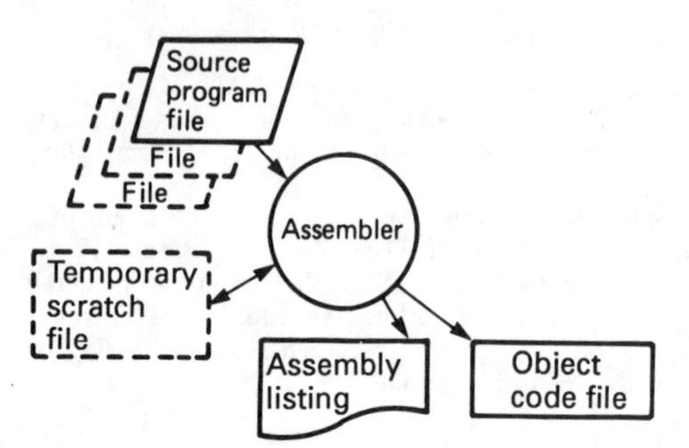

Figure 30.8 Assembler operation summary

30.5 Interrupts

One of the most important features of a microprocessor is its ability to control and act on feedback from peripheral devices such as line printers or machinery controllers. It must be able to sense the operation of the system under its control and respond quickly with corrective commands when necessary.

When conditions that require fast response arise, the system is wired so as to send a signal called an interrupt to the microprocessor. An interrupt causes the microprocessor to stop execution of its main program and jump to a special program, an interupt service routine, that responds to the needs of the external device. The main program resumes when the interrupt service routine is finished.

Important aspects of the 6800 interrupt structure are the stack concept, the use of vectored interrupts, and the interrupt priority scheme provided by the microprocessor logic.

The stack is an area in memory pointed to by the stack pointer (SP) register. The stack has three basic uses:

(1) To save return addresses for subroutine calls.
(2) To move or save data.
(3) To save register contents during an interrupt.

Use of the stack during interrupts is discussed in this chapter.

The 6800 uses four different types of interrupts: reset (RST), non-maskable (NMI), software (SWI), and hardware interrupt request (IRQ). Unique interrupt servicing routines must be written by the system designer for each type of interrupt used, and they can be located anywhere in memory. Access to the routines is provided by the microprocessor that outputs a pair of addresses for the appropriate interrupt. The two locations addressed must contain the address of the required interrupt service routine.

It should be noted that the actual ROM (or PROM) accessed may appear to be at some lower address as long as it also responds to the addresses of the vectors. When one of the four types of interrupts occurs, the microprocessor logic fetches the contents of the appropriate two bytes and loads them into the program counter. This causes the program to jump to the proper interrupt routine. The fetched addresses are commonly called vectors or vector addresses since they point to the software routine used to service the interrupt.

Three of the interrupts (RST, NMI, and IRQ) are activated by signals on the pins of the microprocessor, and the fourth (SWI) is initiated by an instruction. Each of these interrupts have similar, but different, sequences of operation and each will be described.

30.5.1 Reset (RST)

A reset is used to start the program. A LOW pulse on the RESET pin of the microprocessor causes the logic in the microprocessor to be reset and also causes the starting location of the program to be fetched from the reset vector locations.

The $\overline{\text{RESET}}$ line is also connected to any hardware devices that have a hardware reset and need to be initialised, such as the PIA. Grounding of the $\overline{\text{RESET}}$ line clears all registers in the PIA, and the restart service routine reprograms the control and direction registers before allowing the main program to start. The ACIA has no RESET pin and depends entirely on software initialisation. All software flags, or constants in RAM, must also be preset. If the system includes power-failure sensors and associated service routines, additional steps will be needed in the $\overline{\text{RESET}}$ service routine to provide the automatic restart function.

30.5.2 The IRQ interrupt

The IRQ interrupt is typically used when peripheral devices must communicate with the microprocessor. It is activated by a LOW signal on the $\overline{\text{IRQ}}$ pin (pin 4) of the microprocessor. Both the ACIA and PIA have $\overline{\text{IRQ}}$ pins that can be connected to the microprocessor when desired. Even though this line is pulled LOW the $\overline{\text{IRQ}}$ interrupt does not occur if the I bit of the Condition Code (CC) register is SET. This is known as masking the interrupt. The I bit is SET in one of three ways:

(1) By the hardware logic of the microprocessor as a part of the restart procedure.
(2) Whenever the microprocessor is interrupted.
(3) By an SEI (set interrupt mask) instruction.

Once SET, the I bit can be cleared only by a CLI (clear interrupt mask) instruction. Therefore, if a program is to allow interrupts, it must have a CLI instruction near its beginning.

30.5.2.1 Interrupt action

When an interrupt is initiated, the instruction in progress is completed before the microprocessor begins its interrupt sequence. The first step in this sequence is to save the program status by storing the PC, X, A, B, and CC registers on the stack. These seven bytes are written into memory starting at the location in the stack pointer (SP) register, which is decremented on each write. When completed, the SP is pointing to the next empty memory location. The microprocessor next sets the interrupt mask bit (I), which allows the service program ro tun without being interrupted. After setting the interrupt mask, the microprocessor fetches the address of the interrupt service routine from the IRQ vector location and inserts it into the PC. The microprocessor then fetches the first instruction of the service routine from the location now designated by the PC.

30.5.2.2 Nested interrupts

Normally an interrupt service routine proceeds until it is complete without being interrupted itself, because the I flag is SET. If it is desirable to recognise another IRQ interrupt (of higher priority, for example), before the servicing of the first one is completed, the interrupt mask can be cleared by a CLI instruction at the beginning of the current service routine. This allows 'an interrupt of an interrupt,' or nested interrupts. It is handled in the 6800 by storing another sequence of registers on the stack. Because of the automatic decrementing of the stack pointer by each interrupt, and subsequent incrementing by the RTI instruction when an interrupt is completed, they are serviced in the proper order. Interrupts can be nested to any depth, limited only by the amount of memory available for the stack.

30.5.2.3 Return from interrupt (RTI)

The interrupt service routine must end with an RTI (return from interrupt) instruction. It reloads all the microprocessor registers with the values they had before the interrupt and, in the process, moves the stack pointer to where it was before the interrupt. The RTI essentially consists of seven steps that write the contents of the stack into the microprocessor registers and an additional one that allows the program to resume at the address restored to the PC (which is the same place it was before the interrupt occurred). Note that the RTI restores the CC register as it was previously, and interrupts will or will not be allowed as determined by the I bit.

30.6 Development tools

Because the microprocessor (MPU) or microcomputer (MCU) is a programmable device requiring suitably generated machine code to function, development tools are available to generate this code. The range of tools available spans the low-cost evaluation module (EVM), through the medium-priced single-user system (mainly PC-based and having high-level language/assembler capability and in-circuit emulation or ICE) to the top of the range high-performance multi-user software development stations with mass storage, high-level language (HLL) compilers, assemblers and hardware development stations supporting ICE for a range of MPUs and MCUs.

Generally speaking the greater the cost of the development system, the more powerful it is in terms of monitoring editing and debugging facilities; software available; ease of operator interface and flexibility in program storage/retrieval.

Using sophisticated development tools will cut down development time, and EVMs are often used by the first time user to become acquainted with an MPU or MCU.

Development tools are supplied by either the microprocessor manufacturer or some of the instrument manufacturers. The tools from the instrument manufacturers support a range of processors from various microprocessor vendors. Microprocessor suppliers are able to offer comprehensive and up-to-date support for their own and future products. Many software houses offer language support for various devices on a wide range of computer systems.

30.6.1 Use of emulators in the development cycle

Before describing a typical development cycle it is essential to understand what an 'emulator' is and does.

An emulator is called an emulator (and frequently an 'in-circuit emulator') because it replaces the MPU/MCU in the target system, performing the functions of the device it replaces. The emulator is connected to the target by removing the MPU/MCU device from the system and substituting a mating plug which is cabled to the emulator. An emulator system typically consists of a chassis which is cabled to a pod (emulator module) which is in turn cabled to the target system MPU/MCU socket.

Confusion often exists as to the difference between an emulator and a 'simulator'. The emulator, as described above, is a combination hardware/software instrument which endeavours to replicate all functions, including timing, of the target MPU/MCU in its hardware/software environment.

Conversely, a simulator is typically a software tool which tries to provide non-real-time duplication of a processor's instruction set, enabling a software engineer to 'step through' the logic of the engineer's code prior to its execution on the ultimate target system. This implies that a simulator can execute code on a system or processor other than the target MPU/MCU.

In addition to functioning as the replaced MPU/MCU, the emulator provides the development engineer with a window into the operation of the target system. This window allows the engineer to observe what is happening during actual operation to a degree not possible without an emulator.

A fundamental capability of the emulator is to allow the user to exercise the target system by starting and stopping the execution of the target code. The user typically has wide flexibility of start and stop locations and can designate a variety of stopping conditions.

When target execution is stopped, the user may examine or alter target memory or processor registers. The user may also 'step' through target code instructions one at a time and examine the results of the execution. Alternatively, when a stopping condition is specified, the processor is instructed to 'free run', under emulation, until that condition is encountered.

The stop and examine method is an effective mechanism for hardware and software debugging and provides an efficient means of isolating and resolving problems typical in early system development.

Figure 30.9 shows schematically how the host computer, hardware development station, emulator module and target system fit together in a typical development environment.

30.6.2 Development cycle using the Motorola HDS300

Currently MPU/MCU development ranges from the small single-MCU application, requiring code generation of typically 12 kbytes or less, up to multi-MPU systems involving perhaps millions of bytes of code writing.

A small application might typically be written in assembler. The large application would involve high-level languages, an assembler and possibly an operating system (OS). This would involve many programmers with different skills. The individual program modules would then require integration (linking), which would involve yet another set of specialists.

So although there is no 'typical' development cycle certain steps are necessary and therefore common to most applications; it is assumed the system specification is complete.

Development of an application program generally occurs in two distinct phases, creation and debug. At the outset the emulator can be used for testing original system algorithms and input/output (I/O) parameters. Following this the host's editing facilities are used to prepare a source program which is then compiled, assembled, linked and downloaded into the emulator. The application is then debugged using the emulator as described above. Needed changes, if minor, can be made using assembler/disassembler facilities available on the emulator. The latest revision of the software can then be stored on the HDS300's $5\frac{1}{4}$ inch floppy disk which provides greater stand-alone independence and reduces debug time by permitting faster reloads. Major code changes will eventually require recompilation and linking.

An HDS300 system provides resources which allow an application to be configured and testing to begin before all prototype components are available. A clock is available which can be used until the prototype clock is operational. Similarly, emulation memory is available which can be used to replace system memory, and an emulator pod is used in place of the system MPU/MCU. Software can then be exercised on the

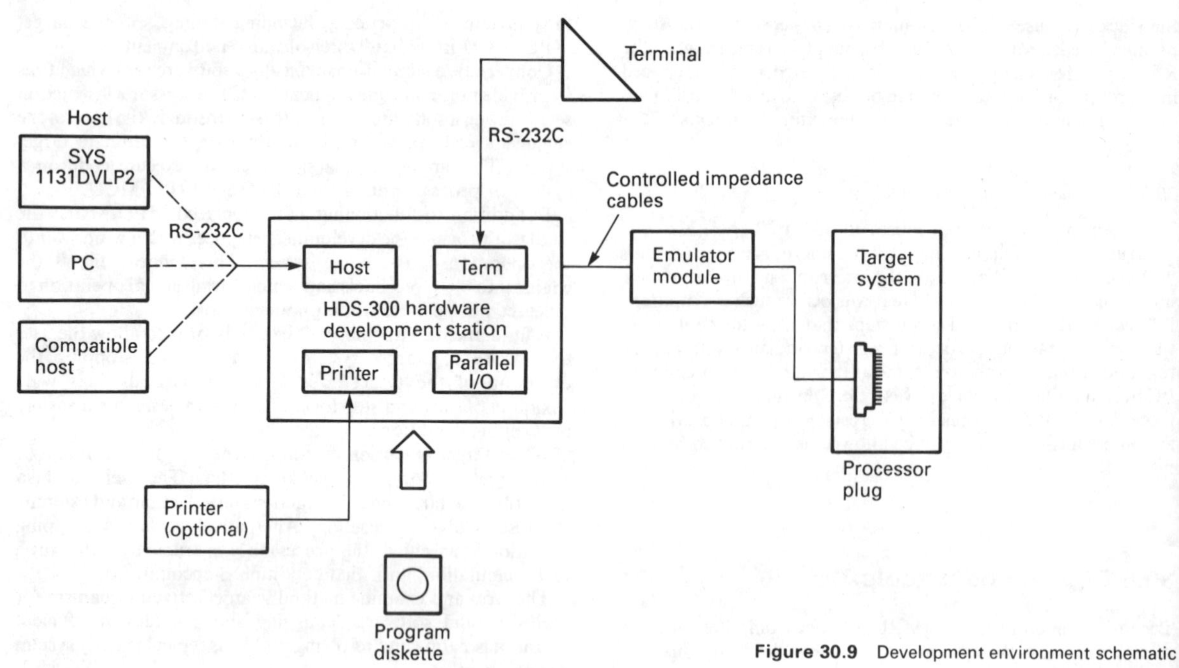

Figure 30.9 Development environment schematic

emulated hardware and needed changes to hardware quickly determined, made and tested until the configuration is fully proven. A complete evaluation of the application system can then be made.

30.6.3 Motorola MC68HC05 EVM evaluation module

At the other end of the scale to the HDS300 is the EVM. The EVM shown in *Figure 30.10* provides a low-cost tool for designing, debugging and evaluating target systems based on the MC68HC05 MCU family.

The MCU devices are emulated by the EVM resident MC68HC05 MCU. Entering data, program debugging and EEPROM MCU programming are accomplished by the monitor ROM firmware via an external RS-232C compatible terminal.

MCU code may be generated using the resident one-line assembler/disassembler or may be generated, for example, on a

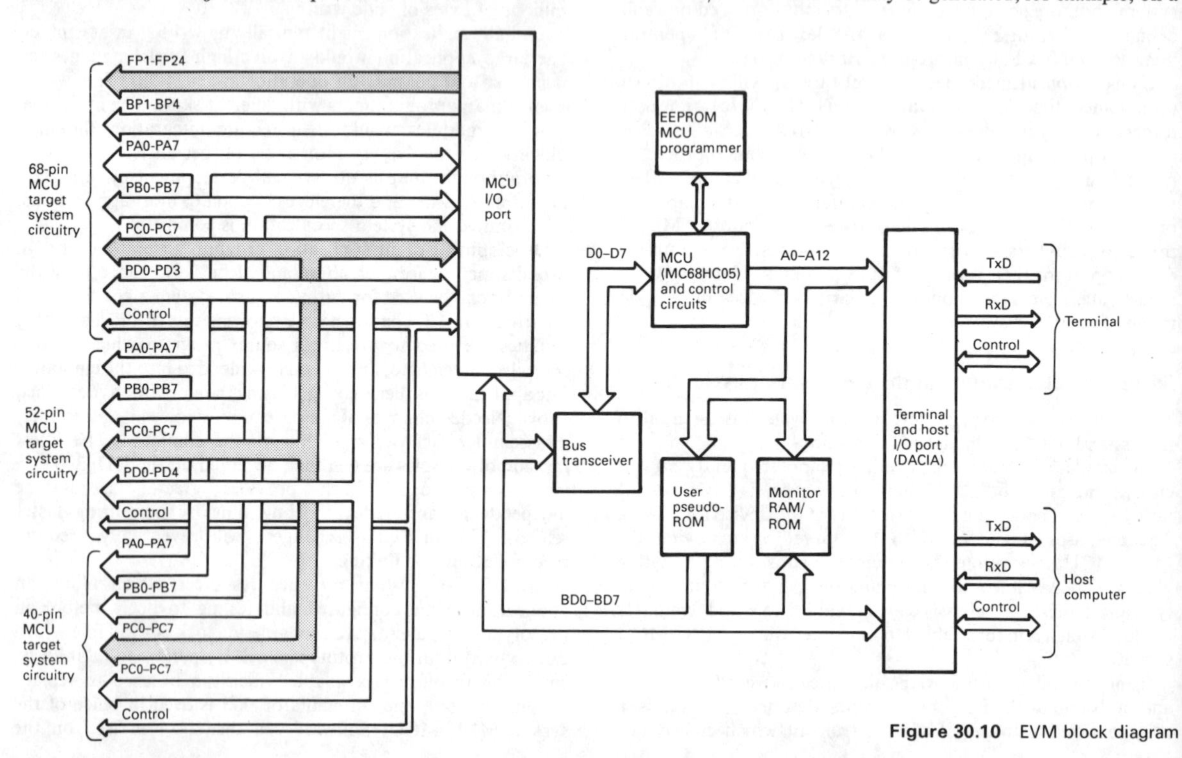

Figure 30.10 EVM block diagram

PC and downloaded to the user program RAM. User code may then be executed using various debugging commands in the monitor. By providing all the essential MCU timing and I/O circuitry, the EVM simplifies user evaluation of the prototype hardware/software product.

30.7 Choosing a microprocessor

The application determines the choice of microprocessor. A high-performance graphics workstation will require the power of a 32-bit processor, whilst a mobile telephone will require a single-chip MCU-type solution. Whereas the workstation manufacturer is primarily concerned with performance, since the chip cost is minimal compared to the overall system cost, the mobile telephone market is highly competitive and price sensitive, so the cost of the processor is very important.

The cost of a device is inversely related to the volume of product used and the age of the device since its first introduction. For prototyping and production volumes of up to about 1000 per annum one would use either microprocessors with program stored in external EPROM or microprocessors with their own EPROM/EEPROM on chip. For greater volumes the cheapest solution is to 'freeze' the software into the silicon by use of on-chip ROM. For specialist high-volume applications, special devices can be designed, providing extra features such as phase-locked loop, analogue-to-digital converters, display driving, power fail backup, etc.

Other considerations are availability of compatible interface circuits; the way the device is packaged, whether dual in line (DIL) for high-volume low-cost, chip carrier for limited physical space, or ceramic for high reliability; the power consumption, which is a function of fabrication technology; availability of software and development tools, which includes good manufacturer support and awareness of the semiconductor vendor's long-term plans to ensure that designs can be sensibly and cost-effectively upgraded; and, finally, existence of second sources to support the prime manufacturer.

30.8 Board-level products

Since their inception, microprocessors have been supplied as board-level products. The system designer may find it more cost effective to integrate these standard boards rather than design his or her own product at the component level. 'Bus' structures are the links between the boards. Examples of such busses are PC bus, S-100 and Multibus. The latest of such busses is VMEbus, which is currently the only standardised bus system, and provides hardware and software compatibility for 8- to 32-bit applications across the 68000 processor family. The greatest advantage of a standard bus is its modularity. To a core system, the designer need only add those software and hardware components required for the application. The Motorola SYS1131DVLP2, shown schematically in *Figure 30.11*, is such an example. The system is used as the host computer for the HDS300 emulator and shows how standard boards (MVMExxx series) along with a standard operating system, in this case the UNIX* SYSTEM V/68, can be combined to produce a high-performance multi-user software and hardware development workstation.

A picture of an MC68020-based VME bus CPU card is shown in *Figure 30.12*.

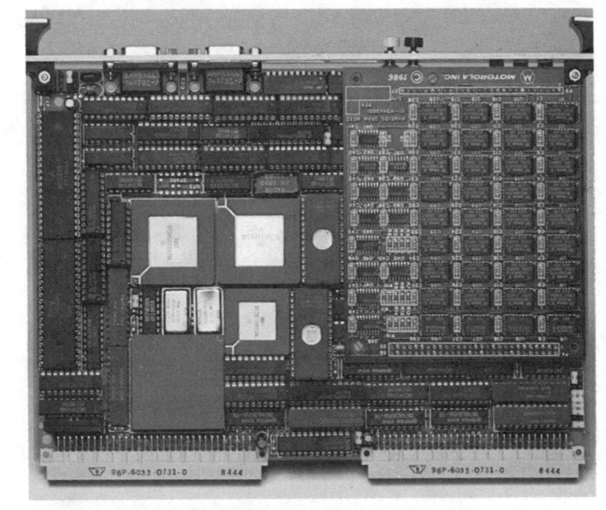

Figure 30.12 MC68020-based VME bus CPU card

*UNIX is a trademark of AT & T.

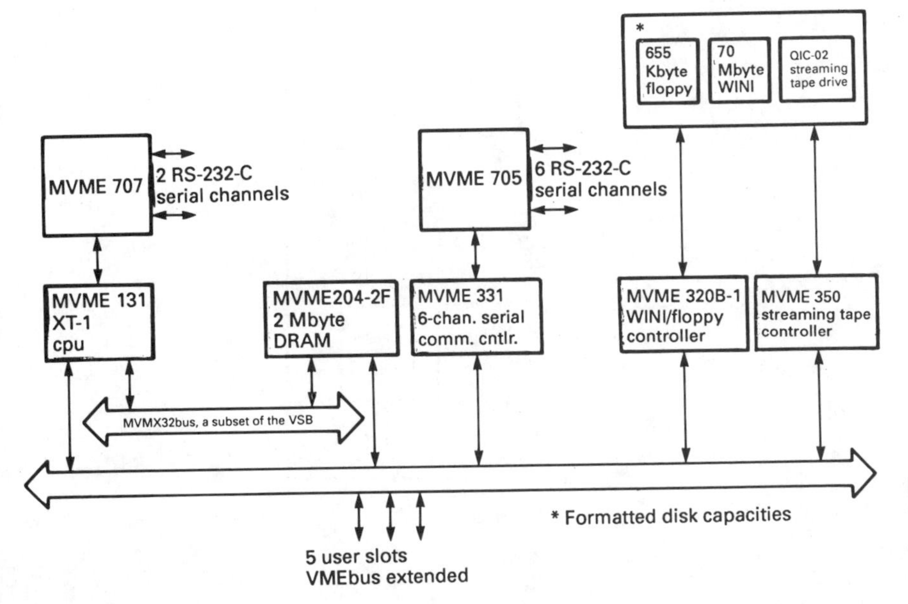

Figure 30.11 SYS1131DVLP2 schematic. (System V/68, HDS300, MVMX 32 bus and VME module are trademarks of Motorola Inc.)

Further reading

BAUER, J. R., 'Alterable microprocessors: tailoring chip design to meet applications', *Digital Design*, August (1982)

CLEMENTS, A., 'An introduction to bit-slice microprocessors', *Electronics & Power*, March (1981)

CUSHMAN, R. H., 'CMOS microprocessor and microcomputer ICs', *EDN*, September (1982)

DETTMER, R., 'Making RISC respectable – the Motorola 88000', *IEE Review*, June (1988)

DONNELLY, W., 'Single chip microcomputers and customised LSI', *Electronics Industry*, March (1982)

GALLANT, J., 'Enhanced microcontroller chips', *EDN*, 21 January (1988)

GRAPPEL, R. D., 'Design powerful systems with the newest 16-bit microprocessors', *EDN*, September (1982)

GRAPPEL, R. D., 'Instruction-set power makes a microprocessor's job easier', *EDN*, November 10 (1982)

HAMILTON, A. *et al.*, 'Lower cost 32 bit processor will open new markets', *Electronic Product Design*, January (1988)

JOHNSON, R. C., 'Operating systems hold a full house of features for 16-bit microprocessors', *Electronics*, March 24 (1982)

JONES, D., 'The MC 88100 RISC processor', *Electronic Engineering*, May (1988)

MOTOROLA, *8-bit Micro and Peripheral Data Book* (1988)

MOTOROLA, *Single Chip Microcomputer Data Book* (1988)

MOTOROLA, *VME Bus Specification CI* (1988)

TILL, J., '32-bit microprocessors', *Electronic Design International*, December (1988)

TITUS, J., 'EDNs 15th annual MUP/MUC chip directory', *EDN*, October (1988)

WILSON, R., '16-bit microcontrollers with a leading role in embedded computing', *Computer Design*, 15 November (1988)

31 Application-Specific Integrated Circuits

J Berry, BSc (Hons)
LSI Logic Ltd

Contents

31.1 Introduction

Integrated circuits destined for one specific application are termed ASICs—application-specific integrated circuits. These devices form one of the fastest growing sectors of the electronic market (see *Figure 31.1*). With well over 100 companies competing in this area, it is also the fastest evolving and most exciting segment. *Figure 31.2* shows the changing technology and maximum complexity associated with the ASIC market. The proliferation of technologies and terms used to describe the various product offerings should not be a barrier to using and taking advantage of ASICs.

This chapter defines what an ASIC is, the potential advantages ASICs offer over other design approaches, the method of design, how to choose an ASIC and the factors that must be considered when selecting both the most appropriate ASIC and the vendor of that product.

31.2 Arrays and cells

ASICs fall into two quite distinct groups—arrays and cells. Arrays, more frequently called gate arrays, are personalised by the final processing steps only. The other group consists of cell-based circuits often called standard cells. They differ from arrays in that all processing steps are used for personalisation. *Figure 31.3* shows the two groups.

31.2.1 Arrays

Arrays were first conceived in the early 1970s when Ferranti introduced their ULAs* (uncommitted logic arrays). These chips

* ULA is a registered trademark of Ferranti

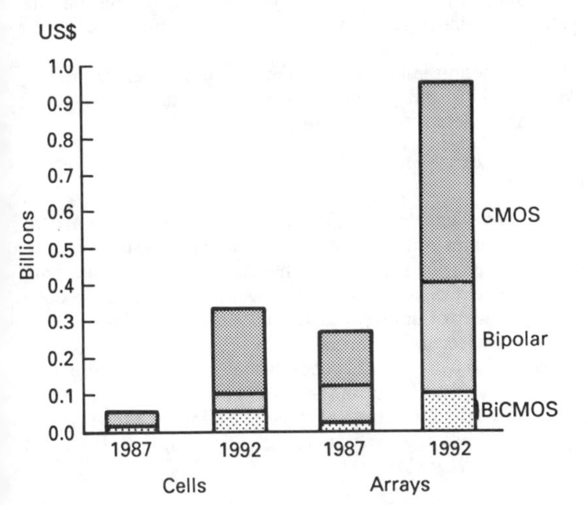

Figure 31.1 European consumption of ASICs (Source: SEMSTAT 1987)

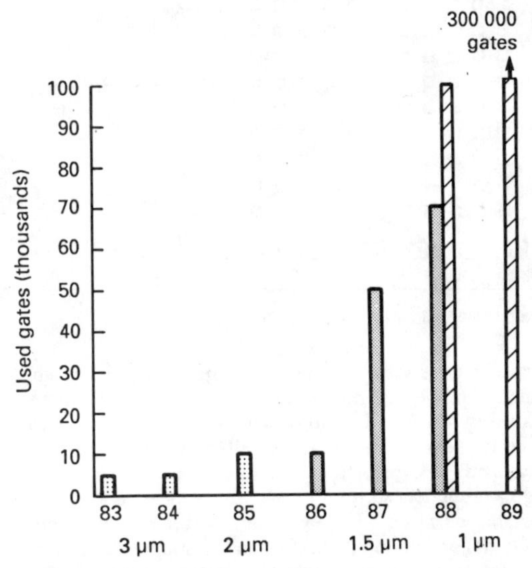

Figure 31.2 Single chip complexity

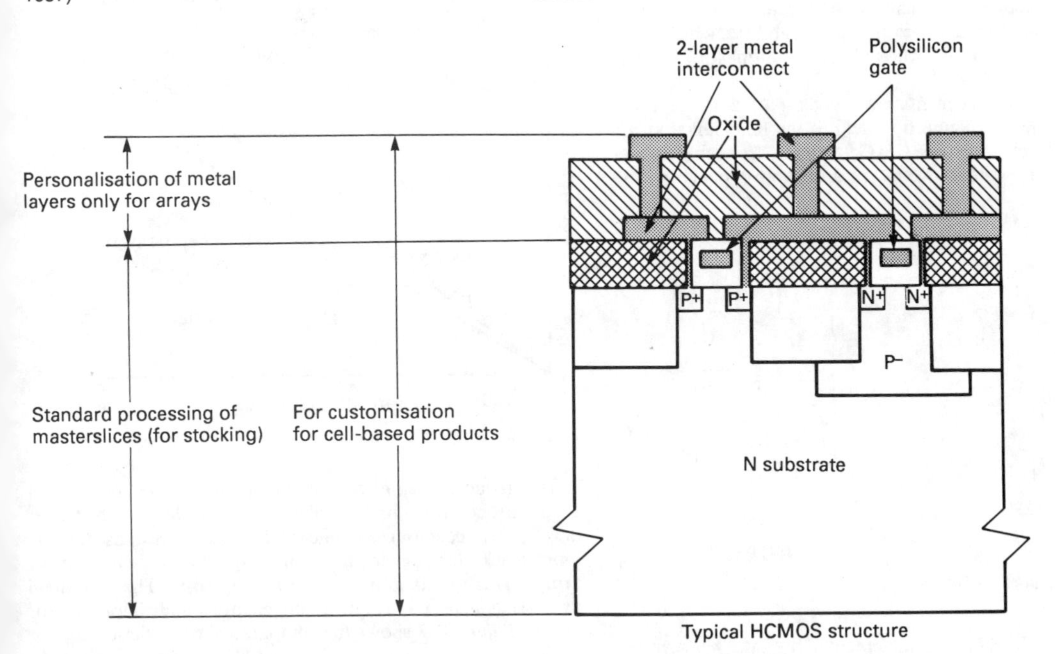

Figure 31.3 Array- and cell-based processing

consist of a regular array of uncommitted (or unconnected) logic elements. A peripheral area surrounds the internal array structure to provide a signal interface and power supply connections. *Figure 31.4* shows this arrangement, unchanged in principle to this day.

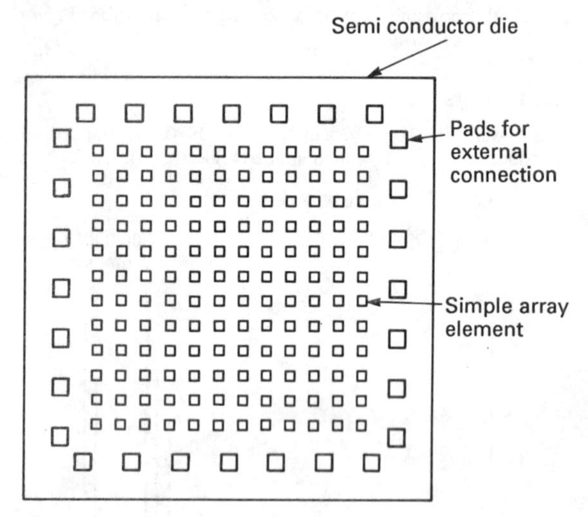

Figure 31.4 Basic array structure

Arrays are processed in high volumes up to the point at which personalisation begins and they are then stockpiled. Economies of scale result from this approach. The limited processing required for personalisation also results in minimal turnround time for array designs.

Peripheral buffers surround the array elements and are used for several purposes. They can be used as power supply connections or as input buffers (also supplying electrical protection for the internal gates) or output buffers. Tristate capability is usually possible, as are bidirectional buffers. Different manufacturers will have specific rules as to possible options, electrical characteristics and the capabilities which may include pull ups, pull downs, and open drain (or source, collector).

The internal array elements each consist of a number of components. The interconnection defines the final function of each element. An example for a CMOS array where each element consists of four transistors (two n-type and two p-type) is shown in *Figure 31.5*.

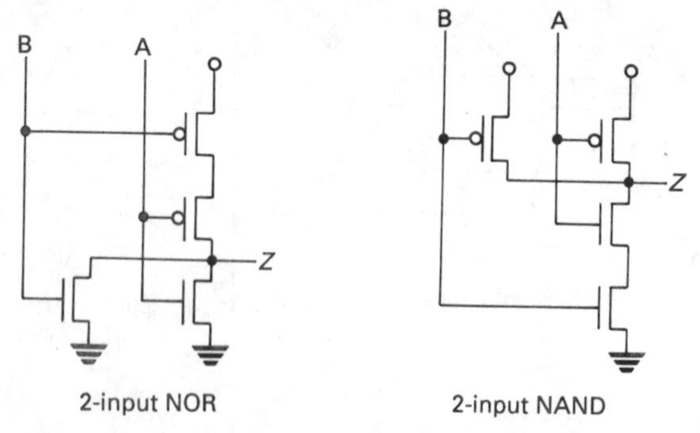

Figure 31.5 Examples of element interconnections

The internal architectural arrangement of the gates has several possibilities. One is shown in *Figure 31.4*, with room around each uncommitted 'gate' on all sides. An alternative is to produce rows or columns of uncommitted gates with dedicated routing channels between them, known as a channelled array.

Routing is achieved in one direction in the 'routing channels'. The other direction is routed across the columns in unused routing slots. Additional routing can be made but at the expense of some gates. This approach to routing can be extended to both directions allowing the sea of gates or Channel-Free* architecture where no dedicated routing channels are included. Clearly gates are lost to routing but the extra gates gained using such an architecture far outweigh these losses. The most important factor is the actual number of usable gates.

Interconnection between components and gates may be achieved using metal, polysilicon or diffusions. The most predictable is metal and this is certainly the most common today. In many cases two layers of metal are used, occasionally three. For designs up to about 50 MHz clock the various interconnection methods generally affect only utilisation (the maximum number of usable gates possible compared to the number available).

Higher performance devices can require separate routing layers for clock and power and need a very full review before designing in detail.

31.2.2 Cells

Cell-based designs do not really differ a great deal from arrays. The most major difference is that all layers are customised as opposed to just the top layers. This allows extra flexibility not possible for arrays. First only used gates are made (arrays usually have a number of unused transistors 'left over'). This allows a continuous spectrum of device sizes.

However, the argument is complicated by being able to alter the size of transistors used in the gates. This produces a gate that uses less area than for arrays. *Figure 31.6* shows the relationship between internal gate area and used gates with both fixed and variable transistor sizes.

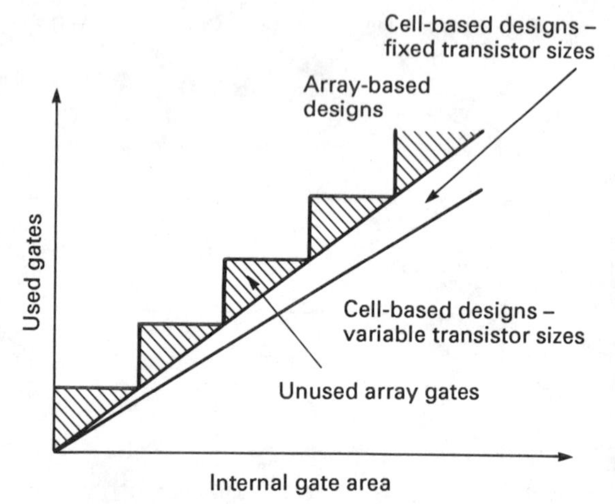

Figure 31.6 Area usage curve

This reduced area of silicon can result in lower production costs, particularly at higher volumes. This is offset by the higher development cost for cells. Since all layers are customised, many more masks (about ten more) are required for cells than for arrays. This results in additional tooling costs. The additional processing steps also result in longer turnround times than for arrays. *Figure 31.7* shows the comparative cost relationship for

* Channel Free is a registered trademark of LSI Logic Corporation

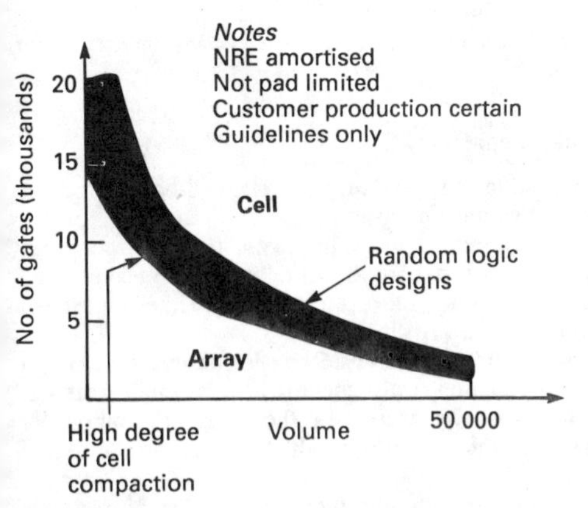

Figure 31.7 Comparative costs of array-based as opposed to cell-based designs

the total spend on a project, totalling the development cost and production costs across the lifetime of the product.

31.3 The reasons for choosing an ASIC

The electronic part of a system can be implemented in a variety of different ways, each with its own set of advantages. ASICs offer advantages in cost, performance, time to market, power consumption, reliability and size.

Cost The cost of manufacturing and maintenance are dramatically reduced because of fewer parts on smaller or fewer boards. This leads to less inventory, assembly and testing. The reduced power consumption also reduces costs of power supplies and cooling requirements whilst increasing reliability. An example of an actual product costing is given in *Table 31.1*.

Table 31.1 ASIC costing example: typical 4000-gate system, 10 000 production volume

	TTL-based system 65 SSI/MSI TTL ICs ($)	*ASIC system one ASIC chip* ($)
Breadboarding	6 000	0
Chip development	0	45 000
PC board (2¢/sq in)	162 000	90 000
Components	390 000	300 000
Assembly ($2/part)	1 300 000	20 000
	1 858 000	455 000

Source: EDN 1984, Advanced Technology Research 1986 and LSI Logic

Size When compared to standard product implementations the ASIC offers a greatly reduced size. Compare the external size of a 30-gate TTL MSI (medium scale integration) part to that of a microprocessor of 1000 or more times the complexity. This can result in significant weight savings as well.

Performance The performance of an ASIC-based system can be dramatically improved over the standard product approach. Removal of buffers is a clear benefit. However, the ability to optimise the design, potentially with added features, also brings improvements to performance. Early access to the next generation of technology and reduction in power consumption increases performance still further.

Time to market The design automation possible with ASICs allows for very rapid design when compared with alternative approaches. The surety of getting good parts back in a short time after design completion (one week is quite possible from some vendors) allows early entry of the final product to the market place. The proprietory features are also well protected, ensuring that the competitive edge is retained for as long as possible.

31.4 The design process

The essential steps of the design process are shown in *Figure 31.8*.

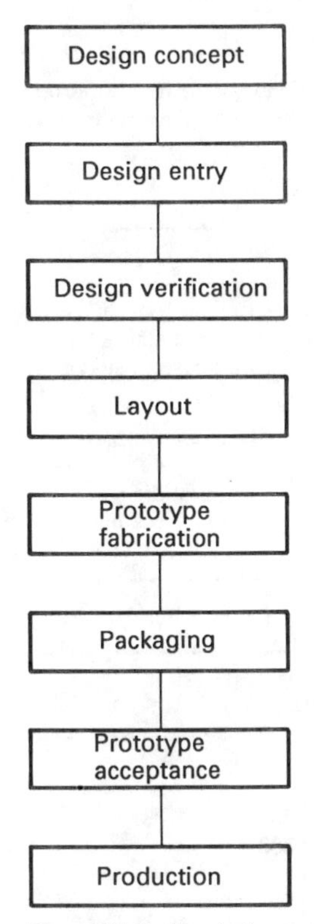

Figure 31.8 The design and manufacturing process

The design concept is arrived at gradually. The method of implementation and technology chosen will affect the feasibility of the project. Software tools such as behavioural simulators allow a very high-level description of the design or entire system to be made. This description may be simulated to check the concept. The model can then be split into appropriate sections and the descriptive detail increased. Since each piece can be worked on separately the design effort is eased and design time reduced. Continuing this process leads to a fully described and optimally partitioned system.

Design entry consists of conversion of the behavioural description to the ASIC vendor's library elements. Ideally the behavioural simulator allows these gate-level implementations to be substituted for their equivalent behavioural block allowing full proof of a correct conversion.

Verification of this design can now proceed. Functionality, performance, testability and package must all be carefully verified against the original specification.

Layout must now be completed, preferably by the ASIC vendor, with full checking after this process against the original specification. Production test patterns are extracted at this stage.

After full approval has been given the prototypes are fabricated in the vendor's facility. The test sequence is then used on the wafers (to determine good devices) before the wafer is sawn into die. Good die are packaged to the required pinout and then fully tested against the production test patterns.

On delivery the prototypes must be very carefully checked. It is only on this final acceptance that all the design details can be released to production.

Vendors each have their own specific flow, but all adhere in general to these steps. Where the sequence is not followed the risk of design failure is greatly increased.

31.5 Selecting the ASIC

With a wide range of different ASICs available it is important that a careful selection is made as to the most appropriate ASIC. This section describes the factors to be considered in that selection process.

31.5.1 General considerations

Select array products first and only move to cell-based products when a clear advantage exists.

Array products provide access to the most recent technology and provide for very rapid product development. When product lifetimes are short and upgrades or redesigns are likely, the lower development cost is a benefit.

Cell-based products are more suitable for designs that have a high-volume production requirement (when the smaller die size allows a lower piece part cost) or the degree of complexity exceeds that available on array products at the time of design definition.

Table 31.2 compares the product types. A list of available products from LSI Logic is included in *Table 31.3* as an example of the devices available from one vendor.

Table 31.2 Product type comparison

	Array products	*Cell-based products*
Description	Metal programmable	Full custom diffused and metalised
	Pre-diffused masterslice	Full custom silicon
Advantages	Fastest development cycle	Highest complexity
	Smallest feature size	Smallest die size
	Lower development costs	Lower piece parts costs

31.5.2 System clock speed

A smaller MOS device geometry will produce shorter gate delays and improve device performance. Thus an approximate, but very rapid, assessment of the technology required can be made simply by using the system clock rate.

Table 31.4 indicates the MOS technology geometries that will support a range of system clock rates. Once the selection of technology has been made a more detailed analysis of critical paths (see *Section 31.5.3*) is required to confirm the choice. The availability of such high-performance CMOS products minimises the design effort needed for high-performance chip design.

Table 31.3 Product range from LSI logic

Device number	*Estimated usable gates*	*Available gate complexity*	*Total pads*	*Max I/O pads*
Channel-Free arrays				
LMA9000—micro array				
LMA9020	700	1968	44	41
LMA9033	1200	3286	58	55
LMA9050	1750	4992	70	67
LMA9072	2500	7238	86	80
LMA9095	3300	9504	98	92
LMA9141	5000	14124	118	110
LMA9190	6700	19000	138	130
LMA9239	8400	23908	154	144
LMA9284	10000	28388	168	158
LCA10000—compacted array				
LCA10026	10000	25740	168	154
LCA10038	15000	37932	204	184
LCA10051	20000	50904	234	214
LCA10075	30000	74970	282	262
LCA10100	40000	100182	326	306
LCA10129	50000	129042	368	348
LCA100K—compacted array plus				
LCA100139	60000	139104	340	320
LCA100188	80000	187748	392	372
LCA100237	100000	236880	436	416
Cell-based				
LCB15—structured cell				
LCB1501	—		20	18
⋮	⋮		⋮	⋮
LCB1560	100000		348	256

Table 31.4 A guide to potential device technology

System clock rate (MHz)	*Largest geometry* (μm) *Drawn*	*Largest geometry* (μm) *Effective*	*Typical relative gate delay* (ns)
<25	3.0	2.4	2.4
<40	2.0	1.5	1.4
<60	1.5	1.1	1.0
<80	1.5	0.9	0.57
<100	1.0	0.7	0.46

The smaller the geometry used the lower the power consumption of the final device. An example is a 33% reduction when 1.5 μm technology is used in place of 2.0 μm technology for an identical function. This can often allow the selection of a lower cost plastic package.

After processing, the effective length is lower than the original drawn gate (see *Table 31.4*). The performance is related to the effective length (L_{eff}) that results after processing.

The use of bipolar transistors in standard bipolar or BiCMOS processes provides for very high performance where designs of 1 GHz or higher are possible. Also different technologies such as SOS (silicon on sapphire) or materials such as GaAs (gallium arsenide) can produce devices of higher performance than CMOS. However, CMOS is by far the dominant ASIC technology.

31.5.3 Critical path performance, design architecture

A circuit will have a path (or paths) that is the limiting factor to performance. The total propagation delay down a path is the sum of the individual cell delays on the path, considering the polarity of the output change. The rise and fall propagation delays of each cell depend on a number of factors, which for estimation purposes can be considered to be technology, device and fanout loading. Fanout loading is the load presented to the output of a gate by all of the gates connected to that output and the connections themselves. For the example in *Figure 31.9* the delays have been calculated for a particular path between two flip-flops for a 0.9 μm CMOS technology.

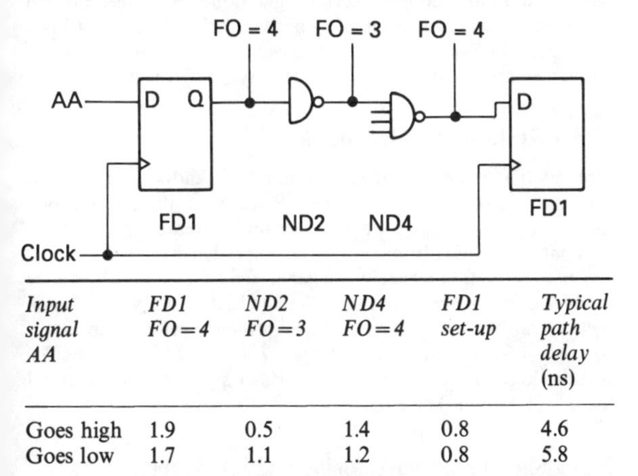

Input signal AA	*FD1 FO=4*	*ND2 FO=3*	*ND4 FO=4*	*FD1 set-up*	*Typical path delay* (ns)
Goes high	1.9	0.5	1.4	0.8	4.6
Goes low	1.7	1.1	1.2	0.8	5.8

Figure 31.9 Calculating critical path delay

The internal elements use standard drive gates in this example. Substituting high drive gates will improve critical path performance. Careful consideration of the design architecture (pipelining, parallel processing) can significantly improve the performance obtainable.

The use of a chip floorplanner is frequently advantageous to provide the designer the ability to pre-place critical hierarchical blocks to give optimum performance with minimum interconnect delays. This high-level approach is particularly useful for larger designs.

31.5.4 Operating conditions

Device performance is affected by the supply voltage, the junction temperature and the processing tolerance. These factors are described in the relevant product data sheets. The processing tolerances are technology dependent, so check these carefully. The effect on delays is global and a simple multiplying factor is applied to the complete delay path. *Table 31.5* summarises these factors for the LSI Logic LMA9000 series.

For example shown in *Figure 31.9*, the worst case commercial path delays would be 8.6 ns (AA goes high) and 10.8 ns (AA goes low) (i.e. the nominal delay × 1.86).

Junction temperature should initially be assumed to be the same as the ambient temperature. However, where timing is marginal a more accurate calculation should be performed including the effect of temperature rise caused by power dissipation.

All these factors (and others) must be taken into account by the design system used, e.g. LDS* for LSI Logic, to ensure the provision of accurate and dependable delay predictions for all paths including the critical ones.

31.5.5 Gate count

Taking a block diagram of the circuit under consideration and partitioning the circuit hierarchically allows a rapid estimation of the gates required. Specific data books and design manuals provide the gate counts for the majority of the smaller elements (e.g. NAND gates and flip-flops) used in a design. Gate counts for many of the larger standard functions, e.g. 29xxx, 82xx and generic functions, may be provided by some vendors. If the design being considered uses standard products, this will be particularly useful, saving a very large amount of design effort and ensuring correct functionality (see *Section 31.5.6*). Having established a gate count for designs consisting solely of random logic, use the following product family dependent guidelines:

(1) *Arrays* Use the available usable gates published in the data sheet to determine the minimum size practicable.
(2) *Cell-based designs* Area calculations are required (see the relevant data sheets or consult the vendor).

Table 31.3 indicates estimated usable gates available for some products. This is a guideline only as the true number is design dependent. A review by the vendor is recommended and some would usually insist upon this design review. Many designs do, however, have elements with a higher density. The following sections examine these elements and the effect they have on design size.

31.5.6 Standard products and compiled functions and cells

Small standard products (i.e. 7474) are generally available in the macrocell library. Medium sized standard products (i.e. 4-bit counters) are frequently available as macrofunctions in the ASIC vendor's library. Medium to large functions (i.e. ALUs, barrel shifters, multipliers and many other industry standard functions) are available in extended libraries as megafunctions. They are provided with complete test patterns greatly easing test program generation for the total design. For Channel-Free arrays these may also be available as metal megacells optimised for size and

* LDS is a registered trademark of LSI Logic Corporation

Table 31.5 Delay multiplier for different operating conditions (LMA9000 series)

Worst case conditions	*Factors*			*Combined multiplier*
	Supply voltage (V)	*Junction temperature* (°C)	*Processing tolerances*	
Commercial	5±5%	−0 to +70	0.6–1.5	×1.86
Industrial	5±5%	−40 to +85	0.6–1.5	×1.96
Military	5±10%	−55 to +125	0.6–1.5	×2.38

performance. This optimisation in size is not practicable where routing channels are used because of the varying channel width with array complexity. For cell-based products these functions may be available as fully diffused megacells.

When a particular element is available as a megacell the effective logic density is significantly increased.

Consult current lists for available standard functions and megacells. *Table 31.6* summarises the types of available logic building blocks. *Table 31.7* gives gate counts for some standard functions.

When a required function or cell is not contained in a library or some customisation is necessary, consider constructing that function from existing cells and functions or compiling a new cell. A range of compilers is available (see *Table 31.8*). These may be used to advantage reducing design size and enhancing performance. The contents of the libraries are constantly enlarged based on the varying requirements of designers. They may thus become the generic functions described earlier.

Table 31.6 Logic building blocks

Macro cells	*Mega cells*
Low-complexity building block	High-complexity building block
Examples: NAND, NOR, FLIP-FLOP	Examples: RAM, ROM, ALUs, UARTs
Fixed transistor interconnection	Fixed and optimised interconnection
Fully characterised	Fully characterised
Macro functions	*Mega functions*
Medium-complexity building block	High-complexity building block
Examples: COUNTERS, ADDERS	Examples: RAM, ROM, ALUs, UARTs
Combinations of macrocells	Combinations of macrocells
Performance layout dependent	Performance layout dependent

Table 31.7 Gate counts for some standard functions

Function	*Description*	*Gate count*
2901	4-bit slice ALU	1006
2910	12-bit microprogram controller	1182
8254	Programmable interval timer	3046
8259A	Programmable interrupt controller	1246
	32-bit carry select adder	643
	16×4 content addressable memory	636
	16×16 two's complement multiplier	2099

Table 31.8 Compilers

Logic compilers	Multipliers, adders, shifters, counters and other logic functions
Schematic compilers	Netlist to schematic conversion
Memory compilers	RAM with multiport options
Logic synthesis interface compilers	ROM with very wide words practicable, conversion of state machine tables, Boolean equations, truth tables, fuse maps and PLAs to optimised logic equivalents
Data path compiler	Data path architecture design and conversion to megafunctions and megacells

31.5.7 On-chip storage requirements

RAM and ROM can be implemented on ASICs and provide fast local memory. Dedicated blocks of memory (megacells) are very dense and fast. Select product type based on the amount of RAM or ROM required. *Table 31.9* shows examples of memory limits on ASIC products, and *Table 31.10* gives some performance examples.

Small amounts of RAM and ROM can be converted to gates using latches and combinatorial logic alone. Sections of ROM can be converted to optimised combinatorial logic using logic synthesis (LLS is one of LSI Logic's compiler tools). Such converted areas can be included on any ASIC product.

31.5.8 Replicated logic blocks

Designs that have a significant amount of random logic that is formed from repeated sections can benefit from the creation of a metal megacell. The repeated section is treated as a single element and typically has a greater logic density.

Since the metal megacells are generated as tiles, the allowance for block interconnect is minimal. The identical layout of each section results in a repeatable performance which can be of significant performance advantage. *Table 31.11* compares the relative logic density for all the various styles of implemented logic on Channel-Free arrays.

31.5.9 Signal pins, partitioning and interfacing

The periphery of every die contains the buffers and associated bonding pads used in packaging. The number of pads a device will need is determined by the sum of input, output and bidirectional signals used in the design and the power and ground requirements of the device. Refer to the device data sheets for specific details but, in general, all inputs and standard drive outputs and bidirects use only one pad.

Table 31.9 Suggested limits of memory on ASIC products

Product		*Maximum bits of RAM*	*Maximum bits of ROM*	*Comments*
Micro array	LMA9000	4K	16K	Metal megacell
Compacted array	LCA10000	8K	64K	Metal megacell
Compacted array plus	LCA100K	32K	128K	Metal megacell
Cell-based	LCB15	72K	512K	Megacell

Table 31.10 Examples of memory performance for LMA9000

Memory block	*Configuration*	*Typical performance* (ns)
RAM	256 × 8	10.0
RAM	64 × 8	9.2
ROM	256 × 8	12.0

Table 31.11 Relative logic densities for Channel-Free arrays

	Percentage
ROM	90
Single-port RAM	75
Multiport RAM	70
Metal megacell	50
Registered logic	45
Typical utilisation	40
Random logic	35

A simple calculation then defines the minimum number of V_{ss} and V_{dd} pads required to adequately support the device for all conditions. When possible, add additional pads. These reduce noise induced by the current power demands of switching outputs since a small resistance exists in the power supply metal. The switching current for a CMOS output buffer may be ten times the maximum static current. It is the dynamic current consumption that must be allowed for. These induced voltages also have the effect of reducing noise immunity on all inputs that share the same power rails. For this reason more advanced products (e.g. Compacted Arrays from LSI Logic) split the power supply rails (three V_{ss} and two V_{dd} power supply rails).

The total number of pads defines the minimum chip size required. *Table 31.12* shows the pads available for the LMA9000 series. It may be necessary to configure additional I/O pads for V_{dd}/V_{ss}, depending on the number and drive of the output buffers. When the number of usable gates is less than that required (pad-limited design), a larger device must be selected. Partitioning the circuit in several different ways can provide an improved solution for multi-chip designs.

The output buffer slew rate can be altered on some vendor's products by appropriate selection of output buffer type (none, moderate or full). This very beneficial feature slows the rate of rise and fall of the chosen output without affecting the steady-state drive capability. This reduces the power requirements (there is a reduction in peak dynamic current) and system noise (less radiated signal noise). Use slew rate control wherever practical.

High drive output buffers will use more pads and require more power supply pads. Reassess the device interface needs, to avoid

Table 31.12 Pads available on LMA9000 series

Device number	*Estimated user gates*	*Available gate complexity*	*Device pads*		
			V_{dd}	V_{ss}	*I/O*
LMA9020	700	1968	1	3	70
LMA9033	1200	3286	1	3	54
LMA9050	1750	4992	1	3	66
LMA9072	2500	7238	2	4	80
LMA9095	3300	9504	2	4	92
LMA9141	5000	14124	2	6	110
LMA9190	6700	19000	2	6	130
LMA9239	8400	23908	4	6	144
LMA9284	10000	28388	4	6	158
LMA9350	12500	34944	4	8	174

pad limited designs. When several designs are required for one system, repartitioning the circuit can radically alter pin count/gate ratios, potentially eliminating pad-limited designs.

31.5.10 Packaging requirements

The package must be able to accept the chosen device. Cavity size available for packages limits the size of device that may be used. The package selected must have sufficient pins for I/O and power supply. Pin grid arrays offer highest pin counts. The package must meet the requirements of the manufacturing process. Pin grid arrays are 'through board', while most chip carriers are surface mounted. The package must suit its operating environment. Ceramic packages are the most robust and can dissipate the most heat, while cavity down packages enable heatsinks to be mounted on them.

Unique requirements can also be accommodated. *Figure 31.10* gives the presently available range of packages. Packaging technology is currently evolving very rapidly and new developments offer many possibilities not currently easily available. The use of multichip assemblies, new substrate materials and electrical connection techniques all offer exciting possibilities for the near future.

31.5.11 Summary

The various factors influencing the choice of ASIC have been addressed above more or less in isolation. In practice, many are interrelated and thus several iterations may be required to obtain the perfect choice. However, unless the design is stretching the limits of technology, it is more important to consider the design from the point of view of how to ensure that the ASIC is a success.

This is particularly true for the first ASIC design which will inevitably be accompanied by justified concern at the potential risk in terms of cost and wasted time and effort. From the company point of view a failure at this point may adversely affect

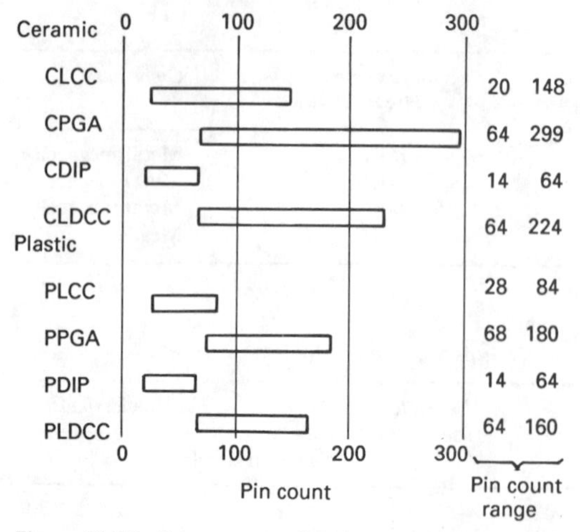

Figure 31.10 Current range of packages

decisions on taking advantage of the ASIC technology available for several years.

It is clear that ASICs are very attractive in terms of their cost, size, power consumption, reliability and the degree of design security offered over alternative solutions. It is not clear what to do in order to produce one. The various horror stories that circulate coupled with over a hundred potential vendors make choices difficult. The potential investment in software and workstations and the concern over repercussions if the design has an error worsen the situation.

It is therefore necessary to use the first ASIC as a learning vehicle. In many respects this is a similar situation to the introduction of microprocessors. Engineers must have hands-on experience to remove confusion and learn new skills. Training in management, manufacturing and procurement is also required.

The approach to use is to choose a major ASIC supplier that you can be certain is doing a large number of designs a year. Check the company's ASIC success record and future stability by asking for ten references and checking the financial situation.

Go directly to the ASIC vendor rather than through a third party to keep communications as direct as possible, to learn from their experience.

Choose the most reliable supplier to keep the chances of success as high as possible. The small cost penalty would be small compared to the cost of failure. The time for high-risk ventures is after the first success.

Select an HCMOS technology. This is the most used, is easily available and yet offers high performance. Also use a medium-sized array for the first design to minimise cost and risk. With a design of 2000–5000 gates the design is simple enough to complete in three to six weeks and yet complex enough to be a valuable learning vehicle. Preferably choose an existing design that has not been implemented with ASICs. Any delays in obtaining the ASIC can be buffered by maintaining existing production.

The package selection should be done in consultation with manufacturing to avoid an unusable device being produced. Purchasing should be involved to check the financial viability of the project and the ASIC vendor.

The ASIC vendor's software will be optimised for the product, putting the responsibility for producing parts to specification on the vendor. Use of 'universal' or 'free' libraries again significantly increases risk of failure. Also, ASIC vendors provide training, usually as a one week class. These provide excellent grounding in ASIC design, testing and the use of the software tools. Go prepared with the questions that must be answered and gain years of experience in just a few days.

It is generally possible to do ASIC designs at the vendor's design centre. Expert advice can be immediately sought and concentration on the design is improved, reducing potential delays. Do not be content with simply completing the logic designs. Check the device performance as exhaustively as possible and most importantly derive production tests before committing the design to silicon. Designing good tests retrospectively is fraught with problems and may prove to be impossible.

The ASIC vendor should also do the layout. It avoids having to spend time (extra delay) in learning all the layout process and the layout will probably be done quicker as well. The potential for performance 'tweaking' by having 'hands on' is usually unnecessary, given today's technology, and is risky.

Checking the design's performance after layout is essential, however, and here software tools can save significant effort. Package pinout should also be checked. Possibly a full design review is an advantage to catch any last problems.

Insisting that all prototypes are fully tested to the production specification by the vendor minimises the risk of delays caused by trying to debug vendor problems. With a test bed prepared and made ready by the designer, the first prototypes can be fully exercised over temperature and voltage.

If all of these steps have been followed the probability is that the ASIC will be a complete success.

Further reading

ANDREWS, W., 'Designers content with an explosion in ASICs', *Comp. Design*, 15 April (1988)

BUTZERIN, T. *et al.*, 'ASIC testing with high fault-coverage', *VLSI Syst. Design*, **9**, No. 9, September (1988)

CORLETT, R., 'Moving towards analog semicustom ICs', *Electron. Syst. Design Mag.*, **18**, No. 10, October (1988)

HARA, D. *et al.*, 'Timing analysis improves efficiency of ASIC design', *EDN*, 26 May (1988)

INGLIS, M., 'ASICs—service *versus* technology', *Electron. Eng.*, February (1988)

KING, H., 'A statistical approach to route estimating', *Electron. Product Design*, May (1988)

LEUNG, S. J. *et al.*, 'A conceptual framework for ASIC design', *IEEE Proc.*, **76**, No. 7, July (1988)

MASTERS, N., 'The basics of ASICs', *Electron. Design Automation*, mid-April/mid-May (1988)

MEYER, E., 'Structured arrays re-enter semicustom arena', *Comp. Design*, 1 January (1989)

MOORE, B., 'Consider the tradeoffs when evaluating linear-semicustom ICs', *EDN*, 4 February (1988)

OSANN, B. *et al.*, 'Compare ASIC capacities with gate array benchmarks', *Electron. Design Int.*, November (1988)

PRYCE, D., 'Semicustom IC's ratings and architectures aid analog- and digital-circuit designers', *EDN*, 15 October (1987)

PRYCE, D., 'Semicustom ICs combine analog and digital functions', *EDN*, 8 December (1988)

RICE, V., 'Analog/digital chips mix problems with promise', *Electron. Business*, 1 February (1988)

SINGER, P. H., 'Fast turnaround for ASIC photomasks', *Semiconductor Int.*, February (1988)

SMITH, T., 'ASICs set the pace in ATE progress', *New Electron.*, April (1988)

WHITE, A., 'The painless path to ASIC design', *J. Semicustom ICs*, **6**, No. 1, September (1988)

32 Electron Valves and Tubes

M J Rose, BSc(Eng)
Mullard Ltd
(Sections 32.1, 32.6.1, 32.8–32.10)

A P O Collis, BA CEng, MIEE
English Electric Valve Co. Ltd
(Section 32.2)

G T Clayworth, BSc
English Electric Valve Co. Ltd
(Section 32.3)

F J Weaver, BSc, CEng, MIEE
English Electric Valve Co. Ltd
(Section 32.4)

P M Chalmers, BEng, AMIEE
English Electric Valve Co. Ltd
(Section 32.5)

E W Herold, BSc, MSc, DSc, FIEE
(Section 32.6.2)

T C Fleming, BSc
Manager, Liddicon Camera Tubes,
English Electric Valve Co. Ltd
(Section 32.7)

Contents

32.10.3 Multistage image intensifier 32/33

32.1 Small valves

32.1.1 Triode

The triode contains three electrodes: a grid is placed between the anode and cathode.

The grid is in the form of a fine wire spiral surrounding the cathode. Because the grid is closer to the cathode than is the anode, small changes in grid voltage V_g have a significant effect on the electron flow and hence on the anode current. If the grid is highly negative with respect to the cathode, all the electrons from the cathode are repelled and no anode current flows. If the grid is highly positive, all the electrons are attracted to the grid and again no anode current flows. If no voltage is applied to the grid, the triode acts as a diode. This is shown by the $V_g=0$ curve in the anode characteristic of *Figure 32.1*.

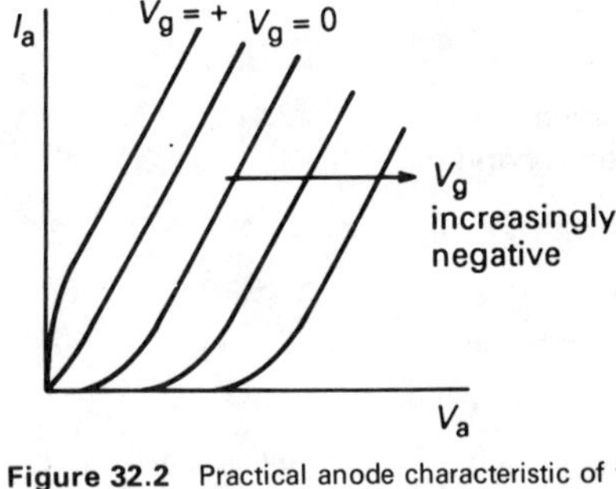

Figure 32.2 Practical anode characteristic of the triode

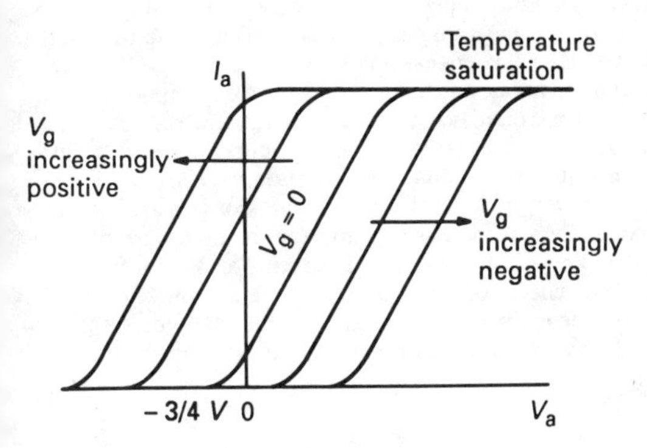

Figure 32.1 Anode characteristic of triode

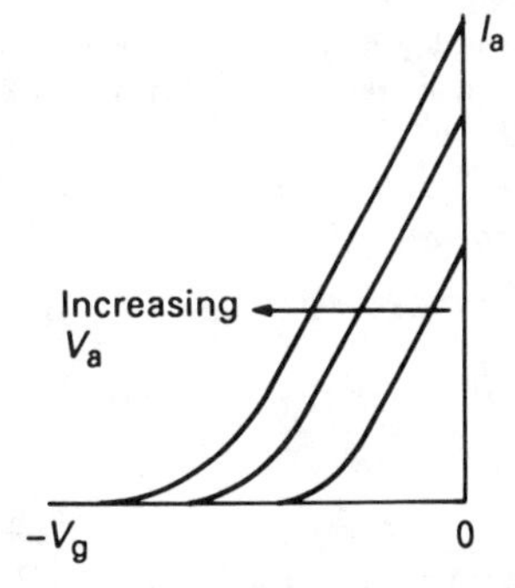

Figure 32.3 Mutual characteristic of the triode

If the grid is slightly positive, the electrons from the cathode are accelerated towards the anode and the triode starts to conduct at an anode voltage more negative than $-\frac{3}{4}$ V. This is shown by the curve to the left of the $V_g=0$ curve in *Figure 32.1*. If the grid is slightly negative, the anode voltage has to overcome the effect of the grid voltage and a larger anode voltage is required before conduction starts. This is shown by the curves to the right of the $V_g=0$ curve in *Figure 32.1*. Because the triode is never operated in circuits with a negative anode voltage, the practical anode characteristic is shown in *Figure 32.2*.

The triode has three principal characteristics. Besides the anode characteristic just described, there are the grid or mutual characteristic, the change of anode current with grid voltage with anode voltage as parameter, and the constant-current characteristic, the change of anode voltage with grid voltage with anode current as parameter. The mutual characteristic is shown in *Figure 32.3* and the constant-current characteristic in *Figure 32.4*.

The slopes of the three characteristics are used to give the small-signal parameters defining the performance of the triode. Because the slopes vary over the length of the characteristics, the value of the parameter is qualified by the operating point at which it applies. The slope of the mutual characteristic defines the mutual conductance g_m of the triode, also called the slope. The slope of the constant-current characteristic defines the amplification factor μ, while the reciprocal of the slope of the anode characteristic defines the anode slope resistance r_a.

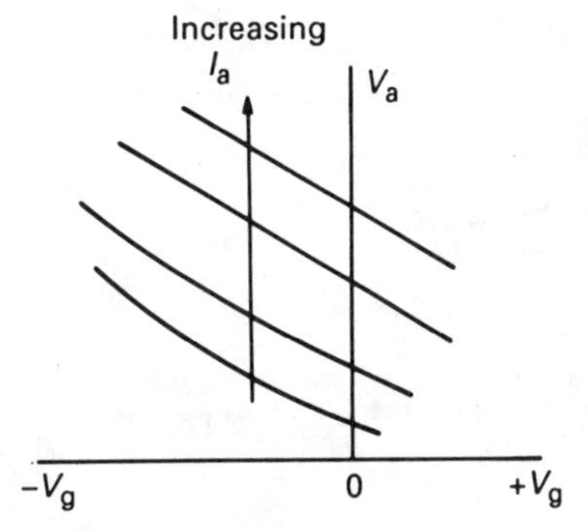

Figure 32.4 Constant-current characteristic of the triode

32.1.2 Tetrode

The performance of a triode is limited by the interelectrode capacitances. The grid-to-anode capacitance can be reduced by placing a second grid, the screen grid, between the grid and the anode. The valve now becomes a tetrode; the circuit symbol is shown in *Figure 32.5*.

The control grid is operated at a low negative voltage, typically up to -10 V, while the screen grid is held at a moderate positive voltage, typically 80 V, to accelerate the electrons from the

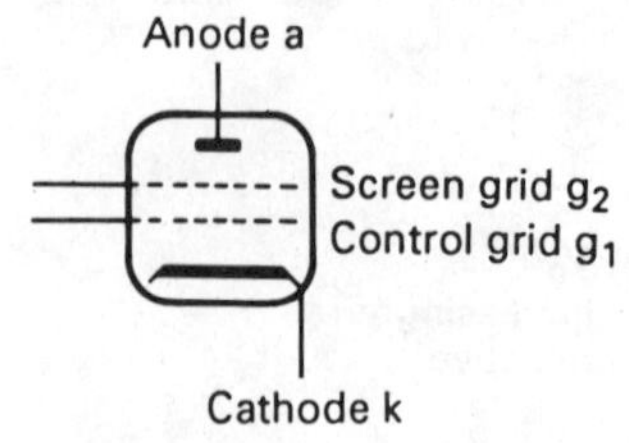

Figure 32.5 Circuit symbol for the tetrode

cathode towards the anode. If the anode voltage is zero, all the electrons flow to the screen grid and a high screen-grid current flows. If the anode voltage is increased but is still lower than the screen-grid voltage, anode current flows and the screen-grid current falls. If the anode voltage is increased further so that it is higher than the screen-grid voltage, the anode current increases further while the screen-grid current is further reduced. This idealised operation of the tetrode is shown by the characteristic of anode current I_a and screen-grid current I_{g2} plotted against anode voltage in *Figure 32.6(a)*.

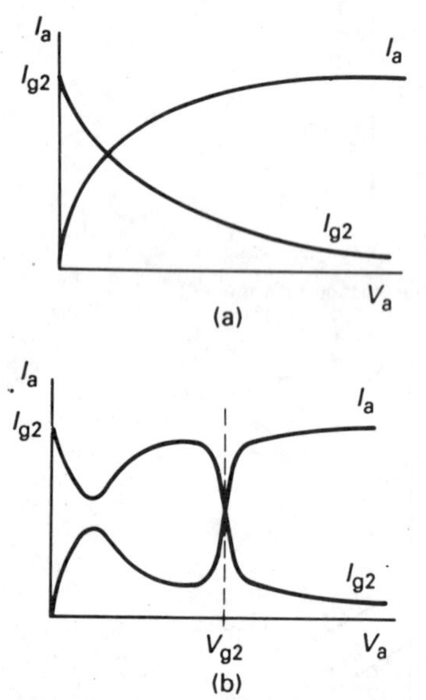

Figure 32.6 Anode current and screen-grid characteristics of tetrode: (a) ideal characteristics neglecting secondary emission, (b) practical characteristics

The operation is ideal because it neglects the effects of secondary emission. The effect of secondary emission is to produce kinks in the characteristic as shown in *Figure 32.6(b)*.

The anode characteristic of the tetrode can be varied by the control-grid voltage V_{g1} to give a family of curves. The mutual characteristic of the tetrode, the variation of anode current with control-grid voltage, with screen-grid voltage constant, is similar to that of the triode.

32.1.3 Pentode

The secondary emission can also be suppressed by inserting a third grid into the valve, the suppressor grid. This grid is wound to a larger pitch than the others, and is placed between the screen grid and anode. The suppressor grid is held at or near cathode potential, and is sometimes connected inside the valve envelope

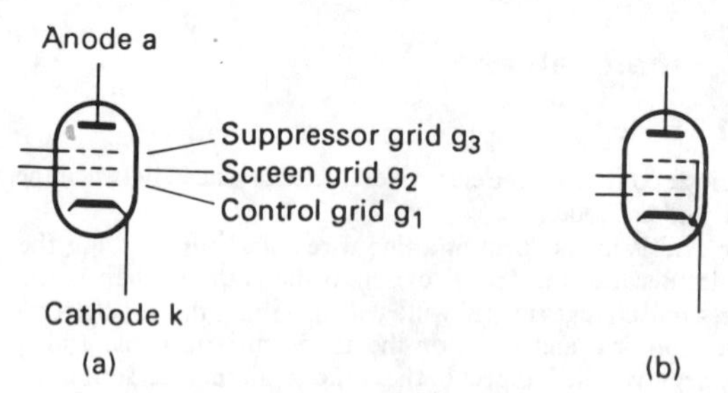

Figure 32.7 Circuit symbol for pentode: (a) separate suppressor grid, (b) suppressor grid connected to cathode

to the cathode. The valve now becomes a pentode; the circuit symbol is shown in *Figure 32.7*.

The control grid is operated at a low negative voltage and the screen grid at a moderate positive voltage as in the tetrode. If the anode voltage is lower than the screen-grid voltage, the screen grid acts as an anode and a high screen-grid current flows. As the anode voltage is increased, the electrons from the cathode flow to the anode because the large pitch of the suppressor grid has little effect on primary electrons. Any secondary electrons from the anode, however, are repelled by the suppressor grid and return to the anode. There is therefore a sharp rise in anode current and a corresponding fall in screen-grid current, as shown in *Figure 32.8*. The effect of control-grid voltage on the anode characteristic is shown in *Figure 32.9*, while the effect of screen-grid voltage on the mutual characteristic is shown in *Figure 32.10*.

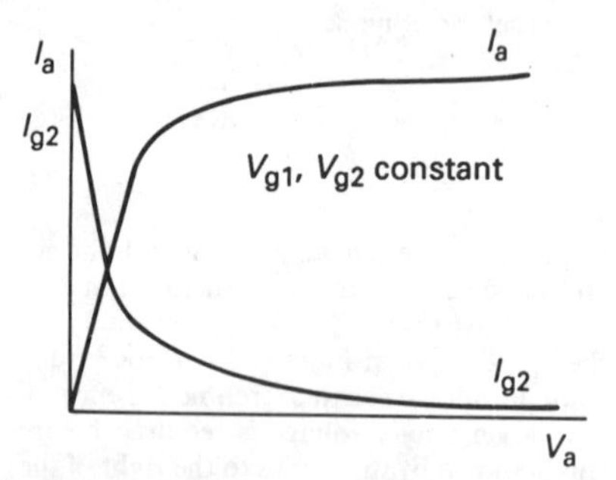

Figure 32.8 Anode current and screen-grid current characteristics of the pentode

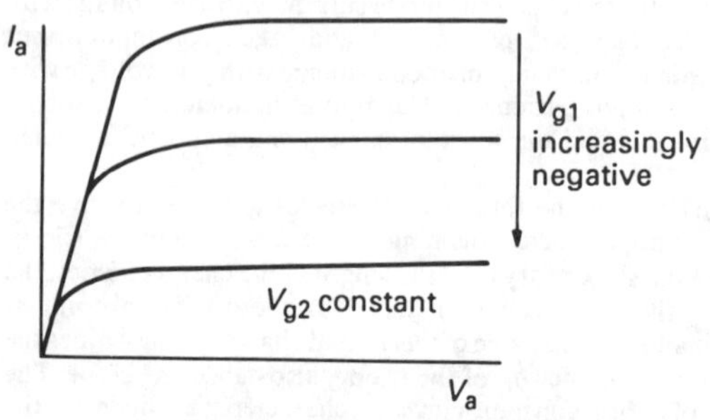

Figure 32.9 Anode characteristic of the pentode with control-grid voltage as parameter

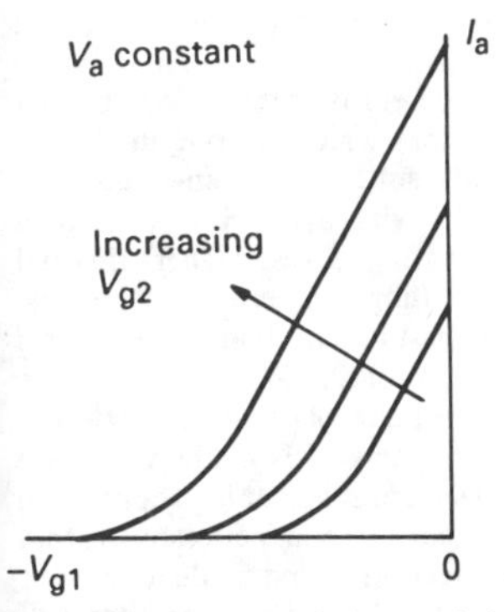

Figure 32.10 Mutual characteristic of pentode with screen-grid voltage as parameter

32.2 High-power transmitting and industrial tubes

Early vacuum tubes had glass envelopes through which the anode was cooled only by radiation, thus limiting the anode dissipation to a few kilowatts before the safe working temperature of the glass was exceeded. The development of a method of making large-diameter glass-to-metal seals overcame this limitation for it was then possible to construct tubes with the anode part of the vacuum envelope so that its external surface could be easily and efficiently cooled.[1] This technique led to the development of the modern high-power electronic tubes and although glass has been replaced by ceramic the principles remain the same. Tubes are available with continuous power outputs up to several megawatts.

In radio and broadcast transmitters, it is now usual to use tetrodes in the power amplifier stages since, with their higher gain, it is often necessary to use only one stage which may be driven directly by solid-state amplifiers. By using the latest design of both transmitter and tube, overall efficiencies better than 75% have been achieved with modulator efficiencies as high as 95%.[2]

In industrial applications such as induction or dielectric heating, triodes are used as power oscillators because they are simple, rugged and reliable, while advances in plasma and high energy atomic physics have led to the production of a range of tubes specially designed for very high power pulse operation both as switches and as r.f. amplifiers.[3–5]

32.2.1 Construction of tubes

A vacuum tube consists of:

(1) The cathode which is heated to a high temperature and is the electron source.
(2) One or more grids which closely surround the cathode and control the flow of electrons to the anode.
(3) The anode on which the electrons are collected.

Each of these electrodes has its own special requirements and particular problems (*Figure 32.11*).

32.2.1.1 Cathode

The properties required of the cathode are that it should be mechanically stable at high temperatures, that it should be a good emitter of electrons and that this emission should be stable and last a long time.[6,7]

In the latest designs of high power tube the cathode is cylindrical in shape and made from a mesh of fine wires of tungsten to which 1–2% of thorium oxide has been added. The

Figure 32.11 (a) A water cooled anode capable of dissipating 120 kW, (b) complete electrode assembly for the BW1185J2 (Courtesy of English Electric Valve Company Limited)

mesh is directly heated to the required operating temperature of 2000 K by passing current through each wire. This design has been developed to overcome the problem of distortion caused by the stresses which occur each time the filament is 'heat-cycled' from room temperature to its working temperature.

To obtain a stable emission the filament is heated in a hydrocarbon atmosphere during manufacture in order to change the outer surface of the wire to tungsten carbide. The thickness of this layer is limited to the equivalent of 20–30% of the cross-sectional area. If the degree of carburisation is greater than this the wires become brittle and there is a risk of them breaking. At the working temperature thorium produced by the reduction of the thorium oxide diffuses to the surface of the carbide to form an emissive layer. During running, some of the thorium is lost by evaporation and some is also lost as the result of bombardment by positive ions which are formed by the collision of high-energy electrons with the molecules of the residual gases left in the tube after evacuation. Therefore the residual gas pressure must be kept as low as possible and the filament must be run sufficiently hot for the rate of thorium diffusion to be fast enough to keep the emissive surface replenished. If this is not done the emission will fail. At the same time, however, carbon from the carbide is also being lost partly by evaporation, and partly by diffusion into the core of the filament wire. When the carbide has completely diffused and no longer exists as a distinct layer, the emission fails.

This process cannot be reversed. The rate of carbon diffusion like that of thorium is a rapid function of temperature, and while the filament temperature must be high enough to maintain the emission it must not be too high or the rapid diffusion of carbon will drastically shorten the life. The importance of this is illustrated by the fact that a 10% rise in filament voltage will increase the operating temperature by about 30 K, doubling the rate of carbon diffusion and halving the life.

32.2.1.2 Grids

The characteristics of the tube are determined by the geometric dimensions of the grid and its distance from cathode and anode.

The grid and anode currents may be calculated using appropriate formulae[8] although computer programs are now often used where either greater accuracy or information on the electron trajectories is required.[9]

The grids must have good mechanical stability at high temperatures but, unlike the cathode, they must have low thermionic emission and low secondary emission.

The grid is heated not only by radiation from the very hot cathode, but when it is driven positive it will attract electrons to itself producing a grid current, which dissipates power (*Figure 32.12(a)*). The combined effect results in the grid being heated to between 1000 and 1500°C and it may itself emit thermionic electrons which are attracted to the neighbouring electrodes (*Figure 32.12(b)*). The impact of the incident or primary electrons in the grid may knock electrons from the surface (*Figure 32.12(c)*).

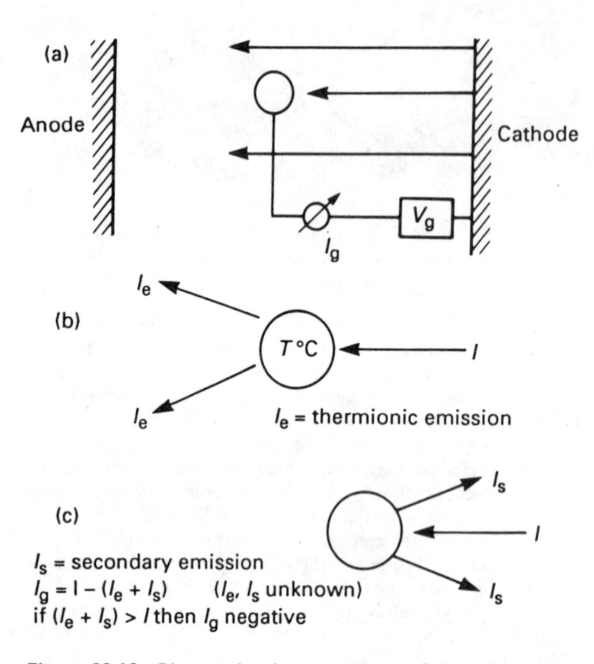

Figure 32.12 Diagram showing components of the grid current

These secondary electrons will be collected on nearby electrodes. The grid current measured externally to the tube therefore has three components: the incident current I, the thermionic current I_e and the secondary electron current I_s, the last two being in the opposite direction from the first.

In general therefore the measured current is less than the incident current and in extreme cases may be negative. The variation from tube to tube, the change during life, plus the possibility of negative currents and hence negative resistance, make circuit design difficult and it is preferable to reduce the effect until it is negligible.

For reasons of mechanical stability and hot strength, metals such as molybdenum and tungsten must be used for making the grid, but these are good thermionic emitters particularly if covered with thorium evaporated from the cathode. A reduction in grid emission can be obtained by coating the grid with a high work function metal such as platinum and this is often done. However platinum has a high secondary emission coefficient. In addition if the grid becomes too hot metallurgical changes occur due to diffusion between the platinum and the core metal, which renders the suppression of thermionic emission ineffective and causes distortion.

One method of solving this problem is to provide a barrier layer between the platinum and core which is done in the so-called K grid.[10] This is made by sintering a thin coating of zirconium carbide onto the grid wire at 1900°C and then platinum plating. The black coating gives higher thermal emissivity thus reducing the operating temperature while the rough surface lessens secondary emission. The final result is a grid which can run with a dissipation of 25 W/cm^2 of surface compared with 7 W/cm^2 for an ordinary platinum coated grid.

In a more recent development[11] the grid is made from pure carbon, thus making use of that element's desirable properties of high hot strength, low thermionic and secondary emission. A thin shell of carbon, the shape of the final grid, is first made by vapour deposition on to a mandrel at 2000°C. The grid apertures are then machined into the shell either by abrasive machining through a mask or by cutting with a numerically controlled laser machine. High-quality grids can be made in this way which can be operated in excess of 50 W/cm^2.

32.2.1.3 Anodes

The anode must have good mechanical strength, good electrical and thermal conductivity and is therefore invariably made of copper (*Figure 32.11(a)*). It can be cooled externally in several ways. When determining the power to be dissipated due allowance must be made for the increase due to mismatching or mistuning of the output and other causes.

32.2.2 Air cooling

This is a very convenient and popular method of cooling but because of the size and weight of the radiator and problems of noise from the air ducting and blowers it is difficult to make a forced air cooled valve with an anode dissipation greater than 50 kW. The temperature distribution over the cooling fins and anode surface can be calculated approximately for a given dissipation and airflow by using a method described by Mouromtseff.[12] The determination of the air flow required is done using the airflow curves given in the manufacturer's tube data such as those shown (*Figure 32.13*). From these the flow and

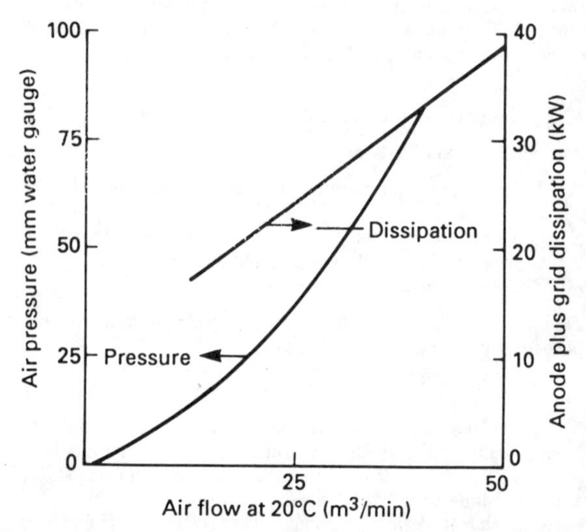

Figure 32.13 Typical air flow characteristics for a forced air cooled anode

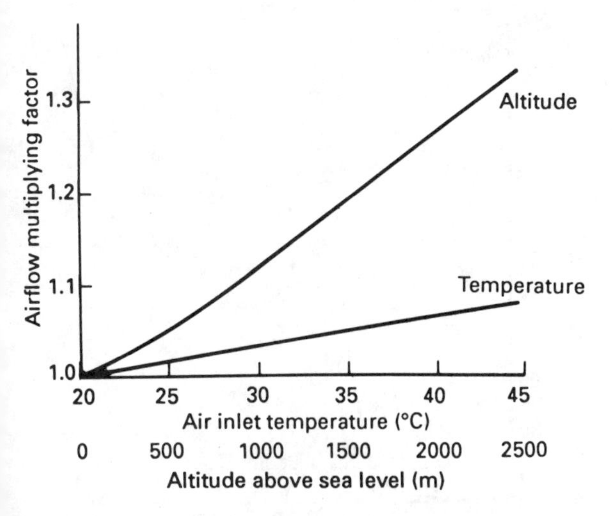

Figure 32.14 Factors to correct air flow for altitude and ambient temperature

pressure head across the radiator are obtained but it may be necessary to correct these for altitude or ambient temperature if these are different from the data (*Figure 32.14*). To the head thus obtained must be added the pressure head in all the other parts of the cooling system including bends, changes in duct size, filters etc. This gives the total pressure head and flow required.[13] Consideration must also be given to the direction of flow. It is recommended that the air should flow from the tube cathode, over the envelope and then through the anode radiator because it means that the cool air will first flow over other components, cooling them and improving their reliability.

The cooling air may be sucked through the cabinet, the inside of which will be at a slightly lower pressure than the outside, when air and dust may leak in. The fan will also be blowing air heated to over 100°C during its passage through the anode radiator. Alternatively air may be blown into the cabinet. The inside pressure will then be greater than outside, which will prevent dust getting in, but the opening of a door or the removal of a panel will severely reduce the cooling of the tube. Once a decision has been made a blower can be selected from data supplied by the manufacturer.

On installation of the tube it is advisable to measure the airflow, and this may be done by commercial instruments or running the tube with a known anode dissipation. The temperature rise of the cooling air is measured and then the airflow can be calculated from the relation:

$$\text{airflow} = \frac{0.173 \times \text{inlet temperature} \times \text{total dissipation}}{\text{temperature rise}} \tag{32.1}$$

where airflow is measured in m^3/min, temperature and temperature rise in K, and dissipation in kW. Once satisfied that the airflow is sufficient the flow switches should be set to switch off all supplies if the flow is reduced. It is good practice to include a temperature switch to detect if the outlet temperature is too high, indicating excessive anode dissipation.

32.2.3 Water cooling

Similar care to ensure adequate flow is equally important with water cooling. However, there are two important additional constraints. Firstly the tube manufacturer will specify a maximum outlet temperature for the water. If this is exceeded it is possible for local boiling to occur, forming a layer of steam over the surface of the anode and severely reducing the heat transfer. A hot spot will form which may cause the anode to distort, or in a severe case the anode may melt.

Secondly, if the quality of the water is not adequate, corrosion and scaling of the anode can occur. The dissolved solids should be less than 30 ppm to prevent scale formation which reduces heat transfer from the anode to the water, restricts the water flow, and may block small water channels used in some tube types. To minimise corrosion the electrical conductivity must also be low, at worst 300 μmho/cm. It is usual to run the anode at high potential above earth, and the cooling water has therefore to be fed through insulating pipes whose length must be about 1 m/kV. Leakage currents flowing through the water will erode the jacket or connecting pipes by electrolysis. Special sacrificial anodes may be fitted in the water pipes and arranged so that they are attacked preferentially, but it is important that these are checked regularly and replaced if necessary. Another solution is to run the anode at earth potential despite the increased circuit complexity. If no suitable water supply is available a closed circuit system with heat exchanger must be installed.

Since the water flow may be measured easily it provides a simple method of determining the anode dissipation and efficiency of the tube. The temperature rise and flow rate are measured and the dissipation is given by:

$$\text{dissipation} = \frac{\text{flow} \times \text{temperature rise}}{15} \tag{32.2}$$

where flow is in l/min and the other quantities as in Equation (32.1).

32.2.4 Vapour cooling[14]

The danger of allowing the water to boil at the surface of the anode has been described above, but it is possible to design an anode where it is quite safe for this to happen. The steam produced is taken via an insulating pipe to a steam condenser, thus utilising the high value of the latent heat of vaporisation to transport the heat (*Figure 32.15*). The first requirement is that adequate water must always be maintained in the boiler and that the level must never drop below the specified minimum level even at full dissipation. To achieve this the pressure in the steam pipe must be cancelled by means of a balance pipe returning to the top

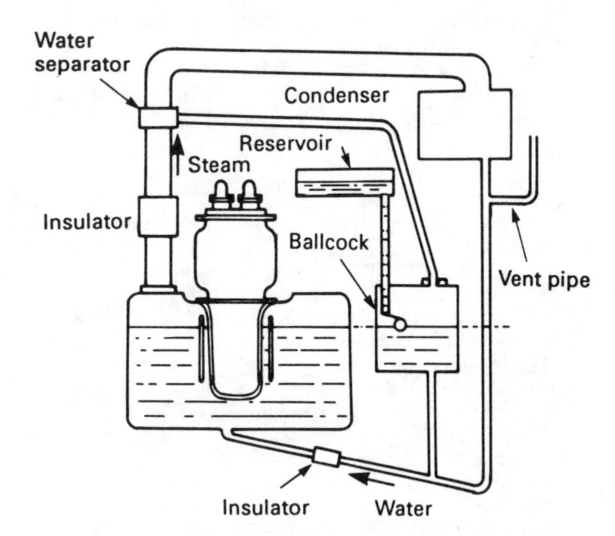

Figure 32.15 Vapour cooling system with external condenser

of the sealed level tank. There is also a pressure fall in the water return pipe due to flow of water back to the boiler and this must be kept as little as possible. Although the flow is only 27 cm^3/(kW min) the return pipe should be of generous size. A temperature switch included in the vent pipe will detect any steam blow off due to condenser failure or excessive dissipation. The system must be filled with distilled or deionised water and it is important that it does not become contaminated with oil or grease as this will result in a severe reduction in permissible dissipation by forming a hydrophobic film over the anode surface (*Figure 32.16*). A simpler system particularly suited for industrial

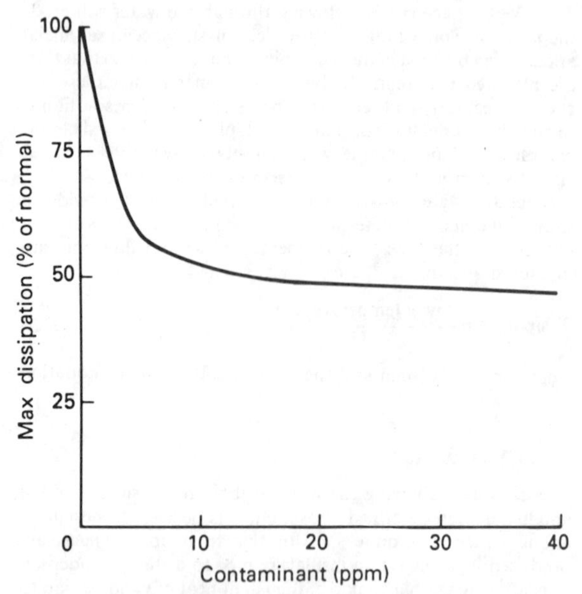

Figure 32.16 Reduction in maximum anode dissipation due to contamination of the water

use where it is often difficult to obtain high quality water is the integral boiler–condenser (*Figure 32.17*). In this water-cooled pipes are built into the boiler itself to condense the steam.

The operation of a vapour-cooled system can be understood by reference to the Nukiyama curve (*Figure 32.18*). This shows the temperature of a flat isothermal metal surface covered with a coolant at its boiling point as a function of the heat flux expressed in W/cm^2. As the flux is increased the temperature of the surface

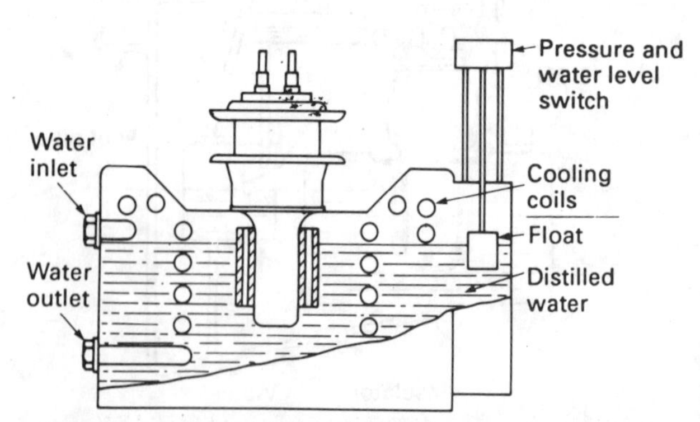

Figure 32.17 Boiler–condenser unit

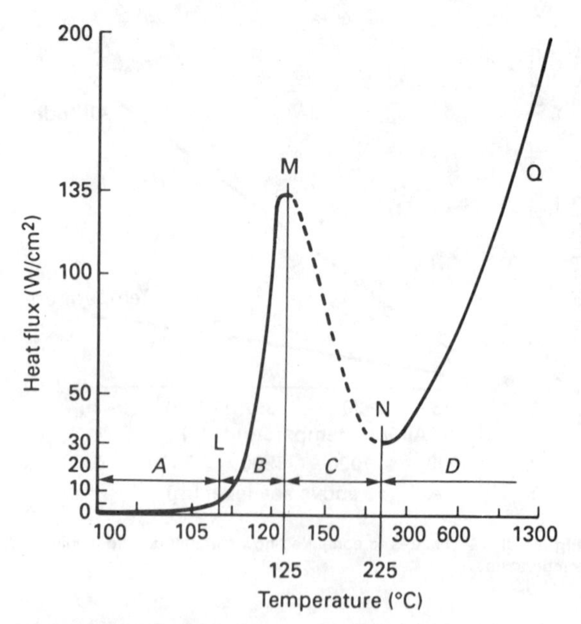

Figure 32.18 Nukiyama curve: variation of surface temperature with heat flow

will follow the curve until point M is reached. Any attempt to increase the heat flux further will force the operating point to jump to Q where the temperature will destroy the anode. Anodes designed for vapour cooling have thick ribs and grooves machined into their surface, and are operated so that, although part of the surface may be in the region MN, the temperature is stabilised by heat conduction to other parts operating in regions ML and A.[14,15] If part of the rib is operating in region A the anode is in thermal equilibrium and an increase in dissipation increases the temperature in that region only.

The Hypervapotron[R] system (trademark of Thomson-Brandt)[16,17] is a recent improvement whereby water-cooled tubes may be run with dissipations up to 2 kW/cm^2, the water inlet temperature being as high as feasible on the equipment using the tube, while outlet temperature can reach up to 90°C at atmospheric pressure or 100°C to 105°C at 1 or 2 bars pressure. The anodes for this system have grooves at right angles to the water flow formed in the surface. Steam formed in the grooves is ejected at high velocity into the water stream where it is rapidly condensed, fresh water rushing back into the groove. This pulsating action prevents the formation of a steam film and allows very high dissipations to be obtained. The variation of the mean heat transfer coefficient with heat flux plays a crucial role in the operation. It has been found that the temperature at the root of the groove tends to be a constant value of 225–300°C, the heat transfer coefficient increasing to maintain this temperature[17] (*Figure 32.19*). All these systems offer the advantages of a smaller heat exchanger the size of which is governed by the average temperature difference between the primary and secondary circuits and since the steam from a vapour-cooled system is at 100°C the heat exchanger can be correspondingly smaller (*Table 32.1*).

32.2.5 Ratings

It is normal practice to express the ratings of high-power tubes as absolute values. This means that the equipment designer has to make sure that no limit is ever exceeded, whatever the variation in

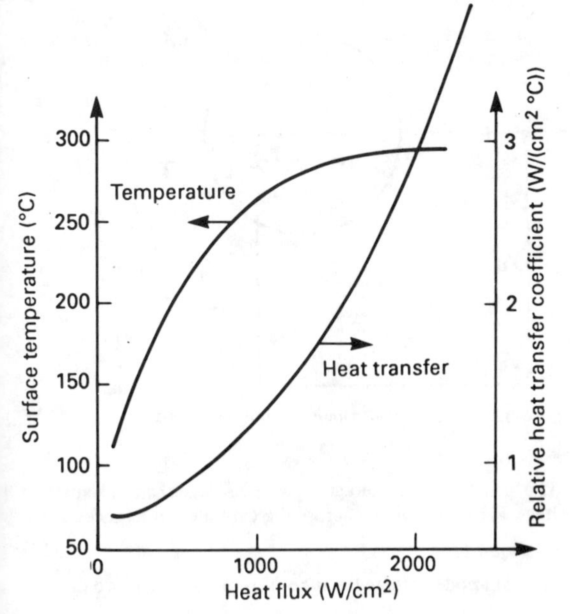

Figure 32.19 Surface temperature and relative heat transfer coefficient for the Hypervapotron system (trademark of Thomson Brandt)

power supply voltage, transients, mistuning, maladjustment or other causes.[18,19]

32.2.6 Calculation of performance

The characteristics of high-power tubes are most conveniently shown by constant-current curves, where the x and y axes are the instantaneous anode and grid voltages respectively, and the plot is the locus of the values of these voltages which give a constant anode or grid current (*Figure 32.20*). For sine waves the instantaneous grid and anode voltages are:

$$v_g = V_g \sin \omega t + V_{bias} \quad (32.3)$$

$$v_a = V_a \sin(\omega t + \pi + \phi) + V_{dc} \quad (32.4)$$

where

V_g = peak r.f. grid voltage

V_{bias} = d.c. grid bias voltage

V_a = peak r.f. anode voltage

V_{dc} = d.c. anode voltage

$\omega = 2\pi \times$ frequency

ϕ = phase angle difference

Coincident values of the voltages v_a and v_g may also be plotted on the constant-current characteristics, and the locus of these points is called the load line, which is in the general case an ellipse. The

Table 32.1 Typical operating conditions for heat exchangers

Method of cooling	*Temperature (°C)*					*Ratio*
	Ambient	*Inlet*	*Outlet*	*Mean*	*Differential*	
Vapour and Hypervapotron R	45	60	100	80	35	1
Water	45	50	70	60	15	2.3

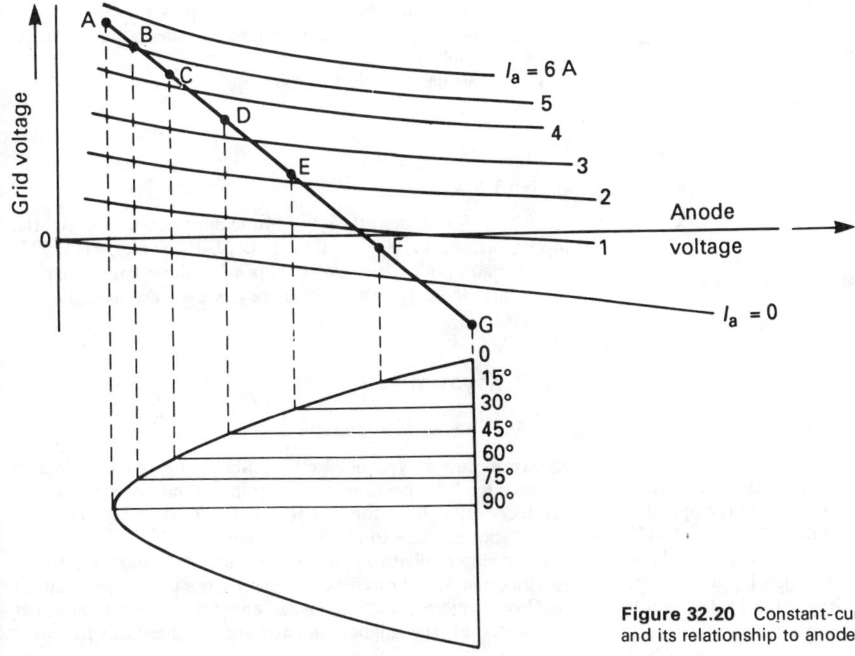

Figure 32.20 Constant-current characteristics showing load line and its relationship to anode and grid voltages

corresponding electrode currents may be read from the characteristics and plotted against time to give the pulses of current taken by the anode and grid. A Fourier analysis may be carried out on these to give the d.c., fundamental r.f. and harmonic components from which the power output and load impedance may be calculated. The analysis can be greatly simplified if the times at which the current values are found are suitably chosen. This is particularly so if the load is a pure resistance, i.e. ϕ is zero in Equation (32.4), when the load line becomes a straight line.

Once the load line has been fixed the values of grid and anode currents at, say, 15° intervals may be read from the curves with the aid of a transparent overlay (*Figure 32.21*).[20,21] Approximate design factors for various operating classes are given in *Table 32.2*.

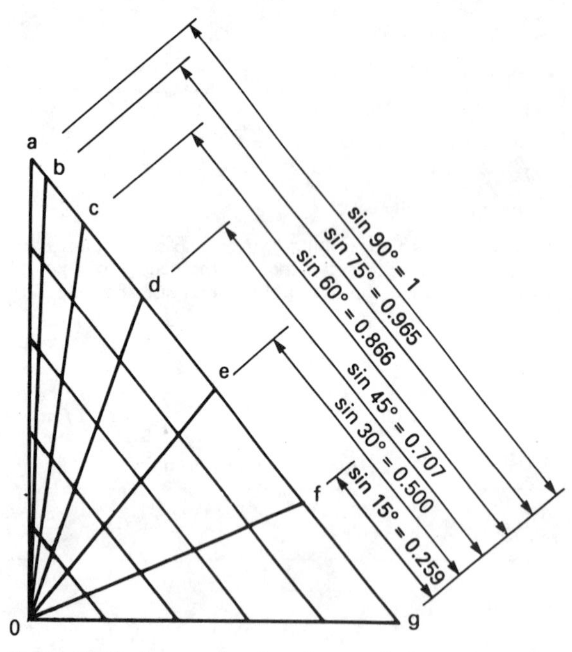

Figure 32.21 Construction of overlay to assist in calculation of tube performance

Table 32.2 Approximate design factors

	Efficiency (%)	*Conduction angle* (degrees)	*Ratio of peak to mean anode current*
Class C	75	120–140	4 to 4.5
Class B	65	180	π
Class AB	60	180	3

32.2.7 High-frequency effects

The time taken for the electrons to travel from cathode to anode is short enough not to have an important effect on the operation of the type of tubes discussed at frequencies below 100 MHz. Special tubes are made which are satisfactory to 1000 MHz or more. However, there is a very important high-frequency effect due to the inductance of the cathode structure inside the tube (*Figure 32.22*).[8]

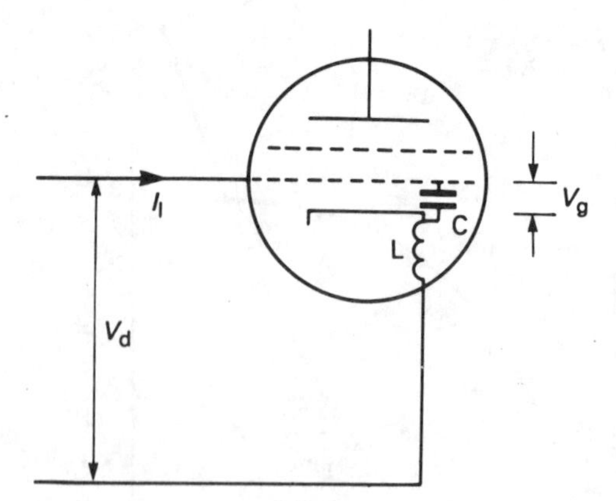

Figure 32.22 Equivalent circuit at high frequencies

The grid cathode voltage V_g will be less than the applied voltage V_i by the voltage across the cathode inductance L:

$$V_i = V_g - j\omega L I_k \quad (32.5)$$

At a fixed anode voltage the tube current may be approximated by

$$I_k = G_m V_g \quad (32.6)$$

Where G_m is the slope expressed in amperes per volt.

The input current I_g to the tube will be given by

$$I_g = V_g \times j\omega C \quad (32.7)$$

Substituting these in Equation (32.5) and rearranging gives the input admittance Y of the tube as

$$Y = j\omega C \frac{1 - j\omega L G_m}{1 + \omega^2 L^2 G_m^2}$$

$$= \frac{j\omega C}{1 + \omega^2 L^2 G_m^2} + \frac{\omega^2 L C G_m}{1 + \omega^2 L^2 G_m^2} \quad (32.8)$$

The second term on the right-hand side of Equation (32.8) represents a conductance and will absorb power. Although this power is not dissipated, being transferred to the output circuit, it must be supplied by the driver.

Typical values for a large tube are:

$L = 5 \times 10^{-9}$ H

$C = 185$ pF

$G_m = 0.3$ A/V

At 30 MHz for example the input conductance is 0.01 mho, i.e. the input resistance is 100 ohm. If the peak r.f. drive voltage is 1200 V then the additional drive power required will be approximately 7 kW, and the drive source must be designed accordingly.

32.3 Klystrons

32.3.1 Types of klystron available

Klystrons are a type of electron tube using the principle of velocity modulation to avoid the transit time problems which produce difficulties for triodes and tetrodes as operating frequencies increase to above 1 GHz.

Klystron oscillators, sometimes called reflex klystrons, produce microwave r.f. energy at power levels up to a few watts at the lower microwave frequencies and up to several hundred milliwatts at the highest microwave frequencies. In some applications klystron oscillators have been replaced by solid-state devices but their inbuilt robust tolerance of supply transients and relatively low cost has recently led to their ousting solid-state devices in some applications.

Klystron amplifiers begin to be useful from about 400 MHz and are available up to around 20 GHz. The majority of applications are below 2 GHz where CW klystron amplifiers can produce power outputs in the range 1–70 kW at gains of 30–40 dB. At frequencies above 2 GHz CW power outputs of up to 25 kW are available. At the lowest frequencies (a few hundred MHz) CW powers of up to 2 MW are obtainable.

Pulsed klystron amplifiers are usually operated above 2 GHz where peak output powers in the range 1–20 MW are available at gains of 30–50 dB. Pulsed amplifiers are less well known at lower frequencies than 1 GHz but tubes are available giving peak output powers in the range 0.5–20 MW.

The klystron was first described by the Varian Brothers[22] in 1939 and by Hahn and Metcalf.[23] The klystron oscillator first appeared in 1940 but successful klystron amplifiers were not available until the early 1950s.

The size and weight range of klystrons is astonishing. A small oscillator klystron might be 10 cm high and weigh 150 g whilst a large CW amplifier klystron could be more than 3 m long and weigh 200 kg. Some special-purpose tubes are much larger and heavier even than this.

32.3.2 Principle of operation

Klystrons use an r.f. voltage to velocity modulate an initially uniform velocity electron beam. An electron beam travels down a metal tunnel which has gaps along its wall across which r.f. voltages are applied. Individual electrons have their velocity shifted by interaction with the r.f. electric fields across the tunnel gaps according to the instantaneous value of the r.f. electric field at the instant when the electron arrives at the gap. Some electrons are slowed, some accelerated, some unaffected according to their arrival time at the gap. In the subsequent tunnel section slowed electrons will be caught by faster electrons and accelerated electrons will catch up slower electrons. Thus the uniform electron beam is converted by velocity modulation into a beam of non-uniform density which may then be used to perform the desired function of either amplification or oscillation.

The tunnel is usually called the *drift tube* and the gaps *interaction gaps*. The r.f. voltages across the interaction gaps are produced by resonant metal boxes called *cavities* suitably coupled to the gaps.

32.3.2.1 Klystron amplifiers

A klystron amplifier having two cavities is shown in *Figure 32.23*. An electron beam (1) is formed by the electron gun (2). The electrons in this beam are accelerated to a constant velocity U_0 determined by the relation:

$$U_0=\left(\frac{2e}{m}V_0\right)^{1/2} \tag{32.9}$$

where

U_0 = electron velocity
e = electronic charge (1.6×10^{-19} C)
m = electronic mass (9.1×10^{-31} kg)
V_0 = beam voltage

The electrons enter the drift tube (3) and continue to travel at velocity V_0 until they enter the input interaction gap (4) which is surrounded by the input resonant cavity (5). When this cavity is adjusted to resonate at the same frequency as that of an r.f. signal coupled into the cavity by means of a small loop antenna an r.f. electric field is developed across the interaction gap parallel to the beam axis.

In each successive half-cycle the electrons crossing the gap are either accelerated or decelerated depending on the phase of the gap voltage. If the instantaneous gap voltage is $V\sin\omega t$, where $\omega=2\pi f$, then the velocity of the emerging electrons will be given by

$$U=U_0\left(1+M\frac{V_1}{V_0}\sin\omega t\right)^{1/2}$$
$$\simeq U_0\left(1+M\frac{V_1}{2V_0}\sin\omega t\right) \tag{32.10}$$

where M is the gap coupling coefficient, dependent on the gap geometry and beam diameter.

The velocity modulated beam enters a field free region (6) called the drift tube and here the fast electrons overtake those slowed down during the preceding half cycle. In this way electron bunches are formed and the beam becomes density modulated.

The first theoretical work on klystron behaviour was published by Webster,[24] who showed that the value of r.f. current on the beam as a function of drift tube length is given by the relation:

$$i_{rf}=2I_0J_1(K) \tag{32.11}$$

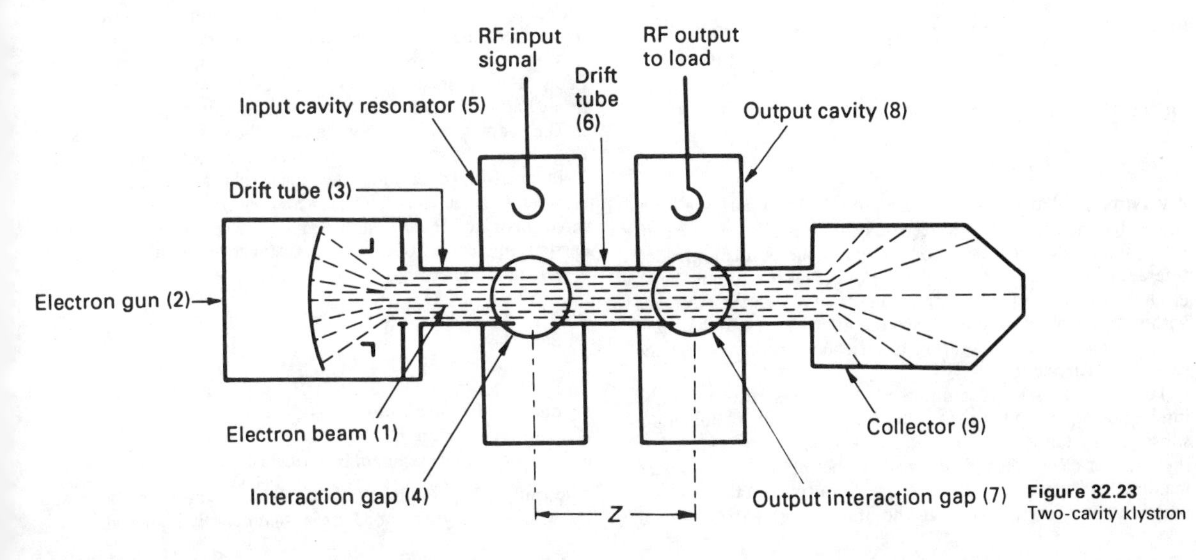

Figure 32.23 Two-cavity klystron

where K, the bunching parameter, is given by

$$K=\frac{\omega z V_1}{2U_0V_0} \quad (32.12)$$

where

I_0 = beam current
J_1 = a Bessel function of the first kind
z = distance measured from the input gap
V_1 = input gap voltage

The maximum value of $J_1(K)$ is 0.58 and occurs when $K=1.84$. It follows that for any given value of V_1 the value of i_{rf} can be made to be a maximum by making z the appropriate length. In theory, therefore, it is possible to achieve very high gain by making z very long, but in practice space charge de-bunching[25] limits the useful value of z.

The output gap (7) is positioned at a distance z which corresponds to the maximum value of i_{rf}. On crossing this gap the bunched beam induces currents in the output cavity (8) which contains a loop antenna which delivers power to an external load via a transmission line.

The voltage developed across the output gap is in anti-phase with the electron bunches. Consequently, more electrons cross the gap during the retarding voltage half cycle than do during the positive cycle. There is, therefore, a nett transfer of energy during each cycle. The electrons are slowed down as they cross the gap and the maximum useful peak voltage is that which just stops the slowest electrons and is approximately the beam voltage V.

The spent beam that emerges from the output gap travels into the collector (9) where the remaining energy is dissipated as heat.

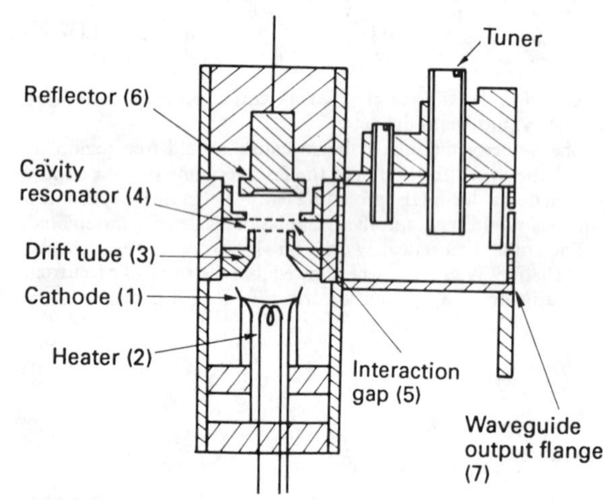

Figure 32.24 Structure of typical reflex klystron

32.3.2.2 Klystron oscillator

A klystron oscillator is shown in *Figure 32.24*. A cathode (1) is heated by a heater (2) and an electron beam is projected via a drift tube (3) into a cavity resonator (4) which surrounds a gridded interaction gap (5). The grids serve to couple more strongly the r.f. electric fields generated by the resonant cavity to the electrons within the beam. Having traversed the interaction gap the beam enters a region of decelerating electric field produced by the reflector electrode (6).

The beam is velocity modulated on its first transit of the interaction gap and density modulation, called bunching, subsequently takes place. The electron bunches are reflected at the reflector electrode and returned to the interaction gap. For maximum r.f. output via the waveguide output (7) the electron bunches must arrive back at the interaction gap so as to experience a maximum r.f. retarding voltage. This requires a drift transit time of $n+\frac{3}{4}$ r.f. cycles, where n is an integer.

Modulation of the electron beam is initiated within the tube by a noise process and oscillation builds up to give full output provided the $n+\frac{3}{4}$ relationship is satisfied. The reflector electrode voltage is adjusted to produce the required relationship and oscillation is sustained. The reflex klystron will, therefore, oscillate in a number of modes determined by the reflector voltage.

Since the bunched beam presents a reactive component to the resonant cavity and the sign of this reactance is dependent upon the phase of the returning bunches, the frequency of oscillation can be adjusted over a limited range by altering the reflector voltage.

The use of a single resonant cavity simplifies mechanical tuning arrangements and most reflex klystron oscillators can be tuned over relatively large frequency ranges with a single control. Electronic tuning is achieved by varying the reflector voltage and as the reflector draws no current very little drive power is required for this type of tuning.

32.3.3 Multicavity klystron amplifiers

Most commercially available klystron amplifiers have at least four resonant cavities and tubes having 5–7 cavities are not uncommon. The two-cavity tube shown in *Figure 32.23* will amplify a single frequency but has a narrow bandwidth. In most real applications for klystron amplifiers significant instantaneous bandwidth is required and this can be achieved by the use of a stagger tuning technique where multiple cavities are tuned to frequencies lower and higher than the centre frequency to build up the required bandpass characteristic.

A typical four-cavity klystron amplifier is shown diagrammatically in *Figure 32.25*. *Figure 32.26* shows an actual four-cavity tube used in a UHF television transmitter. The penultimate and output cavities of the tube can be considered as a power amplifier stage and the earlier cavities (two in this case) as a driver stage for the two-cavity power amplifier stage. The power amplifier stage determines the linearity of the whole tube whilst the driver stage largely determines the gain and bandwidth of the whole tube.

A hot cathode in the electron gun generates an electron beam which is held cylindrical as it traverses the drift tubes by means of the axial magnetic field of a series of coils retained in a supporting structure which also forms a flux return path of low reluctance. The electron gun and collector are external to the magnetic field, though some flux is allowed to link the cathode to ensure correct injection of the electron beam into the drift tube.

Metal cavity boxes surround each of the interaction gaps. Each cavity is mechanically tuned by movable doors and the input and output cavities have loop antennas.

The beam is velocity modulated at the input cavity by a small r.f. voltage. The signal level and hence the electron velocities are insufficient to overcome the space charge repulsion forces between the electrons. Under these conditions the space charge wave theory of Hahn[26] and Ramo[27] is applicable and the r.f. current maximum occurs at a distance of a quarter plasma wavelength z given by

$$z=\frac{U_0}{2\omega_q} \quad (32.13)$$

where

z = distance from the gap
U_0 = initial electron velocity
ω_q = reduced plasma angular frequency

A second, lightly loaded, cavity is positioned at this distance and because of its higher impedance a significantly higher voltage is

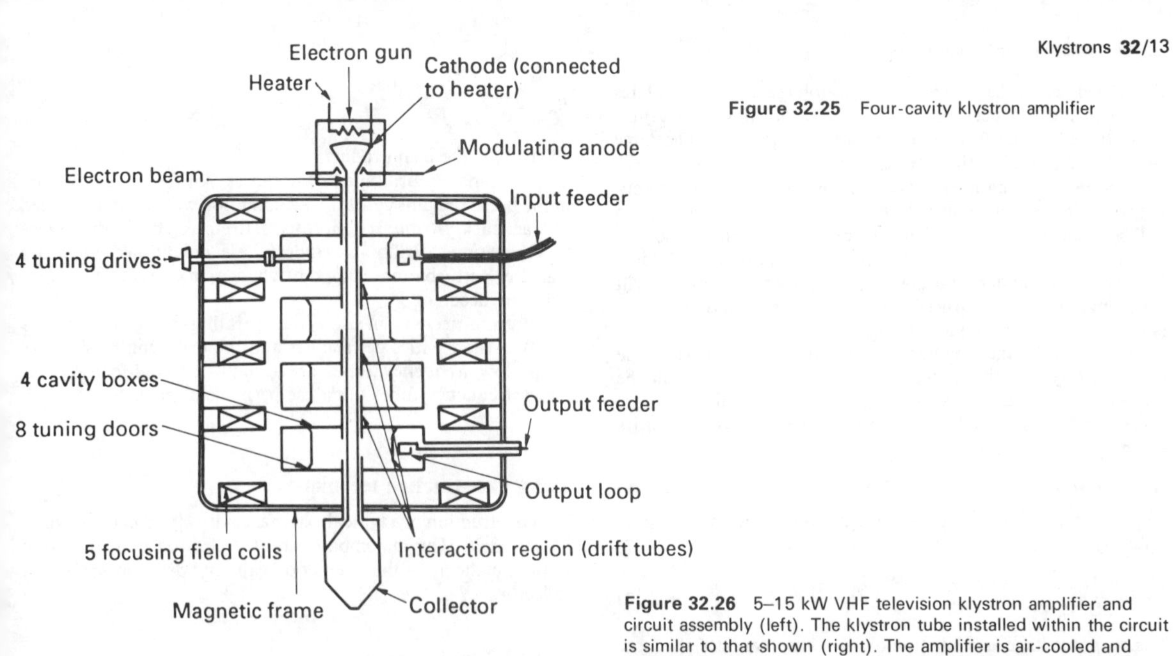

Figure 32.25 Four-cavity klystron amplifier

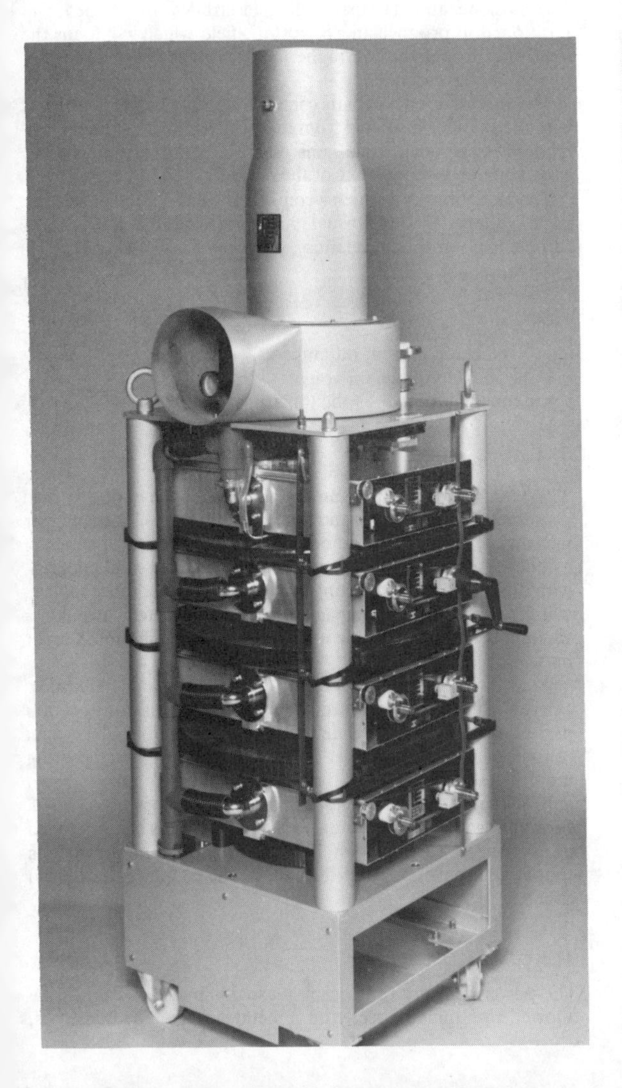

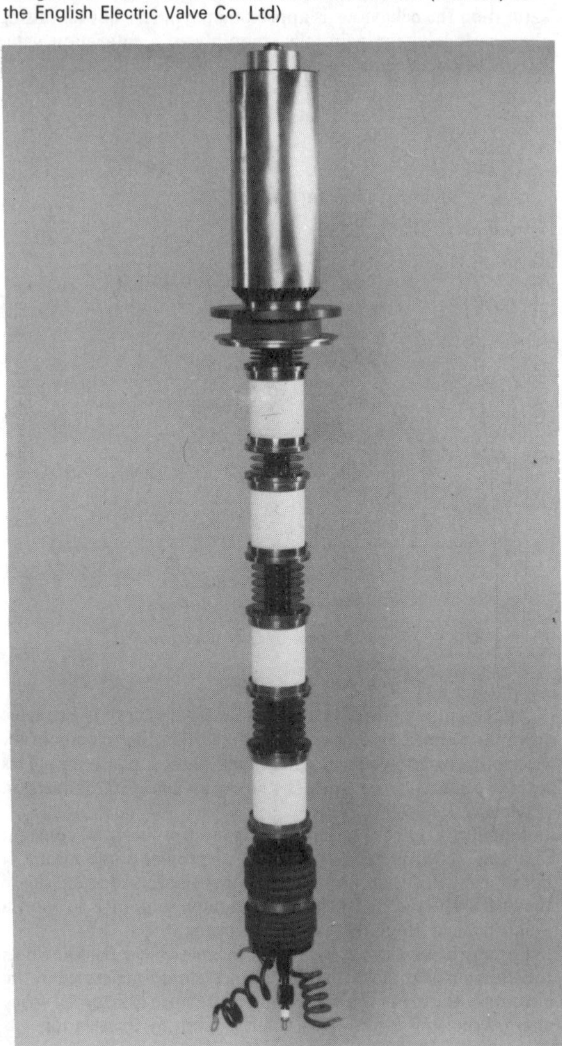

Figure 32.26 5–15 kW VHF television klystron amplifier and circuit assembly (left). The klystron tube installed within the circuit is similar to that shown (right). The amplifier is air-cooled and designed for use in remote, unattended transmitters (Courtesy of the English Electric Valve Co. Ltd)

developed across the second gap. This voltage now remodulates the beam and produces a higher r.f. current maximum at a third gap situated a further 0.25 plasma wavelength away. The same process is repeated at this gap resulting in a higher voltage still. Depending on the number of cavities this process continues until the penultimate gap is reached. By now the signal level at the gap will be large and the space charge forces will be overcome so that the basic two-cavity klystron mechanism will apply. Webber[28] has shown that under large signal conditions the optimum drift tube length is 0.1 of a reduced plasma wavelength and this figure is normally used in klystron design.

In the driver stages the gap voltages are small relative to the beam voltage and the amplification is linear. However, in the output stage the 'drive' level ($V_{\text{pen.gap}}$) is sufficiently high for large signal conditions to apply and the r.f. current at the output gap is given by

$$i_{\text{rf}} = 2I_0 J_1(K_{\text{p}}) \quad (32.14)$$

where K_{p}, the bunching parameter at the penultimate gap, is given by

$$K_{\text{p}} = \frac{\omega z V_{\text{pen.gap}}}{2U_0 V_0} \quad (32.15)$$

The output voltage as a function of drive has, therefore, the form of a Bessel function (*Figure 32.27*). At signal levels well below saturation the behaviour is approximately linear but on raising the power level the gain falls smoothly to a saturation value 6.0 dB below the small signal gain.

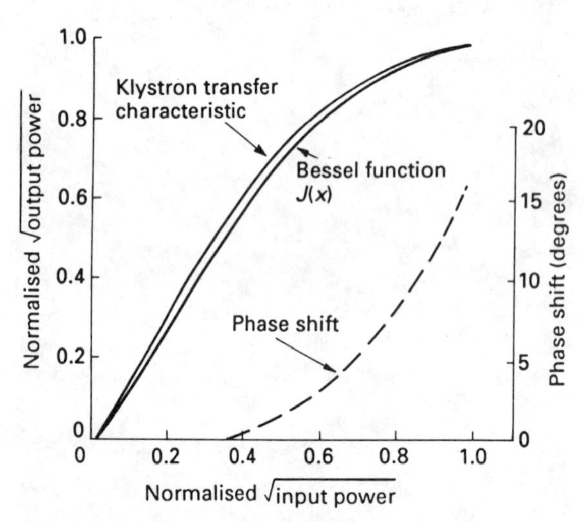

Figure 32.27 Typical klystron transfer characteristic

As the output power increases so kinetic energy is extracted from the beam. This results in a reduction in the velocity of the electrons which introduces a delay and hence a phase shift. This level-dependent phase shift is known as a.m./p.m. conversion (*Figure 32.27*).

Amplifier klystrons require a long electron beam of constant diameter. It is, therefore, necessary to provide some means of constraining the beam which would otherwise spread as a result of the electron forces. By far the most common solution is to apply a magnetic field along the axis of the beam.

For any beam there is a value of magnetic field B_{B}, known as the Brillouin field, at which the centripetal forces generated by the electrons rotating in the magnetic field balance exactly the space charge repulsion forces. This value is given by the relation

$$B_{\text{B}} = \frac{8.32 \times 10^{-4}(I_0)^{1/2}}{r_0 V_0^{1/2}} \quad (32.16)$$

where r_0 is the beam radius.

Unfortunately Brillouin flow requires that the current density in the beam remains constant—a condition which is not satisfied in practical klystrons. It is more usual to employ field values twice to three times Brillouin. Under these conditions the beam diameter perturbations caused by changes in current density are much reduced.

Klystrons decrease in size as the operating frequency increases (because the cavities get smaller and the drift lengths shorter). Large low frequency tubes are usually solenoid-focused but higher frequency tubes often use permanent magnets.

32.3.4 Construction techniques

The construction of a typical external cavity klystron is shown in *Figure 32.28*. The major parts are: the electron gun; the body section, made up of the interaction gaps and drift tubes; and the collector.

32.3.4.1 The electron gun

The cathode and its surrounding electrodes are shaped and positioned to produce an electrostatic field which will focus the electrons through the hole in the anode. A focus electrode usually surrounds the cathode.

The shaped anode, with its central hole, may be made integral with the body or insulated from it in which case it is known as a modulating anode. Some guns employ beam current control grids or electrodes in front of the cathode.

The electron gun geometry controls the beam current for any given voltage. Under normal operating conditions the beam current is given by the relation:

$$I_{\text{beam}} = K V_{\text{A}}^{3/2} \quad (32.17)$$

where

I_{beam} = the beam current (in amperes)
V_{A} = the anode voltage (in volts)
K = a constant known as the (micro)perveance

32.3.4.2 The body

The beam formed by the gun and focused by the magnetic field passes into the drift tubes and the interaction gaps. At the higher microwave frequencies the body usually consists of a series of cylindrical cavities brazed together or milled from solid material, with dividing walls supporting drift tubes. The tuner mechanism, usually an adjustable capacitance plate, is fitted through a vacuum bellows. The whole body is cooled by either liquid or air blowing.

At lower frequencies, particularly between 400 and 1000 MHz, external cavity klystrons are widely used. In this construction part of each cavity is outside the vacuum envelope, which then consists of a ceramic cylinder joined to robust flanges supporting the drift tubes. The external part of the cavity is made in two halves, joining around the ceramic on a plane across which no r.f. current flows. Two opposing doors or cavity walls can be moved transverse to the klystron axis to provide tuning, while a coupling loop can be rotated to adjust the coupling to external circuitry.

32.3.4.3 The collector

The power density of the beam is usually much too high to be allowed to impinge directly on a metal surface. The density is,

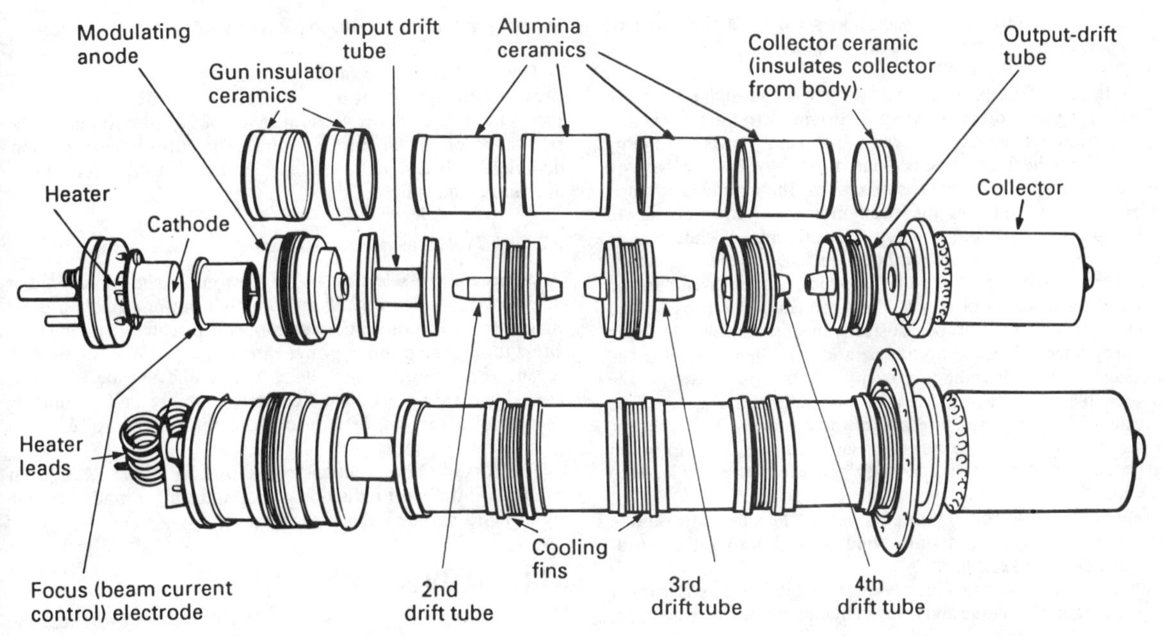

Figure 32.28 Construction of a typical external cavity amplifier klystron (Courtesy of The English Electric Valve Co. Ltd)

therefore, reduced by allowing the beam to spread out on leaving the focusing field beyond the output gap.

32.3.4.4 Cooling

Three methods of heat removal are in common use for high-power klystrons: forced air, liquid and vapour (evaporative) cooling. The first of these is relatively simple but is generally limited to mean beam powers less than 30 kW by the need to keep the collector size reasonable.

High-power tubes may be water or vapour cooled. Water-cooled collectors require large flows and high pressure so the much more efficient vapour cooling technique is usual on the high-power CW tubes.

32.4 Magnetrons

The magnetron is a microwave oscillator with a high efficiency requiring only modest operating voltages for the generation of high power levels. In particular it has found an enduring place in the transmitters of pulsed radar systems and it is with this type of tube that we will be mainly concerned. Other types of magnetron such as the low-power voltage tunable tubes and CW tubes for r.f. heating described more briefly are fully listed in the Further reading section.

The magnetron concept, a cylindrical diode immersed in a uniform magnetic field parallel to its axis, was first reported by A. W. Hull in 1921[29] and was later developed as a high-frequency oscillator.

The development of pulsed radar led to a pressing need for a high-power microwave source in order to improve angular and range resolution and permit the use of physically smaller antennas. The work of Prof. J. R. Randal and Dr H. A. H. Boot[30] at Birmingham University in 1939 satisfied that need with the generation of powers at 3 GHz, several orders of magnitude greater than those previously achieved.

Further war-time development in Britain and the USA provided a profusion of types at frequencies between 1 GHz and 30 GHz operating with output powers ranging from a few kilowatts to several megawatts.[30–33] Methods of mechanical tuning were devised and deployed and a considerable effort was applied, though with less success, to electronic tuning. The post-war years have seen vast improvements in operating stability and life expectancy of magnetrons, while the introduction of the long anode magnetron and the coaxial magnetron, together with the concept of rapid tuning or frequency agility, have further extended the range of application.

32.4.1 Principles of operation

A simplified representation of the main components of the multicavity magnetron in cross-section is given in *Figure 32.29*.

⊕ Magnetic field

$2r_c$ $2r_a$

Cathode Anode structure

Figure 32.29 Main components of the multicavity magnetron

The anode, in addition to being the positive electrode and collector of electrons, also forms a structure resonant near the desired r.f. output frequency. This resonant structure is made up of a number of inter-coupled resonant cavities which provide, in the interaction space, a rotating electromagnetic wave of phase velocity below that in free space so as to give interaction with a rotating cloud of electrons emitted from the cathode.

The magnetic field, which is perpendicular both to the r.f. electric field and to the d.c. electric field (the magnetron is a 'crossed field' device), causes the electrons to rotate round the cathode in the interaction space. The strength of the magnetic field is such that, in the absence of an r.f. field, electrons would be unable to reach the anode.

With an r.f. field present, electrons in a favourable position in the rotating electron cloud will give up energy to the r.f. field and move towards the anode, thereby transforming potential energy arising from the d.c. electric field into r.f. energy. Finally they will be collected on the inner surfaces of the anode structure. Electrons in an unfavourable position in the electron cloud will take up a small amount of energy from the r.f. field and will be quickly returned to the cathode surface where secondary electrons will be emitted, some of which will be useful in the generation of further r.f. power. These returning electrons expend their energy at the cathode surface, an effect known as back bombardment, thus heating the cathode and hence limiting the average power handling capacity. Electrons between the favourable and unfavourable positions described above will be accelerated or retarded into one of these groups. This process, whereby electrons are selected and phased for optimum generation of r.f. power, is known as 'phase focusing' and leads to the high electronic efficiency of the magnetron.

RF output power is coupled from the resonant anode structure, through a vacuum window and into an external waveguide or coaxial line.

The anode voltage at which oscillation can build up is known as the threshold voltage and is given by the following expression:

$$V_{th} = 1.01\left(\frac{r_a}{\lambda}\right)^2 \frac{T\lambda(1 - r_c^2/r_a^2)}{535N} - \frac{4 \times 10^7}{N^2} \quad (32.18)$$

where

V_{th} = the threshold voltage (V)
r_a = the anode bore radius (m)
r_c = the cathode radius (m)
T = the magnetic density (T)
N = the number of resonators
λ = the free space wavelength (m)

The maximum electronic efficiency is given by

$$\eta_{max} = 1 - \frac{1}{NX - 1} \quad (32.19)$$

where

$$X = T\lambda\left(1 - \frac{r_c^2}{r_a^2}\right) \times 4.67 \times 10^{-5} \quad (32.20)$$

32.4.2 Magnetron construction

32.4.2.1 RF structure

The r.f. structure which acts as the magnetron anode is normally made of copper and forms part of the vacuum envelope, thus providing good heat dissipation. However, temperature rise along the vane structure can be a limiting factor on power rating and in high-power CW tubes hollow water-cooled vanes are used. In the case of magnetrons at higher frequencies the power density at the vane tips facing the cathode may be great enough to lead to quite large temperature rises during the operating pulse time.

The r.f. output system is coupled to one resonator of the r.f. structure in conventional types through an appropriate impedance matching system, either waveguide or coaxial line.

In long anode and coaxial magnetrons,[34,35] the r.f. output is symmetrically coupled to the resonator system, in the latter case through a stabilising resonator which can materially improve the frequency pulling and pushing phenomena.

32.4.2.2 Cathode

The cathode which operates in a mixed thermionic and secondary emitting fashion is normally equipped with non-emitting end shields or *hats* to prevent axial spread of the electron stream.

The axial magnetic field required for operation is usually provided by permanent magnets frequently with soft iron pole pieces built through the vacuum envelope in order to reduce the reluctance of the magnetic circuit. The introduction of new materials such as samarium cobalt has led to striking reductions of volume and weight.

32.4.2.3 Typical magnetron structures

A representative selection of magnetrons is shown in *Figures 32.30* to *32.32*. *Figure 32.30* shows a high-power tunable L-band magnetron with conventional strapped r.f. structure giving a 60 MHz tuning range and a power rating of 2.3 MW peak, 3.3 kW mean, with an operating voltage of 39 kV. The tube is vapour cooled, providing excellent frequency stability and permitting the use of a very compact heat exchange system. An external permanent magnet is used.

The 10 kW Q-band magnetron shown in *Figure 32.31* is an example of the great reduction of size and weight made possible by using samarium cobalt magnets.

32.4.3 The 'Dupletron'

Arguably the most innovative development in the marine magnetron field for many years, the 'Dupletron' combines magnetron, duplexer and receiver protection functions in one package, thus eliminating several classes of troublesome interface problem.

Figure 32.30 Vapour-cooled L-band tunable magnetron with anode and cathode assemblies (Courtesy of The English Electric Valve Co. Ltd)

Figure 32.31 Compact 10 kW Q-band magnetron with samarium cobalt magnets

The magnetron is based on the Third Generation Marine Magnetron designs providing power outputs between 3 and 25 kW with good stability and life.

Duplexing is implemented through a modified branched line structure which incorporates a two-stage PIN diode switch/limiter. The switch diode is preactivated by sensing the magnetron input, thus eliminating damaging high-amplitude spikes. Passive receiver protection is provided by self-biasing of the PIN switch diode under incident r.f. power. In both modes of operation the switch is backed up by the second-stage limiter, permitting trouble-free interfacing with the sensitive receivers currently in use. *Figure 32.32* shows an X-band Dupletron rated at 10 kW.

32.4.4 Methods of tuning magnetrons

In many applications it is necessary that the operating frequency of the magnetron be adjustable or variable. In some cases very rapid tuning is required; this is known as frequency agility. Successful tunable magnetrons have been realised using both mechanical and electronic methods.

Figure 32.32 10 kW X-band Dupletron

32.4.4.1 Mechanical methods

Tuning may be accomplished by variation of either the inductance or capacitance of the anode resonator system directly or by tuning a separate resonant cavity coupled to it. In its simplest form the magnetron tuner consists of a conducting ring placed near the end of the magnetron resonator system. When the ring is moved towards the anode the r.f. magnetic field coupling adjacent cavities is modified, the inductance of the resonator is decreased and thus the resonant frequency is increased. The tuner ring may be moved by a simple mechanism passing through a vacuum bellows. The frequency range of the inductive tuner may be increased by attaching conducting pins to the ring; these pins protrude into the resonators of the magnetron anode.

For rapid tuning (frequency agility) the ring may be driven by a transducer capable of rapid movement; both electromagnetic and piezoelectric transducers have been used, the latter being particularly successful in the millimetre wavelength applications. In another manifestation, the spin tuned magnetron, an inductive tuning element takes the form of a toothed ring or tube which rotates in the magnetic r.f. field providing rapid tuning.

Magnetrons with tuners which interact with the electric component of r.f. field, capacitative tuners, have been made. In this case the tuning element may be either a conductor or a dielectric but it must move in regions of high electric field and hence such tuners are prone to arcing when used directly in the anode resonator system.

Coupled cavity tuning has found considerable application in coaxial magnetrons in which the operating frequency may be varied by tuning the stabilising cavity. This has the advantage of separating the tuner from the anode resonator system; it may even be outside the vacuum envelope. In one type of coaxial magnetron frequency agility has been achieved by rapidly rotating dielectric paddles within the external cavity.

32.4.4.2 Electronic methods

Practical electronic tuning of high-power pulsed magnetrons for frequency-agile radar systems has been realised using a multipactor discharge to change the frequency of resonators coupled to the magnetron anode system.

While it is possible to achieve frequency agility using the electromechanical systems described above, these devices suffer from two major limitations. Firstly the variation of frequency with time is essentially cyclic, and although this may be concealed to some extent by varying the radar pulse repetition frequency with time, a modern electron counter measures (ECM) system should be capable of detecting and predicting the inherently cyclic tuning pattern. Secondly the tuning rate is low, leading to a restriction of the pulse to pulse frequency change. Multipactor tuned magnetrons are free from both these limitations permitting improvements in the performance of radar systems to be obtained by reduction of glint, improved subclutter visibility, and better resolution.

Multipactor tuning is achieved by control of a multipactor electron discharge in a number of secondary resonant cavities coupled to the magnetron anode system. The discharge, which is produced by r.f. voltages between cold secondary emitting surfaces, may be inhibited by the application of a switching voltage of 3–4 kV with a negligible current drain if it is applied during the inter pulse period. Thus by switching, say, three tuning elements between 'on' and 'off' states eight discrete operating frequencies can be obtained and the frequency of each operating pulse can be selected at random with a total tuning range of about 1%, 100 MHz at X-band.

The system does not significantly affect the other characteristics of the magnetron and there are only small increases in size and weight. Switching power consumption is very low and variation of power output over the range of frequencies is small, about 1 dB.[36]

32.4.5 Operating characteristics

The major operating characteristics of the magnetron follow from two facts:

(1) There is a sharply nonlinear relation between applied voltage and current through the tube.
(2) It is a self-oscillator.

32.4.5.1 Input characteristics

The input voltage/current relationship results from the fact that no current flows through the device for applied voltages below the threshold value; as soon as the threshold value is exceeded oscillation starts to build up accompanied by a rapid rise in current; the operating voltage rises only slowly above the threshold point. Thus the magnetron input impedance is very high below the threshold point and very low above that point; this fact must be borne in mind when designing magnetron drive circuits. In particular the performance of drive circuits on resistive load will usually differ markedly from performance with a magnetron load.

The performance chart (*Figure 32.33*) shows this nonlinear relation between applied voltage and current for several values of magnetic field, together with contours of constant efficiency. The

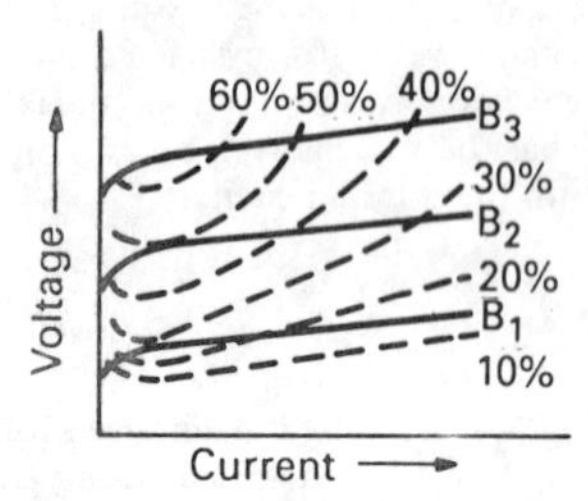

Figure 32.33 Performance chart for magnetron with separate magnet. ——, voltage/current relation for three values of magnetic field; - - - - -, contours of constant efficiency

choice of operating point in this chart may be limited by the onset of various malfunctions in certain regions. It is therefore essential that any deviation from the recommended operating point should be queried with the manufacturer.

In pulsed applications the oscillation must build up from noise levels at the start of each pulse. Thus there is a finite rate of build-up of current and r.f. and hence the rate of build-up of applied voltage must not exceed the recommended value if malfunction is to be avoided.

The front edge of the r.f. output pulse is subject to an inevitable time jitter which is of the order of 5 r.f. cycles in conventional magnetrons but may be an order of magnitude greater in the case of coaxial magnetrons.

32.4.5.2 Output characteristics

The magnetron is a self-oscillator and hence the output power and frequency of operation are affected by the r.f. load impedance; the variation of frequency is referred to as frequency pulling. Output power and frequency are also functions of the input current (the operating voltage remaining nearly constant); the variation of frequency with input current is referred to as frequency pushing.

The effects of both frequency pushing and pulling may be reduced by the use of coaxial magnetrons with their high degree of frequency stabilisation. In particular the low pushing figures obtained by these tubes may in certain circumstances be useful in ensuring minimum r.f. output bandwidth when operating with long duration pulses. The effects of frequency pulling may be reduced by the use of ferrite isolators or circulators. The use of these devices is particularly important when the susceptance of the r.f. load is a rapid function of frequency as in the case of linear accelerators or where the feeder is very long.

Thermal drift of operating frequency occurs as the magnetron warms up or if the ambient temperature changes. In this context the high-power vapour-cooled magnetrons are particularly useful as the temperature of the resonator system is stabilised at the boiling point of the cooling fluid.

32.4.6 Magnetron performance: practical limits

Figures 32.34 and *32.35* indicate the present limits of peak and mean power output in commercially available magnetrons as a function of r.f. frequency.

Available tunability exceeds 10% of the r.f. frequency for slow tuning and 5% for rapid tuning or frequency agility.

Both conventional and coaxial magnetrons have been used successfully in moving target indication radars. In this

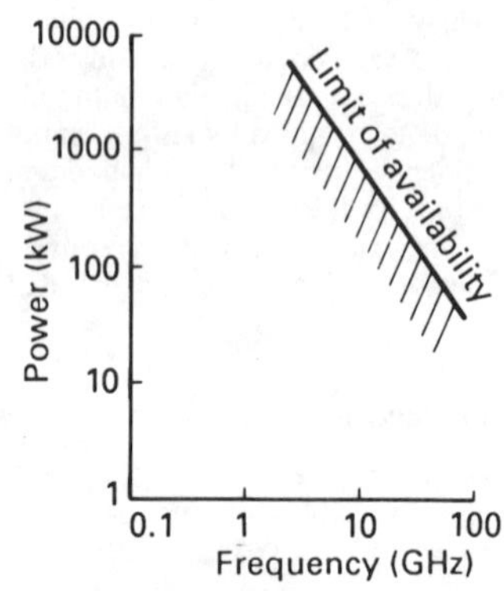

Figure 32.34 Peak output power

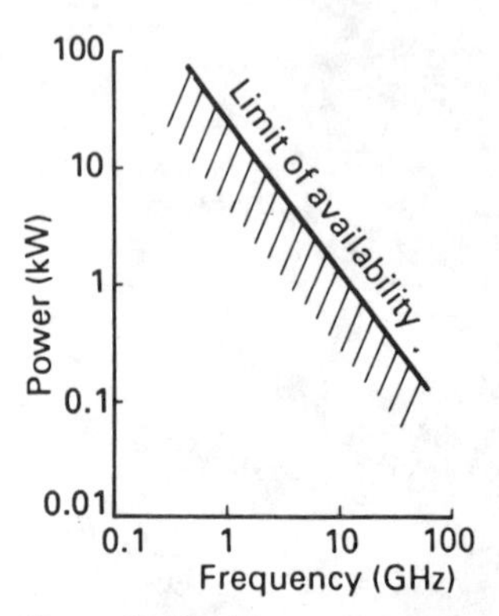

Figure 32.35 Mean output power

application there are demands on magnetron pulse-to-pulse frequency jitter and pulse-to-pulse starting jitter. Rugged magnetrons have been specially developed for airborne and missile applications where the environment can be one of high levels of shock and vibration.

32.5 Travelling-wave tubes

The travelling-wave tube (t.w.t.) is a microwave amplifier capable of amplifying over very wide frequency bands. The amplification process takes place by continuous interaction between an electron beam and an electromagnetic wave propagating along a slow-wave structure. The principle was invented in 1943 by Kompfner[37] who used a simple wire helix as a slow-wave structure. Similar tubes were then developed and first used as microwave amplifiers in microwave relay link systems. During the last 25 years travelling-wave tubes have been continuously developed using other slow-wave structures such as coupled cavities to provide CW output powers of tens of kilowatts and pulse powers of several megawatts with power gains of up to 60 dB. Travelling-wave tubes are usefully employed from u.h.f. to centrimetric wavelengths. However, the original helix slow-wave structure is still one of the most useful due to its great bandwidth. Tubes employing helices have been made with useful amplification properties over a bandwidth greater than two octaves.

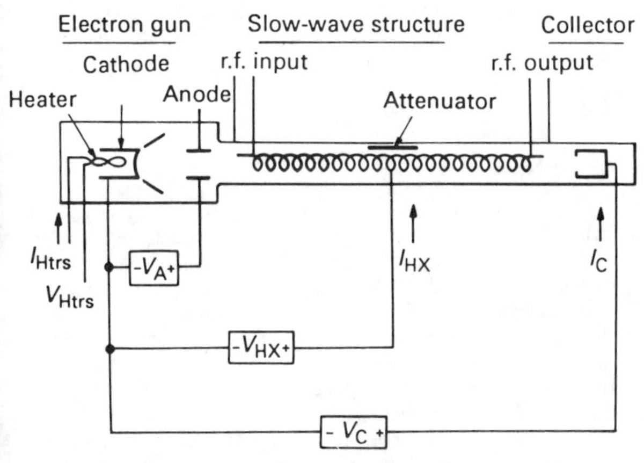

Figure 32.36 Schematic outline of a travelling-wave tube

A travelling-wave tube consists of three main parts as illustrated in *Figure 32.36*:

(1) An electron gun[38] which is capable of producing a beam of electrons.
(2) A slow-wave structure which is in close proximity to the electron beam and which can propagate microwave signals at approximately the same speed as the electrons usually at a fraction of the speed of light. The slow-wave structure has an attenuator region approximately half way along its length to absorb r.f. energy which may propagate in the reverse direction. Without an attenuator the tube could be unstable and self-oscillate.
(3) A collector which is capable of trapping the spent electron beam and dissipating this remaining energy as heat.

An axial magnetic field is required to confine the beam in the structure region and prevent it diverging under the repulsive forces between the electrons. This focusing field can be provided either by a solenoid as in all early tubes or by using permanent magnets. Most travelling-wave tubes with power outputs up to a few hundred watts use a periodic permanent magnet focusing system[39] known as p.p.m. which is lighter, more compact and has less leakage magnetic field than a uniform system.

The velocity of the electron beam in the travelling-wave tube can be adjusted by varying the voltage applied between the cathode and slow-wave structure. The magnitude of the electron beam current is generally controlled independently by varying the voltage on an additional grid or anode in the electron gun.

32.5.1 Amplifier mechanism

At the input and output terminals of the travelling-wave tube there are transitions which allow the r.f. signal to be coupled to the slow-wave structure. The axial component of the r.f. electric field at the start of the slow-wave structure accelerates or decelerates electrons as they enter. This causes a velocity modulation on the beam which gradually gives rise to a density modulation or electron bunching as the beam progresses along the slow-wave structure. As the beam and r.f. wave are in near synchronism the electron bunches induce voltages on the slow-wave structure which reinforce those already present and thereby promote a rapid increase in the r.f. signal. To maintain this interaction the electron beam is initially travelling slightly faster than the r.f. signal wave. However, as the r.f. wave increases the average velocity of the electron bunches is reduced as more energy is extracted from the beam. Finally the electron bunches lose synchronism with the r.f. wave and no further energy is extracted from the beam. At this point the tube is said to be saturated and gives maximum power output. Any further increase of input signal level causes the output to fall. A typical travelling-wave tube input/output or transfer characteristic is shown in *Figure 32.37*.

32.5.2 Gain

The gain depends directly on the length of the slow-wave structure. As the helix or structure voltage is varied above or below the synchronous value the gain decreases rapidly. This

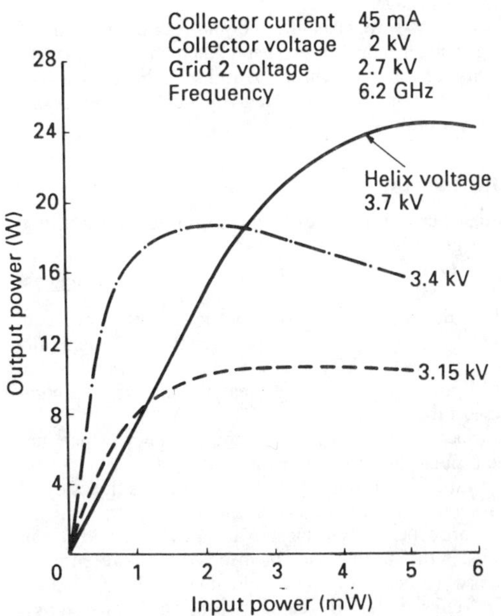

Figure 32.37 Transfer characteristics of a typical N10018 t.w.t. (Courtesy of The English Electric Valve Co. Ltd)

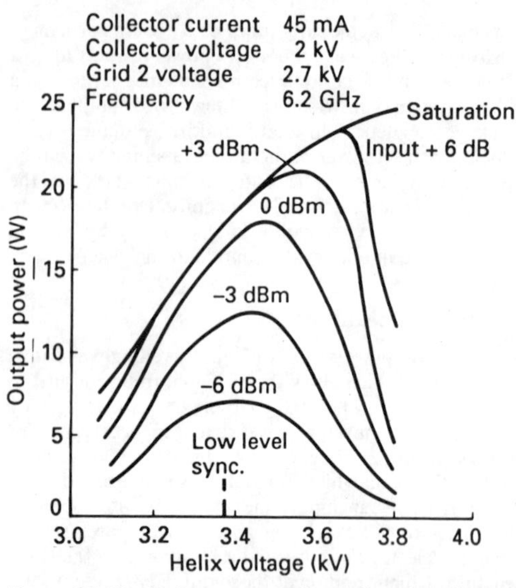

Figure 32.38 Outline power against helix voltage for a typical N10018 t.w.t. (Courtesy of The English Electric Valve Co. Ltd)

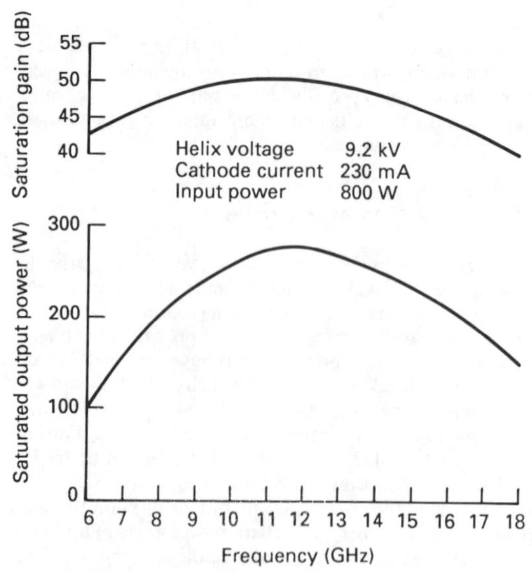

Figure 32.39 Typical performance characteristics of the N10053 t.w.t. (Courtesy of The English Electric Valve Co. Ltd)

characteristic is illustrated in *Figure 32.38*. The gain also increases at a rate proportional to the cube root of the beam current.

The small signal gain G in decibels is given by

$$G = BCN - A - L \tag{32.21}$$

where A, B and L are constants for a given design and have typical values of 10, 40 and 6 respectively. C is Pierce's[40] gain parameter and N is the number of wavelengths along the slow-wave structure:

$$C = \left(\frac{KI_0}{4V_0}\right)^{1/3} \tag{32.22}$$

where K is called the coupling impedance and depends on the geometry of the slow-wave structure, I_0 is the beam current and V_0 is the beam voltage.

For a helix slow-wave structure the phase velocity of the r.f. wave is largely independent of frequency, therefore the wave and beam velocity remain in synchronism and large bandwidths are possible. Typical gain and power/frequency characteristics of a broad-band t.w.t. are shown in *Figure 32.39*.

32.5.3 Efficiency

The highest power which can be obtained from a given travelling-wave tube is given by

$$P_{rf} = kI_0V_0C \tag{32.23}$$

where k depends on the slow-wave structure and electron beam diameters and typically has a value between 2 and 3. Generally the beam efficiencies (100 kC) of the t.w.t.s lie in the range 5% to 10% for low-power tubes and are greater than 35% for high-power pulsed tubes.

One method of improving the beam efficiency which has proved successful is the introduction of a taper[41] in the slow-wave structure towards the output. The r.f. circuit wave is then slowed down at a similar rate to the beam and more energy can be extracted before synchronism is lost. Alternatively the beam can be accelerated in the output region by means of a voltage jump to re-synchronise the beam and wave velocities.

The overall efficiency of a travelling wave can be improved by a factor of 2 or 3 over the beam efficiency by making the cathode-collector potential smaller than the beam voltage and therefore reducing the wasted energy dissipated in the collector. Some travelling-wave tubes used in space applications have overall efficiencies greater than 50%. Such tubes make use of two or three collectors capturing the spent electron beam at progressively lower voltages. The overall efficiency of a travelling-wave tube is

$$\frac{P_{rf}}{I_cV_c + V_{HX}I_{HX} + V_{HRTS}I_{HRTS}} \tag{32.24}$$

32.5.4 Noise

The main sources of noise in a travelling-wave tube arise from the shot noise or current density variations as electrons leave the cathode and secondly the random velocity fluctuations of the emitted electrons. In low-noise tubes[42] the electron gun has several electrodes designed so that the potential profile between the input to the helix and the cathode can be optimised so that minimal noise in the beam is coupled to the helix. Low-noise tubes have noise figures between 6 dB and 10 dB with power outputs of a few milliwatts.

By using a high magnetic focusing field over the cathode and introducing a special beam-forming electrode to produce a divergent electric field at the cathode ultra-low-noise tubes[43,44] are produced with noise figures between 2.5 dB and 5.0 dB.

32.5.5 Linearity

Since travelling-wave tube amplifiers are generally operated in the region of maximum output, nonlinearity of amplitude and phase distortion can occur. In some cases the saturation characteristic is put to good use and the travelling-wave tube may be employed as a limiter. However, in most other amplifier uses where linearity is important the tube is operated 2 dB to 3 dB below the saturated output level.

The conversion of amplitude modulation to phase modulation (a.m./p.m. conversion) occurs in a t.w.t. due to the reduction in the beam velocity as the input signal is increased. At saturation the a.m./p.m. conversion may rise to 5°/dB. TWT microwave link amplifiers are generally designed to have an a.m./p.m. conversion of about 1°/dB at their operating power level.

When several signals are simultaneously amplified by a travelling-wave tube a mixing or intermodulation[45] occurs. This results in intermodulation products (i.p.) at the output with power levels dependent on the relative levels of the original signals. As in the case of a.m./p.m. conversion intermodulation distortion is significantly reduced by operating the tube in the small signal or linear region. With two equal carrier signals operating at saturation the third-order i.p. is typically 10 dB below the carriers.

32.5.6 The construction and types of travelling-wave tube

32.5.6.1 The electron gun

The electron gun shown in *Figure 32.40* produces a controlled diameter beam of electrons from the cathode. The cathode has a

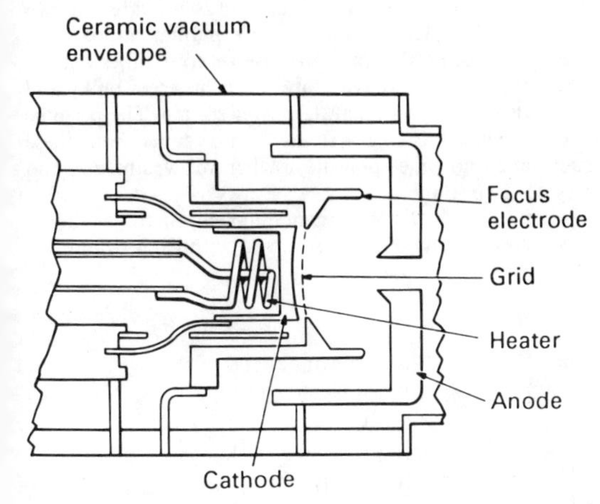

Figure 32.40 The electron gun

surface with a low work function so that electrons are emitted when the cathode is heated. This surface is commonly obtained by depositing a layer of barium and strontium oxides on to a nickel base, stable emission is obtained at temperatures in the region of 750°C. Several other types of cathode construction are in common use but one which is now finding wide application is the barium aluminate cathode. This consists of barium aluminate impregnated into a porous tungsten pellet. The temperature of operation is higher than for the oxide cathodes but higher electron emission densities are possible up to 2 A/cm^2. These cathodes are also less susceptible to ion damage. The operational life of the tube is often governed by cathode life and with care in designs and production this can be in excess of 20 000 hours. The beam current may be switched on and off by modulating the voltage between the slow-wave structure and the cathode. It is also possible to introduce control grids close to the cathode that switch the beam with much smaller voltages.

32.5.6.2 The slow-wave structure

The helix The helix was used in the first t.w.t.s developed and still is commonly used today. The great virtue of the circuit is its capability of producing bandwidths in excess of an octave. This capability has not been equalled by recently developed structures. The helix is however not easily cooled and this generally limits the maximum power output to below 2 kW.

The helix is usually made of tungsten and is wound with great accuracy to give a constant propagation velocity. It is supported by three or sometimes four insulating rods as shown in *Figure 32.41*. In higher-power tubes these are made of beryllia ceramic which having good thermal conductivity keeps the structure cool.

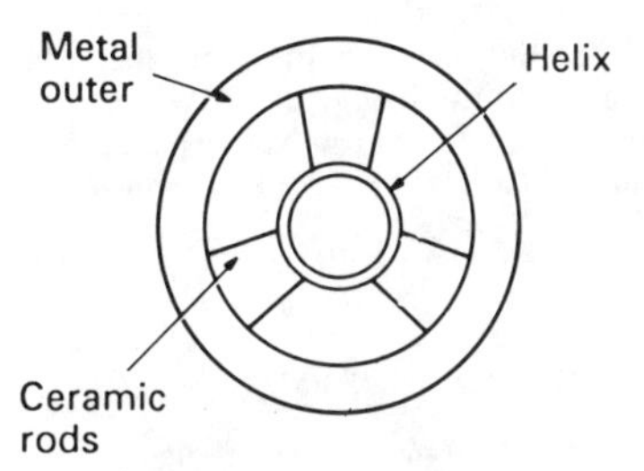

Figure 32.41 The helix structure

For power levels below 20 W the cheaper alumina ceramics or quartz is used.

In low-power non-rugged tubes the outer vacuum envelope is made of glass but with higher power levels a metal cylinder is shrunk or brazed on to the tube rod assembly to ensure a good radial thermal path.

The ring and bar In order to overcome the power limitations of the simple helix the contra-wound helix was investigated and now finds use in the formalised version called the ring and bar structure, *Figure 32.42*. The internal beam diameter is larger than that of a comparative helix and therefore allows beams of

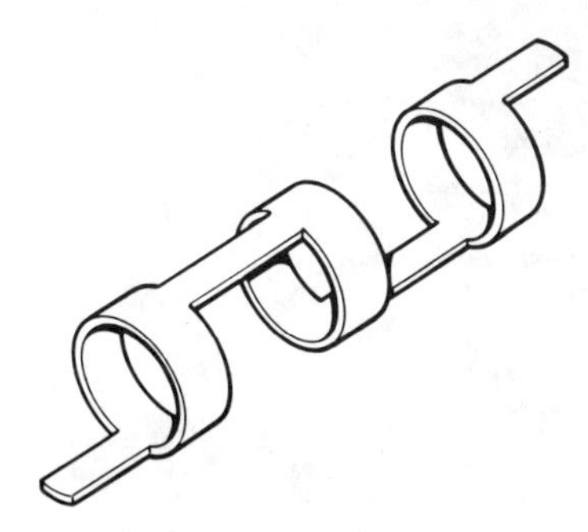

Figure 32.42 The ring and bar structure

lower current density to be used for a given output. The periodic nature of the structure however leads to a reduction in bandwidth. The basic construction is the same as for the helix tubes but power levels of 10 kW have been obtained with 30% bandwidth.

The Hughes structure A series of cavities may be coupled together by slots or loops to form a band-pass structure. These structures are normally made of copper with geometrics such that good thermal paths allow high-power operation. One of the more successful arrangements is the Hughes structure shown in *Figure 32.43*. This structure gives bandwidths of 20% when designed for use in the region of 20 kW CW output power. At higher powers the usable bandwidth is less.

The clover-leaf circuit The clover-leaf circuit has four or six wedges projecting into each pill box cavity and radial slots form

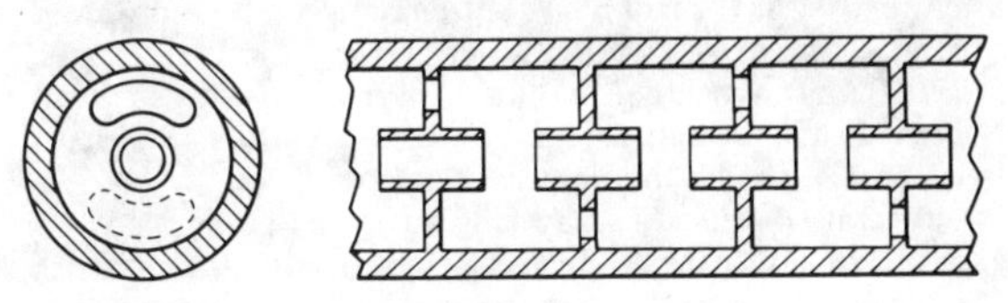

Figure 32.43 The Hughes structure

coupling elements into the adjacent cavities, producing a structure propagating a forward fundamental wave. The structure has found application at output power levels up to 1 MW peak with a bandwidth of 10%. A typical tube[46] is shown in *Figure 32.44*.

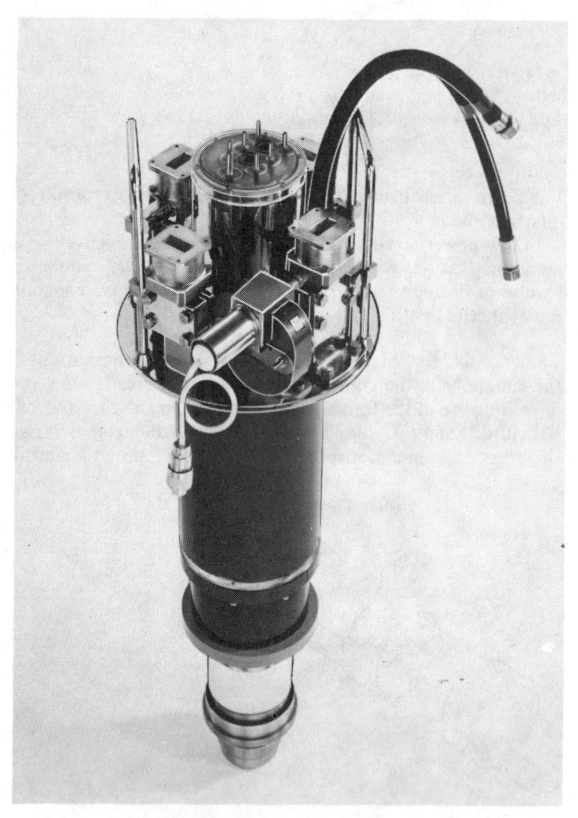

Figure 32.44 A high-power cloverleaf travelling-wave tube type N1061 (Courtesy of The English Electric Valve Co. Ltd)

32.5.6.3 The collector

The collector is an electrode designed to collect the spent beam of electrons after they have left the slow-wave structure. The energy of the electrons is converted into heat on impact with the collector surface and care is taken in the design of collectors for high-power tubes to ensure that internal surfaces of the collector are not overheated. The collector may be cooled by thermal conduction to an external heat sink, by air blown over the collector or by liquid cooling. In tubes where the overall efficiency needs to be high the collector is operated at a reduced voltage relative to the cathode. The electrons are slowed down between the slow-wave structure and the collector and thus have less energy to convert to heat on impact. Operation in this manner is called depressed collector operation.

32.5.6.4 Magnetic focusing

A magnetic field is required to contain the beam as it travels through the slow-wave structure. This field is provided by two main methods—by solenoid or by periodic permanent magnets. The solenoid is used today on low-noise receiver tubes and on some high-power coupled cavity tubes. It is however bulky and heavy and also requires a separate power source. The periodic permanent magnet focusing systems do not suffer from these drawbacks and find wide application wherever weight, size and efficiency are important.

A diagram of a periodic permanent magnet structure is shown in *Figure 32.45*. The polarity is reversed with each cylindrical

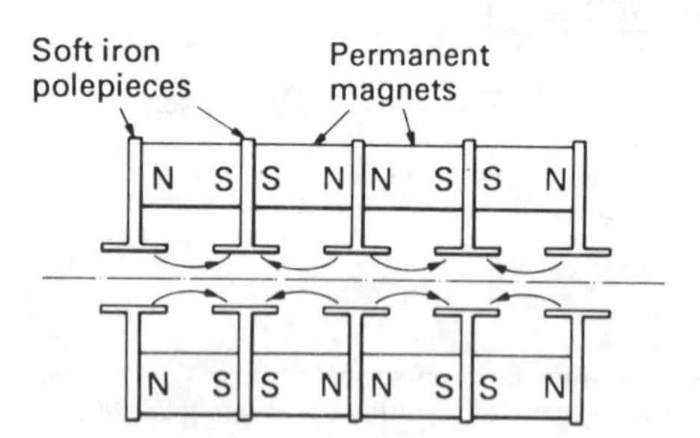

Figure 32.45 Periodic permanent magnet structure

magnet so producing a stack with an alternating magnetic field profile on the beam axis. The focusing system may be an integral part of the tube construction as illustrated in *Figure 32.46* or it may form a separate mount into which tubes are plugged.

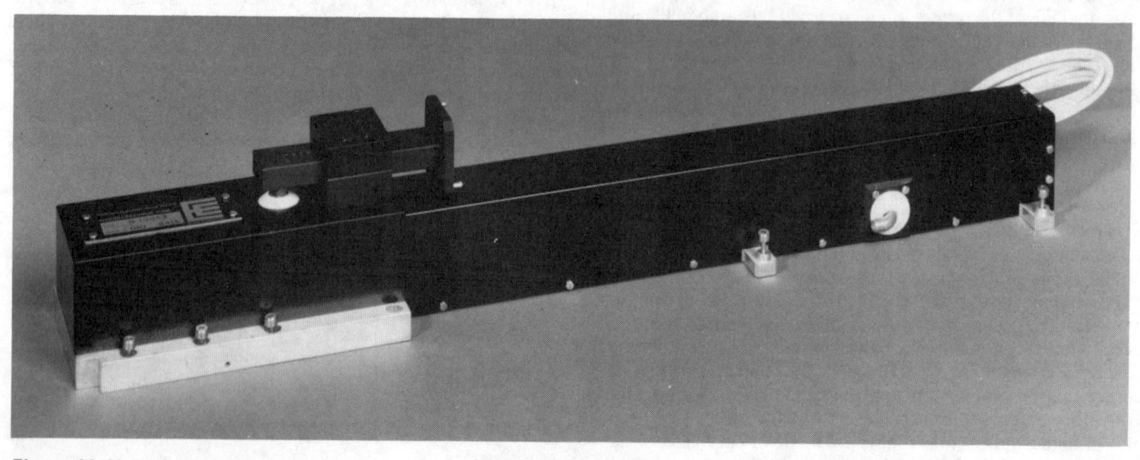

Figure 32.46 A broad-band helix travelling wave tube type N10053 (Courtesy of The English Electric Valve Co. Ltd)

32.5.7 Application of travelling-wave tubes

Travelling-wave tubes are widely used as amplifiers in all types of pulse and CW radars. Where larger bandwidths are required, with greater frequency agility, travelling-wave tubes are used as the final output stage of the radar transmitter as well as in driver stages. Low-noise travelling-wave tubes are used as the first stage in radar receivers. For use in airborne radars where high gain, small size and rugged construction are important the travelling-wave tube finds application. These features also make the travelling-wave tube attractive for space use.

Other applications in the military field include uses in guidance and blind landing systems and control in missile weapon systems. The broad bandwidth of the helix t.w.t. makes it particularly attractive in electronic counter measures (ECM) and jamming applications.

In the civil communications field the t.w.t. is used as a transponder for television broadcasting both on the ground and in satellites. One of the widest uses is in microwave link communication systems where there may be several hundred travelling-wave tubes in one system.

32.6 Television picture tubes

A television picture tube can be regarded as a CRT adapted for the special requirements of displaying a picture. Because a large rectangular picture is to be displayed, a rectangular rather than a circular screen is required to avoid unused screen area. The large display area requires a large deflection angle if the tube is not to be excessively long. This large deflection angle, together with the high beam current when white areas are being displayed on the screen, means that magnetic deflection has to be used instead of electrostatic deflection. It is through considerations such as these that the present-day black-and-white television picture tube has evolved.

32.6.1 Construction of the black-and-white picture tube

The schematic cross-section of a television picture tube is shown in *Figure 32.47*. Electrostatic focusing is used, combined with magnetic deflection. Magnetic focusing was used on some post-war picture tubes, but is not used on present-day tubes.

Electrons are emitted from the heated oxide-coated cathode k, and are brought to a focus between the grid g and the first anode a_1 by the effect of the potentials applied to these two electrodes. It is normal practice to drive the picture tube by modulating the cathode rather than the grid. The d.c. potential of the cathode is determined by the conditions of the video output stage, and this potential is modulated by the video signal. The grid is held at a potential which can be carried to adjust the brightness of the picture on the screen. The d.c. voltage applied to the grid controls the electron flow (beam current) as in any thermionic tube, and so controls the electron bombardment of the screen. When the brightness of the picture has been adjusted, the potential on the grid remains constant. If the video signal were applied to the grid, the grid potential would vary and so alter the point at which the beam was focused. Cathode drive avoids this change of focus, while the varying cathode-to-grid voltage in response to the video signal modulates the electron beam.

32.6.2 Colour television picture tube

In a black-and-white cathode-ray picture tube, a single, pencil-like beam of electrons is accelerated, deflected by a magnetic field into a rectangular scanning pattern and then strikes a phosphor material deposited on the inside of the front glass surface. The high-velocity electrons cause the phosphor to emit white light. There are many inorganic materials that have this light-emitting characteristic, known as cathodoluminescence. The so-called 'white' phosphor is actually a balanced mixture of colour-emitting phosphors. A colour picture tube uses three colour phosphors (red, green and blue emitting) which, if mixed in the right proportion, would also produce white. However, in a colour tube they are not mixed but are deposited on the inside of the front glass surface in many small triplet or triad groups of dots or lines. Instead of a single pencil-like electron beam, the shadow-mask tube uses three such beams, closely spaced. They are accelerated and deflected just as in the black-and-white tube. The three beams converge on each other to meet at the phosphor screen. However, by use of a perforated metal mask close to the phosphor screen, with one perforation for each triplet phosphor group, each beam is cut off or 'shadowed' from two of the phosphor colours and can strike only one of them. Thus, there is one beam that produces only red, one only green and one only blue.

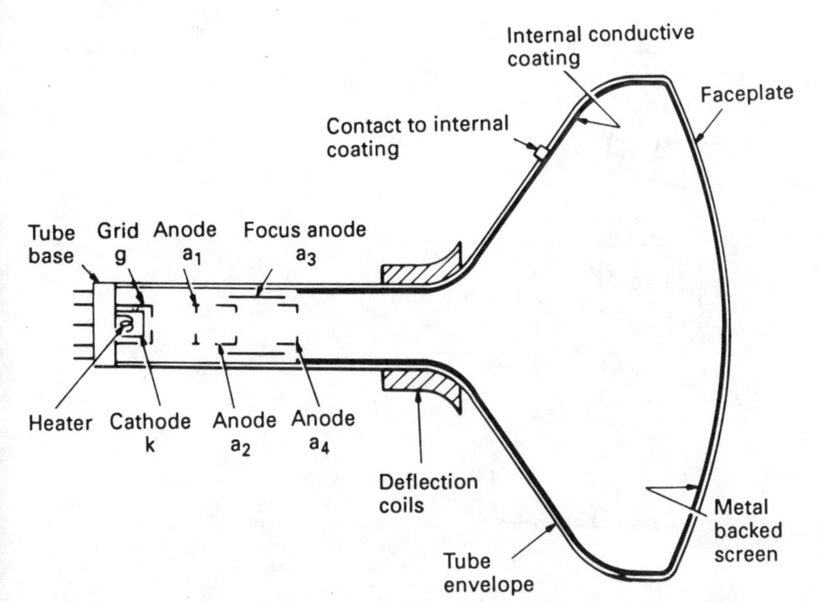

Figure 32.47 Schematic cross-section of black-and-white television picture tube (neck exaggerated for clarity)

With any colour-television system, a receiver can be designed to demodulate or decode the received signal into its three primary component signals—red, green and blue. If each colour component signal is connected to the picture tube to control the corresponding electron beam, separate red, green and blue pictures are produced that appear to the viewer to be superimposed because the eye cannot discern the very small, closely spaced phosphor elements. If all three beams are simultaneously excited, white can be produced, or any colour or combination of colours. Most of the entire gamut of colours distinguishable to the eye are available over a wide range of brightness. Such a colour-picture tube can be used with any system of television.

Figure 32.48 shows schematically a small section of the two most common forms of shadow mask. One form uses small round holes and round phosphor dots, *Figure 32.48(a)*. The three electron beams originate from electron guns in a triangular, or 'delta' arrangement. In the other form, *Figure 32.48(b)*, the mask openings are narrow vertical slits, and the phosphor triplets are narrow vertical lines. In the latter case, the electron guns are on a horizontal line. In both forms, mask and screen are at the same potential and electrons travel in this region in straight lines, as shown. The angle between the beams is exaggerated in *Figure 32.48*; in reality it is only about 1°. The figure does not show the deflecting system or the overall shape of the tube, both of which closely resemble those of a black-and-white tube. However, the figure does show how the shadow effect is used to prevent each beam from striking more than its own colour phosphor. To be noted is that each mask aperture must be exactly in register with a trio of phosphor elements; this is a major problem that makes a colour picture tube so much more difficult to fabricate than its black-and-white counterpart.

32.6.3 Characteristics and limitations

A colour picture tube is characterised by colour fidelity, brightness (luminance), contrast, picture size, resolution and sharpness. Each characteristic usually involves compromise with the ideal because of various limitations, both theoretical and practical.

Colour fidelity is best understood in terms of the CIE colour diagram[47] shown in *Figure 32.49* and derived for an average observer. All visible colours are found by *x*- and *y*-coordinates that lie within the horseshoe-shaped figure. One can think of the *x*-coordinates as one of three colour stimuli, the *y*-coordinates as another and *z*, not plotted, as the third, the sum being unity. Thus, 'white' is in the neighbourhood of $x=\frac{1}{3}$, $y=\frac{1}{3}$, as shown by the central open circle. Extending a radial line out from the white point, the rotational angle determines the hue, and the distance out from white is a measure of the saturation or purity. Colours

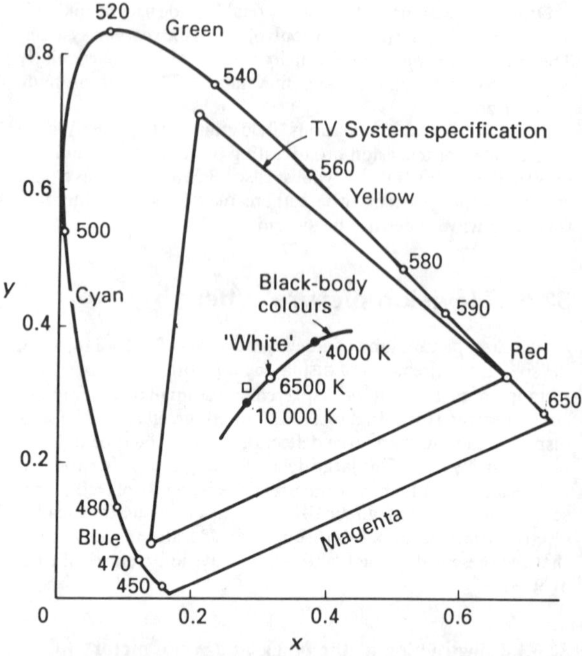

Figure 32.49 The CIE colour diagram. The specified transmitted colour system primaries and white point are shown by open circles. In the USA the receiver 'white' point is often set at the point shown by the square

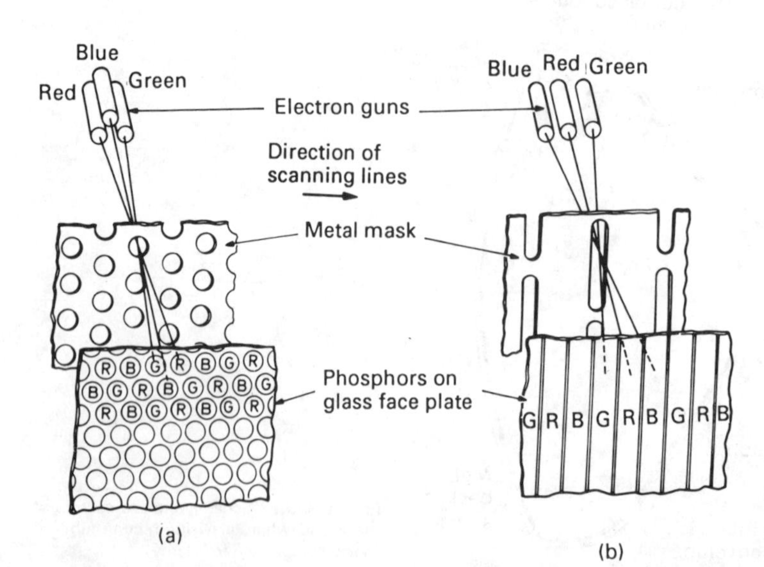

Figure 32.48 Schematic of two shadow-mask systems: (a) with round holes, and guns in a delta configuration, (b) with slit openings and in-line guns. In actual tubes, mask and face-plate are slightly curved

and spectral wavelengths are marked on the figure. A three-colour employs primaries that determine a triangle, and only colours within that triangle are reproduced.

In the colour-television systems used in most of the world, the colour cameras are arranged to produce the three chosen primaries which are approximated by the open circles. A 'white' point is set up, approximately as shown by the central open circle, and actually at $x=0.31$, $y=0.33$. The transmitted colour fidelity is better than that achieved by printing or photography and can be described as excellent. Ideally, the receiver picture tube should have phosphors that match these circles and, in fact, such phosphors are possible but they are not the most efficient. Fortunately, the compromise is hardly noticed by most viewers. Nevertheless, as phosphor research continues, one may expect closer approaches to the ideal, i.e. to the open circles.

A receiver using a colour shadow-mask picture tube, for example with the phosphors indicated by the crosses, can be adjusted to match the transmitted 'white' (open circle) by balancing the three individual beam currents. The open circle represents a *warmer* white than is common in so-called black-and-white tubes, and in the USA many receivers are adjusted to a bluer white shown by the square at $x=0.28$, $y=0.31$. In summary, human colour perception is so adaptive that departures from ideal colour fidelity are well tolerated,[48] and the compromises that are possible are manifold.

The maximum brightness or luminance of a colour picture is primarily determined by the phosphor efficiencies, the electron-beam power in watts per unit area, and the transmission of the shadow-mask and of the faceplate glass of the tube. Bright pictures are highly desirable but there is an increasingly noticeable 'flicker' above a certain brightness, particularly in television systems using 50 field/s picture sequence. Brightness is measured in cd/m^2, also known as *nits* (for those accustomed to older units 1 cd/m^2=0.2918 foot lamberts). The limits to brightness, contrast picture size and sharpness or resolution are interrelated.

The maximum contrast in a picture can be considered as the ratio of the highlight brightness to the brightness of a 'black' portion. The contrast ratio is a maximum when the picture is viewed in a completely dark room, but generally some ambient illumination is preferred. When such an ambient is present, the maximum contrast becomes the ratio of the highlight brightness to the reflected diffused ambient. Phosphors in use have a white body colour, i.e. they are almost fully diffusing. For this reason, means must be employed to minimise the light diffused back to the viewer. In a colour tube, the diffused light will also add white to a pure colour, thereby reducing colour saturation as well as contrast.

There are limits to brightness and, consequently, to contrast that are imposed by the electron-beam power per unit area and the necessary sharpness of the picture. A focused electron beam produces a spot on the screen that gets larger as the current is increased, but the effect is less in shorter tubes with large deflection angles. In a given tube, if an attempt is made to increase the beam current excessively for highlights, the spot gets so much larger that picture sharpness and resolution are lost and the picture appears fuzzy. The beam spot is described as *blooming*. In general, it is desirable to have a small-enough spot to show the scanning lines, although a more important characteristic for a sharp picture is the edge appearance of large-area objects.[49,50]

If all else is equal, large picture sizes are much more pleasing than small ones. The picture shape is fixed by the transmission standards, i.e. rectangular with 4:3 ratio of horizontal width to vertical height. The faceplate is ordinarily made in approximately this shape to conserve space. A practical limit to the size of the picture tube is imposed by cost, and by the awkwardness of length.

32.7 Television camera tubes

32.7.1 Introduction

The central component in the generation of television pictures is the camera tube. A television camera consists of a lens and optical block to direct and focus the light onto the tube, plus control and processing circuitry to operate the tube and amplify/modulate the video output.

The camera tube broadly consists of three elements:

(1) A (generally) planar photosensitive region upon which the incoming light or radiation is focused and absorbed, thereby generating an electronic charge.
(2) A charge storage region usually within the same component as (1), i.e. at the vacuum side of the photosensitive target.
(3) An electron gun which generates a finely focused beam of electrons, which reads off the electronic charge pattern and thus generates the video signal.

The tube generally consists of a cylindrical glass envelope evacuated to a very low pressure. The photosensitive part can either be photoemissive or photoconductive or employ the change in dipole moment due to a temperature increase resulting from radiation absorption. The beam is focused and deflected by electromagnetic and/or electrostatic means and scans the charge storage region sequentially, line by line, restoring the target to its unilluminated state, with this recharging current constituting the video signal. Each tube type varies in the detail of its assembly and the mechanism of signal generation.

Broadcast television began initially with a tube called the iconoscope, but the major tube types contributing to the growth of commercial television in the last 40–50 years are the image orthicon, the vidicon and the lead oxide vidicon.

32.7.2 The image orthicon

The image orthicon was announced in 1946 by RCA[51] and overcame the deficiencies of previous tubes, particularly sensitivity and lag, though it is now almost entirely superseded by the lead oxide vidicon (see *Section 32.7.4.3*).

Seen in *Figure 32.50*, the tube incorporates three major sections:

(1) An image intensifier.
(2) A scanned storage target.
(3) An electron multiplier.

The optical image projected on to the continuous photocathode generates an electron image which is focused by the surrounding solenoid and the electrostatic field in the image section upon the micro-thin glass target. In later tubes this glass has very slight ohmic conductivity[52] compared with the ionic process of early examples. At the target surface secondary electron emission takes place—the secondary electrons being collected by the slightly positive and highly transparent mesh. The loss of secondaries produces a nominal electron gain and a positive charge pattern on the glass of a density varying in proportion to the original pattern of scene brightness. Charging continues in the picture highlight areas until the glass surface slightly exceeds the mesh in potential. The charge pattern is transferred to the opposite target surface for evaluation by the scanning beam. As the target is scanned some beam electrons drift back randomly to the electron gun. Other beam electrons then neutralise the target charges and the then zero potential elements resistively reduce the input face of the target to near zero potential so that a new value of charge may be assumed. The remaining beam electrons are reflected by the zero potential elements roughly along their outgoing path to the limiting aperture of the electron gun. Due to the electron-optical

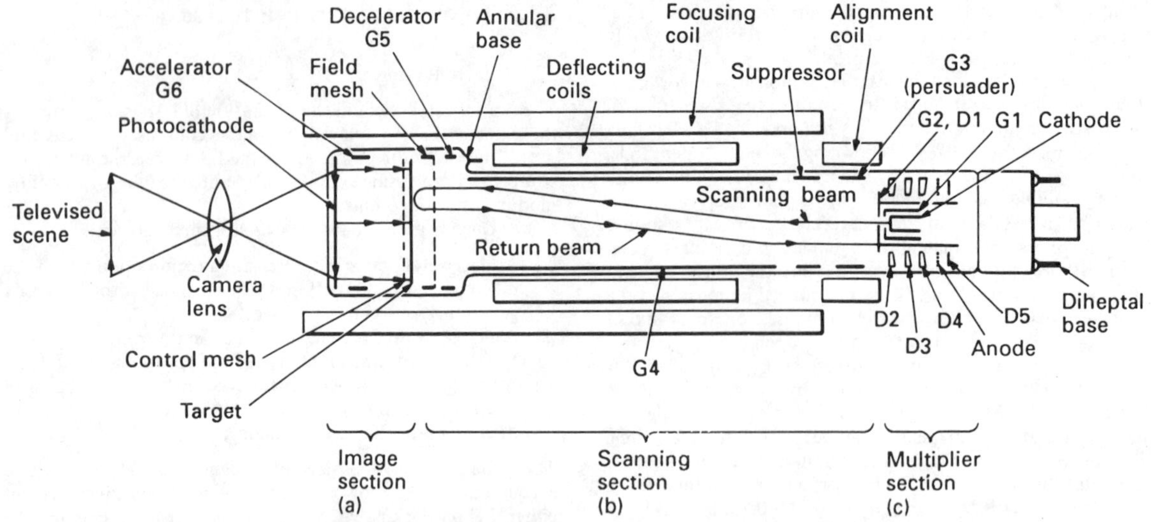

Figure 32.50 Schematic of 3" image orthicon tube. The 4½" version uses the same diameter optical image but incorporates a magnifying electron lens in the image section in order to completely fill the larger target

conditions at the target, the returning electrons impinge on the dynode surface rather than disappear into the aperture. At the dynode, secondary electrons are produced and are guided into the electron multiplier by the negative potential of the persuader. In the five-stage electron multiplier with about 300 V per stage a total gain of about 1000 is achieved to give a final output signal of several microamps—well above the noise level of the head amplifier. The signal amplitude is measured from maximum output corresponding to full beam returned from picture black, to lower values created by the various shades of grey, with a minimum generated by peak white in the original scene. This negative polarity signal is one grave disadvantage of the image orthicon since the maximum return beam from picture black has highest noise content and this is most objectionable in those dark areas of the received picture.

32.7.3 The image isocon

The scanning of the target in an image orthicon gives rise to scattered electrons, in quantity proportional to the scattering charge. By using a selective electrode system, diagrammatically shown in *Figure 32.51*, these can be guided into the electron multiplier and the reflected electrons discarded to form a video signal of correct polarity and with none of the disadvantages of the image orthicon signal. By the end of 1973 such tubes began to assume great importance in low light level television systems due to the possibility of using a lower capacity storage target than is practically feasible in the image orthicon, to produce higher sensitivity without the attendant noise problems.

32.7.4 The vidicon

This name, originally coined by RCA,[53] is now often used as a generic term for all orthogonally scanned camera tubes using a photoconductive target (*Figure 32.52*).

In operation the laminar target suffers a proportionate reduction in resistance according to the local brightness of the incoming light image. The steady potential applied to the transparent conducting backing layer then gives rise to a potential pattern on the vacuum side of the target. When scanned to cathode potential, capacitive coupling across the target layer produces a video output current of varying amplitude proportional to the potential of the element from which it was derived and hence proportional to the elemental brightness of the original scene.

Materials for use as photoconductive targets must satisfy fairly well defined requirements.[54] There must be a maximum conversion of light into available charge carriers and, to achieve this, absorption of the incoming radiation must be high and the effective bandgap of the material should correspond in energy to the longest wavelength in its spectrum. For the visible range a bandgap of less than 2 eV is required. Freedom from trapping centres is also necessary in order that the constituent molecules can return rapidly to the unexcited state.

For satisfactory operation the completed target must have a satisfactory time constant, so imposing upon the basic material a lower limit of resistivity of 10^{12} Ω cm and certain restrictions upon thickness when used as a continuous layer. In this regard adequate thickness is necessary to ensure maximum light absorption but on the other hand it must be limited to minimise charge diffusion, optical scattering and working capacitance.

Various materials can be used as vidicon targets. They include antimony trisulphide, lead oxide and silicon. The latter two materials have low resistivities and must be processed to incorporate blocking contacts in the target layer for satisfactory operation. Details are given below.

32.7.4.1 Antimony trisulphide

Basically of the formula Sb_2S_3, antimony trisulphide has seen extensive use in vidicons.[55] It is the least costly material to process but has limited sensitivity and suffers from movement lag.

As a target layer, antimony trisulphide can be vacuum evaporated to be thin and dense or, in an atmosphere of inert gas such as argon, to a porous film of greater thickness. The operating capacity of the latter is desirably lower as also is its sensitivity. In practice, combinations of hard and porous coatings about 1 μm to 3 μm thick are used. Controlled exposure to air is permissible to allow convenient assembly methods. Antimony trisulphide vidicons have a standing output current or dark current which is temperature dependent when not illuminated, a severe disadvantage in colour cameras. Operational gamma varies from 0.7 at low illumination to 0.4 at higher levels.

Resolving power of early antimony trisulphide vidicons was inadequate and was shown to result from the presence of heavy

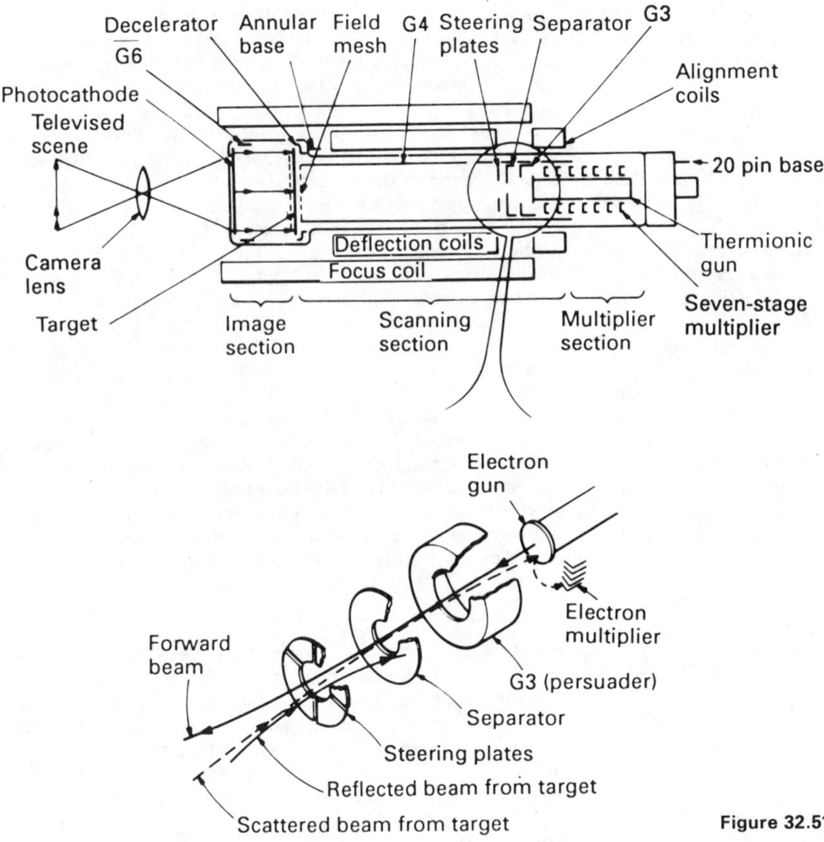

Figure 32.51 Schematic diagram of the image isocon

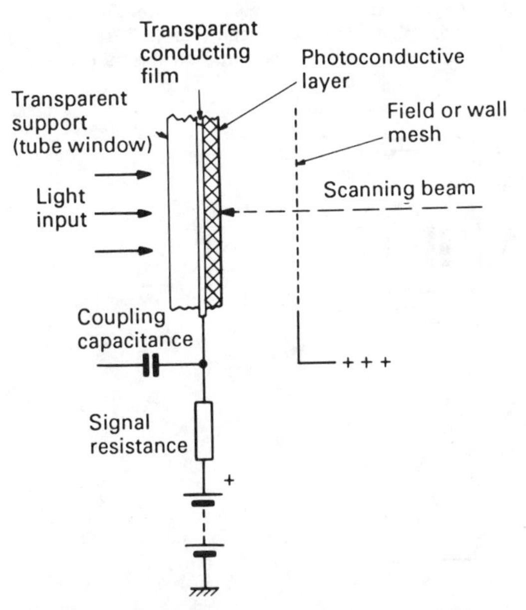

Figure 32.52 Schematic diagram of vidicon target arrangement. The photosensitive layer can have many forms, homogeneously solid or porous. It can also be of sandwich construction combining layers of both types laid down in any sequence

positive ions in the end region of the beam focus electrode near to its terminating field mesh. A considerable improvement is obtained when the latter is electrically separated from the beam focus electrode and the two operated at slightly different potentials, see *Figure 32.53*.[56–58] The separate mesh tube allows a correction to be made for the nonlinear magnetic field at the target end of the focusing solenoid. Furthermore, due to the higher field gradient between the mesh and target, higher beam currents can be used for better control of picture highlights.

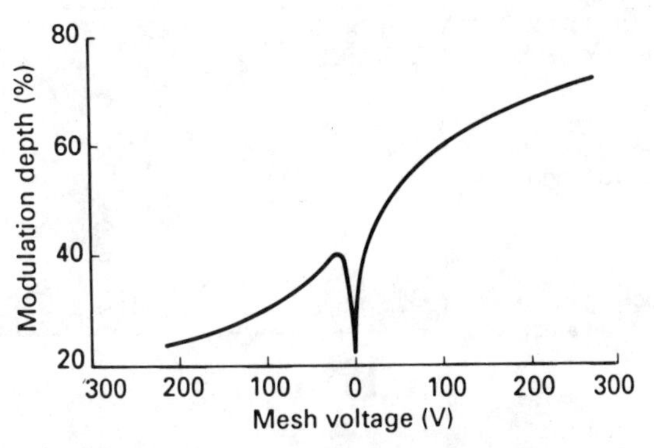

Figure 32.53 Graph showing relation between resolution and mesh potential of a separate mesh vidicon type tube

In spite of several operational drawbacks such as smearing of moving objects, high temperature dependent dark current, etc., the antimony trisulphide vidicon is widely used in a variety of closed-circuit television applications.

32.7.4.2 Silicon

Pure silicon exhibits very high photoconductivity. However, due to its low ohmic resistance, a mosaic of discrete diodes must be formed for it to serve as a camera tube target.[59–61]

One technique is to first heavily oxidise the eventual mosaic surface of an optically worked lamina of single crystal silicon. Then, by photolithographic techniques, the continuous silicon dioxide layer is converted into a fine grid a few microns in pitch—so exposing the base silicon material as a conglomerate of holes. Next a p-type material such as boron glass is fused over the whole surface so that each silicon hole now becomes a p–n junction, surface insulated from its neighbours. To minimise the self-biasing action of the insulating grid when scanned, a metal or reasonably conducting material is applied to each diode surface to reduce the effective bar width of the grid but still maintaining mutual insulation between the diode surfaces. *Figure 32.54* indicates this arrangement.

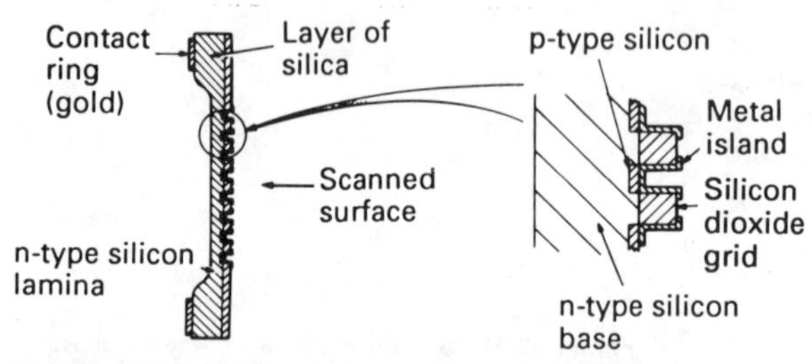

Figure 32.54 Diagrammatic cross-section of a silicon diode array target

The final manufacturing operation is to chemically etch the back of the intended picture raster area to its operational thickness—about 25 μm. Sometimes this surface is given dichroic treatment to enhance a particular spectral sensitivity. The completed target is mechanically mounted in conventional vidicon position with minimum clearance between it and the inside of the tube faceplate. The inevitable small gap is of no optical consequence but does preclude the use of fibre-optic techniques.

In operation, the continuous silicon backing plate is biased some 10–20 V positive with respect to the scanning beam cathode. Scanning the mosaic stabilises it at cathode potential so that each diode assumes a reverse bias condition—the p-type islands just below cathode potential. Behind each one, and probably forming a continuous layer, is a region depleted of current carriers to insulate it from its neighbours.

When illuminated, current carriers, electrons and holes, are photoelectrically produced in the silicon layer. The electrons are attracted by the positive signal plate potential and the holes migrate to charge the p-type islands for evaluation in normal manner by the scanning beam.

The silicon vidicon has very high sensitivity and extends into the near infrared. It is of use in this spectral band for simple surveillance work. Its large-scale use was intended to be for the television telephone where its ability to survive electrical and light overloads and its anticipated long life is of extreme advantage compared with other vidicon-type tubes. The tube is used to a limited extent in the red channel of some vidicon colour cameras.

32.7.4.3 Lead oxide

For broadcast television the lead oxide vidicon under the trade names Leddicon (English Electric Valve Co. Ltd) or Plumbicon (Philips) has become the standard camera tube for colour TV. The target (tetragonal lead monoxide) is structured as a reverse biased PIN diode with p-type region at the free surface of the target, i.e. facing the electron gun. Thus the dark current is very low, approximately 0.5 nA, and does not vary with target voltage as does the antimony trisulphide vidicon. This PIN structure is achieved by incorporating electron traps at or close to the free surface of the target and hole traps at the interface with the transparent signal plate.

Principle of operation The principle of operation is illustrated in *Figure 32.55*. When the tube is scanned in the dark, electrons from the beam land on the target until all the traps are full.

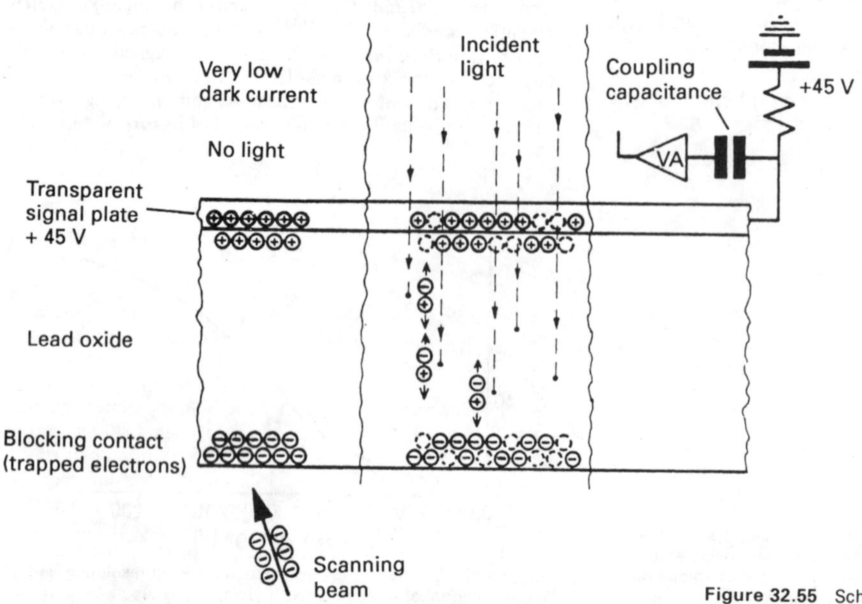

Figure 32.55 Schematic diagram of lead oxide target

Thereafter no further electrons can land and therefore no signal is produced—hence the low dark current. When light is incident on the target it is absorbed at some point, generating an electron–hole pair. Under the influence of the electric field the electron drifts towards the positive signal plate where it annihilates (by recombination) one of the trapped holes. Similarly, the hole drifts to the back of the target where it annihilates one of the trapped electrons. As yet no signal current has been generated but 'vacancies' now exist in the p-type blocking region of the target which relate to the incident light in terms of position and magnitude. As the beam scans the target an electron can now land on each vacant trapping site, and when this occurs an electron instantaneously leaves the signal plate (thus maintaining charge neutrality). This electron leaving the signal plate constitutes the signal current of the tube.

What makes lead oxide most suitable for the high-quality broadcast application is its high sensitivity, fast speed of response to a change in scene (i.e. low picture smear or lag), good resolving power and linear light transfer characteristic. However, several problems arose with early lead oxide tubes and considerable research and development effort over the past 15–20 years has led to significant improvements in performance and diversification in use.

Colorimetry Tetragonal lead monoxide has a bandgap of 2 eV which gives a cut-off wavelength of approximately 640 nm. Since the eye is sensitive to about 700 nm this means that reds were not truly reproduced. This long-wavelength problem was overcome in the 'extended red' tube by doping lead oxide with sulphur to produce a target with a complex structure of lead–oxygen–sulphur. Lead sulphide has a lower bandgap than lead oxide and therefore a higher absorption coefficient for red light, giving improved spectral response and higher red sensitivity. Unfortunately this longer wavelength response is accompanied by a (residual) sensitivity to near-infrared radiation to which the human eye is insensitive and, since many objects such as man-made fibres reflect infrared, colour difelity can again be lost. To overcome this problem an infrared filter is incorporated either in the camera optics or on the face of the tube.

Picture smearing (lag) When an object is moving in a televised scene a 'trail' can appear behind it because of the finite time required by the landed beam in the camera tube to respond to a change in generated photocurrent. This effect is commonly called *lag* and is caused by high-energy electrons in the beam driving the back of the target to below cathode potential, thus increasing the effective resistance of the beam. Lag can be regarded as the time constant associated with layer capacitance and beam resistance in series. It is particularly noticeable at low light levels. A technique to improve this problem of lag is known as light biasing.

The dark current of the tube is increased to about 5 nA by a small amount of illumination either from the camera optics or from a tiny light source incorporated in the tube. This has the effect of stabilising the target at cathode potential, thereby reducing beam resistance and improving lag.

A particular problem with colour TV is that the camera actually produces three pictures—red, green and blue. Differential lag problems can therefore arise between the three channels, and adjustable light bias on each tube helps to minimise this.

Electron gun The most commonly used gun structure is the triode gun as shown in *Figure 32.56*. This has a cathode, control grid G_1 which is held at negative voltage, a first anode G_2, and a wall anode G_3 with a separate mesh G_4. In this gun the beam is extracted from the cathode by field penetration of G_2 as G_1 is made less negative. The beam is focused to a crossover between

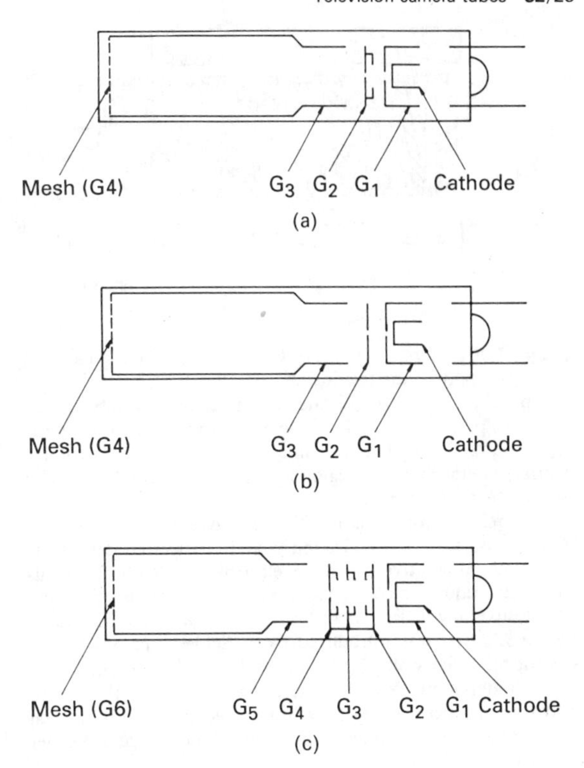

Figure 32.56 Electron gun structures: (a) triode gun, (b) diode gun, (c) tetrode gun (HOP/ACT)

G_1 and G_2. However, electron–electron interactions in the crossover result in a wide energy spread in the beam, giving rise to the high-energy electrons which cause lag problems.

To overcome this, a diode gun structure is now used which by employing a positive (variable) voltage on G_1 with a small G_1 aperture reduces the energy spread in the beam and thus improves lag performance. Another advantage of this gun construction is a steep beam *versus* G_1 voltage slope which permits a high beam to be drawn with just a small excursion of G_1 voltage.

A third structure is designated HOP (highlight overload protection) or ACT (anti-comet tail) and is shown in *Figure 32.56(c)*. This gun incorporates an additional electrode which enables a very high diffuse beam current to be drawn over a short time period.[62]

Comet-tailing A camera tube is normally set up to discharge a fixed signal current, e.g. 600 nA. This means that the tube has sufficient available beam to discharge a light level which will give rise to a signal of 600 nA. The camera iris is then adjusted to generate a typical signal of 300 nA. Therefore, if an increase in light level such as going into a more brightly lit room were to double the photocurrent, the tube could still cope. However, if a very high illumination level occurred, e.g. if the camera panned across a spotlight, the tube has insufficient beam to discharge the generated signal and this would result in beam pulling effects called blooming or (if the highlight moves) comet-tailing.

Two means are available to overcome this. One is to have a fast circuit in the camera which senses the highlight and immediately pulses the electron gun to increase the available beam. This technique is variously known as dynamic beam control (DBC) or comet-tail suppression (CTS), etc. The diode

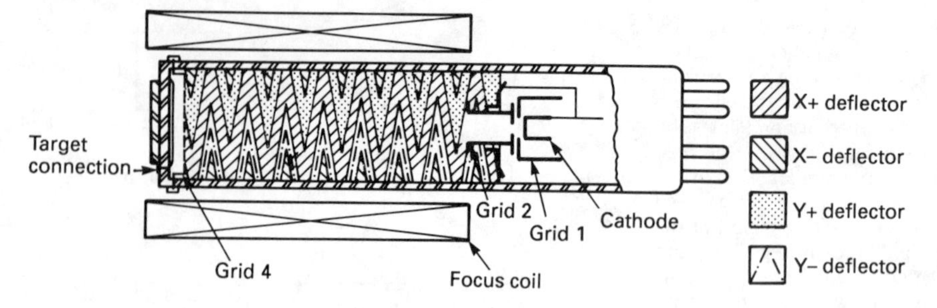

Figure 32.57 Mixed field tube—electromagnetic focus, electrostatic deflection

gun lends itself admirably to this technique although it can and does also apply to the triode gun.

A more elegant but complicated technique uses the tetrode gun (HOP) to pulse a very high diffuse beam up the tube during the short blanked line flyback time. This reduces the voltage excursion in the highlight region to a level which can be handled by the normal beam in read mode.

All three electron guns mentioned above employ magnetic fields generated by the deflection yoke to focus and deflect the beam, but alternative guns use electrostatic fields for focus and/or deflection. The most popular type uses magnetic focusing and electrostatic deflection and is known as a mixed field tube, *Figure 32.57*. It was introduced by K Schlesinger of General Electric for military and industrial use, but with recent advances in computer-controlled manufacturing techniques the mixed field tube has been successfully introduced into the broadcast arena giving improvements in beam landing and corner resolution.

Signal to noise The signal-to-noise ratio is an important parameter in TV camera performance—the higher the ratio the better the overall effective sensitivity of the camera. The noise is evident as random fluctuations, i.e. a grainy, snowflake effect in the TV picture most apparent at low light levels. Most of the noise in a camera is associated with the load resistor and front-end amplifier, with the tube mainly contributing in terms of its stray capacitance to earth. The signal-to-noise ratio is determined by the following equation:

$$S/N \approx 10 \log \frac{I_s^2}{4kTB\left\{\frac{1}{R_T}+\frac{4\pi^2}{3}B^2\left(\frac{C_0^2}{g_m}\right)\right.} \text{ (dB)} \tag{32.25}$$

Where I_s is the signal current, B the camera bandwidth, R_T the load resistor, g_m the transconductance of the pre-amplifier FET, and C_0 the capacitance of the FET plus tube in the deflection yoke. The capacitance of the tube is proportional to the area of the signal plate and target connection ring. Modern tubes incorporate a ceramic signal ring with narrow metallised strips in place of the traditional metal ring. For a typical 25 mm tube this will reduce the in-yoke capacitance by approx. 4 pF, thus improving the S/N ratio by about 4 dB.

32.7.5 Infrared television

Converting infrared radiation or heat scenes into television pictures has many applications, the most dramatic of which is the use of TV cameras by fire services to locate bodies in a smoke-filled room, or to identify the source of a fire. The camera tube used for this application is a pyroelectric vidicon, shown schematically in *Figure 32.58*. The tube has a germanium window designed to transmit 8–14 μm radiation, a triglycine sulphate target and a conventional triode gun. An internal electric dipole in the target changes moment in response to a change in incident radiation, generating a charge on its surface

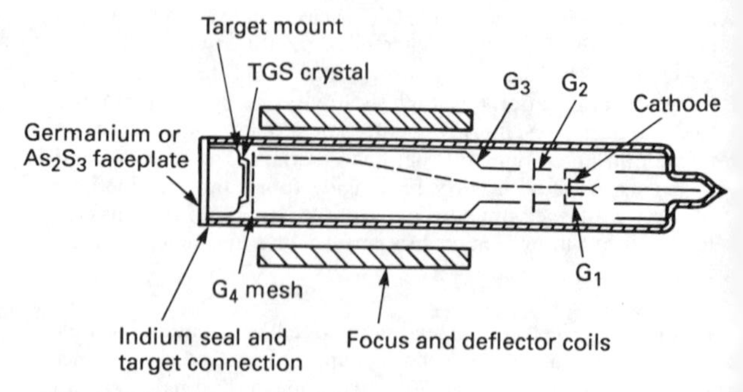

Figure 32.58 Schematic diagram of a pyroelectric vidicon

which is read off by the electron beam much as with the conventional vidicon.

Since the target responds only to temperature changes it is necessary either to pan the camera continuously or to incorporate a rotating chopper in front of the tube in order to achieve TV pictures.

Where shorter wavelengths (1–2.5 μm) are involved for applications that include defect detection in semiconductor crystals, laser profile observation and inspection of old paintings and documents, a lead oxysulphide vidicon has been developed. This is similar to the extended red broadcast lead oxide tube but with the lead sulphide content increased.

32.7.6 Other target materials

A number of other photoconductive materials with particular advantages for particular applications are widely used. In small electronic new gathering (ENG) cameras a selenium–arsenic–tellurium target is used. The tube is called the Saticon[63] and has good resolution performance.

The Chalnicon[64] has a cadmium selenide/cadmium telluride target with high sensitivity and is therefore useful for certain surveillance applications.

The Newvicon tube has a mainly zinc selenide target and is used in surveillance, medical and X-ray intensified systems. It is also used as the tube in home video cameras, with a striped filter on its faceplate to generate the colour pictures. This latter application is currently being superseded by solid-state imagers.

In spite of the progress being made by solid-state imaging, research continues on improving camera tubes. Recently the research laboratory of NHK, the Japanese broadcasting authority, announced a new tube which uses an avalanche effect in the target to give a ten-fold increase in sensitivity. This tube is still very much at the laboratory stage but shows promise particularly for high-definition television.

32.7.7 Low light level camera tubes

These are of two basic forms—either a combination of a conventional type as described above with an image intensifier[65]

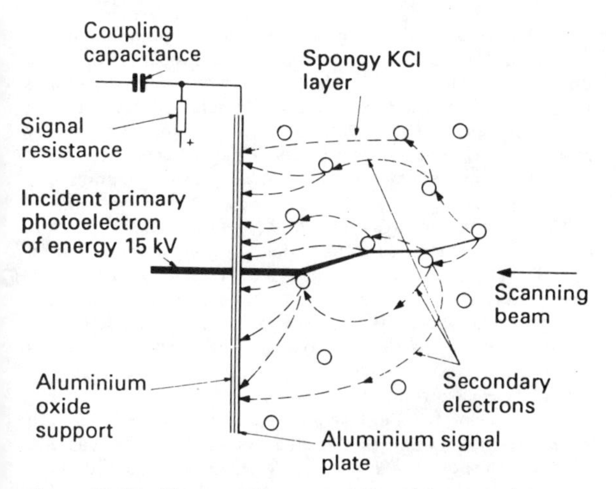

Figure 32.59 Diagrammatic representation of the secondary electron cascade in a target of spongy potassium chloride

or a tube of image orthicon concept but incorporating an intensifier target,[66] *Figure 32.59*. Potassium chloride has been successfully used as an intensifier target.[67] Operated at approximately 15 kV, a cascade of secondary electrons is produced in the thickness of the target from the incoming photoelectrons such that a charge gain of approximately 50 times is achieved. Such tubes carry the generic name of SEC vidicons—the prefix denoting secondary electron conduction.

Arsenic trisulphide,[68] the silicon diode array[61] and zinc sulphide[69] are also used in this manner but in these cases the target conductivity is produced by high-energy electron bombardment. Target gains of 100–500 times are typical of tubes of the electron bombardment silicon target—variously known as SIT (silicon intensifier target) tubes or EBSICON (electron bombarded silicon) type.

In each of the above tubes the video signal is taken from the conducting support for the target mosaic—for the SEC vidicon a mesh-supported aluminium film and, in the case of the EBSICON, the continuous silicon backing layer.

For tube combinations, coupling can be by lens or direct between the output phosphor and fibre-optic window of the image intensifier to the fibre-optic input window of the camera tube. The photocathode of the latter is specially processed to match the spectral output of the phosphor. The increase in sensitivity of the system is, naturally, the gain of the intensifier. Some reduction in resolution is introduced by the fibre optics.

32.8 Photoemissive tubes

Photoemissive tubes are of two types: vacuum tubes and gas-filled tubes. Both types have the same electrode structure sealed in a glass envelope; the difference is that the vacuum type has a high vacuum inside the envelope while the gas-filled type has a low-pressure filling of an inert gas.

The cathode of a photoemissive tube, called the photocathode, is made from a material which emits electrons when illuminated. The photocathode area is large so that the largest amount of radiation is collected and the largest number of electrons emitted. The anode is a thin rod so as not to obstruct the radiation falling on the photocathode. The circuit symbol for a photoemissive tube is shown in *Figure 32.60*.

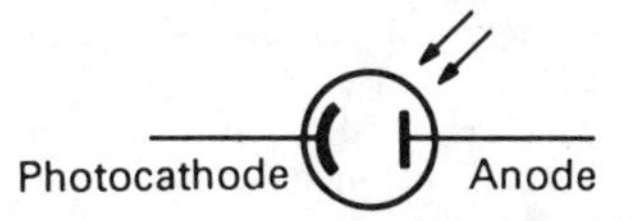

Figure 32.60 Circuit symbol for photoemissive tube

In a vacuum photoemissive tube, if the anode-to-cathode voltage exceeds a value called the saturation voltage V_s, all the electrons emitted when the photocathode is illuminated are attracted to the anode. For a fixed level of radiation, the photocurrent remains constant with increasing anode voltage. As the radiation increases, so a greater number of electrons is emitted from the photocathode and a larger photocurrent flows. Thus a family of anode voltage/current curves with luminous flux as parameter is obtained, as shown in *Figure 32.61*.

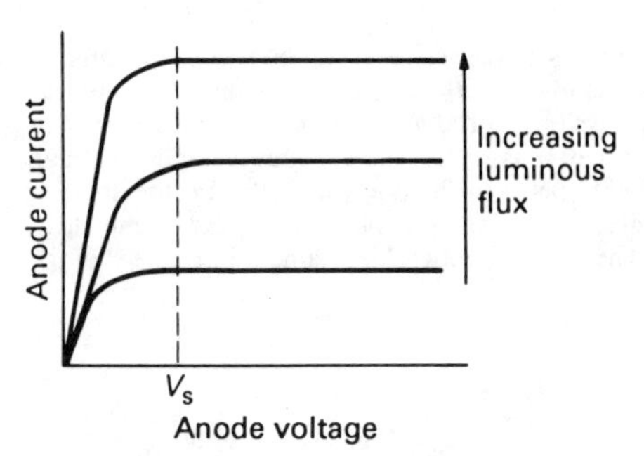

Figure 32.61 Voltage/current characteristic for vacuum photoemissive tube

In a gas-filled photoemissive tube, the photocurrent is initially the same as in a vacuum photoemissive tube up to the saturation voltage V_s. The photocurrent remains reasonably constant with increasing anode voltage until the ionisation voltage v_i is reached, at which voltage the electrons emitted from the photocathode have sufficient energy to ionise the gas filling. There will therefore be an increase in anode current as 'gas amplification' occurs, the gas amplification increasing as the anode voltage is increased. The anode voltage/current characteristic with luminous flux as parameter for a gas-filled photoemissive tube is shown in *Figure 32.62*.

When the photocathode is not illuminated, a small current, the 'dark current', still flows. The magnitude of this current varies

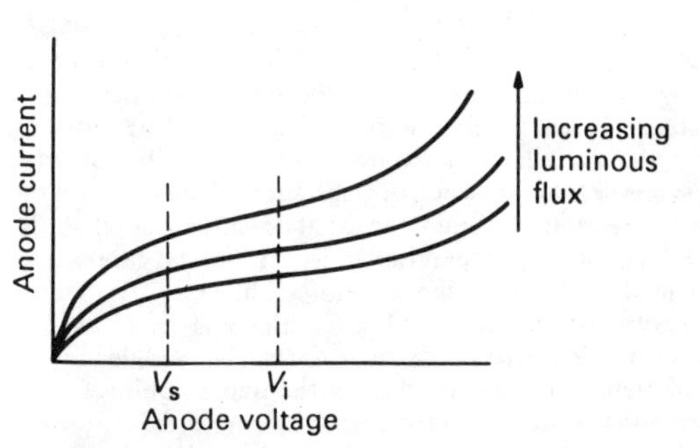

Figure 32.62 Voltage/current characteristic for gas-filled photoemissive tube

with the type of tube, being as low as 0.15 nA typically for a vacuum photoemissive tube and up to 0.1 μA for a gas-filled tube. The spectral response of the tube is determined by the material of the photocathode although this can be modified by using different filters in the window of the tube.

The published data for a photoemissive tube give the spectral response and sensitivity, together with the operating and limiting voltages and currents. The anode voltage/current characteristic may be given with loadlines so that suitable external circuits can be designed.

Photoemissive tubes are used in applications where a current proportional to the incident radiation is required. Many of the applications are similar to those of photodiodes and phototransistors but with the advantages of larger photocurrents being produced and a wider range of spectral responses through the choice of photocathode material.

32.9 Photomultipliers

A photomultiplier is a special form of photoemissive tube in which multiplication of the electrons emitted from the photocathode occurs to produce a larger anode current. The electron multiplication is produced by secondary emission in an electrode structure between the photocathode and the anode.

A simplified cross-section of a photomultiplier is shown in *Figure 32.63*. The photocathode is deposited on the inside of the

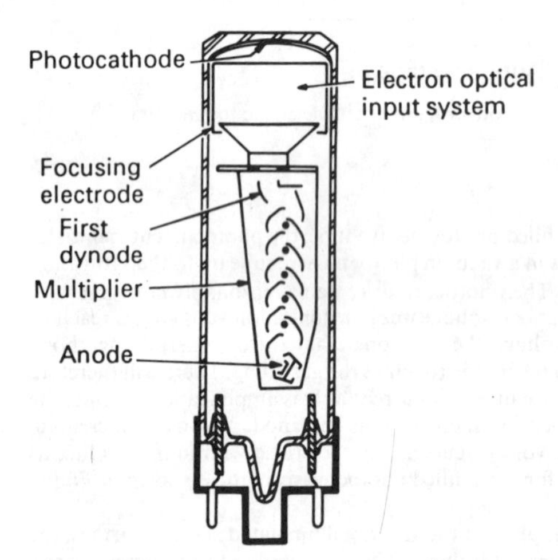

Figure 32.63 Simplified cross-section of photomultiplier

window of the tube. The electrons emitted by the photocathode when illuminated are directed by the electron–optical input system on to the first dynode. Secondary electrons are emitted by this dynode, and these electrons strike the second dynode. Electron multiplication occurs through successive impacts of the secondary electrons through the dynode system so that the anode current of the photomultiplier is considerably larger than the initial current emitted by the photocathode.

The spectral response of the photomultiplier is determined by the photocathode material. The response can be modified by the choice of material for the window of the photomultiplier. It is therefore possible to adapt a photomultiplier to respond to a particular spectrum by the choice of photocathode and window materials.

The transit time of electrons emitted from the photocathode to their striking the first dynode should be independent of the point on the photocathode at which they were emitted. This is particularly important for photomultipliers used in very fast applications. Such differences in transit time are minimised by making the window and photocathode part of a sphere. When a photomultipler is not intended for very fast operation, the window is flat.

The electron–optical system ensures that as many as possible of the electrons emitted from the photocathode reach the first dynode. The focusing electrode is an aluminium ring or a metallised coating on the inside of the tube, and held at the same voltage as the photocathode. An accelerating electrode held at the same voltage as the first dynode accelerates the electrons towards that dynode so that they have sufficient energy to release secondary electrons on impact.

Two forms of dynode structure used in a photomultiplier are the linear cascade and the venetian blind. The linear cascade structure is shown in *Figure 32.64*. Each dynode acts as a

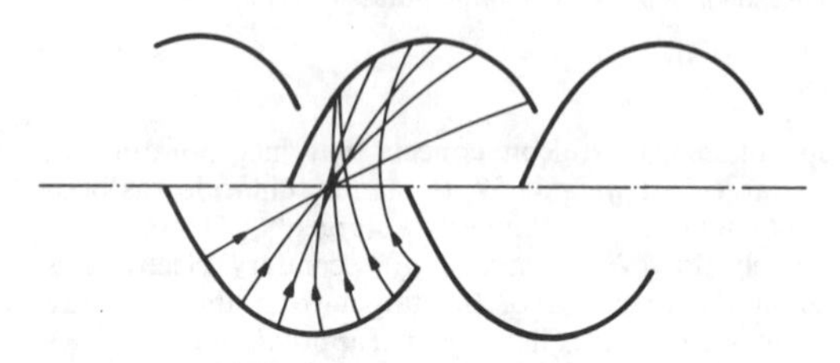

Figure 32.64 Linear cascade dynode system

reflective element, directing the emitted secondary electrons to the next dynode. The voltage between successive dynodes must be large enough to ensure a reasonable collection efficiency and to prevent the formation of a space charge. If a space charge were allowed to form, it would affect the linearity of the anode current with respect to the initial photocathode current. A progressively higher voltage is applied to each dynode through the multiplier towards the anode to obtain the required voltage gradient. The venetian blind dynode structure is shown in *Figure 32.65*. This is

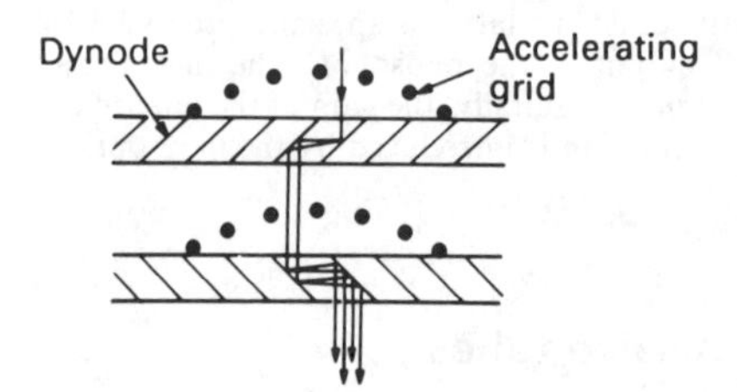

Figure 32.65 Venetian blind dynode system

a regular array of dynodes at 45° preceded by an accelerating grid. All the electrons from the photocathode strike the first dynode to release secondary electrons, which strike the first dynode again to release more secondary electrons, which are attracted to the second dynode. Because of the 'double impact', the venetian blind structure is a relatively slow one and can only be used when the transit time of electrons through the dynode structure is not important.

The material from which the dynodes are made must be one which has a good secondary emission characteristic. Typical materials are magnesium and caesium with a silver oxide layer or copper and beryllium, the latter combination being more widely used as it has a higher stability.

The anode of a photomultiplier is in the form of a grid placed in front of the final dynode. The secondary electrons from the preceding dynode pass through the anode grid to strike the final dynode. The secondary electrons released from the final dynode are then attracted to the anode.

The published data for photomultipliers have some items in common with data for photoemissive tubes already mentioned. The material of the photocathode and of the window are given, together with the spectral response. Luminous and spectral sensitivities under specified conditions are also given.

Typical applications of photomultipliers include their use in scintillation counters, flying-spot scanners, low-level light detectors, photometry, and spectrometry. In scintillation counters, the incident radiation produces small flashes of light in the scintillation material which are amplified by the photomultiplier whose output is used in a counting or detection circuit. In a flying-spot scanner the photomultiplier provides a convenient method of producing an electrical signal proportional to the light transmitted from the spot on the cathode-ray tube through the slide or film being scanned, while the low-level detection, photometry, and spectrometry applications all make use of the high luminous sensitivity and spectral response of the photomultiplier.

32.10 Image intensifier and image converter

The image intensifier and image converter operate on the same principle. An image of the screen to be viewed is focused on the photocathode of the tube. The photocathode emits electrons in response to the incident photons corresponding to the illumination of the scene. The photogenerated electrons are accelerated to strike a luminescent screen. Because each incident photon on the photocathode gives rise to many photons on the screen, a gain in intensity is obtained.

The visible image on the screen is determined by the spectral response of the photocathode. If the response lies within the visible region of the spectrum, and the sensitivity of the photocathode is greater than that of the human eye, an intensified image of a dimly lit scene is obtained. The device is therefore called an image intensifier. If the spectral response of the photocathode lies outside the visible region, say in the infrared part of the spectrum, the image on the photocathode will produce a visible image on the screen. The device now becomes an image converter.

In both types of device, any area of the scene that cannot produce sufficient illumination of the photocathode to release electrons cannot produce an image on the screen, however large the gain of the device.

32.10.1 Proximity image intensifier

The original form of image intensifier, developed in the 1930s and still in use today, is the proximity-focused image intensifier. It consists of a flat photocathode placed close to and parallel with a luminescent screen. The screen is held at a high positive voltage with respect to the photocathode, typically 5–6 kV, and the spacing between the photocathode and screen is approximately 0.6 mm. The electrons released from the photocathode by the incident radiation therefore travel in straight lines to the screen. This system produces a distortion-free image on the screen, and this is the advantage of this type of image intensifier over the more modern types.

32.10.2 Single-stage image intensifier

A simplified cross-section of a single-stage image intensifier tube is shown in *Figure 32.66*. A spherical photocathode is used,

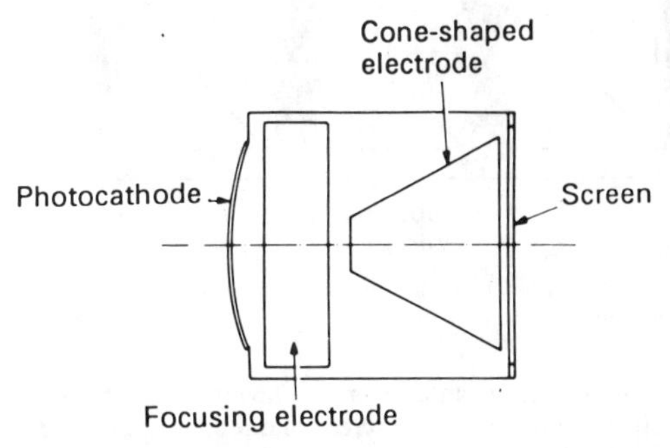

Figure 32.66 Simplified cross-section of image intensifier tube

deposited on the inside face of the window. The emitted electrons are focused by the cylindrical focusing electrode, which is held at the same voltage as the photocathode, at the apex of the cone-shaped electrode. The electrons are accelerated by the cone onto the luminescent screen. Ideally the screen should be spherical with the same radius of curvature as the photocathode, but in practice a flat screen is used. Slight distortion at the edges of the image will therefore occur. The cone-shaped electrode is held at the same voltage as the screen, and the voltage between the photocathode and screen is typically 12–15 kV.

The phosphor used for the luminescent screen is chosen to suit the response of the eyes for direct-viewing systems, or to suit films in systems where the final image is to be photographed. The back of the phosphor on the screen (away from the viewing side) is coated with a layer of aluminium, as in a television picture tube, to increase the brightness of the display. The image with this type of device is inverted with respect to the photocathode. If an upright image is required, however, the input optical system can produce an inverted image on the photocathode.

The gain of an image intensifier is the ratio of the output luminance from the screen to the input illuminance of the photocathode. The gain will depend on the luminous sensitivity of the photocathode and the efficiency of the screen. A typical value for this type of image intensifier is 75.

The published data for this type of image intensifier have elements in common with data for photoemissive tubes. The material, sensitivity, and spectral response of the photocathode are given, combined with details of the luminescent screen. Magnification, resolution, and distortion under specified operating conditions are given. The gain of the device is given, together with a quantity 'background equivalent illumination'. With the device operating but with no input illumination on the photocathode, there will be a background luminance on the screen. This is caused by thermionic or field emission from the photocathode, electron or ion scintillation, or long-term phosphorescence of the screen from previous operation. The background luminance is equivalent to the dark current of photoemissive tubes and photomultipliers, and can be regarded as the noise of the system, determining the minimum input signal that can be amplified by the device.

32.10.3 Multistage image intensifier

Higher gains than can be achieved by the single-stage image intensifier can be obtained by, in effect, connecting several image intensifiers in series. The screen of the first intensifier is used as the input to the photocathode of the second intensifier, and so on.

The schematic cross-section of such a multistage image intensifier is shown in *Figure 32.67*.

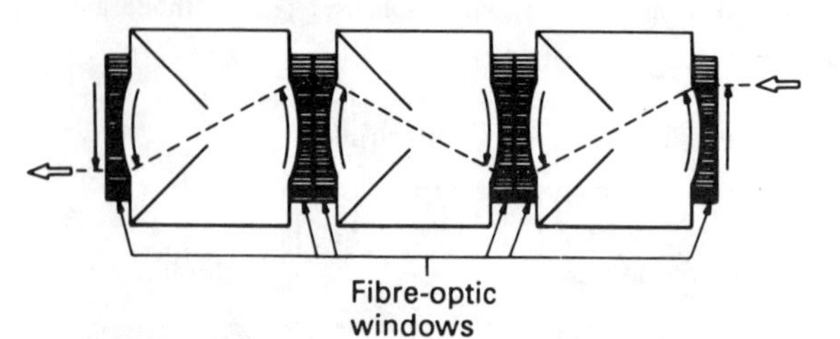

Figure 32.67 Schematic cross-section of three-stage image intensifier with fibre-optic coupling

A three-stage image intensifier is shown, with the input photocathode on the right. The screen of one stage is coupled to the photocathode of the next stage by plano-concave fibre-optic lenses. The radius of curvature of the concave side of the lens on which the photocathode or screen is deposited is chosen to optimise the focusing of the electrons between the photocathode and screen. The plane surface of the lens enables simple coupling between adjacent lenses to be achieved.

32.10.4 Channel image intensifier

A gain comparable to that of the three-stage intensifier in a device of similar size to that of the single-stage intensifier can be obtained from the channel image intensifier. A simplified cross-section of the channel image intensifier is shown in *Figure 32.68*.

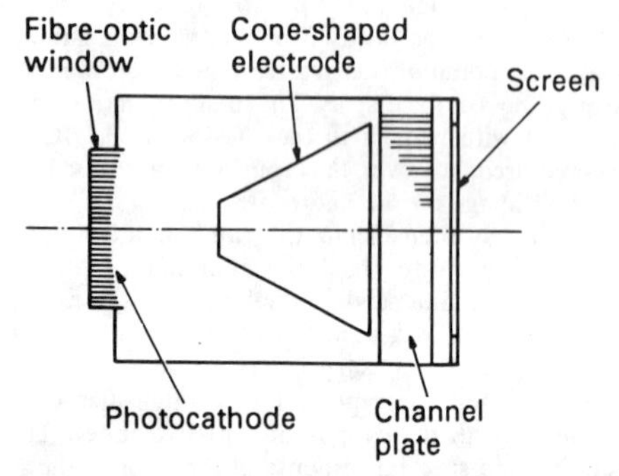

Figure 32.68 Simplified cross-section of channel image intensifier

The high gain results from the use of the channel plate. This may be regarded as a large number of channel electron multipliers packed closely together. A channel electron multiplier is a thin tube, the inside wall of which is coated with a high-resistivity material with good secondary emission characteristics. With a high voltage across the length of the tube, the resistive material on the inside of the tube acts as a continuous dynode. An electron entering the tube and striking the side wall will release secondary electrons. Successive impacts of these secondary electrons result in a large number of electrons emerging from the end of the tube. If the initial electron entering one tube of the channel plate is emitted from a photocathode, and the emerging electrons from the opposite end of the tube are made to strike a luminescent screen, then a high-gain image intensifier can be obtained. The large number of tubes in the channel plate enable all the electrons from the photocathode to be multiplied.

In the cross-section of *Figure 32.68*, the photocathode is deposited on the concave surface of the fibre-optic input window. The electrons emitted from the photocathode are accelerated to the channel plate by the cone-shaped electrode. The channel plate itself is approximately 1 mm thick, and a voltage of 1 kV is maintained across its thickness. The electrons emerging from the channel plate are accelerated to the screen by the high voltage between the plate and the screen, approximately 5 kV. Because of the close spacing of the channel plate and screen, about 0.5 mm, the electrons travel in straight lines to the screen, and so the resolution of the device is high.

The gain of a channel image intensifier is typically 30 000. The gain can be varied by varying the gain of the channel plate, by varying the voltage across it.

The published data for the channel image intensifier are similar to data for the other types of image intensifier. One characteristic that is particular to the channel device is that a curve of gain plotted against channel plate voltage may be included in the data.

32.10.5 Image converter

As mentioned previously, an image converter is a device used to produce a visible image of an 'invisible' scene. The various types of image intensifier described can be used as image converters, the photocathode material being one that responds to radiation outside the visible spectrum. A typical use of an image converter is the viewing of a scene lit by 'invisible light' such as infrared radiation so that the inhabitants of that scene would be unaware of the illumination and of being viewed.

Applications of image converters include military and security uses, as well as astronomical and scientific ones.

References

1 HOUSEKEEPER, W. G., 'The art of sealing base metals through glass', *J. Am. Inst. EE*, **42**, 870 (1923)
2 BRETT, J. E. and MOLYNEUX-BERRY, R. B., 'Pulse width modulator drive for AM broadcast transmitters', *International Broacast Convention* (1982)
3 CLERC, G. and TARDY, M. P., 'The design of high power tetrodes for IRCH applications', *Proc. 14th Symp. Fusion Tech.*, France (1986)
4 STAHL, J. and HOENE, E. L., 'Hochleistungstetroden für die physikalische Forschung', *Vak. Tech.*, **35**, 246 (1986) (in German)
5 FULLER, F. R. and BRUNHART, W., 'A new generation of vacuum tubes for modulator and switch applications', *Int. J. Electronics*, **62**, 837 (1987)
6 SCHNEIDER, P., 'Thermion emission of thoriated tungsten', *J. Chem. Phys.*, **28**, 675 (1958)
7 AMBRUS, J., KEREKES, L. and WALDHAUSER, I., 'The relation between the structure and properties of transmitting tube cathodes', *Tungsram Technische Mitteilungen*, **32** (1977) (in German)
8 SPRANGENBURG, K. R., *Vacuum Tubes*, McGraw-Hill (1948)
9 DUNN, D. A., HAMZA, V. and JOLLY, J. A., 'Computer design of beam pentodes', *IEEE Trans.*, **ED12**, 6 June (1965)
10 PAPENHUIJZEN, P. J., 'A transmitting triode for frequencies up to 900 MHz', *Philips Tech. Rev.*, **19**, 4 (1957/1958)
11 CFTH British Patent 1 206 049
12 MOUROMTSEFF, I. E., 'Temperature distribution in vacuum tube coolers with forced air cooling', *J. Appl. Phys.*, **12**, 491 (1941)
13 PANNET, W. E., *Radio Installations*, Chapman & Hall, London (1951)
14 CFTH British Patent 940 984
15 BEUTHERET, C., 'The vapotron technique', *Rev. Tech. Thomson-CSF*, **24** (1956)
16 BEUTHERET, C., 'A breakthrough in anode cooling systems: the hypervapotron', *IEEE Conf.—Electron Devices and Techniques*, September (1970)

CFTH British Patent 1 194 249
SPITZER, E. E., 'Principles of electrical ratings of vacuum tubes', *Proc. IRE*, **39**, 60 (1951)
British Standards Code of Practice CP1005 (1962)
SARBACHER, R. I., 'Graphical determination of high power amplifier performance', *Electronics*, **15**, 52, Dec. (1942)
THOMAS, H. P., 'Determination of grid driving power in radio frequency power amplifiers', *Proc. IRE*, **21**, 1134 (1938)
VARIAN, R. H. and VARIAN, S. F., 'A high frequency oscillator and amplifier', *J. Appl. Phys.*, **10**, 321, May (1939)
HAHN, W. C. and METCALF, G. F., 'Velocity modulated tubes', *Proc. IRE*, **27**, 106, February (1939)
WEBSTER, D. L., 'Cathode ray bunching', *J. Appl. Phys.*, **10**, 501, July (1939)
WEBSTER, D. L., 'Theory of klystron oscillators', *J. Appl. Phys.*, **10**, 864, December (1939)
HAHN, W. C., 'Small signal theory of velocity modulated electron beams', *GEC Rev.*, **42**, 258 (1939)
RAMO, S., 'Space charge and field waves in an electron beam', *Phys. Rev.*, **56**, 276 (1939)
WEBBER, S., 'Ballistic analysis of a two-cavity finite beam klystron', *Trans. IRE*, **EC-5**, 98 (1958)
HULL, A. W., *Phys. Rev.*, 18–31 (1921)
RANDAL and BOOT, *J. Inst. Elect. Engrs.*, **93**, pt. IIIA, No. 5, 928 (1946)
WILLSHAW, RUSHFORTH, STAINSBY, LATHAM, BALLS and KING, *J. Inst. Elect. Engrs.*, **93**, pt. IIIA, No. 5, 985 (1946)
FISK, HAGSTRUM and HARTMAN, 'The magnetron as a generator of centimetre waves', *Bell Syst. Tech. J.*, 25–167 (1946)
COLLINS, G. B. (ed), *Microwave Magnetrons*, Radiation Laboratory Series, McGraw-Hill (1948)
BOOT, FOSTER AND SELF, *Proc. IEE*, 105B, Suppl. No. 10 (1958)
OKRESS, E. C. (ed.), *Crossed Field Devices*, 2 Vols, Academic Press (1953)
WEAVER, F. J. and STEWART, M., 'Rugged multipactor magnetron for airborne radar system', *IEE Conference Publication* No. 241 (1984)
KOMPFNER, R., *The Invention of the Travelling-Wave Tube*, San Francisco Press (1964)
PIERCE, J. R., *Theory and Design of Electron Beams*, Van Nostrand Reinhold (1954)
STERRETT, J. E. and HEFFNER, H., 'The design of periodic magnetic focussing structures', *Trans. IRE*, **ED-5**, 1, 35 (1958)
PIERCE, J. R., *Travelling-Wave Tubes*, Van Nostrand Reinhold (1950)
SAUSENG, O., 'Efficiency enhancement of travelling-wave tubes by velocity re-synchronisation', *7th International Conference on Microwave and Optical Generation and Amplification, Hamburg*, 16 (1968)
PETER, R. W., 'Low noise travelling-wave amplifier', *RCA Review*, **XIII**, 3, 344 (1952)
CURRIE, M. R. and FORSTER, D. C., 'New mechanism of noise reduction in electron beams', *J. Appl. Phys.*, **30**, 1, 94 (1959)
CHALK, G. O. and JAMES, B. F., 'A wide dynamic range ultra low noise TWT for S-Band', *5th International Conference on Microwave and Optical Generation and Amplification, Paris*, 14 (1964)
KUNZ, W. E., LAZZARINI, R. F. and FOSTER, J. H., 'TWT amplifier characteristics for communications', *Microwave Journal*, **10**, 3, 41 (1967)
CHALK, G. O. and CHALMERS, P. M., 'A 500 kW travelling-wave tube for X-Band', *6th International Conference on Microwave and Optical Generation and Amplification, Cambridge*, 54 (1964)
WRIGHT, W. D., *The Measurement of Colour*, 4th edn, Adam Hilger, Bristol (1969)
BARTLESON, C. J., 'Color perception and color television', *J. SMPTE*, **77**, 1–12 (1968)
HIGGINS, G. C. and PERRIN, F. H., 'The evaluation of optical images', *Photog. Sci. Eng.*, **2**, 66–76 (1958)
MACHIDA, H. and FUSE, Y., 'Gain in definition of color CRT image displays by the aperture grill', *IEEE Conf. Record on Display Devices*, 101–108, Oct. 11–12 (1972)
ROSE, A., WEIMER, P. K. and LAW, H. B., 'The image orthicon—a sensitive television pick-up tube', *Proc. Inst. Radio Eng.*, **34**, 424–432 (1946)
52 BANKS, P. B., 'Improvements in or relating to television camera cathode-ray tubes', *Brit. Pat.* 1 048 390 (1964)
53 WEIMER, P. K., FORGUE, S. V. and GOODRICH, R. R., 'The vidicon photoconductive camera tube', *Electronics*, **23**, 70–74 (1950)
54 TURK, W. E., 'Photoconductive TV camera tubes—a survey', *J. Sci. Tech.* (General Electric Company Ltd), **37**, No 4, 163–170 (1970)
55 FORGUE, S. V., GOODRICH, R. R. and COPE, A. D., 'Properties of some photoconductors, principally antimony trisulphide', *RCA Review*, **12**, No 3, 335–349 (1951)
56 LUBSZYNSKI, H. G., 'Improvements in and relating to television and like systems', *Brit. Pat.* 468 965 (1936)
57 JEPSON, H. B., 'Improvements in or relating to photoconductive devices', *Brit. Pat.* 1 030 173 (1961)
58 DAWE, A. C., 'Special types of vidicon camera tubes', *Industrial Electronics*, November (1963)
59 CROWELL, M. H. and LABUDA, E. F., 'The silicon diode array camera tube', *Bell System Tech. J.*, **48**, 1481–1528 (1969)
60 WOOLGAR, A. J. and BENNETT, E. F., 'Silicon diode array tube and targets', *J.R. Television Society*, **13**, 53–58 (1970)
61 SANTILLI, V. J. and CONGER, III, G. B., 'TV camera tubes with large silicon diode array targets operating in the electron bombarded mode', *Adv. Electronics Electron Physics*, **33A**, 219–228 (1972)
62 DOLLEKAMP, J., 'One-inch diameter Plumbicon camera tube type 19XQ', *Mullard Technical Communications*, **109**, 196–200 (1971)
63 HAOHIRO GOTO, YUKINAO ISOZAKI and KEIICHI SHIDARA, 'New photoconductive camera tube, Saticon', *NHK Laboratories Note No. 170*, September (1973)
64 YOSHIDA, O., 'Chalnicon—a new camera tube for colour TV use', *Japan Electronic Engineering*, 40–44 (1972)
65 NIXON, R. D. and TURK, W. E., 'The image isocon for low light level operation', *J. Soc. Motion Picture Television Eng.*, **81**, 454–458 (1972)
66 GOETZE, G. W., 'Transmission secondary emission from low density deposits of insulators', *Advances in Electronics and Electron Physics*, **XVI**, 145–153 (1962)
67 GOETZE, G. W. and BOERIO, A. H., 'SEC camera-tube performance characteristics and applications', *Advances in Electronics and Electron Physics*, **28A**, 159–171 (1969)
68 SCHNFEBERGER, R. J., SKORINKO, G., DOUGHTY, D. D. and FEIBELMAN, W. A., 'Electron bombardment induced conductivity including its application to ultra-violet imaging in the Schumann region', *Adv. Electronics Electron Physics*, **XVI**, 235–245 (1962)
69 LODGE, J. A., 'A review of television pick-up tubes in the United Kingdom', *Proceedings of Electro-Optic Conference, Brighton, England*, 253–264 (1971)

Further reading

Australian patent 220 414, 26 February (1959)
BAILEY, P. C., 'New lead oxide tubes', *Sound and Vision Broadcasting*, **11**, 19–21 (1970)
BARBIN, R. L. and HUGHES, R. H., 'New color picture tube system for portable TV receivers', *IEEE Trans.*, **BTR-18**, 193–200 (1972)
BARKER, D., 'An experimental investigation of the energy distribution of returning electrons in the magnetron', *Proc. 8th MOGA Conf.*, Amsterdam (1970)
BARTON, D. K., 'Simple procedures for radar detection calculations', *IEEE Trans. on AES*, Sept. (1969)
BECK, A. H. W., *Thermionic Tubes*, Cambridge University Press, London (1953)
BECK, A. H. W., *Thermionic Valves, Their Theory and Design*, Cambridge University Press (1953)
BECK, A. H. W., *Space-Charge Waves*, Pergamon Press (1958)
BENSON, F. A., *Millimetre and Sub-Millimetre Waves*, Iliffe Books (1969)
BIRKENEIMIER, W. P. and WALLACE, N. D., 'Radar tracking accuracy improvement by means of pulse to pulse frequency modulation', *IEE Trans. Comm. Electronics*, January (1968)

BLAHA, R. F., 'Degaussing circuits for color TV receivers', *IEEE Trans.*, **BTR-18**, 7–10 (1972)

BOHLEN, H., 'Properties of a TV transmitter employing grid modulated klystrons', *Proc. 12th Int. TT Symp., Montreux* (1981)

BRODIE, I. and JENKINS, R. O., 'Secondary electron emission from barium dispenser cathodes', *Br. J. Appl. Phys.*, **8**, May (1957)

CHALK, G. O. and O'LOUGHLIN, C. N., *Klystron Amplifiers for Television*, English Electric Valve Co. (1965)

COOPER, B. F. and PLATTS, D. C., 'Frequency agile magnetrons using piezo electric tuning elements', *Proc. European Microwave Conference*, Brussels (1973)

DE HAAN, E. F., 'The Plumbicon, a new television camera tube', *Philips Tech. Rev.*, **24**, 57–58 (1962–63)

DE HAAN, E. F. and WEIMER, K. R. U., 'The beam-indexing colour display tube', *R. Television Soc. J.*, **11**, 278–282 (1967)

DE HAAN, E. F., VAN DER DRIFT, A. and SCHAMPERS, P. P. M., 'The Plumbicon, a new television camera tube', *Philips Tech. Rev.*, **25**, 133–151 (1964)

DOLLEKAMP, J., SCHUT, TH. G. and WEIJLAND, W. P., 'Advances in Plumbicon camera tube design', *Electron Appl.*, **30**, 18–32 (1971)

EDGCOMBE, C. J. and O'LOUGHLIN, C. N., 'The television performance of the klystron amplifier', *Radio and Electronic Engineer*, **41**, 405 (1971)

FIORE, J. P. and KAPLAN, S. H., 'The second-generation color tube providing more than twice the brightness and improved contrast', *IEEE Trans.*, **BTR-15**, 267–275 (1969)

FONDA, G. R. and SEITZ, F., *Preparation and Characteristics of Solid Luminescent Materials*, Wiley/Chapman & Hall, London (1948)

GARLICK, G. F. J., *Luminescent Materials*, Oxford University Press, Oxford (1949)

GITTINS, J. F., *Power Travelling Wave Tubes*, English University Press, Chapters 4 and 5 (1965)

GOLDING, J. F. (ed), *Measuring Oscilloscopes*, Iliffe, London (1971)

HAMILTON, KUPER and KNIPP, *Klystrons and Microwave Triodes*, McGraw-Hill (1948)

HARMAN, W. W., *Fundamentals of Electronic Motion*, McGraw-Hill (1953)

HENDRY, E. D. and TURK, W. E., 'An improved image orthicon', *J. Soc. Motion Picture Television Engineers*, **69**, 88–91 (1960)

HEPPINSTALL, R. and CLAYWORTH, G. T., 'The importance of water purity in the successful operation of vapour cooled television klystrons', *Radio Electric Engineers*, **45**, No. 8, August (1975)

HOLM, K., 'Behaviour of high power tetrodes at high frequencies', *Brown Boveri Rev.*, **74**, 308 (1987)

JAPANESE PATENT, UK 1 328 546, 'An apparatus for modulating and amplifying high frequency carrier waves' (1970)

JEPSON, R. L. and MULLER, M. W., 'Enhanced emission from magnetron cathodes', *J. Appl. Phys.*, **22**, 9, September (1951)

KIND, G., 'Reduction of tracking errors with frequency agility', *IEE Trans. AES*, May (1968)

KLEM, A., 'Delcalix with isocon', *Odelca Mirror*, **9**, 1–4

KLEM, A. and KINGMA, R. V., 'Low light level systems developed by N.V. Optische Industrie de Oude Delft', *Proc. Electro-Optics Int.*, **71**, 304–312 (1971)

KNOLL, M. and KAZAN, B., *Storage Tubes and their Basic Principles*, Wiley, New York (1952)

KONRAD, G. T., 'Performance of a high efficiency high power UHF klystron', *SLAC. Publ. 1896*, March (1977)

LARACH, S. and HARDY, A. E., 'Cathode-ray-tube phosphors: principles and applications', *Proc. IEEE*, **61**, 915–926 (1973)

LATHAM, R., KING, A. A. and RUSHFORTH, L., *The Magnetron*, Chapman & Hall (1952)

LEVERENZ, H. W., *An Introduction to the Luminescence of Solids*, Wiley/Chapman & Hall, London (1950)

MALONEY, C. E. and WEAVER, F. J., 'The effect of gas atmospheres on the secondary emission of magnetron cathodes', *Proc. 7th MOGA Conf.*, Hamburg (1968)

MORRELL, A. M., *et al.*, *Color Television Picture Tubes*, Academic Press, New York (1974)

NARUSE, Y., 'An improved shadow-mask design for in-line, three-beam color picture tubes', *IEEE Trans.*, **ED-18**, 697–702 (1971)

OKRESS, E. C., *Microwave Power Engineering*, 2 Vols, Academic Press (1968)

PARR, G. and DAVIE, O. H., *The Cathode Ray Tube and Its Applications*, Chapman & Hall, London (1959)

PICKERING, A. H., 'Electronic tuning of magnetrons', *Microwave J.*, July (1979)

PICKERING, A. H. and COOPER, B. F., 'Some observations of the secondary emission of cathodes used in high power magnetrons', *Proc. 5th MOGA Conf.*, Paris (1964)

PICKERING, A. H. and LEWIS, P., 'Microwave sources for industrial heating', *Proc. Bradford Conference on Microwave Heating*', October (1970)

PICKERING, A. H., LEWIS, P. F. and BRADY, M., 'Multipactor tuning in pulse magnetrons', *Commun. Int. (GB)*, March (1977)

PIERCE, J. R., *Theory and Design of Electron Beams*, Van Nostrand Reinhold (1954)

PIERCE, J. R. and SHEPHERD, W. G., 'Reflex oscillators', *Bell Syst. Tech. J.*, **26**, 460 (1947)

RAY, H. K., 'Improving radar range and angle detection with frequency agility', *Microwave Journal*, May (1966)

RUDEN, T. E., 'Design and performance of one megawatt 3.1–3.5 GHz coaxial magnetron', *Proc. 9th European Microwave Conf.* (1979)

SHAW, D. F., *An Introduction to Electronics*, 2nd edn, Longman, London (1970)

SIMMS, G. D. and STEPHENSON, I. M., *Microwave Tubes and Semiconductor Devices*, Blackie (1963)

SIMS, C. D. and STEPHENSON, I. M., *Microwave Tubes and Semiconductor Devices*, Blackie (1963)

SKIDMORE, K., 'The comeback of the vacuum tube', *Semiconductor Int.*, **11**, No. 9, August (1988)

SLATER, J. C., *Microwave Electronics*, D. Van Nostrand (1950)

SPANGENBERG, K. R., *Vacuum Tubes*, McGraw-Hill (1948)

SPEAR, B. W. and POWELL, D. E., 'KIMCODE, a method for controlling devacuations of TV tubes', *IEEE Trans.*, **BTR-9**, **1**, 25–31 (1963)

STAPRANS, A., MCCUNE, E. W. and REUTZ, J. A., 'High power linear beam tubes', *Proc. IEEE*, **61**, 299 (1973)

STRINGALL, R. L. and LEBACQZ, J. V., 'High power klystron development at the Stanford linear accelerator centre', *Proc. 8th MOGA Conf.*, Kluver-Deventer (Amsterdam), 14–13 (1970)

TWISLETON, J. R. G., 'Twenty-kilowatt 890 MHz continuous-wave magnetron', *Proc. IEE*, **3**, 1, Jan. (1964)

TYLER, V. J., 'A new high efficiency high power amplifier', *Marconi Review*, **21**, 96, 3rd quarter (1958)

VAN DE POLDER, L. J., 'Target stabilisation effects in television pick-up tubes', *Philips Res. Rep.*, **22**, 178–207 (1967)

VAN DOORN, A. G., 'The Plumbicon compared with other television camera tubes', *Philips Tech. Rev.*, **27**, 1–4 (1966)

VYSE, B. and LEVINSON, H., 'The stability of magnetrons under short pulse conditions', *IEEE Trans.*, **MTT-29**, No. 7, July (1981)

WARNECKE, R. R., CHODOROW, M., GUENARD, P. R. and GINZTON, E. L., 'Velocity modulated tubes', *Adv. Electronics Electronic Physics*, **3**, 43 (1958)

YOSHIDA, S., 'A wide-deflection angle (114°) trinitron color picture tube', *IEEE Trans.*, **BTR-19**, 231–238 (1973)

33 Transducers

J E Harry
BSc(Eng), PhD, DSc, CEng, FIEE
Reader
Department of Electronic and Electrical Engineering
Loughborough University of Technology

Contents

33.1 Introduction

A transducer is a device which is used to convert a measured quantity (pressure, flow, distance, etc.) into some other parameter. Although transducers include devices that convert one form of mechanical input to a different form, for example the manometer for pressure measurement, we are here concerned almost exclusively with electrical transducers which have become almost universally used because of the ease of transmitting and processing the output signal. The ready availability of low-cost electronic systems and increasing degree of automation has resulted in ever increasing applications of transducers in all areas of industry, research and development for direct measurement and process control.

Transducers use a large number of different electrical phenomena, some of which are listed in *Table 33.1*. The different transducers can be classified as active (in which energy conversion occurs) or passive (in which energy is controlled). The overall operation is similar and a complete measurement system normally comprises the transucer or sensor, a signal conditioner or converter, the output of which may be used to supply an indicator or for control purposes, and a supply in the case of passive transducers. Of the many effects that are used in transducers the principal effects used are variation of resistance, inductance, capacitance, piezoelectric effect and thermal effects which are described here.

Table 33.1 Physical effects used in instrument transducers

(1) Active transducers
Electromagnetic
Piezoelectric
Magnetostrictive (as a generator)
Thermoelectric
Photoelectric (photoemission)
Photovoltaic (photojunction)
Electrokinetic (streaming potential)
Pyroelectric
(2) Passive transducers
Resistance
Inductance
Capacitance
Mechanical-resistance (strain)
Magnetoresistance
Thermoresistance
Photoresistance
Piezoresistance
Magnetostrictive (as a variable inductance)
Hall effect
Radioactive ionisation
Radioactive screening
Ionisation (humidity in solids)

33.2 Transducers for distance measurement

33.2.1 Resistance transducers

Resistance transducers using linear or rotating potentiometers for measuring linear or angular movement are both simple and versatile. Simple measurement circuits can be used enabling low-cost systems to be devised. The potentiometer track is manufactured with a linear resistance and may be either wire-wound or continuous using a conducting plastic. The d.c. resistance R of a conductor is given by

$$R=\rho l/A \tag{33.1}$$

where ρ is the resistivity, l the length and A the cross-sectional area.

The potentiometer is usually supplied with an a.c. input, overcoming problems of d.c. bias of the output and the need to amplify a d.c. signal. The a.c. resistance of the conductor is greater than the d.c. resistance due to the skin effect. At high frequencies it is necessary to take into account the increase in a.c. resistance.

The inductance of wound tracks will be higher than for moulded tracks and where high-frequency excitation is used or high-frequency response is required it may need to be taken into account. Both the inductance and the capacitance can normally be ignored when excitation is d.c. or at a low frequency, but capacitive and inductive effects limit the response at higher frequencies.

The resolution of a wire-wound track is limited by the winding density (turns/mm) and the dimensions of the wiper contact. If the wiper is in contact with only one turn at a time then the resolution is R/n, where R is total resistance and n the number of turns. However, for a continuous output the wiper must overlap at least two windings reducing the resolution to $Rn/(n-1)$.

Moulded tracks are normally used enabling very high resolutions up to 0.0035% with a linearity of ±0.2% over distances in excess of 1 m to be obtained. A typical standard resistance of 40 Ω/mm at an operating voltage of 10 V is used. Where required a second parallel track can be incorporated to give two independent electrical outputs.

The conducting plastic track has the advantage of stepless operation and therefore virtually infinite resolution. Imperfect contact between the potentiometer wiper and the track results in contact resistance and hence noise; however, this will only affect dynamic measurements. Wear of the track occurs due to movement of the wiper; however, the operating life is normally in excess of 10^6 cycles.

The accuracy is expressed as a percentage at full scale deflection and therefore the absolute accuracy varies inversely with the deflection. The output voltage signal limits the sensitivity of the system (V/mm) and depends on the voltage supplied to the potentiometer. This in turn is limited by the maximum permissible power dissipation of the potentiometer track, $W=V^2/R$. If the resistance is high the input impedance of the conditioning circuit following the transducer must be high in order to avoid errors due to loading of the track. The error can be calculated from the circuit shown in *Figure 33.1*. The voltage ratio of the unloaded transducer is given by

$$V_o/V_i=1/x \tag{33.2}$$

when the transducer is loaded this becomes

$$\frac{V_o}{V_i}=\frac{x}{1+mx(1-x)} \tag{33.3}$$

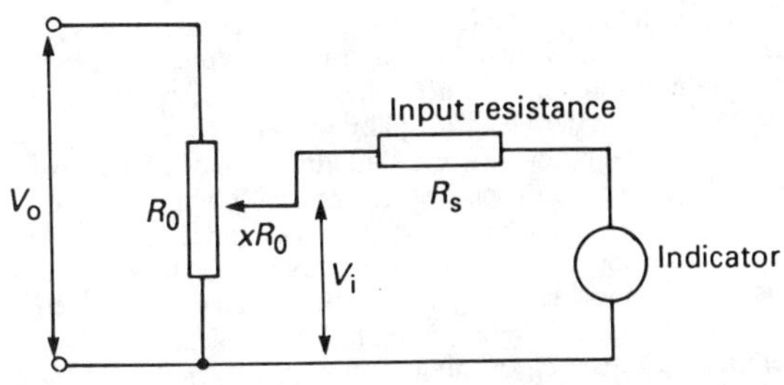

Figure 33.1 Resistance transducer loaded by measuring circuit

where $m = R_0/R_s$ and the error

$$e = \left(1 - \frac{1}{1 + mx(1-x)}\right)100\% \qquad (33.4)$$

The error will be zero at both zero and full-scale deflection of the transducer and will increase to a maximum when the transducer resistance is 50% of the maximum value. In general the load resistance must be of the order of 10 to 20 times greater than the transducer to obtain an accuracy of 1–2% of full scale. The error results in a nonlinear response and to achieve a linear output accurate to within 1% the impedance of the conditioning circuit should be greater than $25R_s$. The principal limitation of potentiometers is the relatively high inertia of the moving parts which limit the rate of response typically to about 0.5 m/s.

33.2.2 Strain gauges

Strain is defined as extension/original length and is dimensionless. Strain gauges utilise the change in resistance of an electrical conductor that occurs when a stress is applied corresponding to a change in resistivity.

The sensitivity of a strain gauge along its axes of measurement can be defined in terms of a gauge factor K:

$$K = \frac{\text{electrical strain}}{\text{mechanical strain}} = \frac{\Delta R/R}{\Delta l/l} \qquad (33.5)$$

For alloy strain gauges $2 < K < 5$ while for semiconductor gauges $K > 100$. The strain gauge will have a conducting path on an axis transverse to the axial path; however, the cross-sensitivity is typically less than 1% for a given axial strain.

The allowable percentage elongation of a gauge depends on the metallurgical condition of the gauge material, the insulating support material, and the adhesive used. It varies from 0.5% for high-temperature applications in which ceramic cements are used up to 5% for polyamide and paper-backed foil gauges.

Strain gauge transducers are extensively used for the measurement of small displacements and, by deduction, of force, pressure and weight. The change in resistance is directly proportional to the strain and is measured with the strain gauge acting as one arm of a Wheatstone bridge. Strain gauges are available in a number of different forms ranging from wire-wound gauges on thin insulating supports to thin film gauges obtained by sputtering and semiconductor gauges.

The measurement of static strain imposes the greatest demands on strain gauge performance and copper nickel alloys are generally used because of their low and controllable temperature coefficients. Nickel iron alloys are used for dynamic measurements, where the larger temperature coefficient is not so important and the higher gauge factor (sensitivity) is an advantage when a high output is required. Nickel chrome and platinum alloys are used where higher operating temperatures are required. Temperature compensation can be achieved by incorporating a dummy strain gauge in the Wheatstone bridge. Temperature limits for dynamic strain measurements are generally much higher than for static conditions.

An important property of a strain gauge material is a very high degree of stability, particularly important for measurement of static strain or in load cells, necessitating the use of high-grade resistance alloy materials, principally nickel chrome alloys and platinum in the form of wire or a thin film, both of which have a low temperature coefficient of resistance. Unbonded strain gauges use fine resistance wire wound in one plane on a former, stress being transmitted to the gauge by an elastic deflection system. Unbonded strain gauges are by nature bulky and cumbersome and are therefore limited in application to areas where they are part of an integral unit such as load cells and accelerometers.

The resistance element of a bonded gauge is attached along its length to the substrate. Various substrate materials are used depending on the application including paper and plastics. The simplest form of bonded strain gauge utilises flat wire formed in a grid (*Figure 33.2(a)*) in the one plane, but this method of construction is relatively bulky. A helically wound grid enables a longer effective strain gauge length to be obtained and consequently higher sensitivity by winding the wire on both sides of the substrate. Very high sensitivities and complex geometrical arrangements are obtainable using photoetched techniques (*Figure 33.2(b)*) to produce miniature strain gauges and composite strain gauges and rosettes to measure strain on more than one axis (*Figures 33.2(c)* and *(d)*). Where high sensitivities are required semiconductor strain gauges formed by deposition can be used.

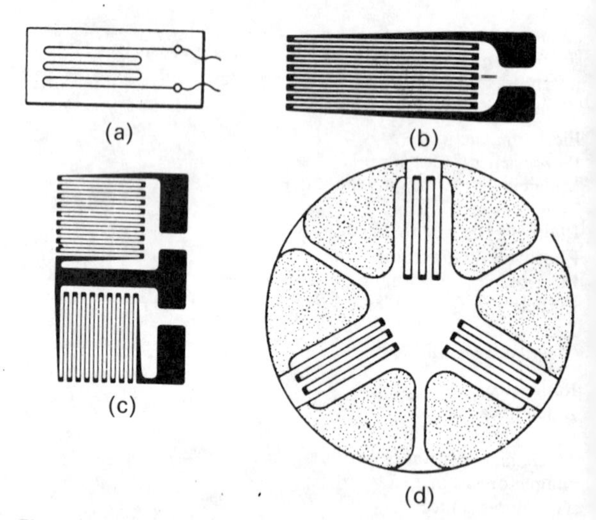

Figure 33.2 Examples of different strain gauges: (a) wire grid, (b) photoetched grid, (c) strain gauge for twin axes measurement, (d) strain gauge rosette

The entire Wheatstone bridge may be incorporated in a single gauge thereby eliminating errors due to temperature variations, etc., and enabling compact transducers suitable for load cells to be constructed.

The attachment of the strain gauge to the assembly for which the strain is to be measured is the critical part of the use of the strain gauge and a wide variety of cements and adhesives have been developed for this purpose. Creep over long periods and shear under high stress must be avoided and the insulation resistance must be high, typically of the order of 10^4 MΩ.

The movement measured in strain measurements is normally very small and as a result the change in resistance of the gauge is correspondingly small and sensitive measuring circuits are required. The most commonly used circuit is the Wheatstone bridge shown in *Figure 33.3*. The bridge is balanced when

$$\frac{R_1}{R_2} = \frac{R_3}{R_4} \qquad (33.6)$$

obtained when no current flows in the null detector. The bridge has the highest sensitivity when the resistors in the bridge arms are equal. Typical values of strain gauge resistance are 100–500 Ω. The bridge is normally balanced by varying a resistance in series with the strain gauge. Since the change in resistance is normally small the effect of temperature variation on the resistance may be

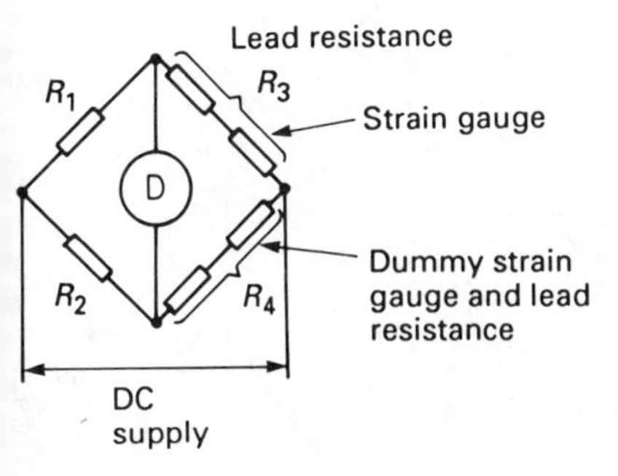

Figure 33.3 Wheatstone bridge with dummy gauge for temperature and lead compensation

significant and it is normal practice to incorporate a dummy strain gauge in the opposite arm of the bridge but which is not subjected to any strain. Any changes in temperature cause a corresponding change in resistance of this gauge therefore minimising the error. The effect of variation in resistivity with temperature will not normally be important for dynamic measurements for which one strain gauge will be adequate. Temperature compensation coupled with increased sensitivity can be obtained where two strain gauges are used, one in tension and one in compression. For example, in the case of a cantilever, or the measurement of torque, the two strain gauges are connected in opposite arms of the bridge, while two active gauges in adjacent arms can be used for measurement of bending strain. The various bridge circuits that are used are referred to as quarter bridge (one strain gauge) half bridge (two strain gauges) and full bridge in which all four arms are strain gauges.

33.2.3 Capacitive transducers

The capacitance between two parallel plates is given by

$$C = \varepsilon_0 \varepsilon_r \frac{A}{d} \tag{33.7}$$

where ε_0 is the permittivity of free space, ε_r is the relative permittivity, A is the effective area of the plates, and d is the separation between the plates. Differentiating gives

$$\frac{dC}{dd} = -\varepsilon_0 \varepsilon_r \frac{A}{d} \tag{33.8}$$

The response of a capacitive transducer is shown in *Figure 33.4*.

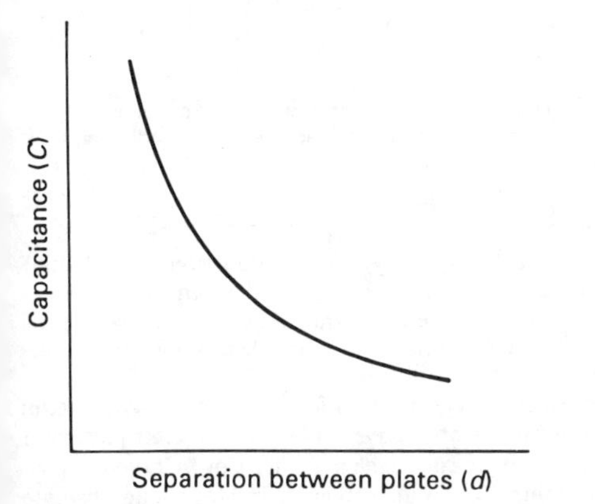

Figure 33.4 Variation of capacitance with separation for parallel plate capacitor

From this it can be seen that the resolution is high when the separation between the plates is small. A number of different capacitor configurations are used for distance measurement some of which are illustrated in *Figure 33.5*. Flat plate

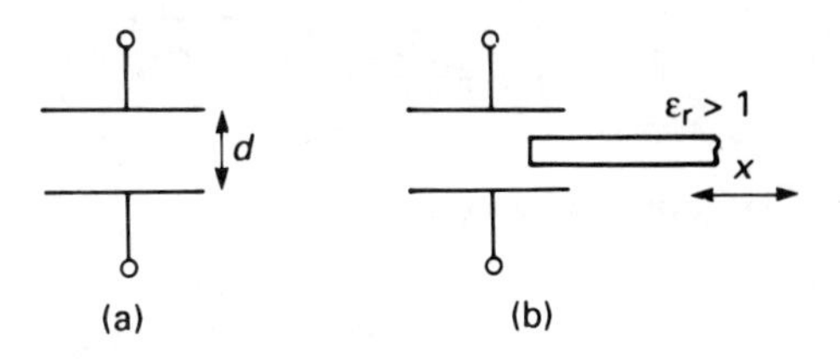

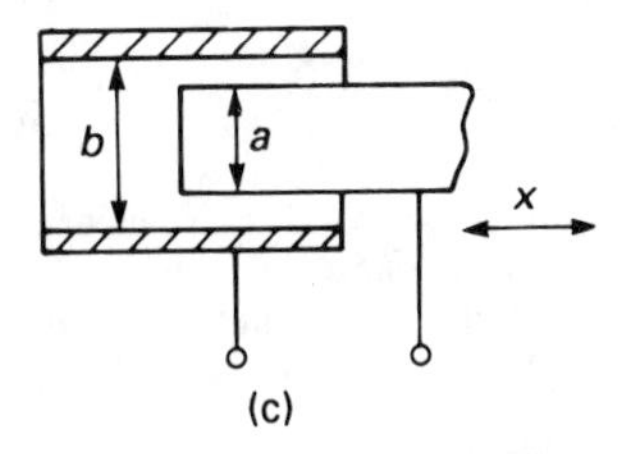

Figure 33.5 Examples of capacitive transducers: (a) parallel-plate, variable separation, (b) parallel-plate, variable permittivity, (c) coaxial tube

transducers are limited to very small but accurate measurement of distance with almost infinite resolution of distances less than 1 mm.

The capacitance of a concentric tube capacitor is given by

$$C = \frac{2\pi\varepsilon_0\varepsilon_r}{\ln(b/a)} \tag{33.9}$$

It varies linearly with distance, and is capable of measurements up to 0.4 m with a resolution of 10^{-4} mm.

Capacitive transducers may also be used for proximity detection in which the capacitance in the ancillary circuit is varied by the proximity of a non-metallic material or by stray capacitance to unearthed metallic objects, and may be used in applications such as counting, etc. The signal conditioning circuit usually uses either amplitude modulation of a high-frequency signal or an a.c. bridge circuit, or measures the change in frequency of a tuned circuit.

33.2.4 Inductive transducers

The inductance of a long solenoid is given by

$$L = \mu_0 \mu_r \frac{AN^2}{l} \tag{33.10}$$

where μ_0 is the permeability of free space, μ_r the relative permeability, A the area of cross-section of the solenoid, N the number of turns and l the length of the solenoid. To achieve adequate sensitivity a core of ferrous material is normally used, typical arrangements being shown in *Figure 33.6*. The inductance may be made to change by varying the reluctance of the magnetic circuit, i.e.

$$L = \frac{\mu_0}{A}\left(\frac{l_i}{\mu_r} + l_a\right) \tag{33.11}$$

where l_i is the length of the path in the magnetic material and l_a the length of the air gap. In practice this is achieved by varying the

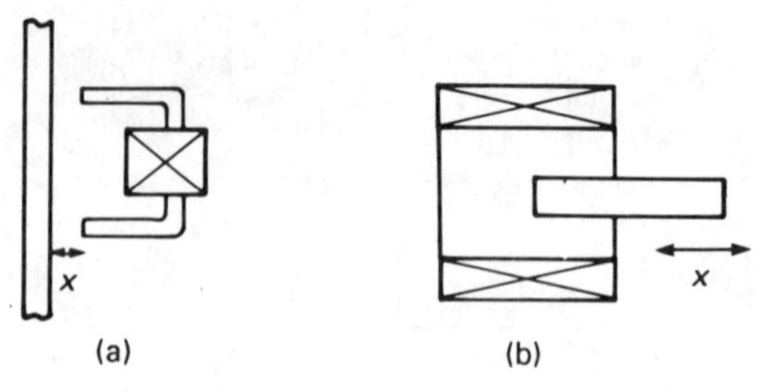

Figure 33.6 Examples of variable reluctance inductive transducers

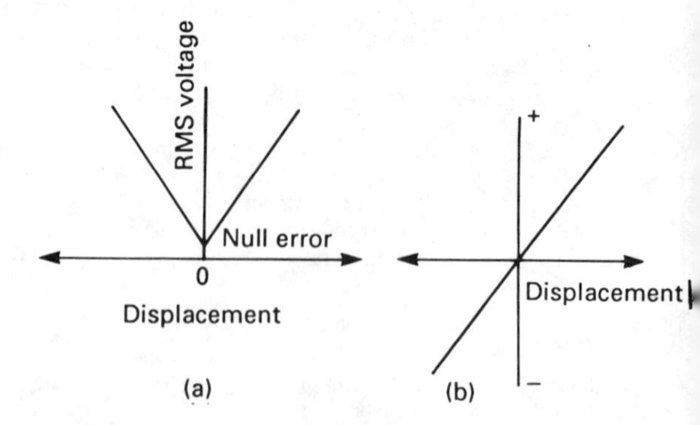

Figure 33.8 Response of LVDT: (a) a.c. r.m.s. output, (b) phase-sensitive rectified output

distance between ferromagnetic core and the coil so as to vary the air gap, i.e.

$$\frac{dL}{dl} = -\mu_0 AN^2\left(\frac{\mu_r}{l_i^2} + \frac{1}{l_a^2}\right) \tag{33.12}$$

Inductive transducers can be used for distance measurement with a high degree of resolution. In one form known as the linear variable differential transformer (LVDT), illustrated in *Figure 33.7*, a core of ferrous material is moved between two coils

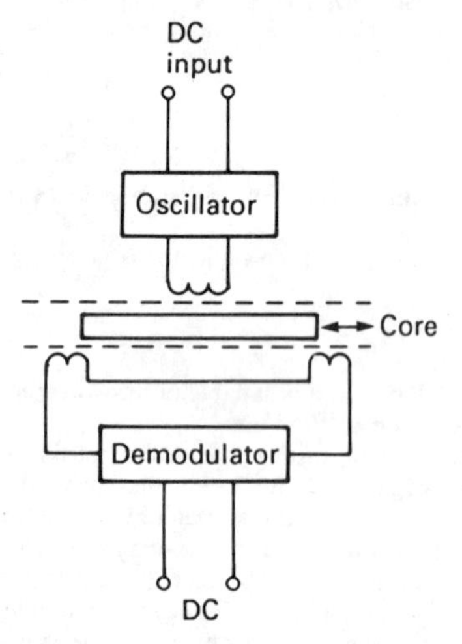

Figure 33.7 Linear variable differential transformer (LVDT) with integral oscillator and demodulator

connected in anti-phase and supplied with a high-frequency excitation voltage. The a.c. input to the primary is normally derived by an ancillary circuit incorporated within the transducer and the secondary output demodulated so that a d.c. output voltage, directly proportional to a displacement of the armature, is obtained. The windings are connected in anti-phase so that the output voltage is at a null when the core is equally positioned in both coils (*Figure 33.8*), although small imperfections in winding, etc., will result in the null voltage being slightly offset from zero. The principal advantage is the absence of contact and virtually infinite resolution. Typical ranges of measurement are from 0.125 mm to 75 mm with a sensitivity of 0.25 mV/mm. A similar construction can be used to measure angular displacement up to 300°.

Inductive sensors may also be used for proximity measurement to give an output signal when a metallic object approaches the front surface of the sensor. The output of the sensor varies non-linearly as the distance from the sensors decreases and sensors of this kind are normally limited to simple systems such as proximity measurement and counting. The inductance is made part of a resonant circuit in which the natural frequency of the circuit is changed by the proximity of the metal object. Similar circuits to those used for capacitive transducers are used.

33.2.5 Optical encoders

Encoders consist of a glass disc in the case of rotary encoders or a strip in the case of linear encoders with accurately generated lines at regular intervals. A narrow beam of light passes through the encoder and relative movement between the encoder and light source generates pulses which are detected by a photoelectric detector and amplified to give a square-wave output. Two signals can be obtained for each output pulse corresponding to the increase and decrease of output, giving twice the resolution. The resolution can be further improved by using a second track (*Figure 33.9*) phase shifted by 90° which automatically gives a count pulse rate of four times the number of lines. The direction of

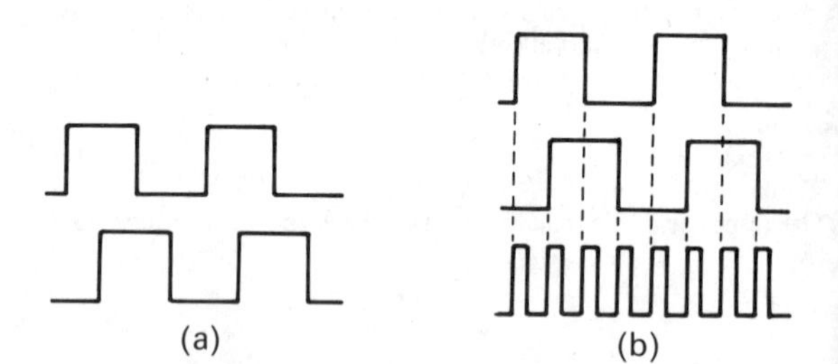

Figure 33.9 Output signals derived from optical encoder: (a) single pulse, (b) pulse output corresponding to leading and trailing pulse edges

movement can be derived by using a direction-sensing logic circuit which differentiates each edge of the signals and checks their relationship with the other track. The overall accuracy is better than ± half bit (1 bit being equivalent to number of lines times four).

The same principle can be used for absolute measurement of distance. However, an absolute encoder has a discrete pattern or code for each line on the disc such as a binary or BCD code which requires decoding. To avoid ambiguity in reading an absolute pattern, cyclic grey codes are normally used in absolute encoders. A grey code is where only one bit changes from line to line. To illustrate the reasons for a cyclic grey code consider the transition from seven to eight in a 4-bit binary code. All four bits change 'simultaneously'. As it is impossible to obtain a truly simultaneous change a momentary erroneous reading could be obtained. To overcome this problem a grey code is generated.

A higher degree of precision can be obtained by the use of Moiré fringes which are formed when two diffraction gratings are superimposed. Instead of two complete plates only one plate and a small part of a plate are required if the plates are tilted. This produces alternating light and dark bands across the grating, illustrated in *Figure 33.10*. These move as the gratings

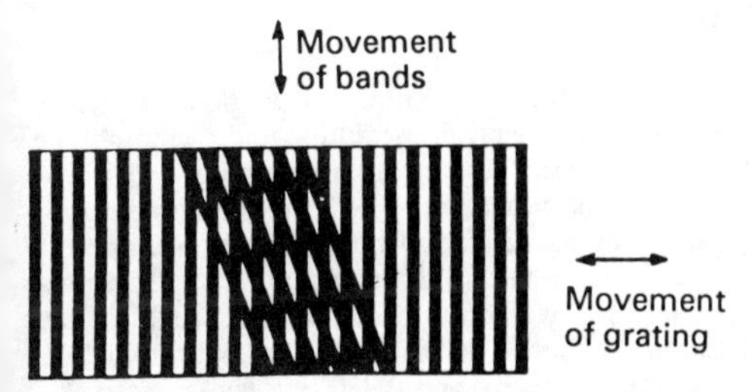

Figure 33.10 Formation of Moiré fringes by crossed gratings

move with respect to each other and can be detected by photocells. The gratings can be made to very high accuracies, a common degree of resolution being 254 lines/mm. Optical encoders enable non-contact measurement of position with very high resolution to be obtained. Resolution up to 1/40 000 of a revolution is possible. Absolute encoders use a second track with associated light source and detector.

Applications of encoders include measurement of distance and velocity in such applications as machine tools, weighing machines, automatic cut to length processes and remote control of position.

The ultimate (at least at present) in precision is obtainable by optical methods. The interferometer which relies on the constructive and destructive interference of light from a coherent source have been used as a laboratory measurement tool for many years. The availability of lasers with depth of coherence of more than a metre has enabled interferometer methods to be taken from the laboratory and these are now in use in such applications as precision machining and calibration. Measurements to accuracies of a fraction of a wavelength are possible enabling measurements of distance over several metres to within 0.1 μm to be made.

33.3 Transducers for velocity measurement

Inductance transducers may be used for measurement of linear velocity which is related to the induced e.m.f., which for a single conductor moving in a magnetic field is given by

$$V = Blu \qquad (33.13)$$

where B is the magnetic flux density, l the length of the conductor and u the velocity of the conductor perpendicular to the magnetic field.

One type of linear velocity transducer uses a magnet coupled to the moving object which moves along the axis of a solenoid coil. This generates an e.m.f. proportional to the velocity but is limited to relatively short path lengths. An advantage of moving the magnet rather than the coil is the elimination of the need for flexible leads. Measurement of rotational velocity is possible using small generators or tachometers (sometimes referred to as tacho generators). The generated e.m.f. is directly proportional to the rotational speed.

The a.c. tachogenerator, essentially an a.c. generator using a rotating permanent magnet to produce an output voltage which is dependent on the rotational frequency, has the advantage that there are no connections required through carbon brushes or slip rings and it is therefore simple and robust. A simple rectifying circuit is required to smooth the output. One of the disadvantages of the a.c. tachogenerator is that the output impedance changes with the frequency which will result in an error unless a high-impedance measuring circuit is used.

The d.c. tachogenerator is similar in construction to the a.c. generator except that the magnet is stationary and the coil rotates. Carbon brushes provide contacts to the rotating commutator and a rectified d.c. output voltage proportional to speed is obtained. The impedance is not dependent on the rotational velocity, and a smoothing circuit is still required to reduce the ripple output.

The drag cup tachometer uses relative motion between the aluminium cup and the rotating magnet to induce eddy currents in the cup which provide an electromagnetic torque proportional to the relative speed. The tachometer can be made direct reading and can also act with a considerable degree of inherent damping which in many cases is an advantage.

Two phase induction generators may be used to provide an output voltage directly proportional to the rotational frequency, at a constant frequency, thereby eliminating some of the problems associated with the permanent magnet a.c. generator. The two windings are wound at right angles to each other, one of which is excited at a constant frequency. Eddy currents are induced in the rotor, which may be either of the squirrel cage or drag cup type. This links the second field coil resulting in an induced voltage which is independent of frequency but varies in amplitude with the speed of the rotor.

33.4 Measurement of force and pressure

Strain gauges (see *Section 33.2.2*) are frequently used for measurement of force either directly in newtons (N) or by weighing (in kilograms, kg). The strain gauges are incorporated in elastic members such as shackles and links (tension), and load cells (compression) connected to half or full bridges. A very high degree of accuracy and stability is required for weighing equipment with accuracies in excess of 0.5%.

Using the piezoelectric effect, the variation of charge with pressure can be used to measure very rapid changes in pressure, torque and acceleration over very large ranges. The construction of a piezoelectric transducer is illustrated in *Figure 33.11*. The

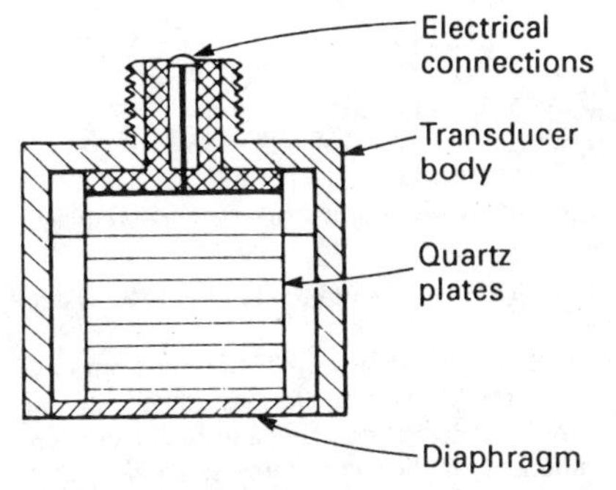

Figure 33.11 Piezoelectric transducer

transducer element may be quartz or a ferroelectric ceramic such as barium titanate. The transducer acts as a capacitor of varying charge whose output is expressed in terms of either charge (as C/g) or voltage (V/g) sensitivity. The transducer has a very high output impedance and therefore it is necessary to use conditioning circuits with very high input impedances, typically of the order of 10^{14} Ω to limit discharge of the capacitor. Since the capacitance is small (typically 20 pF) an amplifier with a high input impedance is necessary. Resolution of 1 part in 10^6 is possible.

Differential pressure transducers (*Section 33.2.4*) using an inductive transducer to measure the deflection of a diaphragm are suitable for use up to about 1 bar. Capacitive transducers, using a pre-tensioned diaphragm, can be used for accurate and rapid measurement of absolute and differential pressure over the range 1 mbar to 10 bar.

Measurement of very low pressure is more difficult. The McLeod gauge traps a sample of the gas in the system which is compressed by a column of mercury of known height. The pressure is related to the volume occupied by the compressed gas. Measurement down to about 10^{-2} Pa is possible but the operation is difficult to carry out automatically. The thermal gauge is suitable for operation over the range 10^{-2} to 10^2 Pa. A resistance heater is inserted in the vacuum; the power dissipation by convection is a function of the pressure and gas. Alternatively the change in resistance may also be used to indicate the pressure.

Ionisation gauges are capable of measuring very low pressure from 0.1 Pa to 10^{-8} Pa. As the gas pressure is reduced through the Paschen minimum the voltage gradient increases and can be used to measure the pressure.

33.5 Accelerometers

Various different forms of accelerometers are based on the acceleration of a known mass, sometimes referred to as the seismic or proof mass, used in conjunction with a displacement transducer for measurement of acceleration. The principle of operation of a seismic accelerometer is defined by Newton's second law:

$$ma = kx \qquad (33.14)$$

in which a is the accelerating force, m the seismic mass, x the deflection of the restoring force, e.g. spring, and k the force due to the deflection. The operating principle is illustrated in *Figure 33.12*. A system based on this alone would result in oscillation at

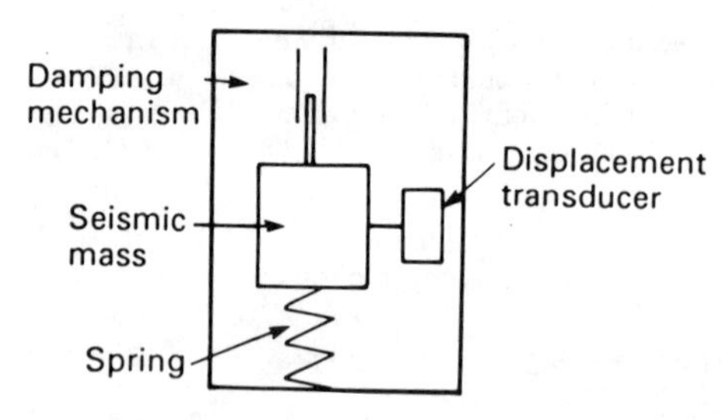

Figure 33.12 Principle of operation of the seismic accelerometer

the resonant frequency of the system and to avoid this the system is critically damped.

Viscous systems are used for damping accelerometers but have the disadvantage that the damping effect is dependent on temperature. This can be overcome by using a drag cup coupled to the moving component in the accelerometer so that as it rotates it cuts the magnetic flux in the gap causing eddy currents to flow in the cup. This results in a force which acts in a direction so as to slow the motion of the cup proportional to the current and therefore the velocity of the cup.

The piezoelectric accelerometer utilises the shear displacement of a piezoelectric crystal (*Figure 33.11*) which generates a capacitive charge. No restoring spring or damping is required and the damping factor which is of the order of 0.01 can be treated as zero over the useful frequency range of the accelerometer which extends to several kHz. Full-scale outputs are obtained for accuraties of from 10^{-10} g to more than 10^4 g, linear over a range of 10^4:1.

33.6 Vibration

Vibration is the oscillatory motion of a body which is characterised by the frequency of oscillation and the amplitude. The oscillation may be sinusoidal or non-sinusoidal with higher harmonics present. Vibration is measured in units of acceleration (m/s^2), velocity (m/s) and displacement (m). Levels of amplitude of vibration vary over a very wide range and an international decibel scale has been agreed for velocity (reference 10^{-3} m/s) and acceleration (reference 10^{-5} m/s^2). The vibration level measured in terms of velocity or acceleration is expressed as

$$\text{vibration level} = 20 \log_{10}(A_1/A_0) \text{ dB} \qquad (33.15)$$

The variation of the amplitude of the measured parameter with frequency is referred to as the vibration spectrum, which in many cases varies uniformly with frequency (response), i.e. no resonance occurs. As the acceleration increases the displacement tends to decrease whilst the velocity remains constant.

The amplitude of vibration can be measured using a seismic mass transducer similar in construction to an accelerometer. If the transducer is vibrated at frequencies greater than its resonant frequency a voltage is induced in the coil by the relative motion between the coil and the magnet, which is proportional to the velocity of the coil with respect to the stationary magnet. The induced voltage is often adequate without further amplification. The piezoelectric accelerometer can be used to measure vibration acceleration.

Vibration displacements can be measured using capacitive or inductive transducers in the same way as distance measurement but are usually limited to the measurement of relative displacement only. Other techniques are occasionally used including stroboscopic methods in which the vibration can be viewed by freezing the motion with a rapidly pulsed lamp and the reed vibrometer in which the effective length of the reed is varied until resonance occurs enabling the frequency to be determined. This method is applicable over the range of 5 Hz to 10 kHz.

33.7 Transducers for fluid flow measurement

33.7.1 Variable area flow meter

The variable area flow meter is illustrated in *Figure 33.13*. The tube is mounted vertically and the fluid flowing through the tube raises the float until the force due to the momentum of the fluid is balanced by the downward force of the float. The variable area flow meter is a simple and effective way of measuring small flows of gases or liquids with a high degree of accuracy and wide range of flow rates. Although normally used as direct indicating flow meters, by using a non-magnetic flow tube combined with a magnetic float the output can be coupled to give an output signal. Variable area flow meters are also used in parallel with orifice plates to measure very high flow rates.

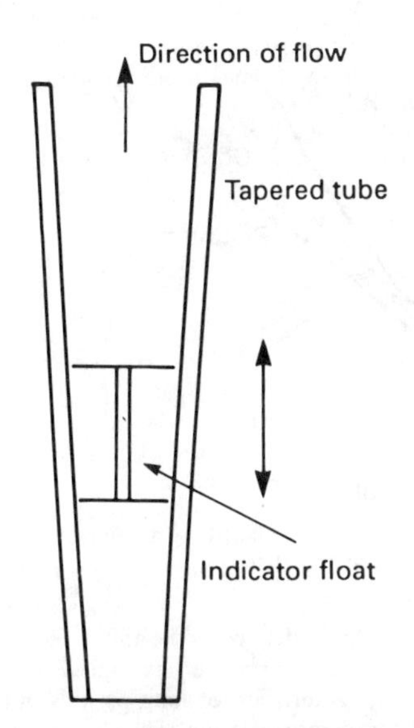

Figure 33.13 Principle of operation of variable area flow meter

33.7.2 Differential pressure flow meters

The rate of flow of a fluid can be deduced from the pressure drop across an orifice. The two principal types of orifice used are the orifice plate, used where the fluid has a high viscosity and/or flow rate, and the Venturi tube for which the total pressure drop is a minimum (*Figure 33.14*). The pressure drop across an orifice may

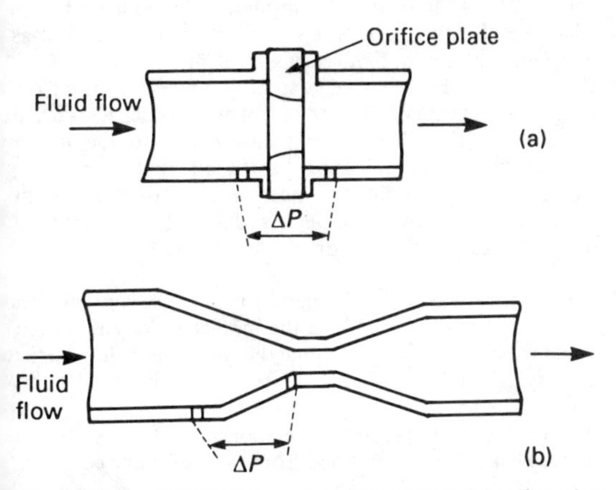

Figure 33.14 Differential pressure flow meters: (a) orifice plate, (b) Venturi tube

be measured with a manometer or with a variable area flow meter or by a differential pressure transducer generally using capacitive transducer elements (*Section 33.2.3*). Flow meters of this kind are capable of operation over flow ranges of up to 3:1 and the square root characteristic produced by orifice meters results in a high sensitivity and accuracy at flow rates above 50% of full scale.

33.7.3 The turbine flow meter

The turbine flow meter uses a turbine wheel in the flow channel which is rotated by the flow. The angular velocity is ideally proportional to the velocity of the liquid and hence the flow rate. The turbine blade itself may be used to generate a voltage by using a paramagnetic material and incorporating a pickup coil in the side walls of the flow meter. Errors may result due to swirling or turbulence of the flow but this can be reduced by incorporating axial vanes at the entrance to the meter. The effects due to viscosity are small for values of Reynolds number greater than about 2000–3000 which set a lower limit to the linear range of the meter. The output of the meter is nominally linear and accurate over a wide range typically ±1% over a range of 20:1.

33.7.4 Magnetic flow meters

The magnetic flow meter relies on the voltage induced in a conductor passing through a magnetic field. The principle of operation is illustrated in *Figure 33.15*. Its application is limited

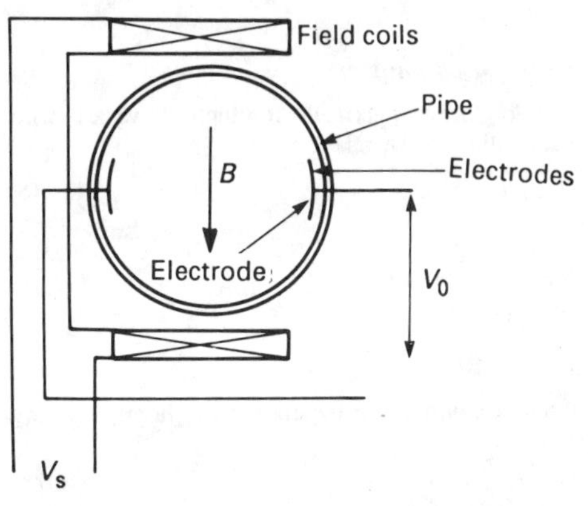

Figure 33.15 Principle of operation of the magnetic flow meter

to liquids with an electrical conductivity above 5×10^{-6} S. No obstruction occurs in the flow line and provided the electrical conductivity is adequate the magnetic flow meter has important applications in metering fluid materials in which pressure drops must be minimised or a substantial particulate material exists such as in slurries. The effect of non-conducting deposits on electrodes can be reduced by using capacitive coupling to the electrodes. The magnetic field is obtained by a.c. excitation of the field coils so as to eliminate the build-up of a polarising potential between the electrodes.

Leakage currents from other sources may also affect the flow indication and, although the error can be reduced by rejecting the quadrature components, the in-phase component may cause an error. Provided this is constant it can be eliminated during setting up of the flow meter; however when it changes the zero setting will need to be changed. This can be overcome by using a pulsed d.c. magnetic field which eliminates problems due to polarisation and errors due to voltage drops from other sources.

33.7.5 Vortex flow meters

The principle of operation of the vortex flow meter is to generate pressure variations to produce a velocity component tangential to the initial axial flow. In one form a bluff body inserted in the pipe produces a vortex shedding effect. This causes a phase shift in an ultrasonic signal which is transmitted across the flow, which is a function of the flow rate. Flowmeters have been developed which use the Doppler shift, due to the scattering caused by solids or bubbles, to provide a corresponding shift in a sound r.f. signal. The technique is suitable for use in applications where adequate scattering signals are present.

33.7.6 Thermal transducers for flow measurement

Alternative methods of measuring fluid velocity include thermistors and hot wire anemometers. Both have an electric current passed through them and rely on the rate of cooling due to the liquid flowing past them which is a function of its velocity. Both methods are capable of accurately measuring instantaneous flow with minimum pressure drop, providing a control signal output.

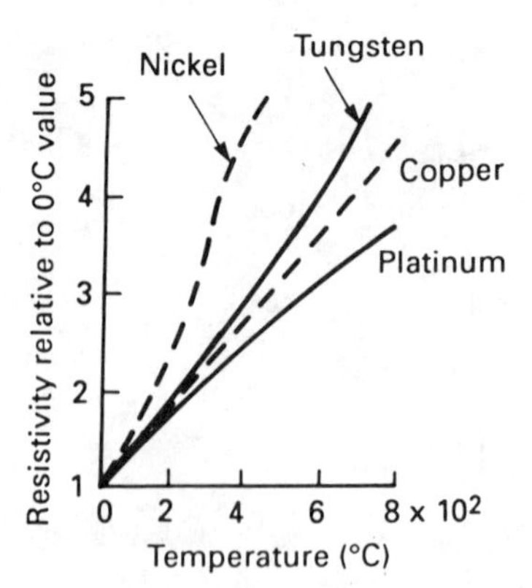

Figure 33.16 Variation of resistivity with temperature of materials used for resistance thermometers

33.8 Transducers for temperature measurement

33.8.1 Resistance thermometers

The resistance of most electrical conductors varies with temperature according to the relation

$$R=R_0(1+\alpha T+\beta T^2+\cdots) \tag{33.16}$$

where

R_0 = resistance at temperature T_0
R = resistance at T
α, β = constants
T = rise in temperature above T_0

Over a small temperature range, depending on the material we may write

$$R=R_0(1+\alpha T) \tag{33.17}$$

where α is the temperature coefficient of resistance.

Important properties of materials for resistance thermometers include a high temperature coefficient of resistance, stable properties so that the resistance characteristic does not drift with repeated heating and cooling or mechanical strain, and a high resistivity to permit the construction of small sensors.

The variation of resistivity with temperature of some of the materials used for resistance thermometers is shown in *Figure 33.16*. Tungsten has a suitable temperature coefficient of resistance but is brittle and difficult to form. Copper has a low resistivity and is generally confined to applications where the sensor size is not restricted. Both platinum and nickel are used extensively because they are relatively easy to obtain in a pure state, but platinum has an advantage over nickel in that its temperature coefficient of resistance is linear over a larger temperature range.

In addition to the temperature range, temperature coefficient and interchangeability characteristics, other important parameters include accuracy, stability, repeatability, rate of response, self-heating effect, insulation resistance and resistance to vibration. Most resistance thermometers adhere to the performance characteristics set out in BS1904 and DIN3760.

The specifications are given based on a typical 100 Ω platinum resistance thermometer.

(1) *Accuracy* The accuracy of calibration is defined as the ability of a thermometer to conform to its predetermined resistance–temperature relationship and is expressed in terms of % of actual temperature reading. The majority of industrial thermometers fall within the 0.1–0.5% range.
(2) *Stability* Stability is defined as the ability of a thermometer to maintain and reproduce its specified resistance–temperature characteristics for long periods of time within its specified temperature range. The drift in the ice-point resistance after 10 000 hours of operation at 600°C must be less than 0.05% (approximately 0.15°C).
(3) *Repeatability* Repeatability is defined as the conformity of consecutive temperature measurements for a thermometer at selected temperatures within its specified range of operation. Consecutive temperature measurements should agree within 0.02% (approximately 0.05°C).
(4) *Time response* The time response is the time required for a thermometer to react to a step change in temperature and reach the resistance corresponding to 63.2% of the total temperature change. The time response can vary from more than 2 s for an industrial encapsulated thermometer to less than 0.5 s for the wafer-type sensing element and as low as 0.2 s for a platinum-film sensing element.
(5) *Self-heating* The heat generated by Joule heating of the resistance element can be a source of error, which is specified as the rise in indicated temperature due to the power dissipated through the sensor over the full range of operating current. The maximum self-heating error over a current range of 0–10 mA must be less than 0.1°C based on a minimum dissipation factor of 100 mW/°C.

Platinum is the material most generally used in the construction of precision laboratory standard thermometers for calibration work and is used for the international thermometer scale from the liquid oxygen point (−182.96°C) to the antimony point (630.74°C).

The resistance–temperature relationship for platinum resistance elements is determined from the Callendar equation

$$T=\frac{100(R_T-R_0)}{R_{100}-R_0}+\delta\left(\frac{T}{100}-1\right)\frac{T}{100} \tag{33.18}$$

where T is temperature and R_T is resistance at temperature T, R_0 is resistance at 0°C, R_{100} is resistance at 100°C, and δ is the Callendar constant (approximately 1.5).

Subsequent errors from the Callendar equation result in temperature differences of less than ±0.1°C at temperature below 500°C. However these differences approach ±1.0°C at 850°C and should be taken into account with industrial thermometers specified for use at high temperatures.

The temperature coefficient of pure nickel is almost twice that of platinum, thus offering the advantage of better sensitivity. Very high purity of nickel is difficult to achieve and different batches of nickel have different temperature coefficients of resistance of the order 0.0062 Ω/°C to 0.00617 Ω/°C. This is compensated for with a coil constructed from a negligible temperature coefficient alloy, such as constantan, connected in series or parallel with the sensing element, and the resultant coefficient and hence the sensitivity is effectively lowered to approximately that of platinum. Construction of nickel resistance thermometers, unlike those of platinum, has been based on a common design. However, the recent widespread use of fully encapsulated platinum thermometers with their superior performance characteristics has replaced nickel in most industrial applications.

Copper of the highest purity is readily available commercially, having a temperature coefficient of resistance slightly higher than that of platinum. The usable temperature range is confined to −200°C to 150°C as copper oxidises at higher temperatures. Construction methods for copper resistance thermometers are very similar to those used in the manufacture of nickel thermometers, but as high-purity copper is easily obtained no compensating resistor is required. The relatively low resistivity of copper necessitates the use of fine wire to avoid excessively large sensors with subsequent slow response times.

The temperature coefficient of resistance is highly linear over the range −50°C to 150°C and this is useful when applied to applications where the measurement of temperature difference is required. In many instances copper themometers with their high degree of reproducibility and interchangeability, linear temperature coefficient of resistance, and relatively simple mode of construction are preferable to the more expensive platinum thermometers for applications near ambient temperature.

33.8.2 Thermistors

The thermistor (temperature sensitive resistor) is a semiconductor whose resistance is a known function of temperature. The variation of resistance with temperature is highly nonlinear, with a high temperature coefficient of resistance of the order of 3–5% per °C compared with 0.4% per °C for platinum. They are normally used up to about 350°C but are available for operation up to 600°C. The temperature coefficient of resistance is normally negative. The resistance, at any temperature T, is given approximately by

$$R_T = R_0 \exp \beta\left(\frac{1}{T} - \frac{1}{T_0}\right) \tag{33.19}$$

where

R_T = thermistor resistance at temperature T (K)
R_0 = thermistor resistance at temperature T_0 (K)
β = a constant determined by calibration

At high temperatures this reduces to

$$R_T = R_0 \exp(\beta/T) \tag{33.20}$$

The resistance is normally high, eliminating errors due to the lead resistance. Thermistors are available in various forms. One form extensively used is a bead of the semiconductor coated with glass for protection. A small size and fast response are obtained. Values of resistance lie between 100 Ω to over 10^7 Ω and bead sizes vary from 0.2 mm to 2.5 mm diameter. The bead may be sealed into a

Table 33.2 Commonly used thermocouples†

Thermocouple	*Maximum continuous operating temperature* (°C)	*Typical output* (μV/°C)‡	*Comments*
Platinum–rhodium			
$Pt–Pt_{87}Rh_{13}$ (Type R)*	1500	12(1600)	Stable; good corrosion resistance
$Pt_{94}Rh_6–Pt_{70}Rh_{30}$	1600	11.6(1600)	
$Pt_{80}Rh_{20}–Pt_{60}Rh_{40}$	1700	4.5(1600)	
Palladium			
$Pt_{90}Ir_{10}–Pd_{40}Au_{60}$	1000	60(800)	Good resistance to corrosion; higher outputs than Pt/Rh alloys
$Pt_{12.5}Pd–Au_{54}Pd_{46}$	1200	35(400)	
Iridium			
$Ir–Ir_{40}Rh_{60}$	2100	6.6(2000)	Fragile at high temperatures
Tungsten–rhenium			
$W–W_{74}Re_{26}$	2700		Operation in inert atmosphere vacuum only
$W_{95}Re_5W_{14}Re_{26}$	2700		
Chromel–alumel (type K)			
$Ni_{90}Cr_{10}–Ni_{94}$, Al, Si, Mn	1300	25(150–1300)	Deteriorates rapidly in H_2S or CO_2; most commonly used up to 1100°C
Iron–constantan (type J)			
$Fe–Cu_{57}Ni_{35}$	800	63(800)	Can be used in oxidising and reducing atmospheres; high output
Copper constantan	350	60(350)	Copper oxidises above 350°C;
Chromel constantan	700	81(500)	very high sensitivity

* Supersedes $Pt–Pt_{90}Rh_{10}$ (type S).
† E.m.f. tables for the following thermocouple combinations are given in BS4937: 1973
Part 1, Platinum/10% rhodium–platinum (type S)
Part 2, Platinum/13% rhodium–platinum (type R)
Part 3, Iron/copper–nickel
Part 4, Nickel–chromium/nickel–aluminium.
‡ Maximum operating temperature is given in brackets.

Table 33.3 Comparison of temperature transducers

Property	*Platinum resistance thermometer*	*Thermistor*	*Thermocouple*
Repeatability	0.03°C to 0.05°C	0.1°C to 1°C	1°C to 10°C
Stability	<0.1% drift in 5 years	0.1°C to 2.5°C drift per year	0.5°C to 1°C drift per year
Sensitivity	0.2 to 10 Ω/°C	100 to 1000 Ω/°C	10–50 μV/°C
Temperature range	−120°C to 850°C	−100°C to 350°C	−200°C to +1600°C
Signal output	1–6 V	1–3 V	0–60 mV
Minimum size	7.5 mm dia. × 6 mm long	0.44 mm dia.	0.4 mm dia.
Linearity	good	poor	good
Special features	greatest accuracy over wide range: highly stable	greatest sensitivity lead effects minimised by high impedance	largest operating range

solid glass rod which can be easily mounted and can be used for measuring liquid temperatures. Discs are made by pressing the thermistor material into round discs up to 25 mm diameter which are sintered. Discs enable higher power dissipation to be achieved and are used for surface temperature measurement where space is not a problem. Washers may be manufactured in the same way to enable mounting on a bolt.

33.8.3 Thermocouples

Various thermocouple materials and methods of construction are used depending on the temperature, environment and sensitivity. Examples of some typical thermocouple materials and conditions of usage are given in *Table 33.2*. Where a continuous output from the temperature sensor is required for indication or control, the thermocouple is the most widely used method of temperature measurement and the chromel–alumel thermocouple is almost solely used today for measuring temperatures up to about 1300°C. Above this temperature, depending on the application, platinum/platinum–rhodium alloy and tungsten/tungsten–rhenium thermocouples are employed. Only a relatively small output of the order of 10–20 μV/°C is obtained and it is necessary to amplify the output. Although over short ranges of temperature the output may be approximately proportional to the temperature, over wide ranges it is nonlinear and linearising circuits are necessary for different thermocouple materials.

The output voltage is measured with respect to the cold junction e.m.f. which must be held constant. This is normally carried out by a built-in temperature-compensated reference voltage. The advent of stable high-sensitivity solid-state amplifiers with built-in linearisation has enabled temperatures to be measured to within 0.1% and has extended the application of thermocouples into many areas where resistance thermometers were formerly used.

33.8.4 Comparison of temperature transducers

A comparison of the advantages and disadvantages of resistance thermometers, thermistors and thermocouples is shown in *Table 33.3*.

The temperature range of thermocouples is the largest and the small mass of the thermocouple junctions means a rapid response time and a sensitive reading at a point.

Reproducibility, stability and accuracy are the main advantages of resistance thermometers. The wire type thermometers have a large surface area useful in area sensing applications. They suffer the disadvantages of limited miniturisation and relatively high cost, but these factors are counterbalanced by linear output which permits the use of less expensive instrumentation. The film-type thermometers are not as yet as accurate, but they offer the possibility of substantial size reductions.

Thermistors combine the small size of thermocouples and are relatively cheap. They suffer from the problems of non-interchangeability of element, nonlinearity of output and poor stability. The most appropriate applications are in small temperature ranges due to their high sensitivity.

Further reading

ABE, O. *et al.*, 'New thick-film strain gauge', *Rev. Sci. Instrum.*, **59**, No. 8, Part 1 (1988)

ANON., 'Progress in smart sensors', *New Electron.*, April (1988)

BENEDICT, R. P., *Fundamentals of Temperature, Pressure and Flow Measurements*, Wiley, New York, p. 58 (1977)

DALLY, J. W. *et al.*, *Instrumentation for Engineering Measurements*. Wiley, New York (1984)

DOEBLIN, E. O., *Measurement Systems, Application and Design*, 3rd edn, McGraw-Hill, New York (1983)

GREEN, D., 'Mechanical guidelines for load cell weighing', *Design Eng.*, December (1987)

HALL, J. A., *The Measurement of Temperature*, Chapman & Hall, London, pp. 13–18 (1966)

HASLAM, J. A., SUMMERS, E. R. and WILLIAMS, D., *Engineering Instrumentation and Control*, Edward Arnold, London (1981)

HICKLENTON, A., 'Precise sensors for fire control', *Test*, October (1988)

JONES, E. B., *Instrument Technology*, Vol. 1, 3rd edn, Newnes–Butterworth, London (1974)

KAJANTO, I. *et al.*, 'A silicon-based fibre-optic temperature sensor', *Eng. Optics*, **1**, No. 4, November (1988)

OLIVER, F. J., *Practical Instrumentation Transducers*, Pitman, London (1972)

ORMOND, T., 'Pressure sensors and transducers', *EDN*, 4 August (1988)

PETERSEN, A., 'The magnetoresistive sensor', *Electron. Components Appl.*, **8**, No. 4 (1988)

WILLIAMS, J., 'Clever techniques improve thermocouple measurements', *EDN*, 26 May (1988)

34 Filters

P J Cottam, BSc
Test Manager
Eltek Semiconductors Ltd

Contents

34.1 Types of filter

The classification of electric wave filters can be done in several ways. They may be grouped in terms of the frequency spectrum against the realisation (*Figure 34.1*), or they may be grouped according to the elements that make up the filters.

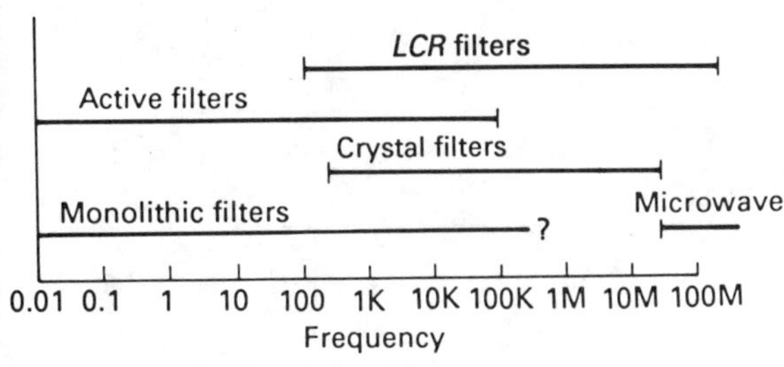

Figure 34.1 Filter frequency guide

There are five basic types of filters used for frequency discrimination in electronic circuits.

(1) The low-pass filter (*Figure 34.2(a)*). This type of filter passes all signals in the frequency band from zero frequency up to the required cut-off frequency.
(2) The high-pass filter (*Figure 34.2(b)*). This type of filter rejects all signals in the frequency band from zero frequency up to the required cut-off frequency.
(3) The band-pass filter (*Figure 34.2(c)*). This type of filter passes all signals in the frequency band defined by a lower and an upper frequency. This is the most common form of filter.
(4) The band-reject filter (*Figure 34.2(d)*). This type of filter rejects all signals in the frequency band defined by a lower and an upper frequency. This type of filter is often used to take out unwanted side tone.
(5) The all-pass filter (*Figure 34.2(e)*). This type of filter is not normally amplitude sensitive, but provides a controlled phase response. It may be used for phase correction in digital transmission systems, phase splitting of signals, or expansion of signals in the time domain.

34.2 Filter design using image parameters

The theory of image parameter design was proposed by O. Zobel in the 1920s, but has in more recent times become less popular as more modern synthesis techniques have been developed.

The basic filter sections used in image parameter design are given in *Figures 34.3–34.6*. These figures give only the simpler elements normally used.

The basic filter circuits are shown in *Figure 34.7*. The basic equations for the image impedances Z_T and Z_π for the $\frac{1}{2}T$ sections are derived by assuming that the network is terminated with impedances that change with frequency in accordance with the following image impedance equations.

Z_T = mid-series image impedance

= impedance looking into the input 1–2 with Z_π across output 3–4

Z_π = mid-shunt image impedance

= impedance looking into the output 3–4 with Z_T across input 1–2

Therefore

$$Z_T=\frac{Z_1}{2}+\frac{2Z_2Z_\pi}{2Z_2+Z_\pi} \tag{34.1}$$

and

$$Z_\pi=\frac{[(Z_1/2)+Z_T]2Z_2}{2Z_2+(Z_1/2)+Z_T} \tag{34.2}$$

This gives

$$Z_TZ_\pi=Z_1Z_2 \tag{34.3}$$

Solving for Z_T and Z_π gives

$$Z_T=(Z_1Z_2)^{1/2}[1+(Z_1/4Z_2)]^{1/2}\,\Omega \tag{34.4}$$

and

$$Z_\pi=(Z_1Z_2)^{1/2}/[1+(Z_1/4Z_2)]^{1/2}\,\Omega \tag{34.5}$$

The general equation for the transfer constant of a network is given as

$$\theta=a+jb \tag{34.6}$$

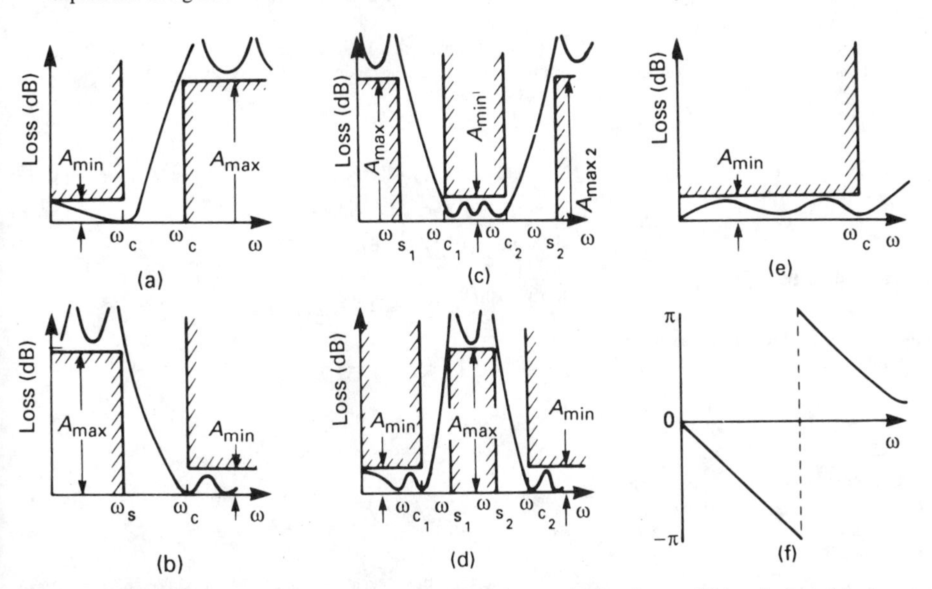

Figure 34.2 Different types of filter: (a) low pass, (b) high pass, (c) band pass, (d) band reject, (e) all pass, (f) all pass

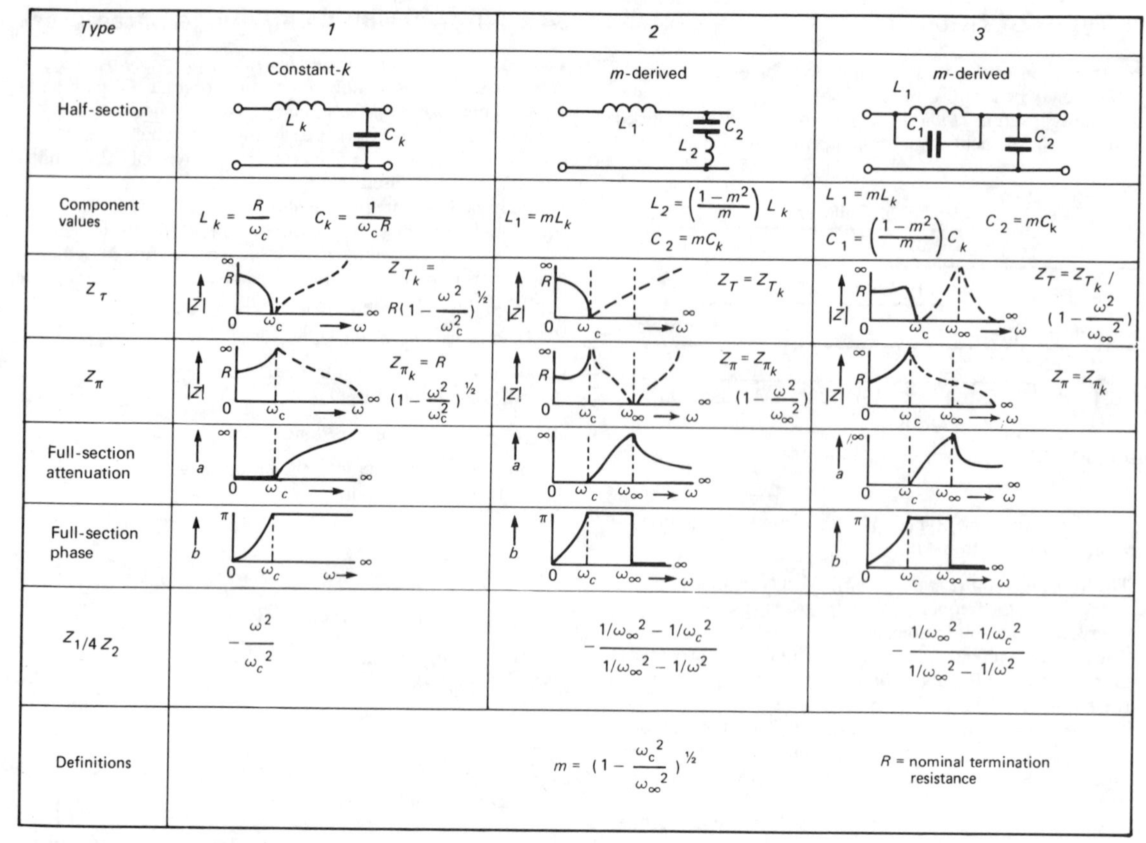

Figure 34.3 Low-pass structures

where a is the image attenuation constant (nepers) and b is the image phase constant (radians).

In the pass band, for a full section, the following equations apply:

$$\cosh\theta = 1 + (Z_1/2Z_2) \tag{34.7}$$

$a=0$ nepers for frequencies giving $-1 \leqslant Z_1/4Z_2 \leqslant 0$, $b=\arccos[1+(Z_1/2Z_2)]$ rad.

In the stop band, for a full section, the following equations apply:

$$a=\text{arcosh}|1+(Z_1/2Z_2)| \quad \text{for } Z_1/4Z_2>0 \tag{34.8}$$

$b=0$ rad

$$a=\text{arcosh}|1+(Z_1/2Z_2)| \quad \text{for } Z_1/4Z_2<-1 \tag{34.9}$$

$b=\pm\pi$ rad

34.2.1 Realisation of a low-pass filter using image parameter design

It is required to design a typical telecommunications low-pass filter with the following requirements:

(1) Cut-off frequency 3.4 Hz.
(2) Peak attenuation 4.5 kHz and 6.5 kHz.
(3) Output load resistance 600 Ω.

We will construct this filter using a constant-k mid-section and an m-derived section with m-derived terminating half-sections. The basic sections are shown in *Figure 34.8*.

(1) Constant-k mid-section (*Figure 34.8(a)*).

$$L_k=\frac{R}{\omega_c}=600/(2\pi\times 3400)=28.1\times 10^{-3}\,\text{H} \tag{34.10}$$

$$C_k=\frac{1}{\omega_c R}=(2\pi\times 3400\times 600)^{-1}=0.078\times 10^{-6}\,\text{F} \tag{34.11}$$

$$a=2\,\text{arcosh}(\omega/\omega_c)=2\,\text{arcosh}(f/3400) \tag{34.12}$$

$$b=2\arcsin(\omega/\omega_c)=2\arcsin(f/3400) \tag{34.13}$$

(2) m-derived mid-section (*Figure 34.8(b)*). We shall use this section to derive the peak attenuation at 6.5 kHz:

$$m=\left(1-\omega_c^2/\omega_\infty^2\right)^{1/2}=\left(1-\frac{3400^2}{6500^2}\right)^{1/2}=0.852 \tag{34.14}$$

$$L_1=mL_k=0.852\times(28.1\times 10^{-3})=23.9\times 10^{-3}\,\text{H} \tag{34.15}$$

$$L_2=[(1-m^2)/m]\times L_k=9.0441\times 10^{-3}\,\text{H} \tag{34.16}$$

$$C_2=mC_k=0.852\times(0.078\times 10^{-6})=0.066\times 10^{-6}\,\text{F} \tag{34.17}$$

$$a=\text{arcosh}\left(1-\frac{2m^2}{(\omega_c/\omega)^2-(1-m^2)}\right)$$

$$=\text{arcosh}\left(1-\frac{1.45}{(3400^2/f^2)-0.27}\right) \tag{34.18}$$

$$b=\arccos\left(1-\frac{2m^2}{(\omega_c/\omega)^2-(1-m^2)}\right)$$

$$=\arccos\left(1-\frac{1.45}{(3400^2/f^2)-0.27}\right) \tag{34.19}$$

(3) m-derived half-sections (*Figure 34.8(c)*). We shall use these sections to derive the peak attenuation at 4.5 kHz:

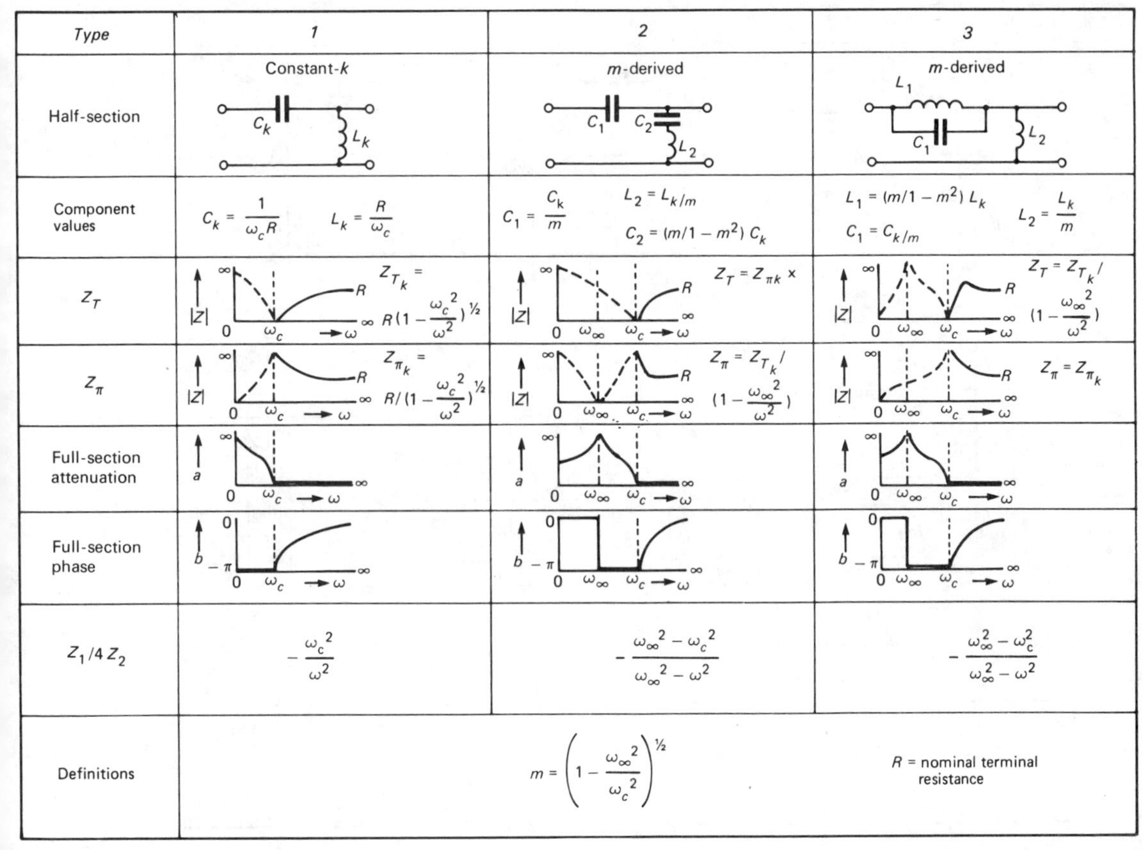

Figure 34.4 High-pass structures

$$m = [1 - (\omega_c^2/\omega_\infty^2)]^{1/2} = 0.655 \qquad (34.20)$$

$$L_1 = mL_k = 0.655 \times (28.1 \times 10^{-3}) = 18.4 \times 10^{-3}\ \text{H} \qquad (34.21)$$

$$L_2 = [(1 - m^2)/m] \times L_k$$

$$L_k = [(1 - 0.655^2)/0.655] \times (28.1 \times 10^{-3}) = 24.5 \times 10^{-3}\ \text{H} \qquad (34.22)$$

$$C_2 = mC_k = 0.655 \times (0.078 \times 10^{-6}) = 0.051 \times 10^{-6}\ \text{F} \qquad (34.23)$$

The final filter is shown together with its response curve in *Figure 34.9*.

34.3 Filter design using synthesis

In modern filter design the image parameter theory is giving way to the design technique which is based on the use of complex polynomials to define the transfer function. All filter synthesis using this approach is done with reference to a normalised low-pass filter, i.e. $\omega_c = 1$ rad/s. Many publications now exist with comprehensive 'look up' tables for normalised low-pass filters. The most commonly used approaches to filter synthesis are the Butterworth, Chebyshev, elliptic and Bessel polynomial approximations.

34.3.1 Butterworth approximation

The well known Butterworth approximation is given by:

$$K(s) = k(s/\omega_c)^n \qquad (34.24)$$

where k is a constant, n is the order of the polynomial, and ω_c is the passband cut-off frequency. This gives the loss function for a low-pass filter as: .

$$H(s) = V_i(s)/V_o(s) = \sqrt{1 + k^2(\omega/\omega_c)^{2n}} \qquad (34.25)$$

The loss in dB is given by

$$A = 10 \lg [1 + k^2(\omega/\omega_c)^{2n}] \qquad (34.26)$$

This response has the flattest possible characteristic at the centre of the passband. At the passband cut-off frequency the loss is given by (again in dB)

$$A = 10 \lg (1 + k^2) \qquad (34.27)$$

In the filter tables this is given as $A_{\max}$. As the frequency ω becomes much greater than ω_c, the loss is given by (dB)

$$A = 20 \lg [k(\omega/\omega_c)]^n \qquad (34.28)$$

This indicates that the loss increases at $6n$ dB/octave.

A graph of the first eight Butterworth responses is given in *Figure 34.10*. The first eight Butterworth polynomials are given in *Table 34.1*. These are derived as follows:

$$|H(s)|^2 = 1 + (-s^2)^n \qquad (34.29)$$

where $\omega_c = 1$ rad, $k = 1$ (hence to denormalise a function we multiply s by $(k^{1/n}/\omega_c)$).

The roots of $|H(s)|^2$ are given by:

$$s_N = \exp\left(\frac{j\pi}{2} \frac{2N + n - 1}{n}\right) \qquad (34.30)$$

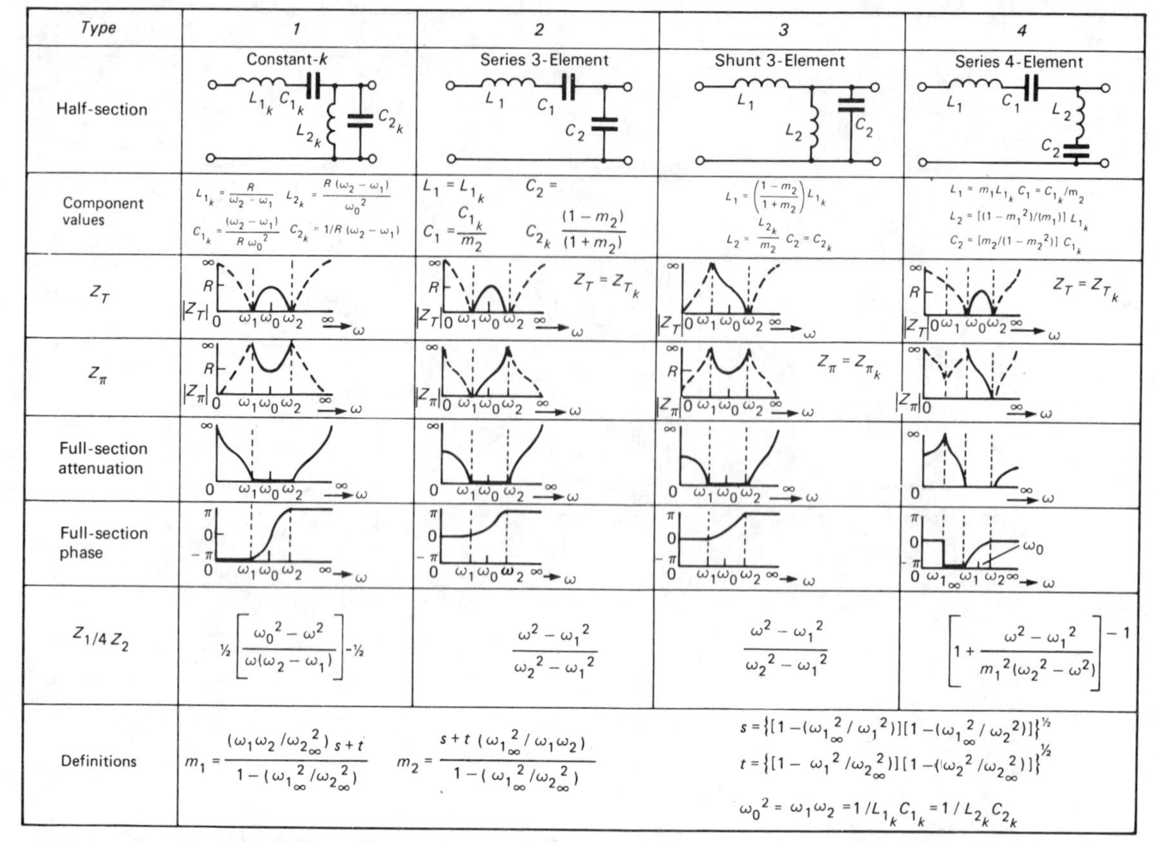

Type
1
2
3
4
Constant-k
Series 3-Element
Shunt 3-Element
Series 4-Element
Half-section
Component values
Full-section attenuation
Full-section phase
Definitions

Figure 34.5 Band-pass structures

where $N=1, 2, \ldots, 2n$. This gives the roots located on a unit circle and equally spaced at π/n rad.

The roots of a fourth-order Butterworth approximation are given in *Figure 34.11*.

34.3.2 Realisation of a Butterworth low-pass filter

The characteristics of the filter are as follows: $A_{\max}=1.0$ dB, $A_{\min}=12$ dB, $f_c=3.4$ kHz, $f_s=4.5$ kHz. From Equation (34.27)

$$k^2=10^{A\max/10}-1 \qquad \text{i.e. } k=0.51 \tag{34.31}$$

The attenuation per octave is given by:

$(12-1)/(4.5-3.4)\times 3.4$ dB/octave $=34$ dB/octave

The order of the filter is 34/6 which equals 5.7, therefore a sixth-order filter is required.

The normalised function is

$$(s^2+0.5176s+1)(s^2+1.4144s+1)(s^2+1.9305s+1)=H(s) \tag{34.32}$$

To denormalise the function we multiply s by $(k^{1/n}/\omega_c)$ and therefore

$$\frac{k^{1/n}}{\omega_c}=\frac{0.51^{1/6}}{2\pi\times 3400}=0.000\,041\,8$$

and

$$H(s)=\frac{[(s^2+12\,382s+1)(s^2+33\,835s+1)(s^2+46\,181s+1)]}{5.723\times 10^8} \tag{34.33}$$

34.3.3 The Chebyshev approximation

The Butterworth approximation provided a response that was maximally flat at $\omega=0$ but got progressively poorer as ω approached ω_c. Also the stopband attenuation is not as good as can be achieved by other types of polynomial approximations. The Chebyshev approximation is a polynomial which can achieve good stopband attenuation. It sacrifices flatness in the passband in favour of equi-ripple, and gives a sharper roll-off in the stopband.

The Chebyshev function is defined as:

$$C_n(\omega/\omega_c)=\begin{cases}\cos(n \arccos \omega/\omega_c) & |\omega/\omega_c|\leqslant 1 \quad (34.34)\\ \cosh(n \operatorname{arcosh} \omega/\omega_c) & |\omega/\omega_c|>1 \quad (34.35)\end{cases}$$

In the passband we aim to use the approximation which gives equi-ripple, and the Chebyshev function can be rewritten as a polynomial:

$$C_{n+1}(\omega/\omega_c)+C_{n-1}(\omega/\omega_c)$$

$$=\cos[(n+1)(\arccos \omega/\omega_c)]+\cos[(n-1)(\arccos \omega/\omega_c)]$$

$$=2\cos(\arccos \omega/\omega_c)\cos(n \arccos \omega/\omega_c)=2\omega/\omega_c C_n(\omega/\omega_c)$$

therefore

$$C_{n+1}(\omega/\omega_c)=2\omega/\omega_c C_n(\omega/\omega_c)-C_{n-1}(\omega/\omega_c) \tag{34.36}$$

For $n=0$

$$C_0(\omega/\omega_c)=1 \tag{34.37}$$

and for $n=1$

$$C_1(\omega/\omega_c)=\omega/\omega_c \tag{34.38}$$

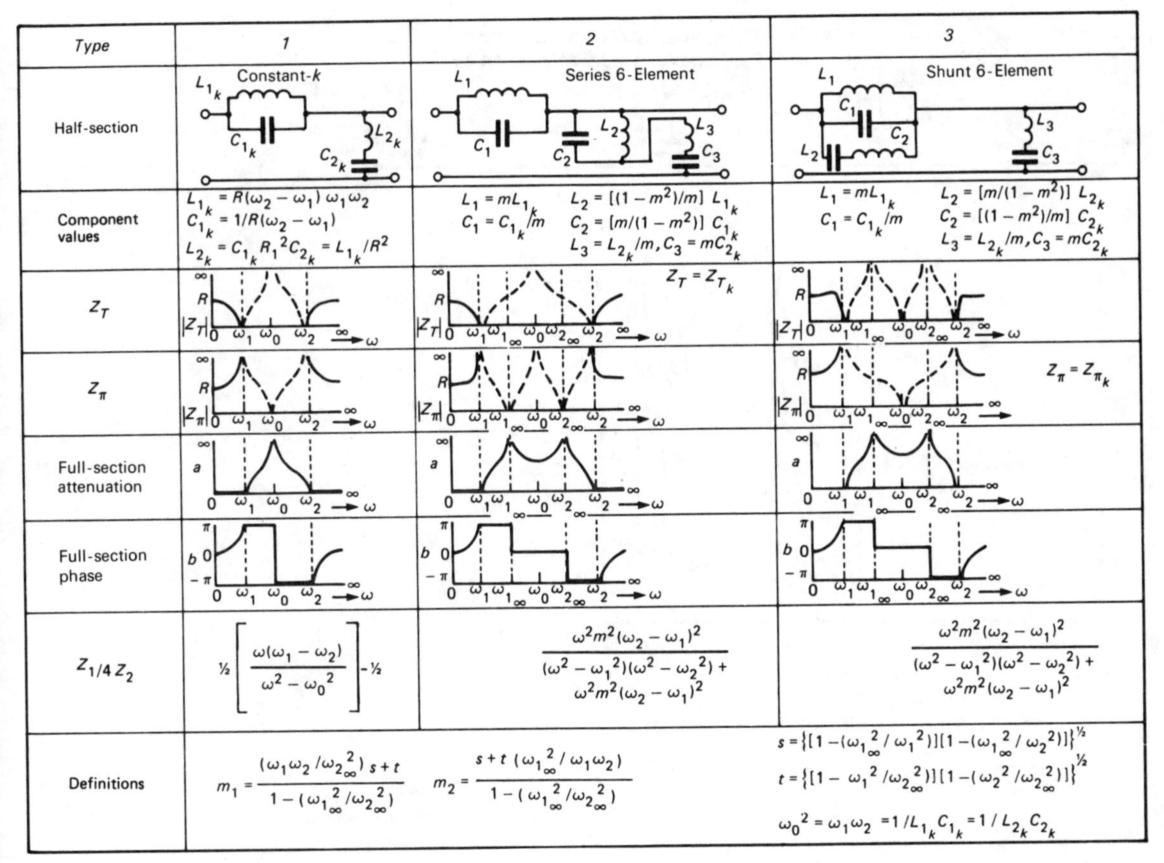

Type
1
2
3
Half-section
Constant-k
Series 6-Element
Shunt 6-Element
Component values
$L_{1_k} = R(\omega_2 - \omega_1)/\omega_1\omega_2$
$C_{1_k} = 1/R(\omega_2 - \omega_1)$
$L_{2_k} = C_{1_k} R_1^2 C_{2_k} = L_{1_k}/R^2$
$L_1 = mL_{1_k}$
$C_1 = C_{1_k}/m$
$L_2 = [(1-m^2)/m]\, L_{1_k}$
$C_2 = [m/(1-m^2)]\, C_{1_k}$
$L_3 = L_{2_k}/m, C_3 = mC_{2_k}$
$L_1 = mL_{1_k}$
$C_1 = C_{1_k}/m$
$L_2 = [m/(1-m^2)]\, L_{2_k}$
$C_2 = [(1-m^2)/m]\, C_{2_k}$
$L_3 = L_{2_k}/m, C_3 = mC_{2_k}$
Z_T
$Z_T = Z_{T_k}$
Z_π
$Z_\pi = Z_{\pi_k}$
Full-section attenuation
Full-section phase
$Z_1/4Z_2$
$\tfrac{1}{2}\left[\dfrac{\omega(\omega_1-\omega_2)}{\omega^2-\omega_0^2}\right]-\tfrac{1}{2}$
$\dfrac{\omega^2 m^2(\omega_2-\omega_1)^2}{(\omega^2-\omega_1^2)(\omega^2-\omega_2^2)+\omega^2 m^2(\omega_2-\omega_1)^2}$
$\dfrac{\omega^2 m^2(\omega_2-\omega_1)^2}{(\omega^2-\omega_1^2)(\omega^2-\omega_2^2)+\omega^2 m^2(\omega_2-\omega_1)^2}$
Definitions
$m_1 = \dfrac{(\omega_1\omega_2/\omega_{2\infty}^2)\, s + t}{1-(\omega_{1\infty}^2/\omega_{2\infty}^2)}$
$m_2 = \dfrac{s + t\,(\omega_{1\infty}^2/\omega_1\omega_2)}{1-(\omega_{1\infty}^2/\omega_{2\infty}^2)}$
$s = \{[1-(\omega_{1\infty}^2/\omega_1^2)][1-(\omega_{1\infty}^2/\omega_2^2)]\}^{1/2}$
$t = \{[1-\omega_1^2/\omega_{2\infty}^2][1-(\omega_2^2/\omega_{2\infty}^2)]\}^{1/2}$
$\omega_0^2 = \omega_1\omega_2 = 1/L_{1_k}C_{1_k} = 1/L_{2_k}C_{2_k}$

Figure 34.6 Band-stop structures

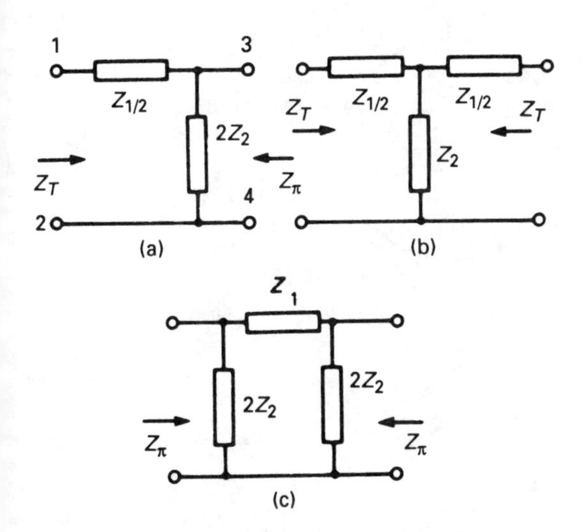

Figure 34.7 Image parameter sections: (a) ½-section, (b) full-*T*, (c) full-π

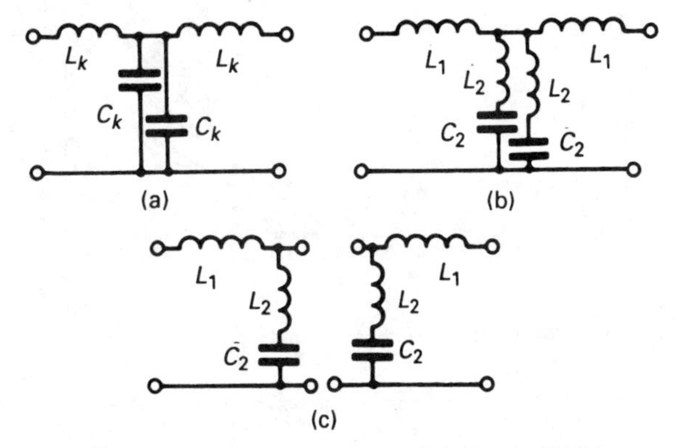

Figure 34.8 Sections for worked example in *Section 34.3.1*: (a) constant-*k* mid-section, (b) *m*-derived mid-section, (c) *m*-derived end section

Then

$$C_2(\omega/\omega_c) = 2(\omega/\omega_c)C_1(\omega/\omega_c) - C_0(\omega/\omega_c) = 2(\omega/\omega_c)^2 - 1 \tag{34.39}$$

$$C_3(\omega/\omega_c) = 2(\omega/\omega_c)C_2(\omega/\omega_c) - C_1(\omega/\omega_c) = 4(\omega/\omega_c)^3 - 3(\omega/\omega_c) \tag{34.40}$$

$$C_4(\omega/\omega_c) = 8(\omega/\omega_c)^4 - 8(\omega/\omega_c)^2 + 1 \tag{34.41}$$

Plots of these polynomial relationships on the (ω/ω_c) axis are given in *Figure 34.12*, for $-1 < \omega/\omega_c < +1$, and show that in the passband the response is equi-ripple. It is also apparent that at $\omega/\omega_c = 1$, $C_n(\omega/\omega_c) = 1$.

The Chebyshev loss function is given by

$$H(s) = V_i(s)/V_o(s) = \sqrt{1 + k^2 C_n^2(\omega/\omega_c)} \tag{34.42}$$

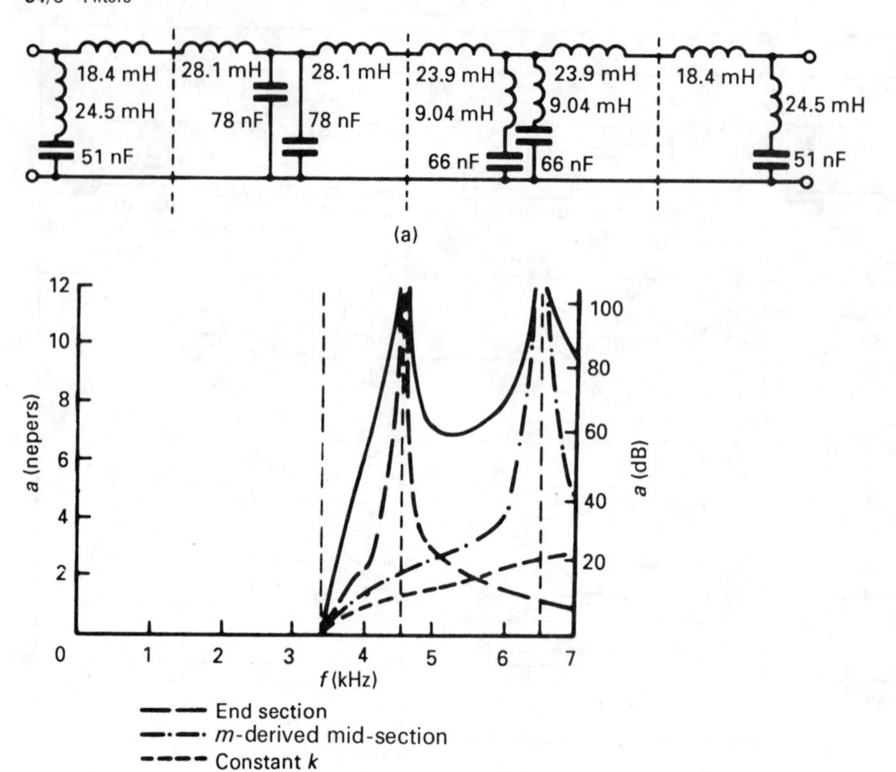

Figure 34.9 Circuit and response of worked example in *Section 34.2.1*: (a) circuit, (b) response

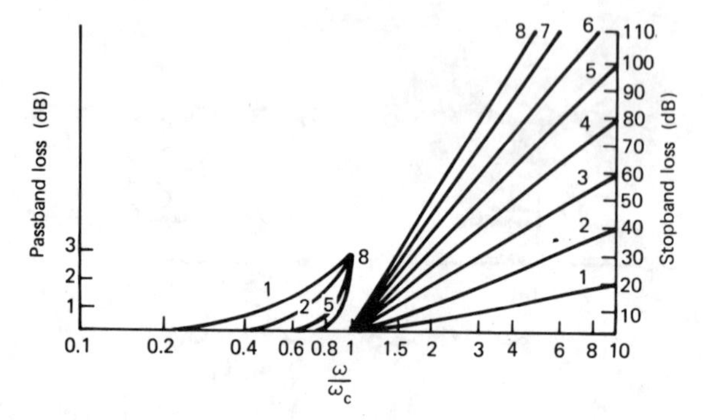

Figure 34.10 Butterworth response

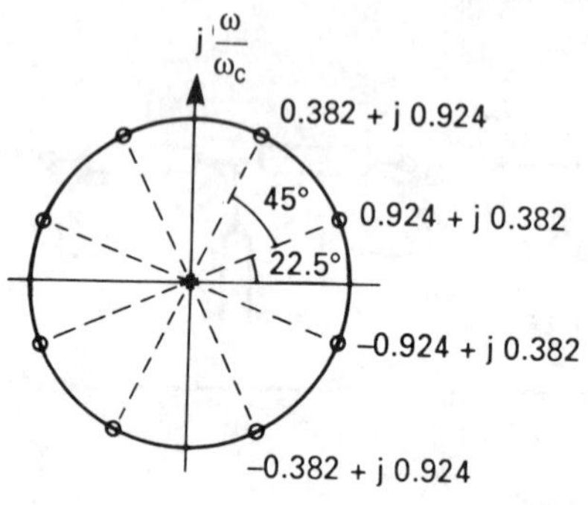

Figure 34.11 Roots of a fourth-order Butterworth

Table 34.1 First eight Butterworth polynomials

n	*H*(*s*)	*H*(*s*) (*factorised*)
1	$s+1$	$s+1$
2	$s^2+1.414s+1$	$s^2+1.414s+1$
3	$s^2+2.000s^2+2.000s+1$	$(s^2+s+1)(s+1)$
4	$s^4+2.1631s^3+3.414s^2+2.6131s+1$	$(s^2+0.7654s+1)(s^2+1.8478s+1)$
5	$s^5+3.236s^4+5.236s^3+5.236s^2+3.236s+1$	$(s^2+0.618s+1)(s^2+1.618s+1)(s+1)$
6	$s^6+3.864s^5+7.464s^4+9.142s^3$ $+7.464s^2+3.864s+1$	$(s^2+0.518s+1)(s^2+1.414s+1)$ $(s^2+1.931s+1)$
7	$s^7+4.494s^6+10.098s^5+14.592s^4+14.592s^3$ $+10.098s^2+4.494s+1$	$(s^2+0.445s+1)(s^2+1.247s+1)$ $(s^2+1.802s+1)(s+1)$
8	$s^8+5.153s^7+13.137s^6+21.846s^5+25.688s^4$ $+21.846s^3+13.137s^2+5.153s+1$	$(s^2+0.390s+1)(s^2+1.111s+1)$ $(s^2+1.664s+1)(s^2+1.961s+1)$

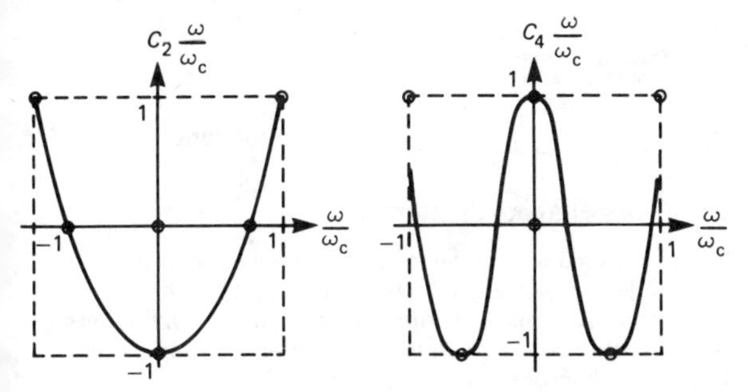

Figure 34.12 Plots of a Chebyshev second- and fourth-order

The loss (dB) is given by

$$A=10\lg\,[1+k^2C_n^2(\omega/\omega_c)] \tag{34.43}$$

At the passband cut-off frequency that loss (in dB) is given by

$$A_{max}=10\lg(1+k^2)=\text{passband ripple} \tag{34.44}$$

As the frequency ω becomes much greater than ω_c, the loss becomes

$$A=20\lg\,[k\,C_n(\omega/\omega_c)]=20\lg\,[k\,2^{n-1}(\omega/\omega_c)^n] \tag{34.45}$$

The Butterworth approximation gave $A=20\lg K(\omega/\omega_c)^n$, therefore the Chebyshev polynomial gives an additional attenuation of $20\lg(2)^{n-1}\sim 6(n-1)$ dB for the same value of n.

Table 34.2 gives the first six Chebyshev polynomials and *Figure 34.13* gives graphs of some of the Chebyshev responses. The multiplying factor in *Table 34.2* is the value necessary to provide a minimum loss in the passband of 0 dB.

From the equation

$$|H(s)|^2=1+k^2C_n^2(\omega/\omega_c) \tag{34.46}$$

it can be shown that the roots of $H(s)$ are $s_N=\sigma_N\pm j\omega_N$, where $N=0, 1, 2, \ldots, 2n-1$. In this expression

$$\sigma_N=\pm\sin\frac{\pi}{2}\frac{(1+2N)}{n}\sinh\left(\frac{1}{n}\operatorname{arsinh}\frac{1}{k}\right) \tag{34.47}$$

and

$$\omega_N=\cos\frac{\pi}{2}\frac{(1+2N)}{n}\cosh\left(\frac{1}{n}\operatorname{arsinh}\frac{1}{k}\right) \tag{34.48}$$

and it can be shown that the roots lie on an ellipse whose equation is

$$\left\{\sigma_N\left[\sinh\left(\frac{1}{n}\operatorname{arsinh}\frac{1}{k}\right)\right]^{-1}\right\}^2+\left\{\omega_N\left[\cosh\left(\frac{1}{n}\operatorname{arsinh}\frac{1}{k}\right)\right]^{-1}\right\}^2$$

as shown in *Figure 34.14*.

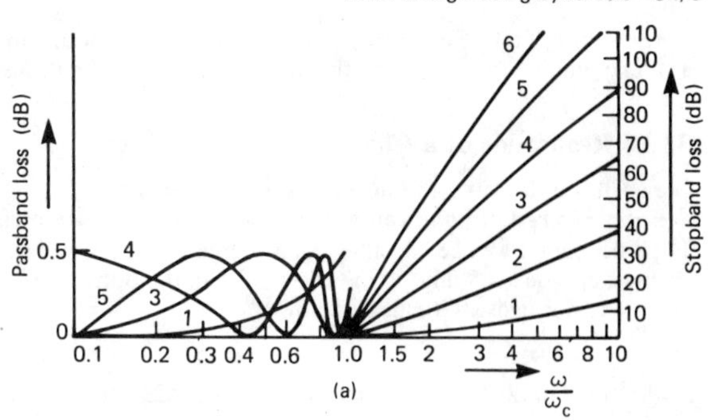

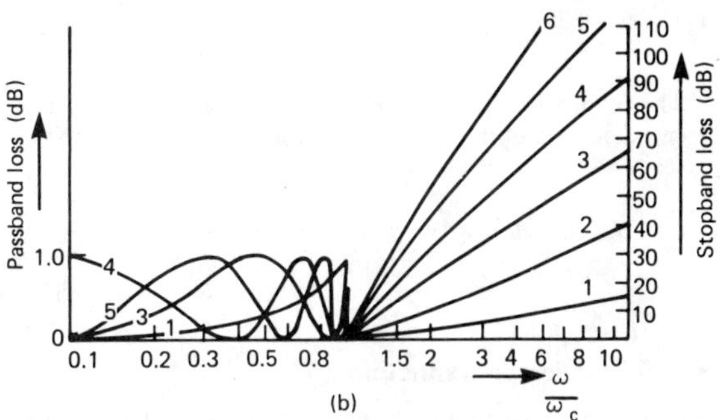

Figure 34.13 Chebyshev responses to sixth order

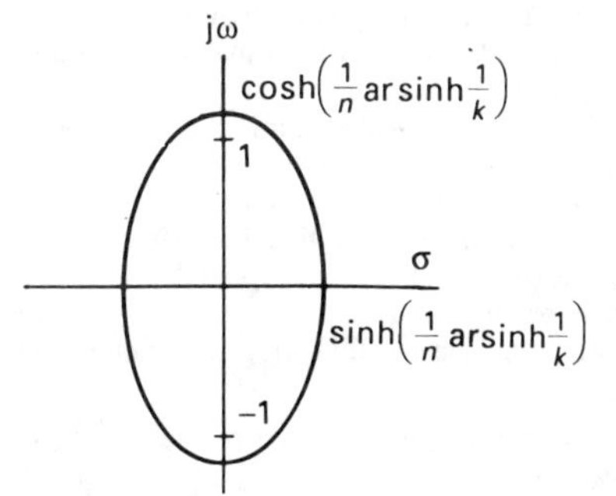

Figure 34.14 Ellipse for roots of Chebyshev polynomial

Table 34.2 First six Chebyshev polynomials

n	A_{max} (dB)	$H(s)$	*Denominator × constant*
1	0.5	$s+2.8628$	2.8628
2		$s^2+1.4256s+1.5162$	1.4314
3		$(s^2+0.6246s+1.1425)(s+0.6265)$	0.7157
4		$(s^2+0.3507s+1.0635)(s^2+0.8467s+0.3564)$	0.3579
5		$(s^2+0.2239s+1.0358)(s^2+0.5863s+0.4768)(s+0.3623)$	0.1789
6		$(s^2+0.1553s+1.0230)(s^2+0.4243s+0.5901)(s^2+0.5796s+0.1570)$	0.0948
1	1.0	$s+1.9652$	1.9652
2		$s^2+1.0977s+1.1025$	0.9826
3		$(s^2+0.4942s+0.9942)(s+0.4942)$	0.4913
4		$(s^2+0.2791s+0.9865)(s^2+0.6737s+0.2794)$	0.2457
5		$(s^2+0.1789s+0.9883)(s^2+0.4684s+0.4293)(s+0.2895)$	0.1228
6		$(s^2+0.1244s+0.9907)(s^2+0.3398s+0.5577)(s^2+0.4641s+0.1247)$	0.0689

The values of the Chebyshev polynomials can now be found in a similar manner to that used in the Butterworth approximation.

34.3.4 Realisation of a Chebyshev low-pass filter

We shall use the same characteristics as in *Section 34.3.2*.

From *Figure 34.13(b)* it can be seen that for an attenuation of 34 dB per octave we need a value of n just greater than 4, and so we must choose $n=5$. This is an order less than the Butterworth filter. The normalised $H(s)$ is as below

normalised $H(s)$

$$=(s^2+0.178\,92s+0.988\,31)(s^2+0.468\,41s+0.429\,30)\times\frac{(s+0{\cdot}289\,49)}{0.122\,83} \quad (34.49)$$

The denormalising multiplier is $(s)\omega_c^{-1}$, the k multiplier having been taken care of in *Table 34.2*. It follows that

$$H(s)=\frac{(s^2+608.3s+11.42\times10^6)(s^2+1592.6s+4.96\times10^6)(s+984.3)}{5.581\times10^{16}} \quad (34.50)$$

34.3.5 Elliptic approximation

Both the Butterworth and the Chebyshev approximations have stopband losses that increase by $6n$ dB/octave for an nth order response. Therefore to get a very sharp roll-off after ω_c needs a filter with a very high order. One way to overcome the use of such complex filters is to provide finite poles of attenuation in the stopband. Butterworth and Chebyshev approximations were both monotonic in the stopband with a pole at infinite frequency. This new type of characteristic is an elliptic function. The most common type of approximation not only gives equi-ripple in the passband, but also gives equi-ripple in the stopband, and this is known as the Cauer approximation.

To define an elliptic response the following parameters are needed:

(1) Frequency transition ratio ω_s/ω_c. This is sometimes given as θ, where $\theta=\arcsin\omega_c/\omega_s$.
(2) Passband ripple A_{max}.
(3) Stopband minimum attenuation A_{min}.
(4) The order of the filter n.

Instead of A_{max} the table will sometimes refer to the reflection coefficient $\rho(\%)$, where

$$A_{max}=-10\lg(1-\rho^2)\,\text{dB}$$

The mathematical proof behind the calculations needed to define the elliptic functions is to be found in R. W. Daniels' text on approximation methods.[1]

The general loss function is given by:

$$H(s)=\prod_{n=2,4,6,\ldots}\frac{(s^2+b_{1_n}+b_{0_n})}{k(s^2+a_{0_n})} \quad \text{even function} \quad (34.51)$$

$$\prod_{n=1,3,5,7,\ldots}\frac{(s^2+b_{1_n}+b_{0_n})(s+c_0)}{k(s^2+a_{0_n})} \quad \text{odd function} \quad (34.52)$$

34.3.6 Bessel approximation

The approximations discussed so far have been concerned with amplitude response, and have taken no account of the associated phase shift. An approximation that gives maximally flat group delay in the frequency domain is the Bessel approximation (sometimes referred to as the Thomson filter).

The ideal delay characteristic is given by the equation

$$V_o(t)=V_i(t-T_0) \quad (34.53)$$

where T_0 is the delay constant.

In the s-plane this equation becomes

$$V_o(s)=V_i(s)\exp(-sT_0)$$

$$H(s)=V_i/V_o(s)=\exp(sT_0) \quad (34.54)$$

If we normalise this function for $T_0=1$ then $H(s)=e^s$. The Bessel function for this is

$$H(s)_n=B_n(s)/B_n(0) \quad (34.55)$$

where

$$B_n(0)=[\overline{1+2\times0}][\overline{1+2\times1}][\overline{1+2\times2}]\ldots[\overline{1+2\times(n-1)}]$$

$B_n(s)$ is the nth order Bessel polynomial which is defined by

$$B_0(s)=1 \quad B_1(s)=s+1 \quad (34.56)$$

and

$$B_n(s)=2(n-1)B_{n-1}(s)+s^2B_{n-2}(s) \quad (34.57)$$

The first six Bessel approximations are given in *Table 34.3* and the first six Bessel characteristics are shown in the graphs in *Figure 34.15*.

34.3.7 Delay equaliser

The Bessel approximation in *Section 34.3.6* gives a maximally flat delay response in the passband but gives a poor amplitude rejection response in the stopband. The alternative approach is to use a filter with the correct attenuation response and then use a phase correcting section to level the delay response in the passband. As the amplitude response of the filter is correct, the amplitude response of the phase correcting section should be flat over both the passband and the stopband of the preceding filter.

This type of phase correction circuit is called a delay equaliser. The transfer characteristic is given by

$$T(s)=\frac{1}{H(s)}=\frac{V_o(s)}{V_i(s)}=\frac{s^2-as+b}{s^2+as+b} \quad (34.58)$$

Table 34.3 The first six Bessel polymomials

n	$H(s)$	*Denominator × constant*
1	$s+1$	1
2	s^2+3s+3	3
3	$(s^2+3.678s+6.459)(s+2.322)$	15
4	$(s^2+5.792s+9.140)(s^2+4.208s+11.488)$	105
5	$(s^2+6.704s+14.273)(s^2+4.649s+18.156)(s+3.647)$	945
6	$(s^2+8.497s+18.801)(s^2+7.471s+20.853)(s^2+45.032s+26.514)$	10395

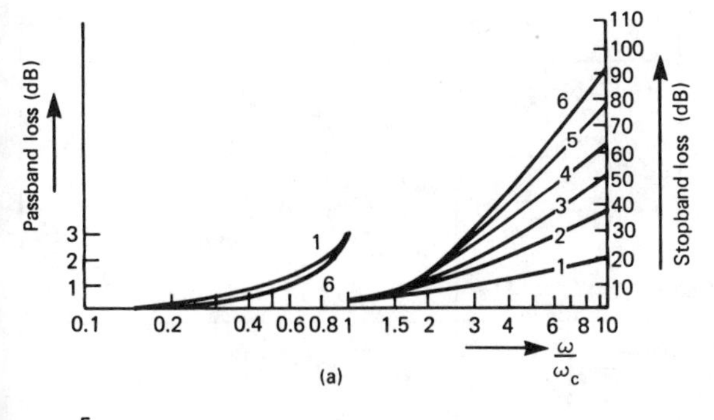

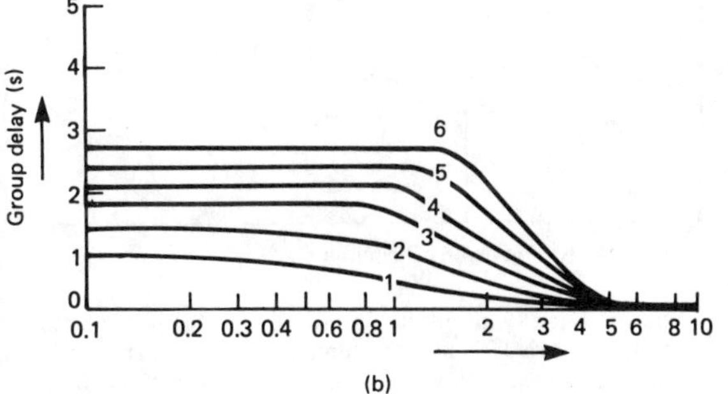

Figure 34.15 Bessel response to sixth order

The gain in dB of this section for all values of ω is

$$20\lg\left|\frac{V_o(s)}{V_i(s)}\right|_{s=j\omega}=10\lg[(b-\omega^2)^2+(-a\omega)^2]$$

$$-10\lg[(b-\omega^2)+(+a\omega)^2]$$

$$=+0\,\text{dB} \quad (34.59)$$

The cut-off frequency is given by

$$\omega_c=\sqrt{b} \quad (34.60)$$

The quality factor for $T(s)$ is given by

$$Q=\sqrt{b}/a \quad (34.61)$$

The delay can then be expressed:

$$\text{delay}=\frac{2s(b+\omega^2)}{(b-\omega^2)^2+a^2\omega^2}\,\text{s} \quad (34.62)$$

The plots of the delay characteristics for different values of Q are given in *Figure 34.16*.

34.3.8 Filter transformations

All the characteristics discussed so far in *Section 34.3* have been with reference to the low-pass characteristic. All the design tables for filter characteristics are given with reference to the low-pass characteristic. We therefore need to know the frequency transforms to convert the low-pass filter to a high-pass, band-pass, and band-stop filter.

34.3.8.1 High pass

Figure 34.17 shows a low-pass characteristic with a lower axis which is the inverse of the frequency axis. If the lower axis is now used as the frequency axis the following factors are apparent.

At $\omega=0$ loss = maximum
At $\omega=\infty$ loss = 0 dB

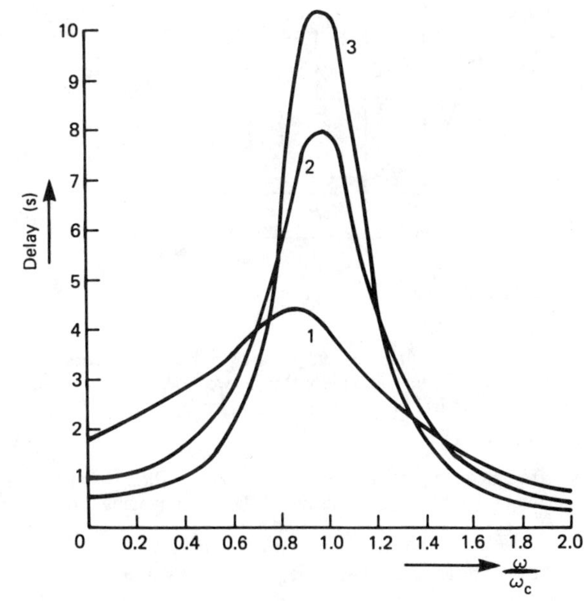

Figure 34.16 Delay equaliser plot for $Q=1, 2, 3$

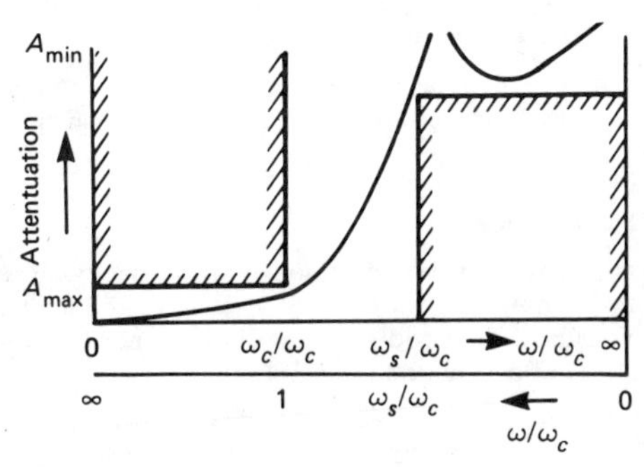

Figure 34.17 Filter transformation—high pass

At $\omega=1$ loss = A_{max}
At $\omega=\omega_s/\omega_c$ loss = A_{min}

This is the characteristic of a high-pass filter. Therefore if we take a low-pass loss characteristic and replace s by $1/s$ we get the complementary high-pass filter.

If the characteristic of a high-pass filter is

$A_{min}=30$ dB, $A_{max}=1$ dB, $\omega_c=1000$, $\omega_s=800$

then the characteristic of a low-pass complement is

$A_{min}=30$ dB, $A_{max}=1$ dB, $\omega_s=1000$, $\omega_c=800$

The normalised low-pass polynomial can now be found. This can be denormalised to a high-pass polynomial by replacing s by ω_c/s.

34.3.8.2 Band pass

Figure 34.18 shows a low-pass characteristic with a lower axis shifted in frequency to realise a band-pass characteristic. The characteristic for the band pass is symmetrical. If the desired band-pass characteristic is not symmetrical then the A_{min} requirement must be made equal on either side to the maximum attenuation, and the frequencies of the stopband corners must be made symmetrical by moving the frequency which is furthest

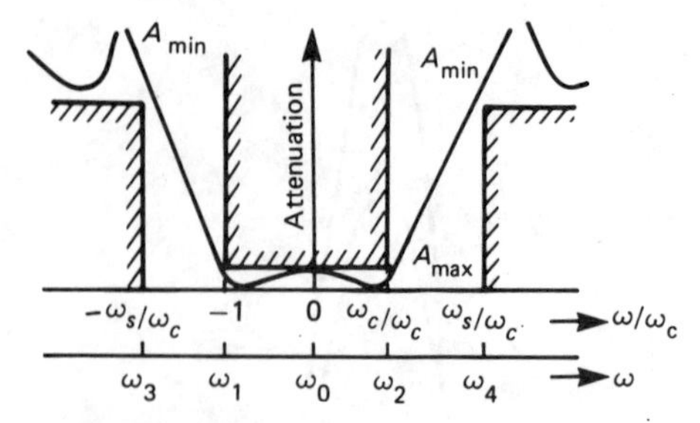

Figure 34.18 Filter transformation—band pass

away from the centre frequency nearer to the centre frequency to regain symmetry. The relationship between the four frequencies of a symmetrical band-pass filter is given by

$$\omega_1\omega_2=\omega_3\omega_4=\omega_0^2 \tag{34.63}$$

The frequency transformation to convert the low-pass to a symmetrical band-pass is accomplished by replacing s by

$(s^2+\omega_0^2)/s(\omega_2-\omega_1)$

To characterise the filters the following low-pass information is required

$$A_{\max},\ A_{\min},\ \omega_c=1,\ \frac{\omega_s}{\omega_c}=\frac{\omega_4-\omega_3}{\omega_2-\omega_1}$$

From this can be realised the normalised low-pass polynomial. This is denormalised and converted to a band-pass by replacing s as above.

34.3.8.3 Band stop

The band-stop characteristic can be derived from the low-pass characteristic by inserting the frequency scale as we did for the high-pass transform. The band-stop transform is therefore the inverse of the band pass, s being replaced by

$$\frac{(\omega_2-\omega_1)s}{s^2+\omega_0^2}$$

To characterise the filter the following low-pass information is required

$$A_{\max},\ A_{\min},\ \omega_c=1,\ \frac{\omega_s}{\omega_c}=\frac{\omega_2-\omega_1}{\omega_4-\omega_3}$$

From this can be realised the normalised low-pass polynomial. This is denormalised and converted to a band pass by replacing s as above.

34.4 Realisation of filters derived by synthesis

34.4.1 Passive realisation

Figure 34.19 shows the impedance functions which can be created using R, L and C combinations. It is apparent that by combining these elements any of the transfer characteristics in *Section 34.3* can be realised.

The usual method of passive realisation is carried out by consulting filter design tables where not only will the polynomial constants be given, but also the normalised values for R, L and C.

Circuit	Impedance
	$(S^2LC+1)/SC$
	$(SL+R)$
	$(SCR+1)/SC$
	$(S^2LC+SCR+1)/SC$
	$SL/(S^2LC+1)$
	$SRL/(SL+R)$
	$R/(SCR+1)$
	$(S/C)/(S^2+S/RC+1/LC)$

Figure 34.19 Passive *RLC* circuits

34.4.2 Active realisation

Because of the high value of inductors necessary for the frequency range below 30 kHz, active realisation has received a great deal of attention. Initially attention was focused on the single amplifier realisation for each section of a filter, but with the arrival of dual, quad and hex operational amplifiers, two- and three-amplifier realisations were produced.

34.4.3 Single and multiple amplifier filters

The advantages of the modern operational amplifier are that it has a very low output impedance and a high input impedance. This makes impedance matching very simple, and so the cascade approach is easily implemented.

34.4.3.1 Sallen and Key filters[2,3]

The basic low-pass circuit shown in *Figure 34.20* has the transfer function

$$T(s)=(\alpha/R_1R_2C_1C_2)\left[s^2+s\left(\frac{1}{R_1C_1}+\frac{1}{R_2C_1}+\frac{1-\alpha}{R_2C_2}\right)+\frac{1}{R_1R_2C_1C_2}\right]^{-1} \tag{34.64}$$

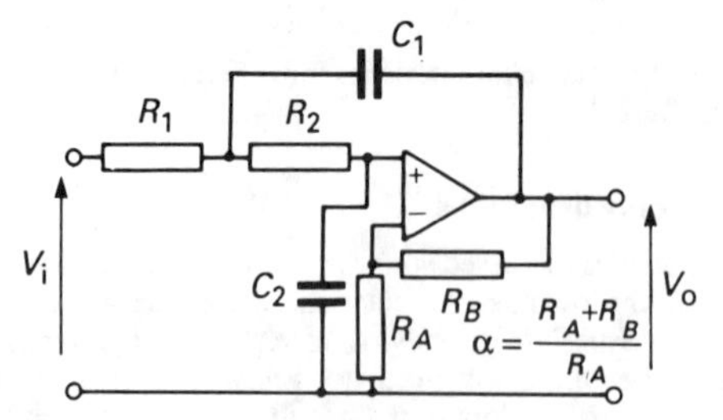

Figure 34.20 Sallen and Key low-pass filters

This is equivalent to the general low-pass function

$T(s)=k\,.\,b/(s^2+as+b)\equiv\omega_c^2/(s^2+(\omega_c/Q_c)s+\omega_c^2)$

where

$b=1/R_1R_2C_1C_2$

$$a=\frac{1}{R_1C_1}+\frac{1}{R_2C_1}+\frac{1-\alpha}{R_2C_2}$$

By comparison we get

$$\omega_c=(1/R_1R_2C_1C_2)^{1/2} \qquad k=\alpha$$

$$Q_c=\frac{\omega_c}{a}=(1/R_1R_2C_1C_2)^{1/2}\left(\frac{1}{R_1C_1}+\frac{1}{R_2C_1}+\frac{1-\alpha}{R_2C_2}\right)^{-1}$$

There are three possible solutions to this. They are as follows.

(1) $R_1=R_2 \qquad \alpha=1$

Then

$$C_1=2Q_c/\omega_c \qquad C_2=(2\omega_cQ_c)^{-1}$$

This has the disadvantage that $C_1/C_2=4Q_c^2$, and this may produce an awkward value for one of the capacitors.

(2) $C_1=C_2=C \qquad R_1=R_2=R$

Then

$$\omega_c=1/RC \quad \text{and} \quad \alpha=3-(Q_c)^{-1}$$

This approach, however, produces a circuit which is more susceptible to component variations than (1). For optimum sensitivity we adopt the Saraga[4] approach in solution (3).

(3) $C_1/C_2=\sqrt{3}\,Q_c \qquad R_2/R_1=Q_c/\sqrt{3}$

Then

$$\alpha=4/3$$

This approach produces the optimum sensitivity to component changes with regard to the stability of Q_c.

The basic high-pass function shown in *Figure 34.21* has the transfer function

$$T(s)=\alpha s^2\left[s^2+s\left(\frac{1}{R_1C_1}+\frac{1}{R_2C_1}+\frac{1-\alpha}{R_2C_2}\right)+\frac{1}{R_1R_2C_1C_2}\right]^{-1} \qquad (34.65)$$

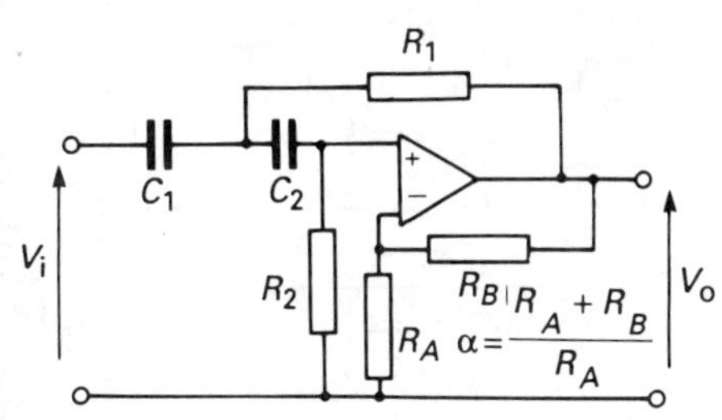

Figure 34.21 Sallen and Key high-pass filters

The method of solving for the component values is exactly the same as for the low-pass function.

By cascading the Sallen and Key high-pass and low-pass sections it is relatively easy to synthesise any Butterworth or Chebyshev filter characteristic which has been derived as in previous sections.

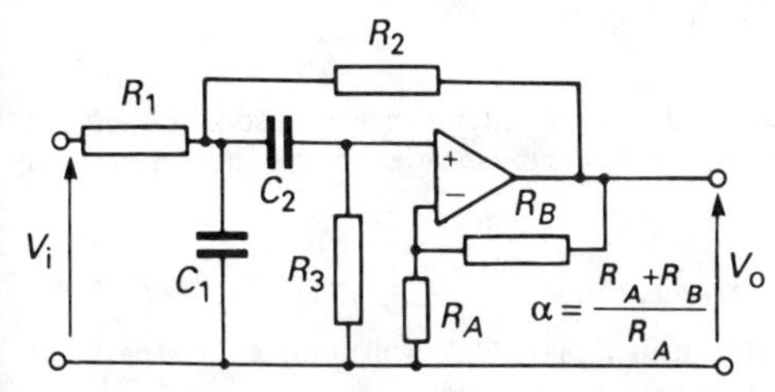

Figure 34.22 Sallen and Key band-pass filters

The basic band-pass function shown in *Figure 34.22* has the transfer function

$$T(s)=\frac{\alpha s}{R_1C_1}\left[s^2+s\left(\frac{1}{R_1C_1}+\frac{1}{R_3C_2}+\frac{1}{R_3C_1}+\frac{1-\alpha}{R_2C_1}\right)+\frac{R_1+R_2}{R_1R_2R_3C_1C_2}\right]^{-1} \qquad (34.66)$$

This is equivalent to the general band-pass function

$$T(s)=Ks[s^2+(\omega_c/Q_c)s+\omega_c^2]^{-1} \qquad (34.67)$$

The usual solution to this is to let

$$R_1=R_2=R_3=R \qquad C_1=C_2=C$$

Then

$$\omega_c=\sqrt{2}/RC \quad \text{and} \quad \alpha=4-\sqrt{2}/Q$$

This leaves

$$K=\frac{\alpha}{R_1C_1}$$

34.4.3.2 Lim filters

The circuit configurations shown in *Figure 34.23* were first proposed by J. T. Lim.[5] The circuit is essentially a balanced twin-*T* network round a unity gain amplifier. The advantage of the twin-*T* approach is that it gives a much lower sensitivity to variations in the amplifier open loop gain than does the Sallen and Key filter.

The design equations which apply to the low pass, high pass, band pass and band stop are:

$$\omega_c=1/CR$$

where

$$C=C_1=C_2=C_3$$

and

$$R=R_2/2=R_1$$

$$\frac{R}{3}=R_5+\frac{R_3R_4}{R_3+R_4}$$

$$Q=\frac{1}{3}\left(\frac{R_4}{R_3}+1\right)$$

The two biquadratic realisations are useful, as they provide us with a means of realising Caver elliptic responses.

The design equations which apply to these filters are

$$\omega_c=1/CR$$

$$C=C_1=C_2+C_4=C_3$$

$$R=R_1+\frac{R_6R_7}{R_6+R_7}=\frac{R_2}{2}$$

$$\frac{R}{3}=R_5+\frac{R_3R_4}{R_3+R_4} \qquad Q=\frac{1}{3}\left(\frac{R_4}{R_3}+1\right)$$

and

$$\frac{\omega_\infty}{\omega_c}=\left[\left(1+\frac{C_2}{C_4}\right)\left(\frac{R_6}{R_6+R_7}\right)\right]^{1/2} \quad \text{for } \omega_\infty>\omega_c \qquad (34.68)$$

and

$$\frac{\omega_c}{\omega_\infty}=\left(\frac{R_6+R_7}{R_7}\right)^{1/2} \quad \text{for } \omega_\infty<\omega_c \qquad (34.69)$$

where ω_∞ is the frequency of maximum attenuation.

The adjustment technique for these filters is worth noting at this point. The adjustment is quite elaborate as it requires three different arrangements of the circuit components.[6] This

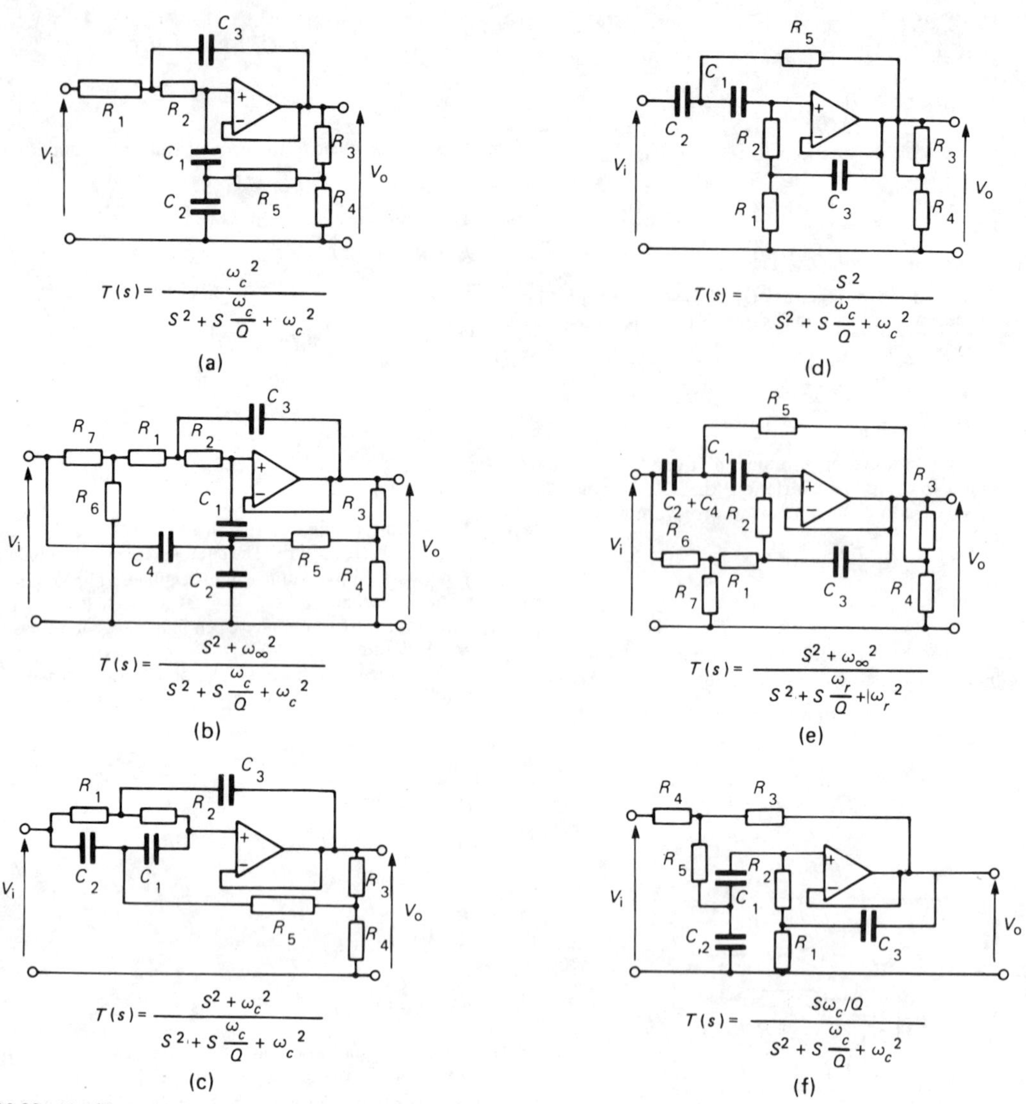

Figure 34.23 Lim filters: (a) Lim low-pass filter, (b) Lim low-pass biquad, (c) Lim band-stop; (d) Lim high-pass, (e) Lim high-pass biquad, (f) Lim band-stop

procedure is shown in *Figure 34.24*. The first two stages are adjusted for a 0° phase shift at ω_c. The third stage of adjustment is to correct for any error in the Q factor of the filter.

34.4.3.3 Wien-bridge notch filter[7]

The typical second-order notch filter response is given by

$$T(s)=(s^2+\omega_c^2)(s^2+s(\omega_c/Q_c)+\omega_c^2)^{-1} \tag{34.70}$$

The circuit shown in *Figure 34.25* gives the response

$$T(s)=K\frac{s^2+(3-R/R_2)\omega_c s+\omega_c^2}{s^2+(3-R_G/(R_G-R))\omega_c s+\omega_c^2} \tag{34.71}$$

where

$$R=\left(\frac{1}{R_1}+\frac{1}{R_2}+\frac{1}{R_G}\right)^{-1}$$

$$K=\frac{RR_G}{(R_G-R)R_2}$$

and

$$\omega_c=\frac{1}{R_cC}$$

To obtain equivalence we require that $R/R_2=3$ and

$$Q_c=\left(3-\frac{R_f}{R_f-R}\right)^{-1}$$

The advantage of this type of filter is that the resonant frequency controlling components are independent of the null-determining components.

34.4.3.4 The three-amplifier biquadratic filter[8]

The circuit shown in *Figure 34.26* will produce low-pass and band-pass functions.

The transfer function between V_i and V_c is given by

$$T(s)V_c/V_i=-(R_1R_4C_1C_2)^{-1}\left(s^2+\frac{1}{R_2C_1}s+\frac{1}{R_4R_3C_1C_2}\right)^{-1}$$

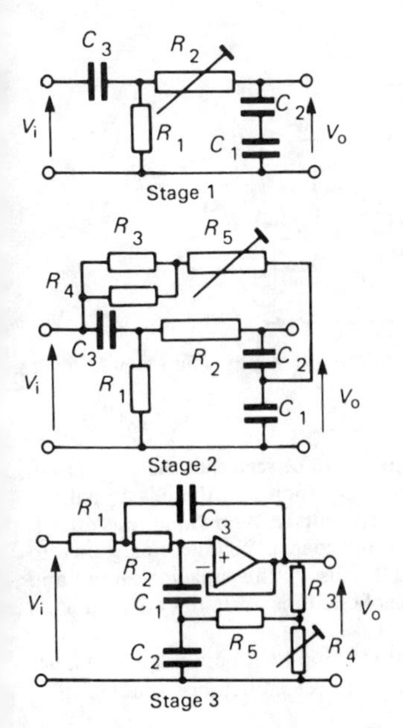

Figure 34.24 Lim low/high-pass filter tuning

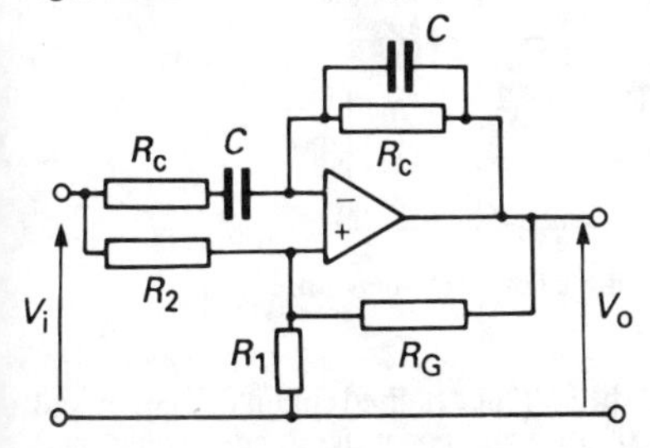

Figure 34.25 Notch filter

$$= T(s)_{V_B/V_i} \quad \text{if } R_5 = R_6 \tag{34.72}$$

which is a second-order low-pass function.

The transfer function between V_i and V_A is given by

$$T(s)_{V_A/V_i} = -s(R_1C)^{-1}\left(s^2 + \frac{1}{R_2C_1}s + \frac{1}{R_4R_3C_1C_2}\right)^{-1} \tag{34.73}$$

which is a second-order band-pass function.

If we now add another amplifier and sum the points V_i, V_A, V_B, we then generate a new transfer function

$$T(s)_{V_D/V_i} = -T(s)_{V_B/V_i}\frac{R_{10}}{R_7} - T(s)_{V_A/V_i}\frac{R_{10}}{R_8} - T(s)_{V_i}\frac{R_{10}}{R_9}$$

$$= \frac{R_{10}}{R_7}\frac{1}{R_1R_4C_1C_2} + \frac{R_{10}}{R_8}\frac{s}{R_1C_1}$$

$$-\frac{R_{10}}{R_9}\left(s^2 + \frac{1}{R_2C_2}s + \frac{1}{R_4R_3C_1C_2}\right)$$

$$\times\left(s^2 + \frac{1}{R_2C_1} + \frac{1}{R_4R_3C_1C_2}\right)^{-1}$$

$$= -K\frac{s^2 + cs + d}{s^2 + as + b} \tag{34.74}$$

the general biquadratic function.

The normal conditions used to design with this filter are to make $C_1 = C_2$, $R_3 = R_4$ and $R_8 = R_{10}$.

A further use of this filter is to realise the delay equaliser function

$$T(s) = -K\frac{s^2 - as + b}{s^2 + as + b} \tag{34.75}$$

This is achieved by making $R_8 = \infty$.

34.4.3.5 Inductance simulation and FDNR simulation

The gyrator is a two-part circuit that inverts an impedance (*Figure 34.27*). K is the gyration impedance multiplier.

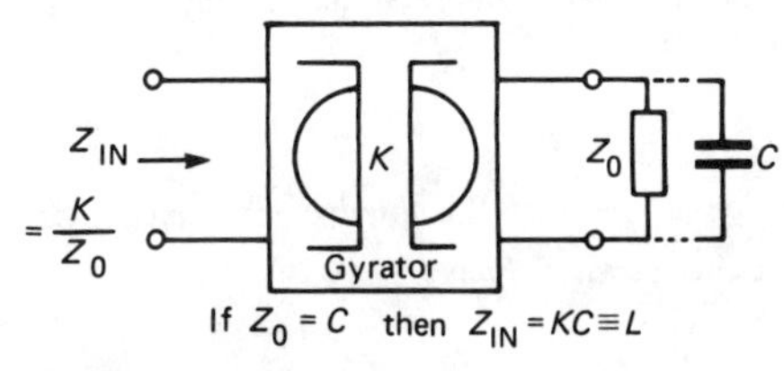

Figure 34.27 Gyrator

Two circuits were proposed for the realisation of the gyrator by Riordan,[9] and Antoniou[10] and Bruton.[11] The circuits are shown in *Figure 34.28*. The input impedance formula for both circuits is the same. It is apparent that if either Z_2 or Z_4 were a capacitor and all other impedances were resistors then the input impedance would be inductive. With this technique values of Q in

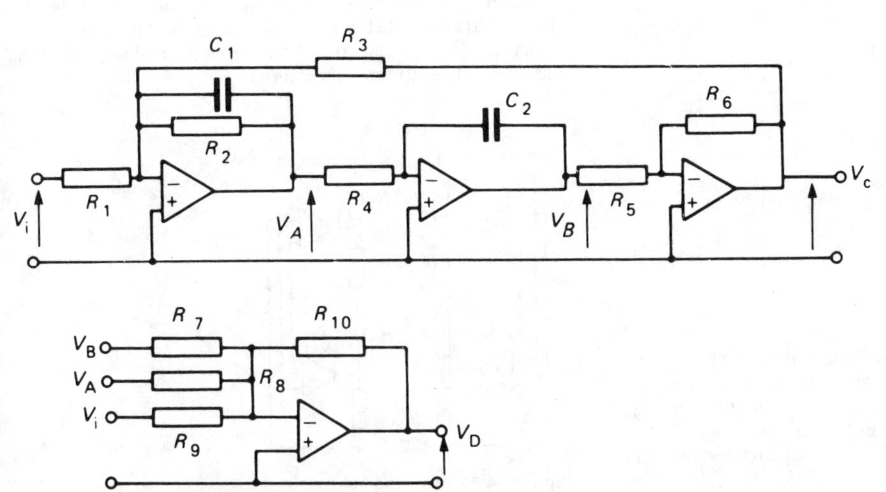

Figure 34.26 The three-amplifier biquadriatic and summing amplifier

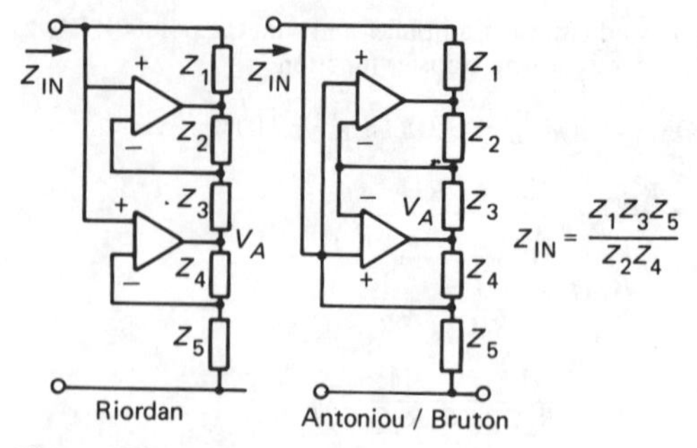

Figure 34.28 Gyrator circuits

excess of 1000 may be obtained, even when using amplifiers with gains as low as 40 dB.

If Z_1 or Z_3 are replaced by a capacitor, and all other impedances are resistors then we generate an input impedance given by the expression

$$Z_{IN}=\frac{R_3}{R_2R_4}\frac{1}{s^2C_1C_3}=\frac{K}{s^2} \quad \text{(symbol = ⊥⊥⊥⊥)} \tag{34.76}$$

This is a frequency dependent negative resistor (FDNR), which varies with the square of the frequency.

When using this type of circuit it should be remembered that the output at V_A is given by

$$V_A=V_i[(Z_1+Z_2)/Z_1] \tag{34.77}$$

This can cause the amplifier to limit without the input voltage approaching the supply lines.

34.4.3.6 Applications of the FDNR and inductance simulations

Consider the circuits shown in *Figure 34.29*.

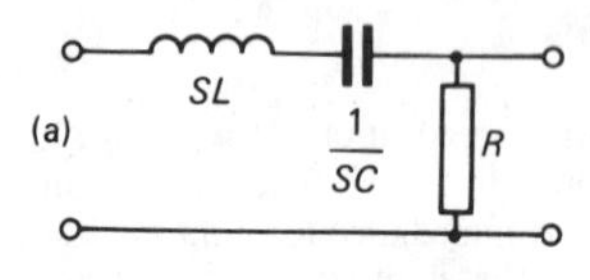

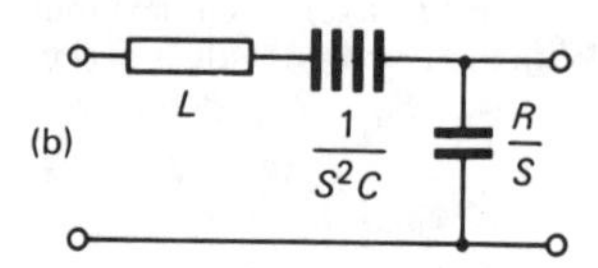

Figure 34.29 Illustration of impedance scaling

$$T(s)=\frac{s(R/L)}{s^2+s(R/L)+1/LC} \tag{34.78}$$

for circuit (a) and

$$T(s)=\frac{s(R/L)}{s^2+s(R/L)+1/LC)} \tag{34.79}$$

for circuit (b).

This technique is known as impedance scaling and holds for all orders of circuits. Therefore, for any *LCR* circuit there are two possible options for the elimination of the inductor.

Consider a typical fifth-order Butterworth/Chebyshev *LCR* network as shown in *Figure 34.30*. Circuit (a) can be converted to circuit (b).

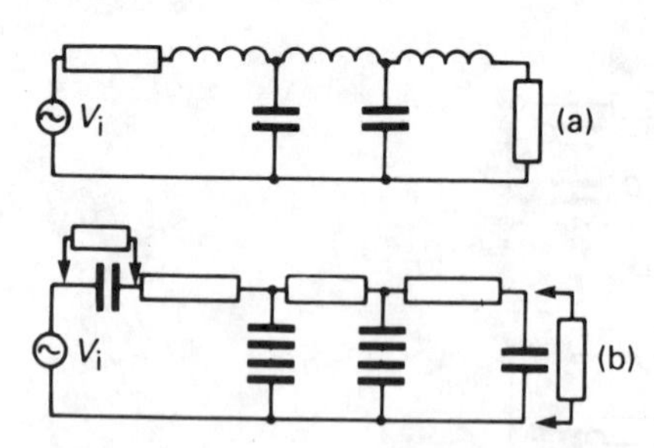

Figure 34.30 Typical fifth-order Butterworth/Chebyshev *LCR* network

From the equivalent circuit it will be seen that the circuits will no longer work down to zero frequency, and there is no path to ground for the amplifier bias currents. A high value resistance is connected across the source and load impedances as shown to overcome these limitations. Because of the capacitive input and output impedances this type of circuit is normally buffered at its input and output.

Now consider a typical fifth elliptic *LCR* network as in *Figure 34.31*. This may be realised by direct inductance simulation.

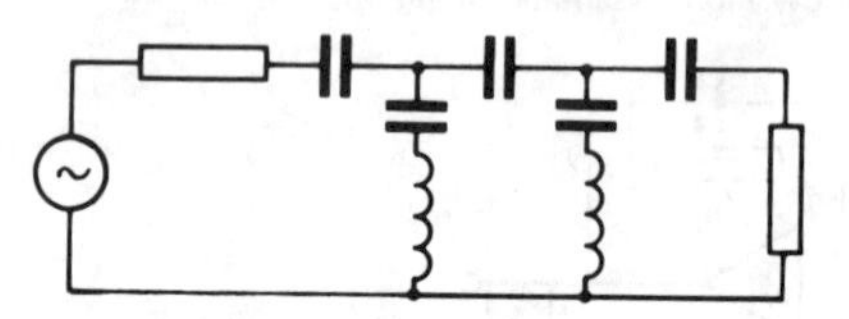

Figure 34.31 Typical fifth elliptic *LCR* network

Therefore, provided the *LCR* normalised circuit can be derived from published tables, the gyrator approach provides a quick and easy means of realising an active filter which will closely follow the theoretical filter characteristic. It should be remembered, though, that a filter is only as good as the components used, and due consideration should be given to the type of capacitors used. The preferred choice of capacitor is the NPO type of ceramic capacitor.

This type of realisation has the disadvantage that it requires two amplifiers to produce a second-order response, but it does produce an active filter with very low sensitivity to component variations.

The problem with the circuits shown in *Figures 34.29–34.32* is that they are grounded. For floating inductance simulation it can be shown that the circuit of *Figure 34.32* realises a floating inductor between terminals 1 and 2.

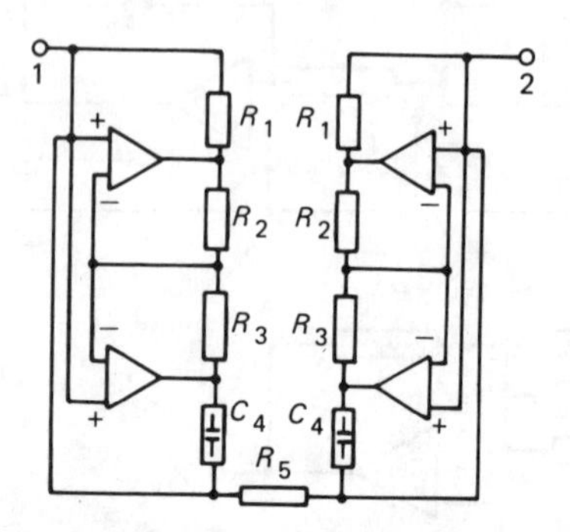

Figure 34.32 Floating inductor simulation

This type of inductance simulation has opened up the way for the production of very low frequency filters as very large values of inductance may be realised with very high values of Q.

34.5 Crystal filters

As the need for more precise control of frequency became necessary, considerable effort was put into developing a frequency conscious component with a very tight frequency selectivity component. The result of this work was the quartz crystal. The crystal structure is shown in *Figure 34.33* and defines the X and Y axes. The Z axis is the optical axis and exhibits no piezoelectric characteristics.

The crystal is cut according to two groups, the X group and the Y group.

For the X group the thickness dimension is parallel to the X axis, and for the Y group the thickness dimension is parallel to the Y axis. The plates are cut at different angles to the Z axis, the angle being classified by the name of the cut AT, BT, MT, etc. The frequency range associated with some of these cuts is given in *Table 34.4*.

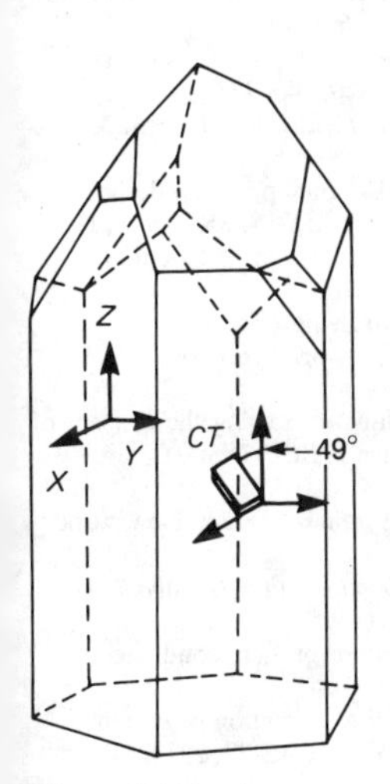

Figure 34.33 Quartz crystals

Table 34.4

Group	*Cut*	*Frequency* (kHz)
X	X	40–20 000
	$5°X$	0.9–500
	MT	50–100
	NT	4–50
Y	Y	1000–20 000
	AT	500–100 000
	BT	1000–75 000
	CT	300–1100
	DT	60–500
	ET	600–1800
	FT	150–1500
	GT	100–550

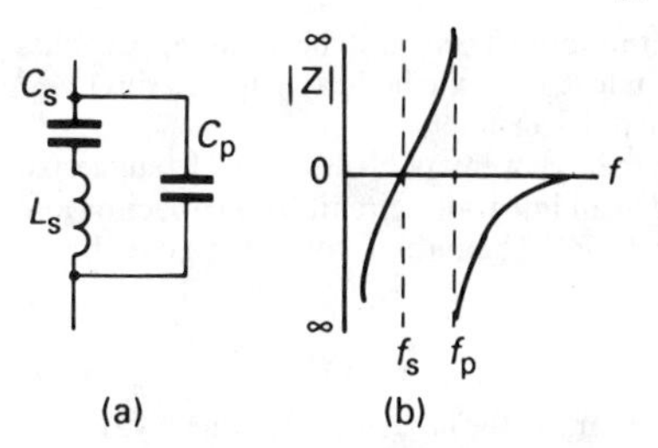

Figure 34.34 Crystal equivalent circuits: (a) circuit, (b) impedance characteristic

The equivalent circuit is given in *Figure 34.34* together with the impedance characteristic. The stability of this circuit is dependent on the Q factor which is the ratio of the reactance to the resistance. The Q of a crystal can lie between 10 000 and over a million, against which the Q of most coils lie between 10 and 500. From the impedance plot it is seen that the crystal has two resonant frequencies, a series and a parallel resonance.

The main use of crystals is in band-pass filters, especially where very narrow bands are required. Two types of band-pass filters are given in *Figure 34.35*.

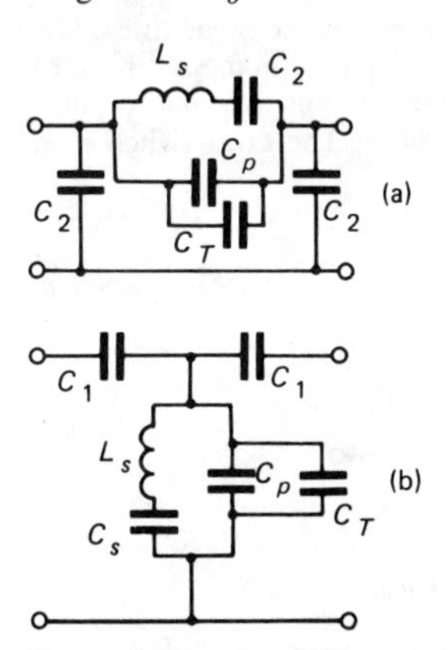

Figure 34.35 Crystal filters: (a) π-section, (b) T-section

For a π network

$$C_1 \approx \frac{1-m^2}{4\pi m f_a R_0} \qquad C_2 \approx \frac{m}{2\pi f_a R_0} \tag{34.80}$$

$$C_s \approx \frac{\Delta f}{2\pi m f_a R_0} \qquad L_s \approx \frac{m R_0}{2\pi\,\Delta f}$$

where f_a is the mid-band frequency $=\frac{1}{2}(f_1+f_2)$, Δf the bandwidth (f_2-f_1), $C_1=C_T+C_P$ and $m=[(f_2^2-f_\infty^2)/(f_1^2-f_\infty^2)]^{1/2}$.

For a T network

$$C_1 \approx \frac{m}{2\pi f_a R_0} \qquad C_2 \approx \frac{m}{\pi(m^2-1) f_a R_0}$$

$$C_s \approx \frac{2m^3\,\Delta f}{\pi(m^2-1)^2 f_a R_0} \qquad L_s \approx \frac{(m^2-1)^2 R_0}{(8\pi m^3\,\Delta f)}$$

where $C_2=C_P+C_T$.

The T and π sections can be cascaded into ladder networks to form wider bandwidth passband filters and tables[12] exist to give different orders of filter with defined ripple in the passband.

34.6 Monolithic approach to filters

The filters described so far have been composed of discrete components. A lot of research has been applied in order to realise these filters in monolithic form. There have been two approaches which have predominated: the switched-capacitance filter[13,14] and the charge coupled device filter.

In 1977 several papers[15] were published which suggested replacing the resistor in an integrator circuit by a capacitor and two switches (*Figure 34.36*). The output voltage is given by

$$V_o = -V_i \frac{C_1}{C_2} \frac{1}{j\omega T}$$

where T is the clock time and ω the input signal frequency. This is equivalent to the conventional integrator using a capacitor and a resistor. The time constant is controlled by the switching frequency and the capacitive ratio, not by the capacitive values. Capacitance ratios are more easily achieved in monolithic form than accurate values of capacitance. Positive and negative integrators are shown in *Figure 34.37*. The extra switches are

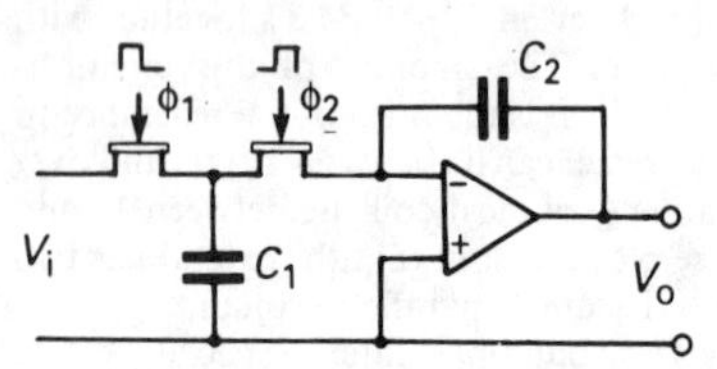

Figure 34.36 Switched-capacitor integrator

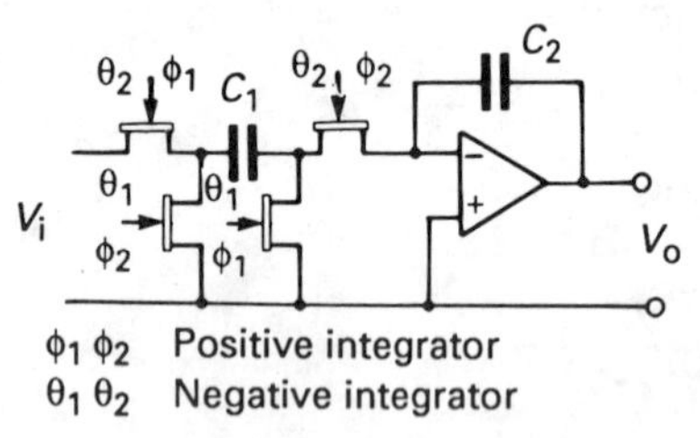

Figure 34.37 Stray capacitor modification

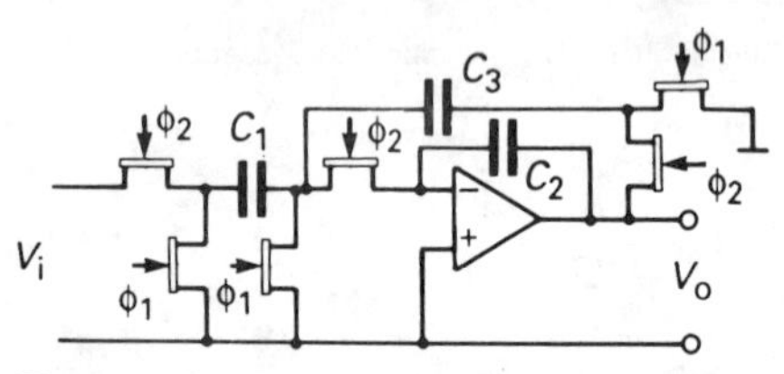

Figure 34.38 Integrator with damping coefficient

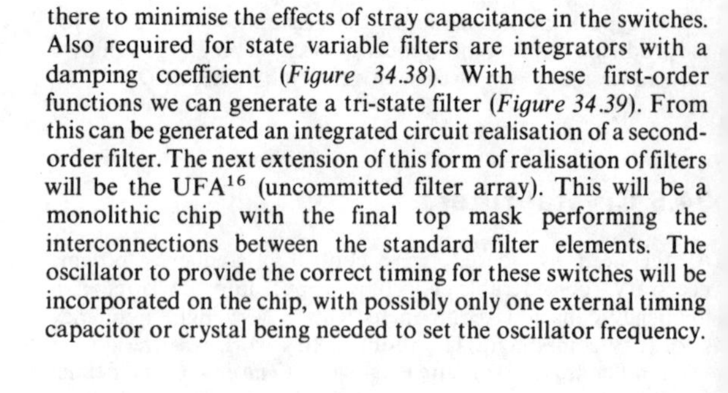
there to minimise the effects of stray capacitance in the switches. Also required for state variable filters are integrators with a damping coefficient (*Figure 34.38*). With these first-order functions we can generate a tri-state filter (*Figure 34.39*). From this can be generated an integrated circuit realisation of a second-order filter. The next extension of this form of realisation of filters will be the UFA[16] (uncommitted filter array). This will be a monolithic chip with the final top mask performing the interconnections between the standard filter elements. The oscillator to provide the correct timing for these switches will be incorporated on the chip, with possibly only one external timing capacitor or crystal being needed to set the oscillator frequency.

References

1 DANIELS, R. W., *Approximation Methods for Electronic Filter Design*, McGraw-Hill, New York (1974)

2 DARYANANI, G., *Principles of Active Network Synthesis and Design*, Wiley, New York (1976)

3 GIVENS, S. D., Application Note AN-40, *Linear and Conversion Applications Handbook*, Precision Monolithics Inc. (1986)

4 SARAGA, W., 'Sensitivity of 2nd order Sallen–Key type active R.C. filters', *Electron. Lett.*, **3**, 442–444 1967)

5 LIM, J. T., 'Improvements in or relating to active filter networks', *British Patent Application 9657* (dated 16 April, 1971)

6 JEFFERS, R. and HAIGH, D. G., 'Active *RC* lowpass filters for FDM and PCM systems', *Proc. IEE*, **120**, 945–953 (1973)

7 DARILEK, G. and TRANBARGER, O., *Electron. Design*, **3**, 80–81 (1978)

8 THOMAS, L. C., 'The biquad: part 1—some practical design considerations', *IEEE Trans. Cct Theory*, **CT-18**, 350–357 (1971)

9 RIORDON, R. H. S., 'Simulated inductors using differential amplifiers', *Electron. Lett.*, **3**, 50–51 (1967)

10 ANTONIOU, 'A realization of gyrator using operational amplifiers and their use in *RC*-active network synthesis', *Proc. IEE*, **116**, 1838–1850 (1969)

11 BRUTON, L. T., 'Network transfer functions using the concept of frequency dependent negative resistance', *IEEE Trans. Cct Theory*, **CT-16**, No. 3, 406–408 (1969)

12 ZVEREV, A. I., *Handbook of Filter Synthesis*, Wiley, New York (1967)

13 LACANETTE, K., *The Switched-Capacitor Filter Handbook*, National Semiconductor Corporation (1985)

14 ANON., *Filtering: A New Concept*, Thomson Semiconducteur (1987)

15 MARTIN, K., 'Improved circuits for the realisation of switched-capacitance filters', *IEEE Trans. Cct Syst.*, **CAS-27**, 237–244 (1980)

16 SEVASTOPOULOS, N., *Application Considerations for an Instrumentation Lowpass Filter*, Application Note 20, *Linear Applications Handbook*, Linear Technology Corporation (1987)

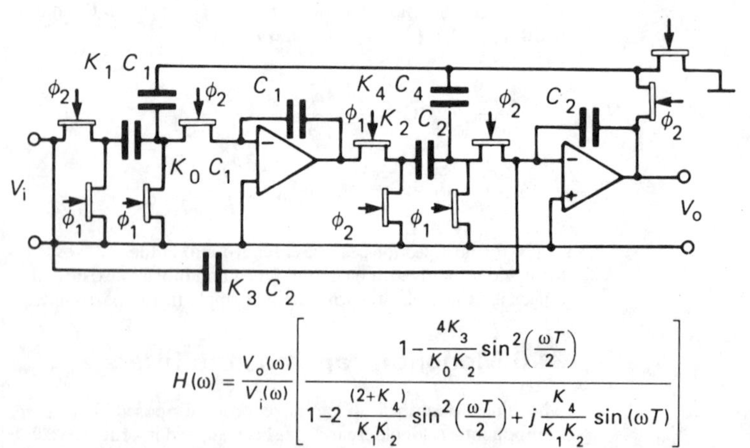

Figure 34.39 Second-order notch filter

Further reading

ASHWELL, R., 'Chip models speed digital filter simulation', *Electron. Product Design*, December (1988)

BALCH, B. F., 'A simple technique boosts performance of active filters', *EDN*, 10 November (1988)

BOWRON, P. and STEPHENSON, F. W., *Active Filters for Communications and Instrumentation*, McGraw-Hill, Maidenhead (1979)

CHEN, C., *Active Filter Design*, Hayden, Rechelle Park (1982)

CHRISTIAN, E. and EISMANN, E., *Filter Design Tables and Graphs*, Wiley, New York (1966)

CHRISTIANSEN, S. and FINK, D. N., *Electronic Engineer's Handbooks*, 2nd edn, McGraw-Hill (1982)

HAJOS, Z., 'Simplified design of wideband bandpass filters', *Tesla Electron.*, **19**, No. 1–2 (1986)

ITT, *Reference Data for Radio Engineers*, 5th edn, Howard W. Sams, New York (1968)

LANE, G., 'Introduction to digital filters', *Electronic Product Design*, Part 1, October; Part 2, November; Part 3, December (1982)

35 Attenuators

V J Green, BSc
STC plc

Contents

35.1 Introduction

An attenuator is a network designed to introduce a known loss when inserted between resistive impedances Z_1 and Z_2 to which the input and output impedances of the attenuator are matched. Z_1 and Z_2 can be interchanged as the source and load with the loss, expressed as a power ratio, remaining the same.

Three forms of resistance network that can conveniently be used are the T section, the π section, and the bridged T section. The equivalent balanced sections are also shown. Also included in this chapter are networks designed to match a source impedance Z_1 to a load impedance Z_2 ($Z_1 > Z_2$) with the minimum possible loss.

35.2 Symbols used

Z_1 and Z_2 are the image impedances of the network, generally: Z_1 is the input impedance and Z_2 the load impedance. N is the ratio of power delivered to the attenuator by the source to the power delivered to the load. K is the ratio of the input current into the attenuator to the output current into the load.

$$\text{Attenuation in dB} = 10\lg N \tag{35.1}$$

$$\text{Attenuation in nepers} = \theta = \tfrac{1}{2}\ln N \tag{35.2}$$

35.3 Ladder attenuators

A ladder attenuator (*Figure 35.1*) allows the input to be switched between shunt arms (S_3, S_2, S_1, S_0). A ladder attenuator can be designed by resolving it into a cascade of π sections (*Figure 35.2*) splitting the shunt arms into two resistors. The last section matches Z_2 to $2Z_1$ with all other sections being symmetrical, matching impedance $2Z_1$. There is a terminating resistor of $2Z_1$ on the first section. Design each section for the loss required between the switch points at the ends of each section.

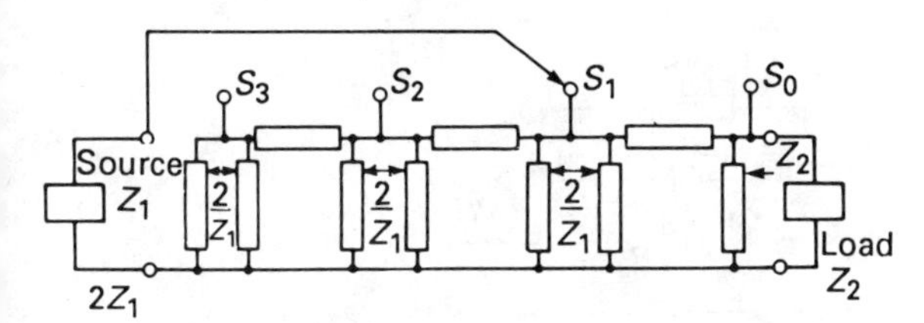

Figure 35.1 A ladder attenuator

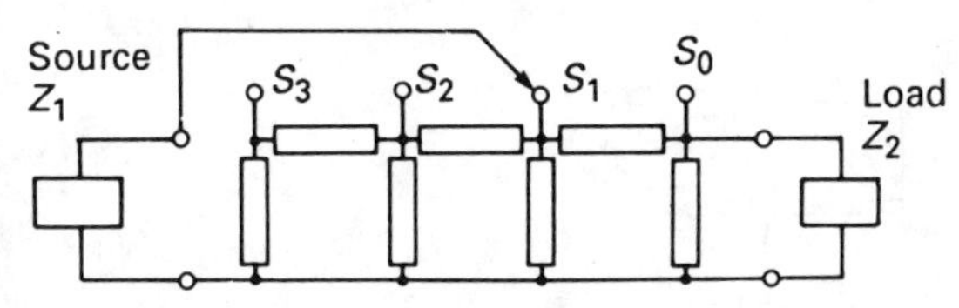

Figure 35.2 A ladder attenuator resolved into π sections

For input connected to S_0:

$$\text{loss in dB} = 10\lg(2Z_1+Z_2)^2/4Z_1Z_2 \tag{35.3}$$

$$\text{input impedance} = Z_2/2 \tag{35.4}$$

$$\text{output impedance} = Z_1Z_2/(Z_1+Z_2) \tag{35.5}$$

For input connected to S_1, S_2, S_3:

loss in dB $= 3 +$ sum of losses of π sections between input and output

input impedance $= Z_1$

35.3.1 Effect of incorrect load impedance on the operation of an attenuator

Let

$Z_2 + \Delta Z_2 =$ actual load impedance terminating the attenuator (ΔZ_2 need not be purely resistive)

$Z_1 + \Delta Z_1 =$ resulting input impedance

$K + \Delta K =$ resulting current ratio.

The relationships between these quantities are:

$$\frac{\Delta Z_1}{Z_1} = \frac{2\Delta Z_2/Z_2}{2N + (N-1)(\Delta Z_2/Z_2)} \tag{35.6}$$

and

$$\frac{\Delta K}{K} = \left(\frac{N-1}{2N}\right)\frac{\Delta Z_2}{Z_2} \tag{35.7}$$

35.4 Symmetrical T and H attenuators

35.4.1 Configuration

Configurations are shown in *Figure 35.3*.

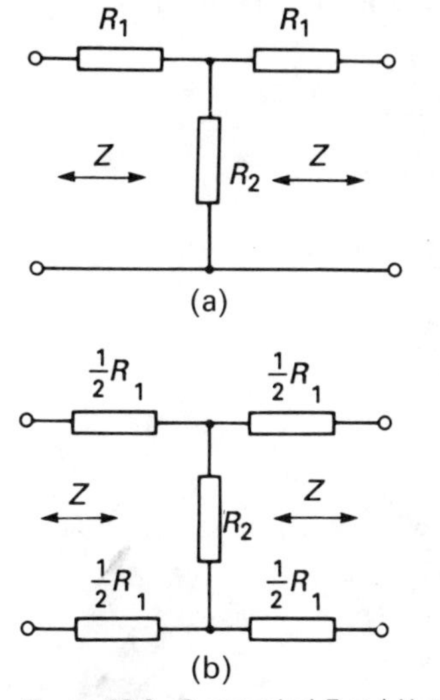

Figure 35.3 Symmetrical T and H attenuators: (a) T attenuator, (b) H attenuator

35.4.2 Design equations

For the T and H attenuators

$$R_1 = Z\tanh(\theta/2) \tag{35.8}$$

$$R_2 = Z/\sinh\theta \tag{35.9}$$

$$R_1 = Z\left(\frac{\sqrt{N}-1}{\sqrt{N}+1}\right) \quad (35.10)$$

$$R_2 = \frac{2Z\sqrt{N}}{N-1} \quad (35.11)$$

35.4.3 Design chart

This is shown in *Figure 35.4*. Values of R_1/Z can be read from the appropriate curve off the left-hand vertical axis. Values of R_2/Z can be read from the appropriate curve off the right-hand vertical axis. Attenuation in dB is on the horizontal axis.

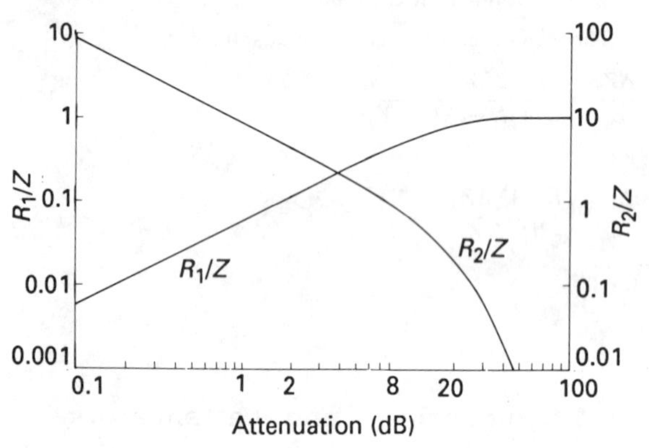

Figure 35.4 *T*-section design chart

35.5 Symmetrical π and 0 attenuators

35.5.1 Configuration

These are shown in *Figure 35.5*.

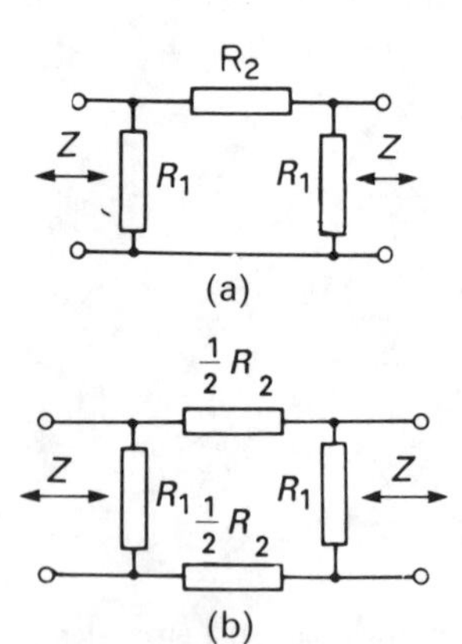

Figure 35.5 Symmetrical π and 0 attenuators: (a) π attenuator, (b) 0 attenuator

35.5.2 Design equations

For the π and 0 attenuators:

$$R_1 = Z/[\tanh(\theta/2)] \quad (35.12)$$

$$R_2 = Z\sinh\theta \quad (35.13)$$

$$R_1 = Z\left(\frac{\sqrt{N}+1}{\sqrt{N}-1}\right) \quad (35.14)$$

$$R_2 = Z\left(\frac{N-1}{2\sqrt{N}}\right) \quad (35.15)$$

35.5.3 Design chart

This is shown in *Figure 35.6*. Values of R_1/Z can be read from the appropriate curve off the left-hand vertical axis. Values of R_2/Z can be read from the appropriate curve off the right-hand vertical axis. Attenuation in dB is on the horizontal axis.

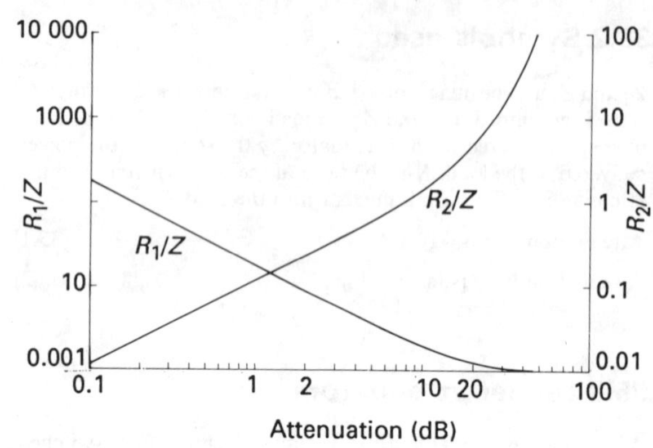

Figure 35.6 π-section design chart

35.6 Bridged *T* or *H* attenuators

35.6.1 Configuration

These are shown in *Figure 35.7*.

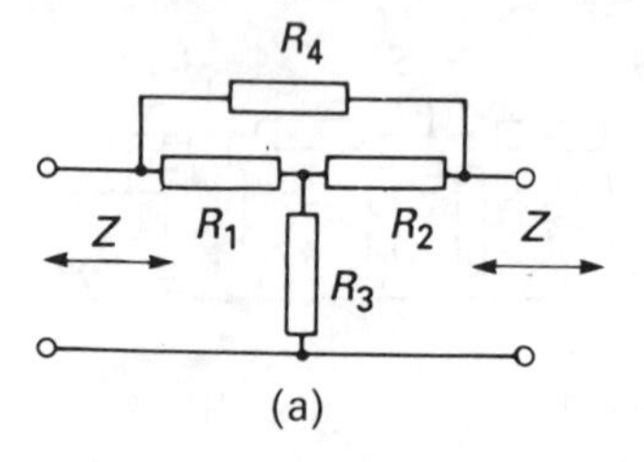

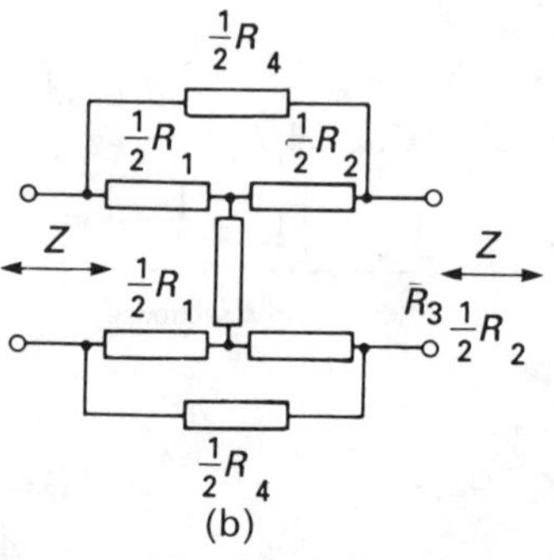

Figure 35.7 Bridged *T* and *H* attenuators: (a) *T* attenuator, (b) *H* attenuator

35.6.2 Design equations

$$R_1 = R_2 = Z \quad (35.16)$$

$$R_3 = Z/(\sqrt{N} - 1) \quad (35.17)$$

$$R_4 = Z(\sqrt{N} - 1) \quad (35.18)$$

35.6.3 Design chart

This is given in *Figure 35.8*. Values of R_3/Z and R_4/Z can be read from the appropriate curve off either vertical axis. Attenuation in dB is on the horizontal axis.

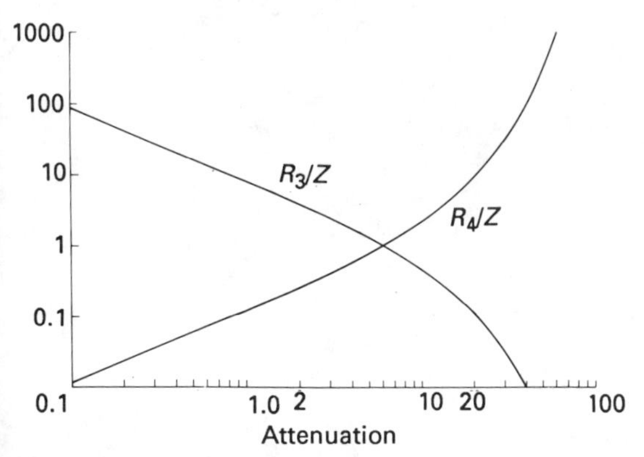

Figure 35.8 Bridged T design chart

35.7 Minimum-loss pads

35.7.1 Configuration

This is shown in *Figure 35.9*.

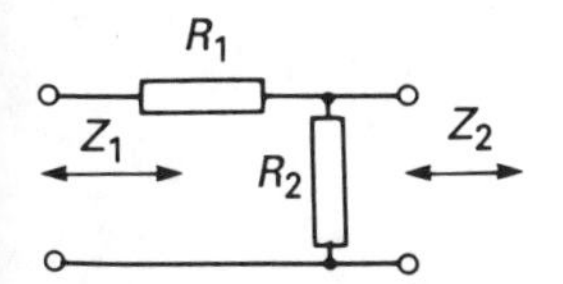

Figure 35.9 Minimum-loss pad

35.7.2 Design equations

$$\cosh\theta = (Z_1/Z_2) \quad (35.19)$$

$$R_1 = Z_1\{[1-(Z_2/Z_1)]^{1/2}\} \quad (35.20)$$

$$R_2 = Z_2/\{[1-(Z_2/Z_1)]^{1/2}\} \quad (35.21)$$

35.7.3 Design charts

Design chart (1) is shown in *Figure 35.10*. The minimum loss in dB can be read from the vertical axis. The ratio Z_1/Z_2 is the horizontal axis.

Design chart (2) is shown in *Figure 35.11*. The value of R_1/Z_1 can be read from the vertical axis. The ratio Z_1/Z_2 is on the horizontal axis.

$$\text{Let ratio } Z_1/Z_2 = M \quad (35.22)$$

$$R_2 = R_1/(M-1) \quad (35.23)$$

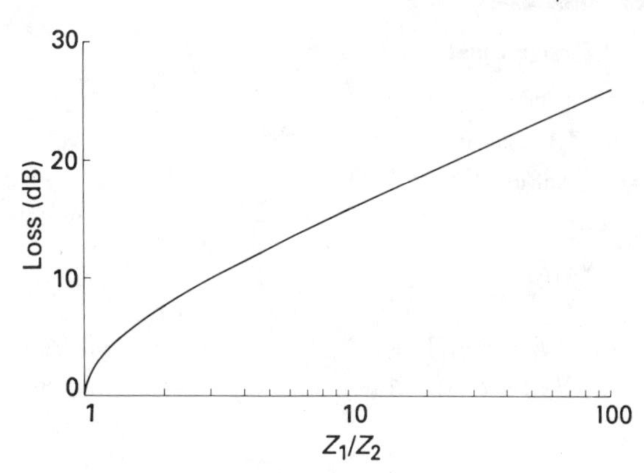

Figure 35.10 Minimum-loss pad design chart (1)

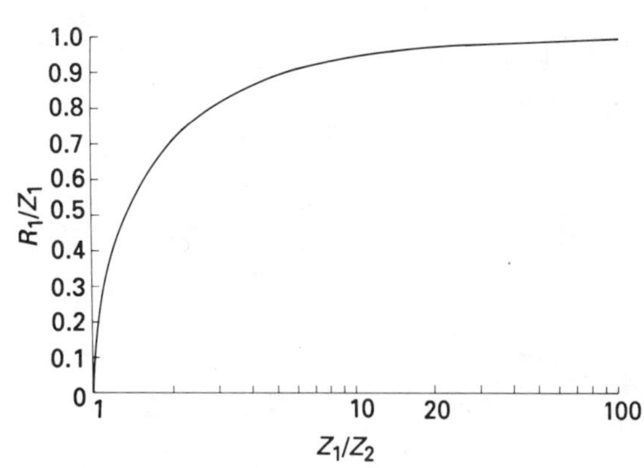

Figure 35.11 Minimum-loss pad design chart (2)

35.8 Miscellaneous T and H pads

35.8.1 Configuration

These are shown in *Figure 35.12*.

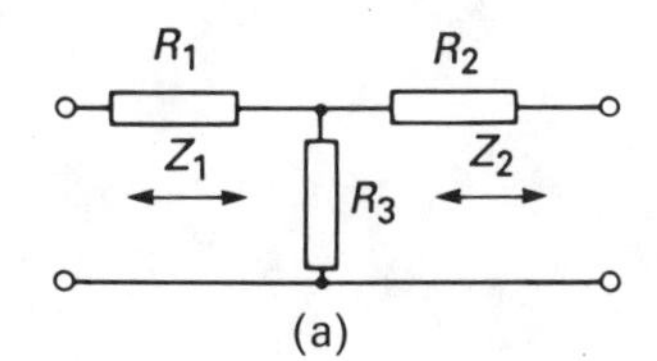

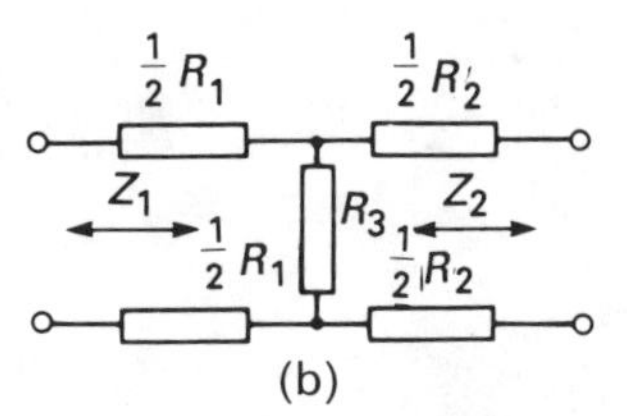

Figure 35.12 T and H pads: (a) T pad. (b) H pad

35.8.2 Design equations

For the *T* and *H* pads:

$$R_3 = (Z_1 Z_2)^{1/2}/\sinh\theta \qquad (35.24)$$

$$R_1 = (Z_1/\tanh\theta) - R_3 \qquad (35.25)$$

$$R_2 = (Z_2/\tanh\theta) - R_3 \qquad (35.26)$$

$$R_3 = \frac{2(NZ_1Z_2)^{1/2}}{N-1} \qquad (35.27)$$

$$R_1 = Z_1[(N+1)/(N-1)] - R_3 \qquad (35.28)$$

$$R_2 = Z_2[(N+1)/(N-1)] - R_3 \qquad (35.29)$$

Part 4

Electronic Design

36

Amplifiers

S W Amos, BSc(Hons), CEng, MIEE
Freelance Technical Editor and Author

Contents

36.1 Introduction

An amplifier is essentially an assembly of active and passive devices designed to increase the amplitude of an electrical signal.

The amplifier may be required to deliver appreciable power output, for example to operate a loudspeaker system. This is an example of a large-signal amplifier and the output signal has significant voltage and current components. In many amplifiers, however, e.g. the r.f. amplifier in a radio or television receiver, the output power is small because one component is small compared with the other. In such small-signal amplifiers the signal is best regarded as a voltage or current waveform. Active devices are by their very nature best suited for operation by a specific type of signal. For example valves and field-effect transistors are controlled by an input voltage and bipolar transistors by an input current. But, irrespective of the type of active device employed, an amplifier can be designed to act as a voltage or current amplifier by suitable choice of input resistance and output resistance. It is, in fact, the ratio of signal source resistance to input resistance and of output resistance to load resistance which determines whether the signal at input or output is best regarded as a voltage or a current. If the ratio is small compared with unity, the signal is best treated as a voltage, and if the ratio is large compared with unity it is more conveniently regarded as a current (*Figure 36.1*).

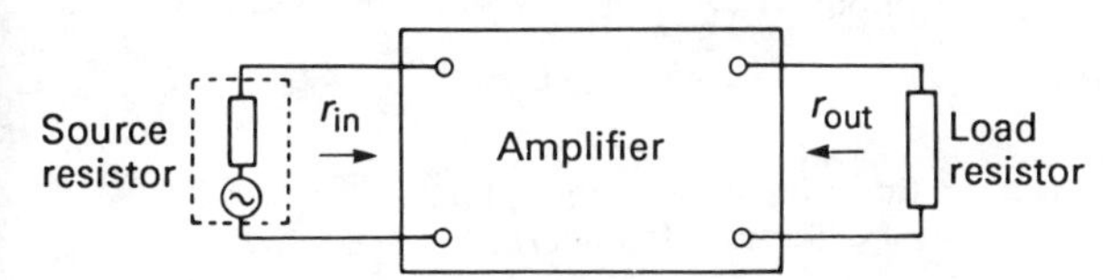

Figure 36.1 External and internal terminating resistances of an amplifier

36.2 Analogue amplifiers

Certain types of signal, notably audio and video signals, are characterised by the fact that at any instant they may have any value within certain limits. They are examples, in fact, of analogue signals, and amplifiers for such signals must accurately reproduce the way in which the input signal varies with time. In other words the output waveform of an analogue amplifier must be a substantially accurate copy of that of the input signal.

The frequency response of an analogue amplifier depends on its purpose. Certain d.c. amplifiers need a response extending from zero to a few kHz. A high-fidelity audio amplifier is likely to have a response level between 30 Hz and 15 kHz which is approximately the frequency range of the average human ear. The upper frequency limit of an amplifier for video or pulse signals is determined by the steepness of the vertical edges in the signals to be reproduced. The steepness of the leading (or trailing) edge of a pulse is measured by the time taken for its amplitude to change from 10% to 90% of its final steady value. This is known as the rise time (or fall time) and is related to the upper frequency limit of the amplifier by the approximate expression

$$f_{max} = (2 \times \text{rise (fall) time})^{-1} \qquad (36.1)$$

Thus if an amplifier is required to reproduce pulses with a rise (or fall) time of 0.1 μs, a response up to at least 5 MHz is needed.

The low-frequency limit of the amplifier is determined by the levelness of the horizontal sections of the signals to be reproduced. The degree of levelness is measured by the amount of sag. The low-frequency extreme f_{min} is given approximately by the relationship

$$f_{min} = \frac{\text{percentage sag}}{100\pi} \times \text{square-wave frequency} \qquad (36.2)$$

Thus to keep the sag in the reproduction of a 50 Hz square wave to less than 2% (representing high-grade performance) requires a low-frequency response extending down to 0.3 Hz!

36.3 Class A, B, C and D operation

Analogue amplifiers are required to reproduce the waveform of the input signal without distortion. Thus the amplifier must have a linear input/output characteristic. The characteristic of many active devices has a linear or near-linear section and the input signal can be arranged to operate on this section as indicated in *Figure 36.2* to give a reasonably undistorted output. This is

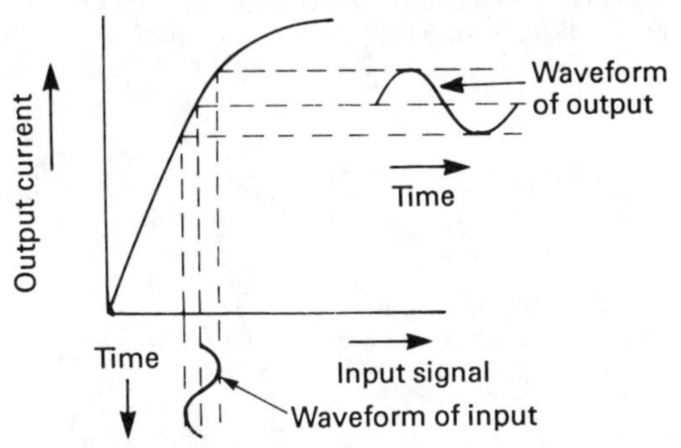

Figure 36.2 Class A operation

known as class A amplification and is extensively used in small-signal amplifiers such as the r.f. and i.f. amplifiers in receivers and the early stages of a.f. and video amplifiers. It is an inefficient form of amplification because the current taken by the device from the supply is independent of input-signal amplitude and remains constant even when the input signal is removed. For a sinusoidal input signal the maximum efficiency (i.e. ratio of output power to power taken from the supply) for a class A stage is 25% and practical values of efficiency are much less.

If the input signal is accommodated at the bottom end of the linear section of the characteristic near the point of output-current cut-off, as shown in *Figure 36.3*, only one-half of each cycle of input signal is reproduced. Linear amplification is, however, possible by using two matched devices in push–pull,

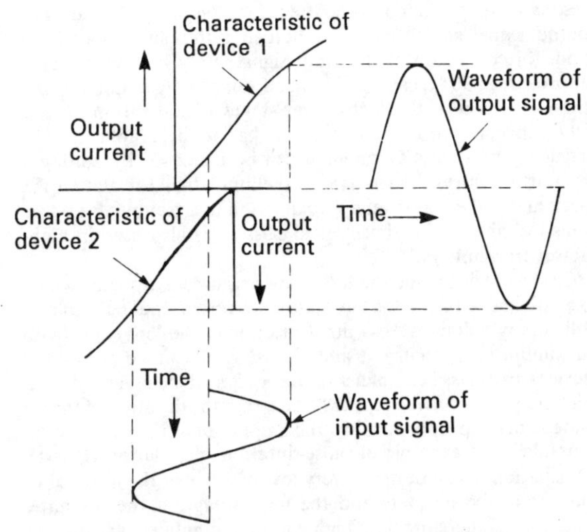

Figure 36.3 Class B operation using a push–pull pair

each biased to cut-off as shown in the diagram. A great advantage of this form of amplification (known as class B) is that the combined current of the two devices taken from the supply is proportional to the amplitude of the input signal and is very low in its absence. Theoretically the efficiency of a class B stage for sinusoidal signals is 78% and figures approaching this can be achieved in practice. This is the form of output stage, using matched bipolar transistors, employed in most modern amplifiers and radio receivers. Neither class A nor class B amplifiers are truly linear, but negative feedback can be used to achieve the required degree of linearity.

There is another mode of operation, known as class C, which can be used only for the amplification of constant-amplitude sinusoidal signals. In this mode only the peaks of the input signal are reproduced as shown in *Figure 36.4*. The output current thus

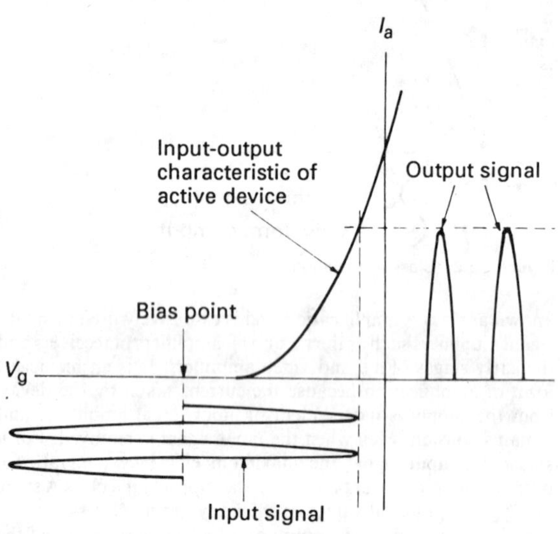

Figure 36.4 Class C operation

consists of a series of pulses at the frequency of the input signal and is not a replica of its waveform. The output current is rich in harmonics but these can be eliminated by use of an *LC* circuit, resonant at the fundamental frequency, as output load. The output signal, generated across the load, is thus an amplified and undistorted version of the input signal. Very high efficiencies, approaching 90%, can be obtained from class C amplifiers and these are used in the high-power stages of radio transmitters where a constant-amplitude signal has to be amplified. The efficiency of a class C amplifier can be increased by adding a proportion of third harmonic to the input signal to flatten the peak and make it a better approximation to a rectangular pulse. This technique, sometimes called class D, is also used in high-power transmitters.

In class D operation the active device is used as a switch which is at all times either in the 'off' state (with zero output current and full supply voltage across the device) or in the 'on' state (with maximum output current and almost zero voltage across the device). In a class D amplifier the device is switched regularly at a high frequency between the two states and the duration of the on state is made proportional to the amplitude of the signal to be amplified—an example of pulse-duration modulation (PDM). Dissipation in the device is very low because of the absence of current in the off state and the low voltage in the on state. Efficiency is therefore very high and such amplifiers are used in the modulators of modern sound transmitters.

36.4 Fundamental amplifying circuits of active devices

The three types of active device—valve, bipolar transistor and field-effect transistor—in their simplest forms have three terminals. The input signal is applied between two of them, of which one must be the base (gate or grid). The output signal is generated between two terminals, of which one must be the collector (drain or anode). It follows that one terminal must be common to input and output circuits and the basic amplifier configurations are classified by this common terminal (*Figure 36.5*).

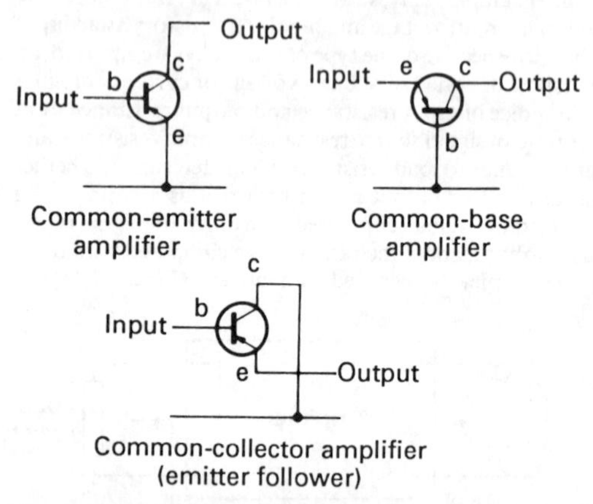

Figure 36.5 Basic forms of bipolar-transistor amplifying circuits

For example, if the common terminal of a bipolar transistor is the base, the circuit is known as the common-base amplifier. It is characterised by very low input resistance, very high output resistance and a current gain of unity. The output signal is in phase with the input signal. Voltage gain can be high. Because the base acts to some extent as an earthed screen between emitter and collector there is little capacitance linking input and output circuits and the amplifier can be used with stability at VHF and UHF.

The common-collector circuit has high input resistance, low output resistance and a voltage gain of approximately unity. Current gain is high. The output signal is in phase with the input signal and, because the gain is unity, is almost equal to it. The output signal thus follows the input signal and the circuit is more popularly known as an emitter follower. The circuit is used where a high input resistance or a low output resistance is required or as a buffer between stages where it is important to minimise unwanted coupling.

The third fundamental amplifying circuit is the common-emitter. The input resistance is low, the output resistance high (though the ratio of the two is much less than in the common-base circuit) and the voltage gain and current gain can both be considerable, so the amplifier can deliver appreciable power. This probably accounts for the fact that this is by far the most used of the three basic circuits. The output signal is inverted with respect to the input signal. A disadvantage of the circuit is that feedback via the capacitance between input and output terminals within the device itself prevents stable r.f. amplification.

36.5 General amplifying circuits

To utilise the wanted part of the characteristic, an active device must be suitably biased. The base of a bipolar transistor needs a

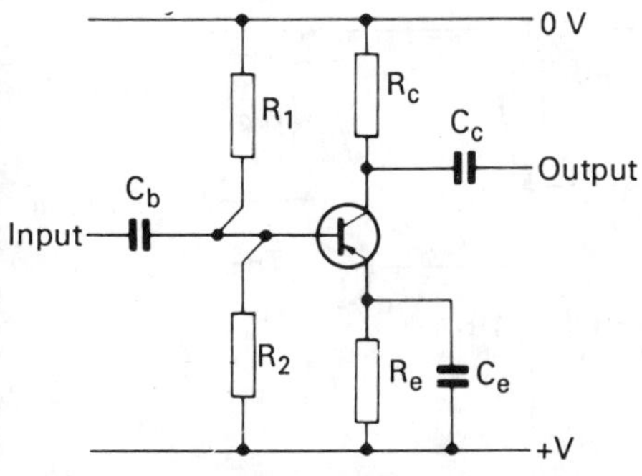

Figure 36.6 Common-emitter stage of amplification with d.c. stabilisation by potential divider and emitter resistor

bias voltage between that of the collector and the emitter, and a potential divider is commonly used to provide it as shown for the class A amplifying stage in *Figure 36.6*. The divider imposes a particular value of voltage on the base and thus defines the emitter voltage (which, for a silicon transistor, is approximately 0.7 V less than the base voltage). The value of the emitter resistor R_e can now be chosen to give the wanted value of mean emitter current. This circuit gives a measure of stabilisation of emitter current because any change in the current brings about a change in base–emitter voltage which offsets the current change. Stabilisation of mean emitter current is important because a number of transistor parameters depend on it, and also because, for a given base bias, emitter current can vary from specimen to specimen of the same transistor type over a range as great as 3:1. This bias circuit can also be used with enhancement-type field-effect transistors which also require a gate bias voltage between the source and drain voltages.

For depletion-type field-effect transistors and valves where the gate voltage lies outside the range of the source–drain voltage, the biasing circuit shown in *Figure 36.7* can be used. Again class

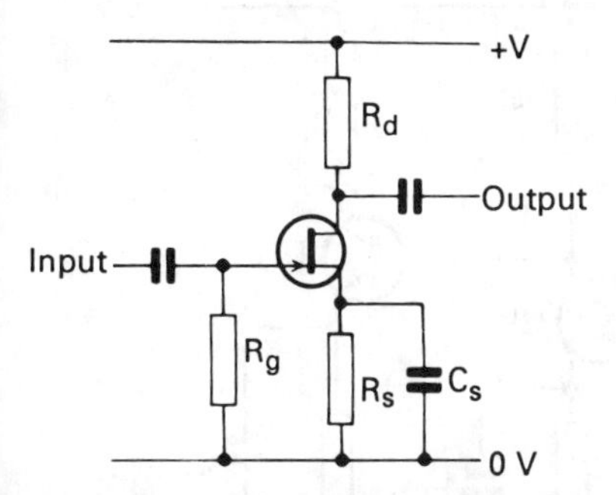

Figure 36.7 A depletion-type junction-gate field-effect transistor with bias and d.c. stabilisation provided by a source resistor

A operation is assumed. The voltage generated across the source resistor R_s by the source current is applied between gate and source, the gate resistor R_g being included to permit the application of the external input signal. The required value of R_s is given by the quotient of the gate bias voltage and the mean source current. This circuit also gives a measure of source-current stabilisation.

36.6 Use of negative feedback

The two biasing circuits just described enable a consistent value of mean current to be achieved, but in spite of this there are inevitably differences in gain between nominally identical circuits because of the variations in characteristics from specimen to specimen of the same transistor type. The application of negative feedback is the usual way of minimising these differences and one method of applying negative feedback is by using an un-decoupled emitter or source resistor. The effect of negative feedback is to reduce gain and improve linearity and it also makes the gain of the stage more dependent on the values of the feedback components and less dependent on the characteristics of the transistor itself. This form of feedback circuit, shown in basic form in *Figure 36.8(a)*, also increases the input resistance

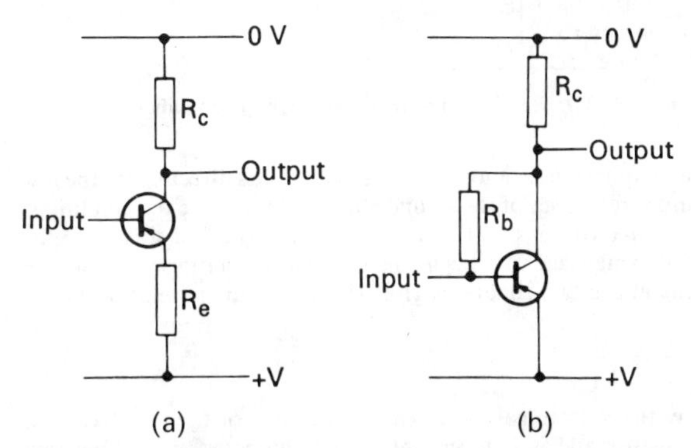

Figure 36.8 Two basic methods of applying negative feedback to a common-emitter amplifier

and the output resistance of the amplifier. The performance of an amplifier with a high input resistance is best expressed in terms of input voltage as mentioned at the beginning of this chapter. The high output resistance implies that the output signal is best regarded as an output current. This type of circuit can thus be treated as a voltage-to-current converter and the ratio of the output current to the input voltage is given approximately by the expression

$$\frac{i_{out}}{v_{in}} = \frac{1}{R_e} \tag{36.3}$$

An alternative method of applying negative feedback to a single transistor is illustrated in *Figure 36.8(b)*. The resistor R_b returns to the input circuit a feedback current proportional to the output voltage. The effect of this type of feedback circuit is, as before, to reduce gain and improve linearity and to make the gain of the stage more dependent on the value of R_b than on the properties of the transistor. However, this form of feedback reduces both the input resistance and the output resistance of the stage. The behaviour of an amplifier with a low input resistance is best explained by assuming the input signal to be a current. The low output resistance implies that the output signal is best regarded as a voltage. This circuit is thus a current-to-voltage converter and the ratio of the output voltage to the input current is given approximately by the expression

$$\frac{v_{out}}{i_{in}} = R_b \tag{36.4}$$

By combining a voltage-to-current converter with a current-to-voltage converter it is possible to produce a two-stage voltage or current amplifier.

36.7 Two-stage voltage amplifier

If the voltage-to-current stage is placed first as indicated in *Figure 36.9*, the two-stage amplifier has a high input resistance and a low output resistance as required in a voltage amplifier. The

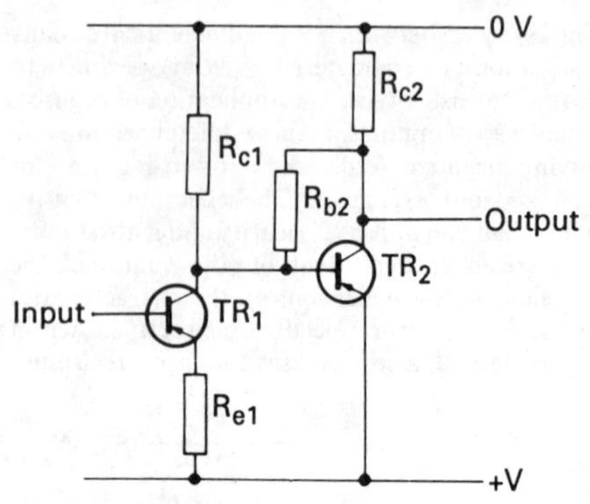

Figure 36.9 Basic form of two-stage voltage amplifier

high output resistance of the first stage feeds directly into the low input resistance of the second stage and these are the conditions required to transfer the current output of the first stage into the following stage with negligible loss. Thus equating i_{out} with i_{in} in the above two equations gives the approximate relationship

$$\frac{v_{out}}{v_{in}}=\frac{R_{b2}}{R_{e1}} \qquad (36.5)$$

i.e. the voltage gain is given by the ratio of the two feedback resistors and is independent of the characteristics of the two transistors.

In practical versions of the voltage-amplifying circuit R_{b2} is usually returned to the emitter circuit of TR_1 as shown in *Figure 36.10*. This has little effect on the performance because the

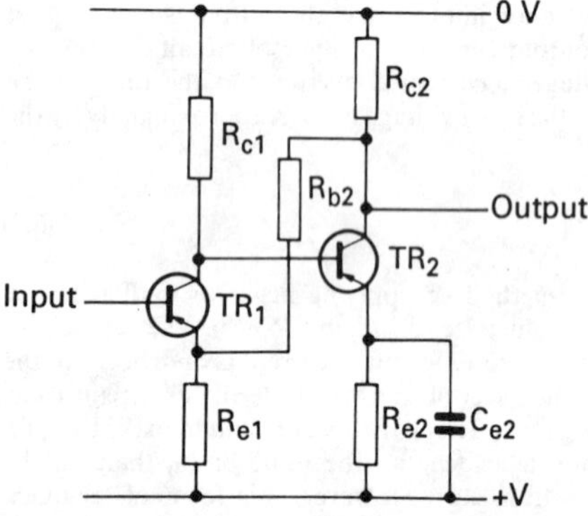

Figure 36.10 Practical circuit for two-stage voltage amplifier incorporating means of ensuring d.c. stability

current injected into TR_1 emitter by R_{b2} emerges with negligible loss from the collector and so enters TR_2 base, R_{c1} being large compared with TR_2 input resistance. R_{b2} and R_{e1} now form a voltage divider across TR_2 with the junction connected to the base via TR_1. This helps in the d.c. stabilisation of the circuit, all that is necessary to complete it being the inclusion of an emitter resistor R_{e2} as shown in *Figure 36.10*. C_{e2} is needed to minimise signal-frequency negative feedback.

36.8 Two-stage current amplifier

If the current-to-voltage converter is placed first, as shown in *Figure 36.11*, the two-stage amplifier has a low input resistance and a high output resistance as required in a current

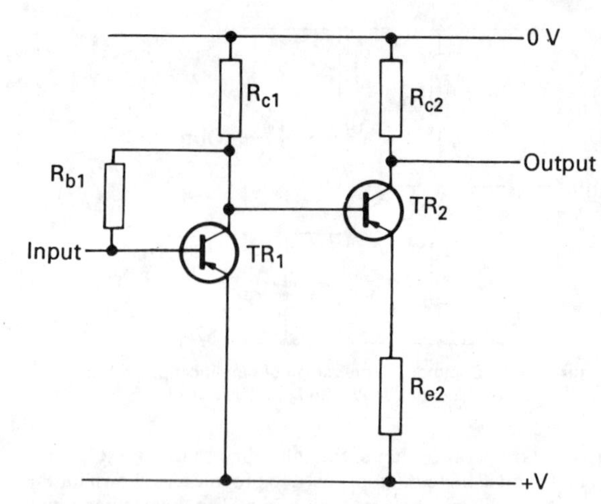

Figure 36.11 Basic form of two-stage current amplifier

amplifier. The low output resistance of the first stage feeds directly into the high input resistance of the second stage. These are the conditions required to transfer the voltage output of the first stage into the input of the second stage with negligible loss. Thus equating v_{in} and v_{out} in the above two equations gives the approximate relationship

$$\frac{i_{out}}{i_{in}}=\frac{R_{b1}}{R_{e2}} \qquad (36.6)$$

i.e. the current gain is given by the ratio of the two feedback resistors and is independent of the characteristics of the two transistors.

In practical versions of the current amplifying circuit R_{b1} is usually fed from TR_2 emitter as shown in *Figure 36.12*. This has

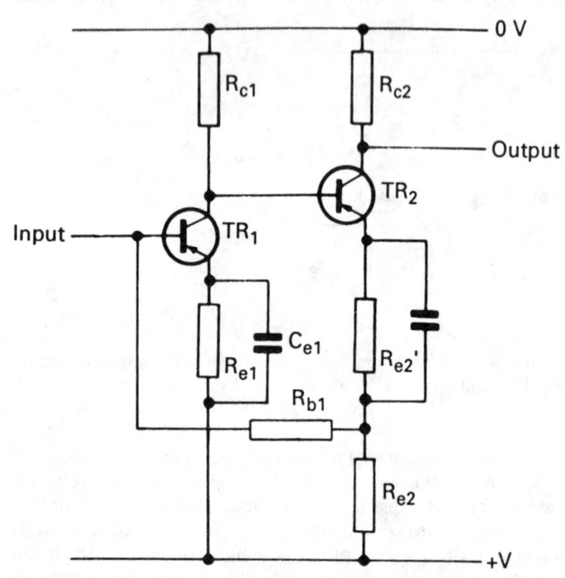

Figure 36.12 Practical circuit for two-stage current amplifier incorporating means of ensuring d.c. stability

little effect on the performance of the circuit because TR_1 collector is connected to TR_2 base and TR_2 emitter follows the signal voltage at the base. To achieve d.c. stability a common technique is to add a resistor $R_{e2'}$ in series with R_{e2} to form a voltage divider to feed TR_1 base via R_{b1}. R_{e1} is then included in TR_1 emitter circuit being decoupled by C_{e1} to eliminate signal frequency negative feedback. $R_{e2'}$ should be decoupled as shown for the same reason.

It is possible to effect economies in components and so produce simpler circuits by using complementary transistors in these two-stage amplifiers. As an example *Figure 36.13* gives the circuit diagram of a complementary voltage amplifier. As before, the gain is given by

$$\frac{v_{out}}{v_{in}} = \frac{R_3}{R_2} \quad (36.7)$$

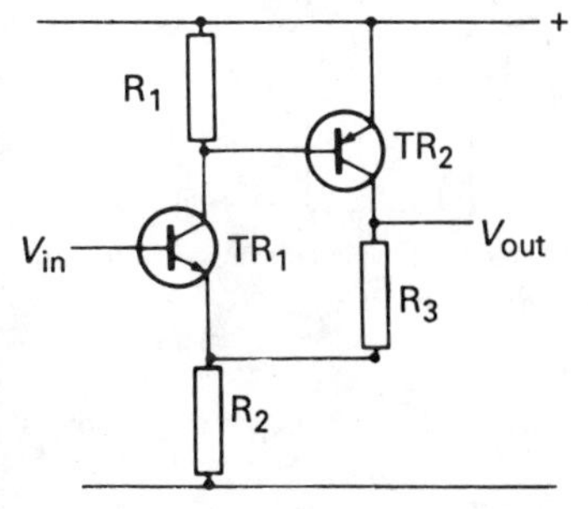

Figure 36.13 Basic form of complementary two-stage voltage amplifier

36.9 DC amplifiers

A d.c. amplifier could consist of a succession of direct-coupled common-emitter stages, each amplifying the output of the previous stage. However, such an arrangement has the great disadvantage that the inevitable slight drift in current in early stages would, after amplification by the following stages, give rise to a serious spurious output. To avoid this it is customary to use long-tailed-pair circuits of the type shown in *Figure 36.14*. Each

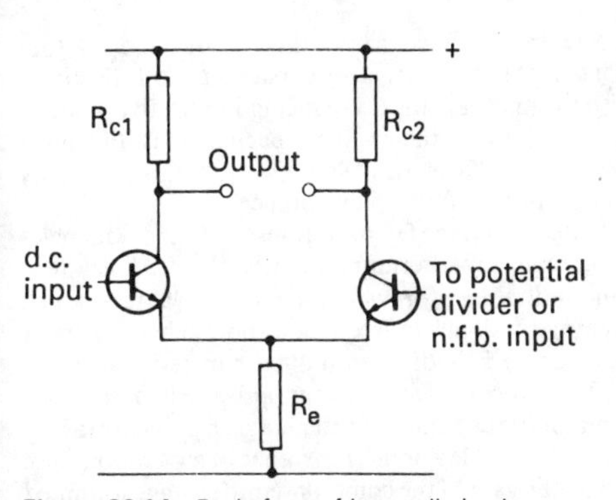

Figure 36.14 Basic form of long-tailed pair

pair consists of two matched transistors with equal collector loads and a common external emitter resistor. Any drift in the collector current of one transistor caused, for example, by a change in ambient temperature is accompanied by an equal change in the collector current of the other transistor which shares the same environment. The output of the pair is developed between the collector terminals and is thus zero for all current changes which affect both transistors equally. In a d.c. amplifier the signal to be amplified is applied to the base of one of the transistors and a negative-feedback voltage can be applied to the base of the other transistor to improve linearity. In long-tailed pairs forming part of an integrated circuit the emitter resistor is usually the collector-emitter path of a third bipolar transistor used to stabilise the mean current through the matched pair.

This technique is used in operational amplifiers. These are d.c. amplifiers with a gain–bandwidth product of, typically, 1 MHz; e.g. they have a gain of 100 up to 10 kHz, 10 to 100 kHz, etc. Such amplifiers are used in analogue computers to carry out arithmetical and other mathematical operations as defined by an external negative feedback loop. For example, by using an *RC* combination with a suitable value of time constant, as shown in *Figure 36.15*, an operational amplifier can be used to perform differentiation.

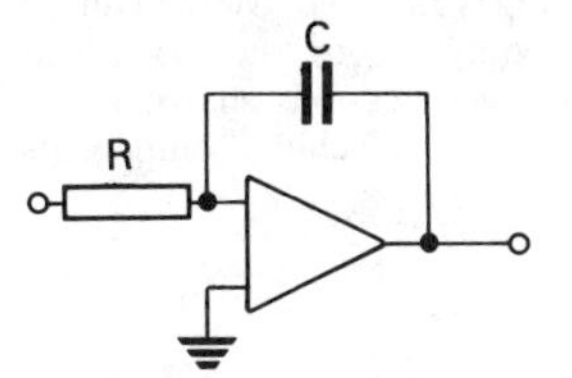

Figure 36.15 Operational amplifier used as a differentiator

36.10 AF output stages

Audio-frequency amplifiers require a frequency response from approximately 30 Hz to 15 kHz, a power output of, say, 30 W and a high degree of linearity. As mentioned earlier the output stage usually operates in class B push–pull, and early examples used the symmetrical circuit shown in *Figure 36.16*, in which a

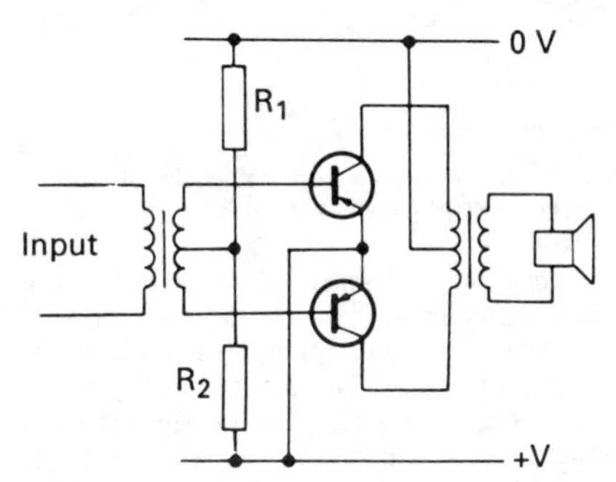

Figure 36.16 Symmetrical push–pull bipolar-transistor output amplifier

transformer with a centre-tapped secondary winding is used for phase splitting and another with centre-tapped primary winding is used to match the output to the loudspeaker. This arrangement was superseded by the single-ended circuit of *Figure 36.17*, which avoids the need for transformers.

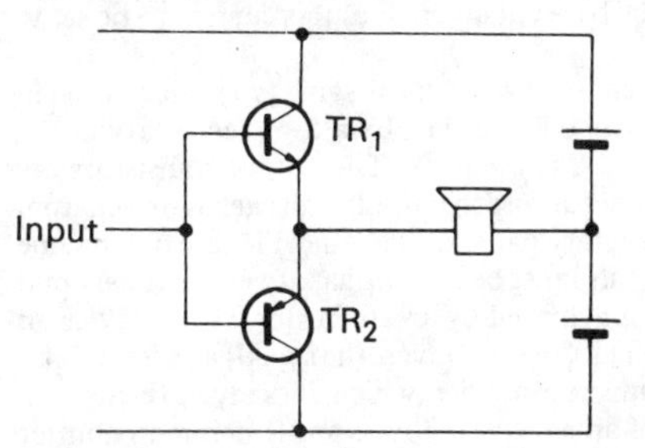

Figure 36.17 Essential features of a complementary push–pull single-ended output stage

A positive voltage applied to the base of a pnp transistor reduces its collector current but the same voltage applied to the base of an npn transistor increases the collector current. If, therefore, the same signal is applied to the bases of a complementary pair of transistors, push–pull operation is obtained without the need for a phase-splitting transformer. The transistors can be connected in series across the supply to form a single-ended output stage as shown in *Figure 36.17*. A signal applied to the common-base input now causes the common-emitter connection to oscillate between the limits of the supply voltage.

A difficulty of the basic circuit of *Figure 36.17* is that the load is in the emitter circuit of both transistors, which therefore operate as emitter followers with unity voltage gain. A large input voltage is needed to deliver maximum output into the load. This disadvantage is overcome by the modifications shown in the circuit diagram of *Figure 36.18*, which also includes components

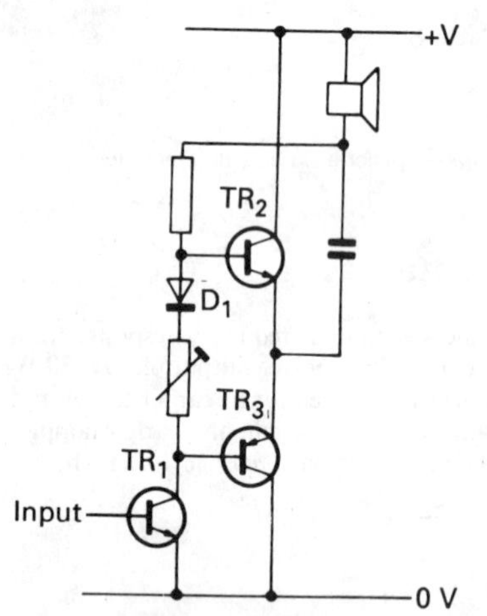

Figure 36.18 Modification of previous circuit to ensure that the output transistors operate in common-emitter mode

for biasing the transistors. The load is now fed from the common-emitter terminal via a capacitor. This enables the other terminal of the load to be connected to the positive supply line so that the junction of load and capacitor can be used as the supply source for the driver transistor TR_1. This arrangement ensures that the signal voltage generated across the collector load resistor for TR_1 is applied between base and emitter of both output transistors. Thus TR_2 and TR_3 operate as common-emitter amplifiers which have much greater voltage gain than the emitter-followers of *Figure 36.17*.

The forward-biased diode D_1 is included to compensate for the changes in base-emitter voltage of the output transistors which occur with alterations in temperature. The diode thus ensures greater constancy in the standing current of TR_2 and TR_3. The preset resistor in the base circuit enables this current to be set to the desired value.

If desired, matched transistors of the same type can be used in the output stage, these being driven by a complementary push–pull pair, as shown in *Figure 36.19*. The output transistors are also connected in series across the supply and each is driven from one of the complementary pairs. As TR_4 and TR_5 are of the same type their input signals must be in antiphase to ensure push–pull operation and this is achieved by arranging for TR_2 to act as an emitter follower (which does not invert the signal) and for TR_3 to act as a common-emitter amplifier (which does invert the signal). The current gain of an emitter follower and a common-emitter amplifier are approximately equal and this ensures that the output pair receive equal-amplitude input signals.

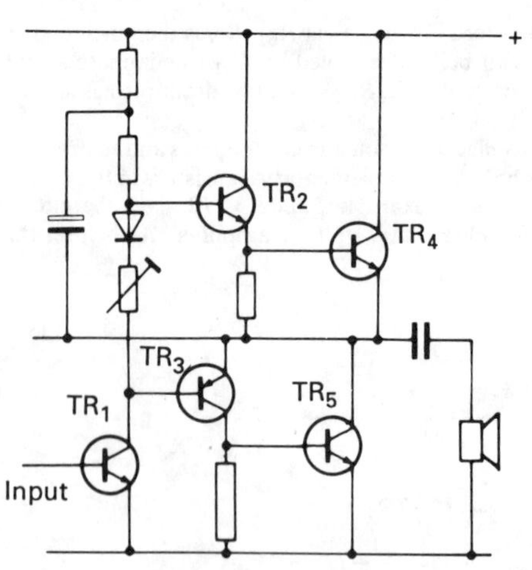

Figure 36.19 Elaboration of the previous circuit to permit the use of similar transistors in the output stage

36.11 Video amplifiers

Video amplifiers need a wider bandwidth than a.f. amplifiers, extending from zero or a very low frequency to, say, 5.5 MHz. Modern silicon transistors have such high cut-off frequencies that these do not cause any limitations in the high-frequency response of amplifiers. Thus two-stage amplifiers of the type described in *Sections 36.7* and *36.8* can be used for video amplification and a particularly suitable circuit can be devised by using complementary transistors in the circuit of *Figure 36.13*. The use of complementary devices simplifies design by permitting direct connection between the collector of TR_1 and the base of TR_2, so eliminating the need for coupling capacitors and extending the low-frequency response. The high-frequency limit of such an amplifier is likely to be determined by shunt capacitance.

Shunt capacitance arises from the input and output capacitance of transistors and by stray capacitance, all effectively in parallel with the load resistor, so reducing its effective value as frequency rises. Of these the input capacitance is the most significant because, by virtue of Miller effect, this can be many times the internal base-collector capacitance.

One way of minimising the fall in response at high frequencies is by the use of frequency-discriminating negative feedback (e.g. by shunting the emitter resistor by a capacitor the value of which is chosen to remove feedback as frequency approaches the upper frequency limit) or by including an inductor in series with the load resistor. But a better way is to minimise Miller effect by limiting the gain of the common-emitter stage, e.g. by arranging for it to feed into the very low input resistance of a common-base stage. In this way the gain of the common-emitter stage is limited to near unity but the common-base stage operates at full gain and the combination (known as a cascode amplifier) has an overall gain equal to that of a common-emitter stage but with a much-improved high-frequency response. The two transistors forming the cascode can be connected in series across the supply as shown in *Figure 36.20*, which uses two field-effect transistors. A cascode is commonly employed as the input stage of a video amplifier because it introduces little noise. Moreover, the high input resistance makes a field-effect cascode particularly suitable for use in a camera-head amplifier because camera tubes have a high, predominantly capacitive output impedance which needs to feed into a high resistance.

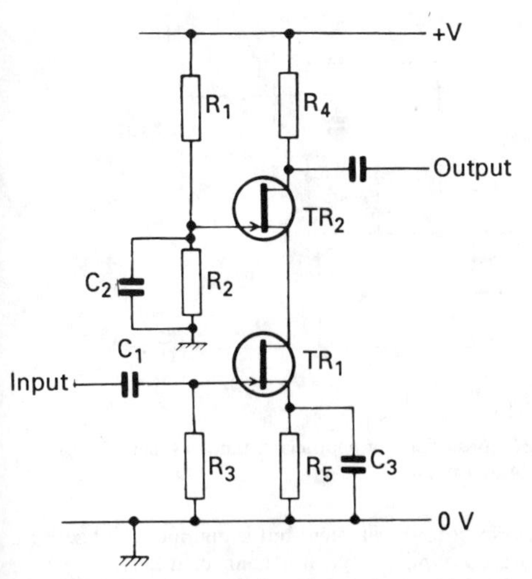

Figure 36.20 A cascode input stage for a wideband amplifier using two junction-gate field-effect transistors

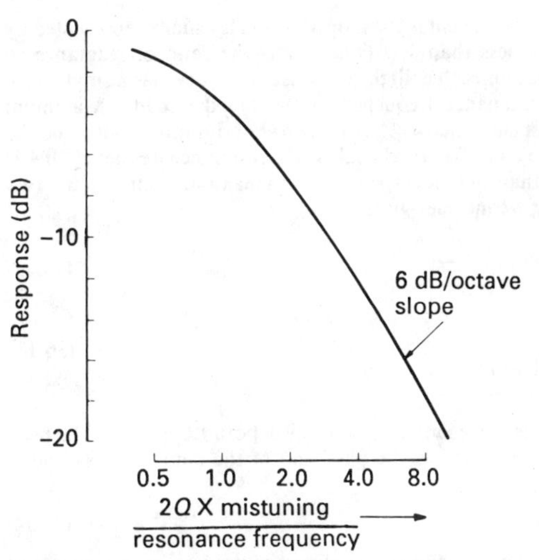

Figure 36.21 Universal curve illustrating the frequency response of a single tuned circuit

36.12 RF amplifiers

So far we have confined our attention to base-band amplifiers, i.e. amplifiers with a frequency response which begins at zero or a very low frequency and extends for several octaves. However, the amplifiers used for r.f. amplification have a restricted bandwidth which is often small compared with the centre frequency. Indeed certain amplifiers, notably those providing the carrier source in radio transmitters, are required to handle a single frequency only. However, those which amplify modulated r.f. signals need a bandwidth which can accommodate the sidebands without significant attenuation. Such a bandwidth is usually obtained by using resonant circuits as loads for the active devices and/or for coupling them.

36.13 Use of single tuned circuits

One form of tuned amplifier which can be used to amplify a.m. signals consists of a succession of *LC* circuits all tuned to the same centre frequency and separated by active devices. The i.f. amplifiers of sound and television receivers are often examples of such amplifiers. The centre frequency of each tuned circuit is determined by the product *LC*: the rate of fall-off of the response on each side of the centre frequency is determined by the effective Q value of the inductor. The relationship between response and mistuning Δf is illustrated in *Figure 36.21*. This shows that the loss is 3 dB when $2Q\,\Delta f/f_c=1$. This gives

$$2\,\Delta f=\frac{f_c}{Q} \qquad (36.8)$$

$2\,\Delta f$ is the frequency difference between the two 3-dB down points and is hence the passband. Thus we have

$$Q=\frac{\text{centre frequency}}{\text{passband}} \qquad (36.9)$$

This is a useful relationship which enables Q values to be assessed for each application. For example the standard centre frequency for the i.f. amplifier of a 625-line television receiver is 36.5 MHz and the bandwidth must be 6 MHz to include vision and sound signals. The required Q value in an *LC* circuit for such an amplifier is thus given by

$$Q=\frac{\text{centre frequency}}{\text{passband}}=\frac{36.5}{6}\approx 6$$

It would be difficult to construct an inductor to have precisely this value of Q and the usual solution to this difficulty is to use coils of higher Q value (say 100) and to damp them by parallel resistance to give the required Q value. The input resistance of transistors is usually sufficiently low at this value of centre frequency that it can conveniently be used to give the required damping. The precise damping effect can be controlled by adjusting the L to C ratio of the tuned circuit, keeping the *LC* product constant to maintain the value of the centre frequency.

As an example a typical value for the input resistance of a bipolar transistor is 2 kΩ. If the transistor base and emitter terminals are effectively in parallel with the *LC* circuit then the input resistance dictates the dynamic resistance R_d of the circuit. R_d is equal to $L\omega Q$ and thus we can calculate the required value of inductance from the relationship

$$L=\frac{R_d}{\omega Q} \qquad (36.10)$$

Substituting for R_d, ω and Q we have that the inductance for the vision i.f. stage is given by

$$L=\frac{2\times 10^3}{6.284\times 36.5\times 10^6\times 6}\,\text{H}=1.4\ \mu\text{H}$$

C can be calculated from the relationship

$$\omega=\frac{1}{\sqrt{(LC)}} \qquad (36.11)$$

Substituting for ω and L we have

$C\approx 14$ pF

In practice it is sometimes more convenient to use values of L and C which give a higher dynamic resistance and to reduce this to the required value by suitable choice of tapping point on the inductor for the base connection of the transistor.

36.14 Use of coupled tuned circuits

A better approximation to the ideal square-topped frequency response required for the amplification of modulated r.f. signals can be obtained by using *LC* circuits in coupled pairs. The response of two identical coupled circuits depends on the coupling between the coils. If the mutual inductance is M, the coupling coefficient is given by M/L and is usually represented by k. If k is less than $1/Q$ (where Q is the ratio of reactance to resistance for each coil) the response is low and has a single peak at the resonance frequency of the tuned circuits. Maximum output occurs when $k=1/Q$, known as optimum coupling and the response has a flattened peak at the resonance frequency. If k is greater than $1/Q$ the response is still a maximum but now has two peaks at frequencies given by

$$f_1 = \frac{f_c}{\sqrt{(1-k)}} \tag{36.12}$$

$$f_2 = \frac{f_c}{\sqrt{(1+k)}} \tag{36.13}$$

where f_c is the resonance frequency of both LC circuits. The peak separation measures the passband of the coupled coils and is given by

$$f_1 - f_2 = f_c k \tag{36.14}$$

Thus the coefficient of coupling required to give a required passband at a particular centre frequency is given by

$$k = \frac{\text{passband}}{\text{centre frequency}} \tag{36.15}$$

As an example consider the i.f. amplifier of an a.m. broadcast receiver. The carrier-frequency spacing on the medium waveband is 9 kHz and this figure is often taken as the bandwidth required. The centre frequency is standardised at 465 kHz and thus the coefficient of coupling required in an i.f. transformer is, from Equation (36.15), given by

$$k = \frac{9}{465} \approx 0.02$$

Thus, if the inductors have a Q value of 50 or less, this value of k corresponds to critical coupling or less and a single-peaked response is obtained. If Q exceeds 50 the response is double-humped and the two peaks become more marked as the Q value is increased.

36.14.1 Means of ensuring stability

If, however, synchronously tuned circuits are used in the input and output circuits of an active device used as a common-cathode (common-emitter or common-source), amplifier instability is likely to occur as a result of the positive feedback via the internal capacitance between the input and output terminals of the device. This can occur even at low radio frequencies and whether the device is a small bipolar transistor in the r.f. or i.f. circuits of a receiver or a large triode valve used in the high-power stages of a broadcast transmitter.

A similar technique can be used in both instances to secure stable r.f. amplification. It is illustrated in *Figure 36.22* as applied to the triode valve. The HT supply for the anode is introduced via a tapping point on the anode-load inductor so that the signal at the top end of the inductor is in antiphase with that at the anode. Thus the signal fed back to the grid via the neutralising capacitor is in antiphase with that fed back via the internal anode-grid capacitance (shown dashed) and can be adjusted empirically to secure stable amplification. This circuit is extensively used at high-power radio transmitters.

An alternative solution is to use a tetrode valve in the output stage of the transmitter. This valve has a second grid, situated between the control grid and the anode, which, if earthed, acts as an electric screen and eliminates the capacitance between input and output, thus preventing positive feedback. To maintain electron emission to the anode, the screen grid must be positively

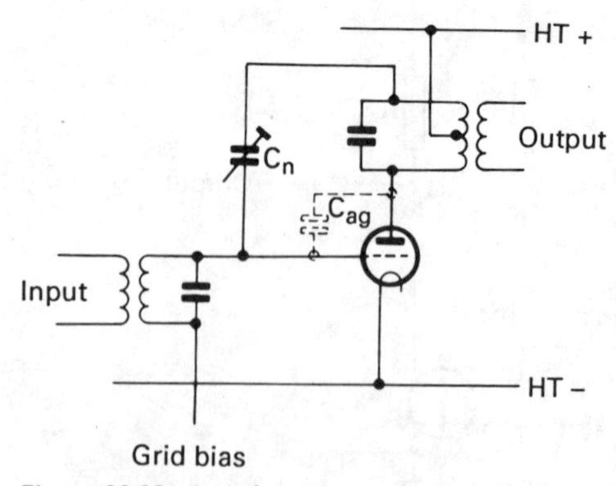

Figure 36.22 Basic form of neutralised-triode r.f. amplifying circuit used in transmitters

biased with respect to the filament but is maintained at zero r.f. potential by a decoupling capacitor connected between screen and filament.

In fact the control grid of a triode valve can be used as a screen if the input signal is applied to the filament as shown in *Figure 36.23* and this common-grid (sometimes called earthed-grid) amplifying circuit is also used in transmitters.

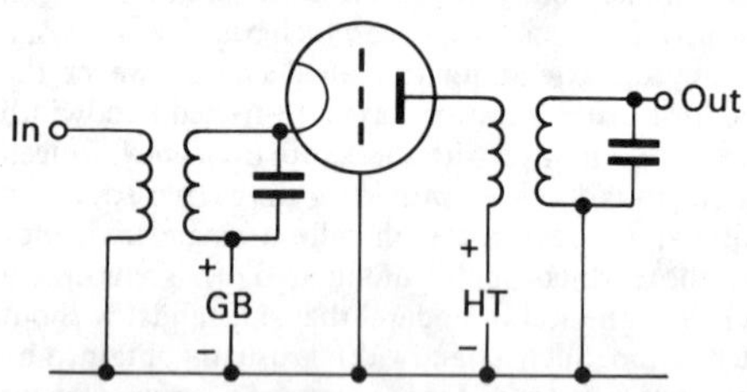

Figure 36.23 Basic arrangement of a triode used as a common-grid amplifier

Similar problems of securing stable r.f. amplification were, of course, encountered in the design of valve receivers and were solved by using r.f. tetrodes and, later, pentodes which had a screen grid between control grid and anode. For bipolar transistor circuits it is more usual to limit the gain of r.f. and i.f. amplifiers (e.g. by using a low tapping point for the input signal to the base) to a value at which the internal feedback cannot cause difficulties.

36.14.2 Cascode amplifiers

At VHF and UHF it is common practice to use a common-base or a common-gate stage as an amplifier and this is often combined with a common-emitter or a common-source stage to form a cascode circuit.

In the circuit of *Figure 36.20*, the cascode is formed of two discrete field-effect transistors, the lower acting as common-source amplifier and the upper as common-gate amplifier. It is significant that there is no external connection to the link between the drain of the lower transistor and the source of the upper. Thus it is possible to replace the two transistors by one dual-gate field-effect transistor as shown in *Figure 36.24* and this form of amplifier is often used in the r.f. stage of a VHF or UHF receiver. The upper gate is connnected to a potential divider across the supply and is decoupled by a low-reactance capacitor, which makes the circuit very similar to that of an r.f. tetrode or pentode used as an r.f. amplifier. In fact the dual-gate transistor is sometimes known as a tetrode transistor.

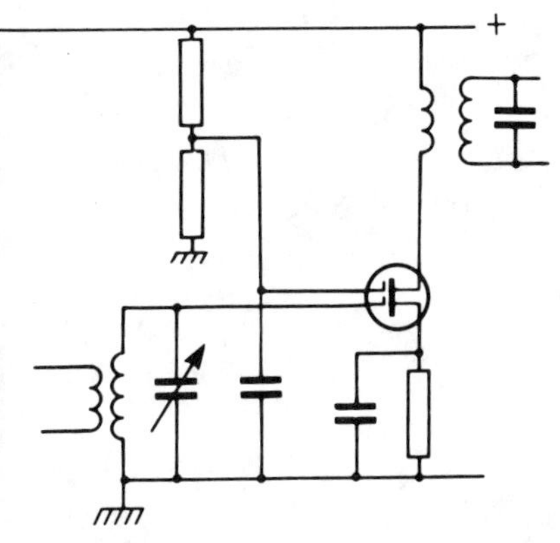

Figure 36.24 A dual-insulated-gate field-effect transistor used as an r.f. amplifier

36.15 Automatic gain control

The gain of the early stages of the i.f. amplifiers of sound and television receivers is usually automatically controlled to combat the effects of signal fading and to ensure that all received signals are reproduced at approximately the same amplitude irrespective of their level at the input to the receiver. Automatic gain control (a.g.c.) is achieved by deriving the d.c. bias for these early stages from the detector output or a post-detector point, modulation frequencies being filtered out, commonly by an *RC* circuit. There are two ways in which the gain of a bipolar transistor amplifying stage can be controlled by a d.c. bias.

In one method the d.c. bias is polarised so as to reduce collector current; this is, of course, the method which was used with valve amplifiers and is known as reverse control. Positive bias is required to control pnp stages and negative bias for npn stages. A circuit using reverse control is illustrated in *Figure 36.25*. This method of applying a.g.c. is not very satisfactory with transistors because increase in bias, as occurs on strong received signals, reduces the signal-handling capacity of the stage when, in fact, an increase in signal-handling capacity is desirable.

Better results can be obtained from the second method of automatic gain control known as forward control. In this method the transistor is forward biased by the a.g.c. signal so that collector current is increased. An essential feature of the circuit, shown in *Figure 36.26*, is the resistor R_c in the collector circuit. As the collector current increases so does the voltage across R_c and the voltage across the transistor falls. With bipolar transistors which are specially designed for this application, the characteristics become more crowded, implying reduced gain as the voltage across the transistor decreases.

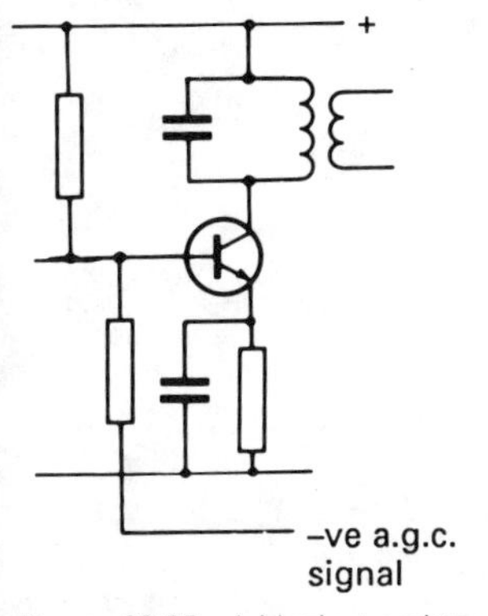

Figure 36.25 A bipolar transistor i.f. stage designed for reverse automatic gain control

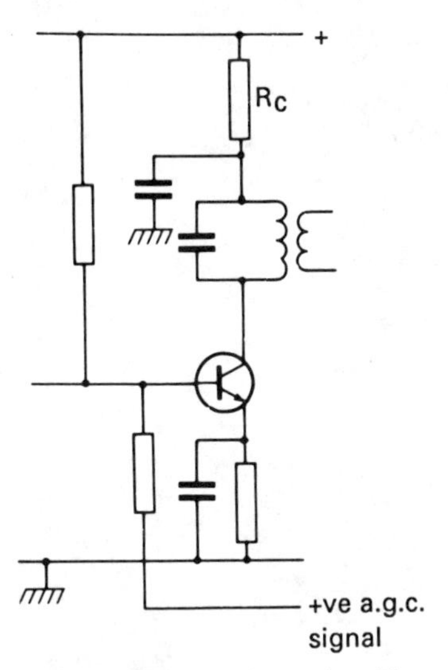

Figure 36.26 A bipolar transistor i.f. stage designed for forward automatic gain control

36.16 Microwave amplification

Modern silicon transistors, used in common-base mode, amplify satisfactorily up to 6 GHz, but their power-handling capacity at such high frequencies is limited and valves are still used in the output stages of high-power transmitters. Examples of these, such as klystrons and travelling-wave tubes, are described in *Chapter 32*.

Waveform Generation

S W Amos BSc(Hons), CEng, MIEE
Freelance Technical Editor and Author

Contents

37.1 Introduction

A wide variety of different waveforms is used in radio and electronics, but the primary constituents of these are the sine wave, the rectangular pulse and the sawtooth or ramp. This chapter surveys methods of generating these basic waveforms.

37.2 Sine-wave generators

Sinusoidal signals have wide applications. They have an obvious use, as a.c. mains, in powering equipment, but are also employed, to name only a few examples, for testing, as carriers in telecommunications systems, for local oscillators in superhet receivers and for bias and erasing in tape recording.

Sine-wave generators, i.e. oscillators, consist essentially of two sections. One determines the frequency of operation and consists of lumped inductance (or resistance) and capacitance at low frequencies, transmission lines at higher frequencies and cavity resonators at microwave frequencies. The other (maintaining) section supplies the energy to keep the frequency-determining section in oscillation and itself requires a d.c. supply.

In many oscillators the maintaining section is an amplifier and the frequency-determining section is used to connect the output of the amplifier to its input to give the positive feedback essential for oscillation. The amplifier input and output connections have a common terminal so that three connections are necessary between the two sections of the oscillator. What distinguishes one type of positive-feedback oscillator from another is the way in which these three connections are derived from the frequency-determining section.

37.2.1 The Hartley oscillator

In the Hartley oscillator (*Figure 37.1*) the three leads are taken from the inductor, a tapping point providing the connection to the common (source) terminal of the field-effect transistor acting as amplifier. The connection is completed via capacitor C_2 which is chosen to have a very low reactance at the operating frequency.

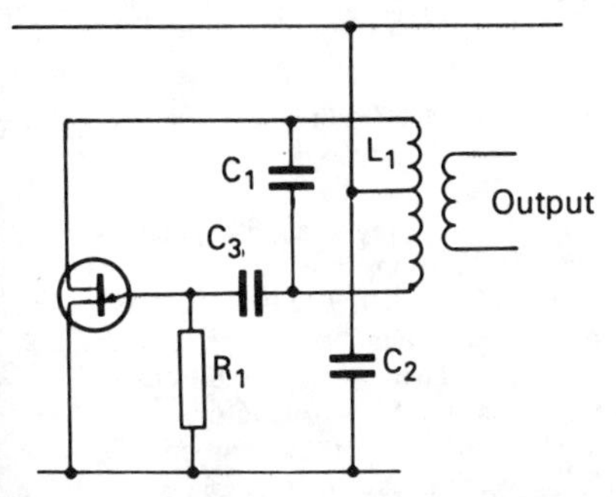

Figure 37.1 A Hartley oscillator circuit using a junction-gate field-effect transistor

C_3 and R_1 provide automatic gate bias. At the moment of switching on there is zero gate bias and the transistor takes considerable drain current. Oscillation thus starts immediately and, as a result, the gate is driven alternately positive and negative with respect to the source. However, when the gate is driven positive the input junction becomes forward-biased and current flows in the gate circuit. This current charges C_3, the polarity of the charge being such as to bias the gate negatively with respect to source. In the final state of equilibrium considerable amplitude of oscillation can result, the gate being driven positive for only a brief period in each cycle. During the remainder of each cycle, when the gate is negative with respect to the source and there is no gate current, the charge on C_3 leaks away through R_1 but, provided R_1 is large enough, very little of it is lost before the next cycle of oscillation restores the charge again. The burst of gate current momentarily applies a low resistance across the tuned circuit and it is this which limits the amplitude of oscillation generated. Normally the amplitude is so large that the transistor is cut off for a large part of each cycle. In other words the amplifier operates in class C. Thus it can be said that the resonant circuit is kept in oscillation by bursts of drain current at the resonance frequency.

For successful results the time constant R_1C_3 should be long compared with the periodic time of the oscillation and the frequency of oscillation is given by the expression

$$f=\frac{1}{2\pi\sqrt{(L_1C_1)}} \tag{37.1}$$

This type of oscillator in which the inductor is centre-tapped is known as the Hartley. The sinusoidal oscillation is set up, of course, in L_1C_1 and the output must be taken from this circuit: a convenient method of doing this is by coupling an inductor to L_1 as shown in *Figure 37.1*.

37.2.2 The Colpitts oscillator

In the Colpitts oscillator the three leads are taken from the capacitive branch of the frequency-determining circuit as shown in *Figure 37.2*. The resistor R_1 is necessary to avoid a short

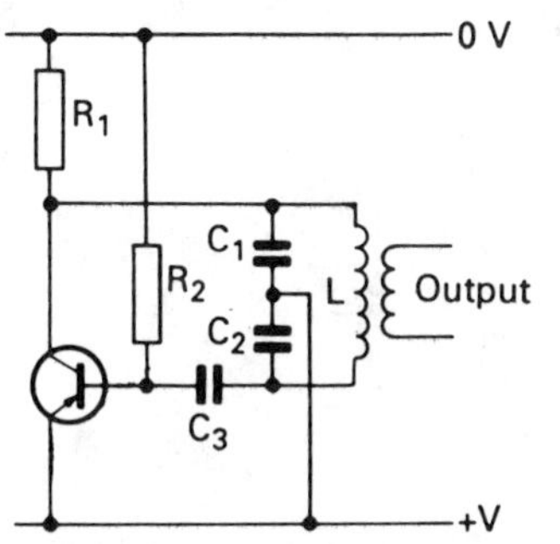

Figure 37.2 A Colpitts oscillator circuit using a bipolar transistor

circuit of the capacitor C_1. As before, C_3 and R_2 provide automatic bias and the transistor operates in Class C so that the sinusoidal output must be taken from a coupling coil as shown.

Oscillation occurs at the frequency

$$f=\frac{1}{2\pi\sqrt{(LC)}} \tag{37.2}$$

where

$$C=\frac{C_1C_2}{C_1+C_2} \tag{37.3}$$

The output is again taken from L–C_1–C_2 by a coupling coil.

37.2.3 The phase-shift oscillator

Generators governed by *RC* circuits can be designed to produce a wide variety of different waveforms and the production of sawtooth and rectangular waves is described elsewhere in the book. However, an *RC* network can also form the frequency-determining element of a sinusoidal oscillator: two examples of such oscillators are described; the first is the *RC* phase-shift oscillator.

In the Hartley and Colpitts oscillators the LC circuit introduced a phase inversion of the signal between its input and output terminals. This, combined with the phase inversion within the amplifier gives the positive feedback essential to sustain oscillation. For a symmetrical wave such as a sinusoid the effect of phase inversion is the same as that of altering the phase by 180°. If therefore we can find a network which shifts the phase by 180° at a particular frequency and if the gain of the amplifier can compensate for the attenuation introduced by the network at this frequency, we have the basis for an oscillator. A single RC network can at best give 90° phase shift and its attenuation is then infinite. However, a network of three RC sections, each introducing 60° phase shift, is a practical possibility: if the resistance values are equal and if the capacitance values are also equal the attenuation introduced by the three-section network is 29 (i.e. this is the value of i_{in}/i_{out}) which can be made good by a single transistor. The circuit diagram of an oscillator operating on such principles is given in *Figure 37.3*. The frequency-determining network is R_1–C_1–R_2–C_2–R_3–C_3 in which $R_1=R_2=R_3=R$ and $C_1=C_2=C_3=C$. DC stabilisation is ensured by the potential divider R_4–R_5 and the emitter resistor R_6. The frequency of oscillation is given by $f=1/2\pi\sqrt{6}\,RC$.

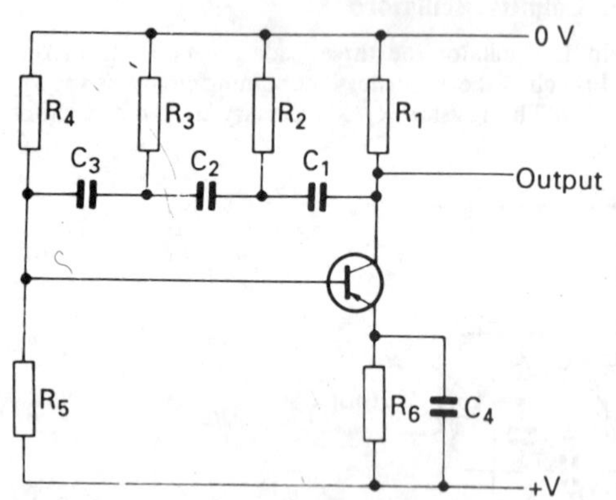

Figure 37.3 Phase-shift oscillator

No means is shown in *Figure 37.3* of limiting the amplitude of oscillation and this is essential to preserve the purity of the waveform generated. It is possible, by adjustment of the value of R_6, to set the emitter current at a value which only just gives sufficient current gain to make up for the attenuation in the RC network, but this is not a good method and it would be better to use an automatic system in which the transistor is biased back when the output amplitude exceeds a predetermined value.

37.2.4 The Wien bridge oscillator

It would be difficult to make a variable-frequency oscillator based on the circuit of *Figure 37.3*. A circuit much better suited to this purpose is the Wien bridge oscillator.

The basic Wien bridge network is shown in *Figure 37.4*. It contains two equal resistors and two equal capacitors. At the frequency for which

$$f=\frac{1}{2\pi RC} \tag{37.4}$$

the network has zero phase shift between input and output, and the voltage attenuation is 3, i.e. $v_{in}/v_{out}=3$. To use such a network as the frequency-determining element in an oscillator the amplifier must also introduce zero phase shift and have a voltage gain of 3. The amplifier must, moreover, have a very high input resistance to minimise any shunting effect on the parallel RC

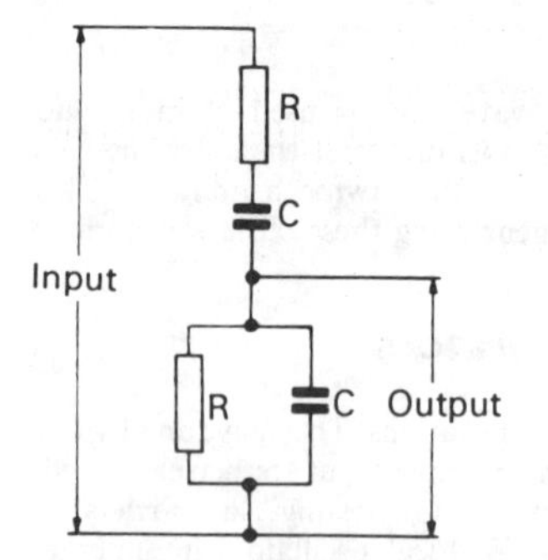

Figure 37.4 Basic Wien bridge network

branch. Two stages at least are required in the amplifier to give the required zero phase shift and it is usual to include a third, an emitter follower, to provide a low-resistance output for the frequency-determining network and for the oscillator itself, as shown in *Figure 37.5*.

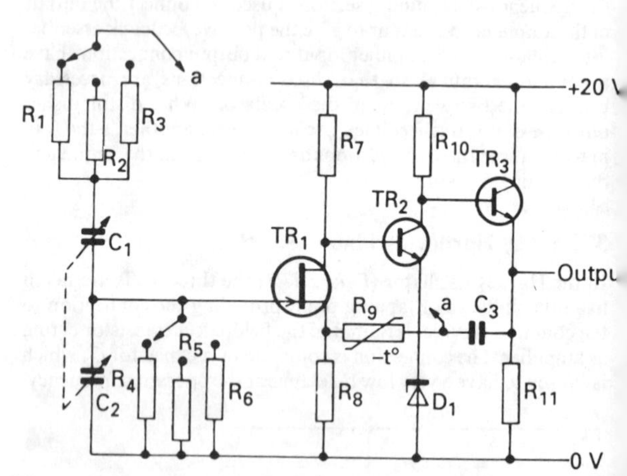

Figure 37.5 A Wien bridge oscillator with three frequency ranges

By using two sections of a two-gang variable capacitor for the two capacitors in *Figure 37.4* it is possible to vary the frequency of oscillation over a range of say 10:1. The two resistors can be varied in decade steps to produce a number of frequency ranges. Three ranges could thus cover the frequency band from 30 Hz to 30 kHz which is suitable for an a.f. test oscillator.

Negative feedback is applied between the emitter of TR_3 (in effect the collector of TR_2 since TR_3 is an emitter follower) and the source of TR_1 via the nonlinear resistor R_9 which has a negative temperature coefficient. This has the effect of maintaining the output of the oscillator substantially constant in spite of frequency or range changes. Immediately after switching on the resistance of R_9 is high because there is no signal in it and hence no heat. The gain of the amplifier is determined by the feedback components and is given by R_9/R_8 and, at the instant of switch on, this is high, much higher than the value of 3 necessary to ensure oscillation. Oscillation therefore begins and builds up rapidly. As soon as the oscillation reaches R_9 via C_3 the resistance of R_9 falls due to the heating effect of the signal in it. The fall continues and the resistance of R_9 settles around a value approximately equal to twice R_8. The type of resistor chosen for R_9 should be such that its resistance will equal twice R_8 with the required value of oscillation amplitude across it.

37.2.5 Negative-resistance oscillators

Not all oscillators require positive feedback for their operation. There is another type in which the frequency-determining circuit is connected to a source of negative resistance. A number of devices have current/voltage characteristics with a negative slope; the dynatron, transitron, unijunction transistor and tunnel diode are four examples. All these devices have a characteristic with the general shape shown in *Figure 37.6*. It is

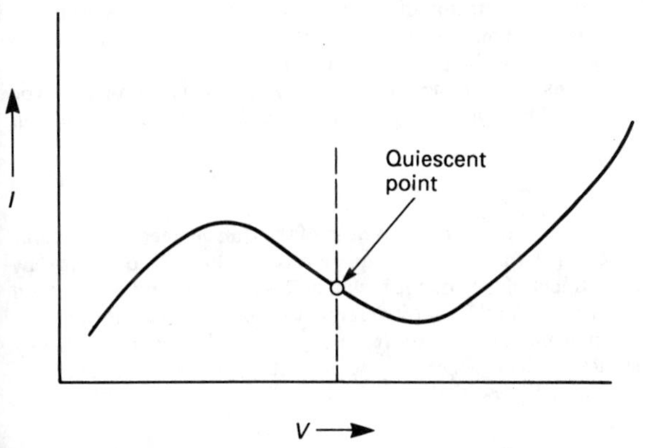

Figure 37.6 A characteristic curve with a region of negative resistance

significant that the region of negative resistance is confined to a particular voltage range and is bounded at both ends by regions of positive resistance. To use a negative-resistance device as an oscillator, the negative resistance must be capable of offsetting the inherent resistance of the resonant circuit used to define the operating frequency. For the shape of characteristic shown in *Figure 37.6*, the region of negative resistance is identified by the voltage and a device with such a characteristic is normally used with a parallel *LC* circuit as the frequency-determining sections. For oscillation to occur the negative resistance must be numerically less than the dynamic resistance of the tuned circuit. The oscillation amplitude then grows until it occupies a voltage range greater than the extent of the negative-resistance region of the characteristic. In fact at the peaks of the oscillation the operating point enters the regions of positive resistance and these apply damping to the tuned circuit, taking power from it and so limiting the amplitude of oscillation. When the amplitude has reached equilibrium the average slope of that part of the characteristic over which the operating point moves during each cycle of oscillation is equal to the dynamic resistance of the tuned circuit. As a result the output amplitude obtainable from a negative-resistance oscillator is limited, and to obtain the maximum the quiescent point must be accurately located at the centre of the negative-resistance region as shown in *Figure 37.6*.

A tunnel diode has a characteristic with a negative-resistance region between voltages of approximately 0.1 and 0.3 V and can be used as an oscillator at frequencies up to 100 GHz.

37.3 Pulse generators

The rectangular pulse, shown in idealised form in *Figure 37.7*, is the fundamental form of signal in computers and digital equipment generally and is also extensively used in television, radar and other equipment; it features in the radiated television waveform.

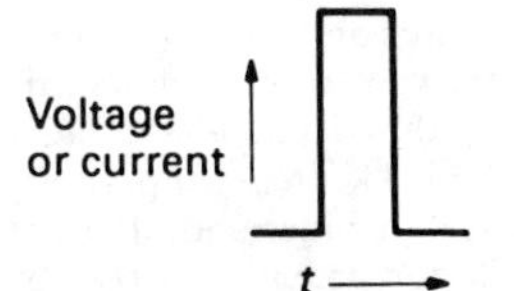

Figure 37.7 Ideal pulse waveform

37.3.1 Limiting circuits

One way of generating rectangular pulses is by shaping other waveforms such as sine waves. For example, if a sinusoidal signal is applied to the circuit of *Figure 37.8(a)* the diode will conduct

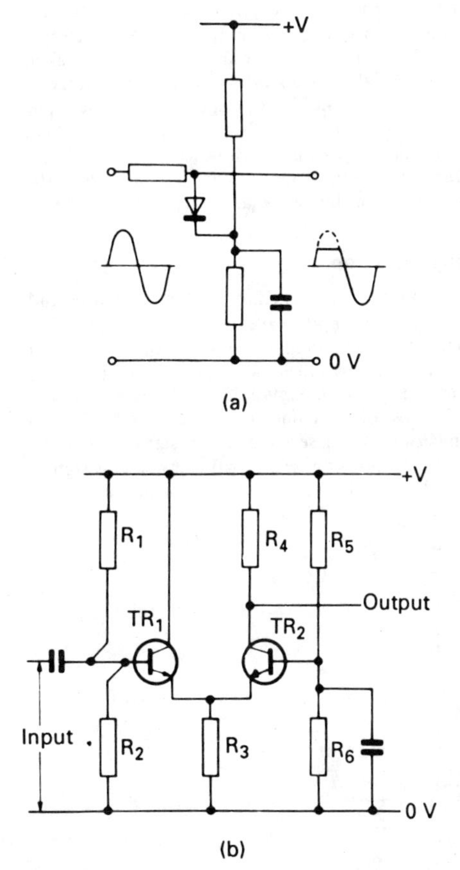

Figure 37.8 Limiting circuits: (a) diode, (b) two-transistor

and hence attenuate all parts of the signal which exceed the diode cathode bias in value, giving an output of the form shown. If the negative-going excursions of this output are eliminated (for example by the use of a second diode) we are left with an approximation to a rectangular pulse: this becomes a nearer approach to the ideal form the greater the ratio of the sine-wave amplitude to the diode bias.

Better results can be obtained by using transistors in place of diodes as limiting devices and the circuit diagram of a two-transistor circuit which limits on positive and negative peaks is given in *Figure 37.8(b)*. The two transistors are emitter-coupled by a common resistor, an arrangement known as a long-tailed pair. TR_1 is an emitter follower driving the common-base stage TR_2. Positive-going signals applied to TR_1 base appear at substantially the same amplitude at TR_2 emitter and, if large enough, cut TR_2 off, causing its collector potential to rise to supply positive value. Negative-going signals, if large enough, cut TR_1 off so that TR_2 takes a steady collector current (determined by the potential divider R_5–R_6 and the emitter resistor R_3) and the collector potential is at a steady value below that of the positive supply rail. Thus a large-amplitude signal at the input to this circuit gives an output at TR_2 collector which alternates between two steady values, i.e. it is a pulse output at the frequency of the input signal.

A simple common-emitter amplifier can, of course, be used as a limiter. By giving it a very large input signal (which, in linear amplifier operation, would be described as a gross overload) the transistor is driven into cut-off on one half-cycle of the input signal and into saturation on the other half-cycle. Thus the collector potential is either at the positive or the negative supply value. The larger the input is made, the shorter is the time taken for the collector potential to switch from one extreme voltage to the other and the better is the shape of the output pulses. The input signal can be sinusoidal or it can be a pulse waveform which has overshoots, ripples or other undesirable features on it. These unwanted features can, with proper design, be removed in the limiting circuit which can thus be regarded as a 'cleaner stage'.

37.3.2 Blocking oscillators

There are other circuits which generate rectangular pulses and do not require any signal input waveform for their operation. These are therefore true rectangular pulse generators and the first example of such a generator is the blocking oscillator. The basic form of the circuit is given in *Figure 37.9*. TR_1 is a common emitter stage with base bias provided by the resistor *R*. There is phase inversion between the base and collector signals and T is a transformer which also gives phase inversion between the signals delivered to the base and collector circuits. Thus a state of

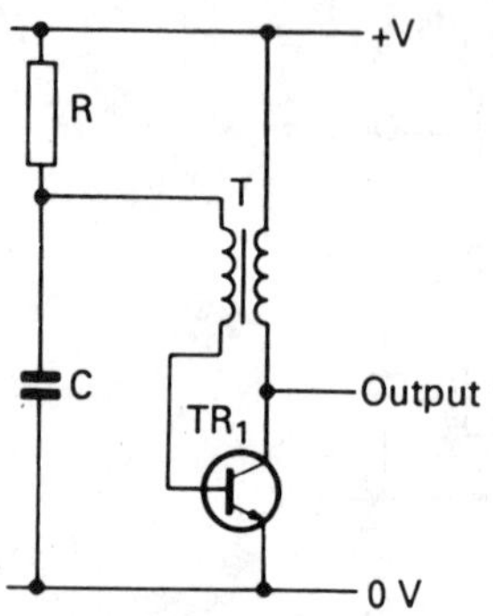

Figure 37.9 Basic circuit for a blocking oscillator

positive feedback exists and this results in oscillation at the resonance frequency of the collector winding, usually the larger of the two windings. One of the aims in the design of blocking oscillators is to provide considerable positive feedback so that the oscillation amplitude builds up very rapidly. As it does so TR_1 takes a burst of base current which charges up the capacitor *C*, the polarity being such as to drive the base negative and to cut the transistor off. This is, of course, the basis of the automatic system of biasing. Ideally TR_1 should be cut off within the first half-cycle of oscillation. *C* now discharges through *R* and TR_1 remains cut off until the voltage across *C* has fallen sufficiently for TR_1 to take base current again. This promotes another half-cycle of oscillation as a result of which TR_1 is again cut off by the negative voltage generated at the junction of *R* and *C* by the burst of base current.

Thus bursts of base current occur regularly and, of course, there are associated bursts of collector current. Negative-going voltage pulses are hence generated at the collector terminal and these can be taken as the output of the circuit. The interval between the pulses is governed by the time constant *RC* so that either *R* or *C* can be made variable to provide a pulse frequency control. The duration of the pulses is a function of the transformer design being primarily dependent on the inductance and self-capacitance of the primary winding.

The natural frequency of a blocking oscillator of the type illustrated in *Figure 37.9* is given by the approximate expression

$$f=\frac{n+1}{RC} \qquad (37.5)$$

where $n:1$ is the step-down ratio of the transformer. It is possible to control the natural frequency of a blocking oscillator by adjustment of the voltage to which *R* is returned and this method is illustrated in the circuit diagram of *Figure 37.10*. The output is taken from the resistor R_2: this avoids the inclusion of any inductive effects which may be present in the output of the circuit shown in *Figure 37.9*.

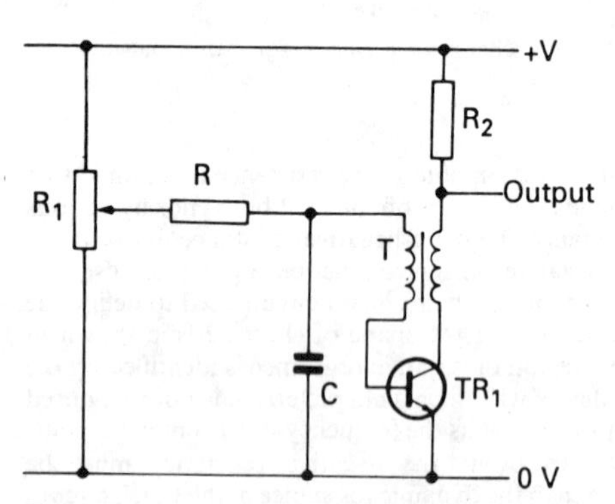

Figure 37.10 Blocking oscillator circuit with frequency control

37.3.3 Synchronisation of blocking oscillators

Although the blocking oscillator does not need an input signal the circuit can readily be synchronised at the frequency of any regularly occurring signal applied to it. The synchronising signal is arranged to terminate the relaxation period earlier than would occur naturally. Thus to synchronise the circuits of *Figure 37.9* or *37.10* a synchronising signal could be in the form of positive-going pulses applied to the base circuit or negative-going pulses applied to the collector circuit. A blocking oscillator intended for synchronised operation should be designed to have a natural frequency slightly lower than that of the synchronising pulses.

If the natural frequency is made slightly below one-half the frequency of the synchronising pulses then the blocking oscillator will be triggered into oscillation by every second synchronising signal, i.e. it will run at precisely one half of the synchronising frequency. This idea can be extended and it is possible to arrange for a blocking oscillator to run at 1/3, 1/4, 1/5, etc., of the synchronising frequency. In other words the blocking oscillator can be used as a frequency divider. Ratios much above 1/5 are a little difficult to achieve with reliability because they demand close control over synchronising-pulse amplitude and natural frequency.

37.3.4 Bistable multivibrators

If the transformer in the circuit of *Figure 37.9* is replaced by any other device which introduces a phase inversion between the signal received from the collector and that fed to the base, then the circuit still has positive feedback and is capable of oscillation. Such a device is a transistor connected in the common-emitter mode and the two-transistor circuit so obtained is known as a multivibrator.

Figure 37.11 shows the circuit diagram of a multivibrator with two direct inter-transistor couplings. The base circuit of each transistor is fed from the collector circuit of the other via a potential divider and the resistor values are so chosen that, when one transistor is conducting, its low collector potential ensures that the other is cut off. Moreover the high collector potential of the cut-off transistor ensures that the other transistor is maintained in the conductive state. For example, TR_1 may be cut off and TR_2 conducting. When the circuit is placed in this state, it will remain in it indefinitely unless compelled to change by an external triggering signal. Such a signal may consist of a positive-going pulse applied to TR_1 base to make it conduct or a negative-going pulse applied to TR_2 base to cut it off. Either form of signal will cause a change of state in both transistors. The circuit will now remain in this new state (TR_1 on and TR_2 off) indefinitely unless compelled to change it by an external signal. Persistent states such as those described are known as stable and the direct-coupled multivibrator thus has two stable states: such a circuit is known as bistable. The changes of state, once initiated by the external signal, are very fast, being accelerated by the positive feedback inherent in the circuit. The transitions from one state to the other can be made even faster by connecting capacitors in parallel with R_2 and R_4 as shown in *Figure 37.12(a)*: C_3 and C_4 are known as speed-up capacitors.

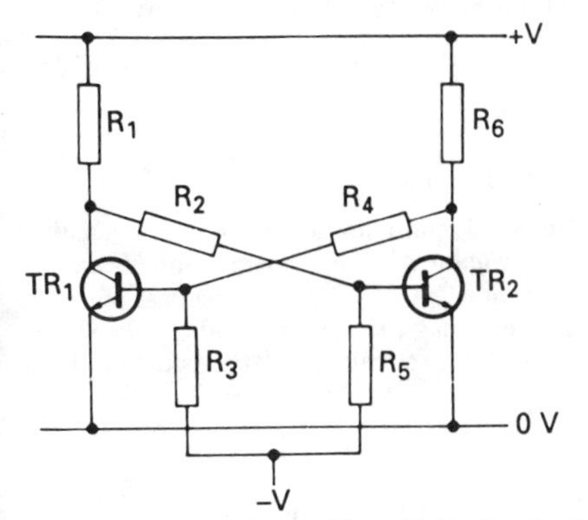

Figure 37.11 Basic bistable multivibrator circuit

One of the principal features of the bistable is its ability to maintain a particular state, i.e. it has a memory. For this reason bistables are extensively used in the stores and registers of computers and similar equipment.

The bistable can also be used as a pulse generator for if it is fed with a regular stream of triggering signals it will change state with each received signal and will thus generate square waves at each collector. If the triggering signals are negative-going blips and if the first cuts TR_1 off then the next must be fed to TR_2 base to cut this off and so on, i.e. the triggers must be directed alternately to the two bases. This routing of triggers can be achieved by a diode gate circuit such as that illustrated in *Figure 37.12(a)*. If TR_1 is cut off the diode D_1 will conduct a negative trigger to TR_1 collector and thus to TR_2 base via the speed-up capacitor C_3. TR_2 is conductive and its low collector potential biases D_2 off so that the trigger cannot reach TR_1 base. Thus the trigger can only affect TR_2 which is cut off by it, causing TR_1 to be turned on. D_2 can now direct the next trigger to TR_1 base. The signals generated at the collectors are thus square waves at half the frequency of the applied triggers as shown in *Figure 37.12(b)*. The signals at the bases are also square waves in antiphase to the collector signals and are of smaller amplitude, but the speed-up capacitors can give some overshoots on the square waves which are illustrated in idealised form.

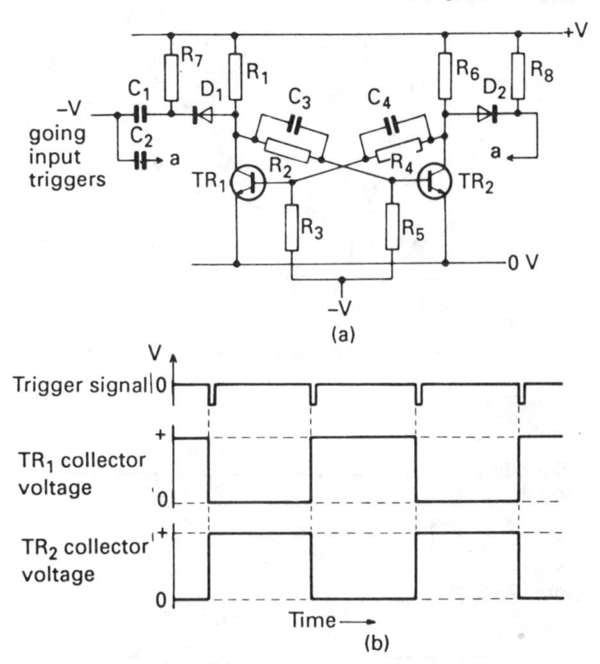

Figure 37.12 Bistable circuit with speed-up capacitors and diode input gate: (a) circuit, (b) associated waveforms

It is significant that the transistors in this circuit are either on (with a substantial collector current) or off (with zero collector current). Except for a very short period at each transition the collector currents never have any values other than these two. The transistors are used, in fact, as switches and the shape of the input/output characteristic, which is of considerable importance in analogue applications, is generally of little consequence in pulse circuits.

37.3.5 Monostable multivibrators

Suppose a multivibrator has one direct coupling and one capacitive coupling as shown in *Figure 37.13(a)*. The introduction of the coupling capacitor C_1 makes a fundamental difference to the behaviour of the circuit because there is now no means of preserving a negative voltage on the base of TR_2 to keep it non-conductive. Because the base resistor R_3 is returned to the positive supply line the circuit will always revert to the state in which TR_2 is on and TR_1 therefore off. This is therefore a stable state like those of the bistable circuit. The circuit can be triggered into the other state (TR_1 on and TR_2 off) but it cannot remain in it and the circuit will automatically return to the stable state without need of external signals to make it do so. The state in which TR_1 is on and TR_2 is off is therefore an unstable state and circuits such as this which have one stable state and one unstable state are known as monostables.

In the stable state TR_2 is on and TR_1 is off. TR_2 has base current and the base potential is near supply negative value. The collector potential of TR_1 is near that of the supply positive line and C_1 is hence charged to the supply voltage. A negative-going

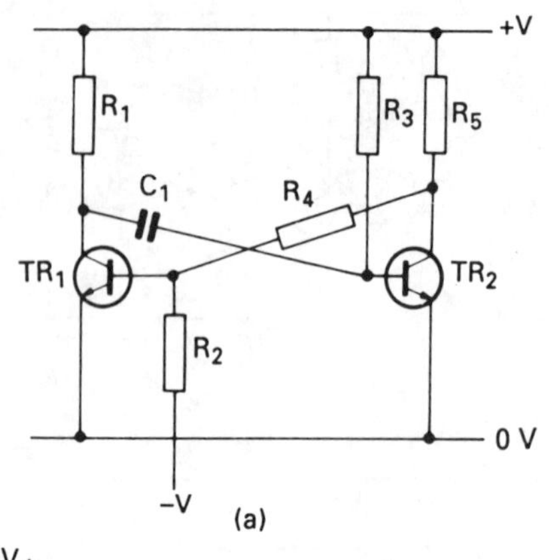

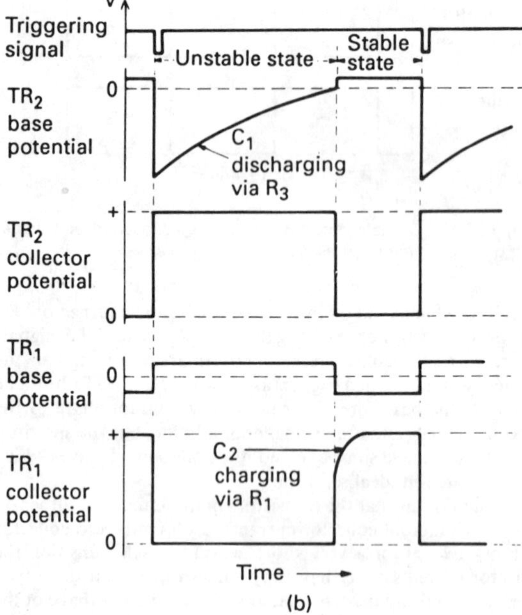

Figure 37.13 Monostable multivibrator: (a) basic circuit, (b) associated waveforms

trigger applied to TR_2 base cuts this transistor off and turns TR_1 on so that TR_1 collector potential falls abruptly to negative supply value, carrying the base potential of TR_2 to a considerable negative voltage. C_1 now begins to discharge through R_3 and as it does so TR_2 base potential rises towards zero. But for the connection to TR_2 base the potential at the junction of R_3 and C_1 would rise to the supply positive value. However as soon as this junction becomes slightly positive with respect to supply negative TR_2 begins to conduct and this initiates another transition, accelerated by positive feedback, which ends with TR_2 on and TR_1 off. The rise in TR_1 collector potential cannot be instantaneous because it is controlled by C_1 which charges via R_1. The waveforms for this circuit are therefore as shown in *Figure 37.13(b)*.

During the unstable state the collector potential of TR_1 is low and that of TR_2 is high. Thus the circuit develops negative-going pulses at TR_1 collector and positive-going pulses at TR_2 collector during this period the duration of which is determined by the time constant R_3C_1. The main application of the monostable is the generation of pulses of a predetermined duration on receipt of a triggering signal. To obtain an estimate of the duration of the pulses we can assume that TR_2 conducts when C_1 has discharged to half the voltage it had across it immediately after triggering. The voltage V_t across a capacitor discharging into a resistance falls exponentially according to the relationship $V_t = V_0\, e^{t/RC}$ where V_0 is the initial voltage. Since $V_t = V_0/2$ we have

$$t = \log_e 2R_3C_1 = 0.6931 R_3C_1 \qquad (37.6)$$

Thus if we require 100-μs pulses $R_3C_1 = 144.3$ μs. R_3 supplies TR_2 with base current in the stable state and if TR_2 is to take, say, 5 mA collector current at this time then the base current should preferably not be less than 100 μA. For a supply voltage of 10 V therefore R_3 should not be less than 100 kΩ. This gives C_1 as

$$C_1 = \frac{144.3}{10^5}\ \mu F = 1.44\ nF$$

37.3.6 Astable multivibrators

The circuit diagram of a multivibrator with two capacitive couplings is given in *Figure 37.14(a)*. There is no means of maintaining a permanent negative voltage on the base of either transistor and this circuit has therefore no stable state. Both states are unstable and the circuit oscillates between them

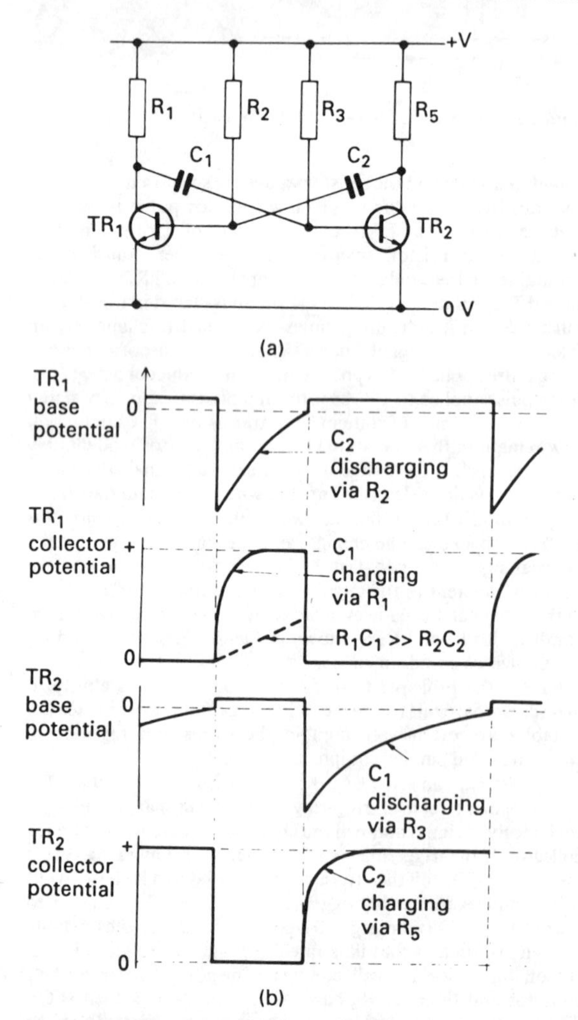

Figure 37.14 Astable multivibrator: (a) basic circuit, (b) associated waveforms

continuously and automatically without need for triggering signals. It is, in fact, a free-running circuit but a multivibrator with two unstable states is generally described as astable.

The waveforms can be deduced from those of the monostable circuit and are given in *Figure 37.14(b)*. The duration of one unstable state is given approximately by $0.6931R_3C_1$ and of the other by $0.6931R_2C_2$ so that the natural frequency of the astable circuit is given by

$$f = \frac{1}{0.6931(R_3C_1 + R_2C_2)} \qquad (37.7)$$

Although the circuit does not require external signals for its operation it can readily be synchronised at the frequency of a regular external signal. Normally synchronisation is achieved by terminating the unstable periods earlier than would occur naturally and therefore positive-going synchronising signals are required at the bases of the npn transistors and the natural frequency of the astable circuit should be slightly lower than that of the synchronising signal.

37.3.7 Emitter-coupled multivibrators

All the multivibrator circuits so far described have included two collector-to-base couplings. It is possible to replace one of these with an emitter-to-emitter coupling: in a direct coupling the emitters are simply bonded and returned to the supply via a common resistor as shown in the monostable circuit of *Figure 37.15*. Alternatively a capacitive emitter coupling can be obtained

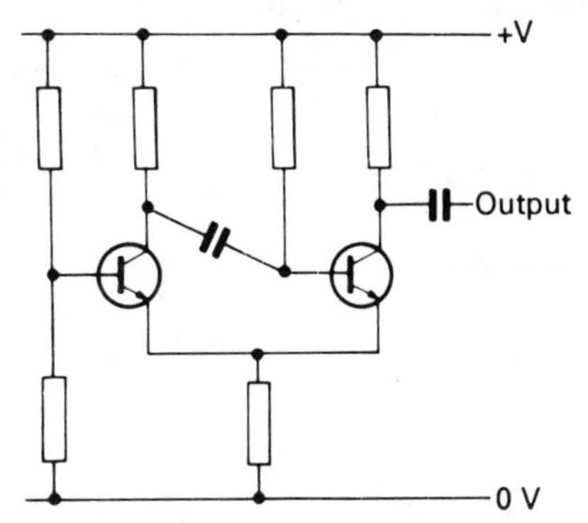

Figure 37.15 Circuit for a monostable emitter-coupled multivibrator

by using individual emitter resistors bridged via a capacitor. An advantage of emitter coupling is that one collector terminal is free, i.e. not involved in the provision of positive feedback, and can thus be used as a convenient output point as suggested in *Figure 37.15*.

37.3.8 Complementary multivibrators

By using a combination of pnp and npn transistors the particularly simple bistable circuit of *Figure 37.16* is possible. When the npn transistor TR_1 is cut off there is no voltage drop across R_1 and therefore the pnp transistor TR_2 is also cut off. The absence of collector current in R_2 ensures that TR_1 remains cut off. Alternatively, if TR_1 is on, there is a large voltage drop across R_1 which biases TR_2 on and the consequent large voltage drop across R_2 keeps TR_1 on. A feature of this circuit therefore is that in one state both transistors are off and in the other state both transistors are on.

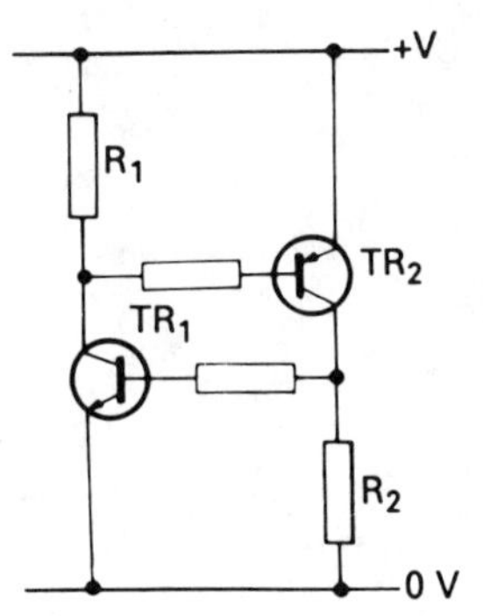

Figure 37.16 Complementary bistable multivibrator circuit

37.4 Sawtooth generators

Sawtooth waveforms are used for electron beam deflection in oscilloscopes, in television transmitting and receiving equipment, in digital-to-analogue conversion equipment and in measuring equipment. Normally the slow rise is required to be linearly related to time for this is the working stroke of the waveform, but the shape of the rapid fall (the return stroke) need not be linear and the shape of this transition is not usually significant.

37.4.1 Production of sawtooth voltages

An approximation to a sawtooth voltage waveform can be obtained from a simple circuit such as that shown in *Figure 37.17*.

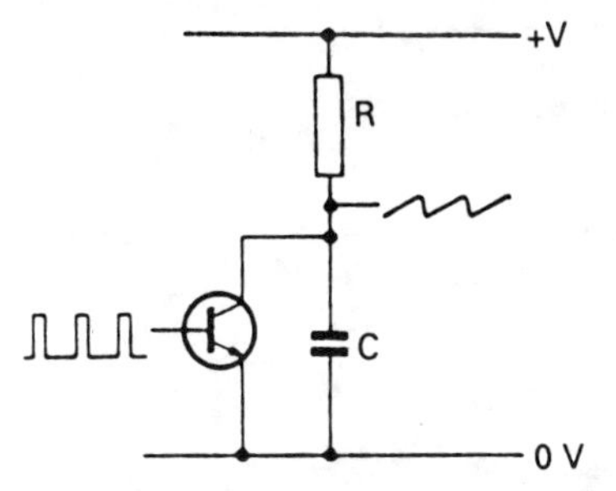

Figure 37.17 Simple discharger circuit for production of sawtooth voltages

The required output is generated across the capacitor C as it charges from the supply via R. The flyback voltage is obtained by discharging C by the transistor which is turned on for the duration of the flyback period. Such a simple circuit has a number of serious limitations: the most important is that the rise of voltage across C is exponential not linear. The departure from linearity is not serious provided that the rise in voltage is restricted to a small fraction of the supply voltage. Normally the performance of the circuit is unsuitable and methods of improving the linearity of the working stroke are necessary. The reason for the lack of linearity is that the rise in voltage across C causes an equal fall in voltage across R and hence a fall in the charging current. Ideally, to achieve linearity, the charging current must be kept constant as charging proceeds.

Two methods of achieving a constant charging current are in common use. The first is the bootstrap circuit illustrated in *Figure 37.18*. An emitter follower is connected across the capacitor C and delivers, at its emitter terminal, a copy of the voltage across

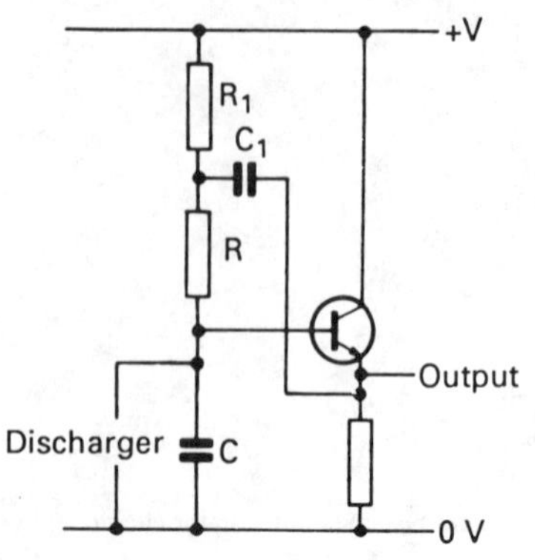

Figure 37.18 Bootstrap circuit

C. The transistor therefore provides a low-resistance output terminal for the sawtooth generator. The emitter follower output is transferred via the long-time-constant circuit R_1–C_1 to the supply point for the charging resistor *R*. Thus as the voltage across *C* rises so does the voltage at the junction of R_1 and *R*. The voltage across *R* remains substantially constant throughout the charging process and thus the charging current is maintained constant.

The second method of achieving linearity is to use a source of constant current in place of the charging resistor *R*. A d.c.-stabilised common-emitter transistor circuit is one possible source of constant current and a circuit using this is given in *Figure 37.19*. The transistor circuit is stabilised by the potential divider R_1–R_2 connected to the base and by the emitter resistor R_3.

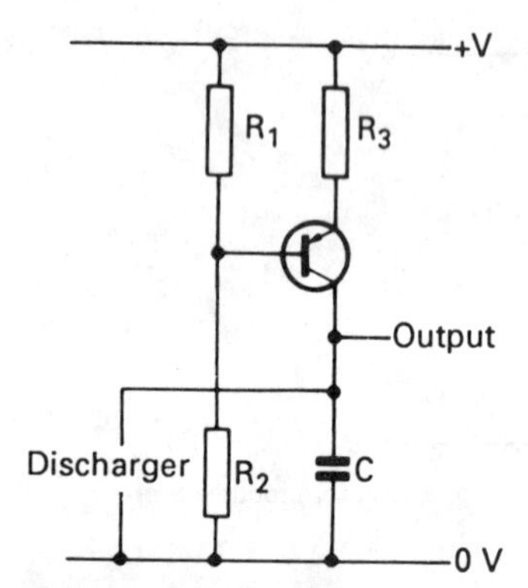

Figure 37.19 Method of achieving linearity using a transistor as a constant-current source

All the circuits described for the production of sawtooth voltages and currents have required a pulse waveform to operate them. Mathematically the circuits have generated the time integral of the input waveform and can thus be described as integrators. Because of the need for input signals such circuits are known as *driven circuits*. By the addition of other components sawtooth generators can be made to provide their own control pulses and can thus generate sawtooths without need for external triggering signals. Such generators are known as *free-running* and a multivibrator can be used as a free-running sawtooth generator.

Fundamentally a multivibrator is a generator of approximately rectangular current pulses and if such pulses are fed to a capacitor a sawtooth waveform is developed across the capacitor. There is no need to introduce an additional capacitor into the circuit for this purpose (indeed in collector-coupled circuits this might affect the positive feedback on which multivibrator action depends) because one of the coupling capacitors can be used.

It has already been pointed out that, when a transistor is cut off in a multivibrator circuit, its collector potential cannot rise instantaneously to supply voltage value because the capacitor coupling the collector to the base of the other transistor must charge during this period via the collector load resistor. Thus in *Figure 37.14(a)* when TR_1 is cut off C_1 charges via R_1. If the time constant R_1C_1 is made long compared with the period of non-conduction of TR_1 (determined by the time constant R_2C_2) C_1 is not fully charged during this period and the voltage generated at TR_1 collector is exponential in form. The initial part of an exponential rise is almost linear as shown in dotted lines in *Figure 37.14(b)*. Thus the condition to be satisfied to obtain such an output is: $R_1C_1 \gg R_2C_2$. The choice of values for R_1 and C_1 is limited because the time constant R_3C_1 determines the period of non-conduction of TR_2 and hence of conduction of TR_1.

37.4.2 Production of sawtooth currents

For television camera and picture tubes a sawtooth current is required in the deflection coils to deflect the beam horizontally at line rate and vertically at field rate. At the line frequency the deflector coils are predominantly inductive and the problem thus is to generate a sawtooth current in an inductive circuit. There is a very easy solution for if an inductor is connected across a constant voltage supply the current in the inductor rises linearly with time. The voltage across an inductor is related to the rate of change of current in it according to the expression $E = -L(\mathrm{d}i/\mathrm{d}t)$. Thus if *L* and *E* are constant then $\mathrm{d}i/\mathrm{d}t$ must be constant. A very simple sawtooth current generator can thus have the circuit diagram shown in *Figure 37.20*. The transistor is held in the

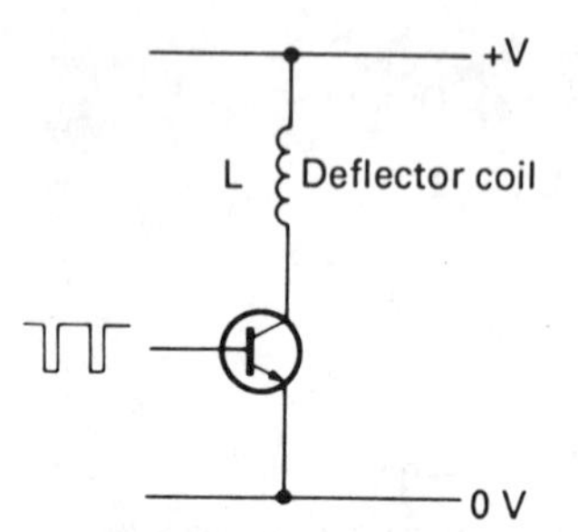

Figure 37.20 Basic circuit for the generation of sawtooth current waveforms

conductive state for the required duration of the working stroke by the positive-going pulse waveform applied to the base. This gives the transistor a low-resistance path between collector and emitter so that, in effect, *L* is connected directly across the supply and current therefore grows in it linearly. At the end of the working stroke the transistor is turned off by the negative-going signal at the base and the current in *L* falls rapidly to zero. This is the very simple basis of the line deflection circuits used in television receivers. In practice the circuits are considerably more complex as illustrated in *Figure 37.21(a)*.

The line scan coils L_1 and L_2 are fed from a line output transformer to ensure that no d.c. flows in the coils (this would produce an undesirable static deflection of the beam) and via the variable inductor L_3 which enables the magnitude of the line-scan current (and hence picture width) to be adjusted. The line-scan circuit is tuned by capacitor C_1: this plays a vital part in the operation of the circuit which will now be described and is illustrated in the curves of *Figure 37.21(b)*.

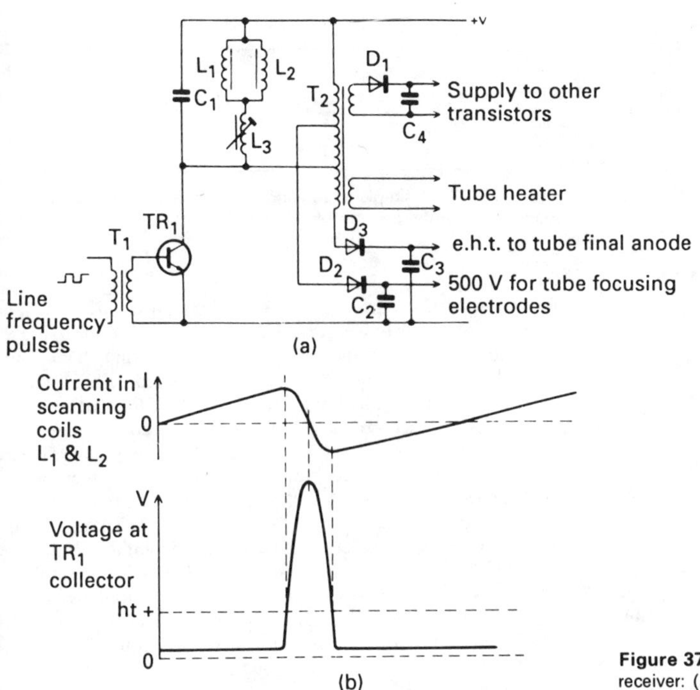

Figure 37.21 More practical line output circuit for a television receiver: (a) circuit, (b) associated waveforms

TR_1 is driven into conduction by pulses from the line oscillator and driver stages and is held conductive while a linearly growing current flows into the coils from the supply. At the end of the working stroke TR_1 is cut off but the current in an inductive circuit cannot cease instantaneously and continues to flow, being taken from C_1 which was previously charged to the supply voltage. C_1 is rapidly discharged by this current which continues to flow and charges C_1 in the reverse direction. As energy flows into C_1 the current in the line-scan coils falls, reaching zero when the voltage across C_1 is a maximum. These exchanges in energy between the scan coils and C_1 are the beginning of free oscillation in the resonant circuit L_1–L_2–L_3–C_1 and in the next stage current begins to flow in the scan coils again but in the reverse direction and this reaches a maximum at the moment when C_1 is again discharged. Scan current continues to flow and C_1 begins to acquire charge again of the polarity assumed initially. When the voltage across C_1 is approximately equal to the supply voltage it forward biases the collector–base junction of TR_1 and this, together with the secondary winding of transformer T_1, provides a low-resistance path from the lower end of L_3 to the negative terminal of the supply. Thus the scanning-coil circuit is again connected directly across the supply. Now however the current in the coils is a maximum and in the reverse direction to that assumed initially. The current therefore falls and the constant voltage across the coils ensures that the rate of fall of current is also constant, i.e. the scan current falls linearly to zero. This current flows into the supply via TR_1 collector–base junction and the secondary winding of T_1. When the current has reached zero the circuit is in the state assumed initially and TR_1 is now turned on again by the external signal to provide another period of linear growth of scan current. With proper design the linear fall and subsequent linear growth of current in the scanning coils combine to produce an uninterrupted linear change of current which constitutes the working stroke. This 'resonant return' circuit is extensively used in scanning systems and is very efficient because almost as much energy is returned by the coils to the supply during the first half of the working stroke as is taken by the coils from the supply during the second half of the working stroke. The transistor TR_1 is turned on only during the second half of the working stroke. The resonant circuit L_1–L_2–L_3–C_1 performs half a cycle of oscillation during the flyback period and this requirement enables suitable values of L_1–L_2–L_3 and C_1 to be calculated.

The rapid change of scan current during flyback causes a high voltage peak to be generated in the line output stage. A sample of this voltage is obtained from a winding on the line output transformer and is rectified by D_3 to provide the e.h.t. voltage for the final anode of the picture tube. The rectifier polarity is such that it conducts during the flyback period and it is commonly a multiplier type to give the high voltage, typically 11 kV, which is required.

A tapping on the primary winding of the line output transformer similarly provides a supply of approximately 500 V for the focusing electrodes of the picture tube. The rectifier is D_2. A secondary winding on the same transformer provides a supply, not rectified, for the tube heater. This is a convenient method of obtaining a low-voltage supply for the heater and avoids the provision of a mains transformer for the purpose.

Another winding on the line output transformer provides a d.c. supply of say 25 V for most of the transistors in the receiver. The rectifier D_1 is arranged to conduct during forward strokes of the line output stage and C_4 provides smoothing. Clearly this 25 V supply cannot be used for the line output transistor itself and the associated driver and oscillator stages and these are usually powered from mains rectifying equipment.

37.5 Waveform shaping circuits

37.5.1 Differentiating and integrating circuits

Bistable and monostable circuits need triggering signals for their action and astable circuits are usually controlled by synchronising signals. It was assumed in the descriptions of these circuits that the external signals had the form of pulses. Certainly pulses can be used for this purpose but it is clear from the descriptions that it is the leading or trailing edge of the pulse which is effective in the triggering or synchronising process: the horizontal part of the pulse is unimportant in this application.

For this reason it is common practice to feed pulses to multivibrators and other circuits via an *RC* circuit of the form shown in *Figure 37.22(a)*. Such an arrangement of series capacitance and shunt resistance is also commonly encountered in the intertransistor coupling circuits of linear amplifiers where the time constant must be so chosen that the signal is transmitted through the *RC* circuit with negligible change in waveform. For a pulse signal a very long time constant would be necessary to transmit the horizontal sections with negligible sag. However the horizontal sections are of no interest if the pulses are used for triggering or synchronising and it is therefore permissible to use a short time constant in the network used to transmit them. The effect of a time constant which is short compared with the pulse repetition period is illustrated in *Figure 37.22(b)*. If the time constant is reduced sufficiently the output becomes simply a succession of alternate positive-going and negative-going spikes. Such a waveform is, of course, quite suitable for triggering or synchronising purposes. Mathematically such a waveform is the first derivative of the input waveform and for this reason the network of *Figure 37.22(a)* is often called a differentiating circuit. Thus the time constants R_7C_1 and R_8C_2 in *Figure 37.12(a)* could both be small compared with the period of the input signal.

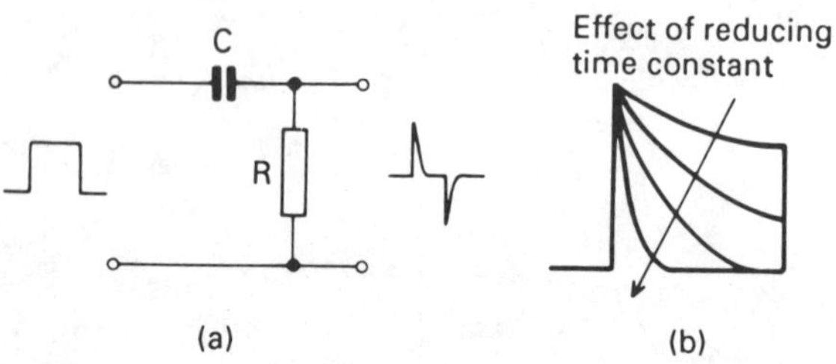

Figure 37.22 A simple differentiating circuit: (a) circuit, (b) the effect on the output waveform of reducing the time constant

The circuit of *Figure 37.23*, on the other hand, gives an output waveform similar to the time integral of the input waveform, provided the time constant *RC* is long compared with the period of the input signal.

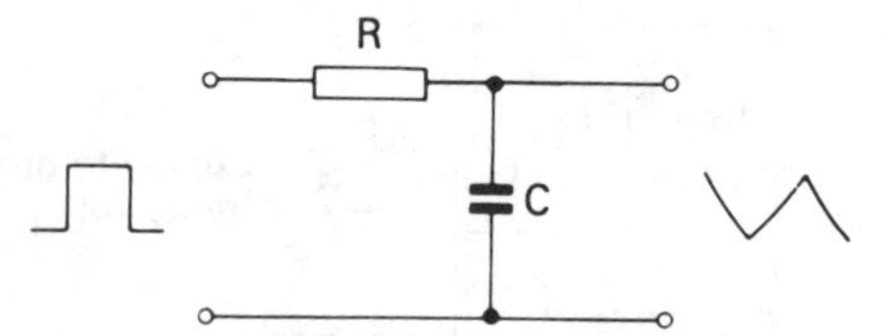

Figure 37.23 A simple integrating network. *RC* must be large compared with the period of the input signal

Modulators and Detectors

S W Amos BSc(Hons), CEng, MIEE
Freelance Technical Editor and Author

Contents

38.1 Introduction

The transmission of information, whether speech, music, vision or data, over long distances requires the use of a carrier channel with a width at least equal to that of the frequency spectrum of the information components. The carrier frequency must have one of its characteristics varied (modulated) by the information and the receiver must contain an information extractor (demodulator) designed to react to the carrier characteristic that is varied and to produce an output which is as close a copy of the original information as possible.

The carrier may consist of a pulsed carrier, whose amplitudes (p.a.m.), positions (p.p.m.) or duration (p.d.m.) are varied by the information as shown in *Figure 38.1*. The pulsed form of carrier

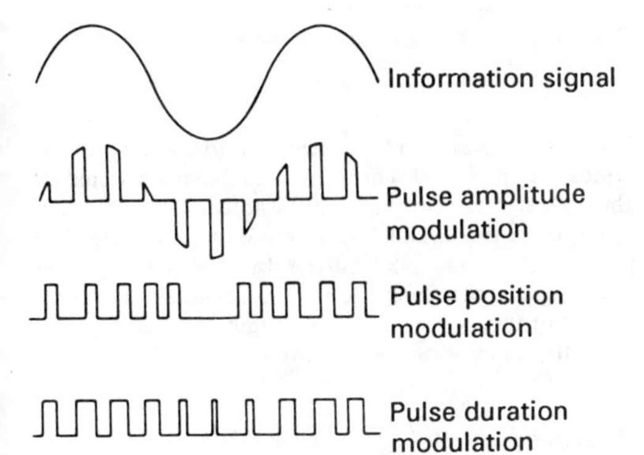

Figure 38.1 Examples of the three types of pulse modulation

operation allows a number of information channels to be transmitted on the same carrier frequency by regularly allocating a pulse to a given channel. For example, eight channels can be accommodated using pulses 1, 10, 19, etc., for channel 1, pulses 2, 11, 20, etc., for channel 2. Pulses 9, 18, 27, etc., are required for synchronising a distribution gate at the receiver. The gate separates the information channels and directs them to their required destinations. Such a system is known as time division multiplex (t.d.m.).

Frequency division multiplex (f.d.m.) is used with continuous carrier in telephony, each speech channel being shifted in frequency before being added to the others occupying different frequency bands. The combined signals modulate the carrier; at the receiver the channels are filtered from each other and their frequency spectra returned to normal.

All the above systems are known as analogue modulation because the information controls the carrier directly. Considerable message-to-noise ratio advantage is gained by converting the amplitude characteristic of the information into a digital form before modulation. This is known as pulse code modulation (p.c.m.). The digital code has to be converted back to its analogue form when it is desired to interpret the information. Time division multiplex is used with p.c.m. when a number of different information channels have to be accommodated.

38.2 Continuous carrier modulation

38.2.1 Amplitude modulation

The carrier amplitude may be represented by $E_c \cos \omega t$ and the modulation by $E_m \cos pt$, so that the a.m. carrier expression is

$$E_c(1+M\cos pt)\cos \omega t \qquad (38.1)$$

where M, the modulation ratio, $\leqslant 1 \propto E_m$.

This is analysed into three components:

$$E_c \cos \omega t + \tfrac{1}{2}E_c M \cos(\omega \pm p)t \qquad (38.2)$$

i.e. the unmodulated carrier and a pair of side frequencies. Information normally consists of several frequencies, and each frequency has its own M proportional to its amplitude (the sum of all M values must not exceed unity) and produces a side frequency pair.

Double sideband (d.s.b.) transmission is the simplest form of a.m. requiring only a very simple unidirectional detector at the receiver and it is therefore used for broadcasting. It requires a transmission channel bandwidth twice the highest frequency of the information components.

To obtain d.s.b. amplitude modulation the information voltage is used to control the gain of a carrier amplifier, and an example with a transistor amplifier is shown in *Figure 38.2*. The

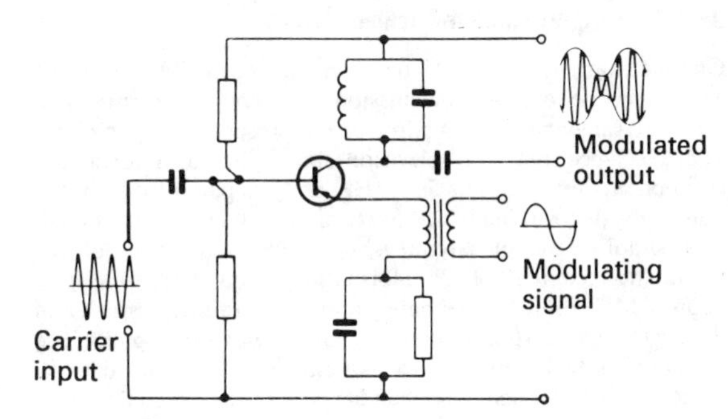

Figure 38.2 A modulated amplifier circuit

information could have been inserted in series with the carrier in the base, and the variation of gain is due to the nonlinear characteristic of the transistor. A similar circuit can be used for an electron tube amplifier. In a triode tube modulation can be achieved by inserting the modulating signal in the grid, cathode or anode circuit. The advantage of grid or cathode modulation is that a relatively low voltage or power is required from the modulator, but modulation envelope distortion occurs at high values of M. Anode modulation requires relatively large power from the modulator but distortion is low up to modulation percentages of 90% ($M=0.9$). Thus broadcasting transmitters using triodes in the modulated amplifier employ anode modulation.

When the modulated amplifier supplies its output direct to the aerial, the system is called high-power modulation; if modulation is carried out before the final r.f. power stage, and envelope distortion is to be avoided, the modulated signal must thereafter be amplified in class B amplifier stages, which have an appreciably lower a.c./d.c. efficiency than the high-power class C modulated amplifier.

A number of techniques have been used to increase the efficiency of transmitters. One is the Doherty amplifier which combines the r.f. amplifying and the modulating functions of the valves used as r.f. amplifiers, so eliminating the need for the large iron-cored components necessary with class B modulators. The Chireix system uses constant-voltage phase-modulated driving signals applied to two saturated power amplifiers feeding a common load, the power output being a function of the phase angle between the input signals. The ampliphase version of this system uses amplitude-modulated drive signals.

Some early sound transmitters used a class A modulator in series with a class C modulated amplifier, but the efficiency was poor. Much higher efficiency is now possible using a modulator operating in class D, a typical p.r.f. being 70 kHz.

In short-wave reception a high signal-to-noise ratio is not always easy to obtain and the intelligibility of signals can be increased by raising the average modulation depth. This is possible using a technique known as trapezium modulation in which bass is attenuated, frequencies above 2 kHz are emphasised and peaks are clipped, a low-pass filter suppressing the harmonics introduced by the clipper.

Television transmitters generally employ grid modulation since it requires much less power than anode modulation—an important consideration with wide-band signals—and nonlinear distortion of video signals is less serious than with speech or music. The synchronising pulse part of the video signal can be operated over the most nonlinear part of the transmitter characteristic and can be predistorted to compensate for the nonlinearity.

38.2.2 Vestigial-sideband transmission

Channel bandwidth should be as small as possible to ensure economic use of the transmission frequency spectrum, and vestigial sideband transmission can represent an appreciable saving of spectrum with television signals having information components up to about 5.5 MHz. Only a part of one of the sidebands, that within about 1 MHz of the carrier, is transmitted; if a simple detector circuit is employed distortion of the modulation content of all information frequencies exceeding about 1 MHz results. Fortunately their amplitudes are small and the degree of distortion is quite acceptable on vision signals. The receiver has to be detuned to half carrier amplitude so as to reduce the low-frequency sideband energy to the original level in relation to the high-frequency sidebands. Vestigial sideband operation is unsuitable for speech or music because of the distortion produced by the simple unidirectional detector used in broadcasting receivers.

38.2.3 Single-sideband transmission

One sideband of a d.s.b. amplitude modulated audio transmission can be removed if a suitable detector is employed in the receiver. The main advantage is the halving of the transmission bandwidth though, if power from the suppressed sideband is transferred to the remaining one, signal-to-noise ratio is increased by 3 dB because halving the bandwidth halves the noise power.

38.2.4 Single sideband with pilot carrier

In a double sideband transmission the ratio of sideband to carrier power is $M^2/2$, which is 50% at 100% modulation. The same is true of s.s.b. when the power from the suppressed sideband is transferred to the other. The carrier power therefore accounts for a very large proportion of the total transmitted power and a worthwhile saving in running costs in a point-to-point communication system is realised by reducing the carrier to a low value. The carrier cannot be entirely eliminated because detection cannot be achieved in its absence. The carrier is restored at the receiver by having an oscillator locked in frequency and phase by the transmitted residual pilot carrier. If almost all the carrier power is transferred to the sideband the transmitted information content can be increased three times relative to the 100% modulation sideband power so that s.s.b. with pilot carrier gives a signal-to-noise ratio approaching 8 dB better than d.s.b. with full carrier (5 dB increase in signal and 3 dB decrease in noise due to halving bandwidth).

A balanced modulator permits the sidebands to be generated and the carrier to be eliminated; it may be a push–pull type r.f. modulator stage with carrier frequency in the same phase to the input base circuits and the information applied in opposite phase as in *Figure 38.3*. If the characteristics of each half stage are identical the carrier component is cancelled at the output but the two sidebands are unaffected. The wanted sideband is selected by passing through a filter rejecting the unwanted.

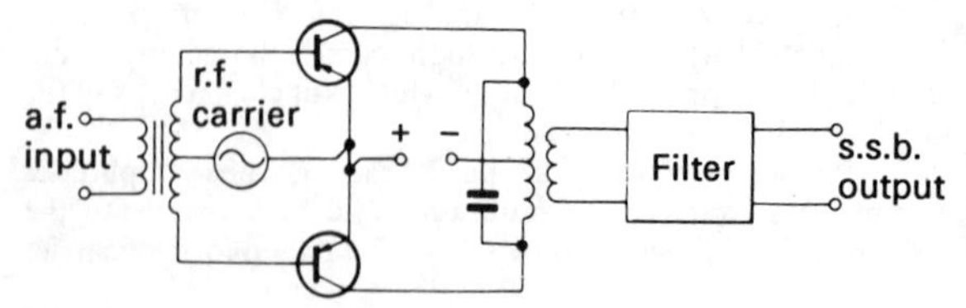

Figure 38.3 A push–pull balanced modulator for carrier elimination

The second sideband channel may be used to carry separate information on the same pilot carrier by means of another balanced modulator and filter; the two independent sidebands and the correct proportion of carrier are combined before power amplification to the information amplitude.

38.2.5 Phase modulation

In phase modulation the phase angle of the carrier is advanced and retarded in proportion to the information amplitude. The mathematical representation is

$$E_c \cos(\omega_c t + \varphi \cos pt) \qquad (38.3)$$

where $\varphi \cos pt \propto E_m \cos pt$.

Like a.m. it can be resolved into carrier and side frequency components but there is more than one pair of side frequencies per information frequency component.

Theoretically there are an infinite number of side frequencies spaced $\pm f_m$, $\pm 2f_m$, etc., from the carrier, but in practice the amplitudes of the higher order side frequencies fall off rapidly if φ is not large.

Phase modulation is produced by variation of the inductive or capacitive element of a buffer amplifier stage following a master oscillator. The variable element may be a varactor diode biased by the information voltage. Phase change must not exceed about $\pm 25°$ if linear modulation is to be achieved, but it can be increased by applying it as an input to a multiplier stage; thus a tripler stage multiplies the phase change by 3. A frequency changer can restore the carrier to its original frequency leaving the multiplied phase untouched.

38.2.6 Frequency modulation

Frequency modulation requires the carrier frequency to be varied in accordance with the information, and the expression for an f.m. carrier is

$$E_c \cos\left(\omega_c t + \frac{f_d}{f_m} \sin pt\right) \qquad (38.4)$$

This is similar to the expression for a phase-modulated wave with f_d/f_m taking the phase of φ. The ratio f_d/f_m is inversely proportional to the modulating frequency, which means that in frequency modulation, for a constant-amplitude modulating signal, the phase shift of the carrier is swept between limits which are inversely proportional to the modulating frequency, whereas in phase modulation the limits are constant. Similarly, in phase modulation, for a constant-amplitude modulating signal, the frequency of the carrier is swept between limits directly proportional to the modulating frequency whilst in frequency modulation the limits are constant. In practice this means that one form of modulation can be converted to the other by including a 6 dB per octave filter in the modulating signal path and, by use of such a filter, the same circuit can be used for the detection of f.m. or p.m. signals.

Frequency modulation is obtained by varying the inductance or capacitance of the tuned circuit of a master oscillator and the modulation must be linear over the range 30 Hz to 75 kHz in transmitters used for stereo sound transmission. (The Zenith-GE system is assumed, using a 38 kHz subcarrier and 19 kHz pilot tone.) Moreover, the average carrier frequency must be kept constant within narrow limits to minimise interference with adjacent channel transmissions. These requirements are met by using balanced reactance modulators which are fed with an a.f.c. voltage in addition to the stereo a.f. signal. In one system the a.f.c. voltage is obtained by comparing a sample of the f.m. oscillator output with two reference frequencies, one 0.5 MHz above the centre frequency and the other an equal amount below it. The input of a discriminator is switched at a subaudio rate between the mixer outputs, and the discriminator output is synchronously switched into two integrators, the outputs of which are compared to obtain the a.f.c. voltage.

38.3 Pulsed carrier modulation

Pulsed carrier modulation is a form of double modulation in which the information is used to modulate a pulse, which is in turn used to modulate the final carrier. It requires a pulse repetition frequency (p.r.f.) at least about 2.3 times the maximum information frequency component. A series of unmodulated pulses can be resolved into a d.c. component and a theoretically infinite number of integral multiples of the p.r.f. When the pulses are amplitude modulated the frequency spectrum can be resolved into a d.c. component and the full range of base-band information frequencies, and sideband pairs of the base-band components associated with each integer multiple of the p.r.f. as indicated in *Figure 38.4*. The side frequency ranges must not overlap if the signal is to be detected without distortion at the receiving point. This cannot occur with a.m. if the p.r.f. exceeds $2f_m$(max).

When the pulse is position or duration modulated, harmonic side frequencies of the pulse repetition integer frequencies ($nf_p \pm mf_m$) are produced but there are no harmonic frequencies of the base band, *Figure 38.4*. Detection at the receiver requires that no overlap shall occur, and if this is probable either the p.r.f. must be increased or the degree of modulation must be reduced. Pulsed carrier modulation is valuable for multiplex operation with t.d.m., but for simplex operation it has no advantages over continuous carrier modulation.

38.3.1 Pulse amplitude modulation

Pulse amplitude modulation is a sampling procedure at the p.r.f. The pulse amplitude may change during the 'on' period in accordance with the information amplitude, or it may remain constant resulting in a stepped envelope version of the original signal. The stepped form is an important stage in the conversion to a digital signal in pulse code modulation. The varying pulse amplitude is used to modulate the carrier amplitude in the normal way. Detection of p.a.m. at the receiving point is achieved by a peak diode followed by a low-pass filter which selects the base band from the pulse spectrum. This type of modulation cannot give any better signal-to-noise ratio than normal amplitude modulation, and it can give a much worse ratio if the receiver is not 'muted' during the pulse off periods.

38.3.2 Pulse position modulation

PPM can be produced by using the information amplitude to frequency or phase modulate the p.r.f. Frequency modulation can be realised by a variable reactance, controlled by the information amplitude, across the p.r.f. oscillator. A limiter in the receiver removes the amplitude noise and allows the signal-to-noise improvement to be achieved. Detection by peak diode followed by a low-pass filter allows the base band frequencies to be recovered. An alternative method is to use a frequency discriminator tuned to the p.r.f.

38.3.3 Pulse duration modulation

Modulation of pulse duration is obtained by varying the level at which a sawtooth voltage derived from the p.r.f. oscillator is sliced. Detection may be by peak diode followed by a low-pass filter, or the pulse duration may be converted to an amplitude variation by using the pulses to control the duration of a ramp voltage.

38.4 Pulse code modulation

With pulse code modulation the information amplitude is divided into a number of levels, each of which is designated by a number; it is regularly sampled and given the number of the specified level nearest the given sample. Normally a binary system of numbering is employed because this is the simplest to achieve electronically, and the digital output consists of a number of pulses per unit time corresponding to the binary number given to the sampled amplitude level. In a binary system the number of levels is 2^n where n is the maximum number of pulses per unit time; thus eight levels require three pulses with zero regarded as a level, whereas 1024 levels require 10 pulses. Since the coded signal represents a number of discrete levels, the reproduced information at the receiving point after decoding will be a stepped copy of the original information, *Figure 38.5*. The steps introduce a spurious signal, and when there are many steps of small amplitude the interference appears as a hiss like random noise superimposed on the original information amplitude, and it is worst when the information amplitude lies midway between two levels. The hiss is termed quantising noise since it is caused by the discrete level quanta; its effect can clearly be reduced by increasing the number of levels, but this requires more digital pulses in a given time interval and therefore a greater transmission bandwidth.

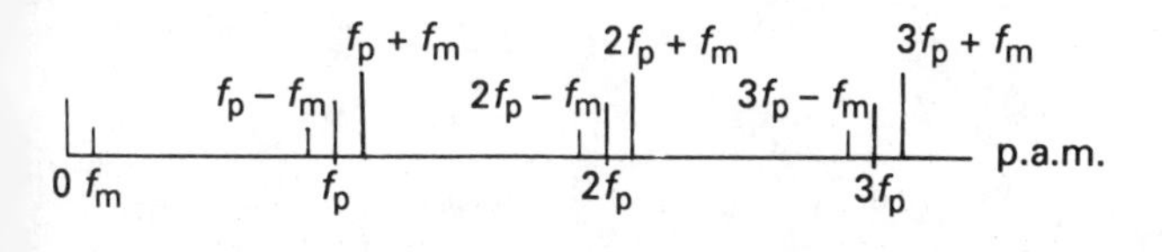

Figure 38.4 Frequency spectrum of p.a.m., p.p.m. and p.d.m. pulses

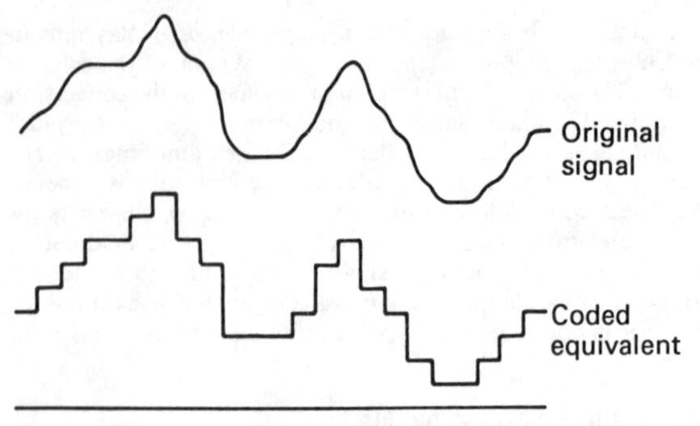

Figure 38.5 Pulse coded equivalent of the original signal

When the number of level steps is small, i.e. when the information amplitude is very small, the step interference changes from a hiss to a seriously distorted signal and this condition is known as granular distortion. In telephonic speech, amplitude variations can be smoothed out by compression, and satisfactory communication can be achieved by a 7-digit binary code (128 levels). The same degree of compression is not permissible for high-quality sound broadcasting, and subjective tests have shown that a 13-digit[2] coding (8192 levels) is required. The 13-digit coding provides a peak signal to r.m.s. quantising noise ratio of 83 dB and a 7-digit 47 dB. It is possible to reduce the audibility of granular distortion by adding a 'dither' voltage to the information before coding. The dither voltage has two components, one a square wave of peak-to-peak amplitude equal to half a level step and at half the sampling frequency, and the other white (thermal) noise at a much lower level (about 4 dB below quantising noise). At the receiver the half sampling frequency square wave is removed by the filter selecting the information frequencies. A slightly improved signal-to-noise ratio is obtained by substituting a pseudo-random signal for the white noise and at the receiver cancelling this by subtracting an identical sequence obtained from another pseudo-random generator synchronised by the transmitted pseudo-random signal.

Conversion of the analogue signal into its digital equivalent involves three processes; sampling, quantising and coding. *Figure 38.6* illustrates this with reference to an 8-level 3-digit coding. A d.c. component must be added to the analogue signal before sampling to make it unidirectional; the combined signal is sampled, *Figure 38.6(a)*, and converted to a p.a.m. signal but the amplitude of each pulse is held constant at the analogue plus d.c. value at the start of the pulse, until the digit number corresponding to the amplitude has been generated, *Figure 38.6(b)*. Quantising involves comparing each pulse amplitude sample with a number of equally spaced reference amplitude levels and either determining to which level its amplitude is nearest or, which is often easier, selecting the level one lower than that which it just fails to reach. Each level is assigned a binary number and the output of the coder provides the binary pulse equivalent of the level determined from the unidirectional p.a.m. analogue signal, *Figure 38.6(c)*.

There are a number of systems for converting the p.a.m. signal into a digital equivalent; one uses a counter coder and another employs successive approximations. A simplified block diagram of the counter coder is illustrated in *Figure 38.7*. The held sample of information amplitude is compared against a ramp voltage of constant slope provided by charging a capacitor at a constant current. When an assessment is to be made a short circuit S across the ramp generator capacitor C is removed and simultaneously clock pulses are fed to a counter. When the voltage across capacitor C equals that of the sample, the comparator sends a pulse which stops the counter. Since the voltage across C increases linearly with time until equality is reached with the sample, the counter registers the binary equivalent of the sample amplitude. The state of the counter is transferred to the shift register via AND gates. The counter is cleared and the ramp capacitor short circuited in readiness for coding the next sample. Shift pulses are fed into the register to release the digital information in time sequence; the digital representation may be transmitted direct over coaxial cable or may be used to modulate a u.h.f. or s.h.f. carrier for transmission over a microwave link.

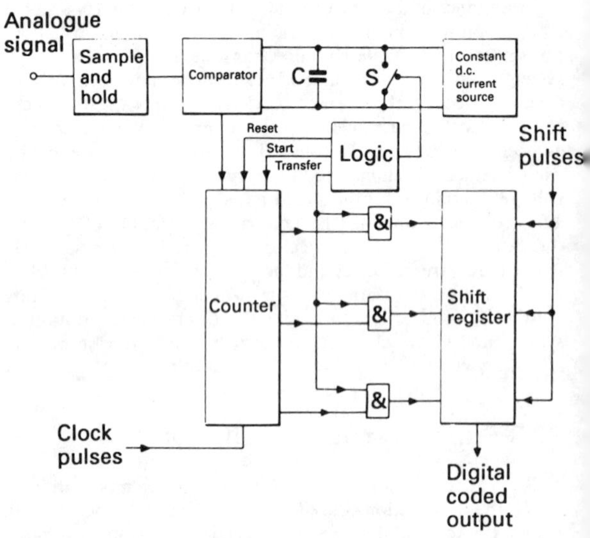

Figure 38.7 Block diagram of counter coder

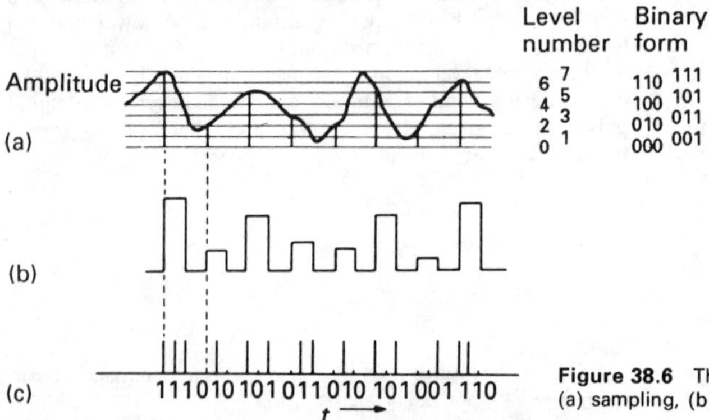

Figure 38.6 The process of pulse coded modulation: (a) sampling, (b) sample and hold signal, (c) coded signal

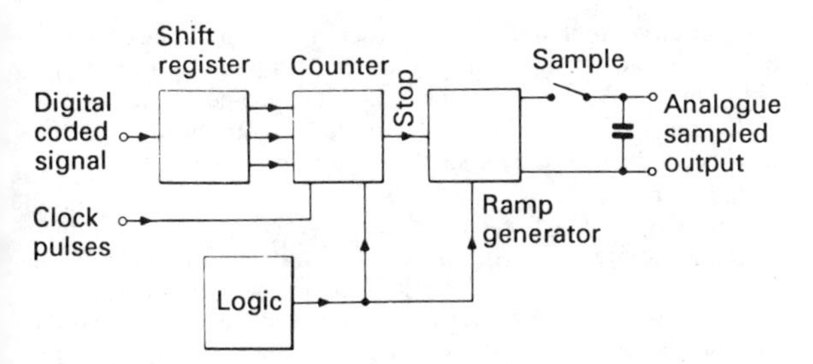

Figure 38.8 Block diagram of a counter decoder

At the receiving point the reverse process may be carried out to convert the digital signal back to its analogue original. A block diagram of the counter decoder is shown in *Figure 38.8*. The coded digital signal is stored in a shift register from which it is transferred to the counter. A ramp generator—a capacitor charged from a constant current—is started simultaneously with the counter which counts back to zero. When all counter outputs register zero, the charging current to the capacitor is cut off, and the voltage across the latter is proportional to the quantised value of the original sample. All these operations are controlled by a sampling oscillator synchronised with that at the sending end, and after each digital conversion the capacitor of the ramp generator is short circuited ready for recharging on the next set of digital pulses. A stepped p.a.m. copy of the original information signal is generated across the capacitor.

The successive-approximation coder compares the sampled analogue signal with a set of successively presented reference voltages corresponding to one digit position in the code. If the maximum reference voltage is less than the sample the comparator gives a logic 1 output; the sample voltage minus the reference voltage is passed to another comparator connected to the next digit voltage reference and if the reference is more than the former the comparator gives a logic 0 output. If the maximum reference voltage is greater than the sample, the comparator registers a logic 0 output and the sample is passed on unchanged to the next reference voltage. This process is repeated until all the digit positions have been defined.

Time division multiplex can readily be applied to allow several p.c.m. signals to share the same transmission channel. With the aid of electronic switches the binary coded samples of each information signal are connected in turn to the transmission system. All information channels are sampled once per cycle of the multiplex operation and extra binary digits called parity bits are added for synchronisation and error detection to complete the whole cycle or frame. The frame synchronising signal maintains the multiplexing and demultiplexing operations in synchronism. The demultiplexer is prevented from responding to a frame synchronising pattern occurring accidentally during the information sequence.

Errors in p.c.m. signals may be detected by transmitting an extra parity bit with the information. The parity bit is inserted at the sending end to bring the total binary digits to an even or an odd number in each transmitted information sample. Failure at the receiver to register an odd or even number (whichever is chosen) of bits in each sample indicates an error.

38.5 Detector nomenclature

According to BS 4727 the function of a detector is to abstract information from a radio wave. The information may be the modulation waveform and the detector is then alternatively termed a demodulator.

The output of an f.m. detector can similarly be a copy of the modulation waveform and such a circuit may also be termed a discriminator. A further complication in the nomenclature of f.m. detectors is that they are sometimes called phase detectors or phase discriminators.

38.6 AM detectors

Any device with a nonlinear transfer characteristic can act as an a.m. detector because it responds unequally to positive and negative half-cycles of the input signal, so giving a modulation-frequency component in the output. An example was the anode-bend detector which used a pentode tube biased near anode-current cut-off. Inevitably such a detector gives considerable harmonic and inter-modulation distortion, and better results can be obtained from diode detectors.

38.6.1 Series-diode detectors

The simplest example of a diode detector is the series-diode circuit shown in *Figure 38.9*. It is similar to a half-wave rectifier circuit and the capacitor C_1 can be called a reservoir capacitor. Operation of the circuit relies on the rapid charging of C_1 through the low-value forward resistance and the subsequent discharge through the high-value diode load resistor R_1.

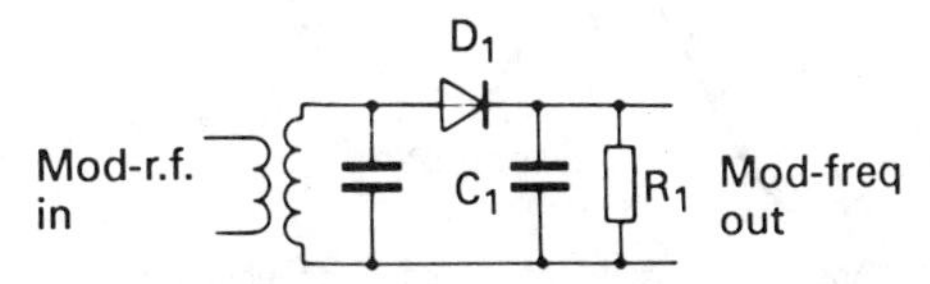

Figure 38.9 The simple series-diode detector circuit (example of a sampling detector)

Diode D_1 conducts during positive half-cycles of r.f. input and charges C_1 to the peak value of the input signal. During negative half-cycles the diode is cut off and C_1 begins to discharge through R_1. The ratio of the time constant R_1C_1 to the period of the carrier is, however, so chosen that very little of the charge on C_1 is lost before D_1 begins to conduct on the next positive half-cycle of input and C_1 is again charged to the peak value. Thus C_1 maintains a positive voltage which keeps D_1 cut off except for the instants when the input signal passes through its positive peaks. In practice the period of conduction is only a small fraction of the positive half-cycle.

Thus the load circuit R_1C_1 is connected to the modulated r.f. source by the low forward resistance of the diode for only a brief fraction of each input cycle and during this time the capacitor voltage is 'topped up' to the peak value of the r.f. input. For the remainder of each cycle the diode is cut off, isolating the load circuit from the r.f. input so that the voltage across R_1C_1 begins a small exponential fall. Thus the diode acts as a switch which is turned on and off by the carrier component of the input signal. This is an example of a sampling process in which the modulated r.f. input signal is sampled once per cycle when it is passing through its positive peak. As the peak value changes as a result of modulation, the voltage across R_1C_1 changes to give a simulation of the modulating signal waveform made up of a number of 'topping up' increases separated by exponential falls. These constitute an r.f. ripple of small amplitude superposed on the modulating-frequency waveform and which is easily removed by an r.f. filter to make the output waveform a good approximation to the modulating signal.

38.6.2 Synchronous diode detectors

Circuits of the type so far considered are used to detect modulated signals in which the carrier is present. They take samples of the positive peaks of the modulated r.f. input and are not affected by variations in the timing or phase of the peaks. To detect carrier-suppressed a.m. signals the detector must be sensitive to the phase as well as the amplitude of the peaks of the input signal; the reason for this will be made clear in the discussion of *Figure 38.10*. Thus the detector must have a reference signal of constant frequency against which it can compare the phase of the modulated r.f. input. To this end the detector is

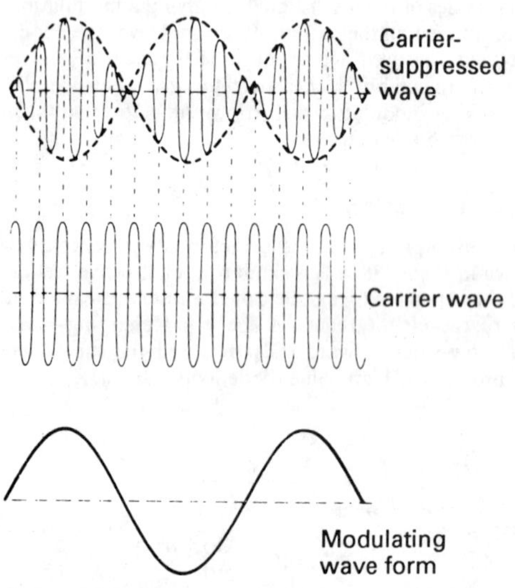

Figure 38.10 Action of a synchronous sampling detector in detecting a carrier-suppressed signal (the dashed lines indicate the sampling periods)

provided with a second input in the form of a constant-amplitude sinusoidal signal synchronised with the (suppressed) carrier frequency of the modulated r.f. signal to be detected.

One possible circuit for a synchronous diode detector is given in *Figure 38.11*. The single series diode of the prototype a.m.

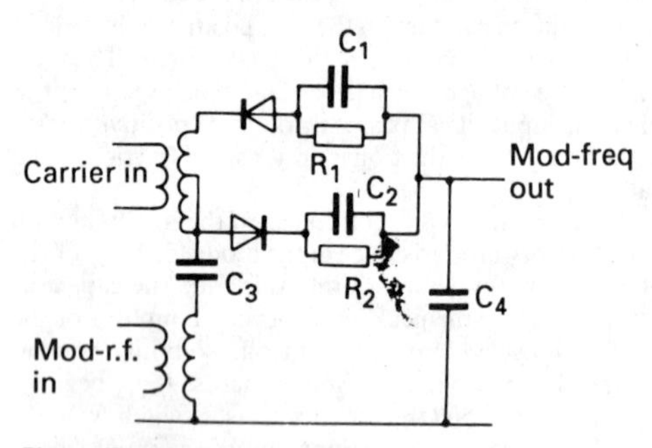

Figure 38.11 Synchronous sampling detector using a diode bridge

detector is replaced by two diodes and a centre-tapped transformer. Both diodes conduct together to produce the low-impedance path which connects the source of modulated r.f. to the capacitor C_4. When the diodes are non-conductive the path is open-circuited and C_4 retains its charge. The diodes must be driven into conduction and non-conduction by the carrier input and not by the modulated r.f. input and thus the carrier input must be large compared with the other input signal. The balanced form of the carrier circuit is adopted to minimise any carrier component which may reach C_4. The time-constant circuits R_1C_1 and R_2C_2 are included as diode loads to ensure that the diodes conduct for only a small fraction of each cycle, i.e. when sampling is required.

The way in which such a detector demodulates a double-sideband suppressed-carrier signal is illustrated in *Figure 38.10*, in which the vertical dashed lines indicate the sampling periods. A non-synchronous a.m. detector, being insensitive to phase, would sample all the positive peaks and would thus produce a grossly distorted output. The synchronous detector operates stricly at carrier-frequency intervals and samples the positive peaks during one half-cycle of the modulating signal and negative peaks during the other half-cycle, thus correctly reconstituting the waveform of the modulating signal. The output has positive and negative swings and, for a symmetrical modulating signal such as a sine wave, has a mean value of zero, i.e. there is no d.c. component as in the output of the prototype non-synchronous series-diode detector.

38.6.3 Shunt-diode detectors

In the circuit of *Figure 38.9* the output of the detector is taken from the reservoir capacitor, but it could alternatively be taken from the diode, the circuit being rearranged as shown in *Figure 38.12* to enable one leg of the output to be earthed. In this version

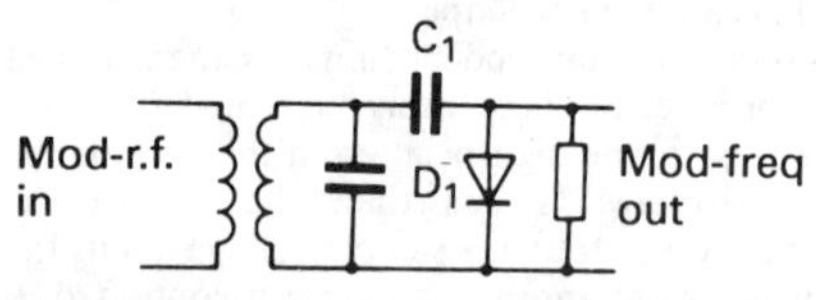

Figure 38.12 The simple shunt-diode detector circuit (example of a clamping detector)

of the circuit, known as the shunt-diode detector, the reservoir capacitor is series-connected, which makes the circuit convenient when d.c. isolation is required between the output terminals and the source of modulated r.f. input.

38.7 FM detectors

Perhaps the most obvious way of detecting an f.m. signal is to convert the frequency variations into corresponding amplitude variations of the carrier which are then applied to an a.m. detector. A number of f.m. detectors operate on this principle. For example, in the slope detector, the centre frequency of the f.m. signal was arranged to fall on a suitable part of the amplitude–frequency characteristic of an *LC* circuit as shown in *Figure 38.13*. The curvature of the skirts of the resonance curve causes distortion, which can be minimised by choice of Q value and resonance frequency but is still serious.

Better results were obtained from the Round–Travis detector by using the push–pull principle. Two similar *LC* circuits were used, one resonant above and the other below the centre frequency. Separate a.m. detectors were used for each *LC* circuit and their outputs were combined (*Figure 38.14*). This detector was abandoned because of the difficulty of tuning the *LC* circuits and because the detector responded to amplitude modulation of the input signal. Modern f.m. detectors have a measure of a.m. suppression.

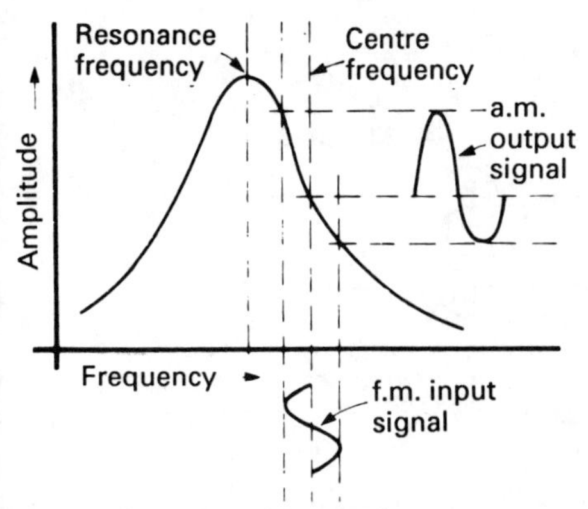

Figure 38.13 Simple f.m. slope detector

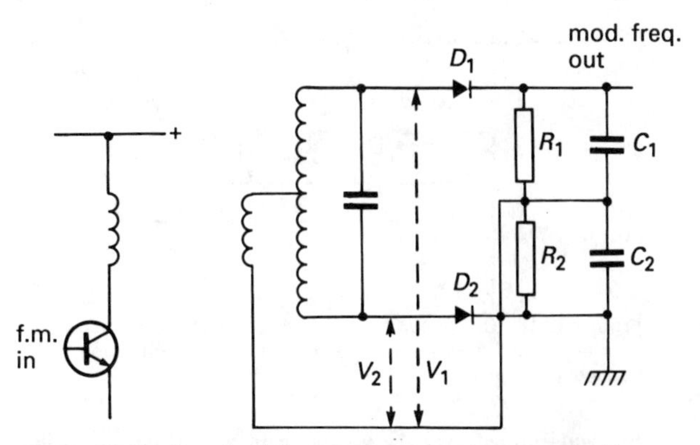

Figure 38.15 One circuit for a Seeley–Foster detector. Note the method used for deriving the diode input voltages V_1 and V_2

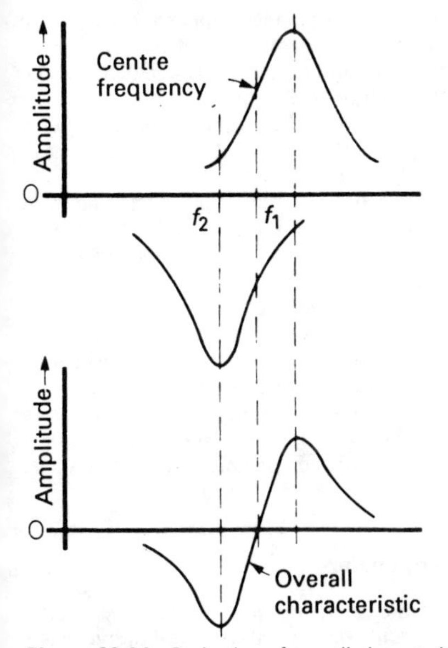

Figure 38.14 Derivation of overall characteristic of the Round–Travis f.m. detector

38.7.1 Seeley–Foster detectors

The tuning difficulty described in the last section was overcome in the Seeley–Foster detector circuit. This f.m. detector uses an arrangement of diodes similar to that of the Round–Travis circuit but the method of providing the diode input signals is different. The method makes use of the phase relationship between the voltage across the tuned secondary winding of a transformer and that across the primary winding. Whether the primary winding is tuned or not, these two voltages are in quadrature when the applied signal is at the resonance frequency of the secondary winding. At frequencies above resonance the secondary voltage lags the quadrature condition to an extent dependent on the frequency deviation and at frequencies below resonance the secondary voltage leads the quadrature condition to an extent depending on the deviation.

If therefore the secondary winding is centre-tapped and if a sample of the primary voltage is injected into the centre tap, as shown in *Figure 38.15*, the voltages V_1 and V_2 at the two ends of the secondary winding vary with frequency in the same way as those from the two tuned circuits in the Round–Travis circuit.

The Seeley–Foster discriminator was extensively employed in early f.m. receivers. Alignment is straightforward, needing only a signal source at the centre frequency, and linearity can be made acceptable. Its chief disadvantage, shared with the Round–Travis circuit, is that it responds to any amplitude modulation of the input signal. Thus to obtain the high signal-to-noise ratio of which an f.m. receiver is capable it is necessary to precede the Seeley–Foster circuit by one or more amplitude-limiting stages to minimise any a.m. content in the received signal.

38.7.2 Ratio detectors

By a simple modification the Seeley–Foster discriminator can be made capable of a useful degree of a.m. suppression. The detector circuit so produced is known as the ratio detector and it is not surprising that it rapidly displaced the Seeley–Foster discriminator. The way in which the ratio detector operates can be approached in the following way.

If one of the diodes in the circuit of *Figure 38.15* is reversed, the net output is the sum of the voltages across the individual diode loads (not the difference as in the Seeley–Foster circuit). Thus, for an input to the circuit at the centre frequency there is a voltage at the combined output approximately equal to the sum of the peak input voltages to the diodes: this compares with zero output from the Seeley–Foster circuit.

If the frequency of the input is displaced from the centre value the output across one diode load increases whilst that across the other decreases and the combined voltage output tends to be independent of frequency and thus of frequency modulation. This combined output is proportional to input signal amplitude and can be used to operate a tuning indicator.

Even though the voltage across (C_1+C_2) is constant (for a given input amplitude) the voltages across the individual reservoir capacitors C_1 and C_2 vary with the frequency of the input signal and either capacitor can be used as the source of modulation-frequency output from the detector as shown in *Figure 38.16*.

To make the circuit capable of a useful degree of a.m. rejection the diode load resistor(s) are given low value(s) so that the tuned circuit feeding the detector is heavily damped. A large value capacitor is then connected across the load resistors to give a time constant approaching one second.

Suppose as a result of amplitude modulation of the input signal there is a momentary increase in the peak amplitude of the signal input to the ratio detector. The voltage across the long time constant diode load circuit cannot instantaneously adjust itself to equal the peak value of the spike and as a result the diodes are driven heavily into conduction and their forward resistance increases the already heavy damping on the tuned circuit thus

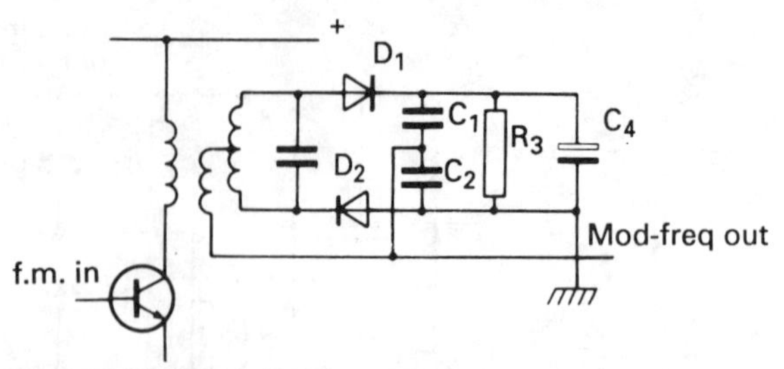

Figure 38.16 One circuit for a ratio detector. R_3 acts as load for both diodes and the voltage across C_2 is taken as modulation frequency output

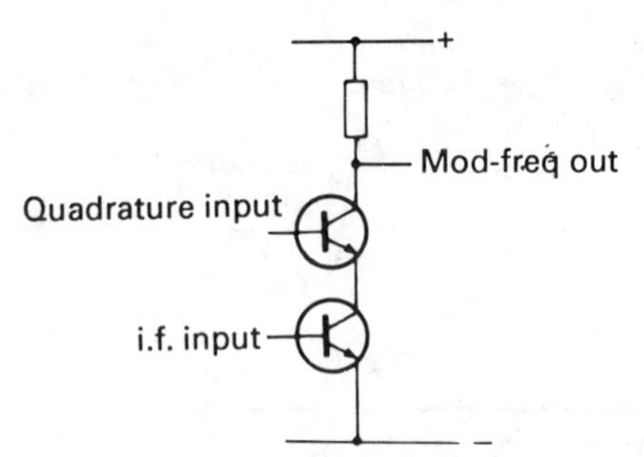

Figure 38.18 Basic form of bipolar-transistor coincidence detector

momentarily reducing the voltage gain of the previous stage, minimising the effect of the spike.

Similarly, if there is a momentary reduction in the peak value of the input signal to the detector, the long time constant network again cannot register the change and the diodes are cut off so removing the damping imposed by the diode load on the tuned circuit. Thus the gain of the previous stage is momentarily increased, offsetting the effect of the change in input signal.

38.7.3 Phase-comparator detectors

This type of detector also makes use of the varying phase relationship between two input signals, nominally in quadrature, such as the voltages across the primary and tuned secondary windings of a transformer.

In phase-comparator detectors the two input signals are limited so as to form rectangular pulses. Limiting may be carried out in separate stages preceding the phase comparator or in the phase comparator itself. The degree of overlap of these pulses varies with the phase difference between the two inputs and determines the output current of the comparator which is therefore a copy of the modulation waveform. The output of the comparator thus depends on the relative timing of the two sets of pulses and is independent of the amplitude of the input signals provided this is sufficient to give satisfactory limiting.

The general form of a phase-comparator detector is illustrated in the block diagram of *Figure 38.17*. In its simplest form a transistor phase comparator could take the form shown in

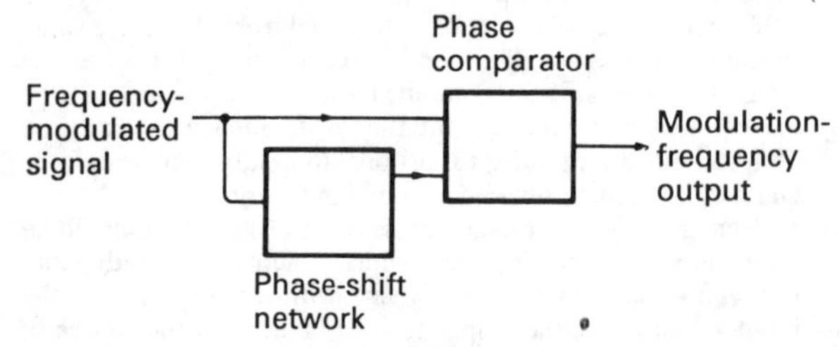

Figure 38.17 General form of phase-comparator detector

Figure 38.18. One of the disadvantages of such a simple circuit is that the output would contain a large component at the input frequency in addition to the wanted modulation-frequency component and in practical forms of phase-comparator detector precautions are taken to minimise this unwanted component.

In integrated circuits, for example, extensive use is made of the push–pull principle and a simplified version of a typical circuit is given in *Figure 38.19*. The output of the i.f. amplifier (also included in the IC) is applied in the form of push–pull pulses to the bases of TR_5 and TR_6 so that when one of these transistors is driven into conduction the other is cut off. The quadrature signal is derived from the i.f. output by use of an external *LC* circuit and associated reactance (one possible arrangement is shown in dashed lines) and is applied also in pulse form to two push–pull pairs TR_1TR_2 and TR_3TR_4 in a balanced circuit which ensures that none of the quadrature component appears between the output terminals. Suppose TR_1 base is driven positive by the quadrature signal at an instant when TR_5 is conductive. The effect is to promote conduction in TR_1 and thus to cut off TR_2, producing a net output between the output terminals. Half a cycle later, when TR_6 is conductive, TR_3 and TR_4 behave similarly and again there is a net output. The duration of these outputs depends, of course, on the extent of the overlap between the i.f. and quadrature inputs and varies with the phase difference between the two inputs. The output can be used as a.f. in an f.m. receiver or for a.f.c. purposes.

TR_7 is included to stabilise the mean current through the detector and is one of the many auxiliary components included in ICs to ensure that the performance is substantially unaffected by variations in ambient temperature or in supply voltage.

A number of ICs designed for use in f.m. receivers incorporate detectors with a circuit similar to that of *Figure 38.19* and they are often described as balanced, symmetrical, quadrature, coincidence or product detectors.

38.7.4 Counter discriminators

These use a principle quite different from those employed in the detectors so far described. If an f.m. signal is rectified the result is a succession of half-sinewave pulses the frequency of which varies according to the modulation. At periods where the pulses are crowded the mean value per unit time is greater than at instants when they are less crowded. This variation in mean value represents the modulation waveform and if the rectified signal is passed through a low-pass filter to suppress all but a.f. the output consists of the wanted modulation-frequency component superposed on a direct component. The change in frequency in the signal radiated from an f.m. broadcast transmitter is, however, very small compared with the centre frequency, typically ± 75 kHz maximum at a carrier frequency of, say, 90 MHz—a variation of less than $\pm 0.1\%$ representing a very small change in mean value of the rectified signal and thus a very small a.f. output from the low-pass filter. The relative change in frequency is greater in the i.f. circuits, ± 75 kHz in 10.7 MHz being approximately $\pm 0.7\%$. It is usual, however, in receivers using pulse-counter discriminators to employ an i.f. of 455 kHz or even lower. At 455 kHz the maximum change in frequency is nearly $\pm 17\%$ which gives a worthwhile modulation-frequency component in the rectified signal.

The signal presented to the low-pass filter must be free of amplitude variations because these would give a spurious output. Moreover all the input pulses must be of identical shape because variations in shape could also give unwanted components in the output. The problem is, therefore, to generate from the i.f. signal a

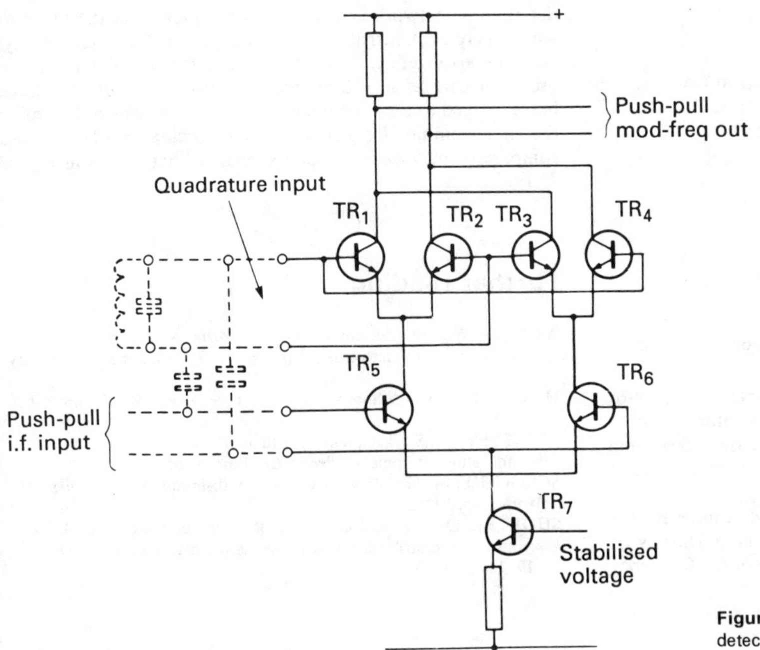

Figure 38.19 Simplified form of coincidence detector used in integrated circuits

series of pulses all of identical form and amplitude, the number per unit time varying according to the modulation.

Early pulse-counter discriminators were fed with square waves from the final limiting stage in the i.f. amplifier. The square wave was differentiated in an *RC* circuit which, as shown in *Figure 38.20*, incorporated diodes to eliminate negative-going blips. The resulting train of positive-going blips was passed through a low-pass filter with a cut-off frequency of, say, 30 kHz. A simple *RC* filter is shown in *Figure 38.20*.

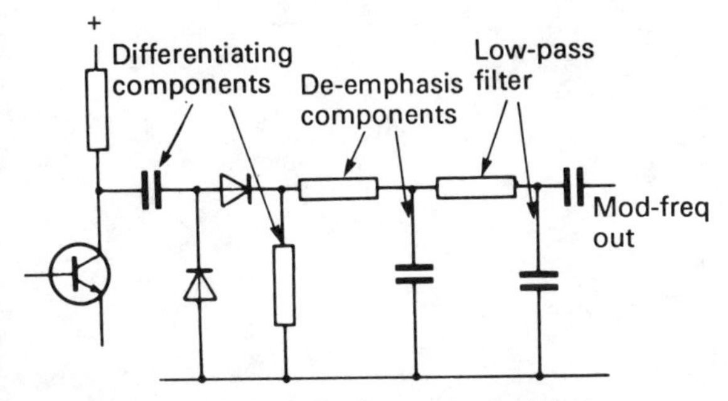

Figure 38.20 Basic form of pulse-counter discriminator

In more recent pulse-counter discriminators the positive-going blips are used to trigger a multivibrator giving, for example, 1 μs pulses which are passed through a squarer stage (to eliminate any overshoots) before being applied to the low-pass filter.

Pulse-counter discriminators are used in applications where linearity is important, e.g. in f.m. rebroadcast receivers and in f.m. deviation meters.

38.7.5 Locked-oscillator discriminators

As the title suggests, this type of f.m. discriminator is based on an oscillator which is synchronised by the f.m. signal so that its frequency follows any changes in that of the input signal. Such a system can be expected to have two useful properties. Firstly the amplitude of the oscillator output can be many times that of the input signal, implying a useful degree of voltage gain. Secondly the oscillator output amplitude is independent of that of the input signal provided this is sufficient to give effective synchronising; in other words the system should give effective amplitude limiting. Thus the oscillator can be used as a source of amplified and amplitude-limited f.m. signals which can be followed by any of the types of discriminator described above. Used in this way, of course, the oscillator is not itself a discriminator but a source of input signal for a discriminator. Circuits of this type were described as early as 1944.

The synchronised oscillator can, however, act as a discriminator. If it operates in class C, taking one burst of current from the supply per cycle of oscillation, the frequency of the bursts follows that of the input signal and so contains a modulation-frequency component which can be used as detector output. For the reasons given in the previous section, however, a low value of intermediate frequency (and hence oscillator frequency) is necessary to give a worthwhile performance from such a circuit.

38.7.6 Phase-locked-loop circuits

In this more recent application of the principle the frequency of the oscillator is controlled not by direct application of the f.m. signal but by a control voltage dependent on the difference between the phase of the oscillator and that of the f.m. signal. The circuit, illustrated in principle in *Figure 38.21*, is so designed that the effect of the oscillator control voltage is to minimise the phase difference between the two signals applied to the phase comparator. Thus the phase of the oscillator is locked to that of the input signal and follows any variations in it. As in the circuits described earlier the output of the phase comparator contains the required modulation-frequency component but the output is usually passed through a low-pass filter to suppress any radio-frequency components.

This type of detector has something in common with those described under 'phase comparator detectors'. Comparison of

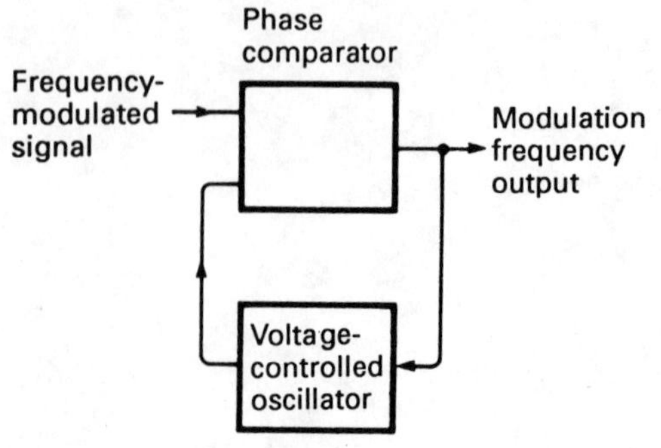

Figure 38.21 Block diagram of phase-locked-loop f.m. detector

Figure 38.7 and *Figure 38.21* shows that in phase-comparator circuits the quadrature input is derived from the other input by use of a phase-shifting network whereas in the phase-locked-loop the second input is derived from the output of the comparator so introducing negative feedback.

The oscillator must be such that its frequency can be readily controlled by a voltage applied to it. It can be a Hartley or Colpitts type in which part of the capacitance of the frequency determining network is provided by a varactor, the d.c. input to which therefore controls the operating frequency. Alternatively, and this arrangement avoids any need for an *LC* circuit, the oscillator can be an astable multivibrator, the control voltage being applied to the resistors of the *RC* circuits which determine the free-running frequency of the oscillator. The phase comparator may have a circuit similar to that shown in *Figure 38.19*.

Further reading

AMOS, S. W., 'AM detectors', *Wireless World*, April (1980)
AMOS, S. W., 'FM detectors', *Wireless World*, January and February (1981)
HERBERT, J. W., 'A homodyne receiver', *Wireless World*, September (1973)
MORTLEY, W. S., 'Frequency-modulated quartz oscillators for broadcasting equipment', *Proc. IEE*, Part B, **104**, 239 (1957)
SCROGGIE, M. G., 'Low-distortion f.m. discriminator', *Wireless World*, April (1956)
SHORTER, D. E. L. and CHEW, J. R., 'Application of P.C.M. to sound signal distribution in a broadcasting network', *Proc. IEE*, **119**, 1442 (1972)

39 Rectifier Circuits

J P Duggan
Mullard Ltd

Contents

39.1 Single-phase rectifier circuits with resistive load

The commonly used single-phase rectifier circuits, and the output voltage waveforms for these circuits when used with a resistive load, are shown in *Figures 39.1* and *39.2* respectively.

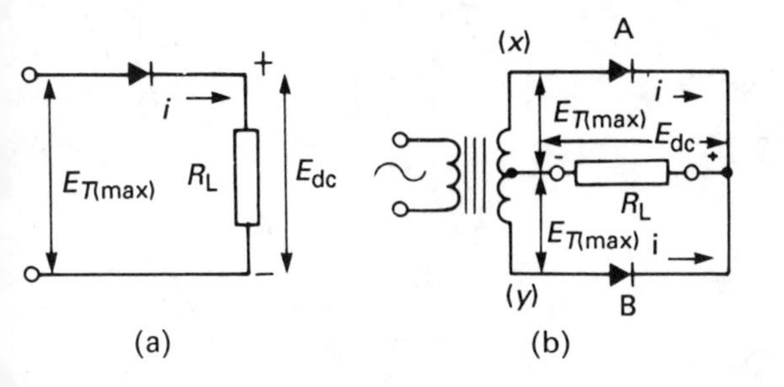

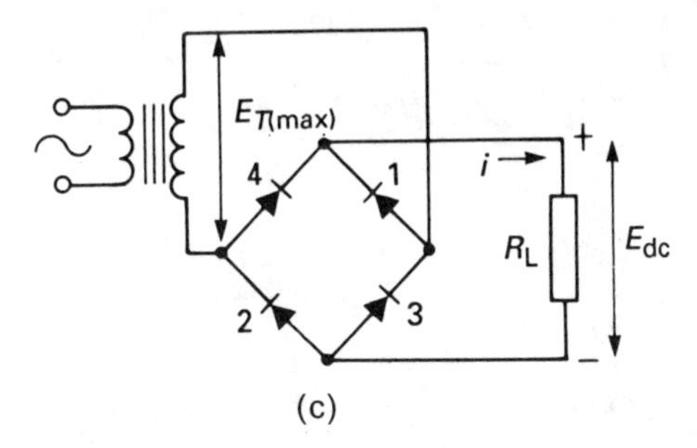

Figure 39.1 Single-phase rectifier circuits: (a) half-wave, (b) full-wave centre-tap, (c) full-wave bridge

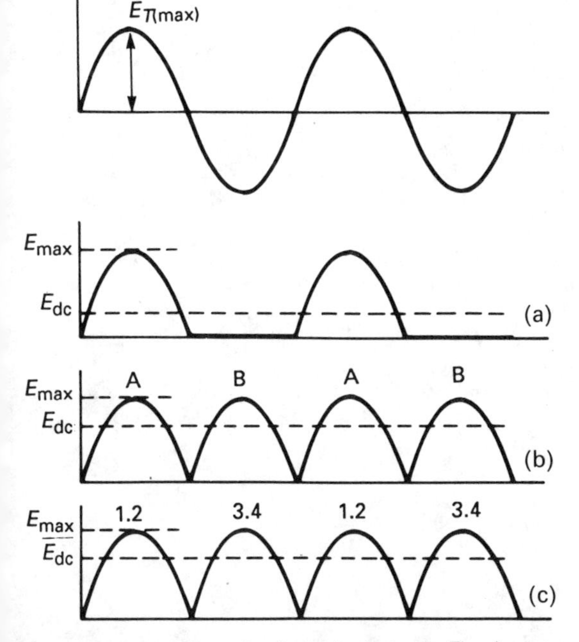

Figure 39.2 Waveforms for single-phase circuits. The sine wave input is shown at the top, and the output waveforms below: (a) half-wave, (b) full-wave centre-tap, (c) full-wave bridge

The secondary input voltage applied to the circuit is sinusoidal and has a crest value $E_{T(max)}$. For all three circuits of *Figure 39.1*, the crest output voltage $E_{max}=E_{T(max)}$.

The half-wave rectifier circuit conducts during the positive half-cycle and blocks during the negative half-cycle of the applied alternating voltage.

In the full-wave centre-tap circuit, the rectifiers are mounted so that rectifier A conducts when point x goes positive, and rectifier B conducts when point y goes positive.

In the full-wave bridge circuit, rectifiers 1 and 2 conduct during the positive half-cycle, and rectifiers 4 and 3 during the negative half-cycle.

The current through the load in each of the three circuits is unidirectional.

If it is assumed that the rectifiers and transformer used are ideal, the performance of any of these circuits can be calculated. The values obtained for each of the above circuits are given in *Table 39.1*.

39.1.1 Percentage ripple

If it is assumed that the amplitudes of the higher harmonics are small compared to that of the harmonic at funcamental frequency f_r, then

$$V_R\% = \frac{\text{fundamental r.m.s. ripple voltage}}{E_{dc}} \times 100 \tag{39.1}$$

From Equation (39.1), the r.m.s. harmonic component at fundamental frequency (which is twice the supply frequency for this circuit) is

$$\frac{4}{3\pi}\frac{E_{max}}{\sqrt{2}}$$

therefore

$$V_R\% = \frac{4}{3\pi}\frac{E_{max}}{\sqrt{2}}\left(\frac{2E_{max}}{\pi}\right)^{-1} \times 100 = 47.2$$

39.2 Single-phase circuits with capacitor input filter

Single-phase half-wave, full-wave, and voltage-doubler circuits are discussed in this section.

39.2.1 Half-wave circuit

The half-wave circuit, *Figure 39.3*, is the simplest rectification circuit giving continuous load current. In the absence of the capacitor C the rectifier will deliver power to the load R_L during the positive half-cycle, and will block during the negative half-cycle. This leads to discontinuous voltage and current in the load.

With the capacitor C in circuit, the capacitor charges to the crest value of the applied voltage on the first positive half-cycle. When the applied voltage falls below the crest value, the capacitor voltage is higher than the applied voltage, and thus the rectifier is reverse biased. The capacitor now discharges into the load until such time as the applied voltage exceeds the capacitor voltage again. The rectifier is then forward biased and charges the capacitor to the crest applied voltage again. The rectifier then ceases to conduct, as previously explained, and the cycle is repeated.

The idealised current waveforms for this circuit, after a steady state has been established, are shown in *Figure 39.4*. The current through the rectifier does not rise instantaneously in practice, because of the time constant formed by the capacitor C and the

Table 39.1 Idealised rectifier circuit performances

		Single-phase			*Three-phase*			
		Half-wave	*Centre-tap full-wave*	*Full-wave bridge*	*Half-wave*	*Full-wave bridge*	*Centre-tap*	*Double-star*
Type of rectifier circuit								
Secondary input voltage per phase			Across BC Across AB		Per phase	Per phase	Per phase	Across CD Across AB Per phase
Output voltage across a–b		$E_{max} = E_{T(max)}$ $E_{rms} = 0.707E_{T(rms)}$	$E_{max} = E_{T(max)}$ $E_{rms} = E_{T(rms)}$	$E_{max} = E_{T(max)}$ $E_{rms} = ET(rms)$	$E_{max} = E_{T(max)}$ $E_{rms} = 1.2E_{T(rms)}$	$E_{max} = \sqrt{3}E_{T(max)}$ $E_{rms} = 2.34E_{T(rms)}$	$E_{max} = E_{T(max)}$ $E_{rms} = 135E_{T(rms)}$	$E_{max} = 0.866E_{T(max)}$ $E_{rms} = 1.17E_{T(rms)}$
Number of output voltage pulses per cycle (N)		1	2	2	3	6	6	6
Output voltage								
E_{dc} in terms of r.m.s. input voltage per phase $E_{T(rms)}$		$0.45E_{T(rms)}$	$0.90E_{T(rms)}$	$0.90E_{T(rms)}$	$1.17E_{T(rms)}$	$2.34E_{T(rms)}$	$1.35E_{T(rms)}$	$1.17E_{T(rms)}$
E_{dc} in terms of r.m.s. output voltage E_{rms}		$0.636E_{rms}$	$0.90E_{rms}$	$0.90E_{rms}$	$0.98E_{rms}$	E_{rms}	E_{rms}	E_{rms}
E_{dc} in terms of peak output voltage E_{max}		$0.318E_{max}$	$0.636E_{max}$	$0.636E_{max}$	$0.826E_{max}$	$0.955E_{max}$	$0.955E_{max}$	$0.955E_{max}$
RMS output voltate E_{rms} in terms of E_{dc}		$1.57E_{dc}$	$1.11E_{dc}$	$1.11E_{dc}$	$1.02E_{dc}$	$1.00E_{dc}$	$1.00E_{dc}$	$1.00E_{dc}$
Peak output voltage E_{max} in terms of E_{dc}		$3.14E_{dc}$	$1.57E_{dc}$	$1.57E_{dc}$	$1.21E_{dc}$	$1.05E_{dc}$	$1.05E_{dc}$	$1.05E_{dc}$
Output current								
Average current per rectifier leg I_0		I_{dc}	$0.5I_{dc}$	$0.5I_{dc}$	$0.33I_{dc}$	$0.33I_{dc}$	$0.167I_{dc}$	$0.167I_{dc}$
I_{rms} per rectifier leg	R	$1.57I_{dc}$	$0.785I_{dc}$	$0.785I_{dc}$	$0.588I_{dc}$	$0.577I_{dc}$	$0.408I_{dc}$	$0.293I_{dc}$
	L		$0.707I_{dc}$	$0.707I_{dc}$	$0.577I_{dc}$	$0.577I_{dc}$	$0.408I_{dc}$	$0.289I_{dc}$
I_{pk} per rectifier leg	R	$3.14I_{dc}$	$1.57I_{dc}$	$1.57I_{dc}$	$1.21I_{dc}$	$1.05I_{dc}$	$1.05I_{dc}$	$0.525I_{dc}$
	L		I_{dc}	I_{dc}	I_{dc}	I_{dc}	I_{dc}	$0.5I_{dc}$
Transformer rating								
Secondary r.m.s. voltage per transformer leg $E_{T(rms)}$		$2.22E_{dc}$	$1.11E_{dc}$ (to centre-tap)	$1.11E_{dc}$ (total)	$0.855E_{dc}$ (to neutral)	$0.428E_{dc}$ (to neutral)	$0.74E_{dc}$ (to neutral)	$0.855E_{dc}$ (to neutral)
Secondary r.m.s. current per transformer leg $I_{T(rms)}$	R	$1.57I_{dc}$	$0.785I_{dc}$	$1.11I_{dc}$	$0.588I_{dc}$	$0.816I_{dc}$	$0.408I_{dc}$	$0.293I_{dc}$
	L		$0.707I_{dc}$	I_{dc}	$0.577I_{dc}$	$0.816I_{dc}$	$0.408I_{dc}$	$0.289I_{dc}$
Secondary volt-amp VA_s	R	$3.48E_{dc}I_{dc}$	$1.74E_{dc}I_{dc}$	$1.23E_{dc}I_{dc}$	$1.50E_{dc}I_{dc}$	$1.05E_{dc}I_{dc}$	$1.81E_{dc}I_{dc}$	$1.50E_{dc}I_{dc}$
	L		$1.57E_{dc}I_{dc}$	$1.11E_{dc}I_{dc}$	$1.48E_{dc}I_{dc}$	$1.05E_{dc}I_{dc}$	$1.81E_{dc}I_{dc}$	$1.48E_{dc}I_{dc}$
Secondary utility factor U_s	R	0.287	0.574	0.813	0.666	0.95	0.552	0.666
	L		0.636	0.90	0.675	0.95	0.552	0.675
Primary voltage per transformer leg (Transformer ratio 1:1)		$2.22E_{dc}$	$1.11E_{dc}$	$1.11E_{dc}$	$0.855E_{dc}$	$0.428E_{dc}$	$0.74E_{dc}$	$0.855E_{dc}$
Primary current per transformer leg (Transformer ratio 1:1)	R	$1.57I_{dc}$	$1.11I_{dc}$	$1.11I_{dc}$	$0.588I_{dc}$	$0.816I_{dc}$	$0.577I_{dc}$	$0.408I_{dc}$
	L		I_{dc}	I_{dc}	$0.471I_{dc}$	$0.816I_{dc}$	$0.577I_{dc}$	$0.408I_{dc}$
Primary volt-amp VA_p	R	$3.48E_{dc}I_{dc}$	$1.23E_{dc}I_{dc}$	$1.23E_{dc}I_{dc}$	$1.50E_{dc}I_{dc}$	$1.05E_{dc}I_{dc}$	$1.28E_{dc}I_{dc}$	$1.05E_{dc}I_{dc}$
	L		$1.11E_{dc}I_{dc}$	$1.11E_{dc}I_{dc}$	$1.21E_{dc}I_{dc}$	$1.05E_{dc}I_{dc}$	$1.28E_{dc}I_{dc}$	$1.05E_{dc}I_{dc}$
Primary utility factor U_p	R	0.287	0.813	0.813	0.666	0.95	0.78	0.95
	L		0.90	0.90	0.827	0.95	0.78	0.95
Fundamental ripple frequency f_r		f	$2f$	$2f$	$3f$	$6f$	$6f$	$6f$
% Ripple = $\frac{\text{r.m.s. fundamental ripple voltage}}{E_{dc}} \times 100$		111	47.2	47.2	17.7	4.0	4.0	4.0
Crest working voltage								
In terms of E_{dc}		$3.14E_{dc}$	$3.14E_{dc}$	$1.57E_{dc}$	$2.09E_{dc}$	$1.05E_{dc}$	$2.09E_{dc}$	$2.42E_{dc}$
In terms of $E_{T(rms)}$		$1.41E_{T(rms)}$	$2.82E_{T(rms)}$	$1.41E_{T(rms)}$	$2.45E_{T(rms)}$	$2.45E_{T(rms)}$	$2.83E_{T(rms)}$	$2.83E_{T(rms)}$

R = Resistive load L = Inductive load f = Supply frequency (Hz)

In the calculation of the above circuit performances, the rectifier forward voltage drop and the transformer impedance have been ignored.
The primary volt-amp rating of the transformer does not take primary magnetising current into account.

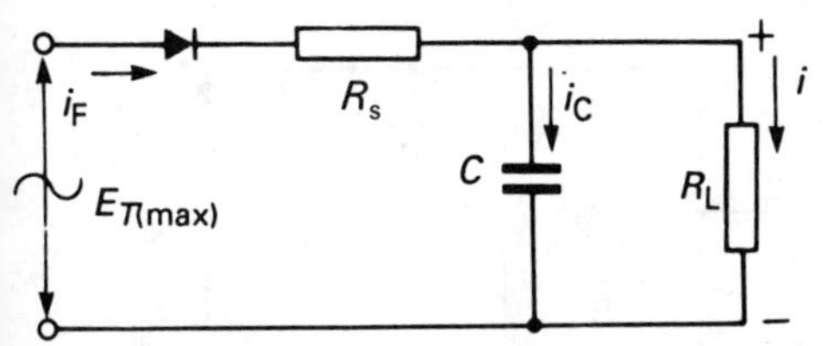

Figure 39.3 Single-phase half-wave circuit

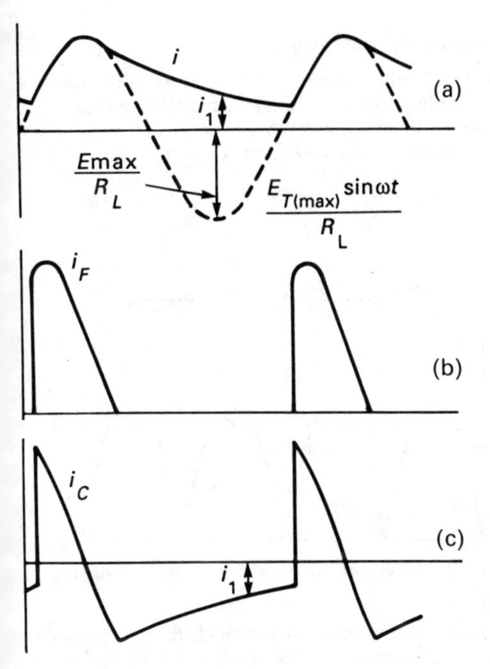

Figure 39.4 Waveforms for single-phase half-wave circuit, after establishment of steady state: (a) load current, (b) rectifier current, (c) capacitor current

source resistance of the supply plus the rectifier and any series resistance.

The capacitor acts as a reservoir, storing up energy during the period that the rectifier conducts. The rectifier current is thus the sum of the capacitor and load currents. The capacitor loses part of its charge during the period that the rectifier is non-conducting, because it discharges into the load during this period. The load current is then equal to the capacitor current i_1. Because of this action, the voltage across the capacitor does not remain constant. The ripple voltage is at the same frequency as that of the applied voltage.

The series resistor R_s is included in the circuit to limit the peak current through the rectifier on initial switch-on.

39.2.2 Performance of half-wave circuit

The charging current from the rectifier to the capacitor flows in pulses which are large in amplitude. The ripple frequency is the same as that of the applied voltage, and an expensive capacitor filter must be used to reduce the ripple to a reasonable value.

If a transformer is used to supply the power to the circuit, then the secondary of the transformer carries unidirectional current each time the rectifier conducts. The transformer has to be rated at the maximum r.m.s. current that flows through the rectifier.

The unidirectional current through the secondary winding of the transformer can lead to core saturation, which in turn leads to increases in magnetising current and hysteresis loss, and introduction of harmonics in the secondary voltage.

The regulation and conversion efficiency of the circuit is low. If a transformer is used, the utility factor of the transformer is also low. Because of the above disadvantages, this circuit is normally only used direct from the mains, and where efficiency is of secondary importance to cost.

39.2.3 Full-wave circuits

There are two types of full-wave rectifier circuit: the full-wave bridge circuit (*Figure 39.5*) and the full-wave centre-tap circuit (*Figure 39.6*). The performance of each is the same, except that, with rectifiers of specified crest working voltage, the d.c. voltage available in a bridge circuit is twice that of the centre-tapped transformer circuit. The voltage and current waveforms for both are shown in *Figure 39.7*.

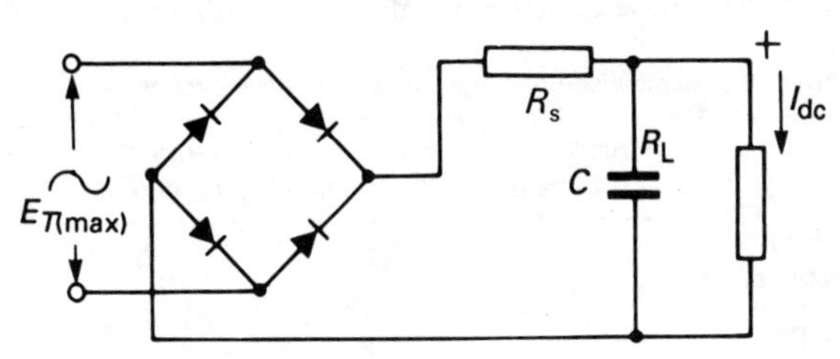

Figure 39.5 Single-phase full-wave bridge circuit

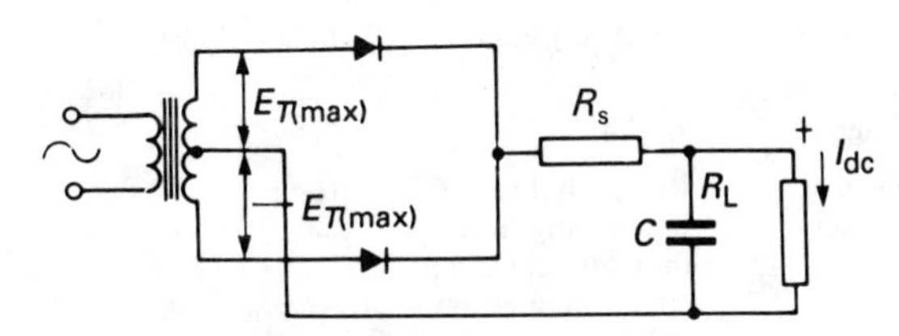

Figure 39.6 Single-phase full-wave centre-tap circuit (also known as two-phase half-wave)

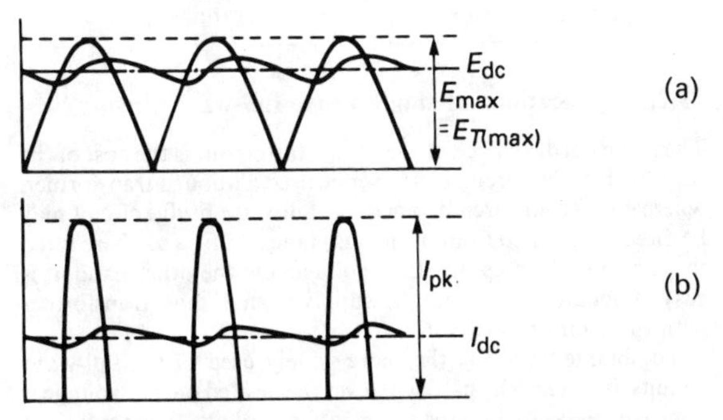

Figure 39.7 Waveforms for single-phase full-wave circuits: (a) voltage, (b) current

In the full-wave bridge circuit the applied alternating voltage is rectified by the bridge, and the output from the bridge is smoothed by the capacitor filter in a similar manner to that described for the half-wave circuit. More efficient smoothing is obtained in this case, because the capacitor maintains the load current for a shorter period, and therefore the capacitor voltage will change by a smaller amount. This means that the d.c. voltage available at the output is greater than that for the half-wave circuit, and the ripple voltage is smaller. The ripple frequency is twice that of the applied voltage.

The centre-tap circuit operates in a similar manner. The rectifiers conduct alternately, and therefore the current flows through each half of the transformer secondary alternately. The rectifiers must withstand a crest working voltage which is equal to the peak value of the applied voltage across both halves of the transformer secondary.

39.2.4 Comparison of single-phase full-wave circuits

The two full-wave circuits are compared in *Table 39.2*.

Table 39.2 Comparison of single-phase full-wave circuits

	Bridge (*Figure 39.5*)	*Centre-tap* (*Figure 39.6*)
No. of rectifiers	4	2
Ripple frequency f_r	$f_r=2f$	$f_r=2f$
Ripple amplitude	Small compared to half-wave circuit	Small compared to half-wave circuit
Smoothing	Relatively easy	Relatively easy
Crest working voltage	$E_{T(max)}$	$2E_{T(max)}$
Conversion efficiency	Relatively high, but slightly lower than for centre-tap circuit because of voltage drop across additional rectifier	High
Transformer	Low transformer secondary volt-amp rating	High transformer secondary volt-amp rating

39.2.5 Applications of single-phase full-wave circuits

The principal drawback of the centre-tap circuit is the cost of the transformer. The circuit can never be used without a transformer, whereas in certain circumstances the full-wave bridge circuit may be directly operated from the mains, if the rectifiers used are rated to withstand the crest working voltage. On the other hand, it is easy to obtain a three-wire d.c. supply from a single transformer with the centre-tap circuit.

The bridge circuit is the more widely used of the full-wave circuits. It is generally used wherever the desired output voltage is approximately equal to the r.m.s. applied voltage. The centre-tap full-wave circuit is used for low-power and low-voltage applications, where low ripple is desired.

39.3 Voltage-doubler circuits

There are two types of voltage-doubler circuit: the symmetrical and common-terminal circuits.

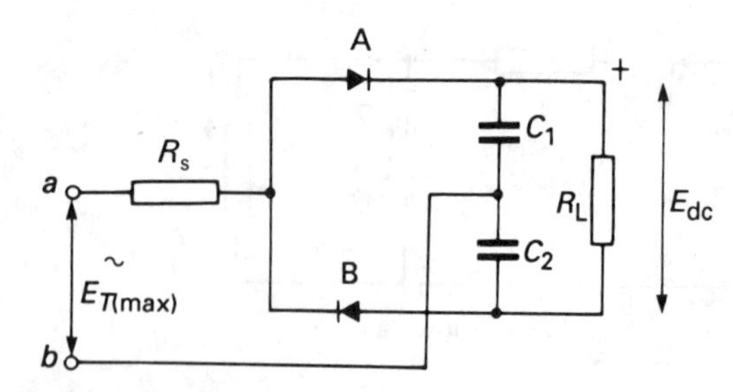

Figure 39.8 Symmetrical voltage-doubler circuit

39.3.1 Symmetrical voltage-doubler

The symmetrical voltage-doubler (*Figure 39.8*) is essentially a combination of two half-wave rectifier circuits, with smoothing filters connected in series, but supplied from the same power source. The output voltage waveform is shown in *Figure 39.9*.

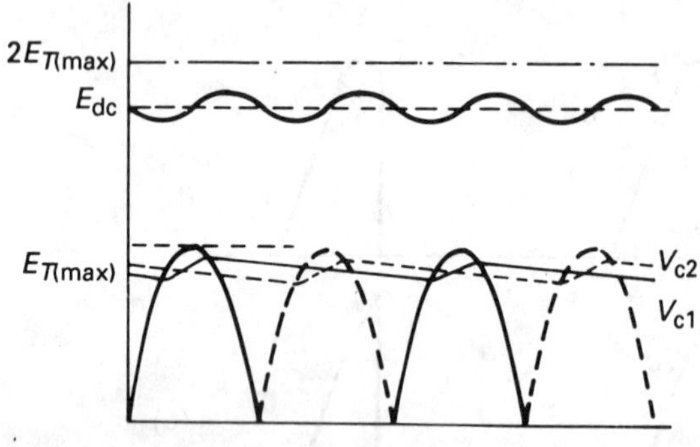

Figure 39.9 Output voltage waveform for symmetrical voltage-doubler

When a is positive, current flows through R_s and rectifier A to charge C_1, whose other terminal is connected to b. When b is positive, C_2 is charged, the return to a being via rectifier B and R_s. Each capacitor charges to the peak applied voltage. The capacitors continually discharge through the load and also act as smoothing elements. The output voltage therefore tends towards twice the peak applied voltage, but cannot achieve it unless the load R_L is disconnected.

The rectifiers must be able to withstand twice the peak applied voltage in the reverse direction. The capacitors must be rated at the peak applied voltage. The ripple frequency of the circuit is twice that of the applied voltage.

39.3.2 Common-terminal voltage-doubler

The common-terminal voltage-doubler is shown in *Figure 39.10*, and the output voltage waveform in *Figure 39.11*.

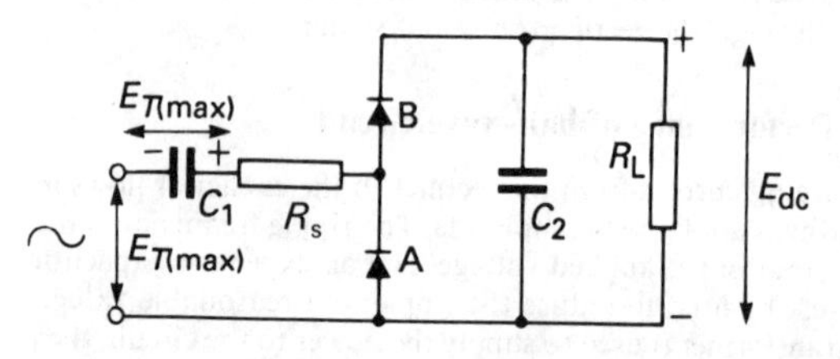

Figure 39.10 Common-terminal voltage-doubler circuit

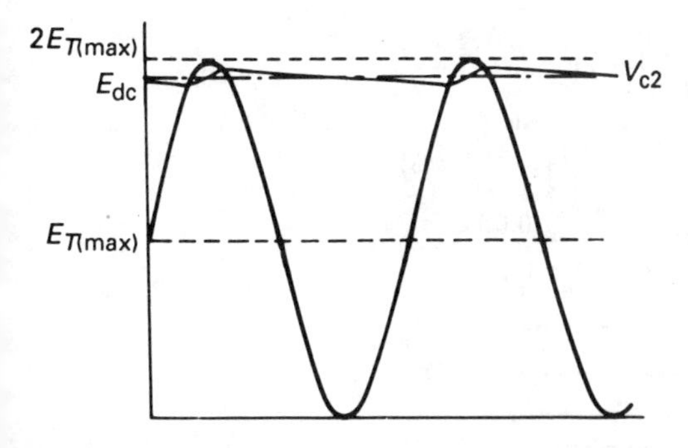

Figure 39.11 Output voltage waveform for common-terminal voltage-doubler

During the first negative half-cycle of the applied voltage, C_1 charges to the peak voltage $E_{T(max)}$ through rectifier A. During the next positive half-cycle, the voltage across C_1 is in series with the applied voltage and aids it to charge C_2 to $2E_{T(max)}$ through rectifier B. Capacitor C_1 loses part of its charge during this process, but charges up to $E_{T(max)}$ again during the next negative half-cycle. The cycle is then repeated.

The voltage across C_2 does not remain constant at approximately $2E_{T(max)}$ because it discharges into the load R_L when rectifier B is not supplying the load current.

The ripple frequency is the same as that of the applied voltage. Capacitor C_2 must be rated at twice the peak applied voltage, and the rectifiers must withstand twice the peak applied voltage.

Applying similar reasoning, it is relatively simple to construct a voltage tripler, a voltage quadrupler, or a circuit with an output voltage which is any other multiple of the peak applied voltage. The essential points to bear in mind are the maximum crest working voltage that the rectifier must withstand, and the rating of the capacitors.

39.3.3 Comparison of voltage-doubler circuits

The two voltage-doubler circuits are compared in *Table 39.3*.

Table 39.3 Comparison of single-phase voltage-doubler circuits

	Symmetrical	*Common-terminal*
Crest working voltage	$2E_{T(max)}$	$2E_{T(max)}$
Ripple frequency f_r	$f_r=2f$	$f_r=f$
Capacitor rating	Rating of C_1 and C_2 must be equal to the peak applied voltage	Rating of C_1 must be equal to the peak applied voltage, and that of C_2 twice the peak applied voltage. C_1 must be rated to carry the r.m.s. load current
Regulation	Poor, but better than for common-terminal doubler	Poor

39.4 Design of single-phase circuits using capacitor input filters

The graphical solution of a capacitor filter rectifier circuit, as put forward by Schade,[1] is presented in *Figures 39.12–39.17*. The peak resistance R_s introduced by Schade to include the peak tube resistance is replaced by the source resistance R_s, which includes the transformer winding resistance, the rectifier resistance, and the series resistance added to limit the initial peak rectifier current.

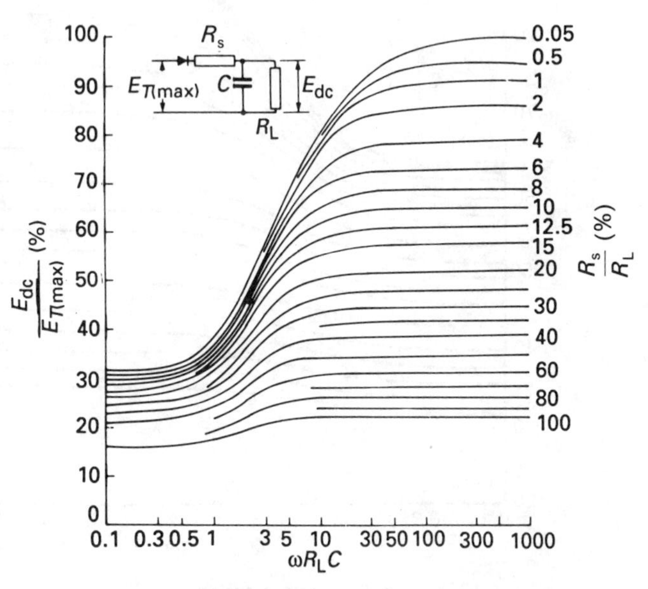

Figure 39.12 $E_{dc}/E_{T(max)}$ % as a function of $\omega R_L C$ for half-wave circuits. C in farads, R_L in Ω, $\omega=2\pi f$

Figures 39.12, *39.13* and *39.14* give the conversion ratio $E_{dc}/E_{T(max)}$ as a function of $\omega R_L C$ for half-wave, full-wave and voltage-doubler circuits respectively. The conversion ratio depends on the value of (R_s/R_L%). For reliable operation the value of $\omega R_L C$ should be selected to allow operation on the flat portion of the curves.

Figure 39.15 gives information on the minimum value of $\omega R_L C$ that must be used to reduce the percentage ripple to a desirable figure. *Figures 39.16* and *39.17* give, respectively, the ratio of r.m.s. rectifier current to average current per rectifier and the ratio of peak repetitive rectifier current to average current per rectifier, plotted as functions of $n\omega R_L C$. These ratios are dependent on the value of R_s/nR_L%.

The transformer leakage reactance has not been taken into account in the design procedure. However, it tends to reduce the peak rectifier current, and therefore assists in limiting the peak current.

39.4.1 Design procedure

The following design procedure is recommended in the design of single-phase silicon rectifier circuits with capacitor input filter.

(1) Determine the value of R_L.
(2) Assume a value of R_s (usually between 1 and 10% of R_L).

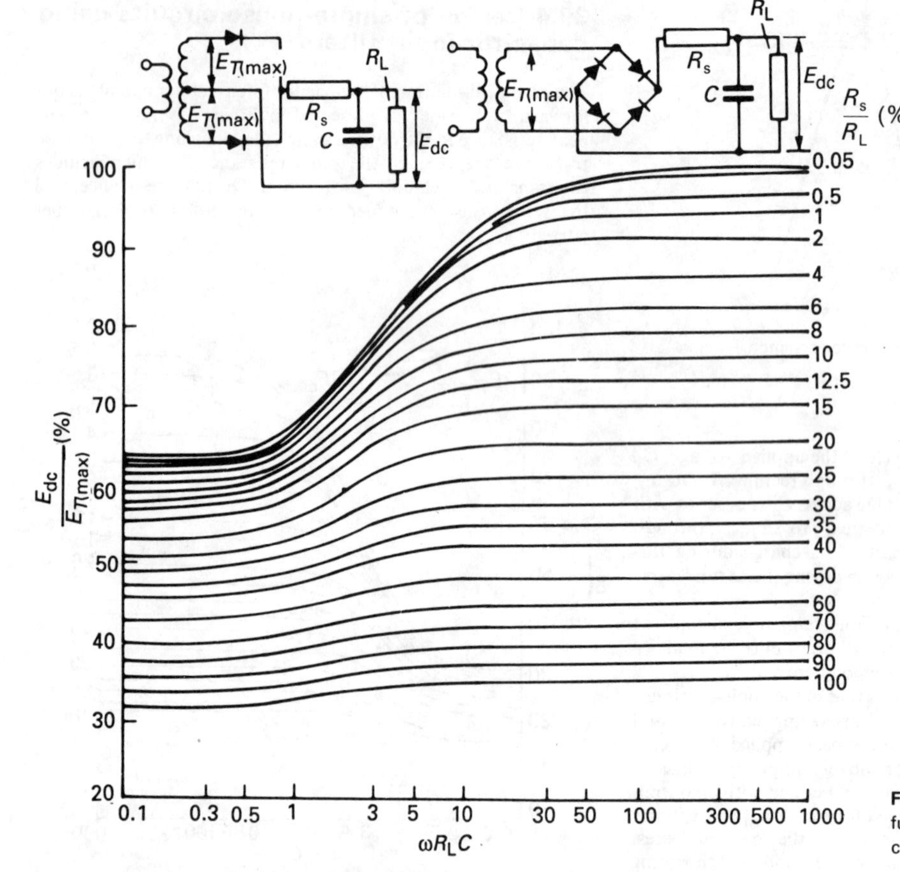

Figure 39.13 $E_{dc}/E_{T(max)}$ % as a function of $\omega R_L C$ for full-wave circuits. C in farads, R_L in Ω, $\omega = 2\pi f$

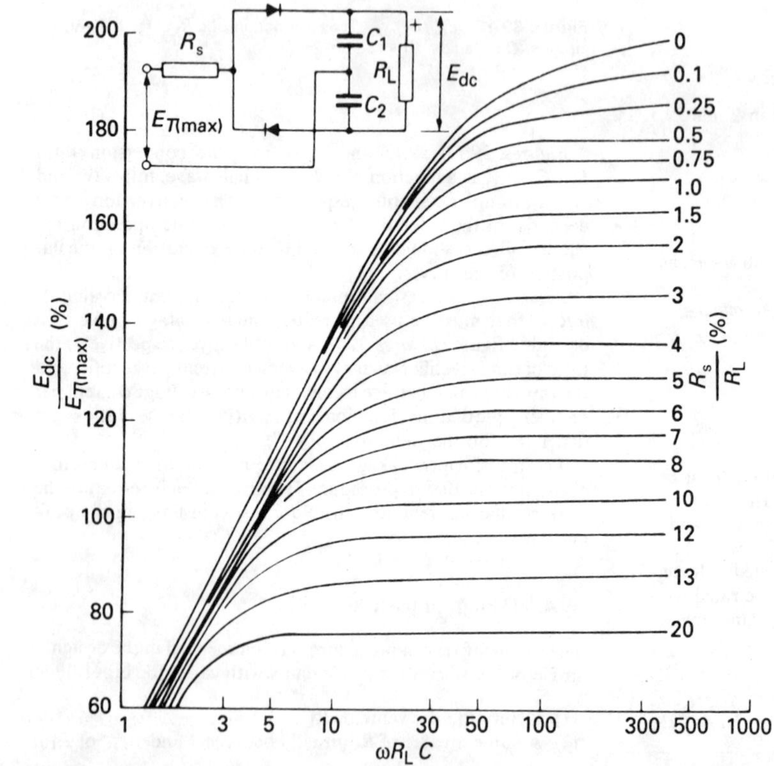

Figure 39.14 $E_{dc}/E_{T(max)}$ % as a function of $\omega R_L C$ for voltage-doubler circuits. C in farads, R_L in Ω, $\omega = 2\pi f$

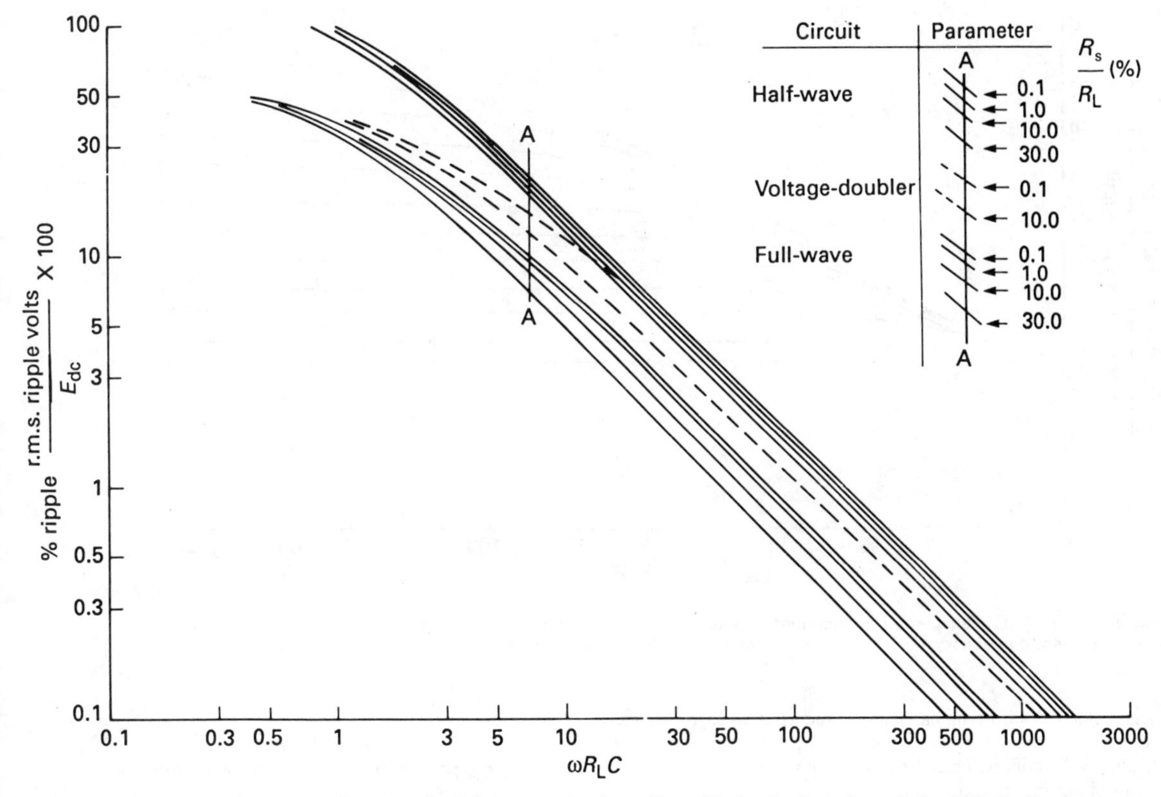

Figure 39.15 Percentage ripple as a function of $\omega R_L C$ for capacitor input filter. C in farads, R_L in Ω, $\omega=2\pi f$ and f is the line frequency

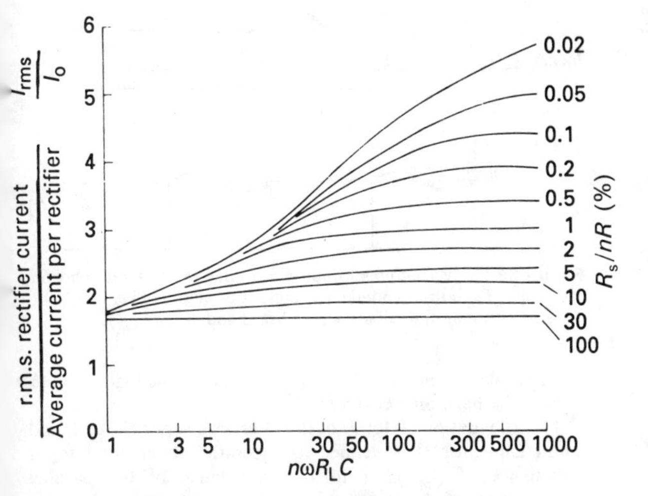

Figure 39.16 The ratio r.m.s. rectifier current/average current per rectifier plotted against $n\omega R_L C$. C in farads, R_L in Ω, $n=1$ for half-wave, $n=2$ for full-wave and $n=0.5$ for voltage doubler

(3) Calculate R_s/R_L %.
(4) From the percentage ripple graph against $\omega R_L C$ (*Figure 39.15*), determine the value of $\omega R_L C$ required to reduce the ripple to a desired value for R_s/R_L % determined in (3). Calculate the value of C required.
(5) From the $E_{dc}/E_{T(max)}$ % against $\omega R_L C$ curves for the appropriate circuit (*Figure 39.12*, *39.13* or *39.14*) determine the conversion ratio for the value of $\omega R_L C$ determined in (4) and R_s/R_L % determined in (3).
(6) Determine the $E_{T(max)}$ and $E_{T(rms)}$ that must be applied to the circuit, using information derived in (5).
(7) Determine the crest working voltage that the rectifiers must withstand.
(8) Determine the r.m.s. current per rectifier from *Figure 39.16*.
(9) Decide on the rectifiers to be used.
(10) Check the peak repetitive current per rectifier from *Figure 39.17*.
(11) Check the initial switch-on current I_{on} given by $E_{T(max)}/R_s$. If the value obtained exceeds that specified for the rectifier, then R_s must be increased and the design procedure repeated.
(12) Design the transformer and adjust the value of R_s accordingly, taking into account the transformer resistance and the forward resistance of the rectifier at the average current.
(13) Check the r.m.s. ripple current through the capacitor.
(14) Design the RC damping circuit as recommended in the published data of the rectifier.
(15) Determine the size of heatsink to be used to allow operation at the desired ambient temperature (from published data).

39.5 Design procedure for single-phase rectifier circuits with choke input filter

The analysis of the capacitor input filter rectifier circuits has shown that, for any high-current conversion, the circuit requires a

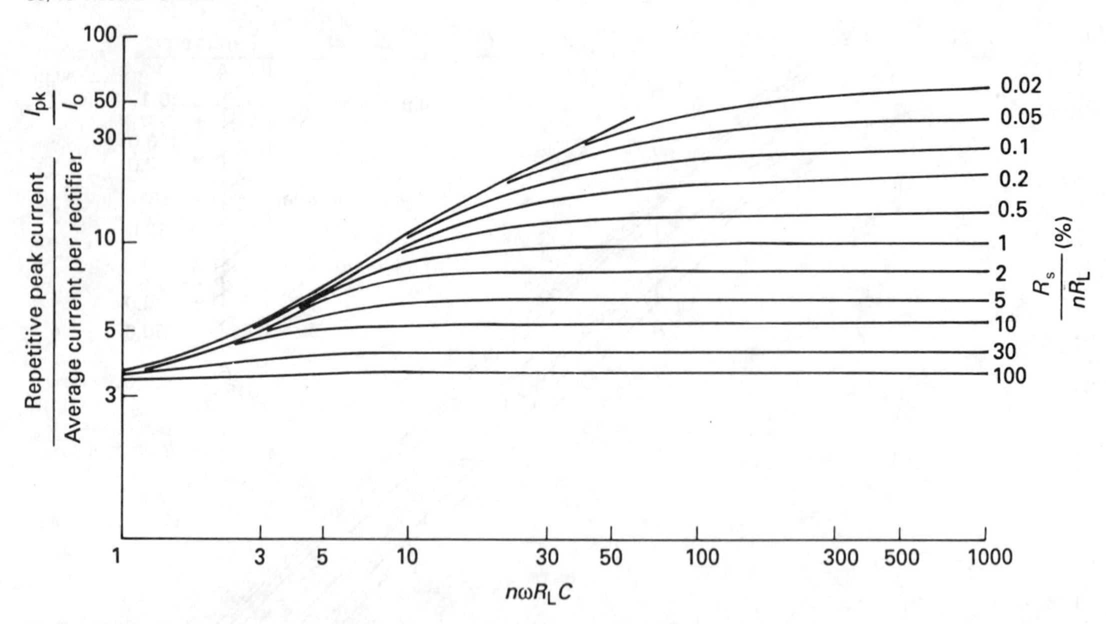

Figure 39.17 The ratio repetitive peak current/average current per rectifier plotted against $n\omega R_LC$. C in farads and R_L in Ω. $\omega=2f$, f is the line frequency. $n=1$ for half-wave, $n=2$ for full-wave and $n=0.5$ for voltage doubler

large value of smoothing capacitor, which has to carry a large ripple current, and large initial and repetitive peak currents flow through the rectifiers. These limitations may be overcome by the use of choke input filters.

The single-phase half-wave circuit (*Figure 39.3*) cannot be used with a choke input filter, as it would require an infinite value of inductance to cause current to flow throughout the cycle.

For the full-wave bridge circuit of *Figure 39.5* and the full-wave centre-tap circuit of *Figure 39.6*, R_s is replaced by a series choke L.

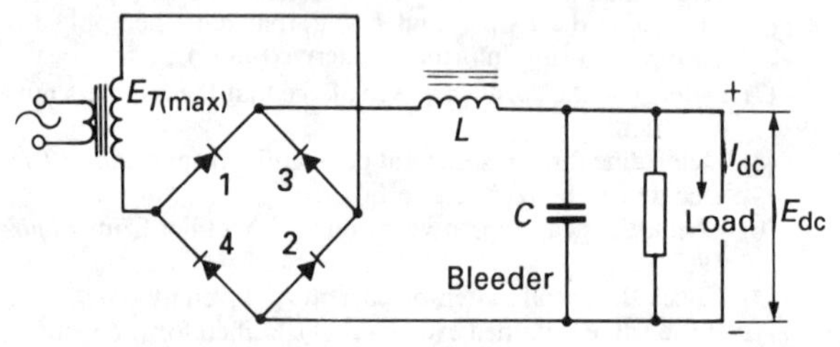

Figure 39.18 Single-phase full-wave bridge circuit with choke input filter

The full-wave bridge circuit with choke input filter is shown in *Figure 39.18*, and the voltage and current waveforms in *Figure 39.19*. The action of the choke is to reduce both the peak and the r.m.s. value of current and to reduce the ripple voltage. The choke input filter circuit, however, requires a higher applied voltage than the capacitor input filter circuit, to produce the same output voltage.

39.5.1 Smoothing circuit

The choke input filter must, ideally, pass only one frequency, which is zero, and attenuate all others. The filter must allow direct current to flow to the load without much power loss, and at the same time present a high impedance to the fundamental and

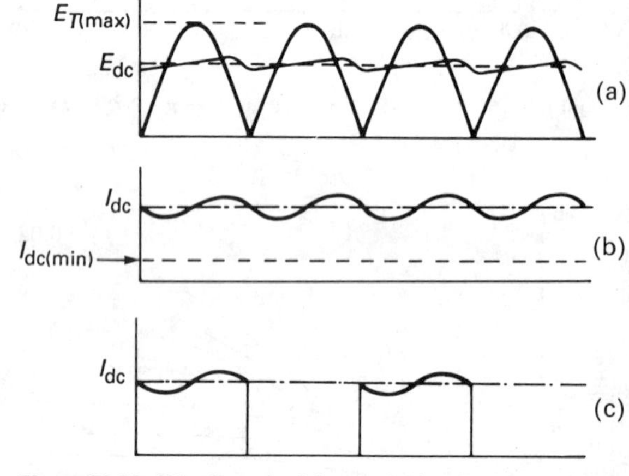

Figure 39.19 Waveforms for full-wave bridge circuit with choke input filter (a) output voltage, (b) current through choke, (c) current through rectifiers 1 and 2 or 3 and 4

other ripple frequencies. The capacitor shunts the load so as to bypass the harmonic currents.

The attenuation factor K of the filter with series choke L and shunt capacitor C is defined as the ratio of the total input impedance of the filter to the impedance of the parallel combination of the shunt capacitor C and load R_L. For the choke input filter to function efficiently, the choke reactance at fundamental ripple frequency f_r should be much greater than its d.c. resistance, and the capacitor reactance much lower than the minimum load resistance. If it is assumed that, the inductance of the choke being L,

$$2\pi f_r L \gg \text{choke resistance } R_L$$

and

$$\frac{1}{2\pi f_r C} \ll R_{L(min)}$$

then

$$K=\frac{2\pi f_r L-(2\pi f_r C)^{-1}}{(2\pi f_r C)^{-1}}$$

therefore

$$K=4\pi^2 f_r^2 LC-1 \tag{39.2}$$

The value of the inductance L used in the circuit must be such as to allow the rectifiers to conduct over one cycle of the fundamental ripple frequency. If the rectifier conducts for a period less than this, then the choke input filter will behave more and more like a capacitor input filter. This will give rise to a higher repetitive peak current through the rectifiers, and will also result in poor regulation.

The use of sufficient inductance allows the rectifier to conduct over the complete cycle; whereas the capacitor input filter allows the rectifier to conduct over only a fraction of a cycle. It follows that, for a given current, the rectifier will switch off before the cycle is completed, for a certain value of inductance. This value is termed the critical inductance L_{crit}.

39.5.2 Output voltage

Consider the single-phase full-wave bridge circuit shown in *Figure 39.18* and the waveforms shown in *Figure 39.19*. The rectified voltage is applied to the choke input filter. This voltage may be expressed as a series containing a d.c. component and harmonic components. The crest value of the output voltage is E_{max}, and it is equal to $E_{T(max)}$ in this circuit.

The rectified voltage can be approximated to a d.c. term plus a harmonic at the fundamental ripple frequency, assuming that the amplitudes of the higher harmonics are negligible. Therefore,

$$e\simeq\frac{2}{\pi}E_{max}-\frac{4}{3\pi}E_{max}\cos 2\omega t \tag{39.3}$$

39.5.3 Critical inductance

From *Figure 39.19* it can be seen that, for the rectifier to conduct throughout the fundamental ripple cycle, the negative-going peak ripple current delivered by the rectifier must not exceed the minimum d.c. current, which occurs with a load of $R_{L(max)}$. Thus

$$I_{dc(min)}=\frac{E_{dc}}{R_{L(max)}}=\frac{2E_{max}}{\pi}\frac{1}{R_{L(max)}} \tag{39.4}$$

If $2\pi f_r L \gg R_L$, and

$$\frac{1}{2\pi f_r C}\ll R_{L(min)}$$

$$\text{peak a.c. current}=\frac{4}{3\pi}E_{max}\frac{1}{2\pi f_r L} \tag{39.5}$$

The critical inductance is reached when the peak a.c. current equals the direct current. That is,

$$\frac{4}{3\pi}E_{max}\frac{1}{2\pi f_r L_{crit}}=\frac{2E_{max}}{\pi}\frac{1}{R_{L(max)}}$$

therefore

$$L_{crit}=\frac{R_{L(max)}}{3\pi f_r} \tag{39.6}$$

For 50 Hz supply frequency and full-wave rectification, $f_r=100$ Hz, so that

$$L_{crit}=\frac{R_{L(max)}}{943} \tag{39.7}$$

Because of the approximations made, it is necessary to use a somewhat higher value of inductance than L_{crit}. In practice, it is found that for reliable and satisfactory operation the optimum value of inductance that should be used is twice the value of L_{crit}.

It is obvious from the nature of the circuit that it is not possible to maintain the critical value of the inductance over all values of load current. This would require an infinite inductance at zero load current. Two methods are available to ensure that current flows throughout the cycle, and that good regulation is maintained over a wide range of load currents: the use of a bleeder resistance or a swinging choke.

39.5.4 Bleeder resistance

A bleeder resistance of a suitable value is connected across the shunt capacitor to maintain the minimum current that will satisfy the critical inductance condition, even when no load is connected. The use of a bleeder will prevent the output voltage from rising to the peak applied voltage in the absence of the load.

39.5.5 Swinging choke

The swinging choke method is based on the fact that the inductance of an iron-cored inductor partly depends on the amount of direct current flowing through it. The swinging choke is designed so that it has a high inductance value at low currents, and this decreases as the d.c. current is increased. The use of such a choke is therefore very satisfactory for maintaining good regulation over a range of load current, and it is also more efficient than the bleeder resistance method.

Since the inductance is continually varying with the load current, the ripple voltage is no longer independent of the load current. When using a swinging choke, it is necessary to ensure that the inductance does not fall to a very low value at the maximum load current, as this will lead to high repetitive peak currents. In practice, the inductance at full load L_F should be such that

$$L_F=2R_{L(min)}/943 \tag{39.8}$$

39.5.6 Ripple current and voltage

If $2\pi f_r L \gg R_L$,

$$\frac{1}{2\pi f_r C}\ll R_{L(min)}$$

and

$$\pi f_r L\gg\frac{1}{2\pi f_r C}$$

then

$$\text{r.m.s. ripple current } I_{c(rms)}=\left(\frac{4}{3}\frac{E_{max}}{\pi}\frac{1}{\sqrt{2}}\right)\frac{1}{2\pi f_r L} \tag{39.9}$$

Since

$$E_{dc}=\frac{2}{\pi}E_{max}$$

$$I_{c(rms)}=\frac{\sqrt{2}}{3}E_{dc}\frac{1}{2\pi f_r L} \tag{39.10}$$

% ripple = % ripple before filtering × $1/K$

From *Table 39.1*, percentage ripple before filtering = 47.2%.

From Equation (39.2), if $4\pi^2 f_r^2 LC \gg 1$ then

$$K \simeq 4\pi^2 f_r^2 LC$$

$$\%\text{ ripple} = \frac{47.2}{4\pi^2 f_r^2 LC} = \frac{1.193}{f_r^2 LC} \quad (39.11)$$

For 50 Hz supply frequency and full-wave rectification, $f_r = 100$ Hz, therefore

$$\%\text{ ripple} = 119.3/LC \quad (39.12)$$

where L is in henries and C is in microfarads.

39.5.7 Minimum value of shunt capacitance

In evaluating the percentage ripple and the attenuation factor of the filter, it has been assumed that the reactance of the capacitor at the fundamental ripple frequency is very much lower than the minimum load resistance. In practice, it is found that satisfactory performance is obtained when the reactance of the capacitor is made less than one-fifth the minimum load resistance. That is,

$$\frac{1}{2\pi f_r C} \leqslant R_{L(min)}/5$$

Therefore

$$C \geqslant \frac{5 \times 10^6}{2\pi} \frac{1}{f_r R_{L(min)}} \mu F$$

$$\geqslant \frac{796\,000}{f_r R_{L(min)}} \mu F \quad (39.13)$$

Because of the nature of the circuit, the capacitor will resonate with the inductor at a certain frequency. At this frequency the output impedance will be greater than the capacitor reactance. Therefore, when a nonlinear loading is applied, precautions must be taken to ensure that the output impedance of the filter is small at the load current frequency.

39.5.8 Additional filter sections

When it is required to reduce the ripple voltage across the load to a very low value, a single-stage choke input filter may require large values of inductance and capacitance, which may lead to an uneconomic filter design. In this case, the same results may be achieved by using a multi-stage filter with small value inductors and capacitors. It can be shown that the optimum smoothing is achieved when all stages are identical.

Figure 39.20 shows the attenuation factor K plotted against $f_r^2 LC$ for one-, two-, and three-stage filters. A suitable arrangement can therefore be selected by studying the filter characteristics. For a K factor between 23 and 160, the two-stage filter is the most economic. For K above 160, a three-stage filter is more suitable.

39.6 Three-phase rectifier circuits

There are many advantages in using a polyphase rectifier system when high-power conversion is required. The object is to superimpose more voltages of the same peak value but in different time relation to each other. An increase in the number of phases leads to the following improvements.

(1) Higher output voltage E_{dc} for the same voltage input.
(2) Higher fundamental ripple frequency and lower amplitude ripple voltage.
(3) Higher overall efficiency.

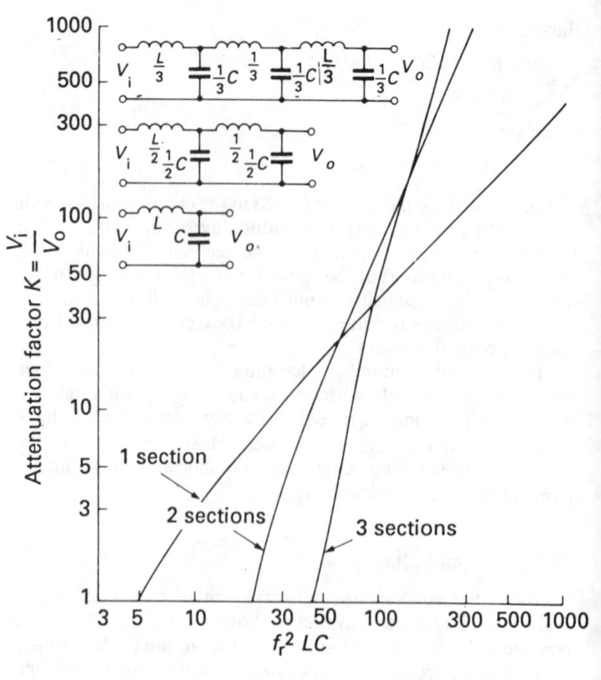

Figure 39.20 Characteristics of choke input filters, L in henries and C in microfarads

39.6.1 Comparison of three-phase circuit performances

Table 39.1 includes the performance of the commonly used three-phase rectifier circuits. In evaluating the results in this table, it has been assumed that the transformer and rectifiers are ideal. The table, however, gives a good indication of the relative merits of the circuits, and may be used to select the best circuit for any particular application. It may also be used for comparing the kilowatts per rectifier available from various circuits. This is best illustrated by an example.

Consider the single-phase and three-phase full-wave bridge circuits, with rectifiers rated at a crest working voltage of 400 V and with a current rating of 20 A. The attainable performances are compared in *Table 39.4*.

From this calculation, it follows that better use of rectifiers is made in the three-phase bridge circuit.

Table 39.4 Comparison of three-phase circuits

	Single-phase bridge	*Three-phase bridge*
Number of rectifiers in circuit from *Table 39.1*	4	6
Output voltage E_{dc}	$\frac{400}{1.57} = 255$ V	$\frac{400}{1.05} = 380$ V
Output current I_{dc}	$2 \times 20 = 40$ A	$3 \times 20 = 60$ A
Power available $E_{dc}I_{dc}$	$255 \times 40 = 10.2$ kW	$360 \times 60 = 22.8$ kW
Kilowatts per rectifier	$\frac{10.2}{4} = 2.55$ kW	$\frac{22.8}{6} = 3.8$ kW

References

1 SCHADE, O. H., 'Analysis of rectifier operation', *Proc. IRE*, **31**, No. 7, 341–361 (1943)

Further reading

CORBYN, D. B. and POTTER, N. L., 'The characteristics and protection of semiconductor rectifiers', *Proc. IEE*, **107**, Part A, No. 33, 255–272 (1960) (Originally Paper 3135U, November 1959)

CROWTHER, G. O. and SPEARMAN, B. R., 'Mains overvoltages: protection and monitoring circuits', *Mullard Technical Communications*, **5**, No. 47, 301–304; **6**, No. 51, 12–21 (1961)

GUTZWILLER, F. W., 'The current-limiting fuse as fault protection for semiconductor rectifiers', *Trans. AIEE* (Part 1, Communication and Electronics), No. 35, 751–755 (1958)

GUTZWILLER, F. W., 'Rectifier voltage transients: causes, detection and reduction', *Electrical Manufacturing*, **64**, No. 12, 167–173 (1959)

ROBERTS, N. H., 'The diode as half-wave, full-wave and voltage-doubling rectifier, with special reference to the voltage output and current input', *Wireless Engineer*, **XIII**, No. 154, 351–362; **XIII**, No. 155, 423–430 (1936)

SAY, M. G., *The Performance and Design of Alternating Current Machines*, Pitman, London (1948) (reprinted 1963)

TOBISCH, G. J., 'Parallel operation of silicon diode rectifiers'. To be published in *Mullard Technical Communications*.

TULEY, J. H., 'Design of cooling fins for silicon power rectifiers', *Mullar Technical Communications*, **5**, No. 44, 118–130 (1960)

WAIDELICH, D. L., 'Diode rectifying circuits with capacitance filters', *Trans. AIEE*, **60**, 1161–1167 (1941)

WAIDELICH, D. L., 'Voltage multiplier circuits', *Electronics*, **XIV**, No. 5, 28–29 (1941)

40 Power Supply Circuits

M Burchall, CEng, FIEE
Head of Engineering
REL Ltd

Contents

40.1 Introduction

A power supply can essentially be considered as a matching device converting energy from a source to that needed by a load usually including some regulation. The source may be a.c. such as 50/60 Hz or 400 Hz mains, in which case rectification is required, or it may be d.c. from a battery, photovoltaic cell, etc., in which case only d.c.–d.c. conversion is required. In most cases where the source is a.c. isolation is required for safety and signal isolation purposes. This isolation is carried out in a transformer working at either the mains frequency or some other internally generated frequency which is usually much higher. In many cases where the source is d.c., isolation is not required and the power supply has to provide only voltage conversion and/or inversion. If isolation is required when operating from a d.c. supply then a transformer is provided with a frequency which is generated internally in the power supply.

In *Figure 40.1* a generalised chain of most of the possible stages from a.c. or d.c. input to d.c. output is shown. After the a.c. input has been isolated and rectified to become d.c. the circuitry becomes common with the d.c. input and all discussion of the following circuit will start from the assumption of an unregulated d.c. input. Some of the circuits such as the linear regulator or the switching regulator can be utilised in different parts of the chain.

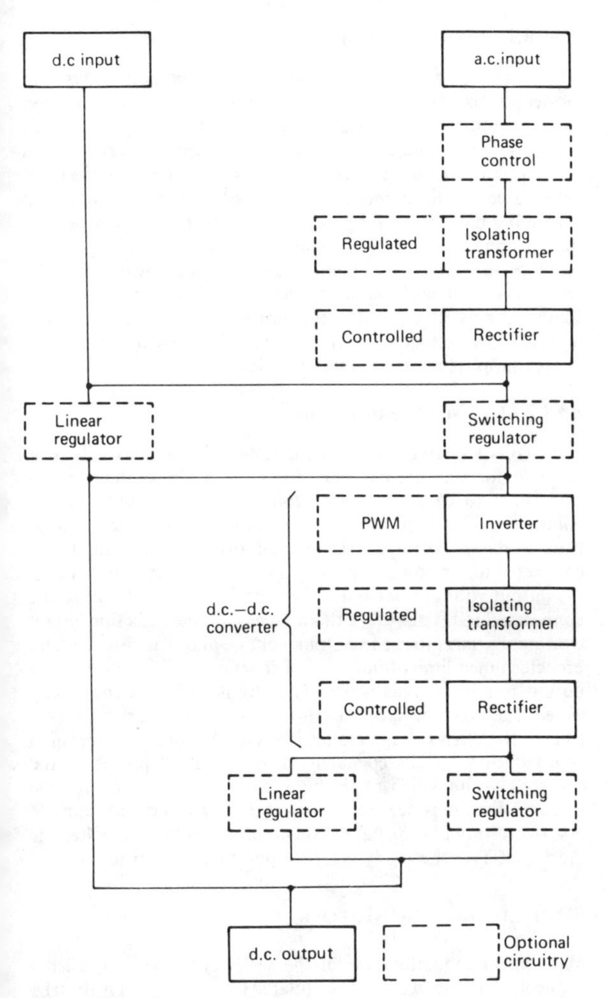

Figure 40.1 Power supply systems

The linear regulator and switching regulator are defined in this context as non-isolated four-terminal networks and the d.c.–d.c. converter as an isolated four-terminal network. There are some exceptions to this general rule in that a d.c.–d.c. converter may be fitted with an autotransformer or have a common path through a d.c. feedback loop but the general principles are still valid.

40.2 Performance

The various factors which affect the performance of a power supply are defined in BS5654 (IEC478).[1] The major items of normal concern are:

(1) Line regulation.
(2) Load regulation.
(3) Temperature regulation.
(4) Ripple and noise.

In a regulated power supply the line and load regulation are controlled by the d.c. loop gain of the feedback loop which can be of any desired level within the bounds of loop stability and bandwidth. Additionally feedforward techniques can be used. Temperature regulation is a function of the quality of the reference voltage, the resistive network sampling the output voltage and the input stages of the feedback amplifier. Ripple and noise are functions, firstly of the type of circuit, e.g. linear or switching and, secondly, the loop gain of the control amplifier at the relevant frequencies and the magnitude and quality of the circuit components.

To a first approximation power supplies can be categorised in four performance brackets (*Table 40.1*).

Table 40.1 Performance of power supplies

Type of power supply	*Total deviation of voltage output (%)*
Unregulated	±10–20
Semi-regulated	±5–10
Regulated	±1–5
Precision	±1

40.2.1 Unregulated power supplies

Unregulated power supplies include all the basic a.c. to d.c. rectification circuits, with or without a transformer, comprising half-wave, full-wave, multi-phase and voltage multiplication circuits. These are described in *Chapter 39* and will not be discussed further except to say that because of their open-loop nature the line regulation approximates to the line voltage variation (usually specified in the range ±6% to ±15%) and the load regulation is highly dependent on the load variation and type of rectification used.

40.2.2 Semi-regulated power supplies

Semi-regulated power supplies fall into two main classes, the phase-regulated thyristor rectified type which is also described in *Chapter 41* and the ferro-resonant transformer[2,3] (CVT) regulated power supply. This is an extension of the unregulated a.c. to d.c. rectifier circuits where the line transformer is replaced by a magnetic regulating transformer which removes most of the input line variation. For fixed load applications this is a very adequate circuit possessing high reliability due to its inherent simplicity. It has been replaced in most applications by the direct-off-line type of switching power supply on the grounds of cost and weight.

40.2.3 Regulated power supplies

Regulated power supplies comprise all types with feedback and/or feedforward to improve line and load regulation. They are either of the linear or switching type or combinations of both and will be discussed in detail in subsequent sections.

40.2.4 Precision regulated power supplies

Precision regulated power supplies are a special group ranging from high-quality general-purpose bench power supplies to the most accurate precision voltage sources which are used for calibration purposes. They are normally linear types with very high feedback loop gains and will not be discussed further.

40.3 Linear or switching power supply

In recent years there has been a significant change in usage away from the linear power supply to the switching power supply[4] as a result of the growth to maturity of the direct-off-line type of switching power supply which is now cost-competitive with linear power supplies down to a few watts. The reasons for this can be summarised as improvements in weight, size and efficiency[5] with some deterioration in the overall performance notably in the area of ripple and noise. For most applications except the most sensitive of linear amplifiers this deterioration is acceptable.

Table 40.2 lists the comparison between linear and switching power supplies for a 5V 20A application. Although the improvement in efficiency from 50% to 75% does not, in itself, look large it leads to a reduction in losses from 100 W to 33 W. It is this reduction, in addition to the reduction in size of magnetic components and capacitors, which enables a switching power supply to be so much smaller than the equivalent linear power supply.

Table 40.2 Comparison between linear and switching power supplies at 5V 20A

	Linear power supply	*Switching power supply*
Efficiency	50%	75%
Power loss (100 W load)	100 W	33 W
Total system power	200 W	133 W
kW/m^3	30	90–180
W/kg	5	15–30

40.4 Protection

It is usual in most power supplies to provide some form of protection against external and internal fault conditions. The most important of these are over-current protection and over-voltage protection.

40.4.1 Over-current protection

If the load current increases beyond its designed level due to some external fault, then, unless the power supply is self-limiting on current (e.g. the shunt regulator, the ferro-resonant regulator and some forms of switching regulators), damage will be caused to the power supply. In the unregulated or phase-regulated supply this protection can be a fuselink, but for most electronic power supplies the protection has to be fast acting and usually consists of extra electronic circuitry which detects the over-current and overrides the voltage control. The three commonest forms of this are:

(1) Current limiting.
(2) Constant current.
(3) Re-entrant or fold-back limiting.

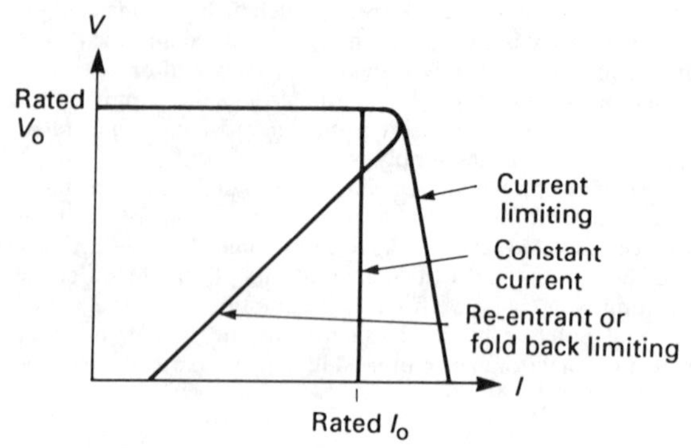

Figure 40.2 Over-current protection

These are shown in *Figure 40.2*. The commonest type of protection is (1). Type (2) is usually found only in precision power supplies with constant voltage constant current (CVCC) characteristics where the constant current performance is comparable with the constant voltage performance. Type (3) is often used on linear series regulators because it reduces the maximum dissipation seen by the series transistors under short-circuit conditions. It is sometimes used with switching power supplies to reduce the magnitude of current flowing into the external circuit fault. It suffers from problems with nonlinear loads (such as tungsten filament lamps) in that it can lock-out at a stable low voltage condition. It can also lock-out when two power supplies are connected in series.

40.4.2 Over-voltage protection

Over-voltages can be caused either internally or externally. It is possible for an internal fault (such as a short-circuit series transistor) to cause a loss of control of the output voltage. Alternatively, it is possible for some external voltage, perhaps from a higher voltage within a multi-voltage system, to be connected to the power supply by an external fault. In either case the output voltage rises and will eventually cause damage to the connected load. To prevent this an over-voltage detection circuit continually monitors the output voltage and if it rises beyond predetermined limits some form of short-circuit is applied to the output terminals. This is usually a thyristor and is commonly called over-voltage crowbar protection. For security of protection the over-voltage circuit is preferably of a two-terminal configuration, i.e. independent of any external power to the control circuits which might be missing at the particular time when protection is needed. *Figure 40.3* shows a typical example of a two-terminal crowbar circuit where the reference voltage is the V_{be} of TR_1 thermally compensated by a thermistor R_3.

40.5 Linear regulators

Most linear regulators are of the series type, i.e. a transistor connected in series with an unregulated d.c. source which absorbs and dissipates the surplus power. The other type of linear

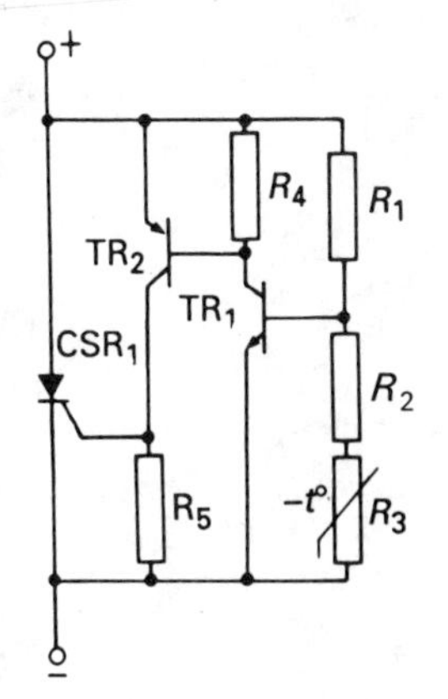

Figure 40.3 Over-voltage crowbar circuit

regulator is the shunt type, but applications for these are mainly confined to low powers because of their inefficiency.

40.5.1 Zener regulator

The zener diode regulator[6] is an open-loop version of the shunt regulator, the simplest form being shown in *Figure 40.4*. To a first approximation the stability ratio S is defined as

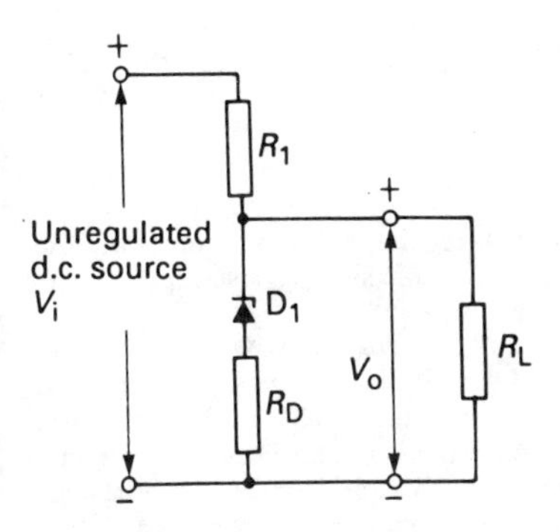

Figure 40.4 Zener diode regulator

$$S = \frac{\%\ \text{input voltage change}}{\%\ \text{output voltage change}} \tag{40.1}$$

$$= \frac{R_1 V_o}{R_D V_i} \tag{40.2}$$

R_D is the incremental resistance of the zener diode at the chosen operating point and is the a.c. or d.c. value depending on the conditions being analysed.

For simple zener diodes where R_D is in the range 5 to 20 Ω values of S of 20 to 50 are typical. Extra stability can be achieved by cascading circuits when the stability ratio increases to the product of the individual stage ratios. With the introduction of IC bandgap references, which have an R_D value of the order of 0.2 Ω, values of S of 500 to 1000 are achievable which are sufficient for most applications.

40.5.2 Shunt regulator

The basic circuit of a shunt regulator is shown in *Figure 40.5*. If the output voltage rises for any reason (such as a reduction in load current) the sampling resistors R_2 and R_3 apply an increasing potential to the base of TR_3. This reduces the potential at the base of TR_2, which turns on transistors TR_2 and TR_1 causing them to draw more current through resistor R_1, counteracting the original tendency for V_o to rise. The output voltage is given by the equation

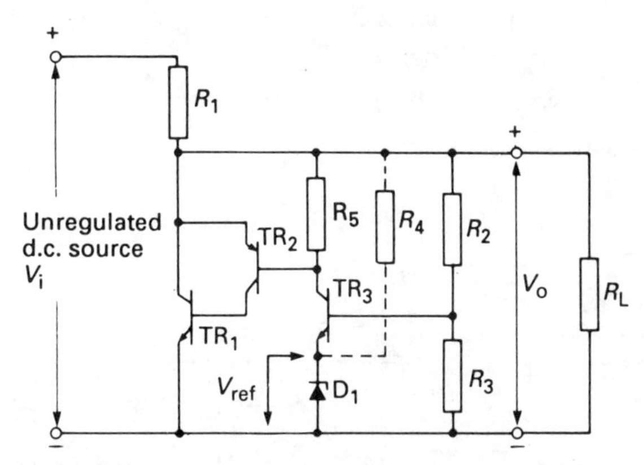

Figure 40.5 Shunt regulator

$$V_o = \frac{R_2 + R_3}{R_3}(V_{REF} + V_{be}) \tag{40.3}$$

Resistor R_4 is usually necessary to provide D_1 with an optimum working current and lower its inherent incremental resistance. D_1 is usually chosen to be around 6.2 V to have a positive temperature coefficient to balance the negative V_{be} coefficient of TR_3. The shunt regulator has one important characteristic not shared by the series regulator in that it can accept current as a sink as well as supply current as a source.

A useful application of this is as an over-voltage protection clamp where the shunt regulator is used to absorb the energy from an over-voltage fault or interference spike in a power system before deciding that the over-voltage is of long duration and applying a crowbar short-circuit in the form of a thyristor. The shunt regulator is also automatically protected against overload currents and needs no extra over-current protection providing the supply resistor R_1 can withstand the extra power of the overloaded condition.

40.5.3 Series regulator

The simplest form of series regulator is shown in *Figure 40.6*. This has a series pass transistor TR_1, a reference source D_1 and a control amplifier TR_2. All other circuits are elaborations of this. As in the shunt regulator, if V_o rises for any reason the potential at the base of TR_2 rises causing it to draw more current through R_1. This lowers the potential at the base of TR_1 causing it to pass less current which counteracts the original tendency for V_o to rise.

The major improvements in this circuit are to increase the current capability of TR_1 by connecting two or three transistors in a Darlington connection, and providing a separate stabilised source for R_1 to provide more gain and independence from variations in the source V_i. This circuit is shown in *Figure 40.7*. This provides a better line and load regulation than the basic circuit. Middlebrook[7] gives a detailed analysis of the

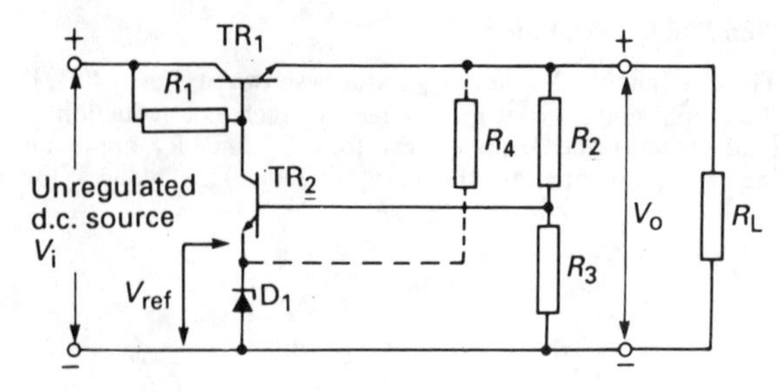

Figure 40.6 Series regulator

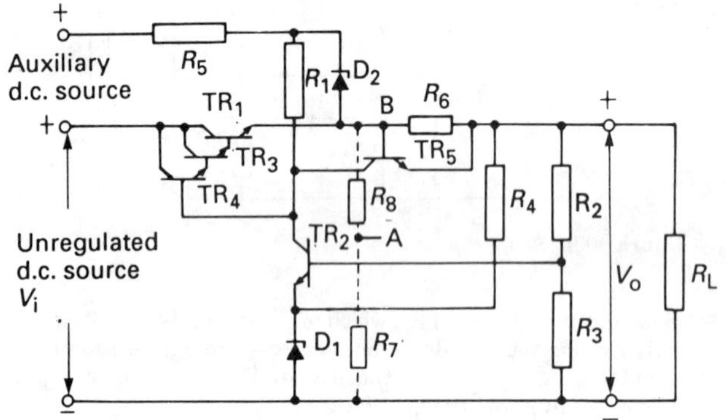

Figure 40.7 Series regulator with current-limiting protection

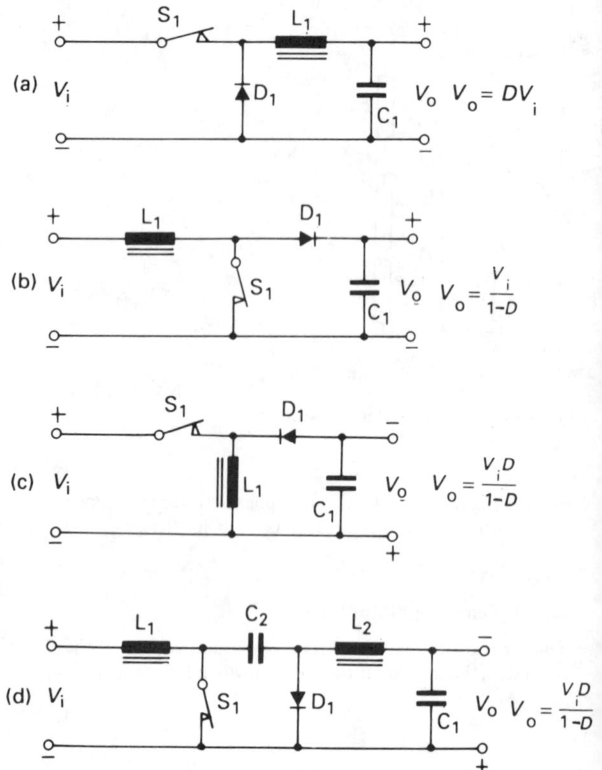

Figure 40.8 Switching regulator: (a) buck, (b) boost, (c) inverting, (d) Euk

performance of this circuit. This circuit also provides current limiting protection. A voltage is developed across R_6 by the load current and when it exceeds the V_{be} of TR_5 it turns TR_5 on and draws current from R_1 diverting it from the series trarsistors. This reduces the output voltage to whatever level is necessary to satisfy the load resistance.

The circuit shown in *Figure 40.7* can be modified for foldback current protection by moving the base of TR_5 from point B to A. Now the voltage developed by the load current has to overcome both the V_{be} of TR_5 and the fraction of the output voltage developed across R_8 by the divider network R_7 and R_8. As the output voltage reduces in the overloaded condition the voltage across R_8 and the amount of current required in R_6 to turn on TR_5 also reduce. This continues all the way down to the short-circuit condition where the load current is determined by R_6 and the V_{be} of TR_5. By the correct selection of R_6, R_7 and R_8 the load current can be made to fold back to any desired short-circuit level.

Series regulators are available in IC and hybrid versions ranging from devices which can provide tens of mA of current through fixed-output TO-3 regulators, to variable three-pin and multi-pin regulators and high power hybrid regulators capable of supplying several amps of current.

40.6 Switching regulators

The two basic switching regulators are the buck (step-down) regulator[8] and the boost (step-up) regulator shown in *Figure 40.8*. They have the generalised property of d.c. voltage conversion where the output voltage can be controlled or varied by modulation of the duty ratio *D*. This is known as pulse-width modulation (PWM). These can be combined as the buck-boost regulator which, by circuit simplification, becomes the inverting regulator; or the boost-buck regulator which becomes the Euk converter.[9] An extensive analysis of various converter topologies and references is given by Severns.[10]

Both the inverting and Euk converters have the extra property of continuous variation of the output voltage above and below the input voltage. The Euk converter also has the benefit of non-pulsating input and output currents and, because the energy storage transfer is by capacitor instead of inductor, can be made smaller and lighter. Capacitors can store many more J/m^3 and J/kg than inductors.

The switching action in *Figure 40.8* can be carried out by the usual switching devices such as a thyristor, GTO[11] (gate turn-off thyristor), bipolar transistor or FET[12] (field-effect transistor).

40.6.1 Buck regulator

A block diagram of a bipolar transistor buck regulator is shown in *Figure 40.9*. This is used to illustrate the principle of pulse-width modulation, which is common to all switching stabilisers and controlled d.c. to d.c. converters. It contains the necessary features of error comparison in amplifier A_1, a ramp generator A_3 (this can be a single or double-sided ramp as convenient) and a modulator in amplifier A_2. Amplifiers A_2 and A_3 need a wide bandwidth to obtain fast switching edges for driving TR_1 with fast 'on' and 'off' transition times particularly at oscillation frequencies above 20 kHz.

If the output voltage increases for any reason the output of amplifier A_1 also increases. This moves the input to amplifier A_2 further up the ramp waveform as shown between *Figures 40.10(a)* and *40.10(b)* and causes the output of A_2 to be high for

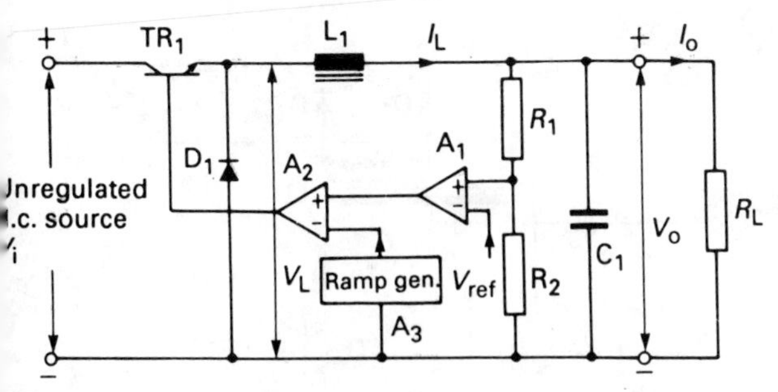

Figure 40.9 Block diagram of a buck regulator

shorter periods of time, thus lowering the duty ratio D of the switch TR_1. This reduces the output voltage to counteract the original tendency to rise.

When transistor TR_1 turns 'on' at time t_0 the current in the inductor rises at the rate

$$\frac{dI_L}{dt}=\frac{V_i-V_o}{L} \tag{40.4}$$

When TR_1 turns 'off' at time t_1 the current in the inductor decays at the rate

$$\frac{dI_L}{dt}=\frac{V_o}{L} \tag{40.5}$$

therefore

$$t_{on}(V_i-V_o)=t_{off}V_o \tag{40.6}$$

where

$$D=\frac{t_{on}}{t_{on}+t_{off}} \tag{40.7}$$

therefore

$$V_o=DV_i \tag{40.8}$$

This is for a theoretically dissipationless system and in practice losses in TR_1, D_1 and L_1 modify this to some extent.

All the above functions of A_1, A_2 and A_3 can be obtained in purpose built ICs.

40.6.2 Boost regulator

The boost regulator will not be examined in detail since the principles are similar to that of the buck regulator, i.e. energy is stored in the input inductor while the switch is closed and released to the output when the switch is open.

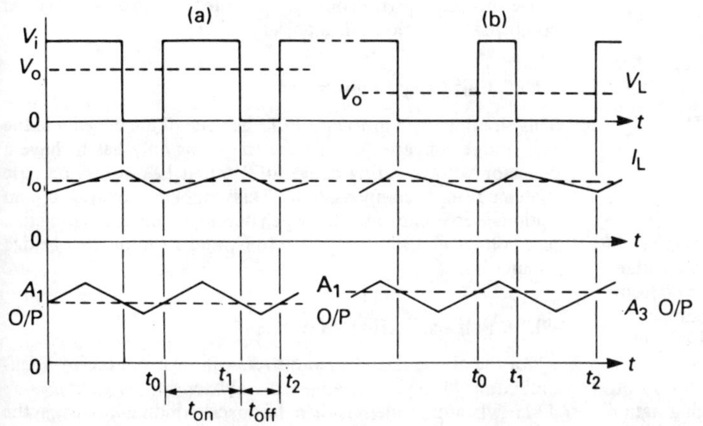

Figure 40.10 Waveforms for a buck regulator: (a) high duty ratio D, (b) low duty ratio D

40.7 DC–DC converters[13–16]

There are five basic types of square wave d.c. to d.c. converter available and these are shown in *Figures 40.11–40.15*. The simplest forms of these are self-oscillating with either voltage or current feedback or both. These give uncontrolled d.c. to d.c. conversion. Externally driven versions can be provided with PWM drive waveforms which give a variable d.c. transformation ratio, in the same manner as the switching regulators, but with input/output isolation.

The two single-transistor converters have poor transformer utilisation since only one half of the B–H loop is used because of the d.c. polarisation. It has been proposed to overcome this by placing a small permanent magnet in series with the magnetic circuit but this has not found any commercial usage so far.

40.7.1 Flyback converter

This is one of the two single-transistor converters and is shown in *Figure 40.11*. The winding polarities are shown by the conventional dot notation. This circuit can be regarded as the isolated version of a boost regulator in that energy is stored (in this case in the transformer primary inductance) during the transistor 'on' time and released to the output during the 'off' time. The extra winding connected through D_1 is only necessary

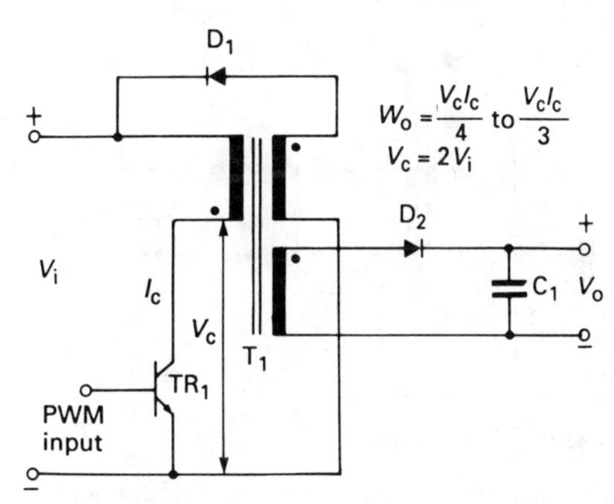

Figure 40.11 Flyback converter

under off-load conditions to return the surplus energy back to the supply and limit the voltage on TR_1 collector to twice V_i.

This circuit is not normally used at powers above 100 W because of the poor utilisation of the peak current capacity of the switching transistor, due to the triangular nature of the collector current waveform, and also because the output capacitor C_1 has to carry ripple currents which are of the same order of magnitude as the output current. These ripple currents can also produce a large output-voltage ripple, not only because of the ramp voltage produced in the capacitance by the charge and discharge currents, but also because of the inherent series resistance and series inductance of the output capacitor. In most practical low-voltage applications the capacitor is of an electrolytic type. The recent availability of ceramic capacitors up to 50 μF capable of carrying large ripple currents has started to revive interest in this circuit because of its inherent simplicity.

The flyback converter is often used for multiple output applications[17] due to the simplicity of coupling extra diodes and capacitors to extra windings. It has superior cross-regulation under these conditions to the forward converter due to the absence of integrating inductors with their attendant resistance.

40.7.2 Forward converter

This is the second of the single-transistor converters and is shown in *Figure 40.12*. In this case power is transferred to the output

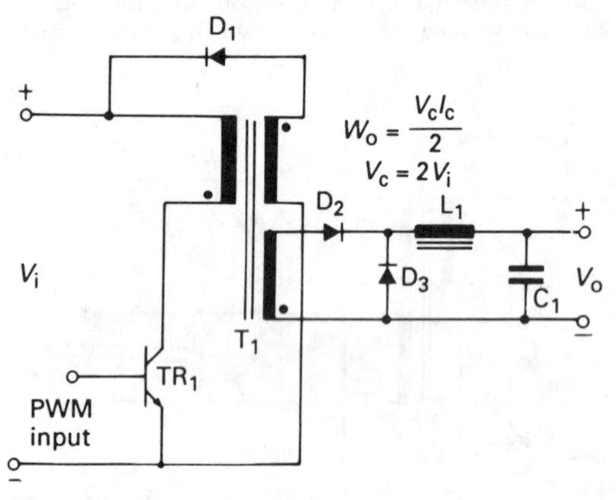

Figure 40.12 Forward converter

winding during the 'on' time of the transistor. The winding connected through D_1 back to the input voltage is now an essential part of the operation and serves to carry the transformer magnetising current while the transformer core re-sets. Since the transistor is now carrying essentially square waves of current the output power can be twice as high as in the flyback converter using a transistor of the same voltage and current ratings. There is no practical limit to the output power of this circuit and multi-transistor and multi-phase versions have been developed up to several kW.

40.7.3 Push–pull converter

This circuit shown in *Figure 40.13* is one of the most popular of the two-transistor converters particularly for low-voltage sources where V_i is below 50 V. This allows transistor collector voltage ratings to be around 100 V giving a good selection of fast switching types. In its self-oscillating form (discussed in *Section 40.7.6*) it is economic and simple. It is not often used where V_i is greater than 300 V, partly because this requires 600 V rating transistors, but also because there is no self-balancing action

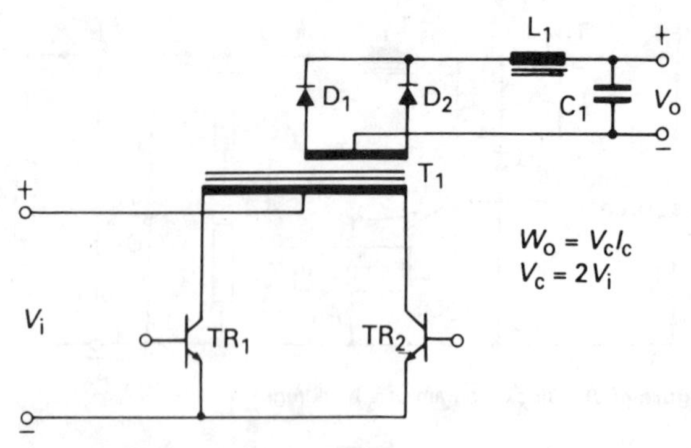

Figure 40.13 Push–pull converter

against asymmetric switching times which cause the transformer flux density to move cycle by cycle into saturation. This saturation is normally followed by the transistor destruction.

40.7.4 Single-ended push–pull or half bridge converter

This circuit, shown in *Figure 40.14*, has two important benefits.

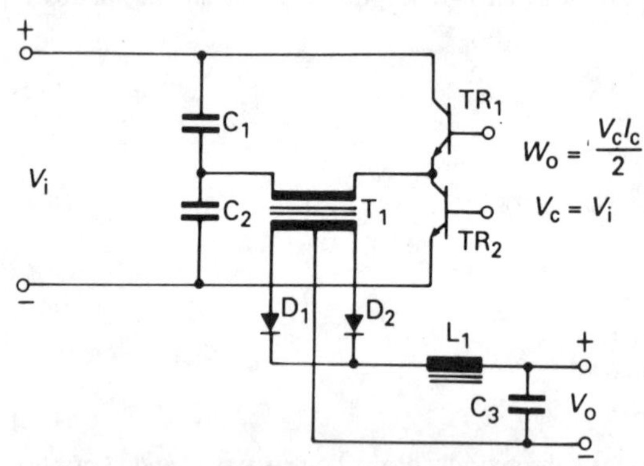

Figure 40.14 Single-ended push–pull or half-bridge converter

Firstly, because the transistors are only switching between the positive and negative of the input source V_i the collector voltage rating can be of the same magnitude as V_i. This is important for direct-off-line power supplies where V_i is the rectified and smoothed mains supply and can reach over 370 V off-load. Secondly, because the output transformer T_1 is connected to the centre point of C_1 and C_2 any direct current flowing in T_1 as a result of asymmetric switching times causes this centre point to move up or down from $\frac{1}{2}V_i$ in such a direction that a compensating action takes place.

40.7.5 Full bridge converter

This circuit shown in *Figure 40.15* has one of the benefits of the half bridge converter in that the transistor only has to have a collector voltage rating equal to V_i but it has no asymmetric switching time compensation. This can be provided by an optional series capacitor C_2 which develops a compensating d.c. bias voltage in a similar fashion to C_1 and C_2 in the half-bridge circuit.

40.7.6 Self-oscillating converters

All of the above d.c.–d.c. converters can be produced in a self-oscillating form. For example the flyback converter of *Figure 40.11* when provided with a feedback winding becomes the

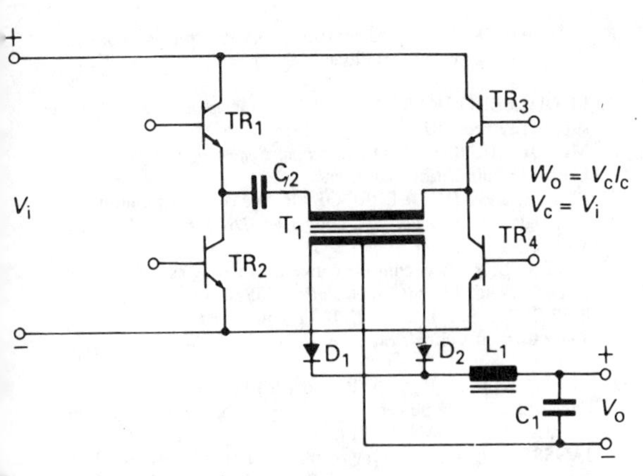

Figure 40.15 Full bridge converter

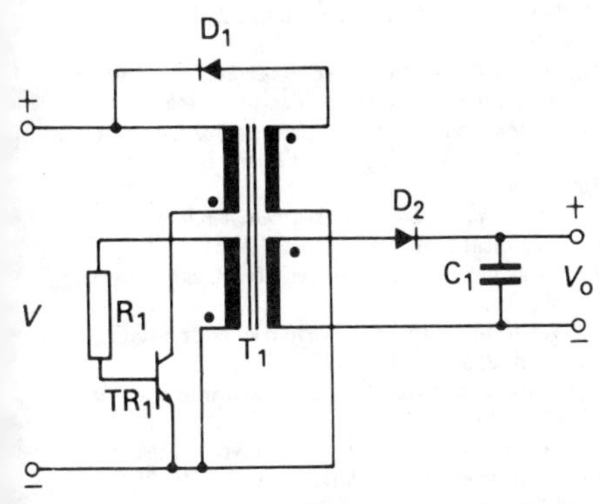

Figure 40.16 Blocking oscillator or ringing choke converter

'ringing choke' or 'blocking oscillator' of *Figure 40.16*. The 'on' period is ended either when the transformer T_1 saturates or when the base current supplied by R_1 is insufficient to maintain the transistor TR_1 in saturation.

Figure 40.17 shows a self-oscillating push–pull converter with both voltage[18] and current feedback. The circuit will work with either type of feedback but has limitations. For example the circuit with voltage feedback tends to be inefficient at partial loads since the drive current has to be designed to be large enough to supply the maximum load condition whereas the circuit with current feedback cannot work at no load since there is no collector current in transformer T_2 primary and, therefore, no available base current. Note that since this is a square wave converter with no PWM the output inductor is not required to integrate the rectified waveform and only a smoothing capacitor C_1 is required.

The oscillation frequency is determined by the saturation of the output transformer T_1 (in the voltage feedback case). This is only practical at low powers since the core losses of a transformer swinging between positive and negative peaks of flux density at a high frequency become unacceptable. At higher powers the frequency is determined by a small auxiliary drive transformer (such as T_2 in the current feedback case) or a small saturating voltage drive transformer.[19] It is also possible for the frequency to be determined by *RC*, *RL* or *LC* circuits in the feedback path or by a saturating base inductor.

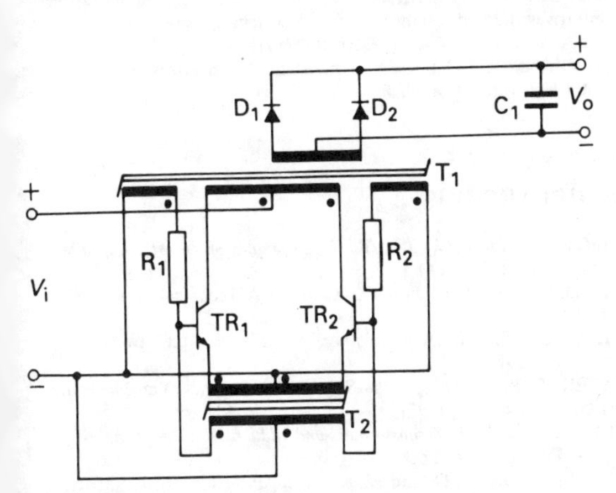

Figure 40.17 Self-oscillating push–pull converter

40.7.7 Non-square wave converters

It is also possible for the d.c. to d.c. conversion process to be carried out in parallel[20,21] or series[22] *LC* resonant circuits. These are similar to those used in d.c. to a.c. inversion except that the oscillation frequency can be optimised at any convenient level instead of at some fixed level, e.g. 400 Hz. There are some benefits to sinusoidal operation mainly in respect of the dV/dt and dI/dt on the switching devices which can be a major source of electromagnetic interference. A typical parallel *LC* converter is shown in *Figure 40.18*.

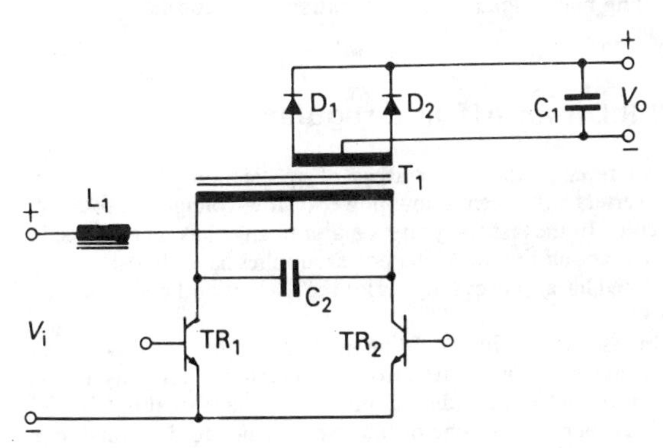

Figure 40.18 Sinusoidal converter

A recent development from these circuits has been the class E converter.[23,24] This circuit aims to improve efficiency by reducing the on-off transition losses of existing square wave and quasi-square wave converters. The class E circuit achieves this by using a shunt capacitor C_1 (including stray capacitance) across the switch TR_1 to delay the rise of switch voltage until the current has fallen to nearly zero at turn-off and by delaying the rise of current at turn-on with series components C_2 and L_2 (including the leakage inductance of transformer T_1). This circuit has been shown to work efficiently up to at least 15 MHz.

The basic circuit is shown in *Figure 40.19*.

Another converter using *LC* circuits in a non-resonant mode but achieving the same results is the Vinciarelli forward

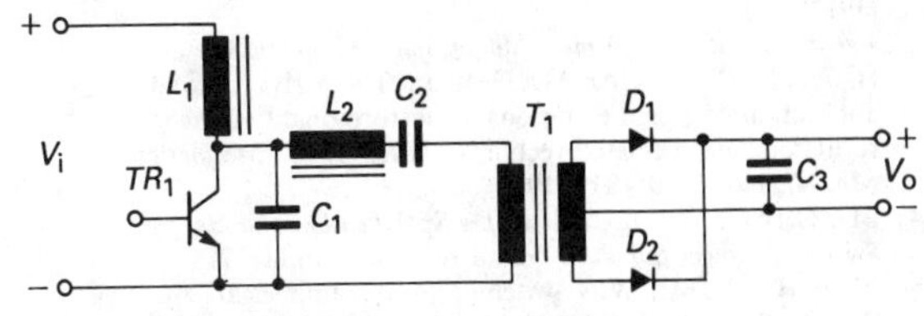

Figure 40.19 Class E converter

converter.[25] DC–DC converters using this circuit are already available commercially working at 1 MHz. The basic circuit is shown in *Figure 40.20*.

Although this circuit has a superficial resemblance to the conventional forward converter of *Figure 40.12*, the introduction of a deliberate amount of leakage inductance in T_1 combined with the correct value of C_1 causes the voltage across TR_1 and the current through TR_1 to adopt semi-sinusoidal wave shapes with nearly zero turn-on and turn-off transition losses.

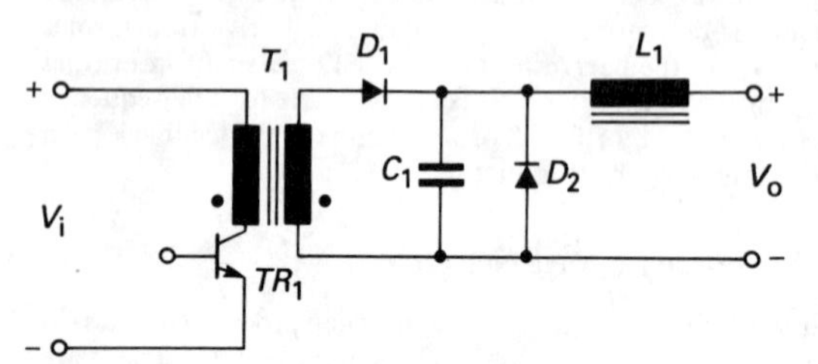

Figure 40.20 Vinciarelli forward converter

Non-square wave converters are particularly useful at very high frequencies, above 100 kHz, where there are practical difficulties in making transformers with low leakage inductance and stray capacity. This is even more so for high voltage power supplies[26] where the winding capacity of the multi-turn output winding makes square wave operation impractical.

40.8 Direct off-line regulators

Until recently the major areas of application of d.c. to d.c. converters have been in low-power or low-voltage (up to 50 V) circuits. In the last few years a major change has taken place in that direct-off-line switching power supplies have almost entirely replaced linear power supplies for the reasons outlined in *Section 40.3*.

In essence a direct-off-line power supply consists of an unregulated supply derived from the a.c. input (normally 115 V, 220 V or 240 V) providing a 'raw' d.c. voltage of about 300 V. This is connected to one of the standard d.c. to d.c. converters detailed in *Section 40.7*, the particular converter being selected according to the voltage, current and power requirements. Regulation against line and load changes can be provided by one of the systems shown in *Figure 40.1*, the commonest being by direct PWM of the d.c. to d.c. converter.[27]

Switching frequencies are normally in the region 20 to 200 kHz although the introduction of the power FET and combinations of FET and bipolar transistors[28] are pushing practical frequencies above 1 MHz, particularly for the non-square wave converters.[29]

References

1 'Stabilised power supplies. D.C. output', BS5654 (1979), IEC478 (1974)
2 *Ferroresonant Transformer Bibliography*, Magnetic Metals Inc.
3 HART, H. P. and KAKALEC, R. J., 'The derivation and application of design equations for ferroresonant voltage regulators and regulated rectifiers', *IEEE Trans. Magnetics*, **MAG-7**, No. 1, 205–211 (1971)
4 BURCHALL, M., *A Guide to the Specification and Use of Switching Power Supplies*, Gould Advance Ltd (1977)
5 BURCHALL, M., 'Why switching power supplies are rivalling linears', *Electronics*, 141–143, September 14 (1978)
6 CHANDLER, J. A., 'The characteristics and applications of Zener (voltage reference) diodes', *Electron. Eng.*, 78–86, February (1960)
7 MIDDLEBROOK, R. D., 'Design of transistor regulated power supplies', *Proc. IRE*, **45**, 11, 1502–1509 (1957)
8 MIDDLEBROOK, R. D., *Switching Regulator Design Guide*, Unitrode Publication No. U-68
9 EUK, S. and MIDDLEBROOK, R. D., 'A new optimum topology switching DC–DC converter', *IEEE Power Electron. Spec. Conf.*, 160–179, June (1977)
10 SEVERNS, R., 'Switchmode converter topologies—make them work for you', *Intersil. Appl. Bull.*, A035 (1980)
11 BURGUM, F., NIJHOF, E. B. G. and WOODWORTH, A., 'Gate turn-off switch', *Electron. Comp. Appl.*, **2**, 4, 194–202, August (1980)
12 CLEMENTE, S., PELLY, B. and RUTTONSHA, R., 'A universal 100 kHz power supply using a single HEXFET', *Int. Rect. Appl. Note*, AN-939 (1981)
13 JANSSON, L. E., 'A survey of converter circuits for switched-mode power supplies', *Mullard Technical Note 24*, TP1442/1 (1975)
14 JANSSON, L. E., 'The design and operation of transistor d.c. converters', *Mullard Technical Communications*, **2**, 17, February (1956)
15 PALMER, M., *The ABC's of DC to AC Inverters*, Application Note AN-222, Motorola Semiconductor Products, Inc.
16 TOWERS, T. D., 'Practical design problems in transistor DC/DC converters and DC/AC inverters', *Proc. IEE*. Pt. B, Suppl. 18, 1373–1383, May (1959)
17 BASELL, M. C., *50 W multiple-output switched-mode power supply*, Mullard Technical Note 33, TP1520 (1975)
18 ROYER, G. H., 'A switching transistor AC to DC converter', *Trans. AIEE*, July (1955)
19 JENSEN, J. L., 'An improved square wave oscillator circuit', *Trans. IRE*, **CT4**, No. 3, September (1957)
20 WAGNER, C. F., 'Parallel inverter with resistance load', *Trans. AIEE*, **54**, 1227–1235 (1935)
21 RIDGERS, C., 'Voltage stabilised sinusoidal inverters using transistors', *Radio Electron. Eng.*, 109–127, August (1967)
22 EBBINGE, W., 'Designing very high efficiency converters with a new high frequency resonant GTO technique', *Powercon 8 Proc.*, A-1, 1–8, April (1981)
23 SOKAL, N. O. and SOKAL, A. D., 'High-efficiency tuned switching power amplifier', US Patent 3 919 656 (1975)
24 REDL, R., MOLNÁR, B. and SOKAL, N. O., 'Class E resonant regulated DC/DC power converters; analysis of operation and experimental results at 1.5 MHz', *IEEE Trans. Power Electronics*, **PE-1**, No. 2, 111–120 (1986)
25 VINCIARELLI, P., 'Forward converter switching at zero current', US Patent 4 415 959 (1983)
26 SCHADE, O. H., 'Radio-frequency-operated high-voltage supplies for cathode-ray tubes', *Proc. IRE*, 158–163, April (1943)
27 WOOD, P. N., *Design of a 5 volt, 1000 watt power supply*, TRW Application Note, No. 122A
28 TAYLOR, B. and FARROW, V., 'A new switching configuration improves the performance of off-line switching converters', *Powercon 8 Proc.*, G-2, 1–7, April (1981)
29 SEVERNS, R., 'The design of switchmode converters above 100 kHz', *Intersil. Appl. Bull.*, A034 (1980)

Further reading

BEDFORD, B. D. and HOFT, R. G., *Principles of Inverter Circuits*, Wiley, New York (1964)
FINK, D. G., *Solid-State Power Circuits*, RCA Technical Series, SP-52 (1971)
FINK, D. G. (ed.), *Electronic Engineer's Handbook*, McGraw-Hill, New York (1975)
HNATEK, E. R., *Design of Solid-State Power Supplies*, Van Nostrand Reinhold, New York (1971)
MCLYMAN, C. W. T., *Transformers and Inductor Design Handbook*, Marcel Dekker, New York (1978)
MIDDLEBROOK, R. D. and EUK, S., *Advances in Switched-Mode Power Conversion*, Teslaco, California (1981)

NIJOFF, E. B. G. and EVERS, H. W., 'Introduction to the series resonant power supply', *Electronic Product Design*, September (1982)
PRESSMAN, A. I., *Switching and Linear Power Supply, Power Converter Design*, Heyden Book Co., NJ (1977)
RODDAM, T., *Transistor Inverters and Converters*, D. Van Nostrand, NJ (1963)
SNELLING, E. C., *Soft Ferrites, Properties and Applications*, Iliffe Books, London (1969)

41 Naturally Commutated Power Circuits

A Clark, DipEE, CEng, MIEE
Electrical Engineering Consultant

Contents

41.1 The principles of natural commutation in rectifier circuits

Natural commutation occurs when the circuit provides the means by which the current is reduced to zero and transferred from one thyristor to another allowing the conducting thyristor to turn off.

The simplest and most widely used method makes use of the alternating voltage waveform to effect the current transfer.

The current in a thyristor can be reduced to zero without interrupting the current in the load by switching on another thyristor connected to a higher-voltage supply source. In *Figure 41.1*, thyristor SCR_2 is triggered whenever voltage V_2 is larger than voltage V_1, the current in thyristor SCR_1 will transfer to thyristor SCR_2 and thyristor SCR_1 will turn off. If the voltages V_1 and V_2 are sinusoidal then everytime voltage V_1 is larger than voltage V_2, thyristor SCR_1 will switch on and resume taking the current.

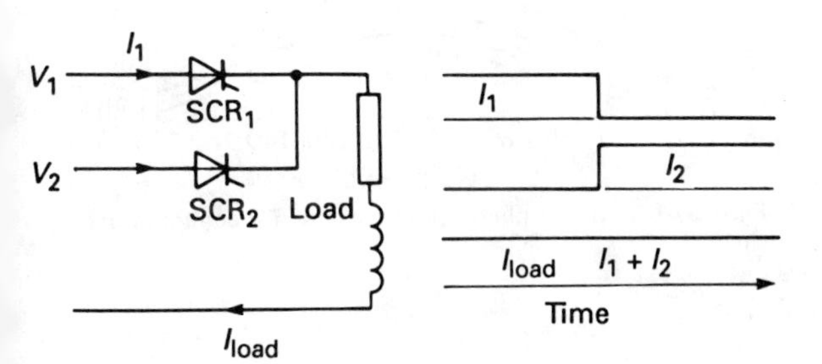

Figure 41.1 Basic circuit for natural commutation

For a simple half-wave circuit as shown in *Figure 41.2*, thyristor SCR_1 is triggered every positive half-cycle of voltage V_1, and the current flows through the load as I_1. When voltage V_2 becomes greater in value, thyristor SCR_2 is triggered and the load current is transferred to it as current I_2. The load voltage is unidirectional and the level is controlled by altering the point at which the thyristors are triggered.

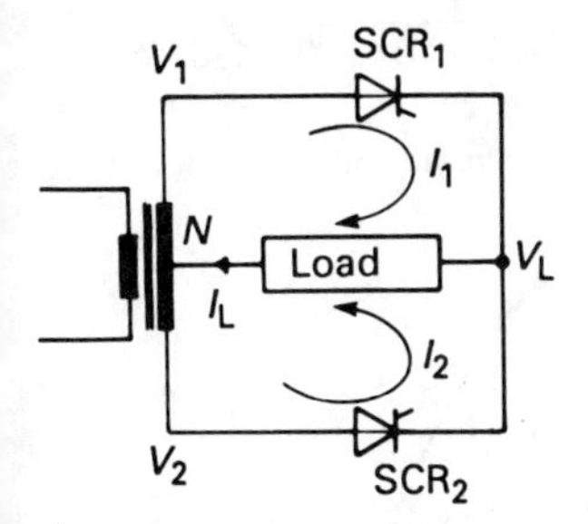

Figure 41.2 Simple half-wave circuit

41.1.1 Waveforms

For a resistive load, the current will cease to flow when the voltage reaches a zero (*Figure 41.3*), but with inductance in the load circuit it will continue for a period after the voltage zero. The period is dependent upon the magnitude of the impedance, and the load voltage will no longer follow the input waveform. As the current becomes continuous the output voltage will remain constant. For a capacitive load, the current will only flow for the period when the d.c. output voltage exceeds the capacitor voltage.

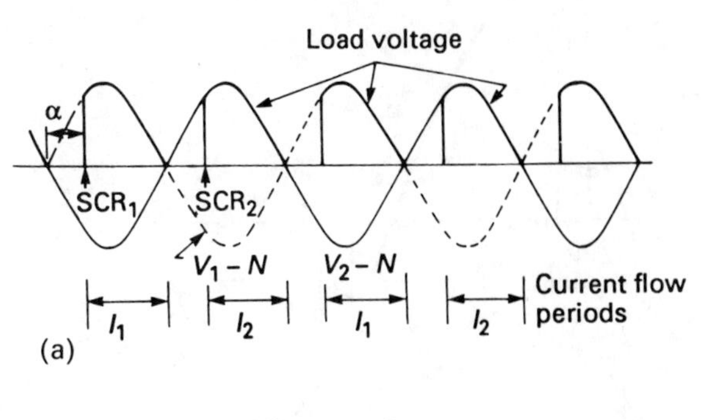

(b)

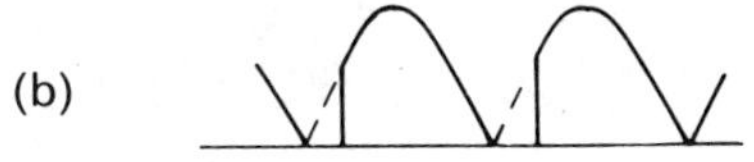

(c)

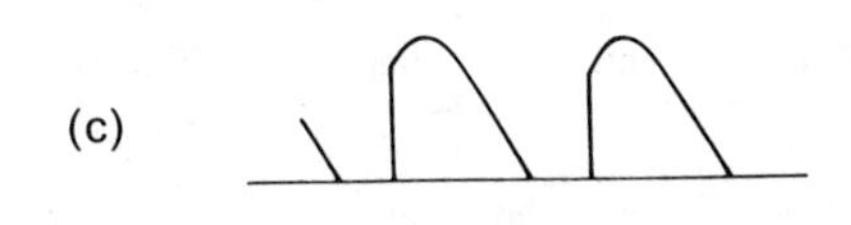

(d)

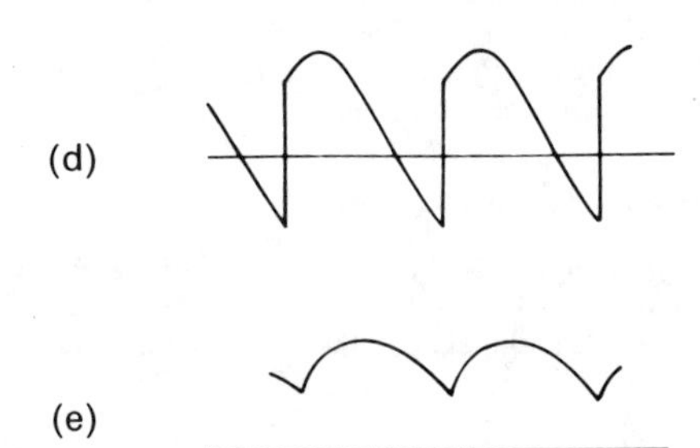

(e)

Figure 41.3 Effect of load inductance: (a) current flow periods, (b) output voltage, resistive load, (c) load current, resistive load, (d) output voltage, inductive load, (e) load current, inductive load

41.1.2 Overlap angle

Due to the stray circuit impedance in each arm, two thyristors can be conducting during the commutation period and this period is called the overlap angle (u). It is dependent on the magnitude of the load current and the value of the transformer leakage impedance. *Figure 41.4* shows the distortion of the a.c. voltage waveform due to overlap.

41.1.3 Regulation

This is defined as the reduction of the output voltage as the load increases. It consists of thyristor and diode voltage drops, and transformer leakage impedance.

41.1.4 AC harmonics

The a.c. current is non-sinusoidal and contains harmonics related to the supply frequency and the circuit arrangement. The

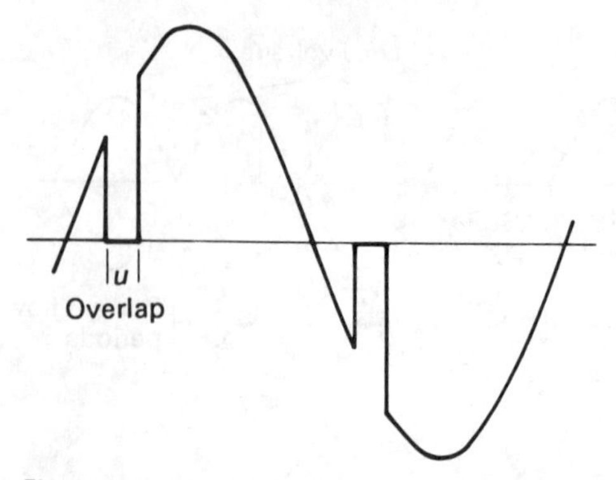

Figure 41.4 Effect of overlap on a.c. voltage waveform

magnitude of these harmonics can be found using Fourier analysis of the waveform.

For smooth d.c. current, and minimum overlap angle, the harmonic content will be constant for all control angles.

$$\text{RMS harmonic} = \frac{100}{m}\,\%\ \text{of fundamental} \qquad (41.1)$$

where m = order of harmonic.

In practical circuits the harmonic content changes with overlap angle and the d.c. ripple content, and changes in circuit impedances will also increase differences.

41.1.5 Power factor of input current

The power factor changes with the delay angle α and the effects of the overlap angle.

For a fully controlled bridge circuit the power factor is given by

$$\cos\varphi = \cos\left(\alpha + \frac{u}{2}\right) \qquad (41.2)$$

The power factor can be measured using the two wattmeter method and is normally defined as the power factor of the fundamental current.

41.2 DC output harmonics

The d.c. waveform of the output for all circuits is easily analysed and the harmonics present in the waveform for each circuit are dependent upon the number of commutations per cycle.

Figures 41.5 to *41.8* illustrate the d.c. harmonic content for continuous d.c. load current.

41.3 Inversion

If the load is sufficiently reactive or has a back e.m.f. such as a d.c. rotating machine, the delay angle can be advanced to 90° where the mean d.c. voltage becomes zero. Providing the current remains continuous the delay angle can be increased beyond 90° and the load voltage will become negative.

With a negative load voltage and the same direction of current flow, the load power is fed back into the supply. The conditions for a single-phase half-wave circuit are shown in *Figures 41.9* and *41.10*. The reverse power flow will continue until the load energy is dissipated.

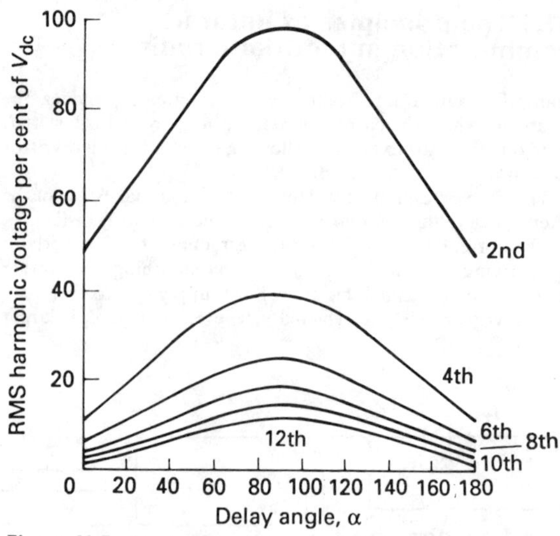

Figure 41.5 Single-phase half-wave and fully controlled bridge circuit

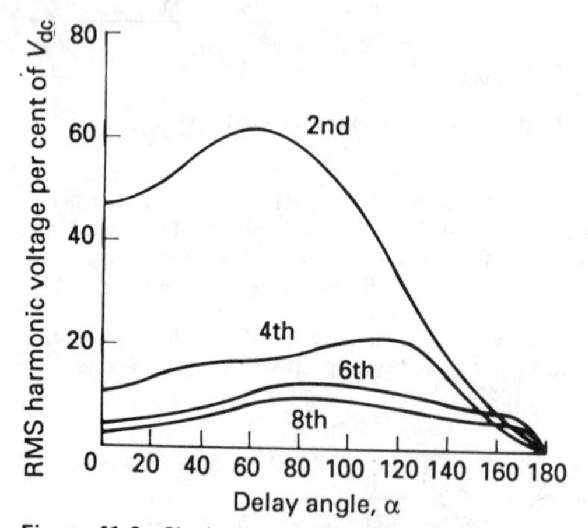

Figure 41.6 Single-phase half controlled bridge

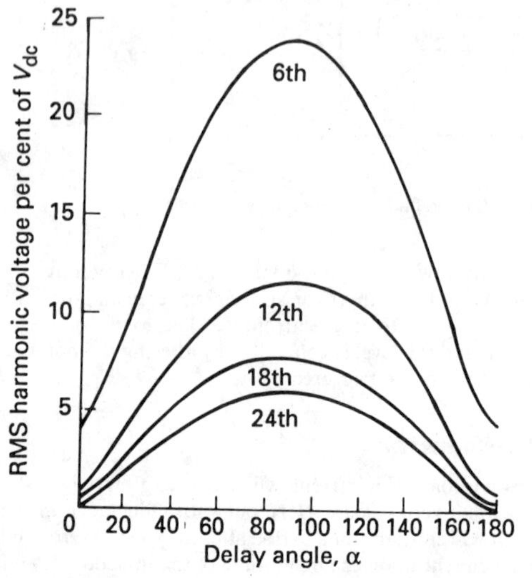

Figure 41.7 Three-phase fully controlled bridge and six-phase half-wave

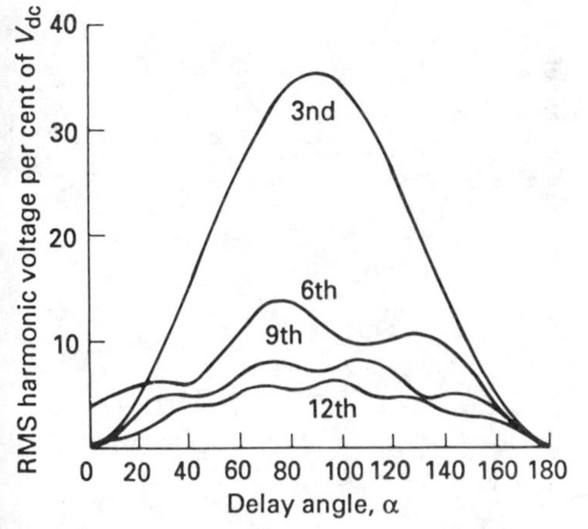

Figure 41.8 Three-phase half controlled bridge

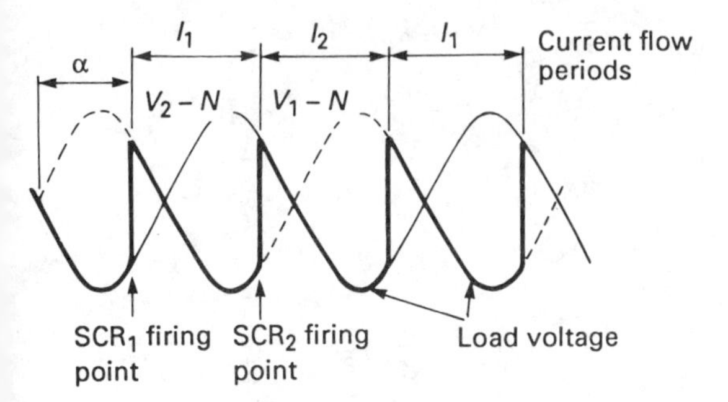

Figure 41.9 Power reversal during inversion

41.4 Circuit parameters for various a.c. configurations

These are shown in *Table 41.1*.

41.5 Typical applications of naturally commutated circuits

A number of circuits are available for controlling a.c. voltages with thyristors. Reverse blocking thyristors may be connected in anti-parallel or a single-phase bridge may be used to rectify the a.c. line current, so that one thyristor can control both halves of the a.c. waveform. This is accomplished by replacing the d.c. load on the single-phase bridge with a short circuit and locating the load in the a.c. line. In some polyphase circuits, the same control of the voltage can be obtained when one thyristor is replaced with a rectifier diode giving a more economical arrangement.

41.5.1 Phase control of the a.c. voltage

When the load is inductive, current flows as a sinusoid which lags the supply voltage by the angle θ, which is a measure of the power factor. If each thyristor is triggered at this angle, the load current will be unaffected. If the triggering angle is made to lag behind θ, the load current will flow in a series of non-sinusoidal pulses of less than 180 electrical degrees duration.

As the angle of phase retard is increased, the pulse becomes increasingly shorter until, at 180 degrees retard, they cease to exist and the voltage across the load is zero. Thus the voltage across the load will be reduced by phase retard in much the same manner as with a resistive load, except that voltage control will take place over a narrower range of triggering angles from θ to 180 degrees. At all triggering angles, the power factor of the load does not depart significantly from the value observed with no

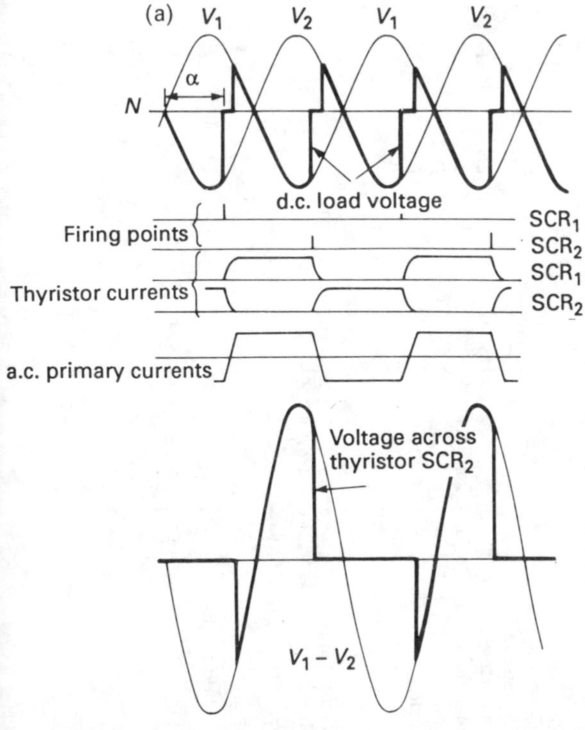

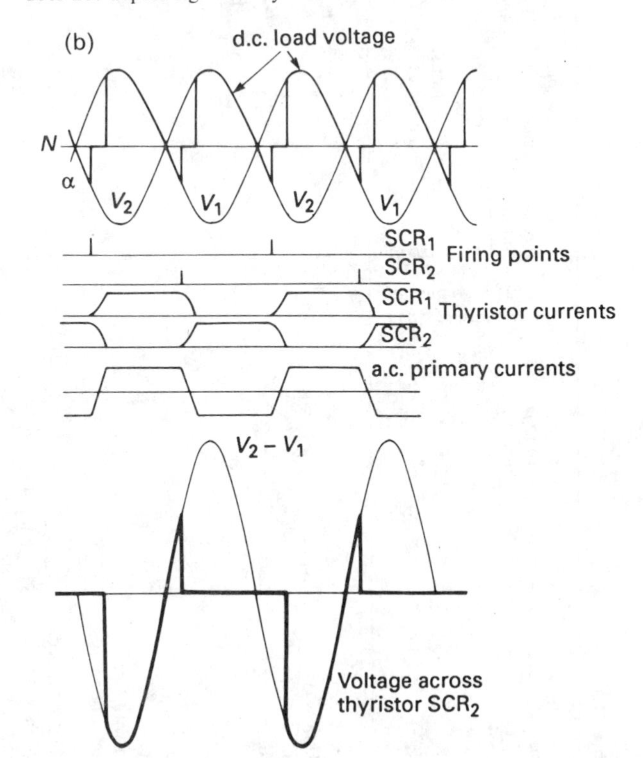

Figure 41.10 Waveforms for single-phase half-wave circuit during inversion, (a) delay angle 120°, (b) delay angle 30°

Table 41.1

Circuit name	Circuit connection	Load voltage waveform	Peak forward voltage on thyristor	Peak reverse voltage: Thyristor	Peak reverse voltage: Diode	Max. load voltage ($\alpha=0$) $E_{dc}=$ av. d.c. $E_{ac}=$ r.m.s. d.c. value	Load voltage (V_S) trigger delay angle (α)	Trigger delay angle range	Maximum thyristor current: I_{AV}	Maximum thyristor current: Conduction angle	Maximum diode current: I_{AV}	Maximum diode current: Conduction angle	Regenerative capability	Ripple frequency of load voltage	Notes
1. Half-wave resistive load	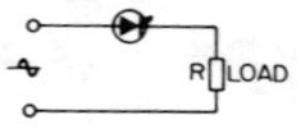	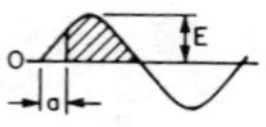	E	E	—	$E_{dc}=\frac{E}{\pi}$ $E_{ac}=\frac{E}{2}$	$E_{dc}=\frac{E}{2\pi}(1+\cos\alpha)$ $E_{ac}=\frac{E}{2\sqrt{\pi}}\left(\pi-\alpha+\frac{\sin 2\alpha}{2}\right)^{1/2}$	180°	$\frac{E}{\pi R}$	180°	—	—	—	f	
2. Half-wave inductive load with free wheel diode	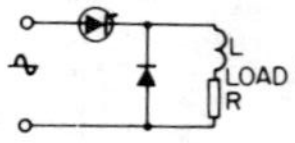	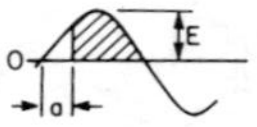	E	E	E	$E_{dc}=\frac{E}{\pi}$	$E_{dc}=\frac{E}{2\pi}(1+\cos\alpha)$	180°	$\frac{E}{2\pi R}$	180°	$0.54\frac{E}{\pi R}$	210°	No	f	
3. Centre-tapped with resistive or inductive load	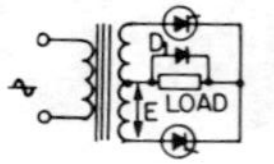	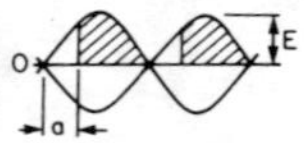	E	$2E$	E	$E_{dc}=\frac{2E}{\pi}$	$E_{dc}=\frac{E}{\pi}(1+\cos\alpha)$	180°	$\frac{E}{\pi R}$	180°	$0.26\frac{E}{\pi R}$	148°	No	$2f$	
4. Centre-tapped with resistive or inductive load with thyristor in d.c. circuit	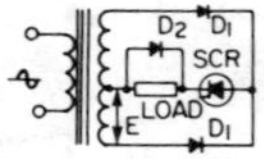	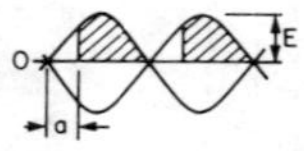	E	0	$2E$ on D_1 E on D_2	$E_{dc}=\frac{2E}{\pi}$	$E_{dc}=\frac{E}{\pi}(1+\cos\alpha)$	180°	$\frac{2E}{\pi R}$	360°	$D_1=\frac{E}{\pi R}$ $D_2=0.26\frac{E}{\pi R}$	180° 148°	No	$2f$	Recovery of thyristor in d.c. load limits frequency
5. Centre-tapped with inductive load	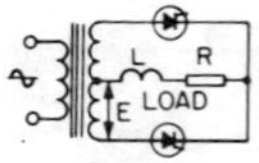	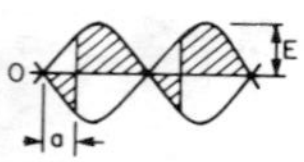	$2E$	$2E$	—	$E_{dc}=\frac{2E}{\pi}$	$E_{dc}=\frac{2E}{\pi}\cos\alpha$ for continuous current	180°	$\frac{E}{\pi R}$	180°	—	—	Yes	$2f$	
6. Hybrid single-phase bridge with free wheel diode	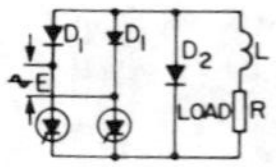	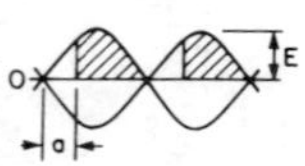	E	E	E	$E_{dc}=\frac{2E}{\pi}$	$E_{dc}=\frac{E}{\pi}(1+\cos\alpha)$	180°	$\frac{E}{\pi R}$	180°	$D_1=\frac{E}{\pi R}$ $D_2=0.26\frac{E}{\pi R}$	180° 148°	No	$2f$	D_2 necessary to ensure turn off on inductive loads

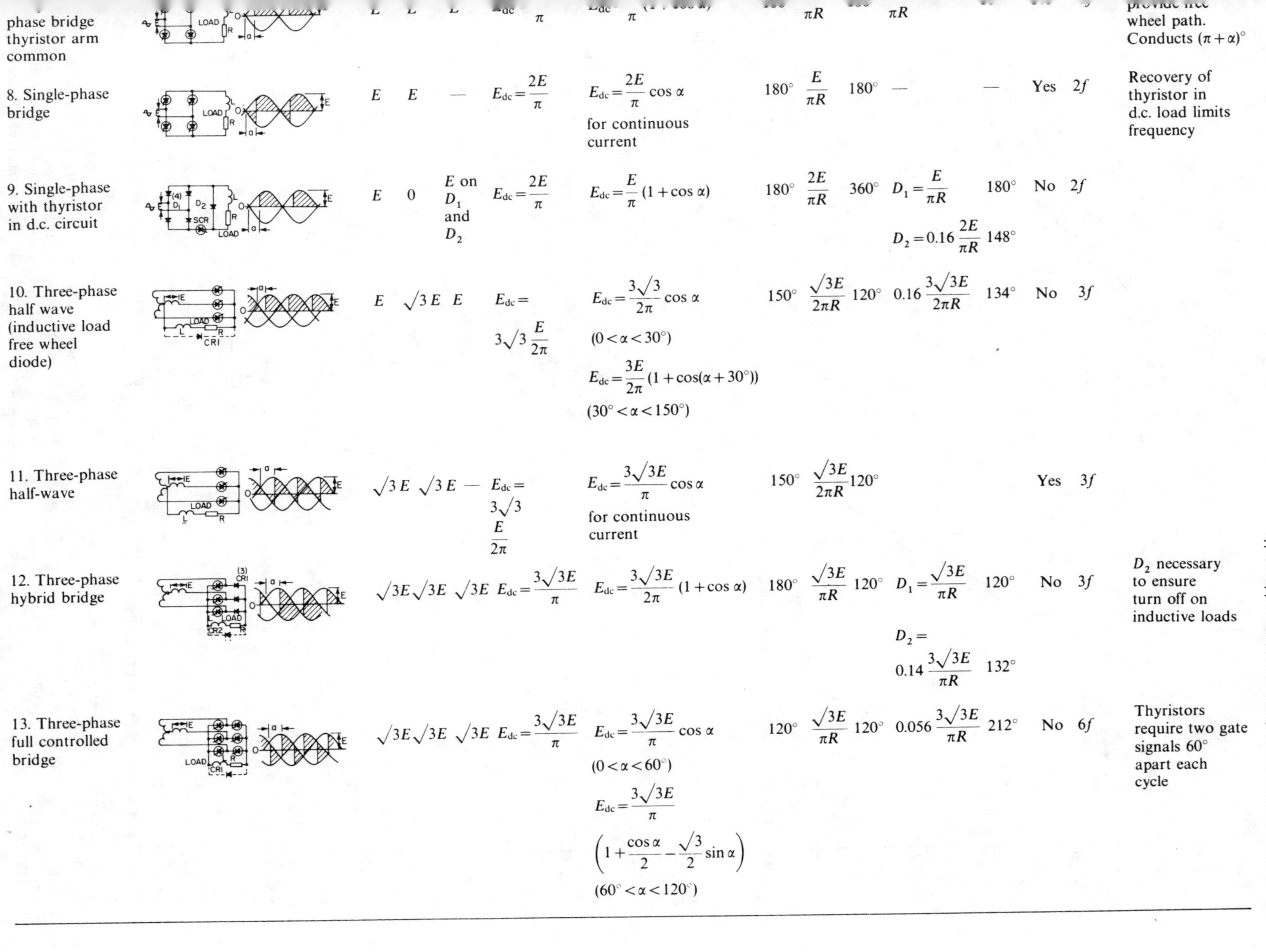

phase bridge thyristor arm common														wheel path. Conducts $(\pi+\alpha)^\circ$
8. Single-phase bridge		E	E	—	$E_{dc}=\frac{2E}{\pi}$	$E_{dc}=\frac{2E}{\pi}\cos\alpha$ for continuous current	180°	$\frac{E}{\pi R}$	180°	—	—	Yes	$2f$	Recovery of thyristor in d.c. load limits frequency
9. Single-phase with thyristor in d.c. circuit		E	0	E on D_1 and D_2	$E_{dc}=\frac{2E}{\pi}$	$E_{dc}=\frac{E}{\pi}(1+\cos\alpha)$	180°	$\frac{2E}{\pi R}$	360°	$D_1=\frac{E}{\pi R}$ $D_2=0.16\frac{2E}{\pi R}$	180° 148°	No	$2f$	
10. Three-phase half wave (inductive load free wheel diode)		E	$\sqrt{3}E$	E	$E_{dc}=3\sqrt{3}\frac{E}{2\pi}$	$E_{dc}=\frac{3\sqrt{3}}{2\pi}\cos\alpha$ $(0<\alpha<30^\circ)$ $E_{dc}=\frac{3E}{2\pi}(1+\cos(\alpha+30^\circ))$ $(30^\circ<\alpha<150^\circ)$	150°	$\frac{\sqrt{3E}}{2\pi R}$	120°	$0.16\frac{3\sqrt{3E}}{2\pi R}$	134°	No	$3f$	
11. Three-phase half-wave		$\sqrt{3}E$	$\sqrt{3}E$	—	$E_{dc}=3\sqrt{3}\frac{E}{2\pi}$	$E_{dc}=\frac{3\sqrt{3E}}{\pi}\cos\alpha$ for continuous current	150°	$\frac{\sqrt{3E}}{2\pi R}$	120°			Yes	$3f$	
12. Three-phase hybrid bridge		$\sqrt{3E}$	$\sqrt{3E}$	$\sqrt{3E}$	$E_{dc}=\frac{3\sqrt{3E}}{\pi}$	$E_{dc}=\frac{3\sqrt{3E}}{2\pi}(1+\cos\alpha)$	180°	$\frac{\sqrt{3E}}{\pi R}$	120°	$D_1=\frac{\sqrt{3E}}{\pi R}$ $D_2=0.14\frac{3\sqrt{3E}}{\pi R}$	120° 132°	No	$3f$	D_2 necessary to ensure turn off on inductive loads
13. Three-phase full controlled bridge		$\sqrt{3E}$	$\sqrt{3E}$	$\sqrt{3E}$	$E_{dc}=\frac{3\sqrt{3E}}{\pi}$	$E_{dc}=\frac{3\sqrt{3E}}{\pi}\cos\alpha$ $(0<\alpha<60^\circ)$ $E_{dc}=\frac{3\sqrt{3E}}{\pi}\left(1+\frac{\cos\alpha}{2}-\frac{\sqrt{3}}{2}\sin\alpha\right)$ $(60^\circ<\alpha<120^\circ)$	120°	$\frac{\sqrt{3E}}{\pi R}$	120°	$0.056\frac{3\sqrt{3E}}{\pi R}$	212°	No	$6f$	Thyristors require two gate signals 60° apart each cycle

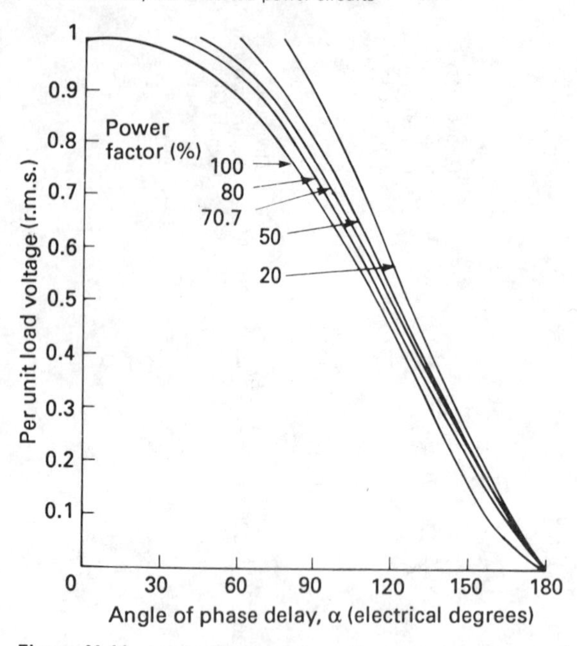

Figure 41.11 Load voltage variation with angle of phase retard

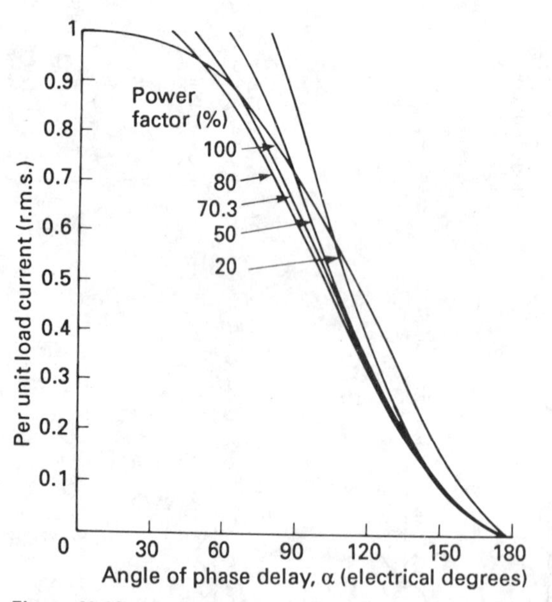

Figure 41.12 Load current variation with angle of phase retard

phase control. Typical transfer characteristics of a.c. phase control circuits are shown in *Figures 41.11–41.13*.

41.6 Gate signal requirements

When thyristors are used to control resistive loads, almost any of the varied forms of triggering circuits may be used with satisfactory results. Synchronization of the triggering pulses may be accomplished from either the line voltage or the voltage across the thyristor.

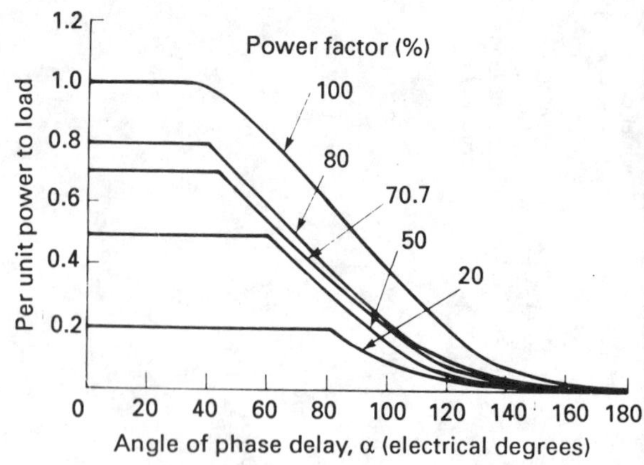

Figure 41.13 Angle of phase retard variation with power to load

However, when the load is inductive, several precautions must be observed in order to achieve optimum performance.

If a triggering circuit is used which produces a narrow spike of gate signal, the charge injected into the thyristor may not be sufficient to maintain the thyristor in the conducting state until the load current has built up to a magnitude larger than the latching current. This may result in mistriggering and erratic control, or no load current whatsoever. One solution is to shunt the inductive load with a small resistive load drawing a current somewhat larger than the maximum value of thyristor holding current. An alternative (and usually more satisfactory) solution is to provide a triggering circuit which produces a square-wave gate signal which lasts from the time at which triggering is initiated until the time when the thyristor conducts a significant amount of current.

For inductive loads, it is mandatory to obtain line synchronisation for the triggering circuit from the line voltage, not from the voltage across the thyristor. If the synchronising signal is obtained from across the thyristor, a large unbalance between the positive and negative half-cycles of current is probable. The result may be overloading of one thyristor, saturation of a transformer core (if a transformer is being controlled), and unstable control. (See *Figure 41.14*.)

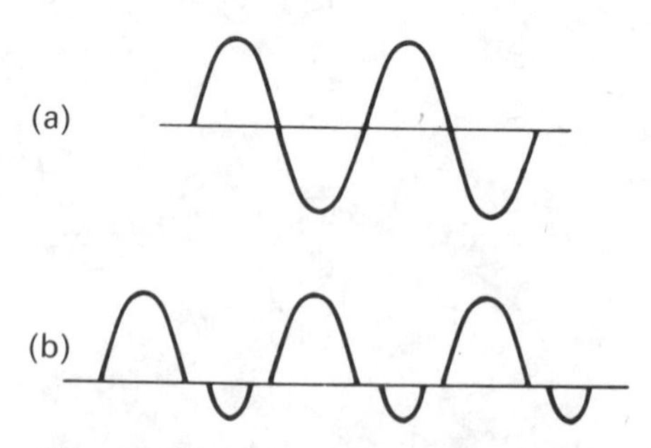

Figure 41.14 Triggering circuit synchronisation: (a) Symmetrical line current when triggering circuit synchronisation is taken from line. (b) Asymmetrical (undesirable) line current which results when triggering circuit synchronisation is taken across *SCR* assembly

For highly inductive loads, triggering the thyristor full on can result in no output if the triggering pulse is so short that it disappears and the thyristor regains its off-state blocking ability before current can flow in the load circuit. To avoid this, it is necessary to make the triggering pulses long enough so that the thyristor will always be able to conduct whenever circuit conditions are right for conduction, once it has been triggered. *Figure 41.15* defines the necessary triggering pulse duration as a function of the load power factor, when the pulse is initiated with zero phase retard.

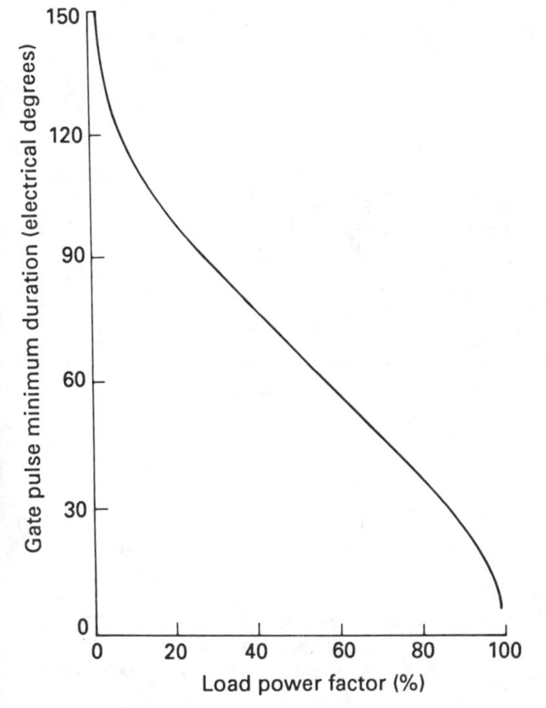

Figure 41.15 Pulse duration variation with load power factor

41.7 Other ways of controlling the a.c. voltage

There are numerous types of equipment or systems which can be controlled readily and advantageously by the use of thyristors as switches (as opposed to using them in the phase control mode for continuous variability of voltage). Of course, many of the switching mode circuits may be readily modified so that they incorporate phase control to provide a fine voltage adjustment to supplement the switching mode of control.

Some of the advantages of the switching mode of operation of thyristors for controlling voltage are as follows:

(1) Switching of load voltage (either from zero to full voltage or from partial to full voltage) is accomplished without mechanical contacts, thus eliminating common maintenance problems which result from burning, pitting and welding of contacts. 'Contact bounce' is also eliminated, thereby reducing radio frequency interference (RFI) caused by the repetitive shock excitation of reactive circuit elements.
(2) 'Zero voltage switching' may be used to essentially eliminate radio frequency interference (RFI) often encountered when voltage is controlled by phase control.
(3) The use of the switching mode of voltage control eliminates the reduction in power factor which inherently occurs when voltage is reduced by phase control.

Many of these switching control circuits will find use in controlling heating elements for ovens, furnaces, hot plates, crucibles, and space heaters. They may also be used for speed controls of squirrel cage and wound rotor induction motors where the motor and/or load inertia is high. Still other types of load, for example resistance welders, stud welders, flashers and flashing beacons, and magnetic hammers or pulsers, demand this type of control as an inherent element in their mode of operation.

Most heating elements have a high thermal inertia and are easily adaptable to control by the switching mode, either by being switched from zero to full voltage or from a low voltage tap to full voltage.

When circuits with a tapped transformer winding are compared to those with a tapped load, it will be seen that the final control result is very similar. However, where the transformer is tapped, the load voltage is initially low and is switched to a higher value. In the tapped load circuit, the initial voltage is high across one section of the load and zero across the other section. After switching, the voltage decreases across one section while increasing across the other. This tapped load 'shunt controller' performs equally well for resistive loads (such as heating elements) and for speed control of wound rotor induction motors, by varying the resistance in the rotor circuit.

Phase control is also very effective (although radio frequency interference filtering may be necessary) using the shunt mode of control. *Figure 41.16* illustrates the control characteristic for both a resistive load and an inductive (70 degree lagging) load. When R_B is small compared to R_A, the range of load current variation is small, but precision of adjustment is very good. When R_B is large compared to R_A, the swing in load current can be very large, but with a reduction in adjustment precision. In either case, the voltage is smoothly adjustable over the design range, and is readily adaptable to automatic control.

In the cases of flashers and beacons, it often becomes possible to substantially lengthen filament life of the lamps by switching from full voltage to a lower transformer tap (reducing the voltage below the incandescence level), thus minimizing the range of filament temperature excursion. A similar improvement in the life of heating elements may also be expected.

To obtain good temperature regulation, or speed regulation, with the switching mode of control, the control system may employ pulse-burst modulation. This mode of control is usually based on a fixed time control period, for example, 20 cycles at power frequency. The number of cycles during this period when the thyristor switch is in conduction is made variable and is adjusted by temperature or speed feedback controls. The switch always conducts for essentially an integral number of cycles, and is either off or conducts at a reduced voltage level for the balance of each period (*Figures 41.17* and *41.18*).

Pulse-burst modulation can be applied to the control of most heating systems and motor drives due to the high thermal or mechanical inertia of these systems. Systems with very high inertia can tolerate relatively long control periods, whereas systems with lower inertia, or requiring a fine control resolution, will demand a short control period.

As mentioned earlier, the use of pulse-burst modulation with thyristor switches will minimise radio frequency interference, but this, by itself, will not completely eliminate it. The fast turn-on of a thyristor when the supply voltage is at any value other than zero, still causes one pulse of shock excitation each time the thyristor is turned on to carry a burst of pulses. However, with thyristor switches, it is possible and often desirable to incorporate zero voltage switching. When controlled in this mode, the thyristors are always turned on at zero voltage (they turn off at zero current), thus eliminating the interference caused by fast switching of heavy currents.

All of the circuits which switch from one tap to another (either tapped transformer or tapped load) may incorporate phase

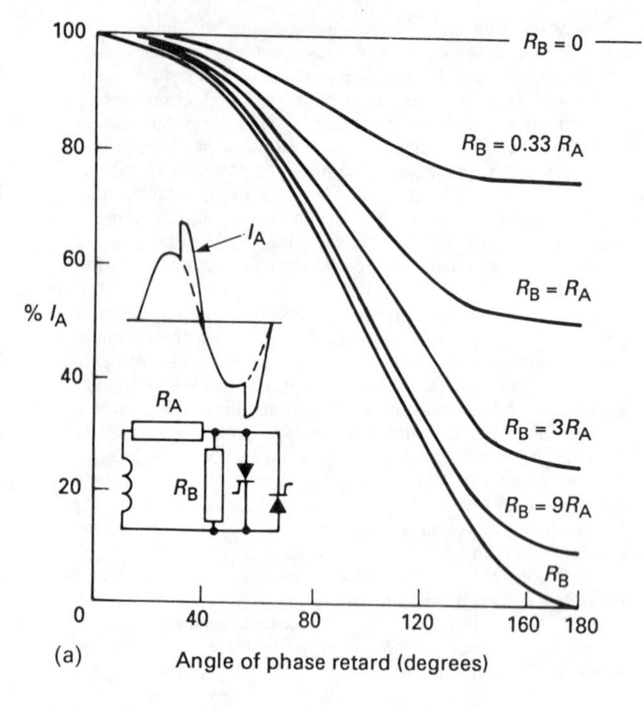

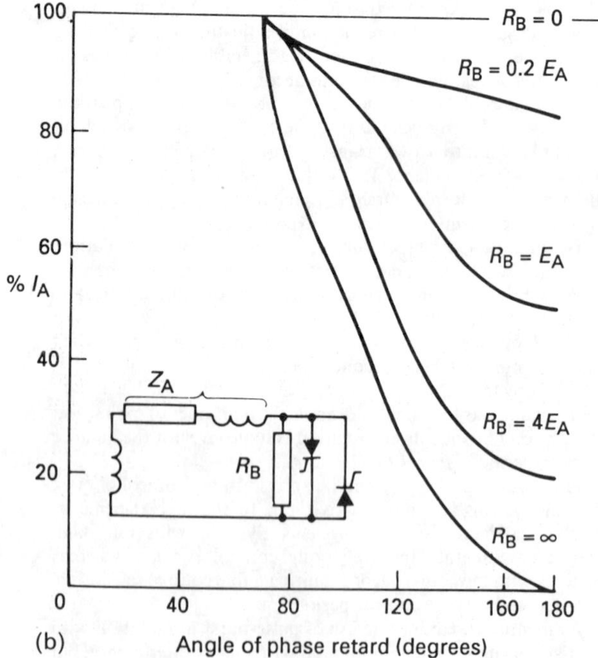

Figure 41.16 *R*C control using shunt thristor: (a) resistive load, (b) inductive load

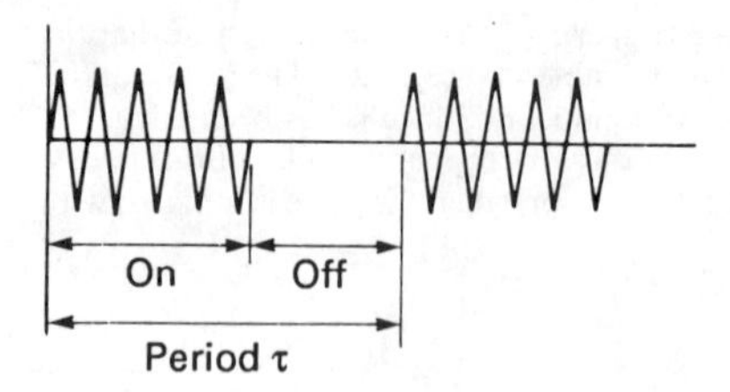

Figure 41.17 Pulse burst modulation voltage control—full on to off

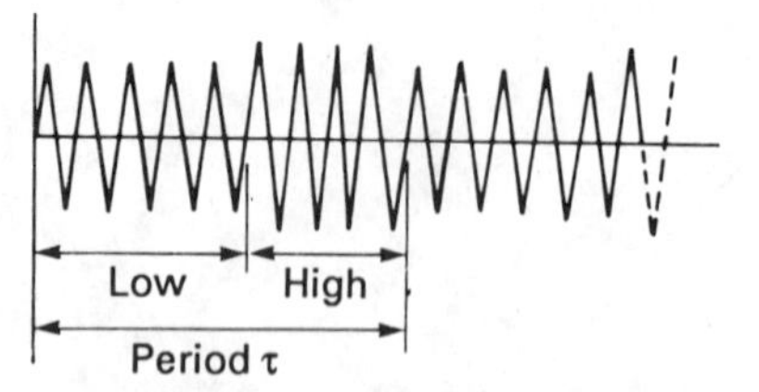

Figure 41.18 Pulse burst modulation voltage control—transformer tap switching

control to supplement the switching type of control to obtain voltage or current regulation of power supplies. Methods of applying phase control have been discussed earlier in this chapter. It is possible to achieve such regulation by phase control alone and eliminate the taps. However, where only a limited range of adjustability is required, the combination of phase control with switching mode operation greatly reduces the peak-to-RMS current ratio in a.c. power supplies. For instance, output voltage can be smoothly varied between the voltages obtained from any two transformer taps by first gating a thyristor connected to the lower voltage tap, and then later in the cycle gating a thyristor connected to the higher voltage tap. This type of voltage control reduces the peak-to-average current ratio. It also minimises the change in power factor as voltage is varied.

41.8 The cycloconverter

A cycloconverter is a means of changing the frequency of alternating power using thyristors which are commutated by the a.c. supply. It can be used as an alternative to a rectifier followed by an inverter.

Cycloconverters utilise any fully controlled, naturally commutated rectifier circuits, using a number of separate circuits in series or parallel. Thyristor ratings will follow the normal practice for mains connected naturally commutated circuits. Anti-parallel operation using two converters in each phase is preferable allowing freedom of choice on the connection of the separate converters employed in the circuit. (Detailed circuit and mathematical analysis is given in References 1, 2 and 3.)

41.8.1 Basic concept of a single-phase circuit

Figure 41.19 shows the circuit discussed in this section. The arms of a single-phase full-wave rectifier circuit are replaced by two pairs of thyristors connected in anti-parallel which can be triggered to produce opposite polarities. When thyristors SCR_1 and SCR_3 are triggered the d.c. output is sensed in one direction, and reversed when thyristors SCR_2 and SCR_4 are triggered instead. By alternatively triggering each pair of thyristors at a frequency lower than the supply frequency a square wave of current is produced and flows in the load. A filter would be required to reduce or eliminate the ripple produced.

A crude sine wave can be synthesised by triggering the individual thyristors at different delay angles.

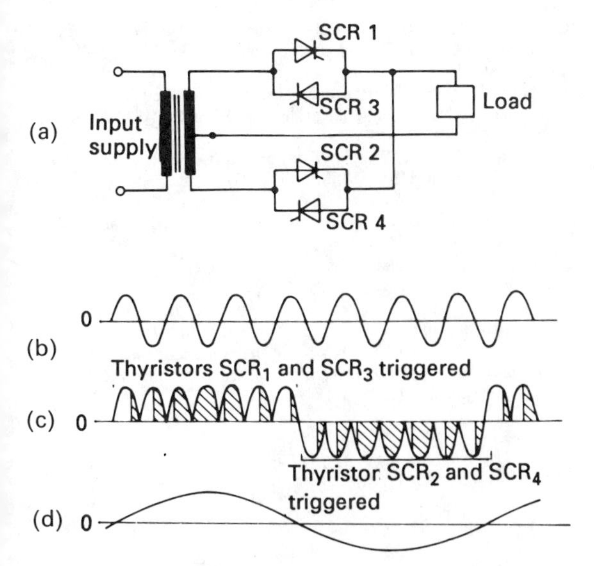

Figure 41.19 Simple single-phase cycloconverter: (a) circuit diagram, (b) input waveform, (c) output waveforms, (d) output waveform after filtering

41.8.2 Practical single-phase circuit

Figure 41.20 shows the circuit discussed in this section. To achieve an improved output waveform shape, two three-phase fully controlled bridges are connected in anti-parallel. This arrangement of the bridges is capable of feeding power to the load

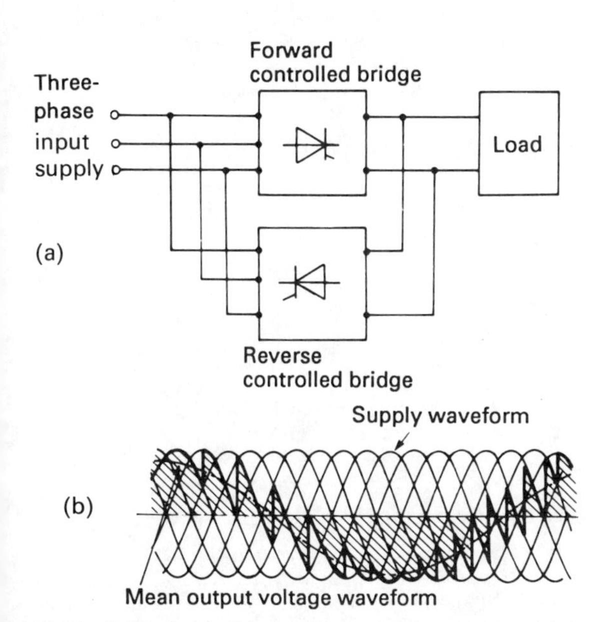

Figure 41.20 Practical single-phase cycloconverter: (a) current diagram, (b) basic output waveform

or returning power to the supply from the load. Both bridges produce full voltage control in either direction and, provided the current is correctly sensed, any combination of output voltage and current is possible.

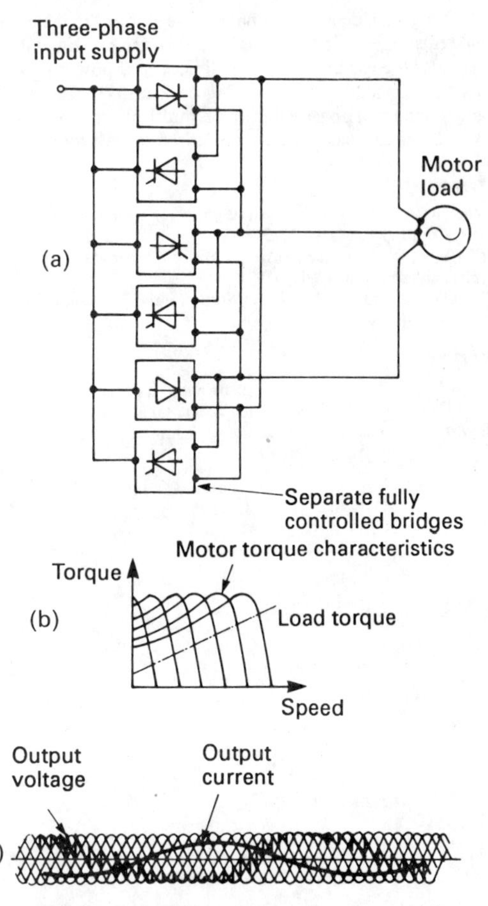

Figure 41.21 Typical three-phase cycloconverter driving a motor load: (a) circuit diagram, (b) load characteristics, (c) output waveforms

41.8.3 Practical three-phase circuit

Figure 41.21 shows the circuit discussed in this section. Three single-phase cycloconverters are usually connected in star or delta to produce a three-phase output, the choice of connection depending on the load characteristics. An example of a delta-connected motor load is illustrated, the circuit producing a wide range of frequency and voltage control to enable the highest torque to be maintained over the speed range. The circuit provides a.c. power in both directions as required by the load.

41.8.4 Characteristics of cycloconverters

(1) A large number of thyristors are required for practical circuits. Three-phase circuits usually need a minimum of 18 devices.
(2) They can only be operated at sub-supply frequency.
(3) Output voltage can be characterised up to maximum cycloconverter voltage. In excess of this level the waveform shape is trapezoidal without voltage control. On motor loads, the input power factor can be improved at the expense of additional harmonic content.
(4) The input power factor is low for most circuits and changes with variation of output frequency. For motor load, output voltage is proportional to frequency to maintain good speed control parameters. There is a low power factor at low speed.

(5) Although cycloconverters have more complicated electronic systems, the output is easily characterised by changing signals to give changes in the direction of power flow from feeding the load to regenerating back onto the supply lines and changes of phase rotation on multi-phase types. Output waveform can easily be characterised to suit load.

References

1 McMURRAY, W., *Theory and Design of Cycloconverters*, MIT Press, Cambridge, Mass. (1972)

2 PELLY, B. R., *Thyristor Phase Controlled Converters and Cycloconverters*, John Wiley (1971)

3 GYUGYI, L. and PELLY, B. R., *Static Power Frequency Changers*, John Wiley (1976)

Further reading

AD'DOWEESH, K. E., 'Microprocessor based a.c. voltage control using GTO thyristors', *Proc. ISMM Symposium Mini and Microcomputers and their Applications*, 29 June–1 July (1987)

BOWLER, P., 'The application of a cycloconverter to the control of induction motors', *IEE Conf. Publication* No. 17, 137–145, November (1965)

HIRANO, M. *et al.*, 'Simultaneous phase and frequency control of a three phase inductor converter bridge', *PESC*, **185**, 24–28 June (1985)

JARL, R. and FRANK, W., 'Thyristor gating ICs', *Power Converters and Intell. Motion* (USA), **13**, March (1987)

MAZDA, F. F., *Thyristor Control*, Newnes-Butterworths (1973)

NISSIERE, C., 'Phase control circuits of triacs', *Elettron. Oggi* (Italy), January (1987)

READ, J. C., 'The calculation of rectifier and inverter performance characteristics', *IEE Proc.*, **92**, 29 (1945)

SALZMAN, T., 'Cycloconverters and automatic control of ring motors driving tube mills', *Siemens Rev.*, **45**, 3–8 (1978)

SCHOTT, W., 'Rectifier converter using thyristors and the TCA 785 integrated phase control', *Siemens Components*, **20**, Nos 4 & 5, August and October (1985)

42 Forced Commutated Power Circuits

F F Mazda, DFH, MPhil, CEng, FIEE, DMS, MBIM
STC Telecommunications Ltd

Contents

42.1 Introduction

The design aspects of d.c. to d.c. converters (choppers) and d.c. to a.c. converters (inverters) are considered in this chapter. The effect of forced commutation circuits, as required by thyristors, is initially ignored so that the results are equally applicable to other forms of switching, such as using power transistors.

42.2 DC–DC converter without commutation

42.2.1 Voltage control methods

Figure 42.1 shows a simplified circuit arrangement of a chopper, and the waveforms for mark-space and frequency control. With

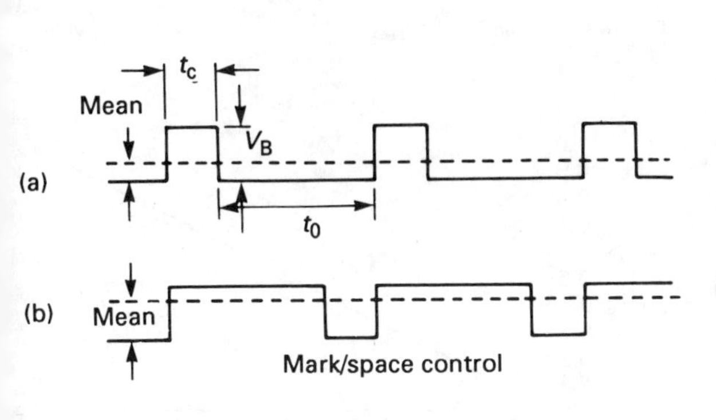

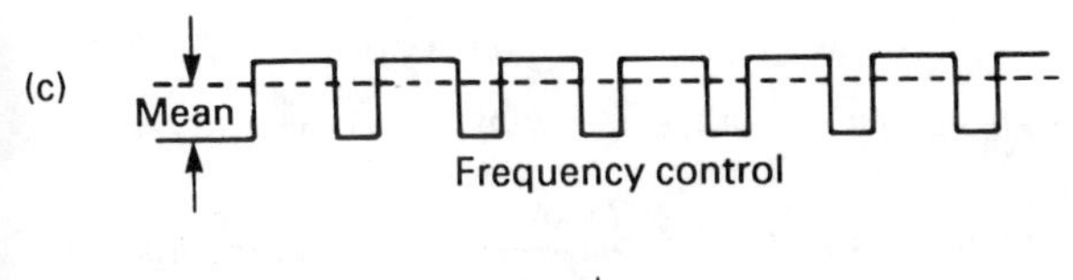

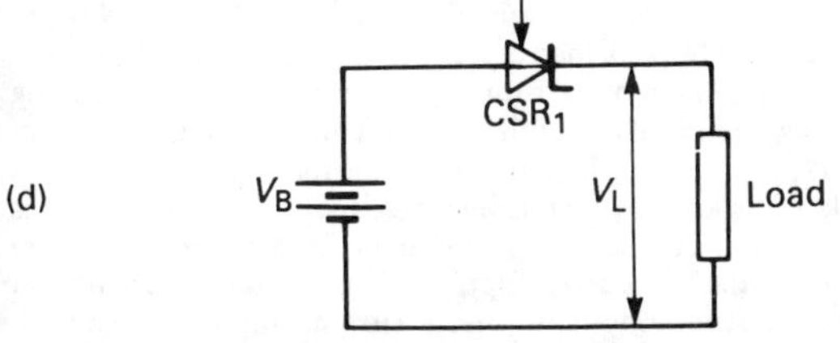

Figure 42.1 Operating principle of a d.c. to d.c. converter (chopper): (a) output waveform, (b) output waveform with mark-space control, (c) output waveform with frequency control, (d) simplified circuit arrangement

thyristor CSR_1 conducting the load voltage is equal to V_B (loss across the thyristor being assumed small). When CSR_1 is non-conducting load voltage is zero. Varying the ratio of thyristor on to off times, either in a variable mark-space or in controlled frequency mode, the mean load voltage is changed, being given by:

$$V_{L(mean)} = V_B \frac{t_c}{t_c + t_0} \tag{42.1}$$

42.2.2 Load and device currents

Figure 42.2 shows an equivalent circuit of a chopper operating into a load of voltage V_F. L, R and C form filter components and

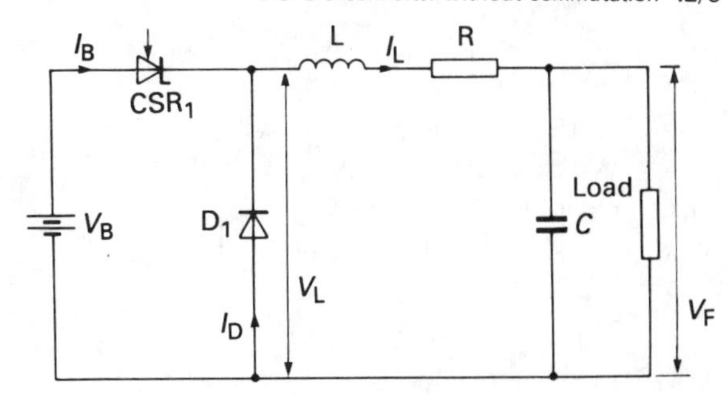

Figure 42.2 Equivalent circuit of a chopper

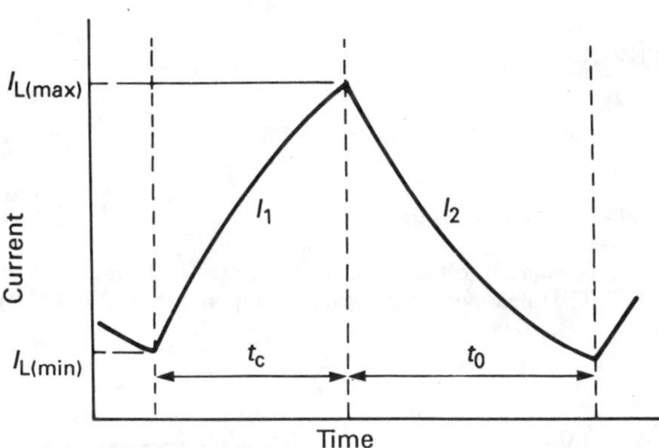

Figure 42.3 Steady-state load–current waveform of a chopper

D_1 is a free-wheeling diode, which carries the inductive load current during the off period of thyristor CSR_1. *Figure 42.3* gives this steady-state current waveform. Time t_c corresponds to the period for which CSR_1 is on and t_0 to that when it is off. The following assumptions are made:

(1) The thyristors and diodes have zero voltage drop when conducting.
(2) The devices have infinite resistance when non-conducting.
(3) Turn-on time of the thyristors is short compared to the switching period. Therefore switching losses can be neglected.
(4) The d.c. source impedance is negligible so that energy can flow in either direction through it without affecting terminal voltage.
(5) V_F is constant during a cycle of operation.
(6) Load current is continuous.

With these assumptions the values of $I_{L(max)}$ and $I_{L(min)}$ can be written from *Figure 42.3* as:

$$\frac{I_{L(min)}}{V_B/R} + \frac{V_F}{V_B} = \exp\left(-\frac{R}{L}t_0\right)\left[1 - \exp\left(-\frac{R}{L}t_c\right)\right] \left[1 - \exp\left(-\frac{R}{L}(t_c + t_0)\right)\right]^{-1} \tag{42.2}$$

$$\frac{I_{L(max)}}{V_B/R} + \frac{V_F}{V_B} = \left[1 - \exp\left(-\frac{R}{L}t_c\right)\right]\left[1 - \exp\left(-\frac{R}{L}(t_c + t_0)\right)\right]^{-1} \tag{42.3}$$

The mean current I_B through thyristor CSR_1 and I_D through diode D_1 can be found as the mean of i_1 over time t_c and i_2 over t_0, respectively.

$$\frac{I_B}{V_B/R} = \frac{t_c}{t_c+t_0} - \frac{L/R}{t_c+t_0}\left[1-\exp\left(-\frac{R}{L}t_c\right)\right]\left[1-\frac{I_{L(min)}}{V_B/R}\right]$$

$$-\frac{V_F}{V_B}\left[\frac{t_c}{t_c+t_0} - \frac{L/R}{t_c+t_0}\left(1-\exp\left(-\frac{R}{L}t_c\right)\right)\right] \quad (42.4)$$

$$\frac{I_D}{V_B/R} = \frac{L/R}{t_c+t_0}\left[1-\exp\left(-\frac{R}{L}t_0\right)\right]\left[\frac{I_{L(max)}}{V_B/R}+\frac{V_F}{V_B}\right]$$

$$-\frac{V_F t_0}{V_B(t_c+t_0)} \quad (42.5)$$

Mean load current I_L is given by:

$$I_L = \frac{I_V + I_D}{V_B/R}$$

or

$$\frac{V_L - V_F}{R}$$

Therefore

$$\frac{I_L}{V_B/R} + \frac{V_F}{V_B} = \frac{t_c}{t_c+t_0} \quad (42.6)$$

Equations (42.3) to (42.6) are shown plotted in *Figures 42.4* to *42.7*. The ripple current, derived from Equations (42.2 and 42.3)

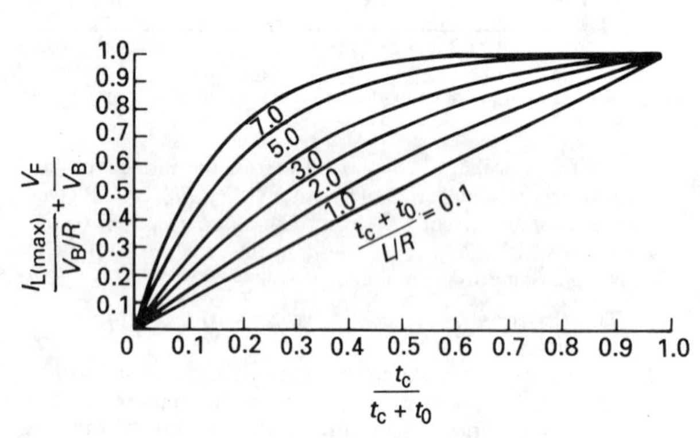

Figure 42.4 Variation of $I_{L(max)}/(V_B/R)+V_F/V_B$ with $t_c/(t_c+t_0)$

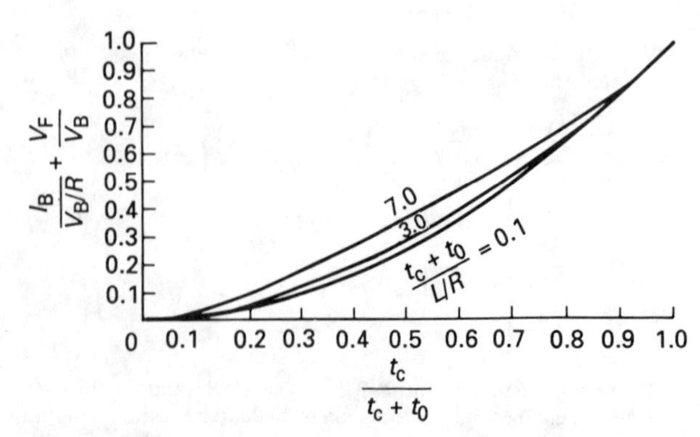

Figure 42.5 Variation of $I_B/(V_B/R)+V_F/V_B$ with $t_c/(t_c+t_0)$

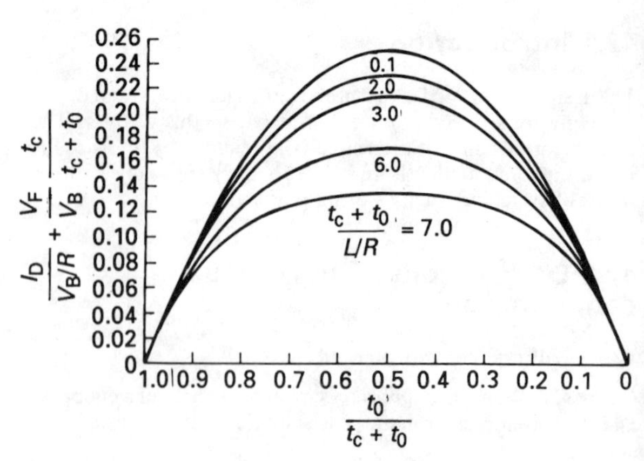

Figure 42.6 Variation of $I_D/(V_B/R)+(V_F/V_B)[t_c/(t_c+t_0)]$ with $t_0/(t_c+t_0)$

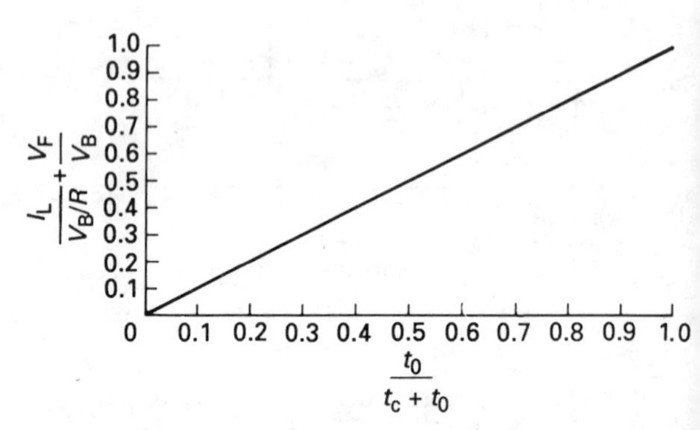

Figure 42.7 Variation of $I_L/(V_B/R)+V_F/V_B$ with $t_c/(t_c+t_0)$

is plotted in *Figure 42.8*. These curves allow device ratings, at any operating frequency, determined by the ratio of $(t_c+t_0)/(L/R)$ to be obtained. The peak load current, $I_{L(max)}$, is that value which has to be commutated in thyristor CSR_1. This varies with operating frequency and pulse width, reaching a maximum value of $(V_B-V_F)/R$ when the thyristor is on all the time. Peak current should be as low as possible for any pulse width. Therefore it is clearly advantageous to operate at higher frequencies. *Figure 42.8* shows that the current ripple is then also minimal and from *Figure 42.5* the thyristor mean current rating is reduced. Although a penalty is paid in terms of increased diode current, a diode is cheaper than an equivalent rated thyristor so that this is

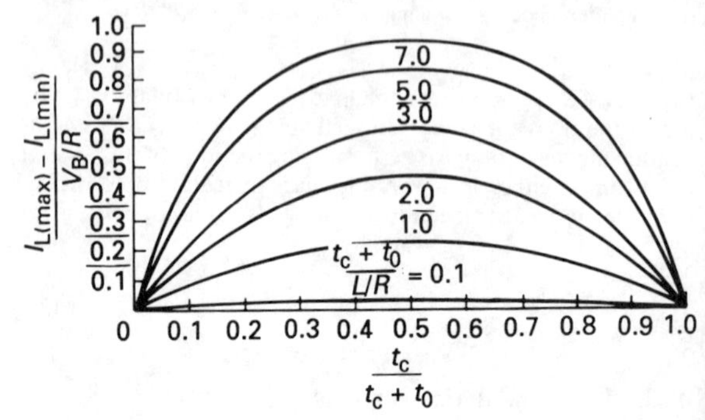

Figure 42.8 Variation of $(I_{L(max)}-I_{L(min)})/(V_B/R)$ with $t_c/(t_c+t_0)$

acceptable. It is interesting to note that ripple current is independent of the value of V_F and reaches a peak at half control width.

42.3 DC–AC converter without commutation

42.3.1 Operating principle

All inverter circuits fall into two broad groups, push–pull and bridge. A push–pull inverter is shown in *Figure 42.9*, the commutation components for the thyristors being omitted.

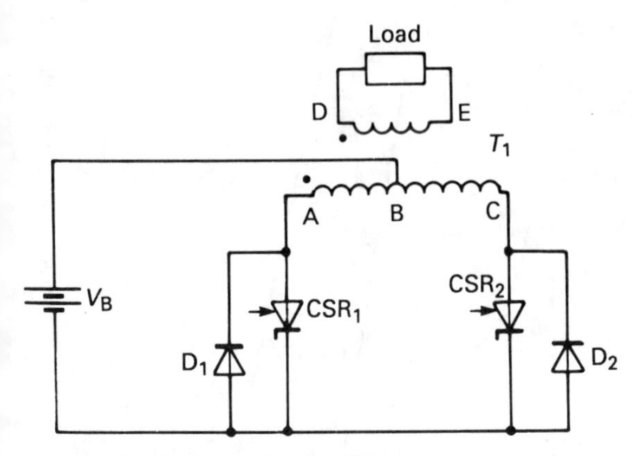

Figure 42.9 Fundamental push–pull inverter

When CSR_1 is turned on, end A of the output transformer T_1 goes negative so that end D of the load is made negative. When CSR_1 is turned off and CSR_2 fired the load polarity reverses. Diodes D_1 and D_2 carry the inductive load current when the thyristors are first turned on. Clearly T_1 is not always essential. It may be possible to make the load itself centre-tapped. For instance ABC could be the stator winding of a single-phase induction motor. In either case it is important to note that when CSR_1 conducts a voltage V_B is impressed across AB. If the two halves of the load are closely coupled, this raises the anode of CSR_2 to a potential $2V_B$ above its cathode. Similarly CSR_1 and D_1 must have a voltage rating of at least $2V_B$.

A bridge inverter without commutation components is shown in *Figure 42.10*. It is evident that such an inverter does not need the load to be centre-tapped or connected via an intermediate centre-tapped transformer. Firing CSR_1, CSR_2 and CSR_3, CSR_4 in pairs gives an alternating load polarity. D_1 to D_4 carry the inductive load current. Although there are twice as many components in a bridge inverter as in a push–pull inverter, the devices now need only be rated to a voltage V_B.

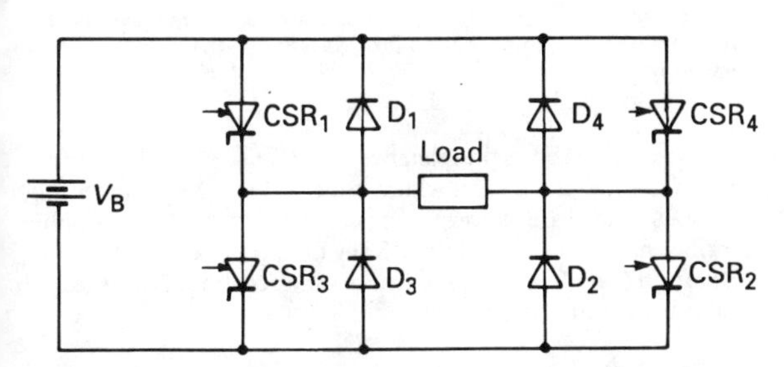

Figure 42.10 Basic bridge inverter

Generally the choice between bridge and push–pull inverters is not difficult. If the load is not centre-tapped, or does not require a transformer, bridge inverters should be used. If the load is transformer coupled, bridge inverters are used at high supply voltages, say above 200 V, and push–pull inverters at lower voltages, which would not necessitate using unduly high voltage rated and expensive devices.

42.3.2 Voltage control methods

There are two instances when voltage control within an inverter is required:

(1) When the output is to be kept at a fixed value in spite of regulation within the inverter, or fluctuations of the supply voltage.
(2) When the output is to be varied in a given manner, e.g. proportional to frequency for variable-frequency drives.

There are several ways in which this voltage control can be achieved. In all cases the a.c. output will be made up of a fundamental component and a band of harmonic frequencies. The various control methods all contrive to reduce the harmonic voltages while avoiding excessive circuit complexity.

Voltage control in inverter circuits is important. All too often the alternatives are not considered during the design stage, and traditional methods such as mark-space (or quasi-square wave) are employed, even though another technique could be much more advantageous.

Perhaps the most popular method of controlling the a.c. voltage is to vary its mark-space ratio, as in *Figure 42.11*. Fourier

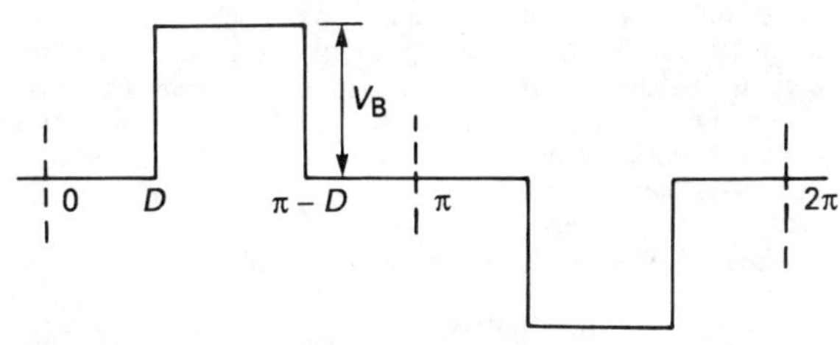

Figure 42.11 Quasi-square waveform

analysis of such a waveform gives the r.m.s. value of the *n*th coefficient as

$$\frac{\%\ \text{r.m.s.}\ (n)}{V_B} = \left[\frac{2\sqrt{2}}{n\pi}\cos nD\right] \times 100 \qquad (42.7)$$

The total r.m.s. voltage of the waveform, including all harmonics, is obtained as

$$\frac{\%\ \text{r.m.s.}\ (T)}{V_B} = \left[1 - \frac{2D}{\pi}\right]^{1/2} \times 100 \qquad (42.8)$$

Equations (42.7) and (42.8) are shown evaluated in *Table 42.1* up to the 15th harmonic. From this table it is evident that the harmonic content of the output increases rapidly as the mark-space ratio of the waveform is reduced. This is illustrated more clearly by *Figure 42.12*. At low voltages various harmonics are almost equal in value to the fundamental, the total harmonic content being about ten times larger. This represents the greatest

Table 42.1 Harmonic composition of a quasi-square wave

	RMS voltage as percentage of d.c. supply								
2D/T	*1*	*3*	*5*	*7*	*9*	*11*	*13*	*15*	*Total*
0.00	90.0	30.0	18.0	12.9	10.0	8.18	6.92	6.00	100
0.02	89.8	29.5	17.1	11.6	8.44	6.30	4.74	3.53	98.0
0.04	89.3	27.9	14.6	8.20	4.26	1.53	0.43	1.85	95.9
0.06	88.4	25.3	10.6	3.20	1.25	3.94	5.33	5.71	93.8
0.08	87.2	21.9	5.56	2.41	6.37	7.61	6.87	4.85	91.7
0.10	85.6	17.6	0.00	7.56	9.51	7.78	4.07	0.00	89.4
0.12	83.7	12.8	5.56	11.3	9.69	4.38	1.30	4.85	87.2
0.14	81.4	7.46	10.6	12.8	6.85	1.03	5.85	5.71	84.9
0.16	78.9	1.88	14.6	12.0	1.87	5.96	6.71	1.85	82.5
0.18	76.0	3.76	17.1	8.80	3.68	8.17	3.34	3.53	80.0
0.20	72.8	9.27	18.0	3.97	8.09	6.62	2.14	6.00	77.5
0.22	69.3	14.5	17.1	1.61	9.98	2.03	6.26	3.53	74.8
0.24	65.6	19.1	14.6	6.89	8.76	3.48	6.44	1.85	72.1
0.26	61.6	23.1	10.6	10.9	4.82	7.40	2.55	5.71	69.3
0.28	57.4	26.3	5.56	12.8	0.63	7.92	2.95	4.85	66.3
0.30	52.9	28.5	0.00	12.2	5.88	4.81	6.58	0.00	63.2
0.32	48.2	29.8	5.56	9.37	9.30	0.51	6.07	4.85	60.0
0.34	43.4	29.9	10.6	4.73	9.82	5.60	1.72	5.71	56.6
0.36	38.3	29.1	14.6	0.81	7.29	8.12	3.71	1.85	52.9
0.38	33.1	27.1	17.1	6.19	2.49	6.91	6.80	3.53	49.0
0.40	27.8	24.3	18.0	10.4	3.09	2.53	5.60	6.00	44.7
0.42	22.4	20.5	17.1	12.6	7.71	3.01	0.87	3.53	40.0
0.44	16.9	16.1	14.6	12.5	9.92	7.17	4.41	1.85	34.5
0.46	11.3	11.0	10.6	9.91	9.05	8.04	6.91	5.71	28.3
0.48	5.65	5.62	5.56	5.47	5.36	5.22	5.05	4.85	20.0
0.50	0.00	0.00	0.00	0.00	0.00	0.00	0.00	0.00	0.00

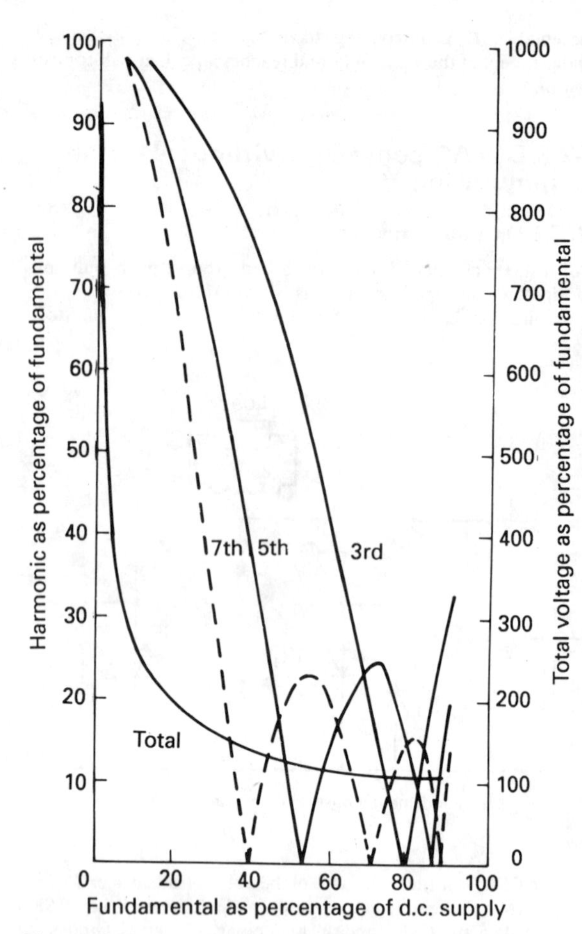

Figure 42.12 Harmonics in a quasi-square waveform

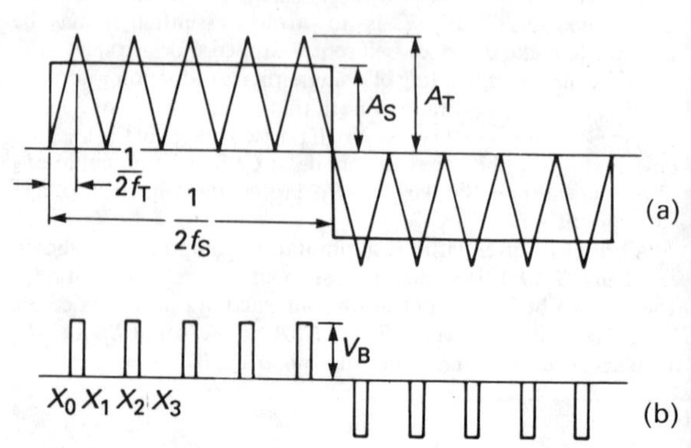

Figure 42.13 Pulse width modulation with a square wave using unidirectional switching: (a) control waveforms, (b) output waveform

disadvantage of the quasi-square voltage control system, and normally limits the output voltage range to between 30 and 90% of the d.c. supply, i.e. a frequency change of 3:1. *Figure 42.11* may also be referred to as a single-pulse, unidirectional wave. It is unidirectional since in any half-cycle the output is either positive or negative, never both. There is also a single pulse in each half-cycle.

A second extension to quasi-square control is to have several equally spaced unidirectional pulses in a half-cycle. This is conveniently obtained by feeding a high-frequency carrier triangular wave and a low-frequency reference square wave into a comparator as in *Figure 42.13(a)*. The output voltage swings between V_B and zero volts according to the relative magnitudes of the two waveforms, as in *Figure 42.13(b)*. The width of the output pulses is determined by the ratio of A_S/A_T and there are as many pulses per cycle as the ratio f_T/f_S. The Fourier coefficient is given by

$$\frac{\%\text{ r.m.s. }(n)}{V_B} = \sum_{M=0,2,4,\dots} \frac{2\sqrt{2}}{n\pi}[\cos nx(M) - \cos nx(M+1)] \quad (42.9)$$

and

$$\frac{\%\text{ r.m.s. }(T)}{V_B} = 100\left[\frac{f_T}{2\pi f_S}\{x(0)-x(1)\}\right]^{1/2} \quad (42.10)$$

The third and seventh harmonics derived from Equations (42.9) and (42.10) are plotted in *Figures 42.14* and *42.15* respectively for $f_T/f_S = 2$, 4, 6, 10 and 20. This shows the improvement in attenuation of the lower harmonics, compared to quasi-square wave control methods, although for higher harmonics it is necessary to go to a larger pulse number to obtain appreciable harmonic reduction.

In *Figure 42.13* a square wave was used as the reference source. A greater reduction in harmonic content would be obtained if a sine wave was used as the reference as in *Figure 42.16*. The output pulses are not of constant width, but provided the values of X_1, X_2, X_3, etc., are known the magnitude of the harmonic coefficient can again be found from Equation (42.9). However, Equation (42.10) is now modified to take account of the unequal pulse widths and is given by

$$\frac{\%\text{ r.m.s. }(T)}{V_B} = \left[\frac{2}{\pi}\sum\{x(M+1)-x(M)\}\right]^{1/2} \quad (42.11)$$

Equations (42.9) and (42.11) are shown evaluated in *Tables 42.2* to *42.5*. Voltage control is again effected by changing the ratio of A_S/A_T. The maximum value of this is limited to 0.98 rather than unity to prevent adjacent pulses from merging into each other. The following factors can be noticed.

(1) Unlike the previous system this method of voltage control results in severe attenuation of frequencies below a certain value. On examining the tables it will be seen that the harmonic numbers with the largest amplitude occur at $f_T/f_S \pm 1$. Therefore, for example, with $f_T/f_S = 10$ the harmonics are largest at the ninth and eleventh. This is logical since the tenth harmonic is the 'carrier' wave itself, and no attempt is made to eliminate it. Quite clearly the higher the ratio of f_T/f_S the more effective this system

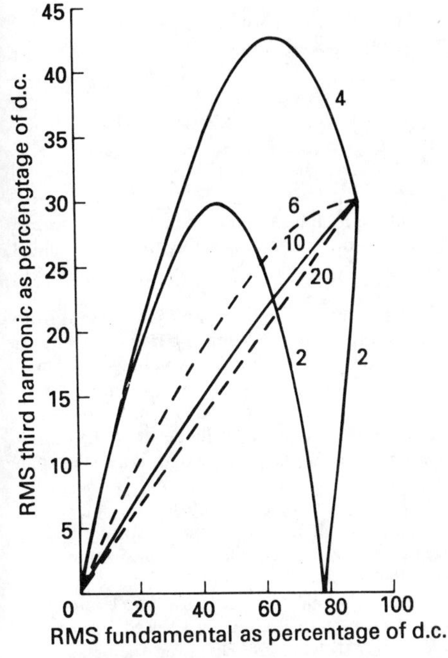

Figure 42.14 Third-harmonic content of a square-modulated unidirectional wave

Table 42.2 Harmonic content of a sine-modulated unidirectional wave with $f_T/f_S=4$

A_S/A_T	RMS voltage as percentage of d.c. supply								
	1	*3*	*5*	*7*	*9*	*11*	*13*	*15*	*Total*
0	0	0	0	0	0	0	0	0	0
0.1	9.93	9.77	9.46	8.99	8.40	7.69	6.88	6.00	26.5
0.2	16.6	15.9	14.4	12.4	9.97	7.32	4.64	2.14	34.4
0.3	23.2	21.1	17.4	12.5	7.18	2.26	1.65	4.13	40.7
0.4	29.6	25.4	17.9	9.19	1.25	4.24	6.48	5.68	46.2
0.5	33.9	27.5	16.9	5.54	3.19	7.27	6.60	2.87	49.5
0.6	40.0	29.5	13.4	1.02	8.41	7.66	2.01	3.52	54.2
0.7	44.0	30.0	10.0	5.36	9.91	5.05	2.41	5.90	57.0
0.8	47.9	29.8	6.01	9.03	9.44	0.95	5.85	5.13	59.7
0.9	51.6	29.0	1.63	11.6	7.08	3.44	6.92	1.61	62.3
0.98	53.4	28.3	0.62	12.4	5.33	5.33	6.38	0.62	63.6

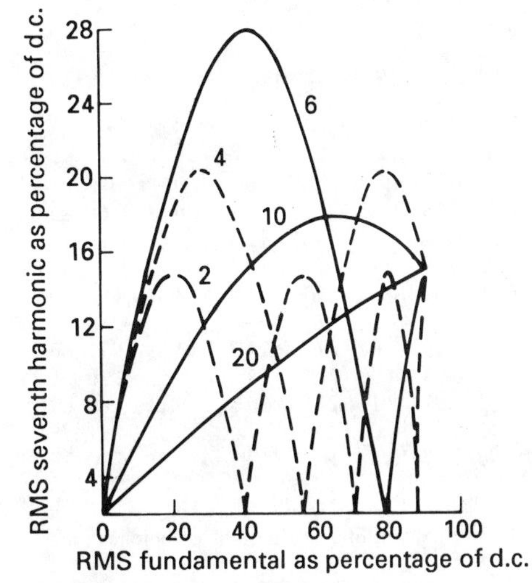

Figure 42.15 Seventh-harmonic content of a square-modulated unidirectional wave

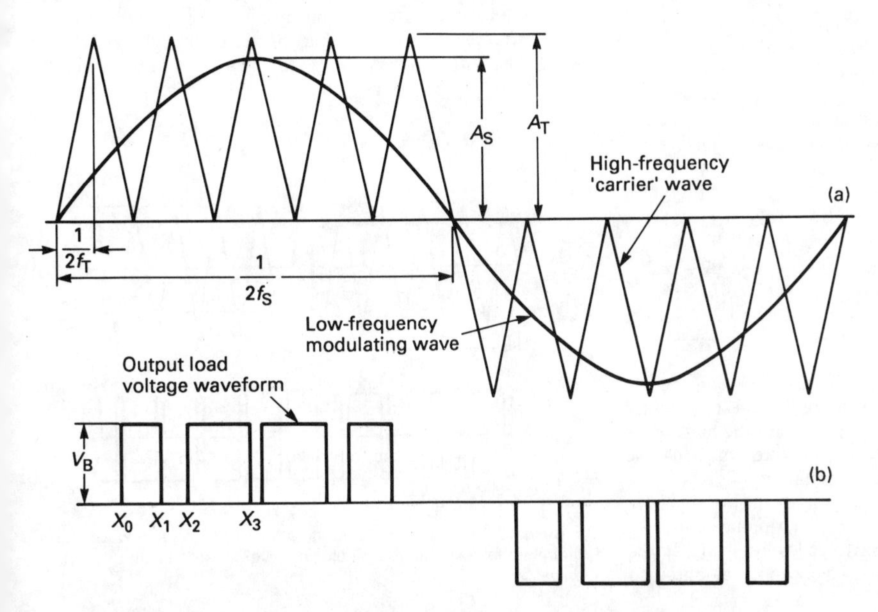

Figure 42.16 High-frequency pulse-width modulation with unidirectional switching using a sine modulating wave: (a) control waveforms, (b) output waveform

Table 42.3 Harmonic content of a sine-modulated unidirectional wave with $f_T/f_S=6$

A_S/S_T	RMS voltage as percentage of d.c. supply								
	1	*3*	*5*	*7*	*9*	*11*	*13*	*15*	*Total*
0	0	0	0	0	0	0	0	0	0
0.1	6.98	0.29	7.20	6.57	0.84	7.20	5.98	1.32	24.0
0.2	14.7	0.60	14.7	13.1	1.60	13.1	10.2	2.09	34.8
0.3	21.2	0.25	19.9	18.0	0.59	14.4	11.4	0.55	41.7
0.4	27.8	0.45	23.9	21.5	0.88	12.7	9.08	0.39	47.7
0.5	34.3	1.46	26.7	23.4	2.17	8.47	3.59	0.38	52.9
0.6	40.8	2.74	28.1	23.1	2.72	3.07	3.48	2.69	57.7
0.7	48.8	5.52	27.2	20.4	2.22	2.73	10.9	6.50	63.0
0.8	55.4	7.28	25.7	15.6	0.32	5.67	13.7	6.41	67.1
0.9	60.8	10.4	22.5	10.8	3.59	5.06	12.6	4.51	70.2
0.98	66.3	13.6	18.7	5.12	7.86	2.74	8.82	0.21	73.1

Table 42.4 Harmonic content of a sine-modulated unidirectional wave with $f_T/f_S=10$

A_S/A_T	RMS voltage as percentage of d.c. supply								
	1	*3*	*5*	*7*	*9*	*11*	*13*	*15*	*Total*
0	0	0	0	0	0	0	0	0	0
0.1	7.26	0.28	0.24	0.88	7.53	6.70	0.43	0.68	25.0
0.2	13.6	0.54	0.12	0.11	13.7	12.3	1.48	0.29	34.6
0.3	20.8	0.95	0.18	0.12	19.9	17.4	2.72	0.27	42.8
0.4	27.5	0.49	0.57	0.02	24.1	21.2	3.63	0.92	49.1
0.5	34.7	0.98	0.69	0.80	27.7	23.0	5.70	0.66	55.2
0.6	42.1	0.60	0.35	3.67	27.8	24.7	6.21	1.71	60.7
0.7	48.9	0.17	0.93	4.90	27.0	23.2	8.51	2.16	65.2
0.8	55.9	1.31	0.91	8.17	24.9	20.8	9.71	2.89	69.9
0.9	62.5	1.43	1.57	11.0	20.7	17.5	10.5	4.11	73.9
0.98	68.5	0.93	1.57	13.4	16.6	12.7	11.9	4.70	77.3

Table 42.5 Harmonic content of a sine-modulated unidirectional wave with $f_T/f_S=20$

A_S/A_T	RMS voltage as percentage of d.c. supply								
	1	*3*	*5*	*7*	*9*	*11*	*13*	*15*	*Total*
0	0	0	0	0	0	0	0	0	0
0.1	7.58	0.68	1.04	0.22	0.33	0.26	0.38	1.26	25.9
0.2	14.9	0.60	0.09	0.50	0.35	0.30	0.62	0.48	36.2
0.3	21.5	0.15	0.15	0.77	0.53	0.52	0.88	0.24	43.7
0.4	28.2	0.03	0.56	0.89	0.00	0.01	1.02	0.55	50.2
0.5	35.3	0.05	0.00	0.41	0.53	0.15	0.47	0.50	56.2
0.6	42.4	0.16	0.05	0.21	0.22	0.13	0.82	0.53	61.6
0.7	49.2	0.10	0.22	0.39	0.29	0.15	0.85	0.97	66.3
0.8	56.8	0.53	0.59	0.26	0.06	0.36	0.28	0.09	71.2
0.9	64.0	0.55	0.02	0.15	0.49	0.21	0.13	0.76	75.5
0.98	69.0	0.26	0.42	0.19	0.10	0.16	0.32	1.25	78.6

becomes. The same statement was made when considering modulation with a square wave, but there the effect of higher carrier frequencies was to keep the proportion of the harmonics constant (and equal to the value for a square wave) as the fundamental was varied, and not to eliminate it.

(2) This system has two disadvantages. Firstly the high inverter frequency required to give effective lower-harmonic reduction leads to smaller efficiencies due to inverter losses. Secondly the maximum output is well below 90% of d.c. supply, as obtained with a square wave. This would limit the maximum operating frequency when running with some types of loads which need to be fully fluxed, as described before.

(3) There is a characteristic increase in total harmonic content, with higher operating frequencies. This is not normally serious since higher harmonics can be more easily filtered out than lower-order harmonics.

In the above discussions all the waveforms looked at have been unidirectional, that is the instantaneous voltage in any half-cycle has been either positive or negative, as in *Figure 42.16*.

Knowing the value of intersection points X_0, X_1, X_2, etc., the r.m.s. of the nth harmonic can readily be obtained by the arithmetic sum of the individual pulses over the 0 to π interval. Therefore

$$\frac{\% \text{ r.m.s. } (n)}{V_B} = \frac{\sqrt{2}}{n\pi}100\Big\{1 - \cos nx(0) + \sum [\cos nx(2M+1) - \cos nx(2M+2)] - \sum [\cos nx(2M) - \cos nx(2M+1)]\Big\} \quad (42.12)$$

The r.m.s. value of the total harmonic is constant, irrespective of the depth of modulation and the operating frequency, and is equal to that of a square wave since there are no zero periods in the output.

For bidirectional switching, even harmonics are absent from the output, apart from the inverter operating frequency and its triple harmonics. For unidirectional switching all even harmonics are missing.

The lower harmonic content with bidirectional switching, when the inverter operating frequency is higher than the output frequency, is less than for a quasi-square wave. It is also less than that obtained with unidirectional switching and is due to the absence of any 'zero' periods in the waveform. For maximum modulation the output reduces to a square wave and the harmonic content is identical to comparable quasi-square and unidirectional systems.

With bidirectional switching the odd-harmonic content generally increases as the operating frequency is raised. The reverse was true for unidirectional switching. However, the presence of the large-valued even harmonics still makes it desirable to operate this system at an inverter frequency of the order of 20 times the desired output frequency.

The square reference wave may of course be replaced by a sine wave. The output is very similar to *Figure 42.17* except that the

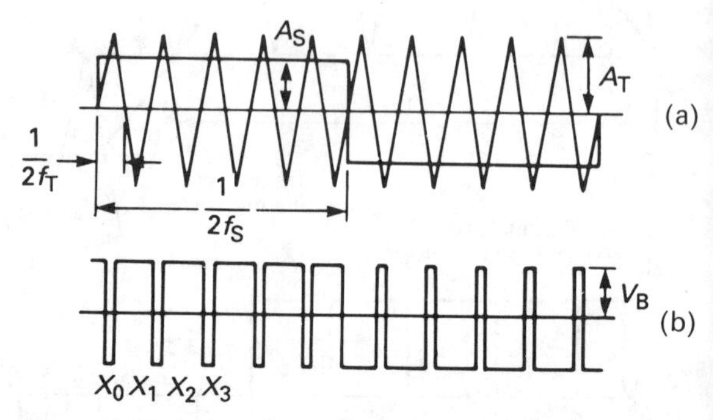

Figure 42.17 Pulse-width modulation with a square wave using bidirectional switching: (a) control waveforms, (b) output waveform

modulation depth varies linearly along the cycle. This waveform will also contain odd and even sine terms, and since $X_0=0$ the r.m.s. voltage of the nth harmonic can be derived from Equation (42.13).

$$\frac{\%\ \text{r.m.s.}\ (n)}{V_B}=\frac{\sqrt{2}}{n\pi}100\left\{\sum[\cos nx(2M+1)-\cos nx(2M+2)]\right.$$

$$\left.-\sum[\cos nx(2M)-\cos nx(2M+1)]\right\} \quad (42.13)$$

The harmonic content of the waveform is very similar to unidirectional switching as in *Tables 42.2* to *42.5*. The harmonic with the largest amplitude is that which occurs close to the chopping frequency f_T, both odd and even harmonics being considered. As an example when operating at $f_T/f_S=6$ the sixth harmonic is very large. For zero modulation depth the fundamental is zero and the sixth harmonic has a value 90% of the d.c. supply, since the output is a square wave at this frequency. As the modulation depth increases the fundamental also increases in value and the sixth harmonic reduces (whereas adjacent even harmonics, i.e. the fourth and the eighth, increase in value). When operating with $f_T/f_S=20$ the harmonic content up to the 15th is very similar to unidirectional methods except that there are now odd and even terms. However, the total harmonic content, which for bidirectional switching is 100% of the d.c. supply irrespective of the modulation depth, is much higher. As in previous methods it is clear that the inverter frequency should be several orders larger than the output frequency for effective harmonic reduction. This is therefore a disadvantage of the system as it can lead to lower efficiencies.

Before leaving this section it would be useful to look at two other techniques for voltage control which have been used specifically as methods for harmonic reduction. The first of these is called staggered phase carrier cancellation. It essentially consists in combining several high-frequency modulated waves in which the carriers are out of phase, whereas the modulating (low-frequency) wave is in phase. This results in a strengthening of the low-frequency and a weakening of the high-frequency carrier.

This system can be extended. For instance combining four waveforms with their low frequencies in phase but carriers phase shifted by 90°, 180° and 270° would result in the carrier and its first, second and third harmonics being eliminated from the output.

The disadvantage of staggered phase-carrier cancellation is the extra hardware needed. Its prime use is in fixed-frequency sine wave inverters where the extra cost can be justified on account of the reduction in size of the output filter.

The second technique worth mentioning is that of waveform synthesis. This is frequently used, especially in larger installations, where several inverters are run in parallel but phase shifted, their outputs being summed by a transformer to produce a stepped waveform with reduced harmonic content. The same effect can be obtained by using a tapped supply. Alternatively, instead of tapping the supply, the primary or secondary of the transformer can be tapped. *Figure 42.18* shows an inverter arrangement with a tapped secondary, the primary being the normal form of a push–pull inverter. The firing sequence of the thyristors is also shown. Provided the tappings on the secondary are such as to give the waveform indicated, it can be calculated that the output contains no harmonics below the eleventh.

42.3.3 Load and device currents

Figure 42.19 shows the load voltage and current waveforms for the inverter of *Figure 42.10*. This also indicates the device conducting periods. Using the same assumptions as made for the

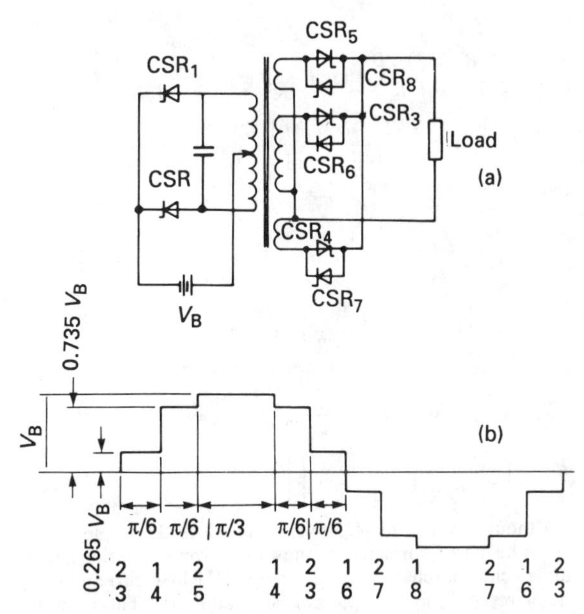

Figure 42.18 Waveform synthesis with a tapped secondary transformer: (a) circuit diagram, (b) load waveform

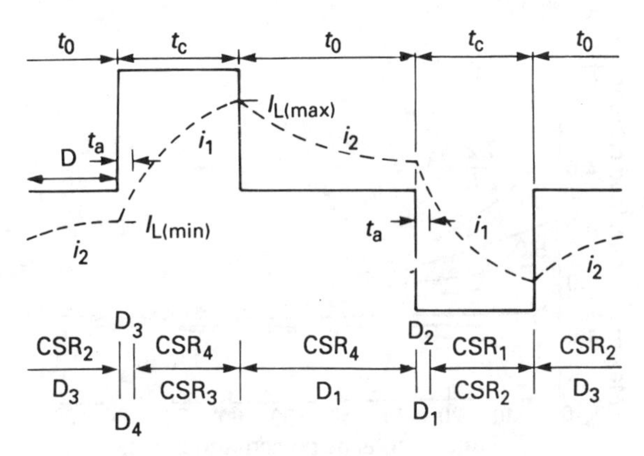

Figure 42.19 Inverter load voltage and current per cycle

chopper, the characteristics of the inverter can be determined as before. If $T=t_c+t_0$ is the chopping period, then:

$$\frac{I_{L(max)}}{V_B/R}=\frac{1-\exp\left[-\frac{R}{L}\left(\frac{T}{2}-2D\right)\right]}{1+\exp\left[-\frac{R}{L}\frac{T}{2}\right]} \quad (42.14)$$

The r.m.s. current rating of the devices is obtained by considering the current and voltages indicated in *Figure 42.19*. It is important to note that one arm of the bridge carries a larger current than the other, and the 'worst case' should be considered. Therefore the rating of the thyristors can be taken as i_1 over t_c-t_a or as the sum of i_1 over t_c-t_a and i_2 over t_2. The latter gives the 'worst case'.

The values of thyristor (I_T), diode (I_D) and load (I_{LR}) r.m.s. currents are given by

$$\frac{I_T}{V_B/R}=\left[\left(\frac{I_{L(max)}}{V_B/R}\right)^2\frac{L/R}{2T}\left\{1-\exp\left[-\frac{T}{L/R}\cdot\frac{4D}{T}\right]\right\}+\left(\frac{I_{TO}}{V_B/R}\right)^2\right]^{1/2} \quad (42.15)$$

where

$$\frac{I_{TO}}{V_B/R}=\left[\frac{L/R}{T}\left\{\frac{T/2-2D}{L/R}-\log P+2P\exp\left(-\frac{T/2-2D}{L/R}\right)-\frac{P^2}{2}\exp\left(-\frac{T-4D}{L/R}\right)-\frac{3}{2}\right\}\right]^{1/2}$$

and

$$P=1+\frac{I_{L(max)}}{V_B/R}\exp\left(-\frac{Rt_0}{L}\right)$$

$$I_D/(V_B/R)=\left[\left(\frac{I_T}{V_B/R}\right)^2-\left(\frac{I_{TO}}{V_B/R}\right)^2+\frac{L/R}{T}\left\{\log P-2P+\frac{P^2}{2}+\frac{3}{2}\right\}\right]^{1/2} \quad (42.16)$$

$$\frac{I_{LR}}{V_B/R}=\left[2\left\{\left(\frac{I_D}{V_B/R}\right)^2+\left(\frac{I_{TO}}{V_B/R}\right)^2\right\}\right]^{1/2} \quad (42.17)$$

Although Equations (42.16) and (42.17) could be plotted, it would be more informative to show the current variations, with fundamental output voltage, as a ratio of the load current, using the curves of *Figure 42.12* for a quasi-square wave. This is shown in *Figures 42.20*, *42.21* and *42.22* and allows peak current and the r.m.s. currents of the thyristor or diode to be determined at any load voltage and r.m.s. current, for a given operating frequency.

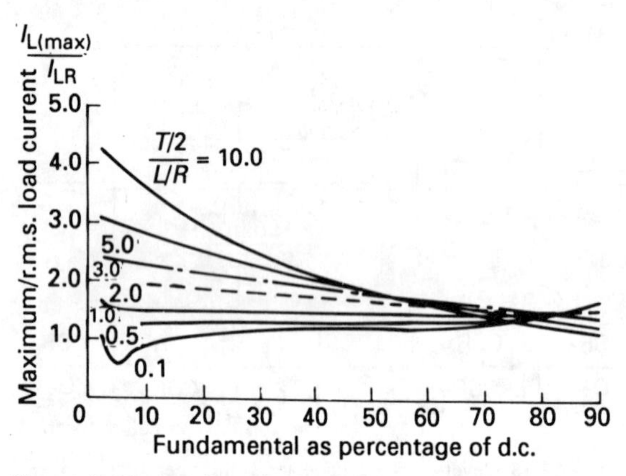

Figure 42.20 Variation of (peak load/r.m.s. load) current with fundamental load voltage

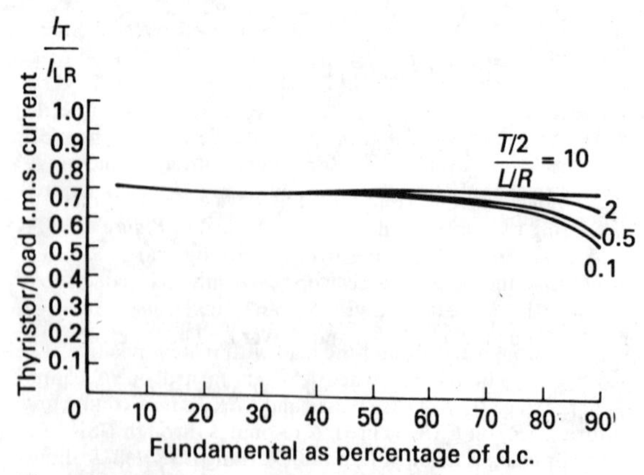

Figure 42.21 Variation of (r.m.s. thyristor/r.m.s. load) current with fundamental load voltage

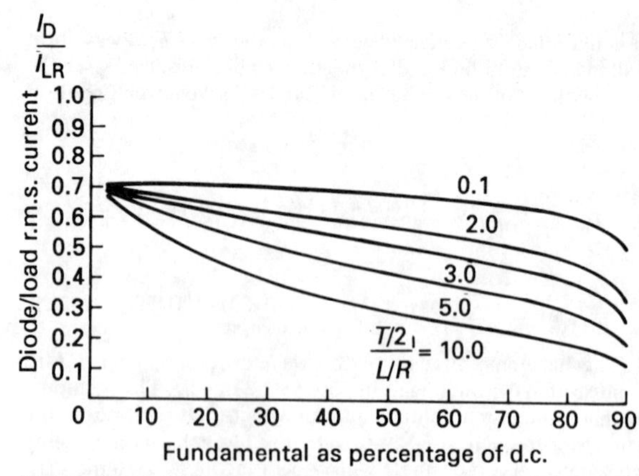

Figure 42.22 Variation of (r.m.s. diode/r.m.s. load) current with fundamental load voltage

Figure 42.20 shows the large increase in commutation current requirements at low frequencies and voltages, illustrating the unsuitability of this control method for low mark-space operation. The thyristor r.m.s. current is almost constant at $1/\sqrt{2}$ times the load current indicating that the devices conduct for approximately half the load current period. The diode ratings also tend to this value at low voltages.

42.4 Commutation in choppers and inverters

42.4.1 Methods of commutation

Although there are many different commutation circuits for choppers and inverters they can be divided into four groupings. These are illustrated with reference to a chopper in *Figure 42.23*.

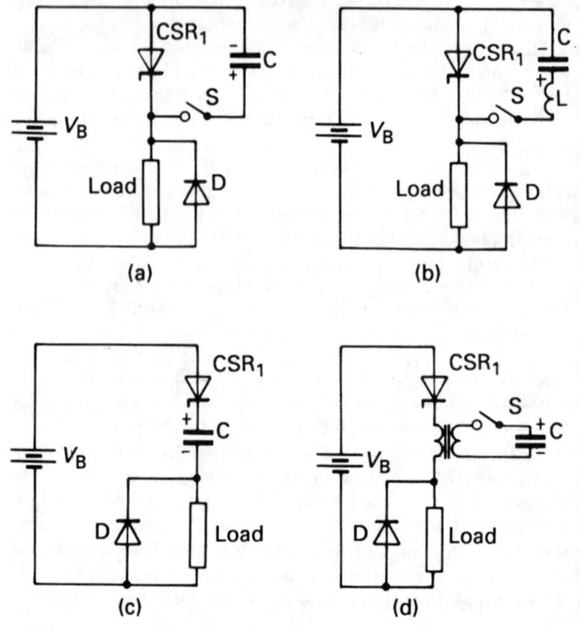

Figure 42.23 Commutation methods: (a) parallel capacitor, (b) parallel capacitor-inductor, (c) series capacitor, (d) coupled pulse

(1) *Parallel-capacitor commutation* The charged capacitor is placed directly across the thyristor to be turned off as in *Figure 42.23(a)*. Switch S forms part of the commutation circuit.
(2) *Parallel capacitor-inductor commutation* An inductor is connected in series with the commutation capacitor as in *Figure 42.23(b)*.
(3) *Series-capacitor commutation* The commutation capacitor is connected in series with the thyristor to be turned off as in *Figure 42.23(c)*.
(4) *Coupled-pulse commutation* The turn-off pulse is coupled to the thyristor through a transformer or auto-transformer as in *Figure 42.23(d)*.

Figure 42.23 indicates only the basic principles of the commutation methods. The auxiliary circuits used to prime the commutation capacitor, ready to turn off the main thyristor CSR_1, are not shown. Typical chopper and inverter circuits in the above four groups are given below.

42.4.1.1 Parallel-capacitor commutation

For the circuit shown in *Figure 42.24* thyristor CSR_2 is fired initially, charging C to voltage V_B with plate b positive. CSR_1 is now turned on. C discharges via CSR_1, D_2, L_1 and resonates, recharging to V_f with plate a positive. For negligible resonant loss V_f is equal to V_B.

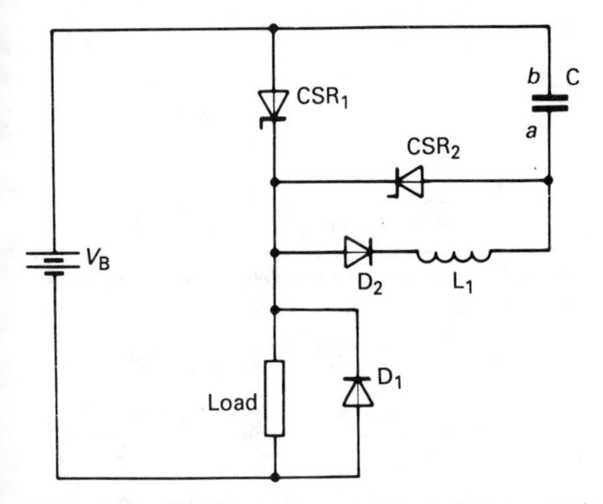

Figure 42.24 Parallel capacitor commutation in a chopper circuit

To turn CSR_1 off CSR_2 is fired. The voltage across CSR_1 is reversed, C discharges through the load (current assumed constant) and recharges with plate b positive. CSR_2 then turns off.

In the circuit shown in *Figure 42.24* assume C to be charged to a voltage V_f. The load current just prior to CSR_2 firing is $I_{L(max)}$ and is assumed constant during the short discharge period of C. If the reverse recovery current of CSR_1 is neglected then:

$$I_{L(max)} = \frac{CV_f}{t_F} \qquad (42.18)$$

where t_F is the time during which CSR_1 is reverse-biased. For commutation to be successful, this time must be greater than the device turn off time at the rate of re-applied forward voltage given by:

$$\frac{dV}{dt} = \frac{I_{L(max)}}{C} \qquad (42.19)$$

Equations (42.18) and (42.19) are applicable to all parallel-capacitor commutated circuits and enable the correct capacitor value to be chosen for a given load and device characteristic. However, the magnitude of V_f will be determined by the auxiliary commutation circuit adopted.

Figure 42.25 shows two examples of commutation in an inverter. In *Figure 42.25(a)* firing CSR_1 charges capacitor C with plate a positive to a voltage $2V_B$. When CSR_2 is fired capacitor C is connected across CSR_1 turning it off. The capacitor now discharges to zero voltage, its stored energy then being dissipated in the L_1–D_1–C–CSR_2 conduction path. After this C charges to $2V_B$ with plate b positive, ready to turn CSR_2 off when CSR_1 is fired.

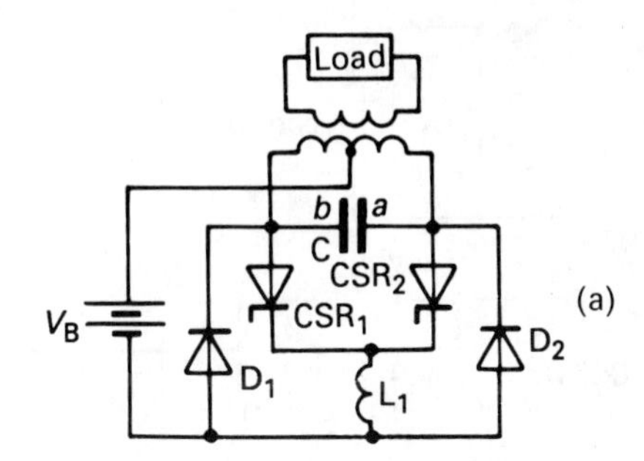

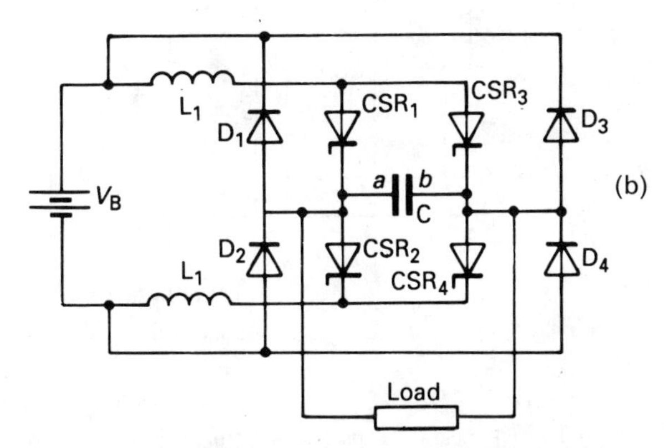

Figure 42.25 Parallel capacitor commutation in inverter circuits: (a) push–pull, (b) bridge

Figure 42.25(b) shows a single-phase bridge inverter circuit. With CSR_1 and CSR_4 conducting capacitor C is charged to V_B with plate a positive. When CSR_2 and CSR_3 are fired to commence the next step of the output, capacitor C is connected across CSR_1 and CSR_4 and turns them off. The circuit is therefore an example of parallel-capacitor commutation. Inductor L_1 prevents the supply from being instantaneously short-circuited during commutation.

42.4.1.2 Parallel capacitor-inductor commutation

Figure 42.26(a) shows a chopper circuit in which C is initially charged through L_1, D_3 and the load with plate a positive. The load cycle is commenced when CSR_1 is turned on. C cannot discharge since D_3 is reverse biased and CSR_2 is non-conducting. To turn CSR_1 off CSR_2 is fired. C resonates through CSR_2 and L_1 and discharges through L_1 and D_3 turning CSR_1 off. D_2 limits the maximum discharge period, even with the load open circuit.

The values of C and L_1 required to commutate a thyristor of

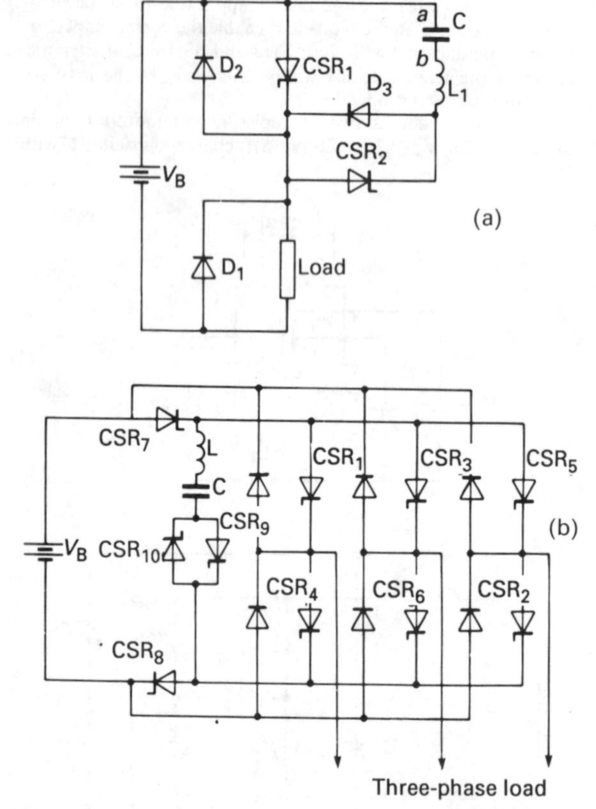

Figure 42.26 Parallel capacitor-inductor commutation: (a) chopper, (b) inverter

turn-off time t_F in parallel capacitor-inductor circuits is given by:

$$C=\frac{2\sqrt{2t_F I_{L(max)}}}{\pi V_B} \tag{42.20}$$

$$L_1=\frac{\sqrt{2t_F V_B}}{\pi I_{L(max)}} \tag{42.21}$$

From this it can be seen that parallel capacitor-inductor commutation is unsuitable for use in systems controlling large currents from low supply voltages. For example, to turn off 250 A from 40 V using thyristors with $t_F=50$ μs would give $C=280$ μF and $L_1=2.5$ μH. In all probability the inductance of the connecting leads would exceed this value.

Figure 42.26(b) shows an inverter circuit which has been developed for aerospace applications. CSR_9 is fired to prime C ready for commutation. When CSR_{10} is now turned on capacitor C discharges via its inductor L, turning CSR_7 and CSR_8 off and so eventually commutating the inverter thyristors. This circuit also has high operating efficiencies.

42.4.1.3 Series-capacitor commutation

Referring to the basic series-capacitor commutator circuit of *Figure 42.23(c)* it is seen that CSR_1 will turn off once C has charged to V_B volts. Before CSR_1 can be refired the capacitor must be reset. There are several ways in which this may be done. In *Figure 42.27(a)* CSR_1 is fired to commence the load cycle. This causes C to charge through L_1 and turn CSR_1 off. CSR_2 is fired which resets C through L_2 after which CSR_2 turns off and CSR_1 can be refired.

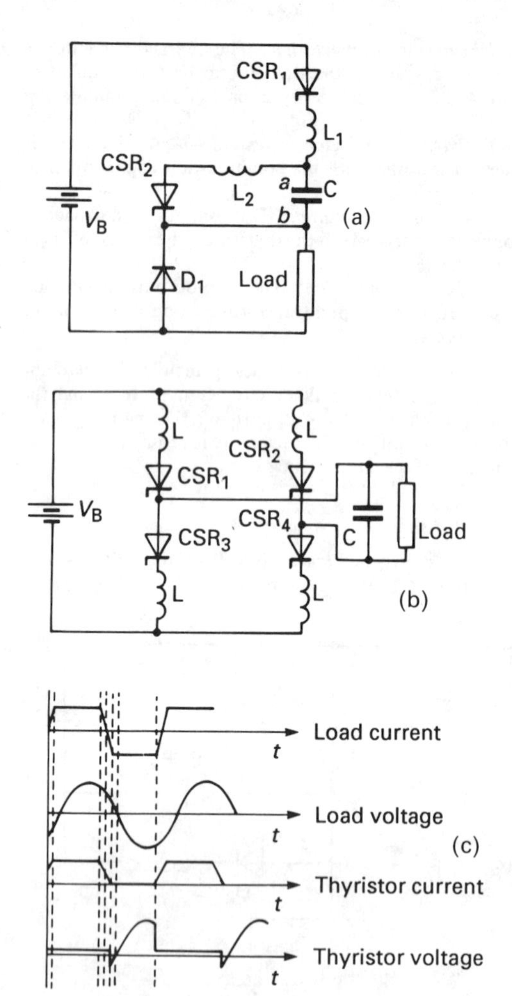

Figure 42.27 Series capacitor commutation: (a) chopper, (b) inverter circuit, (c) inverter waveforms

Series-capacitor commutation circuits have a very limited operating range, due to the capacitor reset time, and are rarely used as choppers. They find much greater application in sine wave inverter circuits. *Figure 42.27(b)* shows a typical circuit with its waveforms (*Figure 42.27(c)*). The inverter is operated at a frequency determined by the oscillatory frequency of the load and C, so that these inverters find most frequent application for producing fixed frequency sine wave output.

42.4.1.4 Coupled-pulse commutation

A chopper circuit using an auto-transformer to couple the turn-off pulse to the thyristors is shown in *Figure 42.28(a)*. CSR_1 is the main thyristor and CSR_2 an auxiliary device used to commutate CSR_1. L_1 L_2 is a tapped auto-transformer. The circuit operation is as follows. CSR_1 is fired to commence the load cycle. This also causes C_1 to charge through L_1 to supply voltage V_B. The clamping action of D_2 prevents C_1 charging to a higher voltage. To turn CSR_1 off CSR_2 is fired. Capacitor C_1 discharges through L_2 coupling a voltage pulse via L_1 to CSR_1 so turning it off. The voltage on C_1 falls to zero after which D_1 conducts, carrying the current due to energy stored in L_2 and the load. CSR_2 turns off when the current in L_2 has fallen to below the device holding value. C_2 is not normally required. By including it CSR_1 can be

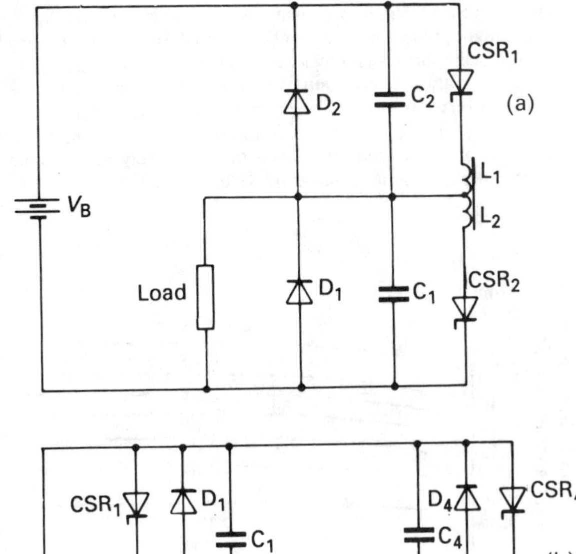

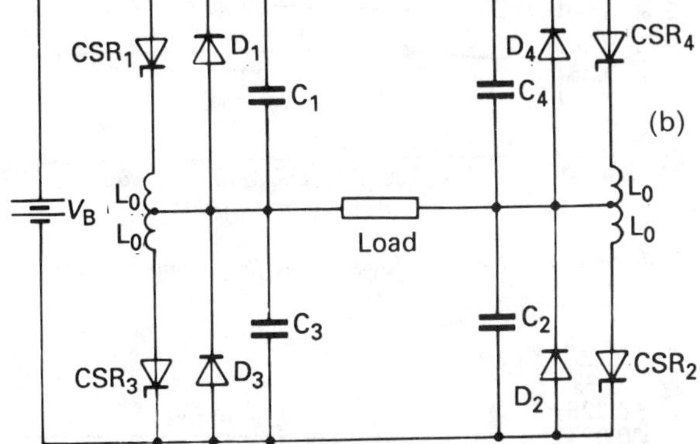

Figure 42.28 Coupled-pulse commutation: (a) chopper, (b) inverter

fired before CSR_2 has turned off since then C_2 discharges through L_1 coupling a pulse to CSR_2 via L_2 and so turning it off. This allows a wider voltage control range.

Assuming a symmetrical system with $C_1=C_2=C$ and $L_1=L_2=L$ the values of L and C can be found as

$$L=\frac{2.76\times V_B\times t_F}{I_{L(max)}} \tag{42.22}$$

$$C=\frac{2.15\times t_F\times I_{L(max)}}{V_B} \tag{42.23}$$

This again shows the unsuitability of this commutation method for low-voltage high-current operation. The circuit of *Figure 42.28(a)* can be extended to give the inverter circuit of *Figure 42.28(b)*, which is perhaps the most reliable and frequently used inverter circuit.

With any device conducting the commutation capacitor in the corresponding half of the bridge is charged to V_B. When the thyristor in that leg is fired the capacitor discharges through half of the auto-transformer, coupling a turn-off pulse to the conducting thyristor via the other half of the transformer. The circuit is versatile in that firing one device automatically commutates the other half of the bridge. It can be used to produce outputs with zeros in the waveform with no additional circuitry. It does, however, suffer from low efficiency, due to commutation losses at higher frequencies, and is therefore not suited for pulse-width modulated voltage control methods. It cannot be used in circuits which have d.c. supply variations to change the output, since the commutation voltage is now affected.

42.4.2 Effect on load and device currents

In this section a typical chopper and inverter are considered along with their commutation components. A method of specifying the device ratings and the value of the commutation inductor and capacitor will also be studied. Although these will differ according to the circuit adopted, the calculation procedure remains unchanged.

In *Figure 42.24*, the mean current rating of diode D_1 is clearly still given by Equation (42.5), its peak rating being $I_{L(max)}$, the current just before CSR_1 turns off (assuming the commutation interval to be short relative to the chopper operating period). The diode voltage rating should exceed $2V_B$ which is obtained when $L_F=1$. Mean current rating of thyristor CSR_1 is I_B, as given by Equation (42.4), plus the resonant current due to C and L_1. This gives a total current

$$I_{CSR_1(mean)}=I_B+\frac{2CV_B}{t_c+t_0} \tag{42.24}$$

The peak current is either $I_{L(max)}$ or the resonant value of $V_B\sqrt{(C/L_1)}$ whichever is greater. The voltage rating must exceed V_B. Assuming C to discharge at constant load current, CSR_1 must have a turn-off time shorter than $CV_B/I_{L(max)}$.

Thyristor CSR_2 carries a current of $I_{L(max)}$ during the charge period of C only. Its mean current rating is $2CV_B/(t_c+t_0)$ although its size is normally determined by its peak current capacity of $I_{L(max)}$. Similarly diode D_2 has a mean rating of $2CV_B/(t_c+t_0)$ but a peak resonant current of $V_B\sqrt{(C/L_1)}$ which can be high. The value of C is fixed by the commutation requirements and L_1 is chosen such that the resonant time $\pi\sqrt{(L_1C)}$ is small compared to the operating frequency (between 5 and 10%).

Figure 42.29 shows the instant of commutation for CSR_3 to an enlarged scale, for the inverter of *Figure 42.28(b)*. Assuming

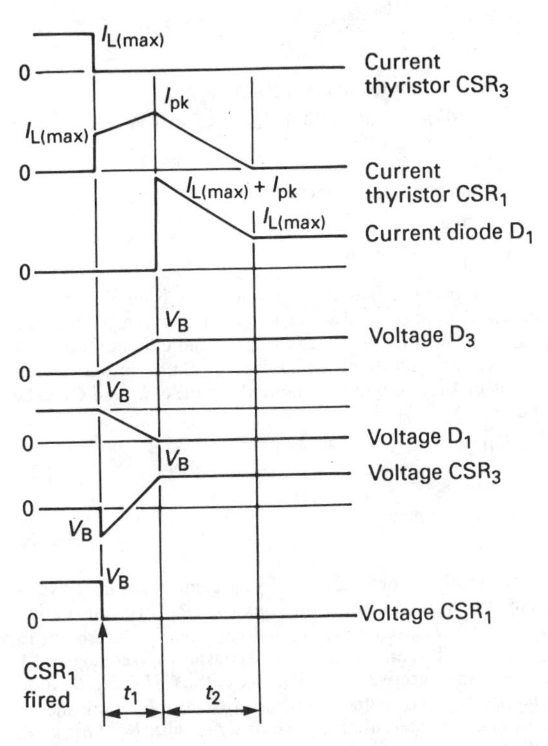

Figure 42.29 Circuit waveforms for *Figure 42.28(b)* during the commutation of thyristor CSR_3

CSR_3 and CSR_4 to be conducting, the load current being at its peak value of $I_{L(max)}$. C_3 and C_4 are at zero voltage (device volt-drops and d.c. resistance of centre-tapped chokes L_0L_0 being neglected). CSR_1 is now fired. If the leakage inductance of the chokes is ignored, current $I_{L(max)}$ transfers instantaneously from CSR_3 to CSR_1, capacitor C_1 discharging to support both this and the load current.

The current through CSR_1 increases, reaching a peak of I_{pk} after time t_1 when C_1 has completely discharged. Energy stored in L_0 now free-wheels through diode D_1 and is dissipated, falling to zero after a further time t_2. Load current is carried by CSR_4 and D_1.

When CSR_2 is fired at a later interval to commutate CSR_4 the load current has decayed to $I_{L(min)}$. Therefore commutation requirements and increased device ratings are not as severe as before. In the 'worst' case, when there are no zero dwell periods in the output voltage waveform, the ratings of all devices are equal. This will be considered here.

If t_F denotes the turn-off time seen by the thyristor being commutated, and W_L is the watts loss per commutation, caused by the dissipation of the energy $\frac{1}{2}L_0(I_{pk})^2$ stored in L_0 then *Figure 42.30* shows the variation of turn-off, peak current and watts-loss

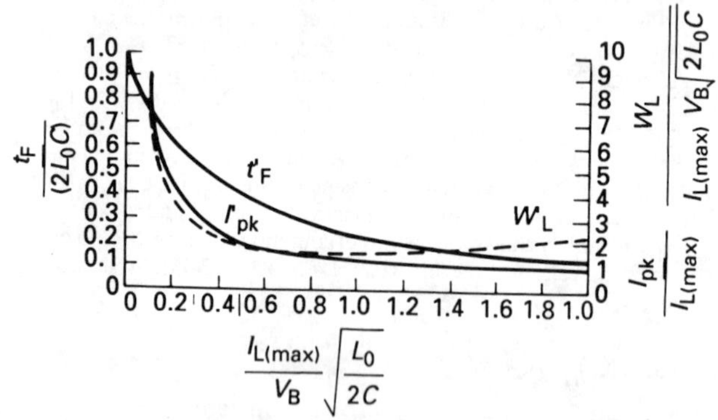

Figure 42.30 Variation of $t_F/\sqrt{(2L_0C)}$, $I_{pk}/I_{L(max)}$ and $W_L \times 2/I_{L(max)}V_B\sqrt{(2L_0C)}$ with $I_{L(max)}\sqrt{(L_0/2C)}/V_B$

factors with commutation factor

$$\frac{I_{L(max)}}{V_B}\left[\frac{L_0}{2C}\right]^{1/2}$$

From this graph it is seen that watts loss is minimum at a commutation factor of 0.8, the variation between 0.6 and 1.0 being slight. Below 0.6 the peak current and commutation loss increases steeply, although available turn-off time also increases. Working on the minimum loss point the value of L_0 and C can be found as:

$$L_0 = \frac{2.76V_B t_F}{I_{L(max)}} \tag{42.25}$$

$$C = \frac{2.15t_F I_{L(max)}}{V_B} \tag{42.26}$$

The contribution to device ratings by the commutation interval is directly dependent on time t_2 in *Figure 42.29*. This is the period required for the current in CSR_1 to decay from I_{pk} to zero due to losses across CSR_1 and D_1 (assumed a constant and equal to ΔV), and due to the 'effective loss' resistance of the CSR_1–L_0–D_1 loop (equal to R_e). To reduce this time to a minimum means introducing an external resistance in series with D_1. This gives greater losses during normal operation, with a subsequent reduction in efficiency. There is no one acceptable solution for all cases. Some inverters may be using devices which are overrated, so that higher r.m.s. currents can be accepted. In others efficiency may not be important so losses across a larger R_e are tolerated. In any case it is always important to ensure that t_2 is less than half a cycle of inverter operation, to prevent commutation failures.

If I_{D_1} and I_{T_1} represent the commutation current, in r.m.s. values, through the diode and thyristor respectively, and T is the inverter periodic time, *Figures 42.31* and *42.32* show plots of

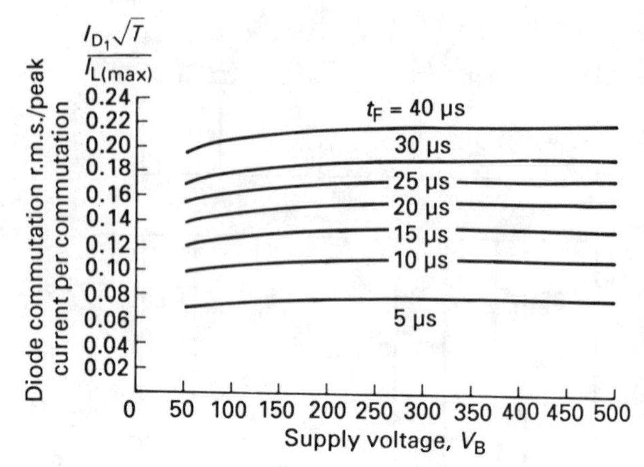

Figure 42.31 Variation of $I_{D_1}\sqrt{T}/I_{L(max)}$ with supply voltage

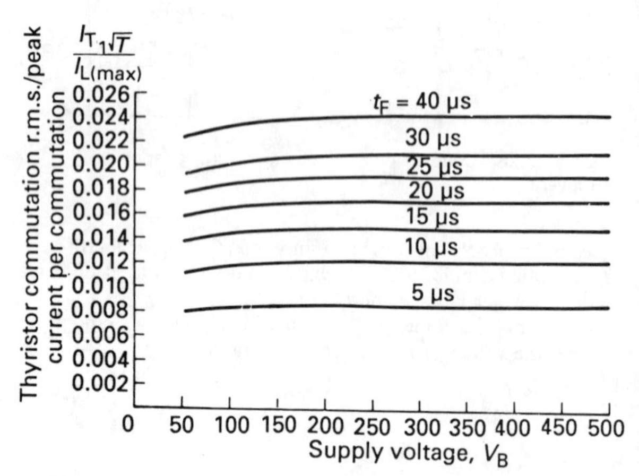

Figure 42.32 Variation of $I_{T_1}\sqrt{T}/I_{L(max)}$ with supply voltage

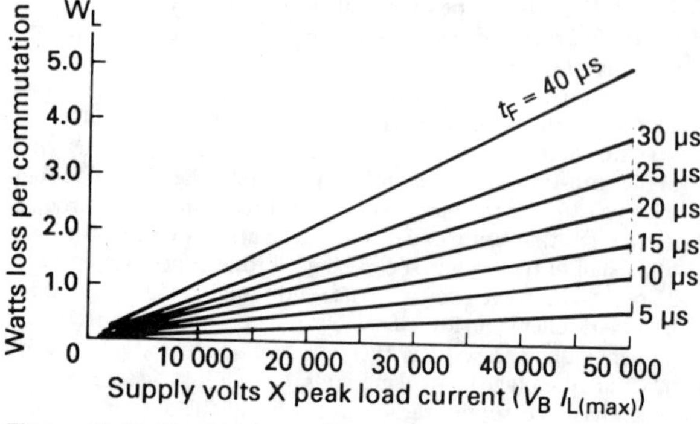

Figure 42.33 Power loss per commutation

$I_{D_1}\sqrt{T}/I_{L(max)}$ and $I_{T_1}\sqrt{T}/I_{L(max)}$ against supply voltage. These are calculated for values of L_0 and C given by Equations (42.25) and (42.26) and for $\Delta V=2.5$ V and $R_e=0.2V_B/I_{L(max)}$. These graphs show the advantage of using fast turn off time devices. The total thyristor and diode ratings are given by the geometric sum of values read off from *Figures 42.21, 42.22, 42.31* and *42.32*. Therefore when operating at an inverter frequency of 10 Hz, $I_{T_1}/I_{L(max)}$ is nearly constant at $0.0248/\sqrt{0.1}=0.079$ whereas *Figure 42.21* gives the thyristor current without commutation as $0.35I_{L(max)}$. The total current is also approximately this value. For 1 kHz operation, however, $I_{T_1}=0.79I_{L(max)}$ and *Figure 42.21* still gives the same value as before so that total thyristor r.m.s. current with commutation is

$$\sqrt{(0.79^2+0.35^2)}I_{L(max)}=0.745I_{L(max)}$$

Therefore this inverter is not suitable for higher-frequency operation. The same conclusions can be derived from *Figure 42.33* which gives the watts loss per commutation. Now, for 40 μs devices and $V_BI_{L(max)}=50\,000$, total energy loss per second at 10 Hz is 200 W whereas at 1 kHz it is 20 kW.

Further reading

BEDFORD, B. D. and HOFT, R. G., *Principles of Inverter Circuits*, Wiley, New York (1964)

CREPAZ, S. *et al.*, 'Electromagnetic components for electronic circuits with forced commutation', *Proc. Int. Conf. Electrical Machines*, 18–21 September (1984)

DA SILVA, E. R. C. *et al.*, 'Forced commutation in thyristor bridge inverters', *Proc. IECON '85*, 18–22 November (1985)

DAVIS, R. M., *Power Diode and Thyristor Circuits*, IEE Monograph No. 7 (1971)

GENTRY, F. E. *et al.*, *Semiconductor Controlled Rectifiers*, Prentice-Hall, Englewood Cliffs, NJ (1964)

GRIFFIN, A. and RAMSHAW, R. W., *The Thyristor and its Applications*, Chapman & Hall (1965)

KAZUNO, H., 'Forced commutation thyristor bridge using pulse transformer', *Elect. Eng. Jpn.* (*USA*), January–February (1986)

MATILAINE, J. L., 'Solid-state 25 kJ/sec capacitor charging modulator', *IEEE Conference Record of 1984 16th Power Modulator Symposium*, 18–20 June (1984)

MAZDA, F. F., *Thyristor Control*, Butterworths, London (1973)

MOREIRA, J. C., 'Analysis, implementation and comparison of forced commutation inverter circuit topologies', *Proc. IECON '85*, 18–22 November (1985)

MURRAY, R. (Ed.), *Silicon Controlled Rectifier Designer's Handbook*, Westinghouse Electric Corporation (1963)

NAKAOKA, M. *et al.*, 'A phase-different angle control-mode PWM high-frequency resonant inverter using static induction transistors and thyristors', *18th Annual IEEE Power Electronics Specialists Conference*, 21–26 June (1987)

PELLY, B. R., *Thyristor Phase-Controlled Converters and Cycloconverters*, Wiley-Interscience, New York (1971)

SEYMOUR, J. (Ed.), *Semiconductor Devices in Power Engineering*, Pitman (1968)

SZPILKA, S. J., 'Induction motor speed control with a compensator in the rotor', *2nd Eur. Conf. Power Electronics and Applications*, 22–24 September (1987)

43 Instrumentation and Measurement

C L S Gilford
MSc, PhD, FInstP, MIEE, FIOA
Independent Acoustic Consultant
(Sections 43.9–43.10)

F F Mazda
DFH, MPhil, CEng, FIEE, DMS, MBIM
STC Telecommunications Ltd
(Sections 43.1, 43.4–43.7, 43.8.8)

R C Whitehead, CEng, MIEE
Polytechnic of North London
(Sections 43.2, 43.4, 43.8.2–43.8.7, 43.8.9)

M J Rose BSc(Eng)
Mullard Ltd
(Section 43.8.1)

Contents

43.1 Measurement standards

No matter what the parameter, eventually all measurements are related back to a common set of standards. In order to ensure correlation between these, measurement standards fall into four levels; international, primary, secondary and working. International standards are defined by international agreement, and are maintained at the International Bureau of Weights and Measures at Sèvres, near Paris. They are not available for general use, but are checked periodically by absolute measurement, in terms of the fundamental unit concerned.

Primary standards are kept in the various National Laboratories of different countries, and are used to calibrate secondary standards which are sent to the laboratories. They are themselves calibrated by absolute measurement. Secondary standards are kept in standards laboratories within industry, and are used to calibrate working standards, which represent the lowest level of measurement standard, being employed to check and calibrate laboratory instruments within that industry.

Standards are generally used to measure five parameters: mechanical, electrical, magnetic, thermal and optical.

43.1.1 Mechanical standards

There are three mechanical standards—those for the measurement of mass, length and time.

43.1.1.1 Mass

The standard for mass in SI units is the kilogram, which was defined in 1889, by the first Conference Générale des Poids et Mesures, as equal to the mass of a cylinder of platinum–iridium alloy which is kept in the Bureau at Sèvres. Secondary mass standards have an accuracy of 1 ppm, whereas working standards have an accuracy of 5 ppm.

43.1.1.2 Length

The SI unit of length is the metre, which was defined in 1983 by the Conference as the length travelled by light in vacuum in a time interval of 1/299 792 458 s. Working standards usually consist of steel gauge blocks having two flat parallel surfaces set at a specified distance apart, and giving an accuracy of about 1 ppm.

43.1.1.3 Time

There are two types of time; elapsed time (i.e. the time between the occurrence of two events) and time-of-day or epoch, which is the time measured from some fixed reference. Atomic standards are generally used to measure time to a very high accuracy. When an atom moves between two energy states E_1 and E_2 it emits radiation of frequency λ given by Equation (43.1), where h is Planck's constant. Caesium, hydrogen and rubidium have been used as atomic standards, and their characteristics are compared in *Table 43.1*.

$$h\lambda = E_2 - E_1 \qquad (43.1)$$

In 1967 the 13th Conference defined the International Second as the duration of 9 192 631 770 periods of radiation corresponding to the transition between two hyperfine levels ($F=4, m_f=0$, and $F=3, m_f=0$) of the ground state ($^2S_{1/2}$) of the caesium-133 atom.

Atomic standards give accurate time scales, but for applications such as navigation correlation with the rotation of the earth is needed. For this measurement apparent solar time can be used, which is based on observing the rotation of the earth about its axis. This time varies due to the elliptical orbit of the earth and the tilt in the earth's axis, and an average is usually taken of all the apparent solar days in a year. This is called a mean solar day, and a mean solar second is equal to a mean solar day divided by 86 400.

In 1956 the International Committee of Weights and Measures defined the second as 1/31 556 925.9747 of the tropical year for 0 January 1900 (equal to 31 December 1899) at 1200 hours. This is known as an ephemeris second. It is determined by observing the motion of the moon, and then referring to lunar ephemeris tables.

43.1.2 Electrical standards

Standards are available for four main electrical parameters: voltage, current, resistance and capacitance.

43.1.2.1 Voltage

The SI unit of voltage is the volt, which is defined as the potential between two points of a wire carrying a current of 1 A, when the power dissipated between them is equal to 1 W. The three main kinds of voltage standard in use are the Weston cadmium cell, the Josephson-effect standard and the zener diode standard.

The saturating Weston cadmium cell is illustrated in *Figure 43.1*. An H glass vessel contains metallic mercury in one limb, which forms the positive electrode, and cadmium–mercury amalgam in the other, forming the negative electrode. A layer of cadmium sulphate crystals maintains the solution in a saturated state. The e.m.f. of the cell is given by Equation (43.2) where $E_{(20)}$ is the value at 20°C, that at any other temperature t being given by Equation (43.3).

$$E_{(20)} = 1.018\,636 - 6.0\times10^{-4}N - 5.0\times10^{-5}N^2 \qquad (43.2)$$

$$E_t = E_{(20)} - 4.6\times10^{-5}(t-20) - 9.5\times10^{-7}(t-20)^2 + 1.0\times10^{-8}(t-20)^3 \qquad (43.3)$$

The saturated cell has a voltage reproducible to within a few microvolts of Equation (43.2), but it has a relatively large

Table 43.1 Comparison of three atomic standards

Characteristics	*Caesium beam*	*Hydrogen maser*	*Rubidium vapour*
Reproducibility	$\pm3\times10^{-12}$	$\pm2\times10^{-12}$	—
Stability (1 s average)	5×10^{-12}	5×10^{-13}	5×10^{-12}
Drift	Low	Low	$\pm1\times10^{-11}$ per month
Resonant frequency (Hz)	9 192 631 770	1 420 405 751	6 834 682 608
Operating temperature	−20°C to +60°C	0°C to +50°C	0°C to +50°C
Approx. weight (kg)	30	400	15
Approx. power required (W)	40	200	40
Atomic interaction time (s)	2.5×10^{-3}	0.5	2×10^{-3}

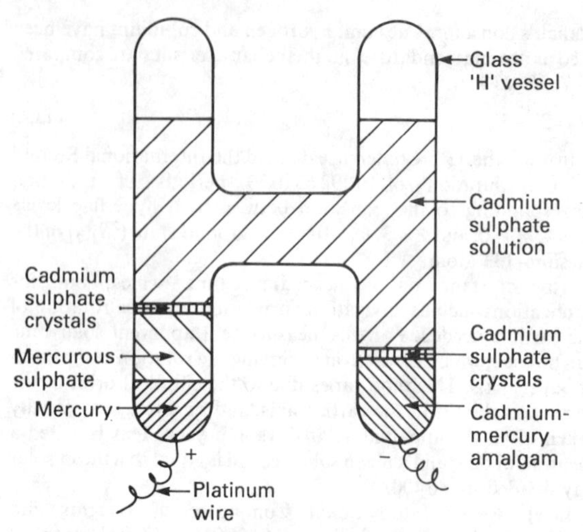

Figure 43.1 Saturating Weston cadmium cell

temperature coefficient of about $-39.4\ \mu V/°C$ at 20°C. It is therefore usually operated in elaborate temperature-controlled environments. The unsaturated cell, which is similar in construction to the saturated cell except that the solution is unsaturated at normal room temperatures, has a lower accuracy than the saturated cell, but is also more stable having a temperature coefficient below $-10\ \mu V/°C$. It therefore needs less complex temperature-controlled environments, making it ideal for use as a working standard. Weston cadmium cells have a high internal resistance and the current drawn from them should not exceed about 100 μA.

Josephson-effect standards consist of a thin film of lead separated by a lead oxide insulating barrier, mounted on a glass substrate and placed in a superinsulated helium Dewar system. The tunnelling junction is excited by a microwave source and is biased by a current source. The Josephson junction produces a voltage given by Equation (43.4), where h is Planck's constant, f is the frequency of the microwave source, e is the electron charge, and N is an integer. The voltage standard has an accuracy of about 0.05 ppm, and is used to calibrate other voltage standard sources.

$$E = Nhf/2e \qquad (43.4)$$

Zener diode voltage standards utilise the sharp, well defined reverse voltage breakdown characteristics and low dynamic impedance of zeners. Alloy junction diodes are used in the range 2.4–12 V, and diffused diodes from 6.8 to 200 V. Usually the zener diode is connected in series with one or more conventional diodes such that the negative temperature coefficient of the diodes compensates for the positive coefficient of the zener. The zener voltage standard, called a transfer standard, is placed in a temperature-controlled oven, held to about 0.01°C. It is used as a laboratory working standard, and has a stability of about 1 ppm/month.

43.1.2.2 Current

The 1960 11th General Conference on Weights and Measures defined the ampere in SI units, also called the Absolute Ampere, as the constant current which, when maintained in two straight parallel conductors of infinite length and negligible cross-section, placed 1 m apart in vacuum, produces a force between them of 2×10^{-7} N/m length. Although the ampere can be measured in absolute terms, using a current balance to determine the force between two current carrying conductors, it is more usual to maintain current standards by using a standard voltage and resistance, and then applying Ohm's law to calculate current.

43.1.2.3 Resistance

The ohm in the SI system is defined in terms of the units of length and mass, and absolute measurements of the ohm are made against a group of resistance standards maintained at the International Bureau of Weights and Measures at Sèvres.

Figure 43.2 shows the construction of a typical standard resistor. The resistive element is made from thin resistance wire having a low temperature coefficient of resistivity, and wound so as to reduce stress. Often the coil is immersed in a bath of oil for stability, and the container sealed against moisture. Errors can arise due to skin effect, and stray inductance and capacitance, when operating at high frequencies. These errors are minimised by limiting the wire diameter, and by winding the coil such that adjacent turns carry current in opposite directions (Ayrton–Perry method of construction). *Figure 43.2* also shows a four-terminal arrangement, where current is fed between the two outer terminals and voltage is read between the inner terminals, so reducing the effects of contact resistance when measuring low valued resistances.

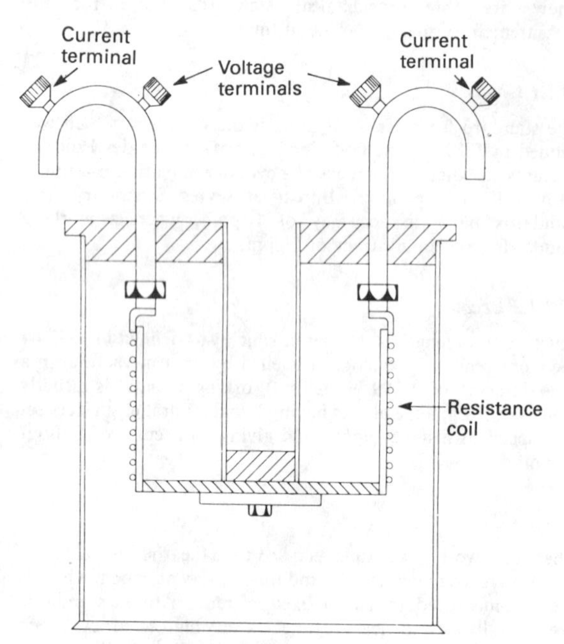

Figure 43.2 Typical standard resistor

43.1.2.4 Capacitance

The farad is the unit of capacitance in the SI system, and is defined as the capacitance between the plates charged to 1 C when there is a potential difference of 1 V between them. Standard capacitors are usually made from interleaved plates of a material such as invar which has a low temperature coefficient of expansion, and are sealed in air or nitrogen environments, which act as the dielectric. Capacitances up to about 1000 pF are obtainable, having an accuracy of 0.02 ppm and a drift of below 20 ppm/year.

Working standards use air dielectric for the smaller sizes, and a solid dielectric for larger valued capacitances.

43.1.3 Magnetic standards

These consist primarily of inductor and flux standards. The SI unit for inductance is the henry, which is defined as the inductance of a closed circuit which produces an induced voltage of 1 V for a current change in the circuit of 1 A/s. Primary standards for inductance consist of single layer coils wound on a stable fused silica former. Working standards are usually of multi-turn construction wound on a ceramic or Bakelite former, or on toroid cores. Inductances from 10 μH to 10 H are available, with stabilities of 100 ppm/year.

The weber is the SI unit for magnetic flux and is defined as the flux which, linking a coil of one turn, produces in it an e.m.f. of 1 V, when the flux decays to zero in 1 s at a uniform rate. Primary flux standards can be obtained by the decay of current through standard inductors. The Hibbert standard uses the effect of a coil moving at a constant rate past a permanent magnet to determine the current induced in the coil, which is proportional to the flux in the air gap.

43.1.4 Thermal standards

The SI unit for thermodynamic temperature is the kelvin, which is defined as 1/273.16 of the value of the thermodynamic temperature of the triple point of water. The triple point of water is the temperature at which ice, liquid water and water vapour are at equilibrium.

The 1927 General Conference of Weights and Measures also defined an alternative temperature scale, called the practical scale, based on the degree celsius. This has two fundamental fixed points, the boiling point of water at atmospheric pressure, equal to 100°C, and the triple point of water at atmospheric pressure, equal to 0.01°C. The primary standard for temperature is the platinum resistance thermometer, whose temperature can be calculated accurately knowing the properties of platinum wire.

43.1.5 Optical standards

The SI unit of luminous intensity is the candela, which is the luminous intensity of a surface of 1/600 000 of a square metre of a black body, at a temperature at which platinum solidifies, and under atmospheric pressure. The primary standard of luminous intensity is a full radiator, black body or Planckian, maintained at the above temperature. Secondary standards consist of special tungsten filament lamps, whose power is adjusted to give an operating temperature at which the spectral power distribution of the lamp matches that of the basic standard.

43.2 Analogue voltmeters

These instruments may be designed for the measurement of d.c. and/or a.c. In addition, resistance measuring facilities may be added. Since these instruments take less power from the circuit under test than is required to operate the moving-coil meter used for the final display, they must incorporate amplifiers. Power sources for the amplifiers may be mains units and/or batteries. The batteries may be of the primary or secondary type, and if they are the latter then charging facilities are usually provided. Provision is made for checking battery voltages using the same moving-coil meters that are used for the display of signal amplitude.

43.2.1 DC measurement

Consider first operation at d.c., for which a basic schematic and simplified circuit are shown in *Figure 43.3*. R_1 is a switch-controlled resistor chain used for range changing. R_2 is first

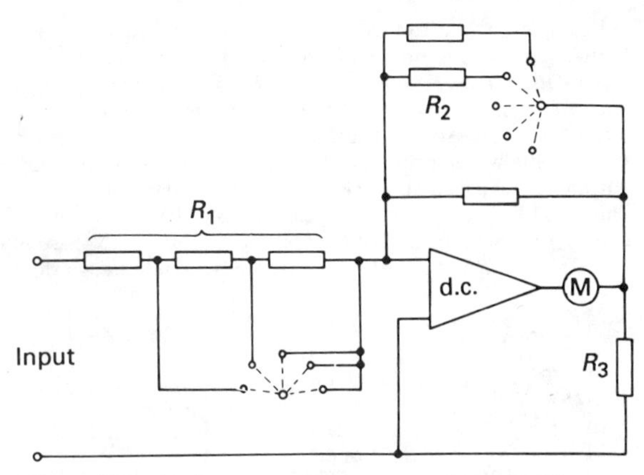

Figure 43.3 A voltmeter employing an operational amplifier

adjusted to standardise the effective supply voltage. The voltage gain of the d.c. amplifier, as modified by negative feedback, is given by R_2/R_1, and the current through the meter M is further controlled by R_3. Typically the value of R_2 is fixed for all except the high voltage ranges at a value of 100 MΩ, while the value of R_1 is switched in accordance with the range in use, so that the overall sensitivity of the complete circuit is 1 μA/V. For ranges above 100 V however this would involve the use of an inconveniently high value of R_1, so for these higher voltage ranges the value of R_1 may be kept fixed at 100 MΩ and reduced values of R_2 be switched into circuit, giving thereby a relatively reduced value of input resistance in terms of Ω/V.

For high values of sensitivity, amplifier stability must be correspondingly high and this is commonly achieved by the use of chopper-stabilised amplifiers. In these, the d.c. signal is modulated to produce a low-frequency signal which is amplified and then rectified. The amplifier should have a narrow bandwidth. For operation at the lowest levels it is advisable to use synchronous detection to make the indication phase sensitive. The signal will produce deflection in one direction only, but noise will produce components in both directions, causing the meter to dither about its correct indication. If a long time-constant circuit is added to the rectifier output, the noise component can be reduced, but at the expense of sluggish operation.

43.2.2 AC measurement

In the measurement of alternating voltages, attention must first be given to the quantity which is to be measured. Although peak-to-peak values may be required for special purposes (especially when the meter is used in conjunction with an oscilloscope), normally these voltmeters are used for the measurement of the root-mean-square values of the fundamental components of waveforms which are nominally sinusoidal, but which may contain small proportions of harmonics, noise, hum and/or other components. The influence of these additional components upon the accuracy of measurement depends upon the measurement circuit employed and sometimes upon the amplitude of the signal. The meter deflection may be proportional to: the arithmetic mean value of the waveform; the peak value; the peak-to-peak value; the root-mean-square (r.m.s.) value. Due notice must be taken of the fact that the relationships between the r.m.s. value of the waveform and the quantities to which the meter readings are proportional are respectively $\pi/2$, $1/\sqrt{2}$, $1/2\sqrt{2}$ and 1. (Manufacturers take account of these factors when calibrating the meter scales.)

Examples of the results of adding an even-order (second) harmonic, and odd-order (third) harmonics to sinusoids are illustrated in *Figure 43.4*. At (*a*) a second harmonic has been added. During any half-period of the fundamental, the arithmetic mean value of the second harmonic is zero. The total arithmetic value of the waveform is thus unchanged when an even-order harmonic of small amplitude is added to the fundamental. This is illustrated by the two shaded areas in (*a*), which as shown are equal for the fundamental and the sum. Variation of the phase of the harmonic relative to the fundamental will not change this.

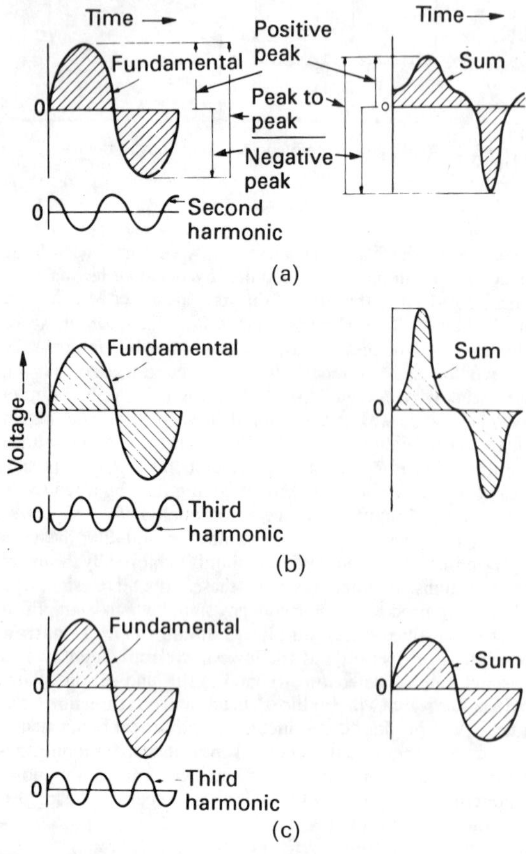

Figure 43.4 Illustrating the influence of added harmonics to a sinusoid: (a) second harmonic, (b,c) third harmonic

The additions of third harmonics to the fundamentals, but with different phase relationships, are shown in (*b*) and (*c*). In (*b*) the sum waveform encloses a smaller area than the fundamental, while in (*c*) the reverse is true. Thus the circuit which responds to the arithmetic mean value of the waveform performs satisfactorily in the presence of even-order harmonics, but less so in the presence of odd-order harmonics, the sign of the error being unknown unless the phase relationship is known.

Consider now the measurement of peak values. Examination of *Figure 43.4*(*a*) shows that for even-order harmonics the sign of the error is dependent upon which peak is measured. Examination of *Figure 43.4*(*b*) and (*c*) shows that while the sign of the error is independent of which peak is measured, the magnitude of the error is dependent upon the phase relationship.

Examination of the peak-to-peak values shows that these are not affected by the addition of small proportions of even-order harmonics, but that positive or negative errors may occur when odd-order harmonics are added.

When root-mean-square operation is employed, the fundamental and harmonic and noise amplitudes are added in quadrature. For instance a 10% harmonic will produce an error of $100\,(\sqrt{1^2+0.1^2}-1)\%=0.5\%$. The error is always positive.

Two typical basic forms of a.c. voltmeters are shown in *Figure 43.5*. In (*a*) the signal is first rectified, then, when necessary, attenuated to keep the amplitude within the operating range of the d.c. amplifier–meter combination. In (*b*), the signal is, when necessary, attenuated, amplified by the a.c. amplifiers, rectified, and a value displayed on the meter. These alternative schemes are now considered in greater detail.

A basic rectifier circuit is shown in *Figure 43.6*(*a*). Initially the following assumptions are made:

(1) Source resistance $R_s \ll$ the load resistance R.
(2) Diode resistance r when highly conductive $\ll$ the load resistance R.
(3) Resistance R_f in low pass filter $\gg$ the load resistance R.
(4) Amplitude of the source is high, e.g. >10 V, so that the diode D acts as an on/off switch.
(5) Time constants CR and $C_fR_f \gg$ the time period of the source t.

If the r.m.s. value of the input voltage is V_i then the capacitor voltage V_c reaches a maximum value of $\sqrt{2}\,V_i$, falling due to the discharge through R during the non-conductive period of the diode to $\sqrt{2}\,V_i-\delta V_c$ where δV_c is the change in capacitor voltage as is shown in (*b*). When $CR \gg t$ then the conductive period of the diode is $\delta t \ll t$, thus the *average* value of V_c is approximately $\sqrt{2}\,V_i-(\delta V_c/2)$. The circuit is nominally peak indicating so that the fractional error, Δ, is as follows:

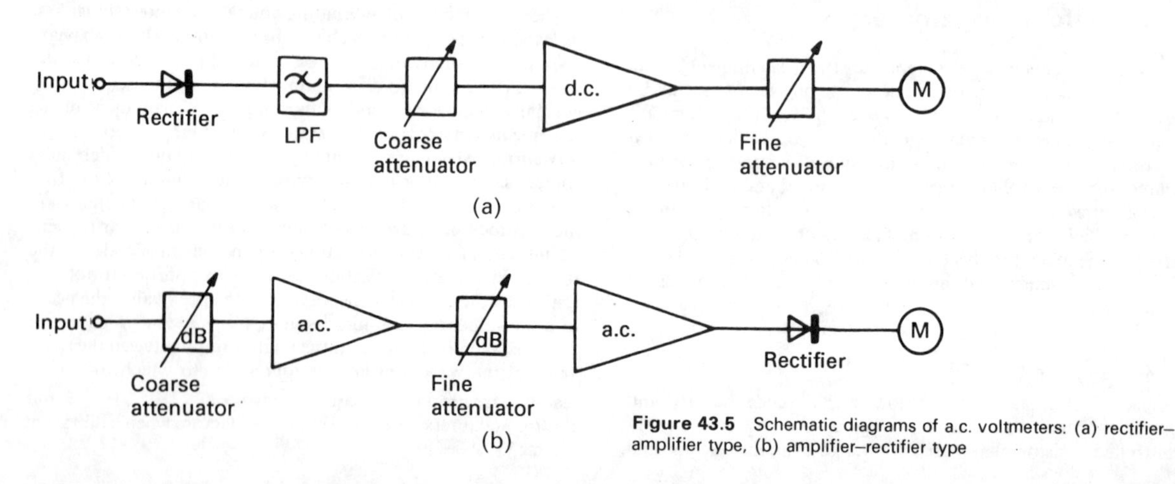

Figure 43.5 Schematic diagrams of a.c. voltmeters: (a) rectifier–amplifier type, (b) amplifier–rectifier type

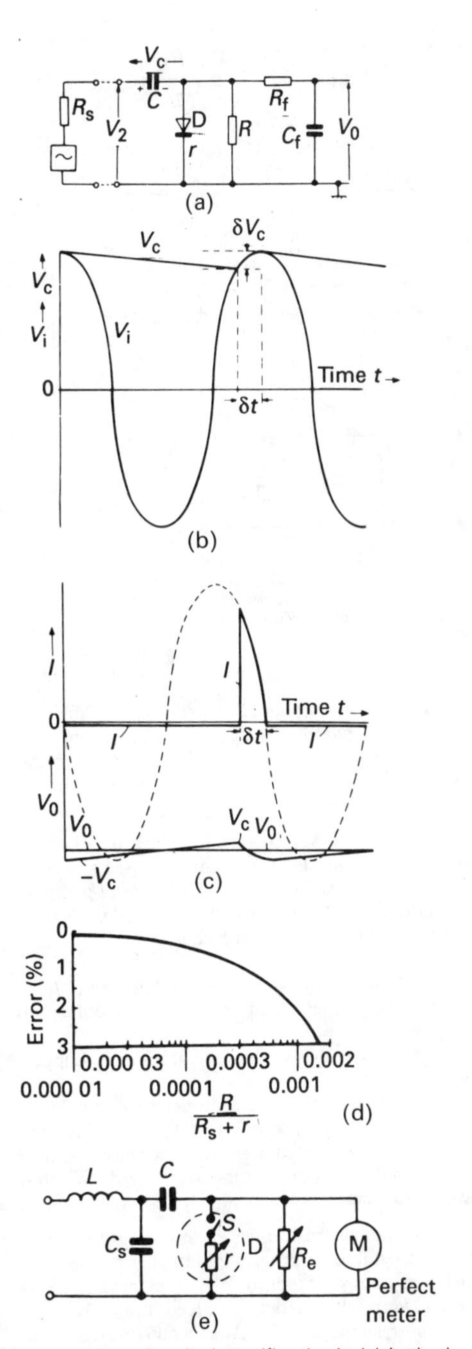

Figure 43.6 The diode rectifier circuit: (a) basic circuit, (b) capacitor and input voltage waveforms, (c) capacitor and circuit current and voltage waveforms, (d) circuit error, (e) equivalent circuit

$$\Delta = \frac{\text{average value of } V_c - \text{maximum value of } V_c}{\text{maximum value of } V_c}$$

$$\simeq \frac{-\delta V_c}{2C_{\text{cmax}}}$$

$$= \frac{-\delta Q}{2V_{\text{cmax}}C} \qquad (43.5)$$

where δQ is the change in capacitor charge.

If the capacitor discharge period approximates to the period t and the capacitor current is I as in (c) then

$$\Delta \simeq \frac{-It}{2V_{\text{cmax}}C}$$

$$= -\frac{t}{2CR} \qquad (43.6)$$

or for design purposes

$$CR = \frac{1}{-\Delta 2f} \qquad (43.7)$$

The required time constant is very much longer than that which is necessary for a similar value of Δ in an amplifier coupling circuit.

When the diode is a thermionic valve, there is hardly any practical limit to the value of capacitance that may be used and hence the minimum frequency of operation. When a semiconductor diode is used, a limit occurs due to the magnitude of the charge that must be passed by the diode when a connection is made to a point where a high value steady potential exists, e.g. at the collector of a transistor or anode of a valve. The output voltage V_o is shown in *Figure 43.6(c)* to be the conjugate of the average value of the capacitor voltage V_c. This is due entirely to the polarity definition which has been adopted for V_c. The capacitor current I is also shown in *Figure 43.6(c)*. During the charge it takes on a shape which is a small portion of a sine wave, and the areas enclosed below and above the axis must be equal. The infinitely sharp wavefront of the current waveform during the charging period implies that the source and diode resistances have zero value. With finite values, the rise-time will of course be finite also, and the peak value of V_c will be less than that of the source e.m.f.

Because the capacitor charging current flows for only a very small fraction of the period t, its peak magnitude must be extremely high in comparison with its average value, and this results in significant errors unless the sum of the source resistance plus diode resistance is extremely small in comparison with the value of the load resistance R. The percentage error, as given by Scroggie, is as shown in *Figure 43.6(d)*. In addition, a combination of high value source resistance with intermittancy of current flow causes distortion of the waveform under investigation. Because of this short duty cycle of the diode, the value of input resistance cannot be specified by any simple figure. Arbitrarily, it could be defined as say 50 times the value of source resistance in connection with which there was a -2% error, but the value would change with the tolerance. However, when the circuit is operated from a resonant circuit of high magnification factor Q, then the input resistance approximates to $R/2$. This is because R is the only component in which there is significant power dissipation at low and medium frequencies, and current is taken from the source only when the instantaneous amplitude is close to $\sqrt{2}V_i$, so that the dissipation in R is $(\sqrt{2}V_i)^2/R$ so that the apparent value of R is $R/2$.

When measuring signals of high amplitude, the output voltage V_o is proportional to the input voltage V_i, so that for this condition the moving-coil meter used for display may carry, say, two scales only and multiplication factors may be used as for a simple milti-meter. For lower amplitudes the signal waveform is accommodated on the curved portion of the characteristic of the diode so that the response changes from peak, through arithmetic mean, to square law, and the period of diode conduction increases. This has the advantage that errors due to harmonics are reduced and the connection of the meter introduces less distortion of the signal under test. One disadvantage, however, is due to the fact that the diode is not operating as an on/off switch, but is conducting moderately for a large proportion of the operating cycle, the dissipation within the diode increases so that the effective value of the input resistance falls. Another disadvantage is that due to the gradual change from linear to square law operation additional meter scales are required such that simple multiplying factors can no longer be applied. The proportional reading accuracy of square-law scales is less than that on linear scales. Some improvement may however be achieved by switching into circuit nonlinear devices to improve the scale law, but such devices tend to be temperature sensitive. The problem of amplifier stability becomes all important at low levels of input signal.

An approximate equivalent circuit for a diode rectifier circuit is given in *Figure 43.6(e)*. Here *S* and *r* represent the diode. At high amplitude, and when driven from low impedance or high *Q* circuits, the duty of cycle of *S* is low as also is the value of *r*. If the amplitude is low and/or the source resistance is high, then the reverse is true. At low frequencies the actual and effective values of R_e are high, and the CR_e product limits the performance at low frequencies. At high frequencies, e.g. above 10 MHz, dielectric losses cause the value of R_e to be reduced drastically. Stray capacitance C_s and stray inductance *L* commonly cause a rise to occur in the response of the circuit at very high frequencies, followed by a progressive fall of response.

Extension of the linear law down to about 100 mV can be achieved by the use of the sharper changes of conductivity of the zener diode using a circuit employing two such diodes of similar voltage rating, one D_1 for rectification, and another D_2 for biasing as in *Figure 43.7*. The output will of course be superimposed upon a steady potential. Due to the absence of hole storage, the circuit is operable up to hundreds of megahertz.

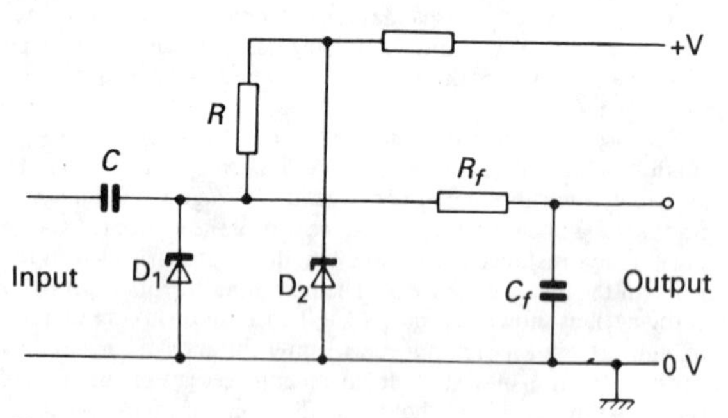

Figure 43.7 A rectifier circuit using zener diodes for operation at low amplitude

A voltage-doubler circuit responding to peak-to-peak values has been shown to have an advantage over the single diode circuit, an example of the former being shown in *Figure 43.8(a)*. An alternative version particularly suitable for use at levels below 1 mV is the balanced circuit of *Figure 43.8(b)*. In order to reduce the noise generated in high value resistors, low value components may be used instead, their effective values being increased by use of positive feedback bootstrapping circuits.

Reference to *Figure 43.5(a)* shows that d.c. amplifiers are needed. For moderate degrees of sensitivity it is necessary either to employ frequent electrical zero setting or else a chopper-stabilised circuit in order to compensate for drift.

In order to minimise capacitance loading of the circuit under test, the diode and other components closely associated with it, are commonly housed in a probe extension from the main unit. In one type a thermostatically controlled heater prevents the operating temperature of the probe from falling below 33°C, because rectification efficiency falls at low temperatures.

The maximum sensitivities which may be achieved using the circuit of *Figure 43.5(a)* are less than those for circuit *(b)*. The former type of circuit is thus only popular for use at very high frequencies for which it would be difficult to design an a.c. amplifier for use in circuit *(b)*.

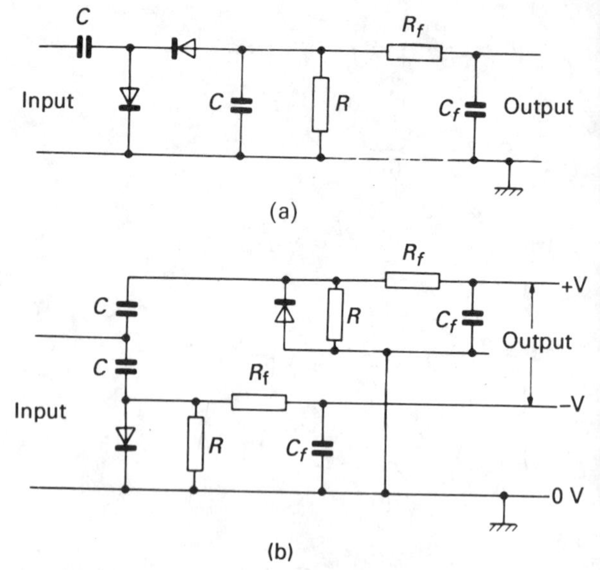

Figure 43.8 Voltage doubler circuits: (a) unbalanced, (b) balanced

For operation below a few tens of megahertz, amplification first and rectification afterwards is to be preferred, as exemplified in the circuit of *Figure 43.5(b)*. The main advantages which this circuit has over the rectifier–amplifier combination are as follows.

(1) An output connection may be taken off just ahead of the rectifier circuit so that the instrument may be used as a wide-band amplifier of adjustable and known value of gain.
(2) The value of the input resistance can be expressed as a simple quantity which does not change with change of amplitude of the signal.
(3) The rectifier circuit always operates at high amplitude, thereby producing high and consistent rectification efficiency.
(4) There is no need for zero setting or the use of chopper-stabilised amplifiers.
(5) Deflection is proportional to the arithmetic mean value of the signal so that the scale law is linear. This minimises the number of scales which appears on the meter face and thus permits scale multiplication factors to be used. Without overcrowding the meter face, a decibel scale may be added.

The detailed design of any instrument in this category centres largely upon the attenuator arrangement. In this connection, steps of 10 dB are almost standard, giving a meter face as shown in *Figure 43.9(a)*. The zero decibel value is normally taken as 775 mV r.m.s., corresponding to 1 mW in a 600 Ω circuit. The scale for a corresponding attenuator switch, which might control either both attenuators or only the fine attenuator, is shown in *Figure 43.9(b)*. Where a single attenuator switch is used, the coarse attenuator may provide steps of 20 dB, the setting of this attenuator changing on *alternate* positions of the switch, while steps of 10 dB are alternately brought in and out of circuit in the fine attenuator. This makes for efficient use of the amplifiers. An alternative arrangement more suitable for use where a very wide range of amplitudes is encountered, or where a probe must be used, is to have separate controls for the attenuators. In this case the coarse attenuator usually has two positions, one marked VOLTS/0 dB and the other marked mV/−60 dB. The fine attenuator will then cover the 60 dB range in steps of 10 dB. Where a probe is provided the coarse attenuator is usually

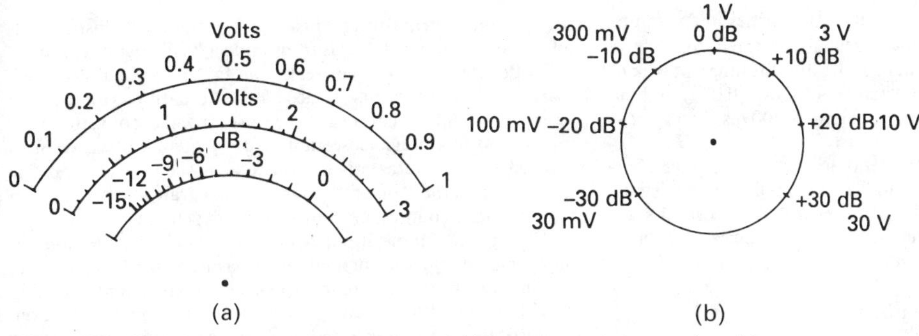

Figure 43.9 Typical voltmeter scales (a) and corresponding range switch engravings (b)

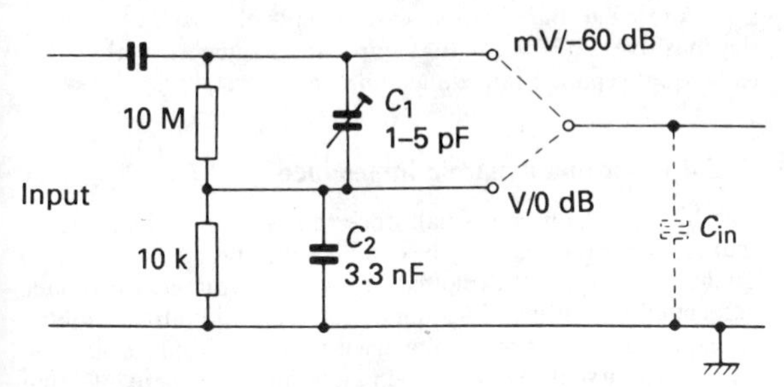

Figure 43.10 A 60 dB attenuator circuit

located within the probe. A basic circuit for such an attenuator is given in *Figure 43.10*.

If the two amplifiers are designed with stages having very high values of input impedance (using say f.e.t. source-followers or compound-emitter followers), then the attenuators may take the form of frequency-compensated potential dividers, rather than constant-impedance networks.

Where the required gain × bandwidth product is small, the first amplifier may be a simple f.e.t. source-follower, feeding a potential-divider-type fine attenuator. Otherwise an f.e.t. input stage and an emitter-follower output are appropriate, preferably with overall feedback. Since the forward resistances of the rectifier diodes change both with changes of current and changes of temperature, they should be fed from a high impedance source. This can be achieved in two ways. First the metering circuit is incorporated into a current negative-feedback loop which thus raises the effective value of the output impedance of the amplifier. Secondly, the output stage of the amplifier may be supplied with constant current load.

43.3 Digital voltmeters

The digital voltmeter provides a digital display of d.c. and/or a.c. inputs, together with coded signals of the visible quantity, enabling the instrument to be coupled to recording or control systems. Depending on the measurement principle adopted, the signals are sampled at intervals over the range 2–500 ms. The basic principles are:

(1) Linear ramp.
(2) Successive approximation/potentiometric.
(3) Voltage to frequency, integration.
(4) Dual slope.
(5) Some combination of the foregoing.

Techniques (3) and (4) are described in *Chapter 28* and (1) and (2) are described below.

43.3.1 Linear ramp

This is a voltage/time conversion in which a linear time base is used to determine the time taken for the internally generated voltage v_s to change by the value of the unknown voltage V. The block diagram, *Figure 43.11(b)*, shows the use of comparison networks to compare V with the rising (or falling) v_s; these

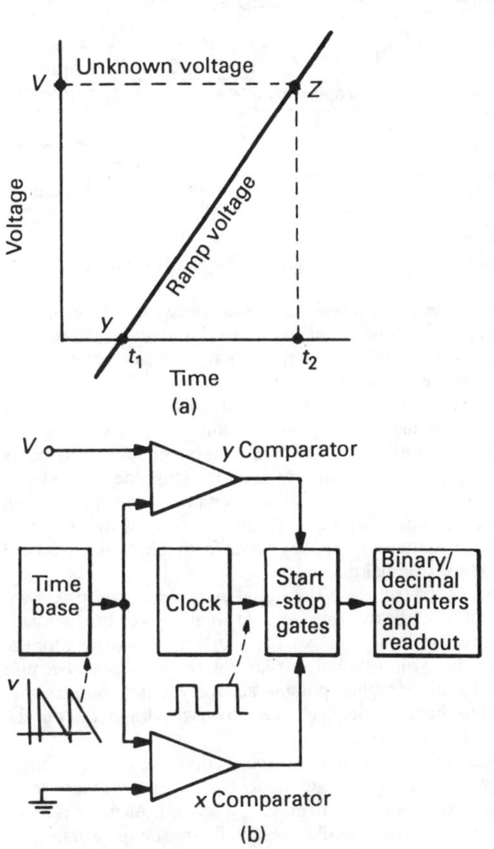

Figure 43.11 Linear-ramp digital voltmeter: (a) ramp voltage, (b) block diagram

networks open and close the path between the continuously running oscillator, which provides the counting pulses at a fixed clock rate, and the counter. Counting is performed by one of the binary coded sequences, the translation networks give the visual decimal output. In addition a binary coded decimal output may be provided for monitoring or control purposes.

Limitations are imposed by small nonlinearities in the ramp, the instability of the ramp and oscillator, imprecision of the coincidence networks at instants *y* and *z*, and the inherent lack of noise rejection. The overall uncertainty is about ±0.05%, and the measurement cycle would be repeated every 200 ms for a typical 4-digit display.

Linear 'staircase' ramp instruments are available in which *V* is measured by counting the number of equal voltage 'steps' required to reach it. The staircase is generated by a solid-state diode pump network, and linearities and accuracies achievable are similar to those with the linear ramp.

43.3.2 Successive approximation

As it is based on the potentiometer principle, this class produces very high accuracy. The arrows in the block diagram of *Figure 43.12* show the signal-flow path for one version; the resistors are selected in sequence so that, with a constant-current supply, the test voltage is created within the voltmeter.

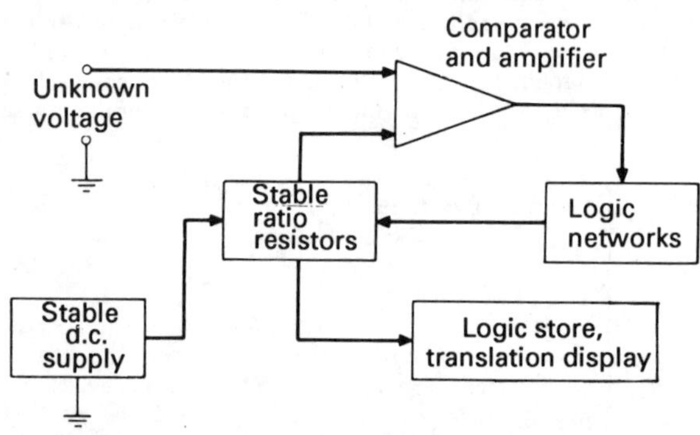

Figure 43.12 Successive-approximation digital voltmeter

Each decade of the unknown voltage is assessed in terms of a sequence of accurate stable voltages, graded in descending magnitudes in accordance with a binary (or similar) counting scale. After each voltage approximation of the final result has been made and stored, the residual voltage is then automatically re-assessed against smaller standard voltages, and so on to the smallest voltage discrimination required in the result. Probably four logic decisions are needed to select the major decade value of the unknown voltage, and this process will be repeated for each lower decade in decimal sequence until, after a few milliseconds, the required voltage is stored in a coded form. This voltage is then translated for decimal display.

It is necessary to sense the initial polarity of the unknown signal, and to select the range and decimal marker for the read-out; the time for the logic networks to settle must be longer for the earlier (higher voltage) choices than for the later ones because they must be of the highest possible accuracy; offset voltages may be added to the earlier logic choices, to be withdrawn later in the sequence; and so forth.

The total measurement and display takes a few milliseconds. When noise is present in the input, the necessary insertion of filters may extend the time to close to a second. As noise is more troublesome for the smaller residuals in the process, it is sometimes convenient to use some different techniques for the latter part. One such is the voltage-frequency principle which has a high noise rejection ratio. The reduced accuracy of the technique can be tolerated as it applies only to the least significant figures.

43.3.3 Digital multimeters

Any digital voltmeter can be scaled to read d.c. or a.c. voltage, current, impedance or any other physical property provided that an appropriate transducer is inserted. The trend with instruments of modest accuracy (0.1%) is to provide a basic digital voltmeter with separate plug-in converter units as required for each parameter. Instruments scaled for alternating voltage and current normally incorporate one of the a.c./d.c. converter units mentioned in a previous section, and the quality of the result is limited by the characteristics inherent in such converters. The digital part of the measurement is more accurate and precise than the analogue counterpart, but is more expensive.

For systems application, command signals can be inserted into, and binary-coded or analogue measurements received from, the instrument through multi-way socket connections, enabling the instrument to form an active element in a control situation.

Resistance, capacitance and inductance measurements depend to some extent on the adaptability of the basic voltage measuring process. The dual-slope technique can be easily adapted for two-, three- or four-terminal ratio measurements of resistance by using the positive and negative ramps in sequence; with other techniques separate impedance units are necessary.

43.3.4 Input and dynamic impedance

The high precision and small uncertainty of digital voltmeters makes it essential that they have a high input impedance if these qualities are to be exploited. Low test voltages are often associated with source impedances of several hundred kilohms: for example, to measure a voltage with source resistance 50 kΩ to an uncertainty of ±0.005% demands an instrument of input resistance 1 GΩ, and for a practical instrument this must be 10 GΩ if the loading error is limited to *one-tenth* of the *total* uncertainty.

The dynamic impedance will vary considerably during the measuring period, and it will always be lower than the quoted null, passive, input impedance. These changes in dynamic impedance are coincident with voltage 'spikes' which appear at the terminals due to normal logic functions; this noise can adversely affect components connected to the terminals, e.g. Weston cadmium standard cells.

Input resistances of the order of 1–10 GΩ represent the conventional range of good quality insulators. To these must be added the stray parallel reactance paths through unwanted capacitive coupling to various conducting and earth planes, frames, chassis, common rails, etc.

43.3.5 Noise limitation

The information signal exists as the potential difference between the two input leads. Each can have unique voltage and impedance conditions superimposed on it with respect to the basic *reference* or *ground* potential of the system, as well as another and different set of values with respect to a local *earth* reference plane.

An elementary electronic instrumentation system will have at least one ground potential and several earth connections—possibly through the (earthed) neutral of the main supply, the signal source, a read-out recorder or a cathode ray oscilloscope. Most true earth connections are at different electrical potentials with respect to each other due to circulation of currents (d.c. to u.h.f.) from other apparatus, through a finite earth resistance path. When multiple earth connections are made to various parts of a high-gain amplifier system, it is possible that a significant frequency spectrum of these signals will be introduced as electrical noise. It is this interference which has to be rejected by the input networks of the instrumentation, quite apart from the removal of any electrostatic/electromagnetic noise introduced by direct coupling into the signal paths. The total contamination voltage can be many times larger (say 100) than the useful information voltage level.

Electrostatic interference in input cables can be greatly reduced by 'screened' cables (which may be 80% effective as screens), and electromagnetic effects minimised by transposition of the input wires and reduction in the 'aerial loop' area of the various conductors. Any residual effects, together with the introduction of 'ground and earth-loop' currents into the system are collectively referred to as *series* and/or *common-mode* signals.

43.3.6 Series and common-mode signals

Series-mode (normal) interference signals V_{sm} occur in series with the required information signal. Common-mode interference signals V_{cm} are present in both input leads with respect to the reference potential plane: the required information signal is the difference voltage between these leads. The results are expressed as *rejection ratio* (in dB) with respect to the input error signal V_e that the interference signals produce, i.e.

$$K_{sm} = 20 \log (V_{sm}/V_e) \quad (43.8)$$

and

$$K_{cm} = 20 \log (V_{cm}/V_e) \quad (43.9)$$

where K is the rejection ratio.

Consider the elementary case in *Figure 43.13*, where the input lead resistances are unequal, as would occur with a transducer

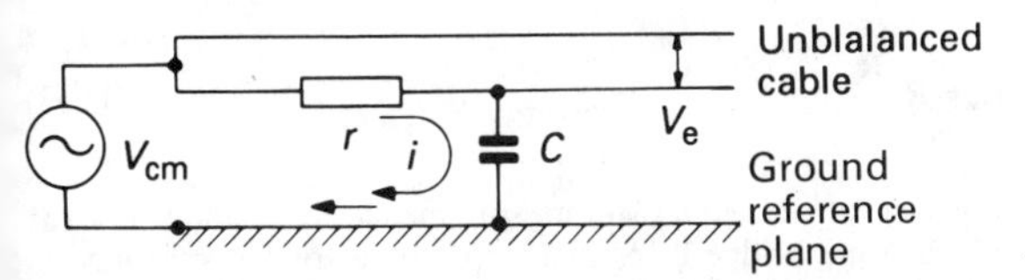

Figure 43.13 Common-mode effect in unbalanced network

input. Let r be the difference resistance, C the cable capacitance, with common-mode signal and error voltages V_{cm} and V_e respectively. Then the common-mode numerical ratio (c.m.r.) is

$$V_{cm}/V_e = V_{cm}/ri = 1/2\pi fCr \quad (43.10)$$

assuming the cable insulation to be ideal, and $X_C \gg r$. Clearly for a common-mode statement to be complete it must have a stated frequency range and include the resistive unbalance of the source. (It is often assumed in c.m.r. statements that $r = 1$ kΩ.)

The c.m.r. for a digital voltmeter could be typically 140 dB (corresponding to a ratio of $10^7/1$) at 50 Hz with a 1 kΩ line unbalance, and leading consequently to $C = 0.3$ pF. As the normal input cable capacitance is of the order of 100 pF/m, the situation is not feasible. The solution is to inhibit the return path of the current i by the introduction of a guard network. Typical guard and shield parameters are shown in *Figure 43.14* for a six-figure digital display on a voltmeter with ±0.005% uncertainty. Consider the magnitude of the common-mode error signal due to a 5 V 50 Hz common-mode voltage between the shield earth E_1 and the signal earth E_2.

Switch S_1 open The a.c. common-mode voltage drives current through the guard network and causes a change of 1.5 mV to appear across r as a series-mode signal; for $V = 1$ V this represents an error V_e of 0.15% for an instrument whose quality is ±0.005%.

Swtich S_1 closed The common-mode current is now limited by the shield impedance, and the resultant series-mode signal is 3.1 μV, an acceptably low value that will be further reduced by the noise-rejection property of the measuring circuits.

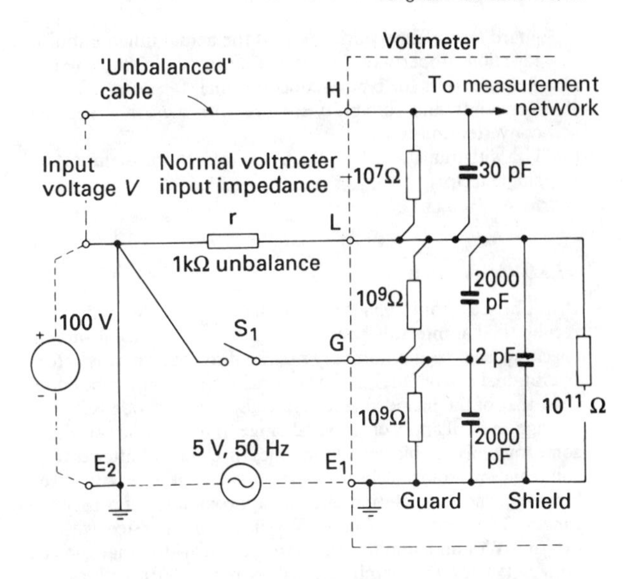

Figure 43.14 Typical guard and shield network for digital voltmeter

43.3.7 Floating-voltage measurement

If the voltage difference V to be measured has a potential difference to E_2 of 100 V, as shown, then with S_1 open the change in potential difference across r will be 50 μV, as a series-mode error of 0.005% for a 1 V measurement. With S_1 closed, the change will be 1 μV, which is negligible.

The interconnection of electronic apparatus must be carefully made to avoid systematic measurement errors (and short-circuits) arising from incorrect screen, ground or earth potentials.

In general it is preferable, wherever possible, to use a single common reference mode, which should be at the zero signal reference potential to avoid leakage current through r. Indiscriminate interconnection of the shields and screens of adjacent components can increase noise currents by short-circuiting the high-impedance stray path between the screens.

43.3.8 Instrument selection

The most precise 7-digit voltmeter, when used for a 10 V measurement, has a discrimination of ±1 part in 10^6 (i.e. ±10 μV), but has an uncertainty ('accuracy') of about ±10 parts in 10^6. The distinction is important with digital readout devices lest a higher quality be accorded to the number indicated than is in fact justified. The quality of any reading must be based upon the time-stability of the total instrument since it was last calibrated against external standards, and the cumulative evidence of previous calibrations of the like kind.

Selection of a digital voltmeter from the list of types given in *Table 43.2* is based on the following considerations:

(1) No more digits than necessary, as the cost increases per digit.
(2) High input impedance, and the effect on likely sources of the dynamic impedance.
(3) Electrical noise-rejection, assessed and compared with *(a)* the common-mode rejection ratio based on the input and

guard terminal networks and (*b*) the actual inherent noise-rejection property of the measuring principle employed.
(4) Requirements for binary-coded decimal facilities.
(5) Versatility and suitability for use with a.c. or impedance converter units.
(6) Use with transducers (in which case (3) is the most important single factor).

43.3.9 Calibration

It will be seen from *Table 43.2* that digital voltmeters should be recalibrated at intervals between 3 and 12 months. Built-in self-checking facilities are normally provided to confirm satisfactory operational behaviour, but the 'accuracy' check cannot be better than that of the included standard cell or zener diode reference voltage and will apply only to one range. If the user has available some low-noise stable voltage supplies (preferably batteries) and some high-resistance helical voltage-dividers, it is easy to check logic sequences, resolution and the approximate ratio between ranges. The accuracy of the 1 V range can be tested with an external Weston cadmium standard cell provided that the cell voltage is known to within $\pm 3\ \mu V$ by recent NPL calibration.

43.4 Bridges

A bridge usually consists of four impedance arms, as in *Figure 43.15*, which are adjusted so that at balance there is no current through the detector. At this point Equations (43.11) and (43.12) are satisfied, where the impedances are assumed to be complex. This basic principle is applied to all the bridges commonly used in instrumentation.

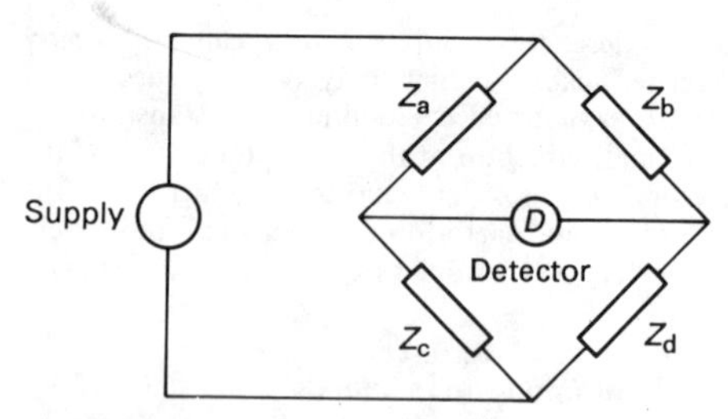

Figure 43.15 An impedance bridge

$$Z_a Z_d = Z_b Z_c \tag{43.11}$$

$$\angle\phi_a + \angle\phi_d = \angle\phi_b + \angle\phi_c \tag{43.12}$$

43.4.1 Resistance bridges

The most frequently used resistance bridge is the Wheatstone bridge, shown in *Figure 43.16(a)*. R_x is the unknown resistor, and resistors R_a, R_b and R_c are adjusted to give a null on the detector. At this point the value of R_x is given by Equation (43.11), which is rewritten in Equation (43.13).

$$R_x = R_c \frac{R_b}{R_a} \tag{43.13}$$

For very low valued resistor measurements, below about 1 Ω, the Wheatstone bridge gives high errors due to the resistance of interconnecting leads and contacts. For these low resistance

Table 43.2 Typical characteristics of digital voltmeters and multimeters

1: Operating principle—
- DS dual slope
- DDS inductive divider + dual slope
- I integration
- ISA mixed I and SA
- SA potentiometer/successive
- R ramp

2: Number of display digits.
3: Operating time.
4: Instrument ranges.
5: Uncertainty, smallest digits to be added in ± ppm of maximum reading.
6: Maximum discrimination, in ± ppm of maximum reading.
7: Parallel input resistance.
8: Parallel input capacitance.
9: Common (or series) mode rejection.
10: Common (or series) mode rejection, 50 Hz.
11: Recalibration period.

1	2	3 ms	4	5 ppm	6 ppm	7 MΩ	8 pF	9 dB	10 dB	11 months
Single-purpose voltmeters										
R	3	500	100 mV–1 kV	5000	1000	10	—	90–10	40	3
I	4	2–200	100 mV–1 kV	300	10	10^5, 10	—	150	150	12
I	6	2–200	20 mV–1 kV	40	10	10^5, 10	—	—	160	6
ISA	7	1000	1 V–1 kV	60	1	10^4, 10	40	—	160	3
DD	7	—	1 V–1 kV	10	1	10^5, 10	—	120	150	3
Small multimeter or panel-meter										
DS	4	200	0.2 V–1 kV 0–20 kHz 200 μA–2A 0.2–2000 kΩ	1000–4000	100	10	110	—	90	12
Modular meters										
SA	5	2–250	0.1 mV–1 kV (d.v.)	200	10	10^4, 10	—	—	100	3
	5	2–250	1 V–1 kV 50 Hz–100 kHz	1000–5000	10	1	100	—	60	3
SA	5	2–250	1 kΩ–10 MΩ	500	10	—	—	—	—	3

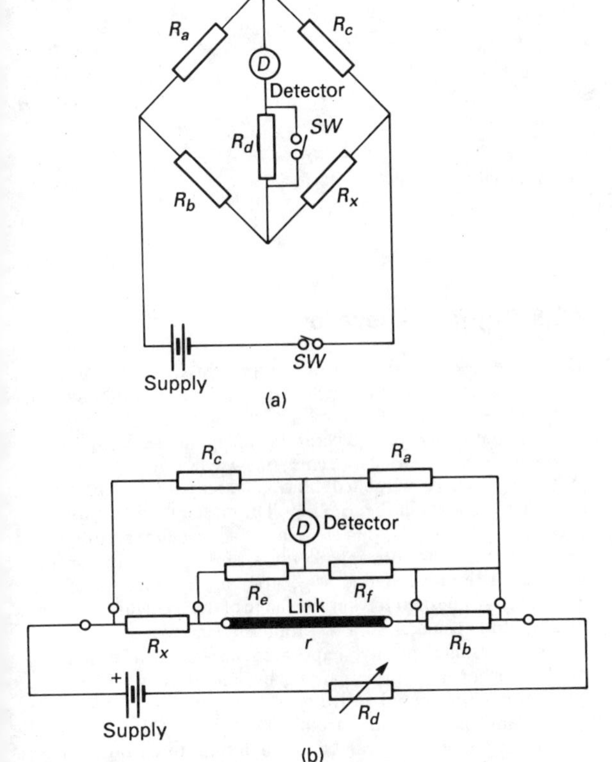

Figure 43.16 Resistance bridges: (a) Wheatstone, (b) Kelvin

measurements a Kelvin bridge, shown in *Figure 43.16(b)*, is preferred, which uses two sets of ratio arms, or a 'double bridge', in which resistors R_e and R_f are employed to balance out the effect of the resistances of connecting links. In the operation of the bridge the values of R_e and R_f are chosen to satisfy Equation (43.14). The bridge is then balanced and link r can be removed without affecting it. Under these conditions the value of the unknown resistor R_x is given by Equation (43.13). A Kelvin bridge can measure resistors in the range 10 $\mu\Omega$ to 1 Ω with an error below 0.02%.

$$\frac{R_c}{R_a}=\frac{R_e}{R_f} \quad (43.14)$$

43.4.2 Inductance bridges

The Maxwell–Wien bridge, shown in *Figure 43.17(a)*, uses a parallel capacitor–resistor combination to balance the unknown inductance L_x. The characteristic of the bridge at null is given by Equations (43.15).

$$\left.\begin{aligned} R_x&=\frac{R_aR_c}{R_b}\\ L_x&=\frac{R_aR_c}{C}\\ Q_x&=\frac{\omega L_x}{R_x}\end{aligned}\right\} \quad (43.15)$$

Although widely used for the measurement of inductors having a Q value of below about 10, the bridge is inconvenient for use with high Q coils since the value of R_b would then need to be impractically large. In these instances the Hay bridge of *Figure 43.17(b)* is preferred, and the bridge equations at null are given by Equation (43.16). Generally the $1/Q_x^2$ term can be omitted, with a resultant error below 1%, since the bridge is used to measure coils with $Q>10$, and this means that the inductance measurement is independent of frequency.

$$\left.\begin{aligned} R_x&=\frac{R_aR_c}{R_b(1+Q_x^2)}\\ L_x&=\frac{R_aR_cC}{1+1/Q_x^2}\\ Q_x&=\omega L_x/R_x=1/\omega R_2C\end{aligned}\right\} \quad (43.16)$$

Figure 43.17(c) shows the Campbell bridge arrangement, which is used to measure mutual inductance of a coil. The resistance and self-inductance of the primary coils are first balanced with both switches in position 2, and then the switches are moved to position 1 and M_a is adjusted to balance M_x. The coil characteristics are given by Equations (43.17).

$$\left.\begin{aligned} M_x&=\frac{M_aR_c}{R_b}\\ L_x&=\frac{L_aR_c}{R_b}\\ R_x&=\frac{R_aR_c}{R_b}\end{aligned}\right\} \quad (43.17)$$

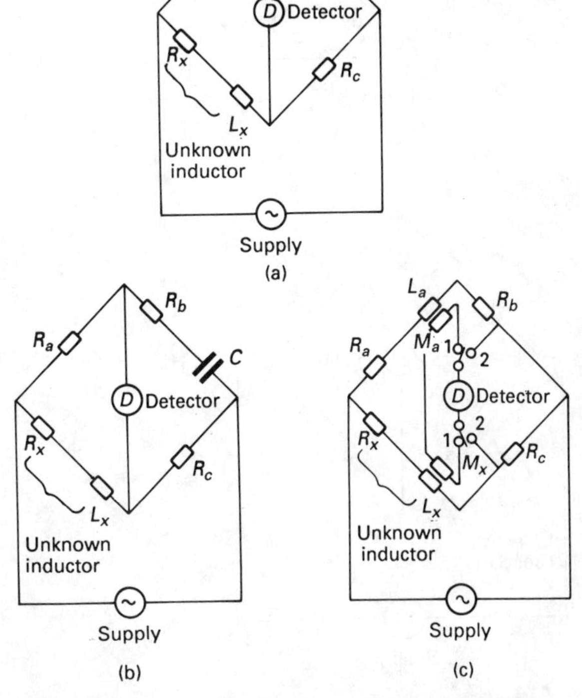

Figure 43.17 Inductance bridges: (a) Maxwell–Wien, (b) Hay, (c) Campbell

43.4.3 Capacitance bridges

The Schering bridge, shown in *Figure 43.18(a)*, is widely used to measure capacitance and dissipation factor. The null equations of the bridge are given by Equations (43.18).

$$\left.\begin{aligned} R_x &= C_b R_c / C_a \\ C_x &= C_a R_b / R_c \\ D &= \omega C_b R_b \end{aligned}\right\} \quad (43.18)$$

An alternative bridge used to measure capacitance is the Wien bridge of *Figure 43.18(b)*, where at balance the values of the unknown capacitance and series loss resistor are given by Equations (43.19). Generally the Wien bridge is more frequently used to measure an unknown frequency f using known values of capacitance and resistance, and Equations (43.19) can then be rewritten as in Equation (43.20).

In a practical bridge C_a and C_x are fixed, and R_a and R_x are gauged so that $R_a = R_x$, with R_b being made twice the value of R_c.

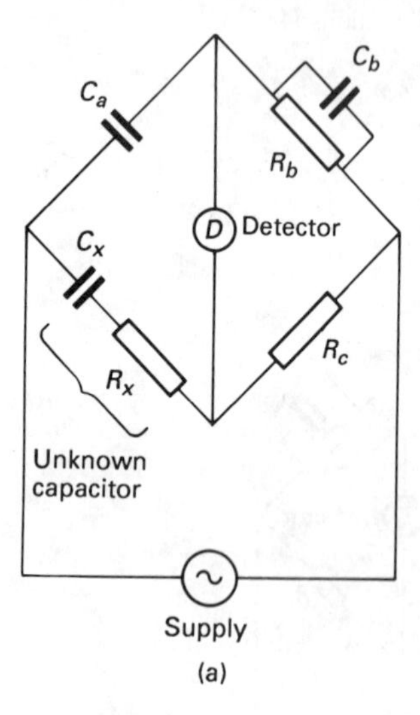

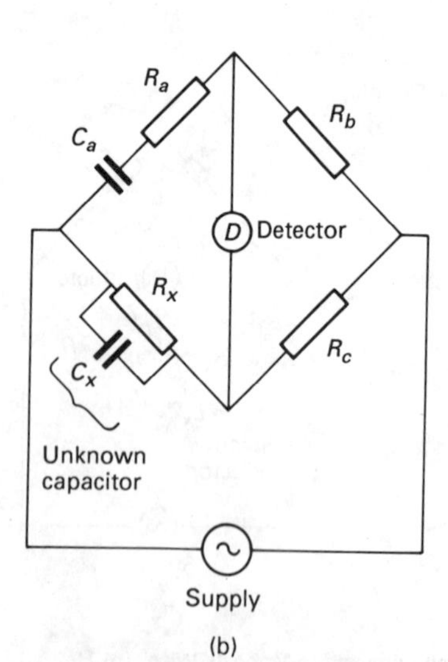

Figure 43.18 Capacitance bridges: (a) Schering, (b) Wien

Under these conditions Equation (43.20) reduces to Equation (43.21).

$$\left.\begin{aligned} \frac{C_x}{C_a} &= \frac{R_b}{R_c} - \frac{R_a}{R_x} \\ C_a C_x &= \frac{1}{\omega^2 R_a R_x} \end{aligned}\right\} \quad (43.19)$$

$$f = 1/2\pi(C_a C_x R_a R_x)^{1/2} \quad (43.20)$$

$$f = 1/2\pi C_a R_a \quad (43.21)$$

43.5 Signal generators

In this section signal generators represent those instruments which generate a waveform whose frequency and amplitude can be varied. This includes sine wave oscillators, square, triangular and sawtooth waveform generators, and pulse generators.

The basic circuits used for generating waveforms are described in *Chapter 37*. A simplified block diagram of a commercial generator is shown in *Figure 43.19*. The master oscillator usually produces sine waves, for a sine wave oscillator. For low frequencies it is not possible to use a tuned *LC* oscillator and integrating circuits are used to generate a triangular waveform, and this is synthesised to alter the shape of the triangular wave as its amplitude changes, and so produce a sine wave.

Generally, different waveshapes are generated in different parts of the system, and these can be brought out in a function generator. Commercial instruments are capable of operating over a range of frequencies from 0.001 Hz to 1 GHz.

The output frequency can be varied in steps by a range switch, and continuously within any step. The oscillator can free-run, or it can be synchronised to an external signal, or it can be f.m. or a.m. modulated by another signal. The amplitude of the output can be varied independently of the frequency.

Sometimes several outputs are available having different impedances for matching. A pulse generator, like a sine wave generator, must also be properly terminated at its operating frequency, with its characteristic impedance, to prevent reflections. If a pulse generator has a rise time of t_r ns then this is equivalent to a sine wave frequency of f_s MHz where

$$f_s = \frac{350}{t_r} \quad (43.22)$$

In selecting a signal generator the quality of the output waveform is an important consideration. For example a sine wave should have minimum distortion; a pulse generator should have clean rise and fall times with no overshoot, ringing or sag; a square wave generator should have equal mark to space periods and often the slope of the edges of the waveform should be variable.

Sine wave generators are often used for testing receivers. Pulse generators are used to test radar and communication systems, for device testing such as the recovery time of transistors, and for circuit analysis such as the transient analysis of an amplifier.

43.5.1 Frequency synthesiser

This is a sine wave generator which uses the phase-locked loop principle to generate a range of output frequencies. *Figure 43.20* shows the operating principle of the instrument.

The input frequency is generated by a stable source such as a voltage controlled crystal oscillator. This is compared in a phase detector whose output consists of $f_2 \pm f_1$. The low pass filter selects the difference frequency $f_2 - f_1$. This is amplified and fed to the voltage-controlled oscillator. This drives f_1 towards f_2 so that they lock together, and only a phase difference exists

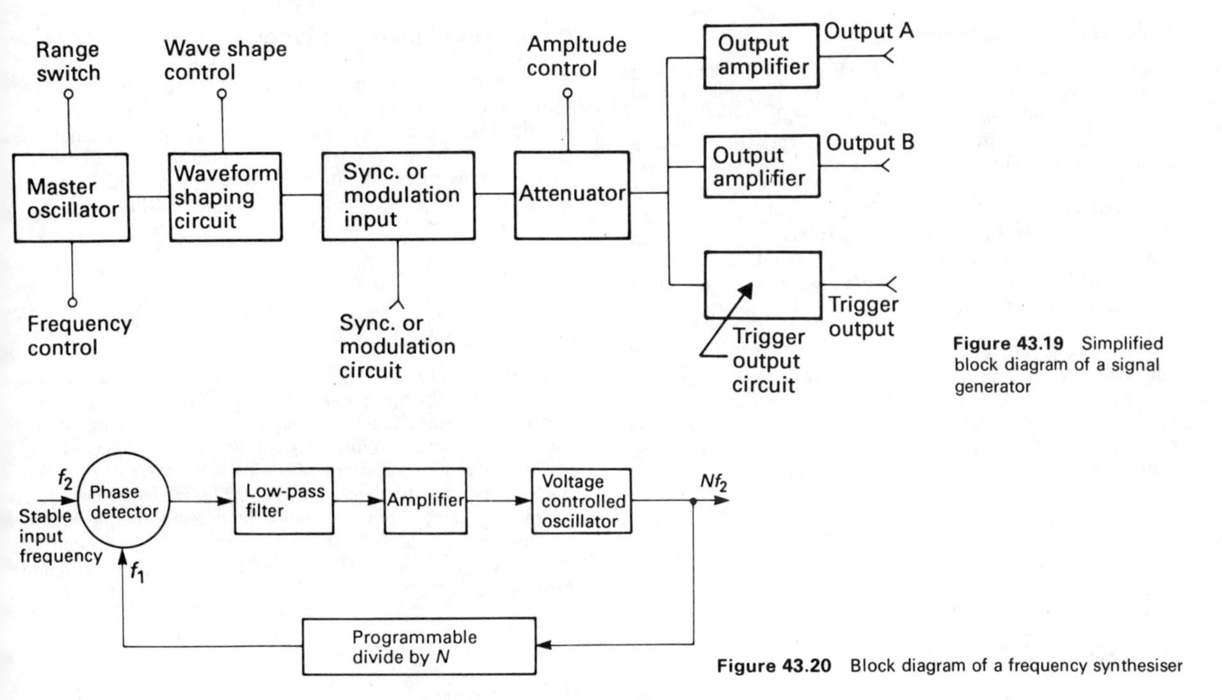

Figure 43.19 Simplified block diagram of a signal generator

Figure 43.20 Block diagram of a frequency synthesiser

between f_1 and f_2 which gives sufficient signal to f_1 to keep it locked to f_2.

To vary the output frequency the value of N is changed in the programmable divider. Usually the input frequency f_2 is obtained from a high frequency oscillator. Since it is expensive to design a programmable divider to run at high frequencies, a fixed count divider, called a pre-scalor, is introduced in series with the programmable divider.

43.5.2 Sweep frequency generator

A sweep frequency generator is usually a signal generator which produces a sine wave output in the r.f. range, whose frequency can be continuously and smoothly varied, usually at a low audio frequency rate. *Figure 43.21* shows a schematic diagram of a sweep frequency generator.

The frequency can be swept electronically or manually. Electronically tuned oscillators include backward wave oscillator tubes. An example of a manual method is the use of a motor driven variable capacitor in a tuned *LC* circuit of an oscillator. The frequency sweeper gives a synchronously varying voltage which can also be used for the horizontal deflection in a cathode ray tube display or an XY recorder. This enables a plot to be obtained giving the response of a device which is fed by the sweep generator.

Manual controls can be used for adjusting the frequency of the master oscillator. The range switch operates in several bands, and the frequency sweeper covers each band usually at 10 to 30 sweeps per second. The level control circuit monitors the r.f. level at some point in the system and holds the power to the load constant as the frequency and load impedance change. This ensures a fixed readout calibration with frequency and prevents source mismatch.

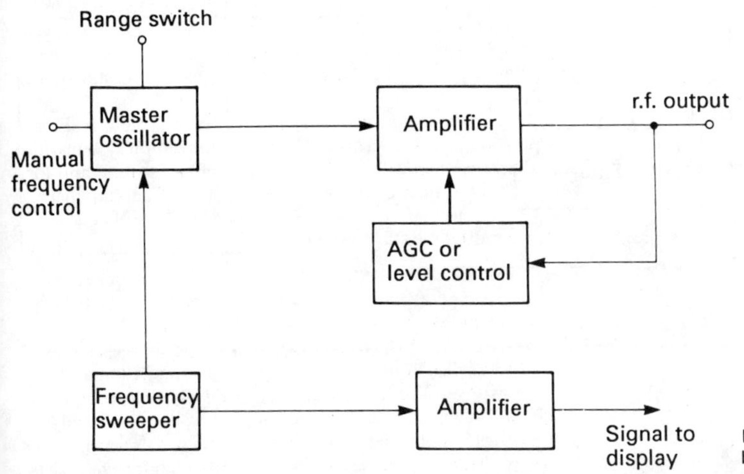

Figure 43.21 Sweep frequency generator block schematic

43.5.3 Random-noise generator

A random-noise generator gives a signal whose instantaneous amplitude varies at random, and contains no periodic frequency component. It has many uses such as testing radio and radar systems for signal detection in the presence of noise, and intermodulation and crosstalk tests in communication systems.

Random noise is usually generated within the instrument by a semiconductor noise diode which gives an output frequency in the band 80 kHz to 220 kHz. This signal is amplified and modulated down to the audio-frequency band. It is then passed through a filter which gives an output noise signal in one of three spectra, white noise, pink noise and USASI noise, as in *Figure 43.22*.

Pink noise is so called by analogy to red light since it emphasises lower frequencies. It has a voltage spectrum which varies inversely as the square root of frequency, and is used in bandwidth analysis. USASI noise is used for testing audio systems since its spectrum approximately equals the energy distribution of speech and music frequencies.

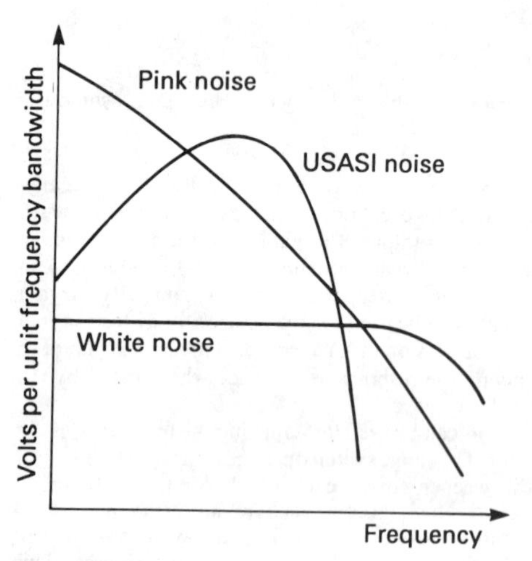

Figure 43.22 Noise spectra

43.6 Waveform analysers

These instruments are primarily used to determine the composition of a waveform. Two types of instruments are described below, harmonic analysers and the spectrum analyser. Harmonic analysers are further divided into harmonic distortion analyser and intermodulation distortion analyser. Transient intermodulation distortion is also described in this section.

43.6.1 Harmonic distortion analyser

This is shown in *Figure 43.23*. The amplifier under test is fed from a signal source having very low distortion. The harmonic output of this source can of course be checked separately by the meter. The output from the amplifier is fed into a notch filter. This is tunable and is designed to cut off the fundamental frequency and to let other signal frequencies through. The voltmeter measures a.c. r.m.s. voltage. Both the total voltage E_T of the harmonics and fundamental, and the harmonic r.m.s. voltage E_H can be read by means of the voltmeter switch. The total harmonic distortion is given by

$$\mathrm{IHD} = \frac{E_H}{E_T} \times 100 \tag{43.23}$$

In practice the meter is calibrated to a percentage scale. It is set to E_T and adjusted for full-scale deflection. When the switch is now moved to E_H the total harmonic distortion can be read off directly on the meter.

There are several types of distortion. Amplitude distortion occurs when the output amplitude is not proportional to its input. When the amplifier is overdriven it will clip and distortion will increase. Crossover distortion occurs in push–pull output amplifiers, and this increases as the test signal level decreases. The distortion meter will also read noise in the system. This noise increases as the signal level decreases so it can be confused with certain types of distortion. Care must therefore be taken to minimise noise.

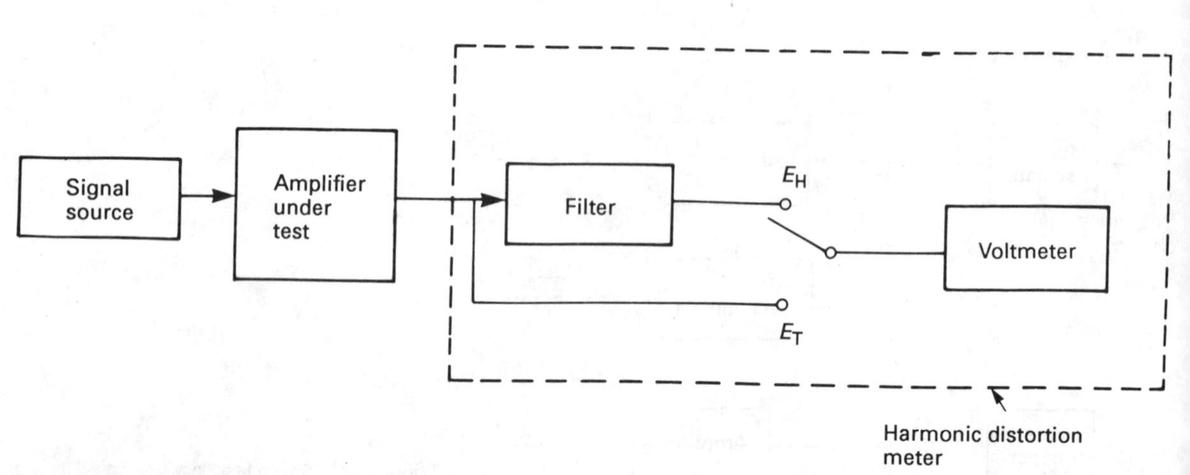

Figure 43.23 Use of a harmonic distortion analyser

43.6.2 Intermodulation distortion analyser

A nonlinear system will generate frequencies in its output which are equal to the sums and differences of the frequencies present in the input signal. When a high frequency f_1 signal and a low frequency signal f_2 are mixed in a linear network the output will contain the two original frequencies only. If the signals are mixed in a nonlinear system, such as an amplifier which has distortion, modulation will occur, and the output will contain at least f_1, f_2, f_1+f_2 and f_1-f_2, and usually also several harmonics and their sum and difference frequencies.

Amplitude modulation is caused by the system distortion as in *Figure 43.24*. The intermodulation distortion is given by

$$\text{IMD} = \frac{C-M}{M} \times 100 \qquad (43.24)$$

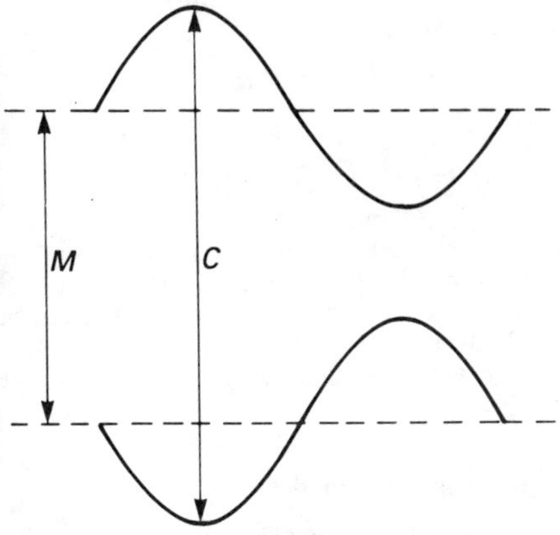

Figure 43.24 Amplifier modulation in a nonlinear system

Figure 43.25 shows how an intermodulation distortion analyser may be used to measure amplifier distortion. The test frequency is usually a two-tone wave in which one frequency is at least 50 times larger than the other. For example 50 Hz and 5 kHz signals may be used in which the 50 Hz signal has an amplitude several times larger than the 5 kHz signal. When the two signals are processed by an amplifier any amplitude nonlinearity will cause the 50 Hz signal to amplitude modulate the 5 kHz signal. The intermodulation distortion amplifier has filters and demodulators for separating and measuring the amplitude modulated signal.

43.6.3 Transient intermodulation distortion

This occurs when there are sudden changes in the input signal and the amplifier cannot respond quickly. The amplifier under test is usually checked by a test signal which is a square wave with a superimposed high frequency sine wave, as in *Figure 43.26(a)*. If transient intermodulation distortion is present then the output from the amplifier will have parts of the high frequency component of the input missing, as at AB in *Figure 43.26(b)*. This can be checked on an oscilloscope.

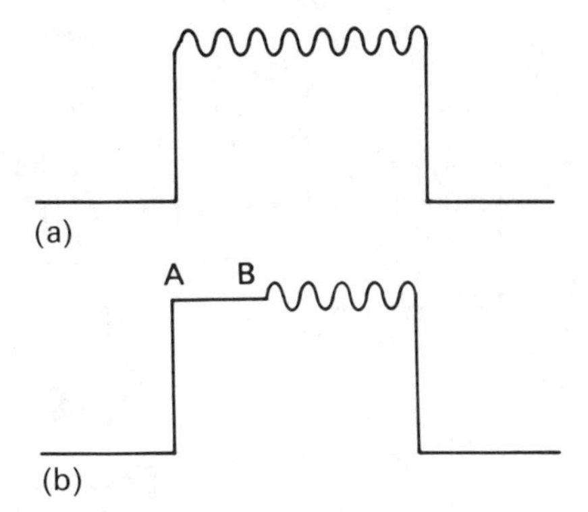

Figure 43.26 Illustration of transient intermodulation distortion: (a) input signal, (b) output signal

43.6.4 Spectrum analyser

Figure 43.27 shows a simplified diagram of a spectrum analyser. The output from the amplifier under test is broken down into its individual harmonics and the frequency and amplitude of each is displayed on an internal CRT display, as in *Figure 43.28*.

The input signal to a spectrum analyser is used to modulate an internal high-frequency oscillator, producing sideband frequencies corresponding to the input frequency. The oscillator is swept by a sweep generator, which also provides the signal for the X plates of the internal cathode ray tube display. The sideband frequencies generated at the mixer are also swept and pass sequentially through the filter. They are then demodulated and applied to the Y plates of the CRT display.

43.7 Counters

The electronic circuits which go into making a counter, and the different types of counters, were described in *Chapter 27*. These circuits will be referred to as the counter module. *Figure 43.29* shows how this module is used in a commercial instrument. The instrument can operate in several modes, such as totalising, frequency measurement, period measurement, ratio measurement and averaging. These modes are selected by a mode select switch and the path which the signals follow in the instrument are varied by internal control circuitry.

The input signals being measured are usually not clean square waves, and may sometimes not have sufficient amplitude to operate other circuits within the instrument. They are therefore amplified (or attenuated if the signals are too large) and sharpened up using waveform shapers and a Schmitt trigger. *Figure 43.30* shows the operation of the circuits. The Schmitt trigger has hysteresis and the input waveform must cross both the upper and lower trigger lines in order to register on the instrument.

In some counters the trigger levels can be varied in order to control the trigger points on the input waveform, and to change the width of the hysteresis band.

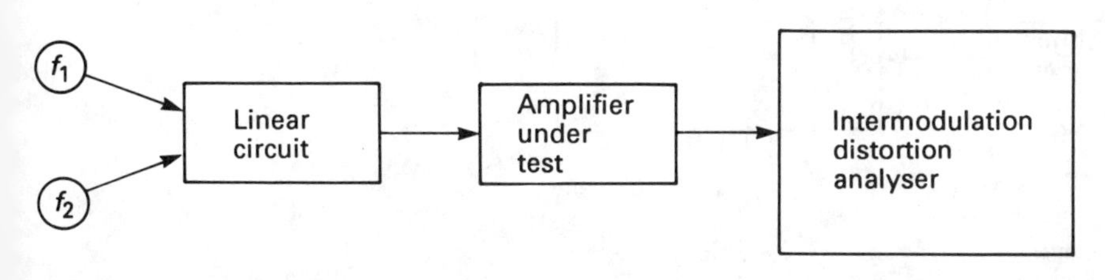

Figure 43.25 Measurement of intermodulation distortion

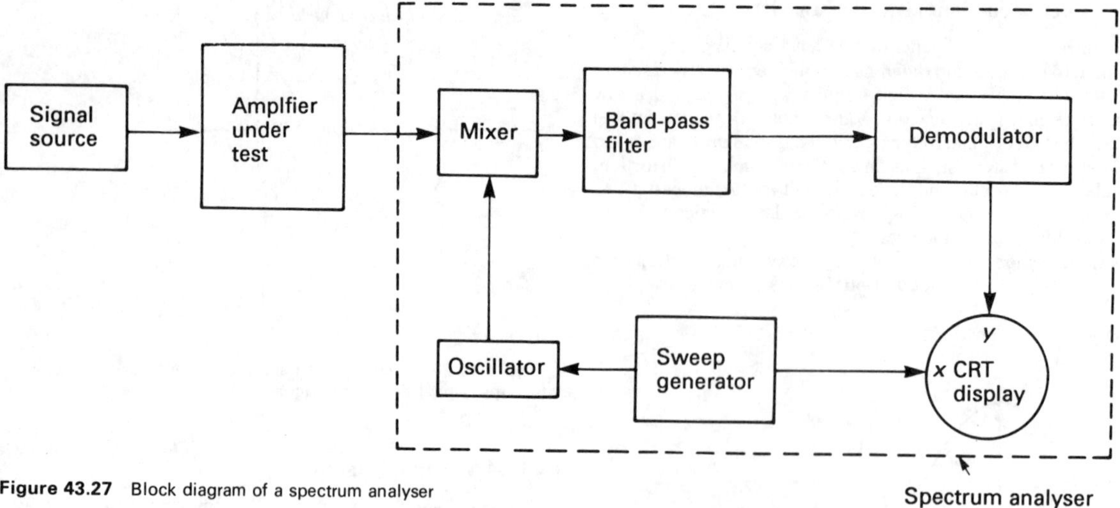

Figure 43.27 Block diagram of a spectrum analyser

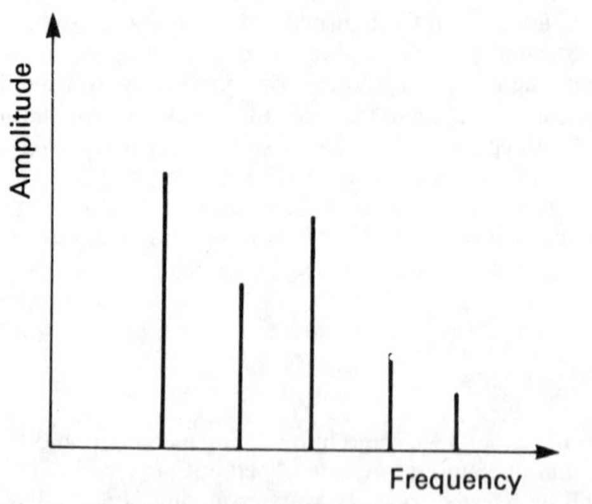

Figure 43.28 Screen display of a spectrum analyser

The counter has a very stable internal crystal oscillator which generates an internal measurement time base. The crystal is usually temperature controlled. Programmable dividers can be used to divide the frequency of the internal time base, or the input signals, before they are used for measurement within the instrument.

43.7.1 Counter measurement modes

The counter usually operates in one of six modes.

43.7.1.1 Totaliser

In this mode of operation the counter simply adds the number of pulses received and displays the total. Counting starts and stops between two periods which are usually determined by external signals, such as switches on the instrument.

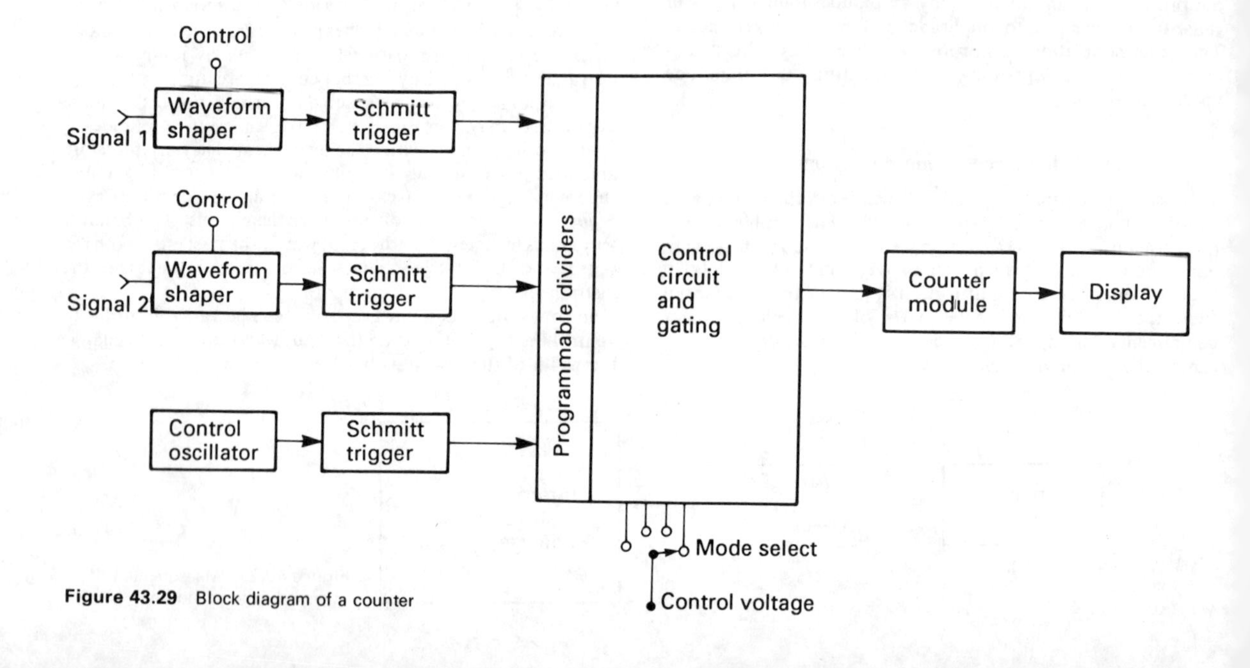

Figure 43.29 Block diagram of a counter

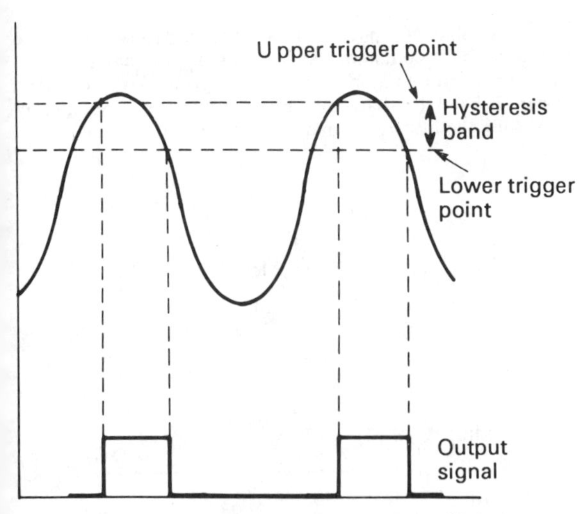

Figure 43.30 Operation of a waveform shaper and Schmitt trigger

43.7.1.2 Frequency measurement

This measures the frequency of the input signal, which is connected to the signal 1 terminal of the instrument. The control gate to the counter module is kept open for a known period of time (a fixed number of internal time base cycles) and the number of input signal pulses is counted. The frequency of the signal is given by

$$f_s = \frac{\text{counter reading}}{\text{time}} \tag{43.25}$$

43.7.1.3 Period and time measurements

A period is the time between two identical points on the input waveform. Now the roles of the signal input and internal time base are reversed. The gate to the counter module is controlled by the input signal and it is kept open for one cycle of the signal. During this time the number of pulses (cycles) from the internal oscillator are counted. The period is given by

$$T_s = \frac{\text{counter reading}}{\text{frequency of time base}} \tag{43.26}$$

It is also possible to use this technique to measure the time between any two points on the input signal, such as its pulse width, by operating the on/off control to the counter module from these points.

43.7.1.4 Ratio measurement

This is used to find the ratio between two input signals. The slower of the two signals is connected to the signal 2 terminal and it replaces the internal time base. The instrument now measures the number of input cycles of signal 1 for a single cycle of signal 2, and this gives the ratio between the two signals.

43.7.1.5 Averaging measurements

This measurement mode is used with the frequency, period or ratio measurements. By using the programmable dividers the counter gate is kept open for several cycles of the control waveform and the average over one cycle is found.

43.7.1.6 High frequency measurement

For measurements of very high frequency, in the gigahertz range, it is usual to divide down the input frequency by using prescalers before counting. This increases the measurement time, and it is also not possible to count a number which is smaller than the divider ratio. An alternative technique is to use a heterodyne counter where the input frequency is mixed with another internal frequency and the difference frequency is selected and measured.

43.7.2 Counter errors

Errors arise in the use of the counter due to three main causes.

43.7.2.1 Time base error

This is caused by a change in the frequency of the internal oscillator. It affects all measurements of frequency and period. The error is constant irrespective of the frequency being measured, for example a 0.001% error will occur whether 1 kHz or 1 MHz is being measured. There are several causes for the time base error:

(1) *Initial error* This is due to faulty setting up during calibration. Counters are most frequently calibrated by comparing them with a standard frequency broadcast.
(2) *Short-term stability error* These result in momentary frequency variations caused by shock, vibration, voltage transients, etc. The effect of this error can be minimised by taking measurements over a long time and using the averaging measurement mode described in *Section 43.7.1*.
(3) *Long-term stability error* This is due to the drift in the crystal oscillator frequency over a long period of time, and is referred to as its ageing rate. As a crystal is temperature cycled, and kept oscillating, its frequency gradually increases with time. The frequency will eventually settle down to a low rate of change, but if the crystal is allowed to cool and start again it will show a jump in frequency of several hertz from its previous value. In many instruments the crystal is kept at a stable temperature inside a temperature-controlled oven. Backup rechargeable batteries are used to keep the oven going if the main power is turned off.

43.7.2.2 Gating error

This occurs when measuring frequency or period and is illustrated in *Figure 43.31*, where, depending on the phase of the input signal relative to the gating waveform, the counter reading can vary by ±1. The error is inversely proportional to the

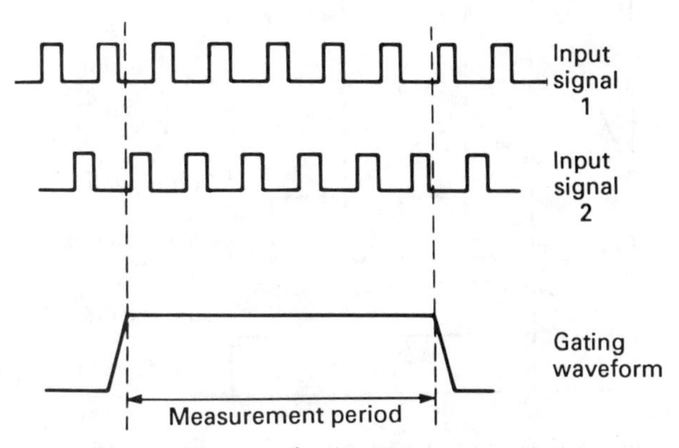

Figure 43.31 Illustration of gating error

frequency being measured, and at low frequencies it can be very large. It is therefore better to take period measurements at low frequencies. Generally if f_i is the frequency of the internal time base then for signal frequencies lower than $(f_i)^{1/2}$ period measurement should be used, and for input frequencies above this value frequency measurements should be used. The gating error is also reduced by using the averaging measurement mode.

43.7.2.3 Trigger error

This is shown in *Figures 43.32–43.34*. The input signal must cross both the trigger level lines of the Schmitt trigger in order to register on the counter. Variation in amplitude, as in *Figure 43.32*, can cause counts to be missed. This can be overcome by amplifying the input signal or by narrowing the hysteresis band and lowering the trigger levels.

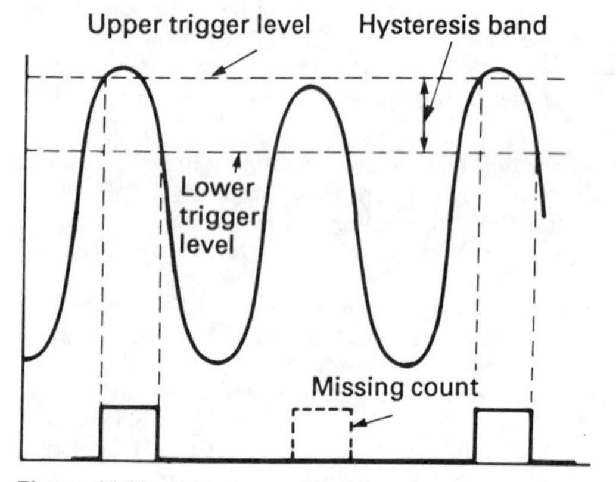

Figure 43.32 Trigger error giving low count

Figure 43.33 illustrates the effect of harmonics on the signal waveform which can result in a counter reading which is too high. This can also occur due to rings or transients on the waveform. It can be avoided by moving the trigger levels so that they do not cross any of the distorted parts of the wave.

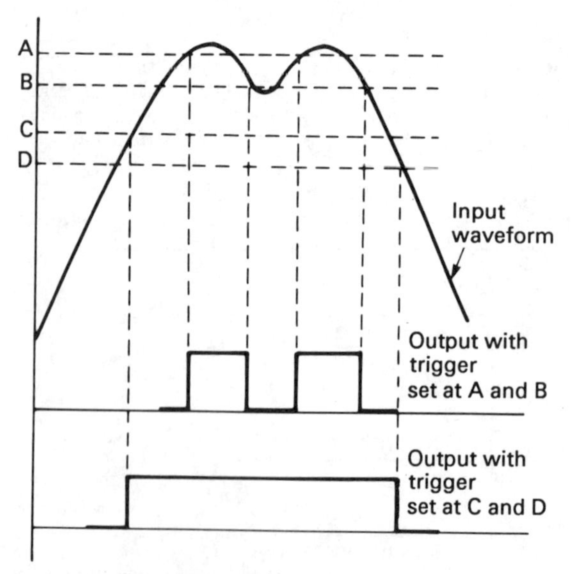

Figure 43.33 Trigger error giving high count

Noise on shallow input signal waveforms can cause jitter in the measurement period, and false readings, as in *Figure 43.34*. Its effect can be reduced by making the signal go quickly through the hysteresis band, by sharpening up the waveform edges, and by narrowing the hysteresis band.

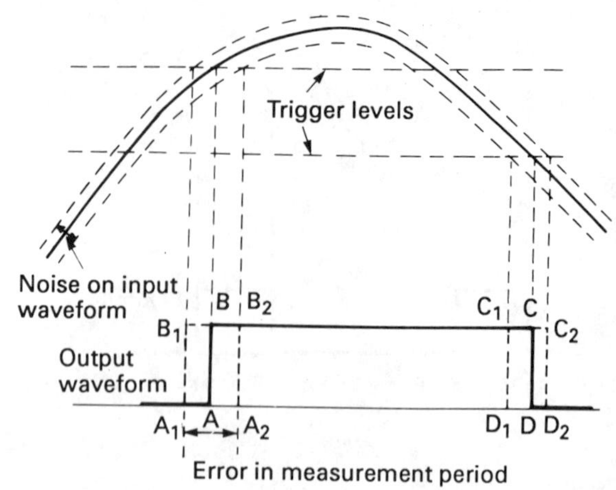

Figure 43.34 Trigger error giving uncertain period of measurement

43.8 Cathode ray oscilloscopes

A cathode ray oscilloscope is an electronic equipment designed to display two-dimensional information on the fluorescent screen of a cathode ray tube in a non-pictorial form. If brightness variations are employed, then to a limited extent it can also present three-dimensional information. The horizontal axis usually represents time and the vertical axis voltage, but the latter may represent one of many quantities such as loudness or magnetic field strength.

The cost of this type of equipment varies considerably and it follows that the range of functions, facilities, performances and techniques employed is also very wide. The high degree of flexibility of the cathode ray tube and the many versions of equipment currently available make it difficult to decide where the boundary line lies between oscilloscopes and various other specialised instruments employing cathode ray tubes as visual display devices. In this section the emphasis is on the medium-priced conventional instrument. Some information is however provided on the more expensive and specialised instruments.

43.8.1 Cathode ray tube

The electrons from a heated cathode can be focused into a narrow beam. If this beam strikes a luminescent screen, a visible indication of the position of the beam is obtained. An electric field applied to the beam will deflect it, and so the screen will show a visible representation of the applied electrical signal. This is the principle of the cathode ray tube.

A schematic cross-section of the cathode ray tube is shown in *Figure 43.35*. The tube can be considered as having three sections: the electron gun, the deflection system and the screen. The electron gun and deflection system are housed in the narrow neck of the tube; the screen terminates the cone part of the tube which is coated on the inside with a conductive layer.

43.8.1.1 Electron gun

The cathode of a cathode ray tube is a cylinder closed at one end.

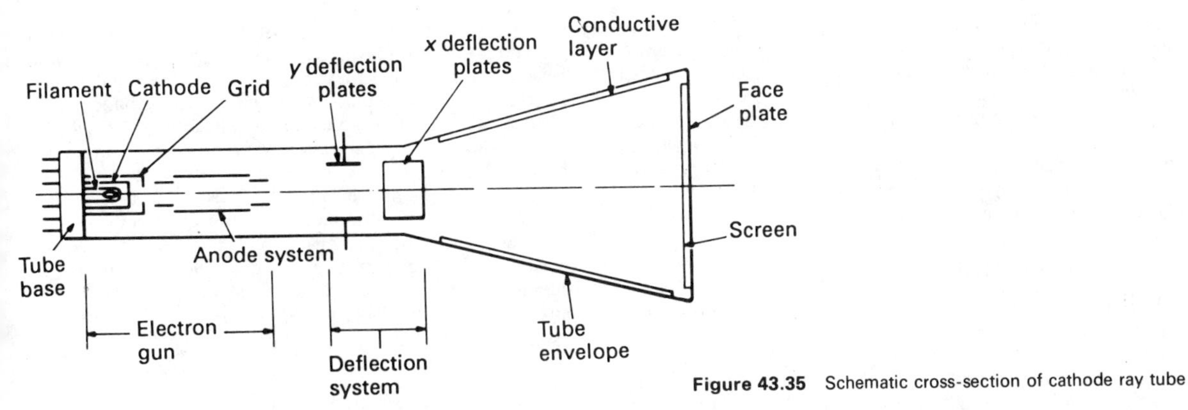

Figure 43.35 Schematic cross-section of cathode ray tube

The outside of the closed end is coated with a mixture of barium and strontium oxides which emits electrons when heated. The cathode is heated by a tungsten filament inside the cylinder, the filament being coated with alumina to insulate it electrically from the cathode. The cathode can then be held at earth potential.

The cathode is surrounded by a second cylinder, again closed at one end, but containing an aperture through which the electrons from the cathode can pass towards the anodes. This second cylinder forms the grid of the cathode ray tube. If the grid is made highly negative with respect to the cathode, no electrons can pass through and no display is obtained on the screen. If the grid is only slightly negative, electrons can pass through. As the brightness of the display depends on the number of electrons striking the screen, and the number of electrons flowing through the tube is inversely proportional to the grid voltage, so this voltage can be used to control the brightness of the display. The range of grid voltage is typically 0 to -50 V. The cutting off of the electron beam by a highly negative grid voltage is called blanking.

A more detailed cross-section of a typical electron gun is shown in *Figure 43.36*. The first anode is in the form of a disc with an aperture in the centre, or a short closed cylinder with apertures at both ends. This anode is held at a constant positive voltage with respect to the cathode, typically 1 to 2 kV. The electric field produced by the first anode brings the electron beam to a focus at point P between the grid and first anode. From this point the beam diverges again, but it is brought to a second focus at a point on the screen by the action of the second and third anodes. The spot on the screen is an image of the electron beam at point P, and so the spot diameter on the screen depends on the beam at P rather than on the cathode diameter or on any aperture in the electron gun. The spot diameter on the screen, also known as the line width, is typically between 0.2 and 0.4 mm.

The second anode is the focus electrode, and is an open cylinder to which a variable voltage about a mean value of typically $+250$ V is applied. The third anode is held at a constant positive voltage similar to that of the first anode and accelerates the electrons towards the screen. The aperture in this anode limits the beam diameter.

The first focus point of the beam, point P, is affected by changes in grid voltage (adjusting the brightness of the display) which may cause defocusing of the spot on the screen. This defocusing can be corrected by adjusting the voltage on the focus anode. Because of the screening effect of the first anode, adjusting the focus voltage does not affect the first focus of the beam.

43.8.1.2 Deflection system

The electron beam in a cathode ray tube can be deflected by a magnetic or electric field. For cathode ray tubes used in television (picture tubes), magnetic deflection is used but for instrument tubes such as those in oscilloscopes electrostatic deflection is used. The beam passes between two parallel plates and a voltage is applied across the plates. The direction in which the beam is deflected and the distance through which it is deflected depend on the polarity and magnitude of the voltage across the plates. To prevent the deflected beam striking the plates, in a practical tube the plates are not parallel but either diverge towards the screen or are shaped to follow the path of the deflected beam. As a result, the deflecting force is not always at right angles to the beam, causing a variation in the force acting over the cross-section of the beam. The spot on the screen may therefore appear elliptical rather than circular at the edges of the screen, an effect called deflection defocusing.

The deflection of the beam in a particular cathode ray tube is related to the voltage applied to the deflection plates by a quantity called the deflection factor. Typical values of deflection factor lie between 4 and 45 V/cm.

Two sets of deflection plates are used in a cathode ray tube, the sets being at right angles to each other. The sets are known as the

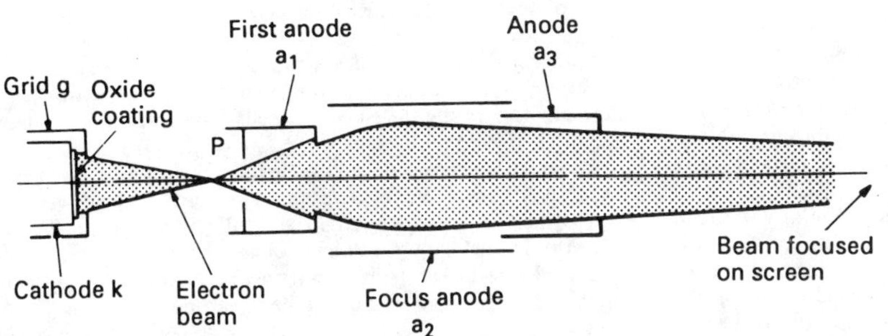

Figure 43.36 Cross-section of electron gun

x- and y-deflection plates, as shown in *Figure 43.35*, referring to the direction in which the spot is deflected on the screen. To prevent interaction between the two deflecting fields, a screen known as the interplate shield is placed between the x and y plates. This shield is usually held at the mean deflection-plate voltage, typically between 0.5 and 2 kV, although sometimes provision is made to vary this voltage to correct distortion of the display. A shield may also be placed around the deflection plates to screen them from external fields.

43.8.1.3 Luminescent screen

The function of the screen of a cathode ray tube is to provide a visible indication of the position of the deflected electron beam at any instant. This is achieved by forming the screen from phosphors, materials that emit visible light when struck by electrons. This effect is called luminescence. Practical phosphors are manufactured from 'host' materials to which are added 'activators'. Typical host materials include the oxides, sulphides, silicates, selenides, and halides of aluminium, cadmium, manganese, silicon and zinc; typical activators are copper, magnesium and silver. The activator prolongs the time for which the light is emitted, but other materials called 'killers' (such as nickel) can be added to reduce the emission time. By choosing the host material and the activator or killer, phosphors of different characteristics can be obtained. The screen of a cathode ray tube is formed by depositing a thin layer of the chosen phosphor on the inside of the faceplate of the tube.

When a point on the screen is struck by the electron beam, the light output from the phosphor rises to a constant level over the build-up time. The level of light output is determined by the energy of the incident beam, proportional to the square of the velocity of the electrons which, in turn, is proportional to the accelerating voltage of the tube. The light output remains constant while the phosphor is bombarded, but decays when the beam is removed. The emission of light when the phosphor is bombarded is called fluorescence and that during the decay is called phosphorescence. A typical light output characteristic is shown in *Figure 43.37*. The build-up and decay times depend on the phosphor used. The decay time is the more important characteristic, and the decay of light from the screen is called the persistence.

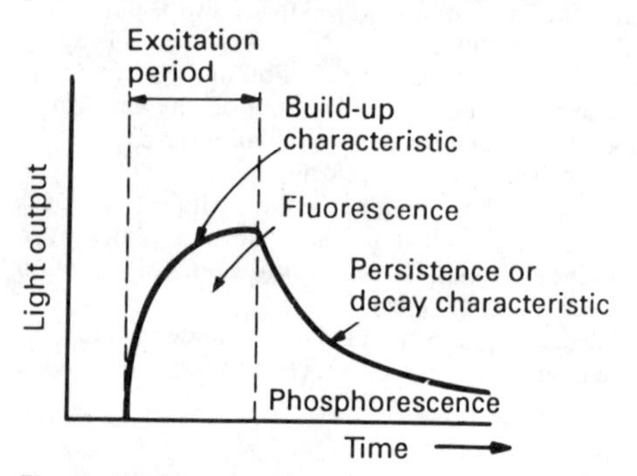

Figure 43.37 Light output characteristic of a phosphor

Screens of difference colours and characteristics can be obtained by choosing the appropriate phosphor. *Table 43.3* shows a classification of phosphors by the fluorescent and phosphorescent colours and the persistence (defined in *Table 43.4*).

When the electron beam strikes the screen, secondary electrons may be released. Some of these secondary electrons may leave the screen to be collected by the conductive layer inside the cone section of the tube (shown in *Figure 43.35*) which is held at the same voltage as the third anode. The ratio of secondary electrons leaving the screen to primary electrons arriving, the secondary emission ratio, depends on the beam energy. At certain energies the secondary electrons may remain on the screen and, because the screen is a poor conductor, a negative voltage builds up and effectively decreases the accelerating voltage on the beam. This effect is called sticking, and results in a reduced light output from the screen.

Sticking can be prevented by depositing a thin layer of aluminium over the phosphor layer and connecting it to the conductive layer on the inside of the tube. Any electrons remaining on the screen are then conducted away. The aluminium backing layer also reflects forward the light from the phosphors that would otherwise shine down the tube and thus be lost.

43.8.1.4 Post-deflection acceleration

To ensure the brightest possible display on the screen, the accelerating voltage on the beam (the voltage on the third anode) should be as high as possible. Unfortunately the higher this voltage, the higher the deflection voltage must be to maintain the deflection factor (each part of the beam remains for a shorter time in the deflection region). There is therefore a limit to the accelerating voltage, and this can affect the brightness of the display for high-speed operation with the type of tube just described, the mono-accelerator cathode ray tube. The beam can, however, be accelerated after deflection to overcome this limitation. A fourth anode is placed between the deflection plates and the screen to form the post-deflection acceleration cathode ray tube.

This fourth anode is provided by the conductive layer on the inside of the tube envelope. The layer is now made in the form of a helix, as shown in *Figure 43.38*. The end of the helix nearest the deflection plates is held at the third-anode voltage, while the end nearest the screen is held at a higher voltage. The ratio of these two voltages is the post-deflection acceleration (p.d.a.) ratio, and for a helical (spiral) p.d.a. tube this ratio is typically 6.

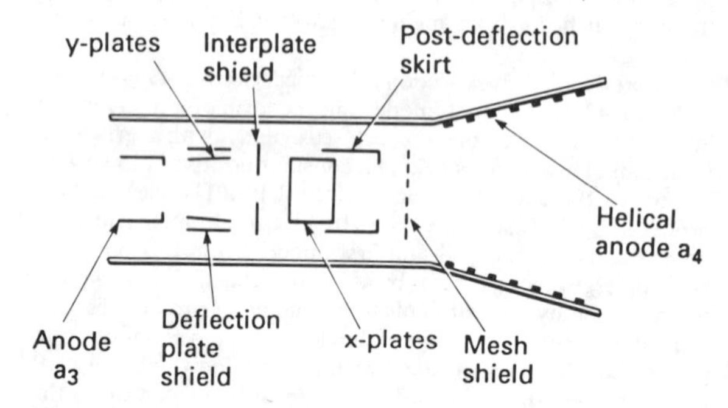

Figure 43.38 Electrode structure of cathode ray tube with post-deflection acceleration

Because the electric field produced by the fourth anode can penetrate into the deflection region, a shield is placed between the deflection plates and the end of the helix. This shield is known as the mesh shield and is shown in *Figure 43.38*.

The electric field produced by the p.d.a. helix without a mesh shield is shown in *Figure 43.39(a)*. The modified field when a mesh shield is incorporated in the tube is shown in *Figure 43.39(b)*. The field now forms a low-magnification diverging lens, and a p.d.a. ratio as high as 10 can now be obtained. The voltage applied to the screen end of the helix can be as high as 10 to 20 kV.

Table 43.3 Designation of phosphors

Pro-electron designation	*Fluorescent colour*	*Phosphorescent colour*	*Persistence*	*Equivalent JEDEC designation*
BA	Purplish blue	—	Very short	—
BC	Purplish blue	—	Killed	—
BD	Blue	—	Very short	—
BE	Blue	Blue	Medium short	P11
BF	Blue	—	Medium short	—
GB	Purplish blue	Yellowish-green	Long	P32
GE	Green	Green	Short	P24
GH	Green	Green	Medium short	P31
GJ	Yellowish-green	Yellowish-green	Medium	P1
GK	Yellowish-green	Yellowish-green	Medium	—
GL	Yellowish-green	Yellowish-green	Medium short	P2
GM	Purplish-blue	Yellowish-green	Long	P7
GN	Blue	Green (infrared excited)	Medium short (fluorescence)	—
GP	Bluish-green	Green	Medium short	P2
GR	Green	Green	Long	P39
GU	White	White	Very short	—
KA	Yellow-green	Yellow-green	Medium	P20
LA	Orange	Orange	Medium	—
LB	Orange	Orange	Long	—
LC	Orange	Orange	Very long	—
LD	Orange	Orange	Very long	P33
W	White	—	—	P4
X	Tricolour screen	—	—	P22
YA	Yellowish-orange	Yellowish-orange	Medium	—

Table 43.4 Persistence of phosphors

less than 1 μs	Very short
1 μs to 10 μs	Short
10 μs to 1 ms	Medium short
1 ms to 100 ms	Medium
100 ms to 1 s	Long
greater than 1 s	Very long

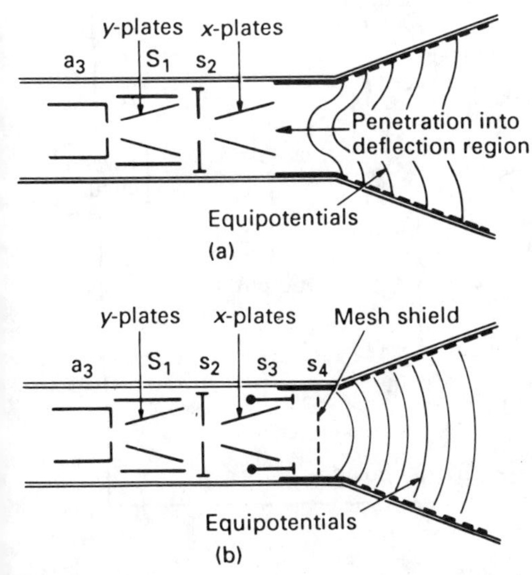

Figure 43.39 Electric field produced by p.d.a. voltage: (a) no mesh shield, (b) with mesh shield

43.8.1.5 Multidisplay cathode ray tubes

It is often useful in an oscilloscope to display two traces simultaneously, particularly if the display is of related waveforms in a circuit. Two electron beams are required in the cathode ray tube to achieve this, each with its own y-deflection system. The two beams may have a common x-deflection system, representing time for example, or can have separate x deflections. Two methods can be used in the tube to obtain the two beams. In one, the double-gun tube, two electron guns are used to provide the two beams. In the other, the split-beam tube, one electron gun is used and the beam is split into two so that each part can be deflected separately.

The splitting of the beam is done by placing a horizontal plate, the splitter plate, between the third anode and the y plates. The splitter plate is held at the same voltage as the third anode, and it is continued along the tube to form a screen between the two y-deflection plates. A deflection voltage applied to the upper plate will therefore operate only on the upper electron beam, while a voltage applied to the lower plate operates only on the lower beam. The two beams can then be deflected simultaneously in the horizontal direction by the x-deflection plates.

Because the beam is split into two, the brightness of the displays on the screen is half that of a single display. This loss of brightness can be a disadvantage when the tube is operating at high frequencies. An alternative method of splitting the beam to overcome this disadvantage is to split it in the electron gun. The third anode has two apertures instead of one, and so two beams emerge.

Post-deflection acceleration can be applied to both beams so the performance of a split-beam tube is comparable to that of a single-beam tube. A problem that may arise in the use of a split-beam tube is that, when the two displays have widely different duty cycles, there will be a considerable difference in the brightness of the displays, which may affect photography of the screen. (The brightness control will affect both beams.) This disadvantage can be overcome by using a double-gun tube.

Because two guns are used in the tube, the brightness and focus of each beam can be adjusted independently. The beams can have a common x-deflection system or separate ones to give two independent displays in one tube. Because there are two electron guns, the double-gun is more bulky than the split-beam tube.

43.8.1.6 Higher frequency operation

There are two limits to the operating frequency of the cathode ray tube previously described. As the frequency of the signal applied to the deflection plates increases, so the reactance of the plates presented to the deflecting signal increases. Each part of the electron beam takes a finite time to pass through the deflection region, the beam transit time, and if the deflecting signal varies over this time, the deflecting force will vary and the deflection of the beam will be less than expected. If the period of the deflecting signal is the same as the beam transit time, the beam will experience equal and opposite deflecting forces and the net deflection is zero.

The limit set by the reactance of the deflection plates depends on the construction of the tube. If the connections to the deflection plates are made by pins in the tube base, the inductance of leads limits the operating frequency to about 10 MHz. If the connections are made through the sidewalls of the tube, and are therefore shorter, the frequency limit is between 50 and 80 MHz. Above this frequency, the transit-time limit starts to be significant.

The operating frequency can be increased further by effectively reducing the beam transit time. This is done by using a series of short deflection plates connected as a transmission line. The deflecting signal is propagated along this line at a velocity equal to that of the electron beam. Each plate deflects the beam, the individual deflections adding to give the final deflection. The frequency limit with this type of tube is about 500 MHz.

Higher operating frequencies can be obtained by using a helical deflection system. The axial propagation velocity of the deflecting signal along the helix is the same as that of the electron beam. Each part of the beam, therefore, remains in a constant field as it travels through the helix but with a steadily increasing deflecting force.

43.8.2 Basic oscilloscope

A block schematic diagram of a typical instrument is given in *Figure 43.40*. The voltage waveform to be examined is applied to the input. A capacitor, which may be shortcircuited by means of a switch, is located in the high-potential input connection. If the waveform is to be examined in its entirety, the switch should be set to the d.c. position. If the signal contains d.c. and a.c. components, but only the latter is to be examined, the switch should be set to the a.c. position.

A variable attenuator Att now follows, the function of which is to enable the required magnitude of the vertical component of the display on the oscilloscope screen to be achieved, with a wide range of possible input amplitudes.

The attenuator is followed by the vertical deflection amplifier Y, the function of which is to deliver a signal of tens or hundreds of volts peak-to-peak to the Y or vertical-deflection plates of the cathode ray tube. Normally a positive rate of change of voltage at the input causes the fluorescent spot on the face of the tube to rise. The Y shift control adjusts the position of the reproduced pattern to the required location in the vertical plane.

The horizontal-deflection amplifier X usually delivers to the X plates a waveform having a constant rate of change of voltage during its active period, causing the spot to move from left to right. The X shift control adjusts the position of the reproduced pattern to the required location in the horizontal plane.

The X amplifier normally derives its input from the sweep circuit. This is triggered to commence its active period of operation during which time the spot on the cathode ray screen travels from left to right. The spot velocity is governed by the setting of the velocity control(s) on the instrument.

An auxiliary output from the sweep circuit is usually available to operate other equipment, thus allowing response/frequency characteristics of circuits to be displayed.

It will be seen that triggering may be effected indirectly from the normal input signal via the Y amplifier or from some external source. When experimenting upon or adjusting a piece of

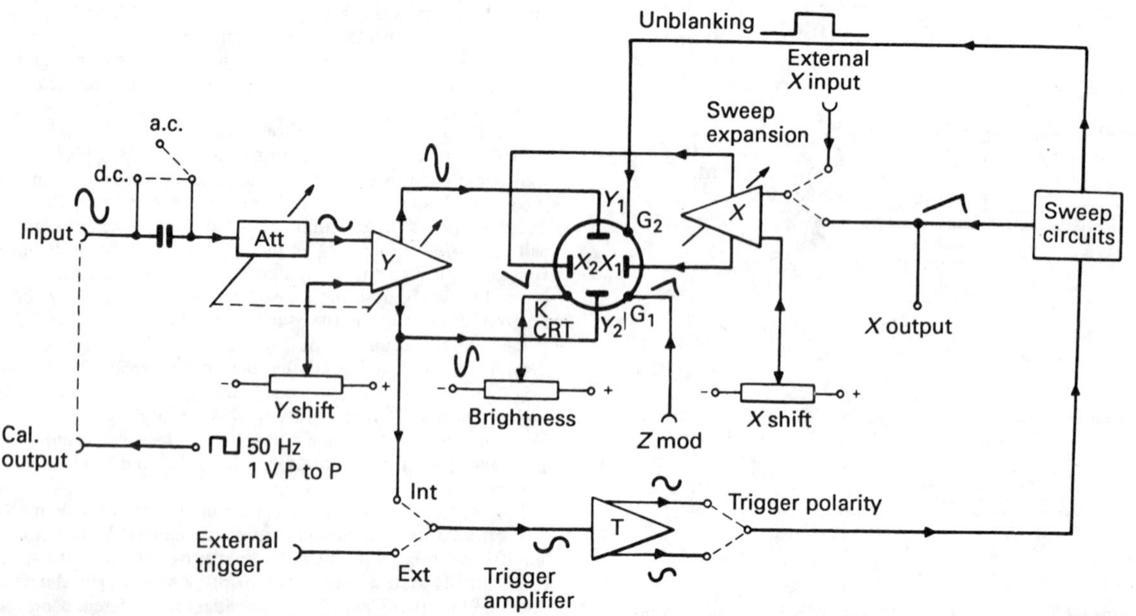

Figure 43.40 Schematic diagram of a typical oscilloscope

equipment under examination it is usually more convenient to trigger the oscilloscope directly from the signal generator in use. Some signal generators have an auxiliary output circuit especially for this purpose, from which a signal is available at a fixed amplitude. This minimises the need for adjustments of the sweep circuits.

Figure 43.40 reveals three further inputs to the cathode ray tube, at the cathode K, and the grids G_1 and G_2. The potential difference between K and G_1 may be varied continuously by means of the brightness control and by means of an external a.c. signal via the *Z* modulation socket. The beam is deflected to prevent it reaching the screen until G_2 receives an unblanking signal which coincides with the active left-to-right movement of the beam.

43.8.3 Oscilloscope amplifiers

The main features of oscilloscope amplifiers which need to be considered are:

(1) Gain
(2) Maximum amplitude of output signal
(3) Bandwidth
(4) Input impedance
(5) Balanced operation

In the case of the *Y* deflection amplifier, the magnitude of the input impedance is kept as high as possible in order to minimise loading of the circuit under test.

The required magnitudes of the output signals from the deflection amplifiers vary from tens of volts to hundreds of volts peak-to-peak. The difference between the *X* and *Y* deflection potentials required for a given cathode ray tube is due to the following causes. The *X* deflection plates are nearer to the screen than the *Y* deflection plates and therefore must bend the beam through a wider angle. In many oscilloscopes provision is made for a longer deflection in the horizontal direction than in the vertical direction, typically in the ratio 4:3. Because of this, the two *X* deflection plates must be further apart than the two *Y* deflection plates which further reduces their deflection sensitivity proportionally.

A moderately priced oscilloscope might have its sensitivity quoted as 50 mV/cm. (More strictly, this is its *inverse* sensitivity.) Instruments having high sensitivity and using tubes of low deflection sensitivities may use amplifiers having gains up to 5000, at which value stabilisation becomes a major problem. In this connection one or more of the following techniques may be employed:

(1) Use of stabilised supply potentials.
(2) Use of a chopper/stabilised amplifier in a feedback network.
(3) Use of balanced stages throughout.

Although the potential required for full screen horizontal deflection is always larger than that which is required for vertical deflection, the amplitude of the available amplifier input potential is normally greater.

In the case of oscilloscopes which are specifically designed for the production of Lissajous figures (*Section 43.8.9*) the *X* amplifier gain should be higher than the *Y* amplifier gain by a factor which is inversely proportional to the relative deflection sensitivities of the cathode ray tube. The number of inverting stages and the manner of the connection to the *X* plates should be such that, in this particular type of oscilloscope, the application of a positive potential to the *X* input causes the spot to move from left to right.

An economically designed amplifier will have its maximum amplitude of output signal and its bandwidth determined almost entirely in the final stage, with earlier stages much more conservatively rated, i.e. these earlier stages will have greater overload capacity and greater bandwidth.

The relative bandwidths of the *X* and *Y* deflection amplifiers need consideration. If the instrument is to be used mainly for Lissajous figures and for the examination of unmodulated waveforms then there is little logic in making the bandwidths different. Most manufacturers, however, take the view that when the *Y* deflection circuit is operating near to its maximum frequency, the instrument is being used either as a peak-to-peak voltmeter, or for viewing only *modulated* waveforms, so that some economy may be effected by providing an *X* amplifier with a lower bandwidth.

The stray capacitances in the circuit however, particularly those of the active devices, effectively shunt the load resistors and thus limit the operation of the amplifiers at the high frequency end of the spectrum. The effective bandwidth of the amplifiers may be extended by the use of one or more of the following techniques:

(1) Stages may have circuit forms which alternate between voltage-amplifier and current-amplifier, e.g. between common-emitting and common-collector types as is shown in *Figure 43.41*.

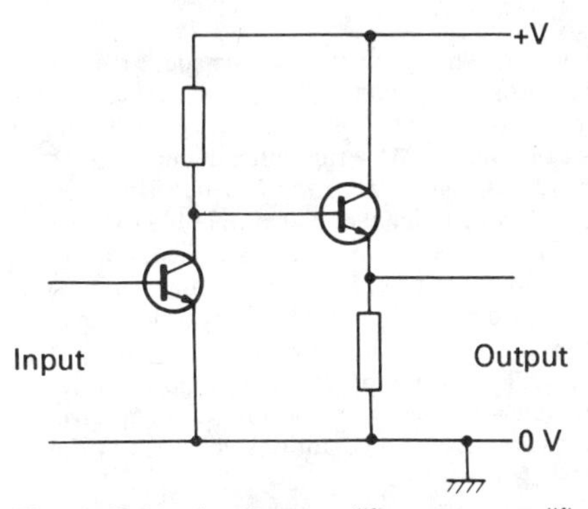

Figure 43.41 A voltage-amplifier current-amplifier combination

(2) A correction inductor may be added to the circuit, its effects offsetting partially the shunting effects of the stray capacitances. An indicator may be connected in series with the load, this being known as shunt correction. Alternatively it may be connected between the output electrode of one active device and the input electrode of the succeeding active device; this is known as series correction.
(3) Emitter (or source) correction circuits may be employed.
(4) Distributed amplifiers may be employed. (These are normally used at frequencies exceeding 20 MHz.)

Some subsidiary (but nevertheless very important) aspects of oscilloscope amplifiers that need consideration are:

(1) Gain control.
(2) Nature of input circuit, balanced or unbalanced. (If the former arrangement is used then the common-mode rejection ratio is important.)
(3) Protection against damage by input signal of excessive amplitude.
(4) Signal delay to permit examination of the leading edge of a pulse.

Y deflection amplifiers are normally provided with continuously variable gain controls which cover the ranges that lie between the discrete steps that are provided by their associated switch controls. The maximum range is usually 1:2.5 and the control may be either a preset or an operational one.

A common requirement is examination of the leading edge of a pulse which is triggering the oscilloscope. To this end it is necessary that a delay circuit be introduced into the *Y* deflection

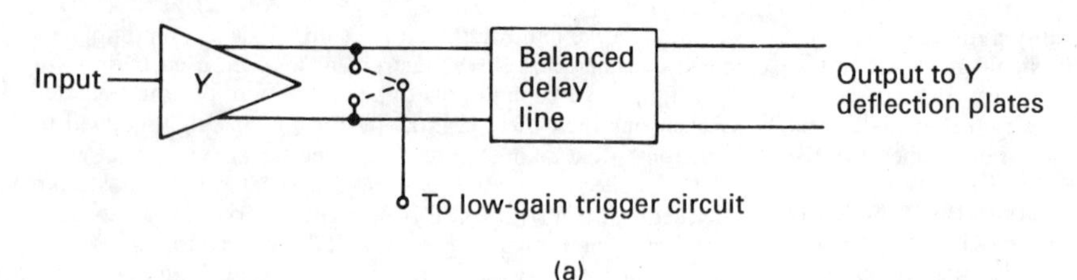

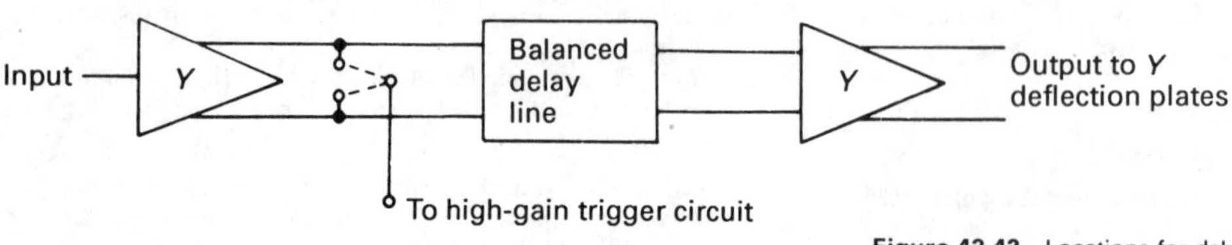

Figure 43.42 Locations for delay lines in: (a) a wide-band oscilloscope and (b) a narrow-band oscilloscope

circuit after the point at which a connection is made to the trigger circuit. The characteristic impedance of a typical balanced delay line is of the order of 200 Ω. For an oscilloscope having a bandwidth of the order of 50 MHz this value of impedance may be suitable for interposition between the *Y* deflection amplifier and the *Y* plates, allowing a signal of high amplitude to be fed to the trigger circuit. However, in an instrument of lower bandwidth, most load impedances have higher values for reasons of current economy and hence the delay line may be located at an earlier stage. The *Y* deflection amplifier is thus split into two sections, the delay line being operated at a lower level of amplitude. This necessitates an increase of gain in the trigger circuit. Schematics for these two conditions are given in *Figure 43.42*.

43.8.4 Attenuators

The main requirements of an attenuator for inclusion in a deflection circuit of an oscilloscope are:

(1) The attenuation must be accurate at all frequencies within the specified band.
(2) The magnitude of the input impedance should be high.
(3) The impedance of the attenuator–amplifier combination should not alter with any change of overall sensitivity. (This is to facilitate the design of probe units and also to maintain a constant loading upon the circuit under test.)

The attenuations, expressed as voltage ratios, are normally in the sequence 1, 2, 5, 10, 20, 50, 100, 200, 500, etc., as far as is necessary. The nine values quoted are adequate for a simple oscilloscope, but further values are necessary for sensitive instruments. In the latter case, variation of overall sensitivity is sometimes achieved by a combination of attenuation adjustment and gain adjustment of the following amplifier, the latter being used at maximum gain only when all attenuation has been switched out. There may be a reduction of bandwidth at maximum sensitivity.

The overall attenuation required is normally provided by the cascade connection of a number of attenuating sections in separately screened compartments. The nine values of attenuation are usually provided by a selection of 1, 2 or 5 in one screened compartment, cascaded with a selection of 1, 10 or 100 in another compartment. An example of a circuit giving unity or 10:1 attenuation is illustrated in *Figure 43.43(a)*.

The input impedance of the amplifier is represented by the dotted C_1. Since its value is likely to be somewhat unstable it may

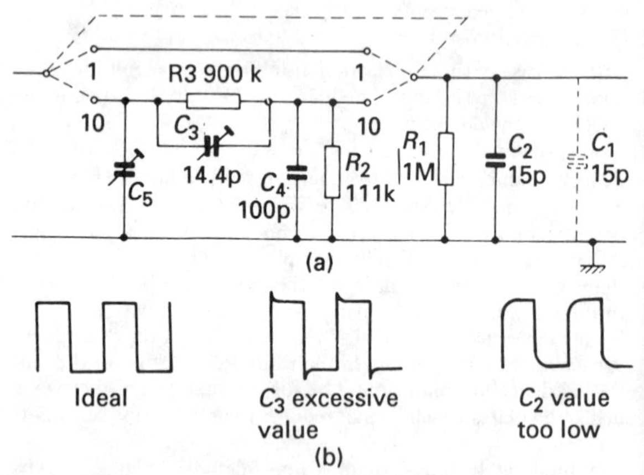

Figure 43.43 Attenuators: (a) basic attenuator circuit and (b) adjustment of high-frequency response

be shunted by a capacitor C_2. The input impedance of the oscilloscope with all attenuators switched out of circuit is thus 1 MΩ shunted by 30 pF, and all these values must be maintained when the attenuators are switched into circuit. The required attenuation at low frequencies is given by

$$\frac{(R_1 \parallel R_2)+R_3}{R_1 \parallel R_2}=10 \quad (\parallel \text{ signifies parallel connection}) \qquad (43.27)$$

while $(R_1 \parallel R_2)+R_3=1$ MΩ as required to maintain constant input resistance. To ensure that this attenuation is maintained over the specified frequency band it is necessary that the time constants of the series and parallel combinations should be identical. Since $(C_1+C_2)/9$ would produce the inconveniently low value of 3.3 pF for the capacitor C_3, a capacitor $C_4=100$ pF is added, so now the equality of time constants can be realised by $C_3R_3=(C_1+C_2+C_4)(R_1 \parallel R_2)$. This produces $C_3=14.4$ pF. The value of input capacitance which has been considered so far is 14.4 pF in series with 130 pF, i.e. 13.5 pF. To maintain a constant value of input capacitance it is therefore necessary to add $C_5=30-13.5=16.5$ pF. Other attenuating circuits are designed on a similar basis.

In the course of production, and possibly at long-term intervals thereafter, the values of C_3 and C_5 (and similar capacitors in other attenuator circuits) should be checked using a square-wave input to the oscilloscope. C_3 will be adjusted with only this attenuator circuit operative. An excessively high value of C_3 will produce overshoot spikes on the displayed waveform, while too low a value will produce prolonged rise times, examples being shown in *Figure 43.43(b)*.

Attenuators are expensive items, so that in the case of oscilloscopes having input circuits that are balanced, and hence requiring the more expensive balanced attenuators, it is usual for the manufacturers to put greater reliance on the use of amplifiers of variable gain and thus to economise on the number of circuits in the balanced attenuators.

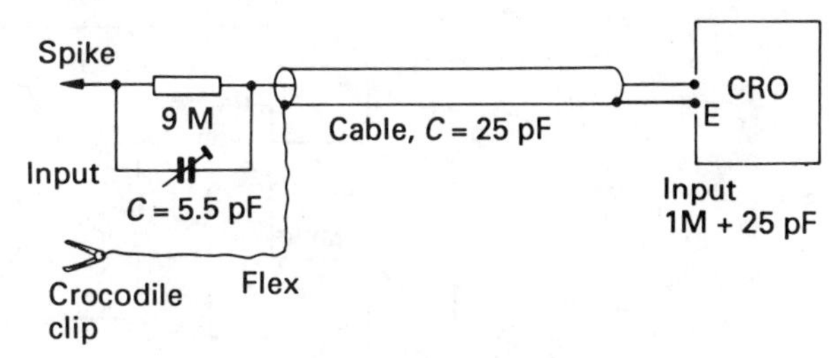

Figure 43.44 Simple probe giving an attenuation of 10

43.8.5 Probes

Probes are designed for use with oscilloscopes and other instruments for one or more of the following purposes:

(1) To isolate the circuit under test from the effects (principally shunt capacitance) caused by normal connecting leads or cable to the oscilloscope and the input circuit of the oscilloscope itself.
(2) To increase or decrease the gain of the measuring system.
(3) To provide detection of an amplitude-modulated waveform.

Unless it is known to the contrary, it is unsafe to take a probe provided by one manufacturer and to use it with another manufacturer's instrument. However, some probes are adjustable in this respect.

The length and type of cable which is provided with the probe are critical and should not be changed. Whereas the capacitance of normal coaxial cable is typically 100 pF/m, special low-capacitance cables employing very thin inner conductors are normally used. These conductors are usually kinked so that the dielectric is mostly air. The inductance of the cable, combined with its capacitance, tends to cause ringing on signals having abrupt transitions, as components are reflected backwards and forwards along the cable. Since such a simple circuit cannot be terminated in its characteristic impedance while maintaining the other properties required, a compromise is reached by using resistance wire for the inner conductor, the resistance value being several hundreds of ohms. Careful design of the cable is necessary if microphony is to be avoided.

Some probes contain only passive components, while others contain active components which must be supplied with power along the interconnecting cable from the main instrument. Probes containing only passive components, and some containing active components also, normally attenuate the signal. In the simpler systems the attenuation is a simple round number such as 2 or 10. Since this may lead to operational errors, more complex system are available which operate when the probe plug is inserted into the socket on the oscilloscope. This action may:

(1) Raise the gain of the vertical deflection amplifier.
(2) Change the circuit of an optical display system using light-emitting diodes or fibre-optic devices, thus indicating a change of sensitivity.

The simplest probe, containing only two components, is shown in *Figure 43.44*. Here the time constant of the probe $5.5\times 10^{-12}\times 9\times 10^{6}=50\times 10^{-6}$ is made to equal the time constant of the cable-plus-oscilloscope input circuit, i.e. $1\times 10^{6}\times (25+25)\times 10^{-12}=50\times 10^{-6}$.

The capacitor in the probe is normally adjusted using a square-wave generator as previously explained in the section on attenuators.

Where much loss of deflection sensitivity cannot be tolerated, where the input impedance of the measuring circuit must be very high, or where an abnormally long cable must be used between probe and oscilloscope, it is usual to use active devices in the probe. An alternative scheme using semiconductors is illustrated in *Figure 43.45*.

Current measuring probes are also available. These are basically current transformers. The current-carrying wire acts as a primary, a split ferrite core has jaws which can be opened to clip over the wire and then closed. A winding on the core acts as the secondary winding. Electrostatic and electromagnetic shielding are necessary. Such probes operate satisfactorily up to about 50 MHz, but performance is less satisfactory below 1 kHz, although frequency-compensated amplifiers may be supplied with the probe.

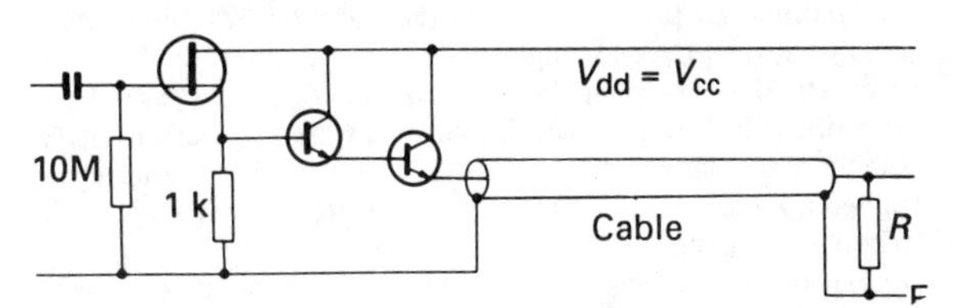

Figure 43.45 An all-semiconductor probe (from *Measuring Oscilloscopes* edited by J. F. Golding, Newnes-Butterworths (1971))

43.8.6 Multiple display facilities

Most multiple display facilities are provided by oscilloscopes specially designed for the purpose; alternatively they are provided by means of additional units used in conjunction with conventional oscilloscopes. In either case the screen size should be large and the aspect ratio preferably 1:1 rather than the 4:3 ratio commonly encountered.

Each channel must be provided with its own sensitivity control(s), its own *Y* shift control, and its own input switch. Some systems have separate horizontal sweep and shift, brightness and focus controls for the two channels, while in some cases they are common.

Multiple displays usually provide dual display facilities, sometimes with the additional facility of displaying either the sum or the difference of two quantities along a single trace. Quadruple display instruments are also available. These are particularly useful for teaching and demonstration purposes, but they are not usually precision instruments. Instruments employing long afterglow cathode ray tubes are available having sweep circuits which operate with repetition frequencies as low as 2 cycles/min. Some four-channel oscilloscopes can be switched to provide two-channel or one-channel operation when required.

The multiplexing facility may be provided either by the use of a special type of cathode ray tube or alternatively by time-division-multiplex electronic circuitry. Several subdivisions of these two techniques are available.

Where a single-beam cathode ray tube is used for multiple display purposes, alternative time-division-multiplex sampling systems are available. Some oscilloscopes use one technique,

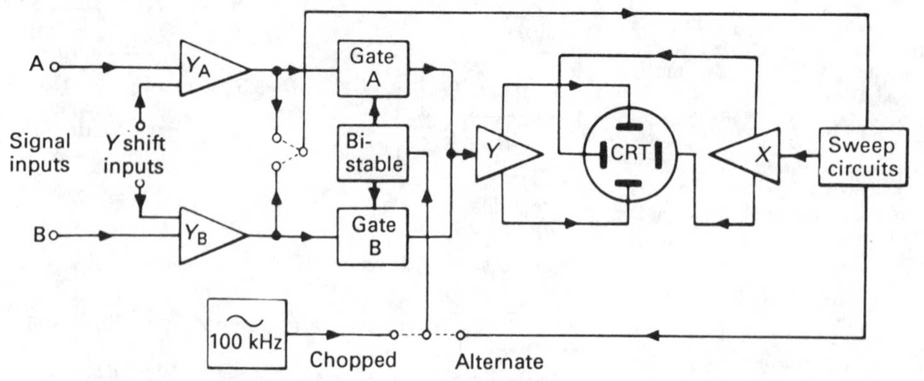

Figure 43.46 A dual-trace oscilloscope

some the other. Some instruments provide facilities for the use of both techniques, the choice being in certain cases left to the user and in other cases governed by the sweep adjustment control.

In one case the switching action in the signal circuits is synchronised with the horizontal sweep, so that each sweep of the spot is devoted entirely to one signal, the various signals being written on to the screen in sequence. This is known as synchronous-mode operation. In alternate mode operation, only two channels are provided. In the other case the switching action is unsynchronised, occurring usually at about 100 kHz. A typical schematic diagram is shown in *Figure 43.46*. In this circuit the gates open alternately under the control of the bistable circuit, so that the signals are displayed in turn. The bistable circuit may be triggered either from the internal sweep circuit or from the 100 kHz internal generator.

Consider as an example two waveforms which are to be examined, a sinusoid being applied to input A and a triangular waveform to input B. To the former, a positive *Y* shift potential is added, producing V_A, and to the latter a negative *Y* shift potential, producing V_B, both being illustrated in *Figure 43.47*. The horizontal sweep waveform is shown as V_X. If alternate mode operation is employed then the input to the common *Y* amplifier will appear as V_{Y_1}. If however chopped operation is employed, then using the chopping waveform V_{CH} the input to the common *Y* amplifier will be as shown at V_{Y_2}. (The vertical transitions of this waveform are so rapid that there is a negligible brightening of the screen in between the two waveform patterns.) The gaps which appear in both waveforms in any one horizontal sweep are filled by successive sweeps, so that irrespective of the switching employed (alternate or chopped), the final result appears as shown at the bottom of *Figure 43.47*.

43.8.7 Sampling

For the examination of waveforms having repetition frequencies between about 100 MHz and 1 GHz, or having rise times of the order of nanoseconds, it is usual to employ a sampling technique, using either a complete sampling oscilloscope or else a sampling adaptor in conjunction with a conventional oscilloscope. A simplified schematic diagram appears in *Figure 43.48* with waveforms in *Figure 43.49*.

The input signal is applied to a trigger generator which delivers a series of pulses, normally at a frequency which is of the order of 30 kHz but at an integral submultiple of the input frequency. (This presupposes that the signal frequency exceeds 30 kHz. If however the signal frequency is less than this, then the sampling frequency is made to equal the signal frequency.)

Each trigger pulse initiates one step in the operation of a staircase generator, the output of which has its steps counted to determine the number of dots per cycle of the waveform which is to appear on the oscilloscope screen in any one horizontal sweep.

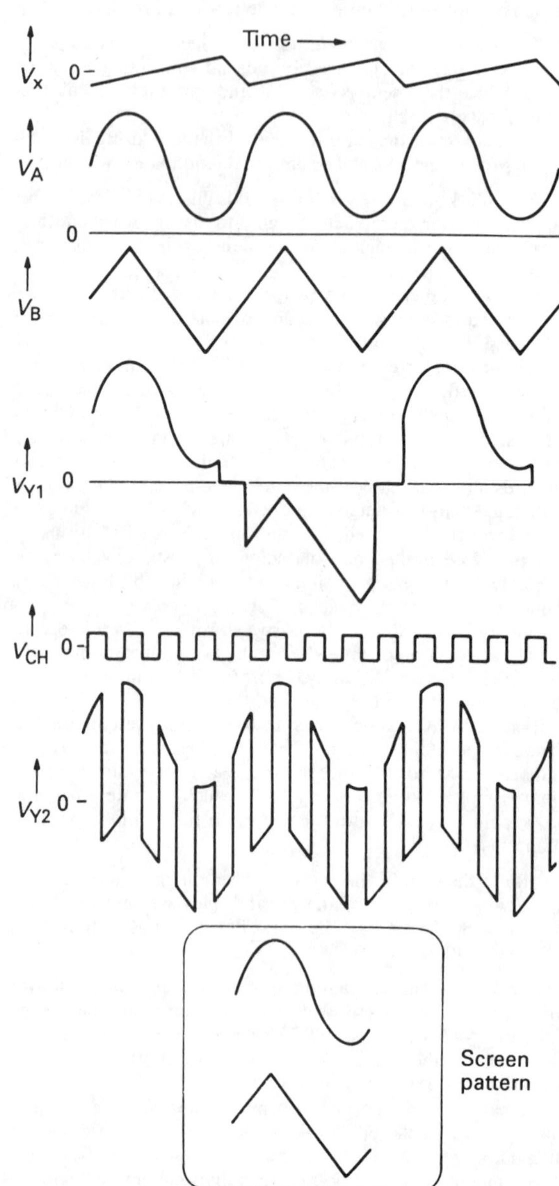

Figure 43.47 Typical waveforms and screen pattern for a single beam dual-channel oscilloscope using either alternate-mode or chopped-mode operation

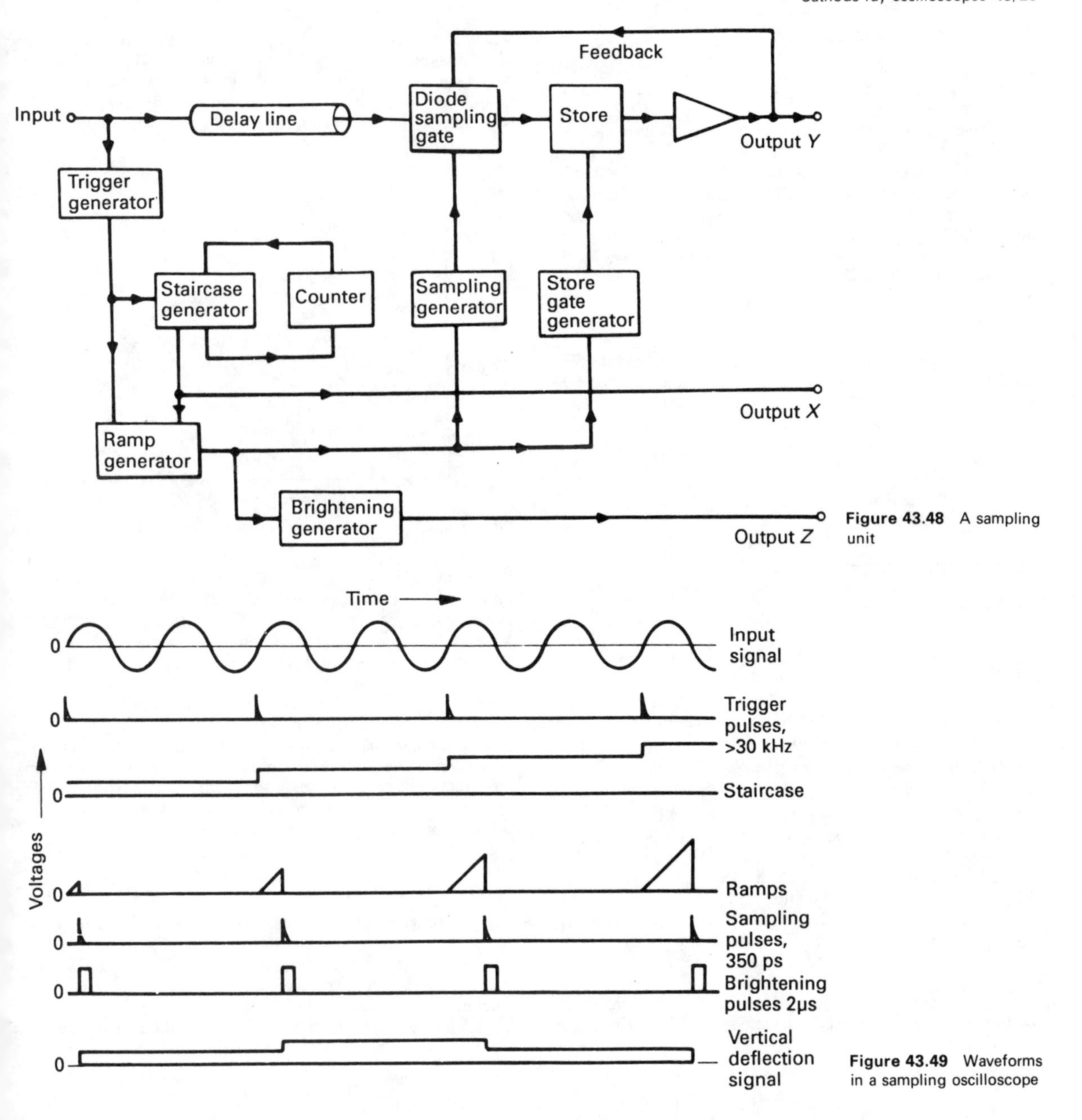

Figure 43.48 A sampling unit

Figure 43.49 Waveforms in a sampling oscilloscope

After the required count, say between 200 and 1000, the counter resets the staircase generator, which then commences the generation of a new staircase.

Each trigger pulse also causes the ramp generator to commence the generation of a very fast ramp of fixed velocity. The duration of each ramp is, however, caused to be proportional to the instantaneous magnitude of the staircase, so that during the staircase cycle the ramps are of progressively increased durations. The *termination* of each ramp now causes the sampling generator to produce a sampling pulse, typically of 350 ps, these sampling pulses occurring progressively later and later in individual periods of the test waveform.

The staircase waveform also causes the spot of the oscilloscope to move horizontally across the screen in a series of rapid movements, and at the same time the spot is brightened by a pulse (typically of 2 μs duration) which coincides in time with the sampling pulse.

The signal now passes via a delay line (producing typically 50 ns delay) to the sampling gate. The function of this delay line is to facilitate the examination of the leading edge of a pulse as explained previously.

Samples of the pulse are taken and stored in a capacitor store, access to which is controlled by the store-gate generator. From the store, an output is taken via an amplifier and feedback path of unity gain, to reverse-bias the diodes in the sampling gate so that during each sampling period the component which is passed into the store is proportional only to the *change* of signal amplitude which has taken place since the last sample was taken. The output from the amplifier is then used to provide vertical deflection of the oscilloscope beam.

43.8.8 Storage oscilloscope

A non-storage oscilloscope needs a periodic signal of a relatively high frequency in order to display a steady trace. The persistence of the screen can vary from milliseconds to a few seconds, depending on the type of phosphor. Storage oscilloscopes are used to display transient events, and also for signals with very low frequencies in which, on a conventional oscilloscope, the first bit of the trace would begin to fade before the end was finished.

Two storage techniques are used in oscilloscopes, analogue and digital. The first uses a special storage cathode ray tube and the second digitises the analogue signal and stores it within the oscilloscope circuitry.

43.8.8.1 Storage tube

Figure 43.50 shows a schematic of one type of storage tube. The flood guns have no associated deflection plates and they flood the entire area of the target. Their function is to maintain the state of the target (write or erase) once this is established by the writing gun. The cathode of the flood guns are at ground potential and they are kept continuously on. The cathode of the writing gun is at a high negative voltage and it can be deflected onto any point on the screen which is to be written into.

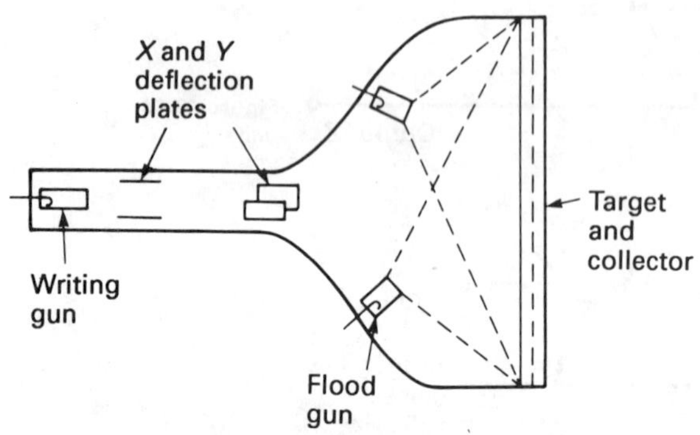

Figure 43.50 Storage tube (schematic diagram)

The screen consists of scattered phosphor particles in which any area can be written into without affecting an adjacent area. This is the target and it is placed on a conductive coated glass face plate which acts as the collector.

When primary electrons from the writing or flood guns strike the target they result in secondary emission. The ratio of the secondary emission current to primary emission current is called the emission ratio. *Figure 43.51* shows a plot of this ratio as the target voltage is changed. The collector is at a positive potential of about V_2. Point A represents the erased position and point C the written position. Assume that the writing gun is off and the target is at point A. The flood guns have insufficient energy to cause the phosphor to move away from A. When the writing gun is switched on it causes a large number of electrons to strike the target and this increases the secondary emissions. From A to B the flood guns oppose the writing gun but beyond B they aid the writing gun in moving the phosphor to point C. The writing gun can now be switched off, the flood guns having sufficient energy to maintain the phosphor in the written position at C.

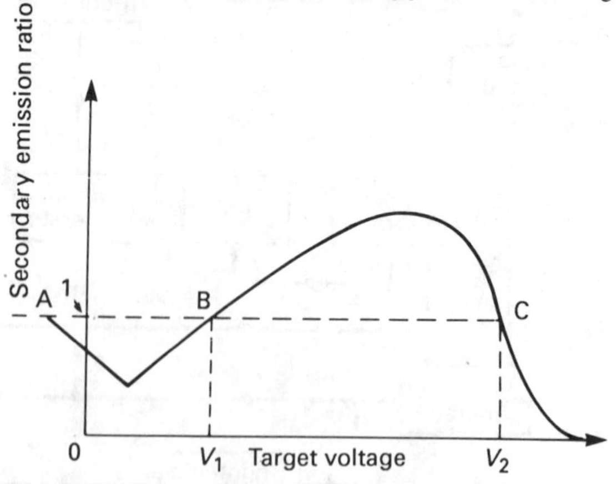

Figure 43.51 Secondary emission due to flood gun

To erase information stored on a storage tube the collector may be pulsed negative. This causes the secondary emission from the target to be repelled back into the target. The target voltage reduces until it reaches the erased point A after which the collector pulse can be slowly reduced to zero.

43.8.8.2 Digital storage

Digital storage oscilloscopes use electronic circuitry to store the signal, and a conventional cathode ray tube which is periodically refreshed by the stored information. *Figure 43.52* shows the schematic of a digital storage oscilloscope. The input signal is amplified and the resulting waveform is sampled and converted to a digital signal, by the analogue-to-digital converter, before

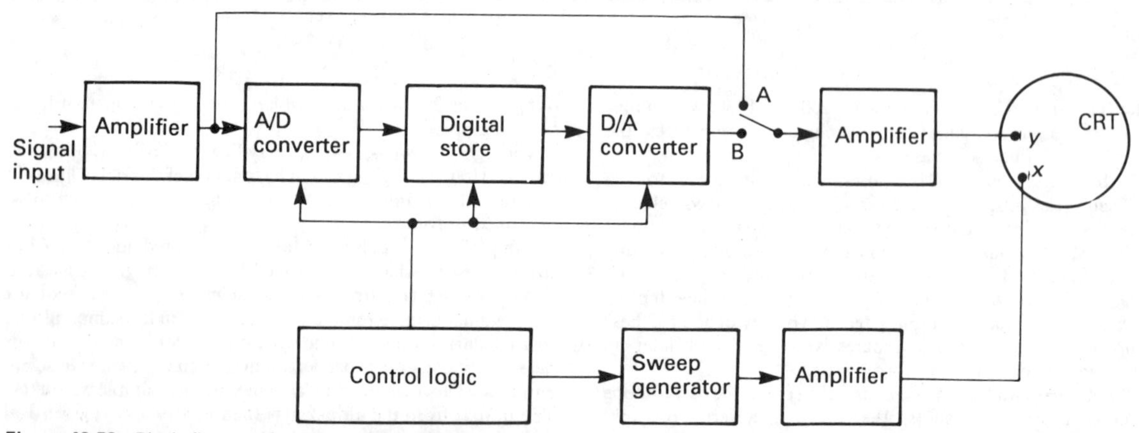

Figure 43.52 Block diagram of a digital storage oscilloscope

being stored. A digital-to-analogue converter changes this stored information back to an analogue signal before it is amplified and fed to the Y plates of the cathode ray tube. When the control switch is set to position A the storage circuitry is bypassed and the instrument behaves like a conventional oscilloscope.

The analogue-to-digital converter is usually the limiting component of a digital storage oscilloscope and its speed determines the frequency response of the instrument. The memory capacity is another important parameter and it determines the resolution of the instrument, and the number of dots used to create a waveform. For example if an 8000 bit memory is available and 8 bits are used to represent a word then the instrument has a resolution of 1 in 2^8, or 1 in 256, and it can use up to 1000 dots to create a waveform. If both channels use the same memory for storage then the number of dots available is halved.

The accuracy of an oscilloscope is usually worse than its resolution, since it includes errors due to resolution of the analogue-to-digital converter and due to nonlinearity of the amplifiers. Modern storage oscilloscopes have resolutions of the order of 0.025% and accuracies of 0.25%. Some oscilloscopes have a dot joining feature which uses simple lines to join between the dots on the waveforms.

The digital storage oscilloscope stores information electronically. This information can therefore be manipulated, such as reproduced on an expanded or reduced time base. It can also be output at different speeds, for example to suit a pen recorder. In this instance the output speed should match the inertia of the pen so that lines between dots are smooth.

43.8.9 Lissajous figures

A less familiar use of the oscilloscope is for the formation of Lissajous figures by a process illustrated in *Figure 43.53*. In this case the Y plates receive their signals in the normal manner but the X plates receive theirs via the external X socket. The internal horizontal sweeping generator and the unblanking operation are inoperative. In the example given, it is shown that the Y plates are receiving a sinusoid while the X plates are receiving a cosinusoid of the same frequency. The two graphs in *Figure 43.53* employ similar time scales. A succession of times, 1, 2, 3, etc., is similarly selected on both graphs and construction lines drawn to indicate

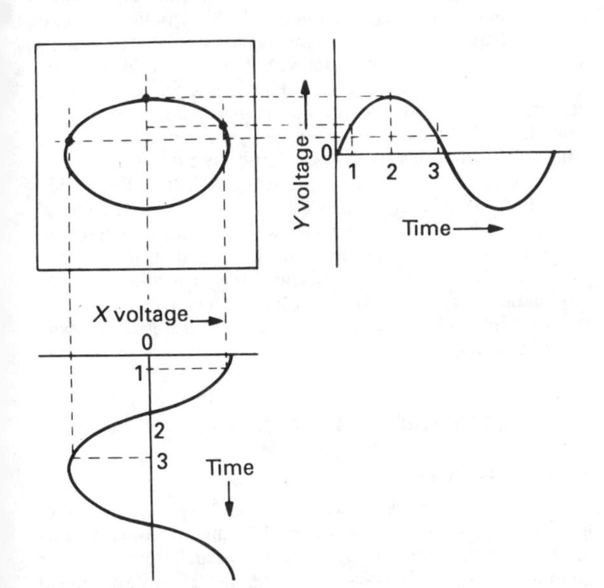

Figure 43.53 Formation of a simple Lissajous figure

the resulting positions of the spot. The result is a pattern which consists of an ellipse having its axes horizontal and vertical, the spot travelling in an anticlockwise direction. Careful consideration shows that two *similar* waveforms would produce a straight line with a positive slope. If one waveform is inverted, the straight line has a negative slope. By employing this principle of construction, the pattern which results may be predicted for any two waveforms of any two repetition frequencies, but in practice the technique is normally restricted to simple waveforms having some simple frequency relationship.

The phase relationship between two sinusoidally shaped waveforms of the same frequency may be determined as follows: First one of the waveforms is applied simultaneously to both X and Y inputs. At least one of these circuits should have a gain control which is continuously variable. The gain controls are now adjusted to produce on the screen a straight line with a slope of either +45° or −45°. (The 45° slope indicates that the deflection sensitivities are identical. The straightness of the line indicates that the deflection circuits have identical phase shift at this frequency.) The preliminary adjustment having been made, the two signals are now fed *separately* to the horizontal and vertical deflection input sockets. The result is likely to be an ellipse, the major and minor axes Z_{maj} and Z_{min} being shown in *Figure 43.54*. The phase difference is now given by:

$$\phi = 2 \arctan \frac{Z_{min}}{Z_{maj}} \tag{43.28}$$

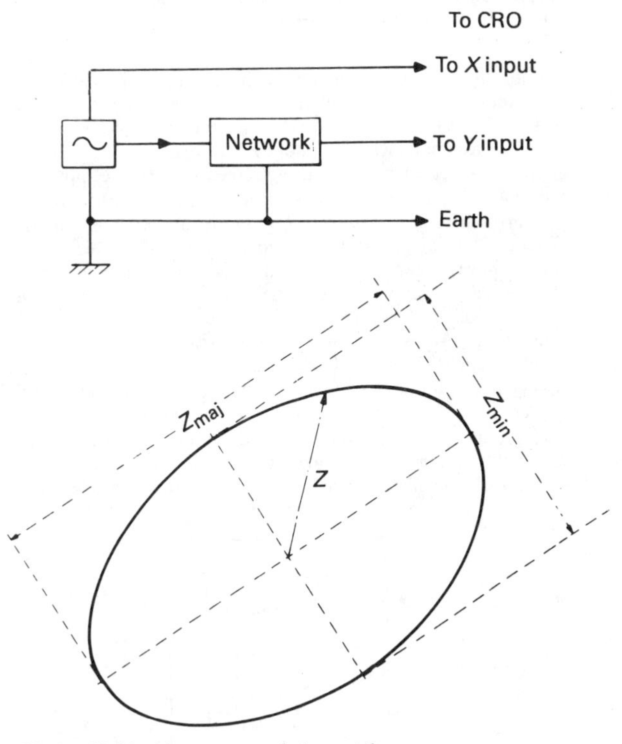

Figure 43.54 Measurement of phase shift

Lissajous figures are also used for frequency comparison purposes, e.g. for checking the operation of frequency multipliers and dividers and for calibrating multifrequency oscillators from single-frequency sources. (It should be noted that the technique can be used only when the relative frequencies are stable to within about 1 Hz.) Examples where $f_X = 2f_Y$ and $f_X = f_Y/3$ are shown in *Figures 43.55(a)* and *(b)* respectively. For any frequency

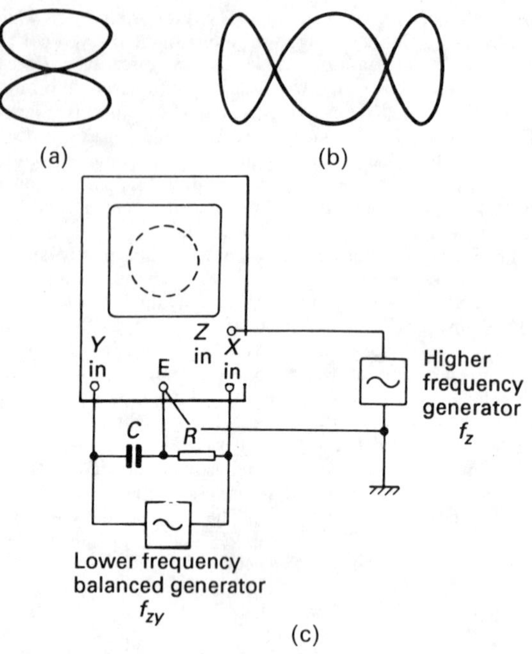

Figure 43.55 Frequency comparison using Lissajous figure: (a) $f_X=2f_Y$, (b) $f_X=f_Y/3$, (c) $f_X=15f_{XY}$

relationship, however, an infinite number of patterns is available depending upon *phase* relationships. Where the frequency relationship is high, e.g. greater than six, it is more convenient to employ a technique which includes Z modulation, the necessary circuit arrangements and resulting screen pattern are shown in *Figure 43.55*(*c*). Here the lower frequency signal is applied from a *balanced* source to a phase-splitting circuit consisting of C and R in series. Assuming that the horizontal and vertical deflection systems have similar sensitivities, the magnitude of the reactance of C should have approximately the same value as the resistance of R. This will produce an ellipse, the axes of which are horizontal and vertical. Adjustment of X or Y sensitivity or C or R will enable a circle to be produced. The higher frequency signal, having an amplitude of several volts, will cause the pattern to be broken up as shown, the number of breaks equalling f_Z/f_{XY}.

Lissajous figures may be employed for the measurement of time delay, using a simple arrangement such as is shown in *Figure 43.56*. It will be realised that whenever the time delay t_d of the network is equal to an odd number of quarter periods of the source, the pattern on the screen will be an ellipse, the axes of which are horizontal and vertical. The most convenient procedure to adopt is to set the generator frequency f to a low value where the delay t_d is only a very small fraction of the generator period $1/f$. The screen pattern should then be a sloping straight line. The value of f is now raised until, for the *first* time, the axes are horizontal and vertical, then $t_d=1/4f$.

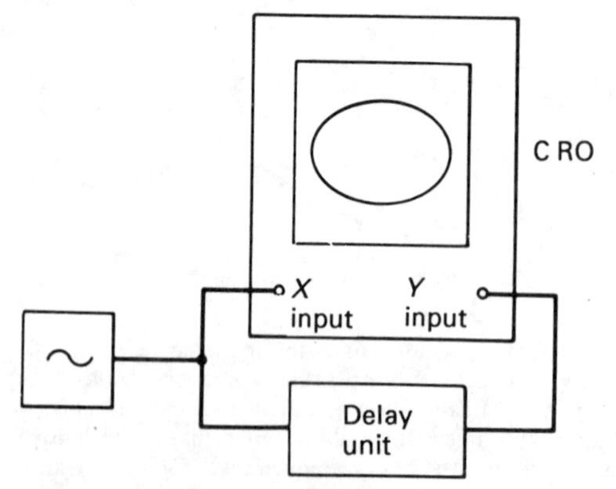

Figure 43.56 Measurement of delay by means of a Lissajous figure

A further use of Lissajous figures is for checking the amplitude modulation of a carrier by means of a trapezium figure. Here the modulated high-frequency signal is applied to the Y deflection circuit (because this normally has higher gain and bandwidth) and the modulating signal is applied to the X deflection circuit. An example is shown in *Figure 43.57*(*a*) where the percentage modulation is given by

$$100\frac{a-b}{a+b} \qquad (43.29)$$

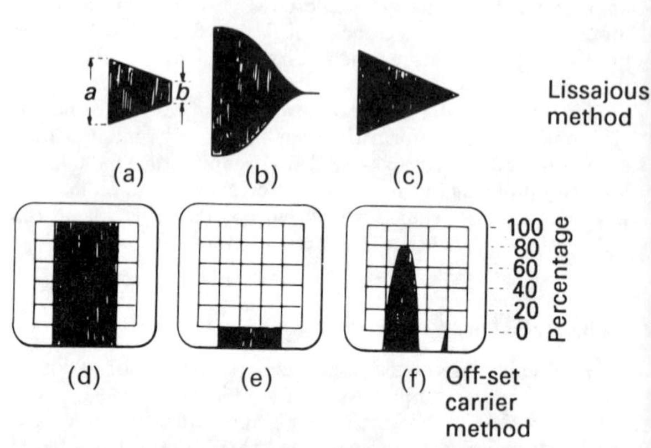

Figure 43.57 Modulation measurements

An example of over-modulation, including peak-amplitude limiting and carrier-cutting, is shown in *Figure 43.57*(*b*). Perfect 100% modulation is illustrated in *Figure 43.57*(*c*).

A more accurate method of measuring modulation is illustrated in *Figure 43.57*(*d*), (*e*) and (*f*). First, the sweep circuit of the oscilloscope is set to operate normally. Then, in the absence of any signal, the Y shift control of the oscilloscope is set to produce a horizontal line to coincide with the bottom line of the oscilloscope graticule. The signal is now applied to the Y deflection circuit in the normal manner, and, while it is *un*modulated, the Y deflection sensitivity (or signal amplitude) is adjusted to produce a pattern which rises by 5 or 10 divisions of the graticule as shown at (*d*). This action calibrates the vertical scale. The Y shift control is now operated to lower the pattern so that its upper extremity coincides with the bottom line of the graticule as shown at (*e*). When modulation is applied, the upper extremity of the pattern indicates the percentage of the modulation, as shown in the example (*f*). It is convenient, but not essential, to synchronise the operation of the sweep circuit with the modulating signal.

43.9 Noise and sound measurements

43.9.1 Fundamental quantities

The words *sound* and *noise* will be treated here as synonymous from the viewpoint of measurement. Two quantities are directly measurable, the sound pressure and the particle velocity, other quantities such as sound power being derived from one or both of these.

It is assumed, unless otherwise stated, that all sound measurements are carried out in air, so that it is necessary for only one of the basic quantities, normally the pressure, to be measured, since pressure and particle velocity in a parallel wave motion are related by the equation

$$p = \rho c v \qquad (43.30)$$

where ρ is the density of the air, c is the velocity of sound in air and v is the particle velocity.

Sound pressures are, for practical purposes, converted into sound pressure levels, defined in decibels (dB) as

$$L = 20 \log_{10}(p/p_{\text{ref}}) \qquad (43.31)$$

where L is the sound pressure level and p_{ref} is a reference pressure of 2×10^{-5} N/m^2.

Alternatively, intensity level (which is the decibel equivalent of intensity, the rate of flow of sound energy across unit cross-section at a point in a sound field) is defined as

$$L_1 = 10 \log_{10}(I/I_{\text{ref}}) \qquad (43.32)$$

where

$$I_{\text{ref}} \text{ is } 10^{-12} \text{ W/m}^2 \qquad (43.33)$$

and

$$I = p^2/\rho c \qquad (43.34)$$

These two reference levels are nearly equal for air at normal temperature and pressure, differing by only about 0.14 dB. For most purposes they are regarded as equivalent.

Subjective measures of sound sensation are derived from objective measurement of pressure level or spectrum pressure level (pressure per unit bandwidth) throughout the spectrum.

43.9.1.1 Measurement of particle velocity

Particle velocity may be measured absolutely by means of a Rayleigh disc,[1] which is a thin disc suspended from a torsion fibre in the sound field. The torque on the disc when held at an angle to the direction of propagation is related to the particle velocity. This method of measurement is unsuitable for use outside a specialised laboratory, and it is therefore better to measure pressure.

43.9.1.2 Measurement of sound pressure

For the measurement of sound pressure a calibrated microphone is needed, together with an amplifier of known gain with rectifier and meter. The microphone may be calibrated by the method of reciprocity,[2] by Rayleigh disc or by comparison with a standard which has been calibrated by one of these methods. The comparison must be carried out in a free-field room (i.e. one in which all surfaces are covered with sound-absorbing material so that only sound directly from the loudspeaker or other source of test sound reaches the microphones).

For calibration of the amplifier, a steady alternating voltage is applied through a calibrated attenuator to its input and the attenuation is varied until the output voltage of the amplifier is equal to the input of the attenuator. *Figure 43.58* shows a circuit for noise measurement embodying these components. The calibrated microphone is connected to the input of the amplifier and the output is rectified and applied to a d.c. meter. The rectification characteristic should give an accurate indication of the r.m.s. value of the signal, which represents the power irrespective of the waveform. British Standard 3489:1962 makes use of an r.m.s. meter obligatory for all noise measurements. To achieve a true r.m.s. reading some form of thermal meter is often used, but simple square-law rectification gives sufficient accuracy even with complex waveforms. The amplitude of the waveform can be measured with the use of a peak-rectifying meter, and the ratio of this to the amplitude of a sinusoid having the same r.m.s. value is known as the crest factor.

As the rectifier is necessarily limited in dynamic range, it is usual to insert switched attenuators into the amplifier, by which the rectifier input and consequently the meter indication, can be brought within a range of 10–20 dB. For the rapid measurement of steady or slowly varying sounds, a calibrated attenuator with 1 dB steps may be used to bring the meter to a reference reading.

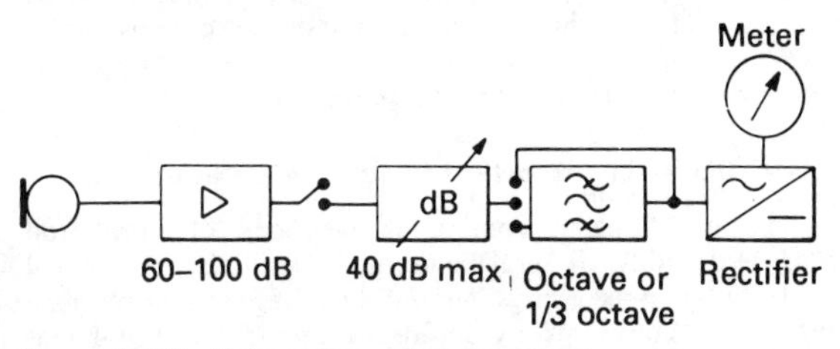

Figure 43.58 Noise and sound measurement

43.9.1.3 Measurement of sound spectra

The spectrum of a complex sound may be measured by inserting band-pass filters into the amplifier chain of a sound-pressure measuring circuit, so as to obtain a series of band pressure levels. The range of audible frequencies is divided into octave or one-third octave bands and filters for the purpose are specified in British Standard 2475:1964. The one-third octave bands, except at very low frequencies, correspond closely to the critical bandwidths of hearing. Thus, measurements in one-third octave bandwidths are used for the evaluation of loudness by the method of Zwicker[3] and also for the great majority of development work on the silencing of machines.

Octave bandwidths are used in the Stevens' method[4] of loudness evaluation and for many applications in connection with industrial and community noise. Noise containing discrete frequency components requires narrow-band heterodyne filters, typically 10% of the centre frequency in width.

43.9.2 Practical equipment for the measurement of sound pressure level

43.9.2.1 Sound level meters

A sound level meter consists of a portable battery-operated noise measuring circuit with an r.m.s. meter calibrated in decibels; most sound level meters are provided with a variety of facilities such as band-pass filters and weighting networks. A typical block diagram is shown in *Figure 43.59*. The requirements of such an instrument are specified in British Standard 5969:1981 in which meters of four different grades of precision are described, replacing the 'precision' and 'industrial' grades of former standards.

The microphone, which is usually of a capacitor or electret type, has an omnidirectional polar diagram in which the off-axis sensitivity deviation is held within specified limits. The main amplifier is followed by an attenuator with switched 10 dB steps by means of which the indicating meter may be brought within

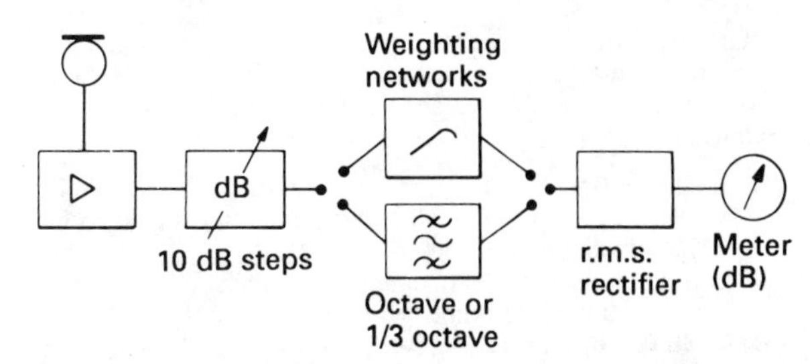

Figure 43.59 Block diagram of a typical sound level meter

the scale for a wide range of microphone inputs. Until recently, the meter scale was usually 10 dB wide with perhaps a small extension at the lower end. Recently it has become possible to provide meters with very much wider scales to facilitate reading of rapidly changing levels. In use, the meter reading is added to the number shown on the attenuator control, which represents the sound pressure level when the meter is at zero.

Weighting networks and other filters are inserted after the attenuators, either built-in and selected by switches or external and connected by input and output terminals.

An overall calibration of sound level meters is carried out by exposing the microphone to a sound generator with a known output. In earlier types of generator, small steel balls were allowed to fall freely on to a metal diaphragm held at a specified distance from the meter. The most common type now is a diaphragm driven into vibration by a stable oscillator and mounted in a tube which can be fitted over the microphone of the meter. In either case, the meter amplifier is adjusted to bring the reading to a designed meter reading.

The gain of the amplifier chain is separately checked by noting the meter reading produced by an inbuilt stable oscillator.

A meter reading of sound pressure level by means of an instrument such as the above, in which equal weight is given to all frequencies within the range of measurement, is not well correlated with the subjective loudness or other subjective effects of the sound. This is because the human ear is relatively insensitive to sounds of very high and, more particularly, low pitch.[5] Weighting networks have therefore been devised for insertion into noise measuring chains by which the low and very high frequency components of complex sounds are progressively attenuated. *Figure 43.60* shows four commonly used networks as

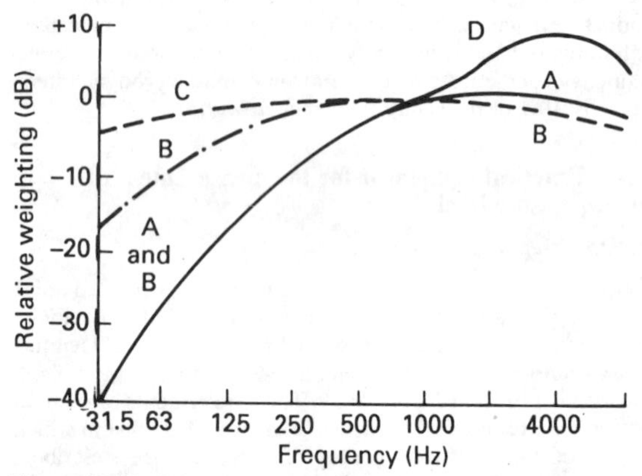

Figure 43.60 Weighting networks (A, B, C, D) (from British Standards 5969:1981 and 5647:1979)

specified in British Standards 5969:1981 and 5647:1979. The *A* network is the most commonly applied and is always provided in sound level meters, being used in connection with road traffic noise, hearing damage risk and assessment of neighbourhood noise.

The proliferation of integrated circuits in the last ten years has enabled the versatility of sound level meters to be increased in several important respects. Facilities now available include the following:

(1) Very fast time constants in conjunction with peak-hold enabling the maximum value of a fluctuating signal to be accurately determined.
(2) Continuous integration or averaging of the acoustic energy input for any desired time.
(3) Real-time display of octave or one-third octave band pressure levels.
(4) AC and d.c. outputs corresponding to the signal reaching the indicating meter, for driving tape recorders or chart level recorders.
(5) Inputs for alternative microphones or for accelerometers for vibration measurement.

43.9.2.2 Chart level recorders

Similar in principle to the measurement of noise levels by measurement of the attenuation necessary to reduce the output of the amplifier to match a reference level, as described in *Section 43.9.1.2*, is the chart level recorder, by which permanent records of level can be made over a range of 75 dB or more. In this instrument, the first of which was marketed by Neumann, a high speed servo system is actuated by the difference between the amplified signal and an arbitrary value. The servo drives the wiper of a potentiometer in the amplifier chain and comes to rest at the balance point. The position of this point is continuously recorded by a pen attached to the wiper arm. The potentiometer is usually designed to give a plot in dB, but any other law may be produced.

For most purposes the chart is driven at constant speed to give a record of level against time, but, by gearing the chart drive to the dial of an oscillator, response/frequency curves can be obtained.

43.9.2.3 Real-time displays of sound spectra

Equipment is now obtainable from several manufacturers, consisting of a noise-measurement chain and a set of narrow band-pass filters which are scanned at short intervals and their outputs displayed on an oscilloscope. The filters are usually one-third octave width or narrower, up to about 400 channels being included in one instance. These instruments are designed as complete systems with a variety of facilities such as the freezing of individual scans as a steady spectrum, the brightening of any selected filter position on the display with a simultaneous readout of its value, or the digital storage of the spectrum of a transient event for subsequent replay. In one instrument, using one-third octave filters, an automatic loudness summation is presented by the method of Zwicker.[3] Time-compression and digital processing may be used to speed up analysis and increase the possible number of channels.

43.9.3 Recording of sound for measurement and analysis

43.9.3.1 Analogue recording

Recording of sound for subsequent analysis may often greatly speed up work on site and also allow short-lived events which cannot be repeated exactly, such as the flyover of an aircraft, to be submitted to frequency analysis by replaying through band-pass filters.

The output of the amplifier stages of a sound level meter are recorded on a magnetic tape recorder, taking the signal from a point after any switched attenuators, filters or weighting networks so that the recording will be able to handle the whole range of levels for which the meter is designed to be used. The range may include, for instance, a calibrating signal of over 124 dB followed by a sound of 30 dB or less.

43.9.3.2 Digital and computer systems for noise measurement

Sound signals, being continuous functions of time, lend themselves readily to digital recording and analysis, either in real time or in digital recordings. The output voltage of a microphone amplifier is converted by means of an analogue-to-digital converter into digital characters which are stored on magnetic tape, with associated information such as time or filter channel. The magnetic tape, usually in the form of a conventional cassette or floppy disk, is subsequently analysed by computer—a process for which the microcomputers now on the market are in widespread use. Other applications of computers and microprocessors to acoustic measurements are dealt with in the section below on general acoustic measurements.

43.9.3.3 Continuous integration of noise power

During the last decade, increasing importance has been attached to the integrated intensity of noise or its average power over a period. The equivalent level L_{eq} of a noise for an averaging time T is defined as

$$L_{eq} = 10\log(T^{-1})\int_0^T (p^2/p_0^2)\,dt \qquad (43.35)$$

and the noise dose is the integral of the energy alone, usually expressed as the percentage of the integral of 90 dB (relative to 10^{-12} W/m^2) for 8 h.

The introduction of microprocessors has enabled sound level meters to produce a direct reading of either noise dose or L_{eq} for such varied purposes as assessing the hearing damage risk in a discotheque or measuring the transmission loss through the facade of a building, using varying traffic noise as the test sound. Reference will be made to other applications below.

43.9.4 Use of equipment for noise measurement

43.9.4.1 Road traffic noise

A-weighted levels are used for all traffic noise measurements whether of the noise emission of industrial vehicles or of the noise climate caused by traffic as a whole, because it is found to correlate more closely with subjective reaction than any other weighting.[6]

The noise emitted by a vehicle is subject to a legal limit, depending on the type of vehicle. British Standard 3425:1967 specifies details of the test site and conditions and of the special sound level meter with a wide scale and no switched attenuators, the settings of which could be questioned in the event of legal proceedings.

Compensation for sound insulation may be claimed by householders living in a climate of excessive traffic noise. The criterion for compensation is that the facade of the house should be subject to a level over 68 dB for more than 10% of the 18 h period from 6 a.m. to 10 p.m. Apparatus for deriving this so-called L_{10} level has been available for many years, the earliest type consisting of a series of mechanical counters triggered at successive 5 dB intervals on the potentiometer of a chart recorder, thus yielding, over a sufficient time, a histogram of time fraction against level, from which various indices can be derived. More recent versions of this equipment give direct readouts of the indices based on digital processing.

43.9.4.2 Machinery noise

The noise output from a machine may be measured in a variety of conditions, e.g. out of doors, reverberant, free-field surveyed *in situ*, or special laboratory. The methods, as specified in British Standards 4196:1981, Parts 1–6, are all basically similar, depending on the measurement of the sound pressure level around the machine and conversion to a total power level attributable to the machine, which may then be used to predict the sound pressure level in any given situation. In every case, the pressure level is sampled in at least three points or by automatic scanning along a line. Corrections are made for the distance of the sampling points from the machine, and the volume and reverberation time of the room.

This procedure is carried out in each octave or one-third of an octave band of frequency within the range of interest. The use of diffusing elements in the room, such as continuously rotating vanes, is recommended to reduce the variation of results.

43.9.4.3 Hearing damage risk

Modern practice in hearing conservation for those exposed to industrial noise at work is summarised in the Department of Employment's *Code of Practice for Reducing the Exposure of Employed Persons to Noise.*[7] It is recommended that exposure to steady sound exceeding 90 dB(*A*) should not be continued for more than 8 h in any day. For varying or intermittent exposures, an integrated level averaged over the whole working period during a day should not exceed 90 dB. Instruments known as dose-meters are available for computing this value of L_{eq} for any working period. Although this code of practice is still in general use, there is a considerable probability that the criterion level of 90 dB will be reduced to 87 or even 85 dB within the foreseeable future. Reductions of this order are being strongly urged in the EEC by many member nations; it is opposed by the UK apparently on economic grounds rather than in the interests of hearing conservation.

43.9.4.4 Industrial noise in residential areas

The application of measurement to the acceptability of noise from industrial premises in a residential area is covered by a complicated set of rules given in BS4142:1967. *A*-weighting is specified, as with road traffic, and the interpretation of level readings must take into account the type of area (e.g. rural, or residential with some industry), the time of day, a notional background level for the area, the actual background level if it can be measured in the absence of the specific interference, the intermittency of the interfering sound and its character.

The above-mentioned British Standard remains in force, but detailed application is much influenced by case law accumulated since the mid-1970s.

43.9.4.5 Aircraft noise

Assessment of community disturbance by aircraft noise requires a knowledge of the number of flyovers as well as the mean level of the peak sound level during individual flyovers. The two factors are combined in an index known as the noise and number index (NNI).

$$\text{NNI} = \text{average level of peak noise} + 15\log N - 80 \qquad (43.36)$$

where N is the number of flyovers per day. The peak level is expressed in *PN* dB, which is a weighted sum of the spectral components having a high correlation with disturbance.[8]

In place of *PN* dB it is now usual to substitute a sound level meter reading with the *D*-weighting (see *Figure 43.60*). Compared with *A*-weighting, this gives prominence to frequencies above 1 kHz.

As with industrial noise in residential areas, the judgments in any situation tend increasingly to take additional factors into consideration, such as the proportion of night flights to the whole and the continental preference for *A* over *D* weighting.

43.10 Acoustic measurements

43.10.1 Definitions

Sound is transmitted from one point in a building to another by various paths through the air or the structure. The point of origin is described as the *source* and the other is usually in a room known as the *receiving room.* Airborne sound transmission is that in which the sound is generated in the air at the first point and received in the air at the second. Sound travelling through the solid material of a building, particularly from a vibrating source in contact with it, is known as *structure-borne sound*. The shortest path of transmission, usually through one or more intervening partitions, is called the *direct path*; sound that travels by indirect paths, often including substantial distances through solid structures, is said to travel by *flanking* transmission.

The sound insulating characteristic of a partition is its *transmission coefficient*. This is the ratio of the sound power radiated into the receiving room to that falling on the source room side. It is usually denoted by τ. The difference in sound power levels is known as the *sound reduction index* (SRI), or *transmission loss* (TL), and is the figure usually quoted. The two quantities are related by the equation

$$\text{SRI} = 10 \log 1/\tau \qquad (43.37)$$

The reduction of sound pressure level between two rooms is known as the *sound level reduction* or *sound level difference.*

43.10.2 Measurement of airborne sound transmission

To measure the sound level difference between two rooms, it is simply necessary to place a sound source in one of them and to measure the difference between the equivalent sound pressure levels in the two rooms. To derive the SRI of the partition, a correction must be applied to the sound level reduction to take into account the absorption of sound in the receiving room. If the sound transmission between the two rooms is entirely by the direct path through the intervening wall, the sound level reduction will depend not only on the SRI of the partition but also on its area and the increase of sound pressure level due to reverberation in the receiving room. The relationship between the SRI and the sound level difference is[9]

$$\text{SRI} = \text{sound level difference} + 10 \log (\tfrac{1}{4} + A/S\alpha) \qquad (43.38)$$

where A is the area of the partition, S the total area of the interior surfaces of the receiving room and α their average absorption coefficient (see *Section 43.10.4*).

43.10.2.1 Standard conditions

Standards for measurement of level reduction and the derivation of SRI in buildings are given in British Standards 2750:1980 Parts 1–8 and 2750:1987 Part 9. In these documents the $\frac{1}{4}$ dB of Equation (43.38) is ignored. This simplification causes little error except when the receiving side is in the open air or a room with very great absorption. A normalising correction is recommended when assessing walls between dwellings or rooms within a dwelling where there may be multiple flanking paths. The correction term becomes $10 \log(10/S\alpha)$ or $10 \log(T/0.5)$ where T is the reverberation time of the room (see *Section 43.10.3*).

To make a full assessment of the SRI of a partition, narrow bands of noise at one-third octave intervals are radiated by one or more loudspeakers into the source room. At each frequency, the sound-pressure level is measured at five microphone positions in each room by one of the methods described in *Section 43.9*, and the mean energy level computed for each room. The microphones should be in positions more than half a wavelength from any wall. The reverberation time also is measured (see *Section 43.10.3*), and the total absorption calculated for substitution in Equation (43.38).

The mean energy level is computed for each room by averaging the power ratios derived from the pressure levels in the five microphone positions and reducing the average to decibels. The measured levels are expressed as decibels relative to any convenient reference, e.g. the multiple of 10 to the next below the lowest, and the power ratios read from a table such as *Table 43.5* (one place of decimals is usually sufficient).

Table 43.5 Power ratios for sound pressure levels

SPL	*Power ratio*	*SPL*	*Power ratio*
0	1.00	7	5.01
1	1.26	8	6.31
2	1.58	9	7.94
3	2.00	10	10.00
4	2.51	11	12.59
5	3.16	12	15.85
6	4.00		etc.

If the spread of the five sound pressure levels (SPL) does not exceed 5 dB, they may be averaged directly, giving results less than 1 dB in error.

Table 43.6 gives the central frequencies recommended for bands of noise used for tests of souna insulation and other acoustic quantities.

The mean SRI of the partition is defined as its average value in one-third octave bands from 100 Hz to 3150 Hz.

Table 43.6 Standard testing frequencies for acoustic measurements

Octave bands (Hz)	*One-third octave bands* (Hz)		
63	50	63	80
125	100	125	160
250	200	250	315
500	400	500	630
1000	800	1000	1250
2000	1600	2000	2500
4000	3150	4000	5000
8000	6300	8000	10000

43.10.2.2 Practical details

Any method may be used for the measurement of the band pressure levels in the two rooms. Readings for all frequency bands and microphone positions may be made in one room before passing to the other. A chart recorder may be used for giving a permanent record of the levels, the source and receiver microphones being switched in alternately. If the dial of the oscillator providing the test signal is driven by gearing from the recorder and the filters are switched by the same mechanism, each filter position gives a pair of levels corresponding to the two rooms.

As five microphone positions are normally necessary in each room to give satisfactory accuracy, one must either go to the expense of a number of matched microphones of high quality or interrupt the measurements to change microphone positions in both rooms. An alternative method is to mount each microphone on a driven rotating arm so that it traces out a circular path along which the pressure level can be sampled.

An important point is that, whichever method is used, checks must be made frequently in measuring the receiving room level to confirm that the wanted signal is sufficiently above the noise level, whether due to electrical or ambient noise, to give reliable measurements. For this purpose the loudspeakers are momentarily switched off and the measured level should fall by at least 6 dB.

For extensive routine measurements at the BBC, digital equipment developed by Moffat[10] has been used, the test signals being played from magnetic tapes which also carry trigger signals for actuating digital processing equipment. The outputs of microphones in the two rooms are then recorded on a second tape which is then processed to yield the sound level reduction.

For laboratory measurement of sound reduction index of partitions and walls the sample must be fixed in an opening in a heavy wall separating two rooms, constituting a so-called transmission suite, in which every precaution has been taken to avoid transmission of sound by any path other than that through the sample partition. Each room must be of adequate size to allow the establishment of a diffuse field and the edge constraints of the sample should be similar to those in its normal situation. It should be rectangular in shape with an area of 10 m^2.

43.10.2.3 Separation of airborne from structure-borne transmission

The measurement of airborne sound transmission by the British Standards method described above gives no information about the path or paths of transmission. These may include structural paths and parallel transmission paths through the partition between the rooms. Part of the energy may be transmitted directly through holes, cracks, thin doors, etc., which will be greatly inferior to the rest of the partition with respect to transmission loss. *Figure 43.61* shows some of the paths by which sound may travel from a source room to an adjacent one. There

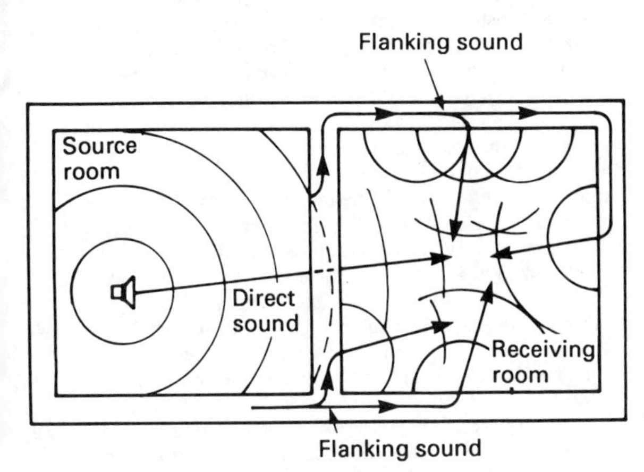

Figure 43.61 Sound transmission between adjacent rooms

are several ways in which the indirect paths may be separated from the principal path for diagnostic purposes or to determine the true sound reduction index of a partition in the presence of other paths even if the energy transmitted through them is the greater.

A non-permeable partition cannot transmit sound except by the bodily movement of its mass and the inherent SRI of a partition can therefore be determined by measuring the velocity amplitude of the surface on the receiving room side with the pressure on the source room surface. The velocity amplitude may be derived from the acceleration amplitude of the surface measured by means of accelerometers attached to it.[11] The velocity amplitude is related to the acceleration amplitude by the equation $v=q/f$, where q is the acceleration and f is the frequency of the sound. Small accelerometers are available in which the sensitive element consists of a steel disc between two discs of piezoelectric material. The acceleration of the whole causes a force difference between the steel and the two piezoelectric discs and the resulting voltage is a measure of the acceleration. The equivalent near-field sound pressure level corresponding to an accelerometer voltage level V relative to 1 V is given by

$$L=V+207-20\log kf \tag{43.39}$$

where f is the frequency and k is the sensitivity in mV/g acceleration.

This equation breaks down below the frequency at which the speed of bending waves in the surface is equal to that of sound in air.

The accelerometer is in other ways also a powerful tool for the tracing of sound and vibration through the structure of a building, and for predicting the near-field sound pressures to be expected in a building subject to vibration.

A second very powerful method is to derive the cross-correlation function between the outputs of microphones in the source and receiver rooms using a comparatively broad band (about an octave) of noise as the signal. A variable time delay is inserted into the source-room microphone circuit and as this is slowly increased the correlation function reaches a maximum value when the delay is equal to the time taken by sound to travel from the source room microphone to the receiving room microphone by any one of the paths. There may be a succession of such maxima, their amplitudes being proportional to the sound pressure levels at the receiving room attributable to the corresponding paths. The transmission losses can therefore be calculated. Details of this method have been given by Goff[12] and Burd.[13]

The radiation of sound in or out of a surface may be reduced by covering the surface with additional flexible layers such as plasterboard supported over a layer of sound absorbing material. BS 2750 points out that this method may be used to reduce the contributions from individual paths of the flanking transmission and so effectively eliminate them from the measurements. This, like the correlation method, is mainly of use in laboratory conditions.

43.10.2.4 Measurement of impact sound transmission

The transmission of sound from impacts within a building cannot be measured in terms of a level reduction. Instead, standard impacts are delivered to a point on the structure and the band pressure level is measured at the standard test frequencies at a number of positions in another room. Most commonly, the blows are delivered to a floor immediately above the receiving room to test the effects of footsteps in producing sound in the room below. The impacts are produced by an impact machine or footsteps machine specified in British Standard 2750:1980 Part 6, consisting of a number of hammer heads each of 500 g weight and a mechanism for allowing the hammers to fall freely in sequence on to the floor at a rate of between five and ten per second. The band pressure levels in the receiving room are plotted against frequency as a spectrum which is compared with a standard form of spectrum. A disadvantage of the tapping machine described above as a representation of footsteps is that the hammers are very much lighter than legs and consequently may give misleading comparisons between structures of widely differing mechanical form, e.g. between a carpet and a floating floor, as means of reducing sound transmission. Cremer and Gilg[14] have found that an electromagnetic shaker gives more reliable results.

43.10.3 Measurement of reverberation time

When a steady sound is radiated into an enclosure and then suddenly cut off, the sound pressure decays according to Franklin's equation[15]

$$P=P_0 e^{-kt} \tag{43.40}$$

where P_0 is the initial pressure amplitude and k is a constant.

The reverberation time of the enclosure is defined as the time required for the sound pressure to fall to 1/1000 of its initial value, i.e. for the level to fall by 60 dB. Thus Equation (43.40) yields

$$L = L_0 - 60t/T \quad \text{(dB)} \tag{43.41}$$

where L, L_0 are the instantaneous and initial levels and T is the reverberation time with a value of $6.9/k$. This is a linear function of time and the term decay curve is usually taken to refer to this form, i.e. the curve of sound pressure level against time.

A straight decay curve will be obtained only in small rooms at the frequency of a strong isolated-room mode; in all other cases it will be modified by beats, rapid fluctuations or changes in general slope as the decay proceeds. Measurement of reverberation time consists of recording or displaying a decay curve and measuring the slope of the best-fit straight line.

43.10.3.1 *Measurement from decay curves*

The block diagram of *Figure 43.62* shows the commonest method of measuring reverberation time by using a chart recorder. A test signal from an oscillator or noise generator is radiated by a loudspeaker into the enclosure under test and received, together with the resulting reverberant sound, by a microphone. The output voltage of the microphone is amplified and fed to a logarithmic chart recorder through band-pass filters which remove noise and harmonics. The chart is started and the sound is cut off so that a decay curve is recorded. This process is repeated at a series of test frequencies, the standard frequencies being those listed in *Table 43.6*. The series is repeated with the microphone in at least four other positions, and in greater numbers in the case of large halls.

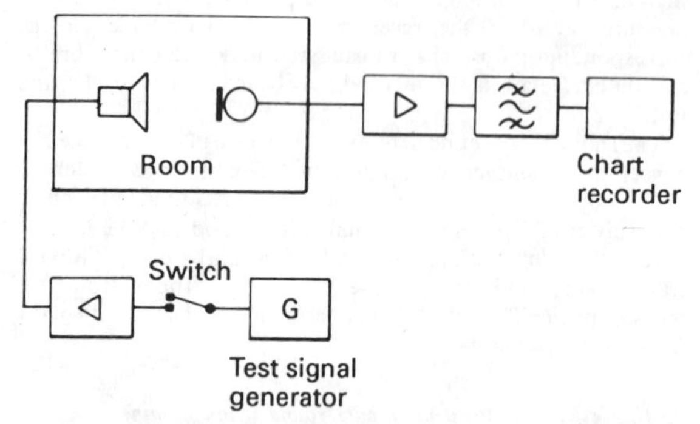

Figure 43.62 Block diagram of equipment for measurement of reverberation time

The slopes of the decay curves are measured or directly converted to reverberation times by means of a calibrated protractor, and averaged for all microphone positions at each frequency.

Figure 43.63 shows the way in which an oscillograph may be used in place of the chart recorder, with advantage for the speed and convenience of measurements.[16]

The on/off switch for the sound source is replaced by an electronic switch which releases short bursts of tone or bands of noise at regular intervals. The end of each burst triggers the time base of the oscilloscope. The microphone signal, after amplification, is converted to its logarithm which is displayed on the oscilloscope as its Y-deflection. The reverberation time is read from the screen with the aid of a rotatable scale. A screen with a phosphor having a persistence of about a second greatly facilitates readings.

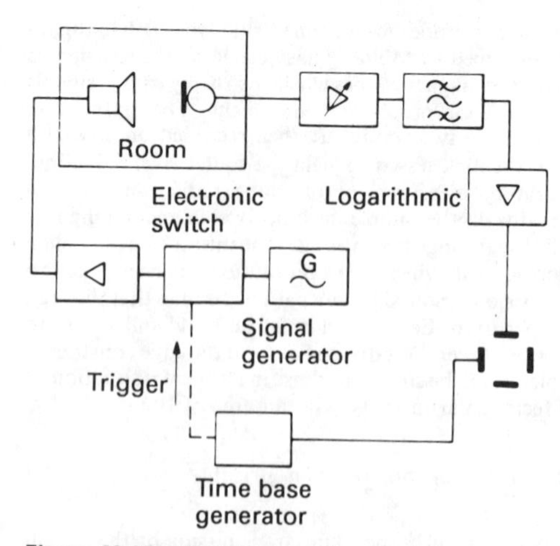

Figure 43.63 Block diagram of reverberation measuring equipment using oscilloscope

43.10.3.2 *Signals for reverberation time measurement*

Owing to the presence of strong standing-wave systems in a room, the shapes and mean slopes of decay curves change rapidly with small variations of frequency. It is therefore essential to use a signal of finite bandwidth so that a number of room modes are simultaneously excited and a decay pattern representative of the frequency region is produced. Random noise may be used, bands of one octave or one-third of an octave being selected in the microphone circuit. A disadvantage is that full use is not made of the power handling capacity of the loudspeaker since only a small fraction of the total radiated power is selected for measurement.

Pistol shots or chords from an orchestra may be used to obtain tape recordings of wide-band decay curves for subsequent analysis through band-pass filters to obtain a graph of reverberation time against frequency. This method is frequently employed to measure the reverberation time of a concert hall in the presence of an audience; pistol shots should not be used if other methods are available as they are found by the author to give unreliable results in comparison with steady test signals.

It has been noted that the decay curves of an enclosure are seldom straight unmodulated lines and therefore some human judgement is required in assessing the best-fit slope. Moreover, background noise and the decreasing gradient, which often occur at the end of a decay, may influence judgement severely. By general agreement, it is now usual to quote the slope between levels 5 and 35 dB below the steady-state level reached before cut-off of the sound.

With either warble tone or bands of noise, successive repetitions of a decay vary noticeably, and a great advantage of the oscilloscope method is that the graticule can be set to agree with the mean slope of a large number of decays which follow one another on to the screen.

43.10.3.3 *Schroeder's method of processing decay curves*

It has been mentioned that a series of decay curves obtained under identical conditions when using finite-band noise such as bands of noise or warble tone will all differ from each other. This is because, although the successive test pulses possess a common spectrum, they represent different functions of time and hence excite the room differently. It is possible to produce identical time functions having a predetermined effective bandwidth by the use of short trains of waves started and stopped at fixed points in their cycles. With such decays, however, it is found that the variability of interpretation is greater than with more usual types of test signals with which a number of successive traces can be superimposed and the average slope estimated.

Schroeder[17] showed that, by using a short tone-burst and integrating the energy in the reverberant signal from present time to infinity, a decay time was derived, representing the ensemble average of all possible decays resulting from a test signal with that particular spectrum. Curves thus obtained are free from the adventitious fluctuations which characterise ordinary decay curves.

In mathematical terms, one plots the function $\int_t^\infty p^2\,dt$ where p is the instantaneous sound pressure and t the time. To obtain this integral, the decay curve can be recorded on magnetic tape which is then reversed and replayed into a squaring and integrating circuit. The output is converted to its logarithm and recorded by a chart recorder. The reverberation time is measured in the usual manner from the slope and divided by two since it represents integrated energy level instead of pressure level.

As an alternative to reversing a tape recording to obtain the integral, the first of two identical decays can be squared and integrated over its whole course and the second continuously integrated and subtracted from the whole integral before recording on the chart.

Thus,

$$\int_t^\infty p^2\,dt = \int_0^\infty p^2\,dt - \int_0^t p^2\,dt \qquad (43.42)$$

Equipment for carrying out this method of measurement has been marketed, but it cannot be regarded as a suitable method for field use.

43.10.3.4 Computer derivation of reverberation time

The intrusion of human judgement into the assignment of decay curve slopes led Moffat and Spring to develop completely objective methods, based on recorded test tapes, for routine use in the BBC studios.

The test tape carried bursts of signals consisting of one-third octave bands of noise and trigger signals. The tape was played into the studio through a loudspeaker and re-recorded, with the added reverberation of the studio. The signals on the tape were subsequently digitised and analysed to extract

(1) The reverberation time of each decay.
(2) The mean reverberation time and variance for each frequency.
(3) The level difference between start of the analysed part of the decay and noise level.

More recently,[18] a portable digital reverberation meter has been developed and put into service by the BBC Research Department, providing direct displays combined with a readout of the corresponding reverberation time. In this equipment the facility of averaging the slopes of successive decays, which was a valuable feature of the oscillograph method described in *Section 43.10.3.1*, is performed automatically when required, the readout settling down to a mean value in the course of a few decays. This instrument is of such versatility and convenience that it can be used equally for the rapid routine assessment of studios and for more involved research measurements. A limited number of them was produced for commercial availability.

43.10.4 Measurements of acoustic absorption coefficients

The absorption coefficient of a material at a stated frequency is the proportion of sound energy incident on it which is absorbed or lost by transmission. This may be quoted for normal or random incidence. The absorption or absorbing cross-section of a finite area of a material is the product of its area and its mean effective absorption coefficient.

43.10.4.1 Measurement by reverberation room

The principle of the reverberation method of measuring absorption coefficient, which yields the random-incidence coefficient, is to measure the reverberation time of a room at a series of frequencies as listed in *Table 43.6*, and then to repeat the measurements after fixing a suitable area of the sample material on to one or more surfaces of the room. The total absorption in the room is calculated from the reverberation time by the formula of Eyring[19] with and without the specimen in place. If $\bar{\alpha}$ is the mean absorption coefficient of all the room surfaces and S their area

$$-\ln(1-\bar{\alpha}) = 0.162V/TS \qquad (43.43)$$

where V is the volume of the room, and absorption $= S\bar{\alpha}$.

According to British Standard 3638:1987, the volume of the room should be between 180 and 250 m^3 and it should be surfaced with hard sound-reflecting surface finishes. The walls and ceiling should be heavy and rigid to avoid absorption at low frequencies by structural vibrations.

The sound should be radiated from a loudspeaker either near one corner of the room or one-third the way along a diagonal of the room, as these positions alone ensure a satisfactorily uniform excitation of all room modes.

The state of diffusion in the sound field should be enhanced by the use of sheets of sound-reflecting material hung from the ceiling. The sizes and orientations of the sheets should be distributed to obtain directional uniformity.

A single sample of 10 m^2 area is recommended for general tests of commercial materials, since this arrangement was found by Kosten[20] to result in the greatest measure of agreement between different testing laboratories. Divided samples, distributed on three or four surfaces of the room, are to be recommended for tests in small areas to promote good diffusion. Subdivision increases the absorption coefficient of most materials especially at frequencies around 500 Hz. It should also be noted that subdivided distributed samples improve the diffusion of a reverberation room to such an extent that hanging sheets are unnecessary.[21] At least five measurements, from different positions of the microphone, should be averaged, and it is an advantage to have two loudspeakers which are used alternately for each microphone position to increase the number of replications at each frequency. Bands of noise or warble tone are suitable as test signals.

It is usual not to make any correction for the loss of absorption of the test room surface when covered by the sample. This effect is kept small by the use of hard materials for the room surfaces, and attempts to make corrections may actually result in larger errors.

43.10.4.2 Measurements by standing-wave tube

Measurements of absorption coefficient by the reverberation method described above requires a large specially treated room, large test samples and facilities for the accurate measurement of reverberation time. A few establishments with a major interest in the development or use of sound absorbers are able to maintain these facilities, but most measurements of coefficients are necessarily carried out by two or three specialist consultant firms. The cost of this service makes it more suitable for checking production prototypes than as an aid for experimental product development. Useful information can, however, be obtained for the latter purpose by measurements of the normal-incidence absorption coefficient in a standing-wave tube.[22]

Figure 43.64(a) shows the principle of this apparatus. The sample, a small disc of 30 to 100 mm diameter, is cut from the material and held in a cap fitting tightly over one end of a tube about 1 m long. The other end of the tube is closed by a loudspeaker diaphragm and the magnet of the loudspeaker is bored centrally to permit the insertion of a probe tube attached to

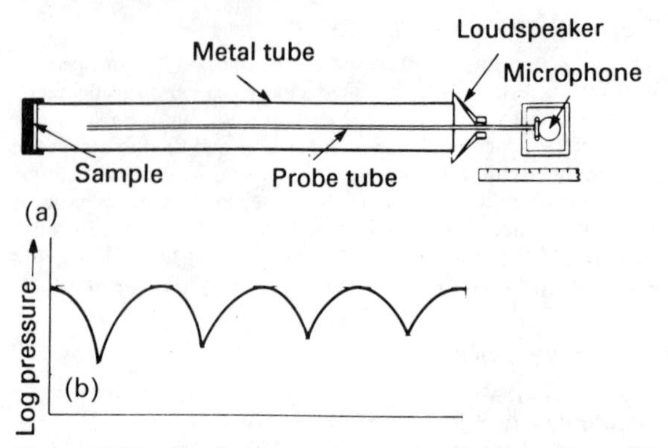

Figure 43.64 Standing-wave tube apparatus for measuring sound absorption coefficient at normal incidence: (a) diagram of equipment, (b) variation of pressure along tube

a microphone. The probe tube is long enough to reach the surface of the sample and the microphone is mounted on a sliding or wheeled carriage so that the sound pressure may be measured at any point along the tube.

With the sample in place and pure tone radiated by the loudspeaker, the sound pressure along the tube shows a series of maxima and minima at intervals of a half wavelength as shown in *Figure 43.64(b)*. The ratio of maximum to minimum pressures diminishes as distance from the sample increases owing to losses in the tube; if its value, extrapolated to the face of the sample, is n, the normal-incidence absorption coefficient is

$$\alpha_N = 4/(n + n^{-1} + 2) \tag{43.44}$$

The theory of this apparatus is given by Beranek.[23]

If the exact distance of the first minimum from the surface of the sample is measured, the real and imaginary parts of the acoustic impedance at the face of the sample can be calculated, and the random-incidence absorption coefficient derived from these parameters for many types of material.[24,25] The calculation depends on the assumption that the impedance at the surface is independent of the angle of incidence. Although this is approximately true for a large proportion of absorbers, there is an element of doubt in the accuracy of the random-incidence coefficients arrived at in this way, and it is essential that final tests on a material to be used in acoustic treatment should be made by the reverberation method.

References

1 KING, L. V., *Proc. R. Soc. A.*, **153**, 17 (1935)
2 COOK, R. K., *J. Res. N.B.S.*, **25**, 489 (1940)
3 ZWICKER, E., *Acustica*, **10**, 304 (1960)
4 STEVENS, S. S., *J. Acoust. Soc. Am.*, **28**, 807 (1956)
5 ROBINSON, D. W. and DADSON, R. S., *Brit. J. Appl. Phys.*, **7**, 166 (1956)
6 MILLS, C. H. G. and ROBINSIN, D. W., *Engineer*, **211**, 1070 (1961)
7 DEPARTMENT OF EMPLOYMENT, *Code of Practice for Reducing the Exposure of Employed Persons to Noise*, HMSO (1972)
8 KRYTER, K. D., *Noise Control*, **6**, 12 (1960)
9 GILFORD, C. L. S., *Acoustics for Radio and Television Studios*, p. 52, Peter Peregrinus, London (1972)
10 MOFFAT, M. E. B., *BBC Research Department Report*, **PH 8** (1967)
11 WARD, F. L., *Proc. 4th Internat. Congress on Acoustics, Copenhagen*, Paper L 11
12 GOFF, K. W., *J. Acoust. Soc. Am.*, **27**, 233 (1955)
13 BURD, A. N., *J. Sound Vib.*, **7**, 13 (1968)
14 CREMER, L. and GILG, J., *Acustica*, **23**, 54 (1970)
15 FRANKLIN, W. S., *Phys. Rev.*, **16**, 372 (1903)
16 SOMERVILLE, T. and GILFORD, C. L. S., *BBC Q.*, **8**, 41 (1952)
17 SCHROEDER, M. R., *J. Acoust. Soc. Am.*, **37**, 409 (1965)
18 WALKER, R., *Proc. Inst. Acoustics Spring Conference*, 1982, Paper C.2.2
19 EYRING, C. F., *J. Acoust. Soc. Am.*, **1**, 217 (1930)
20 KOSTEN, C. W., *Proc. 3rd Internat. Congress on Acoustics, Stuttgart*, **2**, 815 (1959)
21 GILFORD, C. L. S., *Acoustics for Radio and Television Studios*, p. 178, Peter Peregrinus, London (1972)
22 SCOTT, R. A., *Proc. Phys. Soc. London*, **58**, 253 (1946)
23 BERANEK, L. L., *J. Acoust. Soc. Am.*, **12**, 3 (1940)
24 ATAL, B. S., *Acustica*, **9**, 27 (1959)
25 DUBOUT, P. and DAVERN, E., *Acustica*, **19**, 15 (1959)

Further reading

BLASIUS, J. E., 'Selecting digital storage oscilloscopes for test systems', *Test Measur. World*, October (1987)
BURCHAM, T., 'Advances in spectrum analyser intelligence', *Test Measur. World*, September (1987)
JOSELYN, L., 'Standards lab in a box', *Test*, October (1988)
KIBBLE, B. P., 'Redefining the volt and ohm', *IEE Review*, October (1988)
LERMA, J., 'Statistical scheme calibrates counter', *Electron. Design Int.*, December (1988)
MASI, C. G. *et al.*, 'What is the capacitance of this capacitor?', *Test Measur. World*, February (1988)
MASI, C. G., 'Choosing a digital oscilloscope', *Test Measur. World*, October (1988)
MAZDA, F. F., *Electronic Instruments and Measurement Techniques*, Cambridge University Press, Cambridge (1987)
MUSHING, A. *et al.*, 'Performance testing of high speed A/D converters', *Electron. Eng.*, April (1988)
PENSON, G., 'Comparing features on digital oscilloscopes', *Electron. Product Design*, March (1988)
STRASSBERG, D., 'High performance DMMs and calibrators bring standards lab specs to the benchtop', *EDN*, February (1988)
STRASSBERG, D., 'Instruments refine the art of signal generation', *EDN*, November (1988)
STRASSBERG, D., 'High performance DSOs present users with plenty of choices', *EDN*, 22 December (1988)
TAGGART, J., 'The analog view of digital scopes', *Test Measur. World*, June (1988)
TAGGART, J., 'Oscilloscopes and waveform recorders', *Electron. Eng.*, July (1988)
THOMPSON, B. J., 'Voltage, current and resistance calibrators', *Test Measur. World*, March (1988)
WALLING, S., 'Applying arbitrary waveform generators', *Electron. Product Design*, March (1988)
WASHINGDON, D., 'Colour in flat-panel CRTs', *Electron. Comm. Eng. J.*, January/February (1989)

Electron Microscopy

D K Bulgin PhD
Cambridge Instruments Ltd

Contents

44.1 Introduction

The limitations of traditional light microscopy have been well known for many years. The resolution of such microscopes is determined by diffraction effects due to the wavelength of the illumination, and is of the order of 100–200 nm. This corresponds to a useful magnification limit of about ×5000.

In 1924, de Broglie[1] showed wave particle duality and hence that the wavelength of an electron is a function of its energy, $E=h\nu$ where h is Planck's constant. Energy can be imparted to a charged particle by means of an electric accelerating field. Thus, at a sufficiently high voltage, say 50 kV, electrons of extremely short wavelength ($\lambda=0.0055$ nm) can be produced.

Since electrons can also be focused by electrostatic and electromagnetic fields, the potential for use of electrons as a source for microscopy was quickly realised. In addition, electromagnetic lenses are extremely versatile since the focal length of the lens may be changed by altering the current through it.

Two types of electron microscope were originally developed, the transmission electron microscope (TEM) and the scanning electron microscope (SEM) and this led to a combination of these techniques to form the scanning transmission electron microscope (STEM).

The transmission electron microscope is applied to ultra-thin or sectioned material, through which the electron beam is projected to form an image. The whole field of view is illuminated simultaneously and the enlarged image observed on a fluorescent screen incorporated in the microscope.

The scanning electron microscope images solid surfaces. The electron beam is focused to a small spot and is then scanned over the surface. The resultant electron signal is collected and displayed as a brightness modulated image on a cathode ray tube.

44.2 The transmission electron microscope

Transmission electron microscopes were first built in the early 1930s and were soon developed into commercial instruments for research. Modern TEMs have resolutions of the order 0.2 nm, making them powerful instruments in the study of crystalline materials and investigation of the structure of fine particles.

44.2.1 Instrument construction

Figure 44.1 shows a schematic diagram of the main components of a TEM. The electron gun contains a directly heated cathode and a Wehnelt cylinder, acting as a bias shield, mounted on an insulator. The accelerating potential between gun and anode can be varied from about 50 kV to 200 kV depending on the type of instrument, thereby varying the penetrating power of the electron beam. A condenser lens system makes it possible to reduce the cross-section of the beam emitted from the gun and is used to illuminate the area of interest on the sample. This illumination can be varied by adjustment of the condenser lens current for differing working conditions. The current in the objective lens controls the focus of the image on the flourescent viewing screen, and the projector lens is used to vary the magnification, typically ×1000 to ×500 000.

In order to allow the accelerated electrons to reach the fluorescent screen, the electron optics and viewing screen or photographic plate must be maintained in a vacuum. Improved vacuum systems giving working pressures of 10^{-6} to 10^{-7} Torr (mmHg) are obtained using oil diffusion pumps in conjunction with cold traps. This reduces any effect of contamination of the specimen due to residual oil vapours.

The high resolving power of modern TEMs requires that the power supplies for the lenses and high voltage applied to the gun be extremely stable.

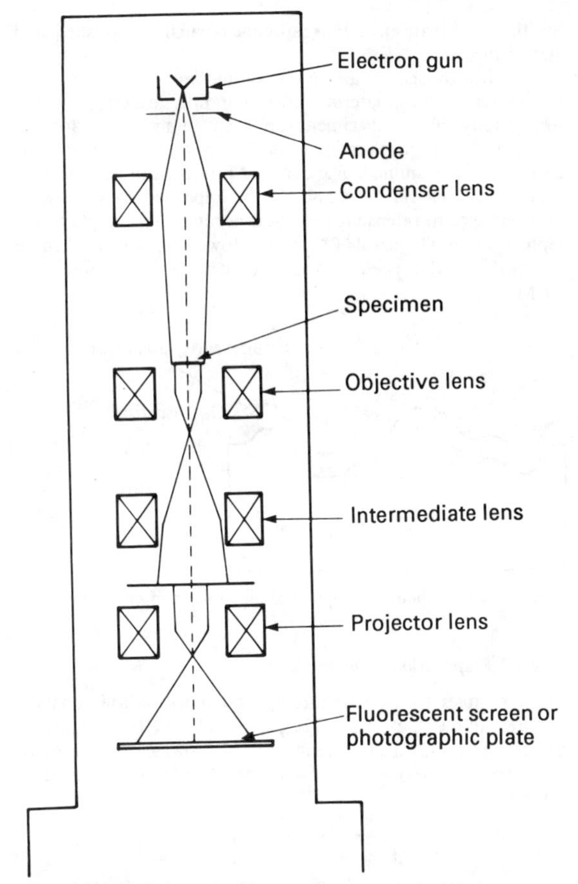

Figure 44.1 Schematic diagram of a transmission electron microscope

Brightness variation in the TEM image is caused by scattering of the electron beam due to different densities within the specimen. The electrons pass through the specimen which must be thin enough to allow the beam to be transmitted, approximately 20–40 nm. When the focus is correct, electrons create a projected image of the sample on the fluorescent screen. Images can be recorded on photographic plate or film by swinging away the fluorescent screen and allowing the image to fall onto the film which is housed within the vacuum environment of the microscope.

44.2.2 Specimen handling in the TEM

Specimen preparation is an important consideration for transmission electron microscopy. Because contrast is formed by electron scattering within the specimen, fine detail will only be observed in very thin specimens. Thick specimens show overlapping of detail from different height levels within the specimen and reduced resolution due to chromatic effects. Scattered radiation is either prevented by the objective aperture from reaching the image plane or merely contributes to the background intensity.

44.2.2.1 Replication

The preparation of replicas of specimens which are opaque to electrons and which cannot be thinned enables their surface structure to be studied in the TEM. Replicas are most widely used for the examination of bulk specimens such as polished and etched metals.

A replica consists of a thin film of material which is electron transparent, the material corresponding exactly to the topography of the specimen surface. One technique for the production of replicas is to vacuum evaporate carbon over the surface to be examined. The contrast produced in the TEM by replicas is often very low and can be improved by shadowing using an electron-dense material evaporated at an angle to the replica surface (*Figure 44.2*). The shadowed replica can then be stripped from the specimen and mounted for examination in the TEM.

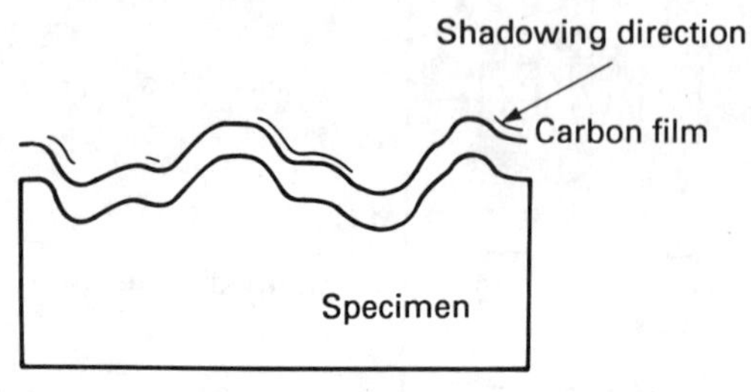

Figure 44.2 Schematic diagram of the replication process

42.2.2.2 Preparation of materials

Thick samples whose structure and composition are constant throughout the thickness can be prepared for TEM by thinning. Controlled removal of material from the surface to a suitable thickness can be obtained by electro-polishing or ion-beam thinning.

44.2.2.3 Embedding and sectioning

This technique is suitable for soft materials, particularly biological tissue. The specimen is embedded in a resin that will remain stable under the electron beam, and thin sections are shaved from the surface using a microtome. Specimens about 40 nm thick can be cut by repeatedly moving the sample past a sharp cutting edge whilst making a small advance towards the knife.

44.2.2.4 Mounting of specimens

Specimens are usually mounted on 3.0 or 2.5 mm diameter metal grids. The grid materials may be of copper or nickel and varying mesh sizes are available. A thin support film may be used to hold the specimen in place on the grid. The grid, complete with specimen, may then be mounted in a special holder and inserted into the vacuum system of the TEM.

44.3 The scanning electron microscope

44.3.1 Construction and performance characteristics

The SEM was postulated in the 1930s by Knoll[2] and Von Ardenne,[3] and serious design study started in 1948 under Oatley in Cambridge, resulting in commercial production in 1965.

A schematic diagram of the SEM is shown in *Figure 44.3*. An electron source is used to provide electrons which are accelerated to a high potential (0.5–50 kV). The beam of electrons is focused by two or three electromagnetic lenses to form a small spot on the sample surface. Double deflection coils housed inside the final lens cause the spot to be scanned over the specimen surface in a raster.

The beam of electrons can interact with the specimen surface in a variety of ways, generating radiations from the sample which

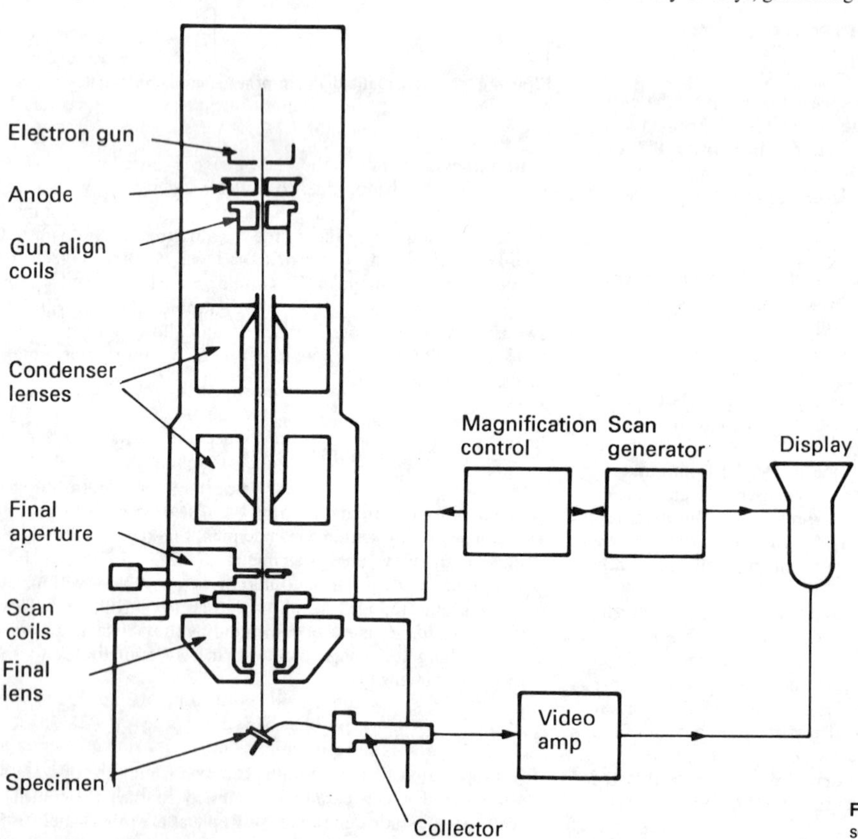

Figure 44.3 Schematic diagram of scanning electron microscope

are characteristic of its composition and topography; a proportion of this radiation is emitted from the surface and can be used to characterise it. Furthermore, as the beam is scanned, the signal level of each characteristic radiation may vary with surface composition and topography. By applying these signals as brightness modulation to a cathode ray tube scanned in synchronisation with the electron beam, images of the surface characteristics can be derived.

The electron source is in general a heated tungsten hairpin, operating over accelerating voltages of 1–50 kV, but alternative sources such as lanthanum hexaboride and field emission sources are also in use, and these give higher brightness. However the alternative sources generally have greater vacuum requirements than that for a tungsten filament, and so ancillary pumping must be provided. The required vacuum level in an SEM is usually provided by an oil diffusion pump backed by a rotary pump, but contamination-free environments are now being required. Liquid nitrogen cooled traps or baffles fitted about the diffusion pump help reduce the contamination. However, oil-free pumps such as turbo-molecular pumps are considered essential to minimise contamination.

The principal imaging mode of the SEM uses secondary electrons. These are generated by electron–atom interactions and have energies sufficient to travel only some 2–20 nm within the substrate material. Thus only those generated close to the surface can be re-emitted and even these are vulnerable to absorption by surface topography.

The spatial resolution attainable in this mode is typically of the order of 3–7 nm. The detector is commonly a positively biased scintillator accelerating electrons into the active area, and thence transmitting a signal via a light guide and photomultiplier to amplifiers and signal processors and finally to a cathode ray tube. Particular properties of the secondary electron image are the large depth of focus available and the flexible magnification range. Magnification derives simply from the ratio of a length scanned on the sample to that scanned on the cathode ray tube, and hence can be varied over a range of typically ×5 to ×300 000 simply by changing the currents through the scanning coils whilst maintaining the scanned area on the cathode ray tube.

Magnification is independent of lens focus and can be zoomed rapidly through its range centring on a fixed point on the specimen surface. Although using a two-dimensional display, the images produced are characterised by their three-dimensional appearance and relative ease of interpretation by non-specialist staff. Images may be viewed at TV rates on the visual display, and interfacing to video recorders is also possible. A wide variety of image processing techniques are available (e.g. differentiation, expanded contrast) which can be used to highlight important features of images to aid interpretation.

Image recording uses a second cathode ray tube and conventional camera system. The record tube is of high resolution (typically 2500 lines) to give optically high-quality pictures. Other imaging signals available in the SEM include the following.

Backscattered electrons Electrons from the primary beam may be scattered through large angles by atomic nuclei and re-emitted with only small losses in energy. Since the backscattering process is dependent on the mean atomic number of the material, the contrast obtained gives compositional detail of the sample rather than topographic detail, provided that the detector is placed symmetrically about the electron beam and the sample surface is normal to the electron beam. If the detector is orientated such that there is a preferred direction for detection, then topographic detail may be enhanced.

Absorbed electrons Under most operating conditions in the SEM, the rate of input of electrons from the beam is greater than the total rate of emission of secondary and backscattered electrons. The resultant charge is usually dissipated by earthing the specimen. However if the sample is connected to a current amplifier rather than earth, it is possible both to measure the absorbed current and to produce a specimen current image. The absorbed current image produced is complementary to the backscattered image. Certain materials exhibit beam induced conductivity effects also, and these are particularly useful in semiconductor research.

X-rays The action of the electron beam to create vacancies in material electronic structure leads to electron transitions accompanied by emission of the appropriate quanta of electromagnetic radiation. In general, this radiation is in the X-ray region of the spectrum, and may be monitored using an appropriate detector. The X-rays so produced are characterised for each element, and this forms the basis of the technique of X-ray microprobe analysis. Detection may be either by the conventional crystal spectrometer (wavelength dispersive systems) or by lithium drifted silicon solid-state devices (energy dispersive system). These give information about the elemental composition of the specimens and can yield qualitative and quantitative analyses.

Cathodoluminescence Some materials exhibit fluorescence in the visible region of the spectrum under electron bombardment as illustrated by the phosphors of cathode ray tubes. The emitted radiation, known as cathodoluminescence, may be used for imaging in the SEM using an appropriate detector. The radiation is in many cases a function of impurity levels within materials and is used both in semiconductor materials research and in many mineralogical investigations.

44.3.2 Specimen handling in the SEM

The necessity of a vacuum environment in the SEM precludes the examination of gases and liquids, and special precautions must be taken with solid materials containing high proportions of liquid or gas. In addition, bombardment by negatively charged electrons ultimately builds up a surface charge and surfaces which are not normally electrically conducting must be rendered so.

Large specimen chambers and specimen stages with wide ranging movements allow samples of up to 175 mm diameter to be examined in a variety of orientations.

Metals may be mounted directly or after cleaning, either chemically or ultrasonically, to remove surface debris. Minerals, ceramics, etc., are coated with a thin (10–20 nm) conductive layer usually of gold, gold–palladium or carbon. This is done by either evaporation or sputtering. Carbon layers are most frequently used where microanalysis is required and no light elements are to be detected. Organic materials are treated similarly, but are generally examined at lower beam voltages to prevent bombardment damage. Biological material contains a large amount of fluid and is generally subjected to some drying procedure such as freeze drying or critical point drying. In some cases quench freezing is used and material is then examined in the frozen state on a cold stage.

Because of the large specimen chamber available, the SEM is particularly well suited to *in situ* sample processing. Behaviour of materials at high and low temperatures, during tensile testing, etc., are studied using videotape to record the dynamic sample changes as they occur.

The SEM is applied in both fundamental and applied research to many problems involving the physical structures of solid surfaces and particularly those requiring greater depth of focus or higher magnification than is obtainable with the light microscope. Other uses include fractography, failure mechanics, the study of powders and compacted materials, tribology and corrosion science. In biology, geology and pathology the SEM is exploited both as a research tool and as a means of presentation of micrographical information to the non-specialist and student. The SEM has found many uses in forensic sciences. It has also gained a wide acceptance throughout the electronics industry, in both research and quality control. Its range of applications in this field is detailed in *Section 44.5*.

44.4 Scanning transmission electron microscopy

STEM has made rapid advances. It is possible to generate variations on STEM by adding scanning attachments to transmission microscopes or transmission detectors to scanning microscopes. However such is the growth of the technique that specially designed STEMs are being built.[4]

Thin sectioned material is used as in the TEM but scanning image collection and processing of the transmitted electrons are retained. Single-atom imaging is obtainable using STEM optimised electron optics. A variety of imaging modes is available for STEM and finds application in materials science particularly in combination with microanalytical techniques.

44.5 Electron microscope applications in electronics

The scanning electron microscope is particularly suited to examination of electronic components and semiconductors. The lack of sample preparation required makes the SEM invaluable in providing rapid examination of failed components and product monitoring which are essential processes in the electronics industry.

The interaction of an SEM beam with a semiconductor sample can generate many phenomena worthy of study, including some that cannot be observed by any other technique. Electron beams are easy to control, to deflect in scan patterns, switch on and off, and to adjust in energy with which they bombard the specimen. Secondly, the alternative operating modes of the SEM quickly answer many of the questions which arise during semiconductor research and production. The use of electron beams in the microelectronics field has grown rapidly in the past ten to fifteen years. Large depth of focus is one of the most important features of the SEM for examination of integrated circuits, allowing detailed study of step coverage at very high tilt angles. As circuits become more complex, with circuit elements reaching the limits of photolithography, the superior resolving power and depth of focus of the SEM has become a necessary part of the evaluation of new circuits, construction techniques and quality control.

A considerable reduction in the size of circuit elements can be made by using an electron beam technique instead of conventional photolithography. The electron beam microfabricator (EBMF) uses an electron beam to write patterns on semiconductor wafers, and the system is computer controlled. The EBMF allows the minimum linear dimension of a circuit element to be reduced from 2 μm (using conventional photolithography) to 100 nm, thereby increasing the available packing density.

44.5.1 Semiconductor investigations in the SEM

Very little sample preparation is needed for examination of semiconductor material in the SEM. Part-processed or completed wafers can be observed in most modern SEMs without breaking and with no preparation. Part-processed wafers can thus be examined at intervals during manufacture to aid quality assessment. Finished devices need to be opened so that the circuit can be seen, and mounted in the SEM with at least one termination grounded.

For the purposes of SEM imaging, a semiconductor device is conductive, even silicon oxide and passivation layers applied to circuits exhibit induced conductivity at the beam energies normally used. Several types of SEM imaging can be used on semiconductor samples to give information which is unobtainable by other techniques.

44.5.1.1 Secondary electron imaging

Imaging using secondary electrons is used to show surface details of topography and also contrast due to differences in atomic number. If a semiconductor wafer is tilted with respect to the incident electron beam, the large depth of focus of the SEM is ideal for monitoring features such as oxide steps formed by multiple diffusions and the integrity and profile of metal tracks (*Figure 44.4*).

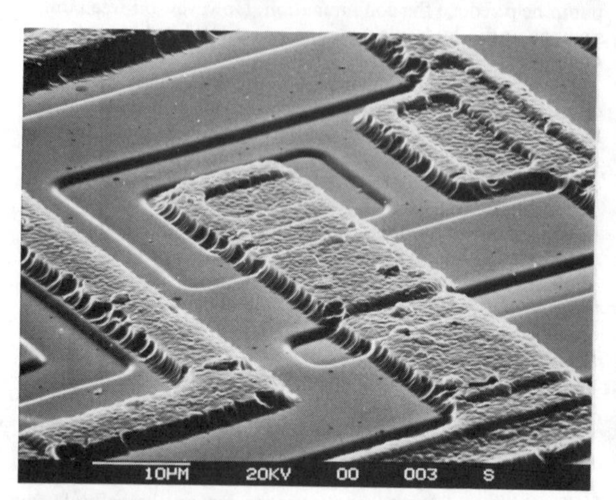

Figure 44.4 SEM micrograph of step coverage on an integrated circuit

The trajectory of secondary electrons leaving a surface will be affected by potential variations in that surface, leading to the phenomenon of voltage contrast. In this mode of operation a semiconductor device can be examined whilst being operated by an external circuit. The potential of any part of the circuit will affect the generation of secondary electrons and therefore the brightness of the resultant image. Thus voltage contrast can be used to observe electrical failure beneath the surface of a device or reverse biased junctions because of discontinuity in contrast. The contrast change is proportional to the value of the applied voltage and therefore it is possible to measure potential at various points in the circuit.

The study of integrated circuits using voltage contrast is often done dynamically, with the circuit under simulated working conditions, e.g. a MOS shift register can be run at high frequency and a beam switching system used to blank the electron beam at some sub-harmonic of the fundamental logic frequency applied to the device. This technique effectively freezes the device in a particular logic state. Dynamic voltage contrast studies can be recorded on videotape, the stroboscopic effect of the beam blanking system effectively slowing down the operating frequency of the device.

Voltage contrast effects are most easily observed if there is no passivation layer over the surface of the device, but it is still possible to observe contrast on passivated devices provided the energy of the primary electron beam is carefully selected with reference to the passivation thickness (*Figure 44.5*(*a*) and (*b*)).

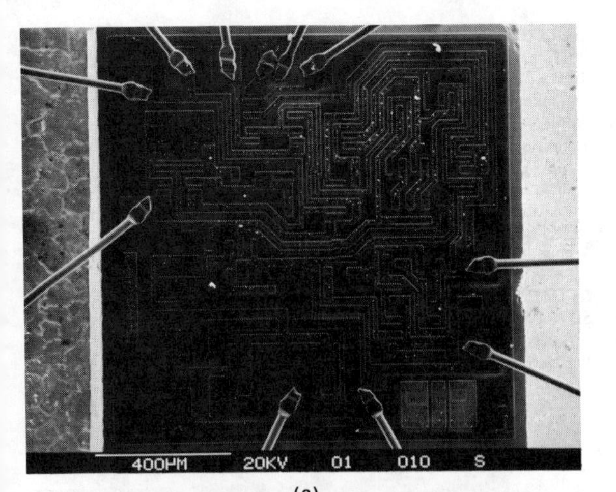

(a)

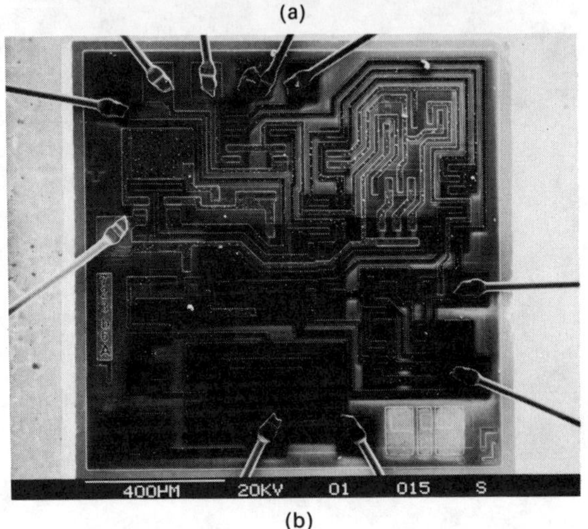

(b)

Figure 44.5 Secondary electron image of standard voltage regulator, + 9 V applied

44.5.1.2 Absorbed electron and conductive imaging

A specimen current amplifier can be used to produce an image by using the absorbed current to modulate the CRT display. The absorbed current image gives contrast which is generally complementary to the backscattered electron image. Whilst the technique can be applied to many types of samples, it is often semiconductor materials which yield most information.

If some pins of a completed semiconductor device are earthed and one or more pins connected to a specimen current amplifier, the specimen current flowing in the device as a result of conductivity induced by the beam can be used to modulate the brightness of the CRT display. The conduction mode depends on the conductivity increase of the target region due to the generation of extra carriers. The primary electrons entering the semiconductor create electron–hole pairs, available for conduction processes until removed by recombination.

The conduction mode (called EBIC, electron beam induced conductivity) can be used to study depletion region boundaries both with and without an externally applied bias (*Figure 44.6*). EBIC imaging is an important technique for monitoring depletion layer spreading, junction breakdowns and oxide pinholes which do not produce compositional or topographic contrast and therefore cannot be observed in any other SEM

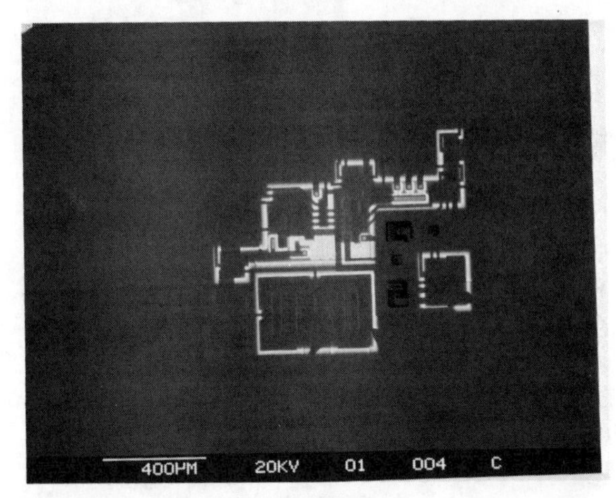

Figure 44.6 EBIC image of voltage regulator, − 0.25 V applied

imaging mode. Since the electron beam will easily penetrate semiconductor materials, careful choice of primary beam voltage allows information in the EBIC mode to be obtained from junctions buried beneath a surface layer. A 10 keV electron beam penetrates about 1 μm of silicon or silicon dioxide, but the penetration depth increases to about 5 μm for a 30 keV electron beam.

The recombination of carriers can also lead to the emission of light photons from the sample. The emission of light under the influence of a primary electron beam is known as cathodoluminescence. Phosphors, light emitting diodes and semiconductor laser materials all exhibit cathodoluminescence which can be detected and used to modulate the CRT display brightness or analysed for wavelength using a spectrometer. Luminescent efficiency of many materials is highly temperature dependent; special-purpose specimen stages available for the SEM allow controlled temperature variation, above and below ambient, during examination.

44.5.1.3 X-ray analysis

Analysis of X-rays produced by interaction of the primary electron beam and the specimen is a powerful technique for characterising the composition of all types of electronic components. Concentrations of elements as low as 0.01% from depths as small as 0.5–1 μm in the specimen surface can be analysed non-destructively on complete components, devices or wafers. X-ray analyses can be displayed in the form of mappings which show the distribution of a chosen element over the surface of a specimen.

The technique is particularly useful in troubleshooting electronic components which have failed due to contamination occurring in production or operating lifetime.

44.5.1.4 Electron channelling

Modification of the operation of the lenses of an SEM allows the beam to be rocked about a point on the surface. The contrast arising from backscattered electrons varies due to the arrangement of atomic planes in the crystal lattice and electron beam channelling patterns are formed. These patterns relate to crystallographic structure and can be used to identify orientation of single-crystal material such as silicon, germanium and gallium arsenide. Selected area channelling patterns can be obtained from grains having a minimum size of about 2 μm.

References

1 DE BROGLIE, L., *Phil. Mag.*, **47**, 466 (1924)
2 KNOLL, M., 'Aufladepotential und Sekundar-emission elektronbestrahlter Oberflachen', *Z. Techn. Phys.*, **2**, 467 (1935)
3 VON ARDENNE, M., 'Das Elektronen raster mikroskop', *Z. Techn. Phys.*, **19**, 407–416 (1938)
4 WIGGINS, J. W., ZUBIN, J. A. and BEER, M., *Res. Sci. Instrum.*, **50**, 403 (1979)

Further reading

GELLER, J. D., 'Semiconductor analysis with Auger electron spectroscopy', *Test Measure. World*, March (1988)
GLAUERT, A. M. (Ed.), *Practical Methods in Electron Microscopy*, North-Holland, Amsterdam (1974 onwards)
GOLDSTEIN, J. I. and YAKOWITZ, H., *Practical Scanning Electron Microscopy*, Plenum Press, New York (1975)
JOHARI, O., *Scanning Electron Microscopy* (Proceedings of Annual Conferences)
JONES, D. R. and WOODWARD, H., 'An overview of IC failure analysis', *New Electron.*, April (1988)
KINAMERI, K. *et al.*, 'A scanning photon microscope for non-destructive observations of crystal defect and interface trap distributions in silicon wafers', *Eng. Optics*, **1**, No. 2, May (1988)
PORTER, M. D., 'IR external reflection spectroscopy: a probe for chemically modified surfaces', *Anal. Chem.*, **60**, No. 20, 15 October (1988)
REIMER, L., *Scanning*, **1**, 3 (1977)
SINGER, P. H., 'The new surface analysis', *Semiconductor Int.*, November (1988)
WELLS, O. C., *Scanning Electron Microscopy*, McGraw-Hill, New York (1974)

45

Digital Design

M D Edwards, BSc, MSc, PhD
Department of Computation,
The University of Manchester
Institute of Science and Technology

Contents

45.1 Number systems

45.1.1 General representation of numbers

In everyday life we normally represent numbers in the decimal radix, or base, for example 1983. Different number systems are employed, however, for other uses; base 12 for counting inches, base 60 for seconds and minutes, base 7 for days of the week and base 3 for feet. In general any number system can represent the integer number N, as

$$N=a_{n-1}r^{n-1}+a_{n-2}r^{n-2}+\ldots+a_1r^1+a_0r^0 \tag{45.1}$$

where

r = radix or base
r^i = digit weighting value
a_i = value of the digit in the ith position, where $0\leqslant a_i\leqslant r-1$. Thus a radix k number system requires k different symbols to represent the digits 0 to $k-1$.
n = number of digits in the representation of the number.

The digits are written in order so that the position of the digit implies the weighting to which it corresponds.

45.1.2 Decimal numbers

We can represent the decimal number, 1983, in the above format as:

$1983=(1\,.\,10^3)+(9\,.\,10^2)+(8\,.\,10^1)+(3\,.\,10^0)$

where $r=10$ and the digit values are either 0, 1, 2, 3, 4, 5, 6, 7, 8 or 9.

45.1.3 Binary numbers

The binary, base 2, number system is used in digital systems and consists of two digits, 0 and 1. Digital systems employ the binary system because it is a straightforward task to decide if an electrical device (or logical element) is either ON (1) or OFF (0). Physical devices exist in one of either two states, e.g. a light bulb is either on or off, a switch is either open or closed. Thus it is logical to use the binary number system in digital systems. Naturally, devices could be constructed to handle numbers with larger bases, but they are ruled out due to their greater complexity.

A number, N, in the binary radix may be represented as:

$$N=b_{n-1}2^{n-1}+b_{n-2}2^{n-2}+\ldots+b_12^1+b_02^0 \tag{45.2}$$

where $r=2$ and the digit values $(b_0\ldots b_{n-1})$ are either 0 or 1 and are known as bits.

Thus the binary representation of the decimal number 1983 is 11110111111 in BINARY NOTATION.

This is equivalent to:

$$\begin{aligned}1983&=(1\cdot2^{10})+(1\cdot2^9)+(1\cdot2^8)+(1\cdot2^7)+(0\cdot2^6)+(1\cdot2^5)\\&\quad+(1\cdot2^4)+(1\cdot2^3)+(1\cdot2^2)+(1\cdot2^1)+(1\cdot2^0)\\&=1024+512+256+128+32+16+8+4+2+1\end{aligned}$$

45.1.4 Conversion from binary to decimal numbers

There are two methods for converting binary numbers to decimal numbers. The first method consists of summing up the powers of 2 corresponding to the 'one' bits in the number. For example:

2^5 2^4 2^3 2^2 2^1 2^0

1 0 1 1 1 0

→ $2=2^1$
→ $4=2^2$
→ $8=2^3$
→ $32=2^5$
—
46

The second method consists of writing the binary number vertically, one bit per line, with the left-most bit on the bottom line. The bottom line is called line 1, the one above it line 2, and so on. The decimal number will be constructed in a column next to the binary number. The procedure to construct the decimal number is to begin by writing a 1 on line 1, the entry on line x consists of two times the entry on line $x-1$ plus the bit on line x (either 0 or 1), the entry on the top line is the answer. This method is known as 'successive doubling' and an example is given below.

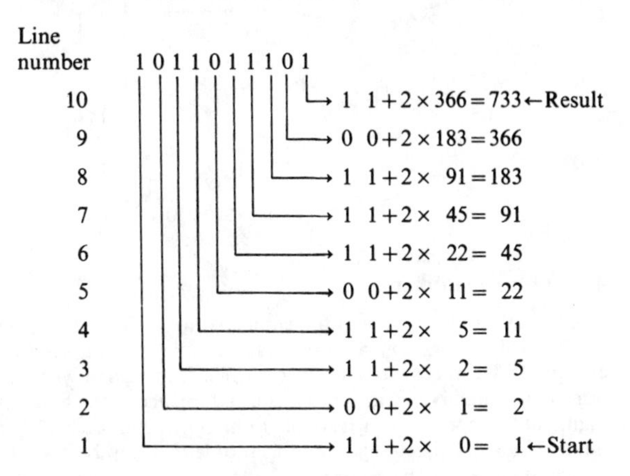

Line number
1 0 1 1 0 1 1 1 0 1
10 → 1 1+2×366=733←Result
9 → 0 0+2×183=366
8 → 1 1+2× 91=183
7 → 1 1+2× 45= 91
6 → 1 1+2× 22= 45
5 → 0 0+2× 11= 22
4 → 1 1+2× 5= 11
3 → 1 1+2× 2= 5
2 → 0 0+2× 1= 2
1 → 1 1+2× 0= 1←Start

45.1.5 Conversion from decimal to binary numbers

There are also two methods for converting decimal numbers to binary numbers. The first method involves subtracting powers of 2 from the decimal number. The largest power of two smaller than the number is subtracted from the number. The process is then repeated on the remainder. When the number has been decomposed into powers of two, the binary number can be assembled with 'ones' in the bit position corresponding to the powers of 2 used and 'zeros' elsewhere. *Table 45.1* gives powers of 2.

Table 45.1 Powers of two

Power of 2	*Decimal*	*Power of 2*	*Decimal*
2^0	1	2^8	256
2^1	2	2^9	512
2^2	4	2^{10}	1024
2^3	8	2^{11}	2048
2^4	16	2^{12}	4096
2^5	32	2^{13}	8 192
2^6	64	2^{14}	16 384
2^7	128	2^{15}	32 768

The second method consists of successively dividing the decimal number by 2. The quotient is written directly underneath the original number, and the remainder, 0 or 1, is written next to the quotient. The quotient is then halved and the process repeated until the number 0 is reached. The binary number can now be obtained directly from the remainder column, starting at the bottom. As an example the decimal number 483 is converted to binary below.

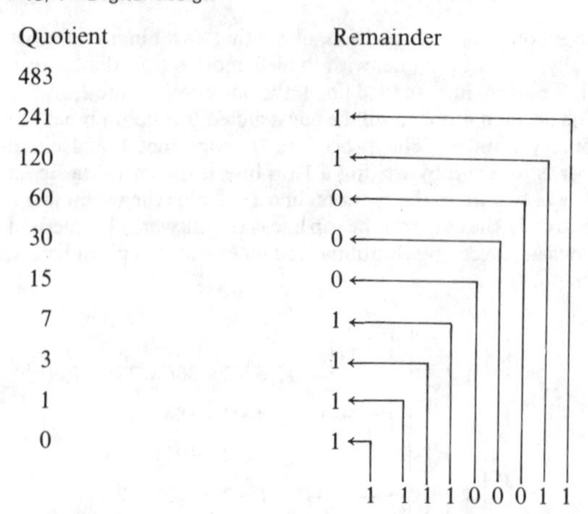

Quotient
Remainder
483
241
120
60
30
15
7
3
1
0
1
1
0
0
0
1
1
1
1
1 1 1 1 0 0 0 1 1

45.1.6 Octal numbers

The binary number system is too cumbersome for human usage especially when large binary numbers containing lots of bits are encountered. It is easy to make errors when reading and writing large binary numbers and hence it is usual to encode the binary digits into a more human readable form. The octal, or base 8, number system helps to alleviate this problem. A number, N, in the octal radix may be represented as:

$$N = l_{n-1}8^{n-1} + l_{n-2}8^{n-2} + \ldots + l_1 8^1 + l_0 8^0 \quad (45.3)$$

where $r = 8$ and the digit values $(l_0 \ldots l_{n-1})$ are either 0, 1, 2, 3, 4, 5, 6 or 7 (see *Table 45.2*).

The octal number system reduces the number of digits when handling binary numbers. To convert binary numbers to octal numbers, the binary digits are divided into 3-bit groups which can be represented by a single octal digit. For example, the binary number

1 1 0 0 1 1 0 1 0 1 0 1 0

is separated into 3-bit groups, starting at the right-hand end of the number and supplying leading zeros if necessary:

0 0 1 1 0 0 1 1 0 1 0 1 0 1 0

The 3-bit groups may be replaced by their octal equivalents:

1 4 6 5 2

and the binary number is converted to its octal equivalent, 14652. Conversely, an octal number can be expanded into a binary number by replacing the octal digits by their 3-bit binary equivalents:

612 = 110 001 010

Octal numbers may be converted to their decimal equivalents, and vice versa, by similar processes used to interchange binary and decimal numbers.

In order to avoid ambiguity between decimal, octal and binary numbers, a subscript of 2, 8 or 10 is used to indicate the radix being used; for example 27(10) = 33(8) = 11011(2).

45.1.7 Hexadecimal numbers

Another convenient number system is the hexadecimal, or base 16, system. The hexadecimal number system, like octal, is used to represent binary numbers in a more readily understandable form. A number $N(16)$, in the hexadecimal radix, may be represented as:

$$N(16) = h_{n-1}16^{n-1} + h_{n-2}16^{n-2} + \ldots + h_1 16^1 + h_0 16^0 \quad (45.4)$$

where $r = 16$ and the digit values $(h_0 \ldots h_{n-1})$ are

0, 1, 2, 3, 4, 5, 6, 7, 8, 9, A, B, C, D, E, F

A table of hexadecimal numbers together with their equivalents is given in *Table 45.2*.

Table 45.2 Decimal, hexadecimal, octal and binary number equivalents

Decimal	*Hex*	*Octal*	*Binary*	*Decimal*	*Hex*	*Octal*	*Binary*
0	0	0	0	8	8	10	1000
1	1	1	1	9	9	11	1001
2	2	2	10	10	A	12	1010
3	3	3	11	11	B	13	1011
4	4	4	100	12	C	14	1100
5	5	5	101	13	D	15	1101
6	6	6	110	14	E	16	1110
7	7	7	111	15	F	17	1111

To convert binary numbers to hexadecimal numbers, the binary digits are divided into 4-bit groups. The 4-bit groups can be represented by a single hexadecimal digit. For example, the binary number,

1 0 1 1 0 1 0 0 1 1 1 0 1

is separated into 4-bit groups, starting at the right-hand end of the number and supplying leading zeros if necessary:

0 0 0 1 0 1 1 0 1 0 0 1 1 1 0 1

The 4-bit groups are replaced by their hexadecimal equivalents:

1 6 9 D (16)

Conversely, a hexadecimal number can be expanded into a binary number by replacing the hexadecimal digits by their binary equivalents:

A 4 F (16) = 1 0 1 0 0 1 0 0 1 1 1 1

Hexadecimal numbers may be converted to decimal numbers, and vice versa, in the usual manner.

45.1.8 Binary addition and subtraction

Binary arithmetic operations, addition and subtraction, are very similar to those in decimal arithmetic. In fact, the rules are much simpler. Consider the addition of two bits, A and B:

A	B	$A+B$
0	0	0
0	1	1
1	0	1
1	1	0 → carry 1

A carry (the next highest power of two) is only propagated if both bits are 'one'.

Binary subtraction follows the same basic rules as binary addition. Consider the subtraction of two bits, A and B:

A	B	$A-B$
0	0	0
0	1	1 → borrow 1
1	0	1
1	1	0

Subtraction is carried out in the normal manner except that a borrow (the next highest power of two) is only propagated if the minuend is 'zero' and the subtrahend is 'one'.

45.1.9 Binary multiplication and division

The remaining two basic arithmetic operations are multiplication and division. There are numerous methods for multiplying and dividing binary numbers, however, only the simplest methods are considered here.

Binary multiplication is similar to decimal multiplication and, again, the rules are simpler. Binary multiplication is performed in the normal manner by multiplying and then shifting one place to the left and finally adding together the partial products. Note that the multiplier digit can only be 0 or 1 and thus the partial product is either zero or equal to the multiplicand. Consider the example below:

```
Multiplicant    1 1 0 1  (13)
Multiplier      0 1 0 1   (5)

      1 1 0 1
      0 1 0 1
      -------
      1 1 0 1
    0 0 0 0
  1 1 0 1
0 0 0 0
-------------
1 0 0 0 0 0 1        (65)
-------------
```

The process of binary division is very similar to that of decimal arithmetic. Binary division is performed by dividing and then shifting the divisor one place to the right and subtracting from the dividend. Note that the division process is simplified to either divide once or not at all. Consider the example below:

```
Dividend:      1 0 1 1 1 1  (47)
Divisor:           1 0 0 0   (8)
                     1 0 1 ←— Quotient (5)
              ____________
     1 0 0 0 ) 1 0 1 1 1 1
               1 0 0 0
               -------
                 1 1 1 1
                 1 0 0 0
                 -------
                   1 1 1 ←— Remainder (7)
                   -----
```

45.1.10 Positive and negative binary numbers

The normal way to represent numbers is to use the 'sign and magnitude' method, for example, +83 and −72. In the binary number system, the most significant bit is used to represent the sign and the remaining bits are used to denote the magnitude. If the most significant bit is '0', the number is considered to be positive and if the bit is '1' then the number is negative.

We will assume that all numbers are 4-bits long for simplicity. Having one bit for the sign reduces the magnitude of the largest numbers that can be represented. If only positive numbers were used then we could represent numbers in the range 0 to 15 ($2^4=16$), but we now have to have approximately half for negative numbers, so the range of numbers is reduced to roughly −8 to +8.

There are three common methods used to represent positive and negative binary numbers; sign and magnitude, one's complement and two's complement.

45.1.10.1 Sign and magnitude

In the sign and magnitude method, the negative number is simply the positive number with the sign reversed. In general, the range of numbers that may be represented in n bits is

$$-(2^{n-1}-1)\leqslant N\leqslant(2^{n-1}-1)$$

Therefore the negative equivalent of an n-bit positive sign and magnitude number, a, is $2^{n-1}+a$.

The sign and magnitude method makes arithmetic complex because there are two values for zero and the binary sum of a positive number and its sign and magnitude negative is non-zero (see *Table 45.3*). For example:

```
6−6≡6+(−6)

 0 1 1 0
+1 1 1 0
---------
1 0 1 0 0=−4 (or +4 if the carry is ignored)
```

The sign and magnitude method is used, for example, in analogue to digital and digital to analogue conversion, but the numbers have to be converted to another form before any arithmetic operations can be performed.

45.1.10.2 One's complement

In the one's complement representation the negative equivalent of a number may be given by:

$$-N=(2^n-1)-N \tag{45.5}$$

where

$$N=(b_{n-1}2^{n-1}+b_{n-2}2^{n-2}+\ldots+b_1 2^1+b_0 2^0)$$

Thus, the negative equivalent of +4 in one's complement may be found by subtracting 4 from 15 where ($15=2^n-1$, $n=4$ bits):

```
  1 1 1 1   (15)
 −0 1 0 0   (4)
 --------
  1 0 1 1   (−4)
 --------
```

The negative representation of a number is the logical complement of the number; that is, replacement of all the zeros in the binary representation by ones and all the ones by zeros. In general, the range of one's complement numbers that may be represented in n bits is

$$-(2^{n-1}-1)\leqslant N\leqslant(2^{n-1}-1)$$

It is possible to perform meaningful arithmetical operations using one's complement numbers. For example, consider the following addition and subtraction sums:

```
(a)    3+2             0 0 1 1
                     + 0 0 1 0
                     ---------
       =5              0 1 0 1
                     ---------

(b)    7−5≡7+(−5)      0 1 1 1
                     + 1 0 1 0
                     ---------
                   (1) 0 0 0 1
          carry     └──────→1
                     ---------
       =2              0 0 1 0
                     ---------
```

In example (b), a carry is generated and in order to produce the correct answer the carry must be added back into the sum. Care must be taken when performing arithmetic operations on one's complement numbers to ensure that numerical overflow does not occur. Overflow is the condition which occurs when the result of an arithmetic operation cannot be correctly expressed. For example, when adding or subtracting 4-bit numbers the result must lie within the range −7 to +7 otherwise the result will be interpreted wrongly. For example:

```
7+6        0 1 1 1
           0 1 1 0
           -------
 =-2       1 1 0 1     (incorrect)
```

The fact that there are two representations of 'zero' in one's complement arithmetic, and addition and subtraction are relatively complex, has made the one's complement representation of negative numbers almost obsolete.

45.1.10.3 Two's complement

In the two's complement representation, the negative equivalent of a number may be given by:

$$-N=2^n-N \quad (45.6)$$

where

$$N=(b_{n-1}2^{n-1}+b_{n-2}2^{n-2}+\ldots+b_1 2^1+b_0 2^0)$$

Thus, the negative equivalent of +5 in two's complement may be found by subtracting 5 from 16; where ($16=2^n$, $n=4$ bits):

```
            1 0 0 0 0    (16)
              0 1 0 1    (5)
ignore →   (0) 1 0 1 1   (−5)
```

The negative representation of a number is the logical complement of the number plus 1. For example:

```
3=0 0 1 1
1's complement  =  1 1 0 0
          +1    +        1
                   -------
                   1 1 0 1  =  −3
```

In general, the range of two's complement numbers that may be represented in n bits is

$$-(2^{n-1})\leqslant N\leqslant(2^{n-1}-1)$$

It is possible to perform meaningful arithmetical operations using two's complement numbers. For example:

```
(a)   4+3                 0 1 0 0
                        + 0 0 1 1
                          -------
       =7                 0 1 1 1

(b)   5−2≡5+(−2)          0 1 0 1
                        + 1 1 1 0
                       (1)0 0 1 1  =  3
                        └ignore the carry

(c)   −3−4≡−3+(−4)        1 1 0 1
                        + 1 1 0 0
                       (1)1 0 0 1  =  −7
                        ignore the carry
```

As with one's complement arithmetic care must be taken to avoid overflow conditions occurring.

The use of two's complement arithmetic is widespread due to the simplicity of the basic arithmetic operations and the fact that there is only one representation of 'zero'.

The three methods for representing positive and negative binary integers are summarised in *Table 45.3*.

45.1.11 Shifting and binary fractions

In addition to the binary arithmetic operations already mentioned there is an additional operation known as the 'shift' operation. The first type of shift operation is known as a logical shift and may be used to multiply or divide a positive number by two. The logical shift left operation, shown below, is equivalent to multiplying by two:

Logical shift left	before	S_3	S_2	S_1	S_0
	after	S_2	S_1	S_0	0
Example:	before	0101	(5)		
	after	1010	(10)		

The logical shift right operation, shown below, is equivalent to dividing by two:

Logical shift right	before	S_3	S_2	S_1	S_0
	after	0	S_3	S_2	S^1
Example:	before	0100	(4)		
	after	0010	(2)		

Table 45.3 Representation of positive and negative binary integers

Representation method	*Representation of +5 and −5*	$-a$	*Zero*	*Max positive*	*Max negative*
Sign and magnitude	+5 0:101 −5 1:101	$2^{n-1}+a$	0:000 1:000	$2^{n-1}-1$	$2^{n-1}-1$
One's complement	+5 0:101 −5 1:010	$2^{n-1}-a$	0:000 1:111	$2^{n-1}-1$	$2^{n-1}-1$
Two's complement	+5 0:101 −5 1:011	2^n-a	0:000	$2^{n-1}-1$	2^{n-1}

The above shift operations will not produce the correct result if two's complement numbers are shifted left or right. Thus, an extra type of shift operation, known as the arithmetic shift, is employed. The arithmetic shift left and right operations also multiply and divide by two respectively, and they maintain the sign of the number. Both arithmetic shifts are illustrated below:

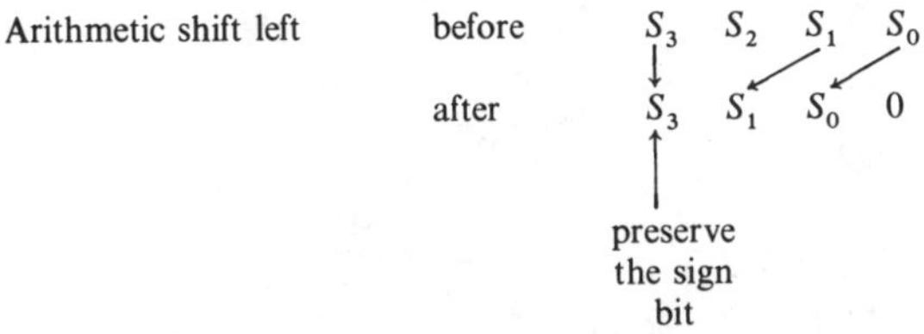

Arithmetic shift left

	before	S_3	S_2	S_1	S_0
	after	S_3	S_1	S_0	0

(preserve the sign bit: S_3 kept in place; $S_1 \to$, $S_0 \to$ moved one place left)

Examples:			
(a)	before	0010	(2)
	after	0100	(4)
(b)	before	1101	(−3)
	after	1010	(−6)

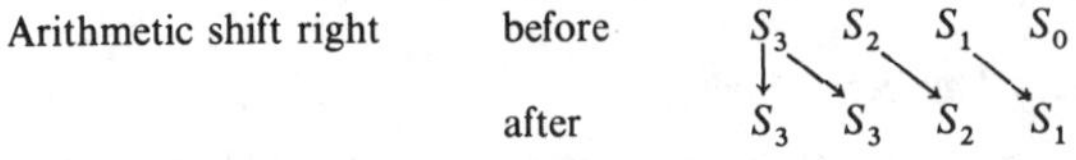

Arithmetic shift right

	before	S_3	S_2	S_1	S_0
	after	S_3	S_3	S_2	S_1

Examples:			
(a)	before	0100	(4)
	after	0010	(2)
(b)	before	1000	(−8)
	after	1100	(−4)

45.1.12 Binary fractions

So far we have only considered integer binary numbers, however, it is also possible to represent binary fractions. A binary number, integer plus fraction may be represented by:

$$N=(b_{n-1}2^{n-1}+\ldots+b_1 2^1+b_0 2^0)$$
$$+(b_{-1}2^{-1}+b_{-2}2^{-2}+\ldots+b_{-m}2^{-m})$$

where the term in the first bracket is the integer part, the + sign is the binary point and the term in the second bracket is the fraction part.

In the decimal number system the fractional part of the number is represented by tenths, hundredths, thousandths, etc., whereas in the binary number system the fractional part is represented by halves, quarters, eighths, etc. For example, the following are valid binary fractions represented in two's complement form:

1.0	=0001.0000	−1.0	=1111.0000
1.5	=0001.1000	−1.5	=1110.1000
3.75	=0011.1100	−3.75	=1100.0100
6.4375	=0110.0111	−6.4375	=1001.1001

45.1.13 Floating point numbers

In many calculations a large dynamic number range is required. It is possible to use fixed point arithmetic to represent large numbers but this incurs the penalty of needing a large number of bits; for example, 20 bits are required to express the decimal number 1 million. Therefore, a system of representing numbers is required where the range of possible numbers is independent of the number of significant digits. The scientific, or floating point, notation may be used, where a floating point number may be expressed in the form:

$$N=f\times b^e$$

where b is the base (or radix), f is the fraction (or mantissa) and e is the exponent.

For example, the decimal number 5280 may be represented in floating point format as 0.528×10^4. The range of the number is determined by the number of digits in the exponent and the precision by the number of digits in the fraction. In floating point format numbers will nearly always have to be rounded to the nearest expressible number. However, the loss of precision is compensated for by a greater dynamic range.

For binary floating point numbers, the number of available bits is divided into a fraction part and an exponent part. The exponent is expressed as a fixed point binary integer and the fractional part as a normal binary fraction. Consider the value 3.125 held in floating point format, base 2, in 16 bits; with 10 bits for the fraction and 6 bits for the exponent:

3.125=011.001 (integer+fraction)

Equivalent to:

	Exponent	*Fraction*
$0.390\,625\times 2^3$	000011	0110010000
$0.781\,25\times 2^2$	000010	1100100000
$0.195\,312\,5\times 2^4$	000100	0011001000

Note that there are several possible ways to represent the number. In order to maximise the number of significant digits in the fraction it is usual to represent all non-zero floating point numbers in *normalised* form. This is achieved by adjustment of the fraction and exponent until the fraction has a non-zero most significant bit; that is, the fraction lies in the range $0.5\leqslant f<1$.

It is usual practice to be able to represent both positive and negative values. However, it is necessary to have both positive and negative exponents, so that negative and normalised numbers of less than 0.5 may be represented. Thus, a 16-bit floating point number may be postulated as:

m.s.b.		l.s.b.
X	X X X X X	X X X X X X X X X X
sign bit 0=+ve 1=−ve	exponent $-16\leqslant e\leqslant 15$	fraction $0.5\leqslant f<1$

Thus, the range of numbers which may be represented is:

$$\pm(0.5\times 2^{-16}\text{ to }0.999\times 2^{15})$$
$$=\pm(7.6\times 10^{-6}\text{ to }3.27\times 10^4)$$

Example:

	Sign	*Exponent*	*Fraction*
$5280=0.645\times 2^{13}$	0	01101	1010010100
0	0	00000	0000000000
$-1.5\times 10^{-5}=0.5\times 2^{-15}$	1	10001	1000000000
$47.0=0.734\,375\times 2^6$	0	00110	1011110000

Any floating point number with a zero fraction represents zero (see above). However, the standard representation of zero (true zero) has a zero fraction and the smallest possible exponent; which means that the exponent should be 10000.

45.2 Basic mathematics of digital systems

In order to design digital systems it is necessary to understand the basic mathematics of the subject, known as Boolean algebra. The relevant mathematical theory is presented so that the engineer may readily appreciate the fundamentals of Boolean algebra and apply it to the design of digital systems.

45.2.1 Boolean algebra

Logic is a tool used in the study of deductive reasoning, also known as propositional logic. In 1854, the mathematician George Boole made a significant advance in this field when he postulated that symbols may be used to represent the structure of logical thought. The work of Boole may be readily applied to the design of digital systems, where a digital circuit may be represented by a set of symbols and the circuit function expressed as a set of relationships between the symbols. The basic symbols used in digital systems are '1' for *true* and '0' for *false*. There are additional symbols used to represent the relationship between the basic symbols. The logical function is the relationship between a set of assertions, which are represented by symbols or letter groups.

Boolean algebra is based on five postulates which may be applied to switching circuits. The five postulates are given below:

Let a symbol X be associated with the state of an element (or circuit), then

(1) $X=0$ if $X \neq 1$ and
(1a) $X=1$ if $X \neq 0$
(This is the definition of a binary variable)

(2) 0 AND 0=0
(2a) 1 OR 1=1

(3) 1 AND 1=1
(3a) 0 OR 0=0

(4) 1 AND 0=0 AND 1=0
(4a) 0 OR 1=1 OR 0=1

(5) NOT 0=1
(5a) NOT 1=0

The basic postulates use three logical relationships which are unique to Boolean algebra, AND, OR, NOT. In addition, a number of theorems may be derived, which are applicable to the algebra of switching circuits. The basic theorems are given below, where the switching variables are represented by the symbols X, Y, Z and may take the value 0 or 1. The AND operator is represented by the symbol '.', the OR operator by '+' and the NOT operator by $\bar{A}$, (*not A*):

(6) $X+0=X$
(6a) $X.1=X$

(7) $1+X=1$
(7a) $0.X=0$

(8) $X+X=X$

(8a) $X.X=X$

(9) $\overline{(X)}=\bar{X}$
(9a) $\overline{(\bar{X})}=X$

(10) $X+\bar{X}=1$

(10a) $X.\bar{X}=0$

(11) $X+Y=Y+X$
(11a) $X.Y=Y.X$

(12) $X+X.Y=X$
(12a) $X.(X+Y)=X$

(13) $Y.(X+\bar{Y})=X.Y$
(13a) $X.\bar{Y}+Y=X+Y$

(14) $X+Y+Z=(X+Y)+Z$
$=X+(Y+Z)$
(14a) $XYZ=(XY).Z=X.(YZ)$

(15) $XY+XZ=X.(Y+Z)$
(15a) $(X+Y).(X+Z)=X+YZ$

(16) $(X+Y).(Y+Z).(Z+\bar{X})$
$=(X+Y).(Z+\bar{X})$
(16a) $XY+YZ+Z\bar{X}=XY+Z\bar{X}$

(17) $(X+Y).(\bar{X}+Z)=XZ+\bar{X}Y$

(18) $\overline{X+Y}=\bar{X}.\bar{Y}$
(18a) $\overline{X.Y}=\bar{X}+\bar{Y}$

A principle of duality exists between the basic theorems; if the values 0, 1 and the operations (.), (+) are interchanged then the alternative representation is obtained (cf 6, 6a; 7, 7a, etc.). The general form of the duality principle is given in Theorems 18 and 18a, and is known as De Morgan's theorem.

The above theorems may be represented in terms of simple electrical switching circuits and the physical representation of some of the theorems is given in *Figure 45.1*.

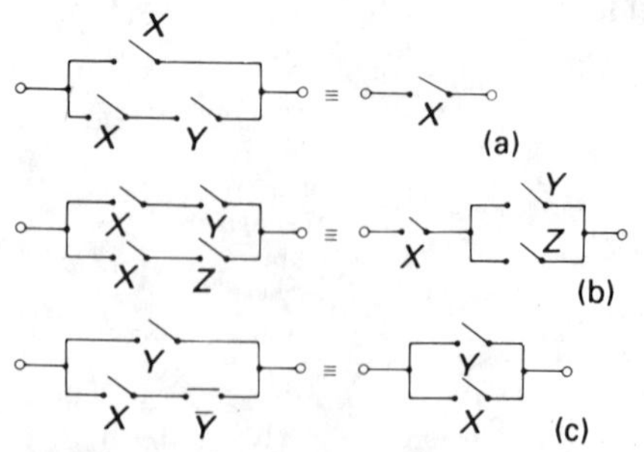

Figure 45.1 Electrical representation of logic theorems (a) $X+X.Y=X$, (b) $X.Y+X.Z=X.(Y+Z)$, (c) $X\bar{Y}+Y=X+Y$

45.2.2 Theorem proving

The theorems of switching algebra may be proved by the use of two techniques; algebraic manipulation and perfect induction.

To prove a theorem (or the equivalence of two logical functions) it is necessary to invoke a subset of the other theorems. Consider the proof of the following relationship: Theorem 15a:

$(X+Y).(X+Z)=X+YZ$

Consider the left-hand side of the equation:

$=(X+Y).(X+Z)$
$=X.X+XZ+YX+YZ$ (by Theorem 15)
$=X+XZ+YX+YZ$ (by Theorem 8a)
$=(X+XZ)+YX+YZ$
$=X+YX+YZ$ (by Theorem 12)
$=(X+YX)+YZ$
$=X+YZ$ (by Theorem 12)

To prove a theorem (or the equivalence of two logical functions), by the method of perfect induction, the following steps must be performed:

(1) Write out all the combinations of the variables in tabular form.
(2) Deduce the result of each expression for all combinations of the variables.
(3) If the result of each expression for all combinations of the variables is the same, then the two expressions are equivalent.

The table of all combinations of the variables and the result of each expression is also known as a *truth table*.

The truth table for Theorem 15a is given in *Table 45.4*.

Table 45.4 Truth table for Theorem 15a

X	Y	Z	$(X+Y).(X+Z)$	$X+YZ$
0	0	0	0	0
0	0	1	0	0
0	1	0	0	0
0	1	1	1	1
1	0	0	1	1
1	0	1	1	1
1	1	0	1	1
1	1	1	1	1

45.2.3 Functions of binary variables

A number of functions may be derived from n binary variables. For n binary variables there are 2^n possible combinations of the variables and 2^{2^n} different functions. Thus, for two binary variables (X, Y) there are four possible combinations and 16 unique functions, as shown in *Table 45.5*.

Table 45.5 Combinations of two binary variables

X	Y	f_0	f_1	f_2	f_3	f_4	f_5	f_6	f_7	f_8	f_9	f_{10}	f_{11}	f_{12}	f_{13}	f_{14}	f_{15}
0	0	0	0	0	0	0	0	0	0	1	1	1	1	1	1	1	1
0	1	0	0	0	0	1	1	1	1	0	0	0	0	1	1	1	1
1	0	0	0	1	1	0	0	1	1	0	0	1	1	0	0	1	1
1	1	0	1	0	1	0	1	0	1	0	1	0	1	0	1	0	1

There are four important functions: f_1 (AND), f_7 (OR), f_8 (NOR), f_{14} (NAND), together with the complement functions f_{12}, f_{10} for X and Y respectively. In general a function may be expressed in terms of the AND/OR operations or its dual in terms of the NAND/NOR/OR operations by application of De Morgan's theorem.

$F = A\,.\,B + C$ (AND/OR)

The following are equivalent to F

$F = \overline{\overline{A\,.\,B}\,.\,\bar{C}}$ (NAND)

$= \overline{(\bar{A} + \bar{B})} + C$ (NOR/OR)

$= \overline{(\bar{A} + \bar{B})\,.\,\bar{C}}$ (OR/NAND)

Thus, any switching system may be expressed in terms of a combination of the AND/OR/NAND/NOR functions (see also *Chapter 27*).

45.2.4 Minterms and maxterms

Logic design problems are normally presented in truth table form. The mathematical description of the function may be extracted from the truth table and then manipulated into a form suitable for implementation by an interconnected set of logic gates. Consider *Table 45.6*.

Table 45.6 Example of a truth table

X	Y	Z	F
0	0	0	1
0	0	1	0
0	1	0	0
0	1	1	1
1	0	0	0
1	0	1	1
1	1	0	1
1	1	1	0

The function F is true (equal to 1) for certain combinations of the variables X, Y, Z. The value of F is false (equal to 0) for the remaining combinations. The function may be described in mathematical terms by taking the logical OR of all the combinations of the variables which produce a 'true' result, thus:

$$F = \bar{X}\bar{Y}\bar{Z} + \bar{X}YZ + X\bar{Y}Z + XY\bar{Z} \quad (45.7)$$

This form of the equation is known as the *sum of products*. The dual of this equation may be obtained by taking the logical OR of all combinations of the variables which produce a 'false' result thus:

$$\bar{F} = \bar{X}\bar{Y}Z + \bar{X}Y\bar{Z} + X\bar{Y}\bar{Z} + XYZ \quad (45.8)$$

By applying De Morgan's theorem to Equation (45.8), we have:

$$\begin{aligned} F &= \overline{(\bar{X}\bar{Y}Z + \bar{X}Y\bar{Z} + X\bar{Y}\bar{Z} + XYZ)} \\ &= (X + Y + \bar{Z})\,.\,(X + \bar{Y} + Z)\,.\,(\bar{X} + Y + Z)\,.\,(\bar{X} + \bar{Y} + \bar{Z}) \quad (45.9) \end{aligned}$$

Thus, Equations (45.7) and (45.9) are identical and the form of Equation (45.9) is known as the *product of sums*.

The sum of products is also known as the *minterm* form of an equation and the product of sums the *maxterm* form. The possible minterms and maxterms for three binary variables, X, Y, Z are given in *Table 45.7*.

Table 45.7 Minterms and maxterms of three variables

X	Y	Z	*Minterms*	*Maxterms*
0	0	0	$m_0 = \bar{X}\bar{Y}\bar{Z}$	$M_0 = \bar{X} + \bar{Y} + \bar{Z}$
0	0	1	$m_1 = \bar{X}\bar{Y}Z$	$M_1 = \bar{X} + \bar{Y} + Z$
0	1	0	$m_2 = \bar{X}Y\bar{Z}$	$M_2 = \bar{X} + Y + \bar{Z}$
0	1	1	$m_3 = \bar{X}YZ$	$M_3 = \bar{X} + Y + Z$
1	0	0	$m_4 = X\bar{Y}\bar{Z}$	$M_4 = X + \bar{Y} + \bar{Z}$
1	0	1	$m_5 = X\bar{Y}Z$	$M_5 = X + \bar{Y} + Z$
1	1	0	$m_6 = XY\bar{Z}$	$M_6 = X + Y + \bar{Z}$
1	1	1	$m_7 = XYZ$	$M_7 = X + Y + Z$

Thus Equations (45.7) and (45.9) may be expressed in their minterm and maxterm forms as:

$$F = (m_0, m_3, m_5, m_6) \quad \text{(minterms)} \quad (45.10)$$

$$F = (M_0, M_3, M_5, M_6) \quad \text{(maxterms)} \quad (45.11)$$

The following relationships exist between minterms and maxterms:

(1) For n binary variables there are 2^n maxterms and 2^n minterms.
(2) The logical OR of all minterms equals 1.
(3) The logical AND of all maxterms equals 0.
(4) The complement of any minterm is a maxterm and vice versa.

45.2.5 Canonical form

If all the binary variables, or their complements, appear once and only once in each term of a logical function then the function is said to be in its *canonical* form. The canonical form of a function is useful in the comparison and simplification of Boolean functions.

To express a function in its canonical form logically multiply (AND) each term by the absent variables in the term, expressed in the form $(A + \bar{A})$.

For example:

$$\begin{aligned} F_1 &= XY + YZ + \bar{X}Z \\ &= XY(Z + \bar{Z}) + YZ(X + \bar{X}) + \bar{X}Z(Y + \bar{Y}) \\ &= XYZ + XY\bar{Z} + XYZ + \bar{X}YZ + \bar{X}YZ + \bar{X}\bar{Y}Z \\ &= XYZ + XY\bar{Z} + \bar{X}YZ + \bar{X}\bar{Y}Z \end{aligned}$$

$$\begin{aligned} F_2 &= XY + \bar{X}Z \\ &= XY(Z + \bar{Z}) + \bar{X}Z(Y + \bar{Y}) \\ &= XYZ + XY\bar{Z} + \bar{X}YZ + \bar{X}\bar{Y}Z \end{aligned}$$

Thus, when expanded into their canonical forms, F_1 and F_2 are equivalent.

45.3 Minimisation of logic functions

A Boolean function may be represented in either a sum of products or a product of sums form. In the design of digital systems the function must be implemented using the least number of logical units (AND, NAND, OR, NOR gates).

It is possible to simplify most Boolean functions by algebraic manipulation. However, for complex functions which involve many variables (greater than five), the simplification process becomes too difficult and tedious. Techniques exist for overcoming this deficiency, which are based on graphical and algorithmic (or tabular) minimisation methods. This section describes both graphical and algorithmic methods for the minimisation of Boolean functions.

45.3.1 Graphical minimisation of Boolean functions

There are numberous ways which may be used to represent a Boolean function graphically. One of the most commonly used approaches is to use a *Venn diagram* which represents variables as areas inside a rectangle, as in *Figure 45.2*.

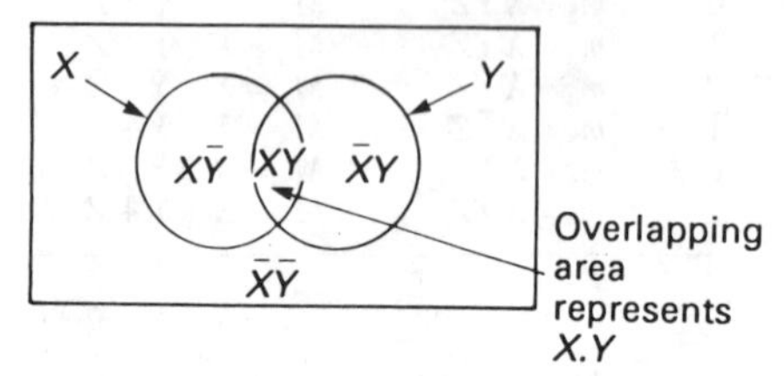

Figure 45.2 A Venn diagram

The rectangle is known as the 'universe of discourse' and the variables are represented by circles inside it. For two variables, the Venn diagram can represent all the possible combinations of the variables and this can be extended to any number of variables, although the diagram becomes too complex to understand for more than four variables.

An alternative representation of Boolean functions may be obtained by the use of 'geometric figures', where variables are represented by nodes, arcs and faces of the geometric figure. Single-, two- and three-variable geometric figures may be drawn, as shown in *Figure 45.3*.

A Boolean function may be simplified by marking each term at an appropriate node and then joining the nodes by the appropriate arcs. The simplified function is then the logical OR of all the arcs. Consider the following example:

$$F=XY\bar{Z}+XYZ+X\bar{Y}Z+\bar{X}\bar{Y}Z \tag{45.12}$$

The terms may be marked on the cube, as shown in *Figure 53.4*, and the four nodes joined by the three arcs.

The four nodes are joined by the three arcs, XY, XZ, $\bar{Y}Z$ and the function thus simplifies to

$$F=XY+XZ+\bar{Y}Z \tag{45.13}$$

However, the function may be further simplified. The XY arc includes the $XY\bar{Z}$ and XYZ nodes and the $\bar{Y}Z$ arc includes the $X\bar{Y}Z$ and $\bar{X}\bar{Y}Z$ nodes; thus, the XZ arc is redundant as it also contains the XYZ and $X\bar{Y}Z$ nodes.

The use of geometric figures to simplify (minimise) Boolean functions is not widely used as the optimum minimisation is not readily apparent and it is difficult to visualise a geometric figure to represent four or more variables.

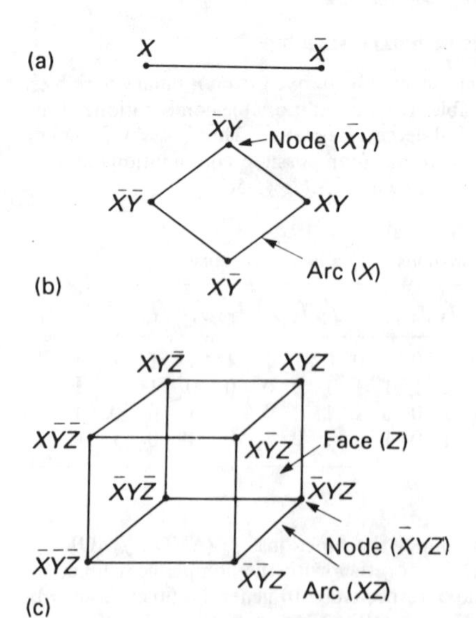

Figure 45.3 Geometrical interpretation of Boolean functions: (a) single variable, (b) two variables, (c) three variables

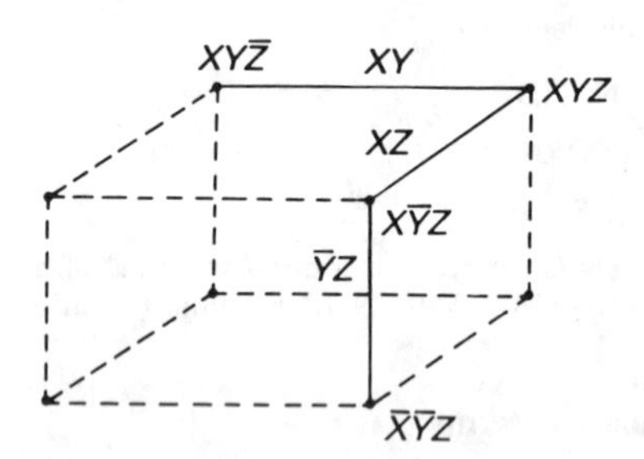

Figure 45.4 Example of a three-variable function

45.3.2 Minimisation using Karnaugh maps

The Karnaugh map method for representing Boolean functions graphically is based on the Venn diagram and geometric figure techniques. An extension to the Venn diagram was proposed by Veitch, where all the possible combinations of n variables are represented as areas inside a rectangle. The areas are arranged in such a manner that a move from one overlapping area to another, in a horizontal or vertical direction, changes only one variable. Thus, all adjacent areas can be related according to the switching theorem $XY+X\bar{Y}=X$. The Veitch diagram for four variables is shown in *Figure 45.5*.

The Karnaugh map is a matrix representation of a Veitch diagram, where each combination of the binary variables is assigned to a square in the matrix. The squares (also known as cells) are assigned binary codes such that each square in a horizontal or vertical direction differs from its neighbour in the value of a single variable only. Thus, terms in a Boolean function may be combined using the relationship $XY+X\bar{Y}=X$. Two, three and four variable Karnaugh maps are shown in *Figure 45.6*.

It must be emphasised that a similar relationship exists between the cells in the top and bottom rows and the leftmost and rightmost columns in the Karnaugh maps.

It is possible to plot Karnaugh maps for n variables, however, the number of cells becomes large above four variables and it is

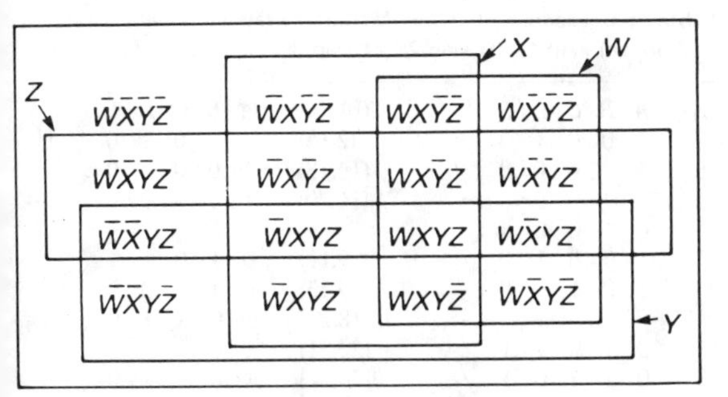

Figure 45.5 Veitch diagram for four variables

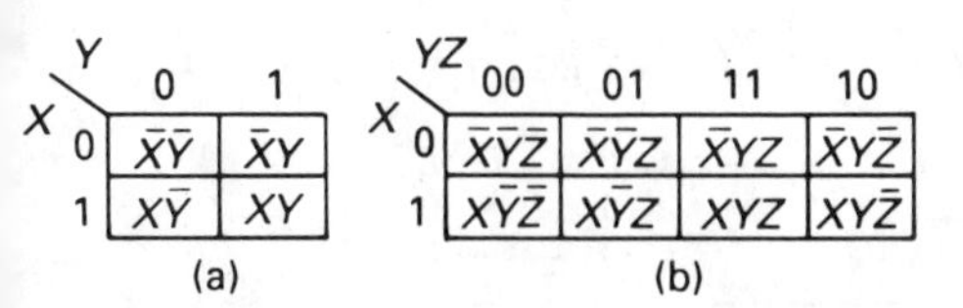

WX \ YZ	00	01	11	10
00	$\bar{W}\bar{X}\bar{Y}\bar{Z}$	$\bar{W}\bar{X}\bar{Y}Z$	$\bar{W}\bar{X}YZ$	$\bar{W}\bar{X}Y\bar{Z}$
01	$\bar{W}X\bar{Y}\bar{Z}$	$\bar{W}X\bar{Y}Z$	$\bar{W}XYZ$	$\bar{W}XY\bar{Z}$
11	$WX\bar{Y}\bar{Z}$	$WX\bar{Y}Z$	$WXYZ$	$WXY\bar{Z}$
10	$W\bar{X}\bar{Y}\bar{Z}$	$W\bar{X}\bar{Y}Z$	$W\bar{X}YZ$	$W\bar{X}Y\bar{Z}$

(c)

Figure 45.6 Karnaugh maps: (a) two variable, (b) three variable, (c) four variable

difficult to see the relationships between cells. Note that for n variables, there are 2^n cells in the Karnaugh map.

Whereas single adjacent cells differ by a single variable, it is possible to group together larger number of cells (known as sub-cubes) which differ by more than one variable but still correspond to the general theorem $XY + X\bar{Y} = X$. A sub-cube may be defined as a set of cells of the Karnaugh map within which one or more of the variables have constant values.

A sub-cube of 2 cells differs by a single variable and thus the variable is redundant. A sub-cube of 4 cells has two redundant variables and similarly a sub-cube of 8 cells has three redundant variables.

To plot a Boolean function on a Karnaugh map, the function must normally be in its canonical sum of products form. If, originally, it is not then it must be expanded into the canonical form by the method described previously. A '1' is then entered into each map cell corresponding to each term in the function. The function may then be minimised (if possible) by grouping together cells to form sub-cubes and thereby eliminating redundant variables. For example:

$$\begin{aligned} F &= \bar{Y}Z + \bar{W}\bar{X}\bar{Y}\bar{Z} + \bar{W}\bar{X}Y\bar{Z} + W\bar{X}\bar{Y}\bar{Z} + W\bar{X}Y\bar{Z} \\ &= WX\bar{Y}Z + W\bar{X}\bar{Y}Z + \bar{W}X\bar{Y}Z + \bar{W}\bar{X}\bar{Y}Z + \bar{W}\bar{X}\bar{Y}\bar{Z} \\ &\quad + \bar{W}\bar{X}Y\bar{Z} + W\bar{X}\bar{Y}\bar{Z} + W\bar{X}Y\bar{Z} \end{aligned} \quad (45.14)$$

The Karnaugh map is shown in *Figure 53.7*.

The sub-cubes enclosing the terms of the function are shown by solid lines and thus by inspection the function may be minimised to yield:

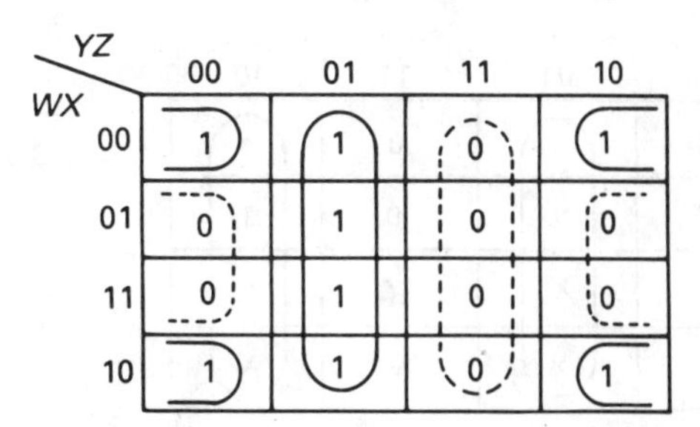

Figure 45.7 Karnaugh map of Equation (45.14)

$$F = \bar{Y}Z + \bar{X}\bar{Z} \quad (45.15)$$

The looped terms are known as *prime implicants*. The cells containing 0s (that is, the terms not in the function) may also be combined as shown by the dotted lines. When reading the result the variables must be inverted and combined in the product of sums form to give

$$F = (\bar{Y} + \bar{Z})(\bar{X} + Z) \quad (45.16)$$

The dotted looped terms are known as *prime implicates*. It should be noted that it is conceptually simpler to combine the 1s rather than the 0s, even though the two results are identical.

When combining cells on a Karnaugh map the following rules should be obeyed:

(1) Every cell that contains a 1 must be included in the minimisation at least once.
(2) The largest possible group of cells must be formed. Only sub-cells containing a power of 2 single cells may be used; that is, groups of 2, 4, 8, ... cells. This ensures that the maximum number of redundant variables is eliminated.

A significant advantage of the Karnaugh map minimisation technique is that it enables 'don't care' conditions to be used in the minimisation process. In some digital systems it is not necessary (or desirable) to specify the output function for all possible combinations of the input variables. This fact may be used to aid minimisation of the function as it is immaterial what value the output function takes for the input variable combination thus it may be assigned the value 0 or 1 in the Karnaugh map. Don't care conditions are assigned the value X (either 0 or 1) in the Karnaugh map. For example, consider the function F given by *Table 45.8*.

Table 45.8 Function F

W	X	Y	Z	F
0	0	0	0	0
0	0	0	1	1
0	0	1	0	1
0	0	1	1	0
0	1	0	0	0
0	1	0	1	1
0	1	1	0	1
0	1	1	1	0
1	0	0	0	X
1	0	0	1	X
1	0	1	0	X
1	0	1	1	X
1	1	0	0	X
1	1	0	1	X
1	1	1	0	X
1	1	1	1	X

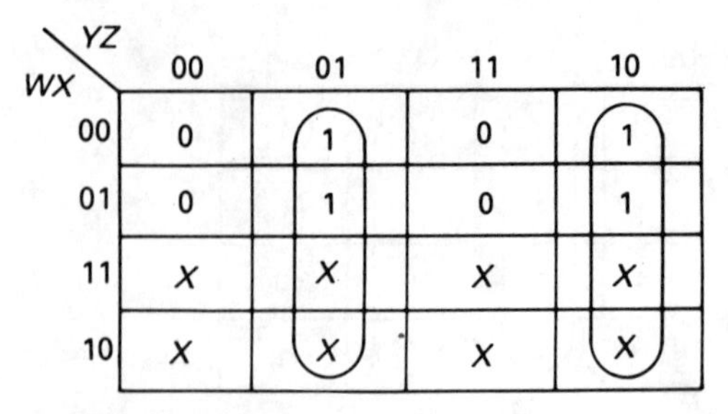

Figure 45.8 Karnaugh map of F defined by *Table 45.8*

The Karnaugh map is shown in *Figure 45.8*.

The function in its minimum form is:

$$F = Y\bar{Z} + \bar{Y}Z \tag{45.17}$$

It is immediately obvious that the use of don't cares has enabled maximum sized sub-cubes to be derived thus eliminating the maximum number of redundant variables.

45.3.3 Tabular minimisation of Boolean functions

The Karnaugh map minimisation technique becomes unwieldy if more than four variables are used. For problems with a greater number of variables a tabular method may be used. The technique most widely used is due to Quine and McCluskey and is based on the technique used in the Karnaugh map method; that is, examining each term and its reduced derivatives, exhaustively and systematically, applying the theorem $X\bar{Y} + XY = X$ at each stage. The result is a minimisation of the function. Although the method is normally performed by hand, it is amenable for programming into a digital computer.

The Quine–McCluskey tabular minimisation method is exemplified below. The following function is to be minimised, expressed in minterm form:

$$F = (2, 9, 11, 13, 15, 16, 18, 20, 21, 22, 23)$$

The first step is to tabulate the terms into groups according to the number of 1s contained in each term as in *Table 45.9(a)*.

The first group contains a single binary digit (2,16) the second group two binary digits (9, 18, 20) and so on. Each term in a group is then compared with each term in the group below it. If terms differ by one variable only, then they may be combined; for example, 00010 is compared with 10010 and found to differ by one variable and the term −0010 (the dash representing the redundant variable) is used to start a new list (*Table 45.9(b)*). Both terms are then 'ticked off' from *Table 45.9(a)* and the comparisons are continued until no more are possible.

The process continues by comparing each term in *Table 45.9(b)*. This time the redundant variables must also correspond.

The process terminates when no terms may be combined. The uncombined terms (unticked list entries) are the prime implicants of the function. Thus, the function F has been minimised to:

$$F = \bar{B}\bar{C}D\bar{E} + A\bar{B}\bar{E} + \bar{A}BE + A\bar{B}C \tag{45.18}$$

During the minimisation process, the repeated terms in a table may be ignored in succeeding tables.

45.4 Sequential systems

A description of combinatorial logic was given in previous sections. A combinatorial logic circuit may be defined as a circuit whose outputs are a function of the present inputs only, as in *Figure 45.9*.

Table 45.9 Example of Quine–McCluskey tabular minimisation method. (a) step 1, (b) step 2, (c) step 3

(a)	A	B	C	D	E	
2	0	0	0	1	0	√
16	1	0	0	0	0	√
9	0	1	0	0	1	√
18	1	0	0	1	0	√
20	1	0	1	0	0	√
11	0	1	0	1	1	√
13	0	1	1	0	1	√
21	1	0	1	0	1	√
22	1	0	1	1	0	√
15	0	1	1	1	1	√
23	1	0	1	1	1	√

(b)	A	B	C	D	E		
(2,18)	–	0	0	1	0		(1
(16,18)	1	0	0	–	0	√	
(16,20)	1	0	–	0	0	√	—
(9,11)	0	1	0	–	1	√	
(9,13)	0	1	–	0	1	√	
(18,22)	1	0	–	1	0	√	
(20,21)	1	0	1	0	–	√	
(20,22)	1	0	1	–	0	√	
(11,15)	0	1	–	1	1	√	
(13,15)	0	1	1	–	1	√	
(21,23)	1	0	1	–	1	√	
(22,23)	1	0	1	1	–	√	

(c)	A	B	C	D	E	
16,18/20,22	1	0	–	–	0	(2
16,20/18,22	1	0	–	–	0	
9,11/13,15	0	1	–	–	1	(3
9,13/11,15	0	1	–	–	1	
20,21/22,23	1	0	1	–	–	(4
20,22/21,23	1	0	1	–	–	

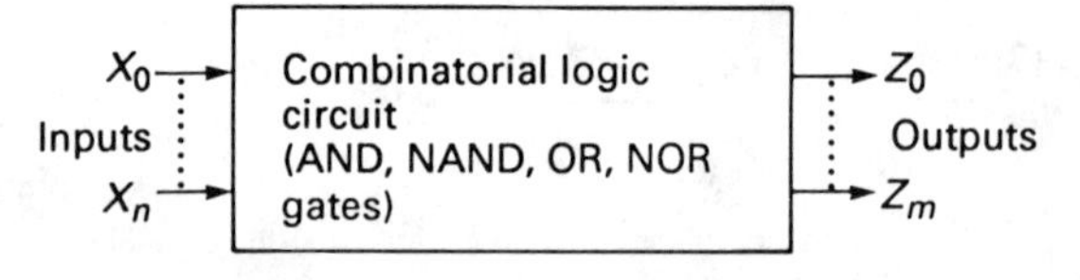

Figure 45.9 Combinatorial logic [$Z=f(X)$]

In order to describe practical digital systems it is often necessary to design circuits whose outputs are a function of the present and past inputs. For example, a combination lock must be aware of the sequence of coded digits which are input to it before the lock will open, and in order to dial a telephone number the digits must be input to the telephone system in the correct sequence. This type of digital system is known as a sequential system.

A sequential system is basically a combinatorial circuit with some of the outputs fed back to the inputs through memory storage elements, as in *Figure 45.10*.

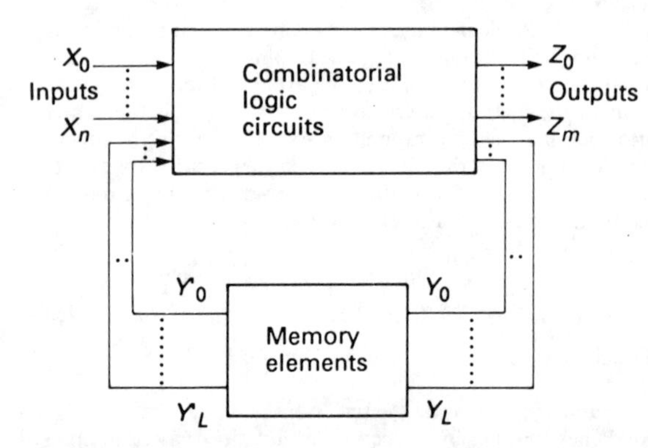

Figure 45.10 Sequential logic [$Z=f(X, Y')$, $Y=f(X, Y')$]

The memory elements are used to remember a particular function of the previous inputs to the combinatorial logic circuit. The values of these variables $Y'_0, \ldots, Y'_L$ are also known as the 'system state' of the sequential system.

There are two types of sequential system—synchronous and asynchronous. In synchronous systems, the inputs, outputs and system state are sampled at regular time intervals. The sampling is controlled by a *clock* which determines the 'frequency' of the system. In general, the inputs, outputs and system state only change on a particular phase of the clock.

In asynchronous systems, the logic circuits proceed at their own 'speed' regardless of any basic regular timing. As a result of asynchronous systems is relatively complex and the operation of such systems is difficult to analyse.

45.4.1 Finite state machines

In order to design sequential systems it is necessary to adopt a rigorous design methodology. It is normal to represent a sequential system by means of a state transition table and a state transition graph. A state transition graph is a graphical representation of a sequential system, where the system state is represented by circles, and the lines connecting the circles represent stage transitions. For example, consider the state transition graph for a 2-bit counter as given in *Figure 45.11*.

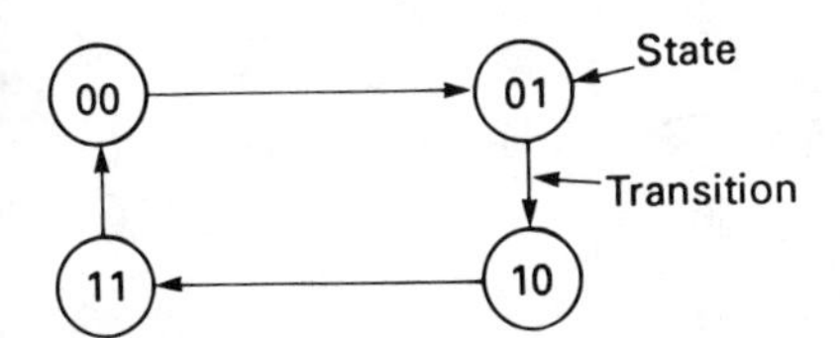

Figure 45.11 State transition graph for a two-bit counter

The state transition graph indicates that the counter counts in the sequence 00, 01, 10, 11, 00...; thus the graph illustrates the operation of the sequential system. The state transition table may be obtained directly from the state transition graph as shown in *Table 45.10*, where *A* and *B* are the system state variables. *Table 45.10* indicates the next state reached when the sequential system is in a particular state. The sequential system may be realised directly from the state transition table. From *Table 45.10*:

$$A' = \bar{A}B + A\bar{B}$$
$$B' = \bar{B}$$

Table 45.10 State transition table for *Figure 45.11*

Present state		*Next state*	
A	*B*	*A'*	*B'*
0	0	0	1
0	1	1	0
1	0	1	1
1	1	0	0

Thus, the sequential system may be implemented as shown in *Figure 45.12* (see also *Chapter 27* for flip-flop description).

The above sequential system is known as a 'finite-state machine' as the system can only exist in one of 2^n states, where *n* is equal to the number of bits in the system state (in this case $n=2$).

It is possible to formalise a general finite-state machine which

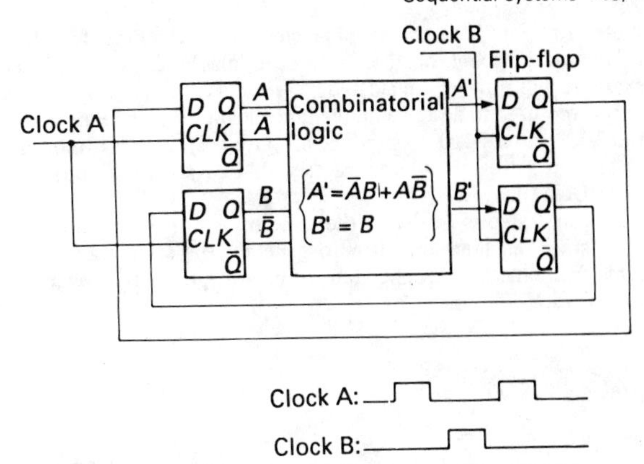

Figure 45.12 Circuit implementation of *Table 45.10*

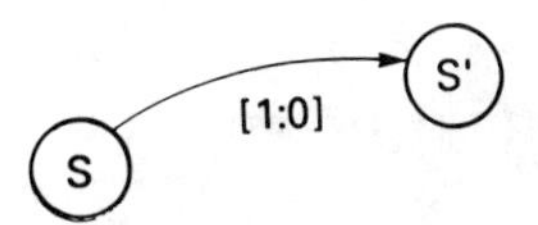

Figure 45.13 A state transition graph

may be used to implement any sequential system. Consider the state transition graph shown in *Figure 45.13*. The circles contain a coded representation of the system state. The transition line, or directed arc, shows the transition between states. The arc leaves one state and terminates on the next state. Alongside the directed arc is displayed the value of any inputs [I] which existed at the beginning of the state transition sequence and also the value of the outputs [O] produced by the state transition sequence. The inputs cause certain 'conditional' stage transitions to occur, which are controlled by the state of the input variables. Thus, the general finite-state machine must be able to generate outputs and state transition sequences which are controlled by inputs and the present system state. The general finite-state machine may be implemented as shown in *Figure 45.14*.

In *Figure 45.14* the following symbols are used:

R = collection of *n* bistables
mf = $f(I, S)$ = machine function
sf = $f(I, S)$ = (state function)
S = system state
I = input variables
O = output variables

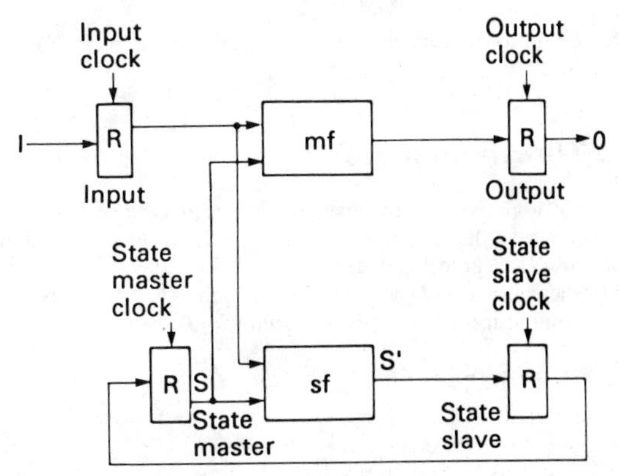

Figure 45.14 General finite-state machine

Note that four non-overlapping clocks are required for the correct sequencing of the finite-state machine. A practical example will now be considered.

It is required to design a finite-state machine which recognises the 3-bit pattern 110, to produce an output Z, in a continuous serial bit stream. A typical serial bit stream input is 1100101011001....

The first step is to derive the state transition graph for the finite-state machine and then to generate the state transition table from the graph. The state transition graph is shown in *Figure 45.15*.

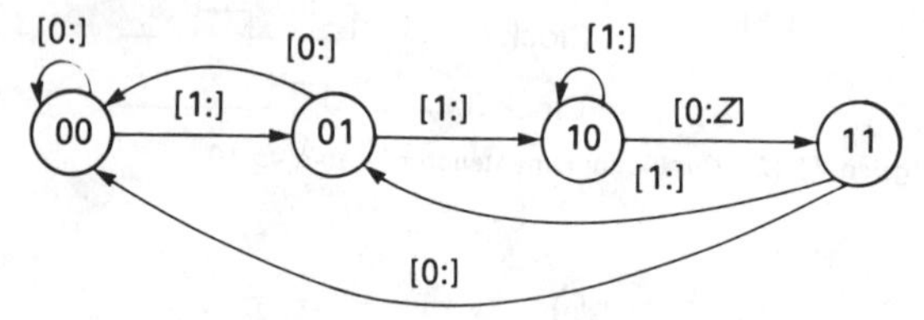

Figure 45.15 State transition graph for the text example

The state transition table is given in *Table 45.11*.

Table 45.11 State transition table for *Figure 45.15*

Input X	Present state A	Present state B	Next state A'	Next state B'	Output Z
0	0	0	0	0	0
0	0	1	0	0	0
0	1	0	1	1	1
0	1	1	0	0	0
1	0	0	0	1	0
1	0	1	1	0	0
1	1	0	1	0	0
1	1	1	0	1	0

The equations for the state function and machine function may be derived from the state transition table:

State function = sf = $f(I, S)$

$$A' = A\bar{B} + X\bar{A}B$$
$$B' = \bar{X}A\bar{B} + X\bar{A}\bar{B} + XAB$$

Machine function = mf = $f(I, S)$

$$= Z = \bar{X}A\bar{B}$$

Thus, the finite-state machine may be implemented as in *Figure 45.16*.

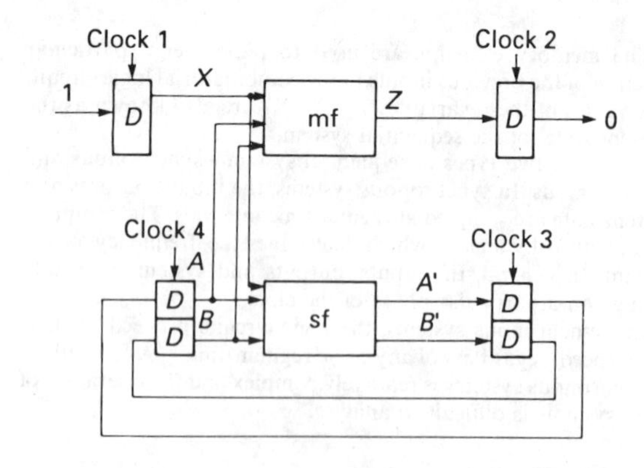

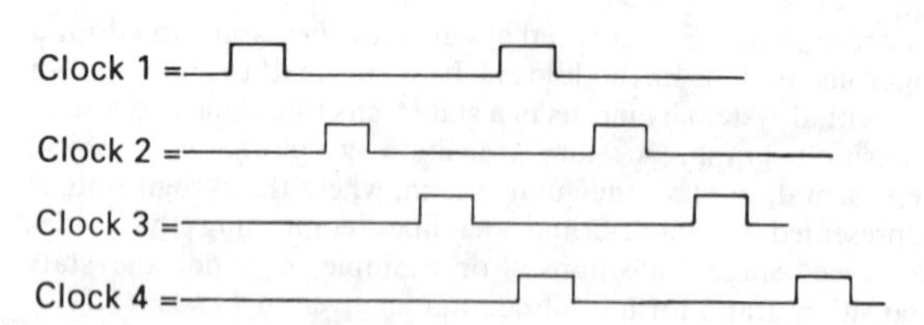

Figure 45.16 Implementation of finite-state machine of the text example

45.5 Threshold logic

Threshold logic is the term given to a special type of Boolean function, where the function is determined by a set of weighted inputs to a threshold logic element.

A Boolean function $F(X)$ of n binary variables is defined to be a threshold function if the following conditions are met:

$$F(X)=1 \quad \text{if} \quad f(X)=\sum_{i=1}^{n} W_i x_i \geq T$$

$$F(X)=0 \quad \text{if} \quad f(X)=\sum_{i=1}^{n} W_i x_i < T$$

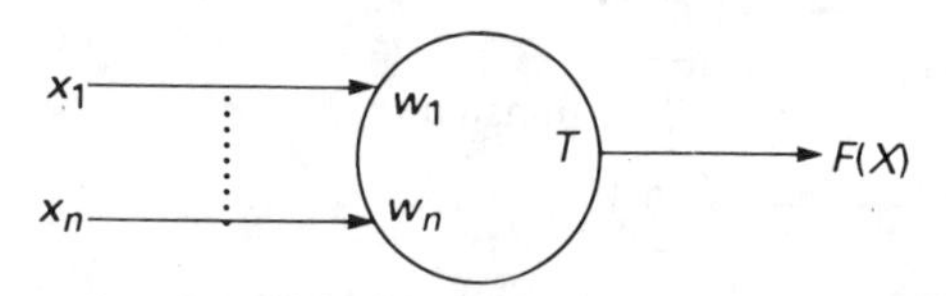

Figure 45.17 Threshold logic element

where

$X=(x_1, x_2, ..., x_n)$
x_i = binary variable
W_i = weight of x_i
T = threshold value
$f(X)$ = algebraic function of X

Thus, the output of a threshold logic element is 'true' if the weighted sum of its inputs is equal to or exceeds a predetermined threshold value, T.

A threshold logic element is a physical device whose inputs consist of n binary variables and whose output is a threshold function of the input variables. A threshold logic element (or threshold gate) may be represented by *Figure 45.18*.

Consider the following threshold function:

$$F = A\bar{C} + A\bar{B} \tag{45.19}$$

Table 45.12 indicates the combinations of the input variables which produce a 'true' output and those which produce a 'false' output.

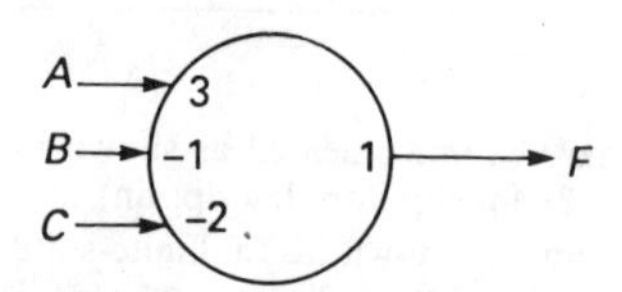

Figure 45.18 Threshold function implementation of Equation (45.19)

Table 45.12 True and false output for Equation (45.19)

	'True' output				*'False' output*		
	A	*B*	*C*		*A*	*B*	*C*
(1)	1	0	0	(4)	0	0	0
(2)	1	0	1	(5)	0	0	1
(3)	1	1	0	(6)	0	1	0
				(7)	0	1	1
				(8)	1	1	1

Thus three input combinations generate a 'true' output and five do not. The problem is to assign weightings to each of the input variables and a value for the threshold such that the terms (1) to (3) produce a 'true' output only. It is normal practice to assign integer weightings to each of the binary input variables and it is possible to have both positive and negative weighting values.

A possible set of weightings, together with the corresponding threshold value, is shown below:

$W_A = 3 \qquad W_B = -1 \qquad W_C = -2 \qquad T = 1$

This is normally expressed in terms of a threshold function, $R[F]$, where:

$R[F] = (3, \ -1, \ -2{:}1)$

Thus, input term (3) produces a 'true' output as the sum of its weighted values exceeds the threshold value $(3+(-1)+0=2>T(1))$ and input term (8) does not produce a 'true' output as the sum of its weighted values is not equal to nor does it exceed the threshold value $(3+(-1)+(-2)=0<T(1))$. The threshold function may be implemented as in *Figure 45.19*.

For n binary variables there are 2^{2n} possible switching functions. This, however, is not the case for threshold functions. In general, there are fewer threshold functions than switching functions; for example, for three binary variables there are 256 possible switching functions but only 104 possible threshold functions. Thus, before a Boolean function may be implemented as a single threshold function, it must be determined if the Boolean function is expressible as a threshold function. If this is the case then the relative weightings of the input variables and the threshold value must be determined for the threshold logic element. This is a difficult process and involves complex mathematics, which is outside the scope of this discussion.

45.5.1 Implementation of threshold logic elements

Threshold logic elements may be physically realised by an operational amplifier and threshold detector circuit, as shown in *Figure 45.19*.

The input variables are either 0 or V volts. The output of the difference amplifier is either a positive or negative voltage depending upon the weighted sum of the inputs (where $w_i \propto 1/R_i$). If negative weightings are required then the input variable should

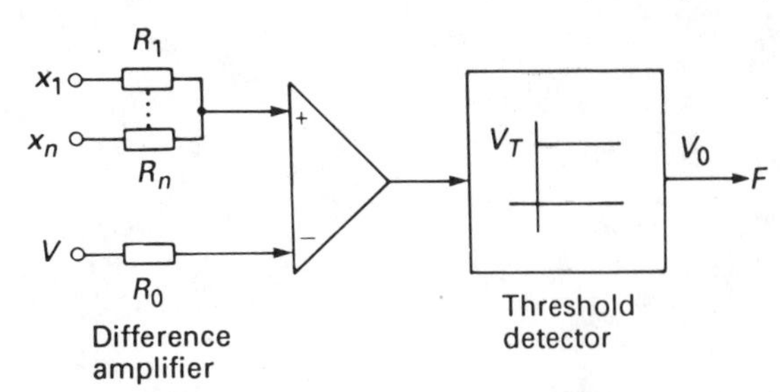

Figure 45.19 Circuit implementation of threshold logic elements

enter the negative input terminal of the difference amplifier. The threshold detector produces an output (V_0 volts) if its input voltage exceeds or is equal to a threshold voltage, V_T; otherwise it produces an output of zero volts. It must be noted that there will be practical limits to the number of inputs to the difference amplifier and the relative sizes of the input variable weightings.

Threshold logic elements may be used as general logic elements as the basic Boolean operations may be readily realised:

Boolean operation $[F]$	*Threshold function* $R[F]$
AND of n variables	$[1, 1, \ldots, 1{:}n]$
OR of n variables	$[1, 1, \ldots, 1{:}1]$
NAND of n variables	$[-1, -1, \ldots, -1{:}(1-n)]$
NOR of n variables	$[-1, -1, \ldots, -1{:}0]$
Inverter	$[-1{:}0]$

Thus, complex Boolean functions may be implemented by a single type of threshold logic element by merely adjusting the weights and threshold value.

The use of threshold logic elements is not widespread at present; however, their usage may become more popular if the threshold logic elements may be implemented without recourse to the use of basic analogue circuitry.

Further reading

ARNOLD, B. K. *et al.*, 'Beyond binary—a multiplier implementation based on multilevel arithmetic', *Electron. Eng.*, January (1989)

GREEN, D., *Modern Logic Design*, Addison-Wesley, Wokingham (1986)

KOPEK, S., 'State machines solve control-sequence problems', *EDN*, 26 May (1988)

LEWIN, D., *Design of Logic Systems*, Van Nostrand Reinhold, London (1985)

McCLUSKEY, E. J., *Logic Design Principles*, Prentice-Hall, Englewood Cliffs, NJ (1986)

MORALES, O. J. *et al.*, 'Beyond binary—an introduction to multilevel arithmetic and its implementation', *Electron. Eng.*, October (1988)

TRESELER, M., 'PDL state machines cut controller design time', *Electron. Design Int.*, November (1988)

46 Software Engineering

R S Hurst
STC Technology Ltd
(Sections 46.1, 46.2 and 46.5)

V A Downes
OVUM Ltd
(Section 46.2)

G H Browton
STC Technology Ltd
(Section 46.3)

P G Hamer
STC Technology Ltd
(Sections 46.4 and 46.7)

C J Tully
Cranfield Information Technology Institute
(Section 46.6)

Contents

46.1 Introduction

If the purpose of engineering is to obtain some product, then the purpose of software engineering is to obtain the software components of that product. Software engineering has attained the status of a branch of engineering; those who call themselves software engineers can be thought of as engineers.

Amongst the characteristics of software there are two, at least, which invite the assertion that software engineering is not true engineering. First, it is soft, not hard, and so shares the complexities and imprecisions of all the 'soft' sciences. It is difficult to measure in useful ways. Second, anyone can write software. Few hardware engineers are daunted by a need to put together a few hundred lines of FORTRAN or BASIC, so where is the engineering need?

The following chapter provides a summary of some of the topics which constitute software engineering as it is studied today. The intent is to introduce a number of ideas, and to point to further sources of reference.

46.2 Requirements and design specification

46.2.1 Requirements analysis

Computer programs have a life-cycle which starts when the first idea for a computerised system is born in the customer's mind and ends when the resulting software becomes obsolete. The program becomes obsolete when it can no longer be modified to meet the user's requirements or when a preferable solution becomes available.

The first stage in this life-cycle is the establishment of the requirements of the customer.[1] In all but the simplest cases, the customer is not the person who will be developing the software. Requirements analysis is, therefore, the process of finding out what the customer envisages that the program will do and then expressing these ideas in a form that can be understood by programmers.

Requirements analysis can be a very complex process. Often, the customer's ideas of what is wanted from the computer are vague and inconsistent. In cases where there are to be several users there are conflicting views of how it should work. The requirements analysis process brings together people who understand the application domain of the proposed system and an analyst who understands the capabilities of computer systems.

The analyst has to establish both the functional and the non-functional requirements of the system. Functional requirements describe the behaviour of the system, such as what outputs can be expected for given inputs. Non-functional requirements deal with qualities such as response time, user friendliness and cost.

Several techniques exist that can be used to help the analyst. These are mainly semi-formal techniques with graphical support that help to identify the data flow expected from the system and activities that are to be performed on the data. The analyst may use these techniques to construct views of the system from the perspective of each of the people who may have an interest in it. Apart from the potential users of the system, this should also cover, for example, those who pay for the development of the system and those who will have to maintain it. These views will then be refined into a single consistent view of the system to be agreed with the customer. The analyst will also advise the customer on the feasibility and analyse the cost of the various requirements. It is essential that the analyst does not allow the customer to foster any illusions concerning what can be achieved for the budget available.

The importance of the requirements analysis phase cannot be over-emphasised. Time taken here in testing out all the implications of what a customer wants is more than compensated for in reduction of the costly and time-consuming re-implementations that are necessary if the delivered program does not turn out to match the customer's expectations. In general, the later a fault is found in the life-cycle, the more expensive it is to correct.

Requirements analysis demands good communication skills and the ability to abstract the important points from a plethora of information.

46.2.2 Specification

The specification is the document that is produced as a result of the requirements analysis. It sets out precisely what the program is to do in a form that can be understood by both the customer and the programmer. It also states the non-functional requirements that are to be placed on the system.

The specification is built up during the requirements analysis stage and should be used as a basis of reasoning about what the customer is asking for and what the system will do. It is essential that the specification be unambiguous—points that may seem clear to the analyst and customer when they discuss them have later to be interpreted by a programmer who has not been party to the earlier discussions.

46.2.2.1 Graphical techniques

Traditionally, specifications have been written in English with, perhaps, some diagrams. An example of this approach is given by SADT.[2] Here the function of the system is described by decomposing it into functional components, identifying for each the required inputs and outputs, and also the controls and mechanisms which support it (*Figure 46.1*).

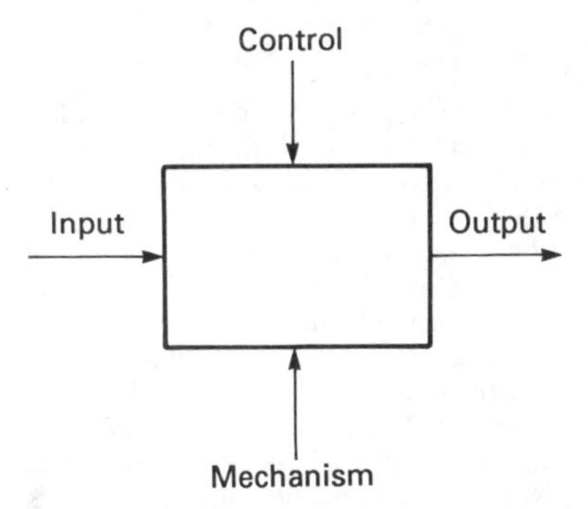

Figure 46.1 SADT

A hierarchy of diagrams is developed, with each diagram analysing one component of the system into a number, generally not more than five, of subcomponents. The subcomponents can, if necessary, be further refined in subsequent diagrams (*Figure 46.2*).

46.2.2.2 Rigorous techniques

Unfortunately, English, however carefully worded, can be open to different interpretations. Take, for example, the statement 'Protective helmets must be worn'. When displayed at the entrance to a building site, this would be interpreted to mean that no-one could go on to the site without wearing a helmet. However, the grammatically equivalent sign 'Dogs must be carried', displayed on the escalators in London Underground stations, does not mean that no-one is allowed on to the escalator without a dog! The potential ambiguity of these sentences is

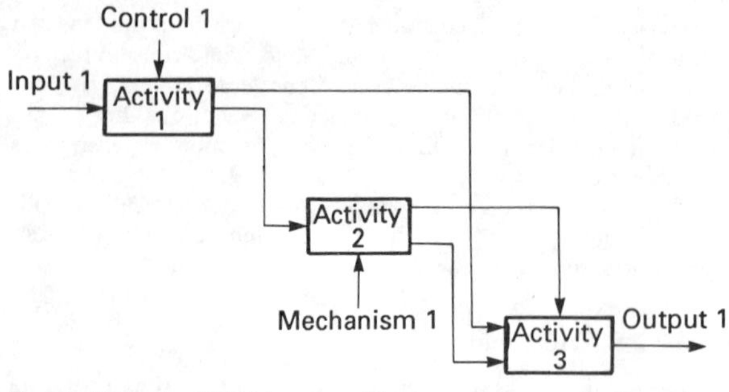

Figure 46.2 An SADT decomposition

resolved by the reader's knowledge of the context in which they are used. Similar ambiguities in program specifications cannot be so easily resolved, since the programmer is not usually familiar with the context in which the customer's application is used.

Formal methods of specification have been developed in recent years to overcome the problems of ambiguity. These are based on mathematics which allow the precise description of properties of data, the operations that can be performed on the data, and the states that the computer system can be in. Being amenable to formal reasoning they give the analyst a powerful tool for deducing properties of the program that are implicitly defined by the customer's requirements.

There are a number of families of formal specification methods, such as:

(1) Modelling languages, such as VDM (the Vienna Development Method[3]) and Z.
(2) Process algebras, such as CSP (Communicating Sequential Processes[4]), CCS (a Calculus of Communicating Systems) and LOTOS (for specification of open system interconnection protocols).

Probably the ones most widely used today are Z, developed at the University of Oxford, and VDM, which originated from IBM's Vienna Research Laboratories. Both of these are limited to the functional aspects of the program.

In addition to providing an unambiguous basis for program design, the specification should also be signed off by the customer as part of the development contract. To do this, customers must be satisfied that they understand what the specification means and believe that it matches their needs. It is in this area that there are difficulties with formal methods, since very few customers are mathematically trained to the degree necessary to understand formal specifications. Two approaches are used to overcome this problem. The first is to annotate the mathematical symbols with structured English text that interprets the specification for the customer. The second is to provide computer-based tools that will animate the specification so that the customer can investigate its functional properties.

Considerable research, development and education is still needed before formal methods can be widely used. Research is needed to extend the mathematics used to cover all aspects of program requirements.

Difficulties currently exist in expressing timing requirements, so important in real-time systems, and concurrency. Research is also needed in theorem-proving techniques which will allow computer-based tools to automate some of the more tedious and error-prone reasoning that is carried out in refining specifications. Development is needed to provide computer aided software engineering (CASE) tools to support requirements analysis and specification. These will include graphical editors to help the analyst, structured editors for formal specifications, and animators that prototype the specified behaviour. Some CASE tools exist now, but significant advances are needed before they match the power that a hardware engineer expects from a computer aided engineering system. Education is also needed, to familiarise programmers and analysts with the mathematics of formal methods.

In the meantime, it is likely that specifications will mainly be prepared as a mixture of diagrams, English text and mathematical symbols. It is also likely that programmers will continue to deliver systems that do not match the customer's requirements, and that a significant proportion of the 'maintenance' stage of the life-cycle will be spent in putting right faults that would never have occurred if more attention had been paid to getting the specification right.

46.2.3 Design

The specification stage in the life-cycle sets down *what* is to be done by the required computer system, but it does not give any indication of *how* the specified behaviour is to be achieved. The design stage sets out to describe how the program will be built.

Basically, design methods are driven by data analysis, operations analysis or object definition. The applications area often determines the most appropriate approach.

46.2.3.1 Functional decomposition

Successful design is based on the maxim of divide and conquer. The details of a significant piece of computer software cannot be held in one person's head at one time so the problem must be broken down into separate pieces, called modules.[5] If the modules are still too large, the process is repeated until the design consists of a series of simple building blocks, each of which can be separately developed and tested.

The first step is to produce a high-level design in which the system is decomposed into these logically distinct modules (*Figure 46.3*). The designer will experiment with different approaches to partitioning the system until he or she finds one that gives the most satisfactory solution. To a degree the merits of the design can be measured. For example: there should be a minimum of dependencies between modules; data that is logically related should be grouped in a single module with the operations that are performed on it; there should be a hierarchical tree structure to the intermodule calls, such that no more than five modules are called at the same level. But the merits of the design are also partly determined, like all good engineering design, by aesthetic factors and an ability to judge these is only built up by experience.

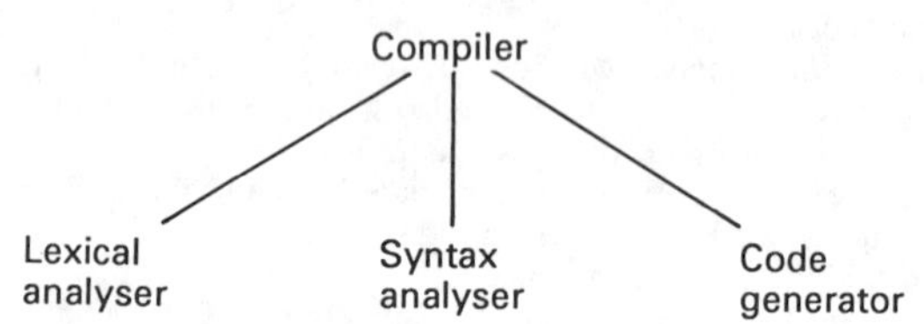

Figure 46.3 Functional decomposition

46.2.3.2 Data flow analysis

Several design methods start from an analysis of the data which is used and generated by the system in order to decide on the structure of the system. In the data flow approach[6] a graph is first constructed of the data flow, described in terms of the processes which act on the data and the stores, sources and sinks of the data (*Figure 46.4*). Then this is transformed into a program hierarchy by associating with each node of the graph three functions: 'get

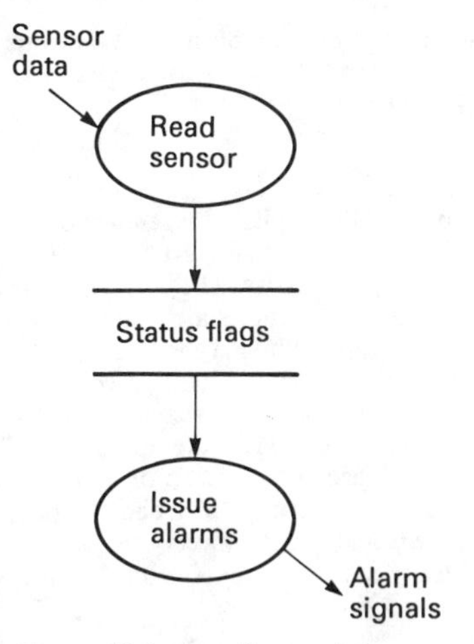

Figure 46.4 Data flow analysis

data', 'transform data' and 'put data'. For analysis of the central 'transform data' function, a functional decomposition approach may still have to be used.

46.2.3.3 Data structure analysis

This approach first looks for a description of the structure of each of the required inputs and outputs of the system and then transforms this model into a model of the necessary system (*Figure 46.5*).

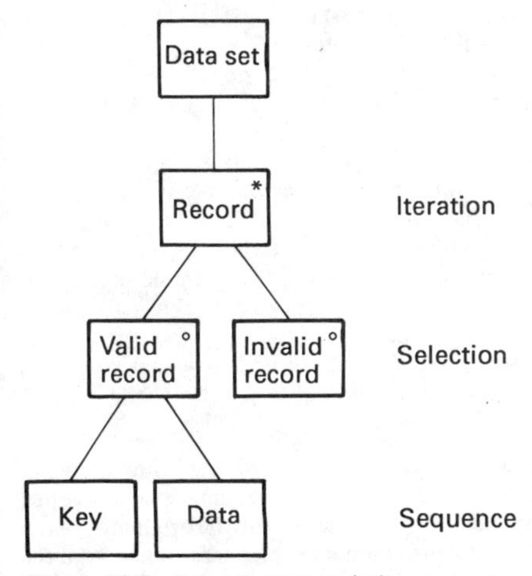

Figure 46.5 Data structure analysis

Such an analysis is used in the Jackson System Development (JSD) method.[7] The method provides solutions to the problems of conflicting data structures (structure clash) where an input structure does not map well on to an output structure.

46.2.3.4 Entity–relationship modelling

This technique[8] is used especially in the development of data bases. The information to be handled is analysed in terms of classes of entities, the attributes which those entities may possess, and the relationships between them (*Figure 46.6*).

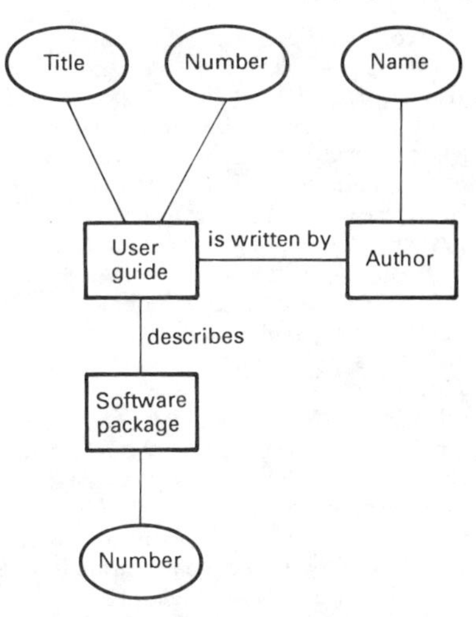

Figure 46.6 Entity–relationship modelling

46.2.3.5 Object-oriented design

This is an approach particularly suited to systems with concurrency. The system is broken down into objects or classes of objects, the actions that can take place on the objects, and those which the objects may require of other objects.[9] Full use of the programming language Ada* calls for design based on object-oriented principles.

HOOD is a proprietory method which is based on an object-oriented design approach which helps with the identification of self-contained modules.

46.2.3.6 Detailed design

After the high-level design is agreed, a process of low-level design is carried out on each module using a program design language. This will set out the structure of the internal working of the module to the point where coding in a particular programming language is a fairly straightforward activity.

46.3 Source program analysis and testing

Part of all engineering activities is the checking that the product meets the customer's requirements. All the processes in software production are subject to error because of the complexity of the process and the abstract nature of the representations used. Until the process of software production can be totally automated, including the capture of the customer's requirements, and those automatic processes proved to be without fault, the need for testing will remain. Certainly the current state of practice in software production does not allow for absolute statements to be made about the quality of the software. A large range of techniques and processes does exist, however, which, properly used and costed, allows statements to be made with a very high level of confidence. In the production of software, the testing of the source code is a vital component of the overall process of checking.

Testing any software product (here taken to be the source code) requires a number of prerequisites. Firstly, the code itself must be available. Secondly, the conditions that define whether,

*Ada is a registered trademark of the US Government—Ada Joint Program Office.

as a result of testing, the code is to be deemed acceptable must be defined. Thirdly, a test plan, which defines the test strategy and techniques to be used and details all the attributes to be tested, is needed. Lastly, a report has to be produced which details the outcome of the tests of the code against the plan and records the decision of the tester with respect to the test conditions. This decision is used by the project and quality management to decide if the code is to be passed as satisfactory or whether it is to be returned to the code producers for rework. This reworking is often called *debugging* (possibly after an original entomologically caused fault). It is not part of the job of testing to carry out that rework but the rework needs to be done and time has to be allowed in the plans for it.

The amount of testing and the level to which it is carried out will vary depending on the cost to the producer of failure in use. This cost of failure depends upon such parameters as the size of the development, the critical nature of the use of the software (for example, software for use in power plants or transport systems needs higher levels of reliability to be demonstrated than software used for payroll production—which needs to be accurate), the users of the software (for example, the software to be used for public domain information services needs to be demonstrably more usable than software used in a mainframe operating system), the cost of maintenance (or change) of the software and the importance to the producer of the customer (varying from life-critical systems, through software packages sold in retail shops to software produced as 'one-offs' in home computing usage).

The selection of techniques to be used in the appropriate testing phases will depend on the resources available for the language, operating system and organisation involved. The selection should be made by the project and quality management of a project at the design stage. That is, the test strategy and test plans for the system should be considered and produced at the same time as the designs for the system.[10,11]

46.3.1 Definitions

As a young discipline, software engineering is still at the phase in its development where neologisms are invented more often than a correct term is used. The following definitions are therefore not universally agreed.

Testing The process of demonstrating and recording that the code which has been written, according to the designs, has been produced with the necessary qualities of completeness, correctness, reliability, functionality, etc., such that the process of code generation may be considered complete.

Checking The collective verb for all activities related to assuring that the product is being prepared in accordance with the customer's wishes and with the quality assurance system.

Verification The process of checking that the product prepared in one phase of the life-cycle is a correct and complete translation from the previous phase.

Validation The process of demonstrating that the product of a software development is a correct and complete translation of the customer's requirements.

Correction Consists of four steps: identifying the error and its location; making the necessary change; verifying and validating the change; recording the change and its impact on documents such as the designs and user manuals.

Fault The manifestation of an error. (Familiarly known as a bug or defect.)

Error A mistake made by a person which may lead to the introduction of a fault into an item.

Failure The inability of a system or component to perform its specified function.

46.3.2 Testing phases

Testing can be divided into a number of phases depending on the size of the system being developed and its importance. For the purpose of this section three broad phases will be described. The first phase is *unit testing* where code modules are tested against their individual designs. Modules are compiled and then tested in harnesses which act as testbeds feeding data to and collecting inputs from the module under test. The second phase is *integration and system testing*, where the modules are collected together and tested as a whole against the system or high-level design. The third phase is *acceptance testing* which demonstrates to the customer that the delivered system meets the specified requirements. These phases are considered in more detail below.

46.3.2.1 Unit testing

This testing of modules in isolation allows the test team to concentrate on the internal feature of the modules, such as the algorithms for handling specific tasks within the program and input and output functions.

The two key questions to be answered in unit testing are 'Does the module work correctly?' and 'Is the module complete?'. The test plans and testing strategy should be directed to answering these questions efficiently and with confidence. The module is to be examined both for its functional performance and for the quality of the implementation. The planning of unit testing occurs with the production of the test plan which should be devised at the requirements specification and high-level design phases. The production of test cases, however, should be done after the detailed design of modules has been carried out as it is then that the designer will be most clear about what the module should do and how it should be tested.

Techniques for unit testing may be divided into two classes, *inspection* and *execution*.

Inspection The first examination of code should be by the programmer. Typically the code will be examined, with a number of values for variables being 'dry run' through the code. This first examination will reveal most of the obvious errors in calculations and logic, but although the programmer knows the code better than anyone else it is very easy to skip parts of the checking because 'I know that bit is right'. Thus there are several techniques which remove this bias—either by machine-driven automation or by the examination of the code by colleagues. A number of these techniques are:

(1) *Structured walkthrough* This is a review of the program module carried out by a group of programmers who are led through the code in a controlled manner by the originator of the code. In this review the programmer explains the function of the module, and how it fits in with the rest of the system, and then shows colleagues the reasoning at each step of the design to code process. The functions of the colleagues are to discuss the translation process of the code from the design, making comments about the style and adherence to programming standards, and to point out any defects in reasoning. As in inspection (see below) this examination is done to help the team achieve its overall quality objectives. This attitude of shared concern for the product will result in better code as each individual accepts the importance of his or her contribution to the whole.

(2) *Fagan inspection* Formal inspection of deliverables, code, designs and manuals is a particularly effective, and comparatively cheap, method of finding errors.[12] Before the

inspection the review team will be given the code and asked to examine it before the meeting. At the meeting the code is read through, preferably by someone other than the programmer, and formally examined against checklists. The outcome is recorded as part of the project history, again by someone other than the programmer. It is the responsibility of the programmer to correct all the errors found in the inspection. Studies have shown that up to 80% of the available errors can be found by properly carried out inspections.[13]

(3) *Control flow analysis* In this technique a flow or directed graph (digraph) is generated from the code. This is rather like making a flowchart after the code has been produced. The digraph can then be examined for errors in the way in which the program is controlled and makes its decisions. The digraph says very little about the calculations, since all statements are reduced to nodes on the graph, but the control flow of decisions and loops becomes clear. Test cases are then devised to exercise all decision points in the graph, such that measurements about the statement coverage can be made, showing that all portions of the code have been tested.

(4) *Data flow analysis* Distinct from control flow analysis is the analysis of the flow of data through a program. The program is run with the code instrumented in such a way that the values and changes of variables are recorded in an execution history. The actions of interest here are the points at which a variable is initialised, referenced, altered and stored. In most programming languages the value of a variable will be undefined if it is referenced before it has been initialised. Other errors which can be detected include the erroneous use of variables after procedure calls and the de-referencing of a variable before it has been used.

Both control and data flow analysis can be performed statically or dynamically, where different errors are detected.

(5) *Compilation* This is often the first of the tests carried out, because it is always available and is an automated process and thus easy and cheap for the tester. The main assumption is that the compiler is correct, and the information given to the tester is that a successful compilation (i.e. one with no errors) infers that the syntax of the code module is correct. This is, of course, far from saying that the module is functionally correct.

(6) *Back to back* This technique involves the production of two independent implementations of a design which are then tested on the same data. The results are compared and any disagreements between the two implementations investigated. This technique finds errors in the implementation and not in the design.

Execution Several techniques exist for execution, as follows:

(1) *Symbolic execution* In this technique the program is 'executed' symbolically. Instead of the value of variables being calculated at each step, an algebraic expression is formulated, reflecting each of the changes applied to input data, until the output is reached. At this point the expression is evaluated with a number of input test data to check if the answers are correct. Though extremely cumbersome, especially when decisions and loops have to be taken into account, symbolic execution has value where the demand for accuracy or reliability warrants the use of a technique akin to program proving.

(2) *Test harnesses or stubs* Two approaches to execution testing of modules are *bottom-up* and *top-down*. In the bottom-up approach, modules are checked in test rigs or harnesses which call the module under test, providing input data and capturing output data. The harness can call upon already tested modules, and can be used to check the output from the module against manually computed answers provided by the tester. The overhead in producing the harness—which itself has to be correct—needs to be recognised. In top-down testing the top-level or main program is tested first, with the subordinate modules being replaced by dummy subroutines called stubs. These stubs provide the calling program with the data expected from them—placed there by the tester. Stubs range in complexity from ones which provide the same answer to every invocation to sophisticated versions which pass a variety of data according to whim and the test case being run. As testing proceeds to integration (see below) the stub modules are successively replaced by the actual implementations as the testing proceeds toward the 'lower' levels of the program.

In both these approaches the problem is to ensure that the harnesses and stubs provide all the range of data needed to carry out the test satisfactorily.

(3) *Test cases* The generation of test input—and the corresponding expected output—is one of the most difficult aspects of testing. To test every single possible input to a system is generally an impossible task. A number of strategies exist for reducing the input space to manageable proportions. Firstly, the system needs only to be tested for those inputs which can occur. The danger here is that the range of inputs which can occur may well be greater than the range of inputs which are supposed to occur. None the less, if it can be shown that the input to a component unit of a software item is constrained, then unit testing of that component can be limited to the bounded input range.

Three approaches may be taken in the generation of test cases for such internal routines. One can generate test cases with *all* the possible values, with values at the *boundaries*, or with *random* values. Boundary cases are chosen by assuming that in the logic of programs it is the values like 0, +1, −1, +2, −2 and plus or minus infinity which are likely to be the cause of errors and that values such as these will be sufficient to check the work of the programmer. Random testing has been shown to be a useful technique in the testing of large systems (that is, it is not generally considered a good technique at unit testing) because the large test data sets do not suffer from the bias of the designer, programmer or tester. The outputs have to be checked against the manually computed values (unless back-to-back testing is used).

Test data need not be numeric. For the generation of any kind of test data, a trade-off has to be made between the total possible input space and the input space which it is feasible to examine, taking into consideration the 'cost of failure'.

46.3.2.2 Integration testing

Integration testing is designed to demonstrate that the interfaces between the previously tested modules function correctly and that the system as a whole functions correctly. Many of the same techniques can be used at integration testing as at unit testing, but often the combinatorial explosion of possible values and paths prohibits the full demonstration of such techniques. The planning for this phase starts during the high level (or architecture) design phase and is enhanced and worked on from that time onward. The key question to be answered is 'Does the system exhibit all the functional and non-functional attributes detailed in the requirements specification document?'. The non-functional requirements will cover such attributes as performance, reliability, security, etc.

The successive integration of tested modules (see above for the discussion on top-down *versus* bottom-up approaches) now takes place with attention placed on those testing techniques which exercise the transfer of data between modules and subassemblies of the system as a whole. The dependence upon the answers provided at this phase will depend critically upon the quality of the unit testing work which precedes it. If the modules are incorrect, the integration testing will take much longer, as the process of tracing errors shown up by failures will have to consider the code inside the modules as well as the control flow governing their execution.

In integration testing (including as it does systems testing for the purpose of this section) tests must also be run to show that the system is usable by the customer, so inspections of the user manuals, examples and error messages must all be included.

46.3.2.3 Acceptance testing

Acceptance testing should be designed by the producer and customer to demonstrate that the former has indeed produced a system which matches the customer's requirements, as expressed in the requirements specification. This phase is typically carried out by the customer, or at their direction, and can range from an informal run-through of the main functions to a full-scale check by testers in the customer's organisation examining the system for all the functional and non-functional attributes laid out in the specification. Typically some part of the payment for production of the system will be retained by the customer until the system has proved itself over a period of sustained use.

46.3.3 Completion of testing

How do we know when testing is complete? The two least helpful criteria are quoted by Myers:[14] 'Stop testing when the planned time has elapsed' and 'Stop when all test cases run without causing errors'. He suggests a number of other criteria, and concludes that testing can be considered complete when the chosen testing techniques at each phase have all been used, when targets of errors found (by module and by system) have been met, and when the curve of the errors found against time show that the rate of error detection becomes acceptably low. These criteria all refer to finding errors. The other aspect of checking, that is the demonstration of validation to the customer, is measured not by defects found but by demonstrating that all the functional and non-functional requirements have been met.

46.4 Simulation

To simulate is 'to duplicate the essence of a system or activity without attaining reality itself'. Computer simulation is often described as a method for understanding, representing and solving complex interdependent problems. It is perhaps truer to say that producing a computer simulation provides a focus and impetus for such activities.

Many real-world disasters[15] could have been anticipated; with hindsight they often appear almost inevitable. Simulation provides a way of learning lessons at a much lower cost. Sometimes this will prevent a disaster; more often it will simply improve a product or system. There are of course no panaceas, and the behaviour of some physical systems is inherently difficult or impossible to predict.[16] The Meteorological Office's inability to predict the damaging storm in the winter of 1987 was due to their over-reliance on their detailed wather model.

46.4.1 Why simulate?

Simulation is used when it enables something to be done more cheaply, more safely, or in a compressed (or accelerated) time frame. Some typical reasons for making a simulation are: to construct prototypes; to reproduce operational behaviour; to gain engineering insight; and to evaluate alternatives and demonstrate trade-offs.

46.4.1.1 Constructing prototypes

Simulation can be used to construct prototypes to explore important properties of products or systems, in particular their dynamic properties. Software engineers use simulation to explore the effectiveness of alternative man–machine interfaces, and get feedback from potential users. Electrical engineers use special-purpose simulation packages to prototype integrated circuit designs. NASA engineers are intending to simulate high-speed air flows as an alternative to using wind tunnels. Materials scientists use simulation to study the long-term effects caused by diffusion of atoms in alloys. An aircraft designer might use a flight simulator to get feedback from pilots on aircraft handling characteristics; this could be especially relevant for high-performance fly-by-wire military planes whose handling characteristics are largely determined by their control systems rather than their inherent aerodynamic characteristics.

All such operator-training models may also be used before the fact to evaluate the acceptability of the process being simulated. So a test pilot might evaluate and optimise the instrument display and handling characteristics of an aircraft before it was built.

46.4.1.2 Reproducing operational behaviour

The prime purpose of being able to reproduce the operational behaviour of an existing product or system is for operator training. The degree of realism provided by such simulations differs greatly. A medical simulation of patients with kidney malfunction might just print out details of the vital signs, answer questions on symptoms, and accept orders about treatment regimens. A significant part of effective simulation is deciding what can be left out.

At the other extreme, aircraft flight simulators probably go further than any other class of simulation in trying to reproduce fully the operator's sensory experiences. The development of the hardware and software required to provide this man–machine interface is a major technical achievement. On the other hand, the actual task of simulating the aircraft's behaviour is not as difficult, as the flying characteristics of the plane are well known, if only as a result of test flights. Flight simulators are cheaper than using real aircraft, but their real benefit is that they enable extensive training to be given in situations that would be too dangerous to reproduce in real life. Note that the aircraft's behaviour in such situations (e.g. engine failure at embarrassing moments) is well known, as it can be studied at safe altitudes.

While few other application areas are as demanding in terms of the man–machine interface as flight simulators, many application areas present greater difficulties in developing a suitable model. For example, emergency situations for chemical plants and nuclear reactors often result in highly unpredictable mixed-phase situations (e.g. water–steam mixtures).

46.4.1.3 Gaining engineering insight

Analytic methods of design are very effective in well understood situations. If failure modes can be anticipated, it is often quite straightforward to predict the circumstances under which they will happen. The difficulty is in gaining sufficient engineering insight to know which failure modes need to be considered. Simulations can provide useful input here. Modelling the heat flow in a reactor vessel and presenting the results visually can indicate potential hot-spots. Similarly, simulating stress build-up can aid the design of lightweight but strong structures. On the other hand, such models may not indicate some failure modes; many box-girder bridges failed or needed strengthening because computer modellers had not anticipated the crumpling failure mode.

46.4.1.4 Evaluating alternatives and demonstrating trade-offs

Often the engineer is aware of design alternatives, either qualitative or quantitative. Simulations can provide a way of evaluating the relative merits of the alternatives. Perhaps the major merit of such an exercise is that it forces the simulator to obtain answers to a large number of questions. Running the simulation with different values of a parameter can indicate the sensitivity of the evaluation to what parameter to be estimated.

46.4.2 Some important topics

There are a number of important topics that the intending simulator should consider. They are as follows.

46.4.2.1 Continuous or discrete event simulation

There are two important questions relating to how the simulation deals with time. The first is if the simulation model is in terms of continuous time or discrete events. The second is if the simulation needs to run in real time.

Continuous models solve mathematical formulae which represent continuous processes, e.g. the flow of liquids through a pipe. In practice the associated differential equations will probably be solved in terms of difference equations, so in the model time will advance in a series of small, equal-sized intervals.

Discrete event simulation is used when the reality being modelled can be represented as a series of discrete events. A major attraction of discrete event simulation is that the computer needs to do very little work except when an event happens, which means that such models can be simulated very efficiently. Obviously the discrete events correspond to the transitions in a finite state model of the customer's state. Developing such a finite state model may be helpful in identifying all the relevant events.

A simulation may need to run interactively in real time because this is an essential aspect of its behaviour (e.g. the flight simulator). Alternatively it may simply be appropriate for it to run non-interactively at a speed where its behaviour can be conveniently observed; this might be slower or faster than real time. On the other hand, non-interactive behaviour can also be usefully presented as statistics; e.g. mean queueing time for the customer, or a histogram of queuing time. In some circumstances it is best to simulate time passing non-uniformly; e.g. a medical simulation would probably skip over periods where the patient's condition does not change (or is not observed).

46.4.2.2 Tabular or graphical presentation

Like any experiment, the success of a simulation is heavily dependent on the skill with which its results are interpreted and presented. Increasingly it is being realised that visual presentation has an important part to play in both the presentation and the interpretation of data.[17] Furthermore it is difficult to over-estimate the importance of visual interactive simulation in understanding the dynamics of the process being studied.

46.4.2.3 Accuracy versus speed

Most simulation models involve a compromise between accuracy and speed. Accuracy encompasses both numerical correctness and the amount of real-world detail which is included. Speed covers both the speed of running the simulation and the speed with which it can be developed. The true art of simulation is in deciding what real-world detail can be left out. Unnecessary detail not only makes the model more complex but, perversely often makes it a poorer model. Decisions on numerical accuracy are more often a simple trade-off between numerical accuracy and simulation speed.

46.4.2.4 Mainframe or micro

The choice of a computer for simulation work represents interesting trade-offs. In general, mainframes are looking increasingly less and less attractive, for several reasons. Firstly, micros frequently have much better facilities for graphical presentation than mainframe terminals. Secondly, many older mainframe simulation packages adopt a batch processing approach and produce comprehensive listings of indigestible data; so are of limited utility. Thirdly, most mainframes are multi-user, so the speed of a simulation will vary according to the load on the computer; this is undesirable if the simulation is being observed as it runs. Fourthly, the speed advantage which mainframes traditionally offered is steadily being eroded. In general, the superior graphics offered by micros is very attractive to those with only moderately computationally demanding simulations.

46.5 Management and control

Much of the task of managing the development of computer software is the same as that of managing any other engineering activity. A team must be formed and motivated, the goals of the activity must be understood and communicated, work must be scheduled and assigned, and costs must be estimated and monitored. Software management calls for the techniques and tools common to all varieties of management.

There are, however, some aspects of software management which are not quite common to management in general, or if they are they are given particular attention in the field of software engineering. A selection of these follows.

46.5.1 Life-cycles and process models

Like most animate and inanimate objects, a software product has a life-cycle; it is conceived, it is brought into existence, it grows and changes, and after a time it dies. It becomes easier to control the creation, change and death of the shoftware if a regular pattern can be observed in the life-cycle, and much attention has been given to different forms of software life-cycle.

46.5.1.1 The waterfall model

The waterfall model[18] became popular in software engineering at the beginning of the 1970s, and it is still the most familiar model of the software development process. It divides the process into a number of phases with the output of each phase becoming the input of its successor. So requirements specification leads into design, then in turn into coding, testing and operational use (*Figure 46.7*).

Adherence to the model provides some useful benefits. It encourages an insistence on completing one phase in the development cycle before the next begins, so the design should not start before the required product is specified and, similarly, the developers should not begin to code the software before its design is completed. In most development projects it is inconvenient to stick rigidly to such a sequence, but consideration of the model will remind the manage of the consequences of the breach of ordering.

In the process of coding, errors will be found in the design; and in the process of unit testing, errors will be found in the code. In these cases, the design and the code must be corrected, and so there is iteration between the phases. With the waterfall model, the idea is that there is iteration only between adjacent phases. It does not allow for interaction between phases which are not successive.

46.5.1.2 The V-model

It is well recognised that to catch an error in the specification of software (as indeed of any manufactured item) at the time the

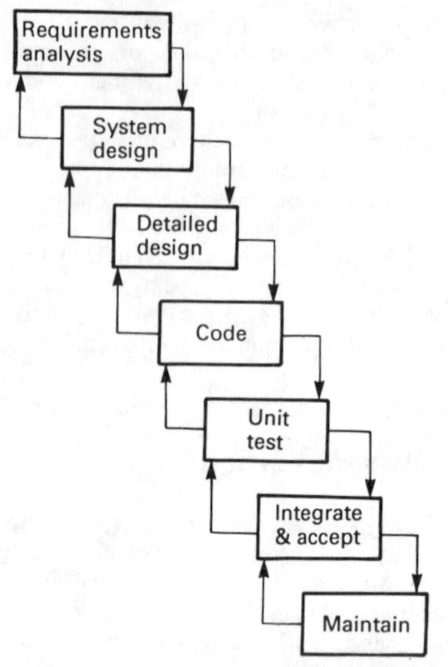

Figure 46.7 The waterfall model

specification is being decided is many times cheaper than to catch the same error when the implemented software is being tested. Therefore it is advisable to verify and validate each representation of the software.

The V-model brings out this continued checking (*Figure 46.8*). It also represents the different phases in the testing of software as it is assembled from its components. Individual modules of code are verified against the detailed design, integrated subsystems are verified against the system design, and the fully integrated product is validated against its specification and its requirements.

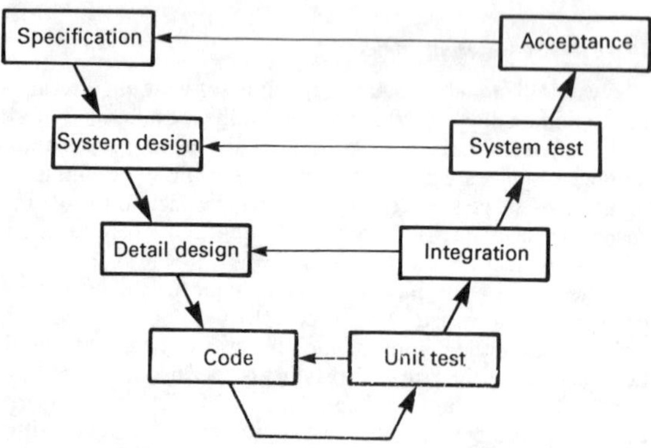

Figure 46.8 The V model

So, in a sense, the output of each phase of development of the code is examined at a corresponding phase of the integration and test of the code. Planning for that phase of integration and test of the time of the development can contribute to validation of the specification or design before it is further refined.

46.5.1.3 The prototyping model

The term 'prototype' is used to refer to any system which demonstrates some characteristic of behaviour of the final product, and which can be made available for examination in advance of the final product. In this sense, a prototype need not exhibit all of the functionality of the final product, and its design may be quite different from that of the final product. The term 'breadboard' might sometimes be more apposite.

The prototyping model calls for the creation of a series of prototypes. These would most typically represent the user interface to the system, but might also allow other issues, such as those of performance, to be investigated.

Later prototypes in the series might be supersets of earlier ones and the union of the set of prototypes might provide a near complete representation of the final product. More often, however, the prototyping model itself is seen as a subset of some other model of the life-cycle, appearing as one or more phases within that model (see *Section 46.5.1.5*).

46.5.1.4 The incremental build model

With this model, a partial version of the final product is produced and made available to its user. The reduced subset does not have all the capability of the final product, but it is based on the same architecture. Then, in turn, each partial version of the product is supplemented by a more complete version until the final product is achieved.

In common with the prototyping model, this approach ensures early feedback from the user, allowing characteristics of the final product to be assessed. It also exploits the possibility that almost 'nobody needs all of what they asked for by the deadline'.[19]

46.5.1.5 The spiral model

The spiral model,[20] which was developed at TRW, takes the view that the development process delivers a series of representations of the required product, and that the way in which each representation is reached from its predecessors is common (*Figure 46.9*).

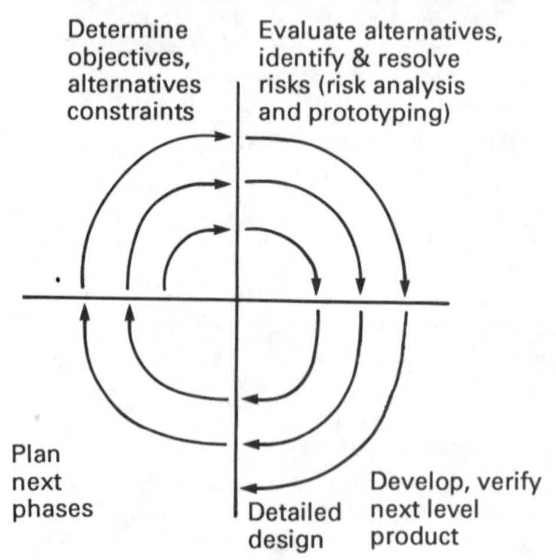

Figure 46.9 Boehm's spiral model

First the objectives, alternatives and constraints of the representation being sought are determined and then the alternatives are evaluated and risks are identified and resolved. The next-level representation is developed and verified, and then the following phase is planned. Each cycle finishes with a review involving the key people concerned with the product, covering both the representations developed during the cycle and the plans for the next one.

46.5.1.6 The contractual model

This is a different style of model (*Figure 46.10*). It does not attempt to represent the phases in the life of a software item or of its development, but it assumes that the project work is analysed by some means into component tasks. Each task now becomes the subject of a contract within the terms of the model, where a contract is a quasi-legal agreement between the person who assigns the task and the one responsible for carrying it out. The contracts may form a hierarchy with the person responsible for a contracted task completing it by letting subcontracts.

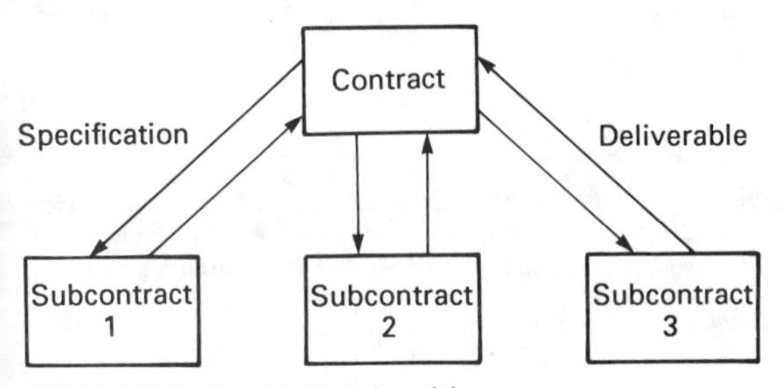

Figure 46.10 The contractual model

A contract specification has three parts: a task specification, which says what the task is; the verification criteria, which say how performance of the task can be checked; and management constraints on the way in which the task is carried out.

Among the benefits given by this model is that it leads naturally into automation of the control of a project. For each task or contract, inputs and outputs are assigned, completion can be detected, and time scales and resource usage can be associated with the task. Mechanisms to support the definition, the letting, the initiating, the monitoring and the assessment of the results of contacts can readily be generated. This model provides the framework of one of the current integrated project support environments.[21]

46.5.1.7 The automation paradigm

There is a considerable pressure for software to be changed. Most software items need to be maintained once they have been created and put into use, and in some areas the investment on such maintenance dwarfs the investment on the creation of new software.

In a sense, the pressure for change is even greater, for much software which is nominally new can be viewed as a variant on existing software. Though the implemented code may be quite distinct, at the level of specification there may be much in common between two software items.

The automation paradigm supposes that it will become possible to automate the generation of the program from its specification (*Figure 46.11*). Clearly, this requires that the specification be expressed formally, and that it include not just the specification of function, but of the other attributes of the product which its user may need to determine. So the specification has to embrace the required performance and the manner of use of the product.

Given such a specification, and given the ability to refine it into executable code, it would become possible to maintain software at the specification level rather than at that of the code. Those errors due to the manual translation of an informal specification into a working program would disappear and the problems of reusability would be reduced, since the commonality of specification can then be exploited.

Such an approach is not yet possible. Fourth generation language systems used in commercial software system generation

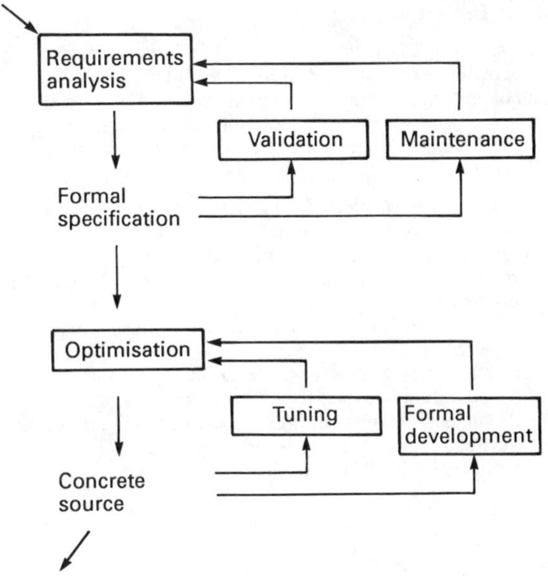

Figure 46.11 The automation paradigm (From Balzer *et al.*[22])

provide some of the advantages but are a long way from fully automated refinements of complete specifications. The automation paradigm does, however, present a goal for the end of the century.[22]

46.5.1.8 The dangers of life-cycles

Though the use of a life-cycle model provides benefits to the manager it also carries risks. Among these are the following, drawn from McCracken and Jackson:[23]

(1) *Inertia* A particular model may not suit all projects and may not fit with the introduction of new practices and tools. The model is no more than a management aid. It should block change only when the change is undesirable.
(2) *Exclusion of the user* It is a common feature of life-cycle models that the end-user appears to play only a minor role. There is a growing trend in some areas of software construction for users to be involved at all stages of development.

46.5.1.9 General process modelling

Life-cycle models provide simple, general-purpose descriptions of the process of developing or evolving a product. They represent structures which the manager may choose to impose on the project. They influence the way it is organised.

Descriptions of the process can also be used by automated management tools. Formal statements of the behaviour of people or of tools needed to carry out an activity of a given type can be created. One use of such descriptions is in planning, as all activities of the described type can be broken down into subtasks in a common way.

Potentially of more value is the ability to execute a process model. For each type of activity a model of the corresponding process is held on the computer, which is then used to prompt the people responsible for carrying out the tasks, and to initiate those which are automated as the necessary inputs become available.

The process model differs from a plan essentially in its generality. The model describes roles whereas the plan describes specific tasks.

46.5.2 Project planning

The usual methods, such as critical path analysis, and tools apply to the planning of a software project just as much as to any other kind of project. A project plan ought however to be more than a dependency network and a schedule. Jensen and Tonies[24] suggest a number of elements of a project plan, including the following:

(1) Technical description of the system, providing an overview of the end goal of the project.
(2) Organisation plan, stating the roles and responsibilities of each element of the organisation. (See, for example, the discussion of chief programmer teams in Brooks.[25])
(3) Methodology, summarising standards, techniques and codes of practice to be used.
(4) Configuration management plan, explaining the configuration and change control procedures.[26]
(5) Documentation plan, listing all documents to be produced.
(6) Test plan, saying what is to be tested, what is to be done when tests are failed, and what the acceptance criteria are.

46.5.3 Measuring

46.5.3.1 Software metrics

A software metric is a unit of measurement of some aspect of a software product or a software process. Examples of metrics include:

(1) *Size* The most common metric of software is the count of the number of program statements or lines of code. A weakness of this measure lies in the uncertainty as to what constitutes a line of code (should comments be counted or not, for example).
(2) *Complexity* There have been a number of metrics which indicate the complexity of the flow of control in a piece of code. McCabe's cyclomatic number[27] is a count of the number of regions in the control flow graph, which is the same as the number of linearly independent control paths in the program (*Figure 46.12*).

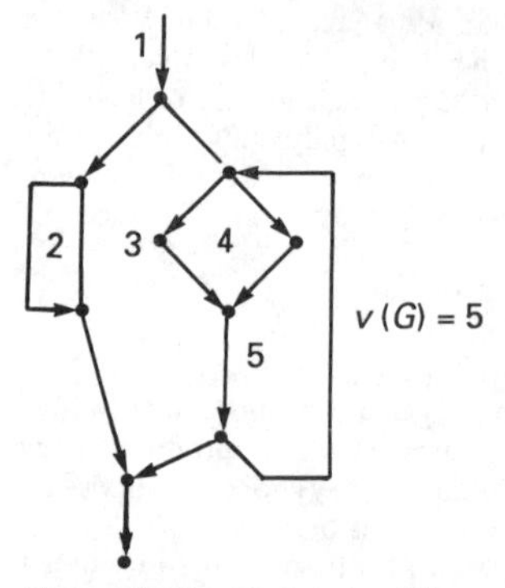

Figure 46.12 McCabe's $v(G)$

Albrecht's metric of function points[28] allocates points to known sources of complexity, such as inputs and outputs. This can offer an alternative measure of size, based more on the scale of the problem than on the particular encoding of its solution.

(3) *Cohesion and coupling* The former values the extent to which the complements of a software item logically belong together, and the latter values the extent to which components of different software items logically belong together. The recommendation is that cohesion should be high and coupling low.
(4) *Test coverage* Any test plan ought to set some target for the extent to which the system is to be tested. To exercise all source statements may be reasonable, to exercise all possible control paths is generally not. Woodward *et al.*[29] introduce the idea of 'linear code sequence and jump (LCSAJ)', and describe tools to assess coverage of testing based on the proportion of those which have been exercised.
(5) *Reliability* Measures of software reliability such as 'rate of occurrence of error reports' or 'mean time between failures' are sometimes used (see also *Section 46.5.3.3*).
(6) *Productivity* Synthetics such as 'lines of code per person-day' are sometimes quoted, but these are often misleading.

Watts[30] lists many more.

46.5.3.2 Cost models

Albrecht's function point analysis mentioned above provides for cost modelling, but perhaps the best known models are Boehm's COCOMO (Constructive Cost Model) and Putnam's SLIM.

COCOMO This comes in three flavours; basic, intermediate and detailed.[31] Basic COCOMO first requires the project to be classed as 'organic', 'semi-detached', or 'embedded'. Organic projects tend to have small experienced teams producing a small product, using a stable environment. Semi-detached projects have a mix of experience in their teams, and may be producing a larger sized product. Embedded projects work under tight constraints and require innovative solutions.

The effort required and the time to complete are then estimated in terms of the probable size of the product (MM = man-months, KDSI = thousands of delivered source instructions, TDEV = development time):

organic:	$MM = 2.4\ KDSI^{1.05}$	$TDEV = 2.5\ MM^{0.38}$
semi-detached:	$MM = 3.0\ KDSI^{1.12}$	$TDEV = 2.5\ MM^{0.35}$
embedded:	$MM = 3.6\ KDSI^{1.20}$	$TDEV = 2.5\ MM^{0.32}$

Intermediate COCOMO improves on these crude estimates by applying multipliers based on:

(1) Product attributes, such as the required reliability.
(2) Computer attributes, such as the performance constraints.
(3) Personnel attributes, such as programmer or designer capability.
(4) Project attributes, such as the degree of use of modern development practices.

Each attribute is scored and a corresponding multiplier is then applied to adjust the basic COCOMO estimates. Detailed COCOMO provides the further refinement that the multipliers are sensitive to life-cycle phases.

In each case, it is necessary for COCOMO to be calibrated to fit the organisation whose projects are being modelled, as no single attribute can be expected to have the same impact in two distinct organisations.

SLIM The Putnam model[32] applies the Rayleigh curve to the manpower profile (*Figure 46.13*). This is given by Equation (46.1):

$$\dot{y}(t) = K * t/t_d^2 * \exp(-t^2/2t_d^2) \tag{46.1}$$

where $y(t)$ is the manpower at time t, K is the total life-cycle effort and t_d is the development time.

46.5.3.3 Reliability models

A reliability model is an important scheduling tool, as it can be used to indicate when a product ought best to be released or how much maintenance effort it requires.

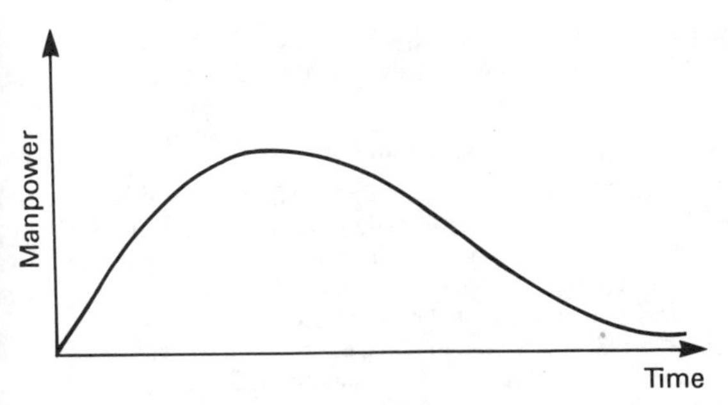

Figure 46.13 The Rayleigh curve

Hardware reliability modelling considers the probability of failure of a component, which is replaced and then carries some probability of failing again. In software reliability modelling it is assumed that the source of a failure can be removed for all time. Different models are therefore appropriate.

Two of the best known are the Jelinski–Moranda[33] and Littlewood–Verall models.[34] The Jelinski–Moranda model supposes that the system starts with a certain number of faults, that these are independent, and that they can be corrected independently. The failure rate is then proportional to the number of errors not yet corrected. The Littlewood–Verrall model supposes that time between failures follows an exponential distribution, parameterised by some variable related to the ability of the programmer and the difficulty of the task.

These and other software reliability models are discussed in Goel.[35]

46.5.4 Configuration management

Though no single life-cycle model necessarily fits all software developments, a common feature of the application of the models is that development is an ordered sequence of operations, where each operation has a tangible output acting as an input to its successor.

This means that a software product can exist in a number of representations at the same time. There may be a specification, a design text, source code for the program and loadable code, each defining the final product. It is a function of configuration management to ensure that each representation is consistent with all of the others.

Software products tend not to be stable; faults are corrected and enhancements are made. Configuration management makes it possible to record the status of any representation of a software item, to be able to say which changes have been applied to it and which have not.[36]

46.5.4.1 Baselines

A baseline is a fixed reference point in the software development. It is a good point from which change can be measured. If the project management is applying a life-cycle model to the development, the work will move through a series of phases, from one phase of the life-cycle to the next. It is conventional to draw a baseline at the end of each of the phases. The end of the phase will be marked by a completion of appropriate documentation or computer files. This tangible output becomes the baseline for the phase.

With most life-cycle models it becomes necessary to re-enter an earlier development phase. (A system design error may be discovered during the detailed design, for example, and so the system design has to be reworked.) If a phase is re-entered, the rework of that phase will generate a change to its baseline. A configuration management system will record established baselines and all changes to those baselines.

There is a difference between work done to establish a baseline and rework done in the same phase of the life-cycle after the baseline is established. The path taken to reach the baseline need not be recorded but each step thereafter is.

46.5.4.2 Configuration identification

Configuration identification establishes the composition of the software item (what other items it is made from) and the documents and files that define it.

Configuration items For all but the most simple software products, the product design activity will include breaking the product down into simpler components. As part of configuration management it has to be decided which components are to be given distinct formal identities and which are not. Those with formal identities are configuration items and are visible to the configuration management system.

Documentation plans Each project should operate a documentation plan identifying all configuration items created or used by the project, and for each configuration item identifying the documents and files which comprise the item. Documentation plans are usually simple and generic.

Parts lists and 'where-used' lists For each configuration item constructed from other configuration items a parts list has to be maintained. This identifies the constituents of the item and also the other items needed for its construction (such as the tools used to build it).

The approximate inverse of the parts list is the where-used list, allowing the consequences of changes to an item to be assessed.

Both parts lists and where-used lists are nominally maintained automatically by configuration management tools.

46.5.4.3 Configuration control

Change control is the administrative mechanism for requesting, preparing, evaluating and authorising change proposals. It is applied both to delivered products and to products under development. Once a baseline has been drawn for a configuration item, all changes to that baseline should be controlled.

The terms used in change control are not consistent across usage, but the following are representative:

(1) *Change request* A request identifies the effect required to be achieved by a change, the advantages and disadvantages of the change, and the urgency of the change as seen by the initiator of the request.
(2) *Change proposal* Proposals are responses to change requests, describing the way in which the change can be implemented and estimating its cost and ramifications.
(3) *Change control board* This is the formal or informal decision making body which determines whether and when a change will be made.
(4) *Change note* The note authorises the change, describes it, and says what other configuration items are affected by it.

46.5.4.4 Status accounting

There is usually a delay between the approval of a change proposal and the completion of implementation of the change. A status accounting mechanism maintains a record of how the product has evolved and how it relates, at any time, to the issued baseline documentation and the set of authorised changes. Status accounting keeps track of changes, recording which parts of the changes have been implemented.

The kinds of data which may be kept include the formal name of each configuration item, the baseline existing for each item, the dates at which baselines were set, who is responsible for the item, what change proposals affect each item, the status of each proposed change, the version number of each item.

In addition, copies of all change requests proposals and notes are held.

46.5.4.5 Configuration auditing

The purpose of auditing in configuration management is to ensure that change to one representation of the product (such as the loadable code) is matched by equivalent change to another (such as the design documentation), and in particular that the change is consistent with customer requirements.

In the later stages of software development several baselines may have been drawn; specifications, design descriptions, code and test packages may all exist. When a change is made it may affect several of the baselines. The auditing function checks that the baselines are in step and that compatible documentation exists.

46.6 Integrated project support environments

A programming environment is a collection of software tools designed to increase the productivity of programmers and the quality of the programs they produce. It provides a way of organising, integrating and controlling the use of such tools.

The term is used by analogy with the widely used term 'working environment', which signifies the facilities and general surroundings provided for workers. The tools in a programming environment are made available to a programmer through a workstation (personal computer or terminal). Since these tools are designed to help with the majority of programming activities, they and the workstation become a very significant part (though not the whole) of the programmer's working environment. *Section 46.6.1* describes such environments.

Although this use of the word 'environment' started in the late 1970s in the context specifically of programming environments, it was not long before it spread. It has since come to describe more extensive collections of tools, which support more activities in the development of software-intensive systems than just programming. In some cases it has been replaced by the more ambitious word 'factory'. These extensions to the original concept of a programming environment are described in *Sections 46.6.2* and *46.6.3*.

Section 46.6.4 provides a small number of basic references, from a literature which is already alarmingly large.

Whatever their scope, environments and factories share a common objective. It is to increase the amount of capital invested per systems/software engineer, to gain returns in productivity and quality comparable to those achieved by capital investment in traditional industry. Only limited gains, however, can be achieved by providing tool support for our current methods of systems and software engineering. The extensive gains we need can only be achieved by the widespread adoption of better methods, supported by adequate tools, organised within environments and factories.

Unlike current methods, it is probable that new advanced methods will not be usable without tool support, and that these complex tools in turn will not be usable unless they are organised within environments. There is thus an intimate relationship between the development of environments and the development of better methods of engineering practice. This is exactly analogous to the way in which industrialisation has gone hand in hand with improving standards of traditional engineering.

46.6.1 Programming environments

The idea of a programming environment first gained currency in the Ada language community, and gave rise to the notion of an Ada Programming Support Environment or APSE. It was based on the recognition (1) that programmers do more than write and test source code, (2) that they need more tools than the ones normally available, like compilers, editors, debugggers, test data generators, and (3) that the many new and existing tools need to be presented in a more coherent and integrated way to achieve economy of effort.

Integration was to be achieved by a common database to be shared by all tools, a common user interface for all tools, and networking workstations together so that members of project teams could communicate easily. These in turn were to be achieved by a high-level operating system, often called a 'kernel', which would manage the storage and movement of information and the invocation of tools, and standards to be adopted by tool writers, so that tools could easily be interfaced to the kernel, often referred to as a 'public tool interface' (PTI).

APSEs belong to a class of language-specific programming environments. This was soon generalised to the idea of language-independent programming environments, which would support programming in a range of languages, sometimes using general tools such as structure editors, operating on formal descriptions of the different languages. By contrast to APSEs, these were often called IPSEs (Integrated Programming Support Environments, a term particularly used in the UK).

By comparison with the more ambitious environments described below, programming environments might be judged rather mundane. Some people might prefer to call them toolkits or workbenches. Nevertheless a lot of advanced work has been, and continues to be, done on the subject of programming support. It includes capturing knowledge about the programming process and embodying it in expert systems. It also includes formal methods for preserving correctness while transforming specifications into executable code. Much of this work is done under the banner of programming environments.

46.6.2 Software engineering environments and factories

From the early 1980s in the UK, the P in IPSE changed its meaning and the term came to stand for Integrated Project Support Environment. At about the same time the term Software Engineering Environment (SEE) came into currency in the USA.

The two terms had very similar meanings. Both signified support for 'programming in the large' rather than for 'programming in the small'. That means shifting the balance of concern from detailed program writing and testing to the whole range of activities involved in the software process, and from relatively small individual programs to very large sets of programs; it also means a concern with the management of software activities.

Not long after, the term 'software factory' (SF) came into currency, mainly in the USA and Japan. The significance of this was a yet further broadening of concern, to include mechanised support for the ancillary activities that surround software development, production and maintenance (such as personnel, finance, distribution), and to include also a concern with the physical workplace of software workers.

The approach usually adopted in the design of software engineering environments or software factories is to make them generic, in the sense that they can be implemented across a range of hardware and operating systems, and can support a range of languages, methods, activities, etc. Indeed, it is often the case that they offer a selection of different toolkits or workbenches, from which users can select those that best fit their ways of working.

The software engineering strategy of the UK Alvey Programme of collaborative R&D in information technology was centred around the IPSE concept, and concurrently a few companies developed commercial IPSEs. (Examples of these include BIS/IPSE from BIS Applied Systems Ltd, Eclipse from Software Sciences Ltd, GENOS from GEC Software Ltd, ISTAR from Imperial Software Technology, and Perspective from Systems Designers plc.[37]) A major project in the Eureka programme of the European Community is the development of a software factory, and a number of individual companies in Japan, Europe and the USA are making serious efforts to develop this idea.

46.6.3 Systems engineering environments and factories

To complete the story, the final recent step has been to look for ways to broaden the scope of environments and factories still further, to support genuine systems engineering of software-intensive systems. The intention here is to incorporate tools for VLSI design and to provide support for the many problems in the design and implementation of complex systems which are not exclusively in the domains of either software or VLSI.

The end goal of the Alvey Software Engineering strategy is to develop what is called an Information Systems Factory (ISF). Similarly, in Europe, under Esprit II, a major project is called the Advanced Systems Engineering Environment. A further project is under way in the USA, entitled Arcadia.

These projects and others seem to agree that the kernel-and-PTI architecture of the early programming environments is inadequate to support their more ambitious goals. There is not yet a clear agreement, however, on the architecture to replace it. It is recognised that among the biggest problems to be solved are the organisation of the very large and very complex information stores necessary to support large systems throughout their lifetimes, protocols which will permit the interoperability of many different tools, and languages for modelling software activities so that they can be effectively described and supported.

46.6.4 Guide to references

Brooks[38] provides a widely admired and provocative discussion of current problems in software engineering, and of the difficulties in finding fundamental solutions. Charette[39] and Sommerville[40] give a good coverage of software engineering environments in their books; Dart *et al.*[41] and Stenning[42] do likewise within the confines of short papers. Hewett and Durham[43] give an excellent (if expensive) market survey. Buxton[44] gives the original and influential definition of broad requirements and architecture for a programming environment, and Lyons and Nissen[45] is a much more detailed discussion of requirements; both works nominally concentrate on the Ada programming language, but they are in fact entirely general in their applicability. Finally, Taylor *et al.*[46] give a statement and discussion of the characteristics of the proposed next generation of environments; it emanates from a group of US academics, but is also representative of advanced thinking in the UK and Europe.

46.7 Languages

To get a computer to do anything it is necessary to develop and perfect (debug) a program describing the operations the computer is to perform. Developing anything but the simplest program is a tricky and error-prone task. The choice of an appropriate programming language, and appropriate methods of program design, can considerably reduce these difficulties.

There is a bewilderingly wide choice of programming languages available to the potential programmer. This section describes the nature of programming languages in general, and indicates which languages are most suitable for particular problem areas.

46.7.1 Types of programming language

The only sort of program a computer can execute is one in machine code; that is, a pattern of ones and zeros in memory corresponding to instructions in the computer's instruction set. Programming languages provide a convenient way of writing a program so that it can be understood by both people and the computer.

There are two main types of programming language; assembly languages and high-level languages.

46.7.1.1 Assembly languages

Assembly languages are the most direct way of programming a computer, short of actually directly setting up the machine code as a bit pattern in the computer's memory. Statements in a simple assembly language are associated one to one with instructions in the computer's instruction set, or with reserving memory for data. More sophisticated assembly languages provide a macro facility, so that commonly used sequences of instructions can be defined once and then included by name. A fragment of the assembly language for the IBM 360/370 mainframe looks like this:

```
X  DC  F'5'  declare a 32-bit variable called X preset to 5
Y  DC  F'7'  declare a 32-bit variable called Y preset to 7
Z  DS  F     declare a 32-bit variable called Z

   L   8,X   load variable X into register 8
   L   9,Y   load variable Y into register 9
   AR  8,9   add register 9 to register 8
   ST  8,Z   store register 8 into variable Z
```

Although this simple example was comparatively straightforward, evaluating a complex formula rapidly becomes cumbersome; and translating between integer and real numbers can be very messy indeed. Obviously anyone programming in assembler needs a detailed knowledge of the computer's instruction set,[47] and programs written in assembler will need to be completely rewritten if the program has to be moved to another type of computer.

Assembly languages can be fun to use, because they provide access to the computer's entire instruction set. This makes it possible, in principle, to write very 'tight' and efficient code. In practice large assembly language programs are often rather inefficient, as the amount of detail required obscures the overall picture from the programmer.

Current industrial opinion suggests that programming in assembler should be avoided if at all possible. If it is absolutely necessary to use assembler (e.g. to use computer functions not accessible from high-level languages), write the bulk of the program in a high-level language and restrict the use of assembler to a few subroutines.

46.7.1.2 High-level languages

High-level languages provide a convenient notation for describing algorithms, without worrying about how they might be expressed in machine code. As a bonus the high-level language support system can detect and flag a number of types of minor mistakes in the program.

The internal structure of most high-level languages (HLLs) fall into the general structure described below.

Variable declarations Variables are used to hold the intermediate results of calculations. In most HLLs a variable can only hold a particular type of value. The two most important types of variable are integers and reals; they might be declared like this:

a : integer; defines a single integer called a.
x : real; defines a single real called x.
y : ARRAY of real; defines ten reals called $y[1]$, $y[2]$, etc. Arrays are invaluable for representing vectors and matrices.

Integers can hold a wide range of integer values, while reals can hold only an approximation to a real number.

Assignment The value of a variable is set by an assignment, which is indicated by a symbol such as :=. Continuing the example above, by assigning values to the variables a and y:

$a := 2;$ sets the integer variable a to 2.
$x := 57.8;$ sets the real variable x to 57.8.
$a := 57.8;$ is illegal, as a cannot hold a real value.
$x := 2;$ is however quite legal, as x can hold an approximation to 2.

Expressions Expressions are written as in normal mathematics, and generally have the expected value.

$a := 4*7-3;$ sets a to 25, i.e. (4 times 7) minus 3.
$x := 1.0-(3.0/3.0);$ in an ideal world this would set x to zero, but actually x will be set to a very small value because of the approximate nature of reals and rounding errors which occur when a computer performs real (floating point) calculations.
$x := \sin(3.14);$ as most HLLs support the normal maths functions.
$a := 2; y[a] := 2.2;$ sets the second element of the array y to 2.2.

Directing control flow In order to write useful programs, it is necessary to be able to make decisions and perform repetitive operations. For example, decisions may be expressed in a simple if ... then ... else form, and repetition using a FOR loop, as shown below:

IF $x<2.8$ THEN $a:=2$ ELSE $a:=3$;
FOR $a:=1$ to 10 DO $y[a]:=1.0$; which sets all ten elements of the array y to 1.0.

Many languages also offer specialised ways of expressing more complicated forms of decision and repetition.

Procedures and functions Procedures allow a sequence of instructions to be defined and then simply used. A procedure to swap the values of two variables might look like this:

```
PROCEDURE swap (x,y);
VAR    x,y,z : real;
BEGIN
       z:=x;
       x:=y;
       y:=z
END;
```

This could then be invoked by simply saying

```
swap (a,b);
```

Notice that the procedure definition is written in terms of formal (i.e. symbolic) parameters, and that the actual parameters that apply for an invocation are given when the procedure is invoked.

Many languages provide built-in functions such as $\sin(x)$; and functions differ from procedures in that they 'return a result'. This enables functions to be used within assignments and expressions, and combined in a variety of ways:

$y := \log(\sin(x)) * \cos(x);$

Input and output All programs require some form of input and output. They may use facilities defined in the language, or a library of 'standard' procedures. Most languages provide sufficient facilities for simple programs.

For some applications, the ability to interface with a database or produce graphical output is becoming increasingly important. Such facilities are rarely provided within a language, but the availability of a suitable library may be important in the selection of a language and its supplier.

46.7.2 Ways of supporting programming languages

There are basically two ways that a programming language can be supported; direct translation to machine code, and interpretation.

46.7.2.1 Direct translation to machine code

In this approach a computer program is used to convert the program as expressed in the programming language to an equivalent program in machine code. If the programming language is an assembly language, this special program is called an assembler. If the programming language is a high-level language, the special program is called a compiler.

Assemblers are quite easy to produce, as the statements in a simple assembly language correspond one to one with instructions in the computer's instruction set, or with reserving memory for data. More sophisticated assembly languages with macro facilities present little extra difficulty to the developer.

A compiler for a high-level language usually detects and flags a number of types of minor mistakes in the program, as well as translating the program into an equivalent machine code program. Compilers can generate very efficient code, comparable to or better than that achieved by most assembler programmers. In addition the program can be easily run on any computer for which a compiler for the chosen language exists.

The picture is not entirely rosy. The first snag is that compilers which generate efficient code can be difficult and expensive to produce. So a compiler may not exist for the desired machine, or it might be rather expensive, or it might not be very efficient. The second snag is that it is not too easy to provide debugging tools that work in terms of the high-level language rather than the generated code.

46.7.2.2 Interpretation

Interpretation is a technique often used to support high-level languages. Instead of the high-level language being converted into machine instructions for a particular computer, it is converted into instructions for an imaginary computer. This instruction set is often chosen to be quite close to the high-level language itself. A small interpreter is then written, which emulates the imaginary computer on the real machine. Generating the code for the imaginary machine is far easier than generating code for a real machine, and the interpreter is usually quite simple to write.

This approach has considerable advantages. Firstly, it is much easier to provide support for language on other machines. Writing an interpreter for a new machine is very much easier than writing an efficient compiler code-generation phase for it. Secondly, as the instruction set for the imaginary computer is much closer to the high-level language than the instruction set for most real computers, it is much easier to provide debugging tools that work in terms of the high-level language.

There is no point in supporting an assembly language by interpretation except in special circumstances, such as during the hardware design and development of a type of computer.

46.7.2.3 Integrated environments

When developing a program with a classical compiler-based support environment, the programmer will repeatedly use an editor to create the program text, run the compiler to generate error messages, reuse the editor to correct the source code, etc. Probably a paper copy of the compiler output will be required so that the error messages can be examined while the program text is being edited. Integrated environments (e.g. Turbo-Pascal) let the programmer concentrate on the editing. The compilation can be invoked from the editor, and the error messages are inserted at appropriate positions in the source text. The editor may also skip directly to the places where the errors occur.

Integrated environments frequently offer additional assistance to the debugging process.

46.7.3 Selecting a language

46.7.3.1 Recognise your needs

As in most activities, the really hard part in deciding on which computer language to use is identifying your real needs. The major needs of the intending programmer are probably good educational material, access to a community with similar problems, and libraries of useful routines.

The best substitute for 20:20 foresight is the advice of experienced users who have the same sort of problems to solve. They will probably have already written (or acquired) all sorts of useful bits of code.

46.7.3.2 Understanding what is available

Without access to experienced users this is difficult, and the advice is very dependent on the sort of computer you are intending to use. If you are going to be using a large mainframe, the local computer centre or system programmer is an obvious contact. They should be able to put you in contact with people with similar problems, or at least explain what is available on the computer and obtainable for it. Probably the choice will be quite limited; traditionally language support on mainframes has been restricted to little more than that available from the manufacturers, and possibly a user group.

If you are going to be using a small computer, especially the IBM PC or a clone, then a very wide range of languages is available, often quite cheaply. Probably the best place to start looking is in the magazines. Shareware, a new method of software publishing and distribution, provides a particularly cheap way to investigate many interesting languages.

Those unfamiliar with programming are strongly urged to learn one of the educational languages such as Pascal or Modula-2 first. Those already familiar with just one language will gain considerable insight by learning another.

46.7.4 Glossary of programming languages by application area

Given sufficient ingenuity and skill it is possible to address almost any problem in almost any programming language. However, workers in various application areas have tended to adopt particular languages. The popularity of a language in an application area tends to be self-sustaining as educational material and libraries of useful code relating to the application area become available.

46.7.4.1 Engineering calculations

Performing scientific and engineering calculations was one of the first areas in which computers were used. FORTRAN (from FORmula TRANslation) was the first high-level language to emerge; and its compilers are often notable for the efficiency of the code they generate. It has gradually evolved over the years, FORTRAN-77 being the latest version currently available.[48] Although not very highly regarded in computer science circles, FORTRAN maintains a pre-eminent position among those performing engineering calculations. Electrical engineers may find the fact that FORTRAN directly supports complex arithmetic, another point in its favour. A factor in FORTRAN's continuing success is existence of a large body of reusable FORTRAN code.

Many engineering calculations are much trickier than might be imagined. Programming apparently simple numerical calculations such as matrix inversion or solving simultaneous equations with a computer actually demands very special skills. The effects of performing the calculations with real, finite precision, floating point arithmetic can often render the results of the computation worthless; and finding an efficient, accurate solution represents an entire field of study called numerical analysis.

Fortunately, the results of this work are often freely available as published FORTRAN programs. A particularly good collection of fundamental numerical methods is presented in Press *et al.*,[49] which includes source code in both FORTRAN and Pascal. Last but by no means least, the FORTRAN linear equation package LINPACK is available at a nominal price from public domain software suppliers. The continuing popularity of FORTRAN is heavily influenced by the availability of such software.

There are few languages which can challenge FORTRAN for engineering calculations. The general-purpose language Pascal[50] is suitable for simpler calculations. Pascal-SC, a recent derivative of Pascal, may be significantly better for numerical calculations. The US DoD-supported language Ada[52] may also be used, and some FORTRAN libraries are being recoded in Ada. APL[53] represents a completely different sort of challenge, being a special-purpose array-processing language; it has a small but dedicated following among engineers. FORTH[54] is another oddity worthy of note, which defies a succinct description.

FORTRAN's main defect is that its use does not encourage the development of good program design and programming style. Although engineers may be wise to do their serious application programming in FORTRAN, they would be well advised to learn an educational language such as Pascal to provide additional insight into the programming process.

46.7.4.2 Data processing

COBOL, the dominant data processing language, was developed to meet the US Navy's data processing needs. It continues to be used in most accountancy and stock-control programs. More recently the so-called fourth generation languages (4GLs) have found a place, especially in areas such as report generation and data entry and validation. Often 4GLs are supported not by a compiler, but by a program which translates them to equivalent COBOL programs.

46.7.4.3 Educational

Several languages have been developed as useful vehicles for teaching the elements of program design and programming. Notable among these are Algol,[55] Pascal[56] and Modula-2,[57] and many excellent books on programming have used these languages for illustrative purposes. Unfortunately several features of these languages, which encourage well structured and thoughtful programming, limit the ability of the languages to tackle some types of application.

46.7.4.4 System programming

System programming includes the writing of operating systems and the production of the utility software needed to keep a computing system operational. Initially systems programming was done in the assembly language of the relevant computer. More recently those developing operating systems have developed languages to meet their needs (e.g. PL/S, S3, PL/M, C).

The only one of these generally available is C, the language used to develop the UNIX* operating system.[57] An evolutionary development C++, although less readily available, is probably to be preferred.[58]

The educational language Modula-2 was developed partly as an exercise in language design and partly to implement the operating system for the Lilith computer.[56] Implementations are becoming readily available.

46.7.4.5 Real-time languages

Computers have always been invaluable in process control and other 'embedded' applications. The languages used have always been rather specialised, not least because the real-time nature of the applications introduces problems not normally encountered in other application areas.

ICI have often used RTL/2 for process control. At one time CORAL[59] was widely used for military and telecommunications applications. The CCITT have developed CHILL[60] for use in telecommunications equipment. However, compilers for these languages are not readily available.

The US DoD developed Ada[61] for use in airborne equipment, partly in response to the discovery that its suppliers were using over 400 different (and obscure) languages for the purpose. Ada has taken considerably longer to implement than originally envisaged, although compilers are becoming available. During the gap between its initial design and availability, some aspects of the Ada language have been questioned.[62]

The language occam[63] has been developed in conjunction with the design of the transputer, and will probably be the language of choice for most people programming this computer.

The more readily available system programming languages are worthy of consideration by those contemplating real-time work; and a radically different alternative is presented by FORTH,[54] which was originally designed to implement the control program for a radio telescope.

46.7.4.6 Declarative languages

Declarative languages are chiefly of interest to computer scientists and workers in artificial intelligence. Programs written in ordinary imperative languages concentrate on describing the algorithm whose execution produces the desired result. Program steps written in a declarative language concentrate on specifying the nature of the required solution. Declarative languages are likely to be of only academic interest to readers of this handbook, not least because they tend to be heavy users of computer resources.

Declarative languages can be divided into relational and functional languages. By far the best known relational language is PROLOG.[64,65] A PROLOG 'program' consists of a series of logical assertions. Given these assertions, the PROLOG 'inference engine' tries to answer questions. For example, and ignoring syntactic details, given the assertions:

Turing is human
Socrates is human
Socrates is Greek

The PROLOG inference engine will answer Socrates to the question

y is human and *y* is Greek?

Functional languages result from an attempt to make programs more like mathematical statements. One effect is to try to replace the use of assignment by functional definition. The best known function language is LISP,[66] which was modelled after the mathematical λ-calculus; a rather daunting introduction to what most people would regard as a list processing language. Another important reference is FP, a language which was developed to illustrate this alternative approach to programming.[67] Hope is another representative of the functional languages.[68]

It is arguably true that where the declarative languages have remained purely declarative they have been found wanting in expressive power. Attempts to increase their expressive power by diluting their declarative style have been a mixed success (look up 'cuts' in the index of a book about PROLOG, or ask a PROLOG programmer).

46.7.4.7 Text and string manipulation

The special problems of text and string manipulation have resulted in a variety of specialist languages. The languages awk and sed are frequently used by programmers of UNIX systems, particularly for tasks such as report generation. Snobol was an early and influential language in this area.[69] Its author has gone on to produce another, more structured, language Icon.[70] Implementations of most of these languages are available.

46.7.4.8 Object-oriented languages

Object-oriented languages represent a different approach to structuring large programs, and the provision of libraries of reusable code.[71] There is still much discussion on terminology.[72]

Simula was probably the first object-oriented language, but was little used outside the simulation community. Smalltalk, which combines a language, an integrated environment, and a large library of useful objects, is used mainly in the construction of rapid prototypes.[73,74]

More recent object-oriented languages include C++, Objective-C and Eiffel. C++ is a 'better C' which includes object-oriented facilities.[58] Objective-C is a Smalltalk/C hybrid which is distributed with a substantial library.[71] Eiffel provides object-oriented facilities in a more Pascal-like style; its developer has published many important and readable papers on aspects of language design.[75]

*UNIX is a trademark of AT&T in the USA and other countries.

References

1 ANSI/IEEE Standard 830-1984, *Software Requirements Specification* (1984)
2 ROSS, D. T., 'Applications and extensions of SADT', *IEEE Computer*, **18**(4), April (1985)
3 JONES, C. B., *Systematic Software Development Using VDM*, Prentice-Hall, Englewood Cliffs, NJ (1986)
4 HOARE, C. A. R., *Communicating Sequential Processes*, Prentice-Hall, Englewood Cliffs, NJ (1985)
5 WIRTH, N., 'Program development by stepwise refinement', *Comm. ACM*, **14**(4), April (1971)
6 YOURDON, E. and CONSTANTINE, L. L., *Structured Design*, Yourdon Press (1975)

7 CAMERON, J. R., 'An overview of JSD', *IEEE Trans. Software Eng.*, **SE-12**(2), February (1986)

8 CHEN, P., 'The entity–relationship model: toward a unified view of data', *ACM Trans. Database Syst.*, **1**(1), March (1976)

9 BOOCH, G., 'Object-oriented development', *IEEE Trans. Software Eng.*, **SE-12**(2), February (1986)

10 ANSI/IEEE Standard 983-1986, *Software Quality Assurance Planning* (1986)

11 ANSI/IEEE Standard 1012-1986, *Software Verification and Validation* (1986)

12 FAGAN, M. E., 'Advances in software inspections', *IEEE Trans. Software Eng.*, **SE-12**(7), July (1986)

13 KITCHENHAM, B. A. and KITCHENHAM, A. P., 'The effects of inspections on software quality and productivity', *ICL Tech. J.* (1986)

14 MEYERS, G. J., *The Art of Software Testing*, Wiley, New York (1979)

15 BIGNELL, V., PETERS, G. and PYM, C., *Catastrophic Failures*, The Open University Press, Milton Keynes (1977)

16 STANLEY, H. E. and OSTROIOSKY, N. (Eds) *On Growth and Form: Fractal and Non-fractal Patterns in Physics*, Martinus Nijhoff, Dordrecht (1986)

17 CHAMBERS, J. M., CLEVELAND, W. S., KLEINER, B. and TUKEY, P. A., *Graphical Methods for Data Analysis*, Wadsworth, Belmont, Calif. (1983)

18 ROYCE, W. W., 'Managing the development of large software systems: concepts and techniques', *Proc. WESCON*, August (1970)

19 GILB, T., 'Deadline pressure: how to cope with short deadlines, low budgets and insufficient staffing levels'. In Kugler, H. J. (Ed.), *Proc. Info. Processing 86*, Elsevier, Amsterdam (1987)

20 BOEHM, B. W., 'A spiral model of software development and enhancement'. In *Proc. Int. Workshop Software Process Software Environments, Software Eng. Notes*, **11**(4), August (1986)

21 LEHMAN, M. M., 'Approach to a disciplined development process—the ISTAR integrated project support environment'. In *Proc. Int. Workshop Software Process Software Environments, Software Eng. Notes*, **11**(4), August (1986)

22 BALZER, R., CHEATHAM, T. E. and GREEN, C., 'Software technology in the 1990s: using a new paradigm', *Computer*, November (1983)

23 McCRACKEN, D. D. and JACKSON, M. A., 'Lifecycle concept considered harmful', *Software Eng. Notes*, **7**(2), April (1982)

24 JENSEN, R. W. and TONIES, C. C., *Software Engineering*, Prentice-Hall, Englewood Cliffs, NJ (1979)

25 BROOKS, F. P., *The Mythical Man-Month—Essays on Software Engineering*, Addison-Wesley, Wokingham (1975)

26 ANSI/IEEE Standard 828-1983, *Software Configuration Management Plans* (1983)

27 McCABE, T. J., 'A complexity measure', *IEEE Trans. Software Eng.*, **SE-2**(4) (1976)

28 ALBRECHT, A. J. and GAFFNEY, J. E., 'Software function, source lines of code and development effort prediction: a software science validation', *IEEE Trans. Software Eng.*, **SE-9**(6), November (1983)

29 WOODWARD, M. R., HEDLEY, D. and HENNELL, M. A., 'Experience with path analysis and testing programs', *IEEE Trans. Software Eng.*, **SE-6**(3) (1980)

30 WATTS, R., *Measuring Software Quality*, NCC Publications (1987)

31 BOEHM, B., *Software Engineering Economics*, Prentice-Hall, Englewood Cliffs, NJ (1981)

32 PUTNAM, L. and FITZSIMMONS, A., 'Estimating software costs', *Datamation*, September, October and November (1979)

33 JELINSKI, Z. and MORANDA, P. B., 'Software reliability research'. In Freiberger, W. (Ed.), *Statistical Computer Performance Evaluation*, Academic Press, New York (1972)

34 LITTLEWOOD, B. and VERRALL, J. L., 'A Bayesian reliability growth model for computer software', *J R. Stat. Soc., Series C*, **22**(3) (1973)

35 GOEL, A. L., 'Software reliability models: assumptions, limitations and applicability', *IEEE Trans. Software Eng.*, **SE-11**(12), December (1985)

36 BERSOFF, E. H., HENDERSON, V. D. and SIEGEL, S. G., 'Software configuration management: a tutorial', *Computer*, January (1979)

37 *The STARTS Guide*, NCC Publications, Manchester (1987)

38 BROOKS, F. P., 'No silver bullet—essence and accidents of software engineering'. In Kugler, H. J. (Ed.) *Information Processing 86*, North-Holland, Amsterdam (1986)

39 CHARETTE, R. N., *Software Engineering Environments: Concepts and Terminology*, McGraw-Hill, Maidenhead (1986)

40 SOMMERVILLE, I. (Ed.), *Software Engineering Environments*, Peter Peregrinus (1986)

41 DART, S. A. *et al.*, 'Software development environments', *IEEE Computer*, **20**(11), November (1987)

42 STENNING, V., 'On the role of an environment'. In *Proc. 9th Int. Conf. Software Eng.*, IEEE Press (1987)

43 HEWETT, J. and DURHAM, A., *Computer Aided Software Engineering: Commercial Strategies*, Ovum Limited (1987)

44 BUXTON, J. N., *Requirements for Ada Programming Support Environments: 'Stoneman'*, US Department of Defense (1980)

45 LYONS, T. G. and NISSEN, J. C. D., *Selecting an Ada Environment*, Cambridge University Press, Cambridge (1986)

46 TAYLOR, R. N. *et al.*, *Next Generation Software Environments: Principles, Problems and Research Directions*, Department of Information and Computer Science, University of California at Irvine, Technical Report 87-63 (1987)

47 MORGAN, C. L. and WAITE, M., *8086/8088 16-bit Microprocessor Primer*, BYTE/McGraw-Hill, Maidenhead (1982)

48 ELLIS, T. M. R., *A Structured Approach to FORTRAN 77 Programming*, Addison-Wesley, Wokingham (1982)

49 PRESS, W. H., FLANNERY, B. P., TEUKOLSKY, S. A. and VETTERLING, W. T., *Numerical Recipes: the Art of Scientific Computing*, Cambridge University Press, Cambridge (1986)

50 GROGONO, P., *Programming in Pascal*, Addison-Wesley, Wokingham (1978)

51 KULISCH, U., ULLRICH, C. and KAUCHER, H. E., *Pascal-SC: a Pascal Extension for Scientific Computation*, Wiley-Teubner (1987)

52 FORD, B., KOK, J. and ROGERS, M. W. (Eds), *Scientific Ada*, Cambridge University Press, Cambridge (1986)

53 IVERSON, K. E., *A Programming Language*, John Wiley, New York (1962)

54 BRODIE, L., *Starting FORTH, an Introduction to the FORTH Language and Operating System for Beginners and Professionals*, Prentice-Hall, Englewood Cliffs, NJ (1987)

55 COLIN, A. J. T., *Programming and Problem Solving in Algol 68*, Macmillan (1978)

56 PLATT, R. and KNEPLEY, E., *Modula-2 Programming*, Renson (1985)

57 KERNIGHAN, B. W. and RITCHIE, D., *The C Programming Language*, Prentice-Hall, Englewood Cliffs, NJ (1978)

58 STROUSTRUP, B., *The C++ Reference Manual*, Addison-Wesley, Wokingham (1986)

59 HALLIWELL, J. D. and EDWARDS, T. A., *A Course in Standard CORAL66*, NCC Publications, Manchester (1977)

60 CCITT, *Introduction to CHILL*, CCITT, Geneva (1980)

61 PYLE, I. C., *The Ada Programming Language*, Prentice-Hall, Englewood Cliffs, NJ (1981)

62 CLARKE, L. A., WILEDEN, J. C. and WOLF, A. L., 'Nesting in Ada programs is for the birds', *ACM SIGPLAN Notices*, **15**(1), 139–45 (1980)

63 INMOS LTD, *Occam Programming Manual*, Prentice-Hall, Englewood Cliffs, NJ

64 CLOCKSIN, W. F. and MELLISH, C. S., *Programming in PROLOG*, 2nd edn, Springer-Verlag, New York (1984)

65 MOREIN, R., 'PD PROLOG', *BYTE*, October, 155–65 (1986)

66 STEELE, G. L., *Common LISP: The Language*, Digital Press, Bedford, Mass. (1984)

67 BACKUS, J., 'Can programming be liberated from the von Neuman style? A fundamental style and its algebra of programs', *Comm. ACM*, **21**(8), 613–41 (1978)

68 BAILEY, R., 'A Hope tutorial', *BYTE*, August, 235–58 (1985)

69 GRISWOLD, R. E., POAGE, J. F. and PLONSKY, I. P., *The Snobol 4 Programming Language*, Prentice-Hall, Englewood Cliffs, NJ (1971)

70 GRISWOLD, R. E. and GRISWOLD, M. T., 'An Icon tutorial', *BYTE*, October, 167–78 (1986)

71 Cox, B., *Object-Oriented Programming: An Evolutionary Approach*, Addison-Wesley, Wokingham (1986)

72 WEGNER, P., 'Dimensions of object-based design', *ACM SIGPLAN Notices*, **22**(12), 168–82 (1987)

73 GOLDBERG, A. and ROBSON, D., *Smalltalk-80: The Language and Its Implementation*, Addison-Wesley, Wokingham (1983)
74 BUDD, T., *A Little Smalltalk*, Addison-Wesley, Wokingham (1987)
75 MEYER, B., 'Genericity versus inheritance', *SIGPLAN Notices*, **21**(11), 391–405 (1986)

Further reading

BATE, G., *The Official Handbook of MASCOT*. Joint IECCA and MUF Committee on MASCOT (JIMCOM), June (1987)
BELL, D. A., 'Issues in relational database performance', *Data Knowledge Eng.*, **3**(1), August (1988)
BOEHM, B. W. *et al.*, 'Understanding and controlling software costs', *IEEE Trans. Software Eng.*, **SE-14**(10), October (1988)
BRANQUART, P. *et al.*, 'Algorithmic languages', *Philips J. Res.*, **43**(3–4) (1988)
DAUGHTREY, T., 'The search for software quality', *Qual. Prog.*, **2**(11), November (1988)
DAVIS, A. H. *et al.*, 'A strategy for comparing alternative software development life cycle models', *IEEE Trans. Software Eng.*, **SE-14**(10), October (1988)
ELLIOTT, J. J., 'Incorporating formal methods into software development practice', *Safetynet*, October/November/December (1988)
FAGAN, M. E., 'Design and code inspections to reduce errors in program development', *IBM Systems J.*, **15**(3), 182–211 (1976)
FALK, H., 'AI techniques enter the realm of conventional languages', *Comp. Design*, 15 October (1988)
FALK, H., 'Software vendors serve up varied palette for CASE users', *Comp. Design*, 1 January (1989)
GEARY, K., 'The practicalities of introducing large-scale software re-use', *IEE Software Eng. J.*, **3**(5), September (1988)
GRETTON-WATSON, P., 'Distributed database development', *Comp. Commun.*, **11**(5), October (1988)
HASS, F. A., 'Common tools are a must', *Comp. Systems Eur.*, January (1989)
HETZEL, W., *The Complete Guide to Software Testing*, Collins (1984)
JOYCE, E. J., 'Reusable software: passage to productivity?', *Datamation*, 15 September (1988)
KILPATRICK, P. *et al.*, 'Software support for the refinement of VDM specifications', *Lecture Notes Comp. Sci.*, **328** (1988)
MARTIN, R., 'Evaluation of current software costing tools', *ACM SIGSOFT Software Eng. Notes*, **13**(3), July (1988)
OULD, M. A. and UNWIN, C. (Eds), *Testing in Software Development*, Cambridge University Press, Cambridge (1986)
PRESSMAN, R. S., *Software Engineering: A Practitioner's Approach*, McGraw-Hill, Maidenhead (1982)
QUESME, P. N., 'Individual and organisational factors and the design of IPSEs', *Comp. J.*, October (1988)
SHEAR, D., 'CASE shows promise but confusion still exists', *EDN*, 8 December (1988)
SPIVEY, J. M., *Understanding Z*. Cambridge University Press, Cambridge (1988)
TERRY, C., 'CASE tools', *EDN*, 28 April (1988)
THOMPSON, B. J., 'Standardizing test data formats', *Test Measure. World*, October (1988)
YU, T. J. *et al.*, 'An analysis of several software defect models', *IEEE Trans. Software Eng.*, September (1988)

47 Digital System Analysis

F F Mazda,
DFH, MPhil, CEng, FIEE, DMS, MBIM
STC Telecommunications Ltd

Contents

47.1 Introduction

This chapter describes the equipment used to analyse digital circuits. These cover the range from the simple logic pulsers and probes, which are used for hardware testing one node at a time, through to the more comprehensive testers such as signature analysers, logic analysers and development systems for software-based products.

47.2 Probes, pulsers and clips

A logic probe is a hand-held device, very similar in shape to a pencil with a very fine point. It detects the voltage level at any point of a circuit to which it is connected. *Figure 47.1* shows the block diagram of a probe. A switch on the probe can be set to measure different logic levels, such as for TTL or CMOS. Usually the maximum voltage is 20 V and input protection circuitry is used to shield the probe from high voltage or current levels.

The probe detects whether the input voltage is above or below the threshold levels for the logic family chosen. Some probes have a single indicator light, and this is on or off for the two logic levels. 'Bad' logic levels cause the light to be at half brightness. Some probes use multicolour indicators to show the different logic states. A pulse stretcher is usually also available which can stretch pulses of down to about 10 ns, so that they can be indicated. Pulse trains up to 50 MHz are also stretched to give an indicator blink at about 10 Hz.

Logic clips are designed to fit over integrated circuits, and they have indicators for each pin of the device. The logic clip draws power from the pin supply, and it contains its own gating logic for finding the supply pins automatically.

A logic pulser is a pulse generator, which is also shaped as a hand-held pen with a fine point. It is capable of inserting voltage and current signals at any node of the circuit, and is usually used in conjunction with logic probes and logic clips. The pulser is capable of producing a high current output, greater than half an ampere, and this is sufficient to override existing logic levels at the nodes to which it is applied. Because the current pulse is applied for less than 300 ns its energy is low, and it does not damage any circuit components. Controls on the body of the pulser enable it to operate in several modes, such as single shot, pulse train or pulse burst.

The logic current tracer is similar to the logic probe in appearance, but it senses pulsing current flow in a circuit, rather than voltage levels. Its narrow tip senses the magnetic field caused by the current, and it can cover a typical range of 1 mA to 1 A. An indicator light turns on when current flow is detected. The device can be used to sense current in multilayer boards, and its sensitivity is adjusted by a control on its side.

Another device, which is mainly used to detect faulty components in a logic circuit, is known as a logic comparator, shown in *Figure 47.2*. It works on the principle of applying an identical series of stimuli to a reference component and the component under test, and indicating a fault if the output response from both is not identical. The reference device is usually plugged into the body of the logic comparator, and a probe clip is available for fixing over the device under test.

47.3 Signature analysis

Signature analysis techniques, and signature analyser equipment, were mainly developed for use in the repair of faulty equipment, especially in the field. When troubleshooting an analogue system the engineer usually has a circuit diagram, on which the voltage levels and waveforms at various nodes are indicated. By tracing through the equipment, and checking the nodes, the source of the fault can be located and repaired.

In digital circuits the signal at the nodes consists of logic ones and zeros. The data stream at the test points can be very complex and the faults may be caused by nodes being stuck at logic one or logic zero, and by timing errors. Analysis can now be done by applying a suitable stimulus to the circuit, and then recording the data output in a convenient form. Signature analysers usually convert the bit stream at the different nodes into a small number of hexadecimal digits, which forms the 'signature' of the circuit at that node. If the circuit diagram is annotated with the signatures expected from a good circuit, the equipment nodes can be tested in turn until the node having the defective signature is found.

Signature analysis should not be confused with transition analysis, which is a long-established method of debugging faulty digital equipment. In this latter method the changes in the logic states at the nodes are counted for a fixed time period and then compared to the value expected from a good circuit. It will be seen in *Section 47.3.2* that signature analysis is much more effective in detecting faults than transition counting.

47.3.1 Principle of signature analysis

The signature analysis technique is based on the use of cyclic redundancy check (CRC) code produced by a pseudo-random binary sequence generator. A pseudo-random binary sequence is a sequence of logic ones and zeros which looks random, but which in practice repeats after a period. A pseudo-random binary generator having a bit size of q will produce a sequence of $(2^q - 1)$ bits before repeating. These will cover all the possible states with q bits, except all zeros.

Figure 47.3 shows a 3-bit pseudo-random sequence generator. The modulo-2 summation gate produces an output of logic zero only when the modulo-2 sum of all its inputs is zero. *Figure 47.4* shows the operation of this circuit, assuming that the input data

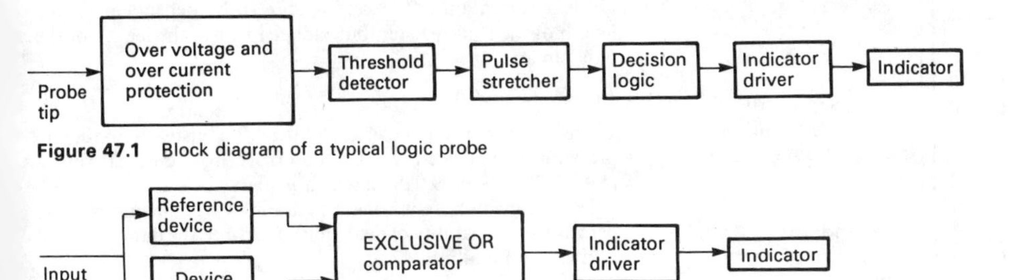

Figure 47.1 Block diagram of a typical logic probe

Figure 47.2 Block diagram of a logic comparator

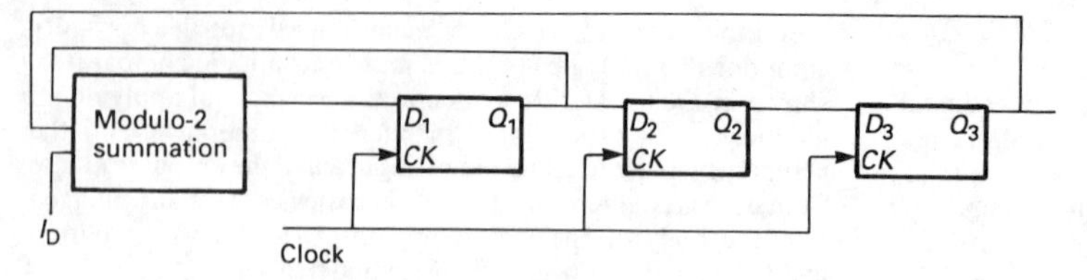

Figure 47.3 A 3-bit pseudo-random binary sequence generator

Clock	D_1	Q_1	Q_2	Q_3	
0	0	0	1	0	Initial state
1	1	0	0	1	Pseudo-random sequence (all possible states except all zeros)
2	1	1	0	0	
3	1	1	1	0	
4	0	1	1	1	
5	1	0	1	1	
6	0	1	0	1	
7	0	0	1	0	
8	0	0	0	1	Repeat

Figure 47.4 Sequence table for the pseudo-random binary sequence generator of *Figure 47.3*, with no data input

I_D is not connected. The initial state is assumed to be $Q_1=0$, $Q_2=1$ and $Q_3=0$. There is feedback from Q_1 and Q_3 to the modulo-2 summation gate, and since both its inputs are zero, its output at D_1 is also zero. Therefore on the first clock pulse Q_1 goes to zero and the states all shift right.

At the end of the first clock pulse one of the inputs to the modulo-2 summation gate is at logic 1, so the output at D_1 is also 1. Therefore on the second clock pulse this logic 1 is fed through to Q_1. The sequence proceeds until after the third clock pulse when Q_1, Q_2 and Q_3 are all at logic 1. Now the output from the modulo-2 summation gate is logic 0 and this is clocked through to Q_1 on the fourth clock.

It is seen by examining *Figure 47.4* that a sequence of (2^3-1) or seven patterns is produced, and that these are random in appearance, but repeat after every seven clock pulses.

If a stream of input data I_D is applied to the pseudo-random binary sequence generator of *Figure 47.3*, then the data sequence is modified depending on the characteristics of the input data. Suppose that the input data consists of a repetitive bit stream 0101111000. Its effect on the pseudo-random binary sequence generator is shown in *Figure 47.5*. In the initial state the data input I_D and feedbacks Q_1 and Q_3 are all at logic 0, so D_1 is also at 0. This is clocked through to Q_1 on the first clock pulse. Now Q_3 goes to logic 1, but since the input data I_D is also at 1 the output at D_1 is at logic 0. This is now fed through to Q_1 on the second clock pulse, which is a different sequence from that of *Figure 47.4*. In fact, checking the sequences of *Figures 47.4* and *47.5*, they are seen to be totally different. The value remaining in the registers Q_1, Q_2 and Q_3 after a specified number of clock pulses, in the present instance nine pulses, is called the signature of the input data bit stream. It can be shown that a single bit change in this data will result in a very different signature, which can therefore be used to measure the performance of the circuit.

The data in signature analysis can be of any length, although the signature length will be determined by the number of bits in the shift register. The data can also be in true or complement state. It is important that, to get consistent signatures on each sample, the system under test is doing the same operation on every sample. This is usually done by using the system under test to provide its own stimulus, and by feeding the system clock into the logic analyser for synchronisation.

Clock	I_D	D_1	Q_1	Q_2	Q_3	
0	0	0	0	1	0	Initial state
1	1	0	0	0	1	Pseudo-random sequence modified by input data
2	0	0	0	0	0	
3	1	1	0	0	0	
4	1	0	1	0	0	
5	1	1	0	1	0	
6	1	1	1	0	1	
7	0	1	1	1	0	
8	0	0	1	1	1	
9	0	1	0	1	1	
–	–	–	1	0	1	Remainder (signature of input data)

Figure 47.5 Sequence table for the pseudo-random binary sequence generator of *Figure 47.3* with data input sequence (I_D) as shown

47.3.2 Error detection

There are two primary requirements in signature analysis; stimulating the circuit under test to produce a data stream, and compressing this data stream to form a signature for the node being measured. As mentioned in the previous section, two main techniques have been used to compress the data stream; transition counting and using a linear feedback shift register to generate a pseudo-random binary sequence. Both these methods are considered further here. The test of any method is how well it detects an error in the data bit stream.

If a data sequence has a bit length of p, and it is fed into a linear shift register with feedback of length q, then the percentage probability P_E of detecting an error in the data sequence is given by Equation (47.1).

$$P_E=[1-K(p-q)(2^{p-q}-1)(2^p-1)^{-1}]\times 100 \tag{47.1}$$

In this equation K is a constant which makes P_E equal to 100% when $p<q$, since in this instance all the data is stored in the shift register and an error will be detected.

If only one error occurs in the input data sequence then it will always be detected, i.e. $P_E=100\%$, since there is never another error bit to cancel the feedback generator. If more than one error occurs, there is a possibility that a later error will cancel out the effect of an earlier one, so both errors go undetected. Under these circumstances the probability of detecting multiple errors is given by Equation (47.2), and is seen to be dependent on the length of the shift register, but independent of the length of the data stream.

$$P_E=(1-2^{-q})\times 100 \tag{47.2}$$

Therefore, for example, if a 16-bit shift register is used, the probability of detecting multiple errors in a data stream is $(1-2^{-16})\times 100=99.998\%$, which is high.

For transition counting, assuming r transitions, the percentage probability of detecting an error in the data sequence is given by Equation (47.3):

$$P_E=\frac{1-\sum_{r=0}^{p}\left[\frac{p!}{(p-r)!\,r!}\right]\left[\frac{p!}{(p-r)!\,r!}-1\right]}{2^p(2^p-1)}\times 100 \tag{47.3}$$

Now for single bit errors, the probability of detecting the error is given by Equation (47.4):

$$P_E = [1-(p-1)(2p)^{-1}] \times 100 \quad (47.4)$$

Therefore for long data sequences there is only a 50% probability that a single bit error will be detected. This compares with 100% probability of detecting single bit errors, for signature analysis using a pseudo-random binary sequence generator.

For multiple bit errors the percentage probability of detecting an error with transition counting varies with the length of the data sequence, p, and is found from Equation (47.4) to vary from about 80% for $p=5$ to 95% for $p=100$. Therefore again the signature analysis technique is superior in detecting data errors.

47.3.3 Signature analysers

A signature analyser is an instrument which is primarily intended for testing and troubleshooting faulty digital equipment. Most modern instruments incorporate other features to avoid the need for multi-instruments. For example, the Hewlett Packard HP 5005A is a signature analyser with a $4\frac{1}{2}$ digit DMM and a 50 MHz counter timer, all working via the same probe. It is therefore possible to check analogue circuits and asynchronous parts, which a signature analyser cannot test on its own.

Figure 47.6 shows the block diagram of a typical multipurpose signature analyser and *Figure 47.7* shows its waveforms in data sample mode. The input data stream is fed via an active probe into the comparator. The high and low threshold setting on this can be programmed by the central controller using the digital-to-analogue converter. This comparator technique digitises the probe readings, and results in a fast set-up time. The threshold levels for the comparator can be varied between ±12.5 V in 50 mV steps, and can be preset to several levels, for example to test TTL, CMOS and ECL circuits.

A single pod also connects into the circuit under test to pick up a clock, and a start and stop pulse which provides the measurement window. Since the same clock is used for the analyser and for the system under test, synchronism is ensured. The active edge of the clock pulse can be selected from the instrument front panel; *Figure 47.7* shows an active rising edge. The clock samples the input data stream at each active edge, and data changes in between clock pulses are ignored.

The maximum clock frequency for a typical instrument, such as the HP 5005A, is 25 MHz with a minimum pulse width of 15 ns in the high or low state. The minimum gate length is one clock cycle, i.e. one data bit between the start and stop pulses. There is no maximum gate length. The data probe has a set-up time, which is the time that the data needs to be present and steady before arrival of the selected gate pulse, of 10 ns. The time for which the data needs to remain steady after occurrence of the clock edge, called the hold time, is zero. The start and stop pulses have a set up time of 20 ns and also no hold time.

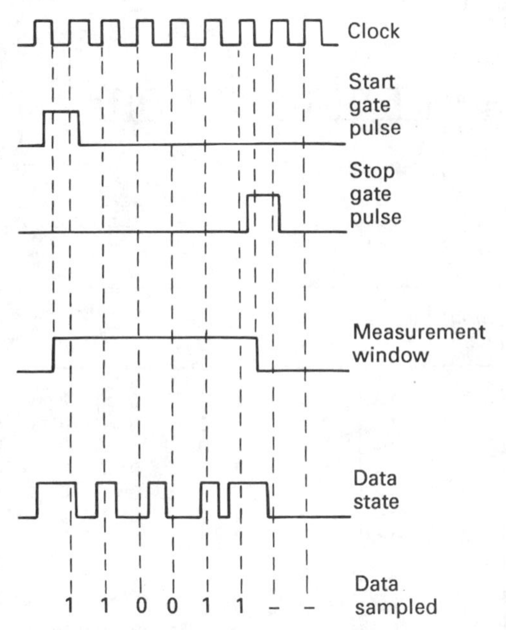

Figure 47.7 Data sampling in a logic analyser

A hold feature on the signature analyser allows one shot signatures to be noted, such as for power-on transients. This will display the signature of the first window only, and will hold this until the reset button is pressed. A flashing 'gate' light on the instrument front panel indicates detection of valid start, stop and clock conditions. If there is a difference between two successive signature readings, a flashing 'unstable' light comes on, and this indicates a possible intermittent fault.

The HP 5005A uses a 16-bit feedback shift register to generate the pseudo-random binary sequence, as in *Figure 47.8*. The tap

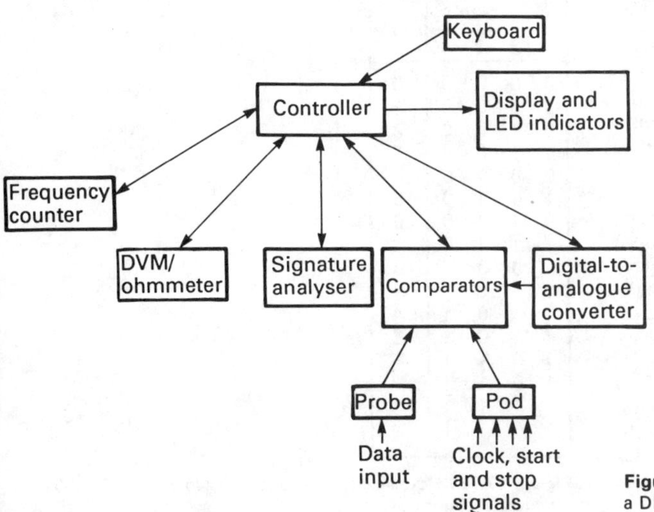

Figure 47.6 Block diagram of a signature analyser incorporating a DMM and a frequency counter

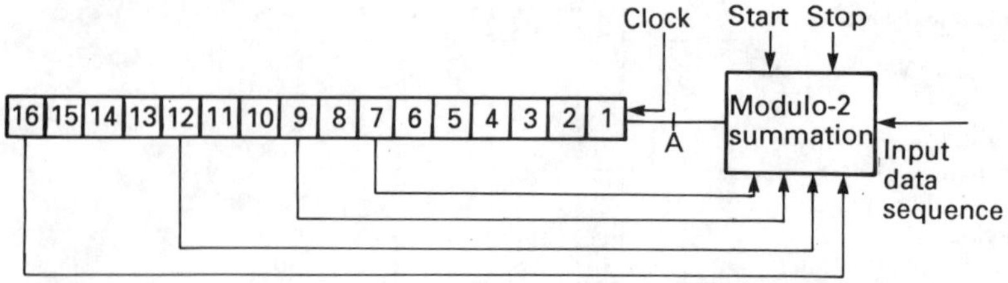

Figure 47.8 Sixteen-bit pseudo-random binary sequence generator

Clock	Input data	Input at A	Shift register															
			1	*2*	*3*	*4*	*5*	*6*	*7*	*8*	*9*	*10*	*11*	*12*	*13*	*14*	*15*	*16*
0	1	1	0	0	0	0	0	0	0	0	0	0	0	0	0	0	0	0
1	1	1	1	0	0	0	0	0	0	0	0	0	0	0	0	0	0	0
2	1	1	1	1	0	0	0	0	0	0	0	0	0	0	0	0	0	0
3	0	0	1	1	1	0	0	0	0	0	0	0	0	0	0	0	0	0
4	0	0	0	1	1	1	0	0	0	0	0	0	0	0	0	0	0	0
5	0	0	0	0	1	1	1	0	0	0	0	0	0	0	0	0	0	0
6	0	0	0	0	0	1	1	0	0	0	0	0	0	0	0	0	0	0
7	1	0	0	0	0	0	1	1	1	0	0	0	0	0	0	0	0	0
8	1	0	0	0	0	0	0	1	1	1	0	0	0	0	0	0	0	0
9	1	1	0	0	0	0	0	0	1	1	1	0	0	0	0	0	0	0
10	0	1	1	0	0	0	0	0	0	1	1	1	0	0	0	0	0	0
11	0	1	1	1	0	0	0	0	0	0	1	1	1	0	0	0	0	0
12	0	1	1	1	1	0	0	0	0	0	0	1	1	1	0	0	0	0
13	0	1	1	1	1	1	0	0	0	0	0	0	1	1	1	0	0	0
14	1	0	1	1	1	1	1	0	0	0	0	0	0	1	1	1	0	0
15	1	1	0	1	1	1	1	1	0	0	0	0	0	0	1	1	1	0
16	1	1	1	0	1	1	1	1	1	0	0	0	0	0	0	1	1	1
17	0	0	1	1	0	1	1	1	1	1	0	0	0	0	0	0	1	1
18	1	0	0	1	1	0	1	1	1	1	1	0	0	0	0	0	0	1
19	1	1	0	0	1	1	0	1	1	1	1	1	0	0	0	0	0	0
20			1	0	0	1	1	0	1	1	1	1	1	0	0	0	0	0
Display reading			9				H				7				0			

Figure 47.9 Sequence table for the 16-bit pseudo-random binary sequence generator of *Figure 47.8*

points on the shift register are chosen to scatter the missed errors as much as possible. Evenly tapped points, for example at 4 or 8 bits, are avoided since most bus-oriented systems, based on microprocessors, tend to repeat patterns at 4- or 8-bit intervals.

Figure 47.9 shows the sequence table for the 16-bit generator of *Figure 47.8* when the input data sequence is 11100001110000111011. Its operation is similar to that shown in *Figure 47.5*. Note the dissimilarity between the input bit sequence and the signature left after the 20th clock pulse. This is displayed using a non-standard hexadecimal code shown in *Figure 47.10*, and forms the signature of the input data sequence.

The DVM/ohmmeter mode can be selected in *Figure 47.6* by I/O signals from the controller. The ohmmeter has several facilities, such as displaying 'open' when the circuit is open circuit, rather than giving an overload display as on most meters. The controller automatically switches to higher ranges when the maximum of any range is reached, and it goes through a self-calibration check every ten readings.

The frequency counter can be used for frequency measurements up to 50 MHz, and for time interval measurements with up to 100 ns resolution. It can be used to test for short-circuits by making time interval measurements. Asynchronous circuits can also be tested by totalizing the pulses between start and stop states. The totalizer can even be used as a form of transition counter.

Bit position				Code
1	*2*	*3*	*4*	
0	0	0	0	0
1	0	0	0	1
0	1	0	0	2
1	1	0	0	3
0	0	1	0	4
1	0	1	0	5
0	1	1	0	6
1	1	1	0	7
0	0	0	1	8
1	0	0	1	9
0	1	0	1	A
1	1	0	1	C
0	0	1	1	F
1	0	1	1	H
0	1	1	1	P
1	1	1	1	U

Figure 47.10 Hexadecimal display code used in the HP 5005A signature analyser

47.3.4 Measurement with signature analysers

When making measurements on a circuit using signature analysers a large number of bit sequences, greater than 20, should be used. It is also often necessary to design the system such that feedback paths within circuits can be broken easily, when testing, to stop faulty sequences being fed back and affecting good nodes. It is then possible to trace the fault back to the defective component.

Microprocessor-based systems are especially suitable for testing using signature analysis. The circuits are designed with a small part of the program memory containing a special diagnostic program, which can be activated to simulate nodes. This causes bit sequence patterns to be generated at various nodes, whose signature can be measured and compared to that of a good circuit.

The microprocessor itself can also be tested by isolating it from its memory, usually by removing test jumpers in the circuit. The microprocessor is then made to free run by its data lines to logic 0 or 1 states, to give an operation such as CLR (clear) or NOP (no operation). The processor now reads the same data at each address, and so free runs around all possible memory addresses. The signatures for the processor bus in this condition can be verified against the expected signature for a good device.

47.4 Logic analysers

The instruments described in earlier sections, logic probes and signature analysers, are primarily used to test electronic circuit hardware. Logic analysers represent a step increase in complexity from signature analysers. Logic analysers were initially primarily intended to test hardware, but with the shift towards microprocessor systems the emphasis is now on using logic analysers for testing software. The change in emphasis from hardware to software has also resulted in a modification of some of the features of a logic analyser. For example, hardware analysers need to run at relatively high speeds, and sampling rates in excess of 500 MHz are often required. However, most microprocessor systems operate at speeds below 20 MHz, so for software analysis lower logic analyser speeds are adequate.

Many of today's systems are based on bus structures, and for effective testing it is necessary to monitor several lines simultaneously. For example, an 8-bit microprocessor would probably have 16 address, eight data and eight control lines, requiring an analyser with 32 inputs. This would be increased to 48 lines for a 16-bit microprocessor. A conventional oscilloscope does not have as many input channels as a logic analyser; several analysers provide up to 72 input channels.

Analogue signals are usually repetitive, and can be shown as a trace on an oscilloscope. Many digital states only occur once during the execution of a microprocessor program. In order to observe these they need to be captured or stored in digital form in memory, and then displayed either as a data stream or as an analogue waveform. Because the signals are all digital the output does not indicate any amplitudes, only logic 0 and logic 1 states.

The size of the memory available with a logic analyser is an important feature, since with multichannel operation, and a high sampling rate, even a large memory will soon be full. Selective triggering is commonly used to conserve memory, by storing data for only that portion of the operation which is of direct interest. *Figure 47.11* shows in simplified form how this works. A trigger word is programmed via the operator panel into the logic analyser. The data input is shifted through the analyser's memory, and is compared with the trigger word. When the data input matches the trigger word, recording of data occurs. Usually this is done by disabling the memory clock after a programmable delay called the trigger delay.

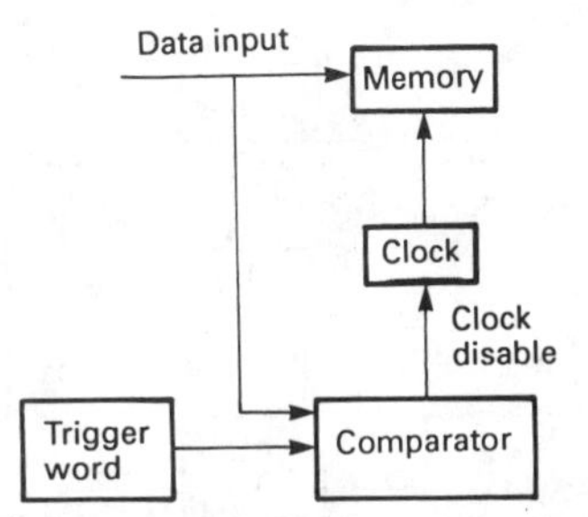

Figure 47.11 Simplified representation of the trigger function within a logic analyser

Logic analysers are complex instruments and it is important that they are designed such that they can be easily used. Usually this is done by setting up the instrument's functions using menus and prompts which appear on the screen. One manufacturer has also provided an instrument with a help button. This can be used to call up onto the screen the entire operating manual for the instrument, which is stored in an internal read-only memory.

47.4.1 Logic analyser construction

A basic block diagram of a typical logic analyser is shown in *Figure 47.12*. The large number of input lines are usually organised into groups onto data pods. The design of the pod is important. Because of the large number of lines involved the pods must be easy to use. Usually they fix onto the body of the integrated circuit (microprocessor) and pick up the signals from its pins. It is also important to avoid ringing and crosstalk, and to have good bandwidth and sensitivity. This is especially important when operating at high clock rates above about 100 MHz. Usually most probes used with logic analysers have a field effect transistor input stage, with an isolating resistor of 1 MΩ and a balancing capacitance of 5 pF. Lower values of resistance and capacitance may be used for higher operating frequencies. Other problems which need to be avoided include skewing, due to different propagation delays of the probe connections.

Usually each pod has some memory associated with it, which communicates with the main backup memory. Most logic analyser manufacturers also provide a family of probes for different microprocessors. These probes plug into the microprocessor socket of the unit under test, and contain special circuits which format the information on the data bus to a convenient form. This allows additional facilities, such as the display of software instructions in mnemonics, to make it easier for debugging.

The data inputs to the pods are compared against a threshold level which sets the logic 0 and logic 1 states. These levels are controlled from the instrument's front panel and it is important that they match the levels which are expected from the type of logic family under test. After passing through the pods the data is compared with the trigger information stored in the pattern and sequence recognition memory. As will be seen in the next section the trigger word can consist of a complex combination of many patterns.

The logic analyser may be operated either on its internal clock or on the clock obtained from the system under test via a data pod. Often the input may consist of many external clocks, all clustered onto one pod. These can be used in any number of logical combinations, and either the rising or falling edge of the clock may be programmed as the active edge. It will be explained in *Section 47.4.2* that both the external and internal clocks are used under different operating mode conditions.

The input to set up the logic analyser is usually via keys, and many instruments use a full ASCII keyboard. Touch-sensitive

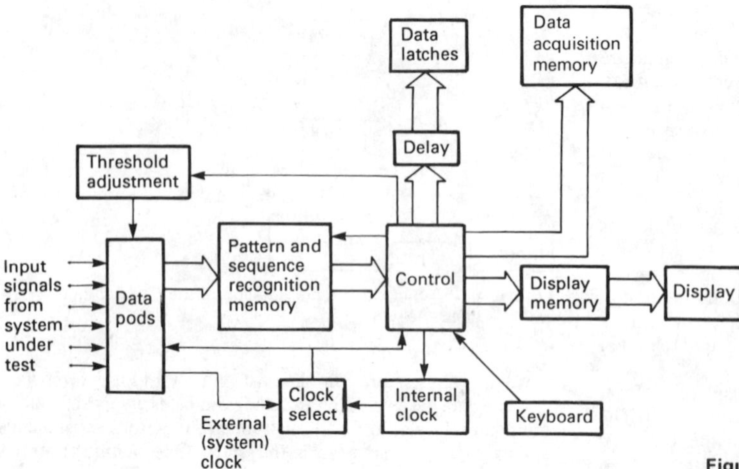

Figure 47.12 Basic block diagram of a logic analyser

keys are also used on the screen to select items such as menus, and soft keys, programmed with short routines, help in making the operator interface easier. The control section, usually based on microprocessors, coordinates the overall operation of the instrument, storing the required data in memory and displaying it in the required format on the display screen, which is usually a cathode ray tube.

47.4.2 Operation of a logic analyser

In this section some of the basic operating features of a logic analyser are introduced. The next section gives simple examples showing the use of the logic analyser.

There are two operating modes for logic analysers, asynchronous and synchronous. In the asynchronous sampling mode the internal clock of the logic analyser is used to sample and transfer the input data to memory. The frequency of the sampling clock should be at least 4–10 times greater than that of the signal being sampled. The output from the asynchronous sample mode is usually shown as *timing diagrams*.

Figure 47.13 shows a simple example, the timing diagram of a 4-bit binary counter. The counter operates on the falling edge of its system clock. If the asynchronous clock operates at a high frequency, then in the asynchronous mode the time difference between sections of the display can be found by showing them on an expanded scale, as in *Figure 47.14*, and noting the time difference based on the asynchronous clock frequency. It should be noted that since the clock and the data are independent there can be a difference of ± 1 clock cycle between successive samples.

In the synchronous sampling mode the clock of the system under test is used as the logic analyser clock. The objective now is to store as many logic states as possible, so a large memory is important rather than a high-speed clock. The recording is made by the clock edge at which the circuit is not triggered, which would be the rising edge of the clock in *Figure 47.13*. This allows for system set-up and hold times. Since logic states are of prime interest in this mode of operation, the timing diagram is not the most appropriate method of displaying the results. A *state list* is preferred, as shown in *Figure 47.15*. The state of the counter after seven clock periods is shown in both binary (BIN) and hexadecimal (HX) format.

Figure 47.16 shows a timing diagram of a bus structured system having eight address and eight data lines. Logic analysers have the capability of displaying all input channels and the full depth of memory simultaneously, so the total system can be studied.

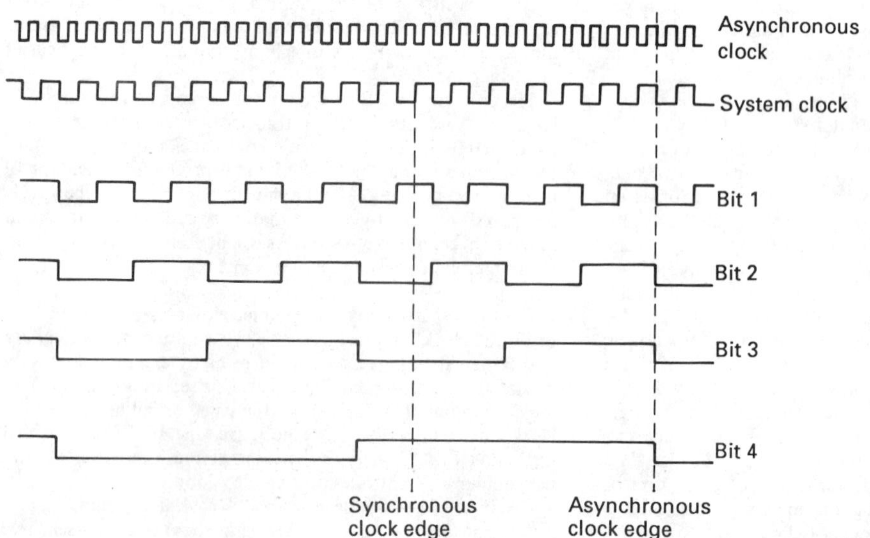

Figure 47.13 Logic waveforms for a 4-bit binary counter

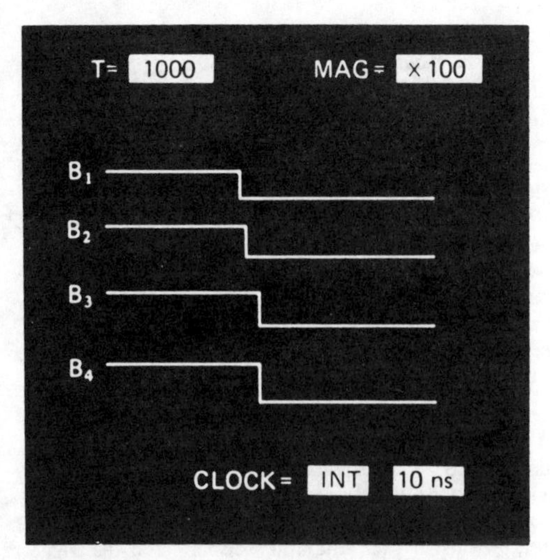

Figure 47.14 Expanded trace of the 4-bit binary counter of *Figure 47.13*

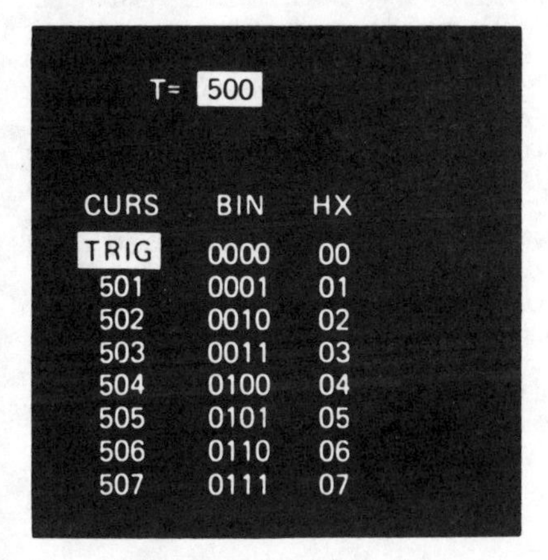

Figure 47.15 State list of the 4-bit binary counter of *Figure 47.13*

Selective areas can then be isolated and magnified for closer scrutiny. In *Figure 47.16* the position of the trigger word is shown by the dotted line *YY*, and the value of the trigger word is given by the T box at the head of the display. The cursor consists of the solid line *XX*, and this can be moved over the screen. The value of the address (A) and data (D) at any position of the cursor can be read from the screen, but for convenience it is also automatically displayed in the A and D boxes at the bottom of the screen.

Time intervals between two points on a timing diagram can also be measured conveniently using the cursor. This is moved to the first point and its position recorded by a marker, and then the cursor is moved to the second point. The elapsed time between these points is displayed in the C box. Also shown on the screen is whether the clock is internal or external, and the clock period. The expansion factor of the display is indicated in the MAG box.

An example of a more complex list than that given in *Figure 47.15*, is shown in *Figure 47.17*. This is sometimes also called a *data list* or a *data domain list*. It is more appropriate for data obtained in synchronous mode, and for microprocessor systems, since it is difficult to show the flow of a microprocessor program as a timing diagram. In a state list the information on the top of the list is usually the data which is recorded first. It is also usual to show only part of the list at a time, and to scroll through the stored memory using the cursor.

The trigger word is given in the T box at the head of the display of *Figure 47.17*. The time interval between sequences can be measured using the cursor and marker, and this is displayed in the C box in terms of the clock period. The channels which are displayed, in this instance A, D and F, can be selected, and also the base in which they are to be shown, the most common being octal (OCT), binary (BIN), hexadecimal (HX) and decimal (DEC). Usually the user would decide after the information had

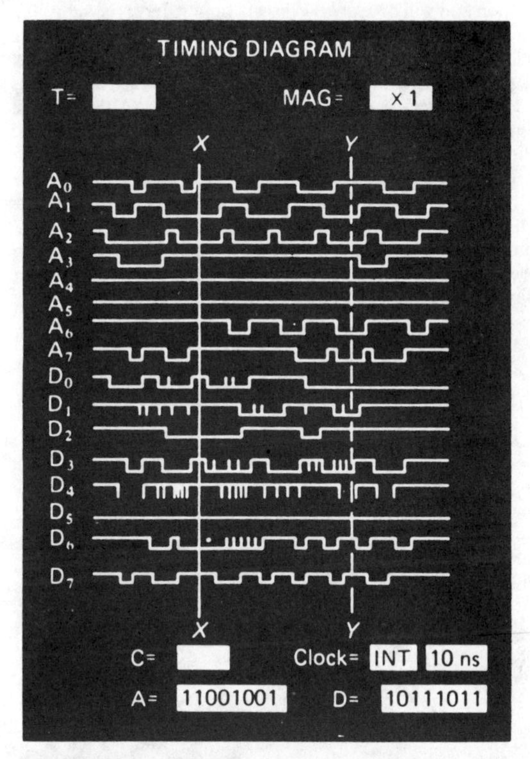

Figure 47.16 A timing diagram

STATE LIST

T= C=

LABEL	A	D	F
BASE	OCT	BIN	BIN
09	01011	11111101	1
10	01004	11111011	0
11	01002	11001010	0
12	01034	10001110	1
13	01201	11010110	0
14	01000	10001111	1
15	01040	10101011	0
16	02002	01100111	1
17	02006	00010010	1
18	02124	10110110	0
19	02126	11001110	1

Figure 47.17 A state list

been recorded whether to display it as a timing diagram or a state list. It is therefore possible to flip from one to the other, and moving the cursor on one display would automatically move it on the other.

In both the synchronous and asynchronous operating modes data is stored in the memory of the logic analyser on an active clock edge only. Therefore short duration glitches which occurred in between these clock pulses would not normally be recorded. To catch these glitches most logic analysers are capable of operating in a latch mode. In this mode, if more than one signal edge occurs between two clock edges, then this will be assumed to be a glitch and will be stored. This will only work if the sampling frequency is greater than the data frequency. The glitches are stored in separate glitch latches, and are displayed in the next clock period. This is illustrated in *Figure 47.18*, which also shows that glitches are lost in a usual sample mode. Although the presence of glitches is displayed in the latch mode the actual position, amplitude or length of the glitch cannot be obtained from the display. However, if a circuit is malfunctioning it is usually sufficient to know that this is being caused by glitches, which can then be suppressed.

Some instruments have a special system just to capture glitches. These are stored in a separate memory and then displayed. The glitches can now be used as a trigger parameter.

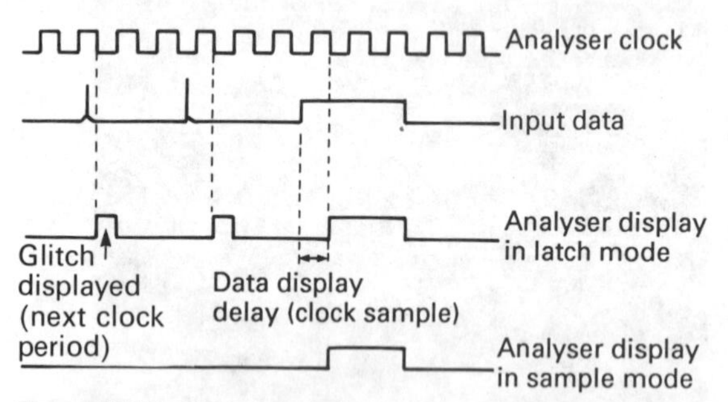

Figure 47.18 Displaying glitches on a logic analyser

The logic analyser is usually set up using a series of menu-driven *format specifications*, an example of which is shown in *Figure 47.19*. In this the start point, clock slope and clock period are input at the top of the screen. The threshold voltages are then assigned to each input data pod, and these may be preset values, such as for TTL and ECL, or they can be adjustable, and the set value is given at the bottom of the page. In this example each pod has eight lines, and a label can be assigned to each line; pods 2 and 3 are labelled address (A), and pod 1 is labelled data (D).

Triggering in a logic analyser is important since it allows the suspect part of the operation to be stored in memory, which is usually limited in size, and it also avoids the need for the engineer to analyse irrelevant data. As mentioned earlier there are three stages to triggering:

(1) Storing pre-trigger information, for example the pre-trigger word and the amount of pre- and post-trigger data, which are determined by the trigger delay.
(2) Searching for the trigger word as the data goes through memory.
(3) Storing the post-trigger data and stopping after the trigger delay.

For hardware systems a single trigger word is usually all that is needed to define the location at which data storage is to start. For software systems, where complex program paths can occur, it is usual to specify a sequential trigger in order to trace a unique program path through the system, and to find passages unambiguously. For example, *Figure 47.20* shows a branching program where it is required to trace the path shown by the dotted line. This is done using a *trace specification* of the form shown in *Figure 47.21*. As for the format specification the label and the base can be selected as required. The count facility allows each item to occur a number of times (usually up to a maximum of several thousand times) before going on to the next item in the sequence. The number of words which must be found in sequence is usually limited to between three and ten. The crosses in any column indicate a 'don't care' state.

Logic analysers are capable of carrying out an analysis of stored data such as histograms of the elapsed time between two events or of the address space used. *Figure 47.22* shows the logic

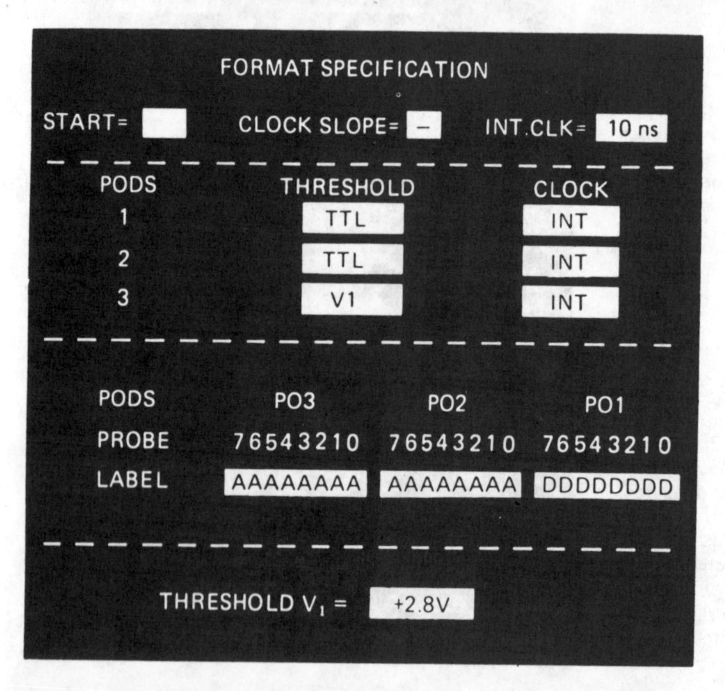

Figure 47.19 Format specification for allocating pods and setting threshold voltages

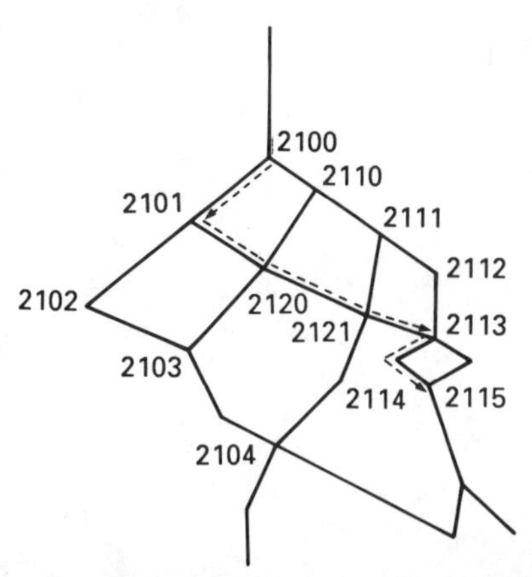

Figure 47.20 Example of a branching program

TRACE SPECIFICATION

LABEL BASE	A OCT	D OCT	COUNT DEC
FIND IN SEQUENCE	2100	XXXX	001
THEN	2101	XXXX	001
THEN	2120	XXXX	001
THEN	2121	XXXX	001
THEN	2113	XXXX	001
THEN	2114	XXXX	001
START TRACE	2115	XXXX	001

TRACE ALL STATES

COUNT OFF

Figure 47.21 Trace specification for the program of *Figure 47.20*

analyser operating in a graph analysis mode. In this mode it can plot any parameter with magnitude on the *y*-axis and sequence of occurrences on the *x*-axis. In *Figure 47.22* the *y*-axis is labelled addresses (A). This means that if the program were to operate in sequence through its address locations the graph would be a straight line. In *Figure 47.22* the program is seen to loop through

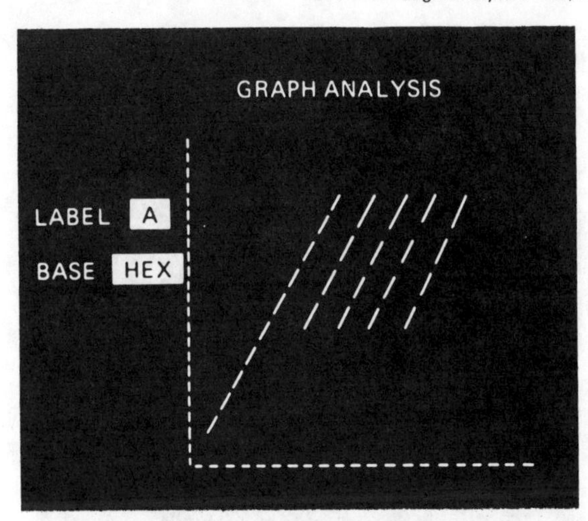

Figure 47.22 A graph of address sequences produced by a logic analyser

a group of addresses, and if this is not expected then this part of the program can be studied in more detail using, for example, a state list output.

47.4.3 Using a logic analyser

Before any measurements are made using a logic analyser, its variables are set up with a format specification menu, and then the section to be recorded is defined within the trace specification menu. As an example, suppose that it is suspected that a fault is occurring within a section of program starting at memory location 00110. *Figure 47.23* shows the trace specification set-up for this. The starting address is specified in A and the D content is unimportant. Triggering starts at the first occurrence of the trigger word (location 00110) hence the count is 1.

The state list corresponding to the specification of *Figure 47.23* is shown in *Figure 47.24*. It is assumed that a special pod for the target microprocessor is in use, so that a mnemonic listing can also be obtained. Examination of this list shows that there is a jump instruction in location 00116, which was not intended and can now be corrected.

An alternative way of locating this discontinuity in the program, which would be easier to spot, especially in a large program, would be to use the graph analysis facility. This is shown in *Figure 47.25*. Each state is represented by a dot (dash shown in the figure), although for a large program several million states can be condensed by first displaying every 1000th state (say) only, and then expanding the scale around the suspect area. *Figure 47.25* shows visually the discontinuity in the program

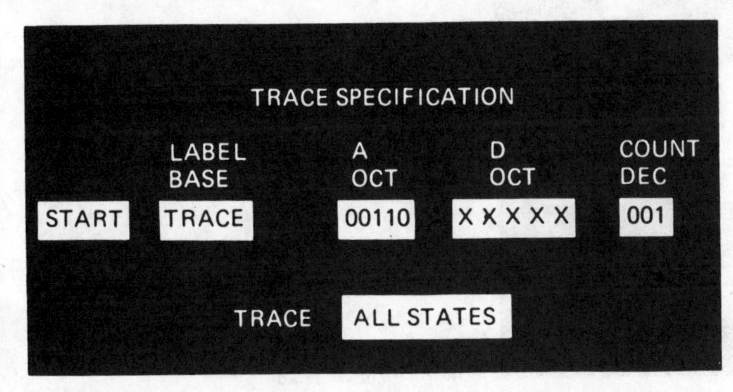

Figure 47.23 Trace specification to print out a section of program starting at location 00110

STATE LIST

LABEL	A	D	
BASE	OCT	OCT	
START	00110	01076	CLA
	00112	00105	TAD X
	00114	01123	CMA IAC
	00116	12621	JMP P
	00210	01076	CLA
	00212	01011	DCA A
	00214	01005	INC B

Figure 47.24 State list corresponding to the trace specification of *Figure 47.23*

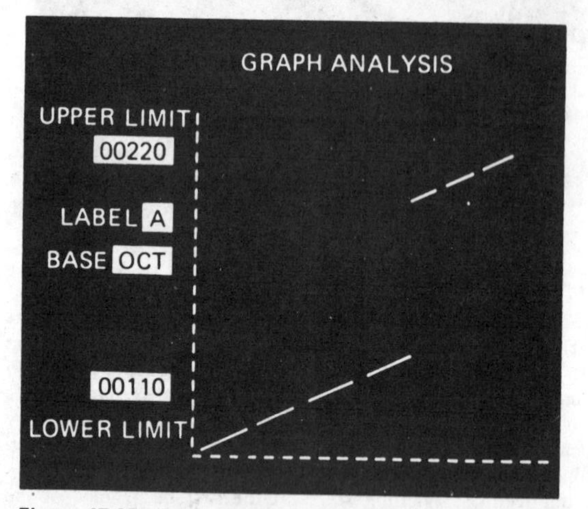

Figure 47.25 Graph analysis for the example of *Figure 47.23*

around address 00116, and this can now be examined using a state list output.

As another example consider *Figure 47.26*, which illustrates the flow diagram for a system used to capture input data. As long as a pulse is not present it waits in a continuous loop of monitoring the input. When a pulse is sensed it is captured and stored. The system then waits 100 ms before operating an output circuit.

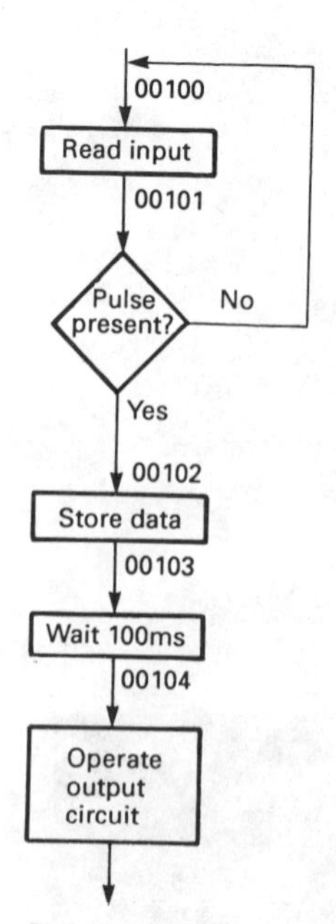

Figure 47.26 A system flow diagram

Figure 47.27 shows the trace specification for *Figure 47.26*, assuming that the sequence shown is to be traced, where the numbers in *Figure 47.26* indicate A locations. Suppose also that the elapsed time is to be indicated so that the count field is programmed with TIME. The time can be absolute (ABS) or relative (REL).

If the time is programmed as REL in *Figure 47.27*, the state list would be as in *Figure 47.28*. The value of A is traced through as specified in *Figure 47.27* and the stored D in each location is also indicated. The time given in this list is the elapsed time between individual states, and it is not accumulative. If the time in *Figure 47.27* was ABS, the state list would show the accumulated time from the trigger word, in *Figure 47.29*, where the start point is at time zero.

Figure 47.30 shows a program with eight nested loops. Suppose that it is suspected that a fault occurs in this program on the 40th occurrence of address 1115 during the 6th execution of the major loop. To find this state the trace specification can be written as in *Figure 47.31*. The count column is now used to

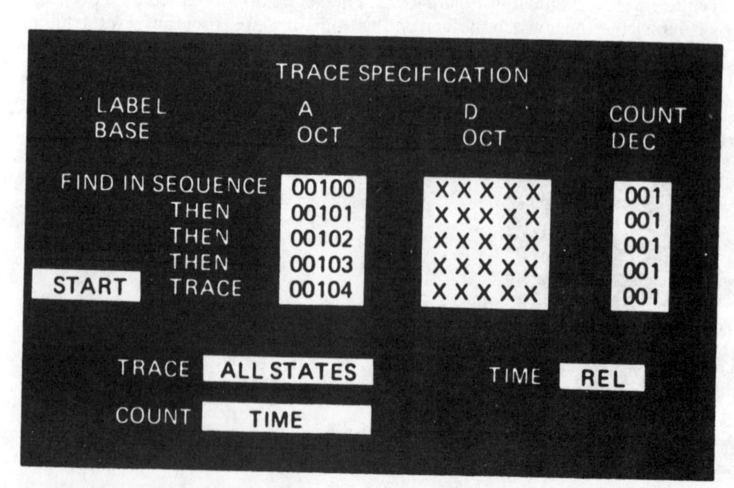

Figure 47.27 Trace specification for the system shown in *Figure 47.26*

STATE LIST

LABEL BASE	A OCT	D OCT	TIME DEC
	00100	01035	–
	00101	00214	35μs
	00102	10123	10μs
	00103	10056	15μs
	00104	01365	100ms

Figure 47.28 State list for the example of *Figure 47.26* using relative time indication

STATE LIST

LABEL BASE	A OCT	D OCT	TIME DEC
	00100	01035	–1.8ms
	00101	00214	–1.765ms
	00102	10123	–1.750ms
	00103	10056	–1.735ms
	00104	01365	0μs

Figure 47.29 State list for the example of *Figure 47.26* using absolute time indication

specify the number of times any of the required addresses is to occur before the analyser is triggered.

The logic analyser is also a useful tool for locating intermittent faults in a system. To do this a good trace of the suspect area is first stored in the internal memory of the analyser, as in *Figure 47.32(a)*. The analyser is then put into the compare mode. In this mode there are several options, such as 'stop if equal' and 'stop if not equal', the former mode being shown in *Figure 47.32(c)*. The

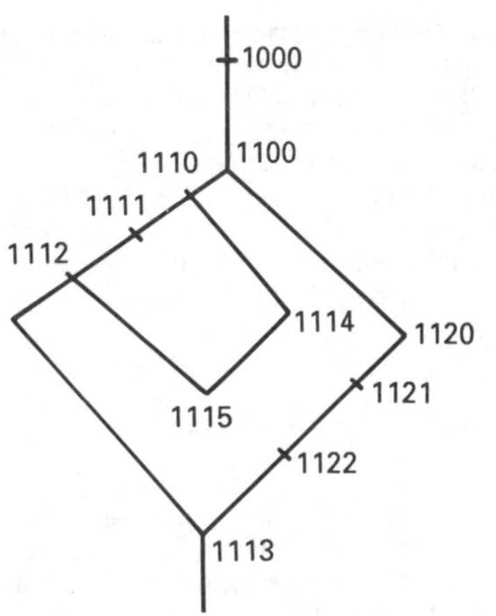

Figure 47.30 Example of a nested loop program

TRACE SPECIFICATION

	LABEL BASE	A OCT	D OCT	COUNT DEC
FIND IN SEQUENCE		1000	XXXX	001
	THEN	1100	XXXX	006
START	TRACE	1115	XXXX	040

TRACE ALLSTATES

COUNT STATE 1115 XXXX

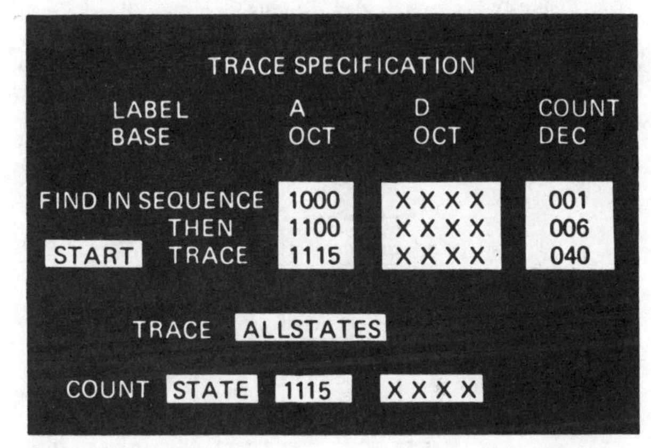

Figure 47.31 Trace specification for the nested loop of *Figure 47.30*

(a) STATE LIST

LABEL BASE	A OCT
START	001100
01	002617
02	713581
03	210456
04	103167
05	001076
06	323567
07	753217
08	325612
09	124700
10	007431

(b) STATE LIST

LABEL BASE	A OCT
START	001100
01	002617
02	713581
03	210456
04	103167
05	001076
06	323567
07	753217
08	325612
09	124700
10	007602

(c) TRACE COMPARE

COMPARE MODE STOP NOT EQ

LABEL BASE	A OCT
START	000000
01	000000
02	000000
03	000000
04	000000
05	000000
06	000000
07	000000
08	000000
09	000000
10	000233

007431 = 0000 0000 0111 0100 0011 0001
007602 = 0000 0000 0111 0110 0000 0010
000233 = 0000 0000 0000 0010 0011 0011

(d)

Figure 47.32 The use of an analyser to locate an intermittent fault: (a) required trace, (b) faulty trace, (c) analyser display in compare mode, (d) derivation of the error code

logic analyser will now continuously cycle through the required section of the program, and will display all zeros as long as the measured trace equals the stored trace. Suppose now that the measured trace momentarily changes to that in *Figure 47.32(b)*, with an intermittent fault occurring which affects item 10. The logic analyser will now display the trace shown in *Figure 47.32(c)*, with a non-zero in item 10, and will stop. The figure shown in item 10 indicates which bits of the two displays in *Figure 47.32(a)* and *(b)* are not equal, as illustrated in *Figure 47.32(d)*.

47.5 Microprocessor development systems

The digital instruments discussed in previous sections have been used primarily to analyse hardware systems, although logic analysers are becoming increasingly accepted for software development. In the present section the techniques used for software development and analysis are discussed, primarily those associated with microprocessor-based systems.

Figure 47.33 shows the flow diagram for product development of a microprocessor-based system (see also *Chapter 30*). The customer requirements are translated into a system specification, and this is followed by system design, which enables the hardware and software contents of the product to be defined.

The hardware is split into modules, usually consisting of one or more printed circuit board assemblies, and these are then specified. Hardware design is initially done on a module basis, and is debugged using computer-based engineering workstations. These enable the schematic diagram to be captured, compiled to check for correctness, and then simulated to check for functional performance and for correct timing. Data required for this analysis is obtained from a database of component characteristics, which is stored within the engineering workstation.

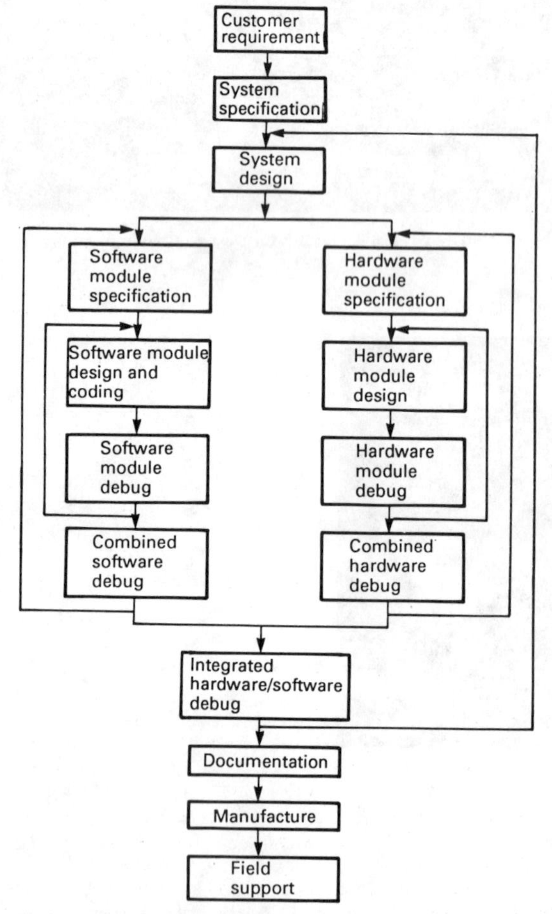

Figure 47.33 Product development flow

The engineering workstation is also used by the test engineer to ensure that the test points provided are adequate for manufacturing test. Test patterns are input to the workstation and the system is run to determine the amount of test coverage. Faults can be simulated at circuit nodes (stuck at 0 and stuck at 1) to see if they are detectable.

Simulation on an engineering workstation eliminates the costly and time-consuming requirement for system bread-boarding, although this can still be done, if required, to verify the operation of critical parts of the circuit (see also *Chapter 52*). When the engineer is satisfied with the design it is passed through a packager, which assigns gates to integrated circuit packages. The design then moves from the circuit design to the physical design phase, where the system is laid out onto a printed circuit board. This may result in a redefinition of package numbers and pins, and this back annotation information is fed back to modify the initial schematic drawing. Conventionally the circuit design and physical design aspects of hardware design have been done on different workstations, usually supplied by different vendors, which has given rise to interface problems. The trend today is to perform all hardware design on a single workstation.

Software development, shown in *Figure 47.33*, follows a similar route to hardware development, but uses a very different set of tools. The individual software modules are initially specified. Very few tools are currently available for this phase, although several are being developed (see *Chapter 46*).

Following module definition is module design, coding and debug. The tools available for these operations are described in *Section 47.5.1*. The software is then combined with the hardware for system test, and it is here that emulation, described in *Section 47.5.2*, is especially useful. Finally documentation for both the hardware and software is prepared for manufacture.

The design phase of a software development project is particularly important since it determines the software quality, reliability and maintainability. This is especially important since about 70% of the cost of software occurs after it has been released, in the form of changes and bug fixes. A well designed and documented software package is much easier to maintain.

Software development costs predominate in the costs of developing a microprocessor-based system, accounting for 60–90% of the overall cost of an average project. The software development is also most likely to cause programme slippage. Time to market is usually the most important consideration for any project. A 50% increase in development costs can usually be expected to reduce the profits obtained from the product by about 5%. On the other hand a 6 month delay to market can reduce profits by as much as 30%. It is because of the importance of software development within a product that much emphasis has been put, in recent years, on improving the productivity of software engineers.

47.5.1 Software development and debug

Several generic software facilities are required for any development system, and these are described in this section. An *editor* is essential in order to enter code and to modify it as required. Editors also perform many tasks which are transparent to the user, such as text compression prior to disk storage to minimise space.

Once the different modules of higher-level language and assembler code have been written they are passed through a utility called a *linker*. The compilers and assemblers (see *Chapter 46*) often do not assign fixed memory addresses to codes, that is they can be reallocated anywhere in memory. The linker's function is to link together these modules into a sequence of memory to produce a single homogeneous program code.

Programs are usually written on development systems, which have the facility for compiling, assembling, linking and then storing the code within the system. After the code has been checked it is transferred from the development system into the memory of the actual target system, so that it can be tested further. This process is known as downloading, and the utility which does it is called a downloader.

The *operating system* provides the interface between the user and the development system, and is the heart of the software development system. Operating systems are usually interactive, and use default conditions to reduce the amount of information which has to be put in by the user. They also tend to be file oriented, in which the user operates on the programs by using their file names.

Operating systems carry out many control tasks, such as calling up compilers and editors, as shown in *Figure 47.34*. The control part of the operating system works off a user source file which specifies items such as memory locations, interrupts, etc.

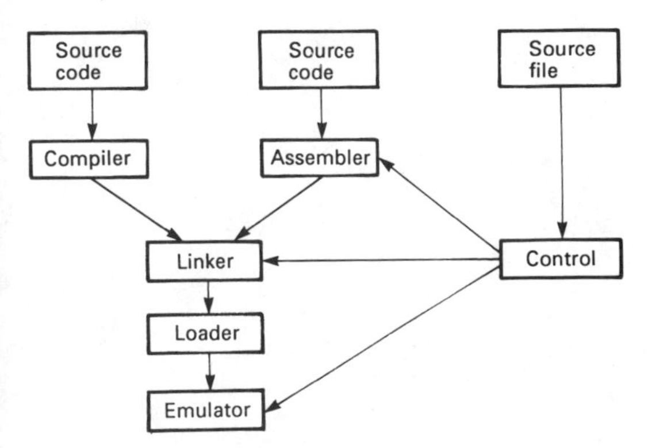

Figure 47.34 The control exercised by an operating system

The control reacts to the user set-up requirements by generating code and command files to automatically handle details which will execute the system program as specified. Examples of this are code generation for initialisation and reset; creating linker command files; and setting up emulator command files, which enable the linked object code to be downloaded and executed in the emulator.

The use of software analysis utilities was introduced in *Section 47.4*. Performance analysis gives an overview of program activity and allows areas of the software to be optimised. Examples of performance analysis utilities available are:

(1) Breakdown of the memory used throughout, which enables more efficient memory allocation and utilization.
(2) An analysis of the time distribution between successive accesses to different memory modules, which enables queuing problems to be identified.
(3) Plot of the time needed to run separate parts of a program including best and worst case conditions. Areas which take more time than expected can be rewritten.
(4) A count of the number of occurrences of different instructions.

47.5.2 Emulation

When testing conventional hardware systems it is possible to understand the operation of the system by studying the circuit diagram. Fault diagnosis of microprocessor-based systems is much more difficult since the circuit performance is determined by its software. Emulation is used in these circumstances. This is done by replacing the circuit under test (also called the target system) by a device that emulates its behaviour. This device is called an emulator, and it can be subject to controlled stimulus in order to study the operation of the target system. This concept of emulation is not new since, for example, when testing hardware circuits it is usual to replace part of the system with external equipment, such as power supplies and function generators, which in effect act as emulators.

When used with microprocessor development systems, the emulator gives control over the target microprocessor and a view of the activities taking place in its registers. The emulator may be part of the development system, so that it is tightly coupled to the overall system, or it can be a stand alone dedicated emulator linked onto the main bus, when softward development is done using a large host mainframe computer.

The target system should work identically with and without the emulator, and this property is known as having transparency. There are two types of transparency. Electrical transparency is when the two systems have the same electrical characteristics, such as operating time and machine cycles, and functional transparency when the same resources, such as memory and interrupts, are used by the two. The emulating microprocessor also needs to be mounted close to the target system to minimise pick-up effects, and it is usually built into an external pod which is plugged in, in place of the target microprocessor. The cables connecting the emulator to the rest of the system should be of good quality and have impedance compensation. Usually the emulating microprocessor needs to be buffered to drive cables and the target system, so performance variances are introduced from the target system, such as extra delays.

Several features are required from an emulator-based development system, to test the target equipment. It should be possible to single step through the program, examining register contents or modifying memory content at each step, and to set special stop points, called breakpoints. These can be software breakpoints, where the program instruction is replaced by an emulator call to stop the program at specific points, or hardware breakpoints, which can also be set by comparing parameters such as memory address or input/output datalines. Hardware breakpoints can also generate triggers for instruments, such as logic analysers and oscilloscopes, to make further readings.

Changes can be made to a program during the debug stage using an emulator. These changes may be patched into the system for assessment. A disassembler facility should be available to observe the microprocessor operation in mnemonics, and an assembler facility for patching. It is also useful to be able to carry out debugging at a higher level language (source code), and to be able to set breakpoints at this level. The complexity of the system is now increased since the emulator needs to track the source code program.

Logic analysers are often teamed with emulators within the development system. The logic analyser is primarily a monitoring instrument, and is very useful in collecting timing data using the asynchronous clock mode. Emulators operate in synchronous mode from the target system clock, and cannot do this effectively. *Figure 47.35* shows an arrangement in which an emulator is used within a development system and is connected to the target system via its microprocessor socket and the emulator interface.

The emulator shown in *Figure 47.35* has access to its internal logic state circuit which provides analysis of functions such as

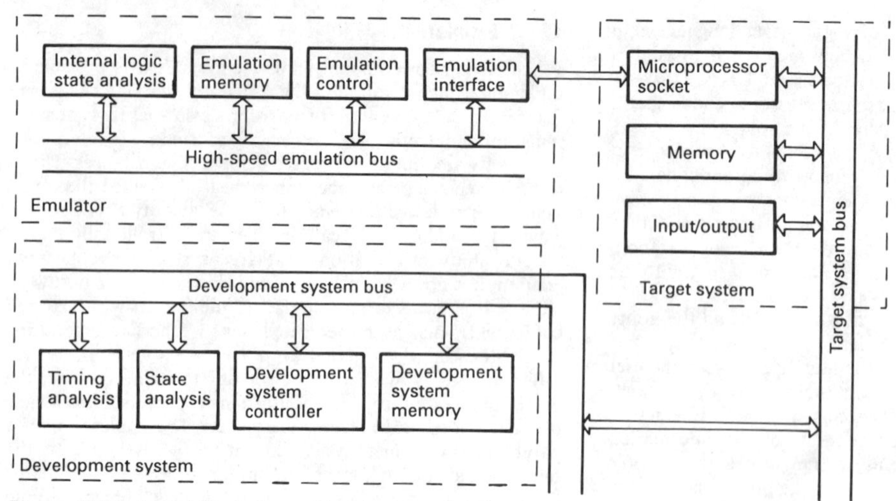

Figure 47.35 Use of an emulator with a development system

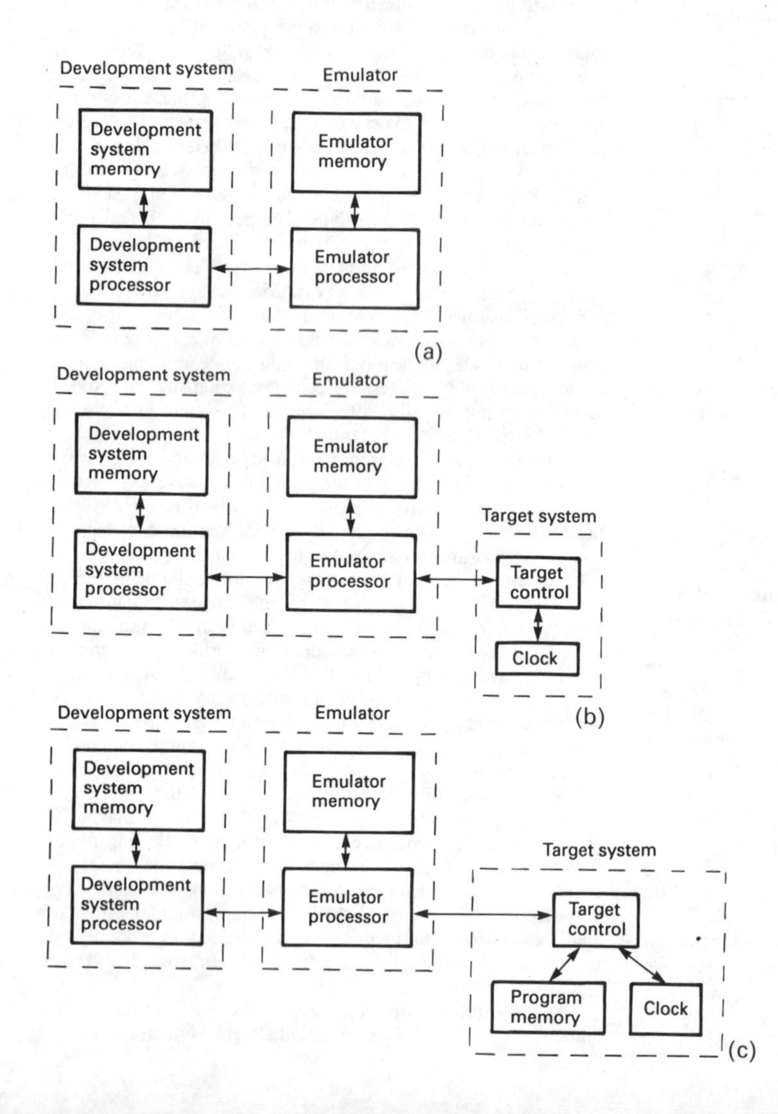

Figure 47.36 Three development modes used for microprocessor-based systems: (a) mode 0, (b) mode 1, (c) mode 2

software performance. It also has access to timing and state analysis within the development system. Software performance analysis does not affect the program execution, so information can be collected and displayed in real time. The emulator memory must be wide enough for the microprocessor address, data, status and control lines, which can be up to 96 bits wide. It must also be deep enough to capture between 1000 and 2000 steps of data. In spite of this memory size, careful setting of breakpoints is essential to optimise use of trace memory.

Figure 47.35 shows separate bus structures for the emulator and the development systems. This allows both systems to operate independently, making it useful for emulation of multiprocessor-based products, for examining logic performance whilst the microprocessor is running, and for updating the emulator trace memory without stopping the processor.

Figure 47.35 shows the emulator connecting into the microprocessor socket of the target system. This in effect replaces the target's microprocessor with the emulator's microprocessor, which is under control of the development system. This form of emulation is known as in-circuit emulation, and is the one most commonly used. It cannot be employed for bit slice processors, since these operate at high speeds and their microcoding feature allows each device to have its own instruction set. In these circumstances memory or ROM emulation is used.

In memory emulation the target's microprocessor is functional, but interface with the emulator is made via the target's program memory. A fast RAM within the emulator replaces the program memory, giving the emulator control over the system's microcode. The emulator can stop the clock at the breakpoints needed. However, although this may be satisfactory for static systems, it is not suitable for dynamic systems, since stopping the clock causes loss of data. If the emulator operates at a much higher speed than the microprocessor cycle time, it can force the processor into a 'do nothing' loop whenever a breakpoint is detected. Therefore the clock is not stopped but the operator can view the internal registers of the system.

The third type of emulation is known as input/output emulation, and is used when the system needs to be tested before the design of the input/output circuit has been completed. The emulator now interfaces with the input/output part of the target system. Pattern and word generator signals are sent to the microprocessor and timing and state analysis is used to check its reponse. The system can be highly programmed; for example, particular signals can be sent to the microprocessor corresponding to particular patterns on the address bus.

Emulation can occur in three modes, as in *Figure 47.36*. In mode 0 the emulator and development system simulate the target system, without any target hardware. The program for the target system is run in real time whilst resident within the emulator's memory. The development system clock is used as the clock for the emulator microprocessor, and the input/output is simulated. Mode 1 emulation is used when some of the target hardware is available. The clock used is that of the target system. Input/outputs can be simulated, or those from the target system used if available. The memory employed may be that within the emulator or the target system, or even a combination of the two. In mode 2 emulation, which is the final hardware–software integration stage, all the hardware is available so the target system's memory, clock and input/output circuits are used. As in the previous two modes the emulator still replaces the target microprocessor, and gives the engineer control over the target system.

Two types of problems can arise when using emulators. In the first the target system operates satisfactorily when the emulator is used, but not when running from its own microprocessor. The most common cause is that the target microprocessor is overloaded, since the emulator is buffered and can therefore support a greater load. Timing problems during power up can also result in this discrepancy in performance. For example, if the power-on reset pulse is too short, it will affect the system when it is running from the target microprocessor, but not when it is running from the emulator, since the emulator execution commences after power-on has occurred.

If the target system works satisfactorily when operating from its own microprocessor, but not when connected to the emulator, the fault could be due to timing problems, caused by buffers, power supply noise, etc. Often these problems also occur when the target system is running off its own microprocessor, but they are masked by other effects and show up as intermittent faults.

After the target has been satisfactorily checked out, using the emulator, the software is programmed into EPROM by transferring the files from the development system to an EPROM programmer. This then forms the program memory for the target system.

Acknowledgement

The content of this chapter has been taken from *Electronic Instruments and Measurement Techniques* by F. F. Mazda, published by Cambridge University Press.

Further reading

FALK, H., 'CASE tools emerge to handle real-time systems', *Computer Design*, 1 January (1988)

FALK, H., 'Development systems evolve toward integrated, host-independent solutions', *Computer Design*, 1 April (1988)

FALK, H., 'C compilers team up with other development tools to ease system design', *Computer Design*, 15 April (1988)

FALK, H., 'Emulators set sights on even faster processor chips', *Computer Design*, 1 May (1988)

GOGESCH, R. S., 'Logic simulators exhibit different levels of device characterization', *EDN*, 13 October (1988)

SCHWEITZER, W., 'Proper glitch capture requires knowledge of logic-analyser limits', *EDN*, 7 January (1988)

SHEAR, D., 'Workstations', *EDN*, 29 October (1987)

SMITH, M., 'The use of emulation', *Electron. Product Design*, April (1988)

STRASSBERG, D., 'PC-based logic analysers', *EDN*, 13 October (1988)

WAAKAAS, E., 'Single board logic analyser for VME bus-based systems', *Electron. Eng.*, October (1988)

WILLIAMS, M., 'A modular approach to microcontroller emulation', *Electron. Eng.*, October (1988)

48

Control Systems

J E Harry, BSc(Eng), PhD, DSc, CEng, FIEE
Reader
Department of Electronic and Electrical Engineering
Loughborough University of Technology

Contents

48.1 Open-loop systems

Control systems are used in all areas of industry. The process may encompass almost any conceivable operation ranging from operation of a machine tool to filling milk bottles. The components comprising a simple control system can be categorised as controller, final actuator or servo, the process and the sensor. The components vary depending on the process and some examples of different components used in different processes are listed in *Table 48.1.* The process parameters are varied by the final actuator. The final actuator may be controlled in turn by a secondary actuator, e.g. the speed of a hydraulic drive may be increased with a motorised valve. The control setting may be varied with the controller which may be mechanical, electromechanical, hydraulic or electronic. The setting (set point) of the controller is fixed by the operative in charge of the process.

Such a system is known as an open-loop system and is illustrated schematically in *Figure 48.1(a).* The sensor indicates the state of the process. Open-loop systems are used in all areas of industry where fluctuations in the process variables occur within acceptable limits or occur slowly and can be corrected by the operator adjusting the set point. If now a feedback signal is obtained from the sensor this can be used to maintain the process within set limits. The feedback path is shown in *Figure 48.1(b).* This now becomes a closed-loop control system and is known as a servomechanism. The action of the controller is now to compare a reference signal corresponding to the set point with the feedback signal, the difference being the error signal E. This is normally amplified and used to restore the process to the set point.

48.2 Closed-loop systems

A simple example of a closed-loop control system is the self-balancing potentiometer illustrated in *Figure 48.2.* Depending on the polarity of the out-of-balance voltage the servomotor varies the tapping position on the slide wire. The error signal tends to zero as the balance position is reached so that in practice a small error exists corresponding to a region either side of the balance position over which no control action occurs known as the dead space. This is reduced by the amplifier. If the gain A of the amplifier is high and the feedback is large so that the amplification is small the sensitivy of the system is increased by a factor $1/A$.

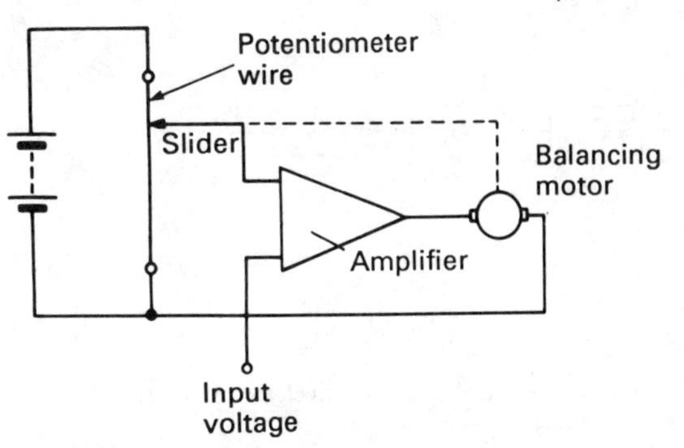

Figure 48.2 Servo-assisted self-balancing potentiometer

48.3 Control equations

Each element of a control system can be typified by its response in terms of the change in output over the input which will be a time-dependent function and can be expressed in terms of first or higher order differential equations.

By taking the Laplace transform of the response the transfer function of a system can be readily determined by combining the transfer functions of the individual elements of the system in

Table 48.1 Examples of elements in control systems

Process	*Control variable*	*Controller*	*Servo*	*Sensor*
Milling	position	on/off	stepper motor	encoder
Oil flow to burner	flow	proportional	d.c. shunt motor	turbine flowmeter
Rolling mill	velocity	proportional and derivative	Ward–Leonard drive	tachometer
Batch furnace for heat treatment	temperature	proportional, derivative and integral control	triac	thermocouple

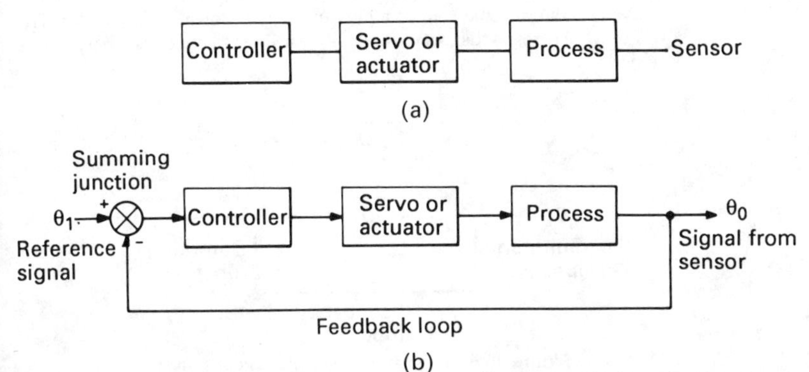

Figure 48.1 Basic components of a control system: (a) open loop; (b) closed loop

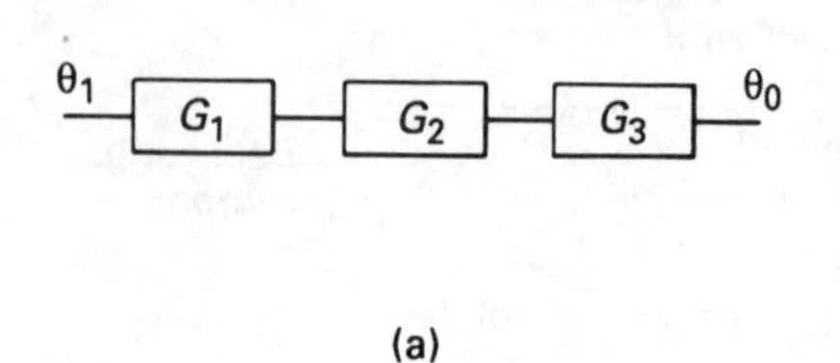

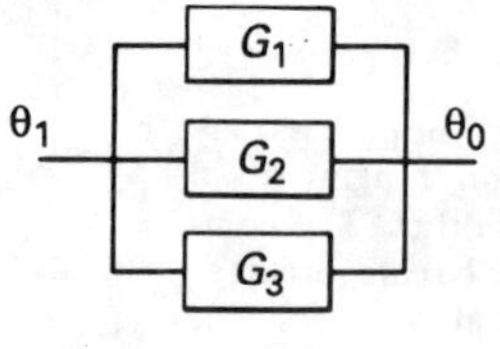

Figure 48.3 Series and parallel connection of system components: (a) series; (b) parallel

Figure 48.3. The transfer function of elements connected in series is equal to their product

$$G=G_1 \cdot G_2 \cdot G_3 \tag{48.1}$$

and in parallel

$$G=G_1+G_2+G_3 \tag{48.2}$$

The servo system can now be represented by the block diagram in *Figure 48.4*.

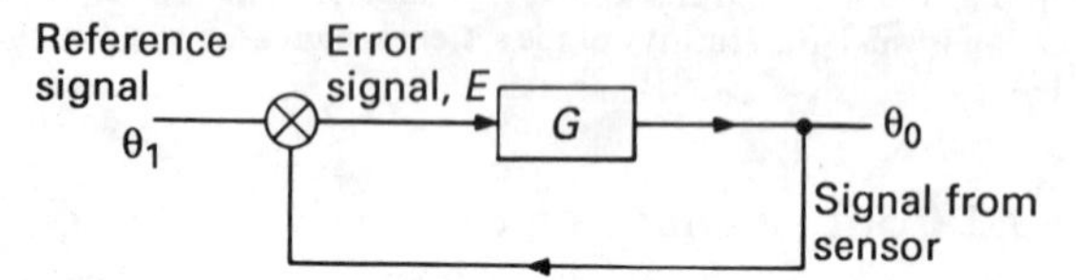

Figure 48.4 Closed-loop control system (servo or batch operation)

The system can be described by the system transfer function, the control ratio and the error function. The system transfer function is given by

$$G=\frac{\theta_o}{E} \tag{48.3}$$

The control ratio is given by

$$\frac{\theta_o}{\theta_i}=\frac{G}{1+G} \tag{48.4}$$

and the error ratio

$$\frac{E}{\theta_i}=\frac{1}{1+G} \tag{48.5}$$

The response of any single control element can be resolved in terms of the damping coefficient ξ, the time constant T, and ω, the critical frequency.

For a first-order system the response is given by

$$T\frac{d\theta_o}{dt}+\theta_o=\theta_i \tag{48.6}$$

and

$$a_2\frac{d^2\theta_o}{dt^2}+a_1\, d\theta_o+a_0\theta_o=a_0'\theta_i \tag{48.7}$$

where

$$T=\frac{2a_2}{a_1} \qquad \xi=\frac{a_1}{2\sqrt{(a_0a_2)}}$$

and

$$\omega_n=a_0/a_2$$

from which it can be seen that

$$T=\xi\omega_n \tag{48.8}$$

By combining the Laplace transforms of each element it is possible to determine the response of the complete system. The Laplace transforms corresponding to various common control system elements are tabulated in *Table 48.2*. The response of the system can be solved analytically for first- and second-order systems. Since the order of the system is equal to the sum of the order of each component in the system, systems with third and higher order responses are often encountered which need to be solved by iterative techniques.

Table 48.2 Examples of control systems and their transfer functions

System or process	*Transfer function*
First-order	$\frac{1}{Ts+1}$
Second-order	$\frac{1}{T^2\xi^2s^2+2T\xi^2s+1}$
Integral controller	$\frac{1}{Ts}$
Proportional controller	K_c
Rate controller	Ts
Linear final control element with a first-order lag	$\frac{K_v}{Ts+1}$

Another example of a simple closed-loop control system is the variable load or regulator system shown in *Figure 48.5*. This corresponds to a process with varying but controlled parameters,

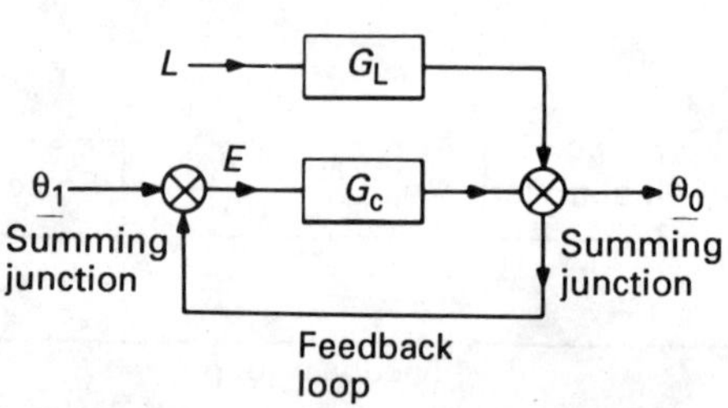

Figure 48.5 Regulator or variable load operation

e.g. throughput. In this case assuming the control level θ_i is fixed the transfer function becomes

$$\frac{E}{L}=\frac{G_2}{1+G} \tag{48.9}$$

$$\frac{\theta_o}{L}=\frac{G_2}{1+G} \tag{48.10}$$

48.4 Control system characteristics

In any closed-loop control system a high degree of precision in terms of the error ratio is required. One factor that limits this is the system stability which decreases as the amplifier gain is increased. If a system becomes unstable it goes out of control which may have disastrous effects.

For any control system there is a small region either side of the set point over which no control action occurs. This may be due to mechanical backlash, thermal inertia, etc. If the dead space is made very small instability may occur. This is known as hunting and occurs where the control element is in effect oscillating about the set point, and is an important consideration in the design of precision control systems.

The stability of a system can be determined if the damping coefficient time constant and critical frequency are known. By varying the relative magnitudes of these parameters the stability can be changed. The cause of instability can be considered in terms of the open-loop frequency response of a system. If delays which exist in the system result in phase lags which are in anti-phase with the input, the output signal is fed back to the input so as to increase the error rather than decreasing it in the same way as positive feedback causes instability in an amplifier.

The magnitude of the delays in a control system is often an important factor governing the stability of the system as well as the degree of precision of control that may be obtained. Delays occur with each element of the system, e.g. sensor controller, actuator and process. Delays also occur in transmission between each element although this will often be small. Depending on the system involved the relative magnitudes of the delays may be important or not. In complex systems the response of individual elements may not be known or the overall response of the entire system may be too complex to analyse. In these cases the response may be measured by determining the response of the system or a model of it to standard input signals. These are usually the response to a continuous sinusoidal input, a ramp function and a square wave input.

The response to a sinusoidal input as a function of frequency enables the amplitude frequency response to be obtained which when plotted on a log scale of frequency is referred to as a Bode diagram. The plot of the phase shift as a function of frequency in polar coordinates is referred to as the Nyquist diagram from which the system stability can be obtained.

48.5 Control modes

The application of control systems can be illustrated by an electric oven operating in an on/off mode as in *Figure 48.6*. The power is supplied until the set point is reached. Delays exist due to the thermal capacity of refractories, the load and the temperature sensor. When the power is switched off as the set point is reached, the temperature of the sensor may continue to rise due to the temperature difference between the sensor and the region close to the heating element, resulting in overshoot, and when the temperature falls through the set point there is a delay before the power is switched on. The delay is known as lag. The relative positions of the sensor and load with respect to the heating element are critical. If the sensor is remote from the heating element, large relatively slow swings of temperature with possible overheating of the load will occur, while if it is close to the heating element rapid oscillations will occur and the load may not reach the required temperature.

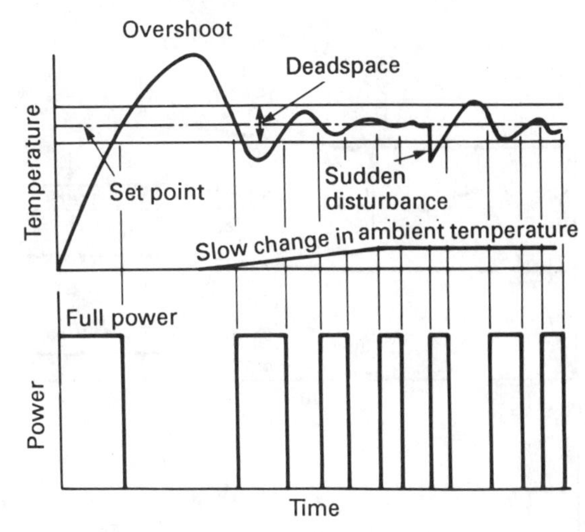

Figure 48.6 Response to on/off control

Oscillation about the set point is inevitable with on/off control since switching occurs only at the set point, when full power is applied. A small difference between the set point and the temperature at which the power is connected often exists and this may be increased to prevent rapid wear of switch contacts.

Proportional action is the basis of continuous control. The power input at any temperature θ within the proportional band is given by

$$W=\frac{\theta-\theta_0}{\theta_2-\theta_1}W_0 \tag{48.11}$$

and the response is

$$p=\frac{e}{b} \tag{48.12}$$

where θ_2 is the temperature of the upper limit of the proportional band, θ_1 at the lower limit, W_0 is the maximum power input at θ_2, e is the error signal and b is a constant for the system.

The power can be varied by controlling the mark/space ratio by relays or contactors, or by using thyristors to give rapid sequence or phase-shift control. The overshoot is reduced by decreasing the power input as the set point is approached so that it is equal to the heat losses when the set point is reached. This is known as proportional control (*Figure 48.7*). The proportional band is confined to only part of the range so that initially full power is available at low temperatures. The position and width of the proportional band can be easily varied.

The position of the set point in the proportional band is preset so that when the set point is reached the power input is exactly equal to the heat losses. This will occur only for the precise conditions existing when the system is set up governed by the position of the set point in the proportional band. Any changes such as in ambient temperature, heat losses and any other factors affecting the power requirements at the set point will result in a change in the power required and hence in the actual control temperature (*Figure 48.7*). The difference is known as offset or droop and is typically up to half the width of the proportional band. The offset can be reduced by increasing the controller gain, i.e. by decreasing the width of the proportional band. However,

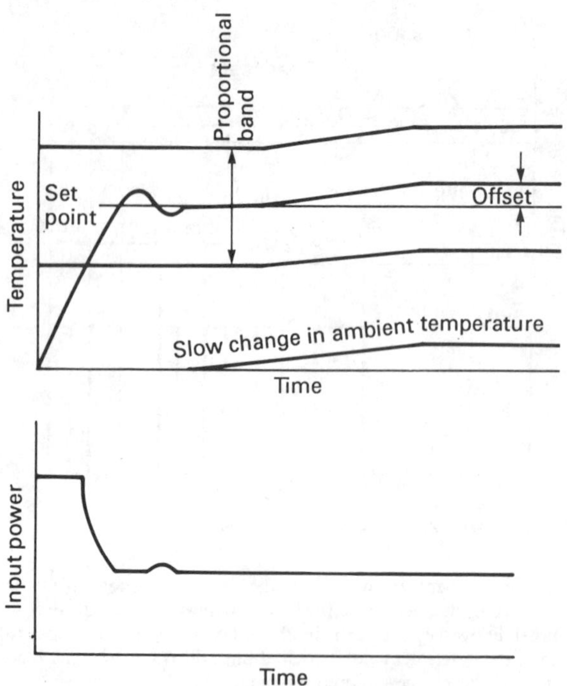

Figure 48.7 Proportional control showing effect of offset

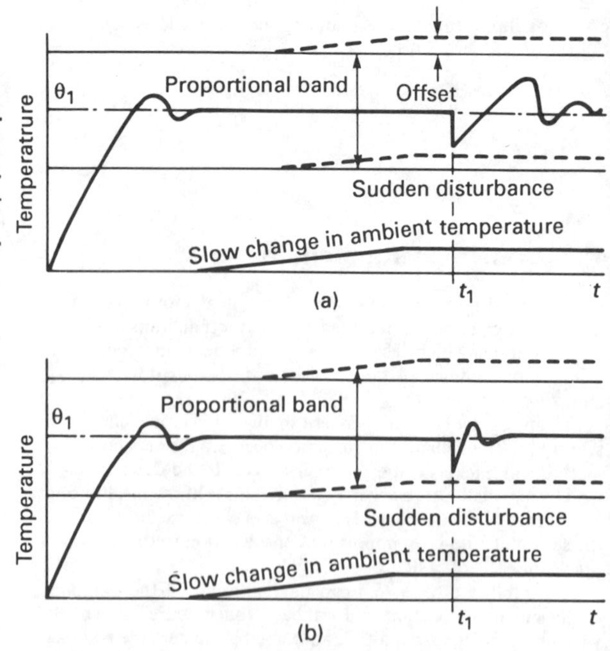

Figure 48.8 Effects of derivative and integral control: (a) proportional and integral control response, (b) proportional, integral and derivative (three-term) control response

this results in increased initial overshoot, reduced stability and limit uncontrolled oscillations equivalent to on/off operation with no dead space. Proportional control alone is used quite extensively for temperature control systems in which a high degree of precision is not required. Fluctuations in the position of the proportional band occur due to changes in the ambient temperature θ_0 and operating conditions in the furnace such as heat losses due to deterioration of the refractories. As a result, the proportional band is offset and the power available at the set point changes since it is critical in governing the control characteristic of the process. Slowly changing fluctuations in variables outside the control system can be reduced by integral control. The function of integral control is to integrate the deviation of the control temperature from the temperature at the set point over a long period and to correct the input power W. The effect of integral control compensating for changes in ambient temperature by restoring the proportional band to its correct level is illustrated in *Figure 48.8(a)* and the response is given by $p=f\int e\,\mathrm{d}t$. The integral time constant can be adjusted to suit the process parameters so that it compensates only for long-term variations and does not respond to small changes.

In any temperature control system, since control only occurs within the proportional band, the integral action may move the proportional band so far up-scale that the set point may be reached before the proportional band. To offset the effect of integral control during start-up, the proportional band may be automatically shifted to a lower temperature.

A sudden fluctuation about the set point can only be slowly corrected by proportional control. This can be compensated for by differential control which responds to fluctuations over only a short period and results in a restoring signal much greater than that due to proportional control alone. The effect of derivative control in response to a sudden change is shown in *Figure 48.8(b)* and is given by

$$p=r\frac{\mathrm{d}e}{\mathrm{d}t} \quad (48.13)$$

The derivative time constant can be adjusted to increase the power input in the proportional band over a period sufficiently long to compensate for a rapid fluctuation.

48.6 Analogue control techniques

48.6.1 The electronic controller

Figure 48.9 Microprocessor-based communicating three-term temperature controller suitable for networking with other controllers (Courtesy of Eurotherm Ltd)

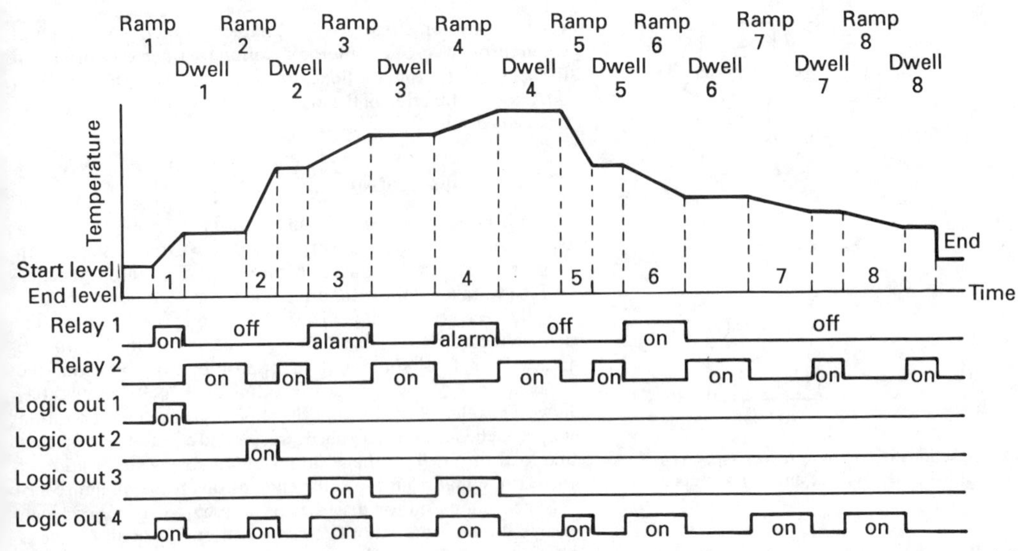

Figure 48.10 Example of an eight-ramp/dwell program utilising a logic interface board with two relays operating and the four logic outputs driving external events

Controllers using microprocessors (*Figure 48.9*) have largely superseded analogue controllers. The use of a microprocessor enables the conventional three-term (PDI) functions to be obtained combined with digital readout of the control variable and error. The controller may also be used to give different formats such as dwell and ramp with the programme state indication. Several controllers can be networked for multizone process control. An example of a complex temperature control programme is shown in *Figure 48.10*.

48.6.2 Analogue computers in control systems

The analogue computer is based on the use of operational amplifiers to carry out arithmetical operations. These include summation, integration, differentiation and scaling. Multiplication and division and nonlinear functions can also be obtained.

Electronic analogue computers may be used to simulate complex systems. Changes in the parameters may be made by varying resistance and capacitance. These may be at preset values or varied according to a preset programme. Analogue computers are usually designed for one specific application although some degree of versatility may be achieved by varying the interconnection of the computer modules.

The analogue computer is capable of dealing with continuous quantities rather than discrete steps and operates in real time. Complex functions can be simulated without the necessity for the large stores and high number of iterative operations required by digital computers. A further advantage is the ability to relate directly the performance of the analogue computer with the process itself and the simplicity of programming.

Notwithstanding these advantages, digital computers have in many areas replaced analogue computers and the falling price of digital computers is likely to increase this trend. The greater accuracy, increased flexibility and the capability of dealing with large numbers of interrelated variables tend to outweigh many of the advantages of analogue computers except for relatively simple applications. Nevertheless, analogue computers still have and will retain an important part in electronic control systems.

48.7 Servomotors

Servomotors are used in an enormous variety of applications ranging from miniature motors used in data recorders to motors with ratings in excess of 1 MW for winding machines used in mines. Both d.c. and a.c. motors are used.

The d.c. commutator motor is the most commonly encountered servomotor drive. Capable of continuous variation in speed from zero to full speed, the torque and power output can be continuously controlled over wide ranges. Although the a.c. squirrel cage induction motor is most extensively used in industry where constant speed drives are required, it is not suitable for applications where wide variations in speed are needed.

48.7.1 Control of speed and torque

Field connections of the d.c. commutator motor are illustrated schematically in *Figure 48.11*. The field may be separately excited, shunt excited, series excited or, a combination of series and shunt excitation (compound excitation). Where very small motors are used the field coil is replaced by a permanent magnet.

The principle of operation of each is similar although the interaction of field and armature currents varies enabling different operating characteristics to be obtained. The simplest machine to consider is the separately excited motor which is extensively used in control systems. The flux from the field coil is proportional to the field current, i.e. $\varphi \propto I_f$. The induced voltage (back e.m.f.) in the armature is proportional to

$$E \propto \frac{N\,\mathrm{d}\varphi}{\mathrm{d}t} \qquad (48.14)$$

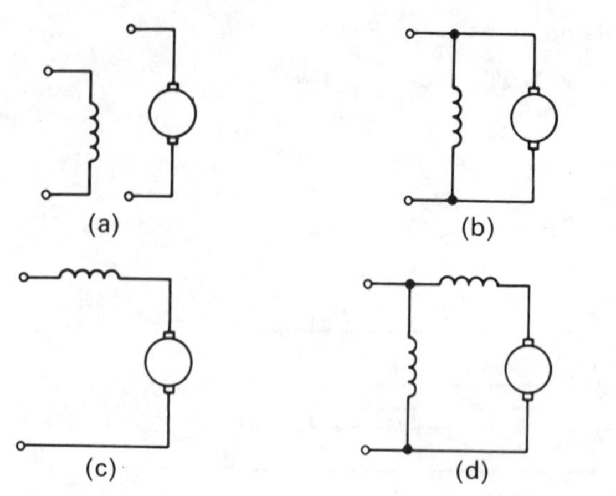

Figure 48.11 Field connections of d.c. motors: (a) separately excited; (b) shunt excited; (c) series excited; and (d) compound excited

which in turn is proportional to $V-I_aR_a$. Hence

$$N \propto \frac{V-I_aR_a}{\varphi} \qquad (48.15)$$

and the torque $T \propto \varphi I_a$.

The rotor should be capable of being rapidly accelerated where the acceleration is given by

$$\frac{d^2\theta}{dt^2}=TgJ \text{ rad/s}^2 \qquad (48.16)$$

where θ is the angular displacement and J is the moment of inertia.

The rotor should be sufficiently damped to reduce overrun where the damping coefficient is given by

$$D=\frac{dT}{dN}gs \qquad (48.17)$$

And the time constant of the rotor should be sufficiently short where

$$t_n=\frac{J}{D}\text{ s} \qquad (48.18)$$

The time required to reverse is approximately $1.7t_n$.

Speed control of motors can be derived directly from the armature current at constant field current. Speed control to an accuracy of 5–15% can be achieved in this way. If armature compensation is used accuracies of $2\frac{1}{2}$ to 5% are possible.

Change of the direction of rotation is obtained by reversing the direction of either the armature current or field current. Split field windings are sometimes used to enable continuous variation of speed through zero. The motor is stationary when the currents in the field are such that the fluxes oppose each other. The speed and direction is a function of the magnitude and direction of the out of balance current.

48.7.2 Position control

The synchro or resolver normally consists of a polyphase stator with a single or polyphase rotor. A single polyphase stator with single phase rotor is shown in *Figure 48.12*. The stator is excited and the voltages in the rotor are a function of its angular position.

If the receiver (which is identical) is also excited by the same source the rotor will be driven by the transmitter until the induced e.m.f. is equal but in opposition to that from the transmitter. A resolution corresponding to about 0.3° is possible, however as the receiver torque decreases as the required position is approached the error signal decreases and effects due to friction and load torque limit the accuracy obtainable. Amplification of the error signal from the rotor enables this to be minimised.

Synchronous motor drives may be used where a very high degree of precision is not necessary and only a small torque is required. The rotational speed is a function of the frequency and power and high frequency supplies may be used. Resolution to about 1/100 of a revolution is possible. A multiple tooth permanent magnet rotor is driven by a two-phase field winding at power or higher frequencies. The two-phase supply can be derived from a single-phase supply as shown in *Figure 48.13* or with a Scott-connected transformer. A 100-tooth rotor operated at 50 Hz has a slew speed of 60 r.p.m.

A stepper motor translates a digital signal into angular movement. Variable reluctance motors use a soft iron rotor in which teeth and slots are cut. A stepper motor is illustrated schematically in *Figure 48.14*. The wound stator has corresponding teeth and slots. When the stator is energised the motor aligns to minimise the reluctance of the air gap. Rotation is controlled and synchronised by input pulses to the stator winding. Variable reluctance motors are capable of very high stepping rates at small stepping angles making them ideally suited for control of position to a high degree of precision. An additional advantage of variable reluctance motor drives is the small compact dimensions that can be achieved. Higher speeds with lower rotor inertia than permanent magnet motors are obtainable, however, the efficiency and power output are lower. Typical step angles vary between 0.9° to 90°.

Another form of construction uses a permanent magnet rotor and wound stator. The rotor rotates until it reaches an equilibrium position opposite to an opposing pole. If the stator windings are excited in sequence the rotor rotates. An alternative construction uses a toothed soft iron rotor and a stator comprising a permanent magnet and control winding. The control winding is used to alter the distribution of the field resulting in stepwise rotation. A feature of this design is the high

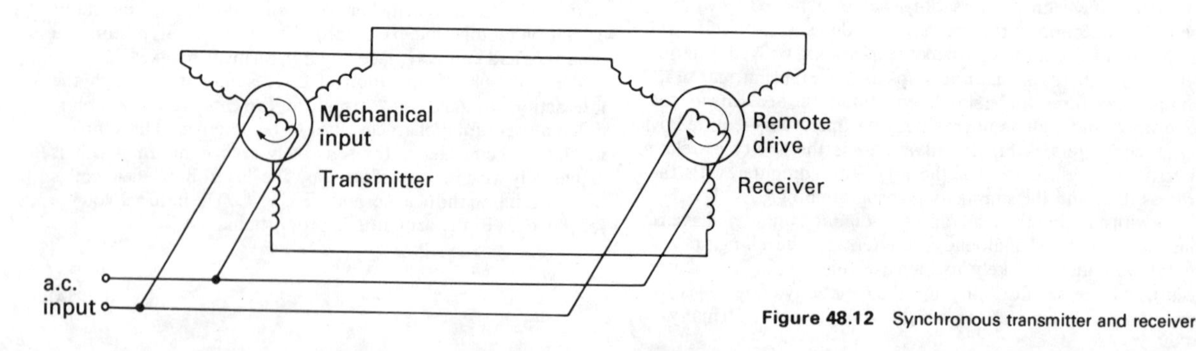

Figure 48.12 Synchronous transmitter and receiver

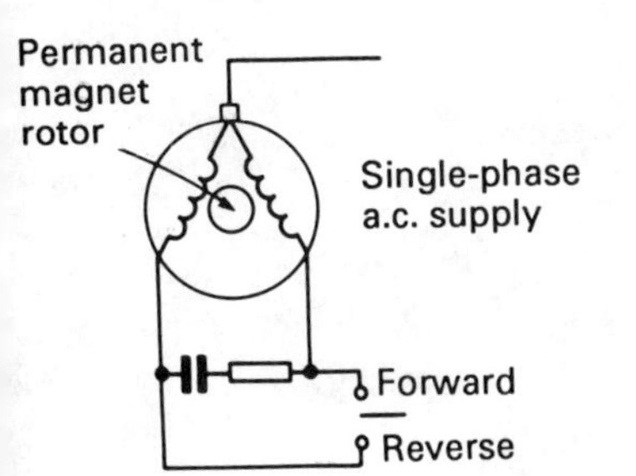

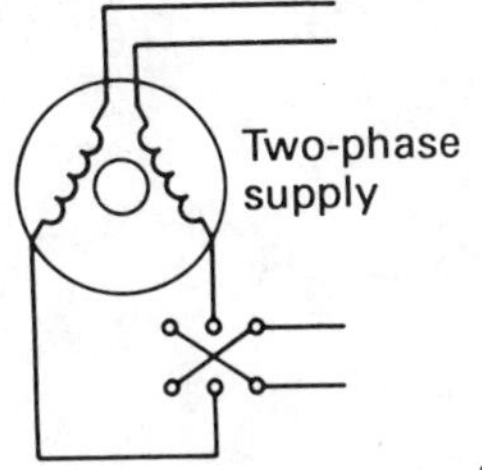

Figure 48.13 Synchronous motor drives

Figure 48.14 Variable reluctance stepper motor

residual torque obtained when the control winding is not energised.

The drive signals for stepper motors are usually obtained from pulses derived from logic circuits operating in sequence. In its simplest form where the motor is operated in on/off mode within its synchronous range the drive circuit is in the form of a transistor switch for each motor coil. This is controlled by a logic circuit to switch the coils in the correct sequence. The complexity of the logic circuit will depend on the number of stator windings. The voltage and current required to drive the motor may vary from a few volts and 50 mA to more than 20 A at 100 V. Stepping speeds from zero to more than 10 000 rev/s are possible.

Further reading

DOEBLIN, E. O., *Control System Principles and Design*, Wiley, New York (1985)

HEALEY, M., *Principles of Automatic Control*, 2nd edn., English Universities Press, London (1970)

IEE, *Electrical Variable Speed Drives*, IEE Conference Publication, Number 93 (1972)

KENJO, T., *Stepping Motors and their Microprocessor Controls*, Clarendon Press, Oxford (1986)

KUO, B. C., *Automatic Control Systems*, 5th edn, Prentice-Hall, Englewood Cliffs, NJ (1987)

LUKE, H. D., *Automation for Productivity*, Wiley, New York (1972)

PIKE, C. H., *Automatic Control of Industrial Drives*, Newnes-Butterworth, London (1971)

RAVEN, F. H., *Automatic Control Engineering*, 2nd edn., McGraw-Hill, New York (1968)

SPEISS, R., 'Hybrid computer aids process control development', *Mini-Micro Systems*, August (1982)

TABAK, D. and KUO, B. S., *Optimal Control by Mathematical Programming*, Prentice-Hall, Englewood Cliffs, NJ (1971)

Antennas and Arrays

A D Monk, MA, CEng, MIEE
Consultant Engineer

Contents

49.1 Fundamentals

49.1.1 Antenna function and performance

The function of an antenna is to provide a transition between a transmission line or a voltage or current feed point, and a plane wave (or more exactly a spectrum of plane waves) propagating in free space. The most important description of the performance of an antenna is its radiation pattern or polar diagram, which quantifies the directional characteristics of the interface with free space provided by the antenna. For the majority of antenna types the most important parameters which may be obtained from the radiation pattern are the peak gain, the beamwidth of the main lobe between half power (3 dB) points, and the levels of the sidelobes relative to the main lobe peak. Other parameters which may be of interest are the operational bandwidth, VSWR or impedance match, front-to-back ratio, and field polarisation properties.

49.1.1.1 Antenna gain

The power gain (generally abbreviated to gain) of an antenna in a specified direction is defined as 4π times the ratio of the power per unit solid angle radiated in that direction, to the net power accepted by the antenna from its generator. If the direction is not specified then it is generally assumed that the 'gain' refers to the maximum or peak value. The mathematical form of this definition is

$$G(\theta, \varphi)=\frac{4\pi\Phi(\theta, \varphi)}{P_0} \qquad (49.1)$$

An alternative but exactly equivalent definition is that the gain is the ratio of the power per unit solid angle radiated by the antenna, to the power per unit solid angle which would be radiated by a lossless isotropic radiator with the same power accepted at its terminals. From this definition the reason for the choice of dBi or 'dB above isotropic' as the unit of antenna gain is apparent.

49.1.1.2 Antenna directivity

The antenna gain G may be expressed as the product

$$G=\eta D \qquad (49.2)$$

where η is the antenna efficiency and accounts for all the losses in the antenna (ohmic loss, transmission line loss, impedance mismatch, etc.), and D is the antenna directivity, which is a function only of the antenna radiation pattern shape. Except in the hypothetical case of a lossless antenna, the gain will always be less than the directivity.

As with antenna gain, antenna directivity may be defined in two exactly equivalent ways, either as the ratio of the radiated power density in the given direction, to the average radiated power density

$$D(\theta, \varphi)=\frac{\Phi(\theta, \varphi)}{\Phi_{av}} \qquad (49.3)$$

or as the ratio

$$D(\theta, \varphi)=\frac{P_t'(\theta, \varphi)}{P_t} \qquad (49.4)$$

where P_t is the total power radiated by the antenna, and $P_t'(\theta, \varphi)$ is the power which would have to be radiated by an isotropic antenna to provide the same power density in the given direction.

From the first definition, the antenna directivity in the given reference direction may be written in terms of the radiated power density in the reference direction Φ_r, and the radiated power density at other angles $\Phi(\theta, \varphi)$.

$$D_r=\frac{\Phi_r}{(P_t/4\pi)}=\frac{4\pi\Phi_r}{\int_0^\pi\int_0^{2\pi}\Phi(\theta, \varphi)\,d\varphi\sin\theta\,d\theta} \qquad (49.5)$$

Thus the directivity may be computed by integrating a measured or theoretical radiation pattern $\Phi(\theta, \varphi)$ over the surface of a sphere.

49.1.2 Field strength and power density relationships in a plane wave

For a plane wave propagating in free space, if the electric and magnetic field strengths E and H are expressed in V/m and A/m respectively, they are related by

$$E=120\pi H \qquad (49.6)$$

where $120\pi\,\Omega \simeq 377\,\Omega$ is known as the impedance of free space.

The power density p expressed in W/m^2 is given in terms of the field strengths as

$$p=E^2/120\pi=120\pi H^2 \qquad (49.7)$$

49.1.3 Radiated power density and field strengths

If a power P_0 is accepted by an antenna whose gain in a given direction is G, then the radiated power density and field strengths in that direction in the antenna far-field at a distance r are

$$p=\frac{P_0G}{4\pi r^2} \qquad (49.8)$$

$$E=\frac{\sqrt{30P_0G}}{r} \qquad (49.9)$$

$$H=\sqrt{\frac{P_0G}{30}}(4\pi r)^{-1} \qquad (49.10)$$

(all units are m.k.s. as before).

49.1.4 Received power

The power received by an antenna at its terminals when illuminated by a plane wave of power density p is

$$P_r=pA_e \qquad (49.11)$$

where A_e is known as the 'effective area' of the antenna, and is related to the antenna gain thus

$$A_e=\frac{G\lambda^2}{4\pi} \qquad (49.12)$$

The received power in terms of the power density and field strengths of the incident plane wave is therefore

$$P_r=\frac{pG\lambda^2}{4\pi}=\left(\frac{E\lambda}{4\pi}\right)^2\frac{G}{30}=30H^2\lambda^2G \qquad (49.13)$$

49.1.5 Aperture efficiency

The gain of a lossless antenna with a uniformly illuminated planar aperture of area A is

$$G_0=\frac{4\pi A}{\lambda^2} \qquad (49.14)$$

The aperture efficiency of an aperture antenna is the ratio of the actual gain to the gain if the same aperture were uniformly illuminated, assuming that the losses are the same in both cases. The aperture efficiency is given in terms of the aperture field distribution f as

$$\eta_a=\frac{1}{A}\frac{|\int_A f\,dA|^2}{\int_A|f|^2\,dA} \qquad (49.15)$$

49.2 Radiation from elemental sources

For the analysis of many antenna types, it is convenient to describe the radiation as occurring from elemental sources disposed across the radiating area or along the radiating length of the antenna. The two basic elemental sources are the electric and magnetic dipoles. The electrical dipole may be considered as an electric current I flowing along a length δL, where δL is much smaller than λ, and has a moment equal to $I\,\delta L$. Similarly the magnetic dipole may be considered as a closed loop of electric current I with an area δA, where the maximum loop dimension is much smaller than λ, and has a moment equal to $I\,\delta A$.

One example of an antenna which can be analysed as a distribution of elemental sources is the linear wire antenna, in which the continuous current distribution along its length is represented as an infinite number of electric dipole sources, each of vanishingly small length and locally constant current. The net radiated field is then equal to the sum or integral of the fields radiated by the elemental sources. Another example would be an aperture antenna such as an electromagnetic horn, with the aperture E and H fields replaced by a combination of magnetic and electric dipoles respectively, by invoking the equivalence principle.

49.2.1 The electric dipole

49.2.1.1 General solution

For a short length δL of electric current I, i.e. an electric dipole, orientated along the z axis of a standard spherical coordinate system as shown in *Figure 49.1(a)*, it may be shown that the components of the radiated electric and magnetic fields in free space are

$$E_r = 60k^2 I\,\delta L\left[\frac{1}{(kr)^2} - \frac{j}{(kr)^3}\right]\cos\theta\, e^{-jkr} \qquad (49.16)$$

$$E_\theta = j30k^2 I\,\delta L\left[\frac{1}{kr} - \frac{j}{(kr)^2} - \frac{1}{(kr)^3}\right]\sin\theta\, e^{-jkr} \qquad (49.17)$$

$$H_\phi = j\frac{k^2}{4\pi} I\,\delta L\left[\frac{1}{kr} - \frac{j}{(kr)^2}\right]\sin\theta\, e^{-jkr} \qquad (49.18)$$

$$E_\phi = H_r = H_\theta = 0 \qquad (49.19)$$

where r is the distance from the dipole to the observation point, and k is the free-space wave number $2\pi/\lambda$. If I is in amperes and δL, r and λ are in metres, then E will be given in volts/metre and H in amperes/metre.

49.2.1.2 Near-field solution

Very close to the dipole, nominally $r<0.01\lambda$, only the E_r and E_θ fields are significant and these become

$$E_r \simeq -j60I\,\delta L\cos\theta(e^{-jkr}/r^3) \qquad (49.20)$$

$$E_\theta \simeq -j30I\,\delta L\sin\theta(e^{-jkr}/r^3) \qquad (49.21)$$

Under this condition these two components are in time phase, with amplitudes simply related

$$E_r/E_\theta \simeq 2\cot\theta \qquad (49.22)$$

49.2.1.3 Far-field solution

Very far from the dipole, nominally $r>5\lambda$, the radial E_r field component vanishes, and E_θ and H_ϕ become

$$E_\theta \simeq j(60\pi/\lambda)I\,\delta L\sin\theta(e^{-jkr}/r) \qquad (49.23)$$

$$H_\phi \simeq (j/2\lambda)I\,\delta L\sin\theta(e^{-jkr}/r) = E_\theta/120\pi \qquad (49.24)$$

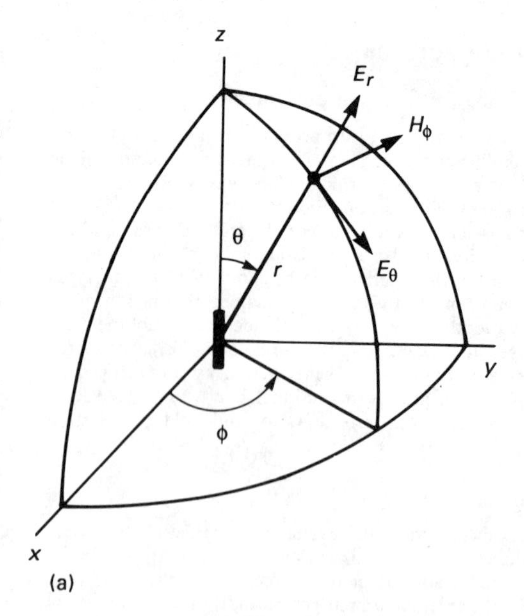

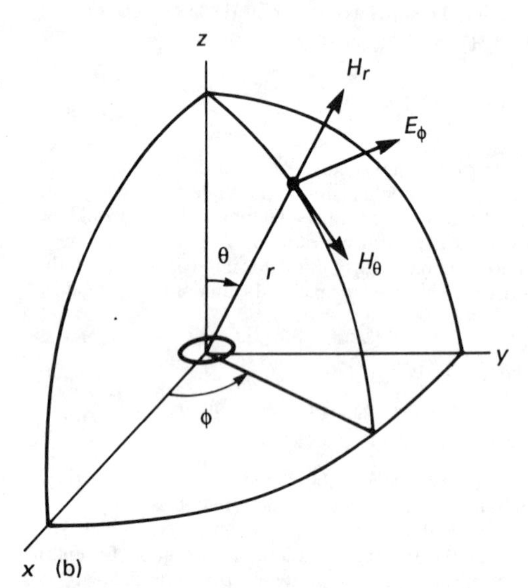

Figure 49.1 Radiation from elemental electric and magnetic dipoles: (a) electric dipole, (b) magnetic dipole

The far-field amplitude is thus proportional to $\sin\theta$, giving the familiar 'figure of eight' voltage pattern in the planes containing the current element, when plotted in polar form. It is also apparent that the far-field amplitude is proportional to $1/r$, so that the power density is proportional to $1/r^2$, giving the inverse square law relationship with distance.

49.2.2 The magnetic dipole

49.2.2.1 General solution

For a small loop area δA of electric current I, i.e. a magnetic dipole, in the xy plane of a standard spherical coordinate system as shown in *Figure 49.1(b)*, it may be shown that the components of the radiated electric and magnetic fields in free space are

$$E_\phi = 30k^3 I\,\delta A\left(\frac{1}{kr} - \frac{j}{(kr)^2}\right)\sin\theta\,e^{-jkr} \quad (49.25)$$

$$H_r = j\frac{k^3}{2\pi} I\,\delta A\left(\frac{1}{(kr)^2} - \frac{j}{(kr)^3}\right)\cos\theta\,e^{-jkr} \quad (49.26)$$

$$H_\theta = \frac{k^3}{4\pi} I\,\delta A\left(\frac{1}{kr} - \frac{j}{(kr)^2} - \frac{1}{(kr)^3}\right)\sin\theta\,e^{-jkr} \quad (49.27)$$

$$E_r = E_\theta = H_\phi = 0 \quad (49.28)$$

where the units and symbols are the same as for the electric dipole.

49.2.2.2 Near-field solution

Very close to the dipole, nominally $r < 0.01\lambda$, only the H_r and H_θ fields are significant, and these become

$$H_r \simeq \frac{1}{2\pi} I\,\delta A \cos\theta(e^{-jkr}/r^3) \quad (49.29)$$

$$H_\theta \simeq -\frac{1}{4\pi} I\,\delta A \sin\theta(e^{-jkr}/r^3) \quad (49.30)$$

so that H_r and H_θ are related in a way similar to E_r and E_θ in the near-field of an electric dipole

$$H_r/H_\theta \simeq -2\cot\theta \quad (49.31)$$

49.2.2.3 Far-field solution

Very far from the dipole, nominally $r > 5\lambda$, the radial H_r field vanishes, and E_ϕ and H_θ become

$$E_\phi \simeq 30k^2 I\,\delta A \sin\theta(e^{-jkr}/r) \quad (49.32)$$

$$H_\theta \simeq \frac{k^2}{4\pi} I\,\delta A \sin\theta(e^{-jkr}/r) = \frac{E_\phi}{120\pi} \quad (49.33)$$

The $\sin\theta$ factor corresponds to the same 'figure of eight' polar diagram as for the electric dipole, in this case in the planes perpendicular to the plane of the loop.

49.2.3 Elemental aperture field source

49.2.3.1 The equivalence principle

The equivalence principle[1,2] is a powerful technique for analysing the radiation from apertures within which the tangential **E** and **H** field distribution is known. Stated briefly, if the field distribution is known across a closed surface surrounding the antenna, then it is possible to define a distribution of electric and magnetic currents, **J** and **M** respectively, across the surface such that the fields everywhere inside that surface are zero

$$\mathbf{J} = \hat{\mathbf{n}} \times \mathbf{H} \quad (49.34)$$

$$\mathbf{M} = \hat{\mathbf{n}} \times \mathbf{E} \quad (49.35)$$

where $\hat{\mathbf{n}}$ is the outwards unit vector normal to the surface. The radiated fields due to the antenna are then exactly equal, except for a change of sign, to the fields radiated by the electric and magnetic currents.

This principle may be applied to a large class of aperture antennas. For example, for an electromagnetic horn, if it is assumed that the field distribution at the aperture has the same form as that of the waveguide modes exciting the horn at its throat, then the radiated fields may be obtained from that aperture field distribution. For a reflector antenna it may be assumed that the fields at the reflector surface due to either illumination by the feed (transmit mode analysis) or by an incoming plane wave (receive mode analysis), are the same as would exist without the reflector present, except that reflection of the wave at the surface cancels the tangential **E** field and doubles the tangential **H** field. The radiation from an element on the reflector surface is then identical to the radiation from the elemental electric current source

$$\mathbf{J} = 2\hat{\mathbf{n}} \times \mathbf{H} \quad (49.36)$$

For an element $dx\,dy$ of a radiating aperture in which the tangential **E** and **H** fields E_x^a and H_y^a are known (see *Figure 49.2*),

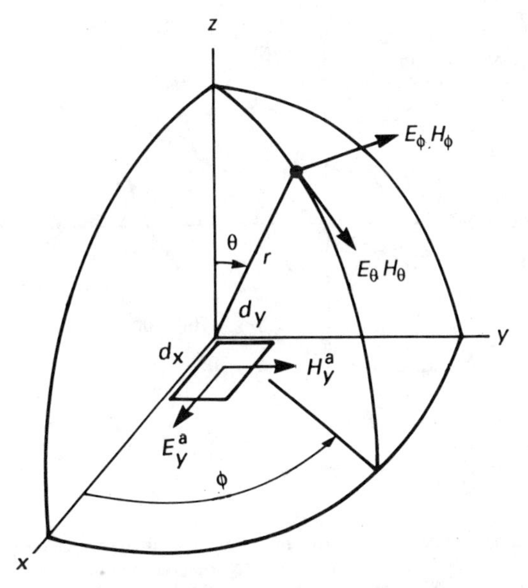

Figure 49.2 Radiation from elemental aperture

the far-field components of the radiated **E** field are

$$E_\theta = -j(E_x^a + 120\pi H_y^a \cos\theta)\frac{\cos\varphi}{2\lambda}\,dx\,dy(e^{-jkr}/r) \quad (49.37)$$

$$E_\phi = j(E_x^a \cos\theta + 120\pi H_y^a)\frac{\sin\varphi}{2\lambda}\,dx\,dy(e^{-jkr}/r) \quad (49.38)$$

For the special case in which the E and H aperture field amplitudes are related by the impedance of free space

$$E_x^a = 120\pi H_y^a \quad (49.39)$$

so that the equivalent current source is a balanced Huygens source, the far-fields become

$$E_\theta = -jE_x^a(1+\cos\theta)\frac{\cos\varphi}{2\lambda}\,dx\,dy(e^{-jkr}/r) \quad (49.40)$$

$$E_\phi = jE_x^a(1+\cos\theta)\frac{\sin\varphi}{2\lambda}\,dx\,dy(e^{-jkr}/r) \quad (49.41)$$

49.3 Antenna polarisation

49.3.1 The polarisation ellipse

The polarisation attributed to an antenna is the polarisation or orientation in space of the field vectors it radiates into space. It has long been a universal convention that antenna or wave polarisation refers to the orientation of the electric or **E** field rather than the magnetic or **H** field. Thus when for example an antenna is referred to as being vertically polarised, the radiated **E** field vector is in the vertical plane and the **H** field vector in the horizontal plane.

For an arbitrarily polarised wave, in a fixed plane normal to the direction of propagation, the locus of the tip of the **E** field vector with respect to time will be an ellipse—the polarisation ellipse. This is shown in *Figure 49.3*. The wave polarisation may

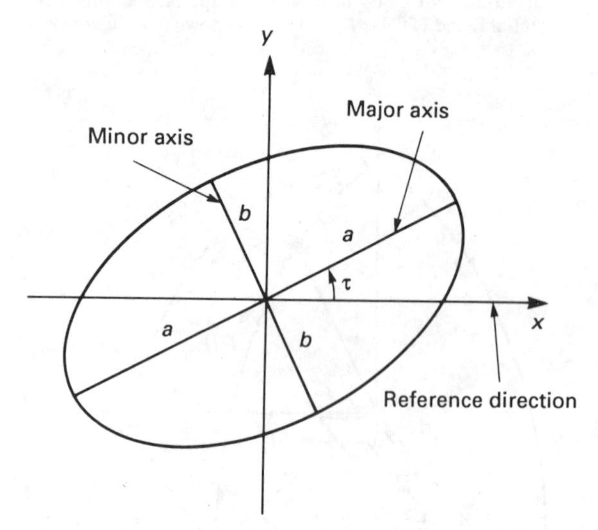

Figure 49.3 The polarisation ellipse

be defined completely by three parameters, all of which are directly related to the geometry of the polarisation ellipse. The first of these is the voltage axial ratio A, which is equal to the ratio of the lengths of the major and minor axes of the ellipse, i.e.

$$A = a/b \tag{49.42}$$

The second is the angle τ between the ellipse major axis and the reference direction, in this case the x-axis. The third is the sense of rotation of the tip of **E** field vector around the ellipse. A clockwise rotation as viewed in the direction of propagation is termed right-hand elliptical polarisation, and rotation anticlockwise is termed left-hand elliptical pollarisation. It is conventional to restrict the axial ratio to the range of values $1 < A < \infty$. A wave polarisation with axial ratio A, which is less than unity, and tilt angle τ is identical to the polarisation with axial ratio $1/A$, which is greater than unity, and tilt angle $\tau \pm \pi/2$. The axial ratio is also often expressed in dB where

$$\text{dB} = 20 \lg A \tag{49.43}$$

so that an axial ratio of unity is 0 dB.

The two special cases are linear and circular polarisation. For linear polarisation the axial ratio is infinity, with the tilt angle τ defining the orientation (the sense of rotation is redundant). For circular polarisation the axial ratio is unity or 0 dB, and the sense of circular polarisation is either right-hand (RHCP) or left-hand (LHCP) (the angle τ is redundant). A circularly polarised wave or field may be considered to comprise two equal and orthogonal linearly polarised components in phase quadrature. The sense of phase quadrature determines the hand of circular polarisation.

49.3.2 Polarisation coupling loss

The power coupling between two antennas or between a receiving antenna and an incident plane wave will depend on the relative polarisation of the two antennas or antenna and incident wave. Maximum coupling with zero polarisation coupling loss will occur when the two polarisations are identical. Zero coupling will occur when the two polarisations are orthogonal.

Quantitatively the polarisation power coupling loss P $(0 \leqslant P \leqslant 1)$ is given by

$$P = \tfrac{1}{2}[1 + \sin 2\theta_1 \sin 2\theta_2 + \cos 2\theta_1 \cos 2\theta_2 \cos 2(\tau_1 - \tau_2)] \tag{49.44}$$

where the two axial ratios are $\tan \theta_1$ and $\tan \theta_2$, and the polarisation ellipse tilt angles (referred to the same reference direction) τ_1 and τ_2. The sign of θ_1 and θ_2 defines the rotation sense or hand.

The maximum and minimum power coupling factors, which occur for $\tau_1 - \tau_2 = 0$ and $\pi/2$ respectively, are

$$P_{\max} = \cos^2(\theta_1 - \theta_2) \tag{49.45}$$

$$P_{\min} = \sin^2(\theta_1 + \theta_2) \tag{49.46}$$

When one polarisation is linear so that $\theta_1 = \pi/2$, the coupling becomes

$$P_{\text{lin}} = \tfrac{1}{2}[1 - \cos 2\theta_2 \cos 2(\tau_1 - \tau_2)] \tag{49.47}$$

which has maximum and minimum values of $\sin^2 \theta_2$ and $\cos^2 \theta_2$ respectively.

When one polarisation is circular so that $\theta_1 = \pi/4$, the coupling is independent of the orientation of the second polarisation ellipse, and is given by

$$P_{\text{cp}} = \tfrac{1}{2}(1 + \sin 2\theta_2) \tag{49.48}$$

49.3.3 Orthogonal polarisations and cross-polarisations

For every polarisation state with axial ratio A and tilt angle τ there is a unique orthogonal polarisation state which has the same axial ratio, tilt angle $\tau \pm \pi/2$ and the opposite sense of rotation. For linear polarisation the orthogonal polarisation is also linear but with the orientation rotated through 90°, and for circular polarisation the orthogonal polarisation is circular of the opposite hand. There is zero coupling between two orthogonally polarised antennas or between a plane wave and an orthogonally polarised receiving antenna.

The concept of polarisation orthogonality is sometimes applied to provide two separate channels from a single antenna, by using two space-orthogonal linear polarisations, or LHCP and RHCP. The most notable example of such an application is frequency reuse in satellite communications systems. A more widely used application of the concept is in a quantitative description of the polarisation purity of an antenna. In any direction in space it is possible to define a 'wanted' and 'unwanted' or 'co' and 'cross' orthogonal polarisation pair, so that the polarisation purity in that direction may be expressed as the ratio of the cross-polar to copolar components of the vector radiation pattern in the same direction.

For a nominally circularly polarised antenna the copolar and cross-polar polarisations are simply LHCP and RHCP in all directions. For a nominally linearly polarised antenna, although in the principal planes the definition of copolar and cross-polar polarisations is obvious, there is some ambiguity in other directions. Ludwig[3] has examined this problem in some detail and has identified three alternative definitions for linear copolarisation and cross-polarisation—Ludwig's first, second and third definitions. The three definitions for an antenna nominally linearly polarised in the y direction are shown graphically in *Figure 49.4* and are described mathematically by

$$\hat{\mathbf{a}}_{\text{co}}^{(1)} = \sin\theta \sin\varphi \hat{\mathbf{r}} + \cos\theta \sin\varphi \hat{\boldsymbol{\theta}} + \cos\varphi \hat{\boldsymbol{\varphi}} \tag{49.49}$$

$$\hat{\mathbf{a}}_{\text{cross}}^{(1)} = \sin\theta \cos\varphi \hat{\mathbf{r}} + \cos\theta \cos\varphi \hat{\boldsymbol{\theta}} - \sin\varphi \hat{\boldsymbol{\varphi}} \tag{49.50}$$

$$\hat{\mathbf{a}}_{\text{co}}^{(2)} = (\sin\varphi \cos\theta \hat{\boldsymbol{\theta}} + \cos\varphi \hat{\boldsymbol{\varphi}})(1 - \sin^2\theta \sin^2\varphi)^{-1/2} \tag{49.51}$$

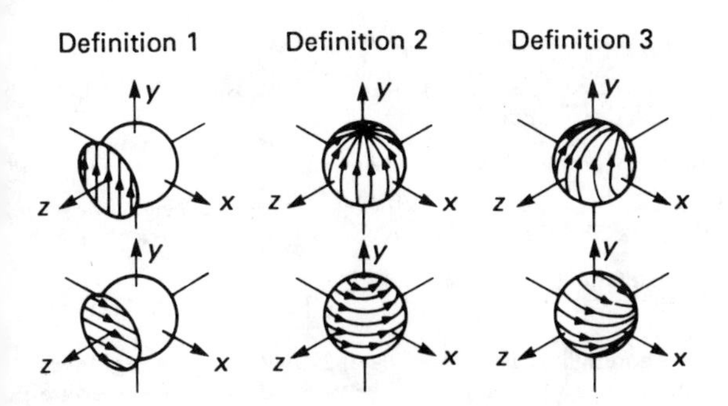

Figure 49.4 Ludwig's three definitions of cross-polarisation. Top, direction of the reference polarisation. Bottom, direction of the cross-polarisation

$$\hat{\mathbf{a}}^{(2)}_{\text{cross}} = (\cos\varphi\hat{\boldsymbol{\theta}} - \cos\theta\sin\varphi\hat{\boldsymbol{\varphi}})(1-\sin^2\theta\sin^2\varphi)^{-1/2} \quad (49.52)$$

$$\hat{\mathbf{a}}_{\text{co}}^{(3)} = \sin\varphi\hat{\boldsymbol{\theta}} + \cos\varphi\hat{\boldsymbol{\varphi}} \quad (49.53)$$

$$\hat{\mathbf{a}}^{(3)}_{\text{cross}} = \cos\varphi\hat{\boldsymbol{\theta}} - \sin\varphi\hat{\boldsymbol{\varphi}} \quad (49.54)$$

where $\hat{\mathbf{a}}_{\text{co}}$ and $\hat{\mathbf{a}}_{\text{cross}}$ are the unit vectors in the copolar and cross-polar directions expressed in terms of the unit vectors $\hat{\mathbf{r}}$, $\hat{\boldsymbol{\theta}}$ and $\hat{\boldsymbol{\varphi}}$ in a standard spherical coordinate system. The copolarised and cross-polarised components of the antenna field $\mathbf{E}$ are

$$E_{\text{co}} = \mathbf{E}\cdot\hat{\mathbf{a}}_{\text{co}} \quad (49.55)$$

$$E_{\text{cross}} = \mathbf{E}\cdot\hat{\mathbf{a}}_{\text{cross}} \quad (49.56)$$

Of the three definitions the third is the most widely used, and has the advantage that a reflector antenna feed having zero cross-polarisation according to Ludwig's third definition will produce parallel field lines in the reflector aperture, and will hence contribute no cross-polarisation to the secondary pattern. It also gives zero cross-polarisation for a balanced Huygens source. The second definition in contrast gives zero cross-polarisation for an electric dipole source, and has the disadvantage that the copolar field line pattern does not become orthogonal to itself when rotated about the z-axis by 90°.

49.3.4 Circular polarisation from orthogonal linear polarisations

Circular polarisation may be obtained by combining two equal amplitude linear polarisations in phase quadrature. Any amplitude imbalance between the two linear components, or departure from the quadrature phase relationship will yield elliptical polarisation with a non-unity axial ratio. This is equivalent to saying that the errors will introduce a circularly polarised component of the opposite hand to that desired, i.e. circular cross-polarisation.

If the two linear polarisation amplitudes are E_1 and E_2 and the relative phase difference is δ, then the ratio of RHCP and LHCP amplitude is

$$\frac{E_R}{E_L} = \left(\frac{E_1^2+E_2^2+2E_1E_2\sin\delta}{E_1^2+E_2^2-2E_1E_2\sin\delta}\right)^{1/2} \quad (49.57)$$

so that pure RHCP is obtained when $E_1 = E_2$ and $\delta = \pi/2$. The voltage axial ratio is given as

$$A = \frac{(E_R/E_L)+1}{(E_R/E_L)-1} \quad (49.58)$$

For the special case where the two linear components are in exact phase quadrature but are imbalanced in amplitude, the net axial ratio is simply equal to the ratio of the linear component amplitudes E_1/E_2.

For the other special case where the two linear components are equal in amplitude but are misphased by an angle Δ from the ideal quadrature relationship so that $\delta = \pi/2 \pm \Delta$, the next axial ratio is given by

$$A = \frac{(1+\cos\Delta)^{1/2}+(1-\cos\Delta)^{1/2}}{(1+\cos\Delta)^{1/2}-(1-\cos\Delta)^{1/2}} \underset{\Delta\to 0}{\simeq} \frac{1+\frac{1}{2}\Delta}{1-\frac{1}{2}\Delta} \quad (49.59)$$

where Δ is in radians.

A set of curves is plotted in *Figure 49.5*, showing how the axial ratio and circular polarisation purity vary with combinations of amplitude and phase imbalance between the linear components.

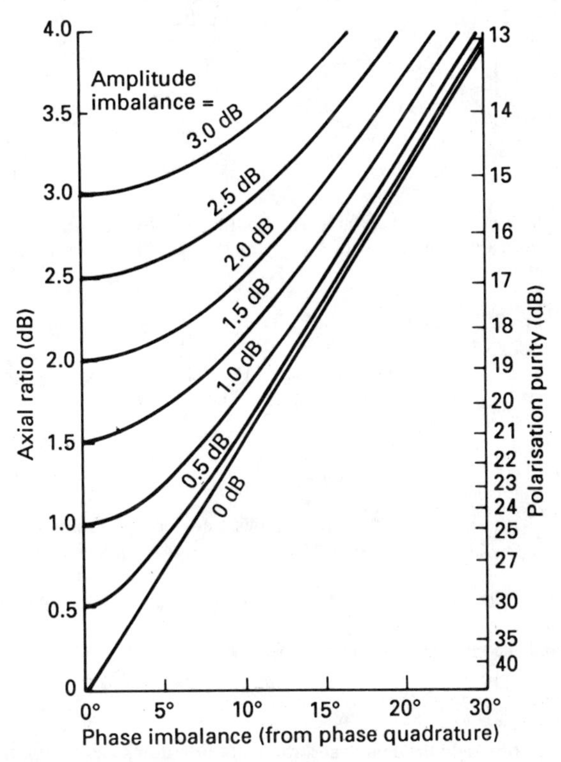

Figure 49.5 Circular polarisation purity versus orthogonal linear component amplitude and phase imbalance

49.4 Linear and planar arrays

Linear and planar arrays comprise elements positioned along a line or across a plane respectively. The excitation function or distribution for a planar array is generally separable in the x and y directions in rectangular coordinates, so that the principal plane patterns are also separable. Because of this, most planar arrays may be analysed in terms of the linear distributions in the two planes.

For an array with a fixed distribution, the main beam position will be fixed in space. If however a linear phase taper is introduced across the array, the main beam will squint, and if the distribution phase can be continuously or nearly continuously varied, then the beam can be continuously scanned in space without physical movement of the array. Such an antenna is known as a phased array, and is widely used in both radar and other applications.

49.4.1 Continuous line source distributions

The spacing between elements in an array is usually of the order of half a wavelength, so that the array pattern is to a fair approximation given by the product of the element pattern and the pattern of a continuous array of isotropic sources with the same aperture extent and distribution as the real array. It is therefore convenient, when comparing the properties of different array distributions, to consider and compare the properties of the equivalent line source distributions.

49.4.1.1 The Fourier transform

For a continuous linear aperture distribution $f(x)$ $(-a/2 \leqslant x \leqslant a/2)$ the far-field radiation pattern is given by

$$g(u)=\int_{-a/2}^{a/2} f(x)\,e^{-j2\pi ux}\,dx \qquad (49.60)$$

where

$$u=\frac{\sin\theta}{\lambda} \qquad (49.61)$$

which may be recognised as the Fourier transform.[4]

For a distribution with even symmetry, i.e. $f(-x)=f(x)$, the transform reduces to

$$g(u)=2\int_{0}^{a/2} f(x)\cos(2\pi ux)\,dx \qquad (49.62)$$

which is known as the cosine Fourier transform. For a distribution with odd symmetry, i.e. $f(-x)=-f(x)$, the transform reduces to

$$g(u)=2j\int_{0}^{a/2} f(x)\sin(2\pi ux)\,dx \qquad (49.63)$$

which is known as the sine Fourier transform. The aperture efficiency of an even symmetrical distribution $f(x)$ is given by

$$\eta_a=\frac{2}{a}\left|\int_{0}^{a/2} f(x)\,dx\right|^2\left[\int_{0}^{a/2} |f(x)|^2\,dx\right]^{-1} \qquad (49.64)$$

and the relative power in the aperture by

$$P=\frac{2}{a}\int_{0}^{a/2} |f(x)|^2\,dx \qquad (49.65)$$

The boresight field $g(0)$ according to Equations (49.60) and (49.62) is given by

$$g(0)=\sqrt{\eta_a P}\,a \qquad (49.66)$$

so that the pattern normalised to $g'(0)=1$ is simply

$$g'(u)=\frac{g(u)}{g(0)}=\frac{g(u)}{\sqrt{\eta_a P}\,a} \qquad (49.67)$$

49.4.1.2 Patterns of common equiphase distributions

In *Table 49.1* the radiation pattern equations and the values of the most important associated parameters for eight common even symmetrical equiphase distributions are tabulated. The beamwidth constant K is the ratio of the 3 dB beamwidth to the aperture width in wavelengths, for large apertures

$$\theta_3 \simeq K/(a/\lambda) \quad \text{(degrees)} \qquad (49.68)$$

The exact relationship valid for large and small apertures is

$$\theta_3=2\arcsin\left(\frac{\pi K}{360a/\lambda}\right) \qquad (49.69)$$

49.4.1.3 Cosine-squared on a pedestal distribution

A widely used linear distribution is the 'cosine-squared on a pedestal' distribution

$$f(x)=e+(1-e)\cos^2(\pi x/a) \qquad (49.70)$$

where $-a/2 \leqslant x \leqslant a/2$, and e is the relative pedestal height or edge taper. The absolute pattern $g(u)$ for this distribution is given by

$$\frac{g(u)}{a}=\frac{\sin(\pi au)}{\pi au}\left[e+\frac{(1-e)}{2}\,\frac{\pi^2}{\pi^2-(\pi au)^2}\right] \qquad (49.71)$$

with a boresight level $g(0)/a=(1+e)/2$. By varying the edge taper parameter e, a varying trade-off between sidelobe levels, aperture efficiency and beamwidth may be obtained. The aperture efficiency is given by

$$\eta_a=\left[1+\tfrac{1}{2}\left(\frac{1-e}{1+e}\right)^2\right]^{-1} \qquad (49.72)$$

and is plotted against the edge taper e together with the beamwidth constant K and the first four sidelobe levels in *Figure 49.6*.

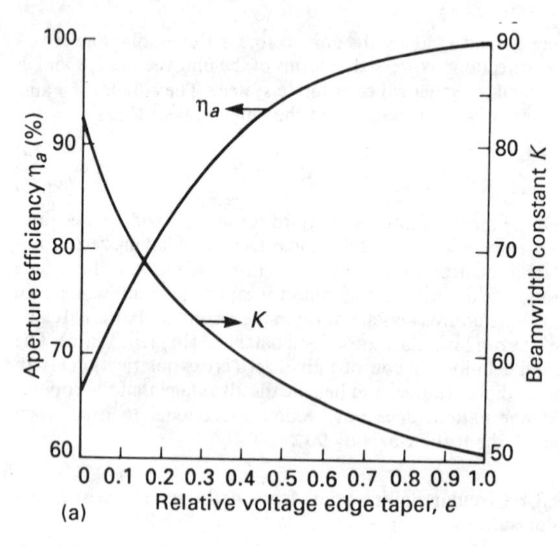

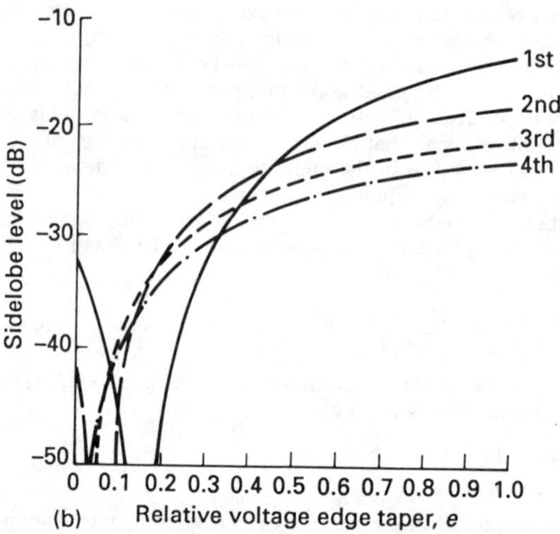

Figure 49.6 Properties of the cosine-squared on a pedestal distribution: (a) aperture efficiency and beamwidth constant, (b) first four sidelobe levels

Table 49.1 Patterns of common equiphase distributions

Distribution type	$f(x)$	$(z=\pi au)$ $g(u)/a$	*Aperture efficiency* η_a	*Relative power (P) in the aperture*	*3 dB beamwidth constant* K	*1st null position (au)*	*1st side-lobe level* (dB)	*2nd side-lobe level* (dB)	*1st side-lobe position (au)*	*2nd side-lobe position (au)*
Uniform	1	$\dfrac{\sin z}{z}$	1 (100%)	1	50.8	1.00	−13.3	−17.8	1.43	2.46
Cosine	$\cos\left(\pi\dfrac{x}{a}\right)$	$\left(\dfrac{\pi}{2}\right)\dfrac{\cos z}{(\pi/2)^2-z^2}$	$8/\pi^2$ (81.1%)	$\frac{1}{2}$	68.1	1.50	−23.0	−30.7	1.89	2.93
Cosine^2	$\cos^2\left(\pi\dfrac{x}{a}\right)$	$\dfrac{1}{2}\dfrac{\sin z}{z}\dfrac{\pi^2}{\pi^2-z^2}$	2/3 (66.7%)	$\frac{3}{8}$	82.5	2.00	−31.5	−41.5	2.36	3.41
Parabolic	$1-\left(\dfrac{2x}{a}\right)^2$	$\dfrac{2}{z^2}\left(\dfrac{\sin z}{z}-\cos z\right)$	5/6 (83.3%)	$\frac{8}{15}$	66.2	1.43	−21.3	−29.0	1.83	2.90
Parabolic^2	$\left[1-\left(\dfrac{2x}{a}\right)^2\right]^2$	$\dfrac{8}{z^2}\left[\dfrac{3}{z^2}\left(\dfrac{\sin z}{z}-\cos z\right)-\dfrac{\sin z}{z}\right]$	7/10 (70.0%)	$\frac{128}{315}$	78.8	1.83	−27.7	−37.8	2.23	3.32
Double cosine	$\left\lvert\sin\left(2\pi\dfrac{x}{a}\right)\right\rvert$	$\pi\dfrac{1+\cos z}{\pi^2-z^2}$	$8/\pi^2$ (81.1%)	$\frac{1}{2}$	53.4	1.00	−7.5	−23.3	1.65	3.89
$1\frac{1}{2}$ cycle cosine	$\cos\left(3\pi\dfrac{x}{a}\right)$	$-\left(\dfrac{3\pi}{2}\right)\dfrac{\cos z}{(3\pi/2)^2-z^2}$	$8/9\pi^2$ (9.01%)	$\frac{1}{2}$	29.7	0.50	+7.6	−9.2	1.40	2.91
Gable	$1-\left\lvert\dfrac{2x}{a}\right\rvert$	$\dfrac{1-\cos z}{z^2}=\frac{1}{2}\left[\dfrac{\sin(z/2)}{(z/2)}\right]^2$	3/4 (75.0%)	$\frac{1}{3}$	73.1	2.00	−26.5	−35.7	2.86	4.92

Table 49.2 Hamming distribution sidelobes

	1st	*2nd*	*3rd*	*4th*	*5th*	*6th*	*7th*
Level (dB)	−44.0	−56.0	−43.6	−42.7	−43.2	−44.1	−45.0
Position (*au*)	2.22	2.79	3.53	4.50	5.49	6.49	7.49

The value of the maximum sidelobe level is minimised for $e=0.08$, and this special case of the cosine-squared on a pedestal distribution is known as the Hamming distribution. The levels and angular positions of the first seven sidelobes are tabulated in *Table 49.2*.

49.4.1.4 The Taylor distribution

It may be shown that the distribution producing the narrowest beamwidth for a specified maximum sidelobe level is that which produces a pattern in the form of a modified Chebyshev polynomial, with all sidelobes of equal amplitude.[5] Such a distribution has become known as the Dolph–Chebyshev distribution. In practical cases this distribution has two disadvantages. Firstly, the distribution tends to require high amplitudes at the edge of the aperture, and secondly for large aperture sizes the directivity will be limited, asymptotically approaching 3 dB above the specified main beam to sidelobe level.

A modified set of distributions was proposed by Taylor,[6] producing equal amplitude sidelobes at the specified level out to the $\bar{n}$th sidelobe, with subsequent side lobes tapering off as $1/u$. The distributions of this form are more realizable in practice, and do not suffer from the gain limitation of the Dolph–Chebyshev distributions.

For a linear aperture of width a, the distribution is given by

$$f(p, A, \bar{n}) = 1 + 2\sum_{n=1}^{\bar{n}-1} F(n, A, \bar{n}) \cos np \tag{49.73}$$

where

$$F(n, A, \bar{n}) = [(n-1)!]^2 \prod_{m=1}^{\bar{n}-1} 1-(n/z_m)^2 \times [(\bar{n}-1+n)!\,(\bar{n}-1-n)!]^{-1} \tag{49.74}$$

$$p = 2\pi x/a \qquad (-a/2 \leqslant x \leqslant a/2) \tag{49.75}$$

$$A = \pi^{-1} \operatorname{arcosh} \eta \tag{49.76}$$

where η is the sidelobe voltage ratio, z_n is the nth pattern zero

$$z_n = \pm\sigma[A^2 + (n-\tfrac{1}{2})^2]^{1/2} \qquad 1 \leqslant n \leqslant \bar{n} \tag{49.77}$$

$$= \pm n \qquad \bar{n} \leqslant n < \infty \tag{49.78}$$

and σ is the pattern stretchout factor (slightly >1)

$$\sigma = n[A^2 + (\bar{n}-\tfrac{1}{2})^2]^{-1/2} \tag{49.79}$$

The resulting pattern is given by

$$g(z, A, \bar{n}) = \frac{\sin \pi z}{\pi z} \prod_{n=1}^{\bar{n}-1} \frac{1-(z/z_n)^2}{1-(z/n)^2} \tag{49.80}$$

where $z = (a/\lambda) \sin \theta$. The 3 dB beamwidth is given by

$$\theta_3 = 2 \arcsin\left(\frac{\lambda\sigma\beta_0}{2a}\right) \tag{49.81}$$

where

$$\beta_0 = 2 \arcsin\left\{\frac{1}{\pi}\left[(\operatorname{arcosh} \eta)^2 - \left(\operatorname{arcosh} \frac{\eta}{\sqrt{2}}\right)^2\right]^{1/2}\right\} \tag{49.82}$$

Comprehensive tables of the aperture distributions for various sidelobe levels and values of $\bar{n}$ are available.[7]

49.4.2 Discrete arrays

For a discrete equispaced array of N elements excited by a voltage or current distribution A_i ($i=1, N$), the far-field pattern is

$$E(\theta) = \sum_{i=1}^{N} A_i \exp[j\pi(s/\lambda) \sin \theta(2i-N-1)] \tag{49.83}$$

where s is the element spacing.

For element spacings in the range $\lambda/2 \leqslant s \leqslant \lambda$ the directivity is given by

$$D = 2\eta_a(N-1)s/\lambda \tag{49.84}$$

where the aperture or illumination efficiency η_a is given by

$$\eta_a = \left(\sum_{i=1}^{N} A_i\right)^2 \left(N \sum_{i=1}^{N} A_i^2\right)^{-1} \tag{49.85}$$

49.4.2.1 The uniform distribution

For a distribution uniform in both amplitude and phase, the far-field pattern is given by

$$E(\theta) = \frac{\sin[N\pi(s/\lambda) \sin \theta]}{N \sin[\pi(s/\lambda) \sin \theta]} \tag{49.86}$$

$$\underset{s/\lambda \sin\theta \to 0}{=} \frac{\sin[N\pi(s/\lambda) \sin \theta]}{N\pi(s/\lambda) \sin \theta} \tag{49.87}$$

where s is the element spacing. For large N therefore the close-in pattern approaches the Fourier transform of a continuous uniform distribution with a −13.3 dB first sidelobe level, and beamwidth constant $K=50.8$.

The directivities or gains for uniform broadside arrays with 2–10 elements are plotted in *Figure 49.7* for isotropic elements, and for short dipoles aligned normal to the array and collinear with the array. In all cases the gain is approximately linear with element spacing for $0.1\lambda < s < 0.9\lambda$, and for isotropic elements the gain is equal to the element number N when $s=\lambda/2$. For both the isotropic and normally orientated dipole elements the gain drops rapidly at $s \simeq \lambda$ due to the emergence of the grating lobe into real space. This effect is largely absent for colinear dipole elements, since the $\cos \theta$ dipole pattern suppresses the grating lobe as it appears in the end-fire mode. The gain of arrays with isotropic or collinear dipole elements are roughly equal for large element numbers, whereas the gain is significantly increased with normally orientated dipole elements.

49.4.3 Beam scanning

An equiphase linear array will produce a broadside main beam normal to the array. The beam may however be scanned away from the broadside direction by an angle θ_0 by applying a linear phase taper of α between consecutive elements to the distribution, where α is given by

$$\alpha = 2\pi(s/\lambda) \sin \theta_0 \tag{49.88}$$

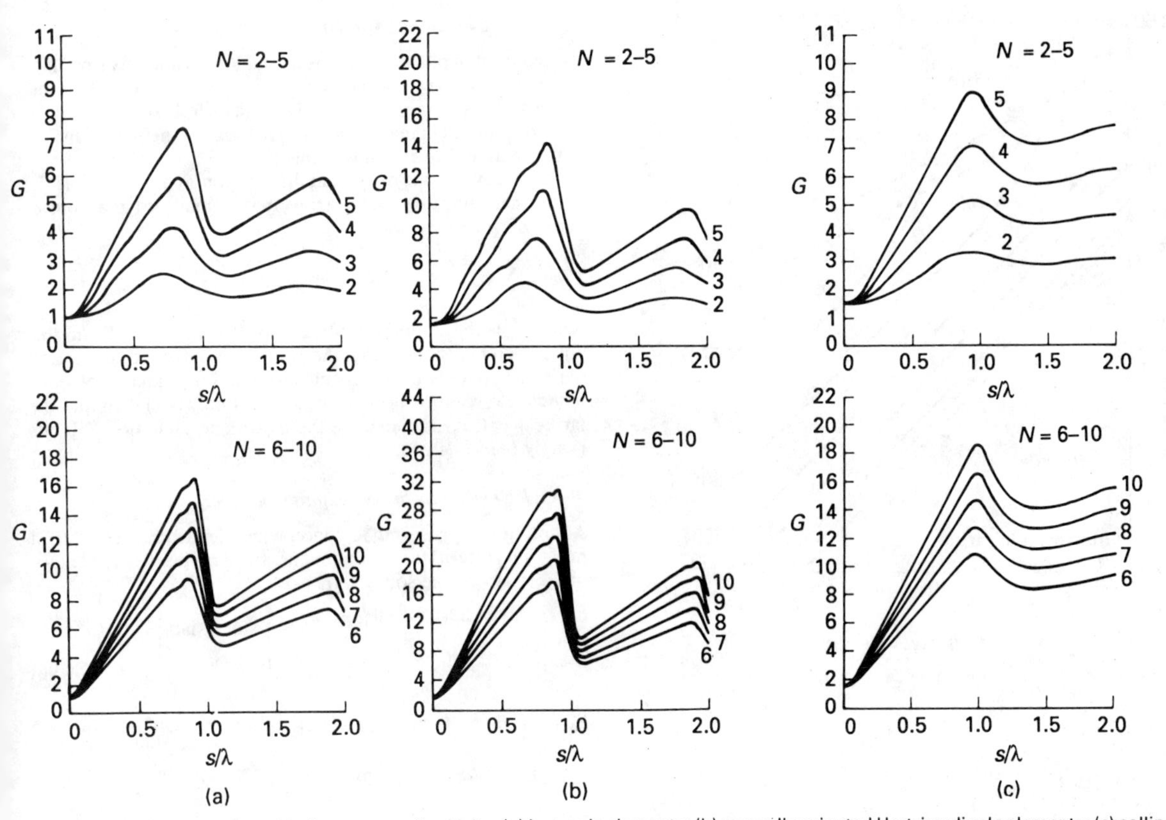

Figure 49.7 Uniform broadside linear array directivity. (a) isotropic elements; (b) normally oriented Hertzian dipole elements; (c) collinear Hertzian dipole elements

where s is the element spacing. The array pattern variable u, normally equal to $\sin\theta/\lambda$, will then be replaced by $\sin(\theta-\theta_0)/\lambda$ with the pattern shape $g(u)$ as a function of this variable remaining the same.

One consequence of scanning the beam away from the broadside direction is that the grating lobe will appear in real space for a smaller electrical spacing.

The spacing just placing the grating lobe peak into real space is given by

$$s/\lambda=(1+|\sin\theta_0|)^{-1} \qquad (49.89)$$

For the broadside case ($\theta_0=0$) this condition is $s/\lambda=1$, and for the end-fire case ($\theta_0=\pm\pi/2$) the condition is $s/\lambda=1/2$.

The other main consequence of beam scanning is that the main beam will broaden as the scan angle is incrreased. For a uniform distribution and array length a, the 3 dB beamwidth is given by

$$\theta_3=\arccos\left(\sin\theta_0-\frac{0.443}{a/\lambda}\right)-\arccos\left(\sin\theta_0+\frac{0.443}{a/\lambda}\right) \qquad (49.90)$$

$$\underset{a/\lambda\to\infty}{\simeq} 0.886\,\frac{\sec\theta_0}{a/\lambda} \quad \text{(at or near broadside)} \qquad (49.91)$$

up to the scan limit where the two beams either side of the array axis start to merge.

At end-fire ($\theta_0=\pm\pi/2$) the beamwidth is given by

$$\theta_3=2\arccos\left(1-\frac{0.443}{a/\lambda}\right) \qquad (49.92)$$

$$\underset{a/\lambda\to\infty}{\simeq} 2\left(\frac{0.886}{a/\lambda}\right)^{1/2} \qquad (49.93)$$

These beamwidth curves are plotted against array length in *Figure 49.8* . To obtain the beamwidths for other distributions, the uniform distribution beamwidth should be multiplied by $K/50.8$, where K is the beamwidth factor in degrees for the distribution in question.

49.5 Circular arrays and loop antennas

To generate useful radiation patterns, it is possible to distribute radiating elements around the circumference of a circle—the circular array. The loop antenna is a special case of the circular array, since it may be considered to comprise an infinite number of infinitesimal elementary dipole sources aligned tangentially to the circumference of the circle. Similarly the annular slot (see *Section 49.7.3*).

It is possible to vary the radiation pattern of a circular array by varying the array excitation function, the array radius and the element orientation. Because of this there are effectively two extra degrees of freedom when compared with the linear and planar array geometries, for which only the array excitation may be varied to attempt to form the desired radiation pattern. An additional advantage of the circular array is its ability to scan a directional radiation pattern about its axis through a full 360° with little or no pattern change.

49.5.1 Element orientation

For a circular array of linearly polarised elements, there are three principal orientations—axial, tangential and radial, as depicted in *Figure 49.9*. For arrays of discrete elements, the axial

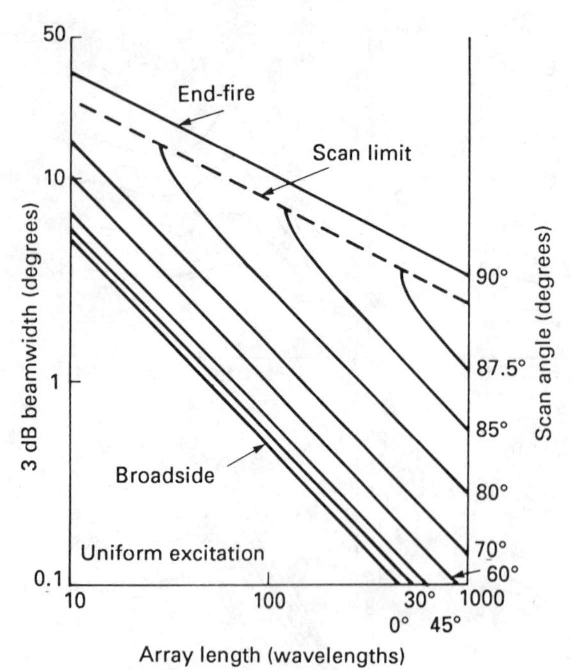

Figure 49.8 Scanned uniform linear array beamwidth

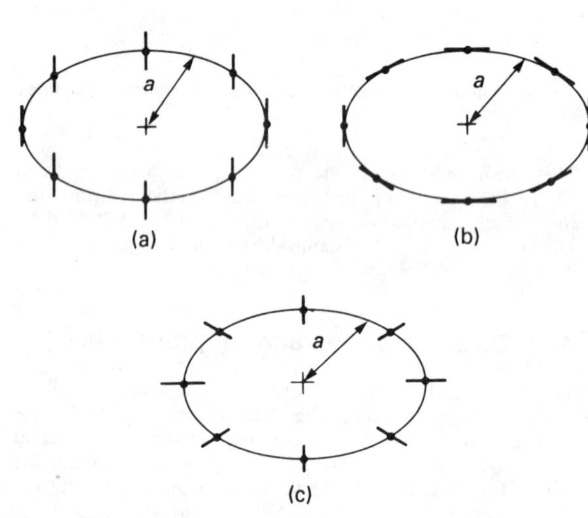

Figure 49.9 Circular array element orientations: (a) axial, (b) tangential, (c) radial

orientation is the most widely used, and the polarisation of the radiated field is everywhere linear and $\hat{\theta}$ directed. The tangential and radial orientations produce elliptical polarisation, and are used, for example, when a circularly polarised radiation maximum is required in the direction of the array axis, which may be achieved with a particular excitation, as will be described later. The analyses for these two orientations are also of use when applied to continuous radiating structures, the circular loop and annular slot already referred to being two examples for the tangential orientation. A centre-fed circular disc would be an example for the radial orientation.

49.5.2 Phase modes for circular arrays

The concept of phase modes in circular arrays not only simplifies the analysis of arrays when excited with particular distributions, but also may be applied directly to the synthesis of radiation patterns. Indeed in many cases the antenna feed network which is actually used directly reflects this.

Any array voltage or current distribution $F(\varphi)$, where φ is the azimuthal angle around the array, may be analysed as a Fourier series of 'phase modes'

$$F(\varphi)=\sum_{-\infty}^{\infty} a_m \, e^{jm\phi} \tag{49.94}$$

where a_m is the complex coefficient for the mth mode. From this equation it may be seen that the phase of the mth mode excitation varies linearly with φ, passing through $2\pi m$ radians for each complete revolution around the array. It may be shown that the far-field pattern radiated by the mth mode excitation exhibits exactly the same property.

49.5.2.1 Continuous array radiation patterns

Adopting the phase mode concept just described, the far-field radiation patterns for continuous arrays of radius a of elementary or Hertzian dipole sources are

$$F_\theta=-\sin\theta J_m(z)\exp[jm(\phi-\pi/2)] \tag{49.95}$$

$$F_\phi=0 \tag{49.96}$$

for axial orientation

$$F_\theta=\frac{m}{z}J_m(z)\cos\theta\exp[jm(\phi-\pi/2)] \tag{49.97}$$

$$F_\phi=jJ'_m(z)\exp[jm(\phi-\pi/2)] \tag{49.98}$$

for tangential orientation

$$F_\theta=jJ'_m(z)\cos\theta\exp[jm(\phi-\pi/2)] \tag{49.99}$$

$$F_\phi=-\frac{m}{z}J_m(z)\exp[jm(\phi-\pi/2)] \tag{49.100}$$

for radial orientation

$$z=2\pi\frac{a}{\lambda}\sin\theta \tag{49.101}$$

These equations apply where the array is aligned in the xy plane in the same spherical coordinate system as for the elemental loop shown in *Figure 49.1(b)* such that at $\varphi=0$ the phase of the excitation voltage or current is zero for each mode. F corresponds to the radiated **E** or **H** field, depending on whether the elemental dipole sources are electric or magnetic respectively. In all cases the $[\exp(-jkr)]/r$ factor is suppressed.

The only cases for which the radiated field is non-zero in the $\theta=0$ direction are the tangential and radial orientations with $m=\pm1$ excitations. In these cases the field in that direction is perfectly circularly polarised, with the hand of polarisation (LHCP or RHCP) changing between the $m=+1$ and $m=-1$ excitations.

The higher the order of the phase mode, the more the radiated energy will be concentrated towards the $\theta=\pi/2$ (array) plane. This is demonstrated in *Figure 49.10*, which shows the gain at $\theta=\pi/2$ (and $\theta=0$ for $m=\pm1$) for tangentially and radially orientated arrays.

49.5.2.2 Discrete array radiation patterns

The continuous array patterns will be closely approximated by those of a discrete array of a finite number of elements, provided two conditions are met.

(1) The circumferential element spacing is a half-wavelength or less, i.e. $2\pi a/s<\lambda/2$, where a is the array radius and s the number of elements.

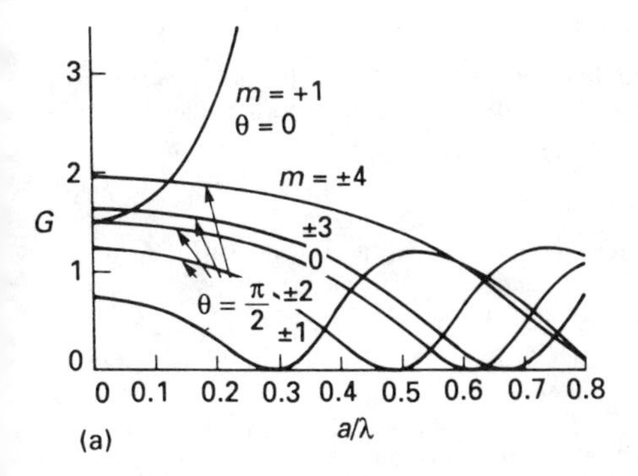

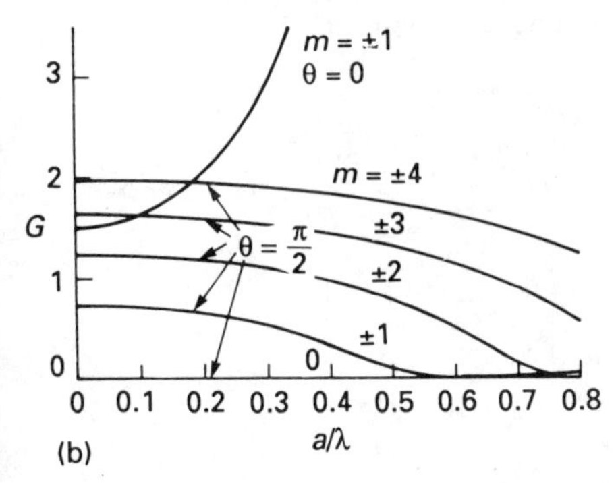

Figure 49.10 Circular array directivity: (a) tangential orientation, (b) radial orientation

(2) The number of elements is much greater than the mode order, i.e. $s>>m$. As a general rule three times the mode order represents a minimum.

If either or both these conditions are not met, the resulting patterns in the azimuthal plane (θ constant) will exhibit amplitude and phase ripple with a dominant period equal to $2\pi/s$, i.e. with one cycle per element. For a discrete array with axially orientated elements the radiation pattern will be modified to

$$F_\theta = -\frac{\sin\theta}{2}\sum_{q=0}^{\infty}(2-\delta_{oq}) \times \{J_{m+qs}(z)\exp[\mathrm{j}(m+qs)(\phi-\pi/2)] + J_{-m+qs}(z)\exp[-\mathrm{j}(-m+qs)(\phi+\pi/2)]\} \qquad (49.102)$$

This may be broken down into the continuous array solution plus an infinite series of perturbation terms consisting of harmonic phase modes. Similar forms apply to discrete arrays with tangentially or radially orientated elements.[8]

49.5.2.3 Array radiation patterns with directional elements

All of the preceding analysis has assumed that the radiating elements are short elemental electric or magnetic dipoles with $\sin\theta$ radiation patterns. When the directional patterns of the elements do not conform to this ideal, the array patterns will be modified from the previously given forms.

One case of interest is that in which the elements are linear dipoles of finite length, so that the element pattern is still symmetrical about its axis, but has a θ dependence other than $\sin\theta$, with a narrower beamwidth. For axially orientated array elements the effect of the element pattern on the net array pattern is straightforward, simply replacing the $\sin\theta$ in Equation (49.95) by the particular element pattern. The basic radiation properties of the array and their dependence on the array radius are unaffected. For the tangential and radial orientations the modifications are more complex,[8] since for these cases the element pattern is not separable from the array pattern. For dipole lengths of less than $\lambda/2$ however the changes to the elemental dipole array patterns are small.

The other case of interest is the axially orientated circular array in which the element pattern is no longer symmetrical about the vertical array axis. Examples of this would be arrays with vertical dipoles backed by reflecting ground planes, or with small horns pointing radially outwards. For moderate element directivity in the φ plane, for example with an element pattern of the form $1+\cos\varphi$, the effects of this directivity are generally beneficial. For a continuous dipole array Equations (49.95) and (49.96) show that, in a given θ direction, the relative radiated field will vary with frequency according to the Bessel function $J_m(z)$. If z is such that $J_m(z)=0$ then it will be impossible to radiate any energy in that direction at that frequency. Elements which are directional in the φ plane tend to 'smooth over' the Bessel function zeros, so that the relative field in any given direction will vary by only a few decibels over a greater than one octave frequency range. This is of great importance for broad-band circular arrays.

49.5.3 Oblique orientation—the Lindenblad antenna

All the preceding arrays have used elements axially, tangentially or radially orientated with respect to the array circle. In the most general case the elements may be arbitrarily orientated with three degrees of freedom—the direction cosines $\cos\alpha$, $\cos\beta$ and $\cos\gamma$ to the $\hat{\mathbf{r}}$, $\hat{\boldsymbol{\theta}}$ and $\hat{\boldsymbol{\varphi}}$ unit vectors respectively in the array or xy plane.

There are several special cases of element orientation apart from the three principal orientations just discussed. One of the more useful of these is with $\alpha=\pi/2$ so that the elements are normal to the radius vector and inclined at an angle γ to the array plane. When excited with the $m=0$ uniform amplitude and phase distribution, and γ is chosen to satisfy

$$\gamma = \pm\arctan[J_1(ka)/J_0(ka)] \qquad (49.103)$$

where k is the free-space wavenumber $2\pi/\lambda$ and a is the array radius, the resulting pattern is azimuthally omnidirectional and circularly polarised—the Lindenblad antenna.[9,10] The relative radiation pattern is of the form

$$F_\theta = \pm[J_1(ka)/J_0(ka)]J_0(z)\sin\theta\cos\gamma \qquad (49.104)$$

$$F_\phi = -\mathrm{j}J_1(z)\cos\gamma \qquad (49.105)$$

where $z=ka\sin\theta$. For small radii ($ka<<1$) these reduce to

$$F_\theta \simeq \tfrac{1}{2}ka\sin\theta \qquad (49.106)$$

$$F_\phi \simeq -\tfrac{1}{2}\mathrm{j}ka\sin\theta \qquad (49.107)$$

so that the field is circularly polarised everywhere in space with the radiation pattern of a short dipole.

49.5.4 Beam cophasal excitation

To produce a pencil beam in the $\theta=\pi/2$ azimuth plane with an axially orientated array of linear elements, the obvious excitation is that which produces equal phase signals from all the elements in the wanted direction—the beam cophasal excitation. For convenience, particularly when scanning the beam electronically, the excitation amplitudes are often uniform. Under these conditions for a continuous array the relative azimuth and elevation plane radiation patterns are

$$F_{\mathrm{az}} = J_0(4\pi a\lambda^{-1}\sin\tfrac{1}{2}\varphi) \qquad (49.108)$$

$$F_{\mathrm{el}} = J_0[2\pi a\lambda^{-1}(1-\sin\theta)] \qquad (49.109)$$

where a is the array radius, φ is the azimuthal angle from the beam peak, and θ is the elevation angle from the zenith (beam

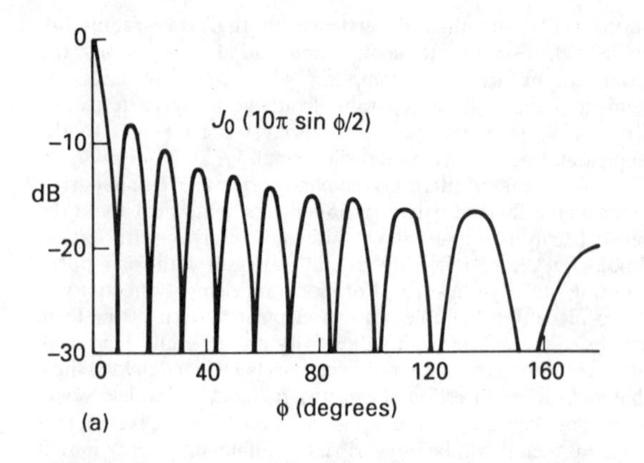

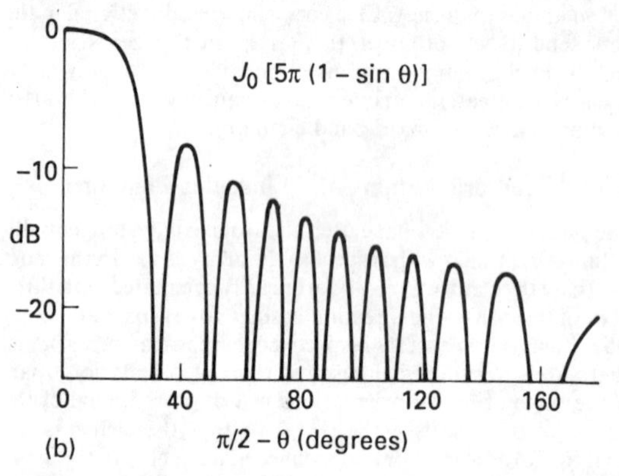

Figure 49.11 Radiation patterns for 5λ diameter beam cophasal circular array: (a) azimuth, (b) elevation

peak at $\theta=\pi/2$). These patterns are plotted for a 5λ diameter array in *Figure 49.11*. The azimuth pattern beamwidth is narrower and the side lobes higher than for a uniform linear array whose length is equal to the circular array diameter, because the geometry of the circular array tends to concentrate the power towards the edges of the aperture. The elevation pattern has the same sidelobe levels as the azimuth pattern, but the main beam is broader because the beam in the elevation plane is effectively formed from the end-fire array mode.

To achieve lower sidelobes it is necessary to excite only the elements along the half circular arc directly visible from the beam maximum direction, and to taper the amplitude from centre to edge. This makes electronic beam scanning more difficult to implement, since the phase *and* amplitude of the excitation to each element must be changed as the beam is scanned.

In all cases the circumferential element spacing must be less than λ/2 to closely approximate the continuous array patterns and avoid the emergence of grating lobes. For elements which are directional in the azimuth plane, the spacing must be made somewhat smaller than what would be allowable for omnidirectional elements.

49.5.5 Loop antennas

An example of a continuous tangentially orientated circular array is the current-fed circular wire loop. In most cases it may be assumed that the current in the loop is uniform in amplitude and phase, so that the excitation function corresponds to the $m=0$ phase mode.

49.5.5.1 Radiation patterns

For a circular loop of radius a aligned in the xy plane of a standard spherical coordinate system, the radiated far-fields are

$$E_\phi = 120\pi^2 I a\lambda^{-1} J_1(2\pi a\lambda^{-1}\sin\theta) \tag{49.110}$$

$$H_\theta = \pi I a\lambda^{-1} J_1(2\pi a\lambda^{-1}\sin\theta) \tag{49.111}$$

where I is the loop current (the $[\exp(-jkr)]/r$ factor is suppressed). For a small loop of arbitrary shape and area A $(<\lambda^2/100)$ the far-field pattern is

$$E_\phi \simeq 120\pi^2 I A\lambda^{-2}\sin\theta \tag{49.112}$$

$$H_\theta \simeq \pi I A\lambda^{-2}\sin\theta \tag{49.113}$$

i.e. the pattern for a short magnetic dipole.

Far-field polar diagrams for circular loops of 0.1, 1 and 5 wavelengths diameter are shown in *Figure 49.12(a)*. *Figure*

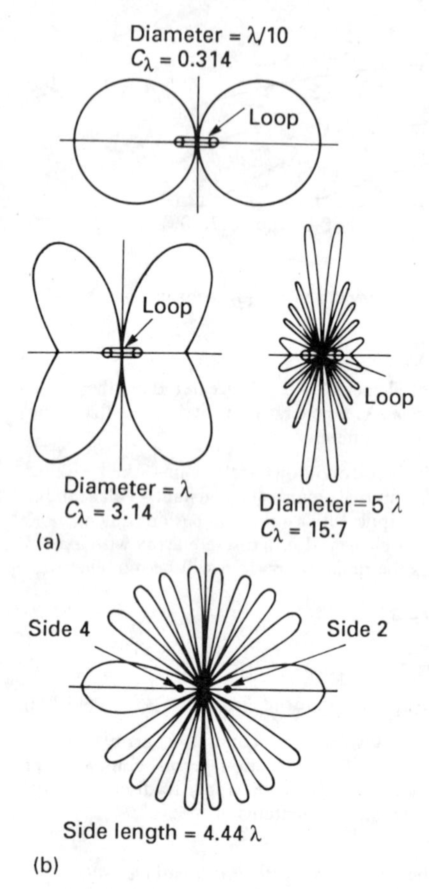

Figure 49.12 Loop radiation patterns: (a) circular loops, (b) square loop

49.12(b) shows the polar diagram for a square loop of side length 4.44λ. Compared with the 5λ diameter (equal area) circular loop pattern, the lobes are of uniform amplitude, rather than decreasing from $\theta=0$ to $\theta=\pi/2$.

49.5.5.2 Directivity

The directivity of a circular loop of radius a is given by

$$G=\frac{4\pi a\lambda^{-1}J_1^2(2\pi a\lambda^{-1}\sin\theta)}{\int_0^{4\pi a/\lambda}J_2(y)\,dy} \quad (49.114)$$

For a small loop ($A<\lambda^2/100$) of arbitrary shape, the directivity in the $\theta=\pi/2$ plane is

$$G\simeq 3/2 \quad (49.115)$$

and for a large circular loop ($2\pi a/\lambda>5$) the peak directivity approaches an asymptotic solution

$$G\simeq 4.25a\lambda^{-1} \quad (49.116)$$

The peak directivity is plotted against loop circumference in *Figure 49.13(a)* showing also the small and large loop approximations.

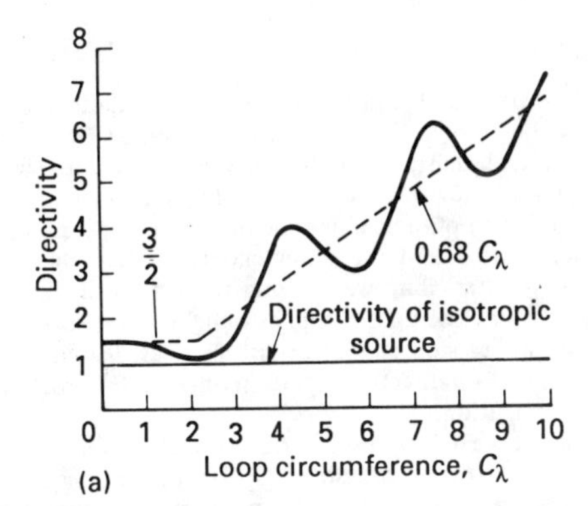

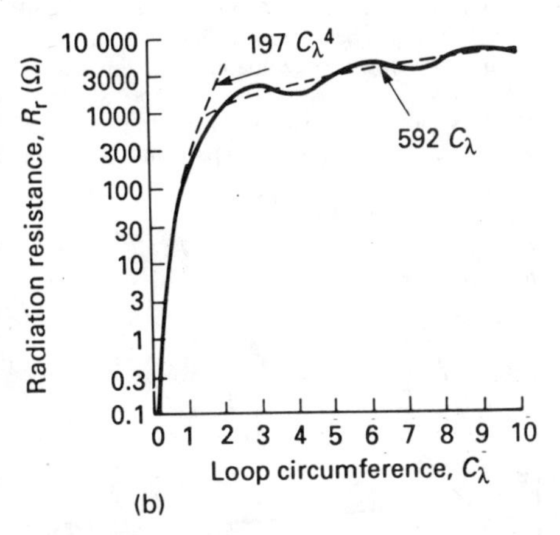

Figure 49.13 (a) Circular loop directivity; (b) circular loop radiation resistance

49.5.5.3 Radiation resistance

The radiation resistance of a circular loop of radius a is given by

$$R_{rad}=120\pi^3 a\lambda^{-1}\int_0^{4\pi a/\lambda}J_2(y)\,dy \quad (49.117)$$

For a small loop of arbitrary shape and area A ($A<\lambda^2/100$) the radiation resistance is approximately

$$R_{rad}\simeq 320\pi^4A^2\lambda^{-4} \quad (49.118)$$

and for a large circular loop ($2\pi a/\lambda>5$) the asymptotic solution is

$$R_{rad}\simeq 3720a\lambda^{-1} \quad (49.119)$$

The radiation resistance of a circular loop is plotted against its circumference in *Figure 49.13(b)*, showing also the small and large loop approximations.

49.5.5.4 Multi-turn loops

If the loop contains more than a single turn, the relative radiation pattern and directivity are unchanged. For a given loop current the radiated fields are N times what they would be for a single turn (N = number of turns), and conversely the induced voltage is N times what it would be for a single turn when illuminated by a plane wave of a given power density. The radiation resistance is raised by a factor N^2.

The applications of multi-turn loops are primarily to raise the radiation resistance of a small loop for matching purposes, and for producing a voltage multiplication effect for maximum sensitivity when feeding a high impedance receiver input.

49.6 Wire antennas

49.6.1 Linear wire antennas (linear dipoles)

A linear wire antenna comprises a straight length of wire whose diameter is much less than its length, fed at some point along its length. A balanced dipole antenna is obtained by feeding the antenna at its centre, which will provide a good non-reactive impedance match to a 50 Ω or 75 Ω transmission line, for a dipole length slightly shorter than a half wavelength. A wire may also be end-fed against earth or a ground plane, which is often a particularly convenient configuration.

49.6.1.1 Radiation patterns

For antenna lengths of up to a few wavelengths, it is a reasonable assumption that the current distribution along the wire is a sinusoidal standing wave with a period equal to one free space wavelength, and with zero current at the free end(s) of the wire. Based on this assumption it is possible to derive the radiated far-field pattern for a centre-fed linear dipole of length L

$$E_\theta(\theta)=j60I\left(\frac{\cos(\pi L\cos\theta/\lambda)-\cos(\pi L/\lambda)}{\sin\theta}\right)\frac{e^{-jkr}}{r} \quad (49.120)$$

where θ is measured from the antenna axis, and I is the current at the point of current maximum. For the special cases of $L=\lambda/2$, λ, $3\lambda/2$, the pattern becomes

$$\underset{L=\lambda/2}{E_\theta(\theta)}=j60I\left(\frac{\cos(\frac{1}{2}\pi\cos\theta)}{\sin\theta}\right)\frac{e^{-jkr}}{r} \quad (49.121)$$

$$\underset{L=\lambda}{=}j60I\left(\frac{\cos(\pi\cos\theta)+1}{\sin\theta}\right)\frac{e^{-jkr}}{r} \quad (49.122)$$

$$\underset{L=3\lambda/2}{=}j60I\left(\frac{\cos(\frac{3}{2}\pi\cos\theta)}{\sin\theta}\right)\frac{e^{-jkr}}{r} \quad (49.123)$$

These patterns are plotted in polar form in *Figure 49.14*. The main lobe in the direction $\theta=\pi/2$ narrows with increasing dipole length, and the secondary anti-phase lobes only appear for lengths greater than one wavelength.

The widely used half-wave dipole has a 3 dB beamwidth of 78° and a gain of 1.64 (2.15 dBi). These values may be compared with the 90° 3 dB beamwidth and 1.5 (1.76 dBi) gain for a vanishingly small Hertzian dipole.

49.6.1.2 Impedance

The radiation resistance of a thin linear dipole of length L at the point of the current maximum is given by

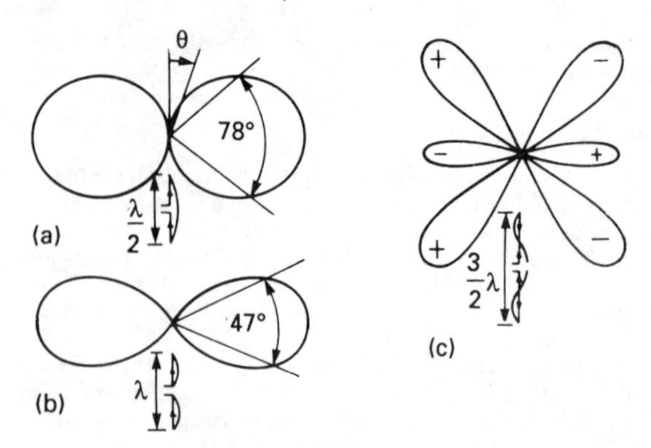

Figure 49.14 Linear dipole patterns. (a) $\frac{1}{2}$ wavelength; (b) full wavelength; (c) $\frac{3}{2}$ wavelength. The antennas are centre-fed, and the current distribution is assumed to be sinusoidal

$$R_{rad}=60\{\gamma+\ln(kL)-C_i(kL)+\tfrac{1}{2}\sin(kL)[S_i(2kL)-2S_i(kL)]$$
$$+\tfrac{1}{2}\cos(kL)[\gamma+\ln(kL/2)+C_i(2kL)-2C_i(kL)]\} \quad (49.124)$$

where S_i and C_i are the standard sine and cosine integrals and γ is Euler's constant (0.5772...). If the dipole is fed symmetrically at the centre the radiation resistance becomes

$$R'_{rad}=R_{rad}\operatorname{cosec}^2(kL/2) \quad (49.125)$$

which predicts an infinite resistance at the dipole centre when its length is equal to an odd number of quarter-wavelengths, and a value of 73.1 Ω for an exact half-wave. For a short ($L \ll \lambda$) centre-fed dipole the radiation resistance is given by

$$R_{rad}\simeq 20\pi^2(L/\lambda)^2 \quad (49.126)$$

The radiation resistance is plotted against dipole length for both feed points in *Figure 49.15*.

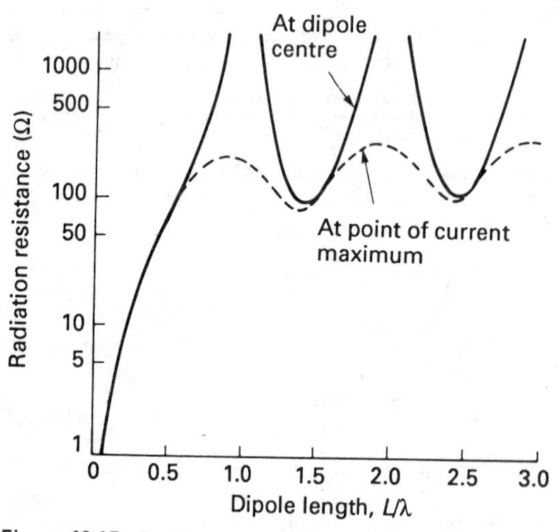

Figure 49.15 Radiation resistance of a thin linear dipole

The reactive component of the antenna impedance is dependent on the radius to length ratio a/L for the wire, and at the current maximum feed point is given by

$$X_a=30\{2S_i(kL)+\cos(kL)[2S_i(kL)-S_i(2kL)]$$
$$-\sin(kL)[2C_i(kL)-C_i(2kL)-C_i(2ka^2/L)]\} \quad (49.127)$$

A resonant design is achieved when this reactance is zero, and this occurs for lengths slightly shorter than a half-wavelength. At resonance the radiation resistance will be in the approximate range 50 Ω–75 Ω, the exact value depending on the wire radius, and such a dipole will therefore present a reasonably good match to a 50 Ω or 75 Ω transmission line.

49.6.2 Linear monopoles

A vertical linear monopole of length L and fed against a ground plane will exhibit identical radiation patterns in the half space above the ground plane to those of a balanced dipole of length $2L$ radiating into free space. Because, for a given drive current, the total radiated power is reduced by a factor of two, the radiation resistance compared to the equivalent dipole will be exactly halved. A quarter-wave monopole for example will have a radiation resistance of 36.6 Ω.

49.6.3 The rhombic antenna

The linear dipole and end-fed antennas just described are members of the class of 'standing wave' antennas. The current along the length of such antennas is formed from two equal waves travelling in opposite directions, such that it is uniform in phase and sinusoidal in amplitude. If one end of a linear wire antenna is correctly terminated and the other end is excited, there will be only a single travelling wave along the length of the wire, propagating at, or close to, the velocity of light in free space. The linear phase progression of the current associated with such a wave will radiate a unidirectional pattern tilted in the direction of the wave propagation.

One of the more widely used forms of wire travelling wave antenna is the horizontal rhombic.[11–13] Its main application is in long distance h.f. communication, and whether excited in free space or a wavelength or so above ground, it has the useful property that the position of the beam peak in the elevation plane may be adjusted by choosing the geometry and dimensions correctly.

The basic rhombic is shown in *Figure 49.16*, orientated in the horizontal or xy plane such that the azimuth pattern beam peak

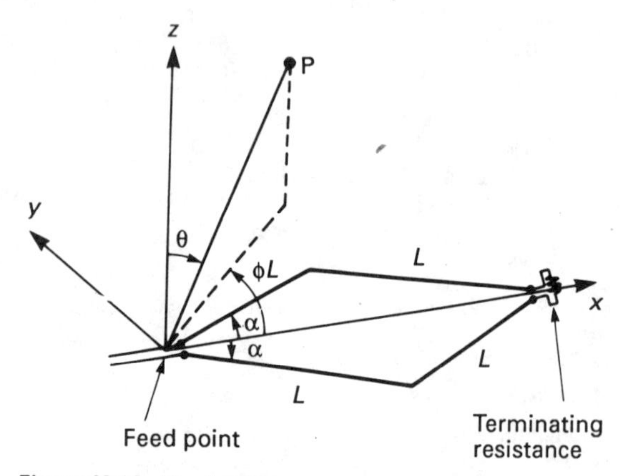

Figure 49.16 The rhombic antenna

is in the direction away from the feed point, along the x-axis. Note the terminating resistor connected at the acute corner opposite to the feed point. For a drive current I the far-field radiation pattern for an isolated rhombic is

$$E(\theta,\varphi)=4(15/\pi)^{1/2}I\sin\alpha\frac{\sin[\frac{1}{2}kL(1-\cos\psi_1)(1-\cos\psi_2)]}{[(1-\cos\psi_1)(1-\cos\psi_2)]^{1/2}} \quad (49.128)$$

where

$$\cos\psi_1 = \sin\theta\cos(\varphi+\alpha) \quad (49.129)$$

$$\cos\psi_2 = \sin\theta\cos(\varphi-\alpha) \quad (49.130)$$

The pattern is expressed in a form such that the radiated power per unit solid angle is equal to $E(\theta,\varphi)^2$. When the rhombic is located at a height h above perfectly conducting ground, the pattern is modified to

$$E'(\theta,\varphi) = 2E(\theta,\varphi)\sin(kh\cos\theta) \quad (49.131)$$

To maximise the radiated field strength at an angle θ in the elevation plane ($\theta = 90° -$ elevation angle above the horizon) it may be shown that the optimum height above ground, arm length and included semi-angle are given by

$$h_{opt} = \lambda/4\cos\theta \quad (49.132)$$

$$L_{opt} = \lambda/2\cos^2\theta \quad (49.133)$$

$$\alpha_{opt} = \tfrac{1}{2}\pi - \theta \quad (49.134)$$

For a typical design the arm length would be in the range 2–7 λ, the height above ground between 1 λ and 2 λ, and the included semi-angle α in the range 15–30°. The input impedance is generally relatively high, say 700–800 Ω, the sidelobes are also rather high at −10 dB or worse, and the 3 dB beamwidth is typically 10–15°.

49.7 Slot antennas

Slots cut in conducting surfaces often provide convenient antennas or antenna elements, since they are 'flush-mounting', and require no protrusion above the ground plane surface. A possible disadvantage is that their radiation patterns are strongly influenced by the size or shape of the ground planes into which they are cut, and the pattern with a ground plane even 10λ across will be significantly different to that of the same slot on an infinite ground plane. A slot may be fed in a number of different ways, either from coaxial line or from waveguide. For example as shown in *Figure 49.13* a small rectangular slot can be fed from a coaxial connected across the narrow dimension of the slot at some point, from the open end of a rectangular waveguide whose axis is normal to the ground plane, or by cutting the slot in either the broad or the narrow face of the rectangular waveguide itself. The method of excitation will have little effect on the radiation pattern of the slot, but will affect primarily the impedance seen by the feeder and the bandwidth. The fourth feed method shown in *Figure 49.17*, and using a coaxial T-fed cavity to excite the slot, offers a relatively wide bandwidth, with a VSWR less than 2:1 over an octave bandwidth.[14]

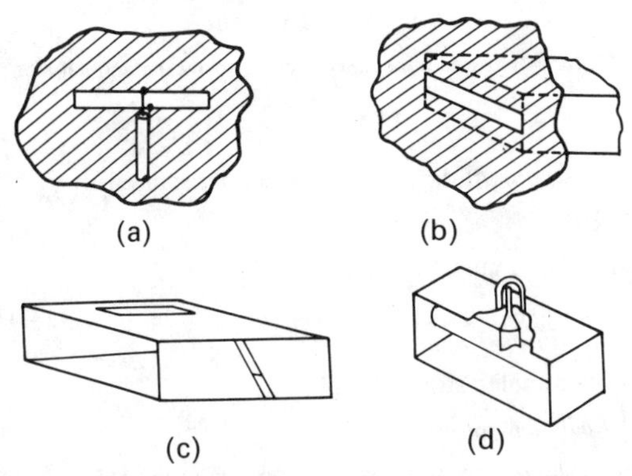

Figure 49.17 Some alternative rectangular slot excitation methods

The simple slot in a ground plane will radiate equally on both sides with a symmetric radiation pattern. In most applications a uni-directional pattern is required, so that the radiation on one side of the ground plane must be suppressed, usually by enclosing the slot on that side with a cavity. Apart from increasing the forward radiated field uniformly by 3 dB, the addition of the cavity will usually have little effect on the radiation pattern. Again the effect will be primarily on the impedance and bandwidth. Suppressing the radiation on one side of the ground plane will double the slot impedance.

49.7.1 Complementary antennas

A useful theory of equivalence between slot and wire antennas based on Babinet's principle has been developed.[15] The basis of this theorem is that any slot antenna has an equivalent wire antenna formed by transposing the areas of conductor and free space. The radiation patterns of these two antennas will be identical except that the **E** and **H** fields will be transposed. The rectangular slot is thus complementary to a linear wire dipole antenna of the same length and width, the former vertically polarized and the latter horizontally polarised, both with the same figure-of-eight pattern in the vertical plane. The annular slot is complementary to a wire loop of the same diameter and width.

This equivalence theory allows the impedance z_{slot} of a slot to be simply related to the impedance z_{comp} of its complementary antenna

$$z_{slot}z_{comp} = \eta_0^2/4 = 3600\pi^2 \quad (49.135)$$

where η_0 is the impedance of free space ($=120\pi$).

49.7.2 The rectangular slot

49.7.2.1 Radiation patterns

The radiation pattern of a slot radiating into free space, and orientated in the xy plane of a standard spherical coordinate system with the **E** field across the narrow dimension parallel to the x-axis, may be computed in two ways. The slot may be converted to its complementary form, and the radiation pattern computed for that using one of the standard forms. Alternatively the field equivalence principle may be invoked and the x-directed electric fields **E** in the aperture replaced by equivalent y-directed magnetic currents **M** from the relationship

$$\mathbf{M} = -\hat{\mathbf{n}}\times\mathbf{E} \quad (49.136)$$

where $\hat{\mathbf{n}}$ is the z-directed unit vector normal to the aperture. The fields radiated by the equivalent magnetic currents are those radiated by elementary magnetic dipole sources, as described in *Section 49.2.2*. The radiated electric fields from a vanishingly small element dx dy of a rectangular slot are thus

$$E_\theta = -jE_x\frac{\cos\varphi}{2\lambda}\,dx\,dy\frac{e^{-jkr}}{r} \quad (49.137)$$

$$E_\phi = jE_x\frac{\cos\theta\sin\varphi}{2\lambda}\,dx\,dy\frac{e^{-jkr}}{r} \quad (49.138)$$

where E_x is the x-directed field in the slot aperture, and k is the free space wavenumber $2\pi/\lambda$. This pattern is uniform in the xz plane, and has a $\cos\theta$ variation in the yz plane. The net field radiated by the slot is found as an integral across the aperture involving the preceding equations for the elemental radiation field. This analysis is only valid if the aperture field in the slot is all or virtually all electric, with little or no magnetic or H component. For this reason the slot width w must be very much less than a wavelength.

The rectangular slot is often made half a free space wavelength long, to provide a resonant design with a purely resistive impedance. For the half-wave resonant slot the radiation pattern in the plane of the slot is the same as the radiation pattern of a half-wave dipole, i.e.

$$E_\phi = \frac{\cos(\frac{1}{2}\pi \sin \theta)}{\cos \theta} \qquad (49.139)$$

As mentioned earlier, the size of the ground plane into which the slot is cut will have a considerable effect on the overall radiation pattern, and the pattern for an infinite ground plane will only be approached for dimensions in excess of 10λ. Most effect will be observed in the plane perpendicular to the slot, and qualitatively the pattern will exhibit a degree of ripple and an amplitude taper in the direction $\theta = \pi/2$ along the ground plane. This is demonstrated in *Figure 49.18*, which shows patterns for a half-wave slot on circular ground planes of various diameters.

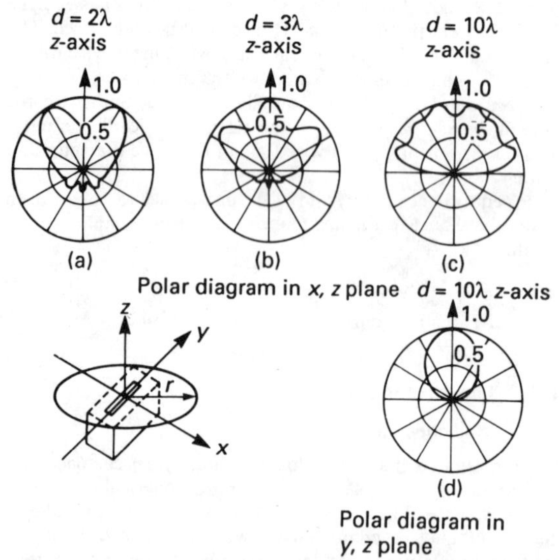

Figure 49.18 Radiation patterns of a narrow half-wavelength slot on a finite circular ground plane

49.7.2.2 Impedance

By using the principle of complementary antennas already described, if a slot is transformed to its complementary antenna and its impedance found, then the slot impedance is given directly by Equation (49.135). This method is applied to half-wave, resonant half-wave, and full-wave rectangular slots in *Figure 49.19*. The complementary cylindrical dipole has the same length as the slot and a diameter equal to half the slot width, since it may be shown that a cylindrical conductor of diameter d is equivalent to a flat strip conductor of width $w = 2d$.[16] It is worth noting that a centre-fed full-wave slot will be almost exactly matched to a 50 Ω coaxial transmission line. Although the centre impedance of a resonant half-wave slot is non-reactive, to match it to a 50 Ω feed line it is necessary to move the feed point to a distance $\lambda/20$ from one end.

Measured slot impedances (*Figure 49.20*) generally confirm the predictions of the preceding analysis, with some slight discrepancies. The most notable discrepancy is that the resonant length appears from the measurements to be independent of the slot width, whereas the complementary antenna analysis suggests that it will be reduced as the slot width is increased.

The radiation conductance of a short rectangular slot of length l may be found, since it may be shown that for the

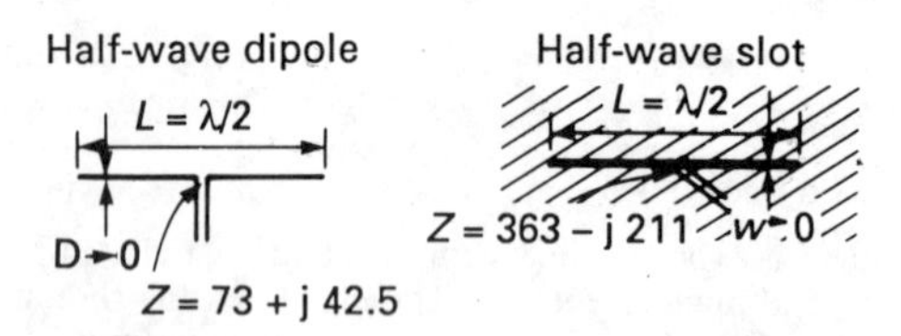

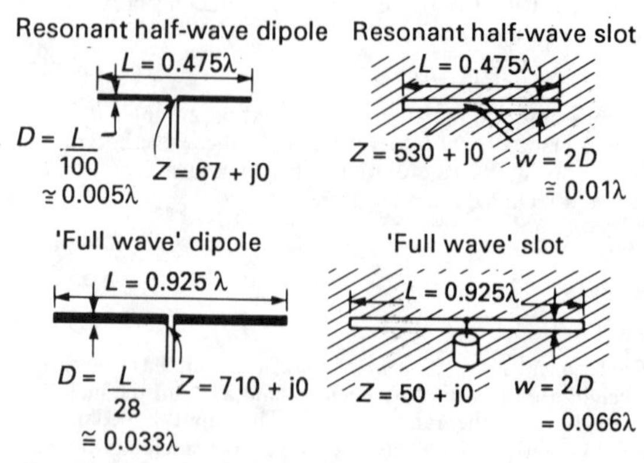

Figure 49.19 Rectangular slot and complementary wire dipole impedances

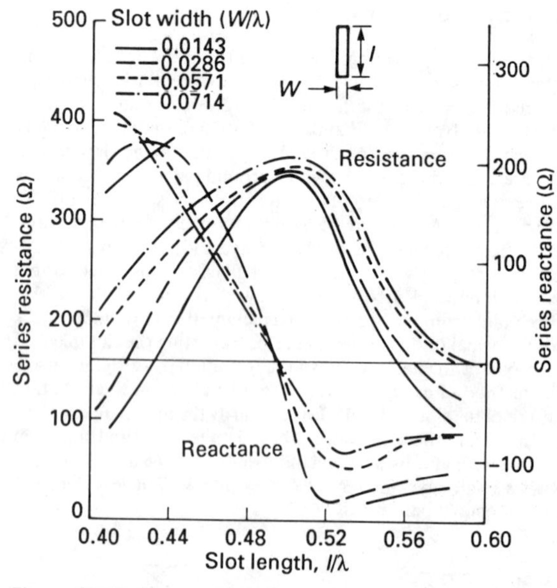

Figure 49.20 Measured impedances for a narrow rectangular slot

complementary short dipole the radiation resistance is

$$R_{rad} = 20\pi^2(l/\lambda)^2 \qquad (49.140)$$

and hence from Equation (49.135)

$$G_{rad} = (l/\lambda)^2/180 \qquad (49.141)$$

49.7.3 The annular slot

49.7.3.1 Radiation patterns

An annular slot in a ground plane may be excited as a radiator by applying a voltage radially across the slot. The radiation pattern is identical to that of the complementary wire loop, with **E** and **H** fields transposed. With the slot in the xy plane of a standard spherical coordinate system, the radiation pattern is symmetrical about the z-axis, and is zero along that axis in the direction $\theta=0$.

The radiation pattern for a narrow circular slot of radius a is

$$E_\theta = j2\pi a V J_1(ka \sin\theta)\, e^{-jkr}/r \tag{49.142}$$

where V is the voltage applied across the slot, k is the free space wave number $2\pi/\lambda$, and J_1 is the first-order Bessel function of the first kind. The radiation pattern is of exactly the same form as that of the circular loop (*Section 49.5.5*), for which patterns are plotted for various radii in *Figure 49.12(a)*. The field along the ground plane in the direction $\theta=\pi/2$ is a function of the electrical radius of the slot, and, for example, for $ka=3.83$ (the first zero of J_1) there is zero radiated field along the ground plane.

For small radii ($a << \lambda$) the equation may be approximated by

$$E_\theta \simeq j2\pi a^2 \lambda^{-1} V \sin\theta\, e^{-jkr}/r \tag{49.143}$$

which is the radiation pattern of a short electric dipole aligned along the z-axis.

The annular slot radiation pattern is affected qualitatively by finite ground plane size, in a similar way to the rectangular slot. The pattern of a small annular slot on a finite ground plane is identical to that of a short stub monopole on the same ground plane (apart from the field transposition).

49.7.3.2 Impedance

The impedance of an annular slot may be computed in the same way as for the rectangular slot. In this case the complementary antenna is a thin wire loop of the same radius.

For a slot fed from a TEM-mode coaxial line with inner diameter and outer diameters equal to the inner and outer diameters, respectively, of the slot, the feed impedance normalized to the coaxial characteristic impedance has been computed.[17] This is plotted in *Figure 49.21* as normalised conductance and susceptance against slot radius, for a number of slot widths. For small radii the conductance is proportional to $(ka)^4$, and the susceptance directly proportional to ka. Although the slot susceptance falls for ka greater than unity, it does not actually become zero for any radius, so that there is no true resonance (cf. the rectangular slot case).

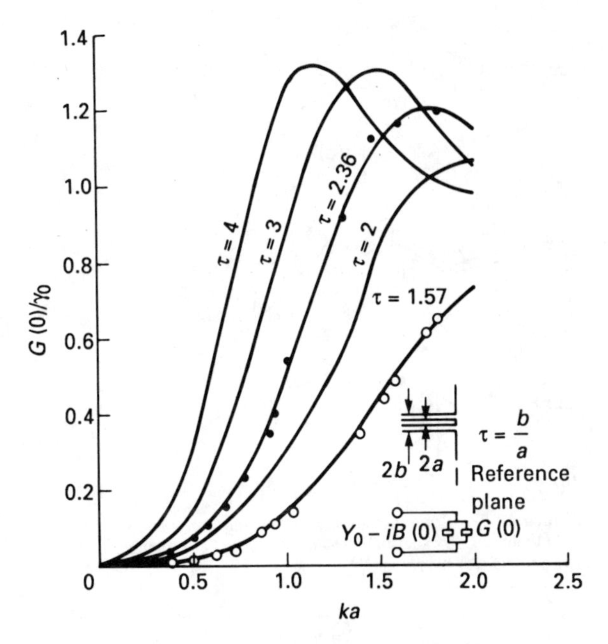

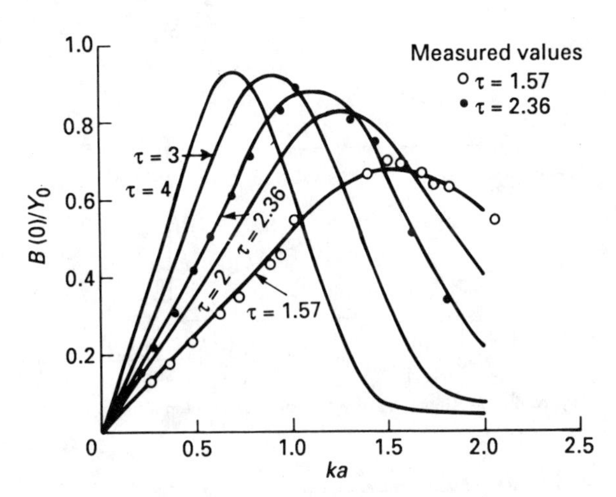

Figure 49.21 Normalised impedance of an annular slot

49.8 Electromagnetic horns

Electromagnetic horns find application as antennas in the frequency range covering UHF to millimetric frequencies, where a moderate gain in the approximate range 10–35 dBi is required over bandwidths of up to an octave (multi-octave bandwidths are also available using special techniques). They can be used as antennas in their own right, for example as earth-coverage satellite antennas, as feeds for reflector antennas, and as accurate and stable antenna gain standards.

Some of the more commonly used horn types are shown in *Figure 49.22*. The pyramidal and conical horns are used to produce beamwidths of the same order in the two principal planes, the sectoral horns produce fan beams which are much narrower in one plane than in the other, and the biconical horn produces a pattern which is omni-directional in the azimuth plane. In nearly all cases the horn is fed from rectangular waveguide (pyramidal and sectoral types) or from circular waveguide (conical and biconical types).

49.8.1 Radiation from open-ended waveguide

The radiation pattern of a horn may be computed from the tangential aperture fields (see *Section 49.2.3*). For a horn with no major discontinuities in the flare it is a reasonable assumption that the fields at the aperture are of the same form as that of the exciting waveguide mode(s) at the throat, with a quadratic phase taper across the aperture due to the wavefront curvature introduced by the flare (see *Figure 49.23*). The edge-to-centre or peak value ψ_0 of this phase error is given quite accurately by the expression

$$\psi_0 = 2\pi\delta_0/\lambda \simeq \tfrac{1}{2}\pi w\alpha/\lambda \tag{49.144}$$

where w is the aperture width in the plane being considered, α is the semi-flare angle in the same plane, and ψ_0 and α are in the same units (degrees or radians).

The effect of the flare-induced quadratic aperture phase error is slight for small to moderate phase errors, certainly over the top 15 dB of the main beam, and it is possible to correct out the phase error with a lens in the aperture. It is therefore useful to consider as a basis the radiation patterns of open-ended waveguides with zero aperture phase error.

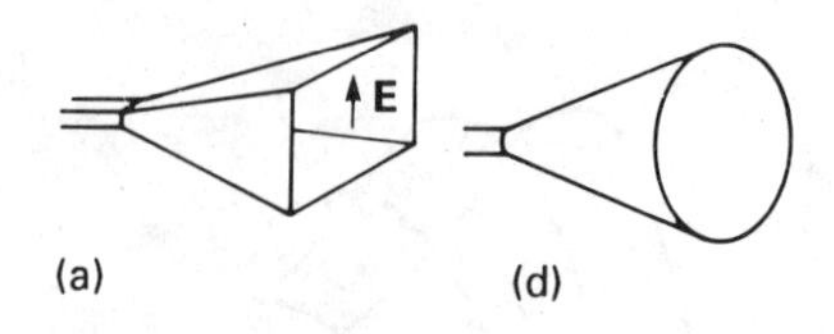

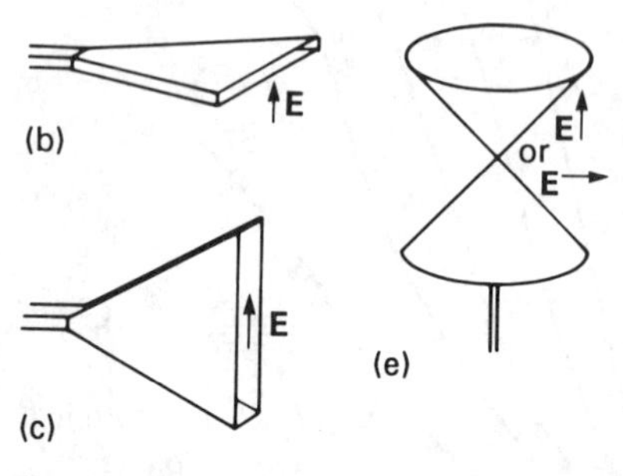

Figure 49.22 Some commonly used electromagnetic horn types: (a) pyramidal, (b) sectoral H plane, (c) sectoral E plane, (d) conical, (e) biconical

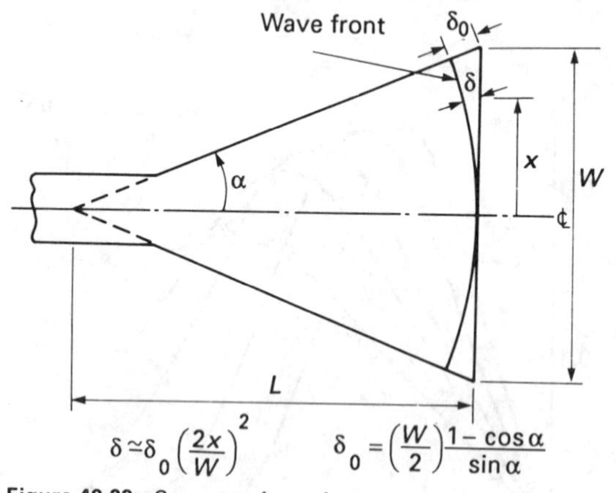

Figure 49.23 Geometry of wavefront curvature in a flared horn

49.8.1.1 Rectangular waveguide

For an open-ended rectangular waveguide of H-plane width a and E-plane height b, the principal plane radiation patterns for the dominant TE_{10} mode are

$$E_E(\theta)=2\sqrt{\frac{\mu}{\varepsilon}\frac{a^2b}{\pi\lambda^2}}\left(1+\frac{\beta_{10}}{k}\cos\theta\right)\frac{\sin(\pi b\lambda^{-1}\sin\theta)}{(\pi b\lambda^{-1}\sin\theta)}\frac{e^{-jkr}}{r} \tag{49.145}$$

$$E_H(\theta)=-\sqrt{\frac{\mu}{\varepsilon}\frac{\pi a^2b}{2\lambda^2}}\left(\frac{\beta_{10}}{k}+\cos\theta\right)\times\frac{\cos(\pi a\lambda^{-1}\sin\theta)}{(\pi a\lambda^{-1}\sin\theta)-(\frac{1}{2}\pi)^2}\frac{e^{-jkr}}{r} \tag{49.146}$$

where k is the free space wavenumber $2\pi/\lambda$, β_{10} is the guide wavenumber for the TE_{10} mode

$$\beta_{10}=k\left[1-\left(\frac{\lambda}{2a}\right)^2\right]^{1/2} \tag{49.147}$$

and r is the distance to the observation point.

The beamwidths between first nulls are given by

$$\theta_n^E=2\arcsin\left(\frac{\lambda}{b}\right) \tag{49.148}$$

$$\theta_n^H=2\arcsin\left(\frac{3\lambda}{2a}\right) \tag{49.149}$$

and for large apertures (narrow beams) the 3 dB beamwidths in degrees are

$$\theta_3^E\simeq 50.8\frac{\lambda}{b} \tag{49.150}$$

$$\theta_3^H\simeq 68.1\frac{\lambda}{a} \tag{49.151}$$

For a given aperture width the pattern is narrower in the E plane than in the H plane, because the aperture field amplitudes in those planes are uniform and cosine-tapered respectively. The sidelobes are higher in the E plane for the same reason.

Equations (49.145) and (49.146) are based on several assumptions. Most notable of these are that the radiation occurs from the fields across the aperture alone and there are no fields outside the aperture or currents flowing on the outside waveguide walls, and that the aperture is matched and there are no higher-order modes generated there. Even though these assumptions are not strictly valid, particularly for small aperture sizes, the results are usefully accurate in practice.

The gain of an open-ended rectangular waveguide or a pyramidal or sectoral horn with small aperture phase error is given by

$$G\simeq\frac{32\ ab}{\pi\ \lambda^2} \tag{49.152}$$

48.8.1.2 Circular waveguide

For an open-ended circular waveguide of radius a, coaxial with the z-axis of a standard spherical coordinate system, and supporting the dominant TE_{11} mode polarised in the y direction, the θ and φ components of the far-field pattern are given by

$$E_\theta(\theta,\varphi)=\frac{\omega\mu}{2}\left(1+\frac{\beta_{11}}{k}\right)J_1(\chi'_{11})\frac{J_1(u)}{\sin\theta}\sin\varphi\frac{e^{-jkr}}{r} \tag{49.153}$$

$$E_\phi(\theta,\varphi)=-\frac{ka\omega\mu}{2}\left(\frac{\beta_{11}}{k}+\cos\theta\right)J_1(\chi'_{11})\times\frac{J'_1(u)}{1-(u/\chi'_{11})^2}\cos\varphi\frac{e^{-jkr}}{r} \tag{49.154}$$

where all the variables are as defined for the rectangular waveguide, together with $u=ka\sin\theta$, χ'_{11} is the first root of $J'_1(x)$ (=1.84118), and β_{11} is the guide wavenumber

$$\beta_{11}=\left[k^2-\left(\frac{\chi'_{11}}{a}\right)^2\right]^{1/2} \tag{49.155}$$

The E_θ pattern respresents the E plane pattern with $\varphi=\pi/2$, and the E_ϕ pattern represents the H plane pattern with $\varphi=0$.

The E and H plane beamwidths between the first nulls are

$$\theta_n^E=2\arcsin\left(\frac{3.832}{2\pi a}\right) \tag{49.156}$$

$$\theta_n^H=2\arcsin\left(\frac{5.331}{2\pi a}\right) \tag{49.157}$$

and for large apertures (narrow beams) the 3 dB beamwidths in degrees are

$$\theta_3^E\simeq 29.4\lambda/a \tag{49.158}$$

$$\theta_3^H \simeq 37.2\lambda/a \qquad (49.159)$$

As with the rectangular waveguide the pattern is narrower in the E plane than the H plane, and the side lobes are also higher in the E plane.

The gain of an open-ended circular waveguide, or a conical horn with small aperture phase error, is given by

$$G \simeq 33.0(a/\lambda)^2 \qquad (49.160)$$

49.8.2 Finite flare angles

The radiation patterns given for open-ended rectangular and circular waveguide will also be representative for pyramidal and conical flared horns, provided the flare is narrow enough to introduce little or no wavefront curvature at the aperture. As a rule of thumb, the edge-to-centre peak aperture phase error ψ_0 given by equation should be less than 20° for this to be satisfied. For quadratic phase errors greater than this, the main beam will broaden, the sidelobes will be increased, the nulls will fill in, and the gain will be reduced. Normalised E and H plane pyramidal horn patterns are plotted for phase errors in the range 0 to 200° in *Figure 49.24*. The pattern changes are similar for the conical horn.

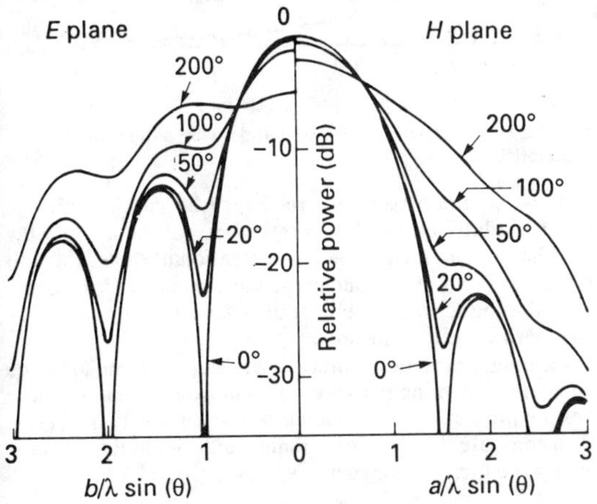

Figure 49.24 Principal plane patterns for a flared pyramidal or sectoral horn with varying aperture phase error

49.8.2.1 Optimum horns

If the axial length L of a flared horn is kept constant, and the flare angle increased, the gain will at first increase due to the increasing aperture size, and then beyond some point start to decrease again as the gain loss due to the increasing quadratic aperture phase error becomes dominant. A horn with the geometry giving the maximum gain for the given flare length is known as an optimum horn.

It is found that for a pyramidal horn the gain maximum occurs when the edge-to-centre path length difference δ_0 is approximately 0.25λ in the E plane and 0.4λ in the H plane (peak phase errors of 90° and 144° respectively). The gain of an optimum pyramidal horn is

$$G \simeq 6.43ab/\lambda^2 \qquad (49.161)$$

where a and b are the H and E plane aperture dimensions.

For an optimum conical horn the edge-to-centre path length difference δ_0 is 0.375λ (peak phase error 135°), and its gain is

$$G \simeq 20.6(a/\lambda)^2 \qquad (49.162)$$

where a is the aperture radius.

The flare semi-angle α is given in terms of the axial length L and edge-to-centre path difference δ_0 as

$$\alpha = \arccos[L/(L+\delta_0)] \qquad (49.163)$$

This equation allows the design of an optimum horn in terms of its length L, using the values of δ_0 just given. Experimentally determined design data together with 3 dB beamwidths for a pyramidal horn are plotted in *Figure 49.25*.

49.9 Reflector antennas

A reflector antenna comprises a feed and one or more reflecting surfaces, configured so that the rays from the feed are redirected in space to achieve a particular radiation pattern. In the most commonly used type—the focusing pencil beam reflector—the feed rays are collimated, producing a narrow pencil beam with a gain considerably in excess of the gain of the feed itself. In other types the reflector profile is chosen to produce a broader secondary radiation pattern of a given shape. In most cases a single reflecting surface is used, however in the Cassegrain and Gregorian types two reflectors are used—one large main reflector, and one much smaller subreflector. Antennas with 'beam waveguide' feed systems use as many as six reflectors, although four of these are effectively used only to replace the main section of the waveguide run to the feed horn.

Despite the advances made in array technology, the reflector antenna remains the most attractive solution as a large aperture antenna in many cases. Its relative simplicity and reliability provide great advantages, and yet very high levels of performance are available, particularly with the benefits of modern design techniques.

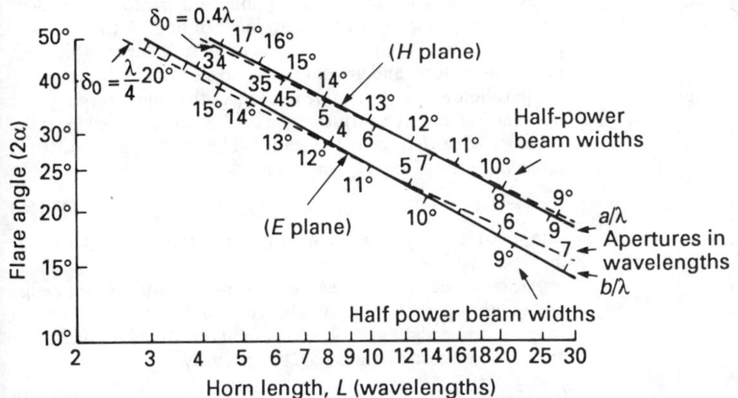

Figure 49.25 Optimum pyramidal horn design data

49.9.1 Focusing pencil beam reflector systems

49.9.1.1 The parabolic reflector

The geometry of the parabolic reflector is shown in *Figure 49.26(a)*. The reflector profile is a parabola defined by

$$y^2 = 4fx \tag{49.164}$$

where f is the focal length, and collimates the diverging rays from a feed placed at the focus into a parallel beam. The reflector is completely defined by its diameter D and focal length, although the focal length is generally expressed indirectly as the ratio f/D, which will in most cases lie within the range 0.25–2.0. The semi-angle α subtended by the reflector at the focus is given in terms of the f/D ratio as

$$\alpha = 2\arctan[(4f/D)^{-1}] \tag{49.165}$$

which is plotted in *Figure 49.27*. The geometrical relationships for a ray emerging from the focus at an angle ε to the axis are

$$\rho = f\sec^2(\tfrac{1}{2}\varepsilon) \tag{49.166}$$

$$y = 2f\tan(\tfrac{1}{2}\varepsilon) \tag{49.167}$$

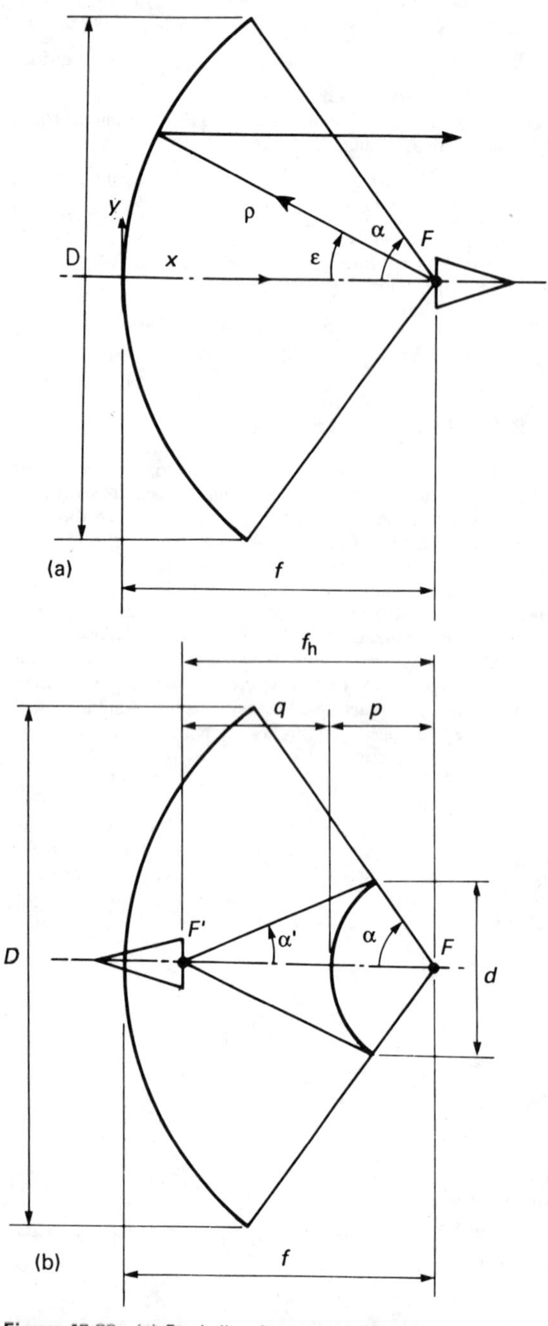

Figure 49.26 (a) Parabolic reflector geometry, (b) Cassegrain (dual reflector) geometry

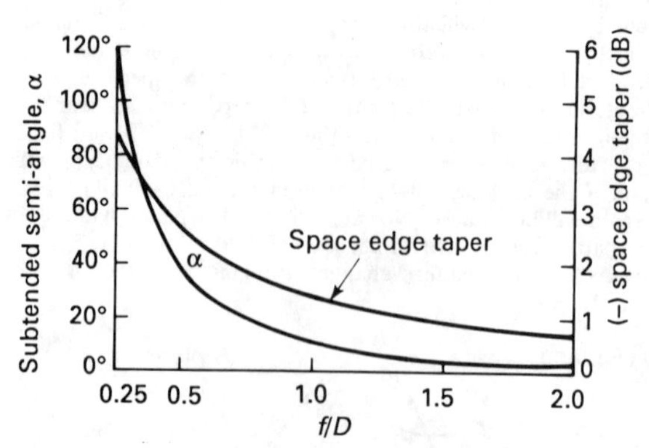

Figure 49.27 Subtended semi-angle and space edge taper for a parabolic reflector

The most common parabolic reflector (the paraboloid) is a surface of revolution formed by rotating the parabola about the axis, and has a point focus. The two-dimensional equivalent with a line focus is parabolic in one plane but unshaped in the other plane, and is used with a line source feed to illuminate the reflector with a cylindrical wave.

The secondary radiation pattern of a reflector antenna may be computed in two principal ways. The more rigorous of these invokes the physical optics principle described briefly in *Section 49.2.3* to compute the electric currents **J** induced in the reflector surface by the incident magnetic field **H**

$$\mathbf{J} = 2\hat{\mathbf{n}} \times \mathbf{H} \tag{49.168}$$

($\hat{\mathbf{n}}$ = unit normal vector), and then to compute the fields radiated into space in the form of an integral over the reflector surface. The second method first computes the fields projected onto the aperture plane of the antenna due to illumination by the feed, and then uses the Kirchhoff principle (Equations (49.37) and (49.38)) to compute the secondary radiation pattern as the Fourier transform of the aperture fields.

For a feed radiating a perfect spherical wave originating at the focus, the fields in the aperture plane of a paraboloid will be uniform in phase. The amplitude distribution of the aperture plane fields will be due to both the amplitude taper of the feed radiation pattern, and an additional 'space taper' introduced by the paraboloid geometry, due to the fact that the diverging rays from the feed must travel further before being collimated, near the aperture edge than near the centre. The space taper (in dB) is given in terms of the ray angle ε as

$$e_{\text{space}} = 20\lg[\cos^2(\tfrac{1}{2}\varepsilon)] \tag{49.169}$$

The space taper at the reflector edge ($\varepsilon = \alpha$) is plotted against f/D in *Figure 49.27*.

The beamwidth of the feed radiation pattern will affect the edge taper of the aperture fields, and hence both the secondary pattern shape and boresight gain. The boresight directivity for a circular aperture reflector of diameter D is given by

$$G = (\pi D/\lambda)^2 \eta_a \eta_s \tag{49.170}$$

where η_a and η_s are the aperture and spillover efficiencies respectively

$$\eta_a=\frac{4}{\pi D^2}\left(\left|\int_0^{2\pi}\int_0^{D/2} f(y,\varphi)y\,dy\,d\varphi\right|^2\right)\times\left(\int_0^{2\pi}\int_0^{D/2}|f(y,\varphi)|^2 y\,dy\,d\varphi\right)^{-1} \quad (49.171)$$

$$\underset{\text{azimuthal symmetry}}{=}\frac{8}{D^2}\left(\left|\int_0^{D/2} f(y)y\,dy\right|^2\right)\left(\int_0^{D/2}|f(y)|^2 y\,dy\right)^{-1} \quad (49.172)$$

$$\eta_s=\left(\int_0^{2\pi}\int_0^{\alpha}|g(\theta,\varphi)|^2\sin\theta\,d\theta\,d\varphi\right)\times\left(\int_0^{2\pi}\int_0^{\pi}|g(\theta,\varphi)|^2\sin\theta\,d\theta\,d\varphi\right)^{-1} \quad (49.173)$$

$$\underset{\text{azimuthal symmetry}}{=}\left(\int_0^{\alpha}|g(\theta)|^2\sin\theta\,d\theta\right)\times\left(\int_0^{\pi}|g(\theta)|^2\sin\theta\,d\theta\right)^{-1} \quad (49.174)$$

where $f(y, \varphi)$ and $g(\theta, \varphi)$ are the fields in the reflector aperture, and feed pattern fields respectively, with other variables as defined earlier.

For a given reflector f/D the aperture efficiency will increase with increasing feed pattern beamwidth, whereas the spillover efficiency will decrease with increasing feed pattern beamwidth. The antenna gain will therefore maximise for a particular value of feed pattern beamwidth, and this will usually occur when the net edge taper is between -8 dB and -12 dB. For most feed patterns the main beam will be approximately Gaussian down to the -12 dB points, so that if a feed pattern edge taper of $-E$ dB is required, the 3 dB beamwidth is approximately given by

$$\theta_3 \simeq 2\alpha(3/E)^{1/2} \quad (49.175)$$

49.9.1.2 Cassegrain (dual reflector) geometry

There are two dual reflector geometries using a parabolic main reflector—the Cassegrain and Gregorian geometries, of which the former is the more widely used. The subreflector of a Cassegrain antenna system (shown in *Figure 49.26(b)*) is hyperbolic in profile and has the effect of creating a virtual focus at F', between the two reflectors.

The subreflector is completely defined by its diameter d, focal length $f_{h'}$ and magnification factor M. The relationships between the various geometrical parameters for the Cassegrain system are

$$M=q/p \quad (49.176)$$

$$e=(M+1)/(M-1) \qquad (e=\text{eccentricity}) \quad (49.177)$$

$$f_h=p+q \quad (49.178)$$

$$\alpha=2\arctan[(4f/D)^{-1}] \quad (49.179)$$

$$\alpha'=2\arctan[(4Mf/D)^{-1}] \quad (49.180)$$

If the main reflector diameter and focal length, together with the subreflector focal length and magnification factor, are already defined, and if, as is usually the case, the subreflector edge is required to subtend the same angle at the prime focus F as the main reflector edge, the subreflector diameter d is then no longer independent

$$d=2f_h/(\cot\alpha+\cot\alpha')^{-1} \quad (49.181)$$

It may be shown that for minimum aperture blockage, when the blockage shadows of the feed horn and subreflector are equal, the subreflector diameter is given approximately by

$$d\simeq\sqrt{2\lambda f} \quad (49.182)$$

49.9.2 Reflector shaping

The basic 'optical' Cassegrain geometry using hyperbolic and parabolic reflector profiles has two disadvantages when applied to microwave antennas. Firstly the illumination of the main reflector will be determined by the shape of the feed pattern (generally amplitude tapered), so that it is not possible to fully maximise the antenna efficiency, except by using special feed types. Secondly internal diffraction effects due to the finite electrical size of the reflectors will come into play, so that the simple geometric or ray tracing analysis will break down, generally reducing the gain below that expected.

To overcome the first of these disadvantages it is possible to synthesise 'shaped' reflector profiles, slightly perturbed from the basic parabolic/hyperbolic pair, which according to geometric optics will produce a uniform aperture field distribution and hence increase the antenna gain.[18,19] This synthesis is generally known as the Williams method. A rigorous analysis of the aperture fields for a geometrically shaped Cassegrain antenna of finite size will show that the field distribution achieved is not exactly uniform, and exhibits amplitude and phase ripple. More sophisticated profile synthesis techniques are available which will further improve the aperture field uniformity and increase the gain by taking account of the internal diffraction effects.[20] Such Cassegrain designs are sometimes termed 'diffraction optimised'.

A modification of both the geometrically shaped (Williams) and diffraction optimised designs is available, in which only the subreflector is shaped and the main reflector is a best-fit paraboloid. Except for large main reflector diameters with geometrically shaped subreflectors, the modified design will provide almost as much gain as a fully shaped system.

To show how the efficiencies of the various shaped designs compare, these are plotted against main reflector diameter in *Figure 49.28*.

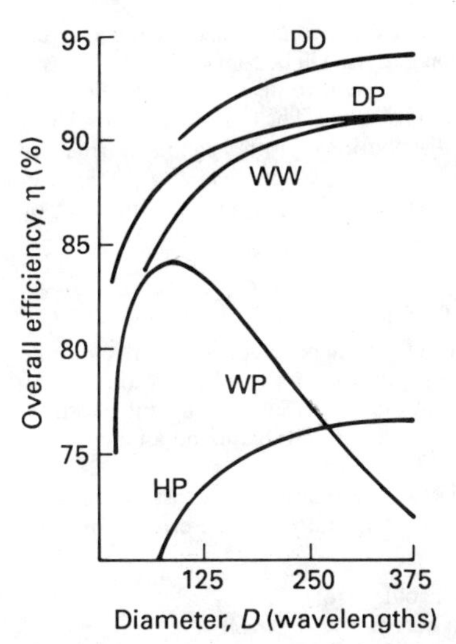

Figure 49.28 Efficiencies for dual reflector antennas with different profiles. HP = hyperboloid/paraboloid; WP = Williams subreflector/paraboloid; WW = Williams subreflector/Williams main reflector; DP = Diffraction optimised subreflector/paraboloid; DD = Diffraction optimised subreflector/diffraction optimised main reflector

49.9.3 Aperture blockage

The aperture of a rotationally symmetric reflector antenna will be blocked by the feed for a front-fed (single reflector) design, and by the subreflector and feed for a Cassegrain (dual reflector) design.

The effect will be to create a central shadow in the aperture which will reduce the antenna gain and increase the sidelobe levels.

If the aperture field distribution is assumed to be circularly symmetric, and parabolic tapered in the radial direction with an edge taper (voltage) e, then it may be shown that the absolute secondary pattern relative to an isotropic source is

$$e(\theta)=2\pi\frac{D}{\lambda}\sqrt{\frac{3}{1+e+e^2}}\left\{2(1-e)\frac{J_2(u)}{u^2}+e\frac{J_1(u)}{u}-\left(\frac{d}{D}\right)^2\left[2(1-f)\frac{J_2(v)}{v^2}+f\frac{J_1(v)}{v}\right]\right\} \quad (49.183)$$

where d and D are the blockage shadow and aperture diameters respectively, and

$$u=\pi D\lambda^{-1}\sin\theta \quad (49.184)$$

$$v=(d/D)u \quad (49.185)$$

$$f=1-(1-e)(d/D)^2 \quad (49.186)$$

For a uniform distribution ($e=1$) Equation (49.183) reduces to

$$e(\theta)=(2\pi D/\lambda)\left[\frac{J_1(u)}{u}-\left(\frac{d}{D}\right)^2\frac{J_1(v)}{v}\right] \quad (49.187)$$

The boresight gain loss due to the aperture blockage is given by

$$\frac{G}{G_{\text{unblocked}}}=\left[1-\left(\frac{d}{D}\right)^2\frac{(1+f)}{(1+e)}\right]^2 \quad (49.188)$$

which for a uniform distribution becomes

$$\frac{G}{G_{\text{unblocked}}}=\left[1-\left(\frac{d}{D}\right)^2\right]^2 \quad (49.189)$$

The blockage gain loss is plotted against normalised blockage diameter and edge taper in *Figure 49.29(a)*.

For small blockage diameters the main effect of the aperture blockage on the radiation pattern will be to increase the levels of the odd sidelobes (1st, 3rd, ...) and to decrease the levels of the first few even sidelobes (2nd, 4th, ...). The first sidelobe level for a parabolic on a pedestal distribution is plotted against normalised blockage diameter and edge taper in *Figure 49.29(b)*. For $d/D>0.3$ it is apparent that the edge taper has little influence on the sidelobe level.

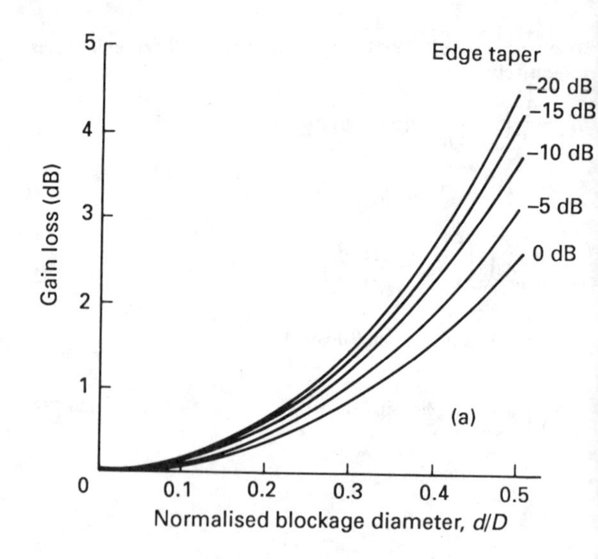

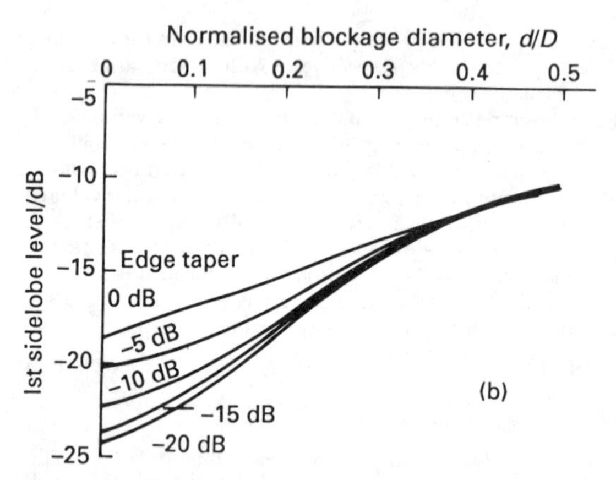

Figure 49.29 Effects of circular central aperture blockage: (a) gain loss, (b) first sidelobe level

49.9.4 Reflector profile errors

If the reflector surfaces of a focusing pencil beam antenna deviate from the required profiles, phase errors will be introduced into the aperture field distribution. The most significant effects of these will be a reduction of the boresight gain and an increase in the sidelobe levels.

For a random distribution of profile errors across the reflector surface, the error distribution may be described by two parameters[21]

ε (the normalised r.m.s. profile error)
c (the error distribution correlation interval)

ε is simply half the r.m.s. path length error, and is related to the r.m.s. profile error measured normal to the surface of, or parallel to the axis of a parabolic reflector (Δ_n and Δ_z respectively) by

$$\frac{\varepsilon}{\Delta_n}=4\frac{f}{D}\left\{\ln\left[1+\left(\frac{D}{4f}\right)^2\right]\right\}^{1/2}\underset{f/D\to\infty}{=1} \quad (49.190)$$

$$\frac{\varepsilon}{\Delta_z}=\left[1+\left(\frac{4f}{D}\right)^{-2}\right]^{-1/2}\underset{f/D\to\infty}{=1} \quad (49.191)$$

where f and D are the reflector focal length and diameter. The correlation interval c is the radius over which the errors remain correlated.

For small errors and small correlation intervals compared to the wavelength and antenna diameter respectively, the boresight gain loss (in dB) is given approximately by

$$\Delta G\simeq 686(\varepsilon/\lambda)^2 \quad (49.192)$$

This is plotted in *Figure 49.30(a)*.

The expected mean side lobe power pattern due to the profiles errors is given by

$$G(\theta)=(2\pi c/\lambda)^2\exp(-\overline{\delta^2})\sum_{n=1}^{\infty}\frac{(\overline{\delta^2})^n}{n\cdot n!}\exp[-(\pi c\sin\theta/\lambda)^2/n] \quad (49.193)$$

where

$$\overline{\delta^2}=(4\pi\varepsilon/\lambda)^2$$

As an example of the profile error tolerances necessary to meet realistic sidelobe specification, *Figure 49.30(b)* plots the maximum permissible r.m.s. error against correlation interval, to meet a widely used satellite earth station antenna sidelobe specification.

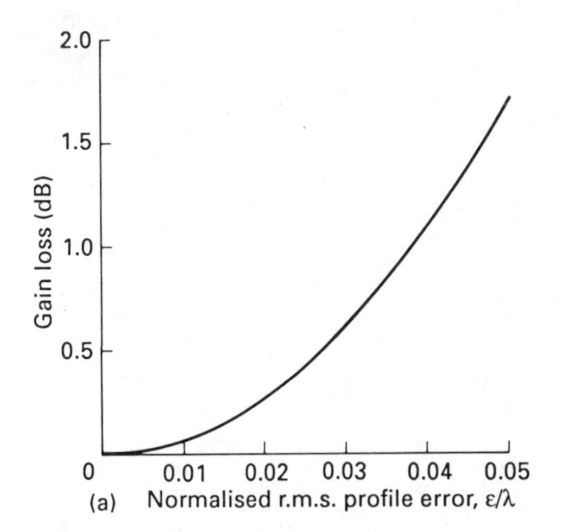

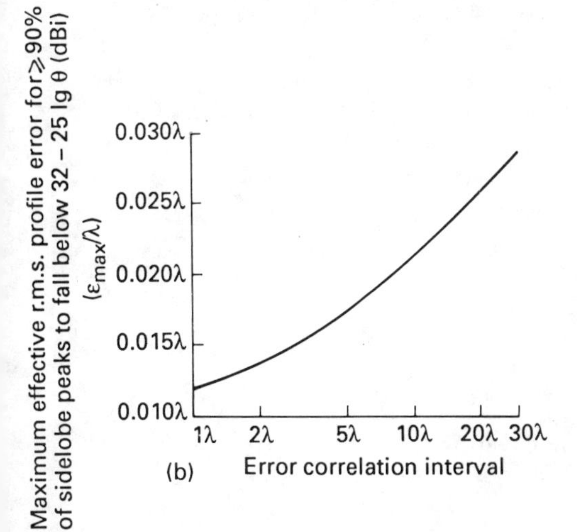

Figure 49.30 Effects of reflector profile errors: (a) gain loss, (b) profile accuracy to meet wide-angle sidelobe specifications

References

1 SCHELKUNOFF, S. A., *Electromagnetic Waves*, Van Nostrand Reinhold, New York (1943)
2 SCHELKUNOFF, S. A. and FRIIS, H. T., *Antennas: Theory and Practice*, Wiley, New York (1952)
3 LUDWIG, A. C., 'The definition of cross polarization', *IEEE Trans.*, **AP-21**, 116–19 (1973)
4 RAMSEY, J. F., 'Fourier transforms in aerial theory', *Marconi Rev.*, **83–9** (1946–48)
5 DOLPH, C. L., 'A current distribution for broadside arrays which optimizes the relationship between beam width and sidelobe level', *Proc. IRE*, **34**, 335 (1946)
6 TAYLOR, T. T., 'Design of line-source antennas for narrow beamwidth and low sidelobes', *IRE Trans.*, **AP-3**, 16–28 (1955)
7 SPELLMIRE, R. J., *Tables of Taylor Aperture Distributions*, TM581, Hughes Aircraft Co., Culver City, California (1958)
8 KNUDSEN, H. L., 'Radiation from ring quasi-arrays', *IRE Trans.*, **AP-4**, 452–72 (1956)
9 LINDENBLAD, N. E., 'Antennas and transmission lines at the Empire State Television Station, Part 2', *Communications*, **21**, 10–14 and 24–6 (1941)
10 BROWN, G. H. and WOODWARD, O. M., 'Circularly-polarized omnidirectional antenna', *RCA Rev.*, **8**, 259–69 (1947)
11 BRUCE, E., BECK, A. C. and LOWRY, L. R., 'Horizontal rhombic antennas', *Proc. IRE*, **23**, 24–46 (1935)
12 FOSTER, D., 'Radiation from rhombic antennas', *Proc. IRE*, **25**, 1327–53 (1937)
13 HARPER, A. E., *Rhombic Antenna Design*, Van Nostrand Reinhold, New York (1941)
14 Radio Research Laboratory Staff, *Very High Frequency Techniques*, McGraw-Hill, New York (1947)
15 BOOKER, H. G., 'Slot aerials and their relation to complementary wire aerials', *J. IEE*, **93**, Part III A, No. 4 (1946)
16 HALLEN, E., 'Theoretical investigations into the transmitting and receiving qualities of antennae', *Nova Acta Regide Soc. Sci. Upsaliensis*, **Ser IV**, **11**, No. 4, 1–44 (1938)
17 LEVINE, H. and PAPAS, C. H., 'Theory of the circular diffraction antenna', *J. Appl. Phys.*, **22**, No. 1, 29–43 (1951)
18 GALINDO, V., 'Design of dual-reflector antennas with arbitrary phase and amplitude distributions', *IEEE Trans.*, **AP-12**, 403–8 (1964)
19 WILLIAMS, W. F., 'High efficiency antenna reflector', *Microwave J.*, **8**, 78–82, July (1965)
20 WOOD, P. J., 'Reflector profiles for the pencil-beam Cassegrain antenna', *Marconi Rev.*, **XXXV**, No. 185, 121–38 (1972)
21 RUZE, J., 'Antenna tolerance theory—a review', *Proc. IEEE*, **54**, 633–40 (1966)

Further reading

BANCROFT, R., 'Accurate design of dual-band patch antennas', *Microwaves and RF*, September (1988)
BLUME, S. *et al.*, 'Biconical antennas and conical horns with elliptic cross-section', *IEEE Trans. (AP)*, August (1988)
PELL, C., 'Phased array radars', *IEE Review*, October (1988)
TESHIROGI, T. *et al.*, 'Multibeam array antenna for data relay satellite', *Electron. Commun. Japan*, Part 1, **75**, No. 5, May (1988)
WU, J. *et al.*, 'A maximum optimisation method for contoured-beam satellite antennas', *ESA J.*, **12**, No. 2 (1988)

50

Noise Management in Electronic Hardware

J M Camarata
Manager, Electronic Design,
Electronics Division,
Xerox Corporation

Contents

50.1 Introduction

50.1.1 Noise susceptibility

Susceptibility of the electronic subsystem to noise is a common concern in the design of electronic hardware. In this chapter some of the important considerations and techniques relative to the control of noise susceptibility are examined.

The rapidly advancing state of art in microelectronics has created many new opportunities for the use of electronics. The proliferation of electronics into new product areas has allowed dramatic product improvements; but it has also brought problems. Often these applications contain relatively noisy environments for the electronics. Consequently, one important area of concern in any product which utilises electronics is the management of electrical noise. The term 'management' is used because noise generation is a practical consequence from the operation of electronic/electrical subsystems. Of course, techniques are used to limit the amount of noise generated consistent with other product objectives. But, for the noise which inevitably remains, proper design (management) is necessary to ensure an acceptable level of performance.

Noise is loosely defined to be any form of electromagnetic energy (radiated or conducted) except for intentional signals along intended signal paths. Noise susceptibility refers to the sensitivity or response of electronic circuits and subsystems to noise. The susceptibility threshold for a particular electronic subsystem under specified operating conditions is normally defined as the point at which functional problems occur.

This chapter focuses on the problem of noise control within and between electronic subsystems and subsystem interactions with the remainder of the machine electrical environment. However, most of the information provided is equally applicable to noise whose origin is external to the machine. Although this information is relevent to any type of equipment, it is oriented towards electronic systems in commercial equipment using digital techniques.

50.1.2 Noise management philosophy

Noise management is an essential part of the design process and must be considered early when design concepts are being formulated. When establishing a noise management strategy, the designer recognises that there are three elements to the problem: noise source, coupling mechanism and receiving circuit/subsystem.

The strategy should address each of these elements to determine the most cost-effective way to prevent harmful interactions. The task for the designer is to recognise and understand potential problems, identify design alternatives and select the most cost-effective combination of techniques. Most of the techniques utilised will be nothing more than good design practice which have little or no cost impact on the product. Occasionally, difficult problems will require special hardware or software solutions which increase product cost. In this chapter, we will only examine the problem relative to hardware design. The total noise management strategy for digital equipment, however, should also include software design. In particular, software filtering and timers, data refresh and recovery techniques, error detection and correction, etc., should be included as appropriate. The number of special hardware solutions required and their cost can be minimised by properly addressing noise management at the beginning and throughout the product design cycle. Solutions that must be patched into an existing design will often carry cost/schedule penalties that could have been avoided.

50.2 Background information

In this section, a brief summary of useful formulas and concepts is presented. For the reader wishing to explore the subject in more depth, there are many good reference books on this material, some of which are listed in the further reading at the end of the chapter.

50.2.1 Waveform analysis

Normally, designers closely monitor and control the time domain characteristics of signal waveforms. However, most circuit and system modelling and analysis relative to the control of interference is done in the frequency domain (i.e. field generation, coupling phenomena, and receptor sensitivity). Understanding the frequency content of potentially disturbing signals greatly aids in the task of interference control.

50.2.1.1 Fourier series

Trigonometric form Any normally encountered periodic waveform of period T may be represented as the sum of a series of sinusoidal components. The components will consist of sinusoids at the fundamental and multiples of the fundamental frequency (harmonics). This is called a Fourier series and is given by:

$$f(t)=\tfrac{1}{2}A_0+\sum_{n=1}^{\infty}(A_n\cos\omega_n t+B_n\sin\omega_n t) \tag{50.1}$$

Where the coefficients A_n and B_n are given by:

$$A_n=2/T\int_{-T/2}^{T/2} f(t)\cos\omega_n t\,dt \tag{50.2}$$

$$B_n=2/T\int_{-T/2}^{T/2} f(t)\sin\omega_n t\,dt \tag{50.3}$$

In Equations (50.1)–(50.3), $\omega_n=n\omega_1$ where $\omega_1=2\pi/T$ and the limits of integration may be any values that are one period apart. Substituting $n=0$ into Equation (50.2) yields A_0 where $A_0/2$ is the average value of the waveform.

The frequency spectrum of a periodic signal consists of discrete line spectra occurring at fundamental and harmonic frequencies. The sine and cosine terms in Equation (50.1) can be combined into the alternate form of Equation (50.4)

$$f(t)=\frac{C_0}{2}+\sum_{n=1}^{\infty}C_n\cos(\omega_n t+\varphi_n) \tag{50.4}$$

where

$$C_0=A_0 \tag{50.5}$$

$$C_n=\sqrt{A_n^2+B_n^2} \tag{50.6}$$

$$\varphi_n=\arctan(-B_n/A_n) \tag{50.7}$$

C_n is the amplitude of the frequency line spectra and φ_n the phase.

Exponential form Rather than compute A_n and B_n, it will often be more convenient to use the exponential form of the Fourier series. This form is given by Equations (50.8) and (50.9)

$$f(t)=\sum_{n=-\infty}^{\infty}D_n^{jn\omega t} \tag{50.8}$$

$$D_n=1/T\int_{-T/2}^{T/2} f(t)\,e^{-jn\omega t}\,dt \tag{50.9}$$

The appearance of negative frequencies in the summation interval of Equation (50.8) is a mathematical requirement associated with complex notation. It should be noted that, in general, D_n is a complex quantity. Also, D_n and D_{-n} are always complex conjugates of one another and $D_n = C_n/2$. The following general symmetry relationships are useful:

(1) If a time function possesses half-wave symmetry, i.e.: $f(t)=f(t+T/2)$, $D_n=0$ for even n.
(2) All D_n are real for even functions.
(3) All D_n are imaginary for odd functions.

(Section 50.2.1.2)

To demonstrate the usefulness of the exponential form of Fourier series, it will be applied to the general rectangular pulse train of *Figure 50.1*.

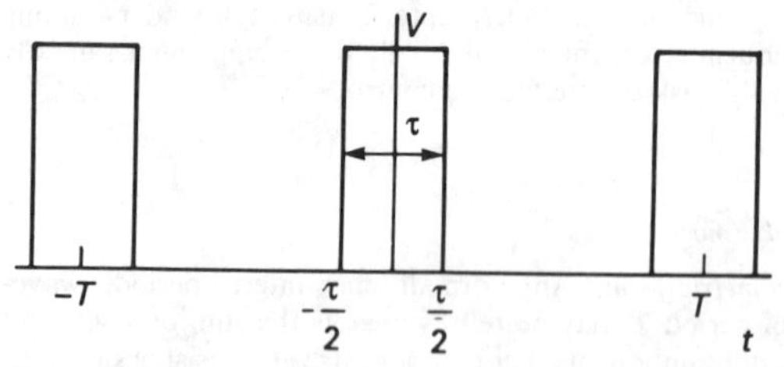

Figure 50.1 Rectangular pulse waveform

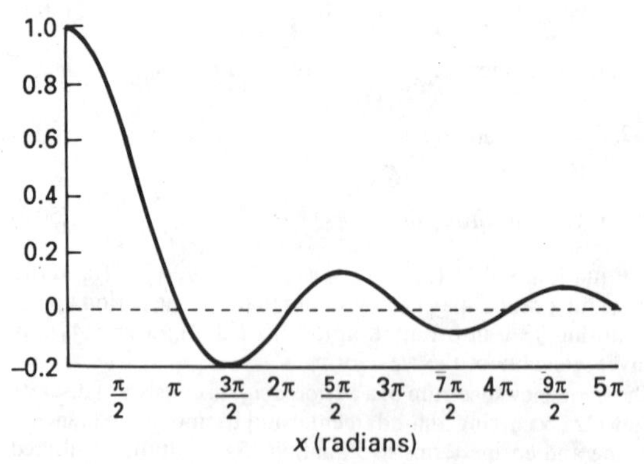

Figure 50.2 sin x/x curve. (From *Electronic Designers Handbook*[1] by R. W. Landee *et al.* Copyright © 1957, McGraw-Hill Book Company. Used with the permission of McGraw-Hill Book Company)

From Equation (50.9)

$$D_n = 1/T \int_{-T/2}^{T/2} f(t)\,e^{-jn\omega_1 t}\,dt = \frac{V\tau}{T}\left(\frac{\sin \frac{1}{2}n\omega_1\tau}{\frac{1}{3}n\omega_1\tau}\right)$$

$$= \frac{V\tau}{T}\left(\frac{\sin n\pi\tau/T}{n\pi\tau/T}\right) = \frac{V\tau}{T}\left(\frac{\sin n\pi f_1\tau}{n\pi f_1\tau}\right) \qquad (50.10)$$

The envelope of the amplitude spectra is of the form sin x/x which frequently occurs in signal analysis. The following characteristics of sin x/x can be immediately deduced:

(1) For x very small ($x \to 0$), $\sin x \to x$ and $\sin x/x \to 1$.
(2) $\sin x/x = 0$ for $x = n\pi$ where $n = 1, 2, 3 \ldots$. For other values of x, $\sin x/x$ oscillates between positive and negative values.
(3) For x very large ($x \to \infty$), $\sin x/x \to 0$. The curve of $\sin x/x$ is plotted in *Figure 50.2*.

The frequency spectrum corresponding to the pulse train of *Figure 50.1* is plotted in *Figure 50.3* for the case of $T/\tau = 4$.

The spectrum consists of discrete line spectra at the fundamental frequency and harmonics. The amplitude of the line spectra is $2|D_n|$. D_n is given by Equation (50.10), and is plotted in *Figure 50.3* as a function of frequency (nf_0). The spectrum envelope is zero at $nf_0 = 1/\tau$, $2/\tau$, e/τ, etc. The number of line spectra between zero crossings of the envelope is equal to $(T/\tau - 1)$.

Equations for the Fourier coefficients for some common waveforms are provided in *Figure 50.4*.

Spectrum envelope approximation Often, one is not interested in obtaining the exact value of the frequency spectrum components but only needs a straight line approximation to the envelope which passes through the maximum values of line spectrum components. Such straight line approximations are shown in *Figure 50.5* on log–log scale for rectangular and trapazoidal pulses. From this graph it can be seen that the low frequency spectrum for these equal area pulses is the same. The first break frequency occurs at $f = 1/\pi\tau$ and the maximum amplitude spectra is falling off at −20 dB/decade. At $f = 1/\pi t_r$ for the trapezoidal pulse the envelope takes another break and it falls off at −40 dB/decade.

50.2.1.2 Even/odd functions

A given function may be even, odd or the combination of even and odd functions. An even function is defined by

$$f(t) = f(-t) \qquad (50.11)$$

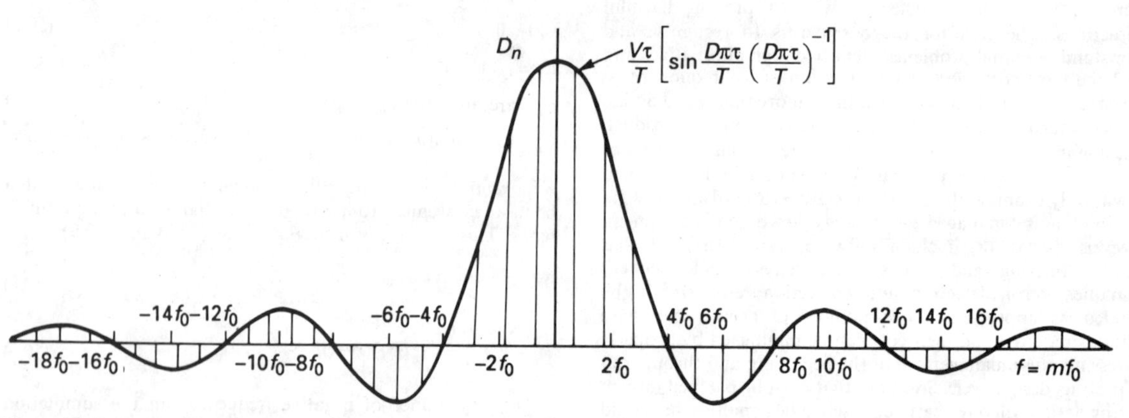

Figure 50.3 Spectral lines for rectangular pulse train with $T/\tau = 4$. (Reproduced from reference 4 with permission)

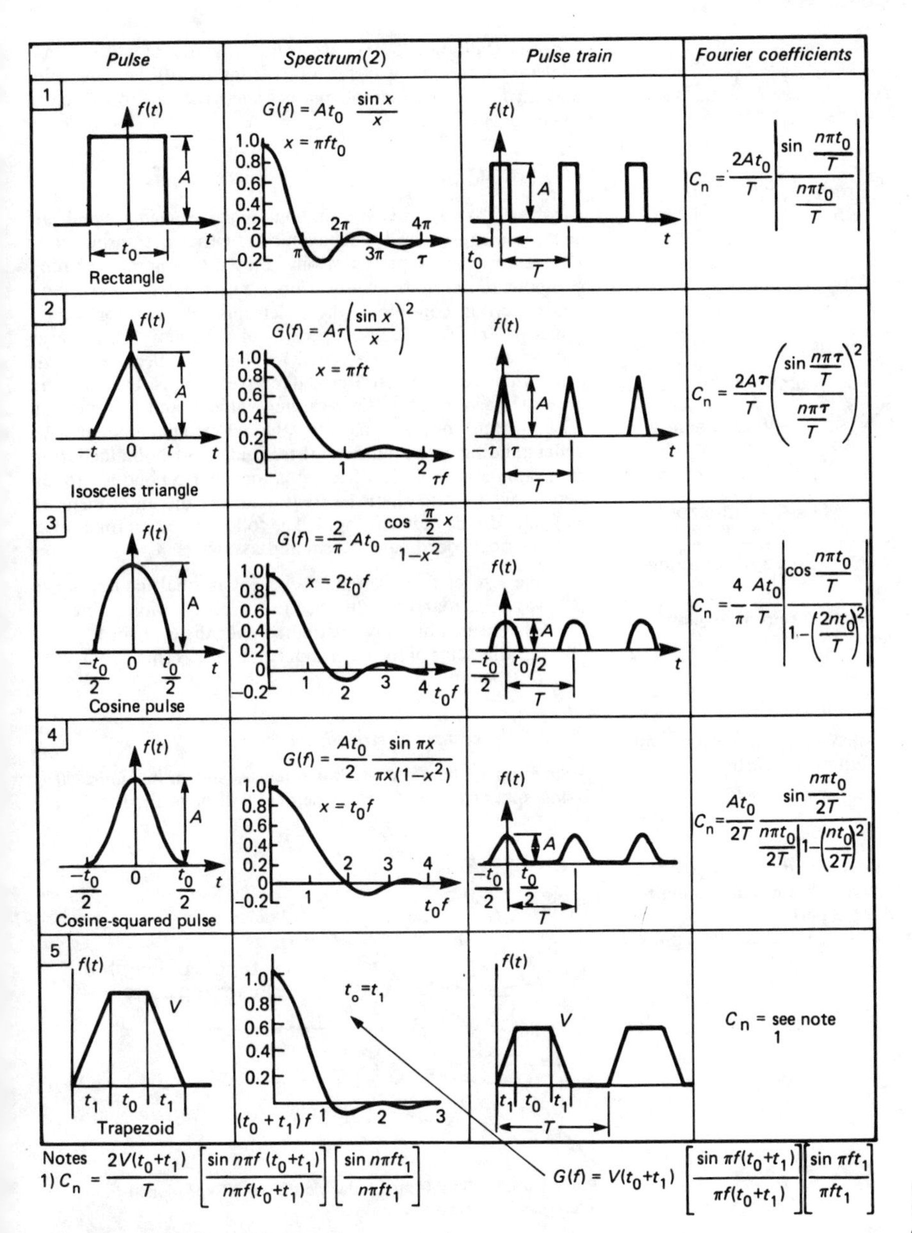

Pulse
Spectrum(2)
Pulse train
Fourier coefficients
1
Rectangle
$G(f) = At_0 \frac{\sin x}{x}$
$x = \pi f t_0$
$C_n = \frac{2At_0}{T} \left| \frac{\sin \frac{n\pi t_0}{T}}{\frac{n\pi t_0}{T}} \right|$
2
Isosceles triangle
$G(f) = A\tau \left(\frac{\sin x}{x}\right)^2$
$x = \pi f t$
$C_n = \frac{2A\tau}{T} \left(\frac{\sin \frac{n\pi\tau}{T}}{\frac{n\pi\tau}{T}} \right)^2$
3
Cosine pulse
$G(f) = \frac{2}{\pi} At_0 \frac{\cos \frac{\pi}{2} x}{1-x^2}$
$x = 2t_0 f$
$C_n = \frac{4}{\pi} \frac{At_0}{T} \left| \frac{\cos \frac{n\pi t_0}{T}}{1-\left(\frac{2nt_0}{T}\right)^2} \right|$
4
Cosine-squared pulse
$G(f) = \frac{At_0}{2} \frac{\sin \pi x}{\pi x(1-x^2)}$
$x = t_0 f$
$C_n = \frac{At_0}{2T} \frac{\sin \frac{n\pi t_0}{2T}}{\frac{n\pi t_0}{2T} \left| 1-\left(\frac{nt_0}{2T}\right)^2 \right|}$
5
Trapezoid
$t_o = t_1$
C_n = see note 1

Notes

1) $C_n = \frac{2V(t_0+t_1)}{T} \left[\frac{\sin n\pi f\,(t_0+t_1)}{n\pi f(t_0+t_1)}\right] \left[\frac{\sin n\pi f t_1}{n\pi f t_1}\right]$ $\quad G(f) = V(t_0+t_1) \left[\frac{\sin \pi f(t_0+t_1)}{\pi f(t_0+t_1)}\right] \left[\frac{\sin \pi f t_1}{\pi f t_1}\right]$

2) Normally the spectrum function ($G(f)$) is multiplied by a factor of 2 to account for the negative frequencies. This is a mathematical requirement and has no physical meaning.

Figure 50.4 Spectrum and Fourier coefficients of some common waveforms. (Reproduced from reference 2 with permission)

The definition of an odd function is given by

$$f(t) = -f(-t) \tag{50.12}$$

Often, a function can be made either even or odd, depending on the choice of the axis. Note that even/odd functions should not be confused with even or odd harmonics which are discussed in the next section. The even or odd harmonic content cannot be affected by the axis choice.

50.2.1.3 Even/odd harmonics

A waveform with a period T will contain only odd harmonics if

$$f(t) = -f(t+T/2) \tag{50.13}$$

Similarly, it will only contain even harmonics if

$$f(t) = f(t+T/2) \tag{50.14}$$

The above conditions are sufficient but not necessary.

50.2.1.4 Fourier integral

As the period of the pulse train increases without limit, the spacing between line spectra ($2\pi/T$) decreases to zero. The discrete line spectrum becomes a continuous frequency spectrum. In Equation (50.9), the ratio D_n/ω is defined as g_n and as $T \to \infty$ this becomes a continuous function $g(\omega)$ given by Equation (50.15).

$$g(\omega) = 1/2\pi \int_{-\infty}^{\infty} f(t)\, e^{-j\omega t}\, dt \tag{50.15}$$

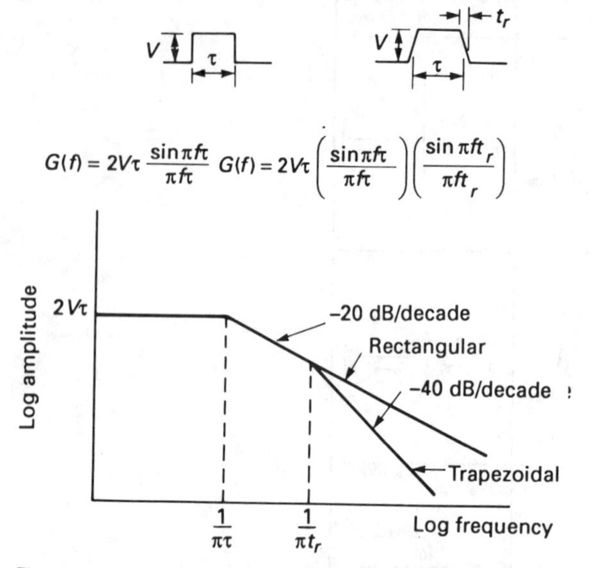

Figure 50.5 Asymptote approximations to spectrum envelope for rectangular and trapezoidal pulses

Also, the summation indicated in Equation (50.8) becomes an integral in the limit and is given by Equation (50.16)

$$f(t)=\int_{-\infty}^{\infty} g(\omega)\, e^{j\omega t}\, d\omega \qquad (50.16)$$

These two equations define the Fourier integral and allow the frequency spectrum for a single pulse to be obtained analogous to the way that the Fourier series did for a periodic waveform.

Applying Equation (50.15) to a single rectangular pulse (*Figure 50.6*) yields

$$g(\omega)=V\tau\left(\frac{\sin \pi f\tau}{\pi f\tau}\right) \qquad (50.17)$$

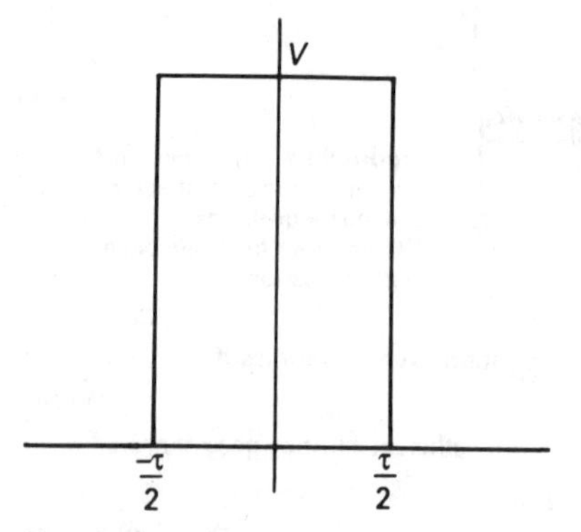

Figure 50.6 Rectangular pulse

Note that the frequency spectrum envelope shape depends only on the pulse's time domain waveform. It is independent of whether the pulse is periodic or non-repetitive. The periodic pulse train has a discrete line spectrum while the non-recurring pulse has a continuous spectrum. It should also be noted that the units for $g(\omega)$ (Equation 50.17) are volts/frequency or volt-seconds while for a periodic waveform (Equation 50.10) they are volts. Spectra for some common pulse waveshapes are provided in *Figure 50.4*.

50.2.2 Field theory

Most electronic hardware contains many elements which are capable of antenna-like behaviour, as either a transmitter or receiver (e.g. cables, printed wiring board interconnect, electronic components, ground or power distribution interconnect, etc.). Under certain conditions, these elements may emit or receive sufficient energy to cause operational problems. This energy transfer takes place via electric, magnetic, or electromagnetic fields which couple the source and receiver circuits. The fields are generated as a result of time-varying voltages and currents that exist within or external to the electronic system. An understanding of the nature of these fields can be obtained from an examination of the field equations corresponding to an elementary electric dipole (current filament) and an elementary magnetic dipole (current loop). The following equations utilise the spherical coordinate system and assume that:

(1) The size of the differential element is small compared to wavelength and the distance to the observation point.
(2) Current is uniformly distributed over the element length.
(3) The diameter of the elemental conductor is small compared to its length.

50.2.2.1 Elementary electric dipole source

Assume an elemental current filament as shown in *Figure 50.7*. Using spherical coordinates, the field will have E_r, E_θ and H_ϕ

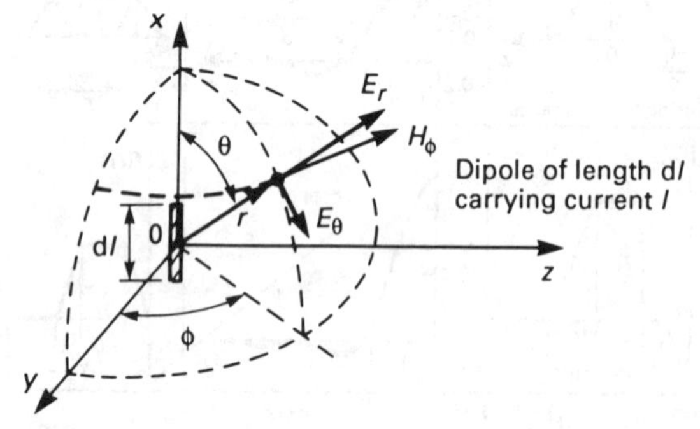

Figure 50.7 Field from an electric dipole. (Reproduced from reference 3 with permission)

vectors. The magnitudes of these vectors are given by Equations (50.18), (50.19) and (50.20)

$$E_r=\frac{I\,dl\,\beta^3\,e^{-j\beta r}}{\omega\quad 2\pi\varepsilon_0}\left[\frac{1}{(\beta r)^2}-\frac{j}{(\beta r)^3}\right]\cos\theta \quad (V/m) \qquad (50.18)$$

$$E_\theta=\frac{I\,dl\,\beta^3\,e^{-j\beta r}}{\omega\quad 4\pi\varepsilon_0}\left[\frac{j}{(\beta r)}+\frac{1}{(\beta r)^2}-\frac{j}{(\beta r)^3}\right]\sin\theta \quad (V/m) \qquad (50.19)$$

$$H_\phi=\frac{I\,dl\,\beta^2\,e^{-j\beta r}}{\omega\quad 4\pi}\left[\frac{j}{(\beta r)}+\frac{1}{(\beta r)^2}\right]\sin\theta \quad (A/m) \qquad (50.20)$$

where

dl = differential length of current element (m)
I = current in element (A)
ω = angular frequency of I (rad/s)
v = velocity of propagation (in free space $v=c=3\times10^8$ m/s)
β = phase constant ω/V

Also,

$$\beta=\frac{2\pi f}{C}=\frac{2\pi}{\lambda}, \text{ where } \lambda=\text{wavelength (m)}$$

ε_0 = permittivity of free space = $1/36\pi \times 10^9$ f/m
r = distance from elemental source to observation point (m)

An examination of Equations (50.18)–(50.20) yields the following information.

(1) For $\beta r << 1$, the higher-order terms dominate with the **E** field components varying as $1/r^3$, $1/r^2$ and the **H** component as $1/r^2$. The field very close to the source is dominated by the $1/r^3$ terms and is called the electrostatic field. This field term corresponds to that of an electric dipole and represents a constant energy field. The $1/r^2$ terms are called the induction field. This total field region where these higher-order terms predominate is called the near field.

(2) For $\beta r >> 1$, E_r becomes insignificant compared to the transverse term (E_θ) and E_θ and H_ϕ vary as $1/r$. This is called the radiation or far field. In this field energy is being propagated from the current dipole.

(3) The transition from the far to the near field is defined as $\beta r = 1$ or $r = 1/\beta = \lambda/2\pi$, approximately at a distance corresponding to one sixth of a wavelength.

50.2.2.2 Elementary current loop

Similar to the previous section, the fields corresponding to the elementary current loop of *Figure 50.8* are given.

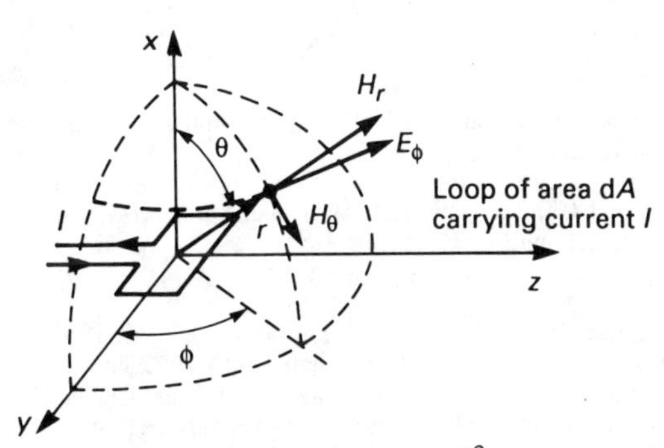

Figure 50.8 Field from a magnetic dipole[3]

The field vectors created by this elementary current loop are H_r, H_θ and E_ϕ whose magnitudes are given by Equations (50.21), (50.22) and (50.23):

$$H_r=\frac{I\,dA\beta^3 e^{-j\beta r}}{2\pi}\left[\frac{j}{(\beta r)^2}+\frac{1}{(\beta r)^3}\right]\cos\theta \quad \text{(A/m)} \tag{50.21}$$

$$H_\theta=\frac{I\,dA\beta^3 e^{-j\beta r}}{4\pi}\left[-\frac{1}{(\beta r)}+\frac{j}{(\beta r)^2}+\frac{1}{(\beta r)^3}\right]\sin\theta \quad \text{(A/m)} \tag{50.22}$$

$$E_\phi\frac{I\,dA\beta^4 e^{-j\beta r}}{\omega 4\pi\varepsilon_0}\left[\frac{1}{(\beta r)}-\frac{j}{(\beta r)^2}\right]\sin\theta \quad \text{(V/m)} \tag{50.23}$$

where dA = differential area of the current loop (m²). All other parameters are the same as defined for the electric dipole.

An analysis of these equations will produce results similar to those obtained for the electric dipole.

50.2.2.3 Wave impedance

At any point in a medium containing an electromagnetic field, the field has an impedance η defined as

$$\eta=\mathbf{E}/\mathbf{H} \tag{50.24}$$

In the far field, this ratio equals the characteristic impedance of the medium which for air is given by Equation (50.25)

$$\eta=\sqrt{\mu/\varepsilon}\simeq 120\pi\simeq 377\ \Omega \tag{50.25}$$

In the near field, the impedance is either higher or lower than 377 Ω depending on the nature of the source. Sources with high voltage and low current generate high impedance fields and low impedance fields result from high current, low voltage sources. For a high impedance field (i.e. electric dipole) the near field impedance is

$$|Z|=\frac{\eta\lambda}{2\pi r}=\frac{1}{2\pi f\varepsilon r} \quad (\Omega) \tag{50.26}$$

For a low impedance field (i.e. current loop), the near field impedance is given by

$$|Z|=\frac{\eta 2\pi r}{\lambda}=2\pi f\mu r \quad (\Omega) \tag{50.27}$$

Figure 50.9 plots wave impedance as a function of distance from the source (normalised to $\lambda/2\pi$). It can be seen that $r=\lambda/2\pi$ represents the transition from the near to far field.

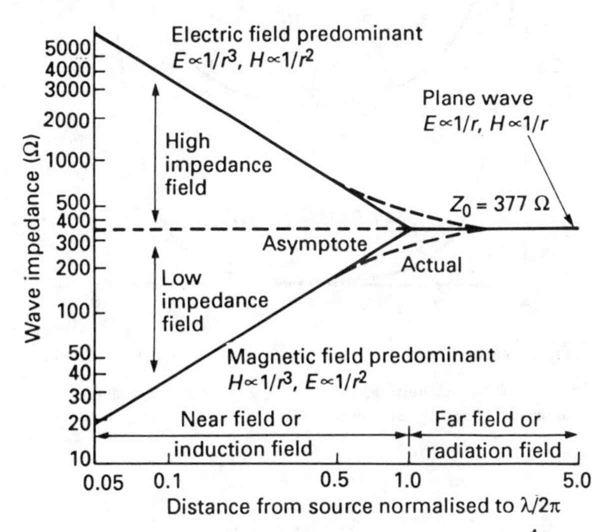

Figure 50.9 Wave impedance against distance from source[4]

50.2.3 Skin effect

A conductor carrying a d.c. current will have the current uniformly distributed across the conductor cross-section. An a.c. current, however, is not uniformly distributed but will tend to concentrate on the conductor surface. As frequency increases, the effect becomes more pronounced until the current flow is confined to a very thin layer on the conductor surface. This phenomenon is called the skin effect and significantly increases the effective resistance of conductors carrying high frequency signals. It is also an important factor relative to shielding effectiveness.

The current density magnitude decreases exponentially with distance from the surface and is reduced to 1/e the surface value at a distance δ (in metres) given by

$$\delta=\sqrt{\frac{1}{\pi f\mu\sigma}} \tag{50.28}$$

where

δ = skin depth
f = frequency
μ = conductor permeability
σ = conductivity (mho/m)

For a copper conductor, $\mu=\mu_0=4\pi\times10^{-7}$, and $\sigma=5.75\times10^7$ mho/m at 20°C and Equation (50.28) becomes

$$\delta=\frac{0.0664}{\sqrt{f}} \tag{50.29}$$

Figure 50.10 shows the current distribution in round solid copper wire of 1 mm diameter (radius a=0.5 mm). *Figure 50.11* shows the depth of current penetration against frequency for some common conductor materials.

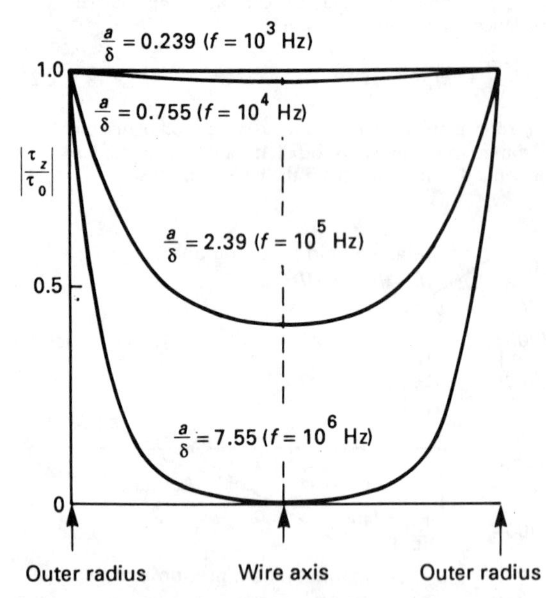

Figure 50.10 Current distribution in cylindrical 1 mm diameter copper wire against frequency[5]

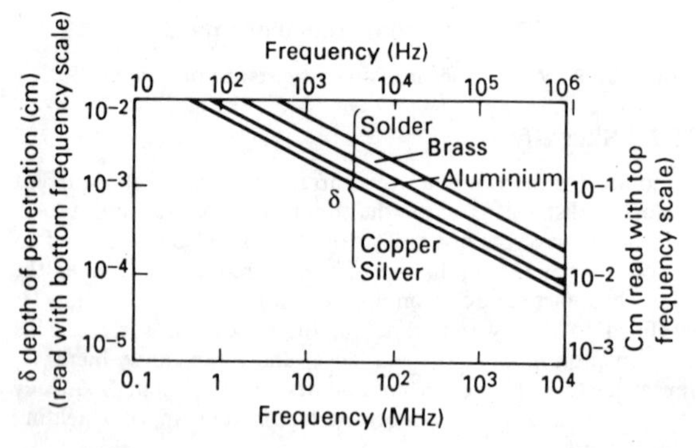

Figure 50.11 Skin effect against frequency for plane conductors[5]

The skin depth, δ, is an important parameter in determining the a.c. resistance of a conductor. Consider a round conductor for which δ is small relative to the conductor radius, a. The equivalent conductor geometry for current conduction may be shown to be a cylindrical shell of thickness δ which carries a uniformly distributed current. The a.c. resistance is proportional to $1/2\pi a\delta$.

The ratio of a.c. to d.c. resistance is

$$\frac{R_{\text{a.c.}}}{R_{\text{d.c.}}}=\frac{a}{2\delta}=\frac{a\sqrt{\pi f\mu\sigma}}{2} \tag{50.30}$$

which for copper becomes

$$\frac{R_{\text{a.c.}}}{R_{\text{d.c.}}}=7.53\,a\sqrt{f} \tag{50.31}$$

50.2.4 Shielding

Shielding is the use of a conducting and/or permeable barrier between a potentially disturbing field source and the circuitry to be protected. Shields may be used to protect interconnect (e.g. cables) or packaged circuit functions. The effectiveness of shielding depends on the characteristics of:

(1) The field (type of field, strength, polarisation, angle of incidence, frequency).
(2) The shielding material (conductivity, permeability).
(3) The physical geometry of the shield (thickness, openings in the shield).

Although the following discussion is strictly true for shields in the form of a large flat sheet, the principles are true in general.

Mathematically, shielding effectiveness (S) for electric and magnetic fields is defined by Equations (50.32) and (50.33) respectively.

$$S=20\lg E_1/E_2 \tag{50.32}$$

$$S=20\lg H_1/H_2 \tag{50.33}$$

in dB, where E_1, H_1 are the incident fields and E_2, H_2 are the attenuated fields.

Shielding attenuates incident fields via two loss mechanisms—absorption (A) and reflection (R). With A and R expressed in dB, the total shielding loss is

$$S=A+R+B \tag{50.34}$$

The B term is a correction factor required if there are multiple reflections within the shield. Normally A is sufficiently high (greater than 10 dB) that B can be neglected. This is especially true for electric fields which have a substantial reflection loss at the incident surface. However, thin shields in the presence of low frequency magnetic fields may require the use of this term. The effect of B reduces the overall shielding effectiveness as indicated in the following expression[4]

$$B=20\lg(1-e^{-2t/\delta}) \tag{50.35}$$

50.2.4.1 Absorption loss

When time-varying electric and magnetic fields impinge on a conducting medium, current flow is induced. In accordance with *Section 50.2.3* (skin effect), the magnitude of current at distance x from the surface (I_x) is given by

$$I_x=I_s e^{-x/\delta} \tag{50.36}$$

I_s = current at surface
δ = skin depth (see Equation (50.28)).

Electromagnetic field attenuation in the medium follows the same relationship which yields

$$A=x/\delta\ \text{(nepers)}=8.69x/\delta\ \text{(dB)} \tag{50.37}$$

From Equation (50.37) it can be seen that absorption loss is directly proportional to the thickness of the material and equals 8.69 dB for each skin depth of distance into the shielding material.

50.2.4.2 Reflection loss

The characteristic impedance of any medium is given by Equation (50.38)

$$Z_0=\left(\frac{j\omega\mu}{\sigma+j\omega\varepsilon}\right)^{1/2} \quad (50.38)$$

In an insulator $\sigma << j\omega\varepsilon$ and

$$Z_0=(\mu/\varepsilon)^{1/2} \quad (50.39)$$

In a conductor $\sigma >> j\omega\varepsilon$ and

$$Z_0=\left(\frac{\pi f\mu}{\sigma}\right)^{1/2}(1+j) \quad (50.40)$$

In *Section 50.2.2*, a field was shown to have an impedance which depends on the electrical distance from the source. The far field impedance is 377 Ω. The near field impedance could be either higher or lower depending on whether the source generator is predominantly electric or magnetic field respectively. When such a field impinges on a conducting barrier, the impedance mismatch will cause some of the incident field to be reflected. The portion that penetrates is attenuated by absorption loss. The penetrating field which reaches the exit surface will undergo another reflection loss. The total reflection loss is given (in dB) by

$$R=20\lg\frac{|Z_W|}{4|Z_S|} \quad (50.41)$$

where Z_W is the impedance of wave just outside the shield and Z_S is the shield impedance.

Equation (50.41) is valid for both electric and magnetic fields. For electric fields, the biggest reflection loss occurs at the incident air/conductor boundary so that very little electric field penetration occurs. Conversely, magnetic fields more readily penetrate the shield material at the incident boundary and suffer large reflection at the exit boundary.

Reflection loss (in dB) in the near field for electric and magnetic fields is given by Equations (50.42) and (50.43) respectively.

$$R_e=320.2-10\lg\left[f^3r^2\left(\frac{\mu_r}{\varepsilon_r}\right)\right] \quad (50.42)$$

$$R_m=14.6-10\lg\left[fr^2\left(\frac{\sigma_r}{\mu_r}\right)\right] \quad (50.43)$$

where

f = frequency (Hz)
r = distance between source and shield (m)
μ_r = relative permeability of shield material (relative to air)
ε_r = relative permittivity of shield material (relative to air)
σ_r = relative conductivity of shield material (relative to copper)

From Equations (50.42) and (50.43) the following facts are evident.

(1) *Electric field*
  (a) Reflection loss decreases 30 dB/decade of frequency.
  (b) Reflection loss decreases 20 dB/decade of distance; loss is greater for shield close to the source.
(2) *Magnetic field*
  (a) Reflection loss increases 10 dB/decade of frequency.
  (b) Reflection loss increases 20 dB/decade of distance; loss is greater for shield farther from source.

Equation (50.42) shows that it is possible to achieve relatively high reflection losses for electric fields. Low frequency magnetic fields, however, have low reflection losses and must depend on absorption loss as the primary loss mechanism.

50.2.5 Interconnect formulae

The distribution of electrical signals and power between physically separated points is a characteristic of all electronic equipment. The interconnect system is the means by which this distribution is accomplished. In general, the interconnect system must handle many different types of signals. Signal frequency content can range from d.c. frequency to hundreds of MHz and power from μW to hundreds of watts. Often these incompatible signals will be in close proximity to each other creating a high potential for interference. The physical/electrical length of the interconnect, effect on signal quality and coupling to other circuits (as either a disturbing source or victim) are factors which the designer must consider.

Depending on signal characteristics, a signal line and its return path may be viewed as a short line (lumped element line) or a long line (distributed parameter transmission line). The important criterion is the electrical length of the line relative to the signal frequency content. If the line is carrying digital signals, a long line is defined as a line whose one-way delay time (t_d) is larger than one half the transition time of the signal (i.e. $t_d > t_r/2$ or $t_f/2$).

Regarding interconnect electrical characteristics, two categories can be defined: self and mutual. Self parameters are those associated with a given signal path itself while mutual parameters concern the coupling between different signal paths. For either type, there can be a lumped or distributed case depending on signal frequencies and line lengths.

The basic electrical parameters of interest are the inductance (L) and capacitance (C) per unit length. For distributed lines, the characteristic impedance (Z_0) and propagation delay (t_d) for a lossless line are given by Equations (50.44) and (50.45)

$$Z_0=(L/C)^{1/2} \quad \text{(in } \Omega) \quad (50.44)$$

$$t_d=v^{-1}=(LC)^{1/2} \quad \text{(in s/m)} \quad (50.45)$$

where

v = velocity of propagation (m/s)
L = inductance (H/m)
C = capacitance (F/m)

Figure 50.12 provides formulae for several useful transmission line configurations.

In *Figure 50.13*, Z_0, L and C are plotted as a function of D/d to illustrate the sensitivity of these parameters to physical spacing.

50.3 Interference mechanisms

When undesirable energy from a source function finds its way into a victim circuit or subsystem, the potential for interference exists. In this section, the primary mechanisms by which this energy transfer can occur are examined. The two predominant types of interference crosstalk are common impedance and induction coupling.

50.3.1 Common impedance coupling

Shared conductive paths between circuits and/or subsystems create the potential for common impedance crosstalk. The two most common places for this to occur are in the power and ground distribution systems. A simple example illustrates how this can occur. Assume a linear amplifier circuit with three stages of amplification implemented as shown in *Figure 50.14*. Further assume that the power handling capacility increases from stage 1 to 3 (i.e. $I_3 > I_2 > I_1$).

From *Figure 50.14* it can be seen that crosstalk between stages is occurring through both the ground and power distribution systems. The input signal to each stage is given by

$$e_{in(n)}=e_{0(n-1)}+e_{gn}=e_{0(n-1)}(1+\varepsilon) \quad (50.46)$$

Transmission line configuration	*Characteristic impedance* (Z_0)	*Capacitance* (f/m)	*Inductance* (H/m)	*Conditions*
Parallel wires (air)	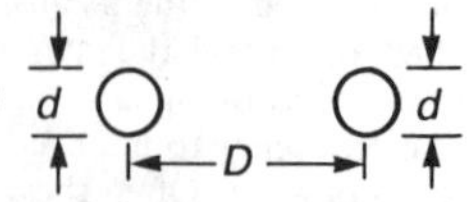 $\frac{(\mu/\varepsilon)^{1/2}}{\pi}\,\text{arcosh}(D/d)$	$\frac{\pi\varepsilon}{\text{arcosh}(D/d)}$	$\mu/\pi\,\text{arcosh}(D/d)$	
Wire over ground	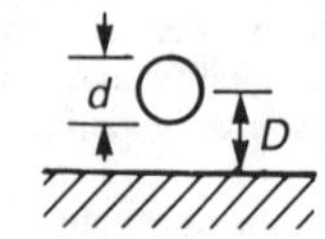 $\frac{(\mu/\varepsilon)^{1/2}}{2\pi}\,\text{arcosh}(2D/d)$	$\frac{2\pi\varepsilon}{\text{arcosh}(2D/d)}$	$\mu/2\pi\,\text{arcosh}(2D/d)$	
Microstrip	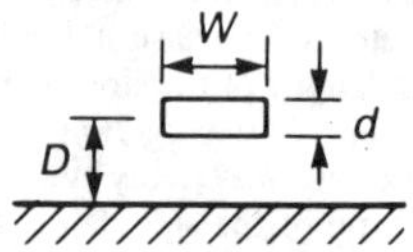 $(\mu/\varepsilon)^{1/2}D/W$	$\varepsilon W/D$	$\mu D/W$	$2D<W$ $d<<2D$
	$\frac{(\mu/\varepsilon)^{1/2}}{2\pi}\,\ln(2\pi D/W+d)$	$\frac{2\pi\varepsilon}{\ln[(2\pi D/W+d)]}$	$\mu/2\pi\,\ln(2\pi D/W+d)$	$2D>>W$
Stripline	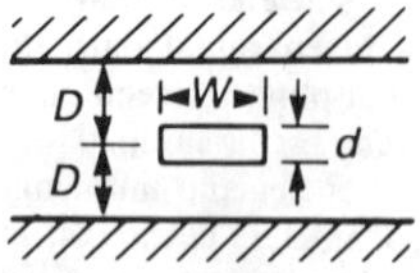 $(\mu/\varepsilon)^{1/2}D/2W$	$2\varepsilon W/D$	$\mu D/2W$	$D<<W$ $d<<D$
Parallel strip	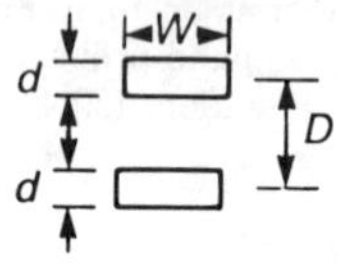 $(\mu/\varepsilon)^{1/2}D/W$	$\varepsilon W/2\pi$	$\mu D/W$	$D<<W$ $d<<D$
	$\frac{(\mu/\varepsilon)^{1/2}}{\pi}\,\ln(\pi D/W+d)$	$\frac{\pi\varepsilon}{\ln[(\pi D/W+d)]}$	$\mu/\pi\,\ln(\pi D/W+d)$	$D>>W$
Coaxial	 $\frac{(\mu/\varepsilon)^{1/2}}{2\pi}\,\ln(D/d)$	$\frac{2\pi\varepsilon}{\ln(D/d)}$	$\mu/2\pi\,\ln(D/d)$	$W>>d$

Figure 50.12 Transmission line formulae[6]

where $\varepsilon = e_{gn}/e_{0(n-1)}$ is the error term, and is indistinguishable from signal. The ground offset error voltage, e_{gn}, contains a component generated from the nth stage itself and a component from each of the stages which follow. The latter terms are particularly serious due to their higher operating currents. In the case of digital circuitry, some relief exists since one is working against a threshold instead of linear gain stages. Here the individual logic stages are roughly identical and there is no difference if the current is flowing towards either stage 1 or stage 3. The power distribution system is subject to the same type of problem, however, and the effect of power distribution noise on circuit performance would need to be characterised for a particular circuit. From *Figure 50.14*, design philosophies for controlling conduction crosstalk coupling are self-evident and can be summarised as follows:

(1) Keep the impedances of the ground and power distribution systems as low as practicable.

(2) Design ground and power distribution systems such that undesirable currents do not flow through current paths which couple noise into critical circuits. This is essentially a task of understanding and managing current flow paths. In particular, preventing currents from high noise sources from flowing through the ground structure of lower power more sensitive circuits. For example, in *Figure 50.14*, connecting the power supply to the high power stage would have currents flowing away from the input stage.

(3) Utilise adequate bypass filtering and decoupling on the power distribution system.

50.3.2 Induction field coupling

In *Section 50.2.2* the field equations for an electric dipole and elemental current loop were examined. The field characteristics as a function of distance from the source allowed two regions to be defined. In the near (induction) field region the field impedance is determined by the source characteristics and field intensity varies as $1/r^n$ ($n=2, 3$). In the far (radiation) field region, field impedance is constant at 377 Ω and field intensity varies as $1/r$. The boundary between these two regions is commonly defined to be approximately 1/6 wavelength ($\lambda/2\pi$) from the source.

At 30 MHz (corresponding to a pulse rise time of about 10 ns), the wavelength in free space is 10 m corresponding to a near/far field transition boundary of approximately 1.8 m. This is still quite a large distance relative to the major crosstalk problems likely to be encountered within most equipment. Consequently, internal machine field coupling interactions are usually from the induction field where the electric and magnetic fields are considered separately. The most likely vehicle for coupling interactions within the equipment environment is the interconnect (signal, power, and ground).

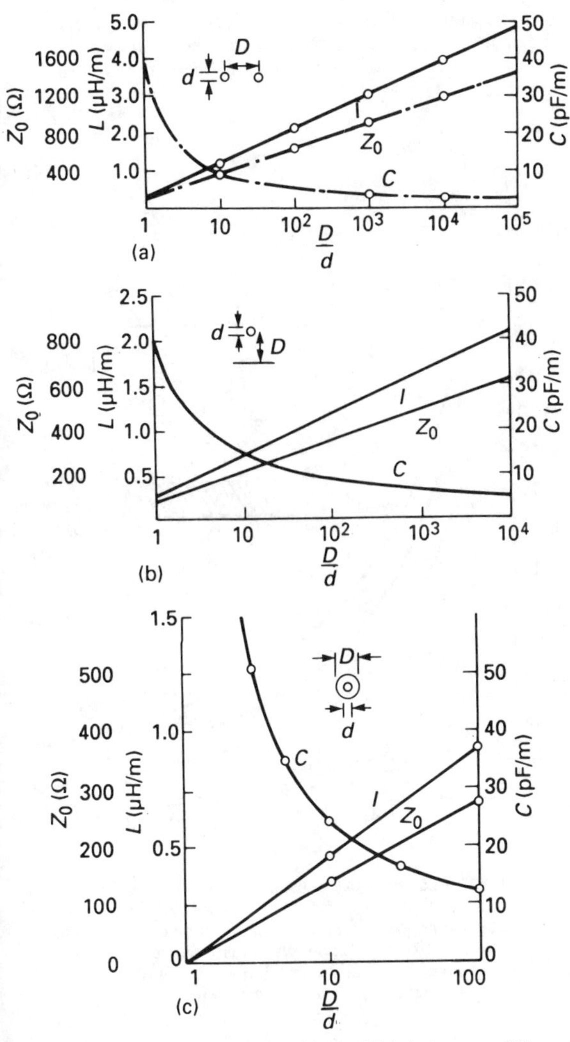

Figure 50.13 Characteristic impedance (Z_0), inductance (L), and capacitance (C) against D/d (all with air dielectric): (a) parallel wires, (b) wire over ground, (c) coaxial cable[1]

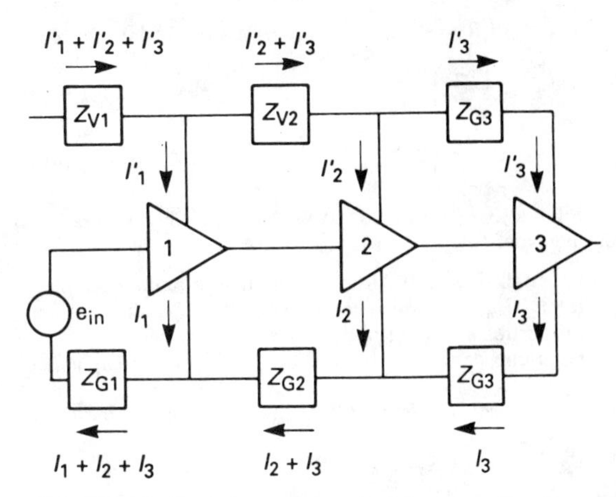

Figure 50.14 Example of common impedance coupling in power and ground distribution

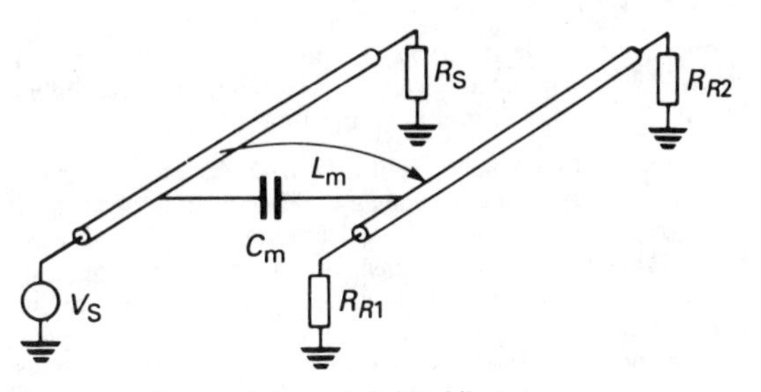

Figure 50.15 Parallel coupled signal lines

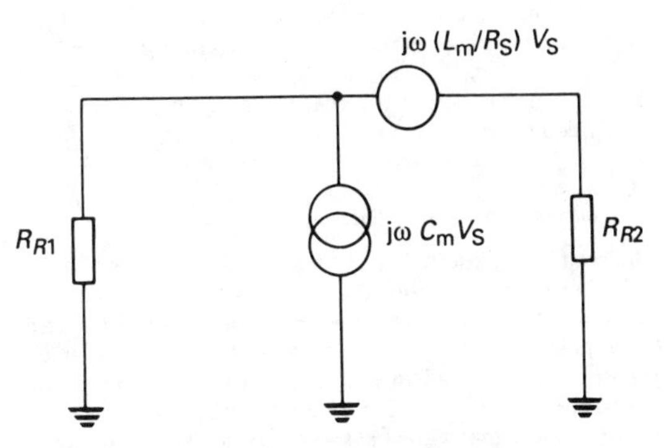

Figure 50.16 Equivalent circuit of victim line

50.3.2.1 Crosstalk equations—electrically short lines

Assume a pair of electrically short parallel wires (two-way delay time less than rise or fall time) over a ground plane which are coupled by mutual capacitance (C_m) and mutual inductance (L_m). Because these are electrically short lines, C_m and L_m are the total lumped mutual capacitance and inductance respectively. One of the wires is actively driven by the generator V_S and the other wire is the receiver or victim circuit (*Figure 50.15*). It is common in wire-to-wire crosstalk situations for the following relationship to exist

$$1/j\omega C_m >> R_{R1},\ R_{R2} >> j\omega L_m$$

The capacitive and inductive coupling may then be considered separately and then combined by superposition to obtain the total effect.

Under these conditions, the equivalent circuit of *Figure 50.16* can be developed. It consists of a series voltage source of magnitude $j\omega L_m V_S/R_S$ from the mutual inductance and a parallel current source of magnitude $j\omega C_m V_S$ from the mutual capacitance. It can be seen that the capacitive and inductive components of crosstalk are in phase at the end closest to the generator and out of phase at the far end. The near and far end crosstalk equations for the receiver line are

$$V_{R1} = C_m \frac{dV_S}{dt}\left(\frac{R_{R1}R_{R2}}{R_{R1}+R_{R2}}\right) + \frac{L_m}{R_S}\frac{dV_S}{dt}\left(\frac{R_{R1}}{R_{R1}+R_{R2}}\right) \tag{50.47}$$

$$V_{R2} = C_m \frac{dV_S}{dt}\left(\frac{R_{R1}R_{R2}}{R_{R1}+R_{R2}}\right) - \frac{L_m}{R_S}\frac{dV_S}{dt}\left(\frac{R_{R2}}{R_{R1}+R_{R2}}\right) \tag{50.48}$$

If $R_{R1}=R_{R2}=R_S=R$, Equations (50.47) and (50.48) reduce to

$$V_{R1}=\tfrac{1}{2}\left(C_m R+\frac{L_m}{R}\right)\frac{dV_S}{dt} \qquad (50.49)$$

$$V_{R2}=\tfrac{1}{2}\left(C_m R-\frac{L_m}{R}\right)\frac{dV_S}{dt} \qquad (50.50)$$

From Equations (50.49) and (50.50), the following observations can be made.

(1) Crosstalk is directly proportional to the mutual coupling terms (L_m, C_m) which for electrically short lines is also proportional to the length of the coupled lines.
(2) High impedance circuits $R>(L_m/C_m)^{1/2}$ favour capacitive crosstalk.
(3) Low impedance circuits, $R<(L_m/C_m)^{1/2}$, favour inductive crosstalk.
(4) V_{R2} is zero for $R=(L_m/C_m)^{1/2}$.

Both capacitive and inductive crosstalk are directly proportional to the rate of change or frequency of the noise generator.

50.3.2.2 Reducing induction field coupling

The observations relative to Equations (50.49) and (50.50) in *Section 50.3.2.1* suggest the following techniques for reducing induction field coupling in interconnecting lines:

(1) Reduce mutual coupling terms, L_m and C_m. This reduction can be achieved by increasing the physical separation of coupled lines, segregating high noise lines from lines associated with sensitive circuits (in cable runs) or, if inductive coupling predominates, minimising loop areas of source and victim circuits especially common area circuits (i.e. use twisted pair, ground plane, etc.). Shield conductors and for microstrip lines interpose grounded line(s) between source and victim lines.

(2) Adjust circuit impedances in accordance with the type of coupling predominating. For capacitive coupling, reduce circuit and line impedances (using close proximity ground planes, etc.) or for inductive coupling increase circuit impedances to reduce currents.

(3) Reduce signal speed, i.e. increase transition times of pulses, use lower frequencies to the extent possible.

50.3.2.3 Coupled transmission line

In high-speed digital equipment the fast transition times of pulse signals often require a distributed transmission line model for the coupled lines. Distributed analysis techniques should be used whenever pulse transition times are less than twice the line's propagation delay. The disturbed line will have two crosstalk components. Forward crosstalk propagates in the same direction as the disturbing signal while backward crosstalk propagates in the opposite direction. Assuming lossless lines terminated in Z_o, the crosstalk equations for near and far ends of the disturbed line are

$$V_{R1}=K_B[V_S(t)-V_S(t-2T_d)] \qquad (50.51)$$

$$V_{R2}=K_f l(d/dt)V_S(t-T_d) \qquad (50.52)$$

$$K_B=[C_m Z_o+(L_m/Z_o)]/4t_d \qquad (50.53)$$

$$K_f=\tfrac{1}{2}[C_m Z_o-(L_m/Z_o)] \qquad (50.54)$$

where

L_m = mutual inductance per unit length
C_m = mutual capacitance per unit length
Z_o = characteristic impedance (Ω)
l = length of coupled region
t_d = delay/unit length

The crosstalk coefficients (K_B, K_f) depend on the geometry of the signal lines and the material properties. *Figures 50.17* and *50.18* provide values of K_B and K_f for different microstrip line geometries.

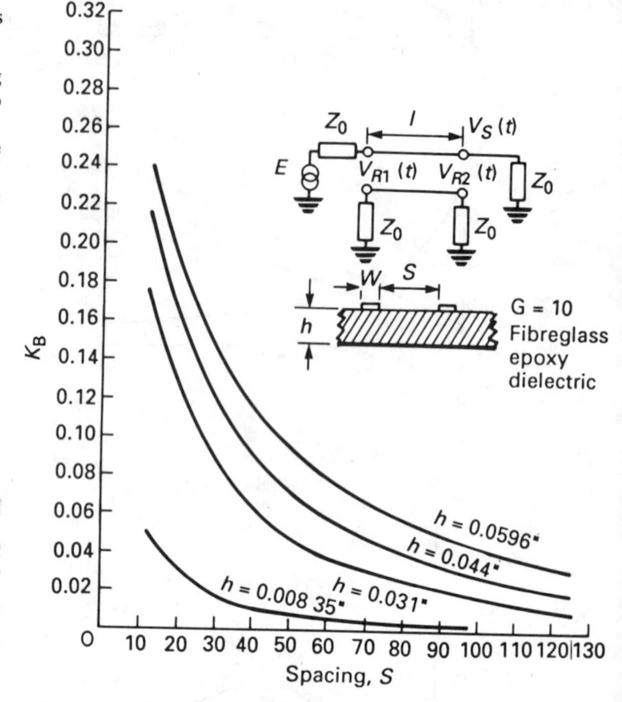

Figure 50.17 Back crosstalk constant against spacing. The following points should be noted. (1) Lines are terminated in Z_o, the characteristic impedance of each line. (2) Curves are valid for $0.0075'' < W < 0.025''$. (3) Copper weight is 2 ounces: 0.0025″ thick. (4) Time delay of line (T_d) is 1.8 nS/ft (~5.4 nS/m). (Reproduced from reference 7 with permission)

From Equations (50.51) and (50.52), the following information can be deduced regarding transmission line crosstalk.

(1) The amplitudes of both the backward and forward crosstalk components are directly proportional to the respective crosstalk coefficient, K_B and K_F.
(2) The amplitude of the forward crosstalk component is directly proportional to the length of the coupled lines (l), and the rate of change of the disturbing signal. The amplitude of backward crosstalk does not depend on the length, l.
(3) The backward crosstalk component is the summation of two parts, each an attenuated replica of the disturbing signal. The first part appears at the same instant as the disturbing signal and is of the same polarity. The second part is inverted and appears after a delay equal to twice the one-way delay of the coupled lines.
(4) The forward crosstalk component waveshape is the derivative of the disturbing signal and it appears one line delay time after the disturbing signal.
(5) The forward crosstalk component is 0 when $Z_o=(L_m/C_m)^{1/2}$. This occurs when the signal lines are embedded in a homogeneous dielectric.

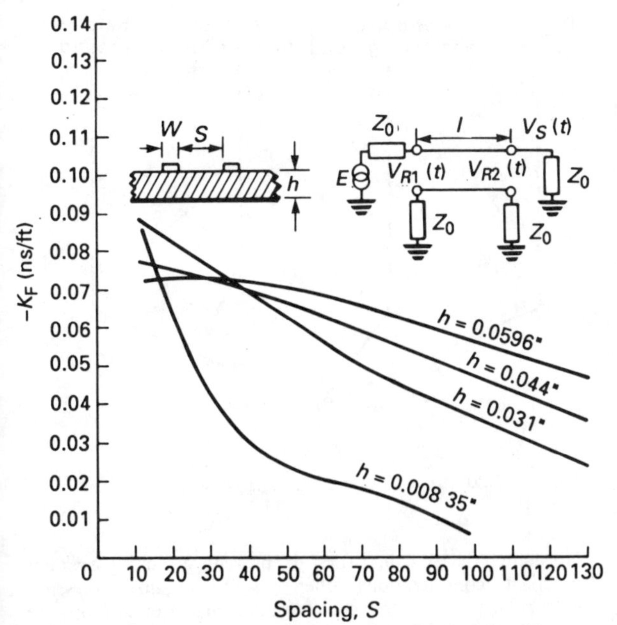

Figure 50.18 Forward crosstalk constant against spacing. Lines terminated in characteristic impedances. Notes (2) and (4) for *Figure 50.17* also apply. (Reproduced from reference 7 with permission)

50.4 Design techniques for interference control

50.4.1 Grounding

A large percentage of noise problems experienced in electronic equipment result from inadequacies in the ground system design. A well considered ground strategy should be established early in the hardware conceptualisation stage of any product.

50.4.1.1 Grounding—objectives/definitions

At the outset, it is necessary to understand that the term 'ground' is not completely descriptive. Different concerns often result in several distinct ground system identifications. Essentially, there are three objectives relative to ground system design. They are:

(1) Safety.
(2) Providing an equipotential reference for electronic circuits/subsystems.
(3) Providing current paths for the control of internal machine crosstalk and emissions into the environment (regulatory limits).

The purpose of safety ground is to prevent shock hazards to operating personnel. Other common names for safety ground are primary ground, green wire ground, hardware ground, a.c. ground, and earth ground.

Signal ground is needed as a common reference for electronic circuitry and as a return medium for currents. Operational considerations (e.g. preventing common impedance crosstalk between incompatible circuits) will often dictate several distinct and separated signal grounds. The grounds within a system are normally electrically connected at some point(s). Occasionally, special performance requirements dictate the use of electrically isolated grounds. This approach tends to be expensive, however, and it is difficult to maintain the isolation at high frequencies.

Other names for signal ground are secondary ground, circuit ground, and d.c. ground.

50.4.1.2 Types of grounds

Safety ground Equipment operating from power mains must be designed such that a fault condition does not cause conducting surfaces (accessible to an equipment user) to become electrically 'hot'. The technique predominantly in use for electrically powered equipment is to provide a safety ground connection to conducting surfaces accessible to the operator. Under normal operating conditions, the current carried by the safety ground conductor is very small and limited by the safety agencies of various countries. Under a fault condition (i.e. short to equipment cabinet), the safety ground conducts the fault current and triggers the circuit breaker or ground-fault-interrupter. In general, safety ground should not be used to perform signal ground functions.

Signal ground Normally, electronic circuits and subsystems require at least one common voltage reference called signal ground. Circuit requirements usually dictate that signal ground closely approximates equipotential characteristics. Because the ground may carry significant current in normal operation, this establishes a limitation on ground impedance. The degree to which the signal ground system must approximate ideal zero impedance conditions is dependent on many factors. For example, the type of circuitry utilised, ground current characteristics, system physical characteristics, and circuit/system susceptibility to ground noise are important considerations.

These factors should be identified early in the design cycle and the knowledge used to establish ground performance specifications. The following sections provide a general framework and guidelines for ground system design.

50.4.1.3 Performance considerations

Performance of the ground system can be defined in terms of the following criteria:

(1) Ground system potential differences (both d.c. and transient).
(2) Conduction crosstalk magnitude.
(3) Susceptibility to external noise sources.
(4) Noise fields emanating from ground system.

Fortunately, these requirements do not normally present conflicting requirements on the ground system design.

Ground system potential differences There are two basic mechanisms for inducing potential differences in the ground system: conduction induced potential differences, i.e. ground current flowing through ground impedance and induction field coupling. Conduction induced offsets can be minimised by keeping the impedance of the ground system and ground currents as low as possible.

The equivalent circuit of a ground system is a series *R*, *L* network. At high frequencies, the a.c. resistance must be used. Both the resistance and inductance of the ground system can be reduced by using ground planes, large diameter wire and minimising physical size. The inductance of the ground system can be kept low by making signal currents return in close proximity to the path from which they came.

Of course, there are practical limitations as to how low the impedance of the ground system can be made. Consequently, current path control is an important technique which must also be utilised.

The primary mechanism relative to induction field induced offsets in the ground system is **H** field coupling into ground loops. The magnitude of the induced voltage is directly proportional to the magnitude of the **H** field, the area of the ground loop and the rate of change of the **H** field.

Common impedance crosstalk In the previous section, conduction-induced potential differences were discussed. These ground potential differences may be particularly troublesome when several circuits share a common ground path. Problems can occur when currents from one circuit induce ground potential differences in another. Apart from reducing ground impedance, the best way to control this problem is to force currents to remain within the ground structure of their associated circuits. By constantly attempting to visualise probable current return paths, the designer can usually manipulate physical layout and impedances to make ground currents follow the desired path. In particular, two approaches which can be used to control ground currents to selected paths are (a) redundant parallel ground returns, and (b) segregated grounds. An example of redundant parallel returns is shown in *Figure 50.19*. The intent

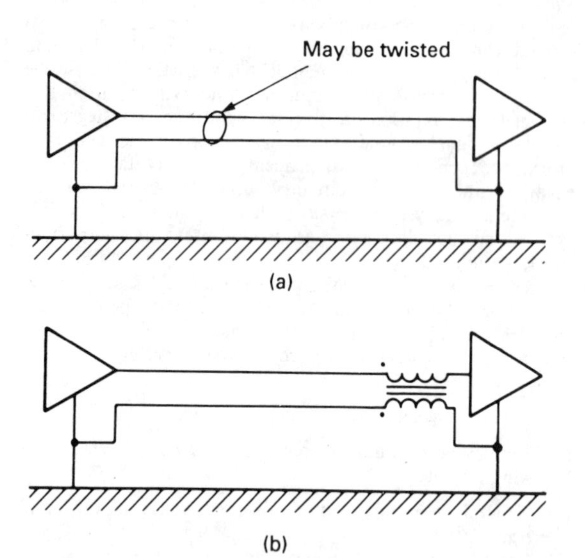

Figure 50.19 Single ended circuit configuration: (a) circuit with redundant current return path, (b) circuit with redundant current return path and Balun transformer

here is to make the redundant return path impedance (primarily inductance) much lower than the ground return path. Accomplishing this will cause a large portion of the return current to use this path. In difficult cases, a Balun transformer can be used to achieve good confinement to the desired path at signal frequencies.

Segregated ground trunks (*Figure 50.20*) are particularly useful when the equipment contains different types of hardware which tend to be incompatible (e.g. electromechanical devices like motors, solenoids, clutches, etc., low-level analogue signals and digital logic circuitry). However, it is often necessary for functions on one ground to communicate with functions on a different ground. Isolated techniques could be used but often aren't because of cost and/or performance issues. If direct coupled circuits are used, signal loop requirements will often require the communicating circuits to use a local ground return. These ground ties introduce ground loops which work against the original intent for ground segregation. They also raise the possibility of additional problems occurring (i.e. circulating currents and induction field coupling).

Ground loops Whenever a circuit/subsystem has more than one path to the common ground point, a ground loop exists.

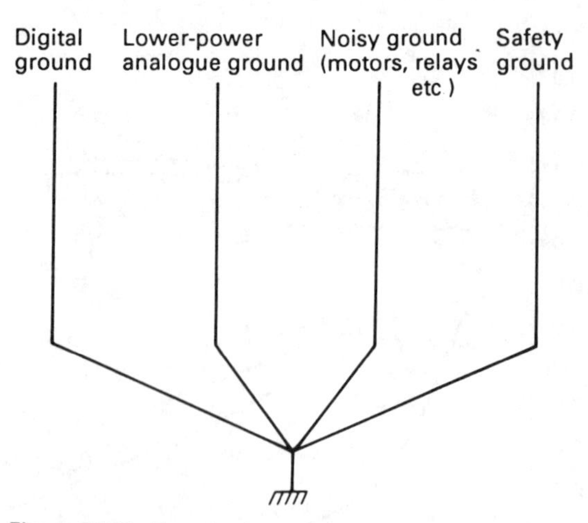

Figure 50.20 Example of ground segregation

Normally, the designer tries to prevent ground loops but practical considerations often result in their presence (particularly in a large piece of equipment). Ground loops are susceptible to **H** field induced voltages and circulating currents. In addition to the effect of the voltage offsets introduced by these circulating currents, they can be a source of noise.

50.4.1.4 System ground philosophies

There are basically four design philosophies for ground systems:

(1) Single-point ground.
(2) Multipoint ground.
(3) Hybrid ground.
(4) Isolated ground.

In general, it is not practicable (or desirable) to utilise a given philosophy at all hardware levels. Unavoidable physical effects (i.e. parasitic coupling) often enter the picture and tend to modify a given grounding philosophy.

The following factors should be considered in the selection of a ground system approach:

(1) Physical size and complexity of equipment.
(2) Frequency content of signals.
(3) Incompatibility between functional elements in the equipment (e.g. high-power drivers and sensitive amplifier circuits).
(4) Sensitivity of functional elements to ground noise (common mode noise susceptibility).
(5) Equipment noise environment.
(6) Communication requirements between circuits/subsystems.

Single-point ground In a single-point ground system (*Figure 50.21*), circuits/subsystems have only one path to a common system ground point. This approach has the advantage of well controlled ground current paths and no ground loops but it also has disadvantages. It is difficult to maintain a low-impedance ground (particularly at high frequency) when the individual ground trunks must all terminate on a common ground point. There is an opportunity for common impedance crosstalk to occur in the trunks with multiple circuit branches. It tends to be physically cumbersome with regard to wiring, especially if an attempt is made to minimise branches. It tends to break down at high frequencies due to parasitic coupling and when it is necessary for circuits/subsystems on different ground trunks to communicate.

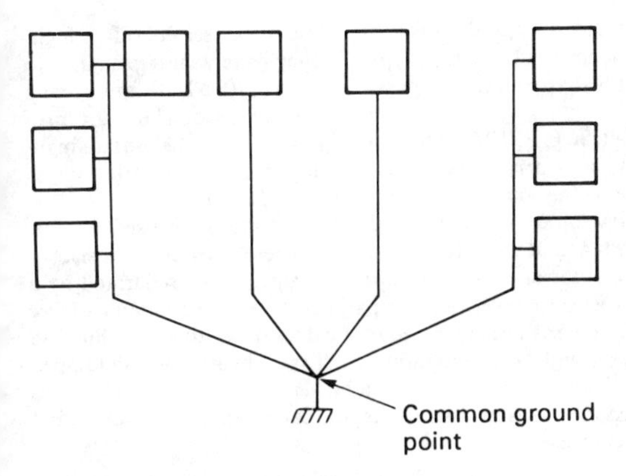

Figure 50.21 Example of single-point ground

Because of the degradation of the single point ground system at high frequencies, it is normally considered to be best at the lower frequencies (i.e. $f < \sim 5$–10 MHz).

Multipoint ground *Figure 50.22* illustrates a multiground system. This approach is usually able to maintain a lower ground

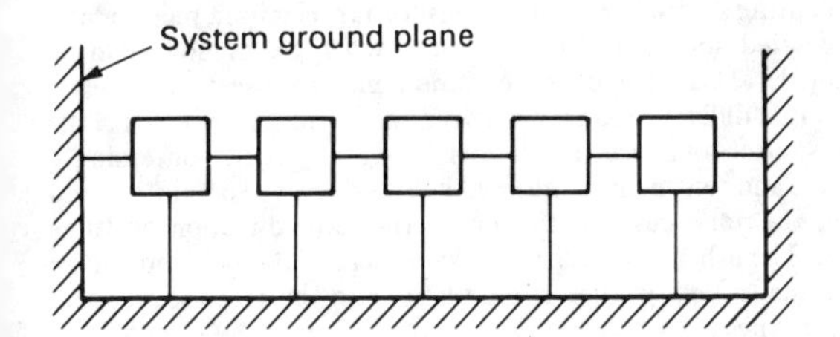

Figure 50.22 Example of multipoint ground

system impedance and thus is used at higher frequencies ($f > 5$–10 MHz). Care must be taken in multipoint ground systems to ensure that the resulting ground loops and common impedance paths do not create interference conditions. These concerns often limit the size of a system using multiground techniques.

Hybrid ground As previously mentioned, a pure single-point or multipoint ground system is normally not practical or desirable. The factors identified in *Section 50.4.1.4* need to be considered when contemplating the ground system design. For example, it will often be convenient to use multipoint ground techniques at the lower packaging levels like the printed wiring board assembly (PWBA). Here the ground loops are small and the multipoint ground structure provides a low-impedance local ground for high-frequency digital circuitry. Depending on the type of equipment, it might be appropriate to extend the multipoint ground structure to a ground of PWBA in a card chassis and then to utilise single-point ground techniques between card chassis.

Isolated ground On occasion, difficult performance requirements require isolated ground systems for subsystems which are communicating.

Figure 50.23 shows three ways that this is commonly accomplished. *Figure 50.23(a)* uses an opto coupler and has response down to d.c. *Figure 50.23(b)* uses transformer coupling which cannot pass d.c. and requires that the signal characteristics (i.e. pulse width, duty cycle) be compatible with the transformer circuit capability. *Figure 50.23(c)* shows a Balun connected

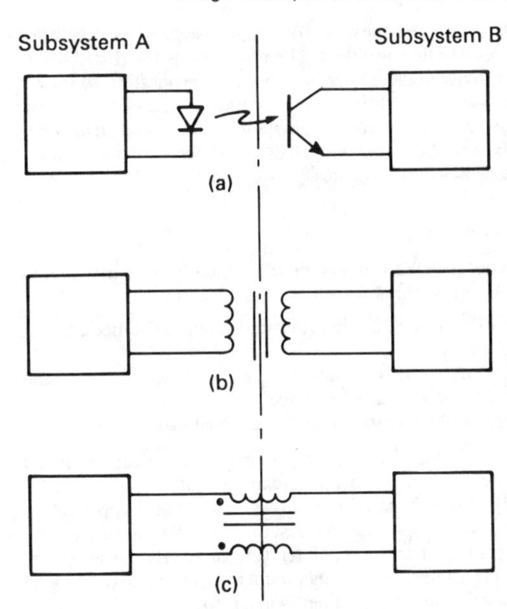

Figure 50.23 Isolation techniques: (a) opto coupler, (b) transformer, (c) Balun transformer

transformer which passes d.c. and thus does not provide low-frequency isolation. It does, however, provide reasonably good high-frequency isolation. An electromagnetic relay is another alternative for achieving isolation if the slower speed can be tolerated.

50.4.2 Power distribution/decoupling

Most electronic hardware requires well regulated d.c. voltages at the point of usage. Normally, these voltages are created from the a.c. mains via a power conversion function. The power conversion may be completely performed within one centralised function or, alternatively, it may be distributed. *Figure 50.24* illustrates these two options. In this discussion only the effect of distribution on power regulation is examined. The technique of

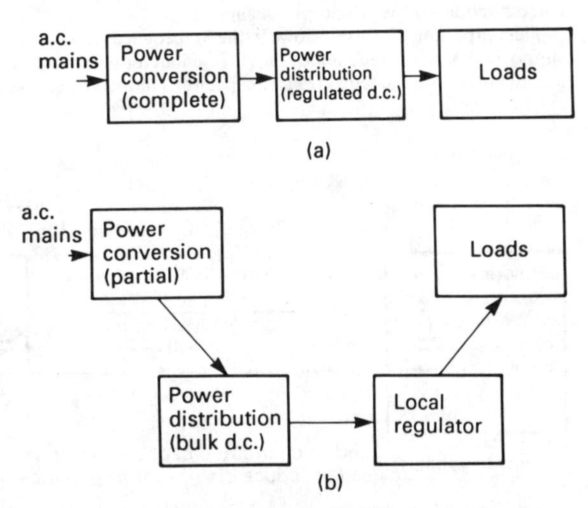

Figure 50.24 Power conversion/distribution approaches: (a) centralised regulation, (b) distributed regulation

Figure 50.24(a) is the one most commonly used and has the more difficult distribution problem. The objective of the distribution system is to distribute power from point-of-generation to point-of-use without the addition of unacceptable noise or losses. The maximum allowable voltage variations (d.c. and transient) should be specified and budgeted to the various distribution system loss/noise mechanisms.

50.4.2.1 Performance considerations

The power distribution system can modify the distributed voltage in any of three ways:

(1) Static d.c. drop due to distribution system resistance and d.c. load current.
(2) Transient voltage changes due to load current and distribution system impedance.
(3) Induction field coupling into the distribution lines.

Static losses in the distribution system are the easiest to predict and control once the maximum load current (I_{LM}) is defined. Generally, it is most satisfactory to use a statistical approach to calculate I_{LM} as opposed to a worst case calculation. In a system of any size, the latter approach tends to be overly conservative. Once I_{LM} is determined, the maximum resistance (R_{DM}) allowed for the distribution system lines is given by

$$R_{DM}=\frac{\Delta V_{d.c.}}{I_{LM}}-R_X \qquad (50.55)$$

where $\Delta V_{d.c.}$ is the allowable d.c. loss and R_X the resistance of other elements in the power distribution path beside wire (i.e. fuses, connectors, etc.). Note that R_{DM} is the total resistance of the distribution lines and includes the resistance of both the source and return lines.

Transient load current changes act on the distribution system impedance to create voltage excursions. In digital equipment, current demand changes/spikes can be associated with logic state changes, capacitive charging, line driving and overlap conduction spikes from logic circuits (i.e. TTL). Distributed filtering with high-frequency bypass capacitors supply charge for high-speed local current transients. The capacitor value (in farads) is calculated as follows

$$C=\Delta I\,\Delta t/\Delta V \qquad (50.56)$$

where ΔI is the current transient magnitude, ΔV the allowable change in voltage at the filtered node and Δt the time interval corresponding to the ΔI current change.

This capacitor effectively lowers the impedance at specific supply nodes. A low loop impedance is essential for good results and requires both the loop area and capacitor's inductance to be kept small. These capacitors should be selected for high-frequency performance. Typically they have values from 0.01 to 0.1 μF. Larger value bypass capacitors (10–50 μF) are usually used to handle the lower frequency variations for a large group of circuits (e.g. PWBA). The general philosophy is to contain high-frequency current transient and current demand variations as close to the source as possible (*Figure 50.25*).

Low-impedance laminated bus-bars are often used for low impedance distribution systems. These systems consist of rectangular cross-section parallel copper strips separated by a thin dielectric layer (e.g. see parallel strip Z_0 formula in *Figure 50.12*). Characteristic impedance values down to a few ohms are possible with this configuration. Physically distributing lumped capacitors along a distribution bus is another way to lower bus impedance within the limits of capacitor and distribution inductance.

50.4.3 Circuit techniques

The net effect of noise on electronic systems is to create extraneous voltages and/or currents. When these voltages and currents interfere with signals, an interference problem exists.

A noise management strategy has two basic thrusts. They are to keep noise out of the electronic system and to utilise circuit techniques which allow satisfactory operation in the noise environment that remains.

Viewing a signal circuit to consist of a transmitter, receiver and associated source and return lines, two types of noise can be defined. The first is called common mode noise and causes a potential difference to exist between the common terminals of the transmitter and receiver circuits. A signal circuit containing a source and return line connected between these two points would have the noise currents flowing in the same direction on these lines. The other noise type is called differential mode and causes currents to flow in opposite directions in the sense source and return lines.

50.4.3.1 Common mode noise

Figure 50.26 shows a circuit operating in the presence of a common mode voltage. This single-ended open wire configuration has very poor performance in the presence of common mode noise. The problem is that the noise voltage appears in the signal loop. It can either detract from a valid signal or be falsely detected as a valid signal. This circuit configuration is very poor relative to common mode to differential mode conversion (the entire V_N appears as a differential mode signal).

Figure 50.27 shows a balanced system using a differential amplifier and twisted pair interconnect. The characteristics of the

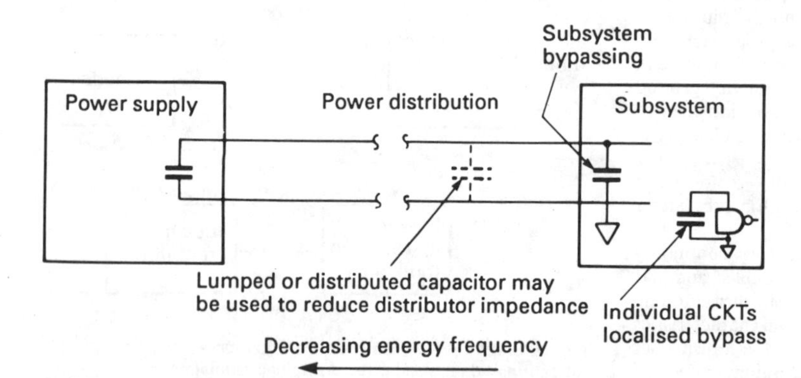

Figure 50.25 Decoupling the power distribution system

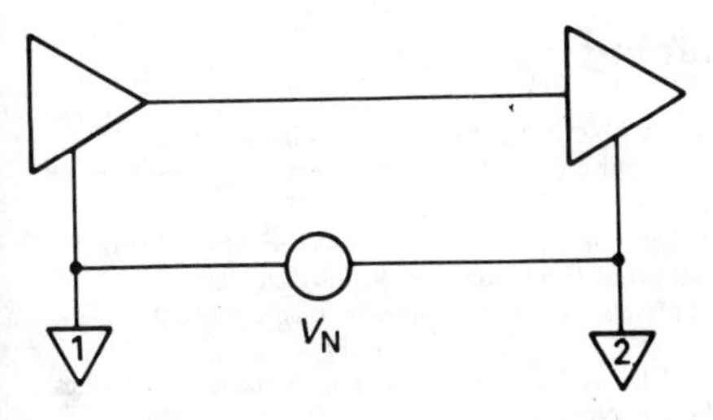

Figure 50.26 Common mode noise on a single-ended circuit

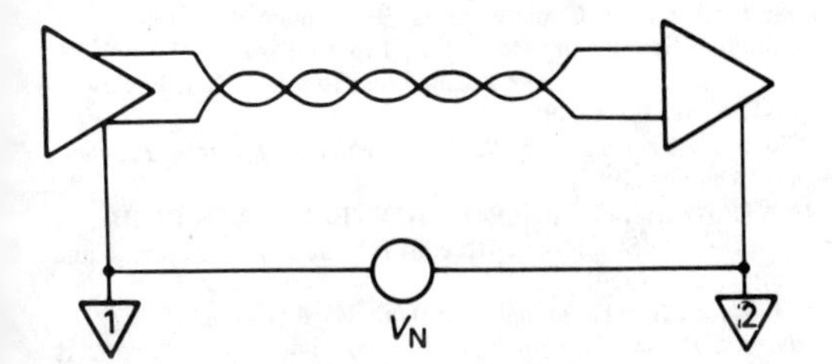

Figure 50.27 Common mode noise on a balanced differential circuit

differential amplifier are to sense the differential signal and reject common mode noise. Because of the balanced structure there is very little common mode to differential mode conversion. Thus, this circuit configuration can tolerate common mode noise quite well.

In *Figure 50.28*, a high impedance current source configuration is used for the transmitting circuit. The current drive signal (and

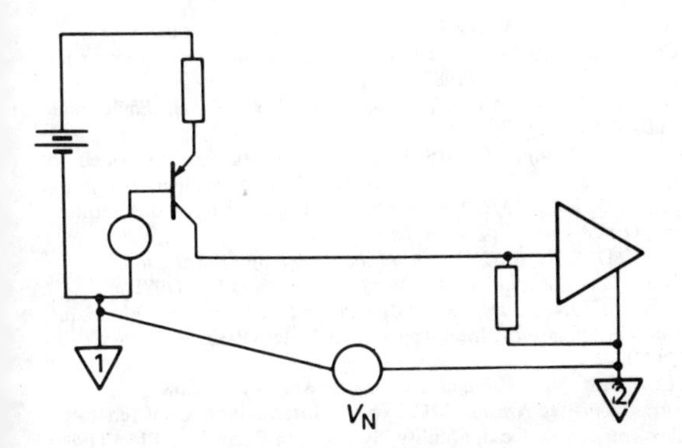

Figure 50.28 Circuit technique to provide common mode rejection

thus the sense voltage) is independent of ground potential difference.

Figure 50.29 shows a single-ended configuration similar to *Figure 50.26* but with coax grounded at both ends used. V_N causes a current (I_S) to flow through the shield which creates a voltage across the shield. Neglecting the resistive component of shield voltage, the transfer function for the error voltage, V_E, generated from the V_N noise source is

$$V_E = V_N \left[\frac{1}{1+(j\omega L_S/R_S)} \right] \tag{50.57}$$

where R_S is the shield resistance and L_S the shield inductance.

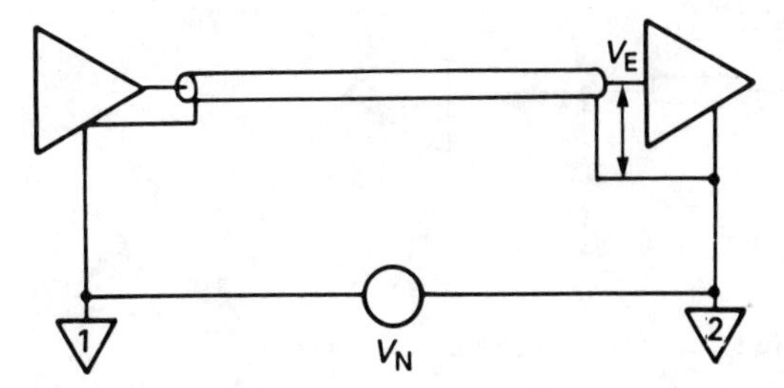

Figure 50.29 Common mode noise on a single-ended system using coaxial cable

The ratio R_S/L_S is called the shield cut-off frequency, f_c^1 where the error signal is reduced to 0.707 of its low-frequency value. From that point on the error voltage decreases at the rate of 6 dB/octave (*Figure 50.30*). Since the values of f_c for most coaxials are quite low (e.g. 0.5–2 kHz), the coaxial cable itself can effectively reject high-frequency common mode noises. The shield resistance drop, however, is not cancelled and care must be taken to keep it small.

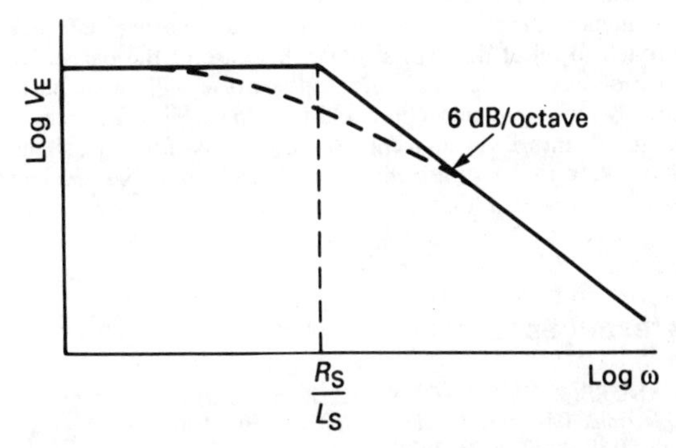

Figure 50.30 Error voltage against frequency for circuit of *Figure 50.29*

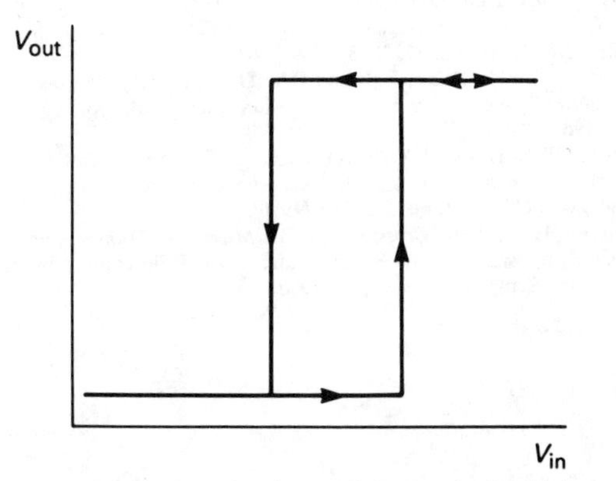

Figure 50.31 Transfer characteristics for circuit with hysteresis

Isolation is another technique often used to reject common mode potential differences (see *Figure 50.23*). However, parasitic coupling across the isolation elements will tend to degrade the isolation at high frequencies and must be evaluated.

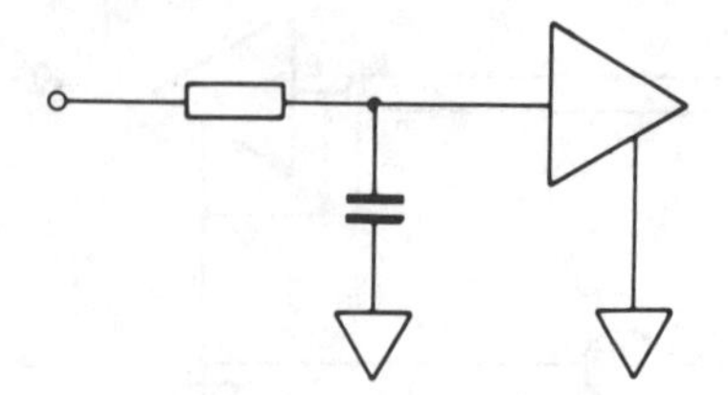

Figure 50.32 Low pass filter on circuit input

50.4.3.2 Differential mode noise

The normal signal mode is differential, thus, this type of noise is particularly troublesome. Amplitude discrimination is the first line of defence against differential noise. The approach is to utilise sense thresholds for circuits and to keep noise amplitudes below these levels. Valid signals, of course, will exceed these thresholds with adequate margin. Input hysteresis (*Figure 50.31*) can further enhance the circuit's noise rejection capability.

Time discrimination (strobe) can also be an effective tool to reject noise (particularly when the noise is synchronised to some machine event). In this approach, the sense circuits are only activated when they are being used.

Frequency discrimination (filtering) is another effective approach to reject differential noise. Because of the derivative nature of most noise mechanisms, the noise will often be of relatively high frequency compared to signals. This fact can be used to advantage through the use of low pass filters at signal inputs likely to be subjected to noise such as cable receiver circuits (*Figure 50.32*).

References

1 LANDEE, R. W., DONOVAN, C. D. and ALBRECHT, A. P., *Electronic Designer's Handbook*, McGraw-Hill, New York, pp. 20.19, 20.20, 20.23 (1957)

2 ENGELSON, M. and TELEWSKI, F., *Spectrum Analyzer Theory and Applications*, Avtech House, p. 66 (1974)

3 EVERETT, W. W. Jr., *Topics in Intersystem Electromagnetic Compatibility*, Holt, Rinehart and Winston (1972)

4 OTT, H. W., *Noise Reduction Techniques in Electronic Systems*, Wiley, New York, pp. 35, 36, 37, 139 (1976)

5 RAMO, S., WHINNEY, J. R. and DANDUZER, T., *Fields and Waves in Communication Electronics*, Wiley, New York, pp. 252, 293 (1965)

6 BELL TELEPHONE LABORATORIES INC., *Physical Design of Electronic Systems: Volume 1, Design Technology*, Prentice-Hall, Englewood Cliffs, NJ, pp. 362, 363 (1970)

7 KAUPP, H. R., *Pulse Crosstalk between Microstrip Transmission Lines*, Symposium Record, Seventh International Electronic Circuit Packaging Symposium, pp. 9, 10 (1966)

Further reading

BUSCHKE, H. A., 'A practical approach to testing electronic equipment for susceptibility to AC line transients', *IEEE Trans. (R)*, October (1988)

CATANI, J.-P., *Effectiveness of Circuits in Common Mode Noise Rejection*, Institution of Electronic and Radio Engineers International Conference on Electromagnetic Compatibility, 1984, p. 251 (1984)

CLEWES, A. B., 'RFI/EMC shielding in cable connector assemblies', *Electronics & Power*, November/December (1987)

D'ARCY, D., 'Compatibility considerations come to prominence', *Electron. Product Design*, December (1988)

GABRIELSON, B., *Future EMC Trends in PC Board Design*, EMC Expo 86 International Conference on Electromagnetic Compatibility, Symposium Record, p. T18.4 (1986)

GRAY, H. S., *Digital Computer Engineering*, Prentice-Hall, Englewood Cliffs, NJ, pp. 273–276 (1963)

HOOPER, N., 'RFI design for 1992 requirements', *Electron. Product Design*, December (1988)

INTERNATIONAL TELEPHONE AND TELEGRAPH CORP., *Reference Data for Engineers*, 4th edn, International Telephone and Telegraph Corp. (1956)

JONES, J. W. E., *Grounding and Inter-Unit Wiring—Some Experimental Results*, Institution of Electronic and Radio Engineers International Conference on Electromagnetic Compatibility, 1984, p. 245 (1984)

KHAN, R. L. and COSTACHE, G. I., *Considerations on Modeling Crosstalk on Printed Circuit Boards*, 1987 IEEE Symposium on Electromagnetic Compatibility, p. 279 (1987)

MORRISON, R., *Grounding and Shielding Techniques in Instrumentation*, Wiley, New York (1977)

NANEVICZ, J. E. *et al.*, 'EMP susceptibility insights from aircraft exposure to lightning', *IEEE Trans. (EMC)*, November (1988)

OSBORN, J. D. M. and WHITE, D. R. J., *Grounding—A Recommendation for the Future*, 1987 IEEE Symposium on Electromagnetic Compatibility, p. 155 (1987)

OTT, H. W., *Digital Circuit Grounding and Interconnection*, 1981 IEEE International Symposium on Electromagnetic Compatibility, p. 292 (1981)

PAUL, C. R. and EVERETT III, W. W., *Printed Circuit Board Crosstalk*, IEEE 1985 International Symposium on Electromagnetic Compatibility, p. 452 (1985)

RYDER, J. D., *Networks Lines and Fields*, Prentice-Hall, Englewood Cliffs, NJ (1958)

VANZURA, E. and ADAMS, J., 'Electromagnetic fields in loaded shielded rooms', *Test & Measurement World*, November (1987)

WHITE, D. R. J., *EMC Handbook*, Vol. 3, Don White Consultants Inc. (1973)

WHITE, D. R. J., *EMI Control in the Design of Printed Circuit Boards and Backplanes*, Don White Consultants Inc. (1981)

WILLIAMSON, T., *Designing Microcontroller Systems for Electrically Noisy Environments*, Intel Application Note AP-125, February (1982)

WILLIAMSON, T., *Designing Computer Systems for Low Susceptibility to Noise*, EMC Expo 86 International Conference on Electromagnetic Compatibility, Symposium Record, p. T14.1 (1986)

ZAIS, A., 'RF shielding becomes more necessary', *Microwave Systems News*, December (1982)

51

Noise and Communication

K R Sturley, BSc, PhD, FIEE, FIEEE
Telecommunications Consultant

Contents

51.1 Interference and noise in communication systems

Information transmission accuracy can be seriously impaired by interference from other transmission systems and by noise. Interference from other transmission channels can usually be reduced to negligible proportions by proper channel allocation, by operating transmitters in adjacent or overlapping channels geographically far apart, and by the use of directive transmitting and receiving aerials. Noise may be impulsive or random. Impulsive noise may be man-made from electrical machinery or natural from electrical storms; the former is controllable and can be reduced to a low level by special precautions taken at the noise source, but the latter has to be accepted when it occurs. Random (or white) noise arises from the random movement of electrons due to temperature and other effects in current-carrying components in, or associated with, the receiving system.

51.2 Man-made noise

Man-made electrical noise is caused by switching surges, electrical motor and thermostat operation, insulator flash-overs on power lines, etc. It is generally transmitted by the mains power lines and its effect can be reduced by:

(1) Suitable r.f. filtering at the noise source.
(2) Siting the receiver aerial well away from mains lines and in a position giving maximum signal pick-up.
(3) Connecting the aerial to the receiver by a shielded lead.

The noise causes a crackle in phones or loudspeaker, or white or black spots on a monochrome television picture screen, and its spectral components decrease with frequency so that its effect is greatest at the lowest received frequencies.

Car ignition is another source of impulsive noise but it gives maximum interference in the v.h.f. and u.h.f. bands; a high degree of suppression is achieved by resistances in distributor and spark plug leads.

51.3 Natural sources of noise

Impulsive noise can also be caused by lightning discharges, and like man-made noise its effect decreases with increase of received frequency. Over the VHF band such noise is only evident when the storm is within a mile or two of the receiving aerial.

Cosmic noise from outer space is quite different in character and generally occurs over relatively narrow bands of the frequency spectrum from about 20 MHz upwards. It is a valuable asset to the radio astronomer and does not at present pose a serious problem for the communications engineer.

51.4 Random noise

This type of noise is caused by random movement of electrons in passive elements such as resistors, conductors and inductors, and in active elements such as electronic valves and transistors.

51.4.1 Thermal noise

Random noise in passive elements is referred to as thermal noise since it is entirely associated with temperature, being directly proportional to absolute temperature. Unlike impulsive noise its energy is distributed evenly through the r.f. spectrum and it must be taken into account when planning any communication system. Thermal noise ultimately limits the maximum amplification that can usefully be employed, and so determines the minimum acceptable value of received signal. It produces a steady hiss in a loudspeaker and a shimmering background to a television picture.

Nyquist has shown that thermal noise in a conductor is equivalent to an r.m.s. voltage V_n in series with the conductor resistance R, where

$$V_n=(4kTR\Delta f)^{1/2} \qquad (51.1)$$

k = Boltzmann's constant, 1.372×10^{-23} J/K
T = absolute temperature of conductor
Δf = pass band (Hz) of the circuits after R

If the frequency response were rectangular the pass band would be the difference between the frequencies defining the sides of the rectangle. In practice the sides are sloping and bandwidth is

$$\Delta f=\frac{1}{E_0^2}\int_0^\infty [E(f)]^2\,df \qquad (51.2)$$

where E_0 = midband or maximum value of the voltage ordinate and $E(f)$ = the voltage expression for the frequency response.

A sufficient degree of accuracy is normally achieved by taking the standard definition of bandwidth, i.e. the frequency difference between points where the response has fallen by 3 dB.

Figure 51.1 allows the r.m.s. noise voltage for a given resistance and bandwidth to be determined. Thus for

$R=10$ kΩ, $T=T_0=17$°C or 290 K

and

$\Delta f=10$ kHz, $V\simeq1.26\ \mu$V

When two resistances in series are at different temperatures

$$V_n=[4k\,\Delta f(R_1T_1+R_2T_2)]^{1/2} \qquad (51.3)$$

Two resistances in parallel at the same temperature (*Figure 51.2(a)*) are equivalent to a noise voltage

$$V_n=[4kT\,\Delta f\,.\,R_1R_2/(R_1+R_2)]^{1/2} \qquad (51.4)$$

in series with two resistances in parallel, *Figure 51.2(b)*.

The equivalent current generator concept is shown in *Figure 51.2(c)* where

$$I_n=[4kT\,\Delta f(G_1+G_2)]^{1/2} \qquad (51.5)$$

If R is the series resistance of a coil in a tuned circuit of Q factor, Q_0, the noise voltage from the tuned circuit becomes

$$V_{no}=V_nQ_0=Q_0(4kTR\,\Delta f)^{1/2} \qquad (51.6)$$

The signal injected into the circuit is also multiplied by Q_0 so that signal-to-noise ratio is unaffected.

51.5 Electronic valve noise

51.5.1 Shot noise

Noise in valves, termed shot noise, is caused by random variations in the flow of electrons from cathode to anode. It may be regarded as the same phenomenon as thermal (conductor) noise with the valve slope resistance acting in place of the conductor resistance at a temperature between 0.5 and 0.7 of the cathode temperature.

Shot noise r.m.s. current from a diode is given by

$$I_n=(4k\alpha T_k g_d\,\Delta f)^{1/2} \qquad (51.7)$$

where

T_k = absolute temperature of the cathode
α = temperature correction factor assumed to be about 0.66

$g_d=\dfrac{dI_d}{dV_a}$, slope conductance of the diode.

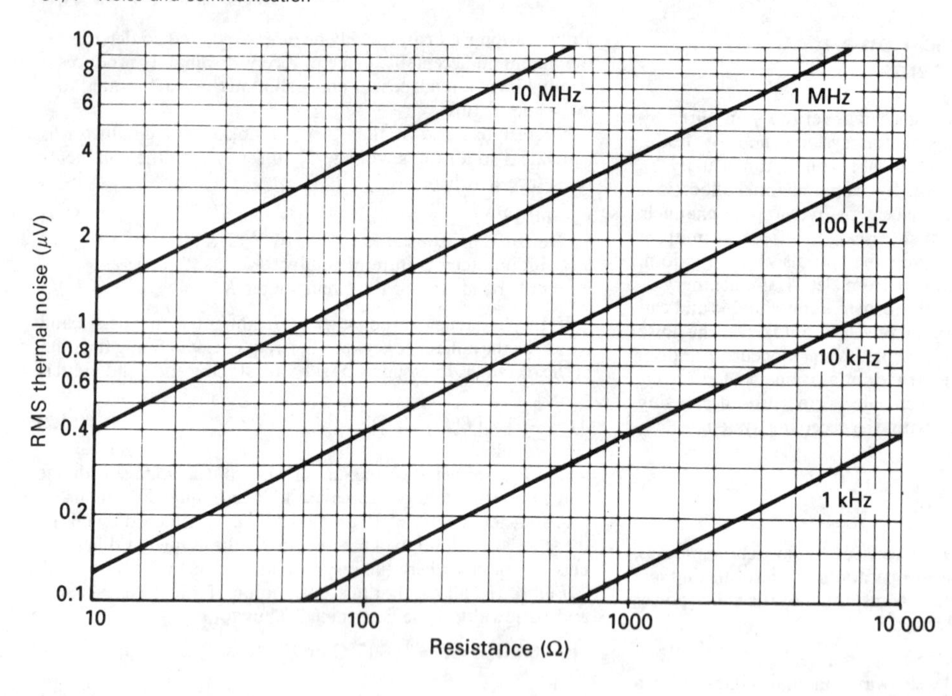

Figure 51.1 RMS thermal noise (kV) plotted against resistance at different bandwidths, $T=290$, $K=17°C$

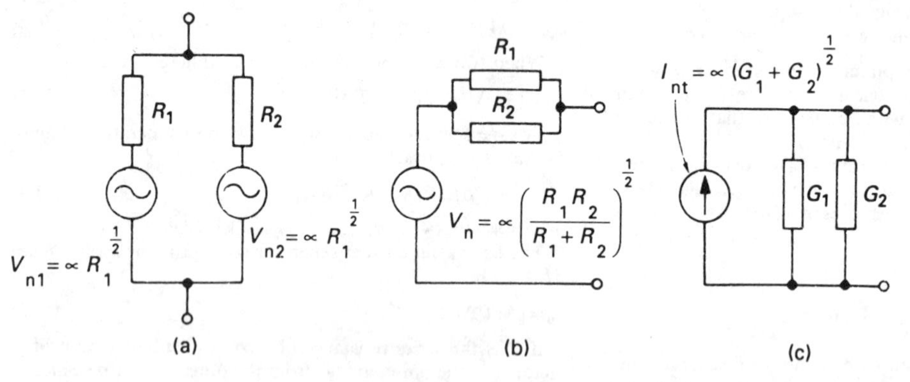

Figure 51.2 (a) Noise voltages of two resistances in parallel; (b) an equivalent circuit and (c) a current noise generator equivalent $\alpha=(4kT\,'f)^{1/2}$

Experiment[1] has shown noise in a triode is obtained by replacing g_d in Equation (51.7) by g_m/β where β has a value between 0.5 and 1 with a typical value of 0.85, thus

$$I_{na}=(4k\alpha T_k g_m \Delta f/\beta)^{1/2} \tag{51.8}$$

Since $I_a=g_m V_g$, the noise current can be converted to a noise voltage at the grid of the valve of

$$V_{ng}=I_{na}/g_m=(4k\alpha T_k \Delta f/\beta g_m)^{1/2}$$
$$=[4kT_0 \Delta f.\alpha T_k/\beta g_m T_0]^{1/2} \tag{51.9}$$

where T_0 is the normal ambient (room) temperature.

The part $\alpha T_k/\beta g_m T_0$ of expression (51.9) above is equivalent to a resistance, which approximates to

$$R_{ng}=2.5/g_m \tag{51.10}$$

and this is the equivalent noise resistance in the grid of the triode at room temperature. The factor 2.5 in R_{ng} may have a range from 2 to 3 in particular cases. The equivalent noise circuit for a triode having a grid leak R_g and fed from a generator of internal resistance R_1 is as in *Figure 51.3*.

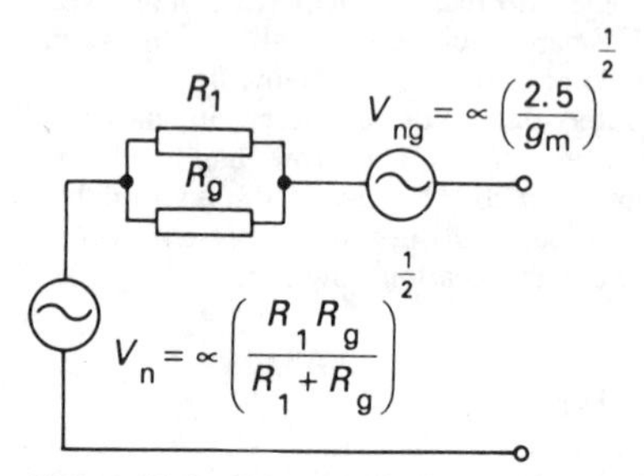

Figure 51.3 Noise voltage input circuit for a valve $\alpha=(4kT\,'f)^{1/2}$

51.5.2 Partition noise

A multielectrode valve such as a tetrode produces greater noise than a triode due to the division of electron current between screen and anode; for this reason the additional noise is known as partition noise. The equivalent noise resistance in the grid circuit becomes

$$R_{ng(tet)}=(I_a/I_k)(20I_s/g_m^2+2.5g_m) \tag{51.11}$$

Where I_a, I_k and I_s are the d.c. anode, cathode and screen currents respectively. I_s should be small and g_m large for low noise in tetrode or multielectrode valves. The factor $20I_s/g_m^2$ is normally between 3 and 6 times $2.5/g_m$ so that a tetrode valve is much noisier than a triode.

At frequencies greater than about 30 MHz, the transit time of the electron from cathode to anode becomes significant and this reduces gain and increases noise. Signal-to-noise ratio therefore deteriorates. Partition noise in multielectrode valves also increases and the neutralised triode, or triodes in cascode, give much better signal-to-noise ratios at high frequencies.

At much higher frequencies (above 1 GHz) the velocity modulated electron tube, such as the klystron and travelling wave tube, replace the normal electron valve. In the klystron, shot noise is present but there is also chromatic noise due to random variations in the velocities of the individual electrons.

51.5.3 Flicker noise

At very low frequencies valve noise is greater than would be expected from thermal considerations. Schottky suggested that this is due to random variations in the state of the cathode surface and termed it *flicker*. Flicker noise tends to be inversely proportional to frequency below about 1 kHz so that the equivalent noise resistance at 10 Hz might be 100 times greater than the shot noise at 1 kHz. Ageing of the valve tends to increase flicker noise and this appears to be due to formation of a high resistance barium silicate layer between nickel cathode and oxide coating.

51.6 Transistor noise

Transistor noise exhibits characteristics very similar to those of valves, with noise increasing at both ends of the frequency scale. Resistance noise is also present due to the extrinsic resistance of the material and the major contributor is the base extrinsic resistance r_b'. Its value is given by expression (51.1), T being the absolute temperature of the transistor under working conditions.

Shot and partition noise arise from random fluctuations in the movement of minority and majority carriers, and there are four sources, namely:

(1) Majority carriers injected from emitter to base and thence to collector.
(2) Majority carriers from emitter which recombine in the base.
(3) Minority carriers injected from base into emitter.
(4) Minority carriers injected from base into the collector.

Sources (1) and (2) are the most important, sources (3) and (4) being significant only at low bias currents. Under the latter condition, which gives least noise, silicon transistors are superior to germanium because of their much lower values of I_{co}.

A simplified equivalent circuit for the noise currents and voltages in a transistor is that of *Figure 51.4* where

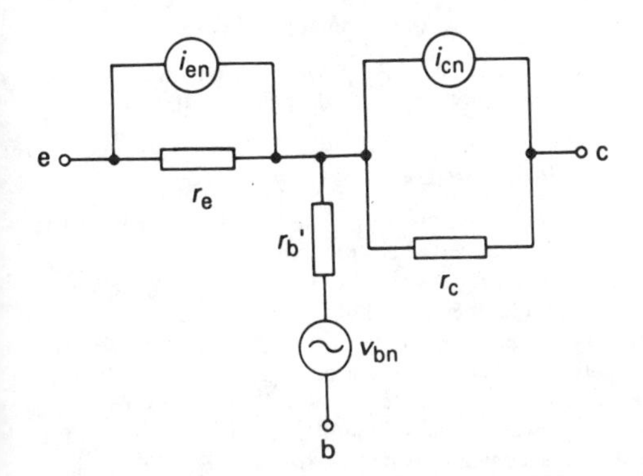

Figure 51.4 Noise circuit equivalent for a transistor

$i_{en} = (2eI_e\,\Delta f)^{1/2}$ the shot noise current in the emitter

$i_{cn} = [2e(I_{co}+I_c(1-\alpha_0))\,\Delta f]^{1/2}$, the shot and partition noise current in the collector

$v_{bn} = (4kT\,.\,r_b'\,\Delta f)^{1/2}$, the thermal noise due to the base extrinsic resistance

$e =$ electronic charge $= 1.602\times10^{-19}$ C

Since transistors are power amplifying devices the equivalent noise resistance concept is less useful and noise quality is defined in terms of noise figure.

Flicker noise, which is important at low frequencies (less than about 1 kHz), is believed to be due to carrier generation and recombination at the base–emitter surface. Above 1 kHz noise remains constant until a frequency of about $f_\alpha(1-\alpha_0)^{1/2}$ is reached, where f_α is the frequency at which the collector–emitter current gain has fallen to $0.7\alpha_0$. Above this frequency, which is about $0.15f_\alpha$, partition noise increases rapidly.

51.7 Noise figure

Noise figure (F) is defined as the ratio of the input signal-to-noise available power ratio to the output signal-to-noise available power ratio, where available power is the maximum power which can be developed from a power source of voltage V and internal resistance R_s. This occurs for matched conditions and is $V^2/4R_s$.

$$F = (P_{si}/P_{ni})/(P_{so}/P_{no}) = P_{no}/G_aP_{ni} \tag{51.12}$$

where G_a = available power gain of the amplifier.

Since noise available output power is the sum of G_aP_{ni} and that contributed by the amplifier P_{na}

$$F = 1 + \frac{P_{na}}{G_aP_{ni}} \tag{51.13}$$

The available thermal input power is $V^2/4R_s$ or $kT\,\Delta f$, which is independent of R_s, hence

$$F = 1 + P_{na}/G_akT\,\Delta f \tag{51.14}$$

and

$$F(\text{dB}) = 10\log_{10}(1 + P_{na}/G_akT\,\Delta f) \tag{51.15}$$

The noise figure for an amplifier whose only source of noise is its input resistance R_1 is

$$F = 1 + R_s/R_1 \tag{51.16}$$

because the available output noise is reduced by $R_1/(R_s+R_1)$ but the available signal gain is reduced by $[R_1/(R_s+R_1)]^2$. For matched conditions $F=2$ or 3 dB and maximum signal-to-noise ratio occurs when $R_1=\infty$. Signal-to-noise ratio is unchanged if R_1 is noiseless because available noise power is then reduced by the same amount as the available gain.

If the above amplifier has a valve, whose equivalent input noise resistance is R_{ng},

$$F = 1 + \frac{R_s}{R_1} + \frac{R_{ng}}{R_s}\left[1 + \frac{R_s}{R_1}\right]^2 \tag{51.17}$$

Noise figure for a transistor over the range of frequencies for which it is constant is

$$F = 1 + \frac{r_b' + 0.5r_e}{R_s} + \frac{(r_b' + r_e + R_s)^2(1-\alpha_0)}{2r_eR_s\alpha_0} \tag{51.18}$$

At frequencies greater than $f_\alpha(1-\alpha_0)^{1/2}$, the last term is multiplied by $[1+(f/f_\alpha)^2/(1-\alpha_0)]$. The frequency f_T at which collector–base current gain is unity is generally given by the transistor manufacturer and it may be noted that $f_T \simeq f_\alpha$, the frequency at which collector–emitter current gain is $0.7\alpha_0$.

Expression (51.18) shows that transistor noise figure is dependent on R_s but it is also affected by I_c through r_e and α_0. As

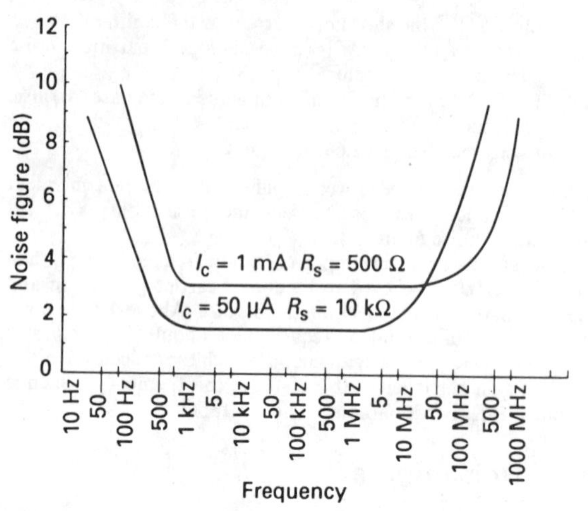

Figure 51.5 Typical noise figure–frequency curves for an r.f. transistor

a general rule the lower the value of I_c the lower the noise figure and the greater is the optimum value of R_s. This is shown in *Figure 51.5* which is typical of an r.f. silicon transistor. Flicker noise causes the increase below 1 kHz, and decrease of gain and increase of partition noise causes increased noise factor at the high frequency end. The high frequency at which F begins to increase is about $0.15f_\alpha$; at low values of collector current f_α falls, being approximately proportional to I_c^{-1}. The type of configuration, common emitter, base or collector has little effect on noise figure.

Transistors do not provide satisfactory noise figures above about 1.5 GHz, but the travelling-wave tube and tunnel diode can achieve noise figures of 3 to 6 dB over the range 1 to 10 GHz.

Sometimes noise temperature is quoted in preference to noise figure and the relationship is

$$F=(1+T/T_0) \tag{51.19}$$

T is the temperature to which the noise source resistance would have to be raised to produce the same available noise output power as the amplifier. Thus if $T=T_0=290$ K, $F=2$ or 3 dB.

The overall noise figure of cascaded amplifiers can easily be calculated and is

$$F_t=F_1+\frac{F_2-1}{G_1}+\frac{F_3-1}{G_1G_2}+\dots\frac{F_n-1}{G_1\dots G_{n-1}} \tag{51.20}$$

where $F_1, F_2, \dots, F_n$ and $G_1, G_2, \dots, G_n$ are respectively the noise figures and available gains of the separate stages from input to output. From Equation (51.20) it can be seen that the first stage of an amplifier system largely determines the overall signal-to-noise ratio, and that when a choice has to be made between two first-stage amplifiers having the same noise figure, the amplifier having the highest gain should be selected because increase of G_1 reduces the noise effect of subsequent stages.

51.8 Measurement of noise

Noise measurement requires a calibrated noise generator to provide a controllable noise input to an amplifier or receiver, and an r.m.s. meter to measure the noise output of the amplifier or receiver. The noise generator generally consists of a temperature-limited (tungsten filament) diode, terminated by a resistance R as shown in *Figure 51.6*. The diode has sufficient anode voltage to ensure that it operates under saturation conditions and anode saturation current is varied by control of the diode filament

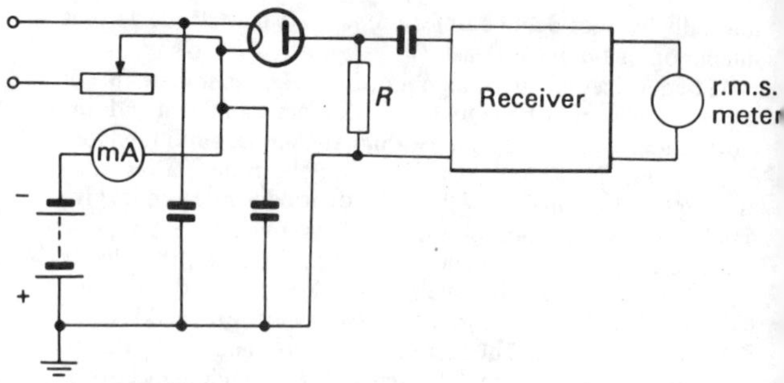

Figure 51.6 Noise figure measurements

current. A milliammeter reads the anode current I_d and the shot noise current component of this is given by $(2I_d e\Delta f)^{1/2}$ where e is electronic charge, 1.602×10^{-19} C. The shot noise has the same flat spectrum as the thermal noise in R, and the meter is calibrated in dB with reference to noise power in R and so provides a direct reading of noise factor. R is generally selected to be 75 Ω, the normal input impedance of a receiver.

To make a measurement, the diode filament current is first switched off, and the reading of the r.m.s. meter in the receiver output noted. The diode filament is switched on and adjusted to increase the r.m.s. output reading 1.414 times (double noise power). The dB reading on the diode anode current meter is the noise figure, since

Noise output power diode off $=GP_{nR}+P_{na}$

Noise output power diode on $=G(P_{nR}+P_{nd})+P_{na}=2GP_{nd}$

Noise figure

$$=10\log_{10}[(GP_{nR}+P_{na})/GP_{nR}]=10\log_{10}P_{nd}/P_{nR}$$

The diode is satisfactory up to about 600 MHz but above this value transit time of electrons begins to cause error. For measurements above 1 GHz a gas discharge tube has to be used as a noise source.

51.9 Methods of improving signal-to-noise ratio

There are five methods of improving signal-to-noise ratio:

(1) Increase the transmitted power of the signal.
(2) Redistribute the transmitted power.
(3) Modify the information content before transmission and return it to normal at the receiving point.
(4) Reduce the effectiveness of the noise interference with signal.
(5) Reduce the noise power.

51.9.1 Increase of transmitted power

An overall increase in transmitted power is costly and could lead to greater interference for users of adjacent channels.

51.9.2 Redistribution of transmitted power

With amplitude modulation it is possible to redistribute the power among the transmitted components so as to increase the effective signal power. As described in *Chapter 38* suppression of the carrier in a double sideband amplitude modulation signal and a commensurate increase in sideband power increases the effective signal power, and therefore signal-to-noise ratio by 4.75 dB (3 times) for the same average power or by 12 dB for the same peak envelope power. Single sideband operation by removal of one sideband reduces signal-to-noise ratio by 3 dB because signal power is reduced to $\frac{1}{4}$ (6 dB) and the non-correlated random noise power is only halved (3 dB). If all the power is transferred to one sideband, single sideband operation increases signal-to-noise ratio by 3 dB.

51.9.3 Modification of information content before transmission and restoration at receiver

51.9.3.1 The compander

A serious problem with speech transmission is that signal-to-noise ratio varies with the amplitude of the speech, and during gaps between syllables and variations in level when speaking, the noise may become obtrusive. This can be overcome by using compression of the level variations before transmission, and expansion after detection at the receiver, a process known as companding. The compressor contains a variable loss circuit which reduces amplification as speech amplitude increases and the expander performs the reverse operation.

A typical block schematic for a compander circuit is shown in *Figure 51.7*. The input speech signal is passed to an amplifier

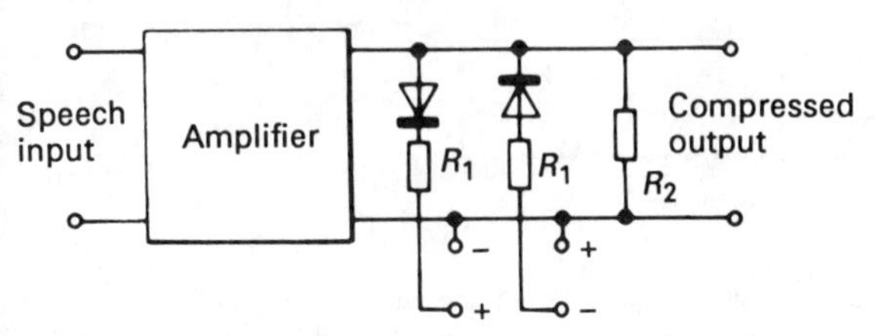

Figure 51.7 A compressor circuit

across whose output are shunted two reverse biased diodes; one becoming conductive and reducing the amplification for positive going signals and the other doing the same for negative-going signals. The input/output characteristic is S shaped as shown in *Figure 51.8*; the diodes should be selected for near identical

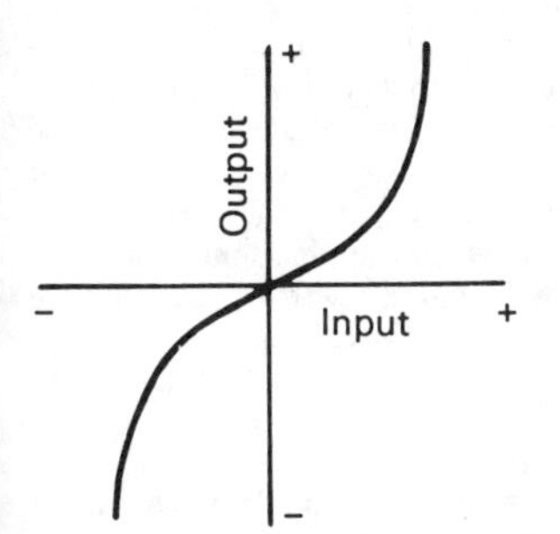

Figure 51.8 Compressor input/output characteristic

shunting characteristics. Series resistances R_1 are included to control the turnover, and shunt resistance R_2 determines the maximum slope near zero.

A similar circuit is used in the expander after detection but as shown, *Figure 51.9*, the diodes form a series arm of a potential divider and the expanded output appears across R_3. The expander characteristic, *Figure 51.10*, has low amplification in the gaps between speech, and amplification increases with increase in speech amplitude.

The diodes have a logarithmic compression characteristic, and with large compression the dB input against dB output tends to a line of low slope, e.g. an input variation of 20 dB being compressed to an output variation of 5 dB. If greater compression is required two compressors are used in tandem.

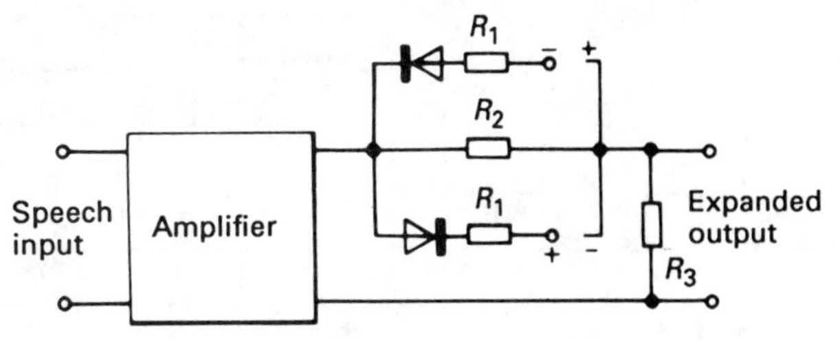

Figure 51.9 An expander circuit

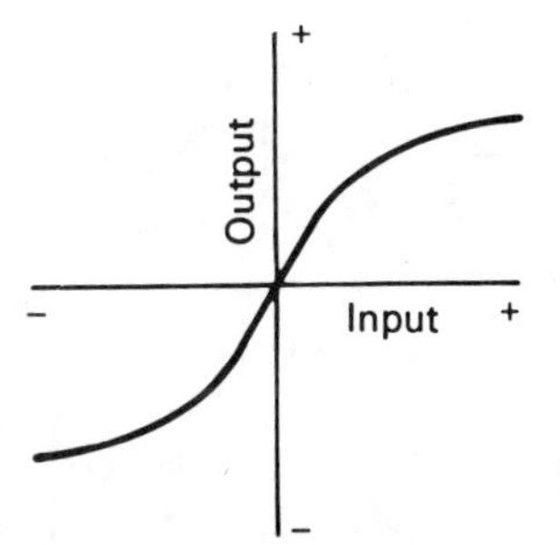

Figure 51.10 Expander input/output characteristics

The collector–emitter resistance of a transistor may be used in place of the diode resistance as the variable gain device. The collector–emitter resistance is varied by base–emitter bias current, which is derived by rectification of the speech signal from a separate auxiliary amplifier. A time delay is inserted in the main controlled channel so that high-amplitude speech transients can be anticipated.

51.9.3.2 Lincompex

The compander system described above proves quite satisfactory provided the propagation loss is constant as it is with a line or coaxial cable. It is quite unsuitable for a shortwave point-to-point communication system via the ionosphere. A method known as Lincompex[2] (linked compression expansion) has been successfully developed by British Telecom. *Figure 51.11* is a block diagram of the transmit–receive paths. The simple form of diode compressor and expander cannot be used and must be replaced by the transistor type, controlled by a current derived from rectification of the speech signal. The current controls the compression directly at the transmitting end and this information must be sent to the receiver by a channel unaffected by any propagational variations. This is done by confining it in a narrow channel (approximately 180 Hz wide) and using it to frequency-modulate a sub-carrier at 2.9 kHz. A limiter at the receiver removes all amplitude variations introduced by the r.f. propagation path, and a frequency discriminator extracts the original control information.

The transmit chain has two paths for the speech signals, one (A) carries the compressed speech signal, which is limited to the range 250 to 2700 Hz by the low-pass output filter. A time delay of 4 ms is included before the two compressors in tandem, each of which has a 2:1 compression ratio, and the delay allows the compressors to anticipate high amplitude transients. The 2:1 compression ratio introduces a loss of $x/2$ dB for every x dB change in input, and the two in tandem introduce a loss of $2(x/2)=x$ dB for every x dB change of input. The result is an almost constant speech output level for a 60 dB variation of speech input. Another time delay (10 ms) is inserted between the compressors and output filter in order to compensate for the control signal delay due to its narrow bandwidth path.

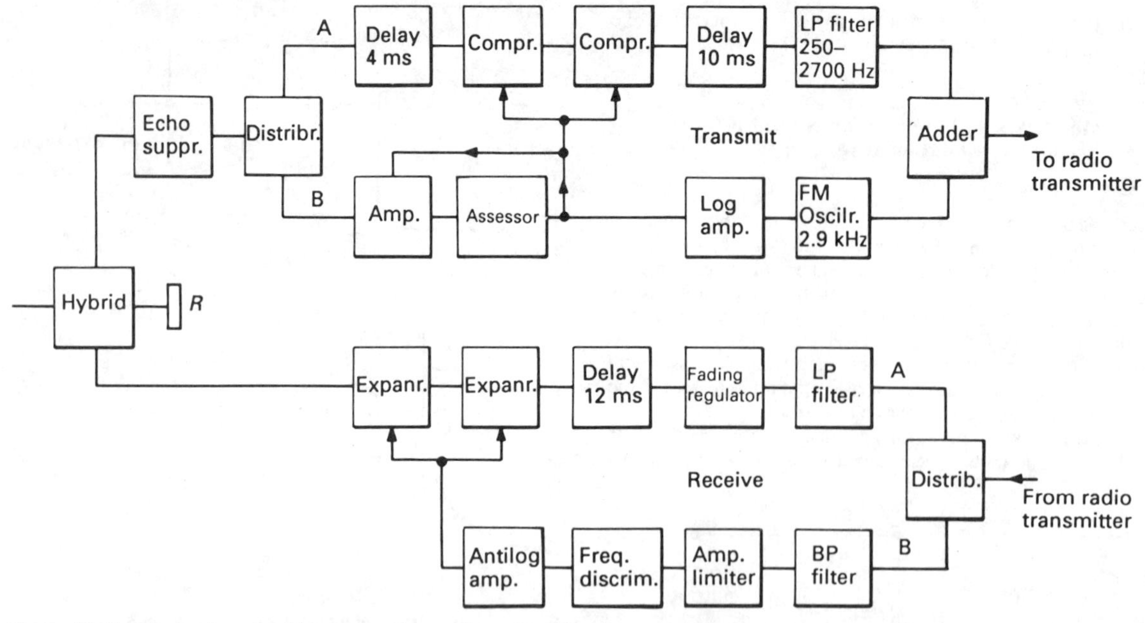

Figure 51.11 Block schematic of Lincompex Compander system for radio transmissions

The other transmit path (B) contains an amplitude-assessor circuit having a rectified d.c. output current proportional to the speech level. This d.c. current controls the compressors, and after passing through a logarithmic amplifier is used to frequency modulate the sub-carrier to produce the control signal having a frequency deviation of 2 Hz/dB speech level change. The time constant of the d.c. control voltage is 19 ms permitting compressor loss to be varied at almost syllabic rate, and the bandwidth of the frequency-modulated sub-carrier to be kept within ±90 Hz. The control signal is added to the compressed speech and the combined signal modulates the transmitter.

The receive chain also has two paths; path (A) filters the compressed speech from the control signal and passes the speech to the expanders via a fading regulator, which removes any speech fading not eliminated by the receiver a.g.c., and a time delay, which compensates for the increased delay due to the narrow-band control path (B). The latter has a band-pass filter to remove the compressed speech from the control signal and an amplitude limiter to remove propagational amplitude variations. The control signal passes to a frequency discriminator and thence to an antilog amplifier, the output from which controls the gain of the expansion circuits. The time constant of the expansion control is between 18 ms and 20 ms.

51.9.3.3 Pre-emphasis and de-emphasis

Audio energy in speech and music broadcasting tends to be greatest at the low frequencies. A more level distribution of energy is achieved if the higher audio frequencies are given greater amplification than the lower before transmission. The receiver circuits must be given a reverse amplification-frequency response to restore the original energy distribution, and this can lead to an improved signal-to-noise ratio since the received noise content is reduced at the same time as the high audio frequencies are reduced. The degree of improvement is not amenable to measurement and a subjective assessment has to be made. The increased high-frequency amplification before transmission is known as pre-emphasis followed by de-emphasis in the receiver audio circuits. FM broadcasting (maximum frequency deviation ±75 kHz) shows a greater subjective improvement than a.m., and it is estimated to be 4.5 dB when the pre- and de-emphasis circuits have time constants of 75 μs. A simple *RC* potential divider can be used for de-emphasis in the receiver audio circuits, and 75 μs time constant gives losses of 3 and 14 dB at 2.1 and 10 kHz respectively compared with 0 dB at low frequencies.

51.9.4 Reduction of noise effectiveness

Noise, like information, has amplitude and time characteristics, and it is noise amplitude that causes the interference with a.m. signals. If the information is made to control the time characteristics of the carrier so that carrier amplitude is transmitted at a constant value, an amplitude limiter in the receiver can remove all amplitude variations due to noise without impairing the information. The noise has some effect on the receiver carrier time variations, which are phase-modulated by noise, but the phase change is very much less than the amplitude change so that signal-to-noise ratio is increased.

51.9.4.1 Frequency modulation

If the information amplitude is used to modulate the carrier frequency, and an amplitude limiter is employed at the receiver, the detected message-to-noise ratio is greatly improved. FM produces many pairs of sidebands per modulating frequency especially at low frequencies, and this 'bass boost' is corrected at the receiver detector to cause a 'bass cut' of the low frequency noise components. This triangulation of noise leads to 4.75 dB signal-to-noise betterment. Phase modulation does not give this improvement because the pairs of sidebands are independent of modulating frequency. The standard deviation of 75 kHz raises signal-to-noise ratio by another 14 dB, and pre-emphasis and de-emphasis by 4.5 dB, bringing the total improvement to 23.25 dB over a.m.

The increased signal-to-noise performance of the f.m. receiver is dependent on having sufficient input signal to operate the amplitude limiter satisfactorily. Below a given input signal-to-noise ratio output information-to-noise ratio is worse than for a.m. The threshold value increases with increase of frequency

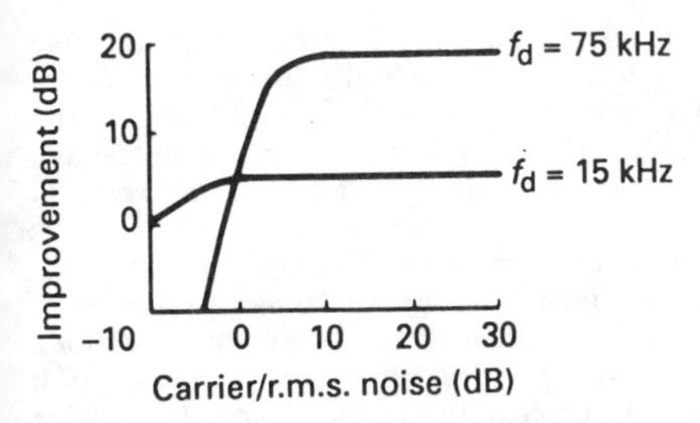

Figure 51.12 Threshold noise effect with f.m. compared with a.m.

deviation because the increased receiver bandwidth brings in more noise as indicated in *Figure 51.12*.

51.9.4.2 Pulse modulation

Pulse modulated systems using change of pulse position (p.p.m.) and change of duration (p.d.m.) can also increase signal-to-noise ratio but pulse amplitude modulation (p.a.m.) is no better than normal a.m. because an amplitude limiter cannot be used.

51.9.4.3 Impulse noise and bandwidth

When an impulse noise occurs at the input of a narrow bandwidth receiver the result is a damped oscillation at the mid-frequency of the pass band as shown in *Figure 51.13(a)*. When a wide bandwidth is employed the result is a large initial amplitude with a very rapid decay, *Figure 51.13(b)*. An amplitude limiter is

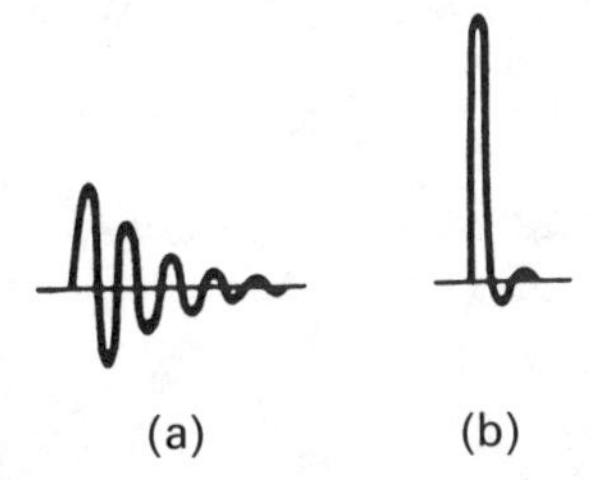

Figure 51.13 Output wave shape due to an impulse in (a) a narrow band amplifier, (b) a wideband amplifier

much more effective in suppressing the large amplitude near-single pulse than the long train of lower amplitude oscillations. Increasing reception bandwidth can therefore appreciably reduce interference due to impulsive noise provided that an amplitude limiter can be used.

51.9.4.4 Pulse code modulation

A very considerable improvement in information-to-noise ratio can be achieved by employing pulse code modulation (p.c.m.).[3] PCM converts the information amplitude into a digital form by sampling and employing constant amplitude pulses, whose presence or absence in a given time order represents the amplitude level as a binary number. Over long cable or microwave links it is possible to amplify the digital pulses when signal-to-noise ratio is very low, to regenerate and pass on a freshly constituted signal almost free of noise to the next link. With analogue or direct non-coded modulation such as a.m. and f.m., noise tends to be cumulative from link to link. The high signal-to-noise ratio of p.c.m. is obtained at the expense of much increased bandwidth, and Shannon has shown that with an ideal system of coding giving zero detection error there is a relationship between information capacity C (binary digits or bit/s), bandwidth W (Hz) and average signal-to-noise thermal noise power ratio (S/N) as follows:

$$C=W\log_2(1+S/N) \qquad (51.21)$$

Two channels having the same C will transmit information equally well though W and S/N may be different. Thus for a channel capacity of 10^6 bit/s, $W=0.167$ MHz and $S/N=63\equiv18$ dB, or $W=0.334$ MHz and $S/N=7\equiv8.5$ dB. Doubling of bandwidth very nearly permits the S/N dB value to be halved, and this is normally a much better exchange rate than for f.m. analogue modulation, for which doubling bandwidth improves S/N power ratio 4 times or by 6 dB.

In any practical system the probability of error is finite, and a probability of 10^{-6} (1 error in 10^6 bits) causes negligible impairment of information. Assuming that the detector registers a pulse when the incoming amplitude exceeds one-half the normal pulse amplitude, an error will occur when the noise amplitude exceeds this value. The probability of an error occurring due to this is

$$P_e=\frac{1}{(2\pi)^{\frac{1}{2}}}\frac{2V_n}{V_p}\exp(-V_p^2/8V_n^2) \qquad (51.22)$$

where V_p = peak voltage of the pulse
and V_n = r.m.s. voltage of the noise.

The curve is plotted in *Figure 51.14*.

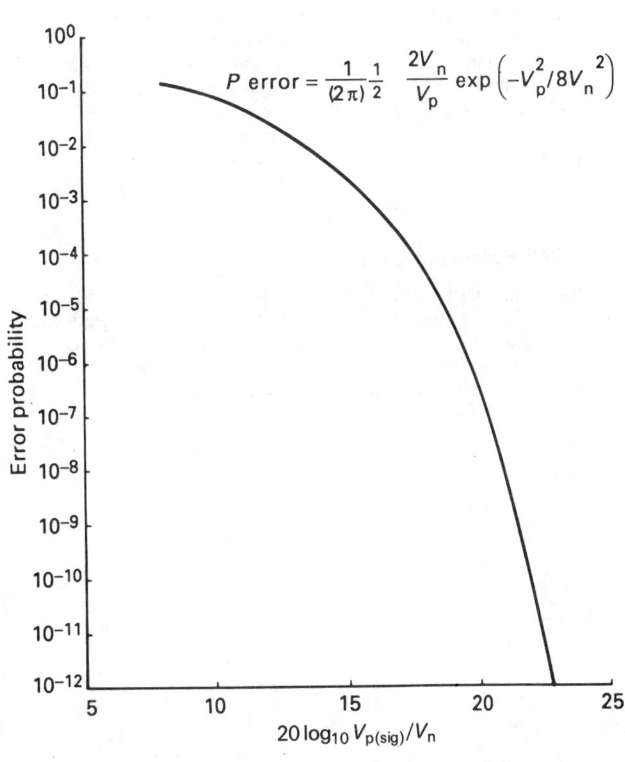

Figure 51.14 Probability of error at different $V_{p(sig)}$/r.m.s. noise ratios

An error probability of 10^{-6} requires a V_p/V_n of approximately 20 dB, or since $V_{av}=V_p/2$ a V_{av}/V_n of 17 dB. In a binary system two pulses can be transmitted per cycle of bandwidth so that by Shannon's ideal system

$$\frac{C}{W}=2=\log_2(1+S/N) \quad \text{and} \quad S/N=3\cong5\text{ dB} \qquad (51.23)$$

Hence the practical system requires 12 dB greater S/N ratio than the ideal, but the output message-to-noise ratio is infinite, i.e. noise introduced in the transmission path and the receiver is completely removed. There will, however, be a form of noise present with the output message due to the necessary sampling process at the transmit end. Conversion of amplitude level to a digital number must be carried out at a constant level, and the reconstructed decoded signal at the receiver is not a smooth wave but a series of steps. These quantum level steps superimpose on the original signal a disturbance having a uniform frequency spectrum similar to that of thermal noise. It is this quantising noise which determines the output message-to-noise ratio and it is made small by decreasing the quantum level steps. The maximum error is half the quantum step, l, and the r.m.s. error introduced is $l/2(3)^{1/2}$. The number of levels present in the p.a.m. wave after sampling are 2^n where n is the number of binary digits. The message peak-to-peak amplitude is $2^n l$, so that the

$$\text{Message (pk-to-pk)/r.m.s. noise} = \frac{2^n l}{l/2(3)^{1/2}} = 2(3)^{1/2} 2^n \quad (51.24)$$

$$\begin{aligned} M/N\ (\text{dB}) &= 20 \log_{10} 2(3)^{1/2} 2^n \\ &= 20n \log_{10} 2 + 20 \log_{10} 2(3)^{1/2} \\ &= (6n + 10.8)\ \text{dB} \end{aligned} \quad (51.25)$$

Increase of digits (n) means an increased message-to-noise ratio but also increased bandwidth and therefore increased transmission path and receiver noise; care must be exercised to ensure that quantising noise remains the limiting factor.

Expression (51.25) represents the message-to-noise ratio for maximum information amplitude, and smaller amplitudes will give an inferior noise result. A companding system should therefore be provided before sampling of the information takes place.

51.9.5 Reduction of noise

Since thermal noise power is proportional to bandwidth, the latter should be restricted to that necessary for the objective in view. Thus the bandwidth of an a.m. receiver should not be greater than twice the maximum modulating frequency for d.s.b. signals, or half this value for s.s.b. operation.

In an f.m. system information power content is proportional to (bandwidth)^2 so that increased bandwidth improves signal-to-noise ratio even though r.m.s. noise is increased. When, however, carrier and noise voltages approach in value, signal-to-noise ratio is worse with the wider band f.m. transmission (threshold effect).

Noise is reduced by appropriate coupling between signal source and receiver input and by adjusting the operating conditions of the first stage transistor for minimum noise figure.

Noise is also reduced by refrigerating the input stage of a receiver with liquid helium and this method is used for satellite communication in earth station receivers using masers. The maser amplifies by virtue of a negative resistance characteristic and its noise contribution is equivalent to the thermal noise generated in a resistance of equal value. The noise temperature of the maser itself may be as low as 2–10 K and that of the other parts of the input equipment 15–30 K.

Parametric amplification, by which gain is achieved by periodic variation of a tuning parameter (usually capacitance), can provide the relatively low noise figures 1.5–6 dB over the range 5–25 GHz. Energy at the 'pump' frequency (f_p) operating the variable reactance, usually a varactor diode, is transferred to the signal frequency (f_s) in the parametric amplifier or to an idler frequency ($f_p \pm f_s$) in the parametric converter. It is the resistance component of the varactor diode that mainly determines the noise figure of the system. Refrigeration is also of value with parametric amplification.

References

1 NORTH, D. O., 'Fluctuations in space-charge-limited currents at moderately high frequencies', *RCA. Rev.*, **4**, 441 (1940), **5**, 106, 244 (1940)

2 WATT-CARTER, D. E. and WHEELER, L. K., 'The Lincompex system for the protection of HF radio telephone circuits', *P.O. Elect. Engrs. J.*, **59**, 163 (1966)

3 BELL SYSTEM LABORATORIES, *Transmission Systems for Communications* (1964)

Further reading

FAZEKAS, P., 'A systematic approach facilitates noise analysis', *EDN*, 12 May (1988)

52 Computer Aided Design

S R Hodge
Racal-Redac Systems Ltd

P A Francis
Racal-Redac UK Ltd

Contents

52.1 The use of computers for electronic design

52.1.1 Historical perspective

Computers have been used to ease the tasks faced by the electronics designer for approximately 25 years. Their use traditionally splits into two distinct areas.

First, computers have long been used for the purpose of simulation (i.e. modelling) of devices and circuits. Almost every engineer at some stage writes a computer program to model the operation of a particular piece of circuitry; this is usually done to establish before a prototype is built that the circuit is likely to behave in the manner expected of it. The need to develop a prototype in software arises from a number of causes; for example, in the case of silicon integrated circuits it is usually a combination of high prototype tooling costs and long waiting times. In the case of some types of analogue circuitry, the need arises through the fact that the complexity or intractability of the mathematics defining the circuit's operation demands the use of numerical methods for its solution.

Software tools which permit the simulation of generalised digital or analogue circuits were developed in the 1970s. For analogue simulation, the best known of these is Spice, originally developed in the University of California at Berkeley, and now available in many different forms from commercial vendors. These simulators perform a straightforward but highly complex function: starting with a description of the circuit to be simulated, usually in the form of an ASCII file containing a netlist, they repeatedly set up and solve a set of simultaneous differential equations which represent the state of the circuit at discrete time intervals defined by the user. In order to do so, they identify the components supplied to them in the netlist and replace them with software models which represent the operation of the component as a mathematical function or set of functions. The simpler components, individual transistors and diodes for example, have models which form part of the simulator itself (the 'primitives' of the simulator) and only require to be parameterised according to part number; the more complex components may be represented by the simulator as assemblages of primitives, usually called functional models, or as 'black boxes' defined by looking only at their terminals, usually referred to as behavioural models. Simulation is examined in more detail in *Section 52.2.1*.

Secondly, computers have been used for at least 20 years to provide an easy and quick method of obtaining a two-dimensional (2-D) graphical record of a design. This may be, for example, a schematic diagram or a physical layout of a piece of silicon or printed circuit board (PCB). Graphical computer systems were initially designed to provide generalised 2-D draughting facilities not for electronic design but for sheet metalwork, mechanical engineering and the like; many of today's electronic computer aided design facilities are adaptations of such 2-D draughting tools to specifically electronic applications.

The very rapid development of integrated circuits (ICs) in the 1970s led to the almost complete abandonment of the manual methods for IC layout (which hitherto had made extensive use of dimensionally stable plastic films such as Rubylith and 'cut and strip' techniques) in favour of 2-D computer-based draughting software. Products became available from a number of vendors who had previously addressed only the mechanical draughting field. The products allowed the user to work on a number of layers simultaneously to design all the patterns necessary for the full definition of an IC as a set of shapes; such products are usually referred to as 'polygon editors' for that reason. Early offerings did not relate any electrical considerations to the physical layouts produced; design rule checking for spacing limit violations and electrical connectivity checking were either performed manually or performed after the initial design had been completed, using off-line batch checking software tools. These provided the user with a list of violations for later correction using the graphical polygon editor.

The early 1980s saw the rapid adoption of workstation-based schematic capture systems by the larger electronics companies who had observed the very high cost of maintaining schematic documentation. Computer systems had the undoubted advantage of speeding up the process of producing correct, neat, reproducible and, above all, easily editable schematics. Engineers tended to adopt these products in advance of drawing offices, perhaps as a result of their background and training. Because the engineer is concerned not only with the accuracy of documentation but also with circuit operation, the vendors of these products were put under some pressure to integrate simulation tools into the schematics capture environment.

It is only in recent years that the parallel paths of CAD development have begun to come together in a useful manner. The integration of simulation tools with schematic capture systems, the development of netlist tools which carry schematic information from the engineer's domain into that of the PCB or silicon layout designer, and the incorporation of software which brings electrical and manufacturing considerations into the physical domain have all been developed in the last few years. *Figure 52.1* shows the various operations carried out in the development of an IC or PCB using the types of CAD tools available to the engineer and layout designer today.

52.1.2 The development of the engineering workstation

Most of the tools used up to about 1980 were designed to run on small mainframes or minicomputers equipped with graphics displays. The operating systems for minicomputer-based systems were almost invariably single user and single tasking; multi-user systems were generally restricted to the mainframes and the emerging superminis. During the late 1970s the so-called intelligent graphics terminal (i.e. a graphics terminal containing its own processor, RAM and ROM on which certain graphics macro functions were performed) allowed systems to be configured in which a number of users could make reasonable use of a shared CPU resource.

By 1980, the cost of CPU power, semiconductor memory and disk storage had fallen to such an extent as a result of rapid technology advances that it became economically possible for each engineer to have a personal CPU resource on his or her desk. The IBM Personal Computer offered the same computing power as the minicomputers of a few years earlier, although its input/output capability was somewhat inferior; for graphics applications, however, this limitation is not a serious one, and indeed the IBM PC and its derivatives are now the most commonly used platforms for both mechanical and electronics CAD.

Advanced CAD software, however, needed more computing power than was available from any machine that relied on physical memory alone. It also needed graphics processing and display resolution which was not at that time available from the MS-DOS* based machines. The distinguishing features of the technical workstations which began to appear in the 1980s were the use of virtual memory, the use of 32-bit processors (either as CMOS microprocessors such as the 68000 family or as bipolar bit-slice microprocessors such as the 2900 family), the use of high-resolution graphics displays, and the use of operating systems such as UNIX† and VMS‡ which permit multiple

* MS-DOS is a trademark of Microsoft Corporation.
† UNIX is a trademark of AT&T.
‡ VMS is a trademark of Digital Equipment Corporation.

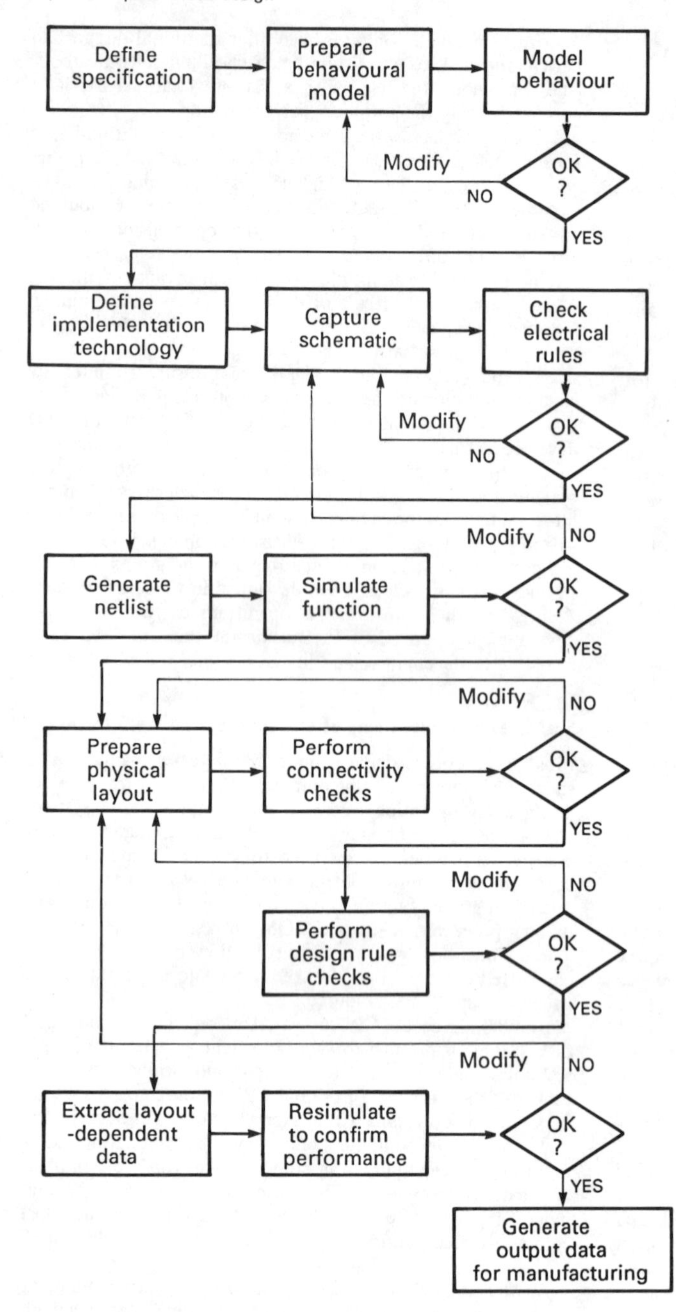

Figure 52.1 Flow of operations for PCB and IC design

processes to run concurrently and permit networking of systems in such a way as to give the user effectively transparent access to data stored at any point.

Today's technical workstations offer a level of performance in terms of sheer CPU power which would have been difficult to envisage only a few years ago. In 1988 4-mips machines are available at less than half the price of a 1-mips supermini in 1984. 300 Mbyte 5.25 inch disks are common, as are 10 Mbit/s communications facilities and full colour graphics with screen resolutions of the order of 1300×1000 pixels. These workstations are now the logical choice for almost all electronics CAD applications and few vendors of CAD software have any remaining justification for failing to adopt them as their standard platforms.

52.1.3 The incentives for adopting CAD systems

There are a number of forces at work in the field of electronics which together form a compelling set of arguments in favour of the adoption of computer aids for design and manufacture.

The accelerating pace of change in almost every field of technology has meant that, over the last few years, product lifetimes have shortened considerably. A product designed 20 years ago could reasonably have been expected to sell competitively, retaining market share and generating useful revenue, for 5–15 years depending on its application; today it is unusual to design for a product lifetime of more than 3 years. The reasons are not difficult to understand. Advances in technology, especially silicon technology, cause designs to become obsolete rapidly on the basis of manufactured cost.

One result of this is that the design cycle can no longer be measured in years; it must be measured in months, sometimes weeks. There is no longer enough time to prepare drawings manually, to order parts and experiment with a series of physical prototypes, or to painstakingly produce PCB masters using opaque self-adhesive tapes on plastic film. Those products that continue to be developed using traditional methods will tend to miss the highly profitable early phases of the product's life. Computer aided design gives a very considerable increase in productivity in design by automating or accelerating those processes which do not require the creative skills of the engineer.

A second effect of the IC revolution is that, because more functionality is being packed into individual components, the level of complexity of a single PCB has reached the point where an engineer can no longer be certain that every aspect of the board's operation has been considered. Some method is needed to formalise the prototype testing process; computer aided engineering, by the use of schematics capture tools and simulation, can be used to develop and test a prototype in software far faster and far more thoroughly than can an engineer working with hardware. In addition, since hardware prototypes offer no easy means of testing to allow for the statistical spread of component tolerances, prototyping in software becomes almost mandatory for today's electronic products. Worst case timing analysis of digital circuits and sensitivity analysis of analogue circuits can ensure that even the most complex of designs can meet its design requirements under any set of operating conditions. This is not to say that physical prototyping has no longer any value and can be eliminated completely, since production engineering departments will almost always need prototypes for such requirements as test fixture and mechanical enclosure development; rather to emphasise the point that a physical prototype can no longer guarantee to satisfy all the verification needs of an electronics engineer.

The physical layout of printed circuit boards has traditionally been a time-consuming and error-prone process. Furthermore, advances in manufacturing technology for fine line routing, multilayer boards and buried vias have made manual methods of preparing camera-ready artwork even more difficult. Graphical, interactive computing systems have been available since the 1960s for PCB layout, but it is only in recent years that they have been widely adopted for general PCB use. Again, they offer considerable productivity advantages over manual methods, but perhaps more importantly they offer better accuracy in terms of the exact mapping of the schematic into its physical implementation. They also allow rigid adherence to the small spacing rules which modern manufacturing technologies allow and which are necessary in order to achieve the high packing densities possible as a result of the advances in integrated circuit technology.

The same rationale lies behind the adoption of computer graphical tools for the design of the integrated circuits themselves; in fact, computer aids were used for IC design before they were used for PCBs, since the adaptation of two-dimensional general-purpose graphics software to IC design followed much simpler rules than its adaption to PCBs; at the same time, the degree of repetition of structure within a typical IC is far higher than that in a PCB, rendering the act of manually laying out an IC design perhaps even less creative and more demanding of intense concentration on geometry than that of manually laying out a PCB.

Computer-based systems, given adequate database structure and access mechanisms, can materially ease the integration of automatic manufacturing tools into the production environment. Numerically controlled profilers and drilling machines are now routinely used for board manufacture, as are automatic pick and place machines for component insertion. These tools also provide improved productivity.

The testing of production PCBs is becoming an issue of addressing rapidly increasing complexity. The time taken for a skilled test technician to troubleshoot a failed board using the tools currently available (logic analysers, signature analysers, logic probes, high-speed oscilloscopes, etc.) and the cost of employing and training test technician staff have reached the point for many companies where automatic testing and fault diagnostics are also becoming mandatory. Computer aided test, involving the use of fault simulation, automatic fault dictionary and test vector generation, and links to automatic functional testers, is an area of electronic design automation which is growing at a very rapid pace.

In addition to the benefits of improved productivity in design, faster and more accurate layout and faster manufacturing and testing, computer aided design enables higher product quality to be sustained. Automatic systems are not subject to the performance variations of humans; as long as they are working and are receiving the materials and data they need, they can continue to deliver identical copies of the master design. Furthermore, since software prototyping and automatic test can systematically examine the operation of the design and of the manufactured product to an extent which an engineer or technician cannot hope to match, electronic design automation makes consistently high-quality product an attainable goal.

52.2 The design process

52.2.1 Computer aided engineering

The name Computer Aided Engineering (CAE) was originally applied to the process of schematic capture and documentation described in *Section 52.2.1.1*. The introduction of such CAE systems was led by a trio of start-up companies (Daisy Systems Inc., Mentor Graphics Inc. and Valid Logic Inc.) in the early 1980s, whose intent was originally to place in the hands of the engineer the tools to produce tidy, accurate schematics. Elimination of the need to use the drawing office and the attendant delays was a motivating factor.

These tools were extensively used for circuit design of both ICs and PCBs, and in addition to being able to produce schematics they offered the ability to output netlists in a form usable for at least functional simulation.

The integration of simulation into CAE was largely driven by the IC designers who, as a result of initiatives in design methodology from several groups, the most notable being that led by Mead and Conway[1] at Xerox Parc, began to use integrated computer-based tools for schematic capture, design verification and analysis and physical layout.

52.2.1.1 Schematic capture

Schematic capture is the usual name given to the process of editing a graphics representation of a circuit diagram and of post-processing the graphical data to produce an output file suitable for driving some sort of plotter to give a permanent paper record. It may also include design rule checking, a facility which checks that electrical rules are not violated by the connectivity defined in the schematic; such electrical rules, for example, may specify the maximum fanout of a device pin, the type of gate to which a pin may legitimately be connected (ECL, TTL, etc.) or the use of wired-OR connections.

Schematic capture tools are usually interactive. In other words, as the schematic is entered, normally by manipulating a mouse, the user is given information relating to what the user has done or is in the process of doing. An example of this is that, when a component is called into the schematic, the system will check to see that the component chosen already exists in the library, and if not will prompt the user to enter the part.

Most schematic capture tools are based on 2-D draughting products and it is common for them to contain no specifically electrical information in the graphics representation of the schematic. It is left to the user to determine where, for example, a connection exists or where two unconnected wires cross. Electrical connectivity is extracted from the schematic only when it is compiled or post-processed to produce a netlist. In these circumstances, it is common practice to combine the design rule checking with the post-processor. Unfortunately this leads to a situation where the user has to restart the schematic capture software in order to edit out the errors detected during compilation.

Advanced schematic capture tools use a different approach. The graphics data contained in the schematic are made to contain all the connectivity information so that the nets exist at the time they are created. By this means, design rule checks can be applied by the graphics software itself, alerting the engineer to errors as they are made.

Today's schematic capture tools also allow for the use of hierarchy in the development of a complex schematic. Hierarchy in essence treats a block of circuitry as if it were another library element, but instead of being a single component it is a collection of components drawn with a particular connectivity. In a schematic where a particular block of circuitry is used a number of times, that block need only be defined and checked once and then multiple instances of it may be called into the schematic. Hierarchical schematic editors allow hierarchical blocks to be fully defined at their own (circuit) level, and to be represented as single entities (e.g. by means of a rectangular outline as in a block diagram) at the level at which they are instanced. Typically a complex schematic which uses hierarchy to simplify its representation will exist on a number of sheets, the top-level sheet containing the overall schematic which references the hierarchical blocks instanced in it and drawn in detail on lower-level sheets.

Digital IC designers normally make extensive use of hierarchy; the nesting of 12 or more levels of hierarchy in a design is not unknown. Hierarchy in PCB designs is not yet so universally adopted, partly because PCBs typically contain far less repetition and partly because until recently hierarchical systems were not available for PCB design.

52.2.1.2 Digital simulation

The simulation of digital designs is a very complex subject which cannot be fully treated here. As an introduction, however, it is instructive to examine the main needs of the engineer in the use of digital simulation and to examine ways in which those needs are met.

Digital simulation divides fairly neatly into three main categories; logic simulation, timing simulation and fault simulation. Each has a specific and complementary purpose.

Logic simulation allows the engineer to simulate the behaviour of a circuit by modelling it as a set of functional blocks. The objective here is to confirm that, viewed from its terminals, the PCB, IC or hierarchical block being designed performs the Boolean function intended. Logic simulation uses the netlist, a library of functions that model the components called in the netlist and a stimulus (usually as an ASCII file defined by the user) to produce an output file which the user examines to confirm circuit operation.

A typical circuit will contain some purely combinatorial logic and some sequential (clocked) elements; logic simulation at its simplest makes the assumption that all combinatorial elements execute their function in zero time, and that all clocked elements take account of their inputs and exhibit changes in their outputs only at the clock edges; it therefore does not make any allowance for element delays, race hazards and so on. It allows the user to determine functionality only for supposedly perfect circuit elements.

The primitive elements for logic simulation are the basic gates (AND, OR, NOT) and the unit time delay element (essentially a D-type).

In practice, a logic simulator that provided only the functionality described here would be of limited use. In fact, most logic simulators go much further than simply modelling Boolean function by means of a two-state (1 or 0) representation of each net.

First, it is necessary to allow for the use of tristate devices and bidirectional buses. Secondly, it is common practice in IC design to wire-OR gate outputs where one gate has the ability to source or sink more current than another and thereby override it; thus the states represented internally in the simulator need to be proliferated to allow for the concept of 'strengths'. One simulator which is used both for IC and PCB design, HHB Cadat 6, actually uses three logic levels at three strengths plus another 12 states, making 21 in all. In interpreting the Boolean functions that define a circuit, the simulator applies a consistent set of logical rules to the 21 states, which avoids the necessity for the user to extend the basic logical definitions of element input/output relationships beyond the normal three states (1,0,X).

Even with this level of detail, logic simulation cannot provide the user with all the information needed to gain confidence that the design will meet its requirements. It is seldom sufficient to know that the circuit designed using idealised Boolean elements functions at some arbitrarily low speed; the designer needs to know that it will function at the clock speed demanded by its engineering specification, and usually with a reasonable margin. Furthermore, it is necessary to know that when the product is manufactured using components which have a statistical distribution of delays, setup and hold times, rise and fall times, etc., it will meet its performance objectives for all combinations of minimum and maximum selections. This need is satisfied by the use of the worst case timing simulator.

It is usual for the logic and timing simulators to be integrated into a single piece of software, since they both use the same netlist, the same library and the same stimulus file. In addition, logic and timing are actually inseparable when asynchronous logic, race conditions, minimum pulse widths and setup and hold times need to be taken into account, as is normal.

Since time delays arise as a result of effects of an analogue nature, the primitive circuit elements used in logic simulation need to be extended in order to contain sufficient information for timing simulation. Simulators normally allow the addition of time delay parameters to logical models, in some cases allowing the delays to be expressed as first-order functions of capacitive loading, fanout or even of temperature and power supply voltage.

So far, the discussion of simulator primitives has been restricted to Boolean functions. In fact, many simulators offer a far wider range of primitives than this; at the more complex end of the spectrum, simulator libraries frequently contain TTL and CMOS MSI and LSI functions and even some commonly used VLSI devices. At the more primitive end, simulators intended to be used for IC design purposes will contain relatively simple models of MOS devices as on/off switches, not only for the obvious purpose of modelling transmission gates but also to model implementations in silicon of Boolean functions which do not exist in a library. They are also likely to contain analogue equivalent circuit models of transistors so that the designer can study in detail the operation of blocks of high-speed circuitry and can use the simulator to make proper and accurate assessments of the required physical sizes of transistors (typically, gate widths and channel lengths of MOS devices).

A digital simulator in which the primitive elements are modelled at the analogue equivalent circuit level is referred to as a *circuit-level* simulator. Similarly, a simulator in which MOS transistors are modelled as switches is referred to as a *switch-level* simulator, and one in which the primitives are gates is called a *logic-level* or *gate-level* simulator. Circuit level simulation is a far more complex task than switch or gate level simulation and for a given piece of circuitry can take 100 or more times as much CPU time, depending on the accuracy required; however, it is often necessary to use circuit-level simulation for speed-critical parts of a design. Hence some simulators offer the capability to define the manner in which simulation is carried out and in which the primitives are defined on a net-by-net or gate-by-gate basis, thus offering the advantage of extreme accuracy where necessary but maintaining most of the speed advantage of gate-level modelling. Such simulators are usually referred to as *mixed mode* or *hybrid* simulators.

It is usual for the output of a logic and timing simulator to be stored on disk as an ASCII file containing columns of data, one column for each input or output, time being on the vertical axis and increasing in the downward direction. Engineers accustomed to using oscilloscopes and logic or timing analysers for physical prototyping find this form of representation difficult to read, and most simulator suppliers now offer a timing analyser form of display of output data, making good use of the graphics performance of the technical workstations (*Figure 52.2*).

Having been satisfied that the circuit is capable of meeting its functional and performance specifications, the engineer needs to determine whether the circuit is testable. This can be achieved in a number of ways, but by far the most universally accepted is through the use of fault simulation.

The fault simulator tests the circuit model, strictly at the logical level, for two possible faults (stuck at 1 and stuck at 0) on each net; in other words, a number of gate-level simulations equal to twice the number of nets is run. The effect of all these runs is observed at the output, and for each error detected at the output a list of possible causes (i.e. net identification and fault type) is developed. This is the so-called *fault dictionary*.

Any fault which does not give rise to an observable change in output either points to a redundancy in the design or to a part of the design which is not testable; thus, at the end of a fault simulation exercise, the designer has a guide to the testability of the design, usually expressed as a fault coverage percentage.

Furthermore, the fault dictionary created during fault simulation now exists as part of the design data associated with that particular design and can be later used by an automatic tester to help the test engineer identify faults down to device or net level.

There have been significant advances in fault simulation techniques, motivated out of the need to improve speed of fault simulation which is self-evidently a CPU-intensive process. Early fault simulators were serial; each pass took only one possible fault into account. However, since the faults are binary in nature and therefore need only 1 bit of the computer's word in order to be fully represented, the development of parallel fault simulators followed rapidly. These allow 16 or 32 fault conditions to be processed simultaneously. In practice, it is common for one of these 16 or 32 to be the 'good circuit' condition in order to have an internal reference. Most of today's fault simulators are parallel.

HHB's Cadat 6 offers the further improvement of concurrent fault simulation. It dynamically identifies net states downstream from the faulted net which have already resulted from the simulation of other fault conditions. The amount of computation needed to produce the same result is thus dramatically decreased; however, the amount of random access memory needed is increased substantially.

52.2.1.3 Behavioural and functional modelling

The terms 'behavioural' and 'functional' are often misunderstood, so it may be useful to review their meanings and their implications for CAD here.

A hierarchical block of circuitry, regardless of the level at which it exists, may be described in a number of ways. At the level at which the block is instanced, it is normal for it only to be described by reference to its external electrical appearance; what happens inside the block is of no consequence. At the level at which it was designed a functional description is used; this description defines the block as an assemblage of smaller entities each of which may itself be either a functional model or a behavioural model.

As an example, take the binary half adder shown in *Figure 52.3(a)*. A behavioural description says that given two inputs A and B, the outputs S (sum) and C (carry) are defined as:

$$S=(A+B) \bmod 2$$

and

$$C=A \text{ and } B$$

The behavioural definition is completed by a statement which specifies the minimum and maximum delays between the input and output pins for the various transitions.

A functional description of the same device will contain information about its internal structure; for example, the half adder may in fact be composed of two-input NAND gates as in *Figure 52.3(b)* or of a CMOS circuit as in *Figure 52.3(c)*. This illustrates the fact that although behaviour can always be deduced from function, function is not fully determined by behaviour. The conversion from behaviour to function involves the elements of creativity and choice on the part of the engineer, since there are always many different ways to produce a single specified behaviour. Behavioural descriptions are generally far more concise than their functional counterparts. It is interesting to note that all basic primitives are of necessity behavioural models, by definition.

Behavioural modelling is used for the purpose of top-down design, especially for ICs. Special languages for behavioural modelling, such as ELLA (designed by the UK Government Royal Signals and Radar Establishment and distributed by Praxis Systems plc) allow hierarchical decomposition of high-level behavioural models into their component parts at successively lower levels of hierarchy, while at the same time verifying that the decomposition accurately represents the model at each level. At some arbitrary level of hierarchy the systems engineer can make a determination that the behavioural blocks can be replaced by functional blocks. By this means, he or she can complete a formally verified design path from a top-level behavioural specification of a system to be detailed implementation of the system at the PCB and custom silicon level.

52.2.1.4 Simulation acceleration

Simulation of large designs is a CPU-intensive task. Frequently, engineers will use batch processing for simulations in order to take advantage of idle resources, for example overnight and at weekends. However, recent advances in workstation technology and in simulation software have given the engineer alternative methods of achieving simulation objectives in acceptable time scales.

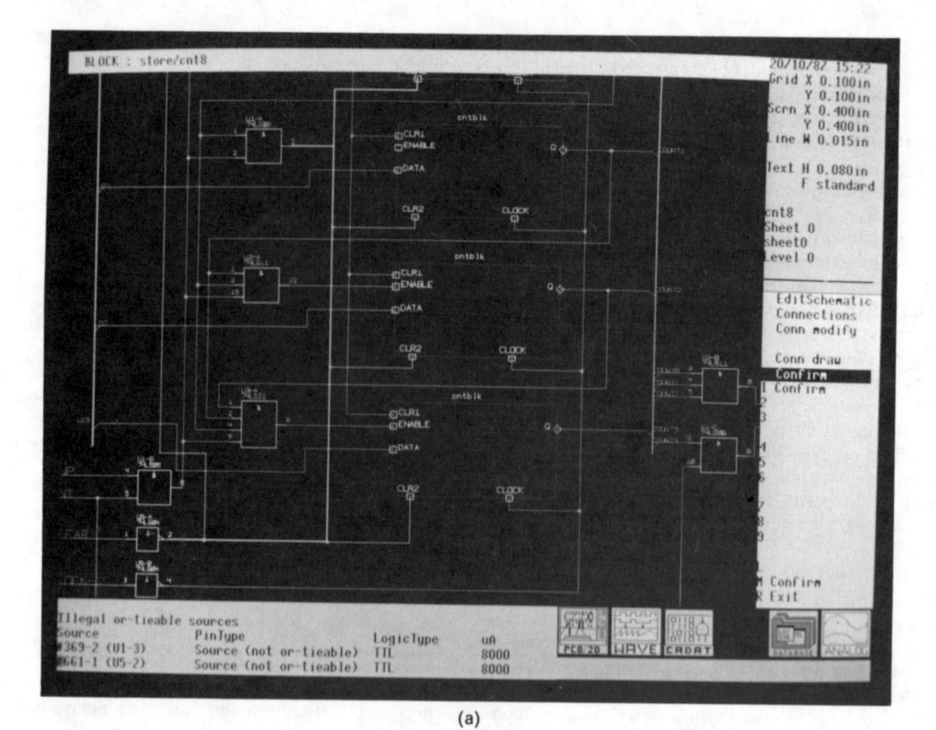

(a)

(b)

Figure 52.2 Graphical performance of workstations: (a) schematics capture, (b) simulator waveform display

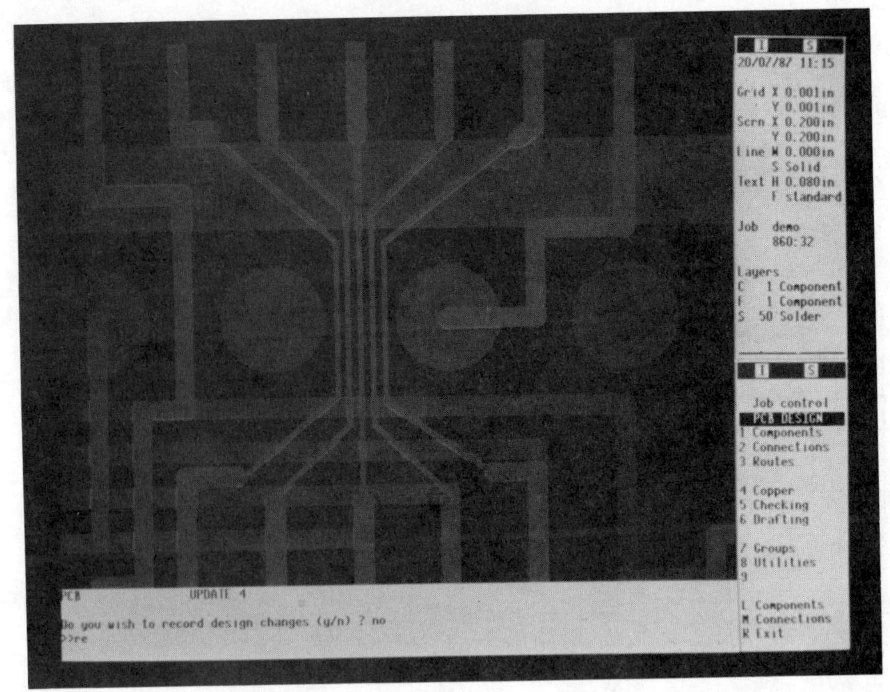

(c)

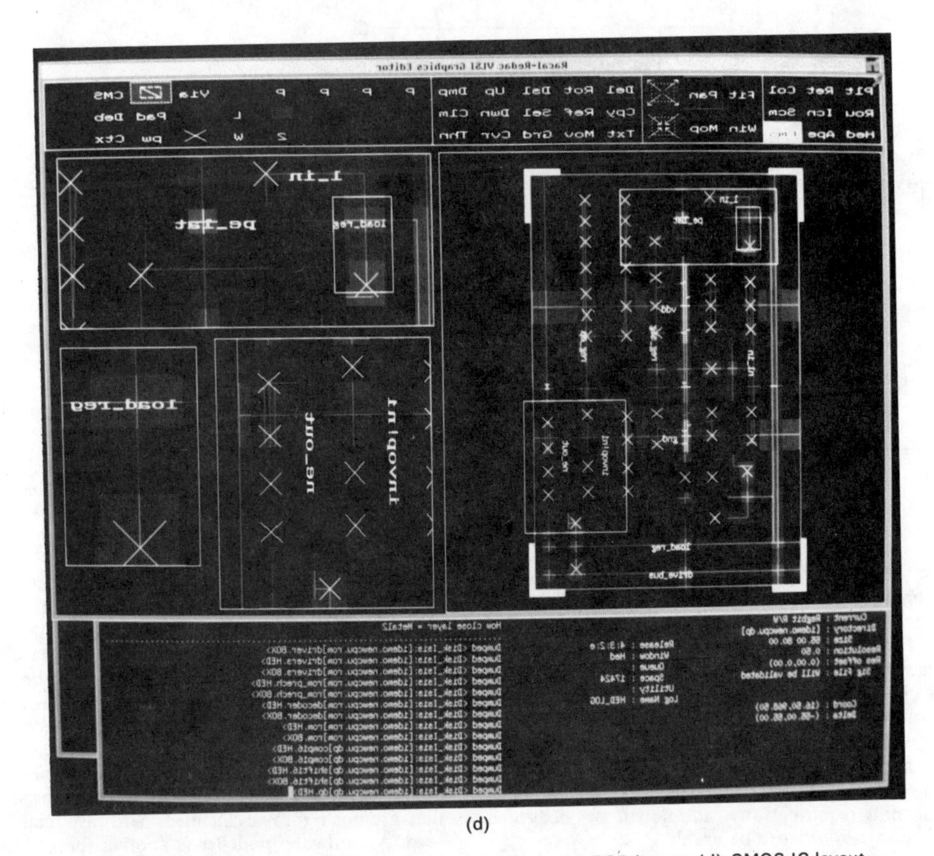

(d)

Figure 52.2 (*cont'd*) Graphical performance of workstations: (c) PCB layout, (d) CMOS IC layout (Courtesy of Racal-Redac Ltd)

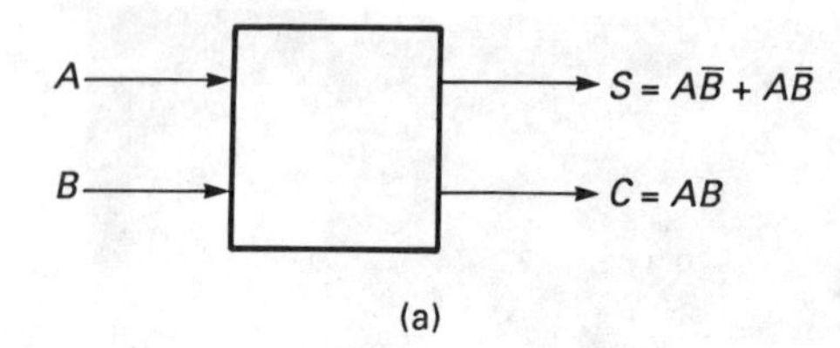

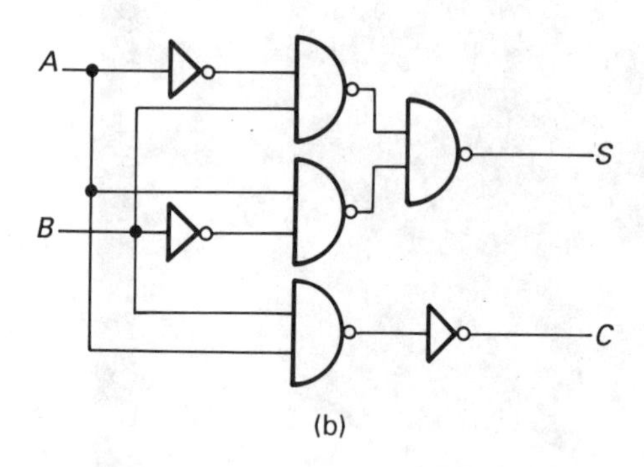

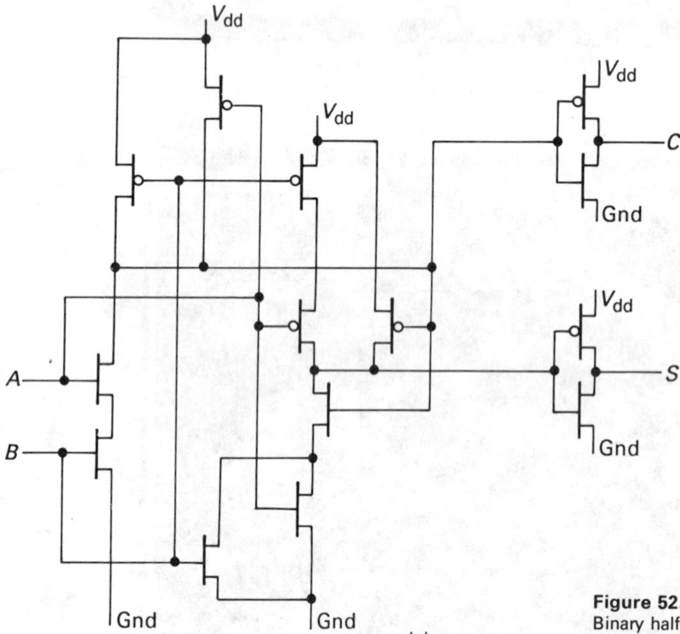

Figure 52.3 Explanation of behavioural and functional models. Binary half-adder: (a) behavioural model, (b) gate-level functional implementation, (c) circuit-level functional implementation

First, a technique exists for fault simulation which makes use of the fact that workstations normally exist in a networked environment. For example, the HHB Cadat 6 fault simulator includes a facility called Local Area Network Acceleration (LANA) which, transparently to the user, automatically schedules blocks of faults to the available idle resources on the network. This makes it possible to use all the computing power at all points on a network for a single simulation exercise driven by a single user seated at just one workstation. It has the clear advantage of not requiring any additional or dedicated computing hardware in order to be usable.

Second, acceleration can be provided by a dedicated resource on the network. This resource usually consists of some form of high-performance non-graphical computer in which the primitive Boolean elements used by the simulator exist as microcoded pseudo-instructions, or even as hardware elements within the CPU. Such accelerators tend to be specific to the particular simulator for which they were designed.

Third, since most PCB designs contain very large scale ICs (VLSI devices) such as microprocessors, memory controllers, UARTs, etc., for which detailed functional models in software are impracticable because of size, speed or the simple fact that they are not rigorously defined, hardware modellers are often used. A hardware modeller is another form of non-graphical, networked computing resource which uses the physical device rather than a software analogue to determine function. It is usual for a hardware modeller to be equipped with a number of slots to take cards, each carrying a different device.

HHB's CATS hardware modeller uses the physical device to determine function and a software shell to determine timing; thus the delays resulting from the physical interconnections to the models do not distort the timing simulation results. A network in which a CATS modeller is installed allows access for logic and timing simulation from anywhere in the network, even allowing the same models to be accessed by different simulation exercises; thus a number of users can make simultaneous use of the one hardware resource.

52.2.1.5 Analogue simulation

The development of analogue simulators has not until recently matched the development of digital simulators in terms of capability, largely because commercial considerations demanded that the bulk of the resource went into the area which covered most of the requirements, i.e. digital design. Most analogue simulators are based on Berkeley Spice, which contains passive and active component models but operates only in the time domain. The need to work in the frequency domain has largely been addressed by specialist active and passive filter design programmes which are not usually considered to be part of the CAD field.

Most board designs in industry are predominantly either purely digital, or digital with some analogue input and/or output. Apart from RF and microwave design, the key need for analogue simulation has been to accommodate line drivers and receivers, clock and clock extraction circuits, analogue-to-digital and digital-to-analogue converters and other self-contained, well defined functions.

Analogue simulators of a fairly general nature are now beginning to make inroads into the area of CAD. One of the most interesting is a product called Saber from Analogy Inc. It not only models electrical characteristics, it also allows modelling, for example of hydraulics, motors, optics and other technologies to which the analogue designer is frequently required to interface electrical designs. Saber also gives a simulation time which increases linearly (sometimes even sublinearly) with increasing complexity, thus overcoming one of the key objections to Spice derivatives, where the simulation time increases as the square of the size of the circuit.

Another requirement is to simulate both analogue and digital circuitry concurrently in the same exercise, while using the output of the analogue simulator as input for the digital simulator and vice versa. This is necessary because in most cases it is not realistic to treat the analogue sections of a design separately from the digital sections; usually, the operation of each area depends on the internal conditions of the other.

RF and microwave simulators tend to be very specialised. As the wavelength of the signals carried becomes a significant fraction of the size of the devices and of the boards on which they are mounted, the physical layout becomes a material source of the passive components of which the product is made. For that reason, r.f. and especially microwave simulators cannot be treated separately from the physical layout tools used to design the ICs and PCBs.

A treatise on the available simulation tools for electronic design is provided by part 2 of a three-part series on computer aided engineering published in *Electronic Design*.[2]

52.2.2 Computer aided design

52.2.2.1 Placement and routing of printed circuit boards

The physical design of a PCB (or, for that matter, of a hybrid or IC) consists fundamentally of two easily understood processes: first, placement (i.e. where to put each individual component) and, second, routing of the metal, polysilicon or diffusion paths which interconnect them.

The starting point for these processes is the netlist, either entered manually or compiled from a schematic capture system; it contains all the component names (library references) and all the necessary connectivity information.

The end-point is a set of data in a form readable by the manufacturing tools to be employed, representing the physical design of a PCB or IC, in which the individual elements are placed and interconnected in accordance with the netlist and with the rules of the manufacturing process. In the case of a PCB, these manufacturing rules are mainly spacing rules determined by photographic resolution and etching accuracy. In the case of an IC, additional effects such as diffusion spreading need to be taken into account, but the principle is essentially the same.

If a PCB or IC is to be designed with a minimal amount of wasted space, placement becomes an extremely complex issue. It needs to take into account the way the product will be routed once the components are placed. However, since no routing exists at the time of placement, any automatic placement software can only take account of a limited subset of the total list of constraints determining placement. Since these constraints may include not only proximity to already connected components but also (in the case of a PCB) thermal and r.f. emission concerns, physical limitations such as height within certain areas of the enclosure, spacing limitations imposed by the physical size of the jaws of any automatic component insertion machine to be used in manufacture and so on, most placement is performed using a combination of automatic and manual interactive methods. Fully automatic placement tools demand a knowledge-based or artificial intelligence system.

Thus, the better CAD tools for placement allow the user to perform the function in a manner which varies according to the type of board being designed and to the priorities attached to the specific placement constraints. The user will normally have available a number of integrated but distinct tools. For example, on a typical design the user may begin by placing manually those components which need to be in specific positions (edge connectors, power transistors, etc.), following that with the use of automatic placement tools to give equally spaced blocks of RAM or ROM devices in a suitable pattern to allow easy routing, followed by auto-interactive placement of random logic devices and other components, and finishing with manual placement of items such as decoupling capacitors and terminating resistors at strategic points on the layout.

As with schematics capture, a system that contains connectivity information at the placement stage can offer considerable advantages over one that does not. If the existence of intended connections can be shown on screen at the time of placement as a set of straight lines linking points to be connected, the user immediately gains some topological knowledge which will help to decide where a component should go; furthermore, if the system contains tools to calculate the total interconnection length for each component, if necessary weighted to allow for critical nets, the user has the knowledge to be able to make a very sound placement decision for each component during interactive placement.

Initial placement usually accounts for less than 20% of the task of designing a PCB. Routing is usually a much more time-consuming process.

Most CAD systems offer a number of routing tools, as they do placement tools, since no single procedure will guarantee a good result in every case. For example, Visula PCB from Racal-Redac contains an interactive manual router plus four automatic routers designed to cope with different applications. One is used to route regular structures such as memory, one to make a quick attempt at routing (primarily for the purpose of making an assessment of the ultimate routability of a given placement), one to perform the bulk of the routing task (as close to 100% routing as time constraints will allow) by means of a 'rip up and re-try' process, and one to apply a final manufacturing pass to a routed layout in order to reduce the number of individual track segments.

The design of a printed circuit board follows a set of rules determined in part by the manufacturing process to be used. Photographic resolution, etching accuracy and soldering technique (among other things) determine the minimum spacing between adjacent, unconnected areas of conductor on the board. Until recently, the development of PCB CAD tools kept pace with the advancing manufacturing technologies as they were adopted; today it is an unfortunate fact that the advanced manufacturing technologies available have overtaken the development of PCB CAD tools and therefore manufacturing capability is increasingly being determined by the capability of the CAD vendors' offerings.

One important reason for this is that the routing algorithms used by most vendors' products are derived from the Lee flood algorithm,[3] which depends upon being able to model the PCB as an array of cells, each of which is individually represented in memory by a single bit which is set once the cell is occupied. Multilayer boards will have a least 1 bit per layer for each cell. Routers using Lee-based algorithms are thus gridded; such routers cannot permit a resolution finer than the minimum grid setting allowed by the word length of the routing software and by the physical memory size of the computer in use. Routing software typically suffers severe speed degradation if the lack of sufficient memory causes paging, hence virtual memory is not a useful alternative to physical memory here. Some attempts have been made to improve this aspect of Lee-based router software but it is becoming evident that few further enhancements are possible.

As an example of the limitation imposed by a gridded router, consider a manufacturing technology which permits 0.060″ pad size and 0.008″ track width and spacing. Such a process allows two tracks between pins on a normal dual-in-line package (DIP) with 0.100″ pin spacing. A gridded system would require a grid no coarser than 0.002″ to accommodate such rules, since the distance from the drill position of the pad to the edge of the first track is 0.038″ as shown in *Figure 52.4*. For a 10″ × 8″ board this implies 5000 × 4000 cells, requiring 20 million memory locations for each layer.

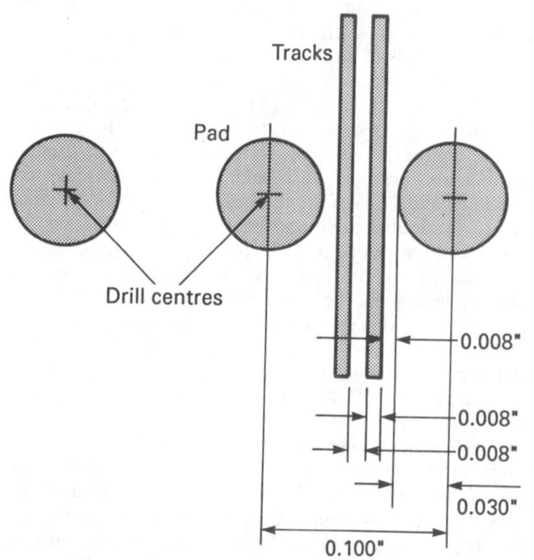

Figure 52.4 Effect of spacing rules on grid size for Lee-based automatic routers

New components, especially surface mount devices (SMDs), compound this problem further. They are frequently manufactured with a pad spacing specified in millimetres but required to share space on a PCB carrying components with pin spacings in imperial units. A 0.100″ pin spacing requires two places of decimals in millimetres (2.54) to avoid accumulation of rounding errors; thus if a metric grid is used it can be no coarser than 20 μm. The situation is even worse if an imperial grid is chosen.

There are routers which are truly gridless; their only resolution constraint exists by virtue of the finite word length (32 bits) used to represent position. The Visula Advanced Router is one such product. With the increasing use of SMDs and mixed metric/imperial units, such gridless routers are becoming increasingly important.[4]

52.2.2.2 Gate array design

Gate arrays are being used in increasing quantities to fulfil the needs for application-specific integrated circuits (ASICs) in a timely and efficient manner. 100000 gate designs are now possible using 1.2 μm CMOS technology, although 100–400 gates seems still to be the common range of complexity.

Gate arrays are supplied by a number of ASIC vendors in the UK, Europe, the USA and Japan. Most vendors regard their place and route capability as being highly proprietary and it appears to be one of the major differentiating factors between suppliers. Partly for that reason and partly because it is not common for gate array users to take overall responsibility for the correct operation of the finished devices, placement and routing are almost always performed by the gate array vendor according to a netlist supplied by the customer. In some cases, for example where specialist arrays are involved or where the user has an in-house gate array foundry, placement and routing may become part of the task facing the electronic engineer or PCB designer.

As far as the user is concerned, the process of designing a gate array is thus normally one of modelling and functional simulation using libraries (and commonly also simulation software) supplied by the gate array vendor. The netlist, together with a functional description, a performance specification, a set of test vectors and some packaging and pinout information, is presented to the gate array vendor as a full specification of the device. The vendor will normally repeat the functional simulation, then perform the physical design process (i.e. placement and routing) from which additional delay parameters (back annotation data) will be extracted for inclusion in a full logic and timing simulation. This will be used to verify the performance specification before manufacture commences. If any discrepancies are found during the process, the customer will be notified.

The placement and routing software used by gate array vendors is generally written to accommodate the style of base layers and library elements specific to the vendors' products. It differs from the type of software used for full custom silicon or PCB design in that the library elements themselves are in fact composed of metallisation patterns only, but in order to be usable they must be placed on the base layers in discrete locations where they make contact through the oxide cuts with the pre-existing silicon artefacts. For a given library element there are many different physical layouts possible; furthermore, each connection to a library element port can be taken from the metallisation pattern at any of a number of different points. Thus placement and routing of gate arrays are inextricably interwoven and there is vast scope for the development of very sophisticated software tools. It is common for gate arrays to utilise less than 50% of the available devices on the base layer, mainly as a result of the inability of current place and route software to make highly efficient assignments automatically.

52.2.2.3 Full custom silicon design

In this context, the words 'full custom' are intended to apply to the manufacturing process rather than the design process. Thus designs that make use of a standard cell library are full custom designs, since all the layers of the finished IC must be manufactured specifically for each design.

Few digital designs these days are produced without at least some use of standard cell libraries, although the tendency in the commodity IC market is still to use polygon editors (essentially, 2-D draughting tools adapted to silicon layout and containing no electrical connectivity in the graphical data) to design for maximum compaction on the chip. The design of an IC using a polygon editor requires the user to have a degree of skill which can only be maintained with constant practice. Since the polygon editor graphical output contains no electrical information until the layout is post-processed to extract connectivity and to check design rule violations, the designer has to be aware, during placement and routing, of all the spacing rules which apply between the same and different layers.

ASIC designers have led the commodity IC houses in the adoption of the latest generation of advanced hierarchical silicon editing tools, having found that the rigorous mapping of electronic design into physical silicon provides them with a secure and speedy route to 'correct by design' silicon, even if it tends sometimes to use more chip area for a given level of functionality than might be possible for a highly skilled designer using a polygon editor. These silicon CAD tools, in which electrical connectivity is maintained and displayed at the same time as the physical representation of the layer information in polygon form, provide the facility, as with connectivity-based PCB systems, of design rule checking as the design is built up on the screen. Racal-Redac's Visula silicon product provides such a facility and has been used by systems designers with little or no IC design experience to design CMOS ASICs with over 250 000 transistors, meeting both the functional and the performance specifications on the first delivered chips. References 5 and 6 are excellent treatises on the design methodology of full custom CMOS VLSI and on the design of CAD tools for VLSI respectively.

52.2.3 Computer aided manufacturing and test

52.2.3.1 Outputs to manufacturing

There are three essential functions carried out in the manufacturing environment which require data and information from the design department. These functions are assembly, test, and the manufacturing control and management system.

All three are areas in which significant investments in automation have already been made by most large companies, but all too often the expected returns are not being realised. The reason is that equipment and software are typically installed in an uncoordinated manner; an attempt to satisfy each function has been made in an isolated manner, leading to the existence of 'islands of information'. Significant manual data entry and data translation work must be carried out in order to gain any benefit from such isolated systems.

A number of electronic design automation systems claim to be capable of providing suitable direct data transfer to manufacturing, but their capability is often more limited than it first appears. The following attributes are required of a system in order for it to be able to deliver the hoped for advantages of automated manufacture across a broad spectrum of possible user needs:

(1) A relational database.
(2) An open software architecture.
(3) Generic, user-definable post-processing.
(4) Comprehensive, standard communications facilities.

The relational database is mandated by the need to be able to extend data and add relations in order to be able to accommodate information needed for manufacturing that was not included at the electronic design stage, as described in *Section 52.3.1*.

The open software architecture is needed because modification of the software itself is often necessary in order to accommodate additions to the data.

By offering generic post-processors to produce drive data, CAD system vendors can provide an economical means of allowing for the multitude of different formats which exist for manufacturing machines. By allowing the post-processor to be modified in an easy and secure manner by users, they remove the dependency of the user on the vendor to supply the tools and, above all, to maintain them when formats change.

Automated assembly plants will generally make use of guided manual or robotic assembly equipment and automatic component insertion or 'onsertion' (a word coined to describe the process of placing surface mounted components on a board or hybrid prior to soldering) tools. For these equipments the data required from the design system will include a list of the components used in the design, relevant data regarding part types, values, positions, heights and pin or lead positions, all in a form suitable for use by the front-end software of the assembly tools. The purpose of transferring data directly between systems is to avoid the laborious and error-prone task of digitsing component positions from PCB artwork.

Test departments generally use in-circuit or functional testing, or both. For in-circuit test equipment, the data requirements are similar to those of the assembly equipment, with the addition of netlist information in the form of node identification for the test points allocated to component pads.

Functional test equipment requires additional information in the form of test vectors and fault dictionaries as described in *Section 52.2.1.2*. This represents a considerable burden, sometimes an impossible task, for the test engineer in cases where the initial design of the product has not taken testability into account. Certain states may arise only as a result of fault conditions not exposed by good circuit simulation, and therefore fault simulation leading to as high a fault coverage percentage as is economically possible should always be a part of the engineer's circuit design function. Often, fault coverage analysis will be used to define test points for the functional test equipment additional to the board's edge connectors. If this function is carried out efficiently, a very few extra points can sometimes be used to increase coverage dramatically for a given total test time, or to shorten the test time for a given desired coverage percentage.

Manufacturing control and management systems need the basic component list information, i.e. a bill of materials, which, in order to be used effectively, will contain component attributes accessible from the design database. This implies extendability of databases. The information needed is likely to include issue numbers, supplier numbers and alternative suppliers, alternative part numbers, stores bin numbers, standard costs, delivery times, mean time between failure (MTBF) figures, and a host of other information of interest for manufacturing and management purposes.

52.2.3.2 Computer integrated manufacturing

The term 'Computer Integrated Manufacturing' is used here to describe a company-wide concept of automation which permeates all the relevant company functions from parts procurement, through manufacturing and test, to the shipment of the final product, including all the management control systems this implicates. The intent is to automate in such a way as to make the goals of high product quality and profitability and short product development times a reality.

No CIM system is available off the shelf from any vendor of electronics CAD tools, nor does any vendor offer the possibility of supplying a standard CIM system in the foreseeable future. Rather, it is the initiative of the user company to inculcate a CIM philosophy, which emphasises the need for integration to be considered in detail before significant commitments are made in individual departments.

Usually, however, some procurement will already have taken place. As a result, a mixture of different systems and standards is likely to evolve, and it is the ability of the different systems to communicate effectively within a heterogeneous environment which is paramount to the selection of tools.

CIM strategies, because of the extent of the applications, require very large investment decisions to be made. It is easy to lose sight of the investment value of the product design and test data in evolving such a strategy; it is this very data, however, which the CIM approach should be designed to protect. *Figure 52.5* gives some idea of the possible scale of data management within a CIM system. Reference 7 offers further explanation of the scope and implementation of CIM systems.

52.3 Architecture of CAD systems

52.3.1 Databases

In a manufacturing environment where CAD tools are routinely used for the design and documentation of electronic products, the value of the data held by the design system frequently outweighs that of the investment in computing hardware and software. Hence the method by which the product data is held by the CAD system, i.e. its database system, is worthy of detailed consideration.

First, duplication of data is highly undesirable. A good CAD system will be based upon a common database which extends across all the applications: engineering, CAD, manufacture and test. It will contain all the library and design data used in every application and in every location, both by the CAD system itself and by the ancillary systems, such as board profilers and drilling machines, automatic component insertion machines, and functional testers.

The common database will enable proper data sharing between processes and will facilitate proper communications between design, physical layout, manufacture and test. The importance of these cannot be underestimated; today's designs are becoming so complex that any requirement for manual

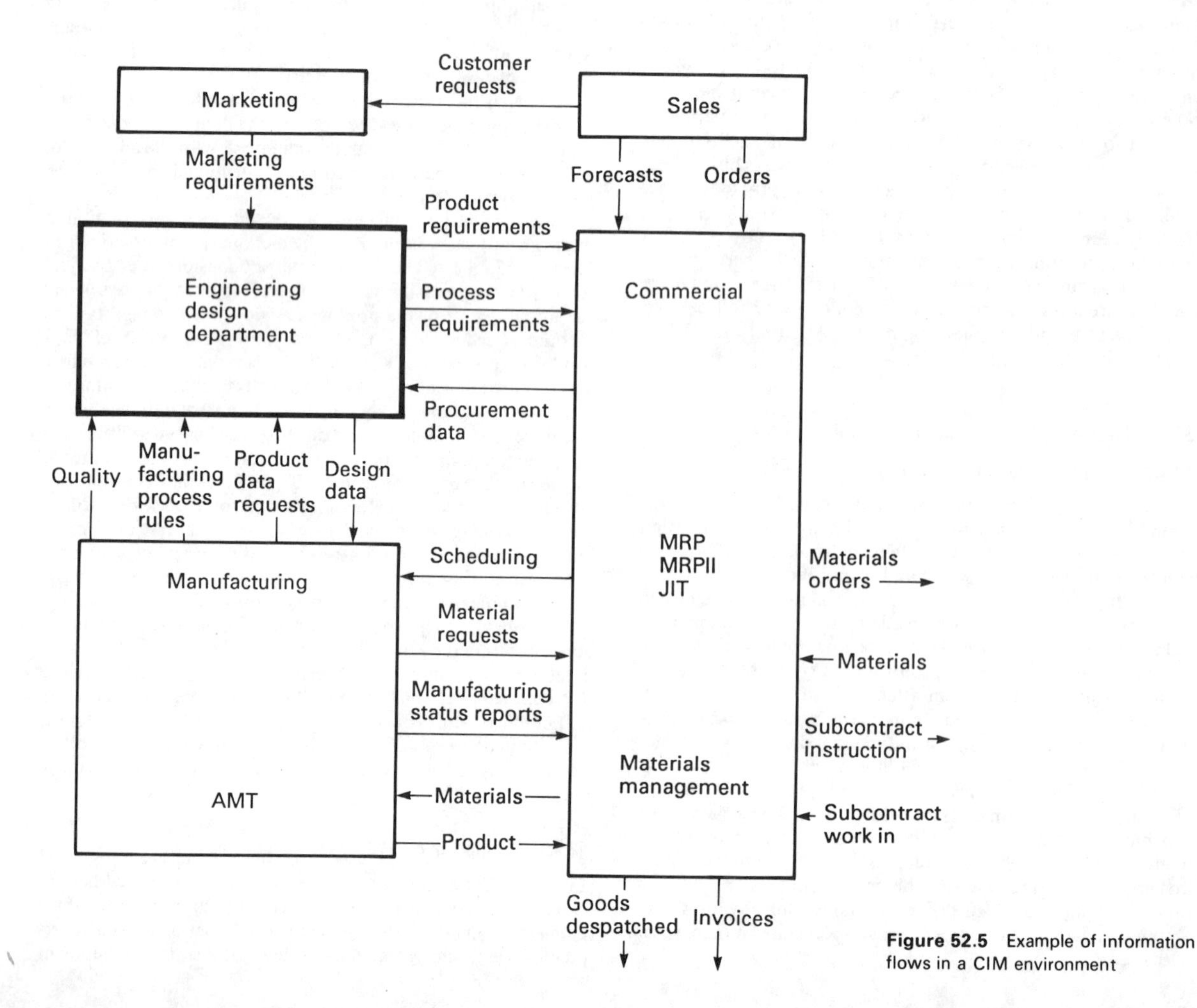

Figure 52.5 Example of information flows in a CIM environment

intervention in configuring data files for transfer from one user to another or from one application to another carries with it risks of error.

The usual mechanism for handling large amounts of data in a systematic manner while retaining fast access is to use a relational database. Commercially supported products such as Informix are designed to run on networked computer systems and can offer standard access mechanisms such as Structured Query Language (SQL) which effectively hide the details of the data management system from the casual user.

52.3.2 Openness and integration facilities

It is commonly desirable to be able to add software developed in house or supplied by a third party to an existing CAD system, and it is instructive to examine the needs for integration and the problems facing the systems manager attempting such a task.

Systems supplied from a single source are generally designed to perform a limited set of tasks, for example CAE and simulation or PCB design. As users begin to realise the value of systems and begin to replace existing work practices with management and control procedures that depend on the use of computer-based tools, there is generally a desire to automate the data interfaces between the various departments which make up the manufacturing operation. These have been dealt with in *Section 52.2.3.1*. However, even within the design environment itself, it is often desirable to be able to interface software tools supplied by third parties.

As examples of this, a common need is to interface to mechanical CAD tools used for package design. Not only do these tools need board 2-D information, they also need parts library access in order to gain knowledge of the heights of components. The interface is not only in the direction of electronics to mechanical CAD; placement of components on a board may be determined by headroom or airflow considerations of which the mechanical CAD system has knowledge.

There are existing and emerging standards for many interfaces, but by no means all. As with all data transfer, the standards cover both the physical communications mechanisms and the format of the data. Standards are dealt with in more detail in *Section 52.3.5*.

In the case where standards do not exist it is important that systems should offer mechanisms for easy access to the data they contain (for example, by the use of Structured Query Language), not only for the purpose of extracting data but also for the purpose of adding data and in particular for the purpose of adding extra fields and relations to data already existing. This last requirement arises as a result of the need to be able to interface software that uses information about library parts and designs and which is not needed by the CAD process, for example, the parts heights and power dissipations alluded to above.

Many CAD products were not initially conceived with this degree of integration in mind, and as a result do not offer the openness which might be considered desirable. Many CAD systems are designed to run on proprietary hardware platforms, or to use proprietary graphics processors or operating systems, or worse, to use proprietary database schemes which are not amenable to extension in order to be interfaced to alien applications and software. Since user's design data is the one part of the CAD investment which grows in value, and since the value depends upon its ability to be accessed and manipulated, such closed systems are rapidly becoming obsolescent.

52.3.3 Software portability and extendability

In recent years the emergence of the technical workstations, the UNIX operating system and the adoption of the C programming language have contributed to the ability of software vendors to make their products hardware independent to a greater degree than hitherto. This has obvious benefits for the software vendors; the software does not have to be written off when the computer manufacturer withdraws his product. However, software portability does not only benefit the supplier. By virtue of the fact that suppliers are no longer concerned with short product lifetimes they are able to invest more in the product's development, with consequent benefits for the user. More importantly, portability means that the software can be available for a new model of computer or workstation within a very short time of its introduction, an important fact when the rapid development of technical workstations is taken into account. While workstation processing power approximately doubles every year, the price of workstations steadily falls.

52.3.4 Distributed applications

The extensive and rapidly growing use of technical workstations for CAD applications has put computing power on the desks of the engineers and designers instead of as previously in a central computing resource under the control of a computer manager. It has also advanced the networking capability of systems; high-speed coaxial and even optical fibre systems offer 100 or more times the bandwidth of the old twisted pair standards such as RS232C and V.24. These networking capabilities are used by the operating systems to provide remote disk access in such a way that the user no longer needs to know where the data physically resides on the system; there is no need for periodically updated local copies to be kept by each user, and common databases do not limit access speed to any significant extent.

CAD applications software can be written in such a way as to make good use of the 'transparent' quality of workstation networks. Some of the tasks undertaken, especially in simulation and in routing, demand very high computing power in order to deliver good results in reasonable time scales. Although workstation power is growing rapidly, applications power requirements seem to be growing at the same rate or even faster.

The main task which can benefit from concurrent processing (i.e. from the use of a number of CPUs operating simultaneously, using the same instructions on different parts of a task) is fault simulation. The local area network acceleration process used by Cadat 6 has been described in *Section 52.2.1.4* and is very effective in reducing time scales for complex simulations, without the need to add applications specific hardware accelerators.

Some tasks are not generally amenable to speed improvements through concurrent processing. Automatic routing is one example of an application which is normally not written to take advantage of concurrency, rather to make use of general-purpose non-graphical high-power computing resources such as reduced instruction set (RISC) machines. By placing such a machine on a network and invoking a management and control system, which allows the application to be accessed from any point, the benefits of high-speed networking can be made to apply even to those tasks which cannot make use of concurrent processing.

It is likely that new router implementations will in fact allow concurrency to be applied to automatic routing tasks, particularly when the Lee routing algorithm is replaced; there is no fundamental reason why automatic routing should remain serial in nature.

52.3.5 Industry standards

In the area of computer aided design generally, certain standards have emerged. Some of these are the result of the determined efforts of international committees and working parties to provide useful standards; others have arisen as a result of less formal processes and have been adopted through popular use.

The following are some of the standards commonly used:

(1) Operating systems:
Berkeley UNIX
Digital Equipment Corporation VMS
(2) Programming languages:
C (especially for graphics applications)
Pascal (mainly for simulation software and functional modelling)
(3) Network communications:
Xerox Corporation Ethernet
(4) Database access:
IBM Corporation SQL (Structured Query Language)
(5) Data transfer formats:
RINF (Redac Interface Neutral Format) (sponsored by Racal-Redac Ltd and supported by Daisy Systems Inc., Mentor Graphics Inc. and Valid Logic Inc.)
EDIF (Electronic Design Interface Format) (sponsored by Motorola Inc., National Semiconductor Inc. and Texas Instruments Inc.)
IGES (Initial Graphic Exchange Specification) (sponsored by the US National Bureau of Standards)
IPC350 (Interconnection and Packaging of Circuits) (US Government sponsored)
MAP (Manufacturing Office Protocol) (sponsored by General Motors Inc. and Boeing Industries Inc.)
TOP (Technical Office Protocol) (sponsored by General Motors Inc. and Boeing Industries Inc.)

References

1 MEAD, C. and CONWAY, L., *Introduction to VLSI Systems*, Addison-Wesley, Wokingham (1980)
2 SCHINDLER, M., 'Demands on simulation escalate as circuit complexity explodes', *Electronic Design*, October, 11–32 (1987)
3 LEE, C., 'An algorithm for path connections and its applications', *IRE Trans. Electronic Computers*, September, 346–65 (1961)
4 FINCH, A. C., MACKENZIE, K. J., BALSDON, G. J. and SYMONDS, G., *A Method of Gridless Routing of Printed Circuit Boards*, Proceedings of the 22nd Design Automation Conference, 1985, pp. 509–15 (1985)
5 WESTE, N. and ESHRAGHIAN, K., *Principles of CMOS VLSI Design*, Addison-Wesley, Wokingham (1985)
6 RUSSEL, G., KINNIMENT, D. J., CHESTER, E. G. and McLAUGHLAN, M. R. (eds), *CAD for VLSI*, Van Nostrand Reinhold, New York (1985)
7 KOCHAN, A. and COWAN, D., *Implementing CIM*, IFS Ltd (1986)

Further reading

CONNER, M. S., 'ASIC simulators', *EDN*, 4 February (1988)
CUTHBERTSON, A., 'PCB thermal analysis on the pc', *Electron. Design Automation*, September/October (1988)
FILSETH, E., 'Spice extensions dynamically model thermal properties', *EDN*, 14 April (1988)
FINCH, J., 'Current trends in fault simulation techniques', *Electron. Product Design*, May (1988)
GABAY, J., 'Circuit simulators conquer new domains', *Comp. Design*, 15 November (1987)
GOERING, R., 'Silicon compilers bridge gap between concepts and silicon', *Comp. Design*, 1 November (1987)
GOERING, R., 'Design tools advance to keep pace with system complexity', *Comp. Design*, December (1987)
GOERING, R., 'A full range of solutions emerge to handle mixed-mode simulation', *Comp. Design*, 1 February (1988)
GOERING, R., 'Logic synthesis tools forge link between behavior and structure', *Comp. Design*, 1 June (1988)
GOODENOUGH, F., 'Complex analog–digital circuits call for fast simulators', *Electron. Design Int.*, December (1988)
HARDING, B., 'Vendors look outside to strengthen their simulation environments', *Comp. Design*, July (1988)
JOHNSON, P., 'Mixed level design tools enhance top-down design', *Comp. Design*, 1 January (1989)
LONG, S. D. *et al.*, 'A team approach to pcb development', *Electron. Product Design*, April (1988)
MORALEE, D., 'Redressing the LSI balance with CAD', *New Electron.*, June (1988)
MOSLEY, J. D., 'PC-based CAE tools', *EDN*, 8 December (1988)
NEILL, T., 'Make the most of your investment', *Eng. Comp.*, January (1989)
QUIRK, K. F., 'Choosing a pc-based CAD package requires careful scrutiny', *EDN*, 8 December (1988)
RIZZATTI, L., 'Worst-case timing analysis ensures board reliability', *Comp. Design*, 15 November (1987)
SCHEIBER, S. F., 'Linking CAD with test generation', *Test Measure. World*, June (1988)
SOUTHARD, J. R., 'CAE software uses algorithms instead of schematics', *EDN*, 4 August (1988)
WADSWORTH, B., 'The role of personal workstations in gate array design', *Electron. Eng.*, November (1987)

53 Television and Sound Broadcasting

K R Sturley, BSc, PhD, FIEE, FIEEE
Telecommunications Consultant
(Sections 53.1–53.7)

P Hawker
Formerly, Independent Broadcasting Authority
(Sections 53.8–53.14)

J D Weston, BSc(Eng)
STC Technology Ltd
(Section 53.15)

C R Spicer, DipEE, AMIEE
BBC Design and Equipment Department
(Sections 53.16–53.21)

J L Eaton, BSc, MIEE, CEng
Consultant (Broadcasting)
(Sections 53.22–53.26)

Contents

53.1 Introduction

Broadcasting is concerned with the generation, control, transmission, propagation and reception of sound and television signals. Sound broadcasting with its relatively small bandwidth presents fewer technical problems than television. Adequate reproduction of speech and music can be achived within a frequency range from about 100 Hz to 8 kHz, but high-fidelity monophonic and stereophonic programmes require a bandwidth from about 30 Hz to 15 kHz. Television broadcasting generates a signal to control the light output of the receiver picture tube and this covers a frequency range of the order of d.c. to 5.5 MHz for 625 line interlaced scanning. This means that the carrier frequency must be much greater than 5.5 MHz.

53.2 Carrier frequency bands

The carrier frequency bands allocated internationally to sound and television broadcasting are shown in *Table 53.1*.

Propagation may be accomplished by means of a ground wave, sky wave or space wave.

Table 53.1 Carrier frequency band allocation

Frequency	*Wavelength*	*Frequency range*	*Purpose*
Low	Long wave	150–285 kHz	Sound
Medium	Medium wave	525–1605 kHz	Sound
High	Short wave	Bands approx. 259 kHz wide located near 4, 6, 7, 9, 11, 15, 17, 21, 26 MHz	Sound
Very high	Band I	47–68 MHz	Television
	Band II	87.5–100 MHz	FM Sound
	Band III	174–216 MHz	Television
Ultra-high	Bands IV and V	470–960 MHz	Television

53.3 Low-frequency propagation

The low frequency range is propagated by the ground wave which follows the contour of the earth; signal strength tends to be inversely proportional to distance, the losses in the ground itself being small, and reception is possible at considerable distances though marred by noise interference. Any sky waves are absorbed in the ionospheric layers.

53.4 Medium-frequency propagation

The medium frequencies are propagated by ground wave during the day, any sky wave being absorbed by the ionospheric *D* layer. The ground wave generates currents in the ground, and the energy loss is greatest over poor conductivity ground such as granite rocks, and is least over sea water, where propagation approaches the inverse square law. The greater the frequency the greater is the loss at a given distance. Curves of ground wave attenuation with distance for sea water and ground of good, moderate and poor conductivity for a 1 kW transmitter and a zero-gain aerial are published by the CCIR.[1] These permit the field strength (referred to 1 μV/m as zero level) of low and medium frequencies to be calculated for distances from 1 to 1000 km and are reproduced in *Figures 53.1* to *53.4*. When using these curves the following points should be especially noted.

(1) They refer to a smooth homogeneous earth.
(2) No account is taken of tropospheric effects at these frequencies.
(3) The transmitter and receiver are both assumed to be on the ground. Height-gain effects can be of considerable importance in connection with navigational aids for high-flying aircraft, but these have not been included.
(4) The curves refer to the following conditions:
   (a) They are calculated for the vertical component of electric field from the rigorous anlysis of van der Pol and Bremmer.
   (b) The transmitter is an ideal Hertzian vertical electric dipole to which a vertical antenna shorter than one quarter wavelength is nearly equivalent.
   (c) The dipole moment is chosen so that the dipole would radiate 1 kW if the Earth were a perfectly conducting infinite plane, under which conditions the radiation field at a distance of 1 km would be 3×10^5 μV/m.
   (d) The curves are drawn for distances measured around the curved surface of the Earth.
   (e) The inverse-distance curve *A* shown in the figures, to which the curves are asymptotic at short distances, passes through the field value of 3×10^5 μV/m at a distance of 1 km.
(5) The curves should, in general, be used to determine field strength only when it is known that ionospheric refelctions at the frequency under consideration will be negligible in amplitude—for example, propagation in daylight between 150 kHz and 2 MHz and for distances of less than about 2000 km. However, under conditions where the sky wave is comparable with, or even greater than, the ground wave, the curves are still applicable when the effect of the ground wave can be separated from that of the sky wave, by the use of pulse transmissions, as in some forms of direction-finding systems and navigational aids.

As an example of the use of the curves, consider a 100 kW transmitter feeding into a 0.4 λ medium wave aerial at 1 MHz. The field strength at a point 50 km away over ground of medium conductivity from a 1 kW transmitter feeding a very short aerial is, from *Figure 53.3*, 53 dB (ref. 1 μV/m). The 0.4λ aerial has a gain of about 1 dB and the transmitter is +20 dB with reference to 1 kW, so that the field strength is 74 dB or 5.5 mV/m. When the signal path conductivity varies, field strength can be calculated by changing from one curve to another. Suppose the receiving point for the above were moved to 100 km with the additional 50 km over ground of poor conductivity. *Figure 53.4* shows that the loss from 50 km to 100 km at 1 MHz is $(40-26)=14$ dB, so the field strength at 100 km would appear to be $74-14=60$ dB (1 mV/m). It has been found in practice that the actual field strength is always less than this and that a more correct value is obtained by determining the field strength with the transmitter and receiver positions interchanged, and then taking the average of the direct and reversed dB values. Thus for the transmitter at the receiving site *Figure 53.4* shows a field strength at 50 km of 40 dB for 1 kW, i.e. a field strength for the 100 kW transmitter and 1 dB gain aerial of 61 dB. The attenuation from 50 to 100 km from *Figure 53.3* is 53 dB to 38 dB, or 15 dB. Hence by this reciprocal method field strength would be 46 dB. Average field strength is $(60+46)/2=53$ dB, and this is taken as the correct estimated field strength.

Sky wave propagation occurs from the ionospheric *E* layer after nightfall due to the disappearance of the absorbing *D* layer. This can be a serious disadvantage because it leads to greatly increased service area often well beyond national boundaries, causing interference in the service areas of distant transmitting stations using the same or adjacent frequencies. Sky wave propagation is subject to considerable fading due to changes in

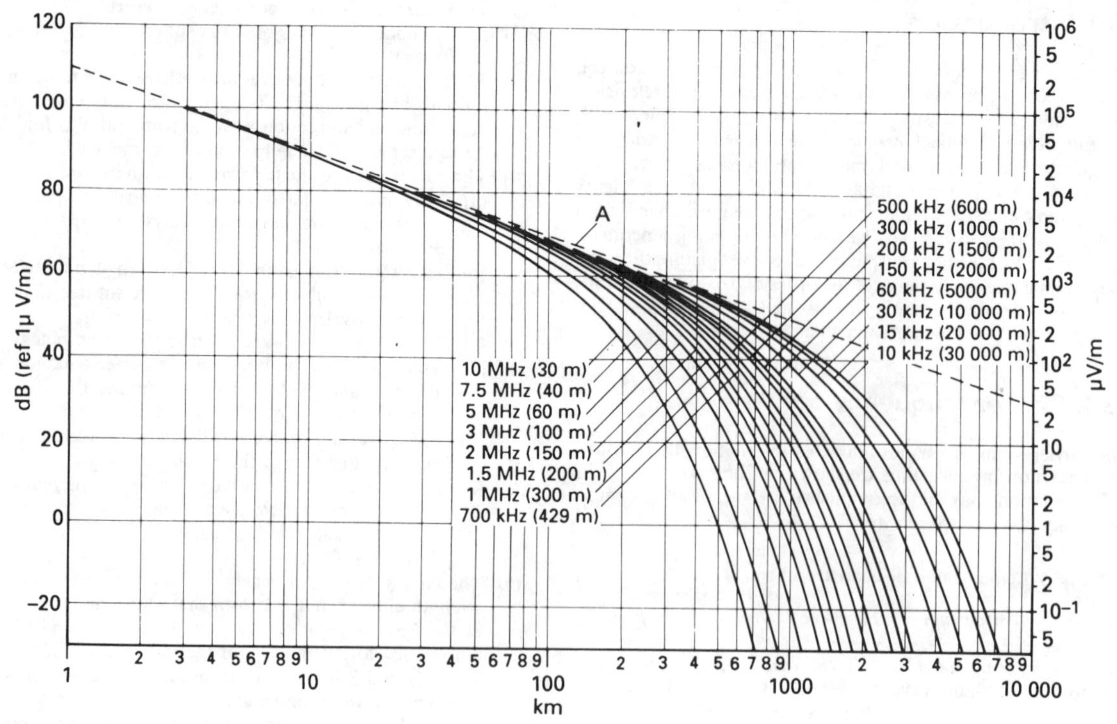

Figure 53.1 Ground-wave propagation curves over sea, $\sigma=4$ mho/m, $\varepsilon=80$ (Courtesy of CCIR)

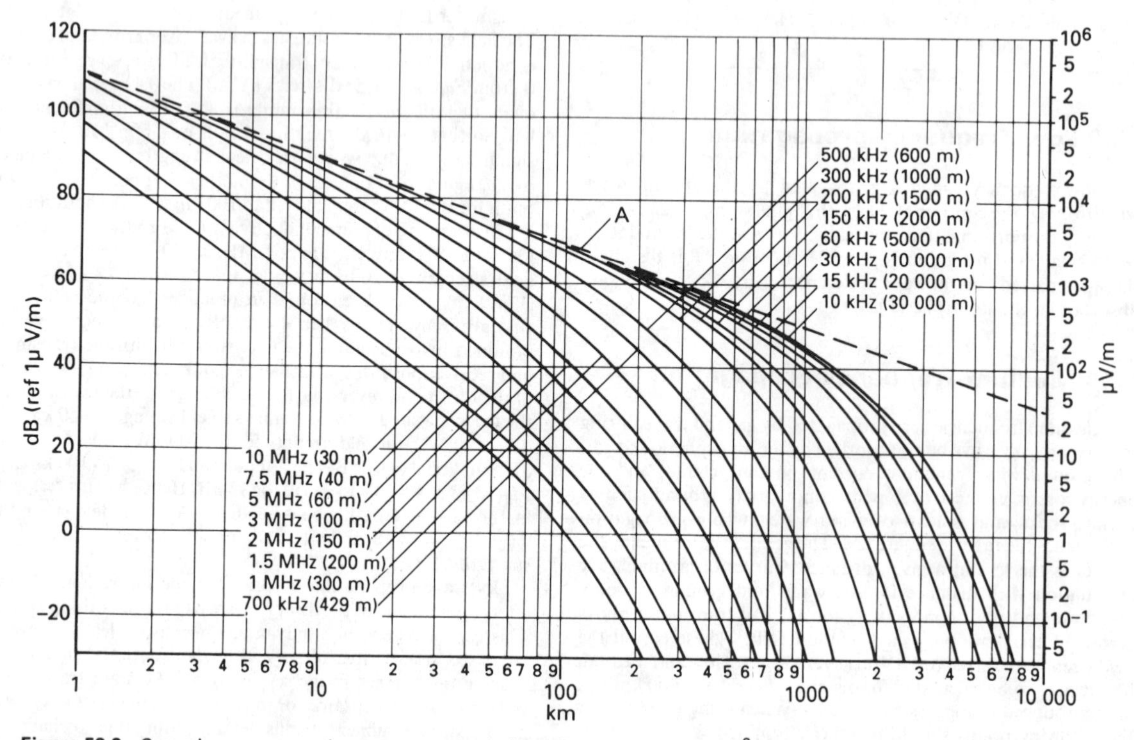

Figure 53.2 Ground-wave propagation curves over land of good conductivity, $\sigma=10^{-2}$ mho/m, $\varepsilon=4$ (Courtesy of CCIR)

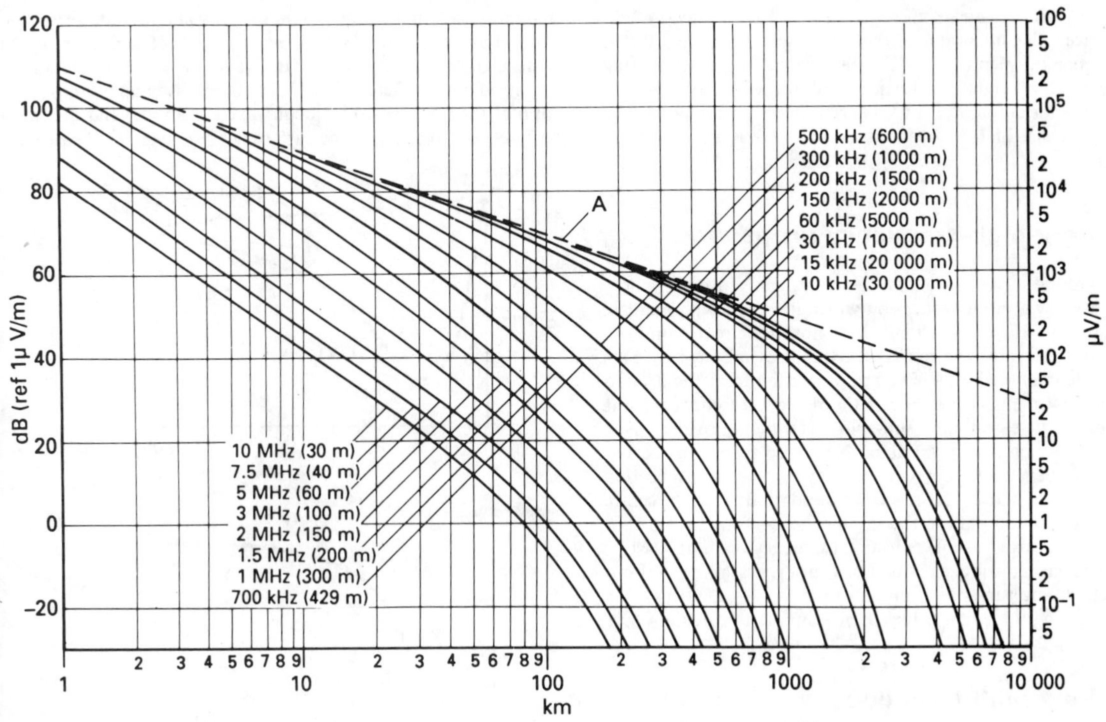

Figure 53.3 Ground-wave propagation curves over land of moderate conductivity, $\sigma=3\times10^{-3}$ mho/m, $\varepsilon=4$ (Courtesy of CCIR)

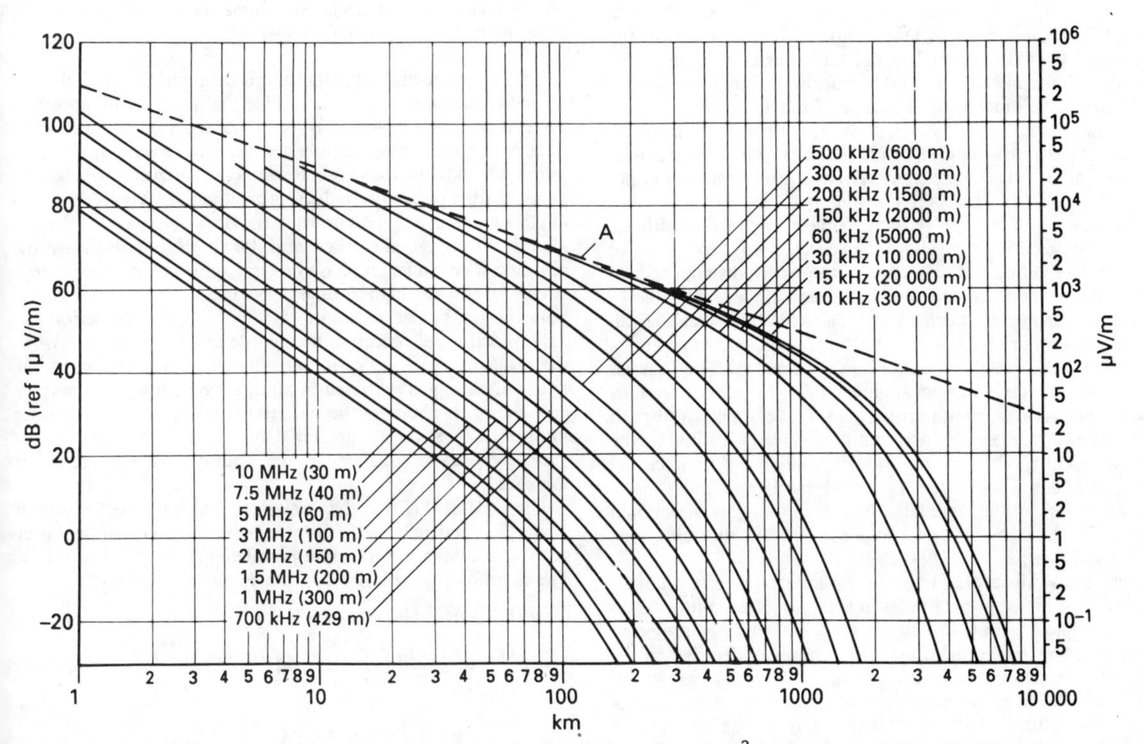

Figure 53.4 Ground-wave propagation curves over land of poor conductivity, $\sigma=10^{-3}$ mho/m, $\varepsilon=4$ (Courtesy of CCIR)

the *E* layer, and at a distance of 300 km to 400 km it may produce a field strength comparable to that of the ground wave. Mutual interference between the two is then greatest and severe fading can occur. The height of a vertical medium-wave aerial largely determines the amount of sky wave reflected by the *E* layer. The greater the height (up to about 0.6 λ) the less the sky-wave reflection.

53.6 Very-high-frequency propagation

As the ground wave from a high-frequency transmitter is rapidly attenuated by ground absorption which increases with increase of frequency, HF propagation is by ionospheric reflection from the *E* and *F* layers. The range of frequencies reflected by each layer is determined by the angle of projection from the ground (the shallower this is the greater the reflection frequency), the time of day, season and sun-spot activity. Above a given frequency, which is very variable, there is no reflection because the wave penetrates beyond the layer. Highest reflection frequency from *E* and *F* layers occurs about midday at the centre of the propagation path in the summer season and at sun-spot maximum. Greatest usable reflection frequency is about 30 MHz from the *F* layer during maximum sun-spot years falling to about 20 MHz during sun-spot minimum years.

Field strength median values are amenable to calculation.[2]

53.6 Very high frequency propagation

Very high frequency propagation[1] is achieved by the space wave; near to the ground energy is quickly lost but the effect disappears with increase of height above ground. Thus, at heights of 2 λ or greater, the space wave is little affected by ground losses. Some bending of the VHF wave occurs in the troposphere, but satisfactory propagation is normally limited to line-of-sight. An intervening hill produces a radio shadow on the side furthest from the transmitter, with much reduced field strength in the shadow. The VHF signal has the advantage of much lower impulsive noise interference and it is normally unaffected by the vagaries of the ionosphere. However, fixed or moving objects comparable or greater in size than the VHF wavelength can act as reflectors to cause interfering wave patterns. A fixed object may increase or decrease the field strength at a given receiving point, depending on the phase relationship between the direct and reflected wave. A moving reflecting object, such as an airliner, causes the phase relationship of the reflected wave to vary, and considerable fading and distortion of the received signal occurs.

With a.m. sound signals the change in field strength due to interference from the reflected wave of a fixed object such as a tall building or water tower would not be serious, but when f.m. is used the phase relationship of the reflected varying carrier frequency is constantly changing and this causes serious distortion (known as multipath) of the sound content. Multipath a.m. television signals produce multiple images (ghosting) on the receiver picture tube due to time delay differences between the received signals. The smaller size of VHF receiving aerials (due to the shorter wavelength) permits the design of highly directional aerials, which can be angled to reduce the undesirable reflections from fixed objects.

The energy from a VHF transmitting aerial is projected at other angles as well as horizontal; some is projected upwards to penetrate the ionosphere and be lost, but some is projected downwards to the ground. Part of the energy is absorbed in the ground at the point of incidence but a good deal may be reflected to cause interference at the receiving aerial in the same manner as reflections from a fixed object. The extent of the interference depends on the amplitude and phase of the reflection coefficient and this is determined by the conductivity and permittivity of the ground at the point of reflection, the angle of incidence to the surface and the polarisation of the wave. Typical examples of the variation of the reflection coefficient of relative magnitude (ρ, the ratio of the reflected to incident wave amplitude) and phase (ϕ) with respect to angle to the surface (ψ) are given in *Figure 53.5*.

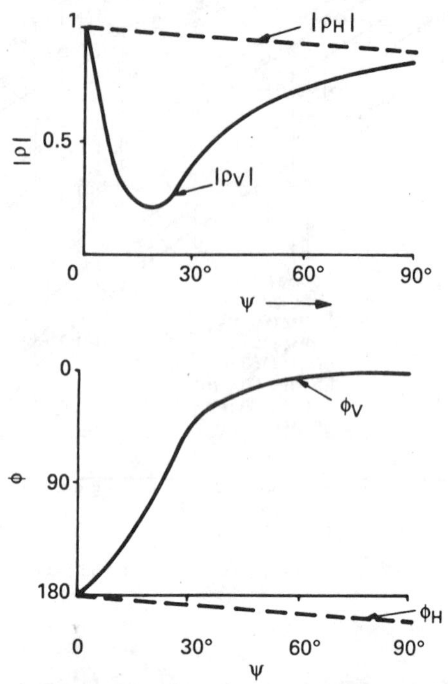

Figure 53.5 Reflection coefficient relative magnitude ρ and phase ϕ at different angles of incidence ψ to the surface (*V* is vertical polarisation and *H* horizontal polarisation)

The reflection coefficient magnitude is maximum at low values of ψ, i.e. at glancing incidence, and for the horizontally polarised wave decreases slowly as ψ is increased. For the vertically polarised wave the magnitude decreases relatively rapidly to a minimum and then rises again towards the horizontally polarised value. With a perfectly conducting ground the magnitude would fall to zero at an angle ψ known as the Brewster angle. At the low angles of surface incidence and for both polarisations the reflected wave suffers a phase reversal, which changes only to a small extent at higher angles of incidence for horizontal polarisation, but decreases rapidly for vertical polarisation.

High values of reflection magnitude are only obtained when the reflecting surface is smooth, for when this is not so the reflected energy is scattered in all directions. A surface may be considered as rough if the variations in height of the surface multiplied by sin ψ are greater than $\lambda/8$, so that a land or sea surface appears rougher as the transmission frequency is increased.

The total field (E_t) at the receiving aerial is the vector sum of direct (E_d) and reflected (E_r) waves with the phase of the latter suitably modified by the extra path length it has had to travel. The extra path length is seen from *Figure 53.6* to be

$$[d^2+(h_t+h_r)^2]^{1/2}-[d^2+(h_t-h_r)^2]^{1/2}$$

$$=d\left\{\left[1+\left(\frac{h_t+h_r}{d}\right)^2\right]^{1/2}-\left[1+\left(\frac{h_t-h_r}{d}\right)^2\right]^{1/2}\right\}$$

$$\simeq\frac{2h_th_r}{d}\quad\text{since } d>>(h_t+h_r)$$

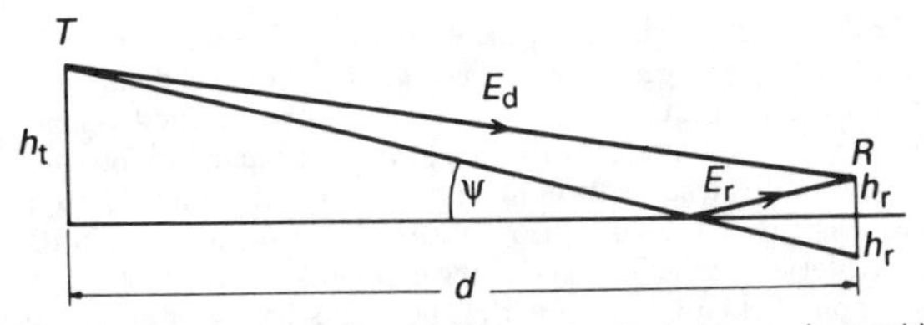

Figure 53.6 Combination of direct and reflected ground ray with VHF propagation (E_d is direct field and E_r reflected field)

This corresponds to a phase angle of

$$\frac{2h_t h_r}{d} \cdot \frac{2\pi}{\lambda} = \frac{4\pi h_t h_r}{\lambda d} \qquad (53.1)$$

Hence the phase angle of the reflected signal at the receiver aerial is ($\phi = 4\pi h_t h_r / \lambda d$) and its amplitude is $k\rho E_d$ where k is the correction factor for the directional characteristic of the transmitter aerial. If $k = \rho = 1$ and $\phi = -180°$ the vector sum of E_d and E_r becomes

$$E_t = 2E_d \sin(2\pi h_t h_r / \lambda d) \qquad (53.2)$$

so that the field at the receiver aerial varies from zero at the ground ($h_r = 0$) through a series of maxima when

$$2\pi h_t h_r / \lambda d = (2n-1) . \pi/2 \qquad (53.3)$$

or

$$h_r = (2n-1)\lambda d / 4h_t$$

and minima when

$$h_r = 2n\lambda d / 4h_t \qquad (53.4)$$

Under the above conditions a transmitting aerial at a height of 300 m would give maximum field strength at a distance of 30 km when the receiving aerial is at a height of 75 m. For low height receiving aerials $2\pi h_t h_r / \lambda d$ is small and $\sin(2\pi h_t h_r / \lambda d) \rightarrow 2\pi h_t h_r / \lambda d$; since the direct wave $\propto 1/d$, the received field strength will tend to be $\propto 1/d^2$.

In practice because of tropospheric refraction it is found that propagation is always greater than line-of-sight by an amount which would be obtained with a ground profile radius of about 1.33 earth radius, and it is usual to use this value in calculations. Temperature inversions with lowest temperatures near the ground can occur from time to time to produce a ducting effect permitting propagation over considerable distances. The effect is more noticeable over sea than over land.

Diffraction of the transmitted wave can occur due to a hill between the transmitting and receiving aerials, and at grazing incidence a loss of about 6 dB in signal strength occurs. The loss increases rapidly when the hill blocks the line of sight. In order to reduce the loss to a low value, the direct path between the two aerials should clear the obstacle by

$$H = \left[\frac{\lambda d_1 (d - d_1)}{d}\right]^{1/2} \qquad (53.5)$$

where d_1 = distance from the receiving aerial to the obstacle.

Thus the summit of a hill halfway between transmitting and receiving aerials spaced apart 30 km should be at least

$$H = \left[\frac{3 \times 30 \times 10^3}{2}\right]^{1/2} = 210 \text{ m} \qquad (53.6)$$

below the direct path if $f = 100$ MHz.

53.7 Ultra-high-frequency propagation

The propagation of ultra-high frequencies follows the same principles as with very high frequencies. The reduction in wavelength means that smaller objects can produce reflections and the absorption from, for example, trees in leaf is greater; on the other hand aerials are smaller and their directivity and gain can be increased.

53.8 Fundamentals of television transmission

For television broadcasting, as for sound radio, the first step is the transmission of a carrier frequency from a transmitter on which the picture information can be radiated and picked up by the receivers. The carrier frequency is selected in accordance with the appropriate channels used in the country concerned. The wanted transmission, provided that the receiver is located within its service area, may be selected by tuning the receiver, although in practice this may be done by means of pre-tuned push or touch buttons, or remote control unit.

The carrier frequency links transmitter to receiver and is modulated with the picture information. For terrestrial television broadcasting the picture or *video* signal is transmitted using *amplitude modulation.* However for space satellite broadcasting, where transmitter power is at a premium, it may be preferable to use FM for transmitting the video signal.

53.8.1 Scanning

Television systems are based on translating a scene into a series of small points, termed *picture elements*, by methodically tracing out or *scanning* the picture, and then transmitting the intensity or light value of each picture element in sequence. In practice the first step in deriving the video signal for a particular scene is to form an optical image on the photocathode of a camera tube. This image is scanned by the electron beam which moves across the image in much the same way as a printed page is read: i.e. from left to right across the top of the image, with a rapid return to the left-hand side of the image, to 'read-off' a second line just below the first, until the entire 'page' has been read. When the bottom of the image is reached the electron beam returns to the initial top left-hand position to read off the next 'page' as shown in *Figure 53.7*.

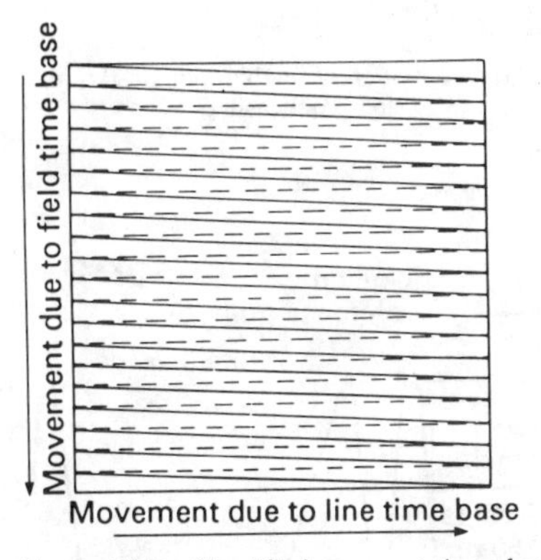

Figure 53.7 Simplified representation of sequential scanning

At the receiver a reproduction of the image is built up on the screen of the picture tube by an electron beam moving in precise synchronism with the beam in the camera tube. The screen glows with an intensity dependent upon the beam current and the glow persists for a short time: this combined with the persistence of vision of the eye gives the viewer the impression that the picture tube face is presenting a complete picture; moreover if the subsequent pictures follow sufficiently rapidly the illusion of movement, as in a cinematograph film, appears natural.

The movement of the camera-tube beam across the optical image, or picture elements, from left to right is termed a line scan, and the rapid return to the left-hand side of the picture the line flyback. In a simple form, supposing a complete picture was transmitted as a series of 100 lines to each picture then it would be a 100-line television system. In practice however certain periods of time are required for synchronising purposes, reducing the number of *active* lines to a lower figure than the theoretical.

A series of still images presented to the eye at the rate of about 10/s provides an illusion of continuous motion but accompanied by pronounced flicker. If the rate is increased to 25/s, flicker is reduced but still noticeable, particularly when the images are bright. A repetition rate of 50/s provides motion virtually without flicker, although for very bright pictures some 60 images are desirable. The rate at which images follow one another is called the *field* or *frame* rate.

53.8.1.1 Interlaced scanning

To transmit a complete high-definition picture at a sequential field rate of 50 or 60/s would require an excessively large bandwidth. An alternative technique is called *interlacing*: instead of transmitting each horizontal line in sequence, alternate lines are scanned first, with the missing lines traced out subsequently as shown in *Figure 53.8*, thus providing two half-scans for each complete picture. Since the eye hardly perceives individual lines,

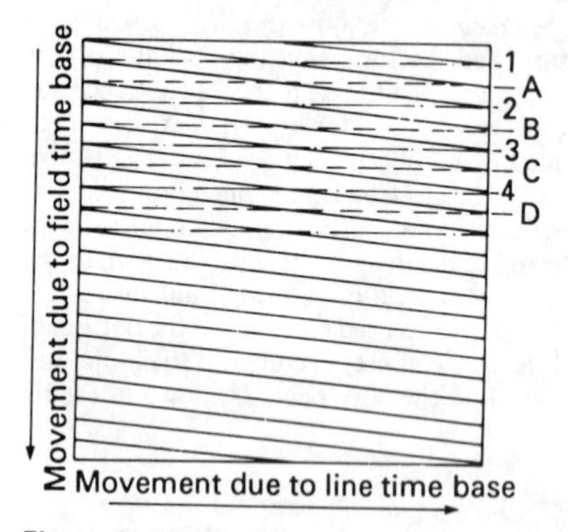

Figure 53.8 Simplified representation of double-interlaced scanning used for all current television broadcasting

the effect is to provide 50 fields or images per second with 25 complete pictures per second, even though the total detail, and hence the bandwidth, is that of a 25 pictures-per-second system. However it is possible with modern techniques to provide roughly equivalent pictures with sequential scanning, though interlacing is universally incorporated in broadcast standards. With interlacing the final line of the first field and the first line of the second field are half lines, the line scan starting or ending at the centre of the top and bottom of the picture.

53.8.2 Video and sync signals

Since the scanning process must be carried out synchronously at both the studio and the receiver it is necessary to provide *synchronising signals* to ensure exact correspondence between the two scanning processes. This is achieved by transmitting timing pulses which are used by the receiver as a reference to which to lock its horizontal and field times bases (it would be feasible in modern practice to provide only one set of timing pulses from which the other set could be derived using digital counting techniques). The way in which the picture signal is combined with the line sync pulses is illustrated in *Figure 53.9* and the waveform of the field sync signals is given in *Figure 53.10*.

The signals required to transmit the proportional brightness of each picture element of a high-definition system are of much higher frequency than for speech or music. As each line is traced out, the output of the camera varies in voltage, changing with changes in light value. If the line consists entirely of white or black or the same shade of grey the output is steady d.c. or 0 Hz. If the line represents a fine network of vertical lines of alternate black and white picture elements the output is at a frequency of several MHz.

It is common practice to term composite picture and synchronising information as *video* signals, but during radio transmission when these frequencies are modulated on a carrier wave, it is termed a *vision* signal.* Video signals are normally distributed as 1 V peak-to-peak signals, across an impedance of 75 Ω.

A high-definition television system thus requires:

(1) Transmission of picture information at frequencies extending to several MHz.

* The AM may be positive or negative as indicated in *Figure 53.10*.

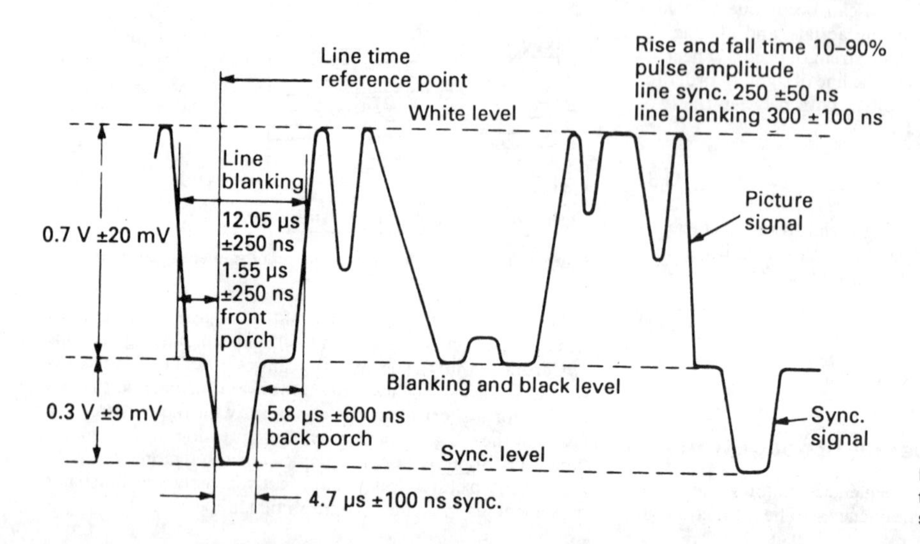

Figure 53.9 The waveform of a typical line showing synchronising signals

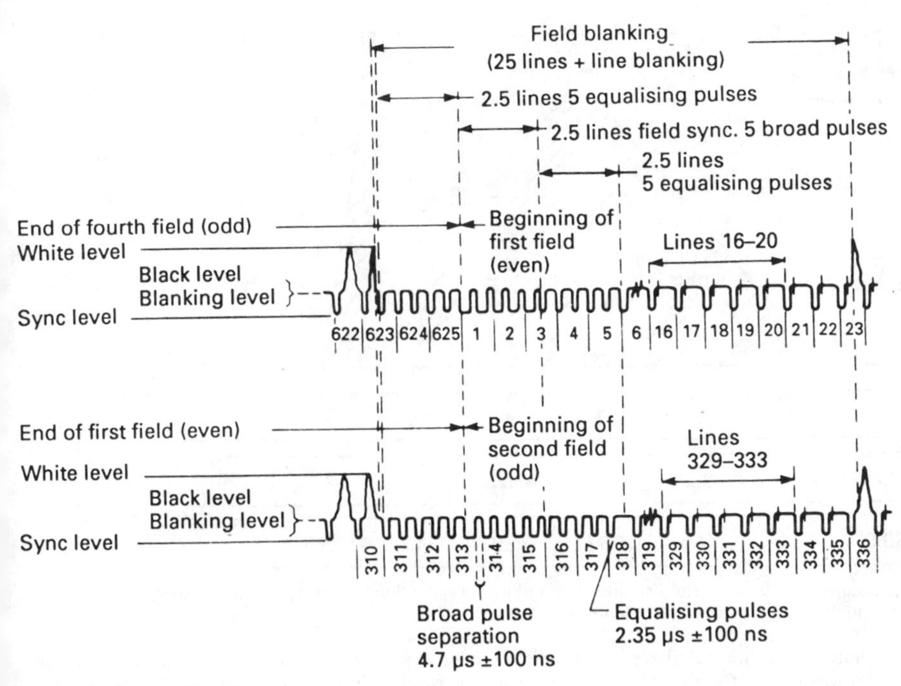

Figure 53.10 Field synchronising and blanking waveforms for a typical signal. Lines 7 to 14 and 320 to 327 have been omitted. Lines 16 to 20 may contain identification, control, or test signals or teletext data. The first and second fields are identical with the third and fourth in all respects except burst blanking

(2) Transmission of synchronising pulses to keep the received time bases in step with those used in the studio; the transmission of these pulses equally requires wide bandwidth.
(3) Transmission of an accompanying sound channel.

The precise way in which these requirements are met and the associated frequencies and tolerances together form a television *standard*. There are a number of different standards in current use; 525-line 30 pictures (60 fields) and several variations of 625-line 25 pictures (50 fields) are the most significant, although 405-line and 819-line standards in the UK and France have been phased out.

53.8.2.1 Bandwidth

The maximum upper frequency of a video signal is governed by the picture content and the scanning standard being used. For the UK System I this is normally taken as 5.5 MHz.

Were conventional double-sideband AM to be used, this would imply a maximum vision bandwidth of $2 \times 5.5 = 11$ MHz. In addition, further bandwidth would be required for the sound signal. In view of the restricted frequency spectrum at VHF and UHF, it has for many years been the practice to reduce the bandwidth of one of the two sidebands to produce *vestigial sideband* or *asymmetric sideband* vision transmission. In System I the full upper sidebands to 5.5 MHz are transmitted but the bandwidth of the lower sideband is restricted to 1.25 MHz (in 625-line System G as used in many countries the lower sideband is restricted even more to 0.75 MHz).

This means that for System I the total vision bandwidth is $1.25 + 5.5$ or 6.75 MHz and a gap of 0.5 MHz is left before the sound carrier, which is thus 6 MHz above the vision carrier, as shown in *Figure 53.11*.

For both Systems I and G in the European region the agreed international UHF channels occupy 8 MHz; in System I the vision carrier is always $+1.25$ MHz from the lower end of the channel; the FM sound carrier $+7.25$ MHz. *Figure 53.11* shows the vision response curve of System I as related to the video signal, the RF channel, and the standard BREMA receiver IF channel. The FM sound is transmitted with a maximum AF signal of 15 kHz and with a peak carrier deviation (corresponding to a 400 Hz tone at a level of $+8$ dBm at the modulator input) of ± 50 kHz. The pre-emphasis time constant is 50 μs.

The use of an asymmetric sideband vision signal results in a degree of quadrature distortion when the signal is envelope demodulated: this form of distortion (which can usually be detected only with some difficulty, even on a test card) can be eliminated by the use of synchronous demodulation.

53.9 Colour television

53.9.1 Development

Demonstrations of low-definition colour television were made by Baird as early as 1928 and later during World War II he was able to demonstrate a system of potentially greater value. The first public service in colour was inaugurated in the United States in 1951, although this system occupied a large bandwidth and could not be satisfactorily received either in colour or black and white except on a receiver designed for the system. It was soon appreciated that to achieve wide acceptability a public colour system needed to be *compatible*, that is the transmission should be receivable as a black-and-white picture on a monochrome receiver, and have *reverse compatibility*, that is allow the reception of black-and-white pictures on the colour receiver. These requirements were carefully studied and a specification for a colour system was drawn up by the National Television System Committee (NTSC) in the United States, and transmissions to this specification began on 1 January 1954. A further important feature of this system was that the colour information is encoded into the video signal in such a way that the transmissions occupy the same bandwidth as the black-and-white transmission. During the period from 1949 onwards, RCA successfully developed the shadow-mask colour display tube which, although difficult to manufacture, was an important and far-reaching advance on earlier colour display systems.

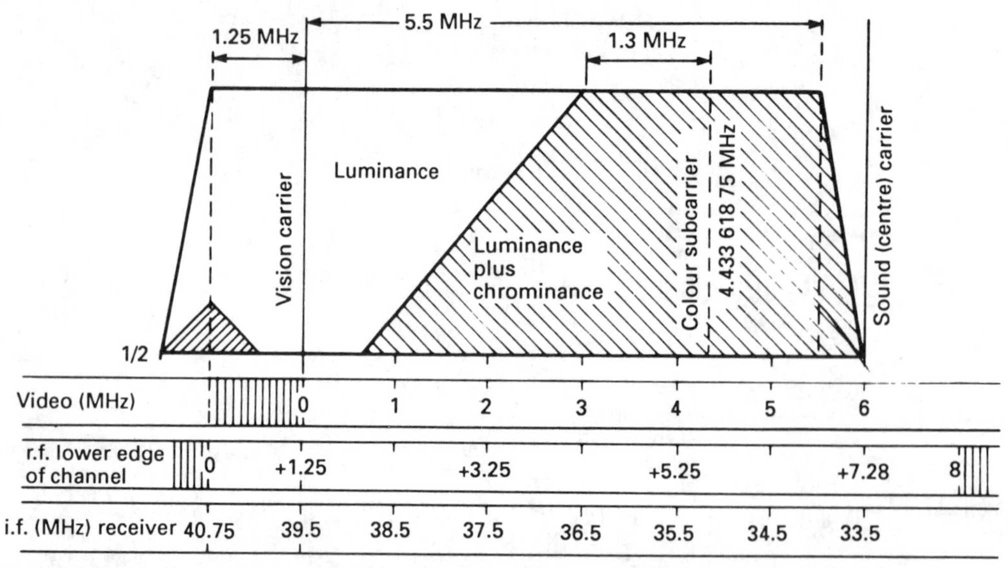

Figure 53.11 The frequency bands occupied by colour picture components and sound signal from an ideal 625-line System I transmitter as related to video, vision, and the IF of a receiver (BREMA standard IF)

The early spread of colour in the United States was relatively slow and some engineers believed this was due to the variable quality of the colour, primarily due to instrumental and transmission shortcomings. It was not until about 1964 to 1966 that demand for colour receivers became widespread.

In Europe the slow growth of American colour television was attributed in some degree to the susceptibility of NTSC signals to relatively small changes of *differential gain* and *differential phase* anywhere within the transmission chain; furthermore, the development in 1956 of quadruplex videotape black-and-white recording appeared to pose particular problems if it was also to meet the more stringent colour requirements. Differential phase refers to changes of phase of the colour subcarrier as a result of changing brightness levels; similarly differential gain is any change of gain with brightness.

As a result alternative encoding systems were developed and studied, though most of the other techniques of NTSC were retained. Two systems proved of particular importance: SECAM (sequential and memory) which introduced the concept of a delay line; and PAL (phase alternation line) which combined the delay line of SECAM with the suppressed subcarrier techniques of NTSC.

All three systems were found to have advantages and disadvantages, and no common agreement could be reached. The outcome was that NTSC is used in North America, Japan, and some other countries; PAL is used in the UK, many European countries, South Africa, Australia, New Zealand, and some others; SECAM is used in France, USSR and East Europe, and some African and Middle East countries. With the improved techniques now available each system is capable of giving roughly similar results, though PAL and SECAM remain less susceptible to transmission errors; SECAM presents rather more problems in the studio, although relatively easy to record on tape. It is possible to transcode between systems and the development of digital standards conversion also allows the signals to be readily converted between the main world standards. The development of fully digital and time-multiplexed analogue systems based on component video rather than composite video is of increasing practical importance for studio and satellite transmissions.

53.9.2 Colour television principles

The sensation of colour derives from the different reactions of the eye/brain system to visible electromagnetic radiation at different frequencies: a colour is what we feel when we look at light of a predominant wavelength: see *Figure 53.12*. The eye is more sensitive to radiation in the middle range of the visible spectrum (green) than to radiation towards the extremities (violet and red, orange and blue).

White light is the reflection of a uniform radiation throughout the visible spectrum; if the brightness is reduced such radiation appears grey; if little or no light is reflected the object appears black. Colour may be experienced by looking directly at a source of radiation or at reflected light.

The eye is less sensitive to the fine detail of colour than to fine detail in black and white: this fact allows colour information to be transmitted more crudely (less bandwidth) than black and white.

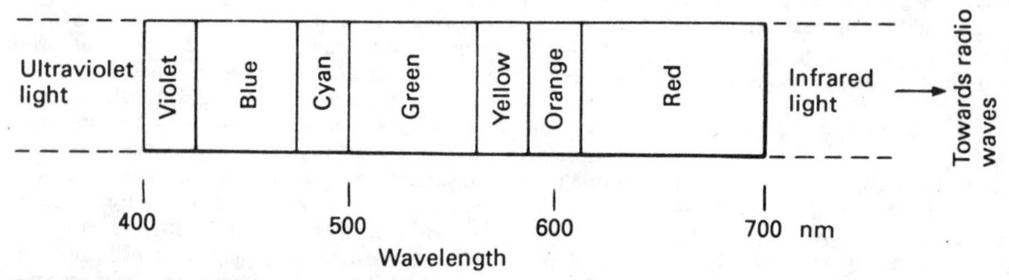

Figure 53.12 The electromagnetic spectrum of visible light. Short wavelengths such as these may be quoted in Angstrom units (10^{-10} m)

The experience of almost any colour can be obtained by adding (*additive mixing*) specific proportions of the three primary colours red, green, and blue. In painting and printing, colour mixing is a *subtractive process* and the primary colours are then red, yellow, and blue.

Any colour can thus be defined by three characteristics: its brightness (*luminance*); its *hue* (dominant electromagnetic wavelength or frequency); and its intensity (*saturation*) which corresponds to its colourfulness.

The standard adopted to define whiteness is known as the *colour temperature* (which determines the amount of blue or green in peak white). The standard adopted in the UK is *Illuminant D* corresponding to 6500 K.

Luminance, the characteristic used in monochrome television, conveys details of the varying levels of brightness of the picture elements. In any compatible colour television system luminance is transmitted, allowing a monochrome picture to be received.

In addition, for colour display, it is necessary to radiate additional information about the *hue* and *saturation* characteristics: such information is termed the *chrominance* (often abbreviated to *chroma*) signal. Saturation is a measure of the intensity or colourfulness of a colour: 0% saturation is entirely grey with a complete absence of colour; 100% saturation has no dilution of the colour with white. An example is that pink and red light may both have the same predominant wavelength (i.e. the same hue) but the red is the more fully saturated colour.

Chrominance information defines the hue and saturation of the picture independently of the luminance, so that theoretically any distortion of the chrominance information does not affect the detail of the picture: a system designed to fulfil this is termed a *constant-luminance* system; in practice some departure from this ideal is to be found in most systems.

To define completely the luminance and chrominance information, a colour television camera analyses the light from the scene in terms of its red (R), green (G) and blue (B) components by means of optical filters and then gamma corrects to take into account differences between camera and display tube characteristics; gamma-corrected signals are termed R', G', B' signals. For closed-circuit sequential colour systems, these three signals may be kept separate, but for a compatible broadcast system the signals are processed to provide the basic luminance signal (Y); in some cameras a fourth pick-up tube was used for operational reasons to obtain a luminance signal although it should be appreciated that the basic R, G, B signals contain all the information required to define luminance, hue and saturation. The green and red signals contribute more to luminance than blue and in practice a matrixing network is designed to the following specification.

$$Y' = 0.3R' + 0.6G' + 0.1B'$$

The four signals, Y', R', G', B' are related mathematically and it is therefore unnecessary to transmit all four. Since for compatibility we require a Y' luminance signal, this is transmitted. Two other signals are obtained by taking the red and blue signals and subtracting from them the luminance signal: that is $(R'-Y')$ and $(B'-Y')$. $(G'-Y')$ can be derived within the receiver by matrixing.

The transmitted signals are thus Y' (i.e. $0.3R'+0.6G'+0.1B'$); $(R'-Y')$ and $(B'-Y')$. These allow us to recover Y', R', G', B' in the receiver.

The current colour systems depend on this process but differ in the way that the three signals are transmitted within a single vision channel. Both NTSC and PAL transmit all three signals simultaneously; SECAM transmits luminance continuously but radiates on any given line only one of the two chrominance signals, changing over between each line. This indicates that in SECAM the vertical colour information is only one-half that of the other two systems, although this is of only minor practical significance since the eye is unable to detect fine colour detail. Delay lines used in SECAM decoders can be more tolerant than those required for PAL.

Although the chrominance signals generated in the colour camera occupy the full range of video frequencies, it is only the luminance signal that requires to be transmitted to this degree of resolution. The ability to resolve fine detail depends on *visual acuity*, and our ability to resolve colour in small details of a picture is inferior to that for corresponding black-and-white or grey pictures. Since the human eye does not resolve colour in small areas there is no need to reproduce this, even for high-quality television pictures. This influences many aspects of colour television, not least the ability to limit the bandwidth of chrominance information relative to that required for the luminance signal; in practice chrominance information in a 625-line system can begin to roll off at about 1.3 MHz. The restriction of chrominance bandwidth makes possible *inband* transmission of chrominance information, although this technique gives rise to some loss of compatibility (dots and crawl being seen on a monochrome picture) and also cross-colour effects on colour reproduction (flaring of patterned jackets is a common example), due to luminance signals appearing in the colour channels.

Basically the information in a monochrome signal is distributed in a series of packets separated by the line frequency, with only little spectrum energy in the gaps between as shown in *Figure 53.13*. By choosing a colour subcarrier frequency (see below) that is accurately placed between multiples of the line frequency, and noting that chrominance modulation energy is similarly in the form of packets it is possible to interleave the basic energy spectra of the luminance and chrominance signals: for 625-line transmission the colour subcarrier frequency is maintained very precisely at 4436618.75 Hz±1 Hz, with a

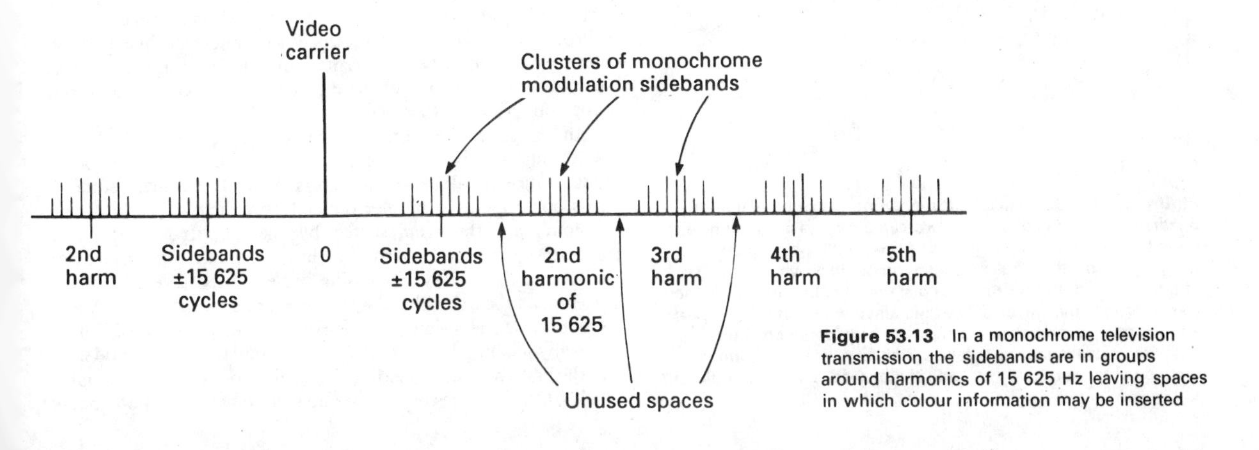

Figure 53.13 In a monochrome television transmission the sidebands are in groups around harmonics of 15 625 Hz leaving spaces in which colour information may be inserted

maximum rate of change of subcarrier frequency not exceeding 0.1 Hz/s.

The relationship between subcarrier and line frequency is

$$f_{sc}=(284-\tfrac{1}{4})f_h+\tfrac{1}{2}f_v \text{ Hz} \quad (53.7)$$

where f_h is the line frequency and f_v is the field frequency.

53.9.3 Chrominance transmission

In the PAL and NTSC colour systems, both of the colour-difference signals are transmitted simultaneously. Both channels of information are transmitted using a single subcarrier frequency with the subcarrier itself suppressed by means of balanced modulators except for a brief synchronising burst (the *colour burst*) which is transmitted following each line synchronising pulse, that is to say it is radiated only during the period known as the back porch of the line pulses as shown in *Figure 53.14*. This brief colour burst is used to correct an accurate

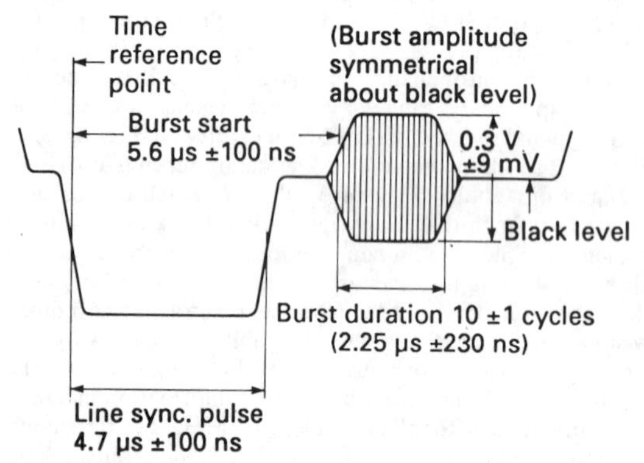

Figure 53.14 The colour burst on a standard level signal (700 mV white level)

reference oscillator in the colour receiver; by this means not only can the subcarrier be suppressed during the transmission of picture information, but it allows the receiver to take advantage of the technique of *quadrature modulation* and to recover separately two channels of colour information using the process known as *synchronous demodulation*: see *Figure 53.15*. One of the

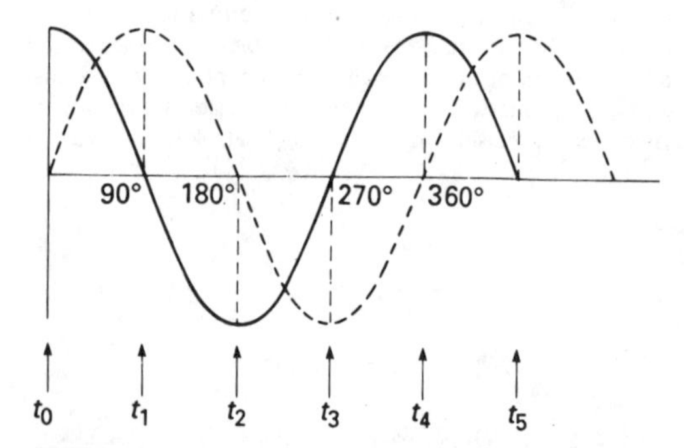

Figure 53.15 The principle of synchronous demodulation of quadrature-modulated signals. Two carrier waves are shown of the same frequency but in quadrature (i.e. with 90° phase difference between them). If each is amplitude modulated by different information and if two demodulators, one synchronised with each carrier, sample the waves only at the times when one is at a peak and the other at null, then the two sets of information can be separately retrieved at the receiver. The solid-line waveform is sampled at times t_0, t_1, t_4, etc. and the dashed-line curve at times t_1, t_3, t_5, etc.

two chrominance signals is shifted through 90° of phase so that the peaks of signal amplitude on one channel occur precisely at the zero crossover points of the other (a similar technique is used in stereo disc recordings).

Since this means that the subcarrier is modulated in both amplitude and phase, the two chrominance signals may be represented by a single phasor of which the length represents the saturation and angular direction (0–360°) of the hue as shown in *Figure 53.16*.

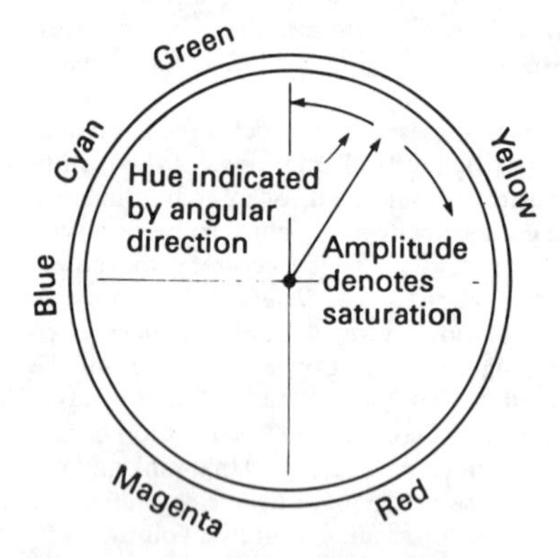

Figure 53.16 In the NTSC and PAL systems any hue can be represented by a phasor having a specific phase angle and an amplitude representing the degree of saturation

The amplitude of the phasor, since it represents saturation, reduces to zero for grey/black tones and is low when representing low saturated colour. This indicates that with PAL and NTSC colour systems little chrominance energy is radiated except for heavily saturated colours, and this feature improves the compatibility of the black-and-white picture.

In the NTSC system the phase of the colour phasors remains constant throughout transmission, the hue information being determined by the degree of synchronism between the studio source and the crystal-controlled reference oscillator for the synchronous demodulators in the receiver. Any disturbance of this synchronism anywhere along the transmission path results in a change of hue, unless corrected or compensated for in the receiver. It was this susceptibility of NTSC to differential phase that led to the development of SECAM and PAL, although many of the colour errors noted in early receivers were due to camera drift and instrumentation faults that have today been largely overcome.

53.9.4 PAL transmission

The PAL system differs from NTSC in that the phase of one of the chrominance signals is reversed during alternate line periods. There are in effect two different colour circles in use, and phase errors induced in one of these colour circles has an equal but opposite effect on the other.

In practice the chrominance signal corresponds to the sideband components of two AM subcarriers in phase quadrature, identified by the letters *U* and *V*, the phase of the *V*-axis component being electronically reversed after every line. Colours are thus represented by the amplitude and phase appropriate for the particular line. The burst of subcarrier following the line synchronising pulse is used to establish the reference subcarrier phase in the receiver and *also* to identify and synchronise the switching of the *V*-axis (*swinging burst*). This is done by making the phase of the colour burst ±135° on odd lines of the first and second fields and even lines of the third and fourth fields. On even lines of the first and second fields and odd lines of

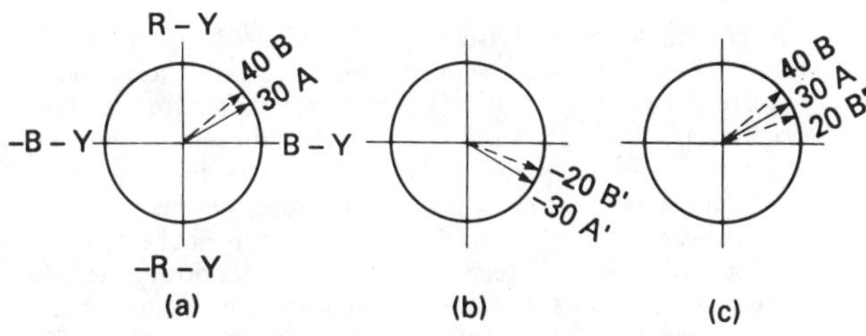

Figure 53.17 Principle of PAL automatic line correction by averaging: (a) hue *A* transmitted with a positive (*R–Y*) component; received as *B* because of a phase shift; (b) same hue transmitted with negative (*R–Y*) component *A'*: received as *B'* due to a phase error equivalent to that at (a); (c) after reversing the polarity of (b) the two received signals *B* and *B'* represent *B* and *B''* which when averaged give the correct hue *A*

the third and fourth fields, the phase of the colour burst is $+225°$. The mean burst phase is held within $180 \pm 2°$ of the reference axis. This allows the PAL receiver to identify whether the *V*-axis ($R' - Y'$) signal is being transmitted in positive or negative form. Receivers incorporating a delay line can provide tolerable colour on signals subjected to as much as 40° differential phase, although every effort is made during transmission to keep radiated errors far below this figure; multipath propagation, however, can induce errors after the signals have left the transmitting antenna.

Between a programme originating source, such as a camera, and the transmitter output, many items of equipment in the signal path are capable of introducing distortion into a colour signal and for this reason broadcasting authorities may specify codes of practice covering studio and outside-broadcast equipment, transmission networks, transmitters, and distinguishing between direct path and worst path conditions, anticipating that, for example, many signals pass more than once through a videotape recorder in the process of production and editing. Such codes normally specify the permissible tolerances of output signal level, nonlinearity distortion (including total phase errors and differential gain), linear distortion, noise and accuracy of synchronising signals.

53.9.5 Insertion test signals

During the field blanking interval, certain of the unused lines may carry additional information, for example, teletext (Oracle/Ceefax) data transmissions, identification, control and test signals. Lines used for such information on the 625-line standard are 16 to 20 inclusive on first and third (even) fields and 329 to 333 inclusive on second and fourth (odd) fields.

Operational teletext transmissions in the UK used lines 15 (328), 16 (329), 17 (330) and 18 (331) and have been further extended.

UK national insertion test signals are normally transmitted on lines 19 (332) and 20 (333).

The UK test signals 1 and 2 are shown in *Figures 53.18* and *53.19*.

Measurements of the test signals allow various distortions to be measured where required throughout the transmission chain. The advantage of the insertion test signal technique is that it allows optimum performance to be achieved without in any way interrupting normal picture transmission. Both manual and automatic monitoring of insertion test signal are possible.

The 2T pulse and 10 μs bar on test signal 1 enable the *K* rating to be found, while measurement of line–time nonlinearity may be made by passing the staircase waveform through a filter of restricted bandwidth. The 10T composite pulse allows chrominance-to-luminance gain and delay inequalities to be assessed.

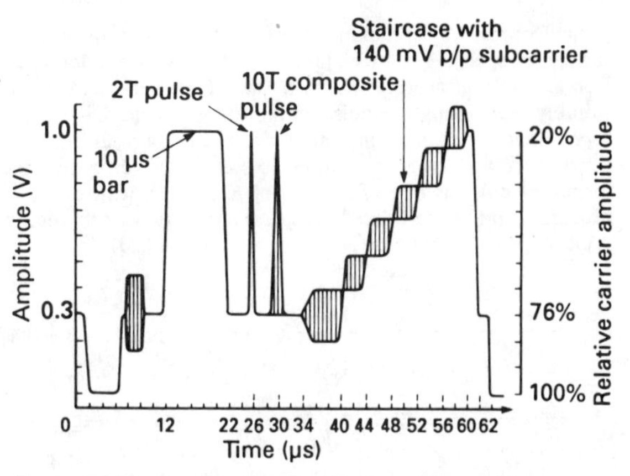

Figure 53.18 Insertion test signal 1 (lines 19 and 332)

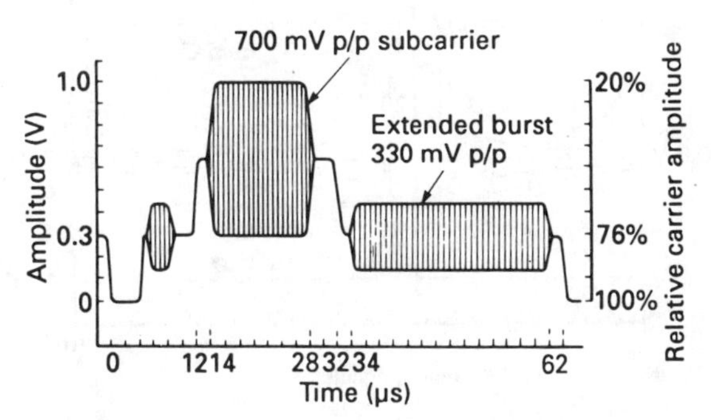

Figure 53.19 Insertion test signal 2 (lines 20 and 333)

Chrominance-to-luminance crosstalk can be measured from the bar waveform on test signal 2. The extended burst on test signal 2 is of constant phase and amplitude and is intended for demodulating the subcarrier on the previous line in order to measure the differential phase.

Table 53.2 illustrates the relevance of measurements made using interval test signals.

Different organisations tend to use the interval test signal facilities in slightly different ways.

An important feature of the test signals is the 2T sine-squared pulse having a half-duration equal to the reciprocal of the nominal bandwidth and the bar which is used both as an amplitude reference for the pulse and also as a means of indicating sag or droop (caused by a falling off in response to lower middle frequencies).

Table 53.2 Measurement using interval test signals

Measurement	*Relevance*
Width of 2T pulse	Resolution
Tilt on 10 μs bar	Smear
Pulse/bar ratio	Amplitude/frequency characteristics
2T echoes and rings	Phase response; echoes
10T pulse	Luminance/chrominance gain and delay inequalities
Chrominance staircase	Differential gain and phase, also luminance linearity
Mini-bar, line 20	Luminance–chroma crosstalk

An accepted way of making waveform measurements involves the use of special graticules in conjunction with oscilloscopes. For example measurements with the 2T pulse are made by suitably positioning the pulse within the graticule. Distortions represented by changes in amplitude, width, or resulting from overshoots and ringing, can be expressed by assessing the displayed pulse as one of a series of *K* ratings, with the least favourable rating considered to represent overall performance. A typical *K*-rating graticule is shown in *Figure 53.20*.

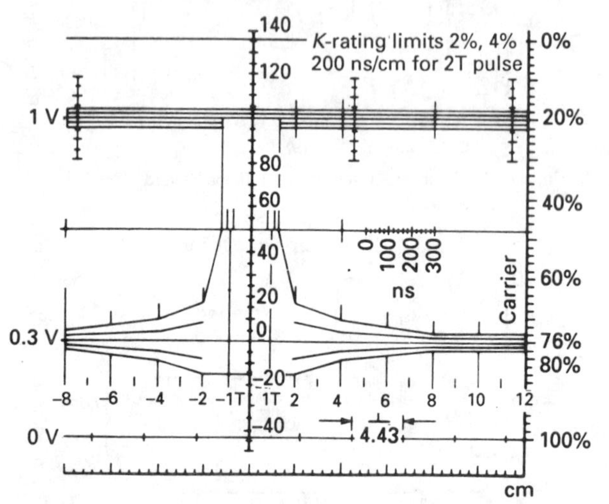

Figure 53.20 IBA *K*-rating graticule

For the 625-line, if a 2T pulse is accompanied by an echo with a delay of 0.8 μs or greater, the equivalent *K* rating is numerically equal to the amplitude of the echo. For shorter delay times the same *K* rating applies to greater amplitudes on a progressive scale, as indicated on the special graticule. Apart from transmitter performance checks, this technique can provide useful information on the quality of a propagation path between stations in assessing multipath reception.

The staircase waveform can be used to determine linearity of the transmission chain.

To facilitate the use of interval test signals for colour transmission a broader pulse, consisting of subcarrier signal, modulated to provide a unilateral sine-squared envelope, was added to permit measurement of luminance-to-chrominance gain and delay inequalities.

The interval test signal system is widely used in Europe for transmitter propagation and link checking and monitoring; it is also used for studio checking although seldom on a continuous basis.

53.9.6 Colour impairments

Although modern colour transmissions provide a high standard of colour fidelity, subjective impairments may occasionally occur. The following notes are based on experience of the Australian Broadcasting Commission during test transmissions prior to the introduction of 625-line (System G) PAL colour television.

(1) *Saturation variations* in received pictures are caused by an incorrect amplitude relationship between the subcarrier burst and the chrominance component of the video signal. Receivers incorporate automatic colour control circuits (ACC) which vary the gain of the chrominance circuits in accordance with the amplitude of the burst on the assumption that any variation in the chrominance level (such as that caused by frequency response) equally affects the burst level. Thus if the ratio of burst to picture is correctly set at the signal source, the saturation of the original signal is maintained at the receiver output. If, however, some variation in frequency response occurs and the burst amplitude only is corrected prior to transmission, the output of the receiver will be incorrect. This problem may not be detected prior to transmission because picture monitors do not usually include ACC circuits. It is essential that the amplitude of burst should always be set accurately at signal sources, that it should never be varied independently of the picture chrominance, and that all sources employ the same standard burst to picture ratio.

(2) *Colour balance* errors may be caused by lack of matching between colour cameras, by the inherent differences in colorimetry between electronic and film cameras, by inherent colour casts in colour film, by poor grading of sections of colour film, or by variations in subcarrier phasing. It is essential for colour matching that all colour monitors be adjusted to the same colour temperature, that the colour scheme of viewing areas be standardised and that a white point reference of the correct colour temperature be available.

(3) *Streak and smear* are the descriptive names given to types of transient distortion and are explanations in themselves of the subjective effect. They are caused by transient distortions of comparatively low frequency and may not always be obvious with pulse and bar when viewed on a picture monitor. A window signal is a suitable test, and as a compromise can be combined with pulse and bar signals by windowing the bar. This fault can be troublesome with long-distance inter-city relay circuits where the effect of a number of small distortions becomes cumulative. It is also a common fault of monitor and receiver circuitry.

(4) *Transient errors* anywhere within the range of video frequencies can cause subjective impairment of pictures. The standard monochrome practice of pulse and bar testing is equally applicable to colour practice, and to explore the response of the chrominance channel a composite pulse is usually added. Other transient test signals are also used for particular applications.

Transient errors may also arise from injudicious use of operational facilities. For example, use of the crispener circuit on a particular colour camera to sharpen the edge detail may produce very pleasing pictures on the transmission monitor of an outside-broadcast van, but can result in an objectionable edge ringing after passing through a mediocre link circuit and transmitter.

(5) *Resolution* may be affected by many faults, from incorrect focusing of optical components to unsatisfactory frequency response of electronic equipment. A small amount of group delay, causing small displacement of the chrominance with respect to the luminance information may also be perceived as a loss of resolution, as may transient errors.

(6) *Crosstalk* in this context refers to the effect of unwanted coupling between two circuits. Usually this is frequency dependent, and increases with higher frequencies. Thus crosstalk becomes a more serious problem with the introduction of colour because the presence of the chrominance channel centred on 4.43 MHz has greatly increased both the HF components of the video signal and their importance. If the crosstalk is synchronous with the wanted signal it will be visible as a faint picture in the background. In the more general case it will cause variations in the colour of at least parts of the picture, usually as a slow pulsing of the colour, as the unwanted signal beats with the chrominance components of the wanted signal.

(7) *Videotape headbanding and velocity errors*. Quad videotape machines have four individual rotating heads which each scan 16 lines of the video signal in turn. Any difference between individual

heads or head amplifiers can result in visible bands across the picture. Any differences in performance between the first line scanned by one head and later lines scanned by the same head will also cause banding. Where the errors affect the amplitude of the chrominance signal (e.g. differences in high-frequency equalisation) there is a change in colour saturation. Where the errors affect the phase of the chrominance signal (e.g. velocity errors) there is a change in colour hue. Sophisticated electronic compensators reduce these errors. These problems are overcome in modern helical scan recorders and by the use of digital timebase correctors.

(8) *Chrominance luminance delay inequalities* cause the chrominance information to be displaced from the luminance information. This fault arises from unsatisfactory phase response at chrominance frequencies. It is often present where older equipment, designed for monochrome, is used.

(9) *Chrominance noise.* In general HF noise is less objectionable than LF noise. However, in colour receivers or monitors the chrominance circuits take the HF noise components from around 4.43 MHz and demodulate them down to low frequencies resulting in a subjective increase in noise when operating in colour. To overcome this either the noise performance of the equipment must be improved and/or notch filters fitted to luminance channels just prior to encoding. Difficulties also occur where the level of lighting is inadequate, e.g. in some outside-broadcast situations. This results in an increase in noise in each camera channel and a consequent increase in chrominance noise.

(10) *PAL ident problems.* The standards for the PAL system lay down that the phase of the subcarrier burst shall be 135° for the first line after the burst blanking sequence. For a continuous picture there is no problem if this is incorrect (225°) because the oscillator in the receiver locks up satisfactorily. Normally there is no problem even if a cut is made between two signals having different burst phasing because the receiver oscillator changes phase and locks to the second phasing very quickly. However, if two such signals are mixed, where the chrominance information of both is compared with the burst of one only, one signal appears in incorrect colours. This situation can occur unless all coders at picture sources operate with the correct ident or where any mechanism can change the PAL ident (e.g. some abnormal videotape conditions).

To check that the PAL ident is correct it is necessary to examine the phase of the burst on the first line after burst blanking and this can be done with a PAL vectorscope with the facility for displaying a single identifiable line (similar to an insertion-test-signal facility). The most convenient line to check is line 6 of field 3.

(11) *Hum and control tones for mains.* The mains supply can be a source of interference to television pictures if any component from it becomes combined with the video signal.

One source of such interference arises from the inter-winding capacitance between the primary and secondary of power transformers. A small current flows via this capacitance to the earth side of the secondary, and then back to the power supply earth via the earth connections of the equipment. It is inevitable that part of this path is common to the signal earth connections, and so a small interfacing signal is injected into the signal paths.

In theory, any hum or other interfering signals below line frequency can be removed by clamp amplifiers, and this method is often used. However, it is desirable to avoid the use of clamp amplifiers where possible as they can add other distortions to the signal. Interference arising from the inter-winding capacitance of power transformers can be overcome by using electrostatic shields and returning all electrostatic shields to mains earth independently of signal earth paths.

(12) *Cross-colour* is an effect whereby luminance information in the region near 4.43 MHz is decoded and appears as LF chrominance information. For example a check pattern on clothing, at a critical distance from a camera, produces coloured effects. One way of minimising this problem is to use a notch filter in the luminance channel just prior to encoding.

53.10 International television standards and systems

Detailed information on the international standards for television broadcasting is given in CCIR Report 624 (ITU Sales Service, Place des Nations, CH-1211, Geneva 20, Switzerland), and summarised in *Tables 53.3* and *53.4*.

53.11 Digital television

For almost two decades the application of digital techniques to video processing and transmission has been under investigation, and progressively used in specialised operational applications since 1971. The ability to make use of picture redundancy and complex coding strategies can provide very large degrees of bit-rate reduction without excessively increasing the harmful effects of bit errors. For composite coded 625-line pictures, transmission bit rates of the order of 34 Mbit/s (17 MHz bandwidth) have been shown experimentally by the BBC, and even less bandwidth may be required for 525-line systems. Digital transmission also provides better energy dispersal.

Table 53.3 International standards for television broadcasting

System	*Lines/fields*	*Channel width* (MHz)	*Video width* (MHz)	*VSB width* (MHz)	*Polarity vision modulation*	*Sound/vision carrier spacing* (MHz)
B	625/50	7	5	0.75	negative	5.5
D, K	625/50	8	6	0.75	negative	6.5
G	625/50	8	5	0.75	negative	5.5
H	625/50	8	5	1.25	negative	5.5
I	625/50	8	5.5	1.25	negative	6.0
K1	625/50	8	6	1.25	negative	6.5
L	625/50	8	6	1.25	positive	6.5
M	525/60	6	4.2	0.75	negative	4.5
N	625/50	6	4.2	0.75	negative	4.5

Table 53.4 Composite colour video signals in terrestrial systems (simplified)

System	*Chrominance signals*	*Chrominance subcarrier* (MHz)	*Chrominance subcarrier modulation*
PAL	E_U, E_V	Approx. 4.43 (3.58 for M/PAL)	Suppressed-carrier AM of two subcarriers in quadrature
NTSC	E_I, E_Q	Approx. 3.58	Suppressed-carrier AM of two subcarriers in quadrature
SECAM	D_R, D_B (line sequential)	Approx. 4.41 (D_R) Approx. 4.25 (D_B)	Frequency modulation alternating one line D_R and one line D_B

Countries using the different systems are as shown in *Table 53.5*.

Table 53.5

I/PAL	Angola,* Botswana, Hong Kong, Ireland, Lesotho, South Africa, UK
B, G/PAL	Albania, Algeria, Australia, Austria, Bahrain, Bangladesh,* Belgium, Brunei, Darussalam,* Cameroon, Denmark, Equatorial Guinea,* Ethiopia, Finland, Germany (FRG), Ghana, Gibraltar, Iceland, India,* Indonesia,* Israel, Italy, Jordan, Kenya, Kuwait, Liberia,* Luxembourg, Malawi, Malaysia, Maldives,* Malta,* Monaco, Mozambique, Netherlands, New Zealand, Nigeria,* Norway, Oman, Pakistan, Papua New Guinea, Portugal, Qatar, Sierra Leone, Singapore, Spain, Sri Lanka,* Sudan,* Sweden, Switzerland, Syria, Tanzania,* Thailand, Tunisia, Turkey, Uganda,* United Arab Emirates, Yemen (AR),* Yemen (PDR),* Yugoslavia, Zambia,* Zimbabwe*
N/PAL	Argentina,* Paraguay,* Uruguay*
D/PAL	China
M/PAL	Brazil
D, K/PAL	Korea (DPR), Romania
B, G/SECAM	Cyprus, Egypt, Germany (DDR), Greece, Iran, Iraq,* Lebanon, Libya,* Mali,* Mauritania,* Mauritius,* Morocco, Saudi Arabia, Tunisia
D, K/SECAM	Afghanistan,* Bulgaria, Czechoslovakia, Hungary, Mongolia,* Poland, USSR, Viet Nam
K1/SECAM	Benin, Burkina Faso, Burundi, Central African Republic, Chad, Congo, Cote d'Ivoire (Ivory Coast), Djibouti,* Gabon, Guinea, Madagascar, Niger, Senegal, Togo, Zaire
L/SECAM	France, Luxembourg, Monaco
M/NTSC	Bermuda,* Bolivia, British Virgin Islands,* Burma,* Canada, Chile, Colombia, Costa Rica, Cuba, Dominican Republic,* Ecuador,* Guatemala,* Haiti,* Honduras,* Jamaica,* Japan, Korea (Republic of), Mexico, Montserrat,* Netherlands Antilles,* Nicaragua,* Panama, Peru, Philippines,* St Christopher & Nevis,* Suriname,* USA, Venezuela

* VHF band only

A digital system is one in which all waveforms are selected from a restricted number, as opposed to the infinite number of shapes and amplitudes of an analogue signal. A digital signal has far cruder and more rugged time relationships and unlike an analogue signal is far less susceptible to differential phase and differential gain distortions. A binary digital system has only two states and so is a 'go/no-go' or 'on/off' system. Analogue systems degrade gradually through all operations; digital systems can show imperceptible degradation until they reach a certain level of errors and then degrade very rapidly.

Digital operations can be of calculable, repeatable, consistent standard whereas analogue systems require accurate adjustment and alignment to achieve optimum results. 'Line-up time' is essentially an analogue situation, whereas an 'on/off' switch may be the controlling factor of a digital system.

Digital systems score where the user needs to store masses of information or perform accurately calculable operations on the information: standards conversion, graphics storage, synchronisation, compression, expansion, noise reduction, image enhancement can all be accomplished more readily in digital form; indeed digital techniques may allow users to do things which cannot be done satisfactorily or at all in analogue form.

Nevertheless such a radical change as an all-digital approach to studio and distribution operations has to be justified on many grounds: economics, performance, reliability, stability and ease of operation amongst them.

The basis of any digital video technique involves quantising the picture into a number of discrete brightness levels, sampling the signal at a repetition rate normally greater than the Nyquist figure of twice the highest frequency component in the signal. Each level is allotted a unique code in the form of the presence or absence of pulses, usually in the form of linear pulse code modulation (PCM). In practice some 256 brightness levels may be used, corresponding to an 8-bit data word. Signal processing is then achieved by arithmetical operations on these data words.

An analogue signal can be converted to digital form, processed and then reconverted to analogue form several times in tandem with little visible degradation. However, if this tandem chain is extended too far it will tend to lead to a marked increase in the quantisation noise caused by the use of a restricted number of amplitude levels. Other impairments in digital video may be due to aliasing, clock jitter, error rates and the impossibility of transmitting perfect pulses. Digital video signals may also call for added complexity for monitoring and measuring the impairments, and some still uncertain cost factors in equipment and maintenance.

The advantages of digital transmission include freedom from the ill-effects of differential phase and differential gain and the ability to regenerate an exact replica of the input data stream at any point in the chain, thus avoiding cumulative signal-to-noise degradation. It was Shannon's communication theory that first unaerlined mathematically the outstanding efficiency of digitally encoded transmission systems.

However it is important to realise that digits do not eliminate all problems. From the earliest days of manual and machine cable telegraphy it has been recognised that the transmission of high-speed pulses within a channel of restricted bandwidth can present severe practical problems, including inter-symbol interference and susceptibility of the error rate to all forms of 'echoes' and multipath propagation.

Digital systems are inevitably subject to quantising noise, which depends upon the number and arrangement of the levels at which the original analogue signal is digitised, and also to aliasing foldover distortion. Aliasing represents spectral components, arising from the process of sapling, not in the original signal; when these fall within the spectrum of the sampled signal they result in foldover distortion. Aliasing can be minimised by effective filtering, although the ease and cost with which such filtering can be accomplished is very much a factor of the sampling rate.

53.11.1 Digital and component systems

Since the early 1970s, digital techniques in video/audio processing and transmission have been under intensive investigation and implementation. This work has led to the first worldwide production standard (CCIR Recommendation 601 of 1982), the implementation of a few experimental and operational all-digital studios, the widespread use of 'stand-alone' digital studio equipment for special effects, noise reduction, standards conversion, etc., and the Nicam 728 dual-channel digital system for television sound.

Initially, the introduction of digital processing was applied to encoded 'composite' waveforms with input and output from each unit in analogue form. However, it was soon appreciated that digital processing offered an opportunity to improve picture quality by eliminating the degradations inherent in the PAL, NTSC and SECAM colour-encoding systems due to 'mixed highs' (frequenty multiplex) insertion of chrominance information within the luminance baseband. Attention was therefore turned to use of 'component' signals with luminance and chrominance signals kept separate, although this involves an extension of the baseband of the analogue signals and consequently higher digital bit rates.

Advantages of handling signals in digital form include the more rugged nature of a digital signal, which requires a much lower signal-to-noise ratio, and its greater immunity to many forms of phase and amplitude distortion and interference. The ability to make use of picture redundancy and complex coding strategies can provide very large degrees of bit-rate reduction without excessively increasing the harmful effects of bit errors. However, once large bit-rate reduction has been applied, it becomes impracticable to use such production techniques as chroma-keying 'downstream' of the bit-rate reduction.

By 1987, all-digital cassette broadcast videotape recorders, based on the EBU/SMPTE D1 digital-component format, had reached the marketplace, making possible all-digital production centres. Such machines can produce more than 20 generations of tape without noticeable degradation, but represent a relatively high-cost system applicable primarily to studios converted for component working. Digital cassette recorders for use on composite signals (D2 format) have also appeared. However, the increased packing density of metal-particle videotapes has led to the successful development of lower-cost analogue-component cassette recorders using $\frac{1}{2}$ inch tape formats (M-2 and Betacam-SP) suitable for both studio and field operation, and is having a greater impact at present than all-digital recording.

CCIR Recommendation 601 (1982) and EBU Tech 3246-E details the specification of the 4:2:2 standard. This forms the fundamental building block of digital television and the basis of a family of standards, with 3:1:1 and 4:1:1 as lower members of the same family. Higher orders are needed for HDTV systems, since the 4:2:2 standard imposes virtually a 'brick wall' filter of 5.5 MHz luminance bandwidth.

CCIR Recommendation 601:

(1) Component signals sampled to 8-bit accuracy. Recommendation 601 also permits 10-bit data words for studio processing.
(2) *Luminance*: 13.5 MHz sample rate, 720 samples per digital active line.
(3) *Colour difference*: 6.75 MHz sample rate, 360 samples per digital active line.
(4) Length of digital active line: 53.33 μs.
(5) Luminance and colour difference samples transferred between equipments at: 27 Mword/s (216 Mbit/s).

Bit rate reduction of 4:2:2 digital bit streams for transmission over contribution and distribution links without significant loss of picture quality or loss of the ability to process the pictures down stream involves sufficient reduction to permit use of the European telecommunications digital hierarchy (e.g. 140 Mbit/s, 70 Mbit/s, 35 Mbit/s). Reduction to 140 Mbit/s or 70 Mbit/s can be achieved without noticeable loss of quality and only small loss of processing capability. Reduction to 34 Mbit/s (4:1 compression ratio) can provide virtually transparent quality but does involve loss of processing capability.

53.11.2 Multiplexed analogue components

The availability of wider channels (19 MHz carrier spacing with 27 MHz overall bandwidth) in the 11.7–12.5 GHz band allotted in Region 1 for direct broadcasting from satellite (DBS) gave an impetus to reconsider the fundamentals of broadcast transmission. On-board limitations of electrical power generation led to the adoption, of FM rather than vestigial sideband AM for the vision signals; consequently with FM noise distribution militating against the use of high-frequency subcarriers.

The MAC/packet system (*Figures 53.21–53.23*) provides a means of broadcasting video in component form, with time separation of luminance and chrominance information, together with high-quality multichannel digital audio on a single carrier. The basic principle is to 'time-compress' analogue video signals by means of digital processing, so that the 64 μs line-period of a 625-line system contains not only the luminance information,

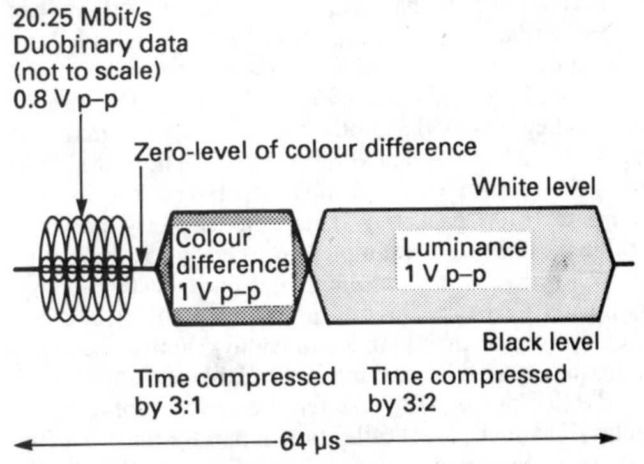

Figure 53.21 Baseband video waveform of a D-MAC television line. The duobinary data burst conveys 206 bits and is followed by time-compressed vision signals

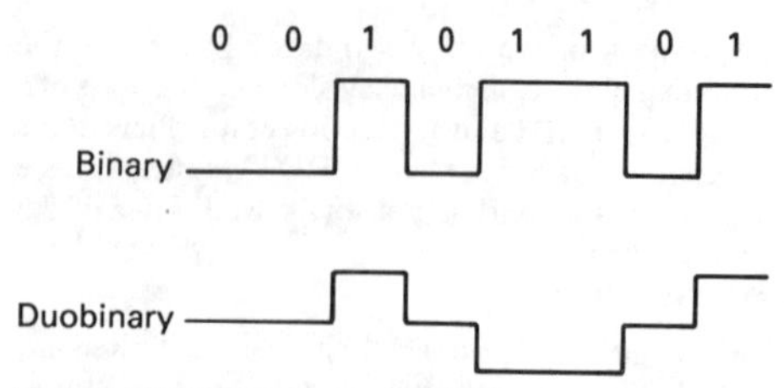

Figure 53.22 Comparison of binary and duobinary coding. After low-pass filtering, the three-level duobinary data signal is time-division multiplexed with the time-compressed vision signal

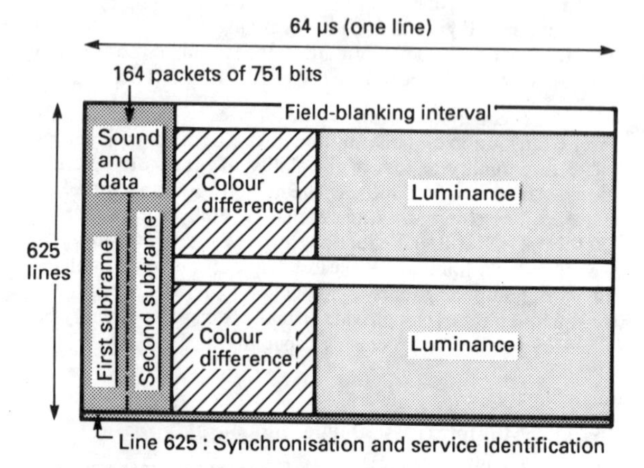

Figure 53.23 Simplified structure of D-MAC/packet frame showing the distribution of vision and 20.25 Mbit/s duobinary data. Organising the data into two subframes allows easy transcoding of one subframe into D2-MAC

one of the colour-difference signals, but also a 10 μs burst of digital data. No conventional synchronisation pulses are transmitted but the data signals include the synchronisation information.

The following parameters apply to the 625-line D-MAC/packet standard. Each line of luminance information is digitally compressed by a factor of 3:2 (52 μs to about 35 μs). The U and V colour-difference signals are reduced by a factor of 3:1 (52 μs to about 17.5 μs). This has the effect of increasing the maximum baseband video frequency from about 5.7 MHz to about 8.5 MHz. The remaining 10 μs periods carry 20.25 Mbit/s duobinary data inserted on to the baseband signal. Each line period thus has 206 data bits with a mean data capacity of about 3 Mbit/s. Line 625 carries only data, with duobinary data inserted on to the baseband signal, with a capacity of 1296 bits. The use of duobinary coding means that both vision and data signals are contained within a baseband of about 8.5 MHz. The digital data are organised into 'packets' of 751 bits with 164 packets in each 625-line frame (4100 packet/s). This data capacity provides considerable flexibility: for example, up to eight 15 kHz audio channels with 32 kHz sampling and near-instantaneous companding (NICAM) of 14 bits per sample to 10 bits per sample, or up to sixteen 7 kHz 'commentary' audio channels. The total data capacity is optionally available for any mix of teletext data, high-quality stereo audio, mono/stereo 'commentary' channels, or for utilisation in part for the provision of analogue wide-screen video information in the MAC enhanced modes or the use of digitally assisted (DATV) vector motion enhancement.

The MAC system also permits the use of 'conditional access' encryption systems offering a high level of security. One possibility is the use of 'double-cut' component rotation with the decoding key transmitted in the digital packets, addressed to individual receivers or groups of receivers.

53.12 Enhanced and high-definition systems

During the 1980s much research work has been devoted to the study and development of electronic television systems capable of providing wide-screen, flicker-free pictures of a resolution comparable with that of 35 mm cinematograph film for broadcast transmission, for electronic cinematography and for high-quality videodiscs. Efforts are being made to establish a single worldwide production standard, to develop transmission systems (broadcast, contribution and distribution links, cable networks) capable of delivering such high-resolution systems, and to develop displays that can do justice to such systems at prices within consumer budgets.

Developments, particularly in large-scale integration and digital electronics, have made such system realisable in the laboratory, with the probability of further improvements in the near future. This means that electronic cameras and videotape recording are becoming the dominant production medium rather than film for television drama, for cinema and videodisc production, and for the production of master tapes for videocassettes.

For large-audience presentation an HDTV system requires:

(1) Image quality directly comparable with that of film.
(2) Elimination of large area and interline flicker.
(3) Elimination of cross-colour defects inherent in composite-coded colour systems.
(4) Wide screen displays.
(5) Economic conversion of material from tape to film and from film to tape, without noticeable degradation of picture quality.

For domestic applications, the HDTV system should preferably be compatible with existing broadcast standards and should be sufficiently flexible to permit advantage to be taken of continuing improvements in display systems, digital signal processing, etc.; it should be suitable for effective down-conversion to conventional broadcast standards, and the very wide video baseband needed for high-definition signals should be amenable to bandwidth reduction techniques.

To reduce flicker and the interline twitter of interlaced scanning, there is a significant advantage in using sequential ('progressive') scanning at frame rates of 50 or 60 Hz, although the bandwidth reduction provided by interlaced scanning cannot readily be accommodated during transmission of sequential scanning; it is however possible, with the aid of frame memories and digital processing, to display an interlaced transmission with progressive scanning.

It has been shown that, as in the cinema, a wide-screen display, viewed at a distance of about three to four times picture height, has significantly more impact on the viewer than pictures with the standard aspect ratio of 4:3 (width/height). Practical HDTV systems have adopted aspect ratios of 5:3 or preferably 16:9.

Although the desirability of a single universal production and transmission standard for HDTV is recognised, the traditional division of the world into 50 Hz and 60 Hz areas (originally due to electricity mains supply practice) may prove impossible to avoid, although a production standard could be devised that would convert to either standard without serious impairment.

As an intermediate step, a number of techniques have been developed to provide 'enhanced' systems that can provide improved-quality wide-screen pictures without increasing the basic scanning rates of conventional systems, and are fully compatible with those systems, suitable for use on satellite channels and, in some instances, for use on terrestrial networks (e.g. Enhanced-MAC).

'Hivision' The NHK broadcasting organisation, in collaboration with Japanese industry, has developed and widely demonstrated a Hivision HDTV system capable of excellent quality on closed circuit. The system, which is not compatible with existing standards, can be transmitted over a single satellite channel by means of a bandwidth-reduction system called MUSE (Multiple Sub-Nyquist-Sampling Encoding). Opinions differ as to the degree of impairment introduced by MUSE.

Hivision parameters are as follows:

Number of scanning lines	1125
Number of active lines	1035
Field rate	60 Hz
Scanning method	2:1 interlace
Aspect ratio	16:9
Samples per active line	1920 for luminance 960 for colour difference

'HD-MAC' A joint European 'Eureka EU95 compatible high-definition television system project' has investigated and proposed a studio standard of 1250/50/2:1/16:9 that is simply related to the present 625-line standard. This would be used for production aimed at the 50 Hz areas. It has been suggested that a single world production standard, with headroom for conversion to the 1125/60/2:1 standard, is required to overcome the problems of the two field-rate standards and to satisfy the needs of cinema-like presentation. A single world production standard meeting this criterion would have the following specification:

Number of active lines	1152
Field rate	50 Hz
Scanning method	Non-interlace
Aspect ratio	16:9
Samples per active line	1920 for luminance 960 for colour difference

53.13 Multichannel sound systems

Although the monaural sound channel of conventional television broadcasting is capable of providing high-fidelity (15 kHz) audio, this has seldom been reflected in the design of domestic receivers. Increasing consumer awareness of the fidelity provided by other media, including pilot-tone stereo radio, stereo cassette tapes and digital CD records, has led to the development of systems for dual-channel and multichannel audio for television capable of providing sufficient separation to permit simultaneous dual-language broadcasting.

The desirable requirements include: general improvement of sound quality; possibility of receiving stereo sound programmes in mono (direct compatibility); possibility of receiving mono programmes on stereo receivers (reverse compatibility); no discernible interference between picture and audio channels; possibility of transmitting either a stereo or two or more separate high-quality mono channels; maximum possible technological compatibility with existing systems and equipment; and receiver decoder costs compatible with perceived benefits to the viewer. For $(L+R)/(L-R)$ stereo a separation of 25–30 dB is considered sufficient. For bilingual transmission, separation of the order of at least 55 dB is required.

A number of different systems have been developed, field tested and brought into use in some countries. The following are selected broadcast systems, but other systems have been used for cable television.

53.13.1 One-carrier systems (with FM subcarriers)

The Japanese FM/FM system (1978) in which the subcarrier is locked to the second harmonic of the line frequency (31.5 kHz for 525-line systems) with a switching (AM) subcarrier at 55 kHz.

The American BTSC/MTS system (Broadcast Television Systems Committee/Multichannel Television Sound) is derived from the pilot-tone FM stereo system used in radio broadcasting but with a pilot frequency (15.734 kHz) locked to the video line frequency and with 'dBx' noise reduction. There is provision for a 15 kHz stereo quadrature channel plus a 10 kHz monophonic 'second audio channel' (sap) on a low-level subcarrier at five times the vision line frequency and, where required, an engineer's order wire ('professional channel') on a subcarrier at six times the vision line frequency.

53.13.2 Two-carrier systems

The West German dual-channel system for stereo or bilingual sound transmission has (for System G) a separate carrier located 5.742 MHz above the vision carrier at -20 dB in addition to the normal sound carrier at 5.5 MHz above the vision carrier at -13 dB. The 242 kHz difference between the sound carrier is an odd harmonic of the vision line frequency. The usual sound FM carrier is modulated with $(L+R)/2$, and the second carrier with only the R signal (or with a second-language or separate audio signal) with a deviation of ± 2.5 kHz. A mode identification signal is provided on a tone-modulated 54.6875 kHz subcarrier (unmodulated for mono, 117.5 Hz AM for stereo, 274.1 Hz AM for separate audio).

The Nicam 728 system was developed initially by the BBC, but is being progressively introduced first by ITV/Channel 4, in some regions from 1989. This uses digital quadrature phase-shift keying to digitally modulate the second carrier located (for System I) 6.552 MHz above the vision carrier with a carrier level of -20 dB with respect to peak vision. The two audio channels (stereo or bilingual) are combined in time division multiplex as a digital bit stream comprising 728-bit frames each of 1 ms duration, giving a bit-rate of 728 kbit/s. The audio signals are sampled at 32 kHz with an initial resolution of 14 bits/sample and near-instantaneous compression to 10 bits/sample in 32-sample blocks. Error protection comprises one parity bit added to each 10-bit sample to check the six most significant bits. The transmission of the conventional mono sound channel is unaffected.

53.14 Teletext transmission

During the 1970s, a data transmission system riding 'piggy-back' on conventional television transmission was developed in the UK and introduced under the service designations Ceefax (BBC) and Oracle (IBA–ITV). An agreed technical specification was introduced in 1974. Teletext transmission systems have since been introduced in a number of countries, and basically similar systems but with rather different technical specifications have also been developed in some countries. Teletext uses the broadcast television signal to carry extra information. These extra signals do not interfere with the transmission and reception of normal programmes. A teletext receiver, a television receiver with additional circuits, is capable of reconstructing written information and displaying it on the screen. The system allows the transmission of very many bulletins of information, and the viewer can choose any one by selecting a three-figure number on a set of controls, usually push buttons or thumbwheel switches. After a short interval the information appears and remains for as long as it is needed.

A major use of teletext is as an information service, but it can also supplement normal television programmes with subtitles or linked pages. The entire signal is accommodated within the existing 625-line television allocation, and so costs nothing in terms of radio frequency spectrum space. It is, effectively, the first broadcasting system to transmit information in digital form.

Teletext pages look rather like pages of typescript, except that they can also include large-sized letters and simple drawings. The standard-sized words can use upper or lower case, and be in any one of six colours—red, green, blue, yellow, cyan and magenta—or white. The shape of the characters is usually based on a 7×5 dot matrix, with a refinement known as character rounding. As many as 24 rows of the standard-sized characters can be fitted on a page, and each row can have up to 40 characters. Each page can carry about 150–200 words.

The specification also allows the option of characters of twice the height of standard characters. Larger-sized characters and also drawings are made by assembling small illuminated rectangles, each one-sixth the size of the space occupied by a standard character, and these too can be in any of the six colours or in white. Part or all of each page can be made to flash on and off (usually once per second), to emphasise any particular item.

The page background is usually black, although the specification allows the editor to define different background colours for part or all of the screen. Also, at the teletext editor's discretion, text can be enclosed in a black window and cut into the normal picture. Furthermore, certain receiver designs allow the whole page (sometimes in white only) to be superimposed upon the picture.

It is potentially possible for the system to carry up to 800 single pages but, because of the way the pages are transmitted, this could mean an appreciable waiting time between page selections. So for the moment not all 800 pages are used at any one time.

Additional circuits are needed in a television receiver to decode teletext. First, data extraction and recognition circuits examine the incoming 'conveyer belt' of teletext data lines, and extract those signals which make up the page which has been selected. The data from these lines is then stored, usually in a semiconductor memory, so that the page can be displayed at the same rate as a normal picture. The binary number codes are then translated to their corresponding characters or graphics patterns. Finally a video raster scan representation of the page is switched onto the screen.

The reception of teletext, like normal television, is susceptible to various kinds of interference. IBA trials, however, indicated that the great majority of those who receive television pictures well are also able to receive teletext.

Figure 53.24 shows one teletext data organisation.

53.14.1 Enhanced graphics

Work is continuing in the UK aimed at defining and specifying broadcast teletext systems in such a way that public broadcast services can be introduced or progressively updated at the lowest possible receiver-decoder costs. Each hierarchical step embraces the properties of lower levels and permits subsequent extensions to higher levels and a fine-line drawing set. Fine-line graphics have the general appearance of alphageometric graphics but demand less 'storage' in the decoder, though with the penalty of providing rather less scope than true alphageometric systems.

The following 'levels' for broadcast teletext systems have been proposed:

Level 1 This is the current operational teletext specification incorporating the joint BBC–BREMA–IBA specification and subsequent minor amendments. Fully operational teletext magazines of news and information have been available in the UK on all three programme channels for a number of years. By 1989 there were over four million teletext receivers in use in the UK, about 20% of television homes. The specification includes optional extensions for linked pages, basic page check word, programme or network label and data for equipment control, including time and data in UTC with local offset.

Level 2 This level offers Polyglot 'C' for multi-language text and incorporates a wider range of display attributes which may be non-spacing.

Level 3 This level introduces dynamically redefined character sets, permitting the use of non-Roman characters such as Arabic or Hindustani, of alphageometric instructions (AGI) in a closed-circuit mode. Tests are shortly to start using similar AGI pages in the operational Oracle magazine.

The efficiency of the system is underlined by its ability to contain within two normal length teletext pages sufficient AGI to provide the following three separate displays:

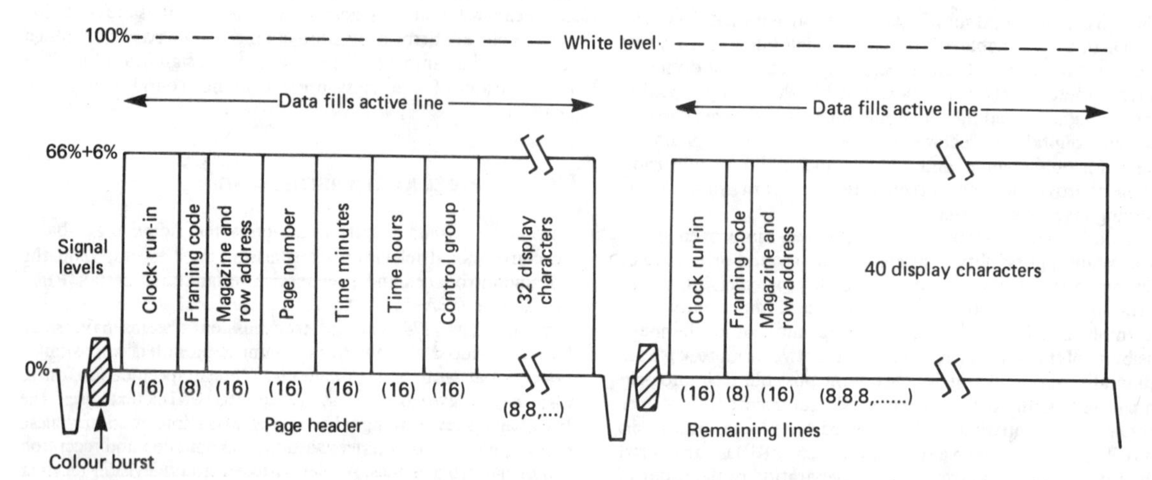

Figure 53.24 Teletext data organisation. (The figures in brackets are the number of binary digits)

(1) Multi-coloured pie chart.
(2) A chessboard with a game in progress.
(3) An engineering announcement incorporating a large IBA logo.

A further one-and-a-half pages of AGI provide enough data for a fully coloured detailed weather map, with symbols representing sunshine and rain clouds, and carrying typical weather forecast captions and titles.

Level 4 This covers the form of alphageometric coding which the IBA is currently developing. Unlike lower levels it requires the use of some computing power within the decoder, and some additional storage capacity.

Level 5 This level covers the transmission of high-definition still pictures. The limitation to the quality of these still pictures, which may be superior to that of the associated television system, becomes that of the receiver display device. Transmissions at this level have included the Oracle-adapted broadcast transmissions of 'Picture-Prestel' during IBC80 (September 1980) and the high-quality still-picture system demonstrated by the BBC.

The level 4 coding as demonstrated by the IBA differs from previously proposed alphageometric systems in two significant ways. Firstly, 'incremental coding' is used which is particularly suited for drawing maps and other irregular outlines and, secondly, a 'delay' command has been introduced. This latter feature is regarded as of particular importance where information is to be transmitted in the form of a series of connected explanatory pages. For example, the IBA system may be explained by a series of 10 pages, the contents of which appear to be drawn as the viewer watches and the complete sequence of pages involves the transmission of only one or two pages of teletext (one or two thousand bytes).

The experimental level 4 system could be introduced into the existing UK system without affecting existing decoders when it becomes economically feasible to build consumer-type decoders having sufficient storage. The system is capable of receiving the currently broadcast level 1 teletext pages, in addition to displaying a series of specially coded pages.

53.15 Cable television

53.15.1 Historical

The use of cables to convey entertainment to the home goes back to the earliest days of the telephone: Gilbert installed it in 1882 and advised Sullivan to do the same to listen to the stage performance of their masterpieces.[3]

In Budapest in 1892 a network carried news services to homes and in the same period the Electrophone Company in London enabled subscribers to listen to theatres, balls and churches.[3]

These rather specialised services did not, however, achieve a wide acceptance and it was not until the advent of the popular demand for broadcasting in the 1920s that wired distribution set off on the path which now seems certain to lead ultimately to the fully integrated communication network carrying any kind of signals—voice, data or vision—to and from each member of the community.

In April 1924 the first commercial radio relay service was started by A. L. Bauling in Koog aan de Zaan in Holland.[3] He was followed independently by Mr A. W. Maton[4] of Hythe, a village in Hampshire, who also started experimenting in 1924 and opened his commercial service in January 1925.

It was upon this foundation that cable television systems grew in Europe, although their rapid growth in North America had a different origin.

53.15.2 Types of systems

In Europe in the early 1930s a substantial business developed in the wired relay of sound radio programmes at audio frequency and the total number of subscribers reached about 1 million in 1950. This business was confined to Europe, mainly Holland and the UK, and developed in large provincial cities, e.g. Hull, Leeds, Newcastle, Nottingham and Bristol. The cost of the cable connection, programme switch and loudspeaker was competitive with the radio receivers of the day and this, with the convenience and reliability of the system, provided the economic base for its growth. With the advent of television in the early 1950s it became necessary for the relay operators, if they were to stay in business, to develop television distribution methods which would, as with sound radio, be competitive with direct reception off-air in the generally good reception conditions of the large provincial cities. The result was the HF multipair system, *Figure 53.25*, which can serve simplified receivers directly, or conventional off-air receivers through an 'inverter'. Since the programmes are carried on physically separate channels these systems may be described as space division multiplex.

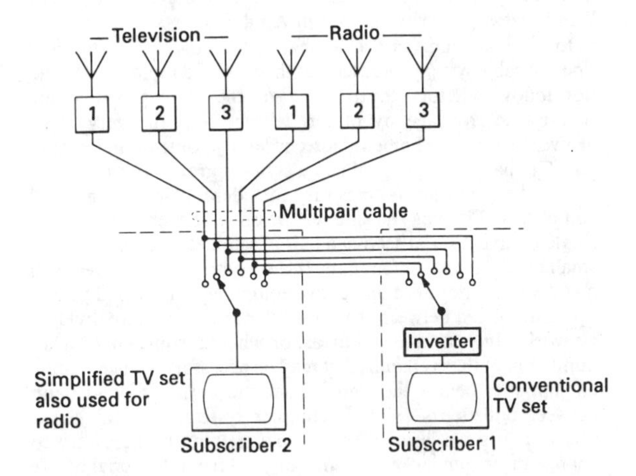

Figure 53.25 HF multi-pair cable television systems

The other type of system was developed, principally in the USA, for different reasons. There is no obligation on the United States broadcasting industry to serve the smaller towns where the available advertising revenue may be insufficient to support a plurality of broadcasting stations. This led to the development of systems which were purely concerned with the improvement of reception of weak signals from distant stations. In these circumstances the system does not have to compete as an alternative means of reception of local transmissions. The most suitable system for this purpose uses a single coaxial cable to which standard television receivers can be directly connected. The different programmes are carried on different frequency channels in a manner exactly analogous to over the air broadcasting (*Figure 53.26*) and may be referred to generally as frequency division multiplex.

The programme material for a cable system could be obtained by reception of normal off-air signals, by direct cable connection from the broadcast authority, or from other sources such as local studios or videotape recorders. There is no technical problem in providing such a signal mix, but in the UK there were restrictions on charging for service. In 1981 the Government issued several licences for pay TV trials and appointed the Information Technology Advisory Panel (ITAP) to study the range of services

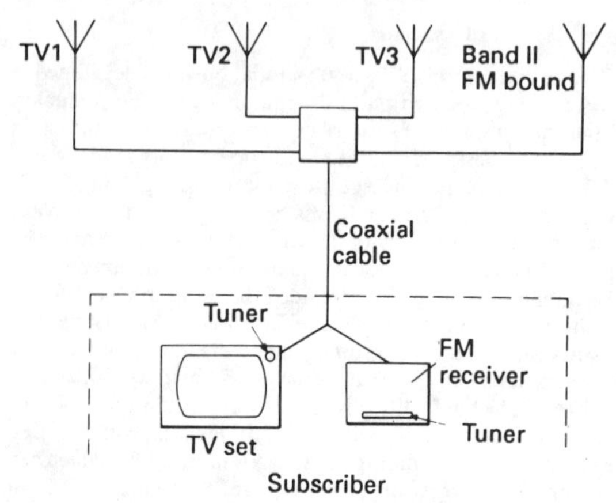

Figure 53.26 VHF coaxial cable television system

which a cable system could offer. 1982 was Information Technology year and the ITAP Report[5] in February was followed by the Hunt Report[6] in September and a Government White Paper[7] policy statement in April 1983.

In the following years a number of franchises to install and operate cable systems were let, but the expected rapid growth did not follow. Factors contributing to the slow growth were nationwide coverage by off-air delivery and the very rapid growth in use of the home videorecorder. By contrast, in Holland and Germany, multilingual cable systems grew quite rapidly.

Wired distribution systems vary considerably in their size and complexity. Perhaps the simplest possible concept is that of a single receiving aerial serving a pair of semidetached houses or a small block of three or four flats. In a very strong signal area such a system may not need any active equipment, the signal picked up being shared between the users by means of passive dividing networks. In weaker signal areas, or where a somewhat higher number of outlets is required, it may be necessary to incorporate an amplifier before the signal from the aerial is divided up between outlets. In still more complex systems, in weak signal areas or to serve large blocks of flats, it may be necessary to change the frequency of the signal derived from the aerial before it is distributed to individual users. This is to ensure that at a television receiver interference does not occur between the signal carried over the wired system and the signal available directly from the broadcasting station.

The most sophisticated wired systems may comprise an elaborate directional aerial system, sited at a high vantage point to gather in weak signals from a distant transmitter; a device to change the frequencies of the incoming signals to other channels more suitable for distribution by wire; and an extensive network of underground and/or overhead cables carrying the signals to many thousands of households in a given locality.

In summary, then, the two basic system types are space division multiplex and frequency division multiplex.[8,9] As will be seen, later optical systems combine these techniques. Another system classification is network structure, e.g. tree and branch (*Figure 53.27*) or mini-hub (possibly with a switch at the hub node) (*Figure 53.28*).

Frequency division multiplex is usually transmitted over tree and branch coaxial cable systems. Amplifiers are used to boost the signal at intervals which depend on the highest frequency being transmitted, the diameter of the cable, and the total number of amplifiers in the chain. At the nodal points the signal is split into the secondary network by an amplifier, normally referred to as a trunk bridger. Amplifiers in the secondary network are called line extenders. The band below 50 MHz is frequently used for upstream traffic. In some installations,

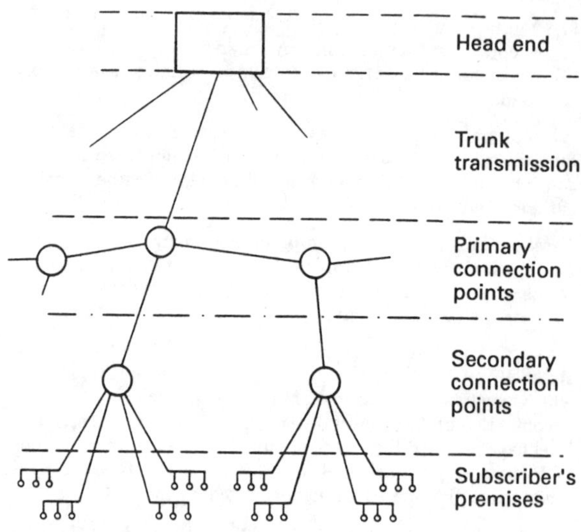

Figure 53.27 Tree and branch structure

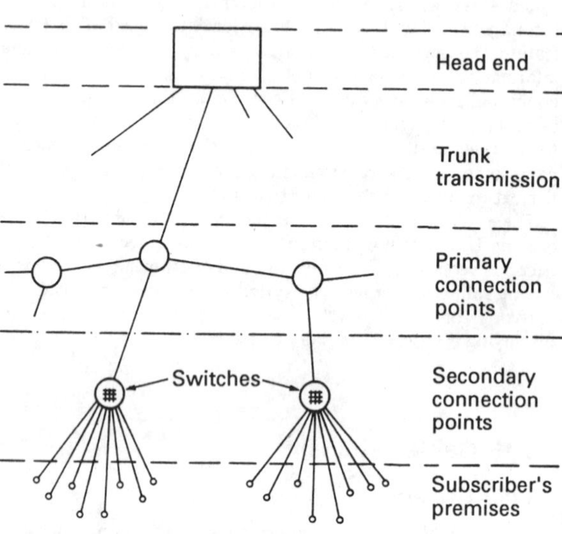

Figure 53.28 Mini-hub structure

especially the larger ones in the USA for instance, trunk transmission is over microwave links, since it is difficult to maintain good quality economically over long cable spans. A backup cable is usually installed in case the microwave link should fail, on the basis that a poor picture is better than no picture in these circumstances. Another alternative to cater for long spans is optical fibre.

In areas of high population density, the final drop cable runs along the wall of terraced houses or flats with the signal being picked off by an asymmetrical tap. The mini-hub structure was introduced initially in areas of lower population density.

53.15.2.1 HF systems

An HF system uses a single channel in the HF band for all programmes and each programme is carried on a separate pair of wires in a multi-pair cable. The vision signals lie in the band between 4 and 10 MHz where cable attenuation is relatively low but the frequency is high enough to avoid difficulties in the demodulation process in the receivers. The result is that a very large number of subscribers can be fed from one amplifier installation; the number ranges from several hundred to as many as 2500. The sound accompanying the vision signal is carried at audio frequency and operates the loudspeaker directly. The sound radio programmes are also carried at audio frequency on separate pairs. More recent installations of this type employed a 12-pair cable of special construction providing capacity for six television programmes with a further six sound radio programmes. In an alternative system a 4-pair cable was used and the sound accompanying the vision signal was carried on an amplitude modulated carrier at approximately 2 MHz, enabling the same pairs to carry a further four sound programmes at audio frequency.

With the increase in the number of channels distributed and the decrease in the cost of standard UHF television receivers HF systems have been abandoned.

53.15.2.2 VHF systems

Modern VHF systems employ wideband amplifiers covering the range 40–450 MHz. Owing to the relatively high cable attenuation at these frequencies, amplifiers must be inserted at fairly frequent intervals. One amplifier may serve up to 150 households and a total system gain up to 800 dB can be achieved with satisfactory performance in respect of noise and cross-modulation, enabling large towns to be served from a single reception site. The system may also carry FM sound programmes in band II exactly as broadcast.

These systems are in use in North America and Europe but following closure of the 405-line system only UHF receivers are used in the UK. Some systems still use VHF for distribution, however, because of the advantage of lower cable attenuation at the lower frequencies. Conversion from VHF to UHF is either performed at the final distribution node and shared between perhaps 100 subscribers, or performed by a separate box next to the TV receiver, especially in cases where pay TV is used and the box also provides a decoding function.

The cable and amplifiers in a VHF system are capable of carrying some 60 channels, and some North American systems use two cables to provide 120 channels. If these signals are delivered to the home receiver there are a number of factors which must be taken into account:

(1) The receiver must have very high selectivity to avoid adjacent channel interference.
(2) Good screening is required to avoid picking up any strong off-air broadcasts in the area.
(3) Radiation from the local oscillator of a receiver connected to the system must be reduced to prevent interference to a nearby receiver tuned to certain off-air frequencies.

53.15.2.3 UHF systems

Modern UHF systems employ wideband amplifiers covering frequencies up to 860 MHz. Because of the high cable attenuation, such systems are limited to areas containing about 200 subscribers. Single standard UHF receivers are used and the programme required is selected by means of the tuner fitted in the receiver. As with the VHF system, FM sound programmes are usually provided in band II exactly as broadcast.

UHF frequencies are unsuited by their nature to large cable systems and are unlikely to find more than limited use in the total system sense. Use of UHF for the final distribution to the subscriber is now becoming standard in the UK.

53.15.2.4 Mini-hub systems

This type of system, introduced in Holland, had the topology of *Figure 53.28*, but without switching at the star nodal point (or mini-hub). They were introduced to:

(1) Avoid the problem of poor receiver selectivity. Transmission from head-end to mini-hub is on adjacent VHF channels. Conversion at the mini-hub to VHF and UHF bands allowed alternate channels to be delivered to the receiver.
(2) Permit later introduction of a switch at the hub.

53.15.2.5 Switched systems

In the old HF systems various switching arrangements were in use. One example is shown in *Figure 53.25*. An alternative was for pairs in the cable to be allocated to individual subscribers to connect them to a switching centre or programme exchange. Channel selection was controlled from the subscriber's premises using either a separate pair or a carrier frequency above the video band in the upstream direction.

Modern systems use the switched star configuration of *Figure 53.28*. Reasons for their introduction are:

(1) To provide long-drop cables in rural areas where cost and performance benefit can be obtained by delivering only the one or two programmes being viewed.
(2) To permit an increase in number of channels on the trunk distribution network without modification to the subscribers' drop.
(3) To provide controlled access to pay channels or alternative services, thus preventing theft of channels.
(4) To reduce the cost of churn (subscriber connection/disconnection).

Channel selection is signalled from the subscriber using an out-of-band data channel. Polling systems are frequently used for this function, especially in tree and branch systems where the upstream capacity is limited.

One method of switching is to use a frequency-agile converter. This is essentially a tuner, similar to that in a television receiver, which selects the required channel and converts it to a suitable frequency for delivery over the subscriber's drop. Examples of this approach are systems by GEC and Cabletime in use in the UK where the subscriber drop operates at UHF, or a system by Texscan in USA using a frequency in the band between 50 and 80 MHz which permits a much longer subscriber drop.

Another method is electronic space switching of the baseband signal as used in the BT system installed in Westminster.[10] This system employs f.d.m. FM for transmission over optical fibre for distribution from the head end to the switch. At the switch the signal is demodulated to baseband video before remodulation to UHF for the final drop, so baseband switching is quite appropriate. The network structure is similar to the mini-hub of *Figure 53.28*, in which each hub has a dedicated trunk.

53.15.2.6 Pay TV

This mode of operation has been introduced to offer a greater range of programming and to enable the network operator to collect a fee for such premium services. The simplest form is payment on a per-channel basis; a number of channels each carrying programmes on a particular subject or group of subjects is offered and subscribers elect to pay a monthly subscription for the channels in which they are interested in the same way that people subscribe to magazines. In the USA the channels are grouped into tiers to facilitate this type of offering; such systems are now widespread. A simple way of implementing pay TV on coaxial systems is to use channels to which ordinary receivers cannot tune and then to supply simple frequency converters to each subscriber. Security against illicit use of converters is a problem, however, and it is more difficult to operate a tiered system. The alternative is to scramble the pay channels and provide the subscriber with a descrambler. The system keeps a record of what each subscriber is allowed access to and as an added security measure might periodically update the descrambling key.

Payment on a per channel basis is satisfactory in many circumstances but it is necessary to achieve payment on a per programme basis if the full potential of pay TV is to be realised. This is because the owners of the most suitable programmes, such as first-run films and important sporting events, will not make them available to television unless they can be sure of receiving their share of the money paid by the public to see their particular programme, as they would from a cinema or sports arena.

Again on coaxial systems unauthorised access is prevented by use of scrambling and the programme is released for view by sending a control signal with an address code for the subscriber in question. The request from the subscriber may be made either by telephone or as the result of interrogation and response signals from a central computer using the upstream signalling channel. These types of system started commercial operation in the USA in 1973.

53.15.2.7 *Optical fibre*

Potentially one optical fibre can carry many hundreds of TV channels, but technological developments such as those in the RACE Programme (see *Section 53.15.3*) will have to be completed before this can be realised economically. Optical fibre cannot be used in the tree and branch subscriber drop because of optical power budget limitations and cost of the electro-optic transducers. In spite of this cost, one of the first applications of optical fibre in cable systems was the subscriber drop of a mini-hub system developed by Times Fibre for use in high-density areas (e.g. blocks of flats).

Optical fibre is much more appropriate for use in trunks, or the even longer-distance super-trunks where degradation of the signal must be avoided. One of the first fibre trunks was installed in the BT Westminster cable system. Using multimode fibre it spans 4 km with no repeaters and carries four frequency-multiplexed TV channels on each of several fibres. Nevertheless an eight-fibre cable has a considerably smaller cross-section than a repeatered coaxial cable carrying the same number of TV channels. Using single mode fibre, spans of 20 km have been achieved. Up to 60 channels per fibre have been reported, using microwave subcarrier multiplexing.[10]

More channels can be added on one fibre by using different optical wavelengths (i.e. very coarse frequency multiplexing in the optical domain). High-density wavelength multiplexing is now being developed. Combination of digitisation and optical multiplexing can provide long-distance low-distortion multichannel transmission.

53.15.3 The future

The future of cable systems will depend on many factors such as:

(1) Regulatory aspects, which differ from country to country.
(2) Public demand, closely bound up with economics.
(3) Alternative means of delivering programmes.
(4) Satellite broadcasting; even low-power satellite signals can now be received in many homes.
(5) Introduction of HDTV, which requires more bandwidth than is available for broadcasting over the air.

In some instances, especially perhaps in countries or areas where off-air programmes are limited or reception is poor, wired broadcasting systems may well continue in their present form. Certainly those who have invested in cable will wish to see a return on capital and this tends to be long term. The more recent installations already provide many services in addition to the type of programming received off-air,[11] and others have built in an ability to up-grade.

There is no fundamental technical reason why we should not have a universal communications network, wideband, two-way and switched, capable of conveying any kind of information, visual, voice or data, from anywhere to anywhere. Besides television and sound for entertainment purposes the range of services which might be provided by such a universal network include:

Telephone and viewphone facilities.
Electricity, gas and water meter reading.
Facsimile reproduction of newspapers, documents, etc.
Electronic mail delivery.
Business concern links to branch offices.
Access to computers.
Information retrieval (library and other reference material).
Computer-to-computer communications.
Special communications to particular neighbourhoods or ethnic groups.
Surveillance of public areas for protection against crime.
Traffic control.
Fire detection.
Educational and training television and sound programmes.

There are many more which could be added to the list. Existing legislation in the UK allows for incorporation of such a wide range of services, but very stringent regulatory requirements and operating practices have to be met if telephony services are to be provided. Future developments will largely depend on non-technical aspects such as political and legal decisions regarding licencing, copyright and other complicated issues. As an example of regulation, in the USA, the Federal Communications Commission has laid down certain rules for the licensing of new wired distribution systems which call for at least 20-television-channel distribution capability, together with some capacity for return communication.

A possible route to the universal communications network in Europe is through the European Commission initiative for Integrated Broadband Communications (IBC) under the heading of Research and Advanced-development for Communications in Europe (RACE). Definition of such a network started in 1986. The overall objective is to provide a total integrated infrastructure for all services. The target is to establish requirements, agree standards, develop the necessary technology, determine network topologies, define the network architecture, and develop suitable products so that commercially viable IBC services can be introduced in a critical minimum number of Member States of the European Community by 1996.

If RACE is successful, the services and facilities currently offered by cable systems will be incorporated into the IBC and the existing cable systems will be gradually replaced at a pace dictated primarily by economic considerations.

53.16 The equipment required for sound broadcasting

The basic equipment required for sound broadcasting would seem relatively simple, namely, a microphone, AF amplifier with volume control, programme distribution to transmitter and a feeder from transmitter to aerial. In practice the technique is much more complicated and involves a smooth change from one programme source to another, the insertion of special effects, e.g. crowd noises, motor car starting, etc., the reduction of the dynamic range of programme volume without destroying artistic values, the recording and reproduction of programmes, the monitoring of output, and the provision of a communication system to coordinate and supervise the distribution to and from studios and to transmitters. A block schematic diagram illustrating a possible grouping of programme requirements is shown in *Figure 53.29*. The programme connection from each source to the control room, from which the composite programme is distributed to the transmitters, is paralleled by a link enabling communication between each source and the distribution centre to be established and to allow a cue feed from the preceding programme to be given when necessary.

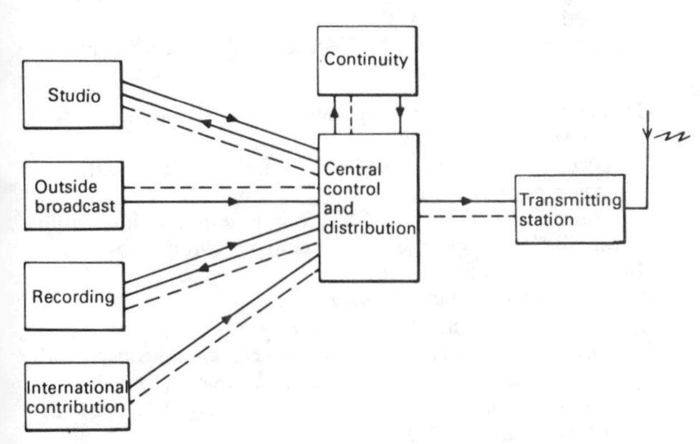

Figure 53.29 Programme collection and distribution (full lines, programme links; dotted lines, two-way communication control links)

53.16.1 The studio

The studio will need to be treated acoustically to give the right 'atmosphere' for performers and listeners, and this will depend on whether it is required for speech, drama or music. Reverberation time, which is time taken for sound to die away to 10^{-6} of its original energy intensity after it has been cut off, is important and should be reasonably constant over the AF range. For speech it should be of the order of 0.2 s and for large concert halls a value of about 1.5 s would be normal. Drama may call for considerable variations from a very low value, less than 0.1 s simulating an open air scene, to a high value, about 3 s simulating a large room with many reflections from relatively bare walls. Artificial reverberation is often used to cover the latter situation.

The range of audio frequencies encompasses about nine octaves and separate studio treatment has to be applied to control the low, middle and high frequencies. Acoustic design is too specialised for inclusion here.

53.16.2 The studio apparatus

The requirements of one particular broadcasting organisation can differ appreciably from another; one may need a relatively simple operation of recorded disc or tape programmes, interspersed with news and advertisements, whereas another may have to cater for live programmes using many performers in large orchestral, chorus and soloist ensembles or in drama productions. Nevertheless there may be a minimum of three microphone channels (rising to perhaps 12), a disc and a tape reproducing channel, together with facilities for adding artificial reverberation or echo, for aural monitoring and for measuring programme volume. An independent channel may be included to permit an emergency announcement to be superimposed on the programme, or a narrator for a drama programme to be inserted.

Outside sources are fed at zero level from the central control room to the studio cubicle, where they can be heard on headphones by using prefade keys.

Talkback may be provided from a microphone in the studio control cubicle to a loudspeaker in the studio so that the producer can issue instructions during rehearsal. During transmission the studio loudspeaker is automatically cut off, but talkback can be provided to artists via headphones. The studio loudspeaker can be used for acoustic effects, which reproduce, for example, crowd noises so that studio background atmosphere is preserved during the insert. A change of background by fading out the studio to insert the effect may destroy an illusion.

Special distortion effects such as the simulation of telephonic speech quality may sometimes be required, and this is achieved by inserting variable frequency-response networks in a channel.

When a discussion is to be broadcast between two or more speakers in geographically distant locations, each must be able to hear the others; unless special precautions are taken 'howlback' is a possibility. This is prevented by feeding back the output of one studio to the other from a point prior to the insertion of the other contributor. The 'clean' feed is supplied on headphones to the latter.

53.16.3 Continuity control

When a programme is normally broadcast nationwide, special steps may be required to maintain it at all normal broadcasting times, and at some point in the main chain before distribution to the transmitters a continuity control, *Figure 53.29*, is inserted. The apparatus is a modified version of the studio equipment with a microphone which can be switched into the circuit by an announcer. The latter can step in and take control with a special announcement or explanation for an over-run of programme or a technical fault, and will also have facilities for inserting a recorded item as a fill-in for an under-running programme.

53.16.4 The main control centre

The chief purpose of the main control centre is to accept programmes from many sources and route them to their destinations, either the studios or the transmitters. A degree of aural and visual monitoring is carried out but no gain control is normally performed except when a channel is being set up with test signals. Communication circuits will be available to all sources originating programme, and there will be facilities for carrying out engineering tests such as those for noise, distortion, frequency response, etc. If the majority of programmes are recorded the distribution network can be made to operate automatically using a memory storage system.

53.17 Stereophonic broadcasting

Monophonic broadcasting suffers from the disadvantage that no spatial sense can be given because the sound source is effectively a single ear. The aim of stereo is to make the sounds appear to come from the position they originally occupied. To do this completely would require many microphone channels with as many loudspeakers. In practice it is not normal to use more than two channels. The illusion of space has to be created by a difference in sound amplitude and phase from two loudspeakers, which should have identical characteristics and operate in phase. Qut-of-phase sources give a diffuse sound image appearing to come from behind the head. An amplitude unbalance of about 20 dB shifts the sound image from the centre line between the loudspeakers to the extreme edge.

The two stereo signals are designated *A* (left-hand side) and *B* (right-hand side), and since monophonic receivers must be able to use the stereophonic signals it is necessary to form an addition of *A* and *B* to give a compatible monophonic version.

A special multiplex system (the GE Zenith system used in many countries including the UK and known as the 'pilot tone' system) is employed to permit both signals to be transmitted on the same carrier and to be separated at the receiver, *Figure 53.30*.

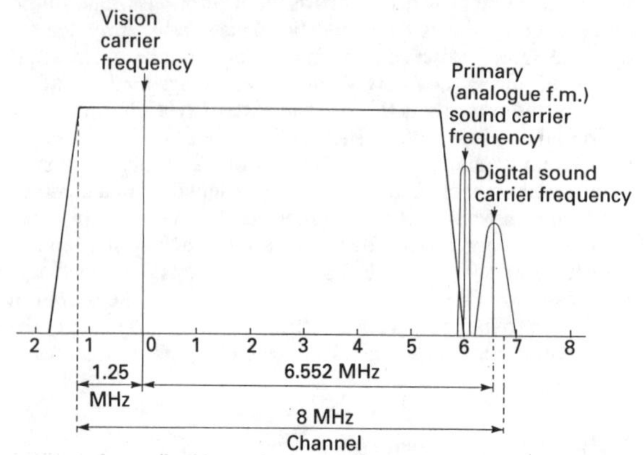

Figure 53.30 The frequency bands occupied by the digital sound signal in relation to that of the picture and primary (analogue, FM) sound signal components of the transmitted signal

The $(A+B)$ signal is left unchanged covering the AF range 30 Hz to 15 kHz but the $(A-B)$ signal together with a subcarrier at 38 kHz is applied to a balanced modulator whose output contains only the amplitude-modulated sidebands covering the range 23 to 53 kHz. A low-amplitude pilot tone at half the subcarrier frequency, i.e. 19 kHz, is added to the $(A+B)$ and the $(A-B)$ sidebands and this composite signal is used to frequency modulate the transmitter. The $(A+B)$ signal is separated at the receiver from the pilot subcarrier (19 kHz) and the $(A-B)$ sidebands. The 19 kHz subcarrier is filtered from the other signals and is amplified and multiplied by two before being applied with the $(A-B)$ sidebands to a detector which extracts the $(A-B)$ signal. Addition and subtraction of the recovered $(A+B)$ and $(A-B)$ signals recovers the original left (A) and right (B) signals. The monophonic receiver rejects the pilot tone and $(A-B)$ sidebands and reproduces only the $(A+B)$ signal.

53.17.1 Stereo sound broadcasting with television

Although television sound is normally monophonic and transmitted using a single frequency-modulated carrier, various means of broadcasting stereo sound with television have been devised. In Europe, two systems have been developed using a secondary sound carrier and provide a stereo service along with compatibility for existing mono listeners. Both systems can be used for dual-language operation.

The analogue dual-carrier system, developed in West Germany, uses a primary sound carrier frequency modulated with the $A+B$ (left + right) signals. A secondary carrier is frequency modulated with the B (right) signal only. For stereo listeners, the right signal is derived directly from the secondary carrier and the left signal derived by matrixing in the decoder. For mono listeners, the $A+B$ signal is obtained from the primary sound carrier in the normal way. If dual-language operation is required, a separate language is modulated onto each carrier, and a control signal is sent to the decoder to control the matrixing.

The NICAM-728 Digital system,[12] developed in the UK, is an approved UK transmission standard for digital stereo sound with terrestial television. Although originally developed for UK System I television, it has also been adapted for use with television Systems B and G.[13]

The $A+B$ mono-compatible signal is transmitted, in the normal way, by frequency modulation of a primary sound carrier. The separate left and right stereo signals are digitally encoded using a near-instantaneous companding system and modulated onto a secondary carrier by differentially encoded quadrature phase shift keying (DQPSK), *Figure 53.30*.

The NICAM-728 system, when used in conjunction with a digital distribution system, makes possible a complete digital-sound broadcasting chain from the studio to the listener. Compatibility with existing mono listeners is maintained as well as the digital encoding being similar to the MAC/packet systems proposed for satellite use.[14]

Characteristics of NICAM-728 used with System I television are:

(1) Primary carrier: +6 MHz and −10 dB (relative to vision carrier).
(2) Secondary carrier: +6.552 MHz and −20 dB (relative to vision carrier).
(3) Data modulation: DQPSK—four-state phase modulation in which each change of state conveys two data bits.
(4) Overall bit rate: 728 kbit/s.
(5) Sampling frequency: 32 kHz.
(6) Initial resolution: 14 bits/sample.
(7) Companding characteristic: near-instantaneous, with compression to 10 bits/sample in 32 sample (1 ms) blocks.
(8) Pre-emphasis: CCITT Recommendation J.17 (6.5 dB attenuation at 800 Hz).

53.18 Distribution to the transmitter

The sound component of television and radio programmes is distributed between the studio centres and the transmitters by means of a network of cable and microwave links. Pulse-code modulation (PCM) techniques[15–17] are replacing older analogue methods of distribution. This provides a higher standard of performance and in the case of stereo radio perfect matching between left and right channels.

Pulse-code modulation ensures that the system is immune to all but the most severe noise and distortion on the bearer circuit. The presence or absence of the pulses may be detected up to a threshold point above which the effect on the audio channel rapidly becomes catastrophic. It can be shown[14] that the threshold for an ideal detector occurs with white noise at a peak signal/RMS noise ratio of approximately 20 dB. Taking this into account a PCM distribution system may be designed so that the audio performance is determined by the coding and decoding equipment, and this performance maintained throughout the complete distribution system.

53.19 Television sound distribution, sound-in-syncs

Television sound can be distributed by the sound-in-syncs (SIS) system.[18] This combines the sound in PCM form with the video signal in such a way as to enable the complete programme to be routed via the same video link. At the transmitter site the signal is split back into separate audio and video signals and transmitted in the usual way to the viewer. The composite SIS signal offers advantages financially due to economies in audio links, and also operationally since the audio cannot become separated from the video between terminal sites.

The SIS system is essentially a time-division multiplex system in which the audio signal is coded in pulse-code modulated form and the pulses inserted during the line-synchronising period of the video waveform (*Figure 53.31*).

The video may be any 625-line colour signal using a field waveform comprising five equalising pulses, five broad pulses

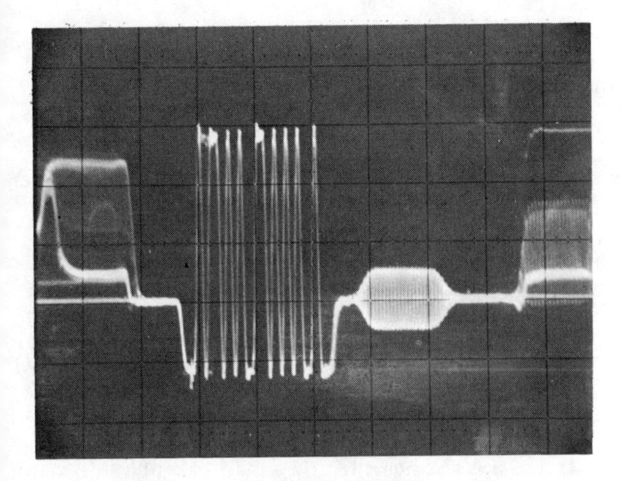

Figure 53.31 Sound-in-syncs waveform showing sound pulses inserted into the line synchronising period

and then a further five equalising pulses. The equipment with modification may be used with 525-line System M video signals.

The audio channel obtained has the following performance:

(1) Frequency response: 50 Hz–13.5 kHz (±0.7 dB reference 1 kHz); 25 Hz–14 kHz (−3 dB reference 1 kHz).
(2) Signal-to-noise ratio: 53 dB CCIR Rec 468-2 quasi-peak meter and weighting network, bandwidth 20 Hz to 20 kHz.
(3) Nonlinear distortion: 0.25% (1 kHz at full modulation).

The quality of the audio channel is independent of the noise and distortion on the bearer circuit up to a threshold point above which the audio channel rapidly deteriorates. The threshold point for SIS is to a large extent determined by the ability to separate the sync pulses from the SIS waveform. This is because the leading edge of the line-sync pulse is used as the timing reference for the insertion of the sound pulses.

Typical distortions permissible on the bearer circuits are:

(1) White noise: 23 dB (peak signal/RMS unweighted noise).
(2) Pulse K rating: 8%.
(3) High-frequency loss: −8 dB at 4.43 MHz.

53.19.1 SIS equipment description

The simplified block diagram of the complete equipment is shown in *Figure 53.32*. The audio input is first compressed in dynamic range by means of a circuit that is basically a fast acting audio limiter. The signal is then sent to an analogue-to-digital converter (A/D) where it is sampled at a frequency equal to twice the video line frequency (31.25 kHz) and each sample converted into a 10-bit binary code. This 10-bit code is then temporarily stored while the next sample is being converted. The two codes are combined to form a 20-bit word, a marker bit added making 21 bits total, and the complete message inserted into the following line sync pulse at a peak-to-peak amplitude of 0.7 V for a nominal 1 V video signal, *Figure 53.33*. The individual pulses are shaped into sine-squared form with a half-amplitude duration of 182 ns, and a spacing of 182 ns between the middle of the pulses.

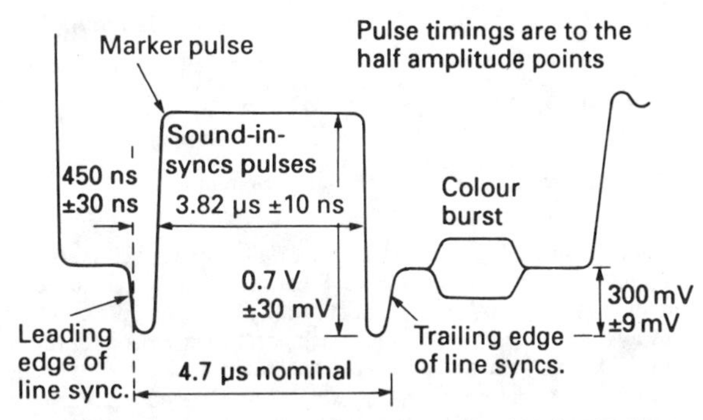

Figure 53.33 Idealised waveform showing combined pulse group in the line-sync period

These pulses have negligible energy above a frequency of 5.5 MHz and so are quite suitable for transmission on a video link. During the field-blanking period alternate equalising pulses are widened to accommodate the pulse groups, *Figure 53.34*. At the decoder the sync pulses are blanked out and a regenerated sync waveform, which is standard in all respects, gated in.

In order to minimise variations in the mean level of the pulse group, the 10-bit words are interleaved and one word

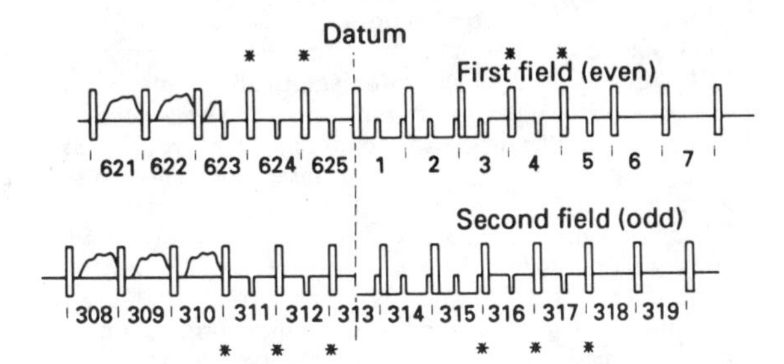

Figure 53.34 Modified field-blanking waveform

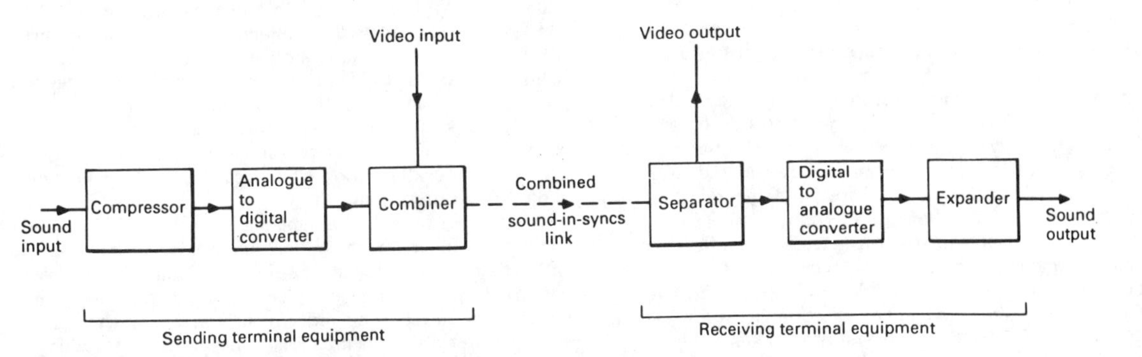

Figure 53.32 Simplified block diagram of the sound-in-syncs system

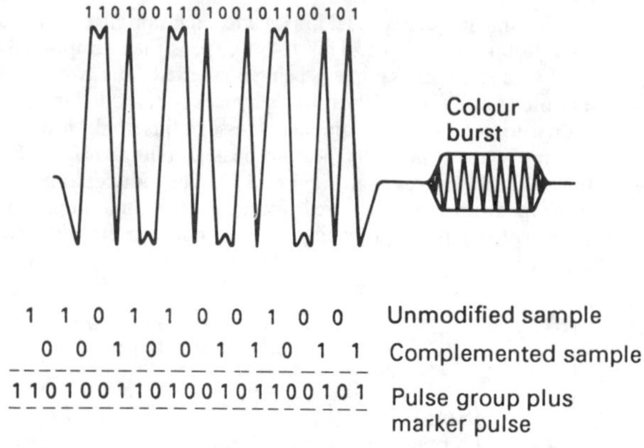

Figure 53.35 Details of pulse group waveform

complemented (logic 1s exchanged for a logic 0 and conversely), *Figure 53.35*. In addition, the least significant bits are arranged to come first so as to keep the most rapidly changing pulses as far away as possible from the video back porch. The reason for this is that, if the video bearer circuit has low-frequency amplitude distortion, variations in mean level during the sync pulse can be impressed on to the back porch, and if a clamp is used these variations may then be transferred to the picture period and produce sound-on-vision effects.

At the receiving terminal the video signal is separated from the combined SIS signal and restored to a standard form. The 21-bit message is sent to a digital-to-analogue (D/A) converter, where it is split into the two 10-bit codes. Each code is then separately decoded. The analogue output of the converter is then fed to the expander where the dynamic range of the audio signal is restored.

53.19.2 Dual-channel sound-in-syncs

With the prospect of stereo sound with television in the UK, a dual-channel sound-in-syncs system[19] has been developed. The DCSIS equipment combines two high-quality audio signals along with a video signal for transmission on a single video link. Two separate and independent sound channels are provided, enabling the possibility of either stereo or dual-language operation.

The audio signals are sampled at 32 kHz and initially encoded to an accuracy of 14 bits/sample. Digital compression to 10 bits/sample is employed using the NICAM-3 near-instantaneous companding system,[20] an established standard format for digital sound transmission.

The data rate for two channels is 676 kbit/s and a total of 44 data bits are necessary in each line-synchronising pulse. In consequence a quaternary (four-level) code is used to cope with the information rate. Unlike mono SIS, the data is non-synchronous with the television line-frequency, and is accommodated by providing the slightly higher gross data capacity of 687.5 kbit/s along with a means of justification.

The data is inserted into most line-synchronising pulses by using a marker pulse and 22 quaternary symbols. Justification is achieved by using, every five or six lines, a marker pulse with only 20 symbols. Line-rate equalising pulses are widened to make room for the data. Each symbol is shaped into sine-squared form with a half-amplitude duration of 182 ns and a symbol spacing of 182 ns. This, like the mono SIS system, ensures that the signal is within the bandwidth limitations of a video circuit.

The audio performance is similar to that obtained using NICAM-3 equipment, and interconnection between systems may be made using HDB3 at a bit-rate of 676 kbit/s. The high-density bipolar (HDB3) code is used to ensure that the signal has no d.c. component and that long strings of zeros are avoided. Normally 1s are coded alternatively as positive and negative pulses, 0s are coded as spaces. If more than three consecutive spaces are to be transmitted the fourth space is transmitted as a pulse but with the same polarity as the previous pulse.

53.20 The 13-channel PCM radio distribution system

The 13-channel PCM system was developed in order to meet the demand for high-quality sound circuits capable of carrying stereo programme material between the studio centres and the main transmitters. All channels provide the same high standard of performance, and any two channels may be used as a stereo pair over long distances without loss of quality. The performance achieved on a single codec is as follows:

(1) Frequency response: 50 Hz to 14.5 kHz ($\pm$0.2 dB with respect to 1 kHz, typically -1.5 dB at 30 Hz and 15 kHz).
(2) Signal-to-noise ratio: 57 dB CCIR Rec 468-2 quasi-peak meter and weighting network, bandwidth 20 Hz to 20 kHz.
(3) Nonlinear distortion: 0.1% (1 kHz at full modulation).

53.20.1 Bit rate

The bit rate was chosen to be 6336 kbit/s. This enables the bit stream to be conveyed on a bearer circuit designed to carry 625-line television signals. The individual pulses are shaped into sine-squared form with a half-amplitude duration of 158 ns, and the pulses are spaced 158 ns apart. These pulses therefore have negligible energy above 6.336 MHz; the bandwidth may be reduced in exchange for a worsening of the immunity to noise and distortion. With a bandwidth up to 6.336 MHz the PCM system can withstand a peak signal/RMS white noise ratio of 20 dB. If the bandwidth is restricted to about 4.5 MHz then the noise immunity becomes approximately 22 dB.

53.20.2 Coding equipment

The simplified block diagram of the coding equipment is shown in *Figure 53.36*.

Each audio input is fed to a limiter via a 15 kHz low-pass filter. If two channels are used as a stereo pair, the limiters are interconnected so that if either limiter operates both change gain by the same amount, thus preventing shift of the stereo image. A 50 μs pre-emphasis network is associated with the limiter. This enables the channel to be used to feed an FM transmitter without an additional limiter at the transmitter site.

Variable emphasis limiters are used and comprise a delay line limiter with a flat frequency response, followed by a 50 μs pre-emphasis network and then finally a variable emphasis network. This arrangement is used to minimise frequent gain changes which can occur when limiters are used with pre-emphasis. When limiting occurs due to high frequencies the limiter introduces a progressive top cut rather than changing the overall gain.

The audio signal passes via a second 15 kHz low-pass filter to the A/D converter. The low-pass filters are designed to pass 15 kHz with little attenuation while frequencies at 16 kHz and above are greatly attenuated.

The A/D converter uses a sampling frequency of 32 kHz. This enables an upper frequency limit of 15 kHz to be achieved without aliasing components causing distortion. Each sample is converted into a 13-bit binary code using a converter with a linear quantising law. In order to protect the system against serious bearer circuit disturbances, a single parity bit is added to each 13-bit word. This provides a check on the five most

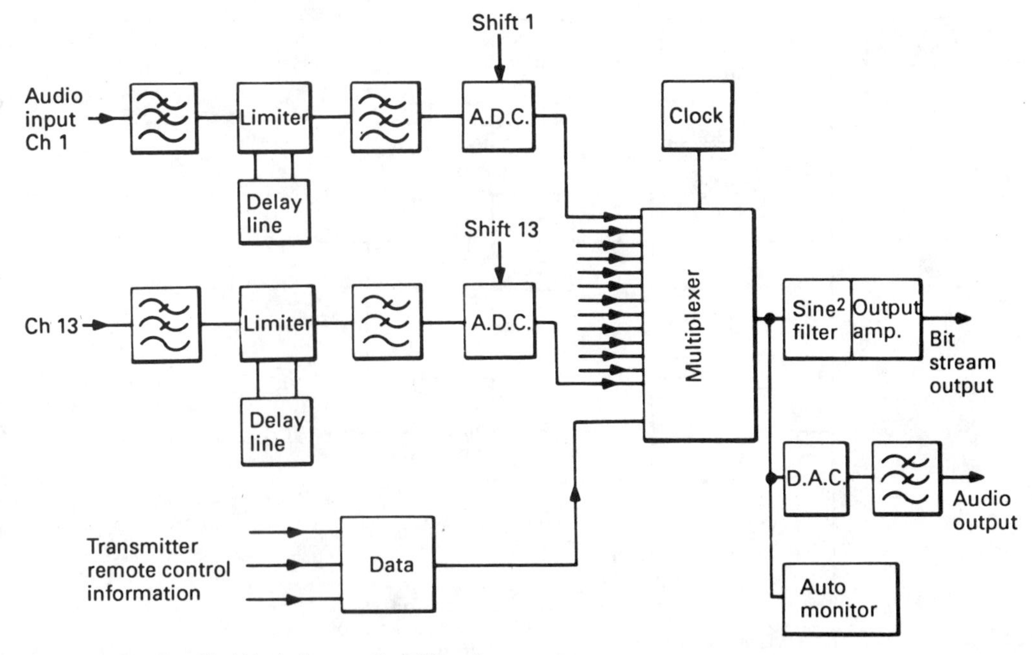

Figure 53.36 Simplified block diagram of a PCM coder

significant bits. If an error is detected, the word is discarded and the previous word substituted in its place. If the error rate becomes high enough to impair seriously the quality of the signal the audio output is muted.

The output of each A/D converter is fed to the multiplexer. The multiplexer carries out the time-division process, combining the 13 channels into a stream of pulses at a bit rate of 6.336 Mbit/s. After pulse shaping into sine-square form the signal is passed to the output. A bit-rate of 6.336 Mbit/s along with a sampling rate of 32 kHz gives a total of 198 bits per frame. The 13 audio channels each use 14 bits including the parity bit, the remaining 16 bits are used as a framing pattern (11 bits) and a data control channel (5 bits).

The data channel provides the facility for remote switching of equipment at decoder sites. The data signal comprises two parts, an address and a switching message. It is possible to switch either equipment at all the decoder sites simultaneously or to send a message to an individual site. The basic capacity of the system is 128 messages but this may be extended if the time to send a message is increased. Typical uses of the data messages are for mono/stereo switching and transmitter remote control.

53.20.3 Decoding equipment

The simplified block diagram is shown in *Figure 53.37*.

The bit stream in the form of sine-squared pulses is fed to the input unit. This unit converts the pulses, which may be noisy and distorted, into definite logic 0 or 1 levels.

The clock regenerator produces clock pulses at 6.336 MHz locked to the incoming bit stream. The synchronisation is achieved by comparing the timing of the framing pattern occurring in the bit stream with a locally generated framing pattern. Any timing error produces a change in a voltage used to control the frequency of the local timing reference.

The regenerated clock pulses are fed to a demultiplexer where groups of bit-rate shift pulses are produced. The shift pulses are used to enter the individual 14-bit pulse groups, present in the continuous bit stream, into the correct D/A converter.

Initially the parity bit is checked and if correct the conversion is allowed to proceed in a D/A converter using the dual ramp-counter technique.

The analogue output from the D/A converter is then fed to a low-pass filter with a cut-off frequency of 15 kHz. This filter removes the unwanted high-frequency components associated with a sampled waveform. The audio signal then passes through an amplifier to the channel audio output of the decoder.

53.21 Digitally companded PCM systems

Multi-channel digital circuits are being increasingly used for telephony with the result that analogue circuits will in the future become scarce. This has resulted in developments designed to use the new digital circuits for high-quality sound programme distribution. Several bit-rates have been recommended by the CCITT,[21] the primary and secondary rates favoured in Europe being 2.048 Mbit/s and 8.448 Mbit/s.

To make the best use of digital circuits, companding techniques are used to conceal the audibility of background noise, i.e. compression at the sending end and expansion at the receiving end. Companding has the effect of varying the noise depending on the signal level, the noise being least at low signal levels and highest at maximum signal level. Analogue companding techniques can be used, but with digital transmission systems it is more usual for the companding to be achieved entirely in the digital domain.

With digital systems the noise is dependent upon the number of bits used to describe each sample of the analogue signal. For high-quality sound transmission, the signal must be quantised to at least 13 bits/sample, and for broadcasting purposes it is becoming more usual to use 14 bits/sample.

Digital companding is used to vary, according to signal level, the effective number of bits per sample used for transmission, so that at low signal levels the signal is quantised to a greater precision than at higher levels. By this means it is possible to obtain up to six high-quality sound channels on a 2.048 Mbit/s bit stream without impairing the subjective quality.

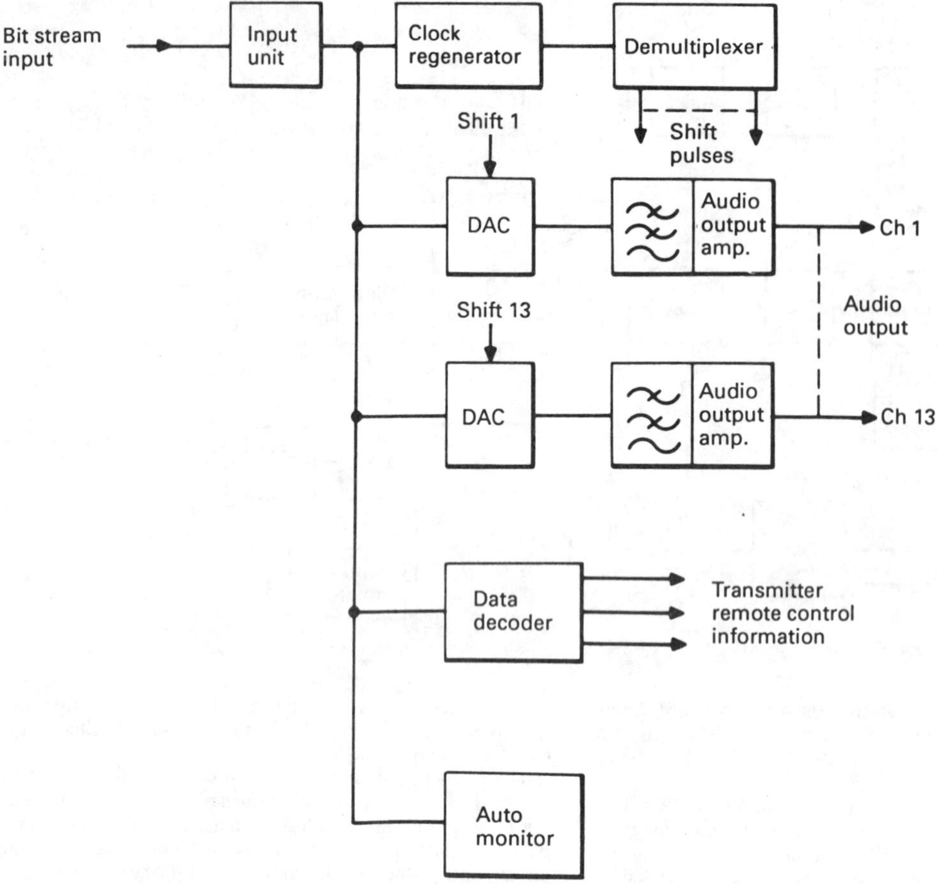

Figure 53.37 Simplified block diagram of a PCM decoder

Digital companding systems, in current use, can be either instantaneous or near-instantaneous.[22] With the instantaneous type, each sample word is changed to another word with fewer bits. The transfer characteristic is a function of the signal level. An example of this type is the A-law often used for telephony and adapted for high-quality sound.

With near-instantaneous companding, the peak value of a block of samples is determined and the whole block coded to an accuracy determined by that largest sample. A scale factor word, indicating the coding range, is transmitted along with the compressed sample words to control the expander.

53.21.1 The NICAM-3 digitally companded PCM distribution system

A system known as NICAM-3,[20,23] using near-instantaneous digital companding, is gradually superseding the 13-channel linear PCM equipment. The bit-rate used for two high-quality sound channels is 676 kbit/s, enabling up to six sound channels and a 1 kbit/s data signalling channel to be obtained on a 2.048 Mbit/s bit stream. Greater numbers of channels can be obtained by multiplexing to give up to 24 channels on an 8.448 Mbit/s bit stream. These signals can be distributed by using digital bearer circuits or circuits normally used for 625-line TV signals.

NICAM-3 encodes the audio samples to 14-bit accuracy and places each block of 32 samples (1 ms) into one of five ranges. The range is determined by the largest sample value found within the block (*Figure 53.38*). This process compresses the samples to 10

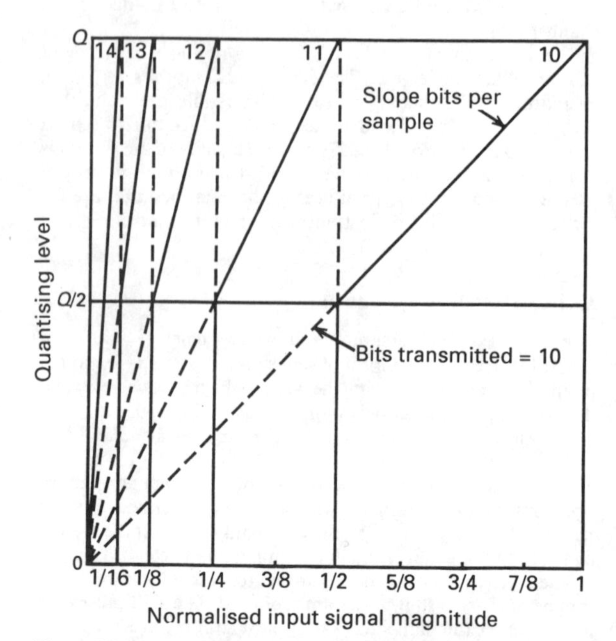

Figure 53.38 NICAM-3 near-instantaneous companding law

bits for transmission by discarding leading zeros and most-significant bits, according to the range. The lowest level samples are, in effect, encoded to a 14-bit resolution and the highest level samples to 10-bit resolution.

The range codes for three blocks of 32 samples are combined into a 7-bit code, and four protection bits are added using a single error correction technique described by R. W. Hamming.[24] The five most significant bits in each sample word are protected by parity so that error concealment by interpolation can be used at the decoder.

The audio channel performance is as follows:

(1) Sampling frequency: 32 kHz.
(2) Initial resolution: 14 bits/sample.
(3) Pre-emphasis: CCITT Recommendation J.17 (6.5 dB attenuation at 800 Hz).
(4) Frequency response: 50 Hz to 14 kHz ($\pm$0.2 dB, reference 1 kHz) typically -1 dB at 40 Hz and 15 kHz.
(5) Signal-to-noise ratio: 63 dB CCIR Rec. 468-2 quasi-peak meter and weighting network, bandwidth 20 Hz to 20 kHz.
(6) Nonlinear distortion: $<0.1\%$ (1 kHz at full modulation).

53.22 Low-frequency and medium-frequency transmitters for sound broadcasting

Figure 53.39 shows a typical schematic for high-power low-frequency (LF) and medium-frequency (MF) transmitters.

The drive unit produces a stable output at the carrier frequency. It often employs a crystal oscillator circuit housed in a temperature-controlled enclosure to achieve maximum stability. Crystal drives can be stable to within one part in 10^9 per month. If greater stability is required a rubidium vapour standard can be employed. This unit makes use of an atomic resonance of rubidium 87 at about 9 GHz. A vapour cell is irradiated with a signal near to the resonance frequency derived from a 10 MHz crystal oscillator. The light from a rubidium spectral lamp also passes through the cell onto a photodetector. When the frequency of the signal equals the resonance frequency the light is absorbed. The output of the photodetector is arranged to control the precise frequency of the crystal oscillator. The unit provides an output at 10 MHz from the crystal oscillator to a stability of about 3 parts in 10^{11} per month. The output is applied to a frequency synthesiser to produce the required carrier frequency.

In high-power transmitters the modulated amplifier is a valve stage employing triodes or tetrodes. It may have a single valve or a pair operating in push–pull. The load for the valve or valves comprises a tuned circuit inductively coupled to the output through a matching circuit. In modern transmitters the output stage is often operated in Class D. A square voltage waveform at the carrier frequency is applied to the grid which drives the valve into saturation. The tuned anode circuit responds to the fundamental component of the train of anode current pulses thus giving a sinusoidal carrier output. If the duration of the pulses is varied in accordance with the audio signal the output will be an amplitude modulated carrier. The fundamental component of a square wave increases according to a sinusoidal law as the mark/space ratio increases from zero, having a maximum at unity mark/space ratio. When fully modulated, therefore, the amplifier can theoretically impose a compression of about 4 dB. As an alternative to control grid modulation, screen grid modulation is possible in tetrodes. With valve circuits output powers of up to 500 kW are typical and greater powers can be obtained by paralleling the outputs of several amplifiers.

For lower output powers, solid-state transmitters are available. The output transistors are operated in a switching mode and since they are either 'on' or 'off' the dissipation is low. In a typical arrangement the power amplifier comprises four transistors in a bridge configuration switched at the carrier frequency across an audio modulated supply rail. These transmitters offer good efficiency (about 50% overall), reliability and easy adjustment of the output power level.

53.23 Short-wave transmitters

Broadcasting on short waves generally calls for regular changes of frequency to match ionospheric conditions, and either a number of crystals must be available to be switched as desired, or a frequency synthesiser must be used. The advantage of the latter is that any frequency can be selected in decade steps from about 4 to 30 MHz, although its stability may be lower than that of the crystal oscillator. The transmitter schematic is similar to *Figure 53.39*. The low-power radio frequency (RF) amplifiers are broadband to accommodate the range of frequencies required and the high-power amplifier can be automatically tuned to a series of spot frequencies; the tuning being under the control of a memory system and a pre-selector switch.

53.24 VHF frequency-modulated sound transmitters

Frequency-modulated transmissions in Band II (88–108 MHz) offer high-quality services with stereo capability. The schematic of a VHF sound transmitter is shown in *Figure 53.40*. In the first stages of the stereo transmitter the incoming left-hand (*A*) and right-hand (*B*) audio signals are passed through pre-emphasis

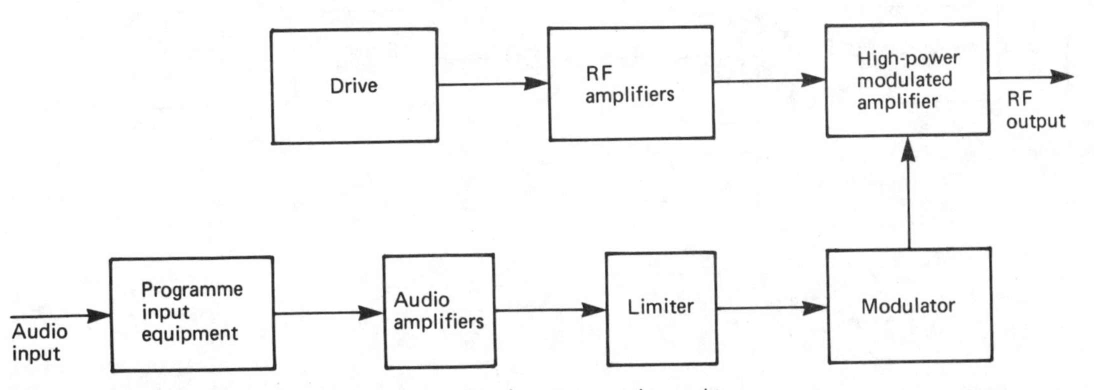

Figure 53.39 Schematic of high-power low- and medium-frequency sound transmitter

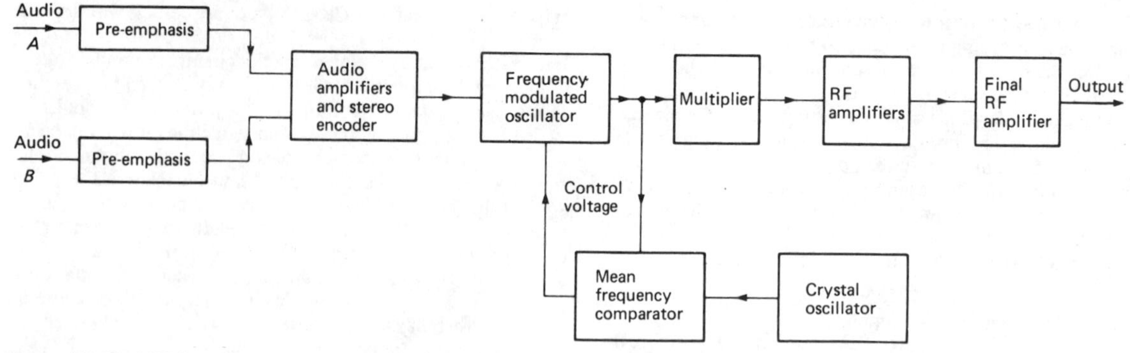

Figure 53.40 VHF frequency-modulated sound transmitter

networks and then applied to a stereo encoder which produces a composite baseband signal prior to modulation. In the 'pilot tone' system the first 15 MHz of the spectrum of the composite signal is occupied by the sum signal $(A+B)/2$. The difference signal $(A-B)/2$ is amplitude suppressed carrier modulated on a subcarrier frequency of 38 kHz and added to the sum signal. A pilot tone at 19 kHz is also added to the composite signal. The multiplier stages following the oscillator increase the frequency deviation by the same ratio as the carrier multiplication. A maximum frequency deviation of ± 75 kHz is usual. The following RF amplifier stages are wideband. Output powers of up to 20 kW are typical and for lower powers all solid-state equipments are available.

53.25 UHF television transmitters

The schematic for one type of UHF television transmitter is shown in *Figure 53.41*. As shown the video signal amplitude modulates the carrier at the final frequency. The diode circuit presents a variable reflection coefficient at the middle port of the circulator thus imposing an amplitude variation on the carrier in accordance with the video signal. An equivalent type of modulator uses a directional coupler with variable reflection coefficient circuits on two ports. The modulation may alternatively be carried out at an intermediate (IF) frequency, the modulator being followed by a frequency changer. The IF modulation method replacing the absorption modulator because precorrection and vestigial sideband (VSB) filtering can be carried out in the IF stages and this gives a better overall performance. *Figure 53.42* illustrates the vision carrier envelope (one line) when modulated with a colour bar signal (PAL system).

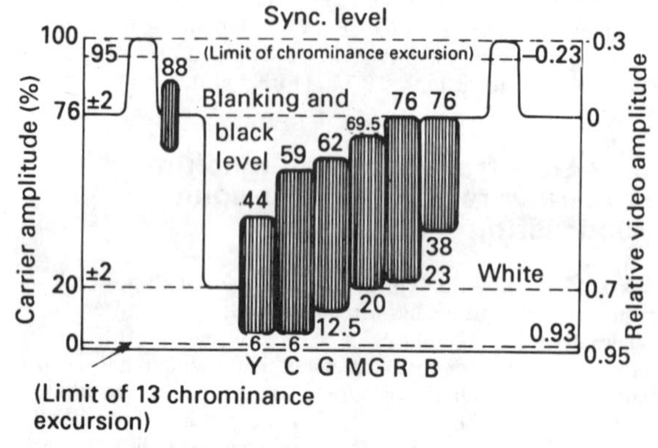

Figure 53.42 Vision carrier envelope (one line) when modulated with a colour bar signal

53.25.1 Television sound transmitter (UHF)

This is a separate transmitter, often employing a kylstron output stage. The type of modulation depends on the television system in use. For system I (in the United Kingdom) frequency modulation is employed on a carrier frequency that is 6 MHz above the vision carrier frequency. The sound carrier power can be between 1/10 and 1/5 of the peak vision carrier power. The outputs of the vision and sound transmitters are combined in a diplexer. Many stations have pairs of sound and vision transmitters operating in parallel which can be switched to single-transmitter operation in the event of a breakdown.

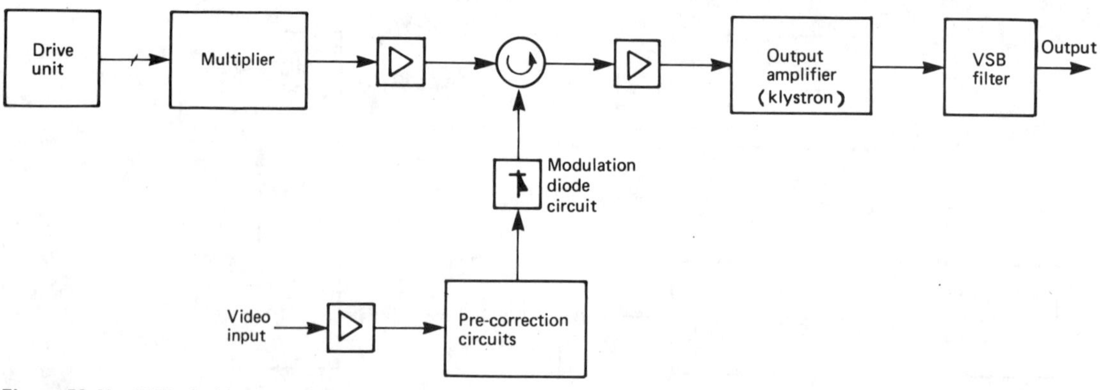

Figure 53.41 UHF television transmitter

53.26 Television relay stations: transposers

Television services in the UHF bands can require a number of low-to-medium-power transmitters to fill 'holes' in the service area of a main transmitter. These relay stations are often of the non-demodulating transposer type in which a complete television channel, comprising sound and vision signals, is received off-air, frequency translated to another channel, and re-radiated. The range of output powers is from a few watts to 1 kW and output amplifiers may be solid-state, travelling-wave tubes or klystrons depending on the power required.

References

1 *CCIR, Documents of the XIIth Plenary Assembly*, Vol. II, 'Propagation', 217–23 (1970)

2 PIGGOTT, W. R., 'The calculation of median sky-wave field strength in tropical regions', *UK Department of Scientific and Industrial Research*, No. 27 (1959)

3 EXWOOD, M., 'Cable television', *Paper to Royal Television Society Convention*, 7 September (1973)

4 COASE, R. H., 'Wire broadcasting in Great Britain', *Economica*, **XV**, No. 59, August (1948)

5 INFORMATION TECHNOLOGY ADVISORY PANEL, *Report on Cable Systems*, HMSO, London (1982)

6 HUNT COMMITTEE, *Report on Policy, Supervisory and Regulatory Aspects Relating to Expansion in the Television Service*, HMSO, London (1982)

7 DEPARTMENT OF INDUSTRY, *The Development of Cable Systems and Services*, Cmnd 8866, HMSO, London (1983)

8 MUDD, L. T., 'Cable television', *Proc. IEE*, **129**, Part A, No. 7 (1982)

9 QUINTON, K., In *Cable Television*, Local Telecommunications 2, J. M. Griffiths (ed.) IEE Telecommunications Series 17, Peter Peregrinus, London (1986)

10 OLSHANSKY, R., 'Microwave multiplexed lightwave systems: a new approach to wideband networks', *Int. J. Digital Analogue Cabled Systems*, **1**, No. 2, April–June (1988)

11 RITCHIE, W. K., 'The British Telecom switched-star cable TV network', *Br. Telecom Tech. J.*, **2**, No. 4 (1984)

12 ELY, S. R., 'The UK system for digital stereo sound with terrestial television', *J. Audio Eng. Soc.*, **35**, No. 9, 653 (1987)

13 BOWER, A. J., 'Digital two-channel sound for terrestial television', *IEEE Trans. Consumer Electronics*, **CE-33**, No. 3, 286 (1987)

14 *Specification of the Systems of the MAC/Packet Family*, Doc. Tech. 3258-E, EBU Technical Centre, Brussels (1986)

15 OLIVER, B. M., PIERCE, J. R. and SHANNON, C. E., 'The philosophy of P.C.M.', *Proc. IRE*, **36** (1948)

16 SHORTER, D. E. L. and CHEW, J. R., 'Applications of pulse-code modulation to sound-signal distribution in a broadcasting network', *Proc. IEE*, **119**, No. 10 (1972)

17 HOWARTH, D. and SHORTER, D. E. L., *Pulse-Code Modulation for High-Quality Sound Signal Distributions: Appraisal of System Requirements*, BBC Research Department Report EL10, 12 (1967)

18 SHORTER, D. E. L., 'The distribution of television sound by pulse-code modulation signals incorporated in the video waveform', *EBU Review*, No. 113-A, February (1969)

19 HOLDER, J. E., SPENCELEY, N. M. and CLEMENTSON, C. S., *A Two Channel Sound-in-Syncs Transmission System, IBC 1984*, IEE Conference Publication No. 240, p. 345 (1984)

20 CAINE, C. R., ENGLISH, A. R. and ROBINSON, J., *NICAM 3—A Companded PCM System for the Transmission of High Quality Sound Programs, IBC 1980*, IEE Conference Publication No. 191, p. 274 (1980)

21 'CCITT, Sixth Plenary Assembly', *Recommendations* C731 and C741, 111–2, p. 425 and pp. 444–46, OCITT, Geneva (1976)

22 CROLL, M. G., OSBORNE, D. W. and SPICER, C. R., *Digital Sound Signals: The Present BBC Distribution System and a Proposal for Bit Rate Reduction by Digital Companding, IBC 1974*, IEE Conference Publication No. 119, p. 90 (1974)

23 CAINE, C. R., ENGLISH, A. R. and O'CLAREY, J. W. H., 'NICAM 3: near-instantaneously companded digital transmission system for high-quality sound programmes', *Radio Electronic Eng.*, **50**, No. 10, 519–530 (1980)

24 HAMMING, R. W., 'Error detecting and error correcting codes', *Bell Syst. Tech. J.*, **XXVI**, No. 2, 147 (1950)

Further reading

CCIR, 'Recommendations and Reports of the CCIR, 1978, XIVth Plenary Assembly, Kyoto', **X**, Rec. 415 (1978)

'Communication Technology for Urban Improvement', June (1971), *Proc. IEEE*, Special Issue on Cable Television, July (1970)

CRAWFORD, D. I., 'High definition television—parameters and transmission', *Br. Telecom Tech. J.*, **5**, No. 4, October (1987)

FORREST, J. R., 'Broadcasting to the future', *Electron. Power*, November/December (1987)

HARDY, A. C., *Handbook of Colorimetry*, MIT, Mass.

HENDERSON, H., *Colorimetry*, BBC Engineering Training Supplement No. 14

HINDIN, H. J., 'Videotext looks brighter as developments mount', *Electronics*, August 25 (1982)

JURGEN, R. K., 'Two-way applications for cable television systems in the 1970s', *IEEE Spectrum*, November (1971)

KLEIN, A. B., *Colour Cinematography*, Chapman and Hall

LUCKIESH and MOSS, *Science of Seeing*, Macmillan

MASON, W. F., *Urban Cycle Systems*, The Mitre Corporation, National Academy of Engineering, Washinfton DC (1972)

MORRELL, LAW, RAMBERG and HEROLD, *Colour Television Picture Tubes*, Academic Press, New York (1974)

MULLARD LIMITED, *30AX Self-Aligning 110° in-line Colour TV Display*, Mullard Technical Note No. 119, Mullard Limited, London (1979)

OHTSUKA, T. *et al.*, 'Digited optical CATV system using hubbed distribution architecture', *IEEE J. Lightwave Tech.*, **6**, No. 11, November (1988)

REPORT OF THE SLOAN COMMISSION, *On the Cable—The Television of Abundance*, McGraw-Hill, New York

SARAGA, P., 'Compatible high definition television', *Electron. Comm. Eng. J.*, January/February (1989)

SIMS, H. V., *Principles of PAL Colour Television and Related Systems*, Newnes-Butterworths, London (1969)

TELEVISION ADVISORY COMMITTEE 1972, papers of the Technical Sub-Committee, Chapter 4, HMSO, London

TSANTES, J., 'AM-stereo technology gains momentum, but no industry standard is in sight', *EDN*, September (1982)

WARD, JOHN E., 'Present and Probable CATV/Broadband Communication Technology', *Report of the Sloan Commission*

WRATTEN, *Colour Filters*, Kodak

WRIGHT, W. D., *Measurement of Colour*, Hilger

Part 5

Applications

54 Communication Satellites

S C Pascall, BSc, PhD, CEng, MIEE
Telecommunications Policies Division,
Commission of the European Communities

Contents

54.1 Introduction

The geostationary satellite concept was first proposed by Arthur C. Clarke who recognised the potential of rocket launches and the advantage of the geostationary orbit.[1] The major uses of communication satellites at present are:

(1) Point-to-point international communications for telephony, data and telegraph traffic and TV program distribution.
(2) Regional and domestic telecommunications and TV program distribution.
(3) Communication with mobile terminals mainly for maritime telecommunication services.

Another use of growing importance is direct television broadcasting and systems for commercial international fixed services.[2]

There are currently three international systems in service and numerous regional and national systems. At international level these are the INTELSAT system, the INTERSPUTNIK system and the INMARSAT system. The INTELSAT system is owned by the INTELSAT Consortium and uses geostationary satellites. The INTERSPUTNIK organisation[3] members are the USSR and its allied countries. The system is based on the use of satellites in highly inclined elliptical orbits as well as geostationary satellites. The INMARSAT system is primarily for maritime mobile communications.

At regional level there are, for example, the ARABSAT system serving Arab countries and the EUTELSAT system serving European countries.

54.2 Frequency bands

Of the bands allocated to communication satellites, those shown in *Table 54.1*, or parts of them, are the ones used or likely to be used for commercial systems.

Table 54.1

Up paths (GHz)	*Down paths* (GHz)
	3.4–4.2
5.85–7.075	4.5–4.8
12.7–12.75 (R2)	10.7–11.7
	11.7–12.2 (R2)
12.5–12.75 (R1)	12.5–12.75 (R1, R3)
12.75–13.25	
14.0–14.5	17.7–21.2
17.3–17.7	
27.5–31	
22.55–23.55 (inter-satellite links)	
32–33 (inter-satellite links)	

R1, R2, R3 refer to the three ITU Regions

In addition the bands 10.7 to 11.7 GHz (in Europe and Africa), 14.5 to 14.8 GHz (except Europe) and 17.3 to 18.1 GHz may be used for up links but only as feeder links to broadcasting satellites. Most of these bands are shared with terrestrial services. When this occurs, there are limits on transmitted powers and the power flux density a satellite may set up at the Earth's surface.[4] In the 3.4–7.75 GHz band the limit is -152 dBW/(m^2 (4 kHz)) below 5° arrival angle, rising to -142 dBW/(m^2 (4 kHz)) at 25° arrival angle and above. In the 12.2–12.75 GHz band, it is 4 dB greater than in the 3.4–7.75 GHz band and in the shared part of the 17.7–19.7 GHz 33 dB greater still, but specified for a 1 MHz rather than 4 kHz bandwidth.

54.3 Orbital considerations

By far the most useful orbit for communication satellites is the geostationary satellite orbit, a circular equatorial orbit at approximate height 36 000 km for which the period is 23 h 56 min, the length of the sidereal day. A satellite in this orbit remains approximately stationary relative to points on the Earth's surface. There are two main perturbations to the orbits of such satellites:

(1) A drift in orbit inclination out of the equatorial plane due to effects of the Sun and Moon. This is at the approximately linear rate of 0.86° a year for small inclinations. If not corrected it causes the satellite to move around a progressively increasing daily 'figure of eight' path as viewed from the Earth.
(2) Acceleration of the satellite in longitude towards one of the two stable points at 79°E and 101°W longitude. This is caused by non-uniformity of the Earth's gravitational field. These perturbations can be corrected (see *Section 54.5*).

54.4 Launching of satellites

The launcher generally places the satellite first into a low-altitude circular inclined orbit and then into an elliptical transfer orbit with apogee at the altitude of the geostationary orbit. At an approximate apogee the apogee boost motor, generally solid fuelled and forming part of the satellite itself, is fired to circularise the orbit and remove most of the remaining orbit inclination. The satellite's own control system is used to obtain and maintain the final desired orbit.

Satellites can be launched using either expendable rockets or the reusable Space Shuttle. Some US launchers suitable for communication satellites, together with their approximate geostationary orbit payload capabilities, are the Thor Delta 3914 (950 kg), 6920 MLV (1415 kg) and 3920 PAM (1247 kg), the Atlas G Centaur (2360 kg) and Titan 34D (4500 kg). Europe has developed the Ariane family of expendable launchers; presently Ariane 3 (2580 kg) is the workhorse for the European Space Agency with the launch of Ariane 4 (4200 kg) imminent and Ariane 5 (8200 kg) being at development stage. Satellites have been launched from the US Space Shuttle, with its various upper stages, in order to reach geostationary orbit, culminating in the Challenger disaster of January 1986 and the withdrawal of the shuttle as a commercial launch vehicle. Other nations have developed their own expendable launch vehicles (ELVs), for example the USSR's Proton (2400 kg), Japan's N_2 (670 kg) and China's Long March (1300 kg).

54.5 Satellite stabilisation and control

Stabilisation of a satellite's attitude is necessary since, for high communications efficiency, directional aerials must be used and pointed at the Earth. Geostationary communication satellites are either spin stabilised or three-axis body stabilised. INTELSAT IV and INTELSAT IVA are spin-stabilised whilst INTELSAT V is three-axis body stabilised.

In a spinning satellite, the body of the spacecraft spins at 30–100 rpm about an axis perpendicular to the orbit plane, but the aerial system is generally de-spun, that is, located on a platform spinning in the opposite direction so that the net effect is a stationary antenna beam relative to the Earth. A reference for the control system is usually obtained primarily by infrared Earth sensors supplemented by Sun sensors. Antenna pointing accuracy of $\pm 0.2°$ or better[5] is obtained through the antenna de-spin control electronics and by occasional adjustments to the direction of the satellite's spin axis.

Body-stabilised designs generally employ an internal momentum wheel with axis perpendicular to the orbit plane. Control about the pitch axis is through the wheel's drive motor electronics, while control about the yaw and roll axes (necessary because of perturbations to the wheel axis direction) may be by gimballing the wheel or by use of hydrazine monopropellant thrusters to correct the axis direction. In any case, thrusters must be used for occasional dumping of momentum.

The orbit of the satellite must also be controlled and this is achieved by ground command of the thrusters. Early thruster systems used monopropellant thrusters in which hydrazine was decomposed at 1300 K. The mass of hydrazine required for a 10-year life, expressed as a percentage of total satellite mass, is approximately 18–25%[6] for N–S station keeping (inclination control) and less than 1% for E–W station keeping assuming a conventional hydrazine system is employed. In a spinning satellite, axial thrusters are used for inclination corrections and radial thrusters, operated in a pulsed mode, for longitude corrections. Methods such as the use of bipropellant, power-augmented electrothermal hydrazine decomposition or electrical propulsion[7,8] exist to improve on the specific impulse of conventional thrusters. In electrical propulsion, a propellant (an alkali metal such as caesium) is ionised by passing it through a heated grid. The ions are then accelerated by an electric field and ejected through a nozzle.

54.6 Satellite power supplies

Silicon solar cells are the accepted source of primary power, except during eclipse when power is maintained by nickel–cadmium cell batteries. Full shadow eclipse occurs on 44 nights in the spring and 44 nights in the autumn. The longest eclipses occur at the equinoxes and last for 65 min (full shadow). The nickel–hydrogen battery is replacing the nickel–cadmium battery now used on operational communications satellites. The nickel–hydrogen battery offers significant advantages in terms of reliability, life expectancy and improved energy density. Research is currently being undertaken on silver–hydrogen batteries.

Spinning satellites have body-mounted solar cells producing at the end of their operating life 9–11 W/kg mass of solar cell array. Body-stabilised satellites using extendable arrays rotated to always face the Sun can give 21–23 W/kg mass. Current research is aimed at increasing deployable solar panel dimensions and reducing weight. Values are reaching 50–60 W/kg and power levels of a few tens of kilowatts. A disadvantage of any extendable array for synchronous orbit missions is the inability to provide full power during transfer orbit when the array is still stored. The solar array mass in each case is of the order of 3–5 kg/m^2 (INTELSAT VI: 59 m^2, 250 kg, 2 kW; TELECOM 1: 16 m^2, 47 kg, 1 kW). High-efficiency solar cells now being developed could cut substantially the weight and area of some satellite solar panels. These cells made with gallium arsenide have efficiency about 4% greater than the best available silicon solar cells.

54.7 Telemetry, tracking and command (TT&C)

Telemetry is used to monitor and evaluate the satellite performance and provide the ground control with data for the operation and failure diagnosis of the satellite. The command function is required for the initiation of manoeuvres by the satellite as part of routine operations, such as to switch paths, in the communications payload or to respond to an emergency by initiating corrective action. Tracking is necessary for the execution of manoeuvres or corrections. TT&C signals generally occupy a narrow channel within the communication band but are outside the communication channels. For example, in the INTELSAT V spacecraft, 6175 MHz is used for the command functions and 3947.5 MHz and 3952.5 MHz for telemetry.

54.8 Satellite aerials

A circular radiation pattern of 17.5° beamwidth is just adequate to cover the area of the Earth visible from a geostationary satellite. This corresponds to a beam edge gain, relative to isotropic, of approximately 16 dB and is generally provided by means of a horn antenna. Spot beam aerials, which may be steerable, are used for increasing the gain and hence the effective isotropically radiated power (e.i.r.p.). On more advanced satellites spot beam aerials are used to permit reuse of the frequency band by relying on the high degree of isolation possible if the beams have adequate angular separation. Spot beam aerials are generally front-fed paraboloids. The beam edge gain of such aerials is approximately 41–20 $\log_{10}$ (beam width in degrees) dB, regardless of frequency.

The aerial types used in INTELSAT satellites to date have been as follows. The INTELSAT I (Early Bird) and INTELSAT II satellites used linear arrays of dipoles (not de-spun) producing toroidal radiation patterns. The INTELSAT III satellite had a de-spun global coverage horn while INTELSAT IV had global horns and two 4.5° spot beam paraboloids all mounted on a de-spun platform which also carried the transponders.[9]

A particular aerial implementation combining large area coverage with frequency reuse between east and west has been used on the INTELSAT IVA satellite. The coverage areas in this case are obtained by composite spot beams formed with multiple feeds illuminating large offset paraboloids, selected feeds being fed in quadrature with others.

INTELSAT V uses six communication aerials, two conventional global coverage horns (6/4 GHz), two hemispherical hemi/zone offset-fed reflectors (6/4 GHz) and two offset-fed spot beam reflectors (14/11 GHz), *Figure 54.1*.[10] Each

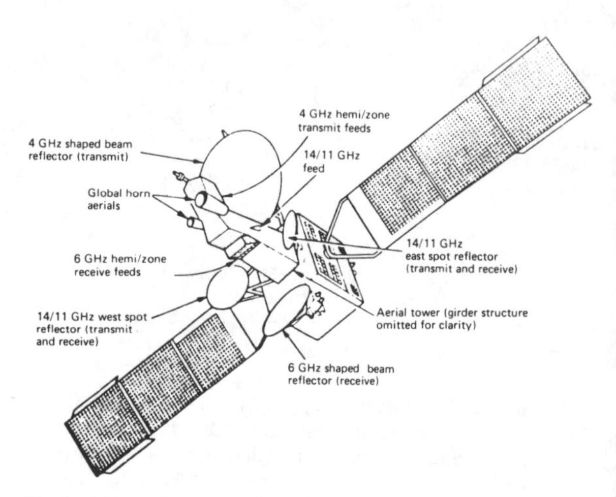

Figure 54.1 INTELSAT V satellite

hemi/zone aerial has 88 feed horns, each feed horn producing a small spot beam that can be set to any amplitude and phase. A combination of these spot beams forms the required footprint shape. The hemi beams are shaped to cover whole continents, such as the Americas, whilst the zone beams, operating in the opposite sense of circular polarisation, cover smaller areas, such as Europe, *Figure 54.2*.

The two 14/11 GHz spot beam aerials are provided to service high traffic areas such as Western Europe and North-East USA. Each aerial receives and transmits into the same spot area. Beam isolation is achieved by beam shaping and linear orthogonal polarisation separation.

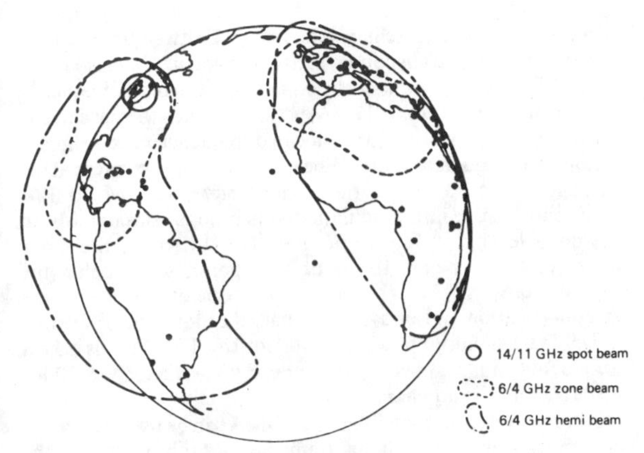

Figure 54.2 Approximate aerial beam converage of INTELSAT V spacecraft for the Atlantic Ocean region

The INTELSAT VI satellite scheduled for launch in 1989 would use ten communication aerials: two hemispherical coverage (6/4 GHz), four zone coverage (6/4 GHz), two global coverage (6/4 GHz) and two spot coverage (14/11 GHz). Antenna reflectors for the 4 GHz transmit and 6.9 GHz receive hemi-zone beams are the most visible features of the communications payload. Their large size arises from the need to generate four co-frequency and co-polarisation zone beams, covering groups of Earth stations, with a minimum interbeam spacing of 2.15°, while maintaining an isolation in excess of 27 dB. An offset feed design is used to eliminate beam blockage and a flat feed array of 146 dual-polarization 'Potter' feed horns is located in the focal plane. Groups of feeds are excited together to produce the required beam pattern, and cancel sidelobes. By this means several independent shaped coverage areas are produced on the visible part of the Earth.

For the two hemi beams, groups of 79 and 64 feeds are activated using the A polarisation feed connectors. To generate the three different ocean sets of zone beam patterns three squarax panels are used. Two fully steerable 14/11 GHz spot beams are produced using offset feed circular reflectors. Global beams at 6/4 GHz are produced using two simple horns.

54.9 Satellite transponders

The essential purpose of the transponder is to translate the received signals to a new frequency band and amplify them for transmission back to the Earth. In early satellites this process was carried out broadband but modern satellites generally separate the signals into a number of channels at least for part of the amplification. Although the INTELSAT IV satellites are nearing the end of their useful life the transponder arrangement used is typical of many current satellites that do not employ frequency reuse and it is therefore the INTELSAT IV repeater[11] that will be described in order to illustrate typical techniques. *Figure 54.3* shows the basic elements. Signals in the band 5.932 to 6.418 GHz from the receive aerial are first amplified in a 6 GHz tunnel diode amplifier (TDA) and then frequency translated by 2225 MHz for further broadband amplification in the 4 GHz TDA and low-level travelling-wave tube (TWT). This part of the transponder subsystem has fourfold redundancy to achieve high reliability.

The signals are then split by a circular filter dividing network into 12 channels of 36 MHz bandwidth and 40 MHz spacing

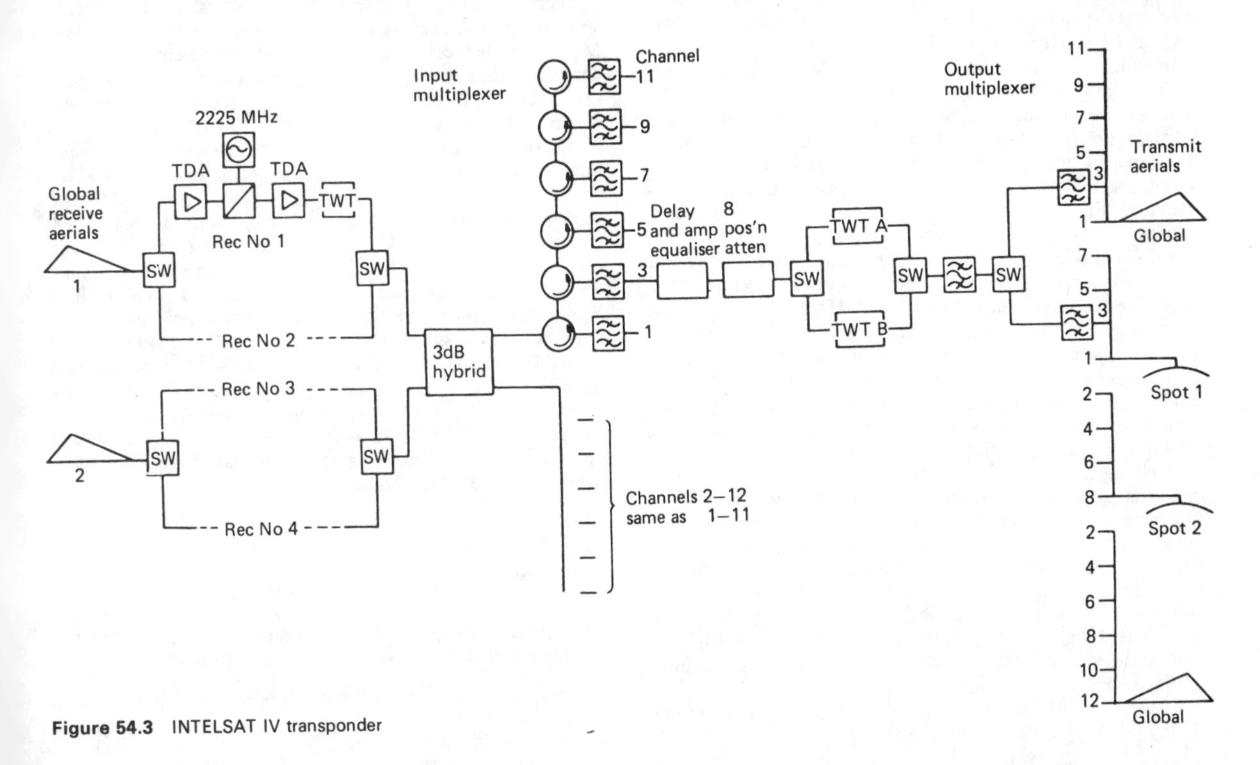

Figure 54.3 INTELSAT IV transponder

between centre frequencies. The channels have gain control attenuators with eight 3.5 dB steps, except channels 9 to 12 where there are only four steps. These are followed by the high-level (6 W) TWT (and standby) and then the switches selecting the transmit aerials. Channels 1, 3, 5 and 7 can be connected to global beam aerial or spot beam aerial No 1, channels 2, 4, 6 and 8 can be connected to the other global beam aerial or to spot beam aerial No 2 and channels 9, 10, 11 and 12 are permanently connected to the global beam aerials. No two adjacent channels are ever connected to the same aerial, which simplifies the output multiplexer filter design.

The principal characteristics of the transponder and aerial subsystem are as follows:

(1) Receive system gain-to-noise temperature ratio $(G/T) - 17.6$ dB/K.
(2) Flux density for TWT saturation -73.7 to -55.7 dBW/m^2.
(3) Transmit e.i.r.p. global beam 22 dBW per channel; spot beam (4.5°) 33.7 dBW per channel.
(4) Receive aerials are left-hand circularly polarised.
(5) Transmit aerials are right-hand circularly polarised.

Other specifications relate to gain stability, gain slope, group delay response, amplitude linearity and amplitude to phase modulation transfer.

The above relates to INTELSAT IV but alternative arrangements are possible. In the European Communication Satellite (ECS), for example,[12] besides operating in the 14/11 GHz rather than 6/4 GHz frequency bands, the receiver employs a parametric amplifier instead of a TDA, FET amplifiers instead of TWT amplifiers and a double conversion transponder is used, the bulk of the amplification taking place in a broadband intermediate frequency amplifier operating at UHF. The channels are connected to one of three 3.7° spot beams or the elliptical (5.2° × 8.9°) European coverage beam referred to as Eurobeam.

There are 12 channels of 72 MHz nominal bandwidth each, with the later models incorporating two 72 MHz channels at 12 GHz to be used for specialised services. Finally, the transponder subsystem is completely duplicated since the satellite employs frequency reuse through polarisation discrimination.

In many modern satellites, the transponder arrangements are further complicated by frequency reuse between spot beams as well as, or instead of, reuse on separate polarisations. Furthermore, the use of more than one set of frequency bands in the same satellite is practised, e.g. 6/4 GHz and 14/11 GHz interconnected with each other. The development of high-frequency FETs fosters the use of single frequency conversion transponder types.

The INTELSAT V spacecraft operates in both 6/4 GHz and 14/11 GHz and utilises these bands in seven distinct coverage areas, five at 6/4 GHz and two at 14/11 GHz, *Figure 54.2*. Reuse of the frequency bands between coverage areas is achieved by means of spatial and/or polarisation isolation.[13]

In all these cases the main complication to the transponders arises through the need to achieve an adequate degree of connectivity between receive and transmit beams, together with flexibility to meet changing requirements.

Satellite transponders are being revolutionised by the use in the satellite of time division switching between beams. A switching capability in the satellite permits interconnections between different up- and down-link beams to be dynamically changed, thus ensuring the desired connectivity and flexibility for re-allocation of capacity. The Olympus[14] satellite, due to be launched in 1988, is equipped with a 4 × 4 satellite switch in the 12/14 GHz band operating in the satellite switch/time division multiple access (SS/TDMA) mode. The satellite switch permits full interconnectivity by means of part-time interconnection of each pair of receive and transmit beams according to expected traffic, so that each beam requires only one receive and transmit chain. The switch matrix connects the output of each receiver to the input of each transmitter sequentially. The switching sequence, or program, is composed on the ground and sent to the satellite by telecommand so it can be changed from time to time.

54.10 Multiple access methods

In a communication satellite system many Earth stations require access to the same transponder subsystem and usually also to the same RF channel in a channellised satellite. Currently frequency division multiple access (FDMA) is used in which the RF carriers from the various stations are allocated frequencies according to an agreed frequency plan. When several carriers are passed through the same TWT the nonlinearity and AM to PM conversion cause intermodulation which adds considerably to the noise level.[15,16] It becomes necessary to reduce the TWT drive level well below the saturation point, so reducing the output power. At the optimum operating level the intermodulation noise is usually half the down-path thermal noise. Typical output power back-off for INTELSAT V is 4.5 dB for global beam channels, 6 dB for hemi/zone channels, 6 dB for 80 MHz spot beam channels.

Alternatively, for digital signals, TDMA can be used,[17,18] each station transmitting its digital traffic in a short burst during the portion of the overall time frame allocated to that station. The frame format of the INTELSAT TDMA system is shown in *Figure 54.4*. A station maintains its correct burst position by applying information provided by the reference station. The reference station derives this information from a comparison of the burst position relative to the reference burst. Initial acquisition is achieved in open loop by stations transmitting a shortened burst at an instant determined from the time at which the reference burst is received and the ranging information supplied from the reference station.

There are other types of multiple access, such as spread spectrum multiple access, but these are reserved for specific uses. Most multiple access systems have fixed assignment of telephone channels, with enough channels allocated to meet the peak demand on each route. For low-capacity routes it may be more economic to use 'demand assignment' systems in which channels from a pool are allotted to pairs of Earth stations as the demand arises. The first system of this type developed for commercial use is known as 'SPADE'.[19] In the INTELSAT system, SPADE uses a 36 MHz transponder bandwidth containing a pool of 800 channels. Individual 4 kHz voice channels are PCM encoded using 8 kHz sampling rate and 7 bits of quantisation. Each digital voice signal is transmitted on a separate carrier using four-phase PSK (phase shift keying) modulation. Systems of this type are referred to as single-channel per carrier (SCPC) systems. The main difference between SPADE and other SCPC systems is that SPADE incorporates a demand assignment system. Pre-assigned SCPC systems using FM modulation with and without companding (SCPC/FM, SCPC/CFM) are widely used and operate at carrier-to-noise values lower than those required by the standard SPADE channel units.

54.11 Transmission equation

The carrier-to-thermal noise power ratio, C/N, in bandwidth B Hz, for either the up path or the down path is given by:

$$C/N = \text{e.i.r.p.} - L - LA + G/T - 10 \log_{10} k - 10 \log_{10} B \text{ dB} \quad (54.1)$$

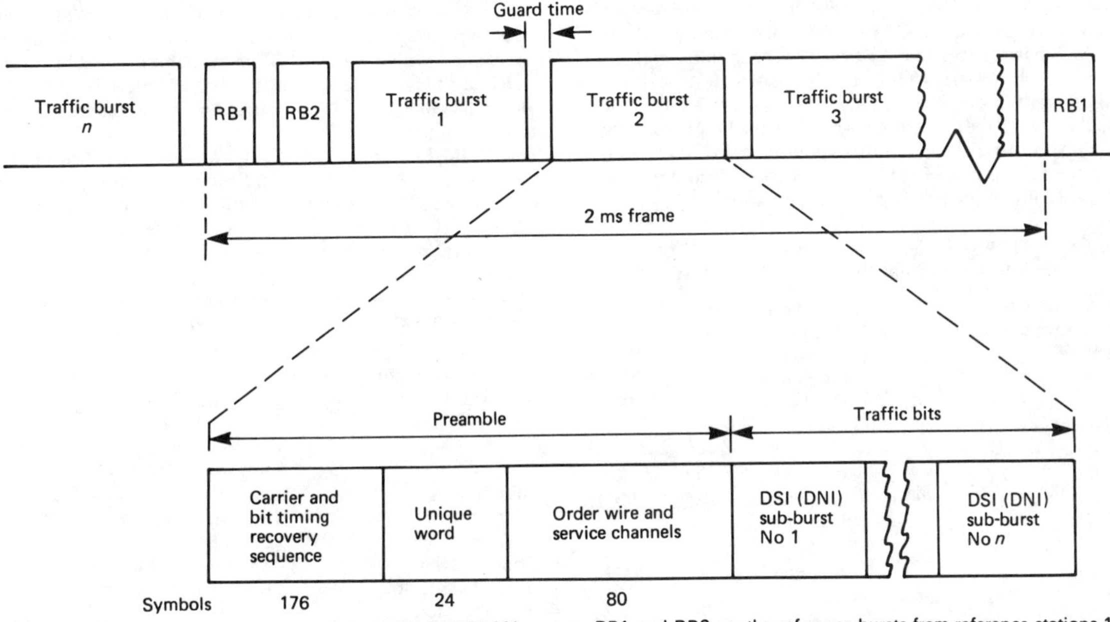

Figure 54.4 Frame and burst format of the INTELSAT TDMA system. RB1 and RB2 are the reference bursts from reference stations 1 and 2 respectively. The drawing is not to scale

where

e.i.r.p. = transmitter power × aerial gain, dB W
L = free-space path loss, $20\log_{10}(4\pi d/\lambda)$, dB
LA = atmospheric loss (0.3 dB at 5° elevation in clear weather at 4 GHz)
G/T = receive system aerial gain-to-noise temperature ratio, dB/K
k = Boltzmann's constant, 1.37×10^{-23} W/(Hz K)

The C/N ratio available in a satellite link is used to determine the channel capacity and performance of the link. Typical contributions to the overall C/N ratio for INTELSAT V are listed in *Table 54.2*.

Table 54.2 Typical INTELSAT V beam edge C/N budgets for a 6/4 GHz hemi-beam and a 14/11 GHz spot beam for multicarrier FM transmission

	Hemi-beam C/N (dB)	*Spot beam* C/N (dB)
Up-path thermal	27.4	26.3
Up-path interference (frequency reuse/spatial and polarisation isolation)	22.5	33.0
Satellite intermodulation	20.6	24.0
Down-path thermal	18.3	19.8
Down-path interference (frequency reuse/spatial and polarisation isolation)	22.5	33.0
Total C/N	14.4	17.5

54.12 Performance objectives

The CCIR recommends performance objectives for international telephone and television connections via satellite. For a telephone channel the most important objective for the system designer is that the psophometrically weighted noise power must not exceed 10000 pW for more than 20% of any month, 50000 pW for more than 0.3% of any month and 1000000 pW for 0.01% of any year.[20] For television the corresponding objective is 53 dB signal-to-weighted-noise ratio for the 2500 km hypothetical reference circuit for 99% of any month measured in the band 0.01 to 5 MHz.[21]

For digital telephony systems bit error rates of 10^{-6} (10 min mean value) 10^{-4} (1 min mean value) and 10^{-3} (1 s mean value) for more 20% of any month, 0.3% of any month and 0.01% of any year, respectively, are recommended by CCIR.[22]

54.13 Modulation systems

Frequency modulation is the most widely used at present. In the case of telephony most traffic is carried in large FM carriers but single-channel per carrier systems are also used, particularly in domestic satellite systems. For the multiple channel per carrier case the channels are first assembled in a frequency division

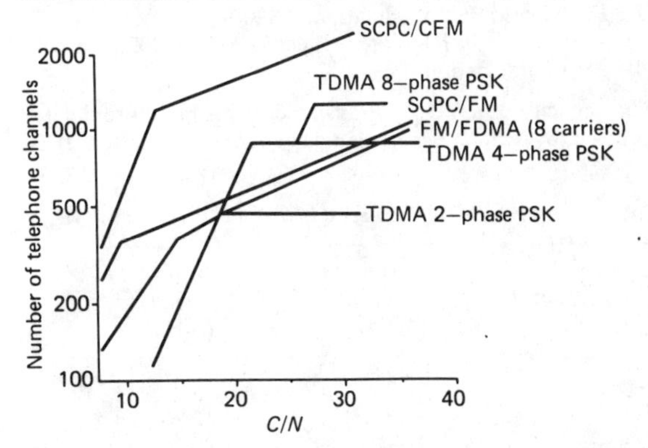

Figure 54.5 Telephone channel capacity in 36 MHz channel

multiplex (FDM) baseband. Modulation indices are chosen according to the available ratio of carrier power to noise density. *Figure 54.5* shows the number of channels (one telephone circuit requires two channels) which can be obtained in a 36 MHz bandwidth as a function of total carrier power to noise ratio. The knee in the curve is the point below which the full 36 MHz cannot be occupied because the C/N ratio would be below threshold, even assuming the use of threshold extension demodulators. Methods of calculating performance, etc., are given in references 23 and 24.

A typical INTELSAT allocation of the 10 000 pW of noise allowed in the telephone channel would be as follows:

Up-path thermal	350	Down-path thermal	2900
Up-path intra-system interference co-channel	585	Down-path intra-system interference co-channel	585
Cross-polarisation	690	Cross-polarisation	690
Satellite intermodulation	1700	Terrestrial interference	1000
		Other satellite system interference	2000
		Margin	500

In the case of television, frequency modulation is again the universally adopted method at present. Reference 24 gives a method for calculating performance.

Table 54.3 gives the modulation parameters used in INTELSAT IVA and V for typical carrier sizes.

Television is normally transmitted in 17.5 MHz bandwidth allowing two 49 dB S/N ratio video signals to be passed through a single global beam transponder. 30 MHz bandwidth, giving 54 dB S/N ratio, is used in certain cases. 20 MHz bandwidth is also now in use on certain INTELSAT transponders with a 41 MHz bandwidth.

The modulation method for digital signals is generally phase shift keying. The theoretical minimum C/N ratio for four-phase PSK, the most commonly preferred system, is 11.4 dB in a bandwidth equal to the symbol rate (half the bit rate) for a bit error rate of 10^{-4}. Practical bandwidth requirements are greater than the symbol rate by typically 20% and the practical C/N ratio is greater than the theoretical minimum value by typically 2 to 3 dB. *Figure 54.5* also shows the capacity obtainable from 36 MHz using PSK modulation for PCM telephony.

The parameters of the TDMA system[17] adopted by INTELSAT are:

(1) Channel encoding at 64 kbit/s (8 bits/sample) with or without digital speech interpolation (DSI).
(2) Four-phase PSK modulation at 120 Mbit/s.
(3) Approximately 1600 channels per 72 MHz (3200 with DSI).
(4) 2 ms frame period.

The capacity of a TDMA transponder can be approximately doubled by employing a digital speech interpolation (DSI) system.[25–27] DSI enables an increase in capacity to be achieved by utilising the quiet periods of a speech channel. Any active input channel is able to switch to any available channel in a pool of satellite channels. Hence, the number of terrestrial channels that can be connected is approximately double the number of satellite channels available.

The FDM/FM and PCM/TDMA systems described above are used mainly for large traffic streams. For thin route applications SCPC operation is frequently adopted. The main parameters of typical SCPC PSK and companded FM systems are as shown in *Table 54.4*.

Table 54.4 SCPC CFM/PSK 4ϕ parameters for SPADE and pre-assigned (INTELSAT standard A station)[28]

Audio channel input bandwidth	300–3400 Hz
Encoding	7 bit PCM $A=87.6$ companding law 8 kHz sampling rate
Modulation	four-phase coherent PSK
Carrier control	voice activated for voice channels
Channel spacing	45 kHz
IF noise bandwidth	38 kHz
C/N at threshold	13.5 dB
Threshold bit error rate	10^{-4}
SCPC/CFM[29] (typical thin route domestic system)	
Audio channel processing	2:1 companding
Modulation	FM
Carrier control	voice activated
Channel spacing	22.5 kHz
Noise bandwidth	18.75 kHz
C/N	16.8 dB for 10 000 pW (subjective equivalent) for a median talker

54.14 Summary of international, regional and domestic satellites

Table 54.5 summarises the main features of communication satellites.

54.15 Earth stations for international telephony and television[30–33]

54.15.1 Aerial system

Several classes of Earth station are currently in use, namely the fixed, mobile and transportable stations. INTELSAT recommends only the standard Earth stations for fixed satellite communication services. Seven standard Earth stations (standard A, B, C, D, E, F and G) are currently authorised by INTELSAT. INMARSAT uses a standard Earth station similar to INTELSAT B as the shore (coast) station in the maritime

Table 54.3 Example of modulation parameters used with INTELSAT IVA and V

Carrier-type No of channels	*Bandwidth allocated* (MHz)	*Bandwidth occupied* (MHz)	*Top baseband limits* (kHz)	*Deviation (r.m.s.) for* 0 dBmO *test tone* kHz	*Overall C/N ratio* (dB)
24	2.5	2.0	108	164	12.7
432	20.0	18.0	1796	616	16.1
972	36.0	36.0	4028	802	17.8

Table 54.5 Main features of communication satellites

Name	*Number successfully launched by end 1981*	*Date of first launch*	*Mass (orbit)* (kg)	*No*	*Transponder characteristics*		*Total telephone channel capacity*
					Bandwidths (MHz)	*Saturation output e.i.r.p.* (dB W)	
INTELSAT							
IV	7	Jan 1971	730	12	36	22, 33.7	8 000 (typ)
IVA	5	May 1975	750	20	36	22, 26, 29	12 000 (typ)
V	2	Dec 1980	1012	27	78, 72, 36 38, 240	23.5, 26.5, 26 29, 41.1, 44.4	24 000 (typ)
RCA	1	Dec 1980	461	24	34	32, 36	28 800
SBS	2	Nov 1980	555	10	43	40–43.7	27 800
ECS	—	Sept 1982	461	9	72	34.8, 40.8	16 000

mobile service, whilst EUTELSAT uses INTELSAT standard C for stations supporting trunking telephony and TV transmission. Domestic systems using leased INTELSAT transponders generally use Earth stations comparable with the standard B. *Table 54.6* lists the standard INTELSAT Earth station parameters.

Modern large Earth stations adopt the Cassegrain construction in which the use of a sub-reflector permits the feed to be located at the centre of the main reflector where it is readily accessible and from where the waveguide runs to the transmitters and receivers are relatively short. Early large systems deployed beam waveguide systems which allow the feed to be located off the moving parts of the aerial close to the low-noise and high-power amplifiers. This eliminates long waveguide runs and awkward flexible or rotary joints.

Aerial steering, generally in azimuth and elevation axes, is necessary because the satellites are not perfectly stationary when considered in terms of the very narrow beamwidth (approximately 0.2° at 6 GHz) of a 30 m aerial. Small beam-pointing adjustments can also be obtained by moving the feed. The satellite beacon is generally used to derive the control signal for the auto-track steering system, a common method being by the detection in the feed system of higher order modes which undergo rapid rate of change as the received signal direction varies near the axis.

54.15.2 Low-noise receiving system[34]

The wide-bandwidth and low-noise temperature of the parametric amplifier makes it the universal choice as the first-stage amplifier. The full satellite extended bandwidth of 575 MHz can be readily covered. Cooling to 20 K or below using closed-cycle cryogenic systems with gaseous helium was used in the past to produce amplifier noise temperatures of the order of 15 K. These have now been abandoned because of their complexity (especially for maintenance, bulkiness and high cost) in favour of thermoelectrically (Peltier)-cooled low-noise amplifiers (LNA), producing noise temperatures of 35 to 55 K (at 4 GHz). The total receiving system noise temperature at 4 GHz, however, is typically 70 K and 90 K for cryogenically and thermoelectrically cooled LNA, respectively at 5° elevation angle. The receiving system noise temperature of 70 K for cryogenically cooled systems is made up of about 25 K due to feed and waveguide losses and 15 K due to the LNA. Uncooled parametric amplifiers with noise temperatures of 50 K (4 GHz) and 120 K (12 GHz) are now widely used in large domestic Earth stations. More recently, FET GaAs amplifiers have found wide application in small domestic and specialised system Earth stations with considerable reduction in cost.

54.15.3 High-power amplifiers

Transmitter powers are relatively modest but the transmitter arrangements are frequently complicated by the need to transmit more than one carrier. In a small or medium size station transmitting only one or two carriers, a klystron of 50–80 MHz bandwidth for each carrier may be the best arrangement. For a large station, klystrons with low-loss combiners (loss from 0.5 to

Table 54.6 INTELSAT Earth station parameters

Standard	*Uplink* (GHz)	*Downlink* (GHz)	*G/T* (*min*) (dB/K)	*Antenna diameter* (m)	*E.i.r.p. range* (dB W)
A	5.85–7.075	3.625–4.2	35	13–18	60–90
B	5.85–7.075	3.625–4.2	31.7	11	60–85
C	14.00–14.50	10.95–11.2 11.45–11.70	37.00	14–18	60–90
D_1	5.925–6.425	3.7–4.2	22.7	5	53–57
D_2			31.7	11	
E_1		10.95–11.2	25	3.5	57–86
E_2	14.0–14.5	11.7–11.95	29	5.5	55–83
E_3		12.5–12.75	34	8	49–77
F_1	5.925–6.256	3.7–4.031	22.7	4.5	63–91
F_2	6.094–6.425	3.869–4.200	27	7.5	60–87
F_3			29	9	59–86
G	6 or 14	4, 11 or 12	NS	NS	NS

NS, not specified

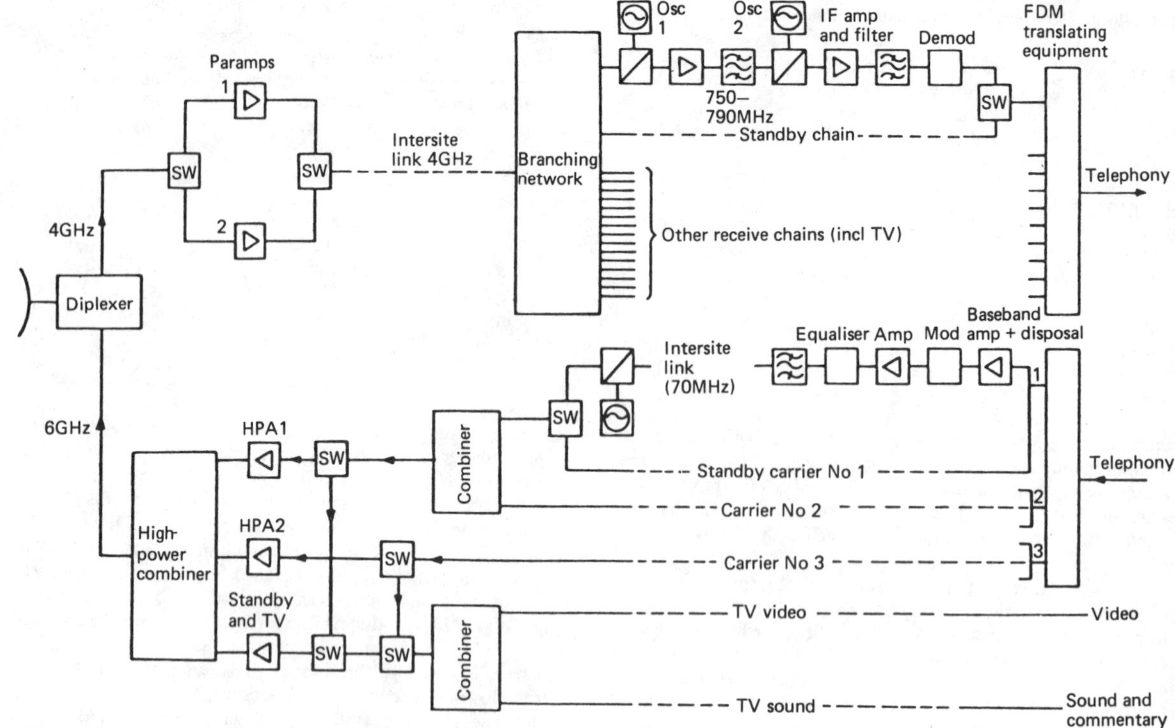

Figure 54.6 Typical Earth station equipment using TWT

4.5 dB) or TWT covering the whole 575 MHz band are used. A TWT may be used with multiple carriers but a large station can still need several TWT, connected through a high-power combining network to the transmit waveguide.

Simple, structure-compensated FET-type solid-state power amplifiers (SSPA) are the usual choice for small stations.

The transmitter power requirements can be judged by the following examples for INTELSAT 6/4 GHz band operation, assuming 60 dB aerial gain.

(1) Global beam carriers: 132 channel, 91 W, 972 channel, 813 W, TV 501 W.
(2) Hemi/zone beam carriers: 132 channel, 129 W, 972 channel, 575 W.
(3) TDMA one carrier per 80 MHz transponder 3200 channel with DSI 316 W.

If two or more carriers are passed through the same TWT or klystron, a back-off of 3–8 dB is necessary to reduce intermodulation products to the allowed levels. At 6/4 GHz band, TWT and klystrons with power outputs of 0.1–2.5 kW and 0.5–5 kW, respectively, are commonly used although TWT with power outputs of 20 kW are commercially available.

54.15.4 Intermediate frequency equipment, combining and branching equipment

On the transmit side the link between the main equipment building and the aerial site is usually at the IF of 70 MHz. Up conversion to 6 GHz is followed by either low-power combining of carriers to be amplified in the same TWT or by intermediate TWT amplifications followed by high-power klystron amplification and combining. On the receive side the frequently large number of received carriers and the need for flexibility make the use of an SHF link from the aerial to the equipment building more convenient. A flexible elliptical waveguide can be used for this.

A large station may need to extract perhaps 30 separate carriers, one for each station from which traffic is to be received. A modern approach to this problem uses stripline branching techniques. The signal path is first broken down to the required number of receive chains, which are all broadband at this stage. A first downconverter, fed with a local oscillator frequency selected to pick out the 40/80 MHz band in which the carrier falls, translates to a 770 MHz IF from which a second downconverter translates to 70 MHz for feeding to the demodulator. For TDMA the first downconverter selects an 80 MHz bandwidth which is translated in the second downconverter to a 140 MHz IF. After equalisation the IF signal is passed to the PSK demodulator. Redundant receive chains and automatic changeover are provided for reliability. Group delay equalisation is carried out at 70 MHz (140 MHz for TDMA), the satellite group delay being normally dealt with by the transmit side equalisation. Carefully designed IF filters are necessary in transmit and receive paths to eliminate adjacent channel interference.

54.15.5 Modulators, demodulators and baseband equipment

The design of frequency modulators follows conventional techniques. The main requirement is for a high degree of linearity.

Demodulators also follow conventional techniques except when the carrier-to-noise ratio is close to the FM threshold (approximately 9 dB). Then threshold extension demodulators[35] are used, the FM feedback demodulator being the commonest of this type.

The telephone channels are assembled into the baseband signals required for the particular transmitted carriers using conventional FDM translating equipment. All these basebands commence at 12 kHz but the band below 12 kHz is used for engineering service circuits and for the energy dispersal signal, a triangular-wave signal which keeps the RF spectrum well spread during the light traffic loading conditions to reduce interference problems. On the receive side those channels destined for the particular station are extracted from the received basebands using FDM translating equipment. They are then reassembled for onward transmission.

Where TDMA is employed, the 120 Mbit/s four-phase PSK signal is demodulated into two 60 Mbit/s data streams which are passed to the TDMA terminal proper. The TDMA terminal extracts the appropriate data bursts containing the relevant telephone channels. These are reformatted as standard 30-channel PCM circuits for onward transmission.

Advances in technology have led to the introduction of small Earth station terminals. Satellite networks, mainly in Europe and the USA, have mushroomed over the last few years. Small terminals located at the user premises can provide live colour video, two-way video conferencing, interactive data and point-to-multipoint data broadcast or collection. Various networks have been set up to provide specialised business services using small terminals with antennae of 1–5 m diameter. *Table 54.6* gives examples of such systems available via INTELSAT satellites. Similar systems operate via EUTELSAT and TELECOM 1. The proliferation of TVRO and VSAT in the USA as a result of telecommunications deregulation has revolutionised the telecommunications scene and a similar phenomenon is expected to occur in Europe following the relaxation of the monopoly of the PTTs.

54.16 Regional and domestic communication satellite systems

Regional and domestic satellites can concentrate all their available radiated power into relatively small areas, thus achieving higher e.i.r.p. compared to international service satellites covering large areas of the globe. Typically, the e.i.r.p. of a global beam is 23 dB W whilst a beam using the same TWT but illuminating mainland USA or Western Europe achieves about 33 dB W. This tenfold increase in incident power can be used to reduce ground segment costs by the use of much smaller aerials than the ones used in international communications, typically 3–9 m.

Some general principles concerning regional and domestic systems can be stated. Such systems are likely to be justified where:

(1) The terrain is very difficult for terrestrial communications.
(2) There are no existing facilities and distances are large.
(3) Terrestrial facilities exist but large volumes of traffic have to be carried over large distances and/or where a wideband service is needed.

Table 54.7 lists basic characteristics of selected regional/domestic service satellites.

ORBITA (USSR)[36–38] The first purely domestic satellite system was the ORBITA network of the USSR based on the use of MOLNIYA satellites placed in a highly elliptical 12 h orbit (538 and 39 300 km at 65.5° inclination) providing mainly TV distribution. The first MOLNIYA (Lightning) satellite was launched in April 1965. A network of ground terminals was set up near big cities with parabolic aerials of 12 m which allowed reception of TV and reception/transmission of voice and telegraph signals. Initially, the frequency band around 800 MHz was used but in 1971 after the launch of MOLNIYA-2 the 6/4 GHz was used. A portable ground station was also deployed which has an aerial of 7 m and works to both MOLNIYA and RADUGA (Rainbow) satellites. The ORBITA system has about 50 Earth stations in the USSR at present.

ANIK (Canada)[39,40] Telesat of Canada with the launch in 1972 of the first of three ANIK-A satellites established the world's second domestic system. The second generation ANIK-B satellites[41] have been added to the original ANIK-A and the first of the third-generation ANIK-C[42] was launched in 1981 (*Table 54.7*). These have been replaced by the fourth-generation satellites ANIK-D. The first of ANIK-D was launched in 1982 and the second in 1984. Contracts have been placed for the fifth-generation dual-band satellites ANIK-E, scheduled for launch in 1990.

WESTAR (USA)[43,44] The first USA domestic satellite, WESTAR I (*Table 54.7*) was launched by the Western Union Telegraph Co. in 1972 followed by WESTAR II in the same year. The WESTAR satellite system provides voice, telegraph and data communications in the USA. The Earth segments consist mainly of five major Earth terminals utilising 15 m aerials located at major cities. Western Union also leases out capacity to other common carriers. The American Satellite Corporation (AMSAT) leases transponders on WESTAR and provides services to its users via about forty 5 to 11 m Earth terminals owned and operated by AMSAT. A WESTAR III was launched in 1979 after having been held in ground storage for 5 years and a WESTAR IV was launched in 1982. A WESTAR V was launched in 1982 and a WESTAR VI is planned for launch in 1988. The latter will deliver higher e.i.r.p. levels and have a greater TWTA redundancy than its predecessors.

RCA SATCOM (USA)[45] RCA satellites SATCOM 1 and SATCOM 2 were launched in 1975 and 1976, respectively (*Table 54.7*). Each satellite has twenty-four 36 MHz transponders with 5 W TWT, operates in the 6/4 GHz band and utilises polarisation diversity for frequency reuse. The SATCOM system services the USA and provides voice, data and TV services to business, media and government. A new data, voice, facsimile, slow-scan TV and teleprint service called '56 plus' is also being offered. SATCOM F4 was launched in 1982 and SATCOM K1 and K2 were launched in 1985. SATCOM K3 is planned for launch in 1989. The F series operates in C band whereas the K series operates in Ku band. The latter are said to be the most powerful domestic satellites to date.

COMSTAR (USA)[46] Comsat General Inc. established the COMSTAR satellite system with the launch of the COMSTAR D1 satellite in 1976 (*Table 54.7*). The system was designed to meet the specific requirements of the American Telephone and Telegraph Co. (AT&T) for long-distance voice communications. The space segment is leased entirely to AT&T and GT&E who operate seven INTELSAT standard A equivalent Earth stations. The satellites employ polarisation isolation frequency reuse in the 6/4 GHz band to provide twenty-four 36 MHz transponders capable of handling 28 800 telephone channels. Three COMSTAR D satellites have been launched to date covering 51 states in the USA.

SATELLITE BUSINESS SYSTEMS (USA)[47] The most ambitious domestic satellite communications system in the USA so far is that of the Satellite Business System (SBS), a partnership of COMSAT General Corporation, Aetna Life and Casualty and IBM. The SBS provides an all-digital private-line switched network for integrated voice, data and video services. The overall system consists of the satellites, Earth stations, facilities for TT&C and a system management function which brings all these together under computer control. The system operates four-phase PSK/TDMA in the 14–11 GHz band (*Table 54.7*).

The space segment comprises three satellites, two operational in orbit (one also acting as a spare) and one ground spare. The first operational satellite was successfully launched in November 1980 and service began in early 1981. Each satellite contains one

Table 54.7 Basic characteristics of selected domestic service satellites

	ANIK-A	*ANIK-B*	*ANIK-C*	*WU WESTAR*	*RCA SATCOM*	*SBS*	*TELECOM 1*	*COMSTAR*	*INDONESIA PALAPA*
Service	Fixed voice	Fixed voice	Fixed voice, TV, data	Fixed voice, TV, data	Fixed voice, TV	Fixed voice, TV, data	Fixed voice, TV, data	Fixed voice	Fixed voice, TV
Frequencies (GHz)	6/4	6/4, 14/11	14/12	6/4	6/4	14/12	6/4, 8/7, 14/12	6/4	6/4
Mass in geostationary orbit (kg)	272	440	567	297	461	555	640	810	200
Coverage	Canada	Canada	Canada	USA	USA	Continental USA	France, Overseas Dept and Territories	USA	Indonesia
No of transponders (bandwidth (MHz))	12(36)	12/6 (36/72)	16(54)	12(36)	24(34)	10(43)	6(36), 2(120), 4(40)	24(36)	12(36)
No of beams	1	1 4/1	4/1	2	2	1	4	4	1
G/T (dB/K)	−7	−6, −1	+1	−6	−5, −10	2 to −2	−8.6 to −4.6, +6.3	−8.8	−7
E.i.r.p. (dB W)	33	36, 47.5	48	33	32, 26	43.7 to 40	27.3–31.5, 47.0	33	32
Modulation	FDM/FM QPSK, SCPC	FDM/FM QPSK, SCPC	FDM/FM QPSK, SCPC	FM QPSK	FDM/FM QPSK, SCPC	QPSK	FDM/FM BPSK, SCPC	FDM/FM	FM/SCPC
Multiple access	FDMA TDMA	FDMA TDMA	FDMA TDMA Re-use	FDMA TDMA	FDMA	TDMA	FDMA TDMA	FDMA Re-use	FDMA
Primary power (W)	300	840	900	300	770	914	920	600	300

active and three spare wideband receivers and ten communication channels. The offset parabolic satellite antennae with maximum gain of 35 dB (at 14/12 GHz) provide shaped beams that cover the contiguous 48 states. The fourth-generation satellites SBS IV, launched in 1984, are more powerful and offer greater capacity than earlier generations. The fifth and sixth generations are scheduled for launch in 1987.

The SBS Earth stations[48,49] which are located in the customers' premises consist of:

(1) RF terminals employing 5.5 m or 7.6 m symmetrically fed parabolic reflectors with Cassegrain feeds.
(2) Burst modems that perform modulation/demodulation functions for the high-speed transmission bursts.
(3) Satellite communications controllers for a series of functions including multiple access control and port adaptor systems for connecting the customers' communications and business equipment to the SBS system.

TELECOM 1 (France)[50,51] The French Government set up its own domestic satellite communications system with the launch in 1984 of TELECOM 1. Three satellites are employed, and the orbital assignments for Telecom F1, F2 and F3 are 8° West, 5° West and E° East, respectively.

TELECOM 1 operates in the 14/12 GHz, 8/7 GHz and 6/4 GHz frequency bands (*Table 54.7*). At 14/12 GHz, coverage is provided over the territory of metropolitan France and neighbouring countries. The six 14/12 GHz transponders provide two different services, namely, an intra-company service for the business communications and a video service from a point of origin towards community viewing receive points. The 8/7 GHz band is used exclusively for French Government communications. A near-global beam provides the required coverage. The 6/4 GHz is used for telephony and television communications between metropolitan France and the Overseas Departments and Territories.

The intra-company communications service includes data, telephony and video conferencing and operates in the TDMA mode. Data rates range from 2.4 kbit/s to 2 Mbit/s. Easy to install, unmanned Earth stations with a 3 to 4 m aerial and 150 W rated transmitter power are installed at or near the business premises subscribing to the service. The satellite and Earth stations employ permanent pooling of transmission capacities, flexibly and dynamically allocated on demand to any system user, according to his varying transmission speed requirements. Five transponders of 36 MHz bandwidth each provide a total of 125 Mbit/s (about 370 channels of 64 kbit/s per transponder).

The video service provides live TV transmissions to cinemas, schools, etc., using small receiving aerials.

Communications between metropolitan France and the Overseas Departments and Territories includes about 1800 voice channels and one TV channel. INTELSAT standard A and B equivalent Earth stations are used in metropolitan France and overseas territories, respectively.

INTELSAT[52] In addition to its global system INTELSAT is currently leasing out about 25 INTELSAT satellite transponders, mostly on spare satellites on a low-tariff pre-emptable basis, to individual signatories for their domestic communications. Countries such as Algeria, Brazil, Chile, Colombia, France, India, Norway, Nigeria, Oman, Peru, Sudan, Saudi Arabia, Spain and Zaire lease whole or partial transponders from INTELSAT for domestic communications. Most domestic systems use 11 to 14 m Earth stations and operate their voice channels in the voice activated SCPC mode. Some countries lease INTELSAT transponders for domestic communications before launching their own domestic satellites. This was done by Indonesia, Australia, China, India and Mexico among others.

In 1986 INTELSAT offered the outright sale of its spare capacity on satellites, available in all three ocean regions, for use only in domestic services. Seven years' use of one transponder cost a mere $3–5 million. To date some 29 transponders have been sold.

PALAPA (Indonesia)[53–55] In 1976, Indonesia initiated a domestic communications service on its PALAPA A1 Satellite (*Table 54.7*). Two in-orbit satellites, one operational and one spare, nearly identical to ANIK-A and WESTAR, form the space segment. Fifty 10 m Earth terminals strategically located on the 13 677-island archipelago (only 1000 islands inhabited) relay satellite traffic to telephone exchanges in the cities and communities served, and interconnect with microwave and cable network for distribution of voice TV and data traffic. The Indonesian Government also leases transponders to neighbouring countries for domestic use.

The first of the second-generation PALAPA B satellites was deployed in 1983. The second was launched in 1987. PALAPA B is twice as big, with twice the capacity and four times the electrical power of PALAPA A.

EUTELSAT[56] Since 1983, the European Telecommunications Satellite Organisation has provided the European regional satellite services using the ECS1 satellites, renamed EUTELSAT-I. The system operates in the 14/11 GHz band using TDMA for voice and data and FM for TV distribution. In addition, EUTELSAT provides specialised business services via small Earth stations using the EUTELSAT-I satellites (flight F2 onwards). EUTELSAT-I F2 was launched in 1984 and EUTELSAT-I F4 was launched in 1987. EUTELSAT's second-generation satellites are scheduled for launch in 1989 onwards. These are more powerful than the first-generation series with improved aerial design and more powerful amplifiers.

ARABSAT[57] The Arab Satellite Communication Organisation was formed in 1976, as a consortium of 22 Arab countries, to supervise constructing, launching and operating a satellite system. The first ARABSAT satellite was launched in February 1985 and the second in June 1985. The system is to provide a variety of communications services such as telephone, low and medium data transmission, multiplexed telex/telegraphy, radio and television distribution and community TV reception. It operates in the C-band (6/4 GHz) using twenty-five 33 MHz transponders. A specific channel is dedicated to community television which operates at 2.5 GHz (S-band) in the down-link.

INSAT[58] The first Indian satellite INSAT-1A, launched on 10 April 1982, had to be de-activated on 6 September 1982 because of failure in orbit and was replaced by the INSAT-1B satellite successfully launched in 1983. This satellite is intended for trunk communications (telephony, data, facsimile, etc.), meteorological Earth observation and data relay, TV broadcasting to community TV sets in rural areas and networking of terrestrial TV transmitters, and finally regional and national networking of radio transmitters.

Aussat[59] Australia's first-generation satellite series comprises three satellites. Aussats 1 and 2 have a total capacity of 30 transponders. Aussat 3 acts as an in-orbit spare. Broadcasting at Ku-band frequencies, the Aussats provide a diverse blend of communication service including direct delivery of TV programming to Australia's outbacks for reception by small dish terminals. Satellite coverage is through four zone beams. There is also a South Pacific beam which covers Papua New Guinea, New Zealand and the Kiribas Islands.

54.16.1 Future regional and domestic systems

Olympus The European Space Agency is planning an experimental regional European satellite, Olympus,[60–62] which will carry mainly satellite multiservice and direct broadcasting payloads. Olympus will be about twice the size of the ECS and is due for launching in 1989.

Astra[56] In 1985, the Duchy of Luxembourg granted a licence to SES (Société Européene de Satellites) to set up a satellite system from the orbital slots assigned to it. The Astra satellite system will initially consist of two medium-powered satellites, one fully operational and the second as a standby. When operational, in 1988, Astra will provide 16 satellite-delivered television channels for reception across Europe using a small rooftop antenna.

TELE-X[63] Sweden, Norway and Finland have decided to go ahead with a satellite system called TELE-X. This is a multi-mission experimental/pre-operational telecommunications satellite which will provide new services for data and video communication and direct TV to users in Sweden, Norway and Finland. The TELE-X platform is based on the platform used in the Franco-German satellite programme TVSAT/TDF-1. The data/video transponders are equipped with 220 W TWTA, one narrow-band transponder with capacity of 500 × 64 kbit/s or 20 × 2 Mbit/s or a combination of these channels and one wideband transponder with capacity of 25 × 2 Mbit/s or 6 × 8 Mbit/s or 2 × 34 Mbit/s or 1 × 139 Mbit/s.

54.17 Satellite systems for aeronautical, land and maritime mobile communications[64–66]

An important advantage of satellite communications over all other modes of long-distance wideband communications is the ability to operate with mobile terminals. A number of satellite systems have been proposed to serve mobile platforms, one of which is the now fully operational 'MARISAT' system.[67] MARISAT was developed to serve mobile maritime users in the three ocean regions, Atlantic, Indian and Pacific, respectively. The system consists of three operational geostationary satellites (launched in 1976), three shore stations and 542 ship terminals.

Each satellite contains one 500 kHz and two 25 kHz channels in the 250/300 MHz bands for use by the US Navy and two 4 MHz transponders for commercial use. One transponder translates shore-to-ship signals from 6 GHz to 1.5 GHz while the other translates ship-to-shore signals from 1.6 GHz to 4 GHz.

The MARISAT communications system provides voice, telegraph and signalling channels. The voice channels are transmitted on an SCPC/CFM/FDMA mode. The telegraph channels are two-phase coherent PSK modulated and use TDM and TDMA for shore-to-ship and ship-to-shore directions, respectively. Each satellite supports one voice and 44 telex channels (low-power mode).

In February 1982, the International Maritime Satellite Organisation (INMARSAT) system replaced MARISAT in providing the space segment for maritime communications. Two European Space Agency MARECS satellites, three INTELSAT V satellites with a maritime services subsystem and an existing MARISAT satellite provide the required three-ocean coverage for INMARSAT.

For economy and continuity of service, INMARSAT has retained the modulation and multiple access methods used by MARISAT. The basic INMARSAT system, which is composed of a standard A ship Earth station and on-shore Coast Earth Station has characteristics which are as follows:

(1) Satellite communication system

	Shore to ship	Ship to shore
Receive G/T (dB/K)	−21 to −17	−17.5 to −21.1
Total e.i.r.p. (dB W)	27–34.2	14.5–20
Transponder bandwidth (MHz)	4–7.5	

(2) Ship terminal characteristics, standard A

G/T	−4 dB/K
Gain	23 dBi
Aerial diameter	1.2 m
E.i.r.p.	37 dB W

(3) Shore Earth station characteristics

G/T	32.2 dB/K
Gain	43–46 dBi (at 60% efficiency)
Aerial diameter	12.8 m
E.i.r.p.	60 dB W (voice), 57 dB W (TDM)

(4) Frequencies used

	Ship to satellite	Satellite to shore
Uplink (MHz)	1626.5–1645.5	6417.5–6425
Downlink (MHz)	1530–1544	4192.5–4200

In addition to the Standard A ship Earth stations, INMARSAT has introduced two other standards: Standard B ($G/T = -12$ dB/K) and Standard C ($G/T = -23$ dB/K). These standards allow for a significant reduction in size and weight of associated terminals. Vocoders, possibly in conjunction with forward error correction techniques such as convolutional encoding, can provide a speech quality with Standard B terminals estimated to be fair to good. The Standard C terminal has been designed as a low-cost message transfer service.

A number of aeronautical satellite communications systems have been proposed in the last decade but, because of international, institutional and economic problems, progress has been slow. Communication with aircraft would probably require a small number of voice channels and perhaps some channels for automatic data transmission as a complement to a radio determination system which would allow closer spacing between aircraft on heavily loaded oceanic routes. Voice channels would probably use narrow-band FM or delta modulation with PSK. The largest single factor influencing system performance is the aircraft antenna which necessarily has very wide beamwidth and hence low gain since it cannot be steerable, except perhaps for rudimentary steering with a phased array or switching between elements. This constraint means that only a handful of channels can be obtained from a satellite which would give thousands of channels to standard INTELSAT Earth stations.

The frequency bands of prime interest for maritime and aeronautical communication satellites[4,68] are as shown in *Table 54.8*.

Table 54.8

	Frequency band (MHz)	
	Space to Earth	*Earth to space*
Maritime mobile satellites	1530–1544	1626.5–1645.5
Aeronautical mobile satellites	1545–1555	1646.5–1656.5
Land mobile satellites	1530–1533	1656.50–1660.50
	1555–1559	1631.50–1634.50

For the links between the satellites and the land-based stations frequencies in the 6/4 GHz bands are used although use of the above-mentioned UHF bands is permitted.

In 1987 the INMARSAT Convention was amended to permit it to offer aeronautical service. Currently trials are being conducted by British Telecom International and British Airways in conjunction with INMARSAT satellites on an aeronautical service over the Atlantic route.

INMARSAT's second-generation satellites are to be equipped with an aeronautical communication payload. This, however, has a limited capacity and would cater for a fraction of the expressed airline needs.

ARINC[69] A USA private company has filed with the Federal Communications Commission (FCC) an application for setting up an aeronautical satellite communications system. This will provide high-quality voice and data to aircraft for the multiplicity of services to serve North American airlines.

54.18 Television broadcasting satellite systems[70,71]

A wide diversity in the technical devices and the performance characteristics of TV broadcasting satellite systems exists. Besides the coverage areas and the frequencies used, the main differences between systems result from the service definition. For example, the Europeans are planning a purely direct to home reception requiring high flux density; others specify community-type reception which requires less powerful satellite transmission.

The WARC 77 drew up a plan[72] in which every country in regions 1 (Europe/USSR/Africa) and 3 (Asia/Australia) was allocated an orbital position with 5 TV channels of 27 MHz necessary bandwidth and specified e.i.r.p. and footprint characteristics. A Regional Administrative Radio Conference (RARC)[73] was held in 1983 to develop a plan and establish technical parameters for the broadcast satellite service in Region 2 (the Americas).

Typical parameters for television broadcasting direct to home and community receivers are given in *Table 54.9*.

At present there are three national television broadcasting satellite systems, namely, CTS/HERMES (Canada/USA),[74] EKRAN (USSR) and BSE (Japan).[75] The main parameters of these systems are listed in *Table 54.10*.

In addition, France and Germany have collaborated on the development of a direct broadcasting satellite system. The first satellite (TV SAT) was launched in November 1987 and will be followed in 1988 by the second, TDF-1. The UK has also embarked on a programme to develop its own direct broadcasting satellite system.

British Satellite Broadcasting (BSB), a private company, was awarded the franchise. The satellite is scheduled for launch in the Autumn of 1989.

The bands allocated to satellite broadcasting[4] are 2.5–2.69 GHz (community reception only) and 11.7–12.5 GHz Region 1, 11.7–12.2 GHz for Region 3 and 12.2–12.7 GHz for Region 2.

Table 54.10 Television broadcasting satellite system parameters

	CTS/HERMES	*EKRAN*	*BSE*
Country	Canada/USA	USSR	Japan
Launch date	1976	1976–1980	1978
Launcher	DELTA-2914	PROTON-D	DELTA-2914
Lifetime (years)	3	Unknown	3
Frequency (GHz)	12	0.7	12
TWTA RF power (W)	200	200	100
Coverage	2.5°	9×10^6 km²	Japan
E.i.r.p. (dB W)	59.5	47.5	55
Aerial diameter (m)	0.6 to 3	YAGI Array	16
G/T (dB/K)	5.8	−6	15.7
C/N_0 (dB Hz)	86.4	87.6	92.5
C/N (dB)	13	13.8	18.5

54.19 Future developments

The trend in the 1990s will be for higher capacity satellites with lower cost per channel making more efficient use of the frequency spectrum and offering better inter-satellite connectivity.

Frequency reuse with multi spot beams[76] and the use of 30–20 GHz frequency bands[77–80] will offer the higher capacity needed.

As the number of satellite beams increases, on-board switching[81] and processing will become increasingly important. Regenerative transponders will demodulate the received digital carrier, reconstruct the signal and remodulate it for transmission to Earth. On-board processing eliminates the up-link noise component of the retransmitted signal making it particularly useful for small terminal links where the up-link noise can be a large portion of the overall C/N.

High transmission efficiency will be achieved by the use of efficient digital coding and processing. Inter-satellite connectivity can be achieved by the use of inter-satellite links (ISL). ISL can interconnect Earth stations operating to different satellites without additional aerials or double hops. This facility will significantly influence the capacity, connectivity, coverage, orbit-utilisation and overall cost of a satellite system.

The Space Shuttle with its 18.3 m by 4.6 m payload bay and 30 000 kg launch capability into low Earth orbit could place in geostationary orbit very large stable platforms that can support complex large antenna and payload configurations serving different missions.[82] However, the grounding of the space shuttle since the January 1986 disaster and its subsequent withdrawal as a commercial launcher has hitherto hampered the realisation of such plans. Alternatives to the platform concept include:

(1) Very large satellites launched by powerful launchers such as Atlas G Centaur or Ariane 5.
(2) Numerous small special-purpose satellites.

Table 54.9 Examples of parameters for FM television broadcast satellite at 12 GHz (Satellite aerial efficiency of 55% is assumed)

Service type	*Receive aerial diameter* (m)	*Receive aerial G/T* (dB/K)	*System noise factor* (dB)	*Overall C/N required* (dB)	*Bandwidth per channel* (MHz)	*Received p.f.d.* (dB W/m²)	*Satellite e.i.r.p.* (dB W)	*Satellite RF power (W) for* 3.4°	1.15°
Home	0.9	6	5.9	14	27	−103	62	631	71
Community	1.8	14	4.2	14	27	−111	54	100	11

(3) Several smaller satellites connected via ISL.
(4) Clusters of satellites[83] interconnected and collocated within the beam of an Earth station.

Various exploratory studies carried out during the past few years have indicated that there was a large unsatisfied demand for specific communications service in the various sectors of business activities which operate small ships, airplanes and land vehicles.

In the field of land mobile communications there exists in Europe extensive advanced Radio Cellular networks, and hence the need for a satellite-based service is less obvious. There are, moreover, other problems such as lack of frequency allocations and the absence of any institutional framework at European and international level.

The provision, at the recent WARC-MOB 87, for a 7 MHz allocation in L-band for a land mobile satellite service goes some way towards alleviating the first problem. However, the realisation of such a service on a European scale would still require the overcoming of the second obstacle. EUTELSAT, in this case, could possibly provide the lead and the necessary institutional framework.

In North America[84] plans are afoot for setting up a land mobile satellite service, subject to approval by FCC. Work is expected to commence in 1988. An operational mobile satellite service would then be available in North America by the 1992.

Information on the following further topics can be found in the references given:

(1) Propagation factors in satellite communication.[85]
(2) Interference with terrestrial or other satellite systems.[86,87]
(3) Efficiency of use of the geostationary orbit.[88]
(4) Effects of propagation delay and echo.[89,90]

References

1 CLARKE, A. C., 'Extra terrestrial relays', *Wireless World*, 305 (1945)
2 EDELSON, B. I. and DAVIS, R. C., 'Satellite communications in the 1980s and after', *Phil. Trans. R. Soc. London*, **289**, 159 (1978)
3 KRUPIN, Y. U., 'The INTERSPUTNIK International Space Communications Organisation', *Radio Telev. (Czechoslovakia)*, **1**, 23 (1978)
4 WARC, 'Final acts of the World Administrative Radio Conference', ITU Geneva (1979)
5 YOSHIDA, N., SHIOMI, T. and OKAMOTO, T., 'On-orbit antenna pointing performance of Japanese communications satellites CS', *AIAA Paper 82-0441* (1982)
6 COLLETTE, R. C. and HERDAN, B. L., 'Design problems of spacecraft for communication missions', *Proc. IEEE*, **65**, No. 3, 342–356 (1977)
7 HAYN, D., BRAITINGER, M. and SCHMUCKER, R. H., 'Performance prediction of power augmented electrothermal hydrazine thrusters', *Technische Universitaet Lehrstuhl für Raumfahrttechnik, Munich, W Germany* (1978)
8 FREE, B. A., 'North–south station keeping with electric propulsion using on-board battery power', *1980 JANNAF Propulsion Meeting*, **5**, 217 (1980)
9 HALL, G. C. and MOSS, P. R., 'A review of the development of the INTELSAT system', *Post Office Elec. Eng. J.*, **71**, 155–163 (1978)
10 EATON, R. J. and KIRKBY, R. J., 'The evolution of the INTELSAT V system and satellite, part 2, spacecraft design', *Post Office Elec. Eng. J.*, **70**, No. 2, 76–80 (1977)
11 JILG, E. T., 'The INTELSAT IV Spacecraft', *COMSAT Tech. Rev.*, **2**, No. 2, 271 (1972)
12 BARTHOLOME, P., 'The European communications satellite programme', *ESA Bull. (France)*, **14**, 40–47 (May 1978)
13 FUENZALIDA, J. C., RIVALAN, P. and WEISS, H. J., 'Summary of the INTELSAT V communications performance specifications', *COMSAT Tech. Rev.*, **7**, No. 1, 311–326 (1977)
14 DINWIDDY, S. E., *Olympus Users' Guide*, ESTEC, Noordwijk, Netherlands, issue 1, December (1983)
15 WESTCOTT, R. J., 'Investigation of multiple FM/FDM carriers through a satellite TWT operating near to saturation', *Proc. IEEE*, **114**, No. 6, 726 (1972)
16 CHITRE, N. K. M. and FUENZALIDA, J. C., 'Baseband distortion caused by intermodulation in multicarrier FM systems', *COMSAT Tech. Rev.*, **2**, No. 1, 147 (1972)
17 INTELSAT, 'INTELSAT TDMA/DSI system specification (TDMA/DSI traffic terminals)', *INTELSAT Document BG 42/65 Rev.*, **1**, June (1981)
18 EUTELSAT, 'EUTELSAT TDMA/DSI system specification', *EUTELSAT Document ECS/C-11-17 Rev.*, **1**, September (1981)
19 EDELSON, B. I. and WERTH, A. W., 'SPADE system progress and application', *EUTELSAT Document ECS/C-11-17*, **2**, No. 1, 221 (1972)
20 CCIR, 'Allowable noise power in the hypothetical reference circuit for frequency-division multiplex telephony in the fixed satellite service', *CCIR Rec. 568, ITU Geneva*, **12**, 39 (1978)
21 CCIR, 'Single value of the signal-to-noise ratio for all television systems', *CCIR Rec. 568, ITU Geneva*, **12**, 39 (1978)
22 CCIR, 'Allowable bit error rates at the output of the hypothetical reference circuit for systems in the fixed satellite service, using pulse-code modulations for telephony', *CCIR Rec. 522, ITU Geneva*, **4**, 61 (1978)
23 HILLS, M. T. and EVANS, B. G., *Telecommunications System Design, Vol. 1 Transmission Systems*, 176–198, George Allen and Unwin (1973)
24 BARGELLINI, P. L., 'The INTELSAT IV communication system', *COMSAT Tech. Rev.*, **2**, No. 2, 437 (1972)
25 TERRELL, P. M., 'Application of digital speech interpolation', *Communs. Int. (GB)*, **6**, No. 2, 22, 24, 26 & 30 (1979)
26 CAMPANELLA, S. J., 'Digital speech interpolation techniques', 1978 National Telecommunications Conf., Birmingham AL USA, *Conf. Record of the IEEE*, 14.1/1.5, December (1978)
27 SEITZER, D., GERHAUSER, H. and LANGENBUCHER, G., 'A comparative study of high quality digital speech interpolation methods', *Int. Conf. on Communications, Toronto, Canada*, 50.2/1.5, June (1978)
28 INTELSAT 'Standard A performance characteristics of Earth stations in the INTELSAT IV, IVA and V systems having a *G/T* of 40.7 dB/K (6/4 GHz frequency bands)', *INTELSAT DOCUMENT BG28/72*, 75, August (1977)
29 FERGUSON, M. E., *FM—The New Single-Channel-per-Carrier Technique*, Technical Memorandum, California Microwave Inc., Sunnyvale CA, USA
30 THE INSTITUTE OF ELECTRICAL ENGINEERS LONDON (UK), *Antennas and Propagation, Part 1 Antennas*, IEE Conf. Publ., IEE, Savoy Place, London WC2, Publication No. 195, April (1981)
31 INTELSAT, 2nd Earth Station Technology Seminar, Athens, Greece, October (1977)
32 LOVE, A. W. (ed.), *Reflector Antennas*, IEEE Press, New York (1978)
33 CLARRICOATS, P. J. B., 'Some recent advances in "microwave reflector antennas"', *Proc. IEE*, **126**, No. 1, 9–25 (1979)
34 KAJIKAWA, M., HAGA, I., FUKUDA, S. and AKINAGA, W., 'Development status of low-noise amplifiers', *AIAA 8th Communications Satellite System Conf., Orlando FA, USA*, 309–316, April (1980)
35 VAN DASLER, G., VAN LAMBALGEN, H. and VAN DAAL, P. M., 'A threshold extension demodulator', *Phillips Telecom Rev. (Netherlands)*, **31**, No. 3, 131–146 (1973)
36 MERCADER, L., 'Direct broadcasting of TV signals from satellites. A description of the Russian systems', *Mundo Electron (Spain)*, No 88, 71–77, September (1979)
37 FISCHER, C., 'Satellite communications systems ORBITA-2', *Radio Fernbehen Elektron. (Germany)*, **23**, No. 17, 548–550 (1974)
38 KANTOR, L. YA., POLUKHIN, U. A. and TALYZIN, N. V., 'New relay stations of the ORBITA-2 satellite communications system', *Elektrosvyaz (USSR)*, **27**, No. 5, 1–8 (1978)
39 ALMOND, J., 'Commercial communication satellite systems in Canada', *IEEE Commun. Mag. (USA)*, **19**, No. 1, 10–20 (1981)
40 ROSCOE, O. S., 'Direct broadcast satellites—the Canadian experience', *Satell. Commun. (USA)*, **4**, No. 8, 22–23, 29–32 (1980)
41 GOTHE, G., 'The ANIK-B slim TDMA pilot project', NTC '80,

IEEE 1980 National Telecommunications Conference, Houston, Texas, USA, 71.4/1-9 (1980)
42 CHAN, K. K., HUANG, C. C., CUCHANSKI, M., RAAB, A. R., CRAIL, T. and TAORMINA, F., 'ANIK-C antenna system', *1980 Int. Symp. Digest*, Antennas and Propagation, Quebec, Canada, 2–6 June 1980, 89–92 (1980)
43 VERMA, S. N., 'US domestic communication system using WESTAR satellites', *World Telecommunication Forum Technical Symp., Geneva*, 6–8 October 1975, 2.4.3/1-6 (1975)
44 SCHNEIDER, P., 'WESTAR today and tomorrow (satellite communications)', *Signal* (*USA*), **34**, No. 3, 43–45 (1979)
45 KEIGLER, J. E., 'RCA SATCOM: an example of weight of optimised satellite design for maximum communications capacity', *Acta Astronautica*, **5**, 219–242 (1978)
46 BRISKMAN, R. D., 'The COMSTAR program', *COMSAT Tech. Rev.*, **7**, No. 1, 1–34 (1977)
47 WHITTAKER, P. N., 'Satellite business system (SBS)—A concept for the 80s', *Policy Implications of Data Network Development in the OECD*, 35–39 (1980)
48 WEISCHADLE, G. M. and KOURY, A., 'SBS terminals demand advanced design', *Microwave Syst. News* (USA), **9**, No. 4, 70 (1979)
49 WESTWOOD, D. H., 'Customer premises RF terminals for the SBS system', *ICC 79 Int. Conf. on Communications*, 6.3/1–5 (1979)
50 FLEURY, L., GUENIN, J. P. and RAMAT, P., 'The TELECOM 1 system', *Echo Rech.* (*France*), No. 101, 11–20 (1980)
51 GRENIER, J., POPOT, M., LAMBARD, D. and PAYET, G., 'TELECOM 1, A national satellite for domestic and business services', ICC '79, *1979 Int. Conf. on Communications, Boston, MA* (*USA*), 10–14 June 1979, 49.5/1 (1979)
52 *INTELSAT News*, **3**, No. 1 (1987)
53 HOGWOOD, P., 'PALAPA—Indonesia to the fore', *J. Br. Interplanet Soc.* (*GB*), **30**, No. 4, 127–130 (1977)
54 SOEWANDI, K. and SOEDARMADI, P., 'Telecommunications in Indonesia', *IEEE Trans. Commun.* (*USA*), **COM-24**, No. 7, 687–690 (1976)
55 TENGKER, J. S., 'Indonesian domestic satellite system', EASCON '76, *Proc. of Eastern Electronics Conf., Washington DC* (*USA*), 11-A to 11-U (1976)
56 LONG, M., *World Satellite Almanac*, Global edition, 21st Century Publishing (1988)
57 AL-MASHAT, A., 'Arabsat system: regional telecommunications programme for the Arab States', *Telecom. J.*, **52-II** (1985)
58 INDIA SATELLITE SYSTEM, *Handbook on Satellite Communications*, Fixed Satellite Services, Geneva (1985)
59 AUSSAT, *Australia's National Satellite System—Network Designer's Guide* (1986)
60 WATSON, J. R., 'L-SAT: A multipurpose satellite for the 1980s and 1990s', *IEE Colloquium on Satellite Broadcasting, London, England, 20 November 1980*, 4/1 (1980)
61 BARTHOLOME, P. and DINWIDDY, S., 'European Satellite Systems for Business Communications', ICC '80, *1980 Int. Conf. on Communications, Seattle WA* (*USA*), 8–12 June 1980, 51.5/1–7 (1980)
62 LORIORE, M., 'Communication satellite payloads: a review of past, present and future ESA developments', Technology Growth for the 80s, *IEEE MTT-S International Microwave Symposium Digest, Washington DC* (*USA*), 28–30 May 1980, 189–191 (1980)
63 GERMAIN, J., 'Telex-X: a multi-purpose communications satellite for the Nordic countries', *Space Commun. Broadcasting*, **2**, No. 3 (1984)
64 IEE, 'Maritime and aeronautical satellite communication and navigation', *Int. Conf. IEE London*, March (1978)
65 CCIR, *Conclusions of the Interim Meeting of Study Group 8* (*Mobile Services*), ITU Geneva (1980)
66 CCIR, *Recommendations and Reports of the CCIR* (*Mobile Service*), **8**, ITU Geneva (1978)
67 LIPKE, D. W. *et al.*, 'MARISAT-A maritime satellite communications system', *COMSAT Tech. Rev.*, **7**, No. 2, 351–391 (1977)
68 *Final Acts of the World Administrative Radio Conference for the Mobile Services* (WARC MOB-87), Geneva (1987)
69 DEMENT, D. K., 'AUSSAT: a dedicated high capacity worldwide aeronautical service'. In: The Royal Aeronautical Society Symposium—Satellite Services for Aviation, June (1987)
70 HMSO, *Direct Broadcasting by Satellite*, Report of Home Office Study, HMSO, London (1980)
71 CCIR Report 215-4, 'Systems for the broadcasting-satellite services (sound and television)', *CCIR*, **11**, 163–185, ITU Geneva (1978)
72 ITU, *Final Acts of the World Broadcasting—Satellite Administrative Radio Conf.*, ITU Geneva (1977)
73 *Final Acts of RARC-83*, Switzerland, July (1983)
74 SIOCOS, C. A., 'Broadcasting satellite reception experiment in Canada using high power satellite HERMES', *IBC 78 IEE Conf. Publ.* (*UK*), **166**, 197–201 (1978)
75 ISHIDA, *et al.*, 'Present situation of Japanese satellite broadcasting for experimental purposes', *IEEE Trans.*, **BC-25**, No. 4, 105 (1979)
76 REUDINK, D. O., 'Spot beam promise satellite communications breakthrough', *IEEE Spectrum*, 36–42 (1978)
77 FUKETA, H., KURAMOTO, M., INOVE, T. and KATO, E., 'Design and performance of 30/20 GHz band Earth stations for domestic satellite communication system', *AIAA Communications Satellite Systems Conf.*, pp. 361–369 (1980)
78 KRIEGL, D. O. W. and BRAUN, H. M., '20/30 GHz satellite business system in Germany', *AIAA Communications Satellite Systems Conf.*, 634–639 (1980)
79 BERRETTA, G., 'Outlook for satellite communications at 20/30 GHz', *ESA J.*, **3** (1979)
80 TIRRO, S., 'Utilisation of the 20/30 GHz spectrum part for high capacity national communications', *Int. Conf., Space Telecommunications and Radio Broadcasting*, Toulouse, France (1979)
81 FORDYCE, S. W. and JAFFE, L., 'Future communications concepts, switchboard in the sky', *Satellite Communications, Feb/March 1978* (1978)
82 MORGAN, W., 'Large geostationary communications platform', *XXX Congress IAF, Munich FR6, September 1979* (1979)
83 VISHER, P., 'Satellite clusters', *Satellite Communications* (*USA*), 22–27 September (1979)
84 SWARD, D. J. and LOK, M. F., 'MSAT System and Service Description', *Land Mobile Services by Satellite Workshop, ESA/ESTEC*, June (1986)
85 IEE, *2nd Int. Conf. on Antennas and Propagation, Part 2, Propagation*, IEE, London, April (1981)
86 JOHNS, P. B. and ROWBOTTOM, T. R., *Communication Systems Analysis*, Butterworths, London (1972)
87 CCIR, 'Methods for determining interference in terrestrial radio-relay systems and systems in the fixed satellite service', *CCIR Report 388-3*, **IX**, 340–357, ITU, Geneva (1978)
88 CCIR, 'Technical factors influencing the efficiency of use of the geostationary satellite orbit by radiocommunication satellites sharing the same frequency bands', *CCIR Report 453-2*, **IV**, 181–206, ITU, Geneva (1978)
89 HUTTER, J., 'Customer response to telephone circuits routed via a synchronous-orbit satellite', *Post Office Elec. Eng. J.*, **60**, part 3, 181 (1967)
90 CAMPANELLA, S. J., SUYDERHOUD, H. G. and ONUFRY, M., 'Analysis of an adoptive impulse response echo cancellor', *Comsat Tech. Rev.*, **2**, No. 1, 1 (1972)

Further reading

CHAYES, A. *et al.*, *Satellite Broadcasting*, Oxford University Press, Oxford (1973)
FEHER, K., *Digital Communications Satellite/Earth Station Engineering*, Prentice-Hall, Englewood Cliffs, NJ (1981)
LOVE, A. W. (ed.), *Reflector Antennas*, IEEE Press, New York (1978)
MARAL, G. and BOUSQUET, M., *Satellite Communications Systems*, John Wiley, New York (1986)
MARTIN, J., *Communications Satellite Systems*, Prentice-Hall, Englewood Cliffs, NJ (1978)
MIYA, K. (ed.), *Satellite Communications Engineering*, Lattice Co., Tokyo (1975)
PELTON, J. N. and SNOW, M. S., *Economic and Policy Problems in Satellite Communications* (*Economic and Political Issues of the First Decade of INTELSAT*), Praeger Publications (1977)
PRITCHARD, W. L. and SCIULL, J. A., *Satellite Communication Systems Engineering*, Prentice-Hall, Englewood Cliffs, NJ (1986)

SCHMIDT, W. G. and LAVEAN, G. E., *Communication Satellite Development Technology*, American Institute of Aeronautics (1976)
SLATER, J. N. and TRINOGGA, L. A., *Satellite Broadcasting Systems Planning and Design*, Ellis Horwood, Chichester (1985)
SNOW, M. S., *International Commercial Satellite Communications (Economic and Political Issues of the First Decade of INTELSAT)*, Praeger Publications (1976)
SPILKER, J., *Digital Communication by Satellite*, Prentice-Hall, Englewood Cliffs, NJ (1977)
UNGER, J. H. W., *Literature Survey of Communication Satellite Systems and Technology*, IEEE Press, New York (1976)
VAN TREES, H. L., *Satellite Communications*, IEEE Press, New York (1979)

55 Point-to-Point Communication

P J Howard, BSc, CEng, MIEE
STC Telecommunications Ltd
(Sections 56.5–56.11)

T Oswald, BSc, MIEE
STC Telecommunications Ltd
(Sections 56.19–56.24)

S F Smith, BSc(Eng), CEng, FIEE
STC Telecommunications Ltd
(Sections 56.1–56.4)

D B Waters, BSc(Eng)
STC Telecommunications Ltd
(Sections 56.12–56.18)

Contents

55.1 Telephone instrument (subset)

The speech transmission elements of a telephone instrument, *Figure 55.1*, consist of a receiver and transmitter (microphone) usually assembled in a common mounting to form a handset and connected to the line through an anti-sidetone induction coil (hybrid transformer) or an equivalent electronic circuit. The complete instrument also incorporates signalling elements, the

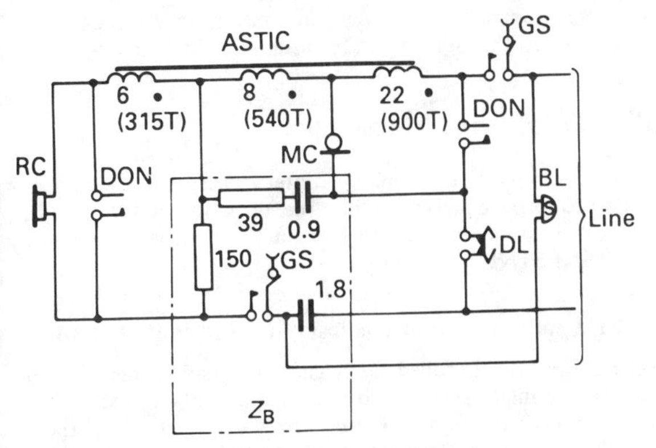

Figure 55.1 Typical telephone instrument circuit

exact form of which depends upon the type of exchange with which it is designed to interwork. Generally they comprise a dial or push buttons and associated gravity switch for signalling to the exchange and a ringer (bell) or an electronic detection circuit and sound transducer for receiving calling (ringing) signals from the exchange.

55.1.1 Transmitter

Most instruments employ a carbon microphone. Sound waves cause the diaphragm to vibrate, varying the pressure exerted on the carbon granules. This produces corresponding variations in the electrical resistance between the granules and hence in the current flowing through them when connected to a d.c. supply.

Power may be derived from a battery or other d.c. supply locally at the instrument but in public systems it is now the universal practice for a single central battery to provide a supply to all lines on the exchange. The a.c. component of the line current produced in this way is known as the speech current and the d.c. component derived from the battery is the microphone feed current (transmitter feed) or polarising current.

The feed to each line is decoupled by individual inductors, often in the form of relay coils, and speech is coupled from one line to another through the exchange switching apparatus by means of capacitors as shown in *Figure 55.2* or through a transformer. The combination of d.c. feed, battery decoupling and speech coupling components is known as a transmission bridge or feeding bridge. The relays providing the battery decoupling also serve to detect line signals to control the holding and releasing of the connection.

The microphone resistance is nonlinear and is subject to wide variations due to temperature, movement of granules and, of course, the effects of sound. Typically it is between 40 and 400 Ω under normal working conditions.

Some telephone instruments employ other types of microphone (e.g. moving coil) which do not themselves require a d.c. feed. Such microphones, however, lack the inherent amplification of the carbon microphone. They are usually supported by a transistor amplifier driven by the line current and designed to operate with similar feeding bridge conditions to carbon microphone instruments.

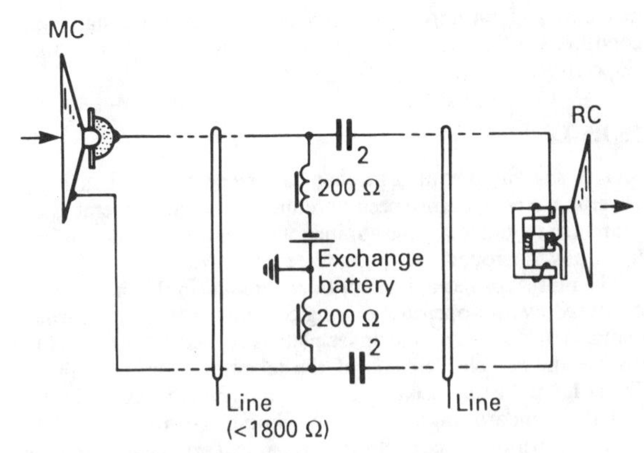

Figure 55.2 Principle of transmission bridge

55.1.2 Receiver

Most types operate on the moving-iron principle employed by Alexander Graham Bell for both transmitting and receiving in his original telephone in 1876. In some cases the diaphragm is operated on directly by the magnets but a more efficient arrangement is shown in *Figure 55.2*, in which the design of armature and diaphragm can each be optimised for its own purpose without conflict of mechanical-acoustic and magnetic requirements. A permanent magnet is included in the simple receiver to prevent frequency doubling due to the diaphragm being pulled on both the positive and the negative half cycle. In the rocking armature arrangement shown in *Figure 55.1*, no movement of the armature would occur at all without the permanent magnet.

Unlike the carbon microphone, the moving iron receiver requires no direct current for its operation and is usually protected against possible depolarisation by a series capacitor.

55.1.3 Anti-sidetone induction coil (ASTIC)

This is a hybrid transformer which performs the dual function of matching the receiver and transmitter to the line and controlling sidetone.

Sidetone is the reproduction at the receiver of sound picked up by the transmitter of the same instrument. It occurs because the transmitter and receiver are both coupled to the same two-wire line. The most comfortable conditions for the user are found to be when he hears his own voice in the receiver at about the same loudness as he would hear it through the air in normal conversation. Too much sidetone causes the talker to lower his voice, reduces his subsequent listening ability and increases the interfering effect of local room noise at the receiving end. The complete absence of sidetone, however, makes the telephone seem to be 'dead'.

The use of a hybrid transformer divides the transmitter output between the line and a corresponding balance impedance. If this impedance exactly balanced that of the line there would be no resultant power transferred to the receiver. In practice there is some transfer to produce an acceptable level of sidetone.

55.1.4 Gravity switch (switch hook)

To operate and release relays at the exchange and thus indicate calling and clearing conditions, a contact is provided to interrupt the line current. This contact is open when the handset is on the rest and closes (makes) when the handset is lifted. Because it is operated by the weight of the handset it is known as the gravity switch. It is also known as a switch hook contact and the terms off-hook and on-hook are often used to describe signalling conditions corresponding to the contact closed or open respectively.

55.1.5 Dial

This is a spring-operated mechanical device with a centrifugal governor to control the speed of return. Pulses are generated on the return motion only, the number of pulses up to ten depending upon how far round the finger plate is pulled.

The pulses consist of interruptions (breaks) in the line current produced by cam-operated pulsing contacts. A set of off-normal contacts operate when the finger plate is moved and are used to disable the speech elements of the telephone during dialling. These dial pulses are usually specified in terms of speed and ratio and the standard values of these parameters are designed to match the performance of electromechanical exchange switching equipment, *Figure 55.3*.

55.1.6 Push-button keypad

For use with exchanges designed for dial pulse signalling there are telephones in which a numerical keypad is used to input numbers to an l.s.i. circuit which then simulates the action of a dial. The output device for signalling to line is usually a relay, typically, a mercury-wetted relay.

For electronic exchanges, and some types of electromechanical exchange, a faster method of signalling known as multifrequency (m.f.) can be used. In this, a 12-button keypad is provided with oscillators having frequencies as shown in *Figure 55.4*. Pressing any button causes a pair of frequencies to be generated, one from each band.

55.1.7 Ringer (bell)

In the idle state, the handset is on its rest, the gravity switch is open and no direct current flows in the line. The ringer is therefore designed to operate on alternating current. This is connected from a common supply at typically 75 V r.m.s. and 25 Hz connected through a 500 Ω resistance at the exchange.

55.2 Telephone networks

Some use is made of telephones interconnected to form self-contained private networks but the majority of the world's telephones are connected to the public switched telephone network (PSTN).

55.2.1 Types of customer (subscriber) installation

Telephones are installed at customers' (subscribers') own premises or made available to the general public at call offices (paystations) with coin collecting boxes (CCB). Each of these installations is normally connected to a local exchange by a pair of wires but in isolated locations radio links may be used. In some cases more than one customer's instrument may be connected to the same line which is then described as a party line. In some

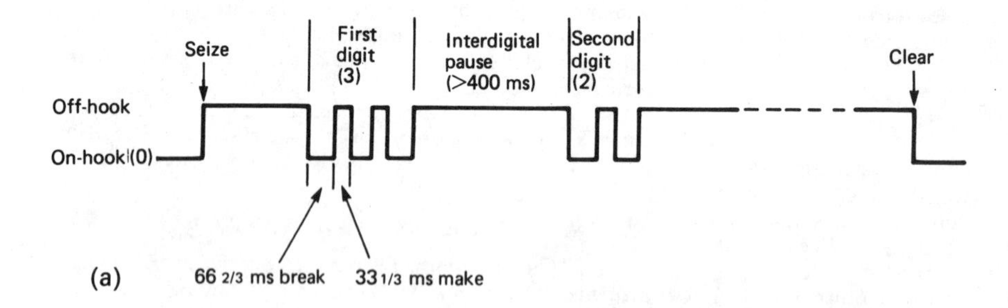

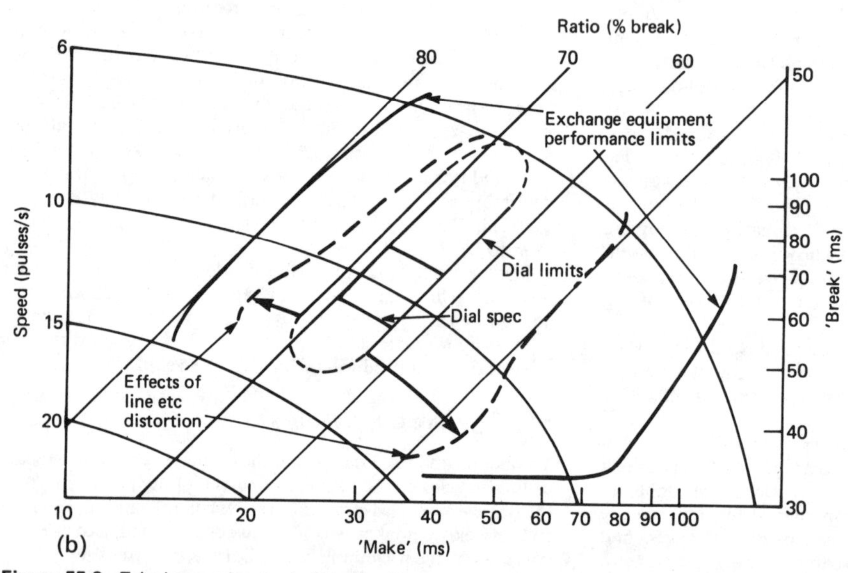

Figure 55.3 Telephone subscriber's line signalling: (a) waveform, (b) performance curves

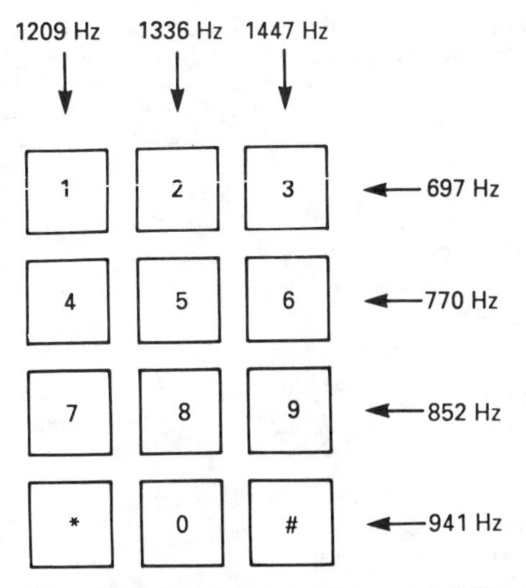

Figure 55.4 Push-button (multifrequency) dialling

countries a party line, especially in remote rural areas, may serve as many as 20 customers with a system of coded ringing signals. In the UK the only party lines are those serving two customers on a system known as shared service. The case of one customer per line is known as exclusive service.

At the customer's premises there can be more than one telephone. Residential and small business users often have two or more instruments with one of several single methods of interconnection known as extension plans. Larger businesses generally have private branch exchanges (PBX) on their premises providing connection between their extensions either under control of a switchboard operator (private manual branch exchange—PMBX) or by dialling (private automatic branch exchange—PABX).

These private exchanges are connected to their local exchange by a group of exchange lines and the public exchange selects a free line within the group (PBX hunting) when a caller dials the first line of the group.

55.2.2 Hierarchy of exchanges

Each local public exchange provides the means for setting up calls between its own customers' lines. It is normally also connected to other local exchanges in the same area (e.g. the same town) by junctions. For calls outside this area, junctions are provided to trunk exchanges which are in turn interconnected by trunk lines enabling calls to be established to customers in other areas. Trunk lines are also provided to one or more international exchanges which enable calls to be established to customers in the national networks of other countries.

A single exchange can sometimes combine two or more of these functions and the larger national networks have more than one level of trunk exchange in their hierarchy (e.g. three levels in the UK, four in the USA). A trunk call in the UK therefore passes through up to six trunk exchanges in addition to the local exchanges at each end.

In some large areas, some local exchanges may be interconnected only by having calls switched through another local exchange, a technique known as tandem switching which can be independent of the trunk switching hierarchy used for calls outside the area.

55.2.3 Signalling

To connect calls through a network of exchanges it is necessary to send data referring to the call between the exchanges concerned.

In the simplest case this consists of loop/disconnect signalling similar to that used on customers' lines, *Figure 55.3*. Each speech circuit consists of a pair of copper wires which have direct current flowing when in use on a call and this current is interrupted to form break pulses corresponding to those produced by the telephone dial. When the called party answers, his exchange sends back an 'answer' signal to the calling exchange by reversing the direction of current flow.

Longer circuits are unsuitable for direct current signalling, e.g. due to the use of amplifiers, multiplexing equipment or radio links. Voice frequency (VF) signalling is then used in which pulses of alternating current are used at frequencies and levels compatible with circuits designed for handling speech currents (typically 600, 750, 2280 and 2400 Hz). This use of frequencies within the speech band (300–3400 Hz) is termed in-band signalling.

An alternative is the use of out-of-band (outband) signalling in which a signalling frequency outside the speech band, e.g. 3825 Hz, is separately modulated on to the carrier in a frequency division transmission system.

In PCM transmission systems, the signals are digitally encoded and certain timeslot(s) are reserved for this purpose separate from the speech timeslots.

These are all channel-associated signalling systems in which the signalling is physically and permanently associated with the individual speech channel even though no signalling information is transmitted during the greater part of the call. The use of stored program control has given rise to a more efficient signalling means known as common-channel signalling. In this, one signalling channel serves a large number of speech channels and consists of a direct data link between the control processors (computers) in the exchanges concerned. Each new item of information relating to a call is sent as a digital message containing an address label identifying the circuit to which it is to be applied. CCITT signalling system No. 7 comprises an internationally standardised set of messages and protocols for common channel signalling.

55.2.4 Manual exchanges

The early exchanges were all manually operated and many are still in service around the world. The exchange equipment consists of a switchboard at which an operator (telephonist) sits. The simplest type is the magneto board on which the caller gains the attention of the operator by using a hand generator to send an a.c. signal which is detected by an electromechanical indicator at the switchboard. An improved design is the central battery (CB) board which uses gravity switch contacts at the telephone to operate a relay at the exchange to light a lamp at the switchboard. In both cases the caller then tells the operator verbally what call he wants to make.

55.2.5 Automatic and automanual exchanges

The term automatic is used to describe exchanges where the switching is carried out by machine under the remote control of the caller who could be either a customer or an operator. Even on these systems there are some calls which customers either cannot or will not set up for themselves. To provide for these, numbers are allocated (e.g. 100) which can be dialled for assistance. Such calls are connected through the automatic exchange to a switchboard where the operator can establish the required connection also through the automatic equipment. These switchboards are called automanual exchanges.

55.3 Automatic exchange switching systems

The first public automatic exchange opened in 1892 at La Porte, Indiana (USA), and used rotary switches invented in 1891 by Almon B. Strowger, an undertaker in Kansas City, after whom the system is named. Since then several systems have been developed. The most successful of these use mechanical devices called crossbar switches or arrays of reed relays. The most recent of these systems use stored program control (s.p.c.) in which the operations to set up each call are determined by software in the exchange processors (computers).

The new generation (digital) systems employ totally solid-state switching as well as stored program control.

55.3.1 Strowger (step by step)

Ratchet-driven switches remotely controlled directly by pulses from the calling telephone are used to simulate the operator's actions on a plug and cord switchboard. The wipers (moving contacts) are connected by short flexible cords and perform a similar function to the switchboard plugs. The place of the jack field is taken by a bank of fixed contacts and the wipers are moved one step from each contact to the next for each pulse, as in *Figure 55.5*, by means of a ratchet and pawl mechanism.

At first, Strowger's exchange provided each customer with his own 100-outlet two-motion selector. The expense of providing one large switch or selector for each customer is now avoided by using a stage of smaller cheaper uniselectors to connect the callers to the selectors as and when they require them. Expansion of the exchange beyond 100 lines is by adding further stages of selectors as shown in *Figure 55.5*.

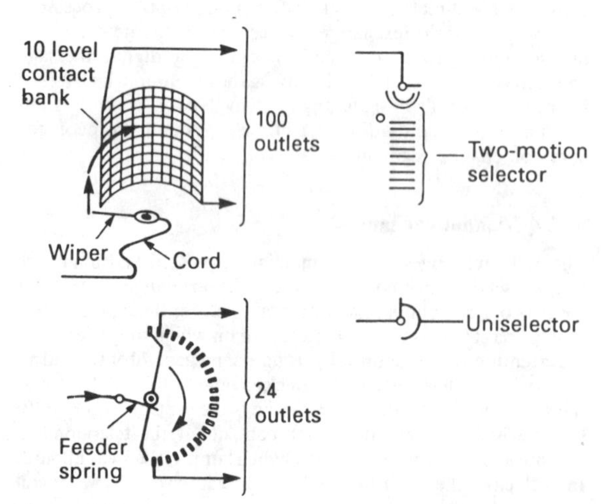

Figure 55.5 Strowger (step-by-step) selectors

The setting up of a call progresses through the exchange stage by stage. Each selector connects the call to a free selector in the next stage in time for the train of pulses to set that selector. The wipers step vertically to the required level under dial pulse control at 10 pulses per second then rotate automatically at 33 steps per second to select a free outlet by examining the potential on each contact. This is step-by-step operation and, when used with Strowger selectors, it needs only simple control circuits which can be provided economically at each selector.

A large city often has hundreds of exchanges with overlapping service areas where it would be unacceptable for callers to have to consult a different list of dialling codes according to which exchange the telephone they happened to be using is connected. To allow the same list of three-digit codes to be used anywhere in the area, director equipment is provided, *Figure 55.7*, which translates the code dialled by the caller into the actual routing digits required.

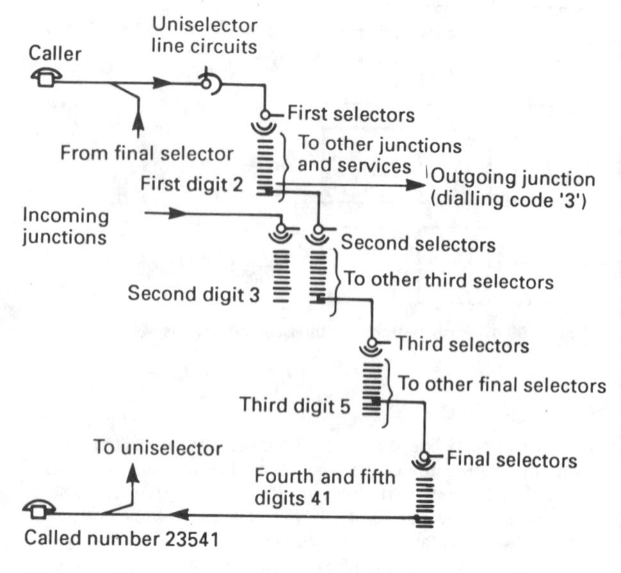

Figure 55.6 Strowger (non-director) exchange trunking

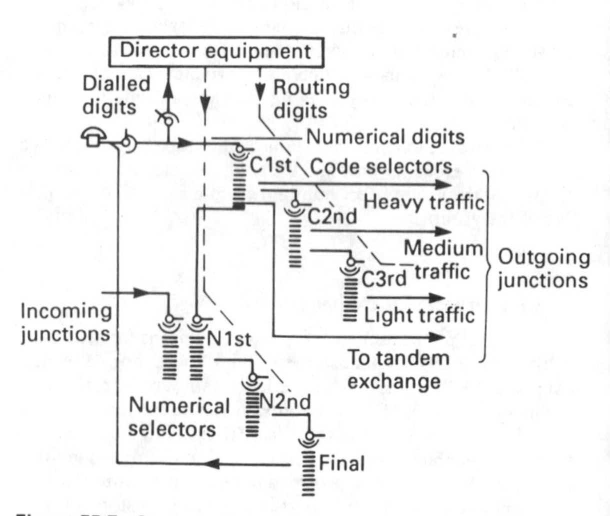

Figure 55.7 Strowger director exchange trunking

55.3.2 Crossbar

The crossbar switch is mechanically simpler than the rotary switches of the Strowger system and can be designed for greater reliability and smaller size. It cannot be directly controlled by dial pulses, however, so more complex control circuits are required which are therefore shared over more than one switch, a method of operation known as common control. Translation facilities like those provided by the director system, but not limited by considerations of mechanical switch sizes, are readily obtained because there is no inherent relationship between number allocation and particular switch outlets.

The switch mechanism consists of a rectangular array of contact sets or crosspoints which can be selected by marking their vertical and horizontal coordinates.

Figure 55.8 shows a much simplified diagram of a single crosspoint and a general view of a complete switch. The flexible selecting finger should be thought of as a stiff piece of wire sticking up out of the paper. The horizontal (select) and vertical (hold) bars move in the direction shown by the arrows when their associated coils are energised. It can be seen that the crosspoint contact will be operated only when the horizontal bar 'selects' before the vertical bar operates. Once operated, the contact remains held by the vertical bar which traps the finger and holds it even when the horizontal bar has been restored. The horizontal bar can be used again to select another crosspoint in the row without disturbing those already in use.

In the most usual form of construction, the movement of the bars is produced by a slight rotation about the pivot in a manner similar to the armature of a telephone relay. The spring finger is attached to the select bar and swings across when the bar rotates.

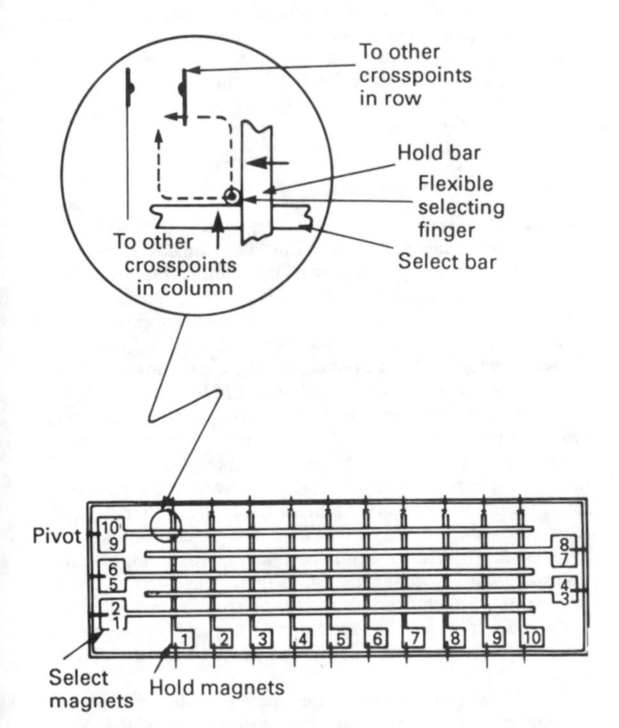

Figure 55.8 Crossbar switch

55.3.3 Crosspoint system trunking

To provide an exchange capable of serving say 10 000 customers it would not be economic either to construct a single large crossbar switch or simply to gang a large number of smaller switches to form a square matrix of 10 000 × 10 000 crosspoints. A more practical arrangement is to connect the switches in two or more stages. To illustrate this *Figure 55.9* shows first a single square matrix of 81 crosspoints providing for nine paths between nine inlets and nine outlets and then how to meet the same requirement with only 54 crosspoints. The advantage of this approach is even greater with larger numbers of circuits and practical crossbar switches having typically 10 × 20 crosspoints or more, rather than only 3 × 3 as shown in the example.

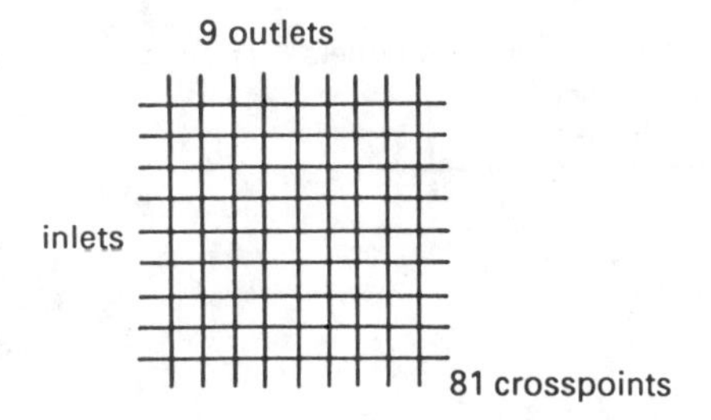

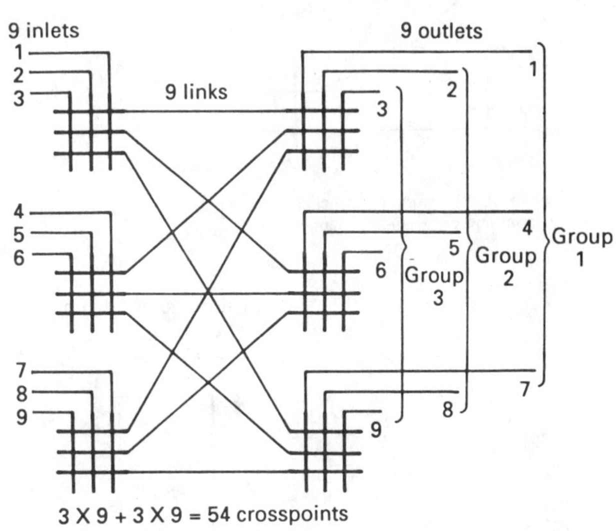

Figure 55.9 Trunking principle of distribution stage for crossbar and electronic exchanges

This simple example provides only for the distribution of calls from the inlets to an equal number of outlets through the same number of links. In practical exchanges it is only necessary to provide as many paths through the switching network as the number of calls which are expected to be in progress at one time. The concentration of lines on to a smaller number of paths or trunks can be achieved using the principle illustrated by another simple example in *Figure 55.10*.

A complete local exchange then consists of a combination of a distribution (group selection) stage and one or more line concentration units as shown in *Figure 55.11*. To establish a call, the line circuit LC detects the loop calling or 'seize' condition (*Figure 55.3*) and signals the identity of the calling line to the exchange control. In analogue systems, the usual method of operating is that the control sets the crosspoints to connect the line to a register which detects the dial (or key) pulses and signals them to the control. In digital electronic systems, the dial pulses are usually detected at the line circuit. In either case the control, which is duplicated or sectioned in some way for security, analyses the digits received and sets the appropriate crosspoints to establish the call.

55.3.4 Analogue electronic exchanges

Typical systems of this sort are TXE2 and TXE4 used in the UK for small (up to 2000 lines) and large exchanges respectively.

This type of system employs a switching matrix having the

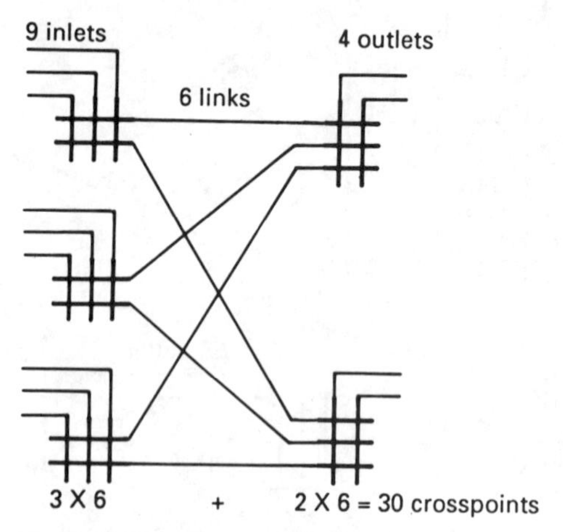

Figure 55.10 Trunking principle of concentration stage for crossbar and electronic exchanges

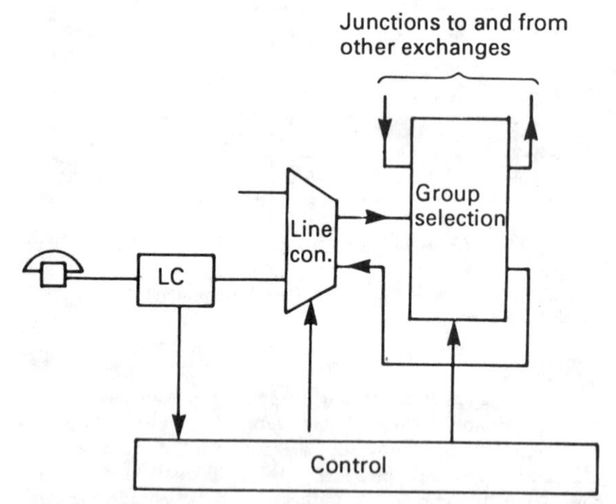

Figure 55.11 Generalised architecture of a complete crossbar or electronic exchange

same structure as crossbar systems (*Figures 55.9*, *55.10* and *55.11*). Public exchanges usually employ reed relays for the crosspoints. These have the advantage of being sealed against corrosion and dirt and they operate in only about 2 ms making them more amenable to electronic control than Strowger or crossbar switches. They also have a separate operating coil for each crosspoint enabling switch sizes to be optimised for system requirements.

The higher speed of operation and the use of electronic control permits greater use of self-checking features to improve reliability and quality of service. The greater use of electronics enables these systems to benefit from advancing technology in that field in terms of size and versatility in the provision of customer facilities.

There are systems using solid-state crosspoints such as thyristors but these are confined mainly to PABX applications where crosstalk, transmission and overvoltage requirements are often less onerous than on public systems.

55.3.5 Space division and time division switching

Exchanges such as Strowger, crossbar and reed relay systems employ space division switching in which a separate physical path is provided for each call and is held continuously for the exclusive use of that call for its whole duration.

Digital and other pulse modulated systems employ time division switching in which each call is provided with a path only during its allocated time slot in a continuous cycle. Each switching element or crosspoint is shared by several simultaneous calls each occupying a single time slot in which a sample of the speech is transmitted. This sample may be a single pulse as in pulse amplitude modulated systems or a group of pulses as in pulse code modulated systems.

The range of exchanges known as System X and System Y recently introduced in the UK are in this category. Speech is pulse code modulated (PCM) and transmitted through the network in 8-bit bytes at 64 kbit/s.

55.3.6 Digital trunk exchange

Digital transmission systems are multiplexed in groups usually of 24 or 30 speech channels on each link. The trunk exchange has to be able to connect an incoming channel in one PCM link to an outgoing channel in another PCM link. Received speech samples must therefore be switched in space to the appropriate link and translated in time to the required channel time slot. *Figure 55.12* shows in simplified form a typical arrangement known as T–S–T (time-space-time). A practical system would have some 1024 time slots per speech store instead of the four shown and a space switch of perhaps 96×96 instead of the 5×5 in the diagram.

The first time switch connects to a cross-office time slot through the space switch, which itself operates in a time division switching mode. The second time switch connects to the time slot required on the outgoing PCM link. The connection illustrated in *Figure 55.12* is from incoming channel 2 to outgoing channel 4 via crosspoint row 3 at cross-office time slot t.

Each time switch operates by storing received speech samples in the order in which they are received and reading them out of store in a different order according to which time slot they are to be connected to.

The arrangement shown carries speech in one direction only. All receive channels are connected to one side of the switch, regardless of whether the circuits are incoming or outgoing in the traffic sense, and all 'transmit' channels are connected to the other side. Each call requires transmission in each direction and these two paths occupy the same crosspoint in the space switch but at different time slots usually 180° out of phase. In the simple example shown time slot $t+2$ would be used for the corresponding opposite direction of transmission.

55.3.7 Digital local exchange

The T–S–T switching network of *Figure 55.12* provides only for distribution, like the crosspoint network of *Figure 55.9*. A complete local exchange needs also a concentration stage. Telephone customers' lines are generally on individual pairs employing analogue transmission. First-generation electronic exchanges have analogue concentration stages, *Figure 55.10*, so that the relatively expensive digital encoding and multiplexing is provided after concentration. Cost trends in electronic components now favour the provision of the conversion on a per line basis so that digital concentration stages can be provided in the form of an additional time switching stage in which the available time slots to the main T–S–T switch are fewer than the total number of input time slots and store locations in the concentrator speech stores they serve. This might consist typically of up to 16 groups of 256 lines, each written cyclically into a 256 word store but with only 256 time slots available for allocation to the acyclic readout from all these stores.

The interface to the analogue customer's line includes certain functions which require voltage and power levels incompatible

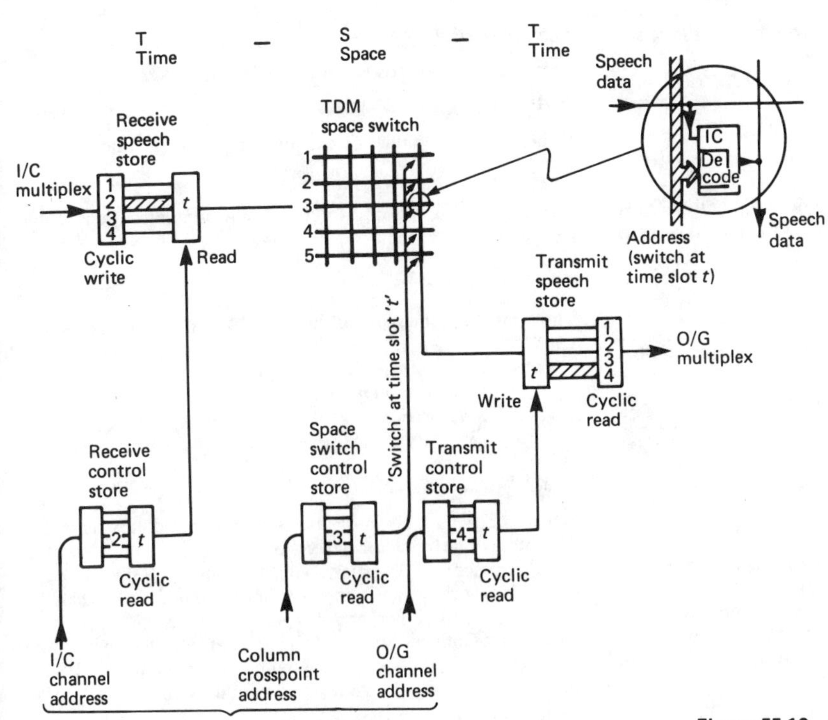

Figure 55.12 Principle of time–space–time switching

with the digital concentrator switch and which therefore have to be provided at the individual line circuit (line card). The line interface functions are generally referred to as the Borscht functions:

(1) Battery feed to line.
(2) Overvoltage protection.
(3) Ringing current injection and ring trip detection.
(4) Supervision (on-hook/off-hook detection).
(5) Codec for analogue-to-digital conversion.
(6) Hybrid for two-wire to four-wire conversion.
(7) Test access to line and associated line circuit.

This stage may be provided in the form of a remote switching unit or concentrator, sited some kilometres from its parent exchange, to which it is connected by a PCM transmission circuit (typically an optical fibre) using a form of common channel signalling.

55.3.8 Integrated networks

The combination of digital exchanges and digital (PCM) junctions and trunk lines constitutes an integrated digital network (IDN) and has advantages of economy and improved transmission quality by avoiding the need for frequent analogue-to-digital conversions.

The availability of end-to-end digital circuits enables data at rates up to 64 kbit/s to be transmitted over the same network as speech without the use of modems. The joint use of the network in this way to provide voice and non-voice services constitutes an integrated services digital network (ISDN).

55.4 Teletraffic

55.4.1 Grade of service

It would be uneconomic to provide switching equipment in quantities sufficient for all customers to be simultaneously engaged on calls. In practice, systems are engineered to provide an acceptable service under normal peak (i.e. busy hour) conditions. For this purpose, a measure called the grade of service (g.o.s.) is defined as the proportion of call attempts made in the busy hour which fail to mature due to the equipment concerned being already engaged on other calls. A typical value would be 0.005 (one lost call in 200) for a single switching stage. To avoid the possibility of serious deterioration of service under sudden abnormal traffic it is also usual to specify grades of service to be met under selected overload conditions.

55.4.2 Traffic units

Traffic is measured in terms of a unit of traffic intensity called the Erlang (formerly known as the traffic unit or TU) which may be defined as the number of call hours per hour (usually, but not necessarily, the busy hour), or call seconds per second, etc. This is a dimensionless unit which expresses the rate of flow of calls and, for a group of circuits, it is numerically equal to the average number of simultaneous calls. It also equates on a single circuit to the proportion of time for which that circuit is engaged, and consequently to the probability of finding that circuit engaged (i.e. the grade of service on that circuit). The traffic on one circuit can, of course, never be greater than one Erlang. Typically a single selector or junction circuit would carry about 0.6 Erlangs and customers lines vary from about 0.05 or less for residential lines to 0.5 Erlang or more for some business lines.

It follows from the definition that calls of average duration t seconds occurring in a period of T seconds constitute a traffic of A Erlangs given by:

$$A = Ct/T \tag{55.1}$$

An alternative unit of traffic sometimes used is the c.c.s. (call cent second) defined as hundreds of call seconds per hour (36 c.c.s. = 1 Erlang).

55.4.3 Erlang's full availability formula

Full availability means that all inlets to a switching stage have access to all the outlets of that stage. Trunking arrangements in which some of the outlets are accessible from only some of the inlets are said to have limited availability.

For a full availability group of N circuits offered a traffic of A Erlangs, the grade of service B is given by

$$B=\frac{A/N!}{1+A+A^2/2!\ldots A^N/N!} \quad (55.2)$$

(This formula assumes pure chance traffic and that all calls originating when all trunks are busy are lost and have zero duration.)

55.4.4 Busy hour call attempts

An important parameter in the design of stored program control systems in particular and common control systems in general is the number of call attempts to be processed, usually expressed in busy hour call attempts (BHCA). Typically a 10 000 line local exchange might require a processing capacity of up to 80 000 BHCA.

55.4.5 Blocking

If two free trunks cannot be connected together because all suitable links are already engaged the call is said to be blocked. In *Figure 55.9*, if one call has already been set up from inlet 1 to outlet 1, an attempted call from inlet 2 to outlet 2 will be blocked. The effects of the blocking are reduced here by allocation of the outlets to the three routes.

Blocking can be reduced by connecting certain outlets permanently to the inlet side of the network to permit a blocked call to seek a new path. Alternatively a third stage of switching may be provided.

It is possible to design a network having an odd number of switching stages in which blocking never occurs. Such non-blocking arrays need more crosspoints than acceptable blocking networks and are uneconomic for most commercial telephone switching applications.

55.5 Analogue transmission by cable

55.5.1 International standardisation

All transmission facilities in public switched telephone networks are planned to achieve the recommended standards of performance set by the CCITT (the International Telephone and Telegraph Consultative Committee of the ITU). In principle, compliance with these recommendations relates only to international connections, but in order to ensure that any call set up from within a country to a point in another country is satisfactory, national connections are operated to the same standards.

The basic transmission qualities recommended by the CCITT include bandwidth, interface levels, noise, return loss and level stability. For specific equipments the CCITT recommends many other parameters to facilitate the setting up of international connections. It is important to recognise that the CCITT is constantly being requested to study aspects of the transmission network and seeks to reflect technological improvements in its recommendation. The current status can be found in the appropriate volumes of the preceding Plenary Assembly. The current issue is the Red Book published by the ITU in Geneva, 1984, of which volume III relates to line transmission.

55.5.2 Terminology

The telecommunication terms used in this section can be explained as follows.

55.5.2.1 Decibel (dB)

This is the unit of power ratios, e.g.

$$P_1/P_2=10\log_{10}(P_1/P_2)\text{ dB} \quad (55.3)$$

If the impedance under discussion is the same it follows that

$$P_1/P_2=V_1^2/V_2^2=20\log_{10}(V_1/V_2)\text{ dB} \quad (55.4)$$

dB is, therefore, the unit for expressing amplifier gains, cable losses, etc.

55.5.2.2 Absolute power

The decibel becomes an absolute unit by reference to a power of 1 mW.

Thus 1 mW corresponds to 0 dBm
1 W corresponds to +30 dBm
1 μW corresponds to −60 dBm
1 pW corresponds to −90 dBm

55.5.2.3 Relative level

When analysing a chain of tandem connected equipments, the level at any one input will depend on prior conditions. However, the equipment in question will have been designed for a particular working level and tolerance range. The concept of relative level, dBr, is a means of expressing the designed operating conditions. Conventionally, a 0 dBm signal at the two-wire audio connection would produce a signal of −10 dBm at a −10 dBr point. Likewise a −15 dBm signal would produce a −25 dBm signal at the same −10 dBr point.

55.5.2.4 Level ratios

It is convenient to express noise levels, pilot levels, etc., relative to the nominal traffic level (i.e. the dBr level) at a point in the transmission path; e.g. a pilot level might be at −10 dBm0, which means 10 dB below traffic signal. Thus a −10 dBm0 pilot measured at a −33 dBr point would read −43 dBm. Similarly, interfering signals and noise can be expressed as dBm0 with respect to traffic levels. Thus an interfering signal of −50 dBm0 would measure −60 dBm at a −10 dBr point.

55.5.2.5 Psophometric weighting

The human ear does not respond equally to all audio frequencies and speech intelligence is primarily conveyed in the midrange frequencies of 500 to 1600 Hz. Thus interfering signals which fall outside this band can be higher than the mid-range frequencies. This property is quantified in the internationally agreed psophometric weighting network. Such a network is always included in voice frequency noise measuring equipment. The effect of the network on randomly distributed noise (white noise) is to reduce its power by 2.6 dB. Also, the noise in a 4 kHz band will be reduced by 1.1 dB when measured in the standard 3.1 kHz bandwidth (0.300 to 3.4 kHz).

Noise results, inclusive of psophometric weighting and bandwidth correction, are expressed as dBm0p, e.g. noise in a 4 kHz bandwidth of −60 dBm0 is equivalent to −63.6 dBm0p in a 3.1 kHz telephony channel.

55.5.2.6 Hypothetical reference circuit

To assist in the allocation of noise between types of equipment a 'hypothetical reference circuit' has been conceived which is representative of a continental call. The circuit is 2500 km long and is made of nine homogeneous line sections, each 280 km long, with various combinations of carrier multiplex equipment at the intermediate points. The noise in any 3.1 kHz channel operating end to end over this circuit must not exceed 10 000 pW0p (−50 dBm0p). 2500 pW0p are allowed in the multiplex equipment and

7500 pW0p for the line. Thus all line transmission plant is designed to achieve a noise performance of not more than 3 pW0p/km. The multiplex equipment noise is allocated between the various translating equipments. For example, the first translation stage, the primary group equipment, is allocated 200 pW0p.

55.5.2.7 Four-wire working and phantom circuits

Local exchange connections to subscribers are two-wire lines carrying both the 'go' and 'return' traffic.

For amplified and carrier circuits, four-wire working is normal, using a separate 'pair' for each direction of transmission.

A third independent 'phantom' circuit can be derived from two pairs as shown in *Figure 55.13*.

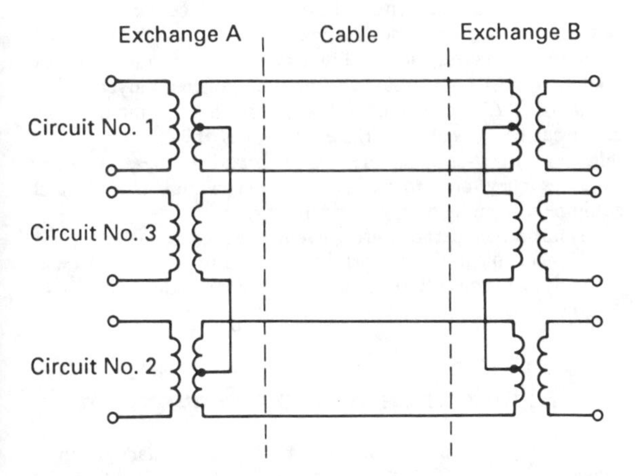

Figure 55.13 Derivation of phantom circuit (No. 3) using centre-tapped line coils

55.5.3 Voice frequency transmission

In addition to subscriber connections, voice frequency circuits are commonly used between nearby exchanges. Loading, the application of coils as a lumped inductance, is normally applied to reduce circuit attenuation and equalise attenuation.

55.5.3.1 Effect of loading

Attenuation is a function of the four primary cable parameters, R, C, L and G (leakance). In modern cables L and G are small and the attenuation approximates to

$$\tfrac{1}{2}R(C/L)^{1/2} \tag{55.5}$$

By increasing L, the attenuation is reduced. The effect of loading is to reduce and flatten the low-frequency response, but attenuation rises rapidly beyond a cut-off frequency, given by

$$f_c=\pi^{-1}(LC)^{-1/2} \tag{55.6}$$

The effective transmission bandwidth is taken as $0.75f_c$, so for a 3.400 kHz circuit f_c needs to be 4.500 kHz. Loading also increases the characteristic impedance to the value

$$Z_0=(L/C)^{1/2} \tag{55.7}$$

(where L and C relate to the loading interval).

55.5.3.2 Standard loading

Standard loading coil inductances are 22, 44, 66 and 88 mH, and a coil spacing of 1830 m is commonly used. *Figure 55.14* shows the loss of 0.63 mm paired conductors of mutual capacitance 45 nF/km, unloaded and with 44 mH and 88 mH loading.

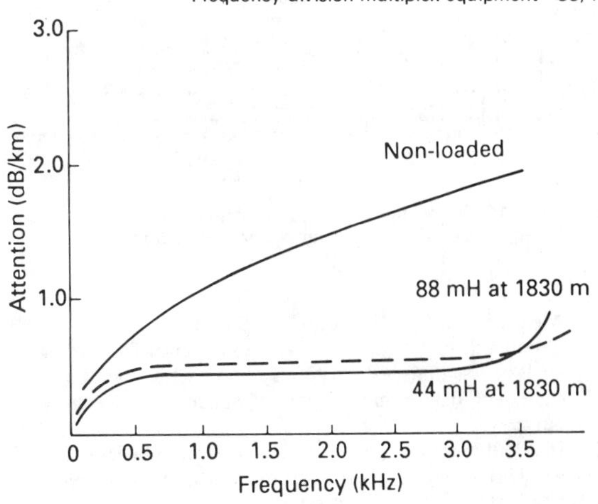

Figure 55.14 Effect of loading: 0.63 mm conductors of mutual capacitance 45 nF/km

55.6 Frequency division multiplex equipment

For the more efficient utilisation of buried copper cables, carrier techniques have been used extensively for very many years in both national and international telephone networks. The particular technique universally adopted is amplitude modulation. However, only one sideband is transmitted and the carrier is suppressed. By this means a maximum number of channels can be contained within a frequency band and the absence of high-level carrier reduces the power loading on the equipment in the transmission path.

At the receiving end the incoming sideband is modulated with a locally generated carrier. Consequently, both the transmitting and receiving carriers have to be generated to a great accuracy, such that the reconstituted voice frequency signal does not differ by more than 1 Hz from the original signal. The internationally defined primary, or basic, group comprises twelve 4 kHz channels assembled into a (60–108) kHz frequency band. Each channel provides an audio path extending from 300 to 3400 Hz.

In some designs a 3825 Hz outband signalling channel is also provided, but this clearly imposes a more severe requirement on the channel separating filters.

55.6.1 Primary group equipment designs

The 12-channel group equipment is by definition the most commonly employed equipment in carrier networks and, hence, design cost effectiveness is particularly important. Consequently, several design approaches have evolved exploiting the economic advantages of technological developments. However, all these designs comply with the international recommendations and can, therefore, be interconnected.

55.6.2 Requirements

The CCITT recommendations (Red Book, Volume III, Rec. G232) are very comprehensive. Key features are as follows:

(1) The average attenuation/frequency characteristic measured through send and receive equipment connected back-to-back shall lie within the mask shown in *Figure 55.15*.

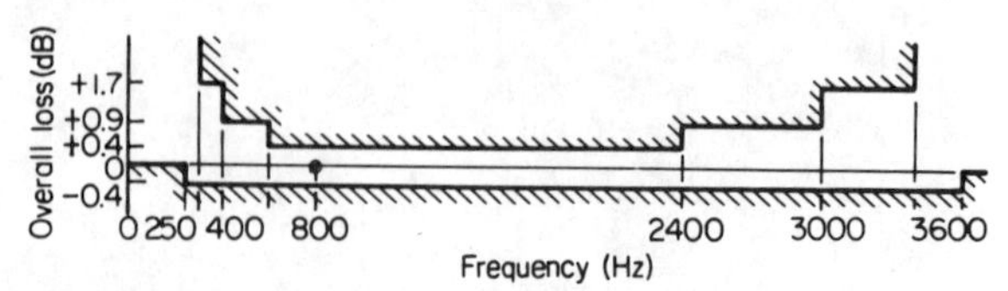

Figure 55.15 Limits for the average variation of overall loss of 12 pairs of equipment of one 12-channel terminal equipment

(2) Measuring at the HF out, no carrier signal in the (60–108) kHz band shall be greater than −26 dBm0. The aggregate shall be less than −20 dBm0, and no carrier signal greater than −50 dBm0 shall be present outside the (60–80) kHz band.
(3) Intelligible crosstalk between channels shall be less than 65 dB. Unintelligible crosstalk from an adjacent channel shall be suppressed by at least 60 dB.

55.6.3 Crystal filter design approach

This design approach exploits the characteristics of quartz crystal filters. This equipment requires the provision of 12 carrier frequencies spaced at 4 kHz intervals from 64 to 108 kHz. The design of each channel unit is identical, apart from the pass band of its crystal filters.

The modulator is switched hard by the carrier signal, effectively by a square-law function. If

$$\text{input signal} = E_1 \sin \omega t \tag{55.8}$$

and

$$\text{carrier signal} = E_2 \sin pt \tag{55.9}$$

then the output signal is given by

$$E_1 \sin \omega t[\tfrac{1}{2} + 2\pi^{-1}(\sin pt + \tfrac{1}{3}\sin 3pt + \ldots)]$$
= modulating signal + upper and lower side bands of the carrier frequency + upper and lower side bands of 3 times the carrier frequency ... (55.10)

An active modulator (namely, a transistor) requires lower carrier powers and provides good linearity over a wide range of signal power levels.

Audio filtering stages prior to the modulator restrict the traffic band to the range 0.3 to 3.4 kHz. The output crystal filters select the lower sideband. In the receive direction, the incoming crystal filter selects the wanted channel and the post-modulator filters suppress the unwanted sidebands and carrier.

Typical crystal filter characteristics are shown in *Figure 55.15*. Clearly the most damaging crosstalk component would be impurity in the carrier supply. Typically, spurious signals which are multiples of 4 kHz need to be 80 dB below carrier level.

55.6.4 Alternative design approaches

One technique exploits the stability, small size and low cost of modern coils and capacitors. In order to reduce the number of designs, use can be made of a pregroup, and to facilitate filter design this pregroup may be first modulated into a frequency spectrum most appropriate to the technology employed.

Figure 55.17 shows one method by which the standard basic group can be derived using these techniques. It will be noted that different carrier supplies are required, and as these are non-standard they are normally generated within the channel equipment from incoming standard carriers.

Several other methods are currently available, including the use of mechanical filters, crystal filters operating at 2 MHz, etc. All equipments meet the internationally agreed requirements and can, therefore, be interworked.

55.7 Higher-order multiplex equipment

The same concepts used in the generation of the primary group of 12 channels in the 60–108 kHz band have been applied to build up a hierarchy of frequency allocations. Thus five 60–108 kHz primary groups are assembled into a 312–552 kHz basic supergroup of 60 channels.

The basic supergroup can in turn be modulated into the 16 supergroup assembly extending from 60 to 4028 kHz. This is a

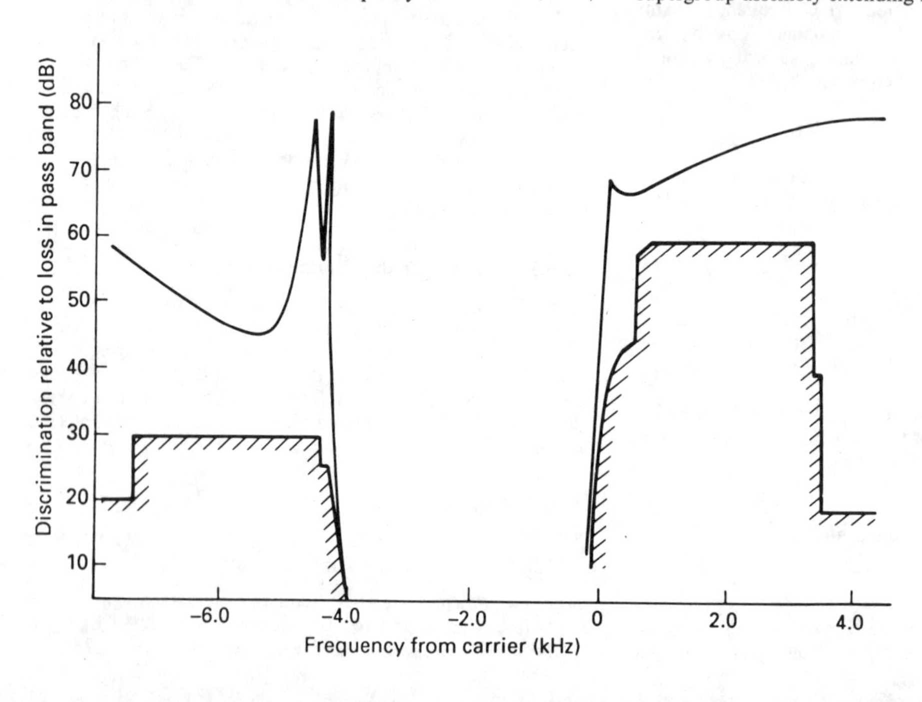

Figure 55.16 Typical channel filter characteristic

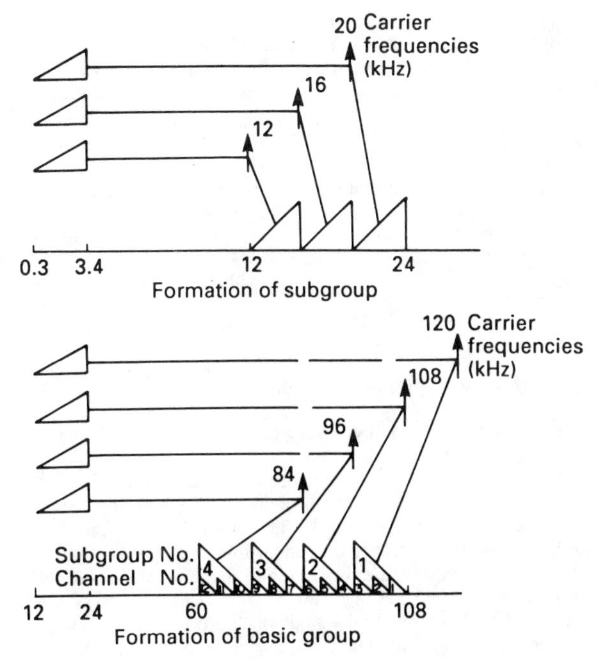

Figure 55.17 Example of basic group formation via a subgroup

commonly used hierarchical level for application to 4 MHz line systems.

The supergroup can also be assembled into a basic master group from 812 to 2204 kHz of 300 channels. Three master groups are assembled into the basic super master group from 8516 to 12 388 kHz of 900 channels. In the United States very similar allocations are used, except that the term master group applies to a 600 channel assembly.

In many countries, including the United Kingdom, it was not advantageous to adopt the master group and super master group levels. Instead a 15-supergroup assembly, extending from 312 to 4028 kHz, is modulated up into the next stage.

The purpose of these hierarchical structures is to provide a modular approach to the assembly of increasing numbers of telephony channels. At most points in the hierarchy, line transmission systems are available to provide interconnections between stations.

55.8 Line transmission systems

The motivation for the development of line transmission systems is to achieve lower cost per telephone circuit. The primary constraints in attaining this objective are the need to provide a transmission bandwidth conforming to one of the hierarchical steps previously discussed, and the need to use one of the internationally accepted cable designs.

A system which requires special translating equipment or a special cable will inevitably attract additional costs cancelling any 'per circuit' advantage. An exception is in the field of submarine cable systems where the unique environmental and commercial factors justify special designs almost on a per system basis. For land lines, cables are installed with a life expectancy of 40 years and the transmission engineer's task is to exploit this medium with ever-increasing efficiency.

55.8.1 Cable designs

Copper pair cables have been used for low-capacity (12, 60, 120 channel) connections, but these techniques are obsolete. Modern FDM transmission systems are designed for use on coaxial pair cables. The coaxial cores standardised by the CCITT are large core, which has a 9.5 mm inside diameter of the outer conductor and a 2.6 mm diameter inner conductor, and small core for which the figures are 4.4 and 1.2 mm respectively. Both these cables are effectively air spaced as the inner is typically supported on relatively widely spaced polyethylene discs.

These standardised coaxial cores can be laid up with other conductors to form a great variety of cable to suit individual administration's needs. Typical designs would comprise some 6, 8 or 12 coaxial cores together with twisted pairs for supervisory use laid up in the interstitial spaces, with perhaps an outside layer of twisted pairs to provide local audio circuits. In some territories such a cable would be armoured and directly buried; in others, and in most urban areas, the armouring is omitted and the cable is installed in cable ducts.

55.8.2 Cable characteristics

The characteristics of these cables for analogue transmission are well defined, and recent work has extended this knowledge further in recognition of the need to apply high-bit-rate digital transmission systems to existing networks.

For FDM applications, the key characteristics are as follows:

(1) Large core cable:
Loss $=2.355F+0.006F$ dB/km at 283 K (10°C)
Impedance $=75\pm3\ \Omega$ at 1 MHz
Temperature coefficient of attenuation $=0.2\%$ per K (°C)
Far end crosstalk = greater than 130 dB

(2) Small core cable:
Loss $=5.32F$ dB/km at 283 K (10°C)
Impedance $=75\pm1\ \Omega$ at 1 MHz
Temperature coefficient of attenuation $=0.2\%$ pper K (°C)
Far end crosstalk = greater than 130 dB

It will be seen that the loss increases with the square root of the frequency. The linear element is very small. A useful approximation is that small core cable has $2\frac{1}{4}$ times more loss. Hence an equipment designed for small-core cable will operate on large-core cable at a spacing $2\frac{1}{4}$ times longer.

Both cables have a characteristic impedance of 75 Ω. The crosstalk performance between cores is excellent and not a practical limitation on system design. Such cables also have good immunity against external interference.

55.9 System design considerations

The principal design considerations are illustrated in the following paragraphs by reference to a 2700 channel system. The bandwidth of this system extends nominally from 300 kHz to 12.5 MHz. The design objective is to provide a stable transmission path with a flat frequency characteristic and negligible noise. This is accomplished by providing amplifying points (repeaters) at regular intervals along the cable.

55.9.1 Traffic path

55.9.1.1 Gain shaping

The recommended repeater or amplifier spacing for a 12 MHz system on 4.4 mm cable is 2 km, corresponding to a loss of about 40 dB at 12.5 MHz and 6 dB at 300 kHz. The usual design technique is to compensate for some of this slope with a shaped negative feedback network, i.e. minimum NFB is provided at the highest frequency, but not less than 15 to 20 dB in order to preserve the virtues that NFB brings.

It is not practical to put the whole cable shape in the feedback path, and additional fixed networks, with zero loss at the top frequency, are used. By splitting these between input and output, good return losses are obtained at low frequencies and buffering to lightning surges from the cable is provided.

A long tandem connection of amplifiers requires equalisation at the terminal stations to compensate for any systematic error accumulating along the line.

55.9.1.2 Noise contribution

The noise contribution of each amplifier must never exceed 6 pW0p in any channel and a design target of 4 pW0p is set to ensure this. This will comprise say two-thirds basic or thermal noise, and one-third intermodulation noise.

Thermal noise contribution = 2.66 pW0p
= −90 + 4.25 = −85.75 dBm0p

Amplifier noise = −139 + NF + gain dBm0p
= −99 + NF dBm0p

Hence the amplifier noise figure must be better than 13 dB at a point of zero relative level. The amplifier cannot be operated at such a level because intermodulation would be too high.

For an output level of −10 dBr, the noise figure would need to be −3 dB. Thus knowing the intermodulation performance and noise figure it is possible to select line levels to give optimum performance. The calculation is complicated because some intermodulation products add on a square law basis. Hence the calculation is most conveniently done with computer and the form of a typical printout is indicated in *Table 55.1*.

The optimum line levels required at the line amplifier outputs can be set by introducing an appropriate network in the transmit terminal (the pre-emphasis network) and taking this slope out at the receiving station (de-emphasis network). The computer program can also be used to test the sensitivity of the design to level errors.

55.9.1.3 Overload performance

In addition to fulfilling the requirements for noise performance the amplifier must also tolerate the equivalent r.m.s. sine-wave power of the peak of the composite multiplex signal. This is given by:

$$P_{eq} = -5 + 10 \log_{10} N + 10 \log_{10}(1 + 15/N^{1/2}) \text{ dBm0} \qquad (55.11)$$

where N in this case is 2700 channels and

$P_{eq} = +30.4$ dBm0.

On a pre-emphasised system, this would apply at the frequency of mean power, typically about 8 MHz where the line level is −17 dBr.

Thus the overload requirement becomes +13 dBm. However, several decibels margin must be added to allow for misalignments and a better specification would be +20 dBm.

It is important to recognise that both the noise and overload requirements have to be met with any secondary lightning protection components connected. These typically comprise high-speed diodes connected across the signal transistors and can contribute to the intermodulation products.

55.9.2 Level regulation

The coaxial cable attenuation is subject to a temperature coefficient of 0.21% per K (°C). Cable is buried at about 80 cm and statistics gathered over many years show a mean annual temperature variation at this depth of ±10 K (°C). Thus a 2 km repeater section of small core cable will change by

$40 \times 0.0021 \times 10 = \pm 0.8$ dB at 12.5 MHz

and by

$6 \times 0.0021 \times 10 = \pm 0.12$ dB at 300 kHz

After a tandem connection of several repeaters this could accrue to an unacceptable level.

It is necessary to compensate for this effect. One solution is to include a temperature sensor in the repeater which controls an appropriately shaped network. This assumes that the local temperature is representative of the whole cable length and that the sensing element has very stable characteristics. Another technique is to send a pilot signal the whole length of the system, and send back a command signal which adjusts the gain of the intermediate amplifiers. The most adaptable solution, however, is to use the line pilot signal to directly control the intermediate amplifiers.

A block schematic of a regulated repeater is shown in *Figure 55.18* and a typical route in *Figure 55.19*. It will be noted that

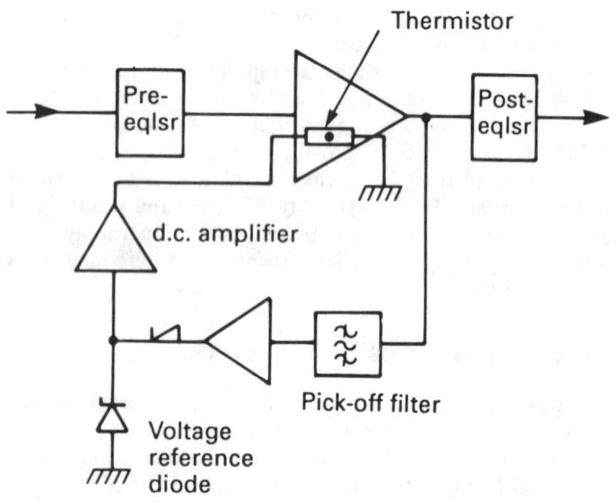

Figure 55.18 Regulated repeater

Table 55.1 Calculated noise performance 500 km route (9.5/2.6 mm cable)

Frequency (MHz)	*Relative level* (dBr)	*Thermal*	*A+B*	*A−B* (pW0p/km)	*A+B−C*	*Total*
0.4	−22	0.092	0	0.003	0.001	0.096
2.3	−22	0.177	0	0.003	0.013	0.193
5.0	−20	0.280	0	0.002	0.029	0.311
7.6	−18	0.396	0	0.001	0.088	0.485
10.4	−15	0.487	0.001	0.001	0.223	0.712
12.3	−12	0.501	0.003	0	0.154	0.658

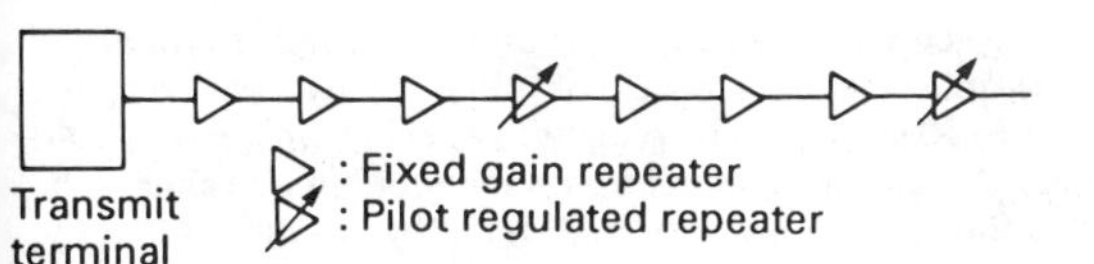

Figure 55.19 Allocation of regulated repeaters

only every fourth amplifier needs be regulated; the rest operate on fixed gain.

A line pilot frequency of 12435 kHz is allocated for this purpose and transmitted 10 dB below the virtual traffic level. It is picked off by a crystal filter in the regulator, amplified and rectified. The d.c. level is compared with a reference signal, the difference signal is amplified and controls the current into a thermistor. The resistance of this thermistor, which changes with current, determines the loss of the Bode network inserted in the amplifier interstage network, or in the negative feedback path. When the resistance is nominal, this network has a nominal loss. If the resistance falls, the loss is reduced; if it rises the loss is increased. The network configuration is such that the loss frequency characteristic is identical to the cable shape.

55.9.2.1 Control ratio

If the pilot level at the amplifier output has changed by 1 dB and the regulator reduces this to 0.1 dB, the control ratio is ten. The control ratio error does not accumulate, because, for example, the next regulator in these circumstances would see a change of 1.1 dB and reduce it to 0.11 dB. The next 1.11 dB reduces to 0.111 dB and so on.

55.9.2.2 Control range

The control range is the range over which the regulator can correct and is determined by the limits of resistance through which the thermistor can be driven and by the network design. Typical practical values are ± 6 dB at 12.5 MHz.

55.9.2.3 Simulation error

In all gain regulating methods the objective is to compensate for the actual cable loss change with temperature. The difference between the characteristic of the compensating network and the cable change is the simulation error. This can be made very small, but nevertheless on a long route it will accumulate in a systematic fashion and require secondary equalisation. Another source of level instability is a change of gain of the repeater with ambient temperature, which can be almost unmeasurable on an amplifier, but significantly accumulative on a long route.

55.9.2.4 Dynamic performance

The dynamic performance of a tandem connection of many pilot gain controlled amplifiers needs analysing as this is a limiting factor on the number of amplifiers, and, therefore, route length, controlled from one station.

The dynamic behaviour can be evaluated in two ways. Firstly, by stepping the send pilot level in say 0.5 dB increments and observing the resultant level change at the far end of the system. The requirement is that the overshoot shall always be less than the input change.

A second method is to amplitude modulate the pilot signal with a low-frequency (e.g. 10 Hz) signal and measure the modulation amplitude at the receive end. This is gain enhancement and a typical requirement is that on the longest regulated section

$$20 \log_{10}(B/A) \leqslant 6 \text{ dB}$$

where A is the amplitude modulation depth applied, and B the modulation depth measured at the receiving end.

55.9.3 Secondary regulation

A second pilot, 308 kHz, is available to control networks compensating for these very small errors. The use of such a network is generally restricted to every hundredth repeater say, but the difficulty in practice is determining the real requirement and then controlling it, when the maximum may indeed not be at 308 kHz. For telephony applications, additional automatic regulating pilots are commonly transmitted within each group or supergroup. These control flat gain networks which is a reasonable assessment of the gain drift which is likely to occur across a limited bandwidth. Another method is periodically to readjust the residual equalisers provided at the ends of a route which notionally compensates for the systematic fixed equalisation error. Equipment is available to do this automatically by means of inter-supergroup pilots, which as implied, are transmitted for maintenance surveillance in the frequency gaps between adjacent supergroups. The solution adopted depends on the lengths of route involved and the maintenance policies of the administration. For route lengths of about 300 to 650 km of large core cable, modern coaxial line equipment has sufficient stability to require no maintenance attention in this respect.

55.9.4 Crosstalk

The most stringent crosstalk requirement stems from the need for the transmission of broadcast programme material over coaxial systems. In this case three or four telephony channels are replaced with a 12 kHz programme modulation equipment.

Assuming unidirectional programme circuits are operating in opposite directions over a 300 km coaxial link, an 86 dB crosstalk margin can be allocated between the two directions of transmission. It is known that crosstalk can add in phase, that is, over narrow bands voltage addition can occur.

Thus, in the 300 km route, with 2 km spaced repeaters there are 150 crosstalk sources voltage adding, i.e. an addition of $20 \log_{10} 150$ or 44 dB. The crosstalk requirement per repeater is then $(86+44)$ or 130 dB. Referring to *Figure 55.20* this means that the signal loss from the output of one amplifier to the input of the

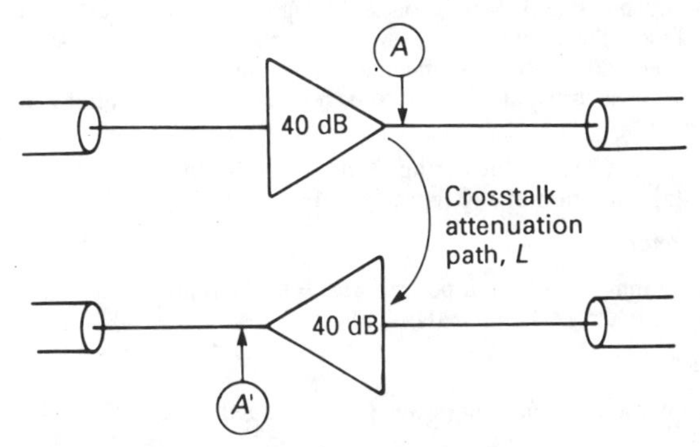

Figure 55.20 Crosstalk. If level $A' = A - 130$ dB then $L = 130 + 40 = 170$ dB

other needs to be 170 dB. Achievement of such a performance consistently in a production environment requires a design in which this need has been recognised from the outset. By careful attention to screening and by minimising common path earth currents, crosstalk values of this order can be realised.

55.9.5 Power feeding

The intermediate line repeaters are installed in underground housings at 2 km intervals. The power for this equipment is fed over the inner conductor of the coaxial cable. The most common technique is to transmit a constant current d.c. as this requires least equipment at the buried repeater point.

Figure 55.21 shows the configuration at a repeater and *Figure 55.22* a complete power feeding section. The spacing between

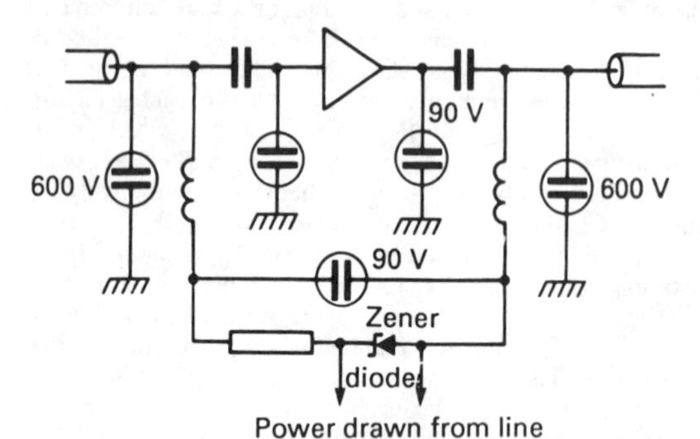

Figure 55.21 Repeater power feed circuit and surge protection components. Voltages indicate nominal breakdown voltages

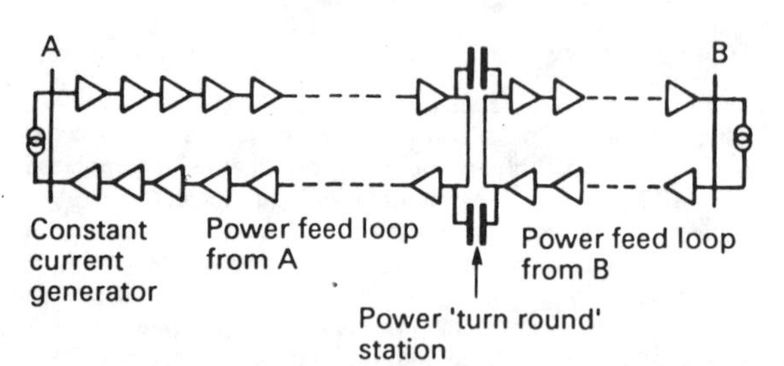

Figure 55.22 Power-feeding sections

power feeding stations can be up to 300 km or more depending on the system and cable type. The power separating filter requirements are calculated on the need to control the effect of the output signal passing back through the power path and adding to the input signal. Such an effect could accumulate along the line and produce an unequalisable ripple.

Suppose the ripple signal is to be restricted to 0.1 dB after 280 km

Maximum level of interfering signal = −38.6 dBm0
Voltage addition of 140 repeaters = 43 dB

Therefore

Maximum contribution per repeater = −82 dBm0
Gain at 12.5 MHz, 0.3 MHz = 40 dB, 6 dB

Then

Suppression required in power path = 122 dB, 86 dB

or

Suppression in filter at 12.5 MHz, 0.3 MHz = 61 dB, 43 dB

At the power turn round point, i.e. where the d.c. power feeding loop is completed, the requirement is the same as the repeater crosstalk requirement. Any extra attenuation can be provided in the turn round connection. The repeater circuits are connected in series and parallel configurations to make best use of the incoming current. In the UK, and some other countries, considerable operating benefit is obtained by restricting the current to 50 mA which is recognised as being inherently safe to personnel. However, it is not uncommon for higher values to be used, although, beyond about 300 mA, the voltage drop in the cable becomes an increasingly dominant factor. A modern repeater consumes less than 2 W. The line voltage drop is of the order of 15 V at 50 mA for each amplifier. With modest sending end voltages, ±350 V say, very long spans between power feeding points can be achieved.

55.9.6 Surge protection

The zener diode shown in *Figure 55.21* is not used as for power stabilisation. Power stability can easily be set by the tolerance on the power feeding equipment constant current source. The purpose of the zener diode is to limit the maximum voltage which could appear across the repeater power input terminals in the event of an induced current surge into the cable.

Surges can result from lightning storms, railway electric traction or electricity power line failures. The problem is to protect the transmission equipment from damage over a very wide range of operational conditions. International agreement has been reached on the level of immunity required (CCITT Rec. K17) although special circumstances may demand higher values. The normal conditions are survival from 5 kV discharges of 10/700 pulse shape, applied transversely to the input and output terminals and longitudinally through the repeater. Additionally, the repeater must withstand 50 Hz currents (10 A for 1 s) applied longitudinally. Several techniques are used to obtain surge immunity of this order. The primary protection comprises gas discharge tubes connected as shown in *Figure 55.21*. These need to be low capacitance designs with very fast (<1 μs) striking characteristics. It is also necessary to ensure that the power feeding current will not maintain the arc once it has been struck. Lower voltage gas tubes are used within the repeater and the pre-and-post attenuation equalisers also buffer the amplifier. Consequently, the components in the equalisers need to tolerate surges. Finally, semiconductor diodes, etc., are distributed within the repeater to absorb residual surges, but these need to be used with care to avoid additional intermodulation of signal products.

55.9.7 Supervisory systems

Analogue line transmission systems are extremely reliable and failure is a rare event. Nevertheless, it is necessary to provide maintenance staff with a means of unambiguously identifying a faulty repeater when the situation arises. The repair team can then be sent to the correct geographical location and replace the equipment in a minimum of time.

In the past, use has been made of the conductors laid up in the interstitial spaces between the coaxial tubes. However, these techniques are not effective on systems with significant numbers of repeaters, or on cables with a small number of interstice pairs. The traffic path can be utilised as the bearer for supervisory information. One such technique is to provide a fixed frequency oscillator at each repeater point, using a different frequency at each location. At the receiving terminal the level of signal received from each oscillator can be used to deduce amplifier conditions along the route. The disadvantages are that a large number of frequencies have to be used at close spacing and the oscillator is therefore expensive. In addition, by virtue of its allocated frequency, the repeater becomes tailored to a location.

An alternative method is to use the same oscillator frequency at all locations and interrogate each oscillator in turn from the receiving terminal. This can be done by stepping an interrogation signal from repeater to repeater as described by Howard.[1]

55.9.8 Cable breaks

The system described above will locate faults in the HF path but, if the cable is cut, power is lost to all repeaters in that power fed section and the oscillator monitoring system is also inoperative. In practice cable breaks are invariably caused by external events which are self-evident in their own right, e.g. road works in the vicinity of telephone cables. Nevertheless, it can be postulated that the break might be due to hidden causes and a location scheme should be provided. A commonly used solution is to provide diodes at each repeater connected across the two cable inner conductors. If the power feeding supply is reversed, these diodes become conducting, the zener diodes in the through path are also forward biassed and hence only drop 1 V. Consequently, a resistance measurement indicates which repeater section is broken. The application of this technique can be extended by grading the resistors connected in series with the transverse diodes.

Location of the actual break point within a repeater section is made by time domain reflectometer measurement on the coaxial core.

55.9.9 Engineering order wire circuit

It is common practice to provide an engineering telephone circuit from end to end of the line system with access at the repeater points. This is particularly useful during system installation and is an important maintenance aid.

Such systems are normally worked as conventional four-wire circuits over loaded interstice pairs. VF amplifiers are provided at the intermediate power feeding stations.

The underground repeater housings are often pressurised and provided with a pressure-sensitive contactor which closes when the pressure falls, indicating the danger of water ingress. These contacts can be paralleled up on the phantom of the speaker circuit. Then any closure can be located by measuring the resistance of the phantom loop.

55.10 Non-telephony services

The transmission equipment described is also designed to cater for non-voice services, e.g. voice frequency telegraph and data modems of various rates. More significant users of bandwidth are programme circuit networks provided for the broadcasting authorities. Television can also be transmitted over FDM line systems. Because the television signals extend down to almost d.c. SSB suppressed carrier modulation is not practical. Instead vestigial sideband modulation is used. For a 12 MHz system the carrier frequency is approximately 7 MHz and the television information extends from about 6.3 to 12.3 MHz. The only change which has to be made to the line system is that it has to be equalised for group delay as well as for attenuation distortion.

55.11 Current status of analogue transmission

The principal characteristics of some analogue line systems currently available are given in *Table 55.2*. Of these the commercially most important are probably the 960 channel and 2700 channel systems. The 10 800 channel (60 MHz) system is in operation in several European countries and a very similar system (Bell L5) is used extensively in the USA. Application of the 40 MHz system is limited and it is not recommended by the CCITT. Technological improvements subsequent to the development of the 12 MHz system made possible the introduction of 18 MHz systems. These provide a 50% increase in channel capacity at the same nominal repeater spacing and their application has been primarily to large coaxial cables.

Table 55.2 Some currently available systems

Bandwidth (MHz)	*Channel capacity*	*Cable* (mm)	*Repeater spacing* (km)	*Notes*
4	960	9.5	9.1	
12	2700	9.5	4.55	Also TV transmission
18	3600	9.5	4.55	Also TV transmission
60	10800	9.5	1.5	Also TV transmission
4	960	4.4	4	
12	2700	4.4	2	
18	3600	4.4	2	
40	7200	4.4	1	Not standardised

Proposals for a larger analogue line system operating at 200 MHz and providing 30 000 channels have been put forward. However, with the change over to digital switching and transmission occurring in all telephone networks, there is no commercial justification for new FDM designs. This position has been reinforced by the emergence of optical fibre technology as a cost-effective means of providing long-distance digital transmission by cable.

Today there is a large installed base of FDM systems of the type listed in *Table 55.2*, which, in conjunction with the FDM multiplex hierarchy discussed in *Section 55.7*, still provides a substantial proportion of national and international telecommunications connections.

55.12 Digital transmission

The past few years have seen a major shift by telecommunication authorities away from analogue techniques for transmission and switching to digital ones. There are a number of economic reasons for this of which the most important are:

(1) The ability to increase the capacity of existing paired transmission cables.
(2) The availability of cheap digital processing.
(3) The ability to carry different types of information, such as speech, data or facsimile, over the same system without mutual interference.
(4) Transmission quality independent of distance and unaffected by digital manipulation such as multiplexing or switching.
(5) The ability to exploit nonlinear media such as fibre optical systems.

Most telecommunication authorities are aiming towards an integrated digital network with digital transmission and switching in which it will be possible to have a digital path from subscriber to subscriber.

The elements of digital transmission are:

(1) Primary multiplex—converting analogue voice or data signals to digital form and multiplexing several channels together.
(2) Higher-order multiplexes—combining lower-speed digit streams to form a higher-speed one for higher-capacitity transmission.
(3) Digital transmission systems of various capacities and using various media.
(4) Coders and multiplexes for non-speech traffic.

55.13 Basic digital techniques

55.13.1 Sampling

An analogue signal can be represented by a series of amplitude samples taken at regular intervals. If the original signal is band limited, so that the sampling frequency is more than twice the highest signal frequency, this band-limited signal may be recovered unimpaired from the sample sequence by ideal low-pass filtering. If the analogue signal is insufficiently band limited, the unwanted signal frequencies are found inverted in the recovered signal, a condition known as aliasing.

For telephone quality speech limited to a maximum frequency of 3400 Hz, a sampling frequency of 8000 Hz (125 μs interval) is standard.

55.13.2 Quantising

Instead of continuously variable amplitudes the samples can be assigned the nearest values on a vertical scale with a finite number of steps; this is known as quantising, see *Figure 55.23*.

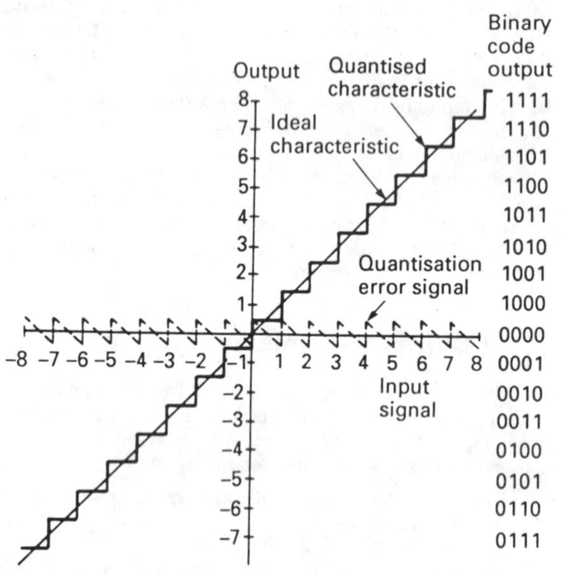

Figure 55.23 Illustration of quantising

The difference between the actual and quantised value is the quantisation error and in use gives rise to a multiplicative, noise-like, error signal known as quantisation distortion (QD). The finer the resolution of the quantised scale the less the QD.

55.13.3 Coding

For transmission the quantised samples can be represented by binary numbers, the number of bits per sample corresponding to the resolution of the quantiser, e.g. N-bit coding would correspond to 2^N quantising levels. In *Figure 55.23*, 16 levels are represented by 4 bits. The resulting bit stream is the pulse-code-modulated signal. Telephone quality speech is typically coded as 8 bits per sample.

55.13.4 Companding

From the description of quantising it can be seen that small amplitude samples are quantised with relatively less accuracy than large ones. Therefore, a low-level signal will have a worse signal to quantising noise ratio than a large one. Ideally the signal to noise ratio would be independent of amplitude. This can be approximated by the use of a nonlinear relationship between the input and output of the coder. The CCITT recommended law for 30 CH systems is:

$$y=\frac{1+\log_{10}(Ax)}{1+\log_{10}A} \qquad A^{-1}\leqslant x\leqslant 1 \tag{55.12}$$

$$y=\frac{Ax}{1+\log_{10}A} \qquad 0\leqslant x\leqslant A^{-1} \tag{55.13}$$

where y is the ideal output, x the input and $A=87.6$. In practice a segmented approximation to this law is used in which the slopes of adjacent segments are in ratios of 2. CCITT recommend 13 segments as shown in *Figure 55.24* for the positive quadrant. The

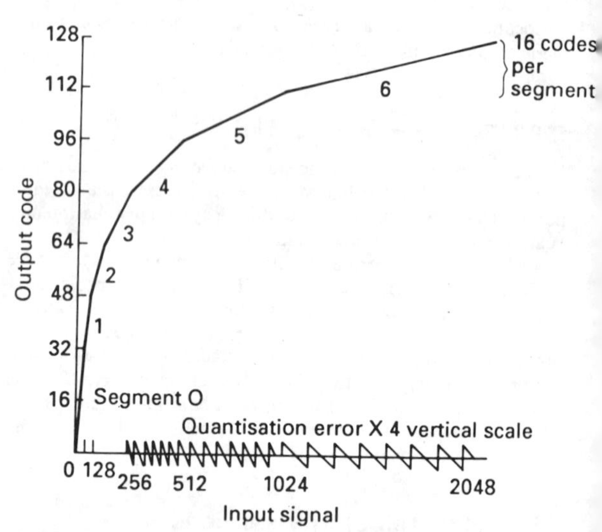

Figure 55.24 Segmented *A* law companding curve

first two segments are colinear. The full range of output codes is 8 bits in 256 levels. 16 of these are allocated to each of the outer segments and 32 to the centre one.

The nonlinear output is said to be compressed. A decoder has a complementary characteristic (expanded) so that the analogue-to-analogue characteristic is linear. The process is known as companding from COMpression and exPANsion.

The anti-aliasing filter, sampling and analogue-to-digital conversion are performed on a per channel basis in a single integrated circuit. The output from these integrated circuits is 64 kbit/s in bursts of 8 bits at an instantaneous rate of 2048 kbit/s. These signals are digitally combined to give the aggregate 32 time slot 2048 kbit/s signal.

55.13.5 Time division multiplexing

If the groups of bits representing successive samples are much shorter in duration than the interval between samples it is possible to use the remainder of the time to transmit other signals. This enables a multiplicity of signals to be carried in one bit stream, so obtaining time division multiplexing.

An arrangement standardised by CCITT is for 32 groups of 8 bits to be transmitted every 125 μs corresponding to the 8 kHz sampling rate of each channel. This gives an aggregate bit rate of $32\times8\times1$ kHz$=2048$ kbit/s for the multiplexed signal.

55.14 Primary multiplexing

A primary multiplex provides conversion between physical signals and their digital representations. Several of these digital signals are multiplexed together by time division multiplexing into a single bit stream.

In order for the multiplexed information in a bit stream to be recovered at the receiver it must be contained within a recognisable frame structure.

For example, the CCITT European frame structure is shown in *Figure 55.25*. It consists of 32 by 8 bit time slots labelled 0 to 31 and repeats at an 8 kHz rate or 125 μs period. Thirty of the time slots are used for speech; these are 1 to 15 and 17 to 31. Time slot 0 is reserved for alignment. That is, it carries fixed bit patterns which can be recognised at the receiver so that the time slot and bit counters in the receiver can be aligned with those of the transmitter (apart from transmission delay) thus enabling the information bits in the frame to be allocated to their correct recive channels.

To enable alignment to be maintained in the presence of digital errors between transmit and receive multiplexes a loss of alignment is not recognised until three successive frame alignment words are corrupted. When alignment has been lost all digit positions are checked until an alignment pattern is found. Confirmation that this is the correct time slot zero (TS0) is achieved by checking for a 1 in digit two of the following TS0 and that the alignment pattern re-occurs in the next TS0.

Time slot 16 (TS16) is reserved for signalling information, that is, information passed between telephone exchanges concerned with the setting up and breaking down of calls on the speech channels. The detailed contents of TS16 depends on the exchange types. For typical electromechanical types it would include seize and release signals and dialling pulses. More modern types of exchange would use multifrequency signalling over speech circuits for numerical information. Stored programme controlled exchanges can use TS16 as a data link between processors (common channel signalling).

Figure 55.26 shows a primary multiplex based on a digital bus with a 32 timeslot 2048 kbit/s structure. A single common card provides bus timing, multiplexing, frame generation and reception and the main 2 Mbit/s HDB3 port. A traffic channel can be allocated to any timeslot.

The two other classes of card are interface cards and processor cards. The interface cards convert between analogue or digital traffic signals and a standard bus format. A typical audio and signalling card uses single channel codecs and high-voltage line condition detectors and generators. For applications such as branching, additional 2 Mbit/s ports can be provided as interface cards. Processor cards act on the bus data to provide per channel functions by common digital processing. Examples of these are tone processors, which detect and generate tones for single and multifrequency signalling, and signalling processors, which convert line conditions to TS16 codes and keep track of call states.

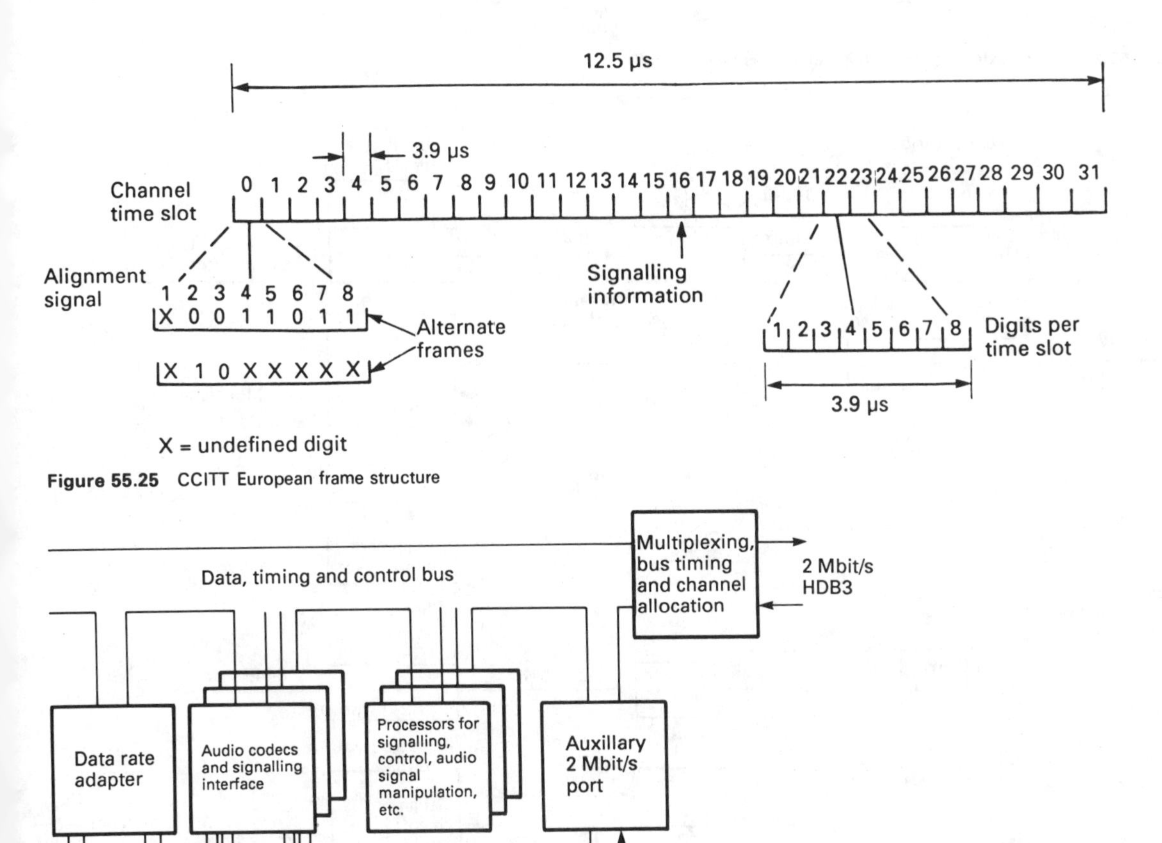

Figure 55.25 CCITT European frame structure

Figure 55.26 A flexible primary multiplexer

55.15 Higher-order multiplexing

Where higher numbers of channels are required the outputs of primary multiplexes can be combined to form higher-order multiplexes. In the CCITT European multiplex hierarchy (see *Figure 55.27*) four primary bit streams are combined to form a secondary multiplex with a bit rate of 8448 kbit/s. Four secondary outputs can be combined to give a third order of 34.368 Mbit/s and four of these combined to give a fourth order at 139.264 Mbit/s corresponding to 1920 speech channels.

The basic technique for each stage of higher-order multiplexing is identical. *Figure 55.28* is a block diagram of a second-order system while *Figure 55.29* shows the frame structure. The four 2048 kbit/s inputs are assumed to be pleisiochronous, that is, with the same nominal rate, but not locked together (the primary multiplex clock tolerance is 50 ppm).

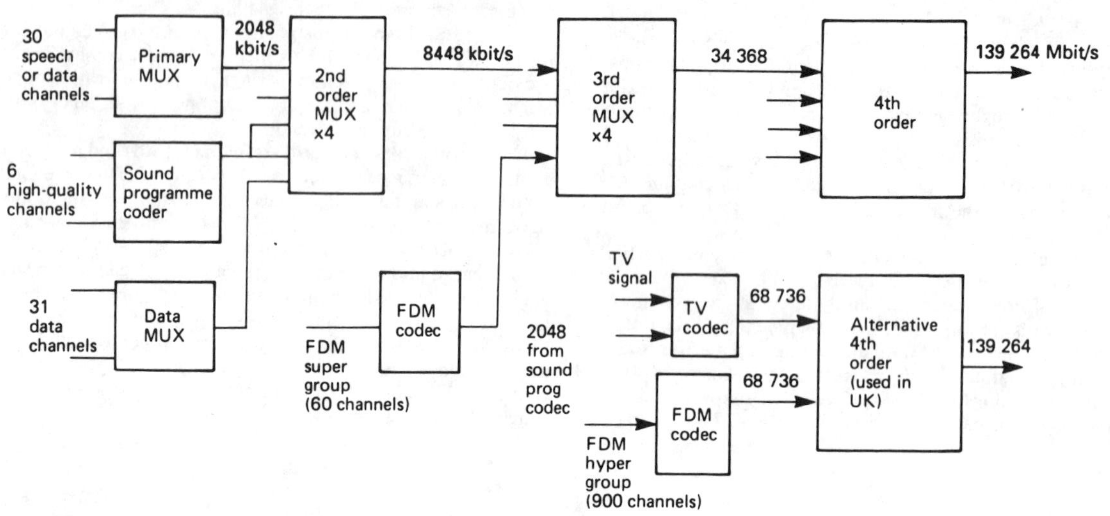

Figure 55.27 Multiplexing hierarchy showing various types of traffic

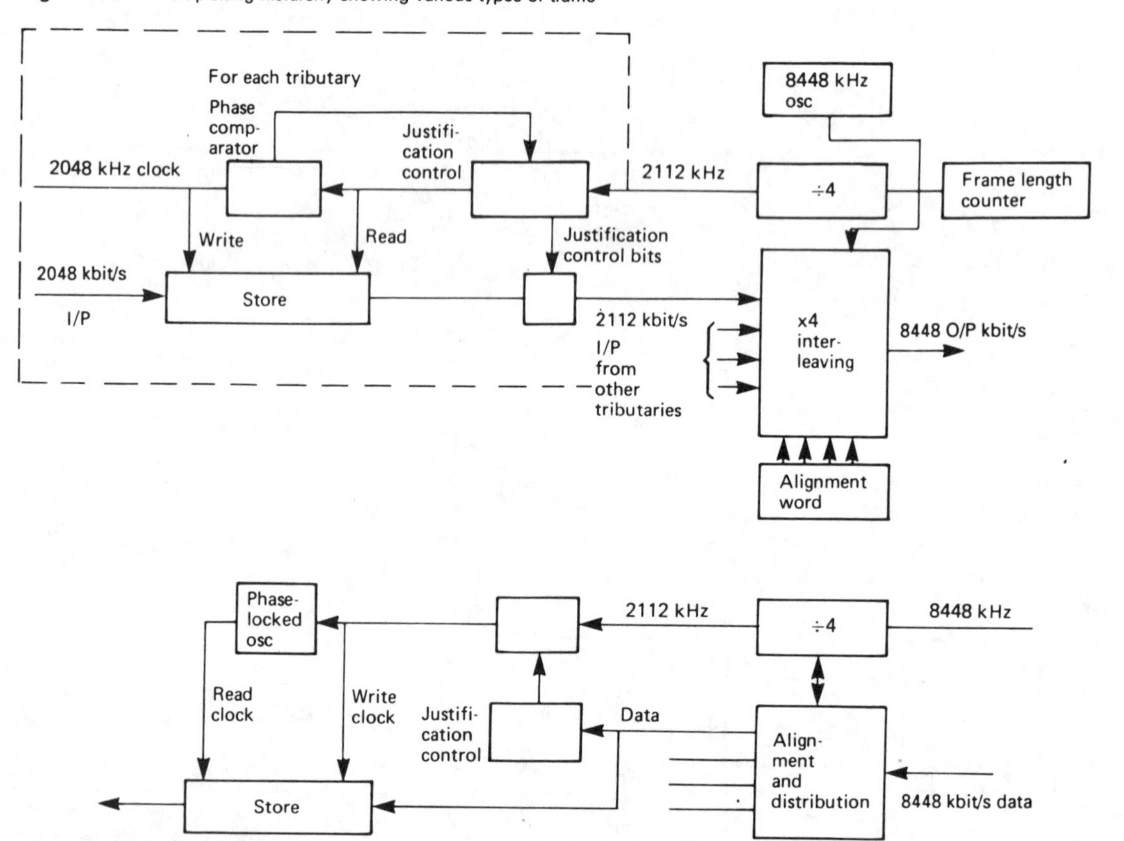

Figure 55.28 Multiplex block diagram

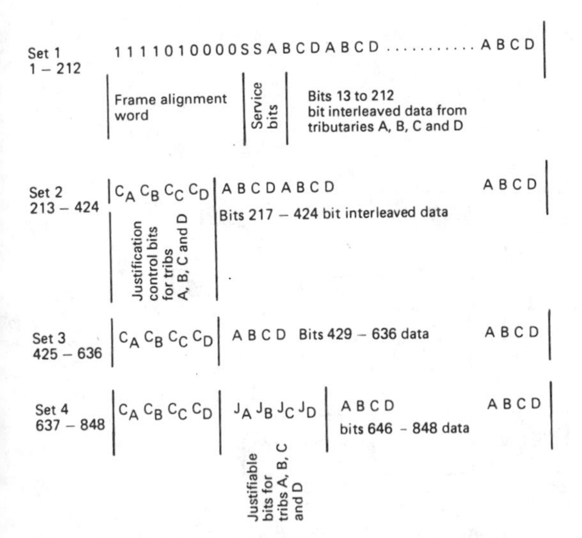

Figure 55.29 Frame structure

In each tributary the 2048 kbit/s data are written into a store. Data is read out from the store at a maximum rate of 206 digits/frame, equal to 2052.2 kbit/s (see *Figure 55.28*) and a minimum rate of 205 digits/frame, equal to 2042.3 kbit/s, the 206th 'justifiable' bit being filled arbitrarily. By comparing the phases of the write and read clocks to the store the choice of using or not using the justifiable digit for data is made so that the mean data rate out of the store exactly equals the input rate. This process is known as justification. Other regular fixed bits are added in the store so that the total digit rate out of the store is 2112 kbit/s. These bits are used for the justification control bits and alignment words.

The outputs from all four tributaries are then bit interleaved to give the final 8448 kbit/s output. The function of the justification control bits is to indicate to the receive demultiplex whether or not the justifiable bit contains valid information.

At the receiver the 8448 kbit/s bit stream is aligned and distributed to the receive tributaries where the the justification control bits are used to remove redundant bits. The resulting bit stream has a clock rate of 2112 kbit/s with clock periods omitted so that the mean rate is exactly the same as that of the original 2048 kbit/s input. A phase-locked oscillator and buffer store are used to smooth out this bit stream to recover the original clock for output. The frame structure of *Figure 55.29* provides more detail of the operation. One justifiable bit is provided per tributary per frame of bits. Three justification control bits, *C* are used per frame. This enables a majority decision to be made at the receiver to protect the process against transmission errors.

Note that no use is made of any information contained in the 2048 kbit/s input data. Thus any 2048 kbit/s bit stream can be multiplexed.

55.16 Digital transmission

A major reason for converting information to digital form is to take advantage of regenerative digital transmission. In addition to employing linear amplification to overcome attenuation a regenerative repeater recognises what the original digital signal would have been and transmits a signal identical to that original. This avoids the build-up of signal imperfections along the length of the route.

Figure 55.30 shows a regenerative system for use with paired cable at say 2048 kbit/s. Power for the intermediate regenerators is provided by a constant direct current (typically 50 mA) which is fed longitudinally over each of the two pairs used for go and return transmission. Transformers at the input and output of each repeater separate and combine the d.c. and signal. Since two pairs are used instead of the 30 pairs which would be required for audio transmission the capacity of the original cable is greatly increased.

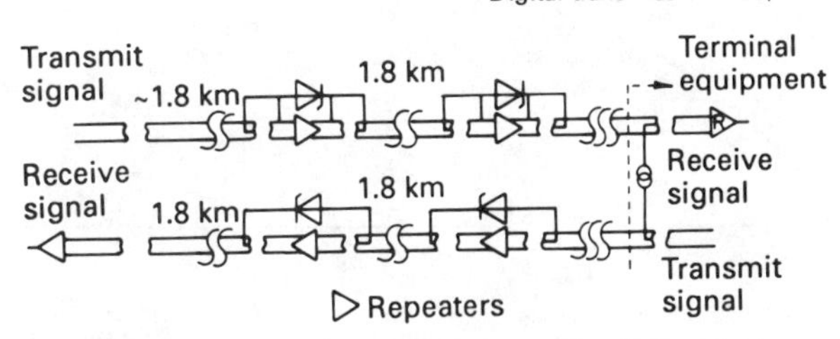

Figure 55.30 Regenerative system for use with paired cable at say 2048 kbit/s

Figure 55.31 shows a block diagram of a paired cable repeater for a three-level signal. The input signal consists of half width pulses transmitted from the previous repeater, attenuated and distorted by the cable pair plus interference from systems on other pairs in the same cable (crosstalk coupled interference). The automatic line-building out (ALBO) section corrects for the effect of various line lengths between repeaters. A passive input equaliser reduces the distortion and the amplifier produces a signal at 'D' the 'decision point' which is the signal to be regenerated. The pulses are approximately full width and the equalised bandwidth up to the digit rate. This signal is split four ways. One path is to the timing extraction where the signal is sliced and rectified to give a spectral component at the digit rate. This is extracted by a tuned circuit and limited to reduced-pattern-induced amplitude variations. The resulting square wave provides a clock to govern the time at which the positive and negative regenerators decide on the presence or absence of pulses.

Two other paths for the amplified signal are to the positive and negative regenerators. The signals are sliced against thresholds to determine whether they exceed the minimum acceptable pulse amplitudes. If these thresholds are exceeded during the sampling instant determined by the timing waveform then a new half width output pulse is generated, the pulse width is governed by the timing waveform. Another path is to a peak detector, which adjusts a variable artificial time to compensate for variations of cable loss with repeated spacing.

Outputs from the positive and negative regenerators are combined in a transformer and sent to line. The block diagram for a coaxial cable repeater would be similar except for the addition of filters at the input and output to separate d.c. power from signal power. Optical repeaters generally use two-level transmission and so have only one regenerator. The input transformer would be replaced by an optical power detector (avalance photodiode) and the output current would drive an optical transmitter, LED or laser. Other forms of timing extraction use phase-locked loops or crystal filters.

Because of the a.c. coupling in the signal path and the need to have frequent signal transitions to maintain the timing waveform non-redundant binary is not suitable for transmission and it is necessary to use a 'line code' with added redundancy to have the necessary properties. Redundancy can be added by using more levels (typically three-level ternary codes) or by increasing the bit rate. A simple code originally used for paired cable systems is AMI (alternate mark inversion), a three-level code in which binary ones are transmitted alternately as +1 or −1. HDB3 (high-density bipolar) is a modification of AMI which avoids lack of timing during long strings of binary zeros. For coaxial systems more efficient codes can be used such as the 4B3T types where blocks of four binary digits are translated to blocks of

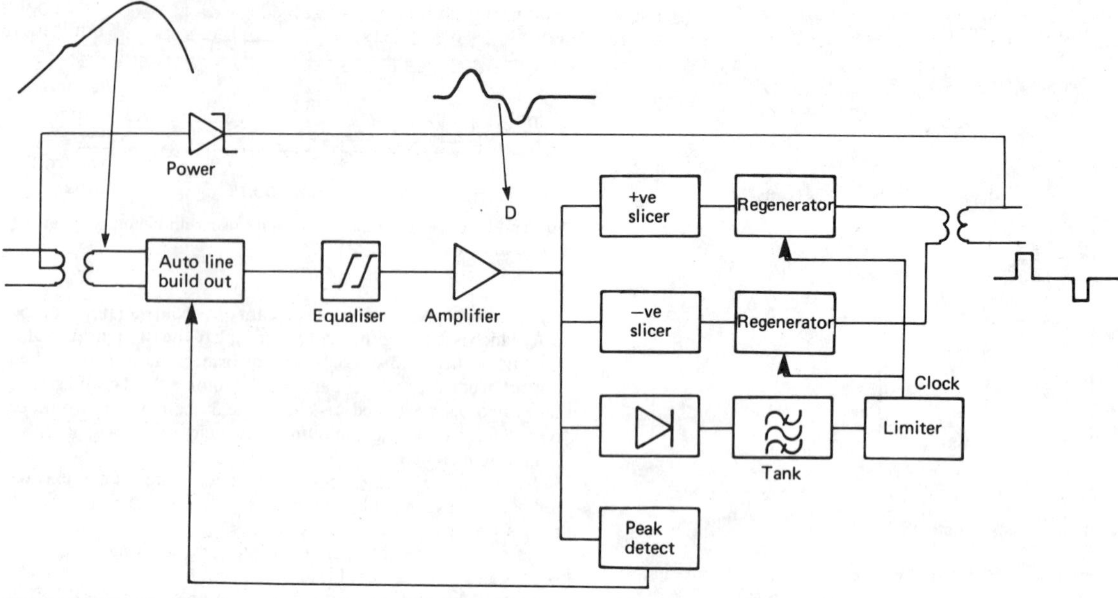

Figure 55.31 Block diagram of a paired cable repeated for a three-level signal

three ternary digits. The reduced symbol rate reduces attenuation and noise bandwidth for a given repeater spacing. Optical systems are restricted to two levels by the nonlinearity of the transducers, binary codes such as 7B8B, where seven information bits are transmitted in 8 bit words, provide sufficient redundancy.

55.17 Coding of other types of information

55.17.1 High-quality sound

A linear coder for broadcast quality sound would require 32 kHz sampling and 14-bit resolution; this would allow four channels in

Table 55.3

System bit rate and medium	*Line code symbol rate*	*Repeater spacing*	*Notes*
2048 kbit/s paired cable	HD83 2048 kbaud	Depends on cable ~1.8 km	Widely used to increase traffic capacity of existing cables
2048 kbit/s paired cable	MS43 (4B3T) 1536 kbaud	Depends on cable ~1.8 km	Allows substantially more systems in existing cables than HDB3 systems
8448 kbit/s paired cable*	MS43 6336 kbaud		Exploits cables originally installed for FDM 24 channel carrier
34*	MS43 27	4 km on 1.2/4.4 mm coax	
139 Mbit/s*	6B4T 692.8	2 km on 1.2/4.4 mm coax	
8448 Mbit/s optical	3B6B	~12 km	Highly redundant code allows cheap terminals and gives auxiliary channel Used for junction systems without intermediate repeaters
34 Mbit/s optical, 139 Mbit/s optical	5B6B	10–100 km depending on fibre and optical devices	

* Not used for new applications. Supplanted by optical systems

a 2048 kbit/s system. Bit rate reduction by *A* law companding can give rise to undesirable distortion. In a system developed by the BBC, NICAM-3, near instantaneous companding is used. Initial analogue-to-digital conversion is 32 kHz sampling and 14 bits, giving an audio bandwidth of 15 kHz. In every period of 1 ms (32 samples) the amplitude of the largest sample is used to determine which of four available linear coding scales will be used. The maximum amplitudes of the four scales are in 6 dB steps and the one chosen is the lowest that can accommodate the largest sample. Each scale has 10-bit resolution, thus the original 14 bits per sample is reduced to 10 giving 320 kbit/s channel. A data channel multiplexed in the 2048 stream indicates to the receiver which scale is required to decode each block of 32 samples. With the reduced bit rate due to compression six sound channels can be included within the 2048 kbit/s rate together with data channels and alignment words.

55.17.2 TV coder

A TV transmission system for a composite PAL signal must handle more than 5.5 Mbit/s bandwidth and have 8-bit linear resolution. It is convenient to lock the sampling rate to three times the colour subcarrier (subcarrier frequency $f_{sc}=4.43$ MHz), that is 13.3 MHz. This gives a bit rate of 106 Mbit/s. Bit rate reduction can be achieved by the use of differential PCM (DPCM). In this technique an estimate is made of the next sample based on previous samples, then instead of sending the next sample with 8-bit resolution the difference between the predicted and actual values is sent with 5-bit resolution. At the receiver the same prediction is made as at the transmitter, and the incoming DPCM sample used to correct this value to reconstruct the original sample. Because small prediction errors are more common than large ones, and because the eye is less critical of prediction errors occurring at large changes of luminousity, it is possible to use a tapered quantising law for the DPCM samples. Thus small differences between predicted and actual signals are coded with finer resolution than large ones. For a TV signal sampled at three times f_{sc} sample 3 before the current one gives a good prediction. The resulting signal has a bit rate of 66.5 Mbit/s; this can be multiplexed with a 2048 kbit/s channel for high-quality sound, etc., and an alignment signal to give a total rate of 68.736 Mbit/s. This occupies half the capacity of a 140 Mbit/s transmission system.

55.18 Submarine telecommunications systems

International telephone circuits were first carried by submarine coaxial cables, having analogue repeaters inserted into the cable at intervals of a few kilometres, in the 1950s. In the following 30 years, over 100 000 km had been installed of British manufacture alone. These systems were designed for a life of 25 years and experience has confirmed this, so that many systems will still be in service at the end of the century.

However, all new systems are expected to use optical fibre cables with digital (PCM) transmission and electro-optic regenerators. A system of this kind first carried international traffic in 1987 and one across the North Atlantic is expected to enter service in 1988. Over 8000 km of optical systems are currently in manufacture, worldwide.

The large bandwidth and low attenuation now achievable on optical fibre, together with the large number of fibres which can be accommodated in a cable, mean that very large circuit capacities are feasible with a spacing between submerged repeaters which is much larger than in analogue systems. Again, with the coming preponderance of digital traffic on internal trunk routes, it is convenient to continue as such onto international connections.

Here, we deal only with submarine systems as they are currently being manufactured; the earlier coaxial systems are comprehensively described in Reference 2.

55.19 Cable

Figure 55.32 shows a typical optical cable as designed in Britain. The optical fibres are coated with an acrylate resin, on manufacture, to protect their surface. The appropriate number

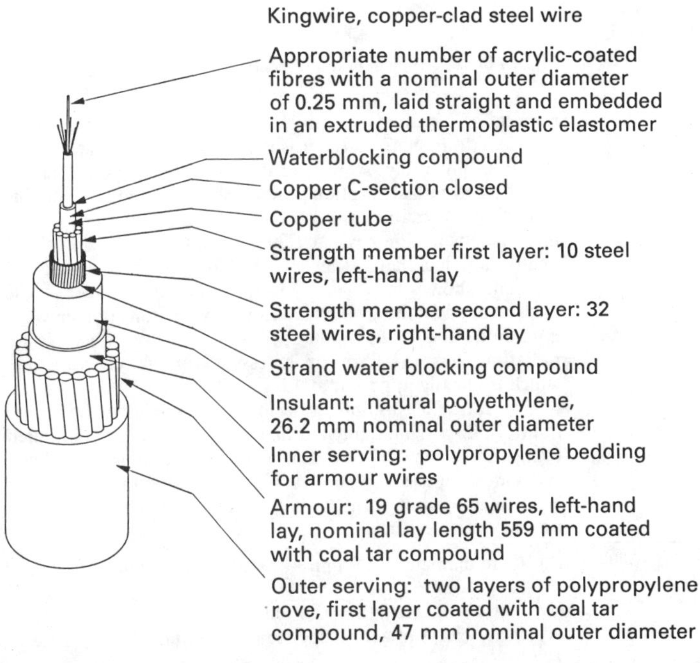

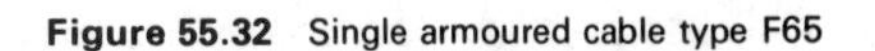

Figure 55.32 Single armoured cable type F65

are laid up straight and then embedded in an extruded thermoplastic. The 'king wire' is included to aid in hauling this package around the factory. The fibre package is then inserted within a copper tube whose purpose is partly to protect the package against hydrostatic pressure and partly to provide a gas-tight enclosure—optical fibres are very susceptible to loss increase due to hydrogen contamination[3] and quite high hydrogen concentrations are possible in some circumstances. Around this tube is a strength member of two layers of high-tensile steel wires with opposite directions of lay. These ensure that no turning moment appears under load, so that the copper tube is not twisted, the cable handles well and the strain for a given load is small. The cable may be subject to rapidly changing load over a long period: appreciable twist of the copper tube could lead to metal fatigue and subsequent loss of hydrogen integrity.[4] A fibre is subject to a static fatigue which is a function of the effects of all the strains that were ever applied to it: if the fibre strain is kept low during installation and service, a less onerous proof test, leading to higher yields, can be applied. At very high pressures, the strength member wires are designed to lock up, so that pressure on the copper tube is relieved. Polythene insulation, so that high terminal voltages can be used to feed current to power the repeaters via the strength member and copper tube in parallel, is applied and is surrounded by armour wires as required. *Figure 55.32* shows fairly light, single-layer armour, used where the risk of damage to the cable is small. In shallow seas, subject to fishing by heavy beam trawlers, the ideal cable is heavy, impenetrable and strong. It may have increased diameter insulation and a double layer of armour, of which the outer layer may have quite a short lay. In the deep sea, the risk of damage from fishing, shipping or abrasion is negligible and a 'lightweight' design is used. Usually this is the same cable up to the polythene outer surface. In some situations there is a risk of damage from marine life, and it is becoming the practice to fit lightweight cable with an outer skin of aluminium or thin steel tape, sheathed with a further protective layer of polythene. This protects the cable against damage from shark bites (which have caused failure in at least one optical cable) and other organisms, such as marine borers. Longitudinal water blocking is used at all interstices, so that, if a cable is cut on the sea bed, it can be repaired without much loss of length from flooding. Any number of fibres, up to about 12, can be supplied in the same cable design.

Fibre for submerged use is always single mode. Usually this is quasi-step index and a typical design has the parameters:

(1) Core diameter = 8.4 μm (germanium doped).
(2) Refractive index difference (Δn) = 0.004.
(3) Fibre diameter = 125 μm (matched cladding).
(4) Cut-off wavelength = 1.15–1.28 μm.
(5) Wavelength of zero dispersion = 1.315 μm.
(6) Attenuation at 1.310 μm = 0.34 dB/km.
(7) Attenuation at 1.55 μm = 0.19 dB/km.
(8) Dispersion at 1.55 μm = 17 ps/(km nm).
(9) Mode field diameter = 9.8 μm.

Such fibres are suitable for high bandwidth (up to about 500 Mbit/s) with Fabry–Perot type lasers—which have a finite line width (wavelength spectrum) and are subject to small jumps of wavelength—at the wavelength of zero dispersion only. *Figures 55.33* and *55.34* show the spectral behaviour of such a fibre, in respect of attenuation and dispersion, as a function of wavelength. In general, the loss is due to Rayleigh scattering in the glass (inversely proportional to the fourth power of the wavelength) with absorption peaks at about 1.2 and 1.39 μm (due to hydroxyl contamination) and infrared absorption above 1.6 μm. The absorption peak at 1.20 μm is quite small, only about one-twentieth of that at 1.39 μm. The large peak between 1.1 and 1.2 μm, shown in *Figure 55.33*, is a consequence of the measurement method which is affected by the transition between multimode and single-mode operation. With this fibre design, radiation from the fibre, on excessive bending, causes a loss which is usually apparent at 1.6 μm and above.

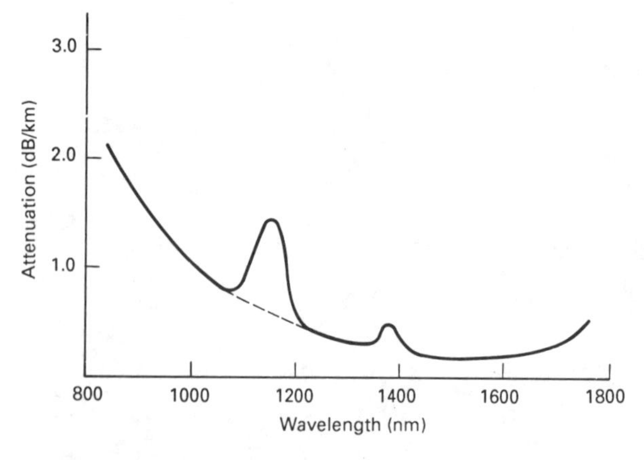

Figure 55.33 Typical spectral attenuation characteristic

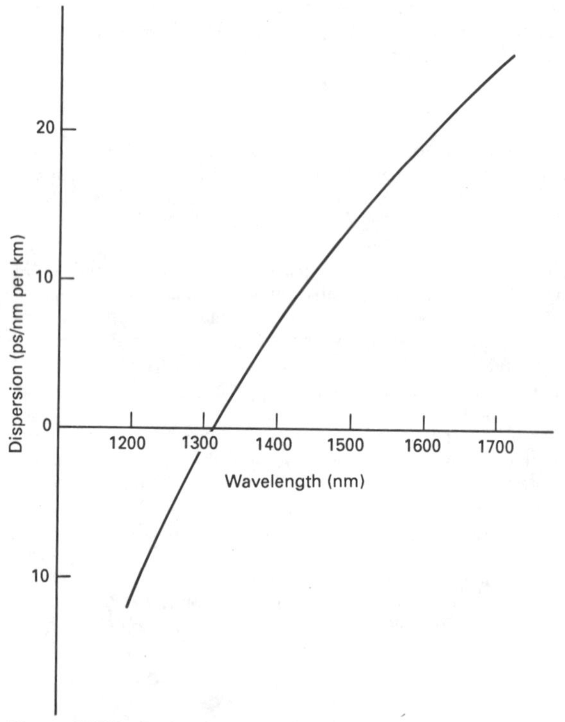

Figure 55.34 Typical fibre dispersion

If operation of this fibre at the lower attenuation at 1.55 μm is not to be dispersion limited at higher bit rates, a laser of limited ('single-line') wavelength width (such as the distributed feedback type) must be used. In some systems, a Fabry–Perot laser has been retained, but with the use of 'dispersion-shifted' fibre of typical characteristics:

(1) Core diameter = 6.9 μm (at base of a triangular profile).
(2) Δn = 0.011.
(3) Cut-off wavelength = 0.84 μm.
(4) Wavelength of zero dispersion = 1.53 μm.

The loss of such a fibre is somewhat greater, due to the higher germanium doping level associated with the higher Δn.

In the structure of *Figure 55.32*, the proof strain requirement for the fibre need be no more than 1%.

55.20 Repeaters

Figure 55.35 shows a block diagram of the apparatus used to receive the input from a fibre, convert the light pulses into electrical ones, amplify, regenerate and retime them, and transmit them as high-level light pulses into the output fibre. In a typical repeater, handling four fibre pairs, there would be eight such regenerators in a single repeater. A regenerator comprises a receiver, an amplifier, a clock extraction circuit, a retiming circuit, a transmitter, a supervisory facility and switching devices which can interconnect fibres or regenerators as required.

55.20.1 Receiver

Most receivers use a transimpedance circuit configuration in which the detector is either a PIN or an avalanche photodiode and the first amplifier uses either a field effect or bipolar transistor or, sometimes, an integrated circuit. A typical PIN/BIP receiver, to operate in a 420 Mbit/s regenerator, would have a (worst case) sensitivity of −32 dBm and a (worst case) overload level of 14 dBm, for a bit error ratio (BER) of 10^{-9}. The sensitivity falls with increasing of bit rate at 1.5–4.5 dB/octave depending on the details of the design.

55.20.2 Amplifier

The function of the amplifier is to:

(1) Amplify the electrical pulses from the receiver to a level suitable for the retiming circuit input—about 0.5 V, peak to peak.
(2) Control the gain so that the output level does not vary with the input level (AGC) over a range of, typically, 40 dB.
(3) Equalise the frequency spectrum, in respect to both level and group delay, to compensate for the frequency response of the receiver and amplifier and to shape the response to give the best compromise between pulse shape and bandwidth. If the fibre has a lowered bandwidth, due to a combination of fibre dispersion with a transmitter of significant spectral width, compensation will be required for this also. A 100% 'raised cosine' shape is popular (relative response $0.5(1+\cos \omega(T/2))$, where ω is angular frequency and T is the time width of a bit). This response gradually falls to zero at the bit rate and above, and produces a pulse with a fairly broad maximum (to minimise the effects of timing errors) and, ideally, no inter-symbol interference.
(4) Provide access to the bit stream for supervisory purposes—typically to provide a signal channel by low-frequency phase modulation of the data rate.

55.20.3 Clock extraction

Bounded codes, such as 7B8B, have been used in submarine systems, but now an unbounded code, 24B1P, is becoming a standard, at least for long-haul systems. This is a code of 24 randomised binary data bits plus one bit to force even mark parity. This gives a near minimum baud rate for a given traffic capacity and provides a good, if low-speed, signalling channel using parity bit violation (PBV). A disadvantage is that the unbounded code, in the NRZ (non-return to zero) format always used on submarine systems, has few transitions so that a high-Q circuit is required ($Q \simeq 800$) to extract a good sine wave from the data. Since the NRZ format has no information at 'clock' frequency, a frequency doubler is required to feed the extraction circuit proper. A surface acoustic wave (SAW) filter is normally used to provide the high Q with the required stability of the corresponding narrow pass band. It should have good phase linearity in the pass band to give a constant time delay, between the transitions of the data and the clock, to minimise timing errors if there are deliberate (FM) or other small variations of the data frequency. A trimmable delay circuit is provided to cause the clock transitions and the nominal data transitions to

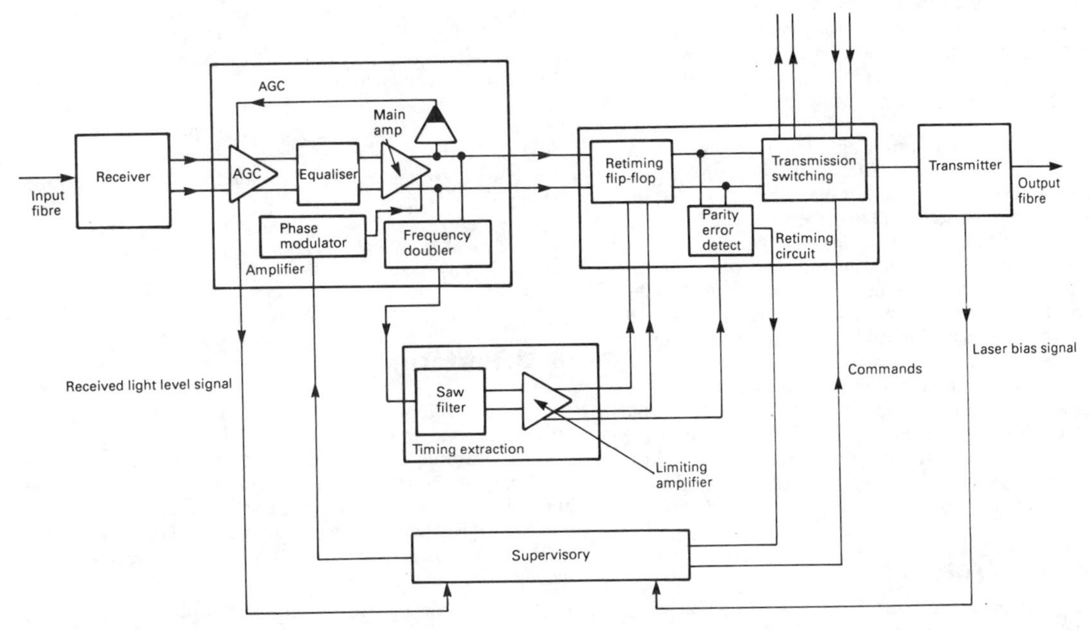

Figure 55.35 Transmission circuit block diagram

coincide. A limiting amplifier is used to drive the retiming circuit to minimise the effects of variations of clock frequency amplitude, due to variations of the data pattern, or whatever.

55.20.4 Retiming

The retiming circuit is a balanced, D-type, master/slave flip-flop, fed by data from the signal amplifier and the steady clock frequency extracted from the data. The product of the flip-flop is a data signal, reshaped and free of imperfections such as noise, and retimed to have accurate transitions.

This circuit contains an even mark parity detector which provides an output, which is a function of the PBV applied to, or inherent in, the data, for the supervisory circuit.

55.20.5 Transmission switching

Circuits are usually provided, at the output of the retiming circuit, to route the data stream to its normal transmitter or in various alternative ways: these can include a spare transmitter (if the normal one is faulty), the transmitter of another sending fibre (to redirect around a fault) and a transmitter of a return fibre (to monitor faults up to that point). These switching circuits are under the control of the supervisory unit, usually on command from the sending station (see *Section 55.20.7*).

55.20.6 Transmitter

Typically this is a semiconductor, GaInAsP, heterojunction laser of the Fabry–Perot type, operating at about the zero-dispersion wavelength of approximately 1.32 μm. The pulses from the timing circuit drive a modulator which operates on the laser current to produce optical pulses which are fed to the fibre. One facet of the laser internal waveguide is coupled to the outgoing fibre; the other facet is coupled to a monitor PIN diode which is part of the control loop of two circuits. One control circuit compares the output voltage of the PIN load with a standard voltage and uses the difference to control the mean power (average of 'ones' and 'zeros' of the data). The other circuit controls the modulation depth of the laser, if the slope of the light output *versus* drive current response should change, e.g. with life. There would also be a total supply current control, to ensure that the proportion of the line current taken by the repeater circuitry does not change with the laser condition. These controls ensure that the laser output is constant during life irrespective of drift of threshold current or whatever. Typical output level is 0 dBm.

Laser development is a fast-moving art so that new designs, without a long previous history, are liable to be used. This, coupled with the facts that laser failure mechanisms (if any) tend to have a low activation energy and that lasers cease to lase at the comparatively low temperature of about 80°C, means that adequate life testing programmes, of sufficient duration, are difficult. For this reason, and not because of any inherent unreliability, it is common, at present, to provide a spare laser which can be alternatively coupled, if necessary, to the output fibre (see *Section 55.20.5*).

For high bit rate operation on non-dispersion-shifted fibre, at the higher wavelength of about 1.55 μm, DCPBH/DFB (double channel buried heterostructure/distributed feedback) lasers are becoming favoured. Because of scattering loss in the DFB structure, the output may be about 2 dB lower than a corresponding Fabry–Perot laser.

55.20.7 Supervisory facilities

Many supervisory methods have been used. The following describes that commonly used in Britain.

Parity of the 24B1P line code is deliberately violated regularly every few frames. This produces a tone at the output of the parity detector which is filtered out and used as a subcarrier, allowing a pulse width modulated channel for outbound signalling. The same tone can be phase modulated in the receive amplifier to provide an inbound channel. On command from the sending station, the following facilities are provided:

(1) Receive light level monitor.
(2) 'Laser health' monitor (i.e. light output at the laser back facet).
(3) Error count. This uses a count of even mark disparity after the command parity violation has ended and reset the counter.
(4) Switching-in of a spare laser module, if fitted.
(5) Out-of-service—transmission switching.

The monitor information (1–3) is signalled back to the sending station by the inbound channel.

An ability to loop back the sending circuit to the receiving circuit is provided so that, on command, comprehensive measurements of the condition of the system up to a particular repeater can be made at the sending station. The same methods can be used to re-route a path via an alternative fibre. If there are several fibres in the cable, this gives a means of redirecting transmission around a fibre or regenerator fault at the expense of reduced total traffic capacity. To facilitate this, the common supervisory unit can be commanded from any regenerator in the repeater and can operate on the same or any other regenerator.

55.20.8 Power feed

Power feed to the repeater is by direct current along the cable centre conductor; 1.6 A is usual. This would be fed from both ends on long routes and the power equipment is largely as for former analogue systems.[2]

Each regenerator in a repeater, of which there would be eight in a four-fibre pair system, is fed from an avalanche diode, whose function is to protect against voltage surges and other current variations, and to provide a semi-standard voltage for control loops (see, for example, *Section 55.20.6*). At about 6.5 V each, the voltage drop per repeater would be about 52 V. The cable resistance is commonly about 0.7 Ω/km, for say 50 km, making the total voltage drop per repeater section about 108 V. For a 100 repeater route, e.g. as may be across the North Atlantic Ocean, the end-to-end voltage is then 11 kV.

Surge suppression measures are simpler than for analogue repeaters, since the signal handling parts of the electronics are no longer connected to the power feed conductor, but the power circuits are given additional protection by gas discharge tubes and suppressor coils.

In some systems, repeaters in the short legs to branching units (see, *Section 55.2.4*) are fed via diode bridge rectifiers, so that the repeater can operate with either direction of feed current.

55.21 Performance

Consider a circuit of 420 Mbit/s traffic capacity which, with the 24B1P line code and some added bits for assembly purposes, gives an operating rate of about 442 Mbaud. The maximum allowable loss would be 32 dB, but, in fact, a cable loss of about 20 dB (corresponding to a section length of about 50 km) is designated. This gives a margin of about 12 dB which, in an 'optical power budget', is divided up into allowances for items such as:

(1) Variability of the loss of fibres as supplied.
(2) Loss at splices, both on installation and for repairs.

(3) Drifts of output power (e.g. drift of the control diode) and receiver sensitivity due, for example, to ageing or temperature.
(4) Minor hydrogen contamination of the fibre.
(5) Margins on sensitivity and overload for unforeseen items, as a matter of prudence.

Systematic jitter can accumulate over a long route—it is proportional to the square root of the product of the number of repeaters, the bit rate, the reciprocal of the clock Q, and the jitter power density per MHz of an individual regenerator; about 0.12 (unit intervals) would be typical of a 100 regenerator route. Note that the clock output can follow the data at low jitter frequencies, so that no timing errors occur in this event; i.e. the regenerators are transparent to low-frequency jitter.

In contrast with an analogue system which, barring a catastrophe, tends to degrade slowly as it grows old, giving notice of impending failure, any digital optical system exhibits what has been called the 'little girl syndrome'—when it is good it is very very good and when it is bad it is horrid! This is a consequence of the large variation of BER with signal-to-noise ratio (SNR): BER can change about two decades for 1 dB change of SNR. It is for this reason that, in submarine systems applications, degradation monitoring, such as laser output and receiver sensitivity measurements, are provided to give advance warning of the need for a repair.

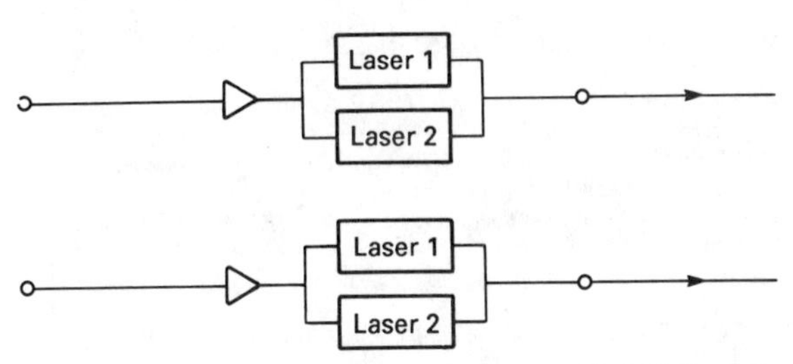

Figure 55.36 Alternative lasers routed by optical hybrid (lines of equal priority)

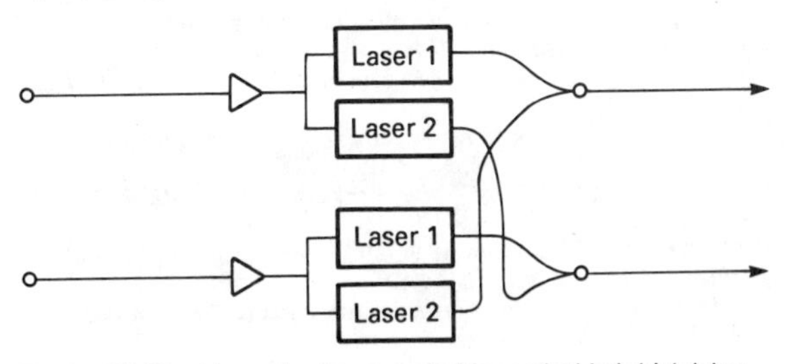

Figure 55.37 Alternative lasers routed by optical hybrid (giving line switching on lines of unequal priority)

55.22 Reliability/redundancy

It will be apparent that an optical digital repeater is much more complex than an analogue one of comparable circuit capacity—but then there are many fewer of them. Also, their use coincides with the coming of age, reliability-wise, of integrated circuits. These generally use the same silicon bipolar technology which was proved by long use in analogue systems. In Britain, a standard eight-cell uncommitted array is used throughout, except, possibly, for a custom-built front end and in the supervisory unit.

There is no reason, then, to doubt that a reliability record similar to that in the past should not be expected in the future. However, bearing in mind such strictures as in *Section 55.20.6*, applicable to new technology, and the large traffic capacity now at risk, methods to increase reliability, using redundant apparatus, are favoured. One method is to use electronic line switching, commanded by the supervisory unit, to enable one to switch a line to any one of the other lines. For protection against laser failure only, the method of *Figure 55.36* has been used, in which a spare laser is commanded by the supervisory unit to take over. The changeover can be by electro-optic switches, which mechanically change the routing of fibres, but a more reliable method, using optical hybrids, is shown. (One must ensure that redundancy switching does not, of itself, increase the hazard to the system.) The method of *Figure 55.36* can be extended to give protection, also, against failures of other apparatus, including fibre failure, by crossing lines, e.g. as in *Figures 55.37* and *55.38*. Because of the hybrid loss (3–3.5 dB), a faily long changeover sequence, say every sixth repeater, would be used.

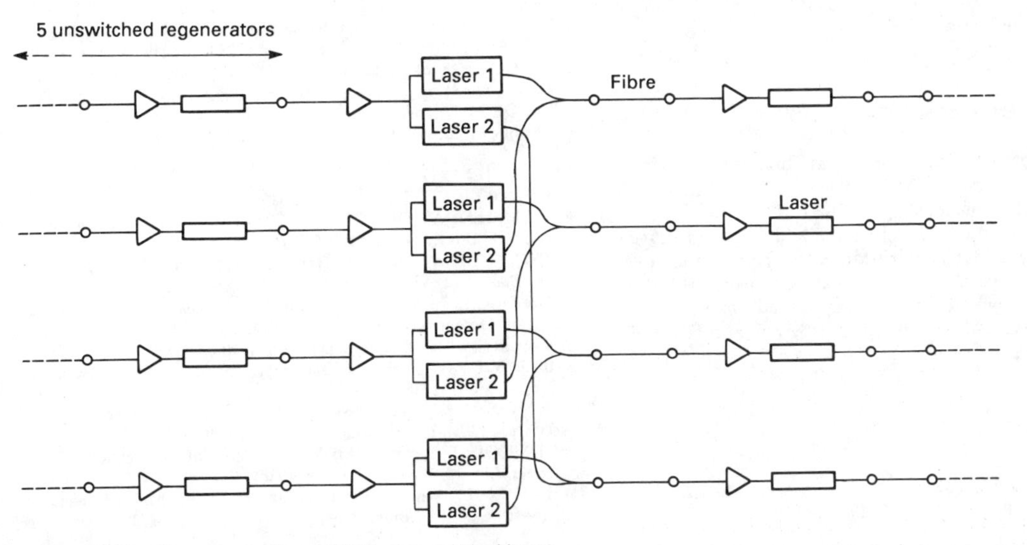

Figure 55.38 Alternative to *Figure 55.36* using switched lasers

55.23 Pressure housings

These generally follow the practice used on analogue systems.[2] Some alterations and improvements have been made, concerned with taking the glass fibres into the internal unit and improving gas tightness (against hydrogen). British practice is as follows.

Glass fibres are taken through the housing bulkhead by glass to metal seals, the fibre being metallised for a short distance. A ceramic seal is used for the power feed (centre conductor) connection, replacing the less gas-tight polythene gland. An additional joint chamber is added to the ends to facilitate fibre jointing (outside of the factory) for which a polythene gland is satisfactory. An articulated device is used at the terminations to give enhanced bend restriction. The steel housing is given a tough, abrasion-resistant coat of epoxy-coal tar resin to ensure that vulnerable parts are insulated from the sea—to avoid corrosion, and to prevent the production of monatomic hydrogen directly on the housing surface by electrolysis.

55.24 Branching

Because there are usually several independent circuits in an optical fibre cable, a long-haul system, say from America to Europe, could have more than one landing point on each continent. These fibres could then branch out to the separate destinations, at a unit in deep water, well off shore, beyond hazard to the cable from shipping or fishing activity. This simplifies traffic administration and adds to the reliability of the system, some of which can still operate if there is a cable fault on one of the shore ends. To avoid waste of capacity, it is usually required to provide traffic between the two or more parts of the same end of the system. Then interchange of sending systems, for the long-distance power feed, is required and a sea earth at the branching unit, to enable power feed, interchangeably, on a local branch must be provided. For this purpose, high-voltage switches, using high-vacuum tube relays, are used. These are highly reliable and have an impressive performance in respect of maximum current and voltage, but are ruined by contamination with traces of a gas at low pressure, such as hydrogen. Since hydrogen is produced by the sea earth of a branching unit (which is cathodically protected by the direction of power feed current against electrolytic corrosion), then, if the earth is close by, it is necessary rigorously to test the vacuum tube seals and/or the housing seals to ensure that there is no degradation of gas tightness after long periods.

55.25 Un-repeatered systems

Because of the low attenuation of optical fibre, it is possible to have long spans without repeaters. We are not concerned with the reliability of submerged plant with these systems, so operating margins can be much lower and less well-proven technology, providing higher performance from more novel components, can be used. This means that section lengths and/or bit rates are much greater than for repeatered systems. Early examples are England–Channel Islands (132 km) and England–Isle of Man (91 km).

References

1 HOWARD, P. J. *et al.*, '12 MHz Line Equipment', *Electrical Communication*, **48**, No. 1/2, 27–37 (1973)

2 MAZDA, F., *Electronics Engineers' Reference Book*, 5th edn, Butterworths, London (1983)

3 PITT, N. J. *et al.*, 'Long term interactions of hydrogen with single-mode optical fibres', *Proc. 10th ECOC*, Stuttgart, 1984, pp. 308–309 (1984)

4 OSWALD, T., 'Submarine optical cables: a phenomenon of resonance in suspended lengths', *IEE Electron. Lett.*, **22**, No. 2, 81–82 (1986)

Further reading

ATKINSON, J., *Telephony*, Pitman, London (1950)

BAX, W. G., '1 pW per km on a 3600 channel coaxial line', *Philips Telecomm. Rev.*, **35**, No. 4 (1977)

BEAR, D., *Principles of Telecommunication Traffic Engineering*, Peter Peregrinus (1976)

Bell Systems Technical Journal, **32**, No. 4, special issue *L3 Coaxial System* (1953)

Bell Systems Technical Journal, **48**, No. 4, special issue *L4 System* (1969)

Bell Systems Technical Journal, **53**, No. 10, Special issue *L5 Coaxial Carrier System* (1974)

BELL TELEPHONE LABORATORIES, *Transmission Systems for Communications* (1965)

BERKELEY, G. S., *Traffic and Trunking Principles in Automatic Telephony*, Benn (1949)

COLBECK, D., 'Public telephone networks', *IEE Review*, January (1989)

Electrical Communication, **48**, No. 1/2, special issue on transmission (1973)

FLOOD, J. E. (Ed.), *Telecommunications Networks*, Peter Peregrinus (1976)

FOX, S., 'Submarine optical fibre cable system applications', *IERE J.*, July–August (1988)

GREEN, J., 'Can centrex conquer the European market?', *Communications*, November (1988)

HALLENBACH, F. J. *et al.*, 'New L multiplex', *Bell Systems Tech. J.*, **42**, No. 2 (1963)

HERMES, W. *et al.*, 'Level regulation of coaxial line equipment', *Philips Telecomm. Rev.*, **30**, No. 1 (1971)

HUGHES, C. J., 'Switching—state of the art', *British Telecom Tech.*, **4**, Nos 1 & 2 (1986)

JEFFREY, B., 'CT2: a new way to talk', *Communications*, November (1988)

KALLGREN, O., 'New generation of line systems', *Ericsson Review*, No. 2, 48–53 (1974)

MALLETT, C. T., 'Digital multiplexing in submarine cable branching units', *Br. Telecom Tech. J.*, October (1987)

MAYO, J. S., 'Communications at a distance', *Mini-Micro Systems*, December (1982)

MILLER, J. R. *et al.*, 'Recent development in FDM transistor line systems', *Post Office Eng. J.*, **63**, Part 4, 234–241 (1971)

NORMAN, P. *et al.*, 'Coaxial system for 2700 circuits', *Elec. Comm.*, **42**, No. 4 (1967)

PROUDFOOT, D. A. *et al.*, 'Digital business services: centrex network evolution', *Proc. Int. Conf. on ISDN*, London, June (1988) Online

RECOMMENDATIONS OF THE CCITT. 'Line Transmission', *Yellow Book*, Vol. 3, ITU, Geneva (1981)

RECOMMENDATIONS OF THE CCITT, 'Protection', *Yellow Book*, Vol. 9, ITU, Geneva (1981)

RECOMMENDATIONS OF THE CCITT. 'Signalling system No. 7', *Red Book*, Vols VI.7 and VI.8, ITU, Geneva (1985)

SCHNURR, L., 'Mobile communications', *Telecommunications*, **16**, No. 11, October (1982)

Siemens Review, **38**, Special issue *Communications Engineering* (1971)

SMITH, S. F., *Telephony and Telegraphy*, 3rd edn, Oxford University Press, Oxford (1978)

TAKAHASHI, K., 'Transmission quality of evolving telephone services', *IEEE Comm. Mag.*, **20**, No. 10, October (1988)

WHITEHOUSE, B., 'The PABX as a hub for data applications', *Comm. Int.*, **15**, No. 10, October (1988)

WILCOCK, L., 'When key systems can provide the perfect business answer', *Communications*, November (1988)

56 Fibre-optic Communication

P Kirkby,
STC Technology Ltd

Contents

56.1 Introduction

The idea of using light as a medium of communication is by no means new: indeed, the technology has now celebrated its centenary.[1] For most of its history, the development of optical communications centred on the use of beams of light, but success was limited. Suitable light sources were a problem until the advent of the laser; practical ranges were limited by atmospheric conditions and beam alignment accuracy.[2] Thus, although the idea was sporadically revived for short-range applications, it was not until the rapidly developing technologies of semiconductor optoelectronics and fibre optics were married that the full potential of optical communications could be realised.

As with most new technologies, interest in fibre optics has centred on the most technically demanding applications: long-distance wideband links for telephony, data and TV signals. However, as is often the case, technological spin-off from the main endeavour has provided a basis for the use of fibre-optic communication techniques in a range of applications that are likely, in total, to represent the dominant market.

The reason for the widening interest in fibre-optic communications is not hard to find since an optical fibre is almost the ideal transmission line. It is made from abundant, low-cost materials, and its bandwidth is larger than that of any other line. It has a low weight per unit length, and it is an electrical insulator. It is immune to electromagnetic interference and, thus, it is also immune to crosstalk from adjacent fibres. It is virtually immune to tapping. Several different types of optical fibre, in various forms, are now available commercially, most having attenuations well below 10 dB/km. To accompany them, connector systems and splicing aids have been developed.

Unlike electrical transmission lines, optical fibres require light sources and photodetectors to interface them with electronic equipment. These are, however, little if any more complicated than the drivers and multiplexers used with wideband cables. A variety of semiconductor generators and detectors have been developed that are very suitable for short- and medium-range applications. Most are in packages designed for easy connection to optical fibres.

It is now possible to assemble from commercially available components optical communications systems that are ideal for use in hostile or hazardous environments, or where large potential differences exist, or where earth loops must be avoided. Applications range from factory instrumentation, through multi-terminal data systems to vehicle control. It is with such practical, local fibre-optic systems that this chapter is concerned.

Optical fibres rely for their operation on the phenomenon of total internal reflection. If a ray of light propagating within a medium of refractive index n_1 approaches the boundary with a second medium of refractive index n_2 at an angle less than a critical angle,

$$\theta_c = \arccos(n_2/n_1) \qquad \text{(Snell's law)} \qquad (56.1)$$

it will not pass through the boundary but will be reflected back into medium n_1, provided that $n_1 > n_2$, as shown in *Figure 56.1*.

Thus, light propagating within a rod of transparent material will be reflected from the walls of the rod if it approaches them at an angle less than the critical angle for the material and the surrounding medium (usually air). Unfortunately, such a simple arrangement is unsuitable for most fibre-optic applications since if two rods touch, as would happen within a cable, light passes from one to the other, *Figure 56.2*. This would lead to excessive losses, crosstalk and lack of security.

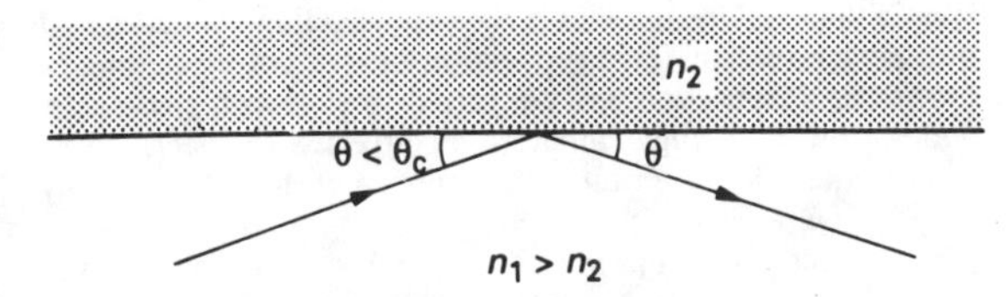

Figure 56.1 Reflection at the boundary of two media

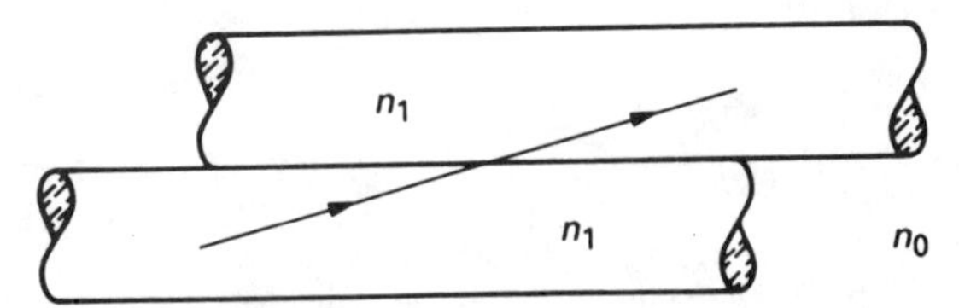

Figure 56.2 Light passing through two single fibres which touch

56.2 Practical optical fibres

In practice, optical fibres for communications purposes are made with a core of higher refractive index than that of the material near the walls, so that they are self-contained. The angle at which light can enter the fibre and be trapped is reduced, but losses due to light escaping and crosstalk are eliminated, and the fibre is almost proof against tapping.

Two basic types of optical fibres are commercially available. The simplest is the step-index fibre which consists of a central homogeneous core surrounded by a cladding layer of lower refractive-index material. Light will be trapped within the core if it approaches the boundary with the cladding at an angle less than

$$\theta_c = \arccos(n_{cl}/n_{co}) \qquad (56.2)$$

where n_{cl} is the refractive index of the cladding and n_{co} the refractive index of the core, *Figure 56.3*. Two fibres of this construction can touch without light escaping from one to the other.

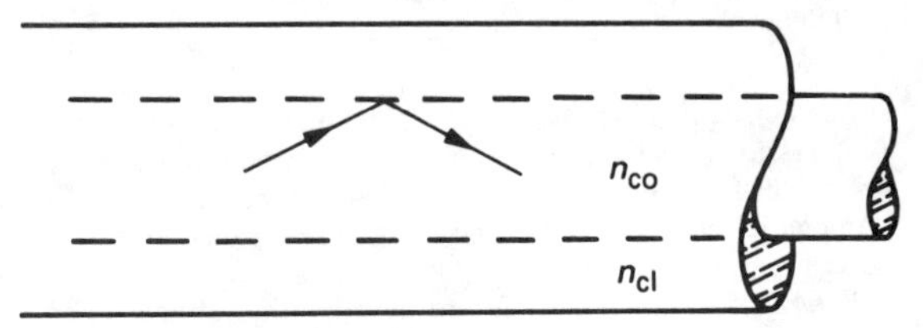

Figure 56.3 Step-index optical fibre

In the second type of fibre, the refractive index of the core decreases with increasing distance from the axis of the core. Again, light will be trapped if the conditions of Equation (56.2) are satisfied. However, due to the refractive-index profile of the core, light crossing the core axis at less than the critical angle is progressively refracted back towards the centre, *Figure 56.4*. This

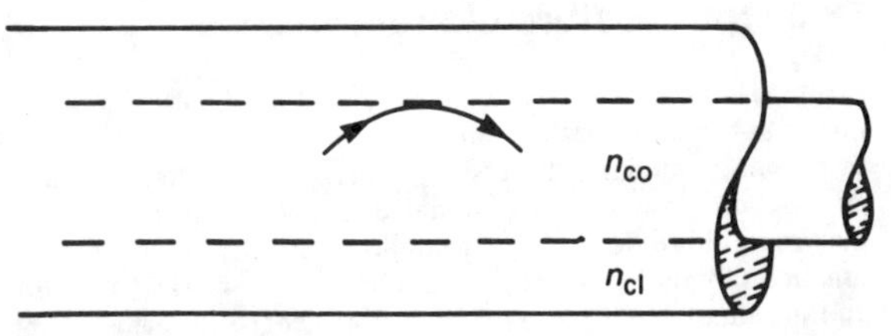

Figure 56.4 Graded-index optical fibre

property of graded-index fibres is called self-focusing. The difference between the refractive index at the axis and at the boundary of the core is about 1%.

Depending on the radius a of the fibre core and refractive indices of core and cladding, light waves propagate through a fibre in one or more modes, much as microwaves in a waveguide.[3] The number of modes N that can be sustained by a fibre at a given free-space wavelength λ_0 is a function of the normalised fibre frequency V:

$$N \simeq \tfrac{1}{2}V^2 \tag{56.3}$$

$$V = \frac{2\pi a}{\lambda_0}(n_{co}^2 - n_{cl}^2)^{1/2} = \frac{2\pi a n_{co}}{\lambda_0}(2\Delta)^{1/2} \tag{56.4}$$

where $\Delta = (n_{co} - n_{cl})/n_{co}$ is the relative refractive index difference.

Figure 56.5 shows the modes present at different V numbers,[4] and the V number as a function of core radius for $\lambda_0 = 900$ nm and n_{co} greater than n_{cl} by 1%. As the radius a and normalised

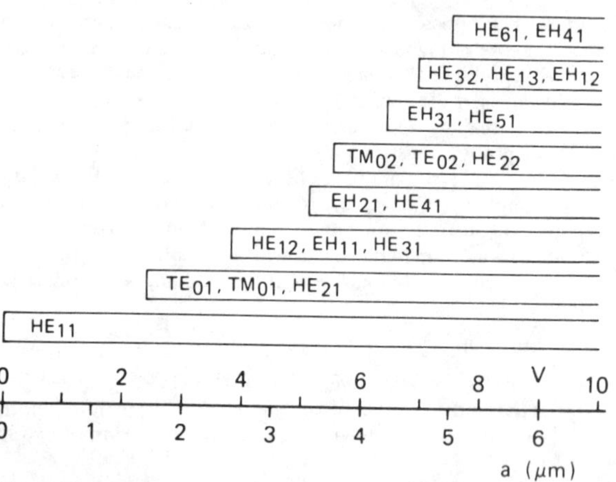

Figure 56.5 Modes present at different values of V and a. $\lambda = 900$ nm, $\Delta = 0.01$, $n_{co} = 1.50$

frequency V of a fibre decrease, so does the number of modes of propagation possible within it. Note that hybrid mode HE_{11} has no cut-off frequency. Thus, for $V < 2.405$ there exists only one propagation mode; the corresponding fibre is called monomode fibre.

Figure 56.6 shows the relation between the refractive index difference and core radius a of this kind of waveguide. For a 1% index difference $a < 1.8$ μm. To avoid such very small core size and thus extreme connector accuracy the index difference is usually reduced to about 0.1%, bringing the maximum core radius to about 5 μm.

On the other hand, when $V > 2.405$, many modes propagate and the fibre is termed multimode. *Figure 56.7* shows cross-sections and index profiles of usual fibre types.

56.3 Optical fibre characteristics

From the point of view of the system designer, the most important characteristics of an optical fibre are material dispersion, modal dispersion, numerical aperture and attenuation. Material and modal dispersion together set the maximum fibre length or maximum repeater spacing as a function of signal frequency or data bit rate. Numerical aperture and attenuation set the maximum fibre length in terms of the available light power.

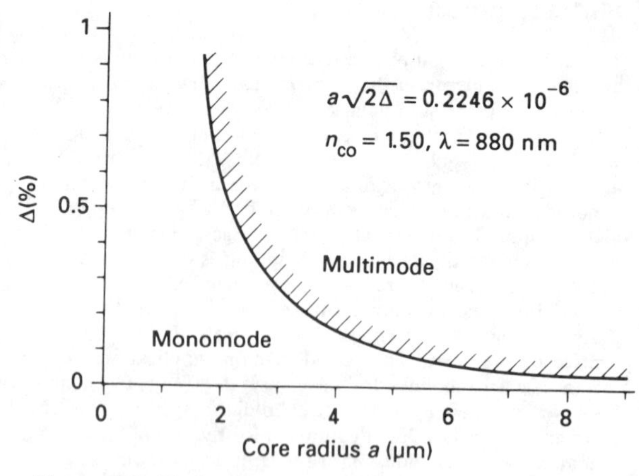

Figure 56.6 The boundary between monomode and multimode fibre operation as a function of refractive index difference

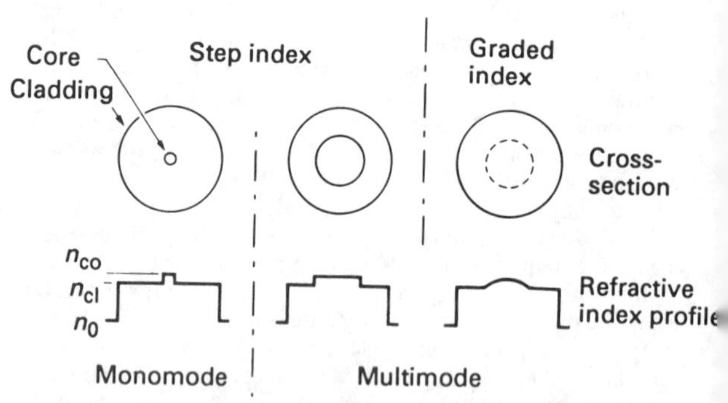

Figure 56.7 Refractive index profiles of the three basic types of optical fibre

56.3.1 Modal dispersion

In step-index fibres, different modes correspond to different angles of reflection from the core/cladding boundary, and, thus, different speeds of propagation through the fibre because path length depends on reflection angle. Since refractive index is the ratio of the velocity of light in a medium to its velocity in free space, c, it follows that phase velocities due to the different propagation modes lie between two limits:

$$\frac{c}{n_{co}} \leqslant v_{ph} \leqslant \frac{c}{n_{cl}} \tag{56.5}$$

where v_{ph} is the phase velocity. The spread of phase velocities results in pulse spreading which, in turn, limits the upper frequency of modulation for a given fibre length.

In practice, due to deviations from the ideal, the degree of pulse spreading in step-index fibres is always less than that predicted by Equation (56.5). Irregularities in the core/cladding boundary lead to mode conversion; fast propagation modes, corresponding to low angles of reflection, are converted to slower, higher-angle, modes; slow, high-angle, modes are converted to faster, lower-angle modes. The net result is a narrowing of the phase velocity spread (*Figure 56.8*).

Perfect step transitions in refractive index cannot be achieved in practice either, so that a continuous transition from n_{co} to n_{cl} takes place about the core/cladding boundary. Within this

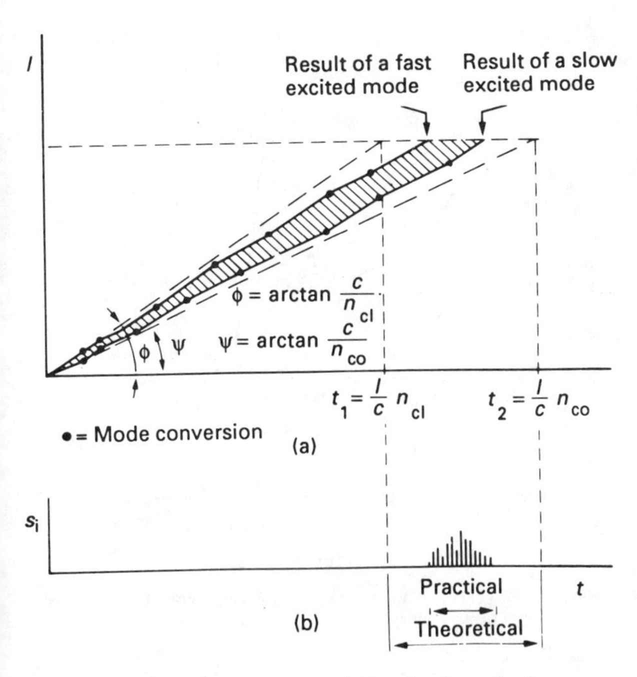

Figure 56.8 (a) Phase velocity and (b) pulse dispersion in step-index fibres. $\phi = \arctan(c/n_{cl})$, $\psi = \arctan(c/n_{co})$

transition zone, some self-focusing similar to that in a graded-index fibre takes place.

Modal dispersion in graded-index fibres is less than that in step-index fibres due to path-length differences being reduced by the self-focusing action. Calculation shows that the phase-velocity spread is least when the core refractive-index profile approximates a parabola.[5]

For step-index fibres, the maximum pulse spread per unit length due to modal dispersion is given by

$$\tau_m = n_{co}\Delta/c \qquad (56.6)$$

In the case of graded-index fibres, the maximum modal dispersion amounts to

$$\tau_m \simeq n_{co}\Delta^2/c \qquad (56.7)$$

Modal dispersion can result in pulse spreads up to 50 ns/km for step-index fibres, but only about 0.5 ns/km for graded-index fibres, both for multimode waveguides.

56.3.2 Material dispersion

Material dispersion is the variation of group velocity with wavelength in a transparent medium whose refractive index depends on wavelength. Pulse dispersion τ_c per unit length as a result of this phenomenon equals:

$$\tau_c = -\frac{\lambda}{c}\frac{d^2n}{d\lambda^2}\Delta\lambda \qquad (56.8)$$

where $\Delta\lambda$ indicates the spectral width of the applied light source. In silica fibres, the behaviour of the refractive index n is such that the material dispersion shows a real zero at a wavelength determined by the chemical composition.

Payne[6] describes a silica fibre with zero material disperson at 1.27 μm. Chang[7] presents a fibre with a different composition showing a zero at 1.55 μm.

Figure 56.9 depicts the material dispersion as a function of

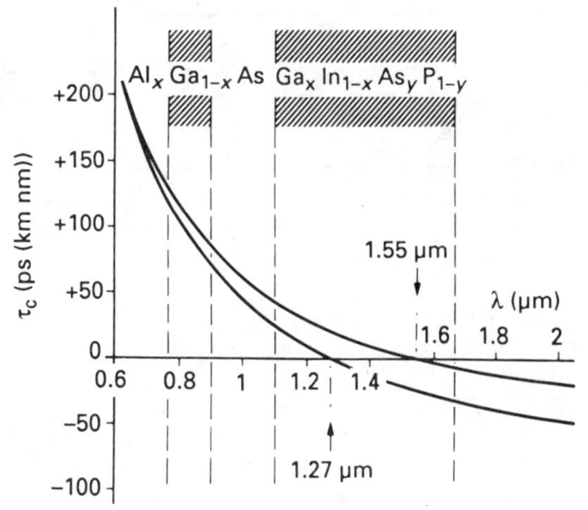

Figure 56.9 Material dispersion τ_c as a function of wavelength for two types of optical fibre

wavelength for these reported fibres. Both curves pass through points of zero material dispersion, offering the prospect of greatly improved optical fibre performance at the longer wavelengths at which the zeros occur. The shaded bands indicate the wavelength regions of two semiconductor light source materials.

56.3.3 Waveguide dispersion

Waveguide dispersion arises principally because of the wavelength dependence of the modal V number. This waveguide dispersion is negligibly small for all modes not close to cut-off. The dispersion contribution from this source is generally not significant and can be disregarded. It is evident that monomode fibres, operated at the zero material dispersion wavelength, behave as an extremely wideband transmission line. At other wavelengths, spectral width of the light source sets the bandwidth of a single-mode fibre. A survey of typical modal and material dispersion data of silica fibres is given in *Table 56.1*.

Table 56.1 Typical dispersion values for silica fibres ($\Delta = 0.01$, $\lambda = 880$ nm)

	Modal dispersion (ns/km)	*Material dispersion* (ps/(km nm))
Monomode	0	60
Multimode		
Graded index	0.5	115
Step index	50	60

56.3.4 Numerical aperture

Since there is a maximum angle at which light can strike the core/cladding boundary and be reflected back, there must also be a maximum angle, called the acceptance angle, at which light can enter the core of an optical fibre and be trapped. This is illustrated by *Figure 56.10*, where it can be seen that, with θ_c as defined in Equation (56.1), and from Snell's law,

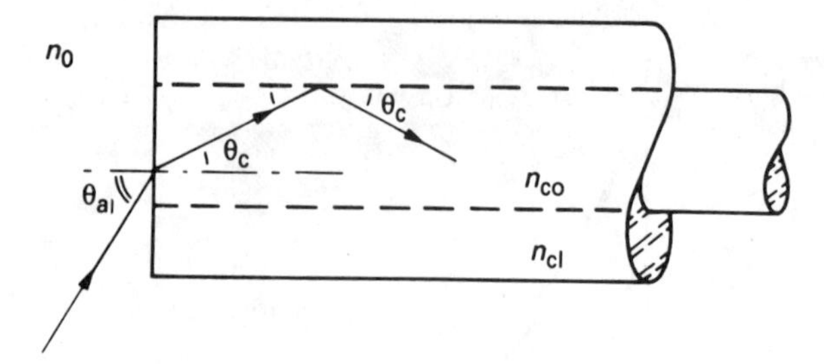

Figure 56.10 Illustration of acceptance angle, $\theta_c = \arccos(n_{cl}/n_{co})$ and $\theta_a = \arcsin(n_{co} - n_{cl})^{1/2}$

$$\frac{\sin \theta_a}{\sin \theta_c} = \frac{n_{co}}{n_0} \quad (56.9)$$

For a fibre-to-air interface, $n_0 = 1$, so

$$\sin \theta_a = n_{co} \sin \theta_c = (n_{co}^2 - n_{cl}^2)^{1/2} \simeq n_{co}(2\Delta)^{1/2} \quad (56.10)$$

Angle θ_a is termed the acceptance angle of the fibre; the quantity $\sin \theta_a$ is the numerical aperture NA.

The numerical aperture determines the power than can be coupled from a light source into an optical fibre. Where the diameter of the emitting area of the source D_e is less than or equal to the diameter of the fibre core D_c, and source and fibre are in contact, the theoretical optical power P_o coupled into a step-index fibre is, assuming a Lambertian source,

$$P_o = \tfrac{1}{4}\pi^2 D_e^2 R_0 (NA)^2 \quad (56.11)$$

When D_e is greater than D_c, so that the emitting area overlaps the fibre core,

$$P_o = \tfrac{1}{4}\pi^2 D_c^2 R_0 (NA)^2 \quad (56.12)$$

where R_0 is the on-axis radiance of the source in both cases.

Equations (56.11) and (56.12) apply only when the light rays lie in a plane containing the axis of the fibre (meridional rays). However, for an extended source, some rays not in a plane with the fibre axis will also be accepted by the fibre (helical rays). This about doubles the power actually coupled[8] for a fibre of $NA = 0.2$.

In practice, source and fibre will not be in perfect contact and there will be a power loss in consequence. In most cases, this loss will be at least 2 dB.

56.3.5 Attenuation

Attenuation, or rather its reduction, is the key to the current success of fibre-optic techniques. Materials research has reduced the attenuation of optical fibres from well over 1000 dB/km to less than 10 dB/km. Specimens have been produced with attenuations approaching 0.2 dB/km. There are two main causes of loss in optical fibres, absorption and scattering.

Absorption is still the principal cause of loss. It is caused by metallic ion impurities in the core material. Many of these ions have electron transition energies corresponding to light wavelengths in the region 0.5 to 2 μm.

Light passing through a fibre will be scattered by inclusions and dislocations in the material that are small compared to the wavelength of the light. Scattering can also be caused by local temperature gradients. Rayleigh scattering due to the molecules of the material themselves is independent of light intensity but varies with $1/\lambda^4$.

Very high grade glasses have Rayleigh scattering losses of about 0.9 dB/km at a wavelength of 1 μm. This must be regarded as close to the practical limit at that wavelength. *Figure 56.11* shows the attenuation of a low-loss monomode fibre as a function of wavelength.[9] The peaks are due to absorption by metallic or hydroxyl ions.

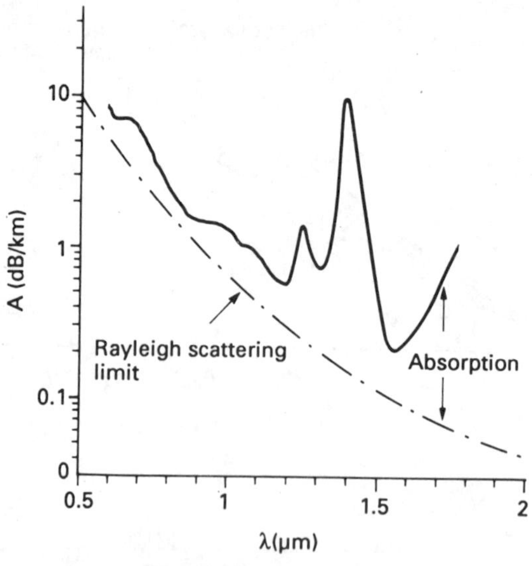

Figure 56.11 The attenuation A of a low-loss monomode fibre as a functional of wavelength

56.3.6 Optical fibre specifications

We are now able to list those optical-fibre properties that must be known if basic quantitative design is to be undertaken. They are:

(1) Dimensions.
(2) Mechanical properties.
(3) Optical attenuation.
(4) Dispersion.
(5) Numerical aperture.
(6) Minimum modulation bandwidth.
(7) Wavelength at which optical characteristics are measured.

By way of example, *Table 56.2* lists the principal characteristics of graded-index optical fibre, type 21183 manufactured by Philips' Glass Division by a chemical vapour-deposition process.[10]

Table 56.2 Characteristics of Philips' graded-index optical fibre type 21183

Bandwidth	⩾800 MHz km
Attenuation (850 nm)	⩽3 dB/km
Pulse dispersion (FWHM)	⩽0.6 ns/km
Numerical aperture	0.21 ± 0.01
Core diameter	50 ± 2 μm
Cladding diameter	125 ± 2 μm

56.4 Light sources

Fibre-optic communications systems require sources of light of adequate brightness capable of being modulated at the desired frequency. The choice, in practice, lies between the semiconductor laser and the light-emitting diode. These devices are attractive in that their drive requirements are compatible with semiconductor practice, they are compact and robust, and, of great importance, they have the long-life and high-reliability potential that characterises well made semiconductor devices. In the case of the LED this potential is well on the way to being realised and useful lives of the order of 100 000 h can now be achieved. *Chapter 23* describes the characteristics of light sources.

56.5 Photodetectors

For the purposes of fibre-optic system design, the principal characteristics of a photodetector are:

(1) Quantum efficiency, the proportion of incident photons converted into current.
(2) Response time, which sets bandwidth.
(3) Internal mutiplication factor, the ratio between primary photocurrent and output current.
(4) Noise.
(5) Practical considerations, such as operating voltage and encapsulation.

For the reasons given in connection with light sources, semiconductor devices are also the preferred choice for detectors in fibre-optic systems.

Table 56.3 lists response time and internal current multiplication factor M for the main types of semiconductor photodetectors. It is apparent that the best available performance for wideband links is currently that offered by silicon PIN and avalanche photodiodes.

Table 56.3 Internal multiplication factor M and response time for various types of semiconductor photodetector

Detector type	*M*	*Response time* (s)
Photoconductor	10^5	10^{-3}
p–n diode	1	10^{-6}
PIN diode	1	10^{-9}
Phototransistor	10^2	10^{-5}
Avalanche photodiode	10^3	10^{-9}
Field-effect transistor	10^2	10^{-7}

Figure 56.12 shows the responsivity and quantum efficiency of various types of semiconductor photodetectors. Silicon is evidently the most suitable material for use with GaAs–AlGaAs light sources. For longer-wavelength sources, InP–InGaAsP seems the most promising material.

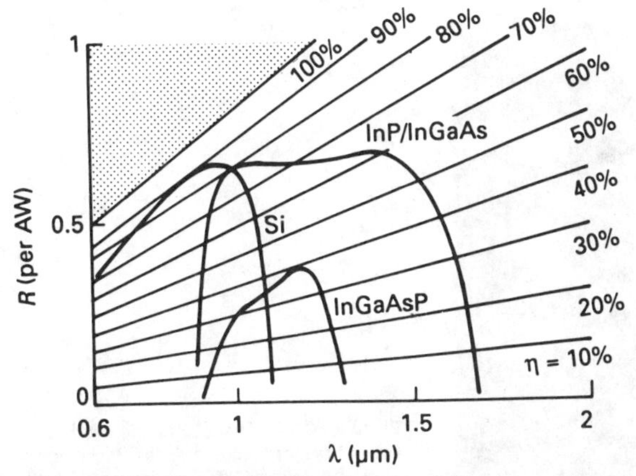

Figure 56.12 Responsivity R and quantum efficiency η of some semiconductor photodetectors in various semiconductor materials

56.5.1 PIN photodiodes

Silicon PIN diodes, *Figure 56.13*, are sufficiently sensitive for most short-range applications. They have the additional advantages of being both comparatively cheap and rugged devices. Moreover, they operate with low bias (5 to 10 V).

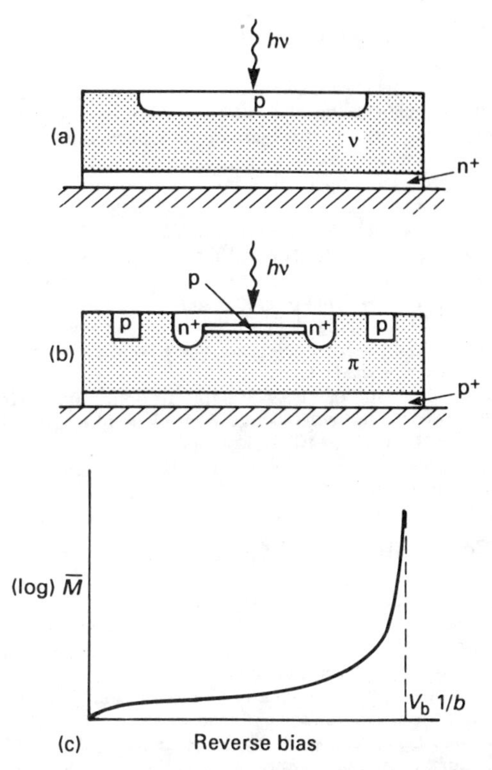

Figure 56.13 Cross-sections of (a) PIN photodiode and (b) avalanche diode photodetectors with (c) typical avalanche multiplication factor M as a function of reverse bias characteristic

For fibre-optic applications, photodetector structures are generally optimised for the LED or laser wavelength to be used. In this way, quantum efficiencies

$$\eta = I_p h\nu / eP_p \qquad (56.13)$$

where I_p is the photocurrent and P_p the optical power, approaching 90% can be achieved in the region 0.75 to 0.9 μm wavelength, with modulation bandwidths of several hundred megahertz.

56.5.2 Avalanche photodiodes

Avalanche photodiodes (APD) combine detection of light with internal amplification by avalanche multiplication. They are operated with a reverse bias close to the breakdown voltage (about 200 V) so that photon-liberated electrons passing through the high field gradient about the junction gain sufficient energy to create new electron–hole pairs by impact ionisation. The amplification of an avalanche diode is strongly dependent on the bias.[11]

Since the process by which gain is achieved is stochastic, excess noise is generated by random gain fluctuations. The excess-noise factor F_e of an avalanche diode is a function of average gain $\bar{M}$,

Figure 56.13. For this reason, there is a value of $\bar{M}$ and, hence, of operating bias, for which detector signal-to-noise ratio is optimum for a given bandwidth, modulation method and optical power combination.

Due to their very high sensitivity, avalanche photodiodes are the preferred detectors in long-distance fibre-optic communication systems using semiconductor light sources.

56.6 Device packages

Optimum coupling between an optical fibre and a laser, LED, or photodetector requires that the end of the fibre be positioned accurately close to the active region of the semiconductor die. This is a job for which the device manufacturer is best equipped, especially since positioning the fibre involves exposing the die to possible contamination.

Light sources and detectors are now available in packages that already incorporate a length of optical fibre one end of which is in close proximity to the active region of the die. Two of these are shown in *Figure 56.14*. That of *Figure 56.14(a)* has a pigtail of fibre about 350 mm long designed to be terminated at an equipment panel or internal bulkhead, where connection to the main fibre system is made.

In the encapsulation of *Figure 56.14(b)* an optical-fibre bar extends to the end of the metal ferrule; this package forms part of a standard BNC, TNC or RIM-SMA optical-fibre connector. Connection may be made to devices in either encapsulation with a loss of about 1.5 dB.

56.7 Coupling optical fibres

In order to assemble a practical fibre-optical communications system, it must be possible to make connections to and between optical fibres. For maintenance purposes, some of the connections will have to be demountable.

In essence, the optical coupling of two fibres is achieved by ensuring that their ends are clean and of adequate optical quality, and then holding them concentrically close together. *Figure 56.15* shows a number of the fibre-to-fibre joint defects that can contribute to connection loss.

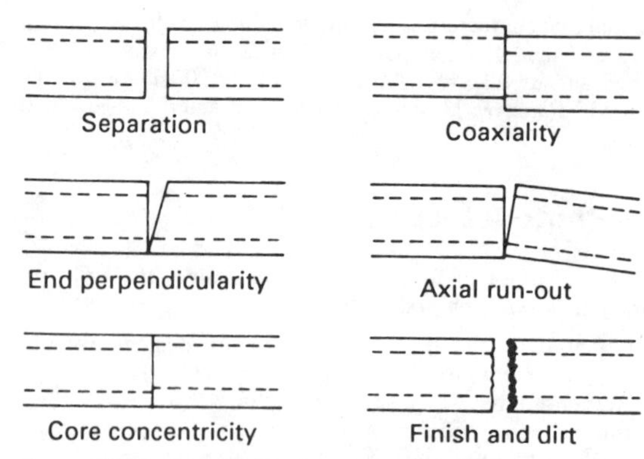

Figure 56.15 Factors affecting losses in fibre-to fibre joints

56.7.1 Fibre-to-fibre connectors

Demountable connectors are available for both single- and multiple-fibre cables. Their essential features are two aligning and holding terminations for the fibres, and a connecting bush. Most commercially available fibre connectors are based on the mechanical arrangements of coaxial types, such as those of *Figure 56.16*. Connectors for thick step-index fibres made from plastic materials are also available (see also *Chapter 18*).

A low-loss connection between graded-index or monomode fibres requires an extremely precise mechanical arrangement. Simple ferrule constructions in standard SMA or DIN housings are generally insufficient.

Figure 56.16 shows two types of fibre-to-fibre connectors suitable for step-index fibres, one for a fibre cable and the other for a single- or double-coated fibre for use inside equipment. Both connectors achieve alignment accuracy by means of a watch jewel at the ends of the male parts.

Welding (fusing) together fibres is an attractive, low-cost solution in cases where demountability is not required, as in buried trunk cables. Losses as low as 0.1 dB can be achieved by this method.[12]

56.7.2 Coupling fibres to active elements

The efficient and reliable coupling of light sources and detectors to system fibres is a crucial part of any fibre-optic link. Discrete optical devices for high-performance telecommunications applications are most commonly packaged in a dual-in-line package with a fibre pigtail 1–3 m long. The fibre pigtail allows the device to be mounted anywhere on a printed circuit board with the connectors mounted suitably for attachment of external cables. Inside the package the fibre is aligned with respect to the laser, LED or detector using soft solders or laser welding techniques. The laser welding techniques have proved much more reliable as creep problems are eliminated. This type of package is available

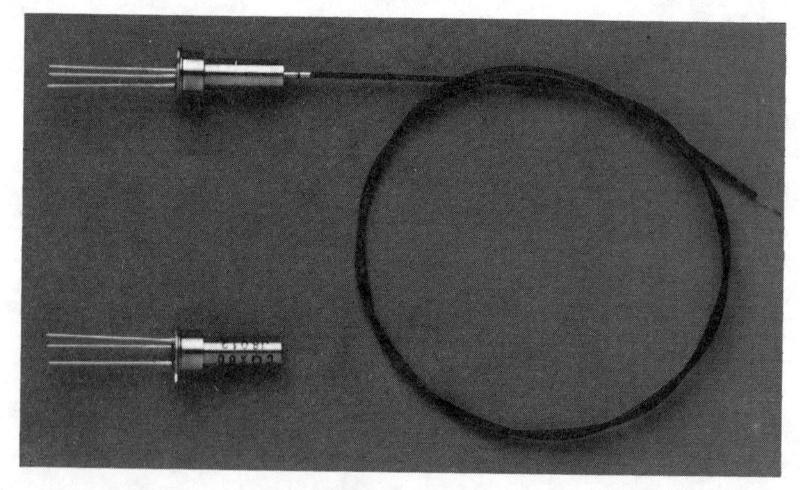

Figure 56.14 Encapsulations used for active fibre-optic devices. (a) Package with fibre 'pigtail', (b) package with short thick fibre which is designed to form part of an active connector

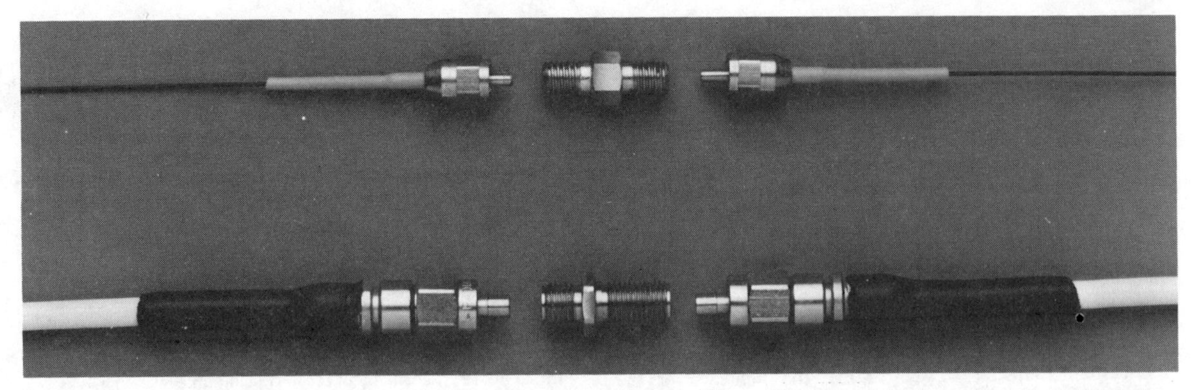

Figure 56.16 Demountable connectors for optical fibres: (a) connector for coated fibres, (b) connector for fibre cable with strain-relieving sleeve

either with or without an internal Peltier cooler to keep the active chip within a specified temperature range.

For slightly less demanding telecommunications links, where thermoelectric coolers are unnecessary, coaxially mounted LEDs and detectors can be used either with a fibre pigtail as shown in *Figure 56.14* or mounted directly into an optical connector. For larger volume short distance applications it is often most economic to incorporate the drive electronics, optical components and connector into one self-contained module to make very compact transmitters and receivers. These links typically require ECL data input levels at up to 200 Mbit/s and give ECL data output with $<10^{-9}$ error rate for multimode link distances up to 3 km. An example of such a link is shown in *Figure 56.17*. Such links are typically used for computer to computer and computer to high-speed peripheral links such as the ICL MACROLAN network and the emerging FDDI (fibre distributed data interface) world standard. Modules are also being developed for the large-volume parts of the telecommunications market, particularly matching the emerging SONNET world standard for 1.3 μm single-mode fibre links at data rates in the less than 200 Mbit/s and less than 600 Mbit/s brackets.

56.8 System design considerations

The essential properties of a fibre-optic communication system are bandwidth and attenuation.

56.8.1 Bandwidth

Assuming Gaussian pulses, the bandwidth can be estimated simply by summing the various system time constants:

$$\tau_{tot}^2 = \tau_s^2 + \tau_f^2 + \tau_d^2 \tag{56.14}$$

Here, τ_{tot} is the overall system time constant, τ_s is the time constant of the light source and driver, τ_d is the time constant of the system fibre and τ_f is the time constant of the detector and receiver input circuit.

The light source time constant is the rise or fall time, whichever is the longer. The optical-fibre time constant is the sum of the time constants due to material and modal dispersion:

$$\tau_f^2 = l^2(\tau_c^2 + \tau_m^2) \tag{56.15}$$

where l is the fibre length, τ_c is the phase delay per unit fibre length due to material dispersion and τ_m is the phase delay per unit length due to modal dispersion.

Once the time constant has been calculated, the overall modulation bandwidth of the system is

$$B = 1/(2\tau_{tot}) = 1/2T \tag{56.16}$$

where T is the minimum pulse length available for digital transmission.

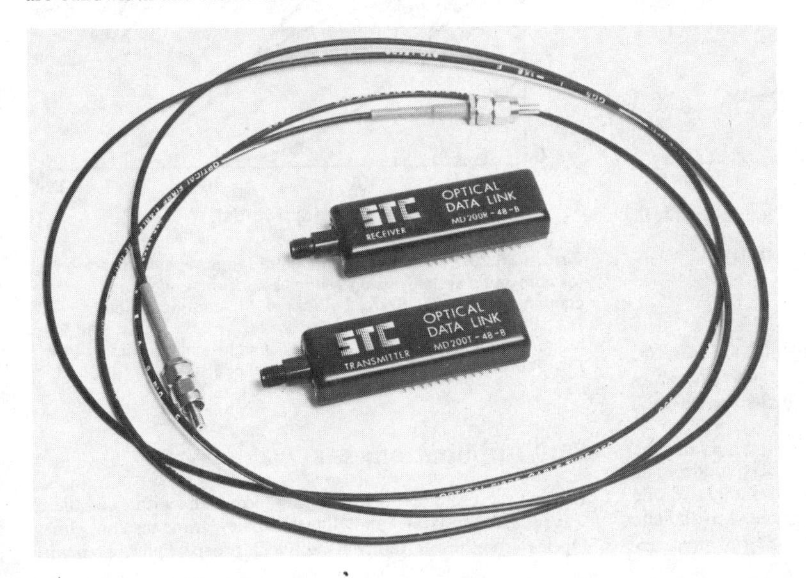

Figure 56.17 Transmitter and receiver modules for an optical data link

56.8.2 Signal attenuation

The maximum permissible signal attenuation in a fibre-optic system is

$$A_{max} = P_{s\,max}/P_{d\,min} \quad (56.17)$$

where P_s is the optical power available from the light source and P_d is the power coupled to the photodetector. A more powerful light source is equivalent to a more sensitive detector.

About 1 to 3 mW of optical power can be coupled into a graded or step-index fibre from a semiconductor laser, depending on coupling efficiency, fibre numerical aperture and laser performance.

The power coupled into a fibre from an LED can be calculated from Equation (56.11) for a step-index fibre. Due to the Lambertian character of the source, the influence of the numerical aperture is much greater than with a laser. Coupled power varies, in practice, from 10 to 200 μW.

The main limitations on the sensitivity of a PIN diode equipped receiver are set by shot and thermal noise. Optimum performance is obtained using a high-resistance transimpedance amplifier, a silicon FET input stage, and a differentiator to compensate for the low-pass characteristic of the input stage.[13] At modulation frequencies above a few tens of megahertz, better results will be obtained with a good bipolar or GaAs FET input stage.

Figure 56.18 gives the sensitivity of a silicon FET equipped PIN diode receiver as a function of a diode load resistor R_L for a bit

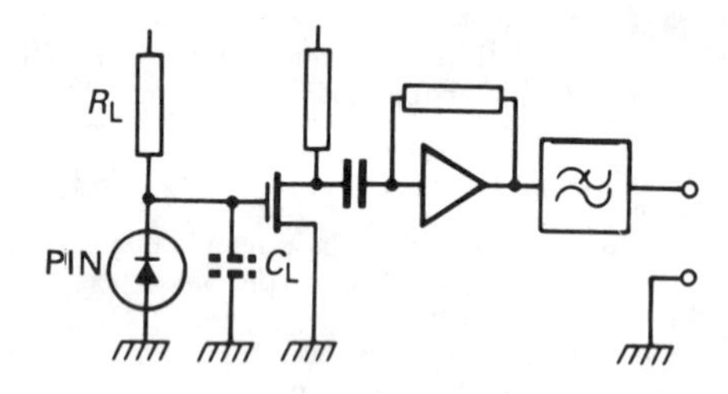

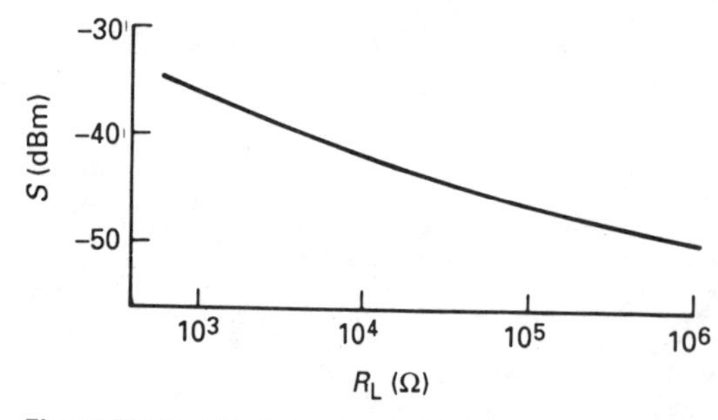

Figure 56.18 PIN diode receiver: (a) circuit diagram, (b) variation of sensitivity S with load resistor R_L

error rate of 10^{-10} at a transmission speed of 20 Mbit/s. An input capacitance of 2 pF and a FET noise equivalent series resistance of 300 Ω are assumed. R_{eq} is the equivalent series noise resistance of the FET.

Due to the internal multiplication factor M, an avalanche photodiode is much more sensitive than a PIN diode. The sensitivity of a receiver equipped with an avalanche diode detector is plotted in *Figure 56.19* as a function of avalanche multiplication factor M. A load resistor of 100 kΩ and an excess

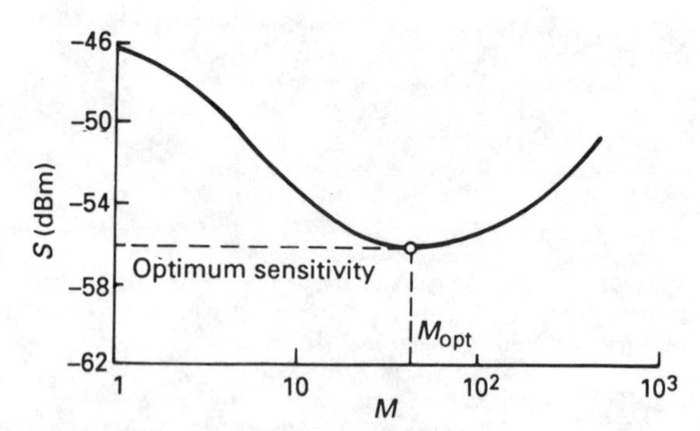

Figure 56.19 Sensitivity of receiver using an avalanche photodiode

noise factor of $2+0.02M$ are assumed; other details are as for *Figure 56.18*.

In order to achieve a bit error rate of 10^{-10} in digital transmission, a signal-to-noise ratio of 22 dB is required. For analogue TV transmission, acceptable picture quality requires a signal-to-noise ratio of at least 45 dB and, consequently, more optical power.

56.8.3 Range

The maximum range of a fibre-optic system is set by either bandwidth or attenuation. The ranges achievable with various source, fibre and detector combinations are plotted as a function of bit rate in *Figure 56.20*.

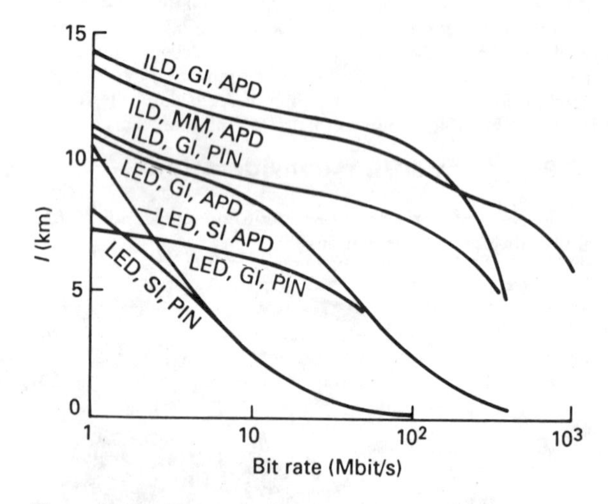

Figure 56.20 Bit rate as a function of range/repeater spacing l for fibre-optic systems using various source–fibre–detector combinations; A=5 dB/km, λ=880 nm. SI, step-index fibre (τ_m=50 ns/km); GI, graded-index fibre (τ_m=0.5 ns/km); MM, monomode fibre; ILD, laser diode; LED, light-emitting diode; PIN, PIN photodiode; APD, avalanche photodiode

56.9 Applications

Optical communications systems compete with millimetre waveguide, coaxial and twisted-pair transmission lines. Optical fibres are certainly attractive in prospect compared with moderate and high-quality coaxial cables. Whether they will replace twisted-pair lines depends on the price of copper.

The extra cost of the optoelectronic interface devices often outweighs the cost advantage of the glass-fibre cable, particularly in short-haul systems. However, an optical system should not necessarily be regarded as a one-to-one replacement for a copper line system. The use of multiplexing techniques to take advantage of the large information-carrying capacity of glass fibre cables can greatly reduce the cost per channel.

In hazardous or noisy environments, glass-fibre cables often do not require the coding and decoding equipment, interference suppressors, heavy shielding or isolation transformers necessary with copper lines. Fibre-optic systems are inherently interference insensitive. Due to their low weight and small size, a given cable duct can accommodate more optical cables than metal ones.

System feasibility can best be judged economically. Compare the value in use of the existing system with the actual cost of an optical replacement, taking all costs, direct and indirect, into account.

Telephone trunk lines and television distribution require levels of performance best achieved with lasers, graded-index fibres and avalanche diode detectors. Particularly on the basis of cost per channel, fibre-optic cables will be cheaper than coaxial cables. In addition, due to the very low attenuation of optical fibres, few or no repeaters will be required.

A 140 Mbit/s optical link has been demonstrated that has a range of 102 km between repeaters, compared to the 1.5 km of an equivalent coaxial system.[14]

The immunity to interference, low weight, compactness and difficulty of tapping make fibre-optics especially attractive for both military and general high-security applications.

Fibre-optic systems are ideal for linking computers to peripherals. Here, speed is the most important property.

Systems using LED, step-index fibres and PIN diode detectors are ideal for monitoring and control purposes in industrial plants, where distances are short and information is transferred at relatively low speeds.

The small size and immunity to crosstalk of optical fibres make them attractive in prospect for internal wiring in such applications as automobiles, data-processing equipment and telephone exchanges.

Using optical fibres, a wideband network to deal with the increasing flow of information to and from the home becomes a practical possibility. One or two fibres could be used to carry TV and radio programmes, telephony, viewdata, telebanking, public utilities metering and similar services. Several projects with this in view are already complete: Hi-Ovis in Japan, BIGFON in the Federal Republic of Germany, DIVAC in The Netherlands, and Biarritz in France. There is now great competition to develop economically viable systems and to introduce them on a large scale. It is widely predicted that by 1991–93 fibre systems will be cost competitive with copper, even for domestic telephone traffic alone.[15,16] Plans are well developed for large-scale implementation in the mid-1990s.

Acknowledgements

The contents have been based on a chapter by H. J. M. Otten in the 5th edition of the *Electronics Engineer's Reference Book*.

References

1 BELL, A. G., 'On the production and reproduction of sound by light', *Proc. Am. Assn Adv. Sci.*, **29**, 115–36 (1881)
2 CADE, C. M., 'Eighty years of photophones', *Brit. Comm. Electron.*, **9**, No. 2, 112–15 (1962)
3 SNITZER, E., 'Cylindrical dielectric waveguide modes', *J. Opt. Soc. Am.*, **51**, No. 5, 491–505 (1961)
4 MILLER, S. E., *et al.*, 'Research towards optical fibre transmission systems', *Proc. IEEE*, **61**, No. 2, 1703–51 (1973)
5 KAWAKAMI, S. and NISHIZAWA, J. I., 'An optical waveguide with optimum distribution of the refractive index with reference to waveguide distortion', *IEEE Trans. MTT*, **16**, No. 10, 814–18 (1968)
6 PAYNE, D. N., *et al.*, 'Zero material dispersion in optical fibres', *Electron. Lett.*, **11**, No. 8, 176–8 (1975)
7 CHANG, C. T., 'Minimum dispersion at 1.55 μm for single mode step index fibres', *Electron. Lett.*, **15**, No. 23, 765–7 (1979)
8 ALLAN, W. B., *Fibre Optics, Theory and Practice*, Plenum Press, New York, 29–33 (1973)
9 MIYA, T., *et al.*, 'Ultimate low loss single mode fibre at 1.55 μm', *Electron. Lett.*, **15**, No. 4, 106–8 (1979)
10 VAN ASS, H. M. J. M., *et al.*, 'The manufacture of glass fibres for optical communication', *Philips Tech. Rev.*, **36**, No. 7, 182–9 (1976)
11 WEBB, P. P., *et al.*, 'Properties of avalanche photodiodes', *RCA Rev.*, **35**, No. 6, 234–78 (1974)
12 FRANKEN, A. J. J., *et al.*, 'Experimental semi-automatic machine for hot splicing glass-fibres for optical communication', *Philips Tech. Rev.*, **38**, No. 6, 158–9 (1978/79)
13 PERSONICK, S. D., 'Receiver design for digital fibre optic communications systems', *Bell Systems Tech. J.*, **52**, No. 6, 843–86 (1973)
14 CAMERON, K. K., *et al.*, '102 km optical fibre transmission experiment at 1.52 μm using an external cavity controlled laser transmitter module', *Electron. Lett.*, **18**, No. 15, 650–1 (1982)
15 STERN, J. R. *et al.*, 'TPON—a passive optical network for telephony', *IEE 14th European Conference on Optical Communications*, Part 1, Pubn no. 292
16 OISHANSKI, R. *et al.*, 'Design and performance of wideband subcarrier multiplexed lightwave systems', *IEE 14th European Conference on Optical Communications*, Part 1, Pubn no. 292

Further reading

ARCHER, J. D., 'Fibre optic communications', *Electron. Product Design*, November (1982)
BOCK, W. J. *et al.*, 'High-pressure fibre-optic leadthrough systems', *Eng. Optics*, **1**, No. 4, November (1988)
CHALLANS, J., 'Connector for optical trunk transmission system', *Electron, Eng.*, September (1982)
DETTMER, R., 'Stretching the fibre', *IEE Rev.*, January (1989)
GOODWIN, A. R. *et al.*, 'The design and realization of a high reliability semiconductor laser for single-mode fibre-optical communication links', *IEEE J. Lightwave Tech.*, September (1988)
HICKLENTON, A., 'Developments in fibre optic sensors', *New Electron.*, 1 September (1987)
LOMBAERDE, R., 'Fiber-optic multiplexer clusters signals from 16 RS—232-C channels', *Electronics*, March (1982)
ORMOND, T., 'Fibre-optic transmitters and receivers enhance data-link performance', *EDN*, 31 March (1988)
PAILLARD, E., 'Recent developments in integrated optics on moldable glass', *SPIE Proc.* **734**, *Fibre Optics* **187**
SUEMATSU, Y. and IGA, K., *Introduction to Optical Fibre Communications*, Wiley Interscience (1982)
TEJA, E. R., 'Light work for high-speed links', *Mini-Micro Systems*, September (1987)
WIESNER, W. and LEWIS, C., 'Designing fibre-optic links', *Electron. Power*, November/December (1987)

The Integrated Services Digital Network (ISDN)

J R Cass
Senior Principal Systems Engineer
STC Telecommunications Ltd

S C Redman, AMIEE
Senior Systems Engineer
STC Telecommunications Ltd

F Welsby, BSc, CEng, MIEE
Head of ISDN Network Product Engineering
British Telecom
(Section 57.9)

Contents

57.1 Introduction

This chapter provides an overview to the principles and operation of the integrated services digital networks (ISDNs) that are being introduced around the world, with particular emphasis on access arrangements.

The key features of the technology employed in ISDN, the role of standards bodies, the differences in ISDN introduction around the world and services offered are examined in this chapter. A brief look is also taken at potential future developments. The discussion of ISDN covers the key elements of telecommunications networks listed below, which ISDN integrates to a degree not previously achieved:

(1) The user
(2) User data
(3) Network services
(4) Transmission techniques
(5) Signalling
(6) Switching
(7) Network management
(8) Power sources
(9) Physical media

The basic concepts of digital telecommunications, and the analogue techniques currently in use today, are discussed in *Chapter 55*. A key feature of ISDN is its dependence on the use of an integrated digital network (IDN), and so such networks are briefly summarised in *Section 57.2*.

It is important to note that digital working is being introduced because it is economic in terms of telephony, which still remains (and will for the foreseeable future) the dominant type of telecommunications traffic. However, the modernisation of the world's telephony networks from analogue to digital working offers an excellent opportunity to converge and enhance telecommunication services in general, and it is being exploited to the full by Network operators.

The effect of enhanced digitally based services on the economic and social environment is potentially great and is not yet known. Telecommunication service availability has been found to have a direct relationship to economic development and therefore *services* are a particularly interesting feature of the introduction of ISDN, and are discussed in this section along with the technical implementations.

57.2 Integrated digital networks

An integrated digital network is one in which digital working is employed for analogue telephony signals (see *Figure 57.1*). Digital working in this context means digital transmission of analogue telephony signals, e.g. speech, digital switching and digital common channel signalling between exchanges. Digital circuits are normally provided via 64 kbit/s channels. Note that the connection between the analogue telephone and the local exchange is still provided by analogue means.

Digital transmission of speech has been studied for some time. The basic technique used generally today in IDNs, i.e. pulse code modulation (PCM) of speech at 64 kbit/s, was developed by STC's Reeves in the late 1930s. PCM was initially exploited using valve technology in specialised military applications. However, its application in public networks became commercially viable with the availability of low-cost semiconductor devices in the 1950s.

PCM was initially employed in inter-local exchange or junction network working as an alternative to the frequency division multiplexing (FDM) techniques then used in trunk networks. PCM was employed as a means of increasing junction capacity, because of its scope for cost reduction as semiconductor technology advanced, and also because the channel-associated digital signalling is provided on a per link basis, as against the individual per-channel signalling arrangements used in FDM systems. PCM signalling therefore became cost effective as channel utilisation rose above a certain figure.

57.3 Role of signalling systems in an ISDN

Signalling is an important element in telecommunication network design, being used to establish network connections at the local switching node, and to control the networks that are transporting the internodal information. A key feature of ISDNs is the powerful signalling systems employed, which utilise highly flexible protocols.

ISDN employs common channel signalling (CCS) which relies on a dedicated channel within the carrier bandwidth to carry the signalling data. Unlike channel associated signalling (CAS), common channel signalling does not time-assign the signalling messages to the individual speech channels; rather, each signalling message contains a bit-coded circuit identifier that associates it with its related circuit or entity.

57.4 ISDN—general aspects

57.4.1 Key features

The CCITT (see *Section 57.5*) defines an ISDN as a network, evolving from a telephony IDN, which provides end-to-end digital connectivity to support a wide range of services, to which users have access via a limited set of standard user–network interfaces (see *Figure 57.2*).

The key features of an ISDN are:

(1) Extension of digital transmission to the customer's premises.
(2) Access to all services by a limited set of connection types.

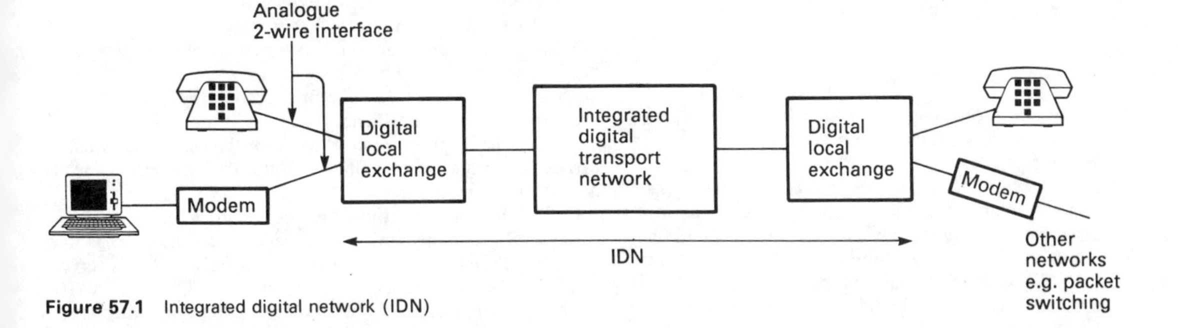

Figure 57.1 Integrated digital network (IDN)

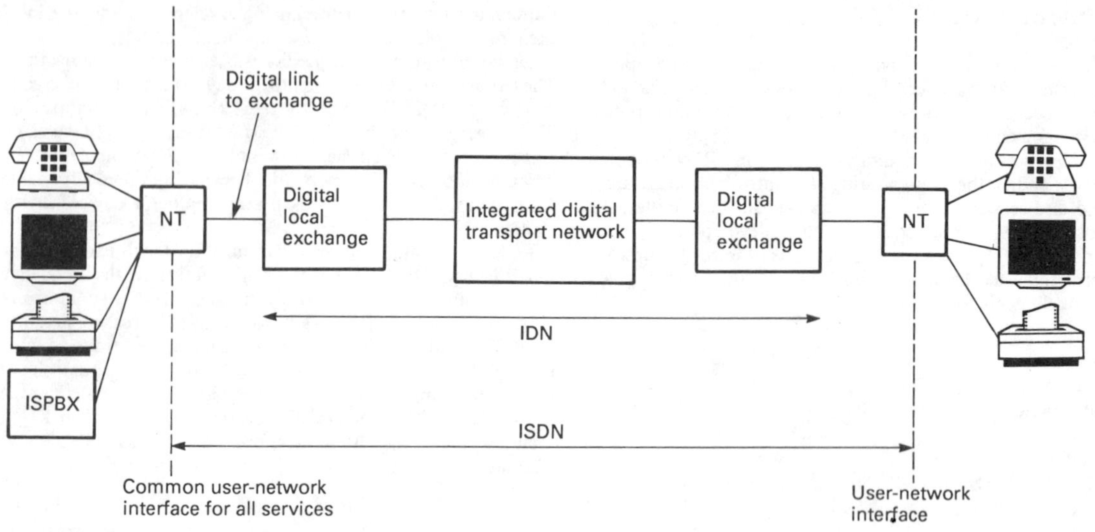

Figure 57.2 Integrated services digital network (ISDN). NT=network termination

(3) Access to all services via a limited set of user–network interface arrangements.
(4) Extension of common channel signalling to the customer's terminal equipment.
(5) The use of a layered protocol structure for the specification of the user's ISDN access.
(6) Circuit switched services (initially) being based on the use of 64 kbit/s channels.
(7) Packet switched connections (initially) being based on a shared use of the D-channel on the ISDN Basic Access.

Note that the term 'B-channel' etc. as used in ISDN means a full duplex communication link.

57.4.2 Types of customer access

Two types of access are currently defined by CCITT recommendations: *basic* and *primary* access.

Basic access provides a user interface at 192 kbit/s and provides two 64-kbit/s channels for user data and one 16-kbit/s channel for signalling and other uses. The remaining bandwidth is used for control and synchronisation of the interface, leaving a net user-signalling data bandwidth of 144 kbit/s. The user interface may be provided in bus or point-to-point arrangements.

Primary access provides a user interface at 2.048 Mbit/s in the UK and 1.544 Mbit/s in the USA, and provides a number of 64-kbit/s channels for user data and one 64-kbit/s signalling channel.

57.5 ISDN standards

57.5.1 International standardisation

A key feature of the ISDN concept is that customer access arrangements and services provision are the subject of international standardisation.

The historical role of standards bodies in telecommunications has been restricted to the creation of common standards for network interworking, e.g. in the specification of transmission system interfaces and internetwork signalling. However, at an early stage in ISDN development, it was recognised that international standardisation of the user network interface, at least as far as basic call control services, had many technical and economic advantages.

User network interface standardisation throughout the world provides the prospect of worldwide terminal portability. The potential large market thereby created for terminal equipment gives the economic benefits of scale by promoting competition between international equipment providers, and encouraging semiconductor manufacturers, on whose technology ISDN relies, to develop the devices required.

The economic benefits offered by international standardisation of ISDNs also are in keeping with the emerging trend in many countries to deregulate national telecommunication markets. International ISDN standardisation, given ISDN's inherent applications for data uses, also conforms with the trend in data communications to move worldwide towards an open systems architecture, following the ISO 7-layer model (see *Chapter 6*). ISDN specifications follow the principles of this model.

In terms of international standardisation, basic access is currently at a more advanced specification status than primary access and therefore various national variants of primary access have been developed, which are more divergent than national variants of basic access.

57.5.2 CCITT

The main worldwide standards body responsible for setting ISDN standards is the International Telegraph and Telephone Consultative Committee (CCITT), a subsidiary body of the International Telecommunications Union (ITU), a United Nations affiliated body based in Geneva.

CCITT standards, known as Recommendations, are updated in four-year cycles and are published in a set of books which all have the same coloured binding, e.g. those published in 1984 are known as the *Red Books*. The final recommendations in each cycle are approved at a Plenary meeting at the end of the cycle. This also sets the programme of work for the next cycle, which is carried out by a number of Study Groups (SGs), each of which is denoted in Roman numerals.

CCITT Recommendations are denoted by a letter and number. ISDN recommendations are denoted as the I-Series and related to ISDN reference user interface configurations, service and network aspects. ISDN is studied in CCITT SGXVIII, although certain aspects are studied in detail in other study groups. The recommendations produced by these other groups are in some cases duplicated as I-Series recommendations.

The I-Series recommendations[1] are grouped as shown in *Figure 57.3* and their detailed structure is given in I110. The key operational aspects of the ISDN customer interfaces are in the I400 Series, which are focused on in this chapter. The first draft recommendations for ISDN basic access were published in the CCITT *Red Book* in 1984. This draft defined the I-Series interface and the signalling procedures for basic call control. Since 1984, work has progressed to refine and consolidate the text on basic access and also to produce a draft for several supplementary services. The results of this work are published in the 1988 CCITT *Blue Book*. The primary access definition is also now defined, but is not as yet in a stable state.

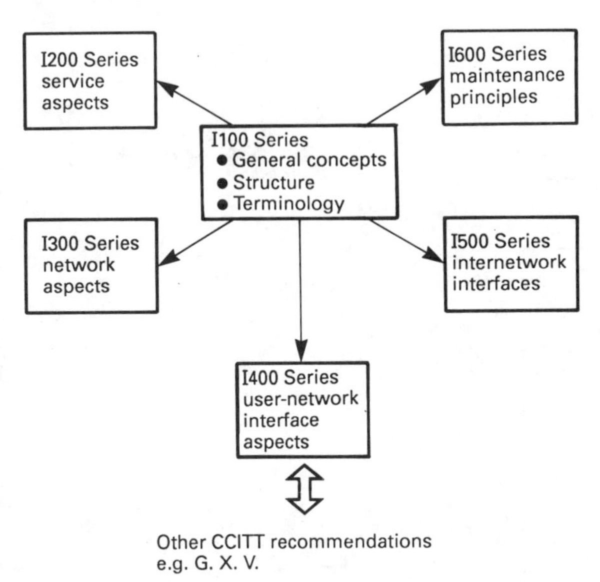

Figure 57.3 CCITT Recommendations (I series) for ISDN

57.5.3 CEPT (Europe)

In Europe, ISDN implementations will be to CEPT versions of the CCITT ISDN standards. The main European telecommunications standards body, the Organisation of European Posts and Telecommunications Administrations (Conference Européenne des Administrations des Postes et des Télécommunications, CEPT) took the CCITT *Red Book* as a base and published its own recommendations for Layers 1, 2 and 3 of ISDN in 1985.[2–4] In the 1984–1988 period four EEC network operators, the Federal Republic of Germany, France, Italy and the UK, produced a harmonised set of agreements for ISDN implementation known as the Quadripartite Agreements, which will form the basis of updated CEPT specifications. These will define the options available within the CCITT *Blue Book* recommendations, and will be standard for implementation within Europe, thereby allowing terminal portability.

In addition, The European Commission has charged CEPT with the task of producing a series of test specifications in an attempt to ensure throughout Europe that there is a commonality of approval for telecommunications equipment that is attached to public networks.

These specifications are known as NETs (Nomme Européennes de Télécommunications) and will be used by all EEC and Scandinavian countries as the basis for type approval, PTT purchasing specification and as a means of obtaining permission to connect. These specifications will be produced, for CEPT approval, by the European Telecommunications Standards Institute (ETSI).

57.5.4 USA

In the USA, ISDN access will be to national standards, developed by the American National Standards Institute (ANSI) accredited T1 committee.[5] These will be based on the *Blue Book* recommendations and will utilise in the case of primary access the 23B+D (1.536 Mbit/s) structure. For regulatory reasons, the NT1 will not necessarily be part of the network and T1 also define the U-Reference point interface.

57.5.5 Japan

Japan has for several years had under trial an ambitious mixed broadband–narrowband ISDN system known as the Integrated Network System or INS, which was not based upon CCITT standards. This pilot system was intended to obtain feedback from users as well as to prove the basic technology. Japan intends that its complete network should support ISDN by the year 2001 and has a trial service employing CCITT-based 2B+D standards operating in several of its cities.

57.5.6 The rest of the world

Published intentions elsewhere in the world are to use the *Blue Book* standards for ISDN introduction in the 1990s and beyond.

57.6 ISDN access in the UK

British Telecommunications (BT), in the absence of firm ISDN primary rate international standards, has developed its own access protocol, Digital Access Signalling System (DASS). Version 2 of this specification (DASS 2)[6] have now been adopted as a UK access British Standard (reference number to be assigned).

Basic access will be provided in the UK via a BT version of the CEPT I-series ISDN protocol.[7] The provision of basic access in the UK therefore requires a translation to be made between I-series basic access and BT primary access DASS 2 protocols.[8] This translation is performed within BT's network by an intelligent primary multiplexer known as an ISDN multiplexer or IMUX, which supports a number of NT1s and associated line transmission systems (see *Figure 57.4*). BT markets ISDN as Integrated Digital Access (IDA), as in *Figure 57.5*, which, apart from being a brand name for ISDN, also reflects a variation of the ISDN concept (see *Section 57.8.4*).

57.7 ISDN architecture and interfaces

57.7.1 Reference architecture and protocol model

The ISDN Reference Architecture as defined by the CCITT is shown in *Figure 57.6*. ISDN customer or user interface functional entities are shown in boxes and interfaces are indicated as crosses. The ISDN protocol architecture conforms to the lower three layers of the Open Systems Interconnection (OSI) model defined by the International Standards Organisation (ISO).

57.7.2 ISDN basic access customer interface

57.7.2.1 General arrangement

The ISDN basic access customer interface is a four-wire arrangement which, at the physical level, supports up to eight terminal

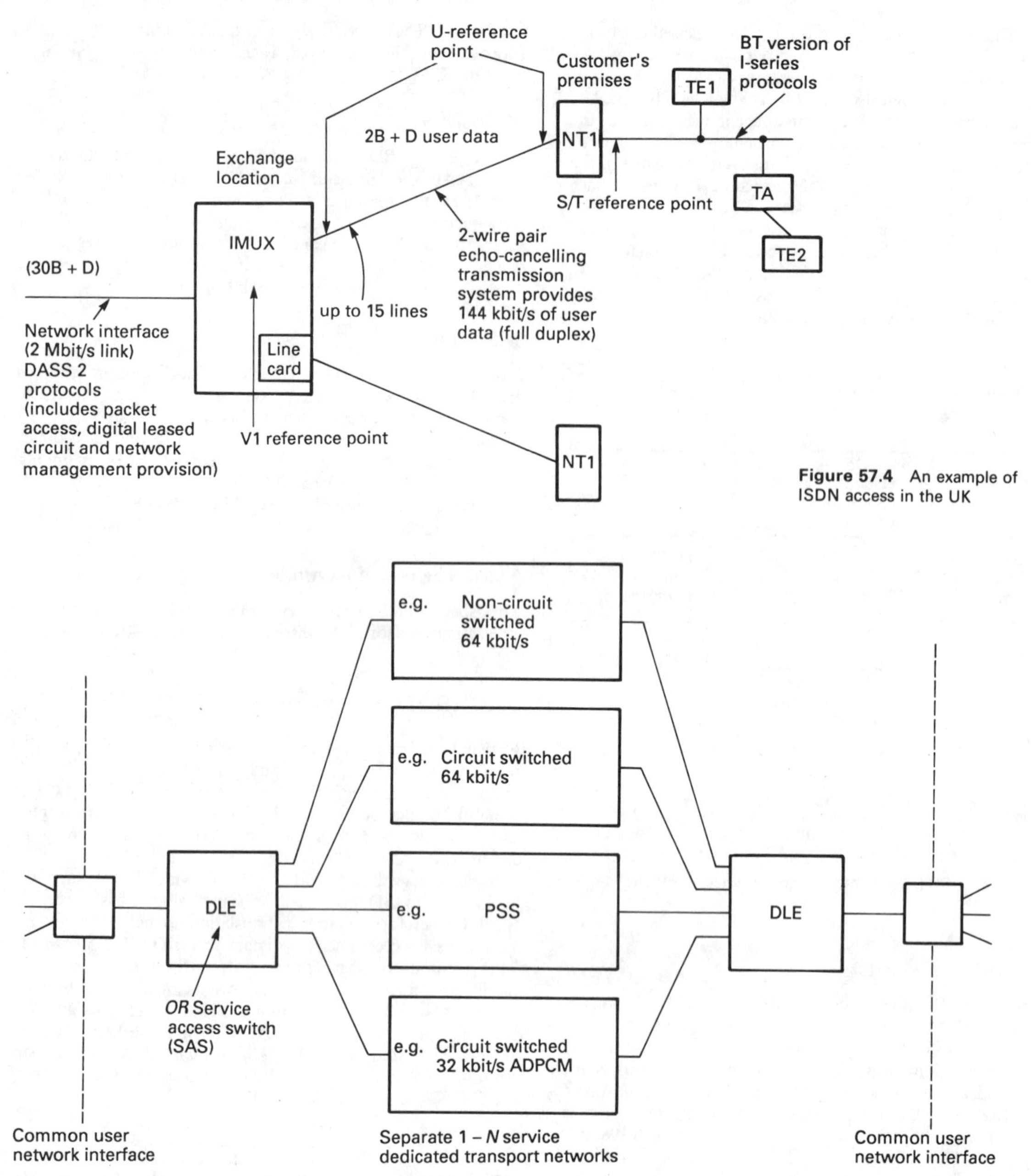

Figure 57.4 An example of ISDN access in the UK

Figure 57.5 Integrated digital access (IDA)

equipments, depending on configuration, and is defined in CCITT Recommendation I420.[9] Terminal equipments (TEs) are connected to the S reference point and may be ISDN TEs which support basic access ISDN (which are known as TE1s) or terminals with existing interfaces such as X21 (which are known as TE2s). TE2s connect to the customer interface via terminal adaptors (TAs).

The customer's interface may be provided via some intermediate processing unit, e.g. an integrated service private automatic branch exchange (ISPBX). In this case the interface between the ISPBX (termed an NT2 in the ISDN reference configuration) and the NT1 is at the T reference point. The S reference point is then on the TE side of the ISPBX, whilst the T reference point then exists between the NT2 and NT1.

Note that the T and S reference points are electrically identical and NT2s and TEs may be mixed on the same interface. The reference point may then be termed an ST reference point. The terminology 'customer interface' or 'S-interface' will be used throughout this chapter to describe T, S/T or S arrangements.

57.7.2.2 Layer 1 features

The layer 1 interface is defined in CCITT Recommendation I430.[10]

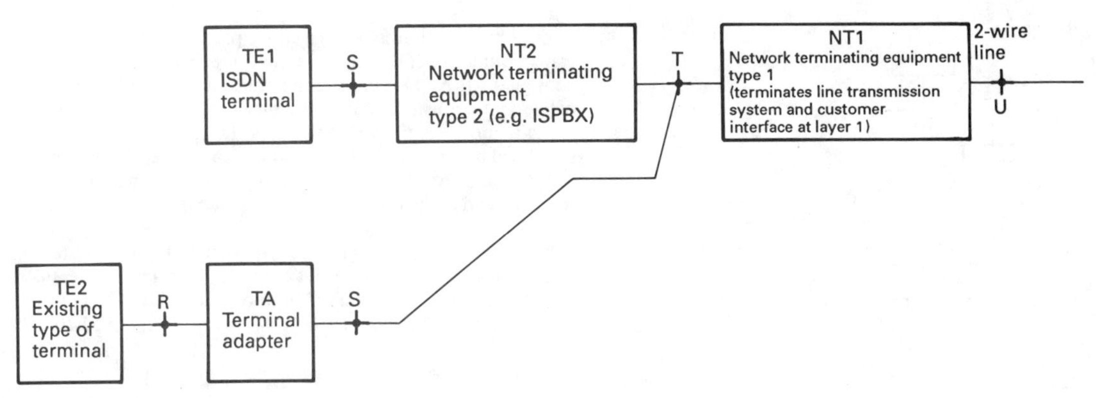

Figure 57.6 ISDN reference architecture

Physical and transmission features The physical access to the S interface is via an 8-way plug/socket arrangement to an ISO standard.[11]

A 192 kbit/s alternate mark inversion (AMI) transmission scheme is used on the S interface. This transports 144 kbit/s of user data and user–network signalling, a layer 1 S interface activation and synchronisation protocol, framing symbols and spare capacity. The S interface frame is shown in *Figure 57.7*. The transmission system allows a variety of S interface configurations to be supported. Depending on the configuration employed, TEs may be located up to 1000 m from the NT1. Some examples are shown in *Figure 57.8*.

The NT1 also supplies a limited amount of power to the S interface, which might be used by a suitably configured terminal to offer emergency telephony service in the event of mains failure.

Access features When the S interface has been successfully activated, a Ready For Data flag is sent to the network by the NT1 in the U interface transmission system's embedded protected operations channel (EPOC). The network indicates that user data is valid from the network by setting a Data Valid flag in the corresponding EPOC.

Up to eight TEs have contention resolved access to the two available B channels and/or the D channel, via a layer 1 protocol applied to the D channel and Echo (or E) channel parts of the S interface frame. TEs may have an equal chance of accessing the D channel (normally the default arrangement) or may be configured in a prioritised manner. The protocol is handled at the NT1 and is not controlled by higher layer signalling entities.

57.7.2.3 Layer 2 features

The higher data link control or HDLC protocol is used at layer 2 (also termed the link layer) to provide a transport mechanism for layer 3 signalling and other D channel data communications between TEs and the network. Note that the NT1 is transparent to link layer information. The HDLC link layer protocol used for ISDN basic access is known as the link access protocol type D or LAPD. LAPD is designed to support multiple terminal installations at the S interface and handles multiple layer 3 entities. Data link layer features are defined in CCITT Recommendation I440/1.[12,13]

Link layer functions The function of the link layer is to deliver across the link frames containing information from the layer above, layer 3, in such a manner that they are error free and in sequence. It also maintains the communication link, and provides a means of detection and notification to the management

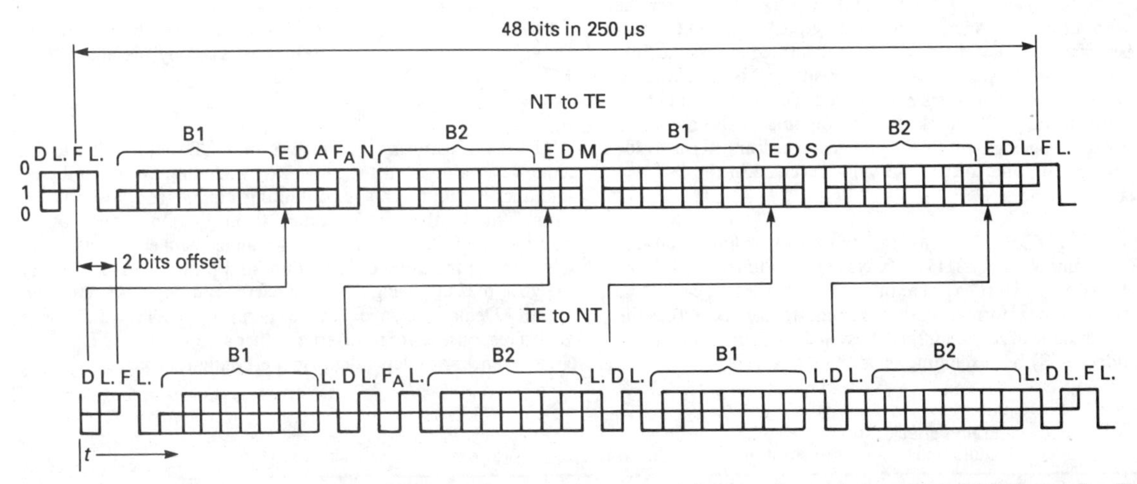

Figure 57.7 Customer interface frame structure. Dots demarcate those parts of the frame that are independently DC-balanced. The F_A bit in the direction TE to NT is used as a *Q*-bit if the *Q* channel capability is applied. The nominal 2-bit offset is as seen from the TE. The corresponding offset at the NT may be greater due to delay in the interface cable and varies by configuration. F, framing bit; L, DC balancing bit; D, D-channel bit; E, D-echo-channel bit; F_A, auxiliary framing bit; N, bit set to a binary value $N=F_A$ (NT to TE); B1, bit within B channel 1; B2, bit within B channel 2; A, bit used for activation; S, reserved for future standardisation; M, multiframing bit

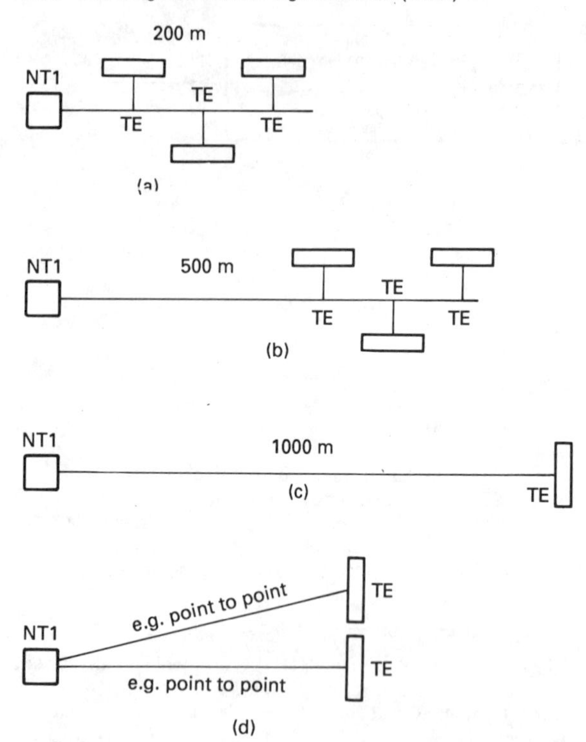

Figure 57.8 Customer interface configurations: (a) short passive bus, (b) extended passive bus, (c) point to point, (d) NT1-star

entity of any errors that occur in transmission, frame format and general operation. It also provides link flow control.

Link layer frame format The information transferred across the link is sent as packets of data known as frames, each of which has the delimiters of a start flag and a finish flag. This is shown in *Figure 57.9*. Note that the information field need not necessarily be present and that the number of bits is application dependent.

Frame check sequence To check that each frame traverses the link without corruption, a frame check sequence (FCS) calculation is performed on each frame to be transmitted across the link, excluding the start and finish flags. The result of this calculation is inserted in the last two octets of the frame. The receive end of the link performs an FCS check on the contents of the incoming frame, excluding flags and FCS, and rejects the frame if this value does not correspond to the FCS value received in the last two octets.

HDLC–use of flags Both the start and the finish flag of a frame have the same value of 01111110. Note that a finish flag of one frame may be the start flag of another. So that the flag cannot be simulated by the information that is being transferred across the link, a technique of zero bit insertion is used on all data (including the added FCS) but excluding flags.

Zero bit insertion is performed by placing in the data at the transmitting end of the link an extra 0 at every occurrence of five contiguous 1s At the receiving end, the first 0 after the occurrence of five contiguous 1s is removed. As the data on the link can never contain more than five contiguous 1s, a flag can never be simulated by the data. This technique also enables seven or more contiguous 1s to be used as an abort signal, thus instructing the receiving end of the link to abort processing of the current frame.

Address field The first and second octet in the frame are the address field, which consists of the service access point identifier (SAPI) octet and the terminal endpoint identifier (TEI) octet. The structure of the address field is shown in *Figure 57.10*.

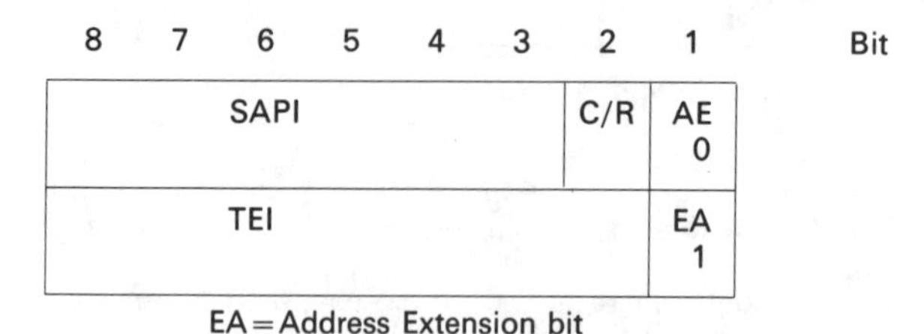

Figure 57.10 Layer 2 address field

Use of the SAPI LAPD identifies the D channel service by the SAPI octet, e.g. signalling has a SAPI value of 0, packet access may have a value of 16, management and maintenance has a value of 63. Considerable spare capacity is therefore available for service enhancement.

TEIs Each TE may consist of a number of logical entities each of which are identified by an addressing mechanism in LAPD known as Terminal Endpoint Identification. Up to 127 TEIs are available for addressing logical entities on any one S interface. An additional TEI value is reserved for broadcast messages to all logical entities. TEIs are assigned as shown in *Table 57.1*.

Table 57.1 TEI assignments

TEI value	*Usage*
0–63	Values pre-assigned to terminals
64–126	Values assigned by the network
127	Broadcast TEI

TEIs 64–126 are assigned as follows. A TE that requires a TEI requests a value from the network by sending an unnumbered Identifier Request message containing a reference whose value is a random number in the range 0–65535. On receipt of an identifier request, the network TEI management entity will either select one of the unused TEI values and return it in an Identity Assigned message using the same reference number, or return an Identity Denied message. A timer is run by the TE which times out if a response is not received. Further actions by the TE in such circumstances are dependent on TE configuration.

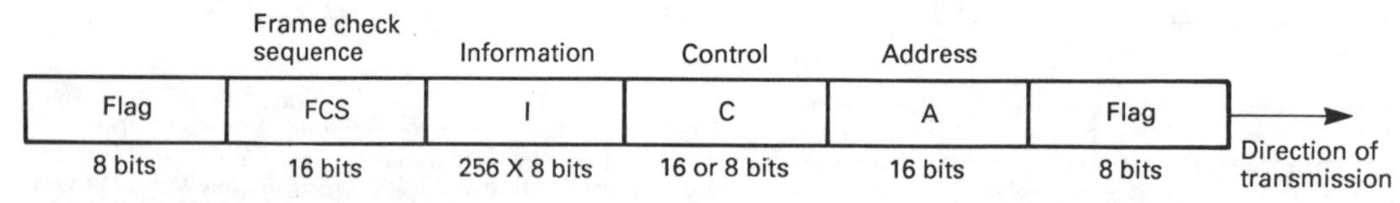

Figure 57.9 Signalling frame format

The link layer utilises the SAPI value of 63 and the TEI broadcast value of 127 in all TEI management procedures.

The network TEI management entity and the TE utilise other layer 2 messages in the TEI management procedure. Identity Check Request and Identity Remove may be sent from the network to the TE, whilst Identity Check Response and Identity Verify may be sent from the TE to the network. The meanings of these messages are self-explanatory. The TEI check procedure is shown in *Figure 57.12*.

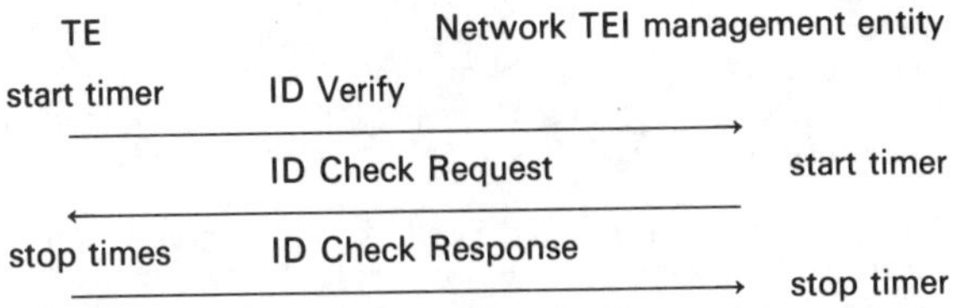

TE — Network TEI management entity

start timer — ID Verify ⟶

ID Check Request ⟵ — start timer

stop times — ID Check Response ⟶ — stop timer

Figure 57.11 TEI check procedure

Control field The control field (see *Figures 57.12* and *57.13* consists of one octet if used for basic working or two octets if used for extended working. Basic working is known as modulo-8 working and employs a 3-bit message sequence counter; extended working is known as modulo-128 operation and employs a 7-bit message sequence counter.

The control field identifies the type of frame, which may be of command or response type. The command frame may be an information frame, supervisory frame, or unnumbered frame. The response frame is either a supervisory frame or an unnumbered frame.

8	7	6	5	4	3	2	1	
N(R)			P	N(S)			0	Information
N(R)			P/F	S	S	0	1	Supervisory
M	M	M	P/F	M	M	1	1	Unnumbered

Figure 57.12 Control single octet (modulo-8)

8	7	6	5	4	3	2	1	
N(S)							O	Information
N(R)							P	2
X	X	X	X	S	S	0	1	Supervisory
N(R)							P/F	2
M	M	M	P/F	M	M	1	1	Unnumbered (1 octet)

N(S) = Transmitter send sequence number
N(R) = Transmitter receive sequence number
S = Supervisory function bit
M = Modifier function bits (function of UI frames)
P/F = Poll/final bit (command/response)
X = Reserved and set to zero

Figure 57.13 Control two octet (modulo-128)

The information frame format is used by all frames that are used to transfer information between layer 3 entities.

The supervisory frame format is used to perform the link control functions such as acknowledgement, request for retransmission and request of suspension of information frames.

The unnumbered frame format is used to provide additional link control functions plus single frame data transfer. These frames do not contain a sequence number in their control field.

Link layer frame exchange The link is established by the terminal sending the unnumbered frame set asynchronous balanced mode extended (SABME). This is acknowledged by the network with an unnumbered frame unnumbered acknowledgement (UA). Information frames may now be transferred across the link and are acknowledged by receiver ready frames (RR)—see *Figure 57.14*.

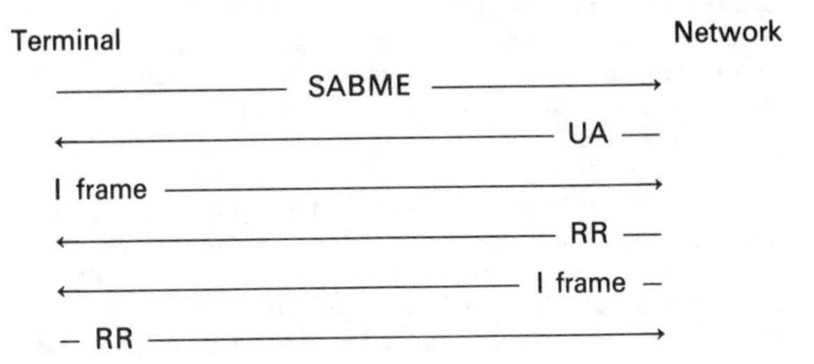

Figure 57.14 Layer 2 frame exchange

57.7.2.4 Layer 3 features

The I series level 3 protocol is used for circuit switched call establishment, control and termination, and is defined in CCITT Recommendation I451 (Q931).[14] The protocol is designed to handle call establishment either in an *en bloc* manner, where all the data to set up the call is contained within one message, or in an overlap manner where the call set-up is performed using a series of messages.

In general, layer 3 signalling may be defined as being of the *functional* or *stimulus* types. The main difference between the two is that stimulus signalling does not necessarily involve any call processing at the terminal, i.e. it allows simple terminal development, and can be used for supplementary services such as calling line identity (CLI) display at the TE. CCITT I series signalling is only defined in terms of functional operation although certain national implementations of ISDN, e.g. in the UK, utilise stimulus procedures for supplementary services.

Frame format All layer 3 messages follow the same basic structure as shown in *Figure 57.15*.

8	7	6	5	4	3	2	1
Protocol discriminator							
0	0	0	0	Length of call reference value			
F	Call reference value						
E	Message type						
Information elements							

F = Call reference flag
E = Message type extension flag (for future use)

Figure 57.15 Layer 3 frame format

The protocol discriminator The function of the protocol discriminator is to indicate the communications protocol used at layer 3 for that frame. At present, I451 defines a single value of 00001000, but the capability exists to allow multiple communication protocols to be used. This feature is used by British Telecom and Telecom Australia to identify proprietary maintenance protocols.

Call reference The next two octets are occupied by the call reference. This is used to identify the messages associated with a particular call. The call reference value is provided by the call-originating side of the interface and its value is assigned at the beginning of a call. It is present in all messages associated with that call for the lifetime of the call. The exception to this rule is a suspended call, which on resumption is given a new call reference by the side that has performed the resumption. A resumed call is identified by the information element 'call identity' within the RESume message. The call reference flag identifies the originator of the call and its purpose is to resolve the problem of the attempts of both ends of the link to allocate the same value. The call originator sets the flag to 0 and the destination to 1.

Message type The message type field is a single octet with an extension indicator. It identifies the message usage which can be divided into four functional groups: call establisment, call information, call clear and miscellaneous. The current list of messages for circuit mode connection in Q931 is shown in *Table 57.2*.

Information elements A message is identified by the message type field and is made up of a series of information elements, each of which is identified by its own particular code and, where appropriate, a length indicator. Variable length information elements have a fixed order within a message, which is determined by descending order of code values. Single octet information elements may appear at any point in a message, in between other elements.

At present there are over 26 information elements identified by CCITT in Q931. Of these, only a few are essential for call control and these are classed as mandatory messages. Other elements are optional, and may or may not be included in a message. Examples of information elements are as follows.

The *bearer capability element* is an information element that contains information that identifies the CCITT I211 bearer service[15] (e.g. 3.1 kHz speech) required from the network, thereby enabling it to handle the call in the appropriate manner. This element is also used by terminals on an incoming call to establish terminal compatibility with the bearer service offered by the network. Note that the originating values of this element may be changed by the network, e.g. A-law speech may be specified at the local part of the network and U-law speech may be specified at the remote end.

Full end-to-end compatibility between terminals may require the use of additional information elements. The *higher layer compatibility* (*HLC*) information element carries information relating to the teleservice offered, e.g. Teletex, Videotex Group 4. It may be monitored by the network (e.g. for tariffing purposes but is transported transparently by it. The *low-layer compatibility* (*LLC*) *information element* is used to identify additional low-layer service attributes, e.g. rate adaption. Again, this information is carried transparently by the network. Note that the values of the HLC and LLC elements are compatible with the bearer capability information element.

The *keypad information element* is used to convey stimulus signalling type information from the terminal to the network. In the reverse direction the *display information element* is used for the same purposes. Both keypad and display information elements may be conveyed across the user network interface within an INFOrmation message, thus having no effect on call control.

Layer 3 frame exchange during a call The message is shown in *Figure 57.16*. The SETUP message sent to the called party is a broadcast message (unnumbered frame). One or more TEIs that fulfil the requirements of this SETUP message may respond to this message. The network is responsible for the decision as to which TEI is connected to the incoming call and for the clearing of the remaining TEIs that responded to the initial SETUP message.

57.7.2.5 Supplementary services

Supplementary services are services additional to normal call control that are provided by the network. These service types are at present usually provided via an operator. They are also extensively available on PABXs and in digital private networks (see *Section 57.7.6*).

A firm specification for the control of supplementary services has yet to be provided by CCITT. In the mean time, ISDN implementors are using the KEYPAD and DISPLAY information elements to offer these facilities.

57.7.3 Primary access

57.7.3.1 General

Primary access is provided as a point-to-point link only. It may be employed to interface to an ISPBX, a basic access multiplexer, or some other call handling device. Primary access is defined in CCITT Recommendation I421.[16]

The two variants of primary access are defined at layer 1 (see below), but primary access signalling is not as well defined as basic access signalling and so various intercept implementations have emerged. These intercept implementations are either based on draft issues of Q921/Q931 specifications or use proprietary ISDN protocols, e.g. the BT-developed DASS 2. These layer 2

Table 57.2 Call control messages

Call establishment	*Call information*	*Call clearing*	*Miscellaneous*
* ALERTing	MODify	* DISConnect	CONgestion CONtrol
* CALL PROCeeding	MODify COMplete	* RELease	INFOrmation
* CONNect	MODify REJect	* RELease COMplete	NOTIFY
* CONNect ACKnowledge	RESume	RESTart	STATUS
PROGress	RESume ACKnowledge	RESTart ACKnowledge	STATUS ENquiry
* SETUP	RESume REJect		
* SETUP ACKnowledge	SUSPend		
	SUSPend ACKnowledge		
	SUSPend REJect		
	USER INFOrmation		

* Messages used in the control of a typical call. Other messages are used for supplementary services and network management

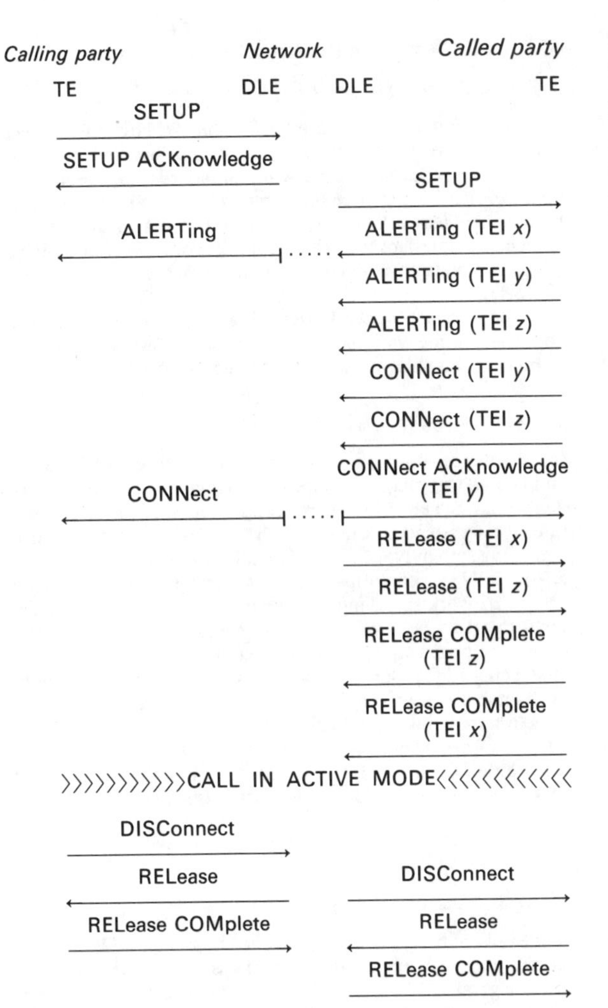

Figure 57.16 Basic call level 3 messages

and 3 features are discussed in this section. Key differences between the intercept standards are the additional mechanisms used for customer line identification when multiplexers are employed.

57.7.3.2 Primary access variants

Two variants of primary access are currently the subject of detailed recommendations by the CCITT, namely access at 2 Mbit/s (used in Europe) and access at 1.5 Mbit/s (used in North America). The interfaces to these access types have differing transmission characteristics and frame structures. Signalling may also be carried in a different manner over these variants of the primary interface, e.g. when used in association with basic access multiplexers.

57.7.3.3 Layer 1 features

The layer 1 primary rate interfaces are defined in I431.[17]

Transmission The layer 1 interface at the primary rate is to CCITT Recommendation G703, for both variants of primary rate, namely 1.5 and 2 Mbit/s.

Frame structure as used for ISDN The 2 Mbit/s frame structure is based on CCITT Recommendation G732 and is composed of 30 64-kbit/s B channels, where a B channel is normally used to transport circuit switched customer data, and one 64-kbit/s D channel that is used to carry all of the signalling information relating to this primary access. Some B channels may be used for other types of traffic in some interface arrangements, e.g. for packet traffic.

A 1.5 Mbit/s frame structure is also based on G732 and is composed of 23 64-kbit/s channels where a B channel is used to carry user information and one 64-kbit/s D channel. Multiplexed interfaces may have different arrangements (see *Section 57.8.2*).

Idle patterns An idle pattern (for example as defined for DASS 2 working) may be applied to a B channel when it is not in traffic. Idle pattern may be used to monitor higher order transmission system traffic levels.

57.7.3.4 Layer 2 features

Both the DASS 2 and I Series variants of primary access use similar layer 2 procedures based on CCITT Recommendation I441. HDLC procedures are used to provide link layer functions. A single Link Access Protocol (LAP) is employed across the link. Customer line identification (CLI) may be carried out in this layer of the protocol.

Maintenance and management procedures (identified by a specific SAPI) are, at this time, in more common usage than the basic access case.

57.7.3.5 Layer 3 features

The layer 3 signalling facilities defined in Q931 apply to both the basic access and primary access configurations.

The layer 3 messages used in non-CCITT applications (e.g. the UK DASS 2) perform similar functions to the layer 3 (Q931) messages, but have different message structures.

57.7.3.6 Customer line identification

The use of a multiplexer to provide a conversion from BRA to PRA has led to several schemes for identifying the customer's line within the signalling system. Some of these schemes are as follows:

(1) A SAPI and TEI are used as part of a 'wrapper' around the existing protocol. The SAPI has a value of 0 and the TEI has the values 1 to 16, which correspond to the customer's lines (Bellcore USA[18]).
(2) A third octet is added to the layer 2 address. This octet is known as the line identifier (LID) and is added to the messages in the direction multiplexer to DLE and removed in the reverse direction (British Telecom, packet traffic[19]).
(3) Tenant group working is an extra information element in the layer 3 protocol and is used the same way that routing information would be used on a PRA to a PABX (Telecom Australia [20]).
(4) Channel-associated signalling over the PRA is accomplished by reducing the number of traffic channels from 30 to 24 and using the remaining six to carry signalling in a time-associated manner. Each of the six channels used for signalling will carry the D channel information for two ISDN customers (Siemens USA[21]).

57.7.4 Alternative access arrangements

Alternative access arrangements are currently defined by CCITT but are not currently supported by Q931 layer 3 signalling procedures. These alternative arrangements are briefly discussed below.

57.7.4.1 *Access at the primary rate*

There are alternative primary rate channel structures defined in the CCITT Recommendations.[22] These are: for the 1544 kbit/s rate a 23 B+E structure and for the 2048 kbit/s rate a 30 B+E structure. The E channel is a 64 kbit/s channel that differs from the D channel above only in that it may carry not only the signalling for its own primary rate interface but also signalling for other primary rate interfaces without an active signalling channel.

57.7.4.2 *Access at other rates*

A third group of primary rate interface structures is the higher rate structures. These make use of H channels that are intended to carry a variety of user information (e.g. fast facsimile, video for teleconferencing, rate adapted up to the H channel rate, or multiplexed 64 kbit/s voice) and signalling, but they do not carry signalling information for circuit switched ISDN. Each H channel has its own associated timing. The following H channel rates are defined in the CCITT Red Book:

H0 channel: 384 kbit/s
H1 channel: 1536 (H11) and 1920 (H12) kbit/s

57.7.5 Packet access to an ISDN

57.7.5.1 *Packet access scenarios*

Packet access to an ISDN is defined in CCITT Recommendation I462 (X31),[23] which states that packet procedures in an ISDN shall be based on CCITT Recommendation X25. Packet access to the customer can be provided by an ISDN basic access over the D channel or the B channel as described below. The form of access is currently known as minimally integrated, i.e. the ISDN provides only an interworking function via an interworking unit (IWU) to the packet switched network. In the future, more highly integrated packet access arrangements will be defined (see *Section 57.8.7*).

57.7.5.2 *Basic access–D channel access*

This method multiplexes the packet traffic with the signalling traffic and is restricted to a data rate of 16 kbit/s. The X25 packets use the same method of access as the signalling traffic. Each packet has a service access protocol indicator (SAPI) code that identifies the frame as being packet traffic, and a terminal endpoint identifier (TEI) used during the communication session to identify the entity to which the packet is directed, in the same way that signalling traffic is transmitted.

Packet call establishment signalling is similar to B channel circuit switched call establishment signalling in that an unnumbered information frame (UI) containing the connection information is broadcast over the customer's line, and the TEIs that can establish a call based on this connection information reply with their TEI values. Contention, call establishment and call control are maintained by the packet handler. This also provides the gateway function to the packet network.

57.7.5.3 *Basic access via the B channel*

Packet access over the B channels using the D channel for establishment of a circuit switched call to the packet handler is another method of access. This approach allows packet working of up to 64 kbit/s but means the loss of a B channel to other traffic.

57.7.6 ISDN and private networks

57.7.6.1 *Status of standards*

The ISDN signalling procedures defined in Q931 do not yet cover working over digital private networks. It is recognised, however, that extension of public network standards to the private application is important in that it would allow ISPBXs to access both types of network in the same way, rather than using different protocols to interface to both types of network. Work to define a common standard is therefore proceeding in CEPT and CCITT (see *Section 57.8.6*).

Digital private network signalling is at present therefore performed using various national or trans-national intercept standard protocols, an example of which is given below.

57.7.6.2 *DPNSS*

In the absence of stable international standards, British Telecom and UK industry have developed a signalling system for use over digital private networks, where access is provided at the primary rate. This system is known as the Digital Private Network Signalling system No. 1,[24] and is similar to DASS 2 (see *Section 57.6*). DPNSS1 conforms with a general requirement of all private network signalling systems in that it is feature-rich, i.e. can support a large number of supplementary services. Currently some 30 supplementary services are supported, with another 10 or so being added each year. It is a mature standard, being used over some 400 networks in the UK. It is also being used over some international digital private circuit connections, and in a number of other countries.

Note that the use of DPNSS1 does not inhibit I series based operation at the user side of the ISPBX (see *Section 57.7.8*).

57.7.7 Rate adaption

57.7.7.1 *General*

At the present time, existing terminal equipments, TE2s in ISDN terminology, usually operate at a rate less than the ISDN bearer rates, e.g. at 9.6 kbit/s, and may be synchronous or asynchronous. Even when the ISDN environment is mature, the situation is likely to continue and therefore rate adaption to the ISDN bearer rates is required. The bearer rates considered in the existing ISDN standards are 16 kbit/s D channel and 64 kbit/s B channel rates. D channel working, by its packet-mode nature, is rate adapted already, and is therefore not considered further in this chapter.

57.7.7.2 *Standards*

The ISDN rate adaption standards have to take into account rate adaption techniques used in existing networks. ISDN CCITT I Series rates adaption recommendations are usually identical to CCITT X and V series recommendations which in turn may be based on other standards, e.g. those generated by the European Computer Manufacturers Association (ECMA).

The base CCITT ISDN rate adaption recommendation is I460 (X57). Other recommendations are: I461 (X30), support by an ISDN of X21 and X21bis data terminals; I462 (X31), support by an ISDN of X25 packet mode terminals; I463 (V110), support of terminal equipments of V series type interfaces. These recommendations also cover the related subject of multiplexing of a number of sub-rate data connections.[23,25–27].

Rate adaption over ISDN is therefore defined in terms of schemes that support existing arrangements or, for new devices, one of the schemes that have been defined specifically for ISDN working. The ISDN layer 3 signalling ensures that compatibility is achieved between both ends of a connection.

One such ISDN specific rate adaption scheme over the B channel is V110, where synchronous or asynchronous data is initially rate adapted into a submultiple of 64 kbit/s which may be 8, 16 or 32 kbit/s. Data is then multiplexed into a 40-bit frame. Spare subchannels within this frame are empty, i.e. there is no reiteration.

Another technique is V120 (formerly known as Vtad) which empoys LAPD techniques over the B channel, and again caters for synchronous or asynchronous data. X31 employs similar techniques.

57.7.7.3 Implementations

In Europe, ISDN terminals (TE1s) and terminal adaptors will generally work to ECMA 102 which is based on CCITT Recommendation V110 (X30). In the USA, V120 is favoured, although in the interim proprietary protocols may be employed.

57.7.8 Access at the user side of an ISPBX

The development of digital PABXs has in general led the development of digital switches used in the public network, and there are a large number of suppliers and digital PABX types. Many of these digital PABXs have an ISDN (and/or digital private network) capability. A digital PABX that is supported by an ISDN is known as an integrated services PABX or ISPBX.

Several types of digital PABX have offered digital access at the user side, in advance of ISDN standards. This digital access offers voice and data capability to proprietary transmission and signalling standards. The CCITT ISDN standardisation process has not, to date, covered basic access at the user side of an ISPBX, and there is no general requirement for this interface to be standardised, as it exists between two pieces of customer equipment.

However, in the interests of terminal compatibility across national and international markets, regulatory authorities may specify that the standard ISPBX–user interface be the ISDN S interface. For example, this has already been proposed by the EEC. It is likely that such regulatory specifications would specify a set of bearer services that would conform with public network capabilities, with proprietary ISPBX–user supplementary services being allowed.

57.8 ISDN access arrangements

Basic access to the ISDN network is either implemented directly onto a digital ISDN local exchange or via some intermediate processing unit such as a multiplexer, concentrator or ISPBX, all of which have a primary ISDN access to the ISDN network. Features of these access types are discussed in this section.

57.8.1 Access direct to a digital switch

Most, if not all, of the types of digital local exchanges that are being introduced worldwide have inherent ISDN capabilities with varying levels of service (and bearer mode) supported. Most of these will offer basic access directly supported by the switch, i.e. the switch will provide a U interface to the customer's line. This interface is not at present the subject of CCITT standardisation and therefore various proprietary transmission schemes will be adopted worldwide.

Primary access will also be supported by digital switches although, in the interim, different protocols will be employed for different implementations. These interim protocols are intercept proprietary standards and will be used at least until a stable CCITT Recommendation for primary rate access is available. A stable primary rate recommendation may not be available until 1996 at the earliest because of the desire to integrate private network signalling into Q931.

Digital exchanges are, in essence, large specialised computers with a large number of input ports controlled by a large amount of software. Their increasing sophistication, particularly in terms of their software content, has meant that their development costs have steadily risen so that the cost of developing a digital exchange, or switch, is now of the order of a billion US dollars. This has in turn meant that the number of switch manufacturers has steadily fallen. The low number of switch manufacturers has increased the volumes of customer accesses to any one type and therefore in the context of ISDN this means that proprietary standards are an attractive proposition in the interim.

The customer, however, requires terminal portability, particularly with regard to basic access TEs, and this was the major reason for the development of ISDN standards by CCITT. To a manufacturer, though, use of proprietary standards means that design changes can be controlled as switches are becoming progressively more complex and redesign requires increasing amounts of development resource. Adherence to the developing CCITT standards is therefore costly in terms of design changes.

This seeming impasse can be overcome by the use of intelligent devices called flexible multiplexers, which act as intelligent front ends to digital switches and interface to the switch in a defined manner. In this way changes are limited to the multiplexer and switch redesign is minimised.

This chapter will not attempt to explain internal digital switch architectures and operation as each type of switch has major differences. Readers should refer to manufacturers' published documentation. The major switch types in use around the world are: System X (GEC/Plessey), E10 (Alcatel), System 12 (Alcatel), EWSD (Siemens), ESS No. 5 (ATT/Phillips), DMS100 (Northern Telecom), AXE10 (Ericsson) and Fetex 150 (Fujitsu). A number of other switches may be encountered that support ISDN in specialised applications, e.g. in rural areas.

57.8.2 Access via basic access multiplexers

An example of this arrangement is shown in *Figure 57.4*. The primary rate access (sometimes termed PRA) multiplexer is used to convert basic rate access (again, sometimes termed BRA in this application) protocols into PRA protocols. Sometimes, concentration of signalling and/or data traffic is provided by the multiplexer. The multiplexer may also be used as a means of conversion from the international standard ISDN signalling (I420) to some national or exchange standard used as the PRA. As discussed in *Section 57.7.3*, customer line identification is a key feature of PRA/BRA multiplexer use.

57.8.3 Integrated digital access (IDA)

Here, rather than all services being carried by one integrated transport network, services are routed via networks optimised for particular types of connection. To the user it appears, however, as a single network (see *Figure 57.5*). Note that this conceptual arrangement is not in conflict with the CCITT ISDN Recommendations. IDA also reflects the role of ISDN in flexible access systems because it is used as a means to obtain managed access to a number of separate voice and data networks.

57.8.4 Transmission aspects of ISDNs

57.8.4.1 Customer benefits

The extension of digital transmission to the customer, which is a major feature of ISDN, provides the customer with a much better quality of service. The digital transmission schemes used are intended to minimise errors and provide loss-free channels. They can be managed by network entities that automatically detect if line quality falls below defined limits. For data use, bit rates are far higher than usually found over analogue circuits and have lower error rates (1 in 10^7 typical). More than one transmission channel is available at the ISDN user interfaces.

57.8.4.2 Primary rate provision

Primary rate access will usually be provided to customers via optical fibres. The transmission techniques employed are discussed in *Chapter 56*. It is possible, however, that provision may be by four-wire copper circuits or via microwave links using established transmission technologies.

57.8.4.3 Basic access provision

Whilst basic access is provided over the S interface transmission scheme, which is relatively undemanding in terms of today's semiconductor technologies, a key technical problem in ISDN provision was providing this interface via the two-wire twisted pair copper lines that connect the customer line to the local exchange. As some 70% of the total investment of the world's telecommunication networks is in the local loop plant it was seen as particularly important to make the best use of this plant until optical fibre is installed as standard.

While it has been known for many years that unloaded two-wire copper cables can transmit signals at frequencies far above those used for PSTN, the design of transmission systems that could support the full duplex transmission of 144 kbit/s of customer data over a two-wire pair, over the planning limits of the world's networks, was probably the key technical challenge faced by ISDN system designers. The key features of such systems are therefore discussed in more detail in this section.

Requirements for 144 kbit/s transmission The requirements for the 144 kbit/s transmission system relate to the cable environment, transmission performance, and constraints on implementation. The transmission system design should be able to work over all cable environments worldwide that conform to national speech planning rules (which are in fact quite similar). These requirements are as follows:

(1) *Two-wire line features*
   (a) The transmission medium is normal twisted pair cable.
   (b) The interface from the network side of the NT1 to the DSLT is assumed to contain no loading coils or loop extension equipment, i.e. only loop cable and protection devices are in series with the transmission system.
   (c) Bridged taps: the transmission method shall be designed to support DSLT to NT1 connections where bridged taps are present, subject to the additional attenuation, etc., introduced by the bridged tap(s) not exceeding the overall attenuation requirements.
   (d) Splices and gauge changes shall be supported.
   (e) The transmission method shall be insensitive to changes in line polarity which may occur on the two-wire loop as a result of line maintenance work by craftsmen.
(2) *Transmission system features*
   (a) The system shall have a design range that exceeds normal speech planning limits (i.e. an attenuation of 46 dB at 100 kHz).
   (b) The system shall support supervisory, maintenance and control action via a protected channel.
   (c) The system shall have in-built CRC-based error monitoring facilities for the B and D channels.
   (d) The system will not necessarily be activated between calls, and therefore fast wake-up is required.
   (e) The system shall support I Series customer interface activation procedures.
   (f) The system shall allow line powering of the NT1.
   (g) The system shall support regenerators and their maintenance. (This requirement impacts on the line powering arrangements.)
   (h) The system shall be implemented such that its cost shall be similar to other low-frequency (LF) local area transmission schemes over copper cable. Due to the complexity of circuitry required to meet the performance requirements, this requirement necessitates system implementation in VLSI technology.
   (i) The system shall interwork with protection devices.

Choice of line transmission scheme Taking into account the requirements listed above, particularly the cost requirement, implementation of the processing functions in a digital semiconductor technology is a key system feature. The use of digital rather than analogue transmission techniques met this requirement best and two types of system were initially considered as possible candidates for implementation. The first class of system, known as burst-mode or ping-pong, squeezed the user data into higher frequency bursts and sent them alternately over the line in each direction. Although technically not too difficult to implement, the use of higher transmission frequencies makes burst-mode system performance more prone to crosstalk and attenuation, and their usable range is limited to about 60% of that required.

The second type of system is based on the use of devices known as echo-cancelling digital hybrids, which transmit and receive data simultaneously, thereby limiting the range of frequencies used. Several subclasses of systems have been proposed, but system designs generally fall into two types, namely binary systems, i.e. those that use two-level codes, and multilevel systems that use three or more level codes. The latter meet the performance requirement of range far more easily than the former, and development worldwide is now mainly based on the use of three- and four-level codes. No CCITT standard is proposed at this time on the coding format to be used, which will depend on field experience with the various system implementations that are now available or proposed. *Table 57.3* compares the near end crosstalk performance of various codes and is collated from data published by system implementors. The SU32 data is based on measured results; other data is based on system computer simulations.

Functional description An example of a line transmission design (the SU32 line code system) is given below. *Figure 57.17* shows

Table 57.3 NEXT performance of various line codes proposed for ISDN two-wire line transmission

		Decision distance/crosstalk power ratio (dB)			
Line code	*Cable insertion loss at 100 kHz*	30 dB	40 dB	46 dB	50dB
SU32		32.14	25.35	21.28	18.56
3B2T		31.52	24.79	20.77	18.04
2B1Q		30.21	23.95	20.25	17.69
DI43		30.64	23.57	19.39	16.65
MMS43		29.72	22.56	18.36	15.65
MDB		29.84	22.30	17.76	14.73
AMI		28.85	21.09	16.60	13.55

Conditions:
(a) Near-end crosstalk (NEXT) attenuation = 50 dB at 100 kHz
(b) Cable = 0.5 mm (24 gauge) copper
(c) Two precursor tap adaptive equaliser with decision feedback equalisation (DFE)

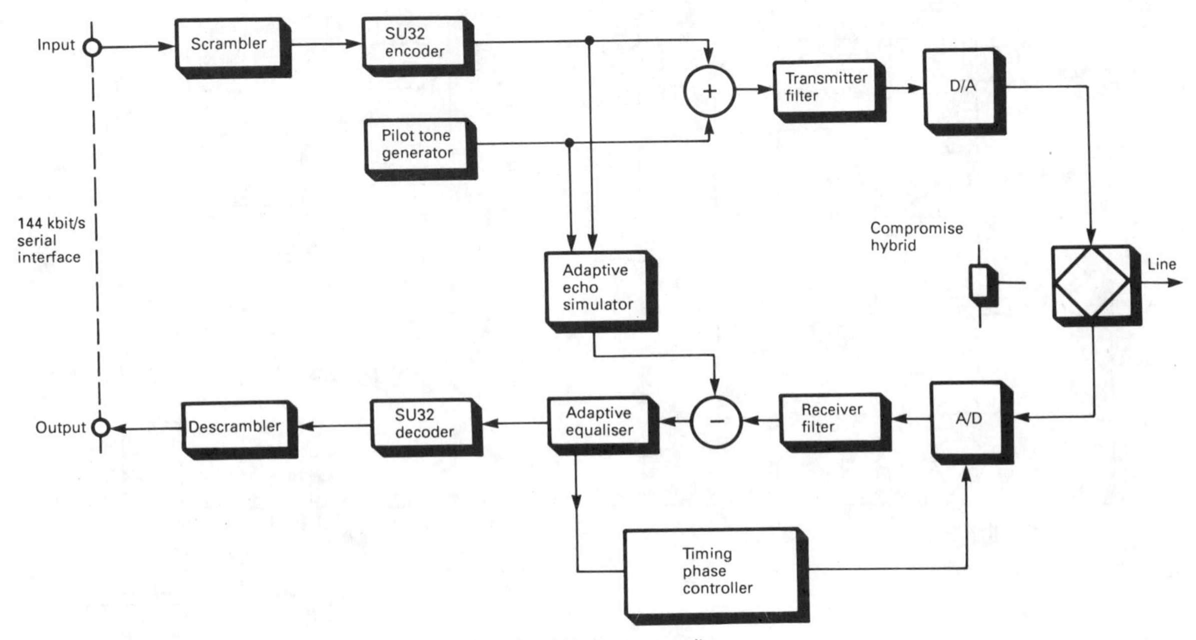

Figure 57.17 U-interface transmission system: main functional blocks (one end)

the main functional blocks contained within the echo-cancelling hybrid arrangement. Both ends of the transmission system in this figure have the same general arrangement, and maintenance, data formatting and S interface functional blocks are not shown.

The transmission method described supports the full duplex transmission of 144 kbit/s of user data as defined in CCITT Recommendation I430. Spare capacity is provided for an embedded protected operation channel (EPOC) which supports various supervisory and maintenance functions. Line errors are detected by data channel CRC checks.

Transmission over the two-wire line is by means of the ternary block code SU32 with an orthogonal timing signal. The baud rate is 108 kbaud/s. An echo-cancelling technique is used whereby the 144 kbit/s of user data to be encoded and transmitted over the two-wire line is scrambled to randomise it and remove cross-correlation between it and the incoming data sequence. The binary data is then coded in an SU32 conditional block coder, filtered and transmitted to line. The reverse occurs in the receive direction.

Frame synchronisation is achieved by the use of a unique sync word. The baud rate of 108 kbaud/s (equivalent to a net line data rate of 162 kbit/s) enables simple support of all system and customer interface clocks. Fast start-up of the system is achieved by the use of a binary handshake procedure which allows separate training of canceller and equaliser.

An orthogonal timing signal is superimposed upon the SU32 line signal in such a way that the superimposed signal contains all the information necessary for symbol sampling. This timing signal does not compromise the performance or efficiency of the line code.

System implementation description An example of a typical line transmission system implemented by STC is shown in *Figure 57.18*. Regeneration, power feed and protection are not shown.

The main components of the system are implemented as a VLSI chipset of two chips, the digital signal processor chip (DSP144) and the analogue line transceiver chip (ALT144). Other components of the system consist of line transformers, power feed circuitry, protection devices and a small number of discrete devices such as crystals and decoupling devices.

The ALT144 contains two sets of pulse density modulation A/D and D/A circuits, S interface line interface devices, and other support elements which include a clock oscillator for the DSP144. It is a full custom bipolar design.

The DSP144 contains all the digital signal processing (DSP) and data formatting logic and is implemented as a semicustom design using standard cells, plus full custom layout for the more regular areas of the chip such as the echo canceller. It is implemented at present using a commercially available 2 μm CMOS process. In order to minimise costs the same two chips are used at both ends of the system. While a few per cent of each chip are unused at each end, this arrangement allows the usual S interface chip to be eliminated thereby enhancing cost-effectiveness. The multimode operation of the chip set also allows regeneration requirements to be accommodated.

57.8.5 ISDN TEs for basic access

A key feature in the exploitation of the network ISDN facilities that are becoming available worldwide is the availability of suitable ISDN terminal equipments and ISDN terminal adapters.

Although it is difficult precisely to classify ISDN terminal functionality, an attempt has been made here to group them into what might be called entry level devices, i.e. those devices which provide existing services in a more advanced way, and more advanced TEs which provide new services not previously generally available.

57.8.5.1 Entry level devices

Current ISDN terminal developments that provide existing services in a more highly featured manner include the following.

Digital telephones The basic telephony function is enhanced by the ISDN transmission and signalling facilities. Transmission of

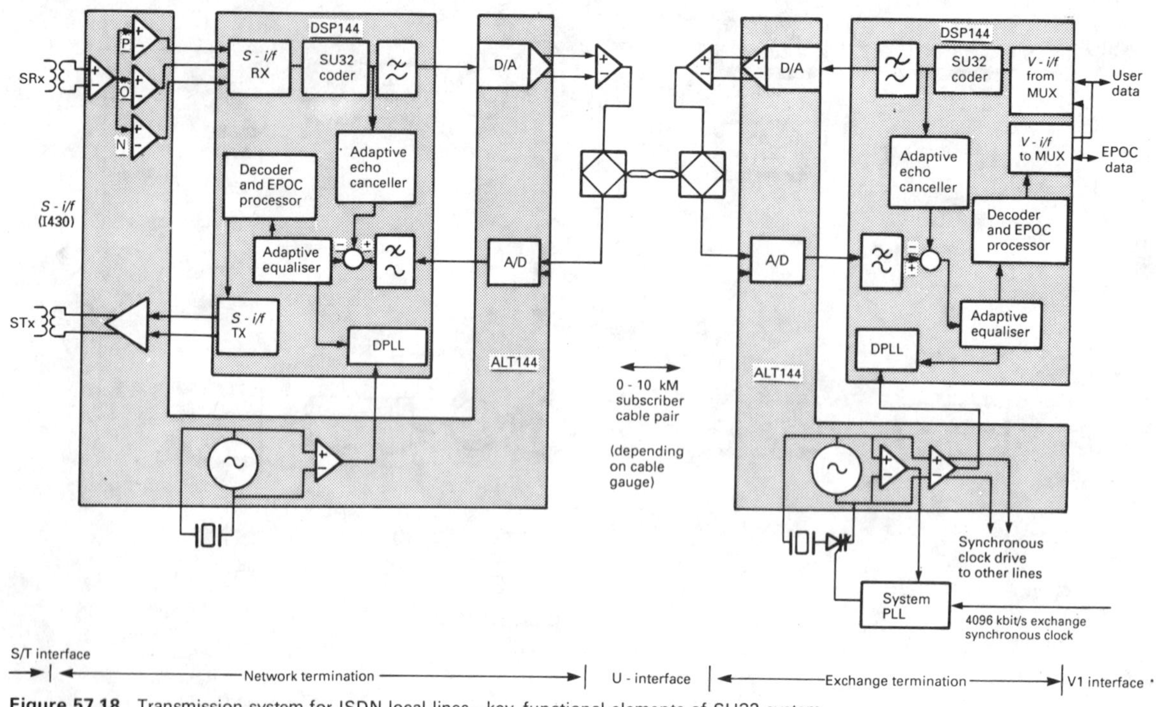

Figure 57.18 Transmission system for ISDN local lines—key functional elements of SU32 system

speech is not dependent on the distance of the customer from the local exchange. A small liquid crystal display (LCD) provides the customer with additional information about outgoing and incoming calls when supplementary services are invoked, e.g. calling line identification (CLI).

Voice/data workstation This unit combines the processing power of a personal computer (PC) and the communications facilities of a telephone, thereby creating a unit with powerful networking facilities. These units would normally be able to make and receive more than one call simultaneously, e.g. voice over B channel, data over D channel, and therefore have many applications in the modern office environment, even when data is transferred at pre-ISDN data rates (e.g. 9.6 kbit/s).

Terminal adapters Most terminals and hosts do not support the ISDN customer interface and therefore terminal adapters are required in order to convert terminal protocols to ISDN formats. Rate adaption, or conversion to analogue modem protocols via a device known as a modec, may also be performed within these units. TAs may be stand-alone devices or add-on boards that plug in extension slots within PCs.

57.8.5.2 Advanced TEs

These can be defined as TEs that take advantage of the unique capabilities offered by the ISDN basic access, namely the 64 kbit/s bandwidth availability and 16 kbit/s D channel. Some examples of this type of TE are the following.

D channel packet access terminals These provide an access for traffic that is of low density, but has a high calling rate, e.g. credit card verification.

Group 4 facsimile (64 kbit/s) Here a high-quality image is transferred via the ISDN in a matter of a few seconds. Data compression techniques are used. When allied to a laser printer, such a system is ideal for remote document retrieval, and is in use for example by the British Library for this purpose.

PhotoVideotex (Picture Prestel) This is similar to videotex, but of colour-slide quality. Access times are of the order provided by existing videotex networks (see *Section 57.9.4*).

Slow scan TV Here an image of a scene is transmitted at reasonable quality levels every second or so, thus allowing remote surveillance to be effected via the public network (see *Section 57.9.5*).

Videotelephones These have been demonstrated at 64 kbit/s, but doubts remain that a cost-effective quality service can be provided at this bearer rate.

LAN bridge These provide a gateway function to a local area network. Extension addresses in the ISDN protocols can be used to identify the LAN entity being addressed (see *Section 57.9.3*).

Digital key system (ISPBX)/multiplexer These perform the sort of functions that are described below for primary access systems, but are obviously limited in the bandwith they can access.

57.8.6 Primary access devices

Devices that access the primary access arrangement so far defined include the following.

ISPBXs Digital private exchanges (PABXs) are being enhanced to support ISDN as well as private network protocols (see also *Section 57.9.2*). These devices are known as integrated services private branch exchanges (ISPBXs), and normally provide access to the ISDN network at the primary rate. Users are offered both local and network ISDN features. The user access to an ISPBX may be via an I Series connection or through some proprietary terminal interface. In some cases an ISPBX may perform the function of a terminal adapter converting non-ISDN protocols into an ISDN format.

Multiplexers Multiplexers connected to a primary access may be basic access multiplexers (see *Section 57.8.2*) or may provide some other data or speech concentration function.

Host computer Here, the host computer provides access to its input ports via an ISDN primary rate interface, possibly via a multiplexer.

57.8.7 Future evolution

ISDN standards have been structured to allow considerable flexibility to be available for future evolution. This evolution will take the form of the additional definition of services and circuit transport modes. Advances in telecommunications technology also have to be catered for, particularly those that relate to increased bandwidth availability at the user–network interface. This section briefly summarises the developments that the authors feel are likely to be important in the future.

57.8.7.1 Basic access

Basic access standards are likely to be enhanced over this CCITT study period. Enhancements will include additional teleservices, supplementary services and more complete support of the various bearer modes.

57.8.7.2 Primary access

It is likely that a draft CCITT Recommendation for primary rate access will be completed over this study period. This will include private circuit applications. A stable specification is not expected until 1996.

57.8.7.3 Other access rates and bearer capabilities

More work will be done on these areas during this study period. Possible other access rates are 384 kbit/s and 704 kbit/s. The authors feel that 384 kbit/s may be developed further as the transmission technology used for 144 kbit/s access is capable of extension to this rate while providing service to the majority of customer lines worldwide, without regeneration. Applications such as videotelephony may be more commercially viable at this rate.

Bearer modes that will be studied include 32 kbit/s (e.g. for mobile services). Both have implications on switch design, as most switches switch at a channel rate of 64 kbit/s.

57.8.7.4 Packet access

This is being studied in detail, the main aim being more fully to integrate packet switching into ISDN networks, e.g. by using the I441/I451 protocols for packet circuit connections (known as the packet control plane operation), access to the ISDN being in the signalling plane. Packet access using the B channel as the data bearer (and the D channel for ISDN access set-up) will also be studied in more depth.

57.8.7.5 Cellular radio

At this time no detailed definition of the role of cellular radio has been made, primarily because existing cellular radio networks are analogue and work to a variety of standards. With the advent of the Pan-Europe Digital Cellular Radio Network in 1991 and beyond, interworking with and operation as part of an ISDN are likely to be studied in more detail. The key technical features to be addressed include the digital cellular radio bandwidth (16 kbit/s) and its effect on the bearer capabilities offered, the high error rates experienced on cellular links, which may require error correction to be employed for data traffic, and the nature of the signalling over the link.

57.8.7.6 Integrated broadband networks

General role of standards International research and study have commenced on customer access networks that integrate telecommunications and wideband services. These studies are based on the use of optical fibre as the transmission medium in the local network and terminating on residential as well as business customers premises. The work being done takes account of the introduction of cable TV networks on a more widespread basis world wide. Cable TV is at present an analogue service with limited switching and signalling capabilities. It is primarily unidirectional, and is based on proprietary standards. The integrated broadband communication networks (IBCNs) being defined rely on the use of digital transmission, switching and signalling techniques.

The CCITT, SGXVIII (ISDN) is examining access structures, services and network requirements. Draft recommendations are to be found within the I Series section of the *Blue Book*. In Europe, a large EEC-inspired programme known as the Research into Advanced Communications in Europe (RACE) initiative is attempting to define standards and develop the required technologies on a pre-competitive basis. This work is being coordinated by CEPT study groups, on behalf of the EEC, and is aimed towards large trial system development by approximately 1995. Both programmes drawn upon work that has previously been done on ISDN access standards.

Access structures The IBCN user data rate in total will be approximately 140 Mbit/s (full duplex), although 2 Mbit/s (approx.) access may be offered as an interim standard. From a transmission standpoint, use of this data rate fits into work being done on business LAN type transmission networks and work being done on synchronous multiplexers, both of which are based on a similar data rate. The 140 Mbit/s rate is also standard world wide, i.e. it is supported by both North American and European type networks.

The precise user interface access structure is not defined at the time of writing but can be expected to have the following general features:

(1) An ISDN interface, i.e. a 2B + D S interface, known as the S0 interface.
(2) An interface at 2 Mbit/s (S1 interface).
(3) Four interfaces at 34 Mbit/s (S2 interface).
(4) One LAN-type interface at about 140 Mbit/s (S3), i.e. S0 + S1 + (4 × S2 or/and S3).

The original intention was to transmit this information in a synchronous frame; however, developments in the study of asynchronous time division multiplexing (ATM) techniques has resulted in general agreement that IBCN will be based on their use for network transport and switching. ATM is a packet-based multiplexing technique in which short packets of data are 'labelled' for transmission over the network. Different types of data can have differing priorities in the transmission queue, e.g. to minimise transmission delay. The technique allows highly efficient network resource utilisation by using this form of dynamic bandwidth allocation. The key technical development of IBCN is the design of the high-speed ATM multiplexers that have to process this data, which will be located in the local and trunk networks of the future.

Services These are at an early stage of definition, but will include ISDN services, videotelephony services and compressed high-quality digital television services.

57.9 ISDN applications

57.9.1 Introduction

Some of the principal features of an ISDN which can be seen to make it attractive in the modern communications environment are:

(1) The availability of data transmission rates, previously obtainable only on private circuits, combined with the ease of use of the public switched telephone network.
(2) For basic rate service, the ability to provide this high data capacity over the local cable network currently used for analogue telephony.
(3) The use of digital techniques end to end, facilitating the automation of call set-up and clear-down procedures within applications.
(4) The availability of ISDN specific supplementary services such as closed user group and access to calling and terminating line identities to improve security of communications.
(5) The ability to connect a number of different terminals on a short passive bus, all accessed by the same directory number (e.g. group 4 facsimile, group 3 facsimile, LAN bridge and a telephone).

These are all features which are more obviously attractive to the business community. Consequently the spread of ISPBXs and the development of LAN gateways and LAN bridges will be key factors in the establishment of ISDN as a ubiquitous service with a flourishing infrastructure of applications development.

57.9.2 ISPBX

The ISPBX provides all the advanced voice and data facilities coming to be expected of a modern digital PBX and is, in addition, connected to the public IDN by one or more primary rate links. Thus ISPBX users can be offered a full range of high-quality voice and/or data services, including access at a basic rate interface.

For a medium-sized company using an ISPBX as the focus of its communications network or a larger company with a private network of interconnected ISPBXs, the introduction of an ISDN offers full facility access to remote users (*Figure 57.19*).

An example of this type of application would be in a bank or building society where there is a clear requirement for both voice and data traffic between the head office PBX and a large number of branch offices. This requirement is currently met with a mixture of digital private circuits to carry high volumes of data normally outside normal office hours, analogue lines for telephony, and possibly dial-up or leased access to the packet switching service for use by a cash dispensing machine.

Using the existing local loop for one analogue telephony line, a basic rate ISDN connection would support a replacement telephone with concurrent access to the packet switching service and would also carry the high-volume out-of-hours data. In addition, configuring the network as a closed user group would increase the security of the communications.

57.9.3 Local area network (LAN) applications

Local area networks linking together personal computers and workstations are becoming quite commonplace in the office environment (see also *Chapter 58*). The basic rate LAN gateway can be used for remote access to the network, and the basic rate

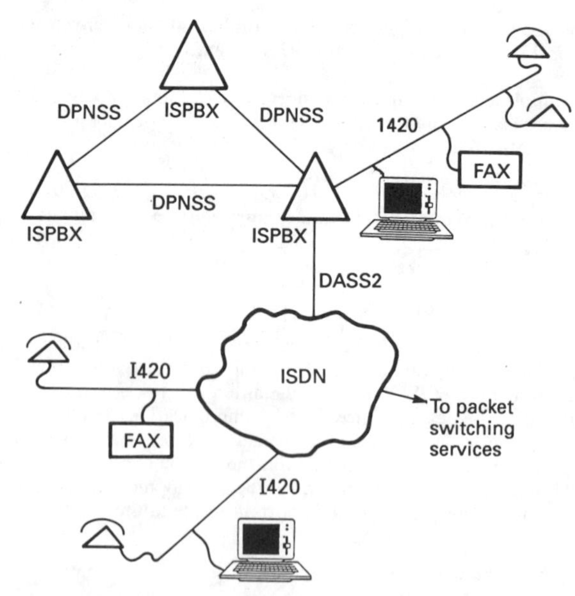

Figure 57.19 ISPBX applications

LAN bridge can be used to link compatible networks economically (*Figure 57.20*).

57.9.3.1 The LAN gateway

The LAN gateway, as the name suggests, provides one or more ports on the LAN for basic rate ISDN connection. Remote users with suitably configured terminals can access the LAN via the ISDN. Levels of access from simple up- or downloading of files to full facility access can be regulated on a call-by-call basis by the security arrangements built into the gateway.

57.9.3.2 The LAN bridge

The LAN bridge is essentially an extension of the gateway concept. The number of circuits connecting two (or more) LANs can be provided on the basis of the inter-network traffic. File server identities are mapped between the LANs, and a user has access to any file server as if it were part of his or her own network.

The principal advantage of using the ISDN for both gateway and bridge connections for a LAN is that the connections are not permanently active. In many instances an intelligent terminal will download an application from the LAN and will thereafter require only infrequent availability of the LAN services, perhaps to access databases or use common facilities such as printers, analogue modems or electronic mail services. If users wish to print a file residing on a remote LAN on their screen or local printer, they will use the same syntax as if it were a local file. Their local equipment will recognise the remote location of the file and will generate a call across the digital network. The call set-up time is such that this activity will be essentially invisible to the user, and the capacity of the link is such that the terminal response will be little different than if it were a local file being accessed. At the end of the print call the link will be cleared and, of course, charges cease. Thus, unlike with a private circuit, the user pays for the link only for the time for which it is active.

57.9.4 PhotoVideotex

PhotoVideotex is a personal computer based application which enables television quality still pictures to be retrieved from a

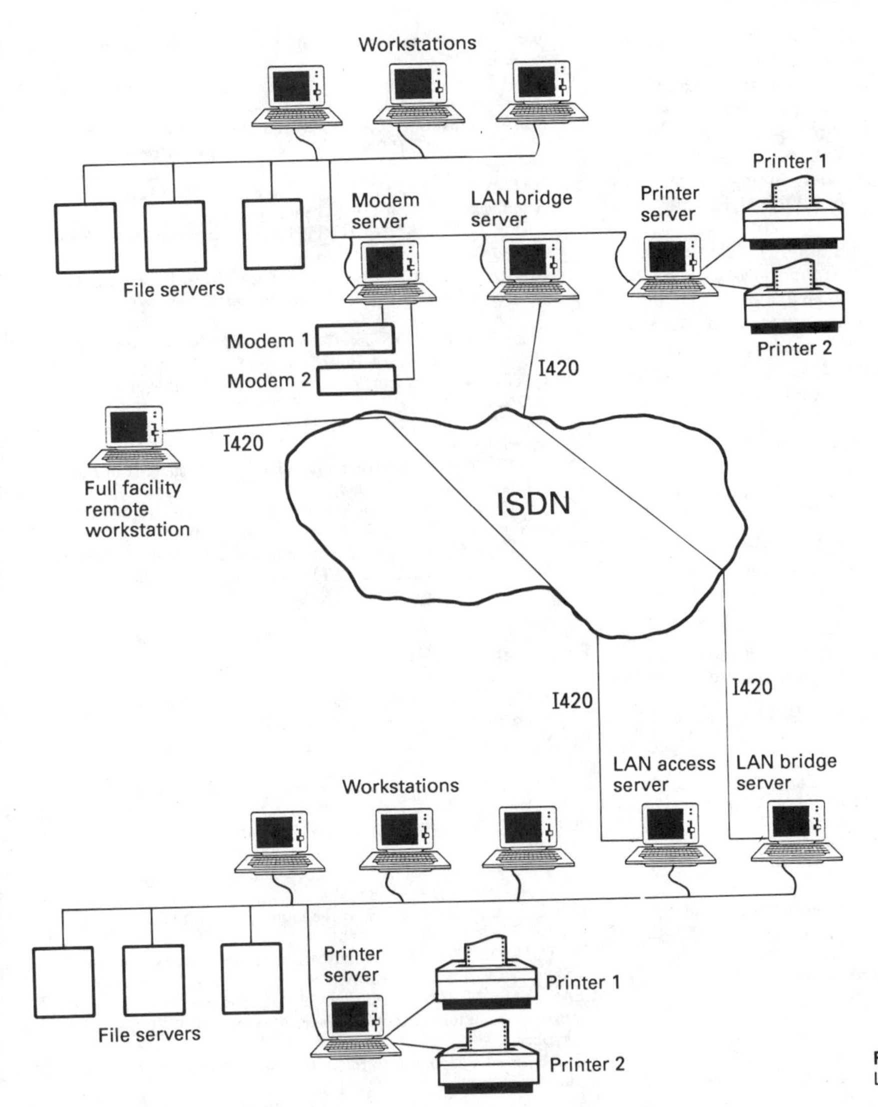

Figure 57.20 LAN bridge and LAN access applications

central database in around 6 s. A service based on this application would enable estate agents throughout the country to access local databases holding high-quality pictures of available housing stock for clients wishing to move into an area. The international origins of the ISDN specifications suggest that such a service could even span national boundaries within the foreseeable future. The plans for international ISDN service also suggest obvious applications for PhotoVideotex services to the travel industry.

PhotoVideotex combined with a conventional database could be of great value in the medical world, enabling a patient's full medical records, including high-quality pictures from radiographs or scans, to be available to any doctor within seconds.

The travel industry is also the source of another novel application of basic rate ISDN used to support slow scan television cameras.

57.9.5 Slow scan TV

The 64 kbit/s capacity of a basic rate ISDN connection enables a new television picture to be produced every 5 s. The existence of a full duplex communication channel provides total control of the remote camera, giving facilities for pan, tilt, zoom and focusing, and even allowing for remote lens cleaning.

Apart from broadcast advertising (posters, newspaper advertisements, television commercials, etc.), the primary medium for promoting holiday resorts in the UK or overseas is through photographs in travel agents' brochures. The advent of ISDN could offer new opportunities to the resorts. Strategically sited cameras covering the most inviting features could be accessed throughout Europe or the USA allowing prospective travellers to be better informed when choosing their destinations.

A more conventional application of slow scan TV technology is in the surveillance of high-security premises, or even traffic monitoring on accident-prone stretches of busy roads or motorways. A simple connection to the local telephone network instead of the time and expense of providing special cables from each camera position to the central control will vastly increase the range of applications for which this solution is economic.

57.9.6 Domestic application

ISDN is new technology which will be constrained in its early days by the costs of the enhancements to the existing telephony centred network to support its particular facilities. The development of specialised terminals to make effective use of those facilities will also be an expensive, if stimulating, exercise. Nevertheless many executives will wish to benefit from the same communications facilities enjoyed in the office whilst they are at home. This, together with the possibilities for home-based working for such services as computer programming and general typing, could provide the introduction of ISDN services into the domestic market.

However, on past performance, there can be little doubt that within a very few years a digital telephone connected to an ISDN line will compare favourably in cost with an analogue telephone and will have the advantage to the customer of the second 64 kbit/s channel and the D channel for access to new packet-based services. When this point is reached the ISDN will fulfil its potential of being the 'PSTN of the 21st century'.

References

1 CCITT, I110 General structure of I-series recommendations, *Blue Book* (1988)
2 *CEPT Recommendation T/CS 46-10,* ISDN user–network layer 1 specification—application of CCITT Recommendation I430
3 *CEPT Recommendation T/CS 46-20,* ISDN user–network layer 2 specification—application of CCITT Recommendations Q920/I440 and Q921/I451
4 *CEPT Recommendation T/CS 46-30,* ISDN user–network layer 3 specification—application of CCITT Recommendations Q930/I450 and Q931/I451
5 ANSI/T1D1.2, *Minimal User–Network Signalling Specification for the ISDN Basic Access Interface*
6 BRITISH TELECOM, *Digital Access Signalling System No. 2 DASS 2 (BTNR 190)*. Part 1, Vol. 1, Core Document
7 BRITISH TELECOM, *BTNR 191–CCITT I-Series Interface for ISDN Basic Access* (British Telecom Version)
8 BRITISH TELECOM, *Digital Access Signalling System No. 2 DASS 2 (BTNR 190)*, Part 3, CCITT I-Series Multiplexor/Concentrator Specification
9 CCITT, I420 Application of I-series Recommendation to ISDN user–network interface: basic user–network interface, *Blue Book* (1988)
10 CCITT, I430 ISDN Basic user–network interface—layer 1 specification, *Blue Book* (1988)
11 ISO, *Draft International Standard DIS 8877*: Specification for interface connector and contact assignments at the ISDN basic access interface located at the reference points S and T
12 CCITT, I440 ISDN user–network interface data link layer—general aspects, *Blue Book* (1988)
13 CCITT, ISDN user–network interface data link layer specification, *Blue Book* (1988)
14 CCITT, I450 (Q930)/I451 (Q931) ISDN user–network interface layer 3 general aspects/specification, *Blue Book* (1988)
15 CCITT, I211 Service aspects of ISDN—bearer services supported by ISDN, *Blue Book* (1988)
16 CCITT, I421 Application of I-series Recommendation to ISDN user–network interface—primary rate user–network interface, *Blue Book* (1988)
17 CCITT, I431 ISDN primary rate user–network interface—layer 1 specification, *Blue Book* (1988)
18 BELL COMMUNICATIONS RESEARCH (BELLCORE), TA-TSY-000397, *ISDN Basic Access Transport Requirements* (Section 7.5)
19 BRITISH TELECOM, D2978—*Specification for Basic Access Multiplexor*, Section 13, D-channel packet access
20 TELECOM AUSTRALIA, TPH 1856, *Australian ISDN: Primary Rate Access Interface*
21 SIEMENS (USA), EWSD, *ISDN User–Network Interface–Layer Protocols for Primary Access*, Document Nos A30808-X3033-X304/5/6
22 CCITT, I412—ISDN user–network interfaces, interface structures and access capabilities, *Blue Book* (1988)
23 CCITT, I462—Support of packet mode terminal equipment by an ISDN, *Blue Book* (1988)
24 BRITISH TELECOM, BTNR188, *Digital Private Network Signalling System No. 1*
25 CCITT, I460 Multiplexing, rate adaption and support of existing interfaces, *Blue Book* (1988)
26 CCITT, I461 Support of X21 and X21 bis based data terminal equipment (DTEs) by an ISDN, *Blue Book* (1988)
27 CCITT, I463—Support of data terminal equipments (DTEs) with V-series type interfaces by an ISDN, *Blue Book* (1988)

Further reading

BAUER, B., 'Private integrated networks: a trigger for public demand', *Telecommuncations*, October (1988)
BLANC, J., 'Efficient ISDN power converters', *Electron. Product Design*, September (1988)
BRAVE, J., 'Will the local loop choke ISDN?', *Data Comm.*, October (1988)
DOMANN, G. H., 'B-ISDN', *IEEE J. Lightwave Tech.*, November (1988)
FREUCK, P. *et al.*, 'ISDN basic access terminal adaptors', In: *Networks for the 1990's*, Reardon, R., ed.
HABARA, K., 'ISDN: a look at the future through the past', *IEEE Comm. Mag.*, November (1988)
KELTON, D., 'The use of digital backbone multiplexers in private ISDN', *Proc. Int. Conf. ISDN*, London, June (1988) online
LEAKEY, D. M., 'Integrated series digital networks: some possible ongoing evolutionary trends', *Comp. Network ISDN Systems*, October (1988)
RALPH, B., 'ISDN and Centrex', *Proc. Int. Conf. ISDN*, London, June (1988) online
SALMONY, M., 'The potential of broadband ISDN', *Comp. Stand. Interf.*, **8**, No. 1 (1988)
STIGTER, J., 'Recent developments in broadband ISDN', *Proc. Int. Conf. ISDN*, London, June (1988) online
TILL, J., 'Sorting out the buses that bind ISDN chips', *Electron. Design Int.*, September (1988)
WRIGHT, R., 'Data networking with ISDN', *Telecommunications*, September (1988)

58 Local Area Networks

J Houldsworth
Manager, Open Systems
International Computers Ltd, UK

Contents

58.1 Introduction

The position on local area network (LAN) standards is now quite firm. Standards for the lower four layers, collectively known as the 'transport function', are in place. Specific combinations of standards in these layers, known as 'profiles', are being selected for LAN system procurement.

Manufacturers must conform to these OSI standards and profiles in order to provide the multi-vendor 'open' local area network environment that users are now beginning to demand.

This chapter describes the Open Systems Interconnections (OSI) architecture, which is the basis for all the networking standards, and the current LAN standards and profiles. It also discusses the future standards programme and the emerging relationship between local area networks and PABXs, which is beginning to receive more attention under a new initiative, known as integrated services PABXs (ISPABX) and integrated services local area networks (ISLN).

58.2 The Open Systems Interconnect architecture—OSI

The classical Open Systems Interconnection architecture (*Figure 58.1*) is a seven-'layer' model in which each layer represents a key function in the system operation. Each layer is labelled with its basic function, which is reasonably self-explanatory. Seven is not a magic number: it was chosen because the organisations involved in its creation agreed that seven is appropriate for achieving a manageable analysis of the functions involved in data communication.

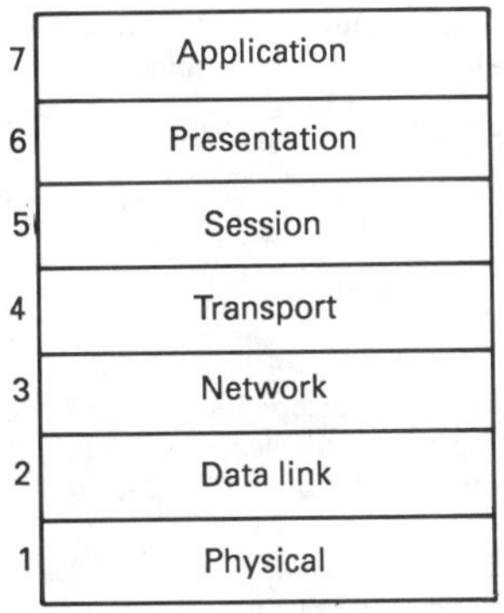

7	Application
6	Presentation
5	Session
4	Transport
3	Network
2	Data link
1	Physical

Figure 58.1 The OSI seven-layer reference model

Alone, this simple model has done more for open systems than anyone could have imagined. Having created the template for the generation of all other standards for interconnection and interworking, the prospect of open systems is now a reality. Indeed, most users now insist that any system that they buy is designed according to the principles of the OSI model.

The seven layers are grouped 1–4 for interconnect (more commonly known as the 'transport' function) and 5–7 for interworking.

58.2.1 The functions of each layer

Physical layer This handles the electrical and mechanical interface to the communications media, and the procedures for activating and deactivating the connection, and converts the data for transmission over the media.

Data link layer This provides the synchronisation and error control for the information transmitted over the physical link.

Network layer This establishes, maintains and terminates connections between end systems, including addressing, routing and facility selection.

Transport layer This provides end-to-end control and information interchange with the level of reliability that is needed for the application. The services provided to the higher layers are independent of the underlying network (see *Section 58.2.2*). The transport layer acts as the liaison between user and network.

Session layer This supports the dialogue between cooperating application processes, binding and unbinding them as communication relationships are needed.

Presentation layer This allows the application process to interpret the information exchanged. Presentation agrees the syntax to be used and the means for local translation.

Application layer This provides a service to support the end-user application process and manage the communication.

58.2.2 Layer services

Each layer is responsible for providing a service to the layer above it and for maintaining a relationship with the equivalent layer in the end-system with which it is communicating.

An OSI 'service definition' is produced for each layer and this defines the service it provides to the next higher layer. The service definition states the mandatory and optional functions that are the responsibility of that layer. It also defines the outline commands and responses, known as 'primitives', which are included in 'protocol data units' (PDUs) which are exchanged between the two layers to invoke the services and respond to requests.

The detailed encoding of the primitives is defined in 'protocol standards', which are produced to specify how the service is met in the required environments. The protocol standards are then implemented in either hardware or software, as appropriate.

The ISO and the CCITT are aligning their service definitions for each of the seven layers.

58.2.3 Simplifying the model

Layers 1–4 (the transport function) contain the interconnection or 'bearer' elements and layers 5–7 (the users of the transport function) contain the interworking elements.

The transport function aims to create a transparent interconnection environment over which the interworking functions in the interworking elements can run independently of the transport media.

The transport layer is the key layer in the model because it forms a neat dividing line between interconnection and interworking. It quarantines the upper three layers from the network, allowing standard applications to run over the various types of local and wide area networks which are needed to suit the specific practical interconnection requirements.

The division at transport layer allows a simple concept of the model to be evolved (*Figure 58.2*) around the two basic layers, 'interconnect' (which joins things together) and 'interworking' (which makes sure that they all understand each other). This figure also indicates the need for gateways in the interworking area. They are required to communicate with systems that use proprietary interworking protocols, but the need will diminish as more manufacturers introduce standard OSI interworking protocols.

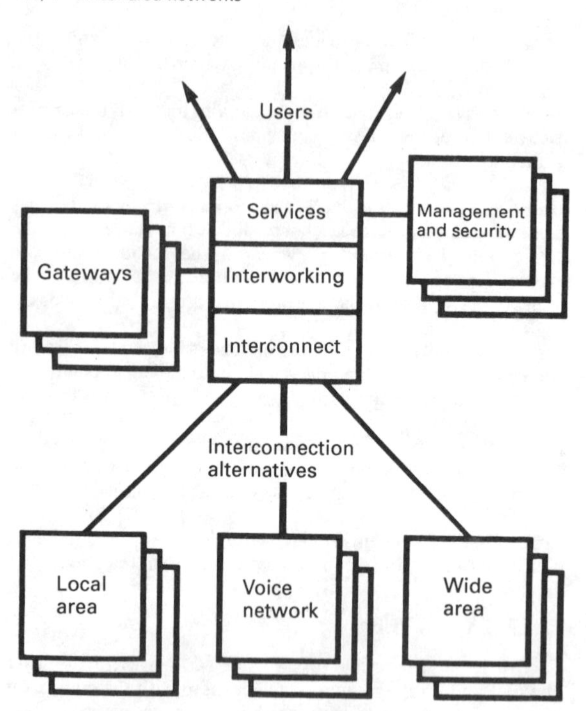

Figure 58.2 Simplified representation of the OSI model

Figure 58.2 is completed by a third grouping, which represents the overlay services that span several layers of the model, such as system management, security and directory control.

58.3 Standards organisations

In the data communication area, two bodies have virtually dominated the standards making process. These are the International Telegraph and Telephone Consultative Committee (CCITT) and the International Standards Organisation (ISO).

Their responsibilities divide, broadly, into the standards necessary for providing a public voice and data network (CCITT) and the standards that provide an end-to-end service over the OSI network. However, there is an increasing overlap in the wide area network environment and the two organisations maintain a close relationship in this area and in the general definition of open system architecture, which is a common template for them both. The CCITT does not produce standards for local area networks, and the ISO is regarded as the key standards committee for this work. However, the introduction of local area networks has also brought new participants into the standards scene.

The first of these is the IEEE in the USA, which is currently leading the technical work on the base LAN standards. The work of the IEEE on LANs is brought to international standards level by the ISO and several of the IEEE standards have already progressed to ISO standard level. The cooperation between the IEEE and the ISO is now so good that the ISO has delegated the printing and distribution of the agreed ISO standards to the IEEE.

The other organisation is the International Electrotechnical Committee (IEC), which became involved in LAN standardisation through its work in specific fields of application, such as process control. They also formed liaisons with the IEEE and, more recently, they have made a joint agreement with the ISO to combine the work on information technology standards under joint ISO–IEC committees. This is bound to accelerate the work.

Another key initiative, which is also accelerating the work, is the selection of preferred combinations of the base standards in the interconnect layers 1–4 and the interworking layers 5–7. These are sometimes referred to as 'functional profiles' or 'functional standards'. Such profiles have already been created for local area networks.

This logical grouping of standards is being exploited by several user groups, including government procurement agencies and the standards organisations. Functional standards are option free and rigidly defined and they can be used by independent test houses for formal conformance testing and certification.

The functional standards activity began in Europe in the SPAG manufacturers Group. The European work has now been adopted by the CEC standardisation committee consortium, CEN/CENELEC/CEPT, which is preparing European standard profiles. General Motors and Boeing have spearheaded an activity to select manufacturing automation protocols and technical office protocols (MAP/TOP). The British and US governments have set up the UK GOSIP and US GOSIP committees to establish preferred government OSI procurement profiles. The US manufacturers have set up the organisation known as COS to establish the preferred US profiles. ICL is the first European company to join COS and has been given a seat on the COS board. ICL is using its joint membership of SPAG and COS to pull the European and US initiatives together. Japan has also set up a similar organisation, called POSI, which has links to COS and SPAG.

The ISO has recently established an activity to put the International seal on the profile work. They are using the other profiling groups as feeder organisations and will produce international standard profiles (ISPs).

The functional standards activity has moved OSI from being a collection of standards to a new role as a procurement tool.

58.3.1 CCITT

The International Telegraph and Telephone Committee (CCITT) is responsible to the International Telecommunications Union (ITU), a treaty organisation formed in 1865 and now a unit of the United Nations.

The CCITT works in 4-year cycles and publishes a complete update of all its regulations following the plenary session which terminates each 4-year cycle. The last 4-year period terminated in 1988.

There are several levels of membership in the CCITT. The first level is the administration, which is the government telecommunications organisation of the member country. Since telecommunications services are provided by the private sector in North America, the State Department in the USA and the Department of Communications in Canada are the members. In the UK the Department of Industry is the official member, but many technical activities are currently coordinated by British Telecom.

The second level of membership covers the private telecommunications carriers which are known as 'recognised private operating agencies' (RPOA). In the USA, ATT and GTE Telenet are example RPOAs. In Canada the Canadian Telecommunications Carriers Association is the sole RPOA.

Although CCITT is principally in support of the telecommunications carriers, industrial and scientific organisations are also given a level of membership to participate in study group activities, but they do not have a right to vote.

58.3.2 ISO and IEC

The International Organisation for Standardisation (ISO) is a voluntary organisation involving the national standardisation activities of each member country. The voting members are the national standards bodies, which draw their membership principally from the manufacturing and users communities. The BSI IST Committee are the participants for the UK; ANSI X3 is the USA participant.

ISO does not work in a structured 4-year cycle and can produce standards at any time. The normal cycle of events is that a Draft Proposal for a Standard (DP) is produced by a subcommittee with input from the member bodies. This is then subjected to a postal ballot and, if 70% of voting members are in favour, it is elevated to the status of a Draft International Standard. It is then subjected to a similar balloting process at the Technical Committee level and then to an ISO council vote for ratification before final publication as an International Standard.

The IEC is also a voluntary organisation, dealing with standards in the field of electrotechnology. Its membership is similar to that of the ISO.

As explained previously, the IEC has been moving into specific application areas and there were liaison difficulties between the ISO and the IEC. A joint agreement has brought the two organisations together in the field of Information Technology Standards under a new Joint Control Committee JTC1. The new grouping will continue to use the ISO balloting procedures described above.

In the new structure, Committee ISO–IEC/JTC1/SC6 deals with the standards in the lower four layers, including all the LAN standards. It has four working groups:

(1) *Working Group 1* Data link layer, developed HDLC and LAN logical link control LLC.
(2) *Working Group 2* Network layer, also maintains close liaison with CCITT providing ISO opinions on work such as X.25.
(3) *Working Group 3* Physical layer, responsible for the physical standards, including the mechanical connectors and electrical signalling.
(4) *Working Group 4* Transport layer, responsible for the various classes of transport standards.

ISO–IEC/JTC1/SC21 deals with the general OSI architecture and the maintenance and enhancement of the architectural reference model—ISO7498. It has several working groups but the key ones in the LAN context are:

(1) *Working Group 1* Reference model.
(2) *Working Group 4* Application and system management.

58.4 Protocol standards

The model caters for a wide variety of environments. It classifies the various communications functions, and each layer is responsible for carrying out its allocated part of the overall process according to the specific environment.

An obvious example is the physical layer, which needs to be adapted to match a variety of environments ranging from cabled local area networks to satellite-connected wide area networks.

The advantage of OSI is that it isolates these differences to the lower few layers so that the higher layers are unaware of the different characteristics of the physical networks, and the same higher layer protocols work on any kind of physical network.

If the implementation in one of the layers provides a reduced service, the rules of OSI permit this to be compensated for in a higher layer to bring the overall service to the standard level.

58.4.1 Physical layer

The physical layer relates to the physical connection to the wide area or local area facility (plugs, sockets and pin allocations) and any conversions necessary for communicating the serial bit stream through the appropriate media. This includes matching the electrical characteristics, sending and reacting to control signals, analogue-to-digital conversions, bit timing, etc.

In the LAN environment the physical connection is referred to as the media access unit (MAU) and the actual connection is very specific to the LAN type.

58.4.2 Data link layer

The data link layer is responsible for deriving a sensible message format from the serial bit stream and providing formatting to allow the remote data link layer to be able to make sense of the returned transmissions. It maintains an intimate dialogue with the communicating data link layer to allow the link to be established in the first place and for the appropriate level of data and link recovery to be effected.

All the current OSI data link control procedures have been derived from a standard which is known as high-level data link control (HDLC). The standard HDLC frame format includes an address field to define the destination of that frame, and a control field to carry the frame number and the number of the last received frame, in addition to control commands and supervisory responses. It closes with a frame checking sequence for error control. The HDLC frame format is shown in *Figure 58.3*.

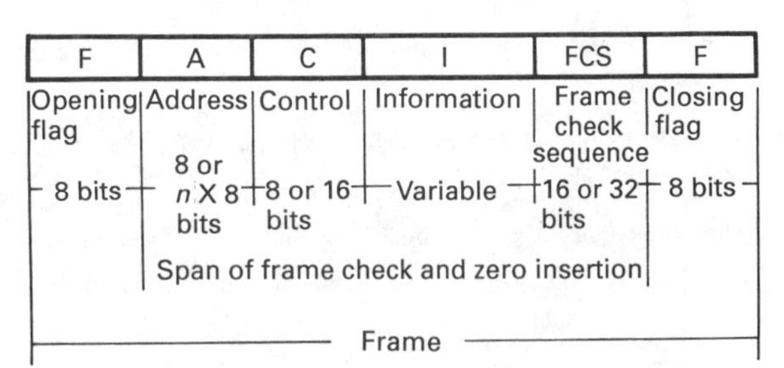

Figure 58.3 HDLC frame format

Each station in a specific dialogue sends frames with sequential sequence numbers and maintains a record of those frame numbers. It monitors the received frames, which contain the sequence number of the last frame the other station has received correctly, and continues to send frames if the sequence number continues to be indexed. It must stop transmission and wait for more responses when the end of its transmission numbering cycle is reached without having received a response for the first frame in that numbering cycle, otherwise it would have two different frames with the same number awaiting a response. This is called the transmit window—eight in the basic version of HDLC, with an option to extend to 128 in special cases. If no response is received (a timeout is used to check this), or if a direct rejection is signalled in the supervisory part of response frame, the station enters a retransmit cycle to effect recovery.

The link layer family for local area networks, known as logical link control (LLC), uses the standard HDLC elements of procedures (with minor detailed variations which must be checked against the specification) but adds a second address to define the source of the data link frame. This is necessary because HDLC originally assumed centralised control, where all transmission is to and from the control station, and LLC must assume free transmission from any station to any other.

58.4.3 Network layer

The network layer deals with the routing of complete messages to the addressed end-systems through intervening networks and with provision of the required facilities and quality of service. The richness of functionality in the network layer depends on the facilities of the intervening network and its routing and control mechanisms.

The Connection Oriented Network Service is matched by the network layer of CCITT X.25 but the emergence of this protocol for use in LANs is only just occurring. The most popular operational standard for the network layer is the 'connectionless' protocol, which is specified in an extension to the network protocol standards series. This allows transmission to take place without first transmitting a LAN frame and receiving a response to confirm that the receiving station is connected. In LAN systems, this is generally regarded as a superfluous action, since no network routing and connection is involved. The use of connectionless network control must be accompanied by the use of class 4 transport (*Section 58.4.4*).

Within an open system the addressed end-system is known as the network service access point (NSAP). A mechanism is being set up by the BSI to allocate OSI NSAP addresses within the UK.

58.4.4 Transport layer

It is vital to provide a standard service from the transport layer, regardless of the richness of the service provided by the individual layers below it. There are several standard classes in the transport layer which can be used to compensate for the different network options, which are provided by layers 1–3.

Some LAN standards groupings permit the omission of the network service altogether and still offer a standard service by choosing the correct transport class.

The transport layer has five classes from which a selection can be made to provide an increasing richness of facility, which compensates for differences in the service offered by the lower three layers. The transport class to be invoked can be negotiated when the connection is established but is usually fixed.

(1) *Class 0* This is the basic class which is for use in networks that have a good error performance and a minimum of interrupts in transmission.
(2) *Class 1* Class 0 plus a recovery feature for errors and interrupts which are signalled by the network layer, avoiding the need to handle these in the application.
(3) *Class 2* The facilities of class 0 but with a multiplexing feature.
(4) *Class 3* A combination of the facilities of classes 1 and 2, providing both error recovery and multiplexing.
(5) *Class 4* Adds the facility to detect errors and out-of-sequence data which are not signalled by the lower layers.

Classes 0 and 2 are for a rich network service environment which is reasonably error free.

Classes 1 and 2 are for networks with a high level of network interrupts.

Class 4 is for use where it is necessary to handle a high level of residual errors which are not reported by the network layer.

58.5 Local area network architecture and profiles

The local area network standards family is mapped on to the model as shown in *Figure 58.4*. Each ring or bus system is specified as a media access control (MAC) service, over which a logical link control protocol (LLC) is used to complete the layer 2 functions.

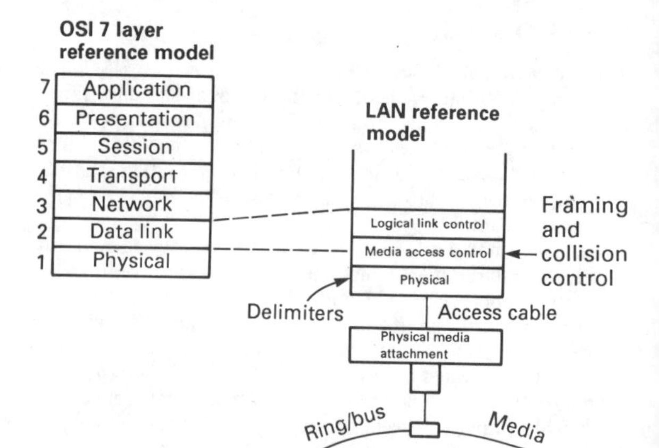

Figure 58.4 Mapping of the LAN standards family on to the OSI reference model

The original target was to produce a single media-independent LLC for all MAC types, but two main options are now contained in the standard; LLC1, which does not include LAN packet acknowledgements, and LLC2 with full acknowledgement and error recovery tools.

LLC1 was created to fit into the scenario known as connectionless, where it is assumed that all the messages pass all the stations and can be extracted by the addressed station. In such systems there is no need to pre-establish a logical connection to the addressed station before sending data; hence the term 'connectionless'.

LLC2 was created to fit into another scenario, known as 'connection-mode', where a logical route is established between the sending and the addressed station before data is transmitted.

Connectionless and connection-oriented solutions may be selected to suit the specific environments. Connectionless is currently the most popular in the LAN environment, whilst connection-oriented solutions (usually X.25) are most often used in WAN environments.

LLC1 and LLC2 are used to form the lower layers of the connectionless and connection-oriented profiles, respectively. The ISO standard permits the use of either LLC1 or LLC2 with the X.25 profile but LLC2 is the conformance standard.

UK GOSIP and CEN/CENELEC are also specifying the X.25/LLC2 profile as an alternative to the connectionless profile. It is quite easy to gateway between the two scenarios, and they may even coexist on the same physical LAN.

A third logical link control variant (LLC3), which provides immediate response opportunity, is being specified by the ISO. This is for use in environments, such as process control, where the return of responses is time critical.

Specific versions of the transport layer (see transport classes, *Section 58.4.4*) are used to complete each layer 1–4 profile. If crossover is required between the scenarios, it is normally implemented at layer 3, but may also be carried out at layer 4 in a 'distributed gateway'.

58.6 LAN media access control standards

Many LAN types have been developed in the commercial world, but the ISO has standardised four media access control (MAC) systems. They divide into two main classes—bus and ring.

The initial standards thrust was within the IEEE in the USA which developed three MAC types. These three standards were adopted in 1983 by the ISO and are about to be published as full International Standards.

A fourth MAC service has been adopted by the ISO, known as the slotted ring. This standard arose from a UK submission, based on the original Cambridge ring system. It has been assigned the reference 8802/7 to avoid confusion with the IEE reference number 802/6, which is assigned to the committee working on municipal area networks (MANs).

The four MAC standards are described in outline below with the three MAC types which originated from the IEEE first.

58.6.1 ISO 8802/3—CSMA/CD

This is a baseband contention bus system. Baseband means that there is no carrier frequency, all transmission is digital. Bus means that all the stations are connected to the same physical media and all of them can monitor each other's transmissions. Contention means that no station is in control of who transmits and all stations are free to contend for the use of the bus medium. A mechanism must be included to resolve the problems which occur when two stations try to transmit simultaneously.

The standard is known as CSMA/CD (carrier sense, multiple access/collision detect), which is a reasonably descriptive label for how the system works.

The basic standard is for a 10 Mbit/s coaxial cable system (a shorthand way of referring to LAN systems) and this standard is known as 10base5 for 10 Mbit/s, baseband and 500 m maximum LAN segment length. There are several alternative enhancements being processed for alternative media. These are all described in *Section 58.8*.

58.6.2 ISO 8802/4—token passing bus

A token-passing system is normally operated over a broadband bus medium. Token passing means that only one station at a time, the station holding the token, can transmit. When the station has completed its transmission it relinquishes the token to the next designated station, which assumes the right to transmit. The token-passing cycle is carried out to a prescribed order, but the order can be changed from time to time.

Broadband means that the transmission is modulated on to a broadband carrier. Transmission is at 10 Mbit/s, modulated on to a coaxial cable using broadband transmitters and receivers at the drop points. The broadband cable carries several carrier frequencies and a 'head end' transceiver ensures that all the 'transmissions' can be received by all the stations.

The token bus system was originally designed for use in the process control industry and is expected to be widely used for shop floor applications in computer integrated manufacturing (CIM).

A media variant, known as 'carrier band', with a single carrier frequency running in cheaper coaxial cable, is available for lower cost distribution from shop floor controllers to the actual control points (cells). This enhancement provides reduced speed transmission at 5 Mbit/s. It has been adopted by the IEEE but it is not yet a part of the 8802/4 specification.

58.6.3 ISO 8802/5—token passing ring

This is another token-passing system, which uses a physical ring configuration operating at baseband. The ring configuration ensures that data passes round the network and, if it is not extracted from the media by the addressed station, it returns to the sending station and this effect is used in control. The token passing discipline is similar to that of the token bus.

The standard currently specifies 4 Mbit/s baseband operation over screened twisted pair wiring with the rings configurable as primary and secondary 'lobes'. Enhancements to increase the speed to 16 Mbit/s are commercially available but this enhancement is not yet in the ISO standard.

58.6.4 ISO 8802/7—slotted ring

Slotted ring is a baseband ring system which uses short, fixed-length LAN packets, defined as 'slots'. The slots are marked 'full' or 'empty' and any station can use any passing slot which is marked empty. This allows all the stations to have a regular access to the LAN without having to wait for messages to be completed before the token is passed.

As in the token ring, the slots pass all the way round the ring if they are not extracted by the addressed station. This is used by a 'monitor ' station to control the use of the LAN.

The slotted ring standard specifies 10 Mbit/s baseband operation over screened twisted pairs. There are currently no plans for enhancements to either the media or the speed.

58.7 Relationship between CSMA/CD LANs and token bus

The most popular of the media access service standards is the CSMA/CD baseband bus, previously known as Ethernet, which is targeted to satisfy the multi-vendor office and data processing markets. Considerable investment has been put into the implementation of this standard by the silicon manufacturers, and there are at least 200 vendors worldwide who supply equipment to this standard.

There are also some new innovations at the physical layer, which allow cheaper cable and transceivers to be used with minor geographical restrictions, which will be described later.

The major thrust behind the token bus system was from the manufacturing process control industry, and in particular from General Motors who started the standards initiative which is widely known as MAP (Manufacturing Automation Protocol). General Motors wanted to procure automation cells from a wide range of suppliers and know that they would all interwork.

A companion thrust for the technical office was started by Boeing who had similar initiatives in their design offices. This became known as TOP (Technical and Office Protocols). TOP specifies the CSMA/CD LAN standard.

The terms MAP and TOP are now completely synonymous and the profiles that the two organisations created are very similar. The ongoing MAP/TOP work is shadowed by organisations worldwide and there is a European MAP/TOP Users Group (EMUG), supported by the interested European manufacturers and users.

It is clear that both token bus and CSMA/CD will coexist in some environments to cater for the specific requirements of combined manufacturing and technical office/data processing systems with gateways between the two subnetworks.

58.8 CSMA/CD configurations

The standard CSMA/CD bus uses a multiple-sheath coaxial cable with a solid central conductor. There are some physical constraints in the installation specification, such as minimum bending radius, but these are not a major problem. A basic bus segment may be up to 500 m in length, and up to 100 transceivers can be inserted in the cable at intervals of 2.5 m or multiples of this distance (*Figure 58.5*). A terminator must be used at each end of each segment to avoid reflections.

The transceivers are attached using a screw-in cable connector, known as a 'bee sting'. The bee sting is screwed in, first to pierce the cable and, as it is screwed in further, to make the appropriate connections to the centre conductor and the sheath. These have the advantage that they do not require the cable to be cut and can be installed whilst the system is operational.

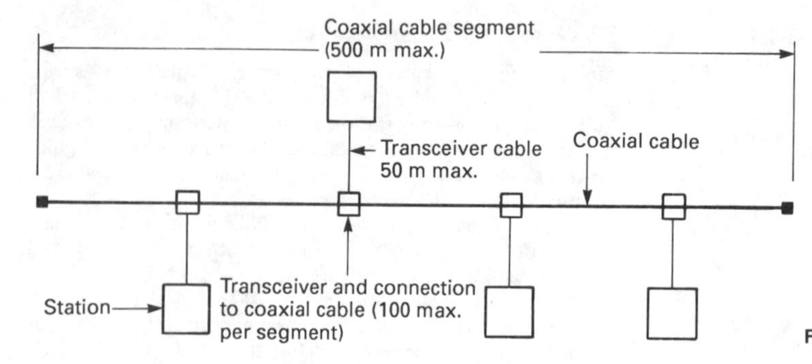

Figure 58.5 CSMA/CD single cable segment

The transceiver drop cable has four twisted pairs, is screened overall and may be up to 50 m in length. The pairs in the transceiver cable carry the transmit and receive signals, a collision detect indication and the power for the transceiver. This is known as the access unit interface (AUI). Passive splitting adaptors are available to allow up to eight units to share a single transceiver.

Typical LAN installations have the main cable and transceivers in cabling ducts, under false floors or in false ceilings, with the transceiver drop cables threaded to the most convenient wall or floor position, where they appear as a 15-pin socket.

Repeaters can be installed to extend the length beyond 500 m, as shown in *Figure 58.6*. They interconnect segments of the LAN so that they operate as a single working cable with common 10 Mbit/s end-to-end timing.

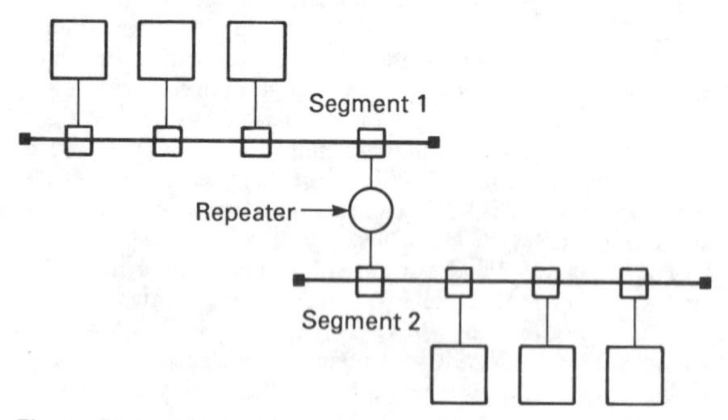

Figure 58.6 CSMA/CD two segments with repeater

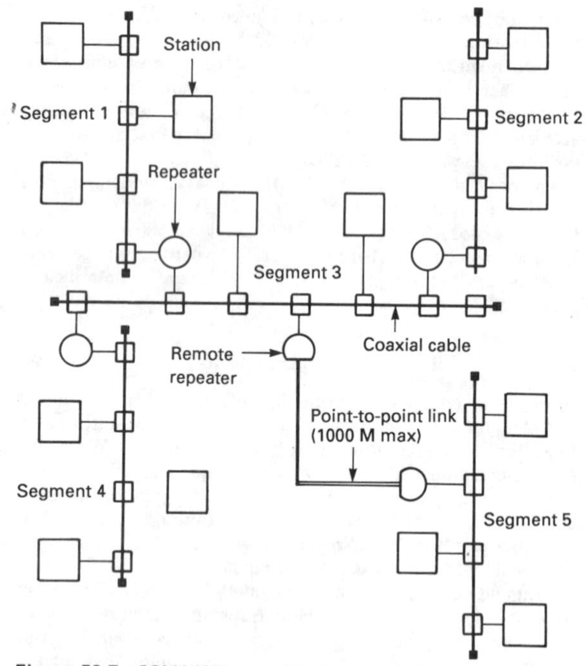

Figure 58.7 CSMA/CD general interconnection

Alternative buffered repeaters, known as MAC bridges, are being standardised to cover all LAN types. They link at the MAC level and share the same data link control. They are already available commercially and are a very useful way of extending the LAN configuration beyond the basic geographic limits. It is possible to interconnect MAC bridges by high-speed digital services to link remote LANs together. Typically these will be the 64 kbit/s and 2048 kbit/s ISDN services.

In a very large LAN system, segment 3 in *Figure 58.7* would be regarded as a 'spinal' segment which could have up to 100 connections to other segments via repeaters, i.e. a maximum of 100 segments, each up to 500 m in length. In a tall building the spinal LAN would run from bottom to top with repeaters to each floor or group of floors.

Figure 58.7 also shows a repeater connection via a fibre-optic cable. This is a part of the 8802/3 standard which is particularly useful if electrical isolation is required in a particular part of the installation, i.e. joining buildings together. The 10 Mbit/s timing is preserved throughout.

An extension to the 8802/3 standard introduces the option of segments of up to 200 M of flexible 50 Ω coaxial cable with no degradation in the performance. This is extremely popular in networks of personal computers (PCs) and office workstations because it allows built-in transceivers, which reduce costs. This standard, known as 10base2, has reached Draft International Standard level. It is often referred to as Cheapernet.

Two later extensions are 10broad36 and 1base5. These are both being processed as draft addenda to 8802/3. Both are available commercially:

(1) 10broad36 defines the use of CSMA/CD over broadband systems, using access unit interface (AUI) compatible connection to the broadband network.
(2) 1base5 defines the use of CDMA/CD at the lower speed of 1 Mbit/s over twisted pair wiring. This is often referred to as Starlan.

Other extensions are being considered for the use of CSMA/CD at 10 Mbit/s over fibre-optic and screened twisted pair wiring. Both are available commercially and they are expected to progress as proposed standards during the next meeting of SC6.

58.9 PABX standards

The IEEE has no interest in standardising the PABX as an alternative local area network. It does not fit into their definition of a LAN, in which all stations interconnect through a single, common media. However, interest is rising due to a new initiative under CEN/CENELEC to standardise integrated services PABXs (ISPABX).

The CCITT is not directly concerned with the use of the PABX as a local area network, but has developed digital subscriber interfaces for the integrated services digital network (ISDN). The ISDN interface (see *Chapter 57*) provides 144 kbit/s which is submultiplexed to provide two 64 kbit/s channels. A number of PABX manufacturers have developed their own digital interfaces to offer combined voice and data transmission over the same physical circuit, but it seems likely that the ISDN digital interface will gradually supersede all these private interfaces.

However, until these standards are widely implemented, the only access method which is genuinely common to all PABX types is the standard voice line—with modems for data transmission.

The future aim is to slot the PABX into the architectural model as an alternative sublayer, to preserve compatibility with the standards at the higher layers. Ultimately, the ISPABX will merge with the integrated services local area network, see *Section 58.10*.

58.10 Future LAN standards—the integrated services LAN

A standard has already been produced for a fibre distributed digital interface (FDDI) by the ISO–IEC committee on peripheral interfacing, JTC/13. This was originally intended for very high speed peripheral connections to mainframes, but has scope for extension.

There is a second version in gestation within the IEEE and ANSI which is specifically designed to support integrated data, voice, image and video transmissions. This is known as FDDI-2 and is expected to enter the ISO–IEC work programme at the next meeting of SC6.

FDDI-2 is becoming known as the integrated services local area network (ISLN). This exciting development will introduce data rates which are an order of magnitude greater than the current 8802 series LANs and bring scope for the true integration of all the information technology services.

Further reading

CLYNE, L., 'Open systems LAN/WAN interworking', *Proc. Internation Conference on Networking Technology and Architectures, Online*, June (1988)

GWILLIM, V., 'Standard performance', *Communications*, November (1988)

HERSKOWITZ, 'Fibre optic LAN topology, access protocols and standards', *Comp. Commun.*, **11**, No. 5, October (1988)

JEFFREE, T., 'A review of OSI management standards', *Comp. Network ISDN Systems*, **16**, No. 1–2, September (1988)

MOULTON, J., 'OSI—an open door for systems integrators', *Mini-Micro Systems*, October (1988)

SENIOR, J.-M. *et al.*, 'Access protocol for an industrial optical fibre LAN', *SPIE Proc.*, **734**, *Fibre Optics*, **187**

SHIMMAN, D., 'Enter the Brouter: an update on linking LANs', *Telecommunications*, November (1988)

'Status of OSI standards', *ACM Comp. Commun. Rev.*, October (1988)

TAYLOR, T., 'The shape of things to come in developing LANs', *Communications*, November (1988)

TOMAS, R., 'Improving security on LANs', *Tech. PC User*, December (1988)

59 Radar Systems

H W Cole
Principal Radar Systems Engineer
Marconi Radar Systems Ltd

Contents

59.1 Introduction

Radar techniques quickly spread into a number of branches different from their original application in early pre-World War II days. The word 'RADAR' is taken generally to be an acronym of USA origin of the phrase *RA*dio *D*etection *A*nd *R*anging. The word 'detection' appears to have been replaced, at the time of its being first coined, by the term 'direction-finding' in the cause of security. The primary radar technique, from which others have grown, is based upon two phenomena:

(1) Radio energy impinging upon a discontinuity in the atmosphere is reflected by the discontinuity.
(2) The velocity of propagation of radio waves is constant.

Exploitation of these phenomena led to transmission of regular pulses of radio energy, range being obtained from measurement of the round-trip time of transmitted pulses. Further pulse transmission and reception engineering led to systems which although radar-based, or inspired, are not true radar, e.g. Oboe, Gee, Omega, etc.—since these are pure distance measuring systems they will not be treated here. A close relative of primary radar, which does not use the reflection of transmitted energy, is, however, so important that it merits separate treatment. This is the secondary surveillance radar (SSR) system wherein pulse transmissions from the ground are received in aircraft, detected and decoded in the aircraft's transponder. The aircraft's transponder then transmits coded pulses back to the ground after a short fixed delay. Thus the operational benefit of primary radar is maintained and many bonuses accrue which are explained later.

59.1.1 History

Radar's history is currently incompletely written. The most prolific historian to date is the generally acknowledged 'father of radar' Sir Robert Watson-Watt.[1–4] He rightly observes the radar principle as being 'often discovered and always rejected'. Early workers, even Heinrich Hertz himself, made observation of the reflection of radio waves from metallic surfaces. In 1922 Marconi introduced publicly the notion of this effect being put to use in the detection and location of the direction of ships. If there is any one point in time at which we can say radar was invented it comes from Sir Robert's memorandum of 27 February 1935. This laid down the basic principle we know as primary radar and also pointed to the need and possibility of SSR by realising how necessary it would be to distinguish 'friend from foe'. Emphasis was placed also on the need for reliable ground–air communications. This memorandum was stimulated by a previous UK Air Ministry request to investigate the possibility of destroying the attacking power of aircraft by radiation. It was concluded that the aim could not be achieved at that time—but in any case the aircraft had first to be located. This would be entirely practicable and Sir Robert elaborated upon this in the second memorandum.

In the few months following February 1935, remarkable progress was made, culminating in detection and ranging of aircraft to half mile accuracy out to nearly 60 miles. Height measurement was also achieved out to 15 miles using range and elevation angle. All this early work was conducted on wavelengths between 50 and 25 m. No measurement of bearing was attempted at this time, it being thought adequate to rely upon independent range measurements taken from a chain of stations of known location. However, around January 1936, the application of receiver DF techniques, using goniometer principles, provided a crude facility which was immediately successful. Similar work was carried out on the European continent, and in the USA. Various reports ascribe the lower level of progress, in Germany for instance, to the German High Command view that the bomber was unstoppable and bombing a very quick way to victory. In the USA, the defence needs were held to be much less urgent than in the UK. This inhibited progress in the early days of radar history.

By 1939 in the UK a radar chain (CH) was established, operating eastwards and ranging from the Orkney Isles to Portsmouth. By September 1941 the chain encircled the whole of England, Scotland and Wales. By this time, the needs and possibilities of radar led to the production of equipments for airborne use, rotatable aerials and the plan position indicator (PPI). In 1940, stimulated by the need for small equipment, Randall and Boot successfully operated their resonant cavity magnetron and thus revolutionised the radar technique. Their device,[5] producing at that time 50 kW peak power at 10 cm wavelength, is now the most widely used generator of microwave power. The use of very short wavelengths made possible the development of small aerials with high discrimination, and equipments with a wide range of power weight and size. The study of radar applications in defence and aircraft navigation continued in peacetime, the latter blossoming into the field of civil aviation electronics, air traffic control, and satellite communications. By the end of World War II, radar engineering had produced many different equipments with differing attributes. Almost immediately the concept of systems engineering took concrete shape and became virtually a separate discipline. The possibility of mixing computer technology into signal and radar data processing has led to the present point of the next revolution in radar.

59.2 Primary radar

59.2.1 Fundamentals

The radar system can be represented in universal form as the functional diagram of *Figure 59.1*.

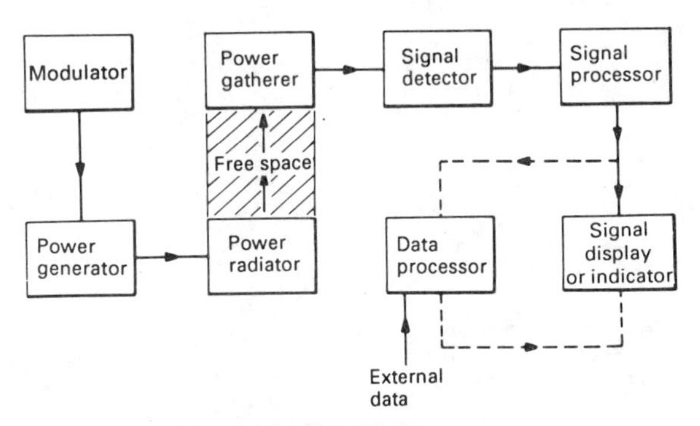

Figure 59.1 Generalised radar system

In any radar system the generated power directed into free space by an aerial will be intercepted by discontinuities in the atmosphere. These are typified as follows:

(1) The ground–air interface.
(2) Hills, mountains, buildings, etc.
(3) Clouds, (precipitating and non-precipitating).
(4) Rain, snow, hail, etc.
(5) Aircraft, ships and vehicles.
(6) Birds, insects, dust clouds, atmospheric discontinuities (i.e. sharp changes of refractive index).

Although some power will be absorbed on impact, most is reflected and some will travel back to the power gatherer, the receiving aerial. The received power is amplified in a receiver, the signal competing with random noise gathered by the aerial and added to that generated at the receiver input (galactic noise, interference signals from electric devices, and receiver noise). After frequency selective amplification, the signal plus noise is passed to a signal processor. Here various characteristics of the numberous types of received signal are exploited to separate the wanted from unwanted, e.g. moving targets from stationary targets, large from small, long from short, etc.

The filtered signals are then passed to a display equipped to register the signals in such a manner as to allow their position to be measured by means of range, bearing and sometimes height scales. The signal processor is being increasingly used to serve a data processor which also accepts external data on targets, situations and various other criteria. Its output can furnish modified display presentations to augment the radar display, and categorise and clarify the total radar-sensed situation.

59.2.1.1 *Monostatic, bistatic and adaptive radars*

Referring to *Figure 59.1* the modulator and power generator constitute the transmitter in all of the above categories. Monostatic systems differ from the bistatic in that a single aerial is used for power radiation and power gathering, *Figure 59.2*. A duplexer is used to separate the transmit and receive functions, both in the time and power amplitude domains. In the bistatic system, *Figure 59.3*, two separate aerials are used, one for the transmit and the other for the receive function. These aerials may sometimes be many miles apart.

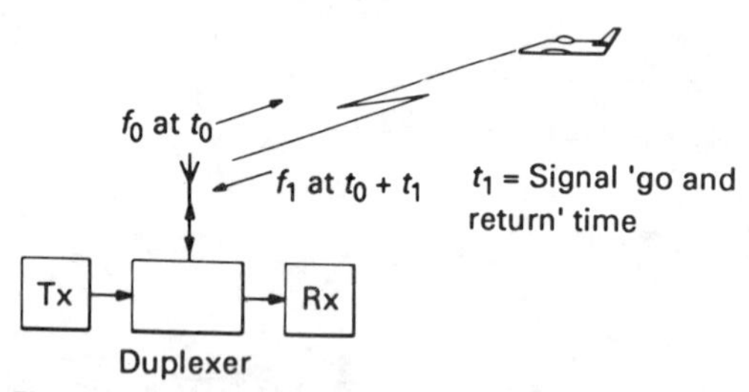

Figure 59.2 Monostatic radar

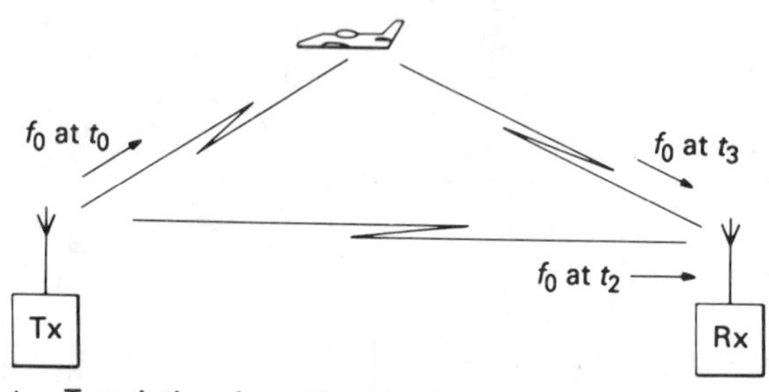

Figure 59.3 Bistatic radar

The range of a reflecting object (aircraft, ship, mountain, etc.) is given by measuring the pulse 'round trip' time from transmission to reception using the leading edges of both transmitted and received pulses as reference.

Round trip time for 1 nautical mile $=12.36\ \mu s$
Round trip time for 1 km $=6.65\ \mu s$

Normally the operating parameters of a radar system are fixed or there can be a few selectable changes during operation, e.g. type of modulation, data gathering rate, power output, receiver sensitivity and selectivity

In the so-called adaptive radar system, parameters of operation (within limits) are varied as a function of the radar's performance on a number of targets. For example, feedback is generated by the processor or data processor to increase information on specific targets by, for instance, directing the aerial to targets in a specific order and governing the dwell time on each to improve data quality; varying the output pulse rate or spectrum. These systems are not yet in general use.

The most commonly used radar system is the monostatic and this section therefore concentrates upon this. Further information on bistatic technique and adaptive radar will be found elsewhere.[6–11]

59.2.1.2 *The radar equation*

There are many forms of the radar equation to be found in the literature. It is therefore considered helpful to develop it from fundamental ideas in order that these various forms, each with their own subtleties and idiosyncrasies, can be better understood. Taking the simple case of free-space performance, the maximum range of radar is a function of:

P_t peak transmitted power
G gain of the aerial in the direction of the target
σ effective reflecting area of the target
λ wavelength of radiation
S_{min} minimum received signal power required to be detectable above system noise
A_r effective area of receiving aerial

Consider a target at range R. The power density at R will be equal to

$$\frac{P_t G}{4\pi R^2} \tag{59.1}$$

The target will intercept this power over its equivalent area of σ and reflect it back over the same distance, R, to the aerial which now becomes the receiver in the monostatic system. Thus the power gathered by the aerial equals

$$\frac{P_t G \sigma A_r}{4\pi R^2 4\pi R^2} \tag{59.2}$$

Now A is related to the aerial gain in the following way:

$$A=\frac{G\lambda^2}{4\pi} \tag{59.3}$$

Maximum gain is obtained when the full aperture of the aerial is available to gather the returned energy and, since the same aerial is used for transmission and reception, Equation (59.2) can be rewritten:

$$P_r=\frac{P_t G^2 \lambda^2 \sigma}{(4\pi)^3 R^4} \tag{59.4}$$

Postulating the maximum range as that achievable when P_r reduces to S_{min}, then

$$R_{max}^4=\frac{P_t G^2 \sigma \lambda^2}{(4\pi)^3 S_{min}} \tag{59.5}$$

The dynamic performance of a radar system cannot, however, be directly calculated by this formula since various statistical factors have yet to be accounted for. One already appears in the notion of S_{min}, since noise is a random phenomenon and is further complicated by the behaviour of σ, the effective target's echoing area. An idea of this can be gained from study of *Figure 59.4* which gives information on the typical values found in practice.

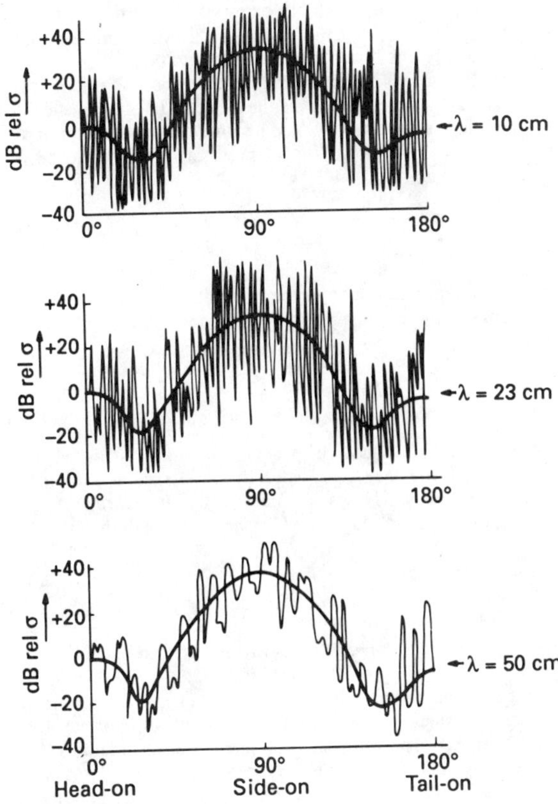

Figure 59.4 Showing typical variation in effective echoing area (σ) of aircraft targets

Table 59.1 Target effective echoing areas

Target	*Attitude*	σ *(typical mean)* (m^2)
Small fighter jet	Head on	2
	Tail on	3
	Side on	500
Large transport jet	Head on	10
	Tail on	15
	Side on	1000

Table 59.1 gives typical values of target effective echoing area. At microwave frequencies there is no extrapolable relationship between wavelength and effective echoing area. The value of σ can change violently over very small azimuth increments at microwave frequencies. As the target attitude to the radar can change even during interpulse periods, there is a range of probabilities that the echoing area will produce a returned signal of a given strength. The distribution of these probabilities is taken variously as Gaussian, Raleigh or exponential. Radar engineers are generally concerned with probability of target detection of the order of 80% and so for all practical purposes the difference in these distributions is negligible. These probabilistic factors, together with others associated with display, operator, atmospheric loss and system loss factors are all combined to modify the final calculation. Two commonly used formulations are those due to L. V. Blake[9] and W. M. Hall.[12]

59.2.1.3 Propagation factor

In all radar systems, theory begins with free-space propagation conditions. These almost always do not pertain since radiations, from and to the aerial via the target and its environment, are seldom via a single path. The most common effect is that of the ground above which the radar aerial is mounted. Consider the situation in *Figure 59.5(a)* where the phase centre of the aerial is at height h above the ground. Energy reaches the distant target by

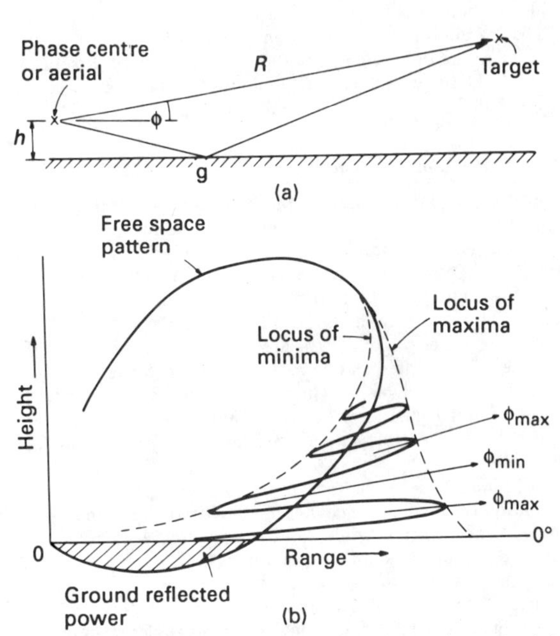

Figure 59.5 (a) Ground reflection mechanism, (b) showing lobes and gaps in vertical polar pattern

the direct ray R and that reflected at g. Over the elevation range of which ϕ is one example, the direct and reflected rays will be alternatively in and out of phase with each other. The free-space pattern of *Figure 59.5(b)* will thus be modulated, a series of lobes being formed. The position of these is governed by h and the wavelength in the following way:

$$\phi_{max} = \frac{\lambda n_1}{4h} \qquad \phi_{min} = \frac{\lambda n_2}{4h} \tag{59.6}$$

where λ is the wavelength, h the height, n_1 are odd integers and n_2 are even integers. Max and min indicate peaks and troughs of lobes respectively.

This model and theory holds for very flat ground or water but becomes a complicated calculation as ground roughness and departures from flat plane increase. Vincent and Lynn[12] have made the problem tractable.

Obviously these lobes are useful providers of extra range but the gaps are less welcome. The modulations can be reduced by several means:

(1) Putting less power into the ground.
(2) Increasing the aerial tilt in elevation.
(3) Intercepting the reflected power and dissipating or scattering it in incoherent fashion.

An interesting alternative is to be found in 50 cm radars used in the 1960s and 1970s by the Civil Aviation Authority for almost all of the UK airways surveillance system. Some are still in use. Here the wavelength and mean aerial height combine to give a long low lobe embracing most of the required long-range airspace. The first null is at an altitude and of such a depth that it is entirely tolerable.

Dependence is placed upon a very flat site, the tolerable roughness being great because of the longer (in the microwave sense) wavelength.

A useful comparison of ground reflection effects across the microwave band is to be found in reference 14. This pattern propagation factor which thus modulates the free-space pattern is also embodied in range calculation.

The magnitude of the lobes and gaps is a function of the effective reflection coefficient of the reflecting surface. Reference 15 treats the subject in great depth and reference 16 gives tables from which specific cases can be calculated.

59.2.1.4 Anomalous propagation

Radar coverage in the vertical plane is usually expressed by a slant range/height diagram drawn for equi-probability of detection of a given effective echoing area and probability of false alarm. A 'standard atmosphere' is generally taken wherein the refractive index of the atmosphere varies with height as the density of the air changes. In such a standard atmosphere containing water vapour, rays are bent downwards as range increases. To counteract this and allow rays in the elevation domain to be drawn as straight lines the range/height diagrams assume an Earth curvature $\frac{4}{3}$ times the normal radius. Naturally local variations in atmospheric condition distort the calculated radar cover diagram.

Sometimes these variations are severe enough to cause a discontinuity in the rate of change of refractive index with height The consequent ray bending can be large enough to equal the Earth's curvature. Under these conditions ('super-refraction') the radar has no horizon and excessive ground ranges can be reached. Apart from difficulties, under these conditions, in obtaining target height data at low elevation angles from height finding radars, normal surveillance radars experience ground clutter and target signals due to transmissions previous to the current one. These are called 'second/third/*n*th-time around' signals. The effect is illustrated in *Figure 59.6*.

In conditions causing anomalous propagation all microwave frequencies are affected. However the onset and degree of effect is later and less severe as wavelength increases. The region of concern can be considered for all practical purposes to be at angles of elevation up to 1°.

59.2.2 Types of radar

Figure 59.1 represents many types of radar; some of the more important of these, starting with the most simple, are now considered.

59.2.2.1 CW radar

Here there is no modulation provided in the transmitter. The movement of the target provides this in its generation of a Doppler frequency. The system is illustrated in *Figure 59.7*. The presence of the target is indicated by the presence of the Doppler frequency of value

$$f_D = \frac{89.4V_r}{\lambda} \quad (59.7)$$

where V_r is the radial speed, λ the wavelength and f_D at 10 cm is approximately 9 Hz/mph (5.6 Hz/(km h)).

Range is not measurable with one c.w. element; but certain modern equipments based on this principle do not need to indicate range, e.g. police speed traps using radar techniques. The technique is widely used for velocity measurements and provided no great range is required, very modest power output can be used, e.g. 10 to 15 W at 10 cm with a small dish aerial of some 90 cm (3 ft) diameter can effectively operate up to 16 to 24 km (10–15 miles) on aircraft.

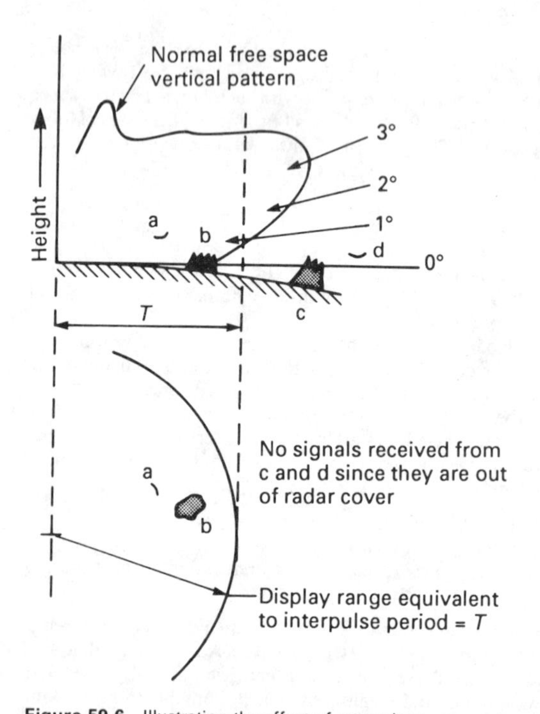

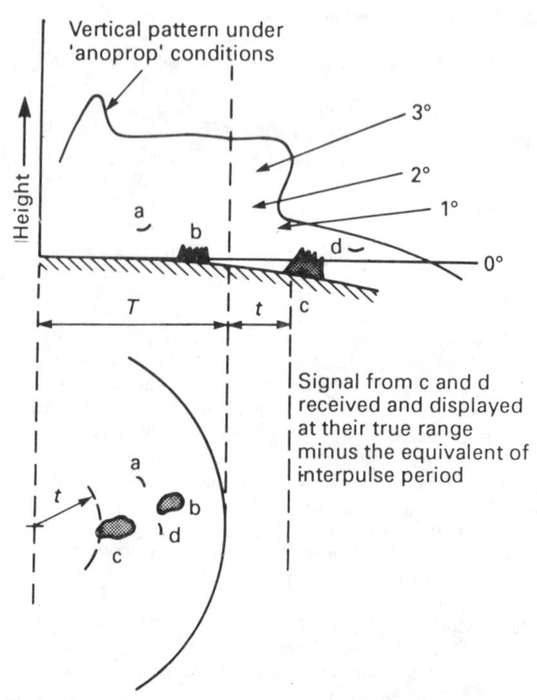

Figure 59.6 Illustrating the effect of anomalous propagation, a and d=aircraft, b and c=ground clutter

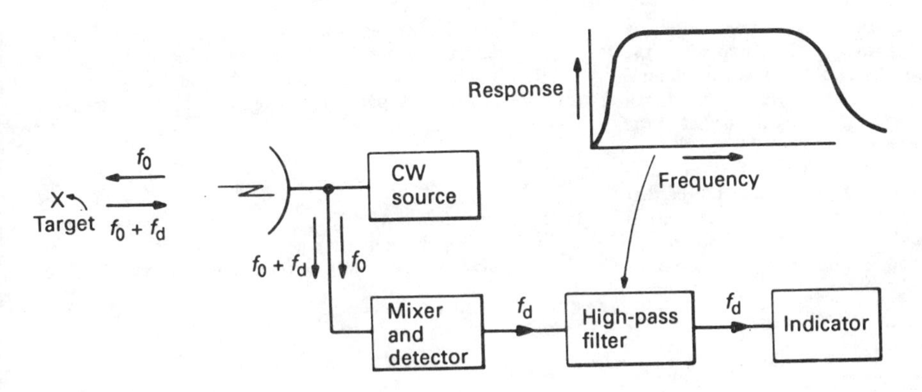

Figure 59.7 Simple c.w. radar

59.2.2.2 AM/CW radar

This variant of the c.w. technique employs two slightly differing c.w. transmissions. With two frequencies f_1 and f_2 a single target at given range will return $f_1 \pm f_D$ and $f_2 \pm f_D$. The Doppler frequency difference can be made small if $f_1 \sim f_2$ is small. However the range domain from the radar is characterised by a phase difference between f_1 and f_2 which is linearly proportional to distance and unambiguous up to phase differences of 2π. By detecting the phase difference of the two almost identical Doppler components of the returned signals the range can be measured, being:

$$R = \frac{C\phi}{4\pi(f_1 - f_2)} \tag{59.8}$$

where R is the range to target, C the velocity of propagation, $f_1 - f_2$ the frequency difference of the two transmitted c.w. and ϕ the phase difference between the two Doppler frequencies.

The complication of dealing with a number of targets simultaneously present in the system is resolved by erecting a number of Doppler filters, each with its own phase measuring element.

59.2.2.3 FM/CW radar

A simple form of this is illustrated in *Figure 59.8*. Here the range continuum has its analogue in frequency deviation from a starting point. Comparison of the frequency of the returned signal and the transmission frequency gives a difference which is directly

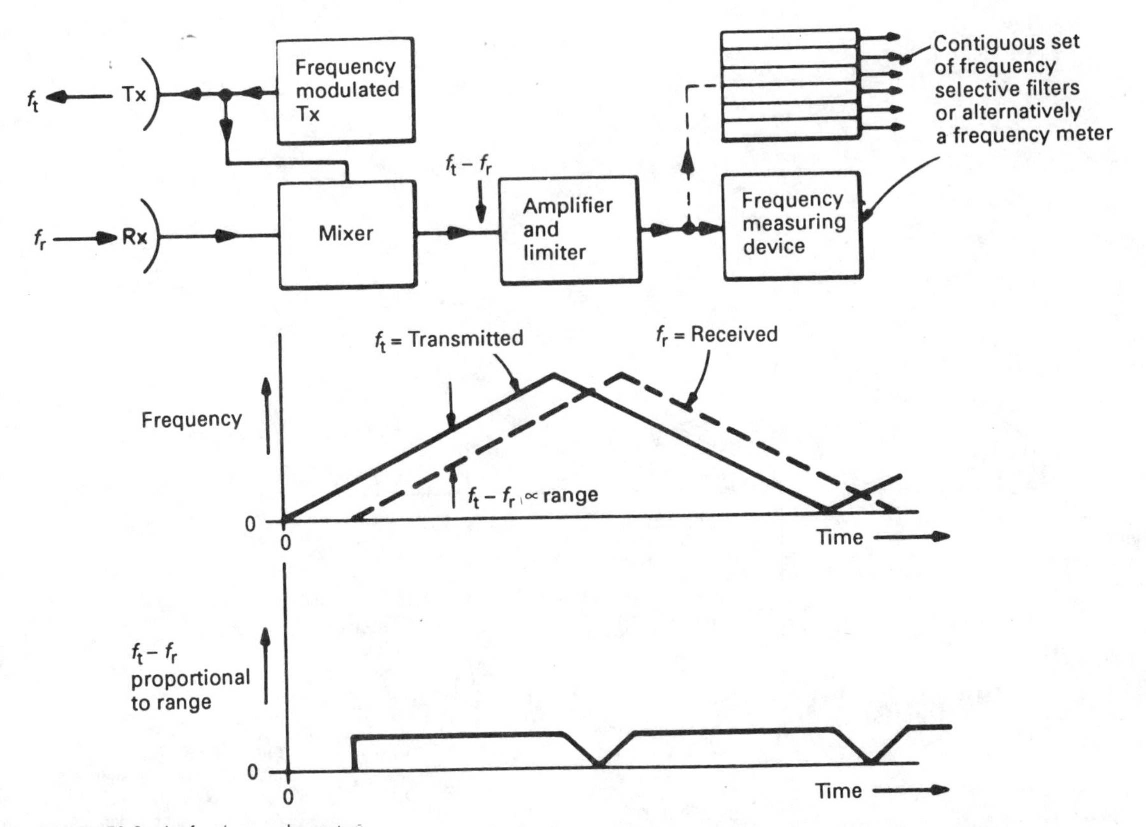

Figure 59.8 An f.m./c.w. radar system

proportional to range. Range accuracy is determined largely by the bandwidth of the frequency measuring cells into which the range scale is broken and the stability of frequency deviations. The technique has been used for a radio altimeter, the Earth being the radar's 'target', and the range being the height of the aircraft above the reflecting surface.[17]

In all the systems described above there was a practical limitation of performance due to the limited c.w. power that could be generated at very high frequencies so that small aerials of high directivity could be used. This limitation of power was eventually overcome by using pulse modulation techniques. In this way very high peak power could be produced within the device's mean power capability. When pulse modulation is used the transit time of the pulses to and from targets is the direct measure of range. A pulse modulated system is illustrated in *Figure 59.9*(*a*) and (*b*).

59.2.2.4 Pulse-modulated Doppler system

In the normal Doppler system, target detection is based upon sensing the Doppler frequency generated by target movement. Straight pulse modulation will detect and indicate both fixed and

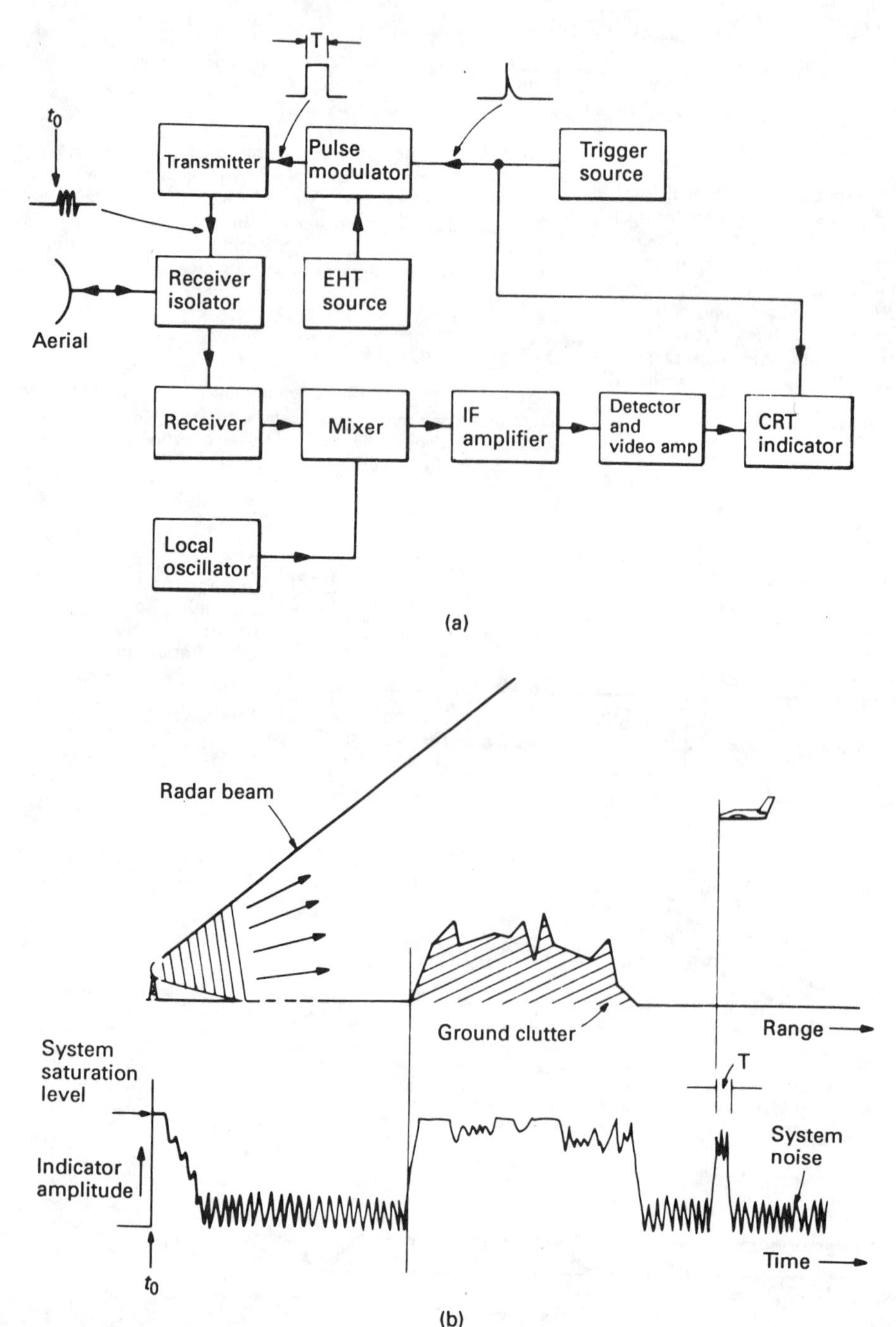

Figure 59.9 (a) Pulse modulated radar system, (b) signals in the pulse modulated radar

moving targets. Bandwidth and other considerations allow the latter to be used as a surveillance radar scanning regularly in azimuth since target dwell time can be short. In order to create a surveillance system that can reject stationary targets at will and preserve moving targets, the pulse modulated system can be made to include the necessary frequency and phase coherence of the Doppler system. This produces the most common form of moving target indicator (MTI) system in use today.

59.2.2.5 MTI systems

Basically these are of two main types:

(1) Self-coherent.
(2) Coherent by phase-locking (commonly called the 'coho-stalo' technique).

Both types operate as follows.

Targets at a fixed range (mountains, etc.) will produce signals which exhibit the same r.f. phase from pulse to pulse when referred to the phase of the transmitted r.f. pulse. Moving targets by the same token, will produce different r.f. phase relationships from pulse to pulse, because of their physical displacement in inter-pulse periods. By use of a phase-sensitive detector these phase relationships can be used in processors to separate fixed from moving targets. The importance of maintaining phase coherence can be seen, for, in order to measure the phase of the received signals relative to that transmitted, a reference has to be laid down.

In the self-coherent system the transmitter is a power amplifier whose r.f. source is crystal controlled by a reference oscillator. The same reference is taken for the receiver's local oscillator. Thus coherence is assured and maintained at i.f. level at which point the phase information is extracted. The system is illustrated in *Figure 59.10*.

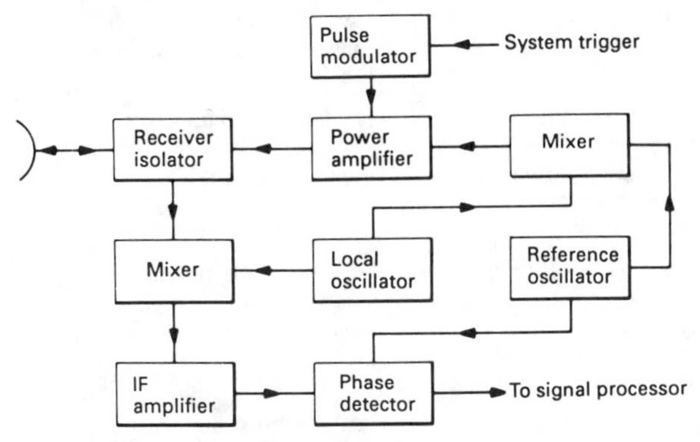

Figure 59.10 Self-coherent pulse radar

In the 'coho-stalo' system, the transmitter is a self-oscillating magnetron whose phase at the onset of oscillation is varying from pulse to pulse. Coherence is achieved as follows.

The receiver has an extremely stable local oscillator (stalo). At each transmission, a sample of the r.f. output power is taken at low level. This, after mixing with the stalo, forms an i.f. locking pulse which is used to prime an oscillator which becomes the coherent reference (the coho). This oscillator is switched off during the time the lock pulse is injected. Upon switching on, the stored energy of the lock pulse starts the oscillation in a controlled manner. Thus the coho preserves the phase reference for a pulse period. The coho is then switched off just prior to the arrival of the next lock pulse and the process is repeated. The system is illustrated in *Figure 59.11*. Signal processing is described in Section 59.2.9.

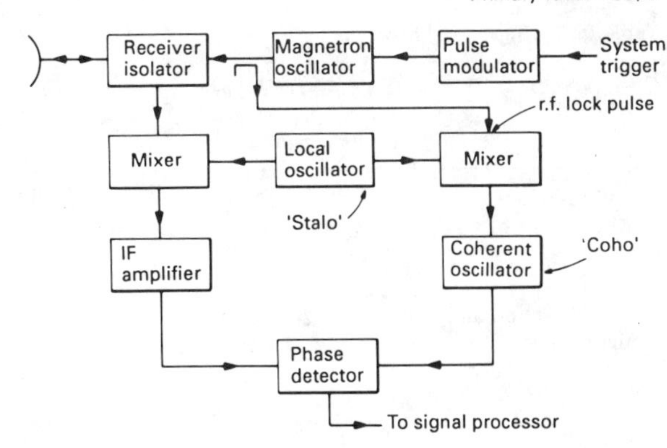

Figure 59.11 Coho-stalo system

59.2.2.6 Chirp

Radar performance is dependent upon the mean power of the system. If this is increased by lengthening the output pulse duration, immediately the range discrimination is reduced (12.36 μs is equivalent to 1 nautical mile in primary radar terms). The *chirp* or pulse compression technique is a means of obtaining high mean power by use of longer pulse output while retaining high range discrimation. It operates as follows.

The transmitter output pulse is frequency modulated, usually linearly, throughout its duration. Received signals which bear this modulation are passed through a pulse compression filter which has a delay characteristic which is frequency dependent. This characteristic is made to be of opposite sense to that used in transmission so that the frequencies occurring later in the transmitted pulse are delayed least. Those occurring early in the output pulse are delayed most. By this means the received energy is compressed in time. If the original pulse duration is τ the output pulse duration from the compression filter is $1/B$ where $B = f_{max} - f_{min}$ used in the frequency modulation. The peak power using this technique is effectively increased by a factor of $B\tau$ which is called the pulse compression ratio or dispersion factor.[7,8]

An unfortunate by-product of this technique is the generation of 'time sidelobes'[9,10,18] and an effect where small signals in close range proximity to large signals are masked until range separation of pulse widths has been established.

The long pulse limits minimum detectable range to the equivalent of the pulse duration (e.g. 60 μs leads to minimum range of about 5 nautical miles). This shortfall is made good by transmitting a short duration pulse with simple amplitude modulation (as in normal radar practice) at a frequency different from any in the chirp regime. A separate receiver and signal processor have then to be added to the system whose total output consists of those from the short pulse transmission followed by those from the chirp system.

59.2.3 Operational roles

Radar's ability to 'see without eyes' has placed the technique at the service of a wide variety of users. A resumé of operational roles with pertinent data is given in *Table 59.2*. It will be seen that a wide range of engineering is required to furnish the requisite hardware.

59.2.4 Transmitters

Across the band of microwaves from 8 mm to 50 cm transmitters of peak power from watts up to 10 MW are found. For the higher powers required, magnetrons are almost exclusively used for

Table 59.2 Operational roles

Operational role	*Typical wave-length*	*Typical peak power and pulse length*	*Deployment*	*Characteristics*
Infantry manpack	8 mm	f.m.–c.w.	Used for detecting vehicles, walking men	Has to combat clutter from ground and wind-blown vegetation etc. Hostile environment. Battlefield conditions
Mobile detection and surveillance	3 cm	20 kW at 0.1 μs to 0.5 μs	Used in security vehicles moving in fog. Also military tactical purposes	Needs high resolving power for target discrimination. Hostile environment. Mobile over rough ground
Airfield surface movement indicator	8 mm	12 kW at 20 ~ 50 ns	Used on airfields to detect and guide moving vehicles in fog and at night	Resolving power has to produce almost photographic picture. Can detect walking men
Marine radar (civil and military)	3 cm 10 cm 23 cm	50 kW at 0.2 μs (see Defence below)	Used for navigation and detection of hazards. Longer wavelength for aircraft detection, i.e. 'floating surveillance' radars	3 cm needs good resolution and very short minimum range performance. All have to contend with sea clutter and ships' movement
Precision approach radar (PAR)	3 cm	20 kW at 0.2 μs	Used to guide aircraft down glidepath and runway centreline to touchdown	Very high positional accuracy required together with very short minimum range capability and high reliability
Airfield control radar (ACR)	3 cm 10 cm	75 kW at $\frac{1}{2}$ μs 0.4 MW at 1 μs	Usually 15 rpm surveillance of airfield area. Guides aircraft into PAR cover or on to instrument landing system	Needs anti-clutter capability. 10 cm has MTI. Good range accuracy. High reliability required
Airborne radar	3 cm	40 kW at 1 μs	Used for weather detection and storm avoidance	Forward-looking sector scanning. Storm intensity measurement system usually incorporated
Meteorological radar	3 cm 6 cm 10 cm	75 kW to $\frac{1}{2}$ MW $\frac{1}{2}$~1 μs	Surveillance and height scanning gives range/bearing/height data on weather. Also balloon-following function	Good discrimination and accuracy. Rain intensity measuring capability
Air traffic control—terminal area surveillance (TMA)	10 cm 23 cm 50 cm	$\frac{1}{2}$ MW to 1 MW 1 μs to 3 μs	Surveillance of control terminal areas. Detection and guidance of aircraft to runways and navigational aids. 60 nautical miles range	Needs good discrimination and accuracy. MTI system necessary. Display system incorporates electronic map as reference
Air traffic control—long range	10 cm 23 cm 50 cm	$\frac{1}{2}$ MW to 2 MW 2 μs to 4 μs	Surveillance of air routes to 200 nautical miles range. Monitoring traffic in relation to flight plans	Has to combat all forms of clutter. MTI essential. Circular polarisation necessary except at 50 cm
Defence—tactical	10 cm 23 cm	$\frac{1}{2}$ MW to 1 MW 1 μs to 3 μs	Mobile or transportable. Used for air support and recovery to base	As for TMA radar plus ability to combat jamming signals
Defence—search	10 cm 23 cm	2 MW to 10 MW 2 μs to 10 μs	Used for detection of attacking aircraft monitoring of defending craft including direction for interception. Usually a height finder operated together with search	As for ATC long range plus ability to combat all forms of jamming. Forward stations report data to defence centre
Defence—3D search	23 cm	5 MW at 60 μS compressed to 0.6 μs	Used for defence search but multiple elevation beams allow automatic calculation of target height	As for defence—search, polarisation is horizontal

wavelengths up to 23 cm. Klystrons and other forms of power amplifiers have been employed at wavelengths of 10 and 23 cm. For 50 cm wavelengths high-power klystrons are universally employed. The advantage of klystrons and power amplifiers is their ability to be driven with crystal-controlled sources. This produces automatic frequency and phase coherence in the system which is called for in all MTI systems. Driven systems confer the ability to change operating frequency rapidly either from pulse to pulse or after a few repetitions of one frequency (burst-to-burst agility). This gives a large measure of protection against jamming—an essential feature of military radar systems.[19]

Magnetrons have to be excited by a very high voltage pulse. This is produced from a source the energy of which can be transferred as a pulse with a controlled rate of rise and fall, at a known time, for a known duration. This is usually effected by a pulse-forming network, the stored energy of which is released to the magnetron by a trigger device via a pulse transformer (usually a thyratron, although hard valve modulators are still used). The released energy, in pulses of some 30–60 kV of up to 10 μs duration, is stored by a charging circuit which can be resonant, for greater efficiency, at the repetition frequency of the transmitted pulses. Care has to be exercised in pulse shaping to avoid magnetron moding and unwanted frequency modulations. Purity of the frequency spectrum is difficult to achieve unless great care in modulator design is exercised.[6] Klystrons are virtually power amplifiers and still need high-voltage pulses to provide the amplification during the required pulse output.

Other methods of providing high peak power in the microwave range are possible but very rarely found in practice; examples are as follows.

High-power klystrons These are available for operation at B (250–500 MHz), L (0.5–2.0 GHz), S (1.5–5.0 GHz) and C (4.0–6.0 GHz) band with output power up to 20 MW at L band. Power gain is typically 40 dB.

High-power travelling-wave tube (TWT) These can be operated at L, S and C band giving output of up to 2 MW peak with gains typically 45 dB. They have very wide bandwidth.

Twystron So called because it uses a klystron-type cathode gun assembly and the TWT technique of slow wave restriction to provide gain of up to 40 dB at S band. It has the very desirable property of some 200 MHz instantaneous bandwidth, allowing frequency agility to be used without transmitter complication.

Crossed-field amplifiers (CFA) These are found for use across the X to C (4.0–11.0 GHz) band and provide typically 10 dB gain. They are a form of amplitron and use cold cathodes.[7,8]

Solid-state devices By use of pulse compression techniques the requisite mean power can be provided by power amplifiers consisting of batteries of transistors whose output is combined in proper phase relationship. This has great advantages in high reliability. No modulator is required, pulses being formed by low-level c.w. input switching.

59.2.5 Receivers

Across the microwave band, receivers almost always use the superheterodyne technique. This is to simplify signal handling in processors, etc. The design aim is to preserve dynamic range to prevent amplitude and phase distortion outside the designer's control. Parametric amplifier techniques are still common which produce noise figures of some 3 dB at 50 cm and 6 dB at 3 cm wavelength.

These commonly employ variable capacitance diodes which 'noiselessly' extract power from a separate microwave c.w. oscillator source called the *pump* and convert it to the signal frequency, thus achieving amplification of signal without introduction of extra noise. The gain of a parametric amplifier of this kind is proportional to the ratio of pump to signal frequency. When integrated into a high-power radar system care has to be used to protect the sensitive diodes from damage due to the leakage of power during transmission.

Very nearly theoretical limits of sensitivity have been reached in some receiver designs (i.e. noise figures approaching 0 dB indicating the receiver operating at the limit of KTB, where K = Boltzmann's constant, T = temperature in degrees absolute and B = bandwidth). This is achieved by super-cooling the receiver input elements, incidentally producing a novel engineering probelm embracing physics, electronics, mechanics, chemistry and refrigeration disciplines. These types of parametric amplifiers are to be found in radio telescopes the powers of which are determined to a large degree by receiver performance.[6] In new radar systems the parametric amplifer is being displaced by a simple transistor 'front end' of very low noise figure.

The radar system sensitivity is given by the sum of all losses between the antenna and receiver plus the receiver noise figure itself; typical figures are shown in *Table 59.3*. In radar performance calculations these losses are entered into the radar equation when deriving S_{min}.

Table 59.3 Typical noise figures

	Noise (dB)
Receiver noise figure	2.5
Rotating joint	0.5
Diplexer (diversity system)	0.25
Duplexer	0.5
Receiver protecting device	0.25
Wave guide and coaxial cable losses	0.25
Total effective noise figure	4.25

59.2.5.1 IF amplifiers

It is common in modern transistorised equipment to find 100 MHz bandwidth easily achieved and bandwidths subsequently restricted to the optimum by the design of a filter at the input or output stages. Dynamic characteristic forming is usually done at i.f. and can produce linear, logarithmic, limiting or compression characteristics fairly readily over 80 dB of dynamic range above noise.

59.2.6 Transmitter–receiver devices

Until very recent times high-power radars used gas discharge tubes as a means of producing isolation between transmitter and receiver. A simple form is illustrated in *Figure 59.12*. When the

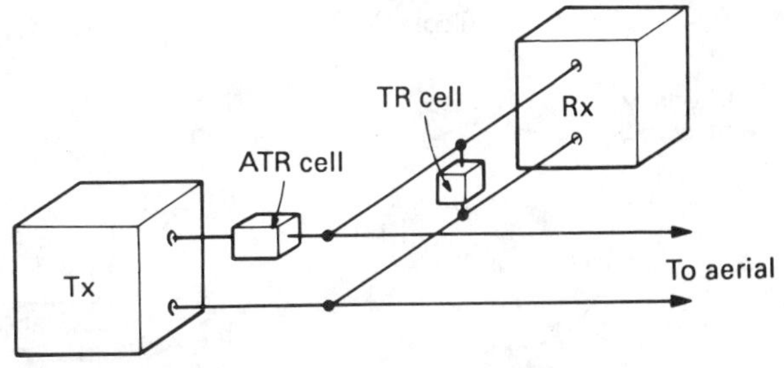

Figure 59.12 Simplified diagram showing receiver protection devices

transmitter fires, the gap in the ATR cell breaks down and allows power to progress to aerial and receiver. Power across the TR cell will cause the cell's tube to strike and impose a short circuit across the receiver input terminals. This short circuit is so placed that matching to the aerial is not disturbed. Upon cessation of the transmission the ATR device will revert to quiescence and thus disconnect the transmitter from the system. The TR cell's short circuit is also removed and thus returned signals can reach the receiver. Disadvantages of this technique are:

(1) Leakage past the TR device into the receiver.
(2) Long recovery time after firing which provides attenuation to received signals. This limits minimum range performance.

These devices are being replaced in modern equipment by high-power-handling diodes using the same technique of open and short circuit lines. They have better recovery times and less insertion loss but produce problems associated with the need to switch bias voltages in synchronism with transmission in order to prevent destruction of the diodes.

59.2.7 Aerials

The gain of an aerial is related to its area. However, the maximum gain is not the only, nor even the prime, consideration. The radar designer attempts to achieve gain in desired directions and thus beam shape in both horizontal and vertical planes becomes extremely important. The beam may be formed in a variety of ways as follows:

(1) Use of groups of radiators (dipoles, unipoles, radiant rods, helices, etc.).
(2) Reflector illuminated by various elements (dipole, horn, poly-rod, linear feed, etc.).
(3) Slotted waveguides.
(4) Microwave lens system.

By far the most common is group 2. There are here many choices of illumination, distribution and reflector shape to permit practically any radiation pattern to be produced.[20]

The beamwidth to the half-power points (θ) in either azimuth or elevation planes is approximately given by

$$\theta = 70\lambda/D$$

where λ is the wavelength and D the physical aperture of antenna in the plane considered, expressed in the same units as λ. The beam is formed at the Rayleigh distance r from the antenna, given by

$$r = fD^2/\lambda$$

where f is a factor not less than unity and up to two. Taking $f=2$ ensures that the transition between near- and far-field effects has been passed.

Further information on antennas will be found in *Chapter 49*.

The advent of small solid-state microwave power devices permits arrays of these to be constructed in large planes of differing proportion (a rectangular plate, a pyramid, a cylinder, etc.). Control of phase and amplitude of the output from each individual device allows beams of varying shape to be formed and for the direction of propagation to be controlled.[19]

59.2.8 Radar displays

Almost exclusively these use cathode ray tubes to represent various related dimensions on its screen, e.g. range and azimuth, range and height, range and relative amplitude. Two types of modulation are used: amplitude and intensity. Typical examples are given in *Figure 59.13*. Of these the most important is the PPI (plan position indictor).

59.2.8.1 The PPI

Although *Figure 59.14* shows magnetic deflection, electrostatic deflection can also be used. The deflection coils, nowadays of sophisticated design with correction for many types of error and aberration due to tube geometry and manufacture, are mounted around the neck of the tube and produce orthogonal fields across the path of the electron beam. Variations of these fields deflect the beam across the face of the tube which it strikes and excites the phosphor on its surface producing an intense glow at the point of contact. This glow is visible through the glass of the screen.

Assume the tube centre is to represent the position of the radar and its edge, the radar's maximum range. The display system generates a *bright-up* waveform synchronised to the radar

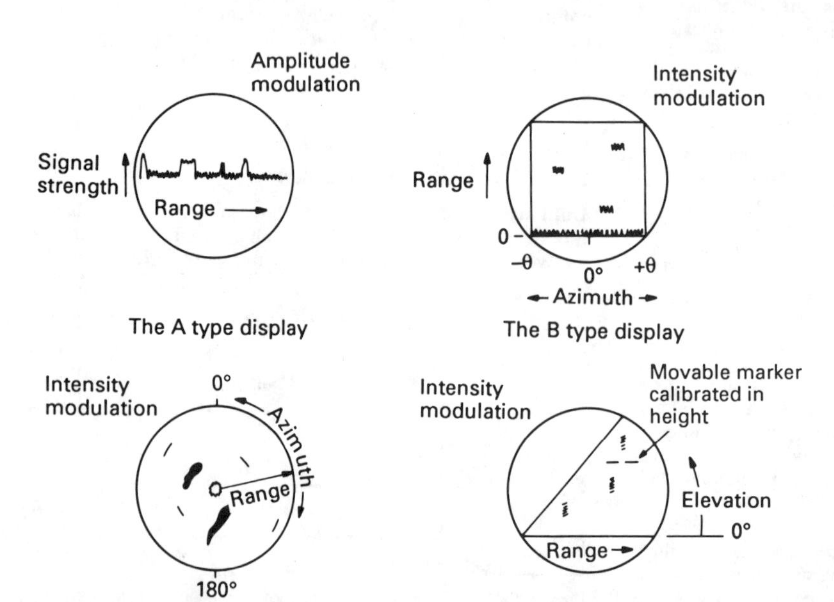

Figure 59.13 Various radar displays

transmission time t_0, and with its amplitude and level set to give a threshold producing very small excitation of the phosphor in the absence of modulating signals. The waveform is also used to govern the period over which the time-base integrators operate. There are identical chains for the X and Y axes and both perform integration of a d.c. voltage expressive of the sine and cosine of the bearing to be indicated. These d.c. voltages can be generated in a number of ways, a common method being a 'sin/cos' potentiometer. The resolver is made to rotate in synchronism with azimuth tell-back elements mechanically geared to the aerial turning system. This may be done, for example, with torque-transmitting rotary transformers ('selsyns') or a servo-drive synchro system, again using rotary transformer technique. As the aerial turns, the values of the resolver's d.c. outputs will change, going through one cycle of maxima and minima per aerial revolution. The action of the integrators will therefore produce X and Y sawtooth waveforms whose amplitude and polarity change in sympathy with the aerial's rotation thus automatically

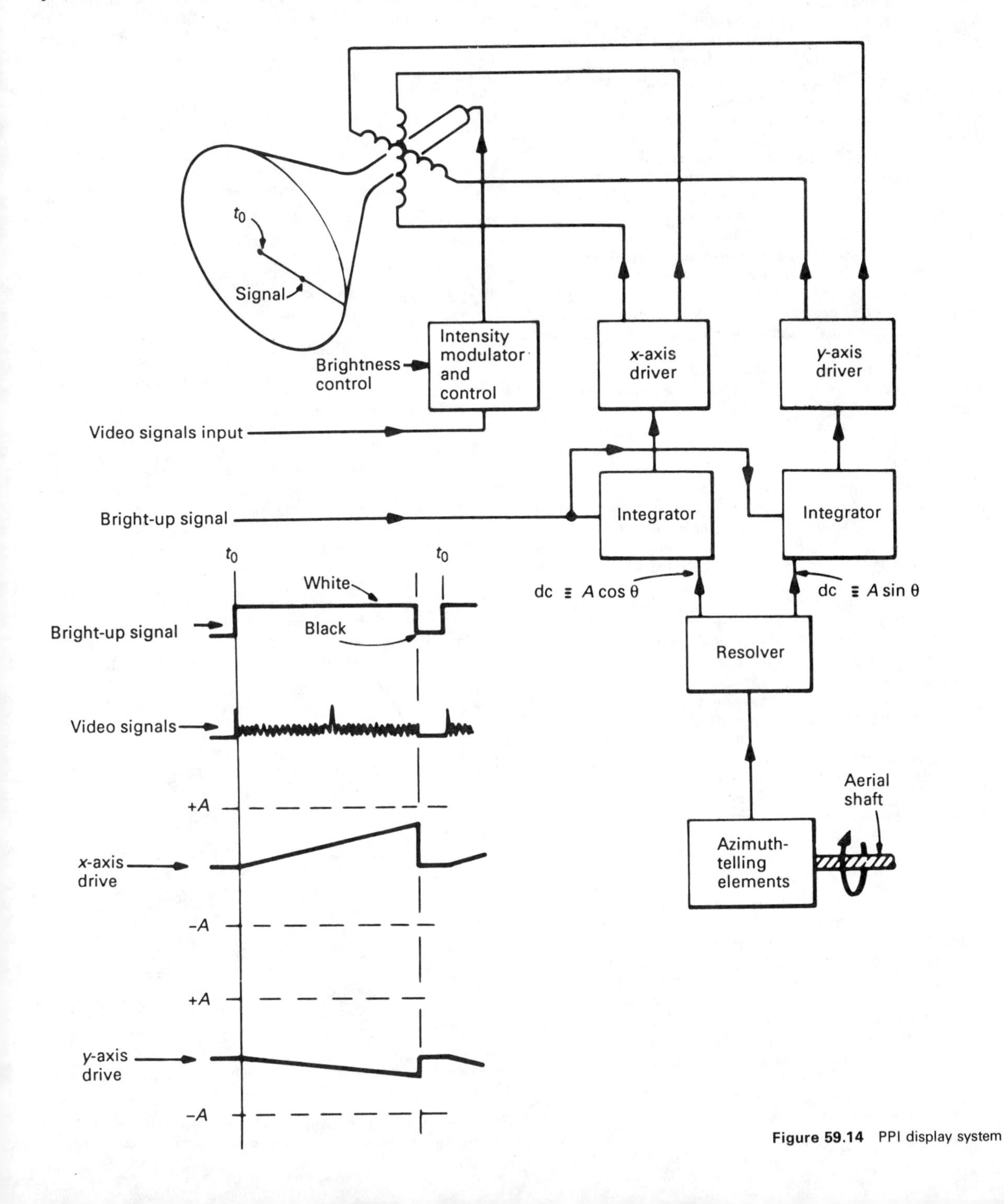

Figure 59.14 PPI display system

reproducing the fields necessary to cause the electron beam to move across the tube at the azimuth of the aerial. By making the X and Y drive waveforms highly linear and governing the overall gain of the deflection system, the tube face can be made to represent areas to calculable scales in a very linear fashion. Since the time base is synchronised with radar transmission time, received signals appear $12.36 \times \gamma$ μs later, where γ is their real range in nautical miles. Signals of detectable amplitude are amplified and used as sources of intensity modulation of the c.r.t. beam. Thus as the radar's narrow beam is rotated in azimuth the signals returned are 'written' on to the tube face. By varying the chemical properties of the phosphor the length of afterglow time can be varied. By this means the display system can add to the radar's fulfilment of its operational role, e.g. if track history is required long afterglow can provide this; if small changes in signal pattern are required in a fast-scan system, a short afterglow is used.

The example in *Figure 59.14* is capable of many variations, e.g. the resolution into X and Y sawtooth deflection waveforms can be done by a rotary transformer element with two fixed orthogonal field coils and the rotor fed with a single fixed amplitude integrated sawtooth. The rotor would be turned in sympathy with the aerial by a servo-driven azimuth transmitting system. Some displays still exist where the deflection coils are rotated around the c.r.t. tube neck; still others are used with a group of coils fixed around the tube which produce a rotating field in sympathy with the aerial rotation. The displays described above are all real-time systems. In non-real-time systems, increasingly to be found, data in the form of symbols or alphanumeric characters are written upon the screen in machine time. Here it is common to find deflection systems consisting of fast and slow elements working in conjunction. In this case no sawtooth integration is needed. The 'slow' deflection coils are made to take upfield values which give the beam a desired position at a desired machine time. The 'fast' deflection coils have analogue waveforms of small amplitude to cause the beam to 'write' desired characters. Typical speeds are 20 μs to move from one tube edge to another and 5 μs per character for data writing. It is unusual to find more than 25 characters (5 × 5) associated with individual targets.

Modern systems utilise digital techniques for both azimuth telling and time-base generation. They may be organised in real or machine time. Two main systems of azimuth telling are to be found, both producing high resolution of the circle into a 13- or 14-bit structure. Digital shaft encoders of magnetic or optical type can produce a multi-bit expression of the aerial's position as a parallel multi-bit word; alternatively, for reporting the position of a continuously rotating shaft (as is common in most search radars), the encoder reports only changes to the least significant digit of the multi-bit word. Integration into the full word is done by a digital resolver at the receiving end of the data link. This produces great economy in data transmission.

59.2.8.2 Raster scan displays

Modern techniques permit range, azimuth and amplitude data on targets to be expressed in digital form. Where this is done, a natural extension to digital displays is possible.

Writing the data into digital stores allows a separate store-reading regime to be organised as a TV-style scan. This in turn permits TV-type displays to be used in place of the traditional PPI. The target data update rate is low (the antenna rotation rate) and the store readout rate is at the much faster TV scan speed. Thus the resultant picture has much higher brightness than the real-time traditional PPI.

Current technology is aiming at 2000 line resolution (already achieved in prototypes), having achieved 1500 in commercially available systems.[21]

59.2.9 The radar environment and signal processing

Wanted signals are always in competition with unwanted signals. The total environment can be appreciated by reference to *Figure 59.15*. This is typical of the general case and illustrates the need to separate the wanted aircraft signals from those simultaneously arriving from, in this radar's role, unwanted or clutter sources. The radar system also suffers total blindness due to shadowing by solid objects and attenuation due to gases, dust, fog and similar particular media, e.g. rain and clouds. The radar aerial attempts to concentrate its radiation in a well defined beam. However there are always sidelobes present, which produce radiation simultaneously with the main beam in unwanted directions. These also gather unwanted signals.

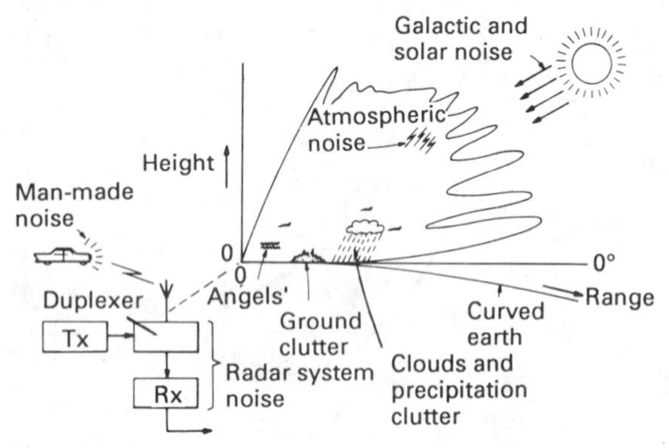

Figure 59.15 The radar environment

Signal processing can be used to discriminate in favour of wanted signals. The basic radar system is also designed to provide some bias in favour of the wanted signals, e.g. by using as narrow a radar beam as possible, by the use of the longest wavelength possible to minimise signals from cloud and rain, by the use of the shortest pulse possible at the highest repetition rate, etc.

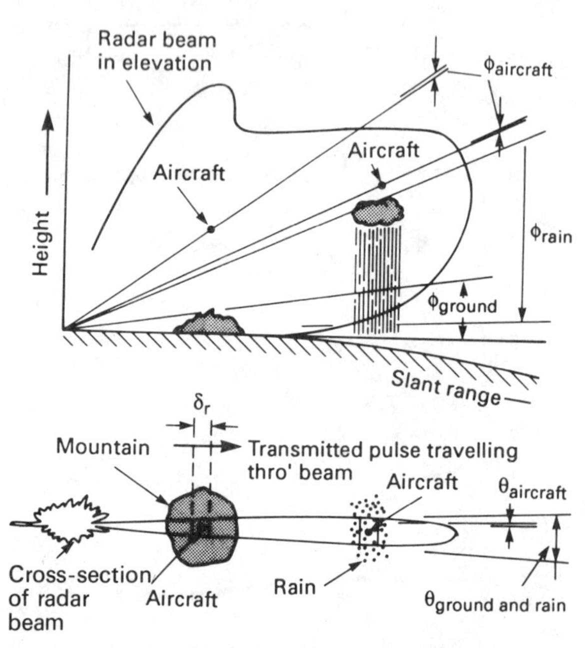

Figure 59.16 The significance of resolution cell. The ground and rain occupy a larger volume of the resolution cell than do aircraft

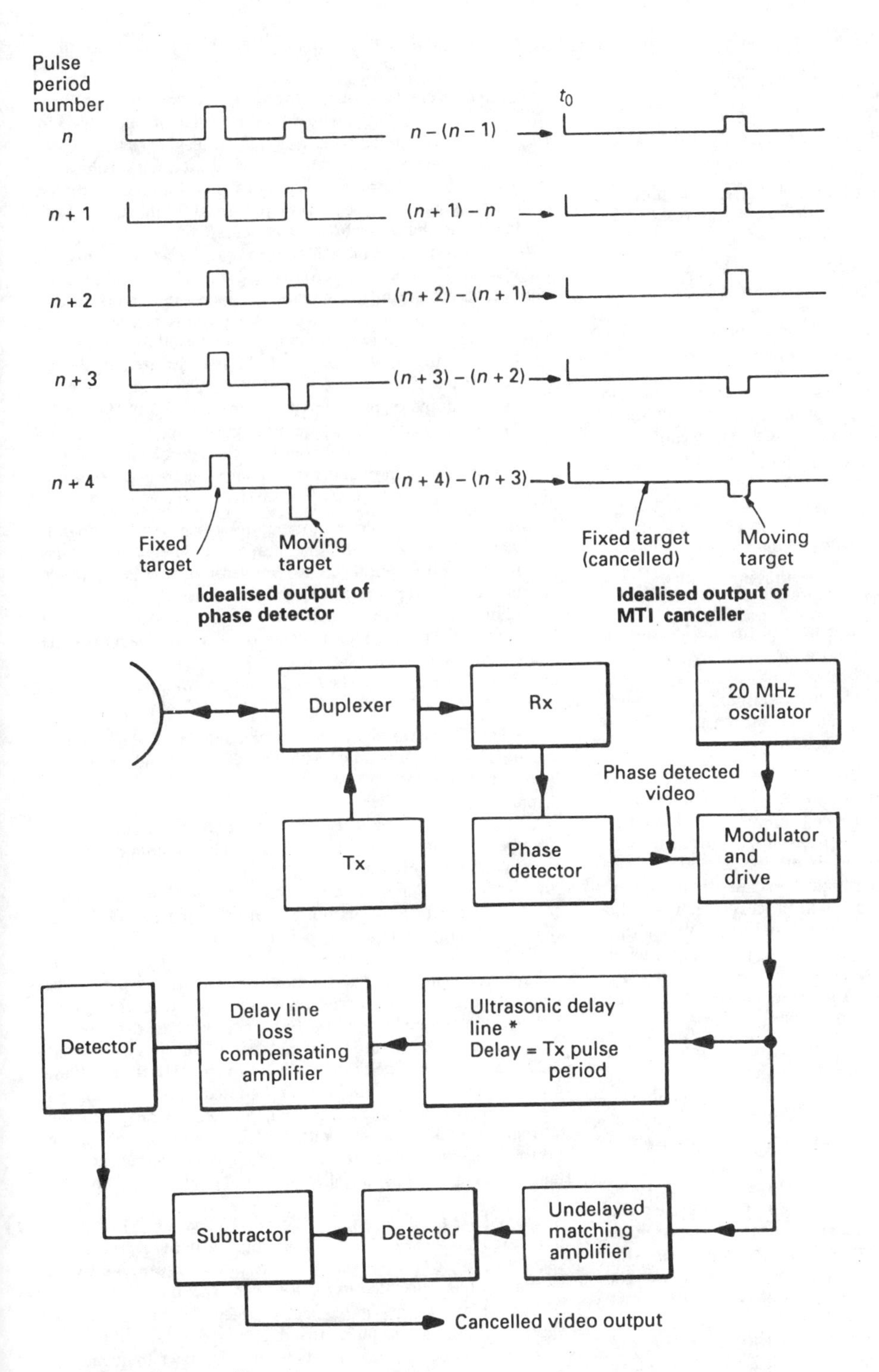

Figure 59.17 (a) The MTI system's phase detector-canceller principle. (b) A simple MTI radar

The radar's resolution capability is largely determined by three parameters:

(1) The azimuth beamwidth.
(2) The elevation beamwidth.
(3) The transmitter pulse length.

This is illustrated in *Figure 59.16*.

The desirability of reducing the size of the resolution cell to improve wanted 'target-to-clutter' ratio can be appreciated by considering the following practical case:

Azimuth beamwidth	1.5°
Target range	60 nautical miles
Pulse duration	3 μs
Elevation beamwidth	10°
(approx. 0 to 70 000 ft at 60 nautical miles)	

The cross-sectional area of the cell is about $\frac{1}{2}$ (nautical miles)^2 and rises to 11 (nautical miles). If the wanted target is an aircraft it is very much smaller than the resolution cell, which can gather unwanted targets (ground clutter, rain and clouds, angels, etc.) throughout its volume.

59.2.9.1 Discrimination against ground clutter

A number of techniques are available ranging from the very simple to the highly sophisticated.

Map blanking An electronic video map of the clutter pattern from a fixed site is used to produce suppression signals which inhibit the radar display, thus producing a 'clean' picture containing only moving targets. This method is not effective with moving targets in regions of clutter and for this reason is seldom used.

Pulse length discrimination (PLD) Here use is made of the difference in the range continuum in the duration between aircraft signals and the general mass of ground clutter. All clutter greater in duration than twice the transmitted pulse length is totally rejected, those up to this limit are displayed, thus preserving aircraft targets and rejecting almost all ground clutter. This is a relatively simple video signal process which has the advantage of providing supra-clutter visibility. Thus, if the wanted signal is of sufficient strength to show above the clutter, it is displayed in clutter regions. A logarithmic receiver is usually employed ahead of the process to increase the dynamic range before signal amplitude limiting takes place.

Moving target indicators (MTI) In this case the processing is closely linked with modulator and transmitter design and is therefore more of a radar system than pure signal processing. The purpose of an MTI system is to discriminate between fixed and moving targets, and system coherence is required (see *Section* 59.2.2.5).

The discrimination is made in the following manner. Consider the radar aerial stationary and illuminating a fixed target, e.g. the face of a mountain. Successive signals received from regular pulse transmissions will be at fixed range R. In the continuum this range can be expressed as $R = n\lambda + d\lambda$, where λ is the wavelength of transmission, n the number of whole wavelengths to the target and d a fraction.

By establishing a phase reference at each transmission and preserving it for a whole reception period it is possible to give coherent meaning to the above expression using a phase-sensitive detector, the output of which for any target is characterised by three factors:

(1) Its occurrence in time relative to transmision (due to the value of $n\lambda$) to the nearest whole wavelength.
(2) Its peak-to-peak amplitude due to the usual radar parameters, e.g. target echoing area, etc.
(3) Its amplitude and polarity within the peak-to-peak range due to the value of $d\lambda$.

In the case cited the output would be as illustrated in *Figure 59.17(a)*. A moving target would produce different values of $d\lambda$ for relatively slow speeds and different values of $n\lambda$ as speed increased. Thus the phase detected output might behave as illustrated. By the storage and comparison technique, successive received signals can be subtracted from each other. Both analogue and digital techniques are employed.

In the example given it will be seen that the signals from fixed targets can be reduced to zero and moving targets will produce non-cancelling outputs. The rate at which the moving targets change their amplitude from pulse to pulse is analogous, and equal in value, to the Doppler frequency referred to in Equation (59.7). A simple pulse Doppler MTI is illustrated in *Figure 59.17(b)*. It has a number of limitations which tend to offset its major advantage of providing sub-clutter visibility (SCV, the ability to see moving targets superimposed upon fixed targets where the moving target is smaller than the fixed). Values of 20 dB for SCV are common in many surveillance radars using this type of MTI. Some of the limitations referred to are as follows:

(1) Blind speeds—where a moving target behaves as though it was stationary, i.e. $n\lambda$ changes by an integer in a pulse period. In this condition the Doppler frequency is a multiple of the pulse repetition frequency of the transmitter.
(2) Blind phases—where a moving target produces the same output from the phase detector whose characteristic is symmetrical in the phase axis.
(3) Phase difference masked by noise or swamped by phase distortion in the receiver.

The system disadvantages can be overcome in varying degrees, e.g. blind speeds by staggered p.r.f. systems; blind phases by dual phase detectors, whose references phases I and Q are 90° displaced.

Other disadvantages are systematic and less tractable. For example 'fixed' targets are seldom found. The 'mountain' referred to above commonly has trees growing on it and wind causes fluctuations which spoil complete cancellation. Also the aerial is rotating in practice and its horizontal polar pattern modulates the target's amplitude again spoiling cancellation.[7,8]

Digital techniques These are now almost always used. The principles described above remain the same. Differences from the analogue technique begin at the output of the phase detector. By quantifying the amplitude and polarity of the phase-detected output at a regular clock rate, typically 1 to 4 MHz, the resultant output is able to be put into a range-ordered store. It remains in store until the next transmission and, range cell by range cell, the current video is compared with that in store. Differences are taken and either used or stored again for second differences to be taken before release as cancelled output. It is a real-time process.

Staggered p.r.f. The simple two-pulse canceller MTI radar illustrated in *Figure 59.17* can be improved by carrying out successive cancellations. The system response for cancellers with regular p.r.f. is illustrated in *Figure 59.18*. The 'blind speeds' are equivalent to a target movement of integral numbers of half wavelengths in an interpulse period. In a typical L-band radar a target radial speed of 150 knots gives a Doppler frequency of about 600 Hz. From Equation (59.7) and *Figure 59.19* it is seen that the relationship is linear. As seen from *Figure 59.18* a regular p.r.f. results in repeated 'blind speeds'. By varing the interpulse period at successive transmissions the overall system response is a compound or average of a number of responses shown in *Figure 59.20*, each slightly displaced because of differences of blind speeds. The first true blind speed will occur when all the zeros of those in the group coincide. Thus if two p.r.f. values were used in

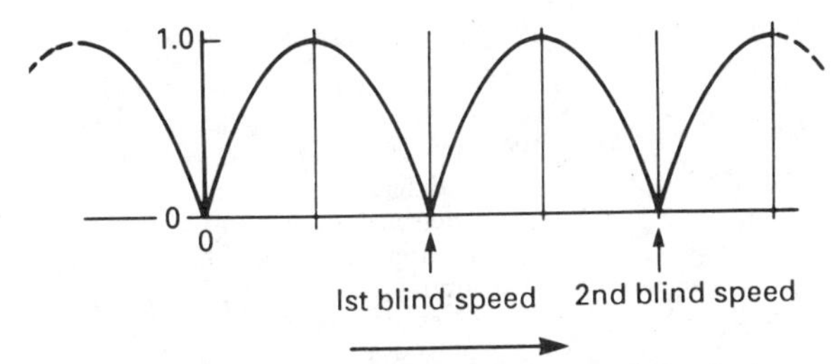

Figure 59.18 MTI response of two-pulse comparator

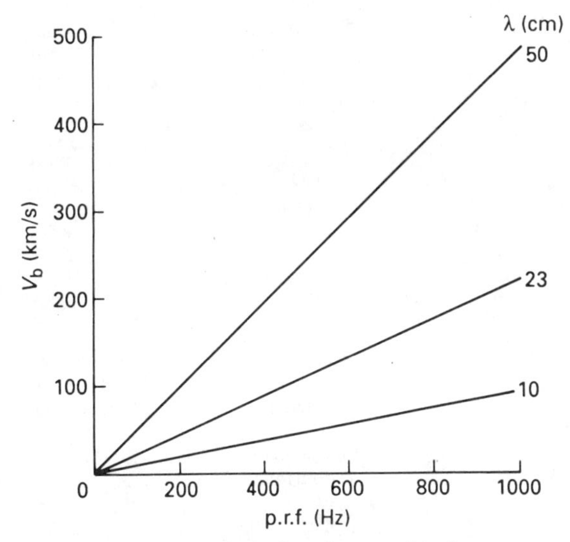

Figure 59.19 MTI blind speeds at three wavelengths. v_b (km/s) = λ (m) × p.r.f. (Hz) × 0.968

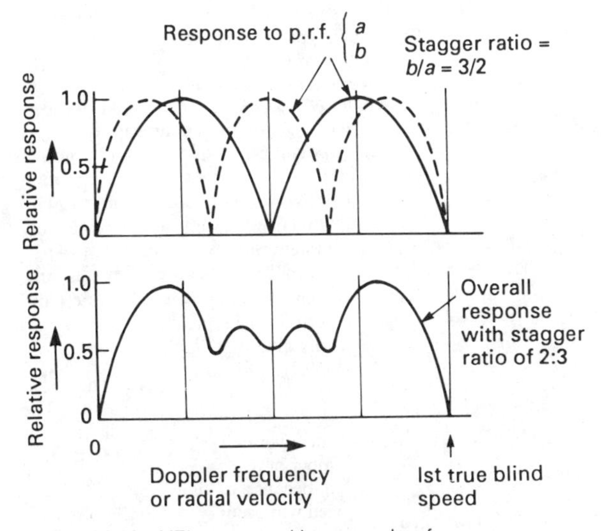

Figure 59.20 MTI response with staggered p.r.f.

the ratio of 4:5 then v_{btrue} is given where $4v_{b4}=5v_{b5}$ where v_{b4} and v_{b5} are the respective blind speeds of the two p.r.f. values. Note that by making the 'stagger ratio' small the value of v_{btrue} increases but vestigial blind speed losses increase.

The response in *Figure 59.18* is a series of half sine waves, resulting from a two-pulse comparison MTI. If a three-pulse comparator is used the response becomes a series of half sine waves squared. A general expression for the response shape of an '*n*' pulse comparison system is:

$$A=\sin^{n-1}\omega t \tag{59.9}$$

where *n* is the number of pulses compared and ωt the equivalent Doppler frequency.

When staggered p.r.f. values are used the general form of the modulus of the overall response is:

$$y=A\sin^{n-1}\omega_1 t+A\sin^{n-1}\omega_2 t+\cdots+A\sin^{n-1}\omega_m t \tag{59.10}$$

where *n* is the number of pulses compared and $\omega_1 t, \omega_2 t, \ldots, \omega_m t$ are the individual responses of each p.r.f. used in the stagger pattern.

Doppler filtering As described above, MTI can be regarded as a time sampling and comparison system. The coherence of receptions with transmissions allows the wavelength to be used as a vernier of the range scale.

The target movement is detected by comparing the phase of returned signals from successive transmissions as well as their range (which changes very little between the samples at the p.r.f. rate).

It happens that the method produces signals whose amplitude changes from the phase detector occur at the Doppler frequency created by the target's movement.

It is possible, using digital techniques, to construct a series of contiguous passband filters in the Doppler domain and by these to discriminate (to the degree governed by the filter bandwidths) between fixed and moving targets. Phase detector outputs form the filter input.[8,22] Such an arrangement is illustrated in *Figure 59.21*. Points of note are:

(1) Each filter has to be followed by some form of TTI (see Integration techniques following). The zero Doppler filter (f_0) output will contain all ground clutter and thus provides targets only when they exhibit super-clutter visibility.
(2) Because the passband characteristics of filters overlap and each has sidelobes, heavy ground clutter can make clutter also appear in filter f_1 output. The same will be true of weather clutter at radial velocities higher than zero.
(3) Because the phase detector cannot distinguish range changes between samples (successive transmissions) of greater than a wavelength, the zero Doppler filter repeats itself at f_D, the sample rate (i.e. the radar p.r.f.). This is directly analogous to 'blind speed'. Thus any target flying at a non-zero blind radial velocity will be masked by the 'alias' of the ground clutter, if at the same range, and not be seen unless it has super-clutter visibility. For this reason it is necessary to operate with staggered p.r.f. as in the pulse comparison system.
(4) The Doppler filter technique has the distinct advantage of giving target visibility in ground and weather clutter where radial velocities of clutter and target are separated by a filter passband width.
(5) Filters formed by digital techniques require samples at a regular rate equal in number to the number of filters. Thus staggered p.r.f. can only be executed in blocks (i.e. if four filters then four samples at one p.r.f. have to be executed before changing to another p.r.f.). This can have the effect of causing half the target data to disappear if the target happens to be at the blind speed of one of the p.r.f. Staggering the p.r.f. disturbs the spectrum of returned signals; this in turn

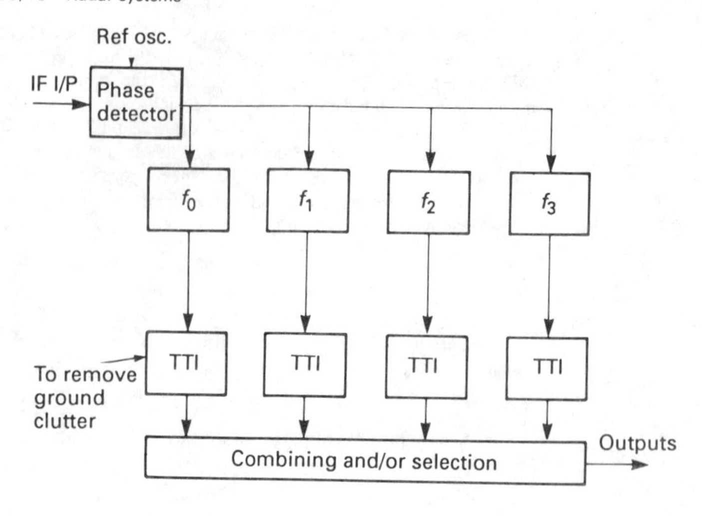

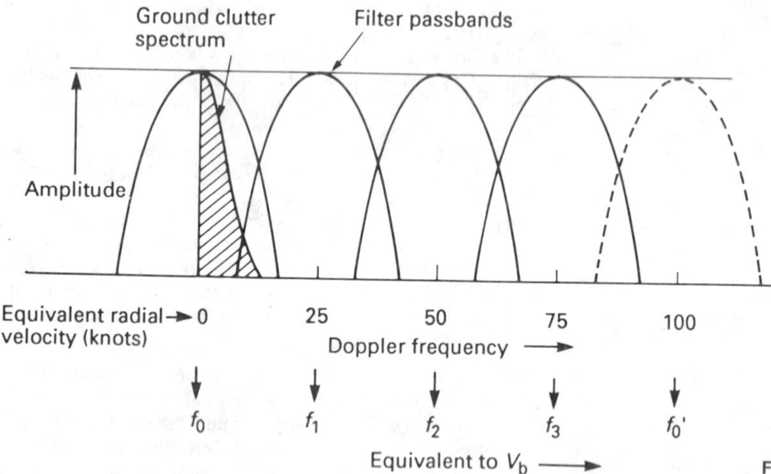

Figure 59.21 A typical four-filter arrangement

degrades the permanent echo cancellation performance. In higher order cancellers the techniques of time varying weights (TVW) is used to combat this.[23]

Data on Doppler filtering techniques can be found in references 24–26.

Integration techniques As far as MTI is concerned ground clutter containing wind-blown vegetation is characterised by exhibiting movement but no displacement over successive aerial revolutions. This is used in the technique of temporal threshold integration (TTI) to remove clutter signals in the following way.

The surveillance area is covered by a large mass of contiguous storage cells (typically 65 000) each of the same azimuth and range size. At successive aerial revolutions the contents of each cell is peak detected and a proportion of the value is stored. Thus over a number of aerial revolutions the true peak value of clutter amplitude would be achieved for non-fluctuating cell contents. The stored value for each cell thus establishes a threshold which must be exceeded if detection is to produce an output. To account for clutter fluctuations a margin is added to the stored values. Thus if an aircraft were to be detected over clutter it requires to exhibit 'super-clutter visibility', i.e. to be of such an amplitude that its addition to the clutter results in exceeding the established clutter threshold plus the margin. The integration time is such that aircraft dwell time in any cell is too short effectively to modify the cell's established threshold.

A variant of this technique performs its integration over batches of cells in the range domain only. The threshold for each cell is now set by the average contents of cells either side of it. It is used in environments which could contain barrage or noise jamming signals which can be rapidly switched on and off. The TTI integration time is many aerial revolutions, therefore any such jamming, having set high thresholds in many cells, when switched off could result in needless desensitisation for long times. In range integration the reaction time is very much less than one aerial revolution.

Adaptive signal processing (ASP) A compound of the principles of digital MTI, Doppler filtering and temporal threshold integration is found in the adaptive signal processor. The principle of operation can be seen from the block diagram in *Figure 59.22*. Phase detected signals are used as input to a four-pulse digital canceller. Staggered p.r.f. gives a Doppler filter passband as shown in *Figure 59.23(b)*. This rejects signals with zero and near-zero Doppler components. Moving weather clutter will still appear at its output and must be rejected. Dual

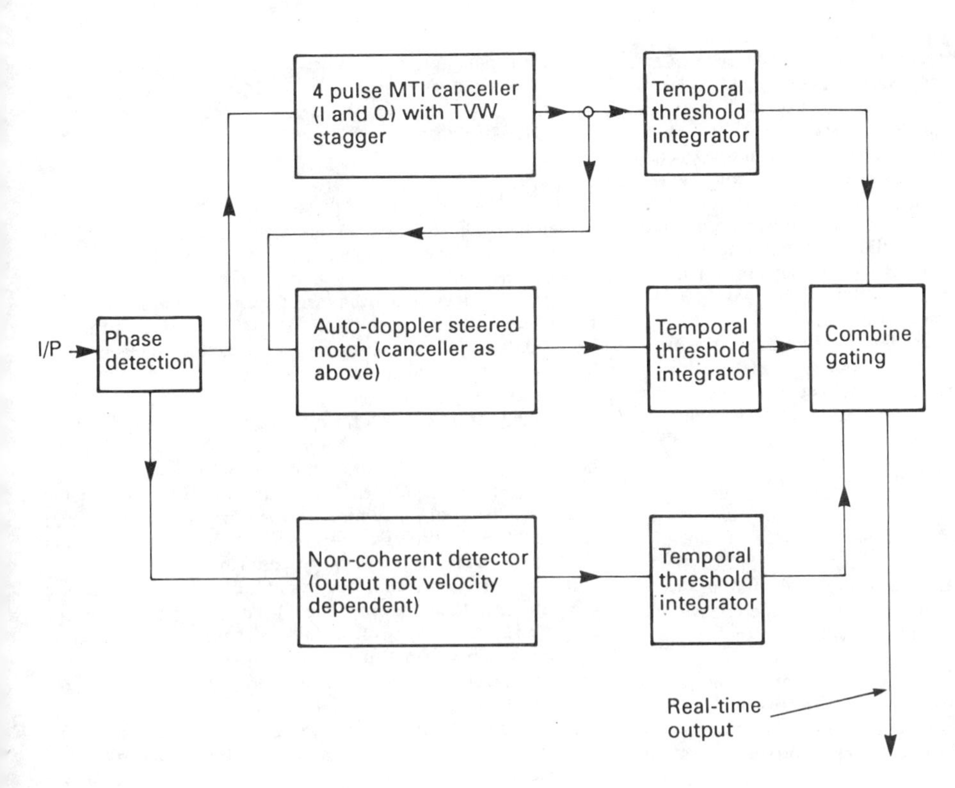

Figure 59.22 The adaptive signal processor

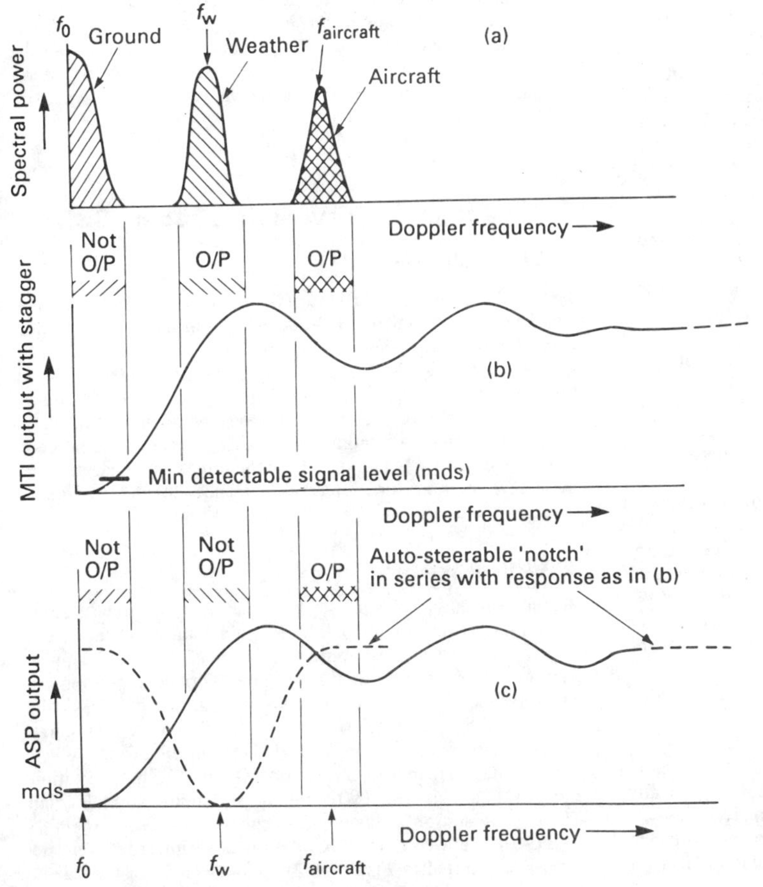

Figure 59.23 Signals and responses in adaptive signal processor (ASP): (a) signal spectra, (b) traditional MTI response, (c) ASP response

phase detectors whose references are 90° phase displaced express the Doppler frequency of the input signal as a phase angle relative to the reference. The measured phase angle is used in a further phase detector to govern the phase shift of the reference oscillation. Subsequent four-pulse cancellation with staggered p.r.f. gives an output characteristic with a rejection 'notch' in the Doppler domain centred upon the measured Doppler frequency. Integration constants are such that aircraft targets of non-zero Doppler components do not cause the phase shift to take place. Thus signals outside the passband of the rejection notch are allowed as output. Simultaneously with the above two detection processes, non-coherent detection is performed.

Temporal threshold integration is applied to each of the three detection processes' outputs and allows any available wanted signal to be gated out in real time. The whole processor thus automatically adapts itself to deal with fixed and moving clutter.

59.2.9.2 Discrimination against interference

PRF discrimination Use is made of the fact that radar transmission is at regular rates, i.e. a known repetition frequency. Using a storage technique similar to that of the MTI system, successive signals are compared. Those due to the station's own transmission will correlate in range, those due to interference will do so only if harmonically or randomly related to the station's p.r.f. Instead of a subtraction process, coincidence gating is used and so wanted signals are released and unwanted interference rejected. In systems not using a fixed p.r.f. use is made of the known time of transmission. Range correlation of wanted targets is used as detailed above, de-correlation again provides rejection of unwanted interference.

59.2.9.3 Discrimination against cloud and precipitation

To a certain extent the PLD technique can be used, but only when clutter is very severe. It is more general to use the circular polarisation (CP) technique. Here the radiation from the radar is changed from linear to circular by either waveguide elements or quarter-wave plates.[27] Raindrops and cloud droplets return nearly equal components in vertical or horizontal planes. When put into the aerial system these components of the signals from rain are of opposite phase and, if of equal amplitude, will cancel. Signals from irregular targets have inequality of vertical and horizontal components and thus do not cancel. However, there will be some loss of wanted target strength (typically 2–3 dB). Rejection of rain signals can be as high as 20 dB, thus discrimination in favour of wanted targets is of the order of 17–18 dB.

The choice of wavelength is critical in achieving discrimination against rain; the longer wavelengths increasingly provide protection due to an inverse fourth power law relating echoing area and wavelength.[6,28]

Many air traffic control radars now incorporate means to preserve weather signal data simultaneously with those in the weather rejection channel. A separate weather receiver and processing channel is required together with means for the aerial feed to develop a 'weather inclusive' port in the circular polarising elements.[29,30] Permanent echoes contained in this 'weather' channel are usually removed by a simple 2-pulse mti canceller.

59.2.9.4 Discrimination against angels

Again, wavelength is an important factor, angels being more apparent at the shortest wavelengths. Angels are now generally recognised as being due to birds, in isolation and in great flocks.[31] Use is made of the target signal amplitude difference between angels and aircraft, angels being smaller in general. Range-dependent signal attenuation or receiver desensitisation is used to reduce angel signals to noise level; the coincident aircraft signals are also reduced at the same time but being stronger are retained above noise.

59.2.10 Plot extraction

Radar signals from given targets exist, in real time, for a few microseconds in periods of milliseconds. From these data the operator reads the signal's centre of gravity as displayed upon the PPI. The data on a given target are repeated each time the radar beam illuminates the aircraft—typically, every 4 to 15 s. Radar data commonly have to be sent over long distances, e.g. from the aerial's high vantage point to a control centre sometimes many miles away. To transfer these data in real time, inordinately large bandwidths must be wastefully employed.

In order to avoid this, and simultaneously to provide positional data in a form more suitable for computer handling, plot extraction equipment is fast emerging and being operationally used. The plot extractor accepts radar signals in real time from the processor, together with the digital expression of the aerial azimuth. The extractor performs logical checks such as range correlation, azimuth contiguity, etc., after signals cross a pre-set threshold and become digital. The extractor derives the range and azimuth centre of all plots which meet the logic criteria and stores these in azimuth and range order ready for release as range and azimuth words in digital form; typically 12 bits for range of 200 miles and 13 bits for 360° of azimuth.

The assembled data are read from store at rates suitable for transmission over telephone bandwidths. By this means 100 to 200 plots can be reported in a few seconds. Computer programmes have been constructed which convert plot data into track data, deriving speed and heading as by-products.[23,32,33] Track data may be fed into PPI-type displays at a high refresh rate (approximately 40 times per second) in machine time, producing a picture which may be viewed in high ambient lighting.

59.3 Secondary surveillance radar (SSR)

59.3.1 Introduction

The need for automatic means of distinguishing friend from foe was foreseen by Sir Robert Watson-Watt in his original radar memorandum. The SSR system, which gives automatic aircraft identity, started its life as the wartime IFF system (identification friend or foe); the term IFF is used universally for the military application of the technique. The civil version is known variously as SSR and, more commonly in the USA, as the radar beacon system (RBS). Both IFF and SSR use the same system parameters these days and no distinction is made in the following text unless necessary.

59.3.2 Basic principles

Regular transmissions of pairs of pulses are made from the ground station at a frequency of 1030 MHz via a rotating aerial with a beam shape narrow in azimuth and wider in elevation. A pulse pair constitutes an interrogation and modes of interrogation are characterised by coding of the separation of the pair, internationally designated P_1 and P_3. Aircraft carry transponders which detect and decode interrogation. When certain criteria are met the transponders transmit a pulse position coded train of pulses at 1090 MHz via an omnidirectional aerial. The ground station receives these replies and decodes them to extract the data contained. This process of interrogation and reply is illustrated in *Figure 59.24*.

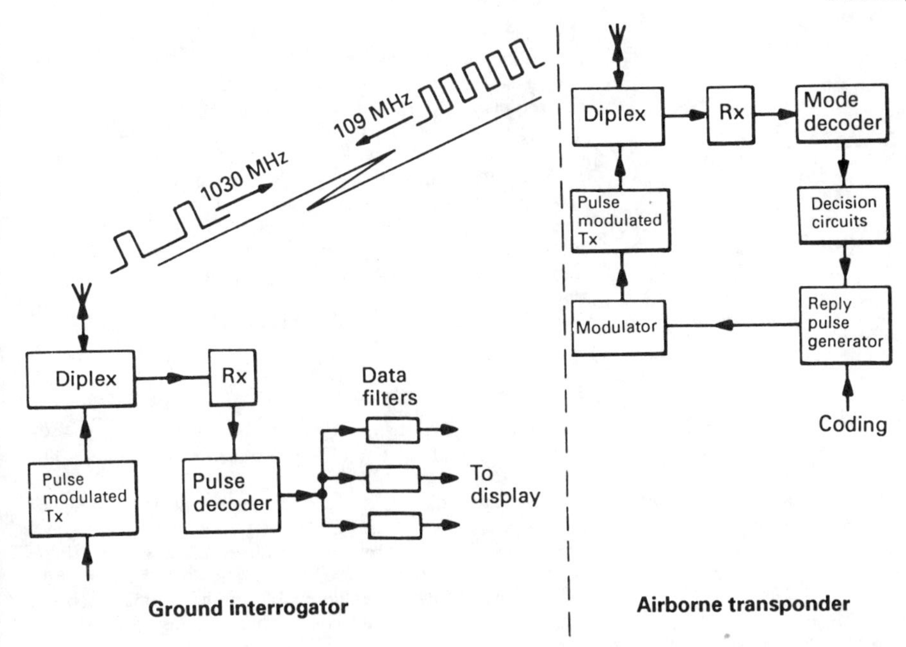

Figure 59.24 THE SSR principle

59.3.2.1 Modes of interrogation

These are shown in *Figure 59.25*. Through the agency of the International Civil Aviation Organisation (ICAO) there is internationally agreed spacing and connotation for the civil modes. Various military agencies treat the military modes in the same fashion:

Military modes:
- Mode 1—Secure
- Mode 2—Secure
- Mode 3—Joint military and civil identity

Civil modes:
- Mode A—Joint military and civil identity (same as Mode 3)
- Mode B—Civil identity
- Mode C—Altitude reporting
- Mode D—System expansion (unassigned)

The dwell time of the ground stations beam on target is generally of the order of 30 ms, dependent upon beamwidth, aerial rotation rate and p.r.f. There is thus time enough to make some 15 to 20 interrogations of a given target. Advantage is taken of this to execute repetition of different modes by mode interlacing. Repetition confers the necessary redundancy of data in the system so that data accuracy is brought to a high level. By this means, for example, the identify of an aircraft and its altitude can be accurately known during one passage of the beam across the target. Hence in one aerial revolution these data are gathered on all targets in the radar cover.

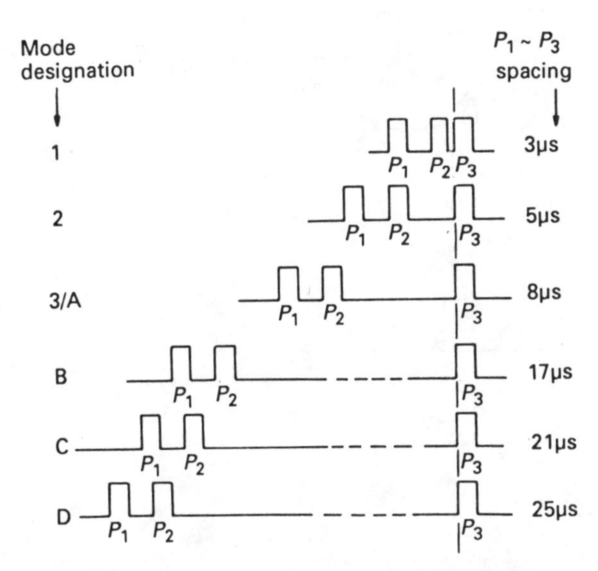

Figure 59.25 Interrogation pulse structure in IFF/SSR systems. Note that $P_1 \sim P_2$ is always 2 μs spacing for all modes. All pulses are 0.8 μs wide

59.3.2.2 Replies

Aircraft fitted with transponders will receive interrogation during the dwell time of the interrogating beam. The transponder carries out logic checks on the interrogations received. If relative pulse amplitudes, signal strength, pulse width and spacing criteria are met, as specified in reference 34, replies are made within 3 μs of the receipt of the second pulse (P_3) of the interrogation pair. Replies are made at a frequency of 1090 MHz. This is 60 MHz higher than the interrogation frequency. In air traffic control terms this provides an important advantage with SSR vis-à-vis primary radar as can be seen from *Figure 59.26* illustrating SSR's clutter freedom.

The pulse train of a reply is constructed, again according to internationally agreed standards, and allows codification of any or all of 12 information pulses contained between two always-present 'bracket' or 'framing' pulses designated F_1 and F_2. The pulse structures possible may be deduced from *Figure 59.27*. The code is octal based; the four groups of three digits are designated in order of significance as groups A, B, C and D. Of the 4096 possible codes, ICAO specifies that code 7700 is reserved for signalling emergencies, code 7600 to signal communication failure and code 7500 to signal 'Skyjack'.

The codification which signals altitude is formulated by a Gilham code pattern giving increments of 100 ft (30.5 m). The

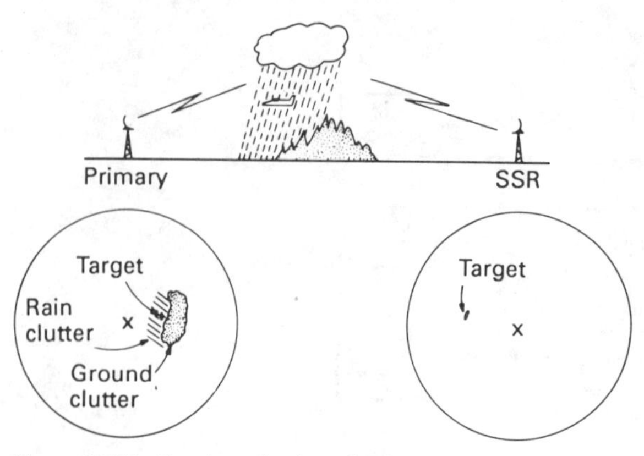

Figure 59.26 The clutter freedom of SSR

tabulation can be found in the ICAO document Annex 10 and covers altitudes from −1000 ft to +100 000 ft (−300 m to +30 000 m).[34]

59.3.2.3 Sidelobe suppression

The interrogation pair P_1 and P_3 is transmitted from an aerial with a radiation pattern narrow in azimuth and inevitably produces sidelobes carrying sufficient signal strength to stimulate replies from aircraft transponders at short and medium range. This impairs azimuth discrimination and causes transponders to reply for longer than necessary, thus generating unwanted interference (fruit) to other stations receiving replies. ICAO specifies a system of interrogation sidelobe suppression (ISLS) to prevent replies being made to interrogations not carried by an interrogator's main beam. It operates as follows. The interrogation pair, P_1 and P_3, both of equal amplitude, is accompanied by a control pulse P_2 of the same amplitude. The aerial is arranged to produce two different horizontal radiation patterns, one of which carries P_1 and P_3 in a narrow beam and the other P_2 at lesser amplitude in the main beam region and greater amplitude in sidelobe regions. A method of achieving this is by means of the 'sum and difference' aerial technique.[35] The control pulse P_2 is always 2 μs after P_1 for any mode of interrogation. The transponder's logic circuits compare the amplitude of P_1 and P_3 with that of P_2. If P_1 and P_3 are less than P_2, no reply is made. If P_1 and P_3 are 9 dB greater than P_2 a reply must be made. Between these limits a reply may or may not be made. We have then the so-called 'grey region' of 9 dB allowed by ICAO. In practice this is currently more like 6 dB with civil transponders and individual transponders can produce quite stable system beamwidths of the order of ±10% of the mean value. The ISLS system is illustrated in *Figure 59.28*.

59.3.2.4 Fruit and defruiting

All interrogations from any ICAO station are made at 1030±0.2 MHz and all replies from all ICAO standard transponders are made at 1090±3 MHz.

Interrogations are made via beams narrow in azimuth; replies are made omnidirectionally. Thus replies to one interrogator station can be received by other stations via their main beam or receiver side lobes (unless receiver side lobe suppression is used). To prevent ambiguous range and bearing information being gathered by these means a technique of reply range correlation is used.

In early equipments before the advent of low-cost digital storage devices each interrogator station was assigned its specific p.r.f. or interrogation rate. A synchronous detector at each station could separate replies to its own interrogator from those made by others. Asynchronous replies are known as 'fruit' and the process of separating them out is known as 'defruiting'.

In modern digital systems incorporating automatic plot and data extraction, the process of range correlation is a central part of the logic necessary for plot detection, thus defruiting is an automatic by-product and does not require the system to include a separate defruiting unit commonly found in early SSRs. Range de-correlation between neighbouring interrogators can now be effected by random staggering of interrogations.

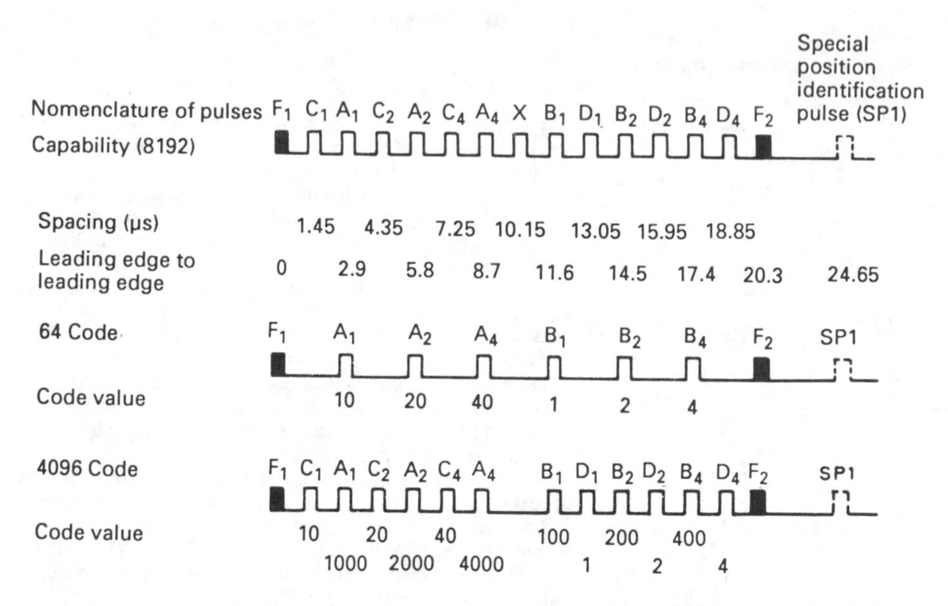

Figure 59.27 The ICAO reply format in SSR

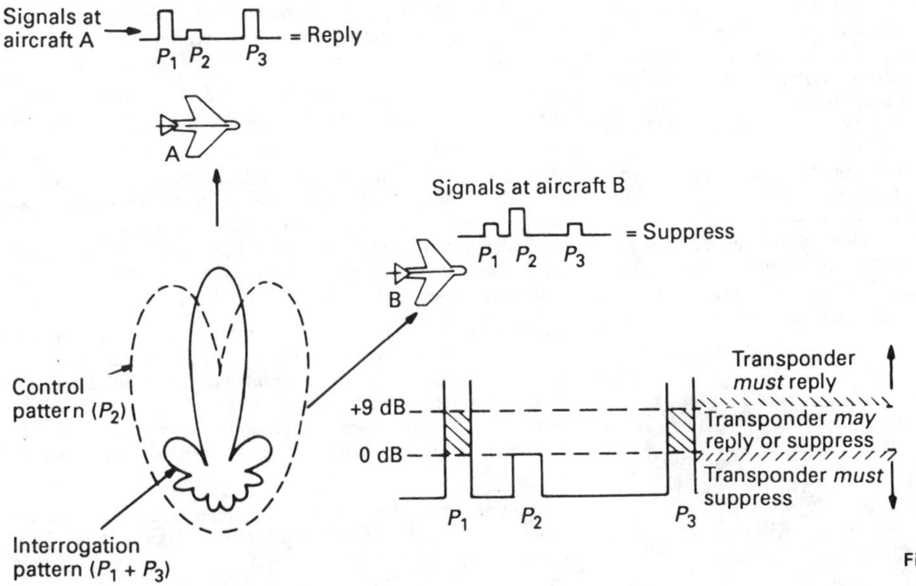

Figure 59.28 The ISLS system

59.3.2.5 *Garbling and degarbling*

If two aircraft on one bearing are close enough in slant range, their replies can overlap because of their simultaneity. This condition is known as synchronous garbling. It is common to find SSR decoders with fast dual registers operated by a commutating clock. These enable all but the most severe pulse masking to be tolerated. In view of the dubiety of data under garbled conditions it is common to find data labelled in such a way that garble conditions are indicated to the operator.

59.3.2.6 *Real-time decoding*

Replies are standardised as shown in *Figure 59.27*. All pulses are 0.45 ± 0.1 μs in duration separated by 1.45 μs and contained within bracket pulses spaced 20.3 μs apart. Thus reply detectors of various sorts can be constructed which sense pulse width, position and spacing. Decoders operating in this manner pass all valid code data to operators who have control units upon which wanted codes may be set by mode/code switches. Every reply contains F_1/F_2 bracket pulses. Decoders are generally arranged to seek coincidence of F_1 and F_2 by delaying F_1 by 20.3 μs. When coincidence is found the decoder produces an output pulse similar to that of a primary radar signal. Thus PPI display responses from SSR appear similarly to the operator, giving range and azimuth data.

59.3.2.7 *Real-time active decoding*

The signals indicating aircraft 'presence' and position can be used by the operator to call out the specific code data of any aircraft reply. By use of a light-pen or real-time gate associated with a steerable display marker, any individual target can be isolated and code data of its replies examined by the decoder. The code structure and the mode of interrogation are indicated to the operator on separate small alphanumeric displays as four octal digits representing the four groups of information pulses referred to above. A similar indicator is used to show automatically reported altitude of targets in decimal form in units of 100 ft (30.5 m) flight levels.

59.3.2.8 *Real-time passive decoding*

Aircraft identity codes can be discovered by use of active decoding as described above. Commonly, identities will be known by either flight planning or radio telephone request. Operators are usually given sets of controls upon which they can set any mode and code combinations, again using the four-digit octal structure. When the control is activated the decoder looks for correspondence between the set values and all replies. When correspondence occurs a further signal is generated by the decoder to augment that due to detection of F_1/F_2. Thus unique identity of a number of targets can be established.

59.3.2.9 *Automatic decoding*

The processes described above operate in real time and require the operator to filter wanted data from that given, by use of separate controls and indicators. Automatic decoding carries out the process of active decoding on all targets continually, storing in digital form all the mode and code data, together with positional data. Using modern fast PPI display techniques these data can be converted into alphanumeric form with other symbols for direct display upon the PPI screen, the position of the data being associated with the position symbol of the relevant target. By this means a fully integrated primary and secondary radar system can be produced. This concept is widely being implemented for air traffic control use.

59.3.2.10 *Modern SSR systems*

Monopulse techniques A new generation of SSR equipments is now in being, incorporating design features which overcome the basic limitations of former systems.[36] A major improvement to azimuth data is achieved by the use of monopulse direction-finding techniques. This allows the azimuth of each and every pulse of every reply to be measured with errors as little as a few minutes of arc (rms); about a four-fold improvement in average performance. The incidence of massive errors ($\pm 1°$) is reduced to very rare occasions. Signal amplitude data is preserved instead of deliberately inhibited. Fast signal processing using 8–20 MHz sampling rates permits individual reply pulse timing (range), azimuth, amplitude and duration to be used in recovery of accurate reply data in very severe garbled conditions.[19]

Large vertical aperture (LVA) antennas A further improvement to both 'traditional' (non-monopulse) or 'monopulse' SSRs has been achieved by replacing the original 'hogtrough' antennas by LVA arrays. The former have a small vertical aperture of about 1½ wavelengths. The resultant beam is very broad in elevation. Consequent ground incident energy produces massive vertical lobing (see *Section 59.2.1.3*).

LVAs are of the order of 6 wavelengths vertical aperture and permit almost perfect 'cosec' shaping in the vertical plane; gain reduces rapidly from its peak (at about +6° elevation) to zero and negative elevation angles. This greatly reduces ground incident energy and hence the vertical lobing effects. False replies created by reflection from vertical surfaces are also dramatically reduced. One such LVA is illustrated in *Figure 59.29* and described in reference 37.

Figure 59.29 Marconi Radar large vertical aperture monopulse SSR antenna type S1095 mounted on top of an ASR type S511. These LVAs are used by the UK CAA to cover UK airspace (Courtesy of Marconi Radar Ltd)

59.3.3 Important distinctions between primary and secondary radar

The following is a brief resumé of the advantages and disadvantages of SSR relative to primary radar.

Positional data Both systems give this, although the quality of data will vary dependent upon the system. Radar stations with both primary and secondary sensors select range or azimuth data from each according to its quality. For instance, primary radar azimuth data would be selected (if available) rather than that from a non-monopulse SSR, whereas the range data on the same target would be preferred from the SSR (because of its generally shorter pulse duration and higher range discretion).

Height data In primary radar systems either a separate height-finding equipment or a multi-beam surveillance system must be deployed. In SSR the height-finding facility is in-built but dependent upon aircraft transponders being fully equipped with height reporting elements (this is not yet mandatory but may become so).

Identity of targets This is inherent in SSR. In primary radar, identity has to established by request of manoeuvre or unreliably inferred from position reports.

Dependence upon aircraft size In SSR all transponders have ICAO standard performance hence the signals from small craft are as strong as those from large aircraft. In primary radar the equivalent echoing area is dependent upon target size and shape. Although both systems suffer from target 'glint', SSR suffers least and can be improved by use of multiple aerials.

Constraints of ground clutter Primary radar has to include clutter rejecting systems with 'in-clutter' target detection capability. This always has a limit which can inhibit target detection in regions of high-level clutter. SSR is completely free from this constraint since the ground station is not tuned to receive on its transmitting frequency; reflected energy is thus rejected.

Constraints of weather clutter The same is true here as for ground clutter, i.e. SSR will not receive energy reflected from weather clutter. Two-way attenuation of microwaves is a factor for consideration in both primary and secondary radar but is less significant in the SSR system which, by ICAO specification, is organised with power in hand and thus more able to cope with losses.

Constraints of angel clutter SSR has complete freedom from angel clutter. This freedom can be achieved in primary radar only by sacrifice of detection of very small targets, or complication of signal processing.

Dependence upon target cooperation SSR is entirely dependent upon target cooperation and is totally impotent unless aircraft carry properly working transponders. Primary radar is dependent only upon the presence of a target within its detection range.

Jamming Frequency agility possible in primary radar systems confers a large measure of protection against jamming. Since SSR operates at two fixed frequencies and with small fractional bandwidths (about 1%) it is relatively easy to jam SSR and IFF systems. The advertised pulse formats and code structures make it easy to generate 'spoof' targets.

Range capability The range performance of SSR is amenable to calculation as with primary radar. A major difference is that SSR range is the compound of two virtually independent inverse square law functions and primary radar range, when a single aerial is used, is an inverse fourth power law function. With SSR it is required to calculate

(1) Maximum interrogation range ($R_{\mathrm{i\,max}}$).
(2) Maximum reply range ($R_{\mathrm{r\,max}}$).

$$R_{\mathrm{i\,max}}^2 = \frac{P_\mathrm{i} G_\mathrm{i} G_\mathrm{r} \lambda_\mathrm{i}^2}{(4\pi)^2 S_{\mathrm{a/c}} L_\mathrm{i} L_{\mathrm{a/c}}} \qquad (59.11)$$

and

$$R_{\mathrm{r\,max}}^2 = \frac{P_\mathrm{r} G_\mathrm{r} G_\mathrm{i} \lambda_\mathrm{r}^2}{(4\pi)^2 S_\mathrm{r} L_\mathrm{i} L_{\mathrm{a/c}}} \qquad (59.12)$$

where

P_i = Peak of interrogator power output
G_i = Gain of the interrogation aerial
G_r = Gain of the aircraft transponder aerial
$S_{\mathrm{a/c}}$ = Signal necessary at aircraft aerial to produce satisfactory interrogation
L_i = Losses between interrogator and its aerial terminals

$L_{a/c}$ = Losses between aircraft aerial and transponder receiver terminals
λ_i = Wavelength of interrogation (29 cm)
P_r = Transponder peak output power
G_r = Gain of aircraft aerial
S_r = Reply signal level necessary at responser (ground receiver) aerial for satisfactory decoder operation
λ_r = Wavelength of reply (27.5 cm)

The following points should be noted relating to the above:

(1) *Interrogator aerial gain* G_i It is usual to find aerials of very small vertical dimension, of the order of 46 cm to 61 cm ($1\frac{1}{2}$ to 2 wavelengths approximately). This produces beams which are very wide in the elevation plane and consequently a great deal of power is directed into the ground. The resultant vertical polar diagram is therefore subject to very large modulation about the free-space pattern by the mechanism described earlier. The system uses vertically polarised radiation.

(2) *Aircraft aerial gain* G_r The aircraft aerial is intended to be omnidirectional. Due to aerodynamic restrictions and varying shape of the airframe, the polar pattern of the aerial is modulated away from the desired shape and, more importantly, subject to shadowing and sometimes total obscuration during aircraft manoeuvring. Average reported gain performance is ± 7 dB about a mean of zero. There is evidence of wide variations over small solid angles of -40 dB and $+20$ dB. For these reasons it is usual to find figures of unity gain assumed in calculations.

(3) *Range performance* If an interrogator is operated at the maximum permitted power ($52\frac{1}{2}$ dBW ERP) and a transponder of nominal performance is taken, maximum interrogation free-space ranges of 300 to 400 nautical miles can be achieved. Allowing the system to have some beamwidth (say equivalent to that at -3 dB relative to the horizontal polar pattern peak), the range is still of the order 200 to 280 nautical miles.

Taking practical values for system loss and ground receiver (responser) sensitivity, reply ranges in excess of 300 nautical miles are obtained. Thus the system has gain in hand on the reply path, ensuring high signal strength of received data.

59.3.4 Future extension of SSR

As indicated above and elsewhere[38–43] traditional SSR systems have their own specific problems and limitations. Prominent among these are over-interrogation and over-suppression of transponders, poor azimuth data, corruption of data by 'fruit', garbling of data, etc. Although monopulse techniques, LVAs and fast processing all produce great improvements, the problem of garbling remains. In the early 1970s a great deal of exploratory work was done to devise ways of overcoming the residual problems of SSR.[44–46] All were based on addressing transponders individually rather than collectively.

In the UK the ADSEL (Address Selective) system was devised; its concurrent counterpart in the USA was called DABS (Direct Addressed Beacon System). Both have now been rationalised into the internationally specified Mode S system[33] and defined in a recent amendment to ICAO Annex 10. To ensure compatibility with current systems the up- and down-link frequencies of SSR are still used in Mode S together with its techniques of sidelobe suppression. The signal formats on up- and down-links are radically changed to create a massive increase in data capacity. The up-link interrogation can contain either 56 or 112 bits per transmission. The down link replies are equally extended. Differential phase shift keying (DPSK) modulation is used in Mode S to improve signal detectability. A highly robust error-correcting system promises the ability to effect very accurate data request and reply in single transactions.

The ability to address individual aicraft presupposes knowledge of the aircraft position. In practice an interrogation station will continue to operate in a surveillance mode albeit at the lower interrogation rate permitted by monopulse direction finding. Tracking on all targets will be performed. A special interrogation categorised as Mode S All Call will result in all Mode S equipped transponders replying with their unique address (one of up to about 16 million possible). The address will be attached to the track data already in store.

When discrete data are required to be passed to or received from individual aircraft, the ground system will know over what azimuth angles to include a specific Mode S interrogation. By this means it will be possible to pass air traffic control instructions and to receive acknowledgements automatically. Other data such as ground or air-sensed weather reports, fuel state, aircraft turn and bank, climb or descent rates can all be had without ground–air–ground radio telephone interchange.

The unique address system obviates two aircraft replying to a given interrogation and so avoids the circumstance of synchronous garbling. Another feature is being implemented with Mode S whereby air-to-air transactions permit suitably equipped aircraft to gain knowledge of the position and track of neighbouring aircraft. This facility is known as ACAS (Airborne Conflict Advisory System) or TCAS (Traffic Conflict Alert Systems). A comprehensive description of the aims and techniques of Mode S are to be found in reference 47 and the latest version of reference 34.

59.3.5 Modern operational methods

Radar art and technology have reached a point where practically any civil operational requirement can be met. In the military sphere electronic countermeasures are developed to great sophistication. As an example of the former, *Figure 59.30* illustrates the way in which primary and secondary radar, display and data handling systems can be combined to give all the information an air traffic control system needs for the safe and expeditious handling of air traffic. Data on selected aircraft are available to operators by simple controls organised in such a way that displays are not cluttered with masses of available data but which is on call at the operator's discretion.

References

1 WOOD, D. and DEMPSTER, D., *The Narrow Margin*, Arrow Books, Revised Illustrated Edition (1969)
2 WATSON-WATT, SIR ROBERT, *Three Steps to Victory*, Odhams Press (1957)
3 WATSON-WATT, SIR ROBERT, 'Radar in War and Peace', *Nature*, **156**, 319 (1945)
4 PRICE, A., *Instruments of Darkness*, McDonald & Jane (1977)
5 RANDALL, J. T. and BOOT, H. A. H., 'Early work on cavity magnetrons', *JIEE*, **93**, 997 (1946)
6 SKOLNIK, M. I., *Introduction to Radar Systems*, McGraw-Hill, International Student Edition (1962)
7 SKOLNIK, M. I., *The Radar Handbook*, McGraw-Hill (1970)
8 NATHANSON, F. E., *Radar Design Principles*, McGraw-Hill (1969)
9 BLAKE, L. V., 'Recent advances in basic radar range calculation technique', *IRE Trans.*, **MIL.5**, 154 (1961)
10 BARTON, D. K., *Radars*, Vol. 3 Pulse Compression, Artech House (1975)
11 GASKELL, S. and FINCH, M., 'Fixed beam multi-lateration radar system for weapon impact scoring', Radar '82', *IEE Conf. Publ.*, No. 216, p. 130
12 HALL, W. M., 'Prediction of pulse radar performance', *IRE*, **44**, 224 (1956)
13 VINCENT, N. and LYNN, P., 'The assessment of site effects of radar polar diagrams', *Marconi Review*, **28**, No. 157 (1965)
14 HANSFORD, R. F., *Radio Aids to Civil Aviation*, Heywood & Co., p. 112 (1960)
15 BECKMAN, P. and SPIZZICHINO, A., *The Scattering of Electromagnetic Waves from Rough Surfaces*, Pergamon (1963)

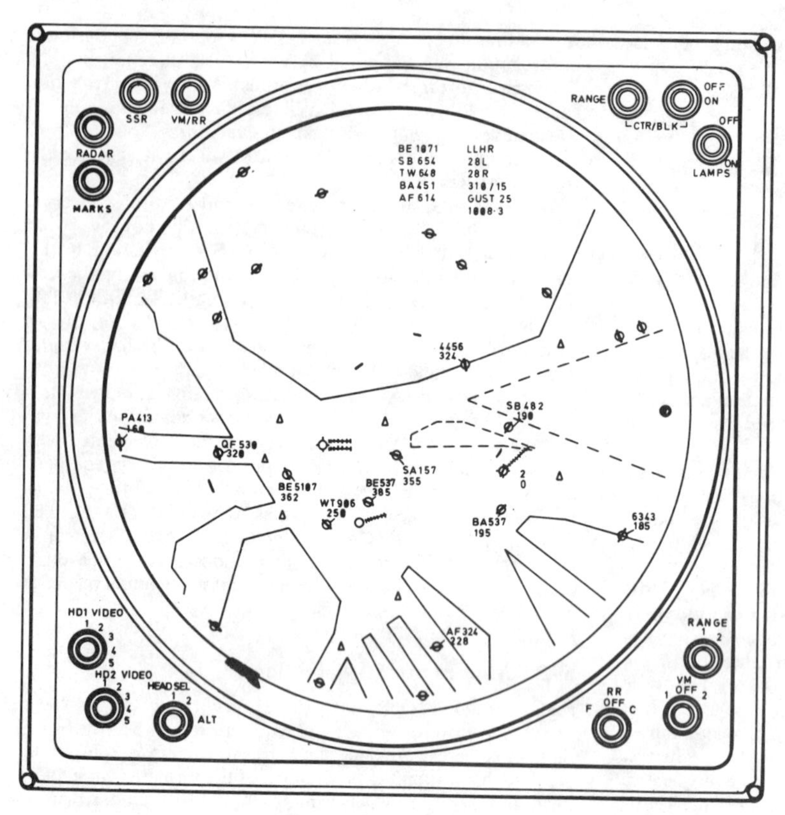

Figure 59.30 Air traffic control system radar display. Example of PPI display capable of presenting real-time signals (arcs) together with machine time data (circle, alphanumerics, video map and symbols) (Courtesy of Marconi Radar Systems Ltd)

16 *Atlas of Radio Wave Propagation Curves for Frequencies Between 30 and 10 000 MHz,* Radio Research Labs, Ministry of Postal Services, Japan (1955)

17 RIDENOUR, L. N., *Radar System Engineering*, McGraw-Hill, p. 143 (1947)

18 BERKOWITZ, R. S., *Modern Radar–Analysis, Evaluation and System Design*, Wiley, chapter 2 (1967)

19 SCANLAN, M. J. B., *Modern Radar Techniques*, Collins (1987)

20 SILVER, S., *Microwave Antenna Theory & Design*, McGraw-Hill (1949)

21 THOMAS, D. H. and HAYMAN, E. E., 'Raster scan radar displays', *IEE Conference Publication* No. 281, *Radar 87*, p. 474 (1987)

22 TAYLOR, J. W. Jr., 'Sacrifices in radar clutter suppression due to compromises in implementation of digital Doppler filters', Radar '82', *IEE Conf. Publ.*, No. 216, p. 46

23 COLE, H. W., *Understanding Radar*, p. 142, Collins (1985)

24 SKOLNIK, M. I., *Introduction to Radar Systems*, McGraw-Hill and Kogakusha, Issue No. 2, para 4.7, pp. 328–331 (1980)

25 FANCY, H., 'Multi-filter MTI system', *IEE Conference Publication* No. 155, *Radar 77*, p. 191 (1977)

26 GROGINSKY, H. L., 'The coherent memory filter'. In: Barton, D. K. (ed.) *Radars*, Vol. 7, p. 239, Artech House (1978)

27 RIDENOUR, L. N., *Radar Systems Engineering*, McGraw-Hill, pp. 81–86 (1947)

28 BARTON, D. K., *Radar Systems Analysis*, Prentice Hall, p. 105 (1964)

29 KLEMBOWSKI, W., 'Detection and recognition of hazardous weather conditions by primary surveillance radar', *IEEE 1985 International Radar Conference Record*, p. 226. IEE Cat. No. 85CH2076-8 (1985)

30 RIDER, G. C., 'A polarisation approach to the clutter problem', *IEEE Conference Publication* No. 155, *Radar 77*, p. 130 (1977)

31 EASTWOOD, SIR ERIC, *Radar Ornithology*, Methuen (1967)

32 HOWICK, R. E., 'A primary radar automatic track extractor', *IEE Conference Publication*, No. 105, 339 (1973)

33 TUNNICLIFFE, R. J., 'A simple automatic radar track extraction system', *IEE Conference Publication* No. 155, *Radar 77*, p. 76 (1977)

34 Annex 10 to the Convention on International Civil Aviation (International Standards & Recommended Practices—Aeronautical Telecommunications) HMSO, **2**, April (1968)

35 SKOLNIK, M. I., *The Radar Handbook*, McGraw-Hill, 38.10 (1970)

36 COLE, H. W., 'The future for SSR, ICAO Bulletin, September (1980)

37 COLE, H. W., 'Messenger—a high performance monopulse secondary system', *GEC Review*, **3**, No. 2 (1987)

38 COLE, H. W., 'S.S.R.—some operational implications in technical aspects', *World Aerospace Systems*, **1**, No. 10, 468, October (1965)

39 ULLYAT, C., 'Secondary Surveillance Radar in ATC', *IEE Conference Publication*, No. 105 (1973)

40 ARCHER, D. A. H., 'Reply probabilities in SSR', *World Aviation Electronics*, December (1961)

41 HERMANN, J. E., 'Problems of broken targets and what to do about them', *Report on the 1972 Seminar on Operational Problems of the ATC Radar Beacon System,* National Aviation Facilities Experimental Center, Atlantic City, NJ, p. 31 (1973)

42 GORDON, J., 'Monopulse technique applied to current SSR', *GEC J. Sci. Tech.*, **47**, No. 3, February (1982)

43 COLE, H. W., 'The suppression of reflected interrogation in SSR', *Marconi Review*, **43**, No. 212 (1980)

44 BOWES, R. C., GRIFFITHS, H. N. and NICHOLS, T. B., 'The design and performance of an experimental selectivity addressed (ADSEL) SSR system', *IEE Conference Publication*, No. 105, p. 32 (1973)

45 STEVENS, M. C., 'New developments in SSR', *IEE Conference Publication*, No. 105, p. 26 (1973)

46 AMLIE, T. S., 'A synchronised discrete-address beacon system', *IEEE Trans. Commun.*, **COM-21**, No. 5, 421 (1973)

47 *Secondary Surveillance Radar Mode S Advisory Circular*, ICAO Circular 174-AN/110, ICAO, Montreal (1983)

60 Computers and their Application

I Robertson
Digital Equipment Corporation

Contents

60.1 Introduction

Although the advent of computers in our everyday lives may seem very recent, the principles of the modern computer were established before the existence of any electronic or electromechanical technology as we know them today, and electronic computers were beginning to take shape in laboratories in 1945.

The work of Charles Babbage, a Cambridge mathematician of the 19th century, in attempting to build an 'analytical engine' from mechanical parts, remarkably anticipated several of the common features of today's electronic computers. His proposed design, had he been able to complete it and overcome mechanical engineering limitations of the day, would have had the equivalent of punched card input, storage registers, the ability to branch according to results of intermediate calculations, and a form of output able to set numeric results in type.

Many purely mechanical forms of analogue computer have existed over the last few centuries. The most common of these is the slide rule, and other examples include mechanical integrators and even devices for solving simultaneous equations.

Much of the development leading to modern electronic computers, both analogue and digital, began during World War II with the intensified need to perform ballistics calculations. The development of radar at this time also provided the stimulus for new forms of electronic circuit which were to be adopted by the designers of computers.

A further development of momentous importance to the technology of computers, as it was for so many branches of electronics, was that of the transistor in 1949. Continued rapid strides in the field of semiconductors have brought us the integrated circuit which allows a complete digital computer to be implemented in a single chip.

60.2 Types of computer

Although there are two fundamentally different types of computer, analogue and digital, the former remains a somewhat specialised branch of computing, completely eclipsed now, both in numbers of systems in operation and in breadth of applications, by digital computers. Whilst problems are solved in the analogue computer by representing the variables by smoothly changing voltages on which various mathematical operations can be performed, in a digital computer all data is represented by binary numbers held in two-state circuits, as are the discrete steps or instructions for manipulating the data, which make up a program.

60.2.1 Analogue computers

An analogue computer consists of a collection of circuit modules capable of individually performing summation, scaling, integration or multiplication of voltages, and also function generating modules. On the most up-to-date systems, these modules contain integrated circuit operational amplifiers and function generators. Several hundred amplifiers are likely to be used on a large analogue computer.

To solve a given problem in which the relationship of physical quantities varying with time can be expressed as differential equations, the inputs and outputs of appropriate modules are interconnected, incorporating scaling, feedback and setting of initial conditions as required, with voltages representing the physical quantities. In this way single or simultaneous equations can be solved, in such applications as engineering and scientific calculation, modelling and simulation.

The interconnection and setting of coefficients and initial conditions required is normally done by means of a patch panel and potentiometers. An analogue computer may also have a CRT display or chart recorder for display or recording of the results of a computation. Where an analogue computer is being used for design work, the designer may choose to observe immediately the effect of changing a certain design parameter by varying the appropriate potentiometer setting, thus altering the voltage representing that parameter. In a simulation application, outputs from the analogue computer may be used as input voltages to another electrical system.

Except in some specialised areas, the work formerly done by analogue computers is now most likely to be carried out on the modern high-speed, cost-effective digital computer, using numerical methods for operations such as integration. Digital computers can work to high precision completely accurately, whereas using analogue techniques there are inherent limits to precision, and accuracy suffers through drift in amplifiers.

60.2.2 Hybrid computers

The analogue part of a hybrid computer is no different in its circuitry and function to a stand alone analogue system. However, the task of setting up the network required to solve a particular problem is carried out by a digital computer, linked to the analogue machine by analogue/digital converters and digital input/output. Thus, the digital computer sets potentiometers and can read and print their values, and can monitor and display or log voltages at selected parts of the analogue network.

Linking the two technologies in this way brings the advantage of programmability to the mathematical capabilities of the analogue computer. Set-up sequences for the analogue circuitry can be stored, and quickly and accurately reproduced at will. Tasks can be carried out in the part of the system best adapted to performing them. In particular, the ability of the digital computer to store, manipulate and present data in various ways is particularly advantageous.

The use of hybrid systems in, for example, the simulation of aircraft and weapon systems, has helped to perpetuate analogue computing as a technique applicable in certain narrow areas.

60.2.3 Digital computers

Digital computers in various forms are now used universally in almost every walk of life, in business, in public service and in the home. In many cases unseen, computers nonetheless influence people in activities such as travel, banking, education and medicine.

There is also a growing range of truly portable computers now available that will fit into briefcases or even pockets. These are primarily used for data collection and onward transmission to a larger computer for detailed analysis and processing. This connection is usually made over the telephone/data network, but can also be achieved by radio transmission back to the host machine.

As can be imagined from the variety of applications, computers exist in many different forms, spanning a range of price, from the smallest personal system to the largest supercomputer, of more than 10 000:1. Yet, there are certain features which are common to all digital computers:

(1) Construction from circuits which have two stable states, forming binary logic elements.
(2) Some form of binary storage of data.
(3) Capability to receive and act on data from the outside world (input), and to transmit data to the outside world (output).
(4) Operation by executing a set of discrete steps or instructions, the sequence of which can be created and modified at any time to carry out a particular series of tasks. This ability to be programmed, with a program stored in the system itself, is what gives great flexibility to the digital computer. Recent advances have enabled changes to be made to a program dynamically whilst it is in operation.

It has also given rise to a whole new profession, that of the computer programmer and the systems analyst, as well as the industry of designing, building and maintaining computers themselves.

60.3 Generations of digital computers

Beginning with circuits consisting of relays, the history of the digital computer can be seen as having fallen into four generations between the 1940s and today.

First generation These were built with valve circuits and delay-line storage, physically very massive, taking up complete rooms, requiring very large amounts of electricity with corresponding heat dissipation, low overall reliability requiring extensive maintenance, often resulting in engineers being on site 24 hours a day. Input/output was rudimentary (teleprinters, punched cards) and programming very laborious, usually in a binary form that the machine could understand without further interpretation.

Second generation These were developed during the 1950s with transistorised circuits. They were faster, smaller and more reliable than the first generation, but still large by today's standards, and had magnetic core main store with magnetic drums and tapes as backup, and line printers for faster printed output. Programming language translators emerged, resulting in the widespread use of assembler-type languages.

Third generation Developed during the mid-1960s, these were heralded by the integrated circuit, allowing more compact construction and steadily improving speed, reliability and capability. The range and capabilities of input/output and mass storage devices increased remarkably. In the software area, high-level languages (e.g. FORTRAN, COBOL, BASIC) became commonplace and manufacturers offered operating system software developed, for example, to manage time-sharing for a large number of computational users, or real-time process control.

Most significantly, a trend of downward cost for given levels of performance was established, and the minicomputer, aimed at providing a few or even one single user with direct access to and control over their own computing facilities, began to gain in numbers over the large, centrally managed and operated computer system per organisation.

Fourth generation While a great many third-generation computer systems are in use, and will remain so for some time, the semiconductor technology of large-scale integration (LSI) has brought complete computers on a chip known as microprocessors, allowing further refinement and enhancement of third-generation equipment.

Semiconductor memory has almost completely replaced core memory, and the continuing reduction in size and cost has brought the 'personal computer', numbered in hundreds of thousands of units supplied, truly within the reach of the individuals in their own home or office.

60.4 Digital computer elements

A digital computer system is a collection of binary logic and storage elements combined in such a way as to perform a useful task. Any computer system, whether a microcomputer based on just a few integrated circuit packages incorporated within a laboratory instrument, or a large data processing system consisting of many cabinets of equipment housed in a specially built air-conditioned computer room, invariably contains a combination of the following parts, described in detail in the following sections and shown in *Figure 60.1*.

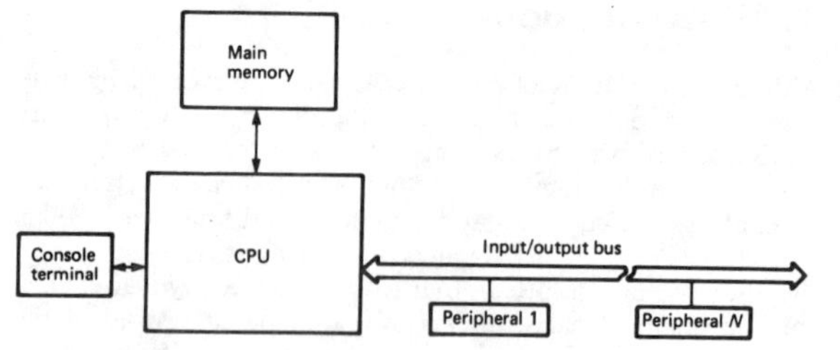

Figure 60.1 Components of a computer system

Central processor unit The CPU is where instructions forming the stored program are examined and executed, and is therefore in control of the operation of the system. Instructions and data for immediate processing by the CPU are held in main memory, which is linked directly to the CPU. This general term covers the units that perform logic decisions and arithmetic, collectively known as the arithmetic and logic unit (ALU).

Input/output This is the structure which provides optimum communication between the CPU and other parts of the system.

Peripherals These are the devices external to the CPU and memory which provide bulk storage, man/machine interaction and communication with other electronic systems.

Bus paths These act as the vehicle for the passing of data and program instructions between all other parts.

60.5 Digital computer systems

In going from microcomputers to large data processing systems there are many differences in complexity, technology, interconnection, capacity and performance. At any given level of system cost or performance, however, there are many similarities between systems from different manufacturers.

Several distinct categories of computer system can be identified. Going from least to most comprehensive and powerful, these are described in subsequent sections.

60.5.1 Hand-held computers

Apart from microprocessor chips incorporated into other equipment and dedicated to performing a fixed application such as control or data acquisition, the smallest type of digital computer system is the hand-held or lap microcomputer. These are typically light and extremely portable, but with a comparatively limited amount of memory and backing storage. They also generally include a flat liquid crystal display screen of between 1 and 16 lines. They are primarily aimed at on-site data capture and problem resolution where communication to more powerful computing facilities is not feasible or desirable.

60.5.2 Personal computers

These are built around a microprocessor chip, normally with a limited amount of memory available to the user, and some simple software stored permanently in read-only memory to allow the user to write programs in a language such as BASIC. A keyboard for entering programs and data, a CRT display or drive circuitry for a domestic television set, storage in the form of an audio tape recorder or small magnetic discs, and a character printer capable of a print rate in the region of 100 characters per second (cps) are the other components of a typical personal computer.

Such a system is capable of a range of data processing and scientific applications, but is essentially capable of only one such task at a given time. Speed of operation is relatively low.

60.5.3 Microcomputers

The first microcomputers were appearing in the mid-1970s and were mainly intended for use by specialist engineers requiring computing power remote from central resources for large and intricate calculations, an example being the Olivetti Pnnn series. However, these were still large and expensive by modern standards and so did not come into general use.

The second half of the decade saw the introduction of mass-produced, and hence cheap, microcomputers, often with an integral CRT, such as the Apple, Nascom and Commodore PET. These were aimed initially at the personal market and great numbers were sold on the basis of video games provided for them. As the novelty wore off, the personal computer market did not explode to the size predicted and the manufacturers had to look to another market sector to finance continued growth and in some cases just to be able to maintain production at current levels. This led to microcomputers being developed for and marketed in the business sector. These are quite different beasts to the personal computers covered in the previous section, with memories and facilities far exceeding those of early minicomputers and in some cases on a par with smaller mainframes.

Many of them have been specially adapted to cater for a specific market sector such as CAD/CAM and graphics use by a single person whilst offering networking and multi-user options as required. Many are covered by the generic term 'workstation' and are intended to permit a person to carry out many different functions without having to leave the desk, examples being Sun Microsystems, Apollo, Computervision and the Digital Equipment Corporation VAXstation/GPX systems.

60.5.4 Minicomputers

Since its introduction as a recognisable category of system in the mid-1960s, with machines such as the Digital Equipment Corporation PDP8, the minicomputer has evolved rapidly. It has been the development which has brought computers out of the realm of specialists and large companies into common and widespread use by non-specialists.

The first such systems were built from early integrated circuit logic families, with core memory. Characteristics were low cost, ability to be used in offices, laboratories and even factories, and simplicity of operation allowing them to be used, and in many cases programmed, by the people who actually had a job to be done, rather than by specialist staff remote from the user.

These were also the first items of computer equipment to be incorporated by original equipment manufacturers (OEM) into other products and systems, a sector of the market which has contributed strongly to the rapid growth of the minicomputer industry.

Applications of minicomputers are almost unlimited, in areas such as laboratories, education, commerce, industrial control, medicine, engineering, government, banking, networking, CAD/CAM, CAE and CIM. There is also a growing use of minicomputers combined with Artificial Intelligence for problem solving that benefits from deduction and backward chaining as opposed to predefined procedural stepping.

With advancing technology, systems are now built using large-scale and very large-scale integrated circuits, and memory is now almost entirely semiconductor. While earlier systems had a very small complement of peripherals, typically a teleprinter and punched paper tape input and output, there has been great development in the range and cost effectiveness of peripherals available. A minicomputer system will now typically have magnetic disk and tape storage holding thousands of millions of characters of data, a printer capable of printing 1200 lines of text per minute, many CRT display terminals (often colour), low-cost matrix printers for local or operator's hard copy requirements, and a selection of other peripherals for specialist use, such as a graphic colour plotter or a laser printer for high-quality output.

60.5.5 Superminis

The word midicomputer, or supermini, has been coined to describe a type of system which has many similarities in implementation to the minicomputer, but, by virtue of architectural advances, has superior performance. These advances include the following:

(1) *Longer word length* The amount of information processed in one step, or transferred in a single operation between different parts of the system, is usually twice that of a minicomputer.
(2) *More comprehensive instruction set* As well as increasing the rate of information handling, the longer word length makes it possible to provide a more comprehensive instruction set. Some common operations, such as handling strings of characters or translating high-level language statements into CPU instructions, have been reduced to single instructions in many superminis.
(3) *Larger memory addressing* This is also a result of longer word length. A technique called virtual memory (see *Section 60.7.11*) gives further flexibility to addressing in some superminis.
(4) *Higher data transfer speeds* on internal data highways, which allow faster and/or larger numbers of peripheral devices to be handled, and larger volumes of data to be transmitted between the system and the outside world.

Despite providing substantial power, even when compared with the mainframe class of system described below, midicomputers fall into a price range below that of the mainframe. This is because they have almost all originated from existing minicomputer manufacturers, who have been able to build on their volume markets, including in most cases the OEM market.

60.5.6 Mainframes

The mainframe is the class of system typically associated with commercial data processing in large companies, where a centralised operation is feasible and desired, and very large volumes of data are required to be processed at high processor speeds, or where a large user base (often in excess of 250 simultaneous users) requires immediate responses during interactive sessions. Today's mainframes, all products of large, established companies in the computer business (except for systems which are software compatible emulators of the most popular mainframe series), are the successors to the first and second generations as described in *Section 60.3*. They inherit the central control and location, emphasis on batch processing and line printers, third and fourth generation programming, and the need for specialised operating staff.

Mainframes are capable of supporting large amounts of on-line disk and magnetic tape storage as well as large main memory capacity, and more recent versions have data communications capabilities supporting remote terminals of various kinds.

Although some of the scientific mainframes have extremely high operating rates (over 100 million instructions per second), most commercial mainframes are distinguishable more by their size, mode of operation and support than by particularly high performance.

60.5.7 Combination technology

There have also been some significant developments in methods of combining computing resources to provide more security and faster processing. These fall into the following categories.

60.5.7.1 Tightly coupled systems

There are two types. In the first instance, certain parts of the system are duplicated and operate in parallel, each mirroring the work performed by the other. This provides security of availability of resource and of data by redundancy, the ultimate being total duplication of every part of the system, with the system designed to continue should one of anything fail. Certain Tandem and Stratus machines are examples of this being applied.

The second technique is to have more than one processor within the same CPU, with each one performing different tasks from the others, but with one having overall control. This provides security of availability should a processor fail, since the system will continue automatically with any remaining processor(s) without any human intervention, albeit with a reduced proessing capacity overall. Certain Digital Equipment VAX 8nnn series fall into this category.

60.5.7.2 Loosely coupled systems

This method employs the technique of sharing all resources across a common group of machines, sometimes known as 'clustering'. Each CPU within the cluster is an independent unit, but it knows of the existence of other members of the cluster. It is generally not necessary to stop processing ('bring down') the cluster in order to add or remove a new member. Each CPU can have its own peripherals attached to it, which it decides whether or not to share with other members. The cluster itself will also own peripherals which are available for use by any member. It is also possible to arrange for communications controllers to monitor the load of each member, to attempt to spread the workload as evenly as possible across the cluster. Should any item, including a CPU, fail within the cluster, processing can be continued on another part of the cluster. Through careful planning of these clusters it is now commonplace to find a logical system (i.e. a cluster) that rarely has to be stopped for hardware maintenance, operating system upgrades, hardware or software failure, etc.

60.6 Central processor unit

This part of the system controls the sequence of individual steps required for the execution of instructions forming a program. These instructions are held in storage, and when executed in the appropriate order carry out a task or series of tasks intended by the programmer.

Within any particular computer system, the word length is the fixed number of binary digits of which most instructions are made up. Arithmetic operations within the CPU are also performed on binary numbers of this fixed word length, normally 8, 12, 16, 24, 32, 36 or 64 binary digits or bits. The CPU is connected via a memory bus, as in *Figure 60.2*, to a section of memory organised as a number of randomly accessible words, each of which can be written to or read from individually. Time for reading one word from or writing one word into main memory is typically 0.2 to 0.6 μs. Each word or location of memory can contain either an instruction or data. Apart from simple systems, some form of magnetic tape or disk memory peripheral is present on a system as file storage and backup to main memory.

Control and timing circuits in the CPU cause instructions and data to be fetched from memory for processing, and write into memory the results of any instructions which require to be stored

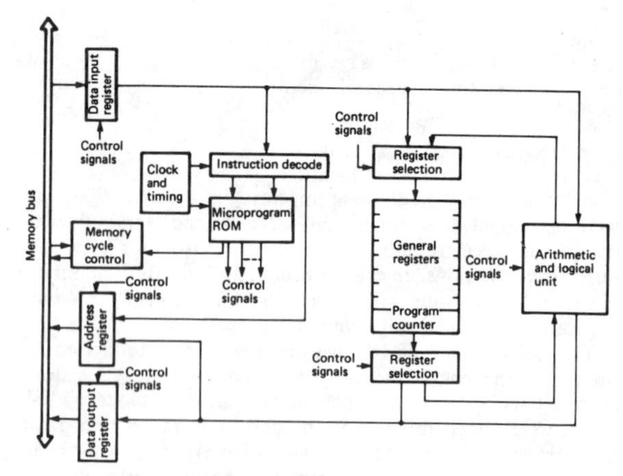

Figure 60.2 CPU block diagram

for further processing. The program counter holds the memory address of the next instruction to be fetched after each instruction has been processed. Frequently, the next instruction is held in the next location in memory and the counter need simply be incremented by one, though some systems work by placing one or more parameters for the instruction immediately after it in memory, thereby causing the next instruction to be displaced further down memory. At other times the sequence of the program dictates that a new value be written into the program counter. Instructions which alter the sequence of a program calculate and insert a new value into the program counter for the next instruction.

In order to start the CPU when no programs are already in memory, the program counter is loaded with a predetermined address by external means, initiated by, for example, a push button or the action of switching power on to the system. The initial program which is entered is a simple, short loader program either held in ROM or hand loading into memory using the front panel switches. Its function is to load a more comprehensive general-purpose loader, which automatically loads user or system programs. This process is known as bootstrapping or booting the system and the initial ROM program is known as the bootstrap loader. However, today most systems perform all of these functions as a result of switching on the power automatically and are ready for use almost immediately, only requesting date and time if they are not held in any form of battery backed up memory.

60.7 Memory

In order to provide storage for program instructions and data in a form where they can be directly accessed and operated upon, all CPUs have main memory which can be implemented in a variety of technologies and methods of organisation.

60.7.1 Memory organisation

Memory is organised into individually addressable words into which binary information can be loaded for storage, and from which the stored data pattern can be read. On some systems, memory is arranged in such a way that more than one word at a time is accessed. This is done to improve effective memory access rates, on the basis that by accessing say two consecutive words, on most occasions the second word will be the one which the CPU requires next. This is generally referred to as interleaved memory.

A memory controller is required between the memory arrays and the CPU, to decode the CPU requests for memory access and to initiate the appropriate read or write cycle. A controller can only handle up to a certain maximum amount of memory, but multiple controllers can be implemented on a single system. This can be used to speed up effective memory access, by arranging that sequentially addressed locations are physically in different blocks of memory with different controllers. With this interleaved memory organisation, in accessing sequential memory locations the operation of the controllers is overlapped, i.e. the second controller begins its cycle before the first has completed. Aggregate memory throughput is thus speeded up.

In some more complex computer systems, all or part of the memory can be shared between different CPUs in a multiprocessor configuration. Shareable memory has a special form of controller with multiple ports, allowing more than one CPU to contend for access to the memory, allocated among them on a cycle-by-cycle basis.

It is sometimes appropriate to implement two types of memory in one system; random access or read/write memory (RAM), and read-only memory (ROM). Programs have to be segregated into two areas:

(1) Pure instructions, which will not change, can be entered into ROM.
(2) Areas with locations which require to be written into, i.e. those containing variable data or modifiable instructions, require to occupy RAM.

Read-only memory is used where absolute security from corruption of programs, such as operating system software or a program performing a fixed control task, is important. It is normally found on microprocessor-based systems, and might be used, for example, to control the operation of a bank cash dispenser.

Use of ROM also provides a low-cost way of manufacturing in quantity a standard system which uses proven programs which never require to be changed. Such systems can be delivered with the programs already loaded and secure, and do not need any form of program loading device.

60.7.2 Memory technology

The commonest technologies for implementing main memory in a CPU are the following.

MOS RAM This technology is very widely used, with the abundant availability, from the major semiconductor suppliers, first of 16 K bit chips, then advancing through 32 K and 64 K chips to 128 K units now available. In the latter form, very high density is achieved, with up to 64 Mbytes of memory available on a single printed circuit board.

Dynamic MOS RAMs require refresh circuitry which automatically at intervals rewrites the data in each memory cell. Static RAM, which does not require refreshing, can also be used. This is generally faster, but also more expensive, than dynamic RAM memory.

Semiconductor RAMs are volatile, i.e. lose contents on powering down. This is catered for in systems with backup storage by reloading programs from the backup device when the system is switched on, or by having battery backup for all or part of the memory.

In specialised applications requiring memory retention without mains for long periods, or battery operation of the complete CPU, CMOS memory can be used because of its very low current drain. It has the disadvantage of being more expensive than normal MOS memory but, where it is essential to use it, circuit boards with on-board battery and trickle charger are now available.

Read only memories Used as described in *Section 60.7.1*, these can be either erasable ROMs or a permanently loaded ROM such as fusible-link ROM.

Bubble memory This has not produced the revolution in memory that it seemed to promise at the start of this decade. It remains at a comparatively higher price-to-performance ratio than semiconductor memory and is not used to any large scale on a commercial basis.

Core memory This remains in some applications but, although it has come down substantially in cost under competition from semiconductor memory, more recently MOS RAMs of higher capacity have been much cheaper and have largely taken over from core memory.

60.7.3 Registers

The CPU contains a number of registers accessible by instructions, together with more which are not accessible, but are a necessary part of its implementation. Other than single-digit status information, the accessible registers are normally of the same number of bits as the word length of the CPU.

Registers are fast-access temporary storage locations within the CPU and implemented in the circuit technology of the CPU. They are used, for example, for temporary storage of intermediate results or as one of the operands in an arithmetic instruction.

A simple CPU may have only one register, often known as the accumulator, plus perhaps an auxiliary accumulator or quotient register used to hold part of the double-length result of a binary multiplication.

Larger word length, more sophisticated CPUs typically have 8 or more general-purpose registers which can be selected as operands by instructions. Some systems such as the VAX use one of its 16 general-purpose registers as the program counter, and can use any register as a stack pointer. A stack in this context is a temporary array of data held in memory on a 'last in, first out' basis. It is used in certain types of memory reference instructions, and for internal housekeeping in interrupt and subroutine handling. The stack pointer register is used to hold the address of the top element of the stack. This address, and hence the stack pointer contents, is incremented or decremented by one at a time as data are added to or removed from the stack.

60.7.4 Memory addressing

Certain instructions perform an operation in which one or more of the operands is the contents of a memory location, for example arithmetic, logical and data movement instructions. In most sophisticated CPUs, various addressing modes are available to give, for example, the capacity of adding together the contents of two different memory locations and depositing the result in a third.

In such CPUs such instructions are double operand, i.e. the programmer is not restricted to always using one fixed register as an operand. In this case any two of the general-purpose registers can be designated either as each containing an operand, or through a variety of addressing modes, where each of the general-purpose registers selected will contain one of the following:

(1) The memory address of an operand.
(2) The memory address of an operand, and the register contents then incremented following execution.
(3) The memory address of an operand, and the register contents then decremented following execution.
(4) A value to which is added the contents of a designated memory location. This is known as indexed addressing.

(5) All of the above, but where the resultant operand is itself the address of the final operand. This is known as indirect or deferred addressing.

This richness of addressing modes is one of the benefits of more advanced CPUs, as, for example, it provides an easy way of processing arrays of data in memory, or of calculating the address portion of an instruction when the program is executed.

Further flexibility is provided by the ability on many processors for many instructions to operate on multiples of 8 bits (known as a byte), on single bits within a word, and on some more comprehensive CPUs such as that of the VAX to operate on double and quadruple length words and also arrays of data in memory.

60.7.5 Memory management

Two further attributes may be required of memory addressing. Together they are often known as memory management.

(1) The ability, particularly for a short word length system (16 bits or less), for a program to use addresses greater than those implied by the word length. This is known as extended addressing. For example, with the 16-bit word length of most minicomputers, the maximum address which can be handled in the CPU is 65 536.

As applications grow bigger this is often a limitation, and extended addressing functions by considering memory as divided into a number of pages. Associated with each page at any given time is a relocation constant which is combined with relative addresses within its page to form a longer address. For example, with extension to 18 bits, memory addresses up to 262 144 can be generated in this way.

Each program is still limited at any given time to 65 536 words of address space, but these are physically divided into a number of pages which can be located anywhere within the larger memory. Each page is assigned a relocation constant, and as a particular program is run, dedicated registers in the CPU memory management unit are loaded with the constant for each page, *Figure 60.3*.

Thus many logically separate programs and data arrays can be resident in memory at the same time, and the process of setting the relocation registers, which is performed by a supervisory program, allows rapid switching between them in accordance with a time scheduling scheme, which is usually based upon resouce usage quota, time allocation or a combination of both. This is known as multiprogramming. Examples of where this is used are a time-sharing system for a number of users with terminals served by the system, or a real-time control system where programs of differing priority need to be executed rapidly in response to external events.

(2) As an adjunct to the hardware for memory paging or segmentation described above, a memory protection scheme is readily implemented. As well as a relocation constant, each page can be given a protection code to prevent it being illegally accessed. This would be desirable for example for a page holding data that is to be used as common data among a number of programs. Protection can also prevent a program from accessing a page outside its own address space.

Both of the above features are desirable for systems performing multiprogramming. In such systems, the most important area to be protected is that containing the supervisory program or operating system, which controls the running and allocation of resources for users' programs.

60.7.6 Virtual memory

Programmers, particularly those engaged in scientific and engineering work, frequently have a need for very large address

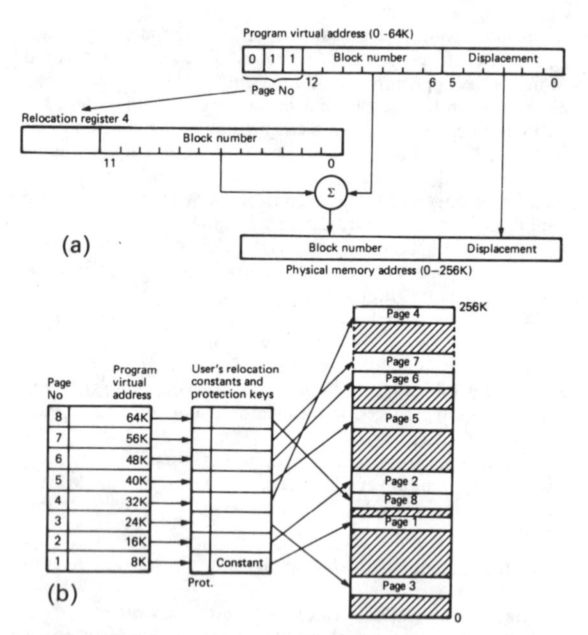

Figure 60.3 Memory management for a 16-bit CPU. (a) Generation of a physical address in the range 0 to 256K by combination of user's program virtual address in the range 0 to 64K with a relocation constant for the page concerned. Memory is handled in 64 byte blocks, with eight relocation registers, giving segmentation into eight pages located anywhere in up to 256K of physical memory. (b) The user's program is considered as up to eight pages, up to 8K bytes each. Relocation constants for that program map these pages anywhere in up to 256K bytes of physical memory. Protection per page can also be specified

space within a single program for instructions and data. This allows them to handle large arrays, and to write very large programs without the need to break them down to fit a limited memory size.

One solution is known as virtual memory, a technique of memory management by hardware and operating system software whereby programs can be written using the full addressing range implied by the word length of the CPU, without regard to the amount of main memory installed in the system. From the hardware point of view, memory is divided into fixed length pages, and the memory management hardware attempts to ensure that pages in most active use at any given time are kept in main memory. All of the current programs are stored on disk backing store, and an attempt to access a page which is not currently in main memory causes paging to occur. This simply means that the page concerned is read into main memory into the area occupied by an inactive page, and that if any changes have been made to the inactive page since it was read into memory then it is written out to disk in its updated form to preserve its integrity.

A table of address translations holds the virtual physical memory translations for all the pages of each program. The operating system generates this information when programs are loaded on to the system, and subsequently keeps it updated. Memory protection on a per page basis is normally provided, and a page can be locked into memory as required to prevent it being swapped out if it is essential for it to be immediately executed without the time overhead of paging.

When a program is scheduled to be run by the operating system, its address translation table becomes the one in current use. A set of hardware registers to hold a number of the most frequent translations in current use speeds up the translation process when pages are being repeatedly accessed.

60.8 Instruction set

The number and complexity of instructions in the instruction set or repertoire of different CPUs varies considerably. The longer the word length, the greater is the variety of instructions which can be coded within it. This means generally that, for a shorter word length CPU, a larger number of instructions will have to be used to achieve the same result, or that a longer word length machine with its more powerful set of instructions needs fewer of them and hence should be able to perform a given task more quickly.

Instructions are coded, according to a fixed format, allowing the instruction decoder to determine readily the type and detailed function of each instruction presented to it. The general instruction format of the VAX is shown as an example in *Figure 60.4*. Digits forming the operation code are first decoded to determine the category of instruction, and the remaining nine bits interpreted in a different way depending into which category the instruction falls.

There are variations to the theme outlined above for CPUs from differing manufacturers, but generally they all employ the principle of decoding a certain group of digits in the instruction word to determine the class of instruction and hence how the remaining digits are to be interpreted.

The contents of a memory location containing data rather than an instruction are not applied to the instruction decoder. Correct initial setting of the program counter, and subsequent automatic setting by any branch instruction to follow the sequence intended by the programmer, ensures that only valid instructions are decoded for execution. In the cases where operands follow the instruction in memory, the decoder will know how many bytes or words to skip in order to arrive at the next instruction in sequence.

Logical and arithmetic instructions perform an operation on data (normally one or two words for any particular instruction) held in either the memory or registers in the CPU. The addressing modes available to the programmer (see *Section 60.7.4*) define the range of possible ways of accessing the data to be operated on. This ranges from the simple single operand type of CPU, where the accumulator is always understood to contain one operand while the other is a location in memory specified by the addressing bits of the instruction, to a multiple operand CPU with a wide choice of how individual operands are addressed.

In some systems such as the VAX, instructions to input data from and output data to peripheral devices are the same as those used for manipulating data in memory. This is achieved by implementing a portion of the memory addresses at the high end as data and control registers in peripheral device controllers. More detail is given in *Section 60.10*.

There are certain basic data transfer, logical, arithmetic and controlling functions which must be provided in the instruction sets of all CPUs. This minimum set allows the CPU to be programmed to carry out any task which can be broken down and expressed in these basic instructions. However, it may be that a program written in this way will not execute quickly enough to perform a time-critical application such as control of an industrial plant or receiving data on a high-speed communication line. Equally the number of steps or instructions required may not fit into the available size of memory. In order to cope more efficiently with this sort of situation, i.e. to increase the power of the CPU, all but the very simplest CPUs have considerable enhancements and variations to the basic instruction set. The more comprehensive the instruction set, the fewer are the steps required to program a given task, and the shorter and faster in execution are the resulting programs.

Basic types of instruction, with examples of the variations to these, are the following.

(1) *Data transfer* Load accumulator from a specified memory location and write contents of accumulator into specified memory location. Most CPUs have variations such as add contents of memory location to accumulator and exchange contents of accumulator and memory location.

CPUs with multiple registers also have some instructions which can move data to or from these registers, as well as the accumulator. Those with 16-bit or greater word lengths may have versions of these and other instruction types which operate on bytes as well as words.

With a double operand addressing mode (see *Section 60.7.4*) a generalised MOVE instruction allows the contents of any memory location or register to be transferred to any other memory location or register.

(2) *Logical* AND function on a bit-by-bit basis between contents of a memory location and a bit pattern in the accumulator. Leaves ones in accumulator bit positions which are also one in the memory word. Appropriate bit patterns in accumulator allow individual bits of chosen word to be tested.

Branch (or skip the next instruction) if the accumulator contents are zero/non-zero/positive/negative.

Many more logical operations and tests are available on more powerful CPUs, such as OR, exclusive OR, complement, branch if greater than or equal to zero, branch if less than or equal to zero, branch if lower or the same. The branch instructions are performed on the contents of the accumulator following a subtraction or comparison of two words, or some other operation which leaves data in the accumulator. The address for branching to is specified in the address part of the instruction. With a skip, the instruction in the next location should be an unconditional branch to the code which is to be followed if the test failed, while for a positive result the code to be followed starts in the next but one location.

Branch or skip tests on other status bits in the CPU are often provided, e.g. on arithmetic carry and overflow.

(3) *Input/output* CPUs like the VAX, with memory-mapped input/output, do not require separate instructions for transferring

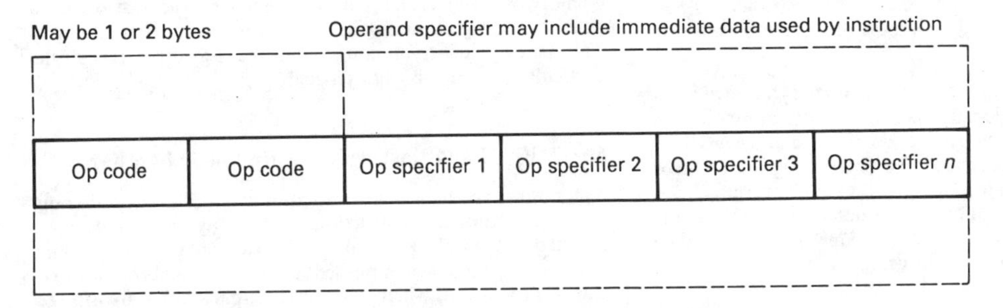

Figure 60.4 General VAX instruction set format

data and status information between CPU and peripheral controllers. For this function, as well as performing tests on status information and input data, the normal data transfer and logical instructions in (1) and (2) are used.

Otherwise, separate input/output instructions provide these functions. Their general format is a transfer of data between the accumulator or other registers, and addressable data, control or status registers in peripheral controllers. Some CPUs also implement special input/output instructions such as:

(a) *Skip if 'ready' flag set* For the particular peripheral addressed, this instruction tests whether it has data awaiting input, or whether it is free to receive new output data. Using a simple program loop, this instruction will synchronise the program with the transfer rate of the peripheral.
(b) *Set interrupt mask* This instruction outputs the state of each accumulator bit to an interrupt control circuit of a particular peripheral controller, so that, by putting the appropriate bit pattern in the accumulator with a single instruction, interrupts can be selectively inhibited or enabled in each peripheral device.

(4) *Arithmetic* Add contents of memory location to contents of accumulator, leaving result in accumulator. This instruction, together with instructions in category (2) for handling a carry bit from the addition, and for complementing a binary number, can be used to carry out all the four arithmetic functions by software subroutines.

Shift. This is also valuable in performing other arithmetic functions, or for sequentially testing bits in the accumulator contents. With simpler instruction sets, only one bit position is shifted for each execution of the instruction. There is usually a choice of left and right shift, and arithmetic shift (preserving the sign of the word and setting the carry bit) or logical rotate.

Extended arithmetic capability, either as standard equipment or a plug-in option, provides multiply and divide instructions and often multiple bit shift instructions.

(5) *Control* Halt, no operation, branch, jump to subroutine, interrupts on, interrupts off, are the typical operations provided as a minimum. A variety of other instructions will be found, peculiar to individual CPUs.

60.8.1 CPU implementation

The considerable amount of control logic required to execute all the possible CPU instructions and other functions is implemented in one of two ways:

(1) With random logic using the available logic elements of gates, flip-flops, etc., combined in a suitable way to implement all the steps for each instruction, using as much commonality between instructions as possible. The various logical combinations are invoked by outputs from the instruction decoder.

(2) Using microcode—a series of internally programmed steps making up each instruction. These steps or micro-instructions are loaded into ROM using patterns determined at design time, and for each instruction decoded, the micro program ROM is entered at the appropriate point for that instruction. Under internal clock control, the micro-instructions cause appropriate control lines to be operated to effect the same steps as would be the case if the CPU were designed using method (1).

The great advantage of microcoded instruction sets is that they can readily be modified or completely changed by using an alternative ROM, which may simply be a single chip in a socket. In this way, a different CPU instruction set may be effected.

In conjunction with microcode, bit slice microprocessors may be used to implement a CPU. The bit slice microprocessor contains a slice or section of a complete CPU, i.e. registers, arithmetic and logic, with suitable paths between these elements. The slice may be one, two or four bits in length, and by cascading a number of these together any desired word length can be achieved. The required instruction set is implemented by suitable programming of the bit slice microprocessors using their external inputs controlled by microcode.

The combination of microcode held in ROM and bit slice microprocessors is used in the implementation of many CPU models, each using the same bit slice device.

60.9 CPU enhancements

There are several areas in which the operating speed of the CPU can be improved with added hardware, either designed in as an original feature or available as an upgrade to be added in field. Some of the more common ones are described below.

60.9.1 Cache memory

An analysis of a typical computer program shows that there is a strong tendency to access instructions and data held in fairly small contiguous areas of memory repetitively. This is due to the fact that loops—short sections of program reused many times in succession—are very frequently used, and data held in arrays of successive memory locations may be repetively accessed in the course of a particular calculation.

This leads to the idea of having a small buffer memory, of higher access speed than the lower cost technology employed in main memory (e.g. disks), between CPU and memory. This is known as cache memory. Various techniques are used to match the addresses of locations in cache with those in main memory, so that for memory addresses generated by the CPU, if the contents of that memory location are in cache, the instruction or data is accessed from the fast cache instead of slower main memory. The contents of a given memory location are initially fetched into cache by being addressed by the CPU. Precautions are taken to ensure that the contents of any location in cache which is altered by a write operation is rewritten back into main memory so that the contents of the location, whether in cache or main memory, are identical at all times.

A constant process of bringing memory contents into cache, thus overwriting previously used information with more currently used words takes place completely transparently to the user. The only effect to be observed is an increase in execution speed. This speeding up depends on two factors, hit rate, i.e. percentage of times when the contents of a required location are already in cache, and the relative access times of main and cache memory. The hit rate, itself determined by the size of cache memory and algorithms for its filling, is normally better than 90%. This is dependent, of course, on the repetitiveness of the particular program being executed.

60.9.2 Fixed and floating point arithmetic hardware

As far as arithmetic instructions go, simpler CPUs contain only add and subtract instructions, operating on single-word operands. Multiplication, of both fixed and floating point numbers, is then accomplished by software subroutines, i.e. standard programs which perform multiplication or division by repetitive use of the add or subtract instructions, which can be invoked by a programmer who requires to perform a multiplication or division operation.

By providing extra hardware to perform fixed point multiply and divide, which also usually implements multiple place shift operations, a very substantial improvement in the speed of multiply and divide operations is obtained. With the hardware techniques used to implement most modern CPUs, however, these instructions are wired in as part of the standard set.

Floating point format (*Figure 60.5*) provides greater range and precision than single-word fixed point format. In floating point

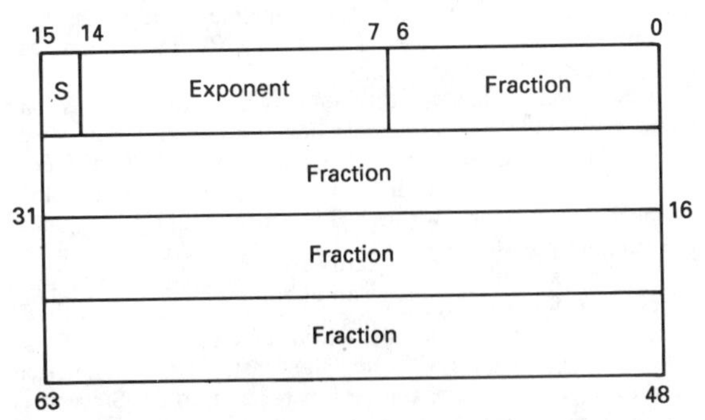

Figure 60.5 32-bit double floating point format

representation, numbers are stored as a fraction times 2 raised to a power which can be positive or negative. The fraction (or mantissa) and exponent are what is stored, usually in two words for single-precision floating point format, or four words for double precision.

Hardware to perform add, subtract, multiply and divide operations is sometimes implemented as a floating point processor, an independent unit with its own registers to which floating point instructions are passed. The floating point processor can then access the operands, perform the required arithmetic operation and signal the CPU, which has been free to continue with its own processing meantime, when the result is available.

An independent floating point processor clearly provides the fastest execution of these instructions, but even without that, implementing them within the normal instruction set of the CPU using its addressing techniques to access operands in memory provides a significant improvement over software subroutines. The inclusion of the FPP into 'standard' CPUs is becoming much more common.

60.9.3 Array processors

Similar to an independent floating point processor described above, an optional hardware unit which can perform complete computations on data held in the form of arrays of data in memory, independent from the CPU and at high speed, is known as an array processor.

These are used in specialised technical applications such as simulation, modelling and seismic work. An example of the type of mathematical operation which would be carried out by such a unit is matrix inversion. The ability of these units to perform very high speed searches based upon text keys has also led to a growing use of them for the rapid retrieval of data from large data banks, particularly in areas such as banking where real-time ATM terminals require fast response to account enquiries from very large data sets.

60.9.4 Timers and counters

For systems which are used in control applications, or where elapsed time needs to be measured for accounting purposes as, for example, in a time-sharing system where users are to be charged according to the amounts of CPU time they use, it is important to be able to measure intervals of time precisely and accurately. This measurement must go on while the system is executing programs, and must be 'real time', i.e. related to events and time intervals in the outside world.

Most CPUs are equipped with a simple real-time clock which derives its reference timing from 50 or 60 Hz mains. These allow a predetermined interval to be timed by setting a count value in a counter which is decremented at the mains cycle rate until it interrupts the CPU on reaching zero.

More elaborate timers are available as options, or are even standard items on some CPUs. These are driven from high-resolution crystal oscillators, and offer such features as:

(1) More than one timer simultaneously.
(2) Timing random external events.
(3) Program selection of different time bases.
(4) External clock input.

The system supervisory software normally keeps the date and time of day up to date by means of a program running in the background all the time the system is switched on and running. Any reports, logs or printouts generated by the systems can then be labelled with the date and time they were initiated. To overcome having to reset the date and time every time the system is stopped or switched off, some CPUs now have a permanent battery-driven date and time clock which keeps running despite stoppages and never needs reloading once loaded initially, with the exception of change to and from Summer Time!

Counters are also useful in control applications to count external events or to generate a set number of pulses, for example to drive a stepping motor. Counters are frequently implemented as external peripheral devices, forming part of the digital section of a process input/output interface (see *Section 60.10*).

60.10 Input/output

In order to perform any useful role, a computer system must be able to communicate with the outside world, either with human users via keyboards, CRT screens, printed output, etc., or with some external hardware or process being controlled or monitored. In the latter case, where connection to other electronic systems is involved, the communication is via electrical signals.

All modern computer systems have a unified means of supporting the variable number of such human or process input/output devices required for a particular application, and indeed for adding such equipment to enhance a system in the field.

As well as all input/output peripherals and external mass storage in the form of magnetic tape and disk units, some systems also communicate with main memory in this common, unified structure. In such a system, for example the VAX, there is no difference between instructions which reference memory and those which read from and write to peripheral devices.

The benefits of a standard input/output bus to the manufacturer are:

(1) It provides a design standard allowing easy development of new input/output devices and other system enhancements.
(2) Devices of widely different data transfer rates can be accommodated without adaptation of the CPU.
(3) It permits development of a family concept.

Many manufacturers have maintained a standard input/output highway and CPU instruction set for as long as a decade or more. This has enabled them to provide constantly improving system performance and decreasing cost by taking advantage of developing technology, whilst protecting very substantial investments in peripheral equipment and software.

For the user of a system in such a family, benefits are:

(1) The ability to upgrade to a more powerful CPU whilst retaining existing peripherals.
(2) Retention of programs in moving up or down range within the family.
(3) In many cases the ability to retain the usefulness of an older system by adding more recently developed peripherals, and in some cases even additional CPU capacity of newer design and technology.

60.10.1 Input/output bus

The common structure for any given model of computer system is implemented in the form of an electrical bus or highway. This specifies both the number, levels and significance of electrical signals and the mechanical mounting of the electrical controller or interface which transforms the standard signals on the highway to ones suitable for the particular input/output or storage device concerned.

A data highway or input/output bus needs to provide the following functions.

Addressing A number of address lines is provided, determining the number of devices which can be accommodated on the system. For example, 6 lines would allow 63 devices. Each interface on the bus decodes the address lines to detect input/output instructions intended for it.

Data The number of data lines on the bus is usually equal to the word length of the CPU, although it may alternatively be a submultiple of the word length, in which case input/output data is packed into or unpacked from complete words in the CPU. In some cases data lines are bidirectional, providing a simpler bus at the expense of more complex drivers and receivers.

Control Control signals are required to synchronise transactions between the CPU and interfaces, and to gate address and data signals to and from the bus. Although all the bits of an address or data word are transmitted at the same instant, in transmission down the bus, because of slightly different electrical characteristics of each individual line, they will arrive at slightly different times. Control signals are provided to gate these skewed signals at a time when they are guaranteed to have reached their correct state.

60.10.2 Types of input/output transaction

Three types of transaction via the input/output bus between CPU and peripheral devices are required, as described below.

Control and status This type of transfer is initiated by a program instruction to command a peripheral device to perform a certain action in readiness for transferring data or to interrogate the status of a peripheral. For example, a magnetic tape unit can be issued with a command to rewind; the read/write head in a disk unit to be positioned above a certain track on the disk; the completion of a conversion by an analogue-to-digital converter verified; or a printer out of paper condition may be sensed.

Normally, a single word of control or status information is output or input as a result of one instruction, with each bit in the word having a particular significance. Thus multiple actions can be initiated by a single control instruction, and several conditions monitored by a single status instruction. For the more complex peripheral devices, more than one word of control or status information may be required.

Programmed data transfer For slow and medium speed devices, for example small floppy disk units or line printers, data input or output one word at a time with a series of program instructions required for every word transferred.

The word or data is transferred to or from one of the CPU registers, normally the accumulator. In order to effect a transfer of a series of words forming a related block of data, as is normally required in any practical situation, a number of CPU instructions per word transferred is required. This is because it is necessary to take the data from, or store it into, memory locations. As a minimum, in a simple case, at least six CPU instructions are required per word of data transferred.

In a system such as the VAX where instructions can reference equally easily memory locations, peripheral device registers and CPU registers, the operation is simplified since a MOVE instruction can transfer a word of data directly from a peripheral to memory, without going through a CPU register. This applies equally to control and status instructions on the VAX, with a further advantage that the state of bits in a peripheral device status register can be tested without transferring the register contents into the CPU.

The rate of execution of the necessary instructions must match the data transfer rate of the peripheral concerned. Since it is usually desired that the CPU continue with the execution of other parts of the user's program while data transfer is going on, some form of synchronisation is necessary between CPU and peripheral to ensure that no data is lost. In the simplest type of system, the CPU simply suspends any other instructions and constantly monitors the device status word awaiting an indication that the peripheral has data ready for input to the CPU or is ready to receive an output from it. This is wasteful of CPU time where the data transfer rate is slow relative to CPU instruction speeds, and in this case the use of 'interrupt' facilities (see *Section 60.10.3*) provides this synchronisation.

Direct memory access For devices which transfer data at a higher rate, in excess of around 20000 words per second, a different solution is required.

At these speeds, efficiency is achieved by giving the peripheral device controller the ability to access memory autonomously without using CPU instructions. With very fast tape or disk units which can transfer data at rates in excess of 1 million bytes per second, direct memory access (DMA) is the only technique which will allow these rates to be sustained.

The peripheral controller has two registers which are loaded by control instructions before data transfer can begin. These contain:

(1) The address in memory of the start of the block of data.
(2) The number of words which it is desired to transfer in the operation.

When the block transfer is started the peripheral controller, using certain control lines in the I/O bus, sequentially accesses the required memory locations until the specified number of words has been transferred. The memory addresses are placed on address lines of the I/O bus, together with the appropriate control and timing signals, for each word transferred. On completion of the number of words specified in the word count register, the peripheral signals to the CPU that the transfer of the block of data is completed.

Other than the instructions required initially to set the start address and word count registers and start the transfer, a DMA transfer is accomplished without any intervention from the CPU. Normal processing of instructions therefore continues. Direct memory access (more than one peripheral at a time can be engaged in such an operation) is, of course, competing with the CPU for memory cycles, and the processing of instructions is slowed down in proportion to the percentage of memory cycles required by peripherals. In the limit, it may be necessary for a very-high-speed peripheral to completely dominate memory usage in a burst mode of operation, to ensure that no data is lost during the transfer through conflicting requests for memory cycles.

60.10.3 Interrupts

The handling of input/output is made much more efficient through the use of a feature found in varying degrees of sophistication on all modern systems. This is known as 'automatic priority interrupt', and is a way of allowing peripheral devices to signal an event of significance to the CPU (e.g. a terminal keyboard having a character ready for transmission, or completion of DMA transfer) in such a way that the CPU is made to suspend its current work temporarily to respond to the condition causing the interrupt.

Interrupts are also used to force the CPU to recognise and take action on alarm or error conditions in a peripheral, e.g. printer out of paper, error detected on writing to a magnetic tape unit.

Information to allow the CPU to resume where it was interrupted, e.g. the value of the program counter, is stored when an interrupt is accepted. It is also necessary for the device causing the interrupt to be identified, and for the program to branch to a section to deal with the condition that caused the interrupt (*Figure 60.6*).

Examples of two types of interrupt structure are given below, one typical of a simpler system such as an 8-bit microprocessor or an older architecture minicomputer, the other representing a more sophisticated architecture such as the VAX.

In the simpler system, a single interrupt line is provided in the input/output bus, on to which the interrupt signal from each peripheral is connected. Within each peripheral controller, access to the interrupt line can be enabled or disabled, either by means of a control input/output instruction to each device separately, or by a 'mask' instruction which, with a single 16-bit word output, sets the interrupt enabled/disabled state for each of up to 16 devices on the input/output bus. When a condition which is defined as able to cause an interrupt occurs within a peripheral, and interrupts are enabled in that device, a signal on the interrupt line will be sent to the CPU. At the end of the instruction currently being executed, this signal will be recognised.

In this simple form of interrupt handing, the interrupt servicing routine always begins at a fixed memory location. The interrupt forces the contents of the program counter, which is the address of next instruction that would have been executed had the interrupt not occurred, to be stored in this first location and the program to start executing at the next instruction. Further interrupts are automatically inhibited within the CPU, and the first action of the interrupt routine must be to store the contents of the accumulator and other registers so that on return to the main stream of the program these registers can be restored to their previous state.

Identification of the interrupting device is done via a series of conditional instructions on each in turn until an interrupting device is found. Having established which device is interrupting, the interrupt handling routine will then branch to a section of program specific to that device. At this point or later within the interrupt routine, an instruction to re-enable the CPU interrupt system may be issued, allowing a further interrupt to be received by the CPU before the existing interrupt handling program has completed. If this 'nesting' of interrupts is to be allowed, each interruptable section of the interrupt routine must store the return value of the program counter elsewhere in memory, so that as each section of the interrupt routine is completed, control can be returned to the point where the last interrupt occurred.

A more comprehensive interrupt system, that on the VAX for example, differs in the following ways from that described above.

Multiple interrupt lines are provided, and any number of devices can be on each line or level.

The CPU status can be set to different priority levels, corresponding to the different interrupt lines. Only interrupts on a level higher than the current priority are accepted by the CPU. This provides a more adaptable way of dealing with a wide range of devices of different speeds and with different degrees of urgency.

When an interrupt is accepted by the CPU, the interrupting device sends a vector or pointer to the CPU on the I/O bus address lines. This points to a fixed memory address for each device, which holds the start address of its interrupt routine and, in the following memory word, a new status word for the CPU, defining its priority level and hence its ability to respond to other levels of interrupt during this interrupt routine. By avoiding the need for the CPU to test each device until it finds the interrupting one, response to interrupts is much faster.

The current value of the program counter and processor status word are automatically placed on a push-down stack when an interrupt occurs. A further interrupt accepted within the current interrupt routine will cause the program counter and status word to be stored on the top of the stack, and the existing contents to be pushed down into the stack. On return from an interrupt routine, the program counter and status word stored when that interrupt occurred are taken from the top of the stack and used by the CPU, allowing whatever was interrupted to continue as before. This can take place for any number of interrupts, subject only to the capacity of the stack. Thus 'nesting' to any level is handled automatically without the need for the programmer to store the program counter at any stage.

60.11 Peripherals

Peripheral devices fall into three categories:

(1) Those which are designed to allow humans to interact with the system by outputting information in the form of readable alphanumeric text or graphics, either on paper or on a display screen, and accepting information from humans

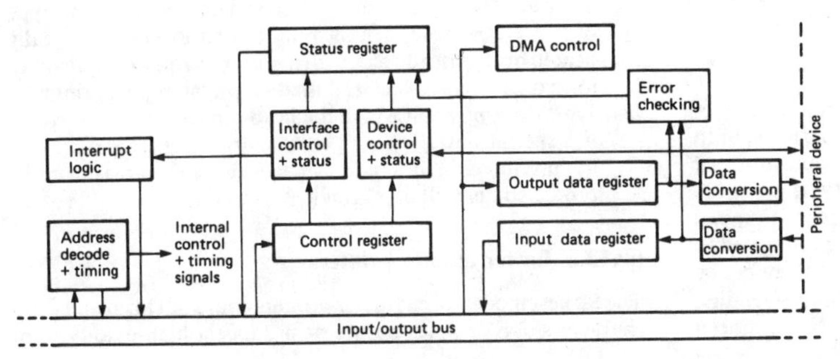

Figure 60.6 Block diagram of peripheral interface

through a keyboard or voice recognition device, or by scanning printed text. The general function performed by devices in this class is sometimes referred to as man–machine interaction.
(2) Those which act as a backup form of storage to supplement the main memory of the system. The most simple of these, now largely superseded, was probably punched paper tape or cards, and the most complex now ranges up to very large disks or other specialised forms of storage, each capable of holding hundreds of millions of characters of information. Peripherals of this type are generally known as mass storage devices.
(3) Interfaces between the computer system and other forms of system or electronic device. Analogue/digital converters, digital input/output and communication line interfaces are good examples.

Either (1) or (3) or both are required in every system. The existence of (2) depends on the need for additional storage over and above main memory. All peripheral devices in a system are connected via the input/output structure (as described in *Section 60.10*) to the CPU, memory, and in some systems to a separate I/O processor.

The throughput rates and flexibility of the input/output structure determine the number and variety of peripheral devices which can be handled in a system before the input/output requirements begin to saturate the system and prevent any processing of instructions being done by the CPU. In deciding on the configuration of a particular system, it is important to analyse the throughput requirement dictated by peripheral devices, to ensure the system does not become I/O bound, and that data from any peripheral device is not lost due to other devices taking too many of the I/O resources.

Historically in the computer industry, independent manufacturers as well as the larger computer systems companies have developed and manufactured peripherals. The products of the independent manufacturers are either bought by system manufacturers for design into their systems or sold by the independent manufacturers direct to users of the more popular computers, with an interface providing compatibility with the I/O bus of the system.

This has fostered the development of many of the widely used, cost-effective peripherals available today, such as floppy disks and printers.

Certain storage devices with removable storage media, where the formats for recording data on the media have been standardised, can be used for exchanging data between systems from different suppliers. This is important where data may be gathered on one system and need to be analysed on a different system more suitable for that purpose. However, due to the very large growth of networking in the 1980s, even between equipment from different manufacturers, the moving of data from one machine to another is most commonly achieved by file transmission, though for very large files magnetic tape is used for this purpose.

60.12 Terminals

A data terminal is essentially a device at which a computer user sits, either to receive data in alphanumeric or graphic form, or to input data through a keyboard or other form of manual input, or both. Terminals range in cost and complexity from a 30 characters per second serial printer to a full graphics terminal with a large, perhaps colour, CRT screen. Most are connected to the CPU by data communications interfaces and can therefore be situated at some distance from the system within a building (LAN), or over a company's private network (WAN), or indeed remote from the computer site communicating over the public telephone network.

60.12.1 Character printers

Only a small proportion of low-speed printers are now supplied with any means of input such as a keyboard. This is because of the widespread use of interactive terminals for data input and the fact that many of these printers are themselves attached in some way to a terminal to act as a slave by producing hard copy output through the terminal. The commonest use for printers with keyboards is as operators' consoles on larger computers. Typical speed is in the range 100–250 characters per second.

The print head consists of seven or more needles held in a vertical plane in the head assembly which is positioned with the needles perpendicular to the paper and spaced a short distance from it, with a carbon ribbon interposed between. Each needle can be individually driven by a solenoid to contact the paper through the ribbon, thus printing a dot. A complete character is formed by stepping the head through five or more positions horizontally, and at each position energising the appropriate solenoids. The head is then stepped on to the position for the next character. When the end of the line is reached, the paper is advanced one line and the print head either returns to the left margin position or, in some faster printers, prints the next line from right to left. This is possible where the printer is provided with storage for a line or more of text, and the characters can be extracted from this store in the reverse order. Throughout speed is improved where this technique is used, by saving redundant head movement.

The 7×5 dot matrix within which this type of printer forms each character allows an acceptable representation of alpha and numeric characters. Better legibility, particularly of lower case characters with descenders, can be achieved by using a larger matrix such as 9×7, i.e. a head with nine needles stepping through seven positions for each character.

Character codes are received for printing and sent from the keyboard in asynchronous serial form, on a line driven by a data communications interface in the computer, whose transmission speed determines the overall printing and keying throughput. Buffer storage for up to one million characters is provided in printers that use serial data communications, to make the most efficient use of communication lines, and in printers with built-in intelligence, to allow look-ahead so that the print head can skip blanks and take the shortest route to the next printable character position.

Character sets can be readily changed by replacing the ROM chip which contains the dot patterns corresponding to each character code, or more usually by sending the character patterns over the network to the printer from the host CPU, often referred to as 'downline loading'. This latter method has the advantage of character selection at any time under program control and without any human intervention.

Other variants of serial dot matrix printer include:

(1) *'Silent' printers*, using special sensitised paper which is marked either by heat or by electrical discharge. In this type of printer, the needles in the print head are either electrically heated or electrically charged, in the appropriate sequence to form characters as the head is advanced along the print line.
(2) *Plotting printers*, in which the head movement is controlled in a special way allowing a simple form of graphic output to be produced. Lines and shaded or solid areas can be produced, as well as normal characters.

60.12.2 Letter quality printers

The earliest models of character printers could not form and print characters of a complex nature or produce a high-quality print image. To overcome this, printers that used a metal wheel, with the characters preformed on the end, were used for occasions where the quality of the image was paramount, e.g. business correspondence. Since character printers have continued to make large improvements in this respect, some offering different quality of print according to requirement and selectable by program control, these letter quality printers have receded in sales and in use because of their higher cost of construction and maintenance.

Another type of serial printer which gives high print quality is the ink jet printer.

60.12.3 Visual display unit

A VDU is a terminal in which a CRT is used to display alphanumeric text or graphics. It is normally also equipped with a keyboard for data input, and occasionally a printer may be slaved to the VDU to produce a permanent paper copy of the information displayed on the screen at a particular time.

The format for the layout of text on the screen is most commonly 24 lines of text (sometimes with a 25th line reserved for system messages to the terminal user), each with up to 80 character positions. Most VDUs allow 132 columns to be displayed in a line, using a special compressed character set. This latter feature is useful for compatibility with computer printouts, which normally have up to 132 columns of print. Displaying of new text on the screen takes place a character at a time, starting in the top left-hand corner and continuing line by line. When the screen is full, the page of text moves up one line allowing a new line to be added at the bottom.

Characters are caused to be displayed either by outputting data from the CPU through a serial communications interface, or by the operator typing on a typewriter-like keyboard.

VDUs are classified depending on which of the following modes of message composition they use:

(1) *Block mode* A full screen of text is composed and, if need be, edited by the operator, and the corresponding character codes held in a buffer store in the VDU. Transmission of the text from the buffer store to the CPU is then done as a continuous block of characters, when a 'transmit' key is pressed.
(2) *Character mode* Each character code is transmitted to the CPU as the corresponding key is pressed. Characters are 'echoed' back from the CPU for display on the screen. This function, plus editing of entered data, is performed by program in the CPU.

To assist with character positioning on the screen, a cursor is displayed, showing the position in which the next character will appear. On most VDUs, its position can be altered by control characters sent by program to the VDU, or by the operator using special keys, so that inefficient and time-consuming use of 'space' and 'new line' controls is unnecessary.

Other functions, some or all of which are commonly provided depending on cost and sophistication, are blinking, dual intensity, reverse video (black characters on a white or coloured background), underline, alternate character sets, and protected areas. Each of these can be selected on a character by character basis. Additionally, the following are attributes of the whole screen: bidirectional scrolling, smooth scrolling (instead of jumping a line at a time), split screen scrolling, enlarged character size, and compressed characters.

The most commonly used colours for displayed characters are white, green and amber, all on a dark background, or the reverse of these where reverse video is available. Almost all VDUs currently produced have keyboards separate from the body of the VDU for ease of use and the comfort of the user, as well as some form of graphics capability ranging from simple graphics to very high resolution colour monitors which produce an image that is difficult to distinguish from a fine grain colour photograph. The level of sophistication on 'standard' VDUs has risen considerably since the start of this decade and high-resolution multi-choice colour terminals are no longer confined to specialist areas such as CAD/CAM, television and engineering.

60.13 High-speed printers and plotters

60.13.1 Line printers

For greater volumes of printed output than can be achieved with serial printers, line printers which can produce a whole line of characters almost simultaneously are available. Using impact techniques, speeds up to 3500 full lines (usually of 132 characters each) per minute are possible. Continuous paper in fan fold form, which may be multi-part to produce copies, is fed through the printer a line at a time by a transport system consisting of tractors which engage sprocket holes at the edges of the paper to move it upwards and through the printer from front to rear. A paper tray at the rear allows the paper to fold up again on exit from the printer.

As well as advancing a line at a time, commands can be given to advance the paper to the top of the next page, or to advance a whole page or line. This is important, for example where pre-printed forms are being used.

Two types of line printer are in common use; drum printers and band printers. Both use a horizontal row of hammers, one per character position or in some cases shared between two positions. These are actuated by solenoids to strike the paper through a carbon film against an engraved representation to print the desired character. In a drum printer, a print drum the length of the desired print line rotates once per print line. In each character position, the full character set is engraved around the circumference of the drum. A band printer has a horizontal revolving band or chain of print elements, each with a character embossed on it. The full character set is represented on the band in this way. To implement different character fonts involves specifying different barrels in the case of a drum printer, whereas a change can be made readily on a band printer by an operator changing bands, or individual print elements in the band can be replaced.

The printer has a memory buffer to hold a full line of character codes. When the buffer is full, or terminated if a short print line is required, a print cycle is initiated automatically. During the print cycle, the stored characters are scanned and compared in synchronism with the rotating characters on the drum or band. The printer activates the hammer as the desired character on the drum or band approaches in each print position.

60.13.2 Laser printers

These are available to meet three different types of printing requirements:

(1) Very high volumes of output at speeds exceeding 200 pages per minute. Normally used by those requiring a constant high-volume printing service since this equipment is expensive to buy, run and service.
(2) Departmental printing requirements, usually consisting of medium to high volumes on an *ad hoc* basis. This equipment would normally be networked to many CPUs and shared by a group of common users. Prints at speeds of up to 40 pages per minute.
(3) Desktop printing. Laser printers small enough to fit on an individual's desk, designed for intermittent low-volume personal printing requirements.

The technology used is common to all three types. The principle used is that of the everyday photocopier, the difference being that the image to be copied is set up according to digital signals received from the host CPU instead of from a photoscan of the document to be copied. The main advantage of this form of output is the clarity and quality of the image printed, which is so good that it is possible not only to print data but also to print the 'form' or 'letterhead' of the paper at the same time, thus avoiding the cost of pre-printed stationery. The disadvantage is that it is not currently possible to print multiple copies simultaneously.

60.13.3 Pen plotters

Another form of hard copy output is provided by pen plotters. These are devices aimed primarily at high-complexity graphics with a limited amount of text. Their uses range from plotting graphs of scientific data to producing complex engineering drawings in computer-aided design applications such as drawings used in integrated circuit chip design.

The plotter has one or more pens held vertically above a table on which the paper lies. These can be of different colours, and as well as being raised or lowered on to the paper individually by program commands, they can be moved in small steps, driven by stepping motors. Plotting in x and y directions is achieved either by having a single fixed sheet of paper on the plotting table with the pen or pens movable in both x and y directions, or by achieving control in one axis by moving the paper back and forth between supply and take-up rolls under stepping motor control, and in the other axis by pen movement. Diagonal lines are produced by combinations of movements in both axes.

With step sizes as small as 0.01 mm, high-accuracy plots can be produced, in multiple colours where more than one pen is used, and annotated with text produced in a variety of sizes and character sets. Supporting software is usually provided with a plotter. This will, for example, scale drawings and text, and generate alphanumeric characters.

60.14 Direct input

Other forms of direct input, eliminating the need for typing on the keyboard of a teleprinter or VDU, are described in this section.

60.14.1 Character recognition

Although the ultimate goal of a machine which can read any handwriting is some way off, the direct conversion of readable documents into codes for direct computer input is a reality. These are used for capture of source data as an alternative to keyboard input, and for processing documents such as cheques. Several types of device exist, with varying capabilities and functions:

(1) Page and document readers, with the capability to read several special OCR founts, plus in some cases lower quality print including hand printing, and hand-marked forms as opposed to written or printed documents. Most character readers have some form of error handling, allowing questionable characters to be displayed to an operator for manual input of the correct character.

    A wide range of capabilities and hence prices is found, from simple, low-speed (several pages per minute) devices handling pages only, to high-speed devices for pages and documents, the former at up to one page per second, with the latter several times faster.

(2) Document readers/sorters which read and optionally sort simple documents such as cheques and payment slips with characters in either magnetic ink or OCR fount. These are geared to higher throughputs, up to 3000 documents per minute, of standard documents.

(3) Transaction devices, which may use both document reading and keyboard data entry, and where single documents at a time are handled.

60.14.2 Writing tablets

Devices using a variety of techniques exist for the conversion of hand-printed characters into codes for direct input to a CPU. The overall function of these is the same—the provision of a surface on which normal forms, typically up to A3 size, can be filled in with hand-printed alphanumeric characters, using either a normal writing instrument in the case of pressure sensitive techniques, or a special pen otherwise.

The benefits of this type of device include:

(1) Immediate capture of data at source, avoiding time-consuming and error-prone transcription of data.

(2) By detecting the movements involved in writing a character, additional information is gained compared with optical recognition, allowing characters which are easily confused by OCR to be correctly distinguished.

60.15 Disk storage

Even though computer main memories are very large when compared to those produced during the 1970s (256 million characters is not uncommon), this is still nowhere near approaching the total memory storage capability required to retain and process all data. Backing storage provides this capability and has the added advantage that it can be copied and stored away from the CPU for security and safekeeping. For backing storage to be most suitable it should be price effective, reliable and easy to exchange and access. It is for this reason that the use of disk storage has grown considerably over the last ten years.

Disks are connected to a CPU by a controller which is normally a DMA device attached to the input/output bus or to a high-speed data channel, except in the case of the slowest of disks, which may be treated as a programmed transfer device. The controller is generally capable of handling a number of drives, usually up to 16 or 32. Having multiple disk drives on a system, as well as providing more on-line storage, allows copying of information from one disk to another and affords a degree of redundancy since, depending on application, the system may continue to function usefully with one less drive in the event of a failure.

A disk controller is relatively complex, since it has to deal with high rates of data transfer, usually with error code generation and error detection, a number of different commands, and a large amount of status information.

Four types of disk drive will be described in the following sections: floppy disk, cartridge disk, removable pack disk and 'Winchester Technology' disk. The following elements are the major functional parts common to all the above types of drive, with differences in implementation between the different types.

Drive motor This drives a spindle on which the disk itself is placed, rotating at a nominally fixed speed. The motor is powered up when a disk is placed in the drive, and powered down, normally with a safety interlock to prevent operator access to rotating parts, until it has stopped spinning, when it is required to remove the disk from the system.

Disk medium The actual recording and storage medium is the item which rotates. It is coated with a magnetic oxide material, and can vary from a flexible diskette of 0.25 Mbyte capacity recording on one surface only (floppy disk) to an assembly of multiple disks stacked one above the other on a single axle (disk pack) holding hundreds of megabytes of data.

Head mechanism This carries read/write heads, one for each recording surface. The number of recording surfaces ranges from only one on a single-sided floppy disk to ten or more for a multi-surface disk pack. In the latter case, the heads are mounted on a comb-like assembly, where the 'teeth' of the comb move together in a radial direction between the disk surfaces.

During operation, the recording heads fly aerodynamically extremely close to the disk surface, except in the case of floppy disks where the head is in contact with the surface. When rotation stops, the heads either retract from the surface or come to rest upon it, depending on the technology involved.

The time taken for the read/write head to be positioned above a particular area on the disk surface for the desired transfer of data is known as the access time. It is a function partly of the rotational speed of the disk, which gives rise to what is known as the average rotational latency, namely one half of the complete revolution time of the disc. Out of a number of accesses, the average length of time it is necessary to wait for the desired point to come below the head approaches this figure. The second component of access time is the head positioning time. This is dependent upon the number of tracks to be traversed in moving from the current head position to the desired one. Again, an average figure emerges from a large number of acesses. The average access time is the sum of these two components. In planning the throughput possible with a given disk system, the worst case figures may also need to be considered.

Electronics The drive must accept commands to seek, i.e. position the head assembly above a particular track, and must be able to recover signals from the read heads, and convert these to binary digits in parallel form for transmission to the disk controller. Conversely, data transmitted in this way from the controller to the disk drive must be translated into appropriate analogue signals to be applied to the head for writing the desired data on to the disk.

Various other functions concerned with control of the drive and sensing of switches on the control panel are performed. On some more advanced drives, much of the operation of the drive and electronics can be tested off-line from the system, allowing fault diagnosis to be performed without affecting the rest of the system.

Information is recorded in a number of concentric, closely spaced tracks on the disk surfaces, and in order to write and hereafter read successfully on the same or a different drive, it must be possible to position the head to a high degree of accuracy and precision above any given track. Data is recorded and read serially on one surface at a time, hence transfer of data between the disk controller and disk surface involves conversion in both directions between serial analogue and parallel digital signals. A phase-locked loop clock system is normally used to ensure reliable reading by compensating for variations in the rotational speed of the disk.

Data are formatted in blocks or sections on all disk systems, generally in fixed block lengths preformatted on the disk medium at time of manufacture. Alternatively 'soft sectoring' allows formatting into blocks of differing length by program. The drive electronics are required to read sector headers, which contain control information to condition the read circuitry of the drive, and sector address information, and to calculate, write and check an error correcting code—normally a cyclic redundancy check—for each block.

Finding the correct track in a seek operation, where the separation between adjacent tracks may be as little as 0.01 mm, requires servo-controlled positioning of the head to ensure accurate registration with the track. All rigid disk systems have servo-controlled head positioning, either using a separate surface pre-written with position information and with a read head only, or with servo information interspersed with data on the normal read/write tracks being sampled by the normal read/write head. Floppy disk systems, where the tolerances are not so fine, have a simpler stepping motor mechanism for head positioning.

60.15.1 Floppy disk

The floppy disk, whilst having the four elements described above, was conceived as a simple, low-cost device providing a moderate amount of random access backup storage to microcomputers, word processors and small business and technical minicomputers.

As the name implies, the magnetic medium used is a flexible, magnetic oxide coated diskette, which is contained in a square envelope with apertures for the drive spindle to engage a hole in the centre of the disk and for the read/write head to make contact with the disk. Diskettes are of two standard diameters, approximately 203 and 133 mm, the latter being more common. The compactness and flexibility of the disk makes it very simple to handle and store, and possible for it to be sent by post.

One major simplification in the design of the floppy disk system is the arrangement of the read/write head. This runs in contact with the disk surface during read/write operations, and is retracted otherwise. This feature, and the choice of disk coating and the pressure loading of the head are such that, at the rotational speed of 360 rpm, the wear on the recording surface is minimal. Eventually, however, wear and therefore error rate are such that the diskette may have to be replaced, copying the information on to a new diskette.

Capacities vary from the 256 kbytes of the earliest drives, which record on one surface of the diskette only, to over 2 Mbytes on more recent units, most of which use both surfaces of the diskette. Access times, imposed by the rather slow head positioning mechanism using a stepping motor, are in the range of 200 to 500 ms. Transfer rates are below 200 kbytes per second.

Another simplification is in the area of operator controls. There are generally no switches or status indicators, the simple action of moving a flap on the front of the drive to load or remove the diskette being the only operator action. The disk motor spins all the time that power is applied to the drive.

60.15.2 Cartridge disk

This type of disk system is so called because the medium, one or two rigid disks on a single spindle, or aluminium coated with magnetic oxide and approximately 350 mm in diameter, are housed permanently in a strong plastic casing or cartridge. When the complete cartridge assembly is loaded into the drive, a slot opens to allow the read/write heads access to the recording surfaces. As well as providing mechanical mounting, the cartridge provides protection for the disk medium when it is removed from the drive.

Drives are designed for loading either from the top when a lid is raised, or from the front when a small door is opened allowing the cartridge to be slotted in. Power to the drive motor is cut off during loading and unloading, and the door is locked until the motor has slowed down to a safe speed. On loading and starting up, the controller cannot access the drive until the motor has reached full speed. Operator controls are normally provided for unload, write protect and some form of unit select switch allowing drive numbers to be reassigned on a multiple drive system. Indicators typically show drive on-line, error and data transfer in progress.

Access times are normally in the region of 30 to 75 ms, aided by a fast servo-controlled head positioning mechanism actuated by a coil or linear motor, the heads being moved in and out over the recording surface by an arm which operates radially. Heads are light weight, spring loaded to fly aerodynamically in the region of 0.001 mm from the surface of the disk when it is rotating at its full speed, usually 2400 or 3600 rpm. Because of the extremely small gap, cleanliness of the oxide surface is vital, as any particle of debris or even smoke will break the thin air gap causing the head to crash into the disk surface. In this rare event, permanent damage to the heads and disk cartridge occurs. Positive air pressure is maintained in the area around the cartridge, in order to minimise the ingress of dirt particles. Care should be taken to ensure cleanliness in the handling and storage of cartridges when they are not mounted in the drive.

The capacity of cartridges is in the range up to 100 Mbytes, with data transfer rates in the region of 1 Mbyte/s. Up to 16 drives can be accommodated per controller and, because of the data transfer rate, direct memory access is necessary for transfer of data to or from the CPU.

60.15.3 Disk packs

The medium used in this type of drive has multiple platters (5 or more) on a single spindle, and is protected when removed from the drive by a plastic casing. When loaded on the drive, however, the casing is withdrawn. The drives are top loading, and unlike cartridge disks, which can generally be rack mounted in the cabinet housing the CPU, are free-standing units.

Other than this difference, most of the design features of disk pack drives follow those of cartridge units. The significant difference is the larger capabilities (up to 1000 Mbytes) and generally high performance in terms of access times (25 to 50 ms) and transfer rates (in the region of 2.5 Mbytes/s).

60.15.4 Winchester drive

So called from a name local to the laboratory in the United States where it was developed, this is a generic name applied to a category of drive where the disk medium itself remains fixed in the drive. The principal feature of the drive, the fixed unit is known as a head–disk assembly (HDA). By being fixed and totally sealed, with the read/write heads and arm assembly within the enclosure, the following benefits are realised:

(1) Contaminant free environment for the medium allows better data integrity and reliability, at the same time having less stringent environmental requirements. Simpler maintenance requirements follow from this.
(2) Lighter weight heads, flying tighter to tolerances closer to the recording surface, allow higher recording densities. Since the disk itself is never removed, instead of retracting, the heads actually rest on special zones on the disk surface when power is removed.
(3) The arrangement of read/write heads is two per surface, providing lower average seek times, by requiring less head movement to span the whole recording area.
    The head positioning arrangement differs mechanically from that of the previously described drives by being pivoted about an axis outside the disk circumference.

Three general types of Winchester drive exist, with aproximate disk diameters of 133, 203 and 355 mm, providing capacities from 25 Mbytes to over 1000 Mbytes. Performance, for the reason described above, can exceed that for disk cartridge or pack drives of corresponding capacity.

The smallest versions of Winchester drive are becoming popular as the storage medium for microcomputers and smaller configurations of minicomputer, offering compact size with very competitive prices and fitting above the top end of the floppy disk range. Operationally, the fact that the disks are not removable from the drive means that a separate form of storage medium which is removable must be present on a system using a Winchester drive. Backup and making portable copies is done using this separate medium, which is usually another type of disk drive, or a magnetic tape system matched to the disk speed and capacity.

60.16 Magnetic tape

Reliable devices for outputting digital data to and reading from magnetic tape have been available for a considerable time. The use of this medium, with agreed standards for the format of recorded data, has become an industry standard for the interchange of data between systems from different manufacturers.

In addition to this, low-cost magnetic tape cartridge systems exist providing useful minimal-cost backup storage for small systems plus a convenient medium for small-volume removable data and the distribution of software releases and updates.

60.16.1 Industry standard tape drive

These allow reels of 12.7 mm wide oxide-coated magnetic tape, which are normally 731 m in length on a 267 mm diameter reel or 365 m on a 178 mm reel, to be driven past write and read head assemblies for writing, and subsequent reading, at linear densities from 800 to 6250 bit/inch. Tapes are written with variable length blocks or records with inter-record gaps in the region of 12.7 mm. Each block has lateral and longitudinal parity information inserted and checked, and a cyclic redundancy code is written and checked for each block. The latter provides a high degree of error correction capability.

The tape motion and stop/start characteristics are held within precise limits by a servo-controlled capstan around which the tape wraps more than 180° for sufficient grip.

Correct tape tension and low inertia is maintained by motors driving the hubs of the two tape reels in response to information on the amount of tape in the path between the two reels at any time. Mechanical buffering for the tape between the capstan and reels is usually provided by a vacuum chamber. This technique, used on modern, high-performance tape drives, has between each reel and the capstan a chamber of the same width as the tape, into which a U-shaped loop of tape of around 1 to 2 m is drawn by vacuum in the chamber. The size of the tape loops is sensed photoelectrically to control the reel motors.

To prevent the tape from being pulled clear of the reel when it has been read or written to the end or rewound to the beginning, reflective tape markers are applied near each end of the reel. These are sensed photoelectrically, and the resulting signal is used to stop the tape on rewind, or to indicate that forward motion should stop on reading or writing.

Three different forms of encoding the data on the tape are encountered, dependent upon which of the standard tape speeds is being used. Up to 800 bit/inch, the technique is called 'non return to zero' (NRZ), while at 1600 bit/inch 'phase encoded' (PE) and at 6250 bit/inch 'group code recording' (GCR) are used. Some drives can be switched between 800 bits per inch NRZ and 1600 bits per inch PE. Very few systems below 1600 bit/inch are now manufactured.

Block format on the tape is variable under program control between certain defined limits and, as part of the standard, tape marks and labels are recorded on the tape, and the inter-block gap is precisely defined. Spacing between write and read heads allows a read-after-write check to be done dynamically to verify written data. Writing and reading can only be done sequentially. These tape units do not perform random access to blocks of data, though those units that permit selective reverse under program control do make it possible for the application to access data other than by sequential read of the the tape. However, this requires prior knowledge by the application of the layout and contents of the tape and is particularly slow and cumbersome, such applications being far better serviced by a disk storage device.

Up to eight tape drives can be handled by a single controller. For PE and GCR, a formatter is required between controller and drive, to convert between normal data representation and that required for these forms of encoding.

Tape drives can vary in physical form from a rack mountable unit that is positioned horizontally to a floor standing unit around 1.75 m in height.

Operator controls for on-line/off-line, manually controlled forward/reverse and rewind motion, unit select and load are normally provided. To prevent accidental erasure of a tape containing vital data by accidental write commands in a program, a write protect ring must be present on a reel when it is to be written to. Its presence or absence is detected by the drive electronics. This is a further part of the standard for interchange of data on magnetic tapes.

60.16.2 Cartridge tape

Low-cost tape units storing in the region of 0.5 Mbyte of data on a tape cartridge are sometimes used for backup storage on microcomputer systems, or for low-volume program loading on larger systems. Cartridge units using block formatted tapes which can search to locate numbered blocks can be used for operations requiring random access, such as supporting operating system software. This is only feasible, however, on a microcomputer system in view of the very slow access times.

Cartridge tape units are normally operable over a serial asynchronous line, and can therefore be connected by any communications interface to a system. Data transfer rates are normally below 5 kbyte/s.

At even lower cost and performance levels, drives using standard cassettes of the dimensions of the normal audio cassette are also found. With some of the personal microcomputers, standard domestic tape recorders are even used as a very simple storage device.

60.16.3 Tape streamer units

The emegence of large-capacity, non-removable disk storage in the form of Winchester technology drives has posed the problem of how to make backup copies of complete disk contents for security or distribution to another similarly equipped system. One solution is a tape drive very similar to the industry standard units described in *Section 60.16.1* but with the simplification of writing in a continuous stream, rather than in blocks. The tape controller and tape motion controls can, therefore, be simpler than those for the industry standard drive. Many modern tape units are able to operate in both block and streamer mode according to operator or program selection, but not on the same tape.

A streamer unit associated with a small Winchester drive can accept the full disk contents on a single reel of tape.

60.17 Digital and analogue input/output

One of the major application areas for minicomputers and microcomputers is direct control and collection of data from other systems by means of interfaces which provide electrical connections directly or via transducers to such systems. Both continuously varying voltages (analogue signals) and signals which have discrete on or off states (digital signals) can be sensed by suitable interfaces and converted into binary form, for analysis by programs in the CPU. For control purposes, binary values can also be converted to analogue or digital form by interfaces for output from the computer system.

In process control and monitoring, machine control and monitoring, data acquisition from laboratory instruments, radar and communications, to take some common examples, computer systems equipped with a range of suitable interface equipment are used extensively for control and monitoring. They may be measuring other physical quantities such as temperature, presssure and flow converted by transducers into electrical signals.

60.17.1 Digital input/output

Relatively simple interfaces are required to convert the ones and zeros in a word output from the CPU into corresponding on or off states of output drivers. These output signals are brought out from the computer on appropriate connectors and cables. The output levels available range from TTL levels for connection to nearby equipment which can receive logic levels to over 100 V d.c. or a.c. levels for industrial environments. In the former case, signals may come straight from a printed circuit board inside the computer enclosure, while in the latter they require to go through power drivers and be brought out to terminal strips capable of taking plant wiring. This latter type of equipment may need to be housed in separate cabinets.

In a similar manner, for input of information to the computer system, interfaces are available to convert a range of signal levels to logic levels within the interface, which are held in a register and can be input by the CPU. In some cases, input and output are performed on the same interface module.

Most mini and micro systems offer a range of logic level input/output interfaces, while the industrial type of input and output equipment is supplied by manufacturers specialising in process control. Optical isolators are sometimes included in each signal line to electrically isolate the computer from other systems. Protection of input interfaces by diode networks or fusible links is sometimes provided to prevent damage by overvoltages. In industrial control where thousands of digital points need to be scanned or controlled, interfaces with many separately addressable input and output words are used.

Although most digital input and output rates of change are fairly slow (less than 1000 words per second), high-speed interfaces at logic levels using direct memory access are available. These can in some cases transfer in burst mode at speeds up to the region of 1 million words per second. High transfer rates are required in areas such as radar data handling and display driving.

60.17.2 Analogue input

Analogue-to-digital converters, in many cases with programmable multiplexers for high- or low-level signals and programmable gain preamplifiers covering a wide range of signals (from microvolts to 10 V), allow conversion commands to be issued, and the digital results to be transferred to the CPU by the interface.

Industrial grade analogue input subsystems typically have a capacity of hundreds of multiplexer channels, low-level capability for sources such as thermocouples and strain gauges, and high common mode signal rejection and protection. As with digital input/output, this type of equipment is usually housed in separate cabinets with terminal strips, and comes from specialised process control equipment or data logger manufacturers.

For laboratory use, converters normally have higher throughput speed, lower multiplexer capacity and often direct cable connection of the analogue signals to a converter board housed within the CPU enclosure. Where converters with very high sampling rates (in the region of 100 000 samples per second) are used, input of data to the CPU may be by direct memory access. Resolution of analogue-to-digital converters used with computer systems is usually in the range 10 to 12 bits, i.e. a resolution of 1 part in 1024 to 1 part in 4096. Resolutions of anything from 8 to 15 bits are, however, available. Where a programmable or auto-ranging preamplifier is used before the analogue-to-digital converter, dynamic signal ranges of one million to one can be handled.

60.17.3 Analogue output

Where variable output voltages are required, for example to drive display or plotting devices or as set points to analogue controllers in industrial process control applications, one or more addressable output word is provided, each with a digital-to-analogue converter continuously outputting the voltage represented by the contents of its register. Resolution is normally no more than 12 bits, with a usual signal range of ±1 V or ±10 V. Current outputs are also available.

60.17.4 Input/output subsystems

Some manufacturers provide a complete subsystem with its own data highway separate from the computer system input/output bus, with a number of module positions into which a range of compatible analogue and digital input/output modules can be plugged. Any module type can be plugged into any position to make up the required number of analogue and digital points.

60.18 Data communications

60.18.1 Introduction

Since 1980 there has been a large-scale growth in the use of data communications between different types and makes of equipment both within a physical location or building and between different buildings situated anywhere on the globe. Even when this communication appears to take place between two points on Earth it has very often actually taken place courtesy of a geostationary satellite positioned in orbit. The creation and maintenance of such 'networks' is now nearly always the role of the Network Manager and his or her staff, a function that is separate from, though working closely with, the traditional computer department.

The requirement for the communication of data is not, of course, new; but what has changed is the basis for that requirement. Previously the only other means available for the transfer of data between machines was to copy by magnetic media (such as tape or disk) or to key in the data again, with the consequent high risk of error and increased time taken. It was seen that data transmission would be faster and more accurate than both of these methods. Interestingly, data communication was not seen as a replacement for the data in hard copy form. Today, more emphasis is being placed on eliminating hard copy transactions.

Large-scale integration and consequent lower costs have made much more readily available very powerful computers which can contain the sophisticated software required to handle complex networks and overcome complex problems such as finding alternative routes for messages when a transmission line is broken. With the computer 'space' thus available, programmers can produce complex programs required to transmit data from a terminal connected to one computer to a program running in another, without the operator being aware of the fact that two computers are involved. Interface devices between the computer and the data networks are becoming increasingly powerful, and are usually microcomputers.

Computers have always been able to communicate with their peripheral devices such as card readers, mass storage devices and printers, but in the 1960s it was not typical for the communications to extend beyond this. Data was transcribed onto 'punching documents' by functional departments within an organisation. The widespread use of interactive VDUs by users at their desks has eliminated almost all of these departments. Even the very large traditional data entry organisations such as the utility companies are now introducing data capture at source using hand-held terminals.

However, in the late 1960s and 1970s the development of both hardware and software technology made it increasingly attractive to replace these terminals with more intelligent remote systems. These systems varied in their sophistication. At one end of the spectrum were interactive screen-based terminals which could interrogate files held on the central computer. Greater sophistication was found in data validation systems which held sufficient data locally to check that, for example, part numbers on a customer order really existed, before sending the order to the computer for processing. Most sophisticated still were complete minicomputers carrying out a considerable amount of local data processing before updating central files to be used in large 'number crunching' applications such as production scheduling and materials planning.

From these systems have grown a whole range of requirements for data communications. We have communications between mainframe computers, between minis, between computers and terminals, between terminals and terminals and so on.

60.18.2 Data communications concepts

Computers communicate data in binary format, the bits being represented by changes in current or voltage on a wire or patterns of light through an optic fibre cable (see *Chapters 55* and *56*). Various code systems have been developed in attempts to standardise the way characters are represented in binary format. One of the earliest of these was the 5-bit Baudot code, invented towards the end of the nineteenth century by Emile Baudot for use on telegraphic circuits. Five bits can be used to represent 32 different characters and whilst this was adequate for its purpose it cannot represent enough characters for modern data communications. Nonetheless Baudot gave his name to the commonly used unit of speed 'baud' which, although strictly speaking meaning signal events per second, is frequently used to denote 'bits per second'.

Nowadays, one of the most commonly used codes is the ASCII code (American Standard Code for Information Interchange). This consists of seven information bits plus one parity (error checking) bit. Another is EBCDIC (Extended Binary Coded Decimal Interchange Code), an 8-bit character code used primarily on IBM equipment.

Within the computer and between the computer and its peripheral devices such as mass storage devices and line printer, data is usually transferred in parallel format (*Figure 60.7*). In

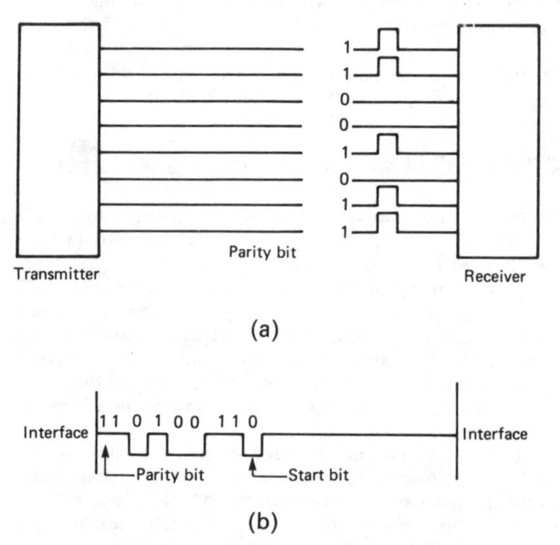

Figure 60.7 Data transmission; (a) parallel, (b) serial

parallel transmission a separate wire is used to carry each bit, with an extra wire carrying a clock signal. This clock signal indicates to the receiving device that a character is present on the information wires. The advantage of parallel transmission is, of course, speed, since an entire character can be transmitted in the time it takes to send one bit. However, the cost would prove prohibitive where the transmitter and receiver are at some distance apart. Consequently, for sending data between computers and terminal devices and between computers which are not closely coupled, serial transmission is used.

Here a pair of wires is used, with data being transmitted on one wire whilst the second acts as a common signal ground. As the term implies, bits are transmitted serially and so this form of transmission is more practical for long-distance communication because of the lower cost of the wiring required. In addition, it is simpler and less expensive to amplify signals rather than use multiple signals in order to overcome the problem of line noise, which increases as the distance between the transmitter and receiver grows. Data transmission frequently makes use of telephone lines designed for voice communication, and since the public voice networks do not consist of parallel channels, serial transmission is the only practical solution.

Parallel data on multiple wires is converted to serial data by means of a device known as an interface. In its simplest form, an interface contains a register or buffer capable of storing the number of bits which comprise one character. In the case of data going from serial to parallel format, the first bit enters the first position in the register and is 'shifted' along, thereby making room for the second bit (*Figure 60.8*). The process continues

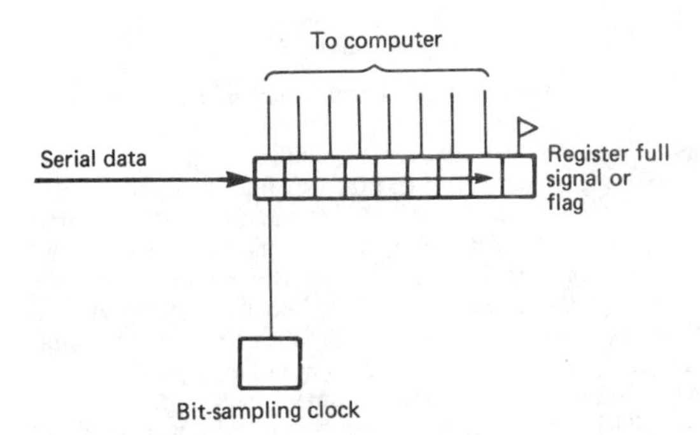

Figure 60.8 Serial-to-parallel interface

until the sampling clock which is strobing the state of the line indicates that the correct number of bits has been received and that a character has been assembled. The clock then generates a signal to the computer which transfers the character in parallel format. The reverse process is carried out to convert parallel to serial data.

This 'single buffered' interface does have limitations, however. The computer effectively has to read the character immediately, since the first bit of a second character will be arriving to begin its occupation of the register. This makes no allowance for the fact that the computer may not be available instantly. Nor does it allow any time to check for any errors in the character received.

To overcome this problem, a second register is added creating a 'double buffered' interface (*Figure 60.9*). Once the signal is received, indicating that the requisite number of bits have been assembled, the character is parallel transferred to the second, or holding, register and the process can continue. The computer now has as much time as it takes to fill the shift register in order to check and transfer (again in parallel format) the character.

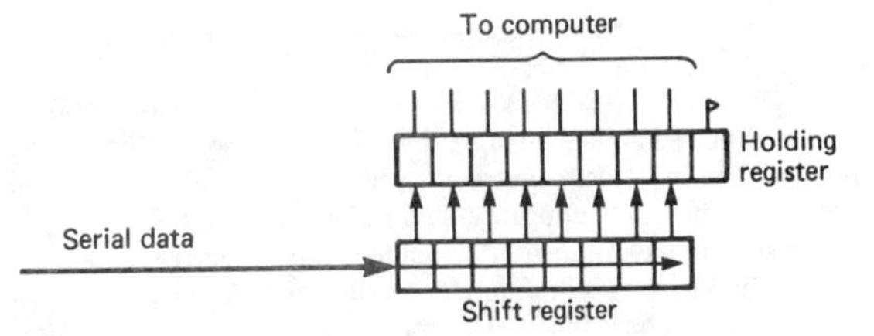

Figure 60.9 Double-buffered interface

60.18.3 Multiline interface

With the development of technology, the transmitter and receiver functions are now carried out by an inexpensive chip and so the major costs in the interface are the costs of the mechanism used to interrupt the CPU when a character has been assembled and the connection to the computer's bus used to transmit the received data to the CPU, or in some cases direct to memory. The interrupt mechanism and the bus interface are not heavily used. Indeed, they function only when a character is received or transmitted. These facilities are shared in a multiline interface, sometimes (though not strictly correctly) known as a 'multiplexer' (*Figure 60.10*). To achieve this the device has several

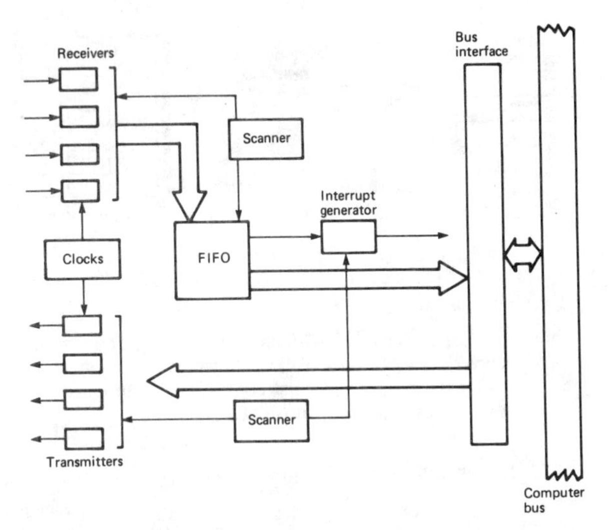

Figure 60.10 Schematic diagram of a multiline interface

receivers and transmitters and a first in, first out (FIFO) buffer for received characters. The receivers are scanned and when a flag is found indicating that a character has been received the character is transmitted into the FIFO buffer, along with its line number. An interrupt tells the CPU that there are characters in the buffer and they are communicated over the bus to the computer. Similarly, a scanner checks the transmitters and, when it discovers a flag indicating that a transmitter buffer is empty, it interrupts the CPU. Typically, the number of lines supported by a multiline interface increases by powers of two for convenient binary representation, 4, 8, 16, 32, 64, 128 and 256 being common.

The economies of scale in such an interface mean that further sophistications can be included such as program selectable formats and line speed, and modem control for some or all of the lines.

However, the term 'multiplexing', strictly speaking, actually refers to the function of sharing a single communications channel across many users (see *Chapter 55*). There are two commonly used methods of achieving this. One is a technique called time division multiplexing. It consists of breaking down the data from each user into separate messages which could be as small as one or two characters and meaningless when taken individually. The messages, together with identifying characters, are interleaved and transmitted along a single line. They are separated at the other end and the messages reassembled. Ths is achieved by use of devices known as concentrators or multiplexers.

The second technique used to achieve this objective of making maximum use of a communication line is frequency division multiplexing. The concept is similar to that of time division multiplexing. It is achieved by transmitting complete messages simultaneously but at different frequencies.

60.18.4 Modems

A significant complication of using public voice networks to transmit data is that voice transmission is analogue whereas data generated by the computer or terminal is digital in format. Thus an additional piece of equipment is required between the digital sender/receiver and the analogue circuit. This device modulates and demodulates the signal as it enters and leaves the analogue circuit, and is known by the abbreviated description of its functions, a modem (*Figure 60.11*). Modems are provided by the common carrier such as British Telecom or by private manufacturers. In the latter case, however, they must be approved by the carrier and must contain or be attached to a device which provides electrical isolation.

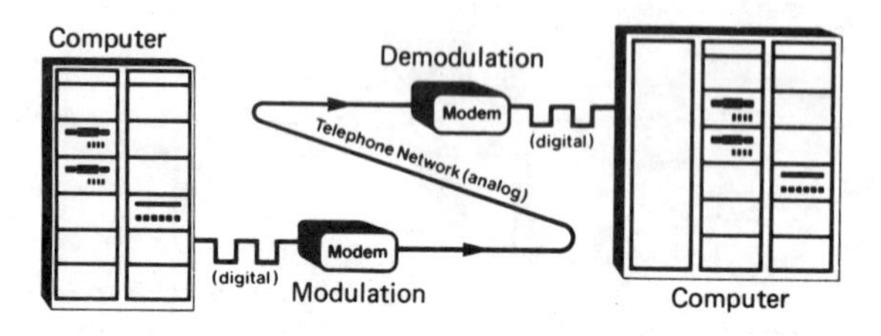

Figure 60.11 The use of modems in a communications link

60.18.5 Transmission techniques

The EIA (Electronics Industries Association) and CCITT (Comité Consultatif Internationale de Télégraphie et Téléphone) publications contain specifications and recommendations for the design of equipment to interface data terminal equipment (computers and terminals) to data communication equipment (modems). The specific EIA standard to which most modem equipment is designed is RS-232-C. The CCITT equivalent of RS-232-C is known as 'V.24—List of Definitions for Interchange Circuits between Data Terminal Equipment and Data Circuit Terminating Equipment'. The EIA/CCITT systems communicate data by reversing the polarity of the voltage; a 0 is represented by a positive voltage and a 1 by a negative voltage.

The signals in the EIA/CCITT specifications are not recommended for use over distances greater than 50 feet. Consequently, the modem and interface should not be more than 50 feet apart, although in practice distances in excess of 1000 feet have been operated without problems.

Different communications applications use one of two types of transmission, asynchronous or synchronous. Slower electromechanical devices such as teleprinters typically use asynchronous (or 'start–stop') transmission in which each character is transmitted separately. In order to tell the receiver that a character is about to arrive, the bits representing the character are preceded by a start bit, usually a zero. After the last data bit and error checking bit the line will return to the 1-bit state for at least one bit time—this is known as the stop bit.

Asynchronous transmission has the advantage that it requires relatively simple and therefore low-cost devices. Since characters can be sent at random times it is also appropriate for interactive applications which do not generate large amounts of data to be transmitted. It is, however, inefficient, since at least two extra bits are required to send eight data bits, and so would not be used for high-speed communication.

In synchronous transmission, characters are assembled into blocks by the transmitter and so the stream of data bits travels along the line uninterrupted by start and stop bits. This means that the receiver must know the number of bits which make up a character so that it can reassemble the original characters from the stream of bits. Preceding the block of data bits synchronisation characters are sent to provide a timing signal for the receiver and enable it to count in the data characters. If the blocks of data are of uniform length, this is all that is required to send a message. However, most systems would include some header information which may be used to indicate the program or task for which the data is destined and the amount of data in the block. In addition, if the messages are of variable length, some end-of-message characters will be required.

Because it does not contain start and stop bits for every character, synchronous transmission is more efficient than asynchronous. However, it can be inappropriate for some character-orientated applications since there is a minimum overhead in characters, which can be high relative to small transmitted block sizes, and the equipment required to implement it is more expensive.

60.18.6 Direction of transmission

There are three types of circuit available for the communication of data and, correspondingly, there are three direction combinations, simplex, half duplex and full duplex. However, it is possible to use a channel to less than its full potential.

Simplex communication is the transmission of data in one direction only, with no capability of reversing that direction. This has limitations and is not used in the majority of data communications applications. It can be used, however, for applications which involve the broadcasting of data for information purposes in a factory for example. In this instance, there is neither a need nor a mechanism for sending data back to the host. The simplex mode of operation could not be used for communication between computers.

Half duplex, requiring a single, two-wire circuit, permits the user to transmit in both directions, but not simultaneously. Two-wire half duplex has a built-in delay factor called turnaround time. This is the time taken to reverse the direction of transmission from sender to receiver and vice versa. The time is required by line propagation effects, modem timing and computer response time. It can be avoided by the use of a four-wire circuit normally used for full duplex. The reason for using four wires for half duplex rather than full duplex may be the existence of limitations in the terminating equipment.

Full duplex operation allows communication in both directions simultaneously. The data may or may not be related, depending on the applications being run in the computer or computers. The decision to use four-wire full duplex facilities is usually based on the demands of the application compared to the increased cost for the circuit and the more sophisticated equipment required.

60.18.7 Error detection and correction

Noise on communications lines will inevitably introduce errors into messages being transmitted. The error rates will vary according to the kind of transmission lines being used. In-house lines are potentially the most noise free since routing and shielding are within the user's control. Public switched networks, on the other hand, are likely to be the worst as a result of noisy switching and transmission.

Whatever the environment, however, there will be a need for error detection and correction. Three systems are commonly used: VRC, LRC and CRC.

VRC, or vertical redundancy check, consists of adding a parity bit to each character. The system will be designed to use either even or odd parity. If the parity is even, the parity bit is set so that the total number of ones in the character plus parity is even. Obviously, for odd parity the total number will be odd. This system will detect single bit errors in a character. However, if two bits are incorrect the parity will appear correct. VRC is therefore a simple system designed to detect single bit errors within a character. It will detect approximately 9 out of 10 errors.

A more sophisticated error-detection system is LRC, longitudinal redundancy check, in which an extra byte is carried at the end of a block of characters to form a parity character. Unlike in VRC, the bits in this character are not sampling an entire character but individual bits from each character in the block. Thus the first bit in the parity character samples the first bit of each data character in the block. As a result, LRC is better than VRC at detecting burst errors which affect several neighbouring characters.

It is possible to combine VRC and LRC and increase the combined error detection rate of 99% (*Figure 60.12*). A bit error can be detected and corrected, because the exact location of the error will be pinpointed in one direction by LRC and the other by VRC.

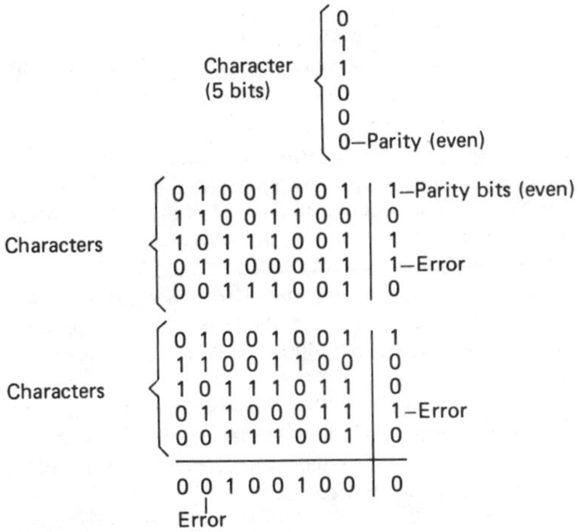

```
                        0
                        1
Character               1
(5 bits)                0
                        0
                        0—Parity (even)

             0 1 0 0 1 0 0 1 | 1—Parity bits (even)
             1 1 0 0 1 1 0 0 | 0
Characters   1 0 1 1 1 0 0 1 | 1
             0 1 1 0 0 0 1 1 | 1—Error
             0 0 1 1 1 0 0 1 | 0

             0 1 0 0 1 0 0 1 | 1
             1 1 0 0 1 1 0 0 | 0
Characters   1 0 1 1 1 0 1 1 | 0
             0 1 1 0 0 0 1 1 | 1—Error
             0 0 1 1 1 0 0 1 | 0

             0 0 1 0 0 1 0 0 | 0
               |
             Error
```

Figure 60.12 VRC, LRC and VRC/LRC combined (with acknowledgements to Digital Equipment Co. Ltd)

Even though the combination of LRC and VRC significantly increases the error detection rate, the burst nature of line noise means that there are still possible error configurations which could go undetected. In addition, the transmission overhead is relatively high. For VRC alone, in the ASCII code, it is 1 bit in 8, or $12\frac{1}{2}$%. If VRC and LRC are used in conjunction it will be $12\frac{1}{2}$% plus one character per block.

A third method which has the advantage of a higher detection rate and, in most circumstances, a lower transmission overhead is CRC, cyclic redundancy check. In this technique the bit stream representing a block of characters is divided by a binary number. In the versions most commonly used for 8-bit character format, CRC-16 and CRC-CCITT, a 16-bit divisor is used. There are no carry overs in the division and a 16-bit remainder is generated. When this calculation has been completed, the transmitter sends these 16 bits—two characters—at the end of the block. The receiver repeats the calculation and compares the two remainders. With this system, the error detection rises to better than 99.99%. The transmission overhead is less than that required for VRC/LRC when there are more than eight characters per block, as is usually the case.

The disadvantage with CRC is that the calculation overhead required is clearly greater than for the other two systems. The check can be performed by hardware or software but, as is usually the case, the higher performance and lower cost of hardware is making CRC more readily available and more commonly used.

Once bad data has been detected, most computer applications require that it be corrected and that this occurs automatically. Whilst it is possible to send sufficient redundant data with a message to enable the receiver to correct errors without reference to the transmitter, the calculation of the effort required to achieve this in the worst possible error conditions means that this technique is rarely used. More commonly, computer systems use error correction methods which involve retransmission. The two most popular of these are 'stop and wait retransmission' and 'continuous retransmission'.

'Stop and wait' is reasonably self-explanatory. The transmitter sends a block and waits for a satisfactory or positive acknowledgement before sending the next block. If the acknowledgement is negative, the block is retransmitted. This technique is simple and effective. However, as the use of satellite links increases, it suffers from the disadvantage that these links have significantly longer propagation times than land-based circuits and so the long acknowledgement times are reducing the efficiency of the network. In these circumstances, 'continuous retransmission' offers greater throughput efficiency. The difference is that the transmitter does not wait for an acknowledgement before sending the next block; it sends continuously. If it receives a negative acknowledgement it searches back through the blocks transmitted and sends it again. This clearly requires a buffer to store the blocks after they have been sent. On receipt of a positive acknowledgement the transmitter deletes the blocks in the buffer up to that point.

60.18.8 Communication protocols

The communications protocol is the syntax of data communications. Without such a set of rules, a stream of bits on a line would be impossible to interpret. Consequently, many organisations, notably computer manufacturers, have created protocols of their own. Unfortunately, however, they are all different and consequently yet another layer of communications software is required to connect computer networks using different protocols. Examples of well known protocols are Bisync and SDLC from IBM, DDCMP from Digital Equipment Corporation, ADCCP from the American National Standards Institute (ANSI) and HDLC from the International Standards Organisation (ISO). The differences between them, however, are not in the functions they set out to perform, but in the way they achieve them. Broadly speaking, these functions are as follows.

Framing and formatting These define where characters begin and end within a series of bits, which characters constitute a message and what the various parts of a message signify. Basically, a transmission block will need control data, usually contained in a 'header' field, text—the information to be transmitted—held in the 'body', and error checking characters, to be found in the 'trailer'. The actual format of the characters is defined by the information code used such as ASCII or EBCDIC.

Synchronisation It involves preceding a message or block with a unique group of characters which the receiver recognises as a synchronisation sequence. This enables the receiver to frame subsequent characters and fields.

Sequencing It numbers messages so that it is possible to identify lost messages, avoid duplicates and request and identify retransmitted messages.

Transparency Ideally all of the special control sequences should be unique and, therefore, never occur in the text. However, the widely varied nature of the information to be transmitted, from computer programs to data from instruments and industrial processes, means that occasionally a bit pattern will occur in the text which could be read by the receiver as a control sequence. Each protocol has its own mechanism for preventing this, or achieving 'transparency' of the text. Bisync employs a technique known as 'character stuffing'. In Bisync the only control character which could be confusing to the receiver if it appeared in the text is DLE (data link escape). When the bit pattern equivalent to DLE appears within the data a 'second' DLE is inserted. When the two DLE sequences are read, the DLE proper is discarded and the original DLE-like bit pattern is treated as data. This is character stuffing. SDLC, ADCCP and HDLC use a technique known as 'bit stuffing' and DECMP uses a bit count to tell the receiver where data begins and ends.

Start-up and timeout These are the procedures required to start transmission when no data has been flowing and recover when transmission ceases.

Line control It is the determination, in the case of half-duplex systems, of which terminal device is going to transmit and which to receive.

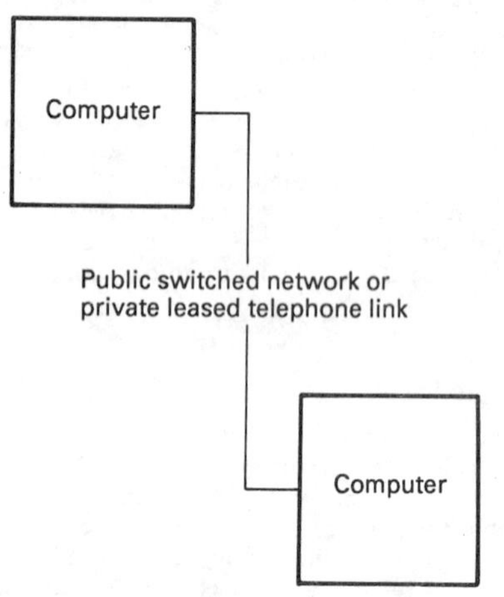

Figure 60.13 A point-to-point link

Error checking and correction As described in the section on error checking techniques, each block of data is verified as it is received. In addition, the sequence in which the blocks are received is checked. For data accuracy all the protocols discussed in this section are capable of supporting CRC (cyclic redundancy check). The check characters are carried in the trailer or block check character (BCC) section.

60.18.9 Computer networks

A communications network may exist within a single site. For the factory environment, many computer manufacturers offer proprietary networks for connecting terminal equipment to circuits based on 'tree' structures or loops. Connections to the circuit may be from video terminals for collection of, for example, stores data, special-purpose card and badge readers used to track the movement of production batches, or transducers used in the movement of industrial processes. Local area networks (LANs) are extensively used in computer networks (see *Chapter 61*).

There are a number of network types.

Point to point This is the simplest form of network and involves the connection of two devices—two computers or a computer and a terminal. If the communication line goes down for any reason then the link is broken, and so it is usual to back up leased lines with dial-up facilities (*Figure 60.13*).

Multipoint As the name implies, Multipoint describes the connection of several tributary stations to one host. It is usual for the host to 'poll' the tributary stations in sequence, requesting messages, and for the network to be based on leased lines. In the case of one 'spur' being disconnected, the tributary station will dial into the host using a port received for the purpose (*Figure 60.14*).

Centralised In this type of network, the host exercises control over the tributary stations, all of which are connected to it. The host may also act as a message-switching device between remote sites (*Figure 60.15*).

Hierarchical A hierarchical structure implies multiple levels of supervisory control. For example, in an industrial environment, special-purpose 'micros' may be linked to the actual process equipment itself. Their function is to monitor and control temperature and pressure.

These 'micros' will then be connected to supervisory 'minis' which can store the programs and set points for the process computers and keep statistical and performance records (*Figure 60.16*).

The next link in the chain will be the 'resource management computers', keeping track of the materials used, times taken, comparing these with standards, calculating replenishment orders, adjusting forecasts and so on.

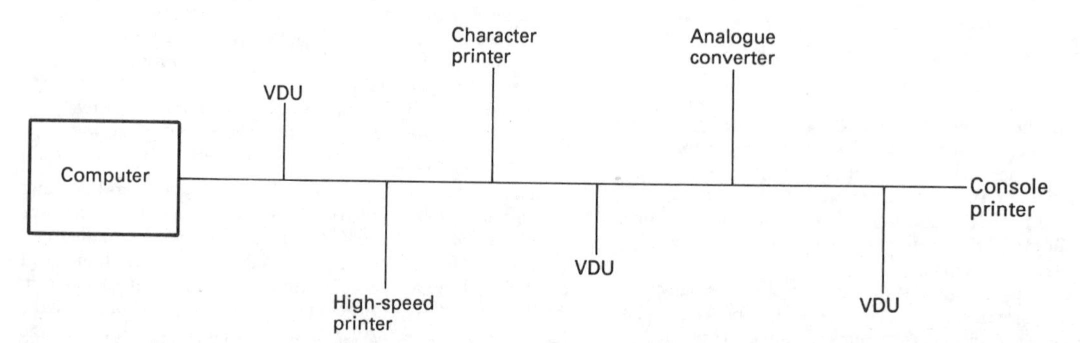

Figure 60.14 A multi-point communications network

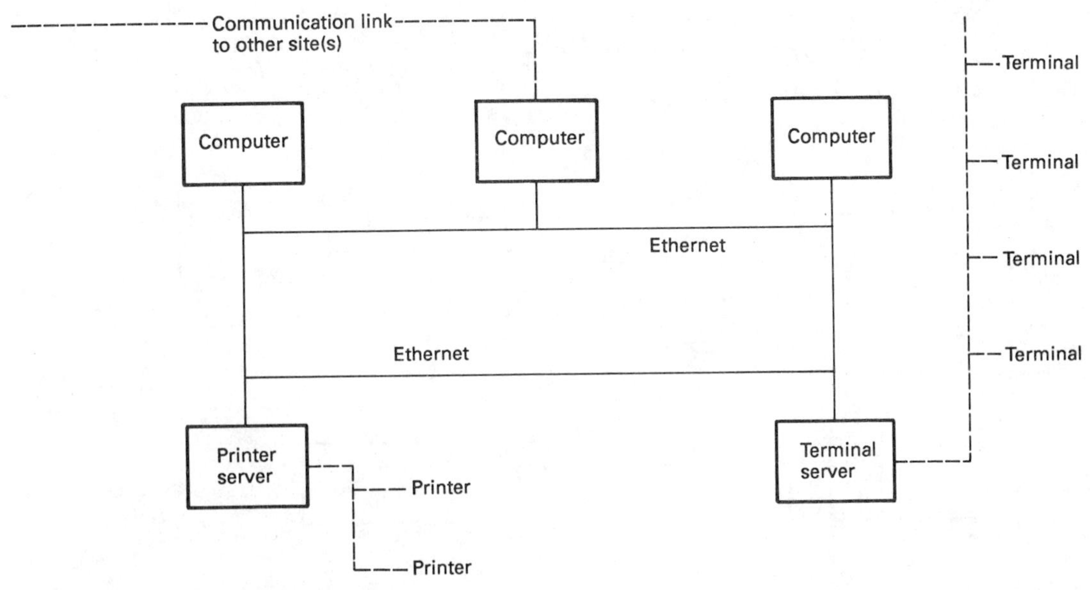

Figure 60.15 A typical centralised single-site network

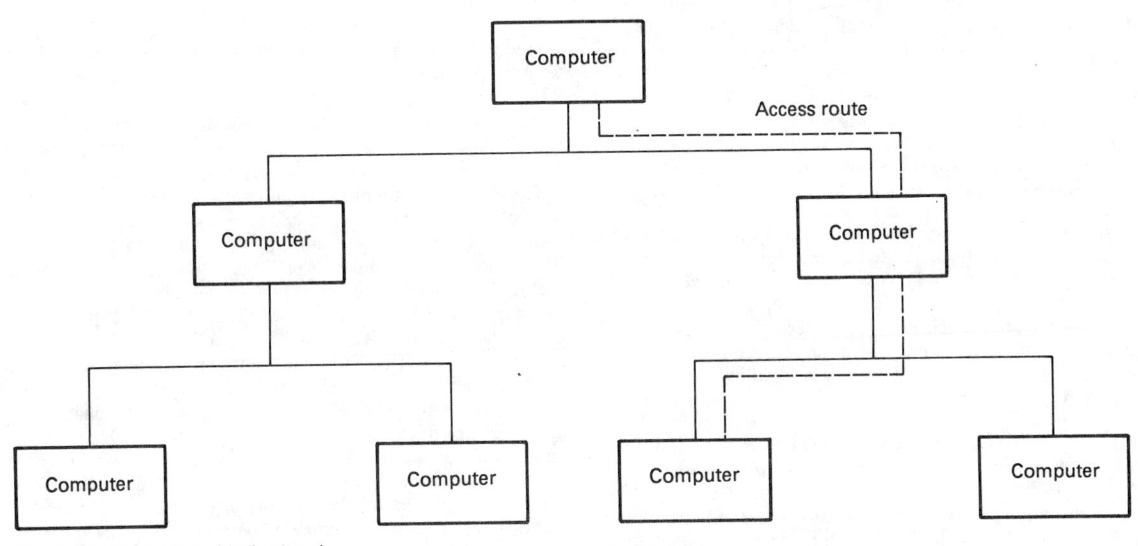

Figure 60.16 A hierarchical network

Finally, at the top of the network, the financial control system records costs and calculates the financial performance of the process.

Fully distributed Here each station may be connected to several others in the network. The possibility then exists to share resources such as specialised peripheral devices or large memory capacity and to distribute the database to the systems that access the data most frequently. It also provides alternative routes for messages when communication lines are broken or traffic on one link becomes excessive (*Figure 60.17*).

However, the design of such systems requires sophisticated analysis of traffic and data usage and even when set up is more difficult to control than less sophisticated networks.

60.18.10 Network concepts

Whatever the type of network, there are a number of concepts which are common.

File transfer A network should have the ability to transfer a file, or a part file, from one node to another without the intervention of programmers each time the transfer takes place. The file may contain programs or data and since different types, and possibly

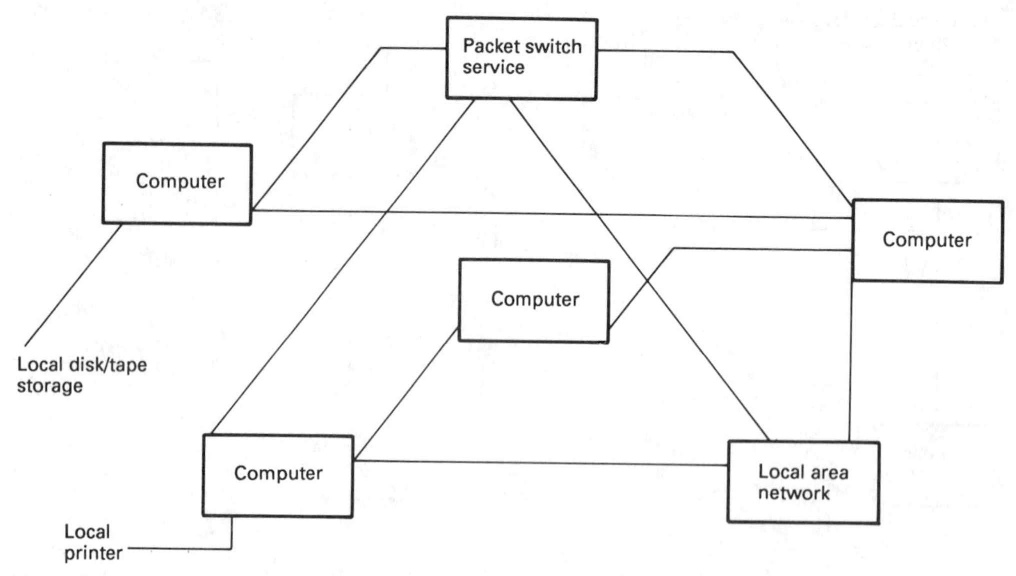

Figure 60.17 A fully distributed wide area network

generations, of computers, and different applications are involved, some reformatting may be required. This requires a set of programs to be written to cover all foreseen transfer requests, and requires a knowledge of all local file access methods and formats.

One good exmple of the need for this is the application known as archiving. It involves the transmission of copies of files held on computer to another system in another location. In the event of original files being lost as a result of fire, the files can be recreated using the archived information.

Resource sharing It may be more cost effective to set up communication links to share expensive peripheral devices than to duplicate them on every computer in the network. For example, one computer may have a large sophisticated flatbed printer/plotted for producing large engineering drawings. To use this the other computers would store the necessary information to load and run the appropriate program remotely. This would be followed by the data describing the drawing to be produced.

Remote file access/enquiry It is not always necessary or desirable to transfer an entire file, especially if only a small amount of data is required. In these circumstances what is required is the ability to send an enquiry from a program (or task) running in one computer and remotely load to the other system. This enquiry program will retrieve the requisite data from the file and send it back to the original task for display or processing. This comes under the broad heading of task-to-task communications.

Logical channels Users of a computer network will know where the programs and data, which they wants to access, exist. They do not want to concern themselves with the mechanics of how to gain acess to them, but expect there to be a set of predefined rules in the system which will provide a 'logical channel' to the programs and data they wish to reach. This logical channel will use one or more logical links to route the users' requests and carry back the response efficiently and without errors. It may be that there is no direct physical link between the users' computer and the machine they are trying to access. In these circumstances the logical channel will consist of a number of logical links. The physical links, in some cases, may be impossible to define in advance, since, in the case of 'dial up' communication using the public switched network, the route will be defined at connection time by the PTT.

Virtual terminal This is a very simple concept. It describes a terminal physically connected to computer A but with access (via A) to computer B.

The fact that communication is via A should be invisible to the user. Indeed, to reach the ultimate destination, the user may unknowingly have to be routed through several nodes.

The use of common systems such as Ethernet and the promotion of common standards such as Open Systems Interconnect have bred a new concept in connecting terminals to computers, with the emphasis placed more on the 'service' that a user requires. Whereas previously users had only to know where the connection was required to and and not how to get there, with Ethernet-based 'servers' they need only to know the name of the service that is required and no longer to specify where it resides. The terminal will be connected to Ethernet through a computer acting as a router. The server will know on which machine or machines the service required is currently available, and needs to know if the service has been moved, whereas the user does not. Furthermore, if the service is available on more than one machine, the server will be capable of balancing the terminal workload given to each machine, all without the user even having to know or be aware of where the service is being provided from.

Many terminal servers are even capable of running more than one terminal to computer 'sessions' simultaneously on the same terminal, enabling the user to switch between them as desired without the host computer thinking that the session has been terminated.

In all of these examples the terminal is considered to be 'virtual' by any of the host machines that it is connected to via the terminal server. This concept and the facilities that it offers are quickly eroding many of the problems associated with previous methods of connecting terminals to computers and the 'switching' and physical 'patching' that were required to connect a terminal to a new machine.

Emulator As the name implies, this consists of one device performing in such a way that it appears as something different. For example, a network designer wrestling with the 'virtual terminal' concept may define that any terminal or computer to be connected to the network should be capable of looking like a member of the IBM 3270 family of video terminals for interactive work and the IBM 2780/3780 family for batch data transfer. In other words, they must be capable of 3270 and 2780/3780 emulation. Indeed, along with the Digital Equipment Corp. VT100/200, these two types of emulation have been amongst the most commonly used in the computer industry.

Routing As soon as we add a third node, C, to a previously point-to-point link from A to B, we have introduced the possibility of taking an alternative route from A to B, namely via C. This has advantages. If the physical link between A and B is broken, we can still transmit the message. If the traffic on the AB link is too high we can ease the load by using the alternative route.

However, it does bring added complications. The designer has to balance such factors as lowest transmission cost versus load sharing. Each computer system has to be capable of recognising which messages are its own and which it is required merely to transmit to the next node in the logical link. In addition, when a node recognises that the physical link it was using has, for some reason, been broken, it must know what alternative route is available.

60.18.11 Network design

Network design is a complicated and specialist science. Computer users do not typically want to re-invent the wheel by writing all the network facilities they require from scratch. They expect their computer supplier to have such software available for rent or purchase, and, indeed, most large computer suppliers have responded with their own offerings.

There are two main network designs in use in the computer industry today.

SNA IBM's Systems Network Architecture (SNA) is a hierarchical network that dominates the many networks hosted by IBM mainframes throughout the world. It is a tried and trusted product developed over a number of years.

DNA Digital Equipment Corporation's Digital Network Architecture (DNA), often referred to by the name of one of its components—DECNET—is a peer-to-peer network, first announced in 1975. Since that time, as with SNA, it has been subject to constant update and development with particular emphasis recently on compliance with OSI.

60.18.12 Standard network architecture

The fact remains, of course, that there is still no standard network architecture permitting any system to talk to any other. Hence the need for emulators. However, the PTTs of the world have long recognised this need and are uniquely placed as the suppliers of the physical links to bridge the gap created by the computer manufacturers. They have developed the concept of public 'packet-switched networks' to transmit data between private computers or private networks.

Public packet-switched networks (PPSNs) Packet switching involves breaking down the message to be transmitted in to 'packets' which are 'addressed' and introduced into the network controlled by the PTT. Consequently, the user has no influence over the route the packets take. Indeed, the complete contents of a message may arrive by several different routes. Users are charged according to line usage and the result is generally greater flexibility and economy. The exception is the case where a user wants to transmit very high volumes of data regularly between two points. In this instance, a high-speed leased line would probably remain the most viable option.

What goes on inside the network should not concern the subscriber, provided the costs, response times and accuracy meet expectations. What does concern the user is how to connect to the network. There are basically two ways of doing this.

(1) If using a relatively unintelligent terminal, the user needs to connect to a device which will divide the message into packets and insert the control information. Such a device is known as a PAD (packet assembler/disassembler) and is located in the local packet-switching exchange. Connection between the terminal and the PAD may be effected using dedicated or dial-up lines.
(2) More sophisticated terminals and computer equipment may be capable of performing the PAD function themselves in which case they will be connected to the network via a line to the exchange, but without the need to use the exchange PAD.

The CCITT has put forward recommendation X25 ('Interface between data terminal equipment for terminals operating in the packet mode in public data networks') with the aim of encouraging standardisation. X25 currently defines three levels within its recommendations:

(1) The *physical* level defines the electrical connection and the hand-shaking sequence between the data terminals equipment (DTE) and the data communications equipment (DCE).
(2) The *link* level describes the protocol to be used for error-free transmission of data between two nodes. It is based on the HDLC protocol.
(3) The *packet* level defines the protocol used for transmitting packets over the network. It includes such information as user identification and charging data.

60.18.13 Data terminal equipment

The basic all-round terminal, the teleprinter, has now been almost completely superseded by the video terminal.

VDUs may be clustered together in order to optimise the use of a single communications line. In this instance a controller is required to connect the screens and printers to the line.

Batch terminals are used when a high volume of non-interactive data is to be transmitted. Most commonly the input medium is punched cards with output on high-speed line printers. As with VDUs it is quite feasible to build intelligence into batch terminals in order to carry out some local data verification and local processing. However, the middle of this decade saw the large-scale introduction of very small but powerful free-standing micros, and these are now in common use as local preprocessors in communication with larger processors at remote sites. The advantage of this method is that the raw data, usually in punched card or magnetic tape format, can be read on to the local machine, verified, reformatted if required, and then transmitted to the central site and its processing initiated; all automatically done by the local micro.

In addition to these commonly found terminals there are a host of special-purpose devices including various types of optical and magnetic readers, graphics terminals, hand-held terminals, badge readers, audio response terminals, point-of-sale terminals and more.

Finally, of course, computers can communicate directly with each other without the involvement of any terminal device.

60.19 Software

60.19.1 Introduction

Software is the collective name for programs. Computer hardware is capable of carrying out a range of functions represented by the instruction set. A program (the American spelling is usually used when referring to a computer program) simply represents the sequence in which these instructions are to be used to carry out a specific application. However, this is achieved in a number of ways. In most cases the most efficient way of using the hardware is to write in a code which directly represents the hardware instruction set. This is known as machine code and is very machine dependent. Unfortunately, it requires a high level of knowledge of the particular type of computer in use, and is time consuming. In practice, therefore, programmers write in languages in which each program instruction represent a number of machine instructions. The programs produced in this 'high-level language' clearly require to be translated into code which can operate upon the computer's instruction set.

It would be possible, of course, to buy computer hardware and then set out to write every program needed. However, this would take a very long time indeed. Most users require their system to perform the same set of basic functions such as reading, printing, storing and displaying data, controlling simultaneous processes, translating program and many others. Consequently, most computers are supplied with pre-written programs to carry out these functions. They fall into three basic categories: operating systems, data management systems and language translators.

60.19.2 The operating system

The operating system sits between the application program designed to solve a particular problem and the general-purpose hardware. It allocates and controls the system's resources such as the CPU memory, storage and input/output, and allocates them to the application program or programs.

Part of the operating system will be permanently resident in main memory and will communicate with the operator and the programs that are running. The functions it will carry out will typically be:

(1) The transfer into memory of non-resident operating system routines.
(2) The transfer into memory of application programs or parts of application programs. In some cases, there is insufficient memory to hold an entire program and so little used portions are held on disk and 'overlaid' into memory as they are required.
(3) The scheduling of processor time when several programs are resident in memory at the same time.
(4) The communication between tasks. For ease of programming, a large program may be broken down into sections known as tasks. In order to complete the application it may be necessary to transfer data from task to task.
(5) Memory protection, ensuring that co-resident programs are kept apart and are not corrupted.
(6) The transfer of data to and from input and output devices.
(7) The queuing of input/output data until the appropriate device or program is ready to accept it.

There are several ways to use the resources of a computer system and each makes different demands on an operating system. The four main distinctions are as follows.

60.19.2.1 Batch processing

This was the original processing method and is still heavily used where large amounts of data have to be processed efficiently. Data is transcribed onto some input medium such as punched cards or magnetic tape, along with some checking data such as column to tab, and then run through the system to produce, typically, a printed report. Classical batch jobs include such applications as payroll and month and statement runs.

Batch operating systems require a command language (often known as JCL—job control language) which can be embedded between the data, and which will load the next program in the sequence. Jobs are frequently queued on disk before being executed and the operating system may offer the facility of changing the sequence in which jobs are run, either as a result of operator intervention or as a result of pre-selected priorities.

Many operating systems are now capable of running multiple batch 'streams' at the same time, and even of selecting which batch streams a particular job should run in. Thus, the person submitting the job is instructing the computer to run it under the best possible circumstances without necessarily knowing in advance where it will be run. This technique is particularly effective in a clustered environment, where batch streams may run across an entire cluster, and the operating system will not only choose the best stream but will also select the best processor for it to run on. Many current JCLs are almost programming languages in their own right, with tremendous flexibility offered to the person submitting the job. However, there is a cost to pay for this flexibility, since the language is translated into machine code at the time of running (this is referred to as an 'interpretive' language), which is much slower than executing a precompiled language. Generally, though, the ratio of instructions to data to be processed is low and this disadvantage is not considered significant.

The advantage of batch processing is its efficiency in processing large amounts of data. The major disadvantage is that once a user has committed a job he or she must wait until the cycle is completed before receiving any results at all. If they are not correct the job must be resubmitted with the necessary amendments. Some operating systems, however, do permit intermediate 'break points' in a job, so that results so far can be obtained, and the job restarted where it suspended without any loss of data. Others allow a batch job to submit data to another batch job for processing, which is very useful if the other batch stream exists to serve a printer, since intermediate results can then be printed without suspending or affecting the running of the original job submitted.

60.19.2.2 Interactive processing

This involves continuous communication between the user and the computer—usually in the form of a dialogue. The user frequently supplies data to the program in response to questions printed or displayed on the terminal, whereas in batch processing all data must be supplied, in the correct sequence, before the job can be run.

A single operator using a keyboard does not use the power of a computer to any more than a fraction of its capacity. Consequently, the resources of the system are usually shared between many users in a process known as *time-sharing*. This should not be apparent to the individual user who should receive a response to any request in two or three seconds. Time-sharing, as the name suggests, involves the system allotting 'time slices', in rotation, to its users, together with an area of memory. Some users may have a higher priority than others and so their requests will be serviced first. However, all requests will be serviced eventually.

Requirements of interactive time-sharing operating systems are efficient system management routines to allocate, modify and control the resources allocated to individual users (CPU time and memory space) and a comprehensive command language. This language should be simple for the user to understand and should prompt the inexperienced operator whilst allowing the experienced operator to enter his commands swiftly and in abbreviated format. To overcome the lack of typing and formatting skills of the average programmer, a new tool has been developed called the 'language sensitive editor' (LSE). This checks what the programmer is keying in as part of a program's coding, as he or she keys it in, for spelling, syntax and format, and highlights any errors at the time of entry.

16.19.2.3 Transaction processing

This is a form of interactive processing which is used when the operations to be carried out can be predefined into a series of structured transactions. The communication will usually take the form of the operators 'filling out' a form displayed on the terminal screen, a typical example being a sales order form. The entered data is then transmitted as a block to the computer which checks the data and sends back any incorrect fields for correction. This block method of form transmission back to the computer is very efficient from a communications perspective, but can be inefficient from the point of view of the terminal operator if there are many fields in error or if the validation of any of the fields is dependent on the contents of other fields on the same form. Some systems, therefore, send back the input character by character and are able to validate any field immediately and not let the operator proceed past a field until it is correct. The options available to the operator will always be limited and he or she may select the job to be performed from a 'menu' displayed on the screen.

Typical requirements of a transaction processing operating system are:

(1) Simple and efficient forms design utilities.
(2) The ability to handle multidrop terminals (terminals 'dropped off' a communications line). This may be achieved by a time-sharing approach or by the central processor 'polling' the terminals in search of a 'message ready' signal.
(3) Efficient file management routines, since many users will be accessing the same files at the same time.
(4) Comprehensive journaling and error recovery. Journaling is the recording of transactions as they occur, so that in the event of a system failure the data files can be updated to the point reached at the moment of failure from a previously known state of the system, usually a regular backup.

60.19.2.4 Real time

This is an expression sometimes used in the computer industry to refer to interactive and transaction processing environments. Here it is used to mean the recording and control of processes. In such applications, the operating system must respond to external stimuli in the form of signals from sensing devices. The system may simply record that the event has taken place, together with the time at which it occurred, or it may call up a program which will initiate corrective action, or pass data to an analysis program.

Such a system can be described as 'event' or 'interrupt' driven. As the event signal is received, it will interrupt whatever processing is currently taking place, provided that it has a higher priority. Interrupt and priority handling are key requirements of a real-time operating system. Some operating systems may offer the user up to 32 possible interrupt levels and the situation can arise in which a number of interrupts of increasing priority occur before the system can return to the program that was originally being executed. The operating system must be capable of recording the point reached by each interrupted process so that it can return to each task according to its priority level.

There are some concepts which are common to most operating systems.

Foreground/background The simplest form of processing is 'single user', either batch or interactive. However, a more effective use of the computer resources is to partition the memory into two areas. One, background, is used for low-priority, interruptable programs. The other, foreground, is occupied by a program requiring faster response to its demands. The latter will therefore have higher priority. The recent increases in both the memory and power of many mini and micro systems has relinquished the use of this technique to the smaller end of the micro range and in particular to home-based PCs.

Multiprogramming This is an extension of foreground/background in which many jobs compete for the system's resources rather than just two. Only one task can have control of the CPU at a time. However, when it requires an input or output operation, it relinquishes control to another task. This is possible because CPU and I/O operations can take place simultaneously. For example a disk controller, having received a request from the operating system, will control the retrieval of data, thus releasing the CPU until it is ready to pass on the data it has retrieved.

Bootstrapping The operating system is normally stored on a systems disk or a ROM. When the computer is started up, the monitor (the memory resident position of the operating system) must be read from storage into memory. The routine which does this is known as the bootstrap.

System generation (sysgen) When a computer is installed or modified, the general-purpose operating system has to be tailored to the particular hardware configuration on which it will run. A sysgen defines such things as the devices which are attached to the CPU, the optional utility programs which are to be included and the amount of memory available and the amount to be allocated to various processes.

It is unlikely that any single operating system can handle all the various processing methods if any of them is likely to be very demanding. An efficient batch processing system would not be able to handle the multiple interrupts of a real-time operating system. There are, however, multipurpose systems which can handle batch and interactive, interactive and real time.

Data management software Data to be retained is usually held in auxiliary storage rather than in memory, since if it were held in memory it would be lost when the system was turned off and, moreover, even with current memory prices, the cost would prove prohibitive. To write and retrieve the data quickly and accurately requires some kind of organisation and this is achieved by data management software. This is usually provided by the hardware manufacturer, although independent software houses do sell such systems which they claim are more efficient or more powerful or both.

60.19.2.4 File organisation

The most commonly used organisational arrangement for storing data is the file structure. A file is a collection of related pieces of information. An inventory file, for example, would contain information on each part stored in the warehouse. For each part would be held such data as the part number, description, quantity in stock, quantity on order and so on. Each of these pieces of data is called a field. All the fields for each part form a record and, of course, all the inventory records together constitute the file.

The file is designed by the computer user, though there will usually be some guidelines as to its size and structure to aid swift processing or efficient usage of the storage medium. With file management systems the programs using the files must understand the type of file being used and the structure of the records with it. There are six types of file organisation, sequential, relative, physical, chain, direct and indexed.

Sequential file organisation Before the widespread use of magnetic storage devices, data was stored on punched cards.

The program would cause a record (punched card) to be read into memory, the information was updated and a new card punched. The files thus created were sequential, the records being stored in numeric sequence. A payroll file, for example, would contain records in employee number sequence.

This type of file organisation still exists on magnetic tapes and disks. However, the main drawback is that to reach any single record, all the preceding records must be read. Consequently, it has application only when the whole file is processed from beginning to end, and random enquiries to individual records are rarely made.

Relative file organisation Relative files permit random access to individual records. Each record is numbered according to its position relative to the first record in the file and a request to access a record must specify its relative number. Unfortunately, most user data, such as part number, order number, customer number and so on, does not lend itself to such a simplistic numbering system.

Physical file organisation Another version of the relative technique is used to retrieve a specific 'block' of data relative to the first block in a file from disk. This is done irrespective of where the actual data records reside in the block and it would be the responsibility of the application program, not the operating system, to separate out individual records (unpacking). Consequently, situations where this method are advantageous are rare, but if the record size equals that of a physical block on disk then this technique offers considerable advantages in speed of retrieval of the data, particularly if the file is in a physically continuous stream on the disk. This type of file is often referred to as a 'physically direct' file.

Chain file organisation This is in effect a file that is required to be read sequentially, but in which not all of the data is available at one time. Earlier file systems did not permit the extension of a sequential file once it was written, and adding data to a file meant reading the whole file, writing it out to a new file as it was read, and then adding the new data on to the end of the new file. To overcome this limitation, the chain file technique was introduced. Each record was written to the file using relative file techniques with the application specifying where each record was to be written to. However, each record contained a pointer to the location of the next record in logical (not physical) sequence in the file, or some method of indicating that there were no more records in the chain (usually a zero value pointer). This then enabled the application program to read the file in sequence irrespective of where the data resided on disk or when it was put there.

The widespread use of sequential files that can be extended, coupled with a vast improvement in database and indexed file techniques, has largely made this technique redundant.

Direct (hashed) file organisation This is a development of the relative file organisation and is aimed at overcoming its record numbering disadvantage. The actual organisation of the file is similar. However, a hashing algorithm is introduced between the user number identifying a particular record and the actual relative record number which would be meaningless to the user. The algorithm is created once and for all when the system is designed and will contain some arithmetic to carry out the conversion (*Figure 60.18*).

This file organisation permits very fast access but it does suffer from the disadvantage that most algorithms will occasionally arrive at the same relative record number from different user record identification numbers, thus creating the problem of 'synonyms'. To overcome this problem the file management software must look to see if the record position indicated by the algorithm is free. If it is, a new record can be stored there. If it is not, a synonym has occurred and the software must look for another available record position. It is, of course, necessary to create a note that this has occurred so that the synonym can subsequently be retrieved. This is usually achieved by means of pointers left in the original position indicating the relative record number of the synonym.

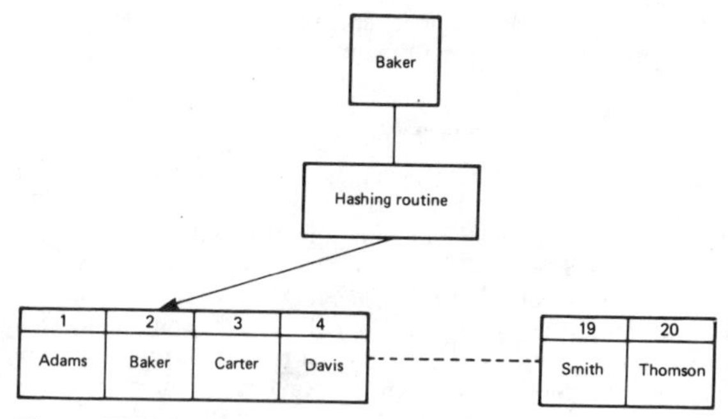

Figure 60.18 Direct (hashed) file organisation

The user numbering possibilities permitted with direct files may be more acceptable to the user since they are not directly tied to the relative record number. However, the need for an algorithm means that these possibilities are limited. In addition the design of the algorithm will affect the efficiency of recording and retrieval since the more synonyms that occur the slower and more cumbersome will be these operations.

Indexed file organisation The indexed method of file organisation is used to achieve the same objectives as direct files, namely the access to individual records by means of an identifier known to the user, without the need to read all the preceding records.

It uses a separate index which lists the unique identifying fields (known as keys) for each record together with a pointer to the location of the record. Within the file the user program makes a request to retrieve part number 97834, for example. The indexed file management software looks in the index until it finds the key 97834 and the pointer it discovers there indicates the location of the record. The disadvantage of the system is fairly apparent; it usually requires a minimum of three accesses to retrieve one record and is therefore slower than the direct method (assuming a low incidence of synonyms in the latter). However, there are a number of advantages.

(1) It is possible to access the data sequentially as well as randomly, since most data management systems chain the records together in the same sequence as the index by maintaining pointers from each record to the next in sequence. Thus we have indexed sequential or ISAM (indexed sequential access method) files.
(2) Depending on the sophistication of the system, multiple keys may be used, thus allowing files to be shared across different applications requiring access from different key data (*Figure 60.19*).
(3) Additional types of keys can be used. Generic keys can be used to identify a group of like records. For example, in a payroll application, employee number 74639 may identify K Jones. However, the first two digits (74) may be used for all employees in the press shop. It is therefore possible to list all employees who work in this department by asking the software to access the file by generic key.

Another possibility is that of asking the system to locate a particular record that contains the key value requested, or the next highest, if the original cannot be found. This is

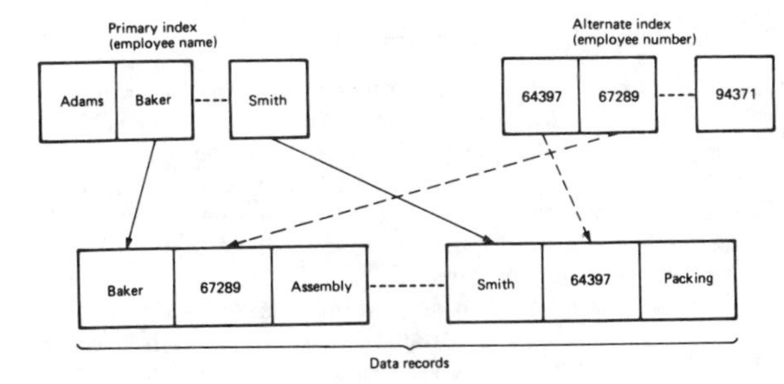

Figure 60.19 Multi-key 'ISAM' file organisation

known as using approximate keys. It is also possible to retrieve records within a given range of keys.

(4) Most computer manufacturers provide multi-key ISAM systems and so the user does not need to be concerned about the mechanics of data retrieval.

60.19.3 Database management

Files tend to be designed for specific applications. As a result, the same piece of information may be held several times within the same system. This has two disadvantages. Firstly it is wasteful of space and effort. Secondly it is very difficult to ensure that the information is held in its most recent form in every location.

It is, of course, possible to share files across applications. However, a program usually contains a definition of the formats of the data files, records and fields it is using. Changes in these formats necessitated by the use of the data within new programs will result in modifications having to be made in the original programs.

The database concept is designed to solve these problems by separating the data from the programs that use it. The characteristics of a database system are: a piece of data is held only once; the data are defined so that all parts of the organisation can use them; it separates data and their description from application programs; it provides definitions of the logical relationships between records in the data so that they need no longer be embedded in the application programs; and it should provide protection of the data from unauthorised changes and from hardware and software. The data definitions and the logical relationships between pieces of data (the data structures) are held in the schema (*Figure 60.20*).

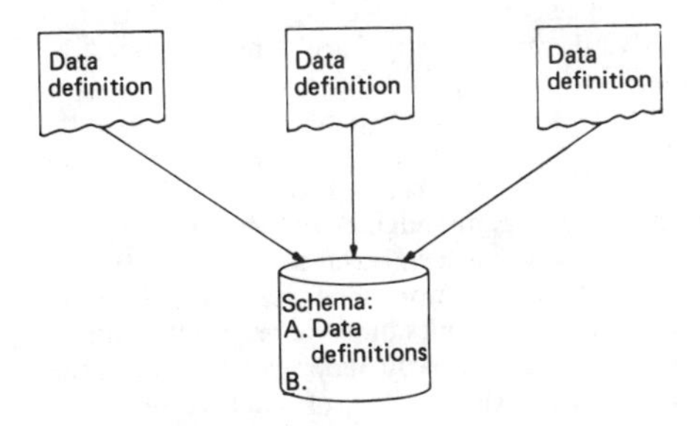

Figure 60.20 The schema

The database is divided into realms—the equivalent of files—and the realms into logical records. Each logical record contains data items which may not be physically contiguous.

Records may be grouped into sets which consist of owner and member records. For example, a customer name and address record may be the owner of a number of individual sales order records.

When an application is developed a subschema is created defining the realms to be used for that application. The same realm can appear in other subschemas for other applications (*Figure 60.21*).

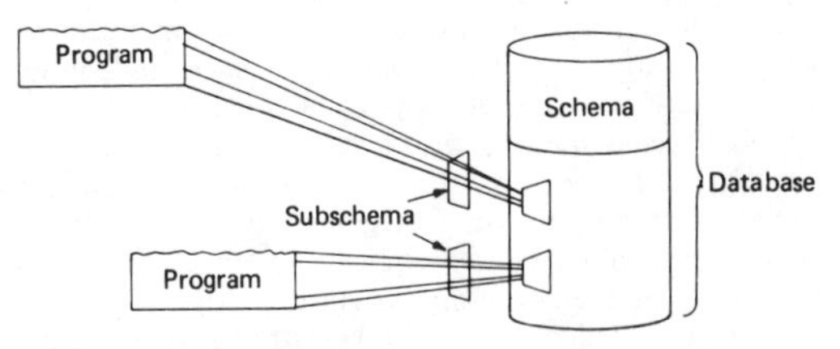

Figure 60.21 The subschema

There are four major definitions of the logical relationships between the data:

(1) *Sequential* (sometimes known as chain) Here each record is related only to the immediately preceding and following records.

(2) *Tree or hierarchical* In this structure each record can be related to more than one record following it. However, records along separate 'branches' are not directly linked with each other and the relationship can be traced only by travelling along the branches.

(3) *Networks* These are the most complex structures. They are effectively groups of trees where records can be related across branches. Any record can, in fact, be related to any other.

(4) *Relational* A relational database stores data in two-dimensional tabular arrays. Each table (file) of the database is referred to as a *relation*, and each row of the table (record) is referred to as a *tuple*. Through the use of normalisation (the successive breaking down of data into groups of two-dimensional arrays where each group's data is functionally dependent upon the group's key) the data is defined in a logical format suitable for use with a relational database. The result is a totally 'flat' file which, using a relational database management system, has the flexibility to dynamically create new relations from extracts of one or more existing relations.

Because, within a database management system, data is separated from the programs that use it, the data is regarded as a corporate asset. Management of this asset is in the hands of a database administrator, who maintains the schema and works with application programmers to define the parts of the database to which they may have access and to help them create subschemas for their particular applications.

Codasyl began to take an interest in developing database standards in 1965 when the database task group was formed. There now exist Codasyl standards for database design.

60.19.4 Language translators

A programming language is a convention comprising words, letters and symbols which have a special meaning within the context of the language. However, programs have to be translated into the binary language understood by computers. The programmer writes a 'source' program which is converted by the language translator into an 'object program'. Usually during this process, checks are made on the syntax of the source program to ensure that the programmer has obeyed the rules. Any errors discovered will be noted, usually in two categories—terminal and warning. Terminal errors are those which will prevent the program running. Warning errors indicate that the translator would have expected something different, but that it may not actually be an error. Program errors are known as 'bugs' and the process of removing them as 'debugging'.

Programs are normally stored in both their original code format, known as 'source', and in their final format, which is understandable and executable directly by the computer, known variously as 'object', 'binary' or 'executable' formats. On disk, programs are often stored as individual files since they can be accessed directly and quickly. However, to store individual programs on individual magnetic tapes would use up too many tapes and have computer operators constantly loading and unloading them. To overcome this limitation, the 'library' principle was introduced. Using this method many different program files are combined into a single file called a library, and facilities exist within the operating system to extract whichever program is next needed from the library without changing the magnetic tape.

The working program is the object program but, when changes have to be made, these will be made to the source program which will then be translated to produce a new object program.

There are two kinds of language translators, assemblers and compilers.

60.19.4.1 Assemblers

An assembler is a language processor designed for use on a particular type of computer. In assembly language there is a one-to-one relationship between most of the language mnemonics and the computer binary instructions, although predefined sets of instructions can be 'called' from the assembly program.

There are four parts to an assembly language instruction:

(1) *Label* This is a name defined by the programmer. When he or she wants to refer to the instruction, this can be done by means of the label.
(2) *Operation code* This will contain a 'call' or an instruction mnemonic. If a call is used the assembler will insert a predefined code during the assembly process. If the programmer used a mnemonic this will define the operation to be carried out.
(3) *Operand* This represents the address of the item to be operated on. An instruction may require one or two operands.
(4) *Comment* This is an optional field used for ease of interpretation and correction by the programmer.

Assembly languages are generally efficient and are consequently used for writing operating systems and routines which require particularly rapid execution. However, they are machine dependent, slow to write and demanding in terms of programmer skill.

60.19.4.2 Compilers

These are used to translate high-level languages into binary code. These languages are relatively machine independent, though some modifications are usually required when transferring them from one type of computer to another. The instructions in the high-level language do not have a one-for-one relationsip with the machine instructions.

Most compilers read the entire source program before translating. This permits a high degree of error checking and optimisation. An incremental compiler, however, translates each statement immediately into machine format. Each statement can be executed before the next is translated. Although it does not allow code optimisation it does check syntax immediately and therefore permits the programmer to correct errors as they occur. These incremental compilers are often referred to as 'interpreters' since they interpret and then act upon each instruction.

60.19.5 Languages

All computers work with and understand instructions in the same format—binary. The content of an instruction to achieve the same objective may well differ from machine to machine, but the instruction will be coded in a binary format. This first format of computer language is called the first generation. It is almost unheard of today to find anyone other than compiler writers who work at this level.

Programming in this first generation of languages was complex, lengthy, skilful and extremely prone to error. It became obvious that improvements were essential and that assistance was required. What better tool to assist than the very computer that was being programmed? If a code system could be developed that was easier to write and read, and which could then be translated by the computer into machine instructions, programming would become easier. These languages formed the second generation of computer languages and are referred to as 'assembly' languages (from the action of the translator of assembling everything together to validate it). Many of the original languages have long since gone or have been replaced by easier and more powerful substitutes, but assembly languages are still the best language for writing programs where flexibility and speed of execution are paramount (such as the operating systems themselves).

Third-generation languages were developed to bring the nature and structure of the instructions much nearer to the programmer's native language, principally to speed up the programming process and to reduce the level of skill required. Since many of these languages were developed in the UK or the USA, the 'native' language was English, and this is still the predominant, almost universal, language of computing today. Some examples of third generation languages are given below.

60.19.5.1 BASIC

BASIC (beginners' all-purpose symbolic instruction code) is an easy to learn, conversational programming language which enables beginners to write reasonably complex programs in a short space of time. The growth in the popularity of time-sharing systems has increased its use to the point where it is used for a whole range of applications from small mathematical problems, through scientific and engineering calculations, and even to commercial systems.

A BASIC program consists of numbered statements which contain English words, words, symbols and numbers such as 'LET', 'IF', 'PRINT', 'INPUT', * (multiply), + and so on.

BASIC was developed at Dartmouth College and while there is a standard BASIC there are many variations developed by different manufacturers.

60.19.5.2 FORTRAN

FORTRAN (formula translation) originated in the 1950s and was the first commercially available high-level language. It was designed for technical applications and its strengths lie in its mathematical capabilities and its ability to express algebraic expressions. It is not particularly appropriate when the application requires a large amount of data editing and manipulation.

In 1966 an attempt was made by ANSI to standardise the FORTRAN language and in 1977 it confirmed FORTRAN 77 as a standard. However, manufacturers have continued to develop their own extensions.

A FORTRAN program consists of four types of statement:

(1) Control statements (such as GOTO, IF, PAUSE and STOP) control the sequence in which operations are performed.

(2) Input/output statements (such as READ, WRITE, PRINT, FIND) cause data to be read from or written to an I/O device.
(3) Arithmetic statements such as * (multiplication), ** (exponentiation), / (division), perform computation.
(4) Specification statements define the format of data input or output.

60.19.5.3 COBOL

COBOL (common business-oriented language) is the most frequently used commercial language. The first Codasyl (Conference of Data Systems Languages) specifications for COBOL were drawn up by 1960, the aims of which were to create a language that was English-like and machine independent.

COBOL is a structured language with well defined formats for individual statements. A COBOL program consists of four divisions:

(1) Identification division, which names and documents the program.
(2) Environment division, which defines the type of computer to be used.
(3) Data division, which names and describes data items and files used.
(4) Procedure division, which describes the processing to be carried out.

The 'sentences' within COBOL can contain 'verbs' such as ADD, SUBTRACT, MULTIPLY and DIVIDE and are readable in their own right. For example, in an invoicing program you may find the line:

IF INVOICE TOTAL IS GREATER THAN 500 THEN GO TO DISCOUNT ROUTINE

As a result, by intelligent use of the language, the programmer can produce a program which is largely self-documenting. This is a significant advantage when modifications have to be made subsequently, possibly by a different programmer.

A general-purpose, structured language like COBOL is not as efficient in terms of machine utilisation as assembly language or machine code. However, in a commercial environment, programmer productivity and good documentation are generally the most important factors. The calculations are usually not complex and therefore do not require great flexibility in terms of number manipulation.

60.19.5.4 PL1

PL1 (programming language one) is primarily an IBM language. PL1 was introduced to provide greater computational capabilities than COBOL but better file handling than FORTRAN.

60.19.5.5 APL

APL (a programming language) was introduced through IBM. It is generally used interactively. At one end of the scale it allows the operator to use the terminal as a calculator. At the other, it enables him or her to perform sophisticated operations such as array manipulations with the minimum of coding.

60.19.5.6 RPG

RPG (report program generator) was introduced by IBM for 360 and System 3 machines originally. A commercially orientated, very structured language, it processes records according to a fixed cycle of operations, although developments have been made in the language to make it more suitable for transaction processing environments.

60.19.5.7 Pascal

This is a strongly typed, block-structured, procedural third-generation programming language, which is becoming more popular. This may be due to its extensive use in educational establishments in recent years.It is now widely used for a variety of differing applications in commercial environments.

Pascal lends itself readily to modern design techniques (including top-down, stepwise refinement, etc.) and has built into it the necessary building blocks to make the most of structured programming. Additionally, because of its strong enforcement of data type, scope and syntax rules, Pascal is a language less prone to programmer errors than many others.

All of the above points assist programmers in writing well designed programs and producing very supportable code.

60.19.5.8 Fourth-generation languages

Fourth-generation languages (4GLs) are the closest yet to native language, enabling a person with very little knowledge or skill in a programming language to write instructions performing complex and lengthy tasks on a computer. They do not generally produce code that is as 'efficient' as previous generations, and the amount of machine resource that is required to support their use is sometimes either misunderstood by those using them or misrepresented by those selling them, and very often both.

Unlike previous generations of programming languages, few 4GLs demand that the instructions given to it need to follow the exact procedure and sequence in which it is required to execute. They are capable of breaking complex English-like statements down into many component instruction parts and then deducing the optimum order in which to carry them out. For this reason they are often referred to generically as 'non-procedural' languages.

These languages are most powerful when used in conjunction with some form of database management system, since they can simplify the interface between the programmer and program by undertaking all necessary communication with the database system. Many of these language/data interfaces have developed so far that the user need not know, and will often be unaware of, how the data is organised or formatted within the system.

Some examples of these languages are Powerhouse, Focus, SQL, Datatrieve and Query.

In broad terms, the increased speed of processing and data handling currently available, and the low cost of memory have reduced the pressures on programmers to code for maximum speed and efficiency. It is frequently more economic to spend more money on hardware than to allow programmers to spend time optimising the performance of their programs.

This, coupled with the shortage of trained programmers has resulted in increased emphasis on simple languages, good program development tools and a general emphasis on programmer productivity.

Further reading

BELITOS, B. and MISRA, J., *Business Telematics (Corporate Networks for the Information Age)*, Dow Jones/Irwin, Homewood, Ill. (1986)

BOND, J., 'New Architectures give supercomputers power to mirror reality', *Comp. Design*, December (1987)

CHANG, S. S. L., *Fundamentals Handbook of Electrical Computer Engineering*, John Wiley & Sons (1982)

CLARK, W. J. *et al.*, 'Electronic memories come in tough little packages', *Comp. Design*, 1 November (1988)

FICHTNER, W. and MORF, M., *VSLI CAD Tools and Applications*, Kluwer Academic, Lancaster (1987)

FREEDMAN, D., 'Optical disks: promises and problems', *Mini-Micro Systems*, October (1982)

GALLANT, J., 'Parallel processing ushers in a revolution in computing', *EDN*, 1 September (1988)
HANNINGTON, S., 'RISC becomes a reality', *New Electron.*, July/August (1988)
HOROWITZ, E., *Programming Languages—A Grand Tour*, Computer Science Press, Rockville, MD (1987)
JUDGE, P., 'Optical computers', *New Electron.*, April (1988)
KENEALY, P., 'Personal computers: a new generation emerges'. *Mini-Micro Systems*, August (1982)
LERNER, E. J., 'Automating programming', *IEEE Spectrum*, August (1982)
LIEBERMAN, D., 'In search of the high-performance controller', *Comp. Design*, 15 September (1988)
MACLENNAN, B. J., *Principles of Programming Languages*, CBS College Publishing, New York (1987)
MAEKAWA, M., OLDEHOEFT, A. E. and OLDEHOEFT, R. R., *Operating Systems, Advanced Concepts*, Benjamin Cummings (1987)
MALLACH, E. G., 'Computer architecture', *Mini-Micro Systems*, December (1982)
MANUEL, T., 'Molding computer terminals to human needs', *Electronics*, June (1982)
MILUTINOUIC, V. M., Computer Architecture, North-Holland, Amsterdam (1988)
ORMOND, T., 'Keyboards enhance their flexibility with cost effective key switch technologies', *EDN*, 23 June (1988)
PLATSHON, M., 'Acoustic touch technology adds a new input dimension', *Comp. Design*, 15 March (1988)
REED, J. S., 'Computer graphics', *Mini-Micro Systems*, December (1982)
STALLINGS, W., *Handbook of Computer Communications Standards*, Vol. 1, *The Open Systems Interconnection (OSI) Model and OSI-Related Standards*, Howard W. Sams, Indianapolis, IN (1987)
STALLINGS, W., *Handbook of Computer Communications Standards*, Vol. 2, *Local Network Standards*, Howard W. Sams, Indianapolis, IN (1987)
SUZUKI, J., 'Recent trends in raster-scan graphic displays', *JEE*, June (1982)
TILL, J., 'Computer system architecture', *Electron. Design Int.*, January (1989)
TORREDO, E. A., *Next Generation Computers*, IEEE Press, New York (1985)
WANG, P. C. C., *Advances in CAD/CAM Workstations—Case Studies*, Kluwer Academic, Lancaster (1986)
WESSELY, H. *et al.*, 'Computer packaging', *Siemens Res. Dev. Rep.*, **17**, No. 5 (1988)
WILLIAMS, T., 'Peripherals and memory systems chase computer price/performance', *Comp. Design*, December (1987)
WILSON, L. B. and CLARK, R. G., *Comparative Programming Languages*, Addison-Wesley, Wokingham (1988)
WILSON, R., 'New CPUs spearhead Intel attack on embedded computing', *Comp. Design*, 15 April (1988)
WILSON, R., 'RISC architectures take on heavyweight applications', *Comp. Design*, 15 May (1988)

61 Videotape Recording

S Lowe
Ampex Great Britain Ltd

Contents

61.1 Introduction

The frequency spectrum occupied by a video signal extends from a few cycles to around 5.5 MHz. This is a range of 18 octaves and well in excess of the 10 octave limit of magnetic recording. Videotape recorders modulate the signal onto a carrier or use digital techniques to overcome the bandwidth limitations. Each system comprises a format which is specified in order to allow tapes to be exchanged between different users and different makes of machine.

Frequency modulation is the major form used, with the addition of amplitude modulation for part of the signal in some types of machine. Digital video recording is used for high-performance machines, especially where multigeneration work is required, as in editing (*Figure 61.1*).

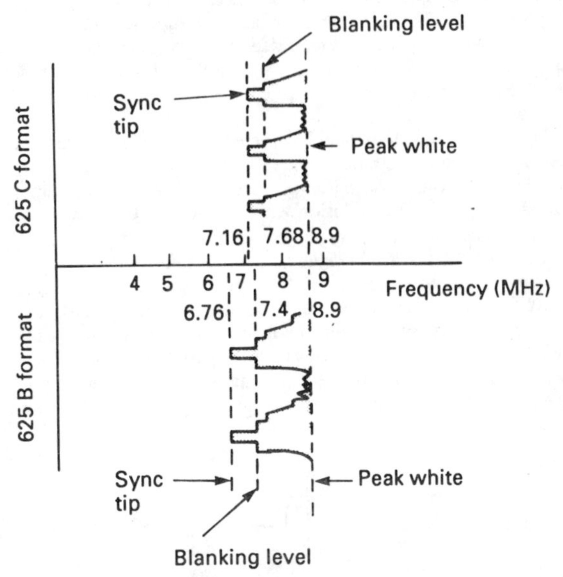

Figure 61.1 Frequencies used in (a) EBU C format, (b) EBU B format videotape recording systems

61.2 Tape transport

In a broadcast-quality machine, the video head has to record and replay signals with sidebands up to 20 MHz. Domestic and industrial machines are limited to around 10 MHz. These high frequencies may easily be recorded and replayed provided that the relative speed of the head across the tape is high enough. The wavelength of the magnetic signal must be large when compared with the size of the magnetic particles and the replay head gap.

Tape is capable of recording wavelengths of 1 μm. The head-to-tape speed is therefore determined by the highest frequency to be recorded and is found to be around 25 m/s (1000 in/s) for broadcast machines and 8 m/s (300 in/s) for domestic and industrial machines. Longitudinal tape transport at these speeds is impractical. Quite apart from the difficulty of moving tape at this speed it would require a reel containing 90 km (56 miles) of tape to record 1 h of programme.

To overcome this problem, videotape recorders have adopted a system of moving video heads which lay down parallel tracks across the tape as it moves from reel to reel. The tape may then move at speeds similar to those of audio recorders. In fact, the audio systems of most videotape recorders are identical to those of audio recorders, with the tape moving at speeds ranging from under 2 cm/s (0.7 in/s) in the case of the domestic Betamax machines to 38 cm/s (15 in/s) for the first broadcast machines.

The first machines used a rotating drum with four video heads equally spaced around it. For this reason they were called quadruplex VTRs. The tape was 2 in wide and the video tracks were laid across the tape almost perpendicular to the edge. Each head recorded a small band of the picture 16 lines high. It was essential to ensure that each head and its associated record and replay channels were identical in performance, or these bands became visible on the screen.

The segmented nature of the recording made it difficult to arrange for slow-motion or still-frame pictures. The next generation of machines, helical scan, used a single head to record a complete television field. This overcame the problem of banding errors, a major restriction on multigeneration work, and allowed still frame or slow motion without the addition of costly field stores.

The helical scan machines lay their tracks at a much shallower angle across the tape. Typically this will be between 2.5° and 5°. The tracks are much longer, up to 400 mm (16 in) and the tape is narrower, 6.25 mm (0.25 in) to 25.4 mm (1 in) (*Figure 61.2*).

61.3 Formats

The various mechanical arrangements have given rise to a number of standard formats. These are internationally agreed track patterns, frequencies and other parameters that ensure that a recording made in one place may be replayed in another.

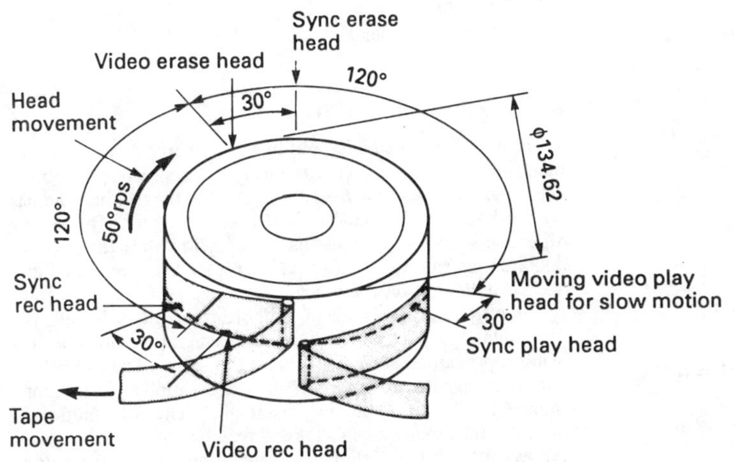

Figure 61.2 C format helical tape path

The first format that achieved worldwide recognition is still in use 25 years or more after its inception. It is the Quad I format which has been mentioned. A Quad II format was specified, and achieved some use. Its main points were that the tape consumption was cut in half by slowing down the capstan and halving the video track widths. The record frequencies were increased and a pilot tone was added which made it possible to monitor the video signal continuously and apply corrections wherever necessary. The output pictures were virtually as good as the input but the operational flexibility of the helical scan machines and their cheapness won out.

The first broadcast helical format was A format, which in a modified form became the now universal C format. One inch wide tape is moved at about 9 in/s past a rotating drum on which are mounted video erase, record and replay heads. For each rotation of the drum the video head lays down or replays the active picture lines of one television field. When the video head reaches the end of a track at the edge of the tape it moves to the start of the next track at the other edge. While it is out of contact with the tape no signals are recorded and a gap, the format dropout, occurs in the video. An optional extra head may be used to record the missing part of the signal.

There are up to four longitudinal audio tracks which are laid along the edge of the tape. A pulse track is also laid along the edge and this is used to control the capstan and drum servo mechanisms of the machine.

At the same time as the C format was introduced, the B format was developed. This has some of the features of both the Quad and C formats. There are two video heads on a 2 in drum. Each television frame is made up of a number of tracks. For the 625/50 PAL machines it is 6 tracks per frame; for the 525/60 NTSC it is 5 tracks per frame (*Figures 61.3* and *61.4*).

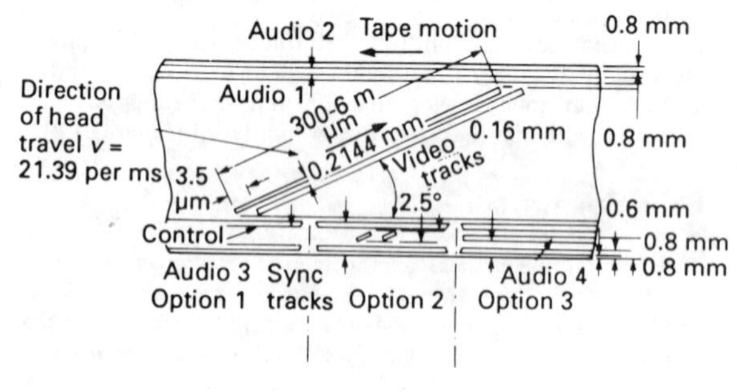

Figure 61.3 Positions of recorded tracks in the three options of C format

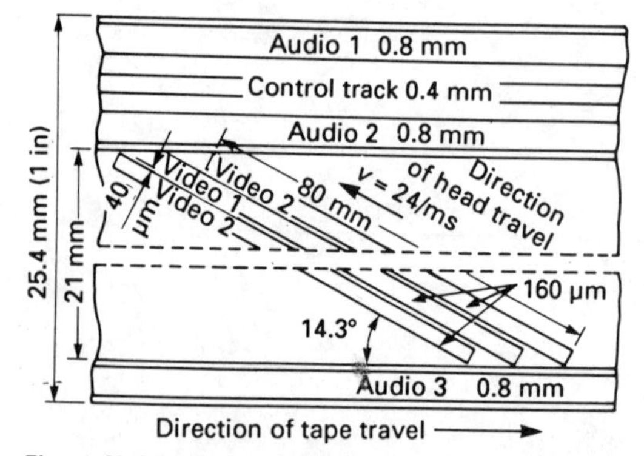

Figure 61.4 B format tape layout

61.3.1 Colour-under formats

After the direct record machines, Quad, B and C format, came a lower cost lower performance series of VTRs suitable for industrial and later domestic use. These VTRs, more often referred to as video cassette recorders (VCRs) because of their universal use of cassettes to hold the tape, used drums with pairs of video heads on them. This allowed each head to record a complete field, but by arranging for the tape to be wrapped more than 180° around the drum the loss of signal between fields is avoided.

Because tape consumption is a major cost in the use of a VTR, these VCRs used lower head-to-tape speeds. This limits the bandwidth that can be recorded to about 3.5 MHz. Since the colour subcarrier for both NTSC and PAL is above this frequency, some way to modify the signal to keep the colour was required. In VCRs the subcarrier has its frequency changed to around 1 MHz. This is then recorded on the tape at a point in the spectrum below the luminance signal. For this reason these machines have become known as 'colour-under' or 'heterodyne' machines (*Figure 61.5*).

Umatic colour-under machines were first used in broadcast for news gathering. Their picture quality was marginal and only second or third generation could be used because of the rapid falloff in colour and resolution with each generation.

The domestic versions of the colour-under machines are the VHS and Betamax VCRs. These machines take an extra step to improve tape economy. Each of the two video heads is angled away from the usual perpendicular azimuth setting. They vary from 5° to 15° depending on the format. This means that each video head will ignore recordings made by its companion. There is no longer any crosstalk between channels and so there is no need to waste tape by leaving an unrecorded guard band between tracks.

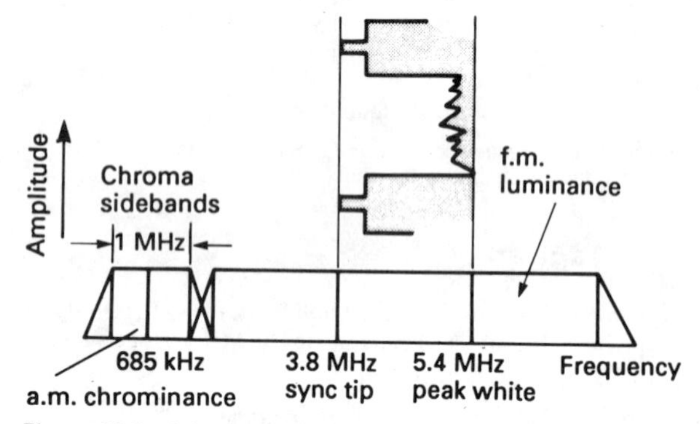

Figure 61.5 Colour-under spectrum

61.3.2 Component formats

With the advent of smaller and lighter cameras it is possible to make one-piece camera recorders (camcorders). Since no cable is needed to feed the signal from the camera to the recorder there is no need to code the colour into a single signal. The colour difference signals and the luminance signal are fed to the recorder and are then recorded on two parallel tracks, luminance on one and chrominance on the other.

There are several versions of component recorders, but all now use similar approaches. The luminance signal is recorded as it is in a direct recorder such as a C format machine. The two colour difference signals are combined and then applied to a record channel similar to the luminance channel. The combination of the colour difference signals takes advantage of the fact that the eye cannot see fine detail in colour. This allows the colour

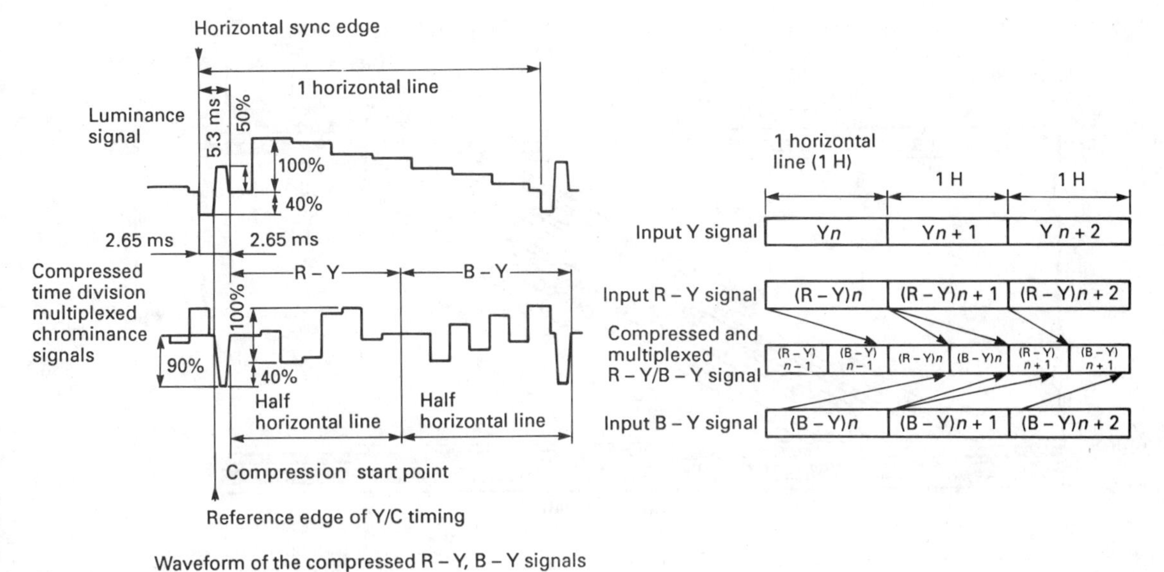

Figure 61.6 Component recording signals

difference signals to be compressed into half the normal line time. The frequency doubles but remains less than the luminance bandwidth. The combined compressed colour difference signals are recorded and replayed by an extra pair of heads on the drum. The transports are based on the VHS and Betamax domestic formats (*Figure 61.6*).

61.3.3 Digital formats

Digital video recorders have been in development for ten years. There are two systems available, D1 and D2. The D1 format uses the component form of video signal while the D2 format uses the composite PAL or NTSC signal.

The basic video signal is converted to an 8-bit code sampled at four times the colour subcarrier frequency. This generates some 208 Mbit/s to which must be added error protection bits and the audio. Unlike analogue recorders, digital recorders can give error-free replay by the use of error-detection systems. This means that once the signal is in the digital form it may be put through many generations without degradation. The machines are therefore ideal for postproduction and high-volume replay operations. The machine can use the error-correction system to monitor the quality of the replay and can warn the operator when a new tape is required (*Figures 61.7* and *61.8*).

61.4 Recording

In the record process the electrical signal is applied to a coil around a core. One side the core arranged to have a small gap, which is in contact with the surface of the magnetic medium. The gap is there to force the magnetic flux into the recording medium, which forms a low-impedance low-reluctance path. The larger the gap the more the pressure for the flux to pass through the recording medium.

The strength of the magnetic flux is greatest at the edge of the gap. A particle of magnetic material will therefore be subject to an increasing flux force as it approaches the gap when the head moves past the tape. There will come a point where the flux is strong enough to realign the magnetic particle and impose the pattern of the signal into the magnetic material. As the particle leaves the area of the trailing edge of the gap in the head it will be subject to decreasing flux levels. At the point where the flux no longer affects the particle alignment a permanent record of the signal will be imprinted into the medium. The point at which this occurs is called the freezing point.

The flux pattern from the head will form a series of cylinders of reducing strength concentric with the gap. Magnetic particles near the surface close to the head will take up alignments which are vertical to the surface. Those particles which are a long way, relatively, into the medium will align themselves parallel to the surface. This gives a combination of perpendicular, vertical and longitudinal recording. The longer wavelengths are recorded deeper into the tape.

The amplitude of the record current in the head coil has a significant effect on the stored magnetic flux strength and the frequency response of the replay. As the record current is increased the signal is moved deeper into the medium, away from the gap. The longitudinal signals are pushed beyond the back surface of the medium and are lost. All the signals are further from the replay gap and so the replay level is reduced. This spacing effect, the increase of distance from the gap to the flux, affects the high frequencies more than the low frequencies. Too high a record current will reduce the high frequencies.

As the record current is increased so the replay signal will increase until a peak is reached. Continued increases in the record level will no longer give an increase in replay level and finally a reduction in replay level will occur. As the record current is varied about this plateau there will be changes in the frequency response of the signal. The higher frequencies carry the colour subcarrier signal. This enables the adjustment of the record current to be fine tuned by measuring the replayed chroma level. On quadruplex machines each of the record replay channels could be matched by ensuring equal chroma replay. This was most critically done by viewing the pictures and therefore had the advantage that special test equipment was not required. For helical machines the correct record current level is best determined by measuring the replayed noise level in the luminance signal. For most machines this occurs at the point where the plateau begins.

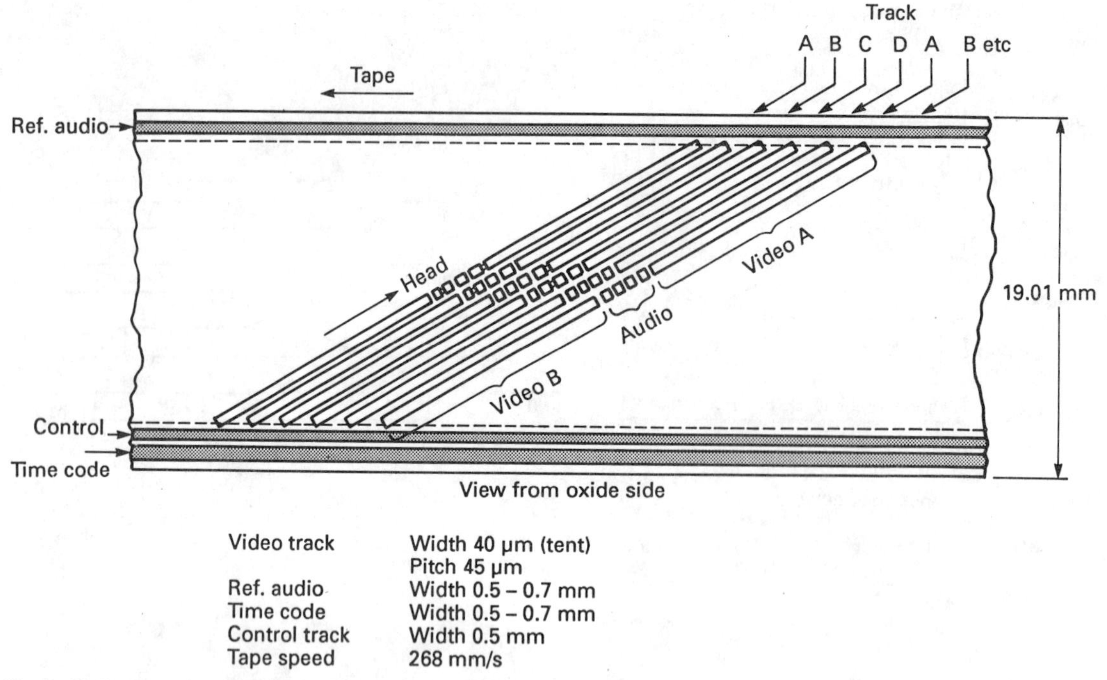

Video track	Width 40 μm (tent)
	Pitch 45 μm
Ref. audio	Width 0.5 – 0.7 mm
Time code	Width 0.5 – 0.7 mm
Control track	Width 0.5 mm
Tape speed	268 mm/s

Figure 61.7 D1 track patterns

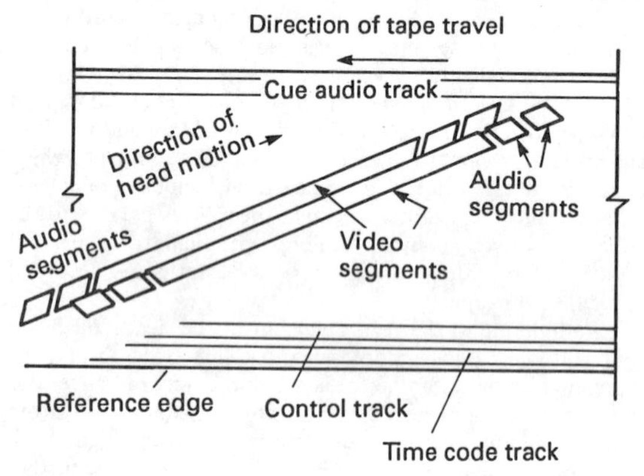

Figure 61.8 D2 track patterns

61.5 Modulation

There are two main reasons for modulating the signal onto a carrier. The first, already mentioned, is bandwidth reduction. The second is the need to make the signal immune to changes in the contact pressure between the head and tape. Fortunately both requirements may be met by frequency modulation.

In direct recording VTRs the input signal is applied to a frequency modulator. Unlike in most FM systems, the carrier and signal frequencies are close and the absence of signal, a black picture, does not imply a constant voltage and therefore frequency. There will still be the line and field synchronising pulses to consider.

It is simpler to consider the FM system in a direct recording VTR as a system in which specific picture levels generate particular frequencies. The various formats are defined by setting a frequency for sync tips, blanking level and peak white. For a typical system these might be 7.16 MHz, 7.68 MHz and 8.90 MHz. These are the chosen frequencies for the 625/50 PAL C format VTR.

A linear characteristic is chosen so that the tolerance on the record modulation frequencies and the replay decoder may be relaxed. This assists in the vital area of interchange between machines.

61.5.1 Moiré

The similarity between the carrier and signal frequencies in the direct record machines gives rise to a picture defect referred to as moiré. The name comes from the fact that its appearance is similar to the moiré fringes seen when two grid patterns of identical pitch are superimposed but slightly offset.

The cause is an interaction between the lower sidebands of the modulated signal and the lower harmonics of the upper sidebands. The problem is particularly severe when saturated colours are present in the picture since these give rise to high levels of colour subcarrier and therefore high energy in the higher orders of sidebands.

The higher order lower sidebands will be placed below zero frequency. This is not a real possibility and so the sidebands are 'reflected' or 'folded' back to the positive side of zero frequency. The reflection introduces a reversal to the phase of the signal. These reflected sidebands then interact with the lower order unreflected sidebands to produce spurious modulations of the signal. The decoder cannot distinguish these spurious signals from the true ones and will incorrectly decode the picture brightness. The result is a characteristic moving moiré pattern of lines on the screen.

61.5.2 Shelf working—low and high band

The carrier frequencies chosen for the original 405 line black and white pictures were 5.9 MHz for black level with peak white at

6.8 MHz and sync tips at 4.28 MHz. There was very little high-frequency energy in the signal and so the sidebands were virtually limited to a single pair. For a picture bandwidth of 4 MHz the lower sideband will occur at 1.9 MHz and so no reflected, or folded, sidebands exist to induce moiré.

With the introduction of colour and the high-energy high-frequency subcarrier, the number of sidebands increased. To reduce the effect of moiré the carrier frequencies were increased so that the second-order lower sideband was above zero frequency and was not folded back. An examination of the effect of increasing the carrier frequencies shows that the disturbance caused by moiré increases slowly to the point where the relevant sideband is no longer reflected. At this point there is a sudden drop in the level of moiré. The plot of moiré against carrier frequency gives a graph with a series of steps or shelves.

Each shelf represents an improvement in moiré performance and gives a band of frequencies which may be specified for sync tip, blanking and peak white. There are three bands which were used for the quadruplex recorders giving rise to the names Low Band, High Band and Super High Band.

The direct recording helical machines use frequencies similar to those of the high-band quadruplex frequencies. Although higher frequencies would give better pictures, the shorter wavelengths involved mean that tape defects become more critical.

61.6 Replay

The mechanism for the replay of a magnetic signal depends on the changing flux in the core of the replay head, which induces a voltage across the coil. The faster the rate of change of the flux the higher the voltage. This means that the higher the recorded frequency the higher the output. For every octave increase in frequency there is a 6 dB increase in output.

The gap in the head has a finite size. Typical head gaps are in the region of 2–3 μm. This compares to the wavelength of the recorded signal at the upper frequency end. There is a point where the heap gap and the wavelength are the same. At this point the average flux across the gap is zero and the output of the head is zero. This is called the extinction frequency. The point at which maximum output is achieved is when a half-cycle of the signal matches the gap size. This is one octave below extinction frequency.

As the frequence falls from the peak output point the output level falls by 6 dB/octave. When the frequency has fallen by 9 octaves the signal level has fallen by 54 dB and the lower frequencies will be lost among the background noise generated by the tape. The maximum range that can be replayed is 10 octaves.

61.7 Time base correction

All VTRs involve mechanical moving parts as an integral part of the system. Inevitably these parts suffer instabilities which give rise to timing and other disturbances in the replayed signal. The stability required by the colour television signal exceeds anything that can be achieved mechanically.

An estimate of the necessary signal timing accuracy may be obtained from the specification of the PAL or NTSC signals. Both have high-frequency subcarriers on which the colour is modulated. To maintain the correct hue the phase of the subcarrier must be kept within ±2°. If we assume a frequency of 4 MHz then one cycle takes 250 ns. The timing accuracy is therefore ±2/360th of 250 ns, approximately 1.4 ns.

In the colour-under formats the luminance signal is not corrected since the picture monitor or home television will use the line sync pulses to control the scan. It is still necessary to restore the subcarrier to its correct frequency and stable phase. The colour decoder in a monitor or receiver relies on an oscillator set by the colour reference burst at the start of each line. The burst-locked oscillator has to maintain its phase from one line to the next and cannot follow rapid changes in phase.

The process of restoring the subcarrier frequency involves mixing the off-tape colour-under signal with a reference oscillator. The reference frequency is varied in sympathy with the tape velocity errors. Both reference and colour-under subcarrier vary at the same rate and so the replayed subcarrier, which is the difference between them, is of constant frequency. This process destroys the vital fixed relationship between line syncs and subcarrier and so the output is not suitable to be fed to the input of a mixer or other broadcast signal chain. A separate time base corrector must be used.

The basic element of a time base corrector is a delay line. The outgoing signal from the VTR demodulator, often referred to as demod out, is fed to the delay line. An output tap is positioned halfway down the delay line. If the demod signal is correctly timed, the tap remains at the halfway point and all the signal is delayed by an amount equal to half the total delay period available (*Figure 61.9*).

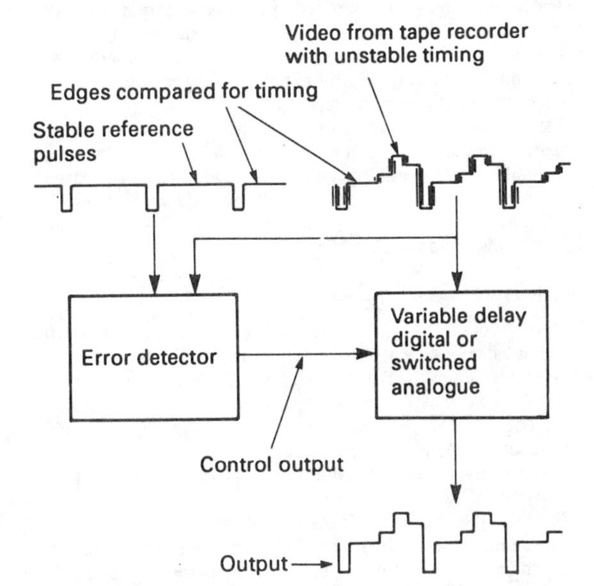

Figure 61.9 Action of a time base corrector

If the demod signal is late, because the video head has slowed or the tape has become distorted, the output tap is moved towards the input end of the line to shorten the delay. The output signal is therefore still correctly timed. If the demod signal is early, delay is added as required.

The most important part of a time base corrector is the timing error detection system. Unless an accurate measure of the error is available, the control to the delay line cannot correct the signal. The reference used for time base correction is the colour burst signal that occurs at the start of each line. There are eight or nine cycles of a high-frequency, and therefore rapid rise time, signal which may be used to compare with the burst on the reference signal fed to the machine.

To achieve the best time base correction results, the signal must be clean. Noise added to the burst will reduce the accuracy of the timing measurement and so in some machines the burst is amplified during recording to improve the signal-to-noise ratio.

Quadruplex machines, with their small drums and short tracks, are subject to small timing errors but with large discontinuities at the switch between heads. The time base correctors were added after the rest of the signal processing and used analogue circuitry. These analogue delay lines needed careful alignment to ensure that the signal amplitude did not vary as the delay changed.

Helical machines do not suffer the sudden jumps in timing because they do not switch heads during the active picture period. However, the longer tracks and larger diameter video head drums are subject to a greater range of timing error. This is especially true when the machines are used in portable form, as in news gathering, or during slow motion replay. Analogue delays are not capable of coping with the large errors, up to 10 or 15 television lines, and so digital delays have been universally adopted.

The demod video signal is first converted to an 8-bit digital signal, 256 levels. This is adequate to prevent the eye detecting the quantising levels under normal picture conditions, although close inspection of a monochrome sawtooth signal will reveal the steps, especially after multigeneration use. The delay can take the form of a random access memory and the major part of the time base corrector (TBC) is concerned with the addressing, writing and reading of the video samples.

The off-tape signals are sampled by a signal generated from the demod signal. This ensures that the video is sampled at regular points along the line. The output from the TBC is controlled by a clock which is generated from the reference signal and therefore reads out the samples at a constant rate.

The demod video is sampled at either three or four times subcarrier frequency. The first TBCs used $3 \times F_{sc}$ because it was not possible to make analogue-to-digital converters work faster. The move to $4 \times F_{sc}$ was made when faster A/D conversion was possible and when the price of memory dropped. At $4F_{sc}$ the data rate is in the region of 150 Mbit/s.

61.7.1 Synchronising pulses

The output of the TBC is usually referenced to the station or edit suite pulse chain. A feed of line and field sync pulses with a colour burst will be fed to the reference of the TBC. The VTR will have to be timed early with reference to station syncs to allow for the normal delay of the TBC, half its correction range. The VTR may use the same reference pulses and set its own video timing early, or the TBC may be used to generate a set of advanced syncs to be used by the VTR.

Since the TBC has a feed of the local station pulses it is normal for the demod video to be stripped of sync pulses which are then replaced by the TBC. In the case of the C format machines the vertical interval pulses will be missing and so the TBC will be required to replace them. A TBC is therefore an essential part of the replay signal chain, even if the timing instabilities could be accepted.

61.7.2 Slow motion

When a helical machine is put into still frame or slow motion, two major changes occur. First, the head-to-tape velocity changes, which alters the line duration; and, second, one field only is replayed, although some machines have a true frame ability which allows both fields to be replayed during slow motion or freeze frame. The TBC has to restore the correct line duration and replace missing lines at the start or end of the field.

The missing lines occur because once the tape movement is stopped, for freeze frame, the video head can no longer make contact with all of the video tracks on the tape. In C format this will mean the loss of $3\frac{1}{2}$ lines for PAL and $2\frac{1}{2}$ lines for NTSC. In fast motion, extra lines will be read and the TBC must discard them.

61.7.3 Velocity errors

If the geometry of the replay machine is not identical to that of the record machine it is likely that the head will travel across the tape at a different speed. Indeed, mechanical instabilities in the system ensure that the head does not have a constant velocity. Part of the function of the TBC is to correct velocity errors. This it does by comparing the actual line timing with the reference line timing.

In first-order velocity compensation a measurement of line duration is made by comparing the timing of two successive line sync pulses. It is assumed that the error will be constant throughout the line, and a sawtooth correction signal is generated. This is a simple model of what actually happens and second-order velocity compensation integrates the errors from three lines to achieve a more accurate curve or error signal. The velocity correction is applied in the TBC by varying the read time at the output of the delay line.

One of the side-effects of a change in velocity is that the off-tape frequency varies. If this happens the luminance level also changes and this can be seen on quadruplex machines where the velocity errors are repeated in bands down the picture. A simple black level clamp at the start of each line will correct this and forms part of the function of the TBC.

61.7.4 Luminance dropout

Where the magnetic medium is damaged or there is dirt on the surface, the replay signal will fall or disappear. The use of frequency modulation reduces the effect of such defects since the picture information is carried in the frequency not the level of the off-tape signal. However, there will come a point where the signal has fallen so low that it cannot be recovered. At this point the demodulator will see only noise, and random voltages will result.

The drop in level can be detected by the replay signal system, and to reduce the visible effect the demodulator output can be clamped to black. At the same time the dropout detector pulse may be fed to an ancillary circuit with the task of concealing the dropout.

61.7.5 Chrominance dropout

The ratio of the sideband energy to the carrier energy determines the chroma level in the replayed signal. A VTR signal system calls for a linearly falling record characteristic. This means that the upper sideband energy is significantly lower than the carrier and lower sideband energies. When the replay head moves away from the tape surface, the shorter wavelength signals attenuate more rapidly than the longer wavelengths. The higher frequencies are lost before the lower ones, which means that as a dropout area is approached the upper sidebands are list first.

The net effect is that the total sideband energy increases with respect to the carrier energy. This causes the chrominance level to increase as the dropout is approached. If the drop in level is not sufficient to result in a full luminance dropout, a brightening of the screen will be seen. This is a chrominance dropout. Such dropouts occur at the start and finish of every luminance dropout, but they are hidden by the concealment system which stretches the dropout pulse to cover the onset and recovery periods.

61.7.6 Dropout concealment

The TBC is fed with the dropout detector pulse from the VTR. When a dropout is present, the TBC will use one or two lines of its delay to recirculate the video from a previous line to conceal the dropout. Extra processing has to be applied to ensure that the correct phase of colour subcarrier is added. If the replacement signal is taken from two lines away, a 180° phase reversal is all that is required and the choice of $4F_{sc}$ sampling makes this easier.

61.8 Editing

Quadruplex recordings, with the video tracks short and perpendicular to the tape edge, could be physically edited. The track pattern could be revealed by an application of iron dust in a volatile liquid. The tape could then be cut and two sections joined with a thin adhesive tape applied to the back surface. It was not possible to change the position of the edit, and the master tape was destroyed in the process. For this reason a system of dub editing was developed and is now the universal method of working.

To make a dub edit, the required scene is replayed from one machine and recorded on another. The two machines are parked, cued, a fixed distance before the edit point. The control for this may be manual but the advent of cheap microprocessors has made computer control universal. Once parked, the machines are started together in play. Once up to speed their outputs are synchronised and at the chosen point the edit machine switches to record.

As it goes from play to record the edit machine must arrange to erase the old tracks in advance of the record head. To achieve this, extra heads have to be placed on the video head drum. They are referred to as flying erase heads.

If the programme is being compiled sequentially, a system known as assemble editing may be used. In this system the new material is added to the old material just as if it were a new recording. The synchronisation is limited to making sure that the join is smooth. If shots are being placed into existing material, the insert edit mode is used.

The position of the video tracks is controlled by a control track which consists of pulses laid on an edge track along the tape. During normal record or assemble editing, a new control track is laid down with the video. In insert editing, the old control trac, is kept to ensure that the new video tracks are laid over the top of the old ones. This ensures a smooth exit from the edit.

The use of computers to control the edit process has made it possible to use cheap VCRs to make the rough edits. This is known as off-line editing. To keep track of the edit points, one of the audio tracks, or on some machines a dedicated track, is used to record a time code signal. This signal numbers each frame of the recording using a 24 h clock system. In addition to the time, the code can carry commands to allow special effects to be triggered automatically during the replay.

Time code edit systems make it possible for one person to control as many as 16 machines, including switchers and picture manipulating systems such as computer graphics and digital effects. Edits may be rehearsed and reviewed with fine tuning made at each stage. The final programme may then be put together automatically using the time code cue list in the computer.

61.9 The next generation

High-definition television is already available in several forms. It can use very large amounts of channel capacity to deliver the improved quality pictures and sound. Videotape recorders able to capture and replay the HDTV signals have been developed alongside the rest of the systems. In general, HDTV VTRs are modified versions of the helical formats we already have. They run at higher speeds and make use of several tracks to record the separate component signals.

Digital machines for the present system are changing the way that postproduction and station operations are performed. Machines require less engineering skill from the operators and are able to perform many more complex activities. A day or more of programming may be loaded into a machine which may then be left to run the network while monitoring the quality of the replay as it goes. The human part of the system is then left to concentrate on the creative aspects of the work.

Further reading

ANDERSON, C. E., 'The modulation system of the Ampex video tape recorder', *J. SMPTE*, **66**, 182–184 (1957)

ARIMURA, I. and SADASHIGE, K., 'A broadcast-quality video/audio recording system with VHS cassette and head scanning system', *SMPTE J.*, **92**, 1186–1192 (1983)

ARIMURA, I. and TANIGUCHI, M., 'A color VTR system using lower-frequency-converted chrominance signal recording', *National Tech. Rep.*, **19**, 205–216 (1973)

BRUSH, R., 'Design considerations for the D-2 NTSC composite DVTR', *SMPTE J.*, **97**, 182–193 (1988)

DARE, P. *et al.*, 'Rotating digital audio tape (RDAT): a format overview', *SMPTE J.*, **96**, 943–948 (1987)

DAVIES, K., 'Formatting and coding the audio in the DTTR', *SMPTE J.*, **96**, 171–176 (1987)

DOLBY, R. M., 'Rotary-head switching of the Ampex video tape recorder', *J. SMPTE*, **66**, 184–188 (1957)

EGUCHI, T., 'The SMPTE D-1 format and possible scanner configurations', *SMPTE J.*, **96**, 166–170 (1987)

ENGBERG, E. *et al.*, 'The composite digital format and its applications', *SMPTE J.*, **96**, 934–942 (1987)

FELIX, M. O. and WALSH, H., 'FM systems of exceptional bandwidth', *IEEE Proc.*, **112**, 1659–1668 (1965)

FIX, H. and HABERMANN, W., 'The signal-to-noise ratio in television tape recording', *EBU Rev.*, **80**, 147–149 (1963)

FIX, H. and HABERMANN, W., 'Recording PAL colour television signals on tape', *EBU Rev.*, **110**, 107–112 (1968)

GINSBURG, C. P., 'Comprehensive description of the Ampex video tape recorder', *J. SMPTE*, **66**, 177–182 (1957)

HEITMANN, J. K. R., 'Development of component digital VTRs and the potential of the D-1 format', *SMPTE J.*, **97**, 126–129 (1988)

MORALEE, D., '30 years of professional video recording', *Electron. Power*, November/December (1987)

SADASHIGE, K., 'Selected topics on modern magnetic video recording technology', *IERE Conf. Proc.*, **26**, 1–26 (1973)

SADASHIGE, K., 'Feasibility study for a new compatible quadruplex video recording standard', *Radio Electron. Eng.*, **44**, 243–249 (1974)

SADASHIGE, K., 'Overview of time-base correction techniques and their applications', *SMPTE J.*, **85**, 787–791 (1976)

SADASHIGE, K., 'An introduction to analog component recording', *SMPTE J.*, **93**, 477–485 (1984)

SADASHIGE, K., 'Transition to digital recording: an emerging trend influencing all analog signal recording application', *SMPTE J.*, **96**, 1073–1078 (1987)

TANIMURA, H. *et al.*, 'A second generation type-C one-inch VTR', *SMPTE J.*, **92**, 1274–1279 (1983)

WILKINSON, J. H., 'A review of the signal format specification for the 4:2:2 component digital VTR', *SMPTE J.*, **96**, 1166–1172 (1987)

ZAHN, H. L., 'The BCN system for magnetic recording of television programs', *Bosch Techn. Berichte*, **6**, 176–188 (1979)

62 Office Communications

J Kempster, BSc(Hons), MBCS
Digital Equipment Co. Ltd
(Sections 62.1, 62.2)

J A Dawson, CChem, MRSC
Rank Xerox Ltd
(Section 62.3)

R C Marshall, MA, FIEE, MInstMC
Protech Instruments & Systems Ltd
(Sections 62.4–62.6)

Contents

62.1 Introduction

A goal for every company is to ensure not only that its internal communications are at a premium, but that it can also effectively communicate with its customers. In practice, however, this is not always achieved. Although many factors prevent good communication, the result is consistent—a waste of time and resources, and a potential loss of competitive edge.

Effective and efficient communication is the timely delivery of useful information between sender(s) and recipient(s). For a business, information may originate from, or may be destined for, both human and computing resources. The need for timely information becomes more acute as markets expand, and hence become more competitive. If a company fails to recognise and respond to this, others will.

Only once information is available to the person who needs it, is a return on investment made. The means by which the information is delivered is therefore vitally important. The information carrier must realise several criteria, including different time zones, geographies, computing environments, user expertise, job family, organisational structures and so on.

Computers are now the key factor in controlling all aspects of office communication. However, paper still remains the most widely used medium for human interface. In this chapter the computer-based communication systems are first introduced, followed by xerography and facsimile, both of which use paper as their communication mechanism.

62.2 Computer-based communication systems

62.2.1 Office systems

Most large computer suppliers now offer 'office systems'. Typically, these are software applications that have been designed to automate tasks that take place in an office environment. Word processing is often quoted as an example.

Since the late 1970s, the number of tasks that have been automated has increased, such that it is now reasonable to expect an office system not only to provide word processing, but also other generic office tools including diary management, electronic mail, calculator, high-quality printing, graphics, and so on. The office system therefore provides its users with an environment through which they may create, extract, manipulate and distribute information.

Generic office tools have helped users to become more efficient. Until recently, however, they have not been closely linked with the data processing applications that support the company's business. Examples include inventory control, customer service, or sales order applications. In other words, the business professional could not easily access the information which should have been available.

The requirement to link together the company's office system with its data processing applications became clear. In this scenario, the professional has direct access to the information that supports the business. This provides greater opportunity for coordinating the company's success in its markets.

As a result, by the mid-1980s, selected suppliers began to offer 'business and office information systems', where the need to tie together office system and data processing applications was realised. The new generation of office systems provides for applications integration, by defining an infrastructure through which information can flow between the office and corporate data processing environments.

62.2.2 Information management

Information management can be defined as the capability of the computer system to store and manage data. Traditionally, information management has been further categorised into: file management, data management, or database management.

Database management is the most sophisticated level of information management. In comparison with data management and file management, it enables data definitions to be stored separately from data processing applications, and also provides for comprehensive data security and integrity. Additionally, techniques for modelling data structures and relationships are offered.

Although database management systems were developed originally for use by central management information services (MIS), they are being used more frequently by departmental users. As a result, suppliers must not only provide for centralised database management, but must also provide a distributed approach to their database management solution.

62.2.2.1 Relationship with the office system

Company information can be stored in a variety of forms. File management tools support access to single, unrelated files, whilst database management tools handle complex relational or CODASYL databases. Despite the method of storage, it is essential that the information flows between the company's data processing and office system environments. Therefore, a fundamental requirement is that the suppliers' office system and information management strategies are complementary to one another.

62.2.2.2 Decision support tools

Access to company information provides only half of the desired solution. Additional tools are required to enable the business professional to manipulate the information once extracted.

One of the objectives of fourth-generation languages (4GLs) is to reduce the expertise required by the programmer. 4GL decision support tools such as report writing, graphic and query facilities provide an environment through which the business professional can manage information. Additional features to simplify use can include on-line tutorials.

Once finalised, information can be filed within the office system and distributed throughout the company using the suppliers' business communication product offerings.

62.2.3 Business communications

The objective of business communications is to ensure that information is available to the people who need it, when they need it. By achieving this goal, business professionals are better able to respond to the demands placed upon them.

Although the supplier's office system should easily communicate information across its other product lines, the real world comprises businesses whose computing purchases are from various suppliers. Even if a company fulfils its computing requirements from a single supplier, it is unlikely that all its trading partners will have made similar purchasing decisions.

The required compatibility of services across all vendors' equipment is resolved only through an adherence to international standards. Solutions that are not based on standards create islands of information resulting from incompatible proprietary solutions. Once compatibility is achieved, a company can concentrate more on its business needs and less on the peculiarities of the supplier's product offering.

During the last decade significant advances have been made in the area of business communications. Three complementary solutions have been defined and developed; electronic mail, videotex and computer conferencing. Each solution enhances one type of interaction that takes place between the information provider(s) and information user(s). Specifically, electronic mail is suited to conveying information from one person to another, or to a defined audience ('one-to-one' communication), videotex to conveying information from one person to a wider audience ('one-to-many' communication), and computer conferencing to conveying information between many people ('many-to-many' communication).

62.2.4 Electronic mail

A company-wide and inter-company mail system provides a common service to unify information providers and users. As a business communication solution, electronic mail carries strategic information, and therefore should not be viewed simply as an incremental application. Rather, it is a competitive tool which if well planned can be a major factor for improving managerial effectiveness.

Electronic mail is an organisational application. It supplies a framework through which people communicate information regardless of the system they use within the company. This framework provides for the reliable, high-speed transmission of information.

Not limited to any industry segment or class of user, various mail interfaces may be designed to support the individual needs of a specific user base. To provide for this requirement the vendor should market user interfaces for its major classes of users, and tools for the design of other interfaces. For example, a financial director may need a preferred interface to give access to a number of profession-specific applications such as a modelling package. A technician may need access to a complex mathematical application.

A company-wide electronic mail system should embrace a number of criteria, including:

(1) The ability to deliver a variety of documents, including image and data.
(2) An electronic forms capability to capture transactions.
(3) Network-wide directories.
(4) Transparent addressing across mail environments.
(5) Support for international standards.

The electronic mail system is an information carrier and should also be available to non-real-time data processing applications. For example, a regression test suite running in Leeds can use the mail system to deliver its results to a product development manager in London. It is important that the vendor's mail strategy allow for electronic data interchange (EDI) application integration.

62.2.4.1 International Standards

The International Standards Organisation (ISO) created a framework to support the different applications and communication technologies across suppliers' systems. The framework, to which most of the major suppliers are now committed, is called the Reference Model for Open Systems Interconnection (OSI) (see *Chapter 58*).

The OSI Reference Model provides peer-to-peer communication between computer applications. In this instance, a personal computer is equal in networking status to a minicomputer or mainframe.

One key application to take advantage of the OSI Reference Model is the X400 Recommendations for Message Handling Systems.

62.2.4.2 X400

In 1984, the International Telegraph and Telephone Consultative Committee (CCITT) completed four years work from which it produced the X400 messaging standards. These recommendations are now the most important messaging standards currently in development.

X400 describes more than just a series of communication protocols. It defines a structural model for electronic mail systems, which describes two layers of service: the message transfer layer (MTL), which handles the envelope services, submission and delivery of the message and its addressing and transfer; and the user agent (UA), which handles the 'content' of the message.

One specific content service, the Interpersonal Messaging System (IPMS) is defined by X400. This is the user's interface to the system. It does not stipulate the appearance of the interface, and so interfaces may be designed to support the needs of various classes of user.

62.2.4.3 Suppliers' implementation

X400 achieves the compatibility of services required between different suppliers' electronic mail implementations.

During 1987, several suppliers confirmed their commitment to X400 by jointly demonstrating their implementations at major exhibitions, including the Hanover Fair and Telecom '87.

At a minimum, the supplier can implement the X400 and OSI communication protocols secondary to its own proprietary solution. Preferably, it will commit fully to the standard by integrating the OSI and X400 structural models into its communication architectures.

The X400 structural model defines the relationship between the supplier's office system and its electronic mail facility. Previously, mail was perceived as a tool to benefit the office system user. Now it is a common resource, available not only through the office system, but also to other computer users and data processing applications.

62.2.4.4 The user

As discussed, people exchange the stragetic information which is used to coordinate a company's success in its markets. It is therefore important that we consider the user.

View X400 as a natural extension to the electronic mail capabilities currently available to the end-user. As such, it need not be seen as a new or disparate technology to which users need migrate. The introduction to X400 should cause minimal (if any) effect to the end-user. This is clearly in the interest of the end-user, whose primary concern is not the technology but its value to the overall company goal.

This can be achieved by using the inherent strengths of the OSI and X400 models.

62.2.4.5 Document handling

Current implementations of X400 support the transmission of IA5 or Teletex character encoded final-form documents. Clearly, in the long term there is the requirement to support the transmission of compound documents.

The suppliers' statement of direction for the support of the ISO proposed Office Document Architecture (ODA) is important. This standard specifies how revisable form compound documents can be supported between various heterogeneous computing environments.

62.2.5 Videotex

Videotex, the second business communication solution, provides for an electronic version of the company library. Similar to a library, it stores information for distribution to a wider audience.

In contrast to electronic mail, each member of the audience does not need to be addressed individually. Rather, the information provider places information into the library, for later consumption by its readers. In this context, the reader may be anyone in the company, or in a specific closed user group, depending on the sensitivity of the information involved.

Unlike its paper equivalent, the electronic library has a far greater opportunity of utilising the company's computing resources to ensure that its information is up to date. For example, as a company's policies change, the electronic videotex library can be used to quickly distribute the new policy statements without the need to reprint costly procedure manuals.

62.2.5.1 Information dissemination

Videotex is concerned with the packaging and subsequent dissemination of information, and it is this definition that links it to the suppliers' electronic publishing strategy. Videotex and electronic publishing receive information from similar sources, utilise the computer network, and package information into a modular format, for example pages. It is only the final distribution medium that changes—electronic publishing uses paper, whilst videotex employs the user's computer terminal.

62.2.5.2 User access

Access to a videotex system is simple. Consider first how a reference book is used: the reader checks the contents page, notes the desired page number, and opens the book at the page. Videotex uses a tree-structured database through which the user is guided by selecting options from menus. The user starts from the first menu page (the top of the page tree) and, through menu selections, branches down the information database until the desired page is presented. Proficient users can learn to bypass menus to arrive directly at the desired page.

With minimal training videotex enables the business professional to retrieve information easily, without the need to understand the complexities of the underlying computer system. Within the engineering world, typical uses of a videotex service include the storage of product specifications, safety procedures, newsletters and service manuals.

62.2.5.3 Relationship with the office system

A strong relationship between the supplier's office system and videotex offering is essential. Business communications is concerned with information flow and, consequently, information stored in the videotex environment must be accessible to the office system user. This relationship not only demands use of the same computer terminal, but also encompasses common command formats and resources including word processors, graphics tools and so on. For the office system user, the videotex service is simply another information source.

62.2.5.4 Infobase

The information database serving the videotex system need not be a single, centralised, physical infobase. Indeed, for many large companies, such a requirement proves limiting. Where desired, infobases can be physically distributed across a number of computers. With this arrangement, the infobases are managed safely, and updated quickly, by the people who provide the information. For the user who continues to see only a single videotex service, the benefit is timely information.

62.2.5.5 Information feeds

A videotex system is not just suited to providing information directly from its infobase. Other dynamic information feeds may be required. These are activated simply by a user requesting a menu option. For example, an engineer requires information on the design status of an electrical component. The component is drawn on the terminal screen from information kept on an optical disk, comments are extracted from the design engineer's working document, and the project timescales are confirmed by an external data processing application. This capability of communicating in a transaction-oriented manner with other information feeds is becoming increasingly important.

62.2.5.6 Infobase tools

The requirement for a high-integrity infobase is satisfied by quality database building tools. The supplier's tools to storyboard, design and maintain infobases are fundamental to a successful videotex service. Other tools can include infobase verification procedures, automatic generation of index menus and page numbers and so on.

62.2.5.7 International Standards

There are a number of videotex presentation standards being used by businesses. The desired format varies across the world, but in summary includes: NAPLPS (North American Presentation Level Protocol Syntax) which is suited to applications requiring high resolution graphics; PRESTEL which was one of the first videotex standards to be developed; and ASCII which is best suited to textual applications.

62.2.6 Computer conferencing

Meetings not only incur travelling and accommodation costs, but are also time consuming to arrange and attend. Potential attendees may not always find the meeting suitably scheduled. Additionally, some meetings do not live up to expectations. Computer conferencing supports the sharing of information between a group of people who have difficulty in meeting face to face. It provides a discussion environment through which topics may be raised, opinions considered, and conclusions reached.

Electronic mail and videotex are unsuitable for this requirement. An electronic mail solution would result in a flood of mail flowing from each attendee wishing to comment on information received from all other attendees. Videotex, although offering a library from which topics may be read, does not provide an environment through which subsequent discussions may be developed. Conference proceedings are stored as text. Contributors use their computer terminal to read previous discussions, and then type their own comments.

Computer conferencing makes full use of the supplier's computer networking capabilities. Conferences can be initiated by any department within a company, and therefore anywhere in the world. Its organisers should have the opportunity to store the proceedings at the most convenient point in their computer network. Therefore, optimal access to the conference by its attendees is dependent on the networking capabilities offered by the supplier.

62.2.6.1 Conference chairperson

The role of the conference chairperson (sometimes known as the moderator) is to ensure that the purpose of the conference is understood, and that it is kept to order. As with face-to-face meetings, the chairperson should hold a position of respect. This will not only confirm the importance of the conference, but also ensure that attention is given to resulting proposals. Other responsibilities extend to determining who is invited to the conference, guiding the conference through to a conclusion, and ensuring that outstanding actions are fulfilled.

62.2.6.2 User access

Attendees enjoy the benefit of participating in conferences at their own convenience. They have the opportunity to maximise their time and effort by contributing to all or some of the conference proceedings, depending on their available time.

A number of features may be available to simplify the attendee's participation in a conference. These include:

(1) Grouping conferences under logical headings.
(2) Defining 'page markers' to enable fast access back to a particular section of a conference.
(3) Requesting information on conference proceedings by date, author and so on.
(4) Updating attendees on all new topics and comments in which they are interested.

62.2.6.3 Body language

Although conference proceedings are stored as text, it is important that body language continue to play a role. As a result, the conference text can include indicators to reflect the mood of the contributor. For example,

CAPITAL LETTERS can signify anger.
:^) can signify humour.
:^ (can signify disappointment.

If the expressions of humour and disappointment remain unclear, turn the page 90° clockwise, and look again!

62.2.6.4 Task groups

Conferencing gives a company the opportunity to maximise the use of its expert resources. Conferences can be used to form task groups to address issues affecting the company goals. For example, an engineering team can benefit by creating a conference to discuss concerns over product design and schedules.

62.3 Xerography

62.3.1 Basic principles

The successful, commercial exploitation of xerography has resulted from the integration of a number of electrostatic process steps. The creation of an electrostatic image, which allows development by charged, pigmented particles, forms the basis of the xerographic process. These particles can then be transferred to plain paper and subsequently fixed to provide a permanent copy. The generation of the electrostatic image in all xerographic copiers depends upon the phenomenon of photoconductivity, whereby a material's electrical conductivity will increase by many orders of magnitude under the influence of light. By initially charging the photoconductor to a high, uniform surface potential, light reflected from the background areas of the original causes charge decay in the corresponding area of the image. Thus, a charge pattern remains on the surface of the photoconductor corresponding to the original document.

The basic steps in the xerographic process are outlined in the following paragraphs and depicted in *Figure 62.1*.

Charging, or sensitisation of the photoconductor to a uniform surface potential of 800–1000 V is usually accomplished by a corotron. Literally a thin wire, stretched between terminals parallel to the photoconductor surface, and driven at approximately 8000 V, the corotron emits a corona of ions which deposit on the photoconductor surface.[1] The charged photoconductor must be kept in the dark to prevent discharge.

Exposure of the sensitised photoconductor to light reflected from the original to be copied generates the required electrostatic

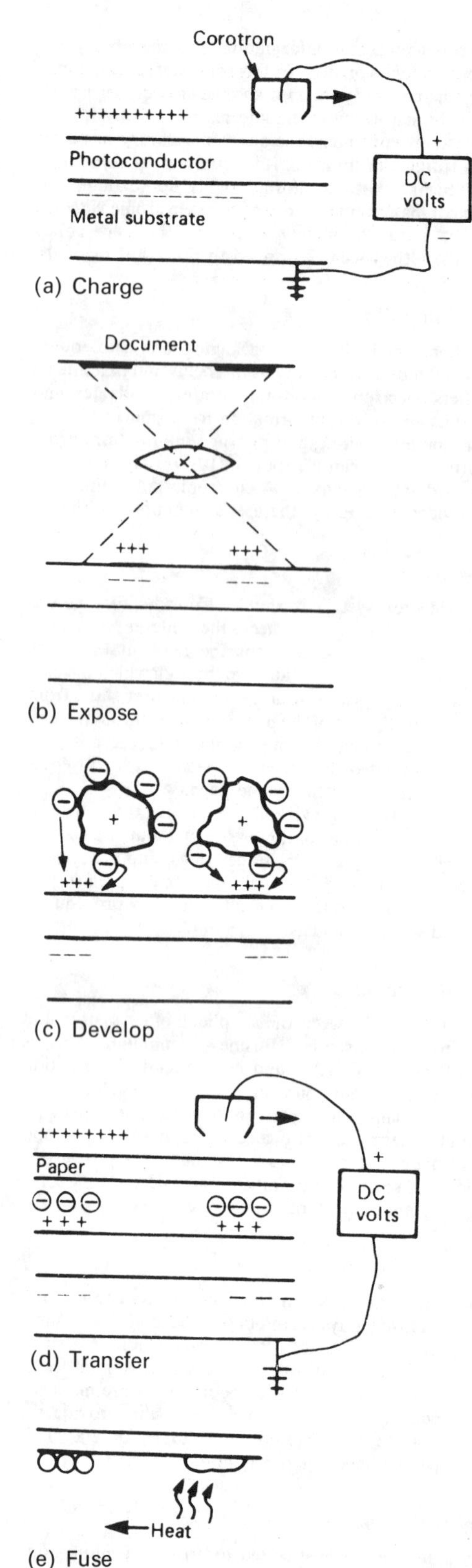

Figure 62.1 The basic steps of xerography

image. Voltage decay occurs via photon absorption by the photoconductor surface with the creation of an electron–hole pair. These separate under the influence of the electrostatic field, the electron neutralising a surface charge, and the hole, transported through the photoconductor, neutralises the corresponding image charge at the photoconductor–substrate interface. Selenium obeys the reciprocity law in that it responds to the product of light intensity and time, regardless of their individual values.[2] Typical light discharge and electrostatic contrast characteristics are shown in *Figure 62.2*.

Xerographic development of the latent electrostatic image renders the image pattern visible. To ensure selective development of the image, the black toner particles are charged to a polarity opposite to that of the photoconductor surface. Toner consists of finely dispersed carbon black in a thermoplastic polymer matrix. The toner particles are small to ensure that reasonable image, edge definition and resolution performance are obtained. Usually transported to the development zone via a carrier, it is the careful selection of carrier–toner pairs that ensures the correct charging, and hence development characteristics.

Transfer of the developed image from the photoconductor surface to plain paper is effected by a corotron. Similar in design and process to the charge corotron, positive charge is sprayed onto the back of the paper, which itself is in contact with the developed photoconductor surface. Sufficient fields are generated to ensure that most, but not all, of the toner will transfer to the paper.

Fusing the image into the surface of the paper is accomplished by heat from a radiant fuser, or a combination of heat and pressure from a fuser roll/backup roll combination. It is this step that will dictate the rheological requirements of the thermoplastic resins used in toner manufacture.

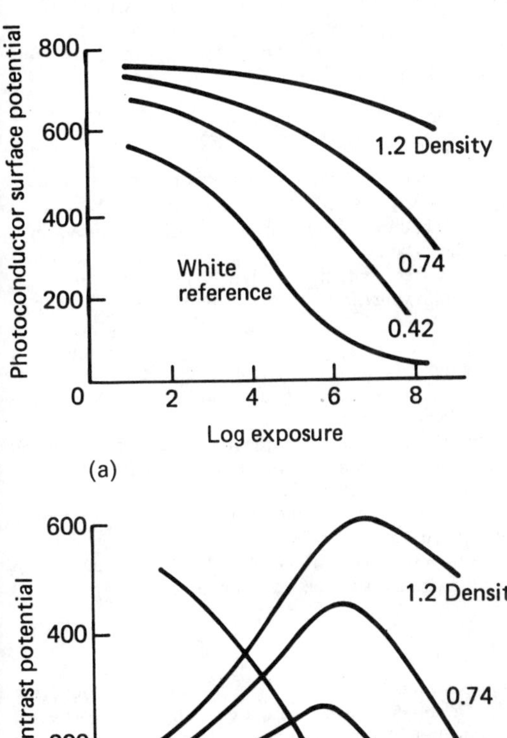

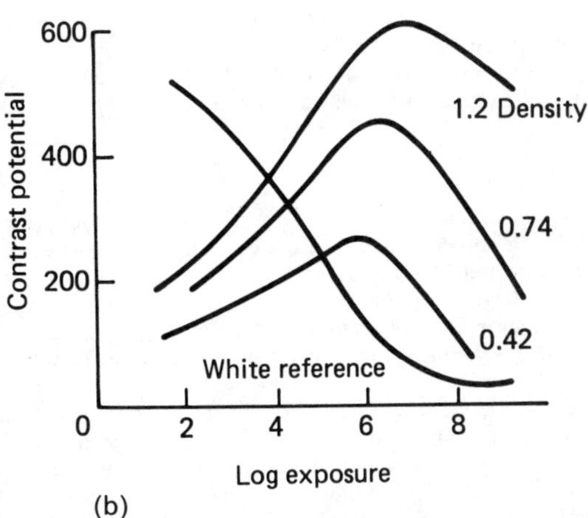

Figure 62.2 Photoconductor characteristics: (a) light discharge, (b) electrostatic contrast

62.3.2 Extension to a dynamic copier

The processes outlined in *Section 62.3.1* allow us to visualise the xerographic steps required to produce a copy via a plate photoconductor in the static mode. Dynamic operation of a copier places many constraints of space and geometry on the elements of process design and this in turn requires significant demands being made on the xerographic developers and photoconductors in use today. The trend towards faster, less expensive and smaller copiers will ensure that technology development will continue for some time to come. A schematic representation of a copier is given in *Figure 62.3*.

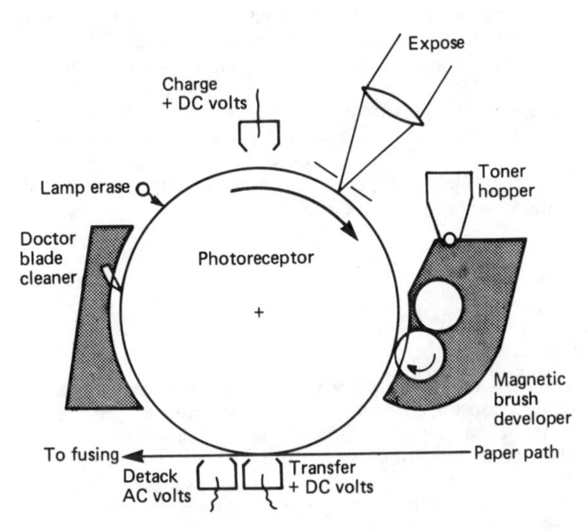

Figure 62.3 Xerographic copier schematic

Charging remains as discussed earlier with the rotation of the photoreceptor now providing the linear motion underneath the corotron. Exposure represents more of a challenge. With stationary platen copiers, the original document is scanned and the reflected light is transmitted via lens, mirrors and exposure slit to the photoreceptor. Some small copiers today incorporate a moving platen and strip optics for cost considerations.

Development of the electrostatic image has been extensively studied following the trend towards faster, smaller copiers.[3] Initial copier designs utilised a cascade development system which poured developer over the rotating photoreceptor. Increasing process speeds demanded developers with smaller particle size (increasing surface area to mass ratio), to increase the toner carrying capacity. Cascading developer over the photoreceptor in the against mode brought some relief for further speed enhancement but the resultant development system remained bulky. The trend towards magnetic brush development systems commenced approximately 15 years ago and immediately offered the advantages of compactness and enhanced developability. The latter, resulting mainly from the fibrous nature of the brush and the density of developer at time of photoreceptor contact, has led to almost all modern copiers using this form of development. A more recent extension of this concept is the single-component magnetic brush copiers used by some Japanese manufacturers and these are discussed in the section on materials.

Apart from the optimisation of currents and geometry, the process steps on transfer (and often detack) are direct analogues of the static environment.

Fusing the toner image onto the paper is one of the major contributors to power consumption within the copier and this has led to the abandonment of the early, inefficient, radiant or oven fusers. Centrally heated polymer-coated, steel or aluminium rolls now effect fusing by a combination of heat and pressure. Release agent fluids are often used on these rollers to prevent toner offsetting from the paper to the polymeric surface.

In the dynamic mode, the photoreceptor has to be returned to a virgin condition prior to reaching the charge corotron and hence embarking on another cycle. If this is not the case, residual voltages will build up within the transport layer of the photoconductor and cyclic problems will rapidly manifest themselves and cause print-to-print variations. Following the transfer of image toner to paper, a residual image (some 5–20% of the developed toner) will remain on the photoreceptor surface. (Total transfer is avoided as it would increase print background levels.) Removal is effected by rotating brushes in some machines, soft webbed fabric in others, but the popular choice for all small and mid-volume copiers today is the so-called doctor blade. A straight-edged polyurethane blade is angled against the photoreceptor and this scrapes the residual toner from the surface. The clean photoreceptor subsequently passes underneath an a.c. corotron or an erase lamp to remove all vestiges of voltage fluctuations prior to charging once again.

62.3.3 Xerographic materials

62.3.3.1 Photoreceptors

The xerographically active part of the photoreceptor is the photoconductor, generally amorphous selenium with one or two minor components, supported on a cylindrical drum of aluminium or a flexible metallic belt. The photoconductor layer is deposited onto the metallic substrates in vacuum coaters. An electrical barrier layer is essential between the 60 μm thick photoconductor and the metallic substrate to prevent charge injection from the latter when the photoconductor has reasonable electron mobility.

Early experimental photoreceptor[2] plates used a variety of photoconductors such as sulphur, anthracene, zinc oxide or sulphide, cadmium sulphide or selenide. The first xerographic copiers were based on extremely pure selenium photoreceptors and these performed adequately at the low cycling speeds. As copiers became faster and smaller, two major improvements were needed in terms of photoreceptor characteristics; greater photosensitivity, and more stable cycling characteristics.

Photosensitivity increases were obtained relatively easily by incorporating small percentage additions of arsenic or tellurium into the deposited film of selenium. Sensitivity increases result from an extension in the photoconductor's spectral response beyond the 550 nm limit of selenium, and enhanced response at smaller wavelengths (see *Figure 62.4*).

Cyclic stability proved more difficult to overcome but is generally improved by halogen dopants at the ppm level in the deposited photoconductor layer. However, careful balancing is required. The presence of chlorine (the most commonly used dopant) at the photoconductor/substrate interface layer reduces the cycle up of residual potentials as desired. However, if the dopant migrates to the outward surface of the photoconductor (often referred to as the generator layer), increasing fatigue

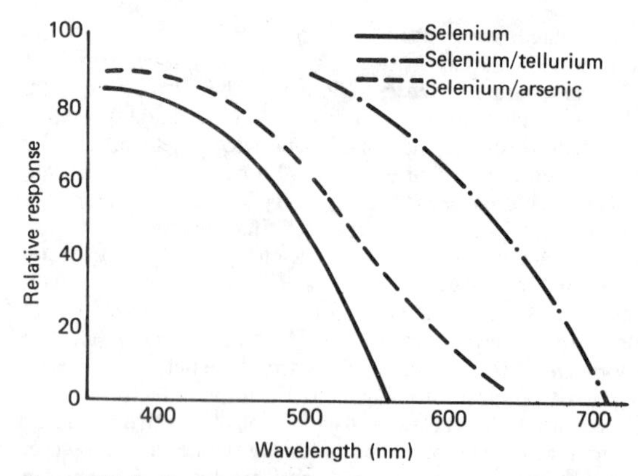

Figure 62.4 Spectral response of typical photoconductors

occurs. In this case, halogen-rich photoconductor can trap electrons in the upper layers and partially discharge the photoreceptor during the charging and expose stations in the copier. The lower apparent charging voltage will manifest itself as loss of density on the final print.

Organic photoconductors are in use today and present their own unique problems but remain a poor second in usage compared with the selenium- or selenide-based photoconductors. The ultimate advantage of cost will probably justify a trend towards organic materials and away from selenium-based photoreceptors.

62.3.3.2 Toners

Dry xerographic toners have not changed appreciably since the introduction of copiers, with an almost universal dependence upon carbon black for pigmentation, finely dispersed in a styrene/methacrylate polymer matrix. Other polymers (polystyrene, styrene/butadiene, polyesters) have been used, the choice often following the desire for different fusing systems to be employed.

Requirements for toner are that it will charge acceptably, to allow development and transfer, it can be cleaned from the photoreceptor surface, and it will fuse readily into the paper. The advent of doctor blade cleaning systems has necessitated additives in the toner to lubricate the photoreceptor surface and prevent blade chatter. Small amounts of zinc stearate are often added to solve this problem.

Toners are melt mixed in either Z-blade mixers or extruders, mechanically crushed to a small size and air-micronised to a final average particle size distribution of 10–15 μm by weight, with virtually all particles being between 5 and 35 μm.

62.3.3.3 Carriers

Performing two basic functions, the carrier acts as a conveying system for the toner, and provides a charging surface to effect triboelectrification. It is this triboelectric charge that ensures selective image development on the surface of the photoconductor. Developers usually contain 1–2% by weight of toner, the development capability being a function of toner concentration (*Figure 62.5*).

In early copiers, xerographic carriers were based on smooth, round sand or glass beads up to 600 μm in diameter. The resultant developer was cascaded over the electrostatic image. Classically, xerographic development is electrostatic field, and not surface potential, dependent and these cascade systems (open development) suffered two major limitations. As there were only fringe fields present, solid areas on originals would not reproduce

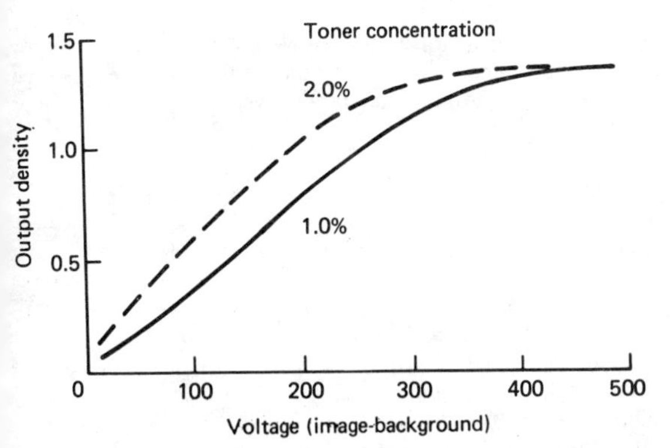

Figure 62.5 Development characteristic as a function of toner concentration

(apart from the edges), and contrast enhancement occurred on development. The imposition of a development electrode above the surface of the photoconductor during the development process significantly improves the solid area performance.[4] The electrode, a positively biased plate in cascade development, or the development roll itself in magnetic brush systems, is set to a value some 100–200 V above the background potential on the photoreceptor, see *Figure 62.6*. Significant gains in latitude are apparent with the electrode in place as background development is now suppressed by a cleaning field and image development is enhanced.

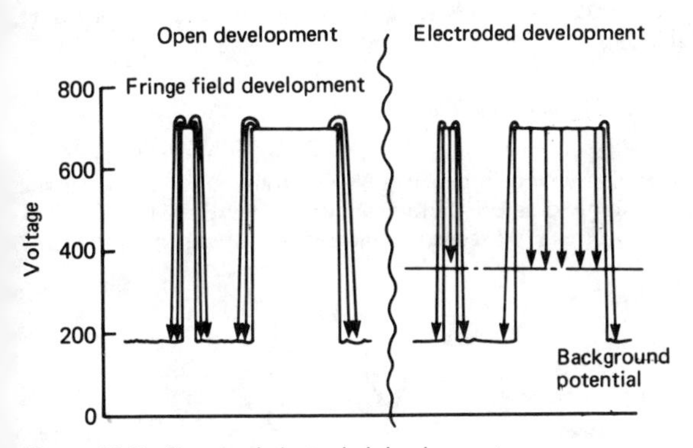

Figure 62.6 Open and electroded development

In order to benefit from the electroded development systems almost universally used today, and to enable increased development speed, carriers have become smaller and more conductive. Steel, or iron shot, in the 80–200 μm size range is now the most common core material in use although magnetite, sponge iron and ferrites have been employed. To produce carriers, the cores are lacquer coated with polymeric films to provide the triboelectric charging interface for the toners. Lacquers have been based on ethyl cellulose but methacrylates are more widely used today. Carrier lacquers and the toner formulation are chosen such that they charge to opposite polarities through physical contact. A triboelectric table (e.g. *Table 62.1*) can be generated such that any material on the list will acquire a negative charge when in contact with a material above it on the list.

Triboelectrification is a poorly understood phenomenon and, in copiers that go faster and faster, has a limitation in terms of time dependence. Incoming toner to the developer housing can be transported to the development zone in less than 5 s. If inadequately charged or wrong sign toner is produced this will develop in background areas. Quantities of such poorly charged

Table 62.1 Triboelectric series[5]

Polymethylmethacrylate
Ethylcellulose
Nylon
Cellulose acetate
Styrene/butadiene copolymer
Polystyrene
Polyvinylchloride
Polytetrafluoroethylene

toner are minimised by design approaches in the developer housing to ensure adequate dynamic mixing. In order for any development system to function, the electrical forces attracting the toner to the electrostatic image must exceed all other forces acting on the toner. Toner deposition occurs by three-body contact (electrostatic image/toner/carrier) or by separation of toner from carrier particles on impact generating a toner cloud within the developer.

62.3.3.4 Single-component developers

A modification of the magnetic brush development system utilises a single-component developer. By loading standard toner polymers with up to 50% magnetite, transport to the development zone is enabled magnetically with the magnetite also acting as the pigment. Charging of the toner particles can be by direct induction from a biased interference bar or by allowing electrostatic induction from the electrostatic image itself. Some problems are immediately apparent with these technologies in that tight control of background and image voltages is needed to maintain development latitude. Also, the requirement for a given conductivity in the toner can diminish the transfer efficiency and lead to low-density prints and high toner consumption rates. Many other development processes are available for specific uses.[3]

62.3.3.5 Paper

The ability to use plain paper has perhaps been one of the major reasons for the success of xerographic copiers. However, the characteristics of the paper are by no means unimportant. Machines that feed sheets of paper often rely on the beam strength of the paper to effect stripping from the photoreceptor or fuser roll. Surface roughness is important in affecting the fixedness (fusing) of the toner image on the copy. Conductivity (especially in humid conditions) can markedly affect transfer.

62.4 Raster scanning

The electronic office brings together two different kinds of technology. On the one hand there is the electronic computer, with its emphasis on data processing, employing a limited set of alphanumeric characters. On the other hand is the photocopier, capable of producing replicas of first-class appearance whether the original be English or Arabic, machine-generated or manuscript. A way has to be found to add this flexibility to work with any sort of image to an electronic system intended for non-specialist use.

This problem was recognised and solved by those pioneers who devised raster scanners to encode and decode written characters and so allow their communication via a telegraph circuit—a process now known as facsimile communication. Today, raster-scanning document readers are being used for character and signature coding and recognition, whilst the raster-scanning output principle allows multi-font character and graphics printers to be designed for computers and word processors. These scanning devices and their associated electronic data processing and communication systems are the subject of this section.

62.4.1 Principles

An image can be dissected into picture elements (pixels or 'pels') for transmission, processing or storage, and reassembled from such elements for display or printing. Such conversion between a two-dimensional image and a serial data stream is usually accomplished by scanning in two orthogonal directions as shown in *Figure 62.7*. This principle is shared with television; the

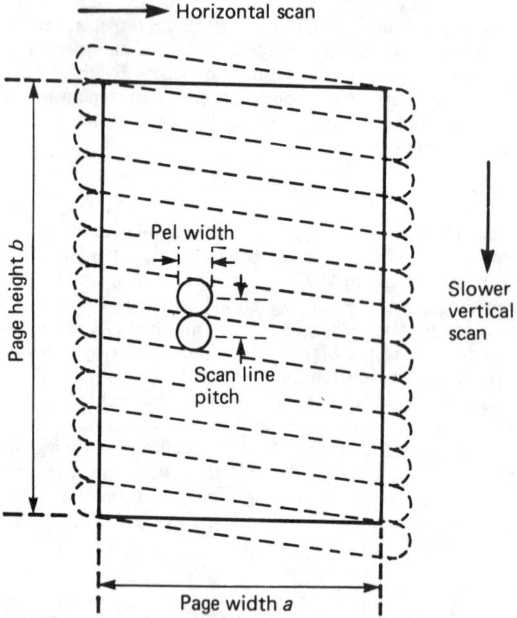

Figure 62.7 The raster scan principle. Horizontal resolution, $r_h \simeq 1$/pel width; vertical resolution, $r_v \simeq 1$/scan line pitch

essential differences lie in the use of paper or film-based input and output media and a frame scanning time increase of several orders of magnitude.

A mechanism that feeds the medium through the machine can also provide scanning in one direction. This has led to a multitude of economical designs in which the horizontal and vertical scanning mechanisms are quite different—a further contrast to TV practice. Very similar raster-scanning mechanisms can be used for image input and output and in many cases this leads to very cost-effective combined mechanisms. A wide variety of configurations are possible and will be discussed later, but first it is necessary to discuss the requirements of various applications.

62.4.2 Resolution

The spatial precision with which an image can be encoded or printed by a raster scanner is commonly measured in lines per millimetre or lines per inch. This contrasts with photographic usage, where line *pairs* per millimetre are normally quoted. The subjective appearance of an image depends on several factors. Firstly, since the scanning methods used for the horizontal and vertical directions are, in general, different, the corresponding resolution figures will also be different. It is found that a worthwhile improvement may result if one dimension has up to twice the resolution of the other. This improvement is useful as it allows quality to be traded for speed by changing just the vertical resolution. *Table 62.2* lists typical resolution specifications for various applications. The performance of the human eye is such that about 8.5 lines/mm can be resolved at 400 mm reading distance.[6]

Table 62.2 Typical resolution figures for raster scanners

Application	*Lines/mm (horizontal × vertical)*
Facsimile transmission: cost of prime importance	3.2 × 2.6
Facsimile, CCITT group 3	3.8 × 3.8
Office systems printer	12 × 12
Office printer with halftones and colour	24 × 24
Graphic arts image correction	80 × 80

62.4.3 Kell factor

Suppose that a black dot is scanned by an aperture of the same width. There is then a chance that the dot will lie partly in the tracks of each of two successive scan lines and so be subsequently reproduced at twice its original width. Averaged over the whole document this effect reduces the resolution by a factor of about 0.7; this factor is known as the Kell factor.[6] It must be applied in the vertical direction (as defined in *Figure 62.7*) of all raster-scan systems. It also applies to the horizontal direction when lines of fixed reading elements such as the photodiode array of *Figure 62.10(d)* are used.

62.4.4 Grey scale

If shades of grey in the original document are to be faithfully reproduced, the system optical density transfer characteristic must approximate to the form shown in *Figure 62.8* curve A. This will require the transmission and perhaps storage of analogue or

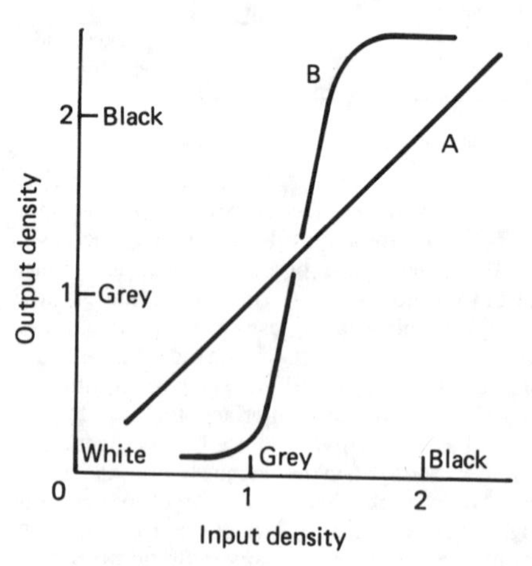

Figure 62.8 Optical density transfer characteristics. A, with grey scale; B, no grey scale

digitised analogue signals. The problem of reproducing analogue grey tones is usually avoided by substituting a pattern of variable-size black dots ('halftones') to give variable percentage cover of the surface. However, if the spatial frequency of these dot patterns is of the same order as that of the raster scan, 'Moiré patterns' will result from the beating of these two frequencies. To avoid this higher resolution must be used. Halftone grey scale print can be produced by raster printers.[7]

If, as is generally the case, shades of grey are not required, then imperfections in both white and black may be removed by nonlinear electronic processing such as that shown by curve B in *Figure 62.8*. This corresponds to a 'high gamma' photographic system. The ease with which this processing can be adjusted to clean up an image is one of the advantages that raster-scanned electronic copiers can offer over page-parallel photocopiers.

62.4.5 Colour

An input scanner may deliberately be made blind to certain colours, such as the boxes that define the layout of a business form, by a suitable optical filter. The more general case of reading full-colour information requires the use of three appropriately filtered photodetectors.[8] Output printing in colour requires separate printing devices for each primary colour. When intermediate shades and combinations of colours are desired, the printer resolution must be such as to permit electronic halftone generation.[7]

62.4.6 Multiple-element scanning

If, as shown in *Figure 62.7*, an image of dimensions $a \times b$ is to be converted into a serial data stream in time t by a single optical detector, and resolutions of r_h and r_v elements per unit length are required respectively in the horizontal and vertical scanning directions, then it follows that the scanning velocity (assuming that no time is wasted) equals abr_v/t. When high performance is required this velocity can become too high for mechanical scanners. Furthermore, the 'exposure time' available for each element of the image equals t/abr_hr_v, which also constrains the design of fast, high-resolution scanners. Both restraints can be relaxed by configurations in which a number of reading or writing elements are in use at the same time, with electronic conversion to or from the serial bit stream. For example, in *Figure 62.10(d)*, the exposure time of each element can approach t/br_v; that is, ar times greater than in the single-detector system.

62.4.7 Drum scanners

If the original document is wrapped around a drum (so that dimension a in *Figure 62.7* is circumferential) then the required scanning pattern becomes a spiral. This may easily be implemented by rotating the drum whilst moving the elemental scanner axially by a screw mechanism as shown in *Figure 62.9*. This simple principle has dominated raster-scanner design since its invention in 1850. Its optical arrangements are simple because the field of view need only cover one pel, and by using one of the direct writing processes to be discussed later, the same scanning mechanism can operate as a raster printer. The principle is therefore of considerable value to facsimile communication. However, the maximum scanning rate is limited by the need for

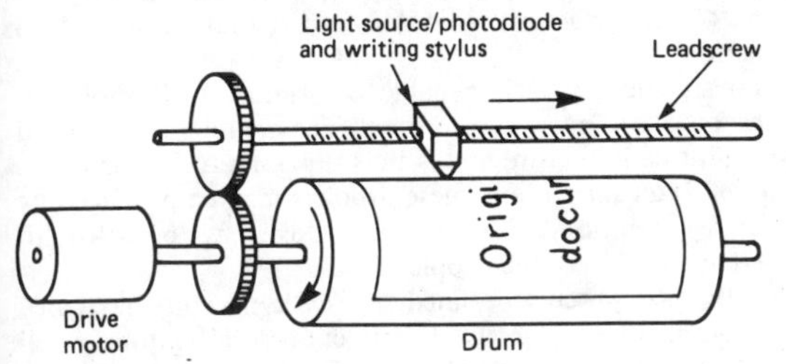

Figure 62.9 Drum scanner principle

precision balancing of mechanical components, and mechanisms must be provided to load, retain and subsequently unload the paper from around the drum.

The rotational speed of the drum must be tightly controlled. It is usual to employ a synchronous motor driven from an electronic power inverter whose frequency is derived from a quartz-crystal source. When varying resolution is to be provided by an adjustable 'gear ratio' between the drum and the leadscrew the latter is turned by a separate stepper motor driven from an adjustable frequency divider from the same source.

62.4.8 Flat-bed scanners

A basic alternative to the drum scanner uses the movement of the paper through the machine to provide the slower 'vertical' scanning action. The horizontal line actually being scanned is kept flat; hence the name 'flat bed'. A great variety of line scanning methods have been proposed for image input and printing, and some of these are discussed in the next section.

62.4.9 Image input scanners

62.4.9.1 Optics

Images are invariably sensed using visible light and a photodetector, with optical filtering added in order to approximate to the response of the human eye. Optical systems are angled so as to avoid the detection of light reflected from the surface of smooth documents. Commonly used light sources include xenon arc lamps and lasers as well as incandescent and fluorescent lamps. It is usually necessary to supply these with smoothed direct current or pulses synchronised to the scanning process in order to avoid modulation of the image. At least in high-resolution or high-speed scanners, the need to obtain sufficient illumination to give an adequate signal-to-noise ratio from the photodetector is a basic design constraint. The designer of lamp power supplies therefore aims to get the maximum possible useful output consistent with the required lamp life.

62.4.9.2 Line scan methods

Figure 62.10 shows a variety of methods by which one line of an input document may be scanned.

In *Figure 62.10(a)* the whole area of the document is illuminated (typically by one or two tubular fluorescent lamps) and the photodetector is arranged to 'see' a single elemental area by the lens. A rotating mirror then scans this elemental field of view across the page. Just as for the drum scanner, precise control of the scanning motor is required.

In *Figure 62.10(b)* the optical arrangement is reversed, and an elemental area of light is scanned across the page and one or more photodetectors arranged to view the whole area. Electronic scanning of a light spot imaged from a cathode ray tube is shown in this example. This is an alternative that can out-perform rotating mirror scanning at high speeds. However, it is bulky, expensive and requires specialised high-voltage and electronic scanning supplies.

It is possible to scan both the light source *and* the photodetector aperture as shown in *Figure 62.10(c)*.

An attractive alternative is to electronically scan the outputs of a line array of CCDs or photodiodes. This has the advantages of multiple-element scanning discussed earlier. Typically, a monolithic array of 1728 elements is used to scan the width of an A4 page. Such elements and their scanning circuits may be integrated onto a single silicon circuit.

If illumination is provided by an array of light-emitting diodes, these may form part of the same assembly, and the optical arrangement may be simplified to produce a very compact system.[9]

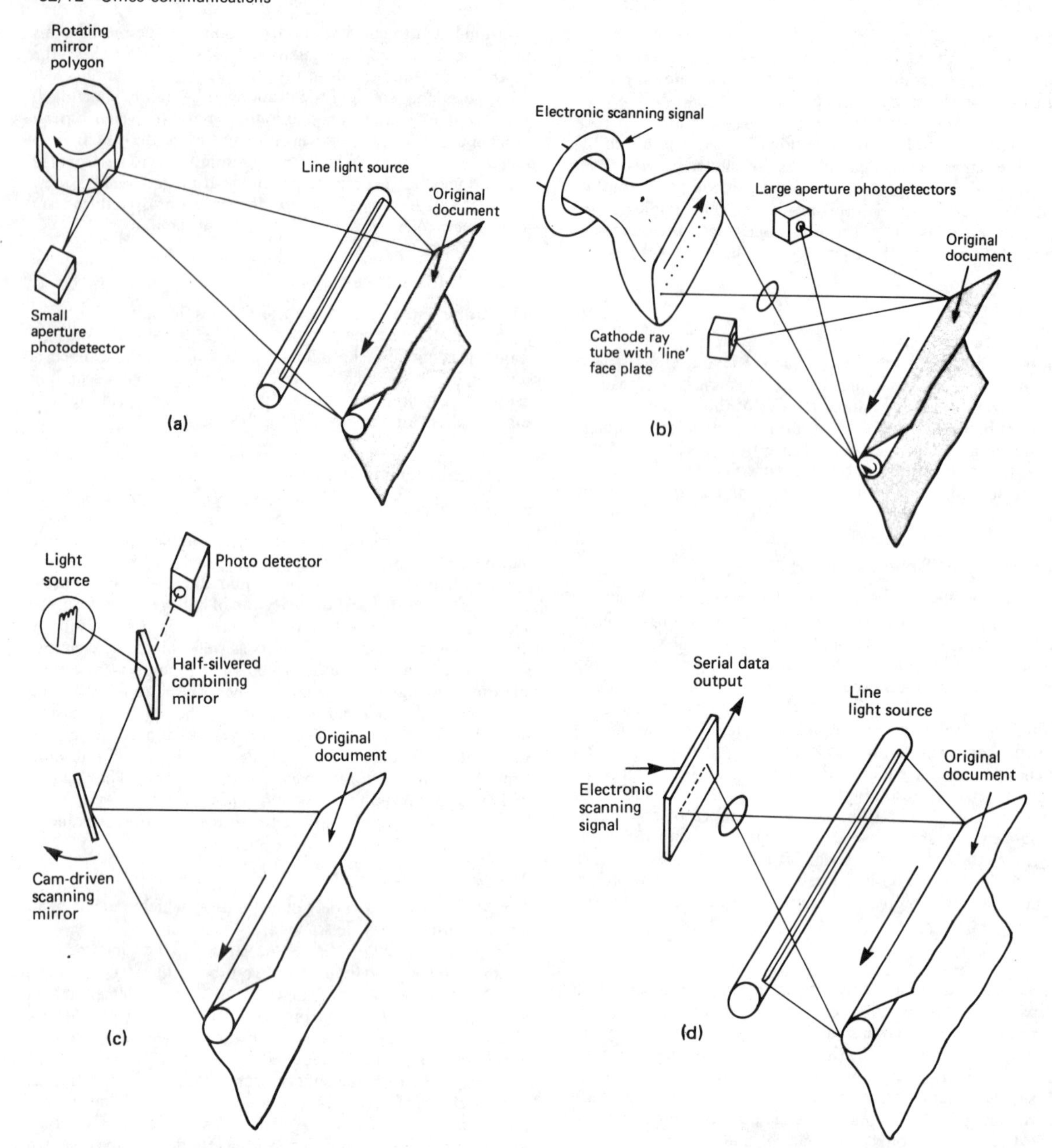

Figure 62.10 Methods of line scanning: (a) flying aperture, (b) flying light spot, (c) flying aperture combined with flying light spot, (d) electronically scanned CCD photosensor array

62.5 Output printers

62.5.1 Direct printers

Figure 62.11 shows examples of writing methods that produce a hard copy by a single process step. Although these are inherently cheap they are difficult to optimise. In particular, most use special papers that look and feel wrong and produce images that may lack permanence.

The electropercussive stylus, the raster-scanned equivalent of the typewriter, is shown in *Figure 62.11*(*a*). It is slow and noisy.

The electroresistive principle shown in *Figure 62.11*(*b*) is widely used with drum scanners, particularly those used in facsimile. Its disadvantage is the ash and smell resulting from the erosion of the surface. It is limited to a resolution of about 4 lines/mm.

Ink jet printing[10,11] is capable of printing high-resolution images onto ordinary plain paper. The liquid ink is forced through a small aperture and is thus broken into fine droplets. *Figure 62.11*(*c*) shows how these droplets may be produced by ultrasonic vibration. Electrostatic fields may be used to accelerate and deflect the droplets.

The thermal process outlined in *Figure 62.11*(*d*) has come into widespread use for cheap printers for calculators, small computers and facsimile machines. Since the transfer of heat to blacken the paper is an inherently slow process, multiple-element

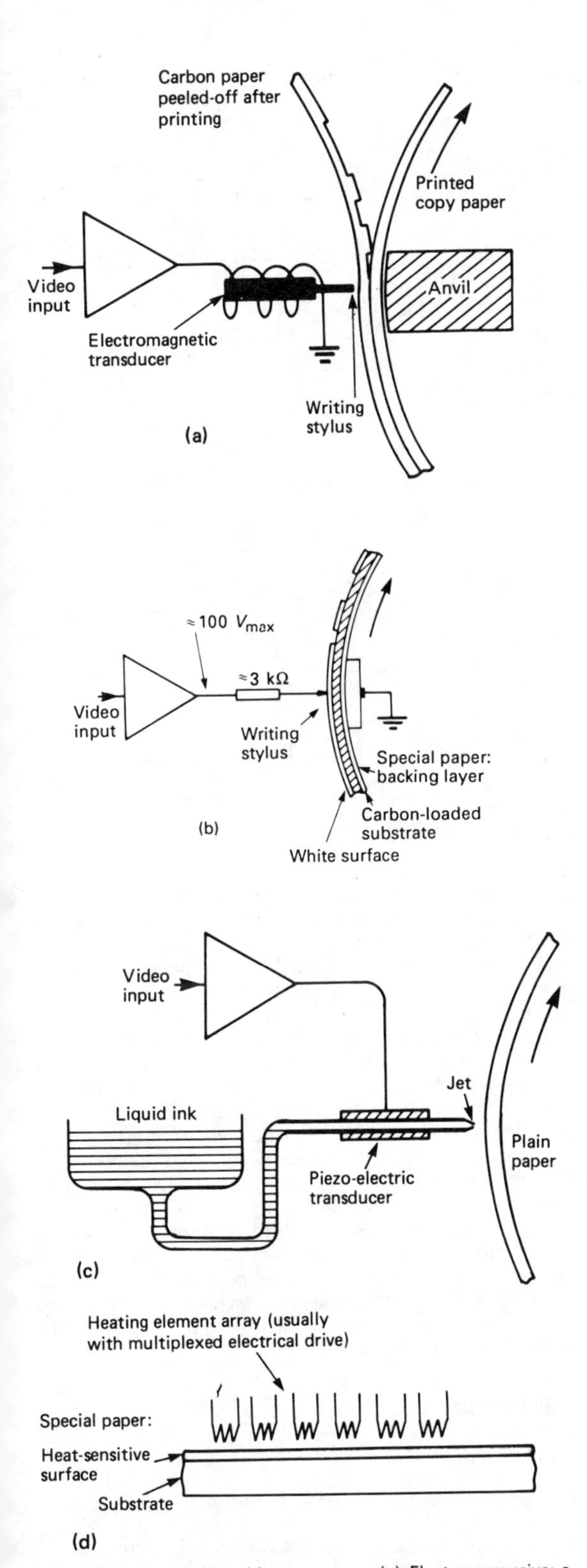

Figure 62.11 Direct writing processes. (a) Electropercussive: a vibrating hammer transfers pigment from carbon paper to the copy paper. (b) Electroresistive: the white surface layer is removed by spark erosion to uncover the black substrate. (c) Ink jet: the ink is held in the tube by capillary action until driven by the shock wave from the transducer. (d) Thermal: the paper coating darkens when the threshold is exceeded

arrays are usual. These arrays use resistors or diodes as heating elements and are energised by some sort of multiplexing arrangement to reduce the number of connections and driver circuits. Resolution is limited by constructional difficulties to about 8 lines/mm but arrays of 1728 elements are in common use.

62.5.2 Electrostatic printers

Electrostatic printers use 'dielectric paper' which acts as a capacitor. A pattern of electric charge corresponding to the image is laid onto the paper surface and this is 'developed' with liquid or powder ink ('toner') and then 'fixed' by heat or pressure just as in a photocopier. *Figure 62.12* shows the operating process of a roll-fed liquid-ink printer widely used as a graphic printer. The 'writing head' consists of an array of styli which are selectively driven to a voltage that ensures ionisation and consequent surface charging of the paper. The ionisation process provides a threshold (no charge is deposited at 400 V, full charge is deposited at 600 V) which allows the use of simple but specialised multiplexing techniques. This same threshold does, however, mean that analogue grey scale printing is hardly possible. 'Ionographic' printers use a similar principle to deposit charge onto a dielectric drum. This charge may then be developed by methods similar to those used in xerographic copiers.

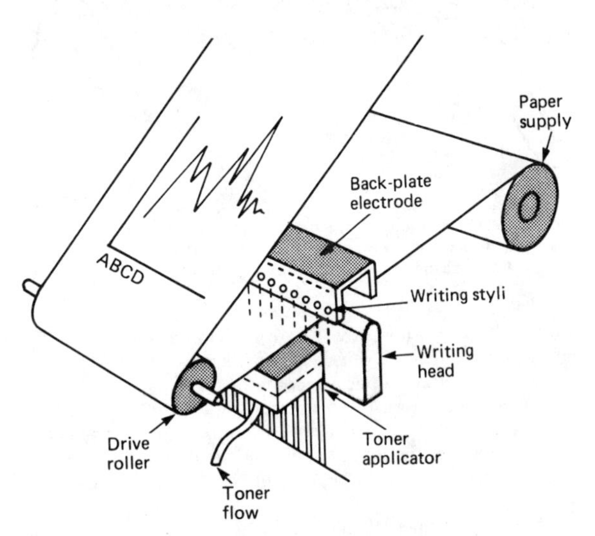

Figure 62.12 The electrostatic writing process (Courtesy of Versatec Inc.)

62.5.3 Optical intermediate printers

Light-sensitive materials such as photographic film and paper can be used as printout media by exposing them to an elemental spot of light that moves in the required raster pattern (*Figure 62.7*) whilst being modulated in intensity by the required video signal. Since subsequent development and fixing will be required the machine process is at least as complex as that needed for an electrostatic printer, but very high resolution and a good analogue grey scale can be obtained.

A crater lamp is a neon glow discharge tube shaped so that light from the crater in its cathode provides a source that can be focused onto an elemental area of the printout medium. *Figure 62.13(a)* shows how this may be applied to a drum scanner. The intensity is modulated by controlling the current and typically 1 to 75 mA might be required. Although this arrangement is very simple the scanning speed is limited by the low light output.

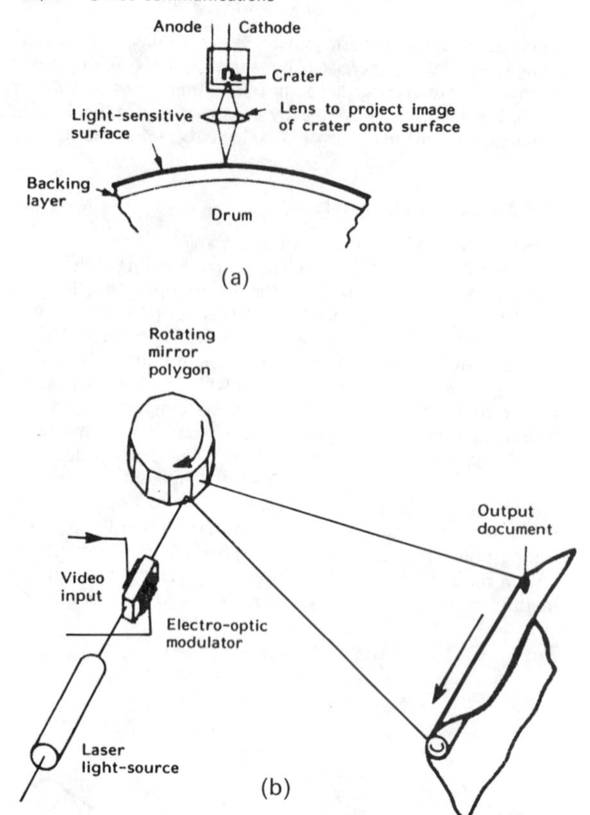

Figure 62.13 Printing arrangements that use light-sensitive copy paper. (a) Crater lamp arranged for use with a drum scanner. (b) Electro-optically modulated laser with rotating mirror scanner

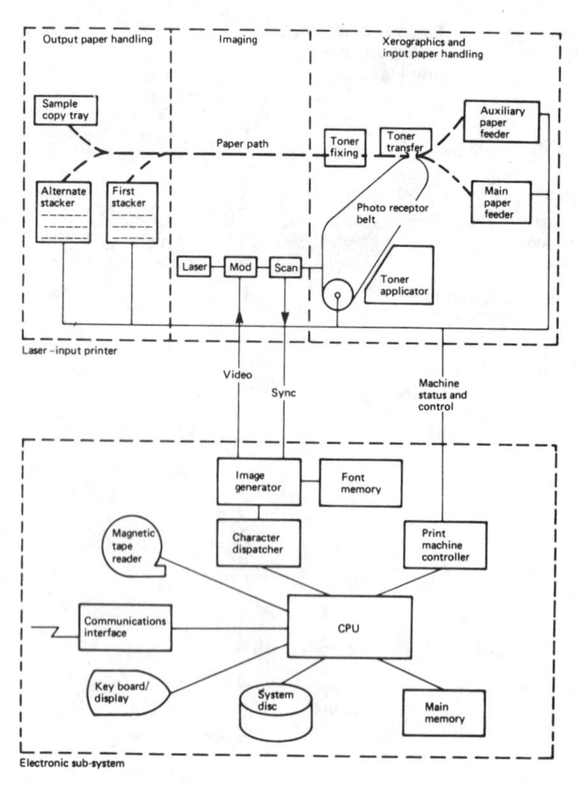

Figure 62.14 Printer system

Figure 62.13(b) shows how light from a laser or xenon arc lamp may be modulated, and caused to scan along a line by a rotating polygon. The electro-optical modulator has a capacitive input impedance and it requires drive circuits similar to those used for electrostatic cathode ray tube deflection. This arrangement closely resembles the input scanner of *Figure 62.10(a)* and the two may be combined to provide an economical input/output scanner.

The input scanner of *Figure 62.10(b)* is also adaptable for output printing, by modulating the electron beam of the cathode ray tube by the video signal, and substituting light-sensitive copy paper for the original document shown.

Photocopiers may be adapted to work as raster-scan printers by using one of the above methods to expose the photoreceptor.[12] The resulting machine is relatively complex but shares the economic benefits of quantity manufacture of the photocopier.

The upper half of *Figure 62.14* shows how the laser scanning arrangement of *Figure 62.13(b)* has been combined with a xerographic duplicator engine to produce a high-quality electronic printer that operates at 2 pages per second.[13,14]

62.5.4 Printer systems

The laser modulator of such a printer must be supplied with some 10 Mbits of video in each half-second page exposure time. To store this whole page in bit-mapped form is inconveniently expensive, so techniques have been developed for storing the information in a more compact form, and translating it into the line-by-line image signal when required.[15] Such compression requires the discarding of combinations that will not be required, or accepting a small statistical risk of being unable to store a desired combination. Alphanumeric characters and features such as vertical and horizontal lines and logos may be stored in a 'font memory' and fetched when required to build up the raster-scanning line.

The hardware required to do this is included in the lower half of *Figure 62.14*, which outlines the electronic subsystem that must be added to the laser-scanned raster printer to make a general-purpose phototypesetter/printer for use in an electronic office system. Input data are received from magnetic tapes produced by other systems or from a communication channel (such as Ethernet), and stored on the systems disk. Job description information, such as format of pages, choice of fonts, details of forms and logos, is prepared and edited by the operator. When a job has been completely specified the print machine is run by the machine controller, whilst the image generator, a specialised hard-wired electronic unit, fetches and assembles the information from the font generator as it is needed to produce the video signal that modulates the writing laser.

62.6 Facsimile

62.6.1 Systems

An input scanner can be used to transmit images over a telecommunications link to a remote printer. This principle has been in use commercially for news photograph transmission[16] since the 1930s. Combined send/receive machines suitable for office use became widely available in the 1960s, and between 1980 and 1987 the number of machines connected to the world-wide telephone network grew from a quarter of a million to two million. Facsimile provides very fast transmission of almost any documentary material without specialist preparation. However, between 2 million and 10 million bits are required for a raster scan of an A4 page; this has to be compared with 20 000 bits for a similar-sized page of ASCII-coded characters. This added transmission burden is somewhat alleviated by the ability to tolerate very high error rates.

The design of facsimile machines has been strongly influenced by the use of the PSTN (public switched telephone network). Firstly, the network was designed for speech, not data. Therefore the power/frequency/time characteristics of the transmitted facsimile signal must be chosen to suit the network. Secondly, transmission time is expensive, so effective data compression and channel modulation methods must be used.[17] Thirdly, since the public network is *switched*, every facsimile machine can, in principle, be connected to every other. Rigorous development and application of international standards is therefore necessary to ensure that this potential for interconnection is not wasted.[18]

Figure 62.15 shows a block diagram of a typical facsimile machine. When data is read from an input document it is first compressed and then modulated onto an audio-frequency carrier prior to being coupled to the line. The receive path is the reverse of this.

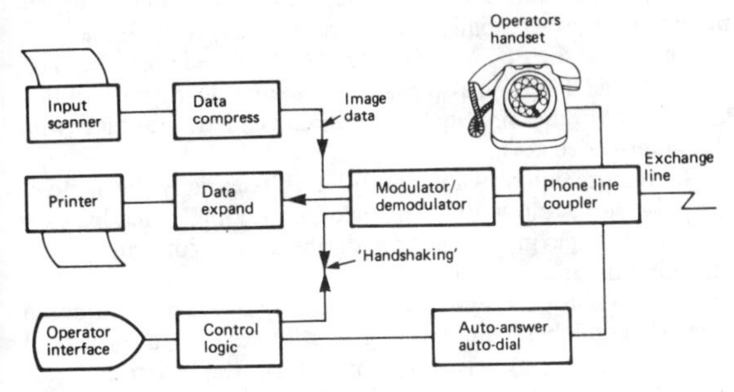

Figure 62.15 A facsimile machine

62.6.2 Handshaking

Control logic in each machine carries out a 'handshaking' process with the other machine designed to ensure that the two machines, which may be supplied by different manufacturers to different specifications, carry out the operator's instructions using the highest possible level of performance that is within their joint capability.

This handshaking is accomplished by an exchange of data, usually employing especially error-proof coding methods to reduce the risk of catastrophic misalignment. When the called machine first comes on line it announces its own capability. The calling machine compares this with its own capability and any special requirements of the operator, such as alternative resolution standards, and instructs the called machine to prepare itself accordingly. The sender then sends a line scan phasing signal. Once the receiver has achieved synchronisation with this signal it sends a 'confirmation for receive' signal. Upon receipt of this confirmation the sender commences image transmission. When this is complete another handshake sequence takes place to confirm safe delivery of the image and arrange for the transmission of the next page, if any.

The more sophisticated machines have automatic facilities for unattended reception and transmission to bridge time differences between users and allow the exploitation of cheaper telecommunication tariffs.

Operating standards of facsimile machines are recommended by the International Telegraph and Telephone Consultative Committee (CCITT), which has over the years defined a succession of 'groups' of increasing performance.[19] The principal characteristics of these groups are summarised in *Table 62.3*, and will be discussed below. It should be nored that many commercial facsimile machines are capable of operating in more than one group, and the necessary decisions are made as part of the 'handshaking' process.

62.6.3 Group 1 operation

Early systems for video transmission by wire used frequency modulation, chosen in the belief that it was the best method of overcoming the distortion of a telephone circuit. Since telecommunications authorities restrict the power in certain frequency bands to avoid interference with control functions,[20] the range 1300 to 2100 Hz was selected for the CCITT group 1 recommendation in 1971. This modulation system limits the maximum data rate, so that at a resolution of 3.8 lines/mm the transmission time of an A4 page is 6 min. Manufacturers therefore provided lower resolution modes to reduce transmission time (e.g. 4 min at 2.6 lines/mm) but these were not standardised. Margin stops are used to reduce the transmission time of short documents but since the recommendations were intended to apply to mechanical scanners whose speed could not be varied no other sort of picture compression was proposed.

62.6.4 Group 2 operation

By 1976 an amplitude modulation system had been developed that offered adequate signal/noise performance at substantially higher transmission rates. Recommendations were prepared to

Table 62.3 A comparison of CCITT facsimile group characteristics applicable to the Public Switched Telephone Network

CCITT 'group'	*A4 page time* (min)	*Horizontal resolution* (pels/mm)	*Vertical resolution* (lines/mm)	*Halftone capability*	*Handshake method* (*T30*)	*Video transmission modulation*	*Picture data compression*
1 (T2)	6	4	3.85	Common option	Tonal	FM Black = 2100 Hz White = 1300 Hz	None
2 (T3)	3	4	3.85	Option	Tonal FSK option	LSB AM Carrier = 2100 Hz	None
3 (T4)	less than 1	8	Std: 3.85 Option: 8	Rare option	FSK (V21) 300 bit/s 1650 Hz +1850 Hz	MPSK Std: (V27ter) 2.4/4.8 kbit/s Opt: (V29) 7.2/9.6 kbit/s	Horizontal: run-length Vertical: optional relative-address Channel: modified Huffman

Note: numbers in parenthesis refer to CCITT recommendations.[21]

standardise the use of a lower-sideband vestigial carrier system using 'alternate phase encoding', in which bandwidth is reduced by arranging that successive periods of full carrier are sent with reversed phase. The recommendation also defined improved 'handshaking' protocols that allow the use of unattended machines.

Some machines in this group employ 'white-space skipping' to reduce transmission time without compromise of quality. When a completely white area is identified a special code is sent and both sender and receiver scan quickly until further data are encountered. This technique can halve the transmission time of a simple business letter.

62.6.5 Group 3 operation

In 1980 agreement was reached on recommendations for a facsimile service that employs developments in digital modem technique, and image processing algorithms, to transmit an A4 page in under 1 min.[22]

The standard modem makes use of multiphase shift keying (MPSK) to provide eight (or alternatively four) modulation levels, so allowing a binary data transmission rate of 4800 bit/s (or alternatively 2400 bit/s) whilst using only an 1800 Hz carrier. An optional modem for use on lines of above-average quality provides data rates of 9600 bit/s or 7200 bit/s by combined multiphase and amplitude modulation.

The handshaking procedures are extended to provide for automatic testing of line quality and subsequent selection of modem data rate from the alternatives described above, together with standardisation of methods for indicating a wide variety of special machine features.

A substantial reduction in transmission time has been achieved by the use of three steps of digital data compression that exploit the correlation between nearby pels. The statistical nature of this process requires raster scan mechanisms that can operate at a rate that varies with the information content of the immediate image area, and so the phasing process used with earlier machines to achieve line synchronisation is no longer appropriate. Run length coding is applied along each horizontal scan line by converting the data to a list of the lengths of each of the black and white sections. Optionally, further lines may be encoded by reference to the previous line 'relative address coding'. Although such coding is not continued beyond four lines, to limit the image degradation due to an error being propagated from one line to the next, it can nevertheless provide about 30% reduction in transmission time. The above two coding processes provide a number of different signals, with widely differing probabilities of occurrence, which have to be sent over the channel to the receiver. 'Channel codes' are therefore chosen so that the most common signals are sent using the shortest codes. This 'modified Huffman' algorithm[23] when combined with the run length coding provides overall compression of data in a typical page[24] by a factor of 5 to 10.

Such coding, and the complementary decoding, must be performed in real time, and typically requires a dedicated microprocessor. Other microprocessors handle machine control and handshaking. Correction of amplitude variation in the reading process, removal of single-pixel transmission errors, and conversion between resolution standards may be provided by custom CMOS large-scale integrated circuits.[25]

To take advantage of cheap-rate transmission tariffs, facsimile machines are commonly equipped with auto-dial as well as auto-answer facilities. Large semiconductor or magnetic bubble memories can be used to enable unattended transmission or retransmission to many destinations: for example, a document may be sent once over a long-distance circuit to a machine that automatically redistributes the information to several other machines in its locality.

Facsimile machines may act as specialised input/output devices for computers, reading and writing graphic and character images locally and at a distance. Those computers that include the necessary modem and graphics hardware can be programmed to provide a complete facsimile function.[26]

62.6.6 The future

Specialised high-performance facsimile systems, such as those used in newspaper production, have been operated over private telecommunication circuits for many years. Typically these can transmit a 300 mm × 450 mm page in 1 min at a resolution of 40 × 40 pels/mm using a 10 Mbit/s channel.[27] However, the exploitation of general-purpose systems that surpass group 3 performance is dependent on the availability of public switched data networks of adequate performance. Since some countries have adopted circuit-switched networks, whilst others use packet switching, group 4 facsimile standards have been built in accordance with the ISO seven-layer architecture.[28] The protocols for this amount to an overhead approaching 50% of the useful data transmitted, so that only the circuit-switched network appears to have the capacity to allow facsimile system performance substantially better than that of group 3 machines connected via the PSTN.[29] However, the detailed protocols T62, T70, T72 and T73[19] do also enable 'mixed-mode' operation; that is, the easy combination of facsimile-coded graphics with character-coded text.

The picture compression algorithm is essentially the same as that of group 3, but with 'relative address coding' extended over the whole page in recognition of the low error rate of the incoming data.

Since text-based systems are universally based on inch-based typewriter pitches, the basic group 4 resolution has been standardised as 200 pels/25.4 mm in both the horizontal and vertical directions. This approximates to the 'optional' resolution of group 3, but can be expected to give improved appearance by virtue of the error-free transmission provided. Mixed-mode terminals must also offer 300 pels/25.4 mm resolution, and may offer 400 pel/mm capability, so providing letter-quality printing in both facsimile and word-processing modes.

References

1 COBINE, J. D., *Gaseous Conductors*, Dover Publications (1958)
2 DESSAUER, A. B. *et al.*, *Xerography and Related Processes*, Focal Press (1965)
3 THOURSON, T. L., 'Xerographic development processes: a review', *IEEE Trans. Electron Devices*, **ED-19**, 495–511 (1972)
4 SCHEIN, L. B., 'The electric field in a magnetic brush developer', *Electrophotography, 2nd Int. Conf.*, SPSE, pp. 65–73 (1974)
5 HENNIKER, J., 'Triboelectricity in polymers', *Nature*, **196**, 474 (1962)
6 PEARSON, D. E., *Transmission and Display of Pictorial Information*, Pentech Press, London (1975)
7 HOLLADAY, T. M., 'An optimum algorithm for half-tone generation for displays and hard copies', *Proc. Soc. Inform. Display*, **21/2**, 185–92 (1980)
8 HUNT, R. W. G., *The Reproduction of Colour*, 3rd Edn, Fountain Press (1975)
9 KOMIYA, K. *et al.*, 'A 2048-element contact-type sensor for facsimile', *IEEE International Electron Devices Meeting*, December 1981, 85CH1708-7, session 13.5, pp. 309–12 (1981)
10 KUHN, L. and MYERS, R. A., 'Ink jet printing', *Scientific American*, **240**, No. 4, pp. 120–8, 131–2, April (1979)
11 KEELING, M. R., 'Ink jet printing', *Physics in Technology*, **12**, 196–203 (1981)
12 BYCKLING, E., 'Laser printing', *Proceedings of Image Science '85*, Acta Polytechnica Scandinavica Applied Physics Series, No. Ph149, pp. 13–20 (1985)

13 SEYBOLD, J., 'Electronic printing and the Xerox 9700', *Seybold Report*, **8**, No. 5 (1978)
14 URBACH, J. C. *et al*., 'Laser scanning for electronic printing', *Proc. IEEE*, **70**, No. 6, 597 (1982)
15 GRAY, R. J., 'Bit map architecture realises raster display potential', *Computer Design*, **19**, part 7, 111–17 (1980)
16 COSTIGAN, D. M., *Electronic Delivery of Documents and Graphics*, Van Nostrand Reinhold, New York (1978)
17 COSTIGAN, D. M., 'Facsimile comes up to speed', *IEEE Communications Magazine*, pp. 30–5, May (1980)
18 JACOBSON, C. L., 'Digital facsimile standards', *ICC '78 Conference Record*, pp. 48.2.1–48.2.3 (1978)
19 INTERNATIONAL TELECOMMUNICATIONS UNION, 'Telegraph Technique', *Red Book*, Vol. 7 (1984)
20 BRITISH STANDARDS INSTITUTION, *Specification for general requirements for apparatus for connection to the British Telecommunications public switched telephone network*, BS6305:1982 (amended 1984)
21 INTERNATIONAL TELECOMMUNICATIONS UNION, *Red Book*, Vol. 8 (1984)
22 HUNTER, R. and ROBINSON, A. H., 'International digital facsimile coding standards', *Proc. IEEE*, **68**, No. 7, 854–67 (1980)
23 FORNEY, G. D. and TAO, W. Y., 'Data compression increases throughput', *Data Communications*, pp. 65–76, May/June (1976)
24 MUSMANN, H. G. and PREUSS, D., 'Comparison of redundancy-reducing codes for facsimile transmission of documents', *IEEE Trans. Communications*, **COM-25**, No. 11, 1425–33 (1977)
25 TADAUCHI, M. *et al*., 'LSI architecture of a facsimile video signal processor', *IEEE International Conference on Communications*, June 1985, 85CH2175-8, session 44.1.1, pp. 1400–4 (1985)
26 LU, C., 'Turning microcomputers into fax machines', *High Technology*, March, 60–1 (1987)
27 SCHMIDT-STOLTING, C., 'The Hell Pressfax system transmits complete pages of newsprint', *Telcom Report*, **9**, No. 4, 224–6 (1986)
28 BRITISH STANDARDS INSTITUTION, *Basic reference model for open systems interconnection*, BS6568:1984 (ISO 7498–1984)
29 BODSON, D. and RANDALL, N. C., 'Analysis of group 4 facsimile throughput', *IEEE Trans. Communications*, **COM-34**, No. 9 (1986)

Further reading

CAWKELL, A. E., 'The state of the electronic office', *Critique*, **1**, No. 2, October (1988)
COLEMAN, V., ERMOLOVICH, T. and VITTERA, J., 'Controlling local area networks', *Electronic Product Design*, October (1982)
FRANGOU, G., 'Networking office automation and applications: an integrated approach', *Proc. Int. Conf. Networking Technology and Architectures, Online*, June (1988)
HINDIN, H. J., 'Dual-chip sets forge vital link for ethernet local-network scheme', *Electronics*, October (1982)
INVERNIZZI, E., 'Information technology: from impact on to support for organizational designs', *MIT Working Paper*, Ser. Mgmt. 1990s, 88(057), September (1988)
LIEBERMAN, D., 'Office automation strives to juggle competing demands', *Computer Design*, November 15 (1987)
RADLETT, A.-L., 'Office document architectures', *Proc. Int. Conf. Networking Technology and Architectures, Online*, June (1988)
SKINNER, C., 'Showing paper the writing on the wall', *Systems Int.*, November (1988)
WANG, F. A., 'Office Automation', *Mini-Micro Systems*, December (1982)

Medical Electronics

D W Hill, MSc, PhD, DSc, FInstP, FIEE
North East Thames Regional Health Authority

Contents

63.1 Introduction

Hospitals, in common with many other large organisations, are making an increasing use of electronic instrumentation and computing. A good indication of the range of transducers, circuitry and applications is given in various publications.[1–3] The purpose of this chapter is to illustrate the broad approach required to obtain a general appreciation of the current role of medical electronics and instrumentation in health care delivery. New developments are covered by abstracting journals[4] and information services.[5] Journals such as *Medical and Biological Engineering and Computing* cover the field in depth, while books[6–8] provide a more detailed account of particular aspects of the subject.

In many cases, but not all, there is a good selection of high-quality apparatus from which to make a choice with a tendency towards 'package' systems and modular arrangements from which a system can be configured to suit an individual application.

An increasing awareness of the high cost of acute beds in hospitals is accelerating health care planners to limit the number of such beds. A consequence of this is the need to provide for better facilities for the care in the community of postoperative patients and also for better diagnostic techniques with a greater availability.

63.2 Diagnosis

Diagnosis is the first task confronting a doctor presented with a patient. At present comparatively little information is used in the general practioner's surgery (office) but in some countries computers are being used to hold patients' records and by decision algorithms and tree branching techniques to assist in arriving at a correct diagnosis.[9,10] It may well happen that a significant contribution to the diagnosis will occur as a result of laboratory analysis of one or more of the patient's body fluids.

An extension of the *ad hoc* diagnostic procedure is the routine bank of tests and questions administered to members of preventive medicine schemes in the form of a multiphasic screening programme which is computer based.[11]

It is evident that the impact of microcomputers on medical instrumentation will be widespread,[12] but it is interesting that the simple mercury-in-glass clinical thermometer still continues to hold its own in spite of the general availability of remote sensing and recording instruments based on thermistors or thermocouples.[13]

Other techniques which may be encountered are as follows.

63.2.1 Radiology[14,15]

One of the first major applications of current electricity in medicine occurred with the generation of X-rays, and X-ray sets now make use of sophisticated electronic control systems both for the accurate positioning of the major components of the imaging system and also for controlling the dose of X-radiation. In district general and teaching hospitals, the department of diagnostic radiology is one of the most expensively equipped, the others being radiotherapy and the pathology laboratories. Traditionally, the high voltage, typically up to 120 kV d.c., applied across the X-ray tube has been derived via a 50 or 60 Hz step-up transformer and valve or solid-state rectifiers. Modern practice is to use a medium frequency of a few kilohertz, which results in a much smaller transformer and increased efficiency of rectification. For the lower powered applications such as mobile X-ray sets, the transformer and rectifiers can be located in the tube head with the X-ray tube. Mobiles are usually motor driven and operate from built-in rechargeable accumulators.

Electro-optical image intensifers are in widespread use. X-radiation having passed through the patient falls on the input window of the intensifier, which provides a smaller but brighter output image for viewing via a television camera and monitor. A contrast medium which is opaque to X-rays is given to the patient—by swallowing to outline the stomach and gut, or by injection to outline the blood vessels of the brain or the chambers of the heart. The contractile action of the cardiac cycle or peristalsis of the gut can be recorded on cine film or video tape. A major advantage of image intensification is that it reduces the dose of X-radiation received by the patient. Sophisticated biplane image intensifier systems are used for heart studies, with contrast medium being injected via a cardiac catheter into individual chambers of the heart. Mobile image intensifiers fitted with a digital memory are widely used in operating theatres during hip pinning procedures.

The visualisation of blood vessels by the injection of a contrast medium has been facilitated by the development of digital subtraction angiography (DSA). An X-ray image of the region of the body in question is produced and stored in the digital memory of the computer via the image intensifier and an analogue-to-digital converter. The contrast medium is then injected using a power injector, and the image is stored when the contrast has reached the blood vessels to be studied. The first image is then subtracted from the second to give a clear image of the vessels. This subtraction approach can eliminate shadows on the image arising from bony structures such as the skull. In conventional X-ray angiography, the contrast medium is injected into a suitable artery. With DSA, in some situations it may be feasible to work with a venous injection.[16,17]

The direct application of digital computing to X-ray techniques occurred with the invention of computer-assisted tomography[18] (CT scanning), in which a beam of X-radiation is directed through the patient in a particular direction and falls onto a bank of detectors usually of a solid-state or high-pressure xenon type. The original CT scanners operated with a translate–rotate motion. The X-ray tube and the detectors were mounted in a gantry and on opposite sides of the patient. The tube and detectors moved sideways under computer control to scan the patient. The gantry rotated through a known angle and the tube–detector combination moved back to rescan the patient. The gantry then rotated through an additional similar angle and the next scan was made.

Each image is digitised and stored in the memory of the CT scanner's computer. From this set of images the computer subsequently reconstructs the X-ray image of that particular 'slice' of the patient. The couch on which the patient lies is then moved on by a known distance, typically in the range 1–10 mm relative to the gantry, and the next slice is scanned. By this means a pre-set series of images of known cross-sections of the patient's body is obtained. The first applications were concerned with the detection of tumours and problems such as brain stem damage and haemorrhaging in the head. CT scanning can distinguish small changes in X-ray density and thus can differentiate between a blood-filled haematoma and a fluid-filled cyst. Dedicated CT head scanners have now been replaced with CT body scanners. A major proportion of CT scanning is concerned with problems of the head and spine and the technique is particularly good at imaging bony structures.

In order to achieve scan times down to 1 s, the X-ray beam may be fan shaped and as many as 1200 detectors may be used. For many applications scan times of 10 s or more may be adequate, but faster scanning eliminates artefacts arising from cardiac, respiratory and peristaltic motions. In such scanners the X-ray tube and bank of detectors continuously rotate around the patient during the period of scanning. The slice approach makes it feasible to visualise a tumour in three dimensions and this has led to the development of radiotherapy planning computers.[19]

With this approach it is possible to superimpose on the CT image an outline of a beam of gamma rays from a radiotherapy cobalt unit or X-rays from a linear accelerator. The nature and position of the beam(s) can be adjusted to optimise the radiation dose received by the tumour.

A technique which is complementary to CT imaging is NMR imaging, known in North America as MRI imaging.[20] It does not use ionising X-radiation but is commonly operated by radiology departments. In contrast to CT, NMR imaging is particularly good with soft tissues but does not image bone.

Nuclear magnetic resonance imaging (or magnetic resonance imaging) does not use X-rays, but requires a large and very homogeneous electromagnet for whole body imaging, together with pulsed coils for systematically perturbing the field of the magnet and a radiofrequency system for exciting protons in the patient's tissues. Conventional wire-wound electromagnets have been used in the range 2×10^{-2} to 0.15 T with cryogenic superconducting magnets in the range 0.35 to 2 T.

NMR imaging depends on the measurement of the response of the protons within the hydrogen atom to a known radio frequency stimulus. When a patient is placed in a magnetic field, all of his protons align with the field and precess around it in a similar fashion to a child's spinning top precessing around the vertical gravitational field. When megahertz radiofrequency radiation from a coil closely fitting the patient is directed into the protons they absorb energy from the irradiating signal in a resonance absorption and change their alignment to the magnetic field. When the radio frequency is switched off, the nuclei radiate their surplus energy to their surroundings at the same resonant frequency. As the protons fall back (relax) to their original alignment, the resonant frequency emitted can be measured. The length of time for this relaxation to occur is associated with the environment of the protons, since the easier it is for them to pass energy to neighbouring atoms the more rapidly they can return to their original state. The intensity of the emitted radiation falls off exponentially with a time characteristic of the environment, the spin-lattice relaxation time.

In order to build up an image of the distribution of protons in body tissues it is necessary to distinguish between different regions of the body having the same resonant frequency. This is accomplished by applying a field gradient across the magnetic field in which the patient lies. Thus the field is slightly weaker on one side than on the other. The region in the slightly lower field will resonate at a lower frequency than that in the stronger field and can thus be distinguished from the higher frequency related to the area of higher field strength. The use of the field gradient facilitates the identification of the location of the particular type of tissue from the frequency of the emitted signal, the amount of water (or proton concentration) from the magnitude of the signal, and the spin-lattice relaxation time (T_1) from the decay of the signal. Different human tissues have their own values of T_1 and NMR can be used as a guide to distinguishing between primary and secondary malignant tumours and benign lesions. The method is not precise owing to the overlapping of values for normal and abnormal tissues. A well illustrated guide to the basic principles of NMR imaging of the human body is given in Reference 21.

Good quality proton images are obtainable with field strengths in the range of 0.3–0.5 T. With field strengths of 1.5–2 T, topical magnetic resonance is possible where an NMR spectrum can be derived from a particular location in the body which is also imaged. Changes in the NMR spectrum can follow changes of a biochemical nature occurring in living tissues, e.g. in the muscles of the forearm following the application of a tourniquet or in a muscle-wasting disease.

NMR imaging offers the advantages of not using X-radiation and of being able to angulate the image slice in any plane. It is, however, slower in use than CT scanning. Typically a CT scanner might handle 25 or more patients per day in contrast to perhaps to 8 for an NMR imager. The cost of an NMR imager is presently more than twice that of a top quality CT scanner.

63.2.2 Nuclear medicine techniques

X-ray imaging techniques have been refined via approaches such as DSA and CT scanning to the point where they are unrivalled for distinguishing fine details in anatomy. However, they cannot follow changes in the function of a particular organ. CT scanning has superseded radionuclide scanning for the localisation of tumours in the brain, but in other organs X-ray images may be made in parallel with isotope tests of functional tissue, e.g. the thyroid and kidneys. In the case of radionuclide imaging, the chosen gamma ray emitting radiopharmaceutical is administered in the form of a suitable radioactively labelled compound which is handled in a desired manner by the organ of interest. Chromium-151 labelled EDTA can be employed to measure the glomerular filtration rate, and iodine-125 labelled inulin to measure the renal plasma flow. A collimated crystal of sodium iodide is placed outside the body and close to the surface above the organ to be studied. As the compound is handled by the organ, gamma rays emitted pass through the collimator if they are travelling in the appropriate direction and strike the crystal, where they produce a scintillation of light which is detected by one or more photomultipliers. The height of the electrical pluse output from the photomultiplier is proportional to the energy of the incident gamma ray and a pulse height analyser can be used to distinguish one radionuclide from another if a dual isotope technique is used. Reference 22 gives an introduction to the methods of nuclear medicine.

Simple single-crystal scintillation detectors have proved to be of great value in renography for the study of renal function, but the use of a large (400 mm diameter, $\frac{3}{8}''$ or $\frac{1}{2}''$ thick) crystal in the form of a gamma camera is now routine for organ imaging, functional studies and the detection of tumour metastases in the skeleton.[23] The radionuclide technetium-99m with a 6 h half-life and a 140 keV gamma ray energy is widely used with gamma camera studies. A typical large field-of-view gamma camera might have 61 or 75 photomultiplier tubes coupled optically to the rear surface of the crystal. A range of collimators (medium energy, high energy, pinhole) can be mounted in front of the crystal (*Figure 63.1*).

Positron-emitting isotopes such as gallium are used in the technique of positron emission tomography (PET).[24] A positron travels no more than a few millimetres in tissue before it is captured by an electron, resulting in the emission of two gamma rays travelling in opposite directions. These can be detected from outside the body and, by reconstructing the lines joining many such gamma ray pairs, a three-dimensional picture of the positron-emitting region can be built up and shown as a series of slices through the patient. PET can quantitatively measure functions such as cerebral and coronary artery blood flow.

Conventional positron emission cameras detect the two gamma rays simultaneously with multiple rings of scintillating crystals and large numbers of photomultipliers. Hence they are expensive and they are used with an on-site cyclotron to produce the necessary radionuclides, such as carbon-11, oxygen-15 and nitrogen-13. A less expensive approach uses multiple lead photon converters added to a proportional gas ionisation counter. Two banks of rotating gamma ray detectors are situated one on either side of the patient. When the gamma rays strike the lead strips, the electrons produced cause ionisation in the gas which is detected on the positively charged anode lines stretched across the frame of the detector. The resulting data is processed to provide three-dimensional images in real time. Tumours as small as 5 mm in diameter can be detected deep in the body.[25]

Figure 63.1 Double-head large field of view scanning gamma camera providing facilities for positron emission computed tomographic images (Courtesy of Siemens Ltd)

Single Photon Emission Tomography (SPET)[26] uses conventional gamma camera instrumentation to measure regional cerebral blood flow using technetium-labelled compounds.

63.2.3 Thermography

Basically, this is a heat-sensing method of imaging which records the temperature distribution patterns from the surface of the patient's body in the region of interest. A television-compatible output format is now adopted, with the line scan accomplished by a rotating polygon mirror and the frame scan by a rocking mirror. A 45° mirror is often placed above the patient and the heat reflected into the scanning system and on to a cooled cadmium mercury telluride detector or a pyroelectric detector.[27,28] Originally, the use of thermography was concentrated on a non-invasive non-ionising radiation method for breast cancer screening. It became clear that it was essential that the subject be thermally in equilibrium with the environment and the interpretation of breast heat patterns is more complicated than was originally envisaged. Progress is being made with the application of computers.[28] Current systems have a resolution of 0.1°C and can produce colour contour displays of the isotherms. Successful applications include the detection of deep vein thrombosis, the effect of drugs on the peripheral circulation and the study of burns.

63.2.4 Ultrasonic imaging

The great advantage of ultrasound over X-rays is that, at the power levels required for diagnostic imaging, ultrasound has not been shown to damage tissues. Thus it has to be of inestimable value in the examination of pregnant women since it has not produced deleterious effects on the fetus. A good general account of the subject is given in reference 29. Static B-scanners were widely used with a hand-held transducer typically operating at 3 MHz and which is moved over the surface above the body region of interest in a raster pattern. Modern systems are capable of producing an image with a grey scale;[30] at least 16 levels are usual. Simple A-scanners with a single transducer serving as both transmitter and receiver and displaying the echoes received from the various tissue interfaces are used with the echoes displayed on a time base to measure any mid-line shift in patients with a suspected head injury.

As the name implies, static scanners cannot be used to observe the real-time motion of body organs, e.g. peristalsis or arterial pulsations or the movements of a fetus. Much recent work in this field has led to the development of four main types of real-time systems: linear array, annular array, rotating crystal and sector scanner.[31] Linear arrays are much used for abdominal and obstetric work, and sector scanners for obstetric and cardiac investigations. Reference 32 discusses the methods and terminology encountered in ultrasound imaging systems. Multi-transducer imaging systems for the heart[33] offer the prospect of reducing the number of patients submitted to invasive cardiac catheterisation procedures for the determination of conditions such as incompetent heart valves.[34]

63.3 Electronic instruments

63.3.1 Cardiovascular instruments

The heart is a vital organ of the body and a substantial effort has been devoted to investigating and controlling its pumping action. It may be necessary to record the electrical potentials generated by the beating heart, the pressures developed, the blood flow and volume pumped, the heart and pulse rates and the heart sounds. The rate can be controlled by a suitable form of cardiac pacemaker and if the heart goes into ventricular fibrillation it can usually be restored to a normal sinus rhythm by means of a defibrillator.

63.3.2 Blood flow measurement

Blood flow can be directly detected and measured by two main methods:

(1) Electromagnetic.
(2) Doppler shift ultrasound techniques.

There are other possibilities such as the use of a laser and the use of thermal flow probes.

63.3.2.1 Electromagnetic blood flow meters

Electromagnetic blood flow probes have to make a snug fit around the vessel in which the flow is to be measured and thus require a surgical exposure of the vessel. The design of the probe and associated cable and connector has to be such that the assembly can be sterilised and is capable of operating while surrounded with body fluids. The principle of operation depends upon Faraday's law of electromagnetic induction, *Figure 63.2*.

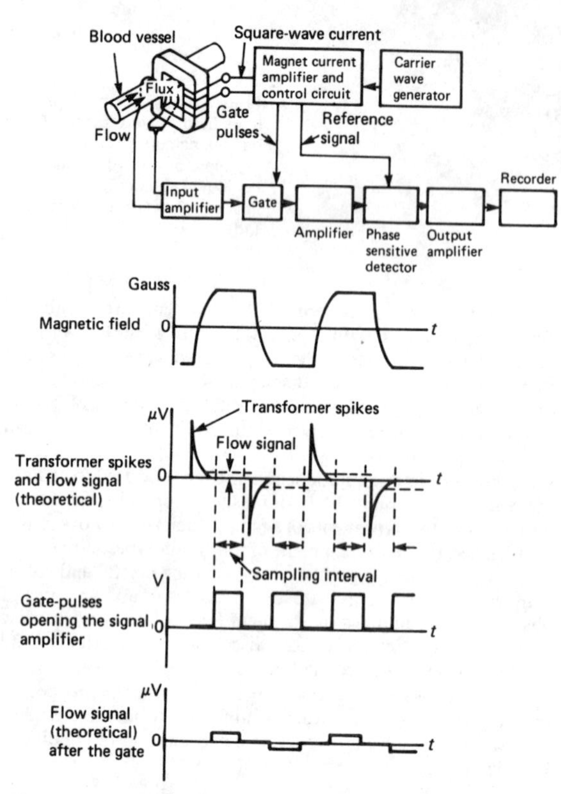

Figure 63.2 Basic principle of operation of a square wave electromagnetic blood flow meter showing the use of gating to eliminate the unwanted transformer spike signals[6]

The direction of flow of electrically conducting blood corresponds with the flow of electric current, and an electric potential whose magnitude is proportional to the velocity of flow is developed in a third plane mutually perpendicular to the plane of flow and the plane of the applied electromagnetic field. Care is required in the probe design because of the small size of the flow signal consequent upon the limited field strength which can be produced without giving rise to a significant heating of the coil. Early flowmeters used a permanent magnet and generated a d.c. flow signal. However, artefacts arising from electrode polarisation changes gave rise to baseline shifts and a.c. systems are now used. The time-dependent magnetic field introduces an unwanted signal which is substantially independent of the blood velocity and is proportional to the rate of change of the field but which is much easier to eliminate than the fluid electrode potentials.[35] This signal arises from both the linkage of flux with the connecting leads to the electrodes and from eddy currents generated in the blood, the vessel wall and the surrounding tissue. It is known as the 'transformer e.m.f.' or 'quadrature voltage'. Both sine wave and square wave excitation of the electromagnet have been employed in practical flowmeters. For a 240 Hz sine wave flowmeter, the wanted flow signal was of the order of microvolts and the unwanted quadrature signal was about 10 times larger.[35,36] A demodulator is employed to extract the flow signal and eliminate the quadrature signal. The electromagnetic flowmeter is able to distinguish forward and reverse flows and to follow the flow pattern during each cardiac cycle. The basic signal is proportional to the blood velocity and thus to the volume flow rate if the vessel's cross-sectional area is known. If the volume flow rate signal is integrated for a known time the result is a signal equivalent to a certain volume of blood. Thus an integrating flowmeter can be calibrated by occluding the normal flow through the vessel surrounded by the flow probe and then injecting a known volume of blood through the probe from a syringe. Flow probes which can be chronically implanted have led to detailed studies of drug action on the circulation.[37] An extractable flow probe which can be placed around the thoracic aorta during surgery and postoperatively withdrawn has been used in man.[38,39]

63.3.2.2 Doppler shift ultrasound

The apparent shift in the frequency when a relative motion occurs between a source of sound and the observer constitutes the 'Doppler effect'.[40] It has been used in the detection of fetal heart movements and in the study of blood flow.

Two piezoelectric crystals in a suitable housing are placed on the surface of the body over the blood vessel or region of interest. The transmitter crystal generates a continuous beam of ultrasound at about 5 MHz and this is reflected back to the receiver crystal from the moving structure in the body. There is an effective increase in frequency when the reflector moves towards the probe and a decrease when it moves away. The change in frequency, the Doppler shift, is usually in the audio range and after electronic detection can be heard on a loudspeaker. Signal processing of the Doppler shift frequency spectrum can be made in real time to indicate the maximum flow velocity and the blood flow pattern. Doppler flowmeters have made an important contribution to the study of occlusive arterial disease.

Lasers are being applied via a microscope for blood velocity studies[41,42] and a thermal dilution probe has been used for the measurement of venous flows.[43,44]

63.3.2.3 Indirect methods

Although indirect methods are not so exact as direct techniques they are most important since they are usually non-invasive and can be applied to people such as outpatients and pregnant women. Plethysmography has proved its worth over the years and is basically a method for measuring the volume changes occurring with time in a particular segment of a limb such as a finger or the calf. In venous occlusion plethysmography, a cuff is applied proximal to the segment and pumped up to a pressure somewhat below the subject's diastolic pressure, i.e. sufficient to occlude the venous outflow from the segment. The arterial inflow continues, causing the segment to swell, and the rate of volume increase is measured and is proportional to the blood flow into the segment. A thin mercury-filled rubber tube can be placed around the circumference of the segment and the increase in its electrical resistance with the swelling is recorded by means of a bridge circuit.[45] The inflation and deflation of the cuff can readily be automated.[44]

Impedance plethysmography usually operates with a constant current technique with frequencies in the range 25 to 11 kHz. Four disposable band electrodes are placed around the circumference of the segment with the current of about 4 mA r.m.s. passed between the outer two electrodes and the resulting voltage change detected at the inner pair.[46] A correlation coefficient of 0.877 has been found in patients[47] between the mercury strain gauge and impedance methods. A block diagram of the commercial Minnesota Impedance Cardiograph is shown in *Figure 63.3*. Impedance plethysmography has been successfully used for the detection of deep vein thrombosis.[48]

Photoelectric plethysmography measures the amount of light transmitted or reflected at the ear, finger or vagina. The amount of light received is influenced by the optical density of blood and varies with the cardiac cycle and the amount of blood present in the capillaries.[49]

63.3.3 Blood pressure measurement

Both invasive (direct) and non-invasive (indirect) methods are well established.[50] Direct measurement requires the cannulation of an artery or central vein, as appropriate, but provides a continuous record of pressure changes. For arterial sites the systolic, diastolic and mean pressures can be measured for each cardiac cycle. Pressure transducers are employed to convert the pressure changes into a corresponding electrical signal. Capacitance change, strain gauge bridge or differential transformer principles have all been used.[51] Currently, emphasis is placed on silicon diaphragm transducers in which a set of silicon strain gauges are laid down on the rear of a silicon diaphragm which deforms slightly under the influence of the blood pressure. Such transducers are mechanically rugged and can be made in a flat format suitable for taping to a wrist. The latest development consists of disposable blood pressure transducers which work on a capicitance change affecting the frequency of an oscillator circuit. A small hollow chamber or 'dome' is located in front of the diaphragm and fitted with connections for two taps to join up with a cannula/catheter for connection to the patient and with a calibration pressure source and a pressurised container of heparinised saline for flushing the transducer dome and cannula. To avoid problems of cross-infection the domes are made from plastic, with the floor of the dome consisting of a slack plastic membrane which sits on the diaphragm of the transducer. The domes can be purchased in packs, are disposable, and also help to electrically insulate the transducer from the patient's blood stream.

In order to obtain a frequency response of some 20 Hz when connected to a cannula, it is important that the chosen transducer has a small volume displacement (0.01 mm^3/100 mmHg). A valuable development has been that of strain gauge microtransducers,[52] which can be mounted at the tip of a cardiac catheter and have a frequency response of several kilohertz which also enables them to record heart sounds. Sophisticated catheters may have two such transducers near the tip to measure the gradient across a heart valve and also a blood flow transducer.

Indirect methods commonly make use of a microphone or stethoscope to detect the onset and cessation of the well known

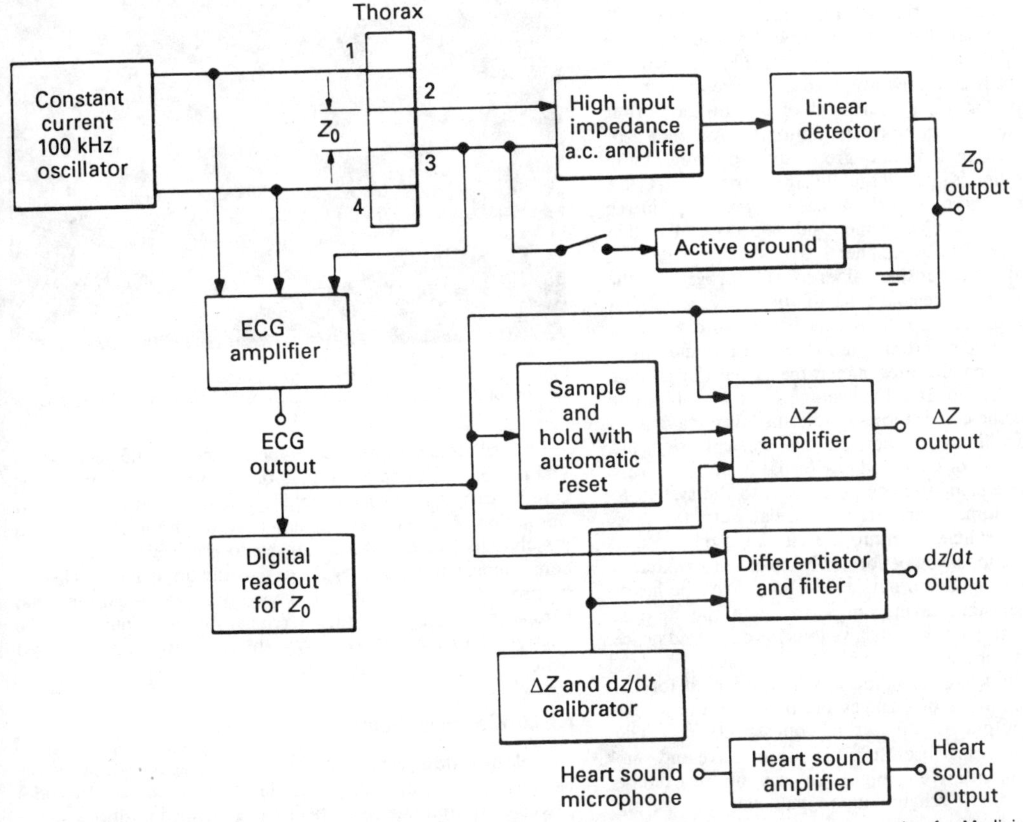

Figure 63.3 Schematic diagram of the IFM/Minnesota impedance cardiograph (Courtesy of Instrumentation for Medicine Inc.)

Korotkoff sounds corresponding to the pressure in the inflation cuff equalling the systolic or diastolic pressures respectively.[53,54] A beam of ultrasound shone on the arterial wall has also been used to detect the onset and cessation of the wall motion at the systolic and diastolic pressures.[55] The operating principle of the ultrasonic arteriosonde monitor is shown in *Figure 63.4*.

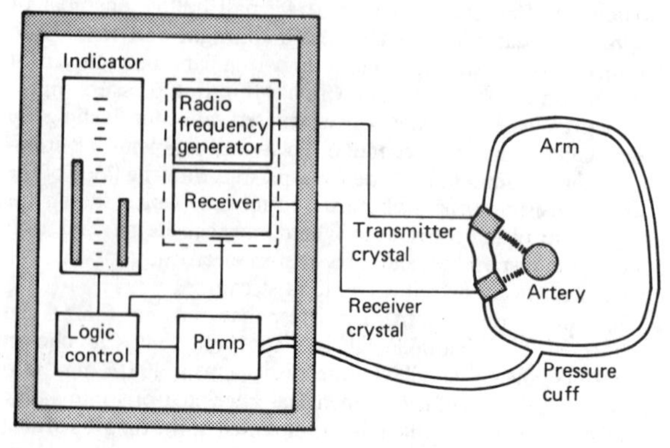

Figure 63.4 Schematic diagram of an ultrasonic indirect blood pressure monitor[6]

63.3.4 Cardiac output and stroke volume

The cardiac output is the volume of blood in litres per minute pumped by the heart. The stroke volume is the volume of blood pumped per beat. In some methods, such as the use of an extractable flow probe[39] or the electrical impedance method, the stroke volume is measured and then multiplied by the heart rate to obtain the cardiac output. In the indicator dilution methods the cardiac output is measured averaged over several beats.

Direct methods for the measurement of the cardiac output require access to the arterial circulation.[13] An indicator substance such as indocyanine green dye, or now more commonly cool saline, is injected rapidly into the right heart via a suitable catheter. It becomes well mixed with the blood in the heart and pulmonary circulation and is subsequently detected in the pulmonary artery or the peripheral arterial circulation by a catheter-mounted optical or thermal sensor. The progress of the indicator past the sensor generates the dilution curve which is a plot of the indicator concentration in the blood plotted against time on a chart recorder. From a knowledge of the amount of indicator injected and the area under the curve the cardiac output can be calculated. It is the average value over the time taken to describe the curve. Fibre-optic transducers have been developed which will monitor both the dye concentration and the oxygen saturation of the blood.[56] For use in intensive care situations, a simple computer can be built in to the system to calculate stroke volume, heart rate and cardiac output.

Iodine-131 labelled human serum albumin can also be used as the indicator, detection being performed by means of an external scintillation counter mounted outside the body over the heart. This method overestimates in comparison with the thermal dilution method, but requires only a venous injection and hence is suitable for use with outpatients.[57]

The electrical impedance method has been adapted for the beat-by-beat monitoring of changes occurring in the stroke volume, cardiac output and myocardial contractility.[58–60] The technique is by no means absolute but is non-invasive and very useful for following relative changes, for example in exercise studies with cardiac rehabilitation patients or during anaesthesia.

63.3.5 Defibrillators

The life-threatening condition of ventricular fibrillation occurs when the contractile action of the fibres (or fibrils) of the heart muscle no longer occurs in a cyclic coordinated fashion. In the majority of incidences the random 'squirming' motion of the fibrillating heart can be converted back into a regular pumping action by the application of a controlled electric shock to the chest wall. If at operation the chest is open, a smaller shock can be applied directly to the heart. Flat 'paddle' electrodes are applied firmly to the chest well, or spoon electrodes are placed around the heart. In a d.c. defibrillator a capacitor is charged up to a maximum energy of 400 J requiring about 7 kV for a 16 μF capacitor. Since irreversible brain damage can occur within a few minutes of the cessation of an effective cerebral circulation, defibrillators are required to be instantly available wherever the patient has collapsed. Lightweight, portable versions powered by rechargeable batteries are available.[13]

A d.c. defibrillator can also be used to reverse the less serious condition of atrial fibrillation. It is now necessary to ensure that the discharge of the smaller shock needed for this application does not coincide with the T-wave of the patient's ECG, since if this occurs the heart might be thrown into ventricular fibrillation. A synchronised defibrillator is used in which the capacitor discharge is initiated after a suitable time delay (typically 25 ms) triggered from the preceding R-wave of the ECG. *Figure 63.5* shows a portable battery-powered defibrillator fitted with a cardioscope and recorder for observing the ECG.

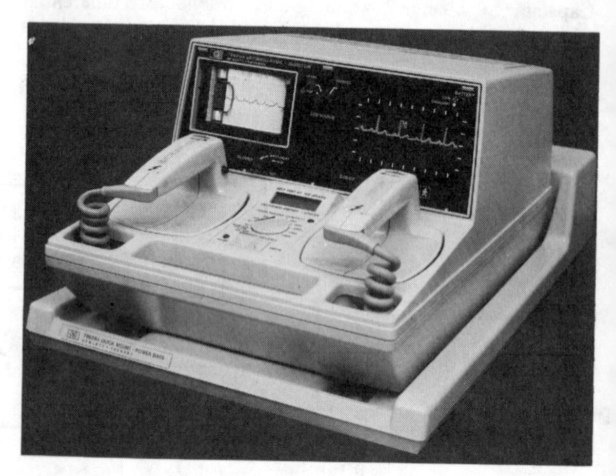

Figure 63.5 Portable battery-powered defibrillator with a built-in cardioscope and ECG recorder (Courtesy of Hewlett Packard Ltd)

Defibrillators have been developed which are sufficiently small and reliable to be implanted in patients who suffer from marked cardiac arrhythmias which could lead to the onset of ventricular fibrillation. The device is implanted in the abdomen with leads attached to two electrodes sutured to the heart. The ECG is continuously monitored by the device and up to 100 shocks can be provided from the stored energy. Whereas a conventional cardiac pacemaker provides a regular series of pulses to the heart, each of some 25 μJ energy, the shock from an implanted defibrillator has an energy of 25 J.

63.3.6 Electrocardiographs

The electrocardiogram (ECG) is the plot against time of the electrical activity of the heart as detected by surface electrodes placed on the limbs of the patient and at defined positions on the thorax. It is capable of giving much information on the history of a 'heart attack' and on the heart rhythm, although it cannot provide information on the pumping capability of the heart except in terms of rhythm. It is a widely used diagnostic technique.

As with all electrical devices connected to the patient and to the mains supply, the question of electrical safety is important and modern ECG recorders are provided with fully floating input circuits which limit the leakage current which can pass through the patient to some 10 μA. Modern circuits have high input impedances in excess of 5 MΩ and common mode rejection ratios better than 60 dB so that interference-free tracings are easily obtainable without mains interference.

Twenty-four hour continuous recording of the ECG (ambulatory monitoring) is of value in determining the cause of transient fainting attacks and rhythm disturbances. Miniature FM tape recorders are now available which can record the ECG, EEG, blood pressure and other physiological signals and can be easily worn by the subject.[61]

In intensive care and coronary care units, automatic detection and classification of arrhythmias in the ECG materially assists prompt clinical intervention and the use of medication. Hard-wired systems are now giving place to software systems.[62,63] Another developing application is in the use of computers for the interpretation of ECGs.[64] Until recently, a minicomputer has been used to analyse a standard 12-lead ECG but recent developments make use of a microcomputer mounted in the ECG trolley.

63.3.7 Cardiac pacemakers

With a normal heart the group of cells forming the natural pacemaker is located by the right atrium and supplies regular impulses, at the heart rate, which travel through the conducting nervous pathways of the heart to cause the ventricles to contract. If the natural pacemaker or the conducting system fails, the ventricles contract spontaneously at a slower rate. This can result in an inadequate cardiac output, particularly under exercise conditions, as can a partial failure of the conducting system (heart block). The missing stimuli can be provided from a fixed rate (70 pulses/min) stimulator or from a stimulator that fires only when it detects that a natural stimulus is absent. Such stimulators are known as cardiac pacemakers and they are electrically connected to the right heart via a flexible lead, the pacing catheter. For short-term pacing for acute situations in hospital an external stimulator is used, but for long-term chronic pacing the device is implanted within the body—usually in the axilla.[65,66]

The circuitry of pacemakers which can respond by inhibiting their stimulus on receipt of an impulse from the heart's own pacemaker is complex and demands great reliability. The microcircuitry is hermetically sealed inside a metal enclosure covered with epoxy resin and an outer layer of silicone rubber to prevent rejection as a foreign body.

Interesting new developments are occurring in rate-responsive pacing. In one approach a crystal sensor in the pacemaker monitors the effect of suitable body motions which correspond with the need to change the heart rate. Other forms of pacemaker measure respiration in order to adjust the pacing rate to the subject's level of physical activity; whilst another system adapts the pacing rate to activity-related changes in the heart's electrical timing.

A new approach to the design of pacing leads involves a steroid drug embedded in a silicone rubber plug placed at the tip of the electrode. The drug is slowly eluted and diffuses into the cardiac tissue adjacent to the electrode and reduces the inflammatory response of the heart to the lead implantation. This makes the heart more sensitive to the pacing stimulus.

Some cardiac patients suffer from tachyarrhythmia, a heart rate which is too fast. Pacemakers exist which can adjust the heart rate back to normal. However, tachycardia can lead to fibrillation, an uncoordinated pumping action of the heart which can rapidly prove fatal. Research is ongoing to design an implantable pacemaker–defibrillator.

In 1986 approximately 250 000 pacemakers were implanted worldwide.

63.3.8 Transcutaneous electrical stimulation

Stimulators strapped to the body and applying electrical stimulation through the skin have been developed for the relief of back pain, sports activity related pain and pain arising from a diminished blood supply to the legs causing ischaemia. In the legs the technique has been reported to be simple, safe, non-invasive and acceptable to both patients and nurses.[67]

63.3.9 Heart sounds (phonocardiogram)

Multichannel physiological recorders usually make provision for the recording of heart sounds via a crystal microphone placed over the praecordium. The sounds are useful for timing events in the cardiac cycle, e.g. the left ventricular ejection time, for obtaining information on the operation of the heart valves and for calculating the 'systolic time intervals' as an index of cardiac performance.[68] Catheter-tip microphones and pressure transducers capable of handling heart sounds are available for sound recording within the heart. Electronic stethoscopes are not in general use.

63.3.10 Pulse or heart rate meters

A pulse rate meter is actuated by the systolic pressure pulse taken from an arterial pressure transducer or from a photoelectric pick-up placed over a surface artery, e.g. at the wrist. A heart rate meter is usually triggered by each R-wave of the ECG. It can happen that the heart does not contract each time its natural pacemaker fires due to the presence of a heart block. The difference between the heart and pulse rates is the pulse deficit. Average rate meters are based upon the use of a leaky integrator but a number of designs exist for beat-by-beat meters.[69]

63.3.11 Respiratory instrumentation

Whilst studies of the heart command much attention, it is clearly of equal importance to be able to study respiratory as well as cardiac mechanics and to quantify the gas exchange in the lungs and the volumes and flow rates together with the composition of the respired gases. In addition to lung function laboratories run by physicians, anaesthetists are also much interested in respiratory measurements.[6] The main areas are pulmonary function studies, respiratory gas analysis, blood-gas analysis and ventilators.

Pulmonary function studies Sophisticated lung function analysers[70] include double spirometers for the measurement of lung volumes and a carbon monoxide analyser for pulmonary diffusing capacity measurement. Simpler spirometers with a timer will measure the peak flow rate, the vital capacity and the forced expiratory volume delivered in one second (FEV_1).

Respiratory gas analysis A wide range of apparatus is available[6] ranging from simple paramagnetic or polarographic oxygen analysers to infrared gas analysers[71] and mass spectrometers[72] capable of monitoring the concentrations of several gases simultaneously on a breath-by-breath basis. A response time of less than 10 ms is required. Developments in infrared-emitting diodes and solid-state sensors have produced compact and rugged CO_2 analysers suitable for mounting on a ventilator.

Blood-gas analysis Modern blood gas analysers can measure the pH, P_{CO_2} and P_{O_2} with a blood sample of only 150 μl and can also calculate derived variables such as the base excess and percentage oxygen saturation. The built-in microcomputer checks the correct functioning of all the components and the availability of consumables. Automatic rinsing and flushing occurs after sampling and automatic calibrations make the analyser available on a 24-hour basis. Surveys of the subject are given in references 73 and 74. Non-invasive continuous monitoring of values which approximate to the arterial oxygen and carbon dioxide tensions can be made with the transcutaneous approach using an electrode heated to 44°C and placed on the skin. Compact combined electrodes for P_{O_2} and P_{CO_2} electrodes are now used for the monitoring of intensive care patients and neonates.[75–77]

Automatic lung ventilators Ventilators are encountered in the operating room for pumping anaesthetic gas and vapour mixtures into the lungs of patients during surgery and in the intensive care unit (ICU).[78] Those for intensive care use tend to be more complicated and incorporate alarm systems to draw attention to mains or gas failure, patient disconnection and obstruction. The flow of gas during the inspiratory phase is usually produced by a constant flow generator and a variety of expiratory phases are available including positive, atmospheric and negative end-expiratory pressure. The latest models incorporate electronic gas flow and pressure transducers, infrared CO_2 analysers and a unit for computing pulmonary mechanics.

Figure 63.6 Microprocessor-controlled central station for a patient monitoring system (Courtesy of Hewlett Packard Ltd)

63.3.12 Patient monitoring systems for intensive care and coronary care units[13]

The aim of patient monitoring systems in the ICU and coronary care unit is to provide a continuous monitoring of one or more important physiological variables such as the ECG and derived values such as the heart rate, and to be able to display and record these as required. The ECG and heart rate are virtually always monitored and in postoperative situations one or more blood pressures may be monitored and the systolic, mean and diastolic pressures derived.[79] The respiratory rate and volume and even the body temperature are not so often monitored automatically but by hand as necessary.

Most systems are modular in construction and are configured for the unit concerned. Facilities for interrogating up to 12 bed stations are normal. Non-fade digital storage type visual displays of the original analogue signals from the patients are shown on a visual display screen together with numerical values for the heart and respiratory rates, temperatures and blood pressures. The use of microprocessors has led to sophisticated R-wave algorithms, digital filtering and a variety of screen formats. *Figure 63.6* shows a modern central station for a microprocessor-based patient monitoring system.

Radio telemetry systems are employed in some units for the continuous monitoring of signals such as the ECG in coronary care units, to which patients are admitted when they are ambulatory after leaving the ICU, and uterine contractions for patients going into labour.[80]

A number of sophisticated minicomputer-based patient monitoring systems have been described,[81] particularly for use with shock patients, but the present trend is to use more limited systems with a built-in microprocessor and good displays.

63.3.13 Instrumentation for obstetric and paediatric investigations

Because it does not involve ionising radiation which can harm the fetus, ultrasound imaging at 3 MHz or 5 MHz has made a great impact in the investigation of problems arising during pregnancy.[82–84] Using a real-time linear array system the beating fetal heart can be detected at 10 weeks. The position of the fetus can be located and fetal head and limb growth monitored throughout gestation. Fetal respiratory movements can be monitored and in some cases the sex of the fetus determined.

The cardiotochograph is a planar pressure transducer which is worn on the maternal abdomen and picks up the fetal heart rate and maternal uterine contractions. Alternatively, the fetal ECG can be picked up from a fetal scalp electrode and the intrauterine pressure monitored via a catheter and pressure transducer. Both signals can be transmitted to a central monitoring station via a radio link[85] to allow the mother freedom of movement.

Premature babies can suffer sudden stoppages of their breathing (apnoea) and several types of apnoea monitor have been developed based on the monitoring of transthoracic electrical impedance changes, chest wall movement by the reflection of microwave radiation or the detection of body movements with the infant lying on a segmented air-filled mattress.

Pulse oximeters have been developed which continuously follow the changes in oxygen saturation from a child's finger or toe and sound an alarm when a significant fall in saturation occurs due to loss of breathing or smothering.[86,87]

63.3.14 Instruments for neurological investigations

The electroencephalograph (EEG) is an 8- or 16-channel recorder for recording on a paper chart the electrical activity picked up from the brain via a montage of electrodes carefully placed on the scalp. It has been widely used in the diagnosis of epilepsy[88] and has a part to play in the diagnosis of brain death.[89] A simplified system for continuously monitoring the cerebral activity of patients in the ICU or during surgery is the cerebral function monitor.[90] The monitoring of intracranial pressure is important to detect swelling of the brain or an accumulation of cerebrospinal fluid in the skull. Infection problems have been minimised by the development of miniature pressure transducers which can be implanted within the skull and communicate via a radio telemetry link, and fibre optic systems are also in use.[91,92]

Whilst the EEG is of help in assessing brain function, neuropathology has benefitted enormously from the advent of the CT head scanner for depicting the extent of cerebral haemorrhage and the presence of lesions.[93]

63.3.15 Applications in physical medicine and rehabilitation

The electromyograph (EMG)[94] for the recording of the electrical activity developed by muscles is widely used for the assessment of muscle-wasting diseases and in monitoring recovery following trauma, and automatic techniques of analysis are being developed.[95] The impedance cardiograph[46] was developed for the cardiac assessment of post-heart-attack patients by non-invasive means and computer-based systems are available for assessing the recovery of post-stroke patients with respect to their performing manual tasks.[96]

Heat treatment of muscles is available in departments of physical medicine via short wave (27 MHz) diathermies or microwave diathermies.[97] The myoelectric control of prostheses and orthoses has not proved easy, but can offer valuable assistance to some categories of amputees.[98]

63.3.16 Instrumentation for ear, nose and throat investigations

The most commonly encountered instrument in audiology departments is the audiometer for determining the acuity of the ear (equivalent to the frequency response of the ear). The application of computers to audiometry is proving valuable.[99,100]

Uncooperative subjects such as babies can have their response to sound tested by the evoked response method in which signal averaging techniques are applied to the subject's EEG to bring out the response of the brain to an applied audio signal.[101]

Hearing aids are in use by many partially deaf people and the requirements for small, cosmetically acceptable units have benefitted from the availability of microcircuit amplifiers and miniature batteries.[102] An interesting new development is the attempt to provide a sense of hearing for the totally deaf by stimulating the auditory cortex of the brain or the cochlea.[103,104] Vibrotactile implanted prostheses are also under trial.[105]

Electronic techniques are not much used in connection with the nose and its function, but techniques have been developed to measure the response to olfactory stimulation. Quantitative assessments of odour sensitivity have been made.[106]

Patients who have lost their vocal cords following a laryngectomy operation require an artificial larynx in the form of a tone generator applied to the throat in order for them to be able to produce understandable audible sounds.[107]

63.3.17 Electrosurgical instrumentation

Modern solid-state diathermy apparatus (electrocautery) consists of a radiofrequency transmitter with an output power typically of 400 W and a frequency of 500–600 kHz.[6,108] In some versions, a pure sine wave output at 600 kHz is available and also two other outputs providing 600 kHz modulated with 76 kHz or 150 kHz. These three waveforms are used for cutting. For coagulation of bleeding points, the 600 kHz sine wave can be modulated with frequencies of 10, 21, 31, 70 or 120 kHz as required, and both cutting and coagulation waveforms can be freely mixed. For general surgical use the output is monopolar, but for delicate work in the brain or nervous system an isolated bipolar output of up to 100 W at 600 kHz is available. Older equipments are based upon valve and spark-gap oscillators which, although robust, radiate many harmonics and have outputs which can be 100% modulated at the mains frequency and will thus play havoc with physiological monitoring systems connected to the patient during surgery. Disposable plate (indifferent) electrodes and plate cable continuity monitors have greatly reduced the risk to the patient of radiofrequency burns arising from a poorly fitting plate electrode or a break in the plate electrode cable.

63.3.18 Psychiatric applications

Electroconvulsive therapy is widely used for the treatment of patients suffering from acute depression and improvements in the instrumentation have been made to provide a closer control of the current levels. The galvanic skin reflex is widely used for monitoring a subject's response to questioning and is based on the measurement of the electrical resistance of the skin.[109] Heart rate meters and biofeedback methods are used in the management of anxiety states and for training in relaxation.[110]

63.3.19 Ophthalmological instrumentation

The movement of the eye generates small electrical potentials which can be recorded on a physiological recording system as the electronystagmogram or the electro-oculogram and these can be used to test the visual field of a patient.[111] Eye movement recording with a non-contact method is of importance in ergonomic and time and motion studies of tasks.[112] Television systems have been designed for the continuous recording of pupil diameter (pupilometry) as a variable in behavioural studies.[113]

The pressure developed within the eyeball is used in the diagnosis of glaucoma and a number of types of tonometer have been designed for clinical screening studies.[114] Potentials developed when the eye is illuminated give rise to the electroretinogram.[115]

63.3.20 Instrumentation for gastrology

Radiopills (endoradiosondes) have been employed to track pressure, temperature and pH changes in the gut and to detect the site of bleeding.[116] However, the complexity of the tracking aerial system has not made their use widespread. The availability of electrodes which can be implanted on the surface of the gut has led to detailed studies of the slow-wave EMG activity of the smooth muscle of organs such as the colon[117] in investigations of peristalsis.

63.3.21 Urological instrumentation

The measurement of bladder pressure, urine flow rate and volume comprise the investigative technique of cystometry used in the study of urinary incontinence and prostatic disease.[118] In studies on incontinence the bladder and rectal pressures together with their difference and the EMGs of the rectal and urethral sphincters may be recorded.[119] Electrical stimulators have been developed to assist in keeping the urethral sphincter closed in patients with incontinence and in helping patients with a neurogenic bladder to micturate.[120] Bladder stones can be shattered *in situ* with an ultrasonic lithotriptor and then removed per-urethrally.[121]

63.3.22 Pathology laboratory instrumentation

Instrumentation in a hospital pathology department is mainly used in the haematology and biochemistry laboratories. The ever-increasing workload has encouraged the evolution of automatic methods, and the advent of microcomputers has made laboratory instruments increasingly 'intelligent' in terms of quality control and ease of operation.

63.3.22.1 Haematology

Much of the automated work of a haematology laboratory is concerned with blood cell counting and sizing and the measurement of the haematocrit (percentage of red cells). There are approximately 5 million cells in 1 ml of whole blood. Red cells (erythrocytes), white cells (leucocytes) and platelets can all be counted in samples of diluted blood passed through an orifice containing electrodes. As each cell passes through the orifice the electrical resistance changes momentarily and a count is recorded. The amplitude and shape of the pulse generated by a cell is related to its volume and can be used to produce a size histogram, calculate the mean cell volume and the haematocrit. Platelets can be counted by a dark field optical scatter method and are important in the blood clotting mechanism. Flying-spot microscope differential cell classifiers working with computer pattern recognition algorithms are starting to automate the tedious classification of blood cell types.[122]

The measurement of the blood clotting time is also important, particularly for anticoagulant therapy. In one approach a reagent is added to the blood in a thermostatted cuvette and the time to form a clot is measured photoelectrically. Other methods use an electromechanical approach with a magnetic stainless-steel ball suspended in an optical path of a vertically oscillating cuvette. The formation of a clot moves the ball out of the light beam. Computer analysis of the clotting time curves can yield details of the clotting mechanisms involved.[123]

The haemoglobin content of a patient's blood sample is important since haemoglobin is the constituent of red cells which by combining with oxygen or carbon dioxide effects their transport in the blood. One laboratory method has required the blood sample to be diluted with potassium ferricyanide to form cyanmethaemoglobin and the absorption at a light wavelength of 540 nm of this compound is a measure of the haemoglobin content. Optical methods are increasingly employed, for example in blood-gas analysers to measure the haemoglobin content.

Computer-based systems are now available for automatic blood grouping.[124]

63.3.22.2 Biochemistry

Analysers capable of automatically analysing a substantial number of specimens fall into two main classes: continuous flow and discrete. In continuous flow analysers, such as the well known Autoanalyser range by Technicon Instruments, up to 20 separate analyses can be performed on one sample such as plasma or urine. The aliquots of the sample fed to each channel are kept separate by interposed air bubbles.

Simple discrete analysers of the 'suck and squirt' type place each sample into a cuvette, add the necessary reagents and then monitor the subsequent chemical reaction, often by measuring a colour change. A new development enables high-speed reactions to be followed in a micro-centrifugal system. The samples and reagents are loaded into a disposable plastic rotor containing perhaps 50 cuvettes. This is spun at a high speed under microprocessor control to mix the sample and reagents and a photometer system measures the colour changes. The arrangement copes well with relatively small batches for different tests rather than the large batches fed through a continuous flow analyser.[125]

Both large continuous flow analysers and centrifugal analysers have their own computers, but off-line computer systems are used in the larger laboratories to provide quality control and to generate laboratory worksheets for the technical staff from the patient test requests and to produce reports for the wards and clinics.[126,127]

An important requirement exists for small, dedicated analysers which can be used out of normal hours in the 'emergency' or 'hot' laboratory for urgent samples such as electrolytes and blood-gas values.

63.4 Safety aspects

In intensive and coronary care units and during major procedures such as cardiac surgery, a large amount of electrical equipment can be connected directly to a patient and most of it will be mains powered. Thus it is important that the patient is not exposed to potentially dangerous currents flowing into or out from him to earth or to equipment. Current practice in Europe is to have all the input circuits fully 'floating', i.e. isolated from earth. It is unfortunate that the human heart is most at risk at mains frequencies (50 to 60 Hz) with respect to the initiation of ventricular fibrillation. Direct current should not be permitted at levels in excess of 10 μA on electrodes and transducers since d.c. is particularly prone to produce burns at electrode–skin interfaces.

The International Electrotechnical Commission has produced its document IEC 601-1 'Safety of Medical Electrical Equipment. Part I: General Requirements' which details three classes of equipment. Class I has protection against electric shock which does not rely only on basic insulation but also has all accessible conductive parts jointed to a protective earth conductor. Class II has double or reinforced insulation and no protective earth. Class III relies for its protection against electric shock on it being supplied only with a safe extra-low voltage (SELV). Having defined the mode of protection, a further classification is possible in terms of application to patients. Type B equipment is class I, II or III equipment with an internal power supply which does not produce more than a specified leakage current and has a reliable protective earth connection. Type BF is type B equipment with an isolated (floating) part for connection to the patient. Type CF is class I or II equipment with an internal power supply and a high degree of protection against leakage currents and a floating part connected to the patient. Type B apparatus is suitable for internal and external application to the patient but not directly to the heart. In that situation type CF should be employed. Type H equipment is class I, II or III equipment with an internal power supply and a degree of protection against electric shock comparable with that of household electrical equipment. Requirements are also specified for anaesthetic-proof apparatus which is safe to use in explosive anaesthetic atmospheres. For type B equipment the maximum allowable leakage current under normal conditions is 0.1 mA rising to 0.5 mA under single fault conditions. In contrast, for type CF apparatus the corresponding values are reduced to 0.01 and 0.05 mA. Thus equipment connected directly to the heart should not pass leakage currents in excess of 50 μA r.m.s. It is possible to classify patients in hospital in terms of their degree of electrical risk.[13] Details of explosive anaesthetic atmospheres are given in reference 78. Details of levels of 50 Hz current which affect the human heart are given in reference 128.

It is most important that when equipment is purchased it should conform to the national safety requirements and that it should be properly maintained throughout its life. There should also be a proper acceptance test and subsequent quality control measurements.

63.5 Lasers

Small, pulsed, ruby lasers originally aroused much attention when embodied in ophthalmoscopes and used for the 'welding' back in place of detached retinas. However, the shock waves associated with the laser light impact can have deleterious effects and current practice is to make use of a continuous-wave argon laser. The argon light is absorbed by blood vessels and these can be sealed off in diabetic retinopathy. Another application concerns the sealing of bleeding gastric ulcers. A fibre-optic catheter has been developed for the latter application.[129] Argon lasers are also suitable for the treatment of some patients suffering from port-wine birthmarks. Continuous-wave carbon dioxide lasers with a power output of 60 W into a jointed mirror applicator have been used for surgery on tumours, organs which are prone to bleed such as the liver, and for some cutaneous conditions.[130,131] Helium–neon lasers are employed in a number of cell counting and sizing instruments and for anemometry with heart valves.[132] It is most important not to look directly at a laser and to beware of scattered light and burns from stray CO_2 laser radiation in the infrared at 10.6 μm. Safety precautions with lasers are discussed in reference 133. A good introduction to medical applications for various types of laser is given in references 43 and 134.

References

1 GEDDES, L. A. and BAKER, L. E., *Principles of Applied Biomedical Instrumentation*. 2nd edn. Wiley-Interscience, New York (1975)

2 JONES, B. E., *Current Advances in Sensors*, Adam Hilger, Bristol (1987)

3 COONEY, D. O. (ed), *Advances in Biomedical Engineering*, Parts 1 and 2, Vol. 6, Marcel Dekker, New York (1980)

4 *Excerpta Medica*, Section 27, Biophysics, Bioengineering and Medical Instrumentation

5 *Topics*, INSPEC Marketing Department, Savoy Place, London WC2

6 HILL, D. W., *Electronic Techniques in Anaesthesia and Surgery*, 2nd edn, Butterworths, London (1973)

7 COBBOLD, R. S. C., *Transducers for Biomedical Measurements —Principles*, Wiley-Interscience, New York (1974)

8 ROLFE, P. (ed.), *Non-Invasive Physiological Measurements*, Vol. 1, Academic Press, London (1978)

9 ZIMMERMAN, J. and RECTOR, A., *Computers for the Physician's Office*, Research Studies Press, Portland, Oregon (1978)

10 REINHOFF, O. and ABRAMS, M. E., *The Computer in the Doctor's Office*, North-Holland, Amsterdam (1980)

11 COLLEN, M. F., 'Data processing techniques for multitest screening and hospital facilities', *Hospital Information Systems*, eds G. A. Bekey and M. D. Schwartz, Marcel Dekker, New York (1972)

12 PINCIROLI, F. and ANDERSON, J. (eds), *Changes in Health Care Instrumentation due to Microprocessor Technology*, North-Holland, Amsterdam (1981)

13 HILL, D. W. and DOLAN, A. M., *Intensive Care Instrumentation*, Academic Press, London (1982)

14 CHRISTENSON, E. E., CURRY, T. S. and DOWDEY, J. E., *An Introduction to the Physics of Diagnostic Radiology*, Lea and Febiger (1978)

15 HILL, D. R. (ed.), *Principles of Diagnostic X-Ray Apparatus*, Macmillan, London (1979)

16 HEMMINGWAY, A. P., 'Intravenous digital subtraction angiography', *Brit. Med. J.*, **295**, 331 (1987)

17 POND, G. D., ORITT, Th. W. and CAPP, M., 'Comparison of conventional pulmonary angiography with intraveous digital subtraction angiography for pulmonary embolic disease', *Radiology*, **147**, 345–50 (1983)

18 EVENS, R. G., ALFIDI, R. J. and HAAGA, J. R., 'Body computed tomography: a clinically important and efficacious procedure', *Radiology*, **123**, 239–40 (1977)

19 SCHERTEL, I., 'Irradiation planning with computed tomography and treatment planning equipment', *Strahlentherapie*, **155**, 757–9 (1979)

20 KITCHENER, P., HOUANG, M. and ANDERSON, B., 'Utilisation review of magnetic resonance imaging: the Australian experience', *Aust. Clin. Rev.*, September, 127–37 (1986)

21 GENERAL ELECTRIC COMPANY, *NMR—An Introduction*, Medical Systems Division, Milwaukee (1981)

22 PARKER, R. P., SMITH, P. H. S. and TAYLOR, D. M., *Basic Science of Nuclear Medicine*, Churchill Livingstone, Edinburgh (1978)

23 LIM, C. B., HOFFER, P. B., ROLLO, F. D. and LICIEN, D. I., 'Performance evaluation of recent wide field scintillation gamma cameras', *J. Nucl. Med.*, **19**, 942–7 (1978)

24 HAUSER, M. F. and GOTTSCHALK, A., 'Comparison of the anger tomographic scanner and the 15-inch scintillation camera gallium imaging', *J. Nucl. Med.*, **19**, 1074–7 (1978)

25 MOORE, G., 'Following the tracks of disease', *Electron. Power*, **33**, 496–8 (1987)

26 ELL, P. J., COSTA, D. C., CULLUM, I. D., JARRITT, P. H. and LUI, D., eds., *The Clinical Application of rCBF by SPET*, Amersham International, Little Chalfont (1987)

27 GARN, L. E. and PENITO, F. C., 'Thermal imaging with pyroelectric vidicons', *IEEE Trans. Electron. Devices*, **24**, 1221–8 (1977)

28 NEWMANN, P., DAVISON, M., JACKSON, I. and JAMES, W. B., 'The analysis of temperature frequency distributions over the breasts by a computerised thermography system', *Acta Thermographica*, **4**, 3–8 (1979)

29 WELLS, P. N. T., *Biomedical Ultrasonics*, Academic Press, London (1977)

30 MATZUK, T. and SKOLNICK, M. L., 'Ultrasound grey scale display on storage oscilloscope system', *Ultrasonics*, **15**, 221–5 (1977)

31 DETER, R. L. and HOBBINS, J. C., 'A survey of abdominal ultrasound scanners: the clinician's point of view', *Proc. IEEE*, **67**, 664–71 (1979)

32 MAGINNES, M. G., 'Methods and terminology for diagnostic ultrasound imaging systems', *Proc. IEEE*, **67**, 641–53 (1979)

33 PEDERSEN, J. F. and NORTHEVED, A., 'An ultrasonic multitransducer scanner for real time heart imaging', *J. Clin. Ultrasound*, **5**, 11–15 (1977)

34 BARNES, R. W., 'Intraoperative monitoring in vascular surgery', *Ultrasound Med. Biol.*, **12**, 919–26 (1986)

35 WYATT, D. G., 'Electromagnetic blood-flow measurement', *IEE Medical Electronics Monographs 1–6*, Ed. B. W. Watson, London, Peter Peregrinus (1971)

36 WYATT, D. G., 'Blood flow and blood velocity measurement *in vivo* by electromagnetic induction', *Med. Biol. Eng. Comp.*, **22**, 193–211 (1984)

37 ASTLEY, C. A., HOHIMER, A. R. and STEPHENSON, R. B., 'Effect of implant duration on *in vivo* sensitivity of electromagnetic flow transducers', *Am. J. Physiol., Heart Circ. Physiol.*, **5**, H508–12 (1979)

38 WILLIAMS, B. T., SANCHO-FORRES, S., CLARK, D. B., ABRAMS, T. D. and SCHENK, W. F. Jr, 'Continuous long-term measurement of cardiac output in man', *Ann. Surg.*, **174**, 357–63 (1971)

39 WILLIAMS, B. T., SANCHO-FORRES, S., CLARK, D. B., ABRAMS, I. D., SCHENK, W. G. Jr. and BAREFOOT, C. A., 'The Williams–Barefoot extractable blood flow probe', *J. Thoracic Cardiovasc. Surg.*, **63**, 917–21 (1972)

40 WOODCOCK, J. P. and SKIDMORE, R., 'Principles and applications of Doppler ultrasound', In *New Techniques and Instrumentation for Ultrasonography*, eds P. N. T. Wells and M. C. Ziskin, Churchill Livingstone, Edinburgh (1980)
41 BORN, G. V. R., MELLING, A. and WHITELAW, J. H., 'Laser Doppler microscope for blood velocity measurements', *Biorheology*, **15**, 363–72 (1978)
42 CARRUTH, J. A. S. and McKENZIE, A. L., *Medical Lasers*, Adam Hilger, Bristol (1986)
43 CLARK, C., 'A local thermal dilution flowmeter for the measurement of venous flow in man', *Med. Electron. Biol. Eng.*, **6**, 133 (1968)
44 RUBINS, S. A., QUILTER, R. and BATTAGIN, R., 'An accurate and rapid inflation device for pneumatic cuffs', *Am. J. Physiol., Heart Circ. Physiol.*, **3**, H740–2 (1978)
45 GREENFIELD, A. D. M., WHITNEY, R. J. and MOWBRAY, J. F., 'Methods for the investigation of peripheral blood flow', *Brit. Med. Bull.*, **19**, 101 (1963)
46 KUBICEK, W. G., PATTERSON, R. P. and WITSOE, D. A., 'Impedance cardiography as a non-invasive method of monitoring cardiac function and other parameters of the cardiovascular system', *Ann. NY Acad. Sci.*, **170**, 724–52 (1970)
47 ARENSON, H. M. and MOHAPATRA, S. N., 'Evaluation of electrical impedance plethysmography for the non-invasive measurement of blood flow', *Brit. J. Anaesth., Proc. Anaesth. Res. Soc.*, September (1976)
48 MULLICK, S. C., WHEELER, H. B. and SONGSTER, G. P., 'Diagnosis of deep venous thrombosis by measurement of electrical impedance', *Am. J. Surg.*, **119**, 417 (1970)
49 SARREL, P. M., FODDY, J. and MCKINNON, J. B., 'Investigation of human sexual response using a cassette recorder', *Arch. Sex. Behav.*, **6**, 341–8 (1977)
50 GEDDES, L. A., *The Direct and Indirect Measurement of Blood Pressure*, Year Book Press, Chicago (1970)
51 COBBOLD, R. S. C., *Transducers and Biomedical Measurements: Principles*, Wiley-Interscience, New York (1974)
52 MILLAR, H. D. and BAKER, L. E., 'A stable ultra-miniature catheter-tip pressure transducer', *Med. Biol. Eng.*, **11**, 86 (1972)
53 GEDDES, L. A. and MOORE, A. G., 'The efficient detection of Korotkoff sounds', *Med. Electron. Biol. Eng.*, **6**, 603 (1968)
54 VAN MONTFANS, G. A., VAN DER HOEVEN, G. M. A., KAREMAKER, J. M. and WIELING, W., 'Accuracy of auscultatory blood pressure measurement with a long cuff', *Brit. Med. J.*, **295**, 354–5 (1987)
55 HOCHBERG, H. M. and SALOMAN, H., 'Accuracy of an automated ultrasound blood pressure monitor', *Curr. Ther. Res.*, **13**, 129–38 (1971)
56 COLES, J. S., MARTIN, W. E., CHEUNG, P. W. and JOHNSON, C. C., 'Clinical studies with a solid state fibre optic oximeter', *Am. J. Cardiol.*, **29**, 383 (1972)
57 HILL, D. W., THOMPSON, D., VALENTINUZZI, M. E. and PATE, T. D., 'The use of a compartmental hypothesis for the estimation of cardiac output from dye dilution curves and the analysis of radiocardiograms', *Med. Biol. Eng.*, **11**, 43–54 (1973)
58 KUBICEK, W. G., KARNEGIS, J. N., PATTERSON, R. P., WITSOE, D. A. and MATTSON, R. H., 'Development and evaluation of an impedance cardiac output system', *Aerospace Med.*, **37**, 377–87 (1966)
59 HILL, D. W. and MOHAPATRA, S. N., 'The current status of electrical impedance techniques for the monitoring of cardiac output and limb blood flow', *IEE Medical Electronics Monographs* 23–27, eds D. W. Hill and B. W. Watson, Peter Peregrinus, Stevenage (1977)
60 LAMBERTS, R., VISSER, K. R. and ZIJLSTRA, W. G., *Impedance Cardiography*, Van Gorcum, Assen, The Netherlands (1984)
61 WINTERBOTTOM, J. T., 'Equiping an ECG department', *Brit. J. Clin. Equip.*, **5**, 176–9 (1980)
62 OLIVER, G. C., NOLLE, F. M. and WOLFF, G. A., 'Detection of premature ventricular contractions with a clinical system for monitoring electrocardiographic rhythm', *Computer Biomed. Res.*, **4**, 523–41 (1971)
63 THOMAS, L. J. Jr, CLACK, K. W. and MEAD, C. N., 'Automated cardiac dysrhthymia analysis', *Proc. IEEE*, **67**, 1322–37 (1979)
64 MACFARLANE, P. W. and LAWRIE, T. D. V., *An Introduction to Automated Electrocardiogram Interpretation*, Butterworths, London (1974)
65 KENNY, J., 'Cardiac pacemakers', *IEE Medical Electronics Monographs* 7–12, Peter Peregrinus, Stevenage (1974)
66 BLOOMFIELD, P. and MILLER, H. C., 'Permanent pacing', *Brit. Med. J.*, **295**, 741–4 (1987)
67 CUSHIERI, R. J., MORRAN, C. G. and POLLOCK, J. G., 'Transcutaneous electrical stimulation for ischaemic pain at rest', *Brit. Med. J.*, **295**, 305 (1987)
68 WEISSLER, A. M., HARRIS, W. S. and SCHOENFIELD, C. D., 'Systolic time intervals in heart failure in man', *Circulation*, **37**, 149 (1969)
69 MASON, C. A. and SHOUP, J. F., 'Cardiotachometer designs', *Med. Biol. Eng. Comput.*, **17**, 349 (1979)
70 PRESTON, T. D., 'Automated lung function tests', *Brit. J. Clin. Equip.*, **5**, 194–7 (1980)
71 HILL, D. W. and POWELL, T., 'Non-dispersive infra-red gas analysis', Adam Hilger, London (1968)
72 PAYNE, J. P., BUSHMAN, J. B. and HILL, D. W., eds, *The Medical and Biological Application of Mass Spectrometry*, Academic Press, London (1971)
73 HILL, D. W., 'Electrode systems for the measurement of blood-gas tensions and contents', *Scientific Foundations of Anaesthesia*, eds C. Scurr and S. Feldman, Heinemann Medical, London (1974)
74 HAHN, G. E. W., 'Techniques for the measurement of partial pressures of gases in the blood', *J. Phys. Elec. Instrum.*, **13**, 470–82 (1980) and **14**, 783–98 (1981)
75 BERAN, A. V., SNIGELAWA, G. Y., YOUNG, H. N. and HUXTABLE, R. F., 'An improved sensor and a method for transcutaneous CO_2 monitoring', *Acta Anaesth. Scand.*, **22**, Suppl. 68, 110–17 (1978)
76 VERSMOLD, H. T., LINDERKAMP, O. and HOLZMANN, M. 'Limits of $tcPO_2$ monitoring in sick neonates', *Acta Anaesth. Scand.*, **22**, Suppl. 68, 88–90 (1978)
77 PARKER, D., 'Sensors for monitoring blood gases in intensive care', *J. Phys. E: Sci. Instrum.*, **20**, 1103–12 (1987)
78 HILL, D. W., *Physics Applied to Anaesthesia*, Butterworths, London (1981)
79 SANDMAN, A. M. and HILL, D. W., 'An analogue pre-processor for the analysis of arterial blood pressure waveforms', *Med. Biol. Eng.*, **12**, 360–3 (1976)
80 VERTA, P. S. and KONOPASEK, F., 'The cardiac disaster alarm', *IEEE Trans. Bio-med. Eng.*, **19**, 248–51 (1972)
81 SHUBIN, H., WEIL, M. H., RALLEY, N. and AFIFI, A. A., 'Monitoring the critically ill patient with the aid of a digital computer', *Comput. Biomed. Res.*, **4**, 460–73 (1971)
82 WELLS, P. N. T. and ZISKIN, M. L., eds, *New Techniques and Instrumentation in Ultrasonography*, Churchill Livingstone, Edinburgh (1980)
83 WELLS, P. N. T., *Biomedical Ultrasonics*, Academic Press, London (1977)
84 WELLS, P. N. T., Ultrasonic imaging: pulse-echo techniques', *IEE Medical Electronics Monographs* 23–27, eds D. W. Hill and B. W. Watson, Peter Peregrinus, Stevenage (1977)
85 FLYNN, A. M., KELLY, J., HOLLINS, G. and LYNCH, P. F., 'Ambulation in labour', *Brit. Med. J.*, **ii**, 591–3 (1978)
86 SOUTHALL, D. P., STEBBENS, V. A., REES, S. V., LANG, M. H., WARNER, J. O. and SHINEBOURNE, E. A., 'Apnoeic episodes induced by smothering: two cases identified by covert video surveillance', *Brit. Med. J.*, **294**, 1637–41 (1987)
87 BARETICH, M. F., 'A tester for respiration rate meters', *J. Clin. Eng.*, **4**, 339–41 (1979)
88 SOREL, I., RONEQUOY PONSOR, M. and HARMANT, J., 'Electroencephalogram and CAT scan in 393 cases of epilepsy', *Acta. Neurol. Belg.*, **78**, 242–52 (1978)
89 SCOTT, D. S., *Understanding EEG*, Duckworth, London (1976)
90 SECHZER, P. H. and OSPINA, I., 'Cerebral function monitor evaluation in anesthesia/critical care', *Curr. Ther. Res. Clin. Exp.*, **22**, 335–47 (1977)
91 LEVIN, A. B., 'The use of a fibreoptic intracranial pressure transducer in the treatment of head injuries', *US J. Trauma*, **17**, 767–74 (1977)
92 BEKS, J. W. F., ABARDA, S. and GIELES, A. C. M., 'Extradural transducer for monitoring intracranial pressure', *Acta Neurochir.*, **38**, 245–51 (1977)

93 DAY, R. E., THOMSON, J. L. G. and SCHOTT, W. H., 'Computerised axial tomography and acute neurological problems of childhood', *Arch. Dis. Child.*, **53**, 2–11 (1978)

94 LOEB, G. E. and GANS, C., *Electromyography for Experimentalists*, University of Chicago Press, Chicago (1986)

95 BOYD, D. C., BRATTY, P. J. A. and LAWRENCE, P. D., 'A review of methods of automatic analysis in clinical electromyography', *Comput. Biol. Med.*, **6**, 179–90 (1973)

96 LYNN, P. A., PARKER, W. R., REED, G. A. J., BALDWIN, J. F. and PISWORTH, B. W., 'New approaches to modelling the disabled human operator', *Med. Biol. Eng. Comput.*, **17**, 344 (1979)

97 GURU, B. S. and CHEN, K. M., 'Hyperthermia in local EM heating and local conductivity change', *IEEE Trans. Biomed. Eng.*, **24**, 473–7 (1977)

98 SHANNON, G. F., 'A myoelectrically controlled prosthesis with sensory feedback', *Med. Biol. Eng. Comput.*, **17**, 73 (1979)

99 MEYER, C. R. and SUTHERLAND, H. C. Jr, 'A technique for totally automated audiometer', *Japan–Scand. Audiol.*, **7**, 105–9 (1978)

100 SAKABE, N., HIRAI, Y. and ITANI, E., 'Modification and application of the computerised automatic audiometer', *Japan–Scand. Audiol.*, **7**, 105–9 (1978)

101 BRACKMANN, D. E. and SELTERS, W. A., 'Electrical response audiometry: clinical applications', *Otolaryngol. Clin. North. Am.*, **11**, 7–18 (1978)

102 WALDEN, B. E. and KASTEN, R. N., 'Threshold improvement and acoustic gain with hearing aids', *Audiology*, **15**, 413–20 (1976)

103 DOBELLE, W. H., STENSAAS, S. S., MLADESOVSKY, M. G. and SMITH, J. B., 'A prosthesis for the deaf based on cortical stimulation', *Trans. Am. Otol. Soc.*, **61**, 157–75 (1973)

104 FEIGENBAUM, E., 'Cochlear implant devices for the profoundly hearing impaired', *IEEE Eng. Med. Biol.*, **6**, 10–21 (1987)

105 DORMER, K. J. and PHILLIPS, M. A., 'Auditory prostheses: implantable and vibrotactile devices', *IEE Eng. Med. Biol.*, **6**, 36–41 (1987)

106 VAN DROGELEN, W., HOLLEY, A. and DOVING, K. B., 'Convergence in the olfactory system: quantitative aspects of odour sensitivity', *J. Theor. Biol.*, **71**, 39–48 (1978)

107 KNORR, S. G. and ZWITMAN, D. H., 'The design of a wireless-controlled intra-oral electrolarynx'

108 GEDDES, L. A., SILVA, L. F., DEWITT, D. P. and PEARCE, J. A., 'What's new in electrosurgical instrumentation?', *Med. Instrum.*, **11**, 355–9 (1977)

109 HALL, S. H., 'A bridge for continuous resistivity observations', *J. Phys. E: Sci. Instrum.*, **9**, 728–9 (1976)

110 WATSON, B. W., WOOLEY-HART, A. and TIMMONS, B. H., 'Biofeedback instruments for the management of anxiety and for relaxation training', *J. Biomed. Eng.*, **1**, 58–63 (1979)

111 ARMON, H., WEINMAN, J. and PELEG, A., 'Automatic testing of the visual field using electro-oculographic potentials', *Docum. Ophthal.*, **43**, 51–63 (1977)

112 HAINES, J. D., 'Non-contacting ultrasound transducer system for eye movement recording', *Ultrasound Med. Biol.*, **3**, 639–45 (1977)

113 USA, S., MURASE, K. and IKEGAYA, K., 'Analysis of the dynamic preformances of a television pupilometer', *Japan J. Med. Electron. Biol. Eng.*, **16**, 177–83 (1978)

114 THORNBURN, W., 'The accuracy of clinical application tonometry', *Acta Ophthalmol.*, **56**, 1–5 (1978)

115 NAKAMURA, Z., 'Human electroretinogram with skin electrode', *Japan. J. Ophthalmol.*, **22**, 101–13 (1978)

116 DELCHAR, I. A. and SMITH, M. J. A., 'An improved method for detecting passive pills', *Phys. Med., Biol.*, **21**, 577–83 (1976)

117 SMALLWOOD, R. H., LINKENS, D. A., KWOK, H. L. and STODDARD, C. J., 'Use of autoregressive modelling techniques for the analysis of colonic myoelectric activity in man', *Med. Biol. Eng. Comput.*, **18**, 591 (1980)

118 KONDO, A., MITSUYA, H. and TORI, H., 'Computer analysis of micturition parameters and accuracy of uroflow meter', *Japan J. Urol.*, **33**, 337–44

119 DOYLE, P. T., HILL, D. W., PERRY, I. R. and STANTON, S. L., 'Computer analysis of electromyographic signals from the human bladder and urethral and anal sphincters', *Invest. Urol.*, **13**, 205–10 (1975)

120 SEIFERTH, J., HEISSING, J. and KARKAMP, H., 'Experiences and critical comments on the temporary intravesical electro-stimulation of the neurogenic bladder in spina bifida children', *GFR Urol. Int.*, **33**, 279–84 (1978)

121 EL FAHIG, S. and WALLACE, D. M., 'Ultrasonic lithotriptor for urethral and bladder use', *Brit. J. Urol.*, **50**, 255–6 (1978)

122 DURIE, B. G. M., VAUGHT, I. and CHEN, Y. P., 'Discrimination between human T and B lymphocytes and monocytes by computer analysis of digitised data from scanning microphotometry', *Blood*, **51**, 579–89

123 FRANK, H. I. I., DREESEN, V. and HENKER, H. C., 'Computer evaluation of performance curves for the estimation of extrinsic circulation factors', *Comput. Med. Biol.*, **8**, 65–70 (1978)

124 GOVAERTS, A. and SCHREYER, H., 'Integration of a Groupamatic MG50 in the routine work of a hospital blood transfusion centre', *Rev. Fr. Transfus. Immuno-Hematol.*, **21**, 715–19 (1978)

125 WILLIAMS, D. L., NUNN, R. F. and MARKS, V., eds, *Scientific Foundations of Clinical Biochemistry*, 2nd edn, Heinemann Medical, London (1978)

126 SIMS, G. E., *Automation of a Biochemical Laboratory*, Butterworths, London (1972)

127 CAVILL, I., RICKETTS, C. and JACOBS, A., *Computers in Haematology*, Butterworths, London (1975)

128 RAFTERY, E. B., GREEN, N. L. and GREGORY, I. C., 'Disturbances of heart rhythm produced by leakage current in human subjects', *Cardiovasc. Res.*, **9**, 256–62 (1975)

129 KIMURA, W. D., GULACSIK, C. and AUTH, D. C., 'Use of gas jet appositional pressurization in endoscopic laser photocoagulation', *IEEE Trans. Biomed. Eng.*, **25**, 218–24 (1978)

130 HALL, R. R., HILL, D. W. and BEACH, A. D., 'A carbon dioxide surgical laser', *Ann. R. Coll. Surg.*, **48**, 181–8 (1971)

131 LABANDER, H. and KAPLAN, I., 'Experience with continuous laser in the treatment of suitable cutaneous conditions', *J. Dermatol. Surg. Oncol.*, **3**, 527–30 (1977)

132 YOGANATHAN, A. A. P., REAME, H. H., CORCORAN, W. H. and HARRISON, E. C., 'Laser-Doppler anemometer to study velocity fields in the vicinity of prosthetic heart valves', *Med. Biol. Eng. Comput.*, **17**, 38 (1979)

133 ELECCION, M., 'Laser hazards', *IEEE Spectrum*, **10**, 32 (1973)

134 KOEBNER, H. K., ed., *Lasers in Medicine*, Vol. 1, Wiley-Interscience, New York (1980)

Index